Limnology

Lake and River Ecosystems

Third Edition

Limnology

Lake and River Ecosystems

Third Edition

ROBERT G. WETZEL

ACADEMIC PRESS
A Harcourt Science and Technology Company

San Diego San Francisco New York Boston London Sydney Tokyo

Cover art from an oil painting by the author, Robert G. Wetzel.

This book is printed on acid-free paper. ♾

Academic Press
A Harcourt Science and Technology Company
525 B Street, Suite 1900, San Diego, California 92101-4495, USA
http://www.academicpress.com

Academic Press
Harcourt Place, 32 Jamestown Road, London NW1 7BY, UK
http://www.academicpress.com

Library of Congress Catalog Card Number: 00-110178

International Standard Book Number: 0-12-744760-1

PRINTED IN THE UNITED STATES OF AMERICA
01 02 03 04 05 06 MB 9 8 7 6 5 4 3 2 1

To Theodore Jones, who took the time to inspire a fledgling
at an early, critical time, and to Carol Ann (né Andree) Wetzel
for her unending understanding and support

Contents

11
THE INORGANIC CARBON COMPLEX

12
THE NITROGEN CYCLE

13
THE PHOSPHORUS CYCLE

14
IRON, SULFUR, AND SILICA CYCLES

15
PLANKTONIC COMMUNITIES: ALGAE AND CYANOBACTERIA

16
PLANKTONIC COMMUNITIES: ZOOPLANKTON AND THEIR INTERACTIONS WITH FISH

17
BACTERIOPLANKTON

18
LAND–WATER INTERFACES: LARGER PLANTS

19
LAND–WATER INTERFACES: ATTACHED MICROORGANISMS, LITTORAL ALGAE, AND ZOOPLANKTON

20
SHALLOW LAKES AND PONDS

21
SEDIMENTS AND MICROFLORA

Preface to the Third Edition

Limnology is currently experiencing a period of introspection. Such criticism is healthy if done constructively and the causes of underlying deficiencies are recognized and addressed. Many problems have arisen, however, in part because of the purported necessity to respond rapidly to governmental and societal demands without an in-depth scientific underpinning. A root cause is our continuing inability to properly educate and train students and the public. Society must recognize that intellectual creativity is essential to excellence in science and that excellence in science is essential to the most effective and cost-efficient management of our resources.

Improvements in science education and training are critically needed. A fundamental requirement of scientific education is to obtain comprehensive background information prior to interpretation and synthesis. One observes an increasing erroneous reliance upon uninformed excuses that old perceptions have little foundation or relevance for modern interpretations. Additional nonsense arguments emerge that too much information exists to integrate. With modern search and organization capabilities, such ignorance of past research is much more than laziness; it represents a deficiency in the teaching of how science is properly conducted. Bibliographic negligence is more than a lack of responsible scholarship because it leads to increasing scientific redundancy and inefficient use of intellectual and financial resources. A result is an increasing tendency to promote old ideas and interpretations under the guise of new and invariably ambiguous "buzz word" terms. These redundant ideas are actively promoted as inspiration among noncritical peers, science writers, and even granting agency administrators who are unfamiliar with the background development of the subdiscipline. Many contemporary topical reviews are incredibly naive, biased, and incomplete, and parts are simply wrong. Such deficiencies in scholarship must be severely condemned and can be countered by more rigorous preparatory study.

There is an increasing tendency, particularly in the United States, to capitulate to the masses of information on any subject and accept, even promote, superficial understanding of ecological subjects. That level of inquiry may be acceptable for the lay public but is not acceptable for professional limnologists, aquatic ecologists, and water resource managers. A superficial level of understanding is unacceptable in all rigorous disciplines that contemplate systems of equal complexity to natural ecosystems. For example, the great strides made in human medicine are the direct result of concerted, systematic evaluation and rigorous experimental investigations of the mechanisms controlling human and pathogen physiology.

In ecology, as in many other disciplines, the "cop out" deference strategy to the information glut is increasingly to specialize in a small area and to then aggregate teams of specialists to attempt to understand how all of the pieces fit and operate together. The present emphasis on interdisciplinary studies is little more than an expansion of the identical process that has been carried on for decades by competent ecosystem scientists. Emphasis, particularly by nonscientist administrators, on forced collaboration, regardless of expertise, either directly or by fiscal coercion is an increasing problem. Such realities of collaborations reinforce the need for participants to be versed on some comprehensive introductory understanding of the whole.

I argue that one cannot manage aquatic ecosystems effectively without understanding how they operate in response to interactions of physical, chemical, and biotic environmental variables. This insistence is analogous to the statement that one cannot effectively manage human health without understanding human physiological and biochemical interactions with environmental variables. Superficial or biased training in limnology can only lead to superficial and biased understanding. There is a need for rigorous limnological training, even at the introductory level. The U.S. National Academy of Sciences (1996) also emphasized that need in the recent thorough evaluation of limnological education.

The first edition of this book, initiated in the early 1970s, resulted from my frustration with the available texts at the time, which, at best, offered interesting but superficial analyses of inland aquatic ecosystems. The initial edition, issued in 1975, was well received as an alternative. The second edition, published in 1983, represented an extensive updating with new findings, appreciable reorganization, and clarifications based on experience in my own classes and constructive inputs from many students and colleagues. Again the book was well received; it has been widely cited as a source work (e.g., ISI Citation Classic) and has been translated into several other languages.

Since the appearance of these editions, a number of other general treatments of limnology have appeared in English, French, German, Spanish, Portuguese, and other languages. Essentially all of these works are directed at providing general overviews of the subject, and some accomplish this objective very well. Some of these books are simple but interesting overviews of the subject that can be assimilated in a long evening of reading, whereas others are more involved and clearly are intended to accompany instructional programs in initial introductory courses in limnology.

Such general books introduce students at college and university levels to the fascinating subject of inland aquatic ecology. These accomplishments assist in fulfilling an enormously important responsibility of educators: the general public, particularly those obtaining higher education, should be aware of the magnitude, general operation, and management problems of inland waters upon which so much of their well-being depends. The information presented in these works need not include details underlying certain basics of the operation and should be presented in an interesting, scientifically correct manner.

There are different levels of training in any subject. The gap is large between general texts suitable for the general public and nonspecialist students and detailed research-level treatments, such as G. E. Hutchinson's four-volume *A Treatise on Limnology.* Somewhere in between is the need for a more detailed *introductory instructional* text for students interested in becoming professionals in this discipline and desiring greater depth of inquiry and understanding. For nearly a decade, among other responsibilities, I have been revising this general limnology text for a third edition for this introductory instructional objective.

Those persons familiar with the past editions of *Limnology* will recognize marked differences in the present *Limnology: Lake and River Ecosystems.* The basic format of previous editions is maintained, although the parts are completely reorganized, modified, and updated with recent advances in understanding. A number of introductory texts have taken the tack of presenting an ecosystem overview initially, in some cases constituting half the text, and then presenting a somewhat hackneyed discussion of the parts of the ecosystems thereafter that may influence the biota within the ecosystem. In *Limnology: Lake and River Ecosystems* I opine first that, apart from the structure of water and a few physical properties, aquatic ecosystem structure, energy flow, and productivity are regulated by biogeochemical cycling. One cannot talk effectively about chemical cycles, for example, without treating microbial mediation of those cycles. Thus, it is essential to comprehend these ecosystem components before discussing the integration of the entire ecosystem. Nonetheless, a very general discussion of aquatic ecosystem structure and productivity appears early in the text, largely in order to introduce the terms used throughout the subsequent topical treatments. True ecosystem integration follows in the individual community treatments and then finally in the important synthesis chapter on organic carbon.

Second, I make major attempts to present data supporting conceptual relationships among biota and the environment in a realistic and unbiased manner. I have been accused of being biased against the roles that animals play in aquatic ecosystem structure and energetic transfer. Nothing could be further from the truth. Evaluation of the data, however, simply does not support some of the entrenched dogma that prevails in contemporary limnological writings, including many textbooks. I do present historical developments of some of those ideas even if new findings demonstrate them to be erroneous. Alternative explanations and hypotheses are then presented. I recognize that some colleagues will have difficulty with these conclusions, although I believe my conclusions accurately represent reality.

Throughout the development of our understanding of limnology, emphasis has been on differences among lakes and rivers, differences among the physical and chemical characteristics of waters of various geomorphological regions, and the diversity of different biota and their growth characteristics. That differentiation among lakes and rivers and their physical, chemical, and biological properties was the impetus behind a half century of lake and river classification studies. A major evolution of my syntheses in *Limnology: Lake and River Ecosystems* is the comparative treatment of topics across lake, reservoir, and river ecosystems. These analyses do indeed indicate differences among the properties of lakes, land–water interface regions, reservoirs, and rivers. Importantly, these analyses also indicate marked commonality in function.

It is essential that we search for commonality, in addition to differences, among these aquatic ecosystems (Wetzel, 2000). Because of the dynamic properties of metabolism, growth, and reproductive capacities of organisms as they respond to the bewildering array of dynamic environmental factors influencing growth and development, variability is enormous and often difficult to fathom. Modeling at the present level of understanding is unable to cope with so many nonlinear variables that are changing rapidly. Although reductionism is essential to provide information on the properties involved, I argue that regulatory generality prevails across the individual species and diversity of processes. Regulation of metabolic rates among different communities is where generality emerges. From both theoretical relationships and management methods of controlling the effects of disturbances, integration of quantitative process rates of metabolism, energy fluxes, and material fluxes is where commonality emerges among highly disparate and complex interacting parameters within ecosystems.

Combining lacustrine, reservoir, and running water limnology in one integrated synthesis volume on this discipline immediately causes manifold increases in the published literature to be evaluated. Indeed, the literature on certain subjects, such as stream invertebrate ecology, is both formidable and intimidating. Yet, amid the maze of individual detail, commonality of processes and regulatory mechanisms emerges. I make no pretense to being comprehensive in coverage, and among the stream and river literature in particular I relied heavily on many thorough reviews of topical subjects.

Saline lakes constitute approximately half of inland waters. Again, chemical and biological characteristics of saline lakes are not treated separately but rather are integrated within topical discussions with fresh waters. The chemistry and biota of saline lakes are fascinating and of great evolutionary significance. The mineral and biotic values of contemporary saline lakes, however, are dwarfed by the collective values of freshwater surface and groundwater repositories to humankind. As a result, treatment here has been subordinate to freshwater ecosystems. Hammer (1986) and Williams (1988, 1996) have summarized well selective characteristics of saline lakes.

Many persons have been generous and helpful with suggestions for improving this edition. In addition to persons cited in the preface to the second edition, numerous colleagues and friends offered constructive criticism on portions of this book: Vernon Ahmadjian, Hartmut Arndt, Michael T. Arts, Arthur C. Benke, Riccardo de Bernardi, J. Marie Boavida, Nina Caraco, Robert Carlson, Perry Churchill, Daniel Conley, G. D. Cooke, Clifford Dahm, Anthony Davis, Stanley Dodson, Eric Espeland, Steven Francoeur, Walter Geller, Chris E. Gibson, Alex Huryn, Frank M. D'Itri, Eileen Jokinen, Klaus Jürgens, Mark Johnson, Susan S. Kilham, Joachim Kleiner, Reiner Koschel, Winfried Lampert, Richard C. Lathrop, Gene E. Likens, S. MacIntyre, Robert P. McIntosh, Jürgen Marxsen, William J. Matthews, R. Michael McKay, J. Melack, Robert E. Moeller, Walter T. Momot, Roland Psenner, Peter H. Rich, Eric E. Roden, Philippe Ross, Mark R. D. Seaward, Arthur J. Stewart, Ruben Sommaruga, Eugene Stoermer, Raymond G. Stross, Keller Suberkropp,

Nancy C. Tuchman, A. Vähätalo, Jack R. Vallentyne, Anthony E. Walsby, Amelia K. Ward, G. Milton Ward, William D. Williams, and Edward O. Wilson. Deborah Cook has patiently sustained my constant barrage of manuscript drafts, tabular materials, endless organization of thousands of references, and requests for help with figure preparations. Her uncompromising attention to detail has assisted me enormously in the preparation of this work. Mark Dedmon also assisted with preparation of many figures. My wife, Carol, continues to exude unending patience and understanding of the siren call of my limnological mistress. Portions of unpublished results cited in the text were supported by subventions of the National Science Foundation. Regardless of the assistance of the persons mentioned and many unnamed, the responsibility for the synthesis and interpretations rests with me.

Robert G. Wetzel

1 PROLOGUE

A basic feature of the earth is an abundance of water, which extends over 71% of its surface to an average depth of 3800 m. Over 99% of this immense hydrosphere is deposited in ocean depressions (Table 1-1). The relatively small amounts of water that occur in freshwater lakes and rivers belie their fundamental importance in the maintenance and survival of terrestrial life.

I. OUR FRESHWATER RESOURCES

Any analysis of inland water resources must address the preeminence of exponential human growth and utilization of fresh water. Humans must be recognized for what they are: an animal whose population growth is in an exponential phase. In spite of its absurdity, a belief prevails that the earth's supply of finite water resources can be increased constantly to meet exponential demands. Fresh waters are a finite resource that can be increased only slightly. For example, desalinization of ocean water requires tremendous energy expenditures for the treatment process and distribution of the product once obtained. That distribution is energetically prohibitive within only short distances from marine sources. Society as a whole, and many freshwater ecologists, have tended to ignore humans, and their use and misuse of fresh waters, as influential factors in the maintenance of lake and river ecosystems. Freshwater utilization is governed by the spiraling relationships in which supply is constantly expanded in response to growing demands. The unfortunate effect of essentially uncontrolled growth is that consumption increases in response to rising supply. Every increase in supply is met by a corresponding increase in consumption, because in contemporary society, voluntary control of consumption is ineffective unless economically advantageous.

II. DEMOTECHNIC GROWTH

The impending environmental problems are not only the result of population growth (Vallentyne, 1972, 1988). They result also from technological growth, both directly in the sense of increased per capita production and consumption and indirectly in that technology has furthered the growth of population and urbanization. This **demotechnic** concept of growth encompasses the combined effects of population in a biological sense as well as of production-consumption in a technological sense. The importance of the concept of demotechnic growth lies in its emphasis on both production and consumption. It describes the cycle in which degradation of the biosphere occurs as a result of utilization of the environment for production and consumption of technological products. Both processes lead to pollution of water resources. In demotechnic growth, attention is correctly focused on all aspects of technological growth, that is, the technological metabolism of humans.

TABLE 1-1 Water in the Biosphere

	Volume (thousands of km^3)	*Percentage of total*	*Renewal time*
Oceans	1,370,000	97.61	3100 years†
Polar ice, glaciers	29,000	2.08	16,000 years
Ground water (actively exchanged)‡	4067	0.295	300 years
Freshwater lakes	126	0.009	1–100 years)§
Saline lakes	104	0.008	10–1000 years)§
Soil and subsoil moisture	67	0.005	280 days
Rivers	1.2	0.00009	12–20 days)¶
Atmospheric water vapor	14	0.0009	9 days

* Modified from Vallentyne (1972), after Kalinin and Bykov. *In:* The Environmental Future, London, Macmillan Publishers, Ltd. Reprinted by permission of Macmillan London and Basingstoke. Slightly different values are given by Shiklomanov (1990), but ratios are similar.

† Based on net evaporation from the oceans.

‡ Kalinin and Bykov (1969) estimated that the total ground water to a depth of 5 km in the Earth's crust amounts to 60×10^6 km^3. This is much greater than the estimate by the U.S. Geological Survey of 8.3×10^6 km^3 to a depth of 4 km. Only the volume of the upper, actively exchanged ground water is included here.

§ Renewal times for lakes vary directly with volume and mean depth and inversely with rate of discharge. The absolute range for saline lakes is from days to thousands of years.

¶ Twelve days for rivers with relatively small catchment areas of less than 100,000 km^2; 20 days for major rivers that drain directly to the sea.

The demands that demotechnic growth have imposed upon freshwater resources are monumental. Although about 105,000 km^3 of precipitation, the ultimate source of the freshwater supply, fall on the land surface per year, only about one-third of it (ca. 37,500 km^3 $year^{-1}$) reaches the oceans as river discharge (Vallentyne, 1972). About two-thirds of the annual water supply is returned to the atmosphere by evaporation and plant transpiration. If the potential water supply of 37,500 km^3 $year^{-1}$ were divided evenly among some 5.5 billion humans now existing on Earth (2000), each person would have potentially 6800 m^3 $year^{-1}$ or 18,680 liters day^{-1}. These values would be halved at a projected population level of 10 billion humans by approximately the year 2050. Even though these quantities seem large in comparison to the human physiological requirement of 2 liters $person^{-1}$ day^{-1}, they are insufficient in view of modern technological demands. Domestic consumption averages 250 liters $person^{-1}$ day^{-1}, the average industrial consumption is 1500 liters $person^{-1}$ day^{-1} in developed countries, and agriculture uses up to several thousand liters $person^{-1}$ day^{-1} in countries with hot, dry climates (Vallentyne, 1972).

Historical patterns of environmental resource utilization have resulted in a series of crises, particularly among populations of developed countries (Fig. 1-1). As the human population increases, the severity of the crises increases. Particular resource-related problems are confronted with various actions by humans (Francko and Wetzel, 1983). Most of these actions are therapeutic adaptations and adjustments rather than true corrective, prophylactic measures. As a result, stress levels increase. The margin between coping successfully with environmental resource limitations and disastrous situations of overexploitation decreases continually. Stability becomes increasingly tenuous.

Vagaries in climate (e.g., Bryson and Murray, 1977) are superimposed upon the immediate demotechnic pressures of resource utilization. Fluctuations in climate have resulted in catastrophic crises both in the availability of resources and effective resource utilization. Rapid climatic changes affect not only the availability of adequate food resources but availability of fresh water as well; the two are inexorably coupled. Although major climatic changes are difficult to predict, such changes are certain and, on the

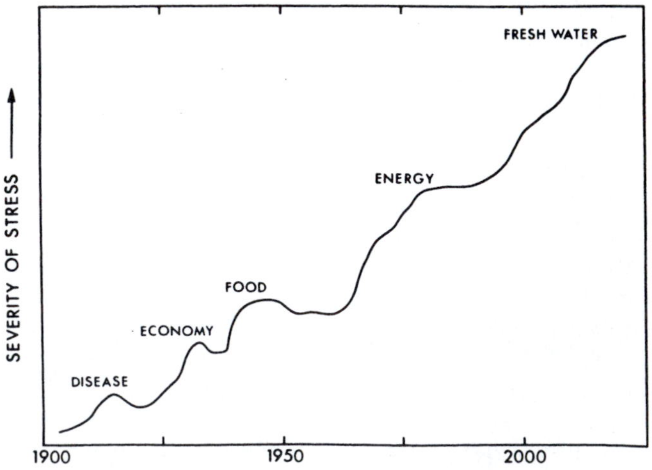

FIGURE 1-1 Stresses of increasing severity imposed upon human populations of developed countries resulting largely from excessive demotechnic growth. (Modified from Wetzel, 1978.)

basis of long-term trends, significant changes are imminent (Silver and DeFries, 1990; Schlesinger, 1991; Firth and Fisher, 1992; Cushing, 1997). In order to cope effectively with these natural and induced climatic changes, we must maintain sufficient flexibility in our use of freshwater resources and an adequate margin below the maximum carrying capacity.

III. HUMAN IMPACT ON FRESHWATER ECOSYSTEMS

The real freshwater supply is in reality much smaller than the potential total because of many factors. First, rainfall is not evenly distributed over land surfaces, and humans themselves are not distributed over land in proportion to water availability. This disparity results in a great expense of resources and energy for distribution systems to move water from places of water abundance to places where it is inadequate to support human activities. Second, total consumption has increased exponentially with demotechnic growth. Expansion of distribution systems to areas of low precipitation, such as for irrigation of semiarid regions, results in disproportionately high use of water because of very high losses by evapotranspiration. Third, potentially the most serious factor stemming from demotechnic growth is the severe degradation from contaminants of water quality. The effect is a severe reduction of the water supply available for other purposes.

Fresh waters of the world are collectively experiencing markedly accelerating rates of qualitative and quantitative degradation (Wetzel, 1992). Certain societies can cope, at least temporarily, with pollution and availability constraints and can even reduce freshwater degradation. In most of the world, however, human population growth continues without significant reduction of rates. Until human growth and consumption is stabilized, one hopes by the mid-twenty-first century, either by intelligence or catastrophes, further losses and degradation of fresh waters can be controlled only partially on a global basis. Control and reversal of degradation requires a proper economic and social valuation of fresh waters. With proper valuation, methods for effective utilization of existing, finite supplies can be applied to agricultural, industrial, and residential uses.

Fresh waters still serve purposes other than water supply, such as recreation, transportation systems, aesthetics, and others. However, the demands of exponential demotechnic growth clearly receive total precedence over uses of fresh waters for other purposes. The most fundamental laws of resource utilization may be recognized by most agencies and industries, but they are not being implemented significantly. The fuel "crisis" of 1973–1974, one of many consequences of demotechnic growth, demonstrated prevailing human behavioral patterns of virtually ignoring all diagnoses or prognostications of impending imbalances. When the crisis finally occurs, the public response is not to seek corrective measures that address the root of the problem, such as stabilization of population and technological growth, but rather to expand the distribution system to acquire more remote sources of energy. Only when disaster is imminent, when risk of survival of a large segment of the population is glaringly obvious to the majority, does a unified global response occur to save humankind from itself.

The remarks above, although pessimistic, accurately assess existing patterns of utilization of our water resources. It is clear that demotechnic growth will continue to impose increasing demands upon freshwater supplies either until inefficient utilization creates a disastrous situation threatening the survival of a major segment of the human race or until the expenditures of energy needed to obtain water exceed tolerable operational levels (Wetzel, 1992). Looking back at the repetitious history of responses to impending environmental disasters, we can be optimistic about the future only until such time as our understanding of the operation of the biosphere, and our knowledge of freshwater ecosystems in particular, is adequate to allow us to recognize the point of irreversibility (Bilsky, 1980; Francko and Wetzel, 1983; Wetzel, 1992). As one reflects on the progress that has been made in limnology since its inception a century ago, it becomes apparent that the time available for understanding fresh waters is disconcertingly limited. We need to intensify study of and time to understand freshwater ecosystems sufficiently to judge their resiliency and capacity for change in response to exponential demotechnic utilization and loading of contaminants. Existing understanding of freshwater ecosystems must be extended to a greater percentage of the population being educated so that this information can be effectively infused into the population at all levels.

It is of the utmost importance, therefore, that we understand the structure and function of freshwater ecosystems. Humans are a component of these ecosystems, and their effects on them will increase markedly until demotechnic growth is stabilized. Emotionalism and alarmist reactions to the momentum of exploitation of the finite biosphere by the technological system accomplish little and, as has been demonstrated repeatedly, are often antagonistic to improvement. Strict conservation and isolation of resource parcels, in the belief that such areas are exempt from technological alterations of the atmosphere and water supply, are naive and likewise contribute little to solution of the overall problem.

Understanding the metabolic responses of aquatic ecosystems is essential in order to confront and offset the effects of these alterations and in order to achieve maximum, effective management of freshwater resources. All waters, of course, cannot be managed directly. Rather, an integration of human growth and utilization with the metabolism of fresh waters is required to minimize detrimental changes. A well-documented effect of human impact upon aquatic ecosystems is *eutrophication,* a multifaceted term associated with increased productivity, simplification of biotic communities, and a reduction in the ability of the metabolism of the organisms to adapt to the imposed loading of nutrients. These conditions lead to reduced stability of the ecosystem. In this condition of eutrophication, excessive inputs often exceed the capacity of the ecosystem to be balanced. In reality, however, the ecosystems are out of equilibrium only with respect to the freshwater chemical and biotic characteristics that are desired by humans for specific purposes. In order to have any hope of effectively integrating humans as a component of aquatic ecosystems, and of monitoring their utilization of these resources, it is mandatory that we comprehend in some detail the functional properties of fresh waters. Only then can we evaluate and predict, with reasonable certainty, the influence human activities will have on the metabolic characteristics of these ecosystems.

IV. THE STUDY OF LIMNOLOGY

Limnology can be defined in several ways, but it is important to recognize that the discipline involves the study of both freshwater and saline inland waters. As noted in Table 1-1 and discussed in detail throughout this synthesis, some 45% of the inland surface waters of the land masses of the world are saline. Although saline lakes are of less practical importance to human activities, they nonetheless are major constituents of our biosphere with a number of unique characteristics (Hammer, 1986; Williams, 1996). The following definitions emphasize several important distinctions:

Limnology *is the study of the structural and functional interrelationships of organisms of inland waters as they are affected by their dynamic physical, chemical, and biotic environments.*

Freshwater ecology *is the study of the structural and functional interrelationships of organisms of fresh waters as they are affected by their dynamic physical, chemical, and biotic environments.* Saline waters (e.g., >0.3‰ or 3 g/liter^{-1}) are excluded from this definition.

Freshwater biology *is the study of the biological characteristics and interactions of organisms of fresh waters.* This study is largely restricted to the organisms themselves, such as their biology, life histories, populations, or, occasionally, communities.

It is important to emphasize that **limnology** correctly encompasses an integration of physical, chemical, and biological components of inland aquatic ecosystems of the drainage basin, movements of water through the drainage basins, and biogeochemical changes that occur en route, and within standing (**lentic**) waters. The lake ecosystem is a system that is intimately coupled with the land surrounding it in its drainage area and its running (**lotic**) waters that transport, and metabolize en route, components of the land to the lake. The analyses and syntheses of topics that follow address standing waters, as in previous editions, but have been expanded to include comparative syntheses of reservoir and stream ecosystems. The limnology of running waters was reviewed masterfully by Hynes (1970), but much new information and understanding has emerged since then. I summarize characteristics of running waters and compare them to those of standing water in this edition. Readers will recognize that this running water limnology is a true subdiscipline of limnology that is correctly receiving well over half of the research attention at the present time.

Many lakes of the world are of glacial origins. Because most limnological research has been concentrated in northern temperate regions, a strong bias occurs in current instruction of limnology by the disproportionate knowledge of natural temperate lakes. Most humans reside in nonglaciated regions in which reservoir and river ecosystems are the predominant surface waters. Understanding of tropical and warm water lake and reservoir ecosystems is increasing rapidly, and I have attempted to integrate their structural and functional differences and similarities into the evaluations of temperate inland waters. Furthermore, the number of human-made reservoirs has increased to the point where very few river systems are not impounded to some extent. Although reservoirs can possess many characteristics that differ from those of lakes, a firm grasp of the dynamics of natural lakes permits a relatively easy transition to an understanding of the more variable and individual characteristics of reservoir ecosystems. Underlying all of these ecosystems are basic metabolic similarities. In this treatment of inland aquatic ecosystems, I attempt to introduce fundamental, functional similarities without becoming mired in the plethora of individual detail.

Minimal background detail is required to appreciate even the rudiments of the discipline of limnology. The selection of material and examples to include in such a synthesis is difficult not only because of individual biases but because there is incomplete understanding of many subjects. I hope that, in the end, the examples presented here are balanced and provide a basic overview of contemporary comparative limnology and a basic minimal understanding of limnology and freshwater ecology at the undergraduate level. A serious major in limnology will realize that he or she needs much greater depth of understanding to comprehend the subject thoroughly. Many of the reference works cited, such as G. E. Hutchinson's classical, perceptive treatise (1957, 1967, 1975, 1993), are only initial introductory summaries of specialized subjects. In the ensuing discussions I often point out forefronts of contemporary limnology as well as gaps in need of intensive investigation.

V. SCIENTIFIC APPROACHES

Hegel (1807) stated, "*Das Wahre ist das Ganze*" (The truth is the whole). Holism alone, however, is inadequate for a comprehensive understanding of limnology. Integration of our knowledge about the individual operational components and environmental factors that regulate the metabolism and productivity of aquatic ecosystems is fundamental. Other treatments of limnology have been deficient in this respect and almost totally directed toward the open-water pelagic communities of standing waters. Other major *operational* components of aquatic ecosystems are largely neglected (i.e., microbial decomposition of dissolved and particulate detritus; the intensive metabolism and biogeochemical cycling that occurs in the littoral zone and adjacent wetlands; metabolism in the sediments; influxes of nutrients and organic matter from outside the lake; the rapidity of nutrient cycling, particularly associated with surfaces). An integrated study of the dynamics of all components of inland aquatic ecosystems requires major alterations in former perspectives on limnology.

Understanding of the causal mechanisms operating in and controlling our natural world is a primary objective in any science, and it is certainly true in limnology because of its premier position for the well-being of humankind. The greater our understanding is, the higher is the probability that we can predict with reasonable accuracy patterns of events within aquatic ecosystems in response to human manipulations and disturbances. How we acquire that understanding is accomplished with a combination of analytical techniques.

1. *Descriptive observations of patterns of biological processes and communities in relation to dynamic patterns of environmental properties.* Such descriptive empirical analyses allow the generation of a hypothesis, that is, a *conceptual* predictive "model" of relationships among observed patterns. These relationships may be compared statistically with correlations of the strength of co-occurrence of two variables. Such regressional correlations are only an extension of descriptions and indicate relationships within a certain statistical probability. The correlations only generate a hypothesis. Correlations do not show the accuracy of relationships, even though the probability of the relationship of patterns may be high, and they do not provide any direct evidence of causality. Comparative correlations can suggest relationships and pattern, but they cannot evaluate or test general patterns of cause and effect. Positive correlations can narrow the search for controlling factors of observed patterns, but spurious and erroneous positive correlations can also obfuscate the analyses.

2. *Experimental examination and evaluation of quantitative response data to selected disturbances imposed on the system.* Great insight about controlling parameters can be gained from experimental manipulation or imposing known, controlled disturbances of specific environmental or community parameters on specific components of the community or ecosystem. Quantitative response data include changes in growth, productivity, reproduction, competition, metabolic adaptations, and other processes of populations and communities as ecosystem components are exposed to the manipulation or disturbance and compared to components not exposed to the treatment.

3. *Application of quantitative predictive models based on experimentally* ***established****, not random, governing variables.* Models allow expansion of *experimentally understood* quantitative relationships, that is, one can insert hypothetical data of various parameters and theoretically estimate system responses to these variables. Models cannot predict anything that was not built into them from the beginning (Lehman, 1986). Even "counterintuitive results" are simply unrecognized consequences of initial assumptions. Models are a tool. Most models greatly overestimate understanding and falsely and naively generate confidence.

Some hypotheses about cause-and-effect regulators of biological phenomena cannot be tested experimentally because the ecosystems cannot be replicated or perturbed. For example, the effects of increased carbon dioxide (CO_2) concentrations in the atmosphere or reductions in ozone concentrations in the stratosphere cannot be replicated at the global scale. However, the effects of increased CO_2 or ozone can be tested experimentally very rigorously at smaller species or community scales, from which strong inferences of cause and effect can be coupled to observations of community changes in biological productivity, diversity, and other factors. These observational comparisons, however, are still correlations of relationships of certain probability, not demonstrations of regulating mechanisms. Although not ideal, such correlations can be highly instructive and useful in the management of ecosystems and aquatic resources even though the underlying causal mechanisms are unknown.

In the ensuing summaries and syntheses of contemporary understanding of the limnology of lakes, reservoirs, and streams, much of the information is presented in a comparative manner based on many, often hundreds, of observations and measurements. Patterns of differences and similarities emerge among the diverse ecosystems. Contemporary comprehension of why inland aquatic ecosystems operate as they do has emanated from experimentally controlled disturbances and manipulations to parts of or to whole ecosystems. These experimental approaches require integration of physiological and biochemical understanding with observed population and community responses to experimental and natural changes (Wetzel, 2000a).

VI. SEARCH FOR COMMONALITY, NOT ONLY DIFFERENCES

Throughout the development of our understanding of limnology, emphasis has been on the differences among lakes and rivers, physical and chemical characteristics among different waters of various geomorphologic regions, and the diversity of different biota and their growth characteristics. That differentiation among lakes and rivers and their physical, chemical, and biological properties was the impetus underlying the half century of lake classification studies. Subsequent comparative analyses attempted to demonstrate functional relationships among different groups but again were less successful than hoped because of the inherent limitations of correlation analyses. Progress has now extended to detailed examination of physiological, biochemical, and molecular variations among biota, commonly coupling these differences to variations in the environmental habitats in which the organisms reside. Some of this continued differentiation of species at the molecular level is most important from evolutionary, biogeographic, and systematic viewpoints, particularly in relation to biodiversity and losses of biodiversity.

Variability is enormous and often difficult to fathom because of the dynamic properties of metabolism, growth, and reproductive capacities of organisms as they respond to the bewildering array of dynamic environmental factors influencing growth and development. Modeling at the present level of understanding is unable to cope with so many nonlinear variables that are changing rapidly. Although reductionism is essential to provide information on the properties involved, regulatory generality prevails across the individual species and diversity of processes (Wetzel, 2000a). Regulation of quantitative rates of metabolism among the different communities is where generality emerges. Consolidation of quantitative process rates of metabolism, energy fluxes, and material fluxes is where commonality emerges among highly disparate and complex interacting parameters within ecosystems. That understanding is essential in terms of both theoretical relationships and in the management of controlling the effects of disturbances.

VII. ALTERED PERSPECTIVES

Fresh waters are biological systems, and biogeochemical processes control the quality of fresh waters. As will be evident in the following pages, we have a reasonable understanding of the "anatomy," that is, the structure, of inland waters; it is essential that we now achieve an understanding of the "physiology," that is, the functional metabolism, of fresh waters and controlling factors of the regulation of that physiology. That understanding is essential for effective management and addressing practical problems of water quality and how water quality is influenced by human-induced changes.

For decades, the operation of aquatic ecosystems has been treated mostly in terms of the higher trophic levels and couplings among largely the animal biota. Many fundamentals of modern ecology are founded on the superbly detailed study of interactions among animals and their supporting resources. This perspective is quite inadequate, however, as we find it essential to evaluate entire ecosystems, not just a small portion of them of potential interest to human culture, for effective management and utilization of the whole. Primary areas of that integrative perspective include: (a) Emphasis on *rates of metabolism and regulation of rates of biogeochemical cycling by that metabolism.* It is essential to understand how species interactions alter rates of energetic flows and storage and of biogeochemical cycling, rather than maintain the prevailing focus on how feeding relationships influence individual population or community growth and reproduction. Of course the processes are related and coupled, but present imbalances of understanding are conspicuous. (b) It will be apparent that *material and energy flows and cycling are regulated largely at the biochemical and microbial levels.* Although many voids exist in understanding at these latter levels, their roles are dominant at the ecosystem level. The ensuing syntheses represent an initial synopsis.

VIII. SUMMARY

1. Over 99% of the immense amount of water of the biosphere occurs in the oceans and polar ice deposits. The turnover time of this water is very long (i.e., the rate of renewal is very slow).
2. Much of the remaining water occurs in inland waters, where renewal times are much shorter than in marine systems (Table 1-1).
3. Finite water resources are being exploited and degraded at an accelerating rate by the activities of humankind.
   a. Demands upon surface and groundwater supplies result from both population growth and the expanding utilization and consumption because of technological growth. This demotechnic growth results in expanded utilization of freshwater and other finite environmental resources. Biospheric degradation results from both biological population expansion and growth of technological production and consumption.
   b. Unless the demands of exponential demotechnic growth upon fresh waters are rapidly controlled, a freshwater crisis at a global level is imminent. The severity of the freshwater crisis will exceed that of past and contemporary resource-utilization problems. The problems are acutely apparent in many parts of the world and will accentuate in the twenty-first century.
4. *Limnology* is the study of the structural and functional relationships and productivity of organisms of inland aquatic ecosystems as they are regulated by the dynamics of their physical, chemical, and biotic environments.
5. Examination of limnological understanding can be approached by an examination of differences in physical and biotic properties among different lake, reservoir, and river ecosystems. However, among the array of dynamic environmental factors influencing metabolism, growth, and reproductive capacities of organisms, functional commonality is found amidst process rates and controls of metabolism, energy fluxes, and material cycling.

2
WATER AS A SUBSTANCE

I. THE CHARACTERISTICS OF WATER

Water is the essence of life on earth and totally dominates the chemical composition of all organisms. The ubiquity of water in biota, as the fulcrum of biochemical metabolism, rests on its unique physical and chemical properties.

The characteristics of water regulate lake metabolism. The unique thermal–density properties, high specific heat, and liquid–solid characteristics of water allow the formation of a stratified environment that controls extensively the chemical and biotic properties of lakes. Water provides a tempered milieu in which extreme fluctuations in water availability and temperatures are ameliorated relative to conditions faced by biota in aerial life. Coupled with a relatively high degree of viscosity, these characteristics have enabled biota to develop many adaptations that improve sustained productivity.

A. Molecular Structure and Properties

The unique properties of water center upon its atomic structure and bonding and the unusual association of water molecules in solid, liquid, and gaseous phases. In equilibrium, the nuclei of a water molecule form an isosceles triangle, with a slightly obtuse bond angle of 104.5° at the oxygen nucleus (Eisenberg and Kauzmann, 1969). The bond length from the center of the oxygen atom to that of each hydrogen atom is 0.96×10^{-8} cm. The nuclei of the molecules are in a continual state of vibration. As the molecule resonates, an electrical dipole moment occurs, such that a complex wave function results in a valence electron density pattern in which electron density is highest near the atoms and along the bonds (Fig. 2-1). The electronic charge of the molecule is not restricted to the planar configuration depicted, but is distributed in multiple directions.

Although most (ca. 56%) water molecules have balanced valances, excitation and ionization do occur. The most common ionized state is characterized by removal of one of the hydrogen atoms, which is charged positively and is functioning essentially as a proton. Higher ionization potentials (energy needed to remove electrons from the molecule) are needed to reach further ionization states of water and occur less frequently. The weak coulombic characteristics of the bonding of hydrogen atoms to the weakly electronegative oxygen atom result in both ionized and covalent states that simultaneously maintain the integrity of water. Water is nearly the only known compound that possesses these characteristics.

The density properties of water that are so germane to limnology and life in water result from the aggregation and bonding characteristics of the water molecules. It is instructive to look first at the structure of ice, whose physical properties are understood better than those of liquid water. Every oxygen atom is at the center of a tetrahedron formed by four oxygen atoms, each about 2.76×10^{-8} cm distant (Eisenberg and Kauzmann, 1969; Horne, 1972). Every water molecule

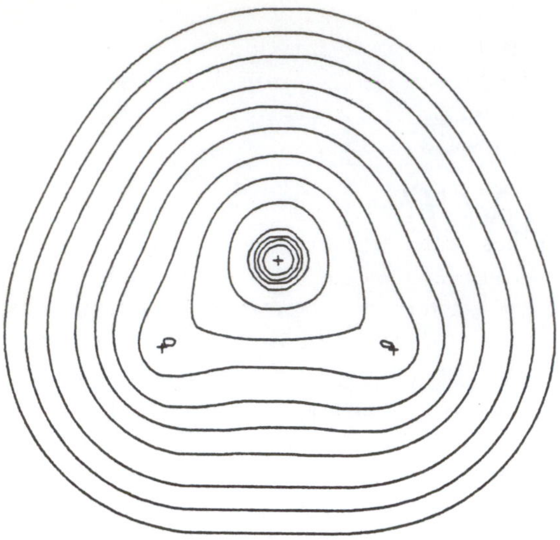

FIGURE 2-1 Contour map of the electronic charge distribution of the water molecule in the plane of the nuclei (positions indicated by crosses). Contours increase in value of atomic units (au) (1 au = 67 e^- per nm^3) from the outermost contour in steps of 2×10^n, 4×10^n, 8×10^n. The smallest contour value is 0.002, with n increasing in steps of unity to yield a maximum value of 20. (Unpublished figure courtesy of Dr. R. F. W. Bader, Department of Chemistry, McMaster University.)

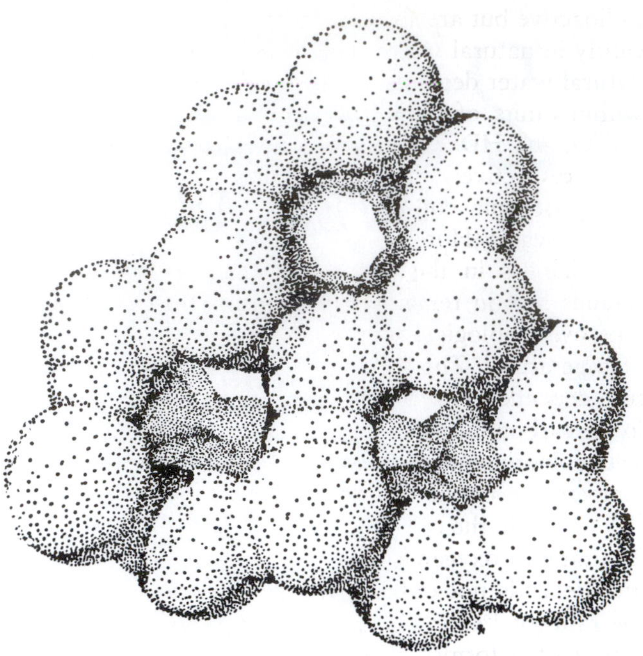

FIGURE 2-2 A diagrammatic representation of an ice crystal showing the van der Waals radii of the atoms and open voids between the aggregated molecules. (From Pimental, G. C., and McClellan, A. L.: *The Hydrogen Bond*. San Francisco, W. H. Freeman and Co., copyright © 1960.)

is hydrogen-bonded to its four nearest neighbors; its O—H bonds are directed toward lone pairs of electrons on two of these adjacent molecules and form two O—H—O hydrogen bonds. In turn, each of its lone pairs is directed toward an O—H bond on one of the other adjacent molecules and forms two O—H—O hydrogen bonds. This arrangement leads to an open lattice in which intermolecular cohesion is great (Fig. 2-2). This structure results in both parallel and perpendicular voids among the molecules and creates an open tetrahedral structure that permits ice to float upon liquid water.

Molecules in liquid water near 0°C experience about 10^{11} to 10^{12} reorientational and translational movements per second, whereas ice molecules near 0°C experience only about 10^5 to 10^6 movements per second. Increasing the temperature of water increases the rate of reorientations and molecular displacements and results in decreased viscosity, decreased molecular relaxation times, and increased rates of self-diffusion. By virtue of their dipolar nature, water molecules interact to form quasi-stable polymers.

We have viewed here water molecules as diffusionally averaged structures both in vibrational time and in space between molecules. However, variations in temperature change the intermolecular distances. Thermal increases result in increased agitation of the molecules, which distorts and breaks down the hydrogen-bonded networks. As the ice melts, increased bond dislocation and rupture occur, which disrupt and fill in the open spaces of the ice lattice structure. The result is an increase in density, and this effect predominates and reaches a maximum at 4°C. As water is heated above 4°C, intermolecular vibrations increase in amplitude and increase interatomic distances. The result is expansion of the liquid and decreased density. It is a characteristic of water that minimum volume, hence maximum density of water, occurs at 3.98°C. At this point, competition between the negative configurational contribution of hydrogen bonding and its positive vibrational contribution is maximized.

B. Isotopic Content of Water

Known isotopes of hydrogen are ^1H, ^2H or D (deuterium), and ^3H (tritium). Tritium is radioactive, with a half-life of 12.5 years, sufficiently short that after its decay to ^3He, most of which is lost from the atmosphere into space, concentrations of tritium in uncontaminated natural water are very low (approximately 1 ^3H atom per 10^{18} ^1H atoms) and do not accumulate. Six isotopes of oxygen are known: ^{14}O, ^{15}O, ^{16}O, ^{17}O, ^{18}O, and ^{19}O. The isotopes ^{14}O, ^{15}O, and ^{19}O are radioactive but are short-lived and do not occur significantly in natural water. The precise isotopic content of natural water depends on the origin of the sample but, within limits of variation, the abundances of $H_2{}^{18}O$, $H_2{}^{17}O$, and $HD^{16}O$ are 0.20, 0.04, and 0.03 mole percent, respectively (Eisenberg and Kauzmann, 1969). Other isotopic combinations are exceedingly rare in natural water (Hutchinson, 1957).

Changes in the ratios of these isotopes in compounds and in remains of organisms formed within a lake over geological time permit an estimation of paleotemperatures. The technique is based on the temperature dependence of fractionation of the oxygen isotopes in the carbon dioxide–water–carbonate system (Stuiver, 1968). During slow precipitation of the carbonate, the temperature of the solution is reflected in the ratio of ^{18}O to ^{16}O in the carbonates of sediments and mollusks. The ratio of oxygen isotopes in the precipitated carbonates depends not only on the temperature but also on the ratio of ^{18}O to ^{16}O isotopes in the water in which carbonate is formed. The measured difference in the ratio of ^{18}O to ^{16}O isotopes between a carbonate fossil and a contemporaneous sample is the result of changes in both temperature and composition of water isotopes. Before paleotemperature determinations can be made, an estimate of the changes in ratio of oxygen isotopes of the water is needed. This measurement is more difficult to make in lakes, where variations over time are greater than in the oceans. However, estimates can be made with reasonable precision.

C. Specific Heat

Specific heat[1] of liquid water is very high (1.0) and is exceeded only by a few substances, such as liquid ammonia (1.23), liquid hydrogen (3.4), and lithium at high temperatures. Other substances of the biosphere, such as many rocks, have much lower specific heats (approximately 0.2). The high specific heat of water, as well as a high latent heat of evaporation (see below), is a function of the relatively large amounts of heat energy required to disrupt the hydrogen bonding of liquid water molecules.

These heat-requiring and heat-retaining properties of water provide a much more stable environment than is found in terrestrial situations. Fluctuations in water temperature occur very gradually. Daily and seasonal extremes in temperature are small in comparison to those of aerial habitats. The high specific heat of water can also have profound effects on climatic conditions of adjacent air and land masses. This ***thermal inertia*** of the hydrosphere (Hutchinson, 1957) occurs on both large and small scales, in relation to the volume of water bodies. Examples are many. The warm Gulf Stream currents of the Atlantic Ocean increase the prevailing temperatures and improve the climate of Western Europe. The prevailing air movements across the Laurentian Great Lakes moderate the climate of Michigan and other states adjacent to and east of the lakes (Leighly, 1942). Milder winters with higher precipitation rates are found in these areas than in the interiors of continents, because the water masses cool and yield heat to the air more slowly than do land masses. Similarly, areas adjacent to large water masses experience moist, cool summer periods. Fall fogs are common over and adjacent to lakes and large rivers, where water vapor from evaporation of the warm lake or river water condenses and stagnates in the cold overlying air. That the lake or river "is steaming" is the description of this phenomenon in common parlance.

The specific heat (thermal capacity) of ice below 0°C is about half (0.5 g^{-1} $°C^{-1}$) that of water and decreases progressively at temperatures below 0°C. The amount of heat required, however, to change ice to liquid water is large (***latent heat of melting*** = 79.72 cal g^{-1}) and is very much larger for disruption of hydrogen bonding in evaporation of water (***latent heat of evaporation*** = 540 cal g^{-1}) or for direct sublimation of ice to water vapor (***latent heat of sublimation*** = 679 cal g^{-1}). Conversely but similarly, a large amount of heat must be lost for the fusion of molecules of 0°C water to ice (***latent heat of fusion*** = 79.72 cal g^{-1}). Because of these properties, large energy inputs are required to melt ice in the spring, and large energy losses are required in order to form ice cover in the winter.

D. Density Relationships

The entire physical and chemical dynamics of lakes and reservoirs and resultant metabolism within them are regulated to a very great extent by differences in density. Throughout the remaining chapters of this book, we will refer repeatedly to the fundamental importance of the unique density properties of water.

The density or specific gravity of pure ice at 0°C is 0.9168 g ml^{-1}, about 8.5% lighter than that of liquid water at 0°C (0.99987). The density of water increases to a maximum of 1.0000 at 3.98°C, beyond which molecular expansion and decreasing density occur at a progressively increasing rate (Fig. 2-3). The density differences are small but highly significant. It is important to examine the difference in density between water at a given temperature and water at a temperature 1°C lower, referred to as the ***density difference per degree lowering*** (Vallentyne, 1957). The magnitudes of this

[1] Specific heat is the amount of heat in calories that is required to raise the temperature 1°C of 1 g mass of a substance.

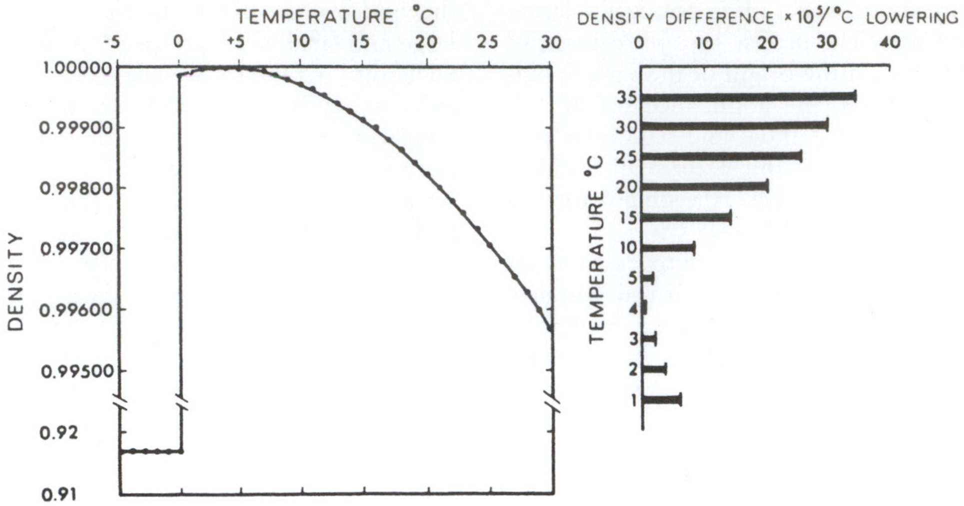

FIGURE 2-3 Density (g ml^{-1}) as a function of temperature for distilled water at 1 atm. The density difference per °C lowering is shown in the right-hand portion of the figure at various temperatures. (Modified from Vallentyne, 1957.)

density difference for water of different temperatures are shown on the right side of Figure 2-3. The density difference per degree lowering increases markedly as the temperature goes above or below 4°C. Physical work is required to mix fluids of differing density, for example, when mixing cream into milk, and the amount of energy input required is proportional to the difference in density. The amount of work required to mix layered water masses between 29 and 30°C is 40 times, and between 24 and 25°C 30 times, that required for the same masses of water between 4 and 5°C.

Density also increases with increasing concentrations of dissolved salts in an approximately linear fashion (Table 2-1). The salinity of most inland waters is within a range of 0.01 and 1.0 g $liter^{-1}$, and usually between 0.1 and 0.5 g $liter^{-1}$. Inorganic salinity of highly saline lakes can commonly exceeds 60 g $liter^{-1}$ and exhibits great seasonal variations (e.g., Rawson and Moore, 1944; International Association of Limnology, 1959; Wetzel, 1964; Eugster and Hardie, 1978; Hammer, 1990; Williams, 1996). In most lakes, however, salinity is very low and varies less than 0.1 g $liter^{-1}$ spatially and seasonally. Salinity-induced variations in density may be small, but they cannot be ignored. Under certain conditions of lake stratification, inorganic salts can accumulate temporarily or permanently (cf. Chap. 6).

TABLE 2-1 Approximate Changes in the Density (g ml^{-1}) of Water with Salt Content

Salinity (parts per thousand, ‰)	*Density (at 4°C)*
0	1.00000
1	1.00085
2	1.00169
3	1.00251
10	1.00818
35 (mean, seawater)	1.02822

After Ruttner (1963).

Salinity also decreases the temperature of maximum density of water at a rate of approximately 0.2°C g^{-1} $liter^{-1}$(‰) increase. The temperature of maximum density of sea water (mean 35 g $liter^{-1}$) is −3.52°C, which does not occur at surface pressures and is below the freezing point (−1.91°C). In most fresh waters, however, changes in the maximum density of water from salinity are very small.

Hydrostatic pressure can be sufficient to compress water and thereby lower the temperature of maximum density. Pressure increases 1 atmosphere per 10 m of depth. The temperature of maximum density decreases about 0.1°C per 100 m of depth (Strøm, 1945). In very deep lakes, temperatures below 4°C are found to be partially, but not totally, related to the relatively high pressures encountered at great depth (Eklund, 1963, 1965; Johnson, 1964, 1966; Bjerke *et al.*, 1979).

E. Viscosity–Density Relationships

The density of water is 775 times greater than that of air at standard temperature and pressure (0°C, 760 mm Hg). The high density of water makes aquatic organisms relatively buoyant against gravitational pull and consequently reduces the amount of energy that an organism must expend to support or maintain its position in the water. Reduction of supporting tissue has occurred in many freshwater animals, particularly among the lower invertebrates, but these adaptations are particularly conspicuous among aquatic vascular plants. Truly submersed angiosperms are good examples; these plants are almost totally limited to fresh waters and have immigrated and adapted to aquatic conditions relatively recently. Modifications have led to marked reductions in vascular tissues, particularly by reduction of the extent of lignification of the xylem and supporting elements, and most of the vascular strands are condensed into a weakly developed central cylinder. Supportive tissues of these hydrophytes develop strongly only in those plants or plant parts that are adapted to aerial existence as surface-floating or emergent forms.

Water viscosity decreases as temperature increases. Water viscosity doubles as the temperature is lowered from 25 to 0°C. The viscosity of water offers approximately 100 times more frictional resistance to a moving organism or particle than air does and is dependent upon the surface exposure area, the speed, and the temperature and chemical composition of the fluid. As will be discussed further on, organisms with locomotion must expend considerable energy to overcome changes in viscosity. Sinking rates and distribution of passive organisms, such as planktonic algae or sedimenting particles, are influenced by density-related changes in viscosity.

F. Surface Tension

The quasi-polymeric bonding properties of liquid water molecules discussed earlier are disturbed at the interface with air. At the interface plane, the molecular attractions are unbalanced and exert an inward adhesion to the liquid phase. The result is an interface surface or film under tension. The surface tension at the air–water interface of pure water is higher than that for any other liquid except mercury. Surface tension decreases with increasing temperature and increases slightly with dissolved salts (Table 2-2).

The surface tension of water is reduced markedly by the addition of dissolved organic compounds. This phenomenon is seen in aquatic habitats wherever dissolved organic matter concentrates (Table 2-2). For example, surface tensions of relatively pristine waters containing few microorganisms differ little from those of distilled water. Bog lakes, which are heavily stained with dissolved organic compounds, exhibit depressions in surface tensions of 6 to 7 dynes cm^{-1} (range 0–20). Where growth of floating algae or of higher aquatic plants is particularly great, the surface tension is depressed by about 20 dynes cm^{-1}. Later on, we will demonstrate that concentrations of dissolved organic compounds are high in more productive waters. Moreover, natural populations of algae and submersed angiosperms secrete large quantities of organic compounds during active photosynthesis, as well as during senescence and lysis. The effects of these organic compounds on surface tension can vary with the dominance of photosynthetic organisms. For example, extracellular organic compounds released from green algae and cyanobacteria lowered surface tensions to a greater extent than those from diatoms (Nägeli and Schanz, 1991). Organic pollutants introduced into fresh waters can also markedly lower surface tensions (Jarvis, 1967; Jarvis *et al.*, 1967).

TABLE 2-2 Surface Tension of Water with Changes in Temperature and Its Depression in Natural Waters under Various Conditions

Temp. (°C)	*Water tension (dynes cm^{-1})*	*Condition*	*Depression of surface tension range (dynes cm^{-1})*
Pure water			
0	75.6	Oligotrophic lakes	0–2
5	74.9	Eutrophic lakes	0–20
10	74.4	Bog lakes	0–20
15	73.5	Lake water with foam	2–9
20	72.7	Near floating-leaved angiosperms	5–20
25	72.0	Near submersed angiosperms	1–2
30	71.2	During a cyanobacterial bloom	0–20
35	70.4	Open sea	<1
40	69.6	Plymouth Sound, near muddy beach	6–20
		Harbor, heavy boat traffic	$15 \geq 20$
Sea water			
ca. 5	75.0		

From data of Adam (1937), Hardman (1941), and Goldacre (1949).

The air–water interface forms a special habitat for organisms adapted to living in or on the surface film. This community is collectively referred to as ***neuston*** and will be discussed in some detail further on (Ch. 8). The surface tension is sufficient to serve as a supporting surface for many organisms of considerable size (e.g., gyrinid and other beetles) and can destroy others, such as cladoceran microcrustacea, that happen to become caught in the interface through wave action and cannot reenter their normal submersed habitat.

II. SUMMARY

1. Water is a unique substance. Its thermal–density properties, high specific heat, and liquid–solid characteristics influence extensively the physical, chemical, and metabolic properties and dynamics of freshwater ecosystems.
2. The molecular aggregation and bonding characteristics of water result in unique density properties that permit water molecules to form quasi-stable polymers.
    a. The maximum density of water (1.00000 g ml^{-1}) occurs at 3.98°C. Molecular expansion and decreasing density occur at progressively increasing rates both above and below 4°C. The density of ice at 0°C (0.9168 g ml^{-1}) is significantly less than that of liquid water at 0°C, and ice consequently floats.
    b. The density difference between water at a given temperature and 1°C lower increases markedly at temperatures above and below 4°C. The amount of energy required to mix water of differing densities increases proportionally to the density differences.
3. The specific heat of liquid water is very high. As a result, thermal conditions within standing waters change slowly, are more stable, and permit a longer growing season than is commonly found in aerial habitats. The heat-absorbing and heat-retaining properties of bodies of water also modify the climate of adjacent land masses.
4. Water viscosity is far greater than the viscosity of air and is inversely proportional to temperature. Aquatic organisms have adapted to the more buoyant aquatic environment and to the resistance water imposed on locomotion; their distribution is often influenced by thermally induced stratification of water masses of differing density and viscosity.
5. High surface tension occurs at the air–water interface. Surface tension decreases with increasing temperature, salinity, and concentrations of organic compounds.

3

RIVERS AND LAKES—THEIR DISTRIBUTION, ORIGINS, AND FORMS

The amount of fresh water on earth is very small in comparison to the water of the oceans, but the fresh waters have much more rapid renewal times (cf. Table 1-1). On a volumetric basis, fresh water is concentrated in large, deep basins of several great lakes. About 20% of surface fresh water is contained in Lake Baikal of Siberian Russia. The number of individual depressions of smaller lakes and reservoirs is extremely large, however, and most of these lakes are located in temperate and subarctic regions of the Northern Hemisphere. Most lakes are shallow and geologically very transient.

Catastrophic events of glacial, volcanic, and tectonic activities have aggregated many freshwater lakes into lake districts. The morphometry and geological substrata of the drainage basins are very important because these properties influence sediment–water interactions, the resultant productivity, and the significance of littoral productivity to that of the river and entire lake ecosystem. Shallow lakes, which have more sediment area per unit of water volume, are generally more productive than deep lakes, and a greater proportion of their total productivity is attributable to communities of the littoral and wetland areas of the land–water interface.

I. DISTRIBUTION OF FRESH WATERS

Inland waters cover less than 2% of the earth's surface, approximately 2.5×10^6 km^2. About 20 lakes are extremely deep (in excess of 400 m), and a significant portion of the world's fresh water is contained in Lake Baikal, Siberian Russia (A = 31,500 km^2; z_m = 1620 m; $\bar{z}$ = 740 m), the deepest lake of enormous volume (23,000 km^3). Like Lake Baikal, nearly all other extremely deep lakes are tectonic or volcanic in origin or have formed from fjords that have subsequently become fresh. Although a few lakes of glacial origin, such as the Laurentian Great Lakes and Great Slave Lake of Canada, have basins of great depth, few exceed a depth of 300 m. Most of the very deep lakes are found in

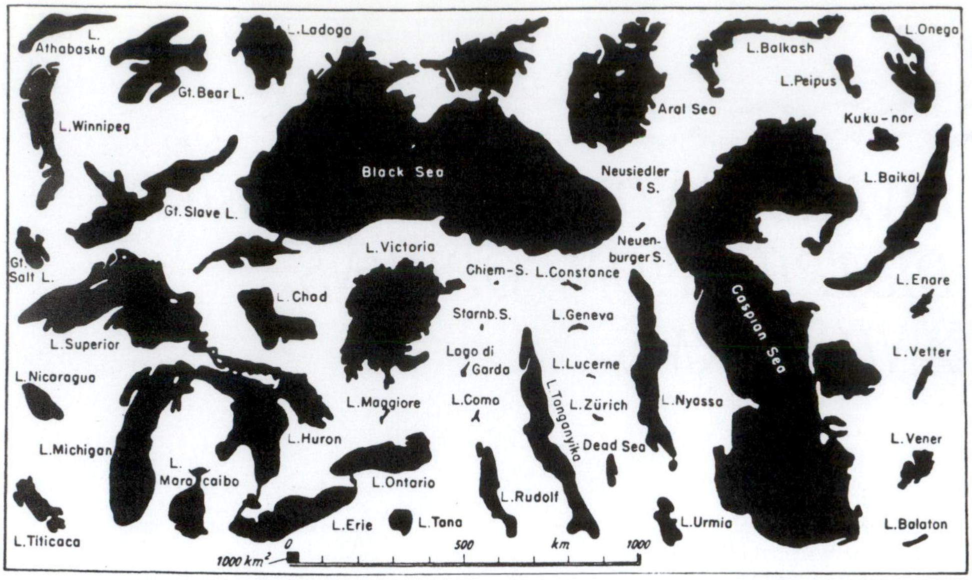

FIGURE 3-1 An approximate comparison of the surface areas of many of the larger inland waters of the world, all drawn to the same scale. The Aral Sea has experienced catastrophic reductions in area (less than half of that depicted here) because of diversions of water for agriculture. (After Ruttner, F.: *Fundamentals of Limnology.* Toronto, University of Toronto Press, 1963.)

mountainous regions along the western portions of North and South America, of Europe, and in the mountainous regions of central Africa and Asia. Lakes Baikal in Asia and Tanganyika in Africa are the only lakes known to have maximum depths in excess of 1000 m and mean depths over 500 m.

On an areal basis, only a few lakes are of very large surface area; most of these are compared in Figure 3-1. If we exclude the Black Sea, which is an inland water really connected to the ocean, the Caspian Sea, a large (436,400 km^2) salt lake, is the largest inland basin separated from the ocean. The Laurentian Great Lakes of North America—lakes Superior, Huron, Michigan, Ontario, and Erie—constitute the greatest continuous mass of fresh water on earth, with a collective area of 245,240 km^2 and a volume of 24,620 km^3. Lake Superior has the greatest area (83,300 km^2) of any purely freshwater lake. Some 253 lakes have a surface area greater than 500 km^2, and nearly half (48%) occur in North America, largely above 40°N latitude

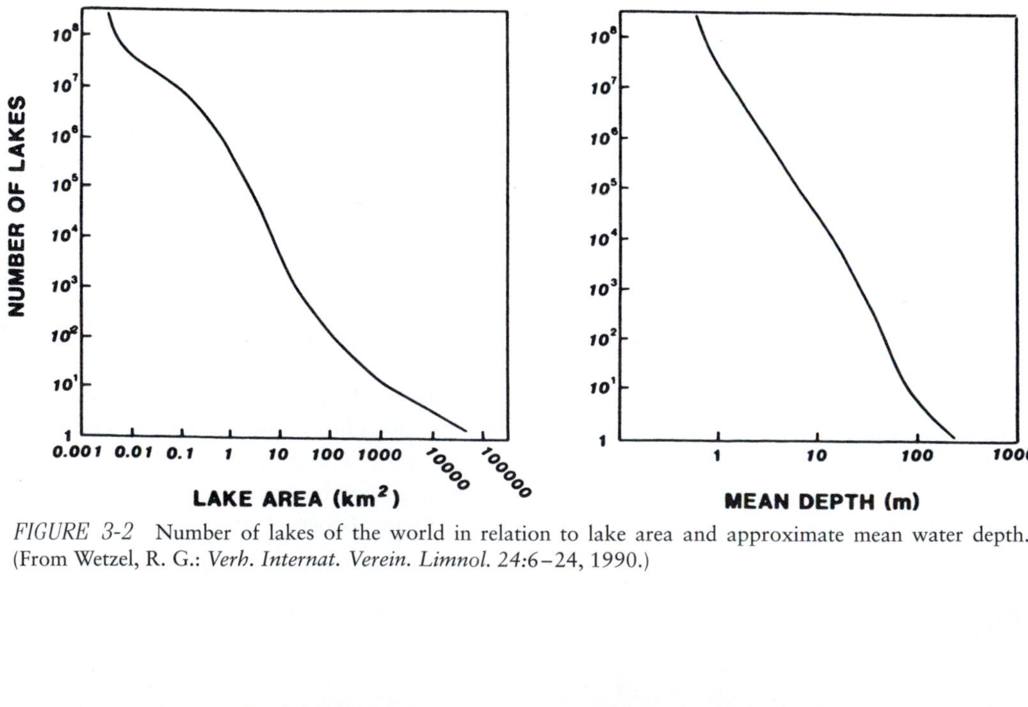

FIGURE 3-2 Number of lakes of the world in relation to lake area and approximate mean water depth. (From Wetzel, R. G.: *Verh. Internat. Verein. Limnol.* 24:6–24, 1990.)

(Herdendorf, 1984). About 75% of these large lakes contain fresh water. The estimated total of these large lakes (179,000 km^3) constitutes over 75% of the total inland surface water of the world.

Although these very large lakes constitute a major freshwater resource, most lakes are much smaller. Most lakes are of catastrophic origin, formed by glacial, volcanic, or tectonic processes. Consequently, many lakes are localized into lake districts in which large numbers of lake basins are concentrated. Glacial activity during the most recent period of major ice advance and retreat created literally millions of small depressions that subsequently were filled with water and modified. Thus large numbers of lakes are found in the Northern Hemisphere, where large land masses of North America and Europe–Asia interfaced with glacial movements. Consequently, numerous small, shallow basins are found throughout the arctic, subarctic, and northern temperate zones (Fig. 3-2).

More recently, humans have created large numbers of reservoirs and ponds (see Chaps. 20 and 26). Most of both natural and artificial lakes are small and relatively shallow, usually <20 m in depth (Fig. 3-2). A vast number of lakes of the subarctic regions are very shallow, with mean depths of about a meter. Because of these morphometric characteristics, a large proportion of the lake volume is exposed to and interacts with the chemical and metabolic processes of soil and sediments. A great proportion of shallow lakes possess properties necessary for the development of sessile littoral flora, which generally results in markedly increased overall productivity.

II. RUNNING WATERS: LOTIC ECOSYSTEMS

Streams and rivers within their drainage basins are central to surface water ecosystems. A number of characteristics differentiate running water from lake ecosystems. Flowing freshwater environments are called ***lotic*** (*lotus,* from *lavo,* to wash) for obvious reasons of unidirectional water movement along a slope in response to gravity. Lotic ecosystems are contrasted to ***lentic*** (*lenis,* to make calm) or lake ecosystems. As already discussed (p. 4), most lakes are open and have distinct flows into, through, and out of their basins. Throughflows, termed the ***water renewal rates,*** are often variable and very slow in lakes but are continuous. The distinction between lakes and running waters focuses on the relative residence times of the water. The importance of variable but continuous and rapid throughput of water and materials contained in it is evident in the biology of most organisms inhabiting running waters. When the energy of flowing water is dissipated, as in the transitional zone of reservoirs, the change to lentic characteristics is rapid.

The directional movement of water is a fundamental property of lotic ecosystems. Dissipation of energy from moving masses of water affects the morphology of streams, sedimentation patterns, water chemistry, and biology of organisms inhabiting them.

Rivers constitute an insignificant amount (0.001 or 0.1%) of the land surface. Only 0.0001% of the water of the Earth occurs in river channels. In spite of these low quantities, running waters are of enormous significance to humans. Erosion moves large amounts of dissolved and particulate matter from the land to the sea. Small streams and rivers often flow to lakes and eventually large river systems. Twenty or so major river systems exceed 2000 km in length and export a large percentage of the eroded materials from major continental land masses (Table 3-1).

TABLE 3-1 Catchment Size and Drainage from Selected Major River Basins[a]

Rivers by continent	*Drainage area (10^3 km^2)*	*Mean annual flow (m^3 s^{-1})*
North America	20,700	191,000
Colorado	629	580
Mississippi	3222	17,300
Rio Grande	352	120
Yukon	932	9100
South America	17,800	336,000
Amazon	5578	212,000
Magdalena	241	7500
Orinoco	881	17,000
Parana	2305	14,900
San Francisco	673	2800
Tocantins	907	10,000
Europe	9800	1,000,000
Danube	817	6200
Po	70	1400
Rhine	145	2200
Rhône	96	1700
Vistula	197	1100
Africa	30,300	136,000
Congo	4015	40,000
Niger	1114	6100
Nile	2980	2800
Orange	640	350
Senegal	338	700
Zambezi	1295	7000
Asia	45,000	435,000
Bramahputra	935	20,000
Ganges	1060	19,000
Indus	927	5600
Irrawaddy	430	13,600
Mekong	803	11,000
Ob-Irtysh	2430	12,000
Tigris-Euphrates	541	1500
Yangtze	1943	22,000
Yellow River (Huang Ho)	673	3300

[a] Data from Szestay (1982).

TABLE 3-2 Drainage Network Characteristics for a Hypothetical Drainage Basin[a]

Order	*Number of streams*	*Average length (km)*	*Average drainage Area* (km^2)	*Average discharge (liters* s^{-1})
1	200,000	0.02	0.00018	0.005
2	65,000	0.03	0.00091	0.025
3	20,000	0.06	0.00414	0.12
4	5500	0.16	0.0129	0.36
5	1500	0.40	0.0906	2.5
6	400	1.0	0.388	11
7	150	2.4	2.20	62
8	40	5.6	9.06	250

[a] After Wetzel and Likens (1991).

III. MORPHOLOGY AND FLOW IN RIVER ECOSYSTEMS

The continual downgradient movement of water, dissolved substances, and suspended particles in streams and rivers is derived primarily from the land area draining into a given stream channel. The hydrological, chemical, and biological characteristics of a stream or river reflect the climate, geology, and vegetational cover of the drainage basin (Hynes, 1970; Oglesby *et al.*, 1972; Beaumont, 1975; Whitton, 1975; Likens *et al.*, 1977, 1995).

A. Drainage Basin

The ***drainage basin***[1] is the area drained by tributary streams that coalesce into a main channel. Several methods have been used for ordering the tributary streams in a drainage network (reviewed by Gregory and Walling, 1973; Gordon *et al.*, 1992). The Horton-Strahler method (Horton, 1945; Strahler, 1952) is widely used. The smallest permanent stream is designated as the first order, and the confluence of two first-order streams (n) creates a second order ($n + 1$) (Fig. 3-3). The order of the trunk stream is not altered by the addition of a stream of lower order. Only when two tributaries of equal order are joined is the order increased. Deficiencies in this popular method of stream ordering, as well as alternatives, are reviewed by Gordon *et al.* (1992). Stream order is negatively correlated with the logarithm of the number of streams within a drainage basin and positively related to the logarithm of stream length (Table 3-2).

Many patterns of drainage networks are observed (Fig. 3-4). Usually surface features (topographic divides) are used to define drainage basins, although subsurface flows may have different boundaries (phreatic divides), particularly in areas underlain with relatively soluble or permeable rocks. Stream length (L_c, cf. Table 3-3) increases in a relatively constant way with drainage basin area (A) as:

$$L_c = jA^m$$

where j and m are derived from many measurements and average about 1.4 and 0.6, respectively. Thus, as drainage basins increase in size, they tend to elongate (Leopold *et al.*, 1964).

B. Water Influx and Movement

Water from all types of precipitation enters the drainage basins and either moves or is temporarily stored. Precipitation may be intercepted by vegetation

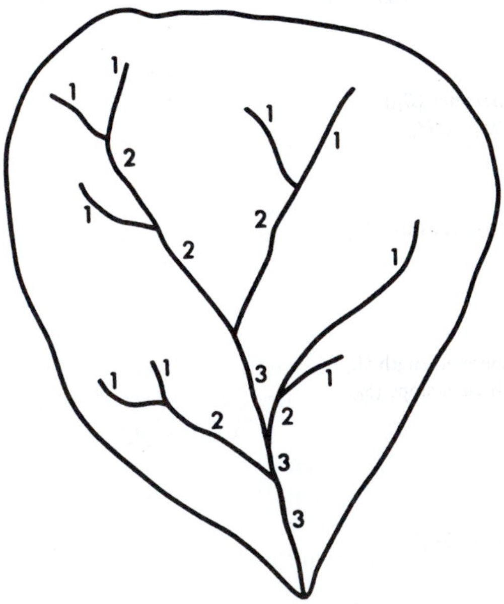

FIGURE 3-3 Stream ordering according to the Horton-Strahler methods. (From Wetzel and Likens, 1991.)

[1] In American usage, ***watershed*** is equivalent to ***drainage basin*** or the European term ***catchment***, all of which refer to the area of land drained by a river system.

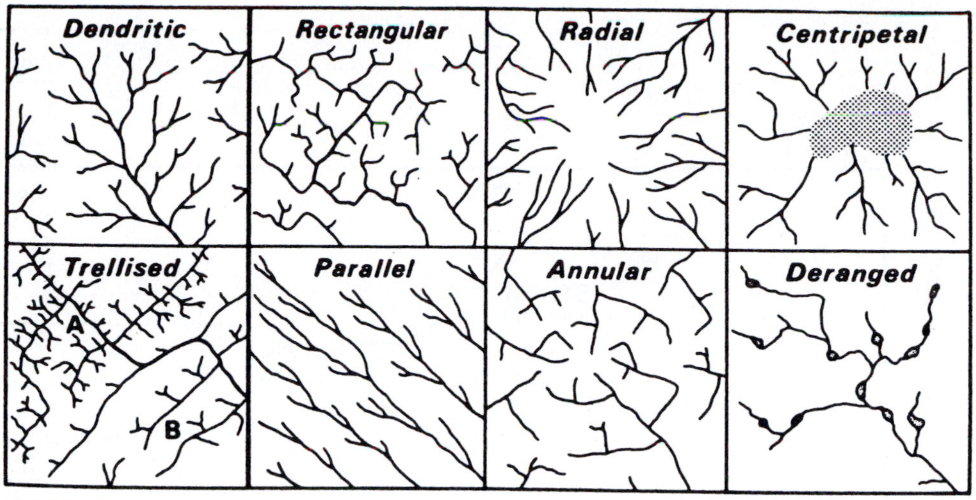

FIGURE 3-4 Common types of drainage networks. (From Gregory, K. J. and Walling, D. E.: *Drainage Basin Form and Process*. 1973. Edward Arnold Ltd., London, by permission.)

TABLE 3-3 Drainage Basin and Channel Characteristics[a]

	Definitions of topographic characteristics
Basin length (L_B)	Straight-line distance from outlet to the point on the basin divide used to determine the main channel length, L_C.
Basin width (W_B)	Average width of the basin determined by dividing the area, A, by the basin length, L_B: $$W_B = A/L_B$$
Basin perimeter (P_B)	The length of the line that defines the surface divide of the basin.
Basin land slope (S_B)	Average land slope calculated at points uniformly distributed throughout the basin. Slopes normal to topographic contours at each of 50- and preferably 100-grid intersections are averaged to obtain S_B. The difference in altitude for the two topographic contours nearest a grid intersection is determined, and the normal distance between these contours is measured.
Basin diameter (B_D)	The diameter of the smallest circle that will encompass the entire basin.
Basin shape (SH_B)	A measure of the shape of the basin computed as the ratio of the length of the basin to its average width: $$SH_B = \frac{(L_B)^2}{A}$$
Compactness ratio (CR_B)	The ratio of the perimeter of the basin to the circumference of a circle of equal area. Computed from A and P_b as follows: $$CR_B = \frac{P_B}{2(\pi A)^{1/2}}$$
Main channel length (L_C)	The length of the main channel from the mouth to the basin divide.
Main channel slope (S_C)	An index of the slope of the main channel computed from the difference in stream-bed altitude at point 10% (E_{10}) and 85% (E_{85}) of the distance along the main channel from the outlet to the basin divide. It can be computed by the equation $$S_C = \frac{(E_{85} - E_{10})}{0.75L_C}$$
Sinuosity ratio (P)	The ratio of the main channel length to the basin length: $$P = \frac{L_C}{L_B}$$

[a] Modified from Winter (1985).

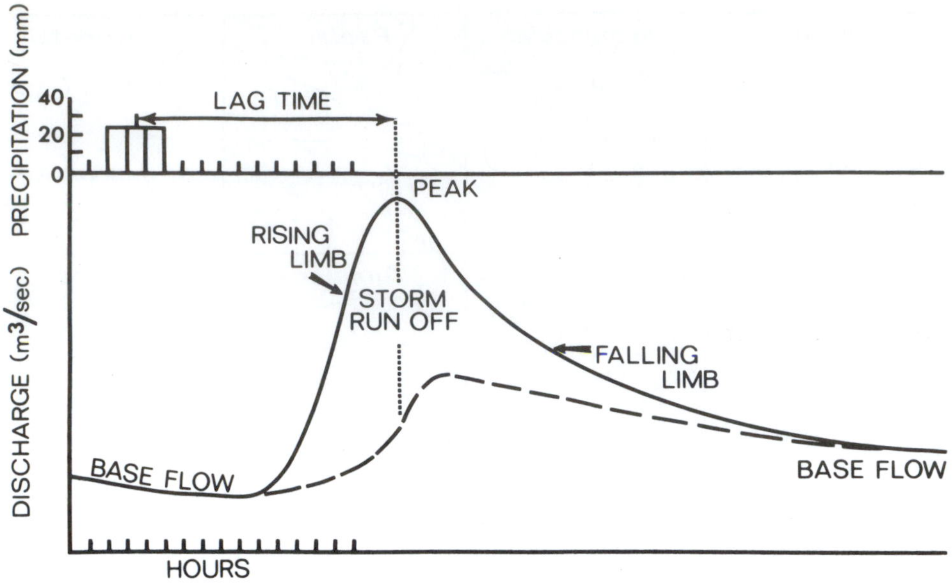

FIGURE 3-5 Stream hydrograph resulting from a single storm precipitation event. (Redrawn from data of Beaumont, 1975.)

and subsequently absorbed and transpired, or it evaporates from the plant surfaces. Some precipitation reaches the litter on the ground and soil surfaces by stem and leaf runoff or by direct soil interception. The rates of absorption of water and movement among soil particles by capillarity are variable with particle size and composition. When water from precipitation or snow/ice melt is added to the soil more rapidly than it can be absorbed, the ***infiltration capacity*** of the soil is exceeded. Excess water will then run downgradient by ***overland flow***.

Most water from precipitation infiltrates the soil. Soils possess variable capacities to store water depending on depth, bulk density, structure, composition, and other factors (Leopold *et al.*, 1964; Todd, 1980). This storage capacity must be exceeded before flow of excess to streams can occur. Water storage capacity is continually made available by losses from evaporation and plant evapotranspiration.

Waterflow through the soil is often channeled by soil structure and voids such as cracks, worm and other animal burrows, and old root channels. Dense or impermeable soil layers impede vertical movement of water and cause ***lateral flow*** of water at intermediate depths of the soil. The chemistry of the water is modified considerably by ionic exchange mechanisms as it passes through soils (Likens, 1984).

The ***surface*** (plane) of the saturated zone of permeable soil is termed the ***water table***. Water held by capillarity in the soil above the water table is called ***vadose water*** and that saturated below the water table, ***ground water***. Ground water flows laterally, intersects stream channels, and provides a moderately stable base flow input to streams. The ***overland flow***, derived from precipitation influxes in excess of soil infiltration capacities, is supplemented by water that infiltrates soils and then flows laterally to the stream channel as ***subsurface storm flow*** to collectively provide the main loading during peak flows or floods (Chorley, 1978; Dunne, 1978).

Stream flow or discharge is the volume of water passing through the cross-sectional area of a stream channel per unit time.[2] Discharge can vary greatly from season to season or within and following a precipitation event. Following a rainstorm or rapid snowmelt period, stream discharge increases to a peak and then decreases (Fig. 3-5). The hydrograph indicates the response of stream discharge to storm events and is useful for comparisons among streams. The hydrograph varies with features of storm events and the drainage basin. The concave slopes of the rising limb reflect the infiltration and storage capacities of the drainage basin. The peak of the hydrograph represents the maximum runoff from the specific inflow precipitation event. The rate of change in the falling limb indicates soil characteristics of flow from storage after precipitation has ceased.

Discharges that exceed the "bankfull" capacities of the river channel are called ***floods***. Floods are essentially random events. The mean annual flood

[2] $Q = Av$, where Q = discharge in $m^3\ s^{-1}$, A = cross-sectional area in m^2, and v = mean velocity in $m\ s^{-1}$.

corresponds to the average value for the largest annual discharge over a series of years. The mean annual flood has a return period of 2.33 years (Leopold *et al.*, 1968). This recurrence interval[3] is the average time interval when the highest discharge of the year will equal or exceed the value of the mean annual flood. Mean bankfull stage recurrence intervals are about 1.5 years.

Stream flow and discharge are directly related to the drainage area. This relationship has been used to predict discharge, particularly for flood events (cf. Strahler, 1964). The relationships between mean annual flood event and size of drainage area are reasonably consistent within different hydrologic regions (Fig. 3-6). Differences reflect variations in elevational gradients and the tendency for drainage basins to elongate as they increase in size.

C. Channel Morphology

The basin or valley trough containing the flowing water is the stream or river ***channel.*** The channel is described physically in terms of length, width, depth, cross-sectional area, slope, aspect, and other parameters (Table 3-3; cf. Leopold *et al.*, 1964; Marisawa, 1968; Wetzel and Likens, 1991; Rosgen, 1996). The channel is usually bordered on one or both sides by a flat area called the ***flood plain.*** Much of the soil of the flood plain is connected hydrologically to the water of the channel. As is discussed later, many chemical changes occur in this water as it moves to the primary channel.

River channels can contain moderate discharges without overflowing the banks. A ***bankfull*** discharge is a flow that fills entirely the cross-sectional stream channel without overflow onto the flood plain. The bankfull stage (Bfs) will be equaled or exceeded on the average of once every 1.5 years (Leopold *et al.*, 1964). Usually, discharges are much smaller than bankfull stages.

River channels rarely flow over an even gradient and are never straight over an appreciable distance. As the energy of flowing water is expended against irregularities in channel morphology, with time there is a tendency to erode the channel to produce a more-even gradient and a more-uniform dissipation of potential energy per unit length of stream. The bending and meandering of the channel approach the theoretical equilibrium of an even energy gradient much more closely than do straight channels. Meandering channels also increase the total volume of the channel and its capacity to transport water over a given distance.

[3] $Tr = \dfrac{N+1}{M}$

where Tr = return period (y), N = number of years of record, and M = rank of an individual event in the set (Leopold *et al.*, 1968).

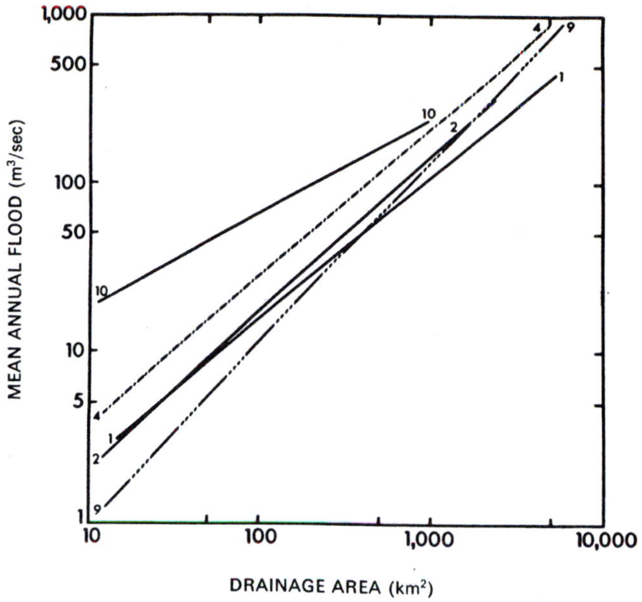

FIGURE 3-6 Relationships between mean annual flood discharge and size of drainage area for rivers in different hydrologic regions of the United States. (Redrawn from data of Tice, 1968.)

Stream and river channels gradually move laterally within the flood plain because of erosion and redeposition of sediment. Old, abandoned portions of river channels are common among most flood plains. The morphology of rivers has been classified and orga.ized on the basis of many criteria (cf. Kellerhals *et al.*, 1976; Kellerhals and Church, 1989). Distinctions can be based on many different criteria of fluvial morphology, such as drainage area, discharge, width–depth ratio, bed material size, channel slope, and others.

Most streams and rivers flow in valleys that exert some lateral and vertical control over the river. Reaches of lotic ecosystems are considered ***constrained*** when the valley floor is narrower than two active stream channel widths (Sedell *et al.*, 1990). Constrained river systems occur where natural geological or human-made features constrict the valley floor and limit lateral mobility and the development of adjacent plant communities. Rivers in valleys without genetic[4] flood plains are also called ***incised*** or ***entrenched.*** Canyons are an extreme case of river incision. Unconstrained river reaches have valley floors that are wider than two active channel widths and lack major lateral geological or artificial constraints (Gregory *et al.*, 1991). Most streams and rivers are of this type and are

[4] A geological term referring to surfaces built or deposited by the present-day river in the course of lateral shifting or flooding (Kellerhals and Church, 1989).

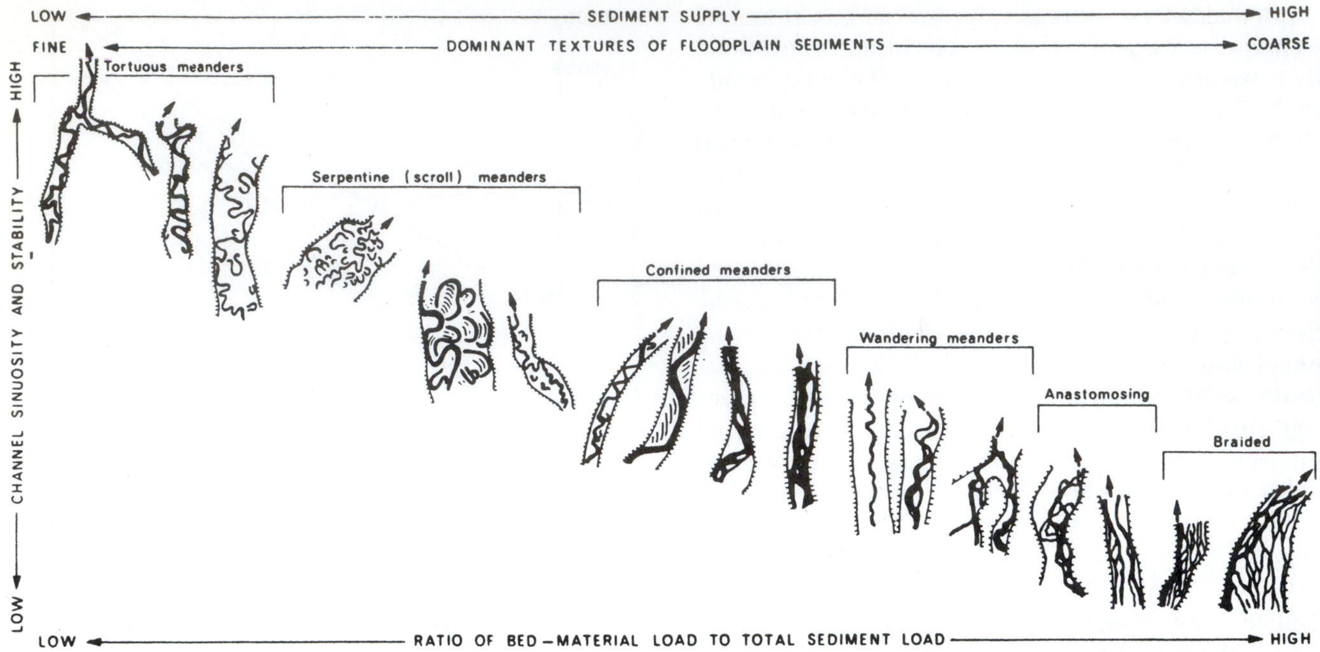

FIGURE 3-7 Form and gradient of alluvial stream channels and the type, supply, and dominant particle sizes of sediments. (Modified from Selby, 1985.)

characterized by active migrations of the channel that expand into extensive flood plains and often form braided channels (Kellerhals and Church, 1989; Pinay *et al.,* 1990; Wissmar and Swanson, 1990). The extent of lateral constraint is relative and quite variable from stream to stream as well as within the same river ecosystem.

The forms of river channels are very diverse (Fig. 3-7) and deviate from four general classes: (a) straight, (b) meandering, (c) braided, and (d) anastomosing. The examples of channel patterns and channel islands depict variations in the great networking of channels and extent of channel anastomosing around channel islands. Channel bars are accumulations of channel bed materials below the average water surface level that often move or are deformed by changing water velocities. Great variations in channel patterns are reviewed in Selby (1985), Kellerhals and Church (1989), Grant *et al.* (1990), Gordon *et al.* (1992), and Rosgen (1994, 1996). It should be rigorously emphasized, however, as correctly noted by Miller and Ritter (1996), that such classification systems function largely as communication tools. Such systems cannot be used effectively to assess stream parameters such as erosion potential, sediment supply, effects of vegetation, and sensitivity to geomorphic disturbances, particularly under different climatic and hydrological regimes.

Although most rivers flow in valleys, exceptions are common. These exceptions include rivers on deltas, fans, or broad plains (Kellerhals and Church, 1989). As discussed further later on (p. 30), deltas are river-deposited continuations of land built by sedimentation out into a body of water. Fans are alluvial, fan-shaped surfaces deposited where a narrow valley emerges onto a broad surface. Broad plains are usually alluvial, lacustrine beds of drained lakes or emergent coastal plains.

IV. GROUNDWATER FLUXES TO LAKES

The dominant inflows to and outflows from lakes occur in surface streams. The sediments of most lake basins have relatively low hydraulic conductivities and lose only small portions of their volume to ground water. Inputs of ground water require the ground water to be elevated above the lake to achieve head pressure, and the soil characteristics must have minimal permeability in connecting strata with the lake basin. Seepage inflow and outflow can fluctuate seasonally depending on the head pressures between the lake and the stagnation point, that is, the point of minimum head pressure along the local flow system boundary (Anderson and Munter, 1981). Groundwater inflows can occur through part of a lake basin, and outflow to recharge the groundwater system through other parts of the lake basin. Alterations in the levels of surrounding water tables or in the water level of the lake can alter the fluxes into or out of the basin. Groundwater inflow rates are more sensitive to seasonal variations in recharge than are outflow rates (Krabbenhoft *et al.,* 1990).

In seepage lakes where groundwater inflows are appreciable, most flow into a lake occurs in the littoral zone whether sediments are present or not (Winter, 1978). If losses of water occur to the groundwater system, in most cases the losses occur in the deeper parts of the lake basin. Water must move through the sediments, thereby considerably reducing the quantity of water lost from the lake compared to that entering the lake in the littoral zone, where sediments of high permeability usually are present. Losses of seepage water through the littoral zone to the ground water are only considerable where no water table mound occurs on the downslope side of the lake. Under these conditions, the water table slopes downward from the shore line, away from the lake. Among glacial terrains, in which most natural lakes occur, seepage from lakes is more common where the water table occurs in end morainal deposits of silty till (Winter, 1981). Much less seepage from lakes occurs in common areas of ground moraine, outwash plains, and flood plains.

A number of analyses have demonstrated the groundwater contributions to water and chemical fluxes of lakes to be negligible (Wetzel and Otsuki, 1974; Malueg *et al.*, 1975; Schindler *et al.*, 1976). In contrast, groundwater contributions to a seepage lake of Minnesota were the largest annual flux of water into (58–76%) and out of (73–83%) the lake during a 12-year study, during which the entire volume of the lake was replaced four times (3-year renewal time) (LaBaugh *et al.*, 1995). Ground water represented as much as half of the annual hydrological input of phosphorus and nitrogen to this lake; the remainder was supplied by atmospheric precipitation.

V. GEOMORPHOLOGY OF LAKE BASINS

The origins of lake basins and their morphometry are of much more than casual interest. The geomorphology of lakes is intimately reflected in physical, chemical, and biological events within the basins and plays a major role in the control of a lake's metabolism, within the climatological constraints of its location. The geomorphology of a lake controls the nature of its drainage, the inputs of nutrients to the lake, and the volume of influx in relation to flushing–renewal time. These patterns in turn govern the distribution of dissolved gases, nutrients, and organisms, so that the entire metabolism of freshwater systems is influenced to varying degrees by the geomorphology of the basin and how it has been modified throughout its subsequent history.

The shape of a lake basin often influences its productivity. Steep-sided U- or V-shaped basins, often formed by tectonic forces, are usually deep and relatively unproductive. In such lakes a proportionally smaller volume of water is contiguous with sediments. Shallow depressions with greater percentage contact of water with the sediments generally exhibit intermediate-to-high productivity.

The following brief résumé on the origin of lakes is based on Hutchinson's (1957) summary of the subject, which was drawn from an array of sources. Particularly seminal was the detailed geomorphologically based evaluation of the origins and classification of lake basins by Davis (1883), which served as a foundation for subsequent treatments. Hutchinson differentiated 76 types of lakes on the basis of geomorphological inception. This detailed classification has been variously modified or amended subsequently (e.g., Horie, 1962 and Timms, 1992, among others) but essentially stands intact. Discussion here is limited to nine distinct groups of lakes, each formed by different processes.

A. Tectonic Basins

Tectonic basins are depressions formed by movements of deeper portions of the earth's crust and are differentiated from lake basins that resulted from volcanic activity. The major type of tectonic basin forms as a result of ***faulting,*** in which depressions occur between the bases of a single fault displacement or in downfaulted troughs (Fig. 3-8). The latter type of basin is referred to as a ***graben,*** and is the mode of origin of many of the most spectacular relict lakes of the world. Foremost among these is Lake Baikal of eastern Siberia, the deepest lake in the world, which has a continuous lacustrine history from at least the early Tertiary period. Lake Baikal, and many other relict lakes, most of which lie in tectonic grabens, are of particular interest because they contain a large number of relict endemic species. For example, of 1200 animal species and at least half that number of plant species known from Lake Baikal, over 80% of those occurring in the open water are endemic to this lake (Kozhov, 1963; Kozhova and Izmest'eva, 1998). Lake Tanganyika of equatorial Africa is a similar deep graben lake (z_m = 1435 m) formed in rift valley displacements along crustal fractures and contains a large number of relict endemic species of plants and animals (Coulter, 1991). Pyramid Lake of Nevada and Lake Tahoe of California–Nevada are familiar examples of American graben lakes.

Tectonic movements causing moderate uplifting of the marine sea bed have isolated several very large lake basins. The relict marine basins of Eastern Europe, which include the Caspian Sea and the Sea of Aral, were separated by the formation of uplifted mountain ranges in the Miocene period. ***Upwarping*** of the earth's

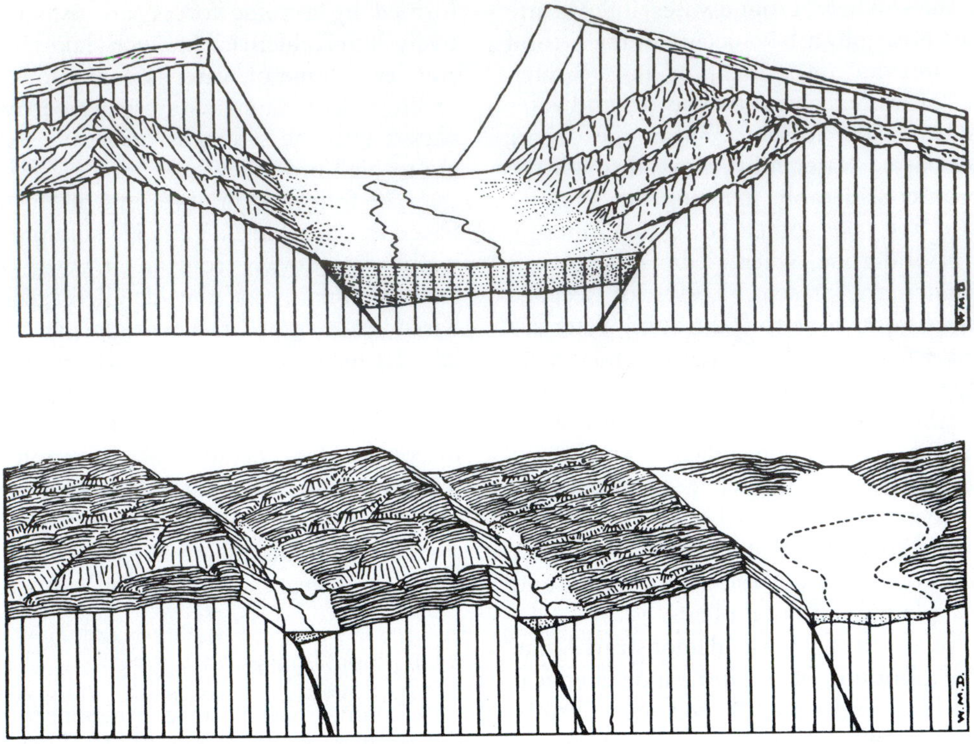

FIGURE 3-8 Tectonic lake basins. *Upper:* In the background, a depressed fault block between two upheaved fault blocks; in the foreground, the same after a considerable period of erosion and deposition. *Lower:* Diagram of the great fault blocks of the northern Sierra Nevada Mountains with the plain of Honey Lake to the east. (From Davis, W. M.: *Calif. J. Mines Geol. 29*:175, 1933.)

crust in lesser degrees also has resulted in the formation of many large lake systems. Lake Okeechobee in Florida resulted from minor depressions (z_m = ca. 4 m) in the sea floor as it uplifted in the Pliocene epoch to form the Floridian peninsula, and it forms one of the largest lakes (approximately 1880 km^2) completely within the United States, exceeded in area only by Lake Michigan. Similarly, Lake Victoria in central Africa resulted from upwarping of the margins of the plateau in which the lake basin lies. This uptilting also created a natural impoundment of the major river system that flowed into the valley. Drainage outlets of the Great Salt Lake basin of Utah were eliminated by uplifted deformations, and a closed lake basin was formed. Upwarping of the earth's crust also contributed, along with the primary glacial scouring activity, to the formation of the Great Lakes of America. As the massive glacial sheet receded, the release of pressure resulted in a significant crustal rebound.

Lake basins are occasionally formed in areas of localized subsidence that result from earthquake activity. Many of these depressions are dry or temporarily contain water, depending on the porosity of the basin material, while others become permanent lakes, usually open basins with outlet drainage.

B. Lakes Formed by Volcanic Activity

Catastrophic events associated with ***volcanic activity*** can generate lake basins in several different ways. As volcanic materials are ejected upward and create a void, or as released magma cools and is distorted in various ways, depressions and cavities are created. If these depressions are undrained, they may contain a lake (Fig. 3-9). Because of the basaltic nature of these lake basins and their drainage areas, which are often very restricted, many lakes associated with volcanic activity contain low concentrations of nutrients and are relatively unproductive.

Small crater lakes are occasionally found occupying cinder cones of quiescent volcanic peaks. However, crater lakes in depressions formed by the violent ejection of magma (Fig. 3-10), or by the collapse of overlying materials where underlying magma has been ejected to create a cavity, termed ***maars,*** are generally small depressions with diameters less than 2 km and result from lava coming into contact with ground water or from degassing of magma. Maars are usually nearly circular in shape and can be extremely deep ($>$100 m) in relation to their small surface area. Basins formed by the subsidence of the roof of a partially emptied magmatic

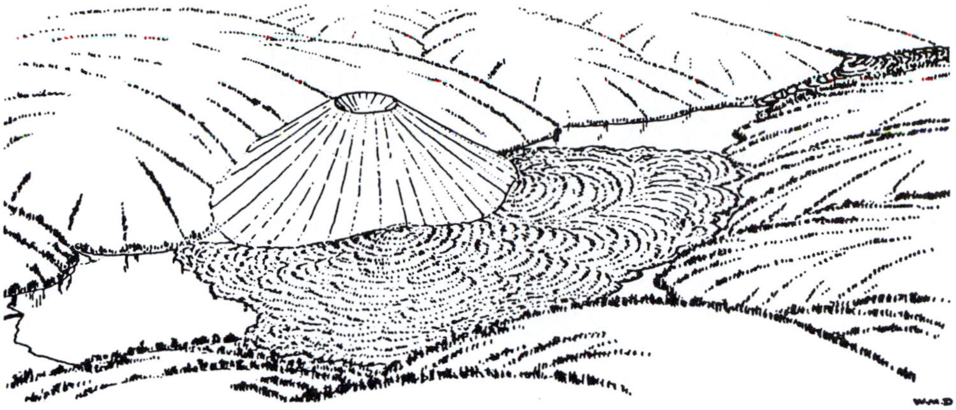

FIGURE 3-9 Volcanic lakes. A caldera lake within the volcanic cone and several lakes within the valleys dammed by lava flows. (From Davis, W. M.: *Calif. J. Mines Geol. 29*:175, 1933.)

chamber are termed ***calderas*** and can be somewhat larger than maars (minimum diameter about 5 km). Among the most spectacular of lakes formed by the collapse of the center of a volcanic cone is Crater Lake, Oregon, with an area of 64 km^2 and a depth of 608 m (seventh-deepest lake in the world). Caldera lakes can be modified in various ways by secondary peaks partially filling the original caldera depressions or by the occurrence of faulting over an emptied magma chamber.

Some volcanic lakes originated by a combination of large-scale volcanic and tectonic processes. In some situations, caldera collapse occurred on such a large scale that extensive portions of the surrounding land subsided in addition to the central portion of the volcano. Such subsidence usually takes place along preexisting fault fractures. Some of the largest lakes associated with volcanic activity were formed in this manner. Many examples of these lakes exist in equatorial Asia and New Zealand (cf. Bayly and Williams, 1973; Larson, 1989).

Lava flows from volcanic activity can form lakes in several ways. As lava streams flow, cool, and solidify, surface lava frequently collapses into voids created by the continued flow of the underlying molten lava. A lake basin may be formed in this manner when the unsupported overstory crust collapses and may be filled when the depression extends below groundwater level. Lava streams also commonly flow into a preexisting river valley and form a dam, behind which a lake can collect (Fig. 3-9). If the dam is of sufficient magnitude,

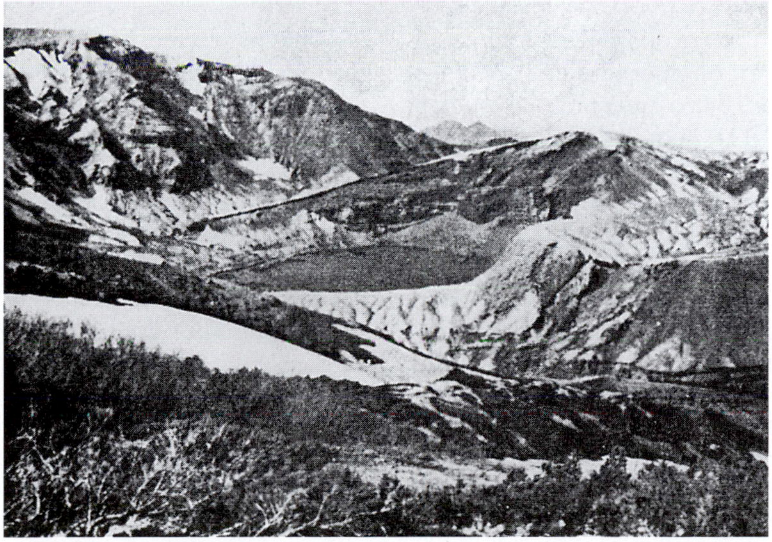

FIGURE 3-10 Lake Okama on Mt. Zao, northern Japan, a caldera lake. (Photo courtesy of I. Saijo.)

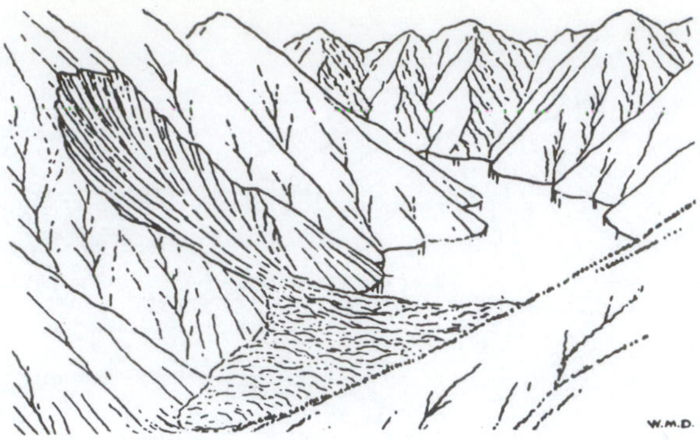

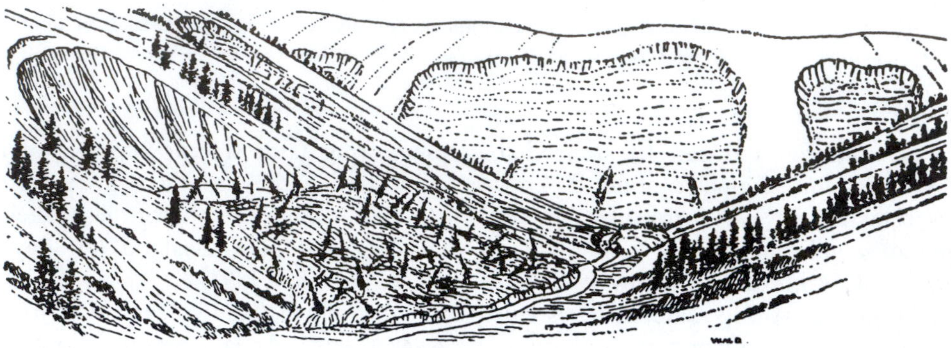

FIGURE 3-11 Lakes formed by a large landslide into a steep-sided stream-eroded canyon (*upper*) and in a hollow behind a recent slide with tilted trees (*lower*). Four mountain landslides are shown in the background. (From Davis, W. M.: *Calif. J. Mines Geol.* 29:175, 1933.)

the entire hydrology of the region can be changed by the resultant reversal of the river system. In some cases, the river flow is forced underground in order to pass the lava obstruction.

C. Lakes Formed by Landslides

Sudden movements of large quantities of unconsolidated material in the form of landslides into the floors of stream valleys can cause dams and create lakes, often of very large size (Fig. 3-11). Such landslide dams may result from rockfalls, mudflows, iceslides, and even flows of large amounts of peat, but they are usually found in glaciated mountains. The landslides usually are brought about by abnormal meteorological events, such as excessive rains acting on unstable slopes. More spectacular slides are occasionally initiated by earthquake activity. Lakes formed behind landslide dams are often transitory and exist only for a few weeks to several months. This short period is due to the fact that unless the slide is very massive, the unconsolidated dam is susceptible to rapid erosion by the effluent of the newly formed lake. Many disastrous floods have resulted from the rapid erosion of the dam material by the effluent, which, once started, can quickly empty the accumulated basin. As with damming by lava flows, if the landslide dam is sufficiently large, the lake can become permanent and reverse the direction of drainage flow from the river valley.

D. Lakes Formed by Glacial Activity

By far the most important agents in the formation of lakes are the gradual but nonetheless catastrophic erosional and depositional effects of glacial ice movements. Land surfaces that are now glaciated include Greenland and several smaller arctic islands, Antarctica, and numerous small areas in high mountains throughout the world. These contemporary glacial activities are small, however, in comparison to the massive Pleistocene glaciation that advanced and receded in four major episodes of activity in the Northern Hemisphere. With the retreat of the last Pleistocene glaciers, an immense number of small lakes were created. Lakes of glacial origin are far more numerous than lakes formed by other processes. The action of glaciers in mountainous regions of high relief usually produces lake basins that are quite different from those that result from the movements of large ice sheets on regions of more mature and gentle relief.

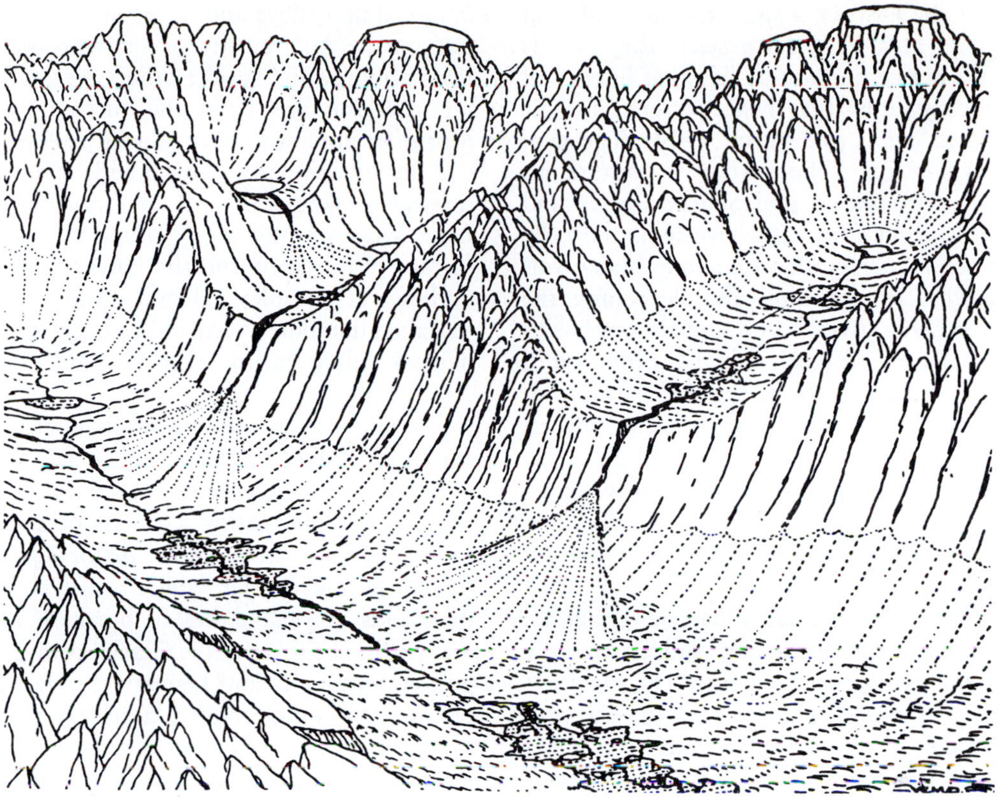

FIGURE 3-12 Several small cirque lakes within a mountain group from which several converging, cirque-headed branch troughs all join the same trunk trough. (From Davis, W. M.: *Calif. J. Mines Geol.* 29:175, 1933.)

A number of lakes, often temporary in nature, occur on the surfaces of, within, or beneath existing glacial ice masses in areas of transitory thaw. For example, numerous meltwater subice lakes several kilometers in diameter occur 3000–4000 m below the Antarctic ice sheet (Oswald and Robin, 1973). In high mountain regions, the fronts and lateral arms of glaciers, as well as terminal morainal deposits supported by the ice, often will function as effective dams in river valleys that originate almost totally from glacial meltwater.

Glacial ice-scour lakes are a vast number of small lakes formed by ice moving over relatively flat, mature rock surfaces that are jointed and contain fractures. The ice-scour lakes are particularly common in mountainous regions where glacial movements have removed loosened rock material along fractures. When the glaciers retreat, the rock basins thus formed fill with meltwater. Such glacial-scour lakes may be found on the upland peneplains of Scandinavia, in the United Kingdom, and in the great Canadian shield region.

A frequently occurring type of ice-scour lake forms in the upper portion of glaciated valleys of mountainous areas, where the valleys are shaped into structures resembling amphitheaters by freezing and thawing ice action (Fig. 3-12). The amphitheaterlike formation is referred to as a ***cirque***, and lakes within such depressions are ***cirque lakes***.[5] Water is held in the cirque depression either by a rock lip above the depression or by morainal deposits. Cirque lakes, which are generally small and relatively shallow (< 50 m), are often found in tandem arrangement within a glaciated trough of a mountain valley, with the higher lake "hanging" above the lower succeeding lake in a stairwaylike fashion. True cirque lakes, which are common to all major mountainous regions, are formed at approximately the snow line in glaciated valleys. When the glaciers extend well below the snow line of constant freezing and thawing, the corrosive action of the ice can form rock basins within the glacial valley. When such valley rock basins form a chain of small lakes in a glaciated valley, resembling a string of beads, they are referred to as ***paternoster lakes***. Where there are mountains adjacent to the sea, as in many areas of Norway and western Canada, ***fjord lakes*** can be formed within narrow, deep basins in glacially deepened valleys.

[5] Sometimes referred to as *tarns*.

The English Lake District, a small group of lakes in a roughly circular area of northwestern England, is worthy of special notice because of the long history of limnological investigations that have centered in this region (reviewed in detail by Macan, 1970). The geological formations of the Lake District consist of an elevated dome of Ordovician and Silurian slates and volcanic intrusions. Rock formations overlying the early uplifted region were later eroded into nine major valleys. In the process of widening and deepening the troughs, glacial erosive activity scoured piedmont basins in each of these valleys. The result was the formation of a series of valleys radiating from the dome, most of which contain one large lake depression and several smaller basins situated in deposits of morainal till.

Where continental ice sheets encountered weak areas in primary basal rock formations, glacial scouring of lake basins occurred on a massive scale in nonmountainous piedmont areas. Great Slave Lake and Great Bear Lake in the central Canadian subarctic region are examples of scouring of preexisting valleys to very great depths by the ice sheet movements. The most impressive example, however, of large rock basins produced by glacial continental ice erosion are the Great Lakes of the St. Lawrence drainage, which collectively form the largest continuous volume of liquid fresh water in the world.

The entire Great Lakes region was covered by continental ice during the glacial maxima. Three important events led to the formation of the contemporary Great Lakes during the retreat of the Wisconsin ice sheet (Fig. 3-13) approximately 15,000 years BP[6] (Hough, 1958). In the early phases of retreat, lakes formed in the Michigan and Huron–Erie basins against the retreating ice lobes and drained southwestward into the Mississippi valley. As the ice sheet repeatedly retreated and advanced over a period of nearly 8000 years, discharge channels from previous scouring were uncovered and altered which resulted in drainage changes to an easterly direction. Another factor that contributed to changes in drainage was the crustal rebound uplifting that took place as the weight of the overlying ice was removed. The flow patterns and configurations of the modern Great Lakes were fixed about 5000 years B.P., except for a lowering of lake levels and diversion of all discharge through the Erie–Ontario basins to the St. Lawrence drainage.

As continental glacial ice sheets retreated in the late Pleistocene, vast amounts of rock debris, moved and incorporated into the ice during former advances, were deposited in terminal moraines and laterally to the lobes of the retreating glacier. These deposits dammed up valleys and depressions in an irregular way and formed lake basins. Some depressions were below later groundwater levels; others were filled by meltwater and drainage from the surrounding topography. ***Morainal damming*** of preglacial valleys, usually at one end but occasionally at both ends, created many of the numerous lakes of the glaciated northern United States. Lake Mendota and many other Wisconsin lakes were formed in this way, and the Finger Lakes of New York underwent complex glacial modification in which morainic damming occurred at both ends of their deep, narrow valley basins (von Engeln, 1961; Bloomfield, 1978).

In regions glaciated by continental ice sheets, a very common process of formation of lakes was associated with the deposition of meltwater outwash left at the border of the retreating ice mass or with blocks of ice buried in this debris. The major features of basins formed in this manner, which accounts for the origin of thousands of small lakes of northern North America, are shown in Figure 3-14. Glacial drift material was deposited terminally and beneath the ice washed out into plains and often contained large segments of ice broken from the decaying glacier. Based on paleolimnological evidence, we know that these blocks of ice, deposited in glacial drift or in the outwash plain, often took several hundred years to melt completely (Florin and Wright, 1969) and resulted in the formation of ***kettle lakes.*** As the ice blocks melted, the morphometry of the resulting depressions was modified variously by the overburden of rock debris (Fig. 3-14c). The ensuing lake basins are highly irregular in shape, size, and slope, corresponding to the irregularities of the original ice blocks. Kettle lakes often exhibit variable underwater relief, with multiple depressions separated by irregular ridges and mounds. These lakes rarely exceed 50 m in depth; their shallow basin is related to the limiting depth of crevasse formation in the fracturing of the terminal glacier.

Most arctic lake basins result from glacial activity, as discussed earlier, or are ***cryogenic lakes*** formed from the effects of permafrost (Hobbie, 1973; Pielou, 1998). The most abundant type of water body found in the Arctic is the shallow pond formed inside of an ice-wedge polygon of raised soil banks that grow above the permafrost from water seepage through cracks in the surface of the ground. Eventually, polygonal networks of ridges are formed that often contain ponds from 10 to 50 m in diameter. Millions of these ponds exist in coastal northern Alaska, Canada, and flat regions of Siberia. The shallow ponds may coalesce into larger ponds, or large amounts of ice deeper in the permafrost may melt, especially if the plant cover is disturbed or destroyed, and thereby form a shallow ***thermokarst lake.***

[6] Before present.

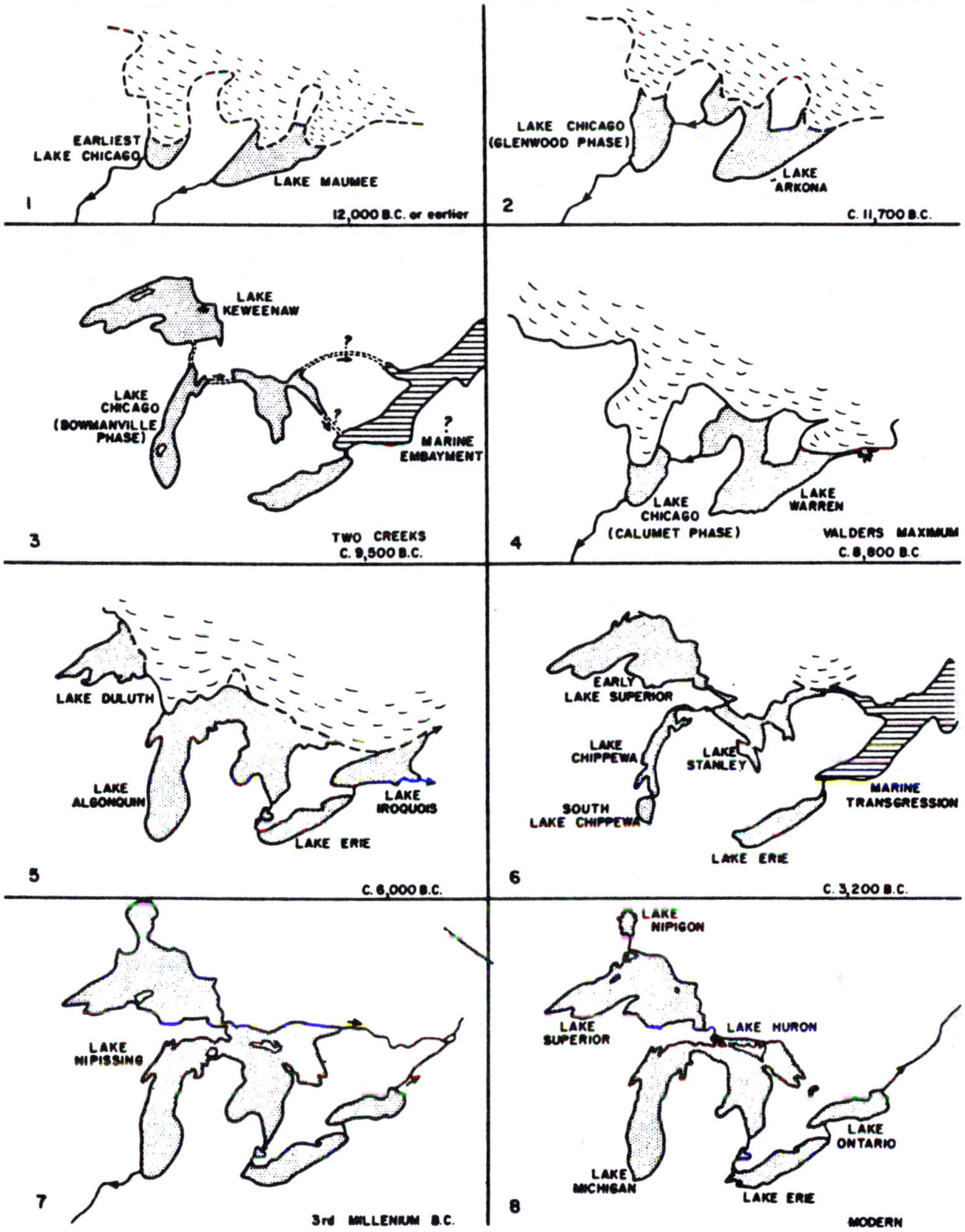

FIGURE 3-13 The history of the Laurentian Great Lakes. (1) Cary substage; (2) Late Cary; (3) Cary-Valders interstadial, low water in eastern basins, marine transgression in Ontario Basin; (4) Valders maximum; (5) post-Mankato retreat; (6) postglacial thermal maximum; (7) Lake Nipissing with triple drainage; (8) modern lakes. (From Hutchinson, G. E.: *A Treatise on Limnology*. Vol. 1, New York, John Wiley & Sons, Inc., 1957, after various sources; see Hough, 1958, for details.)

Some lakes form in the thawed craters of pingos, small conical hills scattered over tundra plains, especially in the North American arctic regions. Water from previously insulated soils can rise by hydraulic pressure through cracks in the permafrost but is overlain with alluvium and tundra vegetation (Pielou, 1998). The water freezes and forms a mound, or pingo. The overburdern soil can rupture and the ice of the depression can thaw to form small ***pingo lakes*** (Likens and Johnson, 1966).

A very large number of arctic thaw lakes are elliptical, with the long axis oriented in a northeast–southwest direction across the prevailing winds. It is apparent, at least among the smaller elliptical lakes of this region of nearly constant wind, that the prevailing system of currents would tend to erode and thaw permafrost at the

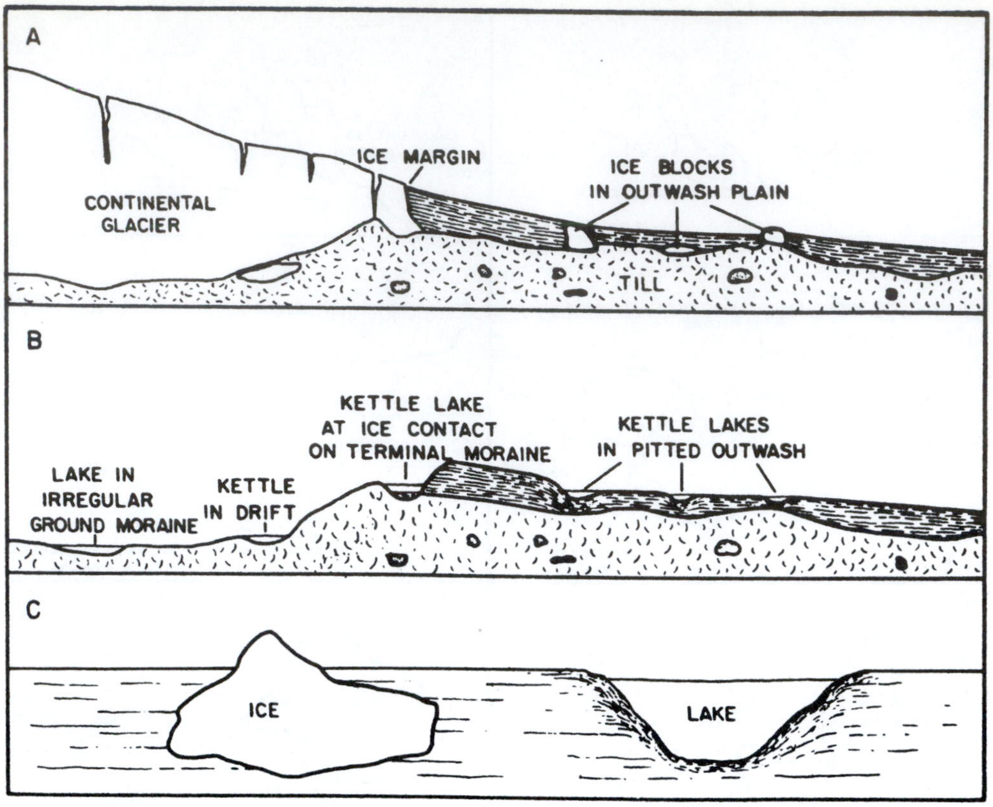

FIGURE 3-14 The formation of various types of kettle lakes. (a) An outwash plain of retreating continental ice containing ice blocks; (b) Lakes formed in the outwash plain and morainal till; (c) Irregular slopes and shelves of a kettle lake formed by deposition of overburden till on an irregular block of ice. (From Hutchinson, G. E.: *A Treatise on Limnology*. Vol. 1, New York, John Wiley & Sons, Inc., 1957, after Zumberge, 1952.)

ends of the long axis of the ellipse lying across the wind (Livingstone, 1954, 1963; Livingstone *et al.*, 1958; Pielou, 1998).

E. Solution Lakes

Lake depressions can be created in any area from deposits of soluble rock that are slowly dissolved by percolating water. Although many rock formations are readily soluble (salts such as sodium chloride [NaCl], calcium sulfate [$CaSO_4$], and ferric and aluminum hydroxides), most ***solution lakes*** are formed in depressions resulting from the solution of limestone (calcium carbonate, $CaCO_3$) by slightly acidic water containing carbon dioxide (CO_2). Solution lakes are very common in limestone regions of the world, notably the karst regions of the Adriatic, especially in the former Yugoslavia, the Balkan Peninsula, the Alps of Central Europe, and in Michigan, Indiana, Kentucky, Tennessee, and particularly Florida in North America.

Solution basins are usually very circular and conically shaped sinks, termed ***dolines,*** which develop from the solution and gradual erosion of the soluble rock stratum. Percolating surface or ground water dissolves limestone most readily at the joints and points of fault fracture through which it drains. Adjacent dolines may eventually fuse to form compound depressions that are more irregular in conformity. Alternatively, the dissolution of limestone frequently occurs in large subterranean caves. Continued erosion of the superstructure by ground water results in its weakening to the point where the roof collapses, which forms a reasonably regular conical doline.

The level of water in the solution basin is often highly variable. Usually, the depressions are sufficiently deep to extend well into the groundwater table and permanently contain water. Other basins that just reach the water table can undergo fluctuations in water level in response to seasonal and long-term variations in groundwater supply.

F. Lake Basins Formed by River Activity

The running waters of rivers possess considerable erosive power that may create lake basins along their courses from elevated land to large lakes or the sea. In the upper reaches of rivers where gradients are steep, excavation by water can produce rock basins that may persist as lakes after the course of the river has been diverted. ***Plunge-pool lakes,*** excavated at the foot of waterfalls, provide a rare but spectacular example of such destructive fluvial action. Several large lakes of the Grand Coulee system of the state of Washington were surely plunge pools shaped by the former interglacial course of the Columbia River system.

A combination of destructive erosional and obstructive depositional processes occurs as rivers flow down more gentle slopes to form many lakes in the river flood plains. Many lakes were formed along large rivers when sediments of the main stream were deposited as levees across the mouths of tributary streams. In this way, the obstruction of the tributary flow continued until the side valley was flooded and a ***lateral lake*** was formed. Lateral lakes are frequently found in tributary valleys along major river systems in all continents, especially in the upper portion of the drainage. The reverse situation of fluviatile dams holding lakes in the main river channel occasionally occurs as a result of deposition by a lateral tributary. If more sediment than the main stream can remove is deposited, the resulting lake may be permanent.

The basins of ***floodplain lakes***[7] contain water throughout the year. When flooded, the flood plain of a river is usually covered continuously with water (Hamilton *et al.,* 1992). As the river level recedes, water drains from the flood plain. Lakes persist in depressions perched above the river or in deep depressions that remain connected with the river.

Three classes of floodplain lakes can be differentiated on the basis of their geomorphological origins. ***Blocked-valley lakes,*** common along the Amazon River, are shaped like drainage valleys and extend inland from the flood plain boundary. Lakes within the boundaries of the fringing flood plain are ***lateral levee lakes*** (Hutchinson, 1957). Two types of levee lakes are common (Hamilton and Lewis, 1990): ***Dish lakes*** are dish-shaped and have low rates of hydraulic throughflow during inundation; ***Channel lakes*** are more riverine, tend to have visible currents during inundation and are frequently in swales among concentric levees of course sediment (scroll bars). Channel lakes have a length-to-width ratio of >5; dish levee lakes have a ratio of <5 (Hamilton and Lewis, 1990).

[7] Floodplain lakes have many regional names. In Australia, such lakes can be termed ***billabongs,*** a word of aboriginal origins that refers to lentic waterbodies on a river flood plain formed by the geomorphic action of the mainstream (Hillman, 1986). In Brazil, *várzea* and *ria* floodplain lakes are differentiated on the basis of suspended particulate and dissolved organic matter loadings from different soil types (Irion and Förstner, 1975).

Where rivers enter the relatively quiescent waters of a lake or the sea, sedimentation results in the formation of deltas, often of very large size. Occasionally, a delta is sufficiently large when entering a long, narrow lake on one side to form a barrier that divides the original lake in half. Much more common, however, are lakes that are formed in the deltas of all major rivers of the world. As the river velocity becomes reduced and sediments are deposited, the water tends to flow around the sediments in a U-shaped pattern where the open end extends seaward. As alluvial deposition extends further seaward, the inward depressions eventually are isolated as shallow lakes, often of large size. Subject to the influence of tides and other water movements of the sea, these ***deltaic lakes*** often receive salt water and frequently are brackish.

As a river meanders within the irregularities of the topography, greater turbulence and erosion occur on the outer, concave side of the river bend, while deposition occurs on the inner, convex side where currents and turbulence are reduced. With time, continued erosion and concavity take place until the U-shaped meander of the river closes in upon itself (Fig. 3-15). The main course of the river cuts a channel through the initial portion of the meander, and levee deposits eventually isolate the loop, referred to as an ***oxbow lake,*** from the river channel. The outer, erosional side of oxbow lakes is usually deeper than the inner, concave side.

G. Wind-Formed Lake Basins

Wind action operates in arid regions to create lake basins by deflation or erosion of broken rock, or by redistribution of sand, and forms ***dune lakes.*** Such lake depressions may be solely or partially the result of wind action, and the water they contain is often temporary and dependent upon fluctuations in climate.

Bayly and Williams (1973) and Timms (1986) differentiate several types of dune lakes based on those found in Australia and New Zealand, although they occur in many other parts of the world as well. Dune barrage lakes form behind sand dunes that are moving inland and blocking a river valley draining toward a coast. These lakes are typically triangular in shape, with the deepest part close to the sand dune, and are found inland in desert regions as well as in coastal areas. If a large amount of deflation occurs so that little or no sand is left on the floor between the trailing arms of parabolic dunes and an impervious rock floor is exposed, a lake may develop. Organic additions from vegetation assist in creating impervious, organically bonded sand–rock in dune depressions. Deflation depressions that are permeable to water can form lakes when they extend below an extensive water table.

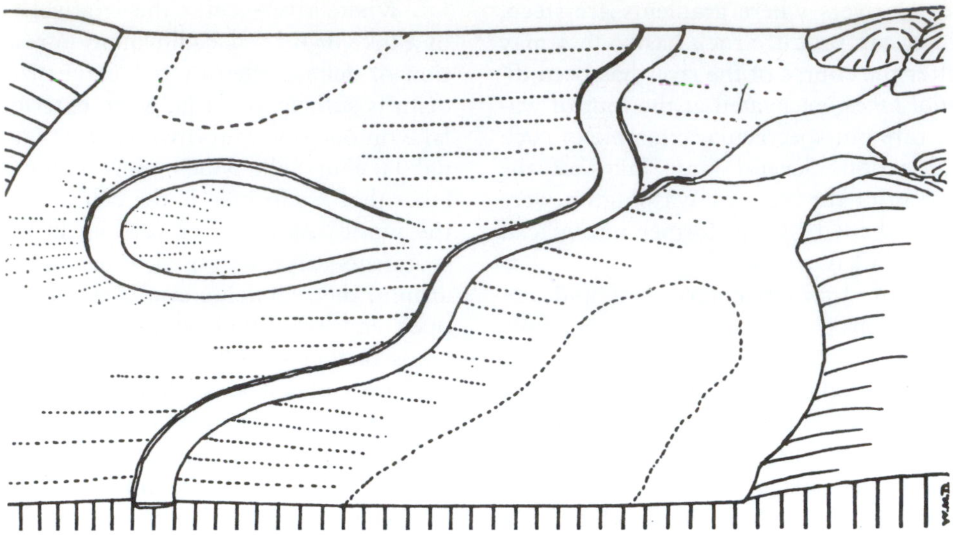

FIGURE 3-15 Diagram of lakes, including an isolated oxbow lake, and sloughs resulting from river activity. (From Davis, W. M.: *Calif. J. Mines Geol.* 29:175, 1933.)

Numerous small, shallow lakes of this type occur along the eastern shore of Lake Michigan among wind-blown depressions of the sand dunes (e.g., Barko *et al.*, 1977). All of these lakes are extremely transitory.

Deflation basins are also common where material is moved and eroded from horizontal strata of rock or clay. The wind-eroded material may be deposited in crescent-shaped mounds downwind or removed completely from the area, which permits the formation of large pans on nearly level areas, often called ***playas.***[8] Deflation basins commonly fill with water during rainy seasons or wet periods and become increasingly saline with evaporation and dry during opposing seasons or dry periods. These ephemeral lakes are common in large portions of Australia, south Africa, endorheic (see p. 46) regions of Asia and South America, and the plains and arid regions of the United States.

H. Basins Formed by Shoreline Activity

When the coast line of a large body of water, such as the sea or a large lake, possesses some irregularity or indentation, the potential exists for the formation of a bar across the depressions to form a ***coastal lake.*** A longshore current flowing along the shore line and carrying sediment will, on encountering a bay, deposit the material in the form of a bar or spit across the mouth of the indentation (Fig. 3-16). Often the spit can eventually separate the bay from the sea or large lake to form a coastal lake.

Marine coastal lakes commonly result from bar formation across the mouths of old estuaries that have been inundated by rising water levels or slight subsidence of the land. Often, river discharge and tidal currents are sufficient to prevent complete separation of the lake from the sea. The result is an alternation between fresh and brackish water in the lake, and the salinity depends upon the ratio of freshwater inputs and saltwater intrusions. Other coastal lakes are completely separated from the sea.

Numerous coastal lakes are found inland, adjacent to large lakes. The formation of these water bodies occurs in analogous fashion, by the deposition of bars across bays and river valleys. Spits are known to have formed on two sides of a lake as a result of current patterns in relation to specific morphometry. The spits may then join in the middle, dividing the lake into two.

I. Lakes of Organic Origin

The array of lakes created by the damming action of plant growth and associated detritus is incompletely known. It is clear that plant growth can be sufficiently profuse to dam the outlet of shallow depressions and alter the drainage pattern of lakes. Such alterations of drainage by plant growth are common in dystrophic lakes (cf. Chap. 25). The drainage patterns of many lakes, particularly in flood plains of tropical river systems, can be extensively altered by the massive growths of emergent and floating aquatic plants (cf. Beadle, 1974). The effectiveness of this method of water

[8] *Playa* is a geological term for flat and generally barren lower portions of arid basins that periodically flood (*a playa lake*) and accumulate sediment. Most basins of playas have multiple origins (see detailed review of Neal, 1975).

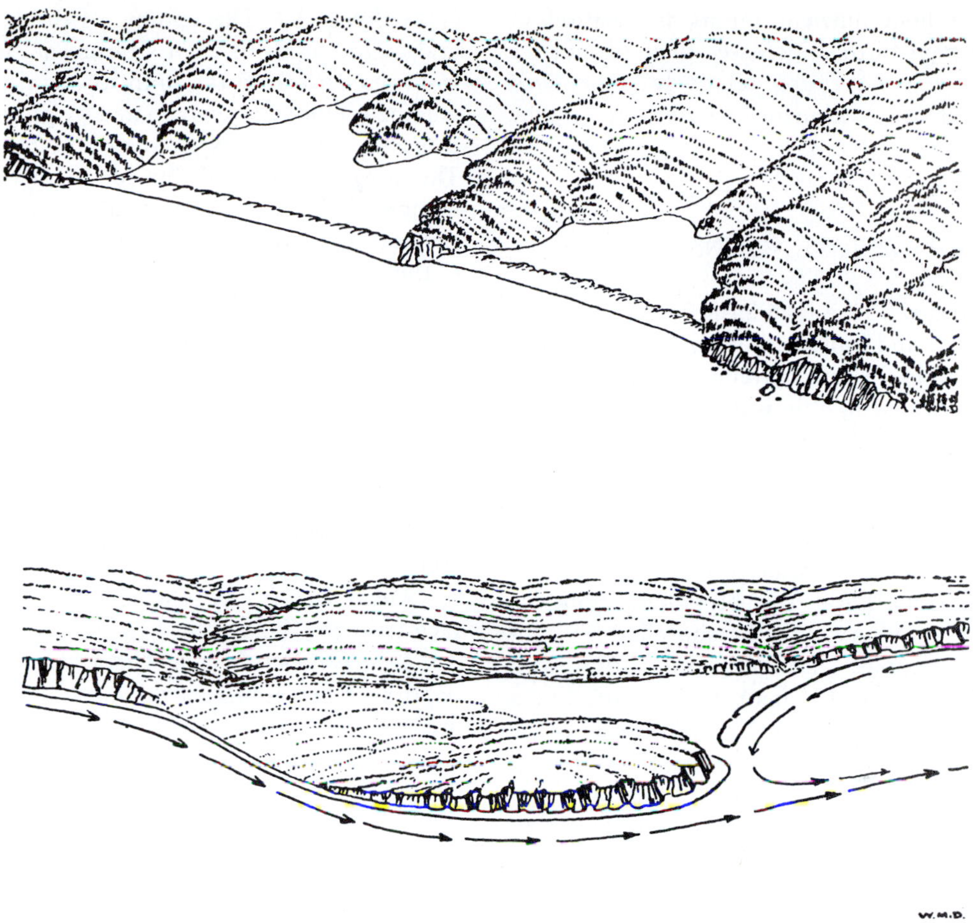

FIGURE 3-16 *Upper:* Coastal lakes formed by enclosure of lagoons by wave and current-built bars on a shore line embayed by slight subsidence. *Lower:* A land-bound island with a bay behind it, closed by a sand-reef beach formed by reverse eddy currents. (After Davis, W. M.: *Calif. J. Mines Geol.* 29:175, 1933.)

retention in flat regions, such as the arctic tundra, is unknown. The photosynthetic activities of attached cyanobacteria and algae as well as other plants can induce massive precipitation of $CaCO_3$ from calcareous waters. In time, these deposits can form barriers and isolate small bodies of water from stream systems (cf. Golubić, 1975).

1. Reservoirs

Two mammals, the beaver and humans, are particularly effective in constructing dams across river valleys to impound water into lakes. The American beaver created numerous large, long-lived lakes, many of which became permanent by means of sediment deposited against the dams. Humans have created artificial lakes by damming streams for at least 4000 years. Only in the last two centuries, however, has this activity become highly significant for the purposes of flood control and the provision of power and water supplies for urban populations.

Reservoirs are being constructed on an unprecedented scale in response to the exponential demophoric demands of humans. Plans for the construction of enormous reservoirs, approaching the area of some of the Laurentian Great Lakes, are underway in Canada and the Soviet Union. Such massive alterations of large drainage systems will result in major modifications in topography and regional climate that are not yet fully recognized or even partially understood. A great need exists for careful planning in the construction of reservoirs because deleterious effects commonly exceed expected benefits (cf. Dussart *et al.*, 1972).

Small, shallow reservoirs, in the form of farm ponds and moderate-sized inundations, have created literally millions of small lakes (see Chap. 20). Particularly characteristic of the morphometry of these lakes is that they are generally shallow and possess large areas where macrophytic vegetation can grow. Such plant life radically alters the productivity of the lake system (Chaps. 18 and 20). Moreover, small reservoirs generally receive high nutrient inputs in relation to their volume, which further increases their productive capacity. Most reservoirs are relatively short-lived because of the extensive sediment load delivered by the influents.

VI. MORPHOLOGY OF LAKE BASINS

The shape and size of a lake basin affect nearly all physical, chemical, and biological parameters of lakes. The forms of lake basins are extremely varied and reflect their modes of origin, how water movements have subsequently modified the basin, and the extent of loading of materials from the surrounding drainage basin.

The morphometry of a lake is best described by a bathymetric map, which is required for the determination of all major morphometric parameters. Such a map is prepared by a survey of the shoreline by standard surveying methods, often in combination with aerial photography. From the map of the shore line a detailed bathymetric map of depth contours must be constructed from a series of accurate depth soundings along intersecting transects. Methods commonly employed for both lake and stream mapping are discussed by Welch (1948), Håkanson (1981), and Wetzel and Likens (2000). Accurate sonar transects, which permit acquisition of much greater detail than was previously possible by manual sounding methods, are now used almost exclusively.

A. Morphometric Parameters

The many morphometric parameters that can be determined from a detailed bathymetric map are discussed at length by Hutchinson (1957). The most commonly used parameters are defined here.

Maximum Length (l). The distance on the lake surface between the two most distant points on the lake shore. This length is the maximum effective length or fetch for wind to interact on the lake without land interruption.

Maximum Width or Breadth (b). The maximum distance on the lake surface at a right angle to the line of maximum length between the shores. The mean width ($\bar{b}$) is equal to the area divided by the maximum length: $\bar{b} = A/1$.

Area (A). The area of the surface and each contour at depth z is best determined by digital integration or planimetry (cf. Welch, 1948; Wetzel and Likens, 2000) or, less precisely, by a grid enumeration analysis (Olson, 1960).

Volume (V). The volume of the basin is the integral of the areas of each stratum at successive depths from the surface to the point of maximum depth. The volume is closely approximated by plotting the areas of contours, as closely spaced as possible, against depth. The integrated area of this curve corresponds to basin volume. Alternatively, the volume can be estimated by summation of the frusta of a series of truncated cones of the strata:

$$V = \frac{h}{3}(A_1 + A_2 + \sqrt{A_1 A_2})$$

where h is the vertical depth of the stratum, A_1 the area of the upper surface, and A_2 the area of the lower surface of the stratum whose volume is to be determined.

Maximum Depth (z_m). The greatest depth of the lake.

Mean Depth ($\bar{z}$). The volume divided by its surface area: $\bar{z} = V/A$.

Relative Depth (z_r). The ratio of the maximum depth as a percentage of the mean diameter of the lake at the surface, expressed as a percentage.

$$z_r = \frac{50 z_m \sqrt{\pi}}{\sqrt{A_0}}$$

Most lakes have a z_r of less than 2%, whereas deep lakes with a small surface area usually have $z_r > 4\%$.

Shore Line (L).[9] The intersection of the land with permanent water is nearly constant in most natural lakes. The shore line, however, can fluctuate widely in ephemeral lakes and especially in reservoirs in response to variations in precipitation and discharge. The length of the shore line can be determined directly or from maps with a map measurer (chartometer or rotometer; cf. Welch, 1948; Wetzel and Likens, 2000).

Shoreline Development (D_L). The ratio of the length of the shore line (L) to the circumference of a circle of area equal to that of the lake:

$$D_L = \frac{L}{2\sqrt{\pi A_0}}$$

Very circular lakes, such as crater lakes and some kettle lakes, approach the minimum shoreline development value of unity. The conformation of most lakes, however, deviates strongly from the circular. Many are subcircular or elliptical in form, with D_L values of about 2. A more elongated morphometry increases the value of D_L markedly, as for example is found in the dendritic outlines of lakes occupying flooded river valleys. Shoreline development is of considerable interest

[9] Two words (*shore line*) as a noun; one (*shoreline*) as an adjective.

because it reflects the potential for greater development of littoral communities in proportion to the volume of the lake.

B. Hypsographic and Volume Curves

The shape of a lake basin can influence biological productivity. In comparative analyses of productivity among lakes, quantitative morphometric analyses of the relationships of lake surface area to the area of sediments exposed to specific volumes of water at different depths permit insights into potential parameters that influence productivity.

The ***hypsographic curve*** (depth–area curve) is a graphical representation of the relationship of the surface area of a lake to its depth (Fig. 3-17). The hypsographic curve represents the relative proportion of the bottom area of the lake that is included between the strata under consideration. The hypsographic curve is, however, only an approximation of the area of the exposed lake bottom because areal measurements are related to the horizontal plane of the lake surface, whereas the actual sediments are slightly sloped and greater in area. The area of the bottom below a square meter of lake surface is larger if the slope of the bottom is appreciable. For example, if the slope of the bottom were 15° from the horizontal, the area of bottom below 1 m^2 at the surface would be 1.04 times greater (Loeb *et al.*, 1983). If the bottom slope were 35°, the benthic area would be 1.22 times greater than the area at the water surface. The hypsographic curve may be expressed in absolute terms (m^2, hectares, or km^2) or as a percentage of the lake area overlying a given depth contour (Fig. 3-17).

The ***depth–volume curve*** is closely related to the hypsographic curve and represents the relationship of

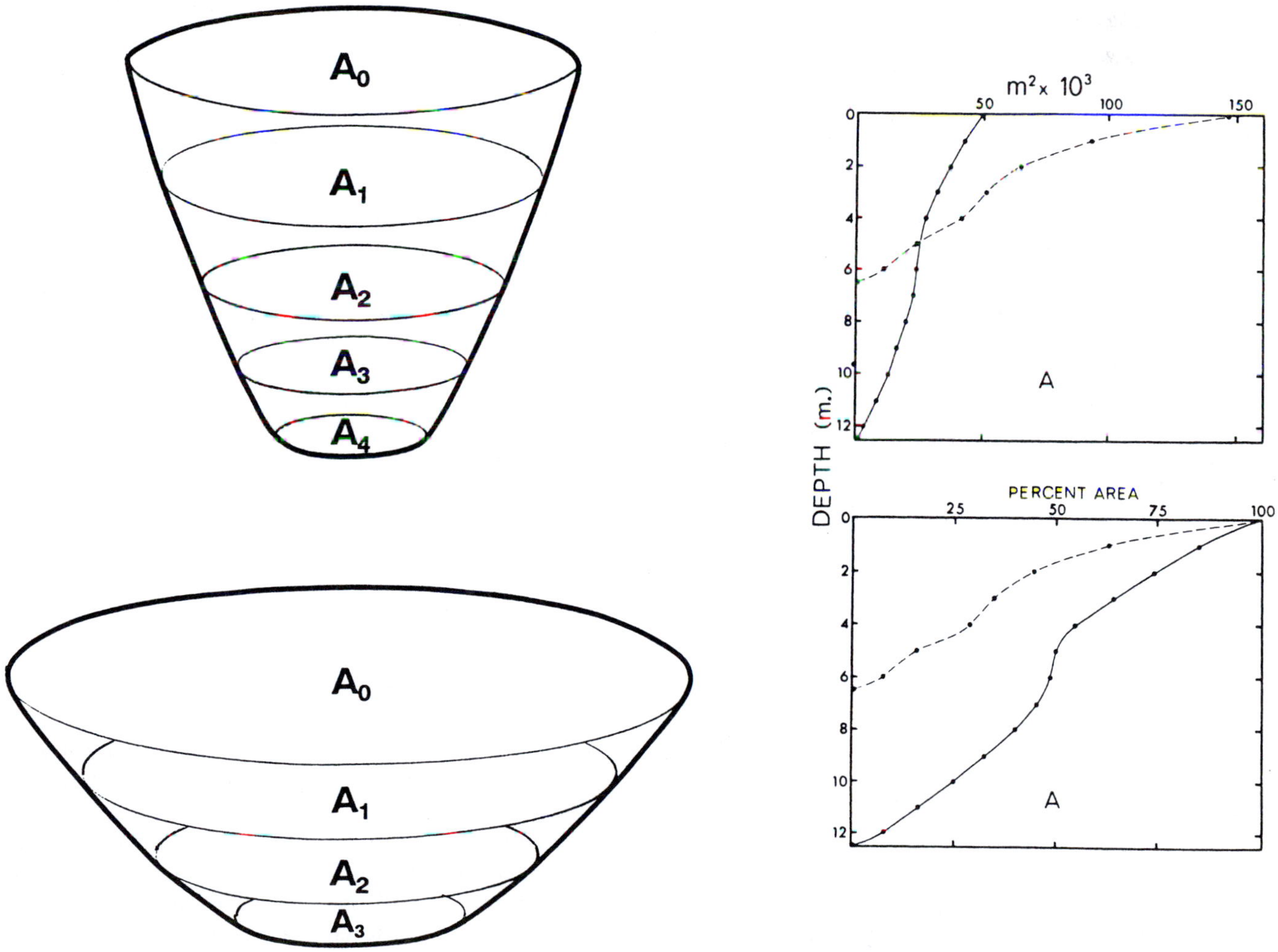

FIGURE 3-17 Hypsographic (depth–area) curves of oligotrophic Lawrence Lake (——) and eutrophic Wintergreen Lake (----), southwestern Michigan. (From Wetzel, R. G., unpublished data.)

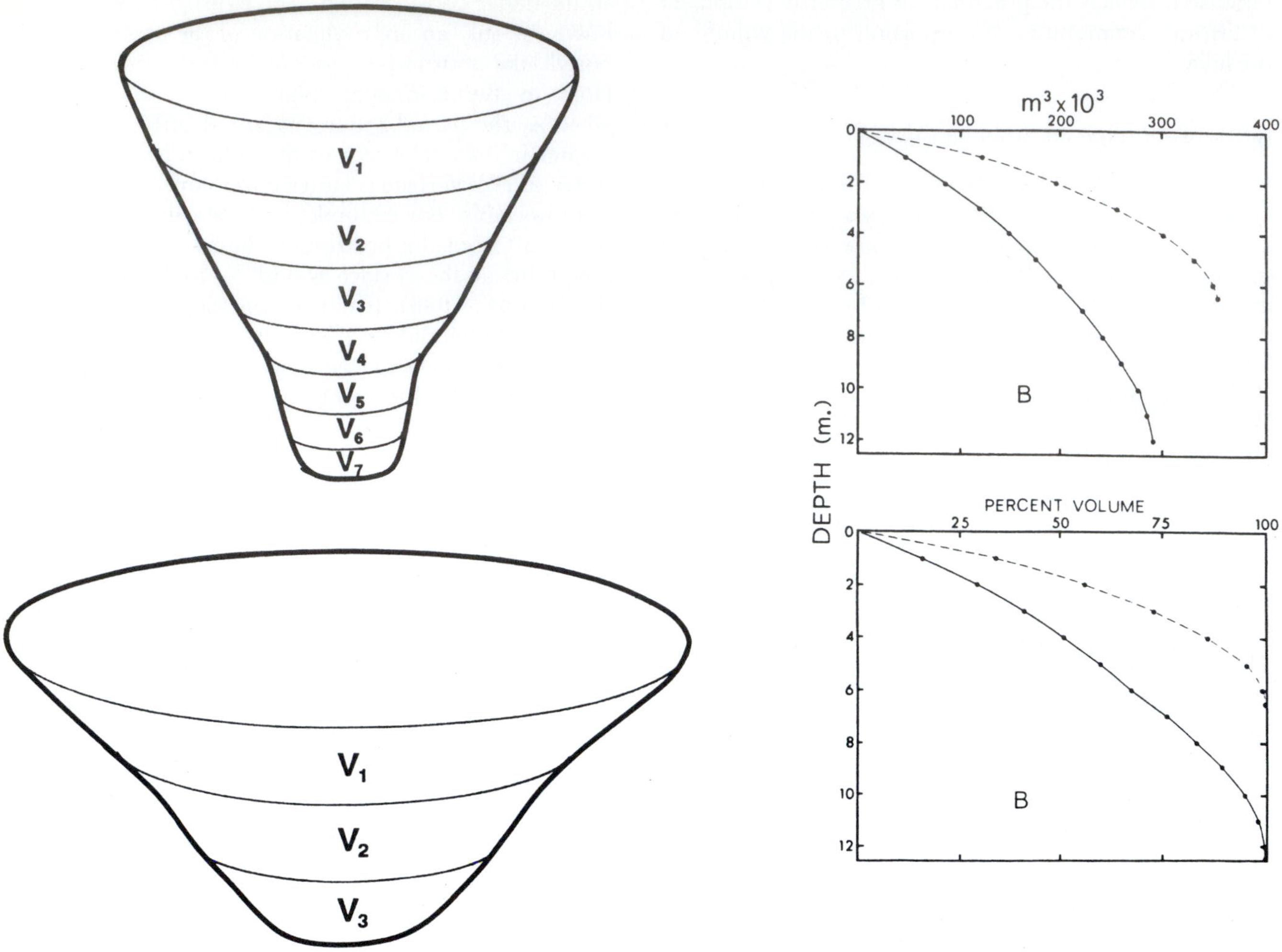

FIGURE 3-18 Depth–volume curves of oligotrophic Lawrence Lake (——) and eutrophic Wintergreen Lake (----), southwestern Michigan. (From Wetzel, R. G., unpublished data.)

lake volume to depth (Fig. 3-18). Similarly, the depth–volume curve can be expressed in actual volume units (m^3 or km^3) or in percentage of total lake volume above a specific depth. The depth–volume curve can be used to estimate readily the approximate mean depth of a body of water, that is, that depth above which 50% of the lake volume occurs.

Among lakes with otherwise comparable conditions, biological productivity is generally greater in those with more sediments in contact with water that receives sufficient light to support photosynthesis (Thienemann, 1927; Strøm, 1933; Rawson, 1955, 1956). The extent of shallow water in a lake is a determining factor in the interrelationships of these zones of photosynthetic production and of decomposition (cf. Ch. 6) and determines the area available for the growth of rooted aquatic plants and associated littoral communities (cf. Chaps. 18 and 19).

The ratio of mean to maximum depth ($z{:}z_m$) is an expression similar to the ratio of the volume of the lake to that of a cone of basal area A and height z_m $[A\bar{z}/\frac{1}{3}z_mA = 3\bar{z}/z_m]$. The ratio $\bar{z}{:}z_m$ thus gives a comparative value of the form of the basin in terms of volume development. For most lakes, the value $\bar{z}{:}z_m$ is >0.33, which is the value that would be given by a perfect conical depression. The ratio exceeds 0.5 in many caldera, graben, and fjord lakes, whereas most lakes in easily eroded rock usually have ratios between 0.33 and 0.5. Very low values of $\bar{z}{:}z_m$ occur only in lakes with deep holes, such as solution or kettle lakes.

Examination of the morphometry of a large number of lakes has shown that the average shape of lake basins approximates an elliptic sinusoid (Neumann, 1959; Fig. 3-19). The elliptic sinusoid is a geometric body whose base is an ellipse, so that planes perpendicular to the base of an ellipse passing through the center

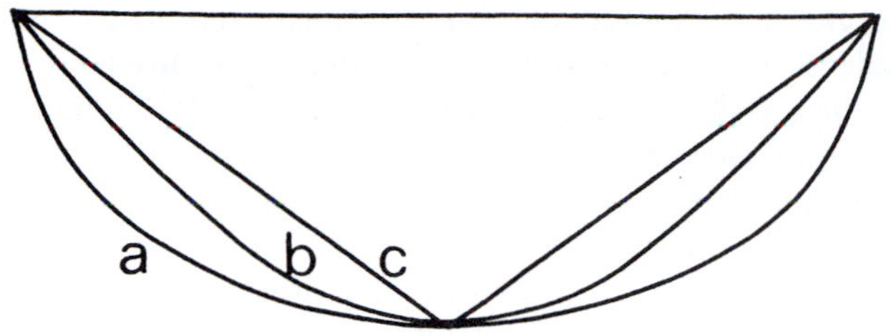

FIGURE 3-19 Vertical cross sections through three forms of lake basins: (a) half an ellipsoid of revolution; (b) elliptic sinusoid; and (c) right elliptic cone. (Modified from Neumann, 1959.)

of the ellipse intersect the surface of the body along troughs of sine curves. The volume of such an elliptic sinusoid is:

$$V = 4\left(1 - \frac{2}{\pi}\right)abz_m = 1.456abz_m$$

where a and b are the half-axes of the lake surface ellipse. Since the area of the lake surface ellipse concerned is πab, $\bar{z} = V/A = 0.464z_m$, the ratio $\bar{z}:z_m = 0.464$. This value is very close to the average value (0.467) of the ratio for over 100 lakes that have been evaluated (cf. also Hayes, 1957; Anderson, 1961; Koshinsky, 1970). Therefore, the elliptic sinusoid serves as a good model for an average lake in which irregularities and submerged depression individuality are not severe. The mean depth of an average lake is slightly less than one-half (0.46) of its maximum depth.

Since the pioneering work of Thienemann, much attention has been focused on the importance of lake morphometry, especially mean depth, to lake productivity in relation to the effects of climatic and edaphic factors (Hutchinson, 1938; Rawson, 1952, 1955, 1956; Edmondson, 1961; Patalas, 1961; Hayes and Anthony, 1964). Mean depth was regarded as the best single index of morphometric conditions, and mean depth exhibits a general inverse correlation to productivity at many trophic levels among large lakes.

A ***morphoedaphic index*** (MEI) was set forth initially to couple empirical log-log relationships of fish productivity (mass/area/time) with abiotic parameters of lake–basin morphometry, expressed as mean depth, volume, and area (Ryder, 1965, 1982). The MEI was a simple ratio, $N/\bar{z}$, where N is a nutrient value or related parameter, and $\bar{z}$ was the mean depth or another appropriate morphometric surrogate. Originally, total dissolved solids and mean depth were used and correlated to some empirical measure of fish yield or of productivity. The ratio was modified extensively with many different parameters for the nutrient component, and the concept was extended to many other biotic measurements among the plankton and benthic organisms. Often, the resulting ratios were misapplied, particularly with self-correlations, and severely criticized for statistical misuse of ratio variables in linear regression modeling (Jackson *et al.*, 1990). These models can be modified to overcome some of the shortcomings to allow certain morphometric parameters of surface area and lake volume to weakly predict, on the basis of total dissolved solids, annual fish yield for moderately deep lakes (Rempel and Colby, 1991). The MEI, however, does not account for the age structure of the fish populations or the recruitment differences within the populations. The production values used are the harvested portion of the older age structures of the populations, taken for commercial or sport purposes, which are not necessarily indicative, and likely marked underestimates, of real production.

The morphoedaphic relationship deteriorates among small lakes and indicates, as will become apparent in later discussions, that regulation of the dynamics of metabolism and productivity in aquatic ecosystems is multifaceted. For example, in shallow lakes various mineral substances of bottom sediments can be resuspended by water movements and cause high turbidity. The suspended materials often include clays brought into lakes and reservoirs from eroding soils, glacial flours, and calcite precipitates in hard waters. Turbidity can persist for long periods of time, reduce light availability, and markedly suppress the productivity of phytoplankton and littoral plants as well as higher trophic levels, regardless of nutrient loading (e.g., Cuker *et al.*, 1990). Thus, the morphometry of the lake basins is only one, although an important, interacting parameter regulating overall productivity but is of very limited value alone for the prediction of ecosystem productivity.

VII. RESERVOIRS

Much of our limnological understanding originates from natural lake ecosystems. Hundreds of thousands of reservoirs, however, have been created by human activities. These artificial water bodies have been created for specific purposes of water management. Examples include water storage, flood control, generation of electrical energy, and recreation.

Reservoirs differ in significant ways from natural lake ecosystems. Study of reservoir ecosystems, however, indicates many functional similarities between artificial and natural lakes. In order to effectively manage and utilize reservoirs, it is important to understand the structural differences between these human-made ecosystems and natural lakes while simultaneously appreciating their functional similarities. Understanding of these structural differences is mandatory for effective management and use of impounded water resources. Failure to recognize basic similarities in metabolic functioning and community interrelationships of the biota will only result in redundancies in reservoir research that can be basically understood from our existing knowledge of processes in natural lakes.

A. Reservoir Characteristics

Reservoirs are most frequently constructed by damming a river valley, within which water accumulates behind the dam. Water released downstream is regulated according to water inputs from the drainage basin as well as uses of the water. As a result of the common linear morphology in a river valley, several distinct physical and biological patterns develop longitudinally along the length of reservoirs. Small reservoirs and artificial ponds are constructed also in valley depressions to which inflows occur by surface flows and/or seepage. In some cases, excavations of soil for construction or mining purposes result in reservoirs or artificial lakes as the excavations cut into water table aquifers (e.g., strip mining lakes, "barrow pit" lakes adjacent to major highways). Among many excavated reservoirs outflows may occur only by subaqueous seepage or evaporation during much of the year. The dam or levees may be breached only occasionally during and following major precipitation events.

Large reservoirs also bring about complex hydrodynamic characteristics with physical and chemical changes that influence aquatic biota in many ways and can be deleterious to the environment and human welfare (e.g., Morgan, 1971; Baxter, 1985). Large impoundments often modify the regions in their vicinities or for large distances below the dam. The reservoir serves as a source or sink for heat, sediments, and solutes that can cause severe effects far downstream from the dam, sometimes for hundreds of kilometers. Diurnal and annual flooding patterns are altered (see Chap. 26), and movements of anadromous fishes are impeded or inhibited.

Because many reservoirs function to reduce river flows, they serve as collectors of riverine sediment loads. As a result, the longevity of reservoirs is often very short, generally less than 100 years. The short-term gains in flood control, hydropower, and other assets must be carefully evaluated against disadvantages. Reservoirs may contribute to the spread of certain diseases that contain waterborne vectors and may reduce the incidence of others. Sometimes serious dislocations of traditional human societies occur.

B. Zonation in Reservoirs

Three distinct zones occur along the longitudinal gradient: a *riverine* zone, a zone of *transition,* and a *lacustrine* zone (Fig. 3-20). Each zone possesses unique and dynamic physical, chemical, and biological characteristics. The riverine zone is often relatively narrow as a result of river geomorphology. Water is usually well mixed. Although water velocities decrease as the water enters the reservoir, advective transport by currents is sufficient to move significant quantities of fine suspended particulates (silts, clays, and organic particulate matter) (Thornton, 1990). High particulate turbidity commonly reduces light penetration and limits primary production within the water of this zone. Loading of organic matter from allochthonous sources is high in proportion to water volume in the riverine zone. High decompositional rates often result with high consumption of dissolved

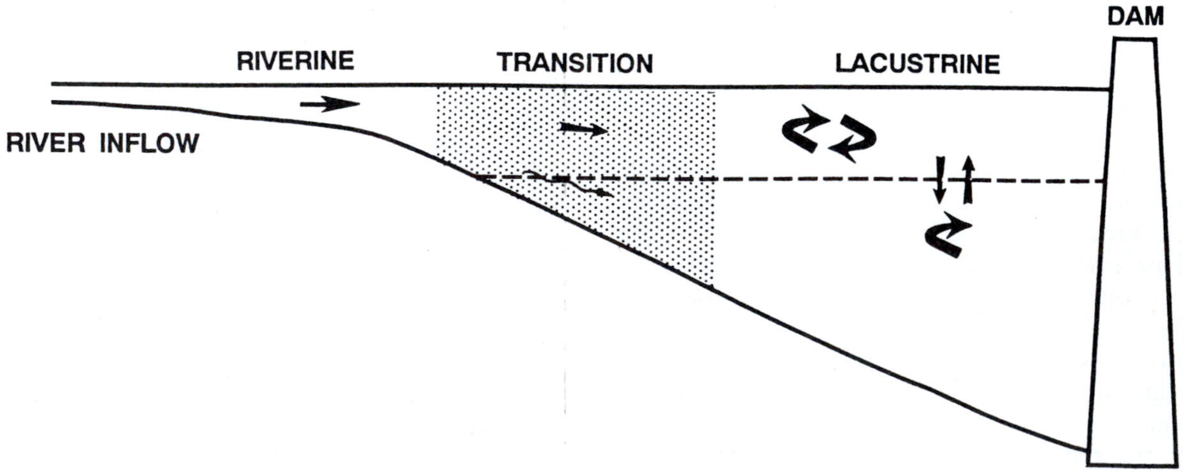

FIGURE 3-20 Generalized zones along longitudinal gradients in reservoirs.

oxygen, but aerobic conditions generally prevail in the shallow, well-mixed riverine zone.

Riverine water velocities decrease as energy is dispersed over larger areas in the transitional zone (Fig. 3-20). An appreciable portion of the turbidity load settles out of the upper water strata as a result. Decreased turbidity results in enhanced depth of light penetration and increased rates of photosynthetic productivity by phytoplankton. Gradually, a shift occurs to an increasing percentage of total organic matter loading from phytoplankton and, in some shallow reservoirs, from rooted vascular plants.

Within the lacustrine zone, characteristics become more similar to lake ecosystems (Fig. 3-20). This portion of the reservoir often stratifies thermally and assumes many of the properties of natural lakes in regard to planktonic production, limitations by nutrients, sedimentation of organic matter, and decomposition in the hypolimnion. Stratification and water movements can be modified or complicated, both spatially and in time, by hypolimnetic or bottom withdrawal of water from the reservoir at the dam, as is commonly the case.

VIII. COMPARATIVE GEOMORPHOLOGICAL CHARACTERISTICS

Comparative analyses of the structural differences in geomorphological characteristics and properties among rivers, reservoirs, and natural lakes can be instructive to point out functional similarities among these different but integrated ecosystems. A comparative summary of relative morphological attributes is presented in Table 3-4.

Rivers are erosional where moderate to rapid flow prevails and are depositional of silt and organic particles in sections of slow current. River channel characteristics are constantly changing in relation to variables of discharge, width, depth, substrata resistance, and transport of sediment.

TABLE 3-4 Comparative Geomorphological Characteristics and Properties among River, Reservoir, and Natural Lake Ecosystems[a]

Property	*Rivers*	*Reservoirs*	*Natural lakes*
Drainage basins	Many small tributaries coalesce into trunk stream; drainage area high in relation to surface area	Usually narrow, elongated lake basin in base or drainage area; area large in comparison to lake area (ca. 100:1 to 300:1)	Circular, lake basin usually central; drainage area usually small in comparison to lake area (ca. 10:1)
Shape	Long, meandering, linear rivers elongate as drainage basins increase in size	Variable, ovoid to triangular	Circular to elliptical predominate
Mean depth	Shallow in headwaters, increases to mouth	Shallow in riverine portions, increasing in lacustrine zones	Moderate to high; average less than 10 m
Depth gradient	Increases from headwaters to mouth	Increases from riverine through transitional to lacustrine zones	Deepest usually remote from shore line
Shoreline erosion and substrata distribution	Extensive, induced by water currents, gravity driven	Extensive in riverine areas by water currents; less from wind-induced currents in lacustrine zones	Localized, induced by wind-generated waves and currents
Shoreline development	Great, astatic	Great, astatic	Relatively low; stable
Sediment loading	High with large drainage basin area	Large with large drainage basin area; flood plains large; deltas large, channelized, gradation rapid	Low to very low; deltas small, broad, gradation slow
Deposition of sediments	Determined by water currents; highly variable with precipitation events	High in riverine zone, decreasing exponentially down reservoir; greatest in old riverbed valley; highly variable seasonally	Low, limited dispersal; relatively constant rates seasonally
Sediment suspended in water (turbidity)	High, variable	High, variable; high percentage of clay and silt particles; turbidity high	Low to very low; turbidity low

[a] Assembled from data of many sources particularly of the syntheses on rivers by Ryder and Pesendorfer (1989) and on reservoirs and lakes by Wetzel (1990b) and many references cited therein.

TABLE 3-5 Comparative Hydrodynamic Characteristics among River, Reservoir, and Natural Lake Ecosystems[a]

	Rivers	*Reservoirs*	*Natural lakes*
Water level fluctuations	Large, rapid, irregular; flooding common	Large, irregular	Small, stable
Inflow	Overland and groundwater runoff; highly irregular and seasonal, less so with large groundwater inflows	Most runoff to reservoir via river tributaries (high stream orders); penetration into stratified strata complex (over-, inter-, and underflows); often flow is directed along old riverbed valley	Runoff to lake via tributaries (often low stream orders) and diffuse sources; penetration into stratified waters small and dispersive
Outflow (withdrawal)	Discharge highly irregular with inflows and precipitation event frequency	Highly irregular with water use; withdrawals from surface layers or from hypolimnion	Relatively stable; usually largely surface water via surface outflow or shallow ground water
Flushing rates	Rapid, unidirectional, horizontal	Short, variable (days to several weeks); increase with surface withdrawal, disruption of stratification with hypolimnetic withdrawal; three dimensional	Long, relatively constant (one to many years); three-dimensional

[a] Assembled from many sources and the syntheses on rivers of Ryder and Pesendorfer (1980) and on reservoirs and lakes by Wetzel (1990b).

Reservoirs are created predominantly in regions where large natural lakes are sparse or unsuitable (e.g., too saline) for human exploitation. In these regions, the climate tends to be warmer than is the case in regions with many natural lakes, resulting in somewhat higher average water temperatures, longer growing seasons, and precipitation inputs that are closely balanced to, or less than, evaporative losses (Wetzel, 1990b). The drainage basins of reservoirs are consistently much larger in relation to the reservoir surface area than is the case among most natural lakes. Because reservoirs are almost always formed in river valleys and in the base of the drainage basins, the morphometry of reservoir basins is usually dendritic, narrow, and elongated (Table 3-5). These physical characteristics affect biological processes in many complex ways, the most important of which are light and nutrient availability. Reservoirs receive runoff water mainly via high-order streams, which results in higher energy for erosion, large sediment-load carrying capacities, and extensive penetration of dissolved and particulate loads into the recipient lake water. Since reservoir inflows are primarily channelized and often not intercepted by energy-dispersive and biologically active wetlands and littoral interface regions, runoff inputs are larger, more directly coupled to precipitation events, and extend much farther into the reservoir lake per se than is the case in most natural lakes (Table 3-5). All of these properties result in high, but irregularly pulsed, nutrient and sediment loading to the recipient reservoir.

Extreme and irregular water level fluctuations commonly occur in reservoirs as a result of flood inflow characteristics, land-use practices not conducive to water retention, channelization of primary influents, flood control, and large, irregular water withdrawals for hydropower generation operations (Table 3-5). Multiplicative effects on loadings result. Large areas of sediments are alternately inundated and exposed; these manipulations usually prevent the establishment of productive, stabilizing wetland and littoral flora. Erosion and resuspension of floodplain sediments augment high loadings from the large drainage basin sources. Surficial sediments are alternately shifted between aerobic and anaerobic conditions, which enhances nutrient release. The reduction or elimination of wetland and littoral communities around many reservoirs minimizes the extensive nutrient and physical sieving capacities that function effectively in most natural lake ecosystems (Wetzel, 1979, 1990a, 1990b). Consequently, shoreline erosion in natural lakes is small and localized to wind-generated currents. As a result, compared to reservoirs the basin morphology of natural lakes is relatively stable.

IX. SUMMARY

1. Some 40% of the total volume of fresh water is contained in the great lakes basins. Most lakes and reservoirs are much smaller, however, and are concentrated in the subarctic and temperate regions of the Northern Hemisphere. Most of the millions of

lakes are small and relatively shallow, usually <10 m in depth.

2. Rivers account for only an extremely small amount (0.0001%) of the water of the Earth but are major transporters of dissolved and particulate matter from the land to the sea.
   a. Water from precipitation enters drainage basins and either moves or is temporarily stored, particularly in soils.
   b. Stream flow and discharge are directly related to the drainage area. Differences are the result of variations in elevational gradients and soil characteristics.
   c. Stream and river channels move laterally within the flood plain because of erosion and redeposition of sediment.
   d. Forms of river channels deviate from four general classes: straight, meandering, braided, and anastomosing. River morphology varies widely with geology, drainage area, discharge, width–depth ratios, river bed material size, and channel slope.
3. Most natural lakes were formed by catastrophic events.
   a. ***Tectonic*** lake basins are depressions formed by displacements of the Earth's crust. Many of the deepest relict lakes of the world, such as *graben lake* basins, were formed by faulting movements. Uplifting of the earth has created a number of lake basins; these basins were often modified by glacial scouring activity.
   b. Lakes can result from ***volcanic*** activity. Small, deep *maar* lakes can form in volcanic cones; the collapse of the roof of partially emptied magmatic chambers can result in large, deep *caldera* lakes within a volcano. Lava flows often dam preexisting river valleys and create isolated lakes.
   c. Temporary or permanent lakes can result from ***landslides*** into stream valleys. Lakes can also form in depressions created behind the landslide material.
   d. The erosional and depositional activity of ***glaciers*** was the most important agent of lake formation. Many lakes were formed from the outwash morainal deposits at the retreating edges of continental glaciers or from melted blocks of ice buried in this morainal debris (*kettle lakes*). In mountainous regions, terminal and lateral morainal deposits can effectively dam river valleys and form lakes. Glaciers often scour lake basins in glaciated valleys of mountainous areas and form amphitheater-shaped depressions termed *cirque lakes*. Several cirque lakes can form in the trough of a mountain valley in a series of connected lakes of successively lower elevation (*paternoster lakes*). Many arctic and subarctic lakes also formed from glacial activity. Some arctic lakes (*cryogenic lakes*) are very shallow and form by water seepage into the permafrost, which, on freezing, forms a polygonic network of ridges that contains subsequent meltwater.
4. Other natural lakes form by gradual events.
   a. ***Solution lakes*** result from sinks, termed *dolines,* formed by the gradual dissolution of rock, such as limestone, along fissures and fractures. Eventually the superstructure is weakened to the point of collapse into the depression.
   b. The erosional and depositional action of ***river water*** can isolate depressions to form lakes (e.g., *plunge-pool lakes* below former waterfalls, *oxbow lakes* of former river channels).
   c. ***Wind erosion*** can form shallow depressions that often contain water temporarily or seasonally (e.g., *dune lakes* in sandy areas and *playa lakes* of flat, arid or semiarid regions).
   d. ***Coastal lakes*** often form along irregularities in the shore line of the sea or large lakes. Long-shore currents deposit sediments in bars or spits that eventually isolate a fresh or brackish-water lake.
   e. ***Reservoirs*** are impoundments created largely by humans by the damming of river valleys. Because of high rates of sedimentation, reservoirs, smaller inundation lakes, and farm ponds are very short-lived.
5. Variation in basin morphology is great; most lake basins approximate an elliptic sinusoid shape. The mean depth of lakes is about half (0.46) of the maximum depth.
6. The morphology of a lake basin has profound effects on nearly all physical, chemical, and biological properties of the lake. Morphometry of lake basins and geological substrates of the drainage basin influence sediment–water interactions and lake productivity, especially extremely productive littoral communities. The greater productivity of small, shallow lakes is usually correlated with the higher water–sediment interface area per water volume (i.e., lower mean depth).
7. Reservoirs, formed by the damming of river valleys, possess three distinct zones along the longitudinal gradient:
   a. A narrow ***riverine zone*** has high advective water transport and high particulate turbidity.
   b. A ***zone of transition*** has decreased water velocities as energy is dispersed over larger areas.

Turbidity general decreases and sedimentation increases.

c. The ***lacustrine zone*** in the region above the dam has characteristics more similar to natural lakes in relation to stratification and morphological interactions.

8. Comparison of geomorphological characteristics among river, reservoir, and natural lake ecosystems focuses upon effects of water movements on erosion and movement of soil, rock, and sediments. Rivers are shallow and shift between erosional and depositional changes in water discharge. Natural lakes are morphologically much more stable than rivers and reservoirs in regard to depth, shoreline development, sediment loading, turbidity, and deposition of sediments.

4 WATER ECONOMY

The water in lakes and rivers is balanced in the basic hydrological relationship in which change in water storage is governed by inputs from all sources less water losses. Water income from precipitation, surface influents, and groundwater sources is balanced by outflows from surface effluents, seepage to ground water, and evapotranspiration. Each of these incomes and losses varies seasonally and geographically and is governed by the characteristics of particular lake basins, their groundwater and river geomorphology, drainage basins, and the climate.

The distribution of water over continental land masses depends on the global hydrological cycle, in which excessive oceanic evaporation is counterbalanced by greater precipitation over land. The hydrological cycle, which can be altered extensively by human-induced changes of surface water systems, determines the distribution of lakes and rivers regionally in relation to the distribution of suitable catchment lake basins.

I. HYDROLOGICAL CYCLE

The hydrological cycle of the earth evaluates the cyclical budgetary processes of water movement. These processes include movement from the atmosphere, inflow, and temporary storage on land, and outflow to the primary reservoir, the oceans. The cycle consists of three principal phases: ***precipitation, evaporation,*** and ***surface and groundwater runoff.*** Each phase involves transport, temporary storage, and a change in the physical state of the water.

A. Evaporation and Precipitation

Evaporation of water into the atmosphere occurs from land, the oceans, and other water surfaces. Primary sites of evaporation include evaporation from precipitation; from precipitation intercepted by vegetation; from the oceans, lakes, and streams; from soils; and from transpiration of plants. Evaporation rates from each of these and other sources of water are determined by an array of dynamic environmental parameters (cf. Meinzer, 1942; Grey, 1970; Gates, 1980; Berner and Berner, 1987; Jones 1992). The mass of atmospheric water vapor, although small in relation to the amount of total global water, is stored as vapor for the least amount of time (average renewal time of 8.9 days) before returning to the earth as rain, snow, sleet, hail, or condensates (dew and frost), either on land or on the oceans.

Precipitated water may be intercepted or transpired by plants, may run off over the land surface to streams (surface runoff), or may infiltrate the ground (Fig. 4-1). Much of the intercepted water and surface runoff (up to 80%) is returned to the atmosphere by evaporation. Infiltrated water may be stored temporarily (average renewal time approximately 280 days) as soil moisture prior to being evapotranspired. Some of the water percolates to deeper zones to be stored as ground water (average renewal time of 300 years). Ground water

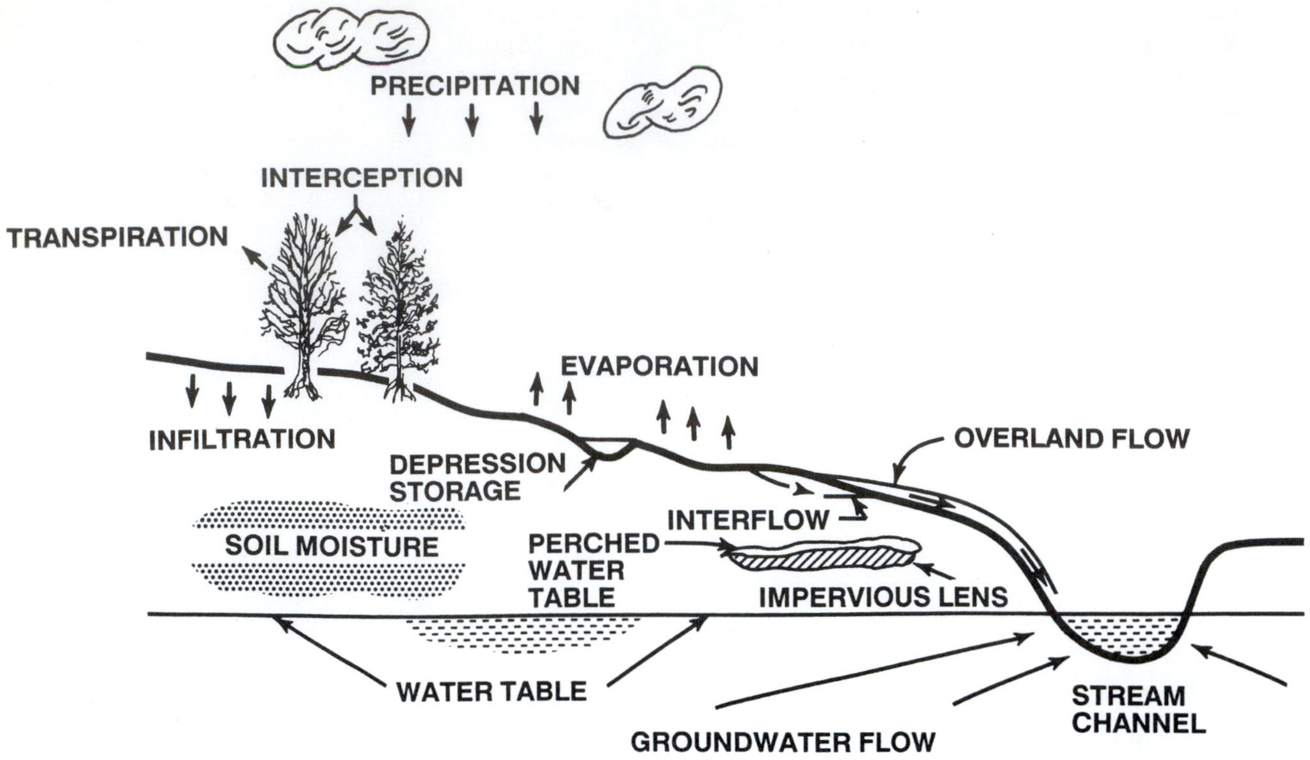

FIGURE 4-1 Major pathways of precipitation and the runoff phase of the hydrologic cycle. (From Wetzel and Likens, 1991.)

moves along flow paths of varying lengths from areas of recharge (precipitation on land, from lake, wetland, or river reservoirs) to areas of discharge. Ground water is actively exchanged and may be used by plants, flow out as springs, or seep to streams as runoff. The runoff phase is exceedingly complex and variable because of the extensive involvement with plants, the extreme heterogeneity of soil structure and composition, and variations in climate.

B. Runoff Flow Processes

The soil and rock of river and lake drainage basins determine pathways and regulate the rates of hillslope runoff of water received as rain and meltwater (cf. review of Dunne, 1978). These pathways of runoff influence many characteristics of the landscape, the uses to which land can be put, and the requirements for effective land management.

When the rate of rainfall or snow meltwater influx exceeds the absorptive capacity of the soil (i.e., the infiltration capacity is exceeded), the excess water flows over the surface as ***overland flow.*** Overland flow is most common in arid and semiarid regions; it also occurs in humid areas where the original vegetation and soil structure have been disturbed or in areas where normally porous soil contains a thin layer of frost and consequently cannot absorb meltwater (Dunne and Black, 1970a, 1970b, 1971).

The storage capacity of the soil is made continually available by evaporation and plant transpiration. Normally, most of the water from precipitation infiltrates the soil. When precipitation is first absorbed by the soil, the water may be stored or it may move by gravity along several pathways within the soil structure and along voids (cracks, worm or other animal burrows, old root channels) toward stream channels. If the soil or rock is deep and of relatively uniform permeability, the ***subsurface water*** moves vertically to the zone of saturation and then follows a generally curving path to the nearest stream drainage channel. This simple pattern of ***groundwater flow*** often is disrupted by irregularities of the base geological structure (Leopold *et al.,* 1964; Davis and DeWiest, 1966; Stephens, 1996; Winter *et al.,* 1998). Water progression can encounter impermeable layers that impede the vertical movement and cause lateral flow. The surface of the saturated zone of permeable soil is called the ***water table.*** Water in the soil above the water table is termed ***vadose water*** and that below the water table, the ***ground water.*** Rates of groundwater flow are generally low and the pathways long; hence much of this ground water contributes to the relatively stable baseflow (see Fig. 3-5) of streams between periods of precipitation. Drainage of water from storms is added to this more uniform groundwater discharge. Although the long route of groundwater flow generally dominates the baseflow of streams, the rate of flow can increase in very permeable rock formations such as limestones and jointed basalts and add significantly to runoff from stormflows to recipient drainage streams.

The sediments of lake basins are generally quite impermeable to rapid inflows from ground water or to seepage outflows. Such connectivity with ground water is usually through shallow sediments near the land–water interface regions, often located in shallow wave-impacted sediments. Lakes, however, and particularly wetlands, can receive groundwater inflows throughout their sediments and have seepage outflows from many different, or nearly the entire, sediment areas. Hydraulic movement through dense organic-rich sediments, particularly of wetlands, is often low but can be accelerated by the uptake, conductance, and evapotranspiration of water by floating-leaved and emergent aquatic plants (Wetzel, 1999a).

When shallow surface soil and weathered rock is permeable and underlaid with relatively impermeable soil horizons, percolating water will be diverted horizontally as ***subsurface stormflow.*** The shallow subsurface pathway of drainage to a stream channel or lake basin is much shorter than the pathway of groundwater flow and generally occurs in soils of high permeability. Therefore, subsurface stormflow can be volumetrically greater than runoff, particularly along steep gradients in narrow drainage valleys.

As subsurface water flows down the gradient of a hillslope, vertical and horizontal ***percolation*** can saturate the soil. Shallow subsurface flow encountering these saturated areas emerges from the soil surface as ***return flow*** and continues to the recipient channel or basin as overland flow. Similarly, direct precipitation onto saturated areas flows over the surface of the soil. These processes assume greater significance in gently sloping, wide valleys with thin soils.

Each of these processes of rainfall or meltwater runoff responds differently to variations in topography, soil, and characteristics of the precipitation events themselves and indirectly to variations in climate, vegetation, and land use. Therefore, runoff flow processes control the volume, periodicity, and chemical characteristics of water entering streams and lake basins. Water that infiltrates the soil and then flows laterally as subsurface stormflows added to overland flow is the main component of peak flows or ***floods*** (cf. Dunne, 1978; Chorley, 1978).

II. GLOBAL WATER BALANCE

More water evaporates from the oceans than is returned via precipitation, whereas on land more water is received via precipitation than is lost by evaporation (Fig. 4-2). Most continental water income originates from evaporation of the oceans. The water of land masses is not uniformly distributed over the major

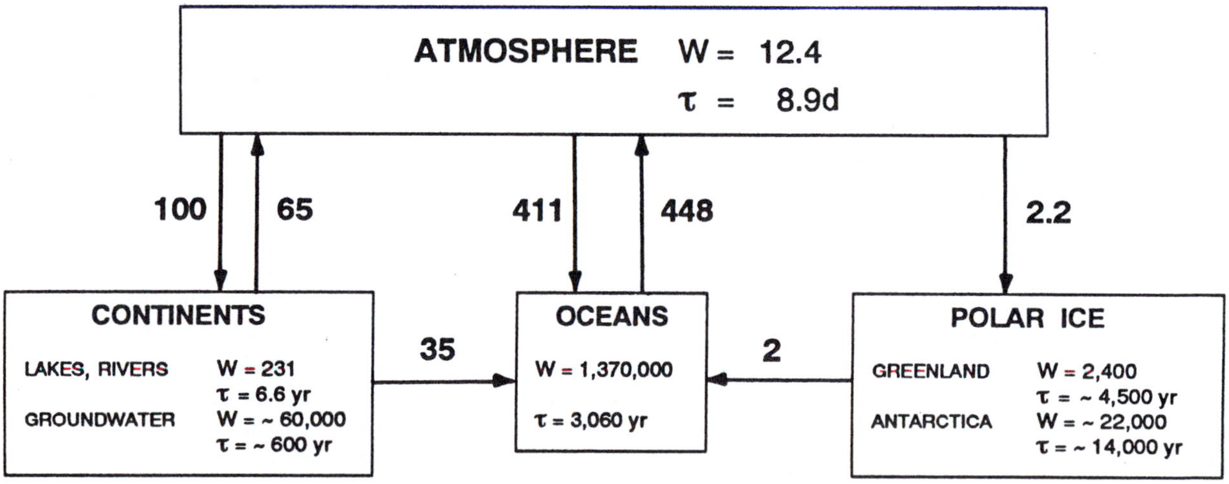

FIGURE 4-2 Global water balance. W = water content in 10^3 km^3, values on arrows = transport in 10^3 km^3 yr^{-1}, and τ = retention time. Estimate of ground water is to a depth of 5 km in the Earth's crust; much of this water is not actively exchanged. (Modified from Flohn, 1973, after L'vovich.)

TABLE 4-1 Water Resources and Annual Water Balance of the Continents of the World[a]

	Europe[b]	*Asia*	*Africa*	*N. America*[c]	*S. America*	*Australia*[d]	*Total*
Area (10^6 km^2)	9.8	45.0	30.3	20.7	17.8	8.7	132.3
Precipitation (km^3)	7165	32,690	20,780	13,910	29,355	6405	110,305
River runoff (km^3)							
Total	3110	13,190	4225	5960	10,380	1965	38,830
Underground	1065	3410	1465	1740	3740	465	11,885
Surface	2045	9780	2760	4220	6640	1500	26,945
Total soil moistening (infiltration and renewal of soil moisture)	5120	22,910	18,020	9690	22,715	4905	83,360
Evaporation	4055	19,500	16,555	7950	18,975	4440	71,475
Underground runoff (% of total)	34	26	35	32	36	24	31

[a] After data from Lvovitch (1973). Data slightly modified in Shiklomanov (1990).
[b] Includes Iceland.
[c] Includes Central America but not the Canadian archipelago.
[d] Includes New Zealand, Tasmania, and Papua New Guinea.

continents (Table 4-1). The total and groundwater runoff are greatest in South America, nearly twice that per area of the other continents.

Until recent times, the global water balance fluctuated very little. Humans have introduced regional fluctuations by extensive environmental modifications for irrigation, industrial, and domestic uses, such as land clearing to obtain more arable land, changes in drainage patterns, and exploitation of groundwater reserves (Flohn, 1973). Water demand for agricultural and industrial purposes is projected to increase the human-induced fraction of continental evaporation from its present nearly 3%–10% soon after the year 2000, and to 50% well before the end of the twenty-first century. The result is an accelerated rate of continental freshwater turnover. Because freshwater supplies are inadequate where demand is greatest, continued expansions at sites of high demand will be curtailed. Large-scale desalinization of sea water, at great energy expense, is a practical contemporary alternative for expansions of freshwater exploitation. The energy costs for redistribution of water, however, restrict desalinization of sea water to coastal regions. Manipulations of surface waters on a massive scale will inevitably lead to irreversible modifications of climate. Modifications of regional climatic conditions have already occurred as a result of extensive alterations of large river systems in which surface waters and evaporation are increased greatly. These modifications are further confounded by those associated with global climate changes induced by human-induced alterations in the gaseous composition of the atmosphere.

Three hydrological regions are recognized among the continental land masses (Hutchinson, 1957). The distribution of lakes is related partly to distribution of lake basins and partly to that of water. ***Exorheic regions,*** within which rivers originate and from which they flow to the sea, contain the major lake districts of the world and most of the lakes. ***Endorheic regions,*** within which rivers arise but never reach the sea, occur between subtropical deserts and the tropical and temperate humid regions. ***Arheic regions,*** within which no rivers arise, are desert areas that occur in the latitudes of the trade winds, between which lies the zone of equatorial rains. Endorheic regions, transitional in nature between the other two, can shift to exorheic or arheic characteristics with relatively small changes in climate.

A. Water Balance in Lake Basins

The water balance of a lake is evaluated by the basic hydrological equation in which the change in storage of the volume of water in or on the given area per unit time is equal to the rate of inflow from all sources less the rate of water loss. Water income to a lake includes several sources.

(a) *Precipitation Directly on the Lake Surface.* Although most lakes, largely in exorheic regions, receive a relatively small proportion of their total water income from direct precipitation, this percentage varies greatly and increases in very large lakes. Extreme examples include equatorial Lake Victoria of Africa, which receives most (>84%) of its water from precipitation on its surface (Kilham and Kilham, 1990; Yin and Nicholson, 1998), and the endorheic Dead Sea, which receives practically no water from direct precipitation.

(b) *Water from Surface Influents of the Drainage Basin.*[1] The amount of the total water income to a lake from surface influents is highly variable. Lakes of endorheic regions receive nearly all their water from surface runoff. The rate of runoff from the drainage basin and corresponding changes in lake level are influenced strongly by the nature of the soil and vegetation cover of the drainage basin. One of the best examples of this effect resulted from the experimental forest cutting and use of herbicides to prevent vegetation regrowth in the Hubbard Brook drainage in New Hampshire (Likens *et al.*, 1967, 1970, 1977; Bormann and Likens, 1979; Likens and Bormann, 1995). Annual streamflow increased 39% the first year and 28% the second year above the values prior to selective deforestation.

(c) *Groundwater Seepage below the Surface of the Lake.* Seepage of ground water is commonly a major source of water for lakes in rock basins and lake basins in glacial tills that extend well below the water table. Sublacustrine groundwater seepage forms the major source of water flow into, and out of, karst and doline lakes of limestone substrata.

(d) *Ground Water Entering Lakes as Discrete Springs.* Sublacustrine springs from ground water occur frequently in hardwater lakes of calcareous drift regions, where the basin is effectively sealed from groundwater seepage by deposits within the basin. For example, ground water largely from springs contributed nearly 40% of the annual water income to calcareous Lawrence Lake, Michigan, in comparison to 10% from precipitation and the remainder from surface runoff in two streams (Table 4-2). Groundwater inputs were closely related to rates of precipitation and evapotranspiration each month.

In ***drainage lakes,*** loss of water occurs by ***flow from an outlet*** (Birge and Juday, 1934), and in ***seepage lakes*** losses occur by ***seepage into the ground water*** through the basin walls. Deposition of clays and silts commonly forms a very effective seal in drainage lakes, from which most or all of the outflow leaves by a surface outlet (e.g., Table 4-2). In seepage lakes the sediments over much of the deeper portions of the basin also often form an effective seal; losses to ground water usually occur from the upper portions of the basin in these lakes. For example, in Mirror Lake most of the large seepage losses occurred within the upper 5 m of sediments (Table 4-2).

[1] The *drainage basin* or *catchment area* (*Einzugsgebiet*; *bassin versant*) is, in American usage, equivalent to *watershed*, the region or area drained by a river. *Watershed*, as used in the United Kingdom, refers to the ridge or crest line dividing two drainage areas and is defined similarly in German (*Wasserscheide*) and French (*ligne de partage des eaux*).

Further losses of water occur directly by ***evaporation*** or by ***evapotranspiration*** from emergent and floating-leaved aquatic macrophytes. The extent and rates of evaporative losses are highly variable according to season and latitude and are greatest in endorheic regions. Lakes of semiarid regions commonly have no outflow and lose water only by evaporation. Such lakes are termed ***closed,*** in contrast to ***open lakes*** that have outflow by an outlet or seepage.

Evaporative losses are modified greatly by the transpiration of emergent and floating-leaved aquatic plants (cf. reviews of Gessner, 1959, and Wetzel, 1999). Rates of transpiration and evaporative losses to the atmosphere vary greatly with an array of physical (e.g., wind velocity, humidity, temperature) and metabolic parameters and structural characteristics of different plant species (Brezney *et al.*, 1973; Bernatowicz *et al.*, 1976; Boyd, 1987; Jones, 1992). Plant growth and evapotranspiration are predominantly seasonal in lake and river wetlands of exorheic regions. In tropical and subtemperate lakes, many of the large perennial hydrophytes grow more or less continually. In most

TABLE 4-2 Annual Water Budget for Lawrence Lake, Michigan, and Mirror Lake, New Hampshire

Source/loss	10^3 m^3	*Percentage of total*
Lawrence Lake:[a]		
Inputs		
Inlet 1	146.6	32.1
Inlet 2	87.5	19.1
Ground water	178.1	39.0
Precipitation	44.6	9.8
Total inputs	456.8	100.0
Outputs		
Outlet	436.5	90.4
Evapotranspiration	46.2	9.6
Seepage losses[b]	0.0	0.0
Total outputs	482.7	100.0
Mirror Lake[c]:		
Inputs		
Runoff	663.1	78.1
Precipitation	182.3	21.9
Total inputs	845.4	100.0
Outputs		
Outlet	409.8	48.5
Evapotranspiration	76.0	9.0
Seepage losses	360.0	42.5
Total outputs	845.8	100.0

[a] After Wetzel and Otsuki (1974).
[b] Indirect evidence indicated that seepage losses were negligible. Subsequent direct measurements have shown that seepage losses are negligible (<0.1 liter m^{-2} day^{-1}) in Lawrence Lake (Wetzel, unpublished data).
[c] After Likens *et al.* (1985).

TABLE 4-3 Comparison of Water Loss from a Stand of the Emergent Aquatic Macrophyte *Phragmites communis* to That of Open Water, Berlin, Germany, 1950[a]

Date	*Evapotranspiration (kg m^{-2} day^{-1})*[b]	*Evaporation (kg m^{-2} day^{-1})*[b]	*Ratio of transpiration to evaporation*
11 May	3.20	3.24	1.0
25 May	2.50	1.44	1.6
27 July	9.82	2.24	4.4
22 August	16.01	2.29	7.0
17 October	2.79	0.79	3.9

[a] Modified from Gessner (1959), after Kiendl.
[b] mm day^{-1} or $\times$ 0.1 = g cm^{-2} day^{-1}.

situations, transport of water from the lake or river to the air is increased greatly by a dense stand of actively growing littoral vegetation, as compared to evaporation rates from open water (Table 4-3 and Wetzel, 1999b). Because most lakes are small and often possess a well-developed littoral flora, these communities contribute significantly to the water balance of many surface waters.

Analytical techniques for a detailed evaluation of the water balance of a lake and its drainage basin are complex and require much work and effort (Sokolov and Chapman, 1974; Winter, 1981, 1995; Wetzel and Likens, 2000). Because of numerous local fluctuations in climate from year to year, analyses should extend over several years. Methodology varies greatly in relation to objectives and is beyond the scope of this work. Many detailed reviews are available on specific hydrological and isotopic methods (see Chow, 1964; Stout, 1967; Gray, 1970; Rodda *et al.*, 1976; Kirkby, 1978; Gordon *et al.*, 1992; Sen, 1995).

III. SUMMARY

1. The hydrological cycle consists of the global processes affecting the distribution and movement of water.
   a. Greater evaporation from the oceans is counterbalanced by greater precipitation onto land masses.
   b. Although the amount of water in the atmosphere is small, its retention time is low, and it cycles on the average every nine days.
   c. Once moved from sites of evaporation, water is returned by precipitation. Much of this water is returned to the atmosphere by evaporation and plant transpiration.
   d. On land, water is absorbed by soil, stored within ground water, and moves by gravity to stream channels and lake depressions. Retention times within groundwater reservoirs are variable and depend on the composition of the soil and rock, slope gradients, vegetation cover, and climate. Groundwater flow rates are generally slow and the pathways long.
   e. Retention times in lakes are generally short (6–7 years on the average).
   f. Human modification of the environment can result in major alterations in the global water balance and the climate.
2. Changes in water storage and retention in lakes result from alterations in the balance between input rates from all sources and rates of water losses.
   a. Water income results from:
      i. Precipitation directly on the lake surface.
      ii. Water from surface influents of the drainage basin.
      iii. Groundwater seepage below the surface of the lake through the sediments or as discrete subsurface springs.
   b. Losses of water from lakes occur by:
      i. Flow from an outlet in the most common ***drainage lakes*** or by seepage through the basin walls into the ground water in ***seepage lakes.***
      ii. Direct evaporation from the lake surface.
      iii. Evapotranspiration from emergent and floating-leaved aquatic plants.

5

LIGHT IN INLAND WATERS

Solar radiation is of fundamental importance to the dynamics of aquatic ecosystems. Nearly all energy that controls the metabolism of lakes and streams is derived directly from the solar energy utilized in photosynthesis. That energy stored in photosynthetically formed organic matter is either synthesized within the lake or stream (***autochthonous***) or within the drainage basin and brought to the lake or stream in various forms (***allochthonous***). Utilization of this energy received within the water or imported from the drainage basin and factors that influence the efficiency of conversion of solar energy to potential chemical energy are fundamental to lake productivity.

In addition to these effects, absorption of solar energy and its dissipation as heat have profound effects on the thermal structure, stratification of water masses, and circulation patterns of lakes. Nutrient cycling, distribution of dissolved gases and biota, and behavioral adaptations of organisms are all markedly influenced by the thermal structure and stratification patterns. Therefore, the optical properties of lakes and reservoirs are important regulatory parameters in the physiology and behavior of aquatic organisms. They deserve detailed scrutiny.

I. LIGHT AS AN ENTITY

The term ***light*** is often confusing, in part because of different usages in absolute physical terms, reactions of visual receptors of animals to light, and responses of plants to light energy. For purposes of this discussion, it is essential that light be treated physically, as part of the radiant energy of the electromagnetic spectrum. Light is energy, that is, something that is capable of doing work and of being transformed from one form into another, but can neither be created nor destroyed. Radiant energy is transformed into potential energy by biochemical reactions, such as photosynthesis, or into heat. Energy transformations are far from 100% efficient in a system such as a lake, and most of the radiant energy is lost as heat.

A. Electromagnetic Spectrum

The electromagnetic spectrum is expressed as units of frequency (cycles per time) and wavelength. At one extreme are cosmic rays of very high frequency (10^{24} cycles per second (cps)) and short wavelength (10^{-14} cm), and at the other are radio and power transmission waves of low

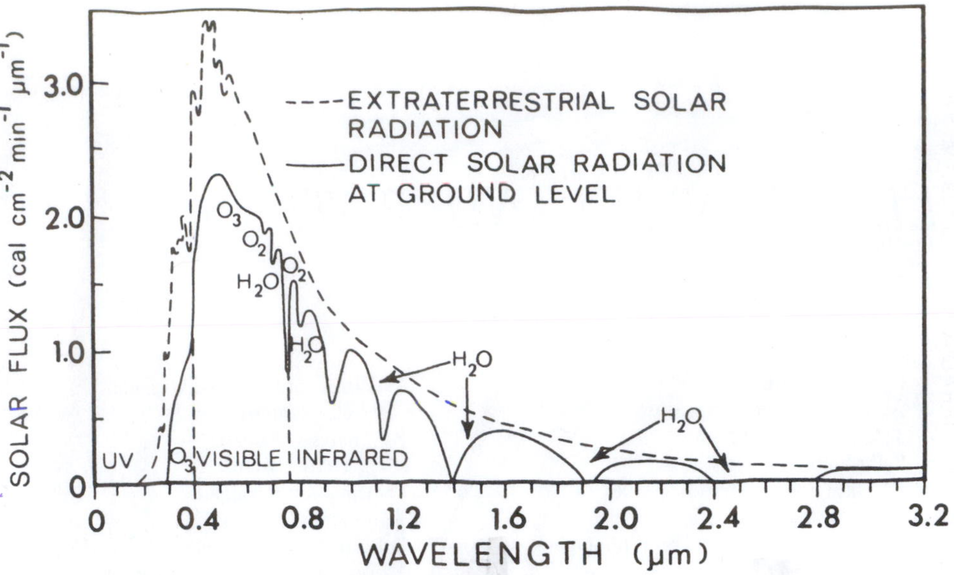

FIGURE 5-1 Extraterrestrial solar flux and that at the surface of the Earth, showing major absorption bands from atmospheric O_2, O_3, and water vapor. (Modified from Gates, 1962.)

frequency (1 cps) and long wavelength (to 10^{10} cm). For all practical purposes, solar radiation constitutes all of the significant energy input to aquatic systems. This solar flux of energy consists of wavelengths of 100 to >3000 nm, from the ultraviolet (UV) to infrared radiation, as it is received extraterrestrially (Fig. 5-1). As solar radiation penetrates and diffuses in the atmosphere of the earth, the energy of certain wavelengths is absorbed strongly and attenuated by scattering. The visible portion of the spectrum, with maximum energy flux in the blue and green portions (480 nm) of the visible range, is only a small amount of the total energy radiated by the Sun (Fig. 5-1). Ultraviolet energy is absorbed largely by ozone and oxygen, and infrared wavelengths are absorbed by water vapor, ozone, and carbon dioxide.

As one evaluates the mechanisms by which life receives light energy, it is important to view light as the radiation of packets of energy termed ***quanta*** or ***photons***. A photon is a pulse of electromagnetic energy; as this energy propagates it has an electric (*E*) and magnetic (*H*) field, with respect to direction of flux and wave characteristics of wavelengths (*l*) and amplitude (*A*) (Fig. 5-2). Hence, light is effectively a transverse wave of energy that behaves as a movement of particles with defined mass. The photon carries energy in a wave conformation.

B. Absorption of Light

Absorption of light energy by atoms and molecules can occur when the electrons of the atoms and molecules resonate at frequencies that correspond to an energy state of a photon. In the collision of an electron and a photon, the electron gains the quantum of energy lost by the photon. It is important to keep this basic photochemical relationship in mind, because the quantum energy imparted by the photon functions in relation to frequency, and each molecular or atomic species has a unique set of absorption characteristics or bands. *Life responds to quantum energy of photons at specific frequencies.*

If the energy distribution of solar flux is plotted against wavelength, as was done in Figure 5-1, the maximum monochromatic intensity of sunlight appears to occur in the blue-green portion of the visible spectrum, an illusion caused by the manner of presenting the data. It is more meaningful to express energy against frequency, where the area under any portion of the curve

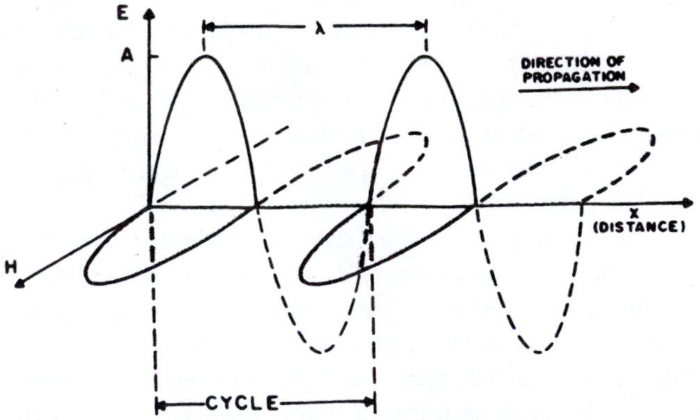

FIGURE 5-2 Instantaneous electric (*E*) and magnetic (*H*) field strength vectors of a light wave as a function of position along the axis of propagation (*x*), showing the amplitude (*A*), wavelength (*l*), cycle, and direction of propagation. (From Bickford, E. D., and Dunn, S.: *Lighting for Plant Growth*. Kent, Ohio, Kent State University Press, 1972.)

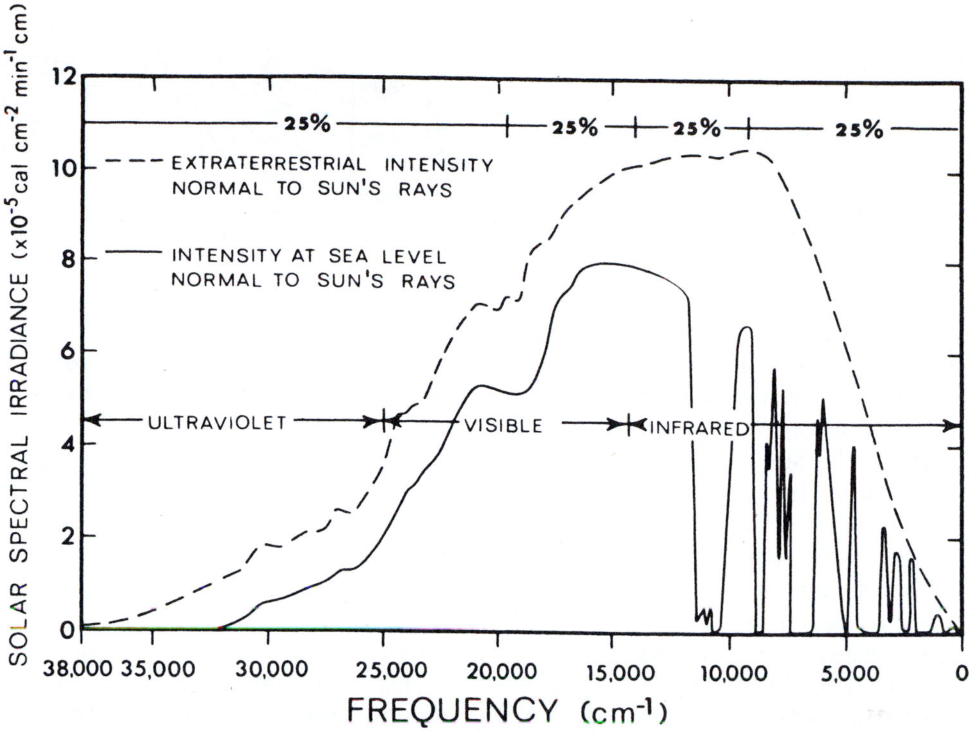

FIGURE 5-3 Solar spectrum at the mean solar distance from the earth as received outside of the earth's atmosphere and at sea level on a surface perpendicular to the solar rays (solar constant of 2.00 cal cm^{-2} min^{-1}). (Modified from Gates, 1962.) Frequency (cm^{-1}) of the abscissa refers to the wavenumber (v^1), or number of wavelengths per cm, and is therefore equal to the reciprocal of the wavelength.

is directly proportional to energy (Fig. 5-3). The energy of the photon in the electromagnetic spectrum is proportional to frequency and inversely proportional to wavelength in accordance with Planck's law:

$$e = hv$$

where:

e = energy of the photon, ergs
h = Planck's constant, 6.63×10^{-27} ergs
v = frequency of the radiation in cps.

When energy is expressed against frequency (Fig. 5-3), the true maximum energy of solar irradiance is found in the infrared at wavelengths somewhat greater than 1000 nm, or 1 mm. The median value of irradiance occurs in the near infrared at a frequency of 14,085 cm^{-1} (= a wavelength of 710 nm), slightly above the visible range. A major portion (29%) of the incoming radiation occurs at wavelengths greater than 1000 nm (frequency $<$ 10,000 cm^{-1}) and 50% beyond the red portion of the visible range. This distribution of energy shifts somewhat as the solar radiation passes through the Earth's atmosphere (Fig. 5-3). The point to be made, however, is that *a large portion of irradiance impinging on the surface of water is in the infrared portion of the solar spectrum and has major thermal effects on the aquatic system.*

Wavelength (l) is a quantitative parameter of any periodic wave motion, not only of light but also of water movements, as will be discussed later on. It is defined simply as the linear distance between adjacent crests of waves and is equal in cm to the speed of light ($c = 2.998\ 10^{10}$ cm s^{-1}) divided by the frequency (v) in cycles per second (cps):

$$l = c/v$$

Wavelength is also often expressed as the wave number (v^1, or k), which is the number of wavelengths per cm, or the reciprocal of the wavelength:

$$v^1 = 1/l$$

The speed of light is reduced in a roughly linear fashion as it passes through transparent materials of increasing density (Table 5-1). (A conversion table for

TABLE 5-1 Speed of Light (589 nm, sodium D-lines)

Medium	*Speed (cm s^{-1})*
Vacuum	2.9979×10^{10}
Air (760 mm, 0°C)	2.9972×10^{10}
Water	2.2492×10^{10}
Glass	1.9822×10^{10}

commonly used units of length and irradiance is given in the appendix.)

II. LIGHT IMPINGING ON WATER

The amount of solar energy that reaches the surface of a lake or other water body is dependent upon an array of dynamic factors. The amount of direct solar energy per unit of time from the Sun, incident upon a surface outside the atmosphere perpendicular (normal) to the rays of the Sun at an average distance of the Earth from the Sun, is referred to as the ***solar constant.***[1] The amount of energy received is a function of the angular height of the Sun incident to the earth and is greatly influenced by latitude and season (Fig. 5-4). The angle of light rays impinging on the water has a marked effect upon the productivity within the water, as will be illustrated repeatedly in subsequent chapters. In equatorial regions, sunlight impinges vertically and leads to relatively constant energy inputs. In temperate and polar areas, however, the sun's angle changes with the sequence of the seasons (Fig. 5-4). Time of day is another factor that influences strongly the solar flux reaching the surface of water bodies, for time of day influences the position of the Sun and the distance that the path of light must travel through the absorbing atmosphere. In polar extremes, for example, direct solar energy decreases to zero for over one-third of the year, and polar waters then receive thermal radiation only from indirect sources.

The absorptive capacities of the atmosphere for solar radiation are governed largely by oxygen, ozone, carbon dioxide, and water vapor, as discussed previously. Moreover, atmospheric transparency can be modified extensively in some regions by both industrial- and urban-derived contaminants. Scattering and absorption also increase in moist air, which is more common on the downwind side of large water bodies. The elevation of a lake and the angular height of the sun both determine the quantity of atmosphere through which the radiation must travel. In summary, the amount and spectral composition of ***direct solar radiation*** reaching the surface of a water body vary markedly with latitude, season, time of day, altitude, and meteorological conditions.

Indirect solar radiation from the sky is largely the result of the scattering of light as it passes through the atmosphere. The extent of scattering is a function

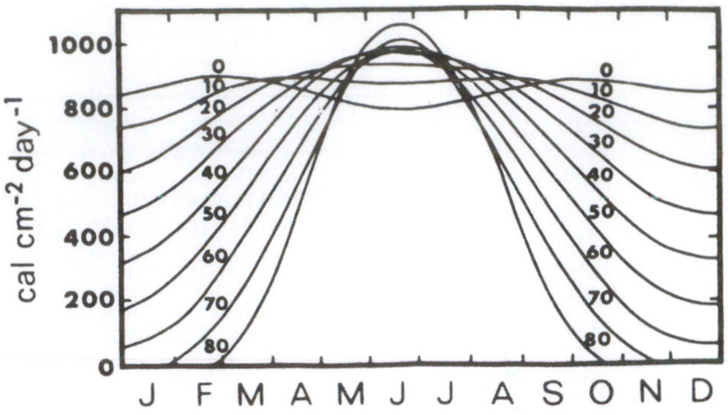

FIGURE 5-4 Daily totals of the undepleted solar radiation received on a horizontal surface for different geographical latitudes as a function of the time of year (solar constant 1.94 cal cm^{-2} min^{-1}). (After Gates, 1962.)

of the fourth power of the frequency, and therefore the UV and shorter wavelength radiation of high frequencies is reduced by about one-fourth as a result of scattering. The result is the blue sky we see directly overhead on clear days. The factors influencing direct solar radiation also influence scattering, but of particular importance are solar height and the atmospheric distance through which the light must pass. The percentage of indirect radiation increases significantly as deviations of the rays from the perpendicular increase, for example, 20–40% at a Sun elevation from the horizontal of 10° in contrast to 8 to 20% at a Sun elevation of 40°.

A. Ultraviolet Radiation and Its Absorption

Reduction of the ozone in the stratosphere from reactions with chemical pollutants has resulted in increases in global ultraviolet radiation (e.g., Kerr and McElroy, 1993; Madronich, 1994). Appreciable increases in the ultraviolet radiation in the past several decades have impacted aquatic ecosystems by enhancing injury to organisms, altering the energy devoted to production of protective pigments, and altering rates of biogeochemical cycling of organic compounds.

Of the ultraviolet radiation (UV) band (100–400 nm), nearly all of UV-C (<280 nm) is absorbed by the stratospheric gases and by the water of aquatic ecosystems. Although relatively little UV-B (280–320 nm) passes through the stratosphere, UV-B is highly energetic and an important photoactivating agent in waters. UV-A (320–400 nm) is less energetic than UV-B but is absorbed less readily in water and penetrates more deeply. Recent measurements *in situ* have demonstrated great variability in the penetration of UV-B and UV-A but much greater penetration than was believed previously (see discussion below).

[1] The solar constant is difficult to measure but recent evidence from satellite instrumentation indicates a value of 1.94 cal cm^{-2} min^{-1} (Drummond, 1971; Gates, D. M., personal communication). Hickey *et al.* (1980), found 1.97 cal cm^{-2} min^{-1} (1376.0 W m^{-2}).

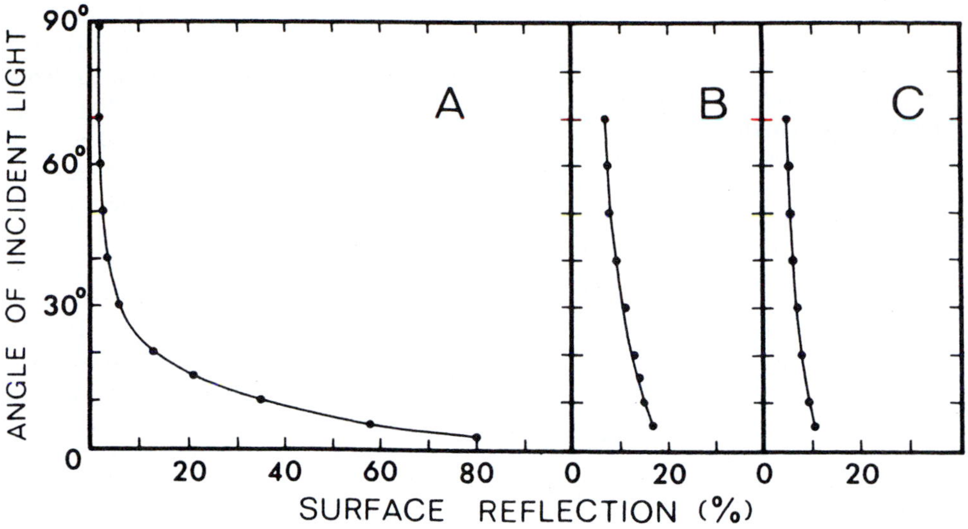

FIGURE 5-5 Surface reflection and backscattering as a percentage of total solar radiation at angles of incidence varying from the horizontal. (A) Clear, cloudless conditions; (B) reflection of diffuse light under moderate cloud cover; (C) heavily overcast conditions. (Generated from data in Steleanu, 1961, and Sauberer, 1962.)

B. Distribution of Radiation Impinging on Water

All of the solar radiation that impinges upon the surface of an inland water body does not penetrate the water. A significant portion is ***reflected*** from the surface and is lost from the system unless it returns to the water after being backscattered from the atmosphere or surrounding topography.

The extent of reflectivity of solar radiation varies greatly with the angle of incidence, the surface characteristics of the water, the surrounding topography, and the meteorological conditions. The ***reflection*** (R) of unpolarized direct sunlight as a fraction of the incident light is a function of Fresnel's law:

$$R = \frac{1}{2}\left[\frac{\sin^2(i-r)}{\sin^2(i+r)} + \frac{\tan^2(i-r)}{\tan^2(i+r)}\right]$$

where i = angle of incidence, and r = angle of refraction. This function states simply that the reflectivity is dependent upon the solar height from the zenith (Fig. 5-5A), that is, the greater the departure of the angle of the Sun from the perpendicular, the greater the reflection will be. Indirect radiation from the sky is also reflected from a water surface but is less affected by solar height (Fig. 5-5B). Under an overcast sky the amount of indirect light that is reflected decreases (Fig. 5-5C). An average reflectance loss value of 6.5% is common, although reflective losses can be reduced further if surrounding topography—mountains, for example—moderates low-angle radiation (Sauberer, 1962).

When the surface of water is disturbed by wave action, reflection increases by about 20% at low angles of incident light (ca. 5°) to approximately a 10% increase at higher angles (5–15°). The difference is small at angles of incidence greater than 15° from the horizontal. Reflection may decrease slightly when the waves are very large and light is exposed to the water surfaces at angles more closely approaching the perpendicular. Ice and particularly snow cover affect markedly the reflectivity of light from the surfaces of lakes. Although data are meager, clear, smooth ice has reflection characteristics similar to those of undisturbed liquid water. Changes in the texture of the ice generally increase reflectivity. Reflection increases markedly with the greatly increased angles and quantity of surface planes of granular ice in the form of snow cover. On the average, about 75% of incident light striking snow is reflected, and under some conditions, as much as 95% can be reflected.[2]

Of the total incident light impinging upon the surface of water, a reasonable average amount that is reflected on a clear, summer day is 5–6%. This mean value increases to about 10% during winter. Qualitatively, light in the red portion of the spectrum is reflected to a slightly greater extent than light of higher frequencies, particularly at low angles of incidence. About half of the total quantity of light leaving the lake is by reflection and half by scattering of light.

[2] The ***albedo***, the ratio of reflected to incident irradiance, is influenced by many factors, such as internal reflections, bubble and particle inclusions, and crystal orientations within the ice and snow, zenith angle, upwelling of scattered light, and other factors (Sauberer, 1962; Adams, 1978; Stewart and Brockett, 1984). Moreover, the distribution of winter ice and snow thickness and condition is often nonuniform over the surface of lakes (Adams, 1978, 1981).

C. Scattering of Light

Scattering of light from the water results in the loss of large amounts of light energy from the lake. This phenomenon is apparent to anyone who has looked down into relatively clear waters where the surface reflection is eliminated. Of the total light energy entering water, portions are absorbed by the water and particles suspended in it, as will be discussed in detail further on, and a significant portion is scattered. ***Scattering of light*** is the result of deflection of quanta by the molecular components of the water and its solutes but also, to a large extent, by particulate materials suspended in the water.

The scattering of light energy can be viewed in a simple way as a composite of reflection at a massive array of angles internally within the water (cf. Mobley, 1994). The energy scattered in all directions within a volume of water varies greatly with the quantity of suspended particulate matter and its optical properties. Volcanic siliceous materials, for example, scatter light much less than suspended particulate matter of lower transparency.

The scattering of light can change significantly with depth, season, and location in a lake and in response to variations in the distribution of particulate matter. When particulate matter is concentrated in the middle zone of great density change (metalimnion) of a thermally stratified lake, either as a result of reduced sinking as the particles encounter increased water densities or as a result of the development of large populations of plankton in certain strata, the scattering of light can increase. Scattering can also increase markedly in areas of the lake where wind-induced currents and wave action agitate and temporarily suspend littoral and shore deposits of particulate matter (Tyler, 1961b). When dimictic or amictic lakes (see Chap. 6) undergo complete circulation, a significant portion of the recent sediments of the lake basin are brought into resuspension (Wetzel *et al.*, 1972; Davis, 1973; White and Wetzel, 1975) and can affect the scattering properties of the lake for an extensive period (weeks). Similarly, the variable influxes of suspended inorganic and organic matter from stream inflows to reservoirs can radically increase the scattering of light nonuniformly within the lake basin. For example, cold, silt-laden, glacial meltwaters can move as subsurface density currents along the metalimnion far out into a lake or reservoir.[3] These alterations in optical properties can be short- or long-lived depending upon the composition and density of the material and the density characteristics of the recipient water at that particular time.

[3] This common phenomenon is called *la bataillere* in the Lake of Geneva, Switzerland.

A significant amount of light can be reflected and scattered from the sediments both in the littoral zone and in shallow areas, as well as from the bottom of moderately deep, clear lakes (Fig. 5-6). The amount of light returned to the water is dependent upon the composition of the sediments; sand sediments or those sediments rich in calcium carbonate (marl; $CaCO_3$) reflect considerably more light than dark-colored sediments of high organic content.

Differential scattering of light also depends on scattering coefficients for different wavelengths, as well as on absorption characteristics for different wavelengths. In very clear water, scattering occurs predominantly in the blue portion of the visible spectrum. As the quantity and size of suspensoids increase, radiation of longer wavelengths is scattered preferentially and greater absorption of the light of high frequencies occurs. Hardwater lakes with large amounts of suspended $CaCO_3$ particles characteristically backscatter light that is predominantly blue-green; lakes rich in suspended organic materials appear more green or yellow.

Diffuse light from scattering and reflecting sources is of obvious importance to organisms that utilize it directly in photosynthesis or indirectly in behavioral responses. In lake ecosystems where light enters the environment unidirectionally from above, diffuse scattered light can form a major supplementary source of energy. The amount of light scattering can easily be one-fourth of the light absorbed by the water. The true values are likely even higher, because most of the existing data on scattering were acquired with unidirectional (2π—flat,

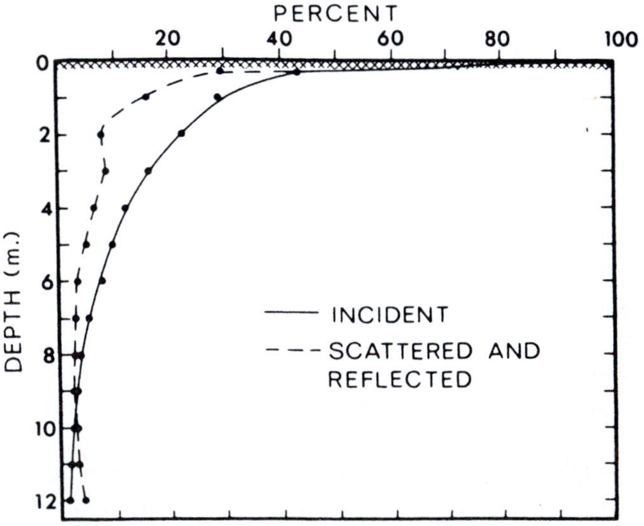

FIGURE 5-6 Comparison of incident light penetration (percentage of total surface light) with depth to that backscattered from concentrations of plankton, especially at 3 m, and from calcareous sediments. Lawrence Lake, Michigan; 17 March 1972; 23 cm cloudy ice.

cosine) instruments. When scattering values obtained with this type of sensor are compared with those obtained with instruments approaching 4π (spherical, scalar) geometry (e.g., Rich and Wetzel, 1969; Smith and Wilson, 1972; Kirk, 1994), higher values are commonly found. Comparisons between cosine and scalar measuring instruments among a range of depths and weather conditions in Loch Neigh, Northern Ireland, for example, showed that cosine values were between 64 and 80% of scalar values (Jewson, 1984). A portion of the scattered light is returned to the surface of the water, and much (80–90%) of that which returns to the surface is lost to the atmosphere. Because of differences in lightwave refraction properties at different frequencies, somewhat less of the red scattered light reaches the surface than the blue.

Several terms are in common usage in relation to light radiation used by photosynthetic organisms. ***Photosynthetically active radiation*** (PAR) is defined as radiation in the 400–700-nm waveband. The photosynthetically active range may extend below 400 nm to wavelengths as low as 290 nm (Halldal, 1967; Klein, 1978), but the amount of subsurface irradiance available at the shortest wavelengths is very small, especially if the water contains significant amounts of dissolved organic compounds (as most inland waters do; see the following discussion). PAR is a general radiation term that is applicable to both energy terms and the preferred photon (quantum) term.

Photosynthetic irradiance is the radiant energy (400–700 nm) incident per unit time on a unit surface (i.e., radiant energy flux density of PAR in units such as W m^{-2}). ***Photosynthetic photon flux density*** is the number of *photons* (quanta) in the 400–700-nm waveband incident per unit time on a unit surface (i.e., photon flux density of PAR in units μmol m^{-2} s^{-1}, formerly μeinsteins m^{-2} s^{-1}).[4]

The total underwater light received by a receptor system, such as that received by an algal cell from all angles, is the optimum measure of radiant energy available for photosynthesis. This realistic value is the ***photosynthetic photon flux fluence rate*** or PPFFR (i.e., photon scalar irradiance or scalar quantum irradiance, Smith and Wilson, 1972), which is defined as the integral of photon flux radiance at a point over all directions about the point. In other words, PPFFR is a measure of the total number of photons of PAR per unit time and area arriving at a point from all directions about the point when all directions are weighted evenly (Smith and Wilson, 1972). Such spherical, 4π quantum sensors are now available commercially and respond equally to all photons in the 400–700 nm range (units in μmol m^{-2} s^{-1}).

Photosynthetic and phototropic responses of organisms are related to the number of quanta of light of specific frequencies (wavelengths) impinging upon biochemical receptor systems. The attenuation of illuminance under water between a spectral range of 350 to 700 nm is not exactly the same for energy units as it is for quanta (Steemann, Nielsen, and Willemoës, 1971; Lewis, 1975). The rate of attenuation of quanta is approximately equal to that of energy in moderately clear water bodies containing intermediate concentrations of dissolved organic carbon (approximately 5 mg $liter^{-1}$). In more transparent waters, energy is absorbed at a slower rate than quanta; the opposite is true in more deeply stained waters containing greater amounts of dissolved organic matter. The divergence in penetration of quanta and energy can increase in certain specific portions of the spectral range and thus affect the utilization by the action spectra of organisms. For example, small differences in the utilization of illuminance during photosynthesis between diatoms, green algae, and cyanobacteria with differing action spectra of photosynthetic pigments occur when the rate of illumination is measured in energy units. The ratio of total quanta to total irradiance energy within the spectral region of photosynthetic activity, however, varies by no more than $\pm 10\%$ and is $2.5 \pm 0.25 \times 10^{18}$ quanta s^{-1} W^{-1} within a number of waters differing in optical characteristics (Morel and Smith, 1974). The ratio can be used to determine accurately the total quanta available for photosynthesis from measurements of total energy; the converse is also true.

III. THERMAL RADIATION IN LAKE WATER

Lake water behaves like a blackbody[5] to radiation of low frequency (wavenumber 14,000 cm^{-1}) and long wavelengths (750–12,500 nm). Because of surface reflection, only about 97% of thermal radiation is emitted into the atmosphere. The atmosphere does not function as a blackbody, but thermal radiation from it is influenced by water vapor pressure and the extent of cloud cover. At night, the net emission of thermal radiation from the surface of a lake, in cal cm^{-2} day^{-1}, approximates [11 × (°K of water − °K of air)] or simply [11 × (temperature of the water − that of the air)] (cf. discussion in Hutchinson, 1957).

[4] The einstein has been used to represent both the quantity of energy in Avogadro's number of photons and also Avogadro's number of photons (Incoll *et al.*, 1977). When used as the quantity of photons, 1 einstein (E) = 1 mole = 6.02×10^{23} photons. The E has now been replaced in the international system (SI) by moles.

[5] A blackbody absorbs all of the radiant energy incident upon it. The thermal radiation (Q) from a blackbody is proportional to the fourth power of the absolute temperature (K): $Q = 8.26 \times 10^{-9}$ $(K)^4$.

A. Net Radiation

The net amount of solar radiation affecting a lake can be referred to as the ***net radiation surplus*** (Q_B):

$$Q_B = Q_S + Q_H + Q_A - Q_R - Q_U - Q_W$$

where

Q_S = direct solar radiation
Q_H = indirect scattered and reflected radiation from sky and clouds
Q_A = long-wave thermal radiation from the atmosphere and from surrounding topography (the latter is usually insignificant)
Q_R = radiation reflected from the lake
Q_U = radiation scattered upward and lost
Q_W = emission of long-wave radiation.

At night, most components are negligible, and the net radiation surplus becomes equal to the long-wave thermal radiation of the atmosphere minus that emitted from the water:

$$Q_B = Q_A - Q_W$$

or, approximately, $Q_B = -11$(temperature of the water − the temperature of the air) in cal cm^{-2} day^{-1}. We will return to these relationships when discussing heat budgets in the following chapter. It should be emphasized, however, that even though the mean value of Q_B may be positive, the lake can be losing heat through evaporation and convective heating of the air (Hutchinson, 1957).

The collective value for the inputs ($Q_S + Q_H + Q_A$) can be measured directly by sensitive Moll thermopile pyranometers. The subject of radiation measurement is treated excellently by Latimer (1972), Bickford and Dunn (1972), Šesták *et al.* (1971), and Coulsen (1975).

IV. TRANSMISSION AND ABSORPTION OF LIGHT BY WATER

The quantity of light energy penetrating the water is dispersed by the mechanisms just discussed and absorbed. The diminution of radiant energy with depth, by both scattering and absorption mechanisms, is referred to as ***light attenuation,*** whereas ***absorption*** is defined as diminution of light energy with depth by transformation to heat (cf. Westlake, 1965). It is important to understand the selective absorptive properties of water, first in pure water and then in waters of differing optical characteristics.

The transmission and absorption of light in water can be approached in several ways. Perhaps the most direct way is to look first at the percentage of transmission or absorption of monochromatic light through given depths of pure water. This ***percentile absorption,*** or Birgean percentile absorption (after E. A. Birge, who used the relationship extensively), is based on the expression

$$\frac{100(I_0 - I_z)}{I_0}$$

where

I_0 = irradiance at the lake surface
I_z = irradiance at depth z, in this case taken as 1 m

In distilled water, the percentile absorption is very high in the infrared region of the spectrum, decreases rapidly in the lower wavelengths to a minimum absorption in the blue, and then increases again in the violet and especially UV wavelengths (Table 5-2). These absorption relationships usually are expressed graphically in linear or, even better, in logarithmic form (Fig. 5-7). Generally about 53% of the total light energy is transformed into heat in the first meter.

The light intensity or irradiance, I_z, at depth z is a function of intensity at the surface (I_0) and the log of the negative extinction coefficient (η) times the depth distance, z, in meters:

$$I_z = I_0 e^{-\eta z}$$

or

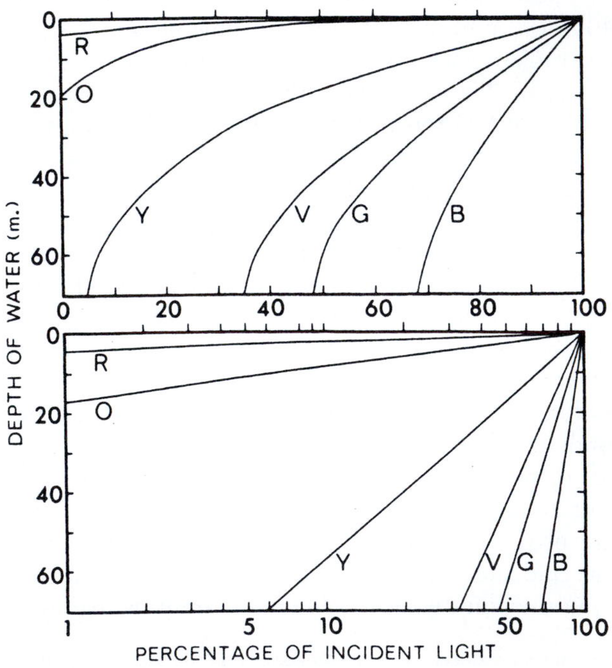

FIGURE 5-7 Transmission of light by distilled water at six wavelengths (*R*–720, *O*–620, *Y*–560, *G*–510, *B*–460, *V*–390 nm). Percentage of incident light that would remain after passing through the indicated depths of water expressed on a linear (*upper*) and a logarithmic (*lower*) scale. (After Clark, 1939.)

TABLE 5-2 Absorption Coefficients (Percentile Absorption) and Extinction Coefficients of Monochromatic Light of Liquid Water at 21.5°C by Laser Optoacoustic Spectroscopy[a]

Wavelength (nm)	*Wave number (v^1) (cm^{-1})*	*Absorption coefficient (10^{-4} cm^{-1}) ("percentile absorption")*	*Extinction coefficient (η m^{-1})*
820 (infrared)	12,200	(91.1)	2.42
800	12,500	(89.4)	2.24
780	12,820	(90.1)	2.31
760	13,160	(91.4)	2.45
740	13,515	(88.5)	2.16
720	13,890	(64.5)	1.04
700	14,285	59.0	0.89
689.2	14,500	47.6	0.646
680.1 (red)	14,700	42.6	0.555
666.5	15,000	38.7	0.489
645.0	15,500	31.4	0.377
624.8 (orange)	16,000	29.6	0.351
605.9	16,500	24.8	0.285
588.0	17,000	12.3	0.131
574.5 (yellow)	17,400	8.1	0.084
546.3	18,300	5.3	0.054
526.2 (green)	19,000	4.0	0.041
512.7	19,500	3.48	0.0354
499.9	20,000	2.33	0.0236
487.7	20,500	1.86	0.0188
473.1	21,000	1.79	0.0181
465.0 (blue)	21,500	2.06	0.0208
454.4	22,000	2.21	0.0224
446.3	22,400	2.38	0.0241

[a] For data from 700 to 446 nm (Tam and Patel, 1979). Values in parentheses are older data determined by 1-m path transmission measurements from James with Birge (1938).

$$\ln I_0 - \ln I_z = \eta z$$

The extinction coefficient (η) is a constant for a given wavelength; approximate values for pure water are given in Table 5-2. This relationship is imperfect in nature because sunlight is not monochromatic, but is instead a composite of many wavelengths.

Direct sunlight rarely enters the water at right angles to the surface, and indirect irradiance is not perpendicular to the surface at any time. Moreover, the natural total extinction coefficient (η_t) is influenced not only by that of the water itself (η_w), but also by absorption of particles suspended in the water (η_p), and particularly by dissolved, colored compounds (η_c). Thus the *in situ* extinction coefficient (η_t) is a composite of these components (Åberg and Rodhe, 1942):

$$\eta_t = \eta_w + \eta_p + \eta_c$$

At low concentrations, the particulate suspensoids have relatively little effect on absorption. With high turbidity, however, the effect is quite significant, particularly at lower wavelengths of the visible spectrum. In detailed analyses of the absorption of lake water and its dissolved components, the particulate fraction is commonly removed by filtration or centrifugation.

A. Effects of Dissolved Organic Compounds

The effects of dissolved organic compounds on the absorption of light energy are very marked and are best introduced by examples taken from the extensive work of James with Birge (1938). In comparison to distilled water, lake water with increasing concentrations of dissolved organic compounds, particularly humic acids, not only reduces drastically the transmission of light but shifts absorption selectively (Table 5-3). Common to all waters is a very high absorption of infrared and red wavelengths, which results in significant heating effects in the first meter of water. At the other extreme, although distilled water absorbs relatively little UV light, even very low concentrations of dissolved organic compounds increase UV absorption greatly. In lakes highly stained with humic compounds, such as Helmet Lake, absorption of UV, blue, and green wavelengths is essentially complete in much less than a depth of 1 m. This relationship of intense absorption of UV light by dissolved organic compounds has been used extensively as a relative assay of their concentrations, as will be discussed in a subsequent chapter on organic matter.

UV-B attenuation depths (z_a = 1% of surface irradiance) ranged from a few centimeters to >10 m

TABLE 5-3 Percentile Absorption of Light of Different Wavelengths by 1 m of Lake Water, Settled of Particulate Matter, of Several Wisconsin Lakes of Progressively Greater Concentrations of Organic Color[a]

Wavelength (nm)	*Distilled water*	*Crystal Lake*	*Lake Mendota*	*Alelaide Lake*	*Mary Lake*	*Helmet Lake*
800	88.9	89.9	90.5	92.4	91.7	93.2
780	90.2	91.3	91.9	93.5	93.0	94.5
760	91.4	93.5	92.6	94.5	94.8	96.0
740	88.5	89.3	91.5	92.7	93.0	96.2
720	64.5	67.6	71.0	78.0	78.0	86.9
700	45.0	50.4	49.7	66.3	70.7	82.5
685	38.0	45.2	42.2	65.7	71.7	86.6
668	33.0	40.3	36.8	65.0	72.3	88.0
648	28.0	37.0	31.9	64.5	75.2	91.2
630	25.0	34.4	28.9	65.8	77.8	94.0
612.5	22.4	32.1	26.3	66.8	80.3	96.0
597	17.8	27.5	22.5	67.0	83.2	97.6
584	9.8	22.0	17.6	67.1	85.7	98.2
568.5	6.0	19.3	14.0	67.6	88.5	98.6
546	4.0	19.2	13.5	70.9	91.6	99.3
525	3.0	19.8	14.1	74.5	94.8	—
504	1.1	20.7	15.2	81.0	97.4	—
473	1.5	21.7	21.7	88.6	99.4	—
448	1.7	23.8	27.8	92.2	—	—
435.9	1.7	24.4	31.0	95.2	—	—
407.8	2.1	28.1	44.3	99.0	—	—
365	3.6	40.0	80.0	—	—	—
Color scale (Pt units)	0	0	6	28	101	264

[a] Selected data from James with Birge (1938).

among a number of lakes (Kirk, 1994; Scully and Lean, 1994; Morris *et al.*, 1995; Williamson *et al.*, 1996; Sommaruga and Psenner, 1997). Much (>90%) of the among-lake variation in diffuse attenuation coefficients (K_d) could be explained by differences in dissolved organic carbon (DOC) concentrations, which directly influenced dissolved absorbance. Throughout the solar UV-B and UV-A range, K_d was well estimated with a univariate power model based on DOC concentration, particularly in waters of low to moderate productivity. The z_a is strongly dependent on DOC concentrations when below 2 mg liter^{-1} and very sensitive to small changes in DOC (Fig. 5-8). In eutrophic lakes, densities of phytoplankton can assume greater importance in UV attenuation (Hodoki and Watanabe, 1998).

UV radiation penetrates more deeply in saline waterbodies than in freshwater ecosystems containing similar concentrations of dissolved organic matter (DOC) (Arts *et al.*, 2000). Over a large number of water bodies in central Saskatchewan, Canada, that ranged widely in DOC concentration (4–156 mg C liter^{-1}) and electrical conductivity (270–74,300 μS cm^{-1}), attenuation of UV-B (280–320 nm) was much greater in fresh waters than in saline waters (Fig. 5-9, *upper*). Attenuation of UV-A (320–400 nm) was also greater in fresh waters than in saline environments (Fig. 5-9, *lower*), although the differences were less pronounced than the situation with UV-B attenuation. The reasons for the differences are not clear, but in shallow saline environments, chromophoric humic substances are very low; most of the DOC is associated with non-

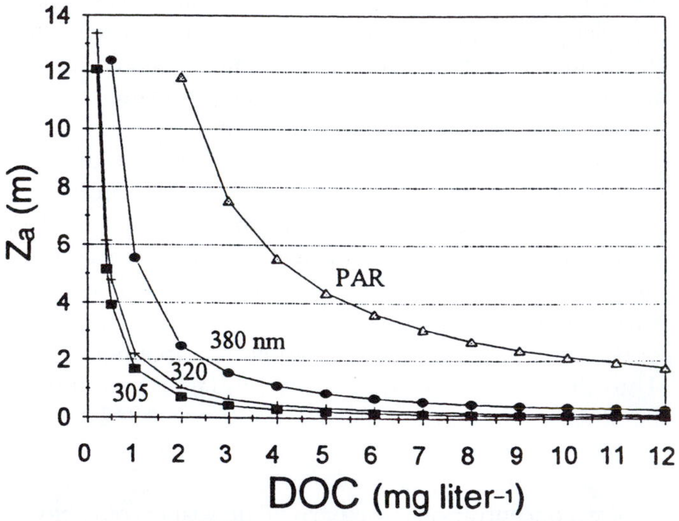

FIGURE 5-8 Relationship between the 1% attenuation depth (z_a = 1% of surface irradiance) and dissolved organic carbon concentration (DOC) based on a survey of 65 glacial lakes in North and South America. Model curves for 305, 320, and 380 nm and photosynthetically active radiation (PAR, 400–700 nm). (After data of Morris *et al.*, 1995, and Williamson *et al.*, 1996.)

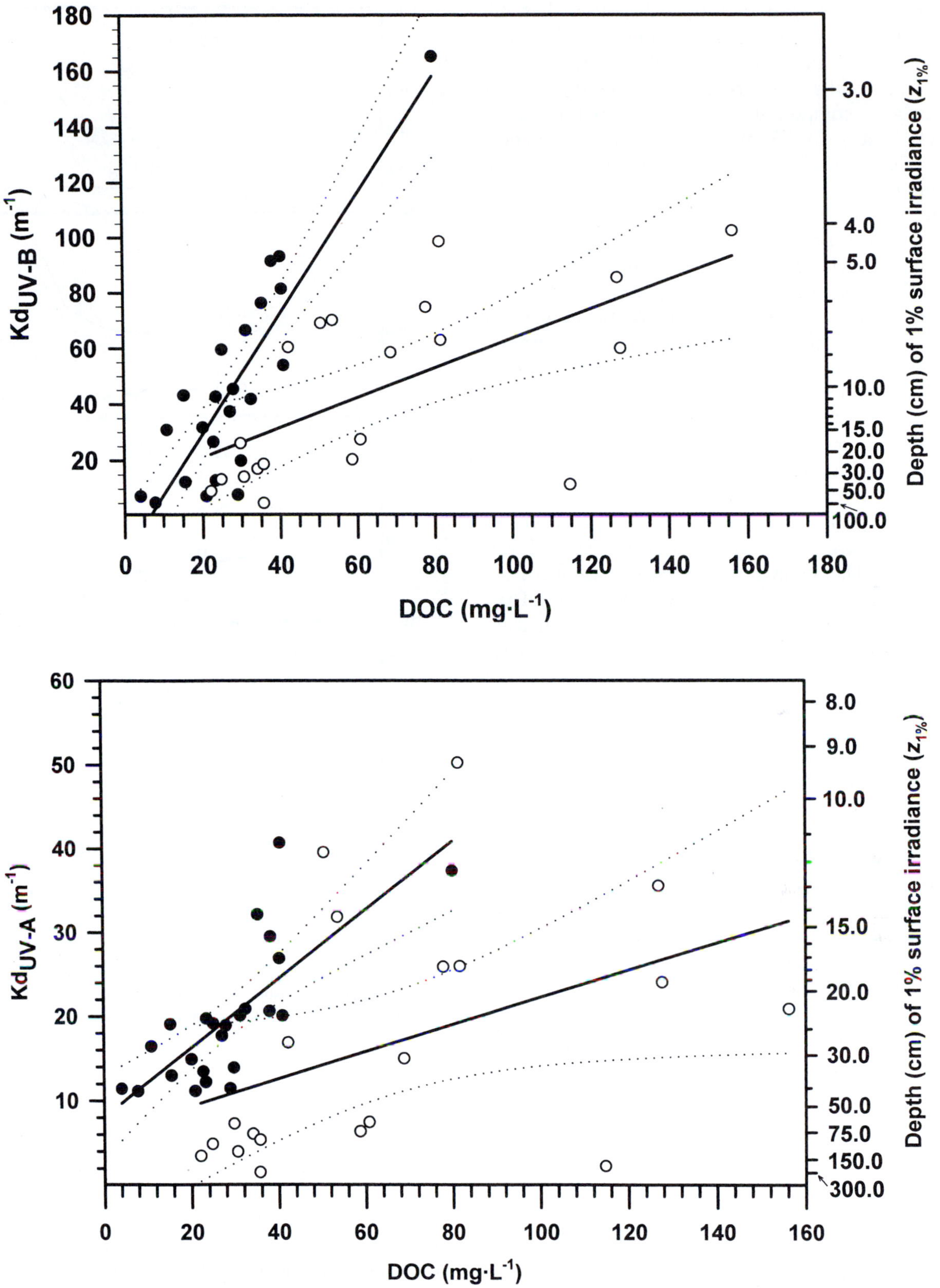

FIGURE 5-9 Vertical attenuation coefficients (K_d, in m^{-1}) and depths ($z_{1\%}$ in cm) at which 1% of the surface irradiance remained as a function of dissolved organic carbon (DOC) concentration for (*upper*) UV-B radiation (280–320 nm) and (*lower*) UV-A radiation (320–400 nm). Filled circles = freshwater lakes; open circles = saline lakes. Dotted lines are the 95% confidence intervals about each regression. (From Arts *et al.*, 2000, by permission.)

chromophoric dissolved organic substances that are quite recalcitrant to microbial degradation. The chromophoric colored substances generally absorb UV radiations that results in partial photolysis to smaller biodegradable compounds or complete photolysis to CO_2 (cf. Chap. 23). Some UV-insensitive non-chromophoric compounds tend to be more stable and can accumulate.

The major effects of dissolved organic matter and particulate suspensoids on absorption of light at varying wavelengths are illustrated graphically in Figure 5-10. The total absorption characteristics in the spectrum (*T*) are compared to the percentile absorption values through 1 m of distilled water (*W*), absorption attributable to particulate suspensoids (*P*), and absorption by lake water that has been filtered to remove particles > 1 mm (*C*), termed *color absorption.* Absorption by dissolved organic compounds ("dissolved color") is selective and greatest in the UV, blue, and green wavelengths. The absorption by dissolved color at the red end of the spectrum is less selective and is probably unrelated to the organic compounds absorbing at lower wavelengths. The extinction coefficients of dissolved color (η_c) increase directly with the color units of the water, which are measured by the relative visual comparisons of the color of the filtered lake water under standard conditions to the color of a specific mixture of platinum–cobalt compounds in serial dilution (discussed further on; cf. also Wetzel and Likens, 1991). In the examples given in Figure 5-10, the water of Crystal Lake indicated zero platinum (Pt) units, Lake George, 24, Rudolf Lake, 50, and that of Helmet Lake, 236 Pt units, which is about the color of weak tea.

Also apparent from this classic work is the relationship that absorption resulting from particulate suspensoids (*P*) is relatively unselective at different wavelengths, particularly at lower concentrations. The η_p functions essentially independently of the η_c but, along with the absorption of water (η_w), η_p values are additive for a particular lake at depth at a given time of year (cf. Åberg and Rodhe, 1942).

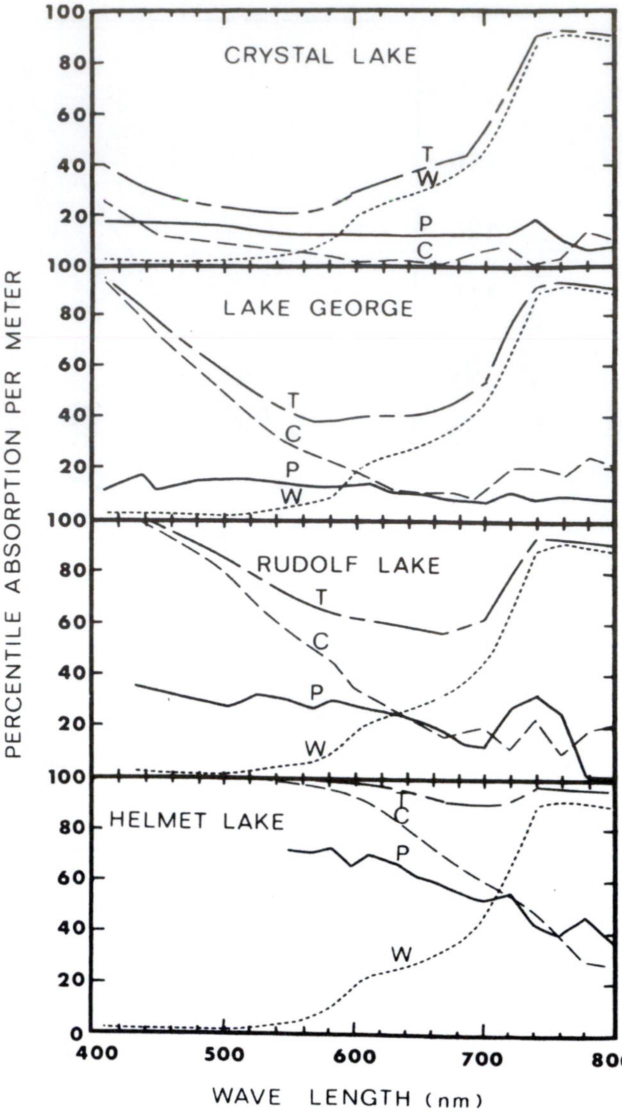

FIGURE 5-10 Percentile absorption of light at different wavelengths passing through 1 m of water of four lakes of northern Wisconsin of increasing concentrations of dissolved organic matter. *T* = total absorption; *C* = absorption by dissolved organic color; *P* = absorption by suspended particulate matter; and *W* = absorption by pure water. (Modified from James with Birge, 1938.)

B. Analysis of Light Transmission

The transmission or absorption within a lake of the total white light from direct insolation and indirectly from the sky has been analyzed in many ways. The vertical extinction coefficient is most commonly determined from the percentile absorption of surface light through depth (Fig. 5-11). The isopleths[6] of these examples of depth–time distribution of light indicate some of the marked fluctuations that are found in natural waters, seasonally and vertically. The composite mean η_t of all depths in Lawrence Lake, an unproductive hardwater lake with rather high concentrations of particulate and colloidal $CaCO_3$ suspensoids, was 0.39 m^{-1} (n = 1746 measured profiles), within an annual range of 0.05 to 1.02. The same value for extremely productive Wintergreen Lake was 1.00 m^{-1} (range, 0.46–1.68). In the former case, the η_t is largely

[6] An isopleth is a line on a graph of a specified constant value showing the occurrence of a parameter as a function of two variables (depth and time in this case). See Wetzel and Likens (2000:26) for details of preparation of these diagrams.

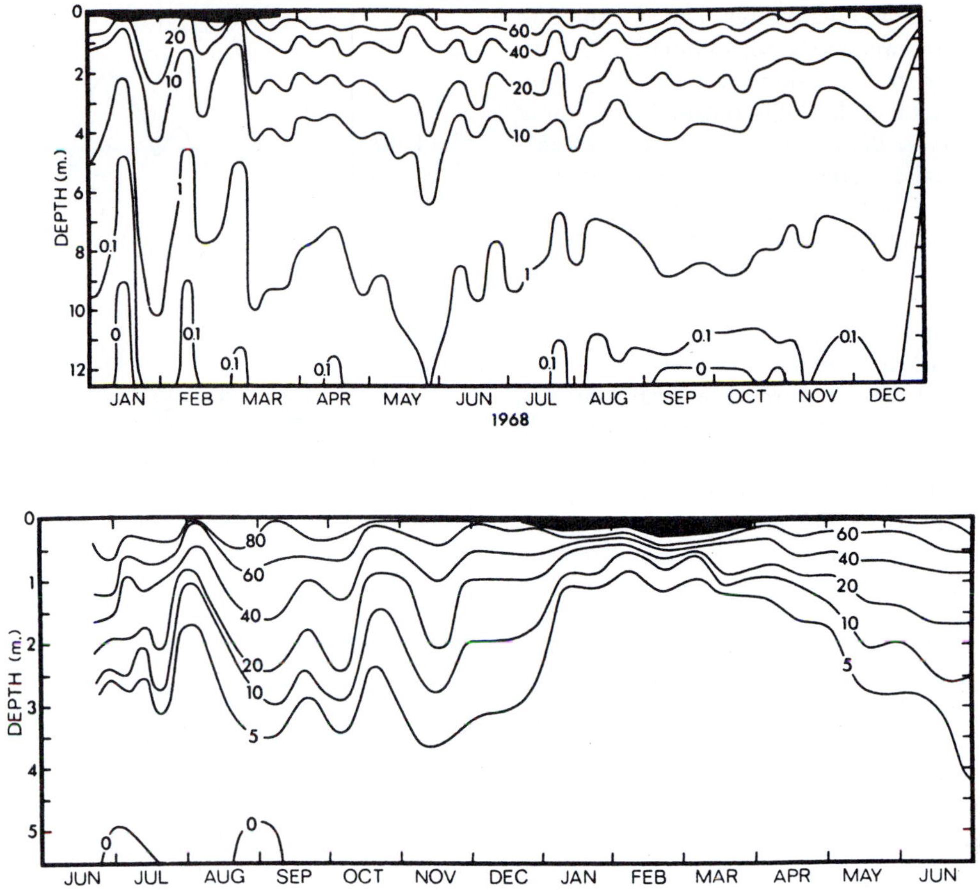

FIGURE 5-11 Isopleths of the percentage transmission of light at the surface with depth and time in unproductive Lawrence Lake (*upper*) and extremely productive Wintergreen Lake (*lower*), southwestern Michigan. (From Wetzel, unpublished observations.)

constant over an annual period, whereas in the latter situation the marked fluctuations in particulate suspensoids of algae are reflected in the mean η_t m^{-1} and mean percentage transmission m^{-1} of the water column (Fig. 5-12).

Calculations of the vertical extinction coefficient are not very reliable in the first meter below the surface because of surface agitation. Average calculations often exclude this region. Direct calculations are made using the formula given earlier, or changed to

$$\eta z = \ln I_0 - \ln I_z$$

The values cited for the vertical extinction coefficient (η_t) obviously represent a composite for all of the wavelengths, each of which is variously influenced by the absorption characteristics of water, particulate

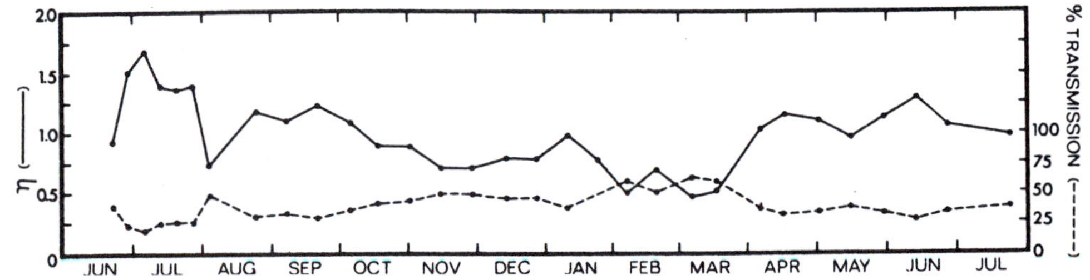

FIGURE 5-12 Average extinction coefficient (η_t) per meter and percentage transmission of light per meter of the water column, Wintergreen Lake, Michigan, 1971–72. (From Wetzel, unpublished data.)

matter, and dissolved organic matter. The η_t values for natural lake waters vary from approximately 0.2 per meter (about 80% transmission m^{-1}) in very clear lakes, such as Crystal Lake, Wisconsin, Lake Tahoe, California, and Crater Lake, Oregon, to about 4.0 m^{-1} in highly stained lake waters or lakes with very high biogenic turbidity. Where turbidity is extremely high, such as in reservoirs near major river inflows under flood conditions (e.g., Roemer and Hoagland, 1979) or in lakes receiving fine materials, such as volcanic ash, that remain in colloidal suspension, extinction coefficients in excess of 10 m^{-1} are not unusual.

As the composite distribution of light transmission with depth is dissected, the spectral selectivity of light reaching deep water should be apparent from the foregoing discussion. It is rather common for the green portion of the spectrum to penetrate most deeply (Fig. 5-13). In very clear waters, the deepest penetration of light is in the blue portion of the spectrum; in Crater Lake, maximum penetration occurred at approximately 469 nm, and in Lake Tahoe at approximately 475 nm (Tyler and Smith, 1970; Smith *et al.*, 1973). In lakes that are darkly stained with dissolved organic matter, practically no light of wavelengths below 600 nm penetrates below 1 m.

The rapid attenuation of light by dense populations of algae or bacteria at certain depths in stratified lakes is common, particularly in the lower depths where certain microorganisms are adapted to low light intensities (Fig. 5-14). Similarly, abiogenic suspensoids

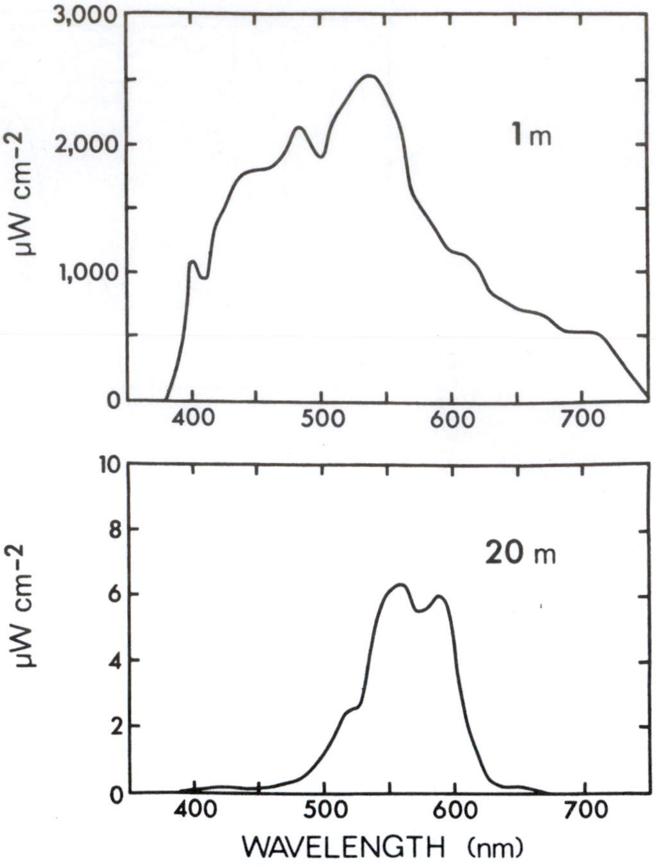

FIGURE 5-13 Comparison of the spectral distributions of energy using a scanning spectroradiometer, Gull Lake, Kalamazoo-Barry counties, Michigan, 16 November 1975 at 1 m (*upper*) and 20 m (*lower*). (From Wetzel and Likens, 1991.)

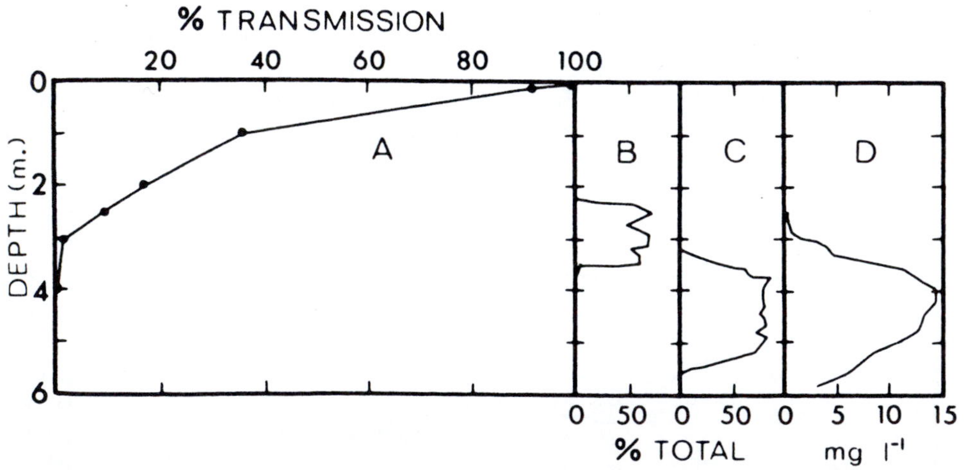

FIGURE 5-14 (A) Percentage transmission of surface incident light with depth in relation to (B) dominating bacterial plate of *Thiopedia* at 2.5–3.5 m and (C) *Clathrochloris* at 3.5–5.5 m (as a percentage of total plankton) and (D) bacteriochlorophyll *d* in Wintergreen Lake, Michigan, 4 July 1971. (From Wetzel, unpublished data, and Caldwell *et al.*, 1975.)

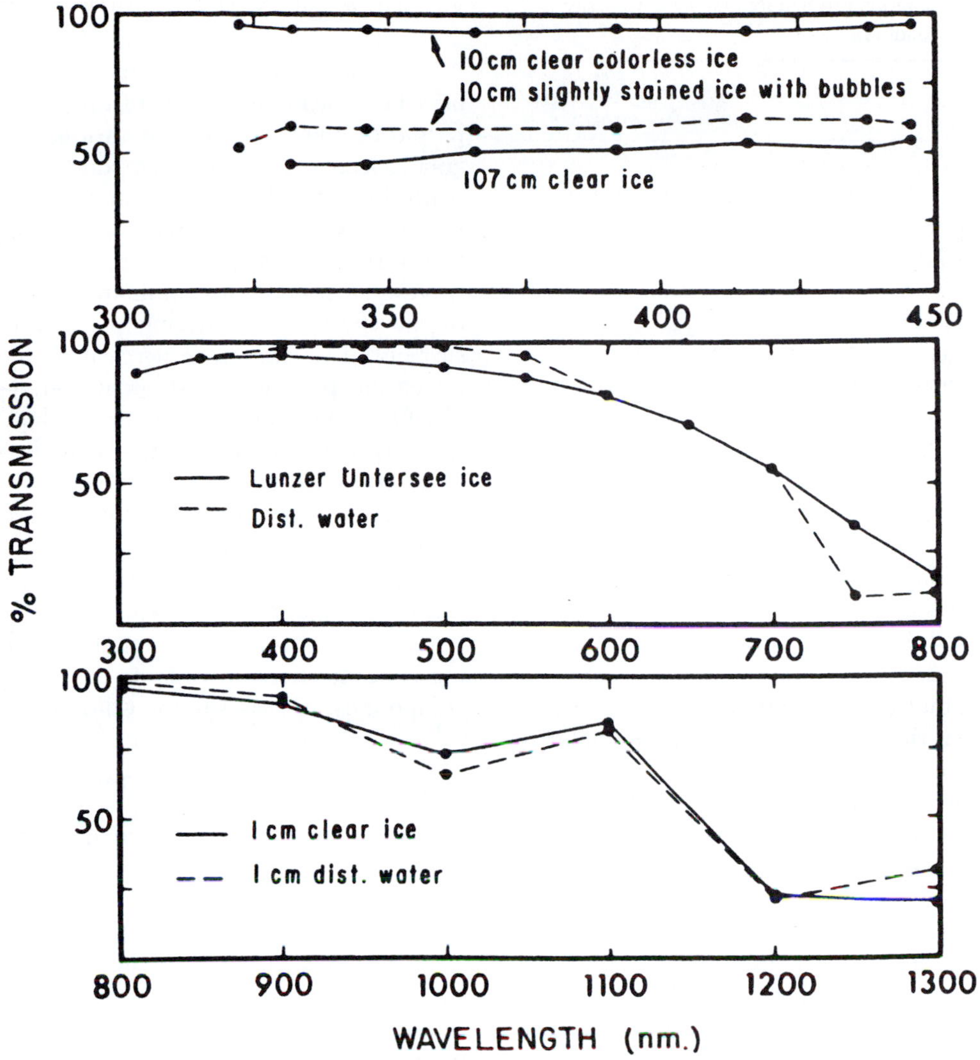

FIGURE 5-15 Percentage spectral transmission of light through ice in comparison to water. (From data of various sources cited in Sauberer, 1950; see also data of Adams, 1978.)

may be brought into a water basin through an inlet at temperatures colder than the surface strata. When this occurs, the suspensoid-laden water will penetrate to deep strata of comparable densities, and the stratified particles can markedly influence the vertical extinction of light.

The characteristics and amount of light penetrating the littoral zone of lakes are greatly altered by the type and extent of development of floating-leaved and emergent higher aquatic plants, as well as by reflection and scattering from the substratum. Since the structure of the littoral macroflora changes seasonally, dense vegetative stands commonly reduce incident light by 50 to over 90% (Gessner, 1955; Szumiec, 1961; Ondok, 1973a, 1973b, 1978; Grimshaw *et al.*, 1997).

V. TRANSMISSION THROUGH ICE AND SNOW

The percentage of light transmission through clear, colorless ice[7] is not greatly different from that through liquid water (Fig. 5-15). Attenuation of light increases greatly, however, if the ice is stained with dissolved organic matter or is cloudy, that is, contains air bubbles or forms irregular crystals upon freezing (Adams, 1978, 1981). Absorption is greatest in the red portion of the photosynthetic spectrum (Adams *et al.*, 1984; Roulet and Adams, 1986; Bolsenga *et al.*, 1991). White

[7] ***Black ice*** is clear and virtually free of gas bubbles; being transparent, it appears very dark over deep water. ***White ice,*** often formed when water saturates lower snow layers, is much less transparent because of internal reflectance and scattering of light.

TABLE 5-4 Light Penetration through Ice and Snow Cover of Lakes under Different Conditions[a]

Ice–snow conditions	*Thickness (cm)*	*Percentage transmission of surface insolation*
Clear ice	43	72
Clear ice	154	23.2
Clear ice with vestige snow	39	53
Clear ice with sediment floc	149	14.8
Milky ice with bubbles	29	54
Wet ice with bubbles	39	41
Translucent ice ("snow ice")	25	11–18
Ice with irregular surface	29	58
Clear ice with 3 cm snow	149 (ice) + 3 (snow)	0.57
New snow	0.5	34
	5.0	20
	10.0	9
	17–20	8.8–6.7
Compacted old snow	17–20	5–1

[a] Selected data from Albrecht (1964) and Bolsenga *et al.* (1991, 1996).

ice exhibits greater red absorption than does snow. Light transmission through black ice is very similar to that through water.

The addition of snow over ice reduces light transmission greatly (Table 5-4). It is common for light to be attenuated to essentially zero at a depth of as little as a meter below the underside of ice and wet, heavy snow. Prolonged periods of heavy snow and ice cover, with attendant severe reduction or elimination of light from the water and from photosynthetic organisms, can have profound effects on the entire metabolism of the lake. In very productive lakes, consumption of dissolved oxygen by catabolic process can exceed production by photosynthesis, which leads to severe reductions in dissolved oxygen or even to total anoxia. Oxygen reduction or anoxia resulting from excessive snow or ice cover occurs commonly in shallow, productive lakes and ponds in temperate latitudes (see Chap. 9) and often leads to the death of many organisms. When fish die under these conditions, this phenomenon is termed ***winterkill.***

Ice and particularly snow are not uniformly distributed over lake surfaces (Adams, 1978; Adams and Lasenby, 1978; Adams and Prowse, 1978; Adams and Roulet, 1982). The thickness of ice and snow can vary as much as a factor of 2 within short distances (< 25 m) and lead to considerable patchiness in both reflectance of light and attenuation of light through ice cover. This patchiness in light distribution beneath ice–snow cover can, in turn, influence horizontal distribution of algal photosynthesis and behavioral responses of planktonic animals.

VI. COLOR OF NATURAL WATERS

The observed color of natural water is the result of light being scattered upward from the water after it has passed through the water to various depths and undergone selective absorption en route. Because molecular scattering of light in the water is a function of the fourth power of the frequency, observed light and therefore color is greater for shorter than for longer wavelengths, and blue dominates in the visible portion of the spectrum. Scattering of light from particulate suspensoids, however, is increasingly less selective with increasing particle size (Hutchinson, 1957). Colloidal $CaCO_3$, common to hardwater lakes and certain glacially fed streams, scatters light in the greens and blues and gives these waters a very characteristic blue-green color. Most of the color of lake waters results from dissolved organic matter and its rapid, selective absorption of the shorter wavelengths of the visible spectrum. As a result, emitted scattered light dominates in the green portion of the spectrum and, with increasing concentrations of organic matter, especially humic compounds, increases in the yellows and reds.

A. Sestonic Color and Turbidity

When the density of particulate matter suspended in the water becomes great, a ***seston***[8] color can be imparted to the water in spite of the relatively nonselective scattering properties of the particles. Suspension of large amounts of inorganic materials, such as clays or volcanic ash, can yield a yellow to brownish-red coloration. Seston color, however, is usually associated with large concentrations of suspended algae or, less frequently, with pigmented bacteria or microcrustaceans. Where cyanobacteria or diatoms occur in great profusion and often accumulate in the surface waters, they may produce blue-green or yellowish-brown colors, respectively. Blood-red color occurs commonly in lakes where conditions are temporarily ideal for massive development of populations of certain dinoflagellate algae, such as *Glenodinium.* At low concentrations of suspended algae, as in the open ocean, the chlorophyll bands of algal pigments have no appreciable influence on water color (Yentsch, 1960). Most of the absorption of light is by the water and dissolved organic compounds. Much absorption and scattering of light by particles is caused by nonchlorophyllous suspensoids (over 65% in a subalpine, oligotrophic lake; Priscu, 1983).

[8] ***Seston*** is a collective term for all particulate material present in the free water. Seston includes both ***bioseston*** (plankton and nekton; see Chap. 8) and ***abioseston*** or ***tripton*** (nonliving particulate material).

Turbidity is also a visual property of water and implies a reduction or lack of clarity that results from the presence of suspended particles or suspensoids. Turbidity usually consists of inorganic particles and originates by erosion of soil of the catchment basin and from resuspension of the bottom sediments (e.g., Nolen *et al.*, 1985). Inorganic turbidity tends to be higher in reservoirs than in natural lakes, in part often associated with larger fetch distances and shallower depths for much of the basin. These inorganic particles often range in size from 0.2 to 2 μm and consist of silicate minerals (mica, illite, montmorillonite, vermiculite, kaolinite) and aluminum and iron oxides (Kirk, 1994).

Suspended particles absorb as well as scatter photons. Although suspensoids usually absorb light substantially only in certain waters, the particles contribute almost entirely to the scattering of solar radiation in natural waters. Light incident upon inorganic particles is reflected externally, can enter the particle to be reflected internally and refracted upon exiting, or be transmitted through the particle and refracted upon leaving. Most of the scattered light will be in a forward direction, within a small angle of the direction of the incident light (Kirk, 1994). The increased scattering of light by turbidity causes enhanced reflection and scattering of light out of the water into the atmosphere.

A particularly common suspensoid in hardwater lakes is associated with colloidal and particulate $CaCO_3$, almost entirely as calcite crystals (cf. detailed discussion in Chap. 11). These suspensions of carbonate become particularly abundant and visually conspicuous as "whitings" in temperate lakes during spring and early summer. During this period, supersaturated conditions are often exceeded because of increasing temperatures and photosynthetic utilization of CO_2 and production of hydroxyl ions, which result in precipitation of particles in the range of 1–20 μm. As these particles settle, turbidity increases markedly, with large increases in backscattering of light, particularly in the blue-green portion of the spectrum (e.g., Weidemann *et al.*, 1985; Effler *et al.*, 1987; Hanson *et al.*, 1990). In certain lakes, such as appropriately named Blind Lake in southeastern Michigan (Schelske *et al.*, 1962), the suspension is so fine and production so intense from supersaturated groundwater sources, that the colloidal suspension renders the water nearly opaque continuously. Light is backscattered with a brilliant blue-green opalescence.

B. Color Scales

Any color (shade or tint) always has two decisive characteristics: color intensity (brightness) and light intensity (lightness) (Albers, 1963). This duality in color intervals results in an extremely subjective ability to discriminate colors. Moreover, visual memory is very poor in comparison with auditory memory. Therefore, the psychophysical nature of reactions of visual organs to light and color has led to several attempts to standardize observations by means of various color scales.

Several color scales have been devised to empirically compare the true color of lake water, after filtration to remove suspensoids, to various combinations of inorganic compounds in serial dilutions. Platinum units[9] is the most widely used comparative scale in the United States. Very clear lake water would yield a value of 0 Pt units, and heavily stained bog water about 300 (cf. Table 5-3). In Europe, the Forel-Ule color scale, involving comparisons to alkaline solutions of cupric sulfate ($CuSO_4$), potassium chromate (K_2CrO_4), and cobaltus sulfate ($CoSO_4$), is commonly used. A strong correlation exists between the brown organic color, which is derived chiefly from decomposing plant detritus, and the amount of dissolved organic carbon in the surface waters (Juday and Birge, 1933). Frequently, color units increase with depth in strongly stratified lakes; this is most likely related to increased concentrations of dissolved organic matter and ferric compounds near the sediments. The subjectivity of color evaluations can be reduced greatly by rather elaborate optical analyses and comparisons with standardized chromaticity coordinates (see Smith *et al.*, 1973).

VII. TRANSPARENCY OF WATER TO LIGHT

Evaluation of the vertical extinction and spectral characteristics of light in natural waters is commonly accomplished *in situ* with modern underwater quantum sensors. An approximate evaluation of the transparency of water to light was devised by an Italian scientist, Secchi (pronounced "sekki"), who observed the point at which a white disk lowered into the water was no longer visible. This method continues to be used widely owing to its simplicity. The Secchi disk transparency is the mean depth of the point where a weighted white disk, 20 cm in diameter, disappears when viewed from the shaded side of a vessel and that point where it reappears upon raising it after it has been lowered beyond visibility (cf. Wetzel and Likens, 1991).

The Secchi disk transparency is essentially a function of the reflection of light from its surface and

[9] 1000 Pt units = the color from 2.492 g potassium hexachloroplatinate (K_2PtCl_6), 2 g cobaltic chloride hexahydrate ($CoCl_2 \cdot 6H_2O$), 200 ml concentrated hydrochloric acid (HCl), and 800 ml water. The color units of filtered water are best examined spectrophotometrically at 410 nm, calibrated against Pt-Co reference solutions (Hongve and Åkesson, 1996).

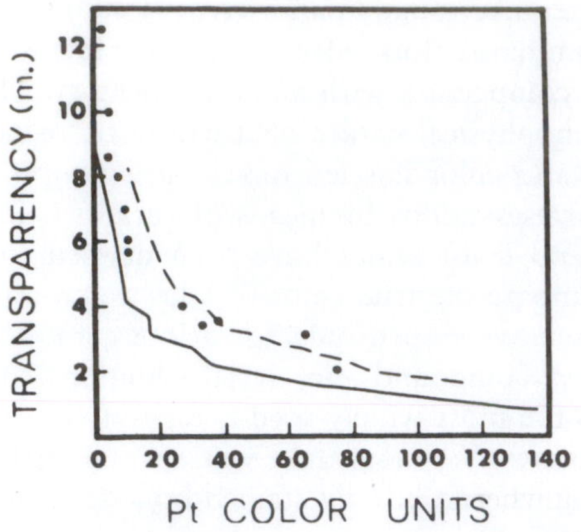

FIGURE 5-16 Relationship between Secchi disk transparency and dissolved color from Swedish lakes (points and dashed line) and 470 Wisconsin lakes (solid line). (From data of Åberg and Rodhe, 1942.)

is therefore influenced by the absorption characteristics both of the water and of its dissolved and particulate matter. Within limits, a parabolic relationship exists between dissolved organic matter and Secchi disk transparency (Fig. 5-16). However, both theoretical analyses and a large number of empirical observations have shown that the reduction in light transmission in relation to Secchi transparency measurements is associated to a greater extent with increased scattering by particulate suspensoids (Štěpánek, 1959; Szczepański, 1968). This correlation is particularly true in very productive lakes and in a very generalized way has been used to estimate the approximate density of phytoplankton populations. Where even moderate amounts of nonalgal turbidity are present, attempts to estimate algal biomass for trophic state classification from Secchi depth data are inappropriate (Lind, 1986). Although the transparency depth is independent of surface light intensity to a significant extent, results become erratic near dawn and dusk (Åberg and Rodhe, 1942), and the Secchi disk should be used preferably at midday.

Observed Secchi disk transparencies range from a few centimeters in turbid reservoirs to over 40 m in a few rare clear lakes. The Secchi disk transparency correlates closely with percentage transmission of light. At the extremes, Secchi disk transparency depths can represent from 1 to 15% transmission (Beeton, 1958; Štěpánek, 1959; Tyler, 1968; and others). The variation is related to differences in the sensitivity of underwater photometers and to the size of suspensoids. In general, the Secchi disk transparency depth corresponds to the depth of approximately 10% of surface light. The relationship between Secchi disk transparency and the extinction coefficient is very good during ice-free periods. The relationship, based on empirical data, of $\eta \text{ m}^{-1} = 1.7/z_{sd}$, where z_{sd} is the Secchi disk transparency depth in meters (Poole and Atkins, 1929), has been shown to be approximately correct in a variety of inland waters (Idso and Gilbert, 1974), although Walker (1980) indicates that $\eta \text{ m}^{-1} = 1.45/z_{sd}$ is more in agreement with results for the sea. Seasonal variations, such as are depicted in Figure 5-17, are common for lakes of the

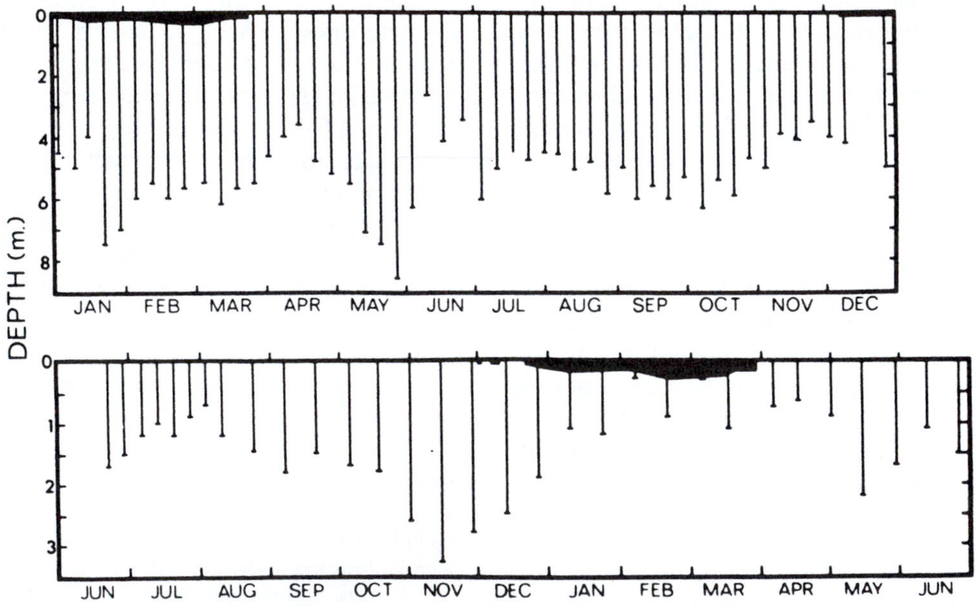

FIGURE 5-17 Annual variations in the Secchi disk transparencies of hardwater Lawrence Lake, 1968 (*upper*), and very productive Wintergreen Lake, 1971–1972 (*lower*), southwestern Michigan. (From Wetzel, unpublished data.)

temperate zone. The examples cited include an unproductive hardwater lake and a highly productive lake with dense algal populations throughout much of the year, particularly in the period of late winter through early summer. Even in the clearest lakes, such as alpine lakes, similar seasonal trends occur.

Colored Secchi disks have been employed in estimations of the spectral distribution of light with depth (Štěpánek, 1959; Elster and Štěpánek, 1967). Within general limits, comparisons of transparencies with a series of colored disks to that of white provides an approximate evaluation of the spectral characteristics of lake waters.

VIII. UTILIZATION AND EFFECTS OF SOLAR RADIATION

Solar radiation is the major energy source driving the productivity of aquatic ecosystems, whether it is incorporated into potential energy by biochemical conversions within the lake directly by its flora or by terrestrial components within the watershed and imported as organic matter to the lake. Algae and macrophytes use between 4 and 9 quanta of light energy per molecule of CO_2 reduced. The photosynthetic receptor is the chloroplast, which consists of a lamellar structure from 1 to 10 μm in diameter and approximately 0.25 μm in thickness. Absorption of light energy is specific for different chlorophyllous pigments: the absorption peaks of the primary pigment chlorophyll *a* are at 640 nm and 405 nm, and those of chlorophyll *b* at 620 nm and 440 nm. Certain plants, such as wheat, have broad responses to quanta of energy with relatively high efficiency of utilization in the green wavelengths, in addition to those of chlorophyll *a*. Other potential adaptations of plants to certain photic environments exist. For example, in the cyanobacterium *Microcystis*, the action spectra indicate that the accessory pigment phycocyanin gains importance when this organism is grown in red light and the cells lose chlorophyll *a*. Other auxiliary pigments supplement the excitation of chlorophyll *a* and improve the efficiency of photosynthesis at long wavelengths.

Photosynthesis of algae is clearly damaged by exposure to UV-B and especially UV-A radiation from natural solar irradiance. Physiological and genetic recovery occurs, and as a result a quasi–steady-state balance between damage and recovery processes is commonly observed (Cullen and Neale, 1994; Karentz *et al.*, 1994). Most species can repair damage to photosystems and DNA and commonly recover during diurnal periods of darkness or repeated and prolonged exposure to UV. Many species produce UV-absorbing pigments; mycosporine–like amino acids are an important and ubiquitous class of such compounds (Vincent and Roy, 1993; Xiong *et al.*, 1997). Many species have a variety of biochemical defenses against toxic end products of UV radiation, such as radical scavenging by carotenoid pigments and superoxide dismutase. Some species avoid intense UV by migration to deeper areas.

The metabolism of epilimnetic phytoplankton can be severely inhibited in the surface water layers. For example, UV-photoinhibition can result in major reductions of photosynthesis (>80%) and of chlorophyll *a* (>60%) when incubated at full natural UV (Bühlmann *et al.*, 1989; Moeller, 1994; Kim and Watanabe, 1993, 1994). UV-A penetrates further than does UV-B, and UV-A photoinhibition was found to be more severe than UV-B alone under natural conditions of UV attenuation.

UV radiation can impact zooplankton and fish directly in shallow water habitats. Direct effects can alter DNA, and UV-B photolysis can generate harmful photochemicals (free radicals, reactive oxygen species) (Hessen, 1994; Siebeck *et al.*, 1994). Although many animals can avoid UV-intense habitats, photoprotective pigments are often developed (cf. discussion on p. 453). Both movements to different habitats as well as development of pigments can alter the susceptibility of prey to predation by fish.

In addition to the adaptations of photosynthetic organisms to the aquatic photic environment, utilization of light by aquatic animals is also specific. The visual receptors of animals have a quantum efficiency of 1, the maximum possible. Among the invertebrates, for example, the opossum shrimp *Mysis* possesses visual receptors with a peak at 515 nm and is found only in deep portions of clear lakes, that is, in an optically blue environment. The water flea *Daphnia* migrates diurnally from deep to surface waters and is adapted for life in changing photic environments with four visual receptor peaks, at 370, 435, 570, and 685 nm (Chap. 16). The dragonfly, which spends much of its life cycle as an immature aquatic nymph, has visual receptors with a peak only in the green (530 nm) and possibly the blue (420 nm) portion of the spectrum (Ruck, 1965). Upon its shifting to the aerial adult stage, an increase in the number of receptor peaks (380, 420, 518, 530, and 550 nm) occurs. Similar situations are found among the fishes; those living only deep in a blue photic environment have maximum reception at approximately 485 nm, whereas shallow-living fishes possess receptors sensitive to longer wavelengths.

Behavioral adaptations to the utilization of light are also common. For example, *Daphnia* uses light as a cue in body orientation while swimming. The long axis of the body is oriented with the vector of maximum light energy, which affects directional swimming of these crustaceans during vertical migrations; hence the distribution of *Daphnia* populations within the lake is light-dependent.

It is now essential that we consider, in addition to these direct physiological effects and utilization responses, the effects of absorption of solar energy and its dissipation as heat on the physical, chemical, and biotic structure of aquatic ecosystems.

IX. SUMMARY

1. Solar radiation is the major energy source for aquatic ecosystems. The productivity and internal metabolism of fresh waters are driven and controlled by energy derived directly from the solar energy utilized in photosynthesis. Photosynthetic biochemical conversions utilizing solar energy occur within fresh waters both directly by aquatic flora as well as indirectly by terrestrial and wetland plants within the drainage basin and imported to the streams and lakes as organic matter. In addition, absorption of solar energy and its dissipation as heat affects the thermal structure, water mass stratification, and hydrodynamics of lakes and reservoirs. These characteristics have marked attendant effects on all chemical cycles, metabolic rates, and population dynamics.
2. Solar radiation can be viewed as pulses of electromagnetic energy, ***quanta*** or ***photons,*** that moves, transversely in a wave conformation of characteristic wavelengths and amplitudes. Absorption of light by atoms and molecules occurs from the resonation of electrons of atoms and molecules at frequencies that correspond to the quantum energy state of impinging light. Each molecular or atomic species, whether photochemical organic compounds in biochemical reactions or inorganic, has a unique set of absorption frequencies at which quantum energy lost by the photon is gained.
3. Much of the incoming irradiance (> 50%) impinging on the surface of water bodies occurs at wavelengths in the infrared portion (> 1,000 nm) of the solar spectrum and has major thermal effects on aquatic systems. Within the visible portion of the spectrum, most of the energy is in the longer wavelengths (reds) of lower frequencies.
4. The amount and spectral composition of solar energy impinging on water bodies is influenced by many dynamic factors.
   a. The angle of light incidence is more perpendicular in equatorial regions, during summer and at midday, than in regions of higher latitude, winter, or at times of the day closer to dawn or dusk. As a result, light must pass through greater distances of atmosphere as the angle of incidence increases from the perpendicular. Selective absorption by oxygen, ozone, carbon dioxide, and water vapor occurs as light passes through the atmosphere, and shorter wavelengths of higher frequencies are rapidly removed (hence the red sky at dawn and dusk).
   b. Indirect solar radiation from the sky results from atmospheric molecular scattering. Short wavelength radiation of higher frequency is scattered more than longer wavelength light (hence the blue appearance of clear sky at midday).
   c. A portion of light reaching water is reflected away from the surface. The greater the departure of the angle of the Sun from the perpendicular, the greater the reflection. When the surface of the water is disturbed by wave action, reflection increases somewhat (10–20%). Cloudy ice and snow greatly increase the amount of reflected light.
5. Of the total light energy entering the water, a portion is scattered and the remainder is absorbed by the water itself, dissolved compounds, and suspended particulate matter. The total diminution of radiant energy by both processes is called ***light attenuation.***
   a. The scattering of light results from deflection of light energy by molecules of water, its solutes, and suspended inorganic and organic particulate materials. Therefore, the extent of this internal reflection varies greatly with depth, season, and the dynamics of inorganic and organic loading in relation to water stratification, productivity and distribution of organisms, and other factors.
   b. ***Light absorption*** is the diminution of light energy with depth by transformation to heat.
   c. The molecular structure of water results in selective absorption of light. Percentile absorption of light by water alone is very high in the infrared and red, lowest in the blue, and increases again somewhat in the violet–ultraviolet portions of the visible spectrum.
   d. Dissolved organic compounds, especially humic compounds, drastically increase absorption of light. Selective absorption is greatest in the infrared and red wavelengths; coupled to the high infrared absorption of water itself, the combined effect is rapid absorption and heating of surface water. Even very low concentrations of dissolved organic compounds absorb greatly the UV, blue, and green wavelengths. Hence, in stained waters,

the dominant wavelengths penetrating to significant depths are in the yellow and red portions of the spectrum.

e. Light absorption by particulate suspensoids, when not in extremely high concentrations, is relatively unselective at different wavelengths.

f. The vertical extinction coefficient (η per meter) is an expression of the exponential attenuation of irradiance at depth in relation to that at the surface. The total extinction (η_t) of natural waters is the sum of absorption by the water itself (η_w), dissolved compounds (η_c), and particles suspended in the water (η_p). Although the mean η_t varies considerably with depth and over a year, the annual mean is fairly representative of the transparency of a lake or reservoir.

g. Although clear, colorless ice absorbs little more light than water, cloudy ice (with air bubbles, irregular crystals) and snow attenuate light severely. Spectral attenuation by clear ice is similar to that of pure water.

6

FATE OF HEAT

The absorption of solar energy by lake water is, as we have seen, influenced by an array of physical, chemical, and, under certain conditions, biotic properties of the water. In addition, headwater areas of low-order streams and the edges of very small lakes can be shaded by terrestrial and floodplain vegetation. These characteristics are dynamic and change seasonally and over geological time for individual river and lake systems. Before discussing the distribution of heat in running waters, it is instructive to evaluate the sources of heat to water in general and its distribution in lakes.

The amount of light energy absorbed by a solution increases exponentially with the distance of the light path through the solution. For light of a wavelength of 750 nm, 90% is absorbed in 1 m, while only 1% transmitted through 2 m of pure water. Absorption is increased markedly by dissolved organic matter. Since much of solar energy is of low frequency in the infrared portion of the spectrum at wavelengths greater than 750 nm, the upper 2 m of lake water absorb over one-half of the Sun's incoming radiation.

The high specific heat of water permits the dissipation of light energy and its accumulation as heat. The retention of heat is coupled with factors that influence its distribution within the river or lake ecosystem: the physical work done by wind energy, currents and other water movements, the morphometry of the channel or basin, and water gains and losses. Resulting patterns of thermal succession and stratification influence in fundamental ways physical and chemical cycles of rivers and lakes, which in turn govern production and decomposition processes.

I. DISTRIBUTION OF HEAT IN RIVERS

The heat content of river water reflects the instantaneous balance among inputs, storage, and outputs. Inputs of heat to a particular section of a river include short-wave solar radiation and long-wave atmospheric and forest radiation from condensation and precipitation, and advection of heat from ground water, from upstream, and from tributary inflows (Walling and Webb, 1992). Heat energy is lost from the water through reflection of solar, atmospheric, and forest radiation, by back radiation from the water, by evaporation, and by outflow from that section of the stream. Energy can be gained or lost by convection and by conduction to or from the atmosphere, stream bed, and banks. Solar, atmospheric, and back radiation terms are dominant in the energy budgets (Brown, 1969). These parameters, and the volume of water, will affect the rates of change of temperatures in streams and rivers.

Temperatures of streams and rivers vary in relation to air temperatures. A strong linear relationship often exists between air and river water temperatures, with some time lags by the water (Smith, 1981; Crisp and Howson, 1982). Departures from this linear relationship are common, as water rarely declines below 0°C even in extreme climates. Melting snow in the spring can keep temperatures well below that of air for weeks or longer, and precipitation on land can result in inflows of surface or ground waters of temperatures quite different from air temperatures for considerable periods (days). Clearly, hydrology is important in modifying the obvious interacting factors of climate and insolation (cf. Hynes, 1970; Ward, 1985) that determine the thermal régimes of running waters.

Even in the tropics, most rivers exhibit diurnal variations in water temperatures, particularly when shallow and exposed to direct solar irradiance. Small and heavily canopied streams exhibit only very small diurnal changes (ca. 2°C d^{-1}). On a large spatial scale, the amplitude of the annual oscillation or harmonic of temperature increases with increasing latitude and altitude (Walling and Webb, 1992). Maximum temperatures often increase progressively from the headwaters to the mouths of river ecosystems. These downstream changes in water temperatures formed the foundation for a widely used river zonation system that will be discussed later in relation to changes in benthic invertebrate and fish communities (Müller, 1951; Ilies, 1953). Within individual sections of a river, temperatures can also vary in response to marginal vegetation cover, groundwater seepage, channel depth and shape, orientation, substratum conditions, and silt content of the water (Walling and Webb, 1992). Submersed aquatic vegetation tends to increase annual mean temperatures and the amplitude of daily fluctuations (Crisp *et al.*, 1982). Within the hyporheic zone of stream sediments, streamwater infiltration occurs at the head of riffles and can alter temperatures of the interstitial water as much as 50 cm deep (White *et al.*, 1987; Crisp, 1990).

Ice formation on rivers is effective in reducing heat losses from the water. Complete ice cover is rare; often, small areas remain open to the atmosphere, particularly over riffles. Ice can form below the surface as well. Crystals of ***frazil ice,*** or slush ice, can form in turbulent, slightly supercooled water. Frazil ice consists of small discs of ice (1–4 mm in diameter and 1–100 μm in thickness) that form at rapids and other areas of open water (Martin, 1981). When moved with the water, frazil ice can scour the substratum and associated organisms. ***Anchor ice***, of much more rigid bonding (Hobbs, 1974), can also form on riffle substrata in shallow water, and impede water flow, increase water levels in pool areas, and cause flooding (Prowse, 1994). When moved by water currents, substrata are picked up with the anchor ice and can be transported large distances downstream. Ice breakup in rivers at high altitudes and latitudes is characterized by large increases in current velocity, stage, water temperatures, concentrations of suspended materials, and scouring of the substrata (Scrimgeour *et al.*, 1994). Major disturbances to biota associated with the substrata result.

Because temperatures within the range acceptable to most biota influence directly rates of metabolism, growth, and reproduction, the cumulative temperature above freezing varies greatly with the length of the growing season. A measure of the cumulative temperature experienced by biota, called ***degree-days***, is the sum of daily mean temperatures above 0°C over a standard period of time, usually a year. Because of the linear relation of water and air temperatures, annual degree-days increase with decreasing latitude at low elevations (Vannote and Sweeney, 1980).

II. DISTRIBUTION OF HEAT IN LAKES AND RESERVOIRS

It is evident from previous discussion that the greatest source of heat to lakes is solar radiation, and most is absorbed directly by the water. Some transfer of heat from the air and from the sediments does occur, but in lakes of moderate depth, this input is small compared to direct absorption. In shallow waters, sediments can absorb significant quantities of solar radiation, and this heat may be transferred in part to the water. However, terrestrial heat is generally very small in comparison to direct absorption of solar radiation by the water. Exceptions include lakes that receive significant percentages of their volume from surface runoff in short periods of time, such as small or shallow reservoirs that have short retention times. Indirect heating also can be significant in lakes that have high-volume inputs from groundwater sources and springs, especially in the case of hot springs and certain volcanic lakes. Condensation of water vapor at the water surface also provides some heat input to the lake from the air.

Some heat is lost from lakes by thermal radiation (cf. Chap. 5). However, because the thermal conductivity of water is very low, heat loss by thermal radiation is predominantly a surface phenomenon, restricted to the first few centimeters of water. Measurable amounts of heat are lost nonetheless by specific conduction both to the air and, to a much lesser degree, to the sediments. Heat is also lost through evaporation. The rate

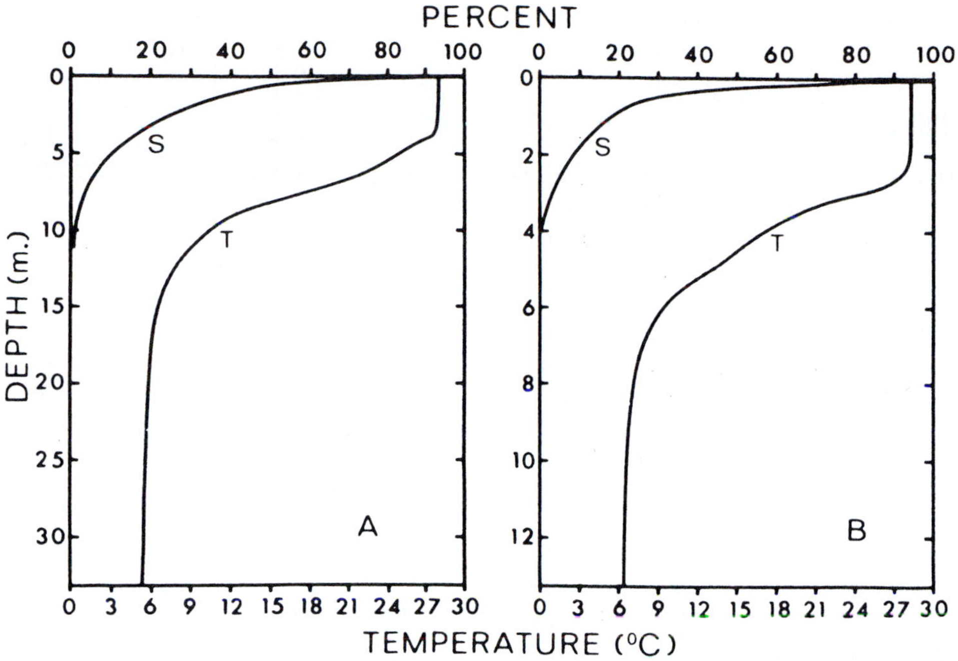

FIGURE 6-1 Vertical depth profiles of the penetration of solar radiation (*S*) and temperature (*T*) in (a) Crooked Lake and (b) interconnected, adjacent Little Crooked Lake, Noble-Whitley counties, Indiana, 18 July 1964. (From Wetzel, unpublished data.) Note differences in depth scales; Crooked Lake has a much larger surface area (79 ha) than Little Crooked Lake (5.3 ha).

of evaporation increases with higher temperatures, reduced vapor pressure, lower barometric pressure, and increased air movement over the water or ice surface and decreases with increasing salinity. When water at 0°C to freezes, 79.7 cal g^{-1} must be released,[1] and, conversely, when ice at 0°C melts to water at 0°C, an equal amount of heat must be absorbed. This latent heat of fusion is among the highest for known compounds. In addition, as ice sublimes to water vapor, 679 cal g^{-1} must be absorbed (see full discussion in Chap. 2). A significant portion of the heat increments to lakes or especially reservoirs can also be lost through outflow. Outflow usually consists of surface waters, and surface waters are often much warmer than underlying water strata.

The inputs and outputs of heat are therefore largely surface phenomena. Solar influx would be expected to dominate in the warmer seasons of the year, and the vertical thermal structure would then approximate the attenuation profile of solar radiation (Fig. 6-1). We might expect that warmer, less dense, and very stable heated water would successively overlie cooler and more dense water. Such is not the case, however, and what is observed (Fig. 6-1) is a relatively uniformly mixed upper portion of the lake that is isothermal often well below the photic zone.

Some convection currents do occur at night when the surface waters cool, become more dense, and sink. Similarly, surface waters can cool during brief shifts in local meteorological conditions, such as under cloudy conditions, increased evaporation rates, cold rain, et cetera, or after seasonal decreases in air temperatures. Surface influents, particularly in reservoirs, also may cool surface waters and induce convection currents. Under most conditions, in typical lakes of temperate latitudes mixing of the surface waters by convection currents is weak (up to 3 m) and insufficient to produce the thermal profiles observed during the heating period of thermal stratification.

Direct absorption of solar radiation accounts for only about 10% of the observed distribution of heat (Birge, 1916). Most of the heat distribution profile results from the action of the wind. As air currents move across the water interface, a frictional wind stress on the surface water generates mixing and currents proportional to the wind velocity (cf. Chap. 7). Very rapid (minutes) changes occur in the microstructure of the temperature profiles of the surface layers. The temperature, and thus the density, is rarely uniform in the surface layer, particularly when strong surface heating is

[1] The correct value is 79.72. The often-cited value of Weast (1970) and earlier editions is in error and is corrected in newer editions of this handbook (personal communication).

occurring or when wind stress is abating (Imberger and Hamblin, 1982; Imberger and Patterson, 1990).

A. Thermal Stratification

It is instructive to begin a discussion of thermal stratification in lakes by examining typical conditions of lakes of the temperate zone that experience strong contrasts in seasonal conditions. Most lakes are concentrated in the temperate and especially northern latitudes, and this permits some degree of generalization. Many exceptions to this general pattern can be found and will be indicated when applicable.

Ice cover on the lake commonly deteriorates in the spring in a relatively slow, progressive way until it is permeated with air columns and/or saturated with water. Warm rains often accelerate this process of ice erosion. The loss of ice cover is usually rapid, often taking place in a few hours, especially if associated with a strong wind. At this time, the water at all depths is near the temperature of maximum density. Slight changes from the temperature of maximum density, either cooling below or warming above 4°C due to vagaries in weather conditions, result in very small changes in density difference per unit change in temperature (cf. Chap. 2). As a result, there is relatively little thermal resistance to mixing, and only small amounts of wind energy are required to mix the water column. In most lakes, the amount of wind energy impinging on the surface is adequate to circulate the entire water column. Circulation, aided to some extent by convection currents induced by cooling at night and by evaporation, can continue for varying periods of time.

The duration of the ***spring turnover*** is governed by many factors. Lakes of small surface area, especially if protected from the wind by surrounding topography or terrestrial vegetation, may circulate only briefly in the spring, often for only a few days. Large lakes, in contrast, often circulate for a period of weeks, and, weather conditions permitting, the temperature of the entire water mass can increase to well above that of maximum density. During this period of mixing in spring turnover, the extent of heating is also a function of the volume of water that must be heated relative to net solar income. In very deep, large lakes such as the Laurentian Great Lakes and Lake Baikal, the heat income can be insufficient to increase significantly the temperatures of the deep water, even if the lake circulates completely. The spring period of circulation in shallow lakes, on the other hand, may allow water temperature to increase to well above 10°C.

As spring progresses, the surface waters of lakes of sufficient depth, usually greater than 6 m, are heated more rapidly than the heat is distributed by mixing.

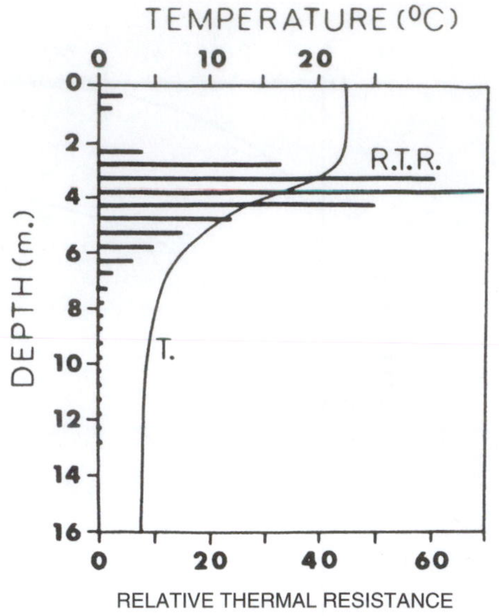

FIGURE 6-2 A summer temperature profile (*single line*) and relative thermal resistance to mixing (bars) for Little Round Lake, Ontario. The relative thermal resistance (R.T.R.) to mixing is given for columns of water 0.5 m deep. One unit of R.T.R. = 8×10^{-6}, that is the density difference between water at 5 and at 4°C. The R.T.R. of the lake water columns is expressed as the ratio of the density difference between water at the top and bottom of each column to the density difference between water at 5 and 4°C. (Modified from Vallentyne, 1957.)

Usually, the rapid surface heating occurs during a warm, calm period of several days. As the surface waters are warmed and become less dense, the relative thermal resistance of mixing increases markedly. A difference of only a few degrees is then sufficient to prevent complete circulation (Fig. 6-2; an identical analysis is given for Lake Mendota by Birge, 1916). From that time onward, the water column is thermally divided into three regions, which are exceedingly resistant to mixing with each other. The initial temperature of the lowest stratum (the ***hypolimnion***) is thus determined by the final water temperature during the spring turnover. The temperature of the hypolimnetic water changes little throughout the period of summer stratification, especially in deep lakes.

The period of summer stratification is characterized by an upper stratum of more or less uniformly warm, circulating, and fairly turbulent water, the ***epilimnion*** (Fig. 6-3). The epilimnion essentially floats upon a cold and relatively undisturbed region, the hypolimnion. The stratum between the epilimnion and the hypolimnion termed the ***metalimnion*** exhibits a marked thermal discontinuity. The metalimnion is defined as the water stratum of steep thermal gradient demarcated by the intersections of the nearby homoiothermal epilimnion and the hypolimnion (graphically depicted in Fig. 6-3). The term ***thermocline*** has been

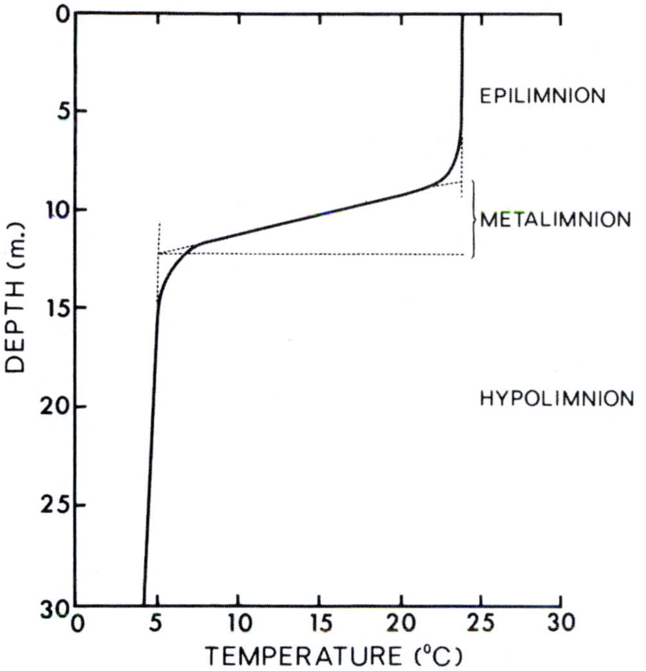

FIGURE 6-3 Typical thermal stratification of a lake into the epilimnetic, metalimnetic, and hypolimnetic water strata. Dashed lines indicate planes for determining the approximate boundaries of the metalimnion (see text).

defined variously but correctly refers to the *plane* of maximum rate of decrease of temperature with respect to depth. An extensive discussion of these terms and their conceptual basis is given by Hutchinson (1957). Terms in wide use that are functionally synonymous with the definition above of the metalimnion include the German *Sprungschicht* and *discontinuity layer* as used in the United Kingdom.

During the transition from spring turnover to summer stratification, the depth of the stratum of greatest thermal discontinuity (usually accepted as a change of >1°C per meter) varies greatly among lakes and from year to year in relation to weather conditions. If the water is being heated rapidly and wind-induced mixing of water is moderately intense but insufficient to circulate the water from top to bottom, the stratum of greatest thermal change will occur deep in the water column (Fig. 6-4). This thermal discontinuity rises rapidly as further heating occurs and density differences increase (see also Fig. 6-9). In contrast, if the weather at this time is relatively calm and hot, strong thermal discontinuities will occur at the surface—in this case, the metalimnion essentially begins at the surface and moves deeper as wind-induced mixing establishes the epilimnion. From that time on, the epilimnion undergoes periods of temporary heating during hot, calm weather and alternates with periods of strong mixing by wind (Fig. 6-4).

The depth of the stratum of steep thermal change is gradually depressed as the upper portion of the hypolimnion slowly gains small amounts of heat throughout the period of thermal stratification (Fig. 6-4 and also particularly Fig. 6-8). It is clear from the early analyses of Ricker (1937) and of Hutchinson (1941) that the heating of a stratified lake proceeds by a combination of direct solar radiation, turbulent conduction, and density currents. The turbulence of the epilimnion is carried into the metalimnion and hypolimnion (Imberger and Patterson, 1990). The conduction of heat by turbulence decreases as the stability of stratification increases throughout the heating season, but within the hypolimnion, heat conduction

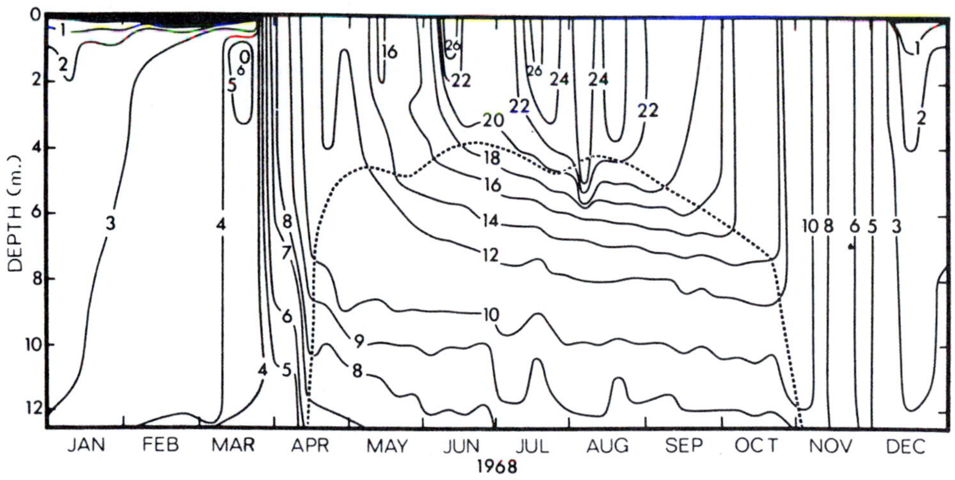

FIGURE 6-4 Depth–time diagram of isotherms (°C) in Lawrence Lake, Michigan, 1968. Dashed line indicates the upper metalimnetic–lower epilimnetic boundary. Ice cover drawn to scale. (Modified from Wetzel *et al.*, 1972.)

varies little with increasing depth. Chemical density currents follow the contour of the basin along the sediments, and as they move to the deepest portions of the basin they can be effective in heat transport. Heat generated by biological oxidations during decomposition is entirely inadequate to account for the observed heat gain by the hypolimnion.

Solar heating of the hypolimnion can occur in lakes that are particularly transparent, such as in alpine situations. In a small, clear mountain lake that is protected from much wind action, an estimated 65–85% of the heating of the upper hypolimnion can be accounted for by direct solar heating (Bachmann and Goldman, 1965; these estimates are likely too high, however; cf. Jassby and Powell, 1975). As might be anticipated, the relative importance of direct solar heating and the transport of heat by turbulence from the upper water strata varies greatly with the conditions of individual lakes. Wind-induced transport of heat to the hypolimnion is usually of greater importance than direct solar heating in most lakes, particularly where light is rapidly attenuated with increasing depth. Solar heating of the hypolimnion, however, can be a significant mechanism in certain clear lakes. Any factor that modifies transparency and penetration of solar radiation can affect thermal structure, seasonal stratification rates, and mixing depths of the epilimnion. Certainly, high concentrations of dissolved organic matter or of high particulate turbidity will increase absorption of infrared; temperatures of highly stained or turbid waters are often several degrees warmer than clear lakes (see Schreiner, 1984). The presence of zooplankton that remove by grazing large portions of the planktonic algae at certain times of the year have been purported to alter transparency, heating, and mixing depths of stratified lakes (Mazumder *et al.*, 1990), although the supporting data could not be separated from other influencing factors.

B. Stability of a Stratified Lake

Stability (*S*) per unit area of a lake is the quantity of work or mechanical energy in ergs required to mix the entire volume of water to a uniform temperature without addition or subtraction of heat (Birge, 1915; Schmidt, 1915, 1928; Idso, 1973). Therefore, during summer stratification, stability expresses the amount of work needed to prevent the lake from developing thermal stratification. More importantly, stability quantifies the resistance of that stratification to disruption by the wind and therefore the extent to which the hypolimnion is isolated from epilimnetic and surface movements.

The stability of a lake is very strongly influenced by its size and morphometry (Table 6-1). The amount of work (*B*) contributed by the wind in distributing heat throughout the lake increases proportionally as the size and volume of the lake decrease, whereas the stability (*S*) or amount of work per unit area to mix a lake to a uniform vertical density is greater in larger lakes. The sum of *B* and *S* is the amount of work per unit area (*G*) needed by the wind to distribute the heat in a lake uniformly at the minimum winter temperature. It should be apparent that the metalimnetic thermal discontinuity represents an effective mixing barrier. The depth of the metalimnion depends primarily on the surface area and can be estimated from knowledge of thermal density differences, morphometry, and maximum wind stress (Gorham and Boyce, 1989). Although not a total barrier, the metalimnion is nearly as effective as the sides of the lake basin. An immense amount of work is required to disrupt it.

TABLE 6-1 Work Required to Distribute Heat by Wind (*B*), to Render the Lake Homoiothermal (*S*), and to Maintain Hypothetical Homoiothermal Conditions Throughout the Heating Season (*G*) for Several Lakes[a]

Lake	*Area* (km^2)	*Maximum depth(m)*	*B* ($g\text{-}cm\ cm^{-2}$)	*S* ($g\text{-}cm\ cm^{-2}$)	*G* ($g\text{-}cm\ cm^{-2}$)
Atitian (Guatemala)	136.9	341	3741	21,500	25,241
Pyramid (Nevada)	532	104	4055	10,255	14,310
Mendota (Wisconsin)	39.2	25.6	1209	514	1723
Amatitlan (Guatemala)	8.2	33.6	752	415	1167
Güija (El Salvador)	44.3	26.0	958	175	1133
Lunzer Untersee (Austria)	0.68	33.7	535	390	925
Lawrence (Michigan)	0.050	12.6	301	208	509
Mirror (New Hampshire)	0.150	10.9	281	135	416
Findley (Washington)	0.114	24.0	175	133	308
Wingra (Wisconsin)	1.370	6.1	110	7	117
Marion (British Columbia)	0.133	4.5	56	11	67

[a] Modified from Hutchinson (1957) and Johnson *et al.* (1978).

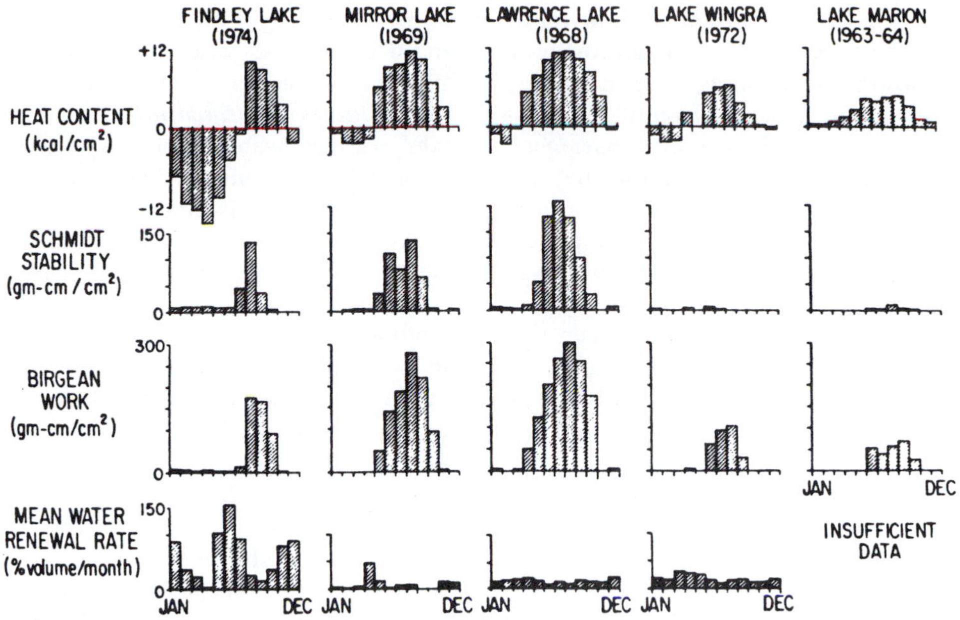

FIGURE 6-5 Monthly distribution of heat content, stability, Birgean work, and mean water renewal rate parameters in five North American lakes. (From Johnson *et al.: Verh. Int. Ver. Limnol.* 20:562–567, 1978.)

An excellent example of the usefulness of thermal energy and stability parameters is seen in the comparative analyses of five North American lakes (Fig. 6-5 and Table 6-1). The five lakes are all small, occur at similar latitudes, and, with the exception of alpine Findley Lake, are at similar elevations (Loucks and Odum, 1978). The maximum and mean depths of the lakes (Table 6-1) and water renewal times[2] differ considerably (Fig. 6-5), both seasonally and among the lakes (Johnson *et al.*, 1978). Marion Lake of this series has a renewal rate of about 76 times per year.

Lakes Wingra and Marion are distinctively unstable and exhibit an order of magnitude less stability (10^1 g-cm cm^{-2}) than lakes Lawrence, Mirror, and Findley (10^2 g-cm cm^{-2}). Lawrence and Mirror lakes are virtually identical with respect to their overall thermal and convective regimes, although Lawrence Lake is more stable than Mirror because of its smaller surface area and greater depth. The latent heat of fusion is the dominant factor in the thermal regime of Lake Findley, which is the coldest lake in this comparison. Winter stability is uniformly low ($S < 10$ g-cm cm^{-2}) for all five lakes, because the presence of ice cover prevents wind work from being exerted on the water column ($B \rightarrow 0$). Advection (horizontal movement of water) dominates the circulation of Lake Marion (approximately 76 renewals per year) compared with Mirror, Lawrence, and Wingra lakes (1–2 renewals per year). Advection is seasonally dominant in Findley Lake (>100% renewal per month during some months). The coincidence of the ice-free period with relaxation of advection (August–November) allows considerable stability to become established.

Density differences that regulate stratification in lakes are not always thermally induced. Differences in salinity can produce a similar stratification, often in combination with temperature differences. The stability so produced can exceed that resulting from thermal density differences and may lead to conditions where disruption of stratification is intermittent or never occurs (see discussion of meromictic lakes below).

C. Loss of Stratification

In late summer and fall, declining air temperatures result in a negative heat income to the lake, and loss of heat exceeds inputs from solar radiation. Surface waters cool and become more dense than underlying, warmer epilimnetic water. As the surface water sinks, it is mixed by a combination of convection currents and wind-induced epilimnetic circulation. The penetration of surface waters into the metalimnion continues as the lake continues to cool. A progressive erosion of the metalimnion from above can be observed as the stratum of thermal discontinuity is reduced and the homoiothermous epilimnion increases in thickness.

[2] ***Water renewal*** is the amount of water required to replace the lake volume during a given time interval (reciprocal water residence time).

On an annual basis, the result is an apparent lowering of the metalimnetic thermocline. The isopleth diagram of Figure 6-4 illustrates lines of equal temperature (isotherms) over time against depth. Depth–time diagrams of quantities or intensities of a given parameter are particularly useful in limnology because they aggregate hundreds and often thousands of data points taken at different depths and times into an annual picture that indicates the seasonal dynamics of physical, chemical, or biological properties of the lake. In the example from Lawrence Lake, it can be seen that stratification was maintained down to a depth of 11 m until late October. Finally, the entire volume of lake water was included in the circulation, and ***fall turnover*** was initiated. The transition from the final stages of weak summer stratification to autumnal circulation usually is abrupt and can occur in a few hours, especially if associated with the high wind velocities of a storm.

Circulation continues with the gradual cooling of the water column. The rate and duration of cooling are highly variable from lake to lake and are dependent upon basin morphometry, the volume of water that must be cooled, and prevailing meteorological conditions (Stewart and Haugen, 1990). For example, in Figure 6-4, Lawrence Lake, with a surface area of 5 hectares (ha) and a volume of 0.292×10^6 m^3, cooled down to and below the temperature of maximum density by early December. Waters of nearby Gull Lake, with an area of 827 ha and a volume of 102.1×10^6 m^3, usually circulate well into January before reaching comparable temperatures. Very large lakes, such as the Great Lakes, which are of similar latitudes to those just mentioned, circulate all winter and rarely reach conditions that permit the formation of an ice cover. In large, very deep lakes—for example, Great Bear Lake of the Canadian Northwest Territories—late summer circulation often does not extend to the lower depths (e.g., below 200 m), because stability at lower levels is sufficiently strong to prevent mixing (Johnson, 1966). The temperature of the maximum density of the water below 200 m is 3.53°C because of hydrostatic pressure. In cold years with slow spring heating, when the upper 200 m of water are cooled to 3.53°C in the autumn, the upper waters reach the same temperature of maximum density as the water strata below 200 m. Circulation then extends to the bottom, which is 450 m in Great Bear Lake.

D. Winter Stratification

As the temperature of the water reaches the point of maximum density (4°C), surface ice can form rapidly with such loss of heat as could occur on a calm, cold night. The difference in water density between 4 and 0°C is, as discussed earlier, very small and results in only a minor density gradient just below the surface. Hence, ***inverse stratification***, in which colder water lies over warmer water, is easily disrupted by a small amount of wind energy. Under stormy weather conditions common at this time of year, it is not unusual for a lake to continue to circulate and cool to well below the temperature of maximum density. Water temperatures of between 3 and 4°C are very common before conditions are such that ice cover can form, as for example is seen in Figure 6-4. Isothermy to less than 1°C has been observed frequently before the formation of ice cover, particularly in large lakes that are subject to much wind action.

After the lake has frozen, the ice cover effectively seals it off from the effects of the wind. Immediately below the ice, water of increasing density at the circulation temperature of the water prior to ice formation underlies water of 0°C. The steep thermal gradient near the undersurface of the ice is referred to as ***inverse thermal stratification***. The term *inverse stagnation* should not be used for this condition because it implies that the water beneath the ice is stagnant and not subject to water movement. Most lakes exhibit considerable water movement under ice (cf. Chap. 7).

Heating of the water under ice may occur throughout the winter to considerable depths (cf. Fig. 6-4). Although clear ice is, as we have seen, quite transparent to solar radiation, the type and amount of snow cover on the ice affect transmission of solar radiation. From the few detailed studies on winter heating, it is clear that most (>75%) is from solar radiation (Birge *et al.*, 1927). The relative contribution of heat stored in the sediments during the summer varies from lake to lake in relation to morphometry, summer heating conditions, and other factors (Hutchinson, 1957). Density currents, generated by heating of water through the ice in the shallow littoral areas, can flow toward the central portion of the basin along the sediments. Thermal–density instability is prevented by diffusion of solutes from the sediments, which raises the density slightly, especially as winter progresses, and accounts for the commonly observed temperatures of above 4°C near the sediments.

It is not unusual to observe apparent thermal–density anomalies under the ice. For example, in Figure 6-4, a cell of water in excess of 6°C was found centered around 1 m in late March in this extremely hardwater lake. The ice at this time was weak and porous from heavy, warm rains. Water very low in dissolved solutes, which had entered through the ice or possibly from the margins of the lake and was running under the ice, was less dense than that of the normal water high in dissolved ions. By this means, solar heating through the ice was able to increase the temperatures in this dilute layer without producing instability. Most of these situations are quite transitory and variable (Kozminski and Wisniewski, 1935) and are likely related in part to air temperatures rising above freezing and advective inflows of warm water from the periphery of the lake (Ellis *et al.*, 1991).

A detailed résumé of the types and characteristics of lake ice is given in Hutchinson (1957), based primarily on the extensive work of Wilson, Zumberge, and Marshall. Two books by Pivovarov (1973) and Ficke and Ficke (1977) summarize in detail the diverse literature on the properties of ice and the thermal characteristics of lakes and rivers under differing conditions of ice cover.

E. Modifications in Stratification

The stratification picture just described, in which a typical temperate lake of moderate size undergoes two periods of mixing, one in the spring and one in the autumn, is called ***dimictic*** and is not always so consistent. Many variations are found in relation to local or regional differences in climate, individual characteristics of the lake morphometry, and movement of the water masses.

A common divergence from the typical summer stratification pattern is the formation of secondary or multiple discontinuity layers (Fig. 6-6). In these situations, which usually last only for a few days or weeks, secondary metalimnia are formed in the epilimnion of the initial stratification during the heating period. These conditions are common when intense heating alternates with periods of extensive mixing. The initial spring metalimnion forms at various depths, as discussed earlier, and stabilizes at a depth characteristic of the lake morphometry and prevailing meteorological conditions. This pattern can be seen in the general limits of the epilimnetic boundary approximately indicated by the dashed line in the example given in Figure 6-4. The mixing of the entire epilimnion can cease

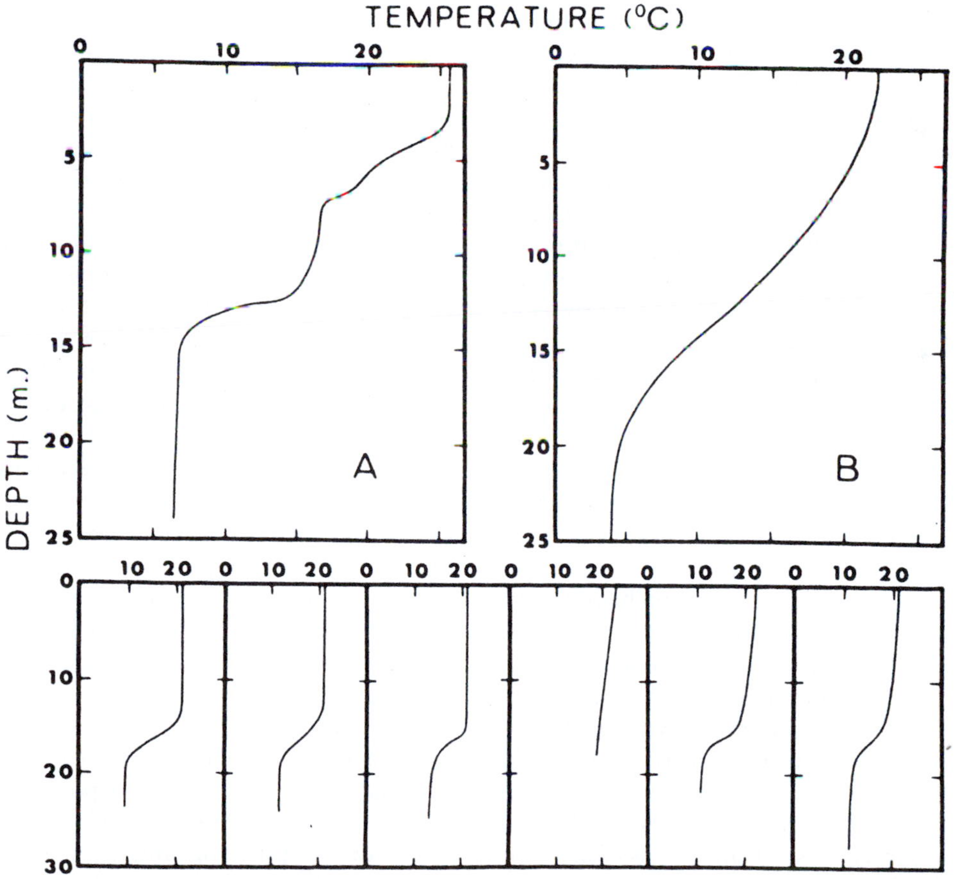

FIGURE 6-6 Variations in summer thermal stratification. (A) Example of multiple thermal discontinuities (generalized from several sources); (B) example of thermal discontinuity greatly reduced by high through-flow volumes (generalized). *Lower series:* Profiles of thermal stratification in submersed depressions of Douglas Lake, Michigan, 10–12 August 1929. (From data of Welch and Eggleton, 1932.)

during a calm, hot period, during which the surface waters can be intensely heated and perhaps mixed by light breezes for only a few meters of depth. In this way, a secondary metalimnion can be formed, overlying the first. The stability of these multiple metalimnia is usually not as great as the stability of the primary deeper one, and they are susceptible to disruption during periods of cooling and strong mixing of the epilimnion. The formation and disruption of these transient discontinuities can occur very rapidly (minutes to hours). Intense solar heating in the morning, wind mixing in the afternoon, and penetrative convective mixing during much of the night is a common pattern of variation in the epilimnion (Imberger, 1985). On a small scale, this process is analogous to normal autumnal erosion of the metalimnion and ensuing circulation.

Thermal stratification can be modified by inflow–outflow relationships, particularly if the volume of influent is large in relation to the volume of the epilimnion. For example, in reservoirs, high inflow from stream discharge, often cooler than the water of the epilimnion, can cause much turbulence and reduce the thermal gradient appreciably (Fig. 6-6b). A similar phenomenon is observed frequently in alpine and northern lakes that receive large flows of glacial or snow meltwater during the later portions of the summer stratification.

Many other variations occur in relation to individual characteristics of lake heating and mixing. Individuality is a prominent characteristic of the observed patterns of thermal structure and is governed strongly by climatic variations, volume of inflow and outflow in relation to the volume of the basin, basin configuration, surface area of the lake, position of the basin in relation to wind action, and other factors. Where the morphometry of the basin is complex, different stratification patterns can be observed within the same basin (see Wetzel, 1966a). Lakes with many submersed basin depressions characteristically exhibit variations in both the position and stability of the metalimnion (Fig. 6-6), which can strongly influence the productivity of these different regions. In the shallow water, depth is insufficient to maintain a typical epilimnion and hypolimnion, but these layers exhibit a reduction in temperature with increased depth. Such diminutive stratification is common in protected shallow areas of large lakes or in entire shallow lakes. Marked horizontal differences in thermal profiles are also observed in steepsided lake basins. For example, thermal microstructures increase as shallow water is approached along a steep lake-bottom slope (Caldwell *et al.*, 1978). These stepped thermal structures result from the mixing activity of internal waves, which are accentuated near the sides of the basin (cf. Chap. 7).

F. Types of Stratification

Density stratification patterns found in many lakes of the world deviate from that described for a typical dimictic situation. Interactions of climate, morphometry, and chemistry increasingly influence stratification and mixing.

F. A. Forel, who is often referred to as the "father of limnology", published extensive monographs on Lac Léman (Lake Geneva), Switzerland, in 1892, 1895, and 1904, that served as a classical foundation in the field for many years. Forel, who proposed a classification of lakes based on their thermal conditions, recognized three types: (1) temperate lakes, which undergo a regular annual alternation of summer and inverse winter stratification between two circulation periods at the temperature of maximum density; (2) tropical lakes, in which the water is never cooled below 4°C and is stratified except for a single period of winter circulation; and (3) polar lakes, in which temperatures never rise above 4°C and the water is inversely stratified except for a single period of summer circulation. However, these descriptions unfortunately suggest that differences depend on latitude, but many exceptions to the classification can be found. An ideal "tropical lake," for example, occurs in the northern temperate region.

Forel's terminology was modified in a number of ways, particularly by Whipple (1898, 1927), Birge (1915), and Yoshimura (1936). Whipple's widely adopted modification of this system subdivided each of the categories into orders on the basis of surface- or deep-water temperatures. According to his scheme, temperatures of bottom waters in first-order lakes remain at 4°C throughout the year. Circulation periods are often absent. In lakes of the second order, water at the greatest depth is near or above 4°C in the summer, and one or two complete circulation periods occur annually between the periods of thermal stratification. Third-order lakes are those in which thermal stratification never occurs, and circulation is more or less continuous, except when the lake is frozen.

As useful as these early categorizations were, it became apparent that alternative groupings were necessary as more information on polar, high-altitude, and tropical lakes became available. A lake classification based on stratification and circulation patterns introduced by Hutchinson and Löffler (1956) and Hutchinson (1957) was widely accepted because it possessed minimal ambiguity. The definitions centered on circulation patterns, as the roots of the names indicate, and refer to lakes that are of sufficient depth to form a hypolimnion. Six lake types were specified:

1. *Amictic*: Perennially ice-covered.
2. *Cold monomictic*: Water temperatures never exceed 4°C, and with only one period of circulation in summer at or below 4°C.
3. *Dimictic*: Circulate freely twice a year in the spring and fall, and are directly stratified in summer and inversely stratified in winter.
4. *Warm monomictic*: Circulate freely once a year in the winter at or above 4°C and are stably stratified for the remainder of the year; not ice-covered.
5. *Oligomictic*: Thermally stratified much of the year but cooling sufficiently for rare circulation periods at irregular intervals; not ice-covered.
6. *Polymictic*: Frequent or continuous periods of mixing per year; not ice-covered. Ruttner (1963) divided polymictic lakes further into (i) ***cold polymictic*** lakes that circulate continually at temperatures near or slightly above 4°C. Such lakes are found in equatorial regions of high wind and low humidity where little seasonal change in air temperatures occurs. At very high altitudes in equatorial regions, cold polymictic lakes gain a significant amount of heat during the day, but nocturnal losses are sufficient to permit complete mixing during the night. (ii) ***Warm polymictic*** lakes are usually tropical lakes that exhibit frequent periods of circulation at temperatures above 4°C. Annual temperature variations are small in equatorial tropics and result in repeated periods of circulation between short intervals of heating and weak stratification, followed by periods of rapid cooling. Under these circumstances, convectional circulation is sufficient, in combination with wind, to disrupt stratification.

These six lake types, based on thermal and circulation characteristics, were generalized in relation to altitudinal and latitudinal distributions (Fig. 6-7). Elevation is relatively unimportant to lake classification however, except at the highest elevations, and many many exceptions to such generalizations exist, particularly at low altitudes, where there is a strong influence of oceanic climate. Figure 6-7 does, however, demonstrate a generally observed geographical distribution of these lake types.

Any classification system is subject to modifications as further information about natural variance is gained. The Hutchinson-Löffler scheme has been criticized by many workers, and various attempts have

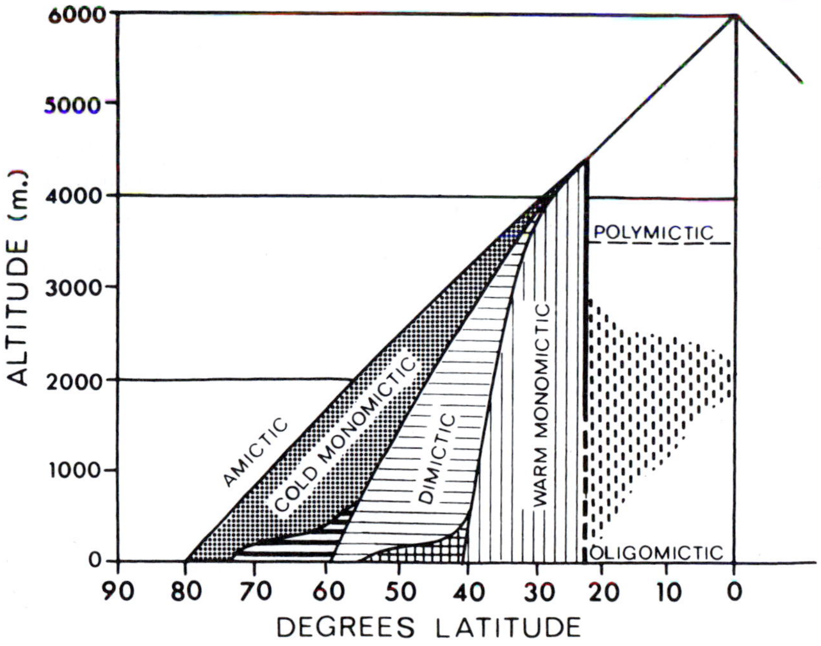

FIGURE 6-7 Schematic arrangement of thermal lake types with latitude and altitude. *Black dots:* cold monomictic; *black-and-white horizontal* bars: transitional regions; *horizontal lines:* dimictic; *crossed lines*: transitional regions; *vertical lines:* warm monomictic. The two equatorial types occupy the unshaded areas labeled oligomictic and polymictic, separated by a region of mixed types, mainly variants of the warm monomictic type (*broken vertical lines*). (Modified from Hutchinson and Löffler, 1956.)

been made for improve and modify it. Lewis (1983a) reviews some of the problems of this system in relation to shallow lakes, separation of meromixis, similarities among tropical lakes, and the rigidity of the 4°C boundary. A modified classification system based upon more recent information was proposed (Lewis, 1983a) and, although somewhat more complicated, does avoid certain intractable problems.

1. Amictic Lakes

Lakes defined as amictic are sealed off perennially by ice from most of the annual variations in temperature. Amictic, perennially ice-covered lakes are rare and largely limited to Antarctica (Goldman, 1972) or, more rarely, to very high mountains. Only a very few cases of such lakes have been recorded in the Arctic, mostly in Greenland, where general conditions necessary for the formation of a permanent ice cover are rare (Hobbie, 1973). It is clear that most of the heating of these lakes is by means of light transmission through the ice and conduction of heat from the surrounding land through the sediments (Ragotzkie and Likens, 1964; McKay *et al.*, 1985). The balance of heat inputs and losses is very constant and results in consistent ice thicknesses on individual lakes of between 3.5 and 6 m.

2. Cold Monomictic Lakes

Lakes with water temperatures never greater than 4°C and ice-covered most of the year with only one period of circulation in the summer at or below 4°C are called *cold monomictic* lakes. The category comprises, for the most part, Arctic and mountain lakes, which, although they may be ice-free for brief periods in the summer, are in frequent contact with glaciers or permafrost. The thermal cycles of Arctic lakes exhibit large variations in relation to their location and summer climate. Shallow lakes are extremely numerous in the subarctic but usually lack sufficient depth to stratify. Water temperatures often rise above 10°C, but never exceed 15°C (Hobbie, 1973). Very deep lakes that become ice-free for brief periods in the summer are truly cold monomictic lakes in which temperatures never reach 4°C. Others, such as Char Lake, Cornwallis Island, Canada (latitude 76°), warm to 4–5°C and are therefore technically dimictic (see below) but do not always stratify. Lake Schrader in the Brooks Range, Alaska, was dimictic and stratified in 1958, with the epilimnion at 10°C and the hypolimnion at 4°C. The following year, summer ice breakup occurred a month later, and the lake did not warm above 4°C; that is it was cold monomictic (Hobbie, 1961).

3. Continuous Cold Polymictic Lakes

Continuous cold polymictic lakes are ice-covered part of the year and ice-free above 4°C during the warm season. During the ice-free period, circulation is continuous or only interrupted by brief diurnal stratification. A great number of shallow north-temperate lakes exhibit this pattern of circulation (e.g., Harvey and Coombs, 1971).

4. Discontinuous Cold Polymictic Lakes

These lakes are ice-covered part of the year and ice-free above 4°C and thermally stratified during the warm season for periods of several days to weeks but with irregular interruption by mixing.

5. Dimictic Lakes

Lakes are called *dimictic* if they circulate freely twice a year in the spring and fall and are directly stratified in summer and inversely stratified under ice cover in winter. Dimictic lakes represent the most common type of thermal stratification observed in most lakes of the cool temperate regions of the world. Their characteristics were described in detail earlier in this chapter. Such lakes also are found commonly at high elevations in subtropical latitudes.

6. Warm Monomictic Lakes

In warm monomictic lakes, temperatures do not drop below 4°C; these lakes circulate freely in the winter at or above 4°C, and they stratify stably in the summer. Warm monomictic lakes are common to warm regions of the temperate zones, particularly those influenced by oceanic climates, and to mountainous areas of subtropical latitudes. Most lakes of the central and eastern portions of North America and the interior of Europe exhibit a distinct continental type of dimictic stratification, whereas a warm monomictic stratification is prevalent in many coastal regions of North America and Northern Europe.

7. Discontinuous Warm Polymictic Lakes

Lakes with frequent or continuous circulation are called *polymictic*. Discontinuous warm polymictic lakes have no seasonal ice cover and stratify for days or weeks at a time but mix more than once per year.

8. Continuous Warm Polymictic Lakes

Mixing in these lakes is more or less continuous, with stratification lasting only for a few hours at a time. No seasonal ice cover occurs, and generally these lakes are tropical with repeated periods of circulation between very short intervals of heating and weak stratification.

This discussion of thermal lake types refers to lakes with sufficient depth to form a hypolimnion. Numerically speaking, when considering the millions of small, relatively shallow lakes that virtually cover the tundra and northern temperate region, it becomes evident that most lakes do not become stratified or become stratified only briefly during ice-free periods. The depth required to become thermally stratified varies so greatly with surface area, latitude, elevation, basin orientation in relation to prevailing wind, depth–volume relations, protection by surrounding topography and vegetation, and other factors that generalizations in this regard would be misleading.

G. Meromixis

The types of stratification discussed thus far describe circulation that occurs throughout the entire water column, as is the case with ***holomictic*** lakes. A number of lakes do not undergo complete circulation, and the primary water mass does not mix with a lower portion. Such lakes are termed ***meromictic*** (Findenegg, 1935; Hutchinson, 1937). In meromictic lakes, the deeper stratum of water that is perennially isolated is the ***monimolimnion***; this stratum underlies the upper ***mixolimnion,*** which periodically circulates. These two strata are separated by a steep salinity gradient, the *stratum* of which is the ***chemolimnion***, the *plane* of density change is called the ***chemocline.***

Although the causes of chemically induced stratification have been treated and named variously, the divisions of Hutchinson (1937) and Walker and Likens (1975) (cf. also Dickman and Hartman, 1979) are particularly appropriate. As pointed out in Chap. 2, a salt concentration of 1 g liter^{-1} increases the density of water by approximately 0.0008. This change in specific gravity is very large in relation to density changes associated with temperature. For example, the density difference between 4 and 5°C is 0.000008, and it would only require 10 mg liter^{-1} of salt to give the same effect in resistance to mixing. The salinity gradient with depth during stratification is almost always much greater than 10 mg liter^{-1}. In most lakes, salinity gradients are insufficient to increase stability to a point where wind energy does not cause holomixis. A large number of lakes, however, do exhibit temporary or permanent meromixis as a result of salinity gradients. In such cases, the term *concentration stability*, in contrast to the thermal stability of holomictic lakes discussed earlier, is more appropriate (Berger, 1955).

1. Ectogenic Meromixis

The condition that results when some external event brings salt water into a freshwater lake or fresh water into a saline lake is called *ectogenic meromixis.* The result in either case is a superficial layer of less dense, less saline water overlying a monimolimnion of saline, more dense water. As might be expected, such situations are found often along marine coastal regions where catastrophic intrusions of salt water from storms associated with unusual tidal activity are fairly common events. In estuarine lakes, such events also occur routinely. A more strict definition of ectogenic meromictic lakes includes only those that are isolated from routine marine influxes of water. Such isolation may be recent or long-standing. An example of the latter is the southern Norwegian Lake Tokke, which most likely has been permanently meromictic for about 6000 years; the isolation of the lake probably took place during a period when the former fjord depression and the surrounding land were elevated some 60 m above sea level (Strøm, 1955).

A most interesting variant of ectogenic meromixis is found in many saline lakes of arid regions or in small depressions that occasionally receive saltwater intrusions (Hudec and Sonnenfeld, 1974). Often, these saline lakes overlie large deposits of such soluble salts as magnesium sulfate and are only a few meters in depth. The infusion of large amounts of fresh water, either naturally during wet periods or artificially as a result of increased irrigation of nearby land, creates a strong meromictic stratification that may persist for many years.

Whatever the causes of this type of meromixis, many of these lakes exhibit marked differences in temperature with depth in which the chemocline is functioning as a heat trap of solar energy. Such ***heliothermic lakes*** contain a solar-heated stratum of warm, saline water beneath a surface layer of cooler, less saline water (cf. review by Kirkland *et al.*, 1983). Much of the sunlight that penetrates the chemocline is transformed to heat, which cannot escape by radiation because water is opaque to infrared and which cannot be released by convection because the specific gravity of the dense water of the monimolimnion below is not significantly decreased by the increasing temperatures. Losses of heat to the air by conduction are slow. As a result, temperatures rise within the chemocline, often to very high temperatures approaching the boiling point.

A striking example is Hot Lake, a shallow (3.5 m) saline body of water occupying a former epsom salt excavation in north central Washington (Anderson, 1958; Kirkland *et al.*, 1983). Solar radiation energy passing through the overlying mixolimnion of fresh water accumulated as heat in the chemolimnion and monimolimnion, where circulation was absent. The resulting temperatures can be in excess of 50°C, and even when the surface of the lake is frozen, a temperature of nearly 30°C can be found at a depth of 2 m (Fig. 6-8). These ***dichothermic*** conditions, in which the monimolimnion is considerably warmer than the overlying water of the chemolimnion, are frequently observed in shallow meromictic lakes (Fig. 6-8 and Kirkland *et al.*, 1983).

Human activities have been known to form meromictic lakes. Creating a connecting channel

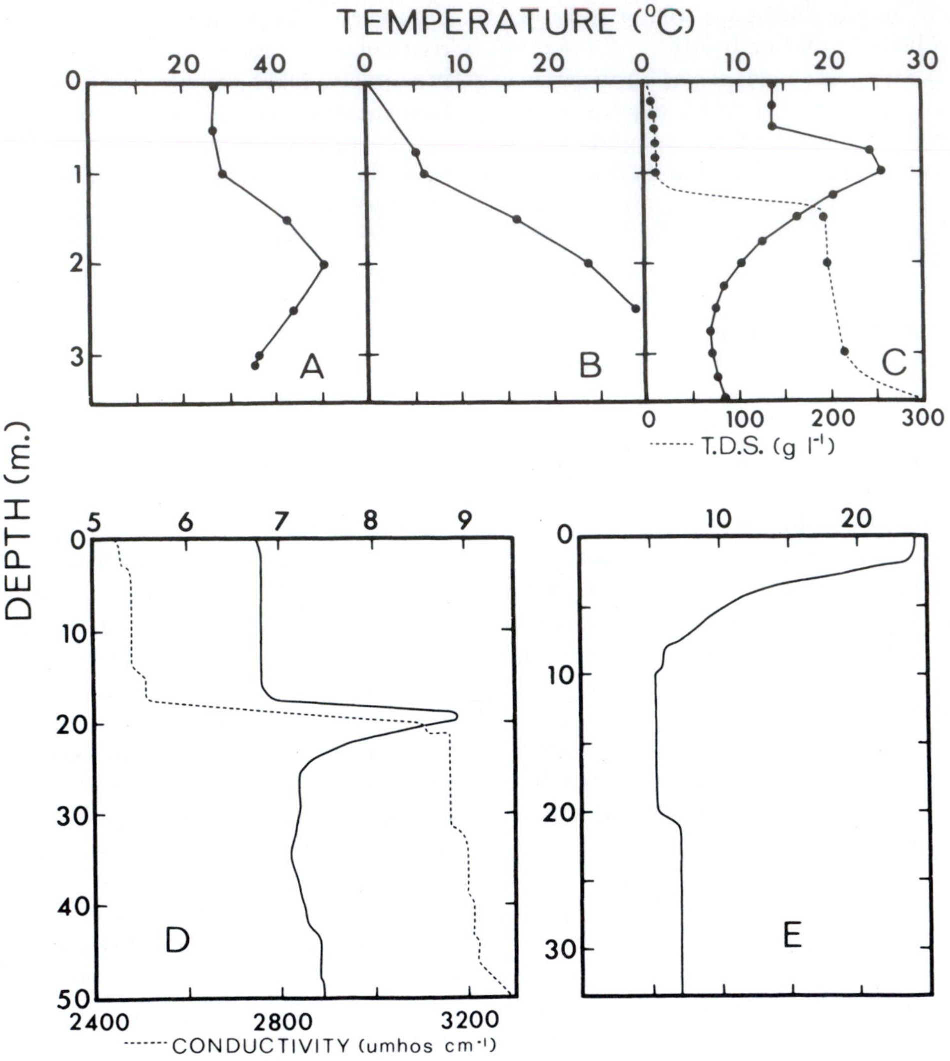

FIGURE 6-8 Thermal discontinuities in meromictic lakes. *Upper series:* Hot Lake, Washington (after Anderson, 1958): (a) 23 July 1955; (b) Heating under thin ice cover, 7 December 1954; (c) Heating in a deeply colored layer, 0.5–1.0 m, 1 May 1956, T.D.S. = Total dissolved solids on evaporation. *Lower series:* (d) Depth distribution of temperature and conductivity, Fayetteville Green Lake, New York, 16 December 1966 (from Brunskill and Ludlam, 1969); (e) Thermal stratification in Ulmener Maar, Germany, 8 August 1911. (Modified from Ruttner, 1963, after Thienemann). The thermal gradient observed earlier in Ulmener Maar has subsequently been modified (cf. Stewart and Hollan, 1975). Note variations in depth and temperature scales.

between a freshwater lake and the sea for purposes of navigation can result in the intrusion of saline water into the lake and cause permanent meromixis. Another event that can form a large number of small meromictic lakes is the introduction of salt caused by runoff from street de-icing (Judd, 1970; cf. also Chap. 8). The excellent analyses of the paleolimnological record of Längsee, Austria, by Frey (1955) clearly show how the clearing of forests surrounding the lake in prehistoric times introduced ectogenic meromixis. The resulting silt-laden runoff flowed into the hypolimnion and initiated a monimolimnion, which was augmented subsequently by salinity inputs of biogenic origin (see further on).

2. Crenogenic Meromixis

Crenogenic meromixis results from submerged saline springs that deliver dense water to deep portions of lake basins. The saline water displaces the water of the mixolimnion. The chemolimnion will stabilize at a depth related to the rate of influx, density differences, and the degree of wind mixing of the mixolimnion. Crenogenic meromixis is really a subclassification of the general group of externally mediated (ectogenic) meromixis (Walker and Likens, 1975).

Examples of crenogenic meromictic lakes are numerous. In Ulmener Maar (Fig. 6-8e), the temperature of the monimolimnion stabilized at 7°C. Meromictic lakes of this type recently have been found in interior Alaska, where subsurface springs introduce saline water into small, deep lakes originating in the thawed craters of ***pingos*** (Likens and Johnson, 1966). Pingos are formed when water, rising by hydraulic pressure through gaps in the permafrost, freezes and uplifts a mound of ice covered by a layer of alluvium. The overburden usually ruptures and the ice of the depression can thaw to form small lakes. In some cases, geothermal and saline waters form complex meromixis with warm, saline waters underlying cooler waters (see Drago, 1989).

3. Biogenic Meromixis

Biogenic or endogenic meromixis results from an accumulation of salts in the monimolimnion, which are usually liberated by means of decomposition in the sediments and from sedimenting organic matter. Often biogenic meromixis is initiated when abnormal meteorological conditions prevent circulation of a normally dimictic lake. Hypolimnetic accumulation in biochemically derived salinity may be sufficient to permit meromixis to persist either temporarily or permanently. Temperature inversions are usually found in the monimolimnion, such as is seen in Figure 6-8d. The high temperatures observed in the chemolimnion are the result of absorption of solar radiation by the dense bacterial plate that occurs at this level. The processes associated with the increasing temperatures in the lower monimolimnion, which are consistently observed, are less obvious. It is clear that the heat of metabolism in decomposition is inadequate to account for such increases (Hutchinson, 1941). When direct solar heating is not possible because of the depth, such heating is likely to be the result of warmer salt-laden water near the sediments that form density currents from overlying strata and descend into the basin along its contours. Heat from the surrounding land and conduction through the sediments is a minor heat source.

Biogenic meromixis is common in lakes that are very deep. For example, nearly all of the extremely deep lakes of the equatorial tropics are meromictic. But biogenic meromixis is also common among lakes of small surface area and moderate depth and that are sheltered, especially in continental regions that experience long, severe winters. ***Partial or temporary meromixis,*** when a normally dimictic lake skips a circulatory period (usually the spring period), is a result of dynamic processes of decomposition in the lower strata of the lake under unusual weather conditions (Findenegg, 1937). For example, after an unusually long and severe winter followed by a rapid thaw and warm conditions, Martin Lake did not undergo spring circulation (Fig. 6-9); the chemically stabilized water of the temporary monimolimnion was carried over from winter accumulation and was augmented throughout the subsequent summer (Wetzel, 1973). Particularly notable is the carryover of low temperatures from the circulation during the preceding fall, which persisted

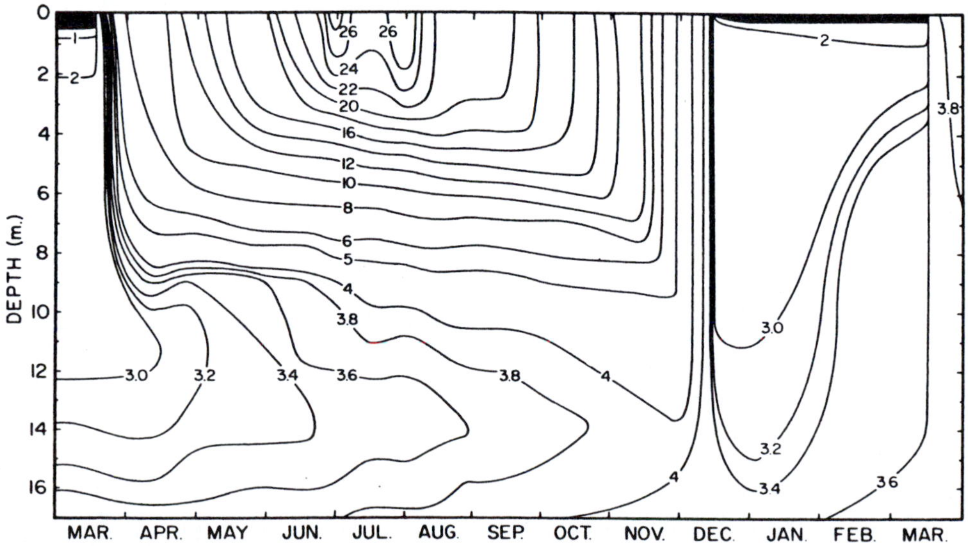

FIGURE 6-9 Isothermal variations (°C) in the major depression of Martin Lake, LaGrange County, Indiana, 1963–64. (From Wetzel, R.: *Hydrobiological Studies 3*:91–143, 1973.)

throughout the summer and fall, prior to the formation of the ice cover. Autumnal circulation in this protected lake did not occur until December, and then only briefly before ice formation. The following spring, a normal complete circulation was observed. This type of temporary, partial meromixis was also found in 2 other lakes in northeastern Indiana, of the 13 investigated during the same period (Wetzel, 1966b). In his classical work on phosphorus cycles and enrichment effects in the small eutrophic Schleinsee, Einsele (1941) found that complete spring turnover occurred, on the average, only every other year. In Lawrence Lake, a small (4.9 ha), deep (12.6 m) hardwater lake of southern Michigan, temporary meromixis occurs about every third year (Wetzel *et al.*, 1972, and unpublished data). Many other examples exist, but these are adequate to emphasize that this phenomenon of intermittent biogenic meromixis is quite common. It is not difficult to envision a combination of circumstances that can shift a lake from temporary to permanent meromixis. Both conditions have profound effects on biota and productivity, as will be discussed later.

Another, quite different cause of endogenic meromixis apparently occurs as a result of precipitation of salts in the upper water strata by "freezing out" from the surface ice layer. The precipitating salts may accumulate in deeper waters in sufficient concentrations to cause meromixis. In Algal Lake, Antarctica, this condition has been termed ***cryogenic meromixis*** (Goldman *et al.*, 1972; cf. also Priddle and Heywood, 1980).

III. THERMAL ENERGY CONTENT: HEAT BUDGETS OF LAKES

Heat content (H, in calories) is a function of the mass (M, in g) of the substance, temperature (t, in °C), and the specific heat (s, in cal g-°C^{-1}). Then,

$$H = M \times t \times s$$

Earlier, we evaluated the thermal radiation in water in relation to gains and losses of heat content (Chap. 5). Because of the high specific heat of water, large volumes of water change temperature relatively slowly. Large lakes moderate local climates and provide longer growing seasons for aquatic biota. This energy stored as heat further alters mixing rates and stratification patterns.

Temperature and density stratification in lakes are dominant regulators of nearly all physicochemical cycles and consequently of lake metabolism and productivity. It is important, therefore, to understand a lake's relative energy content (budget) or thermal capacity[3] both in the amount of heat energy required to develop stratification and the amount of heat a lake must lose in order to overcome density differences.

Heating capacity and energy demand to reach a given state are governed, in the case of a lake, by heat inputs, the volume and accessibility of the water at depth to heat incomes, and heat losses. Although many means of determining the heat content of lakes were developed in the pioneer days of limnology, particularly by Forel, the extensive works of Birge (1915, 1916) on this subject provide the basis for much of the ensuing discussion (cf. also Ragotzkie, 1978).

The heat content of a body of water is the amount of heat in calories that would be released on cooling from its maximum temperature to 0°C, or the amount of heat that has been transferred to the lake to heat it from 0°C to its maximum temperatures at various depths. Of course, the lowest temperatures of many lakes are not near 0°C or even its point of maximum density at 4°C. Hence, the minimum and maximum temperatures of the individual strata are used to determine the heat content of a lake. Although the specific heat of water is unity, which simplifies computations, calculations of the heat budget of a lake require the consideration of changes in volume with depth, since the lower strata contain smaller water volumes and make a smaller contribution to the total heat content than do the upper water layers.

A Birgean heat budget for a typical dimictic lake can be defined as follows (Birge and Juday, 1914; Birge 1915; Hutchinson, 1957):

Summer heat income (**θ_s**) is the amount of heat necessary to raise the temperature of the lake from an isothermal condition at 4°C to the maximum observed summer heat content. This heat income is primarily a result of the heating of the surface layers by direct insolation and the distribution of this heat by wind. It therefore has been called (Birge, 1915) the *wind-distributed heat*, even though not only wind energy is involved in its distribution.

Winter heat income (**θ_w**) is the amount of heat necessary to raise the temperature from its minimum heat content to 4°C. The winter heat income is usually not considered as wind-distributed, even though wind is involved in loss of heat prior to the lake's reaching the minimum temperature during autumnal–early winter cooling. Accurate winter heat incomes must be corrected for the latent heat of ice fusion. This correction can be sizable when the ice cover is considerable.

[3] The use of the word *budget* here is incorrect in terms of its normal definition. Reference actually is to the heat storage capacity of the lake, but the term *budget* is firmly entrenched in limnology.

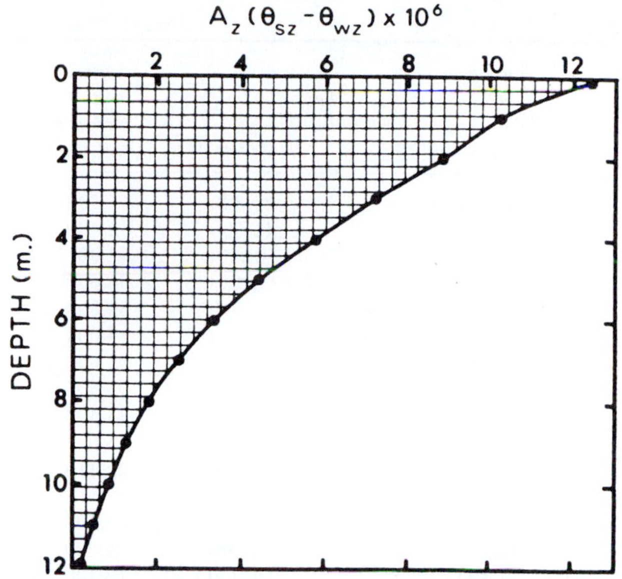

FIGURE 6-10 Curve generated for calculation of the annual heat budget of Lawrence Lake, Michigan, 1968. Calculations resulted in a budget, including a correction for ice, of 10,918 cal cm^{-2} yr^{-1}. (From Wetzel, unpublished data.)

One-cm-thick ice is equivalent to about 66.7 cal cm^{-2} (79.72 cal g^{-1}, where $\rho_{ice} = 0.9168$) or a correction of about 835 cal cm^{-2} for 12.5 cm of ice (see Adams and Lasenby, 1978, and Wetzel and Likens, 2000).

Annual Heat Budget (θ_a) is the total amount of heat necessary to raise the water from the minimum temperature of winter to the maximum temperature of summer. From the standpoint of the whole lake, this refers to the total amount of heat needed between the period of its lowest and its highest heat content.

The heat budget may be calculated by plotting the product A_z ($\theta_{sz} - \theta_{wz}$) against depth z, where θ_{sz} and θ_{wz} are the respective maximum summer and minimum winter temperatures and A_z is the area of the water stratum in cm^{-2} at depth z. The area of the resulting curve is integrated by digitization or planimetry[4] and divided by the surface area of the lake (A_0) because lake heat gains and losses are primarily surface phenomena. To separate the heat budget into summer or winter heat incomes, 4°C is substituted for the respective summer and winter temperatures at depth. A sample computation is given in Figure 6-10 for a small dimictic lake of moderate depth in southwestern Michigan. Note particularly the decreasing contribution of lower water strata to the overall heat budget of a rather typical lake.

[4] The proper use of these ingeniously devised instruments is described in detail in Welch (1948) and Wetzel and Likens (2000).

A large number of annual heat budgets have been computed; about 100 are listed in Hutchinson (1957), and many have appeared in the literature since then. A few exemplary values are given in Table 6-2 to emphasize several points in regard to factors influencing the observed heat budgets. Although most heat budgets have been derived for large, deep lakes, a few generalizations are possible. In temperate lakes that are sufficiently deep to maintain hypolimnetic temperatures at or near 4°C, annual heat budgets are usually between 30,000 and 40,000 cal cm^{-2} yr^{-1}, and most heat (60–80%) is gained during the spring–summer transition. Very few lakes have an annual heat budget in excess of 50,000 cal cm^{-2} yr^{-1}. Notable exceptions include Lake Baikal and Lake Michigan. Annual heat incomes of polar region lakes are small (θ_s values are in the range of 10,000 cal cm^{-2} for large lakes). However, annual values are likely to be very much larger than this when the latent heat of fusion for melting ice is included. The same is true for the apparently low heat budget values of lakes at high altitudes. In the temperate zone, annual heat budgets are strongly correlated with mean depth, area, and lake volumes and rise continuously with lake dimensions, though at a decreasing rate (Gorham, 1964; Allott, 1986). For lakes of a given volume, deeper ones of lesser area take up slightly more heat than shallower lakes of greater area. Dimictic lakes have slightly higher heat budgets than warm monomictic lakes of similar size.

The heat content of a lake can vary significantly (10%) within two or three days; accurate determinations of annual heat budgets therefore require frequent measurements over long periods (Stewart, 1973). Variations in the completeness of vertical mixing from year to year in warm monomictic lakes can cause marked variations in heat storage capacities (Ambrosetti and Barbanti, 1993). Examples of this variation are given in Table 6-3.

The annual income and loss of heat in tropical lakes is very small compared with that of lakes at higher latitudes or altitudes. Also, the mixing patterns are complex, often variable or polymictic in nature. Lewis (1973) demonstrated the weak, variable stratification with the repeated formation of multiple, temporary metalimnia common to many tropical lakes of large size. The minimum yearly heat content, given as negative values in brackets in Table 6-2, represents the heat content above 4°C in lakes that never cool to that temperature. In such lakes at northern latitudes, as in England, these values are very small; they are of course high in equatorial regions.

The stratification patterns and thermal structure of reservoirs with high throughflow volumes are most complicated (Wunderlich, 1971). Where withdrawal of

TABLE 6-2 Heat Budgets of Various Lakes (cal cm^{-2} yr^{-1})[a]

Lake	Surface area (A_0) (km^2)	Maximum depth (z_m) (m)	Mean depth ($\bar{z}$) (m)	Summer heat income θ_s (cal cm^{-2} yr^{-1})	Winter heat income θ_w (cal cm^{-2} yr^{-1})	Annual heat budget θ_a (cal cm^{-2} yr^{-1})
Baikal, Russia	31,500	1741	730	42,300	22,800	65,500
Michigan, USA	57,850	282	77	40,800	11,600	52,400
Washington, Washington	128	65	18	43,000	(−2600)	43,000
Geneva, Switzerland	581.5	310	154.4	36,000	(−23,200)	36,000
Tahoe, California-Nevada	499	501	313	34,800	0	34,800
Green, Wisconsin	29.7	72.3	33.1	26,200	7800	34,000
Pyramid, Nevada	532	104	57	33,600	(−7000)	33,600
Tiberias, Israel	167	50	24	33,500	(−26400)	33,500
Ladoga, Russia	18,150	223	56	18,000	15,300	33,300
Mendota, Wisconsin	39.2	25.6	12.4	~18,250	~5800	24,073
Atitlan, Guatemala	136.9	341	183	22,110	(−288,300)	22,110
Windermere, England						
North Basin	8.16	67	26.0	17,500	(−2900)	17,500
South Basin	10.5	44	17.7	15,680	(−2450)	15,680
Furesø, Denmark	9.9	36	12.3	14,400	2700	17,100
Lunzer Untersee, Austria	0.68	33.7	19.8	12,300	1400	13,700
Lawrence, Michigan	0.049	12.6	5.9	9168	1750	10,918
Schrader, Alaska	13.2	57	33			
1958				9050	—	—
1961				8900	—	—
Chandler, Alaska	15	21	13.5	5760	—	—
Lanao, Philippines	357	112	60.3			
1970				7250	(−121,300)	7250
1971				4500	(−121,300)	4500
Ranu Klindungan, Java	2	134	90	~3410	(~−189,000)	~3410
Hula, Israel	14	4	1.7	~2240	(−1600)	~2240

[a] From numerous sources.

major volumes of water occurs daily and intermittently, the patterns can be further disrupted and altered. Although drawdown usually occurs from the epilimnion, sometimes water is drawn from the hypolimnion. Influent waters also vary greatly in relation to surface runoff, heat content, and retention time in series of reservoirs in tandem on major river systems. Normal Birgean heat dynamics must be modified considerably in order to be applicable to complex reservoir systems. Among the most detailed analyses are those of the

TABLE 6-3 Variations in the Annual Heat Budgets of Several Lakes[a]

Lake	Number of annual observations	Mean depth ($\bar{z}$)	Mean θ_a (cal cm^{-2} yr^{-1})	Range (cal cm^{-2} yr^{-1})
Green, Wisconsin	5	33.1	34,200	32,300–36,400
Geneva, Switzerland	4	154.4	32,325	22,000–40,200
Mendota, Wisconsin	5	12.4	24,073	22,308–25,953
Orta, Italy	8	71.1	22,670	18,809–26,667
Whatcom, Washington	8	45.0	23,000	20,000–26,100
Washington, Washington	13	33.0	21,069	12,000–28,200
Menona, Wisconsin	3	7.7	17,559	17,256–18,041
Waubesa, Wisconsin	3	4.6	11,362	10,948–11,739
Murchison, Tasmania	2	20.6	10,498	10,095–10,900
Valencia, Venezuela	2	19.0	5309	4755–5862
Mize, Florida	3	4.0	4720	3767–6003

[a] From data of Birge (1915), Nordlie (1972), Stewart (1973), Lewis (1983b), Bowling (1990), and Ambrosetti and Barbanti (1993).

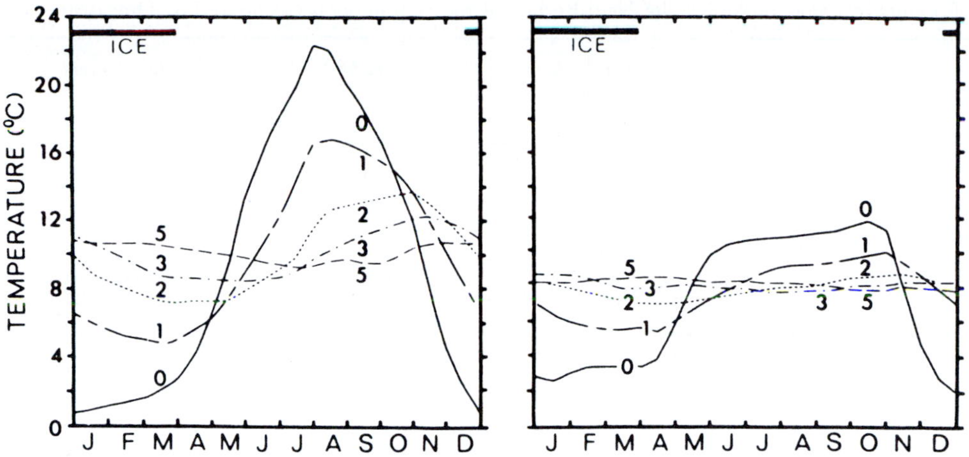

FIGURE 6-11 Annual temperature curves at 0, 1, 2, 3, and 5 m within the sediments of Lake Mendota, Wisconsin (means for 1918, 1919, and 1920). *Left:* Sediments at a water depth of 8 m. *Right:* Sediments at 23.5 m. (Modified from data of Birge, Juday, and March, 1927.)

Slapy Reservoir in Czechoslovakia, where seasonal thermal dynamics have been determined for over a decade (Hrbáček and Straškraba, 1966; Straškraba *et al.*, 1973), and of Lake Powell, Utah–Arizona (Johnson and Merritt, 1979).

A. Heat of Lake Sediments

Sediments of lakes absorb an appreciable amount of heat from the water during the warmer periods of the year and transmit heat to the water during the winter period. Although few detailed studies are available, data from Lake Mendota illustrate this characteristic (Birge *et al.*, 1927). It will be noted from Figure 6-11 that the temperatures of the surface sediments at a water depth of 8 m followed the annual cycle of the epilimnetic water temperatures rather closely. This thermal variation is attenuated with depth in the sediments, so that at a depth of 5 m into the sediments, little seasonal oscillation was observed. A similar relationship was observed in the sediments of the hypolimnion at a water depth of 23.5 m, but the amplitude and heat energy exchange were much less (Fig. 6-11). The heat content of the sediments lagged behind the hypolimnetic water temperature, and the lag duration increased about one month per meter depth of sediment.

This process of transfer of heat from the water to lake sediments during the summer and release from lake sediments into the water during the winter period under ice cover is true on a seasonal average basis. The sediment heat exchange, however, is dynamic and fluxes occur in both directions from and into the sediments (Fang and Stefan, 1996). The direction of heat flux can reverse on shorter, even daily, time scales, particularly in shallow lakes.

Spatial variations in temperatures within the sediments at similar water depths have been found that are apparently due to the type of adjacent land forms of the basin (Matveev, 1964). Shoreline sediments of deep bog lakes were found to be considerably warmer than sediments beneath the open water, but annual variations were largely damped at a sediment depth of 4 m (Likens and Johnson, 1969). Heat flux from the relatively warm shoreline sediments occurred both lakeward and landward. In a shallower bog lake, deep sediments beneath the open water were found to be warmer than those of comparable depth beneath shoreline sediments (Reif, 1969), indicative of a heat flux away from the lake. Differences between the thermal characteristics of these lakes may be related to their maximum depths, because in the shallower lake, absorption of solar energy by the sediments is greater (cf. also Dale and Gillespie, 1977).

Determinations of the heat budget of sediments based on the differences between their minimum and maximum mean temperatures showed that in sediments at a lake depth of 8 m, this value was 2950 cal cm^{-2} yr^{-1}; at 12 m, it was 2200, and at the maximum of 23.5 m, it was 1100 cal cm^{-2} yr^{-1}. An estimated average heat budget of all of the lake sediments was ca. 2000 cal cm^{-2} yr^{-1} for Lake Mendota. As lakes become more shallow, the heat budget of the sediments becomes a greater portion of the whole (Table 6-4). All of the examples of annual heat budgets of lakes cited here (Tables 6-2 and 6-3) are based on water temperatures and ignore sediment temperatures. Although in

TABLE 6-4 Heat Budgets of Lakes in Which the Heat Budget of the Sediments Is Compared to That of the Water[a]

Lake	*Mean depth (m)*	*Heat budget of water* θ_a *(cal cm*$^{-2}$ *yr*$^{-1}$*)*	*Heat budget of sediments* θ_m *(cal cm*$^{-2}$ *yr*$^{-1}$*)*	*Heat budget of lake (cal cm*$^{-2}$ *yr*$^{-1}$*)*	*% Sediments of total heat budget*
Mendota, Wisconsin	12.1	23,500	2000	25,500	7.8
Stewart's Dark, Wisconsin	4.3	7000	730	7730	9.4
Beloie, Russia	4.15	8000	2500	10,500	23.8
Tub, Wisconsin	3.6	8000	970	8970	10.8
El Porcal, Spain	3.0	6068	2962	9030	32.8
Hula, Israel	1.7	2290	1400	3690	37.9

[a] Data from Hutchinson (1957), Likens and Johnson (1969), and Cobelas (1992).

large, deep lakes such a simplifying assumption may be justified, it should be noted that the sediments of shallow lakes are significant to their heat budgets.

IV. COMPARATIVE ANALYSES: THERMAL CHARACTERISTICS OF RIVERS, RESERVOIRS, AND NATURAL LAKES

As discussed earlier, temperatures of streams and rivers vary rapidly on diurnal and seasonal bases, often in close relation to fluctuating air temperatures. Very few rivers are not regulated to some extent by human activities of channel modification or impoundment. Many studies have demonstrated that reservoir impoundment of a river and associated regulation usually increase the mean temperature, reduce or eliminate freezing conditions, depress summer maxima, delay annual temperature cycles, and reduce diurnal temperature fluctuations (Palmer and O'Keeffe, 1989; Webb and Walling, 1993; McRae and Edwards, 1994). These effects persist for a distance of usually several kilometers below the dam; thermal recovery distances showed a positive lognormal relationship with reservoir discharge. In contrast, thermal properties tend to be much more stable in natural lakes, with slower and less marked variations (Table 6-5).

Thermally induced density stratification patterns are rare in river ecosystems and become increasingly complex as water movements slow (Table 6-5). Because

TABLE 6-5 Thermal and Stratification Properties of Rivers, Reservoirs, and Natural Lakes[a]

Property	*Rivers*	*Reservoirs*	*Lakes*
Temperature variations	Rapid, large	Rapid in riverine zones, moderate in lacustrine zone; decreased rates and magnitude of fluctuations	Slow, stable
Thermal density stratification	Rare	Variable, irregular; often too shallow to stratify in riverine and transitional zones; often temporary stratification in lacustrine zone	Common, regular, particularly in dimictic and monomictic lakes
Spatial differences (summer)	Cold headwaters, increasing downstream	Increased mean temperatures; reduced or no freezing	Warm upper strata (epilimnion) and colder lower strata (hypolimnion)
Groundwater effects	High ratio of ground water to surface drainage, decreased temperatures	Relatively small	Significant only in seepage lakes and can reduce temperatures slightly
Tributary effects	Considerable if different from mainstream	Moderate to small	Small, localized to influent area
Shading effects	Considerable, usually seasonal in headwater zones; can enhance thermal stability	Small to negligible	Usually very small or negligible
Ice formations	Transitory	Usually transitory	Persistent; ameliorate variations
Ice scouring effects	Robust, extensive	Usually minor	Localized to windward, near-shore littoral areas

[a] Extracted from numerous sources, particularly Ryder and Pesendorfer (1980) for rivers and Wetzel (1990a) for reservoirs and lakes.

relatively few lakes are strongly influenced by subsurface groundwater influents, the effects of ground water on temperature is much greater in headwater areas of streams than in large rivers, reservoirs, and most lakes. The ice of streams and rivers tends to impact, particularly in scouring effects, and disturb the ecosystem much more than it does in lakes.

V. SUMMARY

1. The high specific heat of water results in dissipation of much of the incoming light energy and its accumulation as heat. In lake systems, the way retained heat is distributed and altered is governed by a number of coupled factors: the physical work of wind energy, currents and other water movements, basin morphometry, and water losses. Resulting patterns of density-induced stratification influence physical and chemical properties and cycles of lakes and, in turn, govern their productivity and decomposition.
2. Temperature is a measure of the *intensity* (not the amount) of heat stored in a volume of water. Heat is measured in calories and is the product of the weight of the substance (in grams), temperature (°C), and the specific heat (cal g-$°C^{-1}$).
3. Temperatures of streams and rivers vary directly with air temperatures, with some time lag by the water. Most rivers exhibit diurnal variations in water temperatures. The size of annual oscillations increases with greater latitude and altitude. Temperatures of streams usually increase downstream.
4. Heat income results from several processes:
   a. Direct absorption of solar radiation; usually the dominant source
   b. Transfer of heat from the air
   c. Condensation of water vapor at the water surface
   d. Transfer of heat from sediments to the water
   e. Heat transfer from terrestrial sources via precipitation, surface runoff, and groundwater inputs
5. Heat losses occur by:
   a. Specific conduction of heat to the air, and, to a lesser extent, to the sediments
   b. Evaporation
   c. Outflow, especially of surface water
6. Although differential heating and cooling of surface waters, as well as influents, can cause convection currents, mixing of surface waters by convection is weak and insufficient to produce the long-term stratification patterns commonly observed in most lakes. Wind energy mechanically distributes most of the heat in a lake.
7. In temperate and other regions, many lakes of moderate depth (approximately > 10 m) exhibit ***thermal stratification.***
   a. During warmer periods of the year, the surface waters are heated, largely by solar radiation, more rapidly than the heat is distributed by mixing. As the surface waters are warmed and become less dense, the ***relative thermal resistance*** to mixing increases.
   b. The lake becomes stratified into three zones (Fig. 6-3):
      i ***Epilimnion:*** An upper stratum of less dense, more or less uniformly warm, circulating, and fairly turbulent water.
      ii. ***Hypolimnion:*** The lower stratum of more dense, cooler, and relatively quiescent water lying below the epilimnion.
      iii. ***Metalimnion:*** The transitional *stratum* of marked thermal change between the epilimnion and hypolimnion.
   c. The ***thermocline*** is the *plane* or surface of maximum rate of decrease of temperature in the metalimnion.
   d. The resistance of thermal- and chemical-density stratification to mixing is a measure of the lake's mechanical stability.
      i. ***Stability*** per unit area of a lake is the quantity of work or mechanical energy in ergs required to mix the entire volume of water to a uniform temperature by the wind without the addition or loss of heat. Stability quantifies the resistance of the stratification to disruption by wind.
      ii. The amount of work by the wind per unit area needed to mix a lake to a uniform vertical density increases as the size and volume of the lake increases.
8. At the end of thermal stratification when loss of heat exceeds heat inputs, surface waters of the epilimnion cool, become more dense, and are mixed with deeper strata of similar density by wind-induced and convection currents. The metalimnion is progressively eroded as the relative thermal resistance to mixing is reduced. Eventually, the entire volume of water is included in the circulation and ***fall turnover*** is initiated. Fall turnover continues with progressive cooling, often to the temperatures of maximum density of 4°C or less.
9. Ice cover forms on the surface under calm, cold conditions. ***Inverse stratification*** of water temperatures occurs under the ice, in which colder,

less dense water overlies warmer, more dense water near the temperature of maximum density at 4°C. Some gradual heating of the water occurs during the winter under ice cover.

10. When the ice cover melts in the spring, the water column is nearly isothermal. If the lake receives sufficient wind energy, as is commonly the case, the lake circulates completely and undergoes ***spring turnover.***
11. Lakes undergoing complete circulation in spring and fall separated by thermal summer stratification and winter inverse stratification are called ***dimictic*** lakes. Dimictic lake types are very common among temperate lakes of moderate size.
12. Many other types of density-related stratification patterns occur among lakes in relation to the interacting effects of climate, morphometry, and chemistry. The types of lakes are classified on the basis of circulation patterns and include lakes of sufficient depth to form a hypolimnion.

Type of lake	*Annual vertical mixing*	*Dominant factors contributing to or preventing vertical mixing*
1. Amictic	No appreciable amount	Permanent ice cover; slow internal mixing
2. Holomictic	Complete, at least once	Predominantly wind energy; convection currents
a. Monomictic	Once	
b. Dimictic	Twice	
c. Polymictic	More than twice	
3. Meromictic	Permanently stratified or interruption of stratification patterns at irregular intervals	Chemically enhanced density stratification
a. Ectogenic	Permanently stratified	1. Surface inflow of (a) fresh water overlying a preexisting saline stratum, (b) saline water underlying a preexisting fresh stratum, or (c) turbidity currents of particulate-laden water. 2. Subsurface inflow of fresh or saline water (= ***crenogenic***)
b. Endogenic	Permanently or temporarily stratified	
i. Biogenic	Permanently stratified	Chemically enhanced density stratification contributed by biological processes with accumulations of bicarbonate or iron/manganese ions in lower stratum and shelter afforded by morphometry of lake basin and surrounding topography
ii. Temporary biogenic	Temporary elimination of complete circulation (spring or fall)	Same factors as biogenic meromixis but borderline conditions where unusual climatic conditions lead to omission of spring or fall circulation and accumulation of hypolimnetic ions
iii. Cryogenic	Permanently stratified	Deep-water accumulation of salts precipitated by freeze concentration from a surface ice layer

13. Water strata of permanently stratified meromictic lakes are named differently but analogously to those of holomictic lakes:
    a. ***Mixolimnion:*** The circulating upper stratum
    b. ***Monimolimnion:*** The deeper stratum of water that is perennially or periodically isolated
    c. ***Chemolimnion:*** The interfacing *stratum* of steep salinity (density) gradient between the mixolimnion and the monimolimnion
14. The heat budget of a lake is a measure of the heat storage capacity under existing conditions of morphometry and the climate to which it is exposed. The annual heat budget is the total amount of heat accumulated between the period of its lowest and its highest heat content. The annual income and loss of heat in tropical lakes is small in comparison to that of lakes at higher latitudes and altitudes. Annual heat budgets of lakes of the temperate zone generally increase with mean depth, area, and lake volume. The sediments of lakes absorb considerable heat from the water during warmer periods of the year and transmit heat to the water during the winter period.
15. The thermal properties of rivers, reservoirs, and lakes are similar but differ considerably in rates of change. Changes in temperature are more rapid in rivers and the magnitude of those differences is usually greater than is the case in reservoirs and especially in natural lakes.

7 WATER MOVEMENTS

I. HYDRODYNAMICS OF WATER MOVEMENTS

As water aggregates into stream channels and flows down elevational gradients, complex turbulent movements occur. The flow is influenced by drag at the air–water interface and at the substratum, where it is affected by the degree of roughness. Velocity of flow in a channel is approximately inversely proportional to the logarithm of the depth. The flows modify the environment markedly and influence gas and nutrient exchanges and substratum properties and generally set the habitat characteristics for the biota. Often, streams and rivers form the primary influents to lakes and modify the hydrodynamics of the receiving basins.

Lakes behave like large mechanical oscillators. They respond in numerous, complex ways to applications of force and are frictionally damped by viscous forces associated with turbulence. Water movements generated by the transfer of wind energy to the water give rise to a spectrum of rhythmic oscillations, both at the water surface and internally deep within the basin. The oscillations and their attendant currents may be in phase or in opposition; their final fate is to degrade into arrhythmic turbulent motions. Each basin has its own set of free modes of oscillation, both surface and internal, which depend on gravity, on basin shape and size, and on the internal distribution of water density. Which of these modes is brought into play during any particular disturbance or storm depends on the duration, the periodicity, and the spatial distribution of the applied force. In other words, the morphometry of the lake basin, the lake's stratification structure, and its exposure to wind are important factors that determine water movement.

Turbulence resulting from water movements is of major significance to the biota and productivity of the lake. Limnological thought is still pervaded by the idea that summer and winter density stratification results in stagnation of lake strata. There is little basis for this view. Water movements are integral components of the aquatic ecosystems, and consideration must be given to their effects on changes in distribution of temperature, dissolved gases and nutrients, and other chemical parameters. Water movements influence not only the

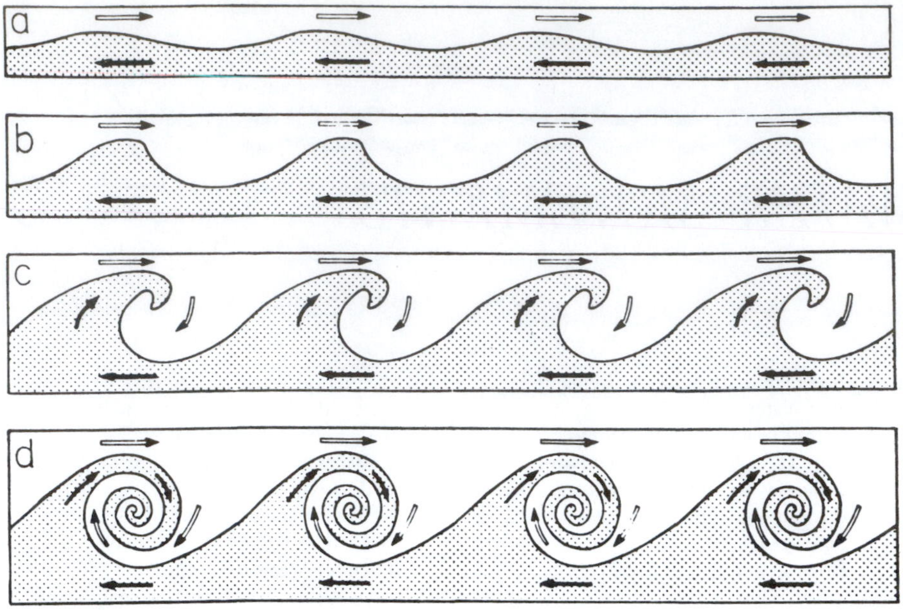

FIGURE 7-1 Stages in vortex formation during shear instability on the interface of a stratified two-layer system of differing densities. (From Mortimer, C. H.: *Mitteilungen Int. Ver. Limnol.* *20*:131, 1974.)

aggregation and distribution of nutrients and food but also the distribution of microorganisms and plankton.

II. FLOW OF WATER

A. Turbulent and Laminar Flow

At sufficiently slow speeds, water flowing in a smooth tube moves along the interface in an apparently orderly manner. If the velocity of movement is increased at the interface, a critical speed is reached above which the smooth and unidirectional laminar flow becomes disordered or turbulent. The transition from laminar to turbulent flow also occurs at the interface of two miscible fluids moving in opposing directions. An example would be opposing horizontal movements between two water strata of differing densities, such as the epilimnion and metalimnion of a lake.

Appreciable experimental work on turbulent shearing stresses (Reynolds stresses) has demonstrated that only a low velocity is needed to shift from laminar flow to turbulent current. The critical velocity is a function of the fluid viscosities and densities and decreases in proportion to an increase in size of the channel or tube. Velocities of only a few mm sec^{-1} can induce turbulent flow, at which point disturbances arising on the interface between the layers will no longer be suppressed by buoyancy (gravity) forces, which tend to keep layers of different densities separated. As a result, laminar flow is almost never found in aquatic systems. It is important that the basic properties of turbulent flow and diffusion be understood in order to appreciate the magnitude of the resultant alterations in distribution of physicochemical parameters and biota.

If a critical velocity difference along a given density interface is exceeded, disturbances grow in amplitude and break into vortices (Fig. 7-1) (Mortimer, 1961, 1974; Smith, 1975; Denny, 1988; Vogel, 1994). Such vortex formation increases the mixing of the two layers by generating a transitional layer, across which there is both a velocity gradient (shear) and a density gradient. This model represents in simplified form what occurs when layers of nearly uniform velocity and density stream parallel to one another, as in the movements in the epi- , meta- , and hypolimnion discussed further on.

If the rate (a) of supply of energy to turbulent vortices or ***eddies***[1] from the shear stresses exceeds the rate (b) at which they have to do work against gravity in disrupting the density stratification, turbulence increases (Richardson, 1925; Mortimer, 1961). Rate *a* is proportional to the square of the shear, while rate *b* is proportional to the density gradient. The magnitude of the ratio b/a determines whether turbulence increases or decreases. In nondimensional form, that ratio is the

[1] The rotating region of fluid, an eddy, refers to an assemblage of shear waves of a spectrum of many lengths or "eddy diameters" (Mortimer, 1974).

Richardson number (R_i) expressed as

$$R_i = \frac{g(d\rho/dz)}{\rho(du/dz)^2}$$

where

g = acceleration of gravity
ρ = density
u = velocity
z = depth

When R_i of a stratified fluid subjected to shearing flow falls below about one-quarter, flow becomes unstable. Or, put in another way, when R_i falls below 0.25, there is a sudden shift in the eddy spectrum from the molecular microturbulence of stable flow toward macroturbulence of the large vortices associated with unstable flow. The shift is accompanied by increases in friction and in mixing *perpendicular* to the direction to the current.

Thus, in a river or lake with stratified shear layers, little mixing and friction between the layers occurs as long as the flow remains stable ($R_i > 0.25$). When flow increases and becomes unstable ($R_i < 0.25$), the stirring action of vortex formation (macroturbulence) increases the interfacial area many times. Mixing is then rapidly completed by microturbulence and molecular diffusion. Soluble components (nutrients) and suspensoids (microalgae, microfauna) are carried along with this dispersion of water into the newly formed, thicker layer in which density and velocity gradients are less than those that were initially present.

The shear instability that results when R_i falls below 0.25 causes a sudden shift in turbulent energy and the layers dissipate and collapse into a complex array of turbulent patches of nearly uniform density (Fig. 7-2). These patches are exceedingly transient in nature, and only the composite final product, after stability is restored, is what is most often observed. The thermal microstratification, commonly observed in detailed studies of the metalimnion (Whitney, 1937; Simpson and Woods, 1970; Baker *et al.*, 1985), is possibly a result of the dissipation of shear instabilities in the thermocline region.

B. Eddy Diffusion and Conductivity

The distribution of turbulent motion must be viewed in a statistical sense, for the distribution of fluid properties such as heat or solute content is a composite of many small random movements. The diffusion models for fluids, developed by Taylor and by Schmidt (Mortimer, 1974), are analogous to those that measure random movement of molecules in molecular diffusion but are obviously on much larger scales.

Heat transfer across a thermal gradient within a layer in a lake may be used to estimate the extent of turbulent transport, because turbulent mixing and heat conduction are quite similar. In a hypothetical layer of water devoid of motion but with a temperature gradient, heat equilibrium is brought about only by conduction. If flow is applied across that gradient and it is turbulent (R_i <0.25), the equalization of temperature is greatly accelerated. Thus, heat (relatively warm water) is transported through a plane perpendicular to the gradient by eddy turbulence. The rate of transport (diffusivity or conductivity) across a fluid plane is the product of the gradient perpendicular to the plane and an exchange coefficient (K_z). If expressed as a substance dissolved in the water across the gradient, it may be written as the gradient concentration per unit volume of water in terms of passage across unit area, g cm^{-2} sec^{-1}.

Thus, the *coefficient of eddy diffusion* (K_z) is a measure of the rate of exchange or intensity of mixing across the plane. It is not a constant but varies with

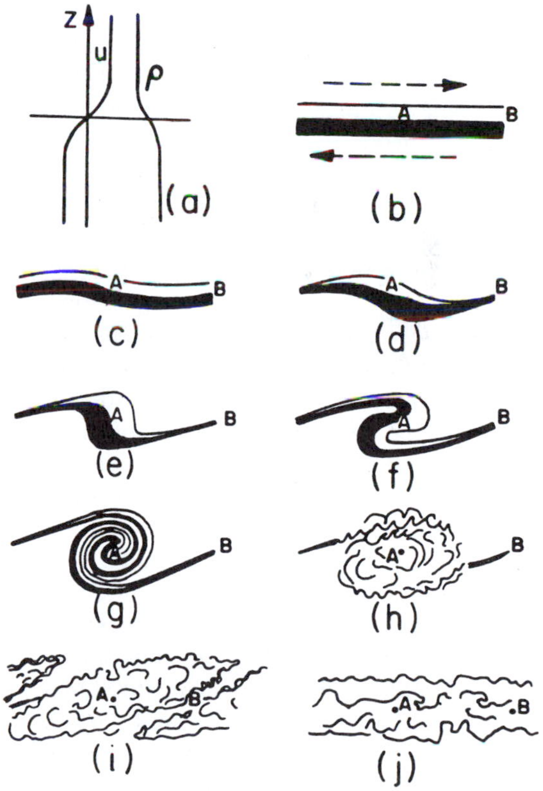

FIGURE 7-2 Growth of shear instability leading to turbulent mixing in a stratified fluid with the velocity and density (ρ) distribution shown in *a*. A and B are fixed points, the arrows indicate direction of flow, and the lines represent surfaces of equal density. (From Mortimer, C. H.: *Mitteilungen Int. Ver. Limnol. 20*:134, 1974, after Thorpe.)

average density, velocity of vertical motion, and mixing length. The horizontal dimensions of lakes are very much larger than the vertical dimensions of stratification; therefore, large differences exist between the magnitudes of horizontal and vertical eddy diffusion coefficients. The concept of an average K_z is subject to many variations related to the distance and is therefore an oversimplification. In reality, dispersion and mixing occur as a result of eddy movements over a wide range of spatial scales in which smaller eddies occur within larger ones. It is necessary to view the turbulent movements and consequent diffusion as taking place across a spectrum of turbulence (the eddy spectrum of Mortimer, 1974).

Turbulent transport causes particles to move apart at right angles to the direction of flow. When particles are close together, their rate of separation is governed by the small eddies of turbulent motion. But as the distance of separation increases, the dispersion of the particles is influenced by eddies of increasing size. Over significant ranges of dispersion, the mean rate at which particles diffuse is proportional to the 4/3 power of the distance between them (Stommel, 1949), a law that has been found to hold true for separation of particles over a range of 10 to 10^8 cm (Olson and Ichige, 1959; Verduin, 1961). The most probable value of turbulent diffusion velocity is in the range of 0.3–1.0 cm sec^{-1}, with a single diffusion velocity of approximately 1.0 cm sec^{-1} being valid for all scales of diffusion processes (Noble, 1961).

The coefficient of eddy diffusion is, for practical purposes, identical to the coefficient of eddy conductivity, which permits, in a general way, estimations of turbulence from changes in temperature. Estimations of eddy conductivity coefficients are useful, as they permit insight into the magnitude of current-induced turbulence in the hypolimnion. The magnitude of K_z and changes observed vertically with stratification, however, depend very much upon the methods of analysis and assumptions about the magnitude of direct solar heating of the upper hypolimnion, convective currents, lateral exchanges of water by internal seiche movements, lateral heating from the sediments, and the accuracy of measurements of small temperature differences (Hutchinson, 1957; Dutton and Bryson, 1962; Lerman and Stiller, 1969; Hesslein and Quay, 1973; Powell and Jassby, 1974; Jassby and Powell, 1975; Di Cola *et al.*, 1977; Imboden and Emerson, 1978; Maiss *et al.*, 1994). A lucid theoretical discussion is given in Denny (1988).

Coefficients of eddy diffusion in the hypolimnion decrease with increasing stability. Therefore, in lakes of large area and depth that have longer fetch for wind exposure, eddy diffusion coefficients of the hypolimnion are greater than in smaller, shallow, stratified lakes (Mortimer, 1941, 1942). In addition, the K_z of a turbulent epilimnion is markedly greater than the K_z in the more stable metalimnion of a stratified lake (Fig. 7-3). This relationship holds seasonally as well. As the gradient of thermal density stratification increases from spring to summer, the coefficient of eddy diffusion decreases, that is, K_z is inversely proportional to stability. In some lakes, particularly high-altitude clear lakes, diurnal heating can markedly alter temperatures of surface strata and form temporary thermal stratification and, with wind-induced mixing, metalimnia (Powell *et al.*, 1984). Marked diurnal changes in eddy diffusivity result. In the deep hypolimnion, K_z tends to increase slightly, especially near the bottom where the effects of other water movements further complicate analyses of turbulent transport (Coleman and Armstrong, 1987). The eddy diffusivity of the hypolimnion of stratified reservoirs increases markedly as hypolimnetic water is discharged at the dam (Sugawa, 1987).

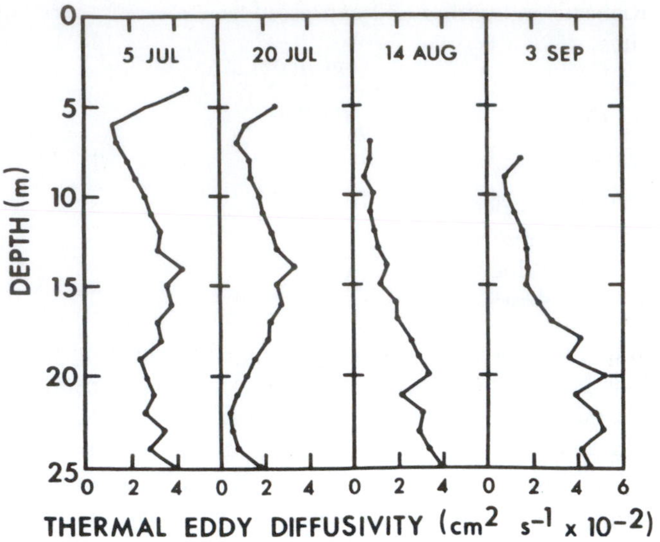

FIGURE 7-3 Selected depth profiles of the vertical thermal eddy diffusivity in Castle Lake, a small cirque lake in northern California, during the summer of 1972. The first value on each date marks the average top of the metalimnion at that date. (Drawn from data of Jassby and Powell, 1975.)

III. IN-STREAM HYDRAULIC MOVEMENTS

Streams and rivers are basically linear channels in which water flows directionally along a declining elevational gradient. The geomorphology and the roughness of the substrata of river channels, however, are variable. Coupled to variable discharge (cf. Chap. 3), water movements are highly dynamic both spatially and over time.

A. Current Velocity

Downstream water movement in a stream channel is referred to as ***current.*** Current is the foremost factor affecting organisms in the channel. The current erodes the channel substrata and determines the extent and type of particle deposition and thus the nature of sediments.

Current velocity, although driven by the gravitational weight of the water moving downslope, is balanced by the frictional resistance of the water against the channel boundaries. Nonturbulent laminar flows are restricted to small but important boundary-layer regions at the substratum–water interface (discussed later). Turbulent flows emanate from the frictional shear forces and propagate logarithmically into the overlying water. Turbulent flows have been separated into hydrodynamic regimes based on the ratio between inertial and viscous forces, the latter related to the particle size of the sediments of the stream bed (Davis and Barmuta, 1989; Carling, 1992).

Current velocity is zero at the substratum surfaces and increases logarithmically within the ***boundary layer*** (Fig. 7-4). The thickness of the boundary layer decreases with the increasing current velocities of bulk water. This stratum is important to biota, particularly microbiota, because, lacking macroturbulence, the transport of gases and nutrients through this stratum are by molecular diffusion processes. Above this layer, current velocities increase and are usually maximum centrally within the channel (Fig. 7-4). Velocities decrease just below the water surface because of frictional shear with the atmosphere. Current velocities are altered by local roughness of the bed materials, and the logarithmic layer usually extends to 10–15% of the channel depth (Carling, 1992). The increase in velocity can extend to close to the surface in shallow streams.

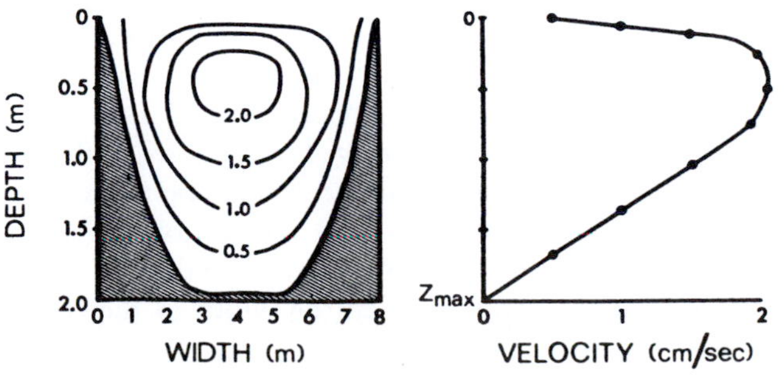

FIGURE 7-4 Idealized current velocity in cm sec^{-1} in a channel cross section (*left*) and in profile at the midpoint of the cross section (*right*). (From Wetzel and Likens, 1991.)

TABLE 7-1 Relationship of Current Velocity to Sediment Composition[a]

Velocity range (cm sec^{-1})	*General bottom composition*	*Approximate diameter (mm)*
3–20	Silt, mud, small organic debris	< 0.02
20–40	Fine sand	0.1–0.3
40–60	Coarse sand to fine gravel	0.5–8
60–120	Small, medium, to large gravel	8–64
120–200	Large cobbles to boulders	> 128

[a] Modified from Einsele (1960).

The shear forces generated by turbulent water movements are transferred to particles and cause them to move along or above the channel substrata. The suspended sediment load is particles that are entrained into the bulk water moving in the channel. The ***bedload*** is particles that are moved downstream in nearly continuous contact with the streambed (Beaumont, 1975). Both movements are, of course, a function of the current velocity. Larger particles may be transported by higher current velocities, and the sediment composition in subsequent sedimentation processes reflects the effects of current in erosion and transport (Table 7-1).

B. Channel Flows

Very few natural river channels are straight. However, small sections of many rivers have the dominant properties of straight channels with relatively uniform flow. Comparing flow in straight vs. the more common meandering channels is instructive. Flow in natural channels is three-dimensional, and because width-to-depth ratios are generally greater than about 13 (Carling, 1992), the effects of the banks are small on the central two-thirds of the flow. However, channel roughness in natural channels causes upwelling and secondary flow cells to develop along the banks (Fig. 7-5). In wide channels, flow may exhibit more than one flow cell, each exhibiting a secondary spiraling flow (Fig. 7-5). Resistance along the channel and banks increases with natural variations in morphometry and protrusions, particularly higher aquatic plants. These variations in resistance can result in secondary flow cells with reverse flow upstream along the bank. The result is alternating scouring and deposition variations in the channel topography at relatively uniform distances (3–10 times stream widths). Current velocities then vary with the roughness and depth variations and result in deeper sections, termed ***pools,*** and shallow sections, called ***riffles.*** Variations in flow resistance and the secondary cells formed by bank shears result

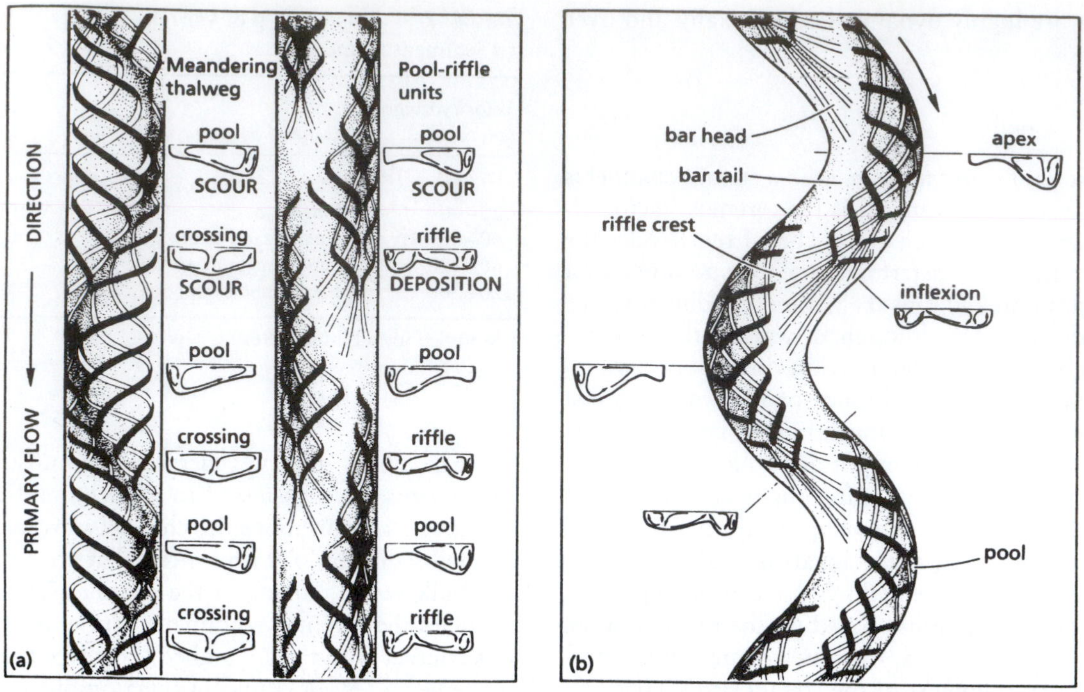

FIGURE 7-5 Models of flow structure in (a) straight and (b) meandering channels. (From Carling, 1992, after Thompson, 1986. Secondary flows and the poolriffle unit: a case study of the processes of meander development. Earth Surface Processes and Landforms *11*:631–641, © John Wiley & Sons Limited. Reproduced with permission.)

in the spiraling flows and shifting within the channel from side to side. As a result, even in relatively straight channels, flow is meandering and erodes the deepest part of the channel, termed the ***thalweg***, in a meandering pathway.

In curved river channels, the fast-flowing surface water of the large cells flows toward the outer bank and is accompanied by downwelling and erosion (Fig. 7-5). Reduced velocities occur on the opposing side and are accompanied by upwelling and deposition and the formation of a ***bar***. Many areas of slowly flowing or relatively quiescent water occur in areas opposite the primary inertial flow regions of the channel. These water masses mix only slowly across a shear zone with the main advective flow (Chatwin and Allen, 1985; Carling, 1992; Rutherford, 1994).

Any obstruction within the stream channel creates shear forces and can introduce turbulence and modify flow patterns and sediment transport in the localized region of the obstruction. Thus, enormously diverse habitats are created, and all are continuously changing with fluctuations in water discharge. This habitat diversity is important to inhabiting biota, particularly as instream flow refugia (Lancaster and Hildrew, 1993) and as sites for accumulation and production of organic matter. These sites can also offer protected habitats of sufficient duration for completion of life cycles before disturbance by high-water spates.

Of many natural obstructions that modify channel flows, two particularly significant agents are wood debris and aquatic plants. Plant materials of many types enter streams either by erosion with water or by direct falling into the channels from adjacent forest trees. Although many leaves, needles, cones, twigs, branches, bark, and rootwads enter streams, most of these are transported downstream until becoming entangled against some larger obstruction, forming a localized ***debris dam***. Retention of leaves and similar materials increases as stream discharge decreases (see Bilby and Likens, 1980; Raikow *et al.*, 1995). Major flow alterations occur from large obstructive debris, such as tree stems and associated root balls of a size greater than ca. 20 cm in diameter that have fallen or been washed as driftwood into the stream or river channel (Maser and Sedell, 1994; Wondzell and Swanson, 1999). Fallen trees extending partially across a channel deflect current laterally, widen the streambed, and increase the hydraulic complexity. Scouring downstream of the obstruction creates pools; backwaters of greatly reduced flow occur upstream from the obstruction along the stream margin. Secondary channels often form around tree obstructions and broaden the width of the stream within the valley.

The growth of aquatic macrophytes in streams and rivers is often in aggregated clumps. Great variations in hydraulic resistance to flow occurs among the many species of aquatic macrophytes adapted to living in flowing water (Dawson, 1978; Haslam, 1978). Long-narrow-leaved species offer the least hydraulic resistance to flow and are most common in streams of moderate discharge, although great variations occur. Water velocities are very low within a plant clump and decline sharply from the edge (<3 cm from the edge) to very low velocities within the plant aggregate (Madsen and Warncke, 1983; Machata-Wenninger and Janauer, 1991; Losee and Wetzel, 1993). Current flows tend to compress the trailing leaves downward and laterally, and water flows around the plant aggregates. Velocities tend to increase between the plants. Markedly reduced flows within the plant clumps lead to increased deposition, often of finer particles than normally occur within the streambed. Faster flows between clumps tends to be more erosive, and particle sizes are larger than within the plant clumps (Marshall and Westlake, 1990). Dense development of aquatic macrophytes can markedly impede stream flows; for example, submersed macrophytes reduced water velocities as much as fourfold during summer in a lowland Danish stream and promoted extensive organic sedimentation (Sand-Jensen *et al.*, 1989).

C. Flows at and within River Sediments

The flow of water over loose sediments may deform the materials and create a variety of bedforms in silt, fine sand, and gravels (Fig. 7-6). Once a critical shear stress is exceeded, ***ripples*** develop in fine sand (<0.6 mm) as crested, triangular, and gentle upstream slopes (Carling, 1992). Ripples move downstream by erosion on the upstream slope and deposition on the steeper downstream (lee) slopes. Heights are less than 4 cm with wavelengths below 60 cm.

At higher water velocities, ripples are replaced by larger ***dunes*** that may be meters in height with wavelengths of tens of meters (Fig. 7-6). Ripples may occur on the upstream slope of dunes. Backflowing turbulence vortices occur on the lee slopes, and the turbulent water above them is termed ***boils***. With further increases in stream velocities, bedload transport increases and dunes are eroded to transitional stages of relatively flat beds (Fig. 7-6). Further increases in current velocities induce water waves with upstream backflows and turbulence. Bedform patterns can be similar among cohesive sediment muds, but the structures are less predictable in relation to flow velocities.

Water moving within a channel encounters substrata variations. These differences induce localized pressure variations in the channel that force water and contained solutes by advection into interstitial spaces within the substratum (Vaux, 1962; Thibodeaux and Boyle, 1987). The penetration of overlying stream water into the substrata varies greatly with the geomorphology of the drainage basin, elevational changes, sediment composition, and particle size, as well as flow characteristics. The saturated interstitial areas beneath the streambed and into the stream banks that contain channel water to the depth to which overlying stream water actively infiltrates by advection is the ***hyporheic zone*** (White, 1993; termed the ***surface hyporheic zone*** by Triska *et al.*, 1989). This advective infiltration zone may extend for several centimeters to well over a meter in depth. Below the surface zone, the ***interactive hy-***

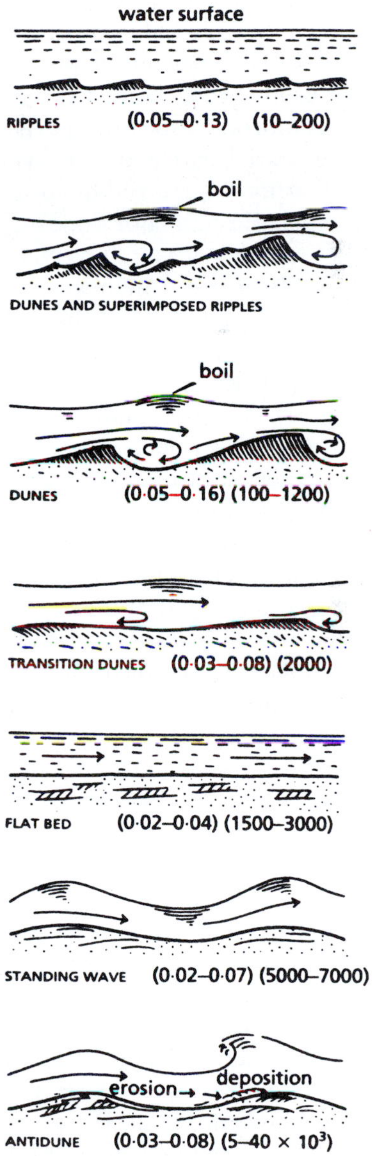

FIGURE 7-6 Sandy bedforms generated as current velocity increases. Parenthetic values refer to a relative friction factor and to the bedload transport rate (ppm). (From Carling, 1992, after Richards, 1982, *Rivers: Form and Process in Alluvial Channels.* Methuen, London, Rutledge. Reproduced with permission.)

porheic zone contains less stream water and more ground water. The boundary between these two hyporheic zones can be considered the hydrological boundary of the stream even though there are active chemical and biotic exchanges between them, because ground water represents water beneath the water table that has not yet been influenced by channel processes (Freeze and Cherry, 1979). The penetration of the interstitial habitat of the hyporheic zone by riverine animals can be viewed as an indicator of spatial and temporal variations in hyporheic boundaries and conditions (Stanford and Ward, 1988).

The hyporheic zone can confer stability to stream ecosystems as potential microbial and invertebrate refuge sites in the alluvium and its sustenance of microbiota important in biogeochemical processes, particularly by storage of nutrients (Grimm *et al.,* 1991; Valett *et al.,* 1994). Geomorphic processes, particularly major floods, alter the spatial extent of the hyporheic zone, however, as well as its volume and hydrologic residence times. For example, during major floods of 50- to 100-year recurrence intervals, water movements and transported large woody debris can rework, disturb, and move sediments of most of the hyporheic volume (Wondzell and Swanson, 1999). The effects on the hyporheic biota would depend on the extent of the hyporheic zone, depth of scour, and species-specific patterns of utilization of the zone.

Stream water penetrates interstitial spaces of sediments to a greater depth on the upstream side of ripples and dunes and less so on the lee side (Grimm and Fisher, 1984; Hendricks and White, 1991). The exchange of dissolved gases and nutrients is consequently much greater in areas of more extensive penetration, which generally increases the distribution and productivity of surface colonizing and interstitial biota in these leading-edge areas.

The sources of water to stream ecosystems vary greatly, but ground water often constitutes the dominant input to base flows. The groundwater dominance often prevails in headwater reaches of streams (Fig. 7-7a). Streamwater penetration into the hyporheic zone of the sediments may be less extensive in these headwater areas. Changes in localized geomorphology and reduction in streambed elevation commonly occur further downstream, with extensive development of riffles and pools (Statzner and Higler, 1986). This flow pattern can result in extensive penetration of surface water into the streambed (Fig. 7-7b), and likely results in a complex of alternating upwellings of ground water and downwellings of stream water through the sediments, as was implicated by the analyses of Godbout and Hynes (1982). Subsurface flow paths do occur in certain coarse substrata beneath the channel, with a slow hyporheic movement (m h^{-1} to m d^{-1}) in straight lines, possibly using relic channels, from one meander to another (Triska *et al.,* 1993). Groundwater sources can constitute a major source of base flows (Williams and Pinder, 1990). In downstream sections of rivers of higher order where the channels widen into flood plains and deltas, elevational gradients and groundwater inputs usually decline. The advective forces of moving water through the sediments also decline. The boundaries between river water and ground water extend laterally further from the river channel as flood plains widen (Fig. 7-7c).

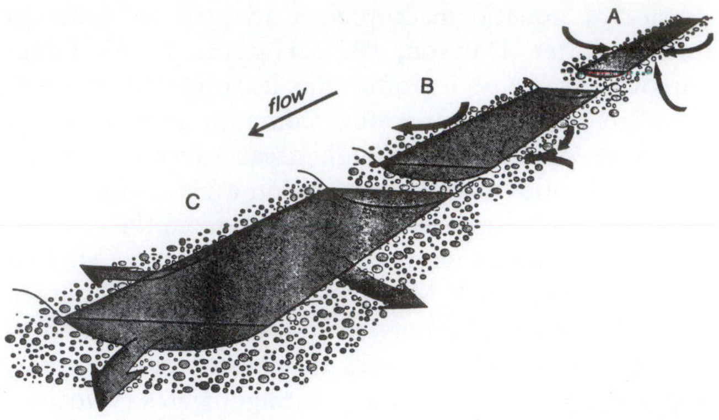

FIGURE 7-7 Directions of water movements within longitudinally connected stream beds. (A) A headwater stream; (B) a mid-reach stream with pool-riffle-pool sequence; and (C) a larger river with a well-developed flood plain. (After White, 1993.)

At bankfull stages (Chap. 3), high discharge can overflow the banks and inundate the adjacent, flat floodplain areas. Thus, the flood plain encompasses areas that are periodically inundated by the lateral overflow of rivers. Overbank flow induces large-scale vorticies into the floodplain areas (Fig. 7-8). The faster-moving water in the main river channel and the slower-moving water on the flood plain produce a shear layer laterally across the system (Knight and Shiono, 1996). The mean velocity decreases from the river to lower values on the flood plain, and the flows are dependent on the average river bank height and the depth of water on the flood plain in relation to the depth of the main river channel. Because of the decreased shear during the later flooding movements of water in the flood plain, the kinetic energy is dispersed over large areas. As a result of shear against the substratum and often many obstructions such as trees, an appreciable sediment load can be deposited from the water. Under very high discharge conditions, materials of the flood plain can be eroded and carried downstream.

The periodicity of inundation of the flood plains covers a wide hydrological spectrum from short (few days) to long duration (several weeks) and from unpre-

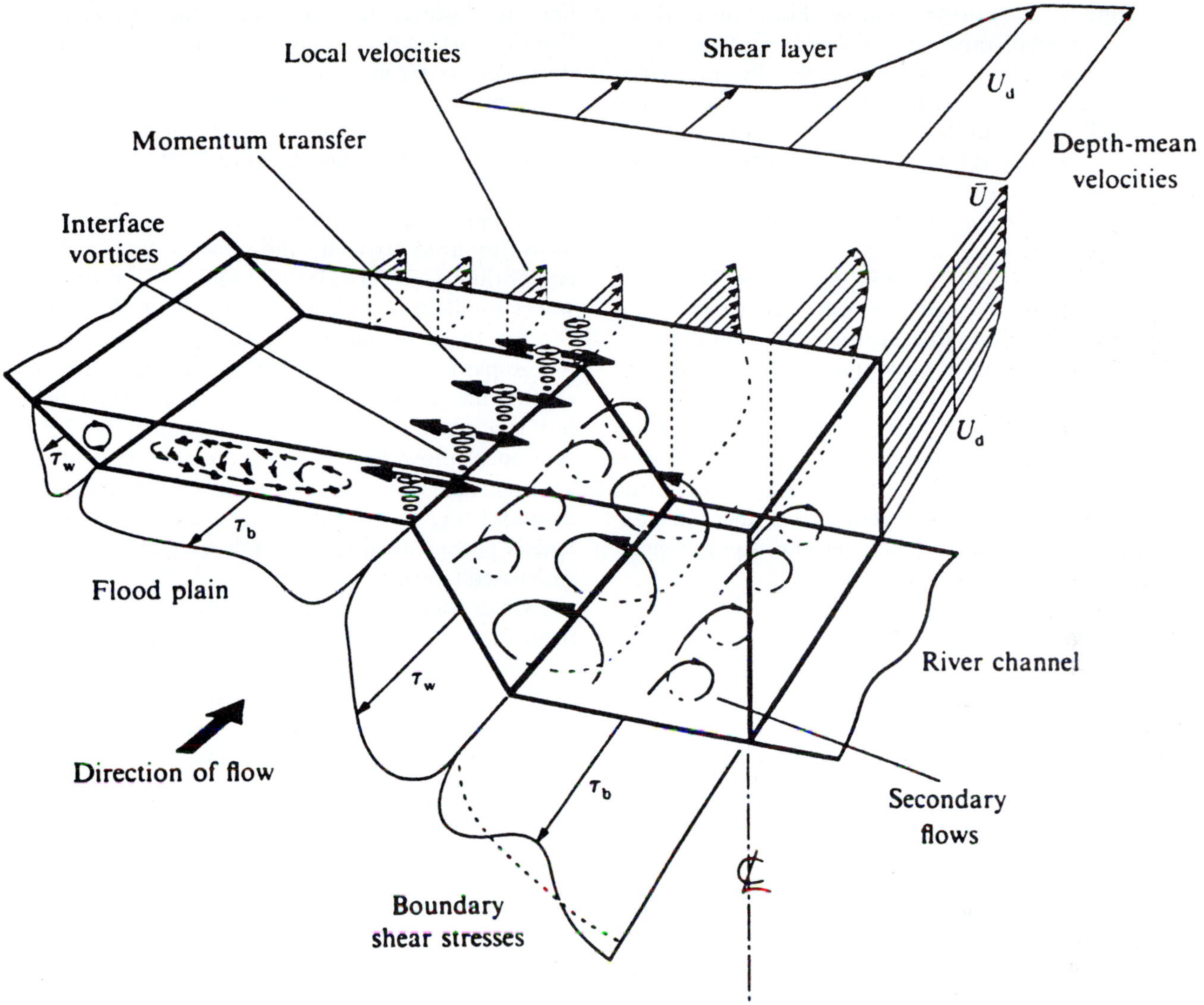

FIGURE 7-8 Hydraulic parameters associated with overbank flows onto and within the flood plain. (After Knight, D. W. and Shiono. 1996, River channel and floodplain hydraulics. In M. G. Anderson, D. E. Walling, and P. D. Bates, eds. Floodplain Processes. John Wiley & Sons, Limited, Chichester. Pp. 139–181. Reproduced with permission.)

dictable to predictable timing (Junk *et al.,* 1989; Bayley, 1995). Among many streams and rivers of the temperate region, flooding is relatively rare and unpredictable, although the long-term frequency is moderately predictable (Chap. 3). Low-order streams have an irregular flood pattern with many peaks that are influenced by local precipitation events. The influence of these events diminishes with increasing size of the drainage basin and become minor in the hydrographs of very large rivers.

The large catchment basins of large rivers offer a strong hydrological buffering capacity, and flooding responses result from (a) seasonality of precipitation and (b) seasonality of evapotranspiration. In tropical and subtropical regions, the hydrographs of large rivers commonly reflect the annual rainy season and typically exhibit only one pronounced flooding event per year. Because of the large size of the rivers and their drainage basins, the downstream flood pulse can persist for long periods of time (weeks to months) after the rainy season has subsided in the upper portions of the drainage system (Junk *et al.,* 1989; Junk, 1997a,b). Many riverine lakes and wetlands can develop in depressions of the flood plain; they become isolated from the main channel for months or years and only become hydrologically connected during periods of high water.

In temperate regions, most rivers have flow régimes altered by dams and channelization. Where medium-sized to large rivers have been unmodified, they also are characterized by low gradients, broad flood plains, and frequent flooding. Flooding in these regions is less predictable and of shorter duration than in regions with large tropical rivers. Because precipitation tends to be less seasonal and more evenly distributed over the year, flooding tends to be controlled primarily by seasonal differences in evapotranspiration (Benke *et al.*, 2000). Runoff and flooding decrease during the summer periods of high losses of water by evapotranspiration.

As rivers and tributaries join, mixing occurs at the confluences. Although many morphological factors influence mixing rates, mixing of water between the confluent rivers is nearly always completed before a downstream distance of 25 channel widths is achieved (Gaudet and Roy, 1995). Turbulence and mixing rates are increased if one of the rivers is appreciably shallower than the other at the confluence zone (Best and Roy, 1991). Shear and scour occur at the junction, with entrainment of the main river water and rapid mixing.

IV. SURFACE WATER MOVEMENTS

A. Surface Waves (Progressive Waves)

The frictional movement of wind blowing over water sets the water surface into motion, producing a wind drift. Wind also sets the surface into oscillation and produces ***traveling surface waves.*** If these waves become large enough to break, their energy flux and dispersion are transferred to the water. Although to an observer looking at a lake or very large river, these surface waves are the most conspicuous periodic characteristic, their limnological importance is largely confined to shore areas. Traveling surface waves are confined to surface layers and cause little displacement of the deeper water masses.

Short surface waves cause the surface water particles to move in a circular path or orbit. In cross section, the path is ***circular*** (Fig. 7-9), with very little significant motion other than slow horizontal translocation (Smith and Sinclair, 1972). In a given wave, water is displaced vertically and returned by gravity to an equilibrium state. Similarly, neighboring short surface waves oscillate periodically in circular fashion. In synchrony, these movements result in a traveling wave whose path may be described as analogous to a point on the rim of a wheel moving along a plane where the hub is at the mean surface of the water (Fig. 7-9). Except in shallow beach areas, the wavelengths (λ) of short surface waves are less than the water depth and, consequently, they are dispersive, that is, they travel at speeds proportional to $\lambda^{1/2}$ of deepwater gravity waves (discussed later).

Of greater interest than the small horizontal movement of short surface waves is the influence such periodic oscillations have vertically, although the height (h) of this vertical oscillation is attenuated rapidly with depth (Fig. 7-9). The decrease of vertical motion with increasing depth can be approximately described as a halving of the cycloid diameter for every depth increase of $\lambda/9$. As we will see shortly, the amplitude or height of surface waves is not directly proportional to wavelength, although a ratio of about 1:20 of h:λ is a common average. If, for example, short surface waves had a λ = 18 m and h = 1 m, the vertical oscillation at a depth of 4 m ($2/9\lambda$) would be 25 cm, at 8 m or $4/9\lambda$, 6.25 cm, and at 18-m depth, 1.95 mm. In other words,

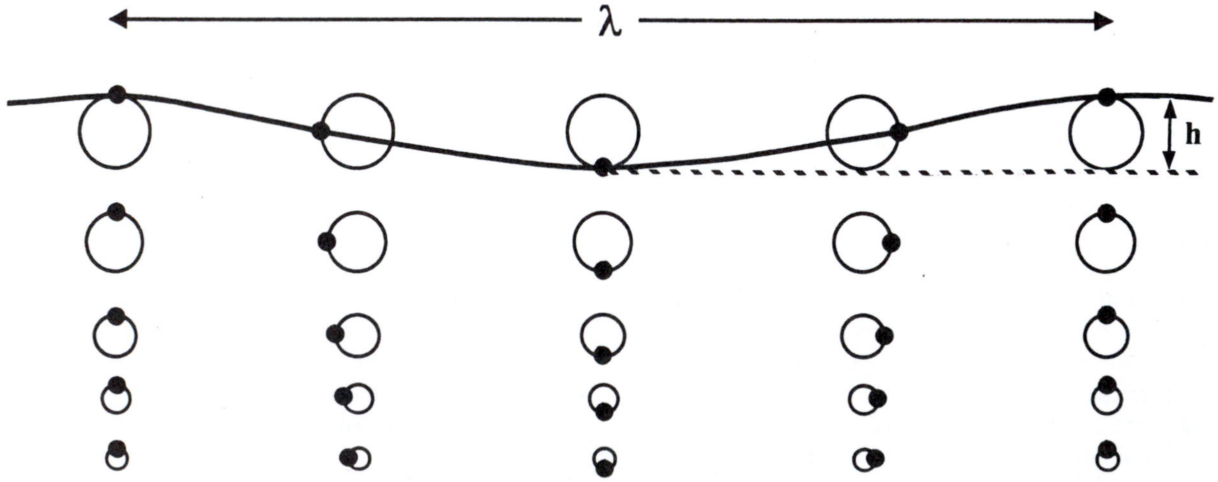

FIGURE 7-9 Circular motion of fluid particles in a sinusoidal wave of length (λ) traveling from left to right on deep water. The height (h) of the circular movement is attenuated with depth by $h/2$, the displacement positive or negative from the equilibrium in a sinusoidal wave. (Modified from Lighthill, 1978.)

the magnitude of vertical displacement at a depth equal to the wavelength will be 1/512 (i.e., $1/2^9$) of that at the surface.

Short surface waves with a wavelength greater than 6.28 cm (2π cm) are referred to as ***gravity waves.*** Waves of a length less than this minimum are called ***ripples*** or ***capillary waves,*** where surface tension effects become appreciable. The ratio of wave height (h) to wavelength (λ) is highly variable within the range of 1:100 to about 1:10. At great height, the crests of the waves sharpen and become less stable. When the angle of the wave height exceeds a $h{:}\lambda$ ratio of about 1:10, the peak collapses and some apical water may be blown off, forming a ***whitecap.*** The occurrence and density of whitecaps increases abruptly as the wind velocity increases from ca. 7 to 8 m sec^{-1} (Monahan, 1969).

For a given wind speed, wave height appears to be nearly independent of depth in small lakes, but in lakes of great area, wave height and length increase with increasing water depth (Hutchinson, 1957). The height of the highest waves observed on a lake appears, without good theoretical explanation, to be proportional to the square root of the ***fetch,*** or distance over which the wind has blown uninterrupted by land. Thus, the maximum height (cm, crest to trough), where x = the fetch distance in centimeters, is expressed as

$$h = 0.105\sqrt{x}$$

For example, the maximum wave height observed in Lake Superior was 6.9 m with a fetch of 482 km (4.82×10^7 cm), which agrees well with a maximum of 7.3 m predicted by this relationship.

Surface wind shear stress on the water surface is often unevenly distributed because of interference by surrounding topography. In the lee of trees, hills, and rocks, surface wind shear stress will remain zero over a length of 6 times the height of the sheltering feature (Ottesen Hansen, 1979). Fully developed wind shear stress and associated growth of turbulent energy at the surface layer requires an additional further distance of 7 times the height. Therefore, the relationship between fetch and surface waves can be greatly modified in small lakes by the surrounding topography.

The waves described in the foregoing paragraphs are "deepwater" or "short" waves, in which wavelength is much less than water depth. Where this condition no longer holds, and the wavelength becomes more than 20 times the water depth, the wave is transformed into a "shallow water" or "long" wave, and the circular motions are transformed into a to-and-fro sloshing, which extends to the bottom of the water column (Fig. 7-10). Hence, the circular orbits found in deepwater waves shift to more elliptical orbits and flatten as they approach the bottom. Shallowwater waves are no longer energy dispersive, because their speed is proportional to the square root of the water depth rather than being determined by wavelength. Therefore, as deepwater waves enter shallowwater areas, their velocity decreases as the square root of depth decreases, and there is a simultaneous reduction in wavelength (Iverson, 1952; Mason,

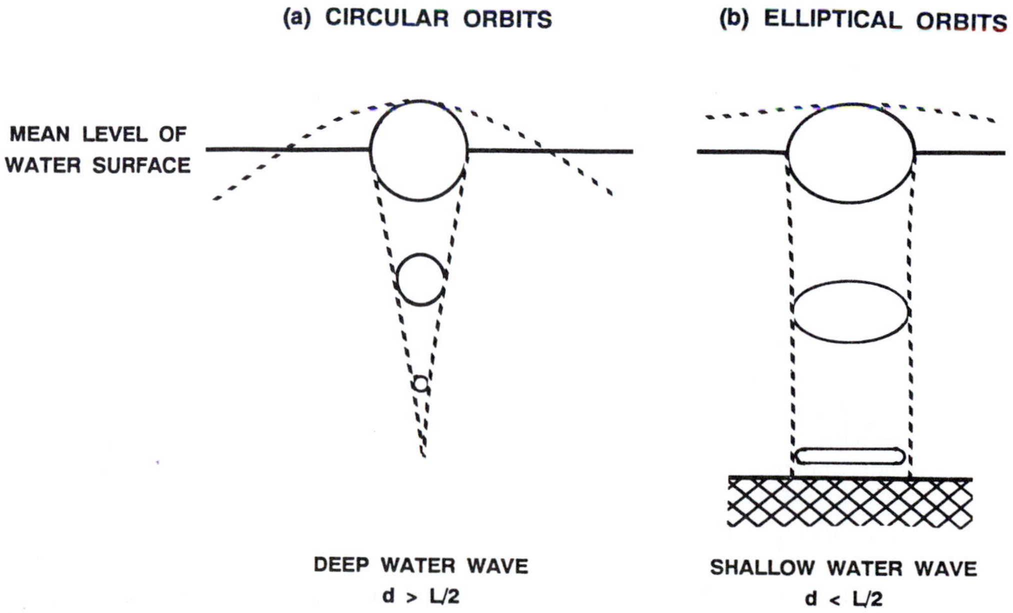

FIGURE 7-10 Orbital motions of water beneath waves. (a) Deepwater waves where water depth (d) is greater than the wavelength $\lambda/2$. (b) Shallow water waves where $d < \lambda/2$. (Modified from Denny, 1988.)

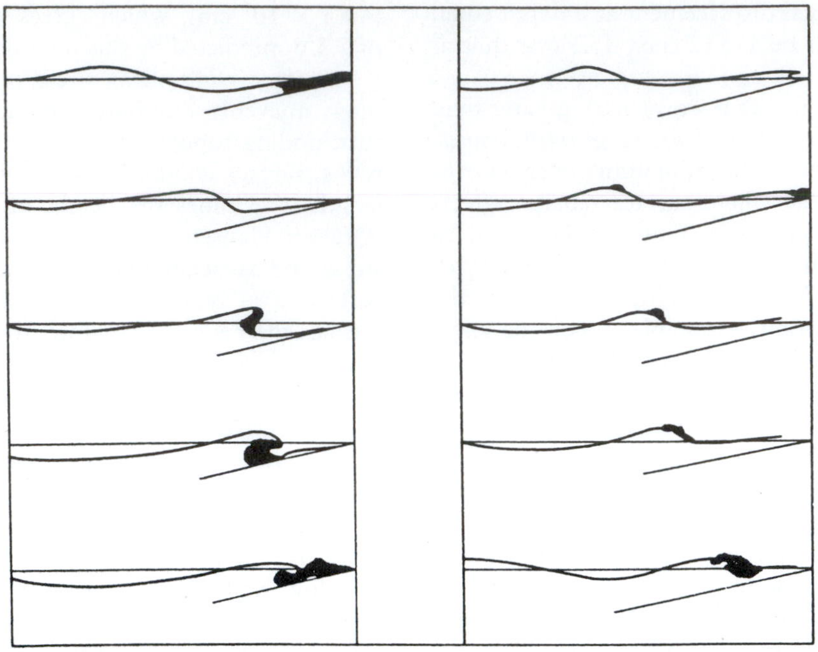

FIGURE 7-11 The breaking of waves on beaches. *Left:* Plunging breaker. *Right:* Spilling breaker. (From Hutchinson, G. E.: *A Treatise on Limnology*. Vol. 1, New York, John Wiley & Sons, Inc., 1957; after Iversen, 1952, and Mason, 1952.)

1952; Hutchinson, 1957). Wave height first decreases slightly, then increases markedly and becomes asymmetrical and unstable. The collapse of water over the front of this now asymmetrical wave is termed a ***breaker*** and occurs in a spectrum of forms between two extreme types (Fig. 7-11). In a ***plunging breaker,*** the forward face of the wave becomes convex and the crest curls over but collapses with insufficient depth to complete a vortex. The crest of a ***spilling breaker*** collapses forward, spilling downward over the front of the wave.

At an early stage of wave development when a fresh wind blows over the water surface, relatively long-lived short gravity waves form in crescent-shaped or "horseshoe" patterns (Dias and Kharif, 1999). These waves are rather steep with sharpened crests and flattened troughs, and they have a front–back asymmetry (Fig. 7-12). The specific horseshoe shape of the wave fronts is always oriented forward.

Where the offshore gradient is low and an abundance of loose sediment exists, the swash from breaking waves can lead to **beach-ridges** of linear mound-shaped ridges of sediment approximately paralleling the shore line (Taylor and Stone, 1996). Beach-ridges are often transient but over a long period of time can prograde with sedimentation and lead to modified shore lines toward the lake basin.

Because in shallow water the wave action extends to the bottom, sedimentation can be prevented in sufficiently shallow water and movement may be too severe for most aquatic plants to grow. In lakes with high rates of sedimentation in the littoral zone, such as in calcareous hardwater lakes where marl banks or benches extend out from the shore line for many meters, recently sedimented material is constantly resuspended and removed to deeper areas. As a result, in suitably shallow water of about a meter, marl benches on the steep sides of these lakes do not increase in height. In a hardwater lake of northern Indiana, for example, the surface sediments of a submersed lakemount island were found to be very old (Wetzel, 1970). The lakemount ceased accretion of sediments abruptly 1 m from the lake surface about 2740 years ago, for more recent sediments were continually moved into deeper parts of the lake.

B. Surface Currents

Currents are nonperiodic water movements generated by external forces. Forces responsible for producing currents include the wind, changes in atmospheric pressure, horizontal density gradients caused by differential heating or by diffusion of dissolved materials from sediments, and the influx of water into a lake. The latter subjects have been treated in other chapters, and the following remarks are therefore concerned primarily with wind-derived surface currents of open-water masses.

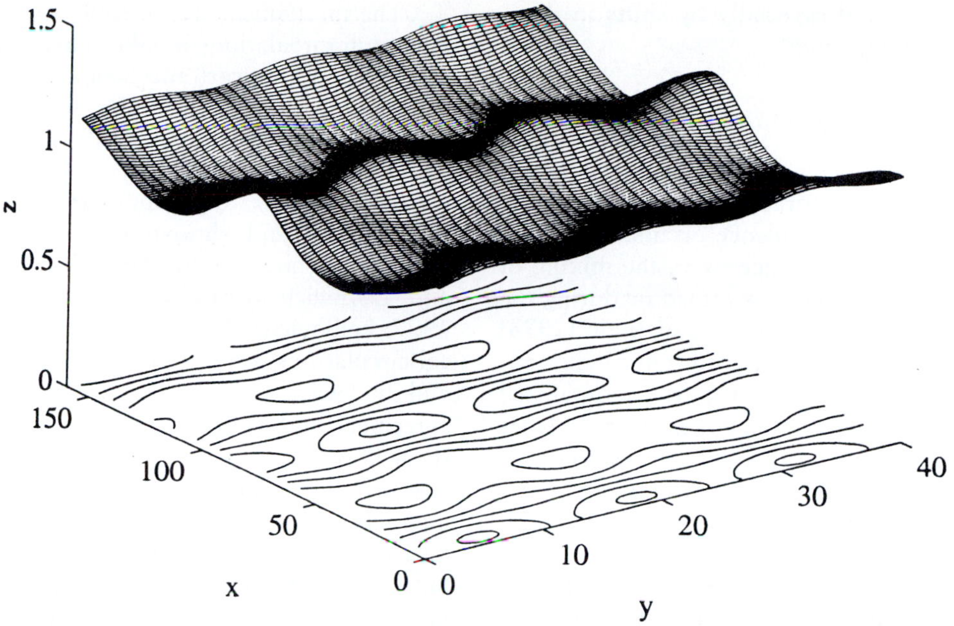

FIGURE 7-12 Crescent-shaped or "horseshoe" patterns of surface gravity waves in their phase of growth with their wave fronts oriented forward. The waves are propagating from right to left. (From Dias and Kharif, 1999, after work of C. Skardrani.)

Geostrophic effects of the deflecting (Coriolis) force due to the Earth's rotation are found in all currents in moderate- to large-sized lakes. In the Northern Hemisphere, surface water currents are deflected to the right relative to the direction of the wind. In the Southern Hemisphere, the Coriolis force causes surface waters to deflect to the left relative to the wind direction.

The combined effects of wind and geostrophic deflection cause surface water to move downwind and to the right in the Northern Hemisphere. Under stratified conditions, there is also a tendency for denser water to collect on the left side of the current and less dense water on the right side of the current. The resulting wind-drift current is deflected 45° from the direction of the wind in a spiral manner (the Ekman spiral) in open waters of large, deep lakes. As the size, and especially the depth, of the lakes decrease, the magnitude of the deflection angle decreases until, in a lake with a depth of less than about 20 m, the angle of declination becomes insignificant. In Lake Mendota, which is of moderate size (39.2 km^2) and depth (25.6 m), a rotation of wind-driven surface currents had an average angle of deflection of 20.6° to the right (Shulman and Bryson, 1961). The depth of frictional influence in Lake Mendota was found to be between 2 and 3.5 m. Stress exerted by the wind on the water surface was found to be a linear function of wind speed.

The wind factor, defined as the ratio of surface water current velocity to wind velocity, is quite variable. In general, the velocity of wind-driven currents is about 2% of the speed of the wind generating them and is largely independent of the height of the surface waves. An average wind factor of 1.3% for the upper half-meter of water was found in Lake Mendota at low wind velocities (Haines and Bryson, 1961). Water velocity in the surface layers increases with wind velocity until a critical wind speed is reached. Above the critical speed, surface water velocity decreases, and the wind factor becomes nonlinear. In Lake Mendota, the critical wind speed was 5.7–6.1 m sec^{-1}.

The surface currents of very large lakes tend to circulate in large swirls or ***gyrals.*** Lake Michigan, for example, is typically exposed to predominantly westerly winds during summer, and its large, conspicuous gyrals are clearly influenced by geostrophic rotational forces (Noble, 1967). In this and other examples, gyral movements are complicated by numerous other water movements related to standing water motions (see later discussion). Inertial currents on a smaller scale also occur in all of the Laurentian Great Lakes, not only at the surface but at all depths and in all seasons (Verber, 1964, 1966). The geostrophic right-hand acceleration of currents also occurs under ice cover. Many complex flow patterns have been observed: straightline, sinusoidal o2 oscillatory, repeated crescent, circular or spiral, and rotary or screw flows. Gyrals in lakes the size of Lake Michigan are not simply the result of geostrophic forces; their direction is strongly modified by large, long waves and especially by shifts in duration of strong prevailing winds.

C. Langmuir Circulations and Streaks

Turbulence generated by surface water movement and wave action is often sporadic and independent of wave direction. Sporadic turbulence occurs, for example, in the dispersion of wave energy in the mixing of the epilimnion when the lake is stratified or during the mixing of the general water mass. Langmuir (1938) demonstrated that under some circumstances the motions associated with turbulent transport are organized into vertical helical currents in the upper layers of lakes (Fig. 7-13). Convection from this vertical motion generates streaks, which are oriented approximately parallel to the direction of the wind. Streaks coincide with lines of surface convergence and downward movement and are marked by windrows of aggregated particulate and surface-active materials. Between the streaks are zones of upwelling. Such ***Langmuir circulation*** occurs in all waveswept lakes of any significant size at wind speeds above 2–3 m sec^{-1}. The streaks are seldom observed at wind speeds above about 7 m sec^{-1}, even though the Langmuir circulations are still occurring (Scott *et al.*, 1969; Ottesen Hansen, 1978), for at high wind speeds, surface turbulence is apparently great enough to disrupt the surface aggregations.

The mechanisms responsible for the generation of Langmuir circulations in lakes have been the subject of much debate (Stewart and Schmitt, 1968; Myer, 1969; Scott *et al.*, 1969; Faller, 1971; Harris and Lott, 1973). It is evident from Faller's (1969, 1978) and Faller and Caponi's (1978) laboratory work that the primary mechanism generating Langmuir circulations probably is the wave and shear-flow interaction mechanisms described in a series of theoretical analyses by Craik and Leibovich (Craik and Leibovich, 1976; Craik, 1977; Leibovich, 1977). In these experiments, the helical circulations were present only when both the wind and the wave patterns existed at the same time, in agreement with the theoretical predictions that both a shear flow due to wind and the Stokes drift of progressive waves (see later) are essential. From this complex interaction, wind energy is converted into organized helical circulations as well as into waves, random turbulence, and the mean shear flow. The resulting Langmuir circulation is one of the primary ways in which turbulence is transported downward and the upper layers of water are mixed.

The Langmuir convectional helices form a series of parallel clockwise and counterclockwise rotations that result in linear alternations of divergences and convergences (Fig. 7-13). In Lake George, New York, Langmuir found the velocity of downward convergence currents to be 1.6 cm sec^{-1} with a wind speed of 6 m sec^{-1}.

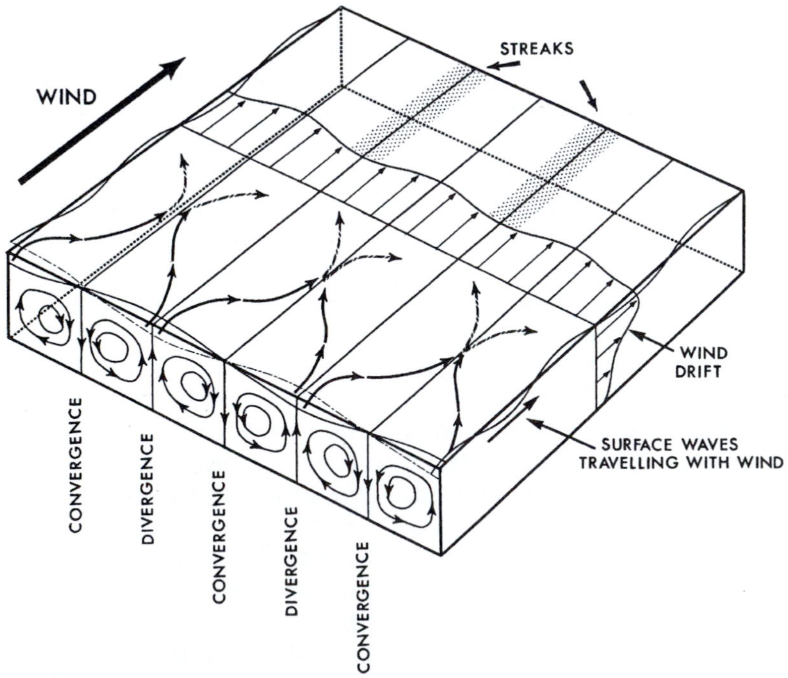

FIGURE 7-13 Diagrammatic representation of the helical flow of Langmuir circulations in surface waters with streaks of aggregated organic matter occurring at lines of divergence. (Based on the experimental work of Faller, 1978, and others.)

At this time, there was a horizontal surface current of 15 cm sec^{-1} and a current of 10 cm sec^{-1} at a depth of 3 m. Downward movement velocities at convergences were about three times the upward current velocities at divergences. Such currents are sufficient to markedly influence the distribution of microflora and fauna suspended in the surface waters of lakes. Algae and zooplankton, with limited or no powers of locomotion, tend to be aggregated in streaks in areas of water divergence (George and Edwards, 1973). The result is a patchiness in horizontal distribution that is important not only in the sampling of the microorganisms for estimates of population size and distribution and in metabolic measurements but also in determining the distribution of predators, which congregate in these zones of higher prey density.

V. INTERNAL WATER MOVEMENTS

A. Metalimnetic Entrainment in Stratified Lakes

Surface waves and Langmuir circulations cause turbulence, but the barrier formed by the density gradient in the metalimnion suppresses the transmission of turbulence into the metalimnion and hypolimnion. However, as wind blows over the surface for a reasonable period of time, wind drift causes water to pile up, with a rise in surface level at the lee end of the lake. The accumulated mass of water is pulled downward by gravity and, when it encounters the more dense water of the metalimnion, it flows back in a direction opposite to that of the prevailing wind (Fig. 7-14). The thermocline, or plane separating the two layers of different density (termed *pycnocline* by Hutchinson, 1957), is tilted in the process. We will return to the vertical movement of this entire water mass in the following section, after discussing the circulation pattern.

Progressive erosion of the metalimnion occurs by ***entrainment***, with a corresponding lowering of the thermocline level (Mortimer, 1961). This erosion, which is characteristic of the autumnal cooling phase, begins when the net heat flux across the water surface changes from downward to upward and ends in the fall overturn. During the early stages of entrainment (Fig. 7-14a), when the return current above the thermocline and the entrained layers below it both are moving upwind, shear turbulence in the metalimnion is low. As long as R_i does not fall below the critical value, the thermocline remains a slippery surface of low frictional shear with little mixing across it. If the shear at the thermocline increases ($R_i < 0.25$), flow becomes unstable and large vortices appear in the return current at the downwind end (Fig. 7-14b, c). These vortices are carried upwind, where they become mixed into the surface drift. This process increases the density gradient at the downwind end until it offsets the local shear to produce neutral turbulence stability, with R_i near 0.25. The resultant whole-basin disposition of isotherms becomes fan-shaped (Fig. 7-14d), and the shape of the metalimnetic density profile varies in different regions of the basin correspondingly.

As the wind velocity decreases in a subsequent calm, the displaced layers slide over each other to redistribute into a new equilibrium. The oscillations so created may or may not be sufficient to create internal instability and waves. The resulting thermal profile retains much of its original form, but the mean metal-

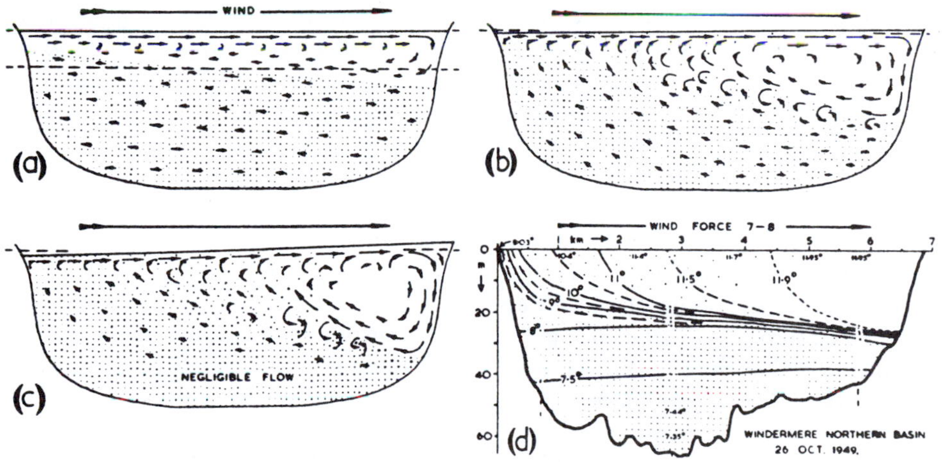

FIGURE 7-14 Stages of wind drift and circulation that led, after about 12 hours of strong wind, to the thermal situation depicted in (d), Lake Windermere, England, in late fall. Broken lines show equilibrium levels of water surface in (a), (b), and (c), and of the thermocline in (a). The initial layer below the thermocline is stippled; speed and direction of flow are roughly indicated by arrows. (From Mortimer, C. H.: *Verhandlungen Int. Ver. Limnol. 14*:81, 1961.)

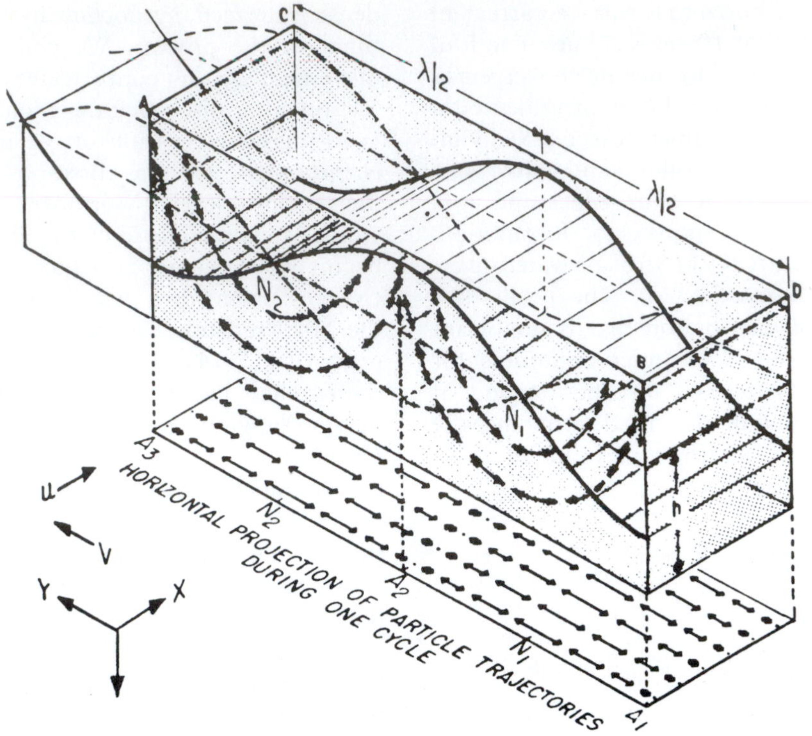

FIGURE 7-15 A long standing wave, without rotation, in a uniform depth (*h*) model. *N*: nodal line; *A*: antinodal line. (From Mortimer, C. H.: *Mitteilungen Int. Ver. Limnol. 20*:157, 1974.)

imnion depth has been pushed to a lower level, depending upon the amount of eddy conductivity. It is important to emphasize that this type of internal water motion is not continuous. Entrainment and metalimnetic erosion are progressive and forced by storms. Erosion can be counteracted by calm periods of heating, but in the fall, when cooling exceeds heating, the erosion continues in stepwise progression with storms. A strong inverse relationship occurs between the magnitude of entrainment and the mean density gradient across the metalimnion, and a strong positive correlation exists between entrainment and basin dimensions (Blanton, 1973).

Entrainment rates are obviously of major importance in the control of the extent of heat intrusion into the metalimnion and nutrient return from the hypolimnion to the epilimnion. A certain degree of compensatory motion occurs in the hypolimnion in a direction opposite that in the metalimnion. Turbulence in the hypolimnion is very much weaker than in the overlying layers but increases in large lakes where space is sufficient for the development of large-scale entrainment.

Surface and epilimnetic water movements and the effects of internal metalimnetic and hypolimnetic currents can be demonstrated very clearly in simple laboratory models of lakes (cf. Mortimer, 1951, 1954; Vallentyne, 1967; Stewart and Hollan, 1984; Stewart, 1988; Wetzel and Likens, 2000). Such models, stratified either thermally or with colored solutions of different chemical densities, are so effective in demonstrating entrainment, seiches, and internal waves (discussed in the next section) that their use is strongly urged in all introductory courses. For example, the upwelling areas of the epilimnion (Fig. 7-14) are seen clearly with tracer dyes in these models. Even more strikingly shown are the vortices of internal turbulence (curved arrows of Fig. 7-14b, c) which, with strong wind, become unstable and break as waves.

VI. WATER MOVEMENTS AFFECTING THE WHOLE LAKE

A. Long Standing Waves

Displacement of the water mass of the whole lake, as illustrated in Figure 7-14, gives rise to rhythmic motions, including both oscillations of the water surface and internal oscillations of isothermal depth. These motions take the form of very long waves, which have wavelengths of the same order as basin dimensions. These long waves are reflected at the basin boundaries and combine into standing wave patterns. The water surface or the thermocline oscillates up and down like a seesaw about a line of no vertical motion (a ***node,*** Fig. 7-15), which coincides with regions of maximum to-and-fro horizontal motion of the water masses. Such standing waves, surface or internal, are referred to as ***seiches,*** a term originally referring to the periodic "drying" exposure of shallow littoral zones. Figure 7-15 illustrates a constant-depth model with two nodes, beneath which the horizontal to-and-fro sloshing of the whole water mass is at a maximum. Seiche motion, particularly the internal variety, has far-reaching effects on the vertical excursion of water masses, much larger than observed at the surface, and on their horizontal motions in the form of alternating currents.

The most common cause of seiches is the wind-induced tilting of the water surface and of the thermocline, as described in the previous section. As the wind slackens and the wind stress is removed, the tilted water surface and the tilted thermocline flow back toward equilibrium. Momentum, however, is great and equilibrium is overshot (Fig. 7-15), which results in a rocking motion about one or more nodal points. No vertical movement occurs at the nodal point, whereas maximum vertical movement takes place at the ends, or ***antinodes.*** The nodal point is in reality a nodal line in this nonrotating rectangular situation running the width of the basin.

1. Surface Seiches

The most conspicuous example of long standing waves is the ***surface seiche,*** which conforms to one or more resonant oscillatory frequencies (free modes) of a particular basin. The free mode of the surface seiche is a barotropic, or surface, wave, which affects the motion of the entire water mass of the lake, whether stratified or not, and attains its maximum amplitude at the surface. The surface seiche contrasts with the baroclinic or internal seiche, discussed further on, which is associated with the density gradient in stratified basins and attains its maximum amplitude at or near the thermocline.

The periodicity of vertical movement at the antinodes is a function of the length and depth of the basin. Particles at the node move back and forth horizontally with the same oscillatory rhythm as the vertical motion at the antinodes. The horizontal oscillation at the nodal line is large when the basin is very long in relation to depth, even though the vertical amplitude of the seiche is small at the antinodes. In a simple rectangular situation, analogous to many lakes in which the length greatly exceeds the mean depth, the period of the uninodal surface oscillation is approximated by

$$t = \frac{2l}{\sqrt{g\bar{z}}}$$

where

t = period (seconds)
l = length of the basin at the surface (cm)
$\bar{z}$ = mean depth of the basin (cm)
g = acceleration of gravity (980.7 cm sec^{-2})

Once surface seiches are set into motion, frictional and gravity damping of the oscillations begins and the water mass gradually returns to equilibrium. The magnitude of damping varies with water depth and the complexity of basin shape. In deep lakes of uncomplicated shapes, for example, weakly damped oscillations may persist with slowly diminishing amplitudes long after the impact of the storm that set them into motion has passed. A few examples in Table 7-2 illustrate the periods of the surface seiche modes, which correspond to values calculated from hydrodynamic theory. In most lakes, the period of the second mode (t_2) is more than half of that of the f)rst mode (t_1).

In long, narrow lakes, transverse seiches are found with periodicities and amplitudes much less than those of longitudinal seiches, which move the length of the longest axis—for example, the conspicuous transverse seiche of 132-min period in Lake Michigan (Mortimer, 1965). Where lake length differs little from breadth, complicated seiches can result when wind direction changes.

Uninodal seiches, as described earlier, are common even in very large lakes. If pressure is exerted on the surface in the center of a basin and released, or periodically exerted and released, binodal surface seiches are generated. Multinodal seiches, with up to 17 nodes, have been observed.

The amplitudes of surface seiches are generally rather small in comparison to internal seiches. For example, in Lake Mendota, the amplitude of a surface seiche was 1–2 mm with a periodicity of 25.8 min, although larger amplitudes of surface seiches (ca. 13 cm) of identical periodicity have been observed in this lake under unusual meteorological conditions (Stewart, 1993). In larger Lake Geneva, Switzerland, the maximum observed surface seiche amplitude was 1.87 m (t = 73 min), although the amplitude damped gradually in two weeks to less than 10 cm. In still larger lakes, such as Lake Erie, where amplitudes of surface seiches can exceed 2 m (t = 14 h), the effects can have both destructive and beneficial results, for example, the flushing of river delta and harbor areas. Surface seiches can occur under ice cover in similar ways.

True lunar tides, the result of gravitational attraction of water masses by the Moon and Sun, have been poorly investigated in lakes. However, it is certain that true tidal movement, even in the largest lakes, is small (a few millimeters). The maximum observed ampli-

TABLE 7-2 Periodicity of Longitudinal Seiches[a]

		Observed period (min)			Observed period (min)
Loch Earn, Scotland	t_1	14.52	Lake Vetter, Sweden	t_1	179.0
	t_2	8.09		t_2	97.5
	t_3	6.01		t_3	80.7
	t_4	3.99		t_4	57.9
	t_5	3.54		t_5	48.1
	t_6	2.88		t_6	42.6

Lake	Length (km)	$\bar{z}$ (m)	First mode period (T_1) (min)
Altausseer, Austria	2.8	34.6	5.3
Garry, Scotland	6.0	15.3	10.5
Mendota, Wisconsin			
NS axis	6.8	12.0	25.6
E axis	9.1	12.8	25.8
Großer Plöner See, Germany	8.7	12.5	27.3
Ness, Scotland	38.6	132.0	31.5
Constance, Germany	66.0	90.0	55.8
Ontario, USA–Canada	311.0	86.0	304.0
Huron, USA–Canada	444.0	76.0	400.0
Erie, USA–Canada	400.0	21.0	786.0

[a] Compiled from data cited in Hutchinson (1957), from various sources, Krambeck (1979), Hamblin (1982), and from Mortimer (personal communication).

tudes in Lake Baikal, Russia, are less than 15 mm, and in Lake Superior, 20 mm (Hutchinson, 1957). All observations are about one-half or less of the theoretical value. The elastic yielding of the Earth to the tide-generating forces presumably accounts for the difference (Mortimer, 1965).

2. Internal Seiches

We have seen that the amplitude of surface seiches is relatively small, even in large lakes, and that their periodicity increases with basin dimensions (Table 7-2). When the lake is stratified, water layers of differing density oscillate relative to one another. Most conspicuous is the successive oscillation of the metalimnion (Fig. 7-16), in which both the period and amplitude of this internal standing wave, or ***internal seiche,*** are much greater than those of the surface seiche (Fig. 7-17).

In a simple, rectangular, troughlike lake or lake model, without considering rotation, one observes the resultant horizontal flow depicted in Figure 7-16. Horizontal flow in this oversimplified stratified situation, which contains only an epilimnion and hypolimnion of differing densities, is maximal at the equilibrium point (node) of the oscillation; it ceases at the point of maximum vertical deflection. Because of the much greater water movement associated with internal seiches, resulting currents that rhythmically flow back and forth in opposing directions are the major deepwater movements of lakes. These currents lead to vertical and horizontal transport of heat and dissolved substances and significantly alter the distribution and productivity of phytoplankton and zooplankton, either directly or indirectly, by means of changes in thermal and chemical stratification (cf. Thomas, 1951).

In the simplest model, consisting of a rectangular basin with a homogeneous "epilimnion" of thickness z_e and density ρ_e and a "hypolimnion" of thickness z_h and density ρ_h, the period (t) of the first internal seiche mode is given by

$$t = \frac{2l}{\sqrt{\dfrac{g(\rho_h - \rho_e)}{\dfrac{\rho_h}{z_h} + \dfrac{\rho_e}{z_e}}}}$$

where

l = length of the basin (cm)
g = acceleration due to gravity (980.7 cm sec^{-2})

It should be noted that this equation is simply an extension of that for the periodicity of a surface seiche in which the density of the upper medium, air, was neglected as very small in comparison to that of water. Although more elaborate formulae have been devel-

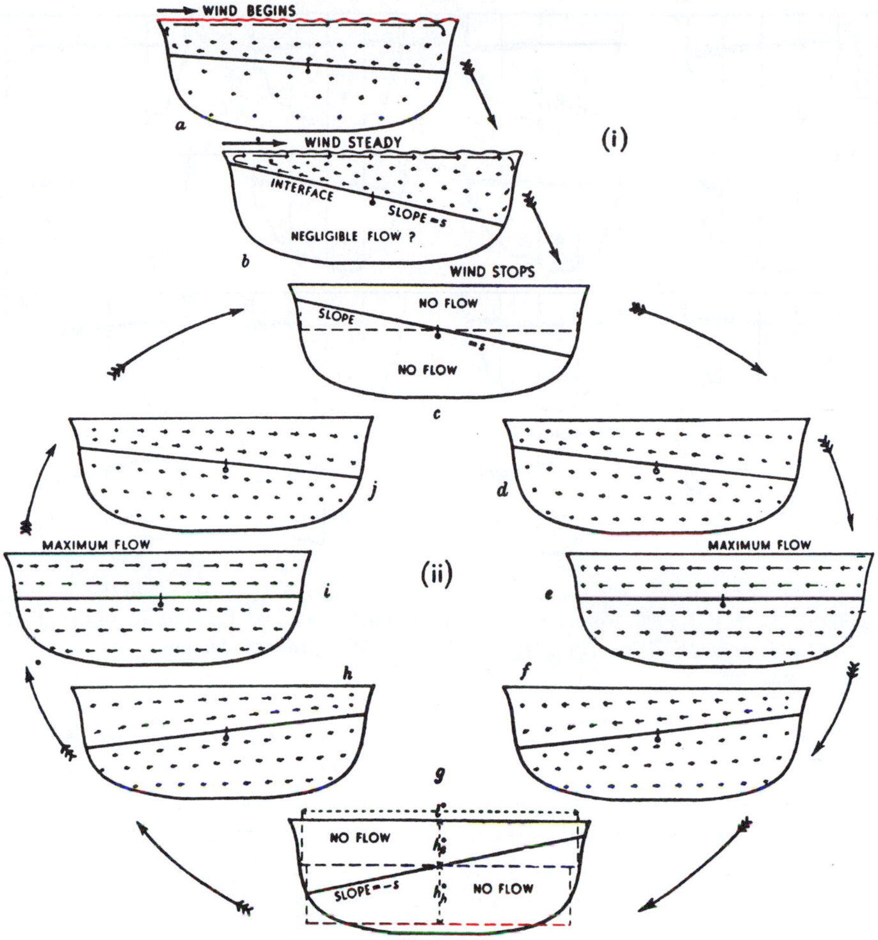

FIGURE 7-16 Movement caused by (*i*) wind stress and (*ii*) a subsequent internal seiche in a hypothetical two-layered lake, neglecting friction. Direction and velocity of flow are approximately indicated by arrows. o = nodal section. (From Mortimer, C. H.: *Proc. Royal Soc. London, Series B 236*:355, 1952.)

oped for continuous density gradients in stratified lakes, this formula yields reasonable estimates of internal seiche periodicity.

The detailed analyses of internal seiches by Mortimer (1952, 1953) permit a few generalizations (Table 7-3). Of all internal waves that are theoretically possible in stratified lakes, the one most commonly set in motion is a uninodal seiche on the metalimnetic boundary (thermocline). In basins of fairly regular shape ranging in length (Table 7-3) from 1.5 km (Lunzer Untersee) to 74 km (Lake Geneva), the uninodal internal seiche along the medial axis always appears as the main resonance. In basins possessing topographic features that may impede a uninodal internal seiche, such as constrictions, islands (Kodomari, 1984), or shallows, as in the southern end of Lake Windermere, the observed period is appreciably longer than the theoretical period. In nearshore areas where local land topography modifies wind fetch, horizontal turbulent mixing rates are altered and often reduced (cf. Lemmin, 1989; Imberger and Patterson, 1990). Multinodal internal seiches form a dominant type of resonance in very large lakes, especially when forced and damped by wind and other short-period disturbances.

Wind energy in the epilimnion dissipatas markedly with increasing depth. Even in large lakes, only about 5% of wind energy flux occurs below about 2 m of depth, but energy dissipation was about eight times higher in the shear zone of the thermocline than in the bottom boundary layer (Kocsis *et al.*, 1998). Energy dissipation increases markedly, particularly where turbulence increases as internal seiches encounter submersed sills in lakes with multidepression hypolimnia.

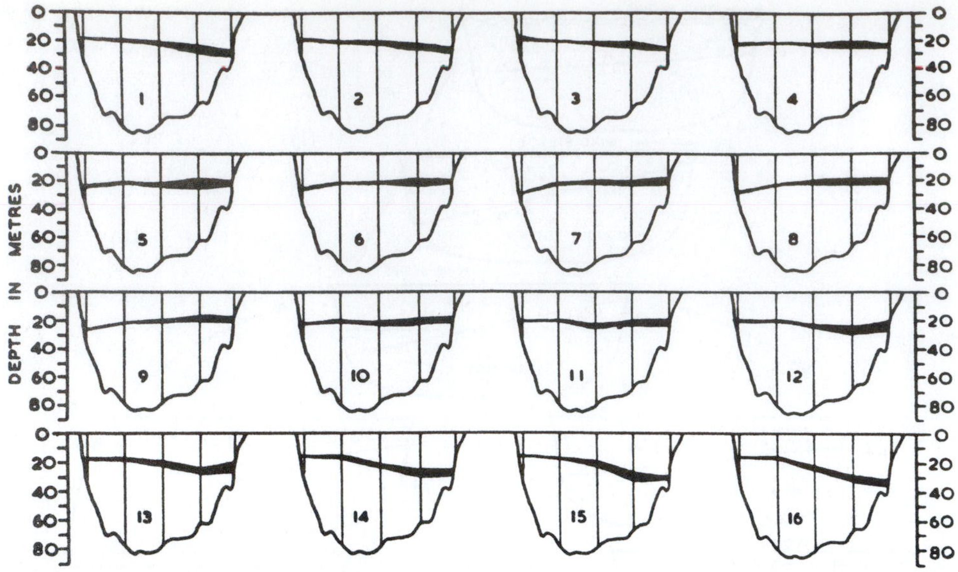

FIGURE 7-17 Successive hourly positions (0100 to 1600 hours, 9 August 1911) of the metalimnion bounded by the 9 and 11°C isotherms on a longitudinal section of Loch Earn, Scotland. (From Mortimer, C. H.: *Schweiz. Zeitschrift f. Hydrologie 15*:94, 1953, after Wedderburn.)

B. Geostrophic Effects on Seiches

Coordinates that appear fixed to an Earthbound observer are in fact rotating relative to coordinates in space (Mortimer, 1955). Inertia causes a moving water mass to attempt to follow a straight line in space, and the resultant track appears curved to the Earthbound observer. Such curvature is equivalent to a small force, the Coriolis force, directed at right angles to the line of motion. This force, which produces a deflection to the right in the Northern Hemisphere and to the left in the Southern Hemisphere, is zero at the equator and

TABLE 7-3 Periodicity of Uninodal Metalimnetic Seiches of Several Lakes[a]

Lake	*Dates*	*Length (km)*	*Calculated period (hours)*	*Observed period (hours)*
Brenda Lake, British Columbia, Canada	May 1991	0.8	1.5	1.7
Lunzer Untersee, Austria	Aug. 1927	1.51	4.0	3.7
Lake Kuttara, Japan	Sept. 1989	2.4	3.6	3.2
Sooke Lake Reservoir, British Columbia	Oct. 1995	4.0	8.0	8.2
Lake Baldegg, Switzerland	Oct. 1979	4.7	9.34	9.3
Lake Chuzenji, Japan	June, Sept. 1982	6.5	12.5	12.0
Windermere, England				
Northern Basin	Aug.–Sept. 1951	6.5	14.4	12–14
Southern Basin	June–July 1951	8.9	23.1	23–25
Loch Earn, Scotland	Aug. 1911	9.6	17.2	16.0
Madüsee, Germany	July–Aug. 1910	13.8	27.3	25.0
Lake Zürich, Switzerland	Aug.–Sept. 1978	29.9	44.0	44.8
Loch Ness, Scotland	Sept. 1904	37.0	62.4	60.0
Geneva, Switzerland	July–Aug. 1942	73.0	96.0	72–108
Kootenay Lake, British Columbia, Canada	July 1976	107.0	—	170.0
Baikal, Russia	Sept. 1914	675.0	1848.0	912.0

[a]From data cited in Mortimer (1953), from various sources, Imboden *et al.* (1983), Hirata and Muraoka (1984), Horn *et al.* (1986), Chikita *et al.* (1993), and Stevens and Lawrence (1997).

maximum at the poles; it is a function of latitude and is proportional to the speed of the current. In sufficiently large lake basins in which frictional forces are small and constraints of the basin sides are negligible, a water mass, once set in motion, follows a circular track, owing to the effects of the Coriolis force. The time taken to complete an "inertial circle" of this type depends only on latitude. If the basin is stratified, this motion may be associated with internal waves, the period of which is referred to as the *period of the inertia oscillation.* For example, in Loch Ness, Scotland, at a latitude of 57°16′, when the velocity of a water mass was 10 cm sec^{-1}, the radius of the circle would be 810 m with a period of the inertia oscillation of 14.2 h (Mortimer, 1955).

Motion in the inertia circle will be encountered only in basins whose width is many times the radius of the inertial circle. In lakes whose width is of the same order or less than that radius, side constraints may be expected to modify the movement. If a lake is stratified and conforms to the simplified two-layered situation in Figure 7-16, then the flow depicted in that figure will be deflected by the Coriolis force onto the right-hand shore (Northern Hemisphere) in the manner indicated in Figure 7-18. This rotating counterpart of the nonrotating conditions of Figure 7-16 has been demonstrated very clearly in Loch Ness (Mortimer, 1955) and subsequently in Lake Michigan, Lac Léman (Bohle-Carbonell, 1986), and other large lakes. In addition to the morphometric parameters of the basin, important control factors include the timing of wind stresses, the time required to tilt the thermocline plane, and the period of the internal seiche.

The amplitude of internal seiches and the currents generated by them are of major importance. For example, in Lunzer Untersee, with a fetch of 1.6 km, an internal seiche with a period of 4 h had an antinodal amplitude of 1 m and generated horizontal currents of 1 cm sec^{-1} at the node. Similarly, currents of 1 cm sec^{-1} were found in Lake Mendota. In larger lakes, for example at the ends of the long basin of Loch Ness and in Lake Michigan, amplitudes of >10 m are common, with horizontal currents of >10 cm sec^{-1} near the nodes.

C. Internal Progressive Waves

The horizontal water movements associated with shearing flow at the metalimnetic interfaces can generate large ***internal progressive*** waves if the Richardson number falls below the 0.25 stability criterion (see Fig. 7-2). Internal waves on the thermocline are roughly an order of magnitude or more larger than waves found on the surface of large lakes. Examples include:

	Period	*Mean amplitude*
Lake Mendota (Bryson and Ragotzkie, 1960)	1.5–7.9 min	0.15 m (max 1.0 m)
Lake Michigan (Mortimer, McNaught, and Stewart, 1968)	3–5 min	1.03 m (max 6 m)
	7 min	0.6 m
	10 min	0.3 m

Internal progressive waves propagate and break much the same as surface waves do (Bryson and Ragotzkie, 1960). Again, these wave movements can be demonstrated clearly in simple lake models containing dyed strata of differing densities (thermally or by salinity) (cf. e.g., the photographs of Mortimer, 1952). The turbulence associated with internal waves is analogous to that at the surface but occurs on a much larger scale and is influential in the transfer of heat and other properties through the metalimnion. Since the large internal waves break at the sides of the basin, their effects, coupled with the vertical movement of the internal seiche on which they move, are particularly significant.

In very large lakes, large circulatory cells of water movement can develop over the troughs of internal progressive waves and extend through the epilimnion to the surface. The result is a series of large, widely spaced convergences and divergences that operate analogously to Langmuir currents in the aggregation and dispersion of suspensoids at the surface. Such aggregations are referred to as ***slicks*** and are much larger and more dispersed than the streaks of Langmuir currents. It is important to note, in addition, that the origins of these circulation patterns are quite different.

VII. OTHER WATER MOVEMENTS

A. Long Surface and Internal Waves

Up to this point, discussion has centered on long standing waves and short progressive waves. In the latter case, the wavelengths (λ), except at the beach, are less than water depth. Consequently, short waves are dispersive, that is, they travel at speeds proportional to $\lambda^{1/2}$ (Mortimer, 1963, 1974). In contrast, ***long waves,*** with a λ long in comparison to and much greater than the water depth, are nondispersive and travel at a speed independent of wavelength. A nonrotating model of such a wave, traveling in water of uniform depth, is illustrated in Figure 7-19. Its speed of progression is gz (in which g is the acceleration of gravity and z is the water depth), and the wave crests are horizontal. A combination of

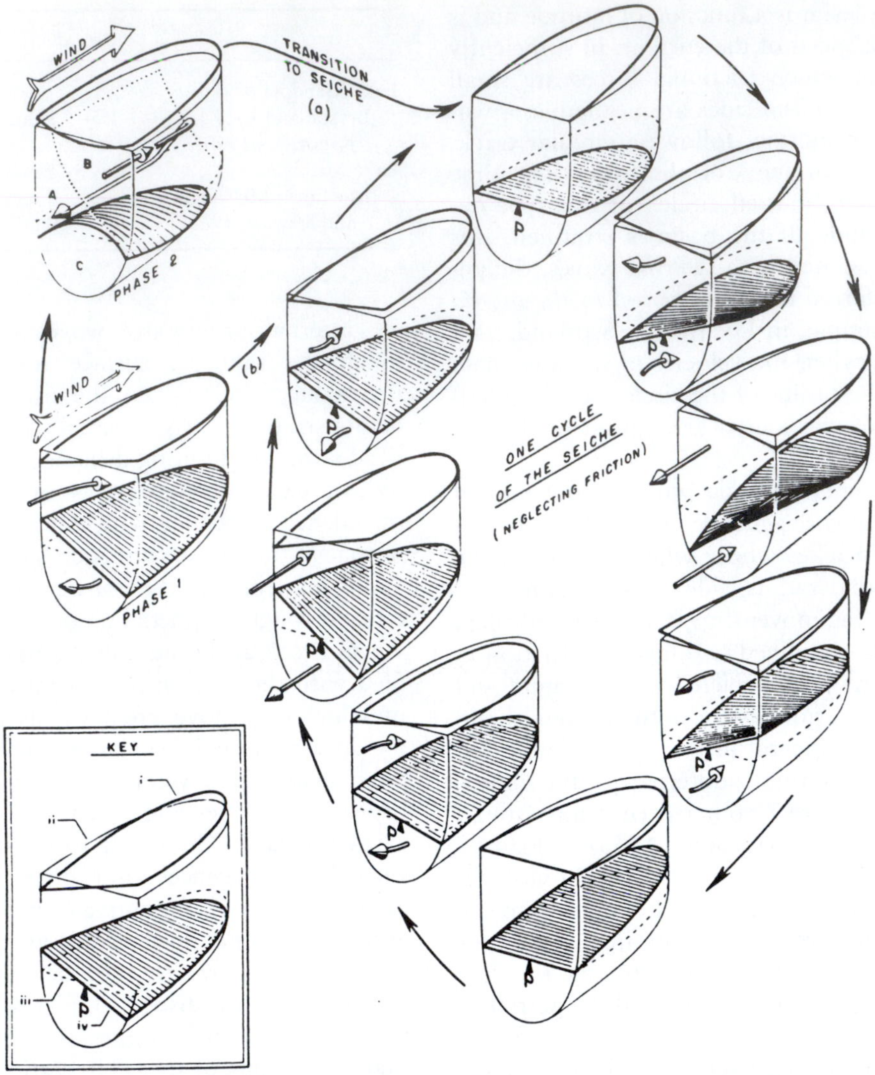

FIGURE 7-18 An internal seiche in a rotating two-layered lake model. In the inset key diagram, (*i*) the oscillating lake surface is shown by a heavy line (this is the surface signature of the internal seiche mode, that is, is not a surface seiche mode); (*ii*) the equilibrium lake surface position is shown by a thin line; (*iii*) the equilibrium interface position is shown by a broken line; and (*iv*) the oscillating interface is shown by a shaded surface.

The two diagrams at upper left illustrate a hypothetical distribution of the layers during the application of the wind stress, which set the seiche in motion. B and A respectively indicate the wind-driven surface and return currents in the upper layer, both deflected to the right by the Coriolis force. C indicates the lower layer, the greater part of which has become displaced out of the half-basin shown.

Eight phases of one oscillation cycle of the first mode internal seiche are shown. Directions of flow in the upper and lower layers are shown by heavy arrows. The nodal point, *P*, the only point of zero elevation change, takes the place of the nodal line in the nonrotating model. Around this point, the internal seiche and its surface (out-of-phase) counterpart rotate counterclockwise. (From Mortimer C. H.: *Mitteilungen Int. Ver. Limnol. 20*:169, 1974.)

two such progressive waves, of equal amplitude and traveling in opposite directions, yields a model of the long waves (seiches) described earlier (Fig. 7-15).

The nonrotating models represented by Figures 7-15 and 7-19 provide simplified descriptions of events in small lakes. When basin dimensions exceed about 15 km, the geostrophic effects of the Earth's rotation must be introduced, and two separate model components emerge to provide the simplest interpretations of long surface and internal waves (Mortimer, 1963,

1974, 1975). Both model components, applicable to channels and basins of constant depth, were developed many years ago in connection with tidal theory and are named after the mathematicians concerned. The first component, associated with a shore line, is illustrated in Figure 7-20. In the ***Kelvin wave*** model, the rightward deflection by the Earth's rotation (Coriolis force) is everywhere exactly balanced by the components of gravity along a wave crest that slopes downward from right to left in the direction of progress (Northern Hemisphere), so that the right hand shore line coincides with maximum wave amplitude. The condition of exact cancellation of the Coriolis force can only be met by an exponential decay of wave amplitude and wave-associated currents along a line normal (perpendicular) to the shore. The result is that the currents are parallel to the shore, which satisfies the boundary condition of zero onshore current component along the shore.

As progressive Kelvin waves travel in opposite directions along a lake basin channel, the long undulations induce currents along the shores (Fig. 7-21). These wave-induced currents are parallel to the direction of the wave progression and the shores of the basin sides. As portions of the water mass rise, others are depressed in a geostrophic balance (Mortimer, 1975). The balance keeps the currents parallel to the shore, with no cross-channel (basin) current at the basin sides. It should be noted that when rotation from geostrophic effects becomes significant in lakes of medium-to-large width, the nodal line in the nonrotating models (comparable to those of small lakes) is replaced by a single midchannel nodal point (*P* in Fig. 7-21).

In very large lakes, such as the Laurentian Great Lakes, the basin width is sufficiently large to permit the long waves to travel without the interference of the shore basin boundaries. These ***Poincaré waves*** are also influenced by Coriolis geostrophic effects, but in this case the wave amplitude does not decrease exponentially away from the shore as in Kelvin waves. That is, Poincaré waves occur in the open water of large lakes, and their reflection generates Kelvin waves

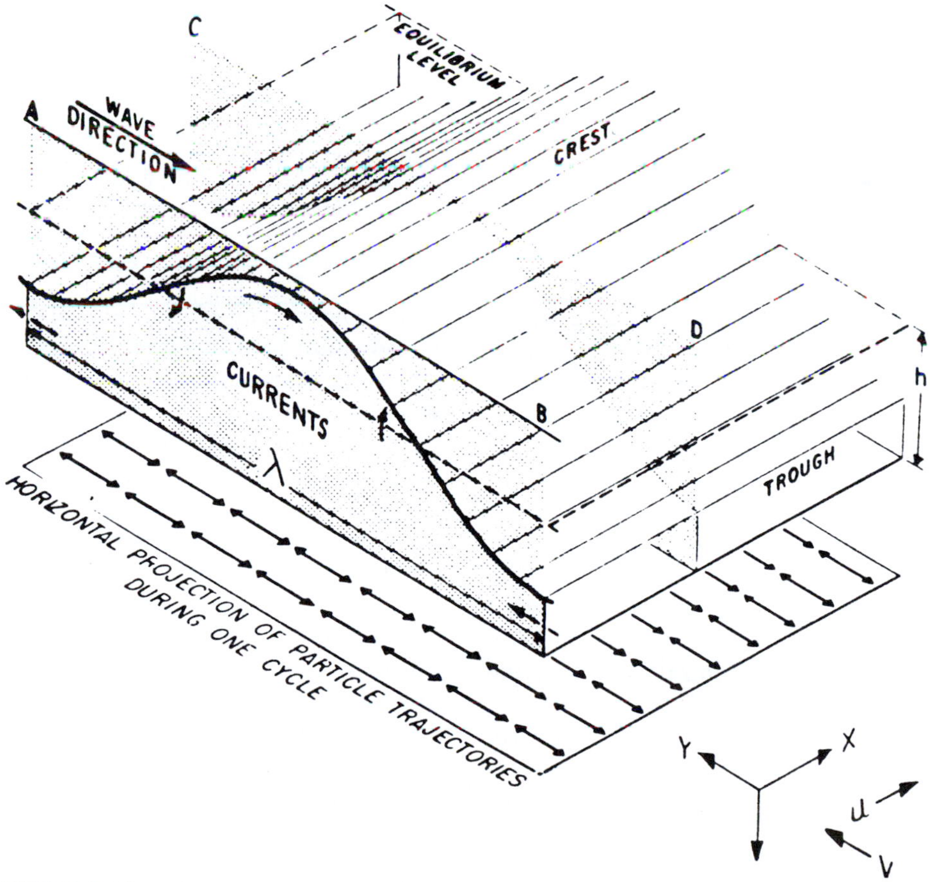

FIGURE 7-19 A long progressive wave, without rotation, in a uniform depth (*h*) model basin. (From Mortimer, C. H.: *Mitteilungen Int. Ver. Limnol. 20*:155, 1974.)

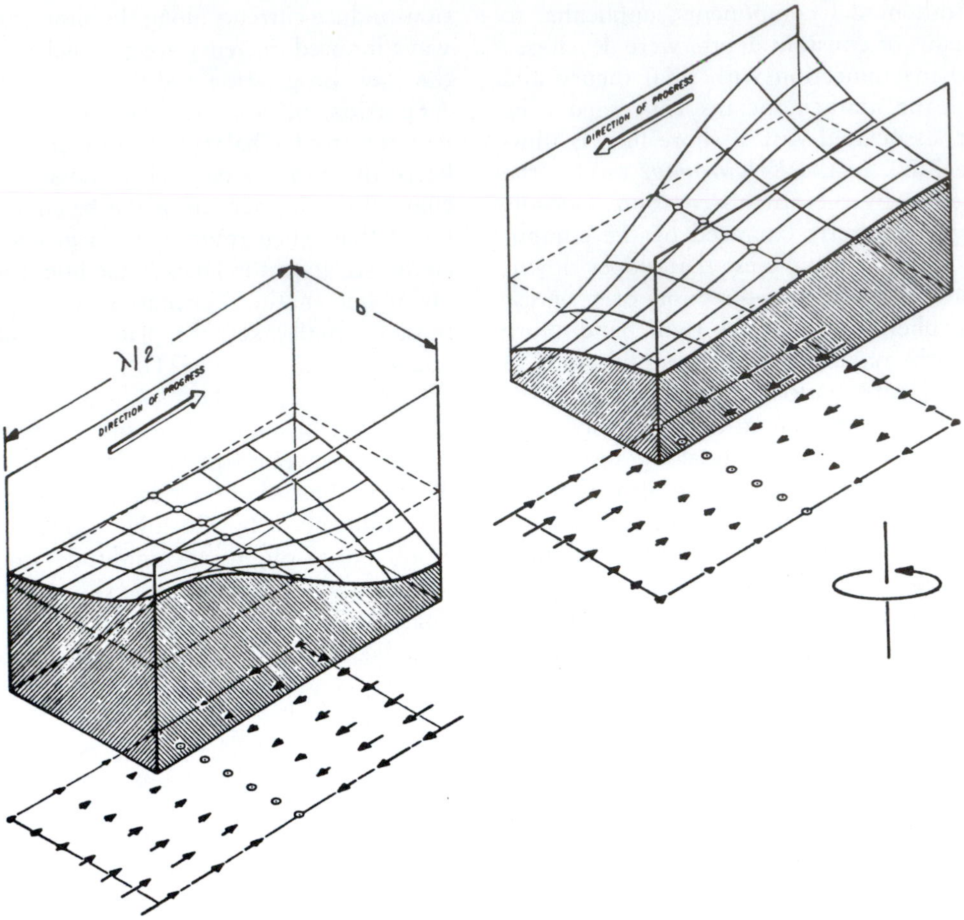

FIGURE 7-20 Long Kelvin waves traveling in opposing directions in straight, rotating (counterclockwise) channels. Half-wavelength portions are shown. Horizontal components of the wave currents are projected as vectors onto the plane below the channels. (From Mortimer, C. H.: *Verhandlungen Int. Ver. Limnol.* 19:65, 1975.)

when encountering shore boundaries. Poincaré waves undulate in a standing wave pattern both across and along the model channel or basin, yielding a cellular pattern of alternating rising and falling hills and valleys with a corresponding cellular pattern of wave-associated currents (Fig. 7-22). Their direction rotates clockwise once every wave cycle. In very large basins—for example, Lake Michigan—the period of observed internal Poincaré waves, equivalent to the Figure 7-22 model, is a little less than the period of motion of the inertial circle. This period depends on latitude and is 17.5 h for the central part of Lake Michigan.

The important difference between the Kelvin wave and the Poincaré wave components is that the Poincaré wave extends with undiminished amplitude right across the channel or basin, whereas the Kelvin wave decreases in amplitude away from the shore and is thereby "trapped" or constrained to travel along the shore. In actual basins we must speak of Kelvin-type and Poincaré-type waves, although the mathematical models illustrated in Figures 7-20, 7-21, and 7-22 (in which the wave surfaces can be visualized either as a water surface or a metalimnetic interface surface) are valid only for constant-depth conditions not met in natural lakes. Nevertheless, even though they are oversimplified, the models do provide useful interpretations of what is observed in large basins (Mortimer, 1963, 1974, 1975).

When the Kelvin wave model is applied to typical conditions of summer stratification in a lake of "medium" or "large" width, for example, Lac Léman and Lake Michigan, with respective widths of the order of 10 and 100 inertial circle radii, Table 7-4 predicts that an internal Kelvin wave traveling along one shore will have decreased to about one-sixth of its initial amplitude on the other shore in Lac Léman and to a negligible amplitude on the opposite shore of Lake Michigan. In the open water of the latter lake, therefore, the Poincaré component should theoretically

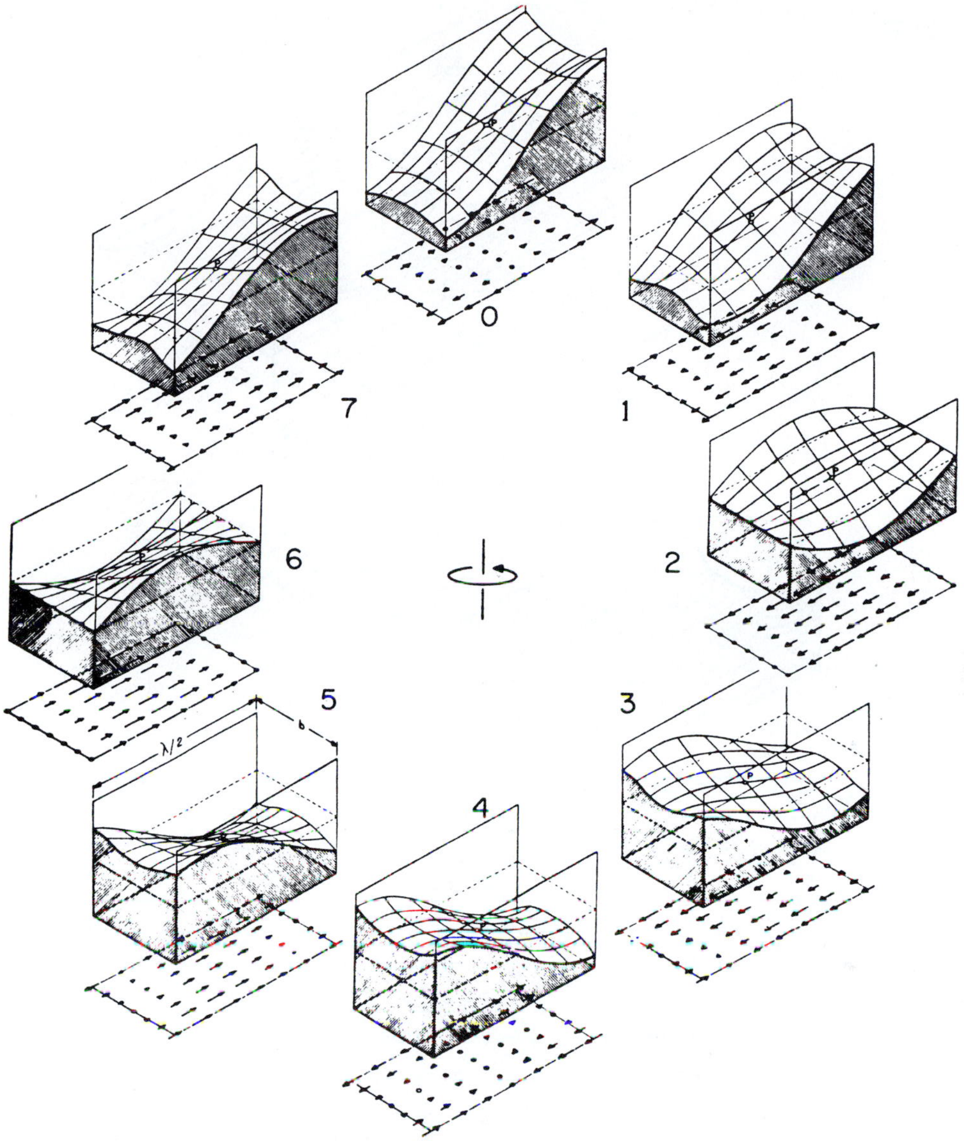

FIGURE 7-21 Successive phases (1/8 cycle) of the long standing wave resulting from combining the two Kelvin waves of Figure 7-20 traveling in opposite directions in a straight, rotating lake basin channel (width of 7 km) of rectangular cross section. The wave topography is at the metalimnion interface, with the upper layer omitted. Horizontal wave current components are projected as vectors onto the plane below the channel. The point of zero elevation change is shown at *P*, around which the wave rotates counterclockwise. (From Mortimer, C. H.: *Verhandlungen Int. Ver. Limnol. 19*:67, 1975.)

dominate the wave pattern, with upwellings and Kelvin-type wave responses restricted to nearshore bands of some 20 km width. This pattern in fact has been observed in Lake Michigan (Mortimer, 1974). One must visualize a transition in the wave-induced current patterns from the cellular, clockwise-rotating, near-inertial periodicity of the offshore Poincaré-type waves in very large lakes and the shore-parallel Kelvin-type currents that increase in amplitude as the shore is approached. This transition is sketched in Figure 7-23. In smaller basins of Lac Léman size, on the other hand, one can predict that the internal, Kelvin-type, long-standing-wave response will dominate. Metalimnetic oscillations with a counterclockwise rotation of the internal wave motion in Lac Léman can be interpreted as an internal Kelvin-type wave progressing counterclockwise around the basin in the manner illustrated in Figure 7-21, with a periodicity of a little over three days (Mortimer, 1963, 1974). As this usually coincides with the periodicity of storm passages, the oscillatory response of Lac Léman is strong. In Lake Michigan, by contrast, the internal Kelvin wave requires about four

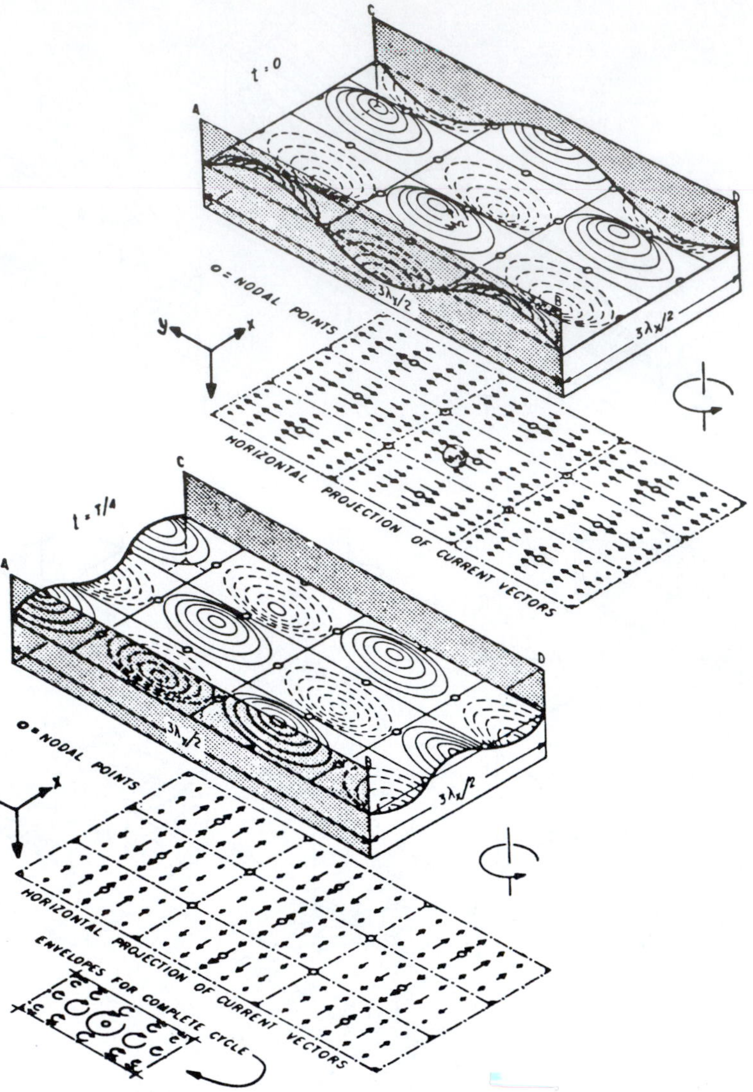

FIGURE 7-22 A standing Poincaré wave in which two phases separated by 1/4 cycle of the oscillation are shown for the cross-channel trinodal case, with a ratio of long-channel to cross-channel wavelengths of 2:1. Current vectors rotate clockwise and attain their maximum values in the center of the cells. The two phases of the figure show the oscillations separated by a quarter period in which, in addition to the clockwise rotation of the vectors, the current directions of one cell remain approximately parallel and opposite in direction to those in neighboring cells. (From Mortimer, C. H.: *Mitteilungen Int. Ver. Limnol.* *20*:17, 1974.)

weeks to complete the circuit of the shores. This cycle is rarely completed before a new storm disturbs its progress.

VIII. CIRCULATION CAUSED BY THERMAL BARS

Another feature of large lakes is a steady circulation set in motion along the shore as a result of density gradients arising when shallow, nearshore waters heat more rapidly than the open water mass (Fig. 7-24). In the illustrated Lake Ontario example, the shallow waters develop stratification, while the main water mass remains in its isothermal (usually below 4°C) mixed winter condition. A narrow transition zone, called a ***thermal bar,*** consisting of a nearly vertical 4°C isotherm, develops between the open water mass and the littoral stratified water. Thermal density differences drive downward flowing currents along the vertical thermal bar into both the lower portions of the inshore and offshore water masses. The density-driven circulation is dependent on the bottom topography,

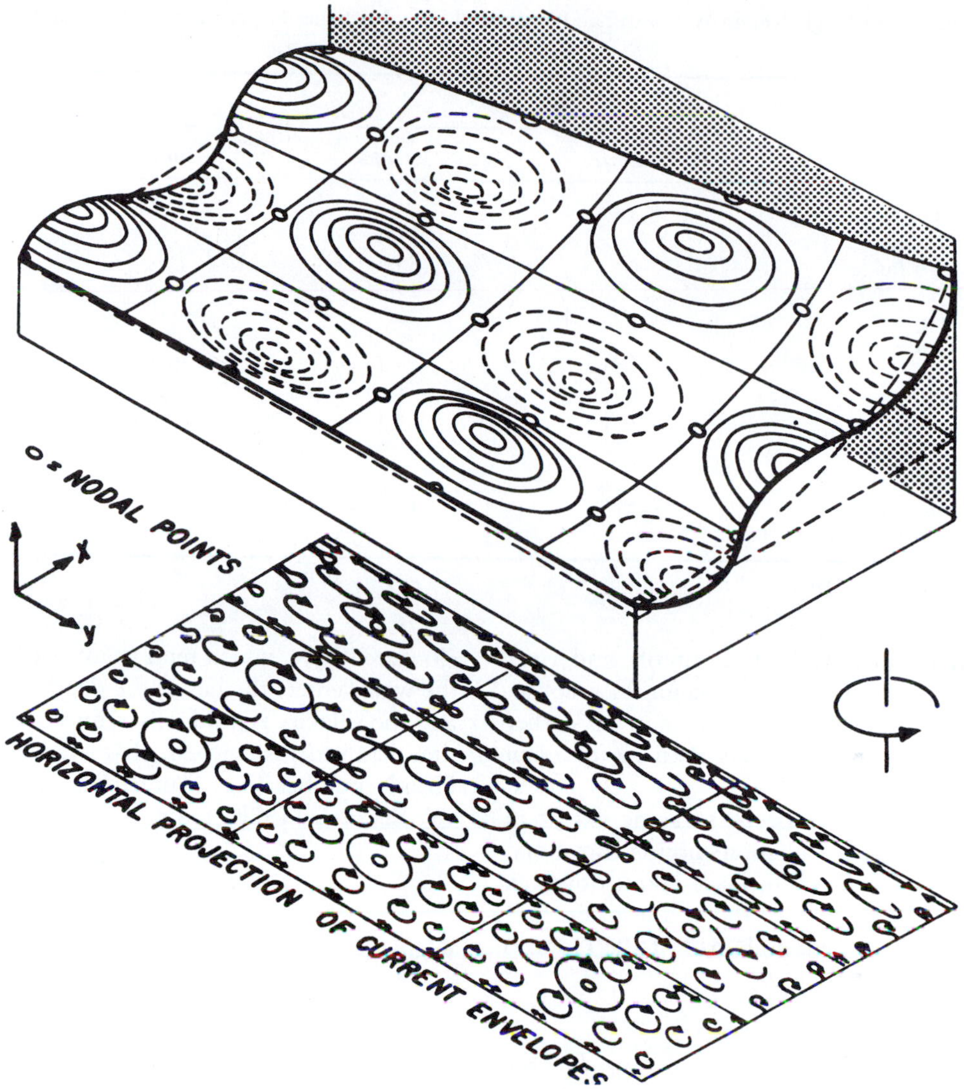

FIGURE 7-23 Combination of a multinodal standing Poincaré wave in a very wide channel model, of which only one side is shown, and a portion of a Kelvin wave traveling along the side. The elevation and current amplitudes associated with the Kelvin wave are at a maximum at the channel side (elevation amplitude, a), and decrease exponentially away from the side in the x-direction, falling to $1/e$ of the on-shore value at a distance d_e from the side. The figure illustrates the transition, in current trajectories, from a nearshore pattern dominated by the Kelvin wave to a pattern dominated by the Poincaré wave at distances $>2d_e$ offshore. (From Mortimer, C. H.: *Large-scale Oscillatory Motions and Seasonal Temperature Changes in Lake Michigan and Lake Ontario*. Special Report no. 12, Center for Great Lakes Studies, University of Wisconsin-Milwaukee, 1971.)

with a more pronounced circulation and considerable descending motions inside the thermal bar zone of convergence in lakes with steep slopes (Malm, 1995). The downward density-mediated currents outside the thermal bar can penetrate to very deep strata or bottom layers and cause appreciable convective mixing (e.g., Lake Baikal; Shimaraev *et al.*, 1993). The Earth's rotation combines with the density gradient to induce a counterclockwise coastal current inside the bar.

The thermal bar moves progressively further from shore as the heat influx continues to warm the larger open water mass. Little mixing occurs between inshore and offshore waters, and a significant portion of inshore water originates from runoff (Spain *et al.*, 1976). Finally, thermal differences between the inshore and offshore regions lessen to the point where stratification of the whole basin occurs (Rodgers, 1966). This mechanism results in temporary isolation of inshore waters

TABLE 7-4 Parameters of Single Kelvin Waves in Lac Léman and Lake Michigan Representative of Nearshore Conditions in Late Summer[a]

	Lac Léman		*Lake Michigan*	
Lake conditions	*Surface*	*Internal*	*Surface*	*Internal*
Inertial period (t_p) in hr	16.6		17.4	
Thickness of upper layer (m)		15.0		15.0
Thickness of lower layer (m)		85.0		60.0
Mean °C of upper layer		19°		20°
Mean °C of lower layer		6°		6°
Density difference $\times 10^{-3}$		1.54		1.74
Mean width (km)	7.9	7.4	102	100
Wave velocity (km hr^{-1})	113	1.59	104	1.60
Offshore distance intervals over which wave amplitude is successively halved	(207)	2.9	(190)	3.1
Percentage of wave amplitude still remaining at opposite shore	98%	17%	70%	$<1 \times 10^{-7}$%

[a] After data cited in Mortimer (1963).

where, when augmented by surface runoff and river discharge, chemical enrichment can occur. As a consequence, increased productivity can occur much earlier in the spring in the inshore regions than in the open water.

Thermal bars commonly occur to some extent in all lakes. In small lakes, the phenomenon is transitory, often lasting only a few days. In large lakes, however, the transition to stratification of the whole basin may take weeks, as seen in the example from Lake Ontario (see Fig. 7-24).

IX. CURRENTS GENERATED BY RIVER INFLUENTS

When a river enters a lake or reservoir, the incoming water will flow into a density layer in the lake that is most similar to its own density; this process is governed by temperature, dissolved material, and suspensoids. Depending on the density differences between the inflowing water and the lake water, three basic types of inflow water movements can result (Fig. 7-25). ***Overflow*** occurs when the inflow water density is less than the lake water density ($\rho_{in} < \rho$) and ***underflow*** results when the inflow water density is greater than the lake water density ($\rho_{in} > \rho$). If the density of inflow is greater than that of the epilimnion but less than that of the metalimnion or hypolimnion, flow enters in a plume at an intermediate depth, and ***interflow*** occurs ($\rho_1 < \rho_{in} < \rho_2$).

A certain amount of turbulent entrainment and mixing nearly always accompanies the flow of water into a lake, given adequate influx velocities. In many lakes, and especially in reservoirs, inflows enter through elongated bays, which tend to inhibit lateral mixing. As the water enters the basin, the depth of inflow increases and the velocity is gradually reduced until a critical section depth (d_0) is reached, characterized the densimetric Froude number (F_0) (Wunderlich, 1971), where mean velocity is divided by reduced gravity:

$$F_0 = \frac{(v/A_0)}{\sqrt{g(\Delta\rho/\rho)d_0}}$$

where

v = flow rate
A_0 = cross-sectional area at the critical section
ρ = density of inflow
$\Delta\rho$ = density difference between the incoming and receiving water

Overflow or underflow initiates at this critical section, with a reduction of flow velocity that may be accompanied by deposition of suspended materials. A critical analysis of density differences and distribution of inflowing water in Lake Biwa, Japan, showed that the inflowing river waters do not mix readily if density differences are reasonably large and flow occurs instead in complex overflow and interflow patterns (Morikawa *et al.*, 1959). Further detailed analyses of these mixing characteristics are given in Imberger and Hamlin (1982) and Imberger and Patterson (1990).

Variations in density differences between incoming water and those of the recipient water body are so great that generalizations are very difficult to make. Discharge also varies widely seasonally, not only in volume but in the accompanying load of dissolved and suspended materials. For example, in alpine situations, the density of river water is greater (cold, high dis-

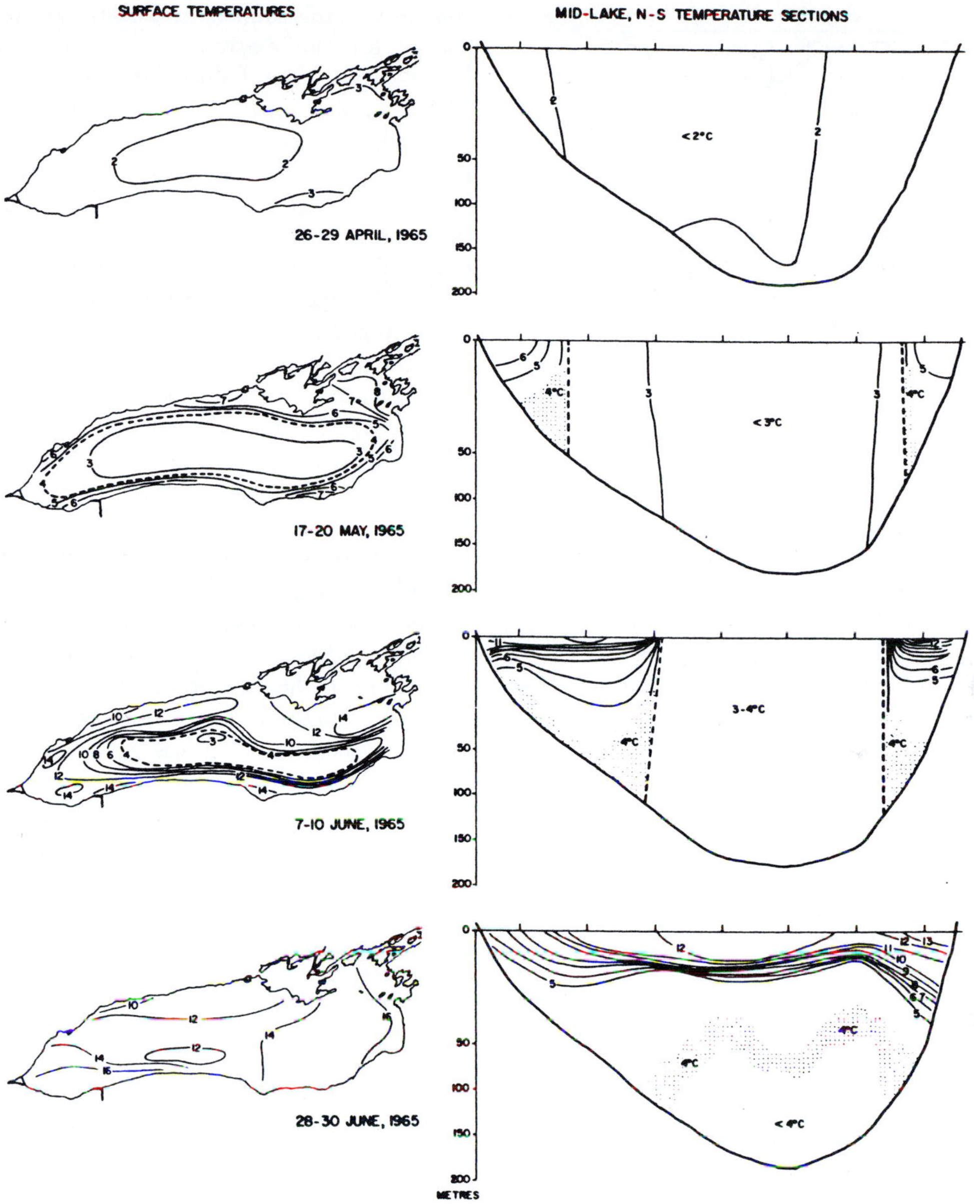

FIGURE 7-24 Formation and progress of a thermal bar in Lake Ontario from winter to full summer stratification. (From Rodgers, G. K.: *Publications Gt. Lakes Res. Div. Univ. Mich.* 15:372, 1966.)

solved and particulate load) than the recipient water and underflow is common, particularly if glacial erosion at the head of the river contributes to the suspended load. During the summer, although alpine river inflow water is still cold, its density is reduced by high-volume dilution with snow and icemelt runoff. In the summer, then, alpine rivers interflow into the metalimnion. This situation has been documented frequently (e.g., Chikita *et al.*, 1985). A particularly striking case is seen in the Bodensee, Germany, where light penetration is abruptly attenuated in the metalimnion over large areas of the lake by the intrusion of river water with a high suspended load (Lehn, 1965).

Geothermal fluids that are abnormally hot and contain high concentrations of dissolved salts create complex density-mediated currents. The hot (59°C) inflows to Lake Aratiatia, New Zealand, initially float on the surface waters (Johnstone *et al.*, 1988). As the influent water cools to a temperature ca. 10°C above that of the lake water, it sinks and forms density currents that rapidly mix with the lake water. Cooler saline or nutrient-rich inflows in other lakes can lead to under-

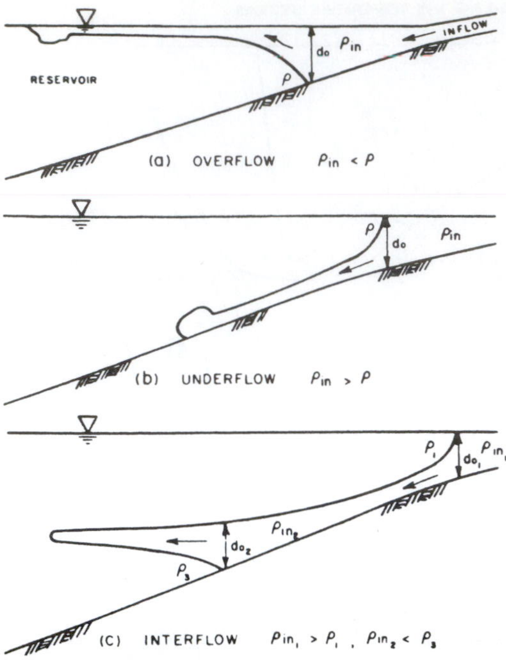

FIGURE 7-25 Types of inflow into lakes and reservoirs. (From Wunderlich, W. O.: The Dynamics of Density-Stratified Reservoirs. *In* Hall, G. E. (ed.), *Reservoir Fisheries and Limnology,* Washington, D.C., American Fisheries Society, 1971.)

flows, disruption of stratification, and alterations of chemical cycling with enhancement of interdependent biological productivity (see Vincent *et al.,* 1991; Gibbs, 1992).

The extent of intrusion and current generation in the receiving lake is also a function of the discharge volume of the river in relation to the volume of the lake. The theoretical retention time of a lake or reservoir, based on the relation of the total influent-outflow to the total volume of the lake, is realized only approximately in most lakes. Retention time varies with the dimensions and shape of the basin, seasonal rates of inflow, and stratification characteristics (Kajosaari, 1966; Englert and Stewart, 1983; Carmack *et al.,* 1986). At high discharge rates, overflow and interflow of rivers may channel across or through the water mass of the stratified basin, whereas at lower discharge rates, river water may penetrate more into the main water mass and may be mixed through more normal circulatory mechanisms. In larger lakes, geostrophic deflection of the incoming water intrusions, surface or interflows, is observed consistently.

X. CURRENTS UNDER ICE COVER

It is clear that the water of a lake under ice cover moves by currents, even in lakes of closed basins with no appreciable inflow and outflow. Much of the evidence for the existence of weak under-ice currents stems from careful direct measurements in which radioactive sodium (^{24}Na) was released at various points beneath the ice and its short-term distribution was measured for several days. Slow horizontal currents, with velocities too low to be measured by mechanical current meters, have been observed in a small ice-covered lake in Wisconsin (Likens and Hasler, 1962) and in a large, permanently ice-covered lake in Antarctica (Ragotzkie and Likens, 1964). Similarly, radioactive tracers have been used to measure slow horizontal movements in a small lake in Nova Scotia (McCarter *et al.,* 1952), in a meromictic lake in Washington (Hutchinson, 1957), and in the monimolimnion of a small meromictic lake in Wisconsin (Likens and Hasler, 1960).

Horizontal displacements of radiosodium were found to be generally asymmetrical in a small ice-covered, closed basin bog lake (Fig. 7-26). Horizontal velocities of at least 10 m day^{-1} (42 cm h^{-1}) were observed near the bottom at the deepest point of the basin, and maximum velocities of 15–20 m day^{-1}. Relatively constant horizontal eddy diffusivities indicated small mixing eddy lengths of $<$1.6 m (Colman and Armstrong, 1983). The vertical component of this motion was between 1 and 3 orders of magnitude less than the horizontal component (Table 7-5). In further experiments, Likens and Ragotzkie (1966) also found a definite rotary movement of horizontal currents, which was cyclonic near the center of the lake and anticyclonic near the perimeter (Fig. 7-27).

These detailed studies and modeling analyses of the Tub Lake data (Rahm, 1983), as well as evidence from a larger lake in Swedish Lapland (Mortimer and Mackereth, 1958), several small Swedish lakes (Bengtsson, 1996), and small arctic lakes (Welch and Bergmann, 1985), all indicate that under-ice water movements are largely the result of convective cells of density currents generated primarily by relatively large quantities of heat flowing from the sediments. The thickness of the bottom current layer is small (ca. 1 dm), however, and the currents low, in the range of 0.1–1 mm s^{-1} (Malm, 1998, 1999). If the temperature of the water into which the heat flows is greater than 4°C, that flow will generate ascending buoyant currents. If, however, the lake has cooled to well below 4°C before the ice cover is established, the heat flux will increase the density of the water in contact with the sediments and cause it to flow as a density current down the sloping sides of the basin into the nearest depression. Chemical exchanges also occur between sediment and water; for example, oxygen is taken up by the sediments and associated microbes. If the lake basin

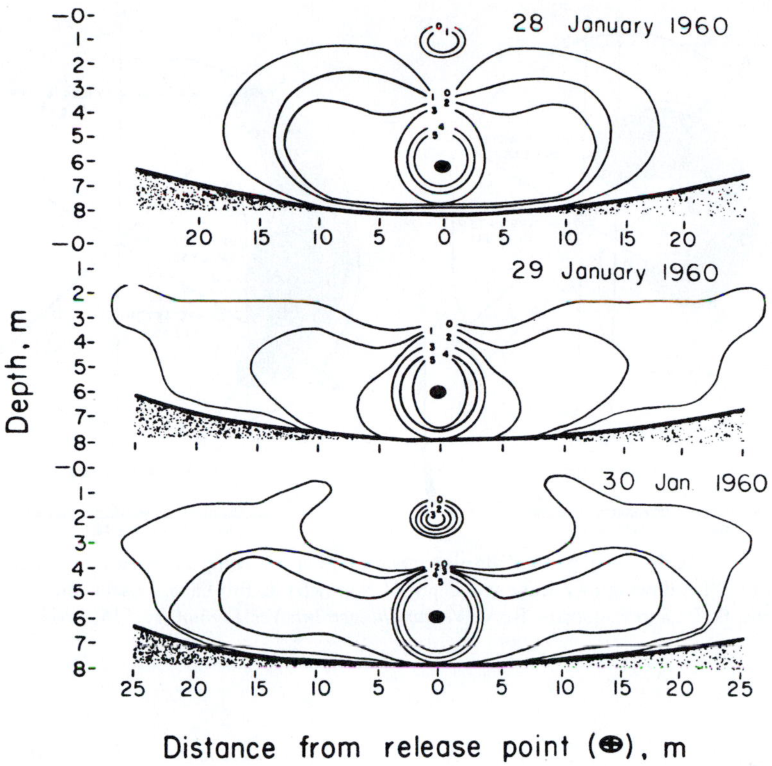

FIGURE 7-26 Diagram of the horizontal and vertical dispersal of ^{24}Na in Tub Lake, Wisconsin, under ice cover 24, 48, and 72 h after release. Integers indicate the logarithmic value of the radioactivity concentration in counts of min^{-1}. (From Likens, G. E., and Ragotzkie, R. A.: J. *Geophys. Res. 70*:2333, 1965. Copyright by American Geophysical Union.)

contains more than one topographically separated depression, the density flows may produce significantly different temperatures and chemical compositions in the bottom waters of these depressions (Mortimer and Mackereth, 1958).

Most of the heat of the sediments is gained during the previous summer period and fall turnover (cf. Chap. 5) and is dissipated slowly. Birge *et al.* (1927) found that about 650 cal cm^{-2} were conducted from the sediment to the water of Lake Mendota during the period of ice

TABLE 7-5 Vertical Motion in Ice-Covered Tub Lake, Wisconsin, Calculated from Measurements of ^{24}Na Dispersal[a]

	Velocity (w) in cm $sec^{-1} \times 10^{-2}$[b]		
Depth (m)	*28–30 Jan. 1960*	*26–27 Jan. 1961*	*27–28 Jan. 1961*
0.5	0	0	0
1.5	0	+0.0132	+0.0378
2.5	+0.0033	−0.0057	+0.0047
3.5	−0.0041	+0.0007	−0.0158
4.5	−0.0233	−0.0498	−0.0508
5.5	−0.0066	−0.0497	−0.0219
6.5	+0.0381	−0.0583	−0.1180
7.5	+0.0661	−0.6990	−1.55

[a] From data given in Likens, G. E., and Ragotzkie, R. A.: *J. Geophys. Res. 70*:2333–2344, 1965. Copyright by American Geophysical Union.

[b] Negative values of *w* indicate downward motion and positive values indicate upward movement.

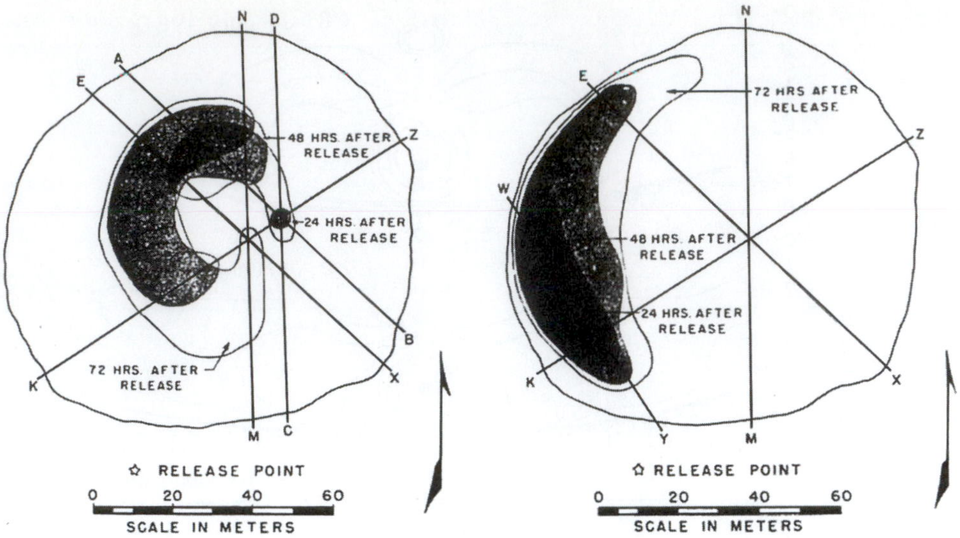

FIGURE 7-27 Maximum horizontal displacement of ^{24}Na following its release at a depth of 5 m (right) and of ^{131}I following its release at a depth of 3 m (left) in Tub Lake, Wisconsin, January, 1962. (From Likens, G. E., and Ragotzkie, R. A.: *Verhandlungen Int. Ver. Limnol. 16*:126, 1966.)

cover (110 days), a heat gain of 5.9 cal cm^{-2} day^{-1}. Thermal gains of 3–4 cal cm^{-2} day^{-1} are common in small temperate lakes during the winter. Tub Lake gained about 3 cal cm^{-2} day^{-1}, and larger Lake Torneträsk, Swedish Lapland, about 1 cal cm^{-2} day^{-1}. The sediment heat source is adequate to explain the observed currents. Biological oxidation during respiration in the sediments represents a caloric equivalent of about 0.04°C cm^{-3}, which is totally inadequate as a heat source.

Geothermal heat flow (approximately 1.23×10^{-6} cal cm^{-2} sec^{-1}) from the interior of the Earth represents only about 0.1 cal cm^{-2} day^{-1} through the sediments of the lake. Ground water that is relatively warm and circulating near the basin probably transmits some heat through the sediments to the lake water. In Wisconsin, this potential heat source was estimated to be about 1 cal cm^{-2} day^{-1}, a significant source of thermal conductivity in addition to that from heat storage from the previous summer. Thus, a major operator of thermally induced convection currents under winter conditions appears to be sediment heat that accumulated during summer.

In addition, oscillating currents can be generated in ice-covered lakes with little river inflow and outflow in proportion to total lake volume. Reciprocating water movements result from such surface seiches and have periodicities that are predicted theoretically from morphometric characteristics of the lake basins (Bengtsson, 1986; Malm *et al.*, 1998; Malm, 1999). The seiche-induced currents are geostrophic, and their magnitudes in Swedish and Russian lakes were in the range of <1–4 mm s^{-1}.

An excess of solar radiation at the lake surface over heat losses is appreciable only at the surface and in late spring; it is absorbed primarily in melting ice and snow. Under certain conditions, some solar radiation penetrates ice cover and warms the water below the ice, which then moves downward in convection currents (0.5–2 mm s^{-1}) because of its higher density (Matthews and Heaney, 1987; Bengtsson and Svensson, 1996). Requirements include high solar radiation inputs, high extinction coefficients, as in dystrophic lakes, to absorb the entering energy, and low water-temperature-induced density differences that form the inverse stratification. Mixing under ice cover can also be induced by turbulent shear flows from river discharges or meltwater (Stigebrant, 1978; Bergmann and Welch, 1985; Kenney, 1991). In large lakes, wind-induced movements of the ice sheet can cause displacements and mixing beneath the ice (Stewart, 1973b).

XI. HYDRODYNAMICS AMONG AQUATIC ECOSYSTEMS

The water of streams and rivers usually moves continuously in a unidirectional flow, except where influenced by tides. Flows are highly irregular seasonally and with precipitation events, as well as with variations in runoff and groundwater inputs (Table 7-6). Current velocities are controlled by the channel morphology and elevational gradient, water volume, type of substratum, and obstructions such as woody debris

TABLE 7-6 Comparative Hydrodynamic Characteristics among River, Reservoir, and Natural Lake Ecosystems[a]

	Rivers	*Reservoirs*	*Natural lakes*
Water-level fluctuations	Large, rapid, irregular; flooding common	Large, irregular	Small, stable
Inflow	Overland and groundwater runoff; highly irregular and seasonal, less so with large groundwater inflows	Most runoff to reservoir via river tributaries (high stream orders); penetration into stratified strata complex (over- , inter- , and underflows); often flow is directed along old riverbed valley	Runoff to lake via tributaries and (often low stream orders) diffuse sources; penetration into stratified waters small and dispersive
Outflow (withdrawal)	Discharge highly irregular with inflows and precipitation event frequency	Highly irregular with water use; withdrawals from surface layers or from hypolimnion	Relatively stable; usually largely surface water via surface outflow or shallow ground water
Flushing rates	Rapid, unidirectional, horizontal	Short, variable (days to several weeks); increase with surface withdrawal; disruption of stratification with hypolimnetic withdrawal; three-dimensional	Long, relatively constant (one to many years); three-dimensional

[a] Assembled from many sources and the syntheses on rivers of Ryder and Pesendorfer (1980) and on reservoirs and lakes of Wetzel (1990b).

and ice. In reservoirs, inflows tend to follow the original river channel (thalweg) and distribute into the water strata on the basis of densities derived from differences in temperature and dissolved salts (Table 7-6). In natural lakes, inflows from surface or groundwater sources tend to be dispersive and small in comparison to the total volume.

Water-level fluctuations in river ecosystems are large, rapid, and irregular, and flooding is common. In reservoirs, water levels are influenced by both the dynamics of river inflows as well as the management of withdrawals from the surface or hypolimnetic strata (Table 7-6; Wetzel, 1990b; Tundisi and Straškraba, 1994). Water-level fluctuations in natural lakes tend to be small and relatively stable. These water-level differences between reservoirs and natural lakes are to a large extent related to the marked differences in the ratios of drainage basin area to lake area, which is very much larger in reservoir ecosystems (Chap. 3).

As a result of these differences in flow and water-level fluctuations, flushing rates in natural lakes are long (one to several years) and relatively constant. Flushing rates of reservoirs are short and variable (days to several weeks) in response to inflow and withdrawal/outflow regimes (Table 7-6). Flushing rates of river ecosystems are comparatively rapid and directional.

XII. SUMMARY

1. Water movements in lakes occur in response to forces, especially wind, that transfer energy to the water. Rhythmic motions (oscillations) result both at the surface of the water and internally, deep within the basin. These motions and their attendant currents may be in phase or in opposition. The ultimate fate of these movements is to degrade into arrhythmic turbulent motions, which disperse the water and the chemicals and organisms within it.
2. As water of a given density moves along an interface, such as the wall of a tube or water of a different density, frictional shearing stresses occur between them. Below a certain speed of flow, water movement is smooth and undisturbed at the interface. This ordered movement is called ***laminar flow.*** As the velocity of movement is increased, the damping effects of gravity are exceeded and vortices occur at the interface, with a decrease in stability. Water movement then becomes disordered in ***turbulent flow,*** in which the frictional increases result in mixing of the two fluids of different density perpendicular to the direction of the current movements. Since the velocities needed to induce turbulent flow are very low (only a few mm sec^{-1}) laminar flow is rarely found in aquatic systems.
3. The coefficient of eddy diffusion (K_z) measures the rate or intensity of mixing (turbulent exchange) across the plane of a density gradient, most often a thermal gradient. K_z decreases markedly in the transition from the turbulent epilimnion into the more stable metalimnion and as the thermal density gradient increases seasonally during sum-

mer stratification. K_z also decreases, that is, there is less turbulent mixing in the metalimnion and hypolimnion of smaller lakes of greater stability than in lakes of large area with longer fetch for wind exposure.

4. Water in streams and rivers flows directionally along decreasing elevational gradients. Frictional shear forces of the water against channel substrata result in turbulent flows. Current velocities are low near the substrata and increase toward the center of the channel. Resistance along the channel and banks results in a spiraling flow pattern that alternates scouring and deposition at relatively uniform distances along the stream channel and result in a meandering channel morphometry. Obstructions (fallen trees, aquatic plants) of any type can modify flow patterns and greatly increase habitat and flow complexity.
5. Flowing water can be forced by advection into the interstitial spaces of stream sediment material, particularly sands and gravel. The zone of sediments into which overlying stream water actively infiltrates by advection is termed the ***hyporheic zone*** and may extend for several centimeters to well over a meter in depth.
6. Movement of air over water sets the water surface into an oscillating motion, called ***progressive*** or ***traveling surface waves.***
   a. The water of surface waves moves in a ***circular*** path (Fig. 7-9). Water is displaced upward and returned to equilibrium by gravity along a circular path (therefore, they are sometimes called *gravity* waves).
   b. The height of the vertical oscillation is attenuated rapidly with depth and decreases by half of the cycloid diameter for every depth increase of 1/9 the wavelength (distance from crest to crest).
   c. While water entrained in surface waves oscillates considerably up and down, horizontal movement is small. The wavelength is much less than the water depth.
   d. When the wavelength of these short surface waves is less than 6.3 cm (2π), the surface waves are called ***ripples*** (capillary waves).
   e. The height of the highest surface waves on a lake is proportional to the square root of the ***fetch,*** that distance over which the wind has blown uninterrupted by land.
7. When surface waves occur near the shore in shallow water, their circular motions are transformed into a to-and-fro sloshing that extends to the bottom of the water column. The wavelength becomes more than 20 times the water depth. As deepwater surface waves enter shallower water, their velocity decreases proportional to the square root of depth. A reduction in wavelength occurs with a marked increase in wave height. With increased height, the waves become asymmetrical and unstable. The collapse or breaking of water over the front of the wave results in a ***breaker*** (Fig. 7-11).
   a. In a ***plunging*** breaker the forward face of the wave becomes convex and the crest curls over and collapses.
   b. The crest of a ***spilling breaker*** collapses forward, spilling water downward over the front of the wave.
   c. Shallowwater wave energy is effective in moving littoral sediments to deeper water and inhibiting the growth of many organisms not adapted to such turbulence.
8. Currents are nonperiodic water movements. Although many external forces contribute to the generation of currents, wind is dominant in driving ***surface currents*** of open water.
   a. The velocity of wind-driven water currents is about 2% of wind speed. The water velocity of surface layers increases in a linear way with wind velocity until a critical wind speed (approximately 6 m s^{-1}) is reached, beyond which the ratio becomes nonlinear.
   b. The Coriolis force from the Earth's rotation creates ***geostrophic effects*** by deflecting the direction of surface currents from that of the wind. Deflection is to the right of wind direction in the Northern Hemisphere and to the left in the Southern Hemisphere. The resulting geostrophic wind-drift deflection is 45° from the wind direction in very large lakes. The angle of declination decreases as the size and especially depth of lakes decrease.
9. ***Langmuir circulations*** are large currents of turbulent motion of the upper strata of lakes that are organized into vertical helices. Langmuir circulations move water and entrained particles in circular cells along cylindrical patterns both vertically and in a helical manner parallel to the wind direction and perpendicular to the lines of surface waves (Fig. 7-13).
   a. Downward movement velocities at cell convergences are about three times greater than upward current velocities at divergences.
   b. Aggregations of surface particles that occur at lines of divergence are called ***streaks.***
   c. Langmuir circulations are generated by surface waves at wind velocities above 2–3 m sec^{-1}. Wind energy is converted through surface wave energy into turbulence; surface instability is

then dispersed downward and drives the convection pattern.

d. Langmuir circulation is a primary way in which turbulent motion is transported downward and the upper layers of water are mixed.

10. Water movements in surface strata cause water to pile up at the downwind end of the lake basin. The epilimnetic water there flows downward by gravity until it encounters the more dense water of the metalimnion and then flows back (upwind) along the epilimnetic-metalimnetic interface.
    a. A portion of the metalimnetic water is entrained (incorporated by mixing) into that of the epilimnion. As a result, the epilimnion can become deeper than was previously the case as the upper portion of the metalimnion is eroded.
    b. If the shear at the thermocline is sufficiently great, resulting in low Richardson numbers ($R_i < 0.25$), turbulent flow and vortices occur in the upwind direction. Internal waves can form at the shear interface.
    c. As entrained water of the metalimnion reaches the upwind end of the basin, especially as the storm event producing the movement ends, some flow of the metalimnion is transferred downward to the hypolimnion. In a similar manner, although on a much smaller scale, movement occurs along the metalimnetic-hypolimnetic density-gradient interface in the windward direction.
11. Wind-induced tilting of the water surface and of the metalimnion (or barometric pressure on one part of a large lake) results in a displacement of more water to one end of the lake than is at the other. When the wind stops as the storm passes, the tilted water strata flow back toward equilibrium. The equilibrium point is overshot, however, by momentum. The resulting rocking motion of the entire water mass of the upper lake strata results in ***long standing waves*** termed ***seiches.***
    a. Seiches have the form of very long wave patterns with wavelengths approximating the dimensions of the basin of many lakes.
    b. The surface of the metalimnion oscillates up and down at the lake ends like a seesaw about a line, the ***node,*** of no vertical motion. Maximum vertical motion occurs at the ends, the ***antinodes,*** often at the ends of the lake basin. Horizontal movement is maximum at the nodal line and nil at the antinodes. The horizontal movements reverse direction with each long oscillation and can create currents of sufficient size to develop instability, vortices, and waves. The seiches return to equilibrium as a result of gravity and frictional damping, with a progressive reduction in the duration of the oscillation periodicity.
    c. ***Surface seiches*** affect the motion of the entire water mass of the lake, whether stratified or not, and have maximum amplitude at the surface.
        i. Horizontal oscillation at the nodal line is large when the basin is very long in relation to depth.
        ii. The amplitude of surface seiches is relatively small (to several meters, maximum) but is sufficient to cause significant flooding and erosion effects of shoreline areas in large lakes.
        iii. The periodicity of surface seiches increases with increasing basin dimensions (see Table 7-2).
    d. ***Internal seiches*** occur when lakes are stratified and the metalimnion is set into oscillation as an internal standing wave.
        i. Both the periodicity and the amplitude of internal seiches are much greater (nearly an order of magnitude) than those of surface seiches.
        ii. Most internal seiches are uninodal, although multinodal seiches occur in very large lakes.
        iii. Since the nodal horizontal oscillations of internal seiches are much larger than those of surface seiches, shearing flow at the metalimnetic interfaces can generate large ***internal progressive waves.*** These internal waves on the metalimnion are about 10 times larger than surface waves, but they propagate and break internally much as surface waves do.
        iv. The geostrophic effects of Coriolis force increase toward either pole north and south of the equator. The geostrophic effects deflect the oscillation patterns and flows associated with internal seiches and cause large-scale twisting undulations of the metalimnion (Fig. 7-18).
        v. The internal currents and waves cause considerable mixing of water, solutes, and organisms across the epilimnetic–metalimnetic interface.
    e. ***Long progressive waves*** of both the surface and metalimnion waters have wavelengths that are much greater than the water depth. Long progressive waves are nondispersive.
        i. As the long waves oscillate in lakes sufficiently large for Coriolis geostrophic effects

to influence motion and intersect the shore line, the rotational effects and decay of wave energy by gravity generate currents that are parallel to the shore and to the progression of the waves (Fig. 7-20). The wave amplitude of these ***Kelvin waves*** decreases exponentially away from the shore.

ii. In very large lakes, the long waves can travel in open water without the interference of shore boundaries. These ***long gravity waves*** (***Poincaré waves***) are influenced by Coriolis geostrophic forces and form a wave pattern of alternating hills and valleys with corresponding cellular patterns of currents (Fig. 7-22).

iii. The amplitude of Poincaré waves increases as the shore boundary is reached and these waves are dispersed into the Kelvin wave configuration (Fig. 7-23).

12. A narrow zone of warm water develops between the shore line and the open water mass in the spring. The shallow nearshore waters heat more rapidly than the open waters. A density-driven circulation perpendicular to the shore consists of two circulation cells, with a zone of convergence, termed a ***thermal bar,*** in between. Little mixing occurs between inshore and offshore waters until the whole lake is thermally stratified.
13. Incoming river water to a lake or reservoir flows into strata of density similar to its own. Inflow water movements are dependent on density differences from temperature, dissolved materials, and particulate suspensoids and vary greatly seasonally and among lake ecosystems.
14. Slow water currents occur in lakes even though the water is sealed from wind by ice cover and there is no appreciable inflow and outflow. Horizontal current velocities tend to be greater than vertical velocities. The currents are thermally induced by convection from heat flowing from the sediments. Most of the heat released from the sediments during winter was accumulated during the previous summer.
15. Comparative analyses of hydrodynamics among river, reservoir, and natural lake ecosystems emphasize the transition from rapid, gravity-mediated, unidirectional and horizontal flushing rates in rivers to long, wind-driven, relatively constant, three-dimensional movements in lakes. Water-level fluctuations shift from large and irregular among streams to relatively small and stable among natural lakes.

8 STRUCTURE AND PRODUCTIVITY OF AQUATIC ECOSYSTEMS

Aquatic ecosystems are open and require a continual input of energy in the form of organic matter. This organic matter is produced almost totally by photosynthesis and is either used by consumers or stored in the ecosystem. Organic matter and energy driving the ecosystem are either produced within the river or lake or are imported from terrestrial or land–water interface communities to the river or lake. Within the aquatic ecosystem, organic matter is degraded physically, as in direct photolysis, utilized by decomposer organisms, or ingested by grazer (metazoan) organisms. Energy and nutrients must continually be replenished in such open systems in which available energy and other resources are constantly being utilized and respired to inorganic compounds and so lost or made unavailable.

Discussion so far has been directed to the basic physical aspects of water itself, how water is distributed to river and lake basins, and the origins of rivers and of lake basins, and how drainage basin morphology is modified with time. We then considered both the penetration of solar radiation and the distribution of heat in various fresh waters and explained how absorbed heat interacts with wind energy to influence water movements, particularly in standing waters.

Now the basic nutrient chemistry of aquatic ecosystems will be evaluated. The biota are inseparably coupled with the dynamics of many chemical elements, in particular the nutrient distribution and their biogeochemical regulation of nutrient fluxes and recycling. Before discussing the major chemical constituents in detail, however, a brief overview of the ecosystem components and how they interrelate is useful.

I. THE DRAINAGE BASIN CONCEPT

Terrestrial and aquatic ecosystems are linked by movements of water and materials through the drainage basin to recipient rivers and lakes. Chemical and biological processes modify the composition of the materials dissolved within and moved by the water. Geological features of the landscape govern the directions of movement and particularly the residence time of water during movement in surface and ground waters toward rivers and lakes. The duration of contact with soil and associated microbiota influences the content of dissolved salts and organic products conveyed in the water. Terrestrial and wetland vegetation in the drainage basin influence the chemical content of both water falling through their leaf canopy as well as water in the soil by selective assimilation and storage within their tissues. The movements and fluxes of a number of these compounds will be treated separately in subsequent chapters.

From drainage basins that are vegetated, such as in forests, usually only about one-third of the incoming precipitation leaves as runoff and groundwater seepage to streams and rivers. Removal of vegetation greatly increases the amount and accelerates the rate of flow of runoff from the landscape. During periods of high precipitation, water in streams frequently exceeds the capacities of the river channels and inundates the flood plains, often for long periods of months. During this inundation, organic matter and many nutrients can be returned to the flooded land and result in highly productive adjacent aquatic habitats.

The drainage basin clearly regulates the characteristics of rivers and lakes within it (Hynes, 1975; Likens, 1984). The geomorphology of the land determines the soil and ionic composition, slope, and, in combination with climate, the vegetation cover. The vegetation and soil composition influence not only the amount of water runoff but both the composition and quantity of organic matter that enter streams and lakes. Finally, as will be reiterated constantly, one cannot study aquatic ecosystems without consideration of human influences. Human pollution and modifications of the environment and climate are now so pervasive that no aquatic environment of the biosphere is unaltered in some manner by these disturbances. Human influences vary greatly and change with time but nonetheless often dominate regulation of biotic productivity and biogeochemical cycling.

II. STREAMS AND RIVERS

The characteristics of stream and river ecosystems, each summarized and integrated in individual topical chapters, are extremely diverse. However, several characteristics distinguish running waters from other inland ecosystems (Hynes, 1970; Minshall, 1988; Calow and Petts, 1992; Allen, 1995; Giller and Malmqvist, 1998).

a. Flow is *unidirectional;* as a result, downstream reaches are influenced by upstream reaches.
b. Running waters are *linear in form* and occupy a very small portion of the total drainage basin area.
c. *Channel substrata and channel morphology tend to be unstable* and undergo constant change as a result of the erosive and molar action of flowing water.
d. *Much of the organic matter entering and supporting metabolism in streams on an annual basis originates from allochthonous sources.* Metabolism at nearly all levels is variably but usually dominantly dependent on organic matter imported from terrestrial sources, and that dependency usually increases with smaller, low-order streams.
e. *Spatial and temporal heterogeneity is high* among many parameters in river ecosystems. The heterogeneity within a river tends to be much more dynamic than in standing waters. Rivers generally possess much greater and more dynamic longitudinal changes in flow, chemical conditions, and biota, particularly in adjacent bank and floodplain areas, than is the case in standing waters. Because streams and rivers respond relatively rapidly to precipitation events within the drainage basin, rapid and large changes occur frequently in loading from the drainage basins and resulting physical and chemical conditions. Flow, dissolved oxygen, and water chemistry are primary regulators of life in flowing water (lotic) ecosystems. Extremes in cyclic variations of these and related parameters set the physiological and behavioral boundaries tolerated by biota.
f. *High variability of physical, chemical, and biological characteristics exists among different streams and rivers* and, as a result, generalizations are more difficult than is the case among many standing waters.
g. Although basic biological functions and growth are similar among biota of lakes and rivers, many of the *stream biota have specialized adaptations to conditions within flowing water.*

III. LAKE ECOSYSTEM CONCEPT

Several early investigators emphasized the functional relationships of organisms within lakes (e.g., Forbes, 1887; Lindeman, 1942). The innovative perceptions of these studies led to intensive evaluation of interrelationships among organisms and changes in specific organism populations in response to alterations in physical, chemical, and biotic properties of the environment. Lakes were initially viewed, incorrectly, as microcosms, functionally isolated from the rest of the landscape and biosphere.

In their time, these analyses were perceptive and they led to extensive investigations of the interrelationships between biotic and physical components within lakes. These studies, which continue to the present time, are of fundamental importance; they have resulted in a comprehensive understanding that is unrivaled by any other area of ecology. Still, within a broad context, these in-lake analyses encompass but a small part of the lake ecosystem.

A number of early workers called attention to the importance of nutrients from the drainage basin in regulating metabolism in lakes. This insight recognized what is commonly accepted now—that the lake ecosystem consists of the lake and its entire drainage basin. Recognition of the importance of the drainage basin, however, is still largely limited to the loading of inorganic and organic nutrients. Comprehensive analyses of the manner in which the terrestrial biota regulate loading to streams and recipient lakes are much more rare (e.g., Likens *et al.,* 1977; Likens, 1985; Likens and Bormann, 1995).

The importance of the catchment area to inputs of both particulate and especially dissolved organic matter to lakes, and of the regulatory influences of this dissolved and colloidal organic matter once it reaches the lake, are finally gaining appreciation and understanding after some 30 years of championing and study by R. G. Wetzel and co-workers. Extremely heterogeneous and productive wetland–littoral areas often lie at the interface between the terrestrial drainage basin and the open-water zone of the lake (Fig. 8-1). These complex wetland–littoral areas are exceedingly important in regulating lake metabolism (Wetzel, 1979, 1990a, 1995). Since a majority of lakes of the biosphere are small and relatively shallow (cf. Fig. 3-2), the metabolically active wetland and littoral components dominate the productivity of most lakes of the world. These complex interface regions are the least understood lake ecosystem component, particularly of very shallow lakes; major aspects will be discussed later (cf. Chaps. 18, 19, 20, and 22).

The effects that terrestrial, wetland, and littoral biota have on the quality and quantity of inorganic and organic loading to a lake can be profound. Water laden with inorganic and organic substances flows from higher elevations to the recipient lake basin both in ground water and in surface streams. Chemical and biological reactions occur en route that selectively modify the quality and quantity of nutrients and organic substances entering the lake. Surface flows often pass through the wetland–littoral complex and can further selectively lose or gain inorganic and organic compounds before reaching the open water.

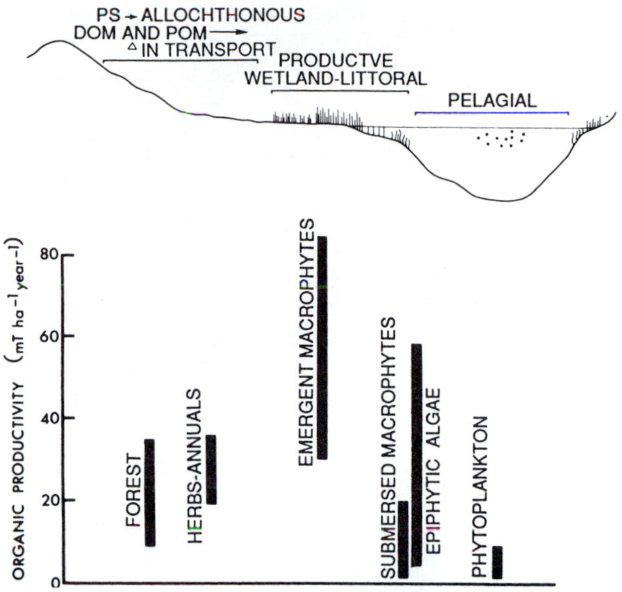

FIGURE 8-1 The lake ecosystem, showing the drainage basin with terrestrial photosynthesis (*PS*) of organic matter, movement of nutrients and dissolved (*DOM*) and particulate (*POM*) organic matter in surface and groundwater flows towards the lake basin and chemical and biotic alteration of these materials en route, especially as they pass through the highly productive and metabolically active wetland–littoral zone of the lake per se (net organic productivity in metric tons per hectare per year).

Finally, appreciable loading of inorganic and organic substances to the atmosphere is common from the industrial and agricultural activities of humans. These compounds reach the drainage basin and lake itself by dryfall and with wet and frozen precipitation. This source of loading is often a significant portion of the total nutrient and contaminant loading to a lake.

A. Lacustrine Zonation and Terminology

The bottom of a lake basin is separable from the free open water, the ***pelagic zone,*** and is further divisible into a number of rather distinct transitional zones from the shore line to the deepest point (Fig. 8-2). The ***epilittoral zone*** lies entirely above the water level and is uninfluenced by spray; the ***supralittoral zone*** also lies entirely above the water level but is subject to spraying by waves. The ***eulittoral zone*** encompasses that shoreline region between the highest and lowest seasonal water levels and is often influenced by the disturbances of breaking waves. The eulittoral zone and the ***infralittoral zone*** collectively constitute the littoral zone. The infralittoral zone is subdivided into three zones in

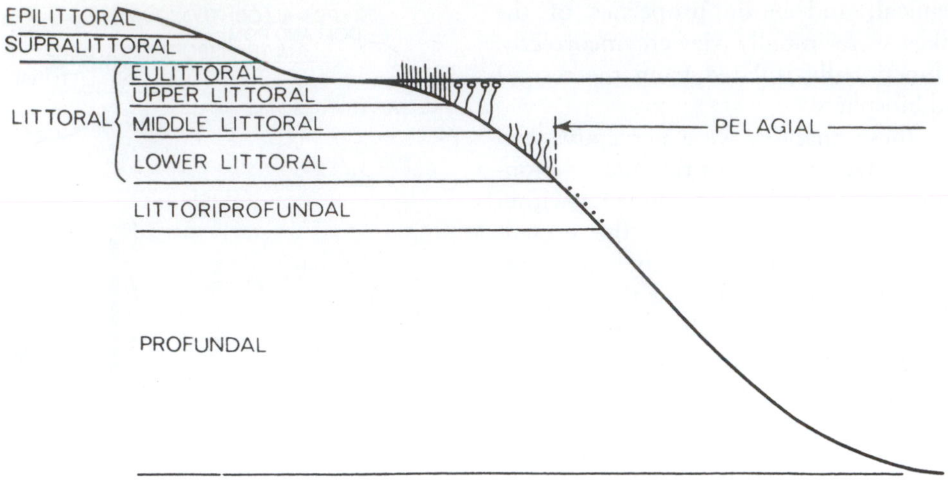

FIGURE 8-2 Lacustrine zonation (see text for discussion). (After Hutchinson, 1967.)

relation to the commonly observed distribution of macrophytic vegetation: *upper infralittoral* or zone of emergent rooted vegetation; *middle infralittoral* or zone of floating-leaved rooted vegetation; and *lower infralittoral* or zone of submersed rooted or adnate macrophytes.

Below the littoral is a transitional zone, the ***littoriprofundal,*** occupied by scattered photosynthetic algae and bacteria. The littoriprofundal zone is often adjacent to the metalimnion of stratified lakes. The upper boundary of the littoriprofundal zone at the lower edge of macrovegetation of the lower infralittoral is usually quite distinct. The lower boundary of the littoriprofundal consists of a gradient of benthic algae and especially cyanobacteria and other photosynthetic bacteria and is less sharply demarcated. The remainder of the sediments, which consists of exposed fine sediment free of vegetation, is referred to as the ***profundal zone.***

The philology associated with bacteria and algae variously attached to substrata in aquatic systems is involved (cf. reviews by Naumann, 1931; Cooke, 1956; Sládečková, 1962; Wetzel, 1964, 1996; Round, 1964a, 1965a; and Hutchinson, 1967). The number of terms is large and many are unnecessarily complex and quite confusing, especially when removed from their original definitions. The subject is rather unrewarding because of the great variation in microhabitats and association of organisms with substrata. A few widely used terms in limnology need to be discussed.

The term ***benthos,*** originating from the Greek word for "bottom," was initially defined broadly to include the assemblage of organisms associated with the bottom or, better, any solid–liquid interface in aquatic systems. Benthos is now nearly uniformly applied to animals associated with substrata.[1] The terms *Aufwuchs* (German: "growth upon"), and to a lesser extent *haptobenthos,* generally connote all organisms adnate to, but not penetrating, a solid surface. Since the term *Aufwuchs* is so broad and ambiguous, it is of little meaning and should be abandoned.

Periphyton, although variously used, usually refers to microbial growth upon substrata. It is a broad term that applies to microbiota living on any substratum, living or dead, plant, animal, or nonliving. Despite being etyomologically imprecise, the word is so entrenched in limnology that its use has been internationally accepted (cf. Wetzel, 1983b). In contrast, ***biofilm*** is a term more frequently used in engineering applications and often refers to predominantly attached heterotrophic bacterial communities. Biofilm, however, is essentially synonymous to periphyton. Reference to more specific subdivisions by complex phrases (e.g., epiphytic periphyton) results in involved and often redundant nomina. A much more explicit manner of expression, and one championed by F. E. Round, is to refer to the organisms with appropriate modifiers descriptive of the substrata upon which they grow in natural habitats. Hence, among the algal communities, one can readily differentiate the following (Fig. 8-3): (a) ***epipelic*** algae as the flora growing on sediments (fine, organic); (b) ***epilithic*** algae growing on rock or stone surfaces; (c) ***epiphytic*** algae growing on macrophytic surfaces; (d) ***epizooic*** algae growing on surfaces of animals; and (e) ***epipsammic*** algae as the rather specific organisms growing on or moving through sand.

[1] *Herpobenthos* as used by Hutchinson (1967) is synonymous with *benthos* when the term *benthos* is unqualified and means "growing on" or "moving through sediment."

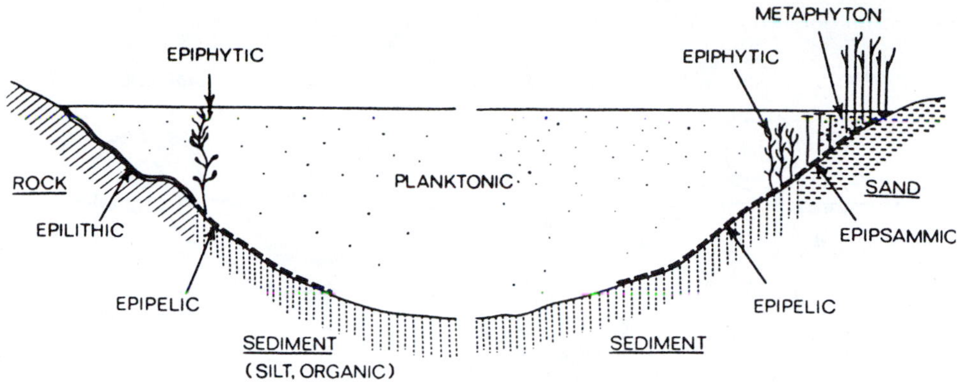

FIGURE 8-3 Major terms of microfloral communities associated with different substrata in inland waters.

The general word ***psammon*** refers to all organisms growing or moving through sand (cf. Cummins, 1962, for a detailed discussion of particle-size differentiation among sediments).

A group of algae found aggregated in the littoral zone is the ***metaphyton,***[2] which is neither strictly attached to substrata nor truly suspended. The metaphyton commonly originates from true floating algal populations that aggregate among macrophytes and debris of the littoral zone as a result of wind-induced water movements. In other situations, the metaphytonic algae can originate from fragmentation of dense epipelic and epiphytic microbial populations. A surprisingly large number of descriptions exist of the clustering of metaphytonic algae and macrophytes into "lake balls," which are densely packed aggregations of algae, plant parts, or both. These balls are formed by the alternating rolling movements of wave action in the littoral zone (Nakazawa, 1973).

The ***phytoplankton*** consist of the assemblage of small plants or photosynthetic bacteria having no or very limited powers of locomotion; they are therefore more or less subject to distribution by water movements. Certain planktonic[3] algae move by means of flagella or possess various mechanisms that alter their buoyancy. Most, however, are "freely floating," a term commonly used when referring to ***plankton*** in general, even though it is obviously somewhat of a misnomer, in that most algae do not float. Most algae are slightly denser than water and sink, or sediment from, the water. Phytoplankton are largely restricted to ***lentic*** (standing) waters and large rivers with relatively low current velocities (cf. Wetzel, 1975b; Wetzel and Ward, 1992). Phytoplankton in streams of moderate flow, which either break loose from the algal communities attached to substrata or enter a stream from the outflow of lakes, are commonly rapidly fragmented or killed by the abrasive action of turbulence and substrata (see, e.g., Chandler, 1937, and Chap. 15). Distinctly macroscopic algae with long filamentous forms usually inhabit parts of the littoral zone. The common groups of phytoplankton, their population and community dynamics, and their productivity have been studied extensively (cf. Chap. 15).

Animals of fresh waters are extremely diverse and include representatives of nearly all phyla. The ***zooplankton*** include animals suspended in water with limited powers of locomotion; they are subject to dispersal by turbulence and other water movements. Like phytoplankton, zooplankton are usually denser than water and constantly sink by gravity to lower depths. The planktonic protozoa and other protists have limited locomotion, but the rotifers, cladoceran and copepod microcrustaceans, and certain immature insect larvae often move extensively in quiescent lake water. These important zooplankton are treated in detail in subsequent chapters (Chaps. 16 and 19).

The distinction between suspended zooplankton having limited powers of locomotion and animals capable of swimming independently of turbulence—the latter referred to as ***nekton***—is often diffuse. Certain zooplankton and early larval stages of fish are initially planktonic but distinctly nektonic in later life stages (cf. Chap. 16).

A number of specialized organisms, the ***pleuston,*** are adapted to the interface habitat between air and water. Numerous macroscopic organisms, such as many of the duckweeds (Lemnaceae) and adult insects (e.g., gerrid hemipterans), are morphologically adapted to the air–water interface. Many species of bacteria and algae,

[2] *Metaphyton* (Behre, 1956) is essentially synonymous with the terms *tychoplankton* and *pseudoplankton* used much earlier by Naumann (1931) and the term *pseudoperiphyton* used by Sládečková (1960).

[3] The word *planktonic,* while etymologically incorrect (see discussion by Rodhe, 1974, and Hutchinson, 1974), is so ingrained in aquatic ecology that change is not desirable.

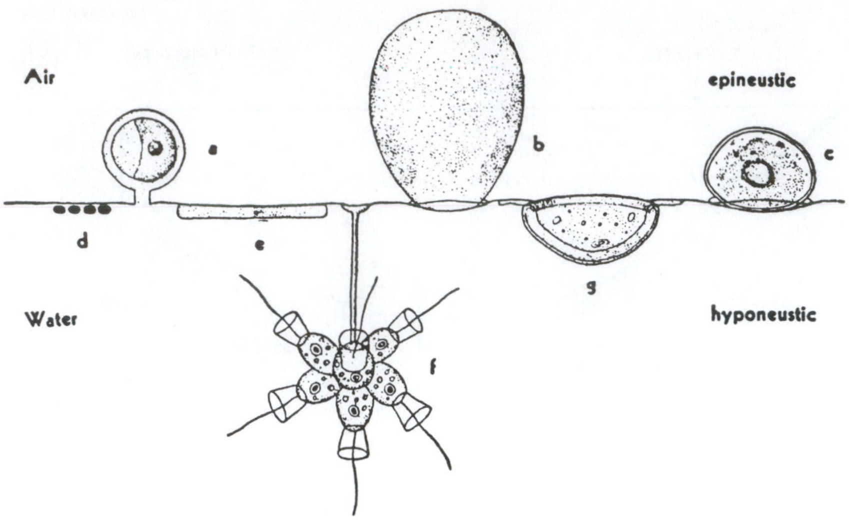

FIGURE 8-4 Exemplary organisms of the neuston at the surface film. Epineustonic: (a) *Chromatophyton* (Chrysophyceae); (b) *Botrydiopsis* (Xanthophyceae); (c) *Nautococcus* (Chlorococcaceae). Hyponeustonic: (d) *Lampropedia* (Coccaceae); (e) *Navicula* (Bacillariophyceae); (f) *Codonosiga* (Craspedomonadaceae); (g) *Arcella* (Rhizopoda). (From Ruttner, F.: *Fundamentals of Limnology,* 3rd ed., Toronto, University of Toronto Press, 1963.)

including some diatoms, chrysophytes, and xanthophytes, are specialized to live at the interface. The microscopic components of the pleuston are collectively termed ***neuston,*** and are separated into those organisms adapted to living on the upper surface of the interface film (the *epineuston*) and those living on the underside of the surface film (*hyponeuston*) (Naumann, 1917; Gladyshev and Malyshevskiy, 1982). Most neustonic organisms live on the upper surface of the interface. A few examples of algal and protozoan neuston are illustrated in Figure 8-4.

The development of neuston is most extensive in sheltered, quiescent waters. In large bodies of water, neustonic algae develop in the pelagic zone during calm periods but are quite changeable and transitory (Nägeli and Schanz, 1990). The algal species of the neuston within the surface-film habitat differ considerably from those of the underlying bulk-water phytoplankton. Similar results have been observed in marine ecosystems (see Zaitsev, 1970; Wandschneider, 1979). Surface films maintain themselves only during periods of very calm conditions for periods of several hours (12–32 h) or days (Fott, 1954; Hühnerfuss *et al.,* 1977).

Conspicuous development of the neuston, particularly of the epineuston, is common in small, quiet waters in the littoral or along the margins and backwater areas of streams. Often, populations of epineustonic organisms become so large that light is reflected from cellular chromatophores, and the water appears to be covered with a dry film of varying coloration. Some neustonic algae, such as *Nautococcus,* release a buoyant, pigmented extracellular layer composed of a hydrophobic mixture of carboxylated polysaccharides and waxes (Pentecost, 1984). A few cladoceran zooplankton and insect larvae are adapted to feed on the neuston from the underside of the interface film.

Very little is known of the physiology and metabolism of the neuston of fresh waters in comparison with that of marine waters (cf. Zaitsev, 1970; Hardy, 1973; De Souza-Lima and Romano, 1983). Although the neustonic productivity and contributions to the total metabolism of fresh waters probably are small in most situations, their roles in shallow lake and wetland ecosystems deserve much further study.

IV. POPULATION GROWTH AND REGULATION

A ***population*** of organisms is a spatially defined assemblage of individuals of one species. A population of a species could broadly include all individuals of that species inhabiting the biosphere, but ecologists generally deal with individuals of a species that occupy, or populate, a more localized area, such as a particular lake or section of flowing water. If movement into and out of the defined area is small, relative to the number of births and deaths occurring within the area, population changes within the area can be studied with reasonable accuracy. The number of individuals per unit area or volume of water, the population ***density,*** is an index of population size at any given time.

Growth among populations and their competitive interactions for available resources are fundamental to nearly all aspects of biological limnology. Analyses of population growth have been the subject of intensive quantitative and theoretical study.[4] Although a detailed analysis is inappropriate here, several fundamental relationships and concepts are discussed, since reference is made to these basic parameters and characteristics in subsequent evaluations of freshwater population and community dynamics.

The growth of a population can be very rapid among smaller organisms. The rate of growth (g) is a geometric expression of the increase of the numbers (N) of the population. The time (t) that it takes a population number to double under ideal conditions is

$$t = \frac{\log_e 2}{\log_e g} = \frac{0.69}{\log_e g} \tag{8-1}$$

Environmental and competitive restrictions quickly limit the size of the populations either by increasing death rate or by decreasing birth rate.

The geometric constant (g, Eq. 8-1) for rate of population change is difficult to use in comparative growth analyses among populations because of differences in population sizes. This problem is mathematically avoided by expressing population growth with an exponential constant, r, where $r = \log_e g$, or $g = e^r$. Calculation of r from population growth data is done by obtaining the difference between the logarithms of the population sizes at the beginning and the end of a period of growth and dividing by the length of the time period over which the observations are conducted:

$$r = \frac{\log_e N_t - \log_e N_0}{t} \tag{8-2}$$

Factor r is an instantaneous, relative measure of population growth rate, in which the percentage rate of increase in a population is determined at every time interval.

The growth rate of a population can be characterized by the logistic equation, which incorporates a term for the intrinsic capacity of a population to grow and a term corresponding to density-dependent limitations upon that growth potential imposed by the competition of others in the population for available environmental resources. Among small, rapidly growing, asexual or parthenogenetic organisms, such as bacteria, many algae, and zooplankton, the differential form of the logistic:

$$\frac{dN}{dt} = rN\left(\frac{K - N}{K}\right) \tag{8-3}$$

[4]The basic discussion of population analyses is presented in many texts on population biology (e.g., Wilson and Bossert, 1971; Hutchinson, 1978; Ricklefs, 1979; cf. also Kingsland, 1982).

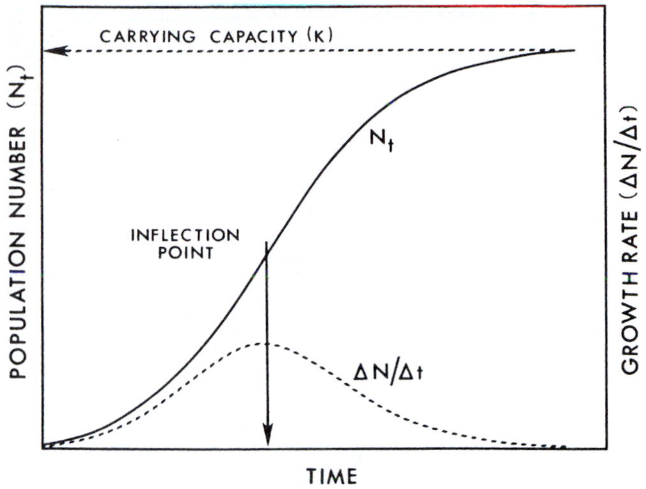

FIGURE 8-5 Common sigmoid growth curve of a population with time (N_t) as it approaches the environmental carrying capacity (K). The growth rate ($\Delta N/\Delta t$) denotes the difference in number of organisms over each time interval. (Modified from Ricklefs, 1979.)

or

$$\frac{dN}{dt} = rN\left(1 - \frac{N}{K}\right) \tag{8-4}$$

can adequately model a population (N) growing from a small number of individuals to an upper-limiting population K, the latter being governed by nutrient supply, space, inhibitory metabolic products, and other density-dependent factors. As the population increases toward the carrying capacity (K), fewer resources are available for reproduction, and reproductive rates decline. As the population density approaches the carrying capacity, $(1 - N/K)$ approaches zero and population growth ceases. The resulting logistic growth curve has a sigmoid shape (Fig. 8-5) and is commonly observed among natural and experimental populations. The basic logistic model discussed here has been variously modified (cf. Hutchinson, 1978) but is commonly found among populations.

A ***life table*** and its resulting ***survivorship curve*** provide the parameters needed for determining the survival and reproductive performance of a population. When a large group (e.g., 1000) of individuals of the same age or nearly the same age, referred to as a ***cohort***, are counted at time intervals, one can obtain an estimate of the number of survivors in each time interval. The fraction (l_x) of the original cohort surviving to time x is plotted logarithmically against x to yield a survivorship curve (Fig. 8-6). Such a survivorship curve, in which the population is followed as it changes with time, is age-specific, since all surviving individuals have similar ages at any time (Hutchinson, 1978). The ***specific death rate*** (q_x) over any time interval x to $x + 1$ is the number of

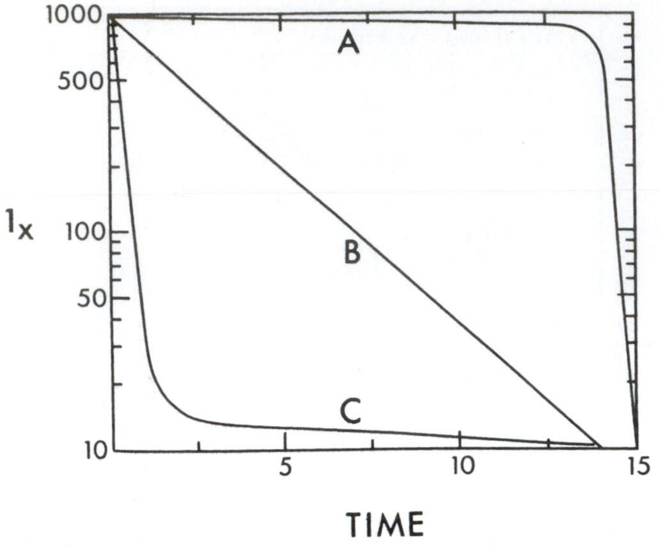

FIGURE 8-6 Idealized survivorship curves of a population (a) with a determined physiological life expectancy (negatively skewed rectangular curve), (b) with a constant death rate (diagonal), and (c) with high juvenile mortality (positively skewed rectangular). (Modified from Hutchinson, 1978.)

individuals dying during that interval ($l_x - l_{x+1}$) divided by l_x. The slope of the survivorship curve at any point corresponds to an instantaneous value of q_x.

Population size tends to oscillate about the equilibrium population level (K) that can be supported by the environment. The rate of increase per individual will decline steadily throughout time as the population grows following the logistic growth equation. The density-dependent decline can result either from a decrease in birth rate (b) or an increase in mortality (death rate, d) as N increases ($r = b - d$). In contrast, density-independent changes in population size can be caused by environmental factors that affect all individuals in a population with equal probability. Density-dependent changes in b or d are particularly important in the regulation of population sizes under uniform environmental conditions. When environmental conditions fluctuate, density-independent constraints to population growth can assume greater importance. Examples of these population changes are discussed in detail in subsequent chapters.

V. COMMUNITY STRUCTURE AND INTERRELATIONSHIPS

Although variously defined, a ***community*** usually refers to a group of interacting populations. Some practical spatial boundaries are most often applied to studies of communities (e.g., planktonic communities) because habitats are diverse and often organisms move from one habitat to another. Throughout subsequent discussions of freshwater biota, the effects of density-dependent and dynamic environmental factors on population and community changes will be evaluated. It is useful to summarize the basic concepts of these relationships and competition among populations of a community in a general way before evaluating specific cases.

A. Ecological Niche

The concept of the ecological niche has had a long history of development and interpretation in ecology. A given habitat consists of a number of interacting environmental gradients. The gradients include both physical and biotic factors, many of which are dynamic and change constantly at differing rates.

In seminal analyses, Hutchinson (1944a, 1957, 1978) formally defined the ***niche*** concept as a certain biological activity space in which an organism exists in a particular habitat. This space is influenced by the physiological and behavioral limits of a species and by the effects of environmental parameters (physical and biotic, such as temperature and predation) acting upon it. Each of these parameters can be ordinated on an axis and can be thought of as a dimension in space. *The fundamental niche, then, can be viewed as an* n-*dimensional space or hypervolume,* with each of its *n*-axes or dimensions corresponding to the range of an environmental parameter over which the organism can exist. Since many physical and biotic factors interact, each species occupies only a portion of its fundamental niche; this portion can be referred to as the species' realized niche.

Exact quantitative definition of a species' niche, realized or fundamental, is impossible because of its infinitely large number of axes or dimensions. In addition, while it is mathematically and statistically possible to describe multidimensional spaces, we can readily understand physical or graphical explanations in only three or fewer dimensions. In many cases, however, only a few dimensions are needed to adequately describe specific niche interactions (cf. Hutchinson, 1978; Morowitz, 1980).

B. Competition

For each species and each resource requirement, a range of possible existence is bounded by extremes of tolerance and bracketing an optimal growth zone. Species differ in resource requirements for optimal growth and reproductive success, but rarely does a species occupy only optimal zones within its requirements. Certain variables (e.g., temperature) are continuous, whereas others (e.g., prey availability) are discrete or discontinuous. Some species have similar requirements along one or more resource axes. When two or more species overlap in terms of resource utilization, competition occurs.

The logistic growth equation can be extended to gain insight into interspecific competition and coexistence of populations in a given habitat over a period of time. In the logistic growth equation (Eq. 8-4), rN is the growth potential of the population in the absence of interspecific competition. The term $1 - N/K$ indicates the change in growth, usually a decrease, caused by competition among individuals of the population (N) with respect to the carrying capacity (K) of the environment. As population growth approaches the capacity (K) of the environment to support it, N/K approaches 1, and the resulting growth potential $(1 - N/K)$ approaches zero.

When two species occur in the same habitat, competition for limiting resources may occur that reduces growth potential and influences the population levels attained. The Volterra equations describe competition between two species:

$$\frac{dN_1}{dt} = r_1N_1\left(1 - \frac{N_1}{K_1} - \frac{\alpha_{12}N_2}{K_1}\right)$$

$$\frac{dN_2}{dt} = r_2N_2\left(1 - \frac{N_2}{K_2} - \frac{\alpha_{21}N_1}{K_2}\right) \qquad (8\text{-}5)$$

The ***competition coefficients*** (α_{12} and α_{21}) indicate the intensity of the effect that each species has on the population growth of the other. The growth rate of each population consequently becomes dependent upon the number of individuals in the populations of both species.

An increase in population size of one species will result in displacement of the other. Numerous combinations of competitive effects have been found. Often, individuals of one species have a greater effect on individuals of a second species than on members of their own (the first) species, and the competitive effect of individuals of the second species on the first is less than the effects on its own (the second) species ($\alpha_{12} > K_1/K_2$ and $\alpha_{21} < K_2/K_1$). When each species has a greater effect on the other species than it does upon its own growth ($\alpha_{12} > K_1/K_2$ and $\alpha_{21} > K_2/K_1$), which species survives will depend upon the initial proportion of the two species in the growing mixed populations (cf. detailed treatment in Hutchinson, 1978).

If each species has a greater effect on its own members than it does on individuals of the other species (i.e., $\alpha_{12} < K_1/K_2$ and $\alpha_{21} < K_2/K_1$), coexistence of the two species can occur. For competitive exclusion of one or more species by a dominant competing species to occur, the environmental medium must be homogeneous, and sufficient time in relation to reproduction must be available. Often, particularly in planktonic environments, conditions change too rapidly for competitive effects to drive a species to extinction. As a result, a number of apparently competing species with similar environmental requirements coexist. Such interactions will be discussed repeatedly in subsequent chapters.

VI. ECOSYSTEM INTERRELATIONSHIPS

In most cases, energy required for growth and maintenance of organisms enters ecosystems as light and is converted to chemical energy by plant photosynthesis (***autotrophy***). Biological communities are supported by and based on the photosynthetic production of organic matter, produced either within a community (autochthonous production) or derived from an external source (allochthonous production) and transported in a dead or decomposing state to a community for utilization (***heterotrophy***).

A. Trophic Structure of Food Cycles

In self-sustaining biological communities, functionally similar organisms can be grouped into a series of operational levels. Each level, which usually consists of many species competing with each other for available resources, forms a ***trophic level.*** The ***trophic structure*** of a community refers to the pathways by which energy is transferred and nutrients are cycled through the community trophic levels.

Photosynthetic organisms are primary producers, and they represent the first level (Λ_1) of the trophic structure.[5] The Λ_1 organisms are eaten by primary consumers or herbivores (Λ_2), which in turn are successively consumed by secondary (Λ_3), tertiary (Λ_4), et cetera, consumers (carnivores). Since energy is lost in each transfer, more than six trophic levels can rarely be sustained. Organisms exist in complex food webs, in which nutrients and energy of one trophic level are utilized by organisms from several different trophic levels. Although greatly oversimplified, Lindeman's (1942) conceptualization is useful for understanding the general patterns of ecosystem structure and energy flow.

The productivity of each trophic level (Λ_n) was considered to be the rate of energy content entering

[5] The notations, which were initially introduced by Hutchinson (unpublished), were later formalized by Lindeman (1942).

(λ_n) the trophic level from the lower one (Λ_{n-1}):

$$\frac{d\Lambda_n}{dt} = \lambda_n + \lambda'_n \tag{8-6}$$

where λ'_n is the rate of energy leaving the trophic level by metabolic dissipation (respiration with dispersion as heat), returned to the environment as excrement or dead organisms, or being passed on to the next trophic level (Λ_{n+1}) as food.

The amount of energy available for metabolism decreases with each increase in trophic level because energy is lost with each transformation. Since organisms expend a considerable amount of energy for maintenance, only a portion of the energy of one trophic level is available for transfer and use by higher trophic levels. Therefore, at equilibrium (Hutchinson, 1978):

$$\lambda_n = -\lambda'_n \tag{8-7}$$

and

$$\cdots \lambda_{n-1} > \lambda_n > \lambda_{n+1} \cdots \tag{8-8}$$

Total biomass and numbers of individuals at each successively higher trophic level also usually decrease, although some exceptions will be discussed later.

Efficiency of energy or material transfer through a trophic level is simply the ratio, expressed as a percentage, of energy or material leaving a level or system to the quantity of energy or material entering the level or system. In the case just discussed, efficiency would equal the ratio of these productivities, λ_n/λ_{n-1} (critically discussed by Kozlovsky, 1968). The ***ecological-efficiency*** is this percentage of transfer of energy from one trophic level to the next. The rate of conversion of solar energy (Λ_0) into chemical energy by photosynthesis (λ_1) by the first trophic level (Λ_1) of primary producers is small, usually less than 1%, because of the great losses of light energy by nonbiological scattering and absorption and dissipation as heat (Chap. 5). The expenditures of energy for maintenance and the losses to the detrital dynamic structure (see Chap. 22) increase with higher trophic levels. As a result, the efficiencies of energy transfer are low from one trophic level to the next (5–15%) and usually decrease as Λ_n increases.

The general trophic-level conceptual view of feeding interactions in aquatic ecosystems has received appreciable criticism because omnivory is so prevalent. In most ecological communities, individuals of a species feed from several trophic levels (cf. review by Persson, 1999). Life-history omnivory is a result of individuals using different resources during different parts of their life cycle. Many of these differences in feeding, particularly during the different stages and sizes of the life cycle of a species, will be detailed in the discussions that follow in this work. Life-history omnivory does not mean that consumer–resource dynamics that are delineated in the trophic level concept become decoupled, but rather that consumer–resource dynamics become coupled in more complex ways.

The presence of size/stage structure means that the utility of the trophic level concept is much less functional in reality, because size-structured interactions cause both horizontal and vertical heterogeneity in food-web interactions (Fig. 8-7). In the simple case of horizontal heterogeneity of food-web feeding interactions (Fig. 8-7, *left,*), several species at each trophic level may lead to covariation among adjacent trophic levels. In the case of vertical heterogeneity (Fig. 8-7, *middle,*), species A is feeding from several trophic levels (omnivory). In the case of both horizontal and vertical

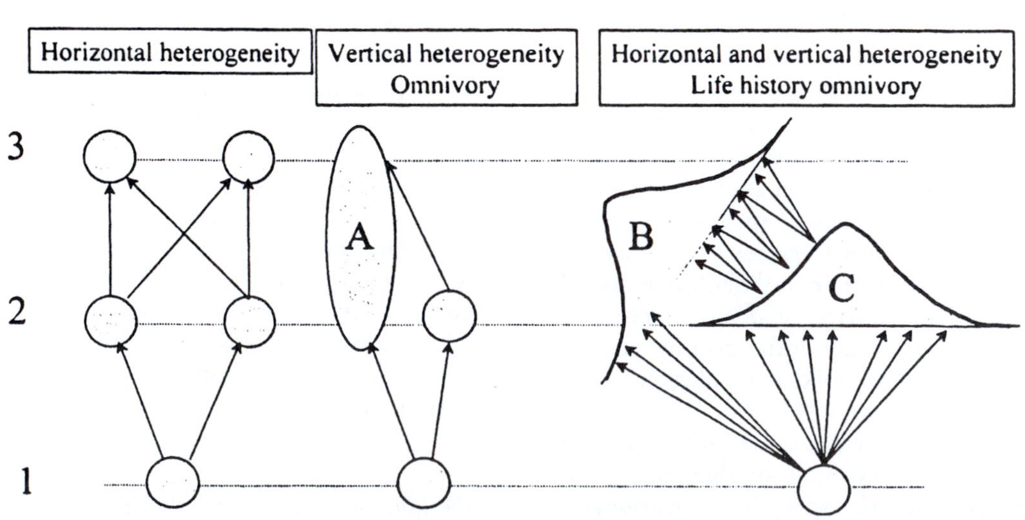

FIGURE 8-7 Three types of heterogeneities in food webs that will give rise to community patterns not expected (see text) from a simple trophic level conceptual approach. (From Persson, 1999.)

heterogeneity (Fig. 8-7, *right,*), species B is feeding from different trophic levels in different parts of its life cycle (life-history omnivory). The distributions in the right-hand figure denote size distributions. Horizontal heterogeneity is present in species C, because individuals of different sizes have different interaction coefficients with the resource being used. Horizontal heterogeneity is present for species B for the same reason as for species C, and, in addition, vertical heterogeneity is present because species B shifts from feeding on the basal resource to feeding on species C with increasing size. Simple linear food-chain models based on direct trophic levels are not realistic.

Another important facet of these trophic structure and energy relationships was that they were developed based almost solely on the ingestion of particulate organic matter (POM) (Wetzel, 1995). Nearly all evaluations of flows of energy within the trophic structures and their variants were specifically orientated along ingestion by animals of particles of organic matter. Excellent understanding was gained from research directed toward the size of POM, morphological aspects of ingestion (e.g., mouth gape, filtration) and avoidance of ingestion (e.g., transparency/visibility, interference by cellular or body projections), behavioral capabilities of organisms for movements within the pelagic zone in relation to refuges and/or escape from predators, nutritional differences in particulate food, and others. Although many of these studies concede that many organisms consume variable amounts of particulate detritus (dead particulate organic matter) and that particulate detritus nearly always dominates over living POM (e.g., Saunders, 1972), quantitative measures of consumption, assimilation, and egestion of detrital POM and its associated attached bacteria are practically unknown.

VII. DETRITUS: DEAD ORGANIC MATTER AND DETRITAL DYNAMIC STRUCTURE

The living biota of lakes constitute only a very small portion of the total organic matter of ecosystems, even in the most productive of freshwater habitats. Most of the organic matter in aquatic ecosystems is nonliving and is collectively referred to as ***detritus.***

The functioning of aquatic ecosystems centers on the cycling of organic carbon between living and nonliving components. Because of the pivotal relationships between the productivity of the living components and the massive amounts of nonliving organic matter, the subject will be discussed in detail later (Chap. 23). Two aspects, however, should be emphasized at this point.

Detritus consists of all nonliving organic matter, in both dissolved and particulate forms. In aquatic ecosystems in general, nearly all of the organic matter consists of dissolved organic carbon compounds (DOC) and particulate organic carbon compounds (POC). The ratio of DOC to POC is usually between 6:1 and 10:1 in both lake and running water ecosystems (Wetzel, 1984). Although living POC of the biota constitutes a very small portion of the total POC, particularly in the pelagic region, the microbial components are important in mediating the fluxes of carbon between the dissolved and particulate phases of detrital organic carbon.

Organic matter originates from photosynthesis within the lake or river ecosystem per se or from the drainage basin. Particulate organic carbon loadings from the drainage basin are relatively small; most organic carbon is transported and imported in dissolved form. Similarly, much of the very high net primary productivity of the wetlands and littoral land–water interface regions of lakes and river ecosystems enters the open–water regions as dissolved organic matter (Wetzel, 1990a; Wetzel and Ward, 1992). This pool of dissolved organic matter is supplemented by the low phytoplanktonic primary production to varying degrees in proportion to the size and geomorphology of the lake and river basins. Clearly, most lakes are small and shallow, and the ratios of organic carbon loadings from allochthonous and littoral sources to those of the pelagic are large (cf. Fig. 3-2; Wetzel, 1990a). The pool of organic matter is variously metabolized, sedimented, or exported.

Very little of either the particulate or the dissolved organic matter emanating from new photosynthetic production is consumed by animals or enters the consumer food chain but instead enters nonpredatory pools of dissolved and particulate detritus (Wetzel *et al.,* 1972; Rich and Wetzel, 1978; Wetzel, 1983, 1995). Because little of this organic carbon does enter the consumer food chain, however, does not mean it is neither metabolized nor important. Not only is this organic carbon the dominant energetic pathway but these microbial and chemical fluxes are essential to the maintenance of nutrient recycling and the metabolic stability of the ecosystem.

Metabolism of particulate and especially dissolved organic detritus from both allochthonous and many pelagic and nonpelagic autochthonous sources dominates both material and energy fluxes. The amount of detrital organic matter is very large, and it is chemically more recalcitrant to biological degradation than is living organic matter. As a result, metabolism of detritus is slow, but the quantity is large and much greater than that resulting from living organic matter that is ingested by anaimals. This slow decomposition provides an inherent ecosystem stability that energetically dampens the ephemeral, volatile fluctuations of higher-trophic-level species populations as they respond to rapidly oscillating factors governing their growth and reproduction (Chap. 23).

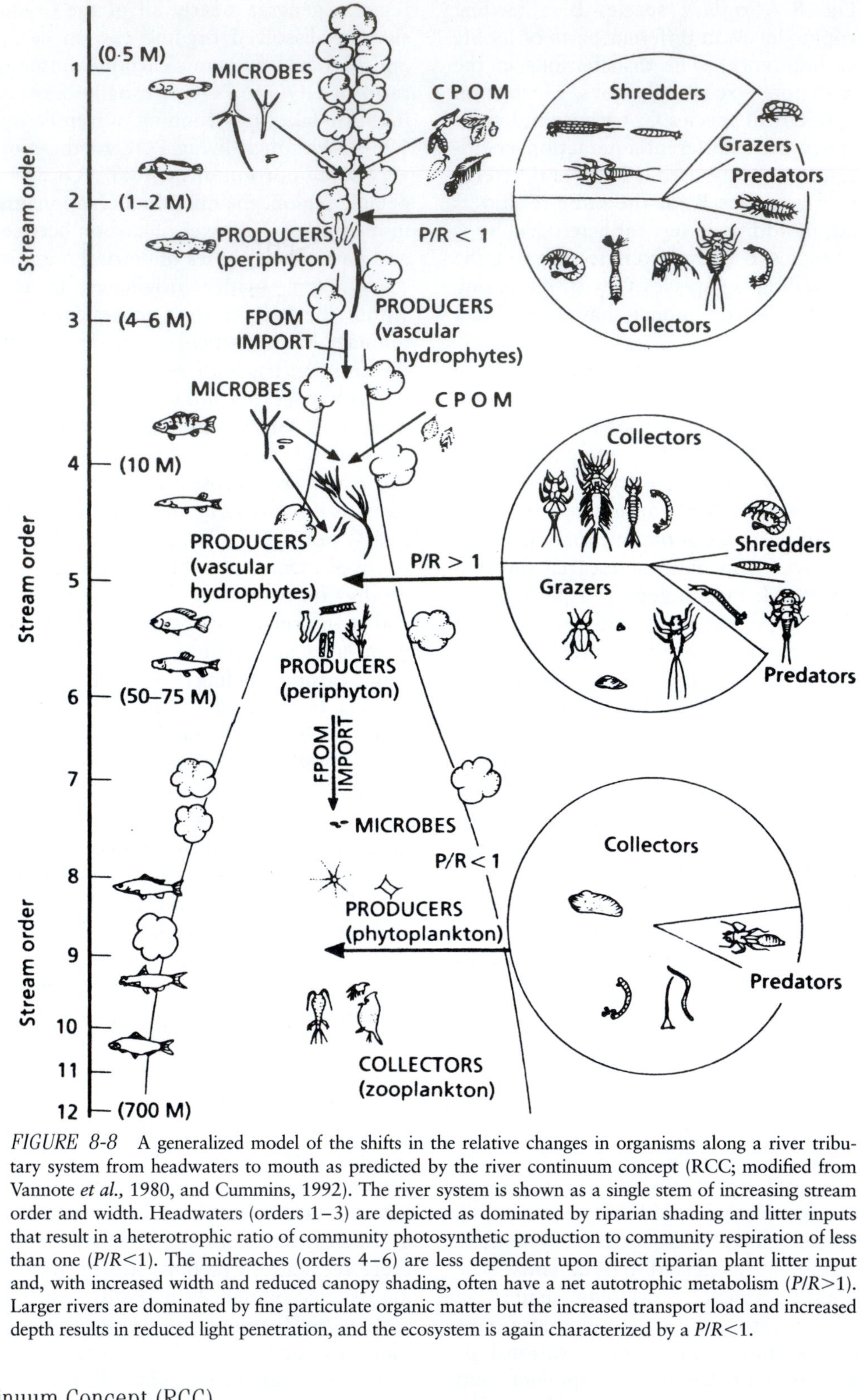

FIGURE 8-8 A generalized model of the shifts in the relative changes in organisms along a river tributary system from headwaters to mouth as predicted by the river continuum concept (RCC; modified from Vannote *et al.*, 1980, and Cummins, 1992). The river system is shown as a single stem of increasing stream order and width. Headwaters (orders 1–3) are depicted as dominated by riparian shading and litter inputs that result in a heterotrophic ratio of community photosynthetic production to community respiration of less than one ($P/R<1$). The midreaches (orders 4–6) are less dependent upon direct riparian plant litter input and, with increased width and reduced canopy shading, often have a net autotrophic metabolism ($P/R>1$). Larger rivers are dominated by fine particulate organic matter but the increased transport load and increased depth results in reduced light penetration, and the ecosystem is again characterized by a $P/R<1$.

A. River Continuum Concept (RCC)

Recognition that the drainage basin was a basic unit for streams and rivers led to the development of the ***river continuum concept*** (***RCC***). The RCC views the entire river ecosystem as longitudinally changing physical templates overlain by biological adaptations along these gradients (Vannote *et al.*, 1980; Cushing *et al.*, 1983; Minshall *et al.*, 1985; Cummins *et al.*, 1995). Geomorphological and hydrological characteristics form the fundamental template along drainage basins upon

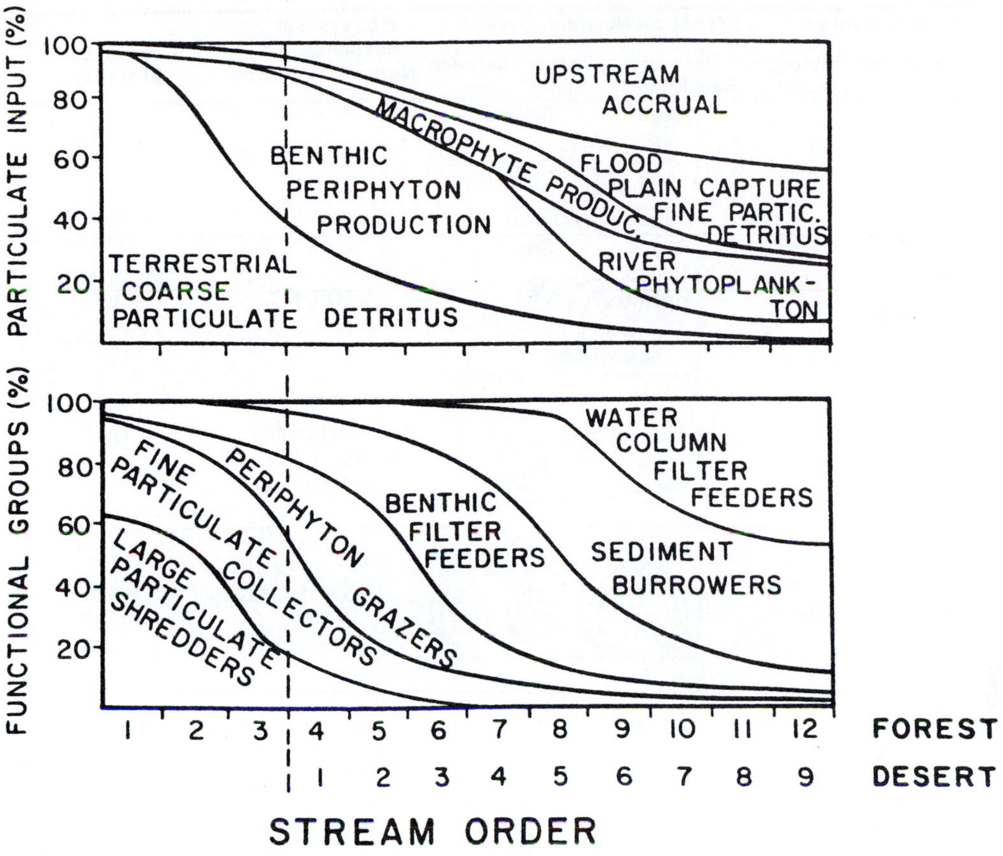

FIGURE 8-9 Commonly anticipated changes in *particulate* organic matter inputs (*upper*) and functional feeding group relationships among invertebrates (*lower*) along a river continuum (after Minshall *et al.*, 1985). The different scales on the abscissa emphasize that different streams enter the continuum at different points. For example, forest streams begin with a strong terrestrial influence with a predominance of allochthonous dissolved and particulate organic matter and detrital macroinvertebrates. In contrast, desert streams, with little if any shading and reduced inputs of allochthonous particulate organic matter, enter the sequence at a position more downstream in the forested stream sequence.

which biological communities become adapted. Both this template and biological communities change along drainage basins from headwaters to river mouth in a reasonably predictable manner (Figs. 8-8 and 8-9).

Generalizations of the RCC focus on seasonal variations of organic matter supply along the continuum (photosynthetic within the stream vs. allochthonous inputs), structure and feeding types of the invertebrate communities, and resource partitioning along the length of the river drainage basin. Heavily forested headwater streams have a leaf canopy that restricts light availability and a very low primary productivity. As a result, these streams exhibit very low primary production: community respiration ratios (P/R) and imports (I) of external organic matter exceed exports (E); thus $E/I < 1$. Such streams are strongly heterotrophic and dependent on imported production of organic matter. Larger rivers less influenced by a canopy of riparian vegetation or streams from desert regions have greater primary production within the river channel. As a result the P/R ratios can be >1, at least at certain times of the year, during which time exports downstream can exceed imports of organic matter from external sources.

The RCC has been modified variously during the past two decades and functions well in a general sense among a variety of river ecosystems. Many of these modifications will be treated in greater detail in subsequent topical chapters. The predictability effectiveness of the RCC varies with changes in physical factors, particularly current velocity, over time. Biological communities have adapted to long-term physical conditions; when human interference alters these properties, particularly over short periods, predictability decreases.

As indicated earlier, materials are displaced downstream, and as a result nutrients are also utilized, stored, released and displaced downstream as they cycle (cf. Chaps. 12 and 13). Recycling of materials within and between water and sediments of streams is controlled by many interacting abiotic and biotic factors, all superimposed on the overall pattern deter-

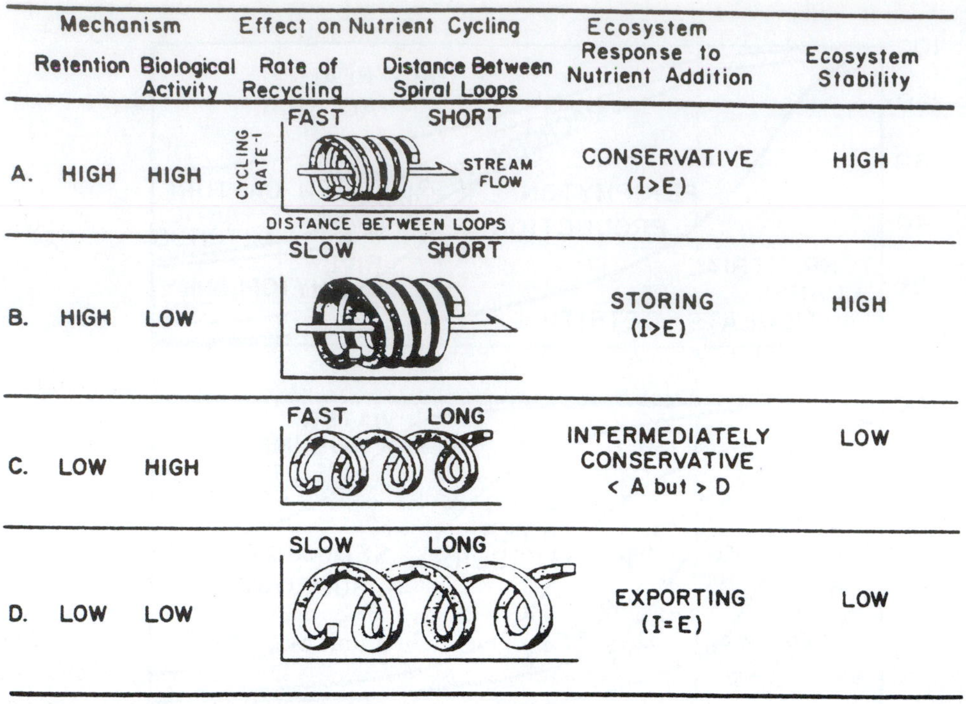
Mechanism
Effect on Nutrient Cycling
Ecosystem Response to Nutrient Addition
Ecosystem Stability
Retention
Biological Activity
Rate of Recycling
Distance Between Spiral Loops
A. HIGH HIGH
FAST SHORT
CYCLING RATE⁻¹
STREAM FLOW
DISTANCE BETWEEN LOOPS
CONSERVATIVE (I>E)
HIGH
B. HIGH LOW
SLOW SHORT
STORING (I>E)
HIGH
C. LOW HIGH
FAST LONG
INTERMEDIATELY CONSERVATIVE < A but > D
LOW
D. LOW LOW
SLOW LONG
EXPORTING (I=E)
LOW

FIGURE 8-10 Hypothetical effects of interactions between distance of downstream movement (velocity × time) and measures of biological activity, such as bacterial metabolism, on organic matter or nutrient cycling in streams of low order (1–4). I = import; E = export. The spiraling rate of recycling reflects biological activity, in which the smaller the diameter of the loop, the faster the rate. Distance between spiral loops represents the extent of downstream displacement of a cycle by the flow of water. The effect of flow can be offset by retention mechanisms so that the higher the retention, the shorter the stream distance between cycles. The quantity of material being cycled is represented by the thickness of the spiraling line. (Modified from Minshall *et al.*, 1983.)

mined largely by flow. The length (openness) of the nutrient cycling along the stream is determined largely by flow and physical retention structures and is modified (size and tightness of loops) by rates of biotic activity in each reach of the stream (Fig. 8-10). The vegetation of the land–water interface along streams greatly influences and modifies productivity within streams and rivers. This vegetation functions in regulation of runoff, nutrient and dissolved and particulate organic matter loadings, as well as the chemical composition of organic matter entering the rivers ecosystems both in dissolved and particulate forms.

B. Diversity

The number of species in a community (species richness, S) increases with the complexity of food webs and with the extent of niche overlap or species packing (i.e., the number of species-niche hypervolumes that a given habitat can contain), especially in the lower trophic levels of food chains. Evaluations of species diversity within a community are complicated by dominant species and an abundance of rare species (Pielou, 1977; Sugihara, 1980). Various indices of diversity have been used, the most common of which account for both relative abundance as well as the number of species. The Shannon diversity index (H') is based on information theory (Shannon and Weaver, 1949):

$$H' = -\sum_{i=1}^{s} p_i \log_2 p_i \tag{8-9}$$

in units per individual per unit volume or area, where p_i is estimated from n_i/N as the proportion of the total population of individuals (N) belonging to the ith species (n_i) and using logarithms to the base 2. In an ecological context, H' measures the diversity in a many-species community.

VIII. PRODUCTIVITY

In any analysis of ecosystem or community productivity, it is important that the values obtained be comparable among different ecosystems, among different community components of the same system, or among different responses of populations to environmental dynamics. The conceptual framework behind productivity historically has been characterized by numerous definitions largely based on agriculture and economics. Perhaps more so than in any other field, in aquatic ecology the questions of productivity have been addressed in detail, and arrays of often incongruous terms and definitions have been proposed. These definitions, many of which are unnecessarily complex, have caused much confusion that has led to ambiguous thinking and expression.

The basic terminology and definitions of the concepts of productivity have been discussed in detail by MacFadyen (1948, 1950), Elster (1954a), Balogh (1958), Davis (1963), Westlake (1963, 1965a), Wetzel (1983, 1995), Benke (1993), and Lucas (1996). Much of the confusion emanates from early concepts that considered productivity as the maximum growth and development of organisms under optimal conditions (Thienemann, 1931); that is, the potential production of organisms or organic matter per unit volume or surface area per unit time (Dussart, 1966). Although the potential of organisms to produce and increase towards infinity may be a useful conceptual framework, in the real world environmental constraints regulate these increases. Optimal conditions for an organism, population, community, or ecosystem can, at best, only be approximated after extensive investigation. Even when the detailed physiological optima for an organism are determined, their applicability to theoretical maximum growth under the dynamics of a nearly infinite number of competitive interactions in natural systems is relegated to the abstract. Such information is much more useful in the interpretation of abiotic and biotic environmental growth controls under *in situ* natural or disturbed conditions.

It is much more meaningful to define the term ***production*** in relation to realized or actual production of organisms, a functional group of organisms in a community, or an ecosystem. Changes in production are related to time and the dynamics of environmental regulatory parameters. However, to generate separate terms for productivity under arbitrarily defined "natural conditions" and under perturbed conditions, as has been variously done, has introduced anthropocentricity and unnecessary complexity. The following definitions of terms represent common usage, and all are in basic agreement with the long history of theoretical discussions (cf. Westlake, 1963; 1965a; Benke, 1993; Lucas, 1996).

A. Terminology and Definitions

Standing crop (also referred to as *standing stock*) is the weight of organic material that can be sampled or harvested at any one time from a given area. Standing crop does not necessarily include the whole population, because certain species or inaccessible parts of the sampled species may be omitted by the sampling procedure. The term derives from agricultural usage of the word *crop,* which is the total weight of organic material removed from a given area over a period of time in the course of normal harvesting practice (usually maximum above-ground foliage or below-ground rooting tissues). The crop is a much less variable measurement than the standing crop, which depends on the time of measurement. For example, a wheat crop is the annual maximum standing crop of above-ground foliage, an alfalfa crop is often the annual total of several standing crops of plant tops, and a sugar cane crop may be a standing crop of tops after nearly two years. Underground organs can constitute a large portion of the total plant mass, and there may be great differences in the standing crop according to the time of sampling.

1. ***Biomass*** is the mass or weight of all living material in a unit area at a given instant in time and is often calculated as the product of mean individual weight (W) and density (N). Biomass has the dimensions of ***mass per unit area or volume*** (e.g., ash-free dry mass m^{-2}). Complications have resulted from the wide usage of standing crop in limnology. When applied to plankton, *standing crop* (or stock) is synonymous with *biomass.* Ambiguities arise when *standing crop,* which refers only to the upper above-ground portions, is applied to aquatic macrophytes; *biomass* includes the entire plant. Biomass evaluations are essential in any analyses of an aquatic plant population or of production dynamics. Because of such inherent differences between the terms, *the measurement of standing crop and use of the term should be abandoned in limnology and only biomass should be used.* If, for some reason, for example, in herbivory, the foliage above ground is specifically of interest, it should be labeled as such: *above-ground biomass.*
2. ***Yield*** is the crop expressed as a rate.
3. Life in ecosystems depends on the flow of energy through the system. ***Production is a flow or flux of mass or energy over time; its dimensions are mass area^{-1} time^{-1} (e.g., g m^{-2} yr^{-1}).*** *Production* thus refers to the increase in biomass of new organic material formed over a period and includes any losses attributable to respiration, excretion, secretion, injury, death, or grazing. Thus, primary production is the quantity of new organic matter created by photosynthesis or the stored energy this material represents. If a photosynthetic organism also uses organic substrates (i.e., is mixotrophic) to supplement photosynthetic metabolism, this additional energy flow is secondary production, even though new organic material may be produced by transformation.

Daily production, weekly production, and annual production are all flows differing only in their specific time units. Natural systems have so many factors causing rapid, frequent, and irregular changes in the instantaneous rates that only average rates can be determined in normal study. It is imperative, as we will stress repeatedly, that the evaluations be done with consideration of the generation times of the organisms under consideration. Weekly evaluations of the productivity of bacteria, for example, are of limited meaning because much of their population dynamics will be excluded; a weekly sampling regimen may be adequate for certain animal components.

In some situations, production may be treated as an integrated accumulation of flows for some time interval. Many secondary production methods are based on first estimating such flow accumulations (Benke, 1993). For example, interval or integrated production (IP) = $N\ \Delta W$, where N = mean density and ΔW = mean increase in weight of an individual over the interval (Table 8-1). Dividing IP by the interval length in days yields daily production (P_d, a flow for the interval), and adding all the IP values for the year gives annual production (P, also a flow). The IP value is an estimate of cohort production of little value in itself, except to serve as an intermediate step in estimating the true flow as annual production. *The annual time period is the only meaningful interval in comparative quantitative analyses of material and energy fluxes at population, community, and ecosystem levels* (Wetzel, 1995).

Losses of production at any ecosystem level occur as a result of nonpredatory losses (respiratory generation of CO_2 and heat, excretion and secretion of dissolved organic materials, and death or injury) and predatory losses (grazing by herbivores or by carnivores). ***Gross production*** (sometimes termed *real production*) refers to the observed change in biomass, plus all predatory and nonpredatory losses, divided by the time interval. Thus, *gross primary production* is the rate of production of new organic matter, or fixation of energy, including that subsequently used and lost during that time interval. ***Net production*** (occasionally termed *apparent production*) is the gross accumulation or production of new organic matter, or stored energy, less losses, divided by the time interval. Sometimes only respiratory losses are subtracted, particularly among evaluations of plant productivity in which losses from processes other than respiration are small. However, all

TABLE 8-1 Terminology and Units Used in Secondary Production Evaluations[a]

Symbol	*Definition*	*Units*	*Type of unit*	*Comments*
W	Individual weight	mg	Mass	Individual weight of an animal at a given time (W_t); often presented as the mean ($\overline{W}$) or difference (ΔW) between sampling dates
N	Density	No. m^{-2}	Storage	Density at a given time (N_t); often presented as the mean ($\overline{N}$) or difference (ΔN) between two sampling dates
B	Biomass	$g \cdot m^{-2}$	Storage	Biomass at a given time (B_t); often presented as a mean over period for which production is calculated (e.g., annual or cohort)
P	Annual production	$g \cdot m^{-2} \cdot yr^{-1}$	Flow	A flow (flux) with time units of per year
P_d	Daily production	$g \cdot m^{-2} \cdot d^{-1}$	Flow	A flow (flux) with time units of per day calculated as $P_d = gB$
IP	Interval production	$g \cdot m^{-2}$	Accumulation (integration)	Integration of production over an arbitrary time period (Δt) (e.g., $IP = gB\ \Delta t$ or $\overline{N}\ \Delta W$); $P = \Sigma IP$ for a year and $P_d = IP/\Delta t$
IP_c	Cohort production	$g \cdot m^{-2}$	Accumulation (integration)	Integration of production over developmental time of cohort (CPI—see below); usually equals annual P when ≤ 1 gen/yr
P/B	Annual P/B	yr^{-1}	Rate	Fraction of biomass which flows per year. Can vary from 0.1 to >200 for aquatic invertebrates
P_d/B	Daily P/B	d^{-1}	Rate	Fraction of biomass that flows per day. Equals daily growth rate, g (see below)
IP_c/B	Cohort P/B	Unitless	Ratio	A true ratio (not a rate) with values usually ranging from 2 to 8
CPI	Cohort production interval	d, yr	Time interval	Time interval from hatching to final size class or emergence
G	Instantaneous biomass growth rate	Unitless	Log of a ratio	$\ln(W_{t+\Delta t}/W_t)$—not a true rate until divided by Δt (see below)
g	Daily instantaneous biomass growth rate	d^{-1}	Rate	Generally computed as G divided by Δt in days

[a] Modified from Benke (1993).

losses should be considered in a true evaluation of net production.

B. Evaluation of Biomass and Production

Numerous parameters have been used in evaluations of the biomass and production of aquatic organisms, including enumeration, volume, wet (fresh) weight, dry weight, organic weight, content of carbon, pigments, energy as heat on combustion, and ATP, and the rates of metabolism of oxygen and carbon dioxide. Often these criteria have been used uncritically, which makes comparability and interpretation difficult (cf. reviews of Lund and Talling, 1957; Strickland, 1960; Westlake, 1965b; Vollenweider, 1969b; Edmondson and Winberg, 1971; Stein, 1973; Wetzel and Likens, 1991; Benke, 1993; among others).

1. Enumeration and Biovolume

Enumeration of organisms per unit volume of water or area is a commonly used method to evaluate microorganisms. Although numbers afford great advantages in permitting qualitative differentiation between species and differentiation of organisms from detrital particles, numbers do not give a true evaluation of biomass because organisms differ greatly in size (cf. Table 8-2). The abundance of each species is better expressed as the biovolume of all individuals (numbers times average cell volume determined from mean dimensions of the cells) per unit volume of water or area of substratum. Among larger organisms, such as zooplankton, benthic fauna, and fish, biomass can be determined using allometric relationships specific to the organisms under study. Volume is a poor measure of aquatic macrophytes because of great variability of internal gas spaces (cf. Chap. 18); thus the relationship between volume and biomass can be extremely variable.

2. Weight (Mass)

Fresh weight, the weight of the organism without any adherent water is, with appropriate precautions, essentially equivalent to wet weight. Owing to the highly variable water content of nearly all organisms, however, wet weight should be avoided. If elaborate precautions are taken, wet-weight analyses can be converted to dry weight on a species basis from a particular environment. Dry weight is variable at temperatures below 100°C; hence, dry weight at 105°C is the international standard. If the very small percentage (usually much less than 1%) of volatile organic constituents lost at 105°C is germane to the analyses, lyophilization (freeze-drying) is the preferred method.

Although dry weight is employed widely in production analyses, organic (ash-free) dry weight, reasonably determined in most situations by the loss in dry weight after ignition at 550°C, is the preferred measure of biomass for larger organisms. The difficulty of separating bacteria, algae, and other small microorganisms from detrital particulate organic matter restricts the application of ash-free dry-weight measurements to larger organisms.

3. Cellular Constituents

Perhaps the most satisfactory means of measuring the biomass of photosynthetic organisms is to oxidize the organic plant material back to carbon dioxide, from which it originated in photosynthetic reduction. The organic carbon content of plants is one of the least variable constituents and falls nearly without exception within the range of 40–60% of ash-free dry weight. The average carbon content among algae is 53 ± 5%, and among aquatic macrophytes, 47% of ash-free dry weight. Exclusion of extraneous organic detritus is not practical at present, and analyses of particulate organic carbon in pelagic samples (cf. Chap. 23) include a very high percentage (usually ≫50%) of nonliving organic material. Algal carbon content usually is estimated from average species content and is extrapolated to natural populations by biovolume measurements of the populations, since the carbon-to-volume relationships are allometric (Mullin *et al.,* 1966; Wetzel and Likens, 2000).

A number of other cellular constituents have been variously employed to estimate changes in population biomass. Except for carbon, however, measurement of biomass by other elements is so complicated, owing to the extreme variability of cellular composition in response to environmental changes, that their use is limited to specific physiological analyses. Although pigment content also varies appreciably with environmental parameters, being able to correct accurately for pigment degradation products in order to measure only the functional pigment content of plants (separate from particulate detritus) permits approximate analyses of composite population dynamics among algae (Wetzel and Likens, 2000).

Several conversion factors have appeared in the literature that allow biomass estimates of one cellular component to be made from another. Even under the most favorable conditions, such factors must be used with utmost caution; under most conditions their use cannot be justified and is best not attempted.

4. Production

Rates of production have been estimated by numerous techniques, most of which are discussed in ensuing treatments of specific groups of organisms. Among aquatic macrophytes, and to a certain extent

TABLE 8-2 Calculated Mean Volumes of Representative Species of Freshwater Plankton Organisms (in μm^3)[a]

Classification	*Volume* (μm^3)
Cyanophyta	
Anabaena flos-aquae (col.)	80,000
Aphanocapsa delicatissima	4
Aphanothece clathrata	10
Aphanothece nidulans	5
Chroococcus limneticus (col.)	400
Chroococcus turgidus (col.)	1000
Coelosphaerium naegelianum (col.)	15,000
Dactylococcopsis smithii (col.)	1500
Gloeocapsa rupestris	18
Gomphosphaeria lacustris (col.)	2000
Merismopedia tenuissima	8
Microcystis flos-aquae	50
Microcystis aeruginosa (col.)	100,000
Oscillatoria limnetica (1 mm)	17,500
Oscillatoria rubescens (1 mm)	30,000
Synechococcus aeruginosus	350
Chlorophyta	
Ankistrodesmus falcatus	250
Chlamydomonas subcompleta	250
Botryococcus braunii (col.)	10,000
Chlorella vulgaris	200
Closterium aciculare	4000
Cosmarium phaseolus	3000
Cosmarium reniforme	30,000
Gloeococcus shroeteri (col.)	5000
Pandorina morum (col.)	4000
Oocystis solitaria	400
Scenedesmus quadricauda	1000
Staurastrum paradoxum	20,000
Tetraedron minimum	40
Ulothrix zonata (1 mm)	6000
Cryptophyta	
Chroomonas nordstedtii	35
Cryptomonas erosa (Michigan)	1000
(Europe)	2500
Cryptomonas ovata	2500
Rhodomonas lacustris (Michigan)	175
(Europe)	200
Rhodomonas minuta	200
Chrysophyta	
Chromulina pyriformis	50
Dinobryon borgei	1500
Dinobryon divergens	800
Dinobryon sociale	800
Mallomonas caudata	12,000
Mallomonas urnaformis	1200
Rhizochrysis limnetica	1200
Uroglena americana (col.)	90,000
Bacillariophyceae	
Amphiprora ornata	17,650
Asterionella formosa (Michigan)	350
(Europe)	700
Cyclotella bodanica	10,000
Cyclotella comensis	400
Cymatopleura solea	80,000
Diatoma vulgare	4350
Fragilaria crotonensis (1 mm)	200,000
Nitzschia gracilis	240
Aulacoseira[b] granulata (1 mm)	60,000
Aulacoseira[b] islandica (1 mm)	80,000
Stephanodiscus astraea	2000
Stephanodiscus hantzscbii var. pusillus	200
Stephanodiscus niagarae	5000
Synedra acus	250
Synedra acus angustissima	1000
Synedra capitata	950
Synedra delicatissima	300
Synedra ulna	50
Tabellaria fenestrata (Michigan)	3000
(Europe)	4000
Pyrrophyta	
Ceratium hirundinella	4000
Gymnodinium fuscum	10,000
Gymnodinium helveticum	20,000
Gymnodinium ordinatum	400
Peridinium cinctum	40,000
Peridinium willei	40,000
Euglenophyta	
Trachelomonas hispida	4200
Trachelomonas volvocina	1800

[a] After Nauwerck (1963), Findenegg (unpublished in Vollenweider *et al.*, 1969) and Wetzel and Allen (unpublished). Numerous values for tropical and other species are given in Lewis (1977) and Hillebrand *et al.* (1999).

[b] Formerly *Melosira*.

attached algae, primary production can be estimated from changes in biomass over time (Chap. 18). Estimates of production rates by planktonic microflora from changes in biomass are much more difficult (Vollenweider, 1969b). A temporal set of biomass measurements results in minimal estimates or underestimations of net production because of losses attributable to grazing, transport, sedimentation, death, and decomposition. Rarely are these parameters sufficiently evaluated to permit analyses of production rates from changes in biomass; exceptions include the investigations by Lund (1949, 1950, 1954), Grim (1952), and Crumpton and Wetzel (1982), among others.

The primary production of phytoplankton has received an extraordinary amount of attention in limnology and has been measured in great detail in a number of aquatic ecosystems. The reasons for the abundance of information on rates of primary production are manifold.

First, phytoplanktonic production represents a major synthesis of organic matter of aquatic systems and can be summarized in the universal equation:

$$6CO_2 + 12H_2O \xrightarrow[\text{pigment receptor}]{\text{light}} C_6H_{12}O_6 + 6H_2O + 6O_2$$

This equation is a gross oversimplification of the complex Calvin–Benson metabolic pathway of photosynthesis of the basic redox reaction:

$$CO_2 + 2AH_2 \xrightarrow{h \cdot v} (HCOH) + 2A + H_2O,$$

in which AH_2 represents a hydrogen donor. Normally, the hydrogen donor is water but, as discussed later, types of donors can include an array of reduced sulfur (e.g., H_2S) or organic carbon compounds among autotrophic and certain other bacteria. In large lakes, phytoplanktonic production can represents a major input of new organic matter and potential energy that drives the ecosystem.

Secondly, as is often the case in ecology, the development of reasonably accurate techniques for the measurement of *in situ* rates of community production has led to their wide application, unfortunately often without any sound rationale as to why the measurements were being made. Only now, after many years of experimentation, is there an adequate understanding of the methodology. The appeal of these techniques lies in being able to measure directly rates of metabolism *in situ;* most other evaluations of production are forced to use indirect methods. A similar plethora of *in situ* measurements of heterotrophic bacterial production is emerging with the employment of direct estimates of rates of synthesis of DNA and other proteins by use of specifically radiolabelled precursors of proteins.

In practice, changes in oxygen production or rates of carbon uptake are usually measured on isolated samples of the natural communities that are incubated for brief periods at the points of collection or under simulated natural conditions aboard ship (cf. Wetzel and Likens, 2000). Certain environmental factors (e.g., temperature and light) are simulated closely; other factors, such as turbulence, nutrient replenishment, and grazing, can differ to varying degrees in the isolated samples. Alternatively, productivity estimates can be made from measurements of changes in oxygen, pH, carbon, or conductivity over short intervals in the natural environment on nonisolated communities. This approach, in addition to possessing numerous complications, yields composite estimates of community microbial metabolism rather than that specifically of the algae or submersed macrophytes.

The light- and dark-bottle techniques for estimating primary production have received wide application. In the oxygen method, samples of phytoplankton populations are incubated at the depths from which they were collected in clear and opaqued bottles. The initial concentration of dissolved oxygen (c_1) can be expected to decline to a lower value (c_2) by respiration in the darkened bottles and increase to a higher concentration (c_3) in the clear bottles, according to the difference between photosynthetic production and respiratory consumption. The difference ($c_1 - c_2$) represents the respiratory activity per unit volume over the time interval of incubation. The difference ($c_3 - c_1$) is equal to the net photosynthetic activity, and the sum $(c_3 - c_1) + (c_1 - c_2) = (c_3 - c_2)$ corresponds to the gross photosynthetic activity. Numerous assumptions are made in the method that can alter the photosynthetic measurements appreciably; for example, respiration rates are not necessarily the same in light and dark, since photorespiration clearly occurs in algae, other processes such as photo-oxidative consumption utilize oxygen separately from apparent respiratory uptake, nonphotolysis of water by bacterial photosynthesis, et cetera. Under many circumstances, these errors are small, but the technique can be considered only as a reasonable estimate. It is probable, however, that these and analytical errors in determination of oxygen concentrations are appreciably less than sampling errors in the analyses of heterogeneous plankton populations in the lake. Large portions of recent methodological works are devoted to detailed discussions of this and the following ^{14}C techniques (see especially Strickland, 1960; Vinberg, 1960; Vollenweider *et al.*, 1969b; Strickland and Parsons, 1972; Westlake *et al.*, 1980; Wetzel and Likens, 2000).

The incorporation of ^{14}C tracer into the organic matter of phytoplankton during photosynthesis has been used as a sensitive measure of the rate of primary production. If the total CO_2 content of the experimental water is known, and if a known amount of $^{14}CO_2$ is added to the water, then determination of the content of labeled carbon in the phytoplankton after incubation permits estimation of the total amount of carbon assimilated. Numerous methodological and physiological problems confront application of the ^{14}C light-and-dark technique; however, with care, most technical problems can be overcome and errors evaluated, for example, respiratory losses of CO_2 and secretion rates of soluble organic products of photosynthesis. Rates of respiration are difficult to evaluate directly with this technique. In many situations, the ^{14}C method yields a measure close to net photosynthetic rates. Comparison of the oxygen and ^{14}C methods under optimal conditions shows close agreement, with a photosynthetic quotient ($PQ = \Delta O_2 / -\Delta CO_2$, by volume) somewhat greater than unity (Fogg, 1963). The photosynthetic quotient varies from near unity when carbohydrates are the principal photosynthetic products, to as high as 3.0 during synthesis of lipids. Assuming a photosynthetic quotient of 1.2 and a statistical significance level of $p < .05$, the smallest amount of photosynthesis that the oxygen light-and-dark technique can measure under ideal conditions is about 10 mg cm^{-3} hr^{-1} (Strickland, 1960). The limit of sensitivity of the ^{14}C method is some 3 orders of magnitude lower, on the order of 0.01 mg cm^{-3} hr^{-1} (Wetzel and Likens, 2000).

Secondary production in fresh waters by invertebrates and vertebrates is much more difficult to estimate accurately than is primary production. Trophic relationships are complex and often change during the life cycle of a species or from one ecosystem to another. The size of animals varies greatly from immature to adult stages, and the diet can also vary greatly throughout the life history. Because most animals are mobile, they actively distribute themselves in response to environmental stimuli, which results in heterogeneous distributions and makes accurate sampling more difficult.

Production measurements of animals are based on estimation of numbers, biomass, and growth rates. The sampling accuracy of changes in population size decreases from zooplankton to benthic organisms to fish. In contrast, growth rates are easier to evaluate in many temperate fishes and in long-lived invertebrates than they are in small, rapidly reproducing zooplankton, other invertebrates, and protists. The basic methods of estimating production of each of these main animal groups will be discussed in subsequent chapters.

When evaluating the population dynamics and production of any animal, estimating food incorporation or ingestion and the utilization of ingested food is imperative. Assimilation means the absorption of food from the digestive system, and the efficiency of assimilation refers to the percentage fraction of ingested food that is digested and absorbed into the body. Assimilation efficiency is not constant and varies greatly with the food quality and rates of food ingestion. Measurements of assimilation are based on the simple relation:

$$\text{Assimilation} = \text{ingestion} - \text{egestion},$$

or

$$\text{Assimilation} = \text{growth} + \text{respiration}$$

Although these relationships are simple, their accurate measurement among natural populations of zooplankton and larger organisms is problematic. The methods employed are discussed critically in Edmondson and Winberg (1971), Winberg (1971), Downing and Rigler (1984), Wetzel and Likens (2000), and Benke (1993). Although ingestion can be measured *in situ* with reasonable sophistication, accurate measurements of egestion rates are often beyond the capacity of contemporary technology. *Growth* is used in a general sense in that it includes the production of eggs or young, much in the same way that respiration includes excretion and other losses as well as normal biochemical respiration.

Assimilation and its efficiency can be approximated with the first of the preceding equations by measuring changes in the radioactivity of the animal after allowing it to feed first on radioactive food and then on nonradioactive food (cf. Sorokin, 1968; Saunders, 1969; Edmondson and Winberg, 1971). Growth and respiration can be estimated by the second equation by measuring the feeding rate (e.g., in calories per time), the growth rate, and respiration. Respiration measured as oxygen consumption can be converted to energy units if the respiratory quotient (CO_2 produced/O_2 consumed) is known.

C. Production/Biomass (P/B) Ratios

Biomass (B) is a measurement of mass for a population present at one point in time in units of mass per unit area (e.g., g m^{-2}) or volume and is therefore a temporary storage of mass or energy. Production (P) is a flow or flux of mass (or energy) per area per time (e.g., g m^{-2} yr^{-1}). The ratio (P/B) of production (P) divided by mean biomass (B) is a rate with units of inverse time (e.g., 1/year). P/B is essentially a weighted mean value of biomass growth rates of all individuals in a population (Benke, 1993). The P/B ratios can be used to estimate the turnover rate of organisms and for an entire trophic level to give a general index of the rate of energy flow relative to the biomass. Determination of P/B values is useful in making comparisons among different trophic levels, as well as among similar trophic levels under different environmental conditions.

Average annual P/B ratios generally decrease with increasing trophic level (Table 8-3) and range over 2

TABLE 8-3 Mean and Range of *P/B* Ratios among Trophic Groups of Freshwater Ecosystems[a]

	Mean	*Range*
Bacteria	141.0	73–237
Phytoplankton	113.0	9–359
Herbivorous Zooplankton	15.9	0.5–44.0
Carnivorous Zooplankton	11.6	1.5–30.4
Herbivorous Benthic Invertebrates	3.7	0.6≥200
Carnivorous Benthic Invertebrates	4.8	1.0–80

[a] After data of Saunders *et al.* (1980), Brylinsky (1980), and Benke (1993).

orders of magnitude (Brylinsky, 1980; Benke, 1993). As would be expected, (a) annual production is higher in short-lived species than in long-lived species, and (b) annual production of a species decreases proportionally as its biomass increases. Therefore, smaller organisms tend to have greater *P/B* values, and *P/B* ratios decrease with increasing organism size and biomass. Production per unit biomass generally increases with decreasing latitude of the aquatic ecosystem since the active growing season is longer, which allows increased numbers of generations per growing season at lower latitudes. Organisms of oligotrophic lakes tend to have lower *P/B* values than those in eutrophic lakes (Brylinsky, 1980; Saunders *et al.*, 1980).

IX. SUMMARY

1. Aquatic ecosystems are open and require a continual input of energy in the form of organic matter.
2. Aquatic ecosystems consist of entire drainage basins. The geomorphology of the land, vegetation and soil composition, and biota, including humans, determine the characteristics of rivers and lakes within the drainage basin.
3. The nutrient and organic-matter content of drainage water from the catchment area is modified in each of the terrestrial, stream, and wetland–littoral components, as well as in the lake or reservoir per se.
4. Productivity is generally low to intermediate in the terrestrial components, highest in the wetland interface region between land and water, and lowest in the open-water portion of the lake.
5. Stream and river ecosystems are distinguished from other inland ecosystems by
   a. Unidirectinal flow
   b. Linear form that occupies little of the total drainage basin area
   c. Unstable and constantly changing channel substrata and morphology
   d. Having most of biotic metabolism supported by organic matter originating from external allochthonous sources that is imported to the stream or river
   e. High spatial heterogeneity that changes rapidly and frequently
   f. High variability and individuality of physical, chemical, and biological characteristics
6. The lake is separated into the open-water *pelagic zone* and the *littoral zone,* the latter consisting of the bottom of the lake basin colonized by macrovegetation. The sediments free of vegetation that lie below the pelagic zone are referred to as the *profundal zone* (Fig. 8-2). The *littoriprofundal zone* is the transitional area of the sediments between the littoral and profundal zones occupied by scattered benthic algae.
7. Various terms have been applied to groups of organisms living within these zones. *Plankton* are small organisms with no or limited powers of locomotion that are suspended in the water; they are subject to dispersal by turbulence and other water movements. Both the small plant plankton, the *phytoplankton,* and animal plankton, the *zooplankton,* are usually denser than water and sink by gravity to lower depths. Organisms with relatively good swimming powers of locomotion are termed *nekton.*
   a. Bacteria and algae growing attached to substrata are collectively called *periphyton* and have been further named in relation to the type of substrata (sediments, rock, plant, animal, sand) upon which they grow (Fig. 8-3).
   b. *Benthos* refers to nonplanktonic animals associated with substrata at the sediment–water interface.
   c. Specialized organisms adapted to living at the air–water interface are called *pleuston.* The pleuston is dominated by microflora collectively termed *neuston.*
8. *A population* is a defined assemblage of individuals of one species. The growth rate of a population is characterized by both its intrinsic capacity to grow and reproduce and the limitations imposed upon that growth potential by the environment and competition of others in the population.
   a. The population growth of many small organisms can be accurately described with the logistic growth equation. As a growing population approaches its environmental carrying capacity, reproduction usually declines because resources become limiting. Under idealized conditions, when a population attains its carrying capacity, reproductive rates decline to zero.

b. Survivorship curves permit an analysis of the survival and reproductive performance of a population at different time intervals.

9. A group of interacting populations forms a *community*.

10. The success of organisms is governed by individual physiologic and behavioral limits and by physical and biotic environmental parameters acting upon them. Each of the regulating parameters forms a dimension; the *niche* refers to an *n*-dimensional resource space, or hypervolume. Each species occupies only a portion of its niche hypervolume.
    a. Species niches can overlap, and overlap in resource utilization can lead to *competition*. The logistic growth equation can be extended to include intra- and interspecific competition and predation and allow more realistic mathematical descriptions of population growth.
    b. In multispecies systems, better differential growth of one species will result in exclusion or displacement of another. Coexistence of competing species can occur where intraspecific competitive effects are greater than interspecific effects, through selective predation, or in cases where the environment changes faster than the time required for competitive exclusion.

11. In biological communities, functionally similar organisms can be grouped into *trophic levels* based on similarities in patterns of food production and consumption. Energy is transferred and nutrients are cycled within an overall ecosystem trophic structure.
    a. The productivity of each trophic level (Λ_n) is the rate at which energy enters the trophic level (λ_n) from the next lower level (Λ_{n-1}).
    b. Since organisms expend considerable energy for maintenance, and because death of an organism routes energy and nutrients into the detrital pool, only a portion of the energy of one trophic level is available for transfer and use by higher trophic levels. Available energy decreases progressively at higher trophic levels, so that rarely can more than five or six trophic levels be supported.
    c. The efficiency of energy transfer from one level to the next is low (5–15%) and often decreases as Λ_n increases.
    d. Life-cycle omnivory, where a species uses different resources during different parts of its life cycle, causes consumer–resource dynamics to be coupled in very complex pathways. These feeding heterogeneities decrease the value of the simple trophic level concept.

12. Living organisms constitute only a very small portion of the total organic matter of ecosystems. Most organic matter is nonliving and is collectively called *detritus*.
    a. Detritus consists of all dead particulate and dissolved organic matter. Dissolved organic matter is about 10 times more abundant than particulate organic matter.
    b. Much of the newly synthesized organic matter of photosynthesis is not consumed by animals, but instead enters the detrital pool and is decomposed.

13. Geomorphological and hydrological characteristics of streams form a longitudinally changing gradient within the drainage basin upon which biological communities of different types become adapted.
    a. This template, referred to as the river continuum concept (RCC), changes in a moderately predictable manner from the headwaters to the river mouth.
    b. As materials are displaced downstream, nutrients and other materials cycle as they are utilized, stored, released, and moved downstream. The rates and distance of nutrient cycling in a spiraling pattern change with different biotic and physical conditions.

14. Biomass is the mass (weight) of living material in a unit area at a given instant of time and has the dimensions of mass per unit area or volume.

15. *Production* is a flow or flux of mass or energy over time and has the dimensions of mass per area formed over a period of time and includes any losses from respiration, excretion, secretion, injury, death, and grazing.
    a. Estimates of primary production by photosynthesis can be obtained directly by following changes in oxygen production or rates of inorganic carbon assimilation.
    b. Secondary productivity by invertebrates and vertebrates is difficult to estimate because of complex trophic relationships, changes in diet during the organism's life history, and animal mobility. Productivity measurements of animals are based on changes in numbers, biomass, and growth rates.
    c. The annual time period is the only meaningful interval in comparative quantitative analyses of material and energy fluxes at population, community, and ecosystem levels.
    d. The ratio of production to biomass (P/B ratio) is a weighted mean value of biomass growth rates of all individuals in a population. The P/B ratio estimates the turnover rates of energy flow relative to biomass; P/B values generally decrease with increasing trophic level.

9 OXYGEN

I. THE OXYGEN CONTENT OF INLAND WATERS

Oxygen is the most fundamental parameter of lakes and streams, aside from water itself. Oxygen dissolved in water is obviously essential to the metabolism of all aerobic aquatic organisms. As a result, the solubility and particularly the dynamics of oxygen distribution in inland waters are basic to the understanding of the distribution, behavior, and growth of aquatic organisms.

The rates of supply of dissolved oxygen from the atmosphere and from photosynthetic inputs, and the hydro-mechanical distribution of oxygen, are counterbalanced by consumptive metabolism by biota and nonbiotic chemical reactions. The rate of oxygen utilization in relation to photosynthesis permits an approximate evaluation of the metabolism of a section of stream or of the lake as a whole.

The resulting distribution of oxygen affects strongly the solubility of many inorganic nutrients. Changes in nutrient availability are governed by shifts from aerobic to anaerobic environments in regions of lakes and of streams. The changes in distribution of nutrients result in rapid growth of many organisms capable of taking advantage of this altered nutrient availability. Population responses may be temporary and transient. If, however, long-term changes in oxygen-regulated nutrient availability are sustained, the productivity of the entire lake or stream can be radically altered. Although oxygen content and dynamics are discussed here largely as a separate entity, its integration with all facets of biotic metabolism will become more apparent in subsequent discussions.

II. SOLUBILITY OF OXYGEN IN WATER

Air contains about 20.95% oxygen by volume, and the remainder is nitrogen, except for a very small percentage of other gases. Because oxygen is more soluble in water than is nitrogen, the amount of oxygen dissolved in water equilibrated with air is greater than nitrogen. The solubility of oxygen is affected nonlinearly by temperature and increases considerably in cold water. The general relationship of oxygen solubility as a function of temperature is given in Table 9-1, although these values are subject to minor modifications as analytical assays improve (extensively discussed in Ohle, 1952; Mortimer, 1956, 1975, 1981; Hutchinson, 1957; Weiss, 1970; and Benson and Krause, 1980). More detailed data are given in the references cited.

The solubility of gases in water is affected by pressure as well as by temperature. Therefore, the

TABLE 9-1 Solubility of Oxygen in Pure Water in Relation to Temperature in Equilibrium with Air at Standard Pressure Saturated with Water Vapor[a]

Equilibrium Temperature (t) (°C)	*Oxygen* (C^*) (*mg liter*$^{-1}$)	*Oxygen* ($C^†$) (μ*gat kg*$^{-1}$)
0	14.621	913.9
1	14.216	888.5
2	13.829	864.4
3	13.460	841.3
4	13.107	819.2
5	12.770	798.2
6	12.447	778.0
7	12.139	758.7
8	11.843	740.2
9	11.559	722.5
10	11.288	705.5
11	11.027	689.2
12	10.777	673.6
13	10.537	658.6
14	10.306	644.1
15	10.084	630.3
16	9.870	616.9
17	9.665	604.1
18	9.467	591.7
19	9.276	579.8
20	9.092	568.3
21	8.915	557.2
22	8.743	546.5
23	8.578	536.1
24	8.418	526.1
25	8.263	516.4
26	8.113	507.1
27	7.968	498.0
28	7.827	489.2
29	7.691	480.7
30	7.558	472.4
31	7.430	464.4
32	7.305	456.6
33	7.183	449.0
34	7.065	441.6
35	6.949	434.4
36	6.837	427.3
37	6.727	420.5
38	6.620	413.7
39	6.515	407.2
40	6.412	400.8

[a] From Mortimer (1981) after data of Benson and Krause (1980). Unit standard concentrations (C^* by volume in mg liter^{-1} or mg cm^{-3} and $C^†$ by weight in μgat kg^{-1}) of oxygen (1 mole = 31.9988 g) in pure water in equilibrium with air at standard pressure (1 atm = 760 mm Hg = 101.325 Pa) saturated with water vapor and with the dry air mole fraction for oxygen taken as 0.20946. To convert mg to cm^3 (ideal gas S.T.P.) or to μgat, multiply mg by 0.70046 or 62.502, respectively. To convert gat to mol, divide by 2. To convert C^* to $C^†$, at temperature $(t)°$ and with the accuracy of this table, multiply C^* by the density of pure water at $t°$.

equilibrium of oxygen of the atmosphere with the oxygen concentration in the water depends on the atmospheric partial pressure of oxygen. Hence, the elevation of the inland water body also influences the concentrations of oxygen. Saturation is usually considered in relation to the pressure at the surface of the lake or river. The percentage saturation of oxygen in water has been treated extensively by Ricker (1934) and Mortimer (1956, 1975, 1981), both of whom have prepared nomograms to aid in such computations. The latter, from Mortimer, is based upon recent data and is recommended for use (cf. Wetzel and Likens, 2000).

The amount of a gas that will remain in dissolved phase is influenced by the atmospheric pressure to which the lake is exposed, by meteorological conditions, and also by hydrostatic pressure exerted by the stratum of water overlying a particular depth. The amount of gas that can be held in water by the combined atmospheric and hydrostatic pressures at a particular depth is called the *absolute saturation* (Ricker, 1934). The actual pressure, P_z, in atmospheres at a given depth is equal to that at the surface, P_0, + 0.0967 times the depth, z, in meters, or

$$P_z = P_0 + 0.0967z$$

The importance of this relationship lies in the amount of pressure required to prevent the formation of bubbles, which rise and escape to the surface. Such bubbles are a composite of many gases, the most common of which are oxygen derived from photosynthesis and methane resulting from anaerobic decomposition under certain conditions. The degree of supersaturation of oxygen necessary for bubble growth increases with depth and hydrostatic pressure. In still water, bubble growth may occur as a result of oxygen tension at depths greater than about 1 m (Ramsey, 1962a). At greater depth, an unusually high degree of supersaturation is required, as well as an absence of turbulence or internal waves that may induce a depth displacement and change in pressure. Very small bubbles can be stabilized by an adsorbed organic film. Such bubbles, on rising to the surface, either rupture or accumulate on the surface as foam (Ramsey, 1962b). Such small bubbles that are coated with adsorbed organic films can also be transported to the surface by turbulent water movements in streams and accumulate as foam on the surface waters of streams. Depending upon turbulence patterns in lakes, oxygen produced at depths greater than 1–4 m can remain dissolved because of the hydrostatic pressure. Oxygen can accumulate to several hundred percent supersaturation relative to the pressure at the surface of the lake and still be below absolute saturation for this gas at depth.

Salinity also reduces the solubility of oxygen in water and must be considered in analyses of oxygen dissolved in inland saline and brackish waters. Gas influx rates are reduced markedly in saline waters by as much as an order of magnitude to water as saline as that of the Dead Sea (Gat and Shatkay, 1991). Bubble entrapment in turbulence from wave action becomes a dominant mode of gas transport in very saline waters. Oxygen solubility declines exponentially with increases in salt content and is reduced by about 20% in normal sea water and much more in hypersaline waters as compared to the amount in fresh water. Tables and a nomogram for oxygen saturation in saline water are given in Green and Carritt (1967) and Weiss (1970) and for hypersaline waters in Sherwood *et al.* (1991, 1992).

The methodology for oxygen analyses is treated elsewhere in great detail (e.g., Welch, 1948, Golterman, 1969; Wetzel and Likens, 2000; Standard Methods, APHA, 1998). The mainstay of oxygen analysis is the Winkler method, based on chemical fixation of the dissolved oxygen and colorimetric titration against reagents of known reaction with changing concentration. Numerous compounds, such as iron, nitrates, and organic matter, significant in polluted waters, can interfere and modifications are needed. In inland saline waters, other chemical methods must be used (Walker et al., 1970; Ellis and Kanamori, 1973). Recent advances in electrode technology now allow reasonably accurate direct measures of oxygen, particularly with microelectrodes, with which nearly instantaneous, continuous measurements are possible (e.g., Revsbech and Jørgensen, 1986).

III. DISTRIBUTION OF DISSOLVED OXYGEN IN RUNNING WATERS

The oxygen content within small, turbulent streams in approximately at, or somewhat above, saturation. Concentrations decline during summer months, as equilibrium solubility declines with increasing temperatures. However, as described later for lake ecosystems, the oxygen content of running waters is modified markedly by chemical and biological processes. Among larger rivers, marked variations in oxygen concentrations occur both spatially and temporally and are often coupled directly to discharge and loadings of organic matter. For example, in backwater areas of reduced current velocities and turbulence, residence times of water are longer and allow for greater alterations of oxygen content from either photosynthetic enrichment or respiratory reductions. Oxygen consumption can be large from direct chemical reactions with dissolved organic compounds of natural or human-related sources or from aerobic respiration of microbial degradation of these compounds. Oxygen demand increases in sites and during times of increased loading of organic matter. For example, during periods of increased leaf fall into heavily canopied streams, increased dissolved organic matter leaching from the leaves can enhance chemical and microbial utilization of dissolved oxygen. Dissolved organic matter leached from soils following precipitation events, or from marsh or swamp areas or loaded from organic sewage sources can similarly cause increased microbially mediated consumption of oxygen (Schneller, 1955; Hynes, 1960; Slack, 1964). In very productive wetland flood plains, as in the Pantanal wetland of the Paraguay River, high rates of organic matter decomposition can lead to high bacterial production of methane (see Chap. 21) (Hamilton et al., 1995, 1997). Bacterial methane oxidation can result in dissolved oxygen reduction or even depletion in backwater areas.

Ground water is often largely or completely devoid of oxygen as a result of chemical oxidations, bacterial respiration, and organic matter as water passes through soils toward stream channels. Groundwater influents are often totally deoxygenated. As anoxic or depleted water passes through hyporheic sediments, reaeration occurs from the stream water. Often, however, only a few millimeters of the upper sediment are well oxygenated.

Photosynthetic activities of algae and macrophytes can increase oxygen concentrations on a diurnal basis, particularly in relatively sluggish streams and rivers (Müller and Weise, 1987; Livingstone, 1991). In nutritionally enriched and productive streams, the magnitude of diurnal oscillations in dissolved oxygen concentrations can increase (Fig. 9-1). Bacterial and plant respiration can be most effective in reducing oxygen concentrations at night. Quantification of the diurnal changes in rates of oxygen production and consumption in a defined section of river has been used as a measure of photosynthetic production (cf. Wetzel and Likens, 2000, and APHA, 1998, for methodology). Oxygen consumption can also occur in lighted reaches of streams and rivers by photochemical oxidation of dissolved organic compounds (cf. review of Brezonik, 1994). Dissolved humic and other organic substances can function as sensitizers of photochemical reactions but also can absorb light and photolyze directly with the consumption of oxygen (e.g., Laane, *et al.*, 1985).

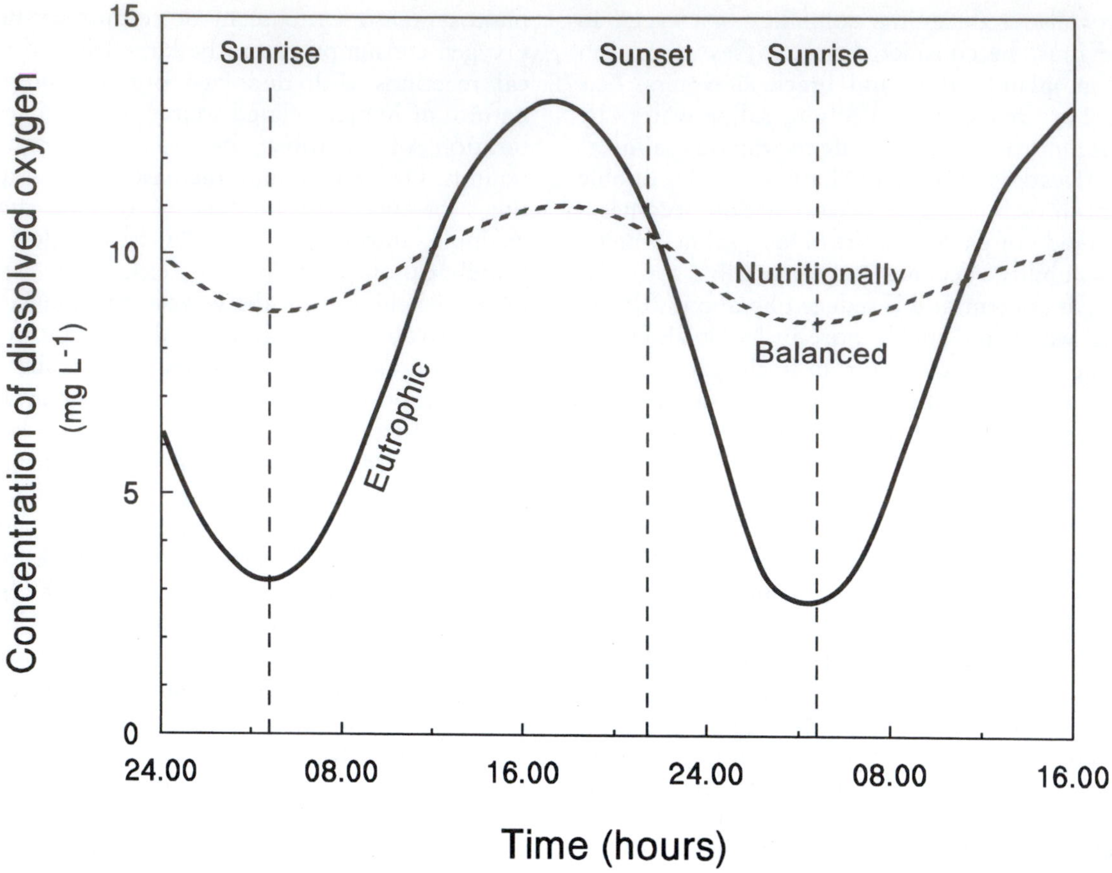

FIGURE 9-1 Diurnal cycle of dissolved oxygen concentrations in a nutritionally balanced and eutrophic stream. (From Walling and Webb, 1994, after data of Gower, 1980.)

IV. DISTRIBUTION OF DISSOLVED OXYGEN IN LAKES

Diffusion of gases into water is a very slow process. To establish an equilibrium between the atmosphere and dissolved oxygen within a reasonable period of time, the water must circulate, as it does during periods of turnover or in the epilimnion of stratified lakes. If the dissolved oxygen concentrations at depth are near saturation, equilibrium at prevailing temperatures and altitudinal pressure is established relatively quickly, usually in a matter of a few days. Very deep lakes, however, require longer periods for complete equilibrium saturation, and saturation may or may not be achieved before thermal stratification effectively terminates circulation for a seasonal interval.

In an idealized lake, the oxygen concentration at spring circulation is at or near 100% saturation, which would be between 12 to 13 mg O_2 liter^{-1} if occurring at about 4°C and at an altitude near sea level (Fig. 9-2). This ideal concentration at 100% saturation is based on physical control of diffusion, mixing, and saturation. Deviations are common, most frequently in the form of a slight supersaturation from photosynthetic activity in which dissolved oxygen is in excess of evasion to the atmosphere. Alternatively, biochemical oxidations can result in undersaturation during conditions where oxygen consumption slightly exceeds inputs from circulation equilibrium mechanisms. The near-surface waters of lakes containing large amounts of dissolved organic matter, such as dystrophic bog lakes or standing waters of wetlands, are often appreciably undersaturated with dissolved oxygen.

A. Orthograde Oxygen Profile

Where the lake is ***oligotrophic*** (low in nutrient inputs with low organic production), oxygen concentrations with depth are regulated largely by physical processes during summer stratification. The oxygen concentration in the circulating epilimnion decreases as the temperatures gradually increase (Fig. 9-2). Accompanying decreasing temperatures in the metalimnion and hypolimnion, the oxygen concentrations increase,

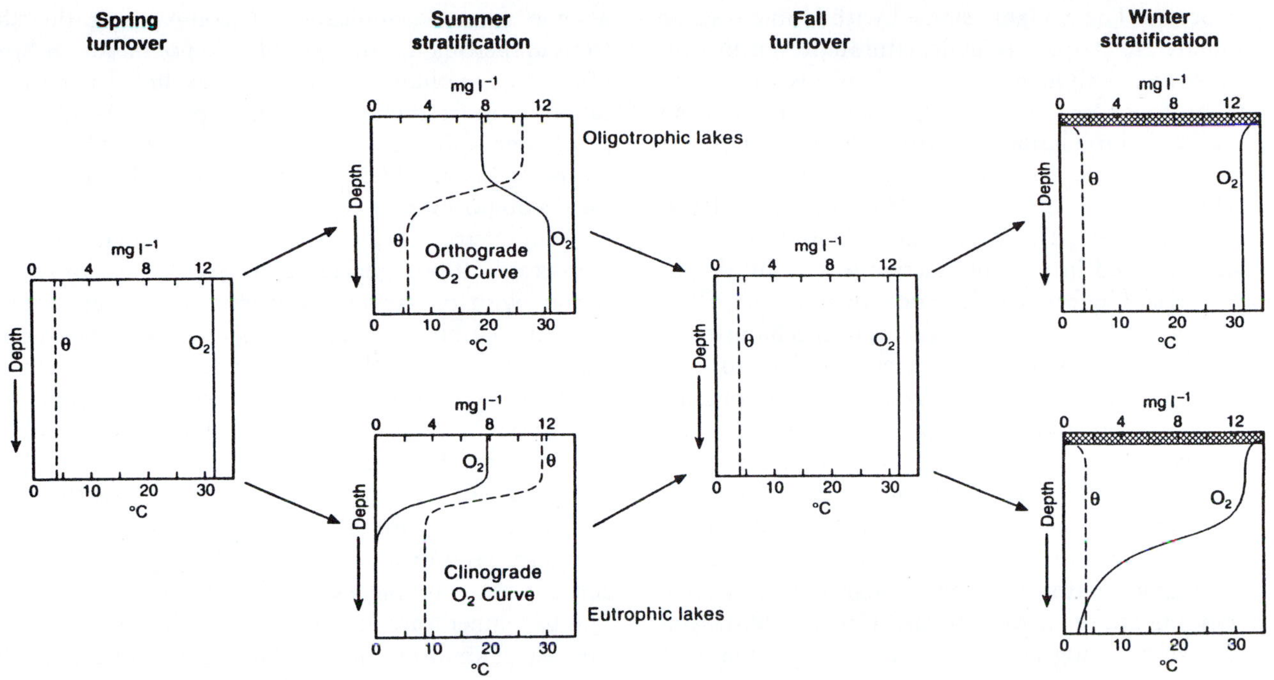

FIGURE 9-2 Idealized vertical distribution of oxygen concentrations and temperature (Θ) during the four main seasonal phases of an oligotrophic and a eutrophic dimictic lake.

that is, they have remained in the hypolimnion essentially at saturation from the period of cold spring turnover just before the summer stratification began. This oxygen profile has been termed (Åberg and Rodhe, 1942) ***orthograde.*** The important characteristic to note is that the percentage of saturation is more or less 100% with increasing depth.

B. Clinograde Oxygen Profile

An idealized orthograde curve is found only in a few extremely unproductive lakes or in moderately oligotrophic lakes during the very early stages of summer stratification. Oxidative processes occur constantly in the hypolimnion, and their intensity is proportional to the amount of organic matter reaching the hypolimnion from the upper zones of the lake. As a result, the oxygen concentrations of the hypolimnion become progressively more reduced and undersaturated. The oxygen content of the hypolimnion of ***eutrophic*** (high in nutrient loading with high organic production) lakes is depleted rapidly by oxidative processes. The resulting profile, in which the hypolimnion is often anaerobic, is termed ***clinograde*** (Fig. 9-2). The hypolimnetic oxygen content of highly eutrophic lakes is often depleted after only a few weeks of summer stratification. The hypolimnion remains anaerobic throughout this stratification period.

Loss of oxygen from the hypolimnion results primarily from biological oxidation of organic matter, both in the water and especially at the sediment–water interface, where bacterial decomposition is greater. Although plant and animal respiration can consume large, often biologically catastrophic, amounts of dissolved oxygen, major consumption from the lake is associated with bacterial decomposition of dissolved and of sedimenting particulate organic matter (Seto *et al.*, 1982). Oxygen consumption in the free water by bacterial respiration is intensive at all depths but in the hypolimnion is rarely offset by the oxygen renewal mechanisms of circulation and photosynthesis that occur in the epilimnion and metalimnion. Oxygen consumption is most intense at the sediment–water interface, where the accumulation of organic matter and bacterial metabolism are greatest. The interface region very rapidly becomes anaerobic during summer stratification, and the hypolimnetic depletion is usually observed first at the lowermost stratum of the hypolimnion. Diffusion of oxygen into this depleted zone from overlying layers occurs slowly. The horizontal distribution of the oxygen profile in the hypolimnion can be modified by vertical turbulence, horizontal translocations, and density currents that move along the basin sediments.

In addition to oxygen consumption by animals, plants, and particularly by bacterial respiration in the open water, chemical oxidation of dissolved organic matter occurs. Lakes highly stained with humic organic compounds are frequently undersaturated even in epilimnetic strata. Although many mechanisms can be involved, at least some of the oxygen uptake results from purely chemical oxidations and from photochemical oxidations induced by ultraviolet light (e.g., Gjessing and Gjerdahl, 1970; Laane *et al.,* 1985; Brezonik, 1994; Wetzel *et al.,* 1995). These chemical oxidation processes are likely masked in very productive waters by the intense bacterial biochemical demands for oxygen.

The relative importance of different mechanisms of hypolimnetic oxygen depletion varies from lake to lake. In large, deep lakes, bacterial respiration of organic matter of phytoplanktonic origin may dominate, and benthic decomposition then will play a minor role. In shallow lakes with relatively large inputs of organic matter from terrestrial, stream, wetland, and littoral sources, decomposition of allochthonously derived dissolved organic matter and benthic particulate organic matter dominate (e.g., Melack and Fisher, 1983). And in highly stained bog lakes that receive large inputs of dissolved humic compounds, chemical oxidations may assume greater significance. Accompanying the shift from an aerobic to an anaerobic hypolimnion, a large, often major volume of the lake is excluded from habitation by most animals and many plants. Another major change is the shift from aerobic to anaerobic bacterial metabolism, which reduces the overall efficiency of the decomposition of organic matter.

Fall overturn begins with the complete loss of summer stratification. During the terminal stages of stratification, with progressive deepening of the epilimnion, the circulation of approximately oxygen-saturated water extends deeper into the hypolimnion. When circulation is complete, oxygen concentrations remain at saturation in accordance with solubilities at existing temperatures (Fig. 9-2).

As ice forms, the exchange of oxygen with the atmosphere ceases for all practical purposes. Based on solubility relationships, one would expect an oxygen concentration profile essentially at saturation in relation to temperature at any given depth. Such would be the case in an ultra-oligotrophic lake where biotic influences are minor. These profiles are very rarely found in

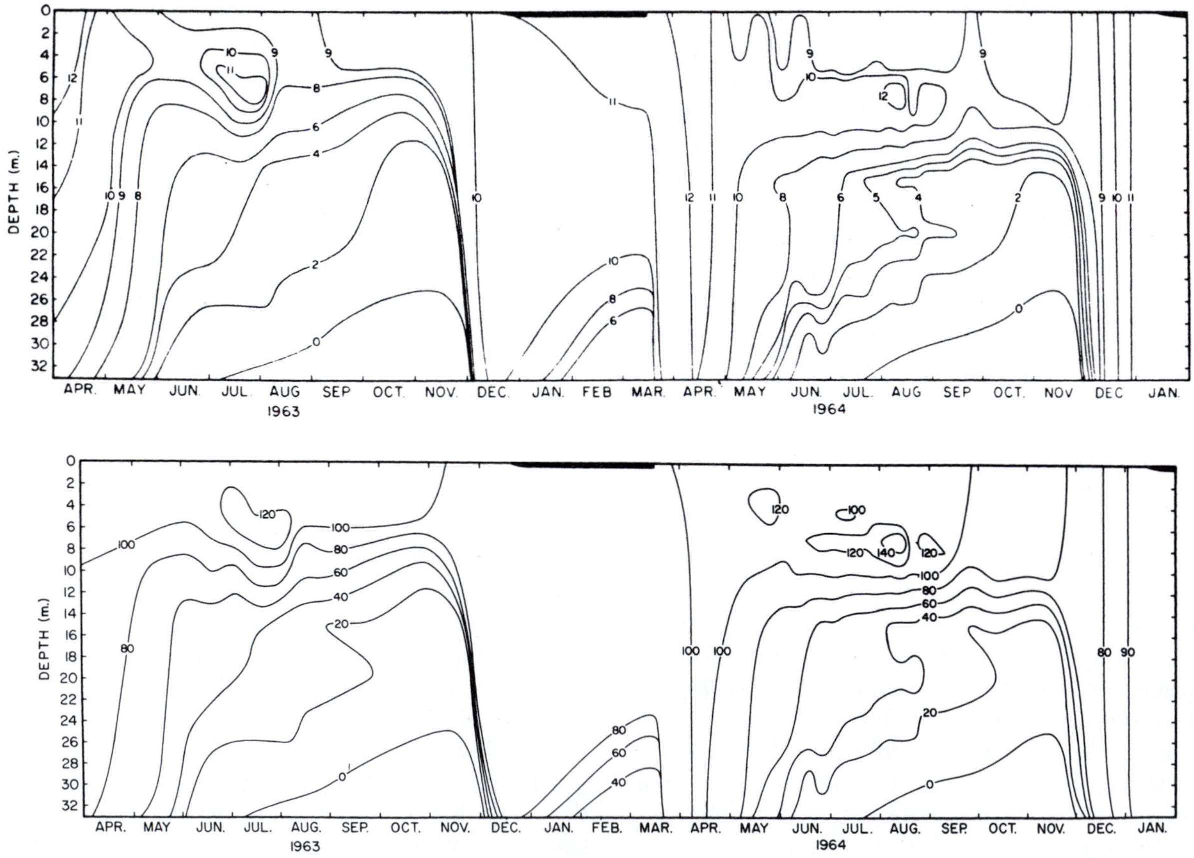

FIGURE 9-3 Depth–time diagram of isopleths of dissolved oxygen concentrations in mg liter^{-1} (*upper*) and percentage oxygen saturation (*lower*), Crooked Lake, Whitley County, northeastern Indiana. Ice cover drawn to scale. (From Wetzel, unpublished data.)

dimictic lakes. Much more frequently observed is a reduction in oxygen concentration with depth, which becomes particularly acute near the sediment.

In eutrophic lakes, photosynthetic production of organic matter, although less than in other seasons, continues throughout the winter and is often vigorous during the later stages of winter ice cover. Light penetration is variable with changing conditions of ice and snow cover, but the photic zone generally is confined to the upper layers. Respiratory utilization and chemical oxidations increase with depth as during summer stratification, although at a lower rate because of the lower temperatures. During winter, the water of the main water mass under ice is commonly colder than 4°C. Water in the littoral areas can be heated slightly through the ice by solar radiation. This warmer, slightly denser water will sink and flow in profile-bound density currents along the sediments to the deeper portions of the basin. Generally, water movement is sufficiently slow so that en route, the oxygen of this water is reduced or depleted as it passes over the sediment–water interface. At the bottom of depressions it displaces water upward and results in a conspicuous depletion at the sediment–water interface, accentuating the decompositional utilization already occurring there.

The stages of the annual cycle just discussed can be illustrated by isopleths of oxygen concentration and percentage saturation for an atypical dimictic lake of moderate productivity (Fig. 9-3). Noteworthy are the high concentrations (near 100% saturation) during the colder periods of spring and fall circulation. Epilimnetic concentrations of oxygen were reduced but, because of reduced solubility in the warmer water, were still at saturation. The high concentrations in the metalimnion were the result of photosynthesis and will be discussed in detail later. The slow, progressive reduction and final depletion of oxygen in the lower hypolimnion is seen in mid- and late summer. The same reduction, but not depletion, is seen under ice cover at the greater depths in Crooked Lake. More extreme changes are seen in the oxygen concentrations of smaller eutrophic Little Crooked Lake (Fig. 9-4), in which the changes were similar to those of Crooked Lake but were greater and more accelerated.

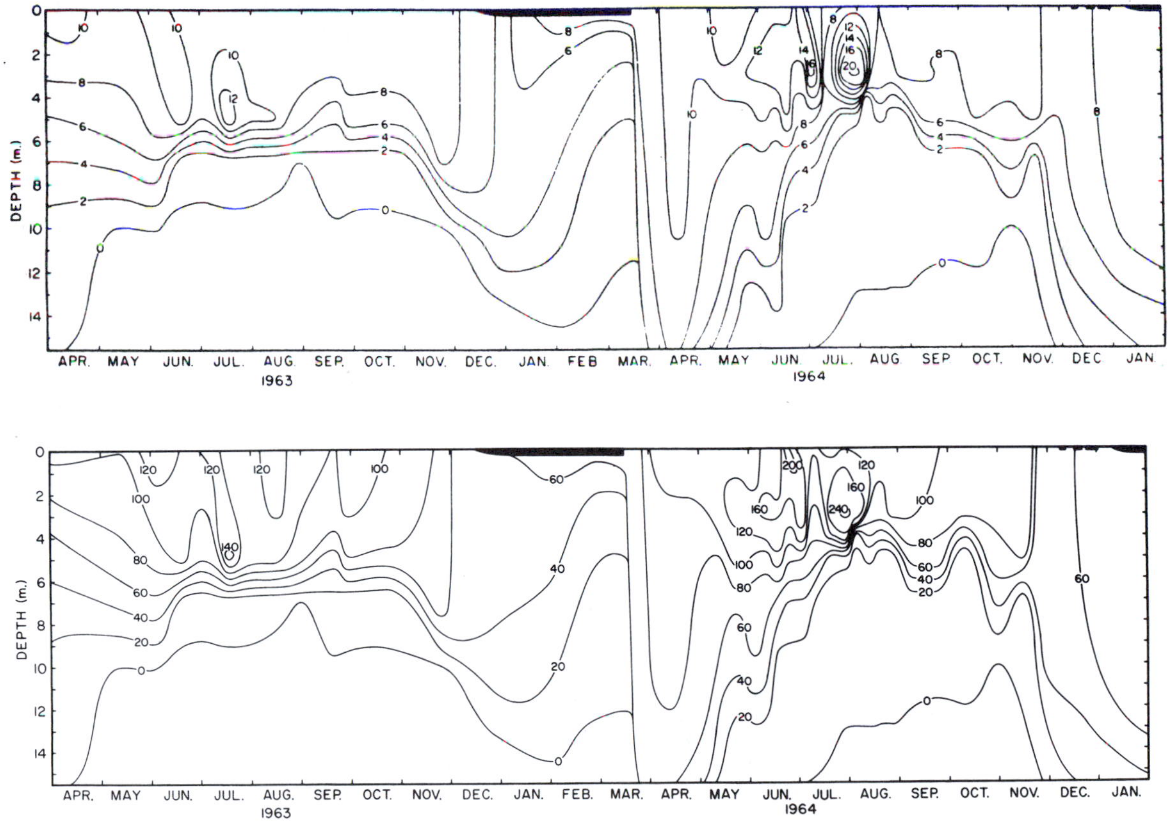

FIGURE 9-4 Depth–time diagram of isopleths of dissolved oxygen concentrations in mg liter^{-1} (*upper*) and percentage oxygen saturation (*lower*), Little Crooked Lake, Whitley County, northeastern Indiana. Ice cover drawn to scale. (From Wetzel, unpublished data.) Isopleth lines appear to merge where gradients are extremely steep; in reality, the lines are separated by small depth–time intervals.

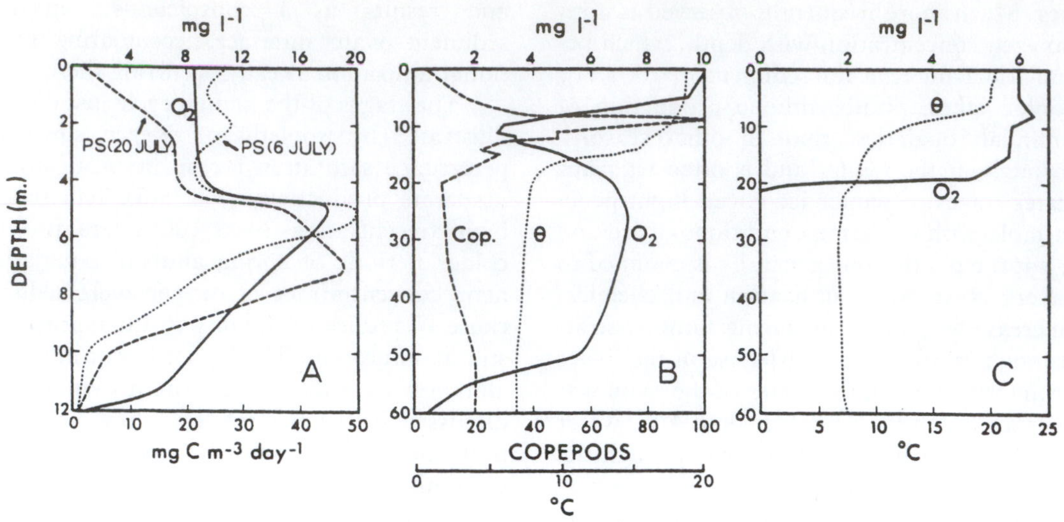

FIGURE 9-5 (*a*) Metalimnetic oxygen maximum, showing a positive heterograde curve, in relation to rates of phytoplanktonic photosynthesis (*PS*), Lawrence Lake, Michigan, 20 July 1971. (From Wetzel, *et al.*, 1972). (*b*) Metalimnetic oxygen minimum, showing a negative heterograde curve, in relation to abundance of copepod microcrustaceans (*Cop.*) and temperature (Θ), Lake Washington, Washington, 18 August 1958. (Drawn from data of Shapiro, 1960.) (*c*) Oxygen concentrations in permanently meromictic Fayetteville Green Lake, New York, 3 September 1935. (Data of Eggleton, 1956.)

V. VARIATIONS IN OXYGEN DISTRIBUTIONS

A. Metalimnetic Oxygen Maxima

Variations in the vertical and horizontal distributions of dissolved oxygen from the general conditions just discussed are often observed. The most common variation is an increase in oxygen in the metalimnion during stratification (Figs. 9-3, 9-4, 9-5, and 9-6); the metalimnetic maximum is referred to as a ***positive heterograde curve*** (Åberg and Rodhe, 1942). As the solubility of the epilimnetic water decreases with increasing summer temperatures, and oxygen consumption in the hypolimnion results in the typical clinograde reduction with depth, the result is an absolute oxygen maximum in the metalimnion, which may be at or above saturation.

Metalimnetic oxygen maxima often are extremely pronounced, with supersaturated values above 200% (Fig. 9-4). Concentrations of nearly 36 mg O_2 liter^{-1} (400% saturation) have been recorded (Birge and Juday, 1911). It is clear that the maxima are nearly always the result of oxygen produced by algal populations that develop more rapidly than they sink out of the lower water strata (Fig. 9-5a). The algae are commonly stenothermal,

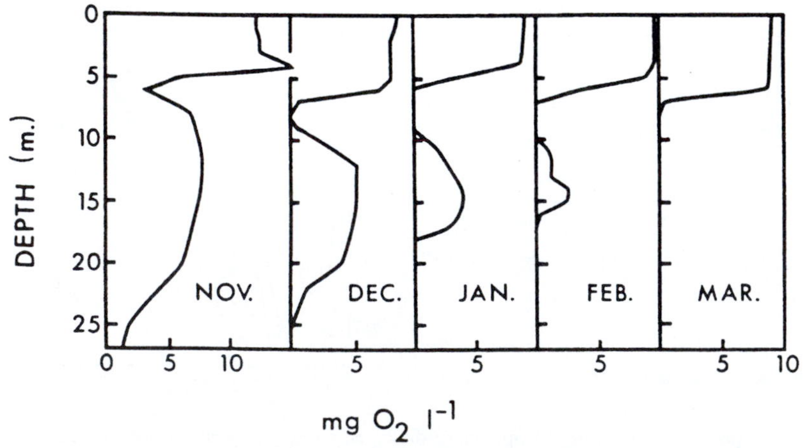

FIGURE 9-6 Depth profiles of dissolved oxygen concentrations during the summer (Southern Hemisphere) in hypereutrophic Lake Johnson, New Zealand, 1970–1971. Similar distributions were observed in the summers of 1969–1970, 1971–1972, and 1974. (Modified from Mitchell and Burns, 1979.)

adapted to growing well at low temperatures and low light intensities but having access to nutrient concentrations that are usually higher in the lower metalimnion than in the epilimnion. Cyanobacteria, especially *Oscillatoria*, are often major contributors to this phenomenon (Eberly, 1959, 1963, 1964; Wetzel, 1966a, 1966b). The depth at which metalimnetic oxygen maxima occur is correlated directly with the transparency of the water (Thienemann, 1928; Yoshimura, 1935). In most lakes metalimnetic maxima are found between 3 to 10 m, but in very clear lakes at depths as great as 50 m.

Extreme metalimnetic oxygen maxima are observed much more frequently in lakes that are strongly stratified. Such lakes commonly have high relative depths,[1] that is, their surface area is small relative to their maximum depth, and they are protected from intense wind action by surrounding topography and vegetation. The average peak in oxygen concentration and relative depth of over 50 lakes that exhibited extreme metalimnetic maxima was 17.2 mg liter^{-1} and 4.30% respectively (Eberly, 1964). Most lakes have a relative depth of <2%.

Metalimnetic oxygen maxima occur in many lakes. When characteristics of morphometry and biota are suitable for their development, the maxima occur consistently from year to year at approximately the same time and depth. For example, in Lawrence Lake, Michigan, the metalimnetic maximum occurred between 4 and 6 m, July through August, for 20 consecutive years of observation (Wetzel, unpublished). The oxygen maxima were superimposed exactly over maxima of photosynthesis by phytoplankton and summer growth maxima of dense beds of rooted aquatic plants on the steep slopes of the littoral zone (Rich *et al.*, 1971). It is likely that water in the littoral zone, enriched with oxygen, dissipates into the metalimnion layer of equal density and contributes to the observed oxygen peak. Ruttner (1963) and Dubay and Simmons (1979) reported a similar situation in small, clear lakes with precipitous slopes, in which the metalimnetic oxygen maximum coincided seasonally with zones of maximum production of submersed macrophytes.

B. Metalimnetic Oxygen Minima

The converse condition, a ***metalimnetic oxygen minimum*** exhibiting a ***negative heterograde curve*** (Fig. 9-5b), is much less frequently observed. Numerous causal mechanisms have been associated with such metalimnetic reductions, and most or all are likely to be in effect simultaneously in situations where such oxygen minima are observed. To single out one mechanism as operational to the exclusion of others (as, e.g., Czeczuga, 1959) can be misleading.

Oxidizable material produced in the epilimnion or brought into the lake from outside the basin continuously sinks. Sinking rates slow when it encounters denser metalimnetic water, which allows more time for decomposition (Birge and Juday, 1911; Thienemann, 1928). Organic matter introduced into reservoirs from storm water can flow, depending on water temperatures and densities, in the metalimnion, as interflow, and contribute to accelerated oxygen reduction in the metalimnion (Nix, 1981). Moreover, decomposition rates are usually higher in the metalimnion, where temperatures are greater than in the colder hypolimnion. As a result, more readily oxidizable organic matter is decomposed at this level, with concomitant consumption of oxygen by bacterial respiration; more resistant organic matter settles slowly into the hypolimnion (Kuznetsov and Karsinken, 1931; Vinberg, 1934; Åberg and Rodhe, 1942). In a review of the subject, Czeczuga (1959) presented some evidence from a lake in which, conversely, decomposition rates were greater in the epilimnion. However, renewal of epilimnetic oxygen and time of residence of sedimenting organic matter in the metalimnion were not clearly delineated. It is clear that a balance between the transparency of the trophogenic zone and the depth of the metalimnion is important in relation to the depth at which photosynthetic maxima occur and in relation to whether oxygen inputs are sufficient to offset decompositional consumption (Ruttner, 1933). It is easy to conceive of a situation in which a pronounced metalimnetic maximum could shift to a minimum within a summer season (Fig. 9-6).

In certain situations, concentrations of massive numbers of zooplanktonic microcrustacea in the metalimnion can contribute to a severe reduction of oxygen. Such respiratory consumption was most likely a major cause of the metalimnetic minima observed in Zürichsee, Switzerland (Minder, 1923), and the same condition can prevail in Lake Washington (Fig. 9-5b), where nonmigrating copepods often develop profusely in late summer (Shapiro, 1960). In the extreme case shown in Figure 9-6, less than 10 percent of the metalimnetic oxygen minimum was accounted for by zooplankton respiration (Mitchell and Burns, 1979). The metalimnetic anoxia was apparently caused mainly by intensive bacterial decomposition.

The basin morphometry of a lake can also contribute to metalimnetic oxygen minima. In cases

[1] Relative depth, or z_r, is an expression of the maximum depth (z_m) as a percentage of the mean diameter and is determined by

$$Z_r = \frac{50 z_m \sqrt{\pi}}{\sqrt{A_0}}$$

where A_0 is the surface area.

where the slope of the basin is gentle and where it coincides with the prevailing metalimnion, a greater area of the sediments with high bacterial utilization of oxygen will be in contact with metalimnetic water than is the case of sediments adjacent to the lower strata. Horizontal mixing and streaming of water from internal water movements is greatest in the metalimnion, where vertical density stability is greatest. As a result, reduction in oxygen content at the sediment–water interface is greater where the slope of the sediments is less. The oxygen reduction in the metalimnion may be sufficient to extend laterally over the entire lake, or, as in the case of Skärshultsjön, Sweden (Alsterberg, 1927), it may occur more strongly in the metalimnion closer to the sediment–metalimnion interface.

Biogenic oxidation of methane by methane-oxidizing bacteria can result in metalimnetic oxygen minima in certain productive lakes (Kuznetsov, 1935, 1970; Ohle, 1958). Methane from anaerobic fermentation in sediments rises from the sediments and is carried to water strata such as the metalimnion, where, with warmer temperatures and greater oxygen, methane oxidation can occur rapidly. In such cases, severe oxygen reduction can occur in a few days if adequate dissolved inorganic nitrogen is available (Rudd *et al.*, 1975, 1976).

C. Distribution in Meromictic Lakes

Extreme clinograde oxygen profiles occur in permanently stratified meromictic lakes (Fig. 9-5c). The permanent monimolimnion receives constant inputs of organic matter or sulfates from saline water, which rapidly deplete any oxygen intrusions.

Temporary (partial) meromixis, discussed in Chap. 6, can lead to highly variable, anomalous oxygen distributions. If a normally dimictic lake is quite productive, a strongly developed clinograde oxygen distribution during both summer and winter stratification would be anticipated, often with anaerobic deeper layers in both seasons. If spring or fall circulation and renewal of oxygen is incomplete before stratification is imposed, the reduced or anaerobic conditions of the lower strata will be carried over to the next stratified period and be further reinforced. Such was clearly the case in Little Crooked Lake in the autumn of 1963 (Fig. 9-4). Mixing was incomplete in Lawrence Lake in the spring of 1968 (Fig. 9-7), which resulted in an accentuated reduction of the hypolimnetic oxygen the following summer and anoxia below 11 m for three months. In normal years with complete spring circulation, as in the following four years (cf. Wetzel *et al.*, 1972), the hypolimnion did not become anaerobic—or only very briefly—just before fall turnover in a half-meter stratum immediately overlying the sediments. In 1973, incomplete mixing occurred again in the spring and resulted in an oxygen distribution pattern similar to that of 1968. Over a 19-year period of study, Lawrence Lake experienced temporary meromixis and enhanced hypolimnetic oxygen reductions every 3–4 years. The chemical nutrient and biotic implications of aerobic and anoxic hypolimnia are very important for the productivity of lakes and will be pursued in detail subsequently.

Oxygen concentrations of many perennially ice-covered lakes of Antarctica are persistently supersaturated. Oxygen carried into the lakes in glacial meltstreams is concentrated as water is removed as ice by ablation and sublimation (Wharton *et al.*, 1986).

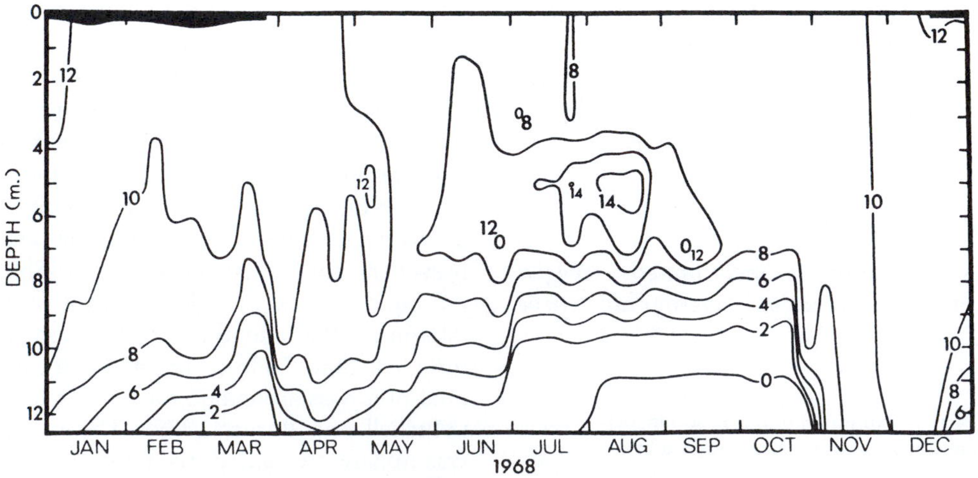

FIGURE 9-7 Depth–time diagram of isopleths of dissolved oxygen concentrations in mg liter^{-1} of Lawrence Lake, Michigan, 1968. Ice cover drawn to scale. (From Wetzel *et al.*, 1972.)

D. Horizontal Variations and Diurnal Cycles

Horizontal spatial variations in the distribution of dissolved oxygen of the open-water pelagic zone of natural lakes are generally small during periods of spring and fall circulation. This situation can change during the productive summer periods and under ice-cover. Littoral aquatic plants often inhabit large areas of more productive lakes, to a depth of 3 to as much as 10 m. Photosynthesis of these macrophytes and of sessile algae growing attached to them and to the sediments can generate large amounts of oxygen. Similarly, large quantities of oxygen are utilized in respiration at night and by littoral bacteria in a zone rich in dissolved and particulate organic matter. The oxygen regime of the littoral zone is usually totally different from that of the open water and experiences marked diurnal cycles (Fig. 9-8). Moreover, the vertical distribution within the vascular plant beds can be highly stratified in accordance with the macrophyte portions that are actively photosynthesizing (often the upper foliage) or those lower parts that are predominately respiring (Fig. 9-8).

The oxygen content of the littoral zone is periodically severely reduced (e.g., Pokorný *et al.,* 1987). Such is the case when large populations of macrophytes (vascular hydrophytes and macroalgae) senesce and die at the end of their growing season. Severe oxygen deficits caused by intense decomposition can persist for several months and may extend from the littoral zone into the open water for some distance (Thomas, 1960). In shallow lakes where submersed hydrophytes grow over the entire basin, decomposition of the macrophytes and attached microflora at the end of the growing season, in late summer when temperatures are still high, can be so intense that the oxygen content of the entire lake may be severely reduced to near anoxia. Such a catastrophic event can result in massive die-offs of many animals; occasionally entire fish populations are lost in such a ***summerkill***.

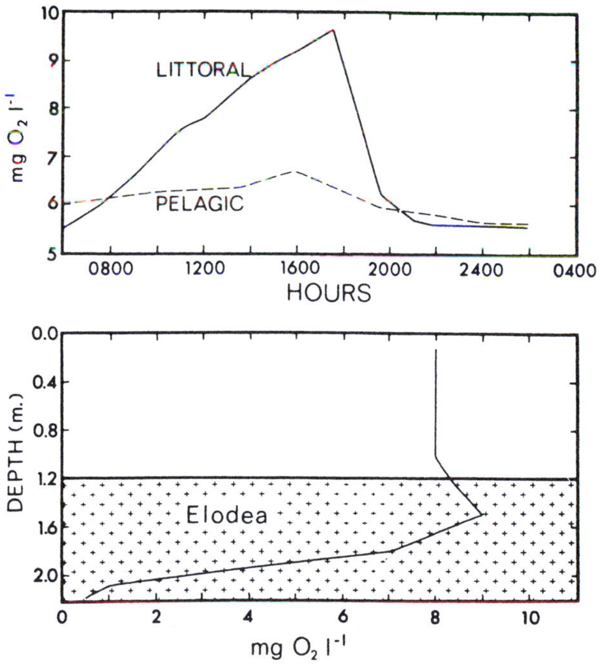

FIGURE 9-8 *Upper*: Changes in dissolved oxygen in the littoral and open-water areas over a diurnal period in eutrophic Winona Lake, Indiana, 9 August 1922. (From data of Scott, 1924.) *Lower*: Vertical stratification of oxygen within the littoral zone of Parvin Lake, Colorado, 9 July 1955, in a luxuriant stand of the submersed macrophyte *Elodea*. (Generated from data of Buscemi, 1958.)

Where lakes are extremely eutrophic, phytoplankton and attached algae often occur in such profusion that they severely attenuate light, and submersed macrophytes are limited to very shallow areas. Under such conditions, diurnal fluctuations in oxygen concentrations of the epilimnion are often just as marked as those of the littoral zone, as shown in Figure 9-8. Changes of 4 to as much as 6 mg O_2 liter^{-1} are seen commonly between early morning darkness and midafternoon (e.g., Lingeman *et al.,* 1975; Pokorný, *et al.,* 1987). Under clear ice cover in such productive waters, oxygen can accumulate and commonly result in diurnal supersaturation. Oxygen analyses in surface waters of productive waters undergoing rapid diurnal changes, when taken only at one time of day, show only one stage of a much more complex situation.

Large horizontal variations in oxygen content are also found in lakes of complex basin morphometry. When a lake consists of many bays or multidepressions, oxygen concentrations are very different in these areas, and each embayment may operate essentially as an individual lake (Welch and Eggleton, 1932; Wetzel, 1966a; Lind, 1987).

In reservoirs, oxygen distribution becomes highly variable horizontally, vertically, and seasonally (Straškraba *et al.,* 1973; and the detailed review of Cole and Hannon, 1990). The complex hydrodynamics of reservoirs are unique in relation to variations in influents, morphometry, draw-down and discharge, and other factors. In addition, seasonal variations in influxes of organic matter from inflowing streams can place varying oxidative demands on the oxygen content of reservoirs.

As a result of the interactions of natural and regulated flow dynamics, temperature, morphology, and community metabolism, a hypolimnetic anoxic zone often develops following the vernal stratification that is common in deep-storage reservoirs. The hypolimnetic anoxic zone develops first in the thalweg of the transition zone during summer stratification and then

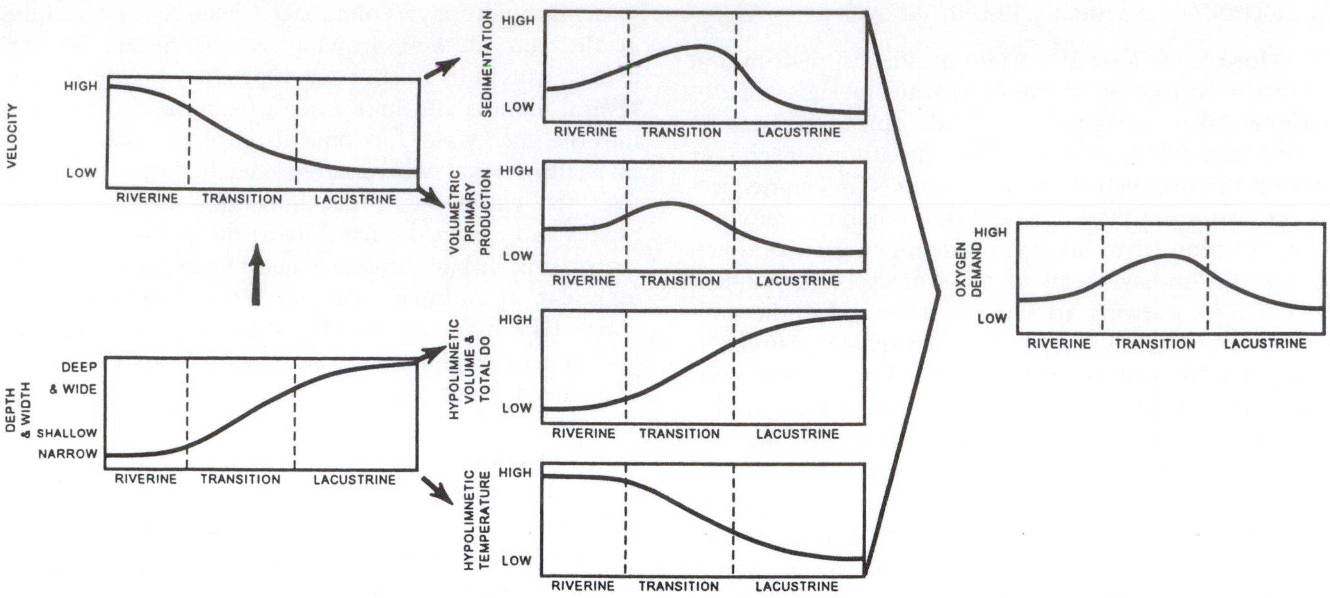

FIGURE 9-9 Relationships of velocity and depth to sedimentation, areal primary production of phytoplankton, hypolimnetic volume and total dissolved oxygen (DO), and hypolimnetic temperature in determining oxygen demand along the length of a reservoir. (Modified from Cole and Hannon, 1990.)

expands horizontally and vertically (Fig. 9-9). The time of onset of hypolimnetic anoxia is regulated largely by meteorological conditions that vary over the year in relation to geographic location. As in natural lakes, the distribution of oxygen and rates of loss are governed by a balance among mixing, inflow intrusions, photosynthesis, and losses by oxidative consumption. The processes are similar among all freshwater ecosystems, but the rates of change in reservoirs are generally much more variable and dynamic than in lakes (Table 9-2).

Horizontal variations in oxygen are particularly conspicuous in the winter under ice cover, especially in the near-surface strata. Fluctuations were noted earlier in response to seasonal changes in photosynthesis (Figs. 9-3, and 9-4). Photosynthesis by algae is strongly suppressed by snow cover on ice, but often accumulated snow cover is very patchy on the surface of windswept ice (Chap. 5, pp. 63ff). The result is variegated photosynthesis and oxygen production below the ice. Heavy snow cover can plunge the lake into virtual darkness; if sustained for several weeks, heavy respiratory demands by decomposition can reduce the oxygen content to very low levels or to anoxia (Greenbank, 1945). The resulting conditions are intolerable to many animals. For example, most fish cannot survive, even at low

TABLE 9-2 Comparison of Dissolved Oxygen Dynamics among Running and Standing Waters[a]

Parameter	*Rivers*	*Reservoirs*	*Lakes*
Dissolved Oxygen	Usually high	Somewhat lower solubilities (higher temperatures); greater horizontal variability with inflow, withdrawal, and POM loading patterns; metalimnetic oxygen minima more common than maxima	Somewhat higher solubilities (lower temperatures); small horizontal variability except in littoral zones; metalimnetic oxygen maxima more common than minima
Oxygen uptake from atmosphere	High	Moderate	Variable to low
Diurnal variations of dissolved oxygen	High, coupled to photosynthesis, respiration, and decomposition	Modest in eutrophic reservoirs	Usually low, can be large in littoral areas

[a] Modified from Ryder and Pesendorfer (1989) and Wetzel (1990a).

temperatures, at less than 2 mg O_2 liter^{-1}. Low oxygen concentrations can lead to selective fish mortality with profound effects on intraspecific population structure (Casselman and Harvey, 1975) and community relationships in subsequent years. Alterations of food-web structure by selective fish mortality in which planktivorous predation is altered can lead to marked differences in zooplanktonic community composition, size structure, and grazing impacts. For example, winterkills eliminated carnivorous fishes of a small eutrophic lake of southern Michigan, which led to domination of the fish community by a small, planktivorous cyprinid minnow (Hall and Ehlinger, 1989). The zooplankton, previously consisting of two large *Daphnia* species, concomitantly shifted to a community of smaller species. Such ***winterkill*** conditions of low winter oxygen concentrations are common in shallow eutrophic lakes and becomes particularly acute at mean depths of <3 m (e.g., Barica and Mathias, 1979).

Empirical models have been developed for shallow eutrophic lakes that allow estimations of the probable risk of rapid oxygen reductions, based on maximum midwinter ammonia–nitrogen and maximum summer phytoplankton development by chlorophyll *a* concentrations, that can lead to summerkill of fishes (Barica, 1984). A model was also prepared to estimate the probability, based on oxygen depletion rates under ice cover, mean depth of the lake, and length of ice cover, of winterkill of fishes. Risk increases greatly in shallow lakes with a mean depth <4 m.

An example of the rapid seasonal changes in oxygen concentrations that can occur can be seen in Green Lake, a small reservoir lake in southern Michigan (Fig. 9-10). In early February, heavy ice and snow cover had severely reduced photosynthesis and the oxygen content of the lake, except in the area immediately adjacent to the small inlet. Three days later, following late winter rains on deteriorating ice, conditions had improved, although horizontal variations in oxygen concentrations still were conspicuous. A week later, oxygen brought in primarily from increased flow of the inlet had virtually obliterated the former distribution.

Methane and hydrogen can be formed in anaerobic sediments of productive lakes in sufficient quantities to bubble up and rise to the surface. Rossolimo (1935) and Kuznetsov (1935) have shown that the methane (CH_4) and hydrogen (H_2) are effectively oxidized by bacteria in water. Under ice cover, this oxidation can cause severe reductions in oxygen content (Fig. 9-11). Generation of these two gases, and oxygen utilization,

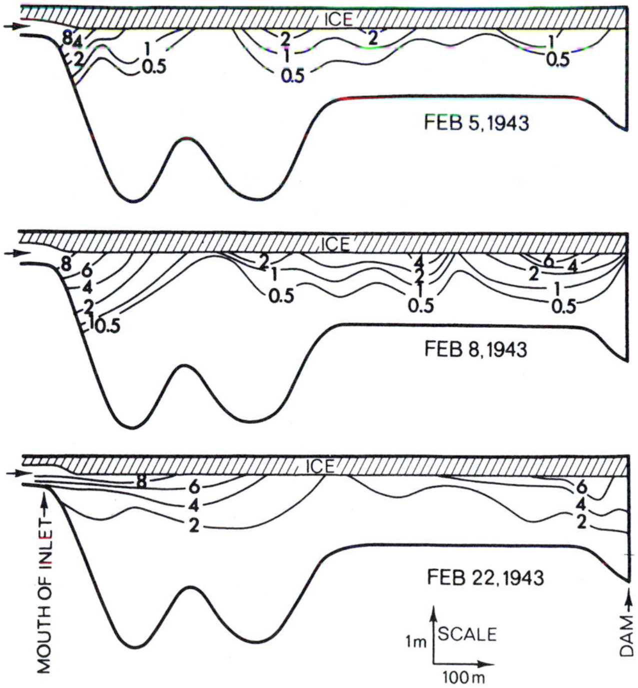

FIGURE 9-10 Horizontal isopleths of oxygen concentrations (mg liter^{-1}) during a two-week period in Green Lake, Michigan, winter 1943. (Modified from Greenbank, 1945.)

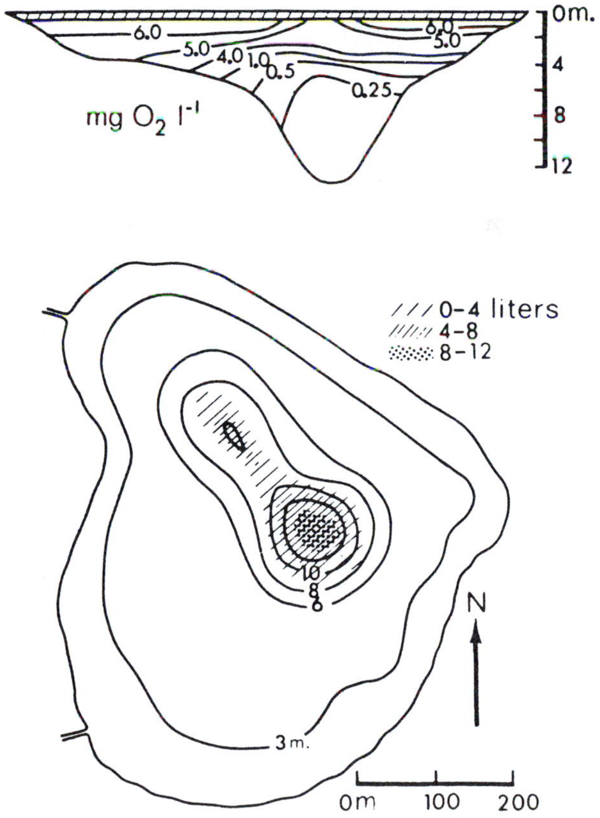

FIGURE 9-11 *Upper*: Isopleths of oxygen concentration in Lake Beloie, Russia, 15 March 1928; *lower*: quantity (liter m^{-2}) of gas, containing 20–24% CH_4, permeating the underside of the thawing ice in late winter. (Modified from Rossolimo, 1935.)

is particularly great over the deeper depressions of the lake basin. The central oxygen depletion is probably accelerated by profile-bound density currents that slowly accumulate in the depression and force overlying water upward. Oxygen concentration in these currents is reduced as they flow along the sediments to the deepest area.

Therefore, the oxygen régime of most dimictic lakes of the temperate zone is governed to a large extent by the quantities of oxidizable matter received during periods of stratification. In lakes that tend toward permanent stratification (either meromictic variants or oligomictic lakes of low elevation in equatorial regions), the oxygen distribution is clinograde and is determined by the duration of the stratification in relation to the inputs of oxidizable organic matter. Lakes containing large quantities of dissolved organic matter, such as bogs and bog lakes, have oxygen concentrations that are appreciably below saturation.

From the numerous examples of oxygen distribution and its control in natural water bodies, the effects of artificial loading of inland waters with pollutional sources of organic matter and other oxidizable materials should be readily apparent. Organic effluents from agricultural activities, sewage, and industry can effectively and rapidly exceed the degradative capacities of lakes and rivers. The resulting effects on aerobic organisms are direct and immediate. Indirect effects on the entire biogeochemical cycling and productivity are more gradual but, as we will see, are more effective in producing generally undesirable conditions.

VI. HYPOLIMNETIC OXYGEN DEPLETION RATES

The rate of loss of oxygen from the hypolimnion during summer stratification increases not only with depth, and hence with decreasing volume of the hypolimnion, but also with time during the stratification period. The difference in amount of oxygen present at the beginning, during, and at the end of stratification below a given depth is referred to as the ***oxygen deficit.*** In its simplest form, the oxygen deficit quantifies a metabolic relationship of the ***trophogenic zone***, the superficial stratum of a lake in which photosynthetic production predominates, with that in the underlying ***tropholytic zone,*** the aphotic deep stratum where decomposition of organic matter predominates. Hence, the amount of organic matter synthesized in the trophogenic zone that rains into and decomposes in the tropholytic zone is reflected in the rate of utilization of hypolimnetic oxygen. Although the dynamics of oxygen utilization are not that simple, the volumetric rate of oxygen consumption in the hypolimnion during summer stratification (oxygen deficit) provides an approximate estimate of the productivity of the lake.

The development of the theory of oxygen deficits as an approximation of lake productivity has a long history since its foundations in the fundamental work of Birge and Juday (1911), Thienemann (1928), and Juday and Birge (1932). Older means of estimating oxygen deficit have serious shortcomings and are only briefly discussed here. All deficit measurements must be used cautiously and only in a general way.

A. Actual, Absolute, and Relative Deficits

The ***actual oxygen deficit*** of the water refers to the difference between the observed oxygen content and the saturation value of a similar quantity of water at its observed temperature, salinity, and atmospheric pressure at the lake surface. A deficiency of this measurement is that it assumes that the water was saturated at the observed temperature during spring turnover and that oxygen intrusions from the trophogenic zone during stratification by currents was negligible. The same criticism can be leveled at the ***absolute oxygen deficit***, the difference between the observed oxygen concentration and the saturation value at 4°C at the pressure of the lake surface (Alsterberg, 1929). The assumption that the oxygen content during spring circulation is at saturation at 4°C is not always so; indeed, more often this is not the case. Therefore, Strøm (1931) proposed and used the ***relative oxygen deficit***, which is the difference between the oxygen content of the hypolimnion during summer stratification and that empirically determined at the end of spring turnover.

Use of the relative oxygen deficit improved estimations of rates of consumption because it took account of different oxygen deficits in layers below the hypolimnetic surface. Thus, consideration was given to the basic property of changing volumes of the strata with depth and to the importance of different hypolimnetic volumes among lakes. Thienemann (1926, 1928) emphasized this point clearly (Fig. 9-12). In comparing two lakes having identical volumes and identical

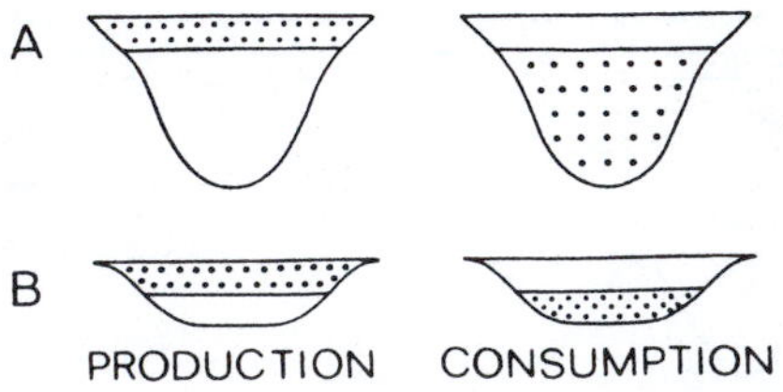

FIGURE 9-12 Diagrammatic representation of two lakes of equal production in the trophogenic zone but of differing volumes of their tropholytic zones, in which oxidative consumption occurs. (Modified from Thienemann, 1926, 1928.)

TABLE 9-3 Sample Computation of the Relative Areal Oxygen Deficit of Gull Lake, Summer 1972[a]

Strata (m)	*Volume (m^3)*	*Mean O_2 (mg liter^{-1})*	*Total O_2 in Strata (metric t)*[b]
Calculations for 7 June 1972:			
15–20	12.910×10^6	9.42	121.612
20–25	5.221×10^6	8.90	46.464
25–30	0.203×10^6	8.68	1.762
30–33.5	0.0428×10^6	8.35	0.358
			170.196
Calculations for 1 August 1972:			76.719 metric t in hypolimnion
Difference:			93.477 *t* or 93.477×10^9 mg

The area of the surface of the hypolimnion at 15 m is $3.4980 \times 10^6 m^2$, or 34.98×10^9 cm^2.

By division, the difference in oxygen or areal oxygen deficit was 2.67 mg cm^{-2} 55 $days^{-1}$ or 0.0486 mg cm^{-2} day^{-1}, or 1.46 mg cm^{-2} $month^{-1}$

[a] Unpublished data of Tague, Lauff, and Wetzel.
[b] 1 mg $liter^{-1}$ = 1000 metric tons km^{-3} = 1 metric *t* $(10^6 m^3)^{-1}$.

trophic status in the trophogenic zones, it would be assumed that the production and influx of organic matter in the trophogenic zone of each would be the same and that similar amounts of oxidizable organic matter would sediment from the trophogenic layer into the tropholytic zones. However the volume of the first lake (*a*) is much larger than that of the second lake (*b*), and the first lake contains a much larger amount of oxygen to be utilized in decomposition. As a result, during a similar interval of time, the oxygen content of the hypolimnion of the shallow lake (*b*) would be reduced more rapidly than the oxygen in the hypolimnion of the deep lake (*a*). One might anticipate a clinograde oxygen curve in the former case and a more or less orthograde curve in the latter, deep lake. Much of the oxygen utilization in the shallow lake may occur after the organic material has settled into the sediments. Turbulence and horizontal currents in the hypolimnion, although small, are usually sufficient to distribute much of the lower deoxygenated water near the sediments into much of the hypolimnetic water. Thus, the relative oxygen deficit, which compensated for differences in hypolimnetic volume with depth, incorporated the intensity of trophogenic production into the consumptive capacity of the hypolimnion or approximate tropholytic zone.

This ***relative areal deficit***, as introduced by Strøm (1931) and modified by Hutchinson (1938, 1957), is the mean oxygen deficit below one cm^2 of hypolimnetic surface and consists of the summation of the individual deficits of a series of layers of decreasing volume with increasing depth. A computation from Gull Lake, southwestern Michigan, illustrates the general approach (Table 9-3). These data consist of measurements of oxygen concentration at several depths; a detailed hydrographic map is essential for accurate estimates of both volume and area. Since the oxygen concentrations did not differ greatly from one depth to the next in Gull Lake at this time, it is possible to find the average concentration in any layer by taking the mean of the concentration at the top of the layer and that at the bottom of the layer. When concentration gradients are steep, simple arithmetic means are not sufficiently accurate; a planimetric integration should then be used (Wetzel and Likens, 1991). This concentration is then multiplied by the volume of the layer to give the total amount of dissolved oxygen in the stratum, and the amounts for each layer of the hypolimnion are added. Thermal and morphometric data show that the surface of the hypolimnion of Gull Lake was 34.98 ha, or 349.8×10^8 cm^2. By division, the difference in oxygen was 2.67 mg cm^{-2} for the 55-day period between 6 June and 1 August, 0.0486 mg cm^{-2} day^{-1}, or 1.46 mg cm^{-2} $month^{-1}$.

Many of the calculations of areal oxygen deficits are from older investigations that make direct comparisons of deficits with various indices of biological productivity difficult. Nonetheless, as would be expected, there is a strong tendency in lakes of moderate depth for the relative areal deficit to increase with increasing productivity. Based on limited data, Hutchinson (1938, 1957) set the ranges among the rather ambiguous general categories of unproductive to productive lakes as: oligotrophic, <0.017 mg O_2 lost cm^{-2} day^{-1}, and eutrophic, >0.033 mg cm^{-2} day^{-1}. Mortimer (1941; Hutchinson, 1957) believes that the limits of 0.025 mg cm^{-2} day^{-1} for the upper limit of oligotrophy and 0.055 mg cm^{-2} day^{-1} for the lower limit of eutrophy are more realistic. These general limits are supported by other criteria for evaluating productivity (e.g., Lasenby, 1975; Burns, 1995).

Oxygen deficits are subject to numerous errors that detract from true representation of an index of trophogenic productivity. Significant inputs of allochthonous particulate and dissolved organic matter can cause large overestimates, while pelagic productivity of moderately to highly stained waters (>10 Pt units) may be underestimated, owing to ultraviolet-light-mediated oxygen consumption by dissolved organic matter (cf. reviews of Leifer, 1988, and Brezonik, 1994). Furthermore, natural ultraviolet photolysis of dissolved humic compounds results in release of numerous small fatty acids that are readily metabolized by bacteria (Wetzel *et al.*, 1995). Although some pelagic and littoral photosynthesis commonly occurs in the upper portion of the hypolimnion in both shallow and deep, clear lakes, this source of error is minimized in larger dimictic lakes with a maximum depth >20 m.

It is further assumed that the organic matter decomposing in the water column and in sediments of the hypolimnion includes most of that produced in the trophogenic zone and that a relatively constant amount sinks into the hypolimnion. Gliwicz (1979) has shown in insightful analyses that the steeper the vertical gradient of water temperature in the metalimnion (greater thermal resistance to mixing), the slower the sinking speed of sedimenting organic particles in the metalimnion. As a result, not only is a larger part of the sedimenting seston prevented from leaving the more turbulent epilimnion and metalimnion but the residence time for nutrients is prolonged in the epilimnion by release from partial decomposition. Consequently, the hypolimnetic relative oxygen deficits are lower in lakes with steeper metalimnetic thermal gradients. Epilimnetic productivity is greater, and respiratory oxidative consumption is largely complete before the organic matter ever reaches the hypolimnion.

Additional critical analyses of areal hypolimnetic oxygen deficits have demonstrated additional relationships to lake parameters (Ohle, 1956; Edmondson, 1966; Lasenby, 1975; Stewart, 1975; Cornett and Rigler, 1979; Charlton, 1980; Stauffer, 1987; Burns, 1995). Areal hypolimnetic oxygen deficits of a number of lakes indicate that (1) the deficit is positively correlated with primary productivity of phytoplanktonic algae; (2) the deficit is inversely proportional to epilimnetic transparency (Secchi disc depth); (3) lakes with higher total phosphorus concentrations have higher oxygen deficits; (4) the deficit is greater in lakes with higher mean summer hypolimnetic temperatures; and (5) oxygen deficits are larger in lakes of greater mean hypolimnetic thickness. The latter correlation, that a lake with a thicker hypolimnion has an oxygen deficit that is greater than a lake with a shallow hypolimnion, is seemingly incongruous with theoretical evaluations. This property may result from differential effects of turbulent oxygen transport across the metalimnion in lakes of varying hypolimnetic depth. Oxygen within the relatively turbulent metalimnion and heat and chemical fluxes into and from the metalimnion increase during windy intervals (Stauffer, 1987). As a result, lakes with shallow hypolimnia can have higher rates of oxygen depletion per unit volume of the hypolimnion but lower rates of oxygen depletion per unit area than do lakes with thicker hypolimnia.

Other parameters influence the rates of hypolimnetic dissolved oxygen depletion (Burns, 1995). Rates of depletion slow when concentrations decline below ca. 2 mg liter^{-1}. Many lakes deoxygenate slowly enough to permit the calculation of several monthly rates before this concentration occurs. An appreciable fraction of the decomposition of both particulate and dissolved organic carbon occurs in the epilimnion and never reaches the hypolimnion (e.g., Mitchell and Burns, 1979). Temperature differences of course affect decomposition rates, and corrections are necessary when comparing deficit results from different lakes. Net photosynthetic oxygen production in the lower metalimnion and upper hypolimnion can offset degradative consumption. River inflows to lakes can transport oxygen and organic matter to the hypolimnion if they underflow the metalimnion after entering the lake (e.g., Gibbs, 1992). Additionally, some of the organic matter settling into the hypolimnion and onto the sediments will not decompose during the stratified period under examination for the deficit calculation.

Oxygen deficit measurements are made because they are relatively easily done from commonly acquired data. However, other methods exist for evaluating lake productivity that are better than hypolimnetic oxygen deficits. Autochthonous productivity can be measured directly, and fluxes of particulate and dissolved organic matter can be evaluated in detail by chemical and biotic assays. Further, the hypolimnetic oxygen deficit method cannot be used effectively in lakes in which the hypolimnion becomes anaerobic during the period of summer stratification. When detailed data on productivity are lacking, however, the oxygen deficit can be informative about the general trophic status of a lake.

Changes in the hypolimnetic oxygen deficit rates over long periods of time can be indicative of overall changes in the productivity of the lake. An example of this is Douglas Lake in the northern part of the southern peninsula of Michigan. Although much is known about the characteristics and biota of this lake, which has been studied for many years, no good estimates of annual productivity exist (Lind, 1978). Analyses of the annual amount of oxygen below the metalimnion showed a progressive, slow decrease between 1911 and 1964 during each year except two (Fig. 9-13). Although considerable scatter in the annual rate of oxygen depletion was found, the rate of change appears to be increasing in recent years (Bazin and Saunders, 1971; Lind, 1978, 1987). The trend reflects an accelerated nutrient input and eutrophication associated with human activity first as a result of deforestation and second as a result of the development of the area for recreational purposes. This pattern has been repeated many times in other lakes, but long-term data are available only rarely. The well-known rapid eutrophication of a much larger lake, Lake Erie, has been followed in a similar way, largely on the basis of losses of hypolimnetic oxygen and benthic fauna (Carr, 1962; Charlton,

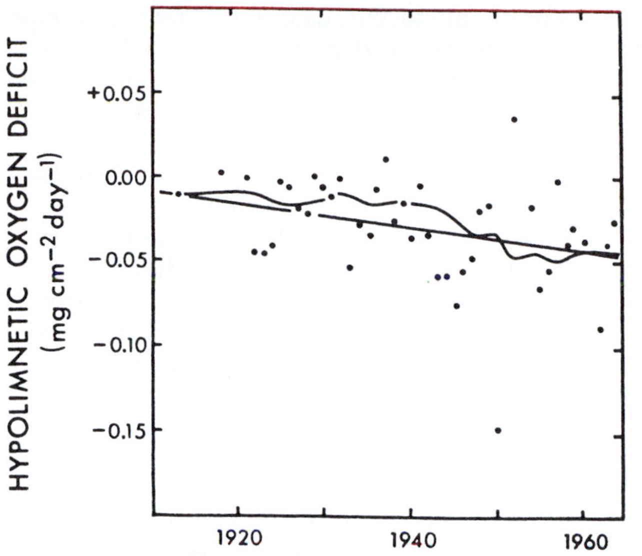

FIGURE 9-13 Change in estimated annual hypolimnetic oxygen deficits over a 53-year period in Douglas Lake, Michigan (negative values in all except two cases indicating a progressively greater rate of hypolimnetic oxygen reduction). The straight line represents the line of best fit by linear regression; the smooth curve represents an exponentially smoothed 1-year forecast at each year. (Modified from data of Bazin and Saunders, 1971.)

1980). Other indices of eutrophication and its reversal will be discussed in other chapters. Hypolimnetic oxygen deficits are but one index and must be used with great caution and careful consideration of differences in hypolimnetic temperatures and thicknesses.

VII. SUMMARY

1. Dissolved oxygen is essential to the respiratory metabolism of most aquatic organisms. The dynamics of oxygen distribution in inland waters are governed by a balance between inputs from the atmosphere and photosynthesis and losses from chemical and biotic oxidations. Oxygen distribution is important for the direct needs of many organisms and affects the solubility and availability of many nutrients and therefore the productivity of aquatic ecosystems.
2. Solubility of oxygen in water decreases as temperature increases. The solubility of oxygen decreases somewhat with lower atmospheric pressures at higher altitudes and increases with greater hydrostatic pressures at depth within lakes. Oxygen solubility decreases exponentially with increases in salt content.
3. Because diffusion of oxygen from the atmosphere into and within water is a relatively slow process, turbulent mixing of water is required for dissolved oxygen to be distributed in equilibrium with that of the atmosphere. Although oxygen content within small, turbulent streams is near saturation, marked variations occur spatially and temporally in larger rivers and are often coupled to variations in discharge and loadings of organic matter. Ground water is often largely or completely devoid of oxygen as a result of chemical and biological oxidations.
4. Distribution of oxygen in the water of thermally stratified lakes is controlled by a combination of solubility conditions, hydrodynamics, inputs from photosynthesis, and losses to chemical and metabolic oxidations.
   a. In unproductive oligotrophic lakes that stratify thermally in the summer, the oxygen content of the epilimnion decreases as the water temperatures increase. The oxygen content of the hypolimnion is higher than that of the epilimnion because the saturated colder water from spring turnover is exposed to limited oxidative consumption. Such a vertical distribution is called an ***orthograde oxygen profile*** (Fig. 9-2).
   b. The loading of organic matter to the hypolimnion and sediments of productive eutrophic lakes increases the consumption of dissolved oxygen. As a result, the oxygen content of the hypolimnion is reduced progressively during the period of summer stratification, usually most rapidly at the deepest portion of the basin where strata of less volume are exposed to the more intensive decomposition in surficial sediments. The resulting vertical distribution is termed a ***clinograde oxygen profile*** (Fig. 9-2).
   c. Oxygen saturation at existing water temperatures returns to all water strata during fall circulation.
   d. The exchange of oxygen with the atmosphere ceases for all practical purposes with the advent of ice formation. The oxygen content and saturation levels are reduced at lower depths in productive lakes, but not to the extent observed during summer stratification, because of prevailing colder water temperatures (greater solubility, reduced respiration).
   e. ***Metalimnetic oxygen maxima*** are often observed, in which the oxygen content in the metalimnion is much greater than, and often supersaturated in relation to, levels in the epilimnion and hypolimnion. The resulting ***positive heterograde oxygen curve*** is usually caused by photosynthetic oxygen production by algae in excess of oxidative consumption in the metalimnion.

f. ***Metalimnetic oxygen minima***, resulting in ***negative heterograde oxygen curves*** (Fig. 9-6), occur less commonly than metalimnetic maxima. Metalimnetic oxygen minima result from a combination of mechanisms. Most often, oxidation during decomposition exceeds oxygen inputs as the rate of organic matter sedimentation from the epilimnion decreases upon encountering the colder, denser metalimnetic water. Respiration by dense populations of microcrustacea in the metalimnion can also contribute to oxygen loss in this stratum.

g. Extreme clinograde oxygen distributions occur in meromictic lakes, in which the monimolimnion may be permanently anoxic.

h. Horizontal variations in oxygen content can be great, particularly in lakes and rivers where the photosynthetic production of oxygen of littoral flora exceeds that of the open-water algae. Rapid decay of massive littoral flora or phytoplankton can result in marked reductions in oxygen content in small, shallow lakes, causing the demise of many animals (called ***summerkill***). Oxygen distribution in reservoirs can vary greatly on an annual basis in relation to differences in organic loading from influents, open-water productivity, and discharge of water from the basin.

i. Under ice and snow cover where light is severely attenuated, photosynthesis augmentation of oxygen content can be eliminated. Respiratory oxidative consumption can reduce oxygen to levels insufficient to support organisms and result in ***winterkill***. Winterkill conditions are common in shallow, productive, temperate lakes and become severe in lakes with mean depths of <3 m.

j. In productive lakes and streams, epilimnetic oxygen concentrations can vary markedly on a diurnal basis. Daily photosynthetic augmentation and nighttime respiratory consumption can exceed turbulent exchange with the atmosphere and can result in rapid (hours) fluctuations between supersaturation and undersaturation.

5. The volumetric rate of dissolved oxygen loss from the hypolimnion during summer stratification, the ***relative areal hypolimnetic oxygen deficit***, has been used as an approximate index of the productivity of dimictic lakes. This method assumes that the organic matter production of the trophogenic zone is reflected in the oxygen consumption that occurs in the hypolimnion. Because of their simplicity of determination, oxygen deficits have often been used uncritically. Recent analyses indicate that differences in hypolimnetic temperatures and volumes (thicknesses) must be considered in comparisons among lakes. Empirical analyses using limited data indicate positive correlations between hypolimnetic oxygen deficits and primary productivity of phytoplankton algae, phosphorus concentrations of the water, mean summer hypolimnetic temperatures, and hypolimnetic thickness, and a negative correlation with epilimnetic water transparency (Secchi disc depth). Used properly and with caution, areal hypolimnetic oxygen deficits can serve to compare productivity approximately for large, deep, dimictic lakes of low to moderate productivity if little allochthonous loading of organic matter occurs.

10

SALINITY OF INLAND WATERS

Salinity is the correct chemical term for the sum concentration of all the ionic constituents dissolved in inland waters, both fresh and saline. Salinity is best expressed as total ion concentration in mg $liter^{-1}$ or meq $liter^{-1}$, which are essentially equivalent as mass or volume in dilute solutions. In saline lakes, the density of the waters diverge significantly from unity, and hence salt lake salinities should be expressed on a mass per mass basis, such as mass total ions per mass of solution (g kg^{-1}), which are temperature independent (Williams 1994). In open lakes with an outlet, the chemical composition of the water is governed largely by the composition of influents from the drainage basin and the atmosphere. Water salinity in closed basins is increased by evaporation and modified by precipitation of salts.

The ensuing summary of the ionic composition of inland waters views salinity along a gradient of concentrations from fresh waters of relatively low salinity to highly saline inland waters with ionic concentrations in excess of those of sea water. Although the volume of saline lakes is nearly equivalent to freshwater lakes (Chap. 1), saline lake ecosystems, though chemically and biologically fascinating, are of distinctly less use to humankind than is fresh water for agricultural, industrial, and domestic purposes, and thus are not given equal evaluation here. Such subordination does not denigrate their relevance, particularly in relation to important biodiversity among their highly specialized biota. Certain minerals are concentrated in saline waters and can be commercially valuable. Detailed summaries of inland saline ecosystems have been made by Eugster and Hardie (1978), Hammer (1986), Williams (1986, 1996), and Vareschi (1987).

I. IONIC COMPOSITION OF SURFACE WATERS

The ionic composition of fresh waters is dominated by dilute solutions of alkalis and alkaline earth compounds, particularly bicarbonates, carbonates, sulfates, and chlorides. The amounts of silicic acid, which occur largely in undissociated form, are usually small but occasionally are significant, particularly in hardwater lakes and streams. The concentrations of four major cations, Ca^{++}, Mg^{++}, Na^{+}, and K^{+}, and four major anions, HCO_3^{-}, CO_3^{-}, SO_4^{-}, and Cl^{-}, usually constitute the total ionic ***salinity*** of the water for all practical purposes. Concentrations of ionized components of other elements such as nitrogen (N), phosphorus (P), and iron (Fe) and numerous minor elements are of immense biological importance but are usually minor contributors to total salinity.

Soft waters refers to waters of low salinity, which are usually derived from drainage of acidic igneous rocks (Hutchinson, 1957). ***Hard waters*** contain large concentrations of alkaline earths, usually derived from the drainage of calcareous deposits.

The salinity of fresh waters is best expressed as the sum of the ionic composition of the major cations and anions in mass or milliequivalents per liter. The quantity ***total solids,*** an estimation of inorganic materials dissolved in water by evaporation to dryness (105°C), is less satisfactory. Combustion of the residue at 550°C yields the ***nonvolatile solids*** per unit mass or volume, as CO_2 from organic carbon is lost. However, $MgCO_3$ and some alkalis and chlorides, which are true nonvolatile solids at normal temperatures, will also release CO_2 at 550°C and cause a considerable underestimate of the nonvolatile solids.

II. SALINITY DISTRIBUTION IN WORLD SURFACE WATERS AND CONTROL MECHANISMS

The salinity of the surface waters of the world is highly variable and depends upon ionic influences of drainage and exchange from the surrounding land, atmospheric sources derived from the rock–soil, ocean, and human activity, and equilibrium and exchange with sediments within the water body. The global mean salinity of river water is 120 mg liter^{-1} (Table 10-1; Livingstone, 1963; Walling and Webb, 1986), a value somewhat lower than Clarke's (1924) average but based on better data that compensate for the large amounts of dilute water draining tropical regions in major river systems. The dissolved composition of rivers is dominated by HCO_3^-, SO_4^-, Ca^{++}, and SiO_2, and 97.3% of global runoff has been classified as the calcium bicarbonate type (Meybeck, 1981). Differences among the continents are in the amounts of calcium and carbonate ions. Low values of these normally dominating ions are found particularly in South America and Australia (Table 10-1).

The three major mechanisms controlling the salinity of world surface water are rock dominance, atmospheric precipitation, and the evaporation–precipitation process (Gibbs, 1970, 1992; Feth, 1971; Kilham, 1975, 1990; Stallard, 1980; Stallard and Edmond, 1981; Meybeck and Helmer, 1989; Eilers *et al.*, 1992; Gibson *et al.*, 1995). Waters of the rock-dominated end of the spectrum are rich in calcium and bicarbonate ions (Fig. 10-1) and are more or less in equilibrium with materials of their drainage basins. Positions within this grouping depend on the climate, basin relief, and particularly the composition of the rock material in the basin (Table 10-2). The chemical composition of low-salinity waters, dominated by sodium and chloride, is influenced by dissolved salts derived from atmospheric precipitation acquired from the ocean. These fresh waters are limited to immediate maritime coastal regions. In very chemically dilute inland lakes with a strong precipitation dominance and little marine influence, appreciable sodium can be derived from weathering (Eilers *et al.*, 1992; Wu and Gibson, 1996). The major ionic composition of these lakes falls outside (lower left, salinity below 10 mg liter^{-1}) the primary envelope of Figure 10-1, which prevails for most inland waters of the world. Inland tropical humid areas of South America and Africa generally have high rainfall of an ionic composition that is either influenced primarily by terrestrial biological emissions

TABLE 10-1 Mean Composition of River Waters of the World (mg liter^{-1})[a]

	Ca^{++}	Mg^{++}	Na^+	K^+	CO_3^{2-} (HCO_3^-)	SO_4^{2-}	Cl^-	NO_3^-	Fe (as Fe_2O_3)	SiO_2	Sum
North America	21.0	5.0	9.0	1.4	68.0	20.0	8.0	1.0	0.16	9.0	142
South America	7.2	1.5	4.0	2.0	31.0	4.8	4.9	0.7	1.4	11.9	69
Europe	31.1	5.6	5.4	1.7	95.0	24.0	6.9	3.7	0.8	7.5	182
Asia	18.4	5.6	5.5	3.8	79.0	8.4	8.7	0.7	0.01	11.7	142
Africa	12.5	3.8	11.0	—	43.0	13.5	12.1	0.8	1.3	23.2	121
Australia	3.9[b]	2.7	2.9	1.4	31.6	2.6	10.0	0.05	0.3	3.9	59
World	15.0	4.1	6.3	2.3	58.4	11.2	7.8	1.0	0.67	13.1	120
Cations (μeq liter^{-1})	750	342	274	59	—	—	—	—	—	—	1425
Anions	—	—	—	—	958	233	220	17	—	—	1428

[a] Selected data from Livingstone (1963) and Benoit (1969).
[b] Values of calcium are likely less, on the average, than Na and Mg in Australian surface waters (Williams and Wan, 1972).

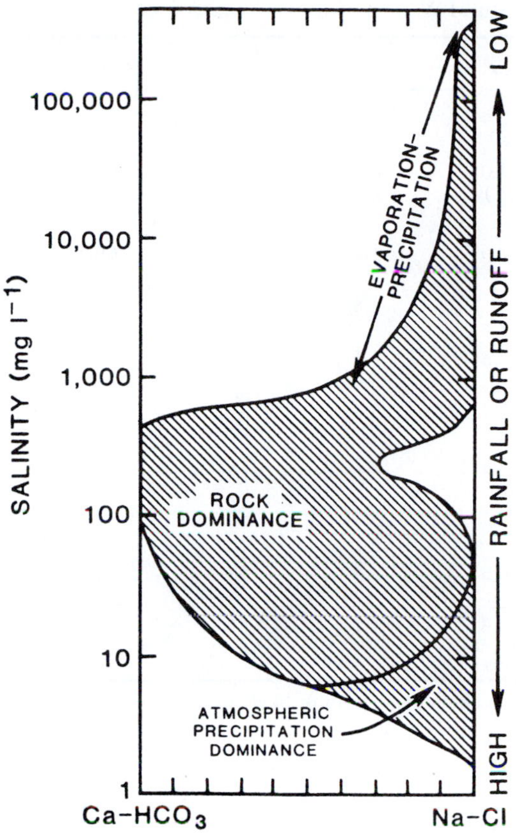

FIGURE 10-1 General processes controlling the ionic ratios and salinity of inland surface waters of the world. (Composite relationships based on the data of Gibbs, 1970; Stallard, 1980; and Kilham, 1975, 1990.)

and particulate dust or, once fallen, is then altered by rapid, intense rock weathering (Kilham, 1975, 1990; Stallard and Edmond, 1982). The third major process that influences the salinity of surface waters is evaporation and fractional crystallization with subsequent sedimentation of mineral salts. Fresh waters extend in a series from calcium- and bicarbonate-rich, to low-salinity waters of rock dominance or to sodium-dominated, high-salinity waters (Fig. 10-1). Rivers and lakes at the extreme saline end of this series are generally located in hot and arid regions. For example, the chemical composition of natural lakes, rivers, and springs in intertropical Africa are controlled primarily by rock dominance via weathering reactions and the evaporation–crystallization process of evaporative concentration and calcite precipitation (Kilham, 1990).

The data discussed above represent average concentrations and general conditions of distribution on a global basis. Within regions, the distribution of salinity of inland waters is highly variable and localized. Hence, some areas or lake districts can be found in which, because of uniform rock formations, the ionic proportions are relatively consistent (Table 10-3). In contrast, in some areas, such as in southwestern Michigan, extremely softwater seepage lakes may be found immediately adjacent to very hardwater calcareous lakes because of complex, varying patterns in the deposition of glacial till.

Over large regions of the temperate zone, dominance by calcium and bicarbonate ions prevails in open lake ecosystems:

$$\text{Cations: Ca} > \text{Mg} \geq \text{Na} > \text{K}$$

$$\text{Anions: } CO_3 > SO_4 > \text{Cl}$$

The relationships are very consistent with those of the average world river water (Tables 10-1 and 10-2), and there is a distinct tendency for the composition of lake waters to approach this average concentration (Rodhe, 1949; Gorham, 1955; Gorham *et al.*, 1983; Meybeck and Helmer, 1989). These salinity relationships led Rodhe (1949) to propose a general standard lakewater concentration, given in Table 10-4 in

TABLE 10-2 Global Average Chemical Composition (μeq liter^{-1}, except as noted) of Unpolluted Rivers and Variations in Composition according to Drainage from Dominant Rock Type[a]

	Global average[b]							
	μeq liter^{-1}	*mg liter*$^{-1}$	*Granite*	*Gneiss*	*Volcanic rocks*	*Sandstone*	*Shale*	*Carbonate rock*
Ca^{++}	670	13.43	39	60	154	88	404	2560
Mg^{++}	259	3.15	31	57	161	63	240	640
Na^{+}	159	3.66	88	80	105	51	105	34
K^{+}	32	1.25	8	10	14	21	20	13
HCO_3^{-}	835	50.94	128	136	425	125	580	3195
SO_4^{2-}	163	7.83	31	56	10	95	143	85
Cl^{-}	86	3.05	0	0	0	0	20	0
SiO_2 (μmole liter^{-1})	173	10.40	150	130	200	150	150	100

[a] Extracted from data of Meybeck and Helmer (1989).
[b] Global discharge-weighted natural concentrations, corrected for oceanic cyclic salts.

TABLE 10-3 Mean Ionic Salinity in Equivalent Percentages of Several Natural Waters

Natural waters	Ca^{++}	Mg^{++}	Na^{+}	K^{+}	CO_3^{2-}	SO_4^{2-}	Cl^{-}
Wisconsin soft waters (Juday, *et al.*, 1938; Lohuis *et al.*, 1938)	46.9	37.7	10.9	4.8	69.9	20.5	9.9
N. German soft waters (Ohle, 1955a)	36.0	14.3	43.0	6.7	42.4	14.1	43.5
Water from igneous rock (Conway, 1942)	48.3	14.2	30.6	6.9	73.3	14.1	12.6
Hubbard Brook, New Hampshire (Likens *et al.*, 1970)	25.2	5.6	67.6	1.6	<5.3	66.3	28.4
Mean sedimentary source material (Hutchinson, 1957, after Clarke, 1924)	53.2	34.0	8.0	4.8	93.8	6.2	—
Swedish rivers, average entire Sweden (Ahl, 1980)	54.9	18.4	22.6	4.1	42.7	39.1	16.5
Ugandan river and lake waters (Viner, 1975)	32.5	26.9	32.1	8.7	62.9	28.0	9.1
Upland Swedish sources (Rodhe, 1948, after Lohammar)	67.3	16.9	13.6	2.2	74.3	16.2	9.5
Hardwater lakes of N.E. Indiana (Wetzel, 1966b)	79.3	14.4	3.3	3.0	60.0	37.9	2.1
Hypereutrophic Sylvan Lake, Indiana[a] (Wetzel, 1966a)	57.3	11.8	26.9	3.4	48.7	34.7	16.6
Average river content of world (Livingstone, 1963)	52.6	24.0	19.3	4.1	67.1	16.3	15.4

[a] A hardwater lake that received large amounts of domestic effluents that had high sodium content, most likely from detergents. Nutrient loading to this lake has subsequently been reduced.

abbreviated form. Deviations from these proportions are common, particularly on the continents of South America and Australia and in drainage basins of igneous source materials and soft waters (see Tables 10-2 and 10-3). Drainage from igneous rock commonly has a salinity of less than 50 mg liter^{-1} and cationic proportions of

$$Ca > Na > Mg > K$$

The proportions of anions in softwater systems shift forward a decrease in carbonates and an increase in halides:

$$Cl \geq SO_4 > CO_3$$

TABLE 10-4 Composition of Swedish Bicarbonate Fresh Waters (mg liter^{-1}) in Relation to Specific Conductance[a]

Salinity	*Ca*	*Mg*	*Na*	*K*	CO_3	SO_4	*Cl*	$\mu S\ cm^{-1}$ (20°C)
10.5	2.5	0.4	0.7	0.3	4.4	1.5	0.7	20
32.9	7.9	1.3	2.2	0.8	13.8	4.7	2.2	60
56.5	13.5	2.3	3.8	1.4	23.7	8.0	3.8	100
92.1	22.0	3.7	6.2	2.3	38.6	13.1	6.2	160
117.3	28.1	4.7	7.9	2.9	49.2	16.6	7.9	200
155.1	37.1	6.2	10.4	3.8	65.2	22.0	10.4	260
180.8	43.3	7.3	12.2	4.4	75.8	25.6	12.2	300
219.7	52.6	8.8	14.8	5.4	92.1	31.2	14.8	360
246.2	59.0	9.9	16.6	6.0	103.2	34.9	16.6	400

[a] From data of Rodhe (1949) and Hutchinson (1957).

Saline lakes form and persist when (a) outflow of water is restricted, as in hydrologically closed basins, (b) evaporation exceeds inflows, and (c) the inflow is sufficient to sustain a standing body of water (Eugster and Hardie, 1978; Hammer, 1986; Williams, 1994). The relative proportions of the major solutes vary greatly among saline lakes. Most saline waters are dominated by Na; very few saline lakes are found in which Ca or Mg is the major cation. Dominating major anions are highly variable, although Cl very commonly predominates. Certain elements, normally present in only trace amounts in surface waters, can occur in very high concentrations in saline lakes (e.g., bromine, strontium, phosphate, or boron) (Wetzel, 1964; Eugster and Hardie, 1978; Hammer, 1978, 1986).

The specific conductance of lake water is a measure of the resistance of a solution to electrical flow. The resistance of an aqueous solution to electrical current or electron flow declines with increasing ion content. Hence, the purer that water is, that is, the lower its salinity, the greater its resistance to electrical flow. By definition, the specific conductance of an electrolyte is the reciprocal of the specific resistance of a solution measured between two electrodes 1 cm^2 in area and

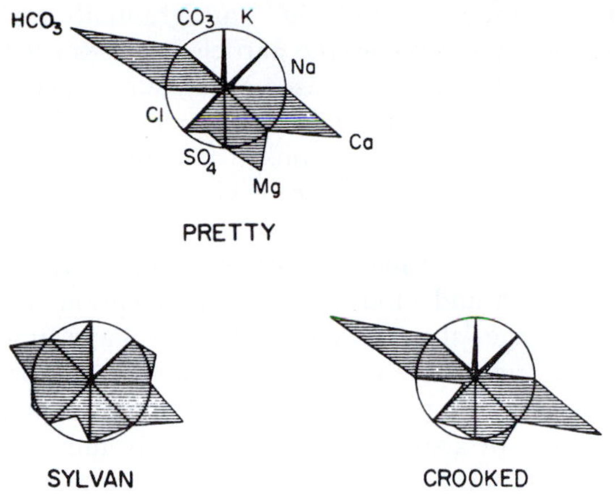

FIGURE 10-2 Ionic polygonic diagrams in which the relative concentrations in equivalent percentage of the major cations and anions are indicated for average concentrations of Pretty, Sylvan, and Crooked lakes, northeastern Indiana. The length of each radius bisecting a sector is proportional to the concentration of a particular ion in equivalent percentage of the total cations or total anions. (After Wetzel, 1966b.)

1 cm apart. The conductance is expressed in μSiemens cm^{-1} (previously μmhos cm^{-1}, the reciprocal of ohms).[1] The temperature of the electrolyte affects ionic velocities; conductance increases about 2% per °C. The international chemical standard reference of 25°C is recommended in all cases, even though 18 or 20°C has been used widely in past limnological studies.

The specific conductance of the common bicarbonate-type of lake and river water is closely proportional (see Table 10-3) to concentrations of the major ions (Juday and Birge, 1933; Rodhe, 1949). Once the concentrations of the major ions are known, changes in specific conductance reflect proportional changes in ionic concentrations (Otsuki and Wetzel, 1974). This relationship is not true, however, for minor constituents of lake water (e.g., N, Fe, Mn [manganese], Sr [strontium], or especially P) (Rodhe, 1951). As would be anticipated, a positive correlation exists between conductance and pH in the intermediate pH range of bicarbonate fresh waters, but this relationship deteriorates among lakes of low salinity and high dissolved organic matter content (Strøm, 1947).

The relative concentrations of the eight major ions of salinity can be shown graphically with ionic polygonic diagrams (Maucha, 1932). A 16-sided regular polygon of standard area is divided into eight equal sectors, the four to the right of the vertical diameter representing the cations and the four to the left the anions (Fig. 10-2). Lines are drawn from the center to bisect each sector. The length of each radius bisecting a sector is proportional to the concentration of a particular ion in equivalent percentage of the total cations or anions. Lines from the radii of the ionic sections are joined to the polygon and shaded. Details of calculation are given in Gessner (1959) and in Broch and Yake (1969).

[1] Specific conductance

$$\mu S\ cm^{-1} = \frac{R \text{ of } 0.00702\ N\ KCl}{R \text{ of sample}} \times 100$$

where R is the resistance in ohms (cf. Rainwater and Thatcher, 1960; Golterman, 1969; Wetzel and Likens, 2006).

III. SOURCES OF SALINITY

A. Weathering of Soil and Rock

The composition of soil and rock and their ion-exchange capacities influence both rates of weathering and ion supply to runoff and percolating water. The adsorption and release of ions is dependent upon (a) the cationic availability, (b) ionic concentrations and proportions in a soil solution or leaching water, (c) the nature and number of exchange sites on the exchange complex of soil particles, and (d) the volume of water in contact with the exchange complex. Four general processes of weathering control ion supply: (a) solution, (b) oxidation-reduction, (c) the action of hydrogen ions, and (d) the formation of complexes (reviewed in detail by Gorham, 1961, and Carroll, 1962).

Ordinary solution not involving hydrolysis or acid weathering is important, primarily in sedimentary deposits rich in soluble salts. Leaching of marine salt deposits results in an enrichment of sodium, potassium, and chlorides relative to other ions in recipient lake and river waters. Relative leaching rates of different ions are time-dependent, and thus marked fluctuations in ionic proportions of runoff water occur during periods of high (flooding) or low rates of runoff because of differences in solubility.

Oxidation and reduction processes primarily affect iron, manganese, sulfur, nitrogen, phosphorus, and carbon compounds in soils. Iron sulfides are common constituents of rocks and water-saturated soils. Oxidation of these sulfides can be a major source of sulfate for natural waters, usually in the form of dilute sulfuric acid, which solubilizes other rock and soil constituents as well. Microbial decomposition of sulfur-containing organic compounds, particularly in woodland soils and peaty bog waters, adds sulfate to natural waters usually as sulfuric acid. Within lakes, oxidation–reduction processes play dominant roles in the sulfur cycle (see Chap. 14).

The action of hydrogen ions from the dissociation of carbonic acid, discussed at length in the following chapter, is of major importance in the weathering of soils and rocks. Concentrations of CO_2 and H^+ ions increase in soils in which intensive microbial decomposition of organic matter occurs. The importance of carbonic acid in weathering is illustrated by the high proportion of bicarbonate ions in most river waters. The proportion is high even in waters of low salinity (< 50 mg liter^{-1}) draining igneous rocks (Table 10-2). Strong acids (H_2SO_4, HNO_3) in rainfall originating from air pollution accelerate the rate of weathering (Likens *et al.*, 1972, 1979; Likens, 1985). Colloidal acids of humic compounds and acid clays can also supply large amounts of H^+ ions for weathering. The "black" waters of tropical rivers draining acidic podzols are colored brown by humic substances originating from incomplete decomposition of forest, flood plain, and wetland/littoral plants. The low pH (3.5–5) of these waters results largely from dissociation of humic acids and the fact that the soils of lowland tropics contain very low levels of bases and of calcium.

Certain soluble organic molecules can chelate or complex ions by molecular bonding, thereby inhibiting these ions from reacting with others and increasing the solubility of the complex. The extent of this process in weathering and the transport of ions to rivers and lakes is probably quite significant but has not been studied adequately.

The exchange of ions between soils and the soil solution is governed by exchange equilibria between hydrogen ions and ions attached to the soil particles. Soil colloids have a finite absorptive capacity, and the exchange rate is highly variable. In humid, well-drained regions, water selectively removes cations from weathering rocks and soils. In areas of limited rainfall and stream activity, an accumulation of sodium and chloride deposited by rain in the soil water can result. The proportion of divalent to monovalent cations in the exchange positions on soil particles is a function of cation concentrations and their relative proportions in the soil water. Sodium ion adsorption to clay particles decreases with increasing concentrations of Ca^{++} and Mg^{++}. Calcium and magnesium are the dominant exchangeable cations in most neutral or alkaline soils. In well-drained situations, the cations added by rainfall will have little effect on exchange leaching from soils or weathering rock. In contrast, when soils are saturated, cations in exchange positions are in equilibrium with those in the soil water and ground water. Cations added by rainwater to the water surrounding clay minerals will establish equilibrium with those in exchange positions, and little weathering or removal of cations will occur. Under arid conditions, salts in soil solutions accumulate, for there is no removal of cations derived from rainfall. Exchange sites contain predominantly Ca^{++} and Mg^{++}; as the soil water becomes saturated with Na^+, the Ca^{++} and Mg^{++} are gradually replaced by Na^+ until all sites on the particle surfaces are filled. This Na^+ can then be released again during periods of rainfall and result in a much higher concentration of Na^+ than of divalent cations. Clay minerals are affected very little by these exchanges and do not weather appreciably.

The salinity of natural waters is influenced further by the depth and mode of water percolation through soils. Seepage lakes receiving relatively large amounts of surface runoff are usually very dilute in comparison to open lakes that drain deeper soil horizons, because the salinity of ground water is generally much higher than that of surface runoff.

B. Atmospheric Precipitation and Fallout

The atmosphere is a significant source of salinity for many dilute fresh waters and for some saline lakes of arid regions. Rainfall carries much of the atmospheric salt to some river and lake waters (cf. reviews of Gorham, 1961; Carroll, 1962; Sutcliffe and Carrick, 1983a, 1983b; and Lesack and Melack, 1991). In less humid regions, dry fallout of salt particles can occur in significant quantities and often exceed those washed down with rainfall. Snow is less efficient in removing atmospheric salts than rain, but at high latitudes, contaminants form an appreciable supply of salinity (e.g., Barica and Armstrong, 1971). All of the major anions in natural waters are cycled in part through the atmosphere as gases as well as in dissolved and particulate form.

The sea is a major source of atmospheric sodium, chloride, magnesium, and sulfate (Table 10-5). Sea

TABLE 10-5 Common Concentrations (mg liter^{-1}) of Major Ions in Continental and Marine Rainfall[a]

Ion	*Continental rain*	*Mean central Amazon rain*[d]	*Marine/coastal rain*
Ca^{++}	0.2–4[b]	0.030	0.2–1.5
Mg^{++}	0.05–0.5	0.013	0.4–1.5
Na^+	0.2–1	0.048	1–5
K^+	0.1–0.5[b]	0.023	0.2–0.6
NH_4^+	0.1–0.5[c]	0.068	0.01–0.05
SO_4^{2-}	1–3[b,c]	0.187	1–3
Cl^-	0.2–2	0.128	1–10
NO_3^-	0.4–1.3[c]	0.192	0.1–0.5
H^+	pH = 4–6	pH 4.9[d]	pH = 5–6

[a] Modified from data of Berner and Berner (1987).
[b] In remote continental areas: Ca^{++} = 0.02–0.20; K^+ = 0.02–0.07; SO_4^{2-} = 0.2–0.8.
[c] In polluted areas: NH_4^+ = 1–2; SO_4^{2-} = 3–8; NO_3^- = 1–3.
[d] From data of Lesack and Melack (1991). Average pH value likely lowered because of abundant dissolved organic acids.

spray carries large amounts of sea water into the atmosphere that, on evaporation, forms salt particles which are capable of acting as nuclei for cloud and raindrop condensation. Atmospheric salinity can be carried for great distances. Although most atmospheric salinity is precipitated with rainfall in the coastal regions, decreasing amounts are carried inland before deposition occurs. Continental rain generally contains more sulfate than chloride; chloride ion concentrations usually increase with proximity to the sea (Table 10-5). In a similar fashion, sodium is the dominant cation of rain, and its concentration decreases in relation to amounts of magnesium and calcium with increasing distance from the sea. The effects of atmospheric transport of ions can be seen in lakes enriched with Na^+ and Cl^- in coastal maritime regions. This enrichment is particularly evident in igneous coastal region mountains, in which many of the maritime lakes of Japan can be found (Sugawara, 1961).

Windblown dust from the soil often contributes salts, especially of calcium and potassium, to rain and snow. This type of transport is the case, for example, in calcareous regions in Sweden. Wind-transported salts from salt pans of lakes in endorheic regions, such as in Russia, western Australia, or the United States, can be moved large distances to drainage basins of other river and lake systems.

A major source of atmospheric salinity is industrial and domestic air pollution. Although numerous ions and particles are emitted into the air, chlorides, calcium, and especially sulfates and nitrates are major contaminants. The magnitude of air pollution on the chemistry of surface waters has now reached such major proportions that global action is required to curtail its effects. Although much of atmospheric pollution returns to the ground in the areas immediately adjacent to the industrial sites of production, sufficient sulfuric, nitric, and hydrochloric acids enter the atmosphere to influence the water of precipitation and surface waters of entire countries or portions of continents. The most infamous example is in Scandinavia, which receives air currents from western and southwestern Europe for much of the year. Contaminants from heavily industrialized regions of the United Kingdom, the Ruhr Valley, and elsewhere caused a 200-fold increase in acidity of rain in a decade (Likens *et al.*, 1972). Values of pH less than 3 were observed, and over large areas of northern Europe, the pH of rain is still less than 4, despite recent reductions of air pollution.

An analogous situation developed in eastern New York and in the smaller eastern states. Acid rain falling in these areas can be traced to industrial origins in the central states and in Canada. In addition to direct effects on the leaching rates of nutrients from plant foliage and soil, plant metabolism is altered and major nutrient inputs to and acidification of surface waters can result in changes in entire drainage basins. In the Hubbard Brook drainage basin of the White Mountains of New Hampshire, sulfate was found to be the principal ion in precipitation (Fisher *et al.*, 1968) and supplied most of the sulfate that was discharged by the streams (Table 10-6). The input of ammonium and nitrate exceeded the discharge of these constituents. The annual deposition of hydrogen ion exceeded that of sulfate and is a major determinant of pH of the waters of that region.

TABLE 10-6 Allocation of Elements (in Percent) within the Hubbard Brook Experimental Forest Drainage Basin[a]

	Ca	*K*	*Mg*	*Na*	*N*	*S*
Source						
Precipitation input	9	11	15	22	31	65
Net gas or aerosol input	—	—	—	—	69	31
Weathering release	91	89	85	78	—	4
Storage or loss						
Biomass accumulation						
Vegetation	35	68	17	2	43	6
Forest floor	6	4	5	<1	37	4
Streamflow						
Dissolved substances	59	22	74	95	19	90
Particulate matter	<1	6	5	3	<1	<1

[a] After Likens and Bormann (1995).

C. Environmental Influences on Salinity

Climate affects markedly the balance between precipitation and evaporation and thus the salinity concentrations of surface waters. This effect has been discussed already in global terms. At a more local level, climatic effects are manifested in a general increase in salinity with decreasing elevation of rivers and lakes. This correlation is because most of the salinity of rain and particulate fallout is deposited at lower elevations.

The salinity of lake waters of closed drainage basins is governed not only by inputs of dissolved ions from runoff but by the fate of these materials upon evaporation (Hutchinson, 1957). Most closed lakes occur in regions with long-term (several years) fluctuating climate and are often exposed to periods of severe aridity. Commonly very shallow, these lakes may evaporate completely or sufficiently to expose large expanses of sediments. Diversions of river inputs by human activities, as in the tragedy of the Aral Sea (Williams and Aladin, 1991), can also lead to the exposure of large areas of sediments of saline lakes. Loss of salts then can occur by wind deflation.

Saline lakes are generally categorized on the basis of dominating anionic concentrations into carbonate, chloride, or sulfate waters (Hammer, 1986; Williams, 1994, 1996). The range of salinity in these lakes is extraordinary, from several hundred to over 200,000 mg liter^{-1} in the Great Salt Lake of Utah and the Dead Sea. Borax Lake of northern California, one of the few shallow saline lakes that has been studied in detail over an annual period, exhibited a nearly fourfold decrease in volume and over a twofold increase in salinity, from less than 28,000 mg liter^{-1} to nearly 60,000 mg liter^{-1} (Wetzel, 1964). An analogous condition was recorded in Lake Chilwa, Malawi, which is typical of many shallow lakes of endorheic regions (Moss and Moss, 1969; Kalk *et al.*, 1979; Hammer, 1986).

Other significant climatic factors influencing salinity are temperature and wind. Temperature influences the rate of rock weathering. Tropical waters, for example, which drain strongly weathered soils, are usually poor in electrolytes, and a large part of their total composition consists of silica. Wind direction and speed may affect the chemical composition of atmospheric precipitation by altering the amount of incorporated sea salinity and sites of deposition inland. Losses of atmospheric salinity are greater in low-elevation, turbulent air masses. The type of vegetation growing on the drainage basin and its requirements for major ions are also influenced by climate. Mineral cycling in tropical perennial forests of dense vegetative cover, for example, differs greatly in higher rates of utilization and leaching of soil nutrients as compared to nutrient cycling in deciduous vegetation of temperate regions.

IV. DISTRIBUTION OF MAJOR IONS IN FRESH WATERS

The spatial and temporal distribution of the major cations and anions of salinity are separable into (1) ***conservative ions,*** whose concentrations within a lake and many streams undergo relatively minor changes from biotic utilization or biotically mediated changes in the environment, and (2) ***dynamic ions,*** whose concentrations can be influenced strongly by metabolism. Of the major cations, magnesium, sodium, and potassium ions are relatively conservative both in their chemical reactivity under typical freshwater conditions and their small biotic requirements. Calcium is more reactive and can exhibit marked seasonal and spatial dynamics. Of the major anions, inorganic carbon is so basic to the metabolism of fresh waters that it is treated in a separate chapter (Chap. 11), and later coupled with organic carbon cycling (Chap. 23). Similarly, sulfate is greatly influenced by microbial cycling and the chemical milieu and is treated separately (Chap. 14), along with iron and silica. Chloride is relatively conservative. The total ionic salinity, composed almost entirely of the eight major ions, is of major importance in osmotic regulation of metabolism and in the distribution of biota and is briefly discussed below.

A. Calcium

Calcium influences the growth and population dynamics of freshwater flora and fauna both directly and indirectly. Calcium is a required nutrient of normal metabolism of higher plants, as well as of prokaryotes (Smith, 1995). A universal requirement for calcium has not been demonstrated for algae but most likely it is required by the green algae and is considered an essential inorganic element of algae. Where demonstrated as essential, calcium is usually needed as a micronutrient. Substitution of calcium by a closely related element, strontium, is readily acceptable to some algae (e.g., *Chlorella, Scenedesmus*) that require calcium and are indifferent to strontium; in other algae, calcium utilization is strongly inhibited by strontium.

The two principal membranes of plant cells, the outer cytoplasmic plasmalemma and the inner cytoplasmic tonoplast, operate in ion transport, one of the basic biological functions of cell membranes (see reviews of Eppley, 1962, and Epstein, 1965). The central vacuole is the main repository of accumulated ions, and both cations and anions may be absorbed from dilute solutions until the vacuolar concentrations vastly exceed external concentrations. Membranes of higher plants have a very low permeability to free ions; ionic exchange is more common among algae. Active transport mechanisms presumably involve high-energy phosphate compounds of phosphorylation. The ion-transporting mechanisms are ion-selective and are often species-specific; the external concentrations of ions and ratios of divalent to monovalent cations can affect transport mechanisms markedly. Calcium is essential for maintenance of the structural and functional integrity of cell membranes.

The distribution of certain algae has been correlated to differing concentrations of calcium. While causality cannot be established definitely for such relationships by singling out one affecting chemical species from an array of simultaneously interacting parameters that regulate growth and distribution, it is clear that calcium is involved indirectly in the metabolism of certain organisms. The desmids are a large algal group that is found in large part in acidic (pH 5–6) waters of low salinity, especially waters of low calcium content. Of the few species of desmids studied, many have a narrow tolerance to calcium concentrations.

Their restriction to soft waters is by no means universal. Species of the larger genera are divisible into those adapted to acidic (pH < 6), calcium-poor (< 10 mg Ca liter^{-1}) waters and those adapted to increasingly alkaline, calcium-rich waters. The detailed studies of Höll on this subject have been verified and are reviewed by Hutchinson (1967). Again, causality between calcium concentration and the metabolism of these algae has not been established through experimental studies, even though evidence for such an interaction is very strong.

A ubiquitous relationship has been found between calcium, calcium-specific regulatory proteins, and cyclic adenosine 3′:5′-monophosphate (cAMP) in both prokaryotic and eukaryotic organisms. These molecules function synergistically to regulate a broad spectrum of important metabolic functions, in particular ion-transport and energy-transduction reactions (Kretsinger, 1979; Marx 1980; Smith, 1995). The relationships between these regulatory molecules in aquatic systems are only beginning to be appreciated. Cyclic AMP is produced and released by aquatic photoautotrophs and bacteria (Francko and Wetzel, 1980, 1981, 1982) and is found in highest concentrations in lake water and plant tissues in midsummer during periods of high photosynthetic demand for carbon and resultant epilimnetic and littoral decalcification (Francko and Wetzel, 1982).

Very few groups of freshwater animals exist in which the distribution of some species has not been related to calcium concentration (cf. Macan, 1961). As among algae and certain macrophytes, in some animal groups most of the species are found in calcareous waters, and their numbers decrease as the concentration of calcium declines. Other species, which are frequently closely related to those in hard, calcium-rich waters, are characteristic of waters poor in calcium. In particular, mollusks, leeches, and tricladian flatworms are divisible into hardwater species (≥ 20 mg Ca liter^{-1}), and those that can tolerate less than this concentration. Analogous but less clear correlations occur among the crustaceans. Although the requirements for calcium in invertebrates are known to some extent (cf. Robertson, 1941), little is known about how small differences in calcium concentrations affect distribution. Attempts to explain the metabolism and distribution of animal species on the basis of single chemical differences have been numerous but largely unrewarding, except where conditions are extreme. Calcium concentrations have been implicated in the aging of rotifers, a process that affects longevity and morphology by accumulation of calcium (Edmondson, 1948). These laboratory findings fit remarkably well into observed distribution and population dynamics of rotifers in lakes of varying calcium concentrations, in which longevity and population success increase in waters of lesser calcium content.

The calcium content of softwater lakes and streams remains well below saturation levels, and these concentrations exhibit minor seasonal variations with depth. The amount of calcium utilized by the biota is usually so small in comparison to existing levels that reduction or depletion by biota cannot be seen in normal analyses. Decomposition processes can lead to some calcium accumulation in the hypolimnion of productive softwater lakes during stratification.

The calcium content of hardwater lakes, however, undergoes marked seasonal dynamics. The changes depicted in Figure 10-3 for a hardwater lake in southern Michigan are quite typical. Between rather uniform concentrations during spring and fall periods of

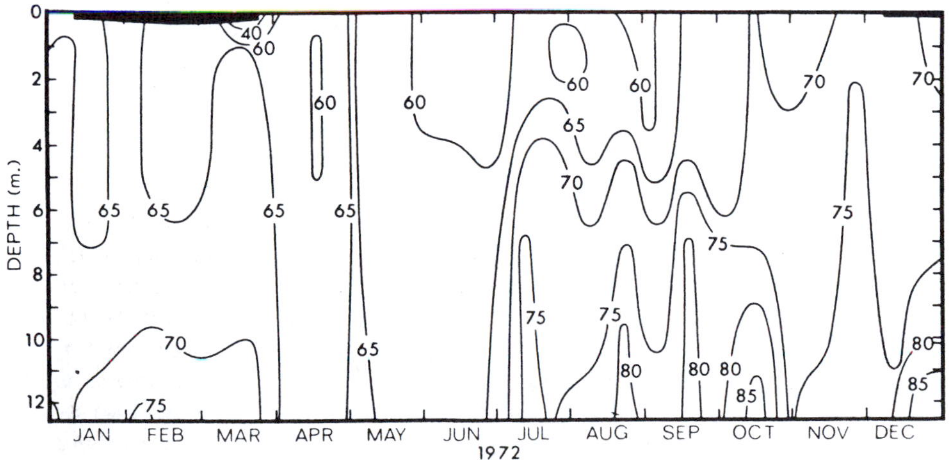

FIGURE 10-3 Depth–time distribution of isopleths of calcium concentrations (mg Ca^{+2} liter^{-1}) of oligotrophic, hardwater Lawrence Lake, Michigan. Opaque areas = ice cover to scale. (From Wetzel, unpublished data.)

circulation, a conspicuous stratification occurs that is repeated annually with only minor variations. Both the levels of calcium (Fig. 10-3) and bicarbonate (Fig. 11-3) decreased markedly in the epilimnion (0–4 m in Figs. 10-3 and 11-3) as a result of precipitation of $CaCO_3$ during the summer months from May through September. Similar losses were observed during the period of winter stratification. In winter, the decrease just beneath the ice is associated in part with dilution by rains permeating the ice and melting ice (cf. Canfield *et al.,* 1983), as well as with increases in photosynthesis just before ice loss. Decreases of concentrations of calcium and of inorganic carbon in the epilimnion and metalimnion are related largely to rapid increases in the rates of photosynthesis by phytoplankton and littoral flora (Otsuki and Wetzel, 1974) and indicate the major role of photosynthetic alteration of dissociation equilibria and induction of epilimnetic decalcification (cf. review of Küchler-Krischun and Kleiner, 1990).

The calcium and bicarbonate budgets of this hardwater lake illustrate relationships among influxes, dynamics within the lake, and outflow over an annual period. Calcium inputs of two spring-fed inlet streams were uniform and contributed from 93 to 98% of the total monthly inputs to the basin during the summer months. After compensation for potential evapotranspiration and precipitation, inputs of calcium from groundwater influxes varied from 2 to 58% of the total monthly inputs. Inputs of bicarbonate of the two inlet streams were also uniform annually and constituted from 82 to 97% of the total monthly influx during the summer period. Contributions from the streams decreased total inorganic carbon input anywhere from 35 to 67% during other times of the year because of the increase in groundwater influxes. The $CaCO_3$ precipitation rate was calculated to be 446 g m^{-2} y^{-1}. This loss of $CaCO_3$ influences the metabolism of hardwater lakes by coprecipitation of inorganic nutrients, such as phosphorus, and selective removal of humic organic acids and other organic compounds by adsorption (Stewart and Wetzel, 1981). Such adsorptive processes and potential biotic effects are discussed under specific sections in the following chapters. The general decalcification processes described here have been observed in many other lakes (e.g., Koschel *et al.,* 1983; Stabel, 1986; Duston *et al.,* 1986; Effler and Johnson, 1987; Küchler-Krischun and Kleiner, 1990).

Epilimnetic decalcification is reflected simultaneously in the distribution of specific conductance of hardwater lakes (Fig. 10-4). Because concentrations of Mg, Na, K, and Cl are relatively conservative, as discussed further on, the specific conductance follows changes in Ca^{+2} and HCO_3^- concentrations in nearly a 1:1 relationship ($r = 0.997$; Otsuki and Wetzel, 1974). A portion of the precipitating $CaCO_3$ is resolubilized in the hypolimnion, which is reflected in both the increased concentrations of Ca^{+2} and the specific conductance in hypolimnetic waters. Some of the $CaCO_3$ is entrained permanently in the sediments and commonly constitutes >30% of the sediments by weight in moderately hardwater lakes (Wetzel, 1970). Adsorption of organic compounds to $CaCO_3$ lowers the rates of dissolution in hypolimnetic strata of reduced pH. Further, calcium complexes with humic acids, especially in the sediments, alter exchange equilibria in the hypolimnion (Ohle, 1955a; Stewart and Wetzel, 1981a; Hering and Morel, 1988).

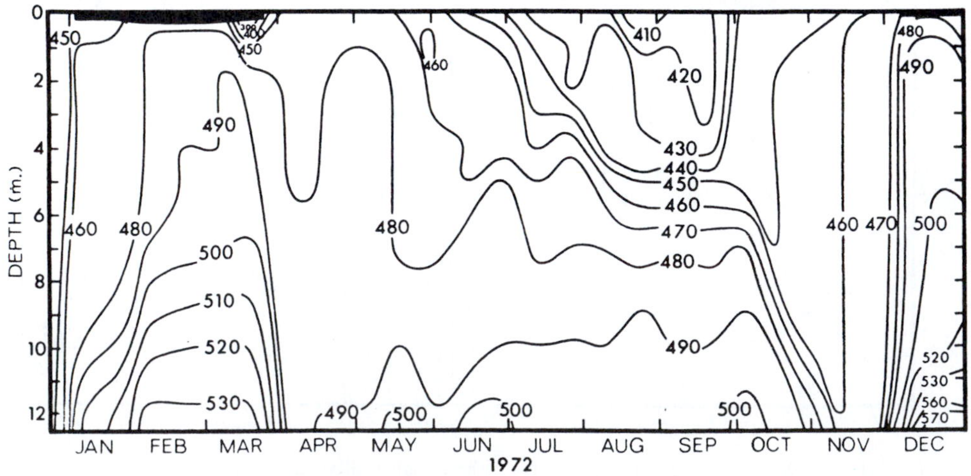

FIGURE 10-4 Depth–time distribution of isopleths of specific conductance (μS cm^{-1} at 25°C) of hardwater Lawrence Lake, Michigan. Opaque areas = ice-cover to scale. (From Wetzel, unpublished data.)

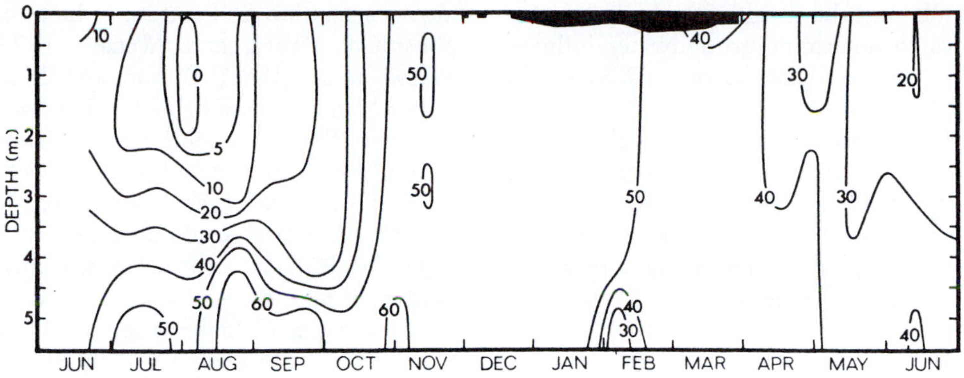

FIGURE 10-5 Depth–time distribution of isopleths of calcium concentrations (mg Ca^{+2} $liter^{-1}$) of hypereutrophic Wintergreen Lake, Kalamazoo County, Michigan, 1971–1972. Opaque area = ice cover to scale. (Wetzel *et al.*, unpublished data.)

Biogenically induced decalcification of the epilimnion reaches an extreme in very productive hard waters. For example, the calcium concentrations of the epilimnion of Wintergreen Lake, Michigan, were reduced from nearly 50 mg $liter^{-1}$ to below analytical detectability in a few weeks (Fig. 10-5). The magnitude of this loss from the trophogenic zone and the resulting increase in monovalent:divalent cation ratios should be recalled in subsequent discussion of the effects of the calcium content and of cationic ratios on species distribution.

B. Magnesium

Magnesium is required universally by chlorophyllous plants for the magnesium prophyrin component of chlorophyll molecules and as a micronutrient in enzymatic transformations, especially in transphosphorylations by algae, fungi, and bacteria. The demands for magnesium in metabolism are minor in comparison to quantities generally available in fresh waters. Magnesium compounds, moreover, are much more soluble than their calcium counterparts. As a result, significant amounts of magnesium rarely precipitate. The monocarbonates of hard waters are usually >95% $CaCO_3$ under ordinary CO_2 pressures (e.g., Murphy and Wilkinson, 1980). $MgCO_3$ and magnesium hydroxide precipitate significantly only at very high pH values (>10) under most natural conditions. The concentrations of magnesium are extremely high in certain closed saline lakes.

Because of magnesium's solubility characteristics and its minor biotic demand, concentrations of magnesium are relatively conservative and fluctuate little both in softwater streams and lakes (e.g., Likens, 1985) and in hardwater streams (e.g., Wetzel and Otsuki, 1974) and lakes (Fig. 10-6). This attribute has been used to

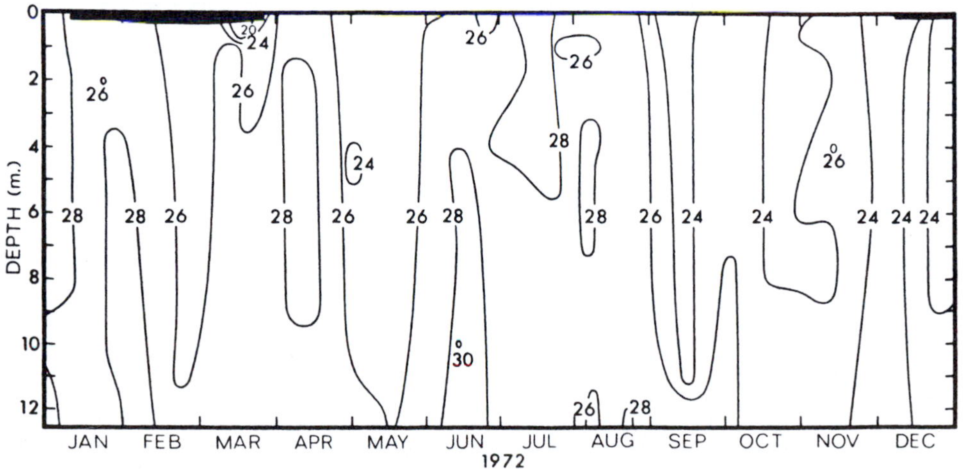

FIGURE 10-6 Depth–time distribution of isopleths of magnesium concentrations (mg Mg^{+2} $liter^{-1}$) of hardwater Lawrence Lake, Michigan. Opaque areas = ice cover to scale. (From Wetzel, unpublished data.)

advantage by employing the magnesium budget of inputs and outflows to determine groundwater influxes to a lake by magnesium mass balance (Wetzel and Otsuki, 1974).

Low available magnesium has been implicated as one of several factors influencing phytoplanktonic productivity in an extremely oligotrophic Alaskan lake (Goldman, 1960). However, such conditions are exceedingly rare in comparison to limitations imposed by the restricted availability of other nutrients.

C. Sodium, Potassium, and Minor Cations

The monovalent cations sodium and potassium are involved primarily in cellular ion transport and exchange. An absolute sodium requirement has been demonstrated in only a few plants. The sodium requirements are particularly high in some species of cyanobacteria (Allen, 1952; Gerloff *et al.*, 1952; Allen and Arnon, 1955) and is required for photosynthesis, bicarbonate transport, intracellular pH regulation, nitrogen fixation and nitrate reduction, and uptake of phosphate (Ward and Wetzel, 1975; Rees, 1985; Maeso *et al.*, 1987; Oleson and Makarewicz, 1990; Valiente and Avendaão, 1993). Potassium and other elements of this series—lithium (Li), rubidium (Rb), and cesium (Cs)—cannot substitute for Na. A threshold level of 4 mg Na^+ liter^{-1} is required for near optimal growth of several cyanobacterial species (Kratz and Myers, 1954), a concentration near mean for numerous hardwater lakes.

Cyanobacteria possess separate systems for the active transport of both CO_2 and HCO_3^-. The active transport system for molecular CO_2 assimilation is so efficient that extracellular CO_2 concentrations are reduced far below the equilibrium values by a transport system that recognizes molecular CO_2 and transports it into the cells (Miller *et al.*, 1991). Cyanobacteria also possess active transport systems for HCO_3^-. The bicarbonate transport system occurs in cells growing at low concentrations of extracellular CO_2, although bicarbonate transport appears to have no effect on CO_2

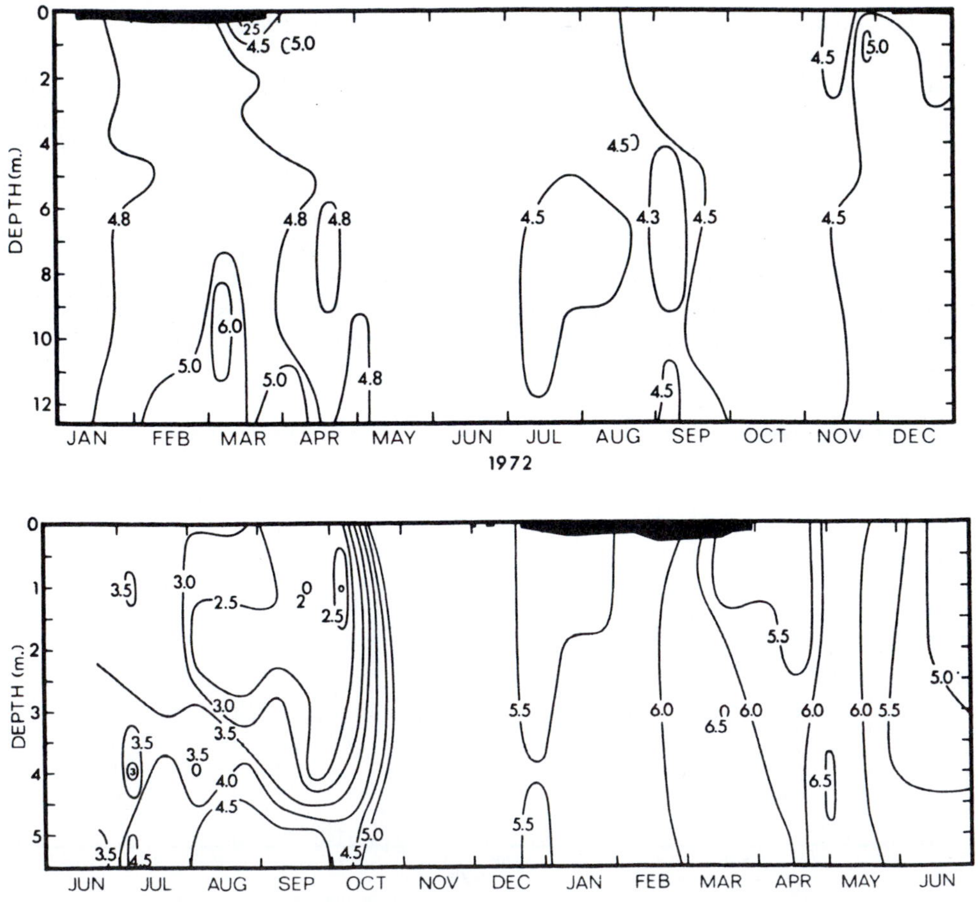

FIGURE 10-7 Depth–time distribution of isopleths of sodium concentrations (mg Na^+ liter^{-1}) of Lawrence Lake (*upper*) and potassium concentrations (mg K^+ liter^{-1}) of hypereutrophic Wintergreen Lake, Michigan, 1971–1972 (*lower*). Opaque areas = ice cover to scale. (From Wetzel, unpublished data.)

transport and CO_2 transport appears to have no effect on bicarbonate transport. One of the HCO_3^- transport systems is dependent on the presence of millimolar concentrations of sodium (Espie *et al.,* 1991). The active transport of CO_2, however, requires only micromolar levels of Na^+ (Miller *et al.,* 1990).

Therefore, sodium, along with many other factors, can influence the development of large populations of the cyanobacteria. Maximal growth of several cyanobacterial species was found at 40 mg liter^{-1}. The enrichment of waters with high levels of sodium and phosphorus, as is the case, for example, in domestic effluents with very high concentrations of both elements from synthetic detergents, was indicated as a potential contributor to effective competition among the cyanobacteria under bloom conditions (Provasoli, 1958; Wetzel, 1966a; Ward and Wetzel, 1975). Sodium and potassium occur in relative abundance as highly soluble cations of numerous salts. Alteration of their concentrations in natural waters is not common, except under conditions of pollution, for example, salinity from industry or domestic sources (Wetzel, 1966a) and animals (Kilham, 1982). Runoff from de-icing salts for roads has resulted in major alterations of the ionic composition of streams and lakes (Hoffman *et al.,* 1981; Sutcliffe and Carrick, 1983a, 1983b; Demers and Sage, 1990; Rich and Murray, 1990; Rosenberry *et al.,* 1999). Alternative de-icing compounds, such as calcium magnesium acetate (Schenk, 1991), are equally effective and much less intrusive biologically.

The spatial and temporal distribution of sodium and potassium in lakes is relatively uniform. Only small seasonal variations are observed, particularly for sodium, in accord with the conservative nature of these ions (Fig. 10-7; see also Stangenberg-Oporowska, 1967). Moderate epilimnetic reductions in potassium concentrations have been observed in productive lakes (e.g., Fig. 10-7) and in fertilized, productive farm ponds (Barrett, 1957). This reduction is related to potassium utilization by the massive algal populations and by submersed macrophytes and their epiphytes (Mickle and Wetzel, 1978a; Barko, 1982). Potassium is actively assimilated into submersed plant tissues with a light-dependent exchange process of reciprocal sodium efflux (Brammer and Wetzel, 1984). Lake sediments are net sources of potassium during summer stratification, and some hypolimnetic enrichment and export occurs in stratified, eutrophic lakes (e.g., Fig. 10-7; see also Stauffer and Armstrong, 1986).

The concentrations of rarer alkaline earth and alkali cations of natural waters vary considerably in relation to the lithology of the drainage basins. Their distribution is discussed by Durum and Haffty (1961), Livingstone (1963), and Cowgill (1976, 1977a,b) in general, and in closed-basin lakes in particular by Whitehead and Feth (1961). General ratios of major to minor cations are given in Table 10-7. Strontium cycling is closely coupled with calcium, and Sr/Ca ratios maintain a nearly constant stoichiometry (Stabel, 1989). Nutritional requirements for these relatively rare elements have not been demonstrated, although they can sometimes substitute for the major cations in metabolic pathways.

TABLE 10-7 Approximate Ratios of Minor Alkaline Earths and Alkalis to the Major Cations[a]

Na/Li	1500	Ca/Ba	400
Na/Rb	3600	Ca/Sr	3000–4500
Na/Cs	31,900	Ca/Be	ca. 40,000
		Ca/Ra	5×10^{10}

[a] From data of Livingstone (1963).

D. Chloride and Other Anions

Earlier discussion of the general distribution of chloride ions in lakes emphasized that chloride is usually not dominant in open lake ecosystems. Streams and lakes near maritime regions, however, often receive significant inputs of chlorides from atmospheric transport from the sea. Pollutional sources of chlorides can modify natural concentrations greatly. For example, the concentration of chloride in Lake Erie increased threefold in 50 years, largely (ca. 70%) from industrial sources, runoff from road salting, and municipal wastewaters (Ownby and Kee, 1967). Chloride is influential in general osmotic salinity balance and ion exchange, but metabolic utilization does not cause large variations in the spatial and seasonal distribution within most lakes (Fig. 10-8). Variations observed in Lake Erie and in many saline lakes are associated with hydrology of the basin and seasonal fluctuations (e.g., Ownbey and Kee, 1967; Wetzel, 1964).

Concentrations of minor halides (Table 10-8) vary somewhat with the lithology within the drainage basin, and higher concentrations often occur in lakes and streams in proximity to marine regions or in those that possess marine rock formations within their drainage basins. Boron is of greater limnological interest because it is a micronutrient required by many algae and other organisms. Concentrations of boron in natural waters are relatively high in comparison to other minor elements. The average concentration in rain and snow is also high (4.7 μg B liter^{-1}) and apparently is of terrestrial origin rather than from the sea (Nishimura *et al.,* 1973). Concentrations reach exceedingly high levels in

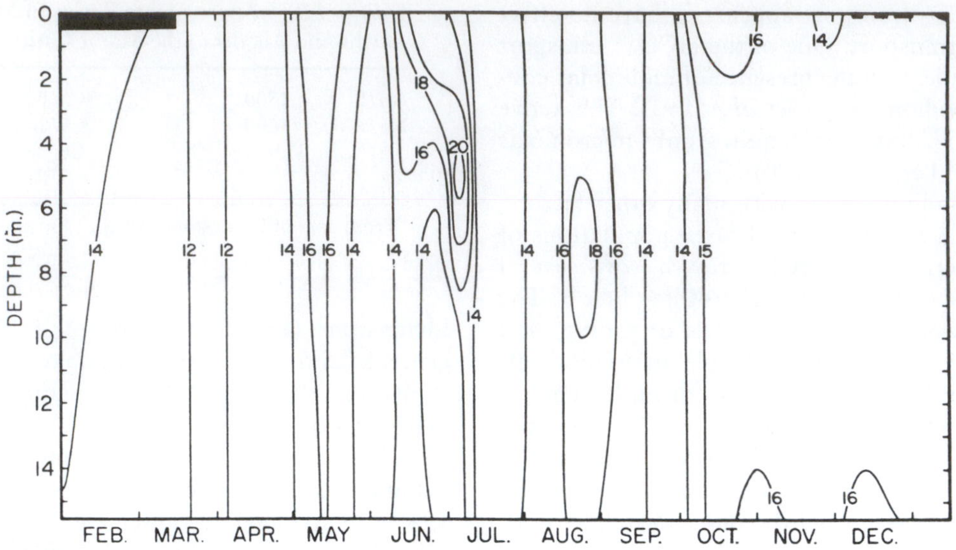

FIGURE 10-8 Depth–time distribution of isopleths of chloride concentrations (mg Cl liter^{-1}) of eutrophic Little Crooked Lake, Whitley County, Indiana, 1964. Opaque areas = ice cover to scale. (From Wetzel, unpublished data.)

certain closed, saline lakes and can approach 1 g per liter. As is the case with most micronutrients, high concentrations are generally toxic to most organisms, as was exemplified by the well-known excessive use of the micronutrient copper as a bioicide. Organisms adapted to high concentrations of boron, to >800 mg B liter^{-1}, were affected little by additions of further borate (Wetzel, 1964). Aquatic angiosperms normally living in waters of very low boron content show a high resistance to large concentrations of boron. Boron additions were found to be stimulatory to photosynthesis to levels of 100 mg liter^{-1}, beyond which inhibitory responses occurred (Baumeister, 1943).

E. Cation Ratios

The ratio of monovalent to divalent cations (M:D) is particularly interesting in relation to the distribution and dynamics of algae and higher aquatic plants in fresh waters. Three major genera of diatoms common to oligotrophic waters, *Fragilaria, Asterionella,* and *Tabellaria,* are stimulated by high levels of calcium (Chu, 1942; Vollenweider, 1950). Increasing levels of potassium permit increased tolerance of these algae to high concentrations of Ca and Mg. Numerous studies, particularly by Provasoli and co-workers (Provasoli, 1958), indicate that the M:D ratio is significant to the observed growth of these algae. Concentrations of Ca and Mg can be manipulated over a wide range as long as the M:D ratio is maintained within reasonably narrow limits for different species. Calcium and Mg are widely interchangeable, and many species are quite tolerant to different Ca:Mg ratios. These experimental results provide excellent confirmation of the much earlier perceptive ecological observations of Pearsall (1922, 1932), who considered a M:D ratio below 1.5 favorable to diatoms and much higher ratios favorable to desmid algae. Diatoms dominate the algal flora of very hard-water lakes with M:D ratios much less than 1.5. This relationship is most strongly correlated with the concentrations of calcium (Shoesmith and Brook, 1983). As the epilimnion undergoes biogenically induced decalcification in early summer in calcareous waters, calcium concentrations following the spring diatom maximum are often halved. It is unknown whether, along with other factors, the resulting shift in M:D ratio, which is concomitant with the shift to mixed populations of predominantly green algae with diatoms, influences phytoplanktonic succession.

TABLE 10-8 Approximate Average Concentrations of Halides and Boron in Natural Fresh Waters[a]

Element	*Average concentration (mg liter^{-1})*	*Chloride ratio*	
Chloride	8.3		
Fluoride	0.26	Cl/F	32
Bromine	0.006	Cl/Br	1400
Iodine	0.0018	Cl/I	4600
Boron	ca. 0.01		

[a] From data of Livingstone (1963).

The effects of cations on the photosynthesis and release of dissolved organic matter were studied on a submersed angiosperm, *Najas flexilis,* a macrophyte that grows poorly in extremely calcareous hardwater lakes (Wetzel, 1969; Wetzel and McGregor, 1968). Concentrations of Ca over 30 mg liter^{-1} and Mg over 10 mg liter^{-1}, both greatly exceeded in hardwater lakes, suppressed rates of photosynthetic carbon fixation and altered the secretion rates of fixed organic carbon as dissolved organic matter. Increasing concentrations of Na above levels commonly found in hardwater lakes increased both the rates of fixation and secretion. The overall effects observed were decreased rates of photosynthetic carbon fixation with decreasing M:D ratios.

Dissolved organic compounds of fresh waters could potentially regulate M:D ratios and indirectly influence rates of photosynthesis (Wetzel, 1968). Sequestering of divalent cations by chelation with amino substances and peptides is well known. Complex formation is also possible by pyrophosphates, binding with macromolecules such as proteins, and the formation of peptized metal hydroxides of humic acids. The mechanisms have not been adequately investigated in relation to controls over algal succession and dynamics, but it is clear that such sequestering mechanisms are additional means of the many by which dissolved organic matter exerts major influences on productivity. This subject is discussed in greater detail in Chap. 15.

F. Ionic Budgets within a Drainage Basin

As we have seen, the temporal and spatial loading, distribution, and fate of ions vary greatly with the lithology, climatic conditions, drainage and limnological characteristics, and biotic activities of soils, streams, and lakes. Examination of a representative example of an ionic budget is instructive to evaluate relative proportions of inputs and fates of ions.

The ionic budget of the Mirror Lake ecosystem within the Hubbard Brook Valley of New Hampshire has been studied many years (Likens, 1986). In this drainage basin, nearly all of the precipitation runoff not lost to evapotranspiration flows through the soil and is collected in streamflow. Precipitation is acidic (pH 4.1) and strongly influenced by air pollution. Cationic loadings in stream water greatly exceed inputs in bulk precipitation, as gains are obtained by weathering, evapotranspiration, biomass, and exchange processes within the soils (Table 10-9). Dissolved cations in stream water draining aggrading forests remained relatively constant despite extremes in hydrologic output and climatic variations. Particulate loadings and outputs were very small and were dependent on storm peaks (Table 10-9). Inputs to the lake and

TABLE 10-9 Average Annual Ionic Budgets (kg yr^{-1}) for Mirror Lake, New Hampshire, 1970–1976[a]

	Ca	*Mg*	*Na*	*K*	*Cl*	*H*$^+$
Inputs						
Precipitation (bulk)	18.3	4.5	19	6.1	88	14.1
Litter	10	1.0	<1	4.6	<1	—
Fluvial						
Dissolved	1943	406	946.0	339.0	882	0.90
Particulate	18	16	21.0	44.0	<1	—
Domestic sewage/ Road salt seepage	?	?	>4[b]	?	>6[e]	—
Total inputs	1980	427	990	394.0	976	15.0
Outputs						
Fluvial						
Dissolved	1881	393	970	380.0	735	0.67
Particulate	0	0	0	—	<1	—
Insect emergence	0.35	0.33	0.35	2.0	0.2	—
Permanent sedimentation	42[c]–117[d]	16[c]–68[d]	32[c]–221[d]	39[c]–205[d]	207[b]	—
Total outputs	1923–1998	409–461	1002–1191	421–587	942	0.67
Change in lake storage	0	0	+4	0	+34	0

[a] Extracted from Likens (1986).
[b] By difference.
[c] Spatially integrated sedimentation since ca. AD 1840.
[d] Extrapolation of precultural deposition rate and chemical content from a sediment depth of 25 cm to 50% of the lake area.
[e] Based on Na value and assumption of NaCl.

outputs from the lake were largely via cations dissolved in the stream water. Regulation of fluxes to the lake can be changed appreciably by disturbances within the drainage basin, such as deforestation and loadings from atmospheric, sewage, road salt leaching, and other sources of pollution. For example, clearcutting forest disturbance resulted in large losses of potassium in particular to stream water (Likens *et al.*, 1994). Of the major cations, potassium was the slowest to recover from clearcutting disturbances.

V. SALINITY, OSMOREGULATION, AND DISTRIBUTION OF BIOTA

A. Origins and Distribution of Freshwater Biota

The salinity of inland waters is generally very low in comparison to that of the sea, although in semiarid regions the salinity of closed-basin lakes occasionally exceeds that of sea water by several times. The distribution of biota in fresh waters has been influenced by a long evolutionary history of physiological adaptations to a wide range of salinities. These mechanisms developed against a background of large differences between the salinity of the environment and that of the cytoplasm or body fluids. The general distribution of the freshwater biota and their tentative origins in terrestrial, freshwater, or marine sources are summarized in the detailed review of Hutchinson (1967). That summary, coupled with reviews of the physiological mechanisms and adaptation of osmotic regulation (Krough, 1939; Beadle, 1943, 1957, 1959; Gessner, 1959; Robertson, 1960; Potts and Parry, 1964), provides an introduction to adaptations to freshwater life. The distribution of biota in brackish water interface regions between marine and fresh waters is summarized in the symposium on brackish waters (1959) and by Remane and Schlieper (1971).

The number of species living in brackish water is very much smaller than the number living in marine regions with similar habitats and much smaller than the number of species in fresh water (Fig. 10-9). However, it should be noted immediately that a paucity of species in this region in no way implies low productivity; productivity of adapted organisms can be exceedingly high. The lowest number of species occurs in salinities of about 5–7‰. The salinity gradient depicted here, with its associated osmotic and ionic properties, is the predominant factor influencing the distribution of biota in brackish waters. In these transitional waters, as in the spectrum of salinities of inland waters, the salinity range occupied by a species depends on the efficiency of the physiological mechanisms by which it is adapted to changes in salinity in the environment.

B. Osmotic Adaptations of Aquatic Plants and Animals

Osmotic regulation functions primarily in the maintenance of a difference in concentrations of ions inside and outside of the cells at appropriate operational physiological levels. The aquatic bacteria and cyanobacteria. Monera (lacking mitochondria or chromoplastids) and the more primitive Protista (algae, fungi, and protozoa with mitochondria and, if they are photosynthetic, with chromoplastids) demonstrate high evolutionary adaptability to changes in salinity (cf. Hutchinson's 1967 description of evolutionary euryhalinity) through relatively small genetic changes. Most freshwater bacteria and cyanobacteria are relatively homoiosmotic and tolerate only a narrow range of salinity but adapt to increasing salinity relatively rapidly by means of genetic change. Extensive adaptive radiation is seen among these groups. Members of the Protista, which are largely single-celled, retain considerable evolutionary euryhalinity and are widely distributed with respect to salinity, although some groups, especially among the green algae, are restricted to fresh water. The contractile vacuole is the primary osmoregulatory organelle among the Protista.

Higher aquatic plants, which are of terrestrial origin, have developed adaptations to fresh water secondarily. Only a few major groups of angiosperms

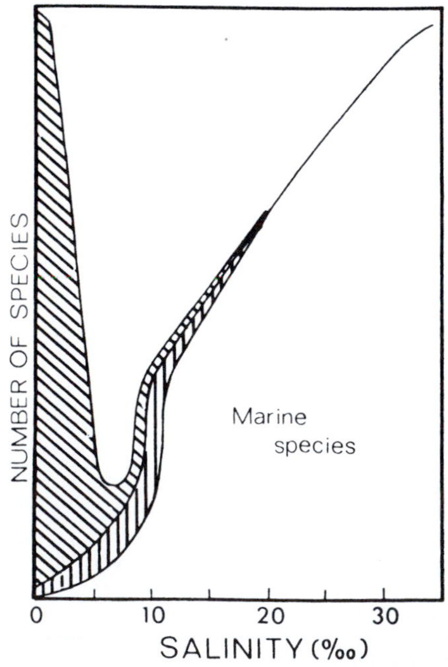

FIGURE 10-9 Number of species in relation to salinity. Diagonal hatching: proportion of freshwater species; vertical hatching: proportion of brackish-water species; lower open area: marines species. (Modified from Remane and Schlieper, 1971.)

have developed extensively in fresh waters, and very few groups extend into saline waters of brackish or marine areas or into hypersaline closed-basin lakes.

Nearly all of the higher freshwater animals originated from the sea, although most aquatic insects are of terrestrial origin. Both in terms of evolutionary and contemporary life in fresh water, osmoregulation is a major problem, for which diverse mechanisms have developed to regulate salt and water content. Adaptation to low salinities by some marine animals has been achieved without osmoregulation. Such ***poikilosmotic*** animals adjust the osmotic pressure of their body fluids to become more or less isotonic with the salinity of the medium. In contrast, a ***homoiosmotic*** animal will tend to retain its initial internal osmotic concentration upon being exposed to modest changes in the salinity of the medium. The general relations between the osmotic pressure among different types of animals, expressed as salinity of the blood versus salinity of the external water, can be visualized in Figure 10-10. The range depicted by area A extends over a wide variation in osmotic pressure of body fluids that is found in brackish-water animals, which tend to be more poikilosmotic at high concentrations and more homoiosmotic at lower salinities. The range of the osmotic pressure curve extends from the most homoiosmotic (a_1) to the most poikilosmotic species (a_2). The lower limits are very variable, represented by the undefined left-hand edge of area A, but all of these species have failed to colonize fresh water.

A few species have succeeded in penetrating fresh water without a renal osmoregulatory mechanism. These brackish-water animals are partially homoiosmotic (area B, Fig. 10-10) and maintain a very high

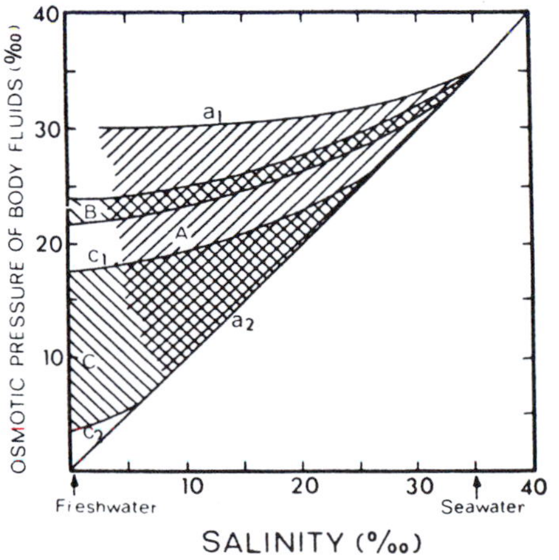

FIGURE 10-10 Relationship between the osmotic pressure, expressed as salinity, of body fluids of brackish-water organisms (A) and freshwater animals (B, C) and the salinity of external water. (Modified from Beadle, 1959.) Relationships of the curves are detailed in the text.

osmotic pressure in hypertonic body fluids by the active uptake of ions from the water. Excretory organs are not involved, since the urine produced is isotonic with the blood.

In most freshwater animals, however, osmotic pressures of the body fluids have decreased to levels equivalent to 5–15‰ salinity to reduce osmotic gradients. These organisms have developed excretory organs that effectively recover ions and produce urine hypotonic to the body fluids. Most freshwater animals therefore effect osmoregulation by active uptake of ions and by a renal mechanism of ion retention. Extremes in osmotic pressures of blood delineate area C of Figure 10-10 by curves c_1 and c_2. The isotonic line along the right-hand edge of area C indicates the upper salinity tolerance limit of most freshwater animals and that they are incapable of hypotonic regulation. Although most freshwater animals are capable of living in water of low salinity, adaptation of body fluids of low osmotic pressure is apparently irreversible, and with few exceptions freshwater organisms are restricted to salinities of <10‰.

Because of the slow diffusion of oxygen in water relative to that in air, movement of water over permeable membranes or tissue surfaces for respiratory needs is almost universal among aquatic animals. The pumping process places high energetic demands on the animals and additionally exposes cellular surfaces to osmotic gradients. Mechanisms for taking up salts against a concentration gradient vary greatly among freshwater animals. In a few organisms, incorporation of salts with the food may be adequate, but more often organs have developed for this purpose. Aside from resorption mechanisms of the excretory organs, which are advantageous energetically, active uptake mechanisms for ions, especially sodium and chloride, are often associated with the respiratory organs (commonly gills) of many invertebrates and vertebrates.

The fauna of extremely saline inland lakes are relatively insensitive to the high salinity of these lakes and to large interseasonal fluctuations in the chemical composition of the water (Beadle, 1969). Most of the animals of saline lakes are of freshwater origin and include particularly representatives of the aquatic insects, phyllopods, copepod and cladoceran crustacea, and rotifers, all of which belong to predominantly freshwater groups. The blood of these saline-inhabiting animals maintains osmotic pressures at levels characteristic of those living in fresh waters. The body surface of these animals exhibits very low permeability, and they possess effective excretory mechanisms for maintaining the body fluids strongly hypotonic to the external medium. Furthermore, the water balance of saline inland waters frequently fluctuates widely; freshwater animals with resting stages capable of withstanding desiccation have a distinct advantage over marine animals, which lack strong development of this characteristic.

VI. SUMMARY

1. Total ***salinity*** of inland waters is usually dominated by four major cations [calcium (Ca^{++}), magnesium (Mg^{++}), sodium (Na^{+}), and potassium (K^{+})] and the major anions [bicarbonate (HCO_3^{-}), carbonate (CO_3^{-}), sulfate (SO_4^{-}), and chloride (Cl^{-})].
2. The salinity of surface waters has a world average concentration of about 120 mg liter^{-1} but varies among continents and with the lithology of the land masses and drainage basins (Tables 10-1 and 10-2).
   a. The salinity of water is a composite of ionic contributions leaching from rock and soil in runoff from the drainage basin, atmospheric precipitation and deposition, and the balance between evaporation and precipitation.
   b. Concentrations of major ions of many surface waters of the world exist in the proportions of $Ca > Mg \geq Na > K$ and $HCO_3–CO_3 > SO_4 > Cl$. In soft waters and in surface waters of coastal regions, Na and Cl often occur in greater equivalent concentrations (Table 10-3).
3. The release of ions as soil and rock weather is controlled by the processes of solution and oxidation–reduction, the action of hydrogen ions, and the formation of organic complexes. A major source of salinity to many dilute fresh waters and certain saline lakes of arid regions is the ionic content of atmospheric precipitation and particulate deposition.
4. Concentrations of the cations magnesium, sodium, and potassium and the major anion chloride are relatively conservative and undergo only minor spatial and temporal fluctuations within a lake from biotic utilization or biotically mediated environmental changes. Calcium, inorganic carbon, and sulfate are dynamic, and concentrations of these ions are influenced strongly by microbial metabolism.
   a. Proportional concentrations of major cations and the ratios of monovalent:divalent cations can influence the metabolism of many organisms, particularly certain algae and submersed macrophytes, as much as absolute concentrations do.
   b. Processes that influence the availability of some cations disproportionately to others (e.g., organic complexing of calcium or biologically induced decalcification of the epilimnion) can indirectly affect seasonal population succession and productivity.
5. The relatively low salinity of fresh waters has influenced greatly the distribution of biota and their long evolutionary history of physiological adaptations for osmotic and ionic regulation in an extremely hypotonic environment.
   a. Although some groups of bacteria and algae are relatively homoiosmotic and can tolerate only a narrow range of salinity, most of the lower flora and fauna are euryhaline, that is, adaptable to a wide range of salinity.
   b. Most higher freshwater animals originated from the sea or from land and adapted to fresh water secondarily. In comparison to marine forms, nearly all of these organisms have reduced osmotic pressures of body fluids and have developed efficient mechanisms for active uptake of ions and renal mechanisms for ion retention.

11

THE INORGANIC CARBON COMPLEX

Inorganic carbon, largely as dissolved carbon dioxide and bicarbonate, is the primary source of carbon for photosynthesis and the generation of organic substances. These organic compounds are generated by cyanobacteria, algae, and higher plants both within the lakes or rivers or externally within the drainage basin and variously imported to the water bodies. These organic carbon-based compounds provide the materials and energy for subsequent metabolism within the ecosystem. Inorganic carbon utilization is balanced by respiratory production of CO_2 by most organisms and by influxes of CO_2 and HCO_3^- from incoming water and from the atmosphere.

The atmosphere and minerals are the sources of inorganic constituents dissolved in fresh water. Precipitation falling on the surface of the Earth as rain or snow contains a variety of gases, aerosols, and dust particles. As the precipitation flows over or penetrates into plants or the soils, water dissolves more gases, particularly carbon dioxide, and provides an acid that reacts with bases of rocks and various mineral substances with which it comes into contact. The water can also lose dissolved carbon to the sediments by precipitation reactions.

Dissolved inorganic carbon is a major constituent of inland waters and can influence many characteristics of gaseous and nutrient availability, as well as serving as the foundation of organic productivity. The loadings of carbon dioxide to and quantities within the atmosphere are increasing progressively, largely from anthropogenic combustion of fossil deposits of organic matter (e.g., Schlesinger, 1991). The dissolved inorganic carbon constituents also influence water quality properties such as acidity, hardness, and related characteristics. Thus, it is essential that the rudiments of dissolved inorganic carbon reactivity be evaluated.

I. THE OCCURRENCE OF INORGANIC CARBON IN FRESHWATER SYSTEMS

A. Carbon Dioxide and Its Solution in Water

The carbon dioxide (CO_2) content of the atmosphere varies with locality and potential enrichment from industrial pollution. The global average is approximately 0.0355% by volume, as of 1991, and is increasing at ca. 2% per year (Machta, 1973; Broecker *et al.*, 1979; Gates, 1993). Carbon dioxide is very soluble in water, some 200 times more than oxygen, and obeys normal solubility laws within the conditions of temperature and pressure encountered in lakes. The amount of CO_2 dissolved in water from atmospheric concentrations is about 1.1 mg liter^{-1} at 0°C, 0.6 mg liter^{-1} at 15°C, and 0.4 mg liter^{-1} at 30°C.

As CO_2 dissolves in water, the solution contains unhydrated CO_2 at about the same concentration by volume (approximately 10 μM) as in the atmosphere (reviewed extensively by Hutchinson, 1957; Kern, 1960; J. C. Goldman *et al.*, 1972; Morel and Hering, 1993; Stumm and Morgan, 1995):

$$CO_2(\text{air}) \rightleftharpoons CO_2(\text{dissolved}) + H_2O$$

Dissolved CO_2 hydrates by a slow reaction (a half-time of approximately 15 s):

$$CO_2 + H_2O \rightleftharpoons H_2CO_3$$

This reaction predominates at a pH of less than 8, with the equilibrium concentration of H_2CO_3 only about 1/400 that of the unhydrated CO_2. Above a pH of 10, $CO_2 + OH^- \rightleftharpoons HCO_3^-$ is the dominant reaction.

H_2CO_3, is a fairly weak acid that dissociates rapidly relative to the hydration reaction:

$$H_2CO_3 \rightleftharpoons H^+ + HCO_3^-$$

$$HCO_3^- \rightleftharpoons H^+ + CO_3^{2-}$$

The pK_1 dissociation value of the first reaction, including both hydrated and unhydrated CO_2 as the undissociated molecule, is 6.43 at 15°C. The pK_2 of the second reaction is 10.43 (15°C).

The bicarbonate and carbonate ions also dissociate to establish an equilibrium:

$$HCO_3^- + H_2O \rightleftharpoons H_2CO_3 + OH^-$$

$$CO_3^{2-} + H_2O \rightleftharpoons HCO_3^- + OH^-$$

$$H_2CO_3 \rightleftharpoons H_2O + CO_2$$

The hydroxyl ions generated in the first two of these reactions result in alkaline waters (above a pH of 7) in lakes and streams that have a naturally high content of carbonates derived from surface and ground water of the drainage basin. As water percolates through soil of the drainage basin, it becomes enriched with CO_2 from plant and microbial respiration. The carbonic acid that forms solubilizes limestone of calcium-enriched rock formations and produces calcium bicarbonate [$Ca(HCO_3)_2$], which is relatively soluble in water, and increases the amount of ionized Ca^{++} and HCO_3^- of the water. As HCO_3^- and CO_3^{2-} increase in hard waters of calcareous regions, common to much of the glaciated temperate region, the pH also increases from the release of hydroxyl ions. In normal hard waters of the midwestern United States, for example, HCO_3^- is the dominant anion (ca. approximately 60%), so that the pH of the water is >8 and [HCO_3^-] >100 mg 1iter^{-1}. Saline lakes of endorheic regions often have a carbonate dominance (CO_3^{2-} >10,000 mg 1iter^{-1} and HCO_3^- >1000 mg liter^{-1} with pH >9.5). Such highly saline carbonate brines, in which total inorganic carbon exceeds several moles per liter, are rare.

Photosynthesis and respiration are two major factors that influence the amounts of CO_2 in water. However, the equilibria of the reactions given here result in the buffering action of alkaline waters, which contain appreciable amounts of bicarbonate. Water tends to resist change in pH as long as these equilibria are operational. An addition of hydrogen ions neutralizes hydroxyl ions formed by the dissociation of HCO_3^- and CO_3^{2-} but more hydroxyl ions are formed immediately by reaction of the carbonate with water. Consequently, the pH remains essentially unaltered, unless the supply of carbonate or bicarbonate ions is exhausted. Similarly, when hydroxyl ions are added they react with bicarbonate ion:

$$HCO_3^- + OH^- \rightleftharpoons CO_3^{2-} + H_2O$$

If the pH of a solution is held constant with buffer reactions, and it is permitted to equilibrate with gaseous carbon dioxide, the total hydrated and unhydrated CO_2 in solution is independent of pH, while the bicarbonate and carbonate concentrations increase with pH in accordance with the pK values. These equilibria are influenced by temperature and by salt concentration (ionic strength). At the salinity of sea water, the pK_1 for HCO_3^- is about 0.5 pH unit and the pK_2 for CO_3^{2-} is about 1 unit lower than in fresh water. Both oceanic and fresh waters are close to equilibrium with atmospheric CO_2. In the marine habitat, the inorganic carbon pool contains about 2 mmol C liter^{-1}, largely as HCO_3^-, a reservoir some 50 times that of the atmosphere. In fresh waters, the total inorganic carbon (ΣCO_2) is much more variable, within a typical range of 50 μmol to 10 mmol liter^{-1}, and more pH dependent. From these dissociation relationships, the proportions of CO_2, HCO_3^-, and CO_3^{2-} at various pH values can be evaluated (Fig. 11-1). Free CO_2 dominates in water at pH 5 and below, while above pH 9.5 CO_3^{2-} is quantitatively significant (Table 11-1). Between pH 7 and 9, HCO_3^- predominates.

B. CO_2 Exchange between the Atmosphere and Water

The diffusion of CO_2 from the atmosphere and the dissociation kinetics of dissolved carbonates are obviously of major importance to photosynthetic organisms

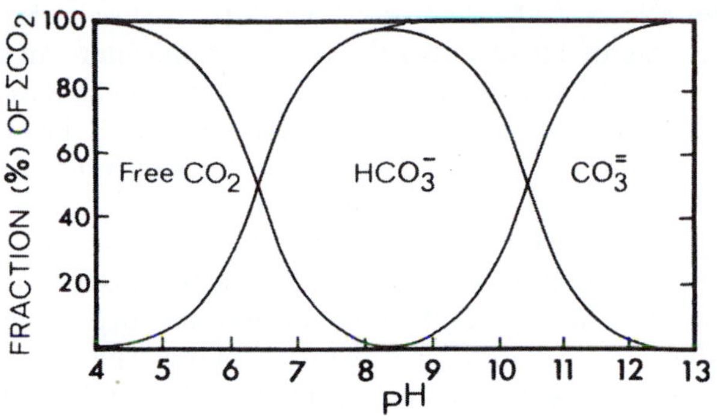

FIGURE 11-1 Relation between pH and the relative proportions of inorganic carbon species of $CO_2(+H_2CO_3)$, HCO_3^-, and CO_3^{2-} in solution. (Slightly modified from Goiterman, H. L. (ed.): *Methods for Chemical Analysis of Fresh Waters,* IBP Handbook No. 8, Oxford England, Blackwell Scientific Publications, 1969.)

dependent on the availability of inorganic carbon for photosynthesis. The magnitude of CO_2 exchange between the atmosphere and water cannot be determined by partial pressure differences alone. Diffusion of atmospheric CO_2 has been examined by utilizing radium-226 and the flux of radon-222 to determine gas transfer rates between the atmosphere and water (Broecker *et al.,* 1965, 1971, 1973). When applied to a very softwater lake of the Canadian Shield of very low ΣCO_2, atmospheric invasion of CO_2 was adequate (0.12 ± 0.06 g C m^{-2} day^{-1}) to account for 30–90% of the carbon fixed by phytoplankton (Schindler *et al.,* 1972). In this sheltered lake, an invasion rate of CO_2 from the air of 17 ± 8 mmol CO_2 m^{-2} day^{-1} was estimated through a hypothetical "stagnant boundary layer" at the surface of about 300 μm in thickness (Emerson *et al.,* 1973). This layer decreased in thickness with increased exchange as wind velocities increased, particularly above 1.5 m sec^{-1}. Consumption of CO_2 by epilimnetic photosynthesis can enhance the flux of

TABLE 11-1 Proportions of CO_2, HCO_3^-, and CO_3^{2-} in Water at Various pH Values[a]

pH	*Total free* CO_2	HCO_3^-	CO_3^{2-}
4	0.996	0.004	1.25×10^{-9}
5	0.962	0.038	1.20×10^{-7}
6	0.725	0.275	0.91×10^{-5}
7	0.208	0.792	2.6×10^{-4}
8	0.025	0.972	3.2×10^{-3}
9	0.003	0.966	0.031
10	0.0002	0.757	0.243

[a] From Hutchinson, G. E.: *A Treatise on Limnology,* New York, John Wiley & Sons, Inc., 1957, p. 657.

atmospheric CO_2 to these soft waters (Weiler, 1974; Wood, 1974, 1977; Emerson, 1975). As noted earlier (Chap. 2), organic substances dissolved in or present on the water surface can also decrease rates of gas exchange and of evaporation.

An extensive survey of the partial pressure of CO_2 in the surface waters from a large number of lakes (1835) of worldwide distribution showed that only a small proportion (<10%) were near equilibrium with the atmosphere and that most samples (87%) were supersaturated (Cole *et al.,* 1994). The average partial pressure of dissolved CO_2 was about 3 times the value in the overlying atmosphere. Data presented for one lake in this paper and many other studies have demonstrated a distinct seasonality to CO_2 evasion to the atmosphere (e.g., Otsuki and Wetzel, 1974; Kling *et al.,* 1991; many others). For example, in a hardwater lake in Michigan that receives high inputs of inorganic carbon as bicarbonate, Otsuki and Wetzel (1974) determined that the partial pressure in two inlet waters and ground water was about 30 times higher than that in the atmosphere. In the outlet, the partial pressure of dissolved CO_2 was 7 times higher during the spring period, and 3 times higher even during the summer period of maximum utilization by photosynthesis and large losses by precipitation and sedimentation of $CaCO_3$. CO_2 in this and most lakes is released into the atmosphere throughout the year.

As discussed later in this chapter, CO_2 from the respiratory processes of decomposition can accumulate to large quantities in the hypolimnia of lakes. Under certain conditions of lake basin morphometry and annual mixing, as in certain deep, tropical crater lakes, CO_2 can accumulate to extraordinarily high concentrations. Decompression of CO_2-saturated water can, with changes in seasonal stabilities of stratification, erupt to the surface at explosive rates (range, 50–90 m s^{-1}) (Zhang, 1996). So much CO_2 is released by this exsolution and evasion to the atmosphere that the dense CO_2 over the lake surface can temporarily cascade over the surrounding terrain. Such massive evasions occurred in Lake Nyos and Lake Monoun of Cameroon, and over 1700 humans and many terrestrial animals died by asphyxiation (Kling *et al.,* 1987, 1991). The sources of the hypolimnetic accumulations of CO_2 and CH_4 were largely biogenic; volcanic sources were minor (Kling *et al.,* 1989).

C. Proportions of Dissolved Inorganic Carbon in Fresh Waters

Carbon is transported continuously by rivers as particulate and dissolved organic carbon (POC, DOC) and as dissolved inorganic carbon (DIC) that functions in weathering reactions. Of the global carbon fluxes to the oceans from the continents, about 18% moves in riverine transport as POC, 37% as DOC, and 45% as DIC (Meybeck, 1993a). Proportions of DIC, POC, and DOC in the rivers of the world are variable; differences emanate from lithology, which controls DIC levels, and differences in relief that control the DOC/POC ratios. DIC is very low in many tropical rivers of humid regions, and there DOC is the dominating form of exported carbon. Within the pelagic zone of lakes and reservoirs, however, the ratio of DOC/POC is ca. 10:1 (Wetzel, 1984).

An average of 80% of the DIC alkalinity of North American river water is attributable to carbonate rock dissolution and 20% to the weathering of aluminosilicate rocks (Morel and Hering, 1993). Of world average river water with higher alkalinity, about 60% of the alkalinity is contributed by dissolution of carbonates, and most of the remainder originates from dissolution of calcium, magnesium, and alumino-silicates. On a global discharge-weighted basis, however, the dissolved content of surface waters is dominated by HCO_3^-, Ca^{+2}, SO_4^{-2}, and SiO_2, and over 97% of global runoff has been classified as of the calcium bicarbonate type (Meybeck, 1993b).

The total inorganic carbon concentration in fresh water depends on the pH, which is governed largely by the buffering reactions of carbonic acid and the amount of bicarbonate and carbonate derived from the weathering of rocks. Carbonates exist as a number of polymorphic and hydrated forms. The most important carbonate of aquatic systems is $CaCO_3$, which occurs in natural waters principally as calcite and the metastable polymorphic aragonite.

The solubility of CO_2 increases markedly in water that contains carbonate. A definite amount of free CO_2 will remain in solution after equilibrium is reached between calcium, bicarbonate, carbonate, and undissociated calcium carbonate. The amount of excess CO_2 required to maintain stability of $Ca(HCO_3)_2$ in solution increases very rapidly with increasing content of bicarbonate in the water derived from carbonates. If the amount of free CO_2 is increased above that required to maintain a given amount of $CaCO_3$ in solution at equilibrium as $Ca(HCO_3)_2$, this aggressive CO_2, as it is termed, will dissolve additional $CaCO_3$.

If a solution of calcium bicarbonate in equilibrium with CO_2, H_2CO_3, and CO_3^{2-} loses a portion of the CO_2 required to maintain the equilibrium (e.g., CO_2 assimilated by photosynthetic organisms), $CaCO_3$ will precipitate until the equilibrium is reestablished by the formation of CO_2:

$$Ca(HCO_3)_2 \rightleftharpoons CaCO_3 \downarrow + H_2O + CO_2$$

The excess CO_2 that is required to maintain large amounts of HCO_3^- in solution at equilibrium can be lost in several ways, and result in massive precipitation of $CaCO_3$. Ground water of limestone regions, heavily enriched with CO_2 from terrestrial decomposition, can release much CO_2 into the atmosphere when it flows to the surface, with resulting precipitation of $CaCO_3$. When such spring water, rich in bicarbonate, surfaces in streams or lakes, all substrata become covered with a dense encrustation of $CaCO_3$. A major cause of the loss of aggressive CO_2 in hardwater lakes and certain streams is photosynthetic utilization of CO_2 by phytoplankton and littoral-submersed macrophytes and epiphytic algae (e.g., Otsuki and Wetzel, 1974; Nebrasova, 1984; Dustin *et al.*, 1986; Stabel, 1986; Küchler-Krischun and Kleiner, 1990). Some of the CO_2 used in photosynthesis is generated by the calcification process, which raises pCO_2 immediately adjacent to the photosynthesizing surface (McConnaughey, 1994). Hardwater lakes rich in bicarbonate commonly undergo massive epilimnetic decalcification during the summer stratification period as a result of photosynthetic removal of CO_2 (a subject discussed later). In the littoral zone of hardwater lakes, the submersed macrophytic vegetation is densely encrusted with $CaCO_3$ precipitated by photosynthetic utilization of CO_2 by the macrophytes and epiphytic algae. The marl encrustations are frequently large and often exceed the weight of the plant (Wetzel, 1960). Cyanobacteria growing attached to substrata in the littoral of lakes and in streams also produce large deposits of carbonates (Golubić, 1973).

In addition to highly dynamic demands for CO_2 from and inputs of CO_2 to fresh water, complex shifts in precipitation and dissolution reactions of carbonate occur spatially and temporally. In alkaline hardwater lakes, often twice the concentrations of calcium and bicarbonate are found than would be expected on the basis of equilibrium with atmospheric pressures of CO_2 (Ohle, 1934, 1952; Wetzel, 1966b, 1972; Otsuki and Wetzel, 1974). The solubility product of $CaCO_3$ is low (0.48×10^{-8}), and $CaCO_3$ can start precipitating from calcareous waters when the pH is sufficiently high in a uniformly buffered system or where equilibria are shifted in microzones associated with sites of active photosynthesis. However, large amounts of inorganic carbon can exist as carbonate and $CaCO_3$ in metastable conditions (House, 1984; Stumm and Morgan, 1995). Once crystallization has started, often in association with a particle such as an algal cell, the rate of crystallization is proportional to the concentrations of calcium and carbonate (Nancollas and Reddy, 1971). A large percentage of the calcite crystals are rhombohedral or dendritic in structure (e.g., Raidt and

Koschel, 1988; Küchler-Krischun and Kleiner, 1990). Although calcite is clearly the most common crystal of $CaCO_3$ formed in lakes and streams, more dense aragonite of differing structure often originating from mollusk shells can dominate in certain lakes, such as seepage, groundwater-dominated marl lakes (e.g., Brown *et al.*, 1992).

The rate of $CaCO_3$ precipitation is slow, unless induced metabolically, as by photosynthesis. The result is a supersaturation with respect to both Ca^{++} and HCO_3^- at concentrations often two to three times that predicted on the basis of equilibria. There is strong evidence that appreciable $CaCO_3$ occurs in stable colloidal form in hardwater lakes (White and Wetzel, 1975). The trophogenic zone in moderately hard Lake Mendota was found to be supersaturated with respect to Ca^{++} and HCO_3^- in all seasons except winter (Morton and Lee, 1968). Extremely hard Lawrence Lake was found to be supersaturated continually (Otsuki and Wetzel, 1974), a situation common where dissolved CO_2 is not in equilibrium with atmospheric CO_2.

The importance of colloidal $CaCO_3$, in addition to larger particulate $CaCO_3$, is poorly understood in relation to indirect effects upon metabolism and flux rates of organic carbon (Wetzel and Rich, 1972). Organic compounds (amino acids, fatty acids, and humic acids) adsorb to particulate and colloidal $CaCO_3$ (Chave, 1965; Suess, 1968, 1970; Chave and Suess, 1970; Wetzel and Allen, 1970; Meyers and Quinn, 1971a; Orlov *et al.*, 1973; Otsuki and Wetzel, 1973; Stewart and Wetzel, 1981). Although such adsorption could be viewed as a scavenging and concentrating process whereby labile dissolved organic carbon from dilute solution is rendered more concentrated for utilization by bacteria, empirical evidence indicates, rather, a chemical competition with the bacteria for the organic substrates. During photosynthetic removal of CO_2, a large fraction of $CaCO_3$ is precipitated by algae and macrophytic vegetation. Frequently, the plant cells serve as a nucleus for particulate $CaCO_3$ formation, which is where organic compounds are being secreted. This association of dissolved organic detrital carbon with $CaCO_3$ is a component of certain freshwater systems that affects the chemical milieu without clearly defined energetic transformations (Wetzel *et al.*, 1972). The organic coatings also reduce the rate of dissolution of sedimenting $CaCO_3$ in lakes and can form a major sink for inorganic and organic detrital carbon in hardwater ecosystems (Wetzel, 1970, 1972; Kleiner, 1990).

D. Alkalinity and Acidity of Natural Waters

Natural waters exhibit wide variations in relative acidity and alkalinity, not only in actual pH values but also in the amount of dissolved material producing the acidity or alkalinity. The concentration of these compounds and the ratio of one to another determine the actual pH and the buffering capacity of a given water. Since the lethal effects of most acids begin to appear near pH 4.5 and of most alkalis near pH 9.5, that buffering can be of major importance in the maintenance of life.

Alkalinity is historically a term that referred to the buffering capacity of the carbonate system in water. Alkalinity is now used interchangeably with ***acid neutralizing capacity (ANC),*** which is the capacity to neutralize strong inorganic acids. Alkalinity in water results from any dissolved species, usually weak acid anions, that can accept and neutralize protons. Because CO_2 is quite soluble and relatively abundant in water in gaseous and dissolved forms, and carbonates are common as primary minerals over much of the Earth, the property of alkalinity of most fresh waters is imparted by the presence of bicarbonates and carbonates, and the CO_2—HCO_3^-—CO_3^{2-} equilibrium system is the major buffering mechanism in fresh waters. Carbonate and hydroxide ions may be present when the pH is very high (Fig. 11-2). Hydroxide, borate, silicate, phosphate, and sulfide are usually present in small quantities in fresh waters, but can be major sources of alkalinity in certain saline inland waters. Dissolved organic anions derived from large amounts of dissolved organic carbon can add alkalinity, particularly among soft waters with limited dissolved inorganic carbon.

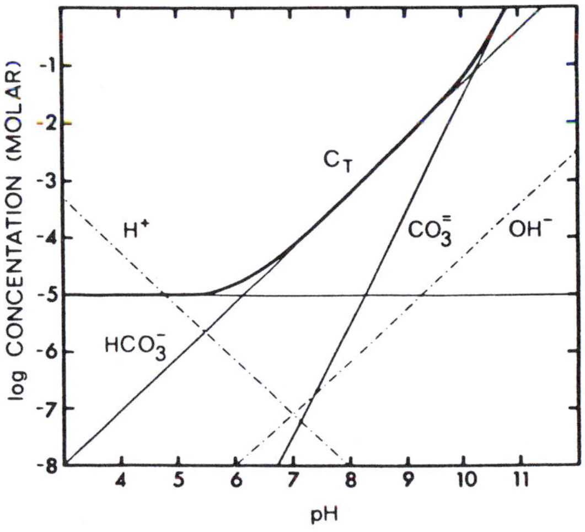

FIGURE 11-2 Model of distribution of carbon species of the carbonate system of natural waters. Pure water is equilibrated with atmospheric CO_2 at a constant partial pressure ($pCO_2 = 10^{-3.5}$ atm, 25°C). The pH can be varied by the addition of a strong acid or strong base, thereby keeping the solution in equilibrium with pCO_2. $\Sigma CO_2 = [CO_2(aq)] + [H_2CO_3] + [HCO_3^-] + [CO_3^{2-}]$. (From Wetzel and Likens, 2000, modified from Stumm and Morgan, 1981, 1995.)

The terms ***alkalinity, carbonate alkalinity, alkaline reserve, titratable base,*** or ***acid-binding capacity*** are frequently used to express the total quantity of base (usually in equilibrium with carbonate or bicarbonate) that can be determined by titration with a strong acid (Hutchinson, 1957). The milliequivalents of acid necessary to neutralize the hydroxyl, carbonate, and bicarbonate ions in a liter of water are known as the ***total alkalinity.*** Alkalinity is numerically the equivalent concentration of titratable base and is determined by titration with a standard solution of a strong acid to equivalency points dictated by pH values at which the alkaline contributions of hydroxide, carbonate, and bicarbonate are neutralized.

The least ambiguous usage of alkalinity is to express values as mass per unit volume; that is, milliequivalents per liter (meq liter^{-1}). In effect, alkalinity measures the proton deficiency with respect to the reference proton level CO_2—H_2O (this topic is reviewed at length by Stumm and Morgan, 1995). Alkalinity is often expressed in milligrams per liter (or parts per million) of $CaCO_3$. This expression assumes that alkalinity results only from calcium carbonate and bicarbonate, which in some lakes (e.g., closed alkaline lakes), implies much greater calcium than actually is present. In moderately hard waters, nearly all of the base is present as bicarbonate, and the term ***bicarbonate alkalinity,*** as mg HCO_3^- liter^{-1}, is sometimes used. The clearest expression, however, is milliequivalents per liter (1 meq liter^{-1} = 50 mg liter^{-1} as $CaCO_3$).

The term ***hardness*** is frequently used as an assessment of the quality of water supplies. The hardness of a water is governed by the content of calcium and magnesium salts, largely combined with bicarbonate and carbonate (temporary hardness) and with sulfates, chlorides, and other anions of mineral acids (permanent hardness). The carbonate hardness can be removed by boiling, which causes precipitation of $CaCO_3$. The fraction of calcium and magnesium remaining in solution as sulfates, chlorides, and nitrates after boiling constitutes the residual noncarbonate hardness. The extent of hardness has been expressed numerically in a remarkably heterogeneous system of scales among different countries (equivalents are given in Table 11-2). For example, one degree of hardness (H°) in the United States equals 1 mg $CaCO_3$ 1iter^{-1}; one German H° corresponds to a concentration of 10 mg lime (CaO) liter^{-1}. Although Höll (1972) has proposed that the international unit be expressed in mval (1 mval = 1 milliequivalent per liter of the material concerned), this terminology has been rescinded by the Système International d'Unités (SI) (see the critical evaluation of the unit Val (= gram equivalent) by Hochmüller and Simoneth (1980a,b).

TABLE 11-2 Various Scales of Hardness of Water[a]

1 German degree of hardness, dH°	= 10 mg CaO liter^{-1} = 7.14 mg Ca liter^{-1} = 17.9 mg $Ca(HCO_3)_2$ liter^{-1}
1 French degree of hardness, French H°	= 10 mg $CaCO_3$ liter^{-1}
1 English degree of hardness, English H°	= 10 mg $CaCO_3^{-1}$ = 0.8° dh
1° dh	= 1.25 English H° = 1.79 French H°
1 French H°	= 0.56 German dH° = 0.7 English H°
1 American degree of hardness	= 1 mg $CaCO_3$ liter^{-1} = 0.056 German dH°
International degree of hardness, mval	= 1 meq liter^{-1} = 2.8 German dH°

[a] After Höll, K.: Water: *Examination, Assessment, Conditioning, Chemistry, Bacteriology, Biology,* Berlin, Walter de Gruyter, 1972. See also Hochmüller and Simoneth (1980a,b).

Acidity is used infrequently as a parameter in limnological investigations. Uncombined carbon dioxide, organic acids such as tannic, humic, and uronic acids, mineral acids, and salts of strong acids and weak bases are usually responsible for the acidity of natural waters. Free CO_2 of most waters is seldom present in large quantities because of its reactions in the carbonate equilibria and exchange with the atmosphere, discussed earlier. In practice, acidity is a measure of the quantity of strong base per liter required to attain a pH equal to that of a solution of sodium carbonate (Na_2CO_3) equivalent to the total inorganic carbon; that is, it is a measure of the active or free CO_2 expressed as meq or mg liter^{-1} $CaCO_3$ rather than as a concentration of CO_2. Mineral acid acidity measures those materials present, other than CO_2, which result in the pH value of water below 4.5. Details on methodology for the evaluation of alkalinity and acidity are given in Golterman and Clymo (1969), Wetzel and Likens (2000), and in the American Public Health Association publication (1998).

II. HYDROGEN ION ACTIVITY

Pure water dissociates weakly to H^+ and OH^- ions. The dissociation constant is very small (10^{-14}), however, and the amounts of H^+ and OH^- present are 10^{-7} g-ions per liter. Natural waters are, of course, not pure, and salts, acids, and bases contribute to the H^+ and OH^- ions in varying ways, depending on the individual circumstances. Since the dissociation constant of water is fixed, the addition of one ion will result in a decrease of the other. The pH is usually defined as the logarithm of the reciprocal of the concentration of free hydrogen ions.[1] The "p" of pH refers to the power (*puissance*) of the hydrogen ion activity. Therefore, more H^+ activity in an acid reaction increases the power from neutrality (10^{-7} or pH 7) to, say, 10^{-4} (pH 4). In more alkaline reactions, H^+ ion activity is decreased from neutrality to, for example, 10^{-10} (pH 10). By definition, pH values cannot be averaged arithmetically, but the average must be estimated from the logarithm of the reciprocals.

The pH of natural waters is governed to a large extent by the interaction of H^+ ions arising from the dissociation of H_2CO_3 and from OH^- ions produced during the hydrolysis of bicarbonate. The pH of natural waters ranges between the extremes of <2–12. Nearly all natural, unpolluted waters with pH values less than 4 occur in volcanic regions that receive strong mineral acids, particularly sulfuric acid. The oxidation of pyrite of rocks and clays in drainage basins can result in sulfuric acid drainage to lakes.

Low pH values are often found in natural waters rich in dissolved organic matter. An appreciable portion (often >50%) of dissolved organic matter occurs as organic acids derived from humic compounds (cf. Chap. 23). Humic compounds are comprised of a large number of complex ligands that have numerous p*K* and charge density (CD, mole sites per mole of carbon) values (Lydersen, 1998). The concentration of organic acids (CD) varies from 5 to 22 μeq mg C^{-1}, often with an average CD of carboxylic acids of ≈10 μeq mg C^{-1}. Recent studies demonstrate organic acids can modify both the acidity of surface waters and changes in strong acid inputs in waters with low or little bicarbonate alkalinity. Organic acids may depress water from 0.5 to 2.5 pH units ANC range of 0–50 μeq liter^{-1} (Lydersen, 1998). Despite these direct additions to acidity, organic acids have a large buffering capacity for strong acids and can reduce pH fluctuations from strong acids in precipitation.

Low pH values are particularly common in bogs and bog lakes that are dominated in the littoral mat by the moss *Sphagnum*. The pH of *Sphagnum* bogs is usually in the range of 3.3–4.5 (0.5–0.03 meq H^+ liter^{-1}). The sources of the H^+ ion activity are several (Clymo, 1963, 1964). Although acidic precipitation resulting from industrial pollution can influence the pH of poorly buffered waters significantly (Likens *et al.*, 1972), in bog areas with no great pollution, the supply of H^+ in rain is unlikely to exceed 30% of the total annual input of H^+. The metabolism of proteins and the reduction of SO_4^- by sulfur-metabolizing bacteria may contribute some H^+ ions to these waters, but their contributions are likely to be small. Although live *Sphagnum* plants secrete organic acids, concentrations of these acids are usually insufficient to account for the observed acidity. Most of the H^+ ion concentrations appear to result from the active cation exchange by the cell walls of *Sphagnum*, during which H^+ is released.

Very high pH values in lakes are usually found in endorheic regions, where lake water contains exceedingly high concentrations of soda (e.g., Na_2CO_3). The range of pH of most open lakes is between 6 and 9. Most of these lakes are the "bicarbonate type," that is, they contain varying amounts of carbonate and are regulated by the CO_2—HCO_3^-—CO_3^{2-} buffering system. Calcareous hardwater lakes commonly are buffered strongly at pH values >8. Seepage lakes and lakes within igneous-rock catchment areas are less well buffered and more acidic, with pH values usually somewhat <7.

[1] It should be noted that measurements of pH involve not the concentration but the activity of the hydrogen ion. One measures the differences of hydrogen ion activity, rapidly being released and incorporated from proton-donor molecules, between unknown solutions and standard buffers of known pH values.

III. SPATIAL AND TEMPORAL DISTRIBUTION OF TOTAL INORGANIC CARBON AND pH IN RIVERS AND LAKES

As will be discussed later, in streams decomposition dominates over in-channel photosynthetic production. High production of CO_2 results. In addition, inflowing water from runoff of soils of both surface and groundwater sources is usually nearly or totally devoid of oxygen but contains large concentrations of CO_2 from bacterial respiration. Spring waters are often anoxic at discharge sites.

Carbon dioxide in high concentrations above equilibrium values is relatively rapidly lost to the atmosphere, particularly in the water turbulence of streams. Respiration within the stream and river water is high, however, and production can exceed evasive losses to the atmosphere and photosynthetic utilization. For example, several Danish streams contained about 8 times more free CO_2 ($pCO_2 = 10^{-2.6}$ atm) than did water in equilibrium with air ($pCO_2 = 10^{-3.5}$ atm) (Rebsdorf *et al.*, 1991). As a result, variations in the CO_2 concentrations are common; higher concentrations and reduced pH can be found in quiescent regions or, times of reduced flows, particularly near wetland areas of high loadings of dissolved and particulate organic matter (Talling, 1958; Minckley, 1961). Changes in alkalinity and pH can also be induced by rapid changes in discharge. For example, alkalinity depressions in stream water can result from dilution by snowmelt water (Molot *et al.*, 1989). Although discharge was positively correlated with alkalinity in circumneutral streams, in acidic streams with little buffering capacity the snowmelt can induce marked short-term reductions in pH and create conditions that can be lethal to certain biota.

In hardwater bicarbonate rivers, such as the Rhine and the Rhone, total CO_2, largely as bicarbonate, increased downstream from the source, mostly as a result of increased organic loadings and decomposition (Golterman and Meyer, 1985). Seasonality was evident in both the bicarbonate concentrations and pH, largely as a result of temperature on the solubility of CO_2. Upon entering reservoirs, many of the features discussed later for lakes apply. However, the complexities of reservoir morphometry, inflow volumes and density-mediated distributions, and retention times make generalizations difficult (cf. Thornton *et al.*, 1990).

Total inorganic carbon (ΣCO_2) is distributed uniformly with depth during periods of circulation in dimictic and monomictic lakes and in shallow lakes of insufficient depth for thermal stratification. The ΣCO_2 content of the water is dependent upon the equilibria established between atmospheric CO_2, the bicarbonate–carbonate system, external loadings, contributions from metabolic respiration, and utilization in photosynthesis. Metabolism is markedly influenced by numerous parameters; in the spring, increasing light and temperature exert major controls on photosynthetic uptake of CO_2, and generation rates of CO_2 from microbial decomposition of organic matter is temperature- and oxygen-dependent.

During the period of thermal stratification, several conspicuous changes occur in the vertical distribution of ΣCO_2. Oligotrophic lakes that exhibit an orthograde oxygen curve and do not possess high concentrations of bicarbonate and carbonate usually also have an orthograde ΣCO_2 curve (Fig. 11-3). Where the dissolved CO_2—H_2CO_3 component constitutes an appreciable part of the total CO_2, as in softwater lakes, the warmer epilimnetic waters contain less CO_2 because of decreased solubility. Photosynthetic utilization may exceed rates of replacement if mixing of the epilimnion is relatively incomplete, but this phenomenon is uncommon in the open water. A slight increase in the ΣCO_2 is often observed in the hypolimnetic water overlying the sediments. The vertical pH distribution exhibits a pattern approximately the inverse to that of ΣCO_2.

The vertical distributions of ΣCO_2 and pH are strongly influenced by various biologically mediated

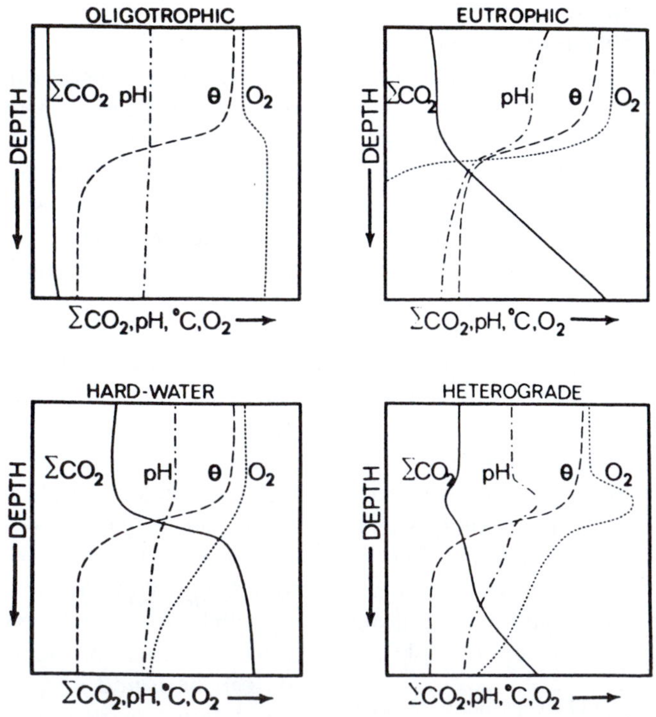

FIGURE 11-3 Generalized vertical distributions of ΣCO_2 and pH in stratified lakes of very low and high productivity, hardwater calcareous lakes exhibiting pronounced epilimnetic decalcification, and a lake with a distinct positive heterograde oxygen curve (θ = temperature).

redox reactions (each of which is treated separately elsewhere). Most conspicuous is the photosynthetic utilization of ΣCO_2 in the trophogenic zone, which tends to reduce CO_2 content and to increase pH, and the respiratory generation of CO_2 throughout the water column and sediments, which tends to decrease pH. In addition to heterotrophic degradation of organic matter, the generation of CO_2 and reduction of pH are augmented by microbial methane fermentation, nitrification of ammonia, and sulfide oxidation. Further, denitrification of nitrate to molecular nitrogen, reduction of sulfate to sulfide, and iron and manganese reduction can result in a net increase in pH and alkalinity (e.g., Kling *et al.*, 1991b; Dillon *et al.*, 1997). The combination of decompositional processes results in an increase in ΣCO_2 of the hypolimnetic waters and a decrease in pH.

As the intensity of decomposition increases in the tropholytic zone, the amount of CO_2 and especially of HCO_3^- increases markedly. The accumulation of ΣCO_2, both free and combined, exceeds the rate of oxygen consumption, and decomposition shifts from aerobic to anaerobic as the hypolimnion becomes anoxic. The origin of increasing concentrations of HCO_3^- in the hypolimnion, especially near the sediments, stems in part from bacterial production of ammonium bicarbonate in the sediments (Ohle, 1952). Ferrous and manganous ions are released as bicarbonates from the sediments when the redox potential is reduced sufficiently under anoxic conditions. In hardwater lakes, part of the $CaCO_3$ sedimenting from the epilimnion undergoes dissolution in the colder, more acidic hypolimnion and results in increases in HCO_3^- concentrations. It is clear, however, as discussed previously, that adsorbed coatings of dissolved organic matter reduce the rates of $CaCO_3$ dissolution. The hypolimnetic increases of cations, especially Ca, usually are delayed somewhat from proportional increases in hypolimnetic HCO_3^-. Under oxidized conditions, Ca^{++} is complexed with the sediments. Under the reducing conditions of anoxia, Ca^{++} is released nearly in proportion to bicarbonate, although some Ca^{++} may be complexed with humic acids (Ohle, 1955; Stewart and Wetzel, 1981; Wetzel, 1990c, 1992b). Hutchinson (1941) demonstrated conclusively that during stratification, the extent of sediment contact per volume of water (and therefore the morphology of the basin) in the metalimnion and hypolimnion is related directly to the development of increasing HCO_3^- as the period of stratification progresses. Therefore, in eutrophic lakes possessing a clinograde oxygen curve, a marked inverse clinograde ΣCO_2 curve can be observed (Fig. 11-3) and pH decreases markedly in the hypolimnion.

The vertical distribution of ΣCO_2 of very hard calcareous lakes shifts seasonally, because photosynthetic utilization of CO_2 in the trophogenic zone occurs rapidly and induces the precipitation of $CaCO_3$ (Fig. 11-4). This epilimnetic biogenic decalcification was described long ago (e.g., Minder, 1922; Pia, 1933), and has been observed frequently. Although precipitation of $CaCO_3$ can be induced by many physical and biotic agents (increasing temperature and bacterial metabolism), photosynthetic utilization of CO_2 by algae and submersed macrophytes is by far the dominant mechanism (cf. Otsuki and Wetzel, 1974). The result is a marked decrease in the ΣCO_2 of the epilimnion by the end of summer stratification (see Fig. 11-3) and a slow progressive increase in the hypolimnion (Fig. 11-4). In the example shown, the hypolimnion became anoxic below a depth of 11 m in September and October, 1968, as reflected in the increasing HCO_3^- in this layer. The pH of the water of the trophogenic zone of unproductive calcareous lakes fluctuates very little seasonally (Fig. 11-5). The bicarbonate buffering capacity

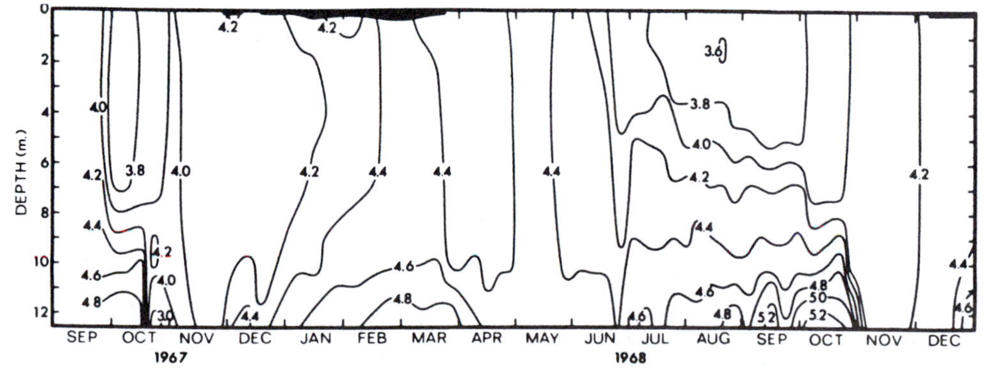

FIGURE 11-4 Depth–time diagram of isopleths of alkalinity, predominately bicarbonate, in meq liter^{-1}, Lawrence Lake, a calcareous hardwater lake of southwestern Michigan. Opaque area = ice cover to scale. (From Wetzel, unpublished data.)

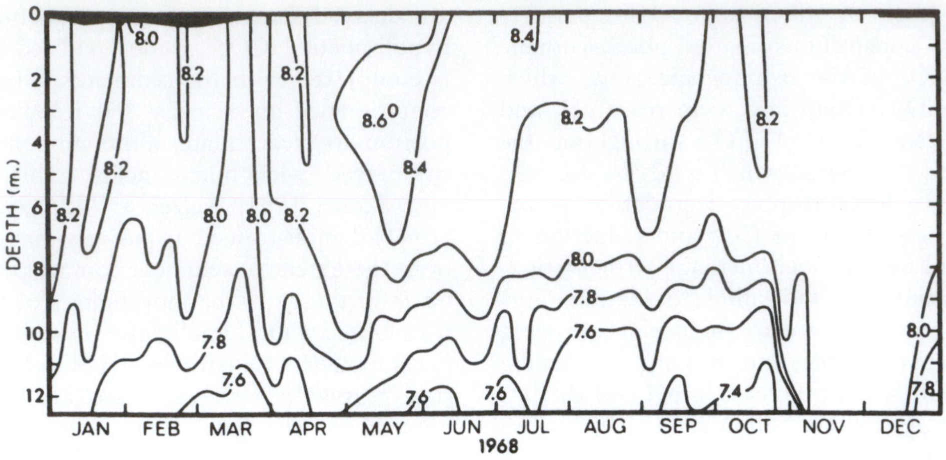

FIGURE 11-5 Depth–time diagram of isopleths of pH in hardwater Lawrence Lake, Michigan. Opaque area = ice cover to scale. (From Wetzel, unpublished data.)

is adequate to compensate for metabolic changes in ΣCO_2. The pH of the hypolimnion in the example progressively decreased throughout the period of stratification. Similar changes also were observed under ice cover, but they were less marked than during the summer period, when rates of production and decomposition were higher.

Vertical changes in ΣCO_2 and pH occur much more rapidly in eutrophic lakes (Figs. 11-6 and 11-7). In the example given in the figures, the hypolimnion below 3 m became anoxic within a month after the onset of thermal stratification and became nearly anoxic under ice cover. The pH values of the epilimnion represent midmorning measurements. Under the conditions of intensive photosynthesis by phytoplankton, the pH can undergo appreciable diurnal fluctuations, often exceeding 10 in the late afternoon and decreasing below 8 during darkness. Biogenic decalcification of the epilimnion also is apparent in this moderately hard hypereutrophic lake (Fig. 11-6).

Where photosynthetic activity in a layer is particularly intense, as might occur in the metalimnion, a positive heterograde oxygen curve is frequently found (see Chap. 9). Commonly associated with this metabolism is a corresponding positive heterograde pH curve and a negative heterograde ΣCO_2 curve (Fig. 11-3, and at 3 m in May of Fig. 11-5). The opposite situation of a negative heterograde oxygen curve, with concomitant ΣCO_2 and pH curves, occasionally is found in strata of intensive respiration, for example, containing plates of bacteria or aggregations of zooplankton (see Nagasawa, 1959).

The total alkalinity of clear softwater rivers and lakes is often very low, particularly those waters of granitic and sedimentary sandstone regions. River and lake water of these areas have relatively poor buffering

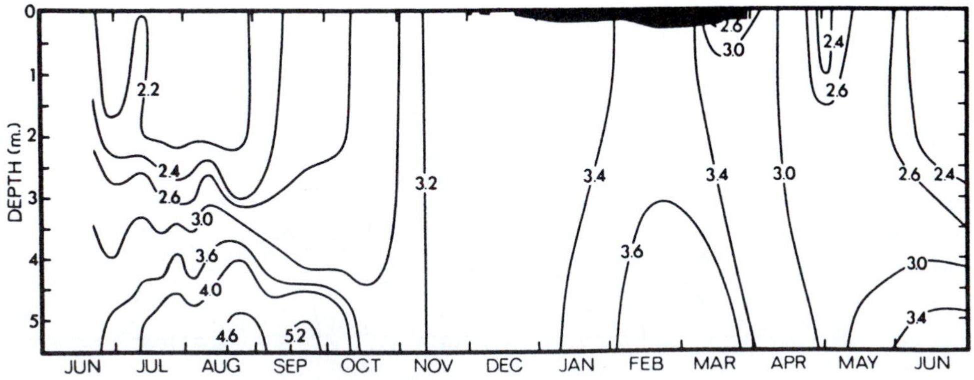

FIGURE 11-6 Depth–time diagram of isopleths of alkalinity, predominately bicarbonate, in meq liter^{-1}, of moderately calcareous, hypereutrophic Wintergreen Lake, southwestern Michigan, 1971–1972. Opaque area = ice cover to scale. (From Wetzel *et al.*, unpublished data.)

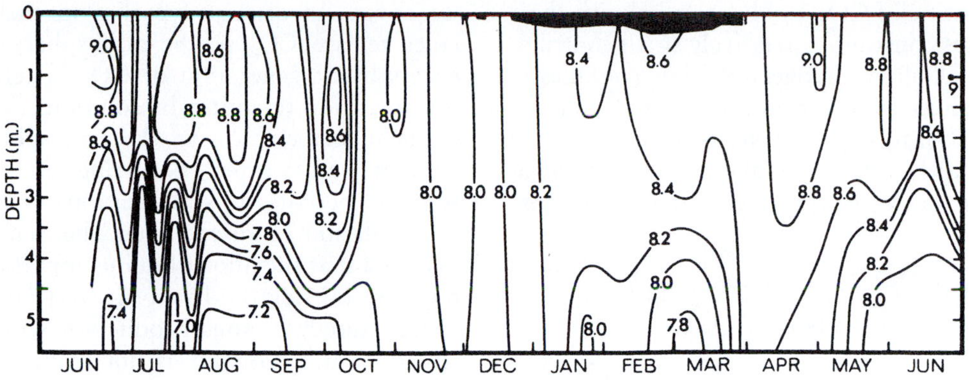

FIGURE 11-7 Depth–time diagram of isopleths of pH in hypereutrophic Wintergreen Lake, Michigan, 1971–1972. Opaque area = ice cover to scale. (From Wetzel *et al.*, unpublished.)

capacities and can be modified quickly by modest inputs of acidic water, such as from rapidly melting snow in spring that had accumulated acidity from atmospheric sources. The literature on the subject of acidification of soils, ground waters, and surface waters is enormous, with thousands of scientific articles and hundreds of books on the subject of sources of natural and anthropogenic acidity and their effects upon the chemistry and biota of fresh waters. Brief reference here is largely limited to effects on the DIC and pH of soft waters.

Atmospheric deposition of particulate and dissolved acid anions has resulted in sustained loads adequate to cause thousands of lakes and streams of low alkalinity to become acidic (e.g., Baker *et al.*, 1991). Major ions in atmospheric deposition that contribute significantly to such acidification are H^+, usually associated with H_2SO_4 and HNO_3, and NH_4^+. Atmospheric H^+ deposition in many areas of sensitive aquatic ecosystems have been in the range of 50–125 meq m^{-2} yr^{-1} (Dillon *et al.*, 1983), although attempts have been made, with mixed success, to reduce these loadings.

In regions where the soils and rock deposits contain high carbonates, acidified precipitation can increase rates of carbonate weathering and result in increased alkalinity and equilibrium pH in softwater seepage lakes (Kilham, 1982). In drainage basins and waters with little buffering capacity, however, the deposition has resulted in loss of acid-neutralizing capacity (alkalinity), decreased pH, increases in sulfate and in dissolved aluminum concentrations, and in marked alterations in ionic speciation and ratios. These acidification changes can persist on a long-term basis or, in less affected areas, chemical changes occur reversibly for short-term periods. In many softwater streams and lakes, for example, melting of acidic snow accumulations can result in rapid (few days), precipitous declines in pH and alkalinity (e.g., Catalan and Camarero, 1993). Lower pH values occurring during late winter and early spring in softwater seepage lakes, on the other hand, were attributed to increases in pCO_2 under the ice because total alkalinity did not change at that time (Kratz *et al.*, 1987).

Much controversy has arisen over the natural versus anthropogenic components of acidification of natural waters and about balances from biologically mediated generation of alkalinity. In remote areas, complex humic and other organic acids, as for example from bog drainage (Schindler *et al.*, 1986), as well as H^+ ions generated from nitrification of NH_4^+, manganese oxidation, ferrous iron oxidation, and sulfide oxidation, all discussed in detail in later chapters, can contribute to acidity and reduction of alkalinity in some lakes (Galloway *et al.*, 1984; Gorham, 1986; Davison, 1987). However, the overwhelming loading of acidity to sensitive softwater surface waters has resulted from anthropogenic pollution from fossil fuel combustion and production of H^+ ionized from several strong acids (Galloway *et al.*, 1984; Morgan, 1991; many others).

Biological generation of alkalinity within lakes has been known for many years (e.g., Hutchinson, 1941; Mortimer, 1941, 1942; Ohle, 1952). Internal alkalinity generation in very softwater lakes is usually dominated by biological reduction of sulfate and nitrate (Baker *et al.*, 1986; Cook *et al.*, 1986; Rudd *et al.*, 1986; Schindler, 1986, 1988; Schindler *et al.*, 1986; Kelly *et al.*, 1987; Giblin *et al.*, 1990). The exchange of base cations and redox reactions, particularly associated with manganese, iron, and sulfate reduction, can also contribute to alkalinity generation (Carignan, 1985; Schiff and Anderson, 1986; Davison, 1987). Anoxia in the water column is not essential, as reduction can occur in surficial sediments, which are nearly always anoxic, with the release of DIC. Despite these sources of in-lake generation of CO_2 and hence alkalinity, these biological sources constitute a relatively small contribution (<7%) to alkalinity budgets of drainage lakes in relation to loadings from the drainage basin (Shaffer and Church, 1989). In seepage lakes or lakes with long hydrological residence times, in-lake processes of alkalinity generation can constitute a greater contribution.

IV. HYPOLIMNETIC CO_2 ACCUMULATION IN RELATION TO LAKE METABOLISM

As in the case of hypolimnetic oxygen deficits discussed earlier, changes in the concentrations of ΣCO_2 in the hypolimnion have been used to indirectly estimate the organic production of the trophogenic zone. The concept is that the accumulation of CO_2 in the hypolimnetic tropholytic zone from decomposition is proportional to the production of organic matter in the trophogenic zone that settles into the hypolimnion. This approach to an estimate of lake metabolism is complicated by inputs of organic matter that do not originate in the trophogenic zone and by losses of carbon from the system that are not represented in CO_2 changes in the tropholytic zone. For many lake systems, the measurement of hypolimnetic CO_2 accumulation can provide a reasonable estimate of the intensity of lake metabolism.

The idea of estimating production from CO_2 accumulation was first suggested by Ruttner in his treatise on tropical lakes (1931). Einsele (1941) compared the CO_2 accumulation in the hypolimnion with several other estimates of production in his critical evaluation of the effects of phosphorus fertilization on productivity of algae and found them to be in agreement. By far the most comprehensive application of the hypolimnetic CO_2 accumulation principle to lakes, and its comparison with other measures of production, especially the oxygen deficit, was done by Ohle (1934, 1952, 1956). The CO_2-accumulation approach has the advantage of being able to follow metabolically mediated changes under both aerobic and anaerobic conditions, whereas the oxygen deficit method is applicable only when the hypolimnion is oxic. More fundamentally, carbon, which is the initial and end product of organic metabolism, is one of the best parameters by which to evaluate productivity.

There are several assumptions underlying calculations of hypolimnetic CO_2 accumulation. First, it is important that the water under consideration be relatively well isolated from the atmosphere and from CO_2 exchange; these conditions may be found in the hypolimnia of small, sheltered lakes with steep thermal stratification gradients or when the entire lake is under ice cover. As discussed earlier in this chapter, most lakes release CO_2 to the atmosphere because of the excess biotic generation of CO_2 in relation to partial pressures for retention in solution. Some seasonality exists in net evasion to the atmosphere, with lower evasion rates often occurring in the productive summer periods of stratification when hypolimnetic accumulation estimates would be made. Second, it is assumed that the allochthonous inputs of organic matter from outside the basin are very small in comparison to autochthonous production within the basin, a situation that probably is not realized very often (cf. Chap. 23). Loss of organic production from the trophogenic zone via the outlet is assumed to be small—a condition which, again, is highly variable in open lakes. A major assumption is that most of the synthesized organic matter of the trophogenic zone decomposes only after it has sedimented to the tropholytic zone. This assumption certainly causes underestimations, because some decomposition of algae, littoral macrophytes, and sessile algae occurs in the warm epilimnetic and metalimnetic layers before reaching the relatively quiescent, cool hypolimnion. The rates of decomposition in the trophogenic zone depend upon many physical and biological factors. Important factors include the proportion of production by the littoral flora vs. phytoplankton, seasonal shifts in the proportion of phytoplankton algae reaching the hypolimnion—for example, silicious diatoms sediment more rapidly than small algae, which are more neutrally buoyant—and water temperature. A significant amount of organic production will be incorporated permanently into the sediments without complete decomposition; the proportion permanently lost in this way generally increases in lakes of greater productivity. Perhaps the greatest problem rests in the assumption of a respiratory quotient ($RQ = CO_2/O_2$) of 0.85, an average figure for the aerobic decomposition of a mixture of carbohydrate, lipid, and protein organic compounds. As will be discussed later (Chap. 23), anaerobic decomposition produces CO_2 utilizing alternate electron acceptors after oxygen has been depleted and often leads to RQ values much greater than 1 (Rich, 1975, 1980; Rich and Wetzel, 1978). Despite these numerous limitations, many estimates of hypolimnetic CO_2 accumulation correlate with general lake productivity and are directly proportional to mean depth of the basin.

Analysis of hypolimnetic CO_2 accumulation requires evaluation of the various components of total CO_2 and their origins. In bicarbonate-poor waters, the difference in CO_2 content at the beginning and after a period of stratification is determined. In bicarbonate-rich waters, the amount of free CO_2 is very small at the onset of

TABLE 11-3 Summary of Method of Evaluating the Hypolimnetic CO_2 Accumulation of a Lake with a Sample Calculation[a]

Calculations	*Lake example*
(1) $NH_4{:}HCO_3 = 18.04{:}61.02$	(1) 1.85 mg NH_4 liter^{-1} present as NH_4HCO_3
(2) $HCO_3 = \frac{(NH_4)(61.02)}{18.04} = (3.38)(NH_4) = \beta$	(2) $\beta = (1.85)(3.38) = 6.25$
(3) $HCO_3{:}CO_2 = 61.02{:}44.01$	(3) $b = (6.25)(0.721) = 4.51$
$CO_2 = \frac{(HCO_3)(44.01)}{61.02} = (0.721)(HCO_3)$	
$b = \beta(0.721)$	
(4) $\alpha = HCO_3$; $\alpha - \beta = x$	(4) $\alpha = 32.4$ mg HCO_3 liter^{-1} $x = 32.4 - 6.25 = 26.15$
(5) $a = \frac{x(0.721)}{2} = x(0.361)$	(5) $a = (26.15)(0.361) = 9.43$
(6) $c = CO_2$	(6) $c = 13.10$ mg CO_2 liter^{-1}
(7) $\Sigma CO_2 = a + b + c$	(7) $\Sigma CO_2 = 9.43 + 4.51 + 13.10 = 27.04$
(8) δO_2 = actual O_2 deficit	(8) mg O_2 liter^{-1} = 0.56
$\frac{CO_2}{O_2} = \frac{44}{32} = 1.375$	°C = 6.7, O_2 at turnover = 12.27 mg liter^{-1}
$\delta O_2\,(1.375) = \delta CO_2$	$\delta O_2 = 12.27 - 0.56 = 11.71$
RQ = 0.85[b]	$\gamma = (11.71)(1.375)(0.85) = 13.69$
$\gamma = \delta CO_2\,(0.85)$	
(9) $\Sigma CO_2 - \gamma = \Delta^1\,CO_2$	(9) $\Delta^1\,CO_2 = 27.04 - 13.69 = 13.35$
(10) $z = \Delta^1\,CO_2(2)$	(10) $z = (13.35)(2) = 26.70$
(11) $\xi = z + \gamma$	(11) $\xi = 26.70 + 13.69 = 40.39$
$\frac{\xi\,(30)}{\text{No. of days}} = \xi$ *per* month	ξ *per* month $= \frac{(40.39)(30)}{120} = 10.1$ mg CO_2 liter^{-1}

[a] Example based on 1.85 NH_4 liter^{-1}; 13.1 mg CO_2 liter^{-1}; 32.4 mg HCO_3 liter^{-1}; 0.56 mg O_2 liter^{-1}; °C = 6.7; period of stratification of 4 months. (After Ohle, 1952.)

[b] An RQ value (CO_2/O_2) of 0.85 is a mean value based on a number of analytical analyses of plant and animal respiration and agrees approximately with some whole closed-lake evaluations (but see discussion of RQ variations in Chap. 23).

hypolimnetic isolation. As stratification is maintained, respiration during aerobic and anaerobic decomposition allows the accumulation of free CO_2 and bicarbonate. In addition, a portion of the bicarbonate is present as "volatile" ammonium carbonate of metabolic origin; additional carbon exists as volatile bicarbonates, half of which are of metabolic origin. Half of the nonvolatile bicarbonate is bound CO_2 of $CaCO_3$, which largely becomes incorporated into the sediment. The sum of these CO_2 inputs is assumed to be an estimate of lake metabolism. A sample calculation is given in Table 11-4 (see Wetzel and Likens, 2000). If CO_2 accumulation is calculated for the entire hypolimnion, and this is divided by the volume of the epilimnion, the quantity termed the *relative assimilation intensity* results. In the example given (Table 11-3),

$$(10.10\text{ mg }CO_2\text{ liter}^{-1}\text{month}^{-1})\left[\frac{4.620 \times 10^6\text{m}^3}{7.442 \times 10^6\text{m}^3}\right] = 6.3\text{ mg }CO_2\text{ liter}^{-1}\text{month}^{-1}$$

This estimate implies that, in the epilimnion of this lake, approximately 6.3 mg CO_2 were photosynthetically combined into carbohydrate per liter of water per month during the period of observation. The CO_2-accumulation method of calculating production usually leads to underestimation because of the confounding factors mentioned earlier. Further, some CO_2 loss from the upper hypolimnion occurs by turbulence, and some bicarbonate is released from the reduction of ferric hydroxide as redox potential decreases during stratification.

The hypolimnetic CO_2 accumulations in several lakes are compared in Table 11-4, along with several other indices of increasing productivity (Ohle, 1955b). The general correlations among various criteria hold. Einsele (1941) applied the CO_2-accumulation evaluation of productivity, along with several other measures, to the trophogenic zone of Schleinsee, Germany, before and after extensive fertilization of the lake with phosphorus (Table 11-5). Schleinsee lends itself well to such analyses because of its limited inflow and outflow.

TABLE 11-4 Comparison of the Relative Assimilation Intensity to Other Indices of Productivity in Lakes of Northern Germany[a]

	Relative assimilation intensity (mg CO_2 liter^{-1} *epilimnion month*$^{-1}$)	*Chlorophyll of Epilimnion* (μg *liter*$^{-1}$)	*Estimate of photosynthetic efficiency of solar radiation* (%)	*Hypolimnetic* O_2 *deficit* (*mg* O_2 *liter*$^{-1}$)
Schaalsee	1.91	1.3	0.76	3.30
Schöhsee	2.13	1.4	0.56	6.96
Zanzen	2.33	1.6	0.96	—
Schmaler Lucin	2.70	1.8	0.80	3.16
Breiter Lucin	4.64	3.1	1.90	2.16
Techiner Binnensee	5.45	3.6	1.65	7.86
Tollensesee	7.04	4.1	2.09	5.49
Edebergsee	14.60	9.7	2.27	10.00

[a] After data of Ohle (1952, 1955b).

Although Einsele's estimations were quite approximate, again the comparison was good. When used in a relative manner, this method is of value.

V. UTILIZATION OF CARBON BY ALGAE AND MACROPHYTES

Algae and submersed aquatic macrophytes require an abundant and readily available source of carbon for high sustained growth. The supply of carbon is regulated by *availability*, a problem of regulation by physical processes of diffusion along concentration gradients within the water and within the communities, and by *physiological assimilation* processes, a problem of transport and biochemical utilization capabilities. The former is discussed elsewhere (Chaps. 7 and 19); discussion here focuses on the utilization of inorganic carbon sources.

Abundant physiological evidence indicates that free CO_2 is most readily utilized by nearly all algae and larger aquatic plants. A number of algae and submersed macrophytes, particularly the mosses, can utilize only free CO_2. Many algae and aquatic vascular plants are capable of utilizing CO_2 from bicarbonate ions when free CO_2 is in very low supply and HCO_3^- is abundant; a few species of algae require HCO_3^- and cannot grow with free CO_2 alone. There is no clear evidence that algae or higher aquatic plants assimilate CO_3^{2-} directly as a carbon source, although it has been implicated in growth at very high pH values (see Felföldy, 1960). The most comprehensive review of the subject is given by Raven (1970, 1984; cf. also Allen and Spence, 1981, Lucas and Berry, 1985).

A. Heterotrophic Augmentation

A large number of algae are heterotrophic, that is, they can remain viable in the absence of light in bacterial-free culture by chemo-organotrophic uptake of dissolved organic compounds (cf. Danforth, 1962;

TABLE 11-5 Estimations of Total Production in the Trophogenic Zone of Schleinsee, Germany, during Summer Stratification (May–September) in the Years Before and in the Year After Application of Phosphorus Fertilization[a]

Basis for calculated production	*kg dry weight* 1935–1937	*kg dry weight* 1939	*Ratio*
Biogenic decalcification	1900	5000	2.6
Ammonia N accumulation	2100	5900	2.8
Hypolimnetic CO_2 accumulation	3800	9600	2.5
Apparent total production	5600	17,200	3.1
Related to surface area	43 g m^{-2}	117 g m^{-2}	2.7

[a] After data of Einsele (1941).

Provasoli, 1963; Stanier, 1973; Droop, 1974; Neilson and Lewin, 1974; Raven, 1984). However, the rates of growth under the experimental conditions used are very low in comparison with normal, light-mediated autotrophic growth, and the levels of substrate concentrations employed are usually exceedingly high, often several orders of magnitude in excess of what would be found in natural waters. The range of organic compounds that support growth in the dark of cyanobacteria is very narrow, because the pentose phosphate pathway is the energy-yielding dissimilatory pathway. Consequently, only exogenous substrates that are readily convertible to glucose-6-phosphate can support growth in the dark. Among obligately photoautotrophic algae, suitable enzymes (e.g., glucose permease) are absent. Investigations conducted at naturally occurring substrate concentrations have consistently shown that bacteria possessing active permease enzyme systems effectively outcompete algae for organic substrates under natural conditions (Wright and Hobbie, 1966; Wetzel, 1967; Hobbie and Wright, 1965; and others). Where active heterotrophic uptake of simple organic substrates is found under natural conditions, carbon assimilation is still low in comparison to the amount assimilated in photoautotrophy, even under very low light conditions. It might be anticipated that any significant augmentation of autotrophy by chemoorganotrophy would be confined largely to the cyanobacteria, which have numerous morphological and physiological similarities to heterotrophic bacteria. This potential heterotrophic augmentation is seen in the cyanobacterium *Oscillatoria,* which grows in massive densities in the microaerobic zone at the metalimnetic–hypolimnetic interface of some productive lakes (Saunders, 1972). Photoheterotrophic growth in the light in the absence of exogenous CO_2 has been shown in two species of cyanobacteria (Ingram, 1973a,b). Studies indicate that respired CO_2 from the substrate oxidation is assimilated by the photosynthetic reactions within the cells. Photoheterotrophy in natural waters was first critically evaluated by McKinley (1977; McKinley and Wetzel, 1979; see also Ellis and Stanford, 1982). Photoheterotrophic contributions to overall carbon cycling were greatest in the morning periods at depths of low light and during spring turnover and late summer stratification. Photoheterotrophic and chemoheterotrophic (not light-mediated) utilization of organic substrates was small in this moderately productive lake in comparison with normal photosynthetic utilization of inorganic carbon.

All of the existing evidence for heterotrophic growth by algae indicates that under natural conditions, photoautotrophy by inorganic carbon uptake overwhelmingly dominates in freshwater systems. Supplementation of photosynthesis by algal heterotrophy is limited to a few specialized conditions. Although probably insignificant for overall plant synthesis, these supplementary processes are potentially important to the survival of some species and in selective competitive interactions.

B. Photosynthetic Inorganic Carbon Use

In most freshwater systems, dissociation rates of ionic CO_2 species and the maintenance of near-equilibrium conditions between atmospheric CO_2 and that of the water are sufficiently rapid that severe inorganic carbon limitation to algal photosynthesis is unlikely in well-mixed pelagic regions, even under conditions of low ΣCO_2. CO_2 limitation can occur, however, in attached algal communities and in submersed angiosperms. Below a pH of 8.5, the theoretical dissociation kinetics of the CO_2—HCO_3^-—CO_3^{2-} complex, derived from pure solutions, agree very well with the concentrations of the ionic species found in natural waters (Talling, 1973). Above pH 8.5, however, the total inorganic carbon dioxide concentrations fall considerably below the values calculated on the basis of the apparent dissociation constants K'_1 and K'_2 of carbonic acid, as derived from measurements of alkalinity (Fig. 11-8). Although the explanation for this phenomenon is not completely clear, it appears to be associated with the presence of a noncarbonate, nonhydroxide alkalinity component resulting from ionized silicate and can be corrected for (Talling, 1973).

Despite the solubility of CO_2 and relatively turbulent water strata of the pelagic trophogenic zone, the photosynthetic carbon supply is not very favorable. Much of the total carbon is in the form of bicarbonate and carbonate, which is not readily available to many aquatic plants. Rapid photosynthesis can rapidly reduce the total DIC and increase pH of the water and shift dissociation equilibria so that dissolved $[CO_2]$ is nearly zero. Diffusion of CO_2 in water along concentration gradients is very slow relative to that in air. Experimental evidence on freshwater photosynthetic utilization of inorganic carbon indicates a strong relationship between physiological availability and the forms of inorganic carbon (Raven, 1970, 1984; Wetzel and Hough, 1973; Hough and Wetzel, 1977; Smith and Walker, 1980; Maberly and Spence, 1983).

The primary enzyme of inorganic carbon fixation in all photosynthetic oxygen-evolving organisms is ribulose-1,5-bisphosphate carboxylase/oxygenase (Rubisco). This carboxylase is involved in fixing some 95% of the carbon assimilated in plants with C_3 metabolism, in which 3-phosphoglycerate is the first major product. Most algae and aquatic macrophytes have C_3 biochemistry, and thus at least 95% of the carbon

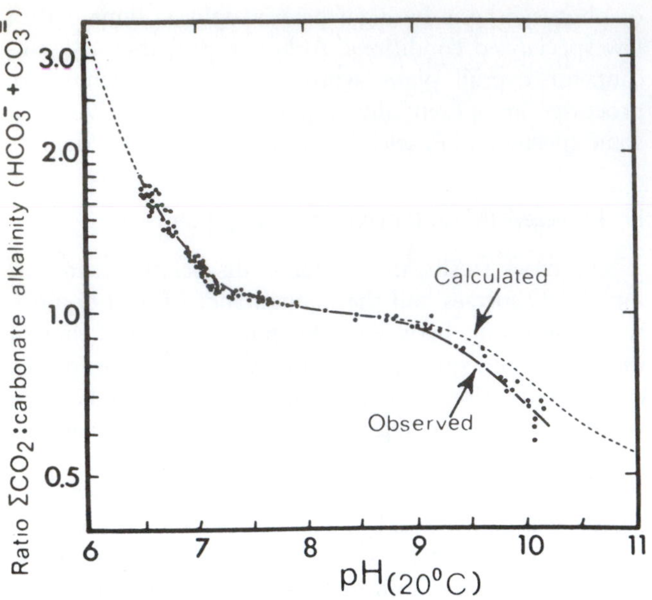

FIGURE 11-8 Variation in the ratio of ΣCO_2 to carbonate alkalinity resulting from HCO_3^- and CO_3^{2-} with pH for samples of water from the surface and bottom (14 m) layers of Esthwaite Water, England. The broken line indicates the relationship calculated from appropriate values of pK_1' (6.38) and pK_2' (10.32). (Slightly modified from Talling, 1973.)

fixed must be supplied as CO_2 at the carboxylase site (Raven, 1984, 1995). The enzyme carboxydismutase is involved in the carboxylation reaction, which produces 3-phosphoglycerate as the first product of carbon fixation, and uses free, unhydrated CO_2 as its immediate substrate.

Carbonic anhydrase, an enzyme that catalyzes the reversible hydration of carbon dioxide, is found universally in photosynthetic cells of plants. Where CO_2 is the carbon species entering the cell, there is no obvious biochemical role for carbonic anhydrase found in plants, although some evidence indicates that this enzyme accelerates the diffusion of CO_2 and transport to sites of biochemical fixation by carboxylases (Raven, 1995). Bicarbonate ions have been implicated as a critical factor in the evolution of oxygen during photosynthesis (Stemler and Govindjee, 1973; Raven, 1984). At high pH values and HCO_3^- concentrations, bicarbonate moves less effectively from binding sites to the chloroplast. When HCO_3^- enters cells, carbonic anhydrase is needed to supply CO_2 to Rubisco. Carbonic anhydrase activity is found in plants that cannot use bicarbonate as well as those that can, although such activity generally increases among algae and certain submersed macrophytes that can utilize bicarbonate (Weaver and Wetzel, 1980; Raven, 1995).

The ability to assimilate bicarbonate ions is variable among planktonic algae, macroalgae, and submersed angiosperms (Raven, 1970, 1984; Allen and Spence, 1981; Maberly and Spence, 1983). When this ability does occur, an additional reaction is needed for HCO_3^- assimilation that is not required for CO_2 assimilation. Active bicarbonate transport followed by HCO_3^- dehydration within the cytoplasm is apparently required and is coupled with a similarly active stoichiometric excretion of hydroxyl ions from the cells. Assimilated bicarbonate ions are used to produce high internal CO_2 pools by action of carbonic anhydrase.

When aquatic plants have similar affinities for CO_2 and the HCO_3^- ion, utilization of bicarbonate generally occurs when the bicarbonate concentration exceeds that of CO_2 by more than 10 times. Free CO_2 concentrations (about 10 micromolar) of most fresh waters and of the sea are in approximate equilibrium with the atmosphere, although many fresh waters contain bicarbonate concentrations far in excess of 10 times that quantity. Equilibrium concentrations of dissolved CO_2, particularly in alkaline hardwater lakes with a pH >8.5, are inadequate to saturate photosynthesis in plants adapted to utilize bicarbonate. As these waters become more productive, and in densely populated littoral zones of less productive lakes, pH is rapidly altered by metabolism on a diurnal basis (pH ranges from a low of 6 to a high of 10 or more per 24 h) and can be associated with reduced carbon fixation and bicarbonate assimilation (e.g., Spencer *et al.*, 1994). Under stagnant conditions, common to heavily colonized littoral zones of lakes, the shift to bicarbonate metabolism, as well as the increased pH, is often associated with severe reduction in CO_2.

Stagnant layers around plant cells (algae) and tissues (macrophytes) clearly have major effects on limiting the assimilation of both CO_2 and HCO_3^- by actively photosynthesizing plants (Smith and Walker, 1980). Even when the plants are exposed to water movements by turbulence or sinking, a residual layer (approximately 10–100 μm thick) parallel to the surface remains. The rate of solute movement across this residual layer approximates that of molecular diffusion. Diffusion of CO_2 and HCO_3^- through these stagnant layers is an important rate-limiting process to both availability of CO_2 and membrane transport of HCO_3^- ions. Mucilage sheaths surrounding algal cells, especially among cyanobacteria, can further reduce the uptake rates of inorganic carbon (Chang, 1980).

Bicarbonate assimilation in media of high pH assumes greater significance in large aquatic macrophytes that have morphologically long diffusion paths. Active transport of HCO_3^- originating from the dissociation of calcium bicarbonate and the secretion of OH^- ions results in the precipitation of $CaCO_3$. Many submersed angiosperms and algae can utilize bicarbonate. Aquatic mosses are able to utilize only free CO_2, and are restricted to waters of relatively low pH and abundant CO_2 (e.g., springs, mountain streams, bogs, or shallow waters) or the lower trophogenic zone where free CO_2 is higher than elsewhere in the lake. CO_2 of the rhizosphere can diffuse into rooting tissues, move to the leaves via internal gas lacunae, and be fixed photosynthetically (Loczy *et al.,* 1983; Steinberg and Melzer, 1983; Wetzel *et al.,* 1985). Although this benthic source of inorganic carbon is widely utilized in submersed macrophytes, the percentage of the total carbon fixed photosynthetically from the rooting tissues is greater among small plants growing in poorly mineralized waters. Moreover, many angiosperms with large intercellular gas lacunae refix the CO_2 of respiration and photorespiration rather efficiently (Hough and Wetzel, 1972; Søndergaard and Wetzel, 1980). The efficiency of refixation and the photosynthetic efficiency of carbon fixation must be highly plastic, related in part to induced shifts to bicarbonate assimilation, and can affect rates of net primary production significantly.

In summary, evidence is available that, among the diversity of concentrations and states of inorganic carbon in fresh waters, there exists a large number of situations where free CO_2, even in equilibrium with the atmosphere, may be inadequate for metabolism at high sustained levels (see especially Talling, 1976). In other cases, free CO_2 may be inadequate as a result of slow diffusion rates and chemical losses from the system. Although it is unusual for inorganic carbon per se to be seriously limiting to phytoplanktonic photosynthetic metabolism under most natural situations, assimilation of bicarbonate at metabolic expense is known and can be induced among plants that have an affinity for both CO_2 and HCO_3^- ions under conditions of low free CO_2 and high bicarbonate concentrations (e.g., Beer and Wetzel, 1981; Goldman and Graham, 1981; Novak and Brune, 1985). Possession of an affinity for bicarbonate is an adaptive advantage in fresh waters and is certainly instrumental in altering competitive succession of algal and certain submersed macrophyte communities.

VI. SUMMARY

1. Most carbon of fresh waters occurs as equilibrium products of carbonic acid (H_2CO_3). A smaller amount of carbon occurs in organic compounds as dissolved and particulate detrital carbon, and a very small fraction occurs as carbon of living biota.
2. The complex equilibrium reactions of inorganic carbon and the distribution of the chemical species of total CO_2 [$\Sigma CO_2 = CO_2 + HCO_3^- + CO_3^{2-}$] are understood in considerable detail.

a. As atmospheric CO_2 dissolves in water containing bicarbonate (primarily associated with the calcium cation), dissociation kinetics of the inorganic carbon of fresh waters below a pH of 8.5 closely follow those predicted by pure dilute solution chemistry.

b. Free CO_2 is very soluble, and in water some becomes hydrated to form carbonic acid:

$$CO_2(\text{air}) \rightleftharpoons CO_2(\text{dissolved}) + H_2O \rightleftharpoons H_2CO_3$$

c. H_2CO_3 is a weak acid and rapidly dissociates to bicarbonate and carbonate:

$$H_2CO_3 \rightleftharpoons H^+ + HCO_3^- \rightleftharpoons H^+ + CO_3^{2-}$$

d. At equilibrium, bicarbonate and carbonate ions dissociate and hydroxyl ions (OH^-) are formed by the hydrolysis of carbonic acid:

$$HCO_3^- + H_2O \rightleftharpoons H_2CO_3 + OH^-$$

$$CO_3^{2-} + H_2O \rightleftharpoons HCO_3^- + OH^-$$

$$H_2CO_3 \rightleftharpoons H_2O + CO_2$$

e. As carbonic acid percolates over rock and through soils, carbonates are solubilized; ionized Ca^{++} and HCO_3^- are released to the water. The dilute solution of bicarbonate of many fresh waters is weakly alkaline because slightly greater concentrations of OH^- ions than H+ ions result from the dissociation of HCO_3^-, CO_3^{2-}, and H_2CO_3.

f. Losses of CO_2 by photosynthetic utilization or additions of CO_2 from biotic respiration tend to change the pH (H^+ ion activity) of the water. The ***buffering action*** of the water, however, tends to resist changes in pH as long as the equilibria of the ΣCO_2 complex are operational. Addition of H^+ ions is neutralized by OH^- ions formed by the dissociation of HCO_3^- and CO_3^{2-}. The pH remains essentially the same as before, unless the supply of HCO_3^- and CO_3^{2-} is exhausted. Similarly, added OH^- ions react with HCO_3^- ions:

$$HCO_3^- + OH^- \rightleftharpoons CO_3^{2-} + H_2O$$

g. If the solution of calcium bicarbonate in equilibrium with CO_2, H_2CO_3, and CO_3^{2-} loses a portion of the CO_2 required to maintain the equilibrium, such as by photosynthetic uptake exceeding replacement of CO_2, the relatively insoluble calcium carbonate (marl) will precipitate:

$$Ca(HCO_3)_2 \rightleftharpoons CaCO_3\downarrow + H_2O + CO_2$$

until the equilibrium is reestablished by the formation of sufficient CO_2. As $CaCO_3$ forms and precipitates, inorganic (e.g., PO_4^{-3} ions) and organic compounds can adsorb to or coprecipitate with the $CaCO_3$ and be carried out of the trophogenic zone of lakes. Metabolic activity can be reduced as a result of nutrient reductions associated with this adsorption and/or coprecipitation.

3. The distribution of ΣCO_2 and pH in surface waters varies both seasonally and vertically in lakes in relation to loading from allochthonous sources, physical conditions, and biotic inputs and consumption.
   a. Although the exchange of CO_2 of the water with that of the atmosphere is rapid and relatively complete in aerated open water of streams and lakes, the spatial and temporal distribution of ΣCO_2 and pH is altered by microbial respiratory metabolism and by photosynthetic consumption, especially in quiescent strata in the pelagic zone of stratified lakes and in productive littoral areas.
   b. Relatively unproductive lakes exhibiting an orthograde oxygen curve generally have an orthograde ΣCO_2 curve.
   c. In productive waters exhibiting a clinograde oxygen curve, an inverse clinograde ΣCO_2 curve is generally found. The pH decreases in the hypolimnion in relation to the increased hypolimnetic concentrations of CO_2 and bicarbonate.
   d. The ΣCO_2 distribution of hardwater calcareous lakes is greatly modified by biologically induced decalcification of the epilimnion.
   e. There is a constant net efflux of excess CO_2 from most lakes of the world; most of this net CO_2 evasion to the atmosphere derives from metabolic respiration of organic compounds.
4. In thermally stratified lakes of low to moderate productivity with minimal inputs of organic matter from outside the basin, the accumulation of CO_2 in the hypolimnion is positively correlated to rates of organic production in the trophogenic zone.
5. Inorganic carbon is a major nutrient of photosynthetic metabolism. However, phosphorus and nitrogen limit photosynthesis more frequently than does inorganic carbon, which occurs in much greater abundance. Because assimilation of CO_2 and HCO_3^- occurs more rapidly than resupply, and conditions of reduced availability result from diffusion resistance at the uptake sites, potential photosynthetic productivity by both algae and submersed macrophytes is not always fully realized. Many algae and vascular aquatic plants have developed compensatory mechanisms to enhance utilization and recycling of respired CO_2.

12 THE NITROGEN CYCLE

Nitrogen is abundant on the surface of the Earth but less than 2% is available to organisms (Galloway, 1998). Reactive nitrogen (defined as N bonded to C, O, or H, as in NO_x, NH_x, organic N) is created largely by biological nitrogen fixation of unreactive nitrogen (triple-bonded N_2) and provides about 90–130 Tg N yr^{-1} (Tg = 10^{12} g) on the continents. Lightning creates only about 1% of reactive nitrogen. Human activities have resulted in the fixation of an additional ~150 Tg N yr^{-1} by energy production, fertilizer production, and cultivation of crops (e.g., legumes and rice) that utilize symbiotic N-fixing microbes. Rates of loss or storage of anthropogenic nitrogen are less accurately known, but clearly nitrogen is accumulating in the environment. Because nitrogen is required for expanding food production for expanding human populations, greater amounts of nitrogen will be converted from unreactive to reactive forms in the future. Much of this reactive nitrogen will enter surface and groundwater ecosystems.

I. SOURCES AND TRANSFORMATIONS OF NITROGEN IN WATER

Nitrogen occurs in fresh waters in numerous forms: dissolved molecular N_2; a large number of organic compounds from amino acids, amines, to proteins and recalcitrant humic compounds of low nitrogen content, ammonia (NH_4^+), nitrite (NO_2^-), and nitrate (NO_3^-). Sources of nitrogen include: (a) precipitation falling directly onto the lake surface, (b) nitrogen fixation both in the water and the sediments, and (c) inputs from surface and groundwater drainage. Losses of nitrogen occur by (a) effluent outflow from the basin, (b) reduction of NO_3^- to N_2 by bacterial denitrification with subsequent return of N_2 to the atmosphere, and (c) permanent sedimentation loss of inorganic and organic nitrogen-containing compounds to the sediments.

A. Nitrogen in Precipitation and Fallout

The amount of influent nitrogen to lakes and their drainage areas from atmospheric sources generally has been considered to be minor in comparison with that from direct terrestrial runoff. However, on closer inspection, and in view of exponentially increasing inputs of nitrogen from atmospheric sources, the amount of nitrogen reaching lakes in this way is often significant to the nitrogen cycle and for productivity. For example, in relatively oligotrophic mountainous regions of granitic bedrock, precipitation is a major source of nitrogen (Likens and Bormann, 1972; Likens *et al.*, 1977). Inorganic nitrogen input by precipitation similarly was found to be a major source of loading to the drainage basin of Lake Tahoe, California–Nevada (Coats *et al.*, 1976; Jassby *et al.*, 1994). Nitrogen from precipitation and direct bulk (particulate) fallout is extremely variable depending on local meteorological conditions, wind patterns, and the location of streams and lakes with respect to industrial and agricultural outputs.

Nitrogen may enter a drainage basin in many forms: as dissolved N_2, nitric acid, NH_4^+, and NO_3^- as NH_4^+ adsorbed to inorganic particulate matter, and as organic compounds, which can occur in either dissolved or particulate phases. No direct relationship exists between the volume of rainfall or snowfall and the quantity of nitrogen influx per area of land or water (Chapin and Uttormark, 1973). Dry fallout can contain as much as 10 times the quantity of nutrients commonly found in rain. The nitrogen content of snow is often much higher than that of rain and can contribute up to half of the total annual nitrogen influx to a stream or lake, even though snow generally constitutes a small proportion of total precipitation. N_2-fixing bacteria occur in rainwater in low numbers (Visser, 1964a), but their contribution to the total input of nitrogen is most likely small in comparison to other sources. In nonpolluted areas, most of the combined nitrogen in the atmosphere is ammonia, much of which originates from the decomposition of terrestrial organic matter (Hutchinson, 1944). Atmospheric ammonia associated with dust particles can be oxidized to nitrate so that precipitation contains both ammonia and nitrate (Hutchinson, 1944, 1975).

The inputs of NO_3^- and NH_4^+ from atmospheric sources average about 0.1 g N m^{-2} yr^{-1} over the continental United States but are highly irregular in distribution. The north central and eastern regions, especially bordering the southern Great Lakes, receive the largest input of nitrogen from precipitation, on the order of 0.3–0.35 g N m^{-2} yr^{-1} (Chapin and Uttormark, 1973). Dry fallout sources increase these values by a factor of 3–4, so that in the Great Lakes region, atmospheric contributions of nitrogen occur at a rate of approximately 1 g N m^{-2} yr^{-1}. Bulk deposition of nitrogen from both wet and dry precipitation in North America now ranges from 0.55 to 1.2 g N m^{-2} yr^{-1} and occurs in approximately equal proportions of NO_3^-, NH_4^+, and dissolved organic nitrogen (Boring *et al.*, 1988; Morris, 1991). Urea can account for nearly half of the water-soluble organic nitrogen compounds deposited from the atmosphere (Timperley *et al.*, 1985). Nitrogen mass-balance techniques applied to a large number of drainage basins and lakes in central Ontario, for example, indicated that on a regional area-weighted basis, 67% of bulk atmospheric total nitrogen was stored or denitrified terrestrially, 12% was denitrified in lakes, 4% was stored in lake sediments, and 17% was exported from the lakes (Molot and Dillon, 1993). Neglecting other nutrients for the time being, this influx alone corresponds to the approximate amounts of nitrogen that are generally required to shift shallow lakes of a mean depth of $<$5m from moderate to high productivity (Vollenweider, 1968). In this region of the country, which also receives high inorganic inputs from runoff of nitrogen-containing sedimentary rock formations, it would be very difficult to control productivity of lakes by limiting nitrogen inputs because of large atmospheric contributions. We will return to this subject in later discussions.

B. Molecular Nitrogen and N_2 Fixation

Although N_2 is not particularly soluble in water, N_2 is usually saturated with respect to surface temperature and pressure in streams and during periods of circulation of lakes (Birge and Juday, 1911). Maximum concentrations are found in winter owing to increased solubility at colder temperatures (approximately 15–20 ml $liter^{-1}$). During summer thermal stratification, the heating of the epilimnion decreases N_2 solubility, while N_2 concentrations in the metalimnion and much of the hypolimnion may be slightly supersaturated as a result of hydrostatic pressure maintaining a higher gas concentration despite the heating in these water strata (Fig. 12-1). A decrease in the N_2 content in the lower hypolimnion above the sediments has been observed (Fig. 12-1), which presumably is related to bacterial fixation of N_2 in productive lakes (Kuznetsov and Khartulari, 1941). Conversely, late in the summer stratification of Lake Mendota, N_2 of the hypolimnion

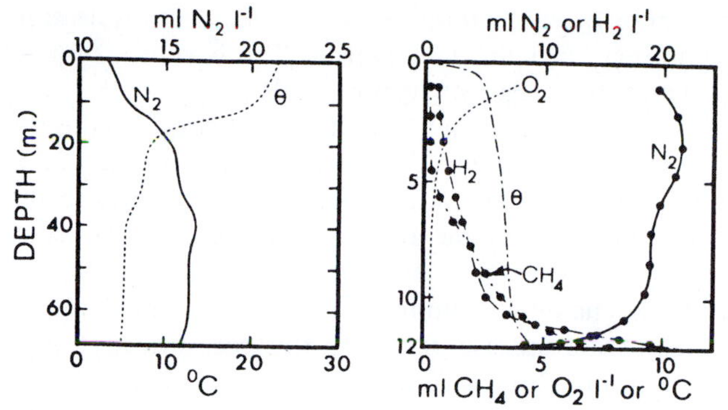

FIGURE 12-1 Vertical distribution of dissolved gases (*left*) in Green Lake, Wisconsin, summer, 1906 (Birge and Juday, 1911) and (*right*) in Beloie Lake, Russia, March 1938. θ = °C. (After Kuznetsov and Khartulari, 1941.)

increased somewhat above what would be expected on the basis of solubility (Brezonik and Lee, 1971); this may be the result of denitrification of nitrate in this lake (see following discussion).

C. Nitrogen Fixation: Cyanobacteria and Other Photosynthetic Bacteria

The importance of *in situ* nitrogen fixation to the productivity of streams and lakes has emerged only within the last two decades, even though the cyanobacteria (formerly termed *blue-green algae*) were implicated in this process many years earlier (Burris *et al.*, 1943). Excellent comparative reviews are given by Hardy *et al.* (1973) and Howarth *et al.* (1988a).

Nitrogen fixation in the open waters of lakes has been correlated strictly with the presence of cyanobacteria that possess heterocysts (Fogg, 1971a, 1974; Riddolls, 1985). Heterocysts are specialized cells that occur singly in most filamentous cyanobacteria, except for the Oscillatoriaceae, and are the sole site of nitrogen fixation in aerobically grown, heterocyst-forming cyanobacteria (Wolk, 1973; Haselkorn and Buikema, 1992). Nitrogen fixation has been found to occur in some unicellular forms that do not produce heterocysts (Fogg, 1974). However, among many species of cyanobacteria, such as *Anabaena* spp., numbers of heterocysts correspond approximately to observed nitrogen-fixing capacity (Horne and Goldman, 1972; Horne *et al.*, 1972, 1979; Riddolls, 1985).

Nitrogen fixation is primarily light-dependent in that the process requires reducing power and adenosine triphosphate (ATP), both of which are generated in photosynthesis. Electrons for the reduction of N_2 are supplied by dinitrogenase reductase in a highly endergonic reaction that requires about 12–15 mol of ATP per mol of N_2 reduced (Fay, 1992). Nitrogen-fixing photosynthetic bacteria can fix limited quantities of N_2 in the dark, but at rates usually <10% of maximum daytime rates (Horne, 1979; Levine and Lewis, 1984; Livingstone *et al.*, 1984; Storch *et al.*, 1990). Determinations *in situ* show a relationship of nitrogen fixation with depth similar to that of photosynthesis (Ward and Wetzel, 1980a). In full sunlight, N_2 fixation is often inhibited at the surface, reaches a maximum some depth below the surface, and shows rapid, nearly exponential decrease with greater depth and associated light attenuation (Fig. 12-2; cf. Chap. 5) (Dugdale and Dugdale, 1962; Goering and Neess, 1964; Horne and Fogg, 1970; Horne, 1979; Lewis and Levine, 1984).

Comparison of N_2—N, NO_3—N, and NH_4—N as nitrogen sources for *Aphanizomenon flos-aquae* (heterocystous) and *Microcystis aeruginosa* (nonheterocystous) at high, low, and variable light intensities revealed that highest growth rates always occurred with NH_4—N, followed by NO_3—N, and then N_2—N (Ward and Wetzel, 1980a). These findings were consistent with those expected from energy expenditures required to assimilate these nitrogen sources (i.e., N_2—N>NO_3—N>

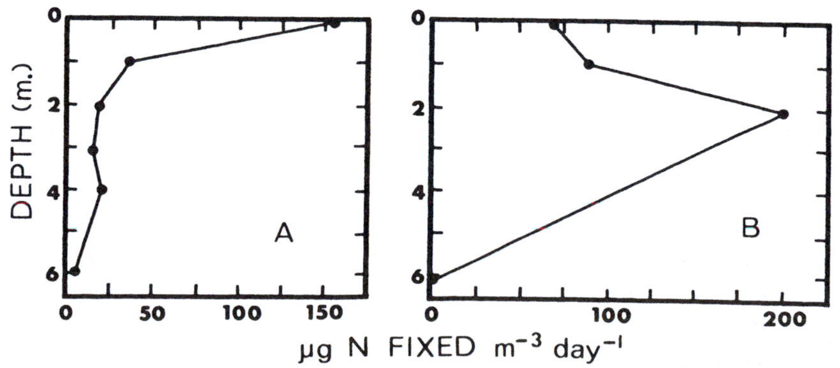

FIGURE 12-2 Variations in nitrogen fixation with depth (a) in Lake Windermere, and (b) in Esthwaite Water, England, 30 August 1966. (Modified from Horne and Fogg, 1970.)

NH_4—N). *Aphanizomenon* would not grow at low light intensities with N_2—N as the sole nitrogen source, although growth could be maintained with either NO_3—N or NH_4—N.

Readily utilizable sources of combined nitrogen suppress synthesis of the nitrogenase complex rather than the activity of any existing enzyme (Fogg, 1971, 1974; Wolk, 1973). Suppression of heterocyst formation by nitrate, even at very high concentrations, is often only partial (Ogawa and Carr, 1969; Ohmori and Hattori, 1972). Similarly, ammonia at low concentrations represses the formation of additional nitrogenase but does not affect activity of extant nitrogenase. In general, then, one would expect an inverse relationship between the rate of nitrogen fixation by cyanobacteria and the concentration of combined nitrogen in stream and lake waters; this is often the case. But because of the carryover of residual nitrogenase activity, the relationship is not always consistent. Thus N_2 fixation by cyanobacteria sometimes may occur at reduced levels in the presence of appreciable quantities of combined nitrogen in the water. The importance of the microenvironment surrounding the cell also should be noted. Molecular N_2 is in higher concentrations and diffuses more readily than either ammonium or nitrate ions. Within the massive mucilage sheaths surrounding many cyanobacteria, a steep gradient could easily develop in which other inorganic nitrogen sources are greatly reduced in comparison to N_2 availability, regardless of the inorganic concentrations of the water (cf. Chang, 1980).

Nitrogen fixation has also been correlated positively with concentrations of dissolved organic nitrogen occurring in the water (Horne and Fogg, 1970; Horne *et al.*, 1972; Paerl, 1985). Algae and cyanobacteria release extracellularly many simple and complex organic carbon and nitrogen compounds. Heterotrophic bacteria colonize common nitrogen-fixing cyanobacteria, such as *Anabaena*, particularly during the maxima of N_2-fixing blooms (Paerl, 1978, 1985; Paerl and Prufert, 1987). These cyanobacterial-heterotrophic bacterial aggregates form internal microenvironments with microzones of redox conditions distinct from those of the surrounding pelagic or benthic habitats. These oxygen-depleted microzones adjacent to the heterocysts protect the highly oxygen-sensitive nitrogenase. Environmental conditions, such as increased water turbulence, that increase diffusion rates of oxygen into the microzones reduce the rates of N_2 fixation. Similarly, reductions of dissolved organic carbon reduce rates of bacterial respiration, consumption of inhibiting oxygen, and rates of N_2 fixation. Many other strategies have evolved to protect nitrogenase from oxygen, such as the heterocyst structure, elaborate sheaths—particularly in nonheterocystous cyanobacteria—rapid changes in respiration, temporary protein inactivation, and production of enzymes to catalyze oxygen radicals (Fay, 1992)

Diurnal rates of nitrogen fixation in open lake water are typically low in the early morning, reach a maximum midday at maximum insolation, and then decline to low rates during the afternoon and evening (Rusness and Burris, 1970; Horne, 1979; Ward and Wetzel, 1980b). The most commonly observed seasonal pattern in temperate lakes is for both the percentage and the absolute rates of N_2 fixation to increase to maximum levels as heterocyst-bearing cyanobacterial populations develop and sources of combined nitrogen are reduced or depleted. High concentrations of total phosphorus are necessary (Lean *et al.*, 1978; Lundgren, 1978; Fiett *et al.*, 1980; Brownlee and Murphy, 1983; Wurtsbaugh *et al.*, 1985). Rates of N_2 fixation decline abruptly as the cyanobacterial populations decrease (cf. Chap. 15 on algal successions). In winter, the N_2 fixation is nonexistent or greatly reduced (Billaud, 1968; Horne and Fogg, 1970; Toetz, 1973). In productive tropical lakes where the periodicity of physicochemical factors and cyanobacteria is less marked, nitrogen fixation rates can extend for longer periods of the year but still exhibit distinct seasonality (e.g., Moyo, 1991).

Evaluations of nitrogen cycles of lakes or streams require estimates of the total nitrogen fixed per annum. Extensive measurements of nitrogen fixation are required because of the marked spatial and temporal heterogeneity of cyanobacterial populations (e.g., Horne and Galat, 1985; Levine and Lewis, 1985; Dudel and Kohl, 1991; Moyo, 1991). Planktonic nitrogen fixation tends to be low in oligotrophic and mesotrophic lakes (generally $\ll 0.1$ g N m^{-2} yr^{-1}) but often is high in eutrophic lakes (0.2–9.2 g N m^{-2} yr^{-1}) (Howarth *et al.*, 1988a,b). Direct estimates of N_2 fixation obtained by measuring $^{15}N_2$ uptake and acetylene reduction by nitrogenase of plankton in oligotrophic lakes, or in lakes of moderate productivity but which contain large pools of combined inorganic nitrogen, are consistently undetectable (Ward and Wetzel, 1980a).

Nitrogen fixation rates and control variables are poorly documented for benthic regions of freshwater ecosystems. Benthic fixation in oligotrophic ecosystems is often dominated by cyanobacteria (Howarth *et al.*, 1988) and tends to increase in cyanobacterial mats of very productive, shallow environments (1–76 g N m^{-2} yr^{-1}). In spite of the relatively small areas of such mats, their nitrogen contributions to the entire ecosystem can be disproportionally large. In moderately productive Lake Windermere, the nitrogen fixation by benthic cyanobacteria growing in abundance on littoral rocks and stones to a depth of 3.5 m was estimated conservatively to be 1 g N m^{-2} yr^{-1} (Horne and Fogg, 1970).

TABLE 12-1 Estimated Annual Rates of Nitrogen Fixation by Phytoplankton and Benthic Algae[a]

Lake	*Area* (km^2)	N_2 *Fixation* (*g N* m^{-2})	*Total N fixed* (*megagrams = metric* t)
Phytoplankton			
Windermere, England			
North basin 1966	8.2	0.037	0.30
South basin 1965	6.7	0.287	1.92
1966	6.7	0.107	0.72
Esthwaite, England			
1965	1.01	0.127	0.13
1966	1.01	0.061	0.06
Clear Lake, California, 1970			
Upper arm	127.0	0.352	361
Lower arm	37.2	0.759	49
Oaks arm	12.5	0.250	50
Total for lake	176.7	0.384	460
Benthic Algae			
Windermere, England	2.9	1.0	0.8
Taharoa, New Zealand			
Exposed beach	—	0.62	—
Submersed beach	—	0.15	—
California stream	—	0.04–0.36	—
Everglades wetland, Florida	>5000	0.42–2.60	—
Range of wetland types	—	0.05–12	—

[a] From data of Horne and Fogg (1970), Horne and Goldman (1972), Horne and Carmiggelt (1975), Lam *et al.* (1979), Bowden (1987), and Browder *et al.* (1994).

For a minimal rocky area of 2.86 km^2, the estimated total annual fixation by the benthic cyanobacteria in Windermere is 0.8 megagrams. In this example, the littoral benthic contribution is considerably larger than the nitrogen fixed by the phytoplankton (Table 12-1). Even though these values are large, the total N_2 fixation by cyanobacteria contributes at most a small proportion (<1%) to the total combined nitrogen income of this lake. An estimate (probably high) for small, eutrophic Chernoye Lake, Russia, indicated that the nitrogen fixation constituted about 13% of the total nitrogen income (Kuznetsov, 1959). In eutrophic Clear Lake, California, biological N_2 fixation, largely associated with cyanobacterial blooms, contributed at least 43% of the nitrogen income, almost as much as NO_3^- from river inflows. Therefore, the importance of nitrogen fixation to the total nitrogen budget is variable. Nitrogen fixation is relatively unimportant as a nitrogen source in many oligotrophic and mesotrophic lakes (<1–10% of total nitrogen inputs). In eutrophic and hypereutrophic lakes of high phosphorus loadings, nitrogen fixation can account for >80% of the annual nitrogen inputs.

A detailed study of the nitrogen fixation in a small subarctic lake of interior Alaska demonstrated that ammonia was more important than N_2 fixation (Billaud, 1968). Of two main algal production periods, the first consisted largely of microflagellates under the ice and depended on ammonia as a nitrogen source. Immediately after the ice melted from the lake, a phytoplankton population composed almost exclusively of the cyanobacterium *Anabaena flos-aquae* developed. During the peak of the *Anabaena* bloom, molecular nitrogen constituted over 25% of the nitrogen assimilated by the plankton. Nitrogen fixed by the plankton in the south basin of Lake Windermere in 1965 amounted to 72% of the amount available as nitrate, although in 1966 and in other water bodies this proportion was less (<0.2–48%) (Horne and Fogg, 1970; Tison *et al.*, 1977; Ward and Wetzel, 1980a). Rates of N_2 fixation are greatly enhanced when the productivity of a lake is increased by phosphorus fertilization (e.g., Lean *et al.*, 1978; Lundgren, 1978).

Nitrogen metabolism is dependent upon the availability of trace quantities of molybdenum and iron for enzymes involved in N_2 fixation, assimilatory nitrate reduction, and denitrification (Miller, 1991). In oxygenated waters, molybdenum occurs primarily in the form of molybdate ion. Uptake follows saturation kinetics by natural phytoplankton assemblages and is light-dependent (Cole *et al.*, 1986). Sulfate anion is very similar to molybdate and is a competitive inhibitor of molybdate assimilation by planktonic algae and bacteria. Because concentrations of dissolved sulfate are several orders of magnitude greater than those of molybdate, sulfate can suppress the availability of molybdate and nitrogen fixation (Howarth *et al.*, 1988b). As sulfate concentrations increase, as in sea water and saline lakes (Marino *et al.*, 1990), the availability of molybdate is further reduced.

Iron is also required as a cofactor for enzyme molybdenum nitrogenase protein activity in nitrogen fixation. Where phosphorus loadings are very high, iron availability can become limiting to N_2-fixation rates (Wurtsbaugh and Horne, 1983; Wurtsbaugh *et al.*, 1985). Combinations of high phosphorus and high dissolved organic matter favor cyanobacterial communities and may be partly related to organic complexation and increased solubility of iron and molybdenum. Under reducing conditions, as in or near sediments, particularly in wetlands and littoral areas, the level of dissolved organic matter tends to be much higher than in pelagic regions (e.g., Mann and Wetzel, 1995). The high level of dissolved organic matter and reducing conditions increase iron solubility and molybdenum stability. The resulting enhanced conditions for N_2 fixation may be counterbalanced by high concentrations of ammonium released from bacterial metabolism in the sediments.

Nitrogen fixation capabilities are widespread throughout other photosynthetic bacteria. As will be discussed in greater detail in Chap. 14, photosynthetic bacteria can develop in great densities in highly structured strata at the interface regions between the aerobic epilimnion or metalimnion and the anaerobic hypolimnion of eutrophic or meromictic lakes, given sufficient light to permit photosynthesis in these strata. N_2 fixation occurs only in the light, and intensive rates occur only under anaerobic conditions among the green and purple photosynthetic bacteria (Kondrat'eva, 1965). N_2 fixation by photosynthetic bacteria occurs simultaneously with the release of molecular H_2 by a noncyclic electron flux resulting from photophosphorylation. The source of these electrons is an exogenous H-donor, such as thiosulfate.

All three families of photosynthetic bacteria of the order Rhodospirillales produce bacteriochlorophyll and carry out anoxygenic photosynthesis (Roberts and Ludden, 1992). The Rhodospirillaceae (Athiorhodaceae), or purple nonsulfur bacteria, are facultative autotrophs but are unable to use elemental sulfur as an electron donor for photosynthesis. The Chromatiaceae (Thiorhodaceae), or purple sulfur bacteria, use elemental and nonelemental sulfur as electron donors for photosynthesis. The Chlorobiaceae (Chlorobacteriaceae), or green sulfur bacteria, use the sulfur of hydrogen sulfide as an electron donor for the photosynthetic reduction of CO_2.

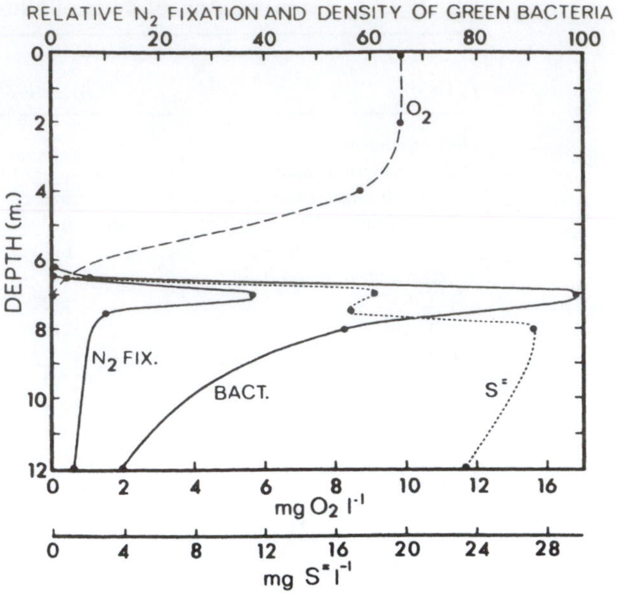

FIGURE 12-3 Nitrogen fixation in a Norwegian meromictic lake, June 1966, in relation to a dense layer of green photosynthetic bacteria, primarily *Pelodictyon,* on the surface of the monimolimnion. (Drawn from data of Stewart, 1968.)

Despite considerable understanding of the biochemical physiology of these organisms, little information is available on the relative quantitative contributions of the nitrogen-fixing capabilities of photosynthetic bacteria in comparison to the cyanobacteria and heterotrophic bacteria in lakes, wetlands, and rivers. In the example given in Figure 12-3, nitrogen fixation is clearly associated with green photosynthetic bacteria, mainly *Pelodictyon,* of the chemolimnion in this meromictic lake. In contrast, the nitrogen-fixation capabilities of photosynthetic sulfur bacteria were not found to contribute significantly to the organic nitrogen loading of Lake Kinneret in Israel (Bergstein *et al.*, 1981). Generalizing from the latter conclusion, however, is not warranted on the basis of existing data.

D. Nitrogen Fixation: Heterotrophic Bacteria

Quantitative information on nonphotosynthetic bacterial nitrogen fixation in lakes is also sparse. Heterotrophic nitrogen fixation is commonly disregarded, on the premise that these nitrogen-fixing bacteria are limited by the low availability of exogenous carbohydrate. One to 25 mg of nitrogen can be fixed per gram of carbohydrate utilized (Stewart, 1969; Hill, 1992). Sufficiently large quantities of soluble carbohydrate rarely are available in natural waters, and there is an intense competition for these substrates by heterotrophic bacteria incapable of N_2 fixation.

TABLE 12-2 Occurrence of Nitrogen-Fixing Bacteria in Water and Sediments of Lakes[a]

Lake type	Number lakes studied	In water (no. ml^{-1})		In sediment (no. g^{-1} wet wt)	
		Azotobacter	*Clostridium pasteurianum*	*Azotobacter*	*Clostridium pasteurianum*
Oligotrophic	5	0	0	0–10	0–10
Mesotrophic	5	0–10	0–10	0–10	0–1000
Eutrophic	10	0–10	1–20	0–10	100–10,000
Eutrophic with high humic content	2	10	1	—	—
Dystrophic with high humic content	3	1–10	0–1	—	—
Eutrophic reservoir	1	5–10	1–5	10–1000	1000–10,000

[a] Modified from Kuznetsov (1970).

The most common heterotrophic N_2-fixing bacteria comprise several species of *Azotobacter* and *Clostridium pasteurianum,* which are found in fair abundance living free in the water, epiphytically on submersed aquatic plants, and in the sediments (Kuznetsov, 1959, 1970). The numbers of these bacteria are generally lowest in the open water, where soluble organic concentrations are low, and tend to increase, with *Azotobacter* being dominant, in bog lakes, which contain high concentrations of dissolved humic organic matter (Table 12-2). Based on scant data from eutrophic Chernoye Lake, Russia, numbers of open-water *Clostridium* increase in the spring just before and after ice loss, decrease to a low in summer, and increase in autumn. This pattern is somewhat analogous to the seasonality of other, non-N_2-fixing bacteria populations that lag slightly behind the productivity pulses of phytoplankton (cf. Chap. 17). In contrast, Niewolak (1972) observed a maximum of *Azotobacter* in spring and summer in Polish lakes.

Azotobacter is found in particular abundance as an epiphyte on submersed aquatic angiosperms and submersed portions of emergent macrophytes. A symbiotic relationship between the *Azotobacter* and the macrophytes is possible since the plants secrete many dissolved organic compounds (Wetzel, 1969) that can serve as substrates for nitrogen-fixing bacteria. The nitrogen released by the bacteria may be utilized by the macrophytes. In the littoral water, populations of the planktonic bacterium *Azotobacter* are higher than those of the open water (J. Overbeck, personal communication). Although few quantitative data are available, the littoral zone may serve as a major site of nitrogen fixation by both heterotrophic bacteria and sessile cyanobacteria. Further investigation is needed.

Large numbers of N_2-fixing bacteria occur in the sediments of lakes; most are concentrated in the upper 2 cm. Bacterial numbers are much higher in productive lakes than in oligotrophic lakes, particularly in the case of *Clostridium* (Table 12-2). Seasonal changes in the numbers of *Clostridium* in the sediments of Chernoye Lake were similar to those observed in the open water: maximum numbers occurred in the fall, followed by low winter populations, that increased to another maximum in the spring. Lowest populations in the sediments were found in midsummer. Similar results were observed for *Azotobacter* populations of sediments in Polish lakes (Niewolak, 1970, 1972).

Several methane-oxidizing bacteria are capable of N_2 fixation (reviewed by Zeikus, 1977; Rudd and Taylor, 1980). The N_2 fixation of these bacteria increases when sources of combined inorganic nitrogen are depleted either within water or at the sediment–water interface where methane is abundant. Although N_2 fixation is physiologically important to the methane-oxidizing bacteria, these nitrogen fixers have not yet been demonstrated to constitute a major source of fixed nitrogen to a lake (Rudd and Taylor, 1980).

In two lakes (Mary and Mize, Table 12-3), N_2 fixation by planktonic heterotrophic bacteria (*Azotobacter*) occurred at a rate several orders of magnitude lower than rates in lakes dominated by the cyanobacterium *Anabaena.* In waters rich in dissolved humic organic compounds (such as in lakes Mary and Mize), cyanobacteria are rare and nitrogen-fixation rates are very low.

E. Nitrogen Fixation: Wetland Sources

In addition to the inputs of nitrogen from drainage runoff from terrestrial sources, lakes and streams often are bordered by dense stands of shrublike trees that fix nitrogen from the atmosphere. Common species of the genera *Alnus* and *Myrica* are nonleguminous angiosperms that form large nodules containing an actinomycetal fungal endophyte at or just below the soil surface. Nitrogen fixation by dense stands of *Alnus* can

TABLE 12-3 Estimates of Fixation of Molecular Nitrogen by Heterotrophic Bacteria and Cyanobacteria in Several Lakes[a]

Lake	*Organisms*	*N_2 Fixation*	*N_2 Fixation in g, whole lake, summer stratification period*
Chernoye, USSR, 1937	*Azotobacter*	1.6×10^{-11} mg cell^{-1} day^{-1}	0.14
	Anabaena scheremetievi	1.12×10^{-9}	13,100
Mendota, Wisconsin, 1960, 1967	Cyanobacterial dominance	0.07–43 μg N liter^{-1} h^{-1}	
Sanctuary, Pennsylvania, 1959	Cyanobacterial dominance	0–6	
Wingra, Wisconsin, 1961	Cyanobacterial dominance	0.005–1	
Smith, Alaska, 1963	*Anabaena* dominance	0–1	
Mary, Wisconsin, June 1968	Heterotrophic bacteria	0.003–0.047	
Mize, Florida	Heterotrophic bacteria		
July 1968		0.083–0.308	
August 1968		0.000–0.083	
Pyramid, Nevada, 1979	Cyanobacterial dominance	2 g N m^{-2} yr^{-1}	

[a] After many sources cited in the text; also Brezonik and Harper (1969).

be as high as 22.5 g N m^{-2} yr^{-1}, most of which enters the plant. Subsequently, leachate from direct leaf-fall or release during decomposition of foliage can reach nearby streams and lakes.

Alnus trees have been implicated as a significant nitrogen source for streams and lakes in glaciated regions of Alaska that are particularly nitrogen-deficient (Goldman, 1960; Dugdale and Dugdale, 1961). In Castle Lake, an alpine cirque lake of northern California, alder (*Alnus*) trees were abundant on only one side of the lake (Goldman, 1961). *Alnus* leaves contained over four times as much nitrogen as those of other deciduous species. Soils, lake sediments, and spring waters draining to the lake from the alder side all contained higher nitrogen levels than those from the non-alder side of the lake. Bioassay of *in situ* rates of photosynthesis by the phytoplankton demonstrated that these nitrogen sources were stimulatory and also that the primary productivity of the lake was significantly higher along the side on which the alders grew.

Cyanobacterial and heterotrophic bacterial N_2-fixation rates can add significant amounts of nitrogen to wetlands. Combined N_2 fixation by microorganisms associated with the aerobic layer of wetland hydrosoils, the anaerobic regions including the rhizosphere of non-nodulated plants, and floating mats of particulate detritus and associated microorganisms often constitute a major source of nitrogen input to wetlands (Buresh *et al.*, 1980). Nitrogen fixation by attached cyanobacteria and heterotrophic bacteria contributed a significant but small (5–10%) portion of the total nitrogen inputs on the flood plain of a tropical riverine lake (Doyle and Fisher, 1994). In another example, the nitrogen fixed annually by heterotrophic bacteria on decaying plant debris of calcareous wetlands ranged from 530 to 2100 mg N m^{-2} and was considerably less (70 mg N m^{-2}) among acidic bog vegetation (Waughman and Bellamy, 1980).

Azolla, a small (1–3 cm) aquatic fern, with global distribution, floats freely on the water surface of many natural and cultivated wetlands. All *Azolla* species contain a symbiotic heterocyst-forming, N_2-fixing cyanobacterium *Anabaena azollae* (Lumpkin and Plucknett, 1980; Peters and Meeks, 1989). These organisms are actively cultivated in rice agriculture to supplement alternative sources of nitrogen fertilization. It is estimated that as much as a quarter of the total nitrogen consumed by humankind emanates from this symbiotic N_2-fixing cyanobacterium–fern association.

II. INORGANIC AND ORGANIC NITROGEN

In addition to atmospheric nitrogen in the form of precipitation, dry fallout, and fixation of N_2 just discussed, a major source of nitrogen income to streams and lakes is from influents, both from surface land drainage and from groundwater sources. Inputs of nitrogen by ground water can be a major part of the annual nitrogen loading in many streams and lakes, especially in regions rich in limestone. In detailing the nitrogen cycle, the forms of nitrogen must be discussed individually and then integrated with rates of biogeochemical utilization and transformations as well as nitrogen losses from the lake ecosystem. An obvious loss is via inorganic dissolved and organic (both dissolved and particulate) compounds in effluents flowing out of lake basins. Further losses occur as a result of permanent interment of partially decomposed biota and inorganic and organic nitrogen compounds adsorbed to inorganic particulate matter in the sediments. Nitrogen can also be lost by volatilization of compounds from

TABLE 12-4 General Relationship of Lake Productivity to Average Concentrations of Epilimnetic Nitrogen[a]

General level of lake productivity	*Change in alkalinity in epilimnion in summer (meq liter^{-1})*	*Inorganic N (mg m^{-3})*	*Approximate average organic N (mg m^{-3})*
Ultra-oligotrophic	<0.2	<200	<200
Oligo-mesotrophic	0.6	200–400	200–400
Meso-eutrophic	0.6–1.0	300–650	400–700
Eutrophic		500–1500	700–1200
Hypereutrophic	>1.0	>1500	>1200

[a] Modified from Vollenweider, R. A.: *Scientific Fundamentals of the Eutrophication of Lakes and Flowing Waters, with Particular Reference to Nitrogen and Phosphorus as Factors in Eutrophication.* OECD Report No. DAS/CSI 68.27, Paris, OECD, 1968, after data of Thomas and Lueschow *et al.* 1970.

the water surface, such as ammonia at high pH and as N_2 formed in microbial denitrification of nitrate.

Combined nitrogen occurs as ammonia (NH_4^+), hydroxylamine (NH_2OH), nitrite (NO_2^-), nitrate (NO_3^-), and dissolved and particulate organic nitrogen. NH_4—N can range from 0 to 5 mg liter^{-1} in unpolluted surface waters, although concentrations are usually low, to well above 10 mg liter^{-1} in anaerobic hypolimnetic waters of eutrophic lakes. The intermediate product, NH_2OH, is rapidly oxidized and occurs only in very low concentrations. Similarly, NO_2—N levels of natural waters are generally very low, in the range of 0–0.01 mg liter^{-1}, although concentrations of up to 1 mg liter^{-1} have been found in the interstitial waters of deep (>90 cm depth) sediments of Lake Mendota (Konrad *et al.*, 1970). Concentrations of NO_2—N increase in the anaerobic hypolimnion of lakes under reducing conditions (e.g., Brezonik and Lee, 1968; Overbeck, 1968) and in streams and lakes receiving heavy organic-matter pollution. Concentrations of NO_2—N are usually low under oxygenated conditions, but significant deepwater maxima (to 10 μg liter^{-1}) have been found in the upper portions of the hypolimnion (Mortonson and Brooks, 1980). Concentrations of NO_3—N range from undetectable levels to nearly 10 mg liter^{-1} in unpolluted fresh waters but are highly variable seasonally and spatially. Organic nitrogen, much of which occurs in forms resistant to rapid bacterial degradation, commonly accounts for more than one-half of the total dissolved nitrogen.

Recently, much attention has been devoted to the importance of nitrogen concentrations of fresh waters in the regulation of algal productivity (e.g., cf. Vollenweider, 1968). Although, as we will see in Chap. 13, phosphorus-cycling rates frequently regulate high sustained productivity, the loading rates of nitrogen are of major importance to the maintenance of high flux rates within the nitrogen cycle. Although there are a number of exceptions, a positive correlation has been found between high sustained productivity of algal populations and average concentrations of inorganic and organic nitrogen (Table 12-4). Mean chemical mass, however, must be used with caution when little consideration is given to the rates of mineralization, microbial transformation, and recycling.

In comparing the distribution of organic and inorganic forms of nitrogen in oligotrophic and eutrophic lake sediments, little correlation was found between concentrations of total or organic forms of nitrogen in sediments and general productivity of the lake (Table 12-5)

TABLE 12-5 Average Distribution of Nitrogen in Some Wisconsin Lake Sediments[a]

	Sediment-N (mg kg^{-1})		*Interstitial water-N (mg liter^{-1})*				
Lake type	*Fixed NH_4*	*Exchangeable NH_4*	*Organic*	*NH_4*	*NO_3*	*Total organic N (%)*	*Acid hydrolyzable (% of total N)*
Softwater, oligotrophic	69	167	2.1	3.8	0.2	1.53	83.1
Softwater, eutrophic	44	415	2.2	9.6	0.3	3.26	80.4
Hardwater, oligotrophic	66	66	—	—	—	0.52	84.4
Hardwater, eutrophic	134	120	2.0	11.4	0.3	0.80	82.0

[a] Modified from Keeney (1973).

(Keeney *et al.*, 1970; Konrad *et al.*, 1970). The percentage of organic nitrogen as hexosamine-N decreased while amino acid-N increased with increasing lake fertility. NH_4—N of interstitial water was somewhat higher in sediments of eutrophic lakes than in those of oligotrophic lakes.

A. Ammonia

Ammonia is generated by the biological dissimilation of nitrate, as will be discussed later in this chapter. Much of the ammonia, however, arises as a primary end product of the decomposition of organic matter by heterotrophic bacteria from the deamination of proteins, amino acids, urea, and other nitrogenous organic compounds. Intermediate nitrogen compounds are formed in the progressive degradation of organic material but rarely accumulate, because deamination by bacteria proceeds rapidly. Phagotrophic protists, particularly bacterivorous flagellates, can accelerate appreciably the regeneration of ammonia by the consumption of bacteria and the release of ammonium ions (Suzuki *et al.*, 1996). Although ammonia is an excretory product of higher aquatic animals, this nitrogen source is quantitatively minor in comparison to that generated by bacterial decomposition.

Ammonia in water is present primarily as NH_4^+ and as undissociated NH_4OH, the latter being highly toxic to many organisms, especially fish (Trussell, 1972). The proportions of NH_4^+ to NH_4OH are dependent on the dissociation dynamics, which are governed by pH and temperature. The approximate ratios of NH_4 to NH_4OH are as follows (Hutchinson, 1957):

pH 6	3000:1
pH 7	300:1
pH 8	30:1
pH 9.5	1:1

Detailed dissociation relationships with pH and temperature are given by Trussell (1972) and Emerson *et al.* (1975). Ammonia is strongly sorbed to particulate and colloidal particles, especially in alkaline lakes containing high concentrations of humic dissolved organic matter.

At high pH values, NH_3 gas can be formed by the deprotonization of NH_4^+. Although few studies have addressed the evasion of NH_3 from fresh waters and littoral areas to the atmosphere, appreciable amounts (range, 10–300 μg N m^{-2} h^{-1}) evade from hydrosoils of lakes and littoral wetlands (cf. review of Schlesinger and Hartley, 1992). The rates of ammonia volatilization from wetlands increase with increasing concentrations of ammonium and pH of the water overlying the sediments, as well as with greater wind velocities and temperatures (Bouwmeester and Vlek, 1981; Jones *et al.*, 1982). On a global basis, volatilization of NH_3 from inland waters and associated wetlands could constitute a significant (e.g., ca. 15%) though not dominating source.

Since NO_3^- must be reduced to NH_4^+ before it can be assimilated by plants, ammonia is an energy-efficient source of nitrogen for plants. The energy necessary to assimilate nitrogen is lowest for NH_4—N and increases for NO_3—N (and N_2—N for N_2-fixing cyanobacteria). Dark assimilation of both NO_3^- and NH_4^+ was about 50% of the assimilation rates at photosynthetically saturating light intensities, which indicated that part of the energy required for inorganic nitrogen assimilation may originate from intermediary metabolism (Priscu, 1984). Among the cyanobacteria, highest growth rates always occurred with NH_4—H as the nitrogen source at many different light intensities (Ward and Wetzel, 1980a,b). Reports of better growth of some algae with NO_3—N as the nitrogen source rather than with NH_4—N may be, in part, the result of the toxicity of NH_4OH at high pH values that can prevail both in culture and *in situ* during periods of high daily photosynthesis in very eutrophic lakes (cf. Rodhe, 1948; Laan *et al.*, 1982). Inhibition of NH_4—N assimilation by natural populations of phytoplankton was observed at high concentrations of NH_4—N, but at low concentrations NH_4—N uptake rates exceeded those of NO_3—N (Toetz *et al.*, 1977; Toetz and Cole, 1980; Berman *et al.*, 1984). NO_3—N uptake can be suppressed by NH_4—N concentrations as low as 2 μg NH_4^+—N $liter^{-1}$ (Priscu *et al.*, 1989). A few algae, such as a species of *Chrysochromulina*, are apparently able to utilize only NH_4^+ as an inorganic N source (Wehr *et al.*, 1987).

The distribution of ammonia in fresh waters is highly variable regionally, seasonally, and spatially within streams and lakes and depends upon the level of productivity and the extent of pollution from organic matter. Although generalizations are difficult to make, the concentration of NH_4—N in well-oxygenated waters is usually low. In the trophogenic zone, NH_4—N is rapidly assimilated by algae and represents the most significant source of nitrogen for the plankton in many lakes (Liao and Lean, 1978). Thus, concentrations of NH_4—N are commonly low in unproductive oligotrophic waters, in the trophogenic zones of most lakes, and in most lakes after periods of circulation (Fig. 12-4). When appreciable amounts of sedimenting organic matter reach the hypolimnion of stratified lakes, NH_4—N can accumulate. The accumulation of NH_4—N greatly accelerates as the hypolimnion becomes anoxic. Under anaerobic conditions, bacterial nitrification of NH_4^+ to NO_2^- and NO_3^- ceases, as the redox potential is reduced to below about + 0.4 V.

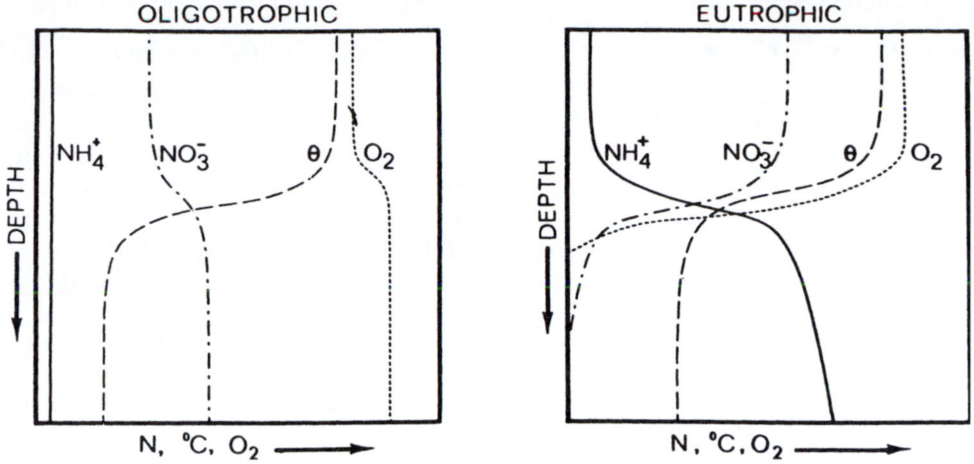

FIGURE 12-4 Generalized vertical distribution of ammonia and nitrate nitrogen in stratified lakes of very low and high productivity.

In the sediments, a large percentage of the ammonia is adsorbed on sediment particles. With the loss of the oxidized microzone at the sediment–water interface under anoxic hypolimnetic conditions, the adsorptive capacity of the sediments is greatly reduced (Kamiyama *et al.*, 1977; Verdouw *et al.*, 1985). A marked release of NH_4^+ from the sediments then occurs. Diffusion transport into the overlying water approximately equals production and release into the interstitial waters of the sediments. These diffusion rates of NH_4^+ into the overlying water can be increased severalfold by the activities of benthic invertebrates such as chironomid larvae, tubificid worms, and bivalve mollusks (Henriksen *et al.*, 1983; Fukuhara and Sakamoto, 1987; Fukuhara and Yasuda, 1989; Svensson, 1997). The ammonium enhancement apparently results more from microwater movements associated with respiratory behavior rather than from direct excretion. Excretion rates of NH_4—N of four species of chironomids and tubificids ranged from 0.33 to 2.87 μg N mg DW^{-1} d^{-1} at 15°C.

If light reaches the sediments in amounts sufficient to support benthic algae, algal and cyanobacterial growth can assimilate NH_4—N from the interstitial water and prevent the flux of NH_4—N from the sediment to the water (Jansson, 1980; Reuter *et al.*, 1986; Risgaard-Petersen *et al.*, 1994; Van Luijn *et al.*, 1995). Because the uptake rates are in part dependent upon photosynthesis, a marked diurnal variation in release rates can occur, with reduced uptake during periods of darkness. These sources of interstitial nitrogen often serve as the dominant inorganic nitrogen source, or supplement N_2 fixation, for the microbiota attached at the sediment–water interface, particularly in nitrogen-deficient lakes and streams. These attached microbial communities can release appreciable amounts of dissolved organic nitrogen to the water above the sediments.

If macrophytes are rooted in the sediments, these plants are capable of absorbing large quantities of nitrogen from the sediments. Much of the nitrogen can be immobilized in the belowground rooting tissues and in the above-sediment foliage tissues. Much of the nitrogen is assimilated as NH_4—N (e.g., Dean and Biesboer, 1985) and can remain for long periods in living tissues and particulate detritus after senescence. Thus, in littoral and wetland regions of freshwater ecosystems, the sediments are a primary storage site for nitrogen, and storage in living and senescent macrophytes is by far the most important component (e.g., Sarvala *et al.*, 1982; Reddy and Patrick, 1984; Bowden, 1987).

B. Nitrification

Nitrification may be broadly defined as the biological conversion of organic and inorganic nitrogenous compounds from a reduced state to a more oxidized state (Alexander, 1965). Of the numerous oxidation and reduction stages outlined in Figure 12-5, initial nitrification by bacteria, fungi, and autotrophic organisms involves (Kuznetsov, 1970):

$$NH_4^+ + 1\tfrac{1}{2}\,O_2 \rightleftharpoons 2H^+ + NO_2^- + H_2O$$
$$(\Delta G_0' = -66.0 \text{ kcal})^1$$

which proceeds by a series of oxidation stages through hydroxylamine and pyruvic oxime to nitrous acid:

$$NH_4^+ \longrightarrow NH_2OH \longrightarrow H_2N_2O_2 \longrightarrow HNO_2$$

[1] Gibbs energy of formation, kcal mol^{-1}.

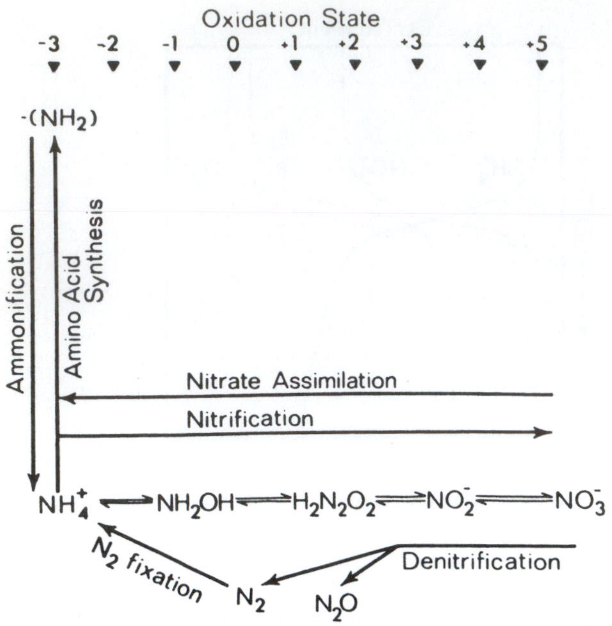

FIGURE 12-5 Biochemical reactions that influence the distribution of nitrogen compounds in water. (After Stadelmann, 1971, and Kuznetsov, 1970.)

The intermediate products are labile to physical and heterotrophic oxidation and are found only rarely in significant quantities relative to other forms of combined nitrogen (cf. Baxter *et al.*, 1973). Much of the energy (total exothermic energy = -84.0 kcal mol^{-1}) released by the oxidation series is used to reduce CO_2 in the formation of organic matter; detailed reactions are discussed by Alexander (1965a), Kuznetsov (1970), and Fenchel and Blackburn (1979).

The nitrifying bacteria capable of the oxidation of $NH_4^+ \rightarrow NO_2^-$ are largely confined to *Nitrosomonas* (Nitrobacteriaceae, order Pseudomonadales), although several other taxa, including methane-oxidizing bacteria, are known to be capable of this process (Alexander, 1965; O'Neill and Wilkinson, 1977). These bacteria are mesophilic, with a wide temperature tolerance range (1–37°C), and grow optimally at a pH near neutrality.

Oxidation of nitrite proceeds further to nitrate by:

$$NO_2^- + \tfrac{1}{2}O_2 \rightleftharpoons NO_3^- \qquad (\Delta G_0' = -18.0 \text{ kcal})$$

Nitrobacter is the primary bacterial genus involved in this oxidation. *Nitrobacter* is somewhat less tolerant of low temperatures and high pH, conditions that can lead to a slight accumulation of NO_2—N. The release of energy for synthesis of organic matter by the oxidation of nitrite is much lower, at -18.0 kcal per mole, than that of NH_4^+ to NO_2^-.

The overall nitrification reactions,

$$NH_4^+ + 2O_2 \longrightarrow NO_3^- + H_2O + 2H^+$$

require two moles of oxygen for the oxidation of each mole of NH_4^+. Although conditions must be aerobic in order for nitrification to occur, these processes will continue until concentrations of dissolved oxygen decline to about 0.3 mg liter^{-1}. Below this concentration, diffusion rates of oxygen to the bacteria become critical. Nitrification is greatly reduced in undisturbed sediments because oxygen is very low or absent (Chen *et al.*, 1972). Some nitrate may diffuse to the water following nitrification in the well-oxygenated surficial sediments of the littoral zone or during periods of circulation of the water and disturbance of sediments (cf. Landner and Larsson, 1972; Laurent and Badia, 1973). Oxidation of ammonia by autotrophic bacteria was found to be of relatively minor importance during the summer stratification of eutrophic Pluss See, Northern Germany (Gode and Overbeck, 1972). Several genera of heterotrophic nitrifying bacteria were found in much higher concentrations in both aerobic and anaerobic water strata. Experimental results suggested that these bacteria were most likely responsible for most of the nitrification that occurred in this lake.

Nitrification is severely inhibited by certain dissolved organic compounds, especially by tannins and their decompositional derivatives (Rice and Pancholy, 1972, 1973). Although this mechanism has been demonstrated only in soil systems, there is a strong possibility that an analogous situation exists in freshwater systems that contain relatively high concentrations of humic materials, for ammonia concentrations are often greater in highly stained lakes than in aerobic waters of other lakes. Therefore, it is likely that rates of nitrification are lower in neutral or alkaline waters containing high concentrations of dissolved humic organic matter. Such was apparently the case in several European waters (Nygaard, 1938; Karcher, 1939). Moreover, nitrification proceeds slowly in acidic waters, such as in acid bogs and acidic bog lakes, where the pH is 5 or less. Nitrate produced in such lakes is probably utilized as rapidly as it is formed, so that most of the time only very low or undetectable quantities are found.

C. Nitrate Reduction and Denitrification

As nitrate is assimilated by algae and larger hydrophytes, it is reduced to ammonia. Molybdenum is required in the enzyme systems associated with this reduction. In a few lake regions, such as granitic mountainous regions, concentrations of molybdenum are extremely low. Rates of carbon fixation by phytoplanktonic algae in lakes of these regions can be increased by additions of molybdenum (Goldman, 1960, 1972). These circumstances are related to the rates of nitrate reduction (Axler *et al.*, 1980) and indicate the coupling of molybdenum to phytoplanktonic growth during portions of the growing season when nitrate concentrations are relatively high.

The assimilation of nitrate and its reduction by green plants are clearly dominant processes in the trophogenic zone of lakes. As much as 60% of the photoassimilated NO_3—N can be excreted as dissolved organic nitrogen compounds, some of which are simple amino acids readily utilized by bacteria (Chan and Campbell, 1978). Nitrate assimilation in oligotrophic lakes may be adequate to reduce the observed concentrations in the trophogenic zone, as depicted in Figure 12-4. Alternatively, NO_3—N assimilative losses may be counterbalanced by nitrification and inflow sources of NO_3—N. In eutrophic lakes, littoral areas, and wetlands, denitrification (see below) is a major process influencing the distribution of NO_3—N. Nitrate assimilation can, however, greatly exceed sources of income and generation, in some cases to the point of reducing NO_3—N to below detectable concentrations (Fig. 12-4).

The ratio of NO_3-N to NH_4-N in fresh waters is variable in relation to natural and pollutional sources of both forms of combined nitrogen. In regions draining calcareous sedimentary landforms, unpolluted waters can have NO_3—N:NH_4—N ratios of 25:1. In other areas where natural sources of NO_3—N are low, the ratio may approach 1:1; where slight-to-moderate sewage contamination or agricultural applications of nitrogen fertilizers influence the waters, ratios in the range of 1:10 are common.

Denitrification by bacteria is the biochemical reduction of oxidized nitrogen anions, NO_3—N and NO_2—N, with concomitant oxidation of organic matter. The general sequence of this process is

$$NO_3^- \longrightarrow NO_2^- \longrightarrow N_2O \longrightarrow N_2$$

which results in a significant reduction of combined nitrogen that can, in part, be lost from the ecosystem if it is not refixed.

Many facultative anaerobic bacteria, particularly of the genera *Pseudomonas, Achromobacter, Escherichia, Bacillus,* and *Micrococcus,* can utilize nitrate as an exogenous terminal H acceptor in the oxidation of organic substrates (Alexander, 1961). The denitrification reactions are associated with the enzyme nitrogen reductase and require, as discussed earlier (p. 209), cofactors of iron and molybdenum. Denitrification operates similarly under both aerobic and anaerobic conditions (Bandurski, 1965).

An exemplary reaction of the oxidation of glucose and the concomitant reduction of nitrate is (Hutchinson, 1957)

$$C_6H_{12}O_6 + 12NO_3^- \rightleftharpoons 12NO_2^- + 6CO_2 + 6H_2O \quad (\Delta G_0' = -460 \text{ kcal mol}^{-1}),$$

and for the reduction of nitrite to molecular nitrogen:

$$C_6H_{12}O_6 + 8NO_2^- \rightleftharpoons 4N_2 + 2CO_2 + 4CO_3^{2-} + 6H_2O \quad (\Delta G_0' = -720 \text{ kcal mol}^{-1}).$$

Approximately as much free energy results as in the aerobic oxidation of glucose by dissolved O_2 ($\Delta G_0' = -699$ kcal mol^{-1}). The denitrification reactions occur intensely in anaerobic environments, such as in the hypolimnia of eutrophic lakes (Fig. 12-4) and in anoxic sediments, where oxidizable organic substrates are relatively abundant.

High concentrations of NO_2^- can occur in large rivers in summer under warm, slow-flowing conditions as a result of dissimilatory nitrate reduction in anaerobic sediments (Kelso *et al.*, 1997, 1999). A range of organic carbon compounds, commonly found in agricultural and domestic pollutants, provide the energy sources in the NO_3^- reduction. Glucose, acetate, and particularly formate, among others, induced high NO_2^- accumulation, caused in part by partial inhibition of denitrification.

A specialized case, of much less general quantitative significance than the heterotrophic denitrification just discussed, is denitrification of nitrate concurrently with the oxidation of sulfur. The process is accomplished by denitrifying sulfur bacteria, particularly *Thiobacillus denitrificans,* that utilize S° or reduced sulfur compounds such as thiosulfate (Kuznetsov, 1970):

$$5S + 6KNO_3 + 2H_2O \longrightarrow 3N_2 + K_2SO_4 + 4KHSO_4$$

$$5Na_2S_2O_3 + 8KNO_3 + 2NaHCO_3 \longrightarrow 6Na_2SO_4 + 4K_2SO_4 + 2CO_2 + H_2O + 4N_2$$

Both processes occur chemosynthetically under dark, anaerobic conditions and yield relatively small changes in free energy. In a similar manner, a chemolithotrophic iron bacterium, *Gallionella ferruginea,* reduces nitrates to N_2 by the oxidation of ferrous ions under anaerobic conditions (Guoy *et al.*, 1984).

The rate of denitrification, as of nitrification, decreases in acidic waters (Keeney, 1973) and is very slow at low temperatures (around 2°C). Optimum rates of denitrification occur well above the temperatures of most natural fresh waters. At high temperatures, the primary product is N_2, while at lower temperatures, nitrous oxide (N_2O) predominates. N_2O is usually rapidly reduced to N_2, but its distribution is quite variable. Although N_2O has not been found in many lakes in appreciable quantities (e.g., Goering and Dugdale, 1966; Kuznetsov, 1970; MacGregor and Keeney, 1973; Kaplan and Wofsy, 1985; Mengis *et al.*, 1996), large N_2O

accumulations were repeatedly observed in strata of greatly reduced oxygen of eutrophic lakes, concomitant with large concentrations of NO_2^- (Yoh *et al.*, 1988, 1990; Yoh, 1992; Mengis *et al.*, 1997). Although nitrification contributed to the accumulations of both compounds to some extent, denitrification was the dominant process of N_2O formation in these lakes. Rates of N_2O and NO_2^- formation were maximal with very low concentrations of dissolved oxygen (ca. 0.1 mg liter^{-1}) and inhibited by reducing conditions in the lower hypolimnion sufficient to reduce iron and sulfide ions (see Chap. 14). In anaerobic littoral and wetland sediments, denitrification dominates as the process producing N_2O, but N_2O to N_2 ratios are greatest at low rates of denitrification (van Cleemput, 1994). Nitrous oxide has been found to be released from aerobic riverine sediments into the overlying water (Wissmar *et al.*, 1987). Higher evasion rates of N_2O, believed to emanate largely from nitrification, occurred from sands of pools (30–695 nM m^{-2} h^{-1}) than from gravel deposits of riffles (65–275 nM m^{-2} h^{-1}). In a series of seven amictic, permanently ice-covered lakes of Antarctica, a detailed study indicates that N_2O maxima were largely a product of nitrification and that denitrification was a sink for this gas in anoxic water (Priscu, 1997). Maxima of N_2O were common in depth gradients where oxygen concentrations and redox potentials were decreasing, and N_2O was nearly absent in anoxic zones of low redox potential. The ice barrier results in an appreciable supersaturation of N_2O, some of which evades to the atmosphere through the brief period of the summer moat of open water at the edge of the ice.

Rates of denitrification by nitrate-reducing bacteria in eutrophic Pluss See, Germany, showed marked seasonal and depth variations (Tan and Overbeck, 1973). Rates of nitrate reduction were particularly high and were correlated with cell numbers during the early portion of summer stratification, before hypolimnetic nitrate concentrations were greatly reduced. High concentrations of oxygen and low levels of nitrate depressed rates of nitrate reduction but had relatively little effect on cell numbers, especially in winter.

Nitrification and denitrification can occur simultaneously. In lake sediments, denitrification of added

TABLE 12-6 Approximate Rates of Bacterial Denitrification in Lake and River Water and Sediments by Direct Measurements[a]

	Rate of denitrification		
Lake	*Water* (*μg N liter^{-1} day^{-1}*)	*Sediment* (*mg N m^{-2} day^{-1}*)	*Source*
Lake 227, Ontario	0–30	15	Chan and Campbell (1980)
Norrviken, Sweden	0.53	100	Tirén *et al.* (1976)
Ramsjön, Sweden	0.65	120	
Smith Lake, Alaska	15	(90)	Goering and Dugdale (1966)
Lake Mendota, Wisconsin	8–26	—	Brezonik and Lee (1968); Keeney *et al.* (1971)
Enriched drainage ditch, Netherlands	—	160	van Kessel (1978)
Lake Kasumiga-ura, Japan	—	3–74[b]	Yoshida *et al.* (1979)
Boyrup Langsø, Denmark	—	57.5	Andersen (1977)
Kvindsø, Denmark	—	34.2	
Eight lakes of Ontario and New York, average summer	—	26.4	Rudd *et al.* (1986)
Fukami-Ike, Japan	0–27	—	Terai *et al.* (1987)
Lake Kizaki, Japan	0–22	—	Terai (1987)
Stream sediments, Ontario			
Without worms	—	50	Chatarpaul *et al.* (1980)
With tubificid worms	—	90	
Littoral sediments,			Chan and Knowles (1979)
Lake St. George, Ontario			
Without submerged macrophytes	—	2.6	
With submerged macrophytes	—	2.2	
Rivers	0	0–116	Seitzinger (1988)
Lakes			Seitzinger (1988)
Oligo mesotrophic	3–27	1.5–19	
Eutrophic	3–500	1–57	
Wetlands	—	0–1000	Bowden (1987); Groffman (1994)

[a] Note: Measurements often use $^{15}NO_3$; other approaches use mass balance techniques, but one can only calculate total lake denitrification by that approach (see text).
[b] Approximate extrapolations from values expressed as rates per weight of sediment and the area sampled by the coring device.

$^{15}NO_3$, was rapid; within 2 h up to 90% of added NO_3—N was reduced, much to $^{15}N_2$ (Chen *et al.*, 1972). Similar results were found in the sediments of several oligotrophic lakes, where denitrification in the sediments was functionally coupled to the supply of nitrate by nitrification (Klingensmith and Alexander, 1983; Dodds, and Jones, 1987). Much of the NO_3—N of lake sediments is incorporated into bacterial organic matter. Keeney *et al.* (1971) found that up to 37% of experimentally added NO_3—N (at the level of 2 mg NO_3—N liter^{-1}) became incorporated into the organic fraction. The remainder was denitrified. Denitrification rates of sediments are 3 to 4 orders of magnitude greater than those of the overlying water (Table 12-6). Denitrification activity within sediments of both lakes and streams is related to the microgradients of reducing conditions (Jones, 1979a,b; Nielsen *et al.*, 1990). In sediments of the littoral zone or those in contact with oxygenated water, denitrifying activity is depressed at the sediment surface and increases at sediment depths (10–15 mm) where reducing conditions increase (electrode potential, E_h, of 210 mV or less; see Chap. 14). Oxygen production by algae attached to the sediments in the light can increase the thickness of the aerobic layer and reduce denitrification in the sediments by as much as 70%. After darkness, oxygen is rapidly consumed by bacterial respiration at rates faster than it diffuses from overlying water, and denitrification again increases as anoxic and reducing conditions return.

D. Dissolved and Particulate Organic Nitrogen

Much of the basic understanding of major nitrogenous fractions and their distribution in lakes has evolved from the classical studies of Wisconsin lakes, especially Lake Mendota (Domogalla *et al.*, 1925; Peterson *et al.*, 1925; Birge and Juday, 1926). Although analytical methodology has improved greatly since these early investigations, their results are still generally valid for many lakes of the temperate region.

The dissolved organic nitrogen (DON) of fresh waters often constitutes over 50% of the total soluble nitrogen. Geographic variation is great, however, in relation to inputs of inorganic nitrogen from natural and artificial sources. Over one-half of the DON is in the form of amino nitrogen compounds, of which about two-thirds is in the form of polypeptides and complex organic compounds and less than one-third occurs as free amino nitrogen (Table 12-7). The qualitative nature of the numerous nitrogen compounds is incom-

TABLE 12-7 Particulate Organic Nitrogen of the Seston and Dissolved Organic Nitrogen of Several Lakes and Streams[a]

	Particulate organic N (μg N liter^{-1})		*Dissolved organic N* (μg N liter^{-1})	
Lake	*0 m*	*20 m*	*0 m*	*20 m*
Mendota, Wisconsin (means)	103	86		
Total soluble N			593	842
Amino N less free acids			170	177
Free amino N			88	88
Peptide N			181	173
Organic nonamino N			187	181
Furesø, Denmark	30–190		440–640	
Ysel, Netherlands	250–1400		590–1840	
Smith, Alaska	80–750			
Bodensee, Germany	10–160		50–150	
Lucerne, Switzerland (surface)	70–390		80–180	
Rotsee, Switzerland (surface)	180–1200		270–660	
Wintergreen, Michigan (surface)	50–2350		500–1320	
Augusta Creek, Michigan	ca. 50–300		90–880	
Lawrence Lake, Michigan				
Pelagic	20–110		80–240	
Surface inlet streams			60–1550	
Ground water			50–650	
Lake Kinneret, Israel				
Epilimnion (3 m)	50–283		89–545	
Hypolimnion (30 m)	55–398		40–343	

[a] After several sources in the text.

pletely known (see reviews of Vallentyne, 1957, and Schnitzer and Khan, 1972). The simple amino acids are substrates that are readily utilized by bacteria; rates of decomposition are high and result in low instantaneous concentrations of free amino acids in fresh waters (Chaps. 17 and 23).

As with organic carbon, the DON of lakes and streams is from 5 to 10 times greater than particulate organic nitrogen (PON) contained in the plankton and seston (Table 12-7) (Peterson *et al.*, 1925; Barica, 1970; Stadelmann, 1971; Manny, 1972a; Manny and Wetzel, 1973, 1974; Serruya *et al.*, 1975; Takahashi and Saijo, 1981; Zehr *et al.*, 1988). The ratios of DON to PON decrease as the lakes become more eutrophic, are closer to 1:1 in the trophogenic zone, and increase in the tropholytic zones. More organic nitrogen is apparently synthesized by small phytoplanktonic algae (<10 μm) per unit cell volume than by larger forms (Manny, 1972b). A significant portion of algal intracellular nitrogen (10–20% in the cyanobacterium *Oscillatoria*) is released extracellularly, mainly as protein and ammonia nitrogen with smaller amounts of nitrite and amino acid nitrogen (Meffert and Zimmerman-Telschow, 1979). Bacterial utilization of these organic compounds, particularly the amino acids, is extremely rapid (Gardner *et al.*, 1987, 1989). Algae and especially cyanobacteria also excrete polypeptides and other organic compounds that are capable of forming complexes with metals, such as iron and copper and with phosphates, and of altering their solubility and physiological availability (Fogg and Westlake, 1955; Murphy *et al.*, 1976; Tuschall and Brezonik, 1980). Similar nitrogenous compounds have been found to be secreted by larger aquatic plants (Wetzel and Manny, 1972); in some situations where the littoral zone is extensively developed, release of DON by macrophytes can form a major source of organic nitrogen to the lake. Furthermore, as aquatic vascular vegetation decomposes, large quantities of organic nitrogen are released (Nichols and Keeney, 1973). Much of this organic nitrogen is absorbed by the sediments and is utilized by attached microbiota, where decomposition can rapidly become limited by inorganic nitrogen, especially under anaerobic conditions (cf. Chap. 23).

Certain microalgae of diverse groups are able to utilize urea as a nitrogen source. Among the phytoplankton of Lake Biwa, Japan, for example, relative assimilation rates were ammonia (average 73.8% of total) > urea (average 20.0%) > nitrate (average 6.2% of total) (Mitamura and Saijo, 1986). The turnover rate for urea varied between 2 and 21 days.

The enzyme urease requires stoichiometric amounts of nickel for catalytic activity. Urease activities and growth of the algae in culture were markedly enhanced by small enrichments of nickel at levels of 0.3–5 μg Ni liter^{-1} (Oliveira and Antia, 1986; Price and Morel, 1991). How applicable these findings are to natural algal populations is unclear.

III. SEASONAL DISTRIBUTION OF NITROGEN

The seasonal changes in nitrogen vary greatly from lake to lake. Several general trends, however, do emerge from the seasonal patterns in oligotrophic and productive lakes. Although many seasonal cycles of inorganic nitrogen have been studied, the cycles of the dissolved and the biota-bound particulate organic nitrogen are often overlooked. The ensuing discussion concerns five lakes: (a) Lawrence Lake, an oligotrophic hardwater lake of southern Michigan that receives high natural inputs of inorganic nitrogen and where N_2 fixation is very low; (b) mesotrophic Vierwaldstattersee (Lake of Lucerne), Switzerland, a lake in which the hypolimnion remains aerobic throughout summer stratification; (c) moderately eutrophic Lake Mendota, Wisconsin; (d) highly eutrophic Wintergreen Lake, southern Michigan, a lake that undergoes normal dimictic circulation; and (e) Rotsee, Switzerland, an extremely eutrophic lake that circulates incompletely in spring and autumn.

The nitrogen inputs to Lawrence Lake, which is located in calcareous glacial till high in nitrates, occur largely as nitrate (Manny and Wetzel, 1982). Concentrations of NO_3—N of two small spring-fed inlet streams varied from 1 to over 20 mg NO_3—N liter^{-1}; groundwater inflow directly into the basin contained lower concentrations (mean 3 mg liter^{-1}), but contributed about 25% to the annual nitrogen income. Ammonia in the influents was much lower (between 0 to 190 μg NH_4—N liter^{-1}), and represented a relatively minor source of nitrogen to this lake. Nitrite concentrations were very small. The total dissolved organic nitrogen (TDON) of the surface and groundwater influents, however, was a significant source and reached levels as high as 1.5 mg liter^{-1}. This TDON was strongly correlated with the metabolic activity of the surrounding marsh vegetation through which the streams flow.

Figure 12-6 contrasts the concentrations of NO_3—N + NO_2—N in mg liter^{-1} (*upper*) and NH_4—N in μg liter^{-1} (*lower*) in seasonal depth–time diagrams. NH_4—N levels of the epilimnion and metalimnion were very low during stratification and increased conspicuously only below 11 m, where anoxic conditions existed in late summer and autumn. This increase was accompanied by a simultaneous denitrification of NO_3 in the lower hypolimnion. NO_2—N

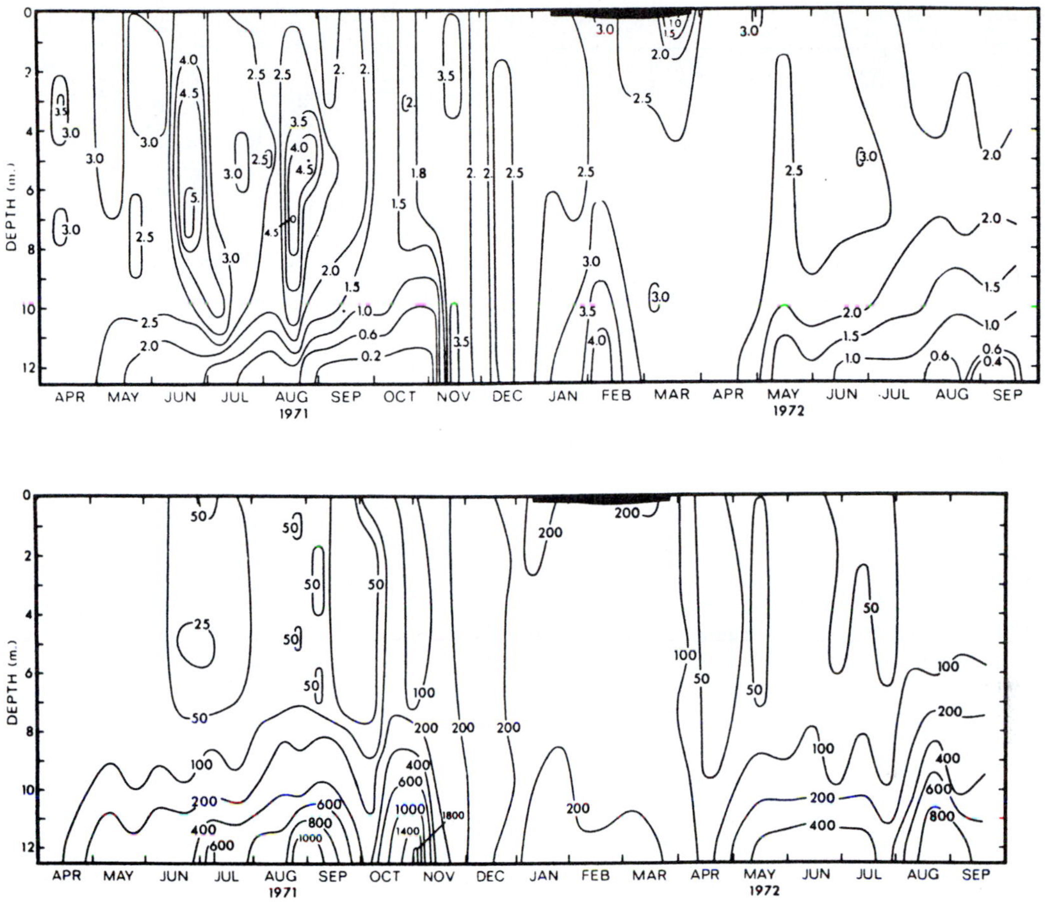

FIGURE 12-6 Depth–time diagrams of seasonal concentrations of NO_3—N + NO_2—N in mg liter^{-1} (*upper*) and NH_4—N in μg liter^{-1} (*lower*), Lawrence Lake, Michigan, 1971–1972. Opaque areas = ice cover to scale. (From Manny and Wetzel, unpublished data.)

concentrations were always very low (< 20 μg liter^{-1}). During turnover in November, and throughout much of the winter period of ice cover, NH_4 concentrations were distributed uniformly with depth at levels over 200 μg liter^{-1}. Nitrification was seen in the greater depths during the winter, and marked decreases in NO_3—N concentrations occurred immediately below the ice, consonant with intensive growth of algal populations. After the spring period of uniform circulation, the pattern was repeated.

The dissolved organic nitrogen (DON) and particulate organic nitrogen (PON) of the open water of Lawrence Lake generally were found to be about equal in concentration (50–1000 μg liter^{-1}). Concentrations of PON closely followed the dynamics of the biomass of the plankton; those of DON were maximal in the summer in the epilimnion and minimal during the winter in the hypolimnion.

The nitrogen budget of a large bay of Vierwaldstattersee was studied in detail by Stadelmann (1971). The nitrate concentrations (Fig. 12-7, *left*) of the epilimnion decreased sharply in the summer because of algal utilization, and this decrease was reflected in the amounts of planktonic particulate organic nitrogen (Fig. 12-7, *right*). NO_2—N concentrations followed a similar seasonal cycle but never exceeded 11 μg NO_2—N liter^{-1}. Ammonia concentrations were always low, usually only in trace quantities; maximum levels (70 μg NH_4—N liter^{-1}) occurred in late summer near the sediments. During this summer period when epilimnetic concentrations of inorganic nitrogen approached zero, cyanobacterial and heterocyst numbers reached their maximum, indicative of a heavy reliance on molecular N_2 fixation. Aerobic hypolimnetic waters of this lake did not undergo significant denitrification.

As lakes become more eutrophic, the processes of assimilation of nitrogen in the trophogenic zone and denitrification in the tropholytic zone intensify. Vertical gradients become extreme. As is exemplified in Wintergreen Lake (Fig. 12-8), nitrate and nitrite are reduced by algal assimilation and denitrification extremely rapidly, from over 1000 μg liter^{-1} to below detectability

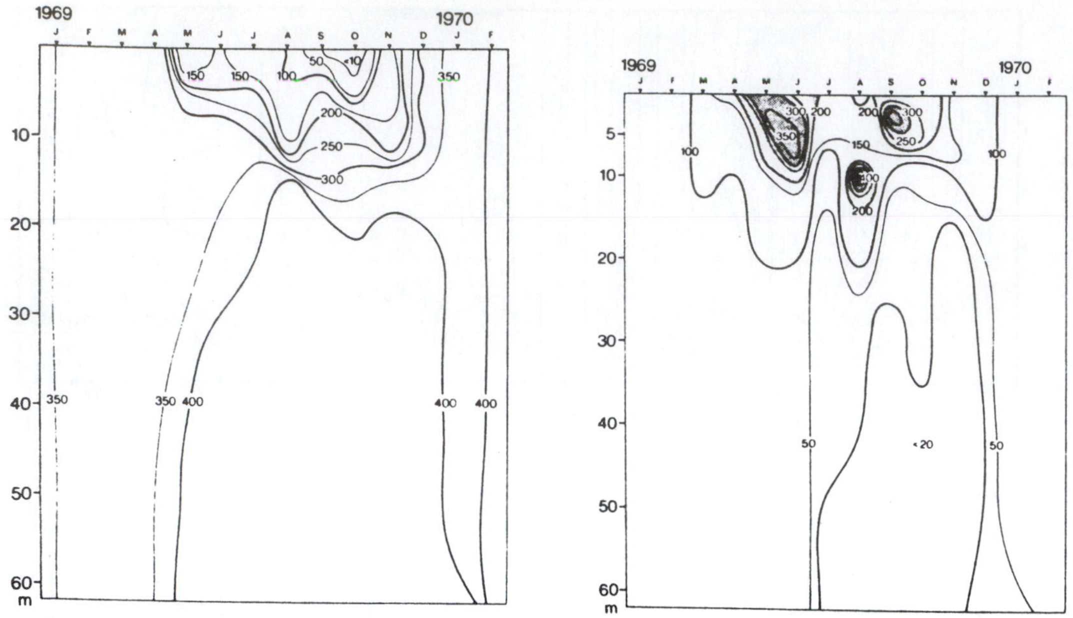

FIGURE 12-7 Depth–time diagrams of the concentrations of nitrate (*left,* in μg NO_3—N liter^{-1}) and particulate organic nitrogen (*right,* in μg N liter^{-1}) in Vierwaldstattersee, Switzerland. (After Stadelmann, P.: *Schweiz. Zeitschrift f. Hydrologie.* 33:1–65, 1971.)

in slightly over a month (Wetzel *et al.,* 1982). Fixation of molecular N_2 was limited to the 0- to 3-m depths in Wintergreen Lake, and very high rates of N_2 fixation by *Anabaena* and *Aphanizomenon* were found only during periods when levels of combined inorganic nitrogen were very low (Duong, 1972; Ward and Wetzel, 1980a). Ammonification and denitrification were particularly intensive in the anaerobic tropholytic zone below a depth of 3 m. As the name of the lake (Wintergreen) implies and as is common in hypereutrophic lakes (Wetzel, 1966a), production continues vigorously beneath the ice and the denitrification process continues, although more slowly, during winter stratification. An analogous situation occurred in the extremely hypereutrophic Rotsee, Switzerland (Stadelmann, 1971). However, in this lake, which is protected from major wind-induced water movements by the surrounding topography, circulation in the spring and autumn was weak and incomplete. The hypolimnion was anaerobic all year long, except for a slight intrusion of oxygen for a week in the spring; this condition was reflected in the high concentrations of ammonia throughout much of the lake most of the year (Fig. 12-9). Hypolimnetic concentrations of nearly 10 mg NH_4—N liter^{-1} are not uncommon near the sediments.

The seasonal distribution of the various forms of nitrogen in eutrophic Lake Mendota (Fig. 12-10) is similar to that commonly found in numerous lakes of temperate regions (Domogalla *et al.,* 1926; Domogalla and Fred, 1926; Barica, 1970). In many cases, autumnal nitrogen minima gradually increase in late fall to maxima in late winter and early summer. Looking at the seasonal processes involved in the surface and hypolimnetic strata over an annual period (Fig. 12-11), the net effects of nitrification and nitrate reduction–assimilation can be seen. In the oxygenated upper waters, nitrate decreases during the spring maximum from increased rates of assimilation by the plankton and nitrate reduction by bacteria. Nitrification decreases rapidly in the tropholytic zone after thermal stratification develops, because the hypolimnion becomes anaerobic. Additional decreases in rates of nitrification in the hypolimnion can occur when nitrifying bacteria suffer losses from grazing of zooplankton (Cavari, 1977). Nitrification ceases and nitrate reduction increases markedly until stratification is disrupted in the fall.

Similar patterns of nitrification and denitrification are found in many diverse lakes. For example, in Lake Titicaca in the high Andes Mountains of Peru–Bolivia, rates of nitrification were highest in the surface mixed layer and lowest in the uppermost portion of the hypolimnion, which is often the stratum of the water column that exhibits very high rates of pelagic decomposition (Vincent *et al.,* 1985). Nitrification activities in the hypolimnion increased somewhat with increasing depth toward the anoxic zone near the sediments, where NH_4^+ concentrations are high as long as some oxygen was present. In many mesotrophic lakes, as stratification decays into autumnal circulation of the water column the accumulated hypolimnetic nitrate-N persists

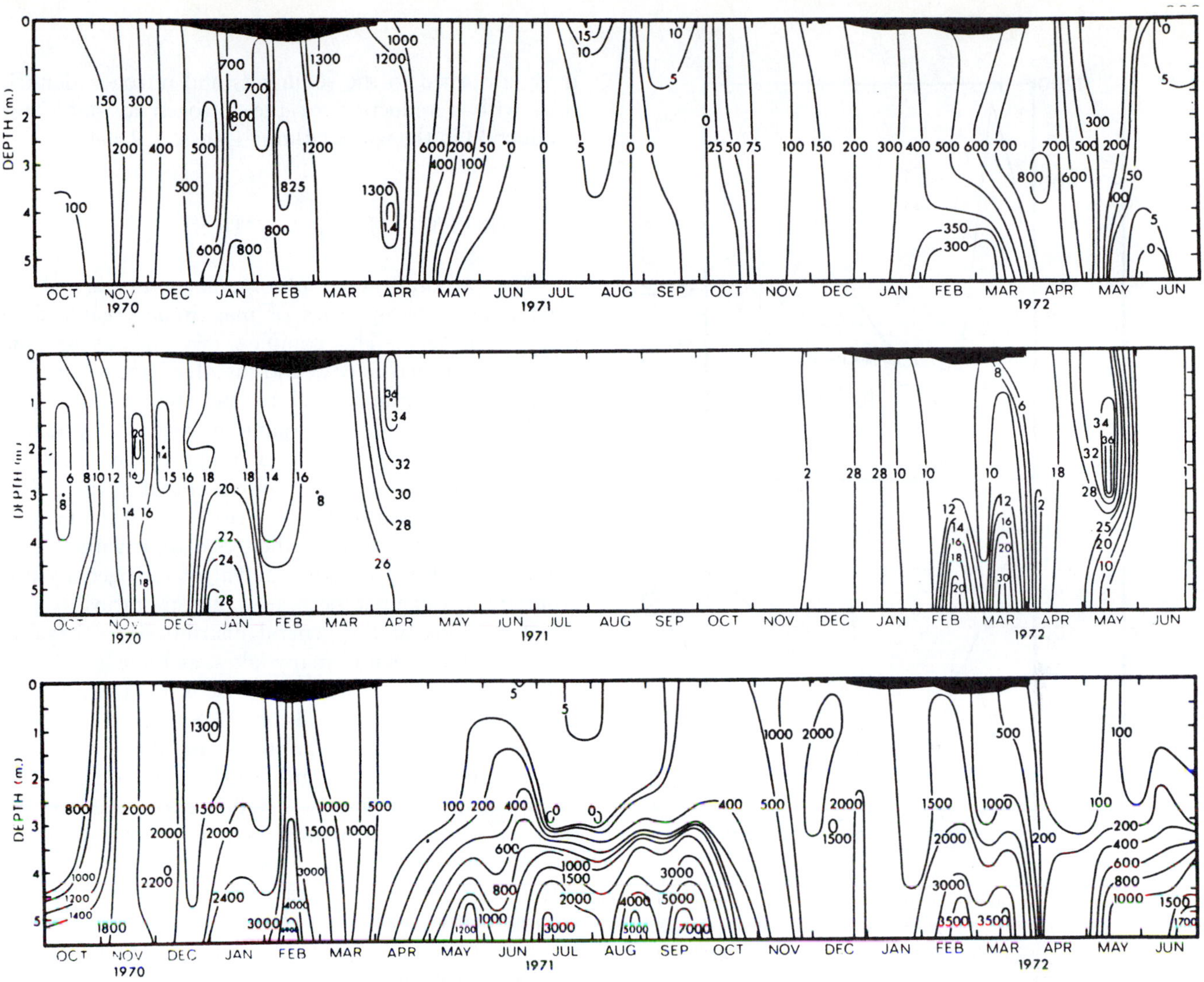

FIGURE 12-8 Depth–time diagrams of the concentrations, all in μg liter^{-1}, of NO_3—N + NO_2—N (*upper*), NO_2—N (*middle*), and NH_4—N (*lower*), Wintergreen Lake, Michigan, 1970–1972. Opaque areas = ice cover to scale. (From Wetzel *et al.*, unpublished data.)

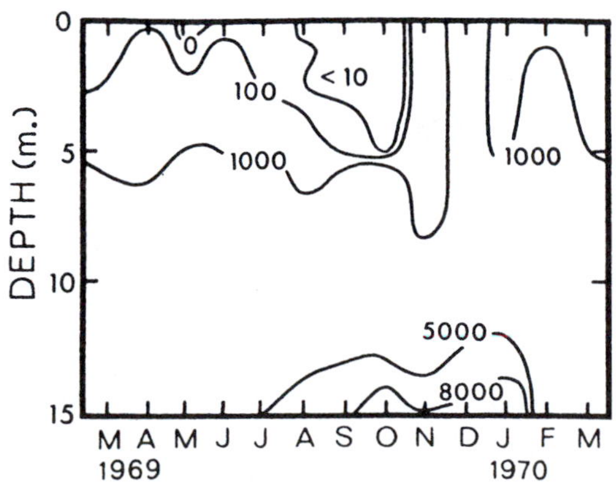

FIGURE 12-9 Depth–time diagram of the concentrations of ammonia (μg NH_4—N liter^{-1}), Rotsee, Switzerland, 1969–1970. (Redrawn from Stadelmann, 1971.)

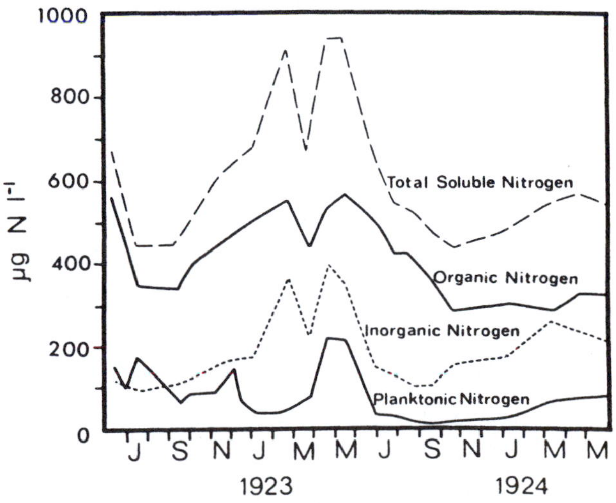

FIGURE 12-10 Average seasonal distribution of forms of nitrogen in Lake Mendota, Wisconsin, June 1922 through May 1924. (Redrawn from Hutchinson, 1957, after Domogalla *et al.*, 1926.)

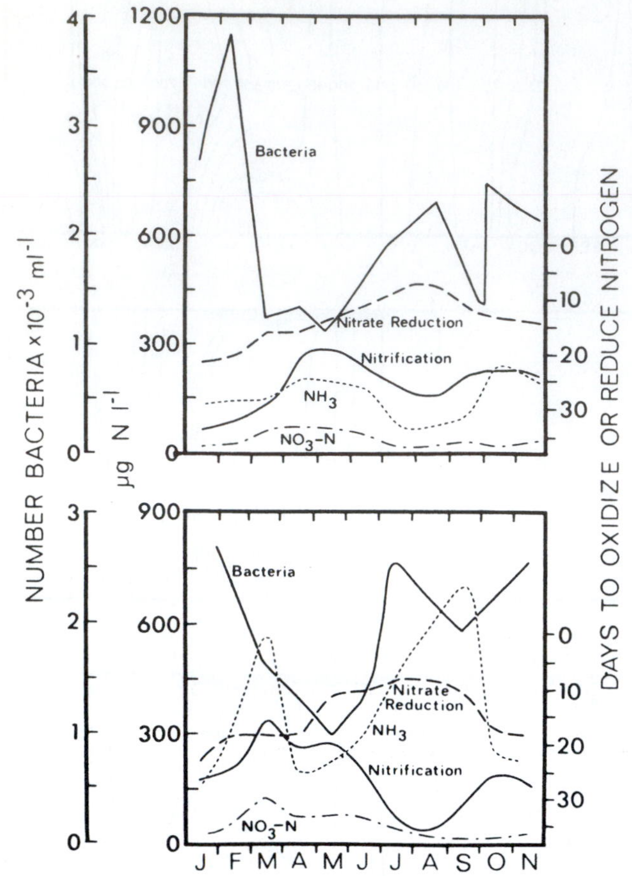

FIGURE 12-11 Rates of nitrification and denitrification, determined by the time required to oxidize or reduce added nitrogen, bacterial numbers, and the concentrations of ammonia and nitrate nitrogen, in Lake Mendota, Wisconsin, 1925. (Redrawn from Domogalla *et al.*, 1926.) *Upper:* surface water of trophogenic zone; *lower:* near bottom water of tropholytic zone.

for some time, often throughout the winter, during periods of lower biotic utilization by the plankton (e.g., Boström, 1981). If the lakes are shallow and not thermally stratified, however, phosphorus can be effectively adsorbed to the sediments and nitrogen denitrified, with a reduction of the nitrogen available for production in the water column (Jansson, 1984).

IV. CARBON-TO-NITROGEN RATIOS

The carbon content of organic matter is on the average at least an order of magnitude greater than that of nitrogen. The complex mixtures of organic compounds in particulate and dissolved phases are decomposed and mineralized to inorganic carbon, primarily as CO_2, and to inorganic nitrogen compounds. The proteolytic metabolism of fungi and bacteria removes proportionately more nitrogen than carbon. The rates of decomposition become slower with the greater chemical recalcitrance of the residual organic compounds, and the selective removal of nitrogen by microbes results in a net increase in the C:N ratios.

If we look at the general distribution of organic carbon and nitrogen in many lakes, as Birge and Juday (1934) did in several hundred Wisconsin lakes, certain trends become apparent (Table 12-8). There are large seasonal and spatial variations in the particulate organic matter in lakes and streams, but most of the organic matter is in the dissolved fraction, and concentrations of dissolved organic matter are very constant (cf. Wetzel *et al.*, 1972). Lakes can be grouped into categories of increasing organic carbon and dissolved organic matter. There is a clear decrease in organic nitrogen with increasing organic carbon content, which results in increasing C:N ratios. This relationship suggests an increase in the refractory nature of the organic compounds in lakes that contain greater amounts of dissolved organic matter. Such is indeed the case when waters containing high concentrations of dissolved organic matter receive increasingly greater proportions of their organic content from organic plant material

TABLE 12-8 Approximate Composition of Organic Matter of Water from Numerous Wisconsin Lakes[a]

Total particulate and dissolved organic carbon (mg liter^{-1})	*Particulate organic matter (mg dry wt liter^{-1})*	*Dissolved organic matter*[b]				
		Dry weight (mg liter^{-1})	*Crude protein*[c] (%)	*Lipid material*[d] (%)	*Carbohydrate* (%)	*C:N ratio*
1.0–1.9	0.62	3.1	24.3	2.3	73.4	12.2
5.0–5.9	1.27	10.3	19.4	1.3	79.0	15.1
10.0–10.9	1.89	20.5	14.4	0.4	85.2	20.1
15.0–15.9	2.32	31.3	12.9	0.2	86.9	22.4
20.0–25.8	2.22	48.1	9.9	0.2	89.9	29.0

[a] After data of Birge and Juday (1934).
[b] Includes colloidal material; particulate matter removed by centrifugation.
[c] Total nitrogen content of organic fraction × 6.25.
[d] Ether extract.

produced in wetland and littoral marsh areas surrounding the water (Wetzel, 1979, 1990a). Material of wetland and littoral origin dominates the dissolved organic matter pool of nearly all small lakes, streams, and particularly bogs and bog lakes.

Organic matter of terrestrial and marsh areas undergoes varying degrees of decomposition prior to and during transport to the lake or stream and during which much of the organic nitrogen can be utilized. Therefore, in general (cf. Hutchinson, 1957), allochthonous organic matter contains about 6% crude protein and has a C:N ratio from 45:1 to 50:1. The dissolved organic matter contains a high percentage of humic acid compounds that are low in nitrogen content (Shapiro, 1957; Schnitzer and Khan, 1972), and these compounds impart a stained brown color to the water. In contrast, autochthonous organic matter produced by decomposition of plankton within the lake contains about 24% crude protein and has a C:N ratio of about 12:1.

Alterations in the protein–carbohydrate–lipid ratios change the C:N ratio as a result of either phosphorus or nitrogen limitation (Vollenweider, 1985). The C:N ratio is thus not only indicative of environmental nutrient availability but also yields an approximate indication of the relative proportions of proteins and nonproteins in the organic matter. One cannot assume that all nitrogen is in proteins, but certainly a majority (ca. 85%) is proteinaceous in the plankton.

Nutrient stoichiometry is important in the regulation of both the species composition and the growth rates of phytoplankton communities. A combination of physiological assays and sestonic ratios is useful in proximate assessments of the nutrient status of phytoplankton (Healey and Hendzel, 1980; Kilham, 1990). For example, a C:N ratio of >14.6 often indicates a severe nitrogen deficiency in phytoplankton, between 8.3 and 14.6 a moderate deficiency, and <8.3 no nitrogen deficiency. The subject of C:N ratios is discussed further in concert with phosphorus in the following chapter (Chap. 13).

V. SUMMARY OF THE NITROGEN CYCLE

A diagrammatic representation of the major nitrogen inputs, transformation pathways, and outputs to a general lake system is given in Figure 12-12. Although numerous pathways are presented, it is obviously a

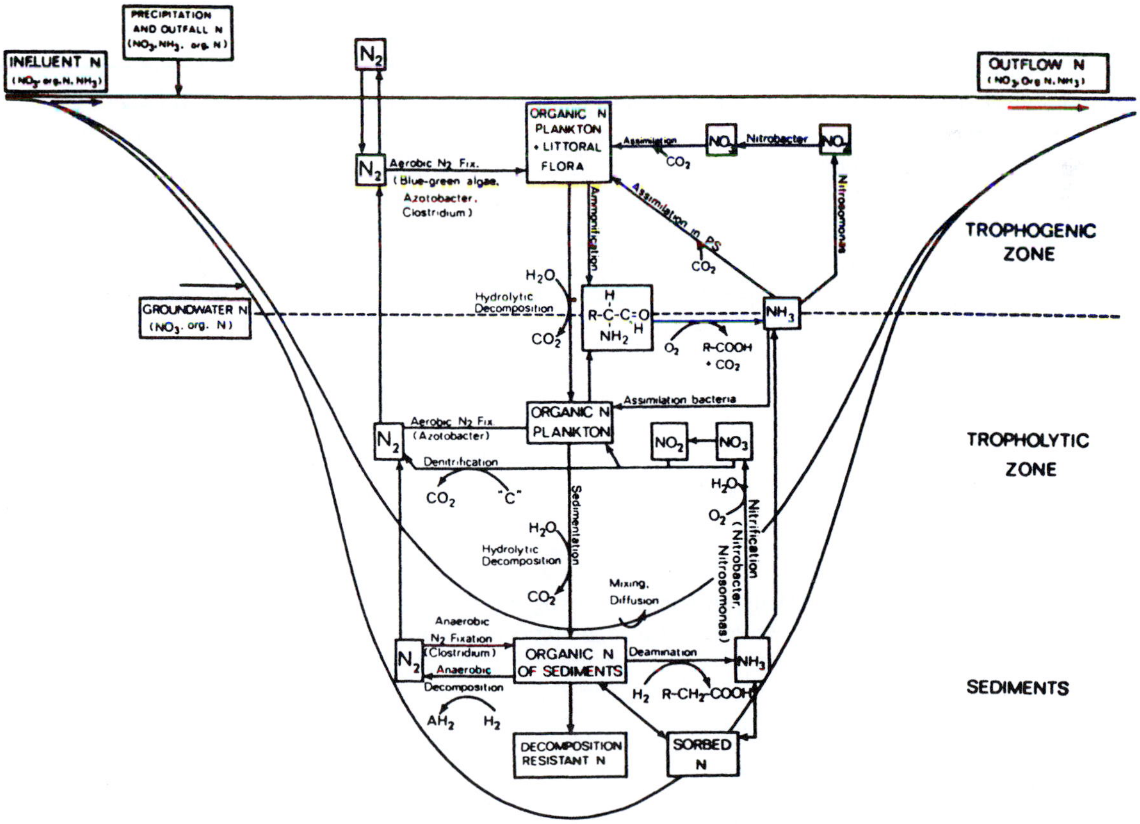

FIGURE 12-12 Generalized nitrogen cycle for fresh waters. PS = photosynthesis. (Greatly modified from Kuznetsov, 1970.)

TABLE 12-9 Approximate Nitrogen Content of Various Components of Bantam Lake, Connecticut[a]

Components	kg N *per lake*	*Percentage*
Lake water	10,700	13
Algae	29,200	33
Vascular aquatic plants	2500	3
Sediment (surface 1 cm)	44,000	51
	86,400	100

[a] Modified from estimates of Frink (1967).

simplification of the complex mechanisms and processes. Numerous analogies in the nitrogen cycle exist between lake and river systems.

The components of the cycle have already been discussed; only a few points need to be emphasized. The processes indicated in the tropholytic zone represent a composite situation under aerobic conditions. Clearly, the obligately aerobic decomposition processes would cease in the anaerobic hypolimnia of productive lakes. The metabolism of the littoral flora, including the vascular macrophytes and their epiphytic associations of bacteria and algae, is not only a major or dominant source of organic nitrogen synthesis in many lakes but can influence significantly the flux of nitrogen from the sediments to the water as well.

Many components and processes are variable seasonally and spatially. For example, inputs of nitrogen from guano of migratory waterfowl that briefly reside on a lake in extraordinary densities (one per m^2) can represent a major input of nitrogen and phosphorus to certain lakes (Manny *et al.*, 1975, 1994; cf. also the extremely detailed review of vertebrate excreta by Hutchinson, 1950). Sewage inputs of organic N, NO_3—N, and NH_4—N to rivers and recipient lakes are often highly pulsed. Similarly, agricultural applications of fertilizers to drainage basin areas are seasonal and can alter nitrogen inputs to lakes and rivers. Nitrogen fixation, normally a minor component of the total nitrogen income, can become a significant driving source of nitrogen to the system at certain times of the year.

The nitrogen dynamics of the sediments are poorly understood. Lake sediments typically contain on the order of 50–200 kg of nitrogen per hectare to a 10-cm sediment depth (Table 12-9), much of which is immobilized and sorbed to inorganic particles (Keeney, 1973). The interstitial water in sediments usually has a much higher concentration of soluble nitrogen compounds, mainly as NH_4—N and organic N, than that of the overlying water. Diffusion rates are exceedingly slow, but mixing of surficial sediments occurs to some extent, even when the lake is strongly stratified, as a result of deepwater water movements, activities of benthic organisms, and loss of gas bubbles from the sediments. At periods of turnover, 1–10 cm of the surface sediment can be mixed and much of it resuspended in the water column (e.g., Wetzel *et al.*, 1972; Davis, 1973; Dillon *et al.*, 1990). Surface sediments are usually oxidized for only a few millimeters in depth, but an oxidized microzone is critical to the solubility and sorption properties of the sediments for ammonia and can fundamentally alter rates of microbial transformations. Nitrogen exchange between sediments and water also varies greatly with sediment composition. For example, the release of nitrogen from sediments of the Rybinsk Reservoir was greatest in silts high in organic matter (Table 12-10). In this reservoir, the anaerobic decomposition of sediment organic matter resulted in considerable loss of nitrogen in the form of N_2.

Because of the functional complexity of the many components of the nitrogen cycle, no direct measurements exist for all processes simultaneously in any lake or stream over an annual period. Where detailed experimentation has dissected individual process rates using ^{15}N isotopic techniques, results from certain times of the year emphasize the major importance of the sediments not only as a large reservoir of nitrogen but as the site of much of the nitrogen metabolism.

Lake Wingra in southern Wisconsin is a shallow, eutrophic lake in which submersed macrophytes constitute a major part of the total lake productivity. An overall average nitrogen content of the various compartments in Lake Wingra was estimated for midsummer

TABLE 12-10 Average Daily Exchange of Nitrogen between the Sediments and Benthic Water Layers in Rybinsk Reservoir, Russia, in mg N m^{-2} day^{-1}[a]

Sediment	*NO_3—N*	*NH_4—N*	*Albuminoid N*	*Organic N*	*Total N*
Sand	+0.14	+1.14	+0.52	−6.00	−4.72
Unflooded soil	−0.44	+0.36	+0.41	+2.13	+2.03
Grey silt (fluvial areas)	−1.20	+21.43	+3.19	+10.98	+31.28
Redeposited peat	−4.3	+1.2	+1.3	+7.1	+4.0

[a] From Kuznetsov (1968).

TABLE 12-11 Estimates of the Nitrogen in Various Compartments in Lake Wingra, Wisconsin, Midsummer[a]

Compartment	Nitrogen (metric tons)	Percentage of total N	
		A[b]	B[c]
Soluble and particulate N in water	3.2	0.5	50.6
Macrophyte N	1.2	0.2	19.3
Sediment, upper 10 cm			
Organic N	138	23.0	—
Interstitial water			
NH_4—N	0.74	0.1	11.5
Organic N	0.19	0.03	3.0
Exchangeable NH_4—N	1.0	0.2	15.6
Sediment, 10–30 cm			
Organic N	460	74.3	—
Interstitial water			
NH_4—N	4.6	0.8	—
Organic N	0.51	0.08	—
Exchangeable NH_4—N	5.4	0.9	—
Total			
Total N in system	615		
Total N excluding sediment organic N and all N below 10 cm in sediment	6.3		

[a] Modified from Isirimah *et al.* (1976).
[b] Total N in system.
[c] Total N excluding sediment organic N and all N below 10 cm in sediment.

conditions (Table 12-11). Much (97%) of the total nitrogen occurred in the sediments as organic nitrogen (column A) but was not readily available for metabolism in the lake system (Isirimah *et al.*, 1976). Neglecting these relatively unreactive nitrogen reservoirs, about 50% of the "available" nitrogen existed in the water column, mostly in dissolved form, 20% in the macrophytes, and 30% in the interstitial water of surficial sediments (column B). Rapid turnover rates of NH_4—N occurred in the water but not in the sediments. NO_3—N turnover was slower in the water than in the sediments, where about 80% of the NO_3—N was rapidly denitrified to N_2.

VI. NITROGEN BUDGETS

Detailed evaluations of the nitrogen cycle that include close-interval quantitative measurements of inputs, metabolic dynamics, and outputs are available for only a few freshwater ecosystems. The cycle is obviously complex and requires accurate analyses of the dynamics of all components for several years. Only a few approximate mass balance analyses of the nitrogen budgets of lakes and streams are available. In these analyses, the rates of bacterial metabolism are obtained indirectly, by assumptions of microbial processing or sedimentation by differences. Nonetheless, the approximate calculated budgets are instructive in characterizing the inputs, outputs, and retention of nutrients by ecosystems.

The nitrogen budget of Lake Mendota, Wisconsin, demonstrates the roughly equivalent contributions from runoff, ground water, and precipitation (Table 12-12). Major losses of nitrogen occur via sedimentation, denitrification, and outflow. Loss by seepage out of the basin was probably small, since most lake basins are well sealed (cf. Wetzel and Otsuki, 1974). In Table 12-13, the nitrogen budget for a large reservoir of complex morphometry accounts for inflows from all river influents and losses by sedimentation and outflow. Although the total nitrogen content of this reservoir decreased in 1962 because of low water levels, inputs of atmospherically derived nitrogen amounted to roughly 10,000 tons per year. About 1300 metric tons entered with rain and snow; the remainder may be attributed largely to biological fixation. Mirror Lake, a kettle lake located in a crystalline granitic bedrock region of the White Mountains of central New Hampshire, receives appreciable (half) nitrogen loading from atmospheric sources (Table 12-14). Primary outputs are via permanent sedimentation and river outflow in approximately equal quantities of inorganic and organic forms. Some 54% of the annual gross sedimentation of organic

TABLE 12-12 Estimated Nitrogen Budget of Lake Mendota, Wisconsin[a]

	Nitrogen income			Nitrogen losses	
Sources	*kg N yr^{-1}*	*%*	*Losses*	*kg N yr^{-1}*	*%*
Municipal and industrial wastewater	21,200	10.4	Outflow	41,300	20.4
			Denitrification	28,100	13.9
Urban runoff	13,700	6.8	Fish catch	11,300	5.6
Rural runoff	23,500	11.6	Weed removal	3250	1.6
Precipitation on lake surface	43,900	21.6	Loss to ground water	[b]	[b]
Ground water			Sedimentation and other[c]	118,850	58.6
Streams	35,900	17.7		202,800	100
Seepage	28,500	14.1			
Nitrogen fixation	36,100	17.8			
Marsh drainage	[d]	[d]			
	202,800	100			

[a] Modified from Brezonik and Lee (1968), with data improvements by Keeney (1972).
[b] Unknown; likely very small in this lake.
[c] By difference between total of other estimated losses and sum of income sources.
[d] Considered significant, but data unavailable.

nitrogen was mineralized in a large, oligotrophic lake in Sweden (Jonsson and Jansson, 1997). About two-thirds of the nitrogen mineralized in the sediments, released to the overlying water, and transported out of the lake as nitrate via the outlet during turnovers in spring and autumn. Only ca. 4% of organic nitrogen remained in the sediments; the remainder was lost via denitrification at about 1 mg N m^{-2} yr^{-1}.

Analyses of nitrogen budgets of some 40 lakes of very different states of productivity indicated that the removal of nitrogen is usually dominated by denitrification, which is directly correlated to organic matter loadings (Hellström, 1996). Where nitrogen fixation is significant, the increase in total nitrogen from N_2 fixation is generally proportional to the concentration of total phosphorus minus a fraction of the estimated nitrogen concentration without any N_2 fixation. General nitrogen budgets of this type leave much to be desired because they provide relatively little insight into the dynamics of internal processing or metabolic controls mechanisms. From an applied viewpoint, however, such mass balance information is useful in relation to nitrogen loading and the effects of this loading on alterations of productivity, as will be discussed in some detail with phosphorus (cf. Chap. 13).

Annual elemental mass balance budgets have also been used to characterize the inputs, outputs, and retention (i.e., the difference between nitrogen inputs and outputs) of stream ecosystems. Hydrological variations can be great in streams and they influence the fluxes. Even among more hydrologically stable streams, interannual variations in discharge can be large (Meyer and Likens, 1979; Meyer *et al.*, 1981; Grimm, 1987). In arid regions, flooding events can cause extreme variations in annual budgets.

Examination of representative nitrogen budgets of streams indicates that total nitrogen storage was similar even though the distributions among storage

TABLE 12-13 Estimated Nitrogen Budget of the Rybinsk Reservoir, Russia, in Metric Tons N per Year[a]

		Time		
	Factors measured	*1 June 1960–1 June 1961*	*1 April 1961–1 April 1962*	*1 June 1962–1 June 1963*
A	Measured nitrogen balance of water mass	+ 12,519	+ 5399	+ 12,170
B	Difference between inflow and outflow	− 180	− 7329	+ 1565
A − B = C	Increase in water mass	+ 12,699	+ 12,728	+ 10,605
D	Precipitated in sediments (20-yr average)	+ 12,500	+ 12,500	+ 12,500
C − D = E	Theoretical balance in water mass	+ 199	+ 228	− 1895
A − E	Nitrogen input from atmosphere and fixation	+ 12,320	+ 5171	+ 14,065

[a] From Kuznetsov (1968).

TABLE 12-14 Average Annual Nitrogen Budget for Mirror Lake, New Hampshire, 1970–1975[a]

	$kg\ yr^{-1} \pm S_{\bar{x}}$
Inputs	
Precipitation	
Inorganic	112 ± 10.7
Organic	7
Litter	13.6
Dry deposition	~15
Fluvial	
Dissolved	
Inorganic	35 ± 7.2
Organic	46
Particulate, organic	9
Total	238
Outputs	
Gaseous flux	?
Fluvial	
Dissolved	
Inorganic	61 ± 8.8
Organic	54
Particulate, organic	11
Net insect emergence	13
Permanent sedimentation	127–139
Total	266–278
Change in lake storage (decrease in NO_3^- storage)	−6

[a] Modified from Likens (1985).

components differed (Table 12-15). Nitrogen storage was largely associated with woody debris (59–80%) in forested streams, whereas in an open desert stream as much as 93% was in algae and autochthonous detritus. Some nutrients are stored in consumer animals, although turnover rates in these organisms are relatively slow. Dissolved organic nitrogen exports are less than inputs and indicated biological utilization associated with mineralization and potentially some sorption and flocculation processes (Triska *et al.,* 1984). Most particulate organic inputs increased in nitrogen concentrations prior to export (Table 12-16).

It has been predicted that ecosystem nitrogen retention would increase as rates of net ecosystem production and biomass accumulation increase and then decline or increase at a slower rate as biomass approaches a steady state (Vitousek and Reiners, 1975; Grimm and Fisher, 1986). At steady state, nutrient inputs would equal outputs, even though the state and forms of the nutrients may change. This relationship has been generally verified in studies that evaluated the patterns of nitrogen retention during successional recovery of algal and macroinvertebrate communities after they were disturbed by flash-flooding events and reduced to virtual zero levels (Grimm, 1987). Biomass and stored nitrogen increased during the early to middle successional stages of recolonization and development and were followed by declines in late stages.

TABLE 12-15 Nitrogen Mass Balances of Five Streams in Percentage of Total in Each Category[a]

	Watershed 10, Oregon	*Beaver Creek Riffle, Quebec*	*Sycamore Creek, Arizona*	*Bear Brook, New Hampshire*	*Mare's Egg Spring, Oregon*
Inputs (% of total)					
Dissolved inorganic N	3	15	16–58	73	(24)
Dissolved organic N	69	67	37–69	11	(24)
Particulate organic N	0	18	9–15	2	43.8
Precipitation and throughfall	2	0.02	0	3	4.2
Litter	19	0.12	—	11	0.2
N_2 fixation	5	0.004	?	?	3.8
Pools retained N (% of total)					
Fine particulate inorganic nitrogen	40	19	0	—	—
Large particulate organic nitrogen	59	80	0	—	—
Producers	0.6	?	86–93	—	—
Consumers	0.2	?	6–14	—	—
Outputs (% of total)					
Dissolved inorganic N	4	15	18–58	84	(21)
Dissolved organic N	74	67	37–72	12	(21)
Particulate organic N	23	18	7–22	3.7	57[b]
Coarse PON	8	0.1	0.1–6	3	
Fine PON	15	18	7–16	0.7	
Emergence	0.2	0.1	0.1–1	?	0.001
Denitrification	—	—	—	—	0.906

[a] Data extracted from Meyer *et al.* (1981), Naiman and Melillo (1984), Triska *et al.* (1984), Dodds and Castenholz (1988), and Grimm (1987).
[b] Includes particulates (16%) and sedimented burial of PON (41%).

TABLE 12-16 Annual Nitrogen Budget for a Small Stream (Unnamed) in the Western Cascade Mountains, Oregon, Heavily Forested with Mature Conifer Trees[a]

	$g\ m^{-2}$		*Sums* $g\ m^{-2}$
Total nitrogen inputs			**15.25**
Hydrologic inputs			**11.06**
Dissolved inorganic N (NO_3—N)	0.50		
Dissolved organic N	10.56	11.06	
Biological inputs			**4.19**
Terrestrial			
Throughfall	0.30	0.30	
Litterfall			
Needles	0.66		
Leaves	0.15		
Cones, bark, twigs, wood	0.17		
Coarse wood debris	0.11		
Microparticulate litterfall	0.18		
Miscellaneous	0.08	1.35	
Lateral movement			
Needles	0.31		
Leaves	0.29		
Cones, bark, twigs, wood	0.85		
Miscellaneous	0.33	1.78	
Aquatic			
N_2-fixation			
Twigs	0.09		
Bark	0.17		
Chips	0.10		
Wood debris	0.34		
Moss	0.06	0.76	
Particulate organic N pool			**11.93**
Fine particulate organic matter			
1 mm–250 μm	0.58		
250 μm–80 μm	0.78		
<80 μm	3.41	4.77	
Large particulate organic matter			
Needles	0.27		
Leaves	0.08		
Cones, twigs, bark, wood	2.18		
Miscellaneous	0.74		
Coarse wood debris	3.80		
Moss	0.07		
Consumers	0.02	7.16	
Total nitrogen outputs			**11.79**
Particulate organic nitrogen		2.53	
Fine particulate organic matter			
1 mm–80 μm	0.28		
<80 μm	1.38	1.66	
Large particulate organic matter			
Needles	0.15		
Leaves	0.03		
Cones, twigs, bark, wood	0.46		
Miscellaneous	0.23		
Coarse wood debris	0.00	0.87	
Emergence of insects		0.02	
Drift of insects		?	
Dissolved organic nitrogen		8.81	
Dissolved inorganic nitrogen (NO_3—N)		0.43	
Denitrification		?	

[a] Data extracted from Triska *et al.* (1984).

A. Nitrogen Loading: Effects of Human Activities

Enrichment of fresh waters with nutrients needed for plant growth occurs commonly as a result of losses from agricultural fertilization, loading from sewage and industrial wastes, and enrichment via atmospheric pollutants (especially nitrate and ammonia). Sufficient information exists about the general responses of many lakes to loading of major nutrients, especially phosphorus and nitrogen, to predict the potential changes in their productivity. The phytoplankton productivity of infertile, oligotrophic lakes is often limited by the availability of phosphorus. As phosphorus loading to fresh waters increases and lakes become more productive, nitrogen often becomes the nutrient limiting to plant growth. Excessive loading of these nutrients permits increased plant growth until other nutrients or light availability become limiting.

The nitrogen loading concepts in relation to increased lake productivity are best discussed simultaneously with phosphorus loading and limitations. Both will be treated in the following chapter.

VII. NITROGEN DYNAMICS IN STREAMS AND RIVERS

A. General Nutrient Dynamics in Flowing Waters

Nutrients move unidirectionally within running waters. Dissolved substances move downstream, may be bound or assimilated for a period of time, and later be released for further movement downgradient. As materials and chemical mass cycle among biota and abiotic components of the stream ecosystem, they are transported downstream, in processes that resemble spirals and have been termed ***nutrient spiraling*** (Webster and Patten, 1979; Newbold *et al.*, 1981; Stream Solute Workshop, 1990). Although upstream movements of nutrients can occur in backflows from eddies, fish migration, and flight of adult aquatic insects, net fluxes are downstream.

Hydrological processes physically move water containing dissolved and particulate components to reactive sites. Exchanges at reactive sites include chemical ionic transformations, sorption and desorption, and metabolically mediated uptake and assimilation by biota (Fig. 12-13). Materials can be transferred from the water column to the stationary streambed. Some of these materials will be retained, utilized by incorporation into living organisms, potentially transferred to other organisms, and subsequently released by excretions or decomposition to the water column and further transported downstream.

Dissolved substances in running waters may be utilized extensively by biota or be reactive abiotically and

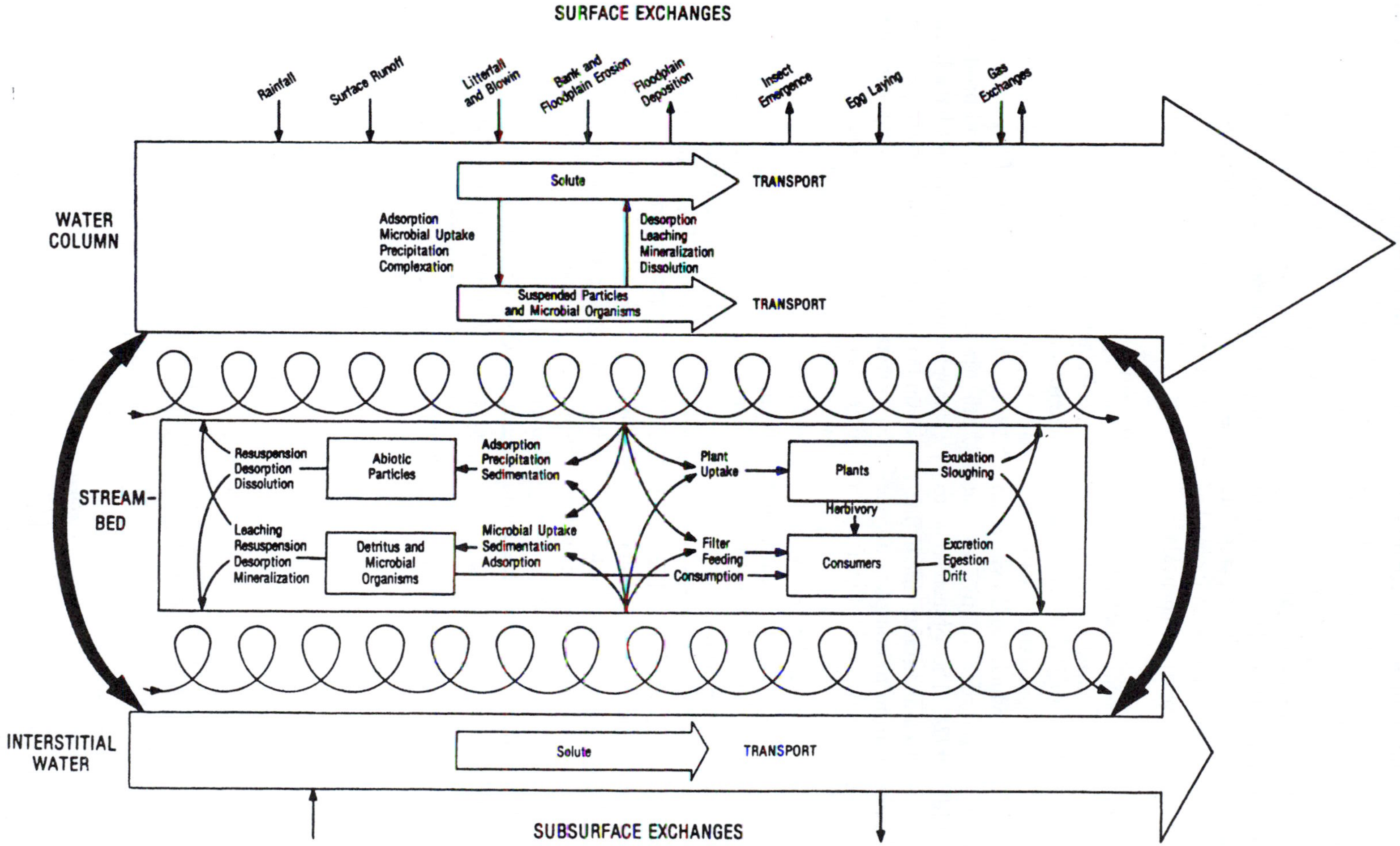

FIGURE 12-13 Solute processes in streams. The two spirals represent the continuous exchange of solutes and particle-bound chemical substances between the water column and the streambed and between the streambed and interstitial water. Materials in the water column and interstitial water are moving downstream, while the streambed materials are stationary. (From Stream Solute Workshop, 1990.)

are termed ***nonconservative.*** Other dissolved substances, termed ***conservative,*** are not generally modified chemically under normal limnological conditions or occur in such abundance that concentrations are not modified substantially by biological removal. Chloride ion is an essential nutrient that is nearly always conservative and is often used as a hydrological tracer. It should be noted, however, that differentiation between conservative and reactive solutes is relative; a solute may be conservative at one time or site and may become reactive at another place or time (Bencala and McKnight, 1987).

A nutrient atom may be used repeatedly as it passes downstream. The rate of utilization and release depends upon physical and biological retentiveness, largely by the microbiota associated with the streambed, that is, the extent to which the downstream transport of the nutrient is retarded relative to that of the water (Newbold, 1992). The objective is to quantify the average downstream distance (S, in meters) for the average nutrient atom to complete a cycle that takes an average time while it moves downstream at an average velocity. The average downstream velocity of the nutrient may be near that of the water in large rivers but is very much slower in streams and rivers where nutrients reside in the sediments and microbiota for a high proportion of the time.

Spiralling length (S) is the average distance a nutrient atom travels downstream during one cycle through the water and biotic compartment. The S equals the sum of distance traveled until uptake ("uptake length," S_W) and the downstream distance traveled within the biota until regeneration ("turnover length," S_B) (Fig. 12-14). The S can be calculated from the downstream nutrient fluxes (mass per unit width of river per unit time in the water (W) component, F_W, and in the biota (B) component, F_B) and the exchange fluxes of biotic utilization (U) of nutrients from the water compartment or regeneration (R) from the biota in mass per unit area per unit time (Fig. 12-14). Details of the methodology, particularly to evaluate the uptake flux rates and release rates by the biotic compartments by use of isotopes of nutrients, are delineated in Newbold *et al.* (1981, 1983).

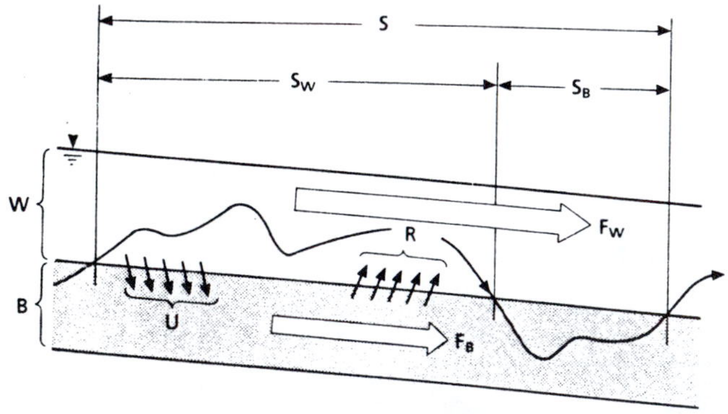

FIGURE 12-14 Spiralling in a river ecosystem of two compartments: water (W) and biota (B). The spiralling length (S) is the average distance a nutrient atom travels downstream during one cycle through the water and biotic compartments. S = the sum of the uptake length (S_W) and the turnover length (S_B) estimated from the downstream nutrient fluxes (F_W and F_B) and the exchange fluxes of uptake (U) and retention (R). (From Newbold, 1992. The Rivers Handbook, vol. 1, Blackwell Science Ltd. Reproduced with permission.)

B. Nutrient Limitations and Retention

Nutrient limitations of primary producers and heterotrophic microbiota in streams and small rivers appear to be uncommon (cf. review of Newbold, 1992). Although cases of phosphorus limitations of both attached algal growth and production and of heterotrophic microbial activities have been determined in a number of streams and rivers, concentrations of soluble reactive phosphorus were consistently <15 μg liter^{-1} and often <5 μg liter^{-1}. Similarly, increased microbial growth is found with inorganic nitrogen enrichments where natural concentrations were $<50-60$ μg liter^{-1}, but rarely above this level. Evidence suggests that in lower-order streams with a greater dominance of biota attached to substrata, retention of nutrients is very high and that these nutrients are intensively recycled within the attached communities (cf. Paul and Duthie, 1989; Wetzel, 1993; Burns, 1998; Chap. 19). Removal of the attached microbial community, as was the case in a desert stream following a flood that eliminated most of the biota, reduced the retention of nitrogen to very low levels, slowed recycling rates, and increased spiralling length (Grimm, 1987). Nitrogen retention increased rapidly as the attached algal and bacterial community reestablished. As rivers become larger, nutrient retention is less and nutrient limitations can become more prevalent. Nutrient loadings to larger rivers, however, are often high because of human activities in the drainage basins. Coupled with high nonalgal particulate turbidity and light reductions, nutrient availability is usually adequate to support microbial productivity within other environmental constraints.

The spiralling length of a nutrient suggests the extent of availability and utilization rates. A shorter spiralling length of one nutrient versus another could imply that the nutrient cycling more rapidly is in greater demand and is possibly limiting the potential growth and productivity of the community. For example, the shorter spiralling uptake length of nitrate nitrogen than that of phosphorus in a small stream of southern Spain was used as evidence of nitrogen limitation under different hydrological regimes (Maltchik *et al.,*

1994). Perhaps a more functional application is to determine the spiralling rates of nutrients among different dissolved, particulate, and animal consumer groups (e.g., Newbold *et al.*, 1983b). Such analyses demonstrate the rapid recycling rates and very short spiralling lengths of the attached microbiota in comparison to abiotic particulate materials and consumer metazoans.

C. Nitrogen Cycling in Running Waters

The inputs, transformations, and cycling of nitrogen in running waters are influenced, as in standing waters, to a large extent by bacterial, fungal, and other microbial metabolism. The transformation processes include both assimilation of nitrogen sources for utilization in synthesis of organic matter and growth, as well as utilization of inorganic and organic nitrogen compounds for energy in oxidation–reduction reactions. These processes have been discussed in detail in this chapter and are largely applicable to running water ecosystems as well as to lakes. Because many of these processes require anaerobic conditions, which are found less frequently in rivers, the subsequent discussion evaluates some of the quantitative differences in flowing waters on these processes. An important initial point of emphasis is that the assumptions that moving waters are fully oxygenated is not always true at even the macrolevel of bulk water volumes in streams and rivers and is often not the case in microenvironments within communities attached to surfaces and in interstitial waters of stream sediments.

D. Dissolved and Particulate Nitrogen in Rivers

The quantities of dissolved inorganic (NO_3^-, NO_2^-, and NH_4^+) and organic nitrogen compounds are highly diverse and variable because of marked differences in the inputs from surface and groundwater sources, particularly as affected by human activities (e.g., Spalding and Exner, 1993), and because of many competing reactions occurring in the nitrogen cycle. Ammonia concentrations tend to be low (7–60 μg N liter^{-1}) in natural river waters, as is the condition in aerobic waters of reservoirs and lakes (Meybeck, 1982, 1993b). Ammonia nitrogen can constitute an appreciable portion (15 to >80%) of the total dissolved inorganic nitrogen (DIN), particularly at low and, among polluted rivers, at very high DIN concentrations. As in lakes, nitrite nitrogen concentrations are very low (<3 μg N liter^{-1}; <1.5% of total dissolved inorganic nitrogen, DIN) among well-oxygenated, unpolluted waters.

Among major rivers, nitrate concentrations range from <25 μg N liter^{-1} in subarctic environments and in Amazonia to ca. 200 μg N liter^{-1} in some temperate rivers (Meybeck, 1982, 1993b). Higher nitrate nitrogen concentrations are found among rivers influenced by agricultural runoff. An average NO_3—N concentration of nearly 100 μg N liter^{-1} is found in natural river waters (Table 12-17). Maximum nitrate concentrations during storm flows were directly related to the magnitude of the storms and resulting high discharge and inversely related to the frequency of storm events (Tate, 1990; Triska *et al.*, 1990a). The activity of terrestrial vegetation of the riparian zones influences the loadings of nitrate and ammonia to the streams; nitrogen concentrations are generally higher during periods of vegetation dormancy or following losses from harvesting or fire (e.g., Likens, 1985; Spencer and Hauer, 1991; McClain *et al.*, 1994). The extensive transformation reactions focused at the upland and stream margins of the riparian zone regulate and diminish transfers of inorganic nitrogen from ground water to stream water.

TABLE 12-17 Distribution of Nitrogen Species (mg liter^{-1}) in Unpolluted Rivers[a]

	Mean (min, max)
Values	
NH_4—N	0.018 (0.005–0.04)
NO_2—N	0.0012
NO_3—N	0.101 (0.05–0.2)
Dissolved organic N (DON)	0.260 (0.05–1.0)
Ratios	
DOC:DON	20 (8–41)
POC:PON	8.5 (7–10)

[a] From many sources but particularly Malcolm and Durham (1976) and Meybeck (1982, 1993b).

Dissolved organic nitrogen (DON) concentrations, although measured less frequently in natural river waters, nearly always constitutes a major part (world average ca. 40%) of the total dissolved nitrogen (Wetzel and Manny, 1977; Meybeck, 1982). In subarctic and humic tropical rivers, DON can constitute over 90% of the dissolved nitrogen. Although dissolved inorganic nitrogen concentrations and discharge can vary widely on a diurnal basis and seasonally, dissolved organic carbon and nitrogen are relatively constant (Manny and Wetzel, 1973). The ratio of dissolved organic carbon to dissolved organic nitrogen is, however, variable (8–41) and averages ca. 20 (Table 12-17).

Particulate nitrogen consists of particulate organic nitrogen (PON), adsorbed ammonia, and organic nitrogen adsorbed to particles (Table 12-17). On a weight ratio basis, the particulate organic carbon to particulate organic nitrogen ratio is relatively constant at 8–10 (mean ca. 8.5).

E. Nitrogen Cycling in Sediments of Flowing Waters

Ammonium sorption to channel and riparian sediments can be extensive. For example, in granular sediments (0.5–2.0-mm grain size) of a third-order stream in California, exchangeable ammonium ranged from 10 μeq 100 g^{-1} of sediment where nitrification and subsurface exchange with stream water were high to 115 μeq 100 g^{-1} in the floodplain riparian sediments where channel water and groundwater mixing and nitrification potential were both low (Triska *et al.*, 1994). Sorbed ammonium was highest during summer/autumn base flow and lowest during winter storm flows. Similar results were found, with somewhat different seasonality, in a small temperate forest stream (Mulholland, 1992) and in the hyporheic zone of desert streams (Holmes *et al.*, 1994; Jones *et al.*, 1995). The riparian zone, when water-saturated, and parafluvial[2] and hyporheic zones are major sources of ammonium and dissolved organic nitrogen to the stream. Once within the stream, nitrogen uptake by attached microbes (bacteria, fungi, and algae) was the primary mechanism controlling spatial and seasonal variations in the water. Diurnal daytime reductions in stream nitrate concentrations at base flow suggest that uptake by photoautotrophs can be an important retentive process (Burns, 1998), although the coupling of photosynthetic oxygen production within the sediment microbial communities to diurnal changes in denitrification are not clear in these environments.

[2] ***Parafluvial zone*** refers to the part of the active stream channel without surface water that is connected hydrologically with the surface stream water.

The physical sorption of ammonia to sediment particles is dynamically coupled to sources from ground water and ammonification and transformations of dissolved inorganic nitrogen by nitrification, denitrification, and nitrate reduction (Fig. 12-15). Experimental studies of ammonium uptake by stream sediments indicated that biotic uptake by attached microflora was quantitatively much greater than physical adsorption (Newbold *et al.*, 1983b; Richey *et al.*, 1985; Aumen *et al.*, 1995). The duration of the storage of ammonium by sorptive processes is variable. Some studies indicate that appreciable retention can occur in summer months and contribute from 12–25% of nitrate released subsequently in winter by nitrification (e.g., Richey *et al.*, 1985), but in larger eutrophic river systems much of the ammonium is rapidly nitrified and exported downstream as nitrate (Lipschultz *et al.*, 1986).

The effectiveness of the denitrification and nitrification processes in the hyporheic zone of stream sedi-

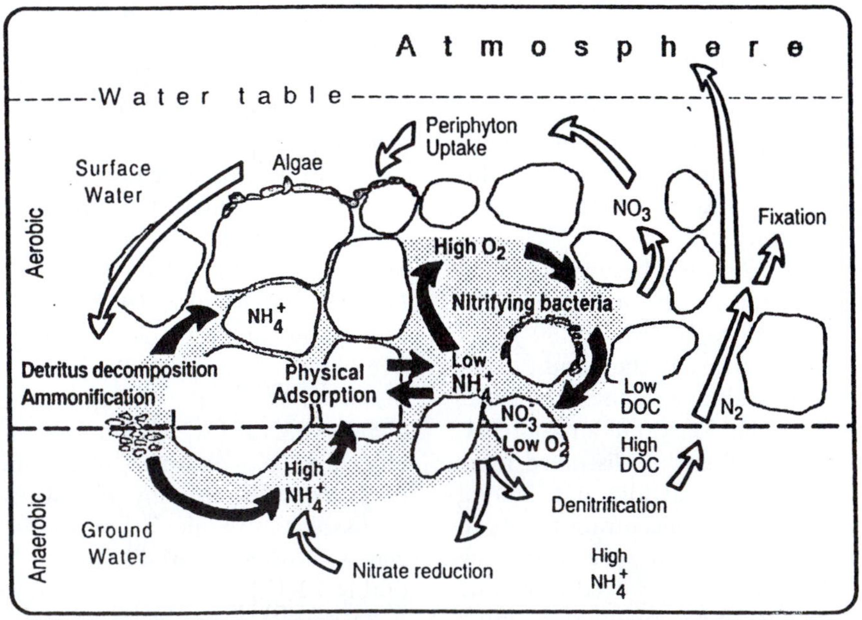

FIGURE 12-15 Linkages among physical sorption of ammonia to sediment particles, sources from ground water and ammonification, and transformations of dissolved inorganic nitrogen by nitrification, denitrification, and nitrate reduction at the groundwater–surface interface of a stream channel. (From Triska *et al.*, 1994.)

ments depends greatly upon the redox conditions within the interstitial water (Fig. 12-15). Flow conditions within the sediment interstices are quite variable spatially because of localized differences in sediment composition and textures. Internal flows also change over time between precipitation events, which lead to variations in both rates of groundwater inflow as well as rates of in-channel flows that penetrate down into the sediments (e.g., Triska *et al.,* 1990b). In organic-rich sediments, the redox profiles can be quite similar to those in lake sediments, in which the zone of nitrification can be restricted to the upper 2–3 mm and intensive denitrification occurs below the stratum of nitrifier activity (e.g., Cooke and White, 1987a,b; Birmingham *et al.,* 1994). Denitrification also occurs in shallow riparian sediments connected to the stream by the hyporheic zone (Duff and Triska, 1990). Denitrification increases with greater distance (10–15 m) from the stream channel were correlated with decreasing oxygen concentrations and increasing concentrations of reduced compounds.

Photosynthetic oxygen production by attached algae can diffuse for several millimeters into the sediments and reduce the rates of denitrification (Christensen *et al.,* 1990). Under shaded conditions or darkness, rates of denitrification increase rapidly as bacterial respiration depletes oxygen in the microzones. Clearly, the processes of nitrification and denitrification can be functioning simultaneously and reciprocally in the sediments, particularly where many microzones of steep redox gradients occur.

VIII. SUMMARY

1. Nitrogen, along with carbon, hydrogen, and phosphorus, is one of the major constituents of the cellular protoplasm of organisms. Nitrogen is a major nutrient that affects the productivity of fresh waters.
2. A major source of nitrogen of the biosphere originates from fixation of atmospheric molecular nitrogen (N_2). The nitrogen cycle is a complex biochemical process in which nitrogen in various forms is altered by nitrogen fixation, assimilation, and reduction of nitrate to N_2 by denitrification. For all practical purposes, the nitrogen cycle of lakes is microbial in nature: Bacterial oxidation and reduction of nitrogen compounds are coupled with photosynthetic assimilation and utilization by algae, photosynthetic bacteria, and larger aquatic plants. The direct role of animals in the nitrogen cycle is certainly very small; under certain conditions, however, their grazing activities can influence microbial populations and nitrogen transformation rates as well as nitrogen utilization rates by photosynthetic organisms. Although the nitrogen cycle of fresh waters is understood qualitatively in appreciable detail, the complex dynamics of quantitative transformation rates have not been clearly delineated, especially at the sediment–water interface, where intensive bacterial metabolism occurs.
3. Dominant forms of nitrogen in fresh waters include (a) dissolved molecular N_2, (b) ammonia nitrogen (NH_4^+), (c) nitrite (NO_2^-), (d) nitrate (NO_3^-), and (e) a large number of organic compounds (e.g., amino acids, amines, nucleotides, proteins, and refractory humic compounds of low nitrogen content).
4. The nitrogen cycle consists of a balance between nitrogen inputs to and nitrogen losses from an aquatic ecosystem.
   a. Sources of nitrogen include (i) nitrogen contained in particulate "dry fallout" and precipitation falling directly on the lake surface, (ii) nitrogen fixation both in the water and the sediments, and (iii) inputs of nitrogen from surface and groundwater drainage.
   b. Losses of nitrogen occur by (i) outflow from the basin, (ii) reduction of NO_3^- to N_2 by bacterial denitrification with loss of N_2 to the atmosphere, and (iii) sedimentation of inorganic and organic nitrogen-containing compounds to the sediments.
5. Microbial fixation of molecular N_2 in soils by bacteria is a major source of nitrogen. In lakes and streams, N_2 fixation by heterotrophic bacteria and certain cyanobacteria is quantitatively less significant, except under certain conditions of severe depletion of combined inorganic nitrogen compounds.
   a. The N_2 content of water is usually in equilibrium with the N_2 of the atmosphere during periods of turbulent mixing. In stratified, productive lakes, concentrations of N_2 may decline in the epilimnion because of reduced solubility as temperatures rise and increase in the hypolimnion from denitrification of NO_3—N.
   b. N_2 fixation by cyanobacteria is usually much greater than fixation by heterotrophic bacteria. In cyanobacteria, N_2 fixation is light-dependent and usually coincides with the spatial and temporal distribution of these microbes. NH_4—N assimilation requires less energy expenditure than NO_3—N, and NO_3—N less than N_2—N. N_2 fixation by cyanobacteria increases when NH_4—N and NO_3—N concentrations decrease in the trophogenic zone.

   c. Bacterial N_2 fixation in wetlands surrounding lakes or adjacent to streams can add significant amounts of combined nitrogen to freshwater ecosystems.
6. Ammonia is generated by heterotrophic bacteria as the primary nitrogenous end product of decomposition of proteins and other nitrogenous organic compounds. Ammonia is present primarily as NH_4^+ ions and is readily assimilated by plants in the trophogenic zone.
   a. NH_4—N concentrations are usually low in aerobic waters because of utilization by plants in the photic zone. Additionally, bacterial nitrification occurs, in which NH_4^+ is oxidized through several intermediate compounds to NO_2^- and NO_3^-.
   b. When the hypolimnion of a eutrophic lake becomes anaerobic, bacterial nitrification of ammonia ceases. The oxidized microzone at the sediment–water interface is also lost, which reduces the adsorptive capacity of the sediments for NH_4—N. A marked increase in the release of NH_4^+ from the sediments then occurs. As a result, the NH_4—N concentrations of the hypolimnion increase (Fig. 12-4).
   c. Bacterial nitrification proceeds in two stages: (i) the oxidation of $NH_4^+ \rightarrow NO_2^-$, largely by *Nitrosomonas* but also by other bacteria, including methane oxidizers; and (ii) the oxidation $NO_2^- \rightarrow NO_3^-$, in which *Nitrobacter* is the dominant bacterial genus involved.
   d. Nitrite (NO_2^-) is readily oxidized and rarely accumulates except in the metalimnion, upper hypolimnion, or interstitial water of sediments of eutrophic lakes. Concentrations are usually very low ($< 100\ \mu g\ liter^{-1}$) unless organic pollution is high.
7. Nitrate is assimilated and aminated into organic nitrogenous compounds within organisms. This organic nitrogen is bound and cycled in photosynthetic and microbial organisms. During the normal metabolism of these organisms, and at death, much of their nitrogen is liberated as ammonia. Additionally, organisms release a variety of organic nitrogenous compounds that are resistant to proteolytic deamination and ammonification by heterotrophic bacteria to varying degrees.
   a. Nitrate (NO_3^-) is the common form of inorganic nitrogen entering fresh waters from the drainage basin in surface waters, ground water, and precipitation. In certain oligotrophic waters in basaltic rock formations, nitrate loading from atmospheric sources, especially if contaminated by human-produced combustion emission products, can dominate nitrogen loading.
   b. Bacterial denitrification is the biochemical reduction of oxidized nitrogen anions (NO_3^- and NO_2^-) concomitant with the oxidation of organic matter: $NO_3^- \rightarrow NO_2^- \rightarrow N_2O \rightarrow N_2$. Nitrous oxide ($N_2O$) is rapidly reduced to N_2 and has rarely been found in lakes in appreciable quantities. Denitrification is accomplished by many genera of facultative anaerobic bacteria, which utilize nitrate as an exogenous terminal hydrogen acceptor in the oxidation of organic substrates. Denitrification occurs in anaerobic environments, such as in the hypolimnia of eutrophic lakes (Fig. 12-4) or in anoxic sediments, where oxidizable substrates are relatively abundant.
8. Dissolved organic nitrogen (DON) often constitutes over 50% of the total soluble nitrogen in fresh waters.
   a. Over half of the DON occurs as amino nitrogen compounds, mostly as polypeptides and complex nitrogen compounds.
   b. The ratios of DON to particulate organic nitrogen (PON) of streams and lakes are usually from 5:1 to 10:1. As lakes become more eutrophic, DON:PON ratios decrease.
9. The distribution of nitrogen in a lake can change rapidly. Examples of the depth–time distributions of the different forms of nitrogen demonstrate that as lakes become more productive from nutrient loading, concentrations of NO_3—N and NH_4—N in the trophogenic zone can be severely reduced and depleted by photosynthetic assimilation. Cyanobacteria with the capability of nitrogen fixation may then come to dominate. In anaerobic hypolimnia, NO_3—N is rapidly denitrified to N_2, which is either fixed or lost to the atmosphere. NH_4—N concentrations accumulate from decomposition of organic matter and release of NH_4—N from sediments under anaerobic conditions.
10. Organic carbon-to-nitrogen ratios (C:N) indicate an approximate state of resistance of complex mixtures of organic compounds to decomposition, because proteolytic metabolism by fungi and bacteria removes proportionally more nitrogen than carbon. Higher C:N ratios commonly occur in residual organic compounds, which are more resistant to decomposition.
    a. Organic materials from allochthonous and wetland sources commonly have C:N ratios from 45:1 to 50:1 and contain many humic compounds of low nitrogen content.
    b. Autochthonous organic matter produced by the decomposition of plankton tends to have higher

protein content and C:N ratios of about 12:1.

c. Alterations in in the protein-carbohydrate-lipid ratios alter the particulate C:N ratio as a result of phosphorus or nitrogen limitation. A C:N ratio of >14.6 often indicates a severe nitrogen deficiency in phytoplankton, between 8.3 and 14.6, a moderate deficiency, and < 8.3, no nitrogen deficiency.

11. Much of the total nitrogen occurs in the sediments in forms that are relatively unavailable for biotic utilization. Of the readily available nitrogen, a majority occurs in soluble form in the water and in the interstitial water of surficial sediments (and in littoral vegetation in shallow, productive lakes). Turnover rates of NH_4-N are rapid in water but slower in the sediments. In contrast, NO_3—N turnover is slower in the water than in sediments, where, under anoxic conditions in eutrophic lakes, NO_3—N is rapidly denitrified to N_2.
12. Increased loading of inorganic nitrogen to rivers and lakes frequently results from agricultural activities, sewage, and anthropogenic atmospheric pollution. In unproductive oligotrophic lakes, phosphorus availability is often the principal limiting nutrient for plant growth. As phosphorus loading to fresh waters increases and they become more productive, nitrogen becomes more important as a growth-limiting nutrient.
13. In running waters, nitrogen is used repeatedly as it passes downstream. The rate of utilization and release depends upon physical and biological retentiveness, largely by the microbiota attached to the streambed. The average distance a nutrient atom travels downstream during one cycle through the water, biotic, and substrata compartments is referred to as the ***spiralling length*** (Figs. 12-13 and 12-14).
14. Nutrient limitations for biotic productivity are uncommon in small rivers and streams, where nutrients are efficiently retained and recycled. As rivers become larger, nutrient retention is less and nutrient limitations can become more prevalent.
15. The processes of nitrogen cycling in the water of streams and rivers are similar to those of lakes and are influenced to a large extent by bacterial, fungal, and other microbial metabolism. Attached bacteria, fungi, and algae are the primary organisms controlling the seasonal spatial and temporal variations within the water. The processes of nitrification and denitrification often function simultaneously and reciprocally in running water sediments, where many microzones of steep redox gradients occur in the hyporheic zone of the streambed.

13 THE PHOSPHORUS CYCLE

I. PHOSPHORUS IN FRESH WATERS

No other element in fresh waters has been studied as intensively as phosphorus. A great body of quantitative data exists on the seasonal distribution of phosphorus in lakes and loading rates of phosphorus from the surrounding drainage basins. Ecological interest in phosphorus stems from its major role in biological metabolism and the relatively small amounts of phosphorus available in the hydrosphere. In comparison to the rich natural supply of other major nutritional and structural components of the biota (carbon, hydrogen, nitrogen, oxygen and sulfur), phosphorus is least abundant and most commonly limits biological productivity.

II. THE DISTRIBUTION OF ORGANIC AND INORGANIC PHOSPHORUS IN LAKES AND STREAMS

In contrast to the numerous forms of nitrogen in lake systems, the most significant form of inorganic phosphorus is orthophosphate (PO_4^{3-}). A very large proportion, greater than 90%, of the phosphorus in fresh water occurs as organic phosphates and cellular constituents in the biota and adsorbed to inorganic and dead particulate organic materials. It is instructive to discuss first the general aspects of the forms and distribution of phosphorus that occur in fresh waters before analyzing the dynamics of exchange between the compartments.

Phosphine (PH_3) is a volatile constituent of the global biogeochemical phosphorus cycle. Generation of phosphine occurs by anaerobic enzymatic reduction of phosphate and has been found in anoxic sediments of fresh waters, wetlands of flood plains and rice fields, and in sewage treatment facilities (Dévai *et al.*, 1988; Gassmann and Schorn, 1993; Gassmann, 1994; Glindemann *et al.*, 1996). In very shallow areas of organic-rich sediments or saturated hydrosoils, phosphine can evade directly to the atmosphere. It has been estimated that very large amounts of phosphorus can be released to the atmosphere from open sewage treatment facilities. Phosphine is reactive under oxic conditions and has not been detected in water. Within sediments, however, concentrations of phosphine have been found in the nanomolar ranges (0–25 nM dm^{-3}). Although phosphine is likely a minor constituent in lake and river waters, evasion to the atmosphere from enriched shallow wetlands and sediments could result in transport to other distant drainage basins and subsequent deposition with precipitation as phosphate.

Total inorganic and organic phosphorus has been separated in various ways for chemical analyses; often, these fractions relate poorly to the way in which phosphorus is metabolized. Perhaps the most important measure is the total phosphorus content of unfiltered water, which consists of the phosphorus in the particulate and in "dissolved" compartments (Juday, 1927; Ohle, 1938). Particulate phosphorus includes (1) phosphorus in organisms as (a) the relatively stable nucleic acids DNA, RNA, and phosphoproteins, which are not involved in rapid cycling of phosphorus; (b) low-molecular-weight esters of enzymes, vitamins, et cetera; and (c) nucleotide phosphates, such as adenosine diphosphate (ADP) and adenosine 5-triphosphate (ATP) used in biochemical pathways of respiration and CO_2 assimilation; (2) mineral phases of rock and soil, such as hydroxyapatite, in which phosphorus is adsorbed onto inorganic complexes such as clays, carbonates, and ferric hydroxides; (3) phosphorus adsorbed onto dead particulate organic matter or in macroorganic aggregations. In contrast to the phosphorus of particulate matter, dissolved phosphorus is composed of (1) orthophosphate (PO_4^{3-}), (2) polyphosphates, often originating from synthetic detergents, (3) organic colloids or phosphorus combined with adsorptive colloids, and (4) low-molecular-weight phosphate esters.

Because of the fundamental importance of phosphorus as a nutrient and major cellular constituent, much emphasis has been placed on its analytical evaluation. Chemical analyses of phosphorus center around the reactivity of phosphorus with molybdate and changes in reactivity during enzymatic and acidic hydrolysis of complex forms of phosphorus compounds as they are converted to orthophosphate. Detailed analyses recognize eight forms of phosphorus, which are differentiated on the basis of reactivity with molybdate, ease of hydrolysis, and particle size (Strickland and Parsons, 1972; Wetzel and Likens, 2000). Four operational categories result: (a) soluble reactive P, (b) soluble unreactive P, (c) particulate reactive P, and (d) particulate unreactive P. However, these operational methods do not necessarily correspond to either the chemical species of phosphorus or to their role in biotic cycling of phosphorus.

Most of the phosphorus data for fresh waters refer to total phosphorus and inorganic soluble phosphorus (orthophosphate), although in more detailed studies, four general fractions have been identified (Hutchinson, 1957). These four fractions are similar to the four operational groups already cited: (a) soluble phosphate phosphorus, (b) acid-soluble suspended (sestonic) phosphorus—mainly ferric phosphate and calcium phosphate, (c) organic soluble and colloidal phosphorus, and (d) organic suspended (sestonic) phosphorus.

Total phosphorus concentrations in nonpolluted natural waters extend over a very wide range from $<1\ \mu g\ liter^{-1}$ to more than 200 mg $liter^{-1}$ in some closed saline lakes. The total phosphorus concentrations of most uncontaminated surface waters are between 10 to 50 μg P $liter^{-1}$. Variation is high, however, and can be related to characteristics of regional geology. Phosphorus levels of fresh waters are generally lowest in mountainous regions of crystalline bedrock geomorphology and increase in lowland waters derived from sedimentary rock deposits. Lakes rich in organic matter, such as bogs and bog lakes, tend to exhibit high total phosphorus concentrations. A few sedimentary coastal areas, such as in the southeastern United States,

TABLE 13-1 General Relationship of Lake Productivity to Average Concentrations of Epilimnetic Total Phosphorus[a]

General level of lake productivity	*Change (reduction) in alkalinity in epilimnion during summer (meq liter^{-1})*	*Total phosphorus (µg liter^{-1})*
Ultra-oligotrophic	<0.2	<5
Oligo-mesotrophic	0.6	5–10
Meso-eutrophic	0.6–1.0	10–30
Eutrophic		30–100
Hypereutrophic	>1.0	>100

[a] Modified from Vollenweider, R. A.: *Scientific Fundamentals of the Eutrophication of Lakes and Flowing Waters, with Particular Reference to Nitrogen and Phosphorus as Factors in Eutrophication.* OECD Report No. DAS/CSI/68.27, Paris, OECD, 1968, after numerous sources.

are rich in phosphatic rock. Lakes with drainage from these deposits have abnormally high phosphorus levels.

In an extremely detailed treatment relating phosphorus and nitrogen to lake productivity, Vollenweider (1968) demonstrated by several criteria that the amount of total phosphorus generally increases with lake productivity (Table 13-1). Although there are a number of exceptions to this relationship, it demonstrates a general principle that is useful when dealing with applied questions of eutrophication. The relationships of total phosphorus content to loading rates of phosphorus from the drainage basin and from lake sediments are discussed in the concluding sections of this chapter.

Separation of the total phosphorus into inorganic and organic fractions in an appreciable number of lakes indicates that most of the total phosphorus is in an organic phase (Table 13-2). Of the total organic phosphorus, about 70% or more is within the particulate (sestonic) organic material, and the remainder is present as dissolved or colloidal organic phosphorus. Rigler (1964) and Lean (1973b) demonstrated that data of former researchers who employed centrifugation and paper filtration methods of fractionation underestimated the importance of the sestonic organic phosphorus. Soluble organic phosphorus includes a significant quantity of phosphorus in a colloidal state. Inorganic soluble phosphorus is consistently very low, constitutes only a few percent of total phosphorus, and, as will be seen further on, is cycled very rapidly in the zones of utilization. The ratio of inorganic soluble phosphorus to other forms of phosphorus of approximately 1:20, or <5%, as inorganic phosphate phosphorus is remarkably constant in a large variety of lakes and streams within the temperate zone. The percentage of total phosphorus occurring as truly ionic orthophosphate is probably considerably <5% in most natural waters (e.g., Prepas and Rigier, 1982; Tarapchak *et al.*, 1982).

The phosphorus distribution within the fractions just discussed is the picture generally observed in the trophogenic zones of lakes. Phosphate, pyrophosphate, triphosphate, and higher polyphosphate anions additionally form complexes, chelates, and insoluble salts with a number of metal ions (Stumm and Morgan, 1996). The extent of complexing and chelation between various phosphates and metal ions in natural waters depends upon the relative concentrations of the phosphates and the metal ions, the pH, and the presence of other ligands (sulfate, carbonate, fluoride, and organic species).

Because phosphate concentrations are generally low, complex formations involving these major cations and various phosphate anions will have little effect on

TABLE 13-2 Fractionation of Total Phosphorus in Lakes Analyzed by Different Techniques of Separation

	Soluble inorganic P		*Soluble organic P*		*Sestonic P*		*Total organic P*		*Total P*
Lakes	(*µg liter^{-1}*)	(%)	(*µg liter^{-1}*)	(%)	(*µg liter^{-1}*)	(%)	(*µg liter^{-1}*)	(%)	(*µg liter^{-1}*)
Northern Wisconsin[a] (Juday and Birge, 1931)	3	13.0	14	60.9	6	26.1	20	87.0	23
Michigan Lakes[b] (Tucker, 1957)	1.5	12.0	5.7	46.9	5.0	41.1	10.7	88.0	12.2
Linsley Pond, Connecticut[c] (Hutchinson, 1957)	2	9.5	6	28.6	13	61.9	19	90.5	21
Ontario Lakes[c] (Rigler, 1964)	—	5.9	—	28.7	—	65.4	—	94.1	—

[a] Centrifugation techniques.
[b] Paper (No. 44 Whatman) filtration.
[c] Membrane filtration (0.5 µm).

the distribution of metal ions but may have marked effects on the phosphate distribution (cf. Golachowska, 1971). Metal ions such as those of ferric iron, manganous manganese, zinc, copper, et cetera, are present in concentrations comparable to or lower than those of phosphates. For these ions, complex formation can significantly affect the distribution of the metal ion, the phosphates, or both. For example, the solubility of aluminum phosphate ($AlPO_4$) is minimal at pH 6 and increases at both higher and lower pH values. Ferric phosphate ($FePO_4$) behaves similarly, although it is more soluble than $AlPO_4$. Calcium concentration influences the formation of hydroxylapatite, $[Ca_5(OH)(PO_4)_3]$. In an aqueous solution lacking other compounds, a calcium concentration of 40 mg liter^{-1} at a pH of 7 limits the solubility of phosphate to approximately 10 μg liter^{-1}. A calcium level of 100 mg liter^{-1} lowers the maximum equilibrium of phosphate to 1 μg liter^{-1}. Elevation of the pH of waters containing typical concentrations of calcium leads to apatite formation (Kümmel, 1981). Moreover, increasing pH leads to formation of calcium carbonate, which coprecipitates phosphate with carbonates (Otsuki and Wetzel, 1972). Sorption of phosphates and polyphosphates on surfaces is well known, particularly onto clay minerals (cf. Stumm and Morgan, 1996), by chemical bonding of the anions to positively charged edges of the clays and by substitution of phosphates for silicate in the clay structure. In general, high phosphate adsorption by clays is favored by low pH levels (approximately 5–6).

Following from these interactions and the distribution of phosphorus in inorganic and organic fractions, the general tendency is for unproductive lakes with orthograde oxygen curves to show little variation in phosphorus content with depth (Fig. 13-1). Similarly, during periods of fall and spring circulation, the vertical distribution of phosphorus is more or less uniform. Oxidized metals, such as iron, and major cations, particularly calcium, can react with and precipitate phosphorus.

Lakes exhibiting clinograde oxygen curves during the periods of stratification, however, possess much more variable vertical distributions of phosphorus. Commonly, there is a marked increase in phosphorus content in the lower hypolimnion, particularly during the later phases of thermal stratification (Fig. 13-1). Much of the hypolimnetic increase is in soluble phosphorus near the sediments. The sestonic phosphorus fraction is highly variable with depth. Sestonic phosphorus in the epilimnion fluctuates widely with oscillations in plankton populations and loadings from littoral areas. Sestonic phosphorus in the metalimnion and hypolimnion varies with sedimentation of plankton, depth-dependent rates of decomposition, and the development of deep-living populations of bacterial and other plankton (e.g., euglenophyceans).

III. PHOSPHORUS CYCLING IN RUNNING WATERS

A. Dissolved and Particulate Phosphorus in Rivers

Dissolved inorganic phosphorus, commonly referred to as orthophosphate or more correctly as soluble reactive phosphorus (SRP) based upon methods used in chemical analyses, averages about 10 μg liter^{-1} worldwide among unpolluted rivers (Meybeck, 1982, 1993b). Total dissolved phosphorus in these waters averages about 25 μg liter^{-1}.

Total dissolved inorganic phosphorus (DIP) of river water often increases by a factor of two- to four-fold

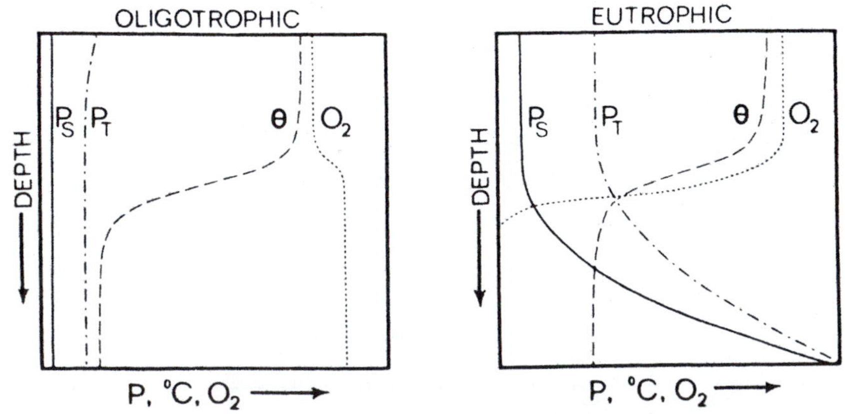

FIGURE 13-1 Generalized vertical distribution of soluble (P_S) and total (P_T) phosphorus in stratified lakes of very low (oligotrophic) and of high (eutrophic) productivity.

during and following increases in discharge from heavy rainfall events or in the early stages of snowmelt. DIP from pollutional sources often increases to levels of 50–100 μg P liter^{-1} from agricultural runoff and to over 1000 μg P liter^{-1} from municipal sewage sources (Meybeck, 1982, 1993b). Where DIP enters streams or rivers at point sources, concentrations decline downstream with net uptake rates in the range of 10–100 mg m^{-2} day^{-1}. As discussed next, uptake occurs both by physical adsorption to benthic substrata and to particulate seston, as well as by assimilation by attached biota. Although uptake of DIP by planktonic seston may be small (ca. 1% of total uptake) in small streams (Newbold *et al.*, 1983a; Mulholland *et al.*, 1990), sestonic uptake increases markedly in larger streams, particularly when dominated by phytoplankton (Newbold, 1992). Some of the seston and adsorbed phosphorus may settle out during base flows and then later be transported downstream during storm-induced high discharge. Exchange from these particles to the water may then occur in these displaced locations.

B. Biological Utilization

Biological uptake of phosphorus by algae, cyanobacteria, bacteria, and larger aquatic plants generally follows Michaelis–Menten kinetics (Chap. 17). Many algae and bacteria assimilate phosphorus at rates more rapid than used for growth. As a result, cells accumulate phosphorus and steady-state growth is saturated by concentrations much lower than the half-saturation constant (where uptake is half its maximum rate). Uptake rates of phosphorus by coarse particulate organic matter and microbiota associated with it commonly reach a maximum in 15–20 days and then decline precipitously (e.g., Mulholland *et al.*, 1984).

The rates of phosphorus uptake correlate with the metabolism of dominant organisms. In highly shaded streams with minimal in-stream photosynthesis, much (60 and 35%, respectively) of the uptake was associated with large (>1 mm) and fine (<1 mm) particles and only 5% with attached algae (Newbold *et al.*, 1983a; Mulholland *et al.*, 1985). In streams where decomposition of leaf detritus was dominant, phosphorus uptake was associated with the attached fungi and bacteria. Much of the phosphorus uptake associated with particulate organic matter is heterotrophic uptake by microbes associated with large and small particles of the benthic detritus (Gregory, 1978; Meyer, 1980; Webster *et al.*, 1991; Suberkropp, 1995). Phosphorus uptake was primarily associated with attached and planktonic algae in streams that were less shaded (e.g., Ball and Hooper, 1963; Petersen *et al.*, 1985). If other growth conditions are adequate, enrichments with phosphorus in the water often result in immediate enhancements of rates of algal photosynthesis and growth (e.g., Hart and Robinson, 1990; Burton *et al.*, 1991; Kjeldsen, 1994). Uptake by macrophytes, particularly by rooted vascular plants that rely largely upon rooting tissues for nutrient uptake, was much less than by attached algae and other microbes.

Average sestonic algal abundance has been positively correlated with total phosphorus and average phosphorus concentrations of many streams at moderate levels of nutrient loadings (Jones *et al.*, 1984; Soballe and Kimmel, 1987; Nieuwenhuyse and Jones, 1996) but becomes increasingly less responsive, as in lakes, to nutrient loadings at higher phosphorus concentrations. Summer mean chlorophyll concentration of seston among temperate streams exhibits a curvilinear relationship with summer mean total phosphorus concentrations (Fig. 13-2). The amount of nutrient loading to a river increases with greater catchment area. The analyses indicated that the phosphorus–sestonic chlorophyll relationship increased some 2.3-fold as the area increased from 100 to 100,000 km^2 (Fig. 13-2). It is important to note that the primary productivity of rivers is generally much less responsive to increased nutrient loadings than is the productivity of lakes. As will be discussed later in the sections on algal productivity, nutrient availability is generally higher in rivers than lakes. Nutrient losses in lakes are greater and large quantities are stored in parts of the ecosystem that are not readily available to the producers.

C. Adsorption and Release

The kinetics of abiotic adsorption and desorption of phosphorus onto and from organic and inorganic particles, such as natural sediment and suspensoid particles, generally comply with Langmuir isotherms. Fine particles (<0.1 mm) account for nearly all of the sorption capacity for phosphorus (0.1–1.0 μg P g^{-1} sediment per μg P liter^{-1} water; Meyer, 1979; Logan, 1982; Stabel and Geiger, 1985). The adsorption capacities of particles saturate and reach quasi-steady state in which uptake and release are about balanced.

The abundance of phosphate is also regulated by solubility reactions, particularly in relation to solid–solution associations with colloids or particles. Phosphorus concentrations in turbid rivers with low calcium concentrations are influenced by a solid ferric hydroxide-phosphate present in colloidal suspensions or on suspended particulates (Fox, 1993). In hardwater streams, the solubility of inorganic phosphorus can decrease as pH increases above 8.5, as is common in areas of intensive photosynthesis (Diaz *et al.*, 1994). Precipitation of inorganic phosphorus as calcium

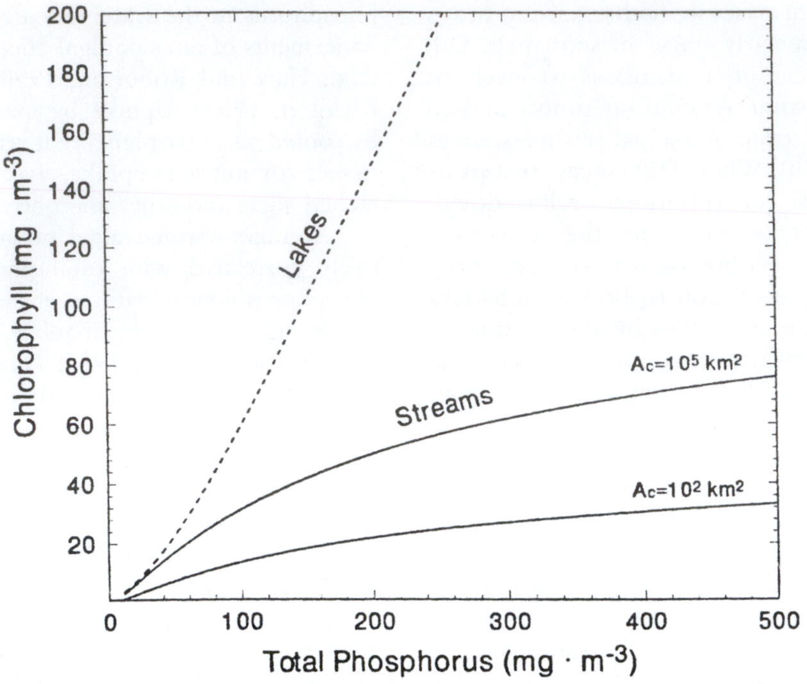

FIGURE 13-2 Phosphorus–sestonic chlorophyll relationships in temperate streams and lakes. Solid curves indicate predicted chlorophyll concentrations in rivers of differing catchment areas (A_C) of 100 and 100,000 km^2. The dashed curve depicts the trajectory of chlorophyll concentration in *P*-limited lakes in which total N:P > 25 (Forsberg and Ryding, 1980). (After Nieuwenhuyse and Jones, 1996.)

phosphate, highly stable as hydroxyapatite, increased to >60% at Ca concentrations approaching 100 mg liter^{-1} and pH values of 9.0 or greater. If calcium carbonate crystallizes under these conditions, phosphorus can be coprecipitated into the crystals and also adsorbed to the surfaces of the carbonates (Otsuki and Wetzel, 1972; cf. Chap. 11). These carbonate deposits are common in travertine encrustations in rivers that receive ground waters supersaturated with calcium and bicarbonates.

Thus, the major divalent metal ions in water, calcium and magnesium, tend to destabilize particles and enhance aggregation and sedimentation rates. In contrast, dissolved organic matter tends to stabilize colloidal particles in stream and lake water, slowing aggregation rates and thereby lowering sedimentation rates (Tipping and Higgins, 1982; Ali *et al.*, 1985). Separation distances and efficacy of van der Waals forces for attraction and aggregation, however, vary greatly with different types and concentrations of natural dissolved organic compounds and with the ionic strengths of charged colloids or particles (O'Melia, 1998).

As particulate organic matter increases, as is common from leaf-fall in late fall and winter in temperate woodland streams, phosphorus uptake can increase as the particulate organic matter increases (Mulholland *et al.*, 1985; Klotz, 1986). It is during this time as well that fungal activity is maximal on particulate organic matter (e.g., Suberkropp, 1995) and bacterial activities are maximal on dissolved organic matter leaching from the particulate organic matter (Wetzel and Manny, 1972b). Phosphorus uptake length, and hence total spiralling length, decreases in winter. These rates are counteracted at other times when detrital particulate organic matter is reduced by high discharge associated with storms.

D. Recycling of Phosphorus in Streams

As discussed earlier in terms of the biota of lakes, phosphorus can be released from biota by excretion in inorganic and organic forms from living microbiota or as the organisms senesce, die, and lyse. Phosphorus can also be released, as discussed earlier, during egestion of consuming animals. As in lakes, the dissolved organic phosphorus compounds in streams are utilized enzymatically appreciably more slowly than is dissolved inorganic phosphorus (reviewed by Newbold, 1992). Because of their slower rates of utilization, an accumulation of phosphorus compounds of high molecular weight (>5000 Daltons) occurs, and these compounds are exported downstream for subsequent utilization (Mulholland *et al.*, 1988).

The spiralling of phosphorus has been measured in only a few streams. In a first-order woodland stream in Tennessee, the spiralling length was 190 m, most of it (165 m) associated with the water movement, and the remainder (25 m) in the particulate components, mostly coarse and fine particulate matter (Newbold *et al.*, 1983a). Less than 3% of the total passed through the invertebrate and vertebrate consumers. Phosphorus uptake lengths ranged from 21 to 165 m and varied inversely with the quantity of detritus on the stream sediments (Mulholland *et al.*, 1985). In a larger river in Michigan, phosphorus uptake lengths varied between 1100 and 1700 m (Ball and Hooper, 1963). Most streams, however, have considerable uptake capacity for phosphorus, with uptake lengths in the range of 5–200 m (e.g., Hart *et al.*, 1991).

In unshaded streams where autotrophic production constitutes an appreciable component of organic matter subsidy of streams, heavy grazing by macroinvertebrate consumers of attached algae and other microbes can reduce the uptake of phosphorus from the water (Mulholland *et al.*, 1983, 1985b). This reduction can result both from decreases in the microbiota as well as increased reduction in sizes of particles. The fine particles of organic matter and phosphorus are then scoured relatively easily and are the predominant form exported downstream (Meyer and Likens, 1979).

Long-term evaluations of phosphorus dynamics in streams are rare. Examination of an average annual phosphorus budget for a forested second-order stream ecosystem of New Hampshire illustrates some of the potential transport and transformation processes (Table 13-3). Phosphorus inputs were dominated by dissolved and fine particulate phosphorus (63%) and falling and windblown litter (23%). Subsurface inflows (10%) and precipitation (5%) were relatively small sources of phosphorus in this ecosystem. The geologic export of phosphorus in the stream water was the only removal vector of consequence. On an annual basis, no annual net retention of phosphorus occurred in the stream. However, over short periods of time, inputs exceeded exports, with phosphorus accumulation. The accumulated phosphorus was exported in large pulses during precipitation-mediated episodes of high-stream discharge. Similar exports have been observed in many other stream ecosystems (e.g., Long and Cooke, 1979; Verhoff *et al.*, 1982; Munn and Prepas, 1986; Wetzel, 1989). For example, in a detailed analysis of daily phosphorus dynamics in a small second-order stream in Michigan passing through a wetland to a lake, between 60 and 80% of the annual PO_4—P from this stream to the receiving lake occurred during three major precipitation events (Wetzel, 1989). Dissolved inorganic and organic nitrogen loadings from the stream, however, were not as strongly coupled to discharge and were controlled to a greater extent by biological retention processes.

TABLE 13-3 Phosphorus Budget for a Forested Second-Order Stream, Bear Brook, New Hampshire[a]

	mg P m^{-2} yr^{-1}		*% of Total*	
Inputs				
Dissolved		346		28
Precipitation	63		5	
Tributaries	152		12	
Subsurface	131		10	
Particulate		900		72
Fine: tributaries	459		37	
Coarse:	441		35	
Litter	283		23	
Tributaries	158		13	
Total Inputs		1246		100
Exports (fluvial):				
Dissolved		242		19
Particulate		1059		81
Fine	807		62	
Coarse	252		19	
Total exports		1301		100

[a] Rounded to whole numbers; modified from Meyer and Likens (1979) and several papers cited therein.

IV. PHOSPHORUS AND THE SEDIMENTS: INTERNAL LOADING

The exchange of phosphorus between sediments and the overlying water is a major component of the phosphorus cycle in natural waters. There is an apparent net movement of phosphorus into the sediments in most lakes. The effectiveness of the net phosphorus sink to the sediments and the rapidity of processes regenerating the phosphorus back to the water depend upon an array of physical, chemical, and biological factors. The correlation between the amount of phosphorus in the sediments and the productivity of the overlying water is modest to weak, and the phosphorus content of the sediments can be several orders of magnitude greater than that of the water. The important factors are (1) the ability of the sediments to retain phosphorus, (2) the conditions of the overlying water, and (3) the biota within the sediments that alter exchange equilibria and effect phosphorus transport back to the water.

A. Exchanges across the Sediment–Water Interface: Overview

The deposition of phosphorus into lake sediments occurs by five mechanisms (Williams and Mayer, 1972;

Boström *et al.*, 1988; Wetzel, 1990a): (1) Sedimentation of phosphorus minerals imported from the drainage basin. Most of this material settles rapidly and is deposited largely in nearshore areas. (2) Adsorption or precipitation of phosphorus with inorganic compounds by three mechanisms: (a) Phosphorus coprecipitation with iron and manganese; (b) adsorption to clays, amorphous oxyhydroxides, and similar materials; and (c) phosphorus associated with carbonates. It is difficult to distinguish direct adsorption of dissolved phosphorus of lake water onto particles in the sediments from diagenetic and transfer processes within surface sediments. (3) Sedimentation of phosphorus with allochthonous organic matter. (4) Sedimentation of phosphorus with autochthonous organic matter. (5) Uptake of phosphorus from the water column by attached algal and other attached microbial communities and to a lesser extent by submersed macrophytes, with eventual transport back to the sediments by translocation and deposition with detritus.

Once the phosphorus is within the sediment in various forms, exchanges across the sediment–water interface are regulated by mechanisms associated with mineral–water equilibria; sorption processes, particularly ion exchange, oxygen and other electron acceptor-dependent redox interactions; and the physiological and behavioral activities of many biota (bacteria, algae, fungi, macrophytes, invertebrates, and fish) associated with the sediments. The sediment–water interface separates into two distinct domains. In all but the upper few millimeters of sediment, exchange is controlled by motions on molecular scales with correspondingly low diffusion rates (Duursma, 1967). In the water, exchange is regulated by much higher and more variable rates of turbulent diffusion (Mortimer, 1971). The exchange rates between various deposits of phosphorus and the interstitial water of the sediments depend on local adsorption and diffusion coefficients and their alteration by enzyme-mediated reactions of the microbiota.

Numerous processes operate, often simultaneously, to mobilize phosphorus from particulate stores to dissolved interstitial phosphorus and then to transport that dissolved phosphorus across the sediment–water interface into the overlying water (Fig. 13-3). Orthophosphate is bound to particles by physical adsorption as well as chemical binding of different strengths (complex, covalent, and ionic bonds). Physical and chemical mobilization includes desorption, dissolution of phosphorus-containing compounds, particularly assisted by microbially mediated acidity, and ligand exchange mechanisms between phosphate and hydroxide ions or organic chelating agents (Boström *et al.*, 1982; Stumm and Morgan, 1996). Microbial biochemical mobilization processes include mineralization by hydrolysis of phosphate–ester bonds, release of phosphorus from living cells as a result of changing environmental conditions, particularly redox, and autolysis of cells.

The internal phosphorus loading to a lake from the sediments depends on hydrodynamic and biotic mechanisms that transport dissolved phosphorus from the sediments to the lake water. Because of the steep concentration gradients of phosphorus between interstitial water and the overlying water, molecular diffusion is a primary transport into the overyling anaerobic water (Fig. 13-3). Currents from wind-induced water turbulence can disrupt gradients and resuspend sediment particles. Disturbance of sediments by benthic invertebrates living on or in the sediments and by bottom-feeding fishes can, when occurring in large densities, cause appreciable bioturbation of sediments. Microbial generation of gases, particularly CO_2, CH_4, and N_2, can accumulate, form bubbles, and rise to the surface. Such ebullition can disrupt gradients and accelerate diffusion of phosphorus upward. The metabolism and growth of plants living on and within the sediments can both suppress and enhance the transport of phosphorus across the sediment–water interface. Because of the major importance of the movement of phosphorus from large accumulations in the sediment to the overlying water, each of these processes is discussed in greater detail in subsequent sections.

B. Oxygen Content of the Microzone

The most conspicuous regulatory features of the sediment boundary are the mud–water interface and the oxygen content. The oxygen content at this microzone is influenced primarily by the metabolism of bacteria, algae, fungi, planktonic invertebrates that migrate to and live within the interface, and sessile benthic invertebrates. Microbial degradation of dead particulate organic matter that settles into the hypolimnion and onto the sediments is the primary consumptive process of oxygen in deepwater areas of lakes. The rate of oxygen depletion is governed by the rates of organic loading to the hypolimnion and by lake or reservoir morphology (cf. Chap. 9). For example, it has been estimated that 88% of the hypolimnetic oxygen consumption in the central basin of Lake Erie resulted from bacterial degradation of algae sedimenting from the trophogenic zone (Burns and Ross, 1971). Decomposition of more labile organic fractions, largely of plant origin, occurs while it is settling to the sediments, and, depending on rates of input and sedimentation, the sediments often receive organic residues that are relatively resistant to rapid decomposition.

Sediment demand for oxygen is high and is governed by the intensity of microbial and respiratory metabolism, slow rates of diffusion, and by the fact

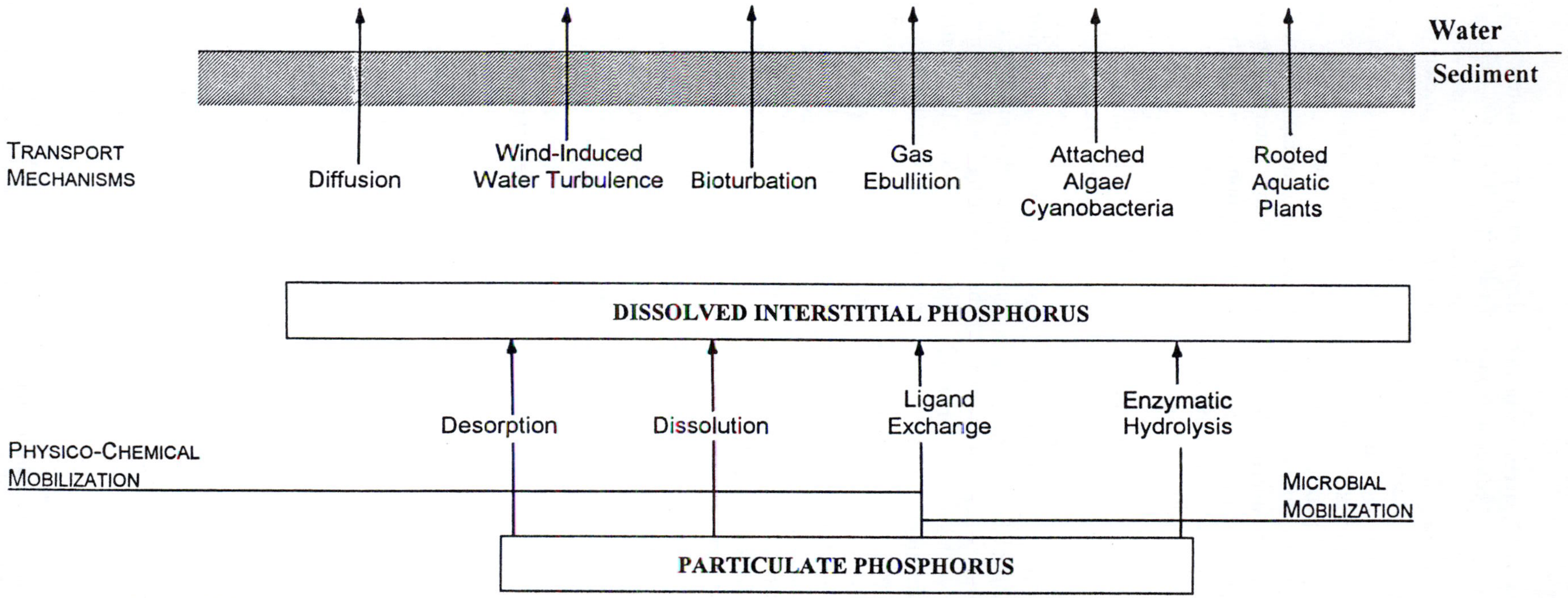

FIGURE 13-3 Processes involved in the mobilization of phosphorus from particulate stores into dissolved states of interstitial water of the sediments and transport across the sediment–water interface into the overlying water. (Modified extensively from Boström *et al.*, 1982, 1988, and Wetzel, 1999c.)

that inorganic elements, such as Fe^{++}, accumulate in reduced form when released into the sediment from decomposing biota. Diffusion regulates transport and is essentially molecular in the sediments, unless the superficial sediments are disturbed by overlying water turbulence. Oxygen from well-aerated overlying water, as in oligotrophic lakes or in more productive lakes at periods of complete circulation, will penetrate only a few centimeters into the sediments by diffusion. Oxygen penetration into the sediments is governed by the rate of oxygen supply to the sediments, turbulent mixing of superficial sediments, if any, and by the oxygen demand per unit volume of the sediment. The superb experimental and observational work of Mortimer (1941, 1942, 1971) has demonstrated the importance of an oxidized microzone to chemical exchanges, especially of phosphorus, from the sediments. At the sediment surface, a difference of a few millimeters in oxygen penetration is the critical factor regulating exchange between sediment and water. These relationships are exemplified by two lakes with organic sediments: The first lake maintained oxygen concentrations at the sediment interface >8 mg liter^{-1} throughout summer stratification, while in the second lake, oxygen levels at the interface decreased to <1 mg liter^{-1}.

In the first situation, illustrated by Mortimer's studies of Lake Windermere, England, oxygen concentration at the sediment surface did not fall below 1 or 2 mg liter^{-1}. Electrode potentials, which approximate composite redox potentials (cf. Chap. 14), of the oxygenated overlying water and surficial sediments to a depth of approximately 5 mm were uniformly high (+200 to 300 mV). Below 40–50 mm in the sediments, the potentials were uniformly low (approximately −200 mV), indicative of extreme reducing conditions and total anoxia (Fig. 13-4). The sediment remained oxidized to a depth of about 5 mm throughout the period of summer stratification. Seasonal differences were observed in the sediment depth at which the transition from high to low potential occurred, but the region of low potential never extended into the water. After five months of stratification, the point of zero mV moved toward the surface of the sediments, to −5 mm from approximately −12 mm at the time of spring turnover and moved downward to −10 mm during fall circulation. The integrity of the oxidized microzone was maintained in a thin but operationally very significant layer during stratification periods. The oxidized microzone was further maintained by diffusion and by turbulent displacement of the uppermost sediments to the overlying water during turnover periods (cf. Gorham, 1958). The effectiveness of the oxidized microzone in preventing a significant release of soluble components from the interstitial waters of the sediments to the overlying water was demonstrated in experimental chambers for over five months (Fig. 13-5, *left*). Phosphorus, in particular, was prevented from migrating upward.

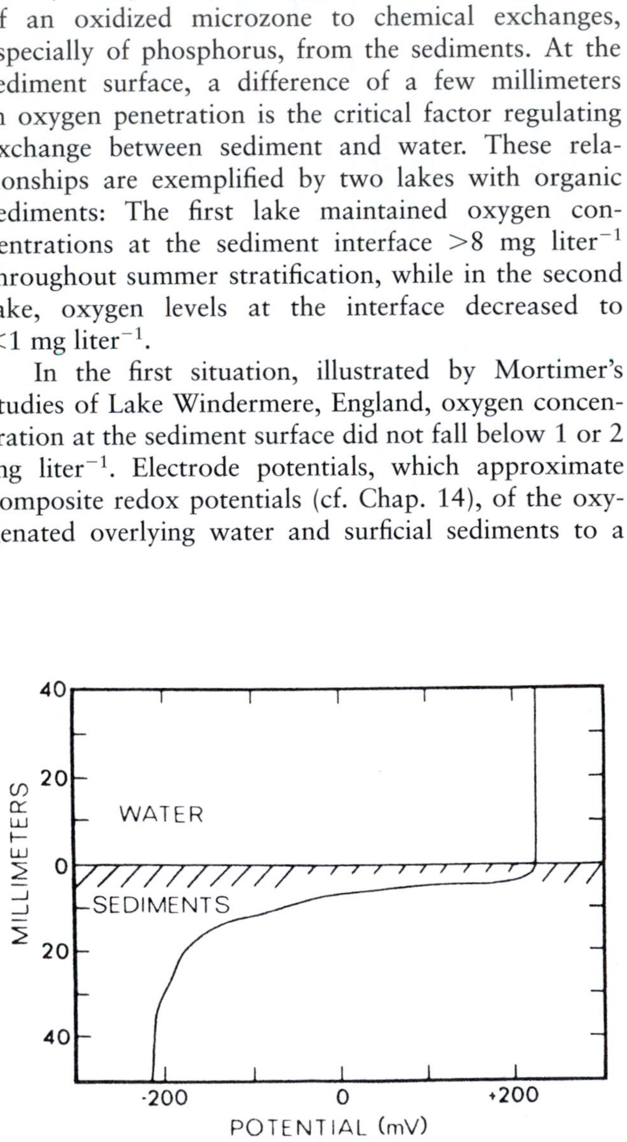

FIGURE 13-4 Diagrammatic profile of composite electrode potentials, not corrected for pH variations, across the sediment–water interface in undisturbed cores from the deepest portion of Lake Windermere before, during, and after stratification. (Based on data of Mortimer, 1971.)

The ability of sediments to retain phosphorus beneath an oxidized microzone at the interface is related to several interacting factors. Much of the organic phosphorus reaching the sediments by sedimentation is decomposed and hydrolyzed (Sommers *et al.*, 1970). Most of the sediment phosphorus is inorganic, for example, apatite, derived from the watershed, and phosphate adsorbed onto clays and aluminum and ferric hydroxides (Frevert, 1979a, b; Detenbeck and Brezonik, 1991; Andersen and Jensen, 1992; Danen-Louwerse *et al.*, 1993). Additionally, phosphate coprecipitates with iron, manganese, and carbonates (Mackereth, 1966; Harter, 1968; Wentz and Lee, 1969; Otsuki and Wetzel, 1972; Boström *et al.*, 1988). Work on Wisconsin lake sediments and the Great Lakes indicated that phosphorus was present in the sediments predominantly as apatites, organic phosphorus, and orthophosphate ions covalently bonded to hydrated iron oxides (Shukla, 1971; Williams *et al.*, 1970, 1971a–c; Williams and Mayer, 1972; Golterman, 1982, 1995). In calcareous sediments of hardwater lakes containing 30–60% $CaCO_3$ by weight, $CaCO_3$ levels were not directly related to inorganic and total phosphorus. These sediments had a lower capacity to adsorb inorganic phosphorus than noncalcareous sediments (cf. Stauffer, 1985; Olila and Reddy, 1993). The observations imply that $CaCO_3$ sorption is less important than iron–phosphate complexes in controlling the

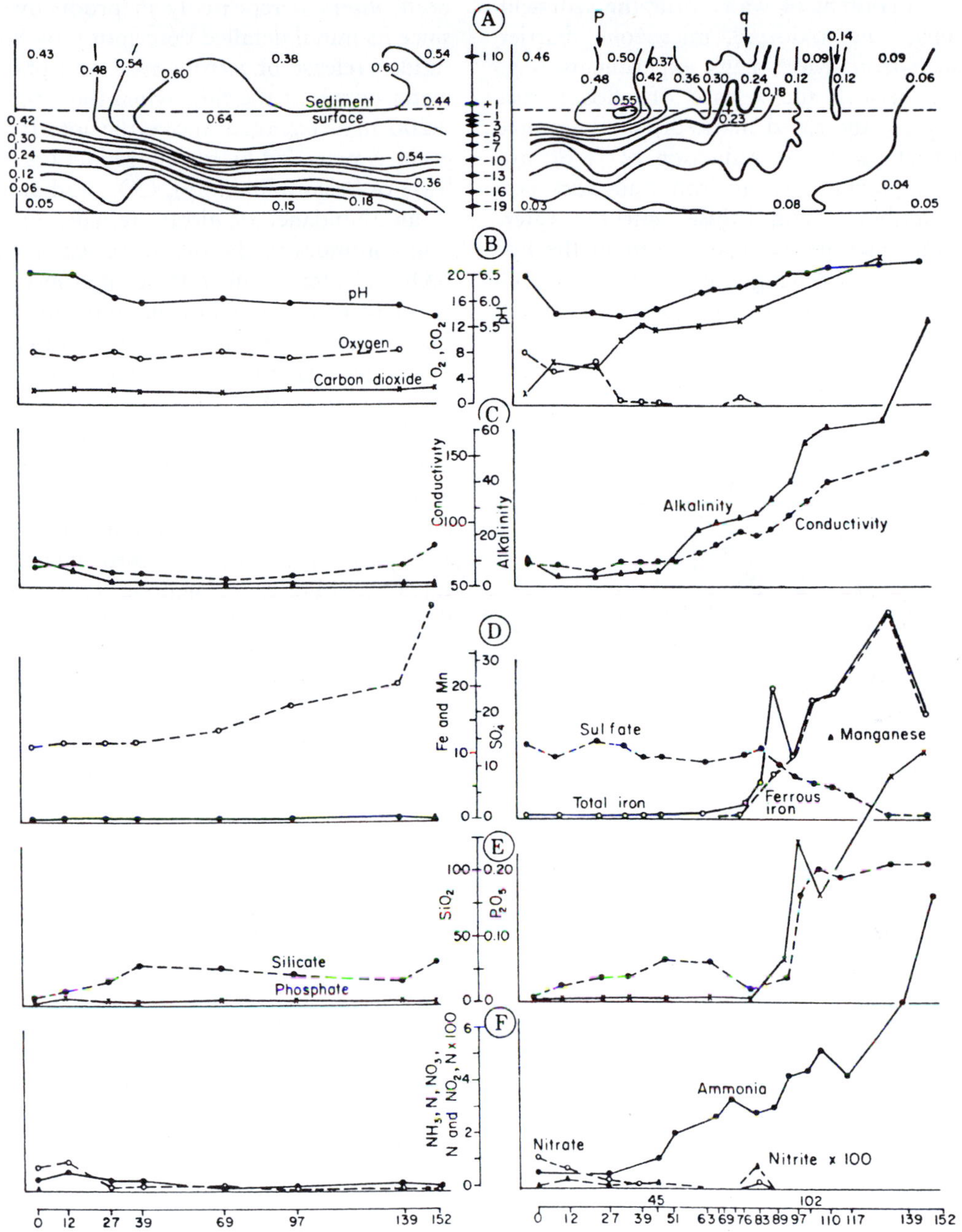

FIGURE 13-5 Variation in chemical composition of water overlying deepwater Lake Windermere sediments over 152 days in experimental sediment–water tanks. *Lefthand series:* aerated chamber; *righthand series:* anoxic chamber. (a) Distribution of redox potential (E_7; Eh adjusted to pH 7) across the sediment–water interface in mm; (b) pH, concentrations of O_2, and CO_2 in mg liter^{-1}; (c) alkalinity expressed as mg $CaCO_3$ liter^{-1} and conductivity, in μS cm^{-1} at 18°C; (d) iron (total and ferrous as Fe) and SO_4 in mg liter^{-1}; (e) phosphate as P_2O_5 and SiO_2 in mg liter^{-1}; (f) nitrate, nitrite ×100, and ammonia, all as N, in mg liter^{-1}. (Mortimer, C. H.: *Limnol. Oceanogr.* *16*:396, 1971, and *J. Ecol.* *29*:280, 1941.)

concentrations of phosphorus in sediments (Andersen, 1975; Frevert, 1980). Although phosphorus exchange by adsorption and desorption within the sediments between sediment particles and interstitial water can be as rapid as a few minutes (Hayes and Phillips, 1958; Li *et al.*, 1972), the rate of transfer across the sediment–water interface depends on the state of the microzone. The oxidized layer forms an efficient trap for iron and manganese (see Chap. 14) as well as for phosphate, thereby greatly reducing the transport of materials into the water and scavenging materials such as phosphate from the water.

As the oxygen content of water near the sediment interface declines, the oxidized microzone barrier weakens. As seen from Mortimer's experiments (Fig. 13-5, right-hand series), the release of phosphorus, iron, and manganese increased markedly as the redox potential decreased. With the reduction of ferric hydroxides and complexes, ferrous iron and adsorbed phosphate were mobilized and diffused into the water. The same general reactions were observed in the hypolimnetic water just overlying the sediments in eutrophic Esthwaite Water (Fig. 13-6), a pattern that has been observed repeatedly in productive dimictic lakes since its initial detailed description by Einsele (1936). A sudden release of ferrous iron and phosphate into the water occurs at the time when the +0.20 isovolt (E_7 = +200 mV) emerged above the interface surface. This event is preceded by nitrate reduction and the slow release of bases (alkalinity), CO_2, and ammonia. As long as an abundance of nitrate remains and nitrate reduction continues in the overlying water (ca. >0.1–1 mg NO_3—N liter^{-1}), no release of iron-bound phosphate occurred from the sediments to the anoxic hypolimnion (Andersen, 1982; Tirén and Pettersson, 1985; Foy, 1986). Manganese is reduced and mobilized at a higher redox potential than iron.

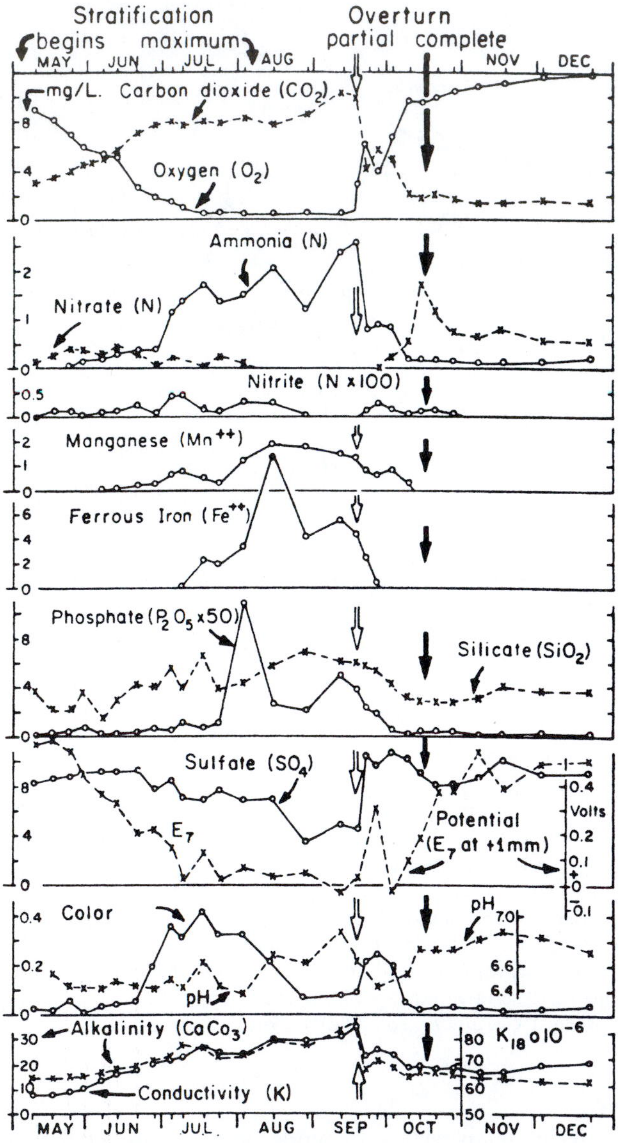

FIGURE 13-6 Seasonal distribution in composition (mg liter^{-1}) and properties of water within 30 cm of the sediments at 14 m in Esthwaite Water, England. Components as in Figure 13-5; color in arbitrary units. (From Mortimer, C. H.: *Limnol. Oceanogr.* 16:387, 1971.)

The introduction of oxygen during autumnal circulation[1] causes ferrous iron to be oxidized and produces a simultaneous reduction of phosphate, part as ferric phosphate, which is less soluble than ferric hydroxide, and part by adsorption onto ferric hydroxide and $CaCO_3$. Although manganese is oxidized more slowly than iron, it nonetheless is effectively precipitated at the time of overturn. Ferrous iron released from the sediments is always in excess of phosphate, and when oxidized, it precipitates much of the phosphate. Some of the ferric phosphate in particulate form may slowly hydrolyze and restore some phosphate to the upper waters and littoral areas (Hutchinson, 1957). Although most phosphate is returned eventually to the sediments, much (>50%) of the hypolimnetic phosphorus upwelled during autumnal circulation is available biologically (Nürnberg, 1985). Often, however, other growth conditions (temperature, light) are not optimal at that time of year.

In very productive lakes where hypolimnetic decomposition of sedimenting organic matter produces anoxic conditions and hydrogen sulfide, some ferrous sulfide (FeS) is precipitated. Ferrous sulfide, like many other metal sulfides, is exceedingly insoluble and forms at a redox potential of about + 100 mV. If large quantities of FeS precipitate, sufficient iron can be removed to permit some of the phosphate accumulated in the hypolimnion to remain in solution during autumnal circulation. The addition of sulfate to a lake in order to increase the bacterial production of hydrogen sulfide (H_2S) and to accelerate the loss of iron has been suggested as a method of fertilizing lakes by regenerating phosphate from the sediments (Hasler and Einsele, 1948).

[1] The normal seasonal erosion in the autumn of the epilimnion and metalimnion with progressive oxygenation of the upper portion of the hypolimnion can be periodically intensified during storm periods, particularly in reservoirs (e.g., James *et al.*, 1990). Stratification can be reestablished subsequently for short periods until the next intensive wind period occurs.

Iron interacts with dissolved humic substances (cf. Chap. 23) to bind phosphorus at acidic to near-neutral pH values, irrespective of redox conditions, in the surficial sediments of certain softwater lakes (Jackson and Schindler, 1975; Francko, 1986). This process should enhance sediment retention of phosphorus in lakes with high loading of iron and humic substances.

C. Phosphorus Release from the Sediments

Because of the obvious importance of phosphorus as a nutrient that often accelerates the productivity of fresh waters, much interest has been devoted to the phosphorus content of sediments and its movement into the overlying water. Lake sediments contain much higher concentrations of phosphorus than the water (Olsen, 1958, 1964; Holden, 1961; Hepher, 1966; Holdren *et al.*, 1977; Boström *et al.*, 1982; and many others). Under aerobic conditions, the exchange equilibria are largely unidirectional toward the sediments. Under anaerobic conditions, however, inorganic exchange at the sediment–water interface is strongly influenced by redox conditions (Frevert, 1979b). The depth of the sediment involved in active migration of phosphorus to the water is considerable (e.g., Carignan and Flett, 1981). In undisturbed anoxic sediments, given sufficient time (2–3 months), phosphorus moved upward readily from a depth of at least −10 cm to the overlying water, regardless of whether the sediments were calcareous eutrophic muds or acidic and peaty in nature (Hynes and Greib, 1970). Comparison of movement in sterile sediments and sediments with anaerobic bacteria showed no significant difference; thus, diffusion predominated.

Phosphorus-mobilizing bacteria, especially of the genera *Pseudomonas, Bacterium,* and *Chromobacterium,* were abundant to at least 15 cm in reservoir sediments (Gak, 1959, 1963). Their abundance and vertical distribution varied with the type of sediments. Low numbers occurred in sandy sediments with small amounts of silt, and the bacteria were concentrated near the interface. Their numbers increased more uniformly with depth in sandy sediments with moderate amounts of organic matter and silts. The greatest numbers of bacteria were found in silts high in organic matter. Hayes and Anthony (1958, 1959) and Anthony and Hayes (1964) examined the relationship of bacterial densities and organic content of sediments in detail and found only a weak correlation in a large number of lakes. Bacterial numbers in the sediments increased proportionally, especially at the interface, with several indices of increasing lake productivity in lakes of neutral or alkaline pH and with low concentrations of humic compounds. In general, however, bacterial production rates are correlated with phosphorus and carbon in the sediments (Eckerrot and Pettersson, 1993). Bacterial biomass in sediments of organically stained acidic bog lakes was also high.

Although bacteria are of major importance in the dynamics of phosphorus cycling in the water, as will be discussed in the following section, their role in expediting phosphorus exchange across the sediment interface is relatively minor in comparison to chemical equilibria processes (Hayes, 1964). Bacterial decomposition is proportional to bacterial densities at the interface and directly related to the general productivity of the lake (Hayes and MacAuley, 1959; Hargrave, 1972). The sediment microflora is important in increasing the concentrations of phosphorus dissolved in the interstitial water of the sediments (Fleischer, 1978). Bacterial metabolism at the interface, however, has relatively little effect on biogenic fixation and removal of phosphate from the overlying water. Using sterilized and natural sediments, it was found that microbial fixation and transport of phosphorus from the water to the sediments amounted to <5% of the total movement under anaerobic reducing conditions (Hayes, 1955; Macpherson *et al.*, 1958; Olsen, 1958; Pomeroy *et al.*, 1965). Under aerobic conditions, bacteria of the interface did significantly increase the microbial transport of phosphorus to the sediments (Hayes, 1955; Hayes and Phillips, 1958), and this loss was related to the amount of microbial phosphorus sedimenting to the interface (cf. also Frevert, 1979a).

Mobilization of sediment-associated P to the overlying water has therefore been ascribed largely to PO_4^{3-} release from Fe(III) oxide as these compounds are reduced by anoxic conditions in surface sediments and the overlying water (Mortimer, 1941; Boström *et al.*, 1982). When release of PO_4^{3-} is compared on a molar ratio to PO_4^{3-} liberated per unit of organic carbon mineralized by bacterial metabolism, the availability of alternate electron-acceptor compounds, such as sulfate (SO_4^{2-}), becomes as important within the interstitial waters of the sediments as the oxidative conditions of the overlying water (Hasler and Einsele, 1948; Sugawara *et al.*, 1957; Caraco *et al.*, 1989, 1990; Roden and Edmonds, 1997). The relative PO_4^{3-} release from sediments can be significantly higher as sulfate concentrations increase from natural or anthropogenic sources, particularly in oligotrophic, softwater lakes.

Iron sulfide formation coupled to sulfate reduction can reduce the abundance of Fe compounds that can complex PO_4^{3-} and thereby promote release of PO_4^{3-} into sediment porewater. When sulfate content is low or absent in anaerobic sediments, microbial reduction generates Fe(II) compounds from microbial reduction of Fe(III) oxide. PO_4^{3-} can be retained effectively with

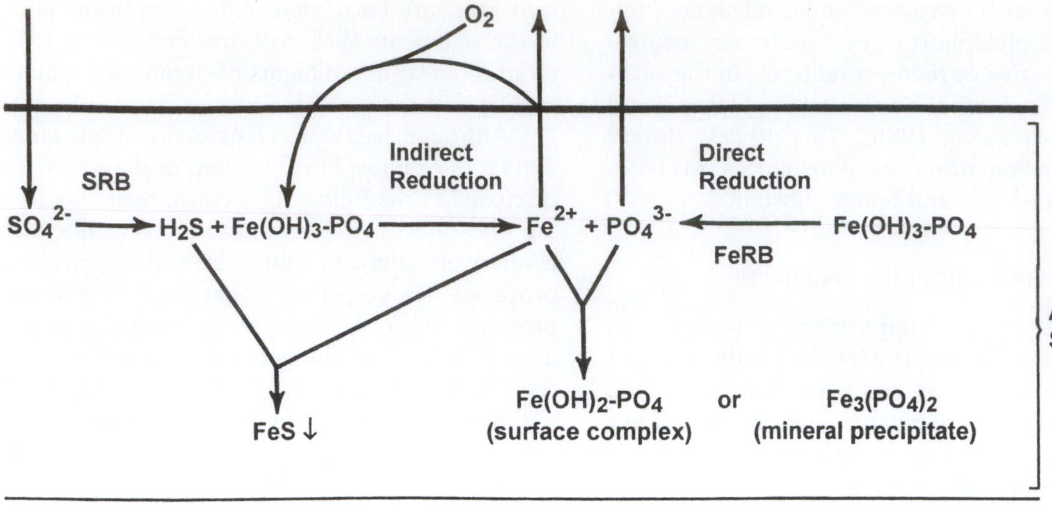

FIGURE 13-7 Interactions of sulfate on the reduction of Fe(III) and release of Fe(II) and phosphate. SRB = sulfate-reducing bacteria; FeRB = iron-reducing bacteria. (Modified from Roden and Edmonds, 1997.)

Fe(II) compounds but Fe-associated PO_4^{3-} is quantitatively released when amorphous Fe(III) oxide of the sediment is converted to iron sulfides (Roden and Edmonds, 1997). Conversion of reactive sediment Fe compounds to iron sulfides by sulfate-reducing bacteria leads to more efficient release of Fe-associated PO_4^{3-} than does direct microbial Fe(III) oxide reduction (Fig. 13-7).

Other processes can contribute to the composite release of Fe-associated PO_4^{3-} from sediments. For example, the uptake of excessive amounts of P by aerobic bacteria (e.g., Fleischer, 1986) and storage as polyphosphates could be rapidly degraded and released with the onset of anaerobic conditions (DeMontigny and Prairie, 1993; Gachter and Meyer, 1993). Because organic sediments are aerobic for only one or a few millimeters, this contribution is likely small in comparison to other mechanisms.

The sorption capacities of Fe(III) oxide decrease as pH levels increase above 6.5 (Stumm and Morgan, 1996). Large increases in pH values can occur within the sediments for several millimeters as a result of diurnal fluctuations in photosynthesis by epipelic algae (Fig. 13-8) and by submersed aquatic macrophytes and associated epiphytic algae. Simultaneous increases in oxygen concentrations (Fig. 13-8) and adsorption to or with photosynthetically induced precipitation of $CaCO_3$ (e.g., Mickle and Wetzel, 1978a) can counteract the availability of desorbed PO_4^{3-}.

The rate of phosphorus release from lake sediments increases (about doubles) markedly if the sediments are disturbed by agitation from turbulence (Zicker *et al.,* 1965). Covering anaerobic sediments with sand or polyethylene sheeting greatly impedes the loss of oxygen in the overlying water and decreases sediment release of phosphorus, iron, and ammonium (Hynes and Greib, 1970).

Algae growing on sediments are able to effectively utilize phosphorus from the sediments (Golterman *et al.,* 1969; Björk-Ramberg, 1986). Moreover, algae suspended in water with various particulate inorganic compounds of extremely low solubility were capable of extracting sufficient phosphorus for growth; without sediment phosphorus sources, the phosphorus content of the water limited algal growth under experimental conditions. The presence or absence of bacteria had little effect on algal utilization of phosphorus. These results stress the importance of extractable phosphates in the sediments if they are agitated into the water column, as in shallow lakes, even though their solubilities may be extremely low.

Some phosphorus release occurs under aerobic conditions from littoral sediments, particularly as temperatures warm above 10–15°C, as concentrations in the overlying water decline below equilibrium levels and as pH increases from intense photosynthetic activity of submersed macrophytes and attached algae (Twinch and Peters, 1984; Drake and Heaney, 1987; Boers and van Hese, 1988; Carlton and Wetzel, 1988; Boers, 1991; James and Barko, 1991). Because the volume of the littoral water is often small in comparison to that of the total lake, the littoral water often cools more rapidly at night than the pelagic zone. Convective flows from the littoral as interflow to the base of the epilimnion can move appreciable quantities of phosphorus from the littoral areas to the pelagic zone.

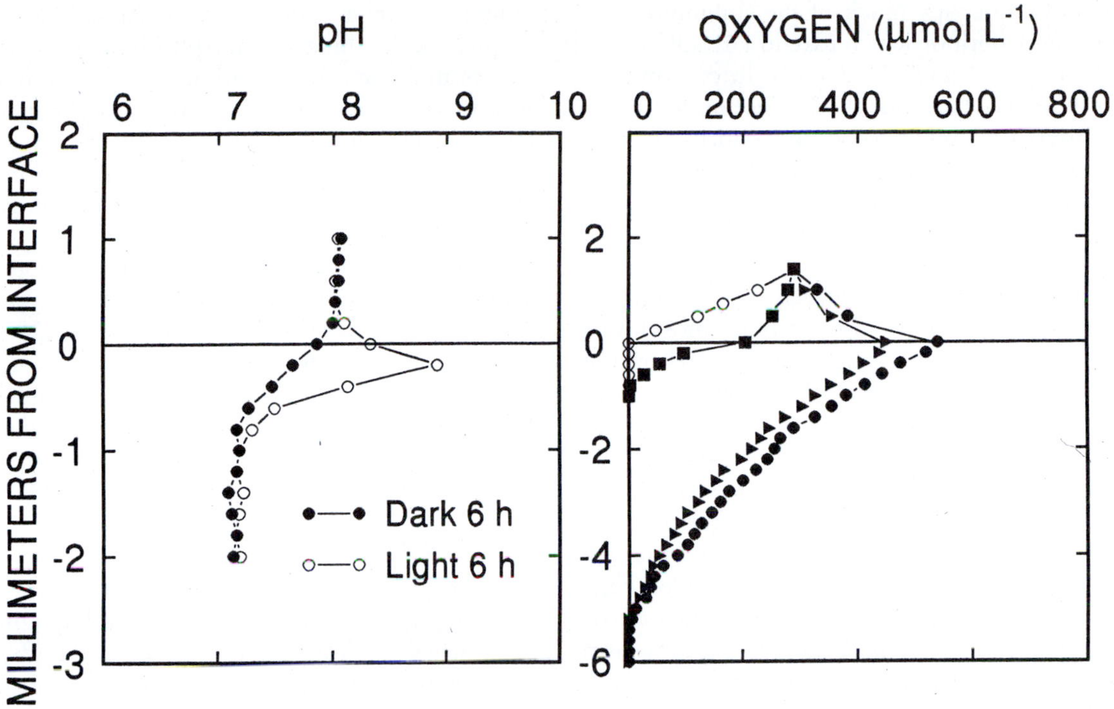

FIGURE 13-8 Distribution of pH (*left*; illumination 30 μmol quanta m^{-2} s^{-1}) and dissolved oxygen (*right*) immediately above and within sediments colonized by epipelic microalgae and bacterial communities. Oxygen microprofiles (*right*) during darkness (○—○), and after 1 h (■—■), 8 h (△—△), and 10 h (●—●) illumination with 10 μmol quanta m^{-2} s^{-1}. (Extracted from data of Carlton and Wetzel, 1987, 1988.)

D. Benthic Algae and Phosphorus Release from Sediments

Thus, oxygen is a primary factor influencing the release of phosphorus from sediments. If the sediments receive light, even at very low intensities (<50 μmol quanta m^{-2} s^{-1}), photosynthesis of epipelic algal communities growing on the sediments can quickly (minutes) produce high, often markedly supersaturated, concentrations of oxygen within the community usually <2 mm in thickness (Fig. 13-8) (Revsbech *et al.*, 1983; Carlton and Wetzel, 1987). This oxygen can diffuse several millimeters into the interstitial water of the supporting sediments at rates greater than it is consumed by bacterial respiration and chemical oxidations. By this mechanism, diurnal changes occur in the oxidized microzone of the sediments from fully oxidized in the daylight hours to fully reducing at nighttime. It was shown experimentally that the rate of phosphorus efflux from sediments is inversely related to the extent of sediment oxygenation and the magnitude of epipelic algal photosynthesis (Carlton and Wetzel, 1988). Although much of the reduced phosphorus efflux is caused by direct chemical redox changes, the microbial community also assimilates and complexes nutrients in organic compounds (cf. Kelderman *et al.*, 1988; Hansson, 1989).

E. Phosphorus Translocation by Migrating Phytoplankton

Several algae and cyanobacteria of lakes exhibit vertical migrations from nutrient-rich sediments or lower water strata to the euphotic trophogenic zone on a seasonal or daily basis. For example, a motile cryptomonad alga and a dinoflagellate migrate vertically from phosphorus-rich strata below the metalimnion at night to surface waters in early morning in sufficient quantities to enrich the epilimnion (Salonen *et al.*, 1984; Taylor *et al.*, 1988). Migrations of a large part of certain benthic-inhabiting cyanobacteria to the epilimnion have been implicated in translocation of benthic phosphorus sufficient to increase phytoplanktonic productivity (Barbiero and Welch, 1992).

F. Phosphorus Cycling Mediated by Aquatic Plants and Epiphytic Microflora

Lakes, reservoirs, and rivers are coupled with their drainage basins by intervening wetlands and littoral zones, the ***land–water interface zones,*** through which much runoff and seepage water flows before entering the main water bodies. The aquatic plants of the interface zones are major biotic components in many, if not most, lakes and in the flood plains of river ecosystems.

These plants often occupy much of the epilimnion and trophogenic zone, both of which extend virtually to the same depth in many small lakes. The littoral macrophytes and their attached microflora function mutualistically as conservative retainers of phosphorus and can modify greatly phosphorus loadings and budgets among aquatic ecosystems (Wetzel, 1990a).

1. Aquatic Plants

Emergent, rooted, floating-leaved and submersed higher aquatic plants take up most of their phosphorus from the interstitial waters of the sediments, where the concentrations are several orders of magnitude greater than in the overlying water (Denny, 1972; Barko and Smart, 1981; Brock *et al.,* 1983). Even though the structural and vascular systems of submersed aquatic angiosperms are simplified and reduced, nutrient uptake is largely by the root–rhizome system from the sediments (cf. Chap. 18). Early studies provided conflicting results. For example, the presence of the macrophytes *Eriocaulon* or *Utricularia* in experimental sediment–water systems increased the movement of radiophosphorus from the water to the sediments (Hayes, 1955). Absorption and translocation of phosphorus, studied in bacteria-free cultures of the macroalga *Chara* common to hardwater lakes, showed all parts of the plants could absorb ^{32}P about equally (Littlefield and Forsberg, 1965). About the same proportion of adsorbed phosphorus was translocated from the apices or from rhizoids to other parts of the plants. Similar results were obtained for other freshwater plants (e.g., Schwoerbel and Tillmanns, 1964; DeMarte and Hartman, 1974).

Rates of phosphorus uptake and excretion by the roots and leaves of submersed macrophytes were found to be dependent on the phosphorus concentrations of the medium (McRoy *et al.,* 1972; Kussatz *et al.,* 1984; Brix and Lyngby, 1985). Because the interstitial sediment phosphorus concentrations are very much greater than those of the water, uptake from the sediments predominates (e.g., Gabrielson *et al.,* 1984; Smith and Adams, 1988). Critical experimental studies of phosphorus uptake both from the sediments and from the water by submersed macrophytes that possessed their normal epiphytic microflora showed, however, that nearly all phosphorus was obtained from the sediments (e.g., Moeller *et al.,* 1988). Nearly all of the nutrient uptake from the water was by the attached microbiota and very little was transferred to the supporting macrophyte.

As foliage matures, a gradual, partial senescence of leaves often occurs, particularly among submersed macrophytes, from which nutrients leach. Some senescing leaves slough off of the living plants and collect among detrital accumulations of the sediments. For example, the leaching of phosphorus from dead macrophytes under sterile conditions was rapid and resulted in a loss of from 20 to 50% of total phosphorus content in a few hours and 65–85% over longer periods (Solski, 1962; Nichols and Keeney, 1973; Landers, 1982). Rates of leaching were greater from roots than from leaves.

Most aquatic angiosperms are herbaceous perennials with multiple cohorts (see Chap. 18). Upon completion of a growth cycle, a major portion (25–75%) of phosphorus of the foliage is translocated to the rooting tissues (cf. review of Granéli and Solander, 1988). Some of this phosphorus store can be permanently interred in senescing root tissues that decompose very slowly in anaerobic sediments.

2. Epiphytic Microbiota

The surfaces of submersed vegetation are colonized with epiphytic microflora. The physiology and growth of attached microflora are coupled to the physical and physiological dynamics of the living substrata upon which they grow. The productivity of epiphytic bacteria and algae is very high (cf. Chap. 19). This community acquires nutrients both from the water within and passing through the littoral zone and from the supporting "host" macrophyte tissues. Although relatively little of the total phosphorus pool within actively growing macrophytes is released, this release can be important to the epiphytes. For example, certain algal species that grow adnate to the macrophyte can obtain over 60% of their phosphorus from the macrophyte (Moeller *et al.,* 1988). Even when phosphorus concentrations are very high in the water, some nutrients are obtained from the macrophyte simply because diffusion within the complex epiphytic community is too slow to meet demands.

The phosphorus of the littoral water is very actively assimilated by the loosely attached epiphytic periphyton, incorporated into the periphyton, and intensively recycled (e.g., Riber and Wetzel, 1987; Wetzel, 1993a). The periphyton, rather than the submersed macrophytes, function as the primary scavenger for limiting nutrients such as phosphorus from the water.

During dormancy phases and senescence of aboveground macrophyte tissues, releases of phosphorus are readily utilized by the periphytic community, which tends to develop profusely during autumn and winter periods. As the senescing macrophytes with their epiphyte communities collapse to the detrital mass near the sediments, much of the released phosphorus is rapidly retained and recycled by epiphytic bacteria and algae (Fig. 13-9) (Wetzel, 1990a, 1993a). Therefore, the attached microflora, particularly that epiphytic on

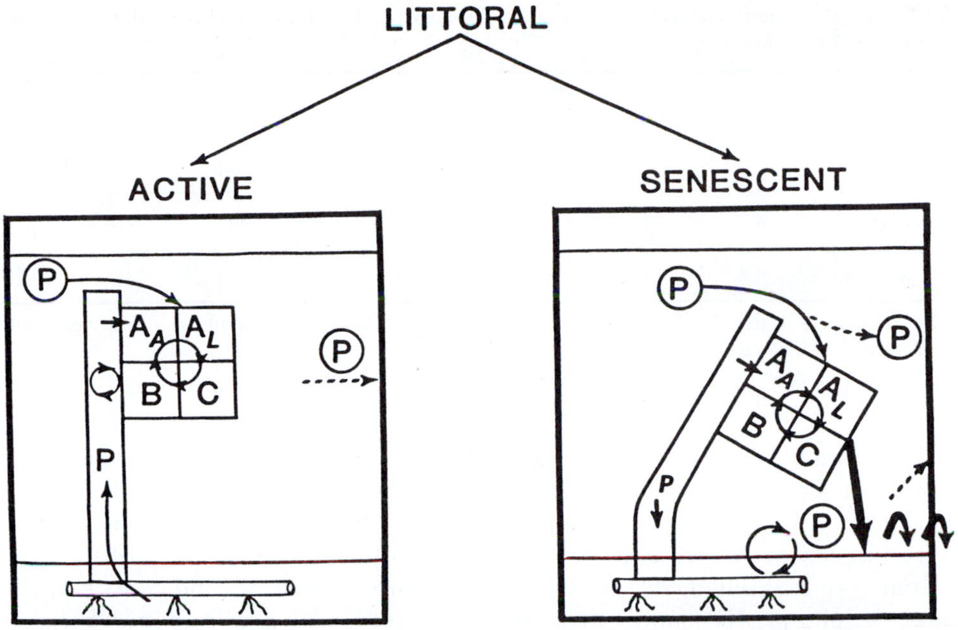

FIGURE 13-9 Fluxes of phosphorus (P) from the sediments to submersed littoral macrophytes and among the epiphytic microflora of the periphyton. A_A = adnate algae; A_L = loosely attached algae, B = bacteria; and C = inorganic or organic particulate detritus, such as calcium carbonate. (From Wetzel, 1990a.)

submersed macrophytes, can function as an effective phosphorus scavenger from the water (e.g., Howard-Williams and Allanson, 1981). The retention capacity is regulated by many environmental parameters and can be exceeded if the loading of phosphorus is very high or the rates of water flushing through this biological sieve is too rapid to allow time for uptake and retention. Rapid water movements through littoral zones, such as by natural storm events (Kairesalo and Matilainen, 1988) or from artificial human-induced runoff (e.g., Adams and Prenkte, 1982), can result in flushing movements of nutrients into the receiving open waters at rates too fast to be retained biologically. Once phosphorus is cycling within the submersed macrophyte–periphyton community, however, it is improbable that much of it will be exported to the open waters.

G. Littoral Phosphorus Fluxes and Loadings

Major sites of phosphorus fluxes in lakes has been examined by dividing them into three general compartments: (a) the pelagic open water and organisms of the epilimnion, (b) the littoral macrophytes and attached microorganisms, and (c) the hypolimnion and sediments. Early tracer experiments employing such compartmental analyses indicated that phosphorus in the epilimnion is extremely mobile (Rigler, 1964). The turnover time of phosphorus in the epilimnion (i.e., the time in which an amount of phosphorus equivalent to the total amount in a compartment leaves that compartment and a similar amount enters it) of a small lake was found to be 3.6 days (Rigler, 1956). Within 20 min of the time it entered, over 95% of the added phosphorus was taken up by the plankton; it had a turnover time of <20 min. In Toussaint Lake, a small acidic lake with a well-developed littoral zone of rooted aquatic plants, the littoral region was the most important contributor to the turnover and retention of phosphorus in the epilimnion; phosphorus was lost to this compartment 10 times more rapidly than to the hypolimnion and sediments and 50 times more rapidly than its loss through the outlet. In comparable experiments in an acidic bog lake in Nova Scotia with extensive developments of littoral *Sphagnum* moss, nearly all of the tracer phosphorus was taken up by the *Sphagnum* and plankton (Coffin *et al.*, 1949; Hayes and Coffin, 1951). Essentially no phosphorus reached the hypolimnion in the short term.

Rate constants of phosphorus loss from the epilimnion cannot be derived from these data. However, reanalyses of the results of several studies indicated that the turnover time of phosphorus in the epilimnion ranged from 20 to 45 days and was inversely correlated with areas of the lakes and estimates of development of the littoral vegetation (Table 13-4). In experimental pond systems with regulated steady-state inputs of phosphorus to and outputs from the epilimnion,

TABLE 13-4 Calculated Rate Constants of Phosphorus Transport in Three Lakes of Differing Littoral Development[a]

Lake	Area (ha)	Littoral vegetation (rank)	P turnover time (days)	Rate constants: k (out of epilimnion)	Rate constants: k (sedimentation from epilimnion)
Toussaint	4.7	1	20	0.05	0.01
Upper Bass	5.8	2	27	0.04	—
Linsley Pond	9.4	3	45	0.02	0.02

[a] Modified from Rigler (1973); lakes are ranked in order of decreasing amount of littoral vegetation, subjectively estimated.

filamentous algae of simulated littoral zones were major sites of phosphorus uptake before it reached the epilimnion (Confer, 1972). The equilibrium exchange between phosphorus inputs to or releases from the littoral flora and from or to the epilimnetic water varies with the amounts of phosphorus input from the drainage basin (Chamberlain, 1968).

Phosphorus uptake by and release from senescing littoral macrophytes and attached algae will vary with the physical constraints of littoral development determined, in part, by basin morphometry and water-exchange patterns. This relationship can shift with seasonal changes in active growth of the littoral macrophytes (e.g., Sarvala *et al.*, 1982). The physiology and growth of attached microflora are coupled to the physical and physiological dynamics of the living substrata upon which they grow. By means of intensive recycling of limiting dissolved nutrients and gases to maintain high sustained growth and biomass, losses are minimized and imported external nutrients can be directed primarily to net growth (Wetzel, 1990a, 1993a). Most aquatic plants are perennial with numerous cohorts and continuous growth and turnover, particularly in regions of moderate climate, and littoral growth is more or less continuous (cf. Chap. 18). Thus, the littoral complex of macrophytes and attached microbiota is often a major component of phosphorus retention and exchange with epilimnetic water, but it is constantly changing in the effectiveness of nutrient retention and loading.

H. Benthic Invertebrates and the Transport of Phosphorus

Evidence for how effectively benthic invertebrates can influence the rates and directions of phosphorus exchange between sediments and the water has been controversial. Clearly, these organisms can enhance rates of transport of phosphorus from the interstitial waters of sediments under certain environmental conditions. In other cases, their influences are minor. Examination of the transport mechanisms related to benthic invertebrate activities on phosphorus fluxes need to address (a) feeding activities of the invertebrates, intestinal decomposition, and excretion; (b) feeding and respiratory behavioral activities that alter redox gradients in sediments and phosphorus solubilities; and (c) movement of organisms from the sediment habitat to other habitats with displaced phosphorus release.

Benthic macroinvertebrates are extremely diverse in their feeding, respiratory, growth, and reproductive behaviors. This diversity contributes to the conflicting, inconclusive, and occasionally uncritical conclusions drawn on the effects of organisms on phosphorus and nitrogen fluxes (cf. reviews of Kamp-Nielsen *et al.*, 1982; Andersson *et al.*, 1988). Simply because phosphorus may be transported from interstitial waters to water immediately above the sediments does not necessarily mean that these compounds will be transported in available chemical forms for use in the trophogenic zones of lakes or rivers. One must retain an ecosystem scale perspective in evaluating these fluxes.

Dominant macroinvertebrate groups include the oligochaete worms, amphipods, bivalve mollusks, and immature chironomid midges. All inhabit and feed in the sediments differently (cf. Chap. 22). Oligocheate worms feed shallowly on surface sediments and do not burrow extensively. Amphipods inhabit and feed on the rich microbiota and detrital particles at the sediment–water interface. Bivalve mollusks move through and disturb surficial sediments and filter particles from the water. The chironomid larvae feed on surficial detrital particles or filter particles from water as they inhabit tubes within the sediments. Their undulating body movements transport water in and out of these tubes.

As populations of benthic invertebrates develop, phosphorus is incorporated into the fauna from the organic material fed upon in the sediments. Adsorption or direct assimilation of inorganic phosphorus is quantitatively insignificant, at least among the microcrustacea (Rigler, 1964). Ingested materials, however, are partially digested and released feces can accelerate the microbial regeneration of nutrients. Direct excretion of dissolved phosphorus, largely as inorganic phosphorus, occurs in small quantities; rates decrease markedly at temperatures below 15°C (e.g., Nalepa *et al.*, 1983; Fukuhara and Yasuda, 1985).

The movement of water by the respiratory and feeding activities of macroinvertebrates is often invoked as a primary means of transporting interstitial water of higher nutrient content from microbial remineralization activities in the sediments to the overlying water. Considerable experimental evidence supports such advective movements of water and particles mediated by benthic animals [***bioturbation*** (Aller, 1982)]. For example, chironomid midge larvae can cause increases in the phosphorus content of the overlying water (Gallepp *et al.*, 1978; Gallepp, 1979; Fukuhara and Sakamoto, 1987). Larger species cause a greater effect. The release of total phosphorus increased approximately linearly (0.3–9.4 mg P m^{-2} d^{-1}) over a range of 0–6585 larvae m^{-2}. In contrast, Davis *et al.* (1975) showed that tubificid worms only slightly influenced phosphorus release and may actually enhance phosphorus deposition at the sediment–water interface. Most of the phosphorus released to the overlying water was as soluble reactive phosphorus, which is readily inactivated chemically under oxidizing conditions or is readily assimilated by bacteria (Johannes, 1964a,b) or algae if light is available. Under oxygenated conditions of the sediment–water interface, only when larval invertebrate densities reach extreme levels (e.g., >100,000 m^{-2}; Lindegaard and Jnasson, 1979; Wisniewski, 1991) do their activities influence the transport of phosphorus across the sediment–water interface.

The role of macroinvertebrate activity at the sediment–water interface is unclear in relation to the transport of phosphorus to the water. Ciliates associated with the sediments are capable of hydrolyzing dissolved organic acids and releasing inorganic phosphate to the water (Hooper and Elliot, 1953). Low oxygen concentrations, however, not only produce an unfavorable environment for most ciliates but also inhibit the release of phosphate by the cells.

Negatively phototactic cladoceran and mysid zooplankton, which migrate to the sediment interface region during daylight, presumably feed actively on the rich microbiota of that region (e.g., Kasuga and Otsuki, 1984). The extent of phosphorus transport to the epilimnion during their subsequent nighttime migration is unknown but worthy of further investigation. Although the experimental addition of snails and ostracod microcrustacea to the sediments altered plankton abundance in ponds, these benthic organisms produced no corresponding changes in uptake rates of phosphorus in open water (Confer, 1972). When benthic invertebrates emerge as adults, they may emigrate from the sediments, thereby transporting phosphorus to other compartments of the ecosystem. For example, in streams a significant upstream migration of phosphorus by fish and invertebrates has been found (Ball et al., 1963a, 1963b). However, this displacement of phosphorus by invertebrates plays only a small part in the overall quantitative cycling of phosphorus in aquatic ecosystems.

I. Contributions of Birds and Fishes to Nutrient Loadings and Cycling

Vertebrate excreta, particularly of birds, are well known to import large quantities of nutrients, especially nitrogen and phosphorus, to inland waters. A particularly comprehensive treatise on the biogeochemistry of vertebrate excretion and global deposits and fluxes has been prepared by Hutchinson (1950). Clearly, nutrient loadings from waterfowl can be large as well as highly variable seasonally in relation to behavioral and migratory patterns of the use of rivers, wetlands, and lakes. A few examples illustrate the potential magnitude of nutrient loadings. In Wintergreen Lake, southwestern Michigan, Canada geese and ducks contributed 69% of all carbon, 27% of all nitrogen, and 70% of all phosphorus that entered the lake from external sources (Manny *et al.*, 1975, 1994). In a small kettle seepage lake of Massachusetts, nearly half the annual loading of phosphorus resulted from two species of gulls (Portnoy and Soukup, 1990). Up to 37% of the annual inputs of nitrogen and 95% of phosphorus of the largest natural lake in France, Grand-Lieu, emanate from large concentrations of resident and migratory birds (Marion *et al.*, 1994). Many other examples exist, but eutrophic conditions uniformly result from such high external loadings. Waterfowl loadings must be incorporated into nutrient load–response models, as proposed by Manny *et al.* (1994), in order to accurately evaluate levels of eutrophication under high bird-use environments.

Fish can impact the distribution of phosphorus and nitrogen in aquatic ecosystems in several ways. As we will see in subsequent discussions, the total amount of phosphorus in fish communities is small in comparison to other inorganic and organic pools in the lake ecosystem (water, littoral zone, sediments, and dissolved and particulate organic matter). However, fish activities can promote and accelerate the transport of phosphorus and nitrogen from sediments to overlying water.

Particularly, the feeding activities by benthic fishes, such as carp and bullheads (*Ictalurus*), can disturb sediments and increase diffusion to overlying waters (e.g., Lamarra, 1975; Keen and Gagliardi, 1981; Schaus *et al.*, 1997). Some phosphorus and nitrogen, obtained from food and detrital sources in the sediments, can be transported into upper strata of lakes and a portion released as soluble phosphorus and ammonia and urea nitrogen with urine and feces. When external loadings to lakes are small, as in dry summer periods, these internal redistributions can be important for planktonic algae and other microbiota (Brabrand *et al.*, 1990). Similarly in streams, fish-derived nitrogen and phosphorus can be moved upstream for large distances by migrating fishes and subsequently released and recycled for use by microbiota, algae, and macroinvertebrates (Schuldt and Hershey, 1995). Feeding history affects the N:P ratios of fish excrements. For example, rates of phosphorus excretion decreased more rapidly after feeding than those of nitrogen, and hence the N:P ratio of excrement increased with time from feeding (Mather *et al.*, 1995).

As discussed earlier, fish predation-induced shifts in the size structure of zooplankton communities can alter rates of sedimentation of seston from the epilimnetic trophogenic zone. Thus, loss rates of phosphorus via sedimenting seston can decline somewhat with a shift by predation to smaller-sized plankton (Mazumder *et al.*, 1989, 1992; cf. also Kairesalo and Seppälä, 1987; Pérez-Fuentetaja *et al.*, 1996). These effects appear to be significant only in very oligotrophic lakes.

Other large vertebrates can influence phosphorus and other nutrient loadings to small waters. For example, hippopotamuses graze on land at night and defecate considerable amounts of leachable phosphorus into lakes during the day (Kilham, 1982).

V. PHOSPHORUS CYCLING WITHIN THE EPILIMNION

The classical studies of Einsele (1941) and many subsequent theoretical and applied analyses of the circulation and fate of phosphorus in the open water of the epilimnion have shown that phosphorus is very rapidly incorporated into planktonic algae and bacteria. Recent study has focused on rates of movement of phosphorus among biologically important forms in lake water. Two broad classes of investigations have resulted: (1) biotically mediated phosphorus-transfer mechanisms and (2) abiotic complexation reactions. The former includes studies on the transfer of phosphorus among seston and various forms of dissolved phosphorus, including the biotically mediated formation of colloidal phosphorus and the enzyme-mediated utilization of dissolved phosphate esters. The latter category includes studies on the sorption and desorption of phosphorus to dissolved humic compounds, colloidal calcium carbonate, and other particles.

The analyses of Rigler and his co-workers have generated a framework that has stimulated considerable further study of phosphorus dynamics (Lean, 1973a, b; Lean and Rigler, 1973; Rigler, 1973). Lean proposed a quantitative, steady-state model of phosphorus exchanges in epilimnetic water in the summer, with the following composition (Fig. 13-10): (a) *particulate phosphorus,* the fraction removed when the water is filtered through a 0.5-μm pore-size filter, which contains the bulk of the phosphorus; (b) reactive inorganic *soluble orthophosphate* (PO_4^{3-}), which has an extremely short turnover time; (c) a *low-molecular-weight* (approximately 250 D) *organic phosphorus compound* (XP); and (d) a soluble macromolecular *colloidal phosphorus* of a molecular weight >5,000,000 D. An exchange mechanism predominates between the inorganic phosphate and the particulate fraction, but some phosphorus is excreted by the microorganisms in the form of the low-molecular-weight compound (XP). Polycondensation of the low-molecular-weight compound (XP) produces the high-molecular-weight colloidal compound. Both fractions, but primarily the latter, release phosphate in the soluble inorganic fraction, which then becomes available to the plankton.

The rate constants shown in Figure 13-10 demonstrate the rapidity of phosphorus dynamics in this algal–bacterial compartment for an exemplary lake in midsummer:

(a) k_1, the uptake of PO_4 = 0.9 relative mass units min^{-1} or 0.002 μg $liter^{-1}$ min^{-1} of total biologically active phosphorus. Some PO_4 is adsorbed to sestonic particulate detritus by physical–chemical processes (Rijkeboer *et al.*, 1991).

(b) k_2, the release of XP, and k_3, the binding of XP by condensation to colloidal P = 0.022 min^{-1}. The excretion of extracellular organic phosphorus compounds is well known among marine and freshwater algae during periods of rapid growth and during senescence (Kuenzler, 1970; Fogg, 1971; Lean, 1973a; Tarapchak and Herche, 1985).

(c) k_4, hydrolysis of colloidal P to PO_4 = 0.0017 min^{-1}. After the rapid formation of colloidal P via XP, in <2 min, subsequent additions of XP displace P from the colloid, again making the soluble inorganic PO_4 form available for uptake. Both colloidal P and XP fractions are not readily utilized as direct phosphorus sources by microorganisms, and they resist rapid attack from enzymes capable of hydrolyzing the most com-

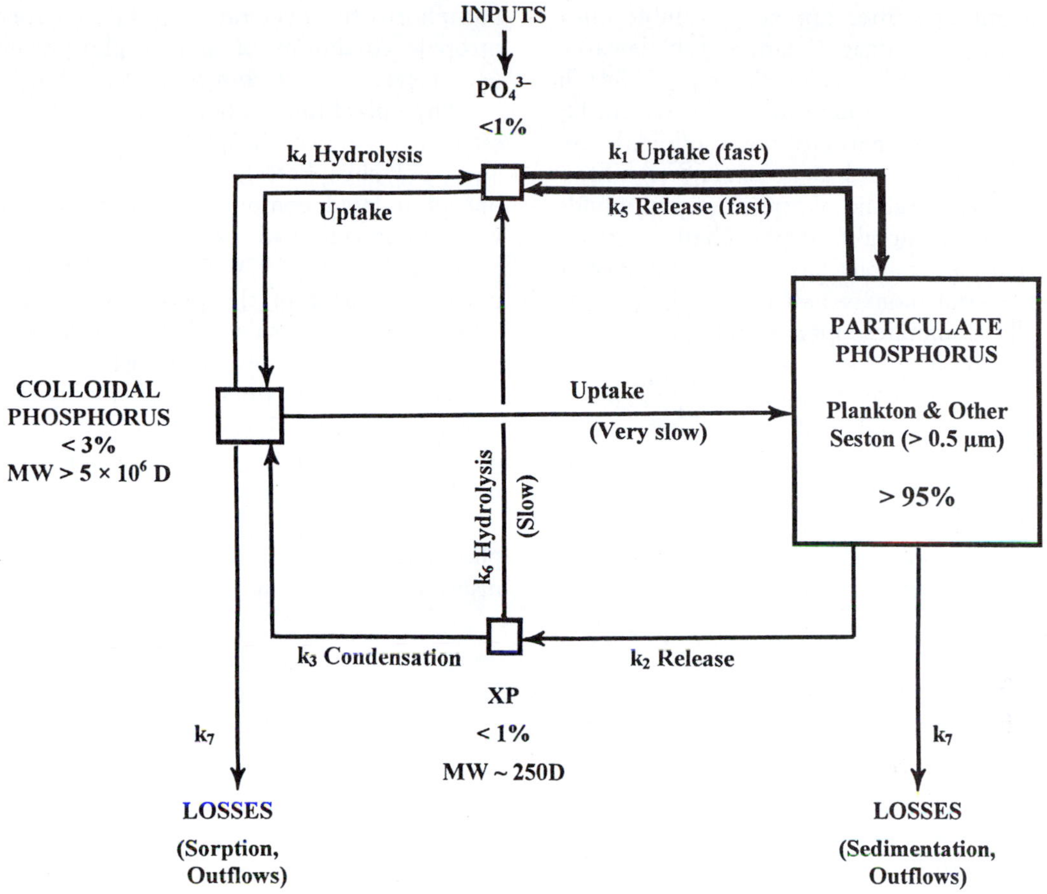

FIGURE 13-10 Phosphorus movement within the epilimnetic openwater zone of lakes, elucidating the exchange mechanisms between the phosphate and the particulate fractions. (Modified and expanded from Lean, 1973a,b.)

mon organic phosphorus compounds (Olsson and Jansson, 1984).

(d) k_5, the release of inorganic PO_4 by the particulate seston, was found to be very much greater than excretion of soluble organic phosphorus (XP). Mass balance calculations showed that k_5 is about 70 times k_2.

(e) k_6, the direct hydrolysis of XP to PO_4, was insignificant in comparison to release via the colloidal form.

(f) k_7, the rate constant for the loss of colloidal phosphorus, is small relative to the other rate constants. However, further aggregation of the colloidal fraction to particulate form or sorption to particulate matter followed by sedimentation represents a phosphorus sink from the epilimnion. Similarly, sedimentation of particulate phosphorus represents a slow but continuous loss from the epilimnion. Thus, during active growth of algae and heterotrophic bacteria, phosphorus must be continuously replaced through input from influents or from the littoral zone or recycled by zooplankton ingestion and subsequent excretion (see following discussion). In experimental systems, much of the colloidal fraction settles out and becomes biologically unavailable after one to five days. Other workers, however, have shown that high-molecular-weight colloidal phosphorus compounds can be used directly as a phosphorus source by algae possessing alkaline phosphatase enzymatic activity (Paerl and Downes, 1978). These rates of particulate loss are congruous with those found in natural waters to which phosphorus additions have been made.

From these initial insights and subsequent studies into the cycling of phosphorus in the open water, it is apparent that sestonic organic phosphorus must be separated into at least two fractions: (a) A rapidly cycled fraction, which is exchanged with soluble forms. In this fraction, phosphate is transferred rapidly through the particulate phase to low-molecular-weight compounds. (b) A fraction of the sestonic phosphorus that is released more slowly. The transfer of colloidal phosphorus material from the phytoplankton to the water appears within minutes after uptake of soluble phosphate by bacteria and algae (Lean and Nalewajko, 1976; Paerl and Lean, 1976; Cembella *et al.*, 1984). In addition to the formation of high-molecular-weight fibrillar and amorphous particles of colloidal size (0.05–0.1 μm), bacteria and algae excrete significant amounts of dissolved organic phosphorus compounds. As discussed later, the uptake of phosphate is greater by bacteria than algae, but since algal biomass is greater than bacterial biomass, algal phosphate incorporation usually dominates uptake and release pathways in the epilimnion.

The uptake and turnover kinetics of phosphorus have been studied in a number of lakes of low to high productivity and at different seasons of the year (Table 13-5). Phosphorus turnover rates are extremely rapid in the summer periods of high demand and relatively low loading inputs but become as much as two orders of magnitude slower during the winter periods. Phosphorus turnover rates are faster under more oligotrophic conditions of greater phosphorus deficiency (e.g., Peters, 1979; Cembella *et al.*, 1983, 1984).

Phytoplankton contain many complex phosphorus esters (e.g., sugar phosphates, nucleotide phosphates, and polyphosphates). These low-molecular-weight phosphorylated compounds can be released to fresh water upon cell death or by active release by microorganisms (Berman, 1970; Kuenzler, 1970). A very large portion (>50%) of the phosphorus content of algal cells can be lost during sedimentation over a period of several days (Otto and Benndorf, 1971). In the Lean model (Fig. 13-10), direct enzymatic hydrolysis of low-molecular-weight XP fractions is presumed to be negligible. In reality, more complex phosphorus regenerative mechanisms likely exist.

Enzyme-mediated hydrolysis of naturally occurring phosphate esters is one of the most plausible mechanisms for phosphorus regeneration in the epilimnion.

TABLE 13-5 Turnover Times of Phosphate in Fresh Waters[a]

Lake	*Turnover times (min)*	*Source*
Lago Maggiore, Italy		
Epilimnion, summer	10–200	Peters (1975)
Epilimnion, winter	200–10,000	
Hypolimnion	1000–100,000	
15 European lakes (epilimnia)	4–74,400	Peters (1975)
Southern Ontario lakes (epilimnia)		Rigler (1964); Planas and Hecky (1984)
Summer	2–14	
Winter	10–10,000	
Lake Kinneret, Israel		Halmann and Stiller (1974);
Summer	11–280	Halmann and Elgavish (1975)
Winter and spring	40–545	
East African lakes	1–1,000	Peters and MacIntyre (1976)
Lake 227, Ontario		Lavine (1975)
Summer	0.4–15	
Winter	120–11,700	
Lake 302, Ontario (summer)	5–26	Levine (1975)
Lake Memphremagog, Quebec		Peters (1979)
Summer	9–26	
Winter	83–770	
Jordan River, Israel (winter)	630	Halmann and Stiller (1974)
Southern Indian Reservoir, Manitoba	150–91,000	Planas and Hecky (1984)
Albertan lakes		Prepas (1983)
Shallow, mixed		
Spring	17–1020	
Summer	2–2160	
Deep, stratified		
Spring	3–42	
Summer	3–16	
Humic lakes, Finland		Jones (1990)
Summer	60–4800	
Winter	100–72,000	

[a] The turnover time is the reciprocal of turnover rate, which is the fraction of phosphate incorporated by the seston per unit time.

Direct enzymatic breakdown of dissolved inorganic phosphorus compounds and polyphosphates to dissolved inorganic phosphorus by relatively nonspecific phosphatase (phosphomonoesterase) activity has been the subject of intensive study. Numerous investigations demonstrate that membrane-bound algal phosphatase activity increases as phosphorus deficiency becomes acute. Increases in alkaline phosphatase activity (APA) under conditions of phosphate limitation, either from increased rates of enzyme synthesis or from derepression of preexisting enzyme, can enhance algal competitive ability by permitting algae to utilize organophosphate or inorganic polyphosphate substrates as alternative phosphorus sources (e.g., Berman, 1970; Jones, 1972; Heath and Cooke, 1975; Jansson, 1976; Stevens and Parr, 1977; Pettersson, 1980; Wetzel, 1981; Jansson *et al.,* 1988; Wynne and Rhee, 1988; Boon, 1994; Newman *et al.,* 1994). Although free, dissolved, alkaline phosphatase is short-lived (Reichardt *et al.,* 1967; Pettersson, 1980), substantial quantities of APA (often in excess of 50% of the total) can be found in dissolved phase (Pettersson, 1980; Wetzel, 1981; Stewart and Wetzel, 1981b; Cotner and Heath, 1988; Rai and Jacobson, 1993). The importance of phosphatase activity to phosphorus cycling in lakes is further confounded because bacterial production of alkaline phosphatase occurs in pelagic environments (Jones, 1972) but need not be induced strictly by phosphate limitation (e.g., Wilkins, 1972; Christ and Overbeck, 1987; Jansson et al., 1988). Several different types of phosphatases are released by higher organisms, such as zooplankton (Wynne and Gophen, 1981; Boavida and Heath, 1984). Their functions, particularly in phosphorus recycling and growth alterations of food sources, are unclear.

The enzyme 5′-nucleotidase catalyzes the hydrolysis of phosphoryl groups from 5′-nucleotides, such as ATP, from the dissolved organic phosphorus pool in natural waters (Ammerman and Azam, 1985, 1991a,b; Tamminen, 1989; Cotner and Wetzel, 1991b; Bentzen and Taylor, 1991). Although the synthesis of the external enzyme alkaline phosphatase, just discussed, is often repressed in high phosphate environments, 5′-nucleotidase is not. Greater 5′-nucleotidase activity has been associated with bacterial-size small microbiota (<1 μm), particularly in environments low in dissolved organic phosphorus concentrations. In high-phosphate environments, most of the phosphate regenerated by 5′-nucleotidase is not assimilated immediately. In comparisons of the different enzyme activities and rates of hydrolysis of different organic compounds, 5′-nucleotidase activity was found to have greater relative importance to phosphorus cycling in an oligotrophic lake than in a moderately eutrophic lake (Cotner and Wetzel, 1991b). The data suggested that nucleotidase activity may be more important to phosphorus regeneration in oligotrophic habitats than phosphatase activity.

Soluble APA and APA associated with algae and nonalgal particulate matter are highly variable seasonally and spatially (Wetzel, 1981; Stewart and Wetzel, 1981b; Cotner and Wetzel, 1991a; Wetzel, 1991). Epilimnetic APA is commonly high during the spring and summer periods of maximal phosphorus demand. During years when spring circulation is incomplete or a lake experiences temporary meromixis, extremely high APA and phosphorus limitations were found in the phytoplankton during spring and summer months (Wetzel, 1981).

Phosphomonoesters (PME) were found in lake water in substantial quantities (up to 55 μg PME-P liter^{-1} in eutrophic Twin Lakes, Ohio; Heath and Cooke, 1975). An inverse relationship was found between PME concentrations and APA during summer stratification. Potential hydrolysis rates of PME of >0.05 mmol per hour by APA were found. These values are similar to those found in marine systems using algal alkaline phosphatase as the hydrolytic enzyme (Rivkin and Swift, 1980). These results, in addition to widely differing chemical analyses of the complex phosphorus pool in lake water (e.g., Lean, 1973a, b; Lean and Nalewajko, 1979; Paerl and Downes, 1978; Downes and Paerl, 1979; Francko and Heath, 1979, 1981, 1982; Boavida and Marques, 1995), suggest that the composition of the soluble phosphorus pool varies greatly among different lake types. Although some lakes may approach the model of phosphorus cycling presented in Figure 13-10, other lakes may possess a phosphorus cycle dominated by the production and biotically mediated hydrolysis of PME.

Physicochemical mechanisms can also influence the cycling of phosphorus and, as a result, affect primary productivity. Certain complex humic materials of both low and high molecular weights contain phosphorus, such as phosphate esters and inositol hexaphosphate (e.g., Anderson and Hance, 1963; Koenings and Hooper, 1976). High-molecular-weight humic compounds can be photoreduced by low-intensity ultraviolet light such as occurs in sunlight (Francko and Heath, 1979; Stewart and Wetzel, 1981a). Orthophosphate adsorbed to ferric-humic compounds can be released by UV-induced photoreduction of ferric iron (Francko and Heath, 1982; Jones *et al.*, 1988). Low-molecular-weight humic compounds can release orthophosphate upon exposure to alkaline phosphatase.

Phosphorus–humic complexes can interact with phytoplankton and bacterioplankton in complex ways. For example, rates of carbon assimilation by phytoplankton were reduced when the algae were exposed to low concentrations of dissolved humic materials of littoral and wetland origin (Stewart and Wetzel, 1982). The effects were species-selective and greatest among smaller (1–5 μm) phytoplankton growing under low light conditions. The production of alkaline phosphatase activity was greatly stimulated by low concentrations of dissolved humic compounds of low molecular weight. Evidence indicates that these humic materials can act as sequestering agents for phosphate ions, organophosphorus compounds, and iron, thereby reducing phosphorus and possibly iron availability to phytoplankton (Jackson and Hecky, 1980; Chow-Fraser and Duthie, 1983; Jones *et al.*, 1988, 1993; de Haan *et al.*, 1990; Münster, 1994; Shaw, 1994). These relationships can be important to both phosphorus cycling and phytoplankton productivity, for major rain events can simultaneously introduce substantial quantities of phosphorus and dissolved humic materials into the pelagial areas of small and moderate-sized lakes.

Additionally noteworthy is that in acidified lakes where aluminum [Al(III)] and iron [Fe(III)] concentrations can be markedly increased, production of acid phosphatases is enhanced (Jansson, 1981). Although the Al(III) and Fe(III) at low pH do not affect the enzymes directly, the metal ions combine with the phosphate group on the substrate and inhibit enzymatic hydrolysis. The organisms may compensate, at considerable energetic cost, for these losses by increased production of phosphatases.

A. Phosphorus Losses by Outflow, Inactivation, and Sedimentation

Accurate nutrient budgeting in a lake or river ecosystem requires quantification of routes of reduced availability for utilization by biota as well as direct losses. Direct ***losses*** by movement with water ***in outflows*** from lakes and rivers occurs in inorganic and organic dissolved and particulate forms. Knowledge of concentrations and discharge volumes allows reasonably accurate evaluations of losses by outflow. Losses by seepage through the sediments, particularly sediments near the shore line, are quite variable (cf. Chap. 4). However, in most lakes, the basin seal is relatively complete and seepage outflows are small in relation to surface outflow. Most of the losses of phosphorus by seepage is via dissolved inorganic and organic compounds.

Sedimentation of nutrients occurs in settling (a) living and dead biota and (b) inorganic particles that are imported to the water from the catchment or are formed within the water and sorb nutrients. As particles settle to the bottom, they can experience partial decomposition and microbial recycling of nutrients to the water or encounter altered chemical conditions that change nutrient sorptive properties of the particles. Once at and within the sediments, periodic resuspension occurs, particularly at times of complete water circulation, during which several centimeters of the surficial sediments can be resuspended into the water column (e.g., Davis, 1973; White and Wetzel, 1975).

Most surface waters of the world are bicarbonate-dominated and many, perhaps a quarter of lakes, are very hard waters with sufficient bicarbonate and calcium to experience periodic precipitation of $CaCO_3$ (cf. chaps. 10 and 11). Precipitation is induced and accelerated by high pH, and these conditions are often associated with microzones adjacent to rapidly photosynthesizing cells of the phytoplankton, attached algae, and submersed macrophytes (Otsuki and Wetzel, 1972, 1974; House and Donaldson, 1986). Much of the phosphorus associated with the $CaCO_3$ is adsorbed onto the crystals; a small portion is incorporated by coprecipitation into the growing crystals. Inhibitors of $CaCO_3$ growth such as magnesium and iron reduce the amount of phosphorus coprecipitated on calcite (House *et al.*, 1986).

The morphology of calcite crystals can change markedly under different environmental conditions (Koschel, 1990, 1997; Raidt and Koschel, 1993). In particular, the complexity of the crystals increases with increasing supersaturation, and they shift from rhombohedral to complex dendritic and columnar crystals (Fig. 13-11). As the crystal complexity increases, the adsorptive surfaces also increase and are potentially more effective in the removal of nutrients and dissolved organic compounds.

Because phosphorus is coprecipitated with and adsorbed to nucleating $CaCO_3$, the process of formation and sedimentation of $CaCO_3$ can markedly alter phosphorus availability to biota. For example, about 35% of the annual total phosphorus removal from the epilimnion of Lake Constance in southern Germany was associated with autochthonous calcite precipitation (Kleiner, 1988), some 25% in eutrophic Wallersee of Austria (Jäger and Röhrs, 1990), and about 30% in hypereutrophic Onondaga Lake of central New York (Effler *et al.*, 1996). In hypereutrophic prairie lakes, very high concentrations of phosphorus are stripped from the epilimnetic waters by periodic precipitation of $CaCO_3$ (Murphy *et al.*, 1983).

B. Phosphorus and Nitrogen Uptake and Recycling by Protists and Zooplankton

As we have seen, the phosphorus content of bacteria is about 10 times that of the algae. Bacteria are

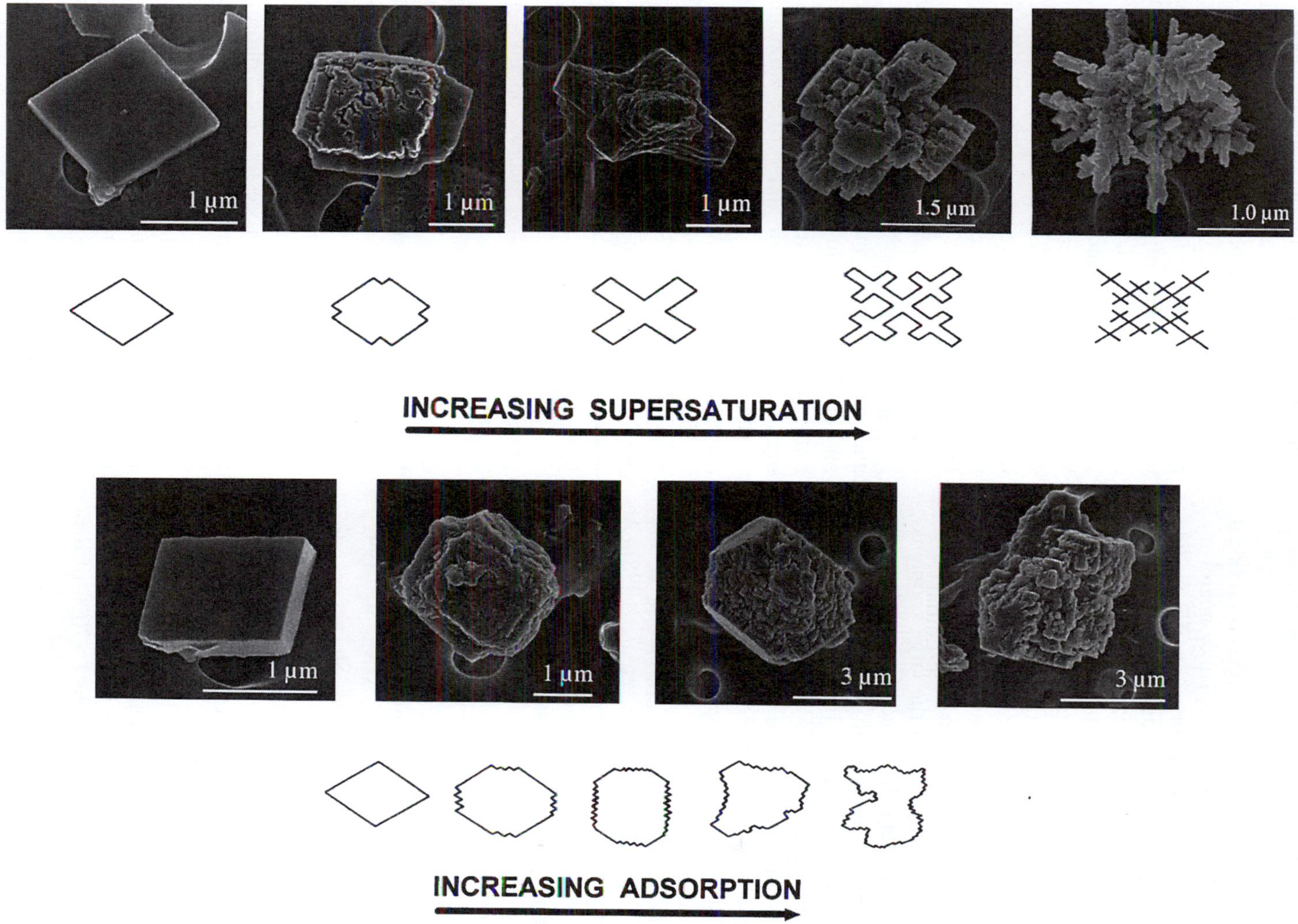

FIGURE 13-11 Calcite crystals from hardwater lakes under differing conditions of chemical saturation. Marked alterations of surface areas and adsorption also occur with changing morphology of the sedimenting crystals. (Photomicrographs courtesy of Prof. R. Koschel, Neuglobsow, Germany.)

commonly responsible for 80% or more of the net uptake of phosphorus (Hessen and Andersen, 1990; Jürgens and Güde, 1990; Vadstein *et al.*, 1993, 1995). Bacteria are often unable to achieve phosphorus saturation and thus compete vigorously with algae for inorganic phosphate.

Many studies have examined the ingestion of algal and bacterial particles by macrozooplankton and the potential regeneration of phosphorus by the feeding activities and from excreted feces, as discussed next. It should be emphasized that such uptake and excretion of phosphorus and other nutrients is a *recycling* process, not a *de novo* source of nutrients. The digestive and metabolic processes alter the storage time of the nutrients already in the ecosystem and potentially render it more available for reuse by other organisms.

Where the uptake and recycling by heterotrophic microflagellates were compared to the large zooplankton, the heterotrophic flagellates feeding on phosphorus-rich bacteria regenerated two-thirds or more of the phosphorus released by grazers (Jürgens and Güde, 1990; Vadstein *et al.*, 1993, 1995). Phosphorus and nitrogen regeneration rates varied, however, among fast- and slow-growing species of heterotrophic flagellates (Eccleston-Parry and Leadbeater, 1995) and were much higher during exponential growth phases than during stationary growth phases (Andersen *et al.*, 1986; Nakano, 1994a; Ferrier-Pagès *et al.*, 1998). A portion of the bacteria senesce and lyse and thereby release nutrients. However, many of the bacteria function as consumers of phosphate, because of their high requirements for phosphorus, and thus serve as phosphorus-rich particles for feeding by flagellates, other protista, and larger zooplankton. The release rates of phosphorus from the bacterivorous feeding of flagellates can account for major recycling sources for phytoplanktonic primary production (e.g., Miyajima, 1992; Rothhaupt, 1992; Nakano, 1994b; Sterner *et al.*, 1995). This recycling by planktonic consumers can be a significant source for the growth of bacteria and algae, particularly late in the summer period of stratification.

The uptake of nutrients, particularly phosphorus and nitrogen, by ingestion of food particles by herbivorous zooplankton and protists can be large at certain times of the year (cf. Chap. 16). During egestion and subsequent leaching from feces, nutrient release occurs in the form of soluble phosphate ions (>70% of dissolved P), ammonium nitrogen (>80% of dissolved N), and some organic phosphorus and nitrogen compounds, particularly urea (usually <10% of dissolved N) (Vargo, 1979; Lehman, 1980a; Mitamura and Saijo, 1980; Madeira *et al.*, 1982; Lenz *et al.*, 1986; Urabe, 1993; Ferrier-Pagès and Rassoulzadegan, 1994; Nakano, 1994a, b; Goma *et al.*, 1996; Ferrier-Pagès *et al.* 1998). The release rates are highly variable, but the rates of N release from crustacean zooplankton (range, 0.2–2 μg N mg DW^{-1} h^{-1}) increased with increasing food abundance, whereas the rates of P release (0.05–0.3 μg mg DW^{-1} h^{-1}) and N:P release ratios were affected directly by the N:P ratio of the food. Release rates of ammonium N were similar per unit biomass for ciliates and phagotrophic flagellates.

When prey food organisms were saturated with phosphorus, release rates from grazing zooplankton were in the range of 1.1–1.5 μg P (mg DW)$^{-1}$ h^{-1} (Olsen and Østgaard, 1985; Lehman and Naumoski, 1985; Jürgens and Güde, 1990; van Donk *et al.*, 1993). When planktonic bacteria and algae are phosphorus-starved and contain less phosphorus, however, release rates from grazers are about 0.05 μg P (mg DW)$^{-1}$ h^{-1}, <5% of the rates under P-sufficient conditions. Detrital particles were also found to contain very low concentrations of phosphorus (Olsen *et al.* 1986).

Appreciable quantities of phosphorus, nitrogen, and carbon remain in the particulate fecal materials and dead zooplankton that settle out of the water column. Loss rates of nitrogen or phosphorus from molting or dead crustacean and rotifer zooplankton were the most rapid (>2% per hour) during the first 24 hr following death (Krause, 1961, 1962; Scavia and McFarland, 1982; Ravera and Gatti, 1993). Thereafter, rates of nutrient and carbon losses decreased but usually >75% was lost in the subsequent six days. Release rates were somewhat faster in the warmer water strata than in the hypolimnion.

The quantitative significance of these sources of nutrient from micro- and macrozooplankton to phytoplanktonic productivity is unclear. Where examined carefully, zooplanktonic excretion of nitrogen likely constitutes only a few percent (<5%) of that required for gross production on an annual basis. Recycling of phosphorus could be appreciably higher, particularly during certain periods of extensive zooplanktonic grazing. The difficulty is that very few quantitative measurements of rates of nutrient regeneration by bacteria, protists, and other processes (e.g., photolysis) have been examined simultaneously with the estimates of release from zooplankton. For example, in eutrophic Lake Loosdrecht in summer, <30% of the phosphorus required by phytoplankton emanated from zooplankton excretion (Oude and Gulati, 1988). Most regeneration of phosphorus resulted from excretion and release from bacteria and algae.

To ignore nutrient regeneration by rotifers and protists as insignificant in comparison to cladocerans is not warranted. Although the individual biomass of the smaller organisms is less by two-thirds to an order of magnitude or more, densities are often several orders of magnitude greater than those of cladocerans. Specific excretion rates are generally higher as biomass decreases (Taylor, 1984). Estimates indicate that microconsumer protists and rotifers can regenerate 50% or more of the total phosphorus and nitrogen that is regenerated by the total zooplankton community (Ejsmont-Karabin, 1983; Lenz *et al.*, 1986; Gulati *et al.*, 1988; Oude and Gulati, 1988; Taylor and Lean, 1991).

Nutrients released by zooplankton are rapidly reassimilated by phytoplankton. Under certain circumstances, when the concentrations of phosphorus in the trophogenic zone are low, phosphorus and nitrogen regenerated by the herbivorous zooplankton can constitute a significant portion of the phosphorus and nitrogen requirements of the algae (reviewed by Lehman, 1980a,b; Gulati *et al.*, 1995). For example, based on nutrient budgets and *in situ* measurements, epilimnetic zooplankton in Lake Washington recycle 10 times more phosphorus and 3 times more nitrogen to the surface-mixed layer during the summer months (June to September) than enters from all external sources combined. Examples of the rates of regeneration of phosphate and ammonia by zooplankton are given in Table 13-6.

The magnitude of nutrient uptake and recycling depends on the nutritional status of the algal cells (Lehman, 1980a; Blažka *et al.*, 1982). Compared with

TABLE 13-6 Regeneration Rates of NH_4-N and PO_4-P by Some Freshwater Zooplankton[a]

Species[b]	Temperature (°C)	μg dry wt animal^{-1}	μg (mg dry wt)$^{-1}$ day^{-1} N	P	Source
Daphnia magna (F)	20–22	250.0	—	0.8	Rigler (1961)
D. rosea (F)	20–22	13.4	—	2.0[c]	Peters and Lean (1973)
D. pulex (F)	15, 20, 25	26.0	5.1	—	Jacobsen and Comita (1966)
D. pulex (F)	20	20	27.9	5.6	Lehman (1980a)
D. longispina (U)	15–22	200–250	—	3.4	James and Salonen (1991)
Daphnia spp. (U)	15	10–48	—	31	Olson and Østgaard (1985)
Mysis relicta (U)	4–7	—	4.5	3.3	Madeira *et al.* (1982)
Diacyclops thomasi	20	4.5	—	5.0	Bowers (1986)
Mostly cladocera (U)	5	1.0	—	2.0	Barlow and Bishop (1965)
	20	0.9	—	4.6	
Mostly copepoda (U)	5	3.6	—	1.9	
	20	2.9	—	6.0	
Rotifers (U)	15–25	—	—	0.10–83	Ejsmont-Karabin (1984)
Euchlanis dilatata	18–19	0.5–1.0	8.8	2.9	Gulati *et al.* (1989)
Mixed cladocera,					
Copepoda, rotifers (U)					
Epilimnion	3–24				Ferrante (1976)
>0.3 mm			—	58.4[d]	
<0.3 mm			—	132.8	
Hypolimnion	4–9				
>0.3 mm			—	120.1	
<0.3 mm			—	170.1	
Mixed epilimnetic zooplankton (U)					
27–28 Sep. '77	Approx. 20		11.9	1.7	Lehman (1980b)
29–30 Sep. '77			12.4	2.0	
30 Sept.-1 Oct. '77			16.7	4.3	
15–16 Sept. '77	Approx. 20		19.2	2.0[e]	
				3.1[f]	
Mixed zooplankton (U)	20	10–70	—	13	Bartell (1981)
	18–20	1–5	—	4.1	Oude and Gulati (1988)
	20	—	6–25	1–7.2	Urabe (1993)
	<15	—	0–85	6–55	Carillo *et al.* (1996)

[a] From the sources cited, especially Lehman (1980b); cf also Taylor and Lean (1982).
[b] Water for the incubations was either filtered (F) or unfiltered (U) to remove phytoplankton.
[c] Total P, including dissolved organic P.
[d] Annual mean excretion rates; values fluctuated 2 orders of magnitude over an annual period.
[e] Soluble reactive P.
[f] Dissolved P, including organic P.

nutrient-sufficient cells, nutrient-deficient algal cells assimilated phosphate and ammonia more rapidly and exhibited lower rates of remineralization.

Zooplankton grazing accelerates rates of phosphorus and nitrogen regeneration. It is also probable that planktonic bacteria, particularly less than 1 μm in size, and very small algae immediately sequester released phosphorus (Dodds *et al.*, 1991). Even though average phosphorus (and often combined nitrogen) concentrations in the epilimnion can be low, the release of nutrients from zooplankton can cause a microheterogeneous patchiness of nutrient concentrations, since the zooplankton distribution is far more clumped than the phytoplankton distribution.

Less clear is the importance of nutrient recycling to the overall rates of algal productivity on an annual basis. Removal of algae and bacteria by grazing zooplankton fluctuates widely on a seasonal and spatial basis (Chap. 16). Moreover, grazing of algae is often size-selective. This selectivity may enhance the availability of nutrients to those algae that are not as effectively grazed. For example, assimilation rates of phosphorus by various dissolved and particulate fractions and different zooplankton species were examined in a mesotrophic lake (Lyche *et al.*, 1996). The microconsumers (<20 μm) exhibited the largest specific phosphorus assimilation (0.28 day^{-1}) and regenerated about 45% ***of assimilated*** phosphorus. The larger microcrustacea, such as *Daphnia,* acted largely as sinks for phosphorus, whereas the cyclopoids regenerated rapidly the small amounts of phosphorus assimilated because of their active feeding on the microconsumers. The zooplankton contributed significantly to sedimentation of phosphorus at this time of the year (late summer) (ca. 1% day^{-1}).

The interactions are furthermore complicated by seasonal patterns of selective predation on zooplankton by planktivorous fishes. As indicated in Chap. 15, planktivorous predation often results in reductions in zooplankton size and therefore grazing rates on phytoplankton. The smaller zooplankton can have higher biomass-specific excretion rates of phosphorus than larger zooplankton (Vanni and Findlay, 1990). Although fish excretion of phosphorus is large per unit biomass, the collective biomass is very small (see Chap. 16). In other studies, rates of phosphorus release from pelagic zooplankton were reduced during June and July when fish predation reached its seasonal maximum and allochthonous supplies were at seasonal lows (Bartell and Kitchell, 1978; Schindler *et al.*, 1993).

Finally, it must be recalled that most of the phosphorus is bound in particulate matter: algae, bacteria, and detritus. Rates of nutrient regeneration from these sources as the phytoplankton senesce and die are poorly understood. Certainly, these organisms with short generation times rapidly recycle nutrients in the trophogenic zone before they settle to lower depths. The relative magnitude of their recycling to that of the zooplankton is not clear, but is variable both spatially, seasonally, and diurnally (Schindler *et al.*, 1993). Zooplankton grazing and concomitant nutrient recycling can enhance rates of primary productivity and algal growth during periods when heavy nutrient demands exceed loading inputs to the trophogenic zone from external sources.

VI. ALGAL REQUIREMENTS FOR PHOSPHORUS

Compounds containing phosphorus influence nearly all phases of metabolism, particularly in the energy transformation of phosphorylation reactions during photosynthesis (Rao, 1997). Phosphorus is required in the synthesis of nucleotides, phospholipids, sugar phosphates, and other phosphorylated intermediate compounds. Further, phosphate is bonded, usually as an ester, in a number of low-molecular-weight enzymes and vitamins essential to algal metabolism.

The importance of phosphorus in algal physiology is of special interest to the limnologist because phosphorus is the least abundant element of the major nutrients required for algal growth in a large majority (ca. 75%) of fresh waters. Furthermore, the most important form of phosphorus for plant nutrition is ionized inorganic PO_4. Although algae can utilize organic phosphate esters, such as glycerophosphates, this ability is variable among species, as is their ability to obtain phosphate groups enzymatically or by release of exoenzymes to the water for catalytic dissociation (see reviews of Krauss, 1958, Provasoli, 1958, and Overbeck, 1962a, 1963; Cembella *et al.*, 1984). Most phosphorus released to the water during active growth of algae is inorganic soluble phosphate and organic esters, which are, in turn, very rapidly recycled. During cell lysis and decomposition, most of the algal phosphorus released is organic and undergoes bacterial degradation (Kraus, 1964).

A. Phosphate Uptake and Light

The uptake of phosphate from the water is influenced by numerous external factors. Among many algae studied in culture, the initial absorption of phosphate and subsequent uptake are greater in the light, especially under CO_2 limitation. The result is a reversible accumulation of cellular phosphate, much of which can be released to the medium. Natural algal populations are more or less synchronized in their productive and growth cycles, and protein synthesis predominates in the initial portion of the daily light cycle (Soeder, 1965). In the green alga *Scenedesmus,* release of phosphate from the algae has been found to be greatest during the latter portion of the light period, both prior to and during cell division in the dark (Overbeck, 1962).

Absorption of phosphate per cell in the light is generally dependent on phosphorus concentration in the medium, within a range that is rather specific for a given algal species. Natural algal and bacterial populations, however, adapt to phosphorus-limited environments by synthesizing uptake enzymatic systems (half-saturation constants) of at least an order of magnitude below those detected in culture conditions (Tarapchak and Herche, 1986). Uptake by phosphorus-deficient cells in the dark is independent of phosphate concentrations (Kuhl, 1962, 1974). When phosphate is supplied in excess, the phosphorus content per cell remains nearly constant; the amount in excess of physiological needs is often incorporated and stored as polyphosphates (Zaiss, 1985). Nitrogen and carbon limitation can lead to intensive polyphosphate production, and such phosphate accumulation is stimulated by calcium (Siderius *et al.,* 1996). When phosphate concentrations become limiting, production of the polyphosphate-forming enzyme, polyphosphate kinase, decreases and the phosphate stores can be degraded in the light.

Phosphorus absorption rates are quite specific to groups of algae and often to species within a genus. Whereas nitrate absorption is independent of phosphate concentrations, optimal growth of many algae occurs at higher concentrations of phosphate when nitrate, rather than ammonia, is the nitrogen source. For example, studies of growth of the natural phytoplankton communities under gradients of nutrients indicated that diatoms were superior competitors for phosphorus and dominated at high silica-to-phosphorus ratios (Kilham, 1986; Tilman *et al.,* 1986). Green algae dominated at low Si:P ratios. In another example, phosphate uptake and transport through the cell membrane, the rate-limiting step, by cyanobacteria is strongly regulated by cation concentrations, particularly Ca^{+2} and Mg^{+2} (Falkner *et al.,* 1980; Budd and Kerson, 1987). Thus, phosphate uptake by cyanobacteria should be favored in hard waters with an abundance of divalent cations.

B. Phosphate Uptake and pH

Many species of algae exhibit optimal growth and uptake of phosphorus within a distinct range of pH of the medium. The pH may alter rates of phosphate absorption by direct effects on the activity of enzymes, on the permeability of the cell membrane, or by changing the degree of phosphate ionization. For example, uptake of phosphorus by the diatom *Asterionella* is greatest at pH values between 6 and 7 (Mackereth, 1953). Uptake rates of phosphorus have been correlated directly with the presence of numerous ions and compounds in the water, such as potassium, availability of micronutrients, and organic compounds. The mechanisms involved are not clearly understood.

C. Phosphorus Concentrations Required for Growth

From an ecological standpoint, the growth of algae in both natural and laboratory cultures exhibits dependency on the amount of available phosphorus and the rate at which it is cycled. Extensive investigations of minimal and maximal phosphorus concentrations, especially by Chu (1943) and Rodhe (1948), grouped freshwater algae into categories according to whether their tolerance ranges fell below, around, or above 20 $\mu g\ PO_4$—P $liter^{-1}$:

a. Species whose optimum growth and upper tolerance limit is below 20 $\mu g\ PO_4$—P $liter^{-1}$ (e.g., *Uroglena*) and some species of the macroalga *Chara*
b. Species whose optimum growth is below 20 μg PO_4—P $liter^{-1}$ but whose tolerance limit is well above that level (e.g., *Asterionella* and other diatoms)
c. Species whose optimal growth and upper tolerance limit is above 20 $\mu g\ PO_4$—P $liter^{-1}$ (e.g., green algae such as *Scenedesmus, Ankistrodesmus,* and many others)

In these studies, phosphorus concentrations of the culture media required for optimal growth were almost always higher than those of the water in natural habitats where the algae were growing. Many explanations were offered for this difference, such as the presence of unknown organic growth factors in the lake but not in the artificial media of the bacteria-free cultures. It is now apparent that the chemical mass of inorganic phosphorus in the water has relatively little relation to growth kinetics. Of overriding importance is the rapidity with which phosphorus is cycled and exchanged between the particulate phosphorus and soluble inorganic and organic phases, as discussed in detail earlier. Secondly, it is of great ecological importance that many algae, when provided with a sufficient supply of phosphorus, can absorb phosphorus in quantities far in excess of their actual needs (Kuhl, 1974). A portion of this phosphorus is lost in normal active growth both as inorganic and organic compounds and is recycled rapidly, at least in part. A large

TABLE 13-7 The Minimal Phosphorus Requirements per Unit Cell Volume of Several Algae or Cyanobacteria Common to Lakes of Progressively Increasing Productivity[a]

Algae	*Minimum P requirement, in μg mm^{-3} cell volume*
Asterionella	<0.2
Fragilaria	0.2–0.35
Tabellaria	0.45–0.6
Scenedesmus	>0.5
Oscillatoria	>0.5
Microcystis	>0.5

[a] From data of Vollenweider, R. A.: *Scientific Fundamentals of the Eutrophication of Lakes and Flowing Waters, with Particular Reference to Nitrogen and Phosphorus as Factors in Eutrophication.* OECD Report No. DAS/CSI/68.27, Paris, OECD, 1968, after numerous sources.

majority (>95%) of the phosphorus is in the particulate phase of algae and dead organic seston. Assimilation of surplus amounts of phosphorus by algae, often referred to as "luxury" consumption, can provide sufficient phosphorus to maintain algal growth in the epilimnion, even though the external concentration of phosphorus may be very low.

Of course, this recycling and utilization of stored phosphorus reserves cannot persist for very long. Losses continually occur from the colloidal phosphorus component, as well as from sedimentation from the particulate phosphorus component and with outflows of both dissolved and particulate components. Inputs of phosphorus to the system are needed continually, either from the littoral zone and turbulent transfer from lower-depth strata and sediments or externally, in order to sustain growth for extended periods. The amount of phosphorus input or loading, then, is important to the sustenance of algal growth and is one of many factors influencing the types of algae that are growing in a particular lake at a particular time of year.

It is therefore more relevant to the question of increasing algal productivity to view phosphorus concentrations in terms of total phosphorus, since most of the phosphorus is bound in the particulate component at any given time. From the few studies available in which common algal species of lakes of differing productivity were studied in relation to phosphorus requirements, the minimum phosphorus required per cell volume can be evaluated (Table 13-7). The colonial diatom *Asterionella formosa* is often found in oligotrophic waters, has a low phosphorus requirement, and can reach maximum cell densities at very low phosphorus concentrations. The quantity of phytoplankton, expressed as cell volume, that may be produced by 1 μg P liter^{-1}, is 2–5 mm^{3} liter^{-1} (Vollenweider, 1968). This quantity of algae is common in lakes of low productivity. *Tabellaria* and *Fragilaria* are diatoms that reach maximum population densities at concentrations of approximately 45 μg PO_4—P liter^{-1}, while *Scenedesmus* needs higher concentrations of around 500 μg liter^{-1}. The cyanobacterium *Oscillatoria rubescens* does not reach its maximum phosphorus content until the initial phosphate concentration of the media used is about 3000 μg PO_4—P liter^{-1}. It is of ecological interest to note that cations stimulate phosphate uptake by the cyanobacterium *Synechococcus* (= *Anacystis*). Calcium ions in particular cause a pronounced reduction in the half-saturation concentration for orthophosphate uptake by increasing the active transport of phosphorus into the cells (Rigby *et al.,* 1980; cf. also Lehman,1976).

Studies on the kinetics of phosphate uptake and growth of *Scenedesmus* in continuous culture have shown that phosphorus uptake velocity is a function of both the concentrations of internal cellular phosphorus compounds and the concentrations of the external substrate (Rhee, 1973). The apparent half-saturation constant (K_m) of phosphorus uptake was 0.6 μM (approximately 18 μg P liter^{-1}), whereas the apparent half-saturation constant for growth was less than K_m by an order of magnitude. Growth was a function of cellular phosphorus concentrations. Internal polyphosphate content appeared to regulate growth rate, particularly in the initiation of cell division. The activity of alkaline phosphatase, the primary enzyme involved in the release of phosphates from polyphosphates, exhibited a relationship that was inversely related to growth rate and was correlated to both polyphosphates and internally stored surplus phosphorus. During phosphate limitation, polyphosphates are mobilized for the synthesis of cellular macromolecules (protein, RNA, and DNA). Similar results were found for species of diatoms (Fuhs *et al.,* 1972) and for green algae and cyanobacteria (Healey and Hendzel, 1975; Senft, 1978).

D. Phosphorus Uptake and Competition by Algae vs. Bacteria

Early studies on the uptake kinetics and competition for phosphorus between planktonic algae and bacteria were inconclusive. For example, bacteria were found to have a lower affinity, in contrast to several other studies, for phosphate than did algae (Table 13-8), but could potentially outgrow the algae because of a more favorable surface-area-to-volume ratio (Fuhs *et al.,* 1972). Competition for phosphate between the alga *Scenedesmus* and the bacterium *Pseudomonas* indicated

TABLE 13-8 Phosphorus-Dependent Growth and Phosphorus Uptake Kinetics of Two Diatoms and Three Bacterial Species[a]

Algae and bacteria	*Mean cell Volume* (μm^3)	*Mean cell surface* (μm^2)	a_0 (10^{-15} *g-atom P*)	μ_m (*doublings per day*)	K_m (10^{-6} *g-atom P liter*$^{-1}$)	V_m (10^{-15} *g-atom P* μm^{-2} *day*$^{-1}$)
Diatoms						
Cyclotella nana	77.5	103	0.95	1.6	0.6	2.0
Thalassiosira fluviatilis	1570	826	12.5	1.6	1.7	7.3
Bacteria						
Corynebacterium bovis	0.71	6	0.19	4.8	6.7	7.7
Pseudomonas aeruginosa	0.41	4.2	0.10	48	12.2	17.9
Bacillus subtilis	0.39	3.9	0.15	12	11.3	12.5

[a] Modified from Fuhs *et al.*: Characterization of phosphorus-limited plankton algae. In Likens, G. E. (Ed.): *Nutrients and Eutrophication,* Milwaukee, American Society of Limnology and Oceanography, 1972.

a_0 = minimum P content per cell.

μ_m = growth rate during unrestricted growth at 22°C, other conditions at or near optimum.

K_m = Michaelis half-saturation constant of uptake mechanism.

V_m = maximum uptake rate for orthophosphate P per unit area of cell surface.

that algal growth was limited in the presence of the bacteria but that growth of the bacteria was little affected by the algae (Rhee, 1972). Competition, seen as cessation of exponential algal growth, was observed some time after the external phosphate concentrations were exhausted. The concentrations of stored internal polyphosphates within the algae decreased to a near-zero critical value at the time of suppressed algal and much faster bacterial growth.

A plethora of recent studies have examined the phosphorus uptake kinetics and specific growth rates of bacterioplankton in comparison to those of phytoplankton. Correlation analyses (Currie and Kalff, 1984a; Currie, 1990) and direct measurements of specific growth rates in relation to phosphorus availability and competition among bacteria and algae (Currie and Kalff, 1984b; Vadstein *et al.*, 1988, 1993; Toolan *et al.*, 1991; Cotner and Wetzel, 1992; Coveney and Wetzel, 1992) indicated the primary points: (a) Bacterioplankton have substantially higher phosphorus requirements than do phytoplankton; (b) Bacterial net consumption of phosphorus is larger (4–10 × higher) than that of the phytoplankton; (c) Bacteria contain much greater amounts of phosphorus (ca. 10×) than algae; (d) Phosphorus as well as or rather than organic carbon commonly limits bacterioplankton; and (e) Bacterioplankton can outcompete algae for phosphorus under a wide range of phosphorus supply rates.

Phosphate is the preferred substrate for uptake into both algae and bacteria (Cotner and Wetzel, 1992). Phytoplankton use both phosphate and dissolved organic phosphorus compounds, particularly at high substrate concentrations. At most natural concentrations, however, uptake of phosphate is completely dominated (>50 to nearly 100%) by bacteria (Currie and Kalff, 1984b; Güde, 1991; Cotner and Wetzel, 1992; Vadstein *et al.,* 1993). The phosphorus contained in the bacteria is then regenerated by death and autolysis of the cells or by feeding activities of heterotrophic flagellates. Because bacteria have high requirements for phosphorus, much of the phosphorus occurs in and is regenerated from these phosphorus-rich particles.

VII. HUMANS AND THE PHOSPHORUS CYCLE IN LAKES

A. Sources of Phosphorus

1. Atmospheric Precipitation

Phosphorus in atmospheric precipitation and particulate material fallout originates from fine particles of soil and rocks, from living and dead organisms, primarily as volatile compounds released from plants and from natural fires and the burning of fossil fuels (Newman, 1995). Phosphorus loadings from the ocean to the atmosphere are very small. Prevailing winds and the water of storms traveling from oceans onto land are usually low in phosphorus content but can be higher in coastal regions (Graham *et al.*, 1979). Generally, the amount of phosphorus in precipitation is less than that of nitrogen. In inland regions, the major source of phosphorus in precipitation is from dust generated over the land from soil erosion and from urban and industrial contamination of the atmosphere. In heavily fertilized agricultural regions, the phosphorus content of precipitation is much higher during the active growing season than in winter. Even in mountainous regions much higher loadings of phosphorus from atmospheric sources were observed in the summer periods (Chapin and Uttormark, 1973; Lewis *et al.*, 1985; Cole *et al.*, 1990). It should be noted additionally that nutrients accumulate in snowpacks and on the ice during winter and can be released rapidly in large amounts during the spring thaw (e.g., English, 1978; Adams *et al.*, 1979).

The phosphorus content of precipitation is generally low (ca. <30 μg P liter^{-1}) in unpopulated regions over land but increases considerably to well over 100 μg P liter^{-1} around urban–industrial aggregations. Atmospheric contributions of phosphorus in rain and dry deposition range from approximately 0.01–0.65 g m^{-2} yr^{-1} (0.1–6.5 kg ha^{-1} yr^{-1}), with most values in the lower portion (ca. 0.2) of that range (Kortmann, 1980; Gibson *et al.*, 1995; Newman, 1995). Atmospheric inputs can represent highly significant loadings when compared for example with the 0.07 g m^{-2} yr^{-1} value that Vollenweider (1968) estimated as a general permissible phosphorus loading rate for lakes. Values of 0.13 g m^{-2} yr^{-1} or higher are considered dangerous from the standpoint of eutrophicational control in lakes with a mean depth of <5 m. Atmospheric loadings of phosphorus can constitute 40% or more of the annual loadings (e.g., Kowalczewski and Rybak, 1981; Psenner, 1984).

2. Ground Water

The phosphorus content of ground water is generally low; average concentrations are about 20 μg P liter^{-1}, even in areas where soils contain relatively large quantities of phosphorus. The low phosphorus content is a result of the relatively insoluble nature of phosphate-containing minerals and the scavenging of surface phosphate by biota and soil particles.

3. Land Runoff and Flowing Waters

In general, the regional chemical characteristics of surface waters are closely related to the soil characteristics of their drainage basins (Keup, 1968; Vollenweider, 1968; Lal, 1998). Soils reflect the regional geological and climatic characteristics. Surface drainage is often a major contributor of phosphorus to streams and lakes. The quantities of phosphorus entering surface drainage vary with the amount of phosphorus in soils, topography, vegetative cover, quantity and duration of runoff flow, land use, and pollution.

The parent rock material from which soil evolves by weathering is extremely variable in phosphorus content, and this variability increases with the thickness and heterogeneity of the stratified soil layers. Basic igneous rock contains relatively little phosphorus as apatite; phosphorus percentages of other rocks are as follows: sandstone is approximately 0.02, gneiss 0.04, unweathered loess 0.07, andesite 0.16, and diabase 0.03. Limestone containing approximately 1.3% P and rare deposits of rock phosphate (10–15% P), in which biotic accumulations either from the organisms themselves or from guano were concentrated, are largely of sedimentary origin (cf. Hutchinson, 1950). Surface layers of soil are relatively rich in organic phosphorus from plant detritus in various stages of decomposition by soil fungi and bacteria. The exchange capacity of soils for phosphorus depends on the composition of the soil and increases with greater quantities of organic and inorganic colloids. Phosphorus is most available and easily leached from soils having a pH of 6–7. At lower pH values, phosphorus combines readily with aluminum, iron, and manganese. Therefore, where drainage basins are acidified from atmospheric pollutional sources, inputs of phosphorus in runoff from acidified soils are reduced (Jansson *et al.*, 1986). At pH 6 and above, progressively greater amounts of phosphate are associated with calcium as apatites and calcium phosphates.

The topography of the catchment basin influences the extent of erosion and subsequent export of nutrients. Flat lands with little runoff and relatively high infiltration rates contribute less nutrient load to runoff than similar lands with steeper gradients (Table 13-9). Further, the relative erosion is influenced markedly by the type of vegetation and use to which the land is put (Table 13-10). Disturbances to the land cover can also have marked effects on nutrient releases to runoff water. For example, phosphorus and nitrogen concentrations of stream waters increased from 5- to 60-fold over background levels within the first two days after forest wildfires impacted the stream catchments (Spencer and Hauer, 1991). Much of the nutrient loading leached from ash. Returns to background concentrations of nutrients required several weeks. The literature on this subject is large, and the reader is referred to the reviews examining phosphorus inputs from runoff by Vollenweider (1968), Biggar and Corey

TABLE 13-9 Concentrations of Nitrogen and Phosphorus in Runoff from Miami Silt Loam of Differing Gradients[a]

	g m^{-2} yr^{-1}	
Gradient (%)	*N*	*P*
8	2.0	0.06
20	4.25	0.2

[a] After data cited in Vollenweider (1968).

TABLE 13-10 Relative Erosion from Soil in Relation to Vegetative Cover and Land Use, Pacific Northwest of United States[a]

Crop or practice	*Relative erosion*
Forest duff	0.001–1
Pastures, humid region, or irrigated	0.001–1
Range or poor pasture	5–10
Grass/legume hayland	5
Lucerne	10
Orchards, vineyards with cover crops	20
Wheat, fallow, stubble not burned	60
Wheat, fallow, stubble burned	75
Orchards, vineyards, without cover crops	90
Row crops and fallow	100

[a] From data cited in Biggar and Corey (1969), after Musgrave.

(1969), Cooper (1969), Keup (1969), Griffith *et al.* (1973), Dillon and Kirchner (1975), Meyer and Likens (1979), Loehr *et al.* (1980), Beaulac and Reckhow (1982), Grobler and Silberbauer (1985), Likens (1985), Lewis (1986), Stibe and Fleischer (1991), Likens and Bormann (1995), and Lal (1998).

An attempt has been made to classify natural and agricultural areas on the basis of the quantity of nitrogen and phosphorus in runoff (Table 13-11). Applications of fertilizers and land management practices, both in agriculture and forestry, will modify and generally increase these values considerably (see, e.g., Clesceri *et al.*, 1986; Skaggs *et al.*, 1994; Jordan and Weller, 1996). Urbanization results in increases in the amount of phosphorus discharged to surface waters in approximately direct proportion to population densities (Weibel, 1969). Phosphorus originating from heavy residential fertilization, storm sewer drainage, and leaves (Cowen and Lee, 1973) can increase phosphorus inputs to surface drainage. For example, in the largest lake of southwestern Michigan, Gull Lake, approximately 24% of the total phosphorus entering this eutrophicating lake was from fertilization of lakeside lawns. The soils of 75% of the lawns, however, were saturated with phosphorus, making additions unnecessary (Moss, 1972b; Moss *et al.*, 1980). In contrast, loadings of phosphorus to a large, undisturbed arctic lake were nearly entirely from stream inputs, largely as dissolved P (Table 13-12). Over two-thirds of the phosphorus left the lake in surface outflows, also largely as dissolved P.

The land–water interface zone where terrestrial ecosystems meet rivers and lakes can alter loadings of both water and nutrients to receiving waters. For example, a land–water "buffer zone" of wetland and littoral vegetation of 20–30 m in width can remove nearly all imported nitrate by denitrification. Rates of denitrification depend on rates of nitrate loadings, carbon availability, and hydrology (Kadlec and Knight, 1996; Fennessy and Cronk, 1997). Plant communities provide a carbon source for denitrifying bacteria, can

TABLE 13-11 Export of Nitrogen and Phosphorus from Soils of Natural and Agricultural Areas

Drainage basin type	*Total dissolved inorganic nitrogen losses, kg N* km^{-2} yr^{-1}	*Total phosphorus losses, kg P* km^{-2} yr^{-1}	*Source*
Undisturbed temperate forest	ca. 200	ca. 2	Hobbie and Likens (1973)
Undisturbed boreal forest	90–160	3–9	Ahl (1975)
Cleared forest, igneous watershed	—	ca. 5	Dillon and Kirchner (1974)
Cleared forest, sedimentary watershed	—	ca. 11	
Cleared forest, volcanic watershed	—	72	
Pasture, low intensity	100–1000	8–20	Ahl (1975); Johnston *et al.* (1965)
Pasture, high intensity	2000–25,000	—	Harper (1992)
Arable, cereals	4000–6000	—	Harper (1992)
Arable, cash crops	4000–< 10,000	—	Letey *et al.* (1977)
Arable, intensive (UK, USA, Netherlands)	—	7–190	Cooke and Williams (1973); Kohlenbrander (1972); Johnston *et al.* (1965); Schuman *et al.* (1973)
Mixed upland (Northern UK)	530–630	ca. 34	Atkinson *et al.* (1986)
Groundwater leachates	427–638	35–93	Reynolds (1979)
Urban runoff	ca. 1000	ca. 100	Harper (1992)
General soil productivity			Vollenweider (1968)
Low	< 500	< 20	
Medium	500–2500	20–50	
High	> 2500	> 50	

TABLE 13-12 Phosphorus Budget for a Deep, Arctic Tundra Lake, Toolik Lake, Alaska[a]

	$mmol\ m^{-2}\ yr^{-1}$		
	Fractional P	*Total P*	*% of Total*
Inputs:			
Stream inflows			
Dissolved P	3.28		70.7
Particulate P	1.12	4.40	24.1
			94.8
Direct precipitation			
Dissolved P	0.09		1.9
Particulate P	0.15	0.24	3.3
			5.2
Total inputs			
Dissolved P	3.37		72.6
Particulate P	1.27	4.64	27.4
			100.0
Exports:			
Sedimentation from water column	1.4–1.7	1.4–1.7	30.1–36.6
Stream outflow			
Dissolved P	2.02		42.1
Particulate P	1.23	3.25	25.6
			67.7
Total outputs			
Dissolved P	2.02		42.1
Particulate P	2.63–2.93	4.65–4.95	56.6–59.2
			100.00

[a] Numbers rounded; from data cited in Whalen and Cornwell (1985)

immobilize many nutrients by assimilation and storage in tissues, and influence nitrification rates by creating oxidized rhizospheres where nitrification can occur (see Chap. 12). Plant removal of nutrients is often, but not universally, seasonal, but denitrification can proceed throughout the year in subsurface hydrology.

Surface runoff receives very large quantities of phosphorus from domestic sewage. Of course, loading of these waters varies greatly with population density, treatment of the sewage for nutrient removal, and points of discharge of effluents. Average values are given in Table 13-13. Industrial inputs, especially those associated with food processing, can be exceedingly high.

Ironically, cleaning detergents are major sources of phosphorus that contribute marked fertilization effects to many fresh waters. Where legally permitted, synthetic detergents include phosphate builders as a major constituent, mainly sodium pyrophosphate and polyphosphates, to complex and inactivate cations of water supplies and permit more effective cleaning action. Until very recently, when detergent phosphorus content was reduced or eliminated, from 7 to 12% of the weight of detergents was phosphorus. The amount of phosphorus used in the production of detergents is staggering; well over 2 million tons are produced annually in the United States alone. Although the percentage of phosphorus in detergents has been reduced somewhat, the phosphorus loading to many water treatment facilities is still extremely high and originates from a source that could be eliminated with relative ease. In a majority of existing treatment facilities, phosphorus has been reduced significantly, but not sufficiently to prevent accelerated productivity in many recipient

TABLE 13-13 Surface Water Loadings of Nitrogen and Phosphorus per Unit Area from Human Excrement and Other Sources Based on an Average of 12 g N Person^{-1} Day^{-1} and 2.25 g P Person^{-1} Day^{-1}[a]

Population density ($persons\ km^{-2}$)	*Nitrogen* ($g\ m^{-2}\ yr^{-1}$)	*Phosphorus* ($g\ m^{-2}\ yr^{-1}$)
50	0.22	0.04
100	0.44	0.08
150	0.66	0.12
200	0.88	0.16
300	1.32	0.24
500	2.20	0.40
1000	4.40	0.80
2500	11.00	2.00
5000	22.00	4.05

[a] After Vollenweider (1968).

lakes. Practical technology for nearly complete removal of phosphorus is available (cf. Vollenweider, 1968; Rohlic, 1969). Where phosphates are banned as a constituent of detergents, the phosphorus loadings to treatment facilities and surface waters decline by 50 to >80% (e.g., Maki *et al.,* 1984; Pallesen *et al.,* 1985; Hoffman and Bishop, 1994).

B. General Eutrophication

The word ***eutrophy*** (from the German adjective *eutrophe*) in general signifies to nutrient-rich and is greatly misused in contemporary dialogue.[2] Naumann (1919) introduced the general concepts of oligotrophy and eutrophy and distinguished them on the basis of phytoplanktonic populations. Oligotrophic lakes contained few planktonic algae and were common in regions dominated by primary rocks. Eutrophic lakes contained more phytoplankton and were common among more naturally fertile lowland regions in which human activity provided an increased supply of nutrients. Although at that time chemical methodology was crude or nonexistent for certain substances, Naumann emphasized that chemical factors, particularly phosphorus, combined nitrogen, and calcium, were primary determining factors. Limnological evidence since that time has shown a strong relationship between the biological dynamics of a lake and its nutritional level (concentrations of nutrients).

Shortly after the turn of the present century, Thienemann (summarized in 1925) found that in alpine and subalpine lakes, midge larvae, largely of the genus *Tanytarsus,* were characteristic of unproductive, deep lakes in which the hypolimnetic water lost little of its oxygen content during summer stratification. Eutrophic lakes (he later adopted Naumann's terms) were shallower and richer in plankton, and the oxygen-reduced hypolimnetic water was dominated by fauna, such as the midge *Chironomus,* which can tolerate very low oxygen concentrations. The extensive studies and conceptualizations of these earlier workers, when coupled with those of numerous others, formed a foundation upon which much regional limnology of the 1930s was based.

A plethora of subsequent limnological studies were directed toward evaluation of the characteristics of different lake types. The terminology for different lake types that developed in the ensuing two decades was indeed phenomenal. Many lake types were differentiated on the basis of indicator species. The basis for many of these descriptive classifications was founded in controlling physical and chemical parameters, while the basis for others was not. The terminology and dichotomies of classification that developed, while instructive, were excessive. Lakes were categorized in nearly every limnological aspect—geomorphological, physical, and chemical—and by indicator species or aggregations of nearly every group of organism from bacteria to fish. Lake types have been based even on the waterfowl that commonly are associated with them. As more lakes were studied, exceptions were found, which led to further splitting and name generation. Early integrative analyses, such as those of Thienemann (1925) and Naumann (1932), became less common as the terminology became more complex. If the details of classifying are carried to an extreme, lakes are so individualistic that one would soon require taxonomic keys for the many lake types (which in fact was proposed by Zafar, 1959).

The concept of trophic state was further elaborated by Naumann (1919, 1929) in which he emphasized the importance of the drainage basin to the functioning of a lake as an ecosystem. He stressed that the quantity, and later production, of phytoplanktonic algae was determined by several factors, particularly phosphorus and nitrogen. This production was translated to greater productivity of the higher trophic levels of the lake biology. Naumann also pointed out the importance of regional variations in algal production in relation to the geology of the drainage basins.

The terms *oligotrophy* for lakes of low production and *eutrophy* for lakes of high production evolved from similar concepts developed for bogs (Weber, 1907; Naumann, 1929, 1932; Hutchinson, 1969; Carlson, 1992; Carlson and Simpson, 1996). Naumann coupled trophic lake types to physical (temperature and light) and chemical (calcium, humic content, nitrogen, phosphorus, iron, pH, oxygen, and CO_2) factors that affected production. The criterion used for classification was production of organic matter, not the factors determining that production (Naumann, 1929). The rationale underlying this criterion was that production of phytoplankton in some of these "disharmonic lake types" was affected to a greater extent by factors other than the primary nutrients phosphorus and nitrogen (Table 13-14). The primary production criteria were combined with a classification system based on hypolimnetic oxygen concentrations in stratified lakes and the presence of benthic macroinvertebrates developed by Thienemann (1921; cf. Rodhe, 1975). Exceptions to the general classification scheme emerged quickly, particularly in relation to the rates of hypolim-

[2]Hutchinson (1973) has written an excellent general review of the subject and the development of the term in limnology. The semipopular book The *Algal Bowl-Lakes and Man* by Vallentyne (1974) complements collections of studies on eutrophication: *Eutrophication: Sources, Consequences, Correctives* (National Academy of Sciences, 1969), and *Nutrients and Eutrophication,* edited by Likens (1972).

TABLE 13-14 Lake Types Based upon Production of Organic Matter by Phytoplankton and Primary Physical and Chemical Determinants of That Production[a]

Lake type	Characteristics
Oligotrophy	Low production, associated with low phosphorus and nitrogen
Eutrophy	High production, associated with high phosphorus and nitrogen
Acidotrophy	Low production, associated with low phosphorus and nitrogen, but also with pH values <5.5
Alkalitrophy	High production, associated with high calcium concentrations
Argillotrophy	Low production, associated with high clay turbidity
Siderotrophy	Low production, associated with high iron content
Dystrophy[b]	Low production, associated with high humic color

[a] After Naumann (1929).
[b] The dystrophic lake type of low P and N but moderate to high humic organic content was described by Thienemann (1921).

netic oxygen depletion that are strongly coupled to temperature and differences in basin morphometry (cf. Chap. 6).

These problems can be avoided if the classification of lakes is made on the basis of production (biomass) rather than trying to incorporate all of the causal variables of limitation and biological structure, as well as biological and abiotic consequences of that production. A more recent development is to use single variables to define the trophic state of lakes. As we see throughout the topical discussions of this book, many variables are evaluated in the context of algal productivity (e.g., phosphorus loadings and concentrations, chlorophyll concentrations, algal productivity, algal biomass, hypolimnetic oxygen deficits, et cetera).

Combinations of many physical and biological variables as regulating trophic state have clouded true cause-and-effect relationships of rates of primary production. As Carlson (1984, Carlson and Simpson, 1996) emphasized, the trophic state concept is important as a critical organizing concept in limnology. The concept represents a continuum of trophic levels that is difficult to quantitatively delineate into subgroups. Organization along a trophic continuum is but one facet of lake typology and uses biomass or rates of production as the original, simplest, and most useful definition of trophic state. Biomass of phytoplankton can be estimated by a number of techniques from pigment concentrations to transparency (Secchi disk analyses). The annual mean rate of production is a much better criterion (cf. Chap. 15) but requires intensive and sophisticated methods for determination. ***Eutrophication*** is simply the alteration of the production of a lake along a continuum in the direction from low to high values, that is, from oligotrophy to eutrophy. The reverse process is termed ***oligotrophication.*** In contrast to annual mean production rate, which is relatively stable for a given lake ecosystem, biological food-web structure is highly susceptible to change along dynamic environmental gradients other than nutrients. No portion of the biological structure other than primary production is recommended as a criterion of the trophic concept.

The grouping of lakes must be left as a spectrum of states of lake metabolism between oligotrophy and eutrophy. Historical studies show that in many small lakes of temperate, glaciated areas, a succession is common from more inorganic sediment, containing oligotrophic-indicator fossils, to a more organic sediment, containing eutrophic-indicator fossils (Hutchinson, 1973). Evidence indicates that once organic sedimentation has become established after the initial oligotrophic phase, a type of trophic equilibrium or reasonably stable steady state occurs. This implies that nutrient inputs from the drainage basin are relatively constant over long periods of time and undergo only minor changes with oscillations in climate and inputs from vegetative cover and erosion. The systems can, and commonly do, regress in productivity (oligotrophication) as the surficial soils become reduced in nutrient content by leaching (cf. Chap. 23). Conversely, productivity can be greatly accelerated by increased nutrient inputs, particularly associated with human activities.

C. The Importance of Nutrient Loading to Aquatic Ecosystems

Great emphasis has been placed on nutrient availability in regulating aquatic productivity. As will be discussed later in treatments of biological productivity, many nonbiological factors other than phosphorus and nitrogen can regulate productivity. The early analyses of differences in productivity among aquatic ecosystems focused upon differences in nutrient concentrations. Increased information gradually demonstrated the rapid dynamics of exchange of nutrients among algae as nutrients were assimilated, utilized, and released during and following growth, reproduction, and death. Still among these studies, however, a prevailing concept was to treat lakes as closed, self-sustained entities. The most rapid change in these viewpoints arose, largely in the late 1960s and 1970s, with full recognition that rivers and particularly lakes are open systems critically dependent upon couplings with their drainage basins.

Phosphorus and nitrogen had been recognized for several decades in agriculture as critical nutrients that often limit plant growth and productivity, and they were similarly recognized in algal productivity. The importance of phosphorus and nitrogen in eutrophication of surface waters was also long recognized, but with the understanding that lakes are open systems it became apparent that the *rates* of loading (i.e., importation) of these nutrients from the drainage basin were critical to trophic conditions. From a management viewpoint, one could also define loadings of nutrients in terms of whether they were acceptable or excessive in terms of the algal productivity generated under the lake conditions of morphometry and water retention times (Vollenweider, 1990). The trophic categories of oligo- , meso- , and eutrophy were gradually analyzed in increasingly quantitative ways, in terms of nutrient concentrations, phytoplankton biomass, chlorophyll concentrations, and water transparency in addition to the classical characterization of hypolimnetic oxygen conditions. These data from many lakes allowed the development of dynamic quantitative modeling for the prediction of probable biotic conditions as a result of nutrient loadings from the drainage basin. A summary of some of these relationships, developed and extensively tested in the 1980s, is presented later.

In more recent times, much attention has been directed to the internal recycling of nutrients within lakes and streams. In particular, internal loadings of nutrients from accumulations in sediments have been examined to determine how this recycling of nutrients supplement those allochthonous sources from the drainage basin and atmosphere. In addition, as is summarized in this chapter, much attention is directed to the recycling of nutrients through food ingestion and alteration by animals. Unfortunately, there has been very little study of the rates of microbial nutrient regeneration, processes that certainly dominate in all aquatic ecosystems. This area of research, though difficult, is a dominant frontier in limnology and is fundamental to effective management of inland aquatic resources.

D. Effects of Phosphorus and Nitrogen Concentrations on Lake Productivity

Although many definitions of lake eutrophication exist, based on an array of conditions associated with increased productivity, the consensus among limnologists is that the term *eutrophication* is synonymous with increased growth of the biota of lakes and that the rate of increasing productivity is accelerated over that rate that would have occurred in the absence of perturbations to the system. The most conspicuous, basic, and measurable criterion of accelerated productivity is an increased quantity of carbon assimilated by algae and larger plants per given area.

Under most lake conditions, the most important nutrient factors causing the shift from a lesser to a more productive state are phosphorus and nitrogen. Typical plant organic matter of aquatic algae and macrophytes contains phosphorus, nitrogen, and carbon approximately in the ratios (Vallentyne, 1974):

1P:7N:40C per 100 dry weight, or

1P:7N:40C per 500 wet weight

If one of the three elements is limiting and all other elements are present in excess of physiological needs, phosphorus can theoretically generate 500 times its weight in living algae, nitrogen 71 (500:7) times, and carbon 12 (500:40) times.

A comparison of the relative amounts of different elements required for algal growth with supplies available in fresh waters illustrates the general importance of phosphorus and nitrogen (Table 13-15). Similar proportional ratios were demonstrated by comparison of the demand among terrestrial plants with the accessible supply of elements from the lithosphere (Hutchinson, 1973). Even though variations in conditions of solubility or availability may at times make very abundant elements (e.g., silicon, iron, and certain micronutrients) almost unobtainable, phosphorus, and secondarily nitrogen, are generally the first to impose limitation on the system. This relationship is emphasized further by consideration of average demand–supply ratios in late winter prior to the spring algal maximum common in lakes of temperate regions and in midsummer during maximum sustained algal productivity (Table 13-16).

Oligotrophic lakes are often limited by phosphorus and contain nitrogen in quantities in excess of demand from growth supported by available phosphorus. As the lakes become more productive, the primary effecting agent is increased loading of phosphorus. As discussed earlier, the instantaneous concentrations of phosphorus usually decrease and are quite variable, but the turnover rate increases markedly. Rates of loss increase, and high algal productivity requires sustained phosphorus loading of the system from allochthonous and littoral sources.

Some studies have indicated that nitrogen rather than phosphorus is limiting to phytoplanktonic productivity of lakes, at least in certain water strata and at certain times of the year (e.g., Elser *et al.*, 1990). Arguments are moot that these findings contradict the general tenet of phosphorus as the foremost macronutrient to limit algal photosynthesis. Combined nitrogen, although found in large quantities in the lithosphere, is largely unavailable to plants. Nitrogen supplies to fresh waters are augmented more readily from external sources (Chap. 12) than are those of phosphorus.

TABLE 13-15 Proportions of Essential Elements for Growth in Living Tissues of Freshwater Plants (Requirements) in the Mean World River Water (Supply) and the Approximate Ratio of Concentrations Required to Those Available[a]

Element	*Average plant content or requirements* (%)	*Average supply in water* (%)	*Ratio of plant content to supply available*
Oxygen	80.5	89	1
Hydrogen	9.7	11	1
Carbon	6.5	0.0012	5,000
Silicon	1.3	0.00065	2,000
Nitrogen	0.7	0.000023	30,000
Calcium	0.4	0.0015	<1,000
Potassium	0.3	0.00023	1,300
Phosphorus	0.08	0.000001	80,00
Magnesium	0.07	0.0004	<1,000
Sulfur	0.06	0.0004	<1,000
Chlorine	0.06	0.0008	<1,000
Sodium	0.04	0.0006	<1,000
Iron	0.02	0.00007	<1,000
Boron	0.001	0.00001	<1,000
Manganese	0.0007	0.0000015	<1,000
Zinc	0.0003	0.000001	<1,000
Copper	0.0001	0.000001	<1,000
Molybdenum	0.00005	0.0000003	<1,000
Cobalt	0.000002	0.000000005	<1,000

[a] After Vallentyne, J. R.: *The Algal Bowl—Lakes* and *Man,* Miscellaneous Special Publication 22, Ottawa, Dept. of the Environment, 1974.

When phosphorus is available in quantities adequate to support metabolism, nitrogen availability can become limiting, particularly to phytoplankton. This limitation is often observed in the two extremes of the trophic spectrum. Under eutrophic conditions of high phosphorus loading, planktonic utilization of combined nitrogen can exceed inputs and literally deplete combined nitrogen supplies in the trophogenic zone in a few days or weeks (cf. Chap. 12). When this occurs, molecular nitrogen fixation by heterocystous cyanobacteria can augment the income of nitrogen, but only at a high metabolic cost. In tropical lakes of low latitudes nitrogen limitation to phytoplanktonic growth can occur (cf. reviews of Salas and Martino, 1991, and of Lewis, 1996). Phytoplanktonic productivity of tropical waters is generally higher than at temperate latitudes (cf. Chap. 15), and higher temperatures can accelerate rates of denitrification and losses of nitrogen to the atmosphere.

TABLE 13-16 Comparisons of the Ratios of Required Concentrations of Inorganic Nutrients to Average Supplies Available in Fresh Waters[a]

Element	*Ratio of demand to available supplies*	
	Late winter	*Midsummer*
Phosphorus	80,000	Up to 800,000
Nitrogen	30,000	Up to 300,000
Carbon	5000	Up to 6000
Iron, silicon	Generally low, but variable	
All other elements	<1000	<1000

[a] After Vallentyne, J. R.: *The Algal Bowl—Lakes* and *Man,* Miscellaneous Special Publication 22, Ottawa, Dept. of the Environment, 1974.

Under oligotrophic conditions, particularly in lakes of mountainous regions and high-latitude waters, phosphorus availability may be adequate to support modest levels of phytoplanktonic productivity, but sources and loadings of combined nitrogen are insufficient to support such productivity. Increasing the availability of nitrogen can result in increased algal photosynthesis until phosphorus again becomes limiting. Enrichments with both nitrogen and phosphorus can stimulate algal productivity markedly (e.g., Holmgren, 1984; White *et al.,* 1986; Elser *et al.,* 1990).

Phytoplanktonic productivity of reservoirs often fluctuates to a much greater extent than in natural lakes (cf. Chap. 15). Because of the frequent variations in inorganic turbidity in reservoirs, the limiting effect of light penetration on photosynthetic activity is more severe than that of nutrients (e.g., Henry *et al.,* 1985).

Nutrient loading to reservoirs from external sources can vary with long-term precipitation patterns and water inflows in comparison to the volumes of the productive zones. If flushing rates are high, turbidity can increase but also nutrient loading and availability can increase (e.g., Hoyer and Jones, 1983; Turner *et al.*, 1983). The availability of combined nitrogen and particularly phosphorus can be shown to intermittently weakly limit photosynthesis of phytoplankton in reservoirs, but light is generally of much greater limitation.

The potential for limitation of eutrophication by inorganic carbon availability has been a subject of much discussion (cf. Chap. 11). In a few exceedingly productive situations, such as sewage lagoons, in which phosphorus and nitrogen compounds are available in excess of any demands, carbon can become limiting to algal growth (Kerr *et al.*, 1972). While there is some evidence for inorganic carbon limitation in exceedingly soft lakes (see, e.g., Allen, 1972), diffusion of atmospheric CO_2 is generally adequate to sustain the carbon requirements of phytoplanktonic populations. It is extremely unlikely that carbon limitation is of significance in the limitation of algal populations in a majority of harder waters, in which dense algae develop under eutrophic conditions.

The importance of phosphorus in comparison to nitrogen and carbon has been particularly well illustrated by large-scale fertilization experiments (Schindler, 1974). Lake 226, located in northwestern Ontario in Precambrian Shield bedrock, was chemically and biologically similar to >50% of the waters draining to the northern Laurentian Great Lakes. The trophogenic zone of Lake 226 was partitioned into two lakes at a constriction in the basin (Fig. 13-12). One basin was fertilized with phosphorus, nitrogen, and carbon, and the other with equivalent concentrations of nitrogen and carbon.[3] The phosphate-enriched basin quickly became highly eutrophic, while the basin receiving only nitrogen and carbon remained at prefertilization conditions (Fig. 13-12). In this phosphate-enriched basin and in several other lakes receiving analogous treatments over a period of several years, algal biomass increased to 2 orders of magnitude over that of lakes receiving only nitrogen and carbon enrichment. Recovery to near prefertilization levels was very rapid when phosphate additions were discontinued.

The amount of evidence for these relationships is so overwhelming, both from experimental and applied investigations, that it is difficult to appreciate how the bitter controversy over the importance of carbon rather than phosphorus as a limiting nutrient in fresh waters developed in Canada in the late 1960s. As a result of a combination of the magnification of a few findings from cyanobacterial cultures and a few rare cases of discovered low carbon availability in natural habitats in which excessive amounts of nitrogen and phosphorus occurred, phosphorus was implied to be of less importance in the acceleration of eutrophication than carbon. So convincing was the misinterpretation of results by a few scientists, and so effective was the exploitation of the situation by the detergent industry and irresponsible journalists, that urgently needed legislation to reduce the effluent loading of surface waters with phosphorus was seriously impeded.

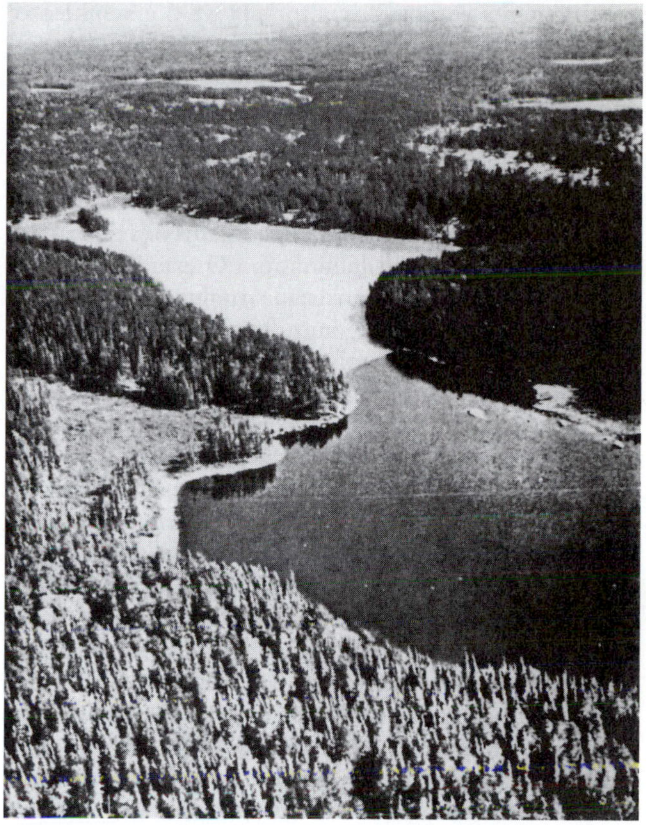

FIGURE 13-12 Lake 226 of northwestern Ontario, which was partitioned at the constriction of the basin, 4 September 1973. The far northeastern basin, fertilized with phosphorus, nitrogen, and carbon, was covered by a dense algal bloom within two months. No increases in algae or species changes were observed in the near basin, which received similar quantities of nitrogen and carbon but no phosphorus. (From Schindler, D. W.: Eutrophication and recovery in experimental lakes: Implications for lake management. *Science, 184:* May 24, 897–899, 1974. Copyright 1974 by the American Association for the Advancement of Science. Photograph courtesy of D. W. Schindler.)

[3]Additions were equivalent to 3.16 g of NO_3—N and 6.05 g of sucrose C per m^2 per year in both basins, made in 20 equal weekly increments. The northeast basin additionally received 0.6 g m^{-2} yr^{-1} of PO_4—P. The N/P and C/P ratios were greater than in treated sewage, but the quantity of P added was not exceptionally high for lakes that commonly receive pollution from domestic sources.

The matter was finally put to rest by the rebuttal of Vallentyne (1970), numerous subsequent investigations, and a national symposium on the subject (Likens, 1972). The effects of the controversy, however, continue to be felt both in research on the subject and in the efforts to reduce the loading of fresh waters with millions of tons of phosphorus each year, so that ultimately lakes will not become, in the words of J. R. Vallentyne, algal bowls.

E. Stoichiometry of Carbon, Nitrogen, and Phosphorus in Particulate Organic Matter

The elemental chemical composition of planktonic particulate organic matter reflects both the planktonic community structure as well as biochemical processes, such as nutrient supply ratios and turnover rates, that occurred in formulation of the composition of the biomass (reviewed in Tilman *et al.*, 1982; Kilham, 1990). For example, phosphorus or nitrogen limitation can lead to alterations in the ratios of protein, carbohydrate, and lipid content of cells, as well as the C:N ratio (cf. discussion in Chap. 12). Hence, the C:N ratio indicates not only characteristics of the nutrient availability but also an approximate evaluation of the relative proportions of cellular proteins and nonprotein structural elements (cf. Vollenweider, 1985).

A relative constancy in the molar ratio of C:N:P of 106:16:1 (or 41:7.2:1 by weight) among marine plankton, termed the Redfield ratio, is generally supported by numerous studies (Redfield *et al.*, 1963; Hecky *et al.*, 1993). The variance in this ratio is small, usually <20%. This constancy has been attributed to the relatively nutrient-sufficient growth conditions of marine plankton and the more homogeneous and stable nature of oceans (J. C. Goldman *et al.*, 1979). In contrast, marked deviations in the sestonic C:N:P proportions occur in lakes. These particulate composition ratios have been coupled to physiological **indicators,** such as rates of growth and productivity, to estimate the nutrient conditions to which phytoplankton of lakes have been exposed. Such relationships of nutrient stoichiometry can also give insights into species composition, growth rates, and successional patterns of phytoplankton (Kilham, 1990; Sommer, 1990).

Elemental cellular stoichiometries of natural phytoplankton communities and the seston can reflect the type and extent of nutrient limitation and availability. Ratios of elements being loaded to a lake are reflected in the elemental composition of the phytoplankton community. For example, N:P ratios in phytoplankton (seston) are strongly correlated with N:P loading rates to lakes (Table 13-17). The C:P and N:P ratios of lake seston are generally higher than the Redfield ratio for marine waters. Phosphorus limitation of algal productivity in lakes tends to be much greater than nitrogen limitation (Healey and Hendzel, 1980; Hecky *et al.*, 1993). Streams, shallow lakes, and reservoirs with short residence times have C:P ratios <350 and N:P ratios <26, whereas lakes with longer residence times (>6 months) differentiate from their inflows typically with C:P >400 and N:P >30. Tropical lakes tend to have relatively high C:N ratios, indicative of potential nitrogen limitations, although the number of lakes sampled was relatively small (Hecky *et al.*, 1993).

TABLE 13-17 Stoichiometric Ratios of Phytoplankton-Dominated Seston of Lakes as Approximate Indicators of Relative Nutrient Limitations[a]

		Degree of nutrient limitation[b]		
Ratio	*Deficiency*	*None*	*Moderate*	*Severe*
C:N	N	<8.3	8.3–14.6	>14.6
N:P	P	<23	—	>23
C:P	P	<133	133–258	>258
Si:P	Si	<20	—	>100
C:Chl *a*	General	<4.2	4.2–8.3	>8.3
APA:Chl *a*	P	<0.003	0.003–0.005	>0.005

[a] After Healey and Hendzel (1980), Kilham (1990), and Hecky *et al.* (1993).
[b] Composition ratios of C:N, N:P, C:P in μmol μmo1^{-1}; C:Chl *a* ratios as μmol μg^{-1}; and physiological ratio of alkaline phosphatase activity (APA):Chl *a* in (μmol μg^{-1})h^{-1}.

Cellular elemental stoichiometry is affected by the availability of nutrients and therefore can be correlated with the extent of nutrient limitations (Table 13-17). An excellent example is observed in the spatial and temporal differences in cellular stoichiometry. Epilimnetic ratios, particularly of N:P, are often higher, indicative of high phosphorus demands in proportion to total inputs, than those of the hypolimnion, and C:P as well as N:P ratios decrease with depth (e.g., Jones, 1976; Gächter and Bloesch, 1985; Tezuka, 1985; Gálvez *et al.*, 1991). These differences with depth are usually related to high phosphorus use in relation to supply, increased nutrient pool size with depth, and respiration of organic carbon as the seston settles.

The ratio of fluxes of nutrients into the dissolved nutrient resource pool (i.e., the supply rates of essential nutrients) can markedly influence the community structure of the phytoplankton (Tilman, 1982). As correctly pointed out by Sterner *et al.* (1992), external sources of nutrient loading can be and have been evaluated with some degree of accuracy, but rates and controls of internal regeneration of nutrients by physical (e.g., resuspension and mixing) and biological (e.g., nitrogen fixation, denitrification, food web regeneration by all heterotrophs, including viruses) processes are much less quantified to levels that allow effective use of predictive models. For example, denitrification was the most important mechanism for reducing N:P ratios in the water column, whereas both nitrogen fixation and sediment resuspension raised N:P ratios (Levine and Schindler, 1992). Sediments tend to be a major nitrogen source, particularly in littoral areas.

The N:P and Si:P supply ratios are fundamental axes along which phytoplanktonic community structures vary (Hecky and Kilham, 1988; Kilham, 1990). Diatoms and some chrysophyte algae have absolute silica requirements. Although total algal biomass is not limited by silica availability, the algal community composition, interspecific competition, and succession can be altered markedly (cf. Chaps. 15 and 19). Species of diatoms with low Si:P requirements, for example, often develop maximally during seasonal periods of lowest Si:P ratios (Kilham, 1984). In another example, many cyanobacteria are capable of nitrogen fixation, which allows these species to maintain high growth rates in habitats low in available combined inorganic nitrogen. The relative competitive capacities and proportions of cyanobacteria in epilimnetic phytoplankton communities are thus indicated by the N:P ratio; cyanobacteria tend to be rare at N:P >29 (Smith, 1983).

Zooplankton and higher animals usually constitute small portions of the total dissolved and seston pools of nitrogen and phosphorus. For example, in an alpine lake in California, zooplankton always constituted <5% and <10% of the pelagic storage compartments for nitrogen and phosphorus, respectively (Elser and George, 1993). However, during brief periods of high zooplankton biomass, the zooplanktonic contributions to total sestonic elemental stoichiometry can be significant. Although the N:P and C:P ratios do not vary greatly intraspecifically among zooplankton, considerable differences occur interspecifically (Watanabe, 1990; Gulati *et al.,* 1991; Sterner *et al.,* 1992; Elser and George, 1993; Urabe, 1995). Zooplanktonic N:P ratios may be altered by the N:P ratio of seston as certain individual species populations of zooplankton, which grow and reproduce most efficiently at a given N:P supply ratio, come to dominate the zooplankton community. Because the biochemical requirements of crustacean zooplankton for a N:P ratio in their tissues is below the ratio of most sestonic assemblages, there is a tendency for crustacean zooplankton to retain P preferentially to N. As a result of this process, nutrient regeneration via feces of zooplankton can be skewed toward nitrogen and thereby enhance phosphorus limitations to phytoplankton during short periods of intense cladoceran grazing (e.g., "clear water phase").

In examining the chemical composition of particulate organic matter, large zooplankton can be effectively separated from smaller particles. However, detrital particles nearly always dominate (>50–80%) over algal and bacterial biomass in the remaining seston (Saunders, 1972a; Uehlinger and Bloesch, 1987; Gálvez *et al.* 1989). If the planktonic microbiota are reproducing rapidly, much of the detritus present can have chemical composition relatively similar to that of the living plankton. Appreciable loadings of particulate organic matter from the structural tissues of terrestrial and littoral plants, however, can result in seston relatively rich in carbon and lesser amounts of nitrogen and lead to high C:N, C:P, and N:P ratios. Although much of the particulate organic matter from external sources tends to be deposited in near-shore regions in natural lakes, importation from the drainage basin is much greater in reservoir ecosystems (Chaps. 15 and 21). In addition, because of the high ratio of drainage basin area to reservoir area, large quantities of inorganic particles (e.g., clay) can be imported to the pelagic zone. These particles of the seston can contain high and variable quantities of nutrients that are undifferentiated from particulate organic matter of biota and can lead to skewed results, requiring care in interpretation. However, in most lakes with water residence times of several months or longer, sestonic composition ratios reflect the general status of the availability of nitrogen and phosphorus for planktonic growth in these waters.

Furthermore, the predictive value of nutrients present in the seston as a reflection of nutrient availability has been extended by the obvious relationship of energy available as light for photosynthesis to nutrient available and incorporated into the seston (Sterner *et al.,* 1997). When phosphorus availability is low, C:P ratios of particulate organic matter will tend to be low and photosynthetic productivity low. Differences in nutrient use efficiency may occur at different light availability, and these ratios will be translated to higher trophic levels feeding upon them. Low light:phosphorus ratios of seston likely indicate simultaneous carbon and organic matter (energy) limitation in other dependent trophic levels, whereas high light:phosphorus ratios suggest phosphorus limitation among several dependent trophic levels.

VIII. PHOSPHORUS AND NITROGEN LOADING AND ALGAL PRODUCTIVITY

When phosphorus is added as a pulse to unproductive lakes or ponds, either experimentally, for purposes of intentional fertilization or in effluents resulting from human activities, the usual response is a very rapid increase in algal productivity (e.g., Einsele, 1941; Maciolek, 1954; Mortimer and Hickling, 1954; Vinberg and Liakhnovich, 1965; Vollenweider, 1968; Correll, 1998). The increased productivity is not sustained as nutrient loadings are reduced, however, but decreases rather rapidly in a few weeks or months to levels near those prior to the addition. Losses in the colloidal fraction and from sedimentation of particulate phosphorus result in continuous reductions of phosphorus from the trophogenic zone. Inputs to the system must be maintained in a continuous or pulsed manner in order to sustain the increased productivity. In other words, steady phosphorus loading of the system is critical to sustaining increased productivity in most lakes of low or medium biological productivity.

Conversely, in order to reduce the productivity of a lake that is receiving a continuous loading of nutrients, algal growth is generally decreased most effectively by reduction of phosphorus inputs to below the level of losses within the lake. Reduction of total phosphorus of the water is the objective, since phosphorus is the major nutrient in greatest demand in relation to supply. Phosphorus is chemically reactive, technologically easier to remove from water than nitrogen, and does not have major reservoirs in the atmosphere.

A. Nutrient Loading

The nutrient loading concept implies that a relationship exists between the quantity of nutrient entering a water body and its response to that nutrient input. The effects of this relationship can be expressed by some quantifiable index of productivity or water-quality parameter (e.g., chlorophyll concentrations and water transparency).

The loading concept was recognized some time ago as applicable to lake changes in response to phosphorus and nitrogen enrichment (e.g., Rawson, 1939, 1955; Ohle, 1956; Edmondson, 1961). Sawyer (1947) suggested that if critical levels of dissolved inorganic nitrogen (300 μg N liter^{-1}) and phosphorus (10 μg P liter^{-1}) measured at the time of spring turnover in Wisconsin lakes were exceeded, nuisance populations of phytoplankton would likely develop during the growing season. Vollenweider (1966, 1968) first formulated definitive quantitative loading criteria for phosphorus and nitrogen and expected trophic conditions in water bodies. He defined boundaries between oligotrophic and eutrophic lakes by relating nutrient loadings to mean depth (as a measure of lake volume) and later refined these relationships (Vollenweider, 1969, 1975). Because phosphorus is commonly the initial limiting nutrient to algae, as discussed earlier, and because the many processes of nitrogen cycling (nitrification, denitrification, nitrogen fixation; Chap. 12) complicate accurate measurements of nitrogen loading of lakes, the loading criteria and models emphasized phosphorus relationships. Attempts to combine phosphorus and nitrogen loadings into lake enrichment models are empirically successful (e.g., Bachmann, 1984).

The loading relationships are all based on the mass balance equation of a substance M between its sources and losses (sinks) in an open system (Vollenweider *et al.*, 1980):

$$\Delta M/\Delta t = I - O - (S - R) \qquad (13\text{-}1)$$

where

$\Delta M/\Delta t$ = storage gain or loss of nutrient M over time Δt
I = external nutrient load
O = nutrient loss by outflow
S = nutrient loss to sediments
R = nutrient regeneration from sediments (internal loading)

While I and O can often be measured directly with accuracy, $(S - R)$ is more difficult to measure and is often obtained by difference. Under steady-state conditions, $\Delta M/\Delta t = 0$. Although in reality steady state does not exist in a lake, lakes that oscillate around relatively constant nutrient loads over periods of several years exhibit a set of trophic characteristics and may be viewed as a tendency toward a repetitive steady state. Often this steady state is defined as the nutrient storage content as measured at spring turnover or as the annual storage content. The models based on mass balance further assume that the nutrient load is instantaneously and completely mixed throughout the lake water (continuously stirred mixed reactor), an assumption that is only partially true for most lakes. The models, consequently, are particularly applicable to a large population of lakes, rather than to describing the specific behavior of an individual lake.

Assuming steady-state conditions, the simple mass balance equation accounts for volumnar loading of phosphorus as well as losses by flushing and sedimentation (Vollenweider and Kerekes, 1980):

$$\frac{\overline{d}[\overline{\mathrm{P}}]_\lambda}{dt} = \left(\frac{1}{\overline{\tau}_w}\right)[\overline{\mathrm{P}}]_i - \left(\frac{1}{\overline{\tau}_p}\right)[\overline{\mathrm{P}}]_\lambda \qquad (13\text{-}2)$$

where

$[\overline{\mathrm{P}}]_\lambda$ = average (total) lake concentration (both dissolved and particulate phosphorus components)
$[\overline{\mathrm{P}}]_i$ = average inflow concentration of total phosphorus
$\overline{\tau}_p$ = average residence time of phosphorus
$\overline{\tau}_w$ = average residence time of water

in which the right-hand terms represent the average rate of supply to and the average rate of loss of total phosphorus from the lake, respectively, and the left-hand term represents the corresponding temporal variations of the average lake concentration. Then, simplifying Equation (13-2):

$$[\bar{P}]_\lambda = (\bar{\tau}_p/\bar{\tau}_w)[\bar{P}]_i \qquad (13\text{-}3)$$

The premise of Equation (13-3) can also be expressed as:

$$[\bar{P}]_\lambda = (1 - R)[\bar{P}]_i \qquad (13\text{-}4)$$

where R = the phosphorus-retention coefficient:

$$R = \frac{[\bar{P}]_i - [\bar{P}]_\lambda}{[\bar{P}]_i} = 1 - \bar{\tau}_p/\bar{\tau}_w \qquad (13\text{-}5)$$

or statistically approximately

$$\bar{\tau}_p/\bar{\tau}_w = \frac{1}{1 + \sqrt{\bar{\tau}_w}}$$

(Larsen and Mercer, 1975; Vollenweider, 1976). An approximate retention coefficient as a function of hydraulic load (q_s) was found to be (Dillon, 1974; Kirchner and Dillon, 1975):

$$R = 0.426e^{-0.271q_s} + 0.574e^{-0.00949q_s}$$

where q_s = areal water load (hydraulic load) to the lake (m yr^{-1}), calculated as the lake outflow volume divided by the lake surface area. These two retention values, while not identical, yield similar results for lakes of mean depths between 10 and 30 m and hydraulic loads between 3 and 90 m yr^{-1} (Vollenweider and Kerekes, 1980). When either of these models was used and data plotted to show the extent of prediction of spring-turnover phosphorus concentration from loading, they described the average statistical behavior of lakes in terms of the relationship between phosphorus load and phosphorus accumulation within the lake (Vollenweider, 1976, 1979; Rast and Lee, 1978). Deviant lakes could be shown to have mechanisms operating to overestimate or underestimate the loading (e.g., high sedimentation rates, high regeneration of phosphorus from the sediments, particularly in lakes that developed anoxic deep waters, Nürnberg, 1984).

B. Biological Responses to Nutrient Loading

Based on these nutrient loading relationships, the loadings to a lake should be related to biological responses such as phytoplankton biomass (e.g., chlorophyll concentrations) and productivity. Early predictions of chlorophyll levels from phosphorus concentrations in lakes at spring turnover or during the growing season (Sakamoto, 1966; Dillon and Rigler, 1974; Jones and Bachmann, 1976; Schindler, 1978) were later extended further with larger worldwide data sets (Vollenweider, 1976; Schindler *et al.*, 1977; Oglesby and Schaffner, 1978; Rast and Lee, 1978; Canfield and Bachmann, 1981). Within certain boundaries, the regression between phosphorus loading and average annual chlorophyll concentrations is approximated by:

$$\overline{[\text{chl. } a]} = 0.55\{[\text{P}]_i/(1 + \sqrt{\tau_w})\}^{0.76}$$

Examples of this relationship are given in Figure 13-13; similar relationships have been found for many other lakes with only minor deviations.

The annual rate of primary productivity of the algae has also been related to the predicted phosphorus concentration (Fig. 13-14). The nonlinearity of the log–log plot results from the light-reducing, self-shading effects of dense algal populations (high biomass) at high levels of productivity (see Chap. 15). This well-known relationship was further reiterated by correlation analyses of Watson *et al.* (1992). The relationship can be represented by

$$\Sigma C(\text{g m}^{-2}\text{ yr}^{-1}) = 7\left[\frac{\{[\text{P}]_i/(1 + \sqrt{\tau_w})\}^{0.76}}{0.3 + 0.011\{[\text{P}]_i/(1 + \sqrt{\tau_w})\}^{0.76}}\right]$$

which operates similarly to the integral of daily photosynthesis and is based on average chlorophyll concentrations and light extinction resulting from turbidity and dissolved organic substances (Vollenweider and Kerekes, 1980).

C. Application of Predictive Models and Data

The evaluation of the trophic status of a lake has great practical importance. Eutrophicational status must be known before remedial corrective measures can be implemented in relation to the desired uses for any lake. Based upon many data obtained during an international program on eutrophication of the Organization for Economic Cooperation and Development (OECD), nutrient load–eutrophication response relationships for lakes and reservoirs were quantified and evaluated (Vollenweider, 1968; Rast and Lee, 1978; Canfield and Bachmann, 1981; Ortiz and Martínez, 1984; and others). The analyses provide a potential basis by which predictions can be made of the changes in water quality that will occur from changes in the phosphorus loadings to phosphorus-limited water bodies. Analyses of over 200 water bodies permitted a general classification of lakes based on water transparency and concentrations of phosphorus, nitrogen, and phytoplankton pigment (Table 13-18). Although any attempt at rigid classification of water bodies is subjective and fraught with exceptions, the general classification is useful.

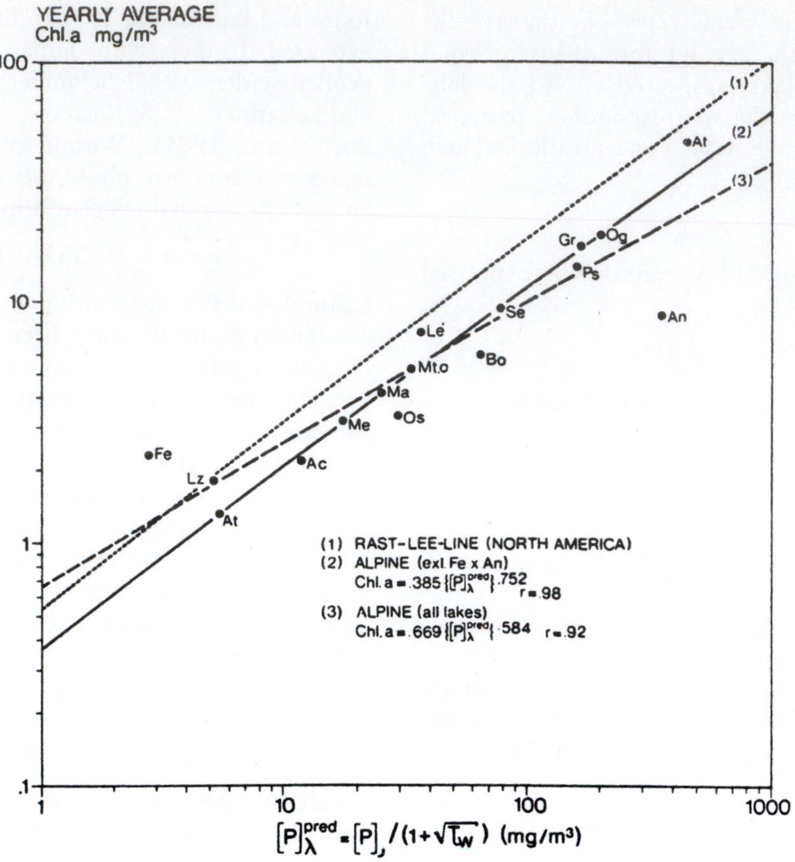

FIGURE 13-13 Statistical relationship between predicted total phosphorus concentration and annual mean chlorophyll *a* concentration for selected alpine lakes. (Modified slightly from Vollenweider, 1979.)

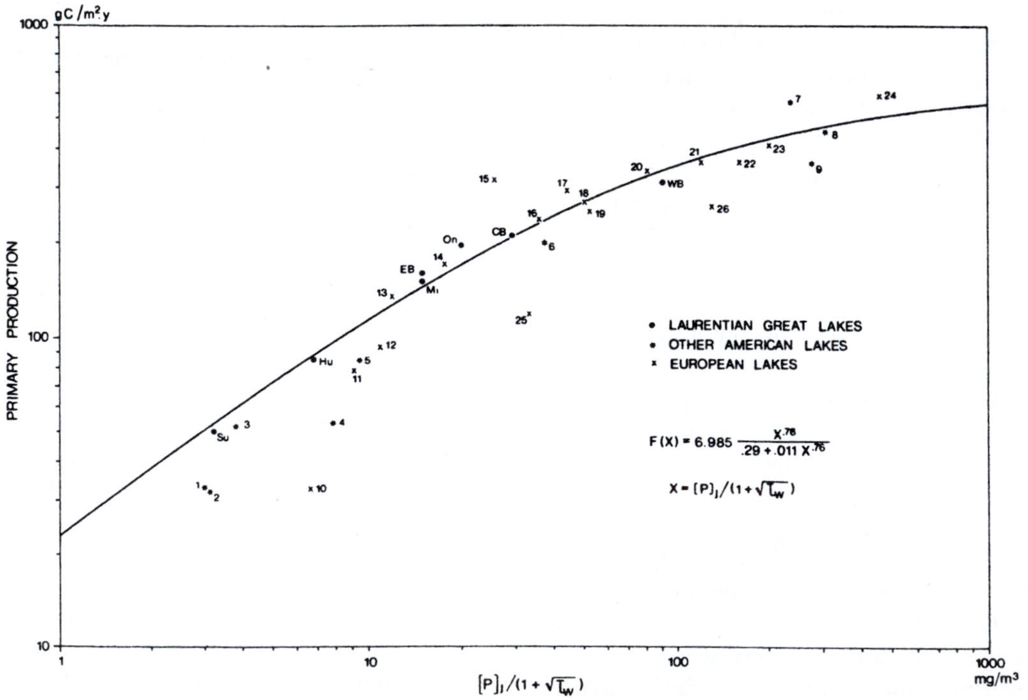

FIGURE 13-14 Annual primary productivity (g C m^{-2} yr^{-1}) as a function of predicted total phosphorus concentration. (From Vollenweider, 1979.)

TABLE 13-18 General Trophic Classification of Lakes and Reservoirs in Relation to Phosphorus and Nitrogen[a]

Parameter (annual mean values)	*Oligotrophic*	*Mesotrophic*	*Eutrophic*	*Hyper-eutrophic*
Total phosphorus (mg m^{-3})				
Mean	8.0	26.7	84.4	—
Range	3.0–17.7	10.9–95.6	16–386	750–1200
N	21	19	71	2
Total nitrogen (mg m^{-3})				
Mean	661	753	1875	—
Range	307–1630	361–1387	393–6100	—
N	11	8	37	—
Chlorophyll *a* (mg m^{-3}) of phytoplankton				
Mean	1.7	4.7	14.3	—
Range	0.3–4.5	3–11	3–78	100–150
N	22	16	70	2
Chlorophyll *a* maxima (mg m^{-3}) ("worst case")				
Mean	4.2	16.1	42.6	—
Range	1.3–10.6	4.9–49.5	9.5–275	—
N	16	12	46	—
Secchi transparency depth (m)				
Mean	9.9	4.2	2.45	—
Range	5.4–28.3	1.5–8.1	0.8–7.0	0.4–0.5
N	13	20	70	2

[a] Based on data of an international eutrophication program. Trophic status based on the opinions of the experienced investigators of each lake. (Modified from Vollenweider, 1979.)

The models permit predictions of the probability of oligotrophic, intermediate, or eutrophic conditions developing in phosphorus-limited lakes in response to various loading regimes (Fig. 13-15). The literature on this subject is very large, for the models have been continuously tested, refined, and improved (e.g., Reckhow, 1979; Chapra and Reckhow, 1981, 1983). As with all modeling efforts of natural and disturbed ecosystems, one must not permit the mathematical manipulations to become an end in themselves (e.g., Pielou, 1981).

Models based on many data permit estimates of permissible loading rates of phosphorus and nitrogen while still allowing tolerable conditions of productivity. Considering a combination of influencing conditions, Vollenweider (1968) estimated very approximate provisional loading rates of nitrogen and phosphorus required to

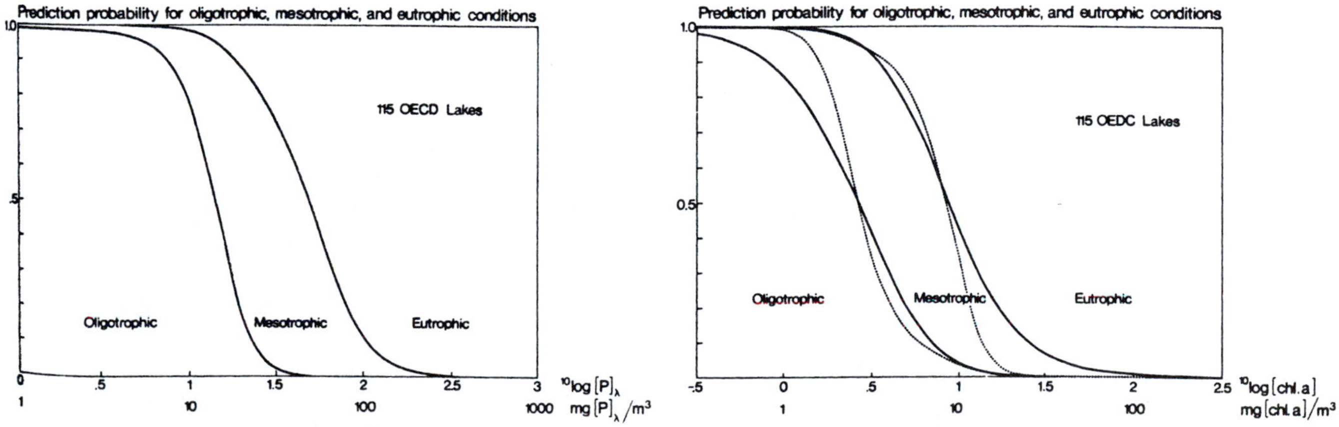

FIGURE 13-15 Probable boundaries of the degrees of trophy of water bodies with differing annual mean values of total phosphorus concentrations (*left*) and chlorophyll *a* concentrations (*right*). (Modified from Vollenweider, 1979.)

TABLE 13-19 Provisional Permissible Loading Levels for Total Nitrogen and Total Phosphorus (Biochemically Active) in g m^{-2} yr^{-1}[a]

	Permissible loading		*Dangerous loading*	
Mean depth (m)	*N*	*P*	*N*	*P*
5	1.0	0.07	2.0	0.13
10	1.5	0.10	3.0	0.20
50	4.0	0.25	8.0	0.50
100	6.0	0.40	12.0	0.80
150	7.5	0.50	15.0	1.00
200	9.0	0.60	18.0	1.20

[a] After Vollenweider, R. A.: *Scientific Fundamentals of the Eutrophication of Lakes and Flowing Waters, with Particular Reference to Nitrogen and Phosphorus as Factors in Eutrophication,* OECD Report No. DAS/CSI/68.27, Paris, OECD, 1968.

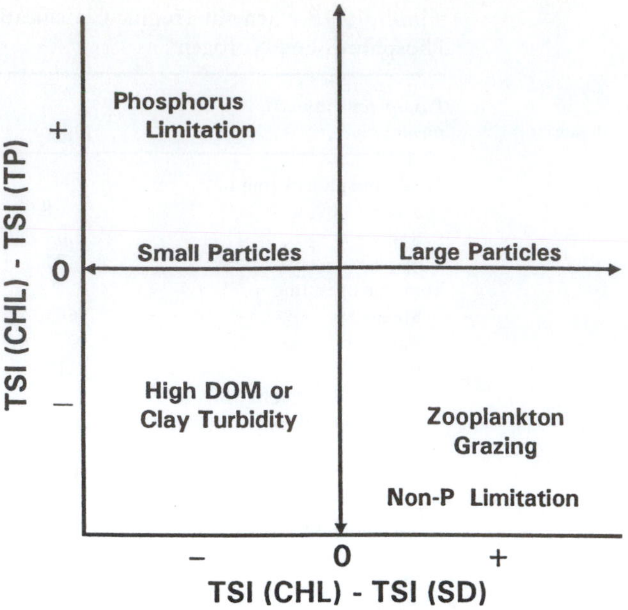

FIGURE 13-16 Potential nutrient-limited and nonnutrient-limited causes for deviation of biomass-based trophic state index (TSI). See text; based upon developments by Carlson (1992).

maintain lakes in a steady state (Table 13-19). Rapid eutrophication is likely to result as loading is increased. Although the predictive models have improved considerably since this analysis appeared, the tolerable loading level of 10 mg P m^{-3} remains unaltered, while the excessive loading level has been increased slightly to 25 mg P m^{-3} (Vollenweider and Kerekes, 1980).

A simple production-based ***trophic state index*** (TSI) uses phytoplankton biomass[4] as a basis for a continuum of trophic states of lakes and reservoirs under both nutrient-limiting and nonnutrient-limited conditions (Carlson 1977, 1980, 1992; Carlson and Simpson, 1996). Algal biomass can be estimated independently by chlorophyll pigment concentrations, Secchi depth, and total phosphorus concentrations. The trophic continuum is divided into units based on a base-2 logarithmic transformation of Secchi depth, with each 10-unit division of the index representing a halving or doubling of Secchi depth. Because total phosphorus is often inversely correlated with transparency, a doubling of the total *P* often corresponds to a halving of Secchi depth transparency. Chlorophyll pigments double every seven units (Carlson, 1980). Derivations of the ensuing linear regression models indicated that total P may be better than chlorophyll at predicting summer trophic state from winter samples, and transparency should be used only if no better data are available:

$$\text{TSI(SD)} = 60 - 14.41 \ln(\text{SD})$$

$$\text{TSI(CHL)} = 9.81 \ln(\text{CHL}) + 30.6$$

$$\text{TSI(TP)} = 14.42 \ln(\text{TP}) + 4.15$$

[4] Although the index was based on phytoplankton biomass, it could likely include macrophytic and periphytic chlorophyll as well.

where SD = Secchi disk transparency (m), CHL = chlorophyll pigment concentrations (mg m^{-3}), and TP = total phosphorus concentrations (mg m^{-3}). Values obtained by the three TSI models should not be averaged. TSI values of <30 are common among lakes and reservoirs of classical oligotrophy, and from 50 to 70 correspond to classical eutrophy. Hypereutrophic conditions are common at TSI values of >70. The TSI changes in some lakes over an annual period, for example, during periods of intensive zooplankton grazing, reduction of nutrient availability seasonally, and with differences in nonalgal turbidity. Deviations of the TSI relationships can be clarified by graphical expression (Fig. 13-16). Values below the zero *X*-axis indicate the likelihood of something other than *P* limitation; points above this axis suggest increasing possibility of *P* limitation. Points to the right of the *Y*-axis indicate transparency is greater than predicted from the chlorophyll index, such as dominance by large cyanobacteria or zooplanktonic grazing and the subsequent relative reduction of smaller particles. Points along the lower left diagonal to the left of the *Y*-axis indicate reduced transparency from nonalgal factors, such as high concentrations of dissolved organic matter or small particles of nonalgal turbidity.

The general trophic classification system for lakes has been extended to streams. Rather than use the chlorophyll of seston, the chlorophyll content of benthic algae has been correlated to total nitrogen and phosphorus content in temperate streams (Dodds *et al.,*

TABLE 13-20 Suggested Boundaries for Trophic Classification of Streams Based on Distributions of Benthic Chlorophyll Content and Nutrient Concentrations[a]

Variable	*Oligotrophic–mesotrophic boundary*	*Mesotrophic–eutrophic boundary*
Mean benthic chlorophyll (mg m^{-2})	20	70
Maximum benthic chlorophyll (mg m^{-2})	60	200
Sestonic chlorophyll (mg $liter^{-1}$)	10	30
Total phosphorus (μg $liter^{-1}$)	25	75
Total nitrogen (μg $liter^{-1}$)	700	1500

[a] After Dodds *et al.* (1998).

1998). The correlations between total phosphorus and chlorophyll for streams are distinctly weaker than is the case for lakes, but a set of proximate trophic categories has been proposed (Table 13-20).

D. Effects of Decreasing Nutrient Loading

The reduction of the productivity of lakes by decreasing phosphorus loading can be very effective. Many examples are given in Vollenweider's (1968) review. Among the most frequently discussed examples is Lake Washington, Seattle, Washington (Edmondson, 1972, 1991). Lake Washington was enriched with increasing volumes of effluent from secondary sewage treatment facilities during the period from 1941 to 1953. Production increased markedly, and the algae became more abundant. Phosphate concentrations also increased proportionally much more than those of nitrate or carbon dioxide. Effluent was diverted away from the lake in 1963, and by 1969, phosphate had decreased to 28% of its 1963 value. Summer chlorophyll concentrations had decreased about as much, but nitrate and total CO_2 fluctuated from year to year at relatively high values. Reduction in phytoplankton and phosphorus continued to a relatively stable level in the 1990s (Edmondson, 1969, 1991; Edmondson and Lehman, 1981).

The rate at which the productivity of a lake will revert toward conditions existing prior to increased loading is variable and depends upon basin morphometry, water chemistry, and the nature of the phosphorus sources (diffuse or concentrated at point sources). The loading of lakes with nitrogen and phosphorus is further influenced by the ratio of the surface area of the drainage basin to that of the lake (Ohle, 1965). Depending on the percentage losses from the land, some of which may be under cultivation, inputs increase roughly in proportion to the "surroundings" ratio.

Upon reduction of loading rates to or below those of the lake prior to accelerated inputs, recovery will require 2–10 yr for lakes of average size and average hydrological replenishment time. The relative residence time, or, inversely, the flow-through rate, of phosphorus is directly coupled to the efficiency of sediment removal and retention capacities in the regulation of phytoplanktonic productivity (e.g., Janus and Vollenweider, 1984; Vollenweider, 1985, 1990). In general, oligotrophic lakes will respond slowly to increasing loading but rapidly to decreasing loads as long as the loading perturbation is of short duration. In contrast, lakes with a long eutrophicational history and large accumulations of nutrients in the sediments have lost resilience and respond more slowly to reductions in nutrient loadings. The rate of recovery, therefore, is much slower in shallower lakes where the "internal loading" of phosphorus from sediment sources is high. In eutrophic Shagawa Lake, Minnesota, for example, where external phosphorus loading was reduced by >70%, predicted reduced equilibrium levels were not achieved (observed levels of phosphorus concentrations after diversion were >100% in excess of those predicted; Larsen *et al.*, 1975). Shagawa Lake is shallow, has an anaerobic hypolimnion, and has an extensively developed littoral zone of submersed macrophytes, all of which contribute to enhanced release of sediment phosphorus. Internal loading from sediment phosphorus sources has been observed in many other shallow lakes, reservoirs, or lakes with anoxic hypolimnia (e.g., Cooke *et al.*, 1977; 1994; Grobler and Davies, 1981; Hillbricht-Ilkowska and Lawacz, 1983; Nürnberg, 1984). The magnitude of internal phosphorus loading can be predicted from the product of an average rate of phosphorus release from anoxic sediments, the surface area of the anoxic sediment, and the period of anoxia. Such internal loading slows the rate of lake recovery implemented by reduction of phosphorus loading from external allochthonous sources. In most lakes, however, the "internal loading" contribution is a relatively small proportion of the external sources (e.g., Bannerman *et al.*, 1974; Edmondson and Lehman, 1981; Smith and Shapiro, 1981; Rast *et al.*, 1983), so that regulation of loading rates can often be effective in controlling phytoplankton productivity.

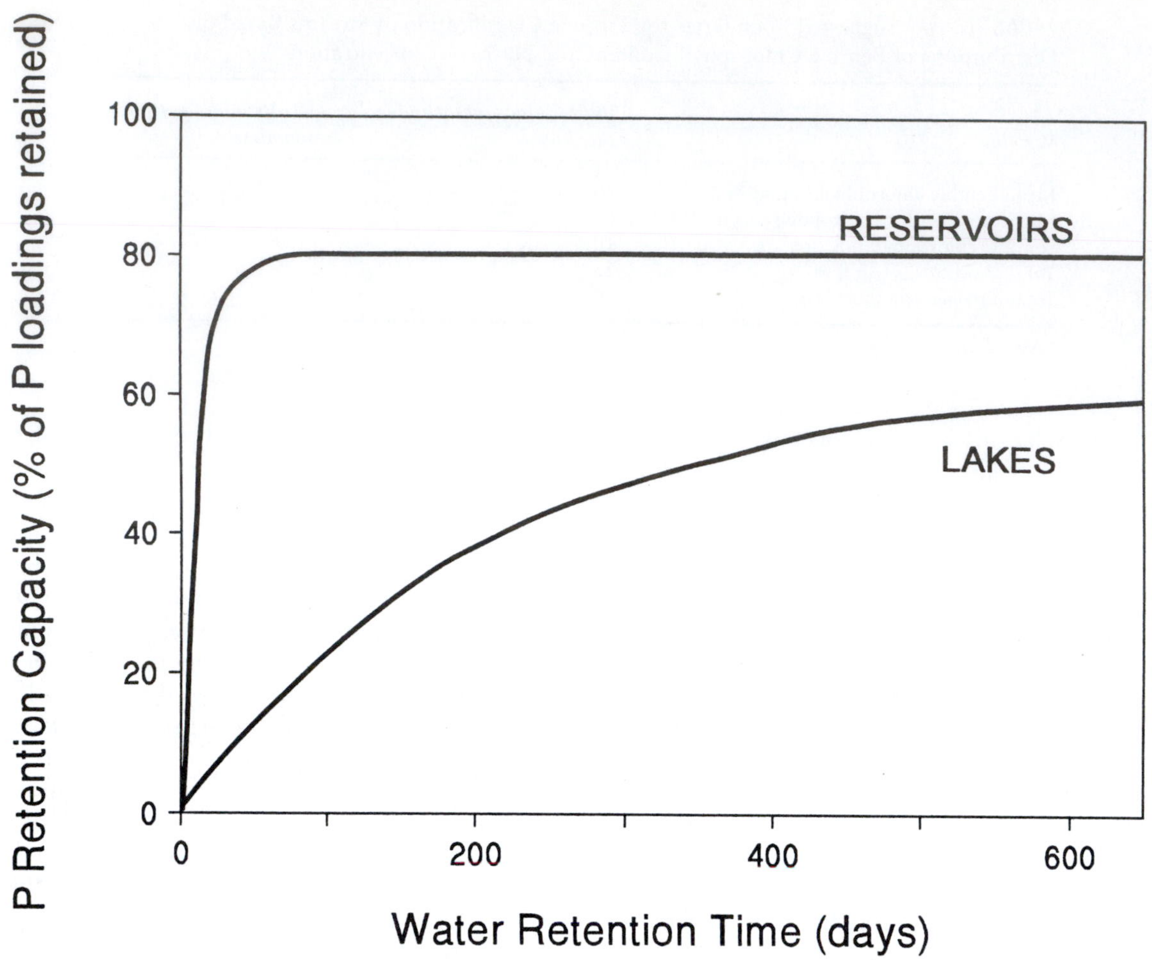

FIGURE 13-17 Phosphorus retention capacity of lakes and reservoirs as a percentage of total phosphorus loadings retained in relation to water retention times. (Modified from Straškraba, 1996.)

Finally, the retention capacities for phosphorus can differ considerably between stratified natural lakes and reservoirs. Much of the difference in phosphorus retention focuses on marked differences in water retention times. Water retention time can be estimated by the number of days needed in a given year to fill the lake or reservoir if it were empty. That water retention can be correlated to the phosphorus retention capacity estimated as the percentage of the total phosphorus loadings remaining and not lost from the water body via its outflow (Straškraba, 1996). Although the phosphorus retention times are similar when comparing water retention times of a year or longer (Fig. 13-17), the water retention times of lakes are much longer (1–7 yr on the average) than those of reservoirs (often <100 days and highly variable). The differences are most pronounced at water retention times <100 days (Fig. 13-16). These differences are related, in part, to more rapid sedimentation of particulate phosphorus and removal of outflow water deep in the stratified waters in reservoirs versus the slower sedimentation of biotic particles and water removal by surface outflows that predominate in lakes.

IX. SUMMARY

1. Phosphorus plays a major role in biological metabolism. In comparison to other macronutrients required by biota, phosphorus is the least abundant and commonly is the first element to limit biological productivity.
2. Many quantitative data exist on the seasonal and spatial distribution of phosphorus in streams and lakes and the loading rates to recipient waters from drainage basins.
   a. Orthophosphate (PO_4^{-3}) is the only directly utilizable form of soluble inorganic phosphorus. Phosphate is extremely reactive and interacts with many cations (e.g., Fe and Ca) to form,

especially under oxidizing conditions, relatively insoluble compounds that precipitate out of the water. Availability of phosphate is also reduced by adsorption to inorganic colloids and particulate compounds (e.g., clays, carbonates, and hydroxides).

b. A large proportion of phosphorus in fresh waters is bound in organic phosphates and cellular constituents of organisms, both living and dead, and within or adsorbed to organic colloids.

c. The range of total phosphorus in fresh waters is large, from <5 μg liter^{-1} in very unproductive waters to >100 μg liter^{-1} in highly eutrophic waters. Most uncontaminated fresh waters contain between 10 and 50 μg total P liter^{-1}.

d. Although concentrations of total and soluble phosphorus of oligotrophic lakes exhibit little variation with increasing depth, eutrophic lakes with strongly clinograde oxygen profiles commonly show a marked increase in phosphorus content in the lower hypolimnion (Fig. 13-1). Much of the hypolimnetic increase is from soluble phosphorus near the sediment–water interface.

3. Dissolved phosphorus in rivers is generally higher than in lakes and often increases markedly following rainfall and snowmelt events on land of the drainage basin.
   a. At low concentrations in streams, phosphorus can be limiting and microbial productivity often increases with increased loading of phosphorus.
   b. In addition to biological uptake, phosphorus abundance is regulated by abiotic adsorption and desorption reactions with organic and inorganic particles. Dissolved organic phosphorus compounds are utilized more slowly than inorganic forms and tend to be exported downstream for considerable distances.
   c. Most of the recycling of phosphorus is associated with microbiota attached to particles. Less than 5% is associated with macroinvertebrates and higher organisms.
4. Exchanges of phosphorus across the sediment–water interface are regulated by oxidation–reduction (redox) interactions dependent on oxygen supply, mineral solubility and sorptive mechanisms, the metabolic activities of bacteria and fungi, and turbulence from physical and biotic activities.
   a. In all but the upper few millimeters of sediment, exchange is slow and controlled by low diffusion rates.
   b. If water above the sediments is oxygenated (approximately >1 mg O_2 liter^{-1}), an oxidized microzone is formed below the sediment–water interface (0 to -5 mm), below which the sediments usually become extremely reducing (Fig. 13-4). The oxidized microzone effectively prevents phosphorus (which is solublized under reducing conditions in the sediments) from migrating by diffusion upward into the water column (Fig. 13-5, *left*).
   c. As the hypolimnion becomes anoxic in productive lakes, the oxidized microzone is lost. The release of phosphate and ferrous iron into the water occurs readily when reducing conditions reach a redox potential (E_7) of about +200 mV (Fig. 13-5, *right*).
   d. Soluble phosphorus can accumulate in large quantities in anaerobic hypolimnia. With the advent of autumnal circulation, ferrous iron is rapidly oxidized and precipitates much of the phosphate as ferric phosphate.
   e. Bacterial metabolism of organic matter is the primary mechanism by which organic phosphorus is converted to phosphate in the sediments and for creating the reducing conditions required for release of phosphate to the water.
   f. Movement of phosphorus from sediment interstitial water can be accelerated by physical turbulence and by biota. If light reaches sediments, photosynthetic activity of epipelic algae can create a highly oxidized microzone in the surficial sediments and regulate phosphorus release to overlying water on a diurnal basis.
      i. Rooted aquatic macrophytes often obtain phosphorus from the sediments and can release certain amounts into the water both during active growth but particularly upon senescence and death. Most of the released phosphorus is retained and recycled by attached algae, bacteria, and other microbiota.
      ii. High population densities of sediment-dwelling invertebrates, such as midge larvae, can increase the exchange of phosphorus across the sediment–water interface.
5. Recent studies of phosphorus cycling in the trophogenic zone have demonstrated that the exchange of phosphorus between its various forms is often rapid and involves numerous complex pathways.
   a. A large portion, often $>95\%$, of the phosphorus is bound in the particulate phase of living biota, especially bacteria, algae, and other microbes.
   b. Organic phosphorus of the seston of the open water consists of at least two major fractions:
      i. A rapidly cycled fraction that is exchanged with soluble forms. In this fraction,

phosphate is transferred rapidly through the particulate phase to low-molecular-weight compounds.

ii. A fraction of dissolved organic and colloidal phosphorus material that is released and cycled more slowly.

c. Phosphorus uptake and turnover kinetics in numerous lakes show that turnover rates are extremely rapid (5–100 min) during summer periods of high demand and low loading inputs. During winter periods, turnover rates become as much as 2 orders of magnitude slower. Phosphorus turnover rates are generally faster under more oligotrophic conditions of lower phosphorus availability.

d. Flagellates, other protista, and zooplankton that feed on bacteria and other seston excrete soluble phosphorus and ammonia. These nutrients are rapidly utilized by algae and bacteria. When supplies of phosphorus are low, this source of recycled phosphorus can be critical to the growth and succession of phytoplankton.

e. Algal requirements for phosphorus are variable among species and can lead to selective advantages of certain algae as phosphorus supplies change seasonally or as phosphorus loadings to a fresh water change over a longer period of time. Algae must compete with bacteria for phosphorus sources; if organic substrates for bacterial growth are high, algal growth may be seriously impeded.

6. Sedimentation of particles results in constant losses of phosphorus from the trophogenic zone. As a result, new phosphorus supplies must enter the ecosystem in order to maintain or increase productivity.

a. Phosphorus enters fresh waters from atmospheric precipitation and from groundwater and surface runoff. The loading rates of phosphorus vary greatly with patterns of land use, geology and morphology of the drainage basin, soil productivity, human activities, pollution, and other factors.

b. When phosphorus is added to unproductive fresh waters, either experimentally or as a result of human activities, the usual response is a rapid increase in algal productivity. Inputs must be more or less continuous, however, to maintain a higher level of productivity.

c. Numerous mass balance models have been developed to predict, on the basis of phosphorus loading and retention times, the anticipated responses of algal biomass and productivity. The models predict a reasonably accurate estimation of permissible phosphorus loading needed to achieve a certain level of productivity if loading is lowered.

d. In certain shallow lakes with greater than average turbulence, large littoral areas, and small, anaerobic hypolimnia, reduced productivity does not always occur as rapidly in response to decreased loading as predicted from the models. In these lakes, phosphorus release from sediment sources ("internal loading") is much greater than values (10–30% of total loading) for deeper, more stratified lakes.

14

IRON, SULFUR, AND SILICA CYCLES

I. BIOGEOCHEMICAL CYCLING OF ESSENTIAL MICRONUTRIENTS

The biogeochemical cycling of essential micronutrients is complex and is regulated to a large extent by variations in oxidation-reduction states, which in turn are mediated by photosynthetic and bacterial metabolism. Many of the reactions among different elemental nutrients are coupled, and the state of one can influence the availability of another. Although the coupled reactions must be viewed simultaneously, the following discussion separates the components as much as possible.

II. OXIDATION-REDUCTION POTENTIALS IN FRESHWATER SYSTEMS

In pure inorganic chemical systems, oxidation-reduction (redox) reactions proceed with a flow of electrons between the oxidized and reduced states until equilibrium is attained. There is a tendency for the reduced phase to lose electrons and be transformed to an oxidized state (e.g., $Fe^{3+} + e^- \rightleftharpoons Fe^{2+}$). Free electrons, however, usually inhibit this process, so large quantities of free ions in reduced and oxidized states can exist together. If electrons are removed, as for example by an immersed platinum electrode, then a transformation of the reduced state to the oxidized state (Fe^{2+} to Fe^{3+}) occurs, and a current of electrons passes through the electrodes. By applying a current (electrons) in the opposite direction, reduction can be induced; the reactions are reversible. Such potentiometric changes in electron flow can be accomplished with any solution containing ionic species, including the composite system of lake water, by drawing off electrons with a calomel[1]–platinum electrode. The resulting measurement, taken against standard conditions, is an expression of the oxidizing or reducing intensity or condition of the solution. The current, referenced against the hydrogen electrode, is termed the *redox potential* or the *electrode potential.*

A. The Redox Potential

Redox potential is proportional to the equivalent free energy change per mole of electrons associated with a given reduction (Stumm, 1966; Morris and Stumm, 1967). Although aqueous solutions do not contain free protons and electrons, it is possible to define proton activity [$pH = -\log(H^+)$] and electron activity [$pE = -\log(e^-)$]. pE is large and positive in strongly oxidizing solutions (low electron activity), just as pH is high in strongly alkaline solutions (low proton activity). Thus, both pH and pE are intensity factors of free energy levels and are not related to capacity or condition (e.g., alkalinity and acidity).

When oxygen is dissolved in water, a redox potential is generated according to the reaction (cf. discussion in Stumm, 1966, and Stumm and Morgan, 1996):

$$H_2O \rightleftharpoons \tfrac{1}{2}O_2 + 2H^+ + 2e^-$$

The pE of water at equilibrium is relatively insensitive to a change in oxygen concentration and extent of saturation, for it is influenced by the fourth root of the partial pressure of oxygen. If the oxygen content is decreased by 99%, the redox potential would be reduced by only about 30 mV. The activity of the hydroxyl ions, however, influences the activity of the hydrogen ions. Therefore, the redox potential is significantly changed by alterations of H^+ and is reflected in the pH. It is customary to express the pE of redox reactions in natural waters at their activities in neutral water at pH 7 rather than at unity. Thus, the redox potential is expressed as E_h or as E_7, in which a correction is made for the change in redox at the pH of the sample to a pH of 7.[2] A change in pH of one unit is accompanied by a change in redox potential of 58 mV. Hence, a common practice is to correct potentials to pH 7 by subtracting 58 mV for every pH unit on the acid side of neutrality and by adding 58 mV for every pH unit on the alkaline side of neutrality. E_h is influenced to only a small extent by temperature, for example, E_h of water = 860 mV at 0°C and 800 at 30°C at pH 7.

The preceding discussion applies to an ideal situation under conditions of equilibrium in which oxidation–reduction systems are reversible (Stumm, 1966). True redox equilibrium is not found in any natural aquatic system because of the extreme slowness of most redox reactions in the absence of appropriate biochemical catalysis. There is, in addition, a continuous cyclic input of photosynthetic energy that disturbs the trend toward equilibrium conditions. In addition to nonequilibrium redox conditions, which are related to the highly dynamic state of biotic activities, electrochemical measurements of redox depend on the nature and composite rates of reactions at the electrode surface. In practice, the redox measurements reflect irreversible electrochemical redox potentials. Thus, in neutral, fully oxygenated water equilibrated with air, redox potentials slightly greater than 500 mV are obtained, which are considerably less than the theoretical E_h of 800 mV.

Few elements (C, O, N, S, Fe, Mn) are predominant reactants in redox processes in natural waters. By the conversion of energy into chemical bonds, photosynthesis produces reduced states (negative E_h) of high free energy and results in nonequilibrium concentrations of C, N, and S compounds (Stumm, 1966). Respiratory, fermentative, and other nonphotosynthetic reactions of organisms tend to restore equilibrium by catalytically decomposing, through energy-yielding redox reactions, the thermodynamically unstable products of photosynthesis. It is through such reactions that nonphotosynthetic organisms obtain a source of free energy for their metabolic demands. The mean E_h is increased by these combined processes. It should be noted that organisms act as redox catalysts by mediating the reactions and transfer of electrons; the organisms themselves do not oxidize substrates or reduce compounds.

Reduced and oxidized iron (Fe^{2+} and Fe^{3+}) is among the most electroactive redox reactants in natural water systems. Other than iron and to a certain extent manganese, the redox components of organic carbon, nitrogen, and sulfur are not electronegative and yield reversible potentials only following enzyme-mediated changes. As a result, redox measurements in natural waters do not lend themselves to quantitative interpretation and comparison. Qualitative, relative comparisons of E_h, representing mixed composite potentials, however, can be extremely instructive, as was demonstrated by the monographic comparison of pH

[1] A calomel electrode consists of a chloride solution (usually KCl), solid Hg_2Cl_2 (calomel), and metallic mercury.

[2] E_h is compared to the hydrogen half-cell at any designated pH; usually it is pH O($[H^+] = 1$). Thus, E_h is usually, but not necessarily always, corrected to neutral pH.

and E_h limits in different environments and among organisms (Baas and Becking *et al.*, 1960).

Redox potentials, as measured with electrodes, cannot be used to predict the dominant redox reactions in anaerobic sediments because of the significant overlap in redox potentials for different reactions. As we will see in the discussion of bacterial metabolism in sediments (cf. Chap. 21), only microorganisms catalyze the oxidation of hydrogen (H_2) coupled to the reduction of nitrate, Mn(IV), Fe(III), sulfate, or carbon dioxide. At steady-state conditions, H_2 concentrations are primarily dependent upon the physiological characteristics of the microorganisms consuming the H_2. Organisms catalyzing H_2 oxidation, with the reduction of a more electrochemically positive electron acceptor, can maintain lower H_2 concentrations than organisms using electron acceptors that yield less energy from H_2 oxidation (Lovley and Goodwin, 1988; Chapelle *et al.* 1996). Each predominant terminal electron-accepting reaction in sediments has a unique range of steady-state H_2 concentrations associated with it: nitrate or Mn(IV) reduction, less than 0.05 nM H_2; Fe(III) reduction, 0.2 nM (range 0.2–0.8 nM); sulfate reduction, 1–4 nM; and methanogenesis, 7–10 nM (range 5–25 nM). Sediments with the same terminal electron acceptor for organic matter oxidation had comparable H_2 concentrations, in spite of variations in the rate of organic matter decomposition, pH, or salinity.

From this discussion, one would anticipate little change in redox potential with increasing depth as long as the water contained dissolved oxygen. Even though a distinctly clinograde oxygen curve may be found, as long as the water is not near anoxia the E_h will remain positive and fairly high (300–500 mV). Such conditions have been found repeatedly in detailed studies of E_h in lake waters of widely diverse types (Kuznetsov, 1935; Hutchinson *et al.*, 1939; Allgeier *et al.*, 1941; Mortimer, 1942; Kjensmo, 1970; Eckert and Trüper, 1993). As oxygen concentrations approach zero and anoxic conditions appear, as in the lower hypolimnion and near the sediments, the E_h decreases precipitously. Similar relationships at the sediment–water interface were discussed in some detail in the previous chapter on phosphorus release from the sediments (cf. Figs. 13-4, 13-5, and 13-6). Within the sediments, reducing conditions prevail, and the E_h declines to 0 mV or below within a few millimeters of the interface (Figs. 13-4, and 13-5). Of the many reducing compounds that contribute to the reductions in E_h, ferrous iron of the sediments is the most important (Fig. 13-5).

Lower redox potentials are generally observed in lake systems containing relatively high concentrations of dissolved humic compounds (Allgeier *et al.*, 1941, Kjensmo, 1970). Humic acids, especially those derived

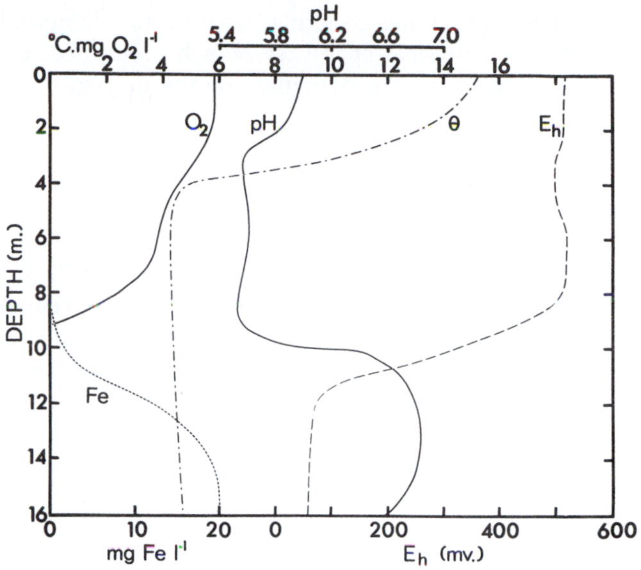

FIGURE 14-1 Vertical distribution of temperature, oxygen, pH, total iron, and redox potential E_h in permanently meromictic Lake Skjennungen, Norway, June 1967. (After Kjensmo, 1970.)

from the moss *Sphagnum* of bogs and bog lakes, have an E_h of about 350 mV (Visser, 1964b), and their reducing properties lead to a metal enrichment by complexing and adsorption to the acid molecules (Szilágyi, 1973; Zimmerman, 1981). The redox potentials of the deep monimolimnia of meromictic lakes are generally extremely low, among the lowest found in natural waters (Fig. 14-1). Extremely high concentrations of soluble total iron are frequently found in the monimolimnia of these permanently stratified lakes of high dissolved organic matter content (Hongve, 1980). In shallow meromictic lakes where the water of the monimolimnion is slowly (over several years) exchanged with the overlying water strata, high monimolimnetic iron concentrations are maintained through the oxidation of ferrous iron in the chemolimnion (Campbell and Torgersen, 1980). Up to 90% of the iron entering the chemolimnion is oxidized and is returned to the monimolimnion by sedimentation and subsequent resolubilization.

III. IRON AND MANGANESE CYCLING IN LAKES

The similarities in chemical reactivity between iron and manganese permit us to discuss them together. Although clear differences exist between the two metals, and although much more is known about the dynamics and cycling of iron, the two metals behave in similar fashion in freshwater systems. A strong interaction exists between the cycling of these metals, especially iron and sulfur. The biogeochemical fluxes of iron and manganese reflect the combined spatial and temporal variations in physical chemistry, which are almost totally controlled by the dynamic conditions regulating bacterial metabolism.

A. Chemical Equilibria and Forms of Iron and Manganese

Iron exists in solution in either the ferrous (Fe^{2+}) or ferric (Fe^{3+}) state. (Detailed reviews are given in Hem and Cropper, 1959; Hem, 1960; Stumm and Lee, 1960; Doyle, 1968a; Millero *et al.*, 1995; and Stumm and Morgan, 1996). Amounts of iron in solution in natural water and the rate of oxidation of Fe^{2+} to Fe^{3+} in oxygenated water are dependent primarily on pH, E_h, and temperature. Ferrous constituents tend to be more soluble than ferric constituents.

Soluble ferrous iron occurs mainly as hydrated Fe^{2+} and hydrated hydroxo ions, the solubility of which is determined largely by the solubility of ferrous hydroxide [$Fe(OH)_2$], ferrous carbonate ($FeCO_3$), and ferrous sulfide (FeS). $Fe(OH)_2$ is exceedingly insoluble within the normal pH range of oxygenated natural waters. The solubility of ferrous iron is generally controlled by the solubility of $FeCO_3$. Even at low pH, carbonate concentrations are usually sufficient to limit solubilization. If water is saturated with $CaCO_3$, $FeCO_3$ is about 200 times less soluble than $CaCO_3$, unless very high pH values (>10) occur. Soft waters with very low concentrations of bicarbonate usually contain somewhat higher concentrations of Fe^{2+}, although most is oxidized to Fe^{3+}. For example, a water free of bicarbonate at pH 7 would contain more than 1000 times as much Fe^{2+} as a water containing 2 meq liter^{-1} alkalinity, an average value for many waters. FeS is also exceedingly insoluble and forms both amorphous and stable, crystalline phases that darken the color of anaerobic sediments (Doyle, 1968b; Davison, 1991). At low redox potentials in the anaerobic hypolimnia of productive or meromictic lakes, bacterial reduction of sulfate to sulfides is common. The resulting excess of sulfide can decrease concentrations of Fe^{2+} through the formation of insoluble FeS.

B. Iron Complexes

The most common species of ferric iron in natural waters is hydrated ferric hydroxide, $Fe(OH)_3$. At equilibrium in the pH range of 5–8, $Fe(OH)_3$ is in the solid state, because its solubility is very low (equilibrium constant approximately 10^{-36} at 25°C). Other insoluble ferric salts are of less significance. For example, phosphate does not influence the solubility of Fe^{3+} when inorganic phosphorus concentration is $<10^{-4}$ mol liter^{-1}, as is usually the case. It is apparent, therefore, that in the absence of organic matter and based only on solubility criteria as mediated by pH and E_h, a number of forms of inorganic iron can exist in natural waters. The simultaneous influence of hydrogen ions and electrons on the equilibria of aqueous iron is illustrated in Figure 14-2. All of the oxygenation reactions of Fe^{2+} to Fe^{3+} are exergonic and capable of supplying energy;

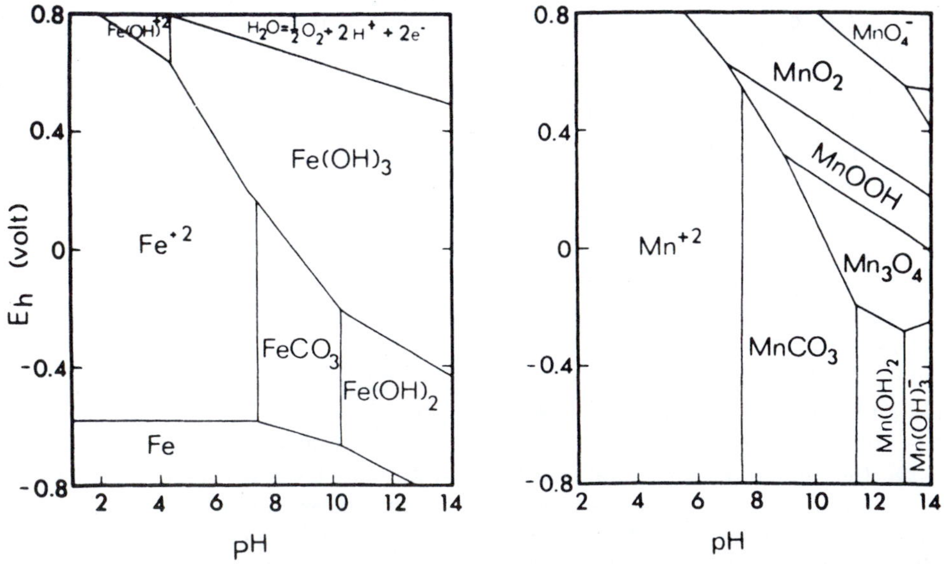

FIGURE 14-2 Approximate distribution of species of iron and manganese in relation to pH and redox potential E_h. Alkalinity is assumed to be equal to 2 meq liter^{-1}. Lines denote points at which the activities of soluble Fe and Mn are 10^{-5} mol liter^{-1}. (Modified from Stumm and Lee, 1960, and Stumm and Morgan, 1996.)

some of these reactions serve as an energy source for microorganisms.

Some of the iron of normal lake water is present as suspensions of flocculent ferricoxyhydroxide, removable by filtration with membranes of a pore size of 0.5 μm (Hutchinson, 1957; Hem and Cropper, 1959; Hem and Skougstad, 1960; Davison and DeVitre, 1992). Under some conditions, a very finely divided precipitate of ferric hydroxide, which has colloidal properties (0.001–0.5 μm, charged), may form (e.g., Tipping *et al.*, 1981; De Haan and De Boer, 1986; Leppard *et al.*, 1988). Colloidal particles of $Fe(OH)_3$ are commonly positively charged, although a negatively charged sol can occur at high pH. Ions in solution, and negatively charged clay particles, organic colloids, and other suspended solids can neutralize the charges on the hydroxide colloidal particles. The uncharged aggregates join to form a rapidly settling precipitate. Other metals, such as copper ions, can be adsorbed by and coprecipitated with the ferric hydroxide precipitate.

Although ferrous iron in solution occurs in extremely low concentrations in oxygenated water, some apparently occurs in particulate and colloidal form. Such suspended ferrous iron is presumably associated with soil and sediment particles. A portion of total iron is contained in the living plankton, but this quantity is a very small part of the whole. The iron content of higher aquatic plants averages 5 mg g^{-1} dry weight, about an order of magnitude greater than the content of terrestrial plants (Oborn, 1960). Plant roots contain a higher proportion of iron than do stems or leaves.

Iron complexes with numerous organic compounds, and this process greatly alters iron solubility and availability to organisms. Many organic bases form strong soluble iron complexes with ferrous and ferric ions (Gjessing, 1964; Millero *et al.*, 1995). Iron enrichment is common in surface waters with a high content of dissolved organic matter, particularly humic and fulvic acids, tannic acids, and other lignin derivatives (Shapiro, 1957; Hem, 1960b; Nürnberg and Dillon, 1993; Shaw, 1994). The intense yellow–brown color of bog waters is, in part, a result of these iron–organic complexes. Natural and synthetic organic compounds, for example, citrate and ethylenediamine tetraacetic acid (EDTA), of high complexing capacity for metals, are commonly used in studies of plant and especially algal nutrition because they can maintain iron available for assimilation. Shapiro (1964, 1966, 1969) showed that complexing of iron with humic derivatives, particularly organic acids of low molecular weight, was by peptization, in which iron is dispersed in a solubilized form of the $Fe(OH)_3$ precipitate by adsorption of the organic acids onto the surfaces of the particles. Some iron is chelated with organic acids by weak chemical bonding, although chelation is not the primary complexing mechanism (cf. Steinberg, 1980). In acidic (pH 4–5) bog lakes, ferric iron may exist not as an inorganic complex but as reactive ferric iron by complexation with colloidal organic acids (Koenings, 1976).

Precipitation particles of iron oxide contain some 30–40% iron by weight (Tipping *et al.*, 1981; Davison and DeVitre, 1992). Carbon content is variable but adsorbed humic carbon can contribute some 4–7% of the total weight. Some particles are negatively charged because of adsorbed humic substances or complexing with COOH and phenolic OH groups of fulvic and humic acids (Tipping, 1981; Schnitzer and Ghosh, 1982; Tipping and Woof, 1983; McKnight *et al.*, 1992; Jones *et al.*, 1993). This complexing, as well as adsorption of phosphate and silicate, would slow flocculation rates. The iron oxide–humic substance complexes can gradually sediment out, of course, and potentially enrich the hypolimnion with humic substances. There is also evidence (Deng and Stumm, 1993) that the reducing power of fulvic acids of the Fe(III)-organic complex can accelerate the shift of the complex to Fe(II).

Photochemical redox reactions of Fe(III) compounds are an important source of Fe(II) and dissolved iron in surface waters. In the absence of organic complexes, iron oxides are solubilized to varying degrees through photodissociation of ferric hydroxy groups at the colloid surfaces (Walte and Morel, 1984). Photochemical redox reactions of Fe(III)-dissolved complexes, dimers, polymers, and precipitates transfer electrons and reduce Fe(III) to Fe(II). Many species of Fe(III) absorb light at wavelengths >300 nm and include species with ligands such as OH^-, H_2O, HO_2^-, HSO_3^-, Cl^-, carboxylates, and O_2^- (Faust, 1994; Sulzberger *et al.*, 1994). Sunlight irradiation of natural waters produced hydrogen peroxide and other radicals in quantities sufficient to affect the speciation of iron.

Natural organic compounds significantly increase the initial rates of dissolution. However, the efficacy of photochemical reductive dissolution of particulate iron varies greatly with stabilization with organic ligands. Organic species may enhance the production of Fe(II) by scavenging hydroxyl and other radicals and thereby decrease the rate of reoxidation of Fe(II). That enhancement is greater in both shallow waters of lakes and streams, particularly in waters of lower pH of 4–6, and is accompanied with an irreversible photooxidation of organic matter (e.g., Collienne, 1983; McKnight *et al.*, 1988).

Another important aspect of photochemical reduction of complexes of humic materials and iron is that phosphorus is commonly adsorbed as well. Irradiance with ultraviolet light resulted in release of phosphate as well as release of reduced Fe(II) in acidic bog water (Cotner and Heath, 1990). The significance of such photochemical release of phosphorus adsorbed to humic–Fe complexes over seasonal scales is unclear.

Total iron found in oxygenated surface waters of pH 5–8 typically ranges from about 50 to 200 μg liter^{-1}, almost none of which occurs in ionic form (Wetzel, 1972). Much higher levels occur in lakes heavily stained with dissolved humic compounds, in acidic volcanic lakes, acidic bog waters, and certain alkaline, closed lakes rich in organic matter.

C. Manganese Complexes

The theoretical thermodynamic redox equilibria of manganese have been studied in some detail (Hem, 1963, 1964; Stumm and Morgan, 1996) and permit an evaluation of manganese species and equilibria under various conditions of oxidation reduction found in natural surface waters. In general, the behavior of manganese in lakes follows these predictions.

Although Mn occurs in several valence states, Mn^{3+} is thermodynamically unstable in aqueous solutions under normal conditions, and Mn^{4+} compounds are insoluble at most environmental pH values. As with ferrous iron, ionic divalent Mn^{+2} occurs at low redox potentials and pH (Fig. 14-2). Some form of oxidized Mn will be in equilibrium with Mn^{2+} under oxidizing conditions of high pH and E_h, and some form of Mn^{2+} may be in equilibrium with manganese carbonate under reducing conditions of low pH and E_h. Supersaturation of both ferrous and manganese carbonate has been observed in anoxic hypolimnia of eutrophic lakes (Verdouw and Dekkers, 1980). The redox equilibrium E_h values are higher and the rates of oxidation slower than for iron; as a result, detectable quantities of Mn are commonly observed longer than comparable quantities of iron under lake conditions. Above a pH of 8.5, an intermediate complex forms in which Mn^{+2} is adsorbed onto manganese oxides. The Mn^{2+} of these oxide complexes can react relatively rapidly with other anions and precipitate as manganous carbonate ($MnCO_3$), manganous sulfide (MnS), and manganous hydroxide ($Mn(OH)_2$). Manganese is also adsorbed to iron oxides and coprecipitates with ferric hydroxide when the pH exceeds 7.

Manganese has a solubility of about 1 mg liter^{-1} in distilled water of E_h of 550 mV and pH 7; solubility decreases markedly with increases in E_h and pH. Manganese forms soluble complexes with bicarbonate and sulfate. Increased bicarbonate activity decreases Mn solubility. At high concentrations, bicarbonate has been found to reduce the oxidation rate of manganese. In a manner analogous to iron, organic molecules can form stable complexes with Mn^{2+}, although their operation in aquatic systems is poorly understood. Manganese occurs in relative abundance in alkaline soils as hydrated oxides, and no doubt organic complexing plays a role in retention of Mn in a complexed dissolved form for transport once redox conditions are altered by microbial and plant metabolism. Igneous rock contains about 0.01% manganese, which is involved in biogeochemical cycling following weathering and subsequent mobilization. Drainage from forest litter, especially from more acidic coniferous forests, is often high in manganese. This observation again points to the probability of the major role of organic complexes in the effective transport of easily oxidized metal ions.

D. Distribution of Iron and Manganese in Lakes and Streams

Under oxidized conditions, as in the epilimnia of lakes and most streams, large amounts of iron are found only in acidic water (pH <3–4), such as in lakes of volcanic origin and influence (Yoshimura, 1936) or runoff streams from strip mining operations. In both cases, organic content is relatively low and acidity usually results from sulfuric acid. When streams that receive acidic mine drainage are exposed to sunlight, photoreduction of ferric iron, likely in colloidal forms, can result in a daytime production of ferrous iron that is nearly four times as great as nighttime oxidation of ferrous iron (McKnight *et al.*, 1988).

The quantity of total iron found in most typical neutral or alkaline surface waters, however, is in the range of 50–200 μg liter^{-1} and is dominated by $Fe(OH)_3$, usually in colloidal and particulate forms, organically complexed iron, and adsorbed sestonic iron in particulate forms. The world average is variable among continents (Table 10-1), and higher than would be generally expected among most oxygenated surface waters of lakes. The range of manganese concentrations (approximately 10–850 μg liter^{-1}) is also highly variable in relation to lithology and drainage of the lake basins (Hutchinson, 1957; Livingstone, 1963; Hongve, 1980). The average quantity of manganese is about 35 μg liter^{-1}, somewhat less than that of iron. The ratio of Fe to Mn in water is generally considerably lower than that of the lithosphere (50:1) and indicates the relative enrichment of Mn with respect to Fe; this is in agreement with the reaction equilibria already discussed.

The vertical distribution of iron and manganese is reflected in the distribution of redox potentials. Concentrations of ionic iron of oxygenated waters of oligotrophic lakes, epilimnia of more productive lakes and of circulating waters are exceedingly low. Manganese is

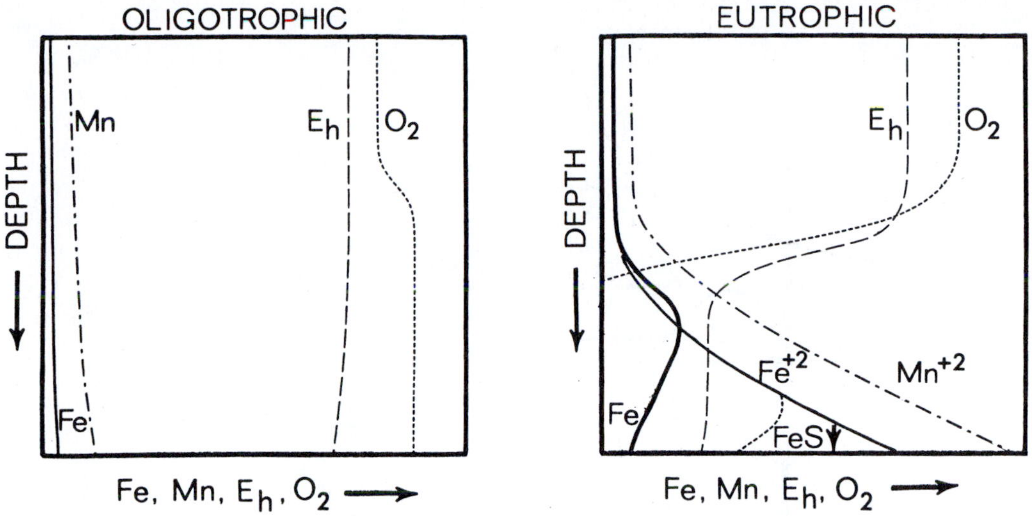

FIGURE 14-3 Generalized vertical distribution of iron, manganese, and redox potential (E_h) in stratified lakes of very low and high productivity.

somewhat more soluble (Fig. 14-3). Ferrous ions diffuse readily from the sediments when redox potentials decline to about 200 mV; migration of Mn^{2+} from the sediments occurs at somewhat greater redox potentials (Robbins and Callender, 1975; Ostendorp and Frevert, 1979). These relationships were demonstrated very well in Mortimer's studies of the sediment interface, both experimentally (see Fig. 13-5) and in a eutrophic lake (see Fig. 13-6). Here it can be seen that the release of Mn precedes that of iron. Released manganese will remain soluble if the oxygen saturation is less than about 50% (Burns and Nriagu, 1976). Iron oxide can form in the absence of particulate matter, but its particle morphology depends upon the presence of particulates. In contrast, the formation of manganese oxide requires living microorganisms to catalyze the oxidation of manganous ions (see later discussion; Ghiorse, 1984; Tipping, 1984; Balikungeri *et al.*, 1985; Tipping *et al.*, 1985).

Thus, a commonly observed seasonal sequence is observed in productive lakes in which the oxic/anoxic boundary migrates from the sediment–water interface well into the hypolimnion (Fig. 14-4). As the hypolimnion becomes more reducing, a progressive shift occurs from the amorphous particulate iron oxyhydroxides to soluble ferrous iron. Where appreciable sulfide is formed by biological reduction of sulfate, ferrous and total iron can be reduced by the formation of and precipitation of insoluble FeS (e.g., Hutchinson, 1957; Cook, 1984). Although the dissolved iron concentrations may be controlled in the lower strata during stratification by FeS precipitation, MnS phases are generally undersaturated and appreciably less than the solubility constants needed for precipitation (Balistrieri *et al.*, 1992). Most of the iron (>90%) reaching the sediments by sedimentation is reduced and redissolved at the sediment–water interface and recycled back into the water column.

Although the seasonal distribution of iron has been studied in some detail (reviewed in Hutchinson, 1957; McMahon, 1969; Davison and De Vitre, 1992), that of manganese has been investigated less extensively (Delfino and Lee, 1968, 1971; Brezonik *et al.*, 1969; Howard and Chisholm, 1975). The general pattern of seasonal distribution is apparent from the examples given of a mesotrophic lake that undergoes hypolimnetic oxygen reduction in the later phases of summer stratification (Fig. 14-5) and of an interconnected eutrophic lake (Fig. 14-6). The eutrophic lake, Little Crooked Lake, is a small, protected eutrophic lake that frequently undergoes partial, temporary meromixis, as was the case in the year shown here when autumnal circulation was incomplete. Of the total manganese, nearly all was in soluble and colloidal form (passing through a 0.45-μm pore size membrane filter) in anaerobic water of the hypolimnion of a small eutrophic lake in New York (Howard and Chisholm, 1975). Less than 13% of the total manganese was in soluble form in the aerobic water strata of this lake.

As decomposition proceeds in the hypolimnion of very productive, thermally stratified lakes, the redox potential of hypolimnetic waters can decline to well below 100 mV. At an E_h below 100 mV, sulfate is reduced to hydrogen sulfide. Hydrogen sulfide is also produced by bacterial decomposition of sulfur-containing organic compounds. Since ferrous iron is released in significant quantities from the sediments at a higher E_h of about

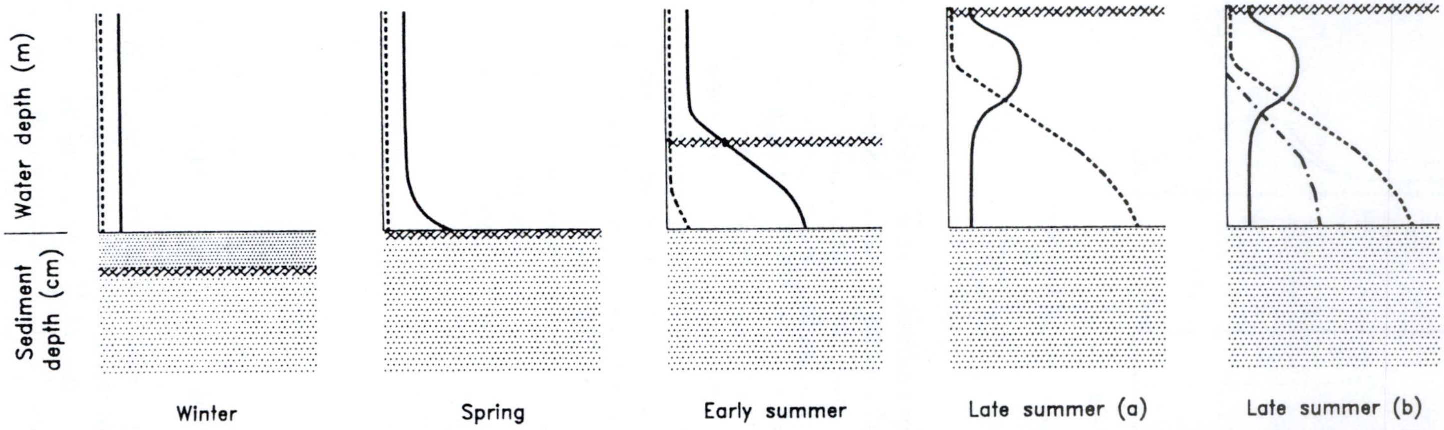

FIGURE 14-4 Changes in the concentration of particulate iron (——) and Fe(II) (- - - -) in the water column of a lake as the oxic/anoxic boundary (xxxx) undergoes seasonal migration. The sediment is represented by light hatching, elevated concentrations of particulate iron in the sediment are represented by darker hatching. The alternative late summer conditions (b) represent marked reducing conditions where sulfide (-·-·-·-) and FeS precipitation can contribute to the alterations of vertical distributions. (Modified from Davison and DeVitre, 1992.)

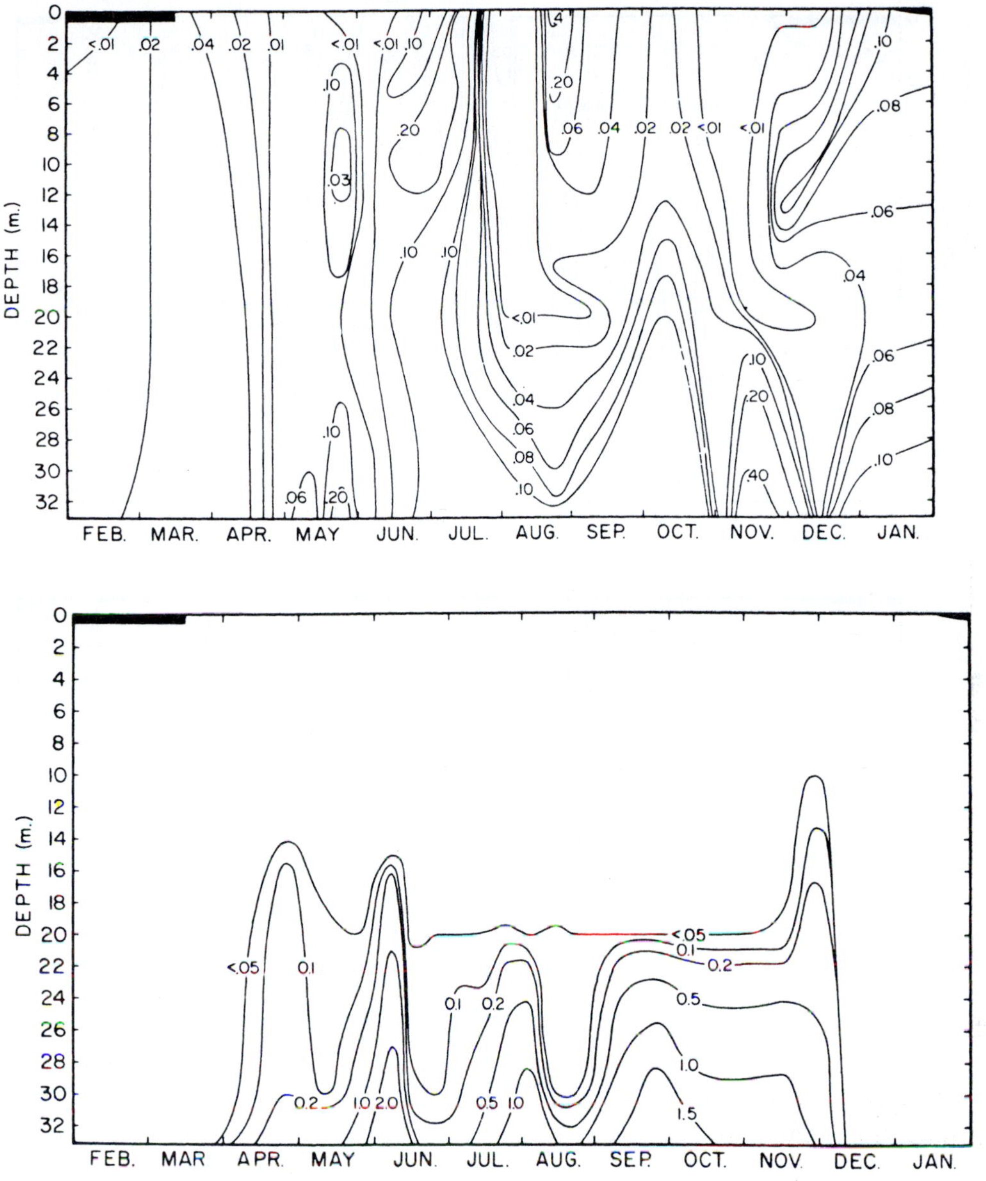

FIGURE 14-5 Depth–time diagrams of isopleths of total iron (*upper*) and manganese (*lower*) in mg liter^{-1} of mesotrophic hardwater Crooked Lake, northeastern Indiana, 1963. (From Wetzel, unpublished data.)

250 mV, much Fe^{2+} is present in the hypolimnion at the time of sulfide formation. The formation of FeS and other metal sulfides (cuprous sulfide, CuS, cadmium sulfide, CdS, et cetera), all of which are very insoluble under normal lake conditions, can result in a significant reduction of iron and other metals toward the end of summer stratification (Davison and Heaney, 1978). Manganous sulfide, on the other hand, is much more soluble and has little effect on the Mn^{2+} concentrations under normal lake conditions.

Iron concentrations in the hypolimnia of softwater lakes can reach very high levels under conditions that prevail in small, deep basins, especially in bog waters receiving high concentrations of humic organic matter. Levels of sulfate are low in such waters, and sulfide concentrations seldom become sufficient to precipitate iron as FeS. Kjensmo (1962, 1967, 1968) demonstrated that the hypolimnetic iron accumulations in protected lakes can reach such levels (>250 mg liter^{-1}) that the salinity gradient becomes adequate to render the lakes permanently meromictic. The precipitation of iron sulfides is inadequate to reduce such high iron concentrations appreciably and contributes to meromictic conditions by allowing iron to accumulate.

Stages along the continuum of declining redox potentials can be divided into four phases of hypolimnetic

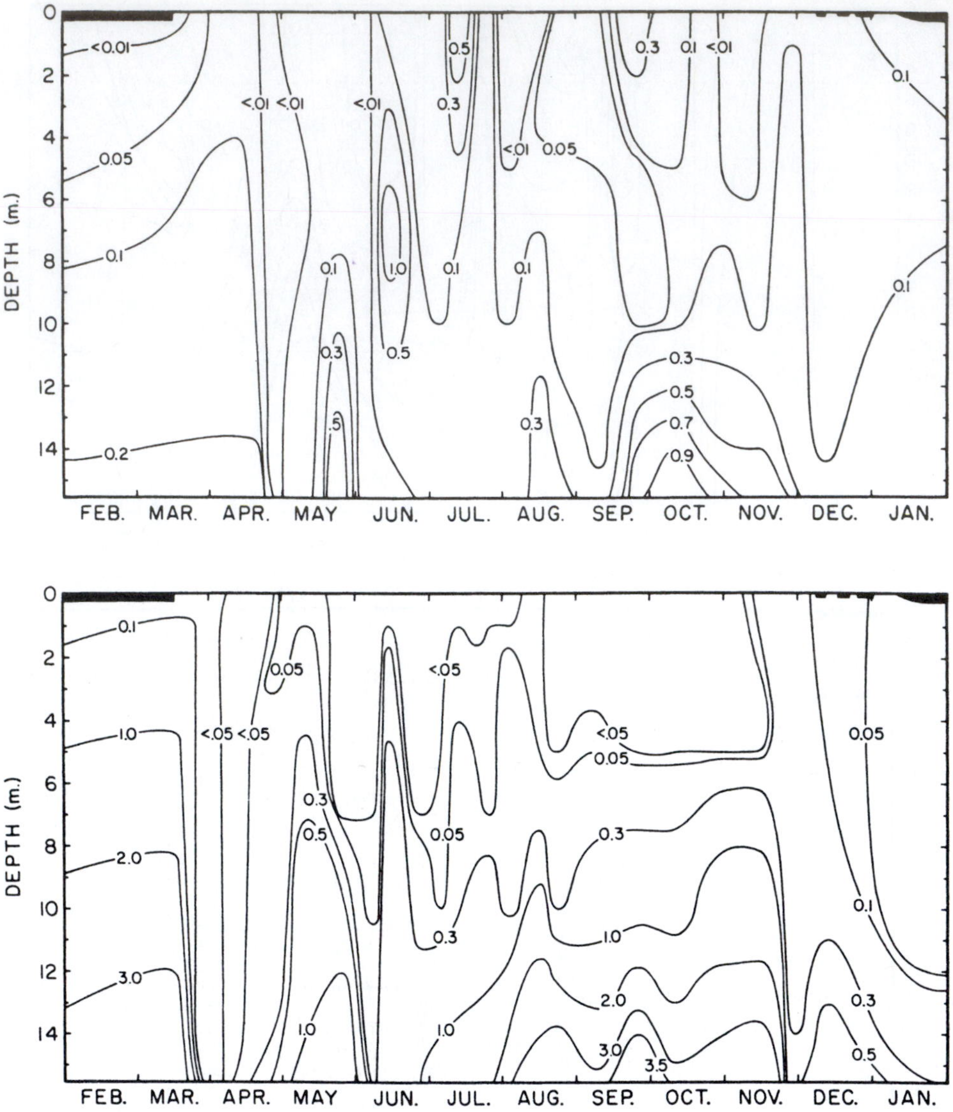

FIGURE 14-6 Depth–time diagrams of isopleths of total iron (*upper*) and manganese (*lower*) in mg liter^{-1} of eutrophic hardwater Little Crooked Lake, northeastern Indiana, 1963. (From Wetzel, unpublished data.)

conditions in stratified lakes of increasing productivity or within a very productive lake during the period of summer stratification (Table 14-1).

E. Cycling of Iron and Manganese

The spatial and temporal distribution of iron and manganese therefore depends upon the balance of many physical, chemical, and biological parameters. Solar radiation and wind affect the hydrodynamic stability of stratification. Biologically mediated chemical processes regulate the redox conditions and the availability of electron acceptors and extent of reduction to allow accumulation of sulfide and subsequent precipitation of metal sulfides.

The mass of particulate iron, for example in eutrophic Esthwaite Water of England, changes little throughout the year (Davison *et al.*, 1980). The large summer increase in the mass of total iron was almost entirely in the soluble and colloidal (<0.7 μm) fraction (Fig. 14-7). Sediment-derived ferrous iron accounted for only 3–18% of the annual iron loading to the lake; most of this sediment-derived fraction was likely oxidized and washed out of the lake rather than returned to the sediment (Fig. 14-7). Most (70–90%) of the iron annually entering the lake accumulated in the sediment (Table 14-2). Redissolution of sinking ferric particles is significant to iron fluxes within the hypolimnion. For example, in Esthwaite Water minimal sinking fluxes were 3–22 μg Fe cm^{-2} day^{-1}. In contrast, much

TABLE 14-1 Changes in the Iron During the Continuum of Declining Redox Potentials in the Hypolimnia of Stratified Lakes of Increasing Productivity[a]

Lake status	$[O_2]$	E_h	Fe^{+2}	H_2S	PO_4^{-3}
Oligotrophic	High (orthograde)	400–500 mV	Absent	Absent	Very low
	↓	↓	↓	↓	↓
↓	Much reduced (clinograde)	400–500 mV	Absent	Absent	Very low
	↓	↓	↓	↓	↓
Europhic	Much reduced (clinograde)	Approx. 250 mV	High	Absent	High
↓	↓	↓	↓	↓	↓
Hypereutrophic	Much reduced (or absent)	<100 mV	Decreasing	High	Very high

[a] After discussion of Hutchinson (1957).

of the total manganese is in organic forms. For example, about 40% of total dissolved manganese was present in organic compounds, mainly with a molecular weight of 1000–50,000 kD in the hypolimnion of a eutrophic Japanese lake (Yagi, 1988).

Within the metalimnion, upward and laterally diffusing Fe(II) oxidizes and coprecipitates a large fraction of soluble phosphorus (Stauffer, 1986; Stauffer and Armstrong, 1986). As the epilimnion deepens in late summer and depresses the oxic/anoxic boundary, reduced iron and manganese of the hypolimnion is progressively oxidized and mixed into the epilimnetic water. The most prevalent form of iron in this fraction is amorphous Fe(III) oxide of small size (<0.5 μm) (Tipping *et al.*, 1982). The concentrations of epilimnetic iron reach maximum near overturn and then slowly decline. Sedimentation of iron oxides likely occurs by flocculation by self-association or by association with other particulate matter. Much iron remains in the water column by resuspension of sedimented materials (Davison *et al.*, 1980).

The cycling and fluxes of manganese are influenced by microbial utilization, particularly in strongly stratified, eutrophic lakes. Although these specialized bacteria are discussed collectively later in this chapter with sulfur, it is instructive to note here forms and fluxes

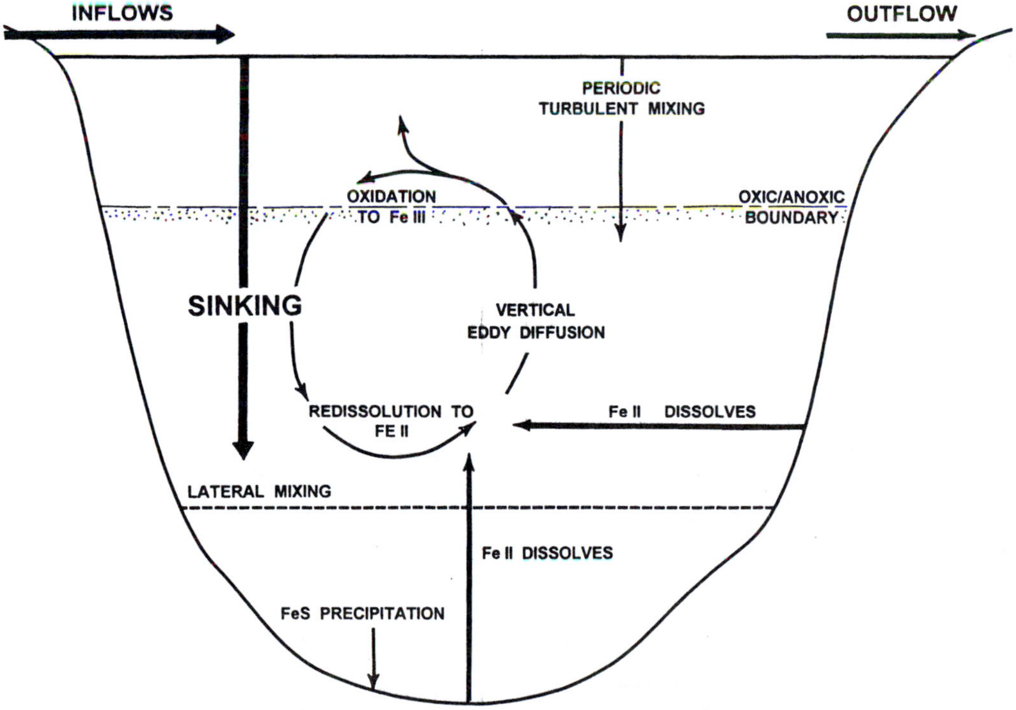

FIGURE 14-7 Model of transport processes of iron in a lake with an anoxic hypolimnion. (Based on Davison *et al.*, 1980.)

TABLE 14-2 Budget of Iron Sources and Sinks, Esthwaite Water, England[a]

	m tons yr^{-1}
Inputs:	
Minimum annual inflows	20
Annual dissolution from sediments	1.8–3.6
Outputs:	
Annual outflow from lake	6.4
Annual accumulation in sediments	15–60
Minimum net loss from water column	14

[a] Extracted from data of Davison *et al.* (1980).

of manganese in a representative eutrophic lake (Fig. 14-8). In Lake Fukami-ike, the boundary stratum between oxic and anoxic waters contained ca. 0.5 mg O_2 liter^{-1} and 0.5 mg S^{-2} liter^{-1} as H_2S (Yagi, 1993). Some 0.08 g Mn m^{-2} day^{-1} was gained in this stratum from epilimnetic sedimentary fluxes and from the anoxic hypolimnion. The largely manganous ions and organically bound Mn of the hypolimnion accumulated from chemical reactions as the redox declined and from manganese-reducing bacteria. DOC concentrations were reduced only in the metalimnion, where some 40, 33, and 27% of the total decrease in DOC was attrib-

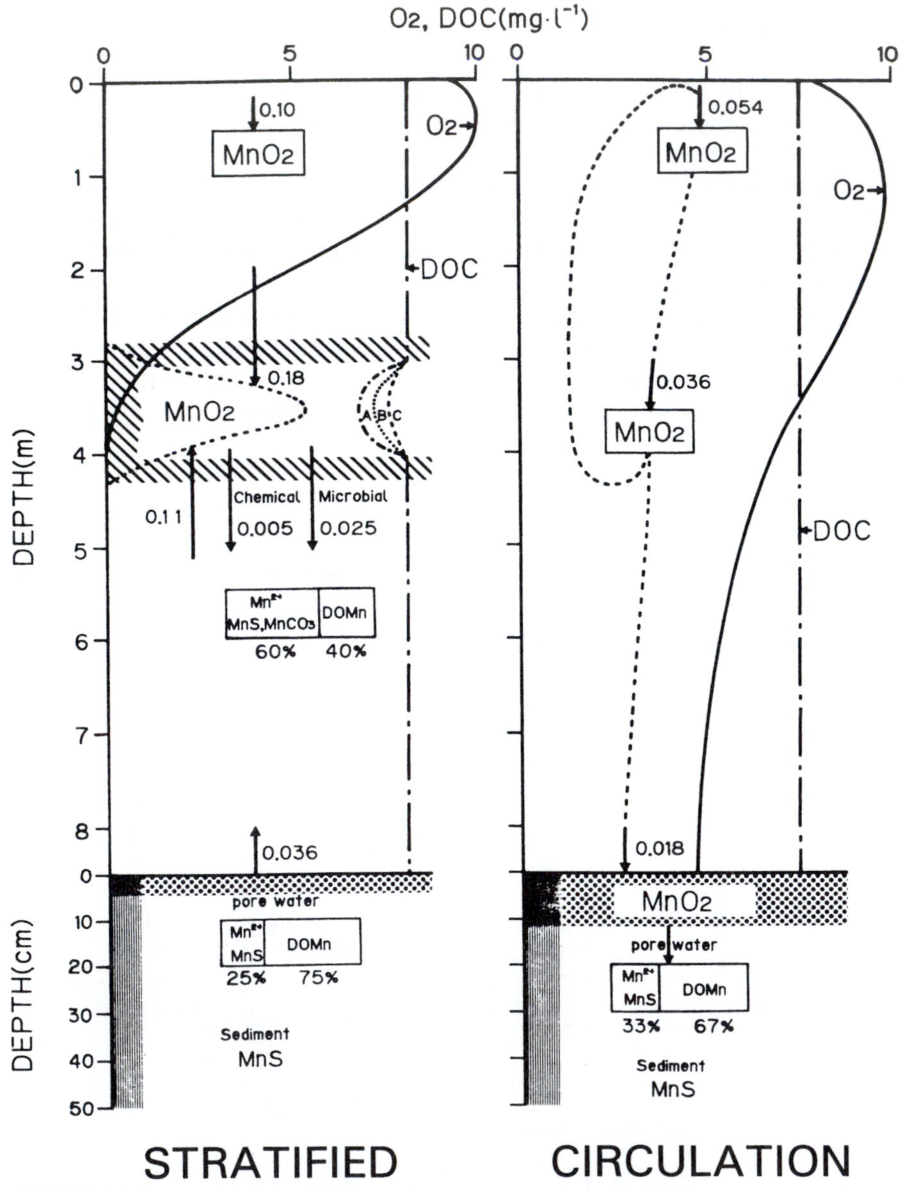

FIGURE 14-8 Manganese cycling and fluxes in relation to dissolved oxygen and dissolved organic carbon in eutrophic Lake Fukami-ike, Japan, during summer stratification and during autumnal circulation. Units of Mn = g Mn m^{-2} day^{-1}. *A*, *B*, and *C* = minimum DOC decrease by Mn-reducing bacteria (40%), the purple nonsulfur bacteria *Rhodopseudomonas paulustris* (33%), and heterotrophic bacteria (27%), respectively. (Modified from Yagi, 1993.)

utable to the Mn-reducing bacteria, a purple nonsulfur bacterium, and heterotrophs, respectively (Fig. 14-8). During lake circulation, most of the Mn was in the form of MnO_2, with appreciable mixing and resuspension in the upper strata. When lakes are acidic or become acidic from anthropogenic causes, Mn tends to be highly conservative, with low rates of retention (e.g., White and Driscoll, 1987).

Fluxes to and mobilization of Fe and Mn from the sediments has already been treated in some detail here and in Chapter 13 in relation to the release of phosphorus from sediments. The cycling and importation of Fe and Mn to the sediments must be evaluated within redox conditions in the lake strata or river system, the settling and focusing (see Chap. 21) of particles, particularly Fe and Mn oxyhydroxides, and the resuspension of surficial sediments. Because sediments of deeper areas tend to be more reducing than those of shallower areas, particulate sedimentation of Fe and Mn oxides tend to be greater in shallow-water sediments (Hamilton-Taylor and Morris, 1985). Hypolimnetic sediments normally exhibit high Mn remobilization rates throughout the year despite a well-mixed, oxygen-saturated water column during autumn and much of winter. Not only are the manganese compounds more soluble than iron, much (40–75%) of the total dissolved Mn in the interstitial waters can be in organic forms of relatively high molecular weight (e.g., two fractions of 1–50 and 50–200 kD; Yagi, 1990).

Iron remobilization from the sediments is regulated by oxygen-mediated redox and rates of sedimenting organic matter (Figs. 13-5 and 13-6). Under the prevailing anoxic conditions of nearly all sediments of lakes, Fe(III) oxides are subject to dissimilatory microbial Fe(III) reduction in which Fe(III) or Mn(IV) are used as an electron acceptor in metabolism. Microbial utilization by Fe(III)-reducing bacteria continues until Fe(III) oxides are depleted. Fe(III)-reducing bacteria are able to outcompete both sulfate-reducing and methanogenic bacteria for organic substrates (Lovley, 1991). High Fe(III) oxide concentrations can be regenerated by rapid oxidation of Fe(II) compounds coupled to oxygen inputs from overlying water, if oxygenated, or in littoral and wetland areas from oxygen released from plant roots (Roden and Wetzel, 1996). Where this oxidation is not available to the sediments, Fe(II) accumulates and migrates into the overlying anoxic water. Radial oxygen loss from roots of aquatic plants (cf. Chap. 18) induces the oxidation of Fe(II) and the formation of appreciable deposits of iron plaque on the roots (e.g., St-Cyr and Crowder, 1989; Kirk and Bajita, 1995; Snowden and Wheeler, 1995; Roden and Wetzel, 1996). Although these deposits have both positive and negative effects on the physiology of the plants, the deposition can result in appreciable oxidation and retention of iron in vegetated sediments.

F. Utilization and Transformations of Iron and Manganese

The metabolic demands for iron and manganese are usually sufficiently low so that the biota do not materially reduce the concentrations of these metals in the environment. Both iron and manganese, however, are essential micronutrients of microflora, plants, and animals (Oborn, 1960b; Wangersky, 1963; Coughlan, 1971). Iron is required in the enzymatic pathways of chlorophyll and protein synthesis, the protein integrity of cell membranes, and in the respiratory enzymes of all living organisms. The function of iron in cytochromes and as the basic component of hemoglobin in higher animals is well known. Manganese and iron are functional components of nitrate assimilation (e.g., Verstreate *et al.,* 1980; Rueter and Ades, 1987), nitrogen fixation (e.g., Wurtsbaugh and Horne, 1983; Rueter, 1988), and photosynthesis and are essential catalysts of numerous enzyme systems in bacteria, cyanobacteria, algae, and animals (Ferreira and Straus, 1994).

Even though the biological requirements for iron and manganese are low, their reactivity, very low concentrations, and restricted availability (particularly of iron) in the trophogenic zone of lakes and in streams suggests that under certain conditions, availability of Fe and Mn can limit photosynthetic productivity. The mechanisms for assimilation of iron from the forms available in oxygenated natural waters are emerging from research. The peptizing and chelating properties of organic acids for iron, which have been found in a large number of lakes (e.g., Shapiro, 1966, 1969), have been shown in a large number of studies to maintain assimilable iron and to enhance the metabolism and growth of many microbiota. Several examples are presented next.

Bacteria, cyanobacteria, and certain algae produce ***siderophores,*** low-molecular-weight glycoproteins that are specific for Fe(III). Under iron-limiting conditions, siderophores are mostly produced extracellularly and excreted into the environment (e.g., Kerry *et al.,* 1988; Benderliev and Ivanova, 1994; Wilhelm, 1995). The iron of these complexes must be transported across the membrane by transport proteins for internal reduction to Fe(II) and assimilation (Guerinot, 1994). Among those microbiota that do not synthesize and release siderophores, iron complexed with natural organic compounds, such as humic acids, can be utilized.

The available iron content of hardwater calcareous lakes is extremely low. For example, the reactive iron of Lawrence Lake, Michigan, seldom exceeded 5 μg liter^{-1}

over an annual period (Wetzel, 1972). In this and similar lakes, it is evident that iron availability is so low that high sustained primary productivity is limited by an effective iron unavailability (Schelske, 1962; Schelske *et al.*, 1962; Wetzel, 1965a, 1966b, 1972). The addition of iron in complex form, either by synthetic chelating or natural complexing organic compounds, resulted in immediate increases in photosynthetic rates. Additions of competing organic compounds were less effective but presumably increased the availability of iron already present in the water. The effectiveness of natural organic compounds from the hypolimnion and amino compounds in maintaining solubility of iron in hardwater lakes has also been demonstrated (Wetzel, 1972). Since these early, detailed studies, iron availability has been implicated, sometimes with marginally convincing experimental evidence, as essential for sustained phytoplanktonic rates of primary production (e.g., Wurtsbaugh and Horne, 1983; de Haan *et al.*, 1985; Chang *et al.*, 1992; Pollingher *et al.*, 1995) and planktonic bacterial productivity (Berman *et al.*, 1993) as well as altering species successions of planktonic algae. It may also be noted that iron can be toxic at very high concentrations, particularly as ferric hydroxide and Fe–humus precipitates on biological and inert surfaces, especially in rivers (Vuori, 1995). Ingestion of sorbed or coprecipitated heavy metals with Fe-oxides can increase the dietary supply of metals to toxic concentrations in the food web.

High concentrations of manganese (>1 mg liter^{-1}) are very inhibitory to cyanobacteria and green algae and can induce marked changes in development and morphology (Gerloff and Skoog, 1957; Patrick *et al.*, 1969; Lorch, 1978). An antagonistic response was demonstrated in which increasing calcium concentrations progressively reduced the inhibitory effects of high manganese. This relationship indicates that high levels of manganese, for example at the time of fall circulation, could be inhibitory to natural populations of cyanobacteria and green algae. Manganese concentrations of <50 μg liter^{-1} were found to inhibit the development of green algae and cyanobacteria in streams and to strongly favor diatom growth. This response by diatoms may be ubiquitous; if so, the moderate levels of Mn and high levels of Ca in hardwater lakes may contribute both to the general dominance of diatoms and to the stimulatory effects of primary productivity in response to small additions of chelated manganese observed by Wetzel (1966b) in these lakes.

In Lake Superior, an apparent synergistic effect between low concentrations of manganese and phosphorus can occur (Shapiro and Glass, 1978). Phytoplanktonic photosynthesis was enhanced much more by low-level enrichments of combined manganese and phosphate than by the addition of either substance alone.

IV. BACTERIAL TRANSFORMATIONS OF IRON AND MANGANESE

The cycling of iron and manganese is largely dictated by the oxidation-reduction conditions of lakes. Although bacterial and photosynthetic metabolism greatly influences these controlling conditions, which indirectly regulate the states of these metals and their fluxes, certain bacteria utilize iron and manganese directly in energetic transformations (Lundgren and Dean, 1979; Ghiorse, 1984; Jones, 1986). These transformations are usually minor, however, in comparison to heterotrophic metabolism of organic substrates in most natural systems.

A. Iron and Manganese Deposition (Oxidation)

Most oxidation of Fe(II) at neutral pH is essentially a chemical process. Clearly, however, bacteria are associated with oxidized iron, and their cellular structures become encrusted with Fe(III). Filamentous forms of iron-oxidizing bacteria are gram-negative rods that form chains bound by sheaths, such as *Sphaerotilus/Leptothrix* (Ghiorse, 1984; Jones, 1986). Prosthecate and appendaged forms (e.g., *Gallionella*) and encapsulated and coccoid forms (e.g., *Metallogenium*, and *Siderocapsa*) add to a diverse group of morphological types. Fe(III) is generally deposited on outer cell surfaces that are usually anionic in charge. Iron- and manganese-depositing bacteria develop at the sites of a redox gradient, usually dominated by the Fe(II)/Fe(III) couple. These redox gradients can be as large as several meters of water strata in the metalimnion and upper hypolimnion of eutrophic lakes or only a few centimeters at oxic/anoxic interfaces, such as iron groundwater seeps into rivers. Chemosynthetic utilization of energy from inorganic oxidations is relatively inefficient, especially in the case of the oxidation of iron and manganese. For example, the oxidation of Fe^{2+} to Fe^{3+} by iron bacteria releases only about 11 kcal per mole Fe.

The cycling of iron and manganese is influenced by two processes (Kuznetsov, 1970; Ghiorse, 1984). First, as pointed out earlier, reduction of the oxidized combined metal occurs under appropriate redox conditions as ferrous bicarbonate or is precipitated and sedimented as a sulfide. Example reactions are

$$Fe_2O_3 + 3H_2S \rightleftharpoons 2FeS + 3H_2O + S$$

$$FeS + 2H_2CO_3 \rightleftharpoons Fe(HCO_3)_2 + H_2S$$

Second, sheathed and stalked bacteria, algae, protozoan flagellates, and some true bacteria precipitate ferric and manganic oxides on their cells. The true iron bacteria occur in iron-rich waters of neutral or alkaline pH. Characteristic reactions of the few chemoautotrophic bacteria that deposit hydroxides and oxides are

$$4Fe(HCO_3)_2 + O_2 + 6H_2O \longrightarrow 4Fe(OH)_3 + 4H_2CO_3 + 4CO_2 + 58 \text{ kcal}$$

$$4MnCO_3 + O_2 \longrightarrow 2Mn_2O_3 + 4CO_2 + 76 \text{ kcal}$$

Some species of *Leptothrix* are facultative iron bacteria that can oxidize both ferrous and manganous salts, whereas *Gallionelia (Spirophyllum)* is restricted obligately to iron. Since at neutral pH and in the presence of oxygen, Fe^{2+} is spontaneously oxidized, iron-oxidizing bacteria are restricted to zones of steep redox gradients, in which they can compete effectively with oxygen for reduced iron. Therefore, the iron bacteria are restricted to the interface regions of iron-bearing rock seeps, swamps, and bogs and to upper hypolimnetic areas, where the redox potential is sufficiently low for reduced iron to occur. Over 220 g of ferrous iron are required to produce 0.5 g of cellular carbon. As a result, much oxidized iron will be precipitated on the sheaths of the bacteria and extruded materials (e.g., Emerson and Revsbech, 1994). Although this iron may not be important from the standpoint of synthesis of organic carbon, it is often important economically, for it causes corrosion and clogging of pipes. These examples refer to strict autotrophic bacteria that satisfy their CO_2 requirements without organic matter; they obtain all required energy from the oxidation of some specific, incompletely oxidized inorganic substance. Iron or manganese thereby enters the cells. Certain facultative autotrophic bacteria (mixotrophic) can develop in water containing only inorganic substances; other facultative autotrophic species are able to utilize organic substances as well. When these organisms occur together and some iron reduction can occur, iron cycling is possible within the attached microbial communities (Emerson and Revsbech, 1994).

Other groups of bacteria involved in the cycling of iron and manganese are heterotrophic. Certain filamentous forms (*Cladothrix,* some *Leptothrix*), in particular, deposit iron and manganese on the cell in sheaths during the process of metabolizing organic compounds. The colonial, coccoid cells or short rods of *Siderocapsa* of the Eubacteriales are a common form occurring at the oxic–anoxic interface zone of hypolimnion–metalimnion, especially in iron meromictic lakes (Dubinina *et al.,* 1973). *Siderocapsa* also is widely distributed in oxygenated zones of streams and lakes (Hardman and Henrici, 1939) and during periods of circulation (Sokalova, 1961). Species of this genus can mineralize humates and increase in numbers coincident with increases in iron humates during periods of high rainfall. *Spirothrix* is another dominant iron bacterium found in both aerobic and anoxic zones of Lake Glubok, Russia, whereas large populations of *Gallionella* developed only at the interface zone of the metalimnion. Over the period of an annual cycle in this lake, the number of iron bacteria reached a very high percentage (19%) of the total bacteria during the later portion of summer stratification. At other times of the year, the percentage was about 5% of the total bacteria. It is apparent that the iron bacteria significantly influence the iron cycle of more productive lakes.

Closely related to *Thiobacillus* (see the next section) is *Ferrobacillus ferrooxidans,* which is abundant in very acidic (pH <3) mine waters, in which ferrous iron is soluble and stable. *Ferrobacillus* oxidizes ferrous carbonate to ferric hydroxide and CO_2:

$$4FeCO_2 + O_2 + 6H_2O \longrightarrow 4Fe(OH)_3 + 4CO_2$$

A few heterotrophic species of iron-oxidizing bacteria of the genera *Sphaerotilis, Leptothrix, Clonothrix,* and *Siderobacter* deposit oxidized manganese along with iron on capsules in their sheaths. Some species of *Metallogenium* obtain part of their energy requirements from the oxidation of manganous oxide (MnO), manganous sulfate ($MnSO_4$), or manganous carbonate ($MnCO_3$) to manganese sesquioxide (Mn_2O_3) and manganese dioxide (MnO_2); others are heterotrophic and, along with several other manganese- and iron-oxidizing bacteria, contribute to the formation of manganese and iron oxides in lake sediments (Oborn, 1964; Perfil'ev and Gabe, 1969; Kuznetsov, 1970; Jaquet *et al.,* 1982). *Metallogenium* is undoubtedly one of the dominant microorganisms involved in the deposition of manganese nodules in lakes (Sokalova, 1961; Sorokin, 1970; Miyajima, 1992).

B. Iron and Manganese Reduction

Microorganisms can catalyze enzymatically the reduction of many metals. Microorganisms have specific metabolic systems for oxidized Fe(III) and Mn(IV) in which the oxidation of organic matter is coupled to metal reduction. Nonenzymatic processes, for example, reduction of Fe(III) by organic compounds and sulfide, are generally of minor significance (Lovley, 1987, 1991, 1995; Nealson and Saffarini, 1994). Mn(IV) reduction by nonenzymatic processes is somewhat more common than that of Fe(III).

During the metabolism, complex organic matter is first hydrolyzed by enzymes that release monomeric compounds such as sugars, amino acids, and long-chain fatty acids (Lovley, 1991). Fermentative bacteria metabolize the sugars and amino acids to short-chain fatty acids (e.g., acetate) and H_2. These fermentative products are then oxidized with the reduction of Fe(III) to CO_2 and water.

Under acidic conditions, several autotrophic bacteria (e.g., *Thiobacillus thiooxidans, T. ferrooxidans*, and *Sulfolobus acidocaldarius*) can oxidize iron disulfides and sulfur (S^0) to sulfate, with Fe(III) serving as the electron acceptor (ZoBell, 1973; Lovley, 1995). For example:

$$FeS_2 + 3\tfrac{1}{2}O_2 + H_2O \longrightarrow FeSO_4 + H_2SO_4$$

$$2FeSO_4 + \tfrac{1}{2}O_2 + H_2SO_4 \longrightarrow Fe_2(SO_4)_3 + H_2O$$

Evidence also indicated that Mn(IV) can also be reduced with S^0 as the electron donor in aquatic sediments.

As will be discussed subsequently in detail (see Chap. 23), a major portion (>60%) of the dissolved organic matter of inland waters consists of humic substances, largely originating from structural tissues of higher plants. Recently, it has been demonstrated that a wide variety of microorganisms capable of Fe(III) reduction were also able to transfer electrons to humic compounds, primarily quinone groups, during the process of oxidation of substrates such as acetate to CO_2, both in water (Lovley *et al.* 1996, 1998; Scott *et al.*, 1998) and in sediments (Roden and Wetzel, 2000). Humic substances can stimulate the reduction of poorly crystalline Fe(III) oxide forms (goethite and hematite) in clay by an electron shuttling between Fe(III)-reducing microorganisms and Fe(III). The transfer of electrons from humic compounds to Fe(III) is an abiotic process, which can occur in the absence of microorganisms. Once oxidized by Fe(III), humic compounds may again accept electrons from humic-reducing microorganisms.

Although Fe(III) reduction is a significant redox process in eutrophic lakes, it contributes to only a small portion of their reducing potential. For example, it has been calculated that Fe(III) and Mn(IV) reduction accounted for <1% of the potential reducing equivalents that enter anoxic hypolimnia as sedimentary organic carbon from primary production (Verdouw and Dekkers, 1980; Davison *et al.*, 1981).

C. Overview

A general summary of these inorganic and bacterial relationships is given diagrammatically in Figure 14-9. The primary distinction between the fluxes of

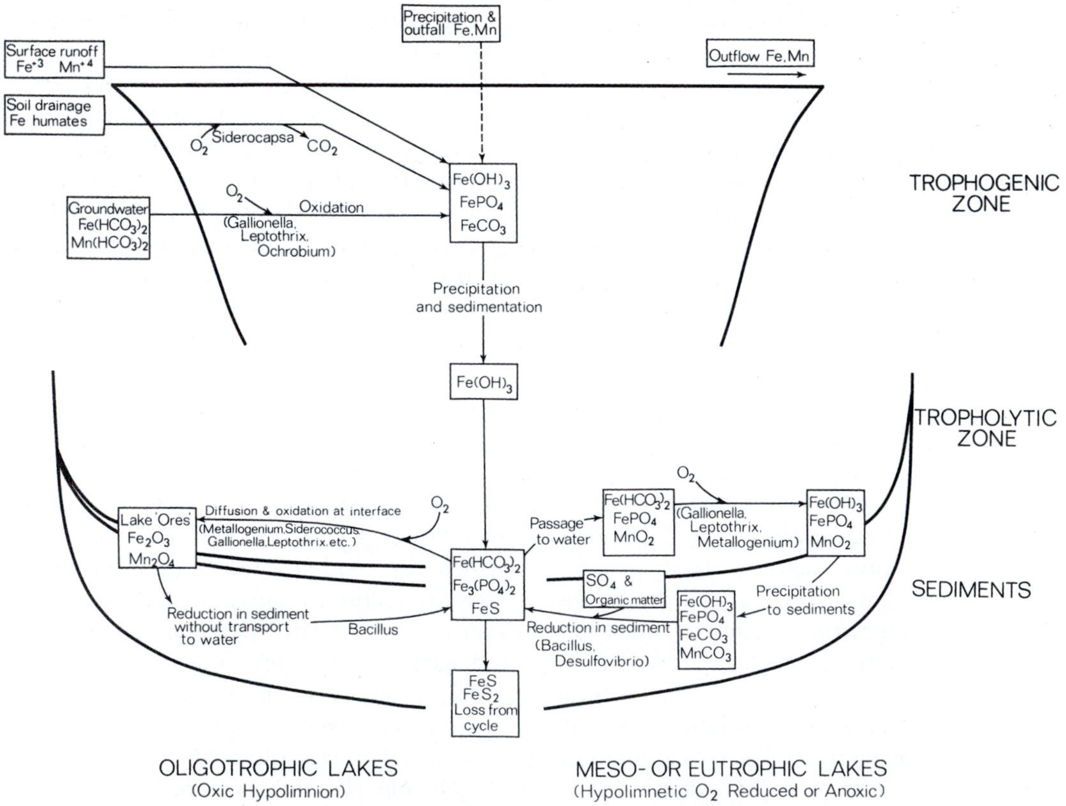

FIGURE 14-9 General iron and manganese cycles in lakes of low and high productivity emphasizing microbial interactions. (Modified from Kuznetsov, 1970.)

iron and manganese in the hypolimnia of lakes of increasing productivity centers on the reduction of redox potential and pH. Our knowledge of the iron and manganese cycles at the bacterial level is largely qualitative owing to the paucity of quantitative information concerning both between-component flux rates and whole-lake metabolism of these two materials. Very little is known about the role of the littoral flora in this cycling. Iron, for example, is translocated from the sediments to the leaves of submersed vegetation (Oborn and Hem, 1962; DeMarte and Hartman, 1974; Wetzel, 1990c). The magnitude of this transport from the sediments to the water in organic-bound phases of plant material as the plants decay is unknown.

V. MINOR ELEMENTS

Minor metallic elements, collectively referred to as ***micronutrients,*** include Fe, Mn, zinc (Zn), copper (Cu), boron (B), cobalt (Co), molybdenum (Mo), nickel (Ni), and vanadium (V), nearly all of which are required for the nutrition of plants and many animals. The universality of an absolute requirement for all of these elements is not clear. In some cases, as with iron and manganese, the essentiality of the element is established. In others, such as vanadium, it is known that one element can substitute for another, for example, vanadium can replace molybdenum in enzymes, and that growth is enhanced by vanadium among certain algae and cyanobacteria (e.g., Meisch *et al.,* 1977; Patrick, 1978; Vaishampayan, 1983; Attridge and Rowell, 1997; Rowell *et al.,* 1998) but not among others (Gerloff, 1963; Holm-Hansen, 1968; Nalewajko *et al.,* 1995). Common physiological functions of these micronutrients are summarized in Table 14-3.

Selenium is another trace element of unknown biochemical significance among aquatic organisms. The dinoflagellate alga *Peridinium cinctum* requires a small amount of selenium (approximately 50 ng Se^{4+} $liter^{-1}$) for optimal growth and reproduction (Lindström and Rodhe, 1978). In a Swedish lake, the concentrations of total and biologically available selenium showed considerable seasonal and vertical variation, which likely resulted from variations in loading from the atmosphere and the drainage basin, as well as algal uptake (Lindström, 1980; cf. Suzuki *et al.,* 1981). Although the potential for regulation exists, the significance of a trace element such as selenium in the regulation of algal growth in natural systems is unclear at the present time.

Most of the information on micronutrient requirements is obtained from studies of physiological deficiencies in cultures, which present an array of experimental difficulties, particularly involving contamination (Weissner, 1962). Requirements are further complicated by observed differences that depend on the composition and concentration of other minerals and an antagonism among these elements. These effects,

TABLE 14-3 Primary Physiological Functions of Micronutrients Generally Accepted as Essential or Potentially Essential for Aquatic Organisms[a]

Element	*Primary functions*
Iron	Electron transport in redox systems of respiration and photosynthesis; enzyme activation, oxygen carrier in N_2 fixation
Manganese	Enzyme activation; electron transport reactions particularly in photosynthesis, detoxification of superoxide radicals; synthesis of secondary metabolites; ribosome structure
Zinc	Membrane integrity; enzyme activation, particularly carbonic anhydrase; gene structure, expression, and regulation; carbohydrate metabolism, anaerobic respiration, protein synthesis; ribosome structure; detoxification of superoxide radicals; phytohormone activity
Copper	Redox reactions of respiration and photosynthetic electron transport; detoxification of superoxide radicals; lignification; haemocyanin in aquatic invertebrates.
Nickel	Iron absorption; nitrogen fixation; several constitutive enzymes, particularly urease; reproductive growth in plants
Boron	Cell wall formation and stabilization; membrane integrity; carbohydrate utilization; pentose phosphate metabolism; lignification; xylem differentiation; stomatal regulation; heterocyst structure and nitrogen fixation in cyanobacteria
Molybdenum	Electron transfer reactions; nitrate reduction and nitrogen fixation (see Chap. 12); sulfate oxidation; protein synthesis
Chloride	Osmoregulation; cation uptake; photosynthesis; reactivity of enzymes
Selenium	Essential growth regulator among certain algae; enhancement of phosphorus metabolism; amino acid and protein synthesis; mitosis; cuticle integrity
Cobalt	Essential for growth among many microbiota, particularly algae; essential component of vitamin B_{12}; potential substitution for zinc
Vanadium	Unclear functions and absolute requirements; enhancement of nitrogenase and nitrate reductase activities; substitution for molybdenum in some algae; phosphorylation
Cadmium	Unclear if a required nutrient in algae; implicated in functions in carbonic anhydrase activity; possible substitution for zinc

[a] Summarized from Lee (1983), Vaishampayan (1983), Keating and Dagbusan (1984), Soeder and Engelmann (1984), Oliveira and Antia (1986), Willsky (1990), Läuchli (1993), Hamilton (1994), Hausinger (1994), Lee *et al.* (1995), Welch (1995), Attridge and Rowell (1997), Kisker *et al.* (1997), Blevins and Lukaszewski (1998), Taylor and Anstiss (1999), and other sources cited in the text.

often found under the conditions of a controlled culture, become much more significant in the inorganic and organic solution of natural waters, which are rendered heterogeneous because of constant seasonal and spatial variations. Hence, one finds a large number of analyses of concentrations of ionic and, less frequently, particulate micronutrients in water that may bear little relationship to actual metabolic requirements. The question of requirements concerns the ***availability*** of the micronutrient within the constraints of other ionic components of salinity, the extent of organic complexing of the micronutrient, and the metabolic demands of varying species. In general, concentrations and availability of micronutrients in most natural waters are adequate to sustain active populations of algae within constraints of light, temperature, and macronutrient availability. There are, however, clear cases in which micronutrients can limit photosynthesis to a degree. A true deficiency of micronutrients is found in some oligotrophic aquatic systems of granitic arctic, alpine, and volcanic areas, in which a paucity of these elements is well known (reviewed by Goldman, 1972). In other situations, such as in hardwater calcareous lakes, the micronutrients are not deficient in the system but are present in forms unavailable for assimilation (Wetzel, 1972). In both situations, the control of productivity can be quite transitory and effective on only certain microbial species, thereby indirectly influencing, for example, algal succession and productivity as well as higher trophic levels.

The importance of dissolved organic matter in regulating micronutrient availability and growth, although long known to be effective in cultures, has been emphasized only recently as a major controlling factor of productivity in lakes (Wetzel, 1968, 1979, 1995; Wetzel and Allen, 1970; Wetzel *et al.,* 1972). Although involved in lake metabolism in many ways (cf. Chap. 23), complexation of micronutrients by organic compounds increases the physiological availability of many micronutrients when the ratio of organic matter to micronutrient concentration is reasonably high. For example, the soluble ionic fraction of total copper along the course of an Asian river was directly proportional to concentrations of dissolved organic carbon (Jun and Bae, 1998). The effects of chelated iron and manganese on algal productivity, discussed earlier, provide only one important example of this process.

It is important to note that the relative concentrations of trace elements in fresh waters can have a significant effect on the competitive abilities of species in algal communities. For example, low concentrations of vanadium and hexavalent chromium favor the development of diatoms; at higher concentrations, green algae and cyanobacteria can dominate the communities (Patrick, 1978). Low concentrations of manganese were correlated with the development of cyanobacteria, whereas diatom communities dominated at concentrations of >40 μg Mn liter^{-1} (Patrick *et al.,* 1969).

Most metallic micronutrients are very toxic when present in excess ionically or when complexed organically to the point where their availability exceeds physiological tolerance limits of biota. The heavy metal toxicity changes markedly with environmental conditions. For example, the inhibitory effect of Cu and of Cd on phosphorus uptake by certain green algae is highly pH-dependent and increases strongly (nearly 200-fold) with increasing pH over the range of 5.5–8.5 probably as a result of competition between H^+ and free metal cations for cellular binding sites (Peterson *et al.,* 1984). Such changes in pH occur commonly on a diurnal basis in shallow productive lakes. An increase in water hardness and concentration of dissolved calcium and magnesium, however, will compete for available binding sites on dissolved organic molecules and reduce the binding of heavy metals. An increase in toxicity of certain metals, such as cadmium, can occur as a result in hard waters even with the same concentrations of dissolved organic matter (Penttinen *et al.,* 1998). In another example, high concentrations of copper (usually as $CuSO_4$) have been used repeatedly as an herbicide to control algal blooms and growth of larger aquatic plants (Hodson *et al.,* 1979; McKnight, 1981). Fogg and Westlake (1955) have demonstrated that polypeptides secreted by cyanobacteria can effectively complex copper ions and reduce toxic effects. These examples indicate many ways in which organic complexing can influence the availability of micronutrients. The effects of micronutrients and their availability to animals are less well understood. It is generally assumed that requirements are met by means of ingestion of and uptake from food. Positive and negative correlations between fluctuations of micronutrient concentrations of the water and animal populations (see Parker and Hazelwood, 1962) yield little insight into this question but suggest that variations in micronutrient content of food can influence, in part, the succession and development of the fauna.

The general average concentrations of the soluble form of minor metallic micronutrients for a range of lakes and streams are given in Table 14-4. However, the ranges of concentration are extreme not only seasonally but particularly in relation to contamination from industrial and other sources of pollution. Some of the concentrations cited in this summary are likely overestimates, particularly from very oligotrophic waters. Contamination from sampling equipment, storage, and analytical instrumentation can be significant, and "metal-clean" sampling can result in appreciably lower

TABLE 14-4 Average Concentrations (μg liter^{-1}) of Soluble Minor Metallic Elements of Natural Waters (Surface)

Water	*Fe*	*Mn*	*Cu*	*Zn*	*Co*	*Mo*	*V*
World surface lakes and rivers (Livingstone, 1963)	Approx. 40	35	10	10	0.9	0.8	Approx. 0.1
Alpine lakes, California (Bradford *et al.*, 1968)	1.3	0.3	1.2	1.5	<0.3	0.4	—
Northern German lakes (Groth, 1971)	31.5	28.6	2.9	6.6	0.05	0.39	—
Northeastern Indiana lakes (Wetzel, 1966b)	15.0	21.3	<2	12.9	<2	30.0	<3
Linsley Pond, Connecticut (Hutchinson, 1957; Cowgill, 1976b, 1977b)	350	140	53	—	0.05	0.19	0.03
South American lakes (Groth, 1971)	533	15.6	1.7	8.7	0.11	0.6	—
Lake Constance, Southern Germany (Hegi, 1976)	10.9	11.7	1.7	6.2	—	—	—
Three Lakes, Victoria, Australia (Hart and Davies, 1981)	417	—	2.4	5.8	—	—	—
Lago Maggiore, Italy (Bando *et al.*, 1981)	81.5	5.5	8.3	39	—	—	—
Tjeukemeer, The Netherlands (de Haan *et al.*, 1990)	227	29	1.3	3.3	0.3	—	—

true concentrations of soluble metals (Nriagu *et al.*, 1993, 1996). Higher concentrations are generally found in nearshore areas and particularly near urban centers and polluted river mouths. Examples are discussed at length in the references given in Table 14-4 and in Chawla and Chau (1969), Mills and Oglesby (1971), Robbins *et al.* (1972), Cowgill (1976, 1977a, 1977b), Hegi (1976), Baccini (1976), Boyle (1979), Leckie and Davis (1979), Martin *et al.* (1980), Nriagu *et al.* (1996), and many others.

A. Distribution of the Minor Elements in Lakes

The distribution of the minor elements among ionic (soluble), organically complexed, and adsorbed fractions, as well as the amounts in living biota, is poorly understood. Quantitative rates of flux between the living and abiotic organic and inorganic phases are even less well delineated; generally, much is inferred from better understood analogous chemical species, such as iron and manganese. Fragmentary evidence on the cycling of the minor elements indicates that these similarities are real and that much of their cycling is comparable.

The concentrations of Cu, Zn, Co, and Mo in ionic solution are usually very small in aerated surface waters. Transport of these trace metals in flowing waters can be partitioned analytically into (1) ionic forms, (2) those complexed in organic materials, (3) those adsorbed to and precipitated onto solids, and (4) those incorporated into crystalline structures (Gibbs, 1973). Although the solubilities of the metals vary somewhat (cf. Groth, 1971; Stumm and Morgan, 1996), in most cases, particularly in lakes, most (>70%) of each adsorbed to biologically produced particulate matter and otherwise associated crystalline solids (Hart, 1982; Wangersky, 1986; Twiss and Campbell, 1998). Much of the remainder is in organic complexes, with very little in solution. Some proportion of the metals removed from the surface waters with settling seston can be solubilized in the deeper water strata by bacterial decomposition of the particles. Thus, many trace metals have distributions in stratified waters that resemble those of the inorganic macronutrients. It should be again emphasized that these distributions are mediated by interacting physical and biological processes, and chemical speciation reactions based on pure chemical analyses in homogeneous systems cannot be applied effectively to natural systems.

A summary of older literature on the cycling of Cu, Zn, Mo, and Co is given in Hutchinson (1957). The detailed analyses of Groth (1971) on Schöhsee in northern Germany highlight the basic phases of the dissolved and particulate fractions over an annual cycle. The total amounts of Co, Mo, and Zn, as well as of Fe and Mn, accumulated in hypolimnetic waters during summer stratification as shown in Table 14-5. The concentrations of Fe and Mn were found to be strongly

TABLE 14-5 Average Concentrations (μg liter^{-1}) of Minor Elements in the Epilimnion and Hypolimnion of Schöhsee, Northern Germany[a]

Stratum	*Mn*	*Fe*	*Cu*	*Zn*	*Co*	*Mo*
Epilimnion (E)	4.5	15	1.0	1.8	0.03	0.21
Hypolimnion (H)	590	425	0.9	1.9	0.07	0.30
Enrichment ratio of H/E	130	28	0.9	1.1	2.3	1.4

[a] Data after Groth (1971).

TABLE 14-6 Average Accumulation of Minor Metallic Elements in the Plankton and Sediment in Comparison to the Average Concentrations in the Epilimnion of Schöhsee, Northern Germany, 1968–1969[a]

Element	*Dissolved in epilimnetic water* (μg liter^{-1})	*In plankton* (μg g^{-1})	*In sediment* (μg g^{-1})
Iron			
Concentration	15	950	58,000
Enrichment factor	1	63×10^3	3900×10^3
Manganese			
Concentration	4.5	130	1600
Enrichment factor	1	29×10^3	355×10^3
Cobalt			
Concentration	0.03	1.1	8.3
Enrichment factor	1	37×10^3	280×10^3
Copper			
Concentration	1.0	60	95
Enrichment factor	1	60×10^3	95×10^3
Zinc			
Concentration	1.8	110	350
Enrichment factor	1	61×10^3	195×10^3
Molybdenum			
Concentration	0.21	4.2	1.4
Enrichment factor	1	20×10^3	7×10^3

[a] Data after Groth (1971).

related to redox conditions, while the primary source of the hypolimnetic accumulation of Co, Mo, and Zn was from release and mineralization of sedimenting organic detritus (cf. also Baccini, 1976). No significant differences in vertical distribution of Cu were found in Schöhsee or the Great Lakes during summer stratification, although moderate increases in hypolimnetic sestonic Cu have been observed elsewhere in eutrophic lakes (Riley, 1939; Baccini and Joller, 1981; Nriagu *et al.*, 1996).

Co, Cu, and Zn, like Fe and Mn, form stable complexes with organic compounds, so that losses of free ions by formation of insoluble hydroxides, sulfides, phosphates, and carbonates can be reduced appreciably (Groth, 1971). Copper is strongly complexed by organic ligands and many of these compounds are likely produced by algae (Becher *et al.*, 1983; Verweij *et al.*, 1989; Xue and Sigg, 1993; Sigg *et al.*, 1995; Jun and Bae, 1998). Greater than 80%, and often nearly all, of total dissolved Co occurs as organic complexes (Qian *et al.*, 1998). Although most of the trace metal micronutrients are in organic complexes, a large fraction of Zn is present in ionic form or weak complexes. Molybdenum, however, exhibits greater mobility than the other ions (Mo > Cu > Zn > Co).

Phytoplankton tend to accumulate minor metallic elements in the order: Fe > Zn > Cu > Co > Mn > Mo (Table 14-6). During decomposition and mineralization of plankton, release to a soluble phase occurs in the order of Mo > Co > Cu > Zn > Fe > Mn, approximately opposite the order of amounts of the elements found to be transported to the sediments by sedimenting organic detritus: Fe > Mn > Co > Zn > Cu > Mo. Coprecipitation of these trace elements with $Fe(OH)_3$ was greatest with Cu, in the order of Cu > Mo > Co > Zn. Therefore, during summer stratification, algal uptake and sedimenting detritus play a major role in the cycling of Co and Zn, but less so for Cu (cf. Reynolds and Hamilton-Taylor, 1992), whereas concentrations of Fe and Mn are regulated largely by redox conditions. Although variations among lakes are great, amounts of minor elements (especially copper) often increase during fall circulation and during winter. Hypolimnetic oxygenation can enhance the release of Cu during aerobic degradation of organic matter in relatively mobile, likely organically complexed forms (Xue *et al.*, 1997). Reduction of Fe and Mn in the sediment and their rapid reoxidation at the sediment/oxic water interface results in an enrichment and deposition of Cu and Zn by freshly formed Mn- and Fe-oxides. Mo is enriched in the sediment and correlates with Mn (Schaller *et al.*, 1997). Similarly, vanadium distributions are highly correlated with iron. Ionic concentrations increase somewhat in winter, but most of this maximum is in the organic fractions (Riley, 1939; Kimball, 1973; Wangersky, 1986). In general, the activity of zinc (Bachmann, 1963) and cobalt (Benoit, 1957; Parker and Hasler, 1969) follows the conclusions summarized above. Molybdenum exhibits much greater mobility than the other minor elements discussed and does not show analogous patterns of distribution (cf. Dumont, 1972; Cowgill, 1977b). In the seasonally anoxic hypolimnion of a eutrophic lake, Mo concentrations declined with depth particularly in the presence of H_2S (Magyar and Moor, 1993). Much of the vertical flux was in the form of particle-bound Mo at 1.1–4.6 μg m^{-2} day^{-1}.

It should be noted that the inputs of many trace elements to fresh waters are increasing. These increases often result from industrial and combustion emissions, which are subsequently deposited onto the drainage basins of rivers and lakes (e.g., Nriagu, 1979; Nriagu and Davidson, 1980). In other cases, the widespread acidification of rain and snow falling on poorly buffered soils and fresh waters can result in high rates of leaching of trace metals. Detailed discussion of metal enrichments above natural concentrations in surface waters is beyond the compass of this synthesis. However, several primary metal pollutants have become so widespread and influential to freshwater biota that some comments are necessary.

B. Cadmium

Cadmium (Cd) occurs in natural fresh waters in the range of ca. 0.01–0.1 μg Cd liter^{-1}, and usually in the high portion of that range in rivers (Martin *et al.*, 1980; Raspor, 1980; Laxen, 1984). Much of the Cd is associated with particles (>0.2 μm) or complexed with dissolved organic macromolecules. As a result Cd availability is often much higher in fresh waters than would be anticipated simply on the basis of inorganic solution chemistry and solubilities. Cadmium releases to the atmosphere are very large (Nriagu, 1980). Natural emissions (e.g., volcanism and soils) are completely dwarfed by anthropogenic emissions from industry and fuel combustion. Concentrations of a few μg Cd liter^{-1} are highly toxic to many organisms, particularly plankton and fishes, primarily by enzyme inhibition (Wong *et al.*, 1980). At very low levels, Cd can function as a metal replacement for Zn in carbonic anhydrase when concentrations of Zn are low (Price and Morel, 1990).

C. Lead

Lead (Pb) is another heavy metal that is widely known for its toxicity to aquatic organisms at relatively low levels. Much of the Pb loadings to drainage basins emanated from the combustion of fossil fuels, particularly from gasoline that contained alkyl–lead additives to improve engine performance. Analyses of lake sediments showed that atmospheric lead deposition increased above background levels from preindustrial uses of Pb in Europe, starting with the Greek and Roman cultures, more than 2600 years B.P. (Renberg *et al.*, 1994). The Pb deposition accelerated throughout the northern hemisphere in the nineteenth and particularly the twentieth centuries, with a deposition maximum in about 1970. In the 1970s, legislation restricted use of lead additives in gasoline and Pb in precipitation has declined in many areas by 97% between 1976 and 1989 (Johnson *et al.*, 1995; Nriagu, 1996). Accumulated Pb in forest soils has declined at much slower rates because of low loss rates to drainage waters. Once in flowing systems, most Pb is associated with dissolved organic substances and adsorbed to particle surfaces (Botelho *et al.*, 1994). Pb content in lake sediments indicates a gradual decline.

D. Mercury

Mercury (Hg) in fresh waters and biota has become of intense interest because of the acute neurological toxicity of this element and widespread findings of elevated concentrations of Hg in fish in lakes remote from direct pollution of surface waters. Recent studies have evaluated the mercury cycle in aquatic ecosystems in some detail (Verta, 1984; Miskimmin *et al.*, 1992; Driscoll *et al.*, 1994, 1995; Henry *et al.*, 1995; Mason *et al.*, 1995; Vandal *et al.*, 1995; Watras *et al.*, 1995; Cai *et al.*, 1996; Baldi, 1997; Lin and Pehkonen, 1999). Atmospheric deposition of Hg to drainage basins and lakes occurs largely as inorganic Hg(0). In oxygenated waters, reactive Hg(0) is converted to Hg(II), which will complex with inorganic ions (e.g., Cl^- and OH^-) and dissolved organic compounds or absorb to particulate matter. Hg(II) can be reduced by bacteria to form Hg(0). Most waters are supersaturated with respect to the solubility of atmospheric Hg(0), and Hg(0) is volatilized to the atmosphere. Within anoxic strata, Hg can form aqueous complexes with sulfide and precipitate as HgS. Sulfate-reducing bacteria of anoxic zones convert Hg(II) to toxic methylmercury (CH_3Hg) and possibly to ethylmercury, which may bind to dissolved organic compounds (particularly humic substances) or be demethylated by microbes. Under anoxic conditions, Hg toxicity is reduced by this complexation and the reaction with sulfides. The reaction of Hg(II) and CH_3Hg can produce stable HgS and the volatile and less toxic dimethylmercury. CH_3Hg accumulates in particulate layers in anoxic strata and is a predominant form of mercury that is assimilated and concentrated in fish. Monomethylmercury tends to be higher in acidified lakes where sulfate concentrations are low and acidic conditions favor CH_3Hg stability. Concentrations of total mercury and CH_3Hg increase directly with increasing concentrations of dissolved organic carbon and with the percentage of nearshore wetlands in the drainage basin. In general, trace and toxic metals tend to be retained more strongly in wetland hydrosoils and detritus than in upland soils (Gambrell, 1994). Mobility of CH_3Hg tends to increase among acidified waters and during warmer periods of low flows in streams and wetlands (Driscoll *et al.*, 1998; Monson and Brezonik, 1998).

E. Aluminum

Acidic precipitation is causing high concentrations of aluminum to be leached from poorly buffered soils in many continental regions (Almer *et al.*, 1978; Cronan and Schofield, 1979). Many inorganic and biotic processes in upland soils can maintain aluminum in nonreactive forms and prevent appreciable loadings with runoff. Sudden acidic precipitation events, however, such as rapid spring snowmelt releases of accumulated acidity from air pollution, can result in marked declines in pH of runoff water to pH 4–5. The result is large increases of reactive Al to receiving streams and lakes. The effectiveness of mobilization of Al from soils depends greatly on the buffering capacities and solubilities of the soils (e.g., Hooper and Shoemaker, 1985; Cronan *et al.*, 1986; Lawrence *et al.*, 1986; Gensemer and Playle, 1999).

Aluminum (Al) chemistry is complex because of the large number of forms occurring in aquatic ecosystems. Within the range of pH and common limnological conditions, Al is a minor dissolved constituent. However, as widespread acidification occurs in poorly buffered surface waters in many areas of the world from atmospheric precipitation of acidic anthropogenic pollution, Al can leach from soils in relatively large amounts, particularly when it is accompanied by an increase of dissolved organic matter leached from the soils (Lorieri and Elsonbeer, 1997). Aqueous ionic Al is highly reactive and is toxic at very low concentrations, can serve as a pH buffer, and is an absorbent of phosphate and organic carbon (Schafran and Driscoll, 1987; Helmer *et al.*, 1990; Gensemer and Playle, 1999). Much of the total dissolved Al is complexed with dissolved organic matter in acidic waters, with a binding capacity that increases from pH 3 to 5, but decreases above pH 5 by formation of Al-hydroxy species (e.g., $Al(OH)_3$). Therefore, in acidic waters (pH <5), concentrations of reactive Al can be 200–500 or more $\mu g\ liter^{-1}$ (e.g., Hongve, 1993). Concentrations of reactive Al are usually <10 $\mu g\ liter^{-1}$ in waters of neutral pH, although some persists in organically complexed and colloidal forms.

VI. THE SULFUR CYCLE

Sulfur is utilized by all living organisms in both inorganic and organic forms. Sulfate is reduced to sulfhydryl (—SH) groups in protein synthesis, with a concomitant production of oxygen that is utilized in oxidative metabolic reactions. Interest in the sulfur cycle of fresh waters, however, extends beyond nutritional demands of the biota, which are almost always met by the abundance and widespread distribution of sulfate, sulfide, and organic sulfur-containing compounds. Decomposition of organic matter containing proteinaceous sulfur and the anaerobic reduction of sulfate in stratified waters both contribute to altered conditions that markedly affect the cycling of other nutrients, ecosystem productivity, and distribution of the biota.

A. Forms and Sources of Sulfur

Sources of sulfur compounds to natural waters include solubilization from rocks, fertilizers, and atmospheric precipitation and dry deposition. At the present time, atmospheric sources, augmented greatly by the combustion products of industry, dominate all other sources.

1. *Weathering*

Sulfate is released during geochemical weathering of rocks and soils containing either sulfides or free sulfur, which are oxidized in the presence of water to form sulfuric acid (ZoBell, 1973):

$$\underset{\text{(pyrite)}}{FeS_2} + 3\tfrac{1}{2}O_2 + H_2O \longrightarrow FeSO_4 + H_2SO_4$$

$$2S + 3O_2 + 2H_2O \longrightarrow 2H_2SO_4$$

These two reactions tend to lower both the pH and E_h, which affects the oxidative weathering reactions of numerous other minerals. Calcium sulfate, which is moderately soluble in water, is a common constituent of sedimentary rocks. As a result, drainage from calcareous regions generally contains higher-than-average concentrations of sulfate (Nriagu and Hem, 1978; Kilham, 1984). As will be discussed further on, bacteria contribute to the oxidation of sulfides and elementary sulfur, both in soil and in water.

2. *Atmospheric Loadings*

Large quantities of reduced sulfur, as hydrogen sulfide, are added to the atmosphere from volcanic gases and biogenic and industrial sources (Kuznetsov, 1964; Kellogg *et al.*, 1972). H_2S undergoes a number of oxidative reactions to become sulfur dioxide (SO_2), sulfur trioxide (SO_3), and sulfuric acid (H_2SO_4). Sulfur dioxide constitutes about 95% of the sulfur compounds resulting from the burning of sulfur-containing fossil fuels. Although the oxidation of SO_2 to H_2SO_4 in air is slow (hours to days), SO_2 is rapidly oxidized to sulfuric acid as it dissolves in atmospheric water. The global cycling of sulfur compounds (Table 14-7) indicates clearly that in industrialized regions, sulfur inputs from human activities are rapidly exceeding inputs from natural sources. In eastern North America, for

TABLE 14-7 Sources, Sinks, and Residence Times of Atmospheric Sulfur Compounds (Units Are 10^6 Tons as Sulfate per Year)[a]

Sources	*10^6 ton yr^{-1}*	*Approximate residence time*
Windblown sea salt SO_4^{2-} in precipitation, 10% of global total deposited on all land	130	
Bacterial and plant production of H_2S (SO_2)	268	0.5–6 days
Man-made production of SO_2 and SO_4^{2-} from fossil fuels (80% deposited on land; 93% of global production in the Northern Hemisphere)	150	0.5–6 days
Volcanic sources (H_2S_2 SO_2, SO_4^{2-})	2	<1 day
Deposition		
Rain over oceans (SO_2, SO_4^{2-})	217	
Rain over land (SO_2, SO_4^{2-})	258	
Plant uptake (SO_2, SO_4^{2-})	45	
Dry fallout deposition (SO_4^{2-})	30	

[a] From data of Kellogg *et al.* (1972).

example, industrial emissions exceed natural ones by a factor of 10 (Table 14-8). The removal processes over land are sufficiently slow (several days) to result in markedly increased concentrations in areas hundreds to thousands of kilometers downwind. In nonindustrial areas, the primary source of sulfate (SO_4^{2-}) in rain and snow is atmospherically oxidized H_2S that is produced along coastal regions by anaerobic bacteria (Jensen and Nakai, 1961). The SO_4^{2-} derived from sea spray is largely returned to the ocean; over land, this source is minor and is generally restricted to coastal lakes (Table 14-8).

The retention of sulfur compounds in soils of the drainage basin and their release to natural waters

TABLE 14-8 Atmospheric Sulfur Budget for Eastern North America[a]

	10^{12} g S yr^{-1}		
Component	*Eastern Canada*	*Eastern USA*	*Total, Eastern North America*
Inputs			
Human-generated emissions	2.1	14	16.1
Natural emissions			
Sea spray	0.06	—	0.06
Terrestrial biogenic	0.06	0.04	0.10
Marine biogenic	0.2	0.4	0.6
Inflow from oceans	0.04	0.02	0.06
Inflow from west	0.1	0.4	0.5
Inflow to United States from Canada	—	0.7	—
Inflow to Canada from United States	2.0	—	—
Total	4.6	15.6	17.4
Outputs			
Wet deposition	3.0	2.5	5.5
Dry deposition	1.2	3.3	4.5
Outflow to oceans	0.4	3.9	4.3
Outflow from Canada to United States	0.7	—	—
Outflow from United States to Canada	—	2.0	—
Total	5.3	11.7	14.3

[a] Modified from Galloway and Whelpdale (1980).

varies with regional lithology, agricultural application of sulfate-containing fertilizers, and atmospheric sources in relation to other sources. In calcareous areas of sedimentary rock, atmospheric contributions can be a small portion of the total. In contrast, in crystalline rock areas of many areas of northeastern North America and of Europe and Asia, wet- and dryfall precipitation supplies nearly all of the sulfate of natural waters (Fisher *et al.*, 1968; Johnson *et al.*, 1972; Kramer, 1978; Dillon *et al.*, 1983; among many others). Much of that atmospherically derived sulfate can be bound in organic compounds of soils of the drainage basin for long periods of time before being released to ground and surface waters.

Such acidic, sulfatic precipitation can increase weathering geochemical processes at rates equal to or greater than the carbonic acid system in these geological regions. The atmospherically derived sulfatic acidity adds to acidity resulting from the organic acids of soils and surface waters. The organic acids can produce ca. 5–10 meq H^+ mg^{-1} organic carbon and lower both the pH of natural waters and the acid-neutralizing capacities of natural waters (Hemond, 1994). Organic acids also increase somewhat the acid–base buffer capacity of water, which can render the pH less sensitive to changes from sulfatic and nitric mineral acid enrichments. Despite these sources, the strong acids associated with sulfate are major contributors to the reduction of acidity in lakes and rivers, even those that contain very high concentrations of natural organic acids (e.g., Gorham *et al.*, 1986; Kerekes *et al.*, 1986; Brakke *et al.*, 1987).

Where soils are sensitive to acidic precipitation, aluminum silicate minerals weather (Dillon *et al.*, 1983):

$$\text{cation–Al-silicate} + \text{acid} + H_2O \longrightarrow H_2SiO_4 + \text{cation} + \text{Al-silicate} + \text{acid anion}$$

Even in noncalcareous terrain, the ionic content of lakes and streams is dominated by HCO_3^- and Ca^{2+}, although in low quantities. The HCO_3^- supplied by the drainage basin is a major component of the acid neutralizing capacity (ANC) of the waters. In surface waters of very low alkalinity, biologically mediated processes add to the ANC by algal uptake of nitrate during photosynthesis, denitrification, and reduction of sulfate and FeOOH. Deposition of strong acids, especially sulfatic, results in the export of cations (Ca^{2+}, Mg^{2+}) and metals (Al, Fe, Mn) and a shift in anion content in the runoff water from bicarbonate to the anions of the anthropogenic acids (SO_4^{2-} and NO_3^-). Because nitrate is readily utilized biologically in the drainage basin, primarily sulfate will increase.

Much of the acidity can accumulate adsorbed to soil particles and in snowfall. Most (ca. 90%) of the total sulfur content is found, however, in the organic matter of the mineral soils (e.g., Houle and Carignan, 1995). Because these organic S pools are often 100–200 times greater than the annual S deposition, the loss of <1% of the organic S pool would be sufficient to explain the long-term releases from soils. Release of the sulfate and associated H^+ ions is often very high during the flushing process associated with commonly occurring rapid snowmelt periods. For example, in a small Canadian Shield basin, flushing of soluble SO_4^{2-} from organic and upper mineral soil horizons during the melt was some four times the amount supplied in meltwater and precipitation (Steele and Buttle, 1994).

Much of the loading of sulfur compounds to streams and lakes, therefore, is in the form of sulfate and soluble organic S constituents (C-bonded S and ester sulfate) (e.g., David and Mitchell, 1985; Lelieveld *et al.*, 1997). Because much of the total S is bound in seston and most of the seston is not mineralized during sedimentation, organic S accumulates and constitutes a major portion (ca. 75%) of the total S of the sediment.

B. Distribution of Sulfur in Natural Waters

The sulfur content of biota ranges from 0.05 to nearly 5% of dry weight in a few bacteria; the average content is 0.2%. The amounts in the biota and detritus, although significant, are generally small in comparison to the inorganic sulfur components of aquatic systems. The distribution and cycling of sulfur therefore involve (a) the chemical species under various conditions, (b) biotic influences on the transformations of sulfur species, and (c) sulfur transport in the system.

The lowest concentrations of sulfate in oxic waters (often slightly <1 mg $liter^{-1}$) appear to be a common feature of numerous African lakes situated in crystalline rock drainage basins (Talling and Talling, 1965). The other extreme is found in sulfate saline lakes (>50 g $liter^{-1}$). The usual range is about 5–30 mg $liter^{-1}$, with an average of about 11 mg SO_4^{2-} $liter^{-1}$. Low levels of SO_4^{2-} were implicated in the limitation of algal productivity in Lake Victoria, Africa (Fish, 1956), but more recently this role of SO_4^{2-} has been shown to be unlikely (Evans, 1961, 1962; Lehman and Branstrator, 1994).

The predominant form of dissolved sulfur in water is sulfate (Fig. 14-10). Nearly all assimilation of sulfur is as sulfate, but during decomposition of organic matter, sulfur is released largely as hydrogen sulfide. Under oxic conditions, H_2S is oxidized rapidly. Therefore, little H_2S would be anticipated in aerated regions of aquatic systems. However, strict application of redox and pH to the evaluation of reactions involving sulfur is difficult because some are chemically slow and medi-

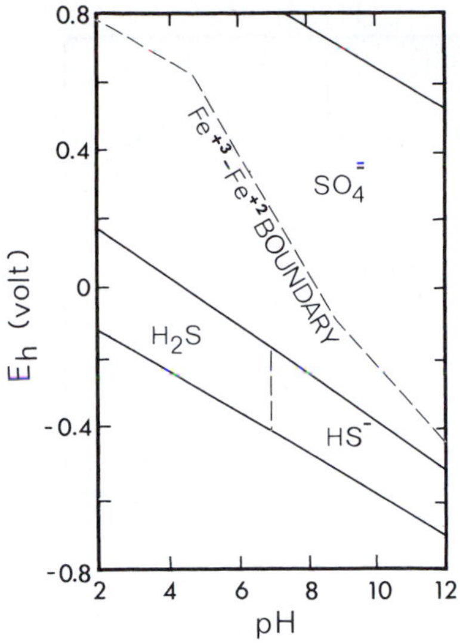

FIGURE 14-10 Approximate redox–pH fields of stability of dissolved sulfur species likely to occur in natural water. (Modified from Hem, 1960, with Chen and Morris, 1972.)

ated by bacterial metabolism. Although SO_4^{2-} and H_2S dominate, HS^- and very low concentrations of S^{-2} occur in strongly alkaline solutions because H_2S, which is very soluble in water, dissociates weakly (k_1 10^{-7}; k_2 10^{-15}) (Hutchinson, 1957; Hem, 1960c). Under certain conditions of low redox and pH, partial oxidation of sulfides occurs and free S^0 may be formed.

Metal sulfides are exceedingly insoluble at the neutral or alkaline pH values commonly encountered in a majority of natural waters. The equilibrium ion activity product of FeS of anaerobic lake sediments is $10^{-17.7}$ (Doyle, 1968b). Therefore, Fe^{+2} released from sediments reacts vigorously with H_2S to form FeS. In an anaerobic hypolimnion, the water must be somewhat acidic in order for appreciable H_2S to accumulate. If the water is alkaline, H_2S will accumulate only after most of the Fe^{+2} has been precipitated as FeS. The removal of sulfide by the release of ferrous ions permits an increase in the migration of other metals from the sediments, such as Cu, Zn, and lead (Pb), which form even more insoluble sulfides than does iron.

Organic volatile sulfur compounds (OVS) of biogenic origins are found in quantities throughout aquatic environments sufficient to contribute to the global atmospheric sulfur cycle (cf. reviews of Caron and Kramer, 1994; Kiene, 1996). Although dimethyl sulfide (CH_3SH) is the most abundant OVS in salt marshes and saline ecosystems, it also occurs by bacterial degradation of dimethylsulfaniopropionate in freshwater ecosystems (Ginzburg *et al.*, 1998). Other sulfide species in addition to H_2S in fresh waters include methanethiol (CH_3SH), dimethyl disulfide (CH_3SSCH_3), carbonyl sulfide (COS), and carbon disulfide (CS_2). OVS are ubiquitous in surface waters and OVS production is independent of sulfate concentrations. Methylated sulfide production often requires methionine, and some of their precursors are readily interconverted during microbial metabolism but are ultimately mineralized. Only a small fraction remains for exchange with the atmosphere.

Because of the high concentrations of organic S in the sediments, these sites of organic enrichment are likely sites of much of OVS production. Concentrations of OVS have been found to be much higher in shallow lakes and wetland/littoral habitats (Richards *et al.*, 1991; Kiene and Hines, 1995; Hines, 1996). As sulfate concentrations, but not dissolved salt concentrations, increase in a gradient of saline lakes, OVS concentrations were several orders of magnitude greater than in fresh waters and saline lakes of low salinity (<7 g $liter^{-1}$) in saline lakes with >20 g SO_4^{2-} $liter^{-1}$ (Richards *et al.*, 1994). As in marine environments, dimethyl sulfide was the dominant species in these saline lakes. OVS compounds are readily degraded in the oxic epilimnion and metalimnion of stratified lakes (Sim *et al.*, 1993) and increase in the hypolimnia of acidified lakes. As a result, volatilization and emissions of OVS from stratified lakes are usually very small and not considered a major sulfur loss mechanism compared to processes such as sulfate reduction in the sediments.

A generalized vertical distribution of sulfate and hydrogen sulfide for stratified lakes is depicted in Figure 14-11. Under oxic conditions, as is the case in many oligotrophic and mesotrophic lakes, and during periods of circulation, H_2S is absent and SO_4^{2-} concentrations change little with depth. Some release of SO_4^{2-} occurs from the sediments, and this increase in sulfate in the hypolimnion can become more pronounced in hypolimnia of mesotrophic or eutrophic lakes in the earlier phases of summer stratification (Fig. 14-12). Reduction of sulfate to H_2S occurs as the redox potential declines to less than about 100 mV as a result of bacterial decomposition. Particularly near the sediments, much of the H_2S reacts with Fe^{2+} ions to form insoluble FeS. In this way, considerable quantities of sulfur can be lost to the sediments (Ingvorsen *et al.*, 1981; Jones *et al.*, 1982). Lakes receiving rich sources of sulfate from inflowing water, such as meromictic lakes of crenogenic formation, often contain immense concentrations of H_2S in their anoxic monimolimnia. Analogous situations occur in certain anoxic stretches of rivers that are polluted with sulfate-rich organic wastes. Effluents from paper-producing industries are a common source of such pollution. Horizontal variations in concentrations of sulfate and sulfides can be large, especially in reservoirs, and are complicated by flow patterns (e.g., Hanušová, 1962).

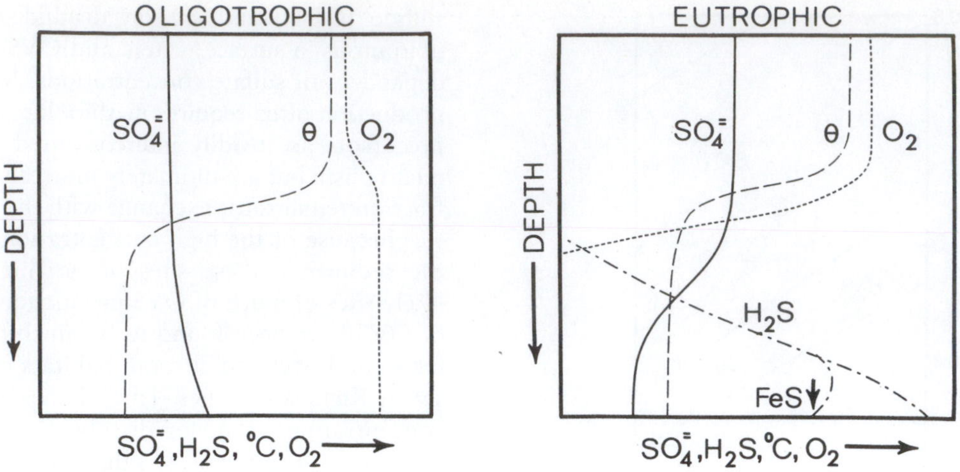

FIGURE 14-11 Generalized distribution of sulfate and hydrogen sulfide in lakes of very low and very high productivity.

The reduction of SO_4^{2-} to sulfide, some of which is lost to the sediments as insoluble metallic sulfides, and oxidation of H_2S to sulfate play a significant role in the modification of conditions for mobilization of phosphate and numerous other nutrients. In an excellent discussion of these reactions and their mediation by various bacterial groups, Ohle (1954) characterizes sulfate as a "catalyzer" of limnetic nutrient cycling.

The sulfur cycle in Linsley Pond, Connecticut, was studied during summer stratification in appreciable detail using ^{35}S-labeled sulfuric acid (Stuiver, 1967). In this eutrophic lake, large quantities of sulfate were lost from the metalimnion and hypolimnion, especially at the sediment–water interface. The rates of reduction of sulfate in the metalimnion and hypolimnion differed little between regions that were anoxic or partly devoid of oxygen but were about 10 times faster than in the fully aerobic epilimnion. Vertical diffusion of sulfate in the metalimnion and upper layers of the hypolimnion was very small during stratification. Horizontal diffusion at a given depth within the lower strata of water was sufficient to transport sulfates to the surrounding sediments, where reduction occurs. Transport of sulfur by sedimenting biological material was negligible and influenced the sulfur cycle of this lake only by providing organic substrates for bacterial metabolism in the hypolimnion and sediments.

This analysis permitted calculation of a total sulfur budget, estimated from the ^{35}S activities of the water and organic compounds (Fig. 14-13). Most of the sulfur was stored as sulfate and sulfide in the water, and as sulfide in the sediments. That fraction utilized by organisms was small and did not materially influence the total cycle. The H_2S in the hypolimnion was oxidized in the epilimnion; escape of H_2S to the atmosphere by diffusion or gas bubbles was low.

Organic sulfur compounds were not studied extensively in this budgetary analysis. Most of the sulfur of the seston occurs as ester sulfates and protein sulfur. The bulk (up to 80%) of sulfur in the sediments of productive lakes consists of organic sulfur compounds (ester sulfates and protein sulfur), and the remainder as pyritic sulfur, acid-volatile sulfides, sulfides dissolved in interstitial water, elemental sulfur, and dissolved sulfates (Table 14-9) (Doyle, 1968a,b; Nriagu, 1968; King and Klug, 1980, 1982a; Mitchell *et al.*, 1981; Smith and Klug, 1981). At least 40% of the acid-soluble sulfide of Linsley Pond was identical to tetragonal FeS (mackinawite).

Sulfur-containing organic compounds are degraded more slowly than other organic compounds. In Lake Mendota, Wisconsin, about 45% of the sulfur precipitated as sulfide was estimated to be derived from mineralization of organic matter, and the remainder (55%) originated from bacterial reduction of sulfates (Nriagu, 1968). Rates of sulfate reduction in the water column were 10^3 times lower than those of the surface sediment and, on an areal basis, accounted for less than 18% of the total sulfate reduction in the hypolimnion during summer stratification (Ingvorsen *et al.*, 1981). Estimates of net mineralization of sestonic sulfur inputs to the sediments in hypereutrophic Wintergreen Lake, Michigan, indicated that only about 45–50% of the total and ester sulfate sulfur inputs, and 75% of the protein sulfur inputs, were mineralized (King and Klug, 1982a). About 3% of the total water column sulfur was permanently lost to the sediments each year.

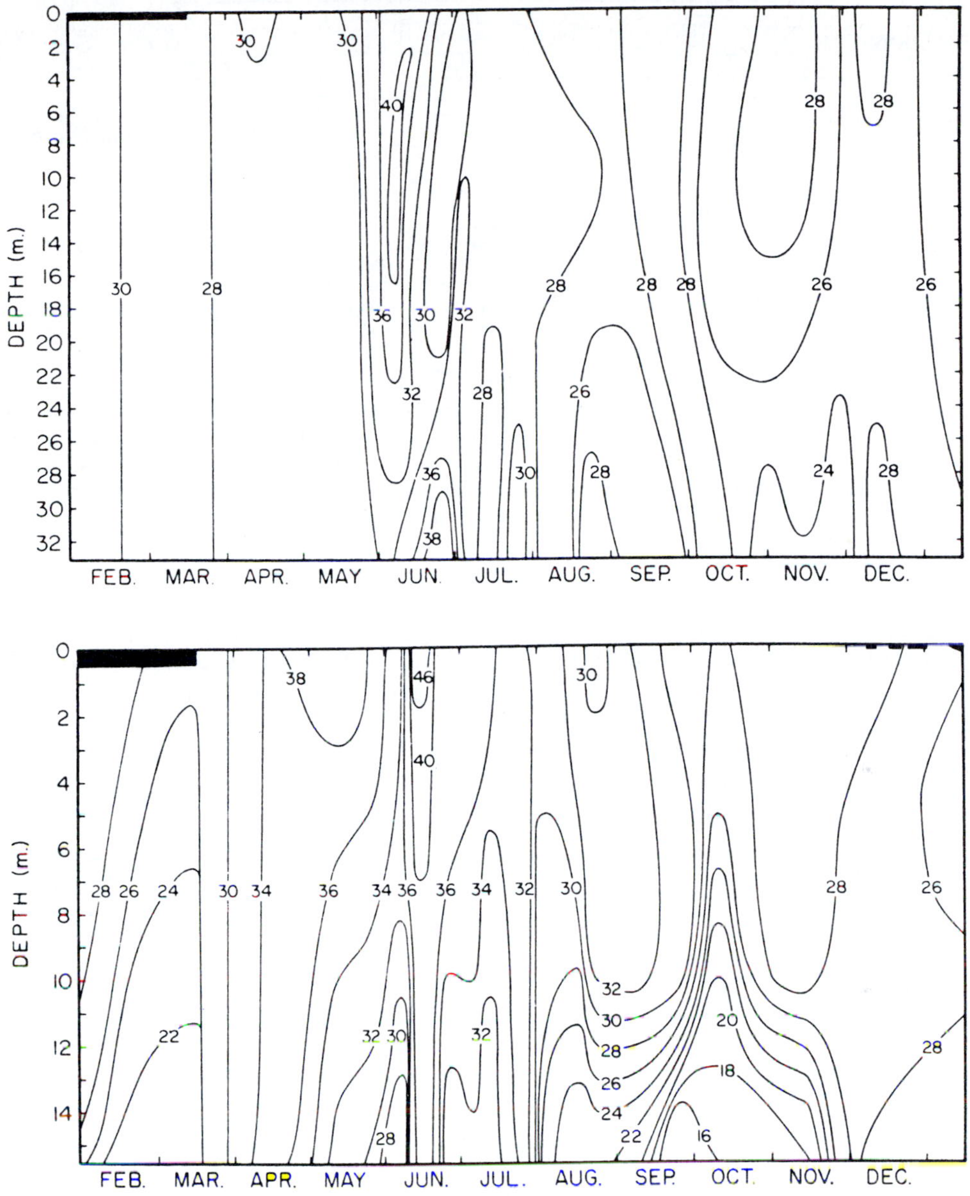

FIGURE 14-12 Depth–time diagrams of isopleths of sulfate concentrations (mg liter^{-1}) of mesotrophic hardwater Crooked Lake (*upper*) and interconnected eutrophic Little Crooked Lake *(lower)*, Noble-Whitley counties, northeastern Indiana. Opaque areas = ice cover to scale. (From Wetzel, unpublished data.)

C. Bacterial Metabolism and the Sulfur Cycle

Sulfate is reduced to the sulfhydryl (—SH) form during the synthesis of proteins by plants and animals. Further reduction of HS^- to H_2S occurs upon decomposition of this organic material by heterotrophic bacterial metabolism (Fig. 14-14). The most important of a large number of protein-decomposing bacteria belong to the genus *Proteus,* which are gram-negative, nonspore-forming rods that usually possess large numbers of flagella (Butlin, 1953). *Proteus* species are particularly active in soil systems. The dominant bacteria involved in proteinaceous decomposition to form H_2S in lakes of varying productivity (Table 14-10) are discussed in Kuznetsov (1970), Zinder and Brock (1978), and Jørgensen (1983). Bacterial densities in the water are 1–3 orders of magnitude lower than in the surface sediments.

A number of bacteria reduce sulfate, sulfite, thiosulfate, hyposulfite, and elemental sulfur to hydrogen sulfide. These ***sulfate-reducing bacteria*** are heterotrophic and anaerobic and use the sulfur compound as a hydrogen acceptor during oxidative metabolism.

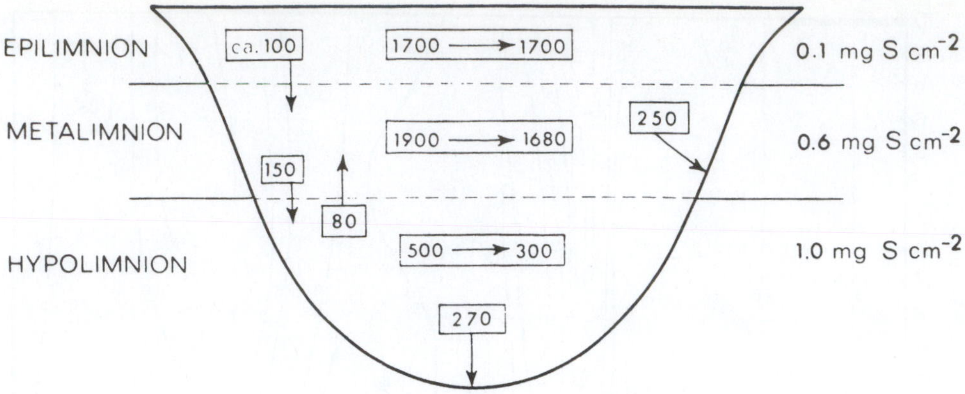

FIGURE 14-13 The total sulfate budget of Linsley, Pond, Connecticut, estimated for an early stage of thermal stratification (*left-hand numbers in central squares*) and final stages (*right hand numbers in central squares*). The other numbers give the transfer rates of sulfate, in kg S, between the different strata. At the right is the total amount of sulfate reduced and stored per cm^2 of sediment during the four-month period. The total amount of dissolved H_2S at the end of stratification was about 15 kg S in the hypolimnion. (After Stuiver, 1967.)

Finally, several groups of bacteria oxidize sulfide to sulfur and sulfur to sulfate.

The ***sulfur-reducing bacteria,*** such as those of the genera *Desulfovibrio* and *Desulfotomaculum,* are obligate anaerobes and derive oxygen from sulfate for the oxidation of either organic matter or molecular hydrogen (Hamilton, 1985; Singleton, 1993):

$$H_2SO_4 + 2(CH_2O) \longrightarrow 2CO_2 + 2H_2O + H_2S$$

$$H_2SO_4 + 4H_2 \longrightarrow 4H_2O + H_2S$$
$$(\Delta G_0' = -60 \text{ kcal mole}^{-1})$$

Although these reactions do not consume oxygen directly (Peck, 1993), the H_2S generated by sulfate-reducing bacteria is readily oxidized and consumes oxygen upon intrusion into, or transport to, aerobic regions. The sulfate-reducing bacteria can reduce sulfite more rapidly, and thiosulfate less rapidly, than sulfate. Colloidal sulfur, but not pure noncolloidal sulfur, is reduced very slowly. Over a concentration range of 20–130 mg SO_4 liter^{-1}, the rate of production of H_2S by bacteria is roughly proportional to SO_4 concentration (Ohle, 1954). The biological oxygen demand of oxidizable organic matter can theoretically be satisfied with about 1.6 g SO_4 g^{-1} by this reduction.

The ***sulfur-oxidizing bacteria*** are commonly differentiated into two groups (Kuenen *et al.,* 1985). The ***chemosynthetic (colorless) sulfur-oxidizing bacteria***

TABLE 14-9 Distribution of Forms of Sulfur in Surface Sediments (0–10 cm)

Sulfur form	*Concentration (moles)*	*Percentage of total*	*Site*	*Percentage ester sulfate of total*
Wintergreen Lake Michigan[a]			Wintergreen Lake, Michigan[b]	
Total sulfur	204	—	Littoral	22.1
Ester sulfate sulfur	65	35	Littoriprofundal	34.2
Pyritic sulfur	21	10	Profundal	32.3
Acid volatile sulfur	8	4	Oneida Lake, New York[c]	80
Elemental sulfur	3	1	South Lake, New York[c]	45
Dissolved sulfide	2	1	Deer lake, New York[c]	79
Dissolved sulfate	0.4	0.2		

[a]Smith and Klug (1981); King and Klug (1982).
[b]King and Klug (1980).
[c]Mitchell *et al.* (1981).

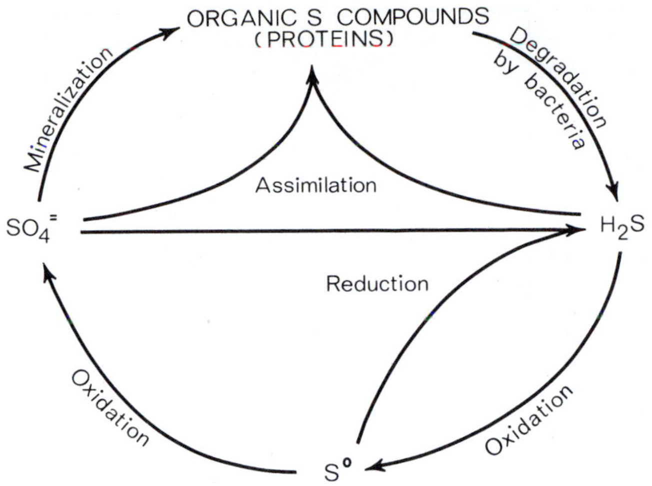

FIGURE 14-14 General sulfur cycle in nature. (Modified from Butlin, 1953.)

are mostly aerobic forms that oxidize H_2S, and are of two types. The first deposits sulfur *inside* the cell:

$$H_2S + \tfrac{1}{2}O_2 \longrightarrow S^0 + H_2O \quad (\Delta G_0' = -41 \text{ kcal mole}^{-1})$$

which accumulates as long as H_2S is available. As sulfide sources are depleted, the internally stored sulfur is oxidized and sulfate is released:

$$S^0 + 1\tfrac{1}{2}O_2 + H_2O \longrightarrow H_2SO_4 \quad (\Delta G_0' = -118 \text{ kcal mole}^{-1})$$

Beggiatoa, a long, filamentous bacterium, and *Thiothrix* are common bacteria that oxidize H_2S with deposition of sulfur intracellularly. These two genera occur in areas where H_2S is being formed (e.g., canals, swamps, and sulfur springs).

By similar reactions, a second type of chemosynthetic sulfur-oxidizing bacteria deposits sulfur *outside* of the cell. This assemblage is represented best by the genus *Thiobacillus*, which oxidizes sulfide, S^0, and other reduced sulfur compounds such as thiosulfate:

$$2Na_2S_2O_3 + O_2 \longrightarrow 2S^0 + 2Na_2SO_4$$

Of the many species of *Thiobacillus*, some (e.g., *T. thiooxidans*) are restricted to acidic waters (pH 1–5), while others such as *T. thioparus* grow optimally at neutral or alkaline pH values. The anaerobe *T. denitrificans* oxidizes thiosulfate in alkaline waters by reduction of nitrate to N_2:

$$5S_2O_3^{2-} + 8NO_3^- + 2HCO_3^- \longrightarrow 10SO_4^{2-} + 2CO_2 + H_2O + 4N_2$$

or

$$5S^0 + 6NO_3^- + 2CO_3^{2-} \longrightarrow 5SO_4^{2-} + 2CO_2 + 3N_2 \quad (\Delta G_0' = -179 \text{ kcal mole}^{-1}).$$

The elemental sulfur-oxidizing bacteria commonly adhere to sulfur granules, continuously utilizing a little at a time in the formation of sulfate.

The other major group of sulfur-oxidizing bacteria is the ***photosynthetic (colored) sulfur bacteria,*** anaerobes that can be divided conveniently into the ***green sulfur bacteria*** (Chlorobacteriaceae) and the ***purple sulfur bacteria*** (Thiorhodaceae). Excellent detailed reviews of the photosynthetic bacteria and their metabolism are given by Gest *et al.* (1963), Vernon (1964), Kondrat'eva (1965), Pfennig (1967), Fierdingstad (1979), Madigan (1988), Kelly (1990), Collins and Remsen (1991), Stolz (1991), and Friedrich (1998). The green sulfur bacteria require light as an energy source and use the sulfur of H_2S as an electron donor in the photosynthetic reduction of CO_2:

$$CO_2 + 2H_2S \xrightarrow{\text{light}} (CH_2O) + H_2O + 2S$$

$$2CO_2 + 2H_2O + H_2S \xrightarrow{\text{light}} 2(CH_2O) + H_2SO_4$$

TABLE 14-10 Predominant Bacteria that Form Hydrogen Sulfide from the Decomposition of Proteinaceous Organic Matter in Different Types of Lakes[a]

Type of lake	*Dominant bacteria*
Oligotrophic	*Mycobacterium phlei, Mycobacterium filiforme*
Mesotrophic	*Bacterium nitriftcans, Pseudomonas liquefaciens, Chromobacter aurantiacum*
Eutrophic	*Pseudomonas liquefaciens, Bacterium delicatum*
Bog lakes of high organic matter	*Pseudomonas fluorescens, Bacillus pituitans*
Saline lakes and estuaries	*Mycobacterium luteum, Micrococcus nitriftcans, Achromobacter halophilum, Flavobacterium halophilum, Bacterium albo-luteum, Vibrio hydrosulfureus*

[a] After data of Kuznetsov (1970).

A few species can utilize molecular hydrogen alone. The green sulfur bacteria, notably the genera *Chlorobium* and *Pelodictyon,* are generally unicellular and nonmotile and produce sulfur granules outside of their cell membranes. At least four bacteriochlorophylls occur in the photosynthetic bacteria that differ from chlorophyll *a* in a primary absorption maximum at higher wavelengths (770–780 nm versus 665 nm for chlorophyll *a*). The green sulfur bacteria can tolerate fairly high concentrations of H_2S, whereas the purple sulfur bacteria are less tolerant of H_2S and grow optimally at high pH values (9.5).

The *purple sulfur bacteria* require light energy for the oxidation of H_2S and other reduced sulfur compounds, especially thiosulfate, to sulfate in the photosynthetic reduction of CO_2. Members of this group are generally large (5–10 μm) and actively motile and deposit free S^0 intracellularly. Sulfide is oxidized to S and SO_4^{2-} by the same reactions described for the green sulfur bacteria. In addition, some of the purple sulfur bacteria are able to grow photoautotrophically, with thiosulfate as the electron donor *(Thiopedia, Thiocapsa, Thiocystis* and *Rhodothece*). Other important genera (*Chromatium, Chlorobium* and *Thiospirillum)* are unable to utilize significant amounts of thiosulfate. Like the green sulfur bacteria, many purple bacteria can utilize hydrogen as the only electron acceptor, simultaneously with an assimilatory sulfate reduction system. In anaerobic environments where iron is abundant, phototrophic sulfur bacteria can use FeS as an electron donor as well as H_2S (Davison and Finlay, 1986; Garcia-Gil *et al.,* 1990; Ehrenreich and Widdel, 1994). Thus, under conditions where light supports phototrophic sulfur bacteria in proximity to combined sources of reduced iron and sulfide, the sulfide is oxidized either directly or indirectly through FeS to sulfate at rates adequate to prevent iron sulfide from forming in the water column.

The Calvin cycle has been shown to be operational in all photosynthetic bacteria studied. Many strains can utilize low-molecular-weight organic substrates, especially fatty acids, as their carbon source, singly or in combination with CO_2. Nearly all green and purple sulfur bacteria require vitamin B_{12} from exogenous sources. This vitamin, as will be discussed further on, is an organic micronutrient that influences photosynthetic productivity under some limnological conditions.

A third group, the ***purple nonsulfur bacteria*** (Athiorhodaceae), is included here among the other pigmented photosynthetic bacteria because of its many metabolic and distributional similarities. The nonsulfur purple bacteria, including *Rhodopseudomonas, Rhodospirillum,* and *Rhodomicrobium,* are facultative photoautotrophs that grow photosynthetically or heterotrophically either aerobically or anaerobically in the dark on organic substrates. Some *Rhodopseudomonas* can utilize thiosulfate anaerobically as a hydrogen donor:

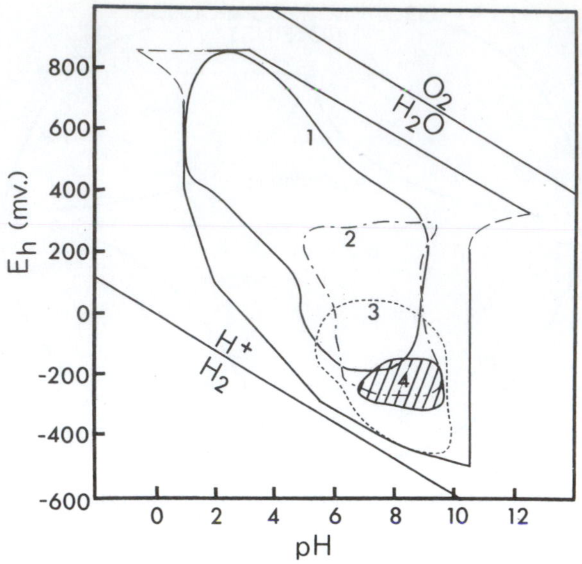

FIGURE 14-15 General E_h–pH environmental limits of (1) chemosynthetic (colorless) sulfur-oxidizing bacteria, (2) photosynthetic purple bacteria, (3) sulfate-reducing bacteria, and (4) green sulfur bacteria, all within the composite distributional field of E_h–pH measurements of habitats of organisms. (After data of Baas Becking *et al.,* 1960.)

$$2CO_2 + Na_2S_2O_3 + 3H_2O \xrightarrow{\text{light}} 2(CH_2O) + Na_2SO_4 + H_2SO_4$$

In this group, sulfur is not stored intracellularly, and hydrogen sulfide inhibits growth.

The occurrence and distribution of the various sulfur-oxidizing or -reducing bacteria are restricted by the redox and pH conditions in relation to oxygen and the state of sulfur compounds (Fig. 14-15). The reducing conditions required by strictly anaerobic photosynthetic sulfur bacteria, for example, must coincide with adequate light of high wavelength before large populations can develop. Often, conditions required for optimal growth of sulfur bacteria occur in stratified lakes as sharply defined layers with steep physical and chemical gradients and result in thin layers or strata of bacterial populations.

Microbial processes involved in the sulfur cycle of lakes are diagrammatically represented in Figure 14-16. The processes on the left-hand side of the figure would be more characteristic of a lake with relatively high concentrations of sulfur in various forms. The gradient between the oxic upper strata and the lower H_2S-rich strata would be steep, but with a diffusion interface zone where both oxygen and H_2S occurred. Those

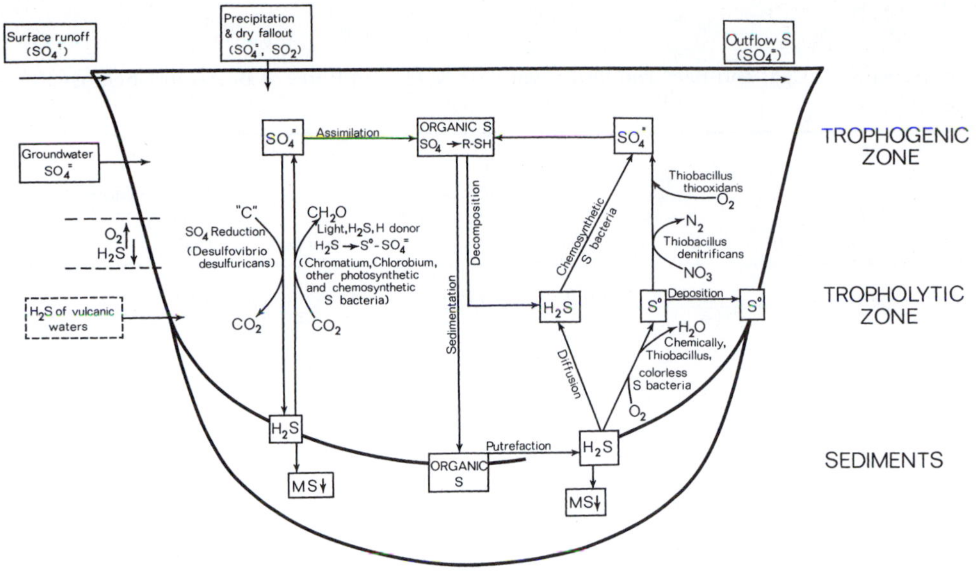

FIGURE 14-16 Composite representation of the sulfur cycle in a lake, with emphasis on the microbiological processes. MS = metallic, primarily iron, sulfides. (Greatly modified from Kuznetsov, 1970.)

processes on the right-hand side of Figure 14-16 are more representative of lakes with lower sulfate content.

The rather specific requirements of the sulfate-reducing bacteria, photosynthetic and colorless sulfur bacteria, and sulfur-oxidizing bacteria would lead one to expect development of massive populations only in localized strata. Such is indeed the case in many situations, particularly in meromictic lakes where gradients within the chemolimnion are steep. Green sulfur bacteria are commonly found in profusion in a thin layer immediately below a dense population of purple sulfur bacteria at the interface of the oxic–anoxic layer of very productive lakes. Light levels at this region are almost always low, usually <10% of intensities at the surface, and rapidly become the dominant limitation to photosynthesis in the bacterial plates. The seasonal development of optimal conditions for the specific groups of sulfur bacteria can be transitory, so that their contribution to the total annual productivity of the lake may be short-lived. Within meromictic lakes, bacterial plates can persist more or less continuously (cf. Abella *et al.*, 1980; Guerrero *et al.*, 1980, 1985; Baker *et al.*, 1985; Overmann *et al.*, 1991).

Although descriptions of the distribution of the sulfur bacteria in aquatic systems are fairly common, and much is known about the physiology of these interesting organisms, little is known of their contribution to the total productivity of lakes. It is clear that at certain periods in productive dimictic and in meromictic lakes, bacterial photosynthesis easily can exceed that of algae and macrophytes (Table 14-11). Most of the data given in Table 14-11, however, are taken from periods when the photosynthesis of bacteria was maximal. Occasionally, when the algal photosynthesis is low, as in the example of Smith Hole Lake, which receives high amounts of allochthonous organic matter, the brief but very productive photosynthetic bacterial development represents a major part of the lake's annual photosynthetic productivity (Wetzel, 1973). In most situations, however, the contribution of bacterial photosynthesis to the entire system over an annual period is small, even though spectacular localized populations may develop periodically.

In several systems, as for example in Lake Gek Gel, an oligotrophic meromictic lake of crenogenic origin, and in the deep, meromictic Black Sea, the most intensive rates of bacterial reduction of sulfates occurred near sediments and in water layers on the littoral slopes at the upper boundary of the anaerobic H_2S zone (Fig. 14-17) (Sorokin, 1964, 1966, 1970). The higher rates of sulfate reduction in these cases were apparently related to higher inputs of organic matter from the littoral zone brought in allochthonously from the drainage basin. Higher H_2S concentrations along these sediment interfaces were then dispersed by weak water currents to the open water (Sorokin, 1970). High rates of sulfate reduction occurred below the zone of most active bacterial chemosynthesis (Fig. 14-17), where again the concentrations of organic substrates were presumably higher.

A similar dichotomous distribution of sulfate-reducing bacteria below the zone of most active

TABLE 14-11 Comparison of Algal and Bacterial Photosynthetic Rates and Rates of Bacterial Chemosynthesis in Several Lakes (PS = Photosynthesis)

Lake	*Units of measurement*	*Algal PS*	*Bacterial PS*	*Bacterial PS/algal PS × 100 (%)*	*Chemosynthesis*	*Chemosynthesis/ total PS × 100 (%)*	*Remarks*
Hiruga, Japan 21 July 66 (Takahashi and Ichimura, 1968)	mg C m^{-2} h^{-1}	29.7	0.9	3.0	4.3	14.1	
Waku-ike, Japan 4 Aug. 65	mg C m^{-2} h^{-1}	19.8	1.8	9.1	—	—	
Kisaratsu, Japan	mg C m^{-2} h^{-1}						
16 Aug. 66		22.1	128	579	3.5	2.3	
2 Sept. 65		12.0	10.6	88	2.5	11.1	
Smith Hole, Indiana Annual period 1963–1964	annual mean						
(Wetzel, 1973)	mg C m^{-2} day^{-1}	194	91	47	6	<10	Massive late summer population of *Chromatium*
14 Aug. 63	mg C m^{-2} day^{-1}	540	5960	1104	—	—	
Wadolek, Poland	mg C m^{-2} day^{-1}						
June 66		32.8	55.3	169	—	—	*Chlorobium* in metalimnion
July 66		144.5	19.4	13	—	—	
Sept. 66 (Czeczuga, 1967, 1968a,b)		24.6	19.1	78	—	—	
Muliczne, Poland	mg C m^{-2} day^{-1}						*Thiopedia* in hypolimnion
19 Aug. 67		478	157	32.8	—	—	
16 Sept. 67		258	136	52.7	—	—	
20 Oct. 67 (Czeczuga, 1968c)		281	28	10.0	—	—	
Belovod, Russia							
July	mg C m^{-2} day^{-1}	500	55	11	15	2.7	
(Sorokin, 1970)	mg C m^{-3} day^{-1}						
	13 m	0	210	—	74	35	
	14 m	0	79	—	44	56	
	18 m	0	2.7	—	16.4	607	
	24 m	0	0	—	5.5	—	
Kuznechikha, Russia June 1972 (Gorlenko *et al.*, 1980)	mg C m^{-2} day^{-1}	160	100	63	—	—	*Chloronema, Chlorochromatium, Chloroplana, Pelochromatium, Thiobacillus*
Arcas, Spain	mg C m^{-3} day^{-1}						
3 Sept. 91 (Camacho and Vicente, 1998)	0.5 m	3.5	0	0	—	—	*Chromatium weissei* in hypolimnion
	7.0 m	13.6	0	0	—	—	
	8.5 m	8.2	0	0	—	—	
	8.7 m	42.8	8.1	18.9	—	—	
	8.9 m	0.0	47.5	100.0	—	—	
Six lakes, Michigan and Wisconsin (Parkin and Brock, 1980)	mg C m^{-2} day^{-1}	276–1510	2–60	0.25–6.3	—	—	Several green and purple sulfur bacterial species
Mahoney, British Columbia (Hall and Northcote, 1990)	mg C m^{-2} day^{-1}	55–233	29–314	17–66	—	—	

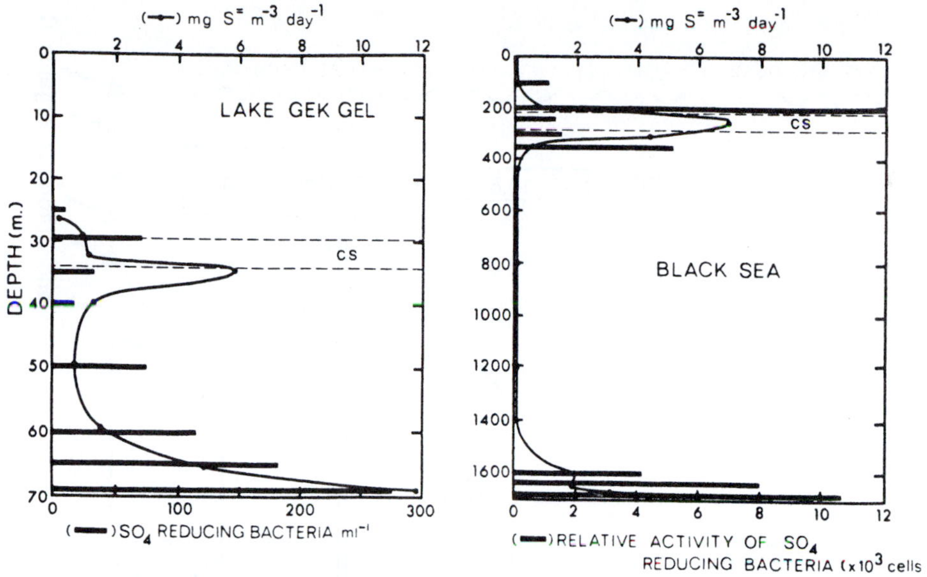

FIGURE 14-17 Rates of sulfate reduction (mg H_2S formed m^{-3} day^{-1}) and the biomass and activity of sulfate-reducing bacteria in meromictic Lake Gek Gel and Black Sea, Russia. CS = zone of active chemosynthetic bacterial metabolism. (After data of Sorokin, 1964, 1970.)

chemosynthesis is seen even more clearly from data of Lake Belovod (Fig. 14-18). The major bacteria in the zone of chemosynthesis were several species of *Thiobacillus,* over which lay a dense population of purple sulfur bacteria. Sorokin has provided evidence on the movements of zooplankton populations in response to changes in bacterial stratification, and studies of feeding indicate that zooplankton, especially the cladoceran microcrustacea, actively feed on the dense populations of these large bacteria.

A detailed study was undertaken of the organic carbon budget and comparative productivity of phytoplankton and phototrophic sulfur bacteria in Mahoney Lake, a small (11 ha), saline meromictic lake in British Columbia (Overmann, 1997). The purple sulfur bacterium *Amoebobacter purpureus* (Chromatiaceae) completely dominated (98%) the microorganisms and concentrated in a 20-cm plate in the monimolimnion between 6- and 7-m depths. Fluxes of organic carbon were calculated from the rates of photosynthesis, sulfate reduction, and sedimentation (Fig. 14-19). Upwelling was a major loss process for *A. purpureus* (80% moved upward to the oxic strata), and sedimentation provided only a small fraction of the carbon substrates of sulfate-reducing bacteria (7.9 of ca. 47 g C m^{-2} yr^{-1}). Only a third of the sulfate reduction within the bacterial plate (13.9 g C m^{-2} yr^{-1}) could be attributed to recycling of *A. purpureus* carbon. Because production and sedimentation of phytoplankton was very

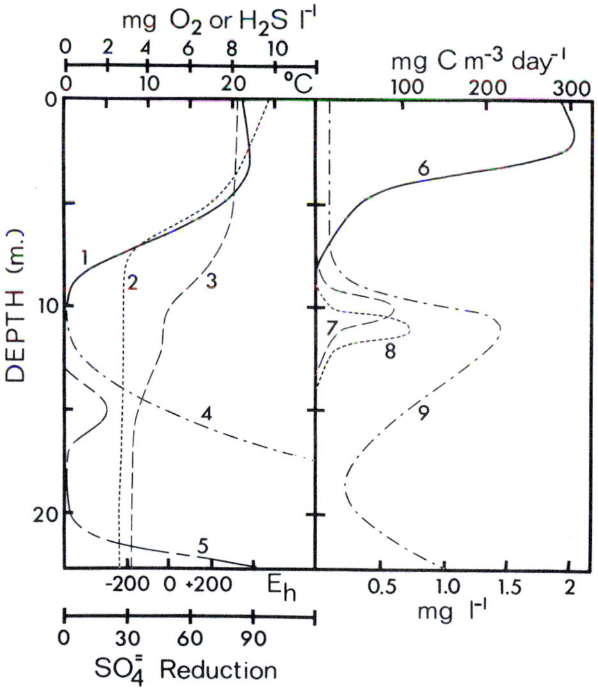

FIGURE 14-18 Distribution of midsummer characteristics and intensity of biological processes in the central depression of meromictic Lake Belovod, Russia. (After data of Sorokin, 1970.) (1) oxygen, mg $liter^{-1}$; (2) °C; (3) E_h in mV; (4) H_2S, mg $liter^{-1}$; (5) rate of sulfate reduction, mg H_2S formed m^{-3} day^{-1}; (6) photosynthesis by algae, mg C m^{-3} day^{-1}; (7) chemosynthesis, mg C m^{-3} day^{-1}; (8) photosynthesis by purple sulfur bacteria, mg C m^{-3} day^{-1}; (9) biomass of bacteria, mg $liter^{-1}$.

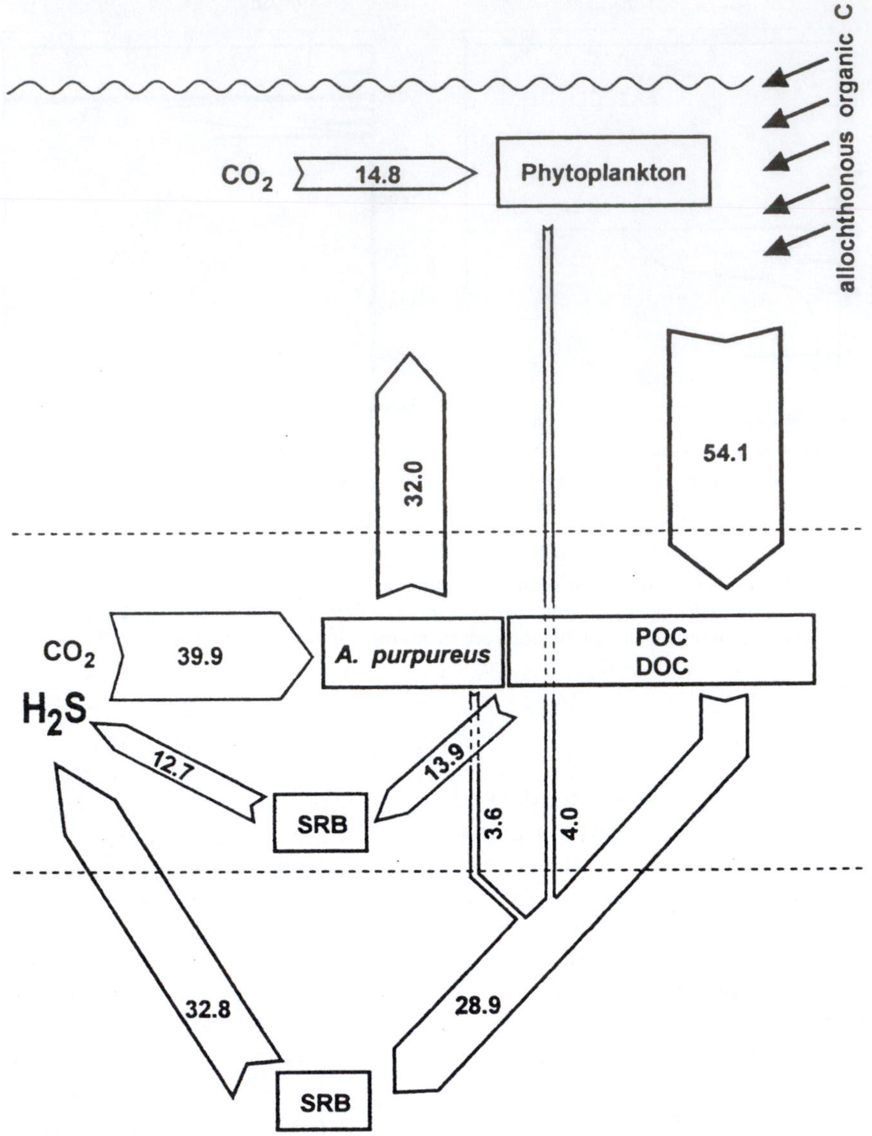

FIGURE 14-19 Carbon cycle (g C m^{-2} of lake surface yr^{-1}) in the anoxic portion (below 6.7 m) of Mahoney Lake, British Columbia, based on rates of photosynthesis, sulfate reduction, sedimentation, and upwelling of the photosynthetic sulfur bacterium *A. purpureus*. Horizontal arrows: photosynthesis; oblique arrows: carbon demand of sulfate-reducing bacteria (SRB); vertical arrows: sedimentation of particulate organic carbon or upwelling of the POC of *A. purpureus*. Dotted lines: upper and lower boundaries of the bacterial place between 6.7 and 6.9 m. (Data from Overmann, 1997.)

small, most of the organic carbon supporting sulfate reduction and generation of H_2S that fueled the phototrophic bacterial production emanated from allochthonous particulate and, especially, dissolved organic carbon (DOC to 90 mg C $liter^{-1}$). In this way, recalcitrant organic carbon from allochthonous sources is converted into readily degradable bacterial biomass using light as the energy source and sulfide as an electron acceptor in the formation of biomass that is then returned to the oxic part of the ecosystem.

D. Sulfur Cycling in the Sediments

Large variability in rates of sulfate reduction within sediments has been observed among different lakes (Table 14-12). Early studies suggested that sediment sulfate concentrations varied little over time, with a relatively constant turnover time for the sulfate (Smith and Klug, 1981). Organic sulfate esters, entering the sediments from settling seston, can serve as a major source of sulfur and be mineralized by an active

TABLE 14-12 Rates of Sulfate and Elemental Sulfur Reduction in Various Sediments

Lake	SO_4^{-2} *concentration* (*mg liter*$^{-1}$)	*Reduction rate* (*mg* H_2S *liter*$^{-1}$ *day*$^{-1}$)	*Source*
Linsley Pond, Connecticut	7.5–17.5	0.8–1.5	Stuiver (1967)
Lake Beloe, Russia	48–205	$2-6 \times 10^{-4}$	Ivanov (1968) in Zinder and Brock (1978)
Gor'kii Reservoir, Russia	32–132	0.04–2.94	
Lake Belovod, Russia	428–625	0.067–0.124	
Wintergreen Lake, Michigan	6–48		Molongoski and Klug (1980a)
Sulfate reduction		0.6–5.8	Smith and Klug (1981)
Elemental sulfur reduction		0.28	

sulfhydrolase system in surficial sediments (King and Klug, 1980, 1982a). These organic sources would augment inorganic reduced sulfur generated from sulfate reduction.

Much of the sulfur in sediments is organic, particularly in eutrophic lakes. In lakes with highly organic sediments and low iron concentrations, a larger portion of the sedimentary iron is present as ferrous sulfides (Giblin *et al.,* 1990). Much of the reduced sulfate is rapidly converted to organic sulfur compounds, and this organic sulfur is more persistent in the sediments than iron sulfides (Rudd *et al.,* 1986). Several recent studies indicate that high rates of sulfide oxidation (pyrite, organically bound sulfur and sulfides) can accompany high rates of sulfate reduction (Marnette *et al.,* 1992; Urban and Brezonik, 1993; Urban *et al.,* 1994). Sulfide oxidation can occur nearly as rapidly as sulfate reduction. Importantly, the formation of organic sulfur in sediments consumes H^+ ions via sulfate reduction in acidified lakes (Anderson and Schiff, 1987). Rates of sulfate reduction in sediments can be significantly enhanced in moderately acidic, organic rich sediments that receive high sulfate inputs from atmospheric or influent sources (Herlihy and Mills, 1985; Schuurkes and Kok, 1988).

VII. THE SILICA CYCLE

Silica (SiO_2) is usually moderately abundant in fresh waters and, although it is relatively unreactive, it is of major significance to diatomaceous algae, chrysophytes, and some higher aquatic plants and sponges. Diatoms assimilate large quantities of silicon in the synthesis of their frustules. Silicon is a major factor influencing algal production in many lakes, and diatom utilization of silica greatly modifies the flux rates of silica in lakes and streams. Availability of silica can have a strong influence on the overall pattern of algal succession and productivity in lakes and streams.

A. Forms and Sources of Silicon

Silicon occurs in fresh waters in two major forms of silicon dioxide or silica (SiO_2). (1) *Dissolved silicic acids* form stable solutions of H_2SiO_4 at much higher concentrations than are encountered in fresh waters (60–80 mg SiO_2 liter^{-1} at 0°C to 100–140 mg SiO_2 liter^{-1} at 25°C at commonly occurring pH values; Krauskopf, 1956). Unreactive silicon is not generally found; polymeric silicon is unstable and depolymerizes rather rapidly (within hours) (Burton *et al.,* 1970). The solution of silica from various rock sources is modified, however, by surface adsorption of silicic acid, which reduces solubility and leads to a general situation in which nearly all natural waters are greatly undersaturated with respect to silica (Tessenow, 1966; Stöber, 1967). (2) *Particulate silica* is found in two forms—that in biotic material, in particular in diatoms and a few other organisms that use large amounts of silica, and that adsorbed to inorganic particles or complexed organically. Silicate complexes with iron and aluminum hydroxides and this process decreases the solubility of silicates in sediments, especially in interstitial waters at pH values above 7 (Ohle, 1964). The solubility of silica is increased by humic compounds and through the formation of iron and aluminum–silicate–humic complexes.

The silica content of drainage to natural waters is less variable than many of the other major inorganic constituents. The world average is about 13 mg SiO_2 liter^{-1}, with relatively little variation among the continents; the average of ground water is somewhat higher than that of surface drainage (Davis, 1964). The major source of silica is from the degradation of aluminosilicate minerals. The greatest concentrations of silica are found in ground water in contact with volcanic rocks; intermediate amounts occur in association with plutonic rocks and sediments containing feldspar and volcanic rock fragments; and small amounts originate from marine sandstones. The lowest amounts of silica are found in water draining from carbonate rocks. Silica forms aggregations and becomes relatively immobile at pH values below 3, but its mobility increases somewhat in the range of pH 4–9. Adsorption is the only significant mechanism of inorganic precipitation; flocculation of colloidal silica could not be demonstrated (Tessenow, 1966).

Carbonic acid originating from dissolved CO_2 reacts with silicates to form carbonates and silica. Concentrations of dissolved silica in soil water increase slightly with higher temperature and decrease with increasing soil pH values (McKeague and Cline, 1963a, b) as adsorption increases within pH 4–9. Above a pH of 10, adsorption decreases sharply. Mineralization of silicates is presumed to be largely or entirely a nonenzymatic hydrolysis (Golterman, 1960). However, silicate bacteria are known to play a role in the weathering of rocks and minerals both in the presence and the absence of organic substrates (Savostin, 1972; Joseph and Bravo, 1990), and benthic-living diatoms are known to attack silicious materials of sediments. Mechanisms of this mineralization are unclear, however.

The silica content of river waters tends to be remarkably uniform and shows little response to change in discharge rates (Edwards and Liss, 1973). This situation is in distinct contrast to other major constituents of river water, which commonly show an inverse relationship between concentration and discharge rates. Although rapid diurnal changes in the silica content of river waters are known and can be associated inversely with division rates and growth of diatoms (Müller-Haeckel, 1965), these and other biological factors are insufficient to explain the relative stability of silica concentrations. An abiological buffering mechanism is apparently also operational, with adsorption reactions between dissolved silica and silica in the solid phase (i.e., in association with hydrated oxides). Adsorption and desorption equilibria can buffer changes in concentration over a period of several days. The effects of dissolved silica and degraded silicates assume much more significance in the oceans, where reactions occur in which silica and alkali metal cations are fixed. Hydrogen ions are released and result in an effective buffering of both pH and silica concentrations (Garrels, 1965; Mackenzie *et al.*, 1965, 1967).

Silica loadings to lakes are usually largely from influent surface waters of rivers. In certain lakes, however, ground waters can be the primary source of silica. For example, even though the short-term, seasonal input of ground water to a small, precipitation-dominated oligotrophic lake in northern Wisconsin was <10% of the annual water budget of the lake, groundwater influents accounted for nearly all of the external silica loading (Hurley *et al.*, 1990).

B. Distribution of Silica in Lakes and Reservoirs

Concentrations of silica within lakes and reservoirs frequently exhibit marked seasonal and spatial variations. Even in oligotrophic waters a conspicuous decrease in silica is often found in the epilimnetic strata during early winter, as well as in the spring during circulation and during thermal stratification (Fig. 14-20). In eutrophic lakes, silica concentrations in the trophogenic zone are commonly reduced to near analytical undetectability. Reductions in silica concentrations within the epilimnetic and metalimnetic zones result in a negative heterograde silica curve against depth (Figs. 14-20 and 14-21). The heterograde silica distribution is clearly associated with intensive assimilation of

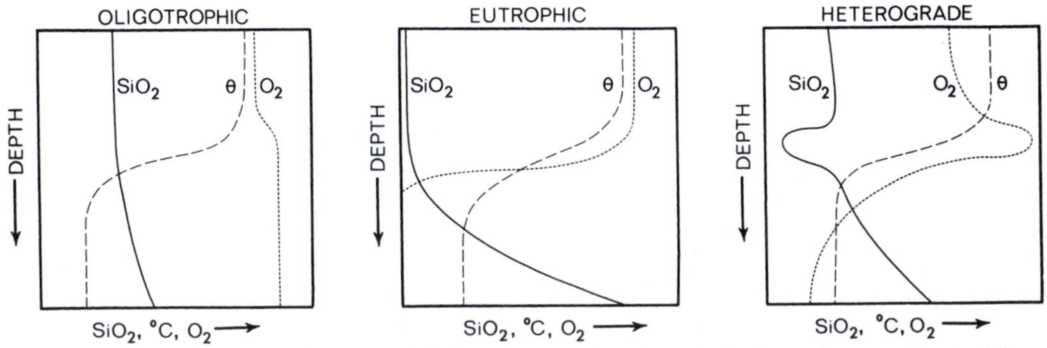

FIGURE 14-20 Generalized distribution of silica concentrations in unproductive and very productive lakes, and in a lake exhibiting a metalimnetic development of diatom algae and a negative heterograde silica curve.

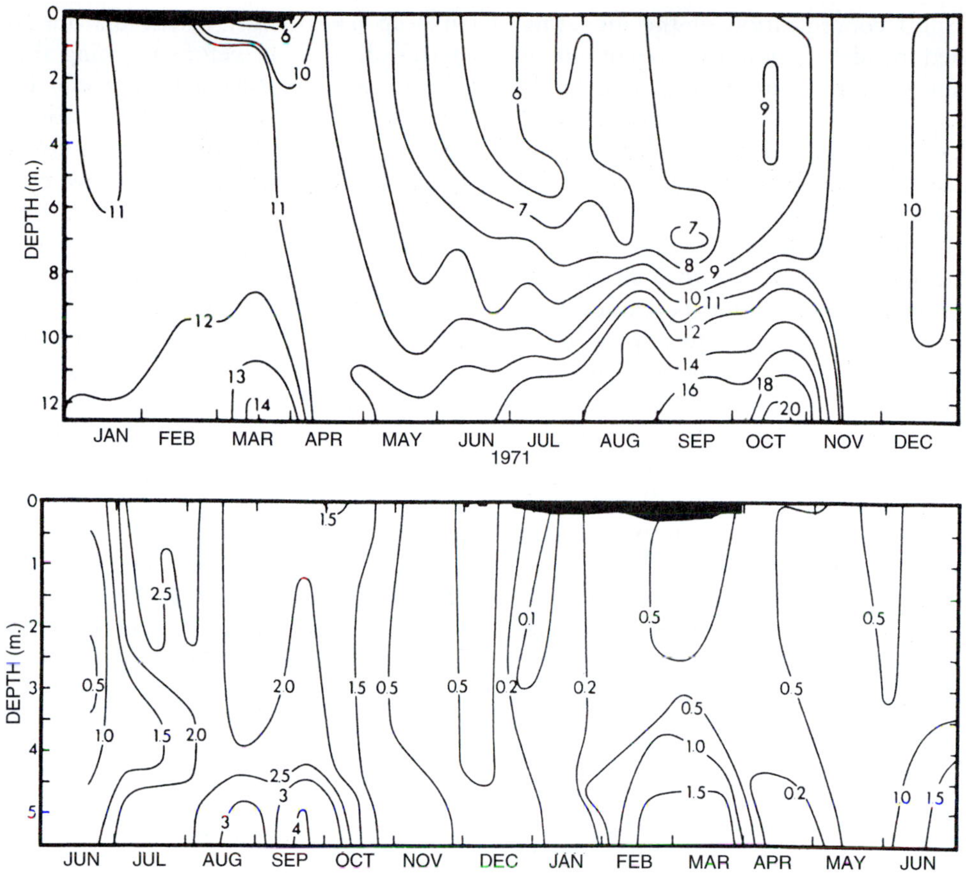

FIGURE 14-21 Depth–time diagram of isopleths of silica concentrations (mg SiO_2 liter^{-1}) in oligotrophic hardwater Lawrence Lake, 1971 (*upper*) and hypereutrophic Wintergreen Lake, 1971–1972 (*lower*), southwestern Michigan. Opaque areas = ice cover to scale. (From Wetzel, unpublished data.)

silica by diatoms, which sediment with their incorporated silica from the trophogenic zone more rapidly than silica is replaced by inputs to the system from surface water and ground water. Other sinks of silica are minor relative to the quantities transported by diatoms. Abiogenic precipitation in the open water is relatively unimportant in the cycle of silica. Dilution by water low in silica can occur, as, for example, by rain percolating through decaying ice (March–April, Lawrence Lake, Fig. 14-21).

The seasonal cycle of silica, demonstrated in Figure 14-21, has been observed frequently. By far the most detailed experimental investigations of mechanisms controlling silica dynamics in lakes is that of Tessenow (1966), from which many of the following statements are drawn. Utilization of silica by diatoms occurs during photosynthesis and increases somewhat in darkness. Adsorption of SiO_2 to dead cells under some conditions can lead to reduction of silica content even below the trophogenic zone. Silica concentrations usually increase in the tropholytic zone of a lake during both summer and winter periods of stratification. The silica gradient becomes more steep in eutrophic lakes that exhibit an anaerobic clinograde oxygen curve. In most stratified lakes, silica concentrations increase in water immediately above the sediments (Fig. 14-21; cf. Conway *et al.*, 1977).

The amorphous silica of diatoms often settles to the sediments. This biochemical condensation of dissolved silica and sedimentation greatly exceeds inputs to the sediments from abiogenic sources and reaches the sediments with variable periodicity. Diatom production is typically greatest in the spring and early winter but often proceeds more or less continuously, particularly in oligotrophic lakes.

Interstitial water is enriched in dissolved silica at concentrations far in excess of those of water entering the lake (Tessenow, 1966; Harriss, 1967). Interstitial concentrations increase as the pH declines below 7, decrease between pH 7–9, and greatly increase above pH 9. Concentrations also increase at higher temperatures within the range found in fresh waters. The dissolved silica of interstitial water is not in equilibrium with amorphous silica, but rather with chemically bound or adsorbed silica. Silica concentration of the interstitial water is controlled by dissolution of ferroaluminum silicate. This complex is formed in the sediments by the reaction of silica from diatoms with aluminum and ferric oxyhydroxides or through the hydrolysis of clay minerals (Nriagu, 1978). Liberation of silica to the overlying water in relatively isolated lake sediments is governed by these equilibria, which are stable only in the presence of the solid phase of adsorbed silica. Sediment material rich in hydrated Al and Fe oxides has a high phosphorus sorption capacity and a very low P concentration (<5 μg liter^{-1}) in the interstitial water under aerobic conditions. Some 25–30% of total P and almost 90% of total Si were found to be bound in mineral lattices in a stable form not participating in biological and chemical transformations (Hartikainen *et al.*, 1996). When the sediments are anaerobic, both P and Si dissolve into the interstitial water. The dissolution patterns suggest that both P and Si are bound to the same components and compete with each other for the sorption sites.

Exchange between sediment and water decreases concentrations of interstitial water and results in greater redissolution from the sediments. Sediment silica release rates increase with rising temperatures (Rippey, 1983), even though in stratified, moderately deep temperate lakes most sediments are continually cold (<10°C). Sediment silica release rates range from ca. 0.5 mmol m^{-2} day^{-1} in oligotrophic lakes to much higher values (1–14 mmol m^{-2} day^{-1}) in eutrophic lakes. Silica regeneration in nearshore regions of shallow sediments, subject to varying seasonal temperatures and high diatom production, can have much higher rates than colder deep sediments (e.g., Quigley and Robbins, 1984). The rate of silica release also varies with differences in silica concentrations between the sediments and the overlying water. Equilibrium is not attained because of the slowness of diffusion (weeks). The difference between silica of interstitial water and in the overlying water is influenced by currents, movements produced by benthic organisms (e.g., larvae of chironomid insects; Tessenow, 1966), and by gas bubbles escaping from the sediments. In unstratified lakes, the concentrations of dissolved silica can increase greatly in the spring from release of silica from the sediments as the temperatures increase among the disturbed sediments (Gibson, 1981).

In lakes dominated by diatom algae, large numbers of sedimenting diatom frustules can accumulate within the sediments and be lost permanently from the system. The extent of this permanent loss depends on rates of diatom productivity, on the morphometry of the lake basin, and upon the percentage of the sediments located in the quiescent waters of the deep hypolimnion. In deep lakes, many of the diatom frustules undergo partial dissolution before reaching the sediments. The dissolution of suspended silica can be accelerated by consumption and fragmentation of diatom frustules by zooplankton, a process that is apparently of significance in both productive shallow ponds and in deep lakes (e.g., Ferrante and Parker, 1978). Silica sedimented from biogenic sources to shallower sediments is returned more rapidly to the overlying and circulating waters. Therefore, the metalimnetic and upper hypolimnetic waters that experience greater movement than the deeper waters are commonly enriched with silica in relation to strata above and below. A resulting positive heterograde silica curve has often been observed (numerous examples are given in Tessenow, 1966). The increased exchange induced by circulation over littoral sediments in shallow lakes can lead to silica concentrations that are sometimes greater than those of drainage inlet sources and can, in addition, result in silica losses from the system by outflow. The silica cycle and economy of most lakes are regulated largely by autochthonous metabolism within the lake, but losses are balanced strongly by allochthonous inputs.

C. Utilization and Role of Silica

Silicified structures occur in many aquatic organisms, but none approaches the importance of the diatoms (Bacillariophyceae). All diatoms are enclosed in a silica wall or frustule in which silicic acid has been dehydrated and polymerized to form silica particles (J. C. Lewin, 1962). The vegetative cells of some species of yellow-brown algae (Chrysophyceae) bear discrete siliceous scales and form cysts with silicified walls. These algae, as well as certain silicoflagellates, however, probably have only an insignificant impact on silica cycling in comparison to the active utilization and more extensive distribution of the diatoms. Similarly, utilization of silicon by certain aquatic macrophytes (*Equisetum*) and siliceous sponges, while of major importance to the organisms *per se* and their development and productivity, is rarely sufficient to alter the quantitative cycling of silica in lakes (cf., however, Conley and Schelske (1993) on the potential additional role of sponge spicules in influencing the silicon biogeochemistry of shallow lakes in Florida).

The succession and productivity of algal populations will be discussed in detail further on, but it is appropriate here to stress the importance of diatoms on the silica cycle and their effects on their own population dynamics. Of all of the aspects of chemical determination of succession and productivity, the relation between diatoms and silica concentrations is among the most apparent. The data, upheld by the detailed investigations of Lund (1949, 1950, 1954, 1955; Heron, 1961; Lund *et al.*, 1963), are irrefutable and have been corroborated by experimental work (cf. Chap. 15).

Phytoplankton in temperate waters usually undergo a spring maximum. This population development may begin beneath the ice as light conditions improve but is most conspicuous during and following spring circulation when the water is relatively rich in nutrients as the winter accumulations are mixed throughout the water column. One or several algal species usually dominates this exponential growth maximum for several weeks, and in a large number of lakes, diatoms constitute the predominant algae of the spring maximum. Increasing light, and to a lesser extent a rise in water temperatures, are major factors initiating the development of diatom populations from smaller residual winter planktonic populations. Circulation and turbulence are much higher in the spring than later in the season and assist in maintaining the relatively dense diatom cells in optimal intensities of light. Other factors influencing the development of succession and development of the algal populations will be treated later (Chap. 15).

The diatom *Asterionella* commonly precedes other diatoms such as *Cyclotella, Fragilaria,* and *Tabellaria* in the spring maximum. Collectively, the spring maximum often declines abruptly as the silica concentrations fall below 0.5 mg liter^{-1} (Fig. 14-22). The same annual pattern is illustrated for Lake Windermere, where it has been demonstrated continuously for over 30 years. Factors such as light intensity, temperature, grazing by zooplankton, fungal parasitism, and changes in other nutrients, especially nitrogen and phosphate, could not be shown to be associated with the decline of the maximum. Experimental results on silica requirements showed that the silica reduction was clearly the major factor contributing to the decline of the diatoms. The diatom succession results from interspecific differences in assimilation efficiencies and the effects of silicon concentration on growth and death. For example, in Lake Constance in southern Germany, silicon depletion in *Asterionella formosa* did not stop cell division but led to the death of most of the population (Sommer and Stabel, 1983). In *Fragilaria* and *Stephanodiscus,* however, cell division stopped but cells did not die and population growth continued at reduced levels after the concentrations of dissolved silica had increased later in the season.

To generalize from these situations would be misleading since, even though this sequence occurs rather commonly, numerous other interacting factors are involved. Silica concentrations and their biogenic reduction from epilimnetic waters, however, are certainly major factors in diatom-community regulation. In lakes where silica levels remain high, even though severely reduced during the summer productive period (e.g., Lawrence Lake, Fig. 14-21, *upper*), the spring diatom maximum persists longer into the summer and is overtaken gradually by a predominance of green algae. In very productive lakes, a maximum diatom peak can be found in the fall and early winter, as, for example, in Wintergreen Lake (Fig. 14-21, *lower*), during which time a competitive advantage to diatoms is apparent until the silica levels decline to very low concentrations (<100 μg liter^{-1}). After spring overturn, diatom growth is very short-lived in this lake, followed by a brief dominance of green algae until inorganic combined nitrogen

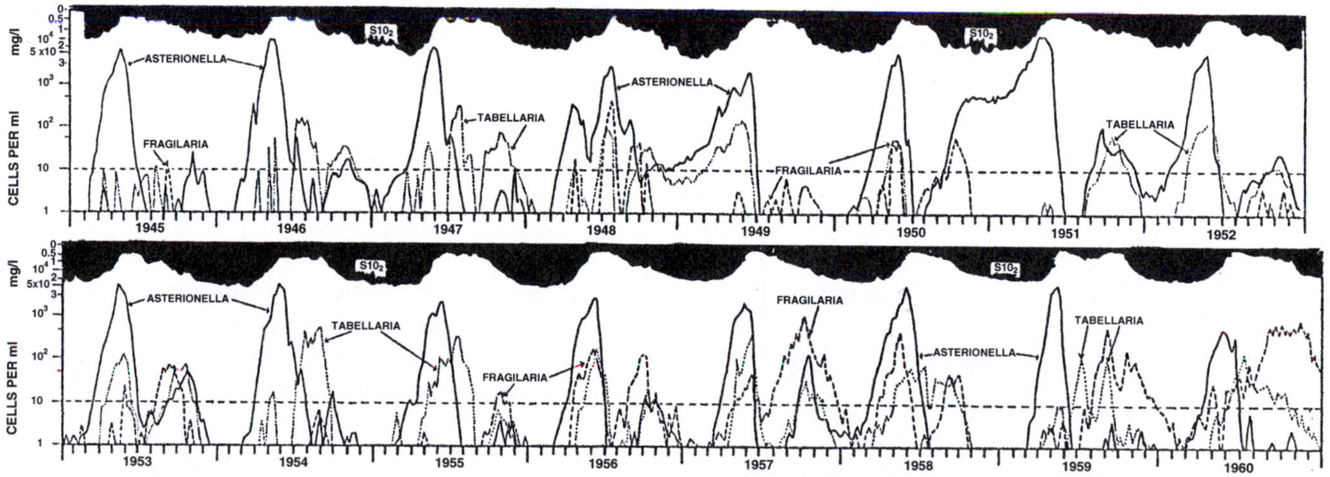

FIGURE 14-22 The periodicity of the diatom algae *Asterionella formosa, Fragilaria crotonensis,* and *Tabellaria flocculosa* in relation to fluctuations in the concentration of dissolved silica, 0–5 m in Lake Windermere, England, 1945–1960. (From Lund, J. W. G.: *Verhandlungen Int. Ver. Limnol. 15*:37, 1964.)

sources are depleted. Nitrogen-fixing cyanobacteria then quickly dominate and persist until the combined effects of increased inorganic nitrogen, reduced light and temperature, and renewed silica reoccur during circulation in late summer and early autumn.

The succession sequence of different species of diatoms can be seen not only within a composite spring maximum but also in the patterns of diatom periodicity as lakes become more productive. In lakes in which silica concentrations are moderate to low (e.g., <5 mg liter^{-1}), progressive long-term enrichment with phosphorus and nitrogen can lead to rapid biogenic reduction in silica levels so that diatoms cannot effectively compete and are replaced by nonsiliceous phytoplankton (Kilham, 1971). A common response to nutrient loading in northern temperate lakes, reservoirs, and even rivers under some circumstances is an increase in diatom growth and biomass. Nutrient enrichments commonly lead to increases in concentrations of total P and total N, but not Si. Such nutrient loading can lead to increased diatom production. The greatly increased sedimentation of diatoms can result in rapid depauperation and even depletion of silica in trophogenic zones. As the reductions of silica from the photosynthetic zones continues over a number of years, more silica is removed to the sediments than is replenished from external sources and regeneration from the sediments (Schelske and Stoermer, 1971; Schelske, 1985, 1988; Schelske *et al.*, 1988; Conley *et al.*, 1993). Under such circumstances, diatoms are gradually outcompeted and excluded by green algae and cyanobacteria during the summer period, and if persisting over long periods of time, permanently as dominant components of the phytoplankton.

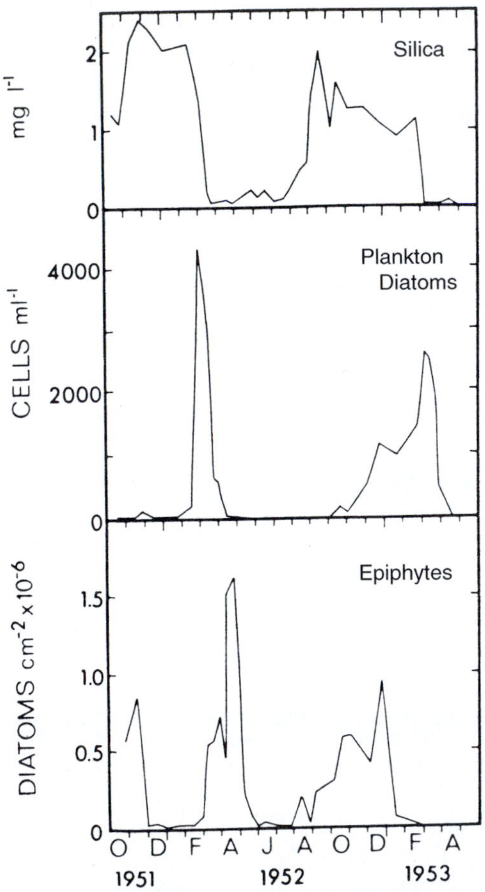

FIGURE 14-23 Variations in numbers of the dominant phytoplanktonic diatom (*Stephanodiscus hantzschii*), silica concentrations at the surface, and epiphytic diatoms per cm^{-2} of *Phragmites* stems at a depth of 20–30 cm, Furesø, Denmark. (After data of Jørgensen, 1957.)

A strong interaction can exist between the population level of littoral diatoms and the development of planktonic diatoms. For example, in Furesø, the spring maximum of the planktonic diatoms, primarily *Stephanodiscus*, reduced the silica concentrations to <40 μg liter^{-1}, levels experimentally demonstrated to inhibit the growth of these algae (Jørgensen, 1957) (Fig. 14-23). An immediate increase in epiphytic diatoms growing on submersed portions of the emergent macrophyte *Phragmites* accompanied this decrease. The *Phragmites* stems were shown to possess large quantities of easily dissolved silica, and their silica content decreased during the development of the epiphytic diatoms.

VIII. SUMMARY

1. The biogeochemical cycling of essential inorganic micronutrients is regulated to a large extent by changes in oxidation-reduction (redox) states, which are governed largely by photosynthetic and bacterial metabolism.
   a. By conversion of light energy into chemical bonds, photosynthesis produces reduced states (negative E_h) of high free energy and nonequilibrium concentrations of carbon, nitrogen, and sulfur compounds. Respiratory and fermentative reactions of nonphotosynthetic organisms tend to restore equilibrium by catalytically decomposing the products of photosynthesis. The nonphotosynthetic organisms obtain free energy for their metabolism from this process.
   b. Complete redox equilibrium conditions do not occur in natural aquatic ecosystems because most redox reactions are slow and consist of a composite of many redox reactions of different reaction rates. In addition, continuously changing inputs of photosynthetic energy occur, which disrupt the tendency toward equilibrium.

   c. The predominant reactants in the redox processes in natural waters are carbon, oxygen, nitrogen, sulfur, iron, and manganese. Under anaerobic conditions, each predominant terminal electron-accepting reaction has an identifiable range of H_2 concentrations, which in sediments allows evaluation of the redox conditions.
   d. Redox remains positive (300–500 mV) as long as some dissolved oxygen (>1 mg $liter^{-1}$) is present. Temperature and pH changes have only minor effects on redox. As oxygen concentrations approach zero, E_h decreases precipitously.
2. Concentrations of ionic iron are exceedingly low in aerated water. Most iron in oxygenated water occurs as ferric hydroxide in particulate and colloidal form and as complexes with organic, especially humic, compounds. The solubility of manganese is considerably higher than that of iron, but it is analogous to iron in the way it reacts.
   a. Under conditions of both low pH and low redox potential (approximately 250 mV), ferrous and manganous ions diffuse from the sediments and accumulate in anaerobic hypolimnetic water of productive lakes (Fig. 14-3).
   b. Under strongly reducing conditions of very low redox potentials (<100 mV), sulfate is reduced to sulfide. Highly insoluble metallic sulfides, particularly ferrous sulfide (FeS), form under these conditions. Therefore, in hypereutrophic lakes, high concentrations of hydrogen sulfide resulting from bacterial decomposition of sulfur-containing organic matter and from reduction of sulfate can lead to significant decreases in dissolved iron (but not the more soluble manganese) as sulfides form during the latter portion of stratification.
   c. Iron complexes with numerous organic compounds, particularly dissolved humic substances, which can increase iron solubility and availability to organisms.
3. Iron and manganese are micronutrients that are essential to freshwater flora and fauna.
   a. Under certain conditions of restricted availability, photosynthetic productivity can be limited by these elements. Manganese is clearly involved in the seasonal succession of certain algal populations.
   b. Certain chemosynthetic bacteria can utilize the energy of inorganic oxidations of ferrous and manganous salts in relatively inefficient reactions involving CO_2 fixation. Other autotrophic and heterotrophic iron-oxidizing bacteria deposit oxidized iron and manganese. These bacteria are restricted to zones of steep redox gradients between reduced metal ions and oxygenated water (Figure 14-9).
4. Quantitative information on the dynamics of other essential metallic micronutrients—zinc, copper, cobalt, molybdenum, vanadium, nickel, and selenium—is limited. The cycling and availability of micronutrients are governed largely by biogeochemically mediated redox processes: organic matter production and degradation, Fe and Mn redox cycling, and sulfide precipitation. These reactions can be altered by metal complexation with organic, particularly humic, substances.
   a. In most natural waters, micronutrient concentrations and availability are usually adequate to meet metabolic requirements within the constraints of light, temperature, and macronutrient availability. A few clear cases of limitations to photosynthetic productivity by micronutrient deficiencies or physiological unavailability have been demonstrated.
   b. The dynamics of copper are strongly affected by redox conditions similar to those of iron. Cobalt, zinc, and molybdenum dynamics are more closely related to microbial metabolism and transport of the seston and to the effectiveness of complexing with organic compounds. Molybdenum exhibits greater mobility than do the other micronutrient ions.
   c. Inputs of many trace elements and heavy metals to fresh waters are increasing as a result of pollution from industrial and combustion emissions to the atmosphere and subsequent deposition via precipitation. Metal enrichments to toxic levels are particularly widespread for cadmium, lead, mercury, and aluminum.
5. Sulfur is nearly always present in quantities adequate to meet high requirements for protein and sulfate ester synthesis. The dynamics of sulfate and the hydrogen sulfide produced by decomposition of organic matter, however, alter conditions in stratified, productive waters that affect the cycling of other nutrients, productivity, and biotic distribution.
   a. Atmospheric sulfur compounds originating primarily from combustion of coal and oil return to land by precipitation and dry fallout of particles. This sulfur constitutes a major global source of sulfur to fresh waters that in many natural waters exceeds inputs from rock and soil weathering and transport in surface runoff and ground water.
   b. Sulfate is the primary dissolved form of sulfur in oxic waters; hydrogen sulfide accumulates in

anoxic zones of intensive decomposition in productive lakes where the redox potential is reduced below about 100 mV.

c. Most of the sulfur in lake water is stored as dissolved sulfate and hydrogen sulfide. Sestonic sulfur-containing proteins and sulfate esters and dissolved sulfides are primary constituents of the sediments. Appreciable quantities of organic volatile sulfur compounds, largely organic sulfides, can evade to the atmosphere from shallow waters.

d. Although oxygen is obtained from sulfate by sulfate-reducing bacteria, the H_2S generated readily oxidizes and utilizes oxygen upon transport or intrusion into aerobic strata.

e. Sulfur-oxidizing bacteria consist of two general types:

   i. Chemosynthetic aerobes that oxidize reduced sulfur compounds and elemental sulfur to sulfate, and

   ii. Photosynthetic sulfur bacteria that utilize light as an energy source and reduced sulfur compounds as electron donors in the photosynthetic reduction of CO_2.

   iii. The redox requirements of the sulfur-oxidizing bacteria, especially those requiring light, are rather specific, and distribution of species is often restricted to zones of steep gradients between anoxic and aerated water strata.

   iv. When conditions are optimal, the photosynthetic bacteria often develop in extreme profusion and may contribute significantly to the annual productivity of lakes.

6. Silica occurs in relative abundance in natural waters as dissolved silicic acid and particulate silica. Much of the particulate Si is biogenic Si associated with living or dead diatom frustules.

   a. Diatom algae assimilate large quantities of silica and markedly modify the flux rates of silica in lakes and streams.

   b. Utilization of silica in the trophogenic zone of lakes by diatoms reduces the epilimnetic concentrations (Fig. 14-20) and induces, along with other factors, a seasonal succession of diatom species that have different assimilation efficiencies for Si and growth rates.

   c. When the concentration of silica is reduced below about 0.5 mg $liter^{-1}$, many diatom species cannot compete effectively with nonsiliceous algae, and their growth rates decline until silica supplies are renewed, usually during autumnal circulation.

   d. Eutrophicational nutrient enrichments of P and N often lead to increased diatom production and removal of Si by sedimenting diatoms at rates faster than renewed by inputs. Diatoms are gradually outcompeted and replaced by algae and cyanobacteria not dependent on Si availability.

15

PLANKTONIC COMMUNITIES: ALGAE AND CYANOBACTERIA

The algae of the open water of lakes and large streams, the ***phytoplankton,*** consist of a diverse assemblage of nearly all major taxonomic groups. Many of these forms have different physiological requirements and vary in response to physical and chemical parameters such as light, temperature, and nutrient regimen. The photosynthetic cyanobacteria, formerly the blue–green algae, constitute a major component of the photoplankton and are discussed throughout this treatment as functionally similar to the planktonic algae. Despite these diversities, both taxonomic and physiological, many algal and cyanobacterial species coexist in the same water volume. However, dominant genera in algal groupings change not only spatially (vertically and horizontally within a lake) but seasonally, as physical, chemical, and biological conditions change in the water body. A general pattern of seasonal succession of phytoplankton has been correlated with environmental factors of many lakes, although the precise reasons for many of these changes are not well known.

Phytoplankton ecology has been one of the most popular areas of aquatic ecology in recent decades and many inroads have been made in understanding algal requirements for light, temperature, and nutrients, buoyancy regulation, competition, productivity, and the effects of predation and parasitism on phytoplankton. However, large voids still exist in our knowledge of the many complex mechanisms that result in the wide array of planktonic algal communities observed worldwide. Further investigations, particularly in the areas of algal–microbial, algal–algal, and algal–herbivore interactions, are greatly needed.

I. COMPOSITION OF THE ALGAE OF PHYTOPLANKTONIC ASSOCIATIONS

A. Morphological and Physiological Characteristics

Before summarizing the algal and cyanobacterial associations and population dynamics of phytoplanktonic communities, a few of the major morphological and physiological characteristics of the primary algal groups should be emphasized. The systematics of algae comprising the phytoplankton are treated well in a number of detailed works but are continually being revised and improved (e.g., Medlin *et al.*, 1993, 1997; Sorhannus *et al.*, 1995; van der Hoek *et al.*, 1995). In this functional physiological and ecological treatment, it is defeating to become engrossed in discussion of whether certain groups are truly algae in a strict sense. In many respects, the cyanobacteria (Myxophyceae) are physiologically similar to the algae. Many motile and colonial flagellates are classified as Protozoa on the basis of certain morphological and reproductive characteristics. The important characteristic, however, is that autotrophic photosynthesis is the primary mode of nutrition and results in a major synthesis of new organic matter. Therefore, these photoautotrophic microbiota are treated collectively with the true algae of the phytoplankton.

The emphasis in the following synthesis is functional, from the standpoint of interactions of physiological characteristics with environmental variables and the effects of these interactions on growth, primary productivity, and resultant population successions within communities and other trophic levels.

B. Pigments

A primary algal characteristic is the presence of photosynthetic pigments—the chlorophylls, carotenoids, and biliproteins. Chlorophyll *a* is the primary photosynthetic pigment of all oxygen-evolving photosynthetic organisms and is present in all algae and cyanobacteria (Table 15-1) and photosynthetic organisms other than the photosynthetic sulfur bacteria (discussed separately in Chap. 14). Chlorophyll *a* has two *in vitro* absorption bands, as discussed on page 67, in the red-light region at 660–665 nm and at lower wavelengths near 430 nm.

Chlorophyll *b*, although common in higher plants, is found only in the green algae, euglenophytes, and certain minor groups (Table 15-1). Chlorophyll *b* is a light-gathering pigment that transfers absorbed light energy to chlorophyll *a* for primary photochemistry (reviewed in Govindjee and Brown, 1974). Maximum absorption bands are approximately 645 nm and 435 nm. Chlorophyll *c*, consisting of three spectrally distinct components, is probably an accessory pigment to photosystem II. Maximum extraction absorption bands are at about 630–635 nm, and in the dominant blue portion of the spectrum at about 583–586 nm and 444–452 nm (Meeks, 1974). Chlorophyll *d*, of no known function, is a minor pigment component found only in certain red algae. A fifth chlorophyll (*e*) has been isolated from a xanthophycean alga, but whether it exists *in vivo* is uncertain (Strain, 1951).

Among the many carotenoids of algae, the carotenes are linear unsaturated hydrocarbons, and the xanthophylls are oxygenated derivatives of carotenes (Goodwin, 1974). As is the case with chlorophyll *b*, light energy absorbed by carotenoids and biliproteins is transferred to chlorophyll *a*, leading to fluorescence and excitation of chlorophyll *a* molecules. β-carotene is the most widely distributed of the carotenes and is replaced by α-carotene only in certain green algae and in the Cryptophyceae. The biliproteins are water-soluble, pigment–protein complexes occurring in the cyanobacteria and, to a lesser extent, in certain cryptophytes and red algae.

C. Cyanobacteria

The cyanobacteria (blue–green algae) have been among the most studied of all planktonic groups. Previously classified as algae in the division Cyanophyta [*cyano* (Greek) = blue–green] or Myxophyceae [*myx* (Greek) = slime], these organisms are true bacteria

TABLE 15-1 Distribution of Photosynthetic Pigments among the Algae and Cyanobacteria[a]

Pigments	*Cyano-bacteria (Myxo-phyceae)*	*Chloro-phyta*	*Xantho-phyceae*	*Chryso-phyceae*	*Bacil-lario-phyceae*	*Crypto-phyceae*	*Dino-phyceae (Pyrro-phyta)*	*Eugleno-phyceae*	*Phaeo-phyceae*	*Rhodo-phyceae*
Chlorophylls:										
Chlorophyll *a*	+	+	+	+	+	+	+	+	+	+
Chlorophyll *b*	–	+	–	–	–	–	–	+	–	–
Chlorophyll *c*	–	–	+	+	+	+	+	–	+	–
Chlorophyll *d*	–	–	–	–	–	–	–	–	–	+
Carotenoids:										
Carotenes										
α-Carotene	–	+	–	–	–	+	–	–	–	+
β-Carotene	+	+	+	+	+		+	+	+	+
γ-Carotene	–	+	–	–	–	–	–	–	–	–
ε-Carotene	–	–	–	–	+	+	–	–	–	–
Xanthophylls										
Lutein	+	+	–	+	–	–	–	–	+	+
Violaxanthin	–	+	+	–	–	–	–	–	+	+
Fucoxanthin	–	–	–	+	+	–	–	–	+	–
Neoxanthin	–	+	+	–	–	–	–	+	–	–
Astaxanthin	–	+	–	–	–	–	–	+	–	
Diatoxanthin	–	–	–	+	+	–	–	–	+	–
Diadinoxanthin	–	–	–	+	+	–	+	–	–	–
Peridinin	–	–	–	–	–	–	+	–	–	–
Dinoxanthin	–	–	–	–	–	–	+	–	–	–
Teraxanthin	–	–	–	–	–	–	–	–	–	+
Antheraxanthin	–	–	–	–	–	–	–	+	–	–
Myxoxanthin	+	–	–	–	–	–	–	–	–	–
Myxoxanthophyll	+	–	–	–	–	–	–	–	–	–
Oscilloxanthin	+	–	–	–	–	–	–	–	–	–
Echinenone	+	–	–	–	–	–	–	+	–	–
Biliproteins:										
Phycocyanin[b]	+	–	–	–	–	+	–	–	–	+
Phycoerythrin[b]	+	–	–	–	–	+	–	–	–	+

[a] From Morris (1967), Meeks (1974), Goodwin (1974), and van den Hoek *et al.* (1995) from numerous sources.

[b] These chromoproteins consist of distinctly different types in the groups in which they occur. Differentiation is based on their absorption spectra (cf. Goodwin, 1974).

with a simple prokaryotic cell structure. Prokaryotic cells lack certain membranous structures, including a nuclear membrane, mitochondria, and chloroplasts. The chloroplasts of other algae and plants originated from cyanobacteria by endosymbiosis. The cells do contain cellular inclusions, which in some cases assume similar functions to eukaryotic organelles such as pigment-bearing lamellae, a plasma membrane, and a nuclear region containing chromosomal material. The cyanobacteria, like most other bacteria, have murein in the cell wall, reproduce by binary fission, and do not divide by mitosis as algae and higher organisms do. However, the cyanobacteria are distinguished from other bacteria by the presence of chlorophyll *a,* which is common to chloroplasts of eukaryotic algae and higher plants and different structurally from bacteriochlorophyll. Cyanobacteria are also able to use water as an electron donor in photosynthesis, which is evolutionarily more advanced than bacterial photosynthesis. Therefore, cyanobacteria possess the enzymatic and pigment capability to photosynthesize under oxygenated conditions as higher plants. In effect, cyanobacteria are structurally and physiologically like bacteria, but they photosynthesize functionally like plants in aquatic systems.

Cyanobacteria occur in unicellular, filamentous, and colonial forms, and most are enclosed in mucilaginous sheaths either individually or in colonies. Most planktonic cyanobacteria consist of members of the coccoid family Chroococcaceae (e.g., *Anacystis* = *Microcystis, Gomphosphaeria* = *Coelosphaerium,* and *Coccochloris*) and filamentous families (e.g., *Planktothrix, Limnothrix,* and *Tychonema,* all formerly classified within the genus *Oscillatoria*), Nostocaceae (e.g., *Anabaena, Aphanizomenon,* and *Nodularia*), and Rivulariaceae (e.g., *Gloeotrichia*).

In filamentous cyanobacteria, cells are arranged end to end to form long trichomes. Trichomes are usually contained within a mucilaginous sheath. Although vegetative reproduction by fragmentation of trichomes is common, some cells are differentiated and specifically involved in reproduction and perennation. Akinetes (nonmotile, resistant cells) form in a number of species as enlarged, thick-walled cells that accumulate proteinaceous reserves in the form of cyanophycin granules (cf. Fogg *et al.,* 1973; Stanier and Cohen-Bazire, 1977; Adams and Carr, 1981). Under favorable conditions, akinetes germinate directly into a trichome or into hormogonia. Hormogonia are short, slightly modified pieces of trichome that fragment from the parent trichome and move away by means of gliding motions to develop into a new filament. In a few species, hormogonia develop within the parent sheath as multiple trichomes. A few other spore structures develop in species that do not form hormogonia; in these species, development continues without resting stages.

Heterocysts are differentiated cells unique to cyanobacteria. Heterocysts can occur in many filamentous species but are absent from those in the Oscillatoriaceae; they are the major sites of nitrogen fixation (Chap. 12). Heterocysts develop from vegetative cells, which form a thick envelope over the cell wall except for a pore at each pole where the heterocyst is connected to adjacent vegetative cells. The envelope provides a barrier to diffusion of gases into the cell. It is through this pore that the plasma membranes are adjoined and the exchange of metabolic products occurs. In contrast to vegetative cells, heterocysts lack phycobilins, which are the primary light absorbers in cyanobacteria, lack an oxygen-evolving photosystem, and show higher reducing activity. The rate of oxygen diffusion into the heterocyst from outside is reduced by waxy layers in the envelope, though sufficient N_2 gets through to support nitrogen fixation (Walsby, 1985). Therefore, the internal environment of the heterocyst is ideal for nitrogen fixation, since the nitrogenase enzyme is inactivated by oxygen. Organic carbon from adjacent cells is transferred into the heterocyst and used as an energy source in nitrogen fixation; reduced nitrogen is, in turn, transferred out of the heterocyst and into vegetative cells (Wolk, 1968, 1973; Adams and Carr, 1981). The interaction of vegetative cells and heterocysts is an example of structural separation of metabolic processes (photosynthesis and nitrogen fixation) in a relatively primitive organism. Nitrogen fixation can also occur in some nonheterocystous cyanobacteria (e.g., *Gloeocapsa*). In this case, photosynthesis and nitrogen fixation may be separated temporally rather than structurally (Gallon *et al.,* 1974). Recent studies indicate that cyanobacteria have circadian rhythms and that the capacity for photosynthesis and nitrogen fixation is regulated by a biological clock, reset by light/dark cues, at the level of gene expression (Golden *et al.,* 1997).

D. Green Algae

The Chlorophyta are an extremely large and morphologically diverse group of algae that is almost totally freshwater in distribution. Most planktonic green algae belong to the orders Volvocales (e.g., *Chlamydomonas, Sphaerocystis, Eudorina,* and *Volvox*) and Chlorococcales (e.g., *Scenedesmus, Ankistrodesmus, Selenastrum,* and *Pediastrum*). Many members are flagellated (2 or 4, rarely more) at least in the gamete stages; in the Zygnematales and the desmids (Conjugales or Desmidiales), the gametes are amoeboid.

Asexual reproduction by vegetative division is common to most of the green algae but is lacking in most of the Chlorococcales and Siphonales. Cell division very often is synchronized so that nuclear and cell division occurs at night. Cell division in colonial species results in enlargement of the colony; new colonies are formed by fragmentation of the colony. Among filamentous species, such as *Spirogyra,* fragmentation of the filament is common. Vegetative formation of single or multiple zoospores within a cell is also common. When liberated from the parent cell, the flagellated zoospores are actively motile and positively phototactic for varying periods of time before losing the flagellae and entering a resting spore stage.

Sexual reproduction in the green algae is diverse. In the simplest case of isogamy, the flagellated male and female gametes are morphologically similar in size and structure. In anisogamy, the flagellated female gamete is larger than the male gamete. Among more specialized algae, oogamy is common, in which union occurs between a large, nonflagellated female gamete and a small, flagellated male antherozoid. Some green algae are sexually monoecious (gametes are derived from the same cell), and others are dioecious (male and female gametes are derived from different cells).

Whereas the planktonic Volvocales and Chlorococcales are ubiquitous in distribution among waters of differing salinity within the normal limnological range, the distribution of most species of desmids of the Conjugales is limited to low concentrations of the divalent cations calcium and magnesium. Although not totally restricted to waters of low salinity (Coesel, 1983; Reif *et al.,* 1983), the desmids are most common and the species diversity is greatest in unproductive soft waters draining land forms developed in granitic or other igneous rocks and especially in waters with a high content of dissolved organic matter. Their abundance and diversity are often greatest in bog waters that drain through deposits of the moss *Sphagnum*. Many desmids are distributed widely, but as a whole they are less cosmopolitan than most unicellular algae (cf. Hutchinson, 1967; Brook, 1981).

E. Yellow–Green Algae

The Xanthophyceae are unicellular, colonial, or filamentous algae that are characterized by conspicuous amounts of carotenoids in comparison to chlorophylls that result in their predominantly yellow–green coloration (Table 15-1). Nearly all of the motile cells possess two flagella, one of which is smooth and longer than the other, "hairy" flagellum (hence the older name, Heterokontae; the Heterokontophyta is now a large phylum that includes many major planktonic algal familes; cf. van der Hoek *et al.*, 1995). A cell wall is often absent; when present, it contains large amounts of pectin and in many species may be silicified. Asexual reproduction is usually by fission, with the formation of zoospores. Sexual reproduction is poorly understood in this group but is most often isogamous.

Most xanthophycean algae are associated with substrata, and many are epiphytic on larger aquatic plants. A few members are planktonic and include common genera such as *Chlorobotrys, Gloeobotrys,* and *Gloeochloris.*

F. Golden-Brown Algae

The chromatophores of the Chrysophyceae often yield a distinctive golden-brown coloration because of the dominance of β-carotene and specific xanthophyll carotenoids in addition to chlorophyll *a*. Most of the chrysophycean algae are unicellular; a few are colonial; and they are rarely filamentous. Flagellation is variable; most often cells are uniflagellate, but some species possess two flagella, usually of equal length. Many species lack a cell wall and are bounded only by a cytoplasmic membrane; others possess a cell surface covered by delicate siliceous or calcareous plates or scales. Vegetative reproduction by longitudinal cell division is most common, especially among the motile unicellular species. Although it has rarely been observed, sexual reproduction is isogamous: The resulting cystlike zygote undergoes reductive division under favorable conditions.

Numerous chrysophycean algae form important components of the phytoplankton. The unicellular species with a single flagellum (e.g., *Chromulina, Chrysococcus,* and *Mallomonas*) are usually very small algal representatives of the nanoplankton (Table 15-2).

TABLE 15-2 Size Classification of Plankton[a]

Term	*Size ranges* (μm)
Femtoplankton	0.01–0.2
Ultraplankton	0.2–20
Picoplankton	0.2–2.0
Nanoplankton	2.0–20
Microplankton	20–200
Mesoplankton	200–20,000

[a] Based on evaluations of Sieburth *et al.* (1978) and the reviews of Hutchinson (1967) and Fogg (1991).

Larger colonial forms such as *Synura, Chrysosphaerella, Uroglena,* and particularly *Dinobryon* are widely distributed and may become major components of the phytoplankton under certain environmental conditions, particularly in temperate, oligotrophic lakes (Hutchinson, 1967). The conspicuous development of populations of certain species of *Dinobryon* and *Uroglena* in lakes of very low phosphorus concentrations (cf. earlier discussion, p. 267) may be related to their abilities to take up phosphate effectively at extremely low ambient concentrations (cf. Lehman, 1976). In contrast, other species of *Dinobryon* and *Synura* have high phosphorus requirements. The phytoplankton of many oligotrophic lakes are dominated throughout the year by chrysophycean algae and secondarily by cryptophytes (e.g., Siver and Chock, 1986; Stewart and Wetzel, 1986). The nanoplanktonic flagellates of the Chrysophyceae and Cryptophyceae totally dominate the phytoplanktonic during the short growing season (60–100 days) of the millions of thermokarst ponds that dominate the northern tundra of Alaska, Canada, and Russia (Sheath, 1986). The chrysophytes are particularly well adapted to low temperatures and light levels (Roijackers, 1986; Sandgren, 1986). Several species of chrysophycean algae, particularly *Dinobryon,* supplement photosynthesis with phagotrophy. These algae can obtain an appreciable fraction of their energy and nutrients by ingesting bacteria (Bird and Kalff, 1986). Although the rates of water clearance are small (e.g., 0.00014 ml day^{-1}), large populations of *Dinobryon* can collectively clear water of bacteria at rates that compete effectively with those of ciliates, rotifers, and microcrustaceans.

G. Diatoms

A most important group of algae of the phytoplankton are the diatoms (Bacillariophyceae), even though most species are sessile and associated with littoral substrata. Their primary characteristic, silicified cell walls, has been discussed in the preceding chapter. Both unicellular and colonial forms are common among the diatoms. The group is commonly divided into the centric diatoms (Centrales), which have radial symmetry, and the pennate diatoms (Pennales), which exhibit essentially bilateral symmetry. The cell wall or *frustule* of diatoms consists of two lidlike valves, one of which fits within the other; the overlapping area of the valves is connected by bands that constitute the girdle. The beautiful structures of the siliceous cell walls of diatoms are complex and used as taxonomic characteristics.

The valves of pennate diatoms exhibit various areas of cell thickenings and dilations. In some species a slit, termed a *raphe,* traverses all or part of the cell wall; in others, a depression in the axial areas of the cell wall, termed a *pseudoraphe,* is found. The four major groups[1] of pennate diatoms are differentiated on the basis of these structures: (a) the Araphidineae, which possess a pseudoraphe (e.g., *Asterionella, Diatoma, Fragilaria,* and *Synedra)*; (b) the Raphidioidineae, in which a rudimentary raphe occurs at the cell ends (e.g., *Actinelia,* and *Eunotia*); (c) the Monoraphidineae, which have a raphe on one valve and a pseudoraphe on the other (e.g., *Achnanthes,* and *Cocconeis*); and (d) the Biraphidineae, in which the raphe occurs on both valves (e.g., *Amphora, Cymbella, Gomphonema, Navicula, Nitzschia, Pinnularia,* and *Surirella*). These divisions are of more than taxonomic interest since distinct nutritional requirements favor the growth of one group over another.

Vegetative reproduction by cell division, usually at night, is the most common mode of multiplication (cf. Werner, 1977). Sexual reproduction occurs periodically when the cells reach a minimum critical size (30–40% of maximum size) through repeated cell division in the course of asexual reproduction (Edlund and Stoermer, 1997). Also required for sexual induction are a set of correct environmental conditions, often species-specific, that may include combinations of light, temperature, nutrients, trace metals, organic growth factors, and osmolarity. After sexual reproduction, size is restored. Amoeboid gametes occur in isogamous reproduction in the Pennales, whereas in the Centrales reproduction is oogamous with flagellated spermatozoids. After fusion of the gametes, an auxospore develops and divides; some of the resulting cells give rise to vegetative cells that are of the largest dimensions of the species.

[1]These groups are differentiated on the basis of morphology but are not necessarily monophyletic (e.g., Medlin *et al.,* 1993; Sorhannus *et al.,* 1995).

H. Cryptomonads

Most of the cryptophycean algae are naked, unicellular, and motile. This class is very small and most of the planktonic members belong to the Cryptomonadineae (e.g., *Cryptomonas, Rhodomonas,* and *Chroomonas*). All members are small and not a great deal is known about their ecology or physiology. The cryptomonads usually are flattened dorsoventrally, with an anterior invagination that bears two equal or subequal flagella. The one or usually two chromatophores contain a variety of pigments (Table 15-1) that can yield a spectrum of apparent colors, including olive green, blue, red, or brown. Reproduction is by longitudinal division; sexual reproduction is unknown in the group.

Cryptomonad algae and other cryptomonadlike microflagellates occur in almost all lakes, regardless of trophic state (Stewart and Wetzel, 1986). The cryptomonads are typically present in low numbers during most of the year but commonly increase intermittently (e.g., Ilmavirta, 1983; Taylor and Wetzel, 1984; Dokulil, 1988). Such increases in abundance often follow the demise of previously dominant "bloom" algal populations. Characteristics of the cryptomonads and other microflagellates include intermittent numerical dominance, high nutritional quality, short turnover times, ability to grow and reproduce at low intensities of light, and effective rapid pulse timing of growth. In combination, these characteristics implicate the cryptomonads and other photosynthetic microflagellates as an ecologically significant internal stabilizing component of planktonic communities.

I. Dinoflagellates

The dinoflagellates (Dinophyceae of the Pyrrophyta) are unicellular flagellated algae, many of which are motile. Although a few species are naked or without a cell wall (e.g., *Gymnodinium* of the Gymnodiniales), most develop a conspicuous cell wall that often is sculptured and bears large spines and elaborate cell-wall processes (the Peridiniales, e.g., *Ceratium, Glenodinium, Peridinium).* In both naked and armored types, the cell surface has transverse and longitudinal furrows that connect and contain the flagella, whose movements create water currents providing for weak locomotion and the disrupting of chemical gradients at the cell surface. Although sexual reproduction occurs, asexual reproduction by the formation of aplanospores (in which a motile phase is omitted) predominates. These asexual resting stages or cysts can undergo considerable periods of diapause (Loeblich and Loeblich, 1984). In the large *Ceratium,* for example, the autumn decline of summer populations in temperate regions results in the production of overwintering cysts. Emergence from benthic cysts can result in an exponential increase in the planktonic cells in the ensuing spring and summer (Heaney *et al.,* 1983; Sako *et al.,* 1985; Pollingher *et al.,* 1993). The distribution of dinoflagellates in relation to major chemical characteristics shows that, while some species are widely tolerant and ubiquitous, especially among species of *Ceratium* and *Peridinium,* most dinoflagellate species have restricted ranges with respect to calcium, pH, dissolved organic matter, and temperature (Taylor and Pollingher, 1987). The relatively large body size (ca. 400 μm) makes these algae relatively inedible to most grazing organisms.

Although many zooplankton change size and form seasonally (Chap. 16), only a few phytoplankton undergo seasonal polymorphism or cyclomorphosis. One example of phytoplanktonic cyclomorphosis is the lengthening of cellular extensions or horns in the dinoflagellate *Ceratium* as temperatures increase from spring into midsummer (Hutchinson, 1967). In *Ceratium,* the conspicuous separation of the cell wall by the transverse furrow into the upper part (epitheca) and lower part (hypotheca) is accentuated further by the development of horns. The epitheca always has one horn; the hypotheca usually has 1–3 horns. As the water warms, the cells of populations exhibit an increase in the length of the horns, a decrease in the width and general size of the cell, and a lessening of the angle of divergence of the hypothecal horns (Fig. 15-1). In midsummer, with even warmer temperatures, the development of a short, fourth horn is common. Field and experimental evidence indicates that these changes are induced primarily by increasing water temperatures. Some experimental evidence points to the fact that the sinking rate is reduced appreciably in less viscous warmer water by the reduced cell dimensions, increased divergence of the horns, and the development of the fourth horn. Presumably, these changes are of adaptive significance to *Ceratium* in that they reduce the rate of sinking out of the photic zone.

J. Euglenoids

Although the euglenoid algae (Euglenophyceae) are a relatively large and diverse group, few species are truly planktonic. Nonetheless, when conditions are favorable, the euglenoids can develop in great profusion. Almost all euglenoids are unicellular, lack a distinct cell wall, and possess one, two, or three flagella that arise from an invagination in the cell membrane. Reproduction occurs by longitudinal division of the motile cell; sexual reproduction has not yet been substantiated.

Although some euglenoids are unpigmented and phagotrophic (able to ingest solid particles) and are best treated as Protozoa, most are photosynthetic and facultatively heterotrophic. Even when phagotrophy constitutes the dominant mode of carbon assimilation among certain nonpigmented members of the Dinophyceae, Chrysophyceae, and especially the Euglenophyceae, none of these algae depends solely on phagotrophy for their major carbon requirements (Droop, 1974). Nutrition is supplemented by the uptake of dissolved organic compounds. Ammonia and dissolved organic nitrogen compounds are the dominant sources of nitrogen among most euglenoid algae.

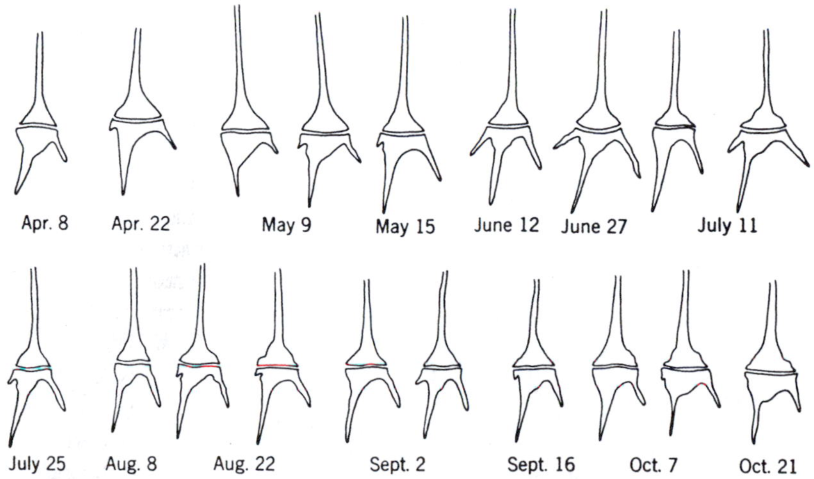

FIGURE 15-1 Seasonal polymorphism in *Ceratium hirundinella* in Kirchbergteich, Darmstadt, Germany. (From Hutchinson, G. E.: *A Treatise on Limnology,* Vol. 2, New York, John Wiley & Sons, Inc., 1967, after work of List.)

Their development in the phytoplankton occurs most often in seasons, strata, or lake systems in which concentrations of ammonia and especially dissolved organic matter are high. For example, detailed seasonal analyses of the phytoplanktonic populations at weekly intervals at each meter of depth for 16 months in hardwater Lawrence Lake in southwestern Michigan showed euglenoid algae developing for only a brief 2-week period immediately prior to autumnal turnover and in a small stratum, 1 m above the deep sediments, that was anoxic during the last 2 months of summer stratification (Stewart and Wetzel, 1986; Taylor and Wetzel, 1988). However, euglenoids are found most often in shallow water rich in organic matter, such as organically polluted lakes (e.g., Yusoff and Patimah, 1994) and farm ponds.

K. Brown and Red Algae

The brown algae (Phaeophyta) are mostly filamentous or thalloid algae, which, as a group, are almost exclusively marine. Only a few genera of this large, primitive group are represented in fresh water; those that are found in fresh water are attached to substrata, such as rocks. No species is planktonic.

The red algae (Rhodophyta) are also very sparsely represented in fresh water, and none is planktonic. A majority of the genera occurring in fresh water are found only there. The thalloid or filamentous species (e.g., *Batrachospermum*) are nearly all restricted to fast-flowing streams of well-oxygenated, cool waters.

II. THE IMPORTANCE OF SIZE: SMALL IS BEAUTIFUL AND PRODUCTIVE

An enormous variation in size is found among the phytoplankton. Early separations and classifications were based largely upon filtering devices, such as fine nets, that were available for capture of the phytoplankton[2]. A more functional basis now underlies size groupings (Table 15-2), in which major environmental parameters controlling growth and succession are coupled to size.

Minute (0.2–2.0 μm) algal and cyanobacterial picoplankton were found to be abundantly and ubiquitously distributed in freshwater ecosystems (Stockner and Antia, 1986; Stockner, 1991). These ***picophytoplankton*** consist mainly of small coccoid cyanobacteria of the *Synechococcus* type, commonly of a size less than 1 μm. These prokaryotes often exceed 300,000 cells ml^{-1} and occupy diverse habitats over a wide range of temperatures, light, salinity, and nutrients. Although less abundant than cyanobacterial picophytoplankton, the eukaryotic *Chlorella*-like picophytoplankton tend to be larger (1.2–2.0 μm) and can dominate in certain extreme habitats, such as acidic or humic-stained dystrophic waters (e.g., Søndergaard, 1991). Colonial picophytoplankton are found in oligotrophic lakes but are likely aggregations of chroococcoid cyanobacteria such as *Synechococcus*.

When the size of planktonic organisms falls below about 20 μm (the ***ultraplankton***, see Table 15-2), the transfer of materials to and from cells is almost entirely by molecular diffusion rather than by the eddy diffusion of turbulence (Fogg, 1991, 1995). The foundational theoretical basis for this size boundary of ca. 20 μm is treated in detail by Karp-Boss *et al.* (1996). A cell radius near 20 μm is needed before sinking or swimming can be expected to increase the flux of nutrients. Macroturbulence is important, however, for moving water parcels containing the ultraplanktonic communities into the light field of the euphotic zone. The ultraphytoplankton are simple, unicellular in form, and have generation times in hours that are similar to those generation times of protistian flagellate predators (Table 15-3). The exchanges by molecular diffusion that small size makes possible, however, allows the establishment of a self-contained, dynamic community of phototrophs, viruses, heterotrophic bacteria, phagotrophic flagellates, and ciliates, in which materials are rapidly cycled with little loss by sedimentation or transfer to higher trophic levels. In such an equilibrium system, population densities are determined by internal dynamics rather than by nutrient inputs.

The larger phytoplankton (***microplankton*** of 20–200 μm; Table 15-2) are not vulnerable to control by flagellates but are inefficient in nutrient uptake and therefore develop prolifically when nutrient concentrations are relatively high. The microphytoplankton are also vulnerable to sedimentation from the photic zone. The availability of sufficient nutrients and transport to light are strongly coupled to macroturbulence (Table 15-3). Water movements are highly variable and, as a result, microphytoplankton must be able to grow and reproduce rapidly when conditions are conducive

[2]Terms were introduced long ago for small plankton components (***bioseston***) of the total particulate material (***seston***). For example, formerly ***nanoplankton*** referred to those plankton passing through older plankton nets of a mesh size of about 50 to 60 μm. The basis for the separations and complex terminology for separations by size are discussed in detail in, for example, Strickland (1960), Hutchinson (1967), and Sicko-Goad and Stoermer (1984). Recent studies have demonstrated the major importance of very small phytoplankton (see Table 15-2). As a result, plankton nets are inappropriate for even qualitative studies of phytoplankton and some, but by no means all, of the early studies of phytoplankton are applicable only to analyses of the largest forms.

TABLE 15-3 Comparison of Functional Characteristics of Ultraphytoplankton and Microphytoplankton[a]

Characteristic	*Ultraplankton*	*Microplankton*
Size	0.2–20 μm	20–200 μm
Main physical process determining nutrient supply	Molecular diffusion	Turbulence, motion relative to water
Form	Simple, unicellular	Elaborate, often colonial
Gas vacuoles in cyanobacteria	Absent	Present
Generation times	Hours	Days
Relative generation time of predators	Equal	Much longer
Variations in natural population densities	$10-10^2$	10^3-10^5

[a] After Fogg (1991, 1995).

(Reynolds, 1984; Fogg, 1991). A cell size of 60 μm or greater is needed to experience substantial gain from macroturbulence (Karp-Boss *et al.,* 1996). Equilibrium is never achieved within the microplanktonic community.

The microphytoplankton separate into two broad types of ecological strategies. Larger algae, such as many of the diatoms, are nonmotile and depend upon mixing to remain poised in the photic zone (Fogg, 1991). Another type, typified by the dinoflagellates and gas-vacuolate cyanobacteria, competes most effectively under relatively quiescent, stable water conditions where motility or buoyancy control allows positioning to optimize access to light and nutrients.

The importance of size relationships among algae and environmental parameters has been emphasized for some time (e.g., Margalef, 1955, 1959). Clearly, there is a general preponderance of small algal species in the phytoplankton of oligotrophic lakes, and the size of dominant algae often increases in more fertile lakes. A number of analyses of the photosynthetic productivity of phytoplankton in which the relative contributions of differently sized algae were determined show that a greater proportion of the total productivity is by smaller forms in less productive waters (Rodhe *et al.,* 1958; Goldman and Wetzel, 1963; Wetzel, 1964, 1965a; Wehr, 1991; others). The contributions of the smaller algae also shifted seasonally; the ultraplankton (<20 μm) can often dominate the productivity, particularly during the winter and spring. More recent studies have examined these relationships in greater detail and generally confirm these results. For example, the picophytoplankton (0.2–2.0 μm) often contributed >60% of the total algal chlorophyll *a* and over 10% of the primary productivity in the trophogenic zone of a mesotrophic upland lake in Wales (Happey-Wood, 1991; Happey-Wood and Lund, 1994) and some 50% of the total primary productivity in Lake Superior (Fahnenstiel *et al.,* 1986).

An inverse relationship is commonly found between total algal biomass and productivity per unit biomass in the phytoplankton. Communities dominated by populations of species with small cells usually have a greater primary productivity per unit algal biomass than communities dominated by populations of large algal species (Findenegg, 1965). Microautoradiographic analyses demonstrated that renewal time (generation time) is related inversely to cell size; cells that are small with a low carbon content tended to have shorter renewal times and were more metabolically active than larger algae (Stull *et al.,* 1973; Amblard, 1988). Some small species of relatively minor contribution to the instantaneous algal community biomass could therefore turn over rapidly and contribute more to the total primary productivity than larger species. The specific rates of population increase for picophytoplankton were found to be low in nature, however, with apparent generation times of several days (Happey-Wood, 1991). Hence, even though the autotrophic picoplankton are ubiquitous, caution is needed to avoid overestimating their contributions to total production.

III. PHYTOPLANKTONIC COMMUNITIES

An outstanding feature of phytoplanktonic communities in lacustrine habitats is the coexistence of a number of algal species. In some cases, one species is found in much greater abundance than others; more often, two or more algal species are codominant in the phytoplanktonic assemblage. A large number of rarer species can always be found among the dominant or subdominant algae. The algae co-occur even though each species has a specific niche based on its physiological requirements and the constraints of the environment. Theoretically, niche overlap can lead to competitive exclusion and should result in dominance by a single species (see Chap. 8). This problem has been formally referred to as the "paradox of the plankton" (Hutchinson, 1961): An apparent multispecific, rather than unispecific, equilibrium occurs in what seem to be physically uniform conditions of the turbulent open water of lakes.

A number of explanations have been advanced to account for species diversity in phytoplanktonic communities that is much higher than would be anticipated from theory and mathematical derivations. First, in order for the relatively slow process of competitive exclusion to occur, the physical conditions must be uniform for a sufficient period of time. If environmental conditions change rapidly, the advantages gained by one species that is a better competitor may not exist long enough to result in the competitive exclusion of other species. Stated another way, differences in the efficiencies of resource utilization among species may be too small for competitive exclusion to occur before conditions change.

Second, competition can be negated partially if commensalism and symbiotic relations exist among species. As discussed later, a number of organic compounds released extracellularly by one alga can influence the metabolism of another species, although most of these compounds are inhibitory and antibiotic in nature. Also, most algae requiring exogenous sources of vitamin micronutrients are motile and small (Chrysophyceae, Cryptophyceae, Euglenophyceae, and Dinophyceae). The differences in organic micronutrient requirements and the release of the excess by species capable of their synthesis would serve as a means of encouraging mixed equilibrium populations.

Greater consumption of one species of algae than another can also promote coexistence, if the species being preyed upon has a much greater competitive advantage. Selective grazing, largely on the basis of algal size, by zooplankton is an important phenomenon and is discussed in some detail in Chap. 16.

Some species of algae are exclusively planktonic. As far as is known, their populations oscillate temporally in abundance, dominating for a period and then becoming extremely rare but nonetheless remaining planktonic (Hutchinson, 1964). Alternatively, some species enter resting stages and thereby leave the competitive arena for a period of time. Some of these species are quiescent in the littoral sediments and later develop sufficiently to form a significant component of the phytoplankton (e.g., Livingstone and Jaworski, 1980). Evidence for benthic algal populations in the littoral zone serving as a "seed" inoculum for some phytoplankton species is sparse (see Chap. 19), but when this occurs, phytoplankton species diversity is increased. This mode of opportunistic expansion into the pelagic region is certainly more common in shallow waters with extensive littoral development.

None of these mechanisms for explaining coexistence are mutually exclusive, and all factors probably operate simultaneously to a greater or lesser degree. Which one dominates at a particular time in a given body of water surely varies. The distribution of phytoplankton is not necessarily as uniform as the concepts of multispecies equilibrium would imply. For example, the phytoplankton of an alpine lake showed a high degree of patchiness for many species, which indicated that the rate of mixing was sufficiently slow relative to the reproductive rate of the algae and allowed communities of many different species to exist simultaneously (Richerson *et al.*, 1970). Under these hypothetical conditions of "contemporaneous disequilibrium," at any given time many patches of water exist in which one species has a competitive advantage relative to the others, but the patches are obliterated frequently enough to prevent exclusive occupation by a single species.

Among saline lakes, phytoplanktonic species diversity is inversely related to lake salinity (Hammer *et al.*, 1983; Blinn, 1993). In general, green algae and a few species of cyanobacteria are dominant; some diatoms are ubiquitous even in very saline lakes (>100 g salinity liter^{-1}).

A. Types of Algal Associations

A great deal of descriptive work has been devoted to the associations of algal species among differing fresh waters. Algae are extremely diverse, and many exhibit a very wide tolerance to environmental conditions found under natural limnological situations. Nonetheless, certain characteristic phytoplanktonic associations occur repeatedly in lakes of increasing nutrient enrichment. Some of the commonly observed major associations are set out in Table 15-4 based on the detailed discussion of Hutchinson (1967). Such categorizations are not very satisfactory because of the wide spectrum of intergradations often observed. Additionally, shifts occur seasonally from one type to another, especially among more productive waters. Even though such characterizations yield little insight into regulating environmental factors, they are useful from the standpoint of general correlations between qualitative and quantitative abundance of the algae and available nutrients.

TABLE 15-4 Characteristics of Common Major Associations of the Phytoplankton in Relation to Increasing Lake Fertility[a]

General lake trophy	*Water characteristics*	*Dominant algae*	*Other commonly occurring algae*
Oligotrophic	Slightly acidic; very low salinity	Desmids *Staurodesmus, Staurastrum*	*Sphaerocystis, Gloeocystis, Rhizosolenia, Tabellaria*
Oligotrophic	Neutral to slightly alkaline; nutrient-poor lakes	Diatoms, especially *Cyclotella* and *Tabellaria*	Some *Asterionella* spp., some *Melosira* spp., *Dinobryon*
Oligotrophic	Neutral to slightly alkaline; nutrient-poor lakes or more productive lakes at seasons of nutrient reduction	Chrysophycean algae, especially *Dinobryon,* some *Mallomonas*	Other chrysophyceans, (e.g., *Synura and Uroglena*); diatom *Tabellaria*
Oligotrophic	Neutral to slightly alkaline; nutrient-poor lakes	Chlorococcal *Oocystis* or chrysophycean *Botryococcus*	Oligotrophic diatoms
Oligotrophic	Neutral to slightly alkaline; generally nutrient poor; common in shallow Arctic lakes	Dinoflagellates, especially some *Peridinium* and *Ceratium* spp.	Small chrysophytes, cryptophytes, and diatoms
Mesotrophic or eutrophic	Neutral to slightly alkaline; annual dominants or in eutrophic lakes at certain seasons	Dinoflagellates, some *Peridinium* and *Ceratium* spp.	*Glenodinium* and many other algae
Eutrophic	Usually alkaline lakes with nutrient enrichment	Diatoms much of year, especially *Asterionella* spp., *Fragilaria crotonensis, Synedra, Stephanodiscus,* and *Melosira granulatea*	Many other algae, especially greens and cyanobacteria during warmer periods of year; desmids if dissolved organic matter is fairly high
Eutrophic	Usually alkaline; nutrient enriched; common in warmer periods of temperate takes or perennially in enriched tropical lakes	Cyanobacteria, especially *Anacystis* (= *Microcystis*), *Aphanizomenon, Anabaena*	Other cyanobacteria; euglenophytes if organically enriched or polluted

[a] After Hutchinson (1967).

A number of phytoplankton indices have been developed, particularly by Thunmark (1945) and by Nygaard (1955), in admirable attempts to quantify more precisely the relationships of rarely occurring and dominant algal species in relation to lake productivity. Their use is revived periodically with limited and varying degrees of success. The indices developed by Nygaard (1949) are based on the ratios or quotients of numbers of species within general groups, such as those of cyanobacteria to desmids, chlorococcalean algae to desmids, centric to pennate diatoms, and euglenophytes to cyanobacteria and green algae. His compound index was the ratio of all species of cyanobacteria and chlorococcalean green, centric diatoms, and euglenoid algae to the species of desmid algae. A general relationship was found between low compound index values of less than 1 and oligotrophic waters. Lowest values occurred in desmid-rich dystrophic lakes enriched by dissolved humic compounds. The values of the compound index increased in more productive waters and approached 50 in highly enriched eutrophic lakes.

Comparisons of these indices with more modern measurements of algal productivity show weak positive correlations, but exceptions are many. Great variations are found among different regions of the world and among lake districts. It is apparent that these indices, while having some value in determining species relationships, are much too superficial in physiological foundation to be of significant use in evaluating productivity among lakes or in elucidating causal mechanisms underlying the composite growth of algae.

IV. GROWTH CHARACTERISTICS OF PHYTOPLANKTON

Analyses and evaluations of seasonal and spatial growth characteristics of phytoplankton are sometimes difficult because of the array of environmental factors involved, the individual physiological properties of each algal species, and the magnitude of change that can occur in both. Clearly, the organisms and environment are both highly dynamic. Some important factors regulating growth and succession are (a) light and temperature; (b) buoyancy regulation, that is, means of remaining within the photic zone by alterations of sinking rates; (c) inorganic nutrient factors; (d) organic micronutrient factors and interactions of organic compounds with inorganic nutrient availability; and (e) biological factors of competition for available resources and predation by other organisms. Each species of algae possesses a range of tolerance to these factors, and population growth proceeds most rapidly at some optimal combination of the interacting factors. The optimal combination of factors required for greatest growth and productivity is probably seldom achieved under most natural conditions. The competitive advantage of one species over another is relative and can change with changes in the physical and biotic conditions that affect growth.

A. Light and Temperature

The ecological effects of light and temperature on the photosynthesis and growth of algae are inseparable because of interrelationships between metabolism and light saturation. Qualitative factors of the selective attenuation of light energy with increasing depth have been discussed earlier (Chap. 5) in relation to photoreceptors and environmental parameters such as dissolved organic matter. Light intensity affects both the rate of photosynthesis and algal growth. Response to light intensity, however, is species-specific, and in many cases, a considerable degree of adaptation to changing light intensities occurs.

The intensity of light required to saturate algal photosynthesis commonly increases as water temperature increases (Fig. 15-2A). Below light saturation, photochemical reactions limit photosynthesis, and these reactions are relatively temperature-independent except at very low temperatures (<5°C) (Wilhelm, 1990). In the example given, exemplified by *Chlorella* (Fig. 15-2A), long-term adaptation to light intensities occurs mainly by changes in the amount of pigment per cell (Steemann Nielsen and Jørgensen, 1968a; Jørgensen, 1969). Cells adapted to high irradiance have a lower chlorophyll *a*

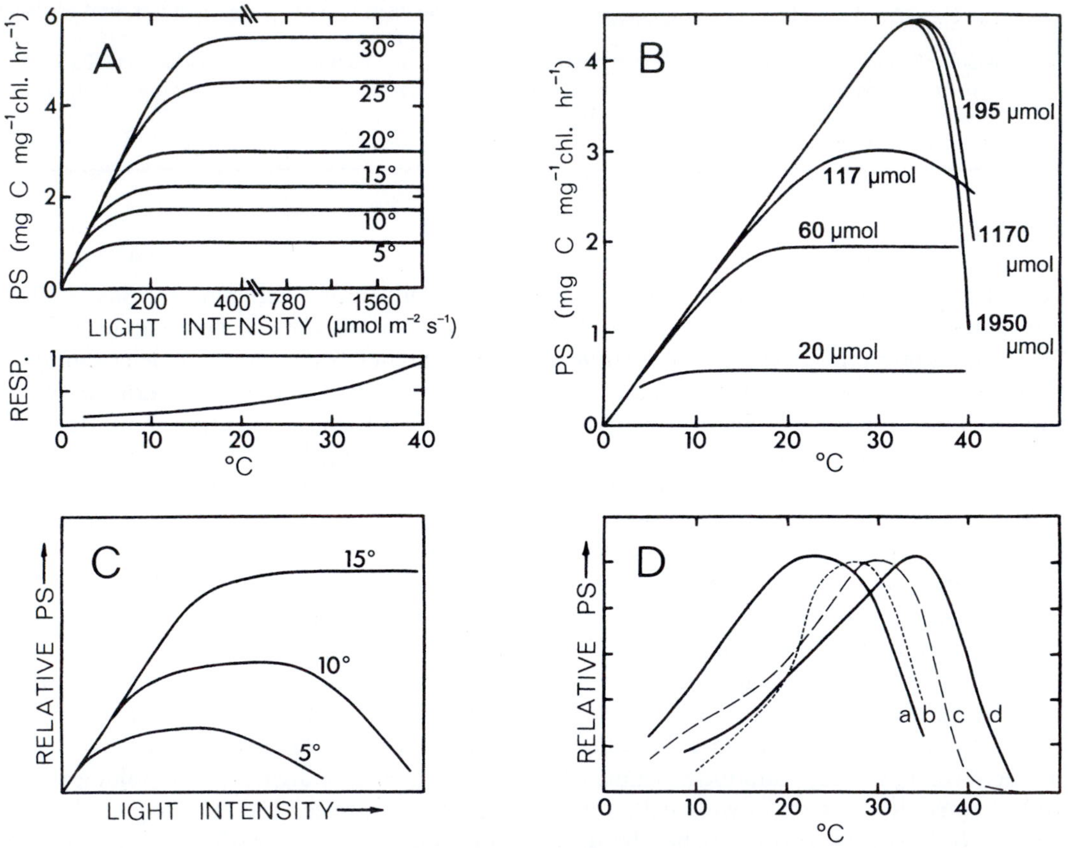

FIGURE 15-2 Interactions between photosynthesis (PS) and light intensity and temperature in plankton algae. (A) Generalized photosynthetic rates (mg C fixed per mg chlorophyll per hour) in relation to increasing light intensities and rates of respiration at different temperatures; (B) commonly observed increasing inhibitory effects of high light intensities that are accentuated at higher temperatures; (C) effects of high light intensity on photosynthetic rates of algae adapted to cold temperatures, as in arctic lakes; and (D) optimum temperatures for photosynthesis of different algae: a. *Synedra,* b. *Anabaena,* c. *Chlorella,* and d. *Scenedesmus.* (After data of Aruga, 1965.)

content per cell than those adapted to low irradiances. In *Chlorella pyrenoidosa,* for example, cells grown at ca. 20 μmol quanta m^{-2} s^{-1} have about 10 times more chlorophyll per cell than those grown at 410 μmol quanta m^{-2} s^{-1}. At high light intensities, but below light inhibition, the rate of photosynthesis per cell was not much greater than at low intensities.

In other algae, such as many diatoms, adaptation to changes in light intensity occurs by changing the light-saturated photosynthetic rate; the chlorophyll content of cells grown at low and high light intensities remains the same. The light-saturated rate of photosynthesis is generally much greater in algae grown under higher light intensities than it is for the same species grown under lower illuminance. Some algae are intermediate between the extreme types discussed here or exhibit a combination of photoadaptive strategies (cf. Prézelin and Sweeney, 1978).

Thus, the early sloping portion of light saturation curves (e.g., Fig. 15-2B) is largely a photochemical response of pigments to increasing irradiance. At light saturation, biochemical enzymatic reactions are rate-limiting and these are regulated by temperature (Davison, 1991). Some algae, such as *Skeletonema,* adapt to lower temperatures by increasing enzyme concentrations; consequently, the same rate of photosynthesis is achieved per given light intensity at high and low temperatures (Jørgensen, 1968; Priscu and Goldman, 1984). In *Skeletonema* adapted to low temperatures, the amount of protein per cell was twice as great at 7°C as at 20°C, and the size of cells was greater at low temperatures. Although rates of respiration are affected little by changes in light intensity, they do increase with increasing temperatures (Fig. 15-2A). Photosynthetic rates of planktonic algae adapted to light and temperature conditions of arctic regions can be nearly as high as those observed in temperate regions during summer under relatively high irradiance (e.g., Steemann Nielsen and Jørgensen, 1968b; Harrison and Platt, 1986).

Light of high intensities is detrimental to many algae and reduces rates of photosynthesis (Fig. 15-2B, C). Photoinhibition occurs when light exceeds physiological saturation and results in an excess of photons that are not dissipated by photosynthetic carbon fixation or by fluorescence or quenching reactions (Long *et al.,* 1994; Falkowski and Raven, 1997). The injury to enzymes and photosystem II can be repaired, and the resynthesis processes are more rapid at higher temperatures. Thus, the repair process is generally faster at higher temperatures, within the limnological range, where photosynthetic rates are higher.

Effects of high light intensities vary with the species in question, and some adaptation occurs with time. On bright days it is very common to observe a distinct depression in rates of *in situ* photosynthesis near the surface. The decrease in photosynthetic rates is associated with photooxidative destruction of enzymes, not chlorophyll (Steemann Nielsen, 1962; Steemann Nielsen and Jørgensen, 1962). The depression of enzymatic processes at high light intensities takes some time to occur. If exposure is of short duration (several hours), reactivation can occur and permanent damage is avoided. A depression of pigment concentrations often occurs at midday on bright days (e.g., Long *et al.,* 1994). Some algae, such as the diatom *Cyclotella,* adapt rapidly (<24 h) to changing light intensities and are not inhibited at very high intensities (Jørgensen, 1964). Other algae, such as natural populations of phytoplankton in Antarctic lakes exposed to continuous light in summer or in Arctic lakes after loss of ice cover in summer, are severely inhibited by high light intensities (Goldman *et al.,* 1963; Hobbie, 1964). In Arctic ponds, the phytoplankton exhibit increasing susceptibility to photoinhibition at cold (2–8°C) temperatures (Fig. 15-2C; Alexander *et al.,* 1980; Hobbie, personal communication). As water temperatures increase during the brief, 100-day period when the ponds are not frozen solid, adaptation to high illuminance occurs.

Ultraviolet light, which mediates a photooxidative alteration of the photosynthetic apparatus, is particularly harmful to planktonic algae. UV light can inactivate photosystem II, disrupt electron transport, and damage pigments, membranes, and thymidines of DNA (e.g., Gessner and Diehl, 1952; van Baalen, 1968; Karentz *et al.,* 1994; Falkowski and Raven, 1997). Detrimental effects are partly reversible if exposure is not too long, such as if plankton exposed to intensive UV irradiance at the immediate surface waters are circulated to deeper waters to which less UV penetrates (see Chap. 5). Resynthesis of damaged proteins and/or pigments can represent a net metabolic loss and affect long-term survival and cell fitness.

Interactions of light and temperature frequently result in one of several possible vertical profiles of photosynthesis in planktonic environments. In some cases, there is a zone of maximum photosynthetic rates at light saturation, which overlies a zone of near-exponential decline of rates with increasing depth (Fig. 15-3A). When light intensities exceed saturation and the photoinhibition threshold of the most productive species in the algal assemblage, a surface photoinhibition of the community photosynthesis often occurs (Fig. 15-3B). The depth at which maximum rates of photosynthesis occur varies with the transparency of the water, which is governed by the concentration of dissolved and particulate organic matter and abiotic turbidity. When densities of phytoplankton, that is, the

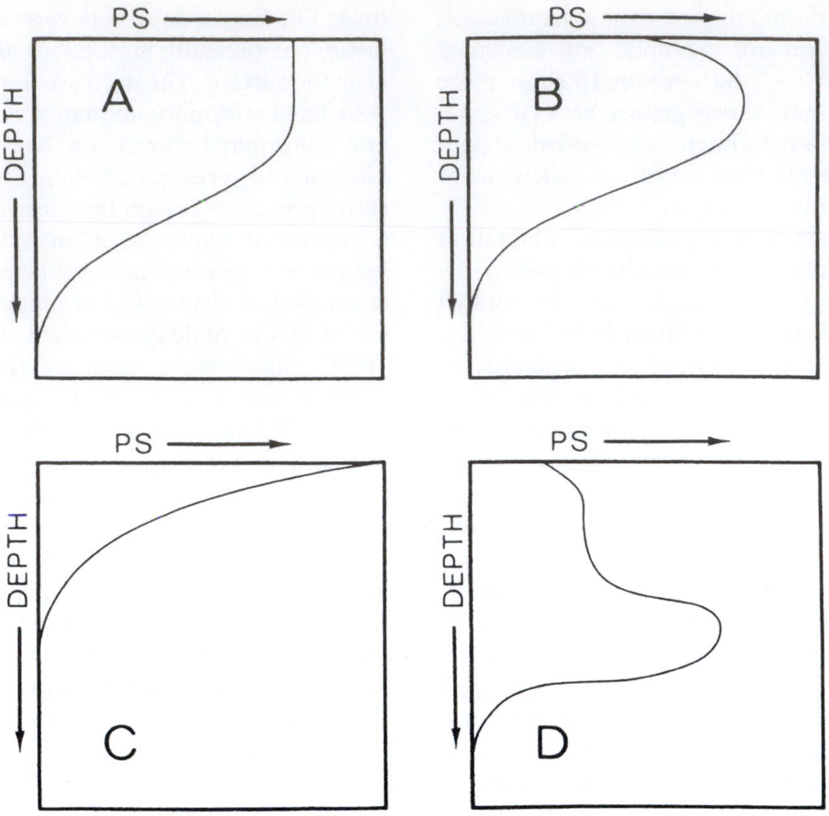

FIGURE 15-3 Generalized variations in vertical profiles of rates of photosynthesis (PS) among phytoplankton. (A) Light saturation in surface waters, without inhibition, underlain with decreasing rates as light is reduced with depth; (B) profile when severe surface photoinhibition occurs; (C) photosynthetic rates when biogenic turbidity is very high; (D) photosynthetic maximum in the metalimnion.

biogenic turbidity, are high, self-shading effects can greatly reduce light penetration and the trophogenic zone is reduced accordingly (Fig. 15-3C). Although surface photoinhibitory effects occur, rapid light attenuation through self-shading of the algal populations reduces the severity of enzymatic damage. In some cases, the algal populations adapted to relatively low light intensities of deeper layers develop greatly and have higher photosynthetic rates than those in the epilimnion (Fig. 15-3D; cf. also Fig. 9-5A).

There is great diversity in tolerances to variations in temperature among the algae (Fig. 15-2D). The minimal temperature at which photosynthesis can occur depends upon the algal species. For many diatoms, the critical temperature is about 5°C; for others, about 15°C (e.g., Rodhe, 1948). In many green algae and cyanobacteria, even higher temperatures are needed for incipient photosynthesis. The cyanobacteria as a group are generally much more tolerant of higher temperatures than are other phytoplankton. Many species of cyanobacteria have higher temperature optima than eukaryotic algae of the same waters. Further, thermophilic algae with optimal growth at temperatures above 45°C are almost exclusively cyanobacteria, and photosynthesis is known to occur at constant temperatures as high as at least 74°C (Castenholz, 1969). Respiration and especially photorespiration increase at high temperatures (Döhler and Przybylla, 1973). In the ensuing discussion of seasonal succession of algal populations, we will see again the frequently observed preponderance of diatoms at colder temperatures and the increasing variety of algae as the waters warm.

B. Flotation Mechanisms and Water Turbulence

The importance of water movements in the transport of particulate organic matter, in this case planktonic algae, has been emphasized at length earlier (Chap. 7). Not only are water movements important in the physical movement of phytoplankton into and out of the photic zone but they are critical in the transport of mineralized matter from lower depths and from the littoral regions to the open water. The degree of turbulence and water movement are obviously critical in the regulation of phytoplanktonic periodicity and production.

The importance of basin morphometry to water movements must be emphasized again in relation to composition of planktonic algae aside from nutrient considerations. An excellent example is seen in the analyses of the spatial distribution of predominantly littoral desmid algae in several lakes in Minnesota (Bland and Brook, 1974). Many desmids occur throughout the year on and among submersed macrophytes of the littoral zone. During spring circulation, many desmid algae are dispersed, particularly to other areas of the littoral, and can colonize newly growing annual macrophytes. In small lakes, circulation in the littoral is not as intense as it is in larger lakes. Larger desmids are carried out of the littoral less frequently in these smaller lakes. The more active circulation of larger lakes, however, transports a variety of desmids into the pelagic regions from the littoral zone, and this input increases the diversity of the planktonic communities.

Various characteristics are found among plankonic algae that improve flotation or reduce sinking rates (Lund, 1959, 1965; Fogg, 1965; Hutchinson, 1967).

a. ***Density.*** The density of most freshwater planktonic organisms is slightly greater than that of the water and causes them to sink when in undisturbed water. Cells of cyanobacteria, which lack sap vacuoles, may have densities in the range of 1.02–1.09 times that of water, but gas vesicles may render them less dense than water (Thomas and Walsby, 1985; Utkilen *et al.*, 1985). Eukaryotic algae, which usually have sap vacuoles, have a density of 1.01–1.03 times that of water.

b. ***Particle Size.*** Spherical particles of a mean diameter of not more than 0.5 mm fall according to Stokes's law: The velocity of sinking varies (i) inversely with the viscosity of the medium (see Chap. 2), (ii) directly with the excess density of the particle over that of the medium, and (iii) directly as the square of the diameter or some appropriate linear dimension (reviewed by Hutchinson, 1967). Thus, in planktonic habitats, velocity of sinking is greatest for relatively large, dense particles in open water (cf. Titman, 1975). It is apparent that sinking out of the photic zone is a distinct disadvantage for phytoplankton. However, movement of a cell through water, either by its own motility or by sinking (or floating) through water has the advantage of disrupting nutrient gradients around the cell and increasing the chances, because diffusion is so slow, of contact with nutrient molecules. Consequently, the disadvantage of sinking out of the photic zone, which is usually fatal, is offset to some extent by the advantage of increased nutrient acquisition.

c. ***Form Resistance.*** Sinking rate is reduced by decreasing size and increasing surface-to-volume ratios (Table 15-5). The sinking rate of an algal cell is almost always less than a sphere of a volume and density equal to those of the alga. The effect of the shape that causes the reduction in this sinking rate is called *form resistance*, the ratio of the sinking rates of the object and a sphere of the same density and volume. Although the rough surface texture of the cell wall or membrane has relatively little effect on sinking, elongation into cylindrical or discoid shapes distinctly decreases settling rates. Projections or more elaborate protrusions such as spines decrease the settling rate by frictional resistance and, if they cause uneven weight distribution, will shift the orientation of the cell during descent.

d. ***Mucilage Production.*** The production of mucilage can alter the sinking rate. The effect on the sinking rate (U) depends on the thickness of the mucilage layer. The mucilage layer can lower the excess density, thereby reducing U, as well as increase the radius (r), which increases U in proportion to r^2 (Walsby and Reynolds, 1980). Thin layers reduce U, but if the layer exceeds a critical thickness, the velocity increases.

Gelatinous sheaths occur in nearly all cyanobacterial phytoplankton, some diatoms, and green algae, particularly most desmids with rather elaborate projections and irregular cell morphology (Boney, 1981). The mucilage generally has a density less than that of the cell, which decreases sinking velocity, but the mucilage layer increases cell size, which increases velocity. The sheath also reduces the efficiency of nutrient uptake by creating an additional diffusional barrier. Mucus secretions can cause diatoms to aggregate, however, when very abundant and accelerate sinking rates as large particles (Smetacek, 1985).

e. ***Gas Vacuoles.*** Gas vacuoles are found in the protoplasm of both cyanobacteria and many other types of bacteria. Each gas vacuole contains numerous hollow, cylindrical gas vesicles, the walls of which are composed of protein. The main cylinder and conical end caps of the gas vesicle are formed by a small hydrophobic protein arranged in ribs. The ribs are bound together by a larger hydrophilic protein, which assembles on the outer surface. Together these proteins form a rigid shell that withstands pressures of several bars. At a certain critical pressure, the structure collapses and no longer provides buoyancy. Cyanobacteria in deep lakes have evolved very narrow, strong gas vesicles that withstand pressures generated at greater depths (Walsby, 1994).

The gas vacuole membranes and the walls of the vesicles are freely permeable to gases, and the gas contained in the vacuoles and vesicles is in equilibrium with gases dissolved in the surrounding water. Water is prevented from entering the gas vacuole by surface tension at the inner surface of the membrane (cf. reviews of Walsby, 1972, 1975, 1994; Fogg *et al.*, 1974; Reynolds and Walsby, 1975).

The gas vacuoles decrease the density of cyanobacteria to below that of water, even though the volume of the vacuoles often occupies <5% of the total cell volume. Cyanobacteria possessing gas vacuoles float toward the

TABLE 15-5 Sinking Rates of Several Freshwater and Marine Algae in Relation to Area:Volume Ratios and Physiological State of Cells[a]

Cell type	*Area:Volume* (μm^2:μm^3)	*Cell size* (μm *diameter or width* × *length*)	*Sinking rate* (*m* day^{-1})
Centric, unicellular			
Thalassiosira nana	0.88–1.20	4.3–5.2	0.10–0.28
T. pseudonana	—	—	0.09
Cyclotelia meneghiniana			
Exponential phase	—	2.0	0.08
Stationary phase	—	2.0	0.24
Coscinodiscus lineatus			
Exponential phase	—	50	1.9
Stationary phase	—	50	6.8
Stephamodiscus astraea			
Before silica depletion	—	—	0.2
After silica depletion	—	—	0.45
Centric, elongate			
Rhizosolenia setigera			
Normal preauxospore	—	6 × 245	0.19–0.44
Spineless preauxospore	0.62–0.75		0.22–0.63
Postauxospore	0.10–0.16	33 × 363	0.79–1.94
Total population	—	—	0.63–0.95
R. robusta			
Exponential phase	—	84	1.1
Stationary phase	—	84	4.7
Centric, chain-forming			
Tholassiosira rotula	0.23–0.29	19–34	0.39–2.10
Skeletonema costatum	0.81–1.01	5 × 20	0.31–1.35
Bacteriastrum hyalinum	0.29–0.33	10 × 30	0.39–1.27
Chaetoceros lauderi	0.19–0.41	20 × 34	0.46–1.54
Aulacoseira[b] *agassizii*			
Exponential phase	—	54.8	0.67
Stationary phase	—	54.8	1.87
Aulacoseira[b] *italica*	—	—	0.86
Pennate, chain-form)ng			
Nitzschia seriata	1.18–1.65	4 × 40	0.26–0.50
Asterionelia formosa			
• Exponential phase	—	25.0	0.20
		—	0.78–1.6
Live	—	—	0.14–0.44
Dead	—	—	0.42–0.95
• Stationary Phase	—	25.0	1.48
			7.6–8.1
Live	—	—	0.17–0.47
Dead	—	—	0.64–0.84
Fragilaria, natural populations			
Germany	—	—	0.5–1.0
England	—	—	0.2–0.65
Fragilaria crotonensis			
Exponential phase	—	—	1.3–1.46
Stationary phase	—	—	7.0–16.6
Tabellaria fenestrata, Natural populations	—	—	0–0.85
Other algae/cyanobacteria			
Staurastrum cingulum			
Exponential phase	—	—	0.38
Stationary phase	—	—	4.5–4.7
Pandorina morum			

(Continues)

TABLE 15-5 (Continued)

Cell type	Area:Volume (μm^2:μm^3)	Cell size (μm diameter or width × length)	Sinking rate (m day^{-1})
Exponential phase	—	—	0.05–0.10
Stationary phase	—	—	0.50–0.70
Mougeotia thylespora	—	—	0.06–2.1
Aphanizomenon flos-aquae	—	—	0.03–0.04
Peridinium cinctum	—	—	7.7
Scenedesmus quadricauda			
Exponential phase	—	8.4	0.27
Stationary phase	—	8.4	0.89

[a] After data of Grim (1952), Smayda and Boleyn (1965, 1966a,b), Pasciak and Gavis (1974), Knoechel and Kalff (1975), Reynolds (1976), Titman and Kilham (1976), Beinfang (1979), Wiseman *et al.* (1983), Gibson (1984), Sommer (1984b), and Gálvez *et al.* (1993).
[b] Formerly *Melosira.*

surface and are not dependent only on turbulence, which is required by other phytoplankton to remain in the photic zone. Changes in gas vacuole:cell-volume ratio determine whether the cells sink or float. The ratio changes as vacuole volume becomes proportionately less of the total volume in exponential growth phases. Gas vacuoles become more abundant when light is reduced and growth rate slows. Rises in turgor pressure of the cells as a result of the accumulation of photosynthate cause a collapse in existing gas vesicles and a reduction in buoyancy (Dinsdale and Walsby, 1972; Konopka, 1980; Walsby and Booker, 1980) but only in cyanobacteria that have relatively weak gas vesicles (Kinsman *et al.*, 1991). Additionally, the "dilution" of newly produced vacuoles during active cell division and the accumulation of dense carbohydrate can reduce buoyancy.

By these mechanisms, cyanobacteria are able to regulate buoyancy and undergo limited vertical migration to poise themselves within physical and chemical gradients favorable to growth, usually toward the bottom of the euphotic zone (Fogg and Walsby, 1971; Takamura and Yasuno, 1984; Klemer, 1986). The population maximum, which results from population growth and movement downward, apparently often occurs as epilimnetic nutrient concentrations are sharply reduced in summer (Klemer, 1978; Konopka, 1981). Metalimnetic populations of planktonic cyanobacteria, particularly of several species of *Planktothrix* (formerly *Oscillatoria*), often develop in strata of low light intensity during the latter portions of summer stratification. Some other filamentous cyanobacteria in the genus *Tychonema* lack gas vesicles and exhibit chromatic adaptation to the low light intensities (Utkilen *et al.*, 1985).

Gas vacuolation can increase under carbon limitation and promote cyanobacterial blooms (Booker and Walsby, 1981), whereas nitrogen limitation decreased vacuolation and contributed to deep population maxima (Klemer *et al.*, 1982; Rijn and Shilo, 1983). Light is an overriding factor regulating buoyancy, however. As with sinking, movement by floating also has the added advantage of disrupting nutrient gradients surrounding actively growing cells. The scattering of light by gas vesicles has been investigated and found to be of little, if any, benefit to cyanobacteria in that they shield the cells from excessively bright light (Shear and Walsby, 1975; Ichimura *et al.*, 1981).

The energetic costs of making gas vesicles to provide flotation have been compared with the combined smaller cost of making flagella and the additional cost of rotating them to provide movement (Walsby, 1994). The gas vesicles provide a cheaper solution in slowly growing organisms like cyanobacteria. In rapidly growing cells, however, they are more expensive because a new set of gas vesicles must be made at each cell division.

Large colonial cyanobacteria occur mainly in turbulent polymictic and epilimnetic waters, since single filaments sink more slowly than helically shaped cells (Booker and Walsby, 1979). As turbulence subsides, these cyanobacteria can adjust their position rapidly. As a result of these movements, large colonial cyanobacteria commonly form phytoplanktonic blooms (cf. Reynolds, 1978). In extreme cases, the blooms of cyanobacteria become so dense on the surfaces that scums or "hyperscums" of thick (several cm), crusted accumulations of cyanobacteria cover the water in wind-protected areas (e.g., Zohary and Robarts, 1990). These surface accumulations form cyanobacterial refuges, which help maintain large planktonic biomass during winter when growth is reduced but have no effect on the perennial survival of the organisms.

f. ***Accumulation of Hydrocarbons.*** The density of some algae is decreased by the accumulation of fats. Some algae, such as *Botryococcus,* can contain lipids up to 30–40% of dry weight and can float because of this accumulation (Fogg, 1965). Usually, however, fat accumulation occurs in senescent cells that are in various stages of degradation.

g. ***Ion Regulation.*** Buoyancy alteration may also be achieved in some marine and brackish-water algae by means of ion regulation (cf. Anderson and Sweeney, 1977). Ionic regulation of the contents of the large sap vacuole in large diatoms and dinoflagellates can make the cells less dense than sea water. K^+ salts replace the denser Na^+ salts and Cl^- the denser SO_4^{2-}. Positive buoyancy can be obtained by this method in sea water but not fresh water.

h. ***Swimming by Flagella.*** Swimming by flagella is common among cryptomonads and dinoflagellate algae. Despite the very small size of many of these flagellated organisms, their locomotion is adequate to move vertically within the water column for 10s of centimeters to several meters daily in response to sinking, shifting light and nutrient concentrations, and possibly predation by zooplankton, especially rotifers (Sommer, 1982, 1986; Arvola, 1984; Gasol *et al.,* 1991; Jones, 1991; Gervais, 1997; Pithart, 1997). Migrational amplitudes and maximal swimming velocities of phytoflagellates increase with body size, but relative amplitudes and relative velocities decrease with body size (Sommer, 1988). Cryptomonads commonly exhibit a regular vertical daytime ascent and a nighttime descent of between 0.5 and 1.5 m. Because the energy costs for flagellar movement in swimming nearly equal or could exceed those benefits gained by movement (Raven and Richardson, 1984), it is assumed that the benefits are in improved nutrient acquisition in heterogeneous environmental conditions of nutrient distribution. If dinoflagellates are moved by water currents into surface waters, motility and phototaxis can be strongly impaired by exposure to ultraviolet radiation (Tirlapur *et al.,* 1993).

C. Inorganic Nutrient Factors

The importance of inorganic macro- and micronutrients, particularly the two major elements, phosphorus and nitrogen, to algal nutrition has been emphasized in preceding chapters on biogeochemical cycling, so reiteration is not necessary here. Facets of algal nutrition and nutrient interactions will be discussed further in the sections on the periodicity and succession of phytoplanktonic populations.

The limiting nutrient concept, originally developed by Liebig well over a century ago as the "Law of the Minimum," has experienced considerable misuse, largely from attempts at oversimplification. Simply paraphrased, the law states that the yield of any organism will be determined by the abundance of the substance that, in relation to the needs of the organism, is least abundant in the environment (cf. Hutchinson, 1973; Talling, 1979). Algal species differ in specific nutrient requirements and optimal ratios of N:P (e.g., Rhee and Gotham, 1980; Tilman *et al.,* 1982; Borchardt, 1996). Thus, in a multispecies algal community, growth rates among different species are likely to be limited by different resources, including differing nutrients. As a result, the concept of single-nutrient limitation does not apply strictly to communities. When such limitation is in effect among dominant species in the community, or a resource such as phosphorus occurs at levels below the minimal threshold requirements of many species of the community, the limitation concept is applicable to community productivity.

Since yield is a result of growth, rate of growth has been substituted for yield in many subsequent analyses, the most important of which is the well-known Monod model for a single nutrient limitation in the growth of microorganisms (cf. Droop, 1973):

$$\mu / \mu_m = S/(K_s + S)$$

when

μ = specific growth rate (increase in biomass per unit biomass per unit time)

μ_m = maximum specific growth rate at infinite external substrate concentration

S = external substrate concentration (mass per unit volume)

K_S = saturation constant when the external substrate concentration results in half the maximal rate of uptake (i.e., S when $\mu = \mu_m/2$)

Irrespective of the internal nutrient concentration, the specific rate of uptake depends upon substrate concentration in a Michaelis function; the rate of uptake increases with increasing substrate concentration to a specific substrate level beyond which no further change in rate of uptake occurs.

Under conditions of steady state with nutrient concentrations that are not limiting, the specific rate of nutrient uptake (u, mass per unit biomass per unit time) equals the product of the specific growth rate (μ) and the cell nutrient quota (Droop, 1973):

$$u = \mu q$$

where q is a demand coefficient or cell nutrient quota (content) in the absence of nutrient excretion [(internal substrate concentration) (mass per unit biomass)]. The specific rate of uptake and the cell nutrient quota have a linear relationship:

$$\mu q = \mu_{m'}(q - k_q) \quad \text{or} \quad \mu/\mu_{m'} = 1 - k_q/q$$

where

$\mu_{m'}$ = maximum specific growth rate at nonlimiting internal substrate concentration,

k_q = the subsistence quota (i.e., the q intercept for zero μ)

The importance of these models, which are confirmed by many studies on planktonic algae (e.g., Goldman, 1977), lies in their applicability to obtaining specific growth values from more easily determined uptake rates (Eppley and Thomas, 1969).

It is important to view nutrient uptake as it really takes place, since uptake of several nutrients occurs simultaneously. Nutrient limitations of photosynthesis can be highly dynamic both spatially and temporally (Wetzel, 1972). Of importance among the many interacting, potentially limiting nutrients is the relative position of limitations within an *intensity* spectrum, which undergoes constant change spatially and temporally. The model of Droop can be expanded to the multinutrient condition, based on the fact that a limiting parameter (in this case μ/μ'_m) is proportional to the product of two or more functions if it is independently proportional to each of them. Therefore,

$$\mu/\mu_{m'} = (1 - k_{qA}/q_A)(1 - k_{qB}/q_B)(1 - k_{qC}/q_C)(\ldots)$$

is a simple polynomial in which the subscripts A, B, C, . . . , refer to various nutrients. In the latter equation, all nutrients are formulated equally; the significance of a particular nutrient in establishing the composite limiting parameter (μ/μ'_m) is determined by the relative magnitude of k_q compared with q. The parameter μ'_m is an abstract proportionality constant in this case, whose value is determined by all factors (physical, nutritional, et cetera) excluded from the equation.

The effectiveness of individual nutrient limitations has been shown in a number of studies (e.g., Droop, 1974, 1977; Rhee, 1974, 1978). Frequently, a relatively sudden shift occurs in the transition from limitation by one nutrient to limitation by another nutrient. These results demonstrate the restricted usefulness of the simple multiplicative polynominal formulation given here as a descriptor of simultaneous limitation of growth or nutrient uptake by two or more nutrients. Although the shift from one nutrient limitation to another can be expressed mathematically (Droop, 1977), the predicted growth rates are likely serious underestimates for a given algal population (Talling, 1979).

1. Resource Competition and Limitations

Nutrient resource competition functions to structure natural communities of freshwater phytoplankton. The Droop equations just discussed describe the dependence of nutrient-limited reproductive growth rates on the cellular content of the limiting nutrient (cell quota) and are applicable to populations grown to steady state, as in a continuous culture, as well as field populations (Sommer, 1991). Although the effects of external nutrient concentrations are difficult to evaluate, particularly for cell quotas (content) of nonlimiting nutrients (Kilham and Hecky, 1988), the Droop model was found to be a superior extension of the Monod model, which relies on ambient concentrations of limiting nutrients for prediction of growth rates (Sommer, 1991).

Growth experiments for algal species under substrate-limiting conditions allow one to obtain μ_m (the maximum growth rate) and K_S (the half-saturation coefficient for growth). These values allow one to estimate the concentration of the limiting nutrient (R^*) required by an alga to grow at a specific mortality or dilution (D) rate (Fig. 15-4), where:

$$R^* = D\,K_S/(\mu_m - D)$$

Species with the lowest R^* for a resource should have the competitive advantage for that resource and dominate (Tilman, 1982). The predicted outcome of competition between two species with two potentially limiting nutrients (Fig. 15-5) defines the isoclines of zero net growth (heavy lines) by the R^* for each species for each nutrient resource. The rate at which each species consumes both resources is represented by a specific consumption vector (diagonal lines), the slope of which is a species' optimal ratio. Species having the lowest R^* values become the competitive dominant under that specific potentially limiting resource. With sufficient information about the physiological capabilities of different algae, the competitive abilities of the different species can be determined in relation to resource availability, as illustrated in the examples of Figure 15-6.

Loss rates experienced by a species can be just as important as growth rates in determining species compositions of phytoplankton communities. Seasonal fluctuations in such mortality rates, as for example by differential susceptibility to grazing, would also direct the succession of populations (Fig. 15-7). Similarly, other environmental factors influencing growth and mortality rates couple to limiting nutrient resources to determine phytoplankton succession. For example, temperature affects mainly the maximal growth rate of a species. Among several species of algae, the species with the lowest R^* should be the superior competitor at a given temperature, and as a result species A should dominate at

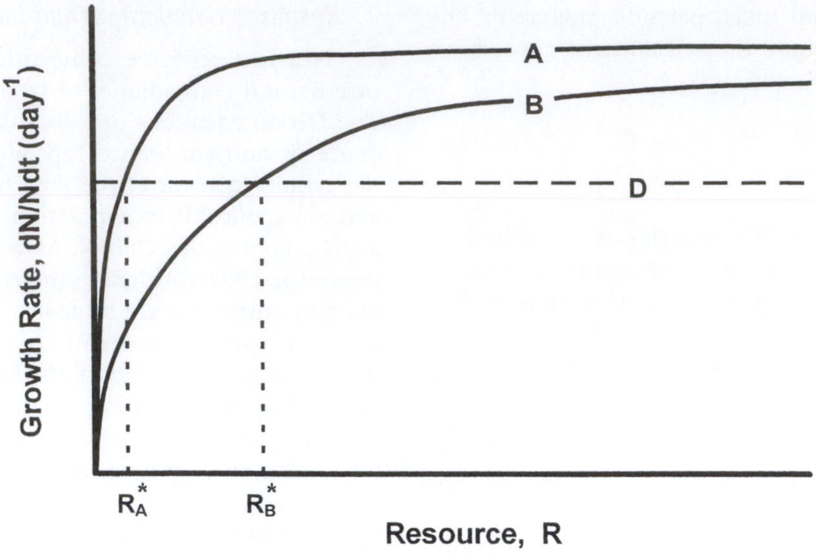

FIGURE 15-4 Monod growth curves for two species of algae (A and B). The R^* value for each species is the equilibrium resource (nutrient) concentration at specific dilution rates (D equals mortality rate at steady state). When the two species A and B compete for a single resource, the species with the lowest R^* for that resource will always win. (Modified from Kilham and Hecky, 1988.)

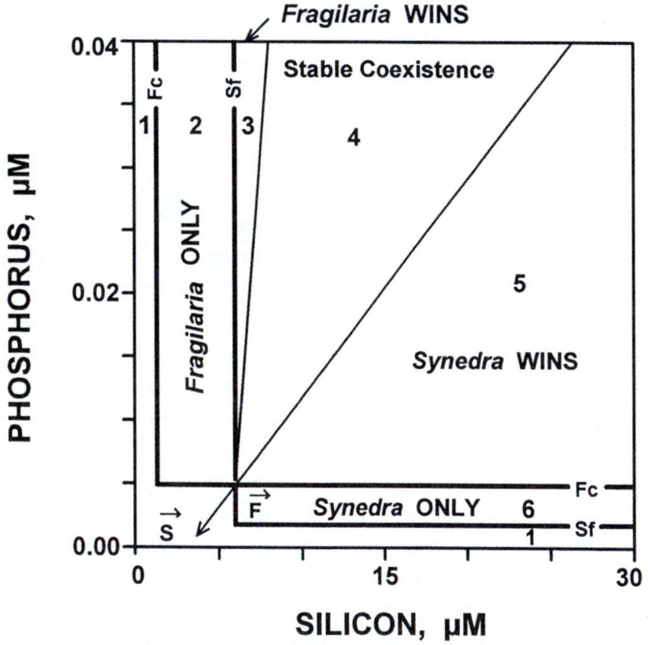

FIGURE 15-5 The predicted outcome of resource (phosphorus or silicon) competition between two freshwater diatoms, *Synedra filiformis* (Sf) and *Fragilaria crotonensis* (Fc). The isoclines of zero net growth (heavy lines) for each species cross at an equilibrium point (•). The consumption vectors (diagonal lines) for each species are labeled S → and F →. Neither species has sufficient nutrients to survive in region 1. In region 2, *Fragilaria* can survive but *Synedra* cannot. In region 3, *Synedra* is outcompeted by *Fragilaria*. Both species stably coexist in region 4. *Fragilaria* is competitively displaced by *Synedra* in region 5. Only *Synedra* can survive in region 6. (Modified from Kilham and Hecky, 1988.)

low temperatures, followed by species B and C as temperature increases, for example, seasonally (Fig. 15-8).

2. r- and K-Selection in Seasonal Phytoplankton Succession

The original concept of two types of selection to which organisms are exposed in natural environments was based on the logistic growth equation.[3] The expanded concepts now group *r*-selection with short-lived, fast-growing species capable of rapid colonization of habitats with abundant resources that are relatively unoccupied with competing species (Kilham and Kilham, 1980; Sommer, 1981; Amblard, 1988). *K*-selection favors organisms that have high competitive abilities by such means as high efficiency of resource utilization and resistance from high losses, such as predation or sedimentation. *r*-selection is commonly associated with early stages of succession following environmental change or disturbances, whereas *K*-selection is typical of more mature, stable stages. A reasonable environmental constancy persisting over some 12–16 generations is needed for the community to reach a more climactic, stable condition (Gaedeke and Sommer, 1986; Holzmann, 1993; Reynolds, 1993). Species diversity tends to increase with greater intervals of time

[3] *r* was the intrinsic rate of growth and *K* the upper asymptote of the growth curve or the "carrying capacity" of the environment for a particular species (MacArthur and Wilson, 1967).

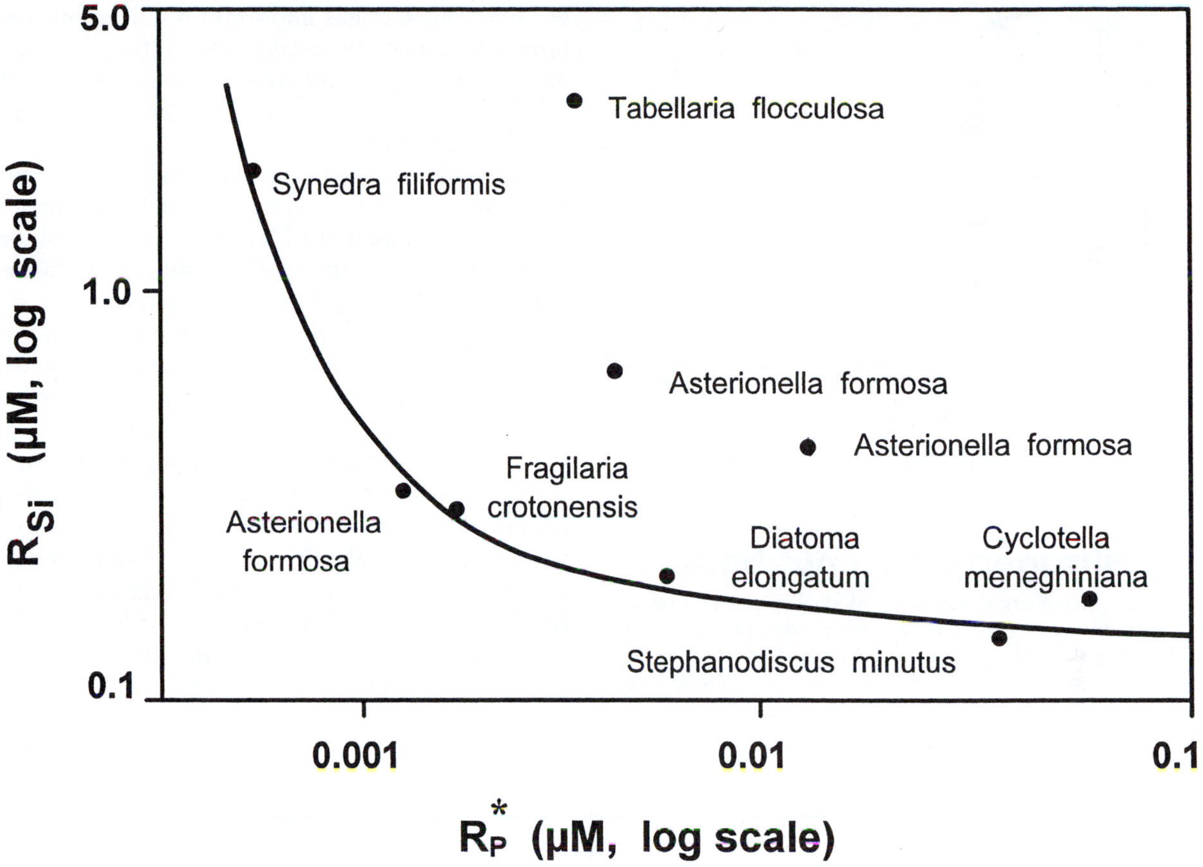

FIGURE 15-6 Differential competitive abilities of freshwater planktonic diatoms at 20°C. *Synedra filiformis,* for example, is the superior competitor for phosphate but the most inferior competitor for silicate. (From Kilham and Hecky, 1988, based on Tilman *et al.*, 1982.)

between disturbance but decrease under very stable conditions where conditions of nutrient limitations and grazing select toward few species.

There is a strong tendency toward balance between fecundity and mortality, where species with rapid growth usually also decline rapidly (Green, 1980; Sommer, 1981). Maximum values occur in the spring in temperate waters, followed by a depression in summer, an increase at the end of summer, and a slow decline in autumn. From spring to late summer, growth classes with the most rapid growth shift to the growth class with the slowest growth, and from the smallest to the largest size class. In the autumn, the developmental sequence commonly reverses.

Species of similar growth class and size class tend to be associated with each other in a community. Therefore, a common sequence is for rapidly growing, small phytoplankton of the spring bloom to shift to a slower-growing algal community with a shift from predominantly *r*-selection to largely *K*-selection. The shift is regulated by both nutrient reductions in the trophogenic zone and early grazing pressures by zooplankton (cf. Kilham and Kilham, 1980). When resources are scarce, the ratio of resource demand (*D*) to resource supply (*S*) tends toward 1, and more *K*-selected organisms are favored (Kilham and Hecky, 1988). When *D*/*S* ratios are considerably <1, more *r*-selected organisms predominate. In the autumn, competition for nutrients declines as the thickness of the trophogenic zone increases, but light and temperature decrease as well. As a result, the biological controls shift to dominance by physical factors.

Complications in simple multiplicative predictive analyses of algal population growth also result when synergistic effects occur between nutrients. An example mentioned earlier was an enhanced rate of photosynthesis resulting from enrichment with both phosphorus and manganese, but not with each nutrient alone (Shapiro and Glass, 1975). As will be discussed later (Chap. 17), among the bacteria, nutrient uptake can be influenced by the previous substrate (nutrient) concentration. If lack of a nutrient previously has limited growth severely, the uptake sites for that nutrient can be reduced. Replenishment of the limiting nutrient may not result in an immediate response of uptake and

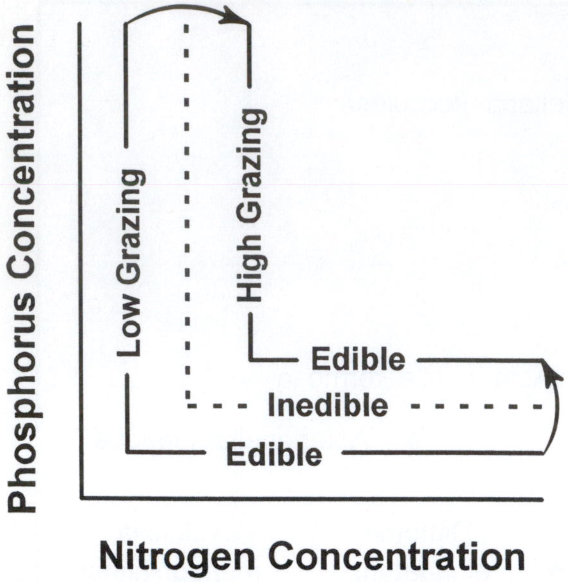

FIGURE 15-7 Hypothetical succession of algal species that results from a shift from low grazing rates to high grazing rates (arrows). The isocline of an inedible species (- - - -) does not vary with grazing pressure, but the isocline of an edible species (——) moves away from the axes as grazing pressure increases. The species with its isocline closest to the axes will be the superior competitor and will dominate. (From Sterner, 1989.)

growth. This fact has important implications for short-term enrichment bioassays since effects of the added nutrient may not be realized in population growth for a considerable period, possibly longer than the duration of the experiment.

In fresh waters of low nutrient concentrations and slow turnover rates, it is metabolically advantageous to be both small and motile. Small size (ultraplankton, <20 μm) increases the ratio of absorptive surface area to cell volume (Table 15-3). Since growth rates among microphytoplankton (>20 μm) are slow and may not be adequate to offset mortality by sedimentation and predation, small size is further advantageous in reducing the settling rate and possibly predation by many grazers. Motility also is effective in disrupting nutrient gradients surrounding cell membranes during active growth. These relationships suggest that one would anticipate a greater percentage of the algal populations of oligotrophic lakes to be small and motile. In most oligotrophic lakes, this relationship is indeed the case. Usually a much larger percentage of the total rates of primary productivity derives from ultraplanktonic algae of oligotrophic lakes than from larger algae. In some extreme cases, all phytoplanktonic primary productivity is by algae <30 μm.

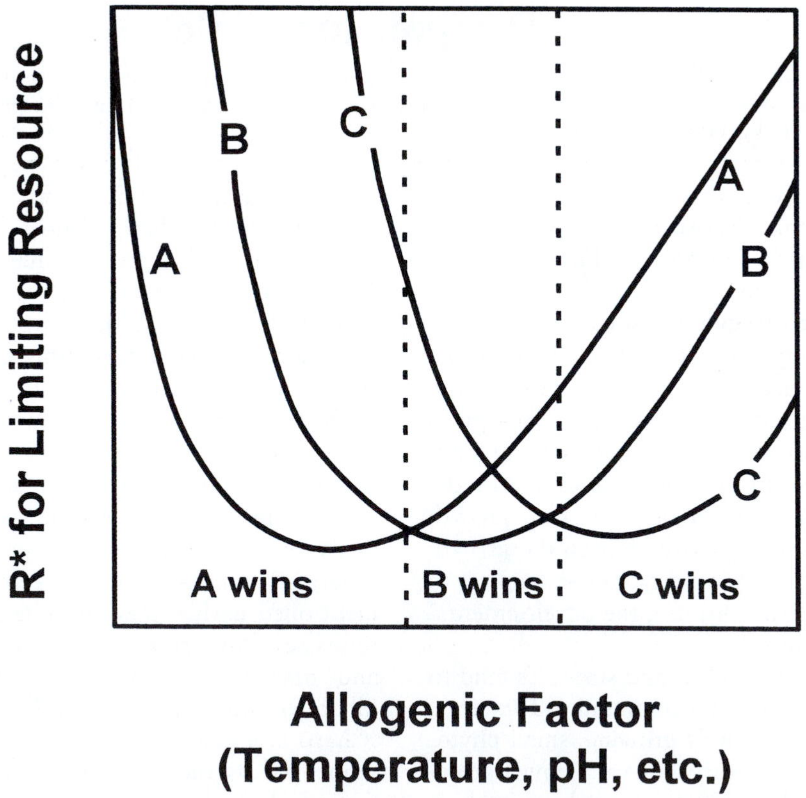

FIGURE 15-8 Competition for one limiting resource in relation to an allogenic factor such as temperature or pH. The species with the lowest requirement (R^*) for the resource at a given level of the allogenic factor should competitively displace the other species. (Modified from Tilman *et al.*, 1982.)

3. Organic Matter and Algal Growth

Dissolved organic matter is intimately coupled with the nutritional requirements of algae for organic growth factors. In addition, the availability of certain organic compounds is dependent upon the quantity and types of dissolved inorganic cofactor compounds that are present. Furthermore, a number of algae can supplement primary photosynthetic autotrophy by the uptake and utilization of organic substrates. Algae capable of synthesizing required organic growth factors (e.g., vitamins) have a distinct competitive advantage over algae that must rely on exogenous supplies synthesized by other microorganisms.

In addition to direct utilization as a micronutrient or utilization as supplementary energy and carbon sources, organic compounds can function as nonessential accessory substances that may stimulate growth of the algae. Growth substances or nonessential organic micronutrients include a large group of hormones known to be produced by algae and effective in growth regulation and cell development (Bentley, 1958a,b; Conrad and Saltman, 1962; Provasoli and Carlucci, 1974). Little is known of the physiological mechanisms of these substances in freshwater algae, although they are probably of some significance in the success and succession of certain more specialized algae under natural conditions.

4. Lipid Production in Algae

Algal lipids are critical in aquatic ecosystems because they provide a high-density source of energy and essential fatty acids (EFA) for aquatic animals. Lipids can also be vectors for the movement of hydrophobic materials, including many organic contaminants (Landrum and Fisher, 1999). Many planktonic and attached algae and cyanobacteria are major synthesizers of lipids. Lipids can constitute 30–50% of the dry mass of lake seston of a size <50 μm in diameter at certain times of the year (Kreeger *et al.*, 1997).

Lipids undergo modest seasonal changes in concentration that are proportional to the percentage of photosynthate allocated to lipids. This relation is directly coupled to light intensity and is proportional to the efficiency of carbon fixation, with greater production in summer than in other seasons (Wainman and Lean, 1992; Arts *et al.*, 1997; Kreeger *et al.*, 1997). In summer, the lipid content as a percentage of production is about 15–30% of total photosynthesis at light saturation (P_m); in winter, the lipid production is 5–20% of P_m. No clear seasonal trend in the lipid percentage of areal photosynthetic production was evident, however, with an annual mean of ca. 17% (Rai *et al.*, 1997; Wainman *et al.*, 1999). This seasonal constancy suggests lipid synthesis is constitutive and not susceptible to "luxury storage," except potentially when the algae can increase lipid synthesis (30–40% of total production) when severely nutrient-limited (e.g., Parrish and Wangersky, 1987; Kilham *et al.*, 1997). Nutrient limitation tends to depress the rate of cell division more than it does of carbon accumulation through photosynthesis. As a result, nutrient limitation should thus result in algae that are rich in carbon storage products as lipids or carbohydrates (Ahlgren *et al.*, 1997; Rai *et al.*, 1997; Olsen, 1999). As noted later, however, essential nutrient limitations can alter the specific production of essential fatty acids in algae, which in turn affects metabolism in both the algae and in the zooplankton that consume them. Light and carbon limitation, however, can limit cell division and should result in a lower level of storage lipids in algae. Also, slow growth in any algal species often leads to a rapid removal of that alga from the plankton through grazing.

Lipids contain some 16 subclasses. Triacylglycerols, in which three fatty acids are esterified to a glycerol molecule, are important intracellular energy stores that are used for generating ATP for respiration and anabolic reactions. Sterols, of similar ring structure, are also neutral lipids associated with membrane structure. Phospholipids, also critical as structural constituents of cellular membranes, are polar and similar functionally to the glycolipids that contain one or more sugar molecules (Parrish, 1999). Glycolipids are major components of the thylakoid membrane in chloroplasts. Nomenclature for fatty acids is based on the total number of carbon atoms (e.g., "18" of 18:2ω6), the number of *cis* double bonds (e.g., "2"), and the position of the first double bond (e.g., "ω6," where the "6" is the position of the first double bond counted from the methyl end of the molecule). Lipids with *trans* double bonds are largely unknown in fresh waters, and only plants can desaturate fatty acids at the ω3 and ω6 positions.

Polyunsaturated fatty acids (PUFA) of algae and cyanobacteria are important biochemical markers because they include some essential fatty acids that animals must assimilate in their diets. The distribution of fatty acid markers in the major groups of freshwater algae and cyanobacteria shows some specificity, but the variations are sufficiently large that fatty acid profiles can be used to distinguish algae at the class level only (Ahlgren *et al.*, 1992; Napolitano, 1999). However, some groups, such as the ω3-PUFAs, are important fatty acid markers of algal biomass because they are virtually absent in most bacteria and terrestrial plants. Among the bacteria, PUFA and fatty acids with a chain length longer than 18 carbon atoms are usually absent in procaryotes. In contrast, eukaryotic algae contain large proportions of fatty acids with 20–22C and 3 to 6 *cis* double bonds. Lipids of terrestrial and higher aquatic plant origins tend to contain longer chain (≈C26) fatty acids than do phytoplankton (≈C16). Thus, the C16/C26 fatty acid ratio in lipids of the particulate organic matter in sediments of lakes can be seen to increase with distance from the land and littoral areas (e.g., Meyers *et al.*, 1984; Boon *et al.*, 1996).

Pronounced differences in fatty acid profiles exist among species and larger taxonomic groups of algae. Dinoflagellates are rich in docosahexaenoic acid (22:6ω3; DHA). Diatoms are rich in eicosapentaenoic acid (20:5ω3; EPA). In general, large quantities of ω3 fatty acids, including EPA and DHA, are far more common in marine than in freshwater algae (Olsen, 1999). Chrysophytes have variable ω3 profiles with either 18:3, 18:4, EPA, or DHA as the dominant ω3 fatty acid. The green algae are more diverse and have variable profiles with 18:3, 18:4, or EPA as the dominant ω3 fatty acids (Ahlgren *et al.*, 1990, 1992, 1998; DeMott and Müller-Navarra, 1997; Desvilettes *et al.*, 1997; Olsen, 1999). Among the cyanobacteria, one group (*Planktothrix, Microcystis*) was similar to green algae, with higher amounts of 18C components of the ω3 acids, whereas the other group (*Anabaena, Spirulina*) contained mostly ω6 acids. The taxonomically diverse flagellates were characterized by high amounts of long-chained (20–22C) polyunsaturated fatty acids. Alterations in growth conditions tend to change the total lipid content rather than specific ratios of fatty acids. Increased lipid content in stressed algae was largely the result of increased saturated fatty acids and ω6 acids, whereas the ω3 acids were relatively unchanged. Branched-chain fatty acids are usually considered a characteristic of cyanobacteria (Mancuso *et al.*, 1990).

There is some evidence that algae, stressed by UV-A (Arts *et al.*, 1999) and/or UV-B radiation, allocate less carbon to lipids and have altered fatty acid profiles (Goes *et al.*, 1994; Wang and Chai, 1994). Relative increases in short-chained, and decreases in polyunsaturated, fatty acids have been observed in phytoplankton exposed to moderate UV (Hessen *et al.*, 1997). Critical membrane fatty acids, such as eicosapentaenoic (20:5ω3) and docosahexaenoic (22:6ω3) acids, can experience peroxidation or reduced biosynthesis under stress of UV radiation.

V. ORGANIC MICRONUTRIENT REQUIREMENTS

Knowledge of the specific requirements of phytoplanktonic algae for growth factors or essential organic micronutrients, especially the vitamins, is extensive (cf. reviews of Provasoli, 1958; Hutner and Provasoli, 1964; Provasoli and Carlucci, 1974). Although most higher plants do not require vitamins (Roth *et al.*, 1996), many algae do; species that require vitamins for growth are termed ***auxotrophic.*** In the case of auxotrophic growth, the requirements for specific organic compounds are low and these compounds do not contribute significantly to the total cell carbon. Only three water-soluble vitamins are known to be required by algae: vitamin B_{12} (cobalamine), thiamine, and biotin. Vitamin B_{12} and thiamine are required alone or in combination by a majority of the auxotrophic algae, and B_{12} is required more often than thiamine. Biotin is known to be required by a few chrysomonads, dinoflagellates, and euglenoid algae.

TABLE 15-6 General Requirements for Vitamins among the Algae and Cyanobacteria[a]

Algal groups	*Biotin*	*Thiamine*	*Vitamin B_{12}*	*Predominant vitamin requirements*
Cyanobacteria	0	0	++	B_{12}
Rhodophyceae	0	0	++	B_{12}
Bacillariophyceae	0	+	++	B_{12}
Xanthophyceae	0	0	0	None
Phaeophyceae	0	0	+	None
Chlorophyceae	0	+	++	B_{12}
Chrysophyceae and Haptophyceae	−	++	+	Thiamine
Cryptophyceae	−	−	+	None
Dinophyceae	+	0	++	B_{12}
Euglenophyceae	−	−	+	None

[a] After discussion of Provasoli and Carlucci (1974).
++ = required in many species; += few species; −= requirement rare; 0 = no known requirement.

There is no strict taxonomic correspondence to the known distribution of the requirement for external vitamins (auxotrophy) among different algae. The differences, however, are sufficient to permit some generalizations. On the basis of relatively few analyses in pure cultures, a significant number of cyanobacteria, diatoms, green algae, and dinoflagellates require exogenous sources of vitamin B_{12}, that is, they are auxotrophic (Table 15-6). Only a few groups show any significant requirements for thiamine. Only the xanthophycean algae exhibit no apparent need for these major water-soluble vitamin growth factors, although in several other groups the requirements are rare (Table 15-6). The cyanobacteria, Chlorophyceae, Xanthophyceae, and Phaeophyceae exhibit the least number of species requiring vitamins. Most of the species in these groups are strictly autotrophic in metabolism. In contrast, a clear predominance of auxotrophic species occurs in the Chrysophyceae, Dinophyceae, Cryptophyceae, and Euglenophyceae.

When algae are unable to synthesize an organic micronutrient, the need for an exogenous source of the correct chemical moiety is obvious. Most of the vitamins are synthesized by bacteria and certain algae and are released to the water during active growth and upon death (Carlucci and Bowes, 1970; Aaronson *et al.*, 1977; Grieco and Desrochers, 1978; Nishijima *et al.*, 1979). Appreciable concentrations of vitamins derived from airborne soil particles, pollen, or active microorganisms (Parker and Wachel, 1971) can enter aquatic ecosystems from precipitation (Table 15-7). Although very low concentrations of organic micronutrients, especially vitamin B_{12}, exist in fresh waters (Table 15-7), these concentrations yield little insight into rates of turnover between supply sources and demand. All assays of the absolute vitamin requirements of algae are necessarily based on bacteria-free cultures and in most cases indicate that the requirements are much lower than observed concentrations in fresh waters.

Studies of seasonal variations in concentrations of vitamin B_{12}, thiamine, and biotin have demonstrated large changes with time (Tal, 1962; Kashiwada *et al.*, 1963; Daisley, 1969; Kurata *et al.*, 1976, 1979a; Cavari and Grossowicz, 1977; Nishijima and Hata, 1977; Parker, 1977). In many cases, concentrations of

TABLE 15-7 Range of Concentrations of Organic Micronutrients Found in Fresh Waters (in μg liter^{-1} = mg m^{-3})

Lake	*Vitamin B_{12} (cyanocobalamin)*	*Thiamine*	*Biotin*	*Niacin (nicotinic acidamide)*	*Pantothenic acid*	*Folic acid*
Linsley Pond, Conn. (Hutchinson, 1967; Benoit, 1957)	0.06–0.075	0.008–0.077	0.0001–0.004	0.15–0.89	—	—
Northern German lakes (Hagedorn, 1971)	—	0.05–12	—	—	—	—
Sagami, Japan (Ohwada and Taga, 1972)	0.005–0.85	0.001–0.38	0.010–0.068	—	—	—
Tsukui, Japan (surface) (Ohwada *et al.*, 1972)						
Dissolved	0.0005–0.0042	0.075–0.436	0.013–0.058	—	—	—
Particulate	0.0019–0.0203	0.031–0.159	0.0005–0.0042	—	—	—
Kasumigaura, Japan (Kashiwada *et al.*, 1963)	0.005–0.028	—	0.0021–0.050	0.30–3.3	0.01–0.26	0.040–0.244
Small Swiss ponds (Clémencon, 1963)	—	<0.001–1.0	<0.001–0.004	<0.001–3.0	<0.01–0.034	<0.01–0.48
Kojima, Japan (Nishijima and Hata, 1979)	0.0021–0.036	0.020–2.30	0.0023–0.1016	—	—	—
Mergozzo, Italy (Kurata *et al.*, 1976)	0.0004–0.0025	0.047–0.334	0.005–0.019	—	—	—
Biwa, Japan (Kurata *et al.*, 1979a)						
Pelagic zone	0.001–0.006	0.008–0.095	0.003–0.015	—	—	—
Littoral zone	0.002–0.018	0.015–0.250	0.004–0.025	—	—	—
Mean of precipitation, Missouri (Parker and Wachtel, 1971)						
April–November	0.001	—	0.004	2.0	—	—
November–March	0.0004	—	0.0008	0.42	—	—

vitamin B_{12} and thiamine, but seldom biotin, were negatively correlated with densities of phytoplankton, particularly small algae (<30 μm). Concentrations of the water-soluble vitamins commonly decrease with increasing water depth. In the surficial profundal sediments, vitamin concentrations are 100–1000 times greater than in the overlying water (e.g., Nishijima and Hata, 1978; Kurata *et al.*, 1979a). Concentrations of vitamin B_{12}, thiamine, and biotin also change diurnally, with higher concentrations occurring in the morning than in the evening (Kurata *et al.*, 1979b). Presumably these differences result from algal utilization in daylight exceeding inputs from synthesizing organisms. Concentrations of the water-soluble vitamins have been found to be consistently higher in the littoral zone than in the open water (Kurata *et al.*, 1979a). The data suggest that epiphytic microorganisms associated with submersed macrophytes are major net exporters of vitamins, especially B_{12}, to the open water.

Concentrations in natural waters are so low that they must be determined by bioassay organisms of known growth responses to concentrations of the micronutrients. Concentrations utilizable by the assay organisms may not be equally utilized by algae. Moreover, evidence indicates that some or many of the natural organic micronutrients may not be readily available for assimilation. Photosynthetic rates and cell numbers of natural phytoplanktonic populations frequently increase markedly by additions of vitamin B_{12} or thiamine (see, e.g., Wetzel, 1965a, 1966b, 1972; Hagedorn, 1971). The concentrations needed, especially in hardwater lakes, to increase the populations in this way are often greatly in excess of those generally known to be required by algae in pure culture. White and Wetzel (1985) have shown experimentally that much vitamin B_{12} is adsorbed to particulate calcium carbonate and is removed from the trophogenic zone by coprecipitation and sedimentation. Other losses are likely. For example, some species of algae produce a glycoprotein exudate that competitively inhibited vitamin B_{12} and suppressed the growth of an algal species that requires B_{12} for competitive species succession (Messina and Baker, 1982). It is doubtful whether the rates of vitamin synthesis generally are limiting to total growth and productivity of phytoplankton of lakes. However, large spatial and seasonal fluctuations occur, and these organic micronutrients probably play a significant role in the succession and competitive success of phytoplankton populations.

VI. HETEROTROPHY OF ORGANIC CARBON BY ALGAE AND CYANOBACTERIA

Photosynthesis dominates metabolism among algae and cyanobacteria. Most algae are obligate ***photoautotrophs:*** Cell carbon is obtained by reduction of carbon dioxide to carbohydrates from transformations in which light energy obtained by light-receptor pigment systems is converted to chemical energy (Table 15-8). The products of photosynthesis (ATP and organic carbon compounds) provide chemical energy as well as organic compounds needed for cellular synthesis. If light is inadequate or unavailable, the cells generally either die after storage compounds are used or they become metabolically inactive in a dormant state until light again becomes available.

In pure culture, an appreciable number of algae assimilate and utilize dissolved organic compounds as a source of carbon and energy both in the dark and in the light. ***Heterotrophy*** or chemo-organotrophy in algae implies the capacity for sustained growth and cell division in the dark; energy and cell carbon are obtained from the metabolism of an organic substrate(s). Heterotrophy in algae occurs by means of aerobic dissimilation, and carbon dioxide may or may not be required. Excellent reviews on the physiology and biochemistry of heterotrophy by algae are given by Droop

TABLE 15-8 Metabolic Processes Utilized by Algae and Cyanobacteria in the Synthesis or Modification of Organic Matter[a]

Metabolic processes	*Energy sources*	*Carbon sources*
Photoautotrophy (photolithotrophy)	Solar radiation	CO_2 only[b]
Photoheterotrophy (photoorganotrophy)	Solar radiation	CO_2 and organic C
Heterotrophy:		
Chemoautotrophy (chemolithotrophy)	Oxidation of inorganic compounds (usually only in certain bacteria)	CO_2 only[b]
Chemoheterotrophy (chemoorganotrophy)	Organic compounds	Organic C only

[a] Modified from Tuchman (1996).
[b] From free CO_2 or HCO_3^-.

(1974), Neilson and Lewin (1974), and Tuchman (1996).

A few algae are ***mixotrophic*** and assimilate carbon dioxide in small amounts simultaneously with organic compounds both in the light and especially in darkness. In other words, photoautotrophy in the light can be supplemented by the assimilation of organic compounds in the dark. In comparison to bacteria and fungi, heterotrophic algae can utilize only a few substrates, such as acetate and related compounds (pyruvate, ethanol, lactate, and higher fatty acids), glycolate, hexose sugars, and amino acids. None of the algae known to grow heterotrophically can do so under anaerobic conditions. The ability to utilize organic substrates requires specific enzymes for transport across cell membranes, and many algae are deficient in these enzymes. In certain species that possess the enzymes, organic substrates still are not utilized because of an apparent impermeability of the cells or an inability to couple their dissimilation to the generation of adenosine 5′-triphosphate (ATP).

It must be emphasized that in most cases in which algal heterotrophic utilization of organic substrates has been demonstrated, concentrations of substrates were very high, often several orders of magnitude greater than those found in most natural waters. Furthermore, culture conditions have been bacteria-free, which eliminates the competitive interactions of bacteria that possess much more efficient active uptake mechanisms. Nearly all analyses *in situ* with natural populations have led to the following conclusions: (a) At naturally occurring substrate concentrations, the low affinity of algae for simple organic substrates makes heterotrophy a relatively inefficient and unimportant process in comparison to photoautotrophy; (b) algae generally cannot compete effectively with bacteria for available substrates (e.g., Wright and Hobbie, 1966). This subject is discussed at greater length later on (Chap. 17).

Although it is obvious that photoautotrophic metabolism is the primary means of synthesis among algae and cyanobacteria in lake systems, it is incorrect to dismiss heterotrophic metabolism in natural populations of algae as unimportant. Indeed, there is evidence that heterotrophy and photoheterotrophy can be instrumental in subtle but important ways in the survival, competition, and succession of algal and cyanobacterial populations. Under conditions of light limitation, as occur deep in the hypolimnion, in turbid reservoirs, under heavy snow and ice cover, at extremely high latitudes at which total darkness exists for a period, or when very dense populations lead to shading, an ability to be facultatively heterotrophic (mixotrophy) is competitively advantageous. Observations of apparently viable and even reproducing phytoplanktonic algae under ice in northern Scandinavia during almost total darkness are incompletely explained (Rodhe, 1955). Some evidence exists that phytoplankton can compete with bacteria for acetate, but not glucose, under low light conditions (Maeda and Ichimura, 1973; Vincent, 1980; Vincent and Goldman, 1980). Certain deepwater cyanobacterial populations, especially *Oscillatoria,* are facultatively heterotrophic under very low light conditions (Saunders, 1972c). Earlier remarks should be recalled on the possible utilization by these populations of extremely low light intensities to generate ATP by photophosphorylation without concomitant CO_2 reduction, which could be potentially important in maintaining viability without significant growth.

If abundant concentrations of organic carbon substrates are present in the environment, the light-independent biochemical reactions of the Calvin cycle may be blocked (Sheen, 1990) and ATP utilized to transport extracellular organic substrates into the chloroplast or cell, in the case of cyanobacteria. True ***photoheterotrophy,*** in which organic substrates serve as a significant source of cell carbon during growth, is more restrictive in that different metabolic pathways are involved. When organic loading is very high, such as in sewage oxidation ponds, algae such as *Chlamydomonas* can assimilate most of the acetate (generated by anaerobic bacteria) by photoheterotrophy, which involves photosynthetic production of ATP and reducing power (Eppley and Macias, 1963). Although photoheterotrophy is generally not of quantitative importance in most natural waters, evidence indicates that it can amount to at least 20% of total inorganic carbon fixation at low light intensities in certain oligotrophic lakes (McKinley and Wetzel, 1979). Two periods of relatively high photoheterotrophic uptake activity were observed: one during spring turnover and a second during the late summer stratified period. The greatest photoheterotrophic contributions to overall carbon cycling occurred during morning-to-midday periods and at intermediate depths of the lower metalimnion. Microautoradiographic techniques have demonstrated selective utilization of organic substrates by algal and bacterial species (cf. Ellis and Stanford, 1982). Slight supporting evidence suggests that glucose can improve survival of algae by photoheterotrophy under light-limited conditions (Amblard *et al.,* 1992). Photoheterotrophy could potentially augment photoautotrophy among algal and cyanobacterial communities attached to surfaces (Tuchman, 1996). Exogenous dissolved organic substrates are certainly more concentrated within these periphytic communities than is the case among phytoplankton. Steep environmental gradients, particularly of light (e.g., Losee and Wetzel, 1983), within the attached communities could enhance photoheterotrophic utilization of organic substrates released by active and senescing microbiota.

VII. OTHER EFFECTS OF DISSOLVED ORGANIC MATTER

The extracellular production and release of antibiotics or growth inhibitors have been suggested in numerous cases from observations of the inhibitory effects of one alga on another in both mixed cultures and natural populations (Proctor, 1957; Krauss, 1962; Lefèvre, 1964; Hellebust, 1974). The chemical composition of these compounds is poorly understood. Inhibitory compounds include peroxides of fatty acids and possibly polyphenolic substances, such as humic acids. Stimulatory compounds include a number of weak organic acids, especially glycolic acid. The effectiveness of antibiotics is governed in part by species specificity, relative concentrations, and rates of bacterial degradation. Such substances may play important roles in the succession of species under natural conditions and are a fertile field for investigation.

Other effects of dissolved organic compounds include a number of ways in which the organic substances affect the availability of inorganic micronutrients (Saunders, 1957; Wetzel, 1968). Chelation of metal ions, an equilibrium reaction between a metal ion and an organic chelating agent, results in the formation of a stable ring structure incorporating the metal ion. Chelation can function in several complex ways in accordance with environmental conditions. The extent to which chelation of metallic ions occurs within a heterogeneous solution, such as lake water, is governed by the bonding characteristics of the organic compounds, the ratio of chelator to metal ions, and stability constants of chelates for different ions. Temperature, hydrogen ion activity, and the concentration and ionic strength of various anions further influence the extent of complexing (Brezonik, 1994). Other organic and inorganic compounds present in fresh waters function as sequestering agents for metals and cations (cf. Chaps. 10 and 14). Complex formation occurs with pyrophosphate, and metals are bound by the complex formation with macromolecules such as proteins (Povoledo, 1961) and by the formation of peptidized metal hydroxides of yellow humic substances (Shapiro, 1964).

Organic complexing of metal ions and major cations can influence the specific composition and succession of algal populations. Examples of some of the mechanisms involved include (a) increases in the availability of inorganically reactive ions, such as iron and manganese (Chap. 14); (b) modifications of membrane permeability and osmoregulation by complexing of cations, which results in changes in monovalent:divalent cation ratios (Chap. 10); (c) reduction of the concentration of a trace metal toxic to a particular organism below the threshold of toxicity by excessive complexing of the metal ions by organic compounds; (d) under certain circumstances increasing the concentration of the toxicant to harmful levels by the complexing and effective completion of organic compounds with a metal ion that is antagonistic to a toxicant; in addition, as discussed earlier (p. 261), evidence exists for the formation of organophosphorus–humic compounds that may reduce phosphorus availability to phytoplankton (Stewart and Wetzel, 1982).

VIII. SEASONAL SUCCESSION OF PHYTOPLANKTON

In spite of the number of pervasive generalizations about the common seasonal succession of phytoplanktonic algal populations in fresh waters, close inspection of existing data shows a great diversity of patterns. Some of the disparity is related to the choice of study methods employed. For example, many of the older analyses were based solely on number of organisms, which is an extremely biased indicator in comparison to biomass because of great differences in size among algae. Furthermore, early investigators often used plankton nets of fairly large mesh size to capture algae through which significant and variable portions of the algae were lost.

Descriptions of the seasonal succession of phytoplankton in north temperate, dimictic lakes frequently include correlations of changes in dominant algal species, biomass, numbers, and rates of photosynthesis with alterations in physical and biotic factors. From these descriptions, a general pattern of phytoplankton succession has emerged. However, many variations on the general succession theme exist, particularly in lakes of other geographical regions (tropical, alpine, and polar). Although correlations are useful in pointing to possible areas of interaction for intensive study, they do not provide specific information on causal mechanisms. Briefly, the temperate pattern of phytoplankton succession involves a winter minimum of small flagellates adapted to low light and temperature, a spring burst of diatom activity and biomass, followed rapidly by a smaller development of green algae, and a transitional lull between spring and summer. Summer populations vary in relation to the trophic status of the lakes but can include either another diatom development in less productive lakes by late summer and early autumn or increases in nitrogen-fixing cyanobacteria in eutrophic lakes.

The discussion that follows integrates a detailed description of phytoplankton succession, correlations with changes in environmental parameters, and recent information on mechanistic explanations of observed patterns. In this latter area, our knowledge is still incomplete.

A. Seasonal Successional Patterns and Periodicity of Phytoplankton

Distinct seasonal patterns and periodicity in the biomass of phytoplankton are observed in polar and temperate fresh waters. Many detailed descriptions of phytoplankton succession have been correlated with changes in environmental parameters, particularly in temperature, light, nutrient availability, and mortality factors such as grazing and parasitism. Because seasonal succession is strongly keyed to meteorological and stratification-mixing processes, patterns in temperate ecosystems differ considerably from those of tropical waters. Each will be discussed separately with the caveat that the processes are similar but the timing can differ considerably.

Generalizations are difficult to make because of the great variability observed among phytoplanktonic numbers and biomass from lake to lake. Several patterns are reasonably consistent, however:

a. The seasonal periodicity of phytoplanktonic biomass is reasonably constant from year to year. If the freshwater ecosystem is not perturbed by outside influences, such as human modifications of the watershed, nutrient loading, et cetera, the seasonal changes in phytoplanktonic populations are similar from year to year.
b. The extent of change in phytoplanktonic numbers and biomass through the seasons is usually very great, on the order of a thousandfold in temperate and polar fresh waters. Seasonal variations in tropical waters are much lower, often as little as fivefold (Fogg, 1965).
c. Maxima and minima of numbers and biomass of phytoplankton often are out of phase with rates of primary production. Primary productivity usually follows closely the annual cycle of incident solar radiation in temperate lakes.

A periodicity of species composition continuously accompanies seasonal changes in total biomass. On an annual basis, algal communities consist of a composite of perennial species (***holoplanktonic***) that persist throughout the year in the plankton and of intermittent species (***meroplanktonic***) that periodically undergo some type of diapause in resting stages. In both cases, but much more markedly in the latter type, an interplay of environmental conditions results in fluctuations in growth and competition among other species. Among the perennial species, population numbers can decrease to extremely low levels, but cells are present in the water in sufficient numbers to re-inoculate the community when growth conditions improve.

The patterns of phytoplanktonic population dynamics discussed below of lakes in the temperate and polar zones are generally in accord with those proposed by Hutchinson (1967), Margalef (1983), Reynolds (1984, 1990, 1997), Harris (1986), Sommer *et al.* (1986), Stewart and Wetzel (1986), and Blomqvist *et al.* (1994). Those patterns of tropical lakes are generally in agreement with the summaries of Lewis (1987, 1996).

B. Dimictic Lakes in the Temperate Zone

The seasonal succession of biomass of major phytoplankton groups in dimictic temperate and arctic lakes is generally separable into eight periods:

1. In ***midwinter,*** low temperatures, high water column stability, reductions of light by snow, moderate nutrient availability, and low zooplankton grazing prevail. The low biomass of phytoplankton is dominated by ultraplankton (<20 μm), particularly by flagellates of the Chrysophyceae and Cryptophyceae (Fig. 15-9). The major structuring environmental factor is light availability, which is low as a result of high reflectance from snow and high attenuation under ice. Concentrations of inorganic nutrients are relatively high, especially near the sediments, from microbial remineralization in excess of utilization and from minimal mixing to strata above.

If nutrients are adequate, growth is limited to species that are adapted to low temperatures and low light irradiance. Population accrual by slow growth often is offset by respiratory losses, secretion of organic compounds, and sedimentation of cells from the limited photic zone under ice cover. In polar regions, inappreciable or zero growth exists under conditions of midwinter near-total darkness and very heavy ice and snow cover. At lower latitudes, within the temperate zone or at latitudes influenced by maritime amelioration of winter conditions, snow and ice are less thick and opaque, allowing more penetration of light.

The community of winter algae beneath ice is usually dominated by small and often motile algae. Cryptophyceans, such as *Rhodomonas* and *Cryptomonas,* are particularly common, as well as pyrrophyceans *(Gymnodinium),* small green algae (*Chlamydomonas*), chrysophyceans (*Dinobryon, Mallomonas, Chrysococcus,* and *Synura*), some diatoms (*Synedra, Tabellaria, Fragilaria),* and euglenophyceans *(Trachelomonas)* (Rodhe, 1955; Verduin, 1959; Wright, 1964; Tilzer, 1972; Maeda and Ichimura, 1973; and Nebaeus, 1984; among many others). The coldwater, low-light-adapted species are evidently quite species-specific in their

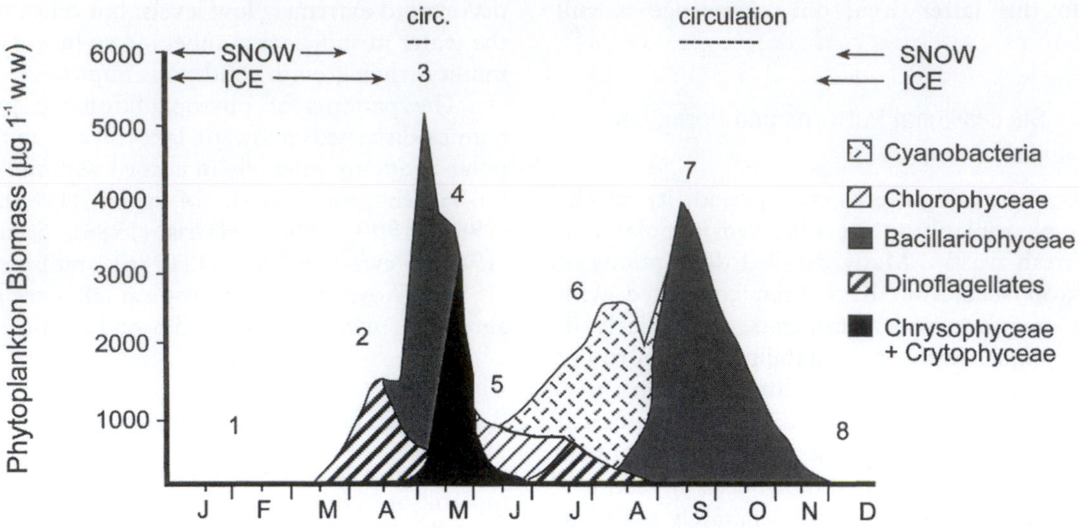

FIGURE 15-9 Model of biomass development and seasonal succession of major phytoplankton groups in a typical temperate lake of moderate productivity. Numbers refer to the eight seasonal periods discussed in the text. (Data from Lake Erken, Sweden; diagram modified from Blomqvist *et al.* (1994.)

tolerances to light quantity and quality and are found in narrow strata in the photic zone beneath the ice cover. For example, over 95% of the phytoplankton occurred in layers of the water column of an ice-covered lake of eastern Massachusetts receiving 0.5–20% of incident radiation (Wright, 1964). As discussed earlier, there is some evidence for mixotrophic growth in which photoautotrophy is supplemented somewhat by heterotrophic assimilation of simple organic substrates under darkness or very low irradiance. However, photoautotrophy dominates the metabolism of these algae, even under exceedingly low light conditions. Changes in light, as occur with snowfall on ice, result in a rapid vertical shift in the distribution and rates of photosynthesis (Fig. 15-10).

Algal blooms occur frequently under winter ice cover in shallow lakes (Nebaeus, 1984). A combination of environmental conditions of light penetrating the ice, particularly clear ice with little snow cover, high nutrient concentrations, influxes of nutrients and other growth-promoting substances from nonfrozen soils, and very low grazing losses can result in algal blooms,

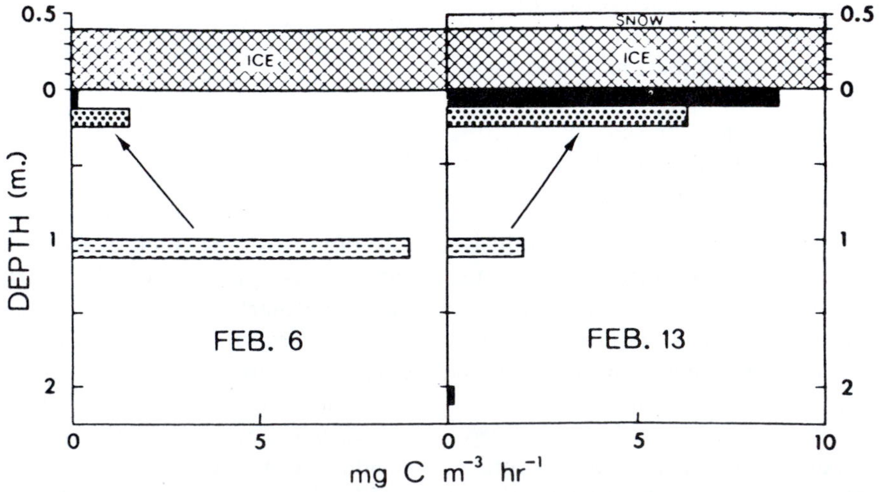

FIGURE 15-10 Rates of primary production beneath ice (6 February) and below ice and snow (13 February). Rectangles with dots are rates obtained when samples taken from 1 m were incubated alongside the surface samples. (Modified from Wright, 1964.)

particularly among the ultraphytoplanktonic (0.2–20 μm) cryptophycean and green algae. Many algae maintain reduced physiological activity under very low light conditions at temperatures of 4°C or less and are able to recover quickly once light and temperature conditions increase (e.g., Hino, 1991).

Rates of primary production under ice cover can constitute a very significant portion of the total annual primary productivity of the phytoplankton. Nearly one-quarter of the annual phytoplanktonic productivity of hypereutrophic Sylvan Lake in northeastern Indiana occurred under the 3-month period of ice cover (Wetzel, 1966a). Similar values have been found since in numerous lakes in the temperate zone, emphasizing that the widely held assumption that winter productivity is insignificant is not universally valid.

2. In ***late winter,*** conditions of low temperature, high water column stability, moderate nutrient availability, and low zooplankton grazing continue, but light availability improves not only with the seasonal (increased intensities and longer photoperiods) but often with less or no snow cover on the ice. Despite the low temperatures, phytoplankton biomass increases rapidly, particularly among the dinoflagellates that can migrate vertically and small centric diatoms (Fig. 15-9).

If temperate lakes are monomictic and circulate throughout the winter without ice cover (e.g., large lakes, subtemperate lakes, lakes at high latitudes but with moderate climate, as in Great Britain), phytoplanktonic growth is very low (e.g., Sommer, 1985). Not only is the seasonal light availability very low but phytoplankton are circulated to water depths of little or no light for long periods of time. Even though nutrient availability is high and grazing is almost nonexistent at these times, low light and movement to sites of low or no light limit photosynthesis. Only when stratification occurs, either under ice or during thermally induced density changes in the spring, can the phytoplankton achieve sufficient duration of stability to develop.

3. During the ***spring circulation,*** temperatures are still low but increasing, the water column is mixing with low stability, light conditions are highly variable but on the average low, and nutrient availability is generally high. Growth and increases in biomass, particularly by diatoms, are very rapid and usually result in the annual maximum of phytoplanktonic biomass (although usually not the maximum productivity, see p. 377). Grazing by herbivorous zooplankton is still low, although in early stages of rapid growth, during this commonly brief period of circulation.

As ice cover is lost in the spring and the season progresses, circulation of the water column results in the mixing of nutrient-laden water from lower depths with surface strata, which have more light. In shallow lakes in which the light penetrates throughout much of the water column, the spring algal maximum is often initiated by coldwater, *polyphotic*[4] species under ice in late winter. These species flourish and develop throughout the period of circulation and may persist for a time after summer stratification begins. The circulation of small lakes is commonly weaker and of shorter duration than it is in larger lakes with larger fetch distances. Among large lakes, net growth of phytoplankton can be significantly suppressed during circulation if mixing occurs at such a rate and to such a depth that the algae are carried out of the photic zone faster than they can multiply. The effects of circulation are accentuated in regions, such as much of the United Kingdom, in which lakes rarely freeze or do so for only brief periods. In these areas, circulation can continue practically all winter, with a significant loss of phytoplankton occurring as a result of outflow and other mechanisms (e.g., Lund *et al.,* 1963).

There is little question that increasing light in the spring is the dominant factor contributing to the development of the spring "outburst," because water temperatures are still low. The spring maximum is frequently dominated by one species, a diatom, such as *Asterionella* discussed earlier (Fig. 14-18), *Cyclotella,* or *Stephanodiscus* (see, e.g., Pechlaner, 1970). Although a lag phase often is not seen before the dominants reach exponential growth phases, it obviously exists and is observed if one samples frequently (e.g., Lund, 1950). In Lake Erken, Sweden, some lag in the response of the algae constituting the spring maximum to the increase in radiation was caused by the necessity for light adaptation, even when the intensities of photosynthetically usable energy were still low (Pechlaner, 1970).

The relative rates of increase[5] of the dominant algae during the spring maximum of exponential growth are usually much less than rates of increase observed in cultures grown under "optimal" conditions of light and temperature (Fogg, 1965). This difference

[4]Adapted to a wide range of light intensities.

[5]The relative rate of increase is a function of the generation time (G), that is, the mean doubling time if the cells divide by two or

$$G = \frac{\log 2}{k'} = \frac{0.301}{k'},$$

when k' = the relative growth constant under exponential conditions, or

$$k = \frac{\log_{10} N - \log_{10} N_0}{t}$$

when N_0 is the initial concentration of cells, N is the final cell concentration, and t the time interval.

is variously attributable to restrictions of light, temperature, nutrients, and losses of cells by sedimentation and other causes.

The combined effects of nutrient limitations and suboptimal temperatures on the growth of diatoms and green algae in cultures showed that the algae required more nutrients as temperatures declined from their optimal temperature range (Rhee and Gotham, 1980, 1981a). The optimum atomic ratio of nitrogen and phosphorus contained in the cells (i.e., the ratio at which one nutrient limitation shifts to another) increased at suboptimal temperatures. The combined effects of nutrient limitation and suboptimal temperature on growth were greater than the sum of the individual effects and were not multiplicative. The larger species, such as major spring diatoms, have generation times of about 4–8 days, which is somewhat longer than those of smaller species under natural conditions. The specific growth rates of the dominant species are considerably higher than those of the entire algal assemblage combined (e.g., Pechlaner, 1970). Continuous cultures simultaneously limited by light and nutrients also showed that the combined effects of these two factors on algal growth were greater than the sum of individual effects and that the effects were not multiplicative (Rhee and Gotham, 1981b). Under both nutrient-limited and nutrient-sufficient conditions, the cell-nutrient quota needed for growth increased as irradiance decreased, indicating that nutrient requirements increase as irradiance decreases.

It should be noted that cold- , low-light-adapted species of algae that are well developed under winter conditions can be severely limited in the spring or develop only in deeper strata in which temperatures and irradiance remain low. Population development in deep strata may result from actual migratory movement, as occurs among the dinoflagellates of alpine lakes, or the development of populations in lower strata through sedimentation of cells from shallower strata, as is the case among some ubiquitous cyanobacteria (e.g., *Oscillatoria rubescens*). Findenegg (1943), and others since, demonstrated the characteristic weak development of *Oscillatoria* in upper layers under winter conditions. Before, during, and after spring circulation, the *Oscillatoria* population weakens in the upper strata and increases in the deep, cold layers of the lower metalimnion and upper hypolimnion.

In high mountain lakes above the tree line, the ultraplankton (0.2–20 μm) dominate the phytoplankton and are autotrophic throughout the long winter (Pechlaner, 1971). These algae are adapted to the low light intensities that penetrate heavy ice and snow cover. Both the concentration of algae and rates of *in situ* photosynthesis shift from the surface layers in winter to deep water strata during spring and summer. Immediately before ice breakup the increasing light induces a downward migration of predominating dinoflagellates.

4. During the ***initial summer stratification,*** temperatures increase rapidly as the water column stability returns, light availability increases, and nutrient availability declines. Phytoplankton biomass declines precipitously in the beginning largely as a result of sedimentation of diatoms. Enhanced sedimentation, photoinhibition, and phosphorus limitation are factors that may slow net diatom accumulation before the onset of silicate limitation (Neale *et al.*, 1991a,b; Poulíčková, 1993). Species-specific differences in mortalities are also known. For example, high death rates of the diatom *Asterionella formosa* in Lake Constance, Germany, in late spring were associated with silicon depletion, whereas the decline of the diatom *Fragilaria crotonensis* was enhanced by fungal parasitism, as discussed later (Sommer, 1984). Rapid growth of small flagellate algae compensates for some of the community losses, but grazing by zooplankton increases rapidly.

5. In the early ***summer "clearwater" phase,*** algal populations decline precipitously under conditions of high temperatures, high water column stability, high light availability, and markedly reduced nutrient availability (Fig. 15-9). Herbivorous zooplankton increase from the hatching of resting stages and from the high fecundity associated with the relative abundance of edible algae. The planktonic herbivores with short generation times (rotifers and ciliates) increase their populations first and are followed by slower-growing species (cladocerans and copepods). Algae <40 μm are particularly vulnerable, and the grazing losses result in community filtration rates that exceed phytoplankton reproductive rates. This "clearwater" phase can result in accelerated microbial nutrient regeneration from zooplankton grazing and can persist for several weeks until less readily grazed and inedible algae develop to significant population sizes.

The decline of the spring maximum of phytoplankton and onset of summer populations in temperate lakes is associated with a complex interaction of physical and biotic parameters. In many straightforward cases, reduction of nutrients in the photic zone of the epilimnion is responsible for slowing the growth of populations of the dominant as well as rarer algae. Since diatoms are often the dominant component of the spring maximum, especially in temperate fresh waters, silica concentrations are often reduced after extensive growth. The reduction of silica concentrations to limiting levels (<0.5 mg liter^{-1}; cf. Kilham, 1975) has been shown repeatedly in many fresh waters but is especially well documented in the northern British lakes, in which maximum concentrations are low (2–3 mg liter^{-1}) (Fig. 15-11; see also Fig. 14-22). Reduction of silica to limiting concentrations occurs in less than two months when turbulence is low and sedimentation of diatom frustules occurs rapidly (Fig. 15-11).

In calcareous lakes in which silica concentrations are naturally high (>10 mg liter^{-1}), silica levels can still be reduced appreciably during the spring and summer in the epilimnion (e.g., Fig. 14-21, *upper*). Utilization by diatoms continues throughout the period of the summer stratification. As hardwater lakes become more productive, the rapidity of silica reduction increases. For example, in extremely eutrophic Wintergreen Lake (Fig. 14-21, *lower*), concentrations are reduced to and below generally accepted inhibiting levels in both winter and spring periods.

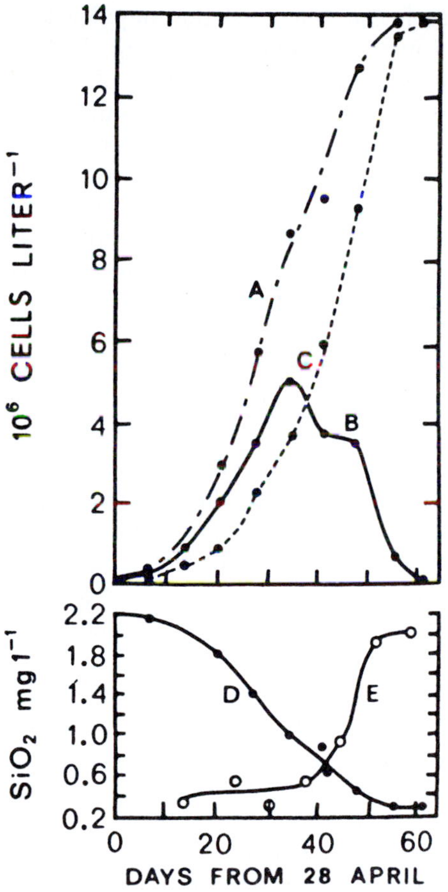

FIGURE 15-11 Production and loss of *Asterionella formosa* in the epilimnion of Lake Windermere, England, from 28 April to 30 June 1947. (A) Cumulative total of cell production computed from silica uptake; (B) mean concentration of cells in epilimnion; (C) cumulative loss of cells from the epilimnion (A minus B); (D) epilimnetic silica concentration; and (E) relative rates of loss of cells from the population. (Modified from Lund *et al.*, 1963.)

The correlations between the observed decline of the diatom maximum with diminishing silica concentrations and the experimental results using bioassays are so distinct that the relationship appears to be predominantly causal. Factors of light intensity, temperature, nitrogen, phosphorus, zooplankton grazing, and fungal parasitism contribute to but are not necessarily instrumental in the decline. Much more complex interactions of both physicochemical and biotic factors undoubtedly exist in the seasonal succession of diatom populations when silica concentrations are not reduced to levels limiting to exponential growth. For example, the decline of the dominant spring diatom *Asterionella formosa* of a small mesotrophic lake was correlated more with combined nitrogen limitation than with depletion of silica (Lehman and Sandgren, 1978; Lehman, 1979).

Much insight into the competitive interactions among dominant spring blooms of algae has been gained from detailed experimentation in cultures (the general methodology of competitive nutrient kinetic analyses is reviewed by Kilham, 1978). Nutrient kinetic parameters of two common spring diatoms, *Asterionella formosa* and *Cyclotella meneghiniana,* showed that maximal growth rates of the two species were not significantly different (Tilman and Kilham, 1976). *Asterionella* was able to grow significantly better, however, at low concentrations of extracellular phosphate, and *Cyclotella* grew better at low concentrations of silica. Using the resource-based competition models (Titman, 1976; Tilman, 1977, 1978) discussed earlier (p. 349), the outcome of nutrient competition between these two species showed that (a) under conditions in which both species were phosphate-limited, *Asterionella* would be dominant; (b) under conditions in which both species were silica-limited, *Cyclotella* would be the competitive dominant; and (c) under intermediate conditions in which the growth rates of each species were limited by different nutrients (*Asterionella* by silica and *Cyclotella* by phosphate), both species would coexist stably. The model was verified for these species both experimentally and in natural populations along nutrient gradients (Kilham and Titman, 1977; Stoermer *et al.,* 1978). These studies indicate that a species is growth-limited by only one nutrient at a time and that where two species are growth-limited by different nutrients simultaneously, they can coexist.

Gradients in nutrients and in the ratios of different nutrients exist both spatially (e.g., from nearshore to openwater areas) and seasonally when utilization or losses exceed inputs. Concentration of a nutrient is important to growth and phytoplanktonic community structure but so is the supply rate of one nutrient relative to other potentially limiting nutrients. Competitive dominance of species populations will develop along nutrient gradients according to the species' maximal growth rates, their half-maximal growth rates (half-saturation constants of nutrient uptake), and their species-specific mortality rates (Kilham and Kilham, 1978; Tilman, 1978, 1980; Kilham and Tilman, 1979). For example, *Asterionella* is a successful competitor at high Si/P ratios, *Fragilaria* can dominate at intermediate ratios, and *Stephanodiscus* grows well when Si/P ratios are low.

6. The ***latter part of summer stratification*** exhibits conditions of high temperatures, high water column stability but a progressive erosion and deepening of the metalimnion, high but declining light availability, and low nutrient availability. Herbivorous zooplankton become food-limited, and their collective biomass declines precipitously from decreased fecundity and greatly increased fish predation. The greatly reduced grazing pressures and increasing nutrient availabilities from processes such as erosion of the metalimnion and parts of the hypolimnion result in a more diverse phytoplanktonic community. Chrysophytes, cryptophytes, and colonial green algae frequently become abundant, and their growth often reduces availability of phosphorus and commonly also combined nitrogen (NO_3^- and NH_4^+) to limiting concentrations.

If silica is abundant, the green algae are often replaced by large diatoms. Commonly, however, silica levels are reduced in the trophogenic zone in late summer. Under these circumstances of silica depletion ($<$0.5 mg SiO_2^- liter^{-1}), dinoflagellates and cyanobacteria dominate (Fig. 15-9). As summer algal growth utilizes combined nitrogen to very low concentrations, filamentous cyanobacteria capable of fixing molecular nitrogen often dominate, a very common sequence in temperate lakes (cf. earlier discussion; Dokulil and Skolaut, 1986; many others). The ability of many cyanobacteria to migrate by buoyancy alterations (cf. p. 347) between lower depths of relative nutrient abundance and upper strata of light abundance (e.g., Gamf and Oliver, 1982), as well as the ability to fix molecular nitrogen from atmospheric sources, results in a superior competitive position in relation to other species of phytoplankton. Certain cyanobacterial species can be recruited from the sediments, particularly during midsummer. For example, a significant portion of the bloom-forming cyanobacterial populations of a small reservoir was gained by migrations from the sediments into the overlying waters (Trimbee and Harris, 1984).

In shallow lakes where inputs of nutrients, such as silica and/or combined nitrogen, from the sediments or external sources exceed microbial utilization, faster-growing chlorophytes or diatoms can become superior competitors compared with the relatively slow-growing cyanobacteria (Hecky *et al.*, 1986; Jensen *et al.*, 1994). Nutrient turnover rates and availability to biota can be very high in shallow lakes, even when inorganic nutrient concentrations are low and the pH high.

The spring maximum of phytoplanktonic biomass is generally short-lived, usually less than one to two months in duration. This maximum often is followed by a period of low algal numbers and biomass that may extend throughout the summer. Among more eutrophic lakes of the temperate region, the summer minimum is often brief and phases into a late summer profusion of cyanobacteria that persists into the autumn, until thermal stratification is disrupted. Populations of phytoplankton are often low throughout the summer in temperate oligotrophic lakes but may develop a second maximum in the autumn period (Fig. 15-9). This second maximum of the autumn, which often consists predominantly of diatoms, is generally not as strongly developed as that of the spring period (e.g., Diaz and Pedrozo, 1993). The decline of the autumnal populations into the winter minimum is often abrupt and more irregular than the decline after the spring development. The limited growing season of lakes of high altitudes and in polar regions often is reflected in a conspicuous, single, summer maximum of phytoplanktonic biomass.

As silica concentrations are reduced in productive lakes, diatom populations are often succeeded by a preponderance of first green algae and later cyanobacteria (e.g., Fig. 15-9). Growth in these eutrophic lakes can be so intense that combined nitrogen ($NO_3^-$and NH_4^+) sources are reduced to below detectable concentrations in the trophogenic zone (e.g., Wintergreen Lake, Chap. 12). When this happens, often by midsummer when the warmest epilimnetic temperatures occur, cyanobacteria with efficient capabilities for fixing molecular nitrogen have a competitive advantage and can predominate (cf. Chap. 12). These lakes require, as a general rule, a reasonably sustained and heavy loading of phosphorus (cf. e.g., Moss, 1973c).

Gradients in nutrient ratios are commonly observed during and after the spring diatom outburst in eutrophic lakes because phosphorus and nitrate concentrations decline at different rates (e.g., Pechlaner, 1970). As a result, the Si/P ratio increases and the N/P ratio decreases markedly. At high Si/P ratios, diatoms can effectively outcompete cyanobacteria (e.g., Holm and Armstrong, 1981). As concentrations of silica are reduced, and later combined nitrogen declines, green algae can compete less effectively with cyanobacteria that possess N_2-fixing capabilities.

Similar relationships have been demonstrated in many lakes (e.g., Reynolds, 1976) and occur on a much larger scale in the Laurentian Great Lakes (Schelske and Stoermer, 1971, 1972; Schelske *et al.*, 1972). Phytoplanktonic algae in the upper Great Lakes are distinctly phosphate-limited. As the loading of phosphorus to the lakes has increased, diatoms reduce the available silica (ambient concentrations about 2–3 mg liter^{-1}) to limiting levels more rapidly with each successive year. The diatoms then are replaced rapidly by green algae and cyanobacteria. Increased loading of trace metals, vitamins, and dissolved organic matter further accelerates the algal succession. Experimental additions of nitrogen and phosphorus to ponds also demonstrated the influence of these nutrients on phytoplankton succession; with nitrogen and phosphorus enrichment, chrysophyte species were replaced by green algae and cyanobacteria (DeNoyelles and O'Brien, 1978).

A series of papers by Moss (1972c, 1973a–c) is instructive in relation to both the seasonal succession and the distribution of phytoplankton. Changes in nutrient concentrations are just as important to seasonal periodicity within a lake as different nutrient concentrations among lakes are important in determining algal distribution. Working with 16 species of diverse but widely distributed phytoplanktonic algae, Moss found little effect of calcium ions above a concentration of 1 mg liter^{-1} among both oligotrophic and eutrophic species. No evidence was found to support the contention that oligotrophic desmids are calciphobic. Certain oligotrophic algae were unable to grow at pH values greater than 8.6, whereas eutrophic species able to utilize both CO_2 and bicarbonate grew well at pH values >9. Temperature optima for growth were similar to those discussed earlier: Diatoms have an extended range into lower temperatures, whereas green algae and cyanobacteria grow well at temperatures >15°C. Temperatures >15°C are common in many temperate waters throughout much of the summer and in tropical waters for the entire year. Temperature and nutrient interactions generally could be verified in experimental bioassays using a variety of natural lake waters ranging from oligotrophic to extremely eutrophic.

7. During ***fall circulation,*** the mixing water is characterized by progressively declining temperatures, low and decreasing light availability, declining grazing by zooplankton, and high nutrient availability from the redistribution from the hypolimnion. Large unicellular and filamentous algae, predominantly diatoms, dominate this long period (often 3–4 weeks) of mixing (Fig. 15-9).

8. During the ***late autumn decline,*** low temperatures, declining light availability, and low nutrient availability results in declining phytoplankton biomass, despite low grazing rates (Fig. 15-9).

C. Tropical Lakes

Tropical lakes are far less numerous than temperate lakes[6] and are dominated by lakes of river origin (Lewis, 1987, 1996). Greater solar irradiance with reduced annual variations result in higher minimum water temperatures. Although smaller thermal discontinuities exist between upper and lower water strata, density differences can be very large and result in high stabilities. The predominantly warm monomictic tropical lakes exhibit great regularity in seasonal mixing, which typically coincides with the hemispheric winter. Although stratification is seasonally persistent, it is less stable than in lakes at higher latitudes. As a result, intraseasonal variations in the mixing and thickness of the mixed strata are much greater than is the case for morphometrically similar temperate lakes (Lewis, 1973; Viner, 1985; Talling, 1986). The high mean temperatures and variable mixing and intrusions into the metalimnion during the stratified period result in high nutrient regeneration rates and, consequently, affect phytoplankton succession and productivity. These intermittent mixing episodes into the metalimnion disturb and reset the successional sequence. Although the successional sequence is essentially the same in tropical and in temperate lakes, more successional episodes occur per year in tropical lakes. On an annual basis, the maximum of phytoplanktonic biomass in tropical lakes often is observed during the winter.

In many tropical lakes surrounded by extensive, nutrient-buffering wetlands and receiving perennial river inflow, total phytoplanktonic biomass and productivity are often both larger and more constant seasonally than those found in temperate lakes. Abrupt changes in abiotic factors (e.g., wind-induced vertical mixing; marked seasonality in rainfall and associated nutrient loading and turbidity) can be more frequent in tropical lakes than in temperate lakes that are stratified for longer periods of time (Talling, 1969, 1986; Lewis, 1978a,b, 1996; Melack, 1979). Episodic changes are often very regular in their seasonality from year to year, and phytoplankton succession within any given episode is similar in tropical and in temperate lakes. Phytoplankton communities are no more complex in tropical lakes than in lakes of higher latitudes (Lewis, 1996).

[6]Some exceptions exist in small areas. For example, in the geologically turbulent and varied archipelago of Indonesia, containing over 11,000 islands, natural lakes are abundant, many of which are large and some are very deep (>250–500 m) (Lehmusluoto *et al.*, 1998).

Phytoplankton diversity is not markedly different than in temperate lakes.

D. Algal Succession and Productivity in Reservoirs

Within reservoirs the irregular dynamics of inflow and rapid, variable flushing rates markedly alter environmental conditions for biotic communities. A reservoir can be viewed as a very dynamic lake in which a significant portion of its volume possesses characteristics of and functions biologically as a river. Often, the riverine portion of reservoirs operates analogously to large, turbid rivers in which turbulence, sediment instability, high turbidity, reduced light availability, and other characteristics preclude extensive photosynthesis despite high nutrient availability (Fig. 15-12). Although the phytoplanktonic primary productivity of riverine sections of reservoirs can be high per unit water volume, the limited photic zone reduces areal productivity, as in large rivers (Wetzel, 1975; Minshall, 1978; Bott, 1983; Wetzel and Ward, 1992). This reduction is only partially ameliorated by turbulent, intermittent recirculation of algae into the photic zone. As turbidity is reduced and the depth of the photic zone increases in the progression through the transitional to the lacustrine reservoir zones, areal primary productivity increases with the resulting greater light penetration and depth of the trophogenic zone (Fig. 15-12). Nutrient limitations, so characteristic to the successional changes of natural lakes of low-to-moderate productivity, can then occur to varying degrees as utilization and losses of nutrients exceed renewal rates.

Light limitation is a dominant control of productivity in most reservoirs, as it is in many productive natural lakes and rivers. In many cases, light limitations in reservoirs result primarily from inorganic clay and silt turbidity. Where light limitations result from high loadings of dissolved organic matter, as in many tropical and subtropical reservoirs, the total and selective spectral light attenuations are quite analogous to those of natural lakes (cf. pp. 56–63).

The succession of phytoplankton communities of the lacustrine portion of reservoirs is generally similar to that found in natural lakes. Disturbance frequencies are much higher in reservoirs than in lakes, with rapid, often irregular and large, changes in water level, flushing rates, turbidity, and mixing (Wetzel, 1990b). In particular, the disturbances to stratification and loadings of inflowing silt that increase turbidity are dominant factors affecting phytoplankton periodicity and succession (e.g., Bettoli *et al.,* 1985; Duthie and Stout, 1986; Ramberg, 1987; Zohary *et al.,* 1996). Variations in flushing rates influence not only nutrient availabilities but also turbidity, and they can overrule the importance of other regulating factors, as is the case in shallow lakes (Bailey-Watts *et al.,* 1990). Green algae and planktonic diatoms are commonly dominant in summer during periods of short residence times, and cyanobacteria develop more strongly when water residence times are longer, such as during dry years (Lepš *et al.,* 1990; Perry *et al.,* 1990). As a result, the phytoplankton communities of reservoirs are disrupted often, which leads to increased diversity with the development of *r*-strategist species whose development and successional sequences are constantly being reset.

IX. MORTALITY OF PHYTOPLANKTON

Growth and productivity of phytoplankton populations are counterbalanced by losses by sedimentation out of the photic zone, viral and fungal parasitism, and grazing by zooplankton. As already discussed in a general way, these mortality losses couple with direct growth factors of temperature, light, and nutrients to determine succession and rates of productivity. Rates of algal sedimentation have already been discussed in relation to environmental factors influencing sinking rates of different groups.

A. Viral, Bacterial, Protozoan, and Fungal Pathogens

Viral pathogens of phytoplankton can lyse many different types of phytoplanktonic algae and cyanobacteria (Saffermann and Morris, 1963; Shilo, 1971; Suttle *et al.,* 1990). Viral particles, in the size range of 0.002–0.2 μm, reduce rates of primary production by as much as 78%. Preliminary data suggest that about 20% of heterotrophic bacteria are infected by viruses and 10–40% of the bacterial community is lysed daily by viruses (Suttle, 1994; Mathias *et al.,* 1995; Wommack and Colwell, 2000). The percentage of eukaryotic phytoplankton lost to viral lysis is likely lower on a daily basis. A diurnal periodicity in viral infectivity is likely if the viruses are exposed to natural sunlight. Viral decay rates increased markedly when exposed to sunlight (Suttle and Chen, 1992).

Viruses exhibit host specificity, and this quality suggested potential application for control of cyanobacterial blooms. Attempts to exploit this specificity for control of cyanobacterial blooms have not been very successful (Jackson and Sládeček, 1970). Cyanophage-resistant mutants can develop quickly among the cyanobacteria. Viral infections of natural blooms of cyanobacteria have apparently led to significant reductions of phytoplankton densities (Daft *et al.,* 1970; Coulombe and Robinson, 1981).

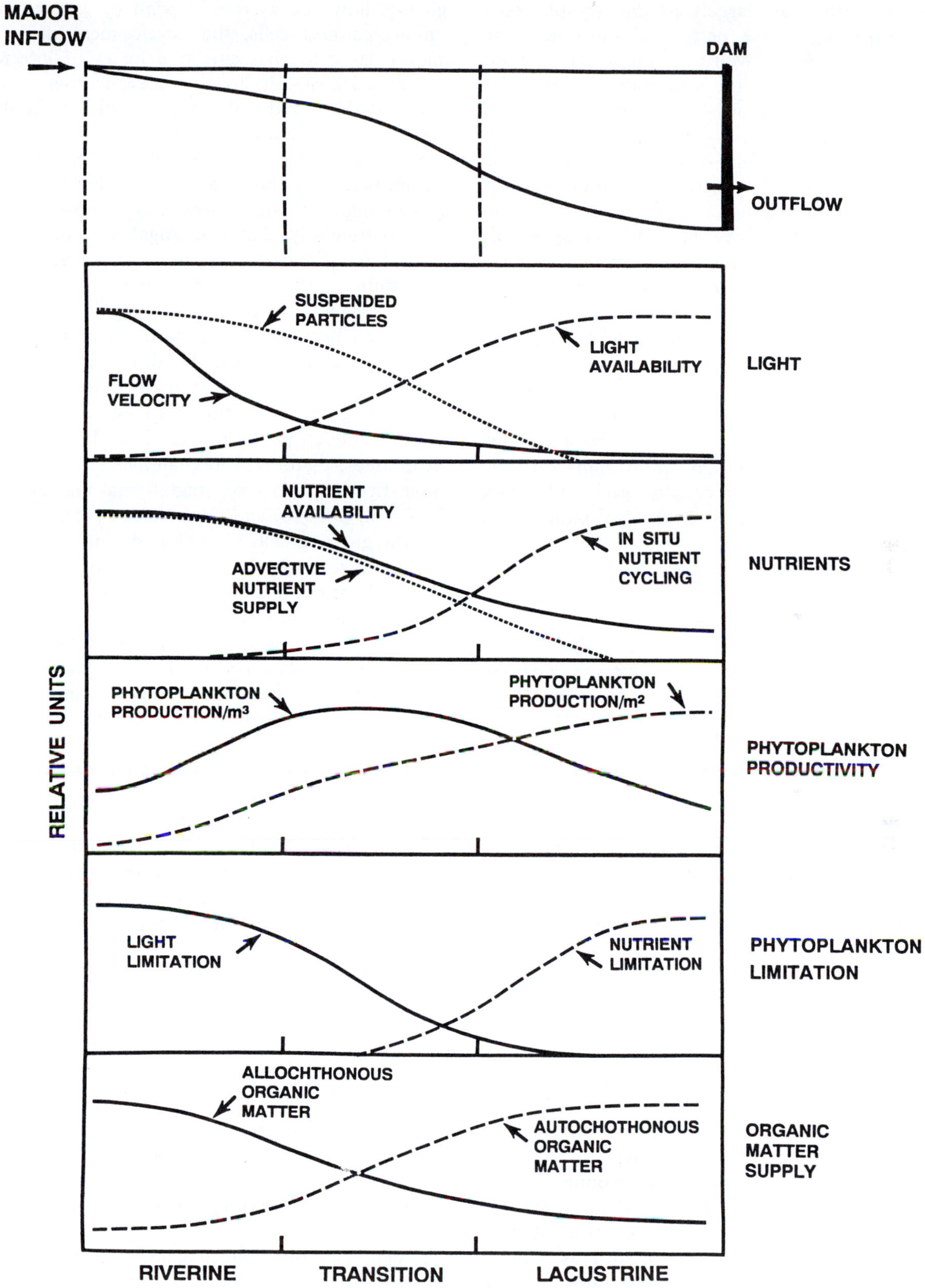

FIGURE 15-12 Commonly observed shifts in flow, turbidity, light, nutrients, phytoplanktonic productivity, and sources of organic matter in the progression through the three major zones of reservoirs. (Modified from Kimmel *et al.*, 1990.)

Bacterial pathogens, largely of the Myxobacteriales, of phytoplankton lyse many species of unicellular and filamentous algae as well as cyanobacteria (Shilo, 1970). Little is known of the significance of bacterial pathogens in the regulation of algal growth, production, and succession.

Many protozoans and related protistian organisms feed directly on algae and bacteria. These processes will be discussed with numerous examples in the following chapter. Some protozoa, however, feed upon algae with parasitic characteristics. Some species of protozoa attach to algae and extract the contents with filiform appendages or digest the contents while remaining outside the cells (Lund, 1965; Canter, 1973; Brabrand *et al.*, 1983). Certain herbivorous protozoans feeding on small colonial algae have been shown to decimate over 99% of certain chlorophycean planktonic algal populations over short periods (7–14 days) (Canter and Lund, 1968). Some of these protozoans, such as *Pseudospora*, consume only certain species of algae. The effects on seasonal succession of algal populations are obvious.

Many algal species are parasitized by aquatic fungi, mostly of the uniflagellate Chytridiales (Chytridiomycetes) that are external parasites and the biflagellate Oomycetes that are internal parasites. In the free-swimming stage of their life cycle, zoospores disperse widely (Canter, 1979; Canter and Heaney, 1984; van Donk, 1989; van Donk and Bruning, 1995). Upon the settling of chytrid fungi on an algal cell, the flagellum penetrates the cell wall and forms an internal rhizoid through which organic contents of the alga are extracted. The fungus then enlarges as a sporangium, which upon maturation produces and releases zoospores. The sporangia of biflagellate fungi are located within the algal host cells and when mature pass from the alga for final zoospore maturation.

Fungal parasitism can influence phytoplankton succession by suppression of maximum population development under the existing environmental conditions. Because most fungal parasites are host specific, infection of one algal species can favor successional development of other competing species (van Donk, 1989). Differences in the composition or amount of extracellular algal exudates likely influence the susceptibility of an algal population to fungal infection by zoospores. However, a number of environmental factors influence both the development of the fungi and their reproductive rates in relation to those of algal host species. Environmental factors, such as light or nutrients, that limit algal reproductive rates may enhance development of a fungal epidemic, as long as the same factors do not limit fungal development (Bruning and Ringelberg, 1987). Losses of host cells can vary greatly, however, as they depend on growth rates of uninfected host cells, the development times of the sporangia of the parasite, and losses of infected and uninfected host cells from processes other than parasitism (Kudoh and Takahashi, 1990, 1992; Bruning, 1991a,b). Temperature variations markedly influence the occurrence and success of phytoplankton parasitism, because at low temperatures development times of sporangia are much slower than those of many algae, particularly diatoms. Algal infection by some chytrid fungi occurs only at low temperatures (Blinn and Button, 1973). Light limitations to host algae could enhance the success of fungal parasitism (van Donk and Bruning, 1995). Nutrient limitations of the algal host could reduce the threshold host densities needed for the development of an epidemic.

Among the best studies of chytrid fungal parasitism, centered in the English Lake District, infection of desmids, diatoms, and cyanobacteria is fairly common. Even though percentage fungal infections can be very high (>70%) and chytrids can parasitize healthy, rapidly growing cells as well as declining senescent algae, parasitism usually does not greatly alter the overall seasonal pattern of periodicity of diatoms and desmids (Canter and Lund, 1948, 1969). Fungal parasitism, however, can have a significant effect on interspecific competition of algae because of the degree of specificity in the algae attacked. Reduction of dominant desmids apparently is influenced to a greater extent by parasitism, whereas the diatom decline, particularly of *Asterionella,* results from a combination of reduced nutrient and parasite interactions. Certain diatom clones react hypersensitively to chytrid parasites, which results in both the death of the host alga and the fungi and causes the demise of the parasite population (Canter and Jaworski, 1979). Parasitism likely increases in eutrophic waters, but little is known of its quantitative significance.

B. Grazing by Zooplankton

Grazing of phytoplankton by metazoan animals, particularly by the microcrustacea, can be a significant factor in the decline of algal populations under certain conditions. As a result, selective feeding contributes to seasonal succession of phytoplankton. Although most of this discussion will be deferred to the subsequent chapter on zooplankton, a few points are appropriate here.

Small phototrophic (e.g., *Cryptomonas*) and heterotrophic microflagellates in the nanoplankton size range (2–20 μm) graze on free-living bacteria and picophytoplankton (cf. many reports cited in Sanders *et al.*, 1985; Stockner and Antia, 1986). Microflagellates are likely major grazers of picophytoplankton and bacteria (cf. Chap. 17). Many ciliates and other microzooplankton (20–200 μm) are far less efficient than flagellates and certain small ciliates in utilization of picoplankton. Although small stages of macrozooplankton can utilize algal picoplankton and some bacteria, such grazing is much less efficient than on larger nanoplankton prey.

Phytoplankton mortality and selective changes in species composition of the phytoplankton algal and cyanobacterial community are highly variable seasonally and spatially within lakes and rivers. As temperatures of temperate waters increase in the spring, reproduction and feeding of zooplankton increase markedly. In some cases, the population and grazing maxima of zooplankton coincide with the decline of algal maxima (e.g., Crumpton and Wetzel, 1982). In other cases, inverse correlations between the two processes are found, or changes in zooplankton populations correlated better with bacterial or detrital concentrations than with those of phytoplankton. Although certain studies indicated that at a few times of the year microcrustaceans could theoretically graze the total volume of the trophogenic zone in a day (cf. Chap. 16), much lower rates (3–25%) are more usual. Zooplankton grazing can have positive effects of increased growth rates of algae by means of nutrient regeneration that could potentially offset the negative effects of grazing mortality (e.g., Elser *et al.*, 1987). A degree of size and species selectivity also exists among the grazing zooplankton. Such selectivity can lead to a competitive advantage by less effectively grazed algal species (often the largest and less edible) that tend to be not or less nutrient-limited (Havens and Costa, 1985; Lehman and Sandgren, 1985). Algal succession can thus be altered by this selectivity within prevailing physical and nutrient availability constraints. In this way, invertebrates and fish possessing size-selective feeding habits on zooplankton can influence zooplankton grazing effectiveness and, in turn, algal succession (e.g., Lampert, 1978; Sterner, 1989).

An example of seasonal succession of phytoplankton and the interaction of growth rates and mortality is seen in detailed studies of a small, oligotrophic temperate lake (Crumpton and Wetzel, 1982). Weekly measurements were made of growth rates, sedimentation rates, and population densities for each of the dominant phytoplankton species and of diurnal zooplankton grazing rates on the algae. The diatom *Cyclotella michiganiana* was the dominant alga through the end of June, at which time *C. comensis* began to increase and became dominant by August (Fig. 15-13). The combined effect of greater growth rates and lower loss rates due to sedimentation of *C. comensis* resulted in its dominance over *C. michiganiana*. In late August, high grazing pressure by herbivorous zooplankton caused rapid declines of both *Cyclotella* species, which were succeeded by the larger green alga *Sphaerocystis schroeteri*. The *Sphaerocystis* species was too large to

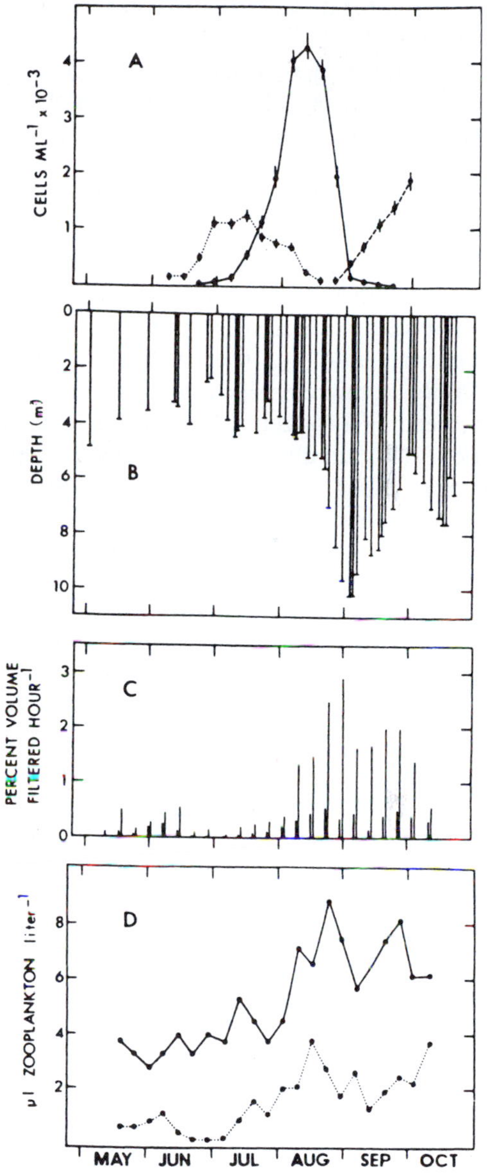

FIGURE 15-13 Dominant phytoplankton, water transparency, zooplankton grazing rates, and zooplankton biomass, Lawrence Lake, Michigan, 1979. (A) Population curves for *Cyclotella michiganiana* (- - - - - -), *Cyclotella comensis* (——), and *Sphaerocystis schroeteri* (- - - -). Error bars denote 95% confidence limits. (B) Transparency (Secchi depth) in meters. (C) Grazing rates as percentage of epilimnion filtered per hour. The first of each pair of bars represents daytime rates; the second represents nighttime rates. (D) Daytime (- - - - - -) and nighttime (——) epilimnetic zooplankton biomass. Dominant zooplankton were copepods (>90%) from May through July and *Daphnia* spp. (>90%) from August through October. (From Crumpton and Wetzel, 1982.)

be fed upon by the zooplankton dominating at that time. The *C. comensis–S. schroeteri* succession clearly resulted from differential mortality rather than from differences in growth rates.

Organic substances released by grazing cladocerans, such as certain *Daphnia*, copepods, and rotifers, can induce rapid morphological changes in certain algae. For example, dissolved substances released from some, but not all, species of *Daphnia*, copepods, and rotifers can induce the formation of large colonies in the polymorphic green alga *Scenedesmus* (Hessen and van Donk, 1993; Lürling and van Donk, 1996, 1997). The substances promoted a shift from unicells ($6-8 \times 4-5$ μm) to 8-cell colonies (40×6 μm), a form that is more grazer resistant and resulted in decreased grazing rates and population growth rates of zooplankton.

Some phytoplankton contain toxic organic substances that selectively alter or inhibit grazing by zooplankton. For example, the cyanobacterium *Planktothrix* contains microcystins and other toxins that are complete feeding deterrents for certain copepods and partial deterrents against grazing by *Daphnia* and other zooplankton (Kurmayer and Jüttner, 1999). Clearly, grazing deterrents can be mediated by chemical defenses rather than only by size and rigidity of the filaments.

Consumption of phytoplankton by zooplankton does not necessarily result in mortality. Some algae are resistant to zooplanktonic digestion and pass intact through the digestive tracts. Algal growth can be stimulated by the acquisition of nutrients or organic growth factors during passage (Porter, 1973, 1975, 1976; Adrian, 1987). The effectiveness of digestion during gut passage can differ depending upon the physiological status of the algae prior to ingestion (van Donk and Hessen, 1993). For example, phosphorus-saturated algae were efficiently digested by *Daphnia*, but P-limited algae passed largely intact through the gut and were likely nutrient enriched.

X. COMPETITIVE INTERACTIONS AND SUCCESSIONAL DIVERSITY

The phytoplanktonic community is composed of a diverse assemblage of algal species populations. Each of these species components and its growth dynamics are influenced by an array of environmental parameters (physical, chemical, and biotic) that undergo constant temporal and spatial variations. Some of the parameters, such as temperature, conservative nutrients, et cetera, undergo slow, periodic oscillations over an annual period. Other parameters, such as light intensity and quality, nutrients in rapid flux, et cetera, exhibit very rapid temporal, physiologically important changes in a matter of minutes or hours. The intensities of these forces greatly affect algal physiology and growth, some with regularity and others in a highly irregular fashion.

It is doubtful that an ideal species hypervolume is ever met under natural conditions. The fit of a species within an array of competing species is relative and changes constantly. The ability of an alga to store a critical nutrient imposes another time factor interaction that stresses the importance of past environmental conditions as well as contemporary ones. On the basis of the number of interacting parameters and ranges of variations tolerated by algae, it is not too surprising that (a) a species may be able to survive indefinitely provided that a suitable combination occurs with sufficient frequency and duration, and (b) the interactions are so complex and variable that, in spite of species overlap in their niche hypervolumes, coexistence can occur even though competitive interactions are in effect. Mathematical formulations of these competitive interactions are set forth in the stimulating theoretical discussions of Greeney *et al.* (1973) and Tilman (1980, 1982).

The diversity of phytoplanktonic algal species has been determined in a number of ways. The best indices of species diversity are probably those that are largely independent of sample size (cf. Chap. 8).[7] Diversity indices are often determined primarily by the proportions of the more common species (equitability) and only secondarily by the number of species, according to derivatives of the Shannon formula (Sager and Hasler, 1969; Moss, 1973d; Hallegraeff and Ringelberg, 1978). Predominance of one or two species results in low diversity values; high values occur when populations of several species each form moderate proportions of the whole. Although data are far from satisfactory because of the variations found, there is a general tendency for species diversity to decrease with increasing fertility of the water. Presumably, the slower growth rates of algae in oligotrophic lakes permit a greater number of species with reasonably similar requirements (high degree of niche overlap) to coexist than would be found in more eutrophic waters.

In eutrophic temperate lakes, the tendency is for algal diversity to increase in summer and decline during winter (Moss, 1973d). However, when the relationship between species diversity and season is examined closely, one can find considerable variability (cf. Hallegraeff and Ringelberg, 1978). Nutrient availability and water temperature may account for much of the

[7]Increasing sample size tends to reveal a greater number of species, as more and more of the rarer species are encountered. Many of the rarer species are of littoral-zone origin and not true members of the phytoplankton (cf. Symons, 1972; Moss, 1973d). Algal species diversity is much greater, commonly by a factor of 10, in the littoral zone than it is in pelagic areas.

observed pattern; during and immediately after spring circulation, nutrient concentrations are high, and progressive depletion of nutrients usually occurs throughout the summer. Relatively few algal species are adapted to the low water temperatures found immediately after ice loss in spring, and these species rapidly come to dominate the algal community. During summer, warmer temperatures are more favorable for growth of a larger number of species but competition for limiting nutrients is likely to increase. Obviously, factors such as parasitism and predation are also involved, but the extent of influence of each of these factors on the overall pattern of succession is variable and poorly studied.

Congruous with these findings are the results of a measure of successional rate, or a seasonal rate of change of species composition. Jassby and Goldman (1974), employing a mathematically more involved but more realistic expression, showed that the rate of change in species composition per unit time decreases precipitously following the spring maximum and midsummer plateaus. Thus, when the community is disturbed by the rapid changes of spring circulation, as in this case, or is perturbed by some catastrophic event (e.g., fertilization by human activities), one would expect rapid changes in species composition and increased successional rates.

The distribution of phytoplanktonic biomass and *in situ* rates of primary productivity has received a great deal of attention and represents perhaps the most thoroughly investigated aspect of lake biota. In contrast, the growth and productivity of phytoplankton of streams and rivers is poorly understood (cf. reviews of Hynes, 1970; Wetzel, 1975b; and Wetzel and Ward, 1992).

The distribution patterns of phytoplanktonic biomass within a lake or river usually change markedly throughout the year, and site-to-site variations are also extreme. Phytoplanktonic productivity is even more variable, for it is influenced by factors that can change within a matter of minutes (e.g., light intensity). However, in spite of these differences, there are some aspects of the distribution patterns that appear to be consistent. In addition, the productive capacity of a planktonic system is bounded by certain constraints beyond which productivity cannot be increased. These considerations are addressed in the following discussion.

XI. PHYTOPLANKTON IN THE GRADIENT ALONG RIVERS, RESERVOIRS, AND LAKES: DIVERSITY AND BIOMASS

A. River Phytoplankton

Many species of river phytoplankton (potamoplankton) reproduce prolifically in rivers and achieve biomass levels of 250 μg chlorophyll *a* liter^{-1} (Friedrich and Viehweg, 1984; Gliwicz *et al.*, 1985; Reynolds, 1988, 1994; Moss and Balls, 1989; Murakami *et al.*, 1992; Reynolds *et al.*, 1994). Diatoms usually dominate in the plankton of larger rivers along with, particularly in summer, a variety of green algae. Many flagellates, chrysophytes, and cyanobacteria are suppressed where currents are vigorous; these groups can increase in significance in areas where currents are reduced. Water movements enhance gaseous and solute exchanges as well as suspension of nonmotile cells. Perenniation of phytoplankton in rivers arises from surviving periphytic and benthic populations, often of backwater areas (Reynolds and Descy, 1996). Thus, much of the phytoplankton community likely derives from attached forms of the substrata (meroplankton) in various stages of growth or dormancy. Alternatively, lake surface water flows to outlets of streams and rivers can serve as a major source of riverine phytoplankton (e.g., Chandler, 1937). Mortality of entering phytoplankton can be high, however, if molar action of rapidly moving water is high.

Successful phytoplankton species in rivers must be adapted, however, to survive frequent fluctuations in irradiance as they are moved by currents erratically through their vertical light field. Often much of the mixed layer is in effective darkness and entrained cells spend more than half their daylight hours at light levels insufficient to support net photosynthesis. Increases in discharge commonly are accompanied by increases in the suspended load of nonliving particulates (clay, silt, and organic particles) derived from the drainage basin as well as within the river from scour. The depth of the euphotic zone within rivers from nonalgal particles decreases sharply as discharge increases (Reynolds, 1988). Under high flow conditions, energy limitations can be as important in regulating the growth of river phytoplankton as is the putative effects of washout downstream. When river ecosystems are disturbed or polluted, reduction or loss of algal diversity is common, particularly through the loss of rare species (Patrick, 1988). Loss of the less common species reduces the resilience of the ecosystem to recover from both natural and human-caused disturbances.

River phytoplankton generally exhibit fast growth rates and hence tend to be smaller in size than phytoplankton of lakes and reservoirs. The generation times of river phytoplankton must be more rapid than the time of displacement downstream. Survival in rivers is aided by a number of water-retentive mechanisms, particularly backwater "dead zones" of reduced currents near and among the banks and flood plains. These backwater areas retain water sufficiently to permit additional cell divisions of suspended phytoplankton. For example, phytoplankton abundance per unit phosphorus did not differ greatly among rivers, reservoirs, and natural lakes when residence times of the water were

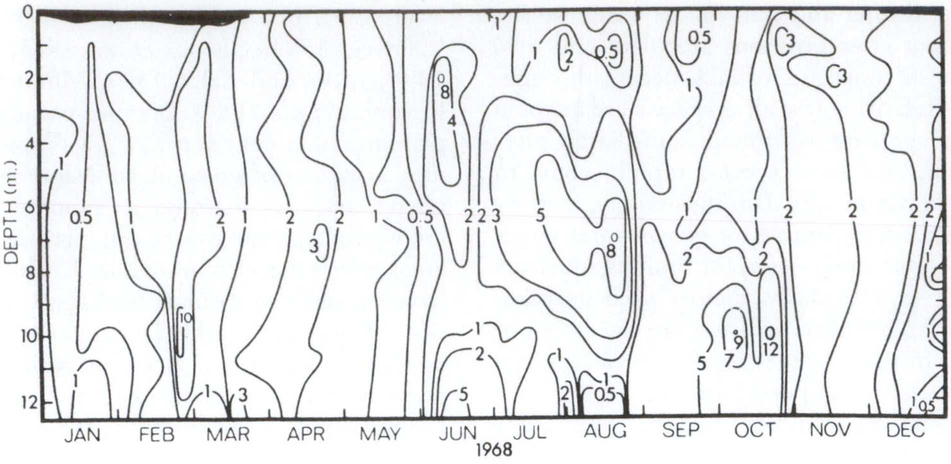

FIGURE 15-14 Depth–time distribution of phytoplanktonic chlorophyll *a* concentrations (mg m^{-3}), Lawrence Lake, Michigan, 1968. Opaque areas = ice cover to scale. Sampled at 7-day intervals at each meter of depth. (From Wetzel, unpublished data.)

similar (Søballe and Kimmel, 1987). In addition, zooplankton in rivers tend to be significantly fewer than in lakes, which would result in less efficient herbivory of phytoplankton in river ecosystems.

B. Vertical Distribution of Biomass in Lakes and Reservoirs

The vertical distribution of phytoplanktonic biomass varies greatly from season to season with shifts in species composition. If the depth–time distribution of phytoplanktonic pigments of an oligotrophic hardwater lake of southern Michigan is taken as an example, conspicuous temporal changes can be seen. In the seasonal depth distribution of chlorophyll *a* concentrations (Fig. 15-14), corrected for pigment degradation products, values were low in winter, increased conspicuously in the spring maximum, and then increased in different strata during the period of summer stratification. Detailed analyses of the phytoplankton community (some 150 major species) revealed that cryptomonads and small green algae predominated in the winter (Taylor and Wetzel, 1988). The composition then changed to predominantly diatoms in the spring and green algae in June and late August during this particular year. The metalimnetic maximum in summer was largely composed of green and non–nitrogen-fixing cyanobacterial populations with lesser percentages of diatoms. The development of a deep chlorophyll maximum (10–11 m) in the later portion of summer stratification was related to very dense populations of euglenoid algae just at the aerobic–anaerobic interface about 1 m above the sediments. Concentrations of chlorophyll *b* were very much lower throughout the year and were irregularly distributed, except in the lower hypolimnion during late summer and autumn. Chlorophyll *c* concentrations were most abundant in algal populations dominated by diatoms. The total composite plant carotenoids, which are somewhat more stable than the chlorophyllous pigments, exhibited a vertical depth distribution seasonally similar to that of chlorophyll *a* (Fig. 15-15).

Pigments must be used with caution as estimates of phytoplanktonic biomass (e.g., Cullen, 1982; Granberg and Harjula, 1982). It is also important to emphasize that a significant portion of the algal cells within the water column is nonviable, that is, is particulate detritus in various stages of decomposition. Chlorophyllous pigments can be corrected for degradation products (e.g., Wetzel and Likens, 2000). Concentrations of these heterogeneous phaeopigments give some insight into the magnitude of nonfunctional pigments in the particulate fractions. Using the same data for Lawrence Lake, Figure 15-16 sets forth the concentrations of phaeopigments with depth over the year. At many times of the year, the phaeopigment concentrations equal or exceed those of chlorophyll *a*. Moreover, the phaeopigment concentrations are often displaced to slightly greater depths and to slightly later times in relation to the functional pigments. The importance of this detrital particulate matter of algal origin to lake metabolism is discussed at length later (Chap. 23). It is important to realize that cell integrity and pigment concentrations can persist for several days after cell death (e.g., Gusev and Nikitina, 1974). Chlorophyll *a* degrades most rapidly upon death and when corrected for more stable phaeopigment degradation products, is an estimate of biomass (cf. Desortová, 1981).

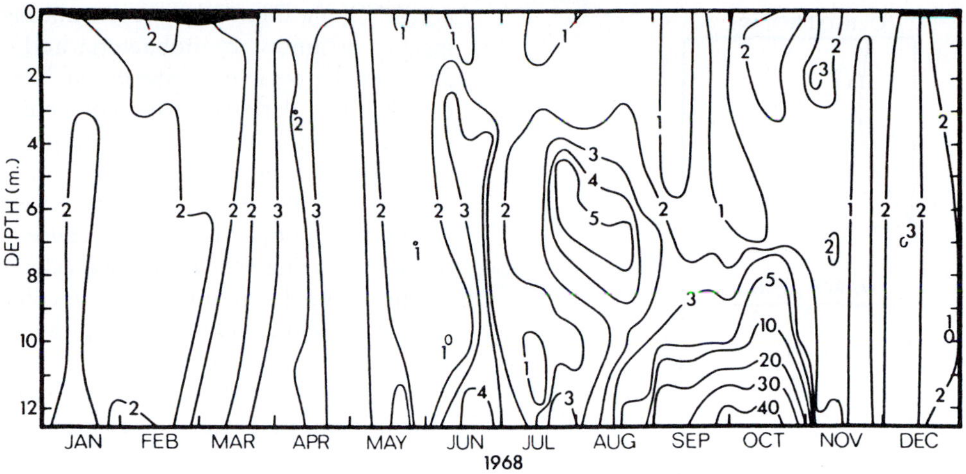

FIGURE 15-15 Depth–time distribution of phytoplanktonic total plant carotenoids (ca. mg m^{-3}), Lawrence Lake, Michigan, 1968. Sampld at meter depths at 7 day intervals; opaque areas = ice cover to scale. (From Wetzel, unpublished data.)

XII. VERTICAL DISTRIBUTION AND MAXIMUM GROWTH IN LAKES AND RESERVOIRS

The pigment–biomass values just discussed are low, which is typical of a rather oligotrophic lake. Considerably lower concentrations, about 1 mg chlorophyll *a* m^{-3}, are not unusual among pristine arctic and alpine oligotrophic lakes. Among alpine oligotrophic lakes and larger oligotrophic lakes in general, high concentrations of chlorophyll are commonly found deep into the upper hypolimnion and into aphotic zones (e.g., Brook and Torke, 1977; Tilzer *et al.,* 1977; Richerson *et al.,* 1978). Analyses have shown that many of the phytoplankton in these deep layers are adapted to low light intensities and that, during circulation periods, aphotic algae can augment phytoplankton growth of the euphotic zone upon reentry into the upper waters. For example, phytoplankton populations can develop in deep waters of some oligotrophic lakes immediately after the ice-free stratification period forms (e.g., Priscu and Goldman, 1983). These photoautotrophic algal communities form a "deep chlorophyll layer" that can persist for months until fall circulation. Redistribution of these algae during the autumnal overturn can

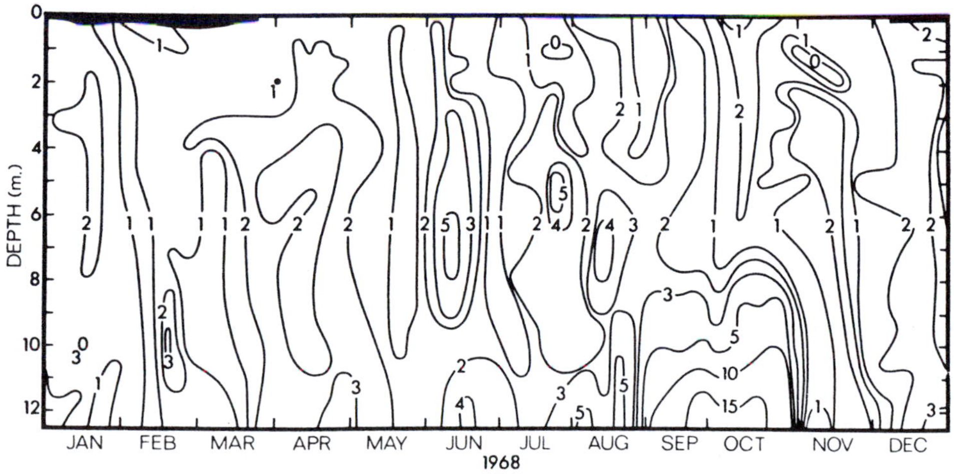

FIGURE 15-16 Depth–time distribution of phytoplanktonic phaeopigment concentrations (mg m^{-3}), Lawrence Lake, Michigan, 1968. Sampled at meter depths at 7-day intervals; opaque areas = ice cover to scale. (From Wetzel, unpublished data.)

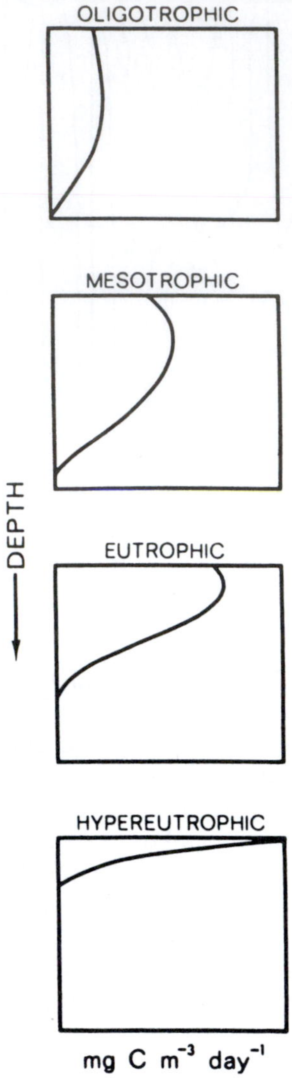

FIGURE 15-17 Generalized increases in productivity of phytoplankton per unit volume of water, with simultaneous reduction of the thickness of the trophogenic zone, in a series of lakes of increasingly greater fertility.

increase algal biomass concentrations of the euphotic zone and potentially increase primary productivity in this zone significantly.

At the other extreme of maximum growth and pigment concentrations, light limitations restrict productivity and the masses of the algae themselves impose controls upon further growth by self-shading. Photosynthetic yield can continue to increase, given adequate nutrient availability, to a theoretical point at which total absorption of photosynthetically active incident radiation occurs. This absorption is distinct from that absorbed by the water itself, by dissolved organic matter, and by abiogenic particulate matter (cf. Talling, 1960, 1971). The theoretical maximum, as indicated by culture analyses, is approximately 1 g chlorophyll per m^3 (1 mg $liter^{-1}$) but varies among algal types, being higher among cyanobacteria and less for green algae and diatoms (e.g., Steemann Nielsen, 1962b). As the algal populations increase biogenic turbidity in progressively more fertile lakes, the productivity per volume of water increases greatly (Fig. 15-17). However, the productivity per square meter of water column does not increase proportionately, because biogenic turbidity of the increasing density of algal cells reduces light penetration. The depth of the trophogenic zone and, hence, the total volume occupied by photosynthesizing algae are consequently reduced.

Therefore, as the composite extinction coefficient of light from biogenic sources increases, the productivity per square meter of water column increases to the maximum imposed by self-shading effects of the algal densities (Fig. 15-18). Increasing the extinction coefficient further, such as by high concentrations of dissolved humic compounds or inorganic particulate and colloidal turbidity, can similarly result in a progressive depression of the maximum possible productivity of phytoplankton.

The highest recorded concentrations of chlorophyll of phytoplanktonic origin emanate from tropical, eutrophic lakes. For example, shallow (2.5 m) but large (250 km^2) equatorial Lake George of western Uganda is a highly productive lake that supports very dense phytoplanktonic populations; these populations consist primarily of cyanobacteria, with very little seasonal change in either species composition or densities. Based on *in situ* estimates of photosynthesis, the predicted concentrations of chlorophyll *a* in this lake stabilized at an algal density of about 500 mg chlorophyll *a* m^{-2} (Ganf, 1974a). Because of the shallow nature of this large lake, frequent daily mixing, and disturbance of the flocculent sediments, light penetration is reduced further by nonalgal components. Therefore, the maximum quantity of chlorophyll *a* actually measured within the euphotic zone is between 230 and 310 mg m^{-2}. However, many of the phytoplankton settling to the sediments are viable and are resuspended each day (Ganf, 1974b), so a more realistic mean algal biomass may be as high as 1000 mg chlorophyll *a* m^{-2}.

Among the maximum recorded values of algal biomass on the basis of chlorophyll content are those from two Ethiopian soda lakes (Talling *et al.*, 1973). In these extremely productive lakes, favored by high temperatures and abundant light, phosphorus, carbon, and other nutrients, the euphoric zone was less than 0.6 m in thickness and the algal populations were limited severely by self-shading. Although concentrations of chlorophyll *a* exceeded 2000 mg m^{-3} in one lake, which appears to be about the maximum possible under natural planktonic conditions (Fig. 15-19), the content of chlorophyll *a* per unit area of the euphotic

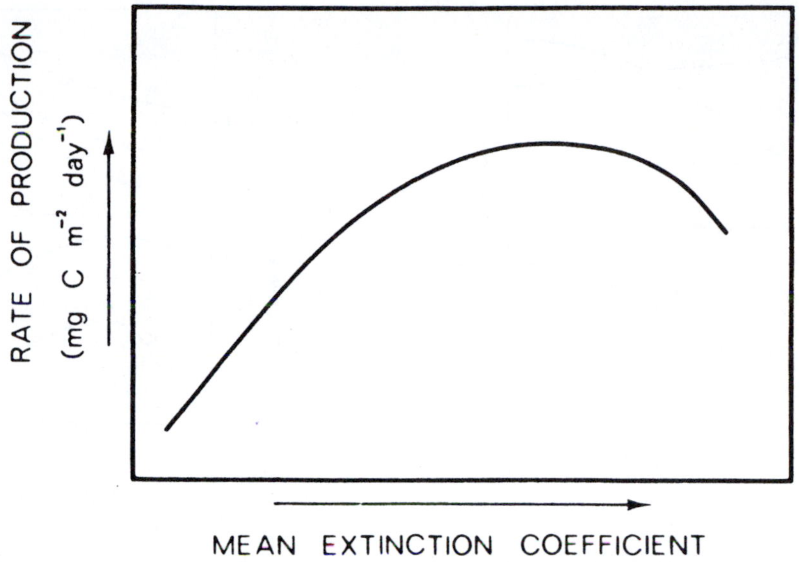

FIGURE 15-18 Generalized maximum possible productivity of phytoplankton in relation to increasing extinction of light, resulting from biogenic turbidity initially, and then combined with increasing effects of extinction by dissolved organic matter and/or particulate and colloidal inorganic matter.

trophogenic zone for these lakes was in the range of 180–325 mg m^{-2}. These values are similar to those maxima (180–450 mg chlorophyll *a* m^{-2}) that have been estimated indirectly or directly to occur in nature (cf. Westlake *et al.*, 1980).

XIII. PRIMARY PRODUCTION OF PHYTOPLANKTON

A. River Phytoplankton

Much of the control of observed primary productivity within running waters has been attributed to the availability of solar insolation reaching the water and its attenuation within the water (e.g., Wetzel, 1975a; Minshall, 1978; Bott, 1983; Wetzel and Ward, 1992; Reynolds and Descy, 1996). Nutrient loadings generally increase as the sizes of the drainage basins and stream order increase, particularly among most river valleys that have been disturbed by agriculture. As a result, nutrient limitations to primary productivity of rivers generally decrease with increasing stream order because of light limitations.

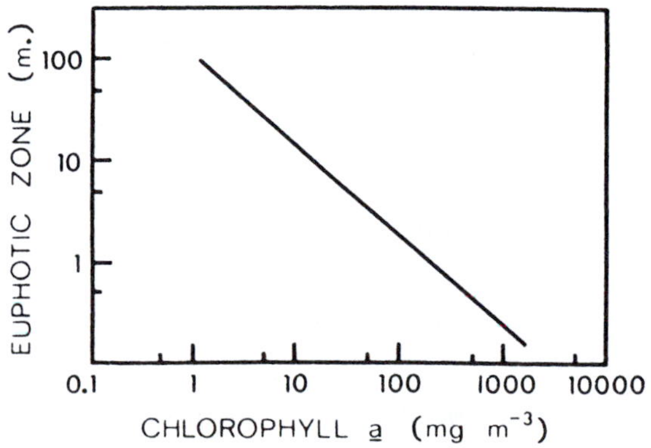

FIGURE 15-19 Generalized relationship of the depth of the euphotic zone in relation to the maximum concentrations of chlorophyll *a* per volume of water. (After data of several authors and Talling *et al.*, 1973.)

Once nutrients are provided from external and internal sources in adequate supply to support maximum rates of photosynthesis, light availability is the dominant factor regulating photosynthetic productivity. Light available *within* river channels is regulated by both the canopy cover of bankside trees and by turbidity (Fig. 15-20). Light availability is often low in small streams and rivers, at least seasonally during the active growing season of streamside vegetation, because of extensive vegetative canopy. Undisturbed rivers usually have low loads of turbidity and high water transparency as size increases downgradient (Fig. 15-20a). Unfortunately, marked alterations in the light quality of most river ecosystems have occurred as land clearance, agriculture, and other disturbances have increased (Fig. 15-20b).

Among relatively nondisturbed ecosystems, primary productivity *within* rivers is directly correlated to stream order (Fig. 15-21a). Much of the increase in stream autotrophy results from increases in production by periphytic algae associated with widespread deposition of sediment surfaces and expansion of shallow substrata with river braiding and meandering. As river ecosystems are disturbed, variance from this general

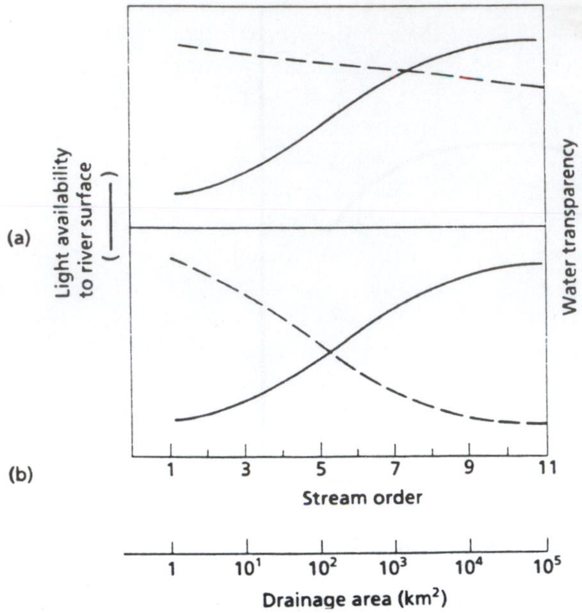

FIGURE 15-20 Changes in patterns of light penetration to the stream surface and water transparency of the stream water along a gradient of increasing stream order and area of drainage basin in (a) preagricultural nondisturbed and (b) present/disturbed lotic ecosystems. (From Wetzel and Ward, 1992.)

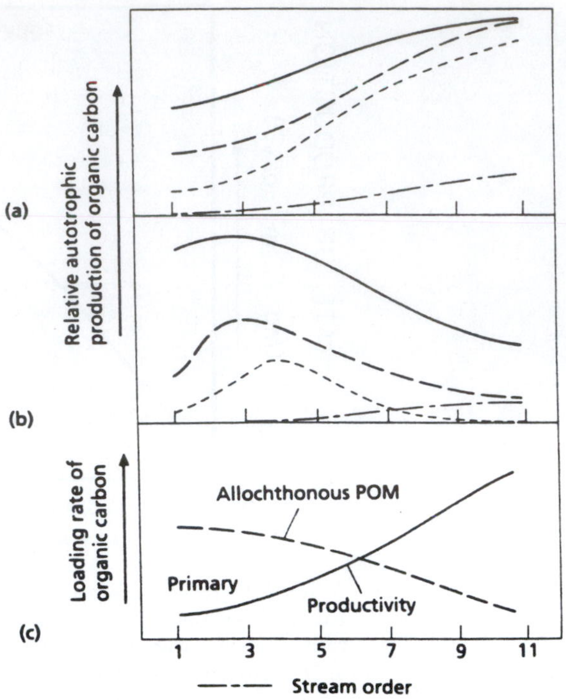

FIGURE 15-21 Changes in patterns of autotrophic production of organic carbon with increasing stream order in (a) nondisturbed and (b) disturbed lotic ecosystems. (c) Longitudinal gradient of organic carbon loading from allochthonous (upland) versus lotic ecosystem primary productivity. (———) = Photosynthetically derived *dissolved* organic matter from wetlands and flood plains; (_ _ _ _) = macrophytes; (- - - -) = periphyton; and (- _ - _ - _) = phytoplankton. (From Wetzel and Ward, 1992.)

correlation is influenced by human deforestation and agriculture (Minshall *et al.*, 1983; Naiman, 1983). Primary productivity within the rivers then generally decreases with increasing stream order because of light limitations (Fig. 15-21b). The contributions by periphyton and submersed macrophytes then decline with increasing size and turbidity.

Therefore, reductions of photosynthetic metabolism *within* the river channel per se are observed commonly along the length of the river with increasing stream order. The elevational slope decreases with increasing stream order, however, and is accompanied by increased deposition of bedloads. Colonization of these bank, wetland, and floodplain areas by higher vegetation results in very high primary productivity. Decomposition of much of the particulate organic matter produced in these sites occurs at or near the sites of production. A large portion of the organic matter synthesized by these high producers is exported to the river water as dissolved organic matter (Fig. 15-21c). As will be discussed later, that dissolved organic matter from allochthonous sources is important to the metabolism and productivity of biota within the streams and rivers.

B. Primary Production of Reservoirs

Reservoir ecosystems are intermediate between rivers and natural lakes in relation to morphology, hydrology, nutrient loadings and cycling, and sources of organic matter. Because of large water level fluctuations from flood control and hydropower operations and light limitations from abiogenic turbidity of suspended clays and silts, attached algal and rooted macrophyte development is restricted. Much of the primary production *within* reservoirs is by phytoplankton. Despite frequent high abiogenic turbidity in reservoirs, the average phytoplankton primary productivity tends to be higher than in natural lakes (Table 15-12). Because of the large drainage basins of reservoirs in comparison to natural lakes, nutrient loadings tend to be much greater. As in all inland aquatic ecosystems, however, much of the organic matter utilized within the ecosystems emanates from external allochthonous sources.

C. Seasonal Rates of Photosynthesis in Lakes and Reservoirs

Rates of *in situ* photosynthesis of the phytoplankton must be evaluated over an annual period to compensate for major differences in the length of active growing season with changing latitude and altitude. Although summer productivity of an arctic or high mountain lake may be just as high as productivity per unit volume of a lake at much lower latitude or elevation, the length of the growing season is considerably less. Since evaluations of phytoplanktonic productivity must be made on an annual basis, it is necessary to look first at the magnitude of these seasonal variations.

Further, it is important to keep in mind the ecological significance of gross versus net productivity. As discussed earlier, rates of the opposing processes of photosynthesis and respiration are very difficult to evaluate in heterogeneous planktonic populations of phytoplankton, bacteria, and zooplankton. The widely used ^{14}C light and dark bottle (Wetzel and Likens, 2000) generally results in values close to net productivity under most *in situ* conditions. This method permits a further evaluation of the extent of extracellular release of soluble organic compounds during growth of the algae. Respiration cannot, however, be evaluated by the ^{14}C method. The sensitive ^{14}C method permits an estimation of the *in situ* rates of photosynthesis approximating net productivity, which is useful since it is this particulate matter that is potentially available for consumption by higher organisms. More elaborate methods, however, are required to ascertain the fate of the total productivity, since most the synthesized organic matter is not consumed by animals but instead enters the detrital food chain (cf. Chap. 23). The less sensitive oxygen-change techniques measure community metabolism, although oxygen production over time is most often the result of phytoplanktonic algae. How light-mediated photorespiration compromises the assumption that dark-bottle estimates of respiration are similar to respiration values in the light have not been evaluated satisfactorily. Similarly, bacterial respiration is assumed to be small in relation to that of the planktonic algae and to be the same in the light as it is in the dark. It can only be concluded that at best, contemporary methods provide estimates of *in situ* rates of photosynthesis that are comparable only in a general way (cf. Peterson, 1980). Nonetheless, the advantages of being able to measure rates directly *in situ* are great and are exceedingly valuable if viewed in the correct perspective.

Several examples of seasonal primary productivity illustrate the types of variations that can be encountered. The changes in productivity of the phytoplankton at depth over the season set out in Figure 15-22 for a hardwater lake of low productivity in Michigan show a common pattern. Rates are low in the winter with periodic minor surges from near-surface populations under ice, particularly toward the end of winter. Rates often decrease during spring circulation, which in this lake for this year was very brief (until early April). Rates of production increased during the spring diatom maximum, entered a low period in May, and then succeeded through a series of pulses through the summer when values were higher than during the spring. Metalimnetic maxima are conspicuous in July and late August prior to a general slow decline in the autumn. These rates of production by phytoplankton are composite values for the community of many species populations, which show a great regularity in periodicity from year to year (monitored continuously for nearly two decades).

When these data at each depth interval are integrated over a square meter of water column, one can

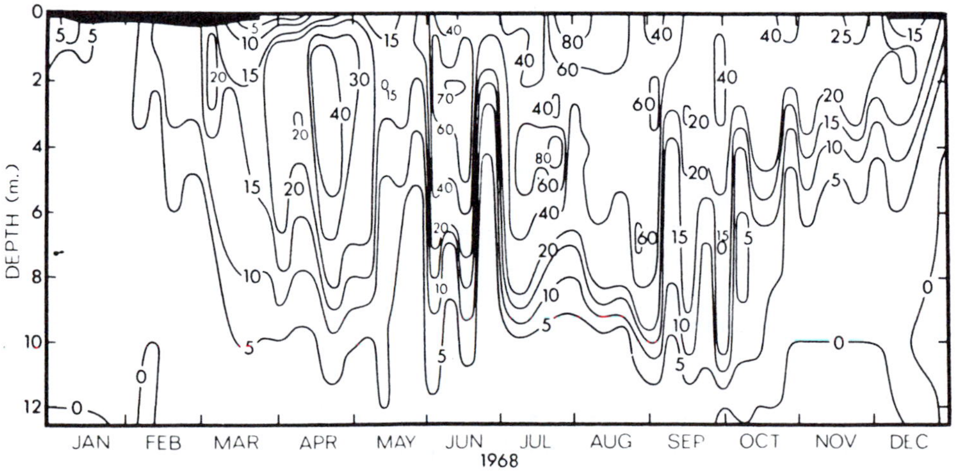

FIGURE 15-22 Depth–time distribution of the *in situ* rates of production of phytoplankton in mg C m^{-3} day^{-1}, Lawrence Lake, Michigan, 1968. Measured at meter depth intervals every 7 days; opaque areas = ice cover to scale. (From Wetzel, unpublished data.)

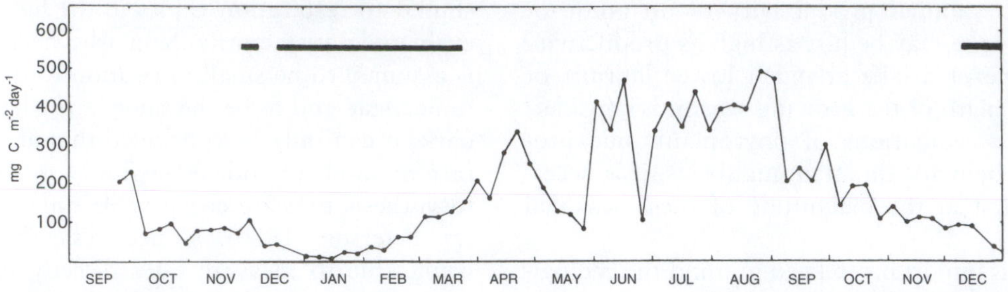

FIGURE 15-23 Integrated areal primary productivity per square meter of the pelagic zone of Lawrence Lake, Michigan, 1967–1968. Bars indicate periods of ice cover. (From Wetzel, unpublished data.)

calculate the rates of *in situ* productivity per surface area of the pelagic zone (Fig. 15-23). It is usual to integrate the areal productivity under such curves for an annual period to obtain the estimate of the total annual productivity by phytoplankton for a year (Wetzel and Likens, 1991). Division by 365 results in an estimate of the mean daily productivity for the lake. Obviously, conically shaped lake basins have much greater volume for phytoplanktonic productivity in upper strata than at greater depth within the trophogenic zone. Therefore, it is advisable to determine the annual productivity for each stratum, say at meter intervals, over the year and integrate these values before summing productivity of each stratum for the whole. In this way, the changes in volume available for productivity are taken into account.

A feature of the total phytoplanktonic productivity and the mean daily productivity for the year is their stability from year to year as long as the system is not perturbed. In spite of variations in the seasonal rates of observed productivity and numerous sources of potential error in measurement and sampling, the values are reasonably constant. Examples of this consistency can be seen in Table 15-9. Some variations are to be expected. In this case, for example, 1972 was a particularly variant year with abnormally high cloud cover and rainfall. Within this type of variation, the annual primary productivity values serve as one of the best available criteria for following changes in the metabolism of lake systems in response to alterations by human activities.

Looking at the primary productivity of phytoplankton of more eutrophic lakes of the same latitude as the previous example, three points are conspicuous (Fig. 15-24). First, the thickness of the trophogenic zone is reduced markedly to the point where most of the productivity occurs in the first 2 m of water. Second, the seasonal succession of productivity is much more irregular and exhibits marked fluctuations in comparison to less productive lakes. As discussed earlier, algal utilization and reduction of critical nutrients (Si, combined N) to levels limiting for certain algal groups progress very rapidly (a few days to weeks) and contribute to the marked oscillations in composite photosynthetic rates. Finally, the data of Figure 15-24 again emphasize the importance of primary productivity under ice and winter conditions, which should not be ignored in contemporary studies.

Because the primary productivity of hypereutrophic Wintergreen Lake is attenuated so rapidly with increasing depth by the self-shading effects of the algae, most of the productivity is restricted to the uppermost portions of the photic zone. Therefore, evaluation of the productivity of the lake on a per-square-meter-of-water-column basis throughout the trophogenic zone is similar to values for the entire lake (Fig. 15-25). Because this lake is shallow and pan-shaped, compensation for the decreasing volume of the productivity in the lower strata only has a small effect on the total lake productivity, that is, the volume of the lower strata within the trophogenic zone is small.

TABLE 15-9 Rates of In Situ Production of Phytoplankton of Lawrence Lake, Michigan, over Several Years

Year	*g C m^{-2} yr^{-1}*	*Mean mg C m^{-2} day^{-1}*
1968	45.43	124.5
1969	41.76	114.4
1970	43.48	119.1
1971	36.13	99.0
1972	29.58	81.0
1973	39.17	107.3
1974	38.87	106.5
1975	43.01	117.8
1976	32.31	88.5
1977	30.93	84.7
1978	29.06	79.6
1979	29.03	79.5
1980	29.74	81.5
1981	39.13	107.2
Mean 14 years	36.26	99.3

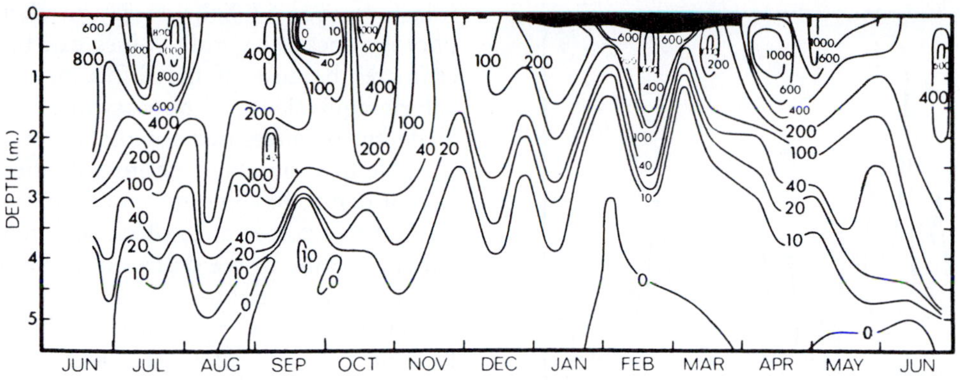

FIGURE 15-24 Depth–time distribution of the *in situ* rates of production of phytoplankton in mg C m^{-3} day^{-1}, Wintergreen Lake, Michigan, 1971–1972. Opaque area = ice cover to scale. (From Wetzel *et al.*, unpublished data.)

Seasonal climatic changes are reduced in tropical regions. The length of the active growing season is extended to the entire year. Monthly irradiances in the tropics are higher than at higher latitudes and result in higher minimum water temperatures, higher mean temperatures for the water column, and smaller thermal discontinuities between upper and lower water strata (Lewis, 1996). Tropical lakes of moderate depth are predominantly warm monomictic and exhibit regularity in seasonal mixing in the winter. Tropical lakes show more intraseasonal variations in thickness of the mixed layer than morphometrically similar temperate lakes. This periodic intraseasonal deep mixing, followed by restoration of a thinner mixed layer, results in a higher efficiency of microbial nutrient regeneration than is the case in cooler, more stable temperate lakes. The combination of efficient nutrient cycling, higher mean temperatures, and greater stability of solar irradiance results in phytoplanktonic primary production about twice as high per unit nutrient as there would be at higher latitudes (Lewis, 1996). Equatorial lakes exhibit the highest values of phytoplanktonic productivity recorded in aquatic ecosystems (e.g., Talling, 1965; Lewis, 1974b, 1996; Westlake *et al.*, 1980).

The successional clock among phytoplankton of tropical lakes is reset frequently because of irregular thickening of the mixed layer of the trophogenic zone. Definite seasonality persists even though it is not nearly as marked as is common in lakes of higher latitudes. Seasonality is reduced in very shallow tropical lakes. In shallow, nonstratified tropical lakes that have high nutrient inputs, the productivity values are not only very high but are relatively uniform throughout the year. An example is Lake George on the equator in Uganda (Talling, 1965; Ganf, 1974). Deeper tropical lakes that stratify thermally, even though weakly so, tend to

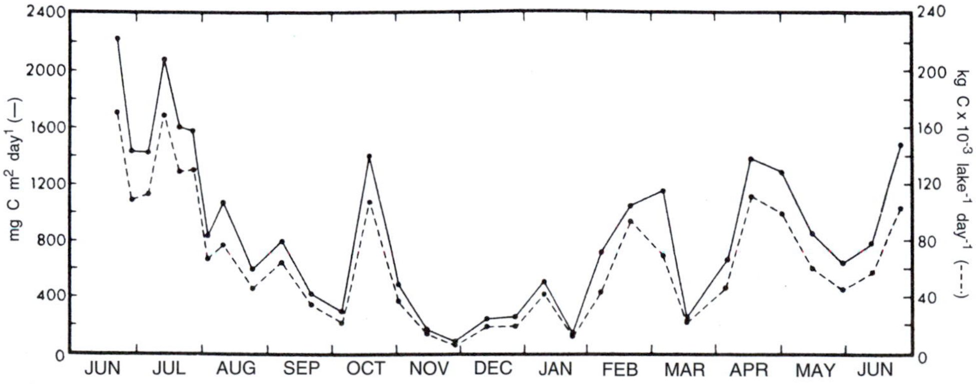

FIGURE 15-25 Integrated primary productivity of the phytoplankton of hypereutrophic Wintergreen Lake, Michigan, 1971–1972, per square meter of the trophogenic zone at the deepest point and for the lake volume with morphometric corrections. (From Wetzel *et al.*, unpublished data.)

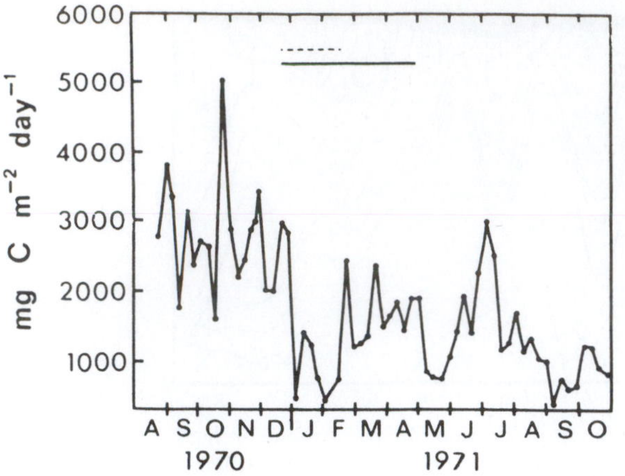

FIGURE 15-26 Net rates of primary production of the phytoplankton of Lake Lanao, Philippines, 1971–1972. The dotted line marks the period of minimum water temperature and deepest seasonal mixing; the solid line marks the period when the lake lacked stable stratification. (Modified from Lewis, 1974b.)

exhibit some periodicity in productivity. For example, in equatorial Lake Victoria of Africa, somewhat higher rates occur in the early months of the year and in June and July, when the breakdown of stratification occurs. The net primary productivity of weakly monomictic Lake Lanao (8°N) in the southern Philippines was relatively high throughout the period of stratification (May–November) but decreased precipitously during the winter period when water temperatures were lowest and deepest seasonal mixing occurred (Fig. 15-26). Respiration rates as a percentage of gross photosynthesis generally are higher (toward 50%) than the average for temperate waters (20–30% of total fixed carbon).

Therefore, when viewing the annual cycles of primary productivity in lakes in relation to latitude, seasonal variation in input of solar radiation at higher latitudes is the general controlling mechanism. Near the equator this seasonal differentiation is muted, and the incidence of vertical mixing is more important in any observed periodicity (Talling, 1969; Lewis, 1996). In some tropical lakes, as in Lake Victoria, seasonal cooling and mixing, regulated chiefly by atmospheric factors like wind and humidity, are decisive events. Nutrient supply has been implicated as a dominant controlling factor of primary productivity of Lake Lanao during stratification (Lewis, 1974b). Nutrient depletion is relieved at frequent intervals by changes in the depth of mixing associated with irregular storms.

D. Efficiency of Light Utilization

The efficiency of light utilization by algae can be estimated in several ways. All methods relate the amount of irradiance available at depth to the rate of transformation of light energy to chemical energy by photosynthesis. Existing values can be viewed only as estimates because of the difficulties in measuring the caloric equivalents of photosynthesis per unit volume at depth, or integrated for the water column, and the variable light inputs of photosynthetically active radiation at depth. In spite of these limitations, it is of great interest to see the range of values encountered in fresh waters because of the desires of some to utilize aquatic systems as potential food sources.

Phytosynthetic efficiencies of light utilization can be defined as an ecological efficiency (ϵ'), in a percentage equal to the amount of photosynthetically stored energy (PSR)[8] in a volume of water $(z_1 + z_2)\text{m}^3$ divided by the photosynthetically available radiation (PAR; 350–700 nm) absorbed or dissipated in the water layer z_1 to z_2, regardless of the variable fraction of PAR absorbed by algal pigments (Dubinsky and Berman, 1976, 1981a; Morel, 1978):

$$\varepsilon' = \frac{\text{PSR}}{\text{PAR}_{z_1} - \text{PAR}_{z_2}} \times 100(\%)$$

The difference in downwelling irradiance flux between successive water planes (z_1 and z_2) approximates light absorbance in the water column and is expressed in calories (or joules) $\text{m}^{-3}\ \text{h}^{-1}$. Backscattered light from lower depths affects the estimate of PAR dissipation in the overlying stratum only slightly. Thus, the photosynthetic efficiency is expressed as a ratio of the caloric equivalent of photosynthesis to radiation inputs.

Photosynthetic efficiency of light utilization by phytoplankton can also be expressed on an areal basis, in which the efficiencies of all the water strata are integrated throughout the euphotic zone and related to the PAR at the surface (cf. Morel, 1978; Dubinsky and Berman, 1981a; Falkowski and Raven, 1997). The areal efficiency permits more direct comparisons with photosynthetic efficiencies in other aquatic, portions of aquatic (e.g., littoral and wetland), and terrestrial ecosystems.

It is well known that the areal efficiencies of light-energy utilization by phytoplankton in aquatic ecosystems are generally much lower than photosynthetic efficiencies of terrestrial systems. Much of the irradiance is selectively and rapidly absorbed by water itself and by dissolved and particulate matter contained in it. The amount of light absorbed by the water and solutes is usually greater than the amount absorbed by plant pigments. Nearly all of the estimated efficiencies are <1%, and the highest values (approximately, 1–2%) reported for phytoplankton are from eutrophic lakes and from

[8]An approximate conversion of assimilated carbon to energy equivalents is 1 mg C assimilated = 9.33 cal, assuming all phytosynthate is glucose (Morel, 1978).

TABLE 15-10 Estimates of Photosynthetic Efficiencies of Utilization of Photosynthetically Active Radiation by Phytoplankton[a]

Ecosystem	*Percentage efficiency*
Lakes	
Tahoe, California–Nevada	0.035
Castle, California[b]	0.040
Finstertaler, Austria[b]	0.068
Oliver, Indiana	0.19
Olin, Indiana	0.23
Walters, Indiana (mean of four basins)	0.26
Pretty, Indiana	0.26
Chad, Chad, Africa	0.26
Crooked, Indiana	0.32
Little Crooked, Indiana	0.33
Martin, Indiana	0.34
Kinneret, Israel	0.35
Smith Hole, Indiana	0.38
Sammamish, Washington	0.42
Wingra, Wisconsin	0.45
Goose, Indiana	0.57
Sylvan, Indiana (mean of three interconnected basins)	0.98
Leven, Scotland	1.76
Rivers	
Spree, Germany	
Beeskow	0.28
Fürstenwalde	0.35
Neu Zittau	0.33
Wernsdorf	0.35
Oceans	0.13–0.16

[a] After data of Wetzel (1966b), Tilzer *et al.* (1975), Dubinsky and Berman (1976), Köhler (1994), and Falkowski and Raven (1997). Lakes are ordered in approximately increasing productivity. Estimates are based on different criteria of conversion; values are only approximately comparable but should be within about 30% (cf. Dubinsky and Berman, 1981b).

[b] Ice-free period only.

tropical areas (Table 15-10). The average light-utilization efficiency of a water column, based on integrated carbon fixation per unit area of water column through the euphotic zone normalized to integrated photon flux, is not constant but increases as the average daily irradiance decreases (Falkowski and Raven, 1997). Similar relationships have been found among aquatic macrophytes (e.g., Mann and Wetzel, 1998). Much of this relationship is attributed to biochemical photoinhibition by excessive light.

Variations in photosynthetic efficiencies are great, both seasonally and vertically within the euphotic zone. As light intensities increase near the surface of the water, both light utilization efficiencies and ratios of photosynthetically stored energy to the amount of PAR absorbed by algal pigments generally decrease, probably because photosynthetic reaction sites become light-saturated (Fig. 15-27; cf. Tilzer *et al.*, 1975; Dubinsky and Berman, 1981a). As irradiance decreases to values below photosynthetic light saturation, photosynthetic efficiencies usually increase. Although high efficiencies of light utilization are common at the bottom of the euphotic zone, photosynthetic rates are so low that these phytoplankton contribute little to the total primary productivity of the lake, since the absolute amount of light energy available at these depths is small. As a result, the overall areal efficiency is increased little.

As the quantity of phytoplanktonic biomass increases, the integral photosynthetic efficiencies generally increase until the maximum levels are restricted by light limitations imposed by self-shading. Therefore, the maximum areal values of photosynthetic efficiency are found in very productive waters containing dense algal populations. Exemplary ranges of photosynthetic efficiency of phytoplankton encountered in lakes of increasing productivity and of rivers are set out in Table 15-10.

It should be emphasized again that the productivity of the littoral zone and surrounding wetlands has usually been ignored as a component of aquatic ecosystems. Littoral and wetland productivity has been critically evaluated in only a few aquatic ecosystems; in every case examined, the combined productivity of

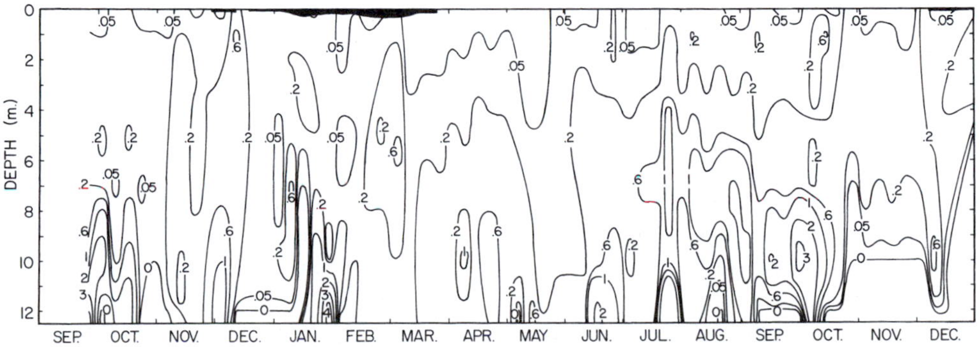

FIGURE 15-27 Depth–time distribution of estimates of the photosynthetic efficiency of utilization of photosynthetically active light income by phytoplankton of Lawrence Lake, Michigan, 1967–1968. Values expressed as per cent, most of which are much <1%. (From Wetzel, unpublished data.)

these sessile plants and their photosynthetic efficiencies of light utilization are much higher than those of the phytoplankton and constitute a major productivity component of most freshwater ecosystems (cf. Chaps. 18 and 19).

E. Extracellular Release of Organic Compounds

Algae release a large number of organic compounds into the water extracellularly. These soluble compounds include glycolic acid, carbohydrates, polysaccharides, amino acids, peptides, organic phosphates, volatile substances, enzymes, vitamins, hormonal substances, inhibitors, and toxins. Literature on this subject is large and has been reviewed by Fogg (1971), Hellebust (1974), Jüttner and Faul (1984), and Carmichael (1986, 1997).

The release of organic compounds extracellularly represents first a significant loss of carbon fixed in photosynthesis. Second, the release of such organic compounds undoubtedly is of much greater importance than is generally recognized and in a number of subtle but significant ways may modify growth, behavior of organisms, and the successional dynamics of algal and other microbial populations. For example, a number of allelopathic interactions have been found among algae, in which organic compounds released by one alga are inhibitors or stimulators to others (e.g., Proctor, 1957; Lefèvre, 1964; Keating, 1977; Wolfe and Rice, 1979; Jüttner, 1981; Gross *et al.*, 1991; Hagmann and Jüttner, 1996). The release of biologically active organic compounds, especially by cyanobacteria, is clearly implicated as one mechanism, among many others discussed earlier, that can influence fine adjustments in species succession and seasonal productivity. Fischerellin is a compound released from a cyanobacterium that can inhibit photosytem II of algae and other microorganisms in a very effective competitive interaction. Production of some of these allelochemicals can be induced when the cyanobacteria experience phosphorus limitations. Although causality is highly probable, quantitative evaluation of organic allelopathy has not yet been demonstrated in detail among natural populations of algae.

Two types of extracellular products are generally recognized (Fogg, 1971): (a) metabolic intermediate compounds of low molecular weight and (b) metabolic end products, usually of higher molecular weight. Glycolic acid, as an intermediate compound in photosynthesis, is known to be released under physiological stress conditions, especially under oligotrophic nutrient conditions and when photosynthesis is light inhibited. Glycolic acid and polysaccharides released during active growth are utilized readily by bacteria (Nalewajko *et al.*, 1980; cf. Chap. 17). Intermediate compounds of respiration that are released include organic acids, organic phosphates, and, to a lesser extent, amino acids and peptides. The liberation of end products of metabolism, usually of higher molecular weight, does not depend on equilibrium, and they are approximately proportional to the amount of growth. Included in this miscellaneous group are carbohydrates, peptides, volatile compounds such as aldehydes and ketones, enzymes, and a number of growth-promoting and -inhibiting substances.

The release of simple compounds (e.g., sugars, amino acids, and organic acids), by actively growing cells probably occurs mainly by diffusion through the plasmalemma (Hellebust, 1974). The rates of release depend on the concentration gradient of the substance across the membrane and the permeability constant of the membrane for the compound. Although active excretion of small compounds is also possible, no convincing evidence for such processes in algae exists. Large molecules, such as polysaccharides, proteins, and polyphenolic substances, probably are excreted by means of more complex processes such as fusion of intracellular vesicles containing the compounds with the plasmalemma. The rates of extracellular release depend on the physiological and environmental factors affecting membrane permeability and on intracellular concentrations (e.g., Mague *et al.*, 1980; Carmichael, 1997).

The rates of extracellular release under *in situ* conditions are quite variable. Although rates have been reported ranging from those equal to the rates of carbon fixation into cellular constituents down to <1% of rates of carbon fixation, on the average most values are <20% (e.g., Søndergaard and Schierup, 1982). For example, in the unproductive hardwater Lawrence Lake, *in situ* rates of release of organic compounds by phytoplankton monitored continuously over nearly two decades ranged from 0.0 to 22.5 mg C m^{-2} day^{-1}, with a mean of 6.5 mg C m^{-2} day^{-1}. The annual mean percentage secretion was 5.7% of net phytoplanktonic primary productivity (Wetzel *et al.*, 1972 and unpublished). Rates of release were greatest in April when production rates were low but increasing (Fig. 15-28). Secretion reached another maximum during the latter portion of summer stratification as primary productivity began to decrease while particulate organic carbon remained high. Rates of release were similar among phytoplankton in rivers. For example, Köhler (1994) found between 2.5 to 5.0% of carbon fixed in photosynthesis was released extracellularly. The mean annual amount of released organic carbon in Lawrence Lake was highest at 1 m and decreased with increasing depth except below 10 m (Fig. 15-28). In these lower strata of very low light and low rates of photosynthesis,

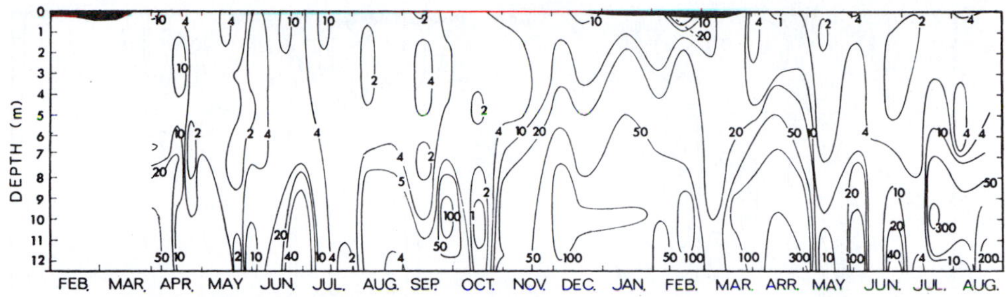

FIGURE 15-28 Depth–time isopleths of extracellular release of organic carbon by phytoplankton as a percentage of rates of carbon fixed photosynthetically, Lawrence Lake, Michigan, 1968–1969. (From Wetzel *et al.*, 1972.)

release rates, although extremely low in absolute value (cf. Fig. 11 of Wetzel *et al.*, 1972), were relatively high in relation to the low rates of primary production occurring in the deeper strata. Therefore, expressed as the mean percentage of extracellular release of all dates and samples, 23.5% of the phytoplanktonic productivity was released to the dissolved phase. Similar results have been found in many other lakes (e.g., Berman, 1976).

As lakes become more productive, the absolute amounts of organic carbon released increase markedly. For example, the annual range in hypereutrophic Wintergreen Lake, Michigan, was between 2 and 100 mg C m^{-3} day^{-1}, with the highest values in the epilimnion (Fig. 15-29). However, because of the extremely high rates of photosynthetic carbon fixation and the rapid seasonal oscillations in photosynthesis within the compressed trophogenic zone, the percentage release values are generally lower than in oligotrophic waters (Fig. 15-30).

High rates of release of extracellular products of algal photosynthesis have been shown to be a function of CO_2 limitation (e.g., by high pH), high population densities, inhibiting light intensities, low light intensities, and low cell densities and growth rates. The high percentage released at both extremes of the light and pH continuum indicates that the release of relatively large quantities of organic carbon compounds is favored by any environmental condition that inhibits cell multiplication but still permits photoassimilation to occur. Membrane permeability alteration or damage is probably the case in situations of high light intensities and other distinctly detrimental conditions. Effects of UV-A and UV-B irradiance on release of organic compounds is unclear (cf. Karentz *et al.*, 1994).

During cyanobacterial blooms, population growth of other algae can be strongly suppressed. Murphy *et al.* (1976) have demonstrated that cyanobacteria, which have high requirements for iron, can release hydroxamate siderchrome compounds, which complex iron and increase its availability, thereby favoring growth of the cyanobacteria. Simultaneously, the hydroxamate chelators have growth-suppressing effects on algae. Certain cyanobacteria, particularly *Microcystis,* produce microsystins and nodularins, a series of some 40 toxic cyclic heptapeptide compounds. Effects are numerous; some are known to inhibit protein phosphatase activity.

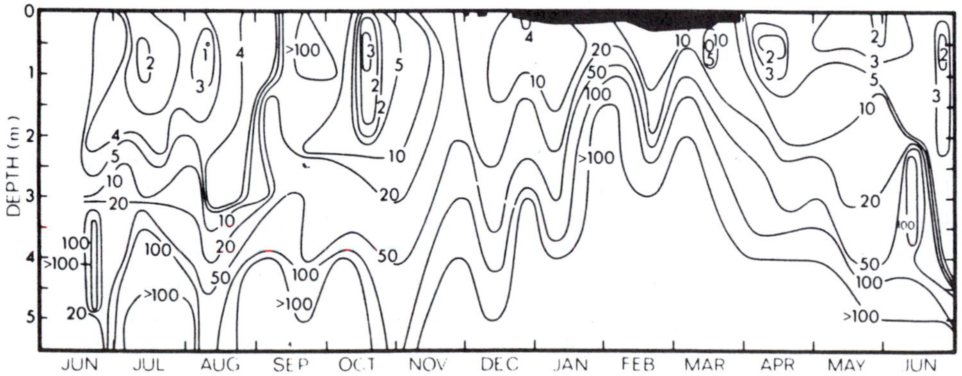

FIGURE 15-29 Depth–time isopleths of the rates of extracellular release of organic carbon by phytoplankton (mg C m^{-3} day^{-1}), hypereutrophic Wintergreen Lake, Michigan, 1971–1972. (From Wetzel *et al.*, unpublished data.)

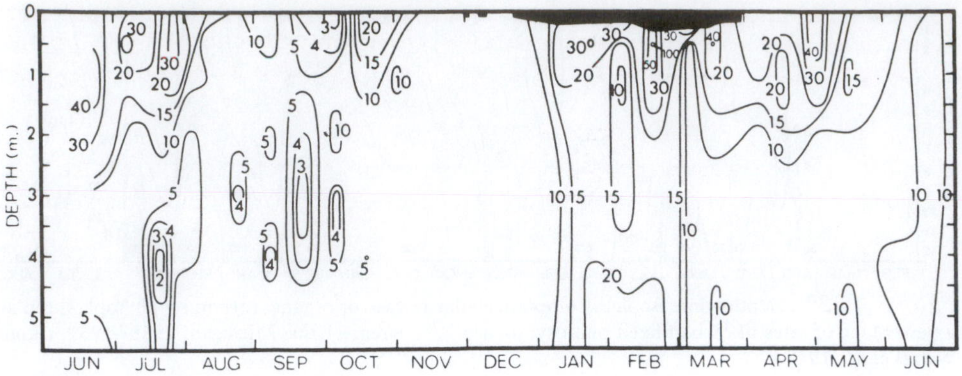

FIGURE 15-30 Depth–time isopleths of extracellular release of organic carbon by phytoplankton as a percentage of rates of carbon fixation, Wintergreen Lake, Michigan, 1971–1972. (From Wetzel *et. al.*, unpublished data.)

This large subject of cyanotoxins is reviewed in detail by Carmichael (1986, 1997). It should be noted that bacterial utilization of excreted organic compounds is extremely rapid (e.g., Nalewajko and Lean, 1972; Herbst and Overbeck, 1978; Nalewajko *et al.*, 1980; Coveney, 1981; Coveney and Wetzel, 1989, 1995). As a result, the simpler low-molecular-weight compounds are utilized and degraded rapidly and large-molecular-weight substances dominate in time because of their much slower rates of degradation by bacteria (cf. Chap. 17).

F. Diurnal Changes in Phytoplanktonic Production Rates

The diurnal periodicity of photosynthesis of phytoplankton is often not proportional to the daily insolation curve. Most studies show maximum photosynthetic rates in the morning hours, with a subsequent reduction at midday; sometimes recovery occurs in the afternoon. This diurnal periodicity is particularly acute at the surface and undoubtedly is partly related to photoinhibition and increased rates of extracellular release of recently synthesized organic compounds. However, much of this surface midday depression dissipates with increasing depth, often in <1 m, as light is attenuated and photosynthesis is undersaturated with light (Vollenweider, 1965; Jones and Ilmavirta, 1978).

Diurnal synchrony of cell division and growth has been well established in cultures of algae (Soeder, 1965), and this synchrony has been observed to take place in the same way in natural populations, particularly among small flagellates, under certain environmental conditions (Heyman and Blomqvist, 1984). As a result, the composite phytoplanktonic populations consist of various combinations of algae in (a) active, strongly anabolic growth stages in which cell numbers are increasing, especially in the morning, and overriding losses by sinking, natural death, and grazing; (b) "neutrally" active stages that are slightly anabolic, not undergoing significant positive or negative changes in biomass; and (c) inactive, catabolically metabolizing stages in which cells in resting, decaying, or degenerating stages are decreasing. Thus, depending on the combination of algae in various physiological stages of population development, and the dominant stage within the composite phytoplanktonic community, diurnal, synchronized, predawn and morning cell division often results in a pronounced photosynthetic periodicity from rapid early morning photosynthesis. For example, cell division in natural populations of the dinoflagellate *Ceratium* is phased to a brief period of the diurnal cycle (Heller, 1977; Heaney and Talling, 1980). The highest frequency of dividing cells occurred late in the dark period, with some carryover into the beginning of the following daylight hours.

It should also be mentioned that photorespiration increases at midday in submersed angiosperms and results in a midday reduction of net photosynthesis (Chap. 18). Photorespiration occurs in algae as well, but little is known as yet of its magnitude and diurnal periodicity under natural conditions.

An excellent example of diurnal biochemical coupling is seen in the bloom-forming cyanobacteria, especially one species of *Anabaena*. During daylight, carbon dioxide fixation in *Anabaena oscillarioides* is greatest in the morning period, while N_2 fixation is maximum during the afternoon hours when oxygen concentrations of the water often become supersaturated (Paerl, 1979; Paerl and Kellar, 1979). High oxygen concentrations can inactivate the nitrogenase enzyme activity of N_2 fixation. Nitrogenase-produced H_2, however, serves to remove excess intracellular oxygen by hydrogenase, leading to the formation of water and ATP (Paerl, 1980). The ATP is used in other metabolic reactions while protecting the N_2 fixation reactions from oxygen.

Vertical migration of phytoflagellates can contribute significantly to observed diurnal periodicity in stratification and photosynthesis of phytoplankton (e.g., Nauwerck, 1963; Tilzer, 1973; Ilmavirta, 1975; Happy-Wood, 1976). In high mountain lakes, for example, the dominant flagellates (e.g., *Gymnodinium*) usually ascend in the afternoon and evening and migrate downward with increasing light intensities (maximum rates approximately 1 m hr^{-1}). Photosynthesis is reduced near the surface during the midday hours as a result of light saturation, photoinhibition, and downward migration, but at greater depths, utilization of light energy usually increases at this time.

G. Horizontal Variations in Phytoplanktonic Biomass and Productivity

It is commonly assumed when measuring phytoplanktonic primary productivity that horizontal variations within the pelagic zone are relatively minor. This assumption is probably valid in many small- to moderate-sized lakes in which variations in different parts of the lakes are relatively small (<10%). In such cases, horizontal variations in phytoplanktonic activity may be masked by larger variation associated with sampling and experimental errors of measurement. Horizontal variations in biomass and productivity of phytoplankton become much more significant, however, in very small lakes, in extremely large lakes such as the Great Lakes, in proximity to the littoral zone of most lakes, and as the morphometric complexity of the lake basins increases, particularly in reservoirs. Even in very productive tropical lakes, phytoplanktonic productivity is often greater in the pelagic zone closer to the perimeter of the lake than in the central portion of the pelagic (e.g., Talling, 1965). In very large lakes, this horizontal disparity becomes more acute (Glooschenko *et al.*, 1973, 1974a, b; Schweizer, 1997). Nutrient loading from specific regions of the drainage basins of such lakes can be sufficient to cause significantly higher, often severalfold higher, rates of primary productivity in nearshore areas or sectors of the pelagic zone.

Phytoplanktonic productivity within the littoral zone or in the pelagic immediately adjacent to it is generally reduced significantly. Within littoral areas of lakes and reservoirs that are colonized by submersed and floating-leaved higher aquatic plants, phytoplankton communities are generally very different from those of the pelagic zone. Overall phytoplanktonic algal productivity is commonly severely reduced because of competition for nutrients by the attached periphytic microorganisms on plant and sediment surfaces (see Chaps. 19 and 21). Additionally, grazing by herbivorous zooplankton is often intensive because many zooplankton have refuge from fish and large invertebrate predation amongst the greatly increased habitat complexity of the higher aquatic vegetation (e.g., Lemly and Dimmick, 1982; Jeppesen *et al.*, 1997).

Horizontal patchiness in phytoplankton distribution has also been observed in relation to wind-induced current patterns (Heaney, 1976; George and Heaney, 1978; Harris *et al.*, 1979; Heaney and Talling, 1980; Camarero and Catalan, 1991; Jones *et al.*, 1995). Light wind stress (approx. 3 m s^{-1}) can cause horizontal flow of surface layers in small lakes. These currents induce upwelling of subsurface concentrations of phytoplankton at the windward end of a lake or downwind transport of algae when concentrated in the surface waters. The direction of the gradient can be readily reversed with a shift in wind direction and circulation patterns (e.g., Jones *et al.*, 1995). Moderate-to-strong wind conditions (>4 m s^{-1}) can cause sufficient turbulent mixing to eliminate heterogeneity within the epilimnion. In large, shallow lakes, wind-induced circulation can cause concentric distribution patterns, with maximum phytoplankton concentrations occurring toward the center (e.g., Ganf, 1974c).

Many lakes and reservoirs have complex morphometry and may contain dendritic valleys with numerous large, long bays. Primary productivity within these relatively isolated bays can be quite different from that of the major basin (Table 15-11). Often the productivity of these isolated arms is higher than that of the primary basin, for they receive proportionally higher loading of nutrients per unit volume.

The primary productivity of phytoplankton of reservoirs is variable seasonally and can vary considerably from year to year as well. During periods of high runoff from the drainage basin, abiogenic turbidity can be very high in the riverine section of the basin that receives the primary influents. In this region, primary production rates often are greatly reduced (Fig. 15-12). As much of the inorganic turbidity settles out of the euphotic zone in the progression downstream along the length of the reservoir toward the dam and outlet, the depth and productivity of the trophogenic zone increase (Fig. 15-12). However, if the retention time of the reservoir is short (i.e., the flow-through rate is high) there may be insufficient time for populations to develop fully before being flushed out of the reservoir. Some reservoirs are so constructed that outflow is drawn from the hypolimnion and lower strata; the increased retention time of surface waters can then lead to a more typically lake-type of phytoplanktonic

TABLE 15-11 Variations in the Annual Primary Productivity of Phytoplankton in Different Sections of Sylvan Lake, Indiana, a Basin of Complex Morphometry[a]

Region of lake	*Annual mean (mg C m⁻² day⁻¹)*	*Annual productivity (kg C ha⁻¹)*
Shallow arm near inlet	1475	5383
Large, isolated bay	1509	5509
Central main basin	1583	5777
Lower arm near outlet	1691	6171
Mean for lake	1564	5710

[a] After data cited in Wetzel (1966a).

community development and primary productivity. Removal of hypolimnetic water, laden with nutrients from sedimenting organic matter and decomposition, can reduce the nutrient loading of the reservoir water.

H. Annual Rates of Primary Production

The number of measurements of the primary productivity of phytoplankton in lake systems is prodigious. Only a few examples have been selected for general comparative purposes in Table 15-12, based on a rather subjective division of lakes into trophic categories. Further examples are listed in Westlake *et al.* (1980).

Among the lowest annual productivity values for phytoplankton of natural freshwater systems are those from Char Lake at latitude 74° in the Canadian Arctic. For this lake, the annual primary productivity of

TABLE 15-12 Comparison of Rates of Primary Production of Phytoplankton in Selected Fresh Waters of Varying Fertility

Ecosystems	*Remarks*	*Mean daily productivity for entire year ($mg\ C\ m^{-2}\ day^{-1}$)*	*Range observed ($mg\ C\ m^{-2}\ day^{-1}$)*	*Annual productivity ($g\ C\ m^{-2}\ yr^{-1}$)*
Lakes				
Oligotrophic lakes				
Crooked, Antarctica (Bayliss *et al.*, 1997)	Deep, permanently ice-covered, ultra-oligotrophic	—	(0–20)	—
Tundra ponds, Alaskan arctic (Alexander *et al.*, 1980)	Mean depth 20 cm; most production by benthic flora	1.6	(0–260)	0.6
Peters, Alaskan arctic (Hobbie, 1964)		2.5	(0–30)	0.9
Schrader, Alaskan arctic (Hobbie, 1964)		19.2	(0–40)	7.0
Char, N.W.T., Canadian arctic (lat. 74°) (Kalff and Welch, 1974)	80% of total production by benthic flora	11.5	0–35	4.2
Meretta, N.W.T., Canada (Kalff and Welch, 1974)	Polluted by sewage	8.5	0–170	11.3
Castle, California (Goldman, pers. comm.)	Deep, alpine	98	6–317	36
Lunzer Untersee, Austria (Jónasson, 1972)	Small, alpine	(123)	—	45
Lawrence, Michigan (Wetzel; see Table 15-9)	Small, hardwater; 14-yr average	99.3	5–497	36.3

(continues)

TABLE 15-12 (Continued)

Ecosystems	*Remarks*	*Mean daily productivity for entire year (mg C m^{-2} day^{-1})*	*Range observed (mg C m^{-2} day^{-1})*	*Annual productivity (g C m^{-2} yr^{-1})*
Ransaren, Sweden (Rodhe, 1958)	Small Lapplandic	—	23–66	—
Lake Superior, USA–Canada (Putnam and Olson, 1961)	Most unproductive of Laurentian Great Lakes	—	50–260	—
Lake Huron, USA–Canada (Vollenweider *et al.*, 1974)	Offshore stations	—	150–700	Approx. 100
Borax, California (Wetzel, 1964)	Large, shallow saline lake (25% of total productivity)	249	10–524	91
Lake Michigan, USA (Vollenweider *et al.*, 1974)	Offshore stations	—	70–1030	Approx. 130
Lake Ontario, USA–Canada (Vollenweider *et al.*, 1974)	Offshore stations	—	60–1400	Approx. 180
Mesotrophic lakes				
Erken, Sweden (Rodhe, 1958	Large, deep, naturally productive	285	40–2205	104
Clear, California (Goldman and Wetzel, 1963)	Very large, shallow	438	2–2440	160
Esrom, Denmark (Jónasson and Mathiesen, 1959)	Large, moderately deep	370	23–422	260
Foresø, Denmark (Jónasson and Mathiesen, 1959)	Large, deep, many macrophytes	462	0–1380	168
Water, Indiana (Wetzel, 1973)	Series of four inter-connected marl lakes			
Basin A		418	102–1395	153
Basin B		210	12–535	77
Basin C		276	30–1048	101
Basin D		437	98–1458	160
Oliver, Indiana (Wetzel, 1973)	Large, deep, marl lake	336	32–775	123
Olin, Indiana (Wetzel, 1973)	Large, deep, marl lake	374	89–996	137
Martin, Indiana (Wetzel, 1973)	Deep, stained marl lake	561	27–1708	205
Pretty, Indiana (Wetzel, 1966b)				
1963		440	68–1850	161
1964		305	6–895	111
Crooked, Indiana (Wetzel, 1966b)	Large, deep, hardwater lake			
1963		469	142–1364	171
1964		359	23–870	131
Little Crooked, Indiana (Wetzel, 1966b)	Small, deep, kettle lake			
1963		618	263–1903	226
1964		598	9–2451	218
Goose, Indiana (Wetzel, 1966a)	Small kettle lake	729	166–1753	266
Lake Erie, USA–Canada (Vollenweider *et al.*, 1974)				
Western stations		—	30–4760	(310)
Central stations		—	120–1690	(210)
Eastern stations		—	140–1440	(160)
Eutrophic lakes				
Wintergreen, Michigan (Wetzel *et al.*, unpublished)	Shallow, extensive nutrient loading	1012	60–2240	369
Frederiksborg, Slotssø, Denmark (Nygaard, 1955)	Shallow, enriched	1030	12–4160	376
Minnetonka, Minnesota (Megard, 1972)	Extremely complex basin, large, deep	(820)	—	(300)

(continues)

TABLE 15-12 (Continued)

Ecosystems	*Remarks*	*Mean daily productivity for entire year (mg C m^{-2} day^{-1})*	*Range observed (mg C m^{-2} day^{-1})*	*Annual productivity (g C m^{-2} yr^{-1})*
Sollerød Sø, Denmark (Steemann Nielsen, 1955)		1430	0–3800	522
Sylvan, Indiana (Wetzel, 1966a; see Table 15-11)	Complex basin, large, shallow	1564	9–4959	570
Lanao, Philippines (Lewis, 1974b)	Large, deep tropical lake	1700	400–5000	620
Victoria, Africa (Talling, 1965)	Large, deep, equatorial lake	1750	1700–3800	640
Dystrophic lakes				
Kattehale Mose, Denmark (Nygaard, 1955)	Very shallow, acidic, peat bog	80	0–400	29
Smith Hole, Indiana (Wetzel, 1973)	Shallow, humic stained	194	24–5960	71
Store Gribsø, Denmark (Nygaard, 1955)	Deep, acidic, humic stained	230	4–680	84
Grane Langsø, Denmark (Nygaard, 1955)	Deep, acidic	248	20–880	91
Reservoirs				
Mean of 10 oligotrophic reservoirs (Kimmel *et al.*, 1990)	Lowest often highly turbid, light-limited	151	67–235	55
Mean of 36 mesotrophic reservoirs (Kimmel *et al.*, 1990)		570	260–940	208
Mean of 21 eutrophic reservoirs (Kimmel *et al.*, 1990)		2019	1125–3975	737
Rivers				
Morgans Creek, Kentucky (Minshall, 1967)	Biomass change	5.5	4–7	20
Walker Branch, Tennessee (Elwood and Nelson, 1972)	Biomass change	9.5	8–11	3.5
Guys Run, Virginia (Hornick *et al.*, 1981)	^{14}C uptake	18	—	6.6
Little Schultz Creek, Alabama (Lay and Ward, 1987)	^{14}C uptake	91.2		33.3
Deep Creek, Idaho (Minshall, 1978)	Oxygen exchange	1180		431
White Clay Creek, Pennsylvania (Bott *et al.*, 1978)	O_2/CO_2 exchange	2220		810

Data approximated in some cases; estimates given in parentheses

phytoplankton is about 4 g C m^{-2} yr^{-1} but the productivity of mosses and epilithic algae increases this figure to about 20 g C m^{-2} yr^{-1} (Kalff and Welch, 1974; Kalff and Wetzel, unpublished). The low productivity results from low light and temperatures as well as from low nutrient availability. For example, Meretta Lake, near Char Lake, receives domestic pollution from a small village in quantities sufficient to increase primary productivity of phytoplankton by five times (Table 15-12), although lower values exist (Table 15-12). To date, all waters analyzed, even amictic, permanently ice-covered lakes, contain a limited algal flora and exhibit some net primary productivity.

A general categorization of the characteristics of lakes between the extremes of very oligotrophic and very eutrophic systems has been attempted many times. The criteria employed, largely resulting from an innate desire of humans for order in the natural world, have

TABLE 15-13 General Ranges of Primary Productivity of Phytoplankton and Related Characteristics of Lakes of Different Trophic Categories[a]

Trophic type	*Mean primary productivity (mg C m^{-2} day^{-1})*[b]	*Phytoplankton density (cm^3 m^{-3})*	*Phytoplankton biomass (mg C m^{-3})*	*Chlorophyll a (mg m^{-3})*	*Dominant phytoplankton*	*Light extinction coefficents (ηm^{-1})*	*Total organic carbon (mg $liter^{-1}$)*	*Total P (μg $liter^{-1}$)*	*Total N (μg $liter^{-1}$)*	*Total inorganic solids (mg $liter^{-1}$)*
Ultraoligotrophic	<50	<1	<50	0.01–0.5		0.03–0.8		<1–5	<1–250	2–15
Oligotrophic	50–300		20–100	0.3–3	Chrysophyceae,	0.05–1.0	<1–3			
					Cryptophyceae,					
Oligomesotrophic		1–3			Dinophyceae,			5–10	250–600	10–200
					Bacillariophyceae					
Mesotrophic	250–1000		100–300	2–15		0.1–2.0	<1–5			
Mesoeutrophic		3–5						10–30	500–1100	100–500
Eutrophic	>1000		>300	10–500	Bacillariophyceae,	0.5–4.0	5–30			
					Cyanobacteria,					
Hypereutrophic		>10			Chlorophyceae,			30–>5000	500–>15,000	400–60,000
					Euglenophyceae					
Dystrophic	<50–500		<50–200	0.1–10		1.0–4.0	3–30	<1–10	<1–500	5–200

[a] Modified from Likens (1975), after many authors and sources.
[b] Referring to approximately net primary productivity, such as measured by the ^{14}C method.

led to classification of lakes on the basis of nearly every parameter, including geomorphology, chemical constituents, nearly every type of organism from bacteria to fish, and even the waterfowl associated with different types of lakes. The most realistic parameter for such categorization, at least in a general way, is one that can be quantitatively determined as a rate of growth and one that integrates the host of environmental parameters controlling the synthesis of organic matter that enters the system. Rates of autochthonous primary production are basic to such an evaluation, and those of the phytoplankton have been proposed many times as the best existing criterion (see, e.g., Rodhe, 1969).

The ranges of primary productivity of phytoplankton commonly associated with oligotrophy and eutrophy are given in Table 15-13, along with several related characteristics of the phytoplankton. Such groupings can only be used in a general way, since there are many exceptions. The relationships, however, predominate in a majority of inland waters. The dystrophic lake category, discussed at length in Chapter 25, is very deviant and variable. Productivity of phytoplankton in dystrophic lakes is frequently attenuated by complex interactions between light, high concentrations of dissolved humic compounds, and limited inorganic nutrient availability.

The primary productivity of phytoplankton is often used as a major criterion for determining the trophic state of a lake or reservoir. The validity of this criterion, however, depends upon the assumption that dissolved and particulate organic matter inputs from littoral and allochthonous sources are small relative to those of the phytoplankton. In many cases on a global basis, littoral productivity and allochthonous organic inputs are much larger than inputs produced by phytoplanktonic productivity (Wetzel, 1990a); in such cases, phytoplanktonic productivity alone is a poor estimator of the trophic state of the lake or reservoir. One of the major objectives of Chapters 18 and 19 is to place these land–water interface components of aquatic ecosystems in proper perspective within the real world. Such analyses do not decrease the importance of phytoplankton to the metabolism and food-web structure of lakes and reservoirs. Rather, the primary productivity of phytoplankton must be viewed as a variably important contributor to overall aquatic metabolism.

XIV. PHYTOPLANKTON AMONG AQUATIC ECOSYSTEMS

Examination of the comparative characteristics of phytoplanktonic communities and their productivity among river, reservoir, and lake ecosystems indicate a number of consistent traits. Phytoplankton diversity is very low in low-order streams but increases in high-order large rivers (Table 15-14). Similarly, the diversity

TABLE 15-14 Comparative Characteristics of Phytoplankton and Their Productivity among River, Reservoir, and Natural Lake Ecosystems[a]

Property	*Rivers*	*Reservoirs*	*Natural lakes*
Phytoplankton diversity	Very low in low-order streams; increasing in high-order large rivers	Low in riverine zones; increasing in lacustrine zone	High diversity in oligotrophic lakes, decreasing in eutrophic lakes; similar in tropical and temperate lakes
Phytoplankton biomass	Very low in low-order streams; increasing in large rivers, although often light-limited	Moderately high in response to high nutrient supply in riverine sections; less in lacustrine zone where nutrients often limiting	Highly variable (to 5 orders of magnitude) in temperate lakes seasonally; much less variable in tropical lakes
Phytoplankton productivity	Low, often light-limited by advective flows and turbidity	Highest in transitional zone; reduced in riverine zone by light availability and often in lacustrine zone by nutrients; marked horizontal gradients; volumetric productivity (P_{max}) decreases from headwaters to dam; areal productivity relatively constant horizontally	Low in comparison to littoral productivity; increasing with moderate nutrient loading; declining at very high nutrient loading; seasonal and vertical gradients predominate; small horizontal gradients; light and inorganic nutrient limitations predominate

[a] Assembled from the syntheses on rivers by Ryder and Pesendorfer (1980) and on reservoirs and lakes by Wetzel (1990) and many sources cited in these papers.

of phytoplankton is low in the riverine portions of reservoirs but often increases in the more stable lacustrine zones with longer retention periods. Phytoplanktonic diversity is distinctly higher in less productive lakes than in eutrophic waters.

The biomass of phytoplankton is very low in low-order streams usually because of light limitations from shading by riverine terrestrial vegetation. Biomass can increase, despite frequent light limitation from abiotic turbidity, in larger rivers where residency times can often exceed those of the algal life cycles. The biomass of phytoplankton is usually low in riverine portions of reservoirs despite nutrient abundance because of frequent turbidity-related light limitations. Phytoplanktonic biomass fluctuates greatly in natural temperate lakes, usually by 5 or more orders of magnitude, on an annual basis, where as biomass in tropical lakes is much less variable.

The productivity of phytoplankton is low, regardless of the optimality of growing conditions, in comparison to the productivity of other photosynthetic organisms in aquatic ecosystems. The spatial distribution of phytoplankton in a dilute aqueous medium and conditions (limited light availability, sedimentation, diffusion limitations of nutrients, and others) that limit productivity have been termed an "aquatic desert" (Moeller *et al.,* 1998). Maximum rates of production of phytoplankton occur under conditions of moderate sustained nutrient availability in which the depth of the euphotic zone is large. The productivity of phytoplankton can dominate in the total productivity of some lakes where the phytoplankton themselves shade out other in-lake producer organisms. The collective productivity of phytoplankton per square meter, however, is low (Table 15-14).

XV. SUMMARY

1. The freshwater phytoplankton are composed of algae of almost every major taxonomic group and the cyanobacteria. Photosynthetic autotrophic metabolism is the sole metabolic pathway for the synthesis of organic matter among most algae. Although certain microbiota possess supplementary nonphotosynthetic means of nutrition, photosynthesis dominates and these organisms can be functionally included with the phytoplankton.
2. Light-energy-absorbing pigments are diverse among the algae. Chlorophyll *a* occurs in all algae and is the primary photosynthetic pigment. Chlorophyll *b* is found only in green algae and euglenophytes, and chlorophyll *c* occurs in many algal groups. Other accessory chlorophylls, of uncertain functions, are restricted to specialized algae. Light energy absorbed by some carotenoids and the biliproteins (phycoerythrin and phycocyanin of the cyanobacteria, cryptomonads, and red algae), as with chlorophylls *b* and *c,* is transferred to chlorophyll *a*.

3. The size of phytoplankton varies greatly. When the size is <20 μm, transfer of material to and from cells is almost entirely by molecular diffusion rather than by eddy diffusion of turbulence (Table 15-3).
   a. The ultraplankton (0.2–20 μm) are simple, unicellular in form, and have generation times in hours. These rates are similar to those of protistian flagellate predators. Small size allows establishment of self-contained, dynamic communities of phototrophs, viruses, heterotrophic bacteria, phagotrophic flagellates, and ciliates, in which materials are rapidly cycled with little loss by sedimentation or transfer to higher trophic levels. Nutrients tend to be recycled rapidly.
   b. The larger microplankton (20–200 μm) are not as vulnerable to predation by flagellates but are inefficient in nutrient uptake and develop prolifically when nutrient concentrations are high. Vulnerable to sedimentation, the microplankton depend on water movements or flotation mechanisms for transport to light and rapid reproduction to succeed. Equilibrium is never achieved within the microplanktonic community.
4. Most phytoplanktonic algae lack or have very limited powers of locomotion. As a result, most phytoplankton are dispersed by turbulent water movements. Since the density of most phytoplankton is greater than that of water, they tend to sink from the lighted, trophogenic zone. Sinking has the advantage of disrupting nutrient gradients surrounding the cells. The disadvantage of sedimenting out of the photic zone is offset to varying degrees by upward movement and turbulent water transport.

   Improved flotation or reduced sinking rates are accomplished in various ways:
   a. Reduction in sinking rate is enhanced by increasing *surface-to-volume ratios*. Deviations in cell morphology from the spherical form by protrusions and projections decrease sinking rates by increasing frictional resistance with the water. Seasonal polymorphism (cyclomorphosis) is rare among phytoplankton but is conspicuous among certain dinoflagellates. *Ceratium* alters its form and reduces its sinking rate as the viscosity of water decreases during warmer seasons (Fig. 15-1).
   b. *Production of mucilage* reduces sinking rates. Gelatinous sheaths are common among cyanobacteria and green algae, including certain desmids and diatoms. The sheaths are less dense than the cell and increase frictional resistance.
   c. *Gas vocuoles* are common in cyanobacteria and decrease the density of the cells to below the density of water. Many cyanobacteria regulate buoyancy and position in the photic zone to poise the populations within vertical physical and chemical gradients favorable to growth.
   d. A few algae reduce density by *accumulation of fats*.
   e. Certain algae can alter density by *regulation of cellular ion content*.
   f. A number of algae move vertically within the water column by *swimming by flagella*. Such movement can improve light and nutrient availability.
5. Coexistence of a number of phytoplankton species is a conspicuous feature of fresh waters. Although a few species commonly dominate a phytoplanktonic assemblage, a number of rarer algae coexist among the dominant species. Many differences in algal physiological characteristics, requirements, and tolerances, as well as seasonal and spatial variations in environmental parameters, permit the apparent multispecific equilibrium to exist for short periods.
   a. Differences in the efficiency of utilization of resources may be too small for competitive exclusion to occur before environmental conditions change.
   b. Partial commensalism or symbiotic dependency (e.g., through production and utilization of organic micronutrients) or both can reduce competition.
   c. Selective herbivory of a more successful species can reduce competition with a less successful species.
   d. Certain algae are meroplanktonic, enter resting stages often in the sediments, and do not compete for periods of varying duration with holoplanktonic (continually planktonic) species.
   e. Many or all of these coexistence mechanisms can function simultaneously.
6. Many environmental factors interact to regulate spatial and seasonal growth and succession of phytoplankton populations.
   a. Light and temperature synergistically affect photosynthesis. Although photosynthetic rates and algal growth are directly related to irradiance intensity, the response to light intensity, especially at light saturation, is temperature dependent (Fig. 15-2) and variable among species. A considerable degree of adaptation to changing light intensities occurs among algae, often by regulation of pigment concentrations per unit biomass.
   b. High light intensities inhibit photosynthesis of many algae; the effects are partly reversible if exposure is not too long (several hours) and phytoplankton are transported to regions of lower light intensities.

c. The vertical distribution of photosynthesis is strongly related to available light. A near-exponential decline with increasing depth underlies a surficial zone of maximum photosynthesis (Fig. 15-3). Photoinhibition by excessive light and ultraviolet radiation is common, with depression of photosynthesis at the surface and a subsurface photosynthetic maximum. As photosynthetic rates per unit volume of water increase in nutrient-enriched waters, the biogenic turbidity resulting from dense algal populations constricts the thickness of the trophogenic zone toward the surface.

d. Algae have definite temperature optima and tolerance ranges, which interact with other parameters to cause seasonal succession. For example, many diatoms can photosynthesize well at cooler water temperatures, whereas the temperature optima of many green algae and cyanobacteria are higher.

e. Major changes in the seasonal succession of phytoplankton in temperate lakes are related to changes in availability of phosphorus, nitrogen, and silica (cf. Chaps. 12, 13, and 14). The growth of a population under conditions of adequate light and temperature is often limited by a single nutrient. Limitation can shift rapidly from nutrient to nutrient as their availabilities change on a diurnal, daily, and seasonal basis.

f. Many algae require, but are unable to synthesize, organic micronutrients, especially vitamin B_{12}. These substances have been found to be actively assimilated by phytoplankton and, at certain times, photosynthesis is enhanced by organic enrichment; they likely play a significant role in succession and competitive success of certain algal populations.

g. Most algae are obligate photoautotrophs in which inorganic carbon is reduced biochemically with light energy. Utilization of organic substrates in the dark (heterotrophy) or in the light (photoheterotrophy) by aerobic dissimilation occurs in certain algae, and especially in the cyanobacteria. Heterotrophic augmentation occurs in some algae, but under most natural conditions of low organic substrate concentrations, bacteria are much better competitors for available substrates.

h. Dissolved organic compounds, largely of terrestrial and wetland plant origin, can influence phytoplankton metabolism by interacting with macro- and micronutrients and influencing their availability, either directly (e.g., complexation) or indirectly (e.g., altering other inorganic regulatory mechanisms).

7. Distinct seasonal patterns and successions of algal populations and biomass are observed in phytoplankton communities.

a. In temperate fresh waters, growth is greatly reduced during winter when both light and temperatures are low. A spring biomass maximum is commonly observed as light conditions improve and often consists predominantly of diatoms and cryptophytes adapted to low light conditions. Lower biomass often occurs during the summer. Among more eutrophic waters with high phosphorus loading, silica concentrations of the trophogenic zone are commonly reduced in the spring to levels inadequate to support large diatom populations. Summer populations of nonsiliceous green algae then flourish until concentrations of combined nitrogen are reduced below replacement. Under these conditions, nitrogen-fixing cyanobacteria have competitive advantages and often proliferate.

b. Summer phytoplankton populations of temperate oligotrophic lakes are often low. A second autumnal maximum, usually predominantly of diatoms, often develops during and after fall circulation. The limited growing season in high altitude and subpolar regions often results in a single summer maximum of phytoplankton biomass. In tropical lakes, total phytoplankton biomass is more constant (maximum often in winter) and larger than in temperate lakes. Frequent changes in abiotic factors, particularly mixing events, occur in tropical lakes that lead to more episodic changes in phytoplankton succession than in stratified temperate lakes. The succession of algae between the episodes, however, is similar in both tropical and temperate lakes.

c. Although variability is large, the general patterns of seasonal succession of phytoplanktonic biomass are reasonably constant from year to year, if the drainage basin and lake are not perturbed.

d. Seasonal changes in phytoplanktonic numbers and biomass are usually very large (approximately a thousandfold) in temperate and subpolar fresh waters. In tropical lakes, seasonal variations are much lower (approximately fivefold).

e. Changes in phytoplanktonic numbers and biomass are often out of phase with rates of photosynthesis. Photosynthetic productivity usually follows more closely the annual cycle of solar irradiance in low or moderately productive lakes, especially in temperate and subpolar areas. Smaller algae with faster turnover rates often predominate at warmer temperatures.

8. Parasitism of phytoplankton, particularly by chytrid fungi and viruses, is common and can increase as algae senesce in aerobic waters. The effects of selective parasites on species succession and algal productivity are unclear.
9. Grazing of phytoplankton by rotifers and microcrusfacea influences algal populations and their succession.
   a. On an annual basis, grazing losses of phytoplankton are usually minor in comparison to losses by sedimentation out of the photic zone.
   b. Grazing losses can be highly significant during certain periods of the year, during which time the algal populations are severely reduced. In the early summer "clearwater" phase, algal populations decline precipitously under conditions of high temperatures, high water column stability, and high light and reduced nutrient availability. During this time, ciliate, rotifer, copepod, and cladoceran zooplankton often increase their populations; algae smaller than 40 μm are especially vulnerable and grazing losses often exceed reproductive rates.
   c. Much of the food ingested by zooplankton is not assimilated, but is egested and enters the detrital pool as dissolved and particulate organic matter.
   d. Selective grazing by microcrustacea, largely on the basis of algal cell size, can alter the seasonal succession of phytoplankton.
   e. Nutrient regeneration, particularly of phosphorus, can be accelerated by zooplanktonic grazing; the increase in nutrient regeneration can enhance the primary productivity of the algae.
10. As nutrient limitations of the phytoplankton of infertile waters are increasingly relieved by nutrient inputs, rates of algal production increase.
   a. The densities of the phytoplanktonic community progressively reduce the available light and decrease the thickness of the trophogenic zone. A point is reached at which self-shading inhibits further increases in productivity, regardless of nutrient availability. Productivity per unit surface area of the freshwater system is usually lower under these hypereutrophic conditions than under less productive conditions where a thicker trophogenic zone exists.
   b. Maximum photosynthetic efficiencies of light utilization by phytoplankton are low, usually $<1\%$ of radiation incident on the water. These efficiencies are considerably less than utilization of light by wetland and terrestrial plants.
   c. An inverse relationship generally exists between algal biomass and productivity per unit biomass in the phytoplankton. Often, small species of relatively minor contribution to the algal community biomass have short generation times and contribute more to the total primary productivity than do larger species.
11. A significant amount ($<20\%$) of the carbon fixed photosynthetically by phytoplankton is released extracellularly as dissolved organic compounds.
   a. Certain low-molecular-weight compounds of intermediary metabolism are released in greater amounts under conditions of stress (e.g., light inhibition, low nutrient availability).
   b. Extracellular dissolved organic carbon released as a percentage of total carbon fixed photosynthetically is usually low (1–5%) in healthy, actively growing phytoplankton. The percentage release increases under very high and very low light conditions and during senescence.
12. Horizontal variations in the primary productivity of phytoplankton can be large. Spatial variations increase in significance in very small lakes, very large lakes (e.g., the Laurentian Great Lakes), in proximity to the littoral zone and inlet areas of most lakes, and as the morphometric complexity of lake basins increases.

16

PLANKTONIC COMMUNITIES: ZOOPLANKTON AND THEIR INTERACTIONS WITH FISH

The animals found in fresh waters are extremely diverse and represent nearly all phyla. The evaluation of their functional roles within aquatic systems requires a balanced understanding of the general modes and timing of growth and reproduction in relation to the availability and utilization of food. The population dynamics and certain important adaptive behavioral characteristics that influence these dynamics regulate the productivity of individual species populations and entire communities. Underlying all evaluations of productivity of the animals are their food or trophic relations with plants and other animals and competitive and predatory interactions that lead to a greater success of one species over another.

I. ZOOPLANKTON

In contrast to the great total biodiversity of animals in freshwater ecosystems, both the phyletic representation and species diversity of zooplankton communities is very much lower in fresh water than in marine habitats (cf. particularly Lehman, 1988). The differences are likely related to the great antiquity, depth, and evolutionary continuity found in the oceans. Even in ancient lakes zooplankton communities are not species enriched.

Planktonic animals are dominated by four major groups: (a) protists that include protozoa and heterotrophic flagellates, (b) the rotifers, and two subclasses of the Crustacea, (c) the Cladocera, and (d) the Copepoda. A few coelenterates, larval trematode flatworms, gastrotrichs, mites, and the larval stages of certain insects and fish occasionally occur among the true zooplankton, if only for a portion of their life cycles.

II. PROTOZOA AND OTHER PROTISTS

Recent studies involving electron microscopy and molecular biology have reshaped understanding of the Protista (Andersen, 1998). Phylogenetically, the Protista represent, at best, a grade, not a clade, and the kingdom Protista is not a natural taxonomic unit and is being abandoned. The Protista does, however, provide an artificial aggregation of predominantly microscopic organisms that is useful as an assemblage for ecological reasons.

Because the primary thrust of the present discussion is concerned with ecology, the protozoa are included among functionally similar phagotrophic protists. These free-living protists include the ciliates, most of the flagellates, and the sarcodines. These protists are eukaryotes, usually unicellular, that possess a nuclear envelope, eukaryotic ribosomal RNA, and endoplasmic membranes (Fenchel, 1987; Laybourn-Parry, 1992; Lee and Kugrens, 1992; Patterson, 1992). These organisms further possess eukaryotic organelles (mitochondria, chloroplasts, and flagella), histones associated with chromosomal DNA, and an ability to perform phagocytosis. Although a few protists can be colonial, they do not form tissues of specialized cells.

Most protozoa and related protists reproduce by binary fission, although multiple fission and sexual reproduction occur in a few free-living forms. Growth and generation times are highly variable among free-living protists. Maximum growth rates are found when food availability is not a constraint; under these conditions, populations generally increase directly with temperature within the limnological range and they decrease as body size increases (Taylor and Sanders, 1991). Many protists produce cysts under harsh environmental conditions (e.g., lack of food and desiccation).

Information on the functional ecology of the protists is emerging rapidly. It is becoming clear that the protistan zooplankton are the most important microbial consumers and have major functions in organic carbon utilization and nutrient recycling. Although the instantaneous biomass of the planktonic protists is small in comparison to those of the rotifers and especially the microcrustaceans, the generation rates of protists are high. The comparative rates of heterotrophic metabolism and productivity among the major groups at the ecosystem levels are presently under intensive investigation. A few examples of population dynamics and productivity estimations are presented next prior to evaluating the importance of these microbial communities to the utilization of organic matter and nutrient recycling.

A. Dynamics of Protistan Zooplankton

1. Flagellates

Flagellates are the most abundant component of the protozooplankton. Autotrophic flagellate species have already been discussed among the phytoplankton, although many are known to be mixotrophic and to feed on bacteria as well as fix carbon photosynthetically (e.g., Sanders and Porter, 1988; Arndt *et al.*, 2000). Common forms include the dinoflagellates (e.g., *Ceratium* and *Peridinium*), chrysomonads (*Dinobryon, Mallomonas,* and *Synura*), euglenids (*Euglena*), volvocids (*Volvox* and *Eudorina*), choanoflagellates (*Astrosiga*), and the diverse large group of heterotrophic flagellates. The heterotrophic groups include heterokont taxa (mainly chrysomonads and bicosoecids), choanoflagellates, kataphepharids, and a number of species, collectively called Protista *incertae sedis*, that cannot be assigned to any of the major groups (Patterson and Larsen, 1991). The heterotrophic flagellates are differentiated into two general groups on the basis of size: (a) heterotrophic nanoflagellates (HNF) below 15 μm and (b) large heterotrophic flagellates within a range of $\geq$15–200 μm (Arndt *et al.*, 2000). Among both temperate and subtropical lakes, dinoflagellates tend to increase in dominance as the pH of the water decreases and can dominate in acidic lakes (Laybourn-Parry, 1992).

2. Ciliates

Many major ciliate genera are represented in freshwater protozooplankton assemblages across the spectrum of trophic states. Of the three major groups, the oligotrichs, particularly *Strombidium* and *Halteria,* are found worldwide in lakes across the trophic spectrum (Laybourn-Parry, 1992). The tintinnid ciliates (order Choreotrichida: *Tintinnidium, Tintinnopsis,* and *Codonella*) are also widely distributed in temperate to tropical regions. Haptorid ciliates (e.g., *Askenasia* and *Mesodinium*) are similarly distributed broadly and abundantly. Although a few ciliates are mixotrophic and supplement nutrition by photosynthesis, most are holozoic and feed on bacteria, algae, particulate detritus, and other protists. A few are carnivorous and feed on small metazoans (see discussion later). Ciliates tend to be more significant components of the zooplankton of eutrophic lakes.

A number of ciliates are common to the zooplankton, although they usually do not dominate except in certain situations (e.g., in very shallow lakes or in the deeper strata of nearly anaerobic hypolimnia). Ciliates can move much more rapidly (200–1000 μm s^{-1}) than other protozoa[1] (0.5–3 μm s^{-1} among those with pseudopodia; 15–300 μm s^{-1} among those with flagella). Such movement contributes significantly to greater dispersal and the higher feeding rates of ciliates.

3. Sarcodines

Sarcodine protozoa are poorly represented in the zooplankton of fresh waters. Even in eutrophic lakes, sarcodines occurred in modest average abundance (Laybourn-Parry *et al.*, 1980). Heliozoans are restricted vertically largely to the epilimnion and metalimnion of stratified lakes. Seasonal variations are great; for example, in Lake Constance abundances ranged from below detection limits in the spring (April to mid-June) to summer maxima of up to 6.6 cells ml^{-1} (Zimmermann *et al.*, 1996). Seasonal mean heliozoan production was, however, equivalent to only about 1% of the combined ciliate and flagellate production in this lake. Testate amoebae, particularly *Difflugia*, are occasionally abundant. For example, populations of the lobosan rhizopod *Difflugia limnetica*, a relatively common planktonic protozoan in several German lakes, increased in early summer and attained a maximum in late summer (Schönborn, 1962). Much of the population then sank as fat globules were metabolized and density increased (Fig. 16-1). Many died and decomposed, some encysted, and others remained active throughout the winter in the littoral sediments. The benthic populations of *D. limnetica* slowly increased and then in late spring became planktonic by reducing their density through formation of fat inclusions and gas bubbles. The tests of this species and some others are covered with minute sand grains; in the planktonic phase, the test may be covered instead with diatom frustules or quartz grains that are circulated into the upper strata (Fig. 16-1). Many other protozoans exhibit this type of meroplanktonic existence, in which only a portion of their life cycle is planktonic.

4. Distributions

Although nearly all protozoa and related protists are aerobic, a majority can grow very well even when oxygen concentrations are low (see, e.g., Bragg, 1960). This microaerophillic ability is conspicuous among the planktonic and benthic ciliates and is attested to by their major development in organic-rich and organically polluted waters (cf. review of

[1] The swimming rates are temperature dependent (e.g., two species of *Loxodes* averaged 270 μm s^{-1} at 10°C and 430 μm s^{-1} at 20°C (Jones and Goulder, 1973).

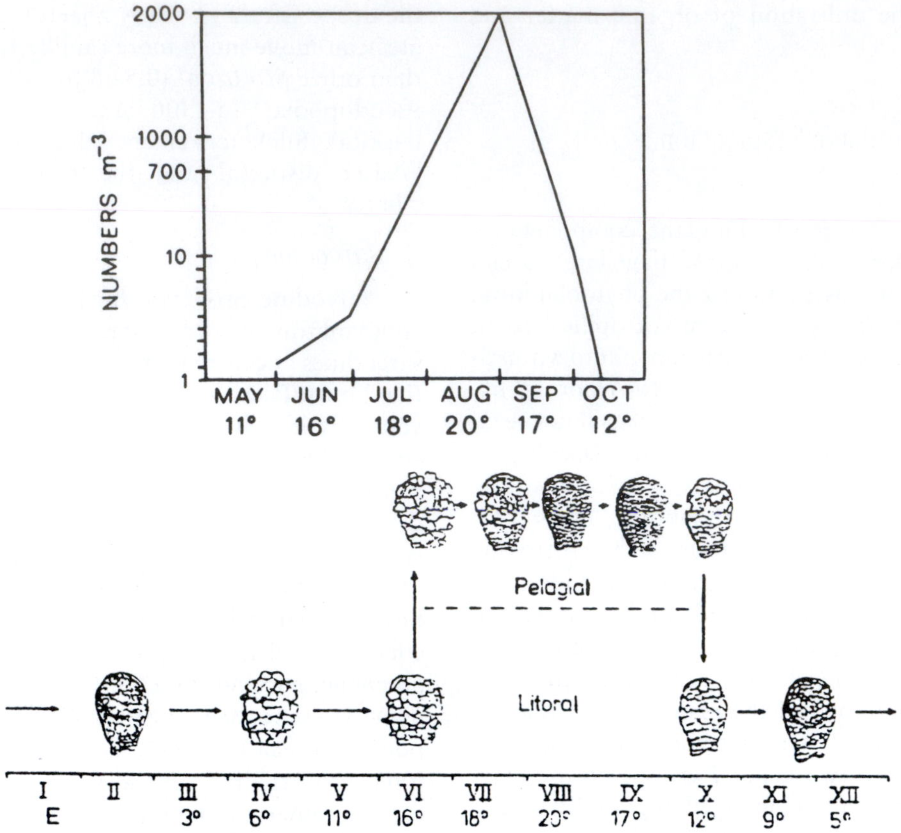

FIGURE 16-1 Changes in populations of the protozoan *Difflugia limnetica* in representative German lakes. *Upper*: Population densities per cubic meter during the planktonic phase. *Lower*: Generalized scheme of morphological changes during the population transitions between the littoral and planktonic habitats. *E* = Ice period. (Modified slightly from Schönborn, 1962.)

Curds, 1982). Various species combinations utilized as indicators of waters of different degrees of organic stress and pollution (cf., e.g., saprobic organism indices, Sládeček, 1973; Foissner *et al.*, 1991, 1992, 1994, 1995; Sládeček and Sládečkova, 1998). Populations of ciliates often develop in water strata greatly reduced in or devoid of oxygen in which bacterial populations tend to be dense, such as in the lower portions of the hypolimnia and the monimolimnion of meromictic lakes.

Four distinct types of ciliated protozoa can occur and be differentially distributed vertically in stratified productive lakes (Finlay, 1990). (1) Specialized anaerobic ciliates (e.g., *Saprodinium*) inhabit anaerobic hypolimnia, utilize particulate organic particles, and have no photosynthetic symbionts, but they do have symbiotic methanogens, which probably remove H_2 as CH_4. (2) Microaerophilic ciliates (e.g., *Loxodes*) are free of symbionts, inhabit dark and microaerobic strata at the metalimnion/hypolimnion interface, and move into lower anaerobic water, where they respire NO_3^-, when the microaerobic zone is illuminated. (3) A third group of ciliated protozoa (e.g., *Frontonia*) lives in the illuminated microaerobic zone of the metalimnion of eutrophic lakes where it benefits from upward diffusion of CO_2 and NH_4^+ that is fixed by its zoochlorellae symbionts. (4) A final group of epilimnetic ciliates (e.g., *Strombidium*) is planktonic and periodically sequesters symbiotic chloroplasts from ingested algae. Distribution of each of these ciliate types is controlled by their biochemical adaptation to anaerobic conditions, the requirements of their symbionts, and their physiological responses to light, oxygen, and gravity. Grazing pressure on anaerobic and microaerophilic ciliates by metazoa has been found to be very low or insignificant (Guhl *et al.*, 1996).

Within lakes that stratify thermally and are moderately productive, two distinct planktonic protozooplanktonic communities develop during summer stratification. An epilimnetic protistan community of obligate planktonic species overlies a hypolimnetic community of migrants from the benthic sediments

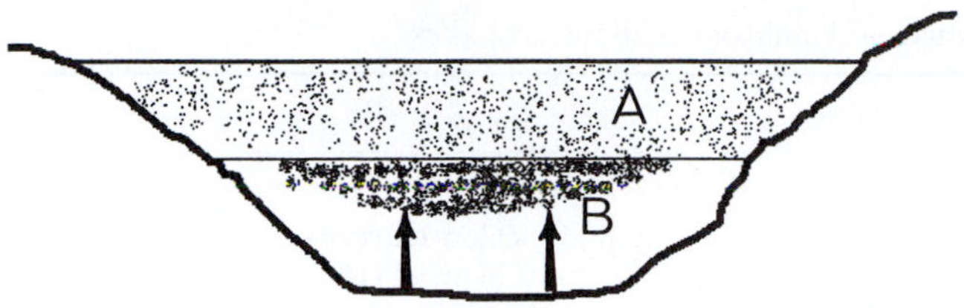

FIGURE 16-2 Distribution patterns of protozooplankton during summer stratification in a eutrophic lake. (a) epilimnetic obligate planktonic species and (b) hypolimnetic community of benthic migrants. (Modified from Laybourn-Parry, 1992.)

(Fig. 16-2). Migration of benthic protozoan communities from the sediments into the deeper strata of the hypolimnion occurred seasonally in response to reduction of dissolved oxygen. For example, two dominant species, *Loxodes* and *Spirostomum*, positioned their populations close to the 1 mg liter^{-1} oxygen isopleth (Goulder, 1974a, 1975; Bark, 1981, 1985; Finlay, 1981). Oxygen is toxic to *Loxodes* and its effect is exacerbated by light (Finlay *et al.* 1986). There is additional evidence that *Loxodes* can respire nitrate as well as oxygen. In another example, in midsummer in large Lake Dalnee, Kamchatka, Russia, flagellate and ciliate protozoans made up a substantial part of the pelagic zooplanktonic community (Sorokin and Paveljeva, 1972). Their maximal biomass was observed during the period of the decline of early summer algal populations and the simultaneous intensive development of bacterial populations. In this lake, protozoan maxima occurred in distinct layers of the water column, generally between 10 and 20 m; their biomass (ca. 3 g m^{-3}) approached the total biomass of other zooplankton at their maximum development. Protozoan utilization of planktonic bacteria in this lake in July constituted a major pathway of energy flux among the fauna in the pelagic zone (discussed further in Chap. 17). Bacterial abundances for protozoan feeding tend to be high at the metalimnetic–hypolimnetic interface (Chap. 17). At shallow sites within the lake basin above the metalimnion, no migration of these species occurs (Laybourn-Parry *et al.*, 1990b).

Mixotrophic ciliates and flagellates are typically distributed within the upper water column in relation to the depth distribution of light. The euphotic zone is shallower among eutrophic lakes, often <5 m, but extends much deeper in less productive lakes. Some flagellates, such as *Ceratium* (see Chap. 15), can migrate extensively to position themselves on a diurnal basis in relation to light, nutrient, and oxic conditions (Reynolds, 1976). Heterotrophic flagellates migrate extensively and often reach very high densities (>140,000 ml^{-1}) in anoxic hypolimnia of eutrophic lakes that contained abundant bacterial communities (Laybourn-Parry *et al.*, 1990b). Abundances of heterotrophic flagellates in most natural plankton assemblages range from 10^2 to 10^6 per liter (Table 16-1) and occur in

TABLE 16-1 Representative Abundances of Planktonic Heterotrophic Flagellates in Fresh Waters

Trophic status/lake/stream	*Flagellates* × 10^3 *ml*$^{-1}$	*Source*
Oligotrophic		
Crooked Lake, Antarctica	0–0.51	Laybourn-Parry *et al.* (1995)
Mountain streams, UK	0.5–1.5	Berninger *et al.* (1991)
Wastwater, UK	2.9 (2.2–3.3)	
Piburger See, Austria	0.75–8.1	Sommaruga and Psenner (1995)
Mesotrophic		
Konstanz, Germany	0.5–8.1	Weisse (1990)
Windermere, UK	16 (13–19)	Berninger *et al.* (1991)
Kinneret, Israel	0.66–1.3	Hadas and Berman (1998)
Black Creek, Georgia, USA	1.5–1.9	Carlough and Meyer (1989)
Ogeechee River, Georgia, USA	1.9	Carlough and Meyer (1989)
Biwa, Japan	0.67–11	Nakano (1994)
Eutrophic		
Oglethorpe, Georgia, USA	1–10	Bennett *et al.* (1990)
Cisó, Spain	0.005–2.8	Jürgens *et al.* (1994)
Priest Pot, UK	59 (56–62)	Berninger *et al.* (1991)
Valencia, Venezuela	0.022	Lewis (1985)
Laneo, Philippines	1.88	Lewis (1985)
Hypereutrophic		
Grosser Binnensee, Germany	0.5–30	Jürgens and Stolpe (1995)
Soda lakes, UK	240–400	Berninger *et al.* (1991)

TABLE 16-2 Representative Range of or Mean Maximum Abundances of Planktonic Ciliates in Lakes

Lake	*Trophic status*	*Ciliates × 10³ liter⁻¹*	*Source*
Temperate			
Lunzer Untersee, Austria	Oligotrophic	1.8	Schlott-Idl (1984a)
Piburger, Austria	Oligotrophic	0.76–12.0	Schlott-Idl (1984b)
		3.5 (0.1–101.4)	Sommaruga and Psenner (1995)
Taupo, New Zealand	Oligotrophic	0.2–2.6	James *et al.* (1991)
Lakes of Quebec, Canada	Oligotrophic	3.3–6.0	Pace (1986)
Nine Ontario Lakes, Canada	Oligotrophic	17.1–37.4	Gates (1984), Gates and Lewg (1984)
Windermere, UK: North basin	Meso/oligotrophic	4.9	Laybourn-Parry (1992)
Windermere, UK: South basin	Mesotrophic	5.4	Laybourn-Parry (1992)
Lakes of Quebec, Canada	Mesotrophic	6.7–8.7	Pace (1986)
18 lakes of Quebec, Canada	Mesotrophic	1.7–51.1	Gasol *et al.* (1995)
Constance, Germany	Mesotrophic	4.7–18.5	Müller *et al.* (1991)
Glubokoe, Russia	Mesotrophic	7.0	Shcherbakov (1969)
Esthwaite, UK	Eutrophic	9.2	Laybourn-Parry *et al.* (1990a)
Okaro, New Zealand	Eutrophic	0.3–10.4	James *et al.* (1995)
Beloe, Russia	Eutrophic	11.0	Shcherbakov (1969)
Lakes of Quebec, Canada	Eutrophic	8.5–21.6	Pace (1986)
Cisó, Spain	Eutrophic	<1–331	Jürgens *et al.* (1994)
Kettle Mere, UK	Hypereutrophic	<1–1530	Bark and Watts (1984)
Subtropical			
Florida lakes	Oligotrophic	10.8	Beaver and Crisman (1982)
Florida lakes	Mesotrophic	27.5	Beaver and Crisman (1982)
Kinneret, Israel	Mesotrophic	3–47	Hadas and Berman (1998)
Florida lakes	Eutrophic	55.5	Beaver and Crisman (1982)
Oglethorpe, Georgia, USA	Eutrophic	<1–200	Pace and Orcutt (1981)
Florida lakes	Hypereutrophic	155.5	Beaver and Crisman (1982)
Tropical			
Tanganyika, Africa	Oligotrophic	8.8	Hecky and Kling (1981)
Laneo, Philippines	Eutrophic	28 (ciliates and amoebae)	Lewis (1985)
Valencia, Venezuela	Eutrophic	218 (ciliates) 111 (amoebae)	Lewis (1985)
General Range			Beaver and Crisman (1989)
	Ultroligotrophic	2.4	
	Oligotrophic	2.3–10.8	
	Mesotrophic	18.0–70.9	
	Eutrophic	55.5–145.1	
	Hypereutrophic	90–215	

much greater abundance, by about 3 orders of magnitude, than the ciliates (Table 16-2).

5. Seasonal Dynamics

Seasonal variations in flagellate and ciliate abundance are variable. Cryptomonad and other microflagellates are commonly numerically dominant across a spectrum of oligotrophic through hypereutrophic lakes (Stewart and Wetzel, 1986). Having high nutritional quality, short turnover times, and ability to grow and reproduce at low intensities of light, these microorganisms function to stabilize variations in nutrient recycling in many spatial and seasonally temporal niches within lakes that are ineffectively utilized by other microbial and small metazoan organisms. For example, in a small oligotrophic lake in Michigan, small microflagellates (<6 μm) and cryptomonads constituted ca. 80% of the total planktonic units throughout the year, but their contributions to algal biomass were nearly always <10% (Taylor and Wetzel, 1988) (Fig. 16-3). Vertical stratification of species occurred but was relatively minor and most conspicuous among certain dinoflagellate and microflagellate species. Predation likely becomes increasingly important in the regulation of the abundance and community composition of protozoans and related protists as lake and river ecosystems become more productive. Where required resources for growth are abundant, the protozoan community is increasingly regulated by predation and correlated with the biomass and species composition of the metazooplankton.

Generally, ciliate abundance is coupled to maximum abundance of phytoplankton in the spring and early summer months (Beaver and Crisman, 1989;

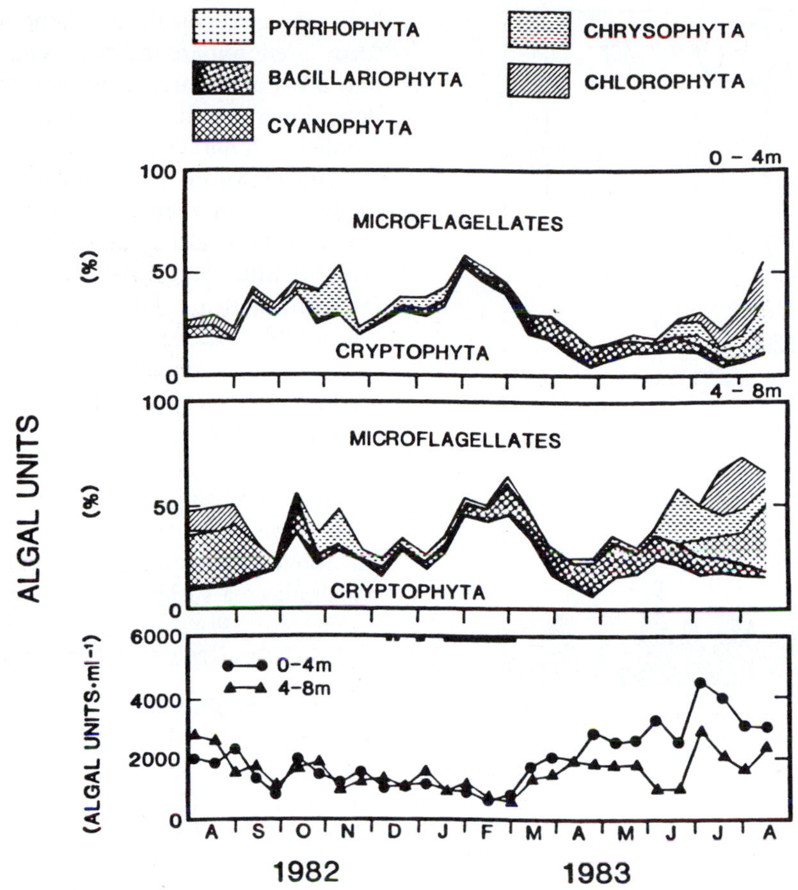

FIGURE 16-3 Percentage microbial units within major phytoplankton and protistan groups in the 0–4-m stratum (*upper*) and 4–8-m stratum (*center*), and total algal unit concentrations (*lower*) in those strata over an annual cycle in Lawrence Lake, Michigan. (From Taylor and Wetzel, 1988.)

Müller *et al.*, 1991; Laybourn-Parry, 1992, 1994). A secondary maximum in the autumn is frequently found in eutrophic and mesotrophic lakes. Species successions over the annual cycle indicate tintinnids are common spring forms in eutrophic and mesotrophic temperate lakes. The tintinnids often decrease during the summer and concentrate in the metalimnion, and later in the autumn increase again in a secondary maximum of lesser magnitude than in the spring.

This seasonal successional pattern among heterotrophic nanoflagellates is exemplified well in Lake Constance (Konstanz), Germany (Weisse, 1991). Nanoflagellate abundance and biomass were lowest in winter and highest in late spring several weeks after the phytoplankton spring maximum (Fig. 16-4). Microflagellate production was balanced through much of the year by grazing by ciliates; predation by larger zooplankton (rotifers and daphnid cladocerans) was significant only during the spring "clearwater" phase (see p. 441). The abundance and seasonal succession patterns of microflagellate communities are controlled both by food resources, particularly in oligotrophic waters, and by predatory losses. Ciliate predation on flagellates is a major regulatory mechanism, especially in more productive waters (see later discussion).

Somewhat analogous seasonal distributions were found in the detailed studies of flagellate, sarcodine, and ciliate communities of Neumühler See, Germany (Mathes and Arndt, 1995). Small heterotrophic nanoflagellates (<15 μm, HNF) were dominant in late summer and in winter, whereas large heterotrophic flagellates (>15 μm, LHF) were dominant in early spring. Ciliates, mostly oligotrichs, formed 50% of protistan biomass on an annual average, whereas HNF (mainly chrysomonads) constituted an average of 29%, LHF (mainly dinoflagellates) 19%, and sarcodines (predominantly heliozoans and testeceans) 2% of the biomass. The protistan zooplankton of these four groups constituted 21% of the annual average total zooplankton biovolume within a range of 3–78% over the year.

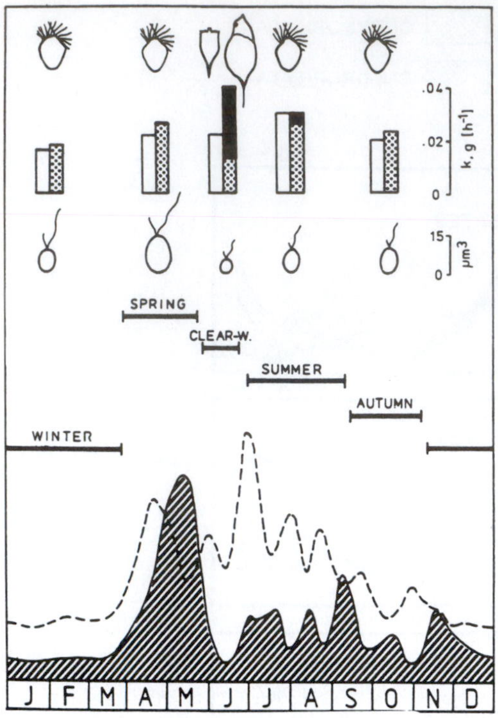

FIGURE 16-4 Seasonal cycle of biomass and cell size of heterotrophic flagellates in relation to bacterial food supply and varying micro- and macrozooplankton grazing impacts. The pictograms in the upper portion denote the major flagellate consumers in the various phases (oligotrioch ciliates, rotifer, *Daphnia*). The bars indicate gross growth rates (k, open bars) and grazing loss rates (g) by ciliate microzooplankton (stippled) or macrozooplankton (filled), respectively. In the lower part, the biomass of bacteria (dashed line) and flagellates (hatched area) are compared. Note that the flagellate biomass is five times that of the bacterial biomass. (From Weisse, 1991; The annual cycle of heterotrophic freshwater nanoflagellates: Role of bottom-up versus top-down control. J Plankton Res. *13*:167–185, by permission of Oxford University Press.)

Among dystrophic lakes that receive very high concentrations of allochthonous dissolved organic matter, microaggregates of particulate detrital material are abundant in the plankton. These aggregates can form foci for bacteria and microflagellates and can likely serve as a significant supplemental food resource, along with bacteria and microalgae, for both microflagellates and ciliated protozoa (Laybourn-Parry *et al.*, 1992, 1994; Carrias *et al.*, 1994). The highly stained waters can attenuate light severely and restrict mixotrophy among protists.

Summer protozooplankton in eutrophic and mesotrophic lakes is dominated by *Strombidium* spp., vorticellids, and *Epistylis* (Laybourn-Parry, 1992). In Lake Tanganyika in equatorial Africa, the biomass of protozoan zooplankton often exceeded that of the phytoplankton in the euphotic zone while the lake was stably stratified (Hecky and Kling, 1981). The dominant protozoan (*Strombidium*) was, however, probably symbiotic with zoochlorellae algae and mixotrophy imposed lower demands on the available food. Mixotrophic protozoa predominate in the summer among all lake types but are particularly common in oligotrophic waters. Both heterotrophic and autotrophic flagellates commonly reach maximum abundances in midsummer.

Among subtropical monomictic lakes that are generally productive and have high average temperatures, ciliate abundances persist at high levels for longer periods of time and often exhibit a bimodal pattern (mesotrophic) or late summer–autumnal maximum (Pace, 1982; Beaver and Crisman, 1989). Some of these increases may be associated with increased nutrient loading and primary productivity during and following mixing in the autumn. Subtropical lakes maintain higher densities of protozooplankton than do lakes of comparable trophic status in the temperate zone (Table 16-2). This increase in productivity is likely related to the longer growing season and higher temperatures, which allow a sustained productivity during the winter.

B. Protistan Growth Rates and Productivity

Phagotrophic flagellates are able to grow at rates comparable to those of autotrophic ultraplankton (0.2–20 μm). Population growth rates of phytoplankton can be <1.0–2.0 day^{-1} under optimal conditions (Chap. 15). The nanoflagellates of a size range of ca. 20–500 μm^{-3} can have generation times of a few hours and reproduce several times per day (Table 16-3). Dinoflagellates (size range ca. 600–50,000 μm^{-3}) rarely have generation times of less than a day, whereas ciliate (size range ca. 2000–150,000 μm^{-3}) generation times are variable in the range of 0.5 to several times per day, although most are about one to two times per day under culture conditions (Table 16-3). Growth rate is inversely proportional to cell size and directly related to temperature (Müller and Geller, 1993; Sherr and Sherr, 1994; Weise, 1997).

Estimates of production rates of protists in lakes and rivers are rare, particularly over an annual period. In a detailed study of the ciliate communities of Lake Constance, Germany, the prostome nanociliates (<20 μm) dominated numerically, while strobiliid ciliates in the size range 20–35 μm contributed most significantly to ciliate production. Ciliate community production[2] was ca. 10–15 g C m^{-2} yr^{-1} in this mesotrophic lake (Müller, 1989). Heterotrophic nanoflagellate production in Lake Constance was higher than that of the ciliates, a direct reflection of the much greater densities of flagellates by a factor of ca. 1000 (Table 16-1) in comparison to ciliates (Table 16-2)

[2]Mean daily production (P) was estimated from the intrinsic growth rates (r_m), generation times of each species, and the mean annual biovolumes (V) of the different taxa, where $P = r_m \cdot V$.

TABLE 16-3 Intrinsic Growth Rates (r_m, day^{-1}) and Minimum Generation Times (h) of Phagotrophic Protists[a]

Group	*Range of temperature*	*Common average r_m (day^{-1})*	*Range of r_m (day^{-1})*	*Generation time (h)*
Nanoflagellates	6–24	4	0.7–6.0	2.8–231
Dinoflagellates	1–27	0.8	0.3–1.3	13–72
Ciliates				
Bacteriovores	4–25	1.5	0.0–6.65	3.6–1200
Algivores	5–22	1.0	0.03–2.1	11.5–800

[a] From data summaries of Müller (1989), Chrzanowski and Šimek (1993), Müller and Geller (1993), Sherr and Sherr (1994), Macek *et al.* (1996), and Jack and Gilbert (1997).

and much greater flagellate growth rates (Table 16-3). The resulting protistan production constitutes a significant part of total zooplankton productivity. For example, in oligotrophic Lake Michigan, heterotrophic protozoan production was equivalent to 40% of the average bacterioplankton production (Carrick *et al.*, 1991, 1992). The phototrophic protozoan production represented 24% of the total planktonic primary production in the epilimnion of Lake Michigan.

C. Protistan Feeding and Trophic Interactions

Protists possess a variety of nutritional mechanisms that include both autotrophy and heterotrophy. The heterotrophic nutritional modes of free-living protists consist of

a. Uptake and assimilation of dissolved organic compounds through the cell plasma membrane (osmotrophy) or by pinching off small-sized vesicles (pinocytosis)
b. Feeding directly on living or dead particulate organic matter (phagotrophy) or by generating water currents to the cell surface where phagocytosis occurs, extending a sucking tentacle (some dinoflagellates), pseudopodial engulfment, or endocytosis within the cytosome (mouth)
c. Metabolic exchange with endosymbionts (a type of mixotrophy)
d. Combinations of the above mechanisms

(Nisbet, 1984; Gaines and Elbrachter, 1987; Sleigh, 1989; Capriulo *et al.*, 1991; Fenchel, 1991; Sanders, 1991a; Taylor and Sanders, 1991; Radek and Hausmann, 1994; Posch and Arndt, 1996).

1. Flagellates

Both phytoflagellates and zooflagellates can gain nutrition by heterotrophic mechanisms. These protists possess the following characteristics: (a) a cytosome in many species; (b) particulate food or nutrient uptake occurring by diffusion, active transport across membranes, or endocytosis; (c) mixotrophy, by which certain species can shift from autotrophic photosynthesis to heterotrophy in the absence of light or practice both simultaneously (e.g., *Ochromonas*); and (d) the use by some species of one or more flagella to create feeding currents to move food particles to the cytostome. Food capture occurs by filtration in many planktonic heterotrophic flagellates. Direct encounter and ingestion is common among benthic flagellates, likely aided by chemoreception.

Certain flagellate protists shift among trophic levels by combining photosynthesis and particle ingestion. Some flagellate photosynthetic algae, commonly chrysomonads, cryptomonads, and dinoflagellates, ingest living or nonliving particles by phagocytosis (Sanders 1991a,b). Conversely, some ciliate and sarcodine protozoa sequester chloroplasts after ingestion of algal cells. The chloroplasts, although no longer dividing, continue to photosynthesize and can contribute to the nutrition of the protist. This endosymbiotic process results in mixotrophy, in which phototrophic and phagotrophic modes of nutrition shift rapidly with changing environmental conditions. For example, the mixotrophic chrysophyte *Poterioochromonas malhamensis* immediately shifts to phagotrophy when abundant bacteria are available (Sanders *et al.*, 1990). Phototrophy was inducible in the light during starvation and is potentially a long-term survival strategy for this mixotrophic alga. The range of size of particles ingested by heterotrophic nanoflagellates extends from common bacteria to ca. 30 μm, but most particles ingested are in the range of 0.2–5 μm (Fig. 16-4).

Feeding rates among flagellates utilizing phagocytosis are highly variable. Representative average values for many lakes and stream are about 15 bacteria per individual per hour within a range of 2–53, or somewhat greater among colonial flagellates (Table 16-4). Even though the ingestion rates of bacteria by

TABLE 16-4 Protistan Grazing Rates on Bacterioplankton or Protists

Ecosystem	*Ingestion rate bacteria protist^{-1} h^{-1}*	*Source*
Flagellates on bacterioplankton		
Lake Oglethorpe, Georgia, USA		Sanders *et al.* (1989)
Individuals	2–53	
Colonial	27–181	
Lake Michigan	17–33	Scavia and Laird (1987)
Lake Vechten, Netherlands	2.2–17	Bloem and Bär-Gilissen (1989) Bloem *et al.* (1989)
Ogeechee River, Georgia, USA	1.7–43	Carlough and Meyer (1990, 1991)
Plußsee, Germany	2.3–15.1	Fukami *et al.* (1991)
Piburger See, Austria	3.4–21.2	Sommaruga and Psenner (1995)
Northern Baltic Sea	2.6–21.7	Kuuppo and Kuosa (1990)
Lake Pavin, France	1.3–92	Carrias *et al.* (1996)
Lake Arlington, Texas, USA	<1–58	Chrzanowski and Šimek (1993)
Lake Cisó, Spain	17–21	Jürgens *et al.* (1994)
Lake Constance, Germany	10–100	Weisse (1990)
Ciliates on bacterioplankton		
Lake Vechten, Netherlands	17.6	Bloem *et al.* (1989)
Lake Oglethorpe, Georgia, USA	34–1276	Sanders *et al.* (1989)
Ogeechee River, Georgia, USA	25–1140	Carlough and Meyer (1990, 1991)
Piburger See, Austria	38–373	Sommaruga and Psenner (1995)
Lake Kinneret, Israel	35–1210	Hadas *et al.* (1998)
Butrón River, Spain	25–4460	Barcina *et al.* (1991)
17 lakes of Norway	60–1000	Stabell (1996)
Lake Pavin, France	6–5910	Carrias *et al.* (1996)
Rimov Reservoir, Czech Republic	360–4200	Šimek *et al.* (1995)
Ciliates on Flagellates		
Lake Cisó, Spain	0.2–1.0	Pedrós-Alió *et al.* (1995)
Lake Constance, Germany	1–42	Weisse (1990)

flagellates are considerably less than those of ciliates (Tables 16-4 and 16-5), the much greater abundance of flagellates than of ciliates renders the flagellate grazing of bacteria dominant under most circumstances. Factors influencing rates include particularly the concentration and type of suspended particles utilized as prey, prey characteristics, especially size, and the velocity and characteristics of feeding currents generated by the flagellate. Feeding rates are directly correlated with temperature as well.

TABLE 16-5 General Characteristics of Suspension Feeding among Different Protists[a]

	Type of protist			
	Nanoflagellate[b]	*Ciliate*	*Ciliate*	*Ciliate*
Food type	Bacteria	Bacteria	Microflagellate	Ciliate
Range of clearance rate (body volumes h^{-1})	5×10^4–10^6	3×10^3–10^4	10^4–5×10^4	5×10^4–5×10^5
Range of clearance rate (ml h^{-1})	2×10^{-6}–10^{-5}	10^{-6}–10^{-4}	10^{-3}–10^{-2}	10^{-2}–5×10^{-1}
Specific protist example	*Ochromonas*	*Glaucoma*	*Paramecium*	*Bursaria*
Clearance rate (ml h^{-1})	10^{-5}	1.3×10^{-5}	10^{-3}	0.43
Minimum filter mesh (μm)	—	0.2	0.4	8
Range of food particle size (μm)	0.2–2	0.2–1	1–5	20–60
Minimum food particle Concentration (mg DW liter^{-1})	0.5	0.8–8	0.08–1.3	0.002–0.15
Body length (μm)	7	35	200	800
Body volume (μm^3)	200	4×10^3	10^5	5×10^7

[a] From data given in Fenchel (1986, 1987).
[b] Termed "microflagellate" in the older literature.

2. Sarcodine Protozoa

Testate and naked amoebae utilize pseudopods to engulf bacteria, cyanobacteria, diatoms, flagellates, other protista, and detrital organic particles as randomly encountered (Canter, 1973; Capriulo *et al.*, 1991). A few amoeboid protozoa (e.g., *Pelomyxa*) harbor endosymbiont methanogenic bacteria (Van Bruggen *et al.*, 1983). Some small amoebae are raptors (Fenchel, 1987).

3. Ciliates

Most free-living ciliates have mouths and feed on particulate food (Fenchel, 1987; Skogstad *et al.*, 1987; Capriulo *et al.*, 1991; Posch and Arndt, 1996). Organisms ingested include bacteria, coccoid cyanobacteria, microalgae, diatoms, dinoflagellates, heterotrophic microflagellates, and other ciliates, as well as detrital organic particles in the range of 0.2–20 μm. Many modes of food acquisition occur: (a) suspension feeding in which oral membranelles may act as filters; (b) active predatory hunting by random encounter or possibly chemoreception; (c) deposit feeding of particles by phagocytosis; and (d) mixotrophy by algal endosymbionts or with functional chloroplasts (e.g., *Paramecium bursarai* with *Chlorella*). The ciliate *Mesodinium rubrum* is completely autotrophic (Lindholm, 1985).

Bacterivory can be very significant for both heterotrophic bacteria and picocyanobacteria at certain times of the year. For example, over a 5-week period in late summer in a eutrophic reservoir, ca. 70% of bacterial production was consumed by heterotrophic flagellates and ca. 20% by ciliates (Šimek *et al.*, 1995). Small (<30 μm) oligotrich ciliates ingested an average of 360–2100 bacteria and 76–210 picocyanobacteria $cell^{-1}\ h^{-1}$ during this period of the year. Such ideal conditions allow high ciliate growth and doubling times of between 24 and 75 h. While only heterotrophic nanoflagellates are able to greatly impact bacterioplankton at low concentrations by grazing, many ciliates are more omnivorous (K. Jürgens, unpublished). Oligotrich ciliates feed on a spectrum from bacteria to nanoalgae and can become important bacterivores, at times exceeding bacterivory of heterotrophic nanoflagellates, which they also eat.

Smaller ciliated protozoans (<30 μm) graze upon bacteria-sized particles (Table 16-5), whereas rotifers (see later discussion) utilize both bacteria and small algae. Larger crustacean zooplankton are primarily phytophagous on small algae (Crisman *et al.*, 1981). Nutrients released when macrozooplankton graze on algae can enhance bacterial growth and thereby provide food for ciliated protozoans and small rotifers. As such, the ciliated protozoans can serve as functional links in freshwater planktonic food chains; they utilize bacteria and very small particulate detritus and provide macrozooplankton with larger particles that can be more efficiently grazed (cf. Porter *et al.*, 1979). There is some evidence that the body size of macrozooplankton (especially cladocerans) is smaller in subtropical than in temperate lakes (e.g., Crisman, 1980). Ciliated protozoans and rotifers become more important in the zooplankton among eutrophic, subtropical lakes.

Two species of benthic ciliates of the genus *Loxodes* were studied in a shallow eutrophic lake in relation to feeding and digestion of algae (Goulder, 1972, 1974a). Feeding rates were low (0.4–13 diatom cells per hour for the larger species), but there was an indication that the protozoan was able to distinguish among three species of the dominating alga *Scenedesmus* and feed selectively. Based on estimates of the feeding rates and the biomass of the algal and protozoan populations, maximum grazing by the ciliates resulted in consumption of <1% of the algal populations per day and had no appreciable effects on them. No evidence was found for interspecific competition between the two species; changes in environmental conditions affected both species in similar ways (Goulder, 1974b, 1980). These species are strongly negatively phototactic and did not migrate from the anaerobic hypolimnion to the oxygenated epilimnion. Most individuals constituting the hypolimnetic population died as the oxygen was depleted.

D. Food Utilization and Changes in Size of Bacteria

Bacterioplankton are significantly grazed by small (<5–8 μm) phagotrophic nonpigmented flagellates. These flagellates often dominate among the heterotrophic planktonic protists. Larger planktonic ciliates can also be important in bacterivory and exhibit high rates of bacterial clearance from water under both natural and experimental conditions (Table 16-4). Phagotrophic phytoflagellates also ingest bacteria at rates comparable to those of nonpigmented flagellates (e.g., Borass *et al.*, 1988; Sanders and Porter, 1988).

Bacterivory by flagellates and ciliates is influenced by temperature as well as the availability of bacterial prey. For example, at 15°C, ciliates filtered 10 times more body volume when bacteria were scarce, but the ingested bacteria were fewer than at high prey densities (Iriberri *et al.*, 1995; cf. also Peters, 1994). Size and shape of prey is important, and clearly prey geometry is the first-order determinant of ingestion through passive mechanical selection by protists. Among both flagellates and ciliates, but particularly among the heterotrophic microflagellates, there is a tendency to ingest larger bacterial cells at a frequency greater than they occur in the bacterial assemblage (Andersson *et al.*, 1986;

Chrzanowski and Šimek, 1990; Gonzalez *et al.*, 1990; Šimek and Chrzanowski, 1992), and some probability exists for chemical discrimination among prey (Verity, 1991). This size-selective grazing by microflagellates likely contributes to the frequently observed decrease in median bacterial cell volume during the summer period of temperate waters.

E. Changes in Bacterial Size and Community Structure Mediated by Interactions among Protists and Larger Zooplankton

Even though the rates of bacterivory by flagellates are smaller than those by ciliates, flagellate numbers are so much greater that their collective grazing effects on bacterioplankton can be much greater than those of the ciliates. Examples of comparative bacterivory in the plankton indicate the importance of flagellates. In contrast to the study of Rimov Reservoir where ciliates accounted for ca. 50% of bacterivory in late summer, in a monomictic reservoir in Georgia, mixotrophic and heterotrophic flagellates dominated grazing (55–99% of the total) at all times of the year (Porter, 1988; Sanders *et al.*, 1989; Bennett *et al.*, 1990). Phagotrophic flagellates are often most abundant in the epilimnion and metalimnion interface during stratification; flagellate ingestion rates tend to be greatest in the epilimnion. Ciliates were responsible for as much as 30% of the bacterivory at some depths in spring and summer, while bacterivorous rotifers were abundant only in summer. The grazing impact of cladoceran crustaceans was generally <1% in this lake, and bacterivory by adult and naupliar copepods was not detected.

Although the cladoceran and copepod bacterivory is generally small, under some conditions it can be significant or may be underestimated, particularly in relation to indirect effects on protists that in turn affect bacterial communities. In an extreme case, where planktivorous fish predation on cladocerans was essentially absent, the cladoceran *Daphnia* dominated the zooplankton community with heavy predation on protists. Under these circumstances, the percentage bacterivory by cladocerans can become a significant part of total bacterivory (e.g., Jürgens *et al.*, 1994a,b, 1997). Similarly, naupliar stages of copepods can graze heavily on protozoan microplankton, picoplankton, and bacteria (Roff *et al.*, 1995; Merrell and Stoecker, 1998). Again, although the naupliar feeding rates are individually small, the collective large numbers of immature copepods can impose both direct and indirect effects on bacterial communities and growth.

Examination of bacterial populations in lakes often demonstrates relatively stable biomass and production distributions. Yet, a number of experimental studies indicate that when predation of larger zooplankton on phagotrophic protists and protozoans occurs, this release of the bacteria from protistan grazing pressure can induce marked alterations of morphology, densities, and community structure of bacterioplankton (cf. reviews of Güde, 1989; Jürgens and Güde, 1994; and Jürgens *et al.*, 1994a,b, 1996, 1997). A major portion of bacterial production can be consumed by nanoflagellates, and moderate densities of copepods and small cladocerans have relatively small effects on the flagellate growth rates and population development (Fig. 16-5). Oligotrichous ciliates are efficient feeders on nanoflagellates, whereas differences among the filter-feeding rotifers, calanoid copepods, and cladocerans were relatively slight. Cyclopoid copepods tend to be relatively inefficient grazers on nanoflagellates. When ciliates or large cladocerans, such as *Daphnia*, are present in dense populations, however, nanoflagellate protists can be markedly suppressed.

Many bacteria can be consumed both by the nanoflagellates or by the metazoans. Depending upon the predators that dominate, the type of predator grazing can induce marked changes in bacterial morphology, usually within 24 h. Under nanoflagellate-dominated bacterivory, a shift to protistan-resistant bacteria in large aggregates and long filaments occurred. When bacteria were exposed to little protistan predation or to metazoan bacterivory, freely dispersed, single-celled, small rod and cocci morphology predominated that are less efficiently grazed by the metazoans. As depicted from laboratory experiments in Figure 16-6, these changes suggest very rapid compensatory responses in bacterial production and formation of size refuges from grazing at both the lower and upper ends of the bacterial size range as potential mechanisms buffering bacterioplankton from large seasonal fluctuations in abundance.

F. Mortality in Perspective

Clearly, *in situ* bacterivory by protistans can be high. For example, the calculated total grazing mortality ranged from 11 to 159% of bacterial cell production in the example just cited. However, theoretical values that protistan grazing could balance bacterial production (Capriulo *et al.*, 1991) are unlikely and require further study. Other sources of mortality are not only poorly understood in nature but rarely measured or even estimated. To assume that all or most mortality of organisms occurs by bacterivory, herbivory, or carnivory in natural ecosystems, as has been commonly done until very recently, is unsound scientifically and grossly misleading (cf. Wetzel, 1995).

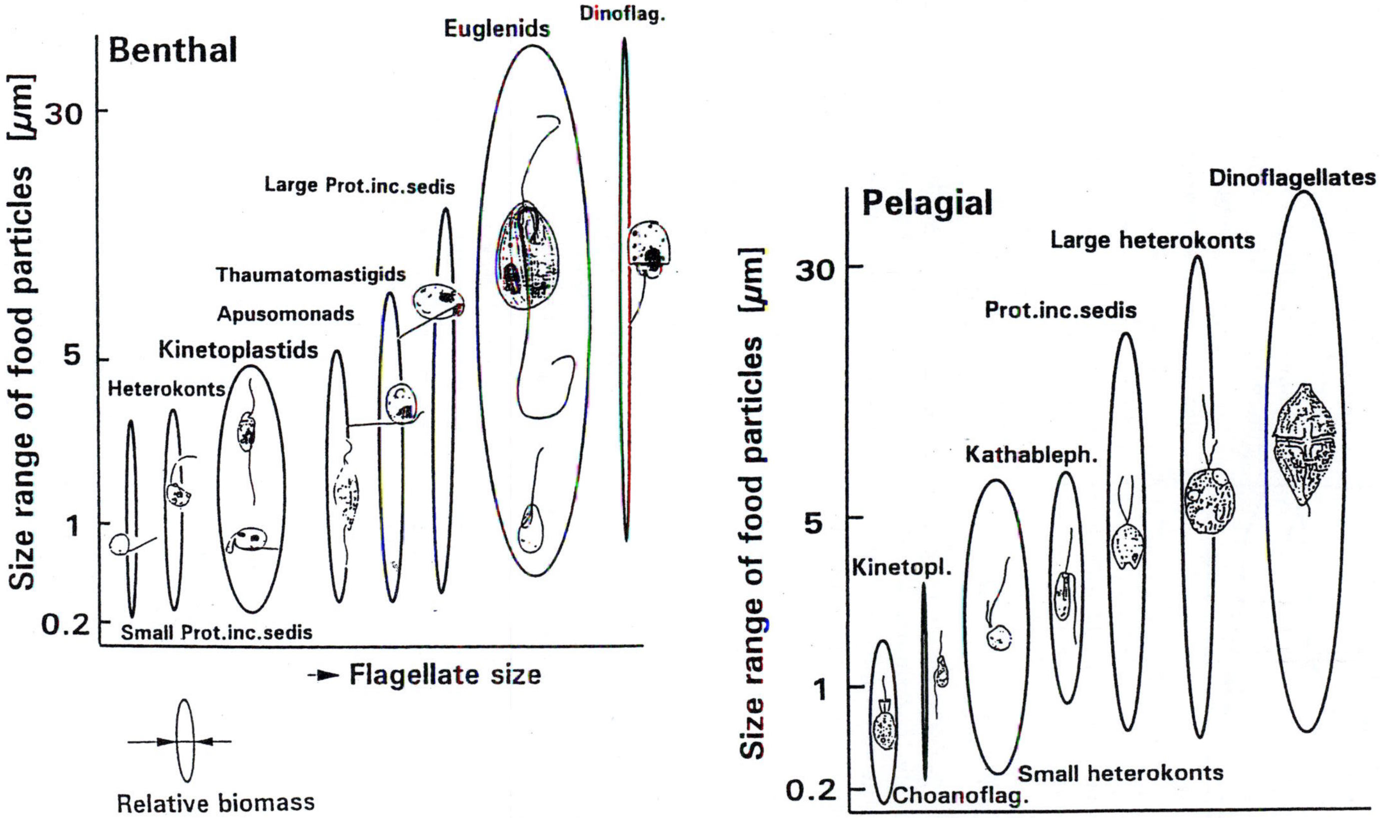

FIGURE 16-5 Comparative food size spectra of important pelagic and benthic flagellate taxa in relation to their mean body size. The relative importance of their biomass contribution is indicated by the width of the elliptical symbols. Prot. inc. sedis = Protista *incertae sedis*. (Slightly modified from Arndt *et al.*, 1999.)

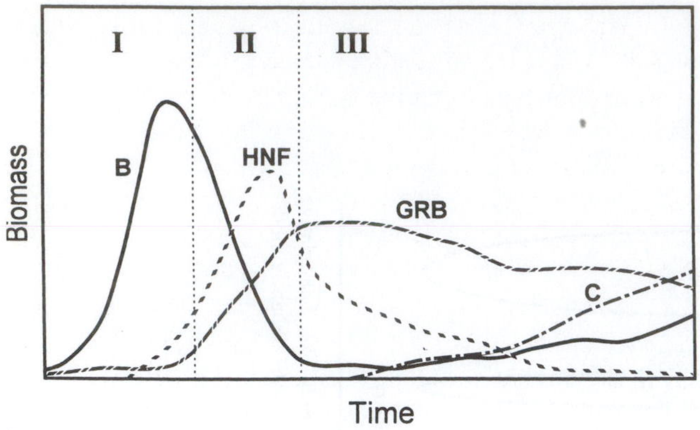

FIGURE 16-6 Model of common microbial succession under conditions of adequate sources of organic substrates for bacteria and with low populations of large zooplankton. *B* = highly edible bacteria; *HNF* = heterotrophic nanoflagellates; *GRB* = grazing-resistant bacteria as aggregates, filaments, and similar forms; *C* = ciliates. (Modified from Jürgens and Güde, 1994.)

Mortality often results simply by termination of biochemical cycles caused by genetic limits of longevity. The common approximate synchronous death of adults of metazoans after reproduction is an example. In addition to losses by simple physiological death, viral lysis can be a major means of bacterial mortality. These statements assume, however, that measures of bacterial productivity are accurate in comparative analyses. Methodological variance is high in sampling, techniques, and metabolic interpretations, and a large number of quantitatively unsubstantiated and likely most erroneous statements of mortality causes and significance litter the scientific literature.

In reservoir ecosystems, rivers can import considerable bacterial biomass from allochthonous sources. Evidence indicates that heterotrophic nanoflagellates consume much of the bacterial production (50–100%) over an annual cycle (Šimek *et al.*, 1999). With increasing trophy of the reservoir ecosystems, however, ciliates (small oligotrichs, vorticellids, and scuticociliates) become as important bacterivores as the nanoflagellates (chrysomonads, bodonids, and choanoflagellates). The bacteria of reservoirs are variably augmented by microbes that are imported from allochthonous terrestrial and wetland sources. It should be noted, however, that apparent genetic variations exist in bacteria from allochthonous versus autochthonous sources (e.g., Iriberri *et al.*, 1994) and that these differences could also lead to variations in consumption by bacterivores.

The nature and quantity of available food have been implicated as major controlling factors in the population dynamics of ciliates. For example, in a shallow pond in Pennsylvania, three population surges of ciliates occurred and were correlated with different nutritional habits (Bamforth, 1958). The small prostomatid *Urotricha* developed rapidly under ice and fed actively on a dense population of chrysomonads. Oxytrichid ciliates appeared as the *Chlamydomonas* algal populations were declining rapidly in spring. Ciliate populations, mainly of the genus *Holophyra*, appeared following spring maxima of phytoplankton and persisted during the summer, coincident with euglenoid algal blooms. Selectivity of algae ingested by ciliates is common (e.g., Takamura and Yasuno, 1983). Food selection is largely a function of the size and shape of the algae, rather than species.

Assimilation of ingested food is of course far from efficient. Little is known of the assimilation efficiencies of consumed prey of protists. Under optimal growth conditions in cultures, the assimilation efficiencies of the protozoan *Amoeba proteus* ranged from 22 to 59% regardless of temperature (Rogerson, 1981). Under ideal growth conditions, assimilation efficiencies as high as 75% have been observed for some protists (Sleigh, 1989). Assimilation efficiencies in nature are likely much lower than this range.

Fecal aggregates are released from protists during feeding. The quantitative variations in such fecal releases certainly varies with prey resource availability. Some particulate egestion values for ciliates are in the range of 20–25% of the volume of algae consumed (Stoecker, 1984). Release of excretory dissolved organic matter by protists is very poorly understood, even though such organic recycling could be highly significant to ecosystem carbon fluxes. General figures cited (Sleigh, 1989) estimate that 30% of food consumed by an actively growing protist is used in respiration, 30% is egested or excreted, and approximately 40% contributes to growth.

Grazing by protists on phytoplankton has been intensively studied, particularly in marine habitats (Capriulo *et al.*, 1991). Phagotrophic ciliate and dinoflagellate ingestion of phytoplankton often equals that of other microzooplankton (copepod nauplii and rotifers). Frequently grazing impacts 20–60% of production of the algal picoplankton (<2 μm in size).

III. TROPHIC RELATIONSHIPS OF PROTISTS IN PELAGIC FOOD WEBS

The functional roles of microbiota in pelagic food webs has received extraordinary study in the past decade. This emphasis is important because clearly a major portion of the pelagic metabolism is occurring within a microbial food web. Before the importance of these microorganisms is evaluated as a support component of the metazoan food web, it is essential to examine the primary characteristics of these pelagic microbial communities.

A. Microbial Food Web

The microbial food web concept was developed in marine studies long ago with the position that heterotrophic and phototrophic microbial metabolism constituted a significant portion of the total metabolism of the ocean (Vernadskii, 1926). With many data from detailed studies of a lake ecosystem, Wetzel *et al.* (1972) developed a general paradigm which stated that bacteria metabolize a significant fraction of total photosynthetic production in natural waters. In particular, that work demonstrated that most of the microbial utilization of primary production occurred in littoral and benthic regions and that only a small portion of the ecosystem primary production entered the metazoan trophic levels (see full discussion in Chap. 23). That same paradigm was developed for pelagic marine waters in a seminal paper by Pomeroy (1974). Many subsequent studies provided data to reinforce the importance of microbes; these were synthesized in the concept of the "microbial loop" (Azam *et al.*, 1983; Sherr and Sherr, 1988).

The ***microbial loop*** is simply a model of the pathways of carbon and nutrient cycling through microbial components of pelagic aquatic communities. In addition to bacterial uptake of nonliving organic matter, many direct links exist among algae, bacteria, and other heterotrophic microbes (Fig. 16-7). By these numerous pathways, fixed organic carbon can be recovered, often into larger-sized microorganisms that may be more available for consumption by larger organisms (Sherr and Sherr, 1988). Metazoans would utilize organisms of the microbial food web by grazing phytoplankton, colorless flagellates, and ciliates >5 μm. In oligotrophic waters that may be dominated by phytoplankton too small (ca. <5 μm) to be effectively

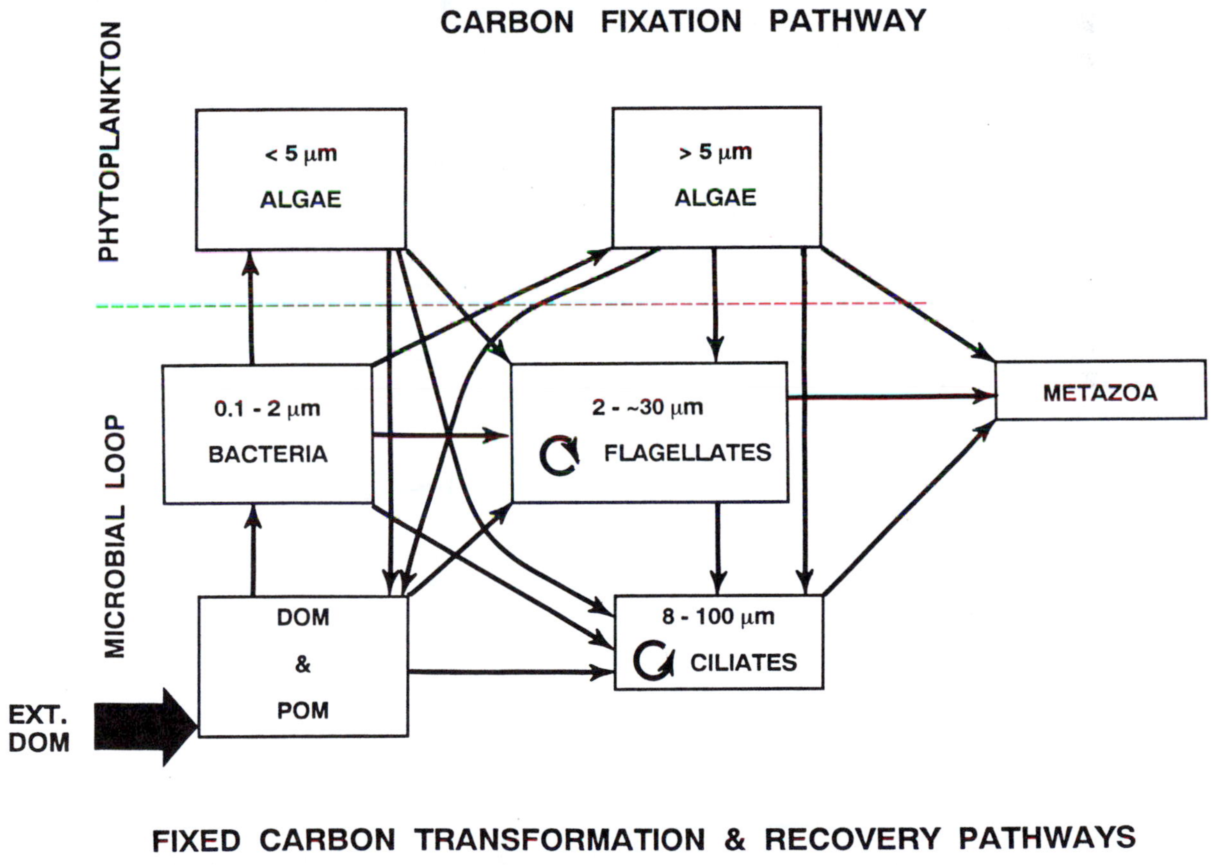

FIGURE 16-7 Trophic interactions within the microbial food web, which is separated here into phytoplankton and "microbial loop" (bacterial and protistan) components. Note the many direct links among heterotrophic and autotrophic microbes, as well as ingestion of phytoplankton by flagellates and ciliates and ingestion of bacteria by mixotrophic algae. The curved arrows in the flagellate and ciliate compartments indicate further predator–prey interactions within these broad classes of organisms. In this model, production of the <5-μm algae is accessible to metazoa only after being transformed into larger protozoan cells. A large portion of the dissolved organic matter (DOM) enters the pelagic from littoral and allochthonous sources in most lakes and rivers. (Modified from Sherr and Sherr, 1988.)

consumed by many zooplankton, the phagotrophic flagellates and ciliates could function as a primary trophic link within the ecosystem.

Under most prevailing conditions, significant fractions, at least 50%, of the dissolved organic matter and nonliving particulate organic materials are metabolized to gases, soluble nutrients, and living bacterial biomass (Fig. 16-7). This bacterial biomass may be living, dormant, senescing and utilized by other bacteria and viruses or may be consumed by bacterivores. The predominant bacterivores are protists, often microflagellates and protozoan ciliates. For example, pelagic bacterivory in eutrophic Lake Oglethorpe, Georgia, was dominated at all times by heterotrophic flagellates, which accounted for 49–81% of grazing on an areal basis and up to 98% of grazing at some depths at certain times of the year (Sanders *et al.*, 1989). Mixotrophic flagellates were major grazers during winter and spring and contributed up to a maximum of 45% of community grazing on an areal basis and to 79% at depths of maximum abundance. In late spring and early summer, ciliate bacterivory constituted as much as 30% at some depths but averaged ca. 11% over the year. Bacterivory by pelagic cladoceran crustaceans was generally <1% of the total and was not detected among copepods.

The bacterivores can also function as decomposers and efficient nutrient remineralizers (Fig. 16-8A) or transfer some of the fixed carbon and nutrients to higher organisms (Fig. 16-8B). In the prevailing conditions (Fig. 16-8A), most of the carbon and nutrients in

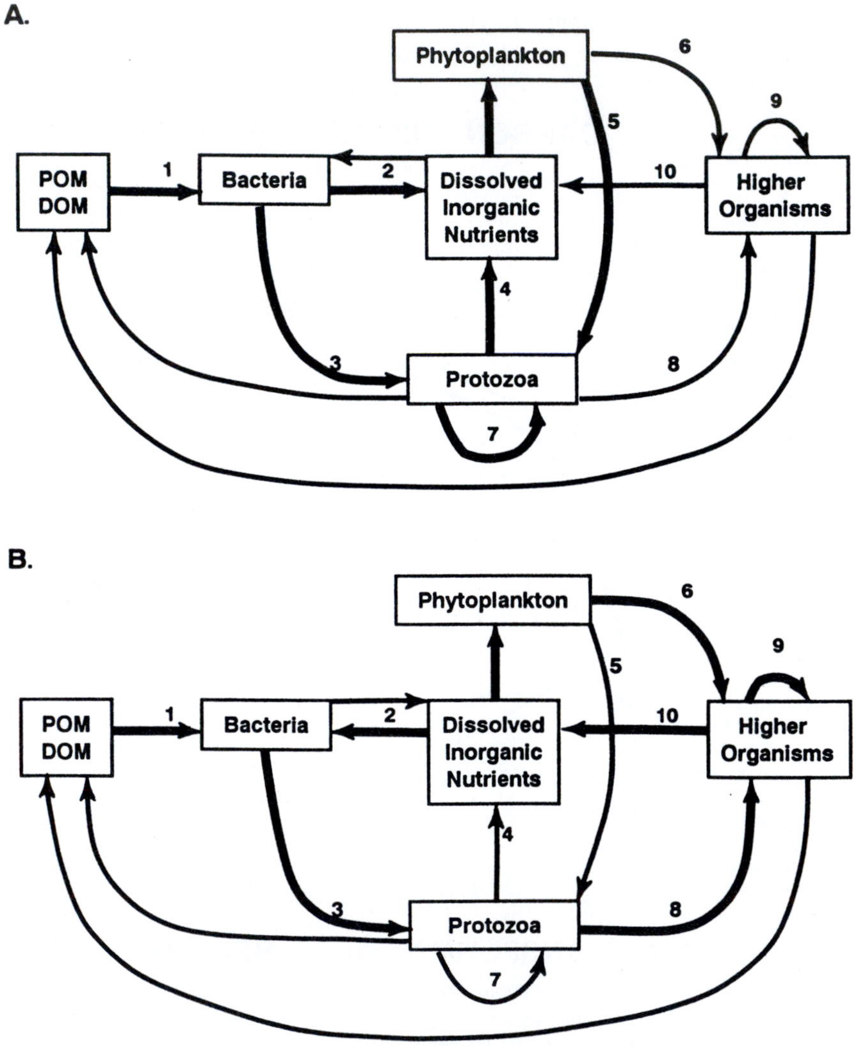

FIGURE 16-8 Idealized models for nutrient cycling through the "microbial loop" of pelagic communities. The thickness of the arrows indicates the hypothesized relative importance of the pathway for nutrient flow. (A) A microbial loop characterized by a high efficiency of nutrient remineralization. (B) A microbial loop characterized by a high efficiency of transfer of nutrients to higher organisms. (Modified from Caron, 1991.)

bacterial substrates are remineralized directly by bacteria (arrow 2) or their protistan consumers (arrow 4) (Caron, 1991). In this case, which is likely often the case, little carbon and nutrients are available to higher organisms as living biomass; hence, the microbial loop functions as both a carbon and a nutrient "sink," that is, the carbon is not directly available to higher animals. As is emphasized later (Chap. 23), however, the carbon and nutrients are not usually lost from the ecosystem but are rather displaced spatially and temporally and recycled later. The size structure, species composition, and possibly the rate of production of the phytoplankton may determine the amount of the composite assemblage that is grazed by the larger zooplankton (arrow 6) or the protozoa/protists (arrow 5). Much complexity within the protistan assemblage (arrow 7) would decrease the quantitative transfer of carbon and nutrients to higher organisms.

In a stronger "link" to the metazoan hypothesis (Fig. 16-8B), carbon and nutrients in organic substrates are utilized by bacteria (arrows 1 and 2) and incorporated into their consumers (arrow 3). As these organisms are utilized by higher organisms (arrow 8), the pathway represents a direct trophic coupling ("link") between dissolved and particulate detrital matter and higher organisms. The large zooplankton then would dominate carbon and nutrient remineralization (arrow 10) and remineralization by the animal microbial loop components would be less (arrow 4). This predation by large zooplankton can impose major regulation of carbon and nutrient fluxes to higher trophic levels. Predation by larger zooplankton is discussed in each of the subsequent sections on the major zooplankton types.

When larger zooplankton are present in abundance, mortality rates on ciliates can be high (Table 16-6), which would allow an appreciable transfer of carbon and energy from the microbial loop to higher trophic levels. Large differences in rates of ingestion of protists have been observed, and these rates are not necessarily proportional to the size of the prey or predator (Wickham, 1995). Although feeding clearance rates of copepods (ml copepod^{-1} day^{-1}) and other large zooplankton do increase in proportion to body length, under natural conditions many types of prey may be available. Predation on ciliates, for example, was appreciably less when copepods had alternative prey species available (Wickham, 1995a). The growth of the ciliate *Stobilidium velox* (43 μm) could be halted in summer by predation of copepods at modest densities (1.6 adult *Epischura* liter^{-1} or 16 *Diaptomus* liter^{-1}) (Burns and Gilbert, 1993). Experimental studies both in the laboratory and in the field demonstrated marked increases in the mortality per day with increases in the size of different species of the cladoceran *Daphnia* (μg per species) (Wickham and Gilbert, 1993; Pace and Vaque, 1994). Heterotrophic nanoflagellates were impacted more severely than ciliates, and the mortality of oligotrich ciliates was more severe than the impact on rotifers. Growth rates of both the flagellates and especially the ciliates were severely suppressed by the large (35 μg) *Daphnia*. Despite this mortality of ciliates, it is well known (e.g., DeBiase *et al.*, 1990) that ciliates alone are insufficient as a food source to support *Daphnia* population growth. The nutritional value of ciliates is variable and represents in part the highly dynamic nutritional values of their prey in relation to season and nutrient availability in the lakes.

B. Organic Carbon Release during Feeding by Protists and Larger Zooplankton

During such feeding and predation, nutrients and dissolved organic compounds are released, either directly during the ingestion and mastication processes, in excretion as dissolved organic matter, and in fecal particles that can leach subsequently into the surrounding water. Nutrients released during protistan and other zooplanktonic feeding are discussed in separate sections (see Chaps. 12 and 13). Practically nothing is known of the release of dissolved organic carbon from

TABLE 16-6 Mortality Rates of Small (< 30 μm) Ciliate Protists by Zooplankton Predators

	Predator density (ind liter^{-1})	°C	*Mortality rate (d day^{-1})*	*Ingestion rates (cells ind^{-1} h^{-1})*	*Source*
Rotifers	20–100	20	0.1–0.7		
On ciliates				32–259	Arndt (1993)
On flagellates				0.5–4.8	
Copepods	10	12–18	0.4–1.6	0–200	Wickham (1995)
				35–100	Burns and Gilbert (1993)
Cladocerans	10	20	0.1–0.8	1–10	Jack and Gilbert (1993a, 1994, 1997)
Larval fish (7-mm *Lebistes*)	0.1	20	0.03	12–125	Jack and Gilbert (1994, 1997)

protistan feeding processes in the microbial food web (cf. review of Nagata and Kirchman, 1997). Clearly, bacterial uptake of dissolved organic substrates is high. As bacteria and pico- (<2 μm) and ultraphytoplankton (<20 μm) are consumed, digested, and remineralized by microflagellates, some of the organic carbon is released. For example, among marine environments, microalgal and bacterial biomass was incorporated into herbivorous/bacterivorous microflagellates with equal efficiency (44%) during exponential growth (Caron *et al.*, 1985). Only 10% of the ingested particulate organic carbon (POC) was released as dissolved organic carbon, while 10% was released as egested POC. Some 70–80% of the organic matter ingested was respired by the microflagellates. Although ammonium regeneration rates of flagellates have been observed to be closely related to the ingested bacterial nitrogen content (e.g., Caron *et al.*, 1988), phosphate regeneration rates depended on bacterial C:P ratios (Nakano, 1994c). P release from grazing by phagotrophic flagellates was most efficient when bacteria were C-limited, a condition that led to the regeneration of 50–90% of the bacterial P (Jürgens and Güde, 1990). When bacteria were P-limited rather than C-limited, regeneration efficiency was variable and often reduced to <20%. These relationships are important because when P availability is low in lakes, regeneration rates are markedly impeded. Conversely, when loading of P increases availability and incorporation into prey, regeneration rates can be accelerated by protistan feeding activities.

Among larger zooplankton, such as *Daphnia*, a significant portion (>15%) of algal carbon ingested is released, largely as feces (Olsen *et al.*, 1986). Of the carbon released, a portion, in the range of 20%, is released as dissolved organic compounds. These compounds are generally relatively small molecules (<10,000 D), some of which consist of nitrogenous compounds and simple amino acids (Riemann *et al.*, 1986; Park *et al.*, 1997; Ferrier-Pagès *et al.*, 1998). The more labile substrates are readily utilized by bacteria. A number of more recalcitrant organic substrates are released as well during zooplankton feeding. These substances are very slowly metabolized but can be photolysed to simple substrates, such as fatty acids, by exposure to natural incident ultraviolet irradiance (e.g., Wetzel *et al.*, 1995).

IV. GENERAL CHARACTERISTICS OF ROTIFERS, CLADOCERA, AND COPEPODS

A. Rotifers

The Rotifera (Rotatoria) is a large class of the pseudocoelomate phylum Aschelminthes, clearly originating in fresh water; only two significant genera and a few species are marine. About three-quarters of the rotifers are sessile and associated with littoral substrates. Approximately 100 species are completely planktonic, and these rotifers form a significant component of the zooplankton. Rotifers form an important group of soft-bodied invertebrates of the plankton. The general characteristics of the group have been treated in some detail by Pennak (1978), Hyman (1951), Hutchinson (1967), Ruttner-Kolisko (1972), Dumont and Green (1980); and Wallace and Snell (1991).

The rotifers exhibit a very wide range of morphological variations and adaptations. In most, the body shape tends to be elongated, and regions of the head, trunk, and foot usually are distinguishable (Fig. 16-9). The cuticle is generally thin and flexible, but in some rotifers it is thickened and more rigid and is termed a *lorica*; the lorica is of taxonomic importance in some groups. The anterior end or corona of rotifers is ciliated; in some species the periphery is ciliated as well. The movement of the cilia functions both in locomotion, especially among planktonic forms, and in movement of food particles toward the mouth. The mouth, although variously located, is generally anterior. The digestive system contains a complex muscular pharynx, termed the *mastax*, and a set of jaws or *trophi* unique to the rotifers that functions to seize and disrupt food particles. Most rotifers, both sessile and planktonic, are nonpredatory. Omnivorous feeding occurs by means of ciliary movement of living and detrital particulate organic matter into the mouth cavity. Predatory species, such as the common *Asplanchna*, are usually large and prey upon protozoa, other rotifers, and other micrometazoa of appropriate size.

Most rotifers are not planktonic, but are sessile and associated with littoral substrata. Population numbers are highest in association with submersed macrophytes, especially plants with finely divided leaves; densities commonly reach 25,000 per liter (Edmondson, 1944, 1945, 1946). Even greater densities are found in the interstitial water of beach sand at or slightly above the waterline (Pennak, 1940). With reduced sites for attachment and presumably less protection from predation, planktonic rotifer populations are much less dense. Densities of planktonic rotifers of 200 to 300 liter^{-1} are common and occasionally reach 1000 liter^{-1}; densities rarely exceed 5000 liter^{-1} under natural conditions.

Several changes characterize the transition from the predominantly sessile to the planktonic life forms (Fig. 16-10). Weight reduction is common as a result of diminution of the lorica and enlargement of body volume with gelatinous materials. Planktonic species tend to have suspension processes and swimming organs in the form of immovable spines or movable setae. A reduction of attachment organs as a result of diminution or total loss of the foot structures also takes place. Adaptations that reduce the sinking rates of reproductive products also occur; for example, attachment of

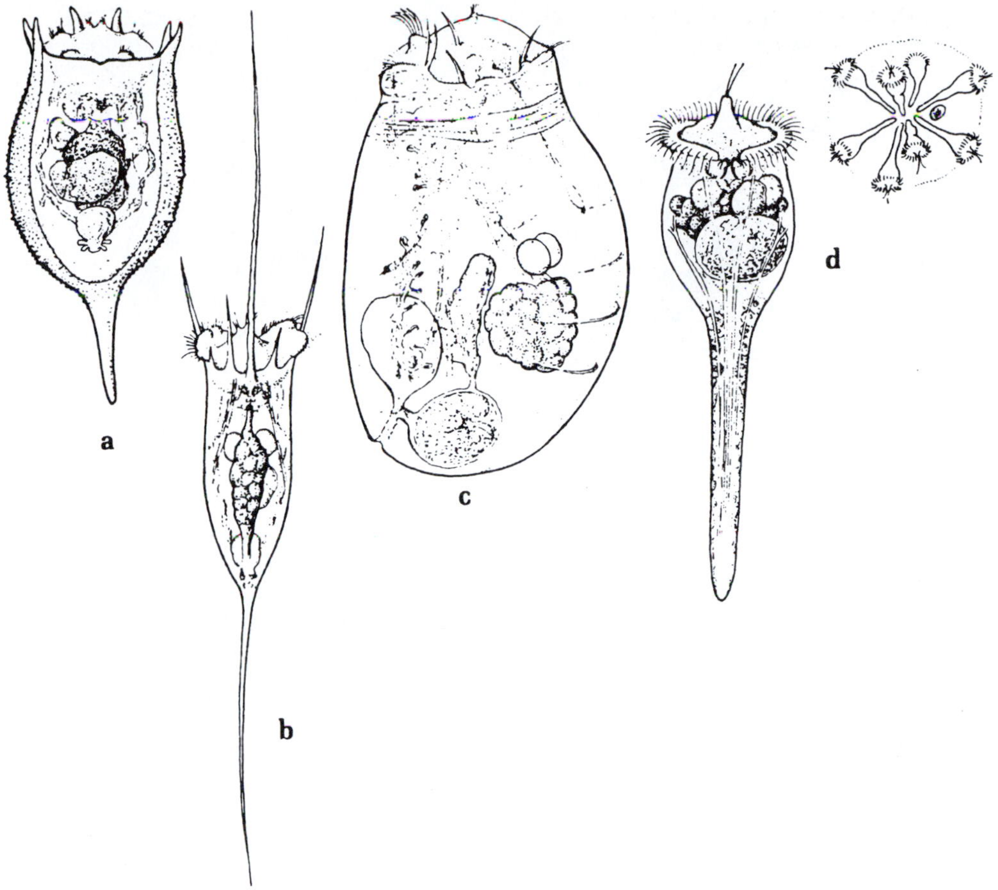

FIGURE 16-9 Exemplary planktonic rotifers: (a) *Keratella cochlearis*; (b) *Kellicottia longispina*; (c) *Asplanchna girodi*; (d) *Conochilus unicornis* singly and in a colony. (From Ruttner-Kolisko, A.: III. Rotatoria. Das Zooplankton der Binnengewässer. I. Teil. *Die Binnengewässer* 26:99–234, 1972, from various sources.)

eggs to the adult, production of lipid-rich eggs that may be extensively ornamented, and vivipary.

B. Crustacean Zooplankton

The crustacean arthropods are almost entirely aquatic; most are marine. Respiration is accomplished through the body surface or gills. The body generally is separated into three distinct regions, but the tendency is toward fusion of abdominal and thoracic segments until, in the Cladocera and Ostracoda, apparent body segmentation has been lost. In many crustaceans, the body bears paired, usually biramous, jointed appendages, and is covered wholly or in part by a carapace.

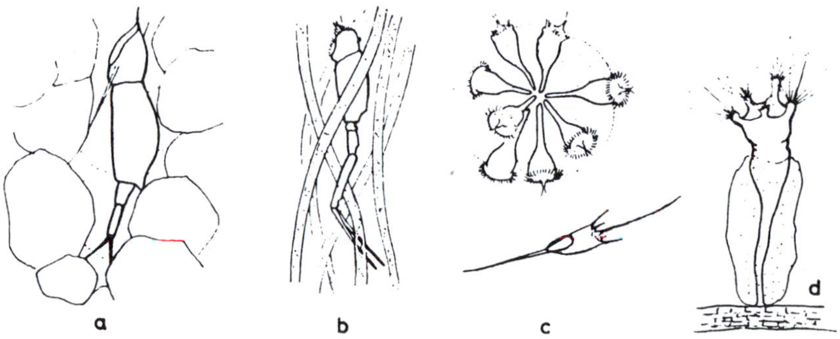

FIGURE 16-10 Exemplary types of rotifers of different habitats. (a) Psammic rotifer (*Bryceella*) among sand grains; (b) a littoral form (*Scaridium*) among algal filaments; (c) planktonic forms (*Conochilus* and *Kellicottia*); and (d) *Collotheca* epiphytic on the stem of a macrophyte. (From Ruttner-Kolisko, A.: III. Rotatoria. Das Zooplankton der Binnengewässer. I. Teil. *Die Binnengewässer* 26:99–234, 1972.)

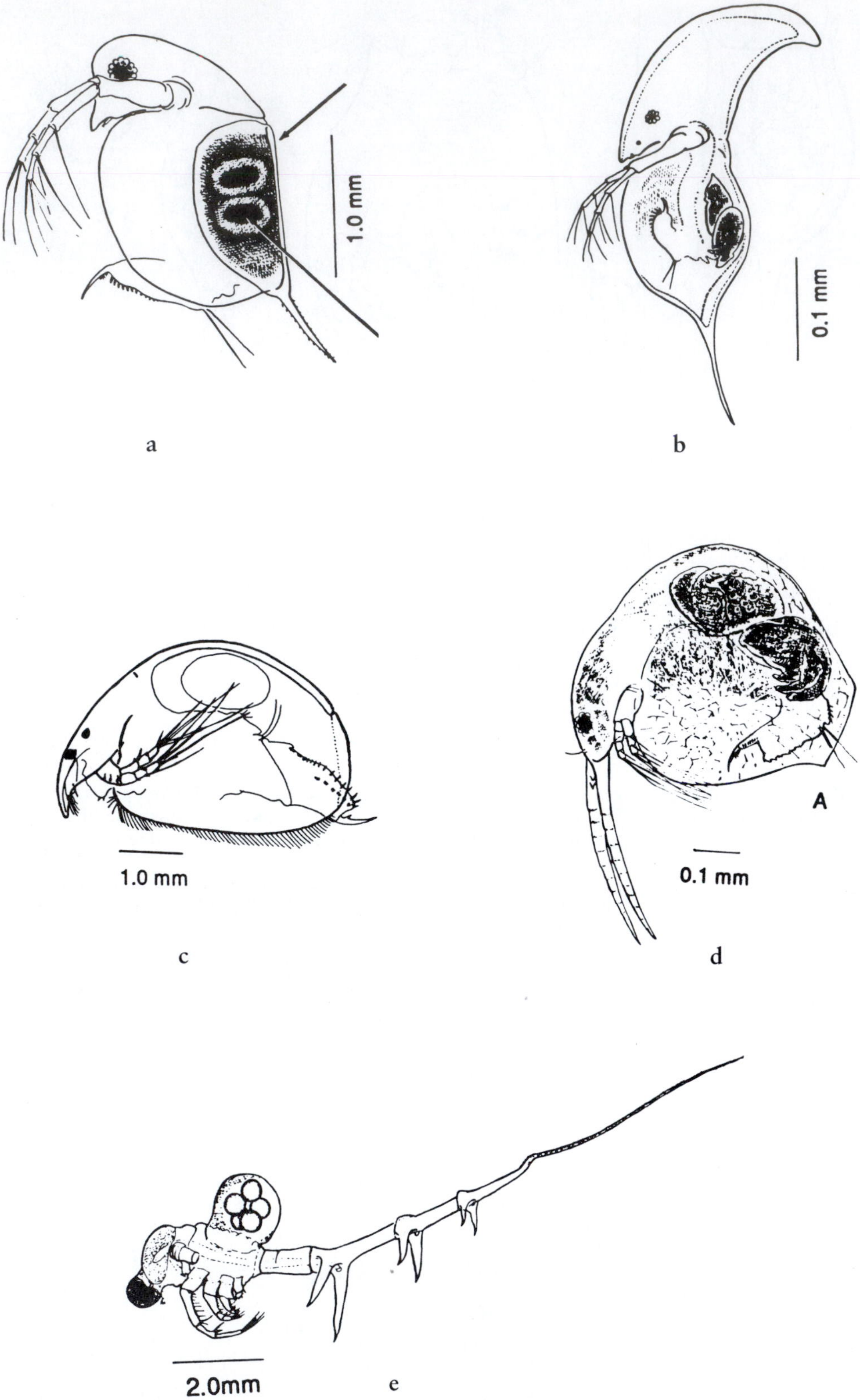

FIGURE 16-11 Examples of common planktonic cladocerans. (a) *Daphnia pulicaria* carrying resting eggs within a protective ephippium case; (b) *Daphnia retrocurva* with an elongated helmet of the carapace; (c) a common littoral species, *Alona bicolor*; (d) *Bosmina* (*Eubosmina*) *coregoni*; and (e) a common predaceous cladoceran *Bytholtrephes cederstroemii*. (From Dodson and Frey, 1991, with permission.)

In fresh water, the truly planktonic Crustacea are dominated almost completely by the cladocerans and Copepoda, to which most of the ensuing discussion is devoted. Only a few insects are planktonic in immature stages; the larvae of *Chaoborus* (Diptera) is a notable example of major importance in the zooplankton that demands special consideration. Most of the Ostracoda are benthic (cf. Chap. 22). A few species of *Cypria* are apparently partly planktonic, but very little is known of their ecology.

The freshwater Branchiopoda (fairy and clam shrimps) are common inhabitants of shallow lakes, particularly of temporary, saline inland waters. All members of this primitive group are distinctly segmented and bear many pairs of swimming and respiratory appendages. The tadpole shrimps (Notostraca) are essentially benthic and most often restricted to shallow, temporary lakes of arid regions. The fairy shrimps (Anostraca), lacking a carapace, and the clam shrimps (Conchostraca), compressed laterally with a bivalved flexible carapace, are common in the plankton in shallow playa lakes and vernal ponds of semiarid regions. The members of the latter groups are bisexual, although parthenogenesis is known to occur in brine shrimp *Artemia* under some conditions. Resistant eggs can be subjected to long periods of desiccation and hatch when wetted again, probably as a result of a reduction in osmotic pressure at the egg surface (Hutchinson, 1967). In semipermanent lakes of more humid regions, hatching and reproductive rates are related strongly to temperature.

1. Cladocera

The suborder Cladocera (Branchiopoda; order Phyllopoda) includes mainly microzooplankton (Dobson and Frey, 1991). With the exception of two species, nearly all the cladocerans range in size from 0.2 to 3.0 mm (Fig. 16-11). All have a distinct head, and the body is covered by a bivalve cuticular carapace. Light-sensitive organs usually consist of a large, compound eye and a smaller ocellus. The second antennae are large swimming appendages and constitute the primary organs of locomotion. The mouthparts consist of (1) large, chitinized mandibles that grind food particles, (2) a pair of small maxillules used to push food between the mandibles, and (3) a median labrum that covers the other mouth parts.

2. Copepoda

The free-living copepods of this order of the class Crustacea are separable into three distinct groups: the suborders Calanoida, Cyclopoida, and Harpacticoida. Although accurate identification is based largely on morphological details of appendages, several general characteristics delineate the major groups (Williamson, 1991). The body consists of the anterior metasome (cephalothorax), which is divided into the head region, bearing five pairs of appendages representing antennae and mouth parts, and the thorax, with six pairs of mainly swimming legs. The posterior urosome consists of abdominal segments, the first of which is modified in females as the genital segment, and terminal caudal rami bearing setae.

The three suborders of free-living copepods can be distinguished by the general structure of the first antennae, urosome, and fifth leg (Fig. 16-12 and Table 16-7). The harpacticoid copepods are almost exclusively littoral, habitating macrovegetation, mosses in particular, and the littoral sediments. Certain species have life histories with a diapause similar to that discussed further on for the cyclopoid copepods. Although the cyclopoid copepods are primarily littoral benthic species, those few members that are predominantly planktonic form major components of the copepod zooplankton,

TABLE 16-7 Some Characteristics of the Three Suborders of Free-Living Freshwater Copepoda[a]

Calanoida	*Cyclopoida*	*Harpacticoida*
Anterior part of body much broader than posterior	Anterior part of body much broader than posterior	Anterior part of body a little broader than posterior
Marked constriction between somite of fifth leg and genital segment	Marked constriction between somites of fourth and fifth legs	Slight or no constriction between somites of fourth and fifth legs
One egg sac, carried medially	Two egg sacs, carried laterally	Usually one egg sac, carried medially
First antennae long, extend from end of metasome to near end of caudal setae, 23–25 segments in female	First antennae short, extend from proximal third of head segment to near end of metasome, 6–17 segments in female	First antennae very short, extend from proximal fifth to end of head segment, 5–9 segments in female
Fifth leg similar to other legs	Fifth leg vestigial	Fifth leg vestigial
Planktonic, rarely littoral	Littoral, a few species planktonic	Exclusively littoral, on macrovegetation and sediments

[a] Modified from the summary of Wilson and Yeatman (1959).

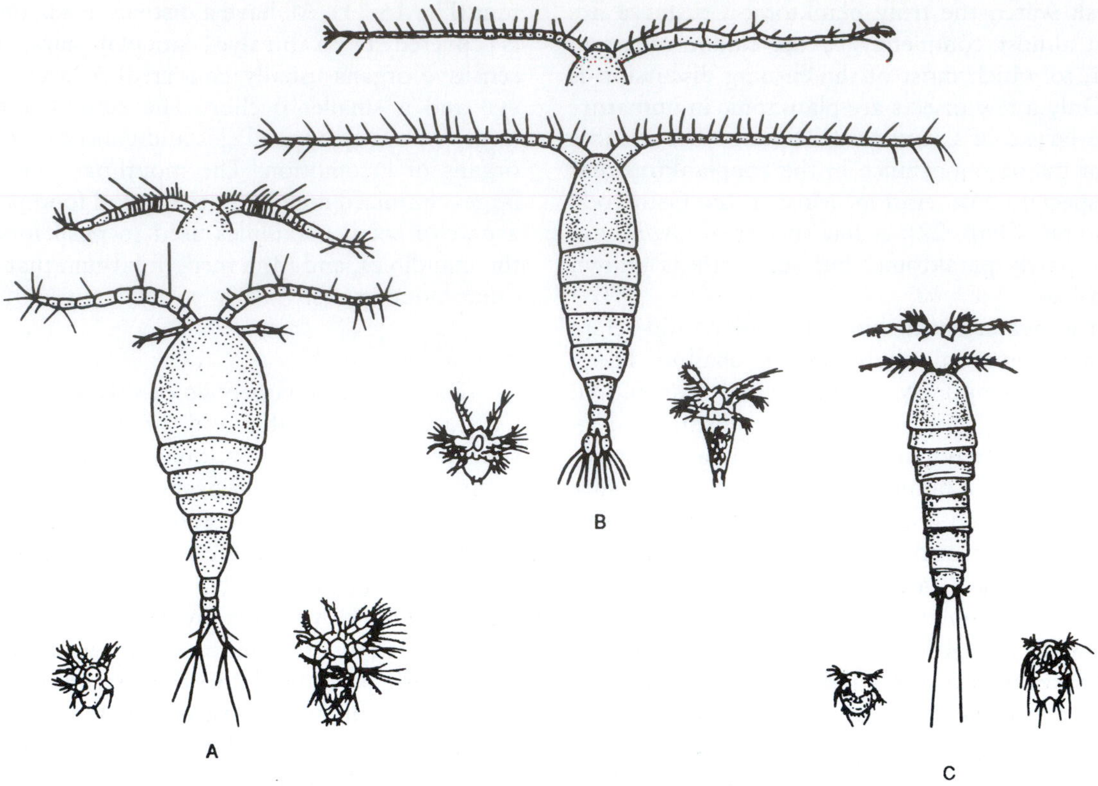

FIGURE 16-12 Diagrams of the three major types of free-living freshwater copepods. Adult females are shown in full, while the adult male first antennae and early- and late-stage nauplii are shown for each group. (A) Cyclopoid; (B) Calanoid; and (C) Harpacticoid. (From Williamson, 1991, with permission.)

especially in small, shallow lakes. The calanoid copepods are almost exclusively planktonic.

V. FOOD, FEEDING, AND FOOD SELECTIVITY

A. Rotifers

The planktonic rotifers feed largely by sedimenting seston particles into the mouth orifice by means of the pulsating action of the coronal cilia (reviewed at length by Pourriot, 1965, and Hutchinson, 1967). The size of the food consumed is quite variable. Most food particles taken are small, less than about 12 μm in diameter, although larger cells, up to approximately 50 μm, are sometimes seized, ruptured, and their particulate parts ingested. Feeding behavior of suspension-feeding rotifers is related to the type of particles, the food size and shape, and food density. Food selectivity clearly occurs in some rotifers and is governed by several rejection mechanisms, either by screening particles or by rejection of particles once they have been ingested (Starkweather, 1980a,b).

Detailed analyses of the feeding behavior of *Brachionus calyciflorus* on varying quantities and mixtures of different types of foods (bacteria, yeast, and algae) demonstrated complex relationships between ingestion rates, food density, and selectivity. Several mechanisms regulate the ingestion of suspended particles (Gilbert and Starkweather, 1977): (a) Cirri of the pseudotrochus may be extended, allowing particles to enter the funnel-shaped buccal field, or they may be bent over to form a screen that prevents even some very small particles (e.g., yeast) from entering the buccal field; (b) Particles in the buccal field may be rejected by ciliary movements before entering the oral canal; and (c) Particles in the oral canal can be rejected by the jaws and pushed back into the buccal field.

Different food types induce different ingestion rates in *Brachionus* (Starkweather and Gilbert, 1977a,b, 1978; Gilbert and Starkweather, 1978). When feeding on yeast, ingestion increased continuously with increasing food-cell densities without the use of pseudotrochal screening. When feeding on certain bacteria and algae, however, ingestion rates were constant over a considerable range of particle densities. The use of pseudotrochal screening increased with larger-sized particles and with increasing particle densities. Screening was also influenced by past feeding history.

Animals selected food of the type they had grown on if exposed to a mixed type of food, and previously starved rotifers used screening less than well-fed animals. The time required for ingested food to pass entirely through the alimentary canal is from 3 to 20 min; this time varies greatly among species and with external conditions.

Some rotifers are raptorial, seizing and ingesting whole prey organisms or drawing in the cell contents after the cell or body wall has been punctured. The largest rotifer, *Asplanchna*, which is predatory on algae, rotifers, and small planktonic crustaceans, has the ability to shift size in response to changes in food size and prey densities, as discussed later (cf. Gilbert, 1980c). Although a large range in size of food particles is consumed by any given rotifer species, in many cases, there is a reasonable separation of rotifer species along a food-particle-size gradient. This separation is congruent with the observed co-occurrence of several species within the pelagial zone of lakes (e.g., Makarowicz and Likens, 1975). It is probable that fairly discrete niches have evolved within the zooplankton, corresponding to specialized utilization of part of the range of available particulate matter. These specializations are adequate to permit co-occurrence without severe competitive interactions.

Food limitations to growth of rotifers vary widely. Threshold food concentrations for rotifers are high in relation to their small size in comparison to other zooplankton (Duncan, 1989). In particular, large rotifer species have high threshold food concentrations that restrict them ecologically to nonpelagic particle-rich environments.

B. Cladocera

Cladocerans usually have five pairs of legs attached to the ventral part of the thorax. The legs are flattened and bear numerous hairs and long setae. Complex movements of these setose legs create a current of water through the valves. This current oxygenates the body surface and forces a stream of food particles anteriorly. The food particles, filtered by the setae, collect in a ventral food groove between the bases of the legs and are impelled forward toward the mouth to be mixed there with oral secretions. The importance of size of food particles in relation to morphological limitations of the filtering apparatus and food selectivity will be discussed later.

The primarily littoral chydorid cladocerans have modified legs that are somewhat prehensile in scraping up larger pieces of detrital material. Feeding by filtration occurs as well. Two other common cladocerans, *Polyphemus* and *Leptodora*, contain members that are predaceous and feed mainly by seizing relatively large particles, such as protozoa, rotifers, and small crustaceans, with their prehensile legs.

C. Copepods

The mouth parts of harpacticoids are adapted for seizing and scraping particles from the sediments and macrovegetation. The food and feeding behavior of the cyclopoid copepods have been studied in detail in two masterful works by Fryer (1957a,b). No filtration mechanisms occur in the free-living Cyclopoida. Feeding is raptorial; plant or animal food particles are seized by mouth parts and brought to the mouth. The maxillules hold and pierce the prey and force particles between the mandibles; intermittent oscillating movements macerate some of the food. Some particles are swallowed intact and are differentially digested. Diatoms tend to be digested, while some green algae, if they are not ruptured, pass through the gut undigested.

Many species of the major cyclopoid genera *Macrocyclops*, *Acanthocyclops*, *Cyclops*, and *Mesocyclops* are carnivorous. The food of these carnivores includes microcrustaceans, dipteran larvae, and oligochaetes, many of which are larger than the copepod that preys on them. Herbivorous cyclopoids include many species of *Eucyclops*, some *Acanthocyclops*, and *Microcyclops*, which feed on a variety of algae ranging from unicellular diatoms to long strands of filamentous species. Carnivorous cyclopoid species tend to be larger than herbivorous species. Random encounter appears to be the dominant mode of finding food in both carnivorous and herbivorous species, which search by discontinuous, irregular movements in the water or over the substratum. Herbivorous species apparently employ gustatory chemoreceptor organs, which may help in food seeking if only to facilitate discrimination between inorganic and organic particles encountered by chance. The filtering rate of *Diaptomus tyrrelli* is reduced in the presence of a predacious copepod *Epischura* (Folt and Goldman, 1981). The alteration in feeding rate is in response to a chemical released into the water by the predator. Presumably, the decreased movement functions as an evasive action taken by *Diaptomus* to reduce predatory success of *Epischura*.

While locomotion in most copepods is in the form of short, jerky swimming movements, the animals being propelled by rapid movement of most appendages simultaneously, swimming is more continuous in the Calanoida copepods. Their gliding movement results from rotary motion of the antennae and mouth appendages. The movements set up small vortical currents that carry particles to the maxillae, which are modified to remove the water. High-speed motion pictures have revealed that calanoid copepods do not strain particles out of the water, but instead propel water past the body by flapping four pairs of feeding appendages (Koehl and Strickler, 1981). The second maxillae actively capture "parcels" of that water containing food particles. Particles are then pushed into the mouth by the first maxillae. Selective feeding also exists among the calanoid copepods. For example, two closely related calanoids, *Diaptomus laticeps* and *D. gracilis*,[3] were found to coexist in the plankton of Lake Windermere, England (Fryer, 1954; cf. Maly and Maly, 1974). The species were separated by size differences, which were correlated with differences in food consumed. The larger *D. laticeps* fed chiefly on *Melosira*, and the smaller *D. gracilis* consumed mainly minute spherical green algae and particulate detritus. Neither species fed upon the then-abundant diatom *Asterionella*. Competition for food by these two calanoid species was virtually nonexistent. Such small differences in feeding habits, based on morphological and behavioral variations, are common in the calanoids and may be sufficient to separate species into different food niches, even though they occupy the same volume of water.

Similar results were found for *Diaptomus* feeding on natural phytoplanktonic populations (McQueen, 1970). Filtration rates were low for cells smaller than 100 μm^3, increased to a maximum of 12.9 ml per animal per day for cells ranging from 102 to 333 μm^3, and remained constant with increased cell volume and decreased cell concentrations.

D. Filtration and Food Selectivity

The entire subject of feeding in Cladocera and in copepods is an area currently under intensive investigation. Recent results indicate not only a number of physical and biotic factors that affect rates of ingestion but also that some organisms possess at least minimal abilities to select food for consumption. The implications of these studies are major not only from the standpoint of effectiveness of food ingestion and differential assimilation by the animals but also because they point out the effects of the consumption on the food populations. This latter subject is most complex. However, it is apparent that at times zooplankton, as a result of direct cropping, can have appreciable effects on phytoplanktonic populations and, through selective grazing, can influence the seasonal succession of the phytoplankton. This area of research has great potential for demonstrating subtle but major community interactions of significant impact on the overall productivity of lake systems.

Filtration of water to remove particulate organic matter is the dominant mode by which most cladocerans collect and ingest food. The ***filtration rate*** of a zooplankter refers to volume per unit time and is defined as the volume of water containing food particles that is filtered by the animal in a given time (cf. Rigler, 1971). This term does not imply that the volume of water passed over the filtering appendages is known, that all particles of any given type have been removed from the water, or that all particles retained by the filtration apparatus have been consumed. The terms *filtration rate*, *grazing rate*, and *filtration capacity* have been used synonymously with *filtering rate*. In contrast, ***feeding rate*** is a measure of the quantity of food ingested by an animal in a given time (Rigler, 1971), measured in terms of number of cells, volume, dry weight, carbon, nitrogen, or some other relevant aspect of the food that is ingested.

E. Suspension Feeding

The size of particles that can be cleared from the water is a function of (a) the morphology of the setae of the moving appendages, (b) physical characteristics of the sieving surfaces and particle movements, and (c) entrapment and concentrations as locomotion of the animal brings particle-laden water to the setae. The morphological characteristics and dimensions of the filter structures and apertures have been examined among many cladocera in particular (e.g., Crittenden, 1981; Geller and Müller, 1981; Brendelberger and Geller, 1985; Ganf and Shiel, 1985; Gophen and Geller, 1985; Hessen, 1985a; Brendelberger, 1991). There is a general direct correlation between the morphology and intersetular spacing of the filtering setae and the ability to retain particles of different sizes. Greater amounts of food particles at low particle concentrations can be collected by increasing filtering rates either by enlarging the area of the filter screens or by increasing appendage beat rate (Lampert and Brendelberger, 1996). Alternatively, over longer periods of time, phenotypic reductions in intersetular distances would reduce mesh sizes.

Discrepancies between sizes of particles retained and morphological apertures and the size selection of particles retained, however, indicated that capture is not a simple mechanical process. In herbivorous copepods, the appendages function more like paddles than filters and move particles along the surfaces for concentration (Brendelberger *et al.*, 1986; Gerritsen *et al.*, 1988). In cladocerans, some flows also occur along the appendages toward the oral groove. Because the intersetular spaces and the resulting Reynolds numbers

[3] Hutchinson (1967) differentiates these species into *Arctodiaptomus laticeps* and *Eudiaptomus gracilis*, respectively.

($\approx 10^{-3}$) are very small, some pressure difference would be needed to force water through the mesh. In cladocerans, a closed filtering chamber allows water to be forced through the apparatus with minimal expenditure of energy (ca. 5% of the total metabolic requirements of the animal) (Brendelberger *et al.*, 1986). Surface charge chemistry likely assists in concentration of particles and the filter apparatus (Gerritsen and Porter, 1982).

Measurements of particle removal are made by observing changes in the number of particles in different size classes caused by grazing over time (cf. Harbison and McAlister, 1980). Caution is needed in extrapolation of laboratory measurements of filtration and ingestion rates among herbivorous zooplankton to grazing rates in natural planktonic environments. Laboratory measurements commonly overestimate *in situ* grazing rates by at least 30–50% because of spatial heterogeneity in natural plankton distributions (Wirick, 1989). The rate of removal of radioactively labeled food particles has also been used to measure zooplankton feeding rates (see review of Rigler, 1971). The latter approach has been extended effectively to natural populations by use of an *in situ* grazing chamber (Haney, 1971) that permits estimates of feeding rates over short periods of time (<10 min) so that ingested food is not excreted during the incubation period of measurement. These methods measure food actually taken into the gut after losses, either through active rejection of food or losses of food particles during the maceration process. Ingestion rates give little insight into assimilation rates of incorporated food, which can be highly variable with food types and diet concentrations.

A number of studies indicate that the rate of feeding stabilizes or decreases as the concentration of food particles increases. For example, in *Daphnia magna*, feeding rate is proportional to concentration of food particles below a critical level (McMahon and Rigler, 1963, 1965). Feeding rate is constant above the incipient limiting concentration of food particles. At concentrations of food above this limiting level, rates of movement of the thoracic appendages that collect the food decrease, movements of mandibles and swallowing of food remain about the same or decrease slightly, and the rate of rejection of food increases (Burns, 1968b, personal communication). For adult *D. rosea* feeding on yeast, the critical concentration for feeding was between 0.75×10^5 and 1.0×10^5 cells ml^{-1} (Burns and Rigler, 1967). Incipient limiting concentrations (cells ml^{-1}) for feeding rate appear to increase with decreasing particle size, although other factors may be involved as well.

Within a given range of food concentrations, filtering efficiency was found to be independent of size of food particles ($0.9–1.8 \times 10^4$ μm^3 for *D. magna*) (McMahon and Rigler, 1963, 1965; see also Kersting and Holterman, 1973; Geller, 1975). At concentrations of yeast cells above 0.25×10^5 cells ml^{-1}, filtering rates of *D. rosea* decreased. The filtering rate of *Ceriodaphnia reticulata* decreased linearly with increasing concentrations of phytoplankton within a size range of 3–24 μm from low to high densities ($10^4–10^6$ ml^{-1}) (O'Brien and deNoyelles, 1974). This relationship was not evident in cultures of *Ceriodaphnia* exposed to low densities ($10^4–10^5$ ml^{-1}) of algae (Czeczuga and Bobiatynska-Ksok, 1970), although the latter methods (very high zooplanktonic densities in a small volume with long incubation periods) do not permit comparison.

Filtering rates have been found commonly to increase with increasing body length (e.g., Peters and Downing, 1984) and with increasing temperatures (Fig. 16-13). Temperatures above a given point result in a decrease in the filtration rate. The temperature of maximum filtration rates differs among species; for *D. rosea*, the maximum was found to be about 20°C (Fig. 16-13), for *D. magna* 28°C, and for *D. galeata mendotae* and *D. pulex* 25°C or above (McMahon, 1965; Burns and Rigler, 1967; Burns, 1969b; Geller, 1975). Brooks and Dodson (1965) have emphasized that the filtering rate should be proportional to the square of the body length in filter-feeding Cladocera. The increase in filtration rate relative to body length, however, varies among species, for example, about to the square in *D. magna* and to the cube of the length in smaller *D. rosea* (Burns, 1969b). Although the area of the filtering setae of these two *Daphnia* species increases approximately as the square of the body length, the filtering area and the filtering rates of *D. rosea* become proportionately greater than those of *D. magna* as the body length increases (Egloff and Palmer, 1971). The difference in the calculated rate of flow of water through the filtering setae, however, remains nearly constant at all body lengths.

The relationship between body size and the maximum size of particles that could be ingested was studied in six species of *Daphnia* and the smaller *Bosmina longirostris* using spherical beads within a size range from <1 to 80 μm in diameter (Burns, 1968a, 1969a). A positive correlation between increasing body size and increasing size of the largest particle ingested was evident for the species assayed (Fig. 16-14). Zooplankton body length, however, may influence the maximal particle size a species can ingest, but it has little influence on the ingestion of smaller particles (Bogdan and Gilbert, 1984; Knoechel and Holtby, 1986). Much selectivity and individual specialization exists, and it is likely unrealistic to predict the diet of a zooplankton species by its size or taxonomic group.

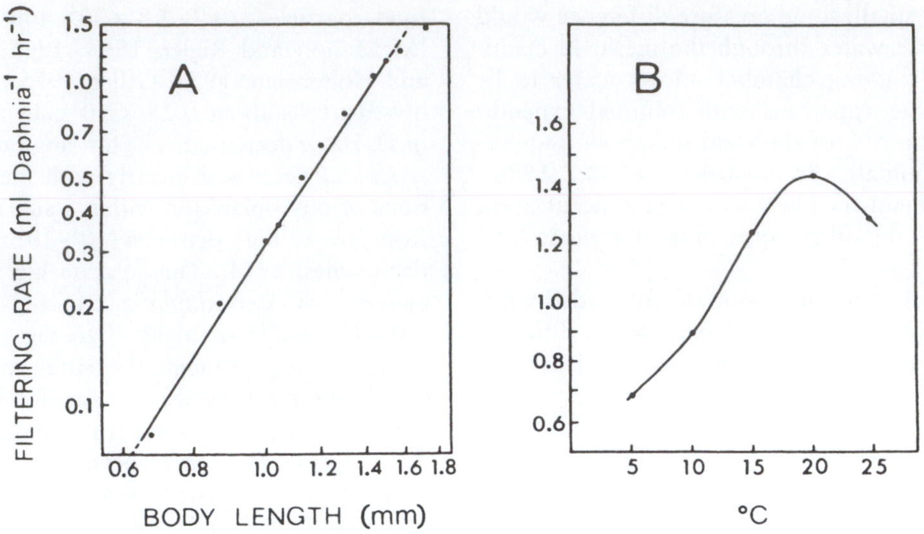

FIGURE 16-13 Relationship between filtering rate and (a) body length at a constant food supply at 20°C, and (b) water temperature (body length 1.65–1.85 mm) of *Daphnia rosea*. (Modified from Burns and Rigler, 1967.)

Extension of these and other results from laboratory feeding studies to natural populations by observations of *in situ* grazing rates has demonstrated some striking relationships. In analyses from small, eutrophic Heart Lake in southern Ontario, seasonal grazing patterns (Fig. 16-15) were found to be similar over a 2-year period (Haney, 1973). At certain depths during periods of the summer, grazing rates exceeded 100% of the water volume per day, but during much of the year, values were less than complete filtration per day and became <10% day^{-1} during the winter (Fig. 16-15; Table 16-8). The lower vertical boundary of zooplanktonic filter feeding was found to be closely defined by the 1 mg liter^{-1} ispoleth of dissolved oxygen concentration, below which filtering rates declined precipitously. There were seasonal patterns in the rates of grazing on different food items. Smaller, bacteria-sized particles apparently were ingested most rapidly in summer, whereas in the autumn the larger-sized algal particles were eaten most rapidly.

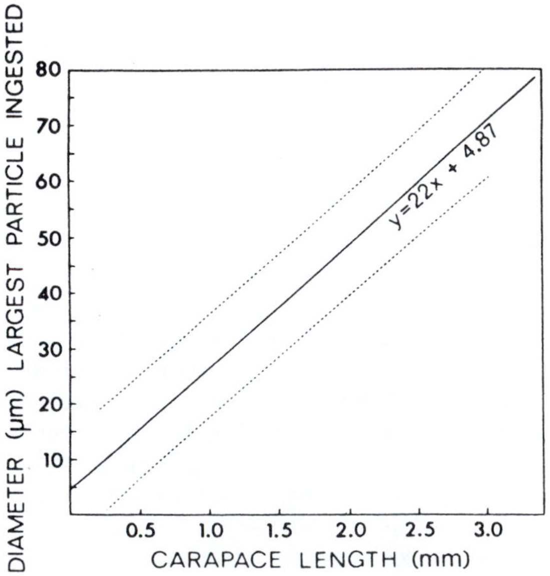

FIGURE 16-14 Relationship between body size and the diameter of the largest particle ingested by seven species of Cladocera. Broken lines equal 95% confidence limits. (Modified from Burns, 1968a.)

Since, as indicated earlier, the grazing rates of cladoceran zooplankton vary greatly with body size, food, and temperature, the *in situ* rates of filtration vary considerably seasonally. The average rates given in Table 16-9, however, show the general relationship to body size. Returning to the example from Heart Lake, the grazing contribution of each species can be estimated by multiplying filtering rates of each species at a given time by its population density at that time. In Heart Lake, *Daphnia rosea* and *D. galeata* were the most important grazers and together accounted for about 80% of the total annual grazing activity by the larger zooplankton (Haney, 1973). Cladocerans (several *Daphnia* species) were also the dominant grazers in moderately productive Lawrence Lake, Michigan (Crumpton and Wetzel, 1982). Grazing of the phytoplankton in this lake became significant only during late summer (Table 16-8; cf. Figure 15-13). Zooplankton filtered <5% of the volume of the epilimnion per day throughout most of the year and had a minor

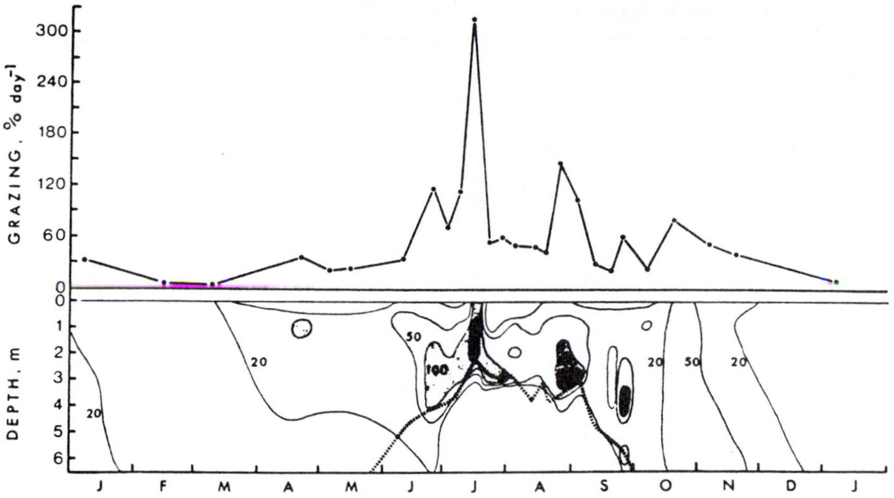

FIGURE 16-15 Seasonal and vertical changes in grazing by the zooplankton community in Heart Lake, Ontario, 1969. Mean grazing rates of the upper figure calculated for the aerobic stratum. Isopleths of grazing rates (lower figure) are at 20, 50, 100, and 200% per day. The broken line is the 1 mg liter^{-1} dissolved oxygen isopleth. (After Haney, J. F.: An in situ examination of the grazing activities of natural zooplankton communities. *Arch. Hydrobiol.* 72:87–132, 1973.)

impact on phytoplankton mortality. In August and September of 1979, however, grazing had a significant effect on both algal productivity and species succession (cf. Chap. 15). During the night, when many cladocerans and copepods migrate to the epilimnetic waters, grazing rates are about five times greater than during daylight hours (Table 16-8; cf. Hart, 1977; Mourelatos *et al.*, 1989). Often, zooplankton encounter strata of reduced oxygen concentrations as they migrate to deep, poorly illuminated water during daylight hours.

TABLE 16-8 Average Grazing Rates of *In Situ* Phytoplankton

		Mean grazing rate (% day^{-1})			
	Total days	*Day*	*Night*	*Total*	*% Total grazing over period*
Lawrence, Michigan[a]					
May	31	0.2	3.0	3.2	4.2
June	30	1.6	3.0	4.6	6.0
July	31	0.4	1.9	2.3	3.0
Aug.	31	4.1	20.2	24.3	31.7
Sept.	30	4.2	21.9	26.1	34.0
Oct.	15	4.2	12.0	16.2	21.1
Heart, Ontario[b]					
Jan.–May	151	—	—	19.2	17.4
Jun.–Sept.	122	—	—	80.1	61.4
Oct.–Jan.	92	—	—	35.2	20.9
Vechten, The Netherlands[c]					
Winter (Dec.–Feb.)	90	—	—	2.1	4.5
Spring (Mar.–Apr.)	61	—	—	9.8	22.0
Summer (May–Sept.)	153	—	—	24	69.0
Autumn (Oct.–Nov.)	61	—	—	5.5	4.5
Annual mean	365	—	—	13	100

[a] After data of Crumpton and Wetzel (1982); see Figure 15-13.
[b] After data of Haney (1973).
[c] After data of Gulati (1978).

TABLE 16-9 Filtering Rates of Various Cladoceran Zooplankters

Species	Animal size range (length in mm)	Average filtering rate (ml animal^{-1} day^{-1})[a]	Source
Daphnia			
D. rosea	1.3–1.6	5.5	Haney (1973)
		3.6	Burns and Rigler (1967)
D. galeata	1.5–1.7	6.4	Haney (1973)
		3.7	Burns and Rigler (1967)
D. parvula	0.7–1.2	3.8	Haney (1973)
D. longispina		2.3	Nauwerck (1963)
D. pulicaria	2.2	14.4[b]	Kasprzak *et al.* (1986)
Ceriodaphnia			
C. quadrangula	0.7–0.9	4.6	Haney (1973)
Diaphanosoma			
D. brachyurum	0.9–1.4	1.6	Haney (1973)
Bosmina			
B. longirostris	0.4–0.6	0.44	Haney (1973)
B. obtusirostris	—	0.56	Persson (1985)
Chydorus			
C. sphaericus	0.1–0.2	0.18	Haney (1973)
Holopedium			
H. gibberum	—	4.4	Persson (1985)

[a] Type of food = *in situ* phytoplankton, in most cases.
[b] Estimated from culture analyses.

Filtering and respiration rates decrease rapidly at oxygen concentrations below 3 mg liter^{-1} (Kring and O'Brien, 1976; Heisey and Porter, 1977).

Filtering rates of the major zooplankton were also studied, although less intensively, in two much less productive lakes in southern Ontario: Halls Lake, a deep oligotrophic lake, and Drowned Bog Lake, a typical *Sphagnum* acidic bog lake (cf. Chap. 25) with high concentrations of dissolved humic matter. The filtration rates of the two dominant zooplankters, *Bosmina* and *Holopedium*, were very different, but the small *Bosmina* formed a dominant percentage of the total grazing in the autumn because of its very large population densities (Table 16-10). In contrast to the bog lake and to eutrophic Heart Lake, in which intense grazing occurred in the upper 3 m, the grazing rates in oligotrophic Halls Lake were extremely low and distributed uniformly throughout the water column. The dominant zooplankter of this oligotrophic lake was the copepod *Diaptomus*, and in general, nonpredatory copepods have much lower grazing rates than the Cladocera (Tables 16-9 and 16-11).

TABLE 16-10 Filtering Rates and Contribution to Total Grazing of Species-Dominant Zooplankton of Acidic Drowned Bog Lake, Ontario, in Early September[a]

	Filtering rates (ml animal^{-1} day^{-1})		Species contribution to total macrozooplankton grazing (%)	
Species	Sept. 1968	Sept. 1969	Sept. 1968	Sept. 1969
Bosmina longirostris	0.46	0.45	85	44.8
Holopedium gibberum	—	9.4	12	46.2
Daphnia parvula	—	1.6	0.1	7.5
Diaptomus oregonensis	—	2.1	2.0	0.9
Diaphanosoma bruchyurum	—	1.2	0.1	0.4

[a] Extracted from data of Haney (1973).

TABLE 16-11 Comparison of Filtering Rates of Various Copepods

Species	*Type of food*	*Particle Concentration* (*1000 × cells ml*$^{-1}$)	*Filtering rate* (*ml animal*$^{-1}$ *day*$^{-1}$)	*Source*
Diaptomus				
D. graciloides	Natural phytoplankton	—	0.3–2.8	Nauwerck (1959)
	Scenedesmus	13.6	4.1	Malovitskaia and Sorokin (1961)
D. siciloides	*Pandorina* and *Chlamydomonas*	—	2.0	Comita (1964)
D. oregonensis	*Chlamydomonas*	1.5–25.0	2.5	Richman (1966)
		25.0–52.0	2.5–1.4	
	Chlorella	52.0–198.0	1.4–0.3	
	In situ phytoplankton	—	1.4–0.00	Haney (1973)
D. pulex	*Aphanizomenon flos-aquae*			
	Single filaments	23	15.1	Holm *et al.* (1983)
	0.8–1.3-mm flakes	20	27.4	
D. gracilis	*Melosira* and *Asterionella*	24.2–52.0	1.92–1.96	
		198.0	0.68	Malovitskaia and Sorokin (1961)
	Chlorella	<30.0	0.61 (5°C) 1.51 (12°C) 2.40 (20°C)	Kibby (1971)
	Scenedesmus	<30.0	0.94 (12°C) 1.32 (20°C)	
	Diplosphaeria	<30.0	1.76 (12°C) 2.54 (20°C)	
	Ankistrodesmus	<30.0	1.61 (12°C) 2.45 (20°C)	
	Carteria	<30.0	0.87 (20°C)	
	Nitzschia	<30.0	1.96 (20°C)	
	Pediastrum	<30.0	0.02 (20°C)	
	Haematococcus	<30.0	2.16 (20°C)	
	Bacteria	<30.0	0.19 (20°C)	
Cyclops				
C. scutifer	Natural phytoplankton	—	0.2–7.3	Persson (1985)
Eudiaptomus				
E. graciloides	Natural phytoplankton	—	0.09–5.4	Persson (1985)
Limnocalanus				
L. macrurus	*Scenedesmus*	—	2.45 (<5°C)	Kibby and Rigler (1973)
	Chlamydomonas	—	1.24 (<5°C)	
	Rhodoturula (yeast)	—	0.1 (<5°C)	
Heterocope				
H. appendiculata	Natural phytoplankton	—	0.2–16	Persson (1985)

In studies in which species of widely different algae were fed to cladoceran and copepod zooplankton at concentrations above critical limitations to feeding, filtering rates decreased markedly with increases in the food concentration (Fig. 16-16). Ingestion rates were calculated as the product of the filtration rate (ml animal^{-1} day^{-1}) and the food concentration (cells ml^{-1}) (Infante, 1973). These results, when coupled with measurements of radioactively labeled carbon from the algae that were incorporated into the animals as an estimate of assimilation, showed marked differences with algal species. The diatom *Asterionella* was ingested and assimilated more readily than *Nitzschia*. Algae with heavy cell walls, such as *Scenedesmus* and *Stichococcus*, were ingested rapidly (Fig. 16-16) but were not utilized well. The Cladocera digested some of the protoplasm of such cells without visible changes in the cell walls, whereas the copepods assayed (*Eudiaptomus* and *Cyclops*) were unable to digest them, and the algae that left the gut were essentially intact. *Staurastrum* and *Cryptomonas* were poorly utilized. Differences also occurred among zooplanktonic species. *Daphnia longispina*, for example, incorporated nearly twice the number of *Nitzschia* that

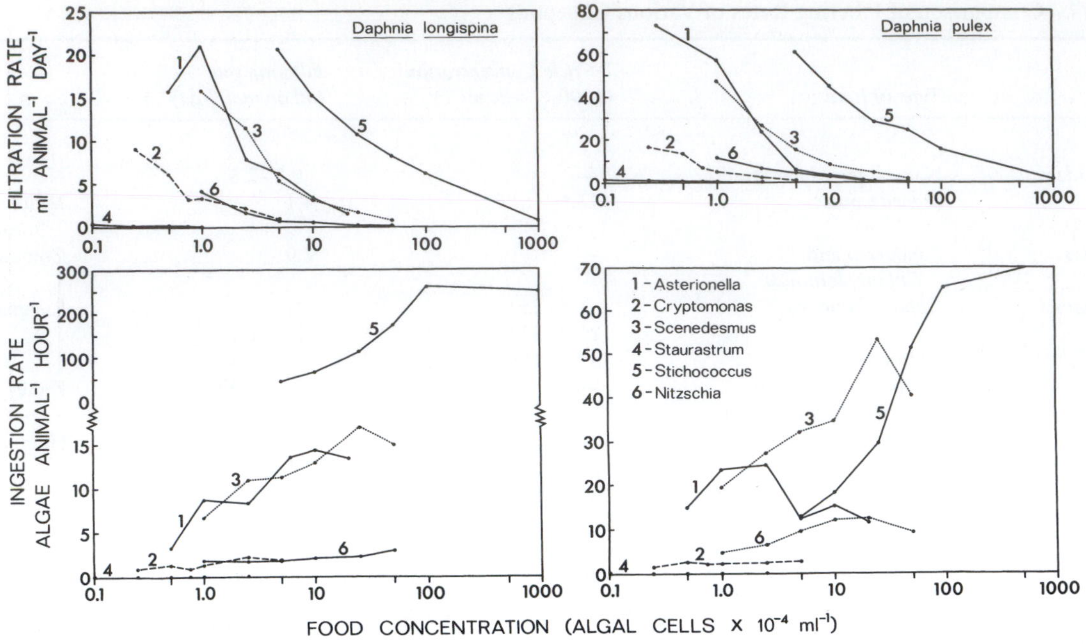

FIGURE 16-16 *Upper:* Filtration rates (in ml animal^{-1} day^{-1}); *lower*: ingestion rates (in cells animal^{-1} h^{-1} as the filtration rate times food concentration in cells ml^{-1}), in *Daphnia longispina* and *D. pulex* exposed to different algal species. The ingestion rate of *Stichococcus* (5) for *D. pulex* is drawn × 0.1 of that found. (Drawn from data of Infante, 1973.)

D. pulex did but only one-half the amount of *Scenedesmus*.

Certain zooplankton feed on a wide variety of algae of different sizes and shapes, with or without sheaths, without any significant selectivity (Gliwicz, 1980; McNaught *et al.*, 1980; Porter and Orcutt, 1980; Kasprzak *et al.*, 1986; Fulton, 1988; Burns and Xu, 1990). Other zooplankton are quite selective in the algal types ingested. Cellular forms are ingested more readily than filamentous or spinuosus forms, and zooplanktonic filtration rates, growth, and survivorship are greater when feeding on cellular forms. Some of the restricted feeding on large or filamentous algae is simply physical reduction of ingestion. For example, the width distance between the margins of carapace valves in filter-feeding cladocerans was significantly smaller when they were exposed to high concentrations of large filamentous algae (Gliwicz and Siedlar, 1980). This reduction in gape width resulted in less ingestion of the large algae in relation to smaller prey.

Evidence also suggests that algae releasing toxic organic compounds are selected against by zooplankton, regardless of suitability of food size and shape, which implies chemoreception (taste). Other zooplankton, often species-specific, feed somewhat selectively on particles impregnated with organic compounds specific to certain algae (DeMott, 1989). Certain cyanobacteria produce and actively release dissolved organic products into the water that reduce thoracic appendage beat rate and filtering rates of cladocera by 50% or more, as well as ingestion rates (Lampert, 1982; Chow-Fraser and Sprules, 1986; Jungmann *et al.*, 1991; Haney *et al.*, 1994, 1995). Such selectivity has been found in certain species of rotifers, copepods, and cladocerans and almost certainly allows for selection of different species of algae of different food quality.

Similarly, when the numbers of grazing Cladocera and copepods were increased experimentally among natural phytoplanktonic populations, small algae such as cryptomonads, certain diatoms, and other nanoplankton decreased, gelatinous green algae such as *Sphaerocystis* increased, and large or unpalatable species such as the cyanobacterium *Anabaena* remained unchanged (Porter, 1973). Although the size of particles grazed is limited by the morphological structure of the setae (Monakov and Sorokin, 1961), large or unpalatable algae result in reduced filtering rates. For example, the ingestion, assimilation, survivorship, and reproduction rates of *Daphnia* that were fed cyanobacteria were lower than those fed green algae (Arnold, 1971). Therefore, even though food species of a usable size may be present, they may be too rare or may be selectively rejected by the grazer at high algal densities. This selection is apparently at least partly related to the relative abundance of particles of different biomass. Selectivity by *Daphnia* increases with increasing particle concentrations until particle consumption is maximized when particles of one size dominate in mixed food (Berman and Richman, 1974; Gliwicz and Lampert, 1990). When concentrations of all particles were approximately equal, selection favored small, abundant particles (e.g., Holm *et al.*, 1983; Hartmann, 1985). Thus, both mechanical and chemical interference in algal selection and ingestion can lead to differential success of algae in lakes of different trophy.

Ingestion of algae does not necessarily result in their utilization and assimilation by the grazers. In some grazer species, algae with durable cell walls, gelatinous sheaths, or masses of colonial cells can pass through the gut intact and remain viable (Porter, 1975, 1976). Other species can utilize broken colonies or partially decayed cells of colonies by breaking them into smaller pieces, while intact colonies pass through relatively unaffected. Increases in the growth rates of certain algae after ingestion and passage through grazers may be the result of several interrelated mechanisms. Nutrients may be obtained from the degradation and digestion of other algae; phosphorus, in particular, is probably a major nutrient that is made more available during this process.

Selective grazing and utilization can remove species or reduce population size in the algal community, which can alter competition for the given resource base. Alternatively, grazer utilization of an algal species can result in enhancement of primary productivity of that species by increased selection for faster-growing genotypes and by increased nutrient supply (Crumpton and Wetzel, 1982).

F. Importance of Food Quality

A number of studies have demonstrated the effects of food quality as well as quantity on assimilation efficiencies. Analyses of *Daphnia pulex* (Richman, 1958) have shown assimilation efficiency to decrease markedly with increasing food concentrations (Table 16-12). Respiratory rates increase considerably with increasing concentrations of algae until a maximum rate is achieved at an algal concentration corresponding to the incipient limiting level for feeding, beyond which filtering rate decreases with increasing food concentrations (Lampert, 1986). Growth as average length, number of young per brood, and total young increased with increasing food concentrations at the levels used. The energy used for growth was nearly constant at all food levels in preadults, because the energy of egestion increased and assimilation efficiency declined as food consumption increased. In adults, however, the major portion of stored energy went into the production of young.

Similar results have been found for a number of other cladocerans (cf. especially Lampert, 1977a,b). A population food threshold is the amount of food necessary for reproduction to offset mortality; a zero change in population size results. As one examines the allocation of food resources into growth and reproduction, slightly more is allocated to reproduction than to body growth among cladocerans (Duncan, 1989). Similarly among rotifers, most food resources were allocated to growth under scarce food resources but shifted largely to increasing egg production under optimal food resource availability.

The individual food threshold is the amount of food necessary for assimilation to balance metabolic losses, which results in a zero change in mass of the individual animal. Maintenance metabolism requires mainly energy, while body growth requires other essential substances, including essential minerals and growth substances in the diet (Sterner and Robinson, 1994; Sterner and Hessen, 1994). For example, algae with a low growth rate resulting from a low availability of phosphorus will contain less phosphorus and be an inferior food for cladoceran growth. Because the carbon content per unit biovolume is similar among both low- and high-growth-rate algal food, both foods are satisfactory as an energy source for maintenance metabolism. Body growth, however, requires essential nutrients, such as phosphorus in this example, which is not uniformly available in food (see stoichiometry discussion in Chap. 15). Therefore, herbivores with high nutrient demands, exemplified by *Daphnia* and phosphorus, appear frequently to be limited not by the food quantity or energy available to them but by the quantity of minerals in their food. Much phosphorus can be associated with bacteria, and as a result, bacteria can be an important source of phosphorus to bacterivorous zooplankton (Hessen and Andersen, 1990).

In addition to the increase in assimilation rates with increasing temperature, several studies show that the assimilation efficiency increases with higher energy content of the food (Schindler, 1968; Pechen-Finenko, 1971). Foods of low caloric value, such as detritus of plant origin in the range of 2–4 cal mg^{-1} dry weight, are assimilated more slowly than algae and bacteria in a caloric range of 5–6 cal mg^{-1}. Although the assimilation efficiency of detritus is low, that of algae and bacteria by zooplankton is variable. The few studies on the subject show that in general algae are assimilated more efficiently (15–90%) than bacteria (3–50%) (Monakov and Sorokin, 1961; Saunders, 1969; Winberg *et al.*, 1973; Hart and Jarvis, 1993; Ojala *et al.*, 1995). The percentage of assimilation varies seasonally with shifts in algal composition and is generally at the lower portion of the range cited. In contrast, when given

TABLE 16-12 Energy Budget of Adult *Daphnia pulex* after 18 Instars (40 Days) of Growth at Four Food Concentrations of *Chlamydomonas*[a]

Food concentration									
Cells × 10^4 *ml*$^{-1}$ *day*$^{-1}$	*mg ml*$^{-1}$	*Energy consumed (cal)*	*Energy of growth and young (cal)*	*Energy of respiration (cal)*	*Energy of egestion (cal; by difference)*	% *Assimilation*	% *Energy consumed as growth and young*	% *Energy assimilated as growth and young*	% *Engery of growth and young as respiration*
25	6.2	6.14	1.07	0.84	4.23	31.1	17.4	56.0	78.6
50	12.4	13.59	1.77	0.94	10.87	20.0	13.1	65.3	53.0
75	18.6	20.74	2.35	1.02	17.37	16.3	11.4	69.8	43.3
100	24.8	29.24	2.93	1.08	25.23	13.7	10.0	73.0	37.0

[a] After data of Richman (1958).

mixed diets of green algae and a photosynthetic bacterium, *Ceriodaphnia* assimilated less of the algae than when fed on algae alone (Gophen *et al.*, 1974). In a related study, live filaments of a cyanobacterium, *Oscillatoria limnetica*, were of lower food quality than detrital particles derived from the cyanobacterium (Repka *et al.*, 1999). Detritus supported growth and reproduction of *Daphnia* at rates comparable to those on a diet of the green alga *Scenedesmus*. Utilization of bacteria was relatively unchanged when the algae were also included in the diet.

G. Utilization of Bacteria by Crustacean Zooplankton

Clearly cladocerans and copepods are capable of ingesting bacteria both in aggregates as well as freely in the water (Horn, 1985; Knoechel and Holtby, 1986; Bern, 1987). Grazing on bacteria by crustacean zooplankton can remove significant (e.g., 25–50%) portions of bacterial production. Removal of the larger zooplankton, such as by planktivorous fish predation, can result in an intensified utilization of bacterioplankton by protistan grazers and rotifers (Riemann, 1985; Geertz-Hansen *et al.*, 1987; Jürgens, 1994b). Regulation of bacterial productivity is much more effective by protistan grazing than metazooplankton (Berninger *et al.*, 1991; Jürgens 1994a).

Many comparative analyses of feeding on natural bacterioplankton and phytoplankton indicate that even at maximal filtration rates bacterioplankton can only be a minor constituent of the diet and are inadequate to support growth and reproduction (e.g., Pedrós-Alio and Brock, 1983; Børsheim and Olsen, 1984; Forsyth and James, 1984; Kankaala, 1988; Sanders *et al.*, 1989; Wylie and Currie, 1991; Urabe and Watanabe, 1991). The ingestion ratios between algae and bacteria, however, exhibit considerable differences in relation to seasonal availability and relative abundances. Bacterial abundance and production, as discussed in detail in Chapter 17, are regulated by nutrient availability. When bacterial abundance and productivity are at low-to-moderate levels, protistan grazing dominates (e.g., Jürgens *et al.*, 1994b; Pace and Cole, 1996). Under highly eutrophic conditions where bacterial abundance is high, rotifer and crustacean zooplankton bacterivory can increase. Bacterial feeding by large cladocerans can be in large part incidental to the process of algal feeding. It should again be noted, as discussed earlier with the algae, that ingestion of bacteria does not necessarily result in mortality. Some bacterioplankton clearly can survive digestion in the guts of zooplankton and gain nutritionally (King *et al.*, 1991).

There are several indications that algae alone do not meet the nutritional requirements of cladocerans (see, e.g., Taub and Dollar, 1968; Sanders *et al.*, 1996). In spite of the moderate assimilation rates by zooplankton of algae, bacteria, and heterotrophic flagellates and ciliates and low assimilation rates of detritus, it is unusual for bacteria and other microheterotrophs to dominate the energy income of zooplankton. Clearly, some nutritional component of algae is needed by cladoceran, copepods, and other zooplankton that is critical to growth and reproduction (see following section on lipids). Moreover, in the few cases in which it has been analyzed in some detail (cf. Chap. 22), detritus can form a major energy resource over much of the year. For macrozooplankton, however, detrital material is of limited value in nutrition. Combined with bacteria, detritus may enable macrozooplankton to survive during periods of algal shortage.

H. Lipids in Zooplankton and Fish

In zooplankton, lipids are dominant energy storage compounds and can constitute a significant portion (to 60–70%) of their dry mass (Arts, 1999b). *De novo* synthesis is very small (Goulden and Place, 1990, 1993); lipids in zooplankton and fish are primarily dietary in origin. As a result, the type and amount of lipid contained within zooplankton tissues are correlated directly with recent feeding activities and food selectivity. The correlation between food type ingestion and body lipid content can be modified by temperature (Arts *et al.*, 1993) and reproduction (Vanderploeg *et al.*, 1992) and result collectively in distinct temporal patterns of lipid content in the zooplankton.

Storage lipids are composed primarily of nonessential lipids and constitute major energy reserves. Some polyunsaturated fatty acids (PUFA) and sterols are essential but are required in trace quantities (Goulden *et al.*, 1999; Olsen, 1999). Essential fatty acids (EFA) include linoleic (18:2ω6), linolenic (18:3ω3), and likely eicosapentaenoic (20:5ω3) acids. The phospholipids, which are abundant in cellular membranes, contain a high percentage of essential PUFA. The absolute amount of phospholipids in tissue of zooplankton and fish is partly under genetic control and relatively constant. In contrast, the triacylglyceride and wax ester contents are highly variable and composition dependent on diet.

Zooplankton are often food-limited during the summer months (Lampert, 1977; Threlkeld, 1979; Tessier, 1986). When food availability is low, lipid reserves are essential to the survival of cladoceran embryos, neonates, and adults (Lampert and Bohrer, 1984; Goulden and Henry, 1985; Goulden *et al.*, 1999). It is important to distinguish the effects of food limitation simply on the basis of reduced total food quantity during summer from limitations in the availability of essential fatty acids (EFA) in the algal food. For example, the low lipid content of the cyanobacterium *Aphanizomenon* provides a totally inadequate food source for normal growth and survival of *Daphnia* and other crustacean zooplankton (Holm and Shapiro, 1984; Arts *et al.*, 1992).

Lipid content of zooplankton has been found to decrease from spring to early summer and to increase again in late summer and autumn (Wainman and Lean, 1990; Goulden *et al.*, 1990; Arts *et al.*, 1992, 1993; Brett and Müller-Navarra, 1997). A number of correlational analyses suggest strongly that the nutritional quality of phytoplankton communities and their essential fatty acid content can be an important factor regulating species and seasonal succession in zooplankton (Ahlgren *et al.*, 1990, 1996; Ahlgren, 1993). For example, in two eutrophic lakes, the total fatty acid content of phytoplankton content varied fivefold during the active growing season, with the lowest values during the summer when cyanobacteria dominated. Highest values occurred during spring and late autumn when diatoms dominated. Essential fatty acids were much more abundant in phytoplankton of the spring and late autumn than during the summer. Specific growth rates of several species of cladocerans were much greater when consuming algae that contained high percentages of essential polyunsaturated fatty acids.

Highly polyunsaturated fatty acids (PUFA) are critical for maintaining high growth, survival, and reproductive rates of zooplankton and fish. High-food-quality algae are rich in such PUFA content, and, conversely, low-food-quality algae are poor in PUFA content (Brett and Müller-Navarra, 1997; DeMott and Müller-Navarra, 1997; Sundbom and Vrede, 1997; Weers and Gulati, 1997). Growth rates of *Daphnia*, for example, can be limited directly by the availability of eicosapentaenoic acid (20:5ω3) within seston used as food (Müller-Navarra, 1995). If fed cryptophyceans or green algae with high PUFA content or if emulsions of PUFAs are added to algal cultures of poor food quality, growth and reproduction rates of zooplankton feeding on these algae can be increased markedly. Fish modify fatty acids such as linolenic (18:3ω3) and linoleic (18:3ω6) acids by elongation and desaturation to form long-chain PUFA such as docosahexaenoic acid (DHA; 22:6ω3) and arachidonic acid (AA; 20:4ω6), respectively. In fish, essential fatty acids, including eicosapentaenoic acid (20:5ω3), DHA, and probably AA, are required to maintain proper growth and development (Bell *et al.*, 1994). For example, deficits in essential fatty acids, in particular DHA, have been shown to affect the number of rods in the photoreceptor population of the eyes of herring (Bell and Dick, 1993) and to hamper the feeding of herring under low light intensities (Bell *et al.*, 1995).

VI. REPRODUCTION AND LIFE HISTORIES

A number of similarities exist among the common life histories and reproductive processes of rotifers and cladocerans. For many successive generations during the main growing season, males are absent and reproduction occurs via parthenogenesis of diploid females. Under certain conditions of environmental stress (e.g., crowding), males develop, meiosis occurs, and sexual (mictic) reproduction occurs, leading to the formation of mictic haploid eggs (Fig. 16-17). If fertilized, the now-diploid eggs ("resting eggs") develop a heavy cell wall and become resistant to environmental extremes.

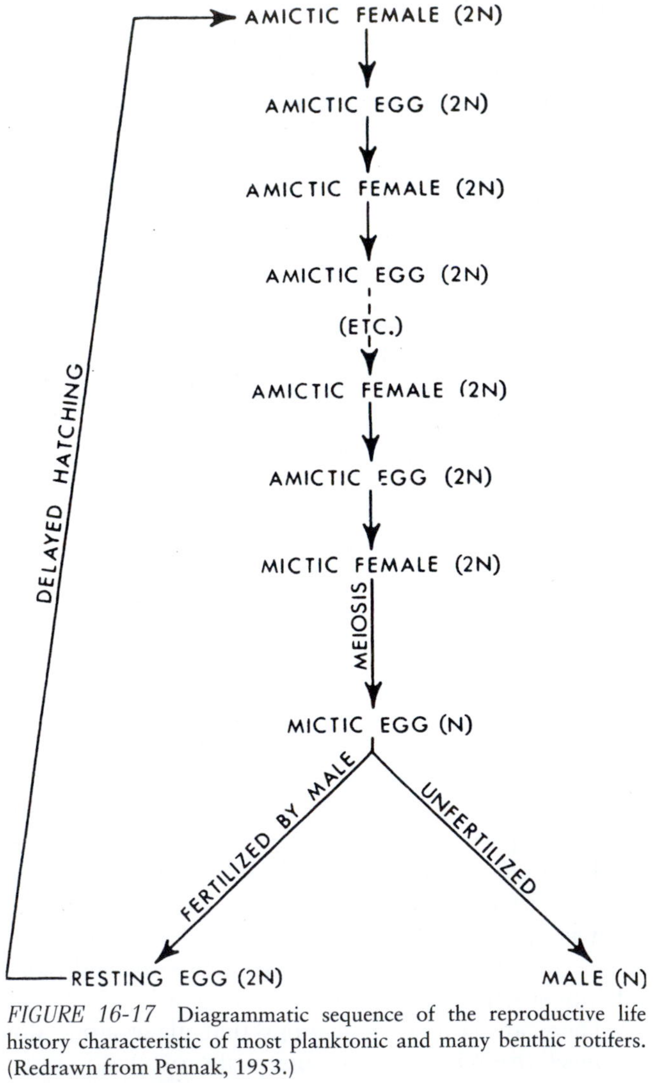

FIGURE 16-17 Diagrammatic sequence of the reproductive life history characteristic of most planktonic and many benthic rotifers. (Redrawn from Pennak, 1953.)

The resting eggs commonly enter diapause, with hatching delayed for long periods, and, upon return of favorable conditions, will develop into amictic diploid females. In contrast, freshwater copepods are bisexual and have a much longer life cycle than do rotifers and cladocerans.

A. Rotifers

The reproductive life history of typical planktonic rotifers, all in the subclass Monogononta, is characterized by a large number of generations in which reproduction is parthenogenetic. These amictic females are all diploid and produce amictic eggs that develop into amictic females (Fig. 16-17). There may be as many as 20–40 of such amictic generations. With an egg-development time of about one day under warm, optimal conditions, and without the need to encounter males for fertilization, a population of amictic females can develop rapidly in 2–5 days.

This nonmeiotic egg production is intermittently broken, often only once or twice per year, by the development of morphologically indistinguishable mictic females. The eggs of the mictic females are produced by normal double meiotic division and are thus haploid (Fig. 16-17). If eggs of these mictic females are fertilized, they develop into thick-walled encysted embryos called resting eggs. The resting eggs undergo prolonged diapause and are highly resistant to adverse environmental conditions. If the mictic eggs are not fertilized, however, they develop rapidly into males. The males are greatly reduced in size and complexity (Gilbert, 1993). The males are short-lived and extremely active and are capable of copulation within an hour of hatching. The hatching of resting eggs can be induced by biochemical factors, but induction under natural conditions is poorly understood. The diapause extends over a period of several weeks or months. Resting eggs accumulate in the sediments and hatch when their environment changes, generally from cold, dark, and anaerobic to relatively warm, light, and aerobic (Gilbert, 1995). The resting eggs always result in parthenogenetic amictic females.

The mechanisms underlying the production of mictic females are unclear, although a number of environmental conditions have been proposed (reviewed by Hutchinson, 1967, and Ruttner-Kolisko, 1972). Although the stimuli are quite species-specific, significant factors include crowding of amictic females in relation to food and accumulation of substances such as pheromones, which are produced by the females. In clones of *Brachionus* reproducing amictically at 20°C, brief exposure to a shock temperature of 6°C resulted in nearly 50% of the clones becoming mictic in <2 days (Ruttner-Kolisko, 1964; Gilbert, 1977).

Asplanchna, the large predatory rotifer, is certainly among the best-studied examples in this regard. Gilbert and co-workers, in a series of superb papers (1967a, 1968, 1972, 1973, 1975, 1976; Gilbert and Thompson, 1968; Gilbert and Birky, 1971; Riggs and Gilbert, 1972; Gilbert and Litton, 1975; reviewed by Gilbert, 1980a), have demonstrated the subtle reproductive controls in this species that have far-reaching implications among the zooplankton. When fed the protozoan *Paramecium*, *Asplanchna* produced only amictic females. Increasing the population densities of the rotifer to extraordinary levels of crowding produced no mictic females, whereas adding low populations of the algae *Chlamydomonas* or *Euglena* to the dense paramecia resulted in the production of a significant number of mictic females. It was also shown that the algal cells had to be eaten to induce the reproductive change in *Asplanchna*; extracellular algal products would not induce the change. It was shown subsequently that the important dietary component from the plants was *d*-α-tocopherol (vitamin E), which induced the transition from parthenogenetic to sexual reproduction and is essential for spermatogenesis or male fertility. The mictic female offspring were also larger and changed in morphology, exhibiting four partially retractile outgrowths of the body wall. These "humps" were elicited by *d*-α-tocopherol within a range of 5×10^{-13} mol (0.2 ng) in a linear fashion to a maximum of 5×10^{-11} mol (20 ng) per female. Tocopherol, as well as cannibalism, also induced a third polymorph, a very large campanulate type.

The adaptive significance of control of sexual reproduction and nongenetic female polymorphisms is complex (reviewed by Gilbert, 1980a). The initial response to tocopherol is a general increase in growth that permits the carnivorous *Asplanchna* to more easily ingest larger, possibly more nutritious, herbivorous prey. Tocopherol can act as an adaptive cue for this growth, for it is synthesized only by photosynthetic organisms and is readily assimilated by herbivores. Upon ingestion of rotifer and crustacean prey, tocopherol is readily assimilated by *Asplanchna* and is transmitted to its embryos without degradation. Tocopherol can promote differential polymorphism in some species of *Asplanchna* through cell enlargement and increased cell division. Population densities of *Asplanchna* increase when herbivorous zooplankton are more abundant; the resulting increase in contacts between males and mictic females facilitates production of resting eggs.

Body-wall outgrowths, which reduce cannibalism by the large individuals, likely evolved with the growth response to tocopherol (Gilbert, 1980a). Two tocopherol-dependent, female morphotypes evolved in some species. The larger, campanulate morphotype is adapted for ingestion of very large prey and, in one species of *Asplanchna*, for parthenogenetic reproduction only. The well-protected cruciform morphotype, with prominent body-wall outgrowths, is adapted for ensuring sexual reproduction and avoiding capture by cannibalistic conspecific *Asplanchna*. Growth responses to tocopherol evolved only in those species that (1) do not depend on the direct utilization of relatively small tocopherol-rich algae and (2) can ingest the copious quantities of zooplankton prey needed for sustaining populations of large individuals.

Other rotifers, such as the sessile bdelloids, none of which are planktonic, reproduce exclusively by parthenogenesis. Resting eggs are unknown, but these adults can survive drying.

The survival and reproductives rate of rotifers are related strongly to the quality and abundance of food as well as to temperature (Edmondson, 1946, 1965; King, 1967; Halbach and Halbach-Keup, 1974; Baker, 1979; Sterzynski, 1979). It is clear that these factors are of major importance in determining seasonal fluctuations in the populations resulting from changes in the balance between increases by reproduction and declines from mortality.

In addition to effects on the rate of development of eggs, temperature obviously influences the rates of biochemical reactions, feeding, movement, longevity, and fecundity. For example, increased temperature reduces survivorship with an age-specific compression of fecundity (Fig. 16-18). Intrinsic rates of increase (r_m) were 0.34 offspring per female at 15°C, 0.48 at 20°C, and 0.82 at 25°C. A number of other factors affect natural populations, but temperature is clearly a major factor affecting birth rate. It is also apparent that species are variable in their responses and habitat restrictions; some are very restricted stenotherms, while others are eurythermal.

A number of studies under both natural and laboratory conditions have shown that food type and quality exert a major influence on the population dynamics of rotifers. For example, in investigations of the dynamics and reproduction of two clones of *Euchlanis dilatata* fed different algal species at various concentrations, King (1967) found several relationships that also have been shown to varying degrees by others. Two genetically identical clones of *Euchlanis* were grown from a single parent by parthenogenesis, so that in succeeding generations, the individuals of one clone, the young orthoclone, were always derived from young parents and those of the other orthoclone from old parents. The clones were fed different concentrations of three algal species, *Chlamydomonas reinhardi, Euglena gracilis*, and *E. geniculata*. Growth rates of individual animals were found to depend on food concentration and not on animal age or algal species. The rate of population increase was related directly to the density of food for concentrations up to 16.4 μg ml^{-1}, beyond which no further changes were observed. In *Brachionus*, filtration rate decreased at high algal densities; this was related to clogging of the filtration apparatus, reduced digestion, or inactivation of the corona by algal toxicants (Halbach and Halbach-Keup, 1974). Within the population, however, growth rates were related to both food species and food concentration (Fig. 16-19). *Euchlanis* fed on *Chlamydomonas* had higher rates of population increase than when feeding upon *Euglena gracilis*. *E. gracilis*, on the other hand, led to higher

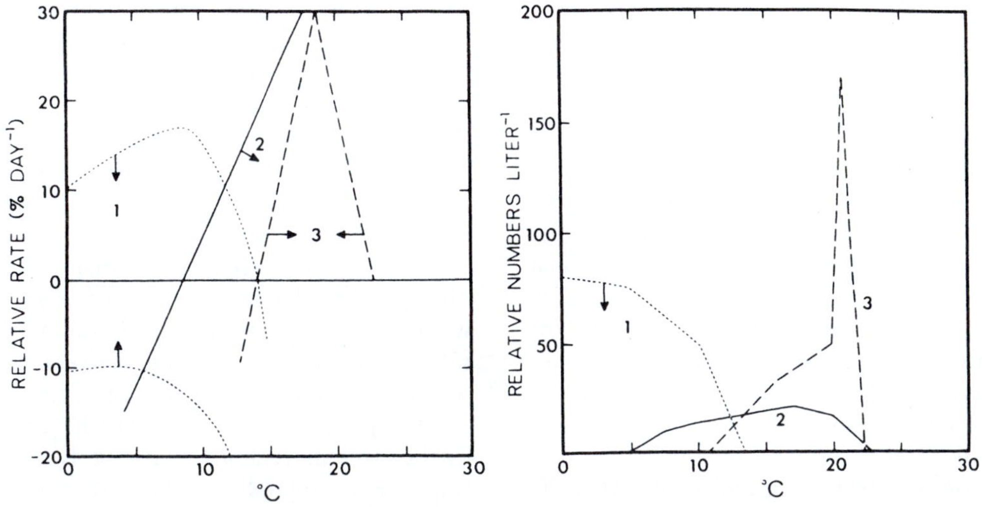

FIGURE 16-18 Range of response (between arrows) in relative rate of change of concentration of rotifers (*left*) and concentration per liter (*right*) in relation to temperature. 1. *Synchaeta lakowitziana*; 2. *Euchlanis dilatata*; 3. *Pompholyx sulcato*. (Drawn from data given in Edmonson, 1946, after Carlin.)

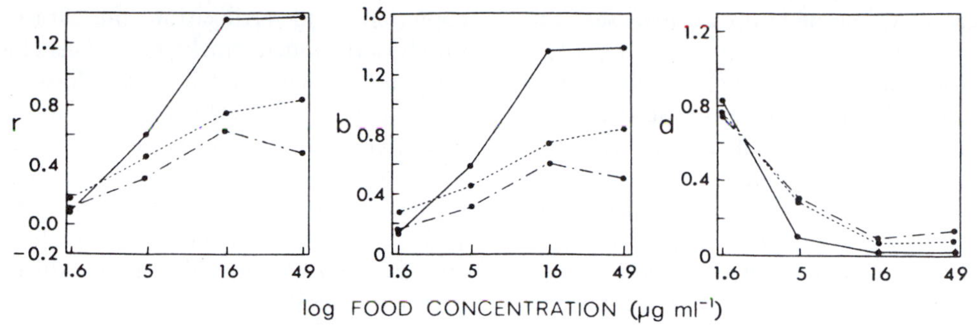

FIGURE 16-19 The instantaneous rate of increase (*r*), birth rate (*b*), and death rate (*d*) of young orthoclones of the rotifer *Euchlanis dilatata* as a function of food level and food species. ________ = *Chlamydomonas reinhardi*; - - - - - - = *Euglena gracilis*; – · – · – · – = *Euglena geniculata*. (Drawn from data of King, 1967.)

rates of population increase than *E. geniculata*. Individuals of the young orthoclone exhibited greater rates of population increase than those of the old orthoclone. The different rates of increase resulted more from corresponding differences in net reproduction (the number of eggs laid by the average female in her lifetime) than from the frequency of egg laying or survivorship. The results, confirmed by others (e.g., Snell, 1980), imply differences in assimilation, differences in the chemical quality of the food, or both.

B. Cladocera

Somewhat analogous to that of the rotifers, reproduction in the cladocerans is parthenogenetic during a greater part of the year. Until interrupted by sexual reproduction, females produce eggs that develop into more parthenogenetic females. The number of eggs produced per clutch varies from two or more in the Chydoridae to as many as 40 in the larger Daphnidae. The eggs are deposited into a brood chamber, or pouch, a cavity dorsal to the body bounded by the valves of the carapace. The eggs develop in the brood pouch and hatch as a small form of the parent. As a result, in contrast to the copepods, there are, with the exception of *Leptodora*, no free-living larval forms among the Cladocera. One clutch of eggs is normally released into the brood pouch during each adult instar.

The number of molts is variable. Four preadult instars are common within a range of 2 to about 8 in some *Daphnia*. Longevity, from the time of egg release into the brood pouch until death of the adult, varies widely with species and environmental conditions (Threlkeld, 1987). Longevity and time between molts are approximately inversely related to temperature. For example, longevity in *Daphnia magna* averaged 108, 88, 42, and 26 days at 8, 10, 18, and 28°C, respectively (MacArthur and Baille, 1929). Longevity also is affected by food availability and commonly increases with a decrease in food consumption, short of starvation, which results in rapid death. Lipid reserves in the Cladocera accumulate in the body coelom when food is abundant and can be used as an energy reserve when food is scarce (Goulden and Hornig, 1980). The quantity of lipid that can be stored by individuals in a species is related to recent feeding activities and food selectivity (see Lipids in Zooplankton and Fish, p. 427). The importance of temperature, however, can be seen in studies of *Daphnia galeata mendotae* (Table 16-13) that were fed differing concentrations of

TABLE 16-13 Instantaneous Rates of Population Increase per Day (*r*) ± S.E. of *Daphnia galeata mendotae* at Different Temperatures and Levels of Food Supply[a]

	Temperature		
Relative food levels	*11°C*	*20°C*	*25°C*
Low ($\frac{1}{4}$)	0.07 ± 0.005	0.23 ± 0.002	0.36 ± 0.015
Medium (1)	0.10 ± 0.003	0.30 ± 0.002	0.46 ± 0.013
High (16)	0.12 ± 0.006	0.33 ± 0.016	0.51 ± 0.006

[a] After data of Hall (1964).

mixed green algae at different temperatures (see Hall, 1964).

In adult cladocerans, the number of molts is more variable than in juveniles and can range from a few to well over 20 instars. Each instar ends with the release of young from the brood pouch, molting, rapid increase in size, and deposition of a new clutch of eggs into the pouch, all of which occurs in a matter of minutes. Therefore, an individual can produce several hundred progeny in a lifespan under favorable growing conditions.

Increases in temperature result in an immediate effect by increasing the rate of molting and brood production, whereas increases in food supply, within limits, affect the rate of development of the population by increasing survivorship and fecundity (i.e., the number of eggs per brood). The latter process is not immediate and this time lag can contribute to population oscillations. When the growth of the population exceeds available food supply, reproduction decreases, with reduced replacement of adults. Reduction of the population below the environmental carrying capacity set by available food results in an increase in the population after a lag period. The extent of population oscillations depends greatly on the duration of the time lag. These relationships have been demonstrated in detailed studies under culture conditions (Frank, 1952, 1957; Slobodkin, 1954), and will be discussed in greater detail later on. When several zooplankton species are competing for limited food resources, the decline of some or all of the populations is responsive to the interactions of food quality and supply, feeding capabilities, and the effect of temperature on reproductive responses of each zooplankton species. Degeneration and abortion of parthenogenic eggs is well known but difficult to quantify (Threlkeld, 1979, 1985). Causes for egg abortion are unclear; egg abortion was found to be inversely correlated with cyanobacterial development. Decline of a population does not necessarily eliminate that species, which may shift to an inactive state by the formation of resting stages.

Parthenogenesis continues until unfavorable conditions arise, whether they are physical, such as temperature reductions, drying, or short day-length photoperiod (cf. Shan, 1974; Pourriot and Clement, 1975; Bunner and Halcrow, 1977; Carvalho and Hughes, 1983), or biotic, that is, induced through crowding by competition for low food supply or a decrease in quality–size of food organisms. As the production of parthenogenetic eggs declines, some of the eggs develop into males. Some females then produce sexual (haploid) eggs, as in the rotifers. Though morphologically similar to parthenogenetic females, sexual females usually produce only one or a few resting eggs. Following copulation and fertilization, the carapace surrounding the brood pouch thickens and encloses the egg(s), forming a semielliptical, saddle-shaped *ephippium* (Zaffagnini, 1987). During the subsequent molt, the ephippium either is shed, as in the daphnids, or remains attached, as in the chydorids. If females carrying resting eggs are not fertilized, the resting eggs are resorbed. One species, the arctic *Daphnia middendorffiana*, is known to be able to produce ephippial eggs parthenogenetically.

The ephippia either sink or float, and can withstand severe conditions such as freezing or drying. It is not unusual to find large accumulations of ephippia along windward shore lines. It is easy to envision entanglement and transport of ephippia by birds to other water bodies. A large percentage of the ephippia hatch under favorable but poorly understood conditions, always into parthenogenetic females. In a laboratory-cultured strain of *Daphnia pulex*, light was required for termination of diapause regardless of temperature or duration of ephippial storage (Stross, 1966). In naturally occurring ephippia of a lake population of the same species, diapause also was broken by light, but prolonged storage in darkness eliminated the need for light. The time required was 5–6 months at 3.5°C and only 3–6 weeks at 22°C. Increases in temperature thus increase the rate of hatching (e.g., Herzig, 1985; Yurista, 1997). Hatching is often irregular over a long period of time, in part related to ample genetic variability as determined from maternal inheritance (Meester and de Jager, 1993).

Males are smaller and only slightly modified in morphology from females (larger antennules and clasping modifications of the first leg for copulation). Mechanisms underlying the periodic induction of males and ephippial females are unclear, although evidence indicates that the stimuli are different (reviewed by Hutchinson, 1967). Male production is correlated with crowding and a rapid reduction of food supply, whereas a constant low food supply simply inhibits reproduction (Slobodkin, 1954). Short-day photoperiods (12 h light and 12 h dark) increased the production of ephippia markedly in *Daphnia pulex*, in contrast to longer light photoperiods, which would be characteristic of midsummer in temperate latitudes (Fig. 16-20). If universal, this photoperiodic response must vary among species.

C. Copepods

Reproduction in free-living copepods is similar in spite of widely varying species differences in sexual behavior and periodicity of breeding. Some species reproduce throughout the year, others only briefly at specific

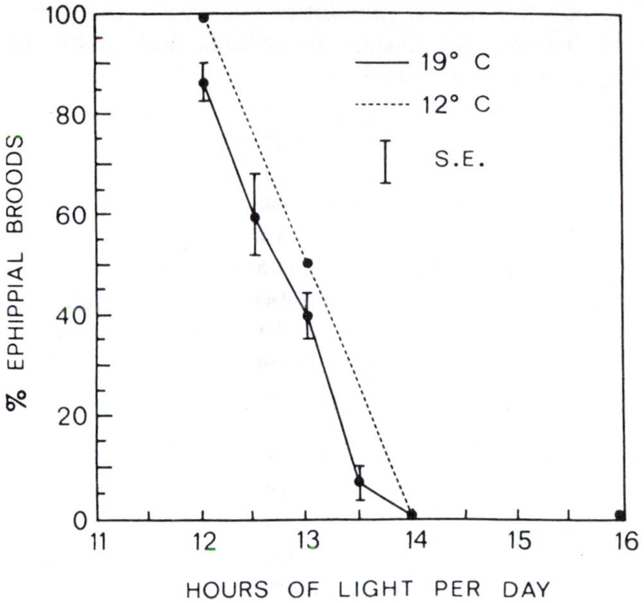

FIGURE 16-20 Influence of photoperiod on the production of ephippial egg broods in *Daphnia pulex* at high population densities. (Modified from Stross and Hill, 1965.)

times of the year. Copepod copulation occurs by a male briefly clasping the female with transfer of spermatophores to the ventral side of the female genital segment. Fertilization may take place immediately or several months after copulation. Fertilized eggs are carried by the female in 1 or 2 egg sacs. The number of eggs in cyclopoid copepods, carried in two lateral egg sacs, is variable; differences observed between populations in different lakes and between seasons, have not been associated clearly with changing environmental conditions. Egg number, up to 72, is generally low in summer and increases in the autumn.

In the calanoid copepods, eggs are carried in a single egg sac and are variable in number (from about 1 to 30). Eggs are differentiated into subitaneous and resting eggs, both being fertilized in many species. Resting eggs usually are dropped to the sediments, where they undergo a period of diapause. Clutch size varies greatly in calanoids; often it is maximal in the spring and the fall, and these two peaks may be separated by a summer minimum. Larger clutch size is correlated with lakes of greater primary productivity and to spring conditions that favor more abundant food supplies. As in the Cladocera, food availability directly influences the clutch size, while temperature determines the rate of egg production and egg development times in copepods (Elster, 1954; Comita and Anderson, 1959; Burgis, 1970; Ceiling and Campbell, 1972; Elmore, 1983; Watras, 1983; Maier, 1989, 1990). The reproductive cycle is apparently not affected by variations in the photoperiod. In this way, a small number of females carrying a few eggs for a short period at high temperatures can produce more young than many females with large egg clutches that are exposed to long periods at low temperatures.

Copepod eggs hatch into small, free-swimming larvae, termed *nauplii*, and then develop by molting through a number of subsequent larval stages. The initial nauplius has three pairs of reduced appendages (first and second antennae, mandibles). The successive naupliar stages, six in all, feed, grow, molt, and acquire further appendages. After six naupliar molts, the next molt results in an enlarged and more elongated form, the first copepodite instar. There are five copepodite stages, during which additional appendages and body segments develop. The sixth and final copepodite stage is the adult. The time required to complete the juvenile stages and period of diapause is highly variable among species and depends upon seasonal conditions (see following discussions).

VII. SEASONAL POPULATION DYNAMICS

A. General Considerations

More useful information on zooplankton population dynamics can be obtained if the stages in the life history of species are analyzed separately than if only the total population is measured.[4] Analyses of egg production and instar dynamics yield information on growth rates and mortality, as well as the duration of each stage and the interval between generations. Instar analyses are easier with copepods than with cladocerans because the naupliar and immature and adult copepodite stages can be distinguished morphologically (cf. Wetzel and Likens, 2000). Among the cladocerans, and to some extent the rotifers, instars can sometimes be evaluated on the basis of size.

Analyses of the population dynamics of zooplankton require sound sampling methods with known statistics on sampling variance and on population heterogeneity within the aquatic system. This subject is outside the purview of the present work and, moreover, has been treated excellently in methods manuals edited by Edmondson and Winberg (1971), Winberg (1971), and Downing and Rigler (1984), and in the reviews by Bottrell *et al.* (1976), Waters (1977), and Benke (1993). Once representative samples of the animal population are obtained, a number of methods have been employed to evaluate population size and fluctuations. Enumeration can be adequate to address certain

[4] The generous counsel of W. T. Edmondson on the population analyses of zooplankton is gratefully acknowledged.

questions concerning population dynamics and interactions, but in order to evaluate secondary productivity, precise knowledge of changes in biomass throughout all stages of the organisms' life cycles is mandatory. Development times must be known and, although they are variable among organisms and environmental conditions, time between samplings must be less than the shortest developmental period. Assuming stable egg distribution, the number of eggs present in a sample of a zooplankton population at some instant in time represents the increment that would have been added to the population over a period of time equal to the duration of development of the eggs (time from laying to hatching). The population would increase by the number hatched during the time period if no deaths occurred.

The population dynamics of rotifers are somewhat simplified because of the very short time between hatching and attainment of reproductive capacity. Analyses of birth and death rates may be estimated from reproductive rates and rates of population change. The reproductive rate can be determined indirectly from the ratio of eggs per female (Edmondson, 1960, 1965, 1968, 1974). The ratio of eggs per female (E) observable in a sample at any given moment can be used to calculate a finite population birth rate (B) as eggs per female per day, by

$$B = \frac{E}{D} \qquad (16\text{-}1)$$

when D is the duration of development of the eggs, or the mean time an individual spends in the egg stage. This relationship assumes that eggs present at a given time will be added to the population during the next period of time, D. If the age distribution of the eggs is stable, the fraction of eggs hatched during a day is $1/D$. If the population is changing size, the value of B is exact only if $D = 1$; the bias introduced by longer durations can be estimated (Edmondson, 1968).

From the reproductive rate B, the instantaneous growth rate of the population, r, can be calculated based on the conventional exponential growth model (cf. Chap. 8), in which positive or negative growth over a short time interval is

$$N_t = N_0 e^{rt} \quad \text{and} \quad r = b - d \qquad (16\text{-}2)$$

in which

N_0 = the size of the initial population at time zero
N_t = the population size at a later time t
r = the growth rate coefficient or intrinsic rate of growth
d = the instantaneous rate of mortality and
b = the instantaneous birth rate (calculated from B)

The effective rate of population increase is the difference between the natural logarithms divided by the time increment ($t - 0$):

$$r = \frac{\ln N_t = \ln N_0}{t} \qquad (16\text{-}3)$$

which is in essence the difference between natality (b) and mortality (d) over the time interval (t).

Alternatively, birth rate, estimated by the egg-ratio method, gives the daily increment to the population. For a population growing by the amount E/D in one day with no deaths, its growth rate would be

$$b = \ln(B + 1) \qquad (16\text{-}4)$$

Here B is estimated at the beginning of the time interval with the assumption that for each female in the population there will be $(E/D) + 1 = B + 1$ females one day later. Assuming no mortality, the population would grow in one day at a rate of

$$\ln(B + 1) - \ln(1) = \ln(B + 1) \qquad (16\text{-}5)$$

Thus by Equation (16-4), a finite per capita birth rate, B, can be used to estimate the instantaneous rate, b. The calculations assume that B is small; under such circumstances both B and the E/D are approximately equal to the instantaneous birth rate b.

As pointed out by Edmondson (1968) and rederived by Paloheimo (1974), for moderately large values of r and bD, the finite per capita birth rate and the egg-ratio/D can diverge considerably (see Lynch, 1982). Further, the relationships are most applicable to planktonic animals that carry their eggs until hatching, at which time free-swimming progeny are liberated, that is, the eggs are subjected to the same mortality as the adults. Paloheimo (1974; Gabriel *et al.*, 1987) has shown that

$$E/D = (e^{bD} - 1)/D \qquad (16\text{-}6)$$

when D is the development time of the eggs, which is algebraically equivalent to

$$b = \ln[(C_t/N_t) + 1]/D \qquad (16\text{-}7)$$

when C_t is the total number of eggs counted at time t, which estimates the instantaneous birth rate b from egg counts C, where C_t/N_t is the egg ratio. This formula assumes that steady-state conditions prevail, which is not usually the case. Development time is temperature dependent, and in analyses of natural populations this primary factor must be estimated from good experimentally determined changes with temperature. Examples of estimates of instantaneous birth rates are given in Table 16-14, in which the C_t/N_t ratio has been held at 1. The egg-ratio method has been used in several detailed population analyses of planktonic rotifers

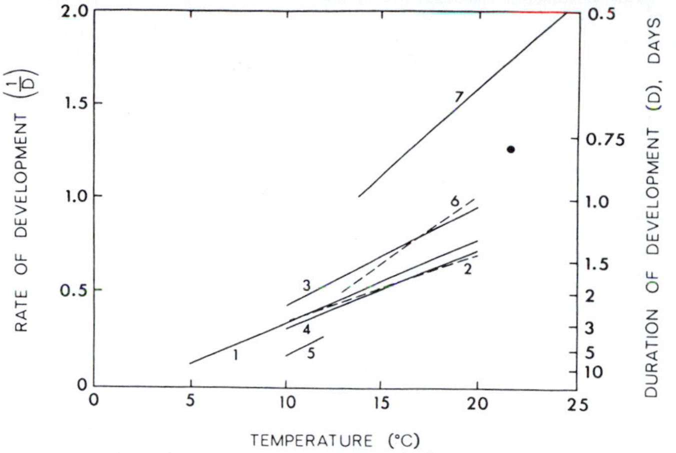

FIGURE 16-21 Average rates of development per day of rotifer eggs in relation to temperature. Line 1, Least-squares regression line for *Keratella cochlearis* and *Kellicottia longispina*; 2. *Polyarthra vulgaris*; 3. *Keratella quadrata*; 4. *Kellicottia bostoniensis*; 5. *Ploesoma truncatum*; 6. *Ploesoma hudsoni*; 7. *Hexarthra fennica*; (•). *Euchlanis dilatata*. (From data of Edmondson, 1960, 1965; and King, 1967.)

(especially by Edmondson, 1960, 1968; Gabriel *et al.*, 1987), from which the following examples are drawn, as well as for cladoceran and copepod zooplanktonic populations to be discussed later.

B. Development Time

The rate of development of eggs, which is required to determine the reproductive rate *B*, has been shown to be a function of temperature among a large number of rotifers and planktonic crustaceans. Examples of this relationship for a number of rotifer species are seen in Figure 16-21. The development time is quite independent of the type and quantitative nutrition of the adult female carrying the eggs (see, e.g., King, 1967; Meier, 1989a, 1990b). The age in cladocerans can be calculated from body size once the relationship of length of juvenile development is known as a function of temperature (Geller, 1987). Embryonic development time for a species of *Diacyclops*, for example, was between 14 days at 2°C and 2 days at 25°C.

The instantaneous growth rate, *r*, is the effective rate of population increase per unit time. A measure of death rate, *d*, is calculated by difference and varies with the abundance of predators.[5] The procedure assumes continuity of birth and death rates during the sampling interval (cf. Keen and Nassar, 1981; Taylor and Slatkin, 1981; Gabriel *et al.*, 1987). The method demands that the sampling interval be short in relation to the life cycle and immaturity time of the animals. Although natural populations do not strictly conform to these assumptions, the discrepancies introduced are usually rather small (see Caswell, 1972; Edmondson, 1974; Paloheimo, 1974; Threlkeld, 1979; Polishchuk and Ghilarov, 1981; Lynch, 1982; Berberovic *et al.*, 1990,

TABLE 16-14 Estimates of Instantaneous Birth Rates at Differing Development Times in Days by Equation (16-7) when the Egg Ratio $C_t/N_t = 1^a$

Development time, d (days)	*Instantaneous birth rate, b*
0.5	1.386
1	0.693
2	0.346
3	0.231
4	0.173
5	0.139
10	0.069

[a] After data of Paloheimo (1974).

[5] In the calanoid copepod *Diaptomus clavipes*, egg size can differ, and larger eggs give rise to larger nauplii, each of which presumably has a greater likelihood of survival compared with smaller nauplii (Cooney and Gehrs, 1980). In some species of cladocerans, lipid reserves deposited in the eggs by the adult depend upon the nutritional status of the adult; neonates with lower lipid reserves (see p. 427) may not survive as well as neonates with extensive lipid reserves (Goulden and Henry, 1982). The extent to which differential survival of newly hatched or neonate zooplankton is influenced by nutritional history of the parent animal is not fully understood and may alter our view of zooplankton life strategies as more data become available.

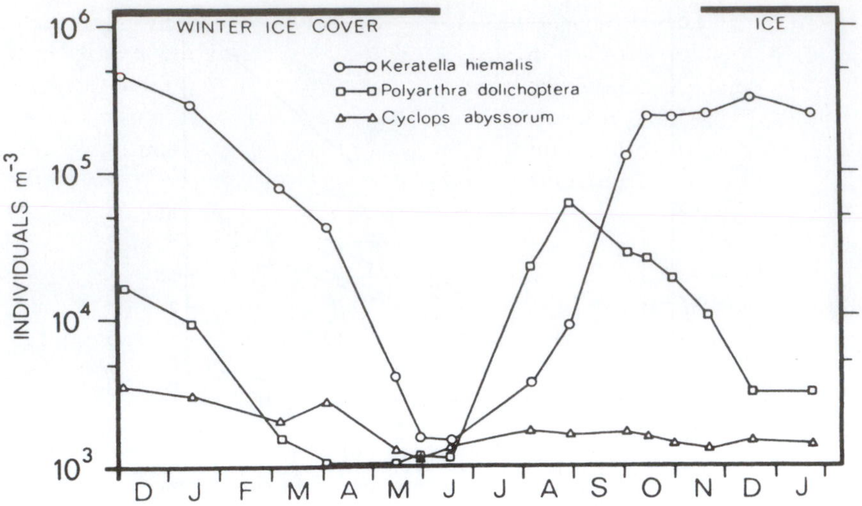

FIGURE 16-22 Seasonal variations in the population densities of three dominating zooplankton species (>99% of the total) of Vorderer Finstertaler See, Central Alps of Austria, 1969. (Modified from Pechlaner *et al.*, 1970.)

for a rigorous discussion of these assumptions). The method is very useful for analyzing factors that control birth and death rates in natural populations of zooplankton.

VIII. ROTIFER POPULATION DYNAMICS

Changes in the seasonal distribution of planktonic rotifer populations are complex and generalizations are difficult to make. A number of perennial species occur that commonly exhibit maximal densities in early summer in temperate regions. Even though most rotifers have wide ranges of temperature tolerances (Berzinš and Pejler, 1989), species are distinctly seasonal and of two general types: (a) cold stenotherms that develop greatest populations in winter and early spring and (b) species that develop maxima in summer, often with two or more maxima, especially in late summer in conjunction with the development of certain cyanobacteria populations. Some of the variations in seasonal succession can be seen in certain well-studied examples from several lakes. Additional examples from several older works are discussed critically in Hutchinson (1967).

The zooplanktonic community of Vorderer Finstertaler See, a deep, alpine drainage lake above the timber line in the Central Alps of Austria, was extremely simple.[6] Two rotifers (*Keratella hiemalis* and *Polyarthra dolichoptera*) and a single copepod (*Cyclops abyssorum*) constituted over 99% of the zooplankton. The species of *Keratella*, known as a ***hypolimnetic cold stenotherm***, developed maximal population densities in midwinter, decreasing conspicuously during spring and early summer (Fig. 16-22). The species of *Polyarthra* is also a coldwater form that tolerates reduced oxygen concentrations and is commonly found in hypolimnia. The *Polyarthra* maximum was seen in midsummer in this lake.

A number of other species of rotifers are similarly cold stenotherms with varying tolerances to reduced oxygen concentrations (Ruttner-Kolisko, 1975, 1980; Elliott, 1977; Herzig, 1980; May, 1980; Laxhuber, 1987). These capabilities permit the rotifers to occupy niches both spatially (metalimnion and hypolimnion) and seasonally (winter throughout much of the water column; spring period) that are unsuitable for other rotifers. Often, bacteria are more abundant in these regions than in the epilimnion and apparently serve as significant food sources. Other rotifers, such as *Notholca*, can feed on the cell contents of large diatoms (e.g., *Asterionella*) and may increase in abundance during the spring diatom bloom.

Studies of the rotifer populations in two small lakes in southern Ontario showed that the seasonal distribution varies considerably among lakes and is synchronous among lakes only in a general way (George and Fernando, 1969). Paradise Lake, for example, is a small (7 ha), shallow (6 m), dimictic lake that exhibits reduced oxygen concentrations in the hypolimnion in late summer but does not become anaerobic. Summer stratification extended from early June through September in the year of the investigation. Of the four

[6] The lake has since been modified drastically by a large hydroelectric project.

species[7] of *Keratella* found in this lake, *K. canadensis* is a cold stenotherm found in the upper strata in early winter. *K. quadrata*, also a cold form, occurred from early winter in the upper layers and moved to the cooler hypolimnion in midsummer before declining. *K. cochlearis* is also a cold stenotherm and exhibited two maxima in this lake, one in winter and another in late spring, but the seasonal distribution of this species varies greatly among different lakes. *K. hiemalis*, known as a winter and early spring species, was common during colder periods but was reduced abruptly as temperatures reached approximately 15°C.

Polyarthra vulgaris, P. euryptera, Conochilus unicornia, Gastropus stylifer, and *Keratella crassa* are largely summer and autumn species, although a considerable displacement and spatial separation of the populations occurred seasonally with depth. The predatory species of *Asplanchna* generally was a minor component of the rotifer community and was relatively uniformly distributed in winter and early summer. The two species of *Kellicottia*, both perennial, were distributed uniformly with depth throughout much of the year; *K. bostoniensis*, however, was restricted to cooler areas during much of the summer. Both species exhibited summer population maxima.

The rotifer community of Lake Constance of southern Germany was examined repeatedly over >50 yr, a period during which the lake experienced gradual eutrophication from oligotrophic to mesotrophic conditions (Walz *et al.*, 1987). A continual quantitative increase in the rotifer populations was observed over this interval, particularly among the epilimnetic populations. Approximately in the middle of this time period, a large *Daphnia hyalina* and a carnivorous species of copepod, *Cyclops*, increased markedly and overwhelmed other cladocerans. Predation by the *Cyclops vicinus* effectively controlled both the *Daphnia* and the large predatory rotifer *Asplanchna*. As predation pressure relaxed in May as the *Cyclops* entered diapause, *Asplanchna* and other rotifers developed rapidly before *Daphnia* populations could recover. Predation on some rotifer species was effective by both *Cyclops* and *Asplanchna*; other species of rotifers were impacted only by *Cyclops*. Similar relationships of *Cyclops* predation on rotifers were found in Lake Michigan zooplankton communities (Stemberger and Evans, 1984).

In a eutrophic reservoir in Georgia, USA, an inverse correlation between the abundance of rotifers and crustaceans was found (Orcutt and Pace, 1984). Rotifers dominated in abundance during the summer months (June–September) and crustacean zooplankton dominated during the remainder of the year.

These examples serve to illustrate some of the variations found in seasonal succession. Variations are very common, and in warm temperate regions very little is known about seasonal succession among rotifers. Interactions among species of rotifers are poorly understood; a notable exception is the predation of *Asplanchna* on *Brachionus*, discussed earlier. Clearly, temperature and food quality and quantity are dominant factors in the regulation of rotifer reproductive rates and population succession.

[7] *Keratella canadensis*, which may not be a valid species, is very similar to *K. quadrata*.

A. Exploitative Competition for Shared Food Resources

The food concentrations for which population growth is zero (i.e., the threshold food level) varies greatly among planktonic rotifer species (Stemberger and Gilbert, 1985). The threshold food concentration was positively correlated to both rotifer body mass and to population growth rates. Thus, the smallest species require less food to attain maximum growth rates; they are better adapted to living in food-poor environments than large species that attain maximum growth in food-rich environments. Under very low food conditions, starvation can occur. Resistance to starvation, that is the ability of a species to persist when energy intake is less than energy expenditure, varies among rotifers and their life stages (Kirk, 1997). Juveniles that allocated little energy to reproduction had greater resistance to starvation than did adults.

Increasing food concentration promotes increasing body growth of adult rotifers, and this direct relationship continues into the reproductive stages until about the laying of the third egg (Walz and Rothbucher, 1991). Relative egg size, volume, and egg–female ratio (fecundity) increase with greater food concentrations and size of adults among many, but not all, rotifer species (Hillbricht-Ilkowska, 1983). High food concentrations enhance growth, shorten juvenile stages, and accelerate reproduction, some of which is seen in increased fecundity. Egg ratios were related to growth rates, however, and were not influenced by the type of algal food available to the rotifers (Rothhaupt, 1990).

Analyses of interactions *in situ* under natural conditions are rare. A detailed regression analysis of the seasonal dynamics of reproductive rates of planktonic rotifers of four lakes in the English Lake District in relation to food abundance verified the results of culture analyses (Edmondson, 1965). The abundance of food at a given time is an imperfect measure of food supply to the feeding organisms; the food consumed during a period of time depends upon the grazers' capabilities, as well as the predators' reproductive rate, which is under the influence of still other environmental variables. Among the three dominant rotifer populations analyzed by Edmondson, reproductive rates were strongly related to temperature, especially in the case of *Keratella*. A small flagellate, *Chrysochromulina*, was of major importance in the success of both *Keratella* and *Kellicottia*. Limited data on bacterial densities showed no significant correlations with reproductive rates of the rotifers. The reproductive rate of *Polyarthra* was directly correlated with the abundance of the small alga *Cryptomonas*.

The significance of food organisms to the rotifers was found to be partly but not wholly related to size. Some food organisms within an acceptable size range were of little significance. The green alga *Chlorella* was consistently not utilized as a food organism. Similarly, *Eudorina* was not eaten and may exert an inhibitory effect. Some rotifer species do not discriminate between feeding upon living and dead algal cells (Bogdan *et al.*, 1980; Starkweather and Bogdan, 1980). *Keratella cochlearis* was found to feed selectively on detritus when given both dead and live algae but, as with several other rotifers, would ingest bacteria, yeast, and algae in mixed diets, although at different rates. Other rotifers (e.g., *Polyarthra*) ingested only algal cells.

Food availability was found to be instrumental in the population dynamics of *Keratella cochlearis* in a eutrophic Danish Lake (Bosselmann, 1979a,b). A small, nonreproducing population occurred in the open water during late autumn and in winter; mortality was low. The existing population started to reproduce and the population was augmented by the hatching of resting eggs. With low mortality in the spring, the population grew exponentially (Fig. 16-23); reproduction was

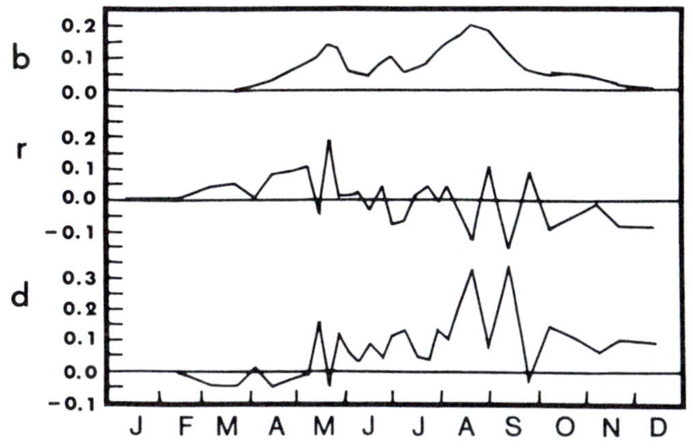

FIGURE 16-23 Estimates of the birth rate (*b*), rate of growth (*r*), and the rate of mortality (*d*) of *Keratella cochlearis* in Lake Esrom, Denmark, 1973. (Modified from Bosselmann, 1979a.)

limited only by low water temperatures. During May, the turnover rate of biomass of this rotifer species was 7.7 days. In early summer, the population was large but constant and had a moderate mortality and a low birth rate because of low food availability. Food availability improved in late summer and, along with high water temperatures, resulted in more rapid biomass turnover rates (5.0 days). Mortality was higher in late summer and the population numbers fluctuated widely. In the autumn, reproduction decreased because of declining temperatures and food availability. Similar direct correlations, with some experimental support, were found between rotifer abundance and food availability in summer in a Wisconsin lake (González and Frost, 1992). In natural rotifer communities of a small lake in Vermont, poor food conditions seasonally induced diapausing amictic eggs in *Synchaeta* that resulted in an obligatory dormant period before resuming development (Gilbert and Schreiber, 1995). Diapausing amictic eggs differ markedly from fertilized resting eggs produced following bisexual reproduction during favorable conditions of food availability (Pourriot and Snell, 1983; Gilbert, 1995). Diapausing eggs are induced immediately after a brief starvation period at relatively little energetic cost. However, a population producing a high proportion of diapausing eggs has a greatly reduced reproductive potential because of lower hatching success subsequently.

A detailed analysis of the zooplanktonic population dynamics and productivity of Mirror Lake, New Hampshire, also demonstrates the spatial and temporal separation of the individual populations (Makarewicz and Likens, 1975, 1979; cf. also Hofmann, 1982). As illustrated in the depth–time productivity data of Figure 16-44, the rotifers were major components of the zooplankton community in this lake. Although copepods constituted 55% of the total zooplankton biomass, they represented only 19% of the zooplankton productivity because of their slow growth over an entire year. In contrast, despite lower biomass, the rotifers accounted for 40% of the annual zooplankton production primarily because of faster growth rates and generation times of 8–10 days (at 17°C). As discussed later (p. 480), the niche hyperspace of the limnetic zooplankton community was subdivided by dispersion of the populations of rotifers, cladocerans, and copepods along gradients of food size, depth, and time. Differences in the life histories in the timing of development of different populations, feeding behavior and food size and types utilized, tolerances of vertical and seasonal variations in temperature and oxygen, and vertical migratory behavior all contributed to minimizing niche overlap and permitting coexistence of species within the annual zooplankton community.

B. Mechanical Interference Competition

Rotifers generally cannot become abundant members of freshwater zooplankton communities in the presence of large ($\geq$1.2 mm) *Daphnia* and copepods even when shared food resources are abundant (Neill, 1984; Gilbert and Stemberger, 1985; Gilbert, 1988; Lair, 1990; MacIsaac and Gilbert, 1991). Commonly, rotifers appear early in the seasonal succession of zooplankton communities and are succeeded by *Daphnia*. As large *Daphnia* populations decline seasonally or are suppressed by planktivorous fish or altered water quality such as lake acidity (Siegfried, 1991), rotifers rapidly dominate. Rotifers are suppressed by large *Daphnia* both through exploitative competition for shared, limiting food resources. Additionally, mechanical interference can occur as rotifers are swept into the brachial chamber of *Daphnia*, where they are immediately killed, mortally wounded, or lose attached eggs before being rejected. In some cases, the rotifers are ingested along with algae. Large rotifers are little damaged by cladocerans but must compete with them for food and are often suppressed by exploitative competition when food is limited. Small cladocerans generally do not mechanically interfere with rotifers.

The larval dipteran midge *Chaoborus* can be a voracious predator on algae and small zooplankton such as rotifers, naupli, and small cladocerans. Impacts on small cladocerans can be significant and can shift the community structure to larger species (e.g., Lynch, 1979; Vanni, 1988). Detailed studies of *Chaoborus* predation on rotifers in several lakes did not, however, demonstrate appreciable selective predation on rotifers, and reproductive output of new individuals could easily exceed predatory losses (Havens, 1990; Rodusky and Havens, 1996).

The predatory rotifer *Asplanchna* releases a soluble organic compound that induces eggs of other rotifers, such as *Keratella*, to develop into larger individuals with longer posterior spines (Gilbert and Stemberger, 1984). This induced polymorphism renders the *Keratella* much less likely to be captured and ingested by the *Asplanchna* and is most effective in reducing predation.

IX. CLADOCERAN POPULATION DYNAMICS

The seasonal succession of the Cladocera is quite variable, both among species and within a species living in different lake conditions. Some species are perennial and overwinter in low population densities as adults (parthenogenetic females) rather than as resting eggs. These species may exhibit one, two, or more irregular maxima. Some perennial species exhibit maxima in surface layers only during colder periods in the spring and in the cooler hypolimnetic and metalimnetic strata during summer stratification. The aestival species that have a distinct diapause in a resting egg stage commonly develop population maxima in the spring and summer when the water is relatively warm. Although one population maximum is general, a second population peak often occurs in the autumn.

Much of the seasonality within cladoceran life histories depends upon the environmental conditions. A comparison of the life history data of an arctic species, *Daphnia middendorffiana*, and a temperate species, *D. pulex*, is instructive. Even though much variation exists among different populations, locations, and seasons, two distinct patterns emerge (Schwartz, 1984). The temperate *D. pulex* matures sooner, at a smaller size, and gives birth to more, smaller progeny in many more adult instars than does the arctic *D. middendorffiana* (Table 16-15). The temperate species also tended to live longer than did its arctic relative that requires more time to mature and produces many fewer broods of large neonates.

Another example of a typical case of temperate pelagic cladoceran population dynamics can be seen in the well-studied zooplankton of Base Line Lake,

TABLE 16-15 Life History Characteristics of Two Species of *Daphnia* from Arctic and Temperate Surface Waters[a]

Characteristic	*Arctic* D. middendorffiana	*Temperate* D. pulex
Time to maturity	10–12 days	5–9 days
Average brood size	4.8–14	3–29.5
Maximum brood size	6–10	70
Number of broods	5	15–30
Size of neonates	0.98 mm	0.69 mm
Longevity	50 days	50–145 days
Maximum size at first reproduction	2.10 mm	1.00 mm
Maximum size	5.90 mm	3.50 mm

[a] Compiled from data in Schwartz (1984) after many sources.

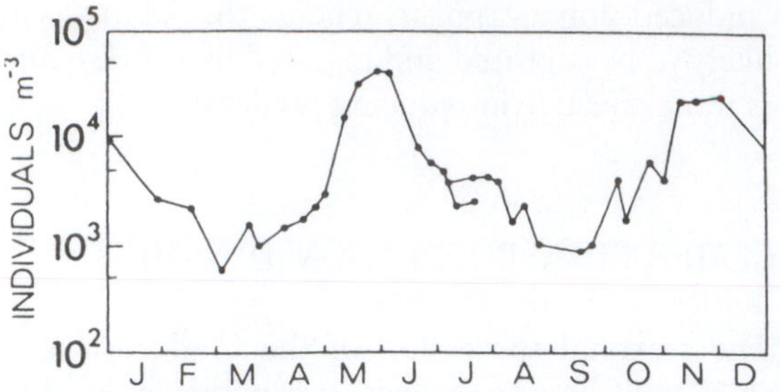

FIGURE 16-24 Population density of *Daphnia galeata mendotae* in Base Line Lake, Michigan, 1960–1961. (Redrawn from Hall, 1964.)

Michigan (Hall, 1964), consisting mainly of *Daphnia*. Two maxima occurred in the population of *Daphnia galeata mendotae*, one in the spring and a second in late autumn (Fig. 16-24). Although a few ephippia were produced, this species, in contrast to *D. pulex* in the same lake, overwintered in the free-swimming stage with little or no reproduction. The initial spring maximum in May–June and that in the fall were preceded by a maximum in brood size (ratio of eggs to mature *Daphnia*) (Fig. 16-25). Degenerate eggs were observed frequently in the brood pouches of *Daphnia* and had no apparent relationship to normal brood size.

Employing the instantaneous rate of increase of the population discussed earlier, the population dynamics of *D. galeata mendotae* were evaluated in Base Line Lake. The instantaneous birth rate (b) was estimated from the finite birth rate of newborn occurring over a time interval in relation to the existing population size. The development time of eggs was determined experimentally in relation to temperature. The instantaneous

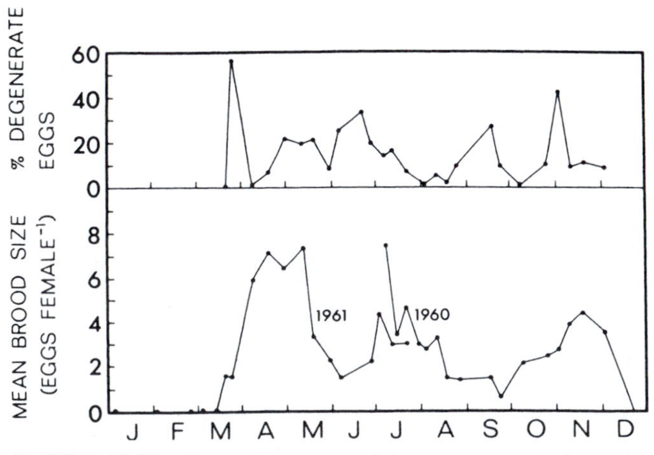

FIGURE 16-25 *Upper*: Percentage of degenerate eggs in the population, *lower*: mean brood size (mean number of eggs per adult) of *Daphnia galeata mendotae*, Base Line Lake, Michigan, 1960–1961. (Redrawn from Hall, 1964.)

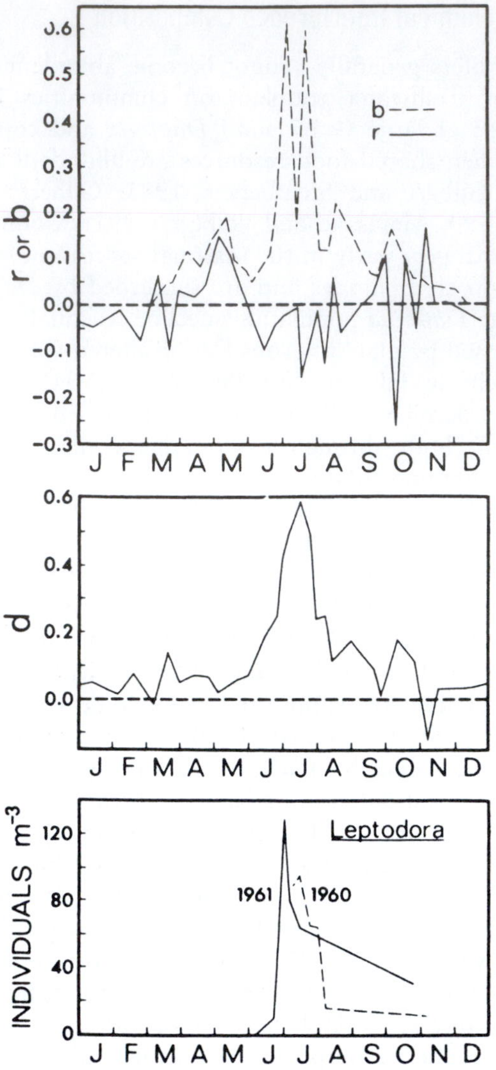

FIGURE 16-26 *Upper*: Instantaneous birth rate (natality), b, and rate of population change (increase or decrease), r, of *Daphnia galeata mendotae*; *middle*: instantaneous death rate, d (b minus r), of *Daphnia galeata mendotae*; *lower*: population density of the predatory *Leptodora*, Base Line Lake, Michigan, 1960–1961. (Redrawn from Hall, 1964.)

rate of population increase, r (see Eq. 16-2), and birth rate were both positive and nearly equal during the spring and fall months (Fig. 16-26). Maximum values of b occurred in the summer months, but r was negative during this period of population decrease because of the increased death rate, d. During the winter, b was zero as reproduction ceased and r was negative as a slow death rate, d, gradually reduced the population until March.

The summer minimum, with a loss rate of 28% per day, apparently was associated with predation on the juvenile stages of *Daphnia galeata* by the large cladoceran *Leptodora*, which reached its maximum in summer (Fig. 16-26). *D. retrocurva* reached its maximum density, greatly exceeding *D. galeata*, in late summer, well after water temperatures exceeded 20°C. Its successful competition for available food in fall was most likely contributory to the reduced birth rate of *D. galeata* at that time. Analyses of fish consumption showed that fish fed extensively on the large *D. pulex* in spring but not significantly on the smaller *D. galeata*. *Daphnia catawba*, which like *D. retrocurva* exhibits one late summer maximum and overwinters with resting eggs, is also known to coexist with *D. galeata* in late summer (Tappa, 1965). The presence of *D. catawba* and its competition for the same food base appeared to be instrumental in delaying the autumnal maximum of *D. galeata*.

In a detailed study of the zooplanktonic populations of shallow ($\bar{z}$ = approx. 1 m), extremely productive Sanctuary Lake in northwestern Pennsylvania, three species of *Daphnia* occurred and exhibited a large spring maximum and a small autumn peak (Cummins et al., 1969), similar to that discussed in the Base Line Lake example. *D. galeata mendotae* dominated these populations and *D. retrocurva* became a codominant in late summer. With the decline of the spring maximum, a massive development of the predacious cladoceran *Leptodora kindtii* occurred and reached a biomass maxima in June in 1 yr and in June, July, and August the following year. Only the large *Leptodora* of from 6 to 12 mm are predaceous and ingest the fluids of their prey. The population dynamics of *Leptodora* were correlated with loss rates of prey; *Leptodora* apparently shifted seasonally to preying on *Daphnia, Ceriodaphnia, Bosmina,* and *Chydorus* and the copepods *Cyclops* and *Diaptomus*. Similar predatory regulation of both small and large cladocerans and copepods nauplii and adults by predation of *Leptodora* has been observed in very different lake environments (e.g., Lunte and Luecke, 1990; Branstrator and Lehman, 1991; Hellsten and Stensen, 1995). Seasonal polymorphism with increased spine length in *Bosmina* often coincides with the heightened predation by *Leptodora* as presumably a defensive mechanism for reduction of predatory success (see discussion on cyclomorphosis, p. 456).

A few cladocerans are distinctly coldwater species and inhabit shallow lakes of northern regions. These species develop large populations only in the early spring and in cooler meta- and hypolimnetic strata in more temperate lakes. These species generally are perennial and tend to tolerate lower oxygen concentrations than the more common warmwater species.

Among cladoceran zooplanktonic populations in general, low populations in winter among perennial species and a near absence in winter among aestival species are common. With increasing food supply from algae, detritus, and bacteria with rising temperatures in spring, cladoceran populations increase from overwintering adults or resting eggs. Temperature increases the rate of molting and brood production, while rising food supply results in increases in the number of eggs per brood. The herbivorous populations can increase exponentially to the point where the community filtration rates crop the phytoplankton at rates that exceed the reproductive rates of the phytoplankton community. As a result of this intensive grazing, phytoplankton biomass can decrease to low levels, with a marked increase in the optical transparency of the water. The "clearwater phase" can persist until less edible algal species populations develop. The herbivorous zooplankton species become food-limited, which results in a precipitous decline in their body weight per unit length and fecundity, both of which quickly translate into declining population densities and biomasses in the summer. Predation by other zooplankton and fish accelerates this decline to very low summer levels.

A shift toward smaller average body size among the surviving crustacean zooplankton is often seen in the summer period of high predation. The larger species of crustacean herbivores are replaced by smaller species and by rotifers. These small species are less vulnerable to fish predation (discussed at length on p. 462ff) and are less affected by interference with their food-collecting apparatus, which can be caused by some forms of inedible algae (Sommer *et al.*, 1986; Christoffersen *et al.*, 1993; Telesh, 1993). As a result, the population mortality of the smaller zooplankton is lower and their fecundity is higher than that of the larger zooplankton species. High predation of both cladocerans and copepods can result also from the mysid malacostracean crustaceans (see Chap. 22). Although differences in life history sequences preclude severe competitive exclusion of cladocerans and copepods under usual natural conditions, when they are introduced to lakes, as has commonly been done for fish management purposes, exploitation of the zooplankton can lead to nearly total suppression of cladocerans and many copepods (e.g., Murtaugh, 1983; Langeland, 1988). Rotifers often proliferate following this release from cladoceran and copepod predation pressures.

Many factors can influence the onset of sexual reproduction in cladocerans (day length, low temperature, flow food availability, and chemical induction from crowding), and distinct species differences are known in response to these factors or a combination of them. In addition, chemical compounds, termed *kairomones*, released by fish induced production of diapausing ephippial eggs in parthenogenetic females of *Daphnia magna* and reduced brood size (Slusarczyk,

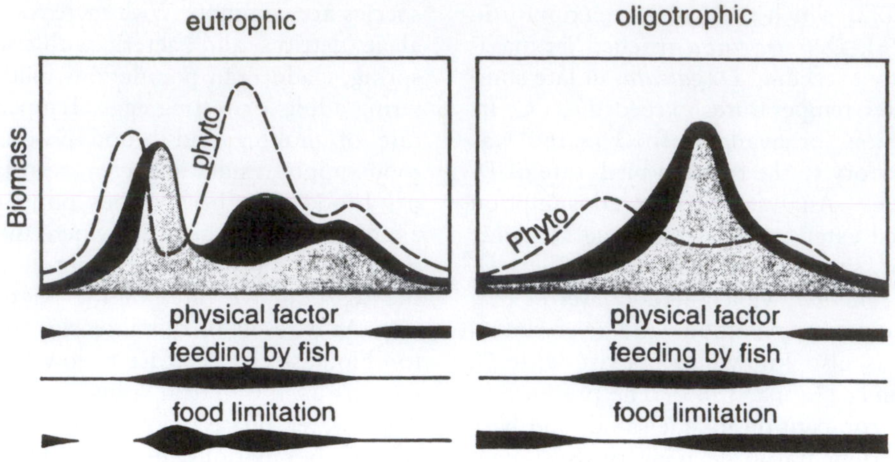

FIGURE 16-27 General model of seasonal succession of zooplankton in typical thermally stratified eutrophic (*left*) and oligotrophic (*right*) lakes of the temperate region. Phytoplankton: dashed line. Zooplankton: small species, dark shading; large species, lighter shading. Black lower bars indicate relative intensity seasonally of factors noted. (Modified from Sommer *et al.*, 1986.)

1995; Pijanowska and Stolpe, 1996). Under heavy predation by fish and low survival probability of parthenogenetic females, ephippia formation in summer can be adaptive by achieving higher fitness by survival than potential immediate reproductive gain via parthenogenesis. Fish kairomones can also alter swimming and population aggregation behavior (Jensen *et al.*, 1998) that may alter vulnerability to predation.

Food availability also influences the size structure of zooplankton communities by age-specific differential mortality. Under conditions of starvation, respiration increases and body weight decreases before death (Threlkeld, 1976). Survival time (t) is greater in larger zooplankton and is a function of body weight (w, in μg dry weight): $t = 2.95w^{0.25}$. Older individuals, however, survive for shorter periods of time than younger zooplankton under nonstarvation conditions. Juveniles tend to be more vulnerable to food limitations than adults (Neill, 1975). The impact of periods of size-related starvation on community size structure may be augmented or reversed by the timing of the onset of food limitation on its members (Threlkeld, 1976) and by the extent to which starvation influences prey utilization by the zooplankton (Williamson, 1980).

Loss of thermal stratification in lakes in autumn and increased nutrient regeneration from deeper strata usually coincides with reduction of light. Large unicellular or filamentous algae, and particularly diatoms, are common in the autumn. Fish predation is generally reduced under declining light conditions and colder temperatures. Zooplankton populations, particularly of certain larger forms, can increase for a brief period. Herbivore biomass rapidly declines because of reduced fecundity from both declining temperatures and food availability. Many cladoceran zooplankton enter diapause stages at this time. These relationships have been set forth in a model, primarily for large cladoceran zooplankton in stratified lakes of the temperate region (Fig. 16-27). Among eutrophic lakes, a spring maximum of small phytoplankton algae is followed by a dominant persisting summer maximum of large, grazing-resistant algae and cyanobacteria. These common phytoplankton maxima of eutrophic lakes are often separated by the "clearwater phase," a very short-lived period when large zooplankton graze phytoplankton rapidly to conditions of acute food limitation and are rapidly replaced by smaller zooplankton species. The phytoplankton "clearwater phase" may persist somewhat longer into the summer, depending on the effectiveness of the grazing of the smaller zooplankton species and nutrient loading, particularly of phosphorus. The collective primary productivity of phytoplankton, however, particularly with smaller species of higher reproductive rates and less biogenic "turbidity" generally is very high during the summer period (cf. Chap. 15). In oligotrophic lakes, the phytoplankton–zooplankton successional process is similar, although highly muted and slower (Fig. 16-27).

X. COPEPOD POPULATION DYNAMICS

A number of cyclopoid copepods are known to enter various periods of diapause, either at the egg stage or in the copepodite stages with or without encystment,

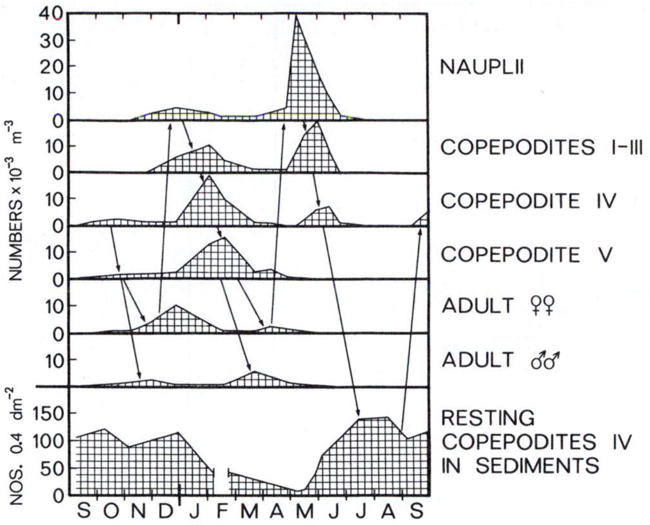

FIGURE 16-28 Development and resting stages of *Cyclops strenuus* in Bergstjern, Norway. Midsummer diapause is indicated by aestivation of copepodites IV in the sediments. (Drawn from data of Elgmork, 1959.)

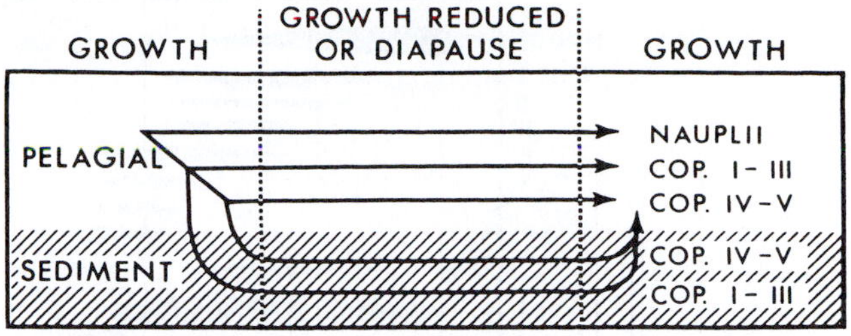

FIGURE 16-29 Generalized life-cycle patterns of limnetic cyclopoid copepods. COP = copepodid instar stages as indicated. (Adapted from Nilssen, 1978.)

upon the sediments (see, e.g., Wierzbicka, 1966; Maier, 1994). Consequently, for reasons discussed later, the annual cycle of the copepod populations may be interrupted by a diapause that persists from one to several months. For example, the resting stage may occur in midwinter, as in *Cyclops bicolor*, or in summer, as is seen in *Cyclops strenuus strenuus* (Fig. 16-28). Nauplii develop from egg-bearing females in this species in a bimodal population during winter and spring. Then, in summer, the larger population of copepodite stage IV aestivates in the sediments for about two months in midsummer before becoming pelagic again in the autumn and developing into adults. In contrast, *Cyclops strenuus abyssorum* exhibits only one effective generation in each year in several lakes in the English Lake District (Smyly, 1973). In the deepest of the four lakes studied in great detail, individuals of this generation hatched from eggs laid in the spring and reached the adult stage in early winter. The adults passed the winter in the plankton and started laying eggs early in the following year. In the three other lakes, most individuals of the spring generation reached the fifth copepodite stage by midsummer and spent the next eight months aestivating in the profundal zone. The copepods left this zone in February or March to return to the pelagic zone, where they became adults and started breeding. Water temperature, oxygen concentrations, light intensity, and day length were believed to be associated with the initiation and termination of this diapause (Smyly, 1973). Co-occurring species of cyclopoid copepods are known to have their maxima and diapause periods at different times. This alternation presumably minimizes competitive interactions for the same food resources.

Although the general pattern of diapause is similar among the cyclopoid copepods, marked variations in the timing of instar stages and their distribution in the water and the sediments have been observed (e.g., Elgmork, 1959; Smyly, 1973; George, 1976; Nilssen and Elgmork, 1977; Vijverberg, 1977; Boers and Carter, 1978; Elgmork *et al.*, 1978; Elgmork and Langeland, 1980; Hansen and Jeppesen, 1992). The basic life-cycle pattern in limnetic cyclopoid copepods can be divided into two parts (Nilssen, 1978; Elgmork and Nilssen, 1978): (a) a period of growth, followed by (b) a period of retarded growth or diapause, the latter resulting from adverse environmental conditions (Fig. 16-29). The intensity of the lines and pathways shown in Figure 16-30 differ in different lake ecosystems. Examples include:

1. In arctic and alpine lakes, diapause as copepodites in the sediments is less common. Adult stages frequently predominate in pelagic waters just above the sediments, perhaps to avoid fish predation. The adults feed heavily on rotifers and their own nauplii during winter.
2. Large differences in life cycles of limnetic cyclopoid copepods occur in temperate environments. In the temperate region, a period of diapause as copepodites IV and V is the rule. In less predictable environments (eutrophic lakes and temporary ponds), many cyclopoid copepods persist during periods of reduced growth in naupliar or advanced copepodid stages in the winter; a relatively small proportion of the population enters copepodid IV–V diapause in the sediments. In more predictable and less productive lakes, during colder periods of reduced or arrested growth the tendency is for most of the population to diapause in naupliar stages or resting stages in the sediments. This response is

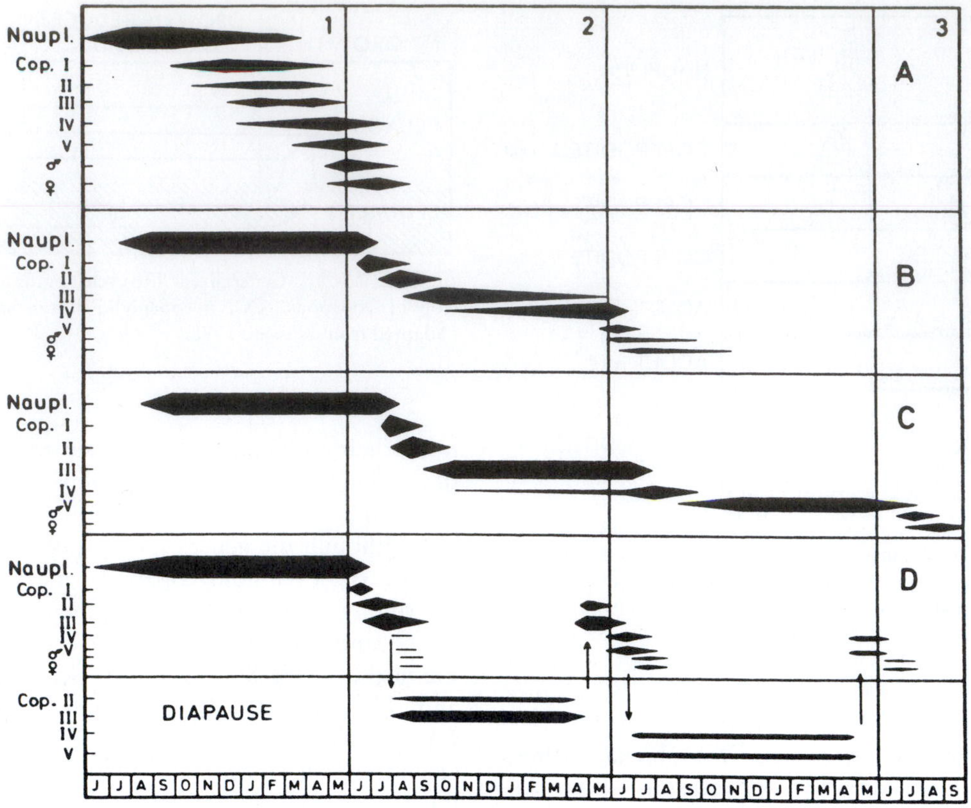

FIGURE 16-30 Variations in the life cycles in the planktonic *Cyclops scutifer*. (A) 1-yr cycle without diapause from a small humic lake; (B) 2-yr cycle without diapause in an oligotrophic lake; (C) 3-yr cycle without diapause in a subpolar lake of Norway; and (D) combined 1-2-3-yr cycle with two periods of diapause in an oligotrophic lake. (From Elgmork, 1985.)

believed to result largely as an adaptation to avoid predation by fish and predatory copepods during periods of slow growth when the population is most vulnerable (Strickler and Twomby, 1975; Nilssen, 1978; Hairston and Munns, 1984).

3. In tropical lakes showing no great seasonal changes in environmental parameters, resting stages have not been observed.

However, adaptability is large and the same species can vary greatly in its life history among different environments. For example, the reproductive period of the planktonic copepod *Cyclops scutifer* can be delayed by retarded development, particularly among the nauplii (Elgmork, 1985). Differences in life cycles ranged from 1 yr to many years (Fig. 16-30). Primary causal factors are focused on differences in temperature and the prolongation was inversely related to food availability. Conversely, when temperatures are high, predation pressures low, and food availability is very high as in shallow hypereutrophic lakes, naupliar stages can be very short and the cyclopoid copepods have several (3–6) generations per year (Hansen and Jeppesen, 1992). This deviation from the normal life cycle has been observed in other eutrophic lakes and can be interpreted as a strategy to optimize the reproductive output (Maier, 1996).

A. Dormancy among Copepoda

Copepod dormancy occurs in various ontogenetic stages as resting eggs, arrested larval development, juvenile and adult encystment, or arrested development of free-swimming nonencysted copepodids or adults (Dahms, 1995). Much study has been devoted to the study of copepod dormancy as resting eggs. Diapause eggs commonly have a thick chorion that serves as protection against digestion by predators (Hairston and Olds, 1984; Marcus and Schmidt-Gegenbach, 1986), desiccation, and bacterial degradation. Species with prolonged diapause tend to be small and tend to be found in inland waters (Hairston and Cáceres, 1996). Although diapause is widespread among the crustaceans, it is particularly prevalent among the copepods.

Diapausing eggs can remain viable in aquatic sediments for decades or longer and have a mortality of ca. 1% per year (Hairston and Olds, 1984; Hairston *et al.*, 1995; Hairston, 1996). Diapausing eggs can serve as a major genetic reservoir for repopulating copepod communities following severe environmental disturbances. Under these conditions, years of good reproductive success can be stored in diapausing eggs in the sediments and allow species to persist during years when reproductive success is poor. The interactions between environmental variations and generation overlap produced by prolonged diapause results in the maintenance of species diversity and genetic variation.

Nonencysted dormant copepodids do not feed and they survive on stored lipids while drifting with the plankton during periods of poor environmental conditions (Fryer and Smyly, 1954; Elgmork, 1980). Encysted copepodids are also common, particularly among the Cyclopoida and Harpacticoida. Dispersal of all of these forms likely occurs only occasionally during major circulation events and is not a viable adaptive strategy (Hairston and Munns, 1984).

Dormancy functions as an energy-saving mechanism that allows individuals to survive transitional periods of harsh, unfavorable environmental conditions (Hairston, 1987; De Stasio, 1989; Hairston *et al.*, 1990; Dahms, 1993; Fryer, 1996). Adverse conditions that have been shown to induce or alter diapause include

a. Reduced temperatures. Production of diapausing eggs has been found to begin earlier in small lakes that cool more rapidly than in large lakes (Hairston and van Brunt, 1994). Rates of emergence from diapause are strongly temperature-dependent (Maier, 1990d).
b. Reduced dissolved oxygen.
c. Photoperiod, particularly long-day lengths.
d. Desiccation, particularly in sediments of reservoirs experiencing fluctuating water levels or of temporary ponds, particularly during the summer periods (e.g., Taylor *et al.*, 1990).
e. Reductions in food availability. There is some evidence supporting the putative adaptive strategy that fourth-instar copepodites entering the sediment during summer would relieve grazing pressures on available food for herbivorous juvenile naupliar stages caused by the grazing of competing cladocerans (Santer and Lampert, 1995). High concentrations of flagellates are required of naupliar development among some cyclopoids. Among lakes that are more productive and have higher food availability, some copepod species do not enter summer diapause.
f. Increased carnivorous predation by invertebrates or fish. Summer diapause in freshwater copepods is often regarded as an adaptation to avoid fish predation.

B. Copepod Community Dynamics

The life cycles of the calanoid copepods are generally somewhat longer than those of the cyclopoids. In the temperate region, most species exhibit prolonged reproductive periods with several generations per year that are indistinguishable from one another. An example of these seasonal dynamics is seen in the populations of *Diaptomus* in a small, shallow beaver pond of southern Ontario (Fig. 16-31). The *Diaptomus* hatched from resting eggs in early May and produced four complete generations, the last maturing in late autumn (Carter, 1974). During spring and summer, generation times were about 4 weeks, while the time was about 6 weeks in autumn generations. Population characteristics were similar over a 3-yr period. In one year, water levels were low, which resulted in increased mean heat content per unit volume; during this year, development time of the generation was reduced and population production increased significantly. Similar annual sequences of the generations were found among the population dynamics of a species, *Eudiaptomus*, in a shallow lake of southern Germany (Maier, 1990b).

A number of studies have demonstrated that most adult cyclopoid copepods are carnivorous and that their predatory activities can play a significant role in the population dynamics of other copepod species. For example, *Mesocyclops* was found to be selectively predacious on copepodites of *Diaptomus* rather than on cladocerans (Confer, 1971). Similarly, adult *Cyclops* preys heavily on nauplii of *Diaptomus* and its own species (McQueen, 1969; Bosch and Santer, 1993). Some 30% of the naupliar recruitment has been observed to be lost by copepod predation and by cannibalism. Juvenile mortality often is very high and is density dependent and can increase conspicuously at high population levels (Elster, 1954).

The coexistence of several congeneric species of copepods in the same volume of water has been a subject of much interest. It is assumed that when coexistence does occur, the species have slightly differ-

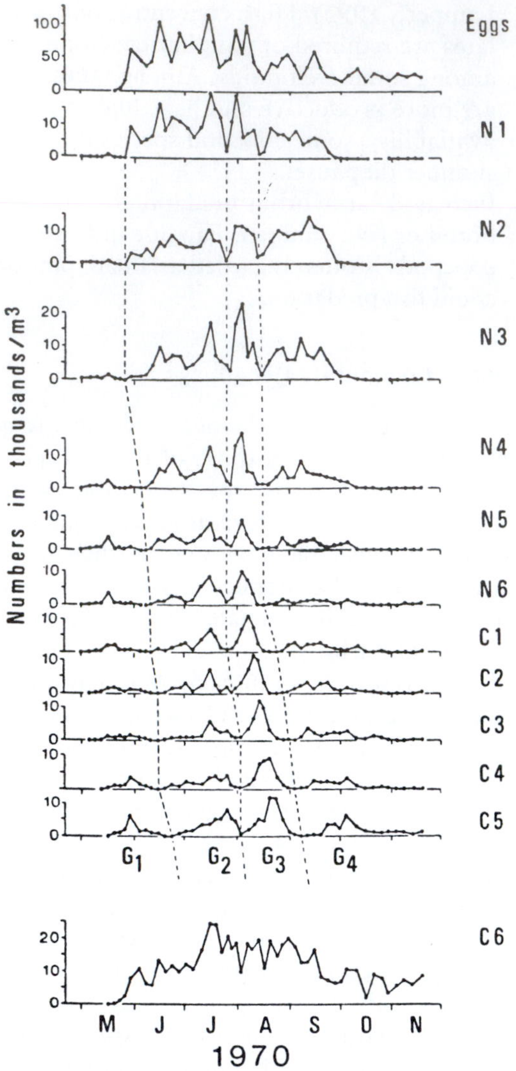

FIGURE 16-31 Seasonal cycles of *Diaptomus reighardi* in Black Pond, Ontario. Broken lines separate the estimated limits of successive generations, G. (From Carter, J. C. H.: Life cycles of three limnetic copepods in a beaver pond. *J. Fish. Res. Bd. Canada 31*: 421–434, 1974. Reprinted by permission of the Journal of the Fisheries Research Board of Canada.)

ing tolerances and optima to existing environmental conditions. Mechanisms promoting coexistence include (a) seasonal separation, discussed earlier; (b) vertical separation in relation to stratification and the distribution of food resources; and (c) size differences in relation to available food particles utilized. All of these mechanisms were implicated as factors permitting the coexistence of three species of *Diaptomus* in an Ontario lake (Sandercock, 1967; see also Hammer and Sawchyn, 1968). *D. minutus* was separated vertically from *D. sanguineus*, and the latter is distinctly smaller than the former species. *D. minutus* and *D. oregonensis* also were separated to some degree by size differences and exhibited different seasonal maxima. *D. sanguineus* and *D. oregonensis* were separated vertically and had different seasonal maxima. Therefore, in this lake, differences in two of the three factors provided the mutual separation among these three congeneric species that is believed necessary to coexistence.

Two common species of predatory cyclopoid copepods, *Acanthocyclops* and *Mesocyclops*, of similar size were found to coexist in similar water strata and have similar seasonal population sequences (Maier, 1990c). Small but significant differences in embryonic and juvenile development rates, feeding rates, and selective predation by fish allowed coexistence. Considerable evidence, reviewed by Hart (1990), exists that stresses the importance of variations in embryonic and juvenile development among copepods. In the generally high predation environment of fresh waters, acceleration of naupliar development potentially reduces the vulnerability of these smaller stages to size-selective tactile predation, while larger copepodid instars are able to reduce the opposing size-selective predation by visual planktivores by virtue of their ability to move quickly and escape fish ingestion.

XI. PARASITISM AND ZOOPLANKTON POPULATION DYNAMICS

The incidence of parasitism among zooplankton is poorly documented, but evidence suggests that population dynamics, species diversity, and productivity can be markedly affected by even moderate parasitic infection. For example, fungal parasitism can induce a high percentage of mortality of copepod eggs and female adults (Redfield and Vincent, 1979; Hoenicke, 1984; Burns, 1985). Subsequent depressions of birth rates can alter population dynamics and function as a predator to increase zooplankton species diversity. A digenetic trematode can infect cladoceran zooplankton and reduce egg production and survival (Schwartz and Cameron, 1993).

XII. ZOOPLANKTON DISTRIBUTION IN RESERVOIRS AND IN FLOODPLAIN LAKES

In contrast to the general distribution in temperate lakes, cladoceran and copepod abundance in reservoirs is markedly influenced by hydrodynamics. The commonly observed distribution of macrozooplankton is (a) low abundance in the riverine portions, (b) maximal abundance and biomass in the transitional zone (cf.

Fig. 15-12), and (c) reduced but somewhat more stable distribution in the lacustrine zone where water residence times are somewhat longer than in the other zones (e.g., Siegfried and Kopache, 1984; Marzolf, 1990). In the transitional zone, deposition of particulate materials and the concentrations of imported silts and clays increases as the load-carrying capacities decrease among reduced currents.

Filter-feeding zooplankton respond to changing quantity and quality of food resources in the shift from imported detrital particles to improved quality of algal and bacterial particles. For example, recruitment, longevity, and fecundity of some species of cladocerans can be reduced during flooding events when abundant silt particles overwhelm algal particles in the water column (Threlkeld, 1986; Kirk, 1992). Other cladocerans are able to grow well in silt-laden water. Although dissolved organic compounds adsorbed onto clay particles can serve as food supplements to particulate microbiota when ingested by cladoceran zooplankton, these sources are likely poor food substitutes for effective long-term reproductive success (Arruda *et al.*, 1983).

Clay and silt turbidity, so common in reservoirs, commonly suppresses zooplankton growth and productivity by direct interference mechanisms and by suppressing food availability from primary productivity as a result of greatly reduced light by shading. Nutrient enrichment to enhance phytoplankton development is usually insufficient to overcome the negative effects of turbidity on higher zooplankton development and productivity. Examples are many. Suspended clay particles suppressed the growth and reproductive rates of some algivorous and bacterivorous ciliates but had minimal effects on others (Hart, 1986; Jack and Gilbert, 1993b). In cladocera, mechanical interference of suspended clay reduces the feeding rate by as much as 70% and severely restricts growth and reproduction (Kirk, 1992). Zooplankton feeding on clay particles may assist in sedimentation and improve water clarity (Gliwicz, 1986c), although the positive correlation observed may simply result from improved algal growth. In contrast, rotifers are more selective in feeding on algae than inorganic particles and are not affected by high concentrations of suspended clay particles, and in turbid reservoirs rotifers are favored over cladocerans (Kirk and Gilbert, 1990; Kirk, 1991; Cuker and Hudson, 1992).

Similarly to irregular and large variations found in reservoirs, physical and biological cycles within floodplain lakes fluctuate widely, particularly in response to the annual rise and fall of the adjacent main river. The productivity of floodplain lakes tends to be very high, particularly in tropical regions, during low water periods. Planktonic crustaceans are commonly abundant during those periods but commonly experience severe declines during high water periods of inundation by the main river when inorganic suspended particles are high and phytoplankton production low (e.g., Twombly and Lewis, 1989; Lansactôha *et al.*, 1993; Junk and Robertson, 1997). Rapid increases in population sizes can result both from high birth rates associated with decreasing turbidity for some period following inundation and from the hatching of resting eggs. Although natality and productivity can be very high for extended periods, high mortality from intense predation by fish and by dipteran *Chaoborus* larvae can be more important in controlling cladoceran abundance than resource limitation in these ephemeral lakes (Twombly and Lewis, 1989).

XIII. ZOOPLANKTON DISTRIBUTION IN TROPICAL FRESH WATERS

Zooplankton of tropical fresh waters do not exhibit greater biodiversity than in more temperate waters. It is believed that tropical zooplankton are mostly opportunists recruited from river backwaters and temporary habitats (Fernando, 1994). Even in ancient lakes, such as Lake Tanganyika, the low biodiversity has likely been regulated by effective year-round continuous fish predation, particularly by young herbivorous Cichlidae (tilapias). Smaller zooplankton, particularly rotifers and protists, predominate (Green, 1993, 1994; Fernando, 1994). Within these large groups, species associations of the tropics are governed largely by temperature and salinity gradients. Large Cladocera are often virtually absent. In tropical lakes and flood plains, the littoral macrophytes offer a partial refuge for many zooplankton. The salinity of the water is restrictive to zooplankton biodiversity. Species richness generally decreases with increasing salinity, particularly above 7 g liter^{-1}, both in tropical and temperate saline waters (Green, 1993; Hammer, 1993; Evans *et al.*, 1996). Those few adapted species, however, can be very effective herbivores in these environments of reduced interspecific competition.

Detailed analyses of zooplankton population dynamics and productivity in Lake Lanao of the Philippines indicate that invertebrate predation by *Chaoborus* is of major importance in the regulation of cladoceran and copepod productivity (Lewis, 1979). At no time do the macrozooplankton approach resource limitations. Similar relationships emerged from analyses of the zooplankton community dynamics and their control mechanisms in a large, warm, monomictic, eutrophic lake of Venezuela (Saunders and Lewis, 1988). Although algal food quality only occasionally suppressed herbivorous zooplankton, carnivorous predation, particularly by the dipteran *Chaoborus* larvae, was a major regulator of mortality. Herbivores coexisted with their predators by several mechanisms: (a) some rotifer species could match predation losses by reproductive increases; (b) copepods under intense predation as adults were able to sustain high losses in the adult stages by recruitment from earlier, less vulnerable developmental stages; and (c) some rotifers and the cladocerans increased in population size only during periods when predator abundance was minimal during the mixing season when the lake was circulating completely.

XIV. ZOOPLANKTON DISTRIBUTION IN FLOWING WATERS

The zooplankton community structure within rivers is generally similar to that of lakes but is commonly dominated by small forms, such as protists, rotifers, bosminids, and juvenile copepods, throughout the year (e.g., Pace *et al.*, 1992; Pourriot *et al.*, 1997; Kobayashi *et al.*, 1998a,b). Zooplankton biomass tends to be negatively correlated with water velocities and positively correlated with temperature and phytoplankton chlorophyll *a* biomass. No evidence was found for significant grazing effects of zooplankton on phytoplankton biomass (Basu and Pick, 1997, 1998). Many microcrustaceans are associated with substrata in benthic and littoral backwater habitats as a means of alleviating downstream movement (Quirós and Cuch, 1990; Richardson, 1991, 1992; Robertson *et al.*, 1995). It is obviously advantageous to remain in a habitat of relatively high food abundance for sufficient time to complete the life cycles. Most dominant planktonic species have relatively rapid life cycles or are predominantly benthic. Microcrustacea and rotifers tend to be moved from prevailing habitats of backwater areas and lower areas of pools during periods of flow exceeding 2.5 cm s^{-1} and exported downstream. Some mortality occurs by net entrapment by polyphagous benthic invertebrate predators, such as certain caddisflies (e.g., Lancaster and Robertson, 1995). It is unclear what proportion of the total zooplankton production is removed by such processes in comparison to many others.

As river ecosystems are regulated with impoundments, zooplankton densities, biomass, and biodiversity usually increase in comparison to unregulated river systems (e.g., Pinel-Alluol *et al.*, 1982). Protists generally increase as a proportion of total zooplankton in higher river orders.

XV. VERTICAL MIGRATION AND SPATIAL DISTRIBUTION

A. Cladocera and Copepods

One of the most conspicuous features of the cladoceran zooplankton, and to a lesser extent the copepods, is the marked vertical migration of these small animals over large distances on a daily basis.[8] There is no question that a light-mediated circadian rhythm underlies at least some cases of vertical migration of microcrustaceans and insect larvae (Dice, 1914; Siebeck, 1960; Ringelberg, 1964; Ringelberg, 1987; Swift and Forward, 1988; Loose, 1993). However, most evidence indicates that changes in light intensity rather than endogenous rhythms trigger diurnal vertical migrations. Movements vary considerably with lake conditions in relation to underwater light characteristics, season, predation pressure, predator presence, and the age and sex of a species. However, certain generalizations do emerge.

Most species migrate upward from deeper waters to more surficial strata as darkness approaches and return to deeper strata at dawn. Maximum numbers are found in the surface layers in nocturnal migration in a single maximum some time between sunset and sunrise. In other cases, twilight migration results in two maxima in surface layers, one at dawn and another at dusk. In a few species, reverse migration occurs with a single surface maximum during the daylight period.

This common pattern of vertical migration may be seen in an example from relatively transparent Lake Michigan (Fig. 16-32). At dusk, the *Daphnia* population migrated from 24 m through the thermal discontinuity of the metalimnion at a rate of over 10 m h^{-1}. Some sinking occurred after the initial ascent, a characteristic common to the twilight type of migration that could be associated with reduced nighttime activity and passive sinking. For example, narcotized *Daphnia galeata mendotae* sink passively between 3.6 and 14.4 m h^{-1}, at a rate inversely related to body length (Brooks and Hutchinson, 1950). The *in situ* populations generally do not sink that rapidly during the night period (usually <0.5 m h^{-1}), which indicates that they are sufficiently active for sinking to only slightly exceed

[8] *Diurnal* refers to an event that occurs in a day (24-h period) or recurs each day. *Diel* (daily) is a more recent term (not yet in most dictionaries) that refers to events that recur at intervals of 24 h or less, with no connotation of either daytime or nighttime (Odum, 1971). In studies of community periodicity, when whole groups of organisms exhibit synchronous activity patterns in the day–night cycle, *diurnal* is often used in a more restricted sense, referring to animals that are active only during the daytime (lighted period) as opposed to others that are active only during the period of darkness (*nocturnal*) and still others only during twilight periods (*crepuscular*).

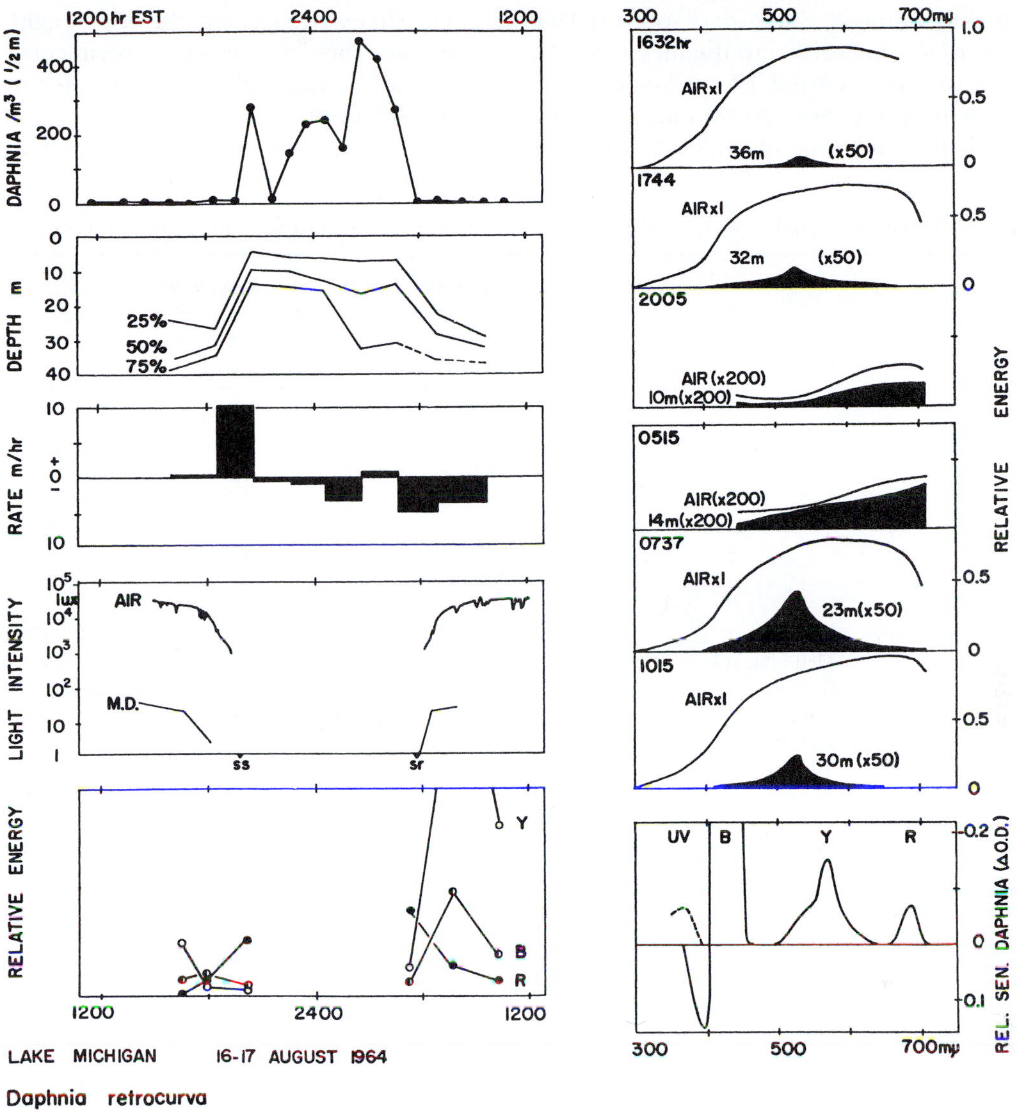

FIGURE 16-32 The vertical distribution and rate of movement of *Daphnia retrocurva* in relation to light quality at depth. *Left, top to bottom*: Number of *Daphnia* m^{-3} at 0.5-m depth; quartile distribution; rate of movement (m h^{-1}); light intensity (lux) in air and at mean population depth (M.D.); and relative energy received by blue (*B*), yellow (*Y*), and red (*R*) photosystems of *Daphnia*. *Right, top to bottom*: Relative energy versus wavelength in air and at mean population depths at different times of the diurnal period; spectral sensitivity of *Daphnia*, expressed as change in optical density (O.D.) of visual pigments extracted with aqueous digitonin upon bleaching, *sr* = sunrise; *ss* = sunset. (From McNaught, D. C., and Hasler, A. D.: Photoenvironments of planktonic Crustacea in Lake Michigan. *Verhandlungen Int. Ver. Limnol. 16*:194–203, 1966.)

upward movements. After another surface maximum before sunrise, the daphnids left the upper waters abruptly about sunrise at velocities of about 5 m h^{-1}. Both the evening ascent and morning descent occurred when light intensity at the mean depth of the population was changing rapidly (Fig. 16-32).

Daphnia has trichromatic vision with spectral peaks of sensitivity in the ultraviolet (370 nm), deep blue–violet (435 nm), yellow (570 nm), and possibly in the red (685 nm) (Fig. 16-32). At the midday depth of habitation, the photoenvironment is predominantly blue–green. The photosystem is most sensitive in the blue and may be best adapted for detecting changes in light intensity at low energy levels (McNaught, 1966). All photosystems may function in the detection of changes in wavelength and intensity at higher levels during daytime vision. The responses to light changes are conspicuous, but there is no evidence that the zooplankters are attempting to remain in a constant photoenvironment.

The vertical migrations also occur under much less transparent conditions of more productive lakes. Under these situations, even though the extent of movement is much less and often only a meter or two, the pattern is conspicuously similar (Table 16-16; Fig. 16-33). The mean morning maximum at the surface occurred 60 min (SD ± 37 min) after sunrise and the mean evening maximum at the surface occurred 19 ± 61 min before sunset. Deviations from the mean pattern increased under less distinct light conditions of moderate to heavy cloud cover. But clearly, the cue is light intensity. The response can be shown experimentally and under abnormal natural light conditions; for example, the response is identical under twilight conditions during a solar eclipse (see, e.g., Bright *et al.*, 1972).

TABLE 16-16 Examples of Estimated Velocity and Distance of Vertical Migration of Several Zooplankters

Group/species	Lake	Maximum observed rate ($m\ h^{-1}$)		Maximum observed migration (m)	Source
		Ascent	Descent		
Rotifera					
Polyarthra vulgaris	Sunfish, Ontario	0.25	0.18	5.6	George and Fernando (1970)
Filinia terminalis		0.65	0.32	4.0	
Keratella quadrata		0.36	0.50	3.0	
K. cochlearis	La Caldera, Spain	1.76	1.18	7.0	Cruz-Pizarro (1978)
Hexarthra bulgarica		0.90	2.05	8.9	
Euchlanis sp.		0.40	0.43	10.0	
Cladocera					
Daphnia galeata mendotae	Michigan	—	—	10	Wells (1960)
	Mendota, WI	1.4	1.4	1.5	McNaught and Hasler (1964)
D. longispina	Grand, CO	—	—	5.8	Pennak (1944)
	Lucerne	—	—	31	Worthington (1931)
	Victoria	—	—	50	
D. retrocurva	Michigan	10.6	5.0	24.3	McNaught and Hasler (1966)
D. pulex	Summit, CO	—	—	2.7	Pennak (1944)
	La Caldera, Spain	1.3	0.6	9.2	Cruz-Pizarro (1978)
D. schoedleri	Mendota, WI	1.4	1.4	1.5	McNaught and Hasler (1964)
Bosmina longirostris	Michigan	19	12	20	McNaught and Halser (1966)
	Silver, CO	—	—	1.8	Pennak (1944)
Polyphemus pediculus	Michigan	3.6	2.0	13	McNaught and Hasler (1966)
Holopedium gibberum		6.9	—	13.7	
Leptodora kindtii		4.2	2.0	8.5	
Copepoda					
Limnocalanus macrurus	Michigan	18.0	9.8	24	McNaught and Hasler (1966)
Diaptomus shoshone	Summit, CO	—	—	8.8	Pennak (1944)
Diaptomus novamexicanus	Castle, California				Redfield and Goldman (1980)
Nauplii		0.0	—	29	
Copepodites		1.3	—	—	
Adults (♂)		3.3	—	—	
Adults (♀)		2.7	—	—	
Cyclops bicuspidatus	Silver, CO	—	—	3.2	Pennak (1944)
Mixodiaptomus laciniatus	La Caldera, Spain				Cruz-Pizarro (1978)
Nauplii		0.4	0.3	6.8	
Copepodites		0.5	0.7	7.6	
Adults (♂)		0.6	0.9	8.5	
Adults (♀)		1.4	0.5	8.8	
Pseudodiaptomus hessei	Sibaya, S. Africa				Hart and Allanson (1976)
Nauplii		8.3	3.9	25	
Copepodites (CI-CIII)		5.5	9.4	40	
Copepodites (CIV-CV)		14.2	10.7	40	
Adults (♂)		16.2	14.9	40	
Adults (♀)		26.0	13.0	40	
Insecta					
Chaoborus trivittatus	Eunice, British Columbia	3	1	9	Swift (1976)
Chaoborus punctipennis	Barbadoes, New Hampshire	0.3–48	—	16	Haney *et al.* (1990)
Chaoborus flavicans	3 German lakes	8.1	6.0	6	Wagner-Döbler (1990)
Amphipoda/Mysids					
Pontoporeia affinis	Michigan	11.7	13.9	40	McNaught and Hasler (1966)
Mysis relicta		18	13	40	
	Michigan	30–48	30–48	76	Beeton (1960)

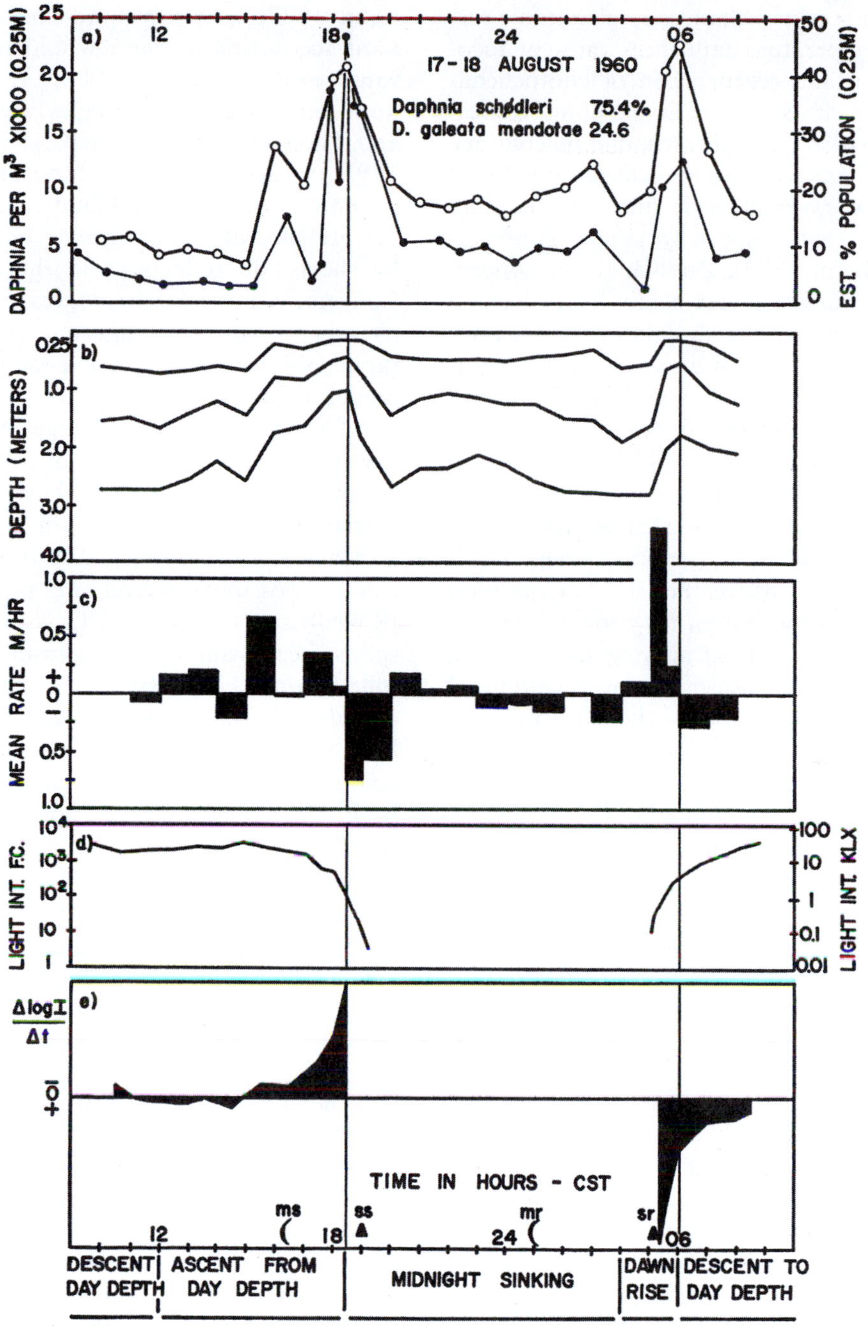

FIGURE 16-33 Diurnal vertical migration of two *Daphnia* populations in Lake Mendota in August in relation to light conditions. (a) Number (●) and estimated population (○) at 0.25-m depth; (b) quartile distribution; (c) net rate of vertical movement (m h^{-1}); (d) light intensity (lux and foot-candles) at 0.25-m depth; and (e) rate of change in the logarithm of the light intensity. *ss* = sunset; *sr* = sunrise; *ms* = moonset; *mr* = moonrise. (From McNaught, D. C., and Hasler, A. D.: Rate of movement of populations of *Daphnia* in relation to changes in light intensity. *J. Fish. Res. Bd., Canada* 21:291–318, 1964. Reprinted by permission of the Journal of the Fisheries Research Board of Canada.)

Variations in vertical migration are great. Generally, young stages have been found to migrate further than adults both in cladocerans and copepods. Differences also occur in the migratory pattern of females and males, especially among copepods. Males tend to be less migratory than females, although the pattern is not consistent. The amplitude and rates of migration vary considerably during the ice-free season. Some of this change may be related to a progressive decrease in oxygen content of lower layers as well as to changing light conditions resulting from seasonal changes in planktonic turbidity. Although temperature influences rates of locomotion, metabolism, and regeneration of photochemical visual pigments, it is minor in affecting vertical migrations in comparison to light. Migrations continue under ice cover and are strongly influenced by light reductions from snow cover. Reverse migration is more common under ice cover, especially among copepods (see, e.g., Cunningham, 1972). Zooplanktonic concentrations are high in surface layers below heavy ice and snow where light levels are low. In lakes of tropical regions, little seasonal variation in diurnal vertical migration has been observed (e.g., Hart and Allanson, 1976).

Evidence for changes in rate of feeding during the diurnal migration of crustacean zooplankton is meager. At certain times of the year in the seasonal succession of zooplanktonic species, the grazing rates are several times greater during the dark period (Haney, 1973; Haney and Hall, 1975; Starkweather, 1975; Crumpton and Wetzel, 1982). At other times, the zooplankton undergo diurnal migration without altering their grazing rates. However, simply because of the massive diurnal movements of major components of the zooplanktonic populations to surface strata at night, a marked increase in grazing pressure on epilimnetic algal, detrital, and protistan populations occurs at night. Certain algae, such as motile unicellular flagellates, can move downward during darkness, and this may represent a survival adaptation to increased nocturnal predation pressure.

Study of the adaptive significance of nocturnal vertical migration has led to a number of hypotheses, all of which probably interact to varying degrees. First, predation by fish and other predators is largely a visual process and requires light. Movement upward into the trophogenic zone in darkness or periods of low light intensities would avoid much of this predation pressure (Zaret and Suffern, 1976; Wright *et al.*, 1980; Zaret, 1980; Lampert, 1989, 1993). Indeed, the evidence for the adaptive significance of this migratory behavior among cladocerans and copepods points largely to a means of reducing mortality from fish predation (Gliwicz, 1986a; Gabriel and Thomas, 1988; Lampert, 1993). Even small amounts of light during the dark period, such as from full moonlight, can increase rates of fish predation and zooplankton mortality (Gliwicz, 1986b). The light of the full moon is of insufficient intensity, however, to suppress zooplankton vertical migration but can alter the amplitude of migration (Dodson, 1990).

Secondary factors influencing vertical migration are food quantity and quality, as well as the physical presence of predators. The food quality of algae is variable diurnally, as is discussed further in the following chapter (see Fig. 17-11). During the day, algae synthesize predominantly carbohydrates; protein synthesis reaches a maximum during the night period (cf. Enright, 1977b). At low food concentrations, increased food acquisition at night would be reproductively advantageous (e.g., Geller, 1986). At high food concentrations such adaptive strategies are less likely (Gabriel and Thomas, 1988a,b; Neill, 1990; Ringelberg *et al.*, 1991). Considerable behavioral flexibility exists under conditions of food availability. Migratory patterns in the presence of predators may be altered substantially by short-term reductions in the availability of food. Conversely, initiation of vertical migration, for example in previously nonmigrating copepods, can be very rapid (<4 h) by *in situ* exposure to an invertebrate predator *Chaoborus*.

Thirdly, although a change in relative light intensity is fundamental to the induction of diurnal vertical migration, in many cases the basic phototactic response is triggered only if a certain threshold concentration of a chemical compound(s), a *kairomone*, released from fish or possibly invertebrate predators, is exceeded (Ringelberg, 1991a,b, 1993; Loose, 1993). This response occurs among microcrustacea and has also been found to induce migration among the phantom midge *Chaoborus* (Dawidowicz *et al.*, 1990; Tjossem, 1990; Kleiven *et al.*, 1996). The chemical composition of the compounds involved is incompletely known but has been characterized as a nonolefinic low-molecular-weight hydroxy-carboxylic acid of intermediate lipophilicity (Parejko and Dodson, 1990; Tollrian and von Elbert, 1994; von Elbert and Loose, 1996). Because no major differences were found in the chemical composition of kairomones from different fish species, and its release and potency is suppressed if the fish were treated with antibiotics, it has been suggested that the source of the kairomones may be from the bacteria associated with fish rather than the fish per se (Ringelberg and van Gool, 1998). Cladocerans such as *Daphnia* have a "memory" for several days once they have been exposed to the kairomones (Hanazato, 1995; Machácek, 1995; Reede, 1995; Ringelberg and van Gool, 1995). Further, life-history traits can be adjusted to different concentrations of kairomones. Age and size at maturity decreased with increasing kairomone levels as a result of reduced growth rate, while the number of eggs increased with smaller offspring.

Fourthly, growth efficiency is somewhat greater at low temperatures, such as are found in the lower strata when summer thermal stratification occurs (McLaren, 1963, 1974). Similarly, feeding rates and efficiency are greater at higher temperatures. According to this differential growth–feeding hypothesis, a significant advantage would be obtained with only one degree of temperature difference between the water strata. Direct experimentation on the growth and reproduction of migrating and nonmigrating *Daphnia* species under simulated food and temperature conditions of diurnal vertical migration did not support the hypothesis that daphnids gain some metabolic advantage from vertical migration (Orcutt and Porter, 1983; Stich and Lampert, 1984). Since the photoperiod would have little influence on advantages conferred by these mechanisms, the adaptations are most coupled with the fitness advantages of reduced predation and higher food quality that are related to the photoperiod.

The energy expenditure in such migrations of microcrustaceans is uncertain but variable among the various swimming and drag characteristics of migrating zooplankton (cf. Vlymen, 1970; Haury and Weihs, 1976; Enright, 1977a; Lehman 1977; Strickler, 1977; Dawidowicz and Loose, 1992a). Among *Daphnia*, the energy expended in swimming is relatively small. Differences in fitness parameters (fecundity, growth) were small, and growth was only slightly less among long-distance migrating populations in comparison to short-distance migrating populations under low availability of food. However, if diurnal vertical migration between warm epilimnetic and cold hypolimnetic waters is induced by predators, zooplankton growth can be reduced significantly (by ca. 60%) because of the time spent in colder waters in predatory avoidance (Dawidowicz and Loose, 1992b; Loose and Dawidowicz, 1994).

Therefore, vertical migration upward during darkness and downward during the lighted period of the day is common behavior among cladocerans and copepods. Such vertical migration occurs primarily among zooplankton that are most visible to fish, particularly of large size and females that are carrying eggs. The amplitude of vertical migration varies as a function of the water transparency and presence and activity of fish. Evidence suggests that fish can release chemical compounds that influence the vertical migratory behavior of zooplankton.

B. Cuticular Pigmentation

A number of zooplankton, particularly copepods, possess a deep red coloration. The red coloration is caused by several carotenoids synthesized from β-carotene present in the algal diet (Ringelberg, 1980). The carotenoid pigment content increases under conditions of high light intensity (e.g., in clear or shallow lakes, especially high altitude lakes) and, among many copepods, increases in summer. Carotenoid content also often increases severalfold in a diurnal rhythm, showing minima around midday and maxima around midnight (Ringelberg, 1980, 1981). Darkly pigmented copepods survived significantly longer than pale copepods of the same species when both were exposed to natural intensities of visible blue light. While carotenoid photoprotection is established among some species, the adaptive significance of irregular carotenoid distribution and diurnal patterns is not obvious, although it is associated with diurnal vertical migration, possible energy utilization from lipids, and feeding periodicity (cf. Hairston, 1976, 1980, 1981; Ringelberg, 1981).

In arctic lakes, the predacious copepod *Heterocope* occurs in distinct dark red and pale green morphs (Luecke and O'Brien, 1981). The dark red form survived better in bright sunlight, but the pale green form was less susceptible to predation by fish that fed visually. In cases where fish predation was intense, the green forms predominated. Vertebrate predators select pigmented calanoid copepods, but common invertebrate predators are not visually selective (Bryon, 1982). In cold environments with short growing seasons, as at high elevations, the advantages of photoprotection from pigmentation at least partially outweigh the costs of selective predation.

Carotenoid pigmentation in cladocerans such as *Daphnia* may be coupled to nutrition and lipid content as well as photoprotection. The presence of cuticular melanin is found in highly light-stressed populations of alpine and arctic zooplankton in shallow habitats (Luecke and O'Brien, 1983; Hebert and Emery, 1990; Hessen, 1993). Even though constituting only 0.03% of body weight in *Daphnia*, melanin can act both directly as a light screen, with more than 90% removal of incident UV radiation, and as a powerful antioxidant to UV-induced oxidants and free radicals. This pigmentation pattern is an adaptation to living in shallow waters exposed to high light intensities and possessing few visual-feeding predators.

C. Avoidance of Shore

Another conspicuous feature of the movements of *pelagic* cladocerans and copepods is a distinct "avoidance of shore." At the shore, the littoral and near-littoral waters of lakes are almost free of pelagic crustaceans. When released within the littoral zone, these animals migrate horizontally away from the shore area if illumination is adequate. This aspect of spatial orientation has been the subject of intensive experimentation (especially Ringelberg, 1964; Siebeck, 1968; Siebeck and Ringelberg, 1969; Daan and Ringelberg, 1969).

The behavioral analyses showed that these crustaceans orient themselves toward the elevation of the horizon and the position of the sun. The direction of migration coincides with the plane of symmetrical light stimuli in the underwater angular light distribution. Near the shore, this plane of symmetry of angular light depends primarily on the position of the elevation of the horizon. With increasing distance from the shore, the effect of the elevation of the horizon on the angular light distribution decreases and finally becomes insignificant, at which point the animals no longer select a directional movement away from shore. The higher the elevation of the horizon, as in mountain lakes, the wider the zone near shore that contains a reduced concentration of pelagic crustacean zooplankton.

Light from the sun and sky incident on water is refracted in the direction of the vertical axis as it penetrates downward into the water. In effect, the angular light distribution operates in an inverted conical fashion, in which light impinging at angles greater than 49° from the vertical is absorbed selectively so rapidly that a light–dark boundary or contrast forms at an angle of about 49° to the vertical axis. The contrast or light gradient is used in orientation of the body axis in swimming. For example, *Daphnia magna* is unable to swim normally or maintain its normal body position when light from every direction is of equal intensity. Normal swimming reappears when a light gradient is introduced. The body and swimming positions are oriented by distinct pairs of ommatidia of the compound eye, which are capable of detecting contrasts. The eye is turned in a dorsal body direction so that the light contrast can be projected on the same dorsal part of the eye. The body is then turned in order to maintain a physiologically fixed angle between the eye axis and the body axis. In the littoral zone, the angular light distribution consists of a dark area along the landward side within the cone of illumination. This distribution results in one of the two contrasts present at 49° in the pelagic being shifted to the vertical, and swimming would be away from the shore (Fig. 16-34). In a plane parallel to the shore line, this contrast difference is absent and the contrasts are the same as in the open water.

Zooplanktivorous fish can shift their distributions on a diurnal basis in relation to the horizontal variations in the distribution of prey. For example, the small planktivorous silverside (*Menidia beryllina*) migrated from the littoral toward the pelagic zone and fed vigorously on zooplankton that had moved horizontally offshore of littoral areas (Wurtsbaugh and Li, 1985). After feeding for a few hours, the fish returned to the littoral before being satiated, perhaps to balance predation losses in offshore areas against foraging gains.

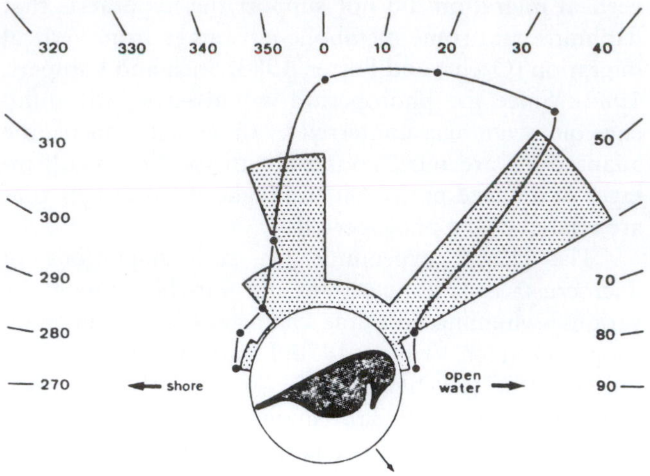

FIGURE 16-34 Orientation of a daphnid zooplankter along a plane perpendicular to the shore. Angular light distribution (solid line), magnitude of contrast expressed as the difference between two succeeding measurements (shaded bars), and the most probable orientation and movement of the daphnid. (From Siebeck, O., and Ringelberg, J.: Spatial orientation of planktonic crustaceans. *Verhandlungen Int. Ver. Limnol.* 17:831–847, 1969.)

D. Rotifers

Although features of diurnal migration are known in some detail for the cladocerans and copepods, relatively little information exists on the daily movements of planktonic rotifers. Commensurate with their relatively small size in comparison with the crustacean zooplankton, the movements of rotifers are relatively slow (Table 16-16). The maximum observed velocity of ascent in *Filinia* of 0.018 cm s^{-1} is nearly equivalent to that of some of the slowest movements among the crustacean zooplankton (George and Fernando, 1970). The maximum velocities of vertical movement, as well as the magnitude, vary among species with descent or ascent and with the season. The general tendency is toward upward movement during the daylight period (e.g., Magnien and Gilbert, 1983), although limited evidence indicates an upward nocturnal migration at one season and the reverse at another season (cf. Pennak, 1944).

No clearcut migration pattern is seen among the rotifers as occurs among the cladocerans and copepods. Presumably, the smaller size of most rotifers removes much of the pressure of visually oriented predation by fishes that probably is associated with nocturnal ascent patterns among Cladocera and some copepods. Analyses of the actual cost of ciliary locomotion among rotifers indicated that it is very inefficient and results in well over half of total metabolism (Epp and Lewis, 1984), much greater that the costs of locomotion among microcrustaceans. The movement of rotifers into the surface waters at midday (e.g., Adeniji, 1978) may, however, reflect an adaptive response to avoid predation by limnetic crustaceans despite the energetic costs.

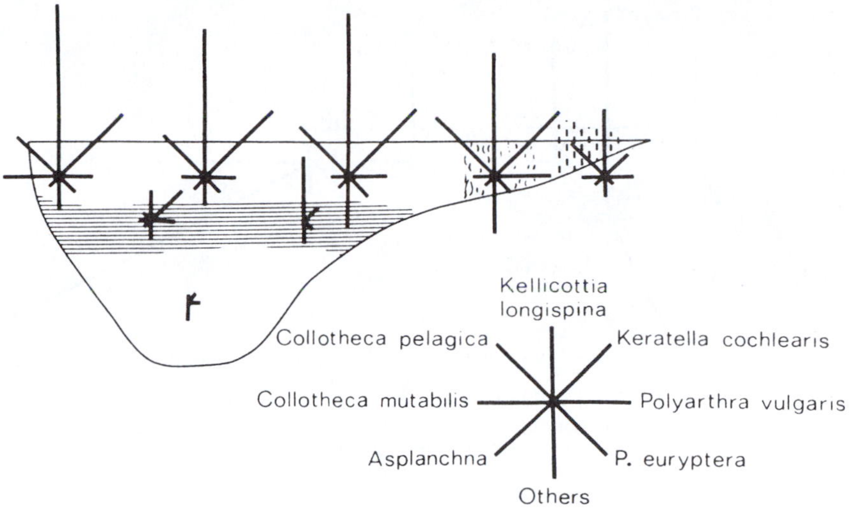

FIGURE 16-35 The relative quantitative distribution of dominant rotifers in percent in different strata along a transect across Skärshultsjön, southern Sweden, 7 August 1950. Horizontal shading indicates the position of the thermal discontinuity. (Modified from Berzinš, 1958.)

XVI. HORIZONTAL VARIATIONS IN DISTRIBUTION

A great deal of horizontal patchiness in rotifer, cladoceran, and copepod zooplankton populations has been found in the pelagic zone of lakes (cf. review of Malone and McQueen, 1983). The four basic types of patches found that are potentially caused by numerous hydrodynamic and biological factors are (a) large-scale, >1 km diameter; (b) small-scale, caused by wind-induced water movements; (c) Langmuir circulation aggregations (cf. pp. 106ff); and (d) swarms, potentially caused by biotic factors. The following few examples illustrate some of the variety found.

Small-scale horizontal variations in the spatial distribution of rotifers is provided by the detailed analyses of zooplankton in a single cross-section across the northern arm of Skärshultsjön, southern Sweden (Berzinš, 1958). The eastern shore of the basin along this 145-m transect is steepsided and drops precipitously to the maximum depth of about 13 m. The western, more gently sloping shore supported a well-developed littoral zone of submersed aquatic plants *Myriophyllum* and *Potamogeton*, grading through waterlilies (*Nuphar*) to the emergent macrophytes *Phragmites*, *Equisetum*, and *Juncus*. The numerous samples were collected within a 2-h period in midafternoon on a single summer day. At this time the lake was stratified, with a freely circulating epilimnion (19°C) to 3 m, below which the metalimnetic thermal discontinuity layer extended to 7 m, at which point the hypolimnion of 8–10°C extended to the bottom. The oxygen content of the hypolimnion was reduced (<2 mg liter^{-1}, <20% saturation) but the hypolimnion was not anoxic.

Within the 41 species of rotifers found, several conspicuous patterns of distribution were noted and are summarized in Figure 16-35, although only a few of the dominant species are considered. The epilimnetic rotifers tended to be more concentrated near the littoral regions on both sides or either side of the basin (*Kellicottia longispina*, *Collotheca mutabilis*, *Polyarthra vulgaris*, and *P. euryptera*). The dominant epilimnetic *Kellicottia* was most concentrated in the open water. Two major species, *Polyarthra major* and *Trichocerca similis*, were restricted to the metalimnion, while several others had distinctly hypolimnetic populations (*Polyarthra longiremis*, *Keratella hiemalis*, and *Filinia longiseta*).

Within the population of *Keratella longispina*, a distinct separation was found according to size. Larger individuals of the population occurred at greater depth, which suggests a greater sinking rate, although this species has some power of vertical movement, which helps it avoid sedimentation to deeper waters. A similar separation according to size was found in the vertical distribution of several species of *Polyarthra*.

Certain pelagic rotifers (e.g., *Asplanchna*), when placed in littoral areas, moved away from the shore toward the open water; others did not demonstrate such horizontal movement (Preissler, 1977). How significant "avoidance of the shore" behavior is in planktonic rotifers is unclear.

In addition to the avoidance of the littoral regions by Cladocera and copepods discussed earlier, a number

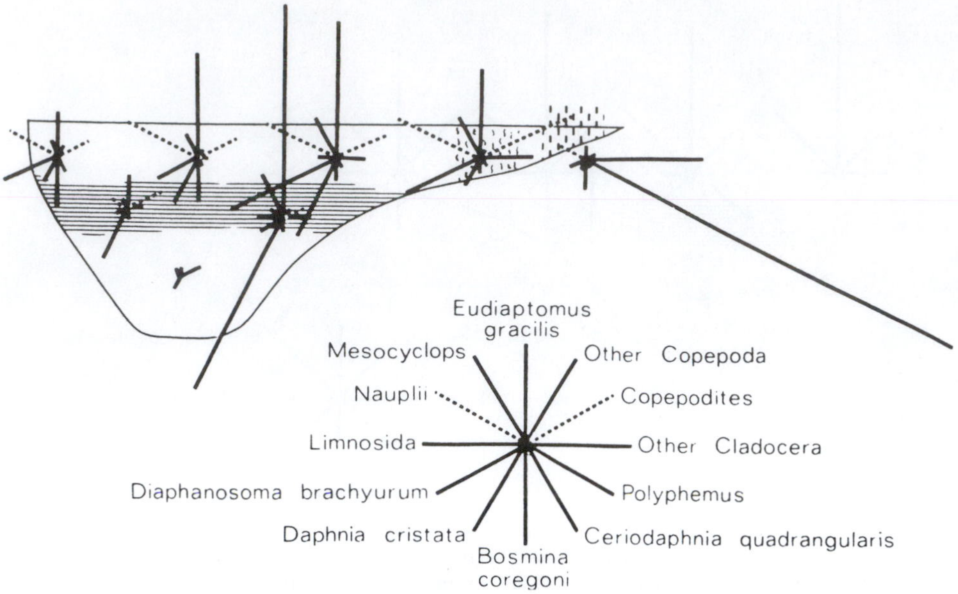

FIGURE 16-36 The relative quantitative distribution of crustacean zooplankton in percentage in different strata along a transact across Skärshultsjön, southern Sweden, midafternoon, 7 August 1950. (Modified from Berzins, 1958.)

of detailed analyses have shown a nonrandom horizontal distribution of zooplankton in the pelagic zone of lakes. An example is seen in the intensive 1-day analysis of Berzins (1958) across Skärshultsjön in Sweden (Fig. 16-36). While *Polyphemus* is distinctly littoral, many of the cladocerans and copepods avoid the littoral and are not uniformly distributed across the pelagic zone. Nonrandom dispersion of phytoplankton and zooplankton within the pelagic zone is in many cases caused by water movements. Microcrustacea tend to accumulate in the epilimnion at the leeward end of the basin and to be rare at the windward end whenever a fairly strong wind persists for an appreciable period of time (see, e.g., Langford and Jermolajev, 1966). Once at the lee side of the basin, weak locomotion is presumably adequate for most of the animals to maintain their depth distribution. Only under exceptionally strong winds and epilimnetic currents can they be swept back along the lower epilimnion to the windward side. Part of the nonrandom dispersion of zooplankton is associated with Langmuir convectional helices whose parallel clockwise and counterclockwise rotations in the epilimnion create linear alternations of divergences and convergences (see p. 106 and Fig. 7-6). Zooplankton clearly concentrate in these areas of upwelling, which occur midway between zones of convergence (George and Edwards, 1973).

This concentration of zooplankton in linear patches affects the feeding behavior of certain fishes. An excellent example is that of the white bass (*Roccus chrysops*), a pelagic fish that, in its younger stages in summer, is largely planktivorous on *Daphnia*. *Daphnia* were found to be concentrated in the linear convergences (McNaught and Hasler, 1961). The schooling white bass were able to locate and feed actively on these concentrations of *Daphnia* and also were keyed to feed most actively in the early morning and evening periods, which coincided precisely with the maximal surface concentrations of the vertically migrating *Daphnia*. Similarly, several species of juvenile cyprinid fishes migrate from the littoral to the pelagic zone only during darkness to feed on zooplankton concentrated in the surface waters by vertical migration (Bohl, 1980; Werner *et al.*, 1980).

XVII. CYCLOMORPHOSIS AND PREDATION AMONG THE ZOOPLANKTON

A. Rotifers

Cyclomorphosis, or seasonal polymorphism, is seasonal morphological variation in successive generations of small aquatic organisms. Cyclomorphosis is a phenomenon that is particularly conspicuous among planktonic Cladocera but is also fairly common among the protozoa, the dinoflagellates (cf. p. 337), and the rotifers. A particularly comprehensive treatment of the subject of cyclomorphosis among plankton is the detailed résumé of Hutchinson (1967). Although it is more difficult to study the genetic continuity of seasonal forms of the same species of rotifers than of the planktonic crustaceans, cyclomorphosis in several rotifers has been analyzed in some depth.

Seasonal polymorphism is defined by the marked change in shape of some part of an organism in relation to some standard dimension of its size. The common changes in growth form among some rotifers include the following. (a) ***Elongation in relation to body width.*** In some species of *Asplanchna*, midsummer populations can be about five times as long as wide, markedly changed from their nearly spherical morphology in late spring. These elongated forms are nearly always sterile and die back and do not reappear until the next spring. (b) ***Enlargement***, with the formation of body-wall outgrowths or humps. As discussed earlier, at least in *Asplanchna sieboldi*, this seasonal change in growth appears to be caused by the tocopherol content of plant-derived food and most likely is an adaptive response to cope with larger-sized food in summer. (c) ***Reduction in size***, usually at higher temperatures in summer, with a disproportionate reduction in length of lorical spines. This form change is common among several species of *Keratella*. Its adaptive significance is unclear, since reduction in spine length increases sinking speed at higher summer temperatures and is compensated for only partly by the reduction in body size. The causal mechanisms for the reduction in spine length are unclear. (d) ***Production of lateral spines.*** The best-studied case of this type of cyclomorphosis is in *Brachionus calyciflorus*, in which the posterior spines elongate in the presence of its major predator, the large rotifer *Asplanchna*. *B. calyciflorus* always has two pairs of anterior spines and one pair of posteromedian spines. A pair of posterolateral spines may or may not be present (Fig. 16-37).

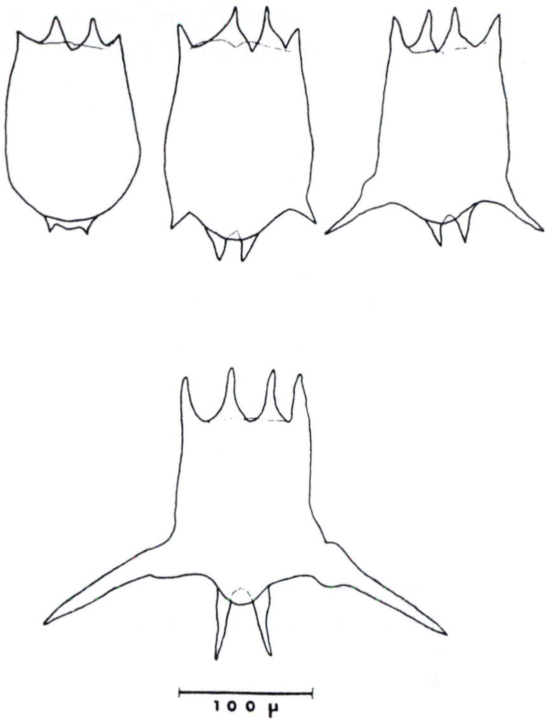

FIGURE 16-37 Spine variation in a clone of the rotifer *Brachionus calyciflorus* induced by increasing concentrations of a proteinaceous substance produced by its predator *Asplanchna*. (From Gilbert, J. J.: *Asplanchna* and posterolateral spine production in *Brachionus calyciflorus*. *Arch. Hydrobiol.* 64:1–62, 1967b.)

Gilbert (1967b; cf. also Pourriot, 1974) has demonstrated that the extension of preexisting anterior and posteromedian spines and the *de novo* induction of large posterolateral spines are caused by a substance released into the medium by the predatory *Asplanchna*. These form changes cannot be explained by temperature induction or by an allometric growth response in which the substance works indirectly on the spines by influencing body size. Changes in body size were found to be independent of posterolateral spine production. The substance was shown to affect the spineform change only at the egg stage before cleavage. The substance released by *Asplanchna* is a relatively thermolabile proteinaceous compound of unknown composition. The shape and movements of long posterolateral spines in *B. calyciflorus* significantly decrease predation by *Asplanchna*. Adult *A. girodi*, which captured about 25% of newly hatched, spineless *Brachionus* with which they made direct contact, were completely unable to capture newly hatched, long-spined individuals. Adult *A. sieboldi* can capture nearly 100% of adult, spineless *B. calyciflorus* contacted, but only about 78% of the adult, long-spined forms. The capture rate drops to <15% of the spined form by young *A. sieboldi*. The spines, however, provide no protection against predation by the copepod *Mesocyclops* (Gilbert, 1980b). A number of analyses of natural populations substantiate the fact that the substance released by *Asplanchna* can induce posterolateral spine production in *B. calyciflorus* populations (Gilbert and Waage, 1967; Green and Lan, 1974). Posterolateral spine lengths were found to be unrelated to water temperature but varied directly both with the density of *Asplanchna* and the presence of threshold quantities of the *Asplancha*-released substance.

The significance of predation upon rotifers varies depending on the species represented in the predator–prey interaction. For example, the predatory copepod *Mesocyclops* feeds effectively on the rotifers *Asplanchna* and *Polyarthra* but not on the rotifer *Keratella* (Gilbert and Williamson, 1978). *Keratella* is released by *Mesocyclops* after it has been captured, since the predator cannot remove soft parts of this rotifer. *Asplanchna* regularly eats *Keratella* but cannot capture *Polyarthra* because of its effective escape behavior. The abundance of *Polyarthra* and *Keratella* in natural communities can be greatly affected by interactions with and between *Asplanchna* and predatory copepods.

B. Cladocera

Seasonal polymorphism in the Cladocera is perhaps more conspicuous than in any other group. The subject has been studied extensively, especially among the genus *Daphnia* (reviewed in detail by Hutchinson, 1967; Kerfoot, 1980a; and Jacobs, 1987).

With increasing water temperatures, light, and food in the spring, the most common seasonal pattern of variation among successive generations in temperate lakes is a ***gradual extension of the anterior part of the head to form a crest or helmet*** (Fig. 16-38). Carapace length, on the other hand, changes little or decreases only slightly between the spring and summer, and then increases again somewhat in autumn. The number of parthenogenetic eggs per individual generally decreases markedly in the strongly helmeted summer forms, although the number of instars increases with higher temperatures. The extent of head development and the change of head shape are extremely variable among species and within the same species under different environmental conditions. In subtropical and tropical waters, cyclomorphosis is weak or does not occur, except when there is a large temperature change between winter and summer (e.g., Zago, 1976; Mitchell, 1978). ***An increase in the length of the tail spine*** also is observed in some species during summer. Cyclomorphosis in the genera *Bosmina, Ceriodaphnia*, and *Chydorus* is much less distinct than in *Daphnia*, and consists of slight reductions in length of the body and of the antennule in summer. In *Bosmina*, changes often occur by means of the formation of transparent dorsal humps, with no increase in length and reductions in antennule length and number of segments; in mucro (caudal spine) length and number of sutures may also occur (Black, 1980a,b; Kerfoot, 1980b).

The mechanisms causing cyclomorphosis in cladocerans have received much study. It is clear that temperature is a primary stimulus affecting the height of the head helmet in *Daphnia* and that it is effective during the middle of embryogenesis. In a number of species, helmet extension does not occur at water temperatures below 10–15°C. Carapace length is maximal in winter, and the ratio of head length to carapace

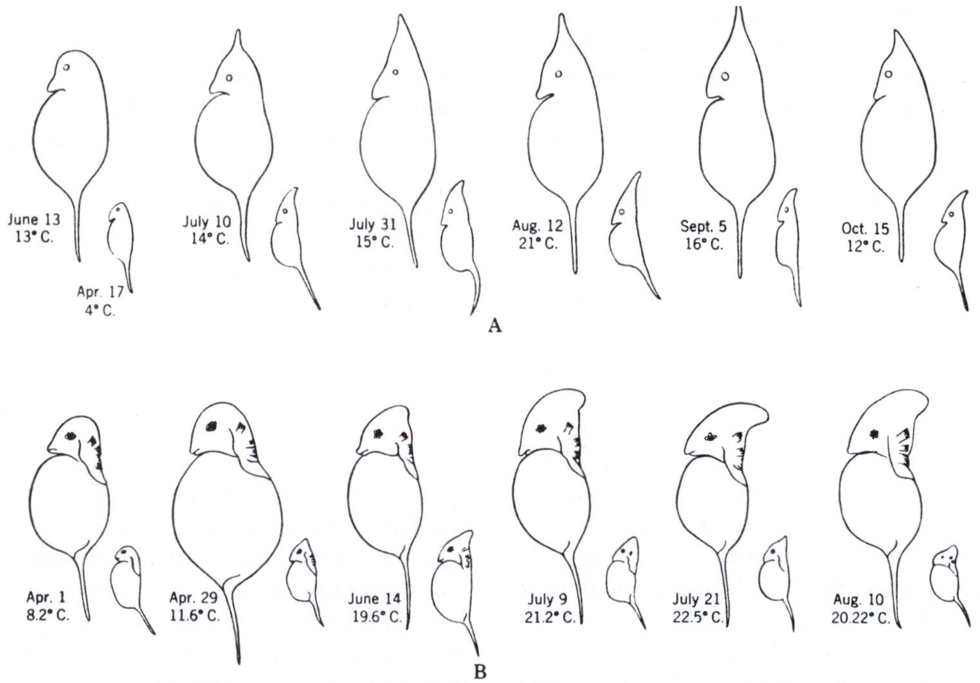

FIGURE 16-38 Cyclomorphosis in (a) *Daphnia cucullata* of Esrom Sø, Denmark and (b) *Daphnia retrocurva* of Bantam Lake, Connecticut. The small individuals at the right are first instar juveniles drawn to the same scale as the adults. (a) From Hutchinson, G. E.: *A Treatise on Limnology*, Vol. 2, New York, John Wiley & Sons, Inc., 1967; (b) From Brooks, John Langdon: Cyclomorphosis in *Daphnia*: *Ecological Monographs* 16:1946. Copyright 1946 by Duke University Press.)

length increases with increasing temperatures. Transfers of animals reared at one temperature to a lower or higher temperature result in a reduction or increase in head length in relation to that of the carapace in later instars. Food supply also affects the specific growth rate and has a relatively greater effect on growth of the head (e.g., Dodson, 1988), but helmet length is more strongly correlated directly with water temperature than with food abundance (Lampert and Wolf, 1986).

Water turbulence also has been shown to be a significant factor in cladoceran cyclomorphosis (see, e.g., Brooks, 1947; Hrbáček, 1959; Jacobs, 1962; Havel and Dodson, 1985). In turbulent cultures and in naturally turbulent environments such as the epilimnion of larger lakes, helmet length relative to that of the carapace is increased significantly in the light in comparison to quiescent situations. Antennal beating rate also increases significantly in turbulent conditions, probably in order to maintain a vertical orientation to light. Helmet development is distinctly more conspicuous in species typically inhabiting epilimnetic turbulent strata (Brooks, 1964). Populations genetically capable of considerable phenotypic variation do not exhibit their extreme phenotype when they live in lower water strata in which growth is slower. Turbulence is virtually ineffective in cyclomorphic induction in the dark (Jacobs, 1961; Hazelwood, 1966). Temperature and turbulence influence relative growth rates from the beginning of the second half of embryogenesis; food becomes effective immediately after birth (Jacobs, 1980). Daphnid cladocerans that are warm, well fed, illuminated, and exposed to turbulence produce the largest helmets. All factors are influential to some extent as long as the animals are capable of growth.

Organic substances released by invertebrate (especially *Chaoborus*) and fish predators can also induce cyclomorphic growth in *Daphnia* more effectively than can changes in temperature, turbulence, and oxygen (Grant and Bayly, 1981; Vuorinen *et al.*, 1989; Hanazato, 1990, 1991; Tollrian, 1990, 1994). Exposure to these chemical cues is most effective at the juvenile growth stages.

A compensatory mechanism exists in daphnid growth: The larger (longer) the helmet resulting from earlier influences, the slower it will grow relative to the body in later instars (Jacobs, 1980, 1987). Certain body proportions are maintained regardless of previous environmental conditions. Since there is no single receptor mechanism for the effective environmental factors, one or more hormones likely interact to regulate specific differential growth.

Morphological variability in cladocerans has been ascribed only to phenotypic plasticity. It is important to note that some cyclomorphic species are, in reality, composed of many genetically differentiated species (e.g., Hebert, 1978a,b; Black, 1980a; Kerfoot, 1980b; Jacobs, 1987). A number of these species often occur in the same community, and seasonal changes in their relative frequencies can cause polymorphic cycles that mimic true cyclomorphosis. Intraspecific genetic variation in head shape likely occurs as well in some species.

C. Copepods

Seasonal polymorphism among adult copepods is minor in comparison with that found among the parthenogenetic Cladocera and rotifers. The most conspicuous feature is a slight inverse relationship of body size to increasing temperature, so that, in the few cases studied, animals of summer populations tend to be somewhat smaller than animals living in colder seasons.

D. Adaptive Significance of Cyclomorphosis

Seasonal polymorphism in cladocerans and rotifers has been assumed to be of major adaptive significance to these organisms. Early investigators assumed that cyclomorphic forms would have greater resistance to sinking because the viscosity of the water decreases at higher epilimnetic temperatures. Cyclomorphosis is confined almost completely to species that inhabit the epilimnion, or at least migrate to it daily. Epilimnetic turbulence, however, decreases the small advantage gained from form resistance (Jacobs, 1967). Such resistance would be somewhat disadvantageous to cladocerans swimming during vertical diurnal migration. Furthermore, there is an energetic and demographic cost to those forms with extreme polymorphic growth, particularly in lowered reproductive rates (Riessen, 1984; Black and Dodson, 1990).

The adaptive significance of cyclomorphic growth appears to center on minimizing two aspects of predatory avoidance. (1) Predation of cladocerans and other large zooplankton by non–filter-feeding fish, which is a visual process (see p. 466). Cyclomorphic growth keeps small the central portion of the body (thorax, abdomen, and basal portion of the head) that is visible to predators. Growth continues by expansion of transparent peripheral structures (Brooks, 1965, 1966; Jacobs, 1966). Thus, when size-dependent planktivory is most intense in summer, the central visible portion of the daphnid body is smallest and the organism is less susceptible to being eaten by fish. Cyclomorphomic organisms are, through selection for high rates of growth in less visible peripheral structures, able to maintain assimilation rates during midsummer without the disadvantage of greater visible size when predation is intense. In the winter, when the rate of predation is negligible and turbulence essentially is eliminated under ice cover, evidence indicates that cessation of molting in juveniles and adults results from the lack of a turbulence cue. With the loss of ice cover, even at low temperatures, molting, growth, and egg production immediately begin from the overwintering population of large females. (2) Reduction of capture success by invertebrate and young fish predators. Many cladocerans of small size (<1.5–2.0 mm) exhibit more cyclomorphic growth than larger species (reviewed by Zaret, 1980). Invertebrate predation by copepods and phantom midge larvae (*Chaoborus*) is more effective on smaller zooplankton (<2.0 mm) (e.g., Havel and Dodson, 1984). The large caudal appendage of cladocerans such as *Bythrotrephes* increases handling time and decreases predation rates by young fish (Barnhisel, 1991).

The lack of cyclomorphosis in copepods in comparison to marked polymorphism in cladocerans and rotifers is likely related to the fundamentally different ways in which these groups have developed different defenses to invertebrate predation (Kerfoot, 1980b). The thoracic legs of copepods are modified for locomotion, while those of cladocerans are modified for filter feeding. Nauplii, copepodite, and adult stages of copepods have evolved precontact defenses, in which they are capable of swift evasive swimming maneuvers (10–35 cm s^{-1}; Strickler, 1975) away from predators. Locomotion in cladocerans by the pair of second antennae results in only feeble, slow escape reactions (e.g., Kerfoot, 1978). Although some rotifers (e.g., *Polyarthra*) have evasive propelling motions, most rotifer locomotion by ciliary movement is slow. Therefore, the cladocerans and rotifers have evolved postcontact defenses, including (a) larger size, (b) elaboration of spines and other body shapes to frustrate handling by predators, (c) hard, chitinized head shield and carapace, (d) protection of eggs and embryos internally in brood pouches, and (e) reduction of development time of immature stages (in contrast to relatively long naupliar and copepodite stages in copepods).

XVIII. FISH WITHIN AQUATIC ECOSYSTEMS

Fish of freshwater ecosystems are enormously important to humans. Although recreational exploitation of fish communities is often a dominant interest among developed countries, throughout much of the world inland fish serve as a major source of protein. Fisheries have developed to maximize utilization of these fish resources. A complex of interrelated problems has evolved as a result. Examples include fish exploitation that often exceeds production capacities, introductions of exotic species, and marked alterations of the trophic structure within lakes and streams.

It is beyond the scope of this synthesis to summarize the biology of fishes and their multifaceted ecological interactions within freshwater ecosystems. Ichthyology, the study of fishes, has a long detailed history that has been summarized, particularly recently, in a number of excellent treatises on the ecology of fishes (e.g., Lagler *et al.*, 1962; Nikolsky, 1963; Lowe-McConnell, 1975; Moyle and Cech, 1988; Wooten, 1990; Gerking, 1994; Jobling, 1994; Diana, 1995; Bond, 1996; Matthews, 1998). The treatments of fish in the present work, however, will be directed to the overall functions of fish in predation on other animals, of how fish feeding ecology can alter the trophic structures, and how fish can influence nutrient cycling within lake and stream ecosystems. Fish contributions to overall production and energy flows will also be coupled to different pathways.

A. Feeding Relationships among Fish

Feeding among fish, as among other animals, is the process for acquisition of both energy and nutrients required for growth and reproduction. The efficiency of this process is determined by both physiological and environmental constraints. Particularly pivotal are the costs of energy expended in acquiring food in relation to the benefits accrued energetically and nutritionally, and how that acquisition translates to the effectiveness of growth and reproduction. Optimal foraging theory predicts that behavioral modifications will be taken by a predator to maximize the returns per unit effort expended in acquiring prey of various qualities.

Nearly all fish use suction feeding to ingest food, by which the fish creates a negative pressure in the buccal cavity and draws into the mouth a stream of water carrying organisms (Gerking, 1994). Intake can also be accomplished by swimming with the mouth and opercles open. When prey is scarce and large, fish commonly feed on individual particles. Conversely, when prey density is high and size is small, fish commonly shift to filtering particles from the water by entrapment on mucus or by mechanical sieving. The filtration apparatus are commonly gill rakers (cf. Fig. 16-39). Interraker spaces are highly variable among fish species and growth stages and influence the size of prey particles acquired.

Olfactory and gustatory chemoreceptors are extremely sensitive among fish and respond to many chemical substances, often at very dilute concentrations (to $<10^{-18}$ M). Vision is critical to the location of prey

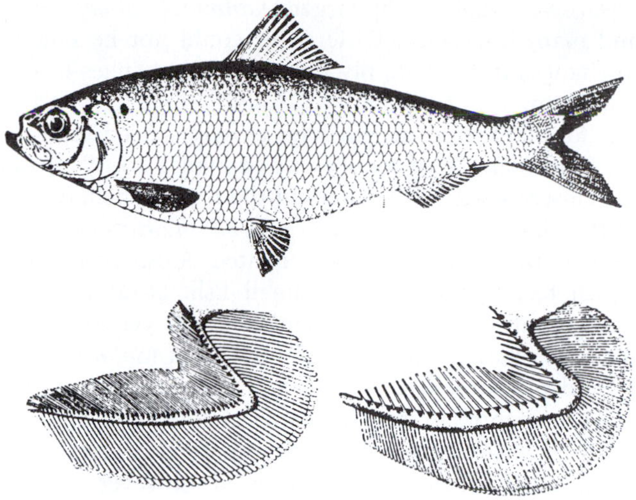

FIGURE 16-39 The alewife, *Alosa pseudoharengus*. *Upper*: A mature specimen, 300 mm in length. *Lower left*: First bronchial arch with closely spaced gill rakers that act as a plankton sieve. *Lower right*: First bronchial arch with widely spaced gill rakers of *A. mediocris*, a species that feeds primarily on small fish. (From Brooks, J. L., and Dodson, S. I.: Predation, body size, and composition of plankton. *Science 150*(3692):28–35, 1965. Copyright 1965 by the American Association for the Advancement of Science.)

among many fish and has led to many behavioral adaptations among prey organisms to seek both spatial and temporal refuges from the visual feeding of fish. The success of visual predation varies greatly with prey size, shape, translucency, and color as well as water clarity, prey behavior, and other factors. Because of the enormous importance of size-biased feeding, by which large prey or large size classes of a single prey species are favored over smaller ones, in food-web structure and energetics in limnology, a separate summary of that subject follows further mention of feeding modes among fishes.

Many fish adapt to a wide variety of food sources and often switch from one food source to another as environmental and food supply conditions change dietary quality and abundance. The species flexibility to switch and to take advantage of the most profitable food source at any particular time, termed *trophic adaptability*, makes it very difficult to accurately group fish into specialist or generalist feeding categories. Specialist fish feed on a restricted diet, whereas generalist fish feed on a broad spectrum of prey species or detrital organic matter. Opportunist fish switch from one food source to another as food populations fluctuate and as their life stages mature. For example, larval fish utilize endogenous sources of nutrient in the yolk sac stages and shift to exogenous mouth ingestion of algae initially as the yolk sac is absorbed and to zooplankton, the latter being selected individually, in the later stages of larval feeding. Mortality among larval fish is very high and some has been attributed to starvation (cf. review of Gerking, 1994). During good reproductive class years, however, abundant fish larvae can severely impact zooplankton, particularly larger species such as *Daphnia* (Mills and Forney, 1983).

A number of fish families are predominantly herbivorous on algae and vascular aquatic plants, or specialize on reproductive parts (fruits, seeds, and flowers) of higher plants. All herbivorous fish ingest some animal food, which may supplement the largely plant diet. These fish commonly possess physical adaptations for mastication of plant materials and partial chemical digestion. Detritivory is common among fish (Gerking, 1994). The diet of adult detritivorous fish consists mainly of decomposed organic matter; living organisms associated with the detritus are also consumed, either actively or fortuitously. Early larvae of detritivores are zooplanktivorous and switch to dead organic matter as they mature.

Diversity of feeding is great among bottom-(benthic) foraging fish. Most species shift among prey types as availabilities change, but certain behavioral adaptations often predominate. Processes include (a) ingesting sediment deposits, (b) scraping periphyton off rock surfaces, (c) grasping invertebrates from surfaces, and (d) crushing mollusks. Extraordinary feeding specializations have evolved among certain fish families, particularly among the nearly 500 endemic cichlid species in East African relict lakes (Greenwood, 1974; Lowe-McConnell, 1987). Fish also feed actively upon moving prey, either those that migrate vertically or horizontally within a lake or are moved by water currents. Zooplankton and some immature insect movements are often coordinated with periods of darkness to avoid heavy fish predation (cf. pp. 488–453). Benthic organisms of streams often are moved downstream by water currents in benthic drift (Chap. 22) and are supplemented by allochthonous invertebrates that fall into the stream from adjacent wetland and land.

Any evaluation of the efficacy of fish predation on the production of benthic invertebrates must be evaluated in the context of total mortality from (a) predation, (b) natural senescence and death, and (c) emigration via emergence, migration, or drift. Fish are clearly capable, under certain conditions, of reducing biomass and production of benthic invertebrates and to markedly alter species composition of prey or even eliminate certain prey by selective predation. Fish predation intensity must be very high, however, to become the main determinant of the zooplankton community, superimposed on lake productivity of food resources (Vanni *et al.*, 1990; Hessen *et al.*, 1995). As in the zooplankton, size selection of benthic prey by fish is common, although not universally a direct correlation. Bluegill or perch predation on large active benthic fauna shifted invertebrate community structure to smaller, less active forms (Hall *et al.*, 1970; Crowder and Cooper, 1982; Post and Cucin, 1984). Increased resource availability by the removal of larger prey can result in increased growth and production of smaller benthic organisms.

B. Predation by Fishes and Size Selectivity of Zooplankton

Thus far in this discussion, competitive interactions among zooplankton populations have centered on differences in spatial and temporal distributions and on food-size requirements that permit coexistence in the same body of water. The succession and competitive success of zooplankton populations can be influenced significantly by predation of other zooplankters, a point which was demonstrated earlier among the rotifers and the large cladoceran *Leptodora*. A large number of studies have demonstrated the importance of planktivorous fish in regulating zooplanktonic populations and in causing a distinct shift favoring survival of species that are smaller in size. In other words, for a number of reasons focusing on energetic efficiencies, predators selectively consume the largest prey possible within their physical and behavioral capabilities and the abundance of prey.

As noted briefly earlier, Hrbáček and collaborators were the first to demonstrate unequivocally the importance of fish in the regulation of the size and species composition of zooplankters (Hrbáček, 1958, 1961, 1962, 1965; Novotná and Korínek, 1966). When large, shallow ponds were stocked heavily with cyprinid fish, the zooplankton consisted of small cladocerans, *Bosmina, Ceriodaphnia*, and rotifers. Water transparency decreased with the predominant development of small nanoplankton algae. Upon selective removal of the fish, the zooplankton changed suddenly to larger cladocerans, particularly *Daphnia longispina*, and rotifer abundance decreased. Transparency increased as the phytoplankton shifted to smaller densities of larger species.

Another series of remarkably simple but definitive studies conducted by these Czechoslovakian workers involved experimental manipulation of portions of the same small lakes by fencing off portions of the littoral from fish and ducks (Straškraba, 1963 1965, 1967; Poštolková, 1967). Within the protected littoral enclosures, submersed vegetation increased and afforded zooplankton greater protection from fish predation than in the open water. Zooplankton biomass increased in the littoral as the small *Chydorus* and other cladocerans were replaced by larger *Daphnia, Simocephalus*, and many copepods. Differences could not be related to changes in food supply but rather to changes in fish predation.

All planktivorous fish have closely spaced gill rakers. The example of the alewife of Figure 16-39 shows the closely spaced gill rakers of the first branchial arch of the planktivorous *Alosa pseudoharengus* in comparison to those of the closely related *Alosa mediocris*, which feeds principally on small fish. Studies of the pharyngeal sieves of the planktivorous yellow perch (*Perca flavescens*) and the rainbow trout (*Salmo gairdneri*) demonstrated that the trout and perch could remove few zooplankters of a size smaller than 1.3 mm (Galbraith, 1967; Wong and Wood, 1972; Nilsson and Pejler, 1973). It must be emphasized, however, that all freshwater planktivorous fish examined so far actively search for and visually select each plankter that they ingest (Brooks, 1968; Seghers, 1975). Some fish (e.g., ciscoes, *Coregonus artedii*; bluegills, *Lepomis macrochirus*) gulp water by constantly opening and closing the mouth while swimming (Janssen, 1980; Werner *et al.*, 1981). These actions ingest aggregations of prey when prey densities are high. However, in the normal respiratory behavior of passing water through the mouth and across the gills, some zooplankters can be collected by gill rakers. Fish such as the alewife are more or less obligately planktivorous, while others such as the trout and perch eat large zooplankters, and if these are unavailable they readily shift to alternate food sources. Size selection of prey appears to be characteristic of both the obligate planktivores, such as some of the alewives, and those species that feed facultatively on the plankton.

The relation between size selection of prey and foraging efficiency was shown in the bluegill sunfish (*Lepomis macrochirus*), a common centrarchid of many temperate waters that selects prey on the basis of size (Werner, 1974; Werner and Hall, 1974; O'Brien *et al.*, 1976; Werner *et al.*, 1981; Mittelbach, 1981). Growth rates increased significantly in direct relation to food size (see, e.g., LeCren, 1958; Hall *et al.*, 1970), and much of this differential growth has been attributed to greater efficiency of foraging with increasing size of prey captured, that is, improved ratio of expenditure of energy or metabolic cost in obtaining food relative to the return (cf. Pyke *et al.*, 1977). In the bluegill, size selection of prey is related to the optimal allocation of time spent searching for and handling the prey. In other words, the size range of prey that maximizes the energy return per unit of energy expended depends on the costs of searching for and handling times of different prey in the environment. Both searching ability and prey-handling efficiency increase with

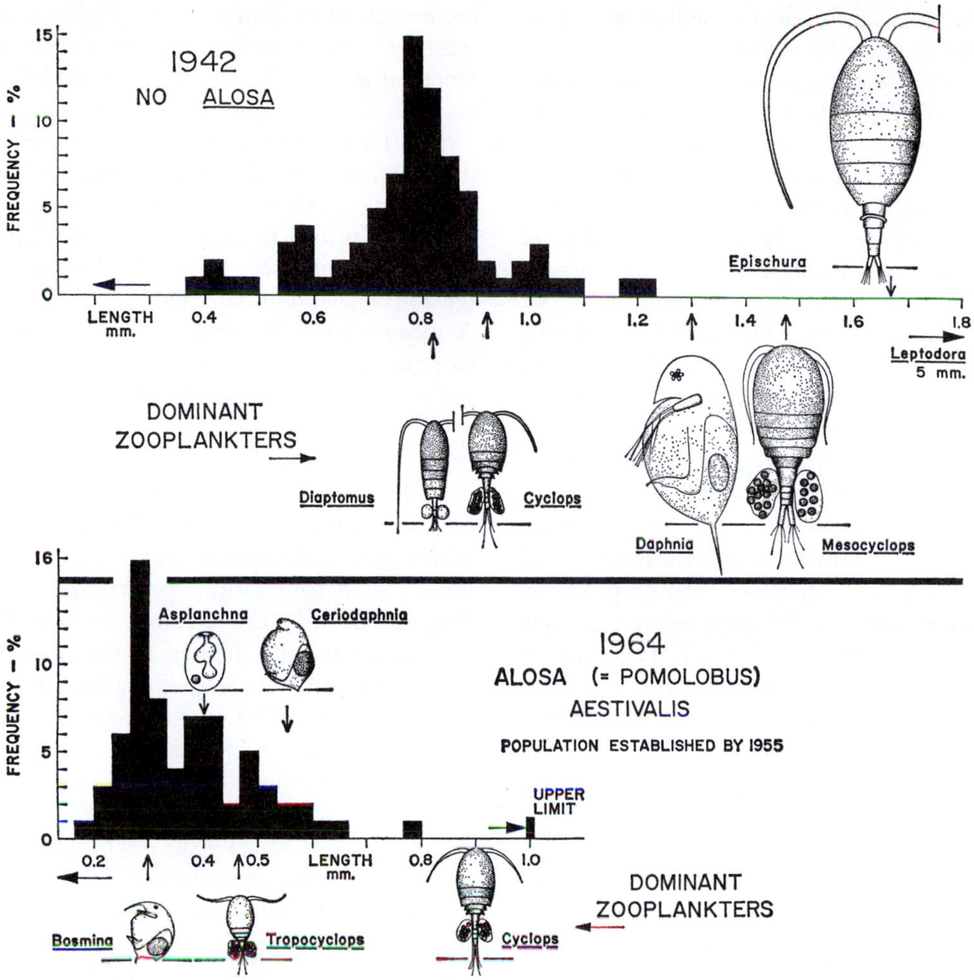

FIGURE 16-40 The composition of the crustacean zooplankton of Crystal Lake, Connecticut, before (1942) and after (1964) a population of *Alosa aestivalis* became well established. Each square of the histograms indicates that 1% of the total sample counted was within that size range. Larger zooplankters are not represented because they were relatively rare. The specimens depicted represent the mean size (length from posterior base lines to the anterior end) of the smallest mature instar. The arrows indicate the position of the smallest mature instar of each dominant species in relation to the histograms. The predaceous rotifer *Asplanchna* is the only noncrustacean included in this study. (From Brooks, J. L., and Dodson, S. I.: Predation, body size, and composition of plankton. *Science 150*(3692):28–35. Copyright 1965 by the American Association for the Advancement of Science.)

increasing fish size (Mittelbach, 1981). The handling time per prey increases exponentially as the ratio of prey size to mouth size increases (Werner and Mittelbach, 1981). At times of low prey abundance, when search time is long, prey of different size are eaten as encountered. As prey abundance increases and search time decreases, smaller-sized classes are eaten less frequently or ignored, such that overall return per time increases (see also Brooks, 1968).

The earlier Czechoslovakian observations on the effects of planktivorous fish on the zooplanktonic species composition were confirmed excellently by a study of the changes in the zooplanktonic populations of Crystal Lake, Connecticut (Brooks and Dodson, 1965). Prior to the introduction of the alewife into the lake, zooplankton included the large calanoid copepod *Epischura, Daphnia*, and the cyclopoid *Mesocyclops*, as well as numerous smaller *Diaptomus* and *Cyclops* copepods (Fig. 16-40). Some 10 yr after the alewife invaded the lake, the larger forms of zooplankters had been replaced and dominants included very much smaller forms, the cladoceran *Bosmina* and two cyclopoid copepods *Tropocyclops* and *Cyclops* (Fig. 16-40). The modal length of the numerically dominant forms had shifted from 0.8 to 0.35 mm in the zooplankton assemblages. Larger forms were found only in the littoral zone and near the sediments, areas that are avoided by the pelagic *Alosa*.

Such size–feeding relationships have been demonstrated in many different environments. Indeed, the excellent early studies have been repeated, with numerous experimental variations in enclosures of highly varying efficacies. For example, the simple zooplankton community structure of high-elevation lakes is strongly influenced by predators present (Starkweather, 1990). Lakes with resident salmonid fish have crustacean zooplankton always smaller than 0.75 mm and substantial rotifer populations, while lakes without fish contained much larger crustaceans (to 2.8 mm) and few or no planktonic rotifers. Such relationships with similar results have been demonstrated often both in experiments where fish and zooplankton predation is modified within enclosures (e.g., Threlkeld, 1988; Mazumder *et al.*, 1990b; Brett *et al.*, 1994) and by predator introductions into or predator alterations within whole-lake ecosystems (e.g., Elser and Carpenter, 1988; Post *et al.*, 1997). The type of planktivore, fish year–class strengths, and seasonal variability in fish feeding all can have marked effects on the altered zooplankton community structure (e.g., Hambright and Hall, 1992; Rudstam *et al.*, 1993). Although the predatory effects on zooplankton were marked, particularly in relation to size, effects on feeding commonly resulted in little change to bacterioplankton abundance or alterations in *total* algal primary production.

C. Buzz Words: The Plague of Fuzzy Concepts and Ambiguous Terminology

Many studies in the last two decades on the selective size–feeding relationships within planktonic communities led to a number of purportedly new hypotheses. All were in reality quite redundant with the studies of several decades earlier presented in summary here. These relationships were studied in a variety of aquatic habitats and provided highly varying degrees of scientific support for the hypothesis that phytoplanktonic algal mass would be influenced by selective control of fish and zooplankton levels within the pelagic food chain. A considerable number of these studies drew conclusions that were unsupported and proposed quite imaginary couplings, all of which persist as an embarrassment to limnology. Among the worst were repeated studies that claimed, without data, that differential predation (unmeasured) was inducing differential grazing (unmeasured) that was causing major nutrient regeneration (unmeasured) and inducing major alterations of photosynthetic productivity and phytoplankton community structure (unmeasured). The purportedly confirmed relationships, with highly variable support, were reiterated with such frequency that a danger emerged where the speculations could be elevated to ecological laws simply over time (e.g., McIntosh, 1980). Early appropriately critical reviews of the data (e.g., DeMelo *et al.*, 1992; Strong, 1992) were effective in forcing more substantive experimental evaluations. Although results of the early foundational studies were generally supported, many of the subsequent overextensions of strengths of trophic interactions were not. A more reasonable and realistic perspective on these trophic interrelationships is now emerging (e.g., Hansson *et al.*, 1998a; Persson, 1999).

1. Trophic Cascades

The term *trophic cascade* originated in studies of marine intertidal organisms (Paine, 1980) and has been widely and ambiguously applied to a number of fresh waters as a foundation for trophic interactions (e.g., Carpenter *et al.*, 1985, Carpenter and Kitchell, 1993). Although formally defined as the propagation of indirect mutualism between nonadjacent levels in a food chain (Schoener, 1993), it is most often applied to well-known pelagic processes, delineated decades ago, of predation by fish and/or invertebrates that sometimes alter the zooplankton community structure and their effectiveness in grazing on phytoplankton. Many studies in enclosures, ponds, and whole-lake experiments demonstrated significant impacts of planktivorous fish on zooplankton that in turn influenced phytoplankton *composition*. Examples of such effects were clearly indicated in the discussions in this chapter. Two important points have emerged from analyses of these interactions, however: (a) The concept of cascading negative trophic interactions is highly simplistic and often fails completely to operate in natural pelagic ecosystems because of a bevy of compensatory mechanisms that emerge almost immediately following predatory alterations (e.g., Mazumder *et al.*, 1990a; McQueen *et al.*, 1992; Ramcharan *et al.*, 1995, 1996; Hansson *et al.*, 1998; Persson, 1999; among many others); and (b) if operational for a period, the "cascading" influence cannot be sustained for long periods under natural conditions without constant disturbance. Trophic cascades are basically a description of one outcome of indirect interactions out of several possible alternatives (cf. Persson, 1999).

Additional ambiguous terms and concepts plague the literature on this subject. "Top-down" and "bottom-up" regulation of some process or property of different trophic levels, usually among the plankton, are terms with definitions nearly as diverse as the number of workers using them. Set forth by McQueen *et al.* (1986, 1989), *top-down* generally refers to predation processes by invertebrates or fish that influence zooplankton community structure that in turn may selectively influence feeding efficacy on seston. *Bottom-up* generally refers to resource regulation of growth and production often beginning with biogeochemical controls of photosynthesis, usually of phytoplankton. By implication, usually only vaguely stated, this resource (nutrient or light) limitation of phytoplankton growth can translate to resource (food) limitations for herbivorous zooplankton and consumers of zooplankton. The hypothesis predicts that "bottom-up" forces are strongest at the bottom of the food chain and "top-down" forces are strongest at the top of the food chain. Organic carbon resources *always* influence the productivity at all levels of the food chain couplings, whereas predatory effects ("top-down" effects) may weakly affect *productivity* at the primary level under some circumstances, but usually does not. Productivity is shifted or displaced to other organisms or to different parts of the ecosystem but is not suppressed greatly or eliminated. Importantly for management implications, however, the shifted productivity may be of a type more "desirable" for human uses of the water, and that characteristic can be an important aspect of management decisions.

2. Biomanipulation

Biomanipulation, a term introduced by Shapiro *et al.* (1975), is a type of biological engineering in which manipulations of biota are used to reduce objectionable algal *types* and biomass in addition to, or to supplant, reductions of nutrient loading. Biomanipulation was originally based on the idea that when the number of planktivorous fish are reduced, the density of large cladoceran zooplankton increases, and their grazing during the summer period can reduce certain species of planktonic algae and reduce the algal turbidity of the water. Planktivorous fish can be reduced by introduction of piscivorous fish or by intensively cropping the planktivorous fish by commercial fishing, as has been successfully done under certain conditions. As a corollary, "biomanipulation" postulates that increased fish planktivory should enhance phytoplankton populations and their development by decreasing grazing by larger zooplankters. Furthermore, manipulation of carnivorous fish predators could alter planktivorous fish populations which in turn could reduce predation on large zooplankton and increase the effectiveness of zooplankton feeding on phytoplankton.

Biomanipulation has often been presented with naive promises as a purportedly inexpensive alternative to control of eutrophication. Many studies have indicated that the presence of large grazers results in an apparent temporary reduction of algal biomass, frequently shifting to irruptions of both ultraphytoplankton, with intensified growth and recycling within the microbial loop, and nuisance blooms of cyanobacteria that are not effectively grazed. If nutrient inputs are controlled, then algal reduction results can be improved. For example, commercial fishing of the planktivorous coregonid fish in Lake Vesijärvi, Finland, has been successfully used as a management tool, coupled with nutrient reductions, improved land use in the drainage basin, improved littoral buffers, and other methods to significantly revert the phytoplankton biomass to acceptable levels for public uses (Kairesalo *et al.*, 1999). This excellent example of lake restoration, however, demonstrates that the fish "biomanipulation" technique requires a continual, sustained removal effort and can only be used effectively in tandem with many simultaneously nutrient reduction and control mechanisms (cf. also Hansson *et al.*, 1998b). Such control of algae cannot be readily realized on a sustained basis in lakes that experience continued high nutrient loadings and multiple usage by humans unless intensive management strategies are constantly applied. Even coupled with aggressive programs to reduce nutrient loading, the primary cause of eutrophication, such manipulations of higher trophic levels and feeding dominance are only of short-lived therapeutic value. Nutrient pools are temporarily displaced but not removed from the ecosystem, and continued nutrient loading will be directed to maximizing primary productivity.

Such alterations of algal composition and biomass distributions by "biomanipulation" are ***not*** "biological control of eutrophication," despite the claims of some workers (e.g., Carpenter *et al.*, 1995; Hansson *et al.*, 1998). Such management processes may temporarily reduce biogenic turbidity in eutrophic lakes for "cosmetic" aesthetic purposes, but it is little more than a temporary displacement of nutrients, particularly of phosphorus, to alternative reservoirs within the lake, some of which will certainly recycle and be reutilized in subsequent photosynthetic production. The proposals (e.g., McQueen *et al.*, 1990, among many others) that such "biomanipulation" will be effective in lakes only when chemical and physical factors are altered to produce algal species compositions that permit strong control of prey by predators are not practical solutions to eutrophication of surface waters. The compensatory capacities of the phytoplankton and coupled microbes of the pelagic ecosystem are multiple, varied, and rapidly responsive and will not allow such a simplistic disturbance to persist for any appreciable period of time if the nutrient resource availability persists for sustained potential productivity. With more sophisticated monitoring and regulation programs in the future, biomanipulation will continue to be a useful, albeit expensive, therapeutic tool to minimize the *effects* of eutrophication.

Terms such as "bottom-up," "top-down," and "biomanipulation" must be used, if at all, with great caution, because they are not only highly ambiguous but they promulgate concepts that are not founded in thermodynamic relationships of nutrient storage and recycling and are used without recognition of the true complexities of the regulation of food-web interactions. Perhaps such simplifications were helpful during the initial development of the concepts; now they inhibit progress toward thorough understanding of the processes involved.

D. Importance of Visibility in Predation

Similar predatory elimination of larger zooplankters, particularly *Daphnia*, from lakes by other planktivorous fish has been shown in many studies since. The planktivorous smelt (*Osmerus mordax*) is particularly effective in this regard (see, e.g., Reif and Tappa, 1966, and especially Galbraith, 1967). However, the relationship of size and planktivore predation is not always that simple. If prey are large but relatively transparent, such as *Leptodora* or planktonic larvae of the dipteran *Chaoborus*, they often are overlooked and predation is reduced (see, e.g., Costa and Cummins, 1972). Furthermore, the slower, steady movements of cladocerans render them more vulnerable to predation than the jerky, irregular movements of copepods.

The importance of visibility in planktivorous predation by fish is demonstrated by the tropical *Ceriodaphnia cornuta*, which shows two distinct polymorphic forms of the same body length within the same lake (Zaret, 1972a,b). One form has pointed, hornlike extensions of the exoskeleton on the head, body, and tail regions with a small area of black pigmentation in the compound eye. The other phenotype is unhorned but possesses a large, pigmented eye. Predation by the dominant planktivore fish, the silverside (Atherinidae; *Melaniris chagresi*) was more intense on the form with the large, pigmented eye. This form of *Ceriodaphnia cornuta* has a superior reproductive potential and a more rapid population growth with greater longevity than that of the horned, small-eyed form. Without fish predation, the large-eyed form can rapidly outcompete the less conspicuous form, but under predation pressure, the form with the small eye, although growing more slowly, can coexist because of reduced visibility to the predator. Similar results were found in the effects of fish predation on *Bosmina* (Zaret and Kerfoot, 1975; Hessen, 1985b). Prey selection was found to be related to the large, black pigmented eye, and body size was of negligible importance. Size-biased feeding by fish that select the larger species of a zooplankton community or larger members of a single species has been documented in at least 20 species of fish (reviewed in Gerking, 1994). Particulate turbidity, particularly clay turbidity common to many reservoirs, reduces visibility and the distance at which predator–prey interactions occur (Abrahams and Kattenfeld, 1997). This constraint will reduce the predation risk on zooplankton by fish in turbid aquatic ecosystems.

As lakes become more eutrophic, a greater proportion of the phytoplankton biomass and productivity often results from large algae (mostly colonial or filamentous; cf. Chap. 15). The larger algae interfere with food collection to a greater extent in larger cladocerans, causing reduced growth and fecundity, than in smaller cladoceran species that feed on small particles (Gliwicz, 1980; Gliwicz and Siedlar, 1980). In this way, larger cladocerans can experience a reduced efficiency of food collection under eutrophic conditions and may consequently be selected against. Such interspecific competition, in addition to size-selective predation, could contribute to the reduction of larger zooplankton.

E. Size-Efficiency Hypothesis

When size selection by fish is not in effect and large zooplankters are present, a common observation is that the smaller-sized zooplankton are not generally found to co-occur with the larger forms. Brooks and Dodson (1965; Brooks, 1968) developed the size-efficiency hypothesis in an attempt to explain the commonly observed inverse relationship between the abundances of small- and of large-bodied herbivorous zooplankton in lakes. According to this hypothesis (Hall *et al.*, 1976):

a. Planktonic herbivorous zooplankton compete for small particles (1–15 μm) of the open waters.
b. Larger zooplankton filter more efficiently and can also take larger particles. This greater effectiveness of food collection leads to relatively reduced metabolic demands per unit mass, permitting more assimilation to go into egg production by the larger herbivores.
c. Therefore, when the intensity of predation pressure by fish is low, the small planktonic herbivores will be competitively eliminated by large forms (dominance of large Cladocera and calanoid copepods).
d. When fish predation is intense, size-dependent predation will eliminate the large forms, allowing the small zooplankton (rotifers, small

Cladocera) that escape predation to become the dominants.

e. When predation pressure is moderate, larger zooplankton species are often kept at reduced populations that allow the coexistence of smaller competitors.

The basic assumptions of the size-efficiency hypothesis imply a complex successional pattern in which optimal body size increases while the range of persisting sizes decreases with decreasing food concentration (critically analyzed by Hall *et al.*, 1976; Zaret, 1980). Vertebrate predation restricts the maximum adult body size of zooplankton, and invertebrate predation may restrict the minimum size. Both of these effects can augment declines of zooplankton productivity caused by food limitations. Minimal food concentrations needed for growth among different species of *Daphnia* decrease as the size of the animals increases (Gliwicz, 1990). As a result, when predation pressure is low, larger cladoceran species should be more successful competitors for food. In contrast, the threshold food concentrations of small species of rotifers are lower than among large species, which is indicative that smaller rotifers are more energetically efficient per unit body mass than large species (Stemberger and Gilbert, 1987).

The size-efficiency relationships among fish and zooplankton, although not fully understood, have major implications on food-web structure and have led to a plethora of research on potential manipulation of food webs by fish introductions ("biomanipulation"). Fish predation clearly can influence species composition and size structure of zooplankton prey. Morphology, physiology, and behavior of prey can all be affected by size-biased feeding, which can lead to significant evolutionary alterations for predator avoidance. The cyclomorphosis among cladocerans, discussed earlier in this chapter, is at least in part a result of predation pressures. Similarly, vertical migration and other evasive behavior among zooplankton are partially induced by predation pressures (O'Brien, 1987; Ringelberg, 1991; Lampert, 1993). The migration behavior of many cladocerans upward as darkness occurs and downward as daylight approaches is most common among large, adult zooplankton, particularly females carrying eggs, that are most visible to fish. That migration occurs only when fish are abundant.

It is erroneous to assume that fish predation on zooplankton is the dominant means of mortality among larger zooplankton size classes (>1.0 mm). Mortality occurs by other, often unclear, factors even under nearly ideal growing conditions (e.g., Boersma *et al.*, 1996). Such factors include simple physiological death, often after resting egg formation, parasitism, microbial infection, invertebrate predation, and physical degradation (e.g., outflow and molar death in flowing systems).

F. Food Limitations in Zooplankton

Food limitation and food competition have not been demonstrated to be strongly in effect in natural populations. However, as discussed earlier, from experimental culture analyses, food limitations clearly can affect the life cycle and reproductive capacities of cladocerans and rotifers (Duncan, 1989). Under food-limiting conditions, body size and mass are reduced, the number of instars or prereproductive period to reproductive stages increases, the duration of development is prolonged, and fecundity is reduced. The threshold concentration at which food limitation occurs is not a constant for a species, but varies with the quality of the food. For example, lipid ingestion from algal food and storage is positively correlated with reproductive success and the threshold for starvation and food limitation (Tessier *et al.*, 1982, 1983). Even under extreme food limitation, some energy is invested into reproduction.

Clearly, food limitations interact with size-selective predation to regulate population dynamics within zooplankton communities. For example, population densities of zooplankton of a shallow temperate lake were affected much more by experimental manipulations of food availability than by manipulations of fish predation. Increases in birth rates and declines in mortality rates resulted from increased phytoplankton availability among cladocerans, some copepods, and rotifers (Vanni, 1987). The densities and size of a few cladoceran populations were reduced by planktivorous fish predation, and size at time of reproduction was smaller. However, flexibility in *community* life history traits allowed the zooplankton to withstand intense size-selective predation by planktivorous fish.

Food competition in nature is uncommon because of (1) the few times of the year when zooplanktonic populations can even approach the potential of completely removing particulate material from the water of certain strata by filtration, as discussed earlier, and (2) the presence of an excess of algae, bacteria, and particulate detrital organic matter beyond what can ever be consumed (most synthesized organic matter is decomposed in nonpredatory pathways; see Chap. 23). Experimental evidence on these food relationships is weak or negative (see, e.g., Sprules, 1972; Dodson, 1974a; Neill, 1975; Gliwicz and Prejs, 1977; Boersma and Vijverberg, 1996) but does not exclude important food-quality interactions.

A much more plausible explanation is again a size-selective predation, but a predation by the larger zooplankters rather than by planktivorous fish. Such predation occurs among the size range of zooplankton that is smaller than can be utilized effectively by fish. A large number of studies have shown that invertebrate predation can be sufficient to eliminate certain smaller species and that this predation, as with some fish, is size-selective (cf. review of Zaret, 1980). Larger copepods, especially, and other predators such as *Chaoborus* and the cladocerans *Leptodora* and *Bythotrephes* are particularly effective in causing significant mortalities of small zooplanktonic species (see, e.g., McQueen, 1969; Anderson, 1970; Dodson, 1970, 1972; Confer, 1971; Confer and Cooley, 1977; Confer and Applegate, 1979; Murtaugh, 1981; Lehman, 1991). While fish planktivory is visual and size-selective, invertebrate predation is more tactile since the eyes of rotifers, crustaceans, and most insects detect light intensity and movements but do not form images. This relationship of size selectivity and tactile responses in consumption suggested to Dodson (1974b) that the shape of prey within the correct size range can influence whether the prey is taken or rejected. Thus, cyclomorphic polymorphism would have an adaptive advantage for prey in both invertebrate predation by morphology and vertebrate predation by reduction of visibility, while still permitting growth.

It should also be noted that in the absence of intense size-selective predation by fish, large cladoceran species could dominate because they possess greater reproductive rates (intrinsic rates of increase, *r*) (Goulden *et al.*, 1978). Even though smaller species mature earlier than larger species, the larger forms have higher fecundities (number of young produced per female per day, *b*), which results in higher rates of increase.

G. Changes in Predation Effects in Shallow Lakes with Macrophytes

The pelagic fish–zooplankton interactions become more complex in shallow lakes (see Chap. 20). Total productivity and particularly fish biomass per unit volume tends to be much higher in shallow lakes than in deep lakes. Similarly, biomass of benthic invertebrates per unit area is higher in shallow lakes, and as a result fish feeding tends to be less dependent on zooplankton prey than in deep lakes (Jeppesen *et al.*, 1996, 1997). Fish tend to shift feeding between zooplankton and benthic animals as changes in prey densities occur seasonally. In pelagic areas of shallow lakes, cladocerans tend to be severely impacted by predation as the effectiveness of vertical migration is reduced. As a result, grazing pressure on phytoplankton is often reduced. Predation of zooplankton by benthic and planktivorous fish is reduced markedly in shallow lakes rich in submersed macrophytes (Jeppesen *et al.*, 1998). The littoral areas serve as a refuge for pelagic cladocerans. As a result, the grazing of both phytoplankton and bacterioplankton among the submersed macrophyte areas can be increased. Much of the nonmacrophyte primary productivity within the littoral areas shifts to the algae and cyanobacteria that are attached to the surfaces of macrophytes and other substrata (see Chap. 19).

The additional presence of piscivorous fish tends to reduce fish planktivory on zooplankton in the pelagic regions of shallow lakes. The presence of piscivores causes small planktivorous fish to make greater use of vegetated habitats and reduced use of pelagic habitats (Turner and Mittelbach, 1990). Planktivory was reduced and larger cladocerans tended to increase in the pelagic areas because of the avoidance of planktivorous fish of predation by piscivores.

XIX. ZOOPLANKTON PRODUCTION

The production rates of specific populations of zooplankton refer to the net productivity or the sum of the growth increments of all specimens of the population. This net productivity excludes maintenance losses (respiration and excretion) and includes the growth increments of the animal itself as well as the biomass produced as gametes and as exuviae during molting. In some larger invertebrates and some vertebrates, the productivity values of specific animal populations are influenced markedly by emigration and immigration. The productivity measurements are complicated when predation causes removal of a significant portion of the population; this loss to the population can be difficult to evaluate accurately.

The methods of estimating the rates of production of a specific population demand an accurate evaluation of the distribution of the organisms, the different stages of development and age, and the generation times. All of these parameters vary among species, seasonally, and under changing environmental conditions and require alterations in methods of sampling, analyses of biomass within each age group, and frequency of sampling. Sampling intervals must be kept shorter than the generation times of the animals.

The manner in which production rates of a specific population are estimated depends on the particular life cycle, reproductive characteristics, and generation times (discussed at length in Edmondson and Winberg, 1971; Winberg, 1971; Edmondson, 1974; Bottrell *et al.*, 1976; Rigler and Downing, 1984; Benke, 1984, 1993). The productivity of a species with a long life cycle and a short period of reproduction is determined relatively easily if individuals are of the same age or if separate cohorts can be recognized and their biomass determined. In such a situation, after the brief period of reproduction, no recruitment occurs to confound estimates of growth and mortality. When reproduction and recruitment to the population are continuous, estimation of production is more involved. In this latter case, which applies to most zooplanktonic populations, cohorts overlap, which makes it difficult to separate changes in their abundance and biomass over time. Consequently, it is necessary to estimate finite birth and growth rates of individuals and evaluate changes in biomass over the life cycle from birth to death. Several simple examples were cited for growth rates of specific cohorts in earlier discussions.

A. *P/B* Ratios

An index of productivity that has become widespread in European and Russian literature in recent years is the P/B coefficient, the ratio of production (P) to biomass (B) (cf. Chap. 8). As indicated earlier, B is a measure of mass for a population present at one point in time in units of mass per unit area or volume and is therefore a temporary storage of mass or energy. P is a flow or flux of mass or energy per area per time. The ratio (P/B) of production (P) divided by mean biomass (B) is a rate with units of inverse time (e.g., 1/yr). P/B is essentially a weighted mean value of biomass growth rates of all individuals in a population (Benke, 1993).

P/B ratios can be useful as an estimate of the turnover of a population and is used for comparative purposes of population growth in response to environmental conditions and perturbations. In the case of a cohort, the cohort P/B ratio is equal to the cohort production divided by the mean cohort biomass (Waters, 1969, 1977; Benke, 1984). The annual P/B ratio is the annual production divided by the mean biomass of the entire 12-month period, even though the generation under study may have been present for less than a year (Wetzel and Likens, 2000). Mean annual biomass is simply the mean of all monthly average biomass values.

In populations with a constant age structure and biomass, a rarity in nature, the P/B ratio is relatively constant. The ratio varies in most situations when age structure, biomass, and growth are discontinuous; what is obtained is an average ratio over an interval of time (e.g., McLaren and Corkett, 1984). Multiplication of the P/B ratio by biomass yields an estimation of production during that interval of time. This ratio is analogous to the turnover rate and is also the reciprocal of P/B, the turnover time, which is the average duration of the life of a species under a given set of growth conditions. Biomass turnover times vary from a few hours for protistan zooplankton to a large range of 1.7 days for rotifers to well over 100 days for large zooplankton under poor growth conditions (Table 16-17). A comparative analysis of the P/B ratios of all major components of the plankton of mesotrophic Lake Constance, southern Germany, showed relatively similar results (Table 16-18). Although heterotrophic plankton biomass was on average twice as large as phototrophic biomass, phytoplankton contributed ca. 69% to the total plankton production in comparison to 19% for microbial plankton (bacteria and protists) and 11% for metazooplankton (Straile, 1998). Thus, autotrophic plankton production was twice the heterotrophic production in the water column (not the lake).

The P/B ratio is an expression of some value in comparing annual productivity among water bodies. However, it masks the large variations that occur within the population dynamics over a year (see, e.g., Straile, 1998). When used indiscriminately, the ratio relates few of the internal population characteristics or, more importantly, of the causal interrelationships leading to the resulting dynamics and productivity. Over brief periods of time, under relatively steady environmental conditions, the turnover rates reflect the relative growth of the populations.

B. Estimates of Zooplankton Productivity

Very few estimates of the annual production of protozoan and other protistan zooplankton communities have been made in lake ecosystems. Nonetheless, the few values given in Table 16-19 indicate that the production by the microzooplankton can be a significant portion of the total and at times exceed that of the larger zooplankton.

Many studies have estimated the productivity of zooplankton (e.g., Morgan *et al.*, 1980). A few examples illustrate the ranges encountered in aquatic ecosystems of differing primary productivity (Tables 16-17 and 16-20). Of many environmental factors regulating the seasonal production cycle of zooplankton, rates of biomass production are set largely by temperature (Shuter and Ing, 1997). Levels of biomass accumulation in zooplankton are set largely by food resource availability and individual body size. In general, a posi-

TABLE 16-17 Examples of the Production of Herbivorous and Predatory Zooplankton Communities of Aquatic Ecosystems

		Biomass (*B*)[a]		Production (*P*)[a]		Biomass turnover time (days)		
Lake type	*Period of investigation*	$g\ m^{-3}$	$kcal\ m^{-3}$	$g\ m^{-3}$	$kcal\ m^{-3}$	*Average*	*Range*	*Remarks and source*
Oligotrophic								
Lake Baikal, Russia	June–July	0.136						Primarily Epischura;
	Sept.	0.43						Moskalenko and Votinsev
	Annual (0–50 m)	—		3.44	—			(1970)
Clear Lake, Ontario	Annual	0.20		3.02	16.435	25	12–333	Herbivorous zooplankton (rotifers, *Holopedium*, *Daphnia*, *Bosmina*, *Diaptomus*); Schindler, (1970)
Lake 239, Ontario	May–Nov.	—		0.61	3.331	22	10–91	
Lake Krugloe, Russia	Annual		0.405	0.94	5.116	29		Arctic lake; Winberg (1970)
Mesotrophic								
Taltowisko Lake, Poland[a]	May–Oct.	0.12		3.04		14.3		29% of total lake area in littoral
Herbivores				4.68	25.43	10.2		zone; Kajak *et al.* (1970);
Predators				0.46	2.50	9.1		Kajak (1970)
Lake Naroch, Russia	Annual	0.07	0.38	1.12	6.11	22.4		Winberg (1970)
Lake Krasnoe, Russia	Annual	0.14	0.76	3.09	16.82	16.5		
Eutrophic								
Lake Mikolajskie, Poland	May–Oct.							Kajak *et al.* (1970); Hillbricht-
Herbivores				6.45	35.09	9.2	4.0–12.5	Ilkowska *et al.* (1970)
Predators				1.32	7.18	25.0		
Lake Sniardwy, Poland	May–Oct.							
Herbivores				3.08	16.78	14.9	8.3–33	
Predators				0.50	2.71	14.3		
Kiev Reservoir, Russia								
Herbivores		0.35	1.9	9.15	49.8	13.9		Winberg (1970)
Predators		0.022	0.12	1.16	6.3			
Severson Lake, Minnesota	May–Oct.							Comita (1972)
Herbivores				2.51	13.55			
Predators				0.11	0.60			

North Pine Dam, Australia	Annual (mean 3 yr)						King (1979)
Rotifers		—	—	—	1.42	—	
Crustaceans		—	—	—	2.60	—	
Total zooplankton		—	—	—	3.87	—	
Lake Kasumigaura, Japan							Nanazato and Yasuno (1985)
Cladocerans	Annual	0.18	—	9.5	—	—	
Lough Neagh, N. Ireland							Andrew and Fitzsimmons (1992)
Rotifers	Annual (3 yr)	0.0065	—	0.146	—	11	
Lake Le Roux, South Africa							Hart (1987)
Herbivores	Annual, 1982	11.8	—	1.78	—	6.7	
	Annual, 1983	24.3	—	2.91	—	8.3	
Dystrophic (high dissolved organic matter)[b]							
Lake Flosek, Poland	May–Oct.						*Sphagnum* bog; high littoral and allochthonous organic inputs; Kajak *et al.* (1970); Hillbricht-Ilkowska *et al.* (1970)
Herbivores				25.68	139.7	6.3	
Predators				1.16	6.3	25.0	
Littoral and pelagic							
Lake Biel, Switzerland	Apr.–Oct.						Vuille (1991)
Littoral planktonic	1987	—	—	7.66	—		
and epiphytic	1988	—	—	3.9	—		
Total zooplankton	1987	—	—	12.6	—		
	1988	—	—	8.4	—		

[a] Estimated using the mean caloric value for microconsumers (Cummins and Wuycheck, 1971).
[b] See discussion in Chapter 25.

TABLE 16-18 Annual Average Production to Biomass Ratios (*P/B*, day^{-1}) and Biomass Turnover Times (days) of Plankton of Mesotrophic Lake Constance, Germany[a]

Plankton component	*Annual average* P/B (*day^{-1}*)	*Biomass turnover time (days)*
Phytoplankton	0.52	1.9
Bacterioplankton	0.13	7.7
Heterotrophic flagellates	0.48	2.1
Ciliates	0.19	5.3
Rotifers	0.13	7.7
Herbivorous crustaceans	0.05	20.0
Carnivorous crustaceans	0.07	14.3

[a] Extracted from data of Straile (1998).

tive correlation exists between the rates of production of heterogeneous phytoplankton communities and of the heterogeneous micro- and macrozooplankton communities among many different reservoirs and natural lakes (Table 16-21; see, e.g., Winberg *et al.*, 1970; Makarewicz and Likens, 1979; Brylinsky, 1980; Richman *et al.*, 1984; Rublee, 1992; Canfield and Jones, 1996). Similarly, horizontal gradients in zooplankton productivity within a lake are often directly correlated to differences in the gradients of primary productivity. However, these statements do not necessarily imply that the herbivorous zooplankters are consuming the algae directly in proportion to their biomass and growth. As was discussed earlier, size, quality, and other factors influence both the ingestion and assimilation of algae. Moreover, much of the algal production enters the detrital pathways of nonpredatory particulate and dissolved organic matter (cf. Chap. 23). Phagotrophic utilization of detrital particulate organic matter with associated microbes is clearly a major route of primary productivity, and that much of the resultant protistan productivity never enters rotiferan and microcrustacean productivity.

Certain algal species are clearly impacted by zooplankton grazing, and many algal species are only minimally impacted by zooplankton grazing. The percentage assimilation of phytoplankton primary productivity by herbivores is nearly always <25% (e.g., Table 16-20).

In a detailed and perceptive study of the relationships between phytoplankton and zooplankton of mesotrophic Lake Erken, southern Sweden, Nauwerck (1963) demonstrated that algal productivity was inadequate to support the herbivorous zooplanktonic

TABLE 16-19 Estimates of Biomass and Production of Protists in the Trophogenic Zone of Lakes[a]

Lake	*Trophic status*	*Seasonal average* (*mg C m^{-3}*)	*Range of production* (*mg C m^{-3} day^{-1}*)	*Source*
Nanoflagellates (NF)				
Lake Michigan, USA	Oligotrophic	—	0.8–8.4	Carrick *et al.* (1992)
Heterotrophic NF			2.68	
Phototrophic NF			3.58	
Microflagellates			2.55	
Lake Biwa, Japan	Mesotrophic	—	1.2–8.4	Nagata (1988)
Lake Constance, Germany (5-yr average)	Mesotrophic	2.9–7.4	1.4–3.5	Weisse (1997)
Ciliates				
Lake Constance, Germany	Mesotrophic	—	1.4–2.1	Müller (1989)
Lake Ontario, USA–Canada	Oligotrophic	0.1–35	0–38	Taylor and Johannsson (1991)
Lake Michigan, USA	Oligotrophic	—	7.83	Carrick *et al.* (1992)

[a] Multiply the mg C figures by 2 in order to approximate mg of biomass (ash-free dry weight) for comparison to data of Tables 16-17 and 16-20.

TABLE 16-20 Examples of Productivity of Herbivorous and Predatory Forms of Zooplankton

Type/species	*Lake/general productivity*	*Period of investigation*	*Production estimates[a]* $g\ m^{-3}\ day^{-1}$	*Production estimates[a]* $kcal\ m^{-2}\ day^{-1}$	*Assimilation % of phyto-plankton production*	*Average biomass turnover time (days)*	*Source*
Filter feeders							
Cladocera							
Daphnia hyalina	Eglwys Nynydd Reservoir, Wales; eutrophic	Annual, 1970	0.57	0.468	1.7	21.3	George and Edwards (1974)
		Annual, 1971	0.32			15.9	
	Lake Constance, Germany; mesotrophic	Season (204 days)	0.07–0.1		6.3	29.3	Geller (1985)
	Ardleigh Reservoir, England; eutrophic	Annual, 1981	0.023			16.3	Mason and Abdul-Hussein (1991)
		Annual, 1982	0.009			19.5	
D. parvula	Severson Lake, MN; eutrophic	Annual	0.010	0.102	0.15	—	Comita (1972)
D. galeata	Lake Constance, Germany; mesotrophic	Season (204 days)	0.07–0.1		4.5	15.7	Geller (1985)
	Lake Esrom, Denmark	Annual		0.329	5.7		Hamburger (1986)
	Lake Esrom, Denmark	Annual	0.0059			10.2	Petersen (1983)
D. galeata mendota	Sanctuary Lake, PA, eutrophic reservoir	May–Nov., 1966	0.407	3.026			Cummins *et al.* (1969)
		May–Nov., 1967	0.030	0.223			
	Canyon Ferry Reservoir, MT, eutrophic	April–Sept.	0.114	0.471	4.4	10.0	Wright (1965)
D. schoedleri	Canyon Ferry Reservoir, MT, eutrophic	April–Sept.	0.227	0.943	8.9	6.7	Wright (1965)
D. longispina	Lake Sevan, southern Russia	Annual	0.006			58.9	Meshkova (1952) in Winberg (1971)
D. lumholzi	Lake Samsonvale, Australia	Annual	0.325			7.6	King and Greenwood (1992)
D. rosea	Pond, Japan	Annual	0.044			11.6	Iwakuma *et al.* (1989)
Bosmina longirostris	Severson Lake, MN	Annual	0.007	0.071		—	Comita (1972)
	Ardleigh Reservoir, England; eutrophic	Annual, 1981	0.0022			18.4	Mason and Abdul-Hussein (1991)
		Annual, 1982	0.0020			18.6	
B. longirostris and *B. coregoni*	Sanctuary Lake, PA	May–Nov., 1966	0.183	1.361			Cummins *et al.* (1969)
		May–Nov., 1967	0.067	0.498			
B. meridionalis	Lake Okaro, New Zealand	Annual	0.007			18.7	Forsyth and James (1991)
Ceriodaphnia reticulata	Sanctuary Lake, PA	July–Nov.	0.031	0.154			Cummins *et al.* (1969)
C. dubia	Lake Okaro, New Zealand	Annual	0.004			6.7	Forsyth and James (1991)
Chydorus sphaericus	Sanctuary Lake, PA	July–Aug., 1966	0.004	0.020			Cummins *et al.* (1969)
		July–Aug., 1967	0.047	0.233			
Alona affinis	River Thames, England	Annual	0.361			11.7	Robertson (1995)
Leydigia leydigi	Benthic populations in sediments,		0.151			15.7	
Diaparalona rostrata	production per m^2		0.173			16.4	
Cladocera	Naroch Lake, Russia	May–Oct.	0.0026	0.117		13.7	Winberg *et al.* (1970)
	Myastro Lake, Russia	May–Oct.	0.015	0.403		10.9	
	Batorin Lake, Russia	May–Oct.	0.033	0.484		10.5	

(*continues*)

TABLE 16-20 (Continued)

Type/species	Lake/general productivity	Period of investigation	Production estimates[a] g m^{-3} day^{-1}	Production estimates[a] kcal m^{-2} day^{-1}	Assimilation % of phyto-plankton production	Average biomass turnover time (days)	Source
Copepods							
Cyclops strenuus	Buttermere, England; oligotrophic	Annual	0.0004				Smyly (1973)
	Rydal Water, England; eutrophic	Anuual	0.0005				
	Grasmere, England; eutrophic	Annual	0.0006				
	Esthwaite Water, England; eutrophic	Annual	0.0017				
	Lake Sevan, southern Russia	Annual	0.0007			79.3	Meshkova (1952) in Winberg (1971)
C. vicinus	Eglwys Nynydd Reservoir,	Annual, 1970	0.13			22.5	George (1976)
	Wales, eutrophic	Annual, 1971	0.12			19.4	
	Lake Constance, Germany; mesotrophic	Annual	0.003			24.3	Wölfl (1991)
Calamoecia lucasi	Lake Otota, New Zealand; oligotrophic	Annual	0.006			11.3	Green (1976)
Boeckella dilatata	Lake Hayes, New Zealand; eutrophic	Annual	0.032			6.4	Burns (1979)
B. minuta	Lake Samsonvale, Australia	Annual	0.27			12.5	King and Greenwood (1992)
Eudiaptomus graciloides	Naroch Lake, Russia	May–Oct.	0.0010	0.044		24.7	Winberg *et al.* (1970)
	Myastro Lake, Russia	May–Oct.	0.0065	0.174		20.2	
	Batorin Lake, Russia	May–Oct.	0.0070	0.104		15.4	
	Lake Esrom, Denmark	Seasonal	—	0.10		6.8	Bosselmann (1975)
		Annual		0.178	3.4		Hamburger (1986)
Mesocyclops edax	Severson Lake, MN	Annual	0.0046	0.045		—	Comita (1972)
Diaphanosoma lecuhten-bergianum			0.0067	0.067		—	
Diaptomus siciloides			0.0342	0.341	0.2	—	
D. connexus	Waldsea Lake, Saskatchewan; mesotrophic	Annual	0.29	—		—	Swift and Hammer (1979)
Argyrodiaptomus	Broa Reservoir, Brazil *furcatus*	Annual	0.021			10.0	Rocha and Mat sumura-Tundisi (1984)
Calamoecia lacasi	Lake Okaro,	Annual New Zealand	0.004			69.9	Forsyth and James (1991)
Arcthodiaptomus (2 species)	Lake Sevan, southern Russia	Annual	0.0014	—		162	Meshkova (1952) in Winberg (1971)
Lovenula africana	Lake Nakuru, Kenya	Annual	0.055–0.080		21.0	21.8	Vareschi and Jacobs (1984)

Rotifers						
Keratella quadrata	Severson Lake, MN	Annual	0.0021	0.021		—	Comita (1972)
K. cochlearis			0.0007	0.0074		—	
Filinia longiseta			0.0011	0.0112		—	
Brachionus sp.			0.0075	0.0752		—	
Polyarthra sp.			0.0010	0.0103		—	
Rotifers	Naroch Lake, Russia	May–Oct.	0.0027	0.120		1.7	Winberg *et al.* (1970)
	Myastro Lake, Russia	May–Oct.	0.0024	0.065		3.9	
	Batorin Lake, Russia	May–Oct.	0.0105	0.156		2.6	
Rotifers	Lake Constance, Germany	Annual, 1984	0.003		3.1	6.7	Pauli (1991)
		Annual, 1985	0.0015			7.1	
Predatory feeders							
Cladocera							
Leptodora kindtii	Sanctuary Lake, PA; eutrophic, shallow	May–Nov., 1966	0.003	0.022		—	Cummins *et al.* (1969)
		May–Nov., 1967	0.013	0.097		—	
	Lake George, NY; deep oligotrophic	June–Aug.	0.006	—		—	LaRow (1975)
Cladocera	Naroch Lake, Russia	May–Oct.	0.0003	0.013		11.3	Winberg *et al.* (1972)
	Myastro Lake, Russia		0.0009	0.023		3.9	
	Batorin Lake, Russia		0.0002	0.003		10.8	
Copepods							
Cyclops sp.	Naroch Lake, Russia	May–Oct.	0.0008	0.034		9.7	
	Myastro Lake, Russia		0.0023	0.062		19.4	
	Batorin Lake, Russia		0.0094	0.140		14.2	
Rotifers							
Asplanchna priodonta	Naroch Lake, Russia	May–Oct.	0.0014	0.061		2.9	
	Myastro Lake, Russia		0.0061	0.163		2.5	
	Batorin Lake, Russia		0.0105	0.156		3.2	
Asplanchna sp.	Severson Lake, MN	Annual	0.0031	0.031		—	Comita (1972)
Synchaeta sp.			0.00009	0.0009		—	
Brachionus dimidiatus	Lake Nakuru, Kenya	Annual	0.109		15.0	2.2	Vareschi and Jacobs (1984)
B. plicatilis			0.288		15.0	1.8	
Insect larvae							
Chaborus punctipennis			0.0001	0.001		—	Comita (1972)
C. flavicans	Pond, Japan	Annual	0.022			15.5	Iwakuma *et al.* (1989)
Mysids							
Neomysis mercedis	Muriel Lake, British Columbia	Annual (2 yr)	0.073			9.4	Cooper *et al.* (1992)
	Kennedy Lake, British Columbia	Annual (2 yr)	0.045			9.3	
Mysis relicta	Lake Mjøsa, Norway	Annual	0.25			50	Kjellberg *et al.* (1991)
	Laurentian Great Lakes, USA	Annual	ca. 0.2			30–45	Sell (1982)
Amphipod							
Pontoporeia affinis	Lake Erken, Sweden	Annual	0.118			31	Johnson (1987)

[a] Conversions estimated using the caloric mean of microconsumers (Cummins and Wuycheck, 1971) when mean depths available.

TABLE 16-21 Correlations between Biomass and Production of Major Components in Fresh Waters[a]

Component	*Sample number* (N)	*Correlation coefficient* (r)
Phytoplankton	27	0.77
Herbivorous zooplankton	26	0.84
Carnivorous zooplankton	25	0.84
Herbivorous benthic organisms	15	0.61
Carnivorous benthic organisms	14	0.72

[a] After data of Brylinsky (1980). All values significant at the 99% level.

productivity.[9] The filter-feeding zooplankton, which accounted for a total productivity of 0.22 g m^{-1} day^{-1} on an annual mean basis, consisted of primarily *Eudiaptomus* (60%), with a turnover time of about 36 days (annual mean; 24 days during the ice-free period). The result, then, is a considerably modified trophic pyramid in which the transfer of energy to higher food levels is short-circuited, to use the words of Nauwerck, via the bacteria and particulate detrital organic matter (Fig. 16-41).

Utilization of bacteria and particulate detrital organic matter was demonstrated in numerous subsequent studies, even though the caloric content of the detrital material is relatively low. In a detailed analysis of carbon pathways among the plankton of a humic lake, most of the particulate organic carbon available for zooplankton was detritus of relatively low nutritional value (Hessen *et al.*, 1990). Ingestion of detrital particles was, however, the major source of food and provided 46–82% of body carbon to the zooplankton. Bacterial organic carbon supplied some 11–42% and phytoplankton organic carbon 6–19% of body carbon to the zooplankton. Pelagic community respiration was high, amounting to some two-thirds of the production.

In another important study, the fate of primary production in the pelagic food web of a mesotrophic lake in midsummer was determined (Lyche *et al.*, 1996). Specific carbon assimilation rates of bacteria were twice those of phytoplankton, and 70% of the bacterial production was fueled by phytoplankton exudates. About half of the gross primary production was lost as excreted dissolved organic carbon and ca. 10% was respired. Only 20% of the net primary production was assimilated by consumers; the remainder was lost as dissolved and particulate detritus or by sedimentation. Of the net bacterial production, a third accumulated as bacterial biomass, a third was assimilated by bacterivores, and a third was lost as detrital dissolved organic matter. Less than 2% of the gross primary production was incorporated into consumer biomass. Clearly, most organic carbon of the microbial food web did not act as a source of carbon to higher consumers but was utilized in the detrital food chain and respired by microbial heterotrophs.

In similar fashion, the greater part of the herbivorous zooplanktonic productivity is not utilized by predatory zooplankton and fish, but instead enters the nonpredatory detrital pathways of decomposition and is partially reutilized by filter feeders as bacteria and detritus. Therefore, Lindeman's classic conceptualization of the generalized lacustrine food-cycle relation-

[9] Certain computational errors have been found in the original paper but these do not alter the general conclusions.

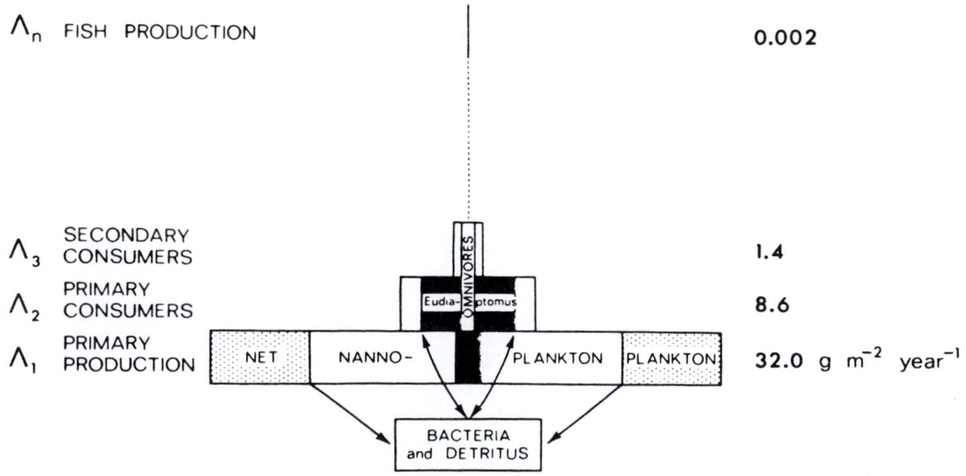

FIGURE 16-41 Diagrammatic representation of the trophic relationships in Lake Erken, Sweden. (Modified from Nauwerck, 1963.) Productivity estimates in g dry weight per m² per year of pelagic zone only.

ships must be modified to emphasize the importance of nonpredatory detrital and bacterial pathways (Fig. 16-42); these pathways were underestimated because inadequate data were available when the cycle was formulated (cf. Rich and Wetzel, 1978). This change does not greatly alter the fundamental contribution of Lindeman, but instead properly emphasizes the major contributions of bacteria and detritus in the trophic relationships of primary consuming animals and major losses of organic matter through decomposition that occur without animal consumption (cf. Chap. 22).

Other comparisons of the proportions of herbivorous zooplankton utilized by predators demonstrate similar results. In detailed studies of nine very different lakes in the Soviet Union, the biomass of predators varied from 6.5 to 52% of the total biomass of zooplankton, averaging about 30% (Winberg, 1970). The productivity of the filter-feeding zooplankters is distinctly higher than that of the predacious zooplankters (Table 16-20). Variation is great, however, particularly in relation to temperature, food quantity and quality, and other factors. Comparisons over two or more years with good sampling show the extent of variation that can occur with the same species in the same lake (Table 16-20; Fig. 16-43). These differences are much greater than can be attributed to sampling and measurement errors. Perhaps the lowest production rates of zooplanktonic populations in a permanent lake are those of polar Char Lake at latitude 74° in the Canadian Arctic (Rigler, 1974). The dominant zooplankters *Limnocalanus macrurus* and *Keratella* produced 261 mg ash-free dry weight per m^2 per year and 3 mg m^{-2} yr^{-1}, respectively. This is equivalent

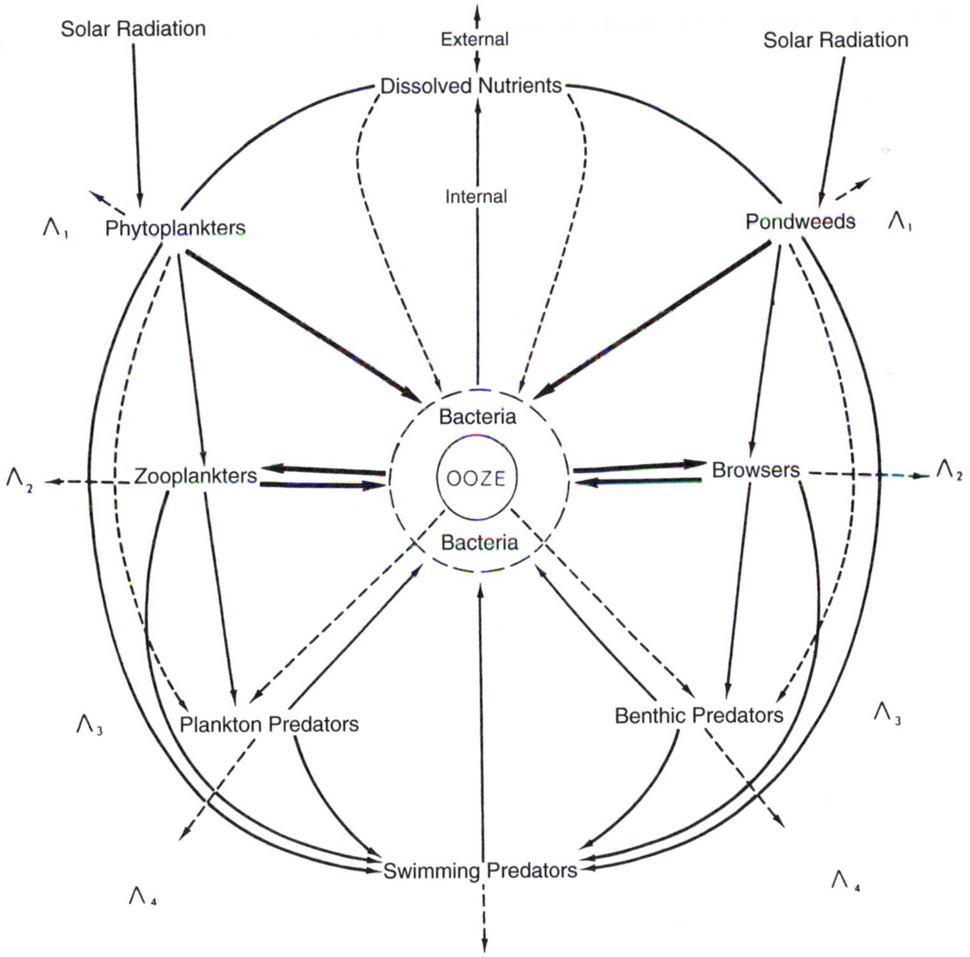

FIGURE 16-42 Generalized lacustrine food-cycle relationships after Lindeman (1942). Added darkened arrows indicate pathways ignored or underestimated in Lindeman's actual calculations, a consequence of accepting existing data on chemical assay of crude fiber in place of direct assimilation or egestion measurements. (From Wetzel, R. G., Rich, P. H., Miller, M. C., and Allen, H. L.: Metabolism of dissolved and particulate detrital carbon in a temperate hardwater lake. *Mem. Ist. Ital. Idrobiol.* 29(Suppl.):185–243, 1972.)

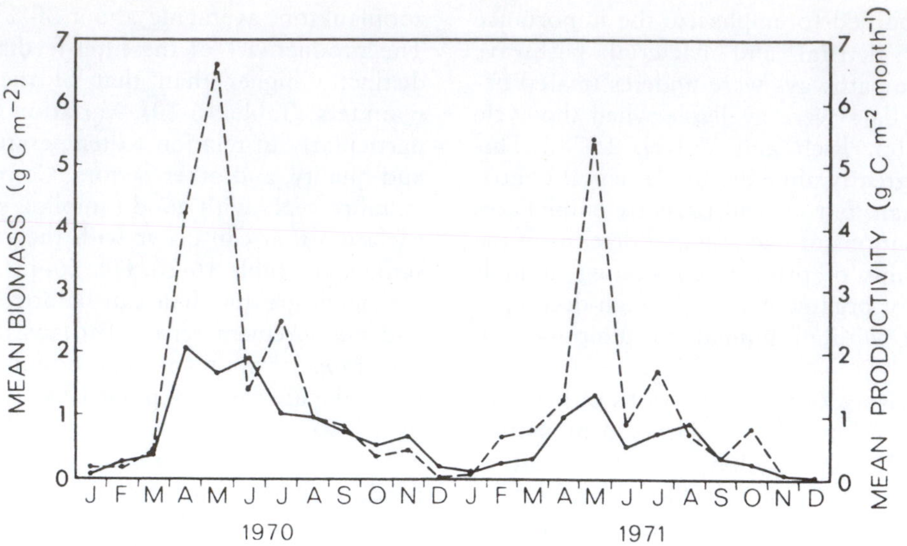

FIGURE 16-43 Seasonal monthly mean biomass (———) and productivity (- - - -) of *Daphnia hyalina* in g C m^{-2} month^{-1}, calculated from daily measurements, of Eglwys Nyndd, Great Britain. (Drawn from data of George and Edwards, 1974.)

to an annual average of about 0.00006 g m^{-3} day^{-1}, an exceedingly low value. In the particular case of *Daphnia hyalina* in a eutrophic reservoir (Fig. 16-43), maximum biomass was found nearly two months before highest numerical density in 1970, but numerical and biomass maxima were coincident the following year (George and Edwards, 1974). Maximum spring biomass occurred during blooms of green algae in late April and early May, when many *Daphnia* survived and grew for some time after reaching maturity. Increases in production coincided with, or lagged somewhat behind, increases in biomass. In the second year, the production was about one-half that of the first year, and the average biomass turnover time also decreased (Table 16-20).

From an intensive analysis of the zooplankton productivity of oligotrophic Mirror Lake, New Hampshire (Makarewicz and Likens, 1979), two major relationships emerged. (a) Rotifers are a major component in energy transfer and intrasystem nutrient cycling. The major importance of rotifers in the transfer of energy is evident in a number of lakes of varying trophic status, from oligotrophic to eutrophic (Table 16-22). One-third of the annual phosphorus flux through the zooplankton of Mirror Lake was by the rotifers. (b) The productivity of the different species of zooplankton, shown in Figure 16-44, indicates how the interacting populations are structured. The population positions or niches are dispersed in relation to complex gradients of time, depth, food, reproduction, and predation (cf., e.g., Zaret, 1975, 1980; Kerfoot and Pastorak, 1978; McNaught, 1978; Seitz, 1980; Kerfoot, 1980a; Sommer *et al.*, 1986; Lehman, 1988). Separation of the niche hyperspace with relatively small regions of species overlap minimizes interspecific competition. Throughout the discussion in this chapter, emphasis has been on the general similarities and often subtle but important differences and separations that occur in feeding, food utilization, reproduction, growth, and behavioral characteristics, the temporal (daily and seasonal) and spatial differences in distribution and growth, seasonal changes in morphology, and various characteristics of predator avoidance and predation effects. All of these factors contribute to the large diversity of population competitive interactions that have evolved to permit coexistence in limnetic zooplankton communities.

The efficiency of energy transformations between components of animal populations is an area of intensive investigation. At present, few cases have been studied in detail among natural populations; much of the

TABLE 16-22 Percentage of Production of Rotifer, Cladoceran, and Copepod Zooplankton in Various Lakes[a]

Lake	*Rotifers*	*Cladocera*	*Copepods*
Mirror, NH	39.8	40.9	19.3
Krivoe, Russia	15.2	36.1	48.7
Krugloe, Russia	19.2	71.8	8.9
Naroch, Russia	43.5	31.2	25.2
Myastro, Russia	23.9	47.6	28.6
Batorin, Russia	29.3	45.5	25.1
239, Ontario	67.2	5.4	27.4

[a] After data of Makarewicz and Likens (1979), Alimov *et al.* (1970), Winberg *et al.* (1970), and Schindler (1970).

TABLE 16-23 Exemplary Estimates of Efficiencies of Food Utilization by Various Animals[a]

Organisms	*% of ingested food utilized in:* Egestion	Assimilation	Respiration	Growth and reproduction	*% of assimilated energy expended in:* Respiration	Growth and reproduction	Source
Zooplankton							
Daphnia pulex	69–86	14–31	4–14	10–17	27–44	56–73	Richman (1958)
Daphnia ambigua	71.8	28.2	17.1	8.4	61	39	Lei and Armitage (1980)
Ceriodaphnia reticulata	—	10.6	1.8	—	—	—	Czeczuga and Bobiatynska-Ksok (1970)
Simocephalus vetulus							Klekowski (1970)
Juveniles (♀)	27.6	72.4	19.5	52.9	26.9	73.1	
Reproducing (♀)	68.3	31.7	11.2	20.5	35.3	64.7	
Leptodora kindtii	60.0	40.0	—	—	92.7	7.3	Cummins *et al.* (1969); Moshiri *et al.* (1969)
Mesocyclops albidus	—	20–75	Approx. 20	Approx. 25	Approx. 50	Approx. 50	Klekowski and Shushkina (1966)
10 Herbivores	52.4	47.6	40.1	7.5	71–82	18–29	Comita (1972)
Benthic animals							
Asellus aquaticus	69.7	30.3	24.7	5.6	81.8	18.2	Klekowski (1970)
Lestes sponsa	63.4	36.6	13.2	23.5	36.0	64.0	Klekowski *et al.* (1970)
Fish							
Phytophagous carp, *Ctenopharyngodon*	86.0	14.0	12.2	1.9	86.0	14.0	Fischer (1970)
Predatory perch, *Perca fluviatilis*	64.2	35.8	16.2	19.5	45.5	54.5	Klekowski *et al.* (1970)

[a] Methods of analysis and experimental conditions vary greatly and are comparable only approximately.

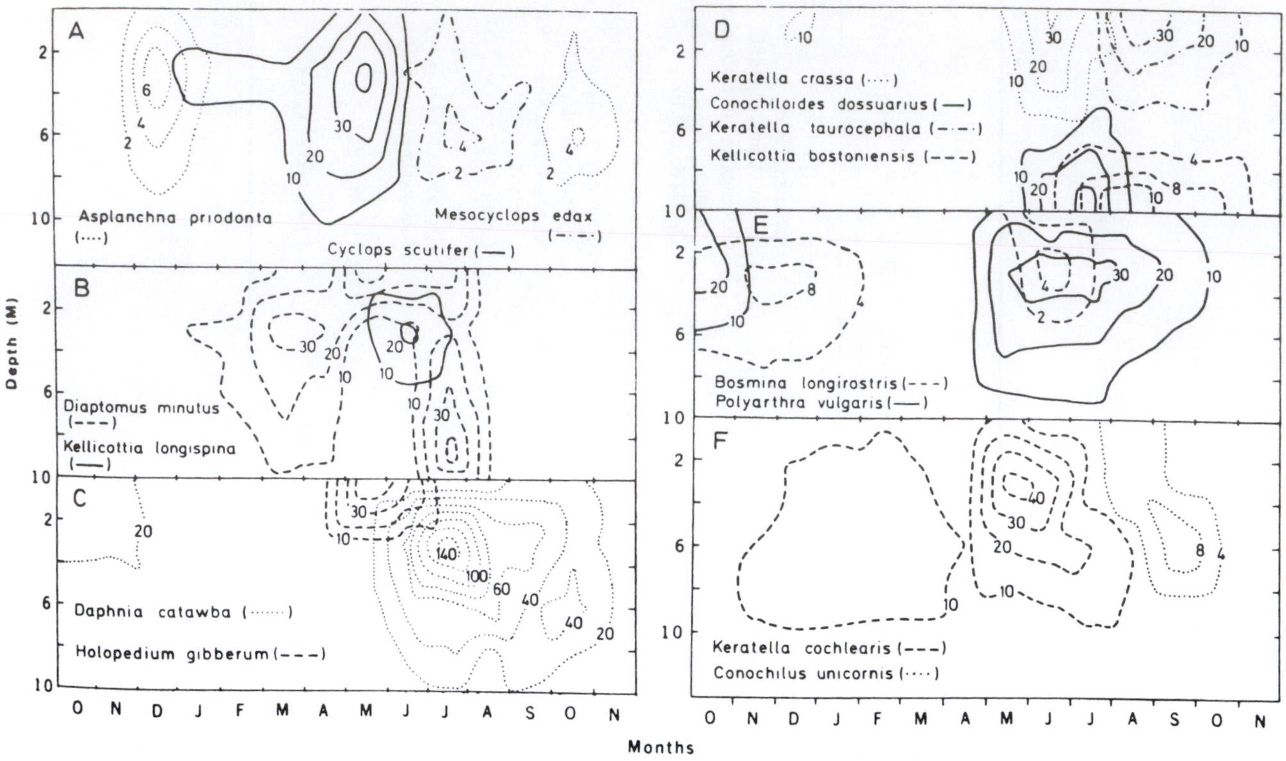

FIGURE 16-44 Productivity response surfaces (from mean production values, μg dry weight liter^{-1} month^{-1}) of zooplankton populations, Mirror Lake, New Hampshire. (a) Predators: rotifer *Asplanchna* and two copepods; (b) macroconsumer herbivores: rotifer *Kellicottia* and copepod *Diaptomus*, (c) microconsumer herbivores: Cladocerans; (d) microconsumer herbivores: rotifers; (e) microconsumer herbivores: rotifer *Polyarthra* and small cladoceran *Bosmina*; (f) microconsumer herbivores: rotifers. (From Makarewitz, J. C., and Likens, G. E.: Niche analysis of a zooplankton community. *Science 190*(4218):1000–1003, 1975. Copyright 1975 by the American Association for the Advancement of Science.)

necessary information on ingestion, respiration, and egestion is determined under controlled laboratory conditions that have been applied variously to natural populations. A certain percentage of the food ingested is not utilized. As a percentage, this egestion rate varies greatly with age, the quantity and quality of food available, and other factors within an observed range of about 30–90% loss (Table 16-23). How much of this egested detrital material is reingested by coprophagy is unclear, at least among the pelagic organisms. Consumption of feces after varying periods of colonization and decomposition by bacteria and fungi is a fairly common phenomenon among benthic fauna of lakes and streams.

The rates of assimilation and respiratory loss for biochemical maintenance also vary widely with environmental conditions (Table 16-23). Assimilation efficiency is generally greater among young animals and decreases with age; average values, mostly determined under experimental conditions better than those *in situ*, are in the range of 10–50%. Assimilation efficiency of course decreases with increasing food concentrations (e.g., Sharma and Pant, 1984). Respiratory costs are high and, as would be anticipated, increase greatly relative to assimilation rates under adverse conditions. For example, the population respiration of the dominant zooplankter of the Canadian Arctic Char Lake, with a maximum water temperature of about 4°C for one month of the year, far exceeded net production (Table 16-24). The ratio of production to assimilation was about 13%.

TABLE 16-24 Mean Production and Respiration of the *Limnocalanus macrurus* Population in Polar Char Lake, Canadian Arctic, over a 4-Yr Period[a]

Production and respiration	*Mean mg m^{-2} yr^{-1}*	*% of total*
Growth	220	11.1
Eggs	5.6	0.3
Exuviae	34.9	1.8
Total production	261	13.1
Respiration	1660	83.4
Assimilation	1990	

[a] After data of Rigler *et al.* (1974).

TABLE 16-25 Utilization and Transfer of Energy from Trophic Levels of Several Freshwater Systems, Expressed in g cal cm^{-2} yr^{-1}[a]

	Lake Mendota, Wisconsin				*Temperate cold spring*			*Silver Springs, Florida*				*Cedar Bog Lake, Minnesota*			*Severson Lake, Minnesota*		
Parameter[b]	Λ_1	Λ_2	Λ_3	Λ_4	Λ_1	Λ_2	Λ_3	Λ_1	Λ_2	Λ_3	Λ_4	Λ_1	Λ_2	Λ_3	Λ_1	Λ_2	Λ_3
Ingestion (I)	118,872	55.4	3.4	0.3	1,095,000	—	—	1,700,000	—	—	—	118,872	19.6	3.4	89,586	57.7	—
Assimilation (A)	501.7	44.3	3.1	0.3	710	2318	242	20,810	3368	383	21	120	16.8	3.1	581.4	27.4	1.2
Egestion (nonassimilation)	118,370	11.1	0.3	0.0	1,094,290	—	—	1,679,190	—	—	—	118,752	2.8	0.3	89,005	30.2	—
Respiration (R)	125.3	16.9	1.8	0.2	55	1746	89	11,977	1890	316	13	30	6.4	1.8	309.1	23.1	0.8
Net productivity ($A-R$)	376.4	27.4	1.3	0.1	655	576	155	8833	1478	67	6	90	10.4	1.3	272.4	4.3	0.4
Production	55.4	3.4	0.3	0.0	(655)	208	—	—	—	—	—	19.6	3.4	0.0	57.7	—	—
Decomposition losses	321.0	24.0	1.0	0.1	—	—	—	—	—	—	—	70.4	7.0	1.3	—	—	—

[a] From data of Juday (1940), Teal (1957), Odum (1957), and Lindeman (1942), as interpreted by Koslovsky (1968), and from Comita (1972).
The lake systems are planktonic only and do not consider littoral and allochthonous inputs, and therefore account for as little as 10% of total ecosystem metabolism.

[b] I = insolation for producers or food ingestion for organisms. Egestion includes nonutilized light and egestion. R includes all energy losses; also urine in higher organisms. Production refers here to that portion of productivity passed on to next trophic level, some of net productivity being dissipated in decomposition, tissue accumulation, and loss from the system.

TABLE 16-26 Phytoplankton Photosynthetic Efficiency and Secondary Production Energy Transfer Efficiencies[a]

Component	*Sample size* (N)	*Mean percentage efficiency*	*Range of efficiencies* (%)
Phytoplankton	93	0.34	0.002–1.0
Herbivorous zooplankton	27	7.1	0.10–27.4
Carnivorous zooplankton	24	1.2	0.17–5.0
Herbivorous benthic animals	17	2.3	0.16–11.1
Carnivorous benthic animals	16	0.3	0.35–1.8

[a] After data of Brylinsky (1980).

Of the production that is utilized for growth and reproduction, usually a small portion is utilized in the diapause stages such as in overwintering eggs. Most of the production enters the nonpredatory pool of detrital organic matter and is decomposed by microorganisms. A variable proportion is utilized by predators. The percentage of the primary consumers to appear as net production of secondary consumers also varies but generally averages around 5–10% or less on an annual basis (e.g., Comita, 1972; Lyche *et al.*, 1996). As was discussed earlier, however, at certain periods of the year, the impact of herbivores and predators can be quite significant, and they can literally decimate populations of the primary producers.

In discussing the roles of zooplankton in the general trophic-level concept (cf. Chap. 8), it should be recalled that trophic levels consist of population groups of different species of organisms that utilize the same mode of nutrition. However, many of the species change their mode of feeding to another trophic level in the course of their life cycles. As was emphasized earlier in this chapter, animal nutrition is strongly dependent upon food density, quality, and many other factors. As a result, the efficiency of food utilization for growth and maintenance varies with the rate of food consumption and population densities. These efficiencies are only reasonably accurate when the populations studied are in a fairly fixed state of equilibrium (Slobodkin, 1960); however, this condition rarely exists among natural populations, since they frequently undergo large fluctuations. Under controlled conditions, metabolic efficiencies of food utilization by particular species have rigorous meanings from a physiological point of view (e.g., Lucas, 1996). Extrapolation of these efficiencies to the complex dynamics of natural systems, however, can only be done in a general way. The evaluation of ecological efficiencies among trophic levels is really only a theoretical way of viewing the interrelationships among organisms. Such an evaluation loses precision as soon as it is extended beyond measurements of solar radiation in studies of primary productivity. Such analyses can be instructive, however, if one can keep in mind the limitations imposed by their simplicity. By way of examples, five freshwater systems are compared in Table 16-25, based on mostly incomplete, rather limited data. In a detailed analysis of the numerous types of efficiency relationships developed by many workers, Kozlovsky (1968) used most of these limited data to illustrate changes taking place in the passage from one trophic level to the next. Some of his conclusions follow, the most important of which is that when the net productivity and assimilation of one level are compared to the net productivity and assimilation of the previous level, the efficiency of transfer is practically constant at about 10% (Table 16-26). In more physiologically based comparisons of food utilization in animals in relation to food ingested, assimilation and respiration increase at higher trophic levels. It then follows that in relation to ingestion and assimilation in animals, production decreases at higher trophic levels.

XX. ZOOPLANKTON AMONG AQUATIC ECOSYSTEMS

Comparative examination of zooplankton communities and their productivity among river, reservoir, and lake exosystems indicate several consistent characteristics (Table 16-27). Although smaller taxa dominate in river ecosystems in comparison to zooplankton in reservoirs and lakes, diversity is large, particularly among the protists and mesozooplankton. The separation of the protistan mesozooplankton from the larger macrozooplankton of rotifers, copepods, and cladocerans becomes most distinct in natural lakes. Zooplankton of rivers and reservoirs tend to rely to a greater extent on detrital particles and associated attached microbes than is the case among zooplankton of natural lakes, where feeding is largely on phytoplankton (Table 16-27). The growth of zooplankton is often limited in rivers and reservoirs of short retention times by water movements, food availability, and relatively high turbidity. The growth of zooplankton in natural lakes is high but fluctuations are strongly influenced by both resource availability and changes in predation pressures.

TABLE 16-27 Comparative Characteristics of Zooplankton among River, Reservoir, and Natural Lake Ecosystems[a]

	Rivers	*Reservoirs*	*Natural lakes*
Zooplankton community structure	Dominated by small forms, such as protists, bosminids, and juvenile copepods; dominant planktonic species have relatively rapid life cycles or are predominantly benthic	Gradient between smaller forms with protistan dominance in riverine portions to common lake forms in lacustrine portions	Strong separation into protistan-dominated mesozooplankton and the macrozooplankton of mostly rotifers, copepods, and cladocerans
Zooplankton community development	Low development; source areas from lakes and floodplain pools; mortality high from molar action and reproduction rates longer than displacement times downstream	Maximum development common in transitional zone; horizontal patchiness high	Vertical and seasonal gradients predominate; horizontal patchiness moderate and coupled to hydrodynamics; relatively high stability and abundance
Zooplankton feeding	Particulate detritus commonly augments algae from planktonic and benthic sources	Particulate detritus, including adsorbed DOM, variably augments phytoplankton as a source of nutrition	Phytoplankton is a predominant food source; consumption of food resources strongly influenced by predator regulation of size of dominating zooplankton
Zooplankton growth	Low; greatest in intermediate- or high-order rivers; limited by unidirectional flow, high turbidity, and low dissolved oxygen	Low to moderate; variable with changing food availability and quality, often limited during periods of high turbidity	Moderate to high; extreme fluctuations mediated by both resource availability and predation pressure variances

[a] Assembled from the syntheses on rivers by Ryder and Psendorfer (1980) and on reservoirs and lakes by Wetzel (1990) and many sources cited in these papers.

XXI. SUMMARY

1. Zooplankton of inland waters are dominated by four major groups of organisms: (a) protists that include protozoa and functionally similar protists, in particular the heterotrophic flagellates, (b) rotifers, and two subclasses of crustaceans, (c) the cladocerans, and (d) the copepods.
2. Protistan zooplankton are the most important microbial consumers and have major functions in organic carbon utilization and nutrient recycling. The instantaneous biomass of planktonic protists is small in comparison to other zooplankton, but generation rates of protists are high.
   a. Flagellates are the most abundant component of the protozooplankton and are often mixotrophic, feeding on bacteria as well as fixing carbon photosynthetically.
   b. Ciliates are largely holozoic and have high feeding rates on bacteria, algae, particulate detritus, and other protists.
   c. Sarcodine protozoa only occasionally reach modest abundance, and their production is usually <10% of combined ciliate and flagellate production.
   d. Many pelagial protozoa are meroplanktonic in that only a portion, usually in the summer, of their life cycle is planktonic. These forms spend the rest of their life cycle in the sediments, often encysted throughout the winter period.
   e. Mixotrophic ciliates and flagellates usually occur in the upper water strata. Migration is common among flagellates on a diurnal basis in relation to light, food, and dissolved oxygen conditions.
   f. Many protists feed on bacteria-sized particles and picophytoplankton and thereby utilize a size class of bacteria, algae, and particulate detritus generally not as effectively utilized by larger zooplankton. Bacterivory by the protozooplankton, particularly the flagellates, is likely a dominant mortality factor for bacterioplankton (cf. Chaps. 17 and 23) and frequently removes 20–60% of the algal picoplankton (<2 μm in size). Assimilation efficiency for growth and reproduction of protists is approximately 40%.
   g. Flagellate and ciliate abundance and production is directly correlated to those of phytoplankton during the winter, spring, and summer. Ciliate predation on flagellates is a major regulatory mechanism, particularly during summer and among more productive lakes.
3. Flagellate growth rates are much greater than those of ciliates. In addition, flagellate densities are greater by a factor of ca. 1000 in comparison to ciliates. As a result the heterotrophic flagellate production equals or exceeds that of ciliates.
4. The "microbial loop" is a model of pathways of carbon and nutrient cycling through microbial components of pelagic aquatic communities (Fig. 16-7). Many direct links exist among algae, bacteria, and heterotrophic protists. Protistan bacterivores can function as decomposers and efficient nutrient remineralizers. Some of the fixed organic carbon can be converted into larger-sized microorganisms that may be consumed by larger metazoans and zooplankton. Under many conditions, the microbial loop functions as both a carbon and nutrient "sink", by which organic matter is respired and is not directly available to higher animals and nutrients are recycled.
5. Although most rotifers are sessile and are associated with the littoral zone, some are completely planktonic; these species can form major components of the zooplankton.
   a. Most rotifers are nonpredatory and omnivorously feed on bacteria, small algae, and detrital particulate organic matter. Most food particles eaten are small (<12 μm in diameter).
   b. A few rotifers are predatory on protozoa, rotifers, and small crustaceans and can alter their size in response to changes in food size.
6. Most cladoceran zooplankton are small (0.2–3.0 mm) and have a distinct head; the body is covered by a bivalve carapace. Locomotion is accomplished mainly by means of the large second antennae.
   a. Most cladocerans feed on particles filtered from the water by means of setae and hairs on five pairs of legs. Particles collect and move in a ventral food groove toward the mouth.
   b. A few cladocerans are predaceous and seize other zooplankton and detrital particles with prehensile legs.
7. Planktonic copepods consist of two major groups, the calanoids and the cyclopoids. These two groups are separated on the basis of body structure, length of antennae, and legs (Fig. 16-12; Table 16-7).
   a. Cyclopoid copepods are raptorial; they seize food particles and draw them to the mouth. Many cyclopoids are carnivorous on other zooplankton; some are herbivorous on a variety of unicellular and filamentous algae. Locomotion is by movement of appendages, which results in short, rapid swimming movements.
   b. Calanoid copepods swim more continuously in rotary motions, which set up currents that carry

particles to modified structures of the maxillae. Particles that are retained are governed by the capture efficiency of the maxillae and differ among calanoid species.

8. Filtration of particles is the dominant means of food collection by rotifers and cladocerans.
   a. *Filtration or grazing rate* is the volume of water containing particles that is filtered by the animal in a given time, regardless of whether or not the particles are retained or ingested. *Feeding rate* is the quantity of food ingested in a given time.
   b. Many zooplankton possess some capacity for selective feeding.
      i. Among suspension-feeding rotifers, food selectivity occurs in some species by several food-rejection mechanisms (screening of particles by cilia or rejection of particles once ingested).
      ii. Among cladocerans, the size of particles cleared from the water is a function of the morphology of the setae of the moving appendages. Feeding rates commonly stabilize or decrease as concentrations of food particles increase.
   c. Filtering rates tend to increase with both increasing body length and increasing temperatures. The size of particles ingested is generally proportional to body size.
   d. The effectiveness of zooplankton grazing varies greatly seasonally and among lakes. Throughout much of the year, zooplankton grazing only filters a small proportion of the water volume (<15% per day), At certain times of the year, grazing can remove large portions of the phytoplankton and can cause marked reduction in phytoplankton productivity.
   e. The effectiveness of zooplankton grazing on phytoplankton is greatest in lakes of intermediate productivity (mesotrophic). In oligotrophic lakes, nutrient availability dominates phytoplanktonic regulation, and in nutrient-rich eutrophic lakes, phytoplanktonic growth greatly exceeds mortality losses by zooplankton.
   f. Algal species succession can also be altered by intensive, selective (usually size-specific) grazing and concomitant regeneration of nutrients. Certain algae can survive gut passage, and their growth can be enhanced by contact with high nutrient levels within the gut of zooplankton.
9. Assimilation efficiency is variable among zooplankton species but is usually <40–50%.
   a. The efficiency of assimilation increases somewhat with higher temperatures and decreases markedly with increasing food concentrations.
   b. Food quality also influences assimilation efficiencies. Rates of assimilation are low when zooplankton are feeding on detritus particles, higher with bacteria, and generally highest when they are feeding on algae of acceptable size and type.
10. Lipids are dominant energy storage compounds in zooplankton (to 60–70% of their mass). Often food-limited in the summer, lipid reserves are essential to survival. Because lipids are largely of dietary origin and differ among species of algal food sources, the essential fatty acid nutritional quality of phytoplankton can be an important regulatory factor in species and seasonal succession of zooplankton.
11. Reproductive rates and life histories of zooplankton are diverse.
    a. Rotifers and most cladocerans reproduce by diploid female parthenogenesis for many successive generations during the main growing season. Under certain conditions of environmental stress, meiosis occurs, which leads to the production of haploid males and females. The fertilized eggs (resting eggs) are diploid and are encased in a heavy cell wall. These resting eggs often enter diapause, and hatching may be delayed for long periods. Under favorable conditions, resting eggs develop into amictic diploid females.
    b. A number of environmental conditions and mechanisms influence the production of sexual mictic females. Reductions in temperature, reduced food availability under crowded population conditions, and reduced availability of various dietary compounds (e.g., vitamin E) can induce sexual production in certain rotifers.
    c. Rates of growth and development of instars in Cladocera are generally proportional to temperature and are often related to food availability. Food supply is, however, of secondary importance to temperature; increased food supply usually increases the rate of population development by increasing fecundity (number of eggs per brood).
    d. Copepods are bisexual. Fertilized eggs are carried externally in egg sacs.
       i. Temperature determines the rate of egg production, and food availability directly influences clutch size.
       ii. Copepod eggs hatch into small, free-swimming larvae (*nauplii*) and molt successively through six naupliar stages. Nauplii then

enlarge and form *copepodites*, and molt through five additional instars before forming an adult. Development time to complete all of the immature stages is far more extended than in the rotifers and Cladocera and varies greatly among species and under different environmental conditions.

12. Analyses of the rates of egg production and the dynamics of instar populations yield information on growth rates, birth rates, mortality, and duration of generations.
    a. The seasonal distribution of planktonic rotifer populations is complex. Many perennial species exhibit maximal densities in early summer in temperate lakes. Other species are more seasonal and include (i) cold stenotherms with greatest populations in winter and early spring, and (ii) species that develop two or more maxima during summer. Certain rotifer species also tolerate low oxygen concentrations, and populations may be separated spatially as well as seasonally.
    b. Threshold food concentrations, at which population growth is zero, vary greatly among planktonic rotifers. High food concentrations and availability enhance growth, shorten juvenile stages, and accelerate reproduction. Predation is important, and generally rotifers cannot become abundant in the presence of large ($\geq$1.2 mm) cladocera and copepods either because of competition for food items or by mortality from mechanical interference during feeding of the larger zooplankton.
    c. The seasonal succession and population dynamics of Cladocera are also variable among species and under different lake conditions.
        i. Some species are perennial (overwinter as adults rather than as resting eggs) and may exhibit one, two, or more irregular maxima.
        ii. Aestival species have a distinct diapause (resting egg stage) and usually develop one population maximum in the spring, and occasionally a second maximum in the late summer or autumn.
        iii. Temperature increases in the spring enhance rates of molting and brood production; increased food supply increases egg number per brood. Shifts in food quality and availability (often size-restrictive), and especially predation by invertebrates and fish, can cause rapid changes in populations of cladocerans. Mortality by predation is normally highest in the summer.
    d. The life cycle of limnetic cyclopoid copepods is separated into two periods: (i) a period of growth followed by (ii) a period of retarded growth or diapause induced by a combination of decreasing water temperatures, photoperiod, reduced food availability, anoxia, and increased predation. The extent to which the population enters a resting stage, and the instar stage at which diapause occurs, is extremely variable. Similarly, the site where the resting stage occurs (water versus sediment) and diapause duration can vary greatly from lake to lake in arctic, alpine, and temperate areas. Diapausing eggs can remain viable in sediments for decades. In tropical lakes, resting stages have not been observed. The seasonal life cycles of the calanoid copepods are usually longer than those of the cyclopoids.
        i. Most cyclopoid copepods are carnivorous and influence the population dynamics of the other copepods by predation. They influence their own dynamics by cannibalism, especially of juveniles.
        ii. Competition between copepods is reduced by (a) variations in timing and duration of diapause, (b) seasonal and vertical (spatial) separation, and (c) differences in utilization of available food particles.
13. Zooplankton of rivers are similar to those of lakes but are dominated by small forms and are negatively correlated with water velocities. As river ecosystems enter reservoir impoundments, protistan and larger zooplankton densities, biomass, and biodiversity usually increase. Clay turbidity, common among reservoirs, often suppresses zooplankton growth and productivity by direct feeding interference and reduced food availability.
14. Many zooplankton, particularly the Cladocera, exhibit marked diurnal vertical migrations.
    a. Most species migrate upward from deeper strata to more surficial regions as darkness approaches and return to the deeper areas at dawn.
    b. Rates of movement over distances in excess of 50 m in certain clear lakes are variable from $<$2 m per hour among rotifers to $>$20 m per hour by certain Cladocera and copepods.
    c. Light intensity is the primary stimulus of vertical migration.
    d. Grazing rates of suspension feeders are usually several times greater during the dark period when they have migrated to upper water strata. The protein content of algal food is commonly higher during the dark period of the diurnal cycle.

e. Therefore, vertical migration upward during darkness and downward during the lighted period of the day is common behavior among cladocerans and copepods. Because fish predation is a visual process, such vertical migration occurs primarily among zooplankton that are most visible to fish, particularly those of large size and females that are carrying eggs. Although changes in relative light intensity are fundamental to the induction of vertical migration, the amplitude of phototactic response varies with the presence and activity of fish. Evidence indicates that fish can release chemical compounds (kairomones) that initiate and influence the vertical migratory behavior of zooplankton.

15. The horizontal spatial distribution of zooplankton in lakes is often uneven and patchy.
    a. Pelagial cladocerans and copepods also migrate away from littoral areas (avoidance-of-shore movements) by behavioral swimming responses to angular light distributions.
    b. In many cases, nonrandom dispersion of zooplankton is caused by water movements, in particular Langmuir circulations and metalimnetic entrainment of epilimnetic water.

16. Seasonal polymorphism, or *cyclomorphosis*, occurs among many zooplankton but is most conspicuous among the Cladocera.
    a. Changes in rotifer growth form include elongation in relation to body width, enlargement, reduction in size, and production of lateral spines. A predator-produced organic substance can induce spine production in certain prey rotifers, which reduces predation success.
    b. Cyclomorphosis in Cladocera often results in extension of the head to form a crest (helmet) with little change or only a slight decrease in carapace length (Fig. 16-38). Caudal spine length often increases.
        i. A combination of environmental parameters has been shown to induce internal growth factors (hormones) that influence differential growth: increased temperature, turbulence, photoperiod, and food enhance cyclomorphosis in daphnid cladocerans. Organic substances (kairomones) released by invertebrate and fish predators can also induce cyclomorphic growth as effectively as changes in temperature, turbulence, and oxygen.
        ii. The adaptive significance of cyclomorphic growth likely centers on reducing predation by allowing continued growth of peripheral transparent structures without enlarging the central portion of the body visible to fish. Small cladocerans that increase size by cyclomorphic growth reduce capture success by invertebrate predators (e.g., copepods).
    c. Cyclomorphosis is lacking in copepods, which, by means of rapid, evasive swimming movements, can defend themselves better from invertebrate predators than can most rotifers and cladocerans.

17. Planktivorous fish can be important in regulating the abundance and size structure of zooplankton populations. Prey are visually selected, in most cases, on an individual basis, although the gill rakers of certain fish collect some zooplankton as water passes through the mouth and across the gills.
    a. Size selection of prey by fish is governed by energy return obtained per unit of energy expended in foraging and by the abundance of prey. When prey are abundant, only larger prey are consumed; as prey abundance decreases, smaller prey are taken.
    b. Planktivorous fish select large zooplankters and can eliminate large cladocerans from lakes. Fish predation intensity must be very high to become the main determinant of the zooplankton community but can be sufficient to restructure the zooplankton community on the basis of size. The restructuring can lead to fewer large zooplankton of greater grazing capacities and reduced grazing mortality of phytoplankton.
    c. When size selection by fish is not in effect, and when large zooplankters are present, smaller-sized zooplankton are generally not found to co-occur with the larger forms. The cause is likely a result of size-selective predation of smaller zooplankton by invertebrates (copepods, phantom midge larvae, and predaceous Cladocera). Phytoplankton grazing is more effective by the larger forms and at certain times of the year and markedly reduces phytoplankton productivity.

18. The production rate (net productivity) of zooplankton is the sum of all the biomass produced in growth, including gametes and exuviae of molting, less maintenance losses from respiration and excretion. Emigration (e.g., outflow losses) and immigration of zooplankton from other aquatic habitats are usually negligible.
    a. A general, positive correlation exists between the rates of production of phytoplankton and of zooplankton (Table 16-20).
    b. Much of autotrophic production is not utilized

by herbivorous zooplankton, but instead enters detrital pathways as nonpredatory particulate and dissolved organic matter. Although particulate detritus has less energy content than living algae, detritus often augments the diet of suspension-feeding zooplankton.

c. The productivity of suspension-feeding zooplankton is higher than that of predaceous zooplankton.

d. The efficiency of assimilation of ingested food is somewhat higher in juvenile stages than in adults but is nearly always <50% in zooplankton. Most of the food is egested and enters detrital pathways, and some may be utilized by the protistan microbial community. Assimilated energy expended in respiration is usually <50%; the remainder is used for growth and reproduction. Assimilation and respiration rates generally increase at higher trophic levels, and production decreases.

17 BACTERIOPLANKTON

Biochemical transformations of particulate and dissolved detrital organic matter by bacteria and fungi are fundamental to the structure and dynamics of nutrient cycling and energy fluxes within aquatic ecosystems. Although substantial gains are being made in our understanding, the entire subject of aquatic microbial ecology is an emerging frontier in limnology. The functioning and stability of aquatic ecosystems is governed by the metabolic transformations of organic matter by bacteria and fungi, yet few areas of understanding in aquatic ecology are in such comparative infancy.

Organic matter is synthesized either autochthonously within the lake, or allochthonously in terrestrial and wetland habitats and is imported to the lake as air and water carries organic matter to the rivers and lake basins. As we will see later, organic matter entering rivers and lakes from the landscape is mostly in dissolved form and is chemically modified by the flora and microflora of streams and wetlands before reaching the lake per se (Chap. 23). A major source of autochthonous organic matter is the phytoplankton. This organic matter is augmented with that synthesized by littoral aquatic plants and sessile photosynthetic microflora (Chaps. 18 and 19). Because most lakes are small and shallow, these littoral sources of autochthonous production often exceed those of the phytoplankton. These composite sources of organic matter form the base upon which all other lake-dwelling organisms depend for nutrients and energy.

Most photosynthetic organisms simply die, add their newly synthesized organic matter to the detrital components of the system, and undergo decomposition. Much of the reduced carbon of organic matter is eventually transformed by microbes to the oxidized state, CO_2. A small and highly variable portion of the total dissolved and particulate organic matter of the aquatic ecosystem, 1–10% on the average, is consumed by animals and is decomposed by their heterotrophic metabolism. In aquatic ecosystems, most organic detritus is metabolized by the bacteria and the fungi.

Decomposition is effective. The organic matter formed within or brought to a lake would quickly fill the basin if it were not decomposed, and the lake would cease to exist. Since decomposition is not complete, residual organic matter remains. Gradually the basin fills in, and the lake is transformed from an aquatic to a terrestrial ecosystem (Chap. 24). The rate of this development, or ontogeny, is governed by the heterotrophic microbial capacity to mineralize organic matter, which is in turn influenced by the geomorphology of the ecosystem and particularly the type (quality) of organic matter synthesized within it or imported to it (Chap. 23).

The ensuing discussion highlights contemporary understanding of the bacterioplankton and their metabolism. As we will see later (Chap. 23), much of the organic matter synthesized within the lake or brought to it from the drainage basin is not decomposed in the water, but instead is decomposed on or in the sediments. The biochemical transformations of organic matter displaced to the sediments are crucial to the operation of freshwater ecosystems and consequently deserve separate treatment (Chaps. 21 and 23).

I. THE ORGANIC CARBON CYCLE

It is perhaps most appropriate to begin discussion with the generalized composite organic carbon cycle of a lake (Fig. 17-1), modeled after the general discussions of Kuznetsov (1959, 1970), and to attempt systematically to provide a quantitative range for the reactions and processes involved among lakes of varying productivity. The sources and general composition of allochthonous and autochthonous organic matter in dissolved and particulate form are treated in some detail in subsequent chapters. Dead organic matter forming the

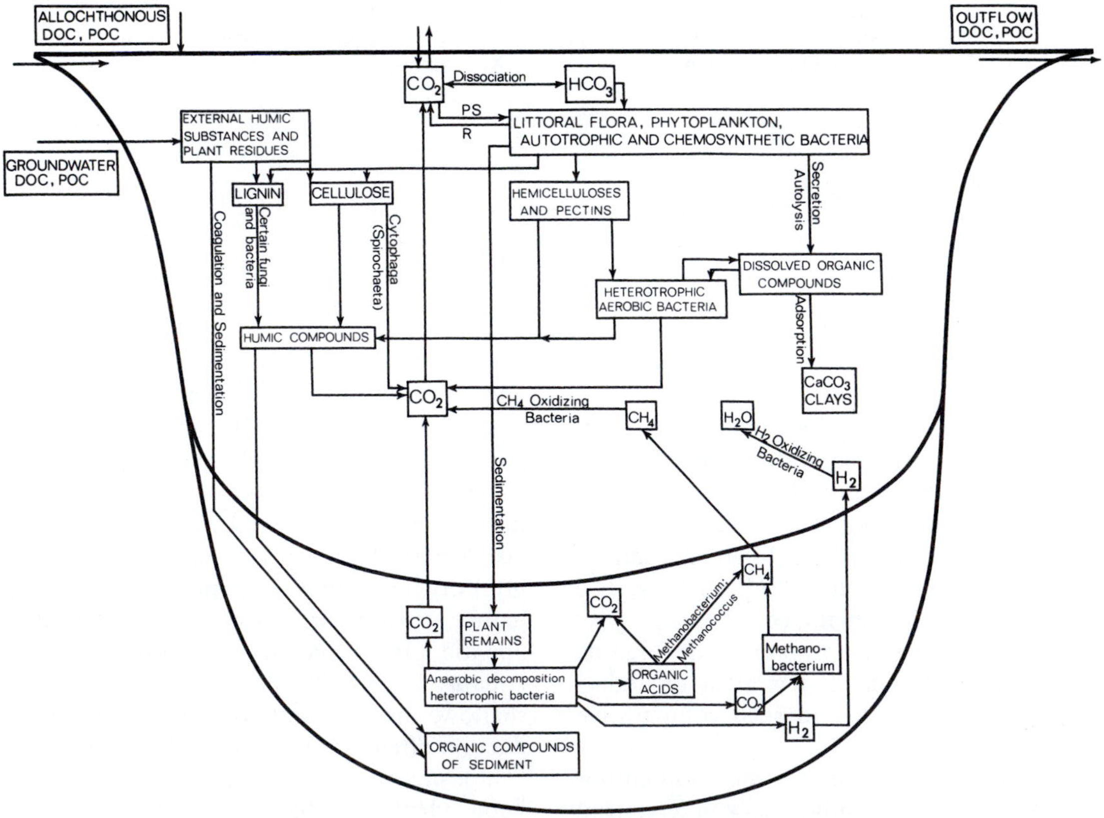

FIGURE 17-1 Simplified organic carbon cycle of a typical freshwater lake. DOC and POC = dissolved and particulate organic carbon; PS = photosynthesis; R = respiration. (Modified from Kuznetsov, 1959, 1970.)

detritus is separable into dissolved and particulate fractions, but it must be emphasized that this demarcation is only an operational separation made by the investigators. From a functional standpoint of the system, the energetic transformations of organic carbon are similar whether they are operating on a large particle or a dissolved organic compound. The rates of enzymatic hydrolysis and transformation differ because of the accessibility of the enzymes to the organic compounds and the chemical structure of the organic compounds.

Bacteria and fungi assimilate dissolved organic compounds, some of which they obtain through enzymatic degradation of particulate organic matter. Only transport of small, low-molecular-weight organic molecules occurs across cytoplasmic membranes in an enzymatic process with permases. Because most (>95%) dissolved organic matter (DOM) is composed of polymeric compounds of high molecular weight, only a small portion of DOM is rapidly utilizable in natural waters (Allen, 1976; Münster, 1985; Münster and Chróst, 1990). However, as will be demonstrated, bacteria can utilize polymeric substrates by a stepwise enzymatic depolymerization and hydrolysis (e.g., Hoppe, 1983, 1991) and by utilization of organic products generated by physical processes such as partial photolysis of the humic polymers (Wetzel *et al.*, 1995; Moran and Hodson, 1998).

Utilization of polymeric substrates can be accomplished by enzymes external to the cells. Most of these enzymes[1] are hydrolases that are bound to the cell surfaces or are in the periplasm of the microbial cells (Chróst, 1991). Monomeric products are cleaved by these enzymes from the ends of the larger molecules.

The decomposition rate of organic substances is greatly dependent on solubility (Vallentyne, 1962). Decomposition will be lowest for those organic compounds or complexes that occur in environmental concentrations exceeding the saturation level in the surrounding water. Degradation rates of soluble organic compounds vary, but for the compound to be preserved and removed from the organic carbon cycle, sedimentation must occur. The prerequisite for sedimentation by gravity is insolubility and aggregation into particles no longer suspended by Brownian motion. Sedimented, relatively insoluble compounds become buried in anaerobic sediments, where rates of degradation are low. Although cellulose can be decomposed very rapidly by anaerobic bacteria under certain conditions (e.g., in sewage digesters and in rumen) in anaerobic aquatic sediments, acidic fermentation products rapidly accumulate and lower the pH sufficiently to reduce or inhibit bacterial metabolism (Brock, 1966). Additionally, temperatures are generally low in many aquatic sediments. Lignin is more resistant to degradation than cellulose, particularly under anaerobic conditions, and both lignin and cellulose form humic substances. The humic materials of the water coagulate to some extent, sink, and form a portion of aquatic sediments.

The primary intermediate products of anaerobic decomposition of sedimented organic matter (especially cellulose) are fatty acids, which are further degraded to CO_2 and hydrogen. Methane is biogenically generated from CO_2 and from hydrogen and then is lost as bubbles or is microbially reoxidized to CO_2 in less reduced strata of overlying sediments or water.

Much of the decomposition of organic compounds occurs in aerobic waters prior to sedimentation to the bottom of the basin. The extent of degradation en route is governed by an array of physical (morphometry, stratification patterns, temperature, and ultraviolet irradiance) and chemical conditions as well as the magnitude and chemical composition of allochthonous and autochthonous inputs of organic matter. Organic inputs to oligotrophic lakes are generally small, and organic matter is exposed during sedimentation to oxic conditions for long distances (greater time). Consequently, degradation of sedimenting organic matter is relatively complete, and organic sediment accumulation is slow. Massive inputs of organic matter in eutrophic lakes result in rapid sedimentation (shorter distances), less volume of aerobic water, and rapid accumulation of organic matter in anaerobic hypolimnia and sediments.

II. DISTRIBUTION OF BACTERIOPLANKTON

Bacterioplankton are often limited by the availability of organic substrates and nutrients, particularly phosphorus. Consequently, one would expect a general increase in bacterial biomass as an aquatic system is increasingly loaded with organic substrates. Evaluation of microbial biomass is commonly done by enumeration and estimation of biovolume by direct microscopic observation (cf. detailed discussions of techniques by Kuznetsov and Romanenko, 1963; Rodina, 1965, 1972; Sorokin and Kadota, 1972; Rosswall, 1973; Romanenko and Kuznetsov, 1974; and particularly Kemp *et al.*, 1993). The culturing of bacteria in media, while useful for species identification in combination with molecular techniques, is extremely selective for specific bacteria under most conditions. These methods underestimate natural heterogeneous bacterial

[1] These enzymes attached to the cells but external to the cytoplasm have been termed *exoenzymes* (Hoppe, 1983), *extracellular enzymes* (Priest, 1984), and *ectoenzymes* (Chróst, 1990).

populations by factors of 10^2–10^5. Direct enumeration by microscopy has the inherent difficulty of discriminating viable cells from detritus and cells that are dead, but recent techniques, such as epifluorescence microscopy with specific fluorescent dyes (Francisco *et al.*, 1973; Porter and Feig, 1980; Labaron *et al.*, 1998; Nobel and Fuhrman, 1998; Weinbauer *et al.*, 1998), have improved accuracy greatly. Biomass of living bacteria can also be separated from nonliving detrital organic matter with a number of specific molecular probes or more generally by measurement of a uniformly distributed cellular constituent, such as adenosine triphosphate (ATP) that decomposes readily on death and can be measured by bioluminescent reactions at extremely low concentrations (Holm-Hansen and Paerl, 1972; Karl, 1993).

In general, the numbers and biomass of bacteria increase with increasing productivity and concentrations of inorganic and organic compounds in lakes (Table 17-1). In spite of great seasonal variations, bacterial numbers and biomass increase from oligotrophic to eutrophic inland waters. The highest levels have been observed in tropical alkaline, saline lakes (e.g., Kilham, 1981) and in eutrophic reservoirs, probably because the large reservoirs studied tend to occur on rivers that receive both industrial and municipal wastes. Numbers of bacteria are markedly lower in acidic dystrophic lakes, which contain high concentrations of humic matter. In a similar manner, production of bacteria as calculated from changes in biomass over short time periods increases with increasing phytoplanktonic productivity of lakes. The average generation time of bacteria decreases

TABLE 17-1 Numbers, Volumes, and Biomass of Bacteria in Inland Waters of Differing Productivity[a]

Habitat	*Number* (10^6 ml^{-1})	*Biovolume*[b] (μm^3)	*Biomass*[b] ($g\ m^{-3}$)	*Reference*
Oligotrophic				
Schirmacher Lakes, Antarctica	0.018–0.035	0.034	—	Ramaiah (1995)
Lake Fryxell, Antarctica	0.54–4.5	0.13–0.39	—	Konda *et al.* (1994)
Ace, Antarctica	0.13–7.28	—	—	Bell and Laybourn-Parry (1999)
Zelenetskoye, Russia (1970, 1971)	0.175	0.265	0.06	Kuznetsov (1970)
Baikal, Russia	0.20	—	—	
Krivoye, Russia (1968, 1969)	0.67	0.43	0.21	
Onezhskoe, Russia	0.29	—	—	
Ladozhskoe, Russia	0.35	—	—	
Pert, Russia	0.13	—	—	
Lawrence Lake, Michigan	2–6	0.09–0.35	0.28	Coveney and Wetzel (1995)
Mirror Lake, New Hampshire	0.5–7	0.12	—	Ochs *et al.* (1995)
Toolik Lake, Alaska	0.1–3.1	0.056–0.26	0.01–0.04	Hobbie *et al.* (1983, 1999)
Mesotrophic				
Krasnoye, Russia (1964–1970)	0.70	0.43	0.30	Kuznetsov (1970)
Naroch, Russia (1968–1970)	0.64	0.50	0.32	
Dal'nee, Russia (1970, 1971)	1.50	0.67	1.00	
Sevan, Russia (1952, 1962, 1966)	0.39	1.12	0.43	
Glubokoe, Russia (1932)	1.2	—	0.97	
Ladoga, Russia (1977–1989)	0.2–1.2	—	—	Kapustina (1996)
10 shallow lakes of Florida (mostly mesotrophic)	1.4–10.5	—	—	Crisman *et al.* (1984)
Lake Biwa Japan				
North Basin, mesotrophic	1.6–3.4	0.02–0.20	—	Nagata (1984)
South Basin, eutrophic	5.9–8.6	—	—	
Eutrophic				
Drivyaty, Russia (1964)	1.84	0.76	1.40	Kuznetsov (1970)
Myastro, Russia (1968–1970)	2.20	0.50	1.10	
Batorin, Russia (1969, 1970)	6.40	0.50	3.20	
Beloe, Russia (1932)	2.23	—	—	
Chernoe, Russia (1932)	4.00	—	—	
B. Krivoe, Russia (saline)	12.3	—	—	
Lake Constance, Germany	1–10.0	—	0.016–0.18	Güde *et al.* (1985)
7 years	0.5–10.8	—	—	Simon *et al.* (1998)
Himon-ya Pond, Japan	4.1–19	—	—	Konda (1984)
Lake Valencia, Venezuela	0.1–1.4	—	1.0	Lewis *et al.* (1986)
Four African saline lakes	3.7–360	—	—	Kilham (1981)

(Continues)

TABLE 17-1 (Continued)

Habitat	Number (10^6 ml^{-1})	Biovolume[b] (μm^3)	Biomass[b] ($g\ m^{-3}$)	Reference
Reservoirs				
Rybinsk, Russia (1964–1968)	1.70	0.60	1.00	Kuznetsov (1970)
Bratsk, Russia (1965–1972)	0.85	0.90	0.77	
Kiev, Russia (1967–1968)	4.10	0.84	3.35	
Kremenchug, Russia (1968)	3.50	0.60	2.10	
Kakhov, Russia (1968)	4.00	0.47	1.90	
Dneprodzerzhin, Russia (1968)	3.40	0.65	2.20	
Kashkorenskoe, Russia	7.8–57.9	—	—	
Rybovodnye, Russia (pond)	1.0–40.0	—	—	
Kramet-Niyaz Russia (pond)				
Without fertilization	2.0–6.0	—	2.0–6.0	
With fertilization	5.0–20.0	—	5.0–25.0	
Lake Arlington, Texas	11.0	0.160	0.172	Chrzanowski (1985)
Bluestone, West Virginia	0.15–8.55	—	—	Perry *et al.* (1990)
Pareloup, France	2.2–6.3	0.08–0.45	0.03–0.10	Lavandier (1990)
Dystrophic				
Chernoe, Russia	1.07	—	—	Kuznetsov (1970)
Piyarochnoe, Russia	0.43	—	—	
Serpovidnoe, Russia	0.1–0.5	—	—	
Lake Botjärn, Sweden	1.2–7.3	0.18–0.20	0.15–0.29	Johansson (1983)
Lake Vitalampa, Sweden	0.7–8.2	0.17–0.20	0.15–0.32	
Loch Ness, Scotland	0.23–0.71	—	—	Laybourn-Parry *et al.* (1994)
Rivers				
Kuparnk River, Alaska	0.3–2.7	0.1–3.1	—	Hobbie *et al.* (1983)
Ogeechee River, Georgia	8–75	0.06–0.18	0.02–1.14	Edwards (1987)
Black Creek, Georgia	7–110	0.02–0.09	—	
Average values				
Oligotrophic	0.50	0.2–0.4	0.15	Saunders *et al.* (1980)
Mesotrophic	1.00	0.4–1.2	0.70	
Eutrophic	3.70	0.5–0.9	2.30	

[a] Often, summer values in the trophogenic zone.
[b] Biovolumes of bacteria tend to decrease seasonally as water temperatures increase (Chrzanowski *et al.*, 1988) or with intensified bacterivory (see text and Chap. 16). Bacterial cell carbon content can be approximated by a conversion factor of 5.6×10^{-13} g of C μm^{-3} (Bratbak, 1985).

in more eutrophic lakes. A positive relationship has been demonstrated between the average rates of phytoplanktonic productivity and bacterial numbers (Table 17-2)[2] (cf. Overbeck, 1965; Godlewska-Lipowa, 1975, 1976; Jones, 1977; Straškrabová and Komárková, 1979; Overbeck, 1994).

[2] The term *saprophytic bacteria* (zymogenous, copiotrophic), as used in these and similar studies, refers to those bacteria that grow on organic-enriched agar, such as nutrient agar or casein–glucose agar. These techniques are highly selective but reflect bacteria that grow on high concentrations of easily assimilable organic substances. Saprophytes generally require organic substrates at high concentrations for appreciable development. In contrast, *oligotrophic bacteria* develop at minimal organic substrate concentrations (1–15 mg C liter^{-1}), even though they can grow on richer media (Kuznetsov *et al.*, 1979). As such, oligotrophic bacteria are conceived as those bacteria adapted to uninterrupted nutrient limitation (Poindexter, 1981). Such bacteria possess physiological and morphological characteristics that maximize their ability to gather nutrients across steep spatial gradients and to utilize nutrients conservatively.

A. Seasonal and Vertical Distribution

The seasonal distribution of bacterial populations is highly variable from lake to lake and from year-to-year within a lake. A few generalizations, however, can be advanced even from population data determined with sampling frequencies that are far greater than the generation times of the bacteria. Although direct correlations are often observed between the numbers of phytoplanktonic algae and heterotrophic bacteria both seasonally (Fig. 17-2) and vertically with depth (Fig. 17-3), careful examination of this correlation among many lakes has shown that it is weak at best and erroneous for rivers. Bacterial numbers and biomass are commonly high in the epilimnion, decrease to a minimum in the metalimnion and upper hypolimnion, and increase in the lower hypolimnion, especially if anoxic (e.g., Niewolak, 1974; Jones, 1978; Kato and Sakamoto, 1979, 1981; Chróst and Rai, 1994; Coveney and Wetzel, 1995; Simon *et al.*, 1998b). Vertical profiles of bacterioplankton and

TABLE 17-2 Relationship between Average Rates of Phytoplanktonic Primary Production and Bacterial Numbers at 1 m Depth in Four Lakes of Northern Germany[a]

Lake	*Primary productivity (mg C m^{-3} day^{-1})*	*Saprophytes (no. ml^{-1})*	*Total bacteria (× 10^6 ml^{-1})*
Schöhsee	16	300	0.5
Schluënsee	17	350	0.5
Grosser Plöner See	35	650	1.0
Plußsee	57	1000	1.1

[a] Modified from Overbeck (1965).

phytoplankton are usually more strongly correlated than they are seasonally.

The primary assumption is that the release of extracellular dissolved organic compounds from phytoplankton are approximately proportional to their biomass and production and that bacterial productivity is largely limited by organic substrate availability (e.g., Cole *et al.*, 1988). As will be delineated, correlations of this type can be highly misleading, and regulating physical and biotic factors have been shown to be much more complex and dynamic. There is no doubt that the heterotrophic metabolism of the bacteria depends to a significant extent on organic substrates produced by

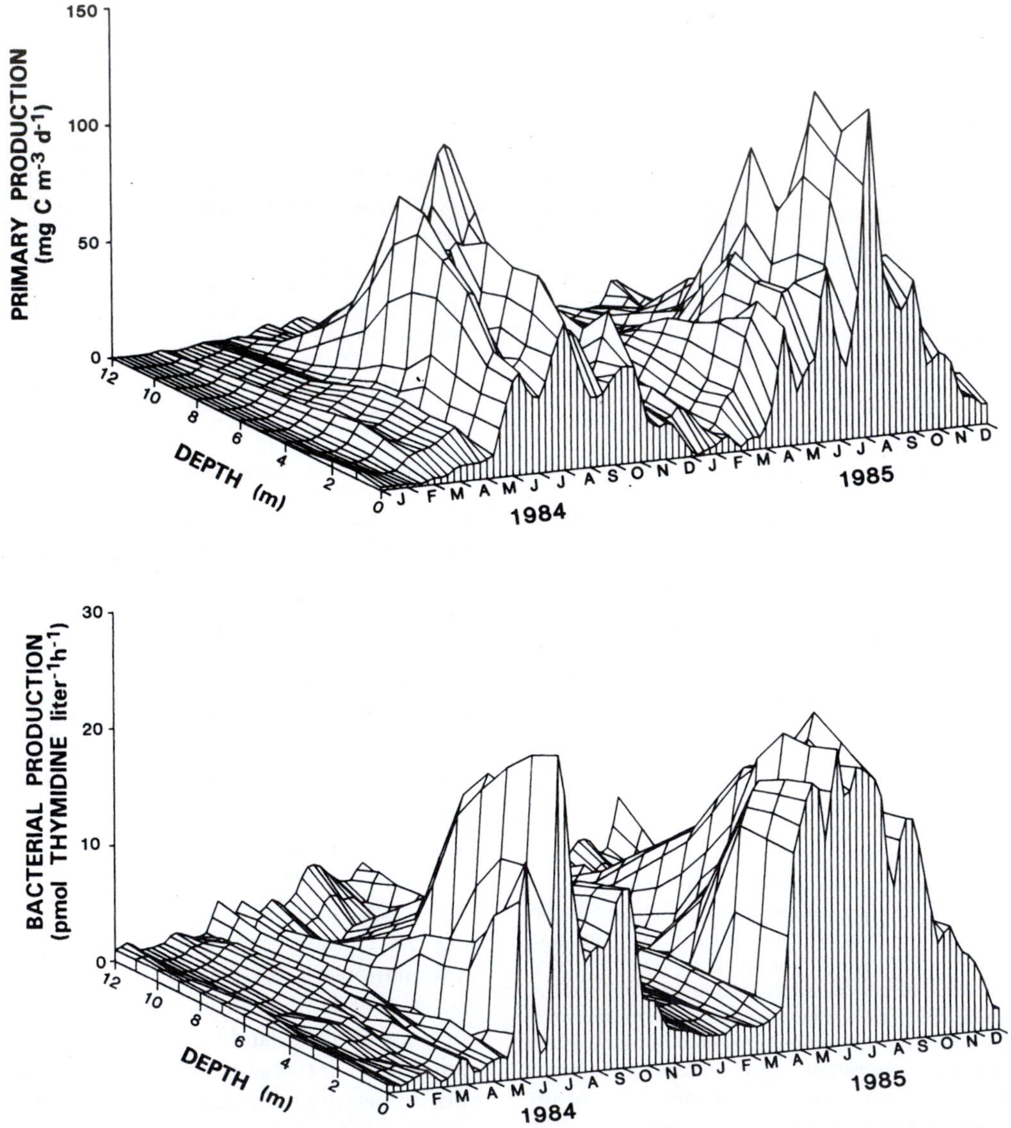

FIGURE 17-2 Comparison of simultaneous measurements of production of phytoplankton (*upper*) and bacterioplankton (*lower*) in oligotrophic–mesotrophic Lawrence Lake, southern Michigan, over a 2-yr period. Bacterial production was determined from *in situ* rates of conversion of radiolabeled thymidine to DNA of the bacteria. (From Coveney and Wetzel, 1995.)

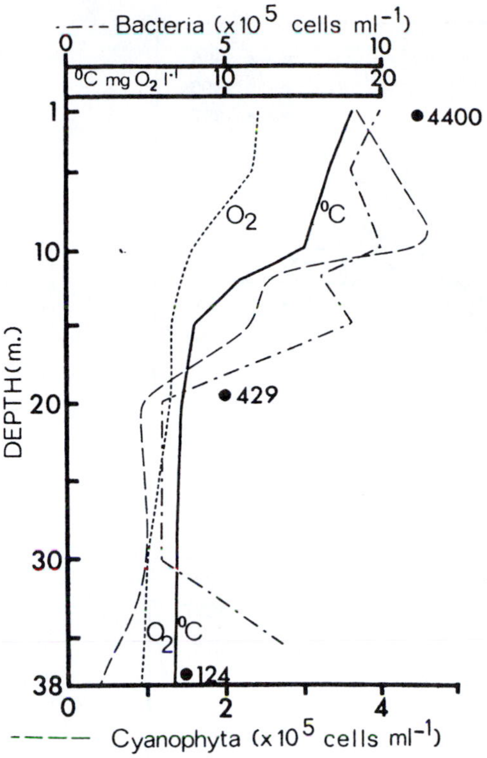

FIGURE 17-3 Vertical distribution of heterotrophic bacteria in relation to planktonic algae in Schluënsee, northern Germany, 18 August 1965. Circles = saprophytes ml^{-1}. (After data of Overbeck, 1968.)

the phytoplankton, as will be described in detail. Often, increases in bacterial populations are delayed and follow phytoplankton maxima by many days or weeks (Straškrabová and Komárková, 1979). More frequent sampling has demonstrated that the vertical distribution of bacteria in dimictic lakes can change very rapidly (Rasumov, 1962; Saunders, 1971). For example, the rapid changes observed in Figure 17-4 became particularly acute in mid-Júne, following a phytoplanktonic maximum; bacterial numbers increased sharply, then decreased again in a few days. The rapid inversions in vertical distribution of the waxing and waning bacterial populations (Fig. 17-4) indicate that many short-term variables and processes control bacterial abundance, distribution, and productivity (Saunders, 1971; Saunders *et al.*, 1980). Correlations have been found between pulses of bacteria and changes in sources of organic matter, such as death and decomposition of phytoplankton and of littoral plants, as well as changes in precipitation and runoff of organic matter from allochthonous sources, nutrients, temperature, mixing, parasitism, and grazing by protists and zooplankton (e.g., Lane, 1977; Tanaka *et al.*, 1977; Goulder, 1980; Coveney and Wetzel, 1989, 1992, 1995). However, correlations alone yield little insight into the mechanisms controlling the apparent interaction; they only establish hypotheses on possible controlling factors.

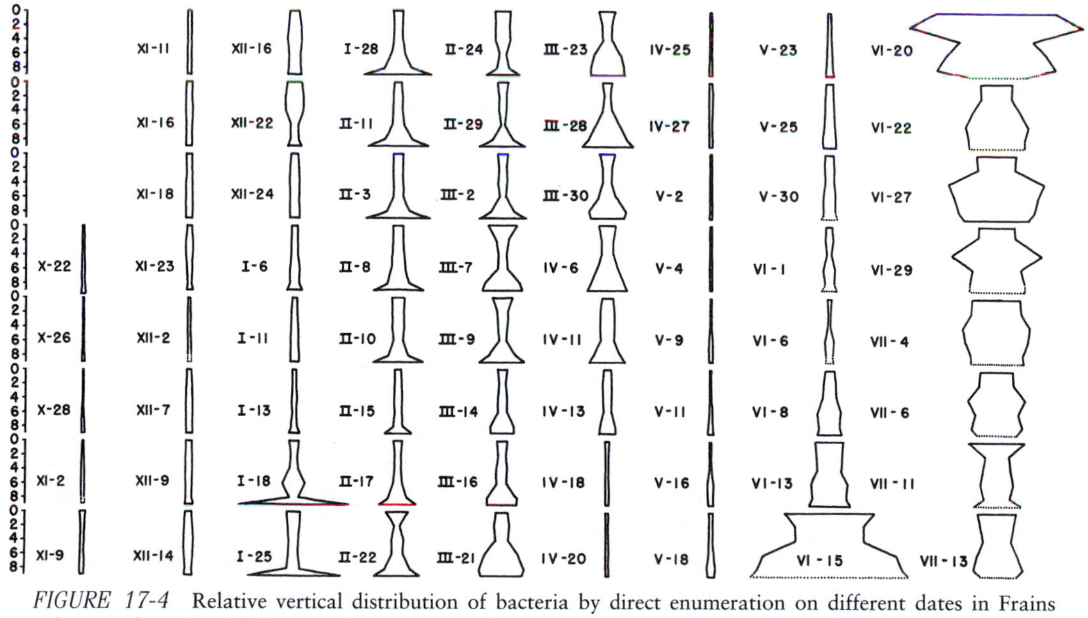

FIGURE 17-4 Relative vertical distribution of bacteria by direct enumeration on different dates in Frains Lake, southeastern Michigan, 1959–1960. Ordinate: depth in meters; abscissa: relative numbers. (From Saunders, G. W.: Carbon flow in the aquatic system. *In* Cairns, J., Jr. (ed.) *The Structure and Function of Fresh-Water Microbial Communities,* Blacksburg, Va., Virginia Polytechnic Institute, 1971.)

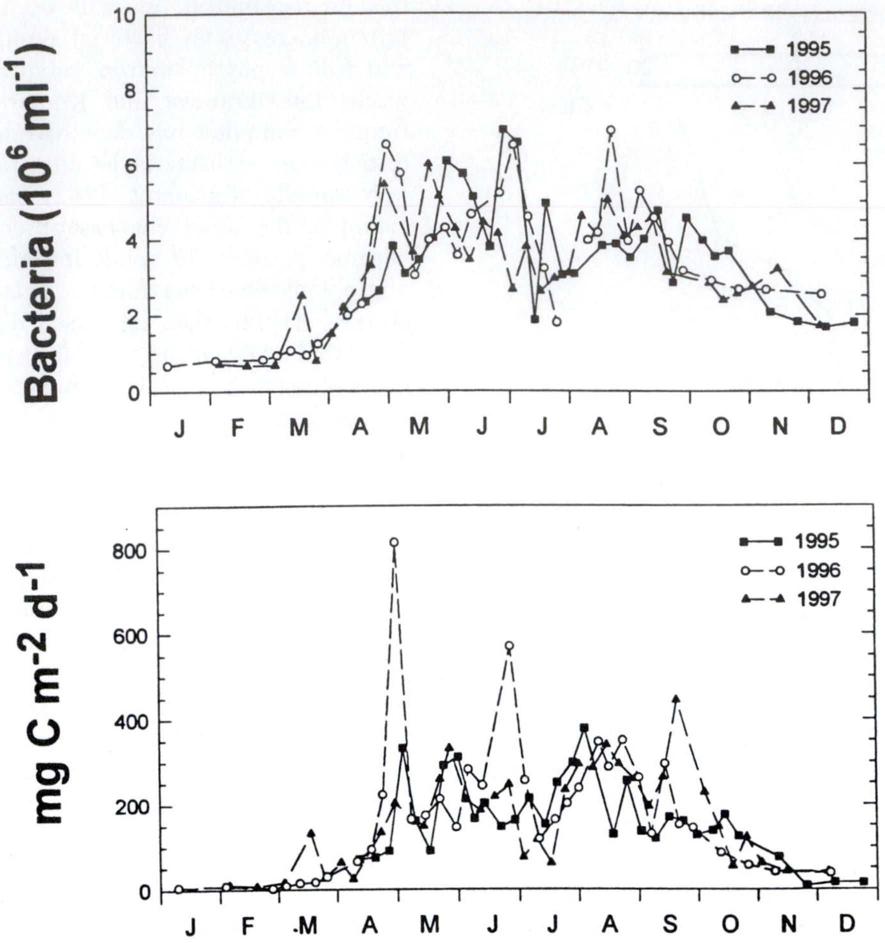

FIGURE 17-5 Seasonal changes in epilimnetic (0–10 m) bacterial numbers (*upper*) and in bacterial production of the trophogenic zone (0–20 m) (*lower*) in Lake Constance, southern Germany. (Modified from Simon *et al.*, 1998b.)

Planktonic bacterial biomass and production is generally lower during winter than during summer in temperate lakes and reservoirs; this relationship is correlated with low winter temperatures and reduced loading of particulate and dissolved organic substrates from autochthonous (phytoplankton and littoral plants) and allochthonous sources when the soils are frozen. A significant portion of the planktonic bacterial community may be physiologically dormant when conditions are cold and limited organic substrates are available (e.g., Stevenson, 1978; Wright, 1978).

Before discussing potential parameters that regulate bacterial biomass and production, it is instructive to examine a detailed analysis of seasonal dynamics and potential controls of bacterioplankton abundance and production in Lake Constance (Simon *et al.*, 1998b). This lake was eutrophic in the 1980s but has been undergoing marked reductions in nutrient loadings in the 1990s and is experiencing oligotrophication. Lake Constance is now a large temperate, mesotrophic lake with bacterioplankton properties generally representative of many lakes. Seasonal changes in bacterial production (Fig. 17-5) generally coincide with those of bacterial biomass (Fig. 17-5) and ascend from low winter values to increases in the spring and summer before declining again in the autumn and winter. Within this general annual cycle, however, a number of rapid changes in both bacterial biomass and production have been observed, and these changes have been coupled to a bevy of physical and biotic variables.

The generic model set forth by Simon *et al.* (1998a) suggests that growth of heterotrophic bacterioplankton is limited by the availability of organic carbon substrates because of reduced winter levels of phytoplankton productivity and release of extracellular organic carbon (Fig. 17-6). As we will see, however, allochthonous loadings of dissolved organic substrates can be high in the winter and readily utilized by many bacteria.

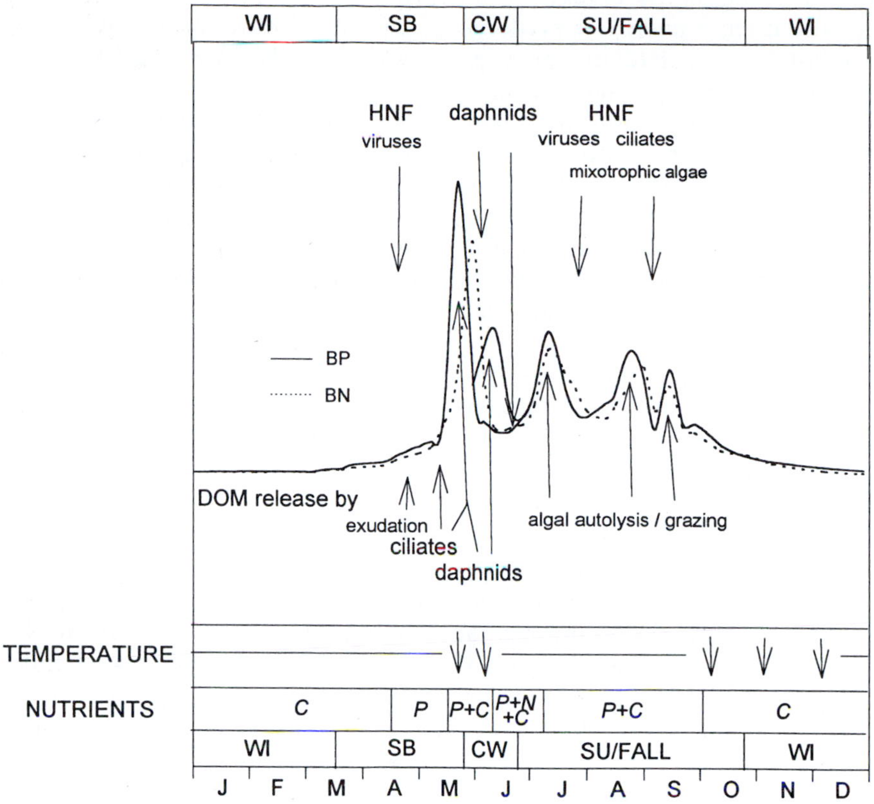

FIGURE 17-6 Idealized patterns of seasonal dynamics and regulation of bacterioplankton abundance and production in temperate, mesotrophic Lake Constance, southern Germany. *BN* = bacterial numbers; *BP* = bacterial production; *WI* = winter; *SB* = spring increase in phytoplankton abundance; *CW* = "clearwater" phase, which is highly variable from year-to-year depending on the relative abundance of daphnid zooplankton and their effectiveness in grazing of phytoplankton algae; *SU/FALL* = summer and autumn periods. Upward arrows = net positive effect, and downward arrows = net negative effects on bacterioplankton. (Modified from Simon *et al.,* 1998a.)

Temperatures are low and also control bacteria to a certain extent, but many bacteria are adapted to low temperatures (5–10°C), particularly in the spring, and capitalize upon the early phases of spring increases in production of phytoplankton. During the spring algal bloom, phytoplankton production enhances availability of extracellular organic substrates, and the supply of organic substrates is enhanced by intensified grazing activities on the algae by protists and zooplankton (Fig. 17-6). Dissolved carbohydrates and amino acids are available and utilized in large amounts at this time (Hanisch *et al.,* 1996; Simon, 1998), but bacterial growth can be limited by the intense algal and bacterial demands for inorganic nutrients, mainly phosphorus limitations. Although cladoceran daphnid zooplankton feed primarily on algae, as they decimate the spring algal bloom, increasing, very short-lived impacts on bacterioplankton occur. Most limitation results from mortality of bacteria from viral lysis, often exceeding 50%, and from grazing by protistan protozooplankton (see discussion in Chap. 16 and later in this chapter). Viral lysis and grazing activities by zooplankton on both algae and bacteria enhance the regenerated supply of dissolved organic substrates and subsequent growth of bacteria. During the summer, the same combination of biotic and abiotic factors influence bacterial growth rates and production, but usually in a more balanced manner. Gradually, as the seasons progress toward winter, declining temperatures and organic carbon substrates reduce production of bacterioplankton.

Controversy prevails concerning whether bacterioplankton are freely planktonic or attached to suspended inorganic or organic detrital particulate matter. Often most bacteria are observed to be freely planktonic, particularly in oligotrophic waters (e.g., Riemann, 1978; Geesey and Costerton, 1979; Saunders *et al.*, 1980; Johansson, 1983; Lewis *et al.*, 1986). However, bacteria are also often found to be predominantly attached to detrital particulate matter (e.g., Hutchinson, 1967; Paerl, 1975; Lind and Dávalos-Lind, 1991), but marked seasonal variations have been observed (e.g., Kirchman, 1983). For example, in lakes or reservoirs with high, persistent in-lake suspended clay–organic aggregates, <20% of the bacterioplankton may be free in the water. Microbial attachment to particles may be in response to enhanced availability of adsorbed organic substrates relative to substrate concentrations existing in the water.

Horizontal spatial variability in biomass and production of bacterioplankton is usually low in the central pelagic regions of lakes that are not morphometrically complex or disturbed by high effluent loading from point sources (e.g., Jones, 1977; Zmyslowska and Sobierajska, 1977; Jones and Simon, 1980). Closer examination of horizontal variations of bacterial production has demonstrated consistently that production increases markedly along a gradient from the central portions of the basins toward littoral areas. This gradient has been found in all lakes examined, including the very large great lakes (Scavia and Laird, 1988). Loading of both nutrients and organic substrates from the land and littoral producers certainly enhances bacterial productivity. Patchiness in bacterial distribution is found also upon critical sampling and analysis of large-scale differences in organic loading from human effluents (e.g., Palmer *et al.*, 1976; Rao and Burnison, 1976). Sinking rates of individual bacteria are so low (approximately 1 mm day^{-1}; Jassby, 1975) that horizontal differences are usually coupled to differences in water turbulence and variations in availability of nutrients and organic compounds (e.g., phytoplankton, littoral sources and influent sources of organic substrates).

Neuston bacterioplankton in the surface films of inland waters is poorly understood; most studies are of marine waters (Norkrans, 1980). The potential for enrichment of organic substrates and nutrients from atmospheric sources suggests that bacteria adapted to this habitat of high ultraviolet irradiance may grow more prolifically than true bacterioplankton. However, turbulence commonly disperses surface communities frequently. Comparisons of the metabolic activities of neuston bacteria of the surface film with those of bacterioplankton several centimeters below the air/water interface indicated that assimilation and respiratory rates can be higher in the neuston, particularly if the waters are oligotrophic (Crawford *et al.*, 1982; Hermansson and Dahlbäck, 1983). In more eutrophic situations, however, respiration rates were about the same in both habitats.

III. CONTROL OF BACTERIOPLANKTON BY RESOURCE AVAILABILITY AND ENVIRONMENTAL FACTORS

The growth and development of heterotrophic bacteria usually is limited by availability of some resource such as organic substrates for carbon, energy, or nutrients, particularly phosphorus and nitrogen. Second, environmental parameters, particularly temperature and ultraviolet light can affect significantly bacterial growth. Finally, predation upon bacteria by animals can alter the growth, development, and production of bacteria. Each of these regulators should be examined in relation to observed spatial and temporal patterns of growth and production of bacterioplankton.

A. Temperature

A number of studies have shown that bacterioplanktonic growth is positively correlated with temperature, particularly at relatively low temperatures (<10–15°C) (Scavia and Laird, 1987; Morris and Lewis, 1992; Coveney and Wetzel, 1995; Ochs *et al.*, 1995; Felip *et al.*, 1996; Tibbles, 1997; Simon and Wünsch, 1998). As was discussed earlier (Chap. 6), a large volume of lakes in the lower metalimnion and hypolimnion rarely warm to >12–15°C at any time of the year in the temperate regions. Hence, in the lower part of the water column temperature can directly control bacterial growth by a simple Q_{10} relationship for biochemical reactions. Above these temperatures, bacterial growth is less strongly correlated to temperature differences, and presumably in the seasonally warmer upper part of the water column, temperature-adapted bacterial communities develop and are limited to a greater extent by other parameters and resources, particularly organic substrates and nutrients.

Bacterial development can increase under conditions of thermal bar development in large lakes (e.g., Moll and Brahce, 1986; Kapustina, 1996). Although temperatures are cold, they are several degrees warmer than the open pelagic region beyond the thermal bar, and these temperature differences in the lower range are significant to increased bacterial metabolism. Nutrient availability from littoral runoff may also increase within the thermal bar zones as well.

B. Extremes in pH Values

High rates of photosynthesis by phytoplankton in eutrophic lakes or by submersed macrophytes and attached algae can induce high pH values, often in excess of pH 10 on a diurnal basis (cf. Chaps. 11 and 22). Such values are particularly common in eutrophic waters of intermediate bicarbonate buffering capacity, and this chemical composition dominates surface waters globally. Bacterioplankton production commonly decreases under conditions of high pH (ca. ≥10) (e.g., Jeppesen *et al.*, 1997). Such reductions are likely the result of reduced efficiencies of enzymatic activities at high pH ranges. Bacterioplankton production has been observed to be similar under acidic and neutral to slightly alkaline conditions (Bell and Tranvik, 1993). When softwater lakes that have been acidified from anthropogenically induced atmospheric pollution are neutralized by applications of lime, bacterioplankton production can increase and be more susceptible to disturbance by protistan grazing and viral mortality.

C. Nutrient Availability

Restricted nutrient availability can often limit the growth and productivity of bacterioplankton. Indeed, the pelagic environment can be referred to as a "desert" in relation to the availability of organic substrates for carbon and nutrients, particularly phosphorus, for the bacterioplankton. Studies of the seasonal dynamics of bacterioplankton often suggest relative constancy of cell densities over time within a general seasonal increase during the warmer, more productive periods (e.g., Fig. 17-5). As indicated later, appreciable mortality occurs by viral lysis and protistan grazers. The limited availability of reduced carbon substrates is often invoked as the primary restriction to bacterioplanktonic growth, in part evolving from the cross-system, generally positive correlation between bacterial production and phytoplankton biomass (Cole *et al.*, 1988; White *et al.* 1991). However, an increasing number of analyses indicated that phosphorus, rather than or in addition to organic carbon, was the critical nutrient regulating the growth of bacterioplankton (e.g., Coveney and Wetzel, 1992; Morris and Lewis, 1992; Vrede, 1996; Watanabe, 1996; Simon *et al.*, 1998).

The requirements of bacterioplankton for inorganic nutrients, particularly phosphorus and nitrogen, had been usually assumed to be met either through the utilization of organically bound fractions or through high-affinity uptake systems for inorganic forms. As indicated in the earlier detailed discussions on phosphorus utilization and competition by algae and bacteria (see Chap. 13, pp. 268–269), uptake of inorganic phosphorus in phosphorus-limited aquatic ecosystems is dominated by bacterioplankton because they have much higher phosphorus requirements than do phytoplankton and their uptake systems have greater affinities than those of algae.

Phosphorus commonly limits bacterioplankton. For example, the addition of inorganic phosphorus was shown to stimulate the production of bacterioplankton diluted in filtered lake water (Toolan *et al.*, 1991). Additions of organic nutrients showed those organic substrates alone did not increase specific growth rates[3] of bacterioplankton of an oligotrophic lake, but the addition of inorganic phosphorus alone increased specific growth rates (Coveney and Wetzel, 1992). Bacterioplankton can accumulate phosphorus above that required to meet metabolic demands ("luxury uptake"). That accumulation is much higher when nitrogen availability is low (N:P <40:1) than when nitrogen was abundant relative to phosphorus (N:P > 40:1) (Chrzanowski and Kyle, 1996; Chrzanowski *et al.*, 1996). These results indicated that phosphorus, rather than organic carbon, can limit bacterioplankton growth and that direct competition occurs between phytoplankton and bacterioplankton for inorganic phosphorus. Although phosphate is the preferred substrate for uptake into both bacteria and algae, enzyme-mediated hydrolysis of natural organic phosphate-containing compounds, such as phosphate esters and nucleotides such as ATP, can form a major source of phosphorus for bacterioplankton when inorganic phosphorus availability is reduced (cf. Chap. 13, pp. 260ff).

Inorganic nitrogen is much less frequently limiting to bacterial growth than is phosphorus (e.g., Elser *et al.*, 1995a,b; Jansson *et al.*, 1996). In three temperate lakes of low productivity, nitrogen fluxes into bacterial-size fractions of the seston relative to the N/P ratio of the bacteria indicated a threshold N/P ratio (ca. 22:1, by atoms); below this value, bacteria assimilated added nitrogen, and above it nitrogen was released. Limitation by phosphorus or nitrogen can shift rapidly seasonally, even diurnally, and with changes in the composition of the microbiota. During periods when phosphorus availability increases, it is common to find enhancements of bacterial production from inorganic nitrogen enrichments of the plankton (e.g., Wang *et al.*, 1992; Wang and Priscu, 1992; Simon *et al.*, 1998a). The enhanced bacterial productivity may be direct, in response to utilization by the bacteria, or indirect, from the stimulation of phytoplankton growth and their release of organic substrates that in turn enhance bacterial growth.

Many researchers espouse that one element is more limiting than another and that there can be a shift from one nutrient limitation to another seasonally. Such discussions are governed by scale and can quickly become circular and pointless. If a nutrient such as phosphorus is acutely scarce in an aquatic ecosystem, the growth

[3] *Specific growth rate (SGR)* is calculated from endpoint bacterial densities or bacterial biovolume (initial, D_0; final, D_T) as $(\ln D_T - \ln D_0)/T$, where T is incubation time.

and productivity of many components, from bacteria to plants to large animals, can become chronically nutrient-limited (e.g., Moeller *et al.,* 1998). In many ecosystems, particularly in the pelagic zones of lakes, the many different kinetic rates for nutrient uptake exist among the diverse biota of hundreds of coexisting species of the plankton. At any one time, the composite dominating nutrient limitation may be one element, and with changes in loadings, utilization, and losses, the composite dominating nutrient limitation can shift to another element. Under transient growth conditions, it is possible to demonstrate multiple nutrient-limited growth (Egli, 1995). Under natural conditions, however, usually such a shift requires several weeks, but theoretically it could occur over very short time periods (hours) and oscillate back and forth. What is generally studied and reported is a collective, composite, community nutrient limitation.

Examination of the vertical distributions of bacterioplankton also supports the general scenario that nutrient availability is a common limitation in the epilimnetic waters of stratified lakes and decreases in the hypolimnion. Bacterial cell size is commonly observed to be smaller (greater surface area to biovolume ratio) in epilimnia than in hypolimnia. For example, bacteria from anoxic hypolimnia were found to be 2–10 times larger than those of oxic waters, and cell size was independent of temperature (Cole *et al.,* 1993). A related observation was that bacterioplanktonic abundance and biomass (product of mean cell size and bacterial abundance) were also markedly greater in anoxic hypolimnia than in oxic epilimnia.

IV. DECOMPOSITION OF DISSOLVED ORGANIC MATTER

The decomposition of organic matter by bacteria is governed by many factors, particularly chemical characteristics such as the structure of the compounds and biological parameters such as the synthesis of enzymes capable of hydrolyzing the organic substances. The instantaneous concentrations of labile organic compounds, such as simple carbohydrates and amino acids, are very small, indicative of their rapid assimilation and rapid rates of turnover. The remainder of the dissolved organic matter of inland waters is much more recalcitrant to decomposition and is dominated by humic and fulvic acids, largely of higher plant origins. The labile compounds, however, are important because of their high energy, relatively simple chemical bonding and structure, and relatively high level of nutrient per total carbon content. Insight into the complex differential utilization of these relatively labile organic compounds, such as amino acids, is gained from several detailed analyses of the amino acid and carbohydrate cycling in lakes.

It has long been argued that labile organic substrates occur at "threshold" concentrations (Jannasch, 1970) and dissolved free amino acids and carbohydrates usually occur at nM concentrations. This common observation led to the assumption that bacterioplankton are often living under chronic substrate limitation and to the erroneous assumption that the large residual pool of dissolved organic matter is largely unavailable for bacterial utilization. Nonetheless, the recalcitrant pool of organic matter is used slowly, and much of the bacterial metabolism is strongly coupled to the sources of simple substrates and their rapid turnover rates.

A. Amino Acids and Related Compounds

Concentrations of dissolved free amino acids (DFAA) of lake water and the interstitial water of sediments generally are very low ($<10\ \mu g\ liter^{-1}$; range 0.0017–2.4 nM) at any given time (Brehm, 1967; Gardner and Lee, 1975; Münster, 1993; Simon, 1998a), although they can increase to considerably higher levels in eutrophic pond waters (Zygmuntowa, 1972). All of the protein amino acids as well as several nonprotein amino acids have been found. Over an annual period and in several different types of waters, concentrations of serine, glycine, and alanine were relatively high, those of aspartic acid and threonine occurred in less abundance, and those of all remaining amino acids and glutamic acid were very low. Maximum epilimnetic concentrations commonly occur in the winter, especially just below the ice, and during the summer period of maximum algal production. Concentrations in the metalimnion are generally very low but may increase somewhat in the hypolimnion. Just above the sediments, concentrations of amino acids often decline again. Many of the dissolved amino acids were found to be adsorbed to colloidal carbohydrates liberated by the microflora. These colloids are most abundant in the epilimnion during the summer and are reduced to extremely low concentrations in the hypolimnion, where intensive bacterial metabolism occurs.

High-molecular-weight peptides of the free water and interstitial water of the sediments were associated with humic substances and increased in surface waters during autumnal influxes of allochthonous plant material (Brehm, 1967). These compounds and, additionally, mucopeptides of cyanobacterial and heterotrophic bacterial origin undergo relatively slow bacterial degradation. Amino acids are adsorbed selectively by these humic substances of dead particulate matter and by the cell

TABLE 17-3 Dissolved Organic Nitrogen Compounds in the Epilimnion of Two Lakes of Northern Germany, 27 June 1967[a]

Dissolved organic nitrogen compounds	*Schuënsee (mesotrophic)*	*Plußsee (eutrophic)*
Extracellular amino compounds		
Free amino acids	39 μg liter^{-1}	71 μg liter^{-1}
Peptides	226	250
Colloids	118	414
Glucosamines	31	126
Total	414 μg liter^{-1}	861 μg liter^{-1}
Amino-*N* as a percentage of the total extracellular organic nitrogen	32.2%	23.7%
Amino acids and amino sugars as a percentage of the total dissolved organic matter	10.3%	5.5%
C:N of dissolved organic matter	7.5:1	10.7:1

[a] From data of Gocke (1970).

walls of both heterotrophic and cyanobacteria. These polypeptides remain relatively undegraded after autolysis of algae. In contrast, certain amino acids, especially aspartic and glutamic acids, are released very rapidly during autolysis or grazing of phytoplankton and are assimilated immediately by heterotrophic bacteria. Other amino acids are released through autolysis more slowly and result in a higher relative concentration in the dead plankton. Bacterial decomposition of plankton occurs first on the autolytically soluble products. Most of the cell decomposition occurs in the epilimnion; more resistant cell components are degraded more slowly after the soluble autolytic substances are utilized.

A more recent detailed study with markedly improved analytical techniques was performed on mesotrophic Lake Constance (Simon, 1998a). The spatial and temporal distributions of dissolved free amino acids (DFAA) and dissolved combined amino acids[4] (DCAA) were determined over an annual period and coupled to rates of bacterial biomass and productivity. Concentrations of DFAA ranged from <20 to 480 nM, with highest values at the end of the spring maximum of the phytoplankton development. In contrast, concentrations of DCAA ranged from 610 to 3755 nM of amino acid equivalents, and highest concentrations of DCAA occurred in middle to late summer. The low C:N ratios of the amino compounds (2.8–3.3) emphasize the excellent substrate quality of these compounds.

Measurements with individual amino acids of uptake kinetics demonstrated that those in least abundance were assimilated most rapidly (<20 h), and conversely that those compounds found in higher concentrations exhibited longer (30–90 h) turnover times (the time required for the plankton remove and metabolize all the substrate without replacement) (Hobbie *et al.*, 1968). These results are reinforced by Gocke's (1970) demonstration of the rapid bacterial decomposition of free amino acids (reduced to 2% within 6 days) and peptides (to 13% in 6 days) by natural bacterial populations, while amino–colloidal complexes were relatively resistant to degradation and accumulated. The differences in rates of degradation of amino components are reflected in the absolute concentrations found (Table 17-3). Similarly, in the study of dissolved amino acid utilization in Lake Constance, the turnover times for DFAA fluctuated greatly seasonally and ranged between 2.6 and >50 days, with the shortest times occurring during the spring bloom of phytoplankton (Rosenstock and Simon, 1993; Simon, 1998a). From rates of DFAA incorporation into bacterioplankton per rate of bacterial production, it was estimated that between 20 and 40% of the bacterial biomass production was met by the carbon of these amino compounds. This and other studies (e.g., Gardner *et al.*, 1987) indicate bacterial production can be limited by substrate availability at certain times of the year.

Present understanding indicates that both dissolved free and combined amino compounds constitute important substrates for bacterioplankton growth. Simple organic nitrogen compounds are utilized more effectively in more productive lakes (faster turnover; lower instantaneous concentrations as a percentage of the total organic nitrogen). This observation suggests that the enzymatic systems of bacteria of eutrophic lakes are

[4] DCAA were derived from the hydrolyzable amino acids derived from total dissolved proteinaceous compounds minus DFAA, expressed as amino acid equivalents.

adapted or keyed to respond to a much greater variety of substrates than those of oligotrophic waters exposed to conditions of less frequent substrate encounters. The oligotrophic lake species, upon receipt of an influx of a specific substrate for which it does not possess an active enzymatic uptake system, must adapt to that substrate (cf. Hollibaugh, 1979). Hence, the observed assimilation rates are initially low but can increase markedly as adaptation occurs. Those species of eutrophic waters, which are exposed more frequently or continuously to a variety of substrates, respond to and assimilate the substrates immediately.

Little is known about utilization rates of more complex compounds, such as amino sugars. Amino sugars, particularly *N*-acetylglucosamine, have been widely observed at concentrations of 10–100 μg liter^{-1} (Nedoma *et al.,* 1994). Because analytical assays for direct evaluation of these compounds are difficult, the ability of natural microorganisms to use such compounds can be inferred indirectly from the presence of specific extracellular enzymes capable of liberating the compounds from polymer organic matter. Such enzyme activities, such as *β*-*N*-acetylglucosaminidase, have been found to be tightly coupled to bacterial activity in many different lakes and rivers. It is also noteworthy that bacteria attached to particles had a much higher (>100 times) specific activity of aminopeptidases than that of free-living bacteria (Debroas, 1998). The densities of seston particles, both dead and living phytoplankton containing protein, and attached bacteria tend to increase in eutrophic waters.

Expectedly, therefore, the highest rates of detrital proteolysis and aminopeptidase activity are found associated with the highest phytoplankton productivity both spatially and temporally over seasons and on a daily basis. For example, the highest enzymatic and proteolytic activities are commonly in the trophogenic zone, seasonally following phytoplankton blooms, and diurnally from noon to sunset (e.g., Halemejko and Chróst, 1986).

B. Carbohydrates, Simple Organic Acids, and Lipids

The distribution of carbohydrates and simple organic acids has been investigated in several lake systems (see, e.g., Saunders, 1963; Walsh, 1965a,b, 1966; and particularly Weinmann, 1970; Münster, 1984; Hanisch *et al.,* 1996; Meon and Jüttner, 1999). As with simple amino compounds, the instantaneous concentrations of individual organic acids and carbohydrate compounds are very low (usually <30 μg liter^{-1}) and represent a small portion of the total dissolved organic matter (cf. Hama and Handa, 1980; Münster, 1993). Free monosaccharides and oligosaccharides are assimilated readily by bacteria and occur at very low instantaneous concentrations (<10 μg liter^{-1}; 1–100 nM, Moen and Jüttner, 1999). Dissolved polysaccharides, consisting of numerous hexoses, pentoses, and methylpentoses, and especially phenolic and cresolic compounds, occur in greater abundance. Various organic acids, particularly acetic, levulinic, formic lactic, citric, oxalic, malic, glycolic, and short-chain fatty acids, have also been found to be generated in appreciable quantities by partial photolysis by natural ultraviolet light of humic substances (see later discussion below and Chap. 23).

Detailed evaluation of the vertical distribution of dissolved carbohydrates in spring in eutrophic Plußsee, northern Germany, found moderate concentrations of total dissolved carbohydrates (<0.5–800 μg C liter^{-1}), with higher values in the epilimnion and decreasing values with depth (Münster, 1984, 1985, 1991). Similar vertical distributions were found in Lake Constance (Hanisch *et al.,* 1996). In Plußsee on one spring day, about 75% of the total dissolved carbohydrate occurred as monosaccharides and 10–15% as disaccharides; on a diurnal basis, maxima were observed about 2–4 h after the photosynthetic maximum of the phytoplankton in the surface waters (Münster, 1984, 1991).

Examination of seasonal changes in total dissolved carbohydrates (monosaccharides, disaccharides, cellobiose) in mesotrophic Lake Constance, Germany, showed relatively low concentrations (130–400 μg C liter^{-1}) at all times with relatively small changes in concentrations seasonally (Hanisch *et al.,* 1996). Concentrations of dissolved free carbohydrates were always below analytical detection limits. Rates of utilization of total dissolved carbohydrates by bacterioplankton ranged between 0.5 and 3.4 μg C liter^{-1} hr^{-1}, with the highest rates occurring at the maximum and decline of the spring bloom of phytoplankton. Turnover times of total dissolved carbohydrates (ratio of concentration to rate of utilization) varied between 3.6 and 21 days during the spring and summer periods (not assayed during other seasons).

Another significant source of these compounds is the secretion products of phytoplankton (Chap. 15) and littoral flora (Chap. 18), autolysis of the plants and microflora, and intermediary products of microbial degradation. In bacteria-free algal cultures, carbohydrates, especially mono- and disaccharides, and total dissolved organic matter increase in proportion to algal growth. In the presence of a natural bacterial flora, the extracellular products are degraded rapidly (hours). This reduction is selective. The labile carbohydrates, organic acids, and amino acids decrease much more rapidly than the total dissolved organic matter (Weinmann, 1970; Gocke, 1970; Poltz, 1972; Münster, 1993; Hanisch *et al.,* 1996).

TABLE 17-4 Average Percentage of Lipid Fractions in the Net Plankton, Sedimenting Particulate Organic Matter, and Sediments of Several Lakes of Northern Germany[a]

Fractions	*Net plankton (>20 μm)*	*Sedimenting particulate matter*	*Sediment*[b]
Total organic matter per dry weight	83.1%	36.7%	16.0%
Total lipids per organic matter	22.3	9.0	6.8
Total fatty acids per total lipids	61.2	32.5	12.1
Triglycerides of total lipids	30.8	8.0	1.8

[a] After Poltz (1972).
[b] Grosser Plöner See only; mean values; values greatest near surficial interface and decreased progressively within first meter.

Although relatively little is known about the distribution and metabolism of lipid compounds in fresh waters, the detailed investigations of Poltz (1972) of lakes of northern Germany demonstrate relationships that probably occur frequently. Total lipid, fatty acid, and triglyceride content declines rapidly when cells die and settle out of the trophogenic zone; concentrations of these materials in the sediments are much lower than in living net plankton (Table 17-4). Lipid content variations of net plankton, sedimenting particulate organic matter, and lake sediment were relatively large, because of changes in lipid composition among the phytoplankton (see pp. 353–354) and zooplankton (see pp. 427–428). Vertical migration and population oscillations both contributed to fluctuations in short-term lipid content in these particle types. Changes in the lipid content and fatty acid compositional patterns of the plankton over an annual period were not as great, however, because of opposing changes in lipid content of the phyto- and zooplankton. The relative amount of short-chained fatty acids, especially of the 16:1 isomer, increased from summer to winter so that the mean chain length of the fatty acids declined. The same fatty acids were found in the triglycerides of the plankton and in the total fatty acid fractions, but the relative concentrations of highly unsaturated fatty acids in the triglycerides were lower.

Decomposition of organic matter and fatty acids was most rapid under the warmer, aerobic conditions of the epilimnion. The mineralization of the planktonic particulate organic matter of the pelagial zone was rapid and usually complete (>85%). The intensity of decomposition of triglycerides was greatest (>98%) in the epilimnion and followed the general sequence of degradation: triglycerides > total fatty acids > lipids > total organic substances (Poltz, 1972). Rates of decomposition decreased markedly in the hypolimnion. Only small amounts of the substances produced in the plankton were found settling to the sediments: 4 to 10% of particulate organic matter, 1–2.5% of lipids, 0.1–0.5% of total fatty acids, and <0.1% of triglycerides. After deposition, further degradation rates were most rapid at the surface interface zone of the sediments, although the rates were slower than those occurring in epilimnetic waters during the process of sedimentation.

Much less is known about the production, utilization rates, and losses of dissolved fatty acids in aquatic ecosystems. Many long-chain lipophilic compounds are attached to particulate organic carbon. Short-chain dissolved fatty acids are commonly directly proportional to the release of organic compounds from photosynthesizing phytoplankton (Fogg, 1983). For example, glycolic acid, a product in the competition of oxygen with CO_2 as a substrate in the ribulosebisphosphate carboxylase reaction, can be excreted by algae rather than metabolized to serine or glycine. Released glycolic acid can show a distinct diurnal variation in the trophogenic zone of a eutrophic lake with maxima (ca. 50–100 μg C as glycolic acid liter^{-1}) during the daylight periods of maximum photosynthesis and declining to minima (ca. 1 μg C liter^{-1}) during darkness (Münster, 1991). Because glycolic acid is an excellent substrate for bacterioplankton, the observed diurnal distribution is likely a balance between production and release in excess of bacterial utilization during the day and rapid utilization as an energy source for respiration and transport of other carbon substrates across membranes during darkness.

These analyses were limited to the pelagic zone of lakes and consequently do not reflect the massive amounts of particulate organic matter that can reach the sediments of many lakes when the littoral flora dominates production. The sediment interface is clearly the site of the most intensive respiratory metabolism in many lakes (see Chap. 21), and much of the organic matter undergoing degradation there originates from littoral and allochthonous sources. Part of the lipid derivatives originate from the bacteria, algae, and fungi; these derivatives may be intermediate metabolic products formed during the degradation of this accumulated organic matter (cf. Schulz and Quinn, 1973).

C. Humic Compounds and Their Photolysis to Simple Organic Substrates

As much as 80% of dissolved organic matter (DOM) in inland waters is composed of organic acids that originate largely from higher aquatic and terrestrial plants. Of these organic acids, some 30–40% are composed of aromatic carbon originating from structural plant tissues (Malcolm, 1990). Concentrations of organic acids are commonly in the range of 4–8 mg liter^{-1} and can exceed 50 mg C liter^{-1} in wetlands, floodplain waters, and interstitial waters of hydrosoils (Wetzel, 1984; Mann and Wetzel, 1995).

The high concentrations of DOM in inland surface waters clearly result from high photosynthetic productivity associated with lake and river ecosystems, particularly associated with wetland and littoral regions, as well as large loadings of DOM from decomposition of plant materials within the drainage basin (Wetzel, 1990a, 1992). Similarly, the high photosynthesis and decomposition within the extensive floodplain marginal areas of river ecosystems also serve as origins of DOM to river runoff water (Wetzel and Ward, 1992). More labile compounds of these heterogeneous organic mixtures are selectively degraded by microbiota as the water is transported along the gradient from land through the littoral to pelagic regions. Residual organic compounds contain high concentrations of the relatively recalcitrant humic substances that originate largely from the partially degraded plant structural tissues (Frimmel and Christman, 1988; Purdue and Gjessing, 1990).

Humic compounds are structurally complex (phenolic linkages; see detailed discussion in Chap. 23) and tend to have long residence times in lakes and streams. In general, their degradation by aquatic microflora proceeds slowly. A portion of the humic materials present in a lake is truly dissolved, but under certain conditions, humic substances can aggregate into colloids or flocs. High-molecular-weight dissolved humic materials readily adsorb to particulate matter (Davis and Gloor, 1981), which may further alter their rates of degradation. Rates of glucose mineralization are also very low in tropical lakes that receive high loading of dissolved humic materials (Rai, 1978, 1979).

Despite the frequently stated erroneous opinion that humic substances are poorly metabolized, in reality total dissolved organic matter, of which 50–80% is made up of humic substances, of most surface waters does not increase even though sources constantly import large quantities to the pelagic pool. Spatial concentrations of total dissolved organic matter with depth change relatively little over seasons (e.g., Wetzel *et al.*, 1972). Losses that balance these influxes are occurring by many mechanisms: microbial decomposition, photolysis, aggregation and sedimentation, and outflows.

Inputs of allochthonous dissolved organic matter to surface waters are directly proportional and coincident with rainfall (see Chap. 23). Bacterioplankton growth and productivity are often more strongly correlated with the inputs of allochthonous dissolved organic carbon (DOC) from the catchment than with phytoplankton productivity in small lakes (e.g., Wetzel and Otsuki, 1974; Mickle and Wetzel, 1978b; Hessen, 1985; Coveney and Wetzel, 1992, 1995), very large lakes (Arvola and Kankaala, 1989; Bertoni and Callieri, 1989; Laybourn-Parry *et al.*, 1994), and rivers (McKnight *et al.*, 1993). Tranvik (1990) found that at least half of bacterioplankton growth utilized DOC of humic substances of a molecular weight between 1,000 and 10,000 D. Rates of uptake and turnover of polymeric and monomeric aromatic dissolved organic compounds were nearly as efficient and rapid as those of simple amino acids (Münster *et al.*, 1999). The ratio between aromatic carbon and amino acid carbon uptake varies between 0.2 and 5.2 (mean 2.8; see Table 17-5). Although these recalcitrant compounds are used less efficiently than more simple substrates, the abundance of the more complex humic substances is so large proportionally that the collective slower rates of mineralization compensate in the large pool available.

Haan (1972) demonstrated that a species of *Arthrobacter* was able to grow slowly on humic fractions and that it was able to hydrolyze the phenolic ether linkages. Studies of the degradation of dissolved humic compounds in heavily stained Finnish lake waters also provided evidence that these compounds, although relatively resistant, are mineralized slowly (Ryhanen, 1968; Sederholm *et al.*, 1973). Enrichments with inorganic nitrogen and phosphorus accelerated the rate of degradation. More recent studies have confirmed that bacterial degradation of dissolved organic matter of humic-rich waters can be prodigious, although these high-molecular-weight and phenolic-dominated compounds are not utilized directly at rates comparable to those of simple extracellularly released DOC (e.g., Tranvik, 1988, 1989, 1990; Sundh, 1992; Amon and Benner, 1996; see later discussion on bacterial productivity).

Studies of the effects of a low-molecular-weight humic fraction on the growth of a species of *Pseudomonas* showed that the presence of the humic fraction caused an increase in the cell yield and in the

TABLE 17-5 Comparison of Approximate Rates of Turnover of Organic Substrates by Natural Bacterioplankton in Rivers and in Lakes of Increasing Productivity Based on the Range of Maximum Phytoplanktonic Photosynthetic Rates of Carbon Fixation (P_{max})

Habitat	*Substrates*	P_{max} *(mg C* m^{-3} *day*$^{-1}$*)*	T_t *(h)*	*Source*
Lakes				
Oligotrophic				
Lapplandic lakes, Sweden (summer)	Glucose, acetate	1–30	>10,000	Rodhe *et al.* (1966)
Lawrence, Michigan (annual)	Glucose	1–80	40–300	Wetzel *et al.* (1972)
	Acetate		10–120	
Tupé, Brazil (annual)	Glucose	—	0.5–>30,000	Rai (1978, 1979)
Mesotrophic				
Crooked, Indiana (annual)	Glucose	63–110	80–470	Wetzel (1967, 1968)
	Acetate		20–350	
Klamath, Oregon (summer)	Water, glucose	—	220	Harrison *et al.* (1971)
	Acetate	—	250	
	Sediments, glucose		2.25	
	Acetate		0.75	
Gravelly Pond, Massachusetts (summer)	Glycolate	—	60–200	Wright (1975)
Kizaki, Japan	Glucose	—	12–58	Kato and Sakamoto (1983)
	Amino acids	—	3–13	
Kinneret, Israel	Glucose	—		Cavari *et al.* (1978)
0–10m		—	88	
20–40m		—	133	
Janauari, Brazil (annual)	Glucose	—	15–900	Rai (1978)
Four Várzea lakes, Brazil (annual)	Glucose	—		Rai (1979)
Low water		Higher	40–101	
High water		Lower	105–>10,000	
Erken, Sweden				
Summer	Glucose	40–130	10–100	Hobbie (1967)
Winter	Glucose	2–20	100–1,000	
Lake Mekkojärvi, Finland				
Epilimnion	Leucine	—	48 (9–178)	Münster *et al.* (1999)
	Hydroxybenzoic	—	38 (28–48)	
Hypolimnion	Leucine	—	35 (1–47)	
	Hydroxybenzoic	—	55 (35–68)	
Eutrophic				
Wintergreen, Michigan (summer)	Glucose	80–120	<1–20	Saunders (pers. comm), Wetzel (unpublished)
Little Crooked, Indiana (annual)	Glucose	190–205	36–232	Wetzel (1967, 1968)
	Acetate		24–190	
Duck, Michigan (annual)	Glucose	10–320	8–50	Miller (1972)
	Acetate		4–40	
Pamlico River (estuary), North Carolina	Amino acids	—	1.5–26	Hobbie (1971)
Lötsjön, Sweden				
Summer	Glucose	<100	0.4–5	Allen (1969)
Winter	Glucose	<20	20–300	
Upper Klamath, Oregon (summer)	Glucose	—	2.4	Wright (1975)
	Acetate	—	2.3	
	Glycine	—	8	
	Glycolate	—	26	
Three Polish lakes, Poland	Palmitic acid	—	7–18	Chróst and Gajewski (1995)
Plußsee, Northern Germany (annual)	Glucose	20–150	6–202	Overbeck (1975)
	Glucose	—	126–263	Chróst (1984)
	Acetate	—	150–290	
Streams/Rivers				
Breitenbach, Germany	Glucose	—	14–464	Marxsen (1980)
	Acetate	—	3–224	
Rohrwiesenbach, Germany	Glucose	—	7–450	
	Acetate	—	10–950	
Two Australian river waters	Glucose	—	2–7	Boon (1991)
	Acetate	—	4–6	
	Glutamic acid	—	5–6	

cell yield per unit of respiratory consumption of oxygen as compared to growth in organic media lacking the fraction (Haan, 1974). These results suggest that the stimulating effect of low-molecular-weight humic material on bacterial metabolism operated by coupling to the metabolism of less recalcitrant organic substrates. Similar results were found by Stabel *et al.* (1979) and Geller (1983). Cometabolism of many very recalcitrant compounds (e.g., herbicides) is well known (Horvath, 1972; Hulbert and Kraviec, 1977) and is probably an important mechanism for the degradation of the more recalcitrant compounds of fresh waters. The mechanisms of such cometabolism deserve greatly increased investigation and probably represent a major mechanism for biological utilization of these compounds.

It is well-known that ultraviolet irradiance (UV) can photolyse portions of proteinaceous and humic macromolecules (e.g., Manny *et al.*, 1971; Zepp *et al.*, 1981; Geller, 1986; Amon and Benner, 1996). Detailed chemical analyses of these transformations showed that small organic fractions, particularly numerous small fatty acids—such as acetic, formic, citric, pyruvic, malic, levulinic, among others—were generated by photolysis of the humic substances (Wetzel *et al.*, 1995; Moran and Zepp, 1997). Even before these photolytically generated simple organic substrates were identified, many studies demonstrated that DOM exposed to natural UV radiation exhibited immediate stimulation of and sustained bacterial growth (e.g., Stewart and Wetzel, 1981, 1982b; Wetzel *et al.*, 1995; Lindell *et al.*, 1995, 1996; Moran and Zepp, 1997). Different plant species yield very different fatty acids upon natural photolysis of DOM from different stages of their long-term decomposition (Boon *et al.*, 1982; Wetzel, 2000c).

Photolysis of natural humic and related substances by natural irradiance is unquestionably a major physical process that can generate large quantities of simple, readily utilizable organic substrates for bacterial metabolism (see further discussion in Chap. 23). The question is how significant this photolysis is to sustained bacterial metabolism and productivity in surface waters. As we have discussed (Chap. 5), UV-B and UV-A are attenuated rapidly with depth, and rates of attenuation are directly proportional the content of DOC. Short-term photoinhibitory effects of UV irradiance to planktonic bacteria and algae are common in oligotrophic waters (Herndl *et al.*, 1993, 1997; Karentz *et al.*, 1994; Sommaruga *et al.*, 1997). Convective and turbulent mixing of the surface layers can move bacterioplankton to depths where photoenzymatic repair of DNA leads to efficient recovery of bacterial activity. Humic substances that were modified by surface irradiance are also moved to photoprotected regions and are readily utilized by bacterioplankton. Humic substances not only alter the kinetics of photochemical reactions that lead to photolytic degradation but in addition, by utilization of some of the energy, these recalcitrant DOC compounds lessen the photoinhibitory effects of the UV radiation. Exposure of bacterioplankton and phytoplankton to low intensities of UV irradiance enhanced bacterial activity by 20–>100% in a few hours, much of which may have resulted from possible injury to algae and the increased release of organic substrates from the algae. The primary effect of photolysis of humic substances, however, is to potentially generate hundreds of μg liter^{-1} of simple fatty acid substrates for bacterial utilization. Even if only 10% of the large pool of dissolved humic substances is photolysed in the surface half-meter of water, these compounds can be circulated into photoprotected strata of the epilimnetic waters and be readily utilized by bacterioplankton, particularly during periods of darkness. Recent evidence also indicates that portions of photosynthetically active radiation (PAR) also photodegrade humic substances (see Chapt. 23).

D. Utilization Rates of Specific Dissolved Organic Compounds

The relative rates of degradation of the composite pool of hundreds of organic compounds that constitute total dissolved organic matter (DOM) by a heterogeneous composite community of hundreds of species of heterotrophic bacteria are difficult to quantify in natural waters. In natural systems, each species of bacteria potentially has different abilities to assimilate specific substrates. Measurements of total community production rates of the composite heterotrophic bacterial communities are now reasonably estimated by rates of incorporation of amino precursors into bacterial DNA and proteins (e.g., Kemp *et al.*, 1993; Wetzel and Likens, 1999). Measurements of specific substrate assimilation by bacteria, however, are fraught with methodological problems (cf. Bell, 1980; Bell and Sakshaug, 1980; Wetzel and Likens, 1991, 2000). The kinetics of organic substrate utilization by heterogeneous planktonic populations, determined by the uptake rates of radioactivity-labeled organic compounds at substrate concentrations that occur naturally, are instructive in a relative if not absolute way.

The uptake of inorganic ^{14}C-labeled carbon by natural populations over short intervals of time has been used widely as a measure of rates of carbon fixation by photosynthesis. Parsons and Strickland (1962) employed labeled organic substrates to measure uptake by natural heterotrophic populations in the dark by the relationship

$$v = \frac{cf(S_n + A)}{C\mu t}$$

when

v = rate of uptake (mg C m^{-3} h^{-1})
c = radioactivity of the filtered organisms (cpm, counts per minute)
f = a correction for isotopic discrimination (1.05 or 5% slower uptake of ^{14}C, which has a greater mass than ^{12}C)
S_n = *in situ* concentration (mg $liter^{-1}$) of the organic substrate
A = concentration (mg $liter^{-1}$) of added substrate (labeled and unlabeled)
C = cpm from 1 μCi of ^{14}C-labeled substrate on the radioassay instrumentation used
μ = quantity of ^{14}C added to the sample
t = incubation time (hours)

This equation assumes that natural substrate concentrations (S_n) are much less than A. When the uptake of a solute is mediated by a transport system located on or in the cell membrane, the rate of uptake can be described by Michaelis-Menten kinetics, when

$$v = \frac{V_{max}(S)}{K_t + S}$$

v = velocity at a given substrate concentration S
V_{max} = maximum velocity, attained when uptake sites are continually saturated with substrate
K_t = transport constant, by definition a measure of the affinity of the uptake system for the substrate. Also called K_m, the Michaelis constant, the substrate concentration at which the velocity is one-half of the maximum velocity V. By definition, $v = V_{max}/2$.

As developed by Wright and Hobbie (1966; cf. also Allen, 1969), this nonlinear uptake relation over substrate concentration can be transformed into a linear relationship by the Lineweaver–Burk equation to yield

$$\frac{C\mu t}{c} = \frac{(K_t + S_n)}{V} + \frac{A}{V}$$

With this equation, data from uptake measurements from algae and bacteria at low substrate concentrations can be plotted as $C\mu t/c$ vs. A, giving values for (K_t + S_n) and V (Fig. 17-7). The negative intercept on the abscissa is equal to (K_t + S_n), and the reciprocal of the

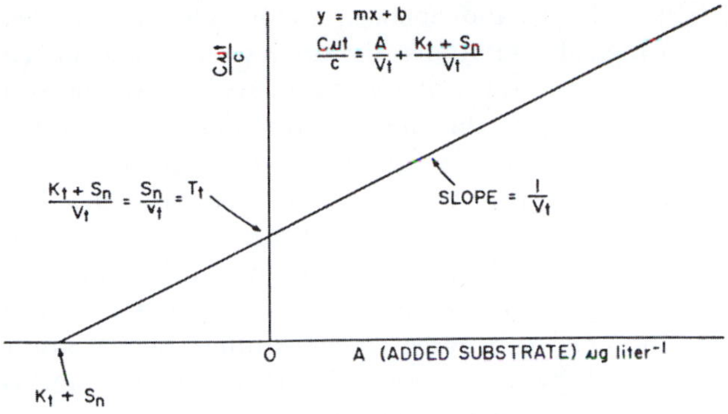

FIGURE 17-7 Graphical analysis of bacterial uptake at low organic substrate concentrations following Michaelis–Menten enzyme kinetics. A plot of $C\mu t/c$ against increasing substrate concentrations, S, illustrating derivation of (1) maximum natural substrate concentrations, $K_t + S_n$, as μg $liter^{-1}$, (2) maximum velocity, V, as μg $liter^{-1}$ h^{-1}, and (3) turnover time for substrate regeneration (T_t) in hours. (After Allen, H. L.: Chemo-organotrophic utilization of dissolved organic compounds by planktonic algae and bacteria in a pond. *Int. Rev. ges. Hydrobiol.* 54:1–33, 1969.)

slope is V, the maximum rate of uptake. The ordinate intercept is equivalent to the turnover time (T_t), which is the time required for complete removal of the natural substrate by the microflora (Hobbie, 1967). The (K_t + S_n) approximates the maximum natural substrate concentration (S_n) if K_t is very small, as is often, but not always, the case. These relationships assume that a constant rate of regeneration of the organic solute is occurring *in situ*, or that steady-state conditions exist. An appreciable portion of ^{14}C organic substrate is respired rapidly by the microflora, and corrections for this loss must be made (Hobbie and Crawford, 1969; Wetzel and Likens, 2000).

In contrast to the already discussed nonlinear active uptake velocities at low substrate concentrations, the uptake of organic compounds by natural planktonic populations at high substrate concentrations (approximately >0.5 mg $liter^{-1}$) does not exhibit rate-limitation kinetics or saturation of uptake sites. Passive uptake velocity continually increases with rising substrate concentrations. The slope of the response line is constant (K_d), derived from diffusion kinetics, and has been used to estimate diffusion uptake by natural populations of algae and bacteria.

Analyses of utilization rates of organic substrates under *in situ* conditions with natural populations have been variously applied and usually employ only a few simple substrates such as glucose, other sugars and isomers of glucose, acetate, glycolate, and amino acids. Only a few analyses have been extended over an annual period (Hobbie, 1967, 1971; Wetzel, 1967; Allen, 1969; Overbeck, 1975; Rai, 1978, 1979; Robarts,

1979; Gillespie and Spencer, 1980). Despite inherent physiological limitations, the analyses do yield insight into utilization rates of specific dissolved organic substrates. Foremost, they demonstrate that dark diffusive algal uptake of simple organic substrates for use in heterotrophic growth at low natural substrate concentrations is almost always <10% of active permease-mediated uptake by bacteria. Stated in another way, the algae are generally ineffective in competing for available organic substrates at substrate concentrations maintained by active bacterial heterotrophic uptake (see discussion in Chap. 15, pp. 356ff).

Maximum velocities of uptake are quite variable among lakes and different organic substrates but generally occurred after spring or following late summer algal maxima. Rates decrease by about an order of magnitude during winter in temperate lakes. Concentrations of substrates such as glucose, acetate, and amino acids remain low throughout the year, for the rate of inputs to the labile dissolved organic pool is balanced roughly by the rate of removal by bacteria. Patterns of vertical distribution of uptake velocities tend toward maxima during and following algal maxima in the trophogenic zone, minima in the lower metalimnion–upper hypolimnion, and marked increases in rates near the sediments. Within the sediments, utilization velocities of simple organic solutes are several orders of magnitude greater than in the overlying water (Harrison *et al.*, 1971).

Utilization rates of simple substrates generally increase with greater phytoplankton productivity in an approximately direct relationship (Table 17-5). Great variations are observed in these rates both seasonally and vertically with changes in depth. Uptake rates of simple carbohydrates generally increase with greater bacterial densities and biomass and with increasing concentrations of total inorganic nitrogen and total phosphorus (e.g., Bowie and Gillespie, 1976; Spencer, 1978). Glucose assimilation rates by planktonic bacteria were found to be nearly an order of magnitude greater in littoral areas where submersed macrophytes were abundant than in the pelagic areas (Gillespie and Spencer, 1980). Heterotrophic utilization of glucose by neustonic bacteria at the surface air–water interface was found to be much less than by planktonic bacteria (Dietz *et al.*, 1976). These data may be viewed as general *direct* measurements of the expected positive correlation between increased bacterial metabolism of ubiquitous simple substrates and waters with larger inputs of organic matter. Methodological problems, such as competitive inhibition of one substrate on the uptake of another (Burnison and Morita, 1973), substrate affinities (Button, 1978; Moaledj and Overbeck, 1980), variable substrate uptake in relation to phosphorus availability (Overbeck and Tóth, 1978), and variable corrections for respiratory losses of CO_2, preclude a more detailed evaluation of interactions.

V. CONTROL OF BACTERIOPLANKTON BY BIOTA

As we have seen above, communities of bacterioplankton have specific growth rates on the order of 1 day^{-1}. Although appreciable diurnal variations in growth are seen and reasonably explained, there is relatively little seasonal variation in bacterial abundance, which suggests that growth and morality are relatively balanced. Although few investigators would disagree that resource limits to growth are approximately balanced by mortality losses, differences in opinion arise concerning the relative importance of parasitism, predation, UV lysis, and other potential causes of death. Supporting data are few but are improving for inland water ecosystems.

A. Dormancy and Natural Mortality

As pelagic conditions alter, bacteria either adjust physiologically or by means of rapid genetic modifications to the new environment, die, or enter a state of dormancy. Statements that fissile, vegetative microbes do not age, senesce, and die, but rather are only replaced by two young individuals (e.g., Postgate, 1976) are thermodynamically, biochemically, genetically, ecologically, and intuitively unreasonable. Alternatively, it is often argued that the physiological state of a significant portion of the bacterial community in most aquatic environments is dormant (e.g., Stevenson, 1978). Although there is repeated evidence for the formation of smaller cells, microcysts, and similar diminutive cells under nutrient, substrate, and other environmental limitations, physiological data supporting a dormant state are sparse. As indicated earlier, microautoradiographic analyses of the percentage of natural bacteria that assimilate specific organic substrates nearly always demonstrate uptake by a small (<50%) portion of the composite community of many different bacterial species. Inability or inactivity for assimilation of one substrate does not necessarily indicate inability or inactivity for others, however.

Mortality of bacterioplankton in the absence of grazing by protists and zooplankton is usually <50% per day (e.g., Servais *et al.*, 1985; Pace, 1988; Wetzel, 1995). Much of that loss is likely associated with simple senescence under starvation conditions, particularly if exposed to high light, as well as parasitism.

TABLE 17-6 Number of Viruses or Phages and Ratio of Number of Viruses to Bacteria (V/B) in the Pelagic of Fresh Waters Determined by Direct Enumeration by Transmission Electron Microscopy (TEM) or Epifluorescent Light Microscopy (ELM)

Habitat	*Viruses/ml*	*V/B*	*Preparation*	*Source*
Plußsee, Germany	2.5×10^8	40	TEM	Bergh *et al.* (1989)
	$>10^8$	—	TEM	Demuth *et al.* (1993)
	1.6×10^7 to 3.7×10^8	2–50	ELM	Witzel *et al.* (1994)
Lake Kalandsvannet, Norway	1.9×10^7 to 2.0×10^8	24–60	TEM	Heldal and Bratbak (1991)

B. Viral Mortality

The bacterioplankton community is made up of many strains, all exhibiting varying population dynamics and growth rates. Changes in abundance can be rapid, in the range of 0.2–0.6 day^{-1} (Tuomi *et al.*, 1997). Mortality of bacterioplankton from viral infection can be very high. Viruses have been observed to be very abundant, as, for example, ca. $2.5–3.7 \times 10^8$ ml^{-1} in eutrophic Plußsee, Germany (Bergh *et al.*, 1989; Witzel *et al.*, 1994). Average virus-to-bacteria ratios of 50:1 have been observed with a high direct correlation between the abundances of viruses and bacteria (Table 17-6). If viruses do not encounter a bacterial or algal host, viral decay rates among natural viral populations occur, during which viral nucleic acid is released from the capsid (Bratbak *et al.*, 1994; Wommack and Colwell, 2000). The rate of this inactivation of phages is typically on the order of 0.1–2 day^{-1}, although much longer rates can occur before instability sets in (see Chap. 23).

Based on *in situ* observations and experimental results, most estimates indicate that viral phages lyse about one-fourth (range, 2–24%) of the bacterial production per hour (Heldal and Bratbak, 1991). On the average, 50 mature virus particles (range, 10–300) are released from each killed bacterium. Phage-induced mortality of bacterioplankton in Lake Constance, Germany, was estimated to vary between 1 to 24% of total mortality over an annual period (Simon *et al.*, 1998). Other viral mortality estimates of bacterioplankton are in a similar or higher range (11–66%) for lakes, coastal waters, and rivers (e.g., Suttle, 1994; Fuhrman and Noble, 1995; Mathias *et al.*, 1995; Wommack and Colwell, 2000).

C. Grazing by Protists and Larger Zooplankton

Bacterivory by protists and larger zooplankton is clearly a major and often dominant mortality factor within natural bacterioplankton communities. The feeding effectiveness on bacterioplankton and their importance in the nutrition, growth, and productivity of both protistan protozooplankton and larger zooplankton were discussed at length in Chapter 16 (pp. 405–406; 427). Many protists feed on bacteria-sized particles and picophytoplankton (<2 μm) and thus utilize a size class of bacteria, algae, and particulate detritus that is not as effectively utilized by larger zooplankton. This protistan bacterivory, particularly by flagellates, is a dominant mortality factor of bacterioplankton. At this point, the questions are directed to the effects of grazing on the development, growth, and productivity of the bacterioplankton under different conditions.

Although crustacean zooplankton can consume bacteria in seemingly large quantities (Table 17-7), the rates are relatively low in comparison to the bacterial biomass present. In comparison to other bacterivorous grazers over an annual period, on average grazing by daphnids accounted for 9–12% of bacterioplankton mortality; ciliates, 14–19%; and heterotrophic nanoflagellates, 52–68% in Lake Constance (Güde, 1986; Simon *et al.*, 1998). An attempt was made to generalize the relative impacts of different groups of bacterivores on total bacterioplankton productivity (Table 17-8), but variance is very large seasonally and among lakes of differing chemical composition, dissolved organic matter, and other parameters. These values are likely overestimates because the true rates of viral mortality is conservatively estimated and not known directly.

Bacterivory values greatly in excess of 100% are frequently found in the literature. Such values may be real during particular instantaneous measurements but obviously could be sustained only for very short periods before supply is suppressed and feeding efficiencies decline precipitously. Furthermore, many of the estimates of feeding are based on the likely erroneous assumption that bacterivory rates are equal among extant protists and that their physical presence and

TABLE 17-7 Grazing Rates by Large Zooplankton on Bacterioplankton

Ecosystem	*Ingestion rate bacteria individual^{-1} h^{-1}*	*Source*
Rotifers		
Anuraeopsis fissa	500–920	Ooms-Wilms *et al.* (1995)
Keratella cochlearis	150–230	
Conochilus unicornis	150	
Brachionus angularis	820–910	
Filinia longiseta	5100–7950	
Cladocerans		
Bosmina coregoni	2650	Ooms-Wilms *et al.* (1995)
Bosmina longirostris	3100	
	38–82	Jürgens and Stolpe (1995)
Chydorus sphaericus	5700	Ooms-Wilms *et al.* (1995)
Daphnia cucullata	20,000	
D. magna	475–2780	Jürgens and Stolpe (1995)
D. hyalina	170–900	

enumeration allows a direct estimate of community bacterivory. Careful analyses among several different habitats indicated that rarely are more than half of the bacterivorous nanoflagellates actively feeding at any given time (González, 1999). The fraction of active protistan grazers is time-dependent and reaches a saturation point in a few minutes, depending upon the environmental conditions. Percentages of actively grazing nanoflagellates appear to be lower (ca. 25%) in oligotrophic ecosystems in comparison to those (ca. 40–60%) of more eutrophic pelagic systems.

Bacterivory is dominated by heterotrophic nanoflagellates. Ciliates and rotifers rarely impact bacterioplankton communities significantly by ingestion. Even though *Daphnia* and other cladoceran zooplankton are relatively inefficient bacteriovores, at times seasonally, as for example during the "clearwater phase," very large populations of cladocerans can be effective in reducing the bacterioplankton. Moreover, among very eutrophic lakes where bacterial specific growth rates and production are enhanced by high nutrient loading, the total bacterivory by cladocerans can increase relative to protistans. Many examples of these relationships have been observed (e.g., Pace *et al.,* 1990; Chrzanowski and Šimek, 1993; Pace and Cole, 1996; Simon *et al.,* 1998b).

VI. BIOTICALLY RELEASED DISSOLVED ORGANIC MATTER

Dissolved organic compounds are released by organisms during active growth and by autolysis upon senescence and death.

TABLE 17-8 Potential Bacterivory in Lakes Expressed as a Percentage of the Bacterioplankton Production Based on Abundances of Zooplankton and Bacteria and Literature Values of Specific Clearance Rates[a]

Organisms	*Clearance rate (μl individual^{-1} h^{-1})*	*% of bacterial production*
Heterotrophic nanoflagellates	0.6–29	3–70
Ciliates	0–0.53	1–19
Rotifers	0.5–10	<7
Cladocera		
Daphnia spp.	12.5–450	1–12
Other cladocerans	10–330	<10
Copepoda	4.6–158	6–12

[a] Composite from Sanders *et al.* (1989), Tranvik (1989), Chrzanowski and Šimek (1993), and Simon *et al.* (1998a,b).

TABLE 17-9 Average Percentage Loss of Total Substance of Different Aquatic Organisms Immediately and 24 h after Death, Aerobic, 20° or 25°C

Organisms	*Immediate loss* (%)	*Loss after 24 h* (%)	*Source*
Planktonic green algae	11	33	Krause (1962)
Planktonic diatoms	11	15	
Phytoplankton	4–34	11–43	Hansen *et al.* (1986)
Leaves of aquatic plants			
Callitriche hamulata	9	16	Krause (1962)
Scirpus subterminalis	35	—	Otsuki and Wetzel (1974b)
Scirpus acutus	—	4	Godshalk and Wetzel (1978b)
Najas flexilis	—	20	
Myriophyllum heterophyllum	—	15	
Nuphar variegatum	—	23	
Carex rostrata	7	—	Danell and Sjöberg (1979)
Equisetum fluviatile	1	—	
Leaf litter			
Acer platanoides	—	80–90	Gunnarsson *et al.* (1988)
Ulmus americana	9	15	Kaushik and Hynes (1971)
Alnus rugosa	3	5	
Mixed zooplankton	15	52	Krause (1962)
Cladocera *(Daphnia)*	29	—	
Copepods *(Cyclops and Diaptomus)*	8	—	
Sediment-living worms (*Tubifex*)	5	68	
Small fish (*Lebistes*)	7	28	

A. Biotically Released Dissolved Organic Matter: Autolysis

In the past two decades, great enthusiasm arose for explanations of complex food-web structures and interrelationships. The underlying themes of these excellent studies elucidating material fluxes among biota focused on animal predation and predatory impact relationships within the communities. Emerging from these collective studies was a highly erroneous assumption that much of the production was ingested and the most organic matter was utilized by heterotrophic metazoans. In reality, most organic production simply senesces, lyses, and is decomposed by heterotrophic microorganisms (Wetzel, 1995). Longevity and senescence are genetically programmed, and although the temporal sequences can be modified by environmental parameters, genetic and biochemical senescence and death is finite. This area of intensive contemporary research among higher plants and animals is almost totally neglected among aquatic organisms and deserves greatly accelerated study because it is a fulcrum for the detrital dominance of metabolism in freshwater ecosystems.

As organisms, including bacteria and viruses, senesce and die, ***autolysis*** of cytoplasmic cellular contents by deterioration of cellular membranes rapidly releases significant amounts of dissolved organic matter (DOM). This release is followed by rapid utilization of energy- , carbon- , and nitrogen-rich substrates by bacteria. The loss of total substance is very high immediately after "death" in all organisms (Table 17-9). Although such release is high among higher organisms, such is stated to be so, without evidence, for bacteria on the premise that bacteria simply enter states of starvation and dormancy (Güde, 1988; cf. review of Morita, 1982). It is certainly unlikely that autolysis would occur at the same rate as growth, but it must be emphasized that although bacterivory can be and is often dominant, mortality can occur by other means as well, and these means are usually never examined simultaneously with grazing.

The autolytic release of DOM initially by algae, aquatic plants, and zooplankton and its progressive decomposition thereafter has been studied in a few cases (Botan *et al.*, 1960; Krause, 1964; Otsuki and Hanya, 1968, 1972a,b; Golterman, 1971; Otsuki and Wetzel, 1974; Godshalk and Wetzel, 1978a–c; Hansen *et al.*, 1986). Data from these studies illustrate the general relationships believed to prevail. Organic carbon of dead green algal cells decomposing under aerobic conditions declined about 55% in 5 days (20°C) and decreased more slowly thereafter. Cell morphology changed little during the first 50 days, indicating that although the cellular contents were decomposed quickly after death, the cell wall was relatively resistant to microbial degrada-

TABLE 17-10 Changes in Composition of Dissolved Organic Products during the Aerobic Decomposition of *Scenedesmus* Cells at 20°C[a]

Time (days)	*Organic C (mg liter^{-1})*	*Organic N (mg liter^{-1})*	*C:N ratio*
0	213	22	11
35	184	21	10
185	87	17	6

[a] From Otsuki and Hanya (1972a).

tion. Organic nitrogen of the cells was reduced by about 70% during the first 30 days; thereafter, decomposition of both organic C and N was slower. Dissolved organic C and N released by autolysis, and subsequently from intermediate bacterial products of decomposition, are degraded rapidly at first; rates of degradation then decrease with time (Table 17-10). Similar results were found with the decomposition of submersed vascular aquatic plants (Chap. 21).

B. Utilization of Extracellularly Released Dissolved Organic Carbon

Dissolved organic compounds also originate by secretion (extracellular release) from actively growing algae (discussed in Chap. 15, p. 382ff.) and larger aquatic plants (Chap. 18). In general, extracellular release of dissolved organic carbon from phytoplankton increases nonlinearly with increasing productivity and becomes a very low percentage (<5%) of total photosynthesis in eutrophic lakes (see Chap. 15, pp. 382ff.). The quantities and qualitative composition of extracellularly released organic matter are variable and depend upon the species and their physiological status under existing environmental conditions. Simple organic acids, sugars, and more complex carbohydrates, amino acids, peptides, pigments, and enzymes (see Chap. 15, pp. 382–384) are among the compounds known to be liberated during active growth. Aquatic bacteria possess uptake systems of high substrate affinity and low substrate specificity.

Most of these secreted compounds are assimilated rapidly by the microbial flora, so their concentrations in the water would always be expected to be low. Some extracellular products, such as polypeptides, are more resistant to rapid degradation and can form a more significant portion of the total dissolved organic matter at any given time. Extracellular release of organic compounds during active growth represents an elimination of metabolic products and is a relatively small loss of synthesized organic matter for the plants. Demarcation between true physiological loss of photosynthate from release of dissolved organic compounds during cellular senescence and autolysis is not distinct; certainly both occur simultaneously with rapidly changing rates of release in natural heterogeneous algal and bacterial communities containing hundreds of species at any given time and location.

Dissolved organic compounds are also released by secretion from zooplankton and by leaching of their feces. For example, a significant amount (up to 17%) of the algal carbon ingested by cladoceran zooplankton can be immediately lost as dissolved organic carbon as the algae are damaged during feeding (Lampert, 1978a). More than 10% of the algal particulate carbon removed from suspension by grazing can be transformed to dissolved organic carbon by these processes. Compounds released through these processes can be of significance to the planktonic bacteria, since substrate availability in pelagic waters is usually limiting.

Extracellular organic products of phytoplankton are utilized in different ways (Nalewajko, 1977). Clearly, different phytoplankton release different compounds extracellularly. For example, low-molecular-weight compounds (<1 kD) predominated those released by diatoms, whereas cyanobacteria released high-molecular-weight compounds (Cho *et al.,* 1997). Low-molecular-weight substances are assimilated rapidly by bacteria or cyanobacteria, essentially simultaneously as rapidly as excreted (Nalewajko *et al.,* 1980; Chang, 1981). Binding of low-molecular-weight compounds to colloids or to the surfaces of particulate matter may also occur and can result in abiotic removal of these substances from solution through subsequent sedimentation. One would anticipate marked differences in assimilation rates among bacterioplankton based on the phytoplankton predominance at any particular time or spatial distribution.

Estimates of bacterioplanktonic utilization of secreted organic compounds from natural populations indicate rapid rates of assimilation. For example, in two eutrophic lakes in southern Sweden as well as in natural mixed planktonic communities from an oligotrophic lake in Michigan, from 28 to 80% of the released dissolved organic carbon was recovered in bacterial particulate fractions (Coveney, 1981; Coveney and Wetzel, 1989). First-order rate constants for utilization of extracellularly released organic carbon exhibited a mode of 0.05–0.15 per day. An average of 50% (range, 33–78%) of gross uptake of extracellularly released organic carbon was respired. Variations in utilization were positively correlated with temperature. As a result of this utilization, an appreciable portion of the dissolved organic carbon released from photosynthesizing algae is returned to the pelagic food web as microbial biomass, despite high losses to respiration.

Similar results have been found in many freshwater and marine pelagic ecosystems (e.g., Saunders and Storch, 1971; Meffert and Overbeck, 1979: Larsson and Hagström, 1979; Bell and Satzshang, 1980; Chróst and Faust, 1983; Brock and Clyne, 1984; Jones and Cannon, 1986; Feuillade *et al.,* 1988; Chrzanowski and Hubbard, 1989; Baines and Pace, 1991; Nürnberg and Shaw, 1998). As the quantities of recalcitrant organic substrates of total DOC increase, for example, in lakes and rivers receiving high loads of humic substances, the importance of the DOC released extracellularly from algae decreases markedly as substrates for bacterioplankton (Jones and Salonen, 1985; Miller, 1987; Sundh and Bell, 1992). In summary, it is estimated that under most circumstances, less than half of the carbon required for growth of bacterioplankton results from extracellular release of organic substrates from phytoplankton.

VII. DECOMPOSITION OF PARTICULATE ORGANIC DETRITUS

Kinetic evaluations of aerobic decomposition of algae by bacteria approximate first-order reaction rates during the initial 30 days of degradation. Decomposition of cellular nitrogen and the production of dissolved organic nitrogen demonstrate that dead algal organic nitrogen is separable into labile and rocalcitrant constituents according to their relative resistance to bacterial degradation. Additional dissolved organic nitrogen is generated by bacteria through reassimilation of mineralized nitrogen. The general sequence of decomposition of dead algal material, the major organisms involved in this degradation, and the simultaneous regeneration of organic nitrogen are indicated in Figure 17-8. Following proteolysis by means of various enzyme systems, amino acids are deaminated by deaminases to ammonia; the ammonia then proceeds through a series of nitrification stages to hydroxylamine, nitrite, and nitrate. Organic nitrogen is synthesized during the processes of nitrate reduction, denitrification, and nitrogen fixation (see Chap. 12), each of which varies with numerous environmental parameters. In turn, lysis of the microbial flora generates additional dissolved organic nitrogenous compounds.

Decomposition rates of organic matter are quite different under anaerobic conditions. For example, after 60 days of anaerobic decomposition of green algal cells at 20°C, 30% of algal cell carbon was transformed into dissolved organic carbon (DOC) and 20% was mineralized to CO_2; 50% remained as particulate matter. Of the algal cell nitrogen, only 8% was transformed to soluble form, 48% was mineralized, and 44% remained in particulate form after 60 days. The DOC in solution consisted largely of yellow organic

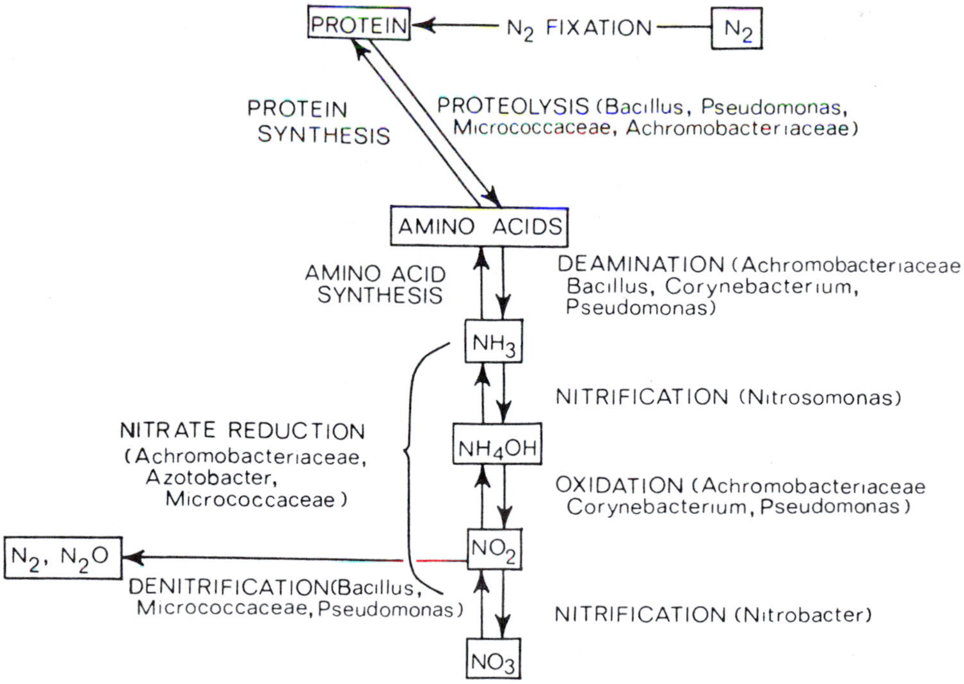

FIGURE 17-8 Decomposition and utilization stages of nitrogenous organic matter of aquatic organisms. (Modified from Botan *et al.,* 1960.)

acids that were resistant to further degradation. The decomposition rates and production of dissolved organic compounds from dead algal cells were only one-fourth as great under anaerobic as under aerobic conditions, and the rate constant of cell carbon and nitrogen decomposition decreased to less than one-half under anoxic conditions. Similar results have been found for the decomposition of aquatic angiosperms under anaerobic conditions (cf. Godshalk and Wetzel, 1978a–c).

The rates of decomposition under both aerobic and anaerobic conditions are strongly influenced by temperature (usually higher rates with increasing temperatures in the natural range), pH (commonly lower at acidic pH values), the major cation concentrations and cationic ratios, and other factors. The rates of cycling and utilization of the numerous dissolved organic intermediate products of bacterial and fungal degradation, such as amino acids, keto acids, various fatty and nonvolatile organic acids, phenolic derivatives, mercaptans, and others, are highly variable (see, e.g., Krause *et al.*, 1961).

Experimental studies on the transformation of artificially generated radioactive algal detritus under conditions approaching a natural community are instructive in understanding the relationships involved (Saunders, 1972). The living components of the water constitute a very small portion of the total organic matter (Fig. 17-9); most organic matter occurs as dissolved organic matter (DOM) and as detrital particulate organic matter (POM). For example, in the surface water of a productive lake, the summer POM averaged 57% of the total seston (inorganic and organic; Table 17-11). The phytoplanktonic dry weight ranged from between 5 and 25% of the organic particulate detritus most of the time and only occasionally exceeded 40–50%. Similar results have been shown for the seston composition of Lake Erie (Leach, 1975) and several Russian lakes (Winberg *et al.*, 1970). In the autumn of Lake Constance, Germany, dead organic seston exceeded microbial biomass by a factor of 4–10 (Simon and Tilzer, 1982). Considering that the POM constitutes only about 10% of the total detrital organic matter (ca. 90% as DOM), and that living organic matter makes up only a small fraction of the particulate component, it is clear that dissolved and particulate organic detritus is a major component of the organic mass of aquatic ecosystems.

TABLE 17-11 Ratio of Dry Weight of Organic Particulate Detritus to Phytoplankton and Organic Particulate Detritus as a Percentage of Seston, Frains Lake, Michigan[a]

Time period	*Ratio of detrital POM to phytoplankton*	*Organic particulate detritus as a % of seston*
21 Jan. 1967	5.7	—
27 Apr. 1967	4.6	45
3 May 1967	6.6	72
9 May 1967	16.9	65
16 May 1967	15.1	67
23 May 1967	2.3	21
30 May 1967	12.0	58
6 June 1967	4.5	57
14 June 1967	4.9	61
20 June 1967	8.4	68
27 June 1967	4.8	62
22 June 1968	1.3	44

[a] After Saunders (1972a).

Decomposition of the particulate organic detritus generated from planktonic communities of surface lake water occurs more slowly, at rates on the order of 10% per day (Table 17-12). Similar rates of decomposition of planktonic algae in an oligotrophic lake ranged between 0.5% day^{-1} (winter) to 6% day^{-1} in summer (Cole *et al.*, 1984). Radiolabeled dissolved organic carbon derived from the particulate organic detritus amounted to about 1% of the initial radioactivity after one day; these values were about 2–6 times greater than the amount of dissolved organic matter released extracellularly by phytoplankton in the trophogenic zone.

Decomposition rates of detrital particulate organic matter (POM) of course are influenced by an array of compositional and environmental factors. Certainly differences in the organic composition of algal cell

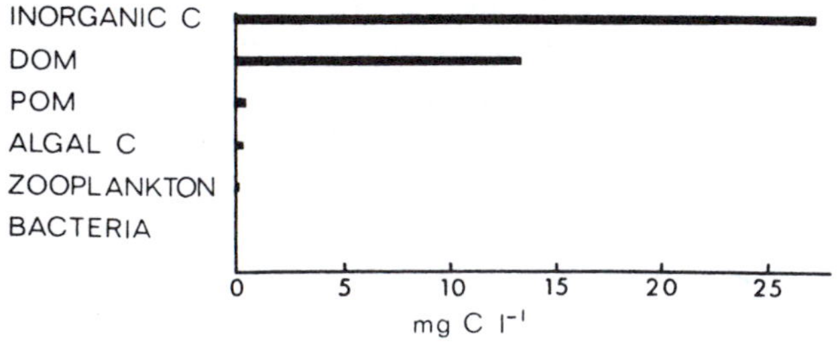

FIGURE 17-9 Distribution of carbon in inorganic and organic fractions at 1 m in Frains Lake, Michigan, 22 July 1968. (Modified from Saunders, 1972.)

TABLE 17-12 Decomposition of Fine Particulate Organic Matter in Lake Water[a]

Substrate	*Percentage loss per day*
Phytoplankton	14
Mixed dead phytoplankton and detritus	5–20
Detritus (algal)	8
Large filamentous algae	7–25
Zooplankton	1–33

[a] After data of Krause (1959), Saunders (1972a), and Pieczyńska (1972a).

walls lead to differences in degradation rates. For example, POM of dead cyanobacteria decomposes much more rapidly than that of green algae and diatoms; desmid algal POM is especially resistant (Mills and Alexander, 1974; Gunnison and Alexander, 1975). The concentrations and species diversity of bacteria also are involved. Among environmental parameters, temperature and oxygen are paramount, and availability of inorganic nutrients, especially nitrogen, can also be an important variable.

Bacterial decomposition of organic detritus (soluble or particulate) can be considered simply as the rate of change between bacteria and detrital substrates, compounds, or particles and can be expressed as a biomolecular second-order reaction (Saunders, 1972b):

$$\begin{array}{c}(m_1, m_2, \cdot\cdot\cdot)(T,[N], n_1, n_2, \cdot\cdot\cdot)\\ \downarrow \qquad \downarrow\\ \dfrac{-d[S]}{dt} = C_1[S] \times C_2[B]\end{array}$$

when

$[S]$ = concentration of detrital substrate
$[B]$ = concentration of bacteria
T = temperature
$[N]$ = concentration of inorganic nutrients
$m_1, m_2, \ldots; n_1, n_2, \ldots$ = other variables
C_1, C_2 = coefficients

The rate of decomposition of detrital substrates is a function of their concentration and of the enzymatic activity of the bacteria dispersed in the water or attached to surfaces of detrital particles (Saunders, 1972b). The coefficients C_1 and C_2 are obviously highly dynamic and variable operators on the detrital–bacterial interactions. Control variables that operate on the substrate coefficient C_1 include various biodegradability factors (Alexander, 1965b, 1975), examples of which are many. A structural characteristic of the substrate molecule or of the particle surface parameters (i.e., area, particle size, and adsorption sites) can inhibit the enzymes. Certain enzymes can be inactivated by adsorption to surfaces of minerals or colloids or inhibited by phenolic and polyaromatic substrates or their derivatives (e.g., Boavida and Wetzel, 1998). Substrates may be rendered inaccessible not only by surface characteristics but also by transport of the substrates to microenvironments in which conditions preclude or greatly reduce bacterial or enzymatic activities.

Control variables that operate on the bacterial component C_2 include obvious parameters such as temperature, T, and inorganic nutrient concentrations, $[N]$; as well as other essential growth factors or biologically utilizable terminal electron acceptors. The presence of biologically generated organic inhibitors, microbially formed inorganic inhibitors, high salt concentrations, temperature extremes, acidity, or other environmental conditions outside the range suitable for microbial proliferation can all inhibit bacterial growth (Alexander, 1965b, 1975). A microbial community lacking the appropriate enzyme system may not be able to metabolize certain compounds, or the substrate may not be able to penetrate into cells that possess the appropriate enzymes. Thus, the decomposition rates of organic substrates in natural systems are a function of the physical and biological structure at any instant in time, as well as of the substrate characteristics and physiological state of the microflora and their enzymatic systems.

Within this general framework, it is frequently desirable to know the flux rates of organic carbon among the different components of the system. As we have seen earlier (Fig. 17-9), the system structure is dominated by two very large pools of nonliving carbon, the inorganic and dissolved organic carbon. Algae and particulate organic detritus contain much less carbon, zooplankton and other fauna contain even less; and bacteria contain about 10% of that of the zooplankton (Saunders, 1971). This approximate general distribution of the carbon mass of the trophogenic zone of lakes bears little relation to carbon flux rates, however.

A 1-day intensive study undertaken to evaluate relationships between the carbon structure of a lake's pelagic system and various metabolic flux rates indicates that the microflora control the movement of carbon (Fig. 17-10). The biota utilize the carbon of the various pools within the limits of the environmental conditions. The environmental variables change constantly and result in continually changing rates of

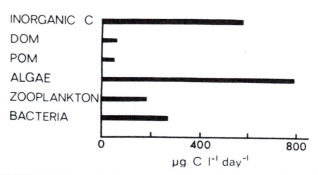

FIGURE 17-10 Assimilation rates of carbon in μg C liter^{-1} day^{-1} by different microbial components at 1 m in Frains Lake, Michigan, on 22 July 1968. (Modified from Saunders, G. W.: Carbon flow in the aquatic system. *In* Cairns, J., Jr. (ed.) *The Structure and Function of Fresh-Water Microbial Communities,* Blacksburg, Va., Virginia Polytechnic Institute, 1971.)

transfer among the carbon pools. The changes in carbon concentrations and metabolic rates in the surface waters over a 24-h period during the summer can be generalized in a theoretical way in Figure 17-11. Photosynthetic rates of algae are coupled with light quantity. Usually, maximum photosynthetic rates are skewed toward the morning period of high light intensity and less photoinhibitory and photorespiratory reductions of photosynthetic efficiencies. Photosynthetic rates decline precipitously with decreasing light in the evening (Saunders and Storch, 1971). Coupled with the cycle of extracellular release of dissolved organic compounds by algae, bacterial assimilation rates increase and reach maximal levels about 3 h following the secretion maximum. The division activity of bacterial populations is highest during the night and early morning, when the output of organic substances by the algae is low, whereas volume growth is highest during the second half of the day, when release of photosynthetic algal products is highest (Krambeck, 1978). These general relationships were supported by diurnal measurements of bacterial productivity (Simon, 1994). The highest rates of bacterial biomass production were found in the evening during the spring phytoplankton bloom. The ratio of protein to carbohydrate of the algae also shifts diurnally; maximum carbohydrate levels are reached at the conclusion of the daily photosynthetic period. Zooplanktonic grazing rates in the epilimnion can increase markedly at dusk with a rapid diurnal migration of many zooplankters to the trophogenic zone, and thus impact both bacterial and particularly algal growth. Ingestion and assimilation rates of algae by many zooplankton, particularly *Daphnia,* are greater in the night than during the day. Thus, we can envision a series of coupled oscillating reaction systems among the major components of the pelagic system in the trophogenic zone, that respond to an array of environmental variables.

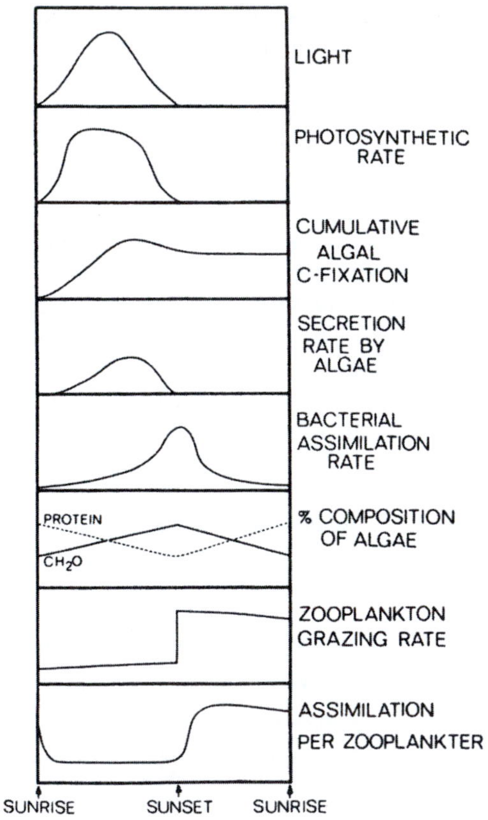

FIGURE 17-11 Generalized diagrams of the daily cycles of various processes and concentrations in the upper trophogenic zone of a lake system in summer. (Modified from Saunders, G. W.: Carbon flow in the aquatic system. *In* Cairns, J., Jr. (ed.) *The Structure and Function of Fresh-Water Microbial Communities,* Blacksburg, Va., Virginia Polytechnic Institute. 1971.)

A. Sedimentation and Decomposition of Pelagic Particulate Organic Matter

The vertical distribution of bacterial numbers and activity in stratified lakes commonly shows a maximum in the trophogenic zone, a minimum in the metalimnion, and an increase in the lower hypolimnion. The decomposition of sedimenting organic matter produced in the epilimnion follows a similar distribution. A number of devices have been used to measure the amount of tripton, the term for all nonliving suspended matter (cf. methods discussed in Edmondson and Winberg, 1971; and Bloesch and Burns, 1980). Containers of various types are suspended at several depths to entrap sedimenting materials over a period of one to several weeks, depending on the sedimentation rates. After a period of collection, kept as short as possible to reduce complicating interferences, chemical analyses of the collected seston permit an approximate indirect evaluation of decomposition-induced changes in the organic matter that have occurred during sedimentation.

Most (>75%) soluble organic matter released by secretion and autolysis of phytoplankton decomposes rapidly to a major extent at the site of its generation and release in the trophogenic zone. The particulate organic detritus undergoes slower degradation. During stratification periods in lakes of moderate depth, 75 to >99% of the particulate organic matter synthesized in the trophogenic zone is decomposed in the water column by the time it reaches the sediment interface (Table 17-13). In Lawrence Lake, Michigan, an annual mean of 88% of the sedimenting particulate organic carbon was decomposed by the time it reached the lower epilimnion (5 m; Wetzel *et al.,* 1972). Little additional decomposition occurred on further hypolimnetic sedimentation; an annual mean of 10% of planktonic

TABLE 17-13 Percentage of Organic Carbon Produced by Phytoplanktonic Primary Production That was Mineralized by Bacterial Decomposition per Area at Depth and at the Bottom of the Water Column in Three Lakes of Northern Germany[a]

	Schluënsee	*% Mineralization*			*Schöhsee*	*% Mineralization*			*Plußsee*	*% Mineralization*	
Dates	*Depth (m)*	*Per area*	*Total*	*Dates*	*Depth (m)*	*Per area*	*Total*	*Dates*	*Depth (m)*	*Per area*	*Total*
9 June–6 July	20	58.8		11 April–	10	52.8		11 May–	5	93.1	
	30	33.5	97.3	18 May	20	30.7	83.5	30 May	15	5.0	98.9
	40	5.0							25	0.9	
6 July–16 Aug.	20	73.7		18 May–	10	75.1		30 May–	5	90.6	
	30	12.2	97.6	4 July	20	16.9	92.0	23 June	15	7.6	99.5
	40	11.7							25	1.3	
16 Aug.–7 Sept.	20	77.6		4 July–	10	26.8		23 June–	5	89.8	
	30	9.8	98.5	10 Aug.	20	54.0	80.8	8 Aug.	15	9.1	99.7
	40	1.5							25	0.9	
7 Sept.–5 Oct.	20	87.8		10 Aug.–	10	75.3		8 Aug.–	5	83.7	
	30	4.2	98.8	5 Sept.	20	18.8	94.1	16 Sept.	15	13.5	99.7
	40	6.8							25	2.6	
		Mean =	98.1%								
				5 Sept.–	10	17.2		16 Sept.–	5	70.0	
				10 Oct.	20	77.2	94.4	4 Nov.	15	26.5	99.1
									25	2.5	
										Mean =	99.4%
				10 Oct.–	10	12.9					
				17 Nov.	20	80.7	93.6				
				17 Nov.–	10	8.8					
				5 Dec.	20	74.3	83.1				
						Mean =	88.8%				

[a] After Ohle (1962).

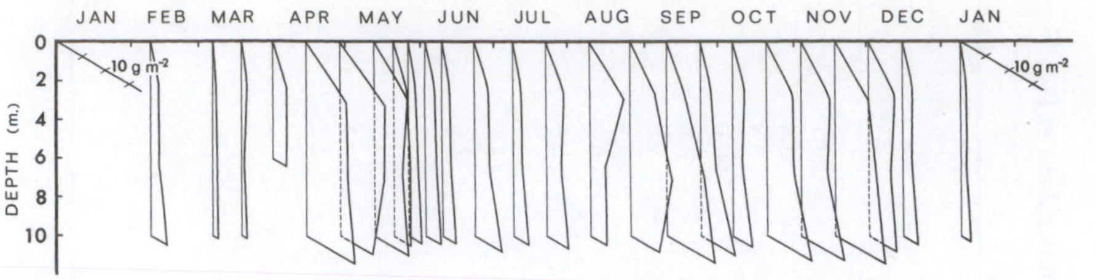

FIGURE 17-12 Seasonal changes in the organic content of sedimenting seston (g m^{-2}), Lawrence Lake, Michigan, 1972. (From White and Wetzel, unpublished.)

particulate organic carbon of the trophogenic zone was found at 10.5 m. Similar results were obtained by Lawacz (1969) using different methods.

Even though the degradation of particulate organic matter is fairly complete with increasing depth, seasonal variations are great (Fig. 17-12). The increasing concentrations of sedimenting organic matter with depth must be considered in relation to the decreasing area of each stratum as the basin morphometry constricts with depth. This relationship is illustrated diagrammatically in Figure 17-13 for Schluënsee; in this lake, the funneling effects of the basin morphology increased the concentration of organic carbon per unit area, even though the total amount of synthesized particulate organic matter was greatly reduced (>97%) by decomposition at a cumulative depth of 40 m (Ohle, 1962). As the size of the basin decreases, the significance of the funneling effects of concentration increases.

As lakes become shallower and the importance of littoral production in relation to that of the phytoplankton increases, a greater proportion of the synthesized organic matter is shifted to the sediments. We would expect decomposition to be more complete in the water column of larger, oligotrophic lakes because of the amount of time required for decomposition of the more recalcitrant materials of algal cells (cf. Jewell and McCarty, 1971) and the distance and time through which a particle travels during sedimentation. Evidence supports these expectations.

For example, in a shallow (z_m 6.5 m), hypereutrophic lake in southern Michigan, the high primary productivity of phytoplankton, the relatively short distances that algal particulate organic matter must fall to reach the sediments, and the close proximity of the anaerobic hypolimnion (anaerobic 4 m) to the photic zone resulted in high sedimentation of incompletely decomposed algal particulate matter to the sediments (Molongowski and Klug, 1980b; cf. also Gasith, 1975, 1976). The quantity of organic seston reaching the sediments was not only high in this shallow lake but was relatively uniform throughout the thermally stratified summer period of high phytoplankton productivity. The residual sedimenting seston had a high protein content (50–60% of that of phytoplankton) and, as a result, a low C/N ratio (4–6). In contrast, in deeper lakes, proteinaceous materials in sedimenting seston are rapidly degraded in the upper water strata (Brehm, 1967; Matsuyama, 1973; Lastein, 1976). The carbohydrate fraction of the sedimenting seston reaching the sediments of this shallow lake consisted primarily of glucose (19.2%),

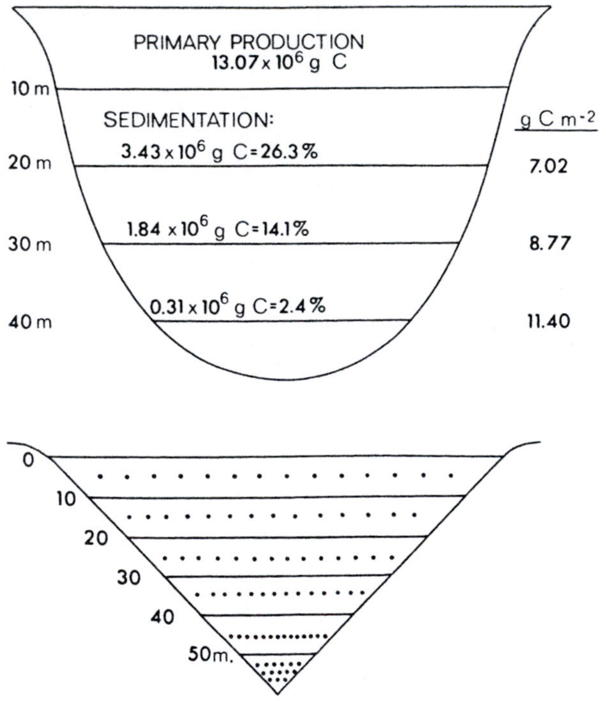

FIGURE 17-13 Schematic diagrams of (*upper*) the relationships between planktonic primary production and sedimentation of organic particulate carbon in Schluënsee, Germany, 16 July–16 August 1960, and (*lower*) the morphometric funnelling effect on sedimentation in lakes. (Modified from Ohle, 1962.) The numbers to the right of the upper figure (g C m^{-2}) indicate the concentrating effect of basin morphometry on the quantity of particulate carbon per unit area in each of the strata at increasing depths. The percentage of the particulate carbon produced in the trophogenic zone and sedimenting to each stratum (inside the diagram) is markedly decreased by decomposition at each step.

other soluble monomers (12.4%), and soluble polysaccharides (43.4%). Cellulosic polysaccharides represented only 7.2% of the total carbohydrate fraction, which indicates that most of the carbohydrates were of algal origin, even though this lake has a massive littoral development that covers over a third of the lake area. As will be discussed later (Chaps. 21 and 23), much of the littoral particulate plant production decomposes on or within the sediments within the littoral zone.

VIII. PRODUCTIVITY OF BACTERIOPLANKTON

Comprehensive evaluations of planktonic bacterial productivity have increased markedly in the last decade, largely as a result of greatly improved radiotracer methodology. Several of these studies yield sufficient detail for functional analyses of carbon fluxes. Although many of the estimates of bacterial productivity will be discussed in Chap. 23 on the organic carbon cycling in aquatic ecosystems, a few statements are appropriate here in relation to the discussion of the distribution of pelagic bacteria and their growth.

A. Planktonic Bacterial Productivity

Bacterioplankton live in natural environments that are relatively hostile to rapid growth—literally a pelagic desert. Natural bacteria grow at slower rates than their maximum potential and may be limited by organic carbon substrate or nutrient availability, as well as physicochemical factors such as temperature. Even under relatively optimal growing conditions in natural pelagic environments, average generation times are often several days. The metabolic and physical factors regulating growth vary so rapidly in natural environments that even under the best of circumstances our measurements of bacterial growth and production are meager composite estimates of the true values. These generation rates may slow to average doubling times of 100–200 days or even more during winter periods when resources as well as temperatures are reduced (Moriarty and Bell, 1993). Clearly, pelagic bacteria communities are usually not growing at maximum rates much of the time. Individual species populations within communities compete inefficiently for resources and survive in a starved state. Both organic substrate and nutrient (especially phosphorus) availability are critical limitations under many circumstances in lakes. Slow growth rates and long periods in stationary phase and starvation are normal conditions for bacteria in the pelagic environment.

The efficiency of bacterial metabolism and growth is highly variable. Although it is theoretically possible for bacteria to convert organic substrates to biomass with high efficiency (>85%) under optimal conditions (Payne and Wiebe, 1978), such high rates are never realized under natural conditions. Growth is rarely in exponential phase and with many resource and physical limitations, growth efficiencies [bacteria production/(bacterial production + bacterial respiration)] are nearly always <50%, and mostly <30% (e.g., Schroeder, 1981; Pomeroy and Wiebe, 1988; Cole and Pace, 1995).

The rates of biomass turnover of bacterioplankton is a useful parameter to evaluate differences in spatial and seasonal dynamics. The ratio of production to biomass (P/B ratio) or turnover time of actively metabolizing bacteria was comparable to the generation time of actively metabolizing cells (Simon, 1988). The biomass of small (0.2–1.0 μm) free-living bacteria usually turned over significantly faster (higher P/B ratios) than that of large (1.0–3.0 μm) bacteria throughout the water column (Fig. 17-14). The turnover rates of attached bacterial biomass exhibited large fluctuations. When large amounts of decaying organic particles were present, turnover rates of attached bacteria were similar to those of large free-living bacteria and shorter than those of small free-living bacteria, particularly in aphotic zones (Fig. 17-14). Bacterial production is usually greater in anoxic hypolimnetic water of stratified lakes over a range from oligotrophic to eutrophic, despite lower hypolimnetic temperatures (Cole and Pace, 1995). Although the growth efficiency of anaerobic bacteria is about the same as that of aerobic bacteria, the relatively high rates of bacterial production in anoxic waters resulted from larger bacterial populations growing more slowly than the aerobic bacteria.

Once the organic matter is converted from dissolved to particulate phases, bacterial biomass is subject to mortality as indicated earlier—natural genetically programmed senescence and death with autolysis, viral parasitism, and particularly grazing by protists and other zooplankton. The methodology of assaying production rates allows some separation of the grazing by zooplankton, but in general the observed production rates represent net production rates after losses from respiration, autolysis, parasitism, and grazing. In addition, a significant proportion of the bacterioplankton community can be in a dormant state. The percentage of the total bacterial community that is dormant can also change rapidly under different conditions. For example, in Lake Constance, Germany, during the spring bloom of phytoplankton when bacterial production reached it maximum, 60–80% of all free-living bacteria were metabolically active (Simon, 1987, 1988). During this period of abundant dead algae and other detrital particles, a high proportion of the detrital particles were colonized with metabolically active attached bacteria.

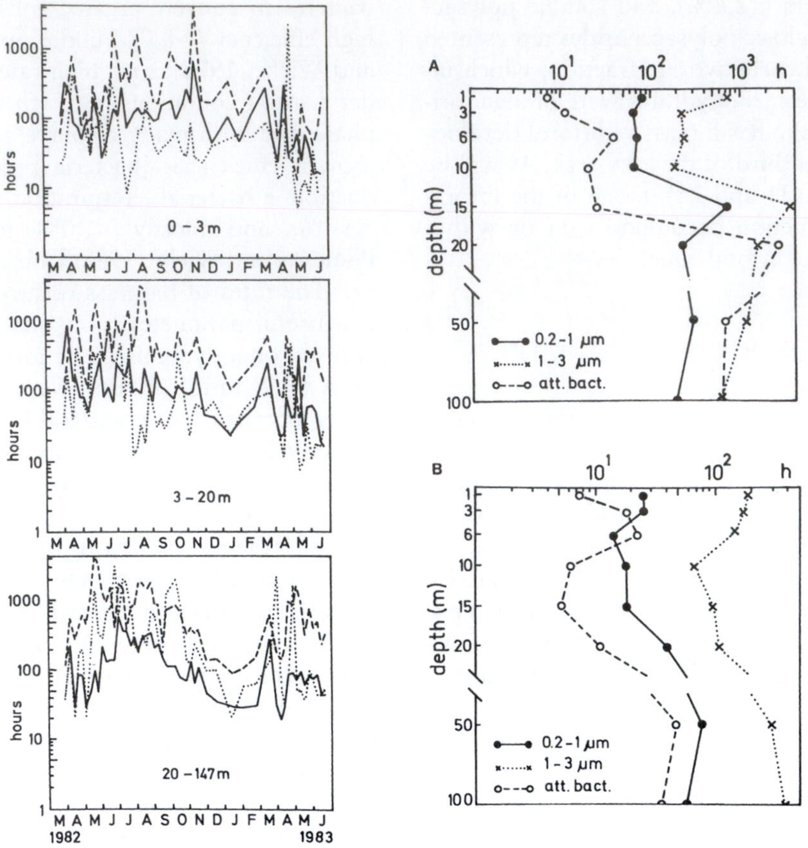

FIGURE 17-14 Bacterial turnover times (*P*/*B* ratio of biomass/production) in hours (hr) of small (0.2–1.0 μm; ———) and large, free-living (1–3 μm; - - - -) and attached (· · · · · ·) bacteria in mesotrophic Lake Constance, Germany. Curves = mean values from the water strata at the depths indicated. In the depth profiles, (*a*) = 19 July 1982 during a diatom bloom in the epilimnion; (*b*) = 16 May 1983 during a spring phytoplankton bloom. (From Simon, 1988.)

These detrital particles in various stages of decomposition can aggregate and be colonized by microbial consortia. Microenvironmental conditions within such detrital–microbial consortia can be characterized by reductions in dissolved oxygen and steep redox gradients (Paerl and Pinckney, 1996). These microgradients affect rates of organic degradation and nutrient regeneration and are certainly very different from those of the immediately surrounding pelagic water. On an annual basis, however, most (>75%) of the bacteria were small (0.2–1 μm) and free-living, made up over two-thirds of the bacterioplankton productivity, and were most important in recycling of organic matter. Larger (1.0–3.0 μm) free-living bacteria constituted about half of the total bacterial biomass but had much slower turnover times than the smaller forms. During periods of stratification, turnover rates are generally much faster, by an order of magnitude, in the trophogenic zone than in the tropholytic zone.

Examples of annual mean production of bacterioplankton across a spectrum of lakes of different nutrient availabilities show an approximate direct correlation with increasing production of phytoplankton (Table 17-14). It must be emphasized repeatedly that this bacterial production is often a very small portion of total bacterial metabolism; most bacterial metabolism occurs in the sediments and littoral regions of lakes and rivers (see Chap. 21). Although the annual production of the bacterioplankton is usually appreciably less than the estimated annual production of phytoplankton, in many lake and most river ecosystems inputs of organic matter from allochthonous and littoral sources greatly exceed those generated by the phytoplankton. Although the summed respiration of heterotrophic consumers (predominantly microbes) cannot exceed organic inputs to an ecosystem, the total organic carbon assimilated by heterotrophic consumers and their production can exceed the organic inputs to

TABLE 17-14 Estimates of Average Annual Production of Bacterioplankton in Comparison to Rates of Phytoplankton Production in Aquatic Habitats of Differing Productivity

Habitat	*Average annual bacterial production mg C m^{-2} day^{-1}*	*Average annual phytoplankton production mg C m^{-2} day^{-1}*	*Ratio of bacterial to phytoplankton production (%)*	*Methods for estimating bacterial production*[a]	*Source*
Oligotrophic					
Mirror, New Hampshire	18.2 (13.8–26.0)	101	18	TdR, others	Jordan and Likens (1980); Ochs *et al.* (1995)
Eastern Brook Lake, California	37.2	136.5	25	TdR	Thomas *et al.* (1991)
Lawrence, Michigan					
1984	186.3	126.0	148	TdR	Coveney and Wetzel (1992, 1995)
1985	468.5	153.4	305	TdR	
Toolik Lake, Alaska	8.2–22	—	—	Leu	Hobbie *et al.* (1999)
Lake Michigan, USA	652 (summer)	457 (annual)	—	TdR	Scavia and Laird (1987)
Mesotrophic					
Biwa, Japan (0–20 m, north basin)	100–1200	—	30	FDC, TdR	Nagata (1987)
Awassa, Ethiopia	470	—	10–30	FDC, TdR	Gebre-Mariam and Taylor (1990)
Constance, Germany (8-yr of trophogenic zone only)	73–186	534–822	9–29	TdR	Simon *et al.* (1998)
Eutrophic					
Nesjøvatn, Norway	160	580	25	FDC	Vadstein *et al.* (1989)
Tystrup, Denmark	279	622	45	FDC	Riemann (1983)
Arlington Reservoir, Texas	342	1611	17	TdR	Chrzanowski and Hubbard (1988)
Mendota, Wisconsin					
1979	244	1792	27–62	FDC	Pedrós-Alió and Brock (1982)
1980	592–1123	2197	27–51	FDC	
Norrvikan, Sweden (summer only)	24–850	~1500	~50	TdR	Bell *et al.* (1983)
Hartbeespoort Reservoir, So. Africa	7.2–168	400–>10,000	2	TdR	Robarts and Wicks (1990)
Plußsee, Germany	228 (64–590)	512	42	TdR, Leu	Chróst and Rai (1994)
Frederiksborg Slotssø, Denmark	342	1096	31	TdR	Søndergaard (1993)
Buresø, Denmark	252	356	70	TdR	
Mikolajskie, Poland	~2500	~1500	166	TdR	Gajewski and Chróst (1995)
Søbygård, Denmark (8-yr mean)	106	3250	1.6–5.5	TdR	Jeppesen *et al.* (1997)
Humboldt, Saskatchewan	20–422	975	42–67	TdR	Robarts *et al.* (1994)
Xolotlán, Nicaragua	600–1200	—	—	TdR	Erikson *et al.* (1998)
Stream/River/Wetland					
Breitenbach, Germany	440 (190–670)[b]	444[b]	100	FDC, Leu	Marxsen (1988, 1999)
Pelagic Sites, Okefenokee Swamp, Georgia	88	—	—	TdR	Murray and Hodson (1985)

[a] FDC = frequency of dividing cells; TdR = tritiated thymidine incorporation into DNA; Leu = tritiated leucine incorporation into protein.
[b] Largely benthic bacterial and algal production.

the ecosystem simply because organic carbon is recycled, although inefficiently, by consumers (Strayer, 1988).[5] However, sources of organic matter, particularly dissolved, from external allochthonous and littoral areas are known to supplement greatly and often dominate over those produced by the phytoplankton. In addition, storage of organic matter can greatly alter the rates of respiration, utilization, and recycling.

The horizontal spatial distribution of productivity of bacterioplankton within a lake or river is markedly influenced by differences in the availability of sources of organic substrates and nutrients. For example, bacterial production rates at coastal sites of Lake Erie were found to be 25–50 times greater than those at the offshore sites (Hwang and Heath, 1997; similarly in Lake Michigan; Scavia and Laird, 1987). Percentage grazing of bacterioplankton by protists was much more extensive in offshore regions than was the case in the inshore areas of much higher bacterioplankton productivity.

B. Chemosynthesis

Microbial production by chemosynthesis occurs to a marked extent in layers having contact with the anaerobic zones of an aquatic system, especially in boundary layers between anaerobic and aerobic zones. As indicated in earlier discussions of the cycling of several elements, the anaerobic processes of decomposition of organic matter provide reduced inorganic compounds that serve as energy substrates for the chemoautotrophic bacteria. Chemosynthetic secondary production becomes significant primarily in steep gradients of redox potential (Sorokin, 1964a, 1965, 1970). Outside of these layers, chemosynthesis is very low in relation to total heterotrophic bacterial production (Romanenko, 1966; Jordan and Likens, 1980). Chemosynthesis by bacteria is normally very low in the pelagic of streams, infertile waters, and lakes of intermediate productivity and becomes a significant contribution to the whole only in productive or meromictic lakes exhibiting steep redox gradients.

The ratio of dark CO_2 fixation (almost all by heterotrophic metabolism of bacteria; see Gerletti, 1968; Romanenko, 1973) to photosynthetic fixation of CO_2 and bacterial chemosynthesis generally is small in oligotrophic waters. For example, the 14-yr average of dark CO_2 fixation was 13.3% of phytoplanktonic carbon fixation in the light in a small Michigan lake (Table 17-15). This percentage is much less for the lake, about one-tenth, if the littoral photosynthesis of this lake is considered in addition to the carbon fixation of the phytoplankton. The ratio of photosynthetic to dark CO_2 fixation decreases in the transition to planktonic eutrophy and hypereutrophy. For example, chemosynthetic bacterial fixation rates of CO_2 in eutrophic Plußsee, northern Germany, fluctuated between 0.4 and 30.0% of the total bacterial productivity over an annual period (Overbeck, 1979). The total carbon flux through planktonic bacteria in oligotrophic Mirror Lake was between 16 and 43% of phytoplankton net production, or 11–31% of all autochthonous and allochthonous organic carbon inputs. In lakes that are deeper than Mirror Lake (z_m, 10 m), planktonic bacterial production probably accounts for a greater proportion of the total decomposition simply because particulate detritus would have a longer residence time in the water column. With further transition of the system or in lakes with a predominance of productivity by emergent macroflora and associated attached and littoral microflora, the relative contribution of dark CO_2 fixation to the DOC pool apparently decreases (cf. Wetzel, 1979).

TABLE 17-15 Dark CO_2 Fixation in the Pelagic Zone of Lawrence Lake, Michigan, Integrated for the Water Column at the Central Depression, 1968–1981[a]

Years	*g C m^{-2} yr^{-1}*	*Mean mg C m^{-2} day^{-1}*	*% of light CO_2 fixation*
1968	4.7	12.9	10.4
1969	10.2	27.9	24.4
1970	8.2	22.4	18.8
1971	5.0	13.8	14.0
1972	3.6	9.9	12.2
1973	1.8	4.9	4.5
1974	0.7	1.9	1.8
1975	3.2	8.9	7.5
1976	3.7	10.2	11.5
1977	3.4	9.3	11.0
1978	3.5	9.6	12.1
1979	6.2	17.0	21.4
1980	5.9	16.2	19.8
1981	6.5	17.8	16.6
Average (14 years)	4.7	13.0	13.3

[a] Data of Wetzel (unpublished). Based on individual measurements at each meter of depth (0–12 m) at weekly and biweekly intervals over the 14-yr period.

IX. COMPARISON OF BACTERIOPLANKTON AMONG AQUATIC ECOSYSTEMS

General characteristics of the bacterioplankton among river, reservoir, and natural lake ecosystems are set forth in Table 17-16. Such analyses are difficult, however, because of the paucity of extensive, long-term

[5] The growth efficiency figures used in the theoretical discussion by Strayer (1988) are much greater than would be found generally under natural conditions.

TABLE 17-16 Comparison of Characteristics of Bacterioplankton among River, Reservoir, and Natural Lake Ecosystems[a]

	Rivers	*Reservoirs*	*Natural lakes*
Bacterioplankton abundance	High to very high depending on sources of organic substrates and nutrients	Moderate to high; greater proportion of bacteria attached to seston particles	Low; increasing with increasing nutrient loading and productivity of phytoplankton
Bacterioplankton productivity	Low and increasing with stream order; heavy reliance on allochthonous dissolved organic matter and floodplain primary production	Low to moderate; often particle-associated; commonly higher in riverine portions than in lacustrian sections	Low; greater rates in trophogenic zone (epilimnion), decreasing in metalimnion, increasing in hypolimnion; relatively small seasonal changes because of mortality
Respiration	High	Moderate	Low
Mortality	High, from protistan predation, viral parasitism, UV photolysis, and molar action	Moderate, largely from protistan predation and viral parasitism	High; about half from protistan, especially nanoflagellate, predation, 20–40% from viral parasitism, and the remainder from autolysis and photolysis

[a] Gleaned from many individual refereces cited in this chapter and Wetzel (1990b).

data. The abundance and productivity of bacterioplankton are obviously strongly coupled to the availability of energy-rich, readily hydrolyzable organic substrates and nutrients, all of which are highly dynamic and variable among all aquatic ecosystems. Mortality of bacterioplankton is poorly understood but is dominated by protistan, particularly nanoflagellate, predation and viral parasitism.

X. SUMMARY

1. Much decomposition of particulate and dissolved organic matter occurs by means of planktonic bacteria in the aerobic pelagic waters prior to sedimentation of particulate detritus to the bottom of the basin. Rates of degradation are governed by an array of conditions: quality (chemical composition) and quantity of organic matter, physical parameters (e.g., temperature, stratification patterns, and basin morphometry), and chemical parameters (e.g., availability of terminal electron acceptors, particularly oxygen, and inorganic nutrients).
2. Numbers, biomass, and productivity of bacterioplankton generally increase with increasing photosynthetic productivity of fresh waters.
   a. A close correlation often exists between seasonal changes in biomass of phytoplankton and of heterotrophic bacterioplankton. Bacteria increase rapidly in response to enhanced rates of metabolism of phytoplankton.
   b. In thermally stratified lakes, bacterial biomass and productivity are commonly highest in the epilimnion, decrease to a minimum in the metalimnion and upper hypolimnion, and increase in the lower hypolimnion.
   c. Seasonal and vertical distributions in bacterial populations can change very rapidly (e.g., a few days). Seasonal changes, however, generally follow low values during colder periods and increase to maxima in spring and summer.
   d. Bacterioplankton are predominantly free-living (>75%), but in reservoirs and other environments where concentrations of suspended particles are high, bacterial attachment to particles is higher and can dominate (>80%).
3. Growth of bacterioplankton is controlled by resource availability and environmental factors.
   a. Bacterioplanktonic growth is positively correlated with temperature, particularly at low temperatures below 10–15°C). Above this temperature other factors regulate growth to a greater extent.
   b. Restricted nutrient availability, particularly of phosphorus, often limits the growth and productivity of bacterioplankton. Inorganic nitrogen is much less frequently limiting to bacterial growth than is phosphorus. Under transient growth conditions, multiple nutrient limitations are possible, but collective composite community nutrient limitations are usually relatively stable for periods of several weeks.
4. Decomposition of organic matter by bacteria is governed also by chemical characteristics of compounds. Simple organic substrates (amino acids, mono- and oligosaccharides, simple organic acids, and short-chained unsaturated fatty acids) are assimilated and mineralized by planktonic bacteria in aerobic waters more rapidly than larger, more complex soluble organic compounds.
   a. Dissolved organic substrates are utilized more effectively (faster turnover times) in warm, more productive, nutrient-rich fresh waters than they are in oligotrophic systems at lower temperatures.
   b. Both dissolved free and combined carbohydrate, amino, and lipid compounds are important substrates for bacterioplankton growth.
   c. The dynamics and utilization of carbohydrates and simple organic acids are known in some detail for lakes. The ratios of concentration to rate of utilization (turnover times) of total carbohydrates were found to be in the range of 3–20 days in the summer in temperate lakes.
   d. As much as 80% of dissolved organic matter is composed of organic acids that originate largely from higher aquatic and terrestrial plants. Many of these degradation products form humic compounds that often possess phenolic linkages that are more recalcitrant to degradation than simple substrates and have longer residence times in fresh waters. Despite their complexity and relatively high molecular weight, these humic compounds can serve as substrates for more than half of the bacterioplankton production. Humic compounds are often cometabolized with simpler substrates in a coupled metabolic process. Partial photolysis of humic substances by ultraviolet light can generate large quantities of simple, readily utilizable fatty acids and related substrates and support bacterial growth.
5. Dissolved organic compounds released by phytoplankton and other biota during active growth and upon senescence and death (autolysis) undergo rapid initial decomposition; declining mineralization at progressively slower rates of degradation then follows.
   a. Upon death, large amounts (5–35% of the total biomass) of organic matter are released as soluble organic compounds.

b. Extracellular release of dissolved organic substrates from phytoplankton can form a major source (to 95% at certain times) of carbon and energy for bacterioplankton.
c. Usually >75% of the soluble organic matter released by secretion and autolysis of phytoplankton is decomposed in the trophogenic zone. An average of 50% of the gross uptake of extracellularly released organic carbon is respired by the bacteria after assimilation.
d. In the pelagic trophogenic zone, decomposition of the algal-dissolved organic matter can operate in a coupled oscillating manner (Fig. 17-11). Photosynthetic rates are coupled to light, often with high rates in the morning and increased extracellular release of dissolved organic compounds in the afternoon. Bacterial assimilation and growth attain maximum rates slightly after (about 3 h) the maximum of algal product release and during the early period of darkness.

6. Resource limits to growth are approximately balanced by mortality and other losses.
   a. The importance of natural, genetically regulated physiological death and of dormancy under conditions of environmental stress is likely high but is not understood in bacterioplankton communities.
   b. Viral parasitism and mortality results in a major loss of bacterioplankton. Usually at least 25% of bacterial mortality results from viral lysis.
   c. Bacterivory by protists, especially nanoflagellates, and other zooplankton is a dominant mortality factor in natural bacterioplankton communities. Usually about 50% of bacterial mortality results from predation, mainly by nanoflagellates.
7. Decomposition rates of particulate organic detritus approximate first-order reaction rates during the initial stages of degradation and then decline as more recalcitrant compounds accumulate.
   a. Most pelagic organic matter occurs as dissolved organic matter. The particulate organic matter, most of which is nonliving, usually constitutes <10% of the total organic matter.
   b. Decomposition of fine particulate organic matter occurs at rates from 1 to about 15% per day in warm, aerobic lake water and is much slower than degradation of dissolved organic matter released by secretion and autolysis.
   c. The rapidity and completeness of decomposition of particulate organic matter is about the same under anaerobic conditions but can decline, as in anoxic hypolimnia of productive lakes, if the temperatures are very low.
   d. Nitrogen-containing amino and proteinaceous compounds are generally utilized more rapidly than carbohydrate-based compounds. As a result, organic C/N ratios generally increase with time.
   e. As particulate organic matter sinks from the trophogenic zone in stratified lakes, the extent of decomposition before reaching the sediments is controlled by many factors.
      i. In oligotrophic to moderately productive lakes of moderate depth (>10–20 m), 75–99% of the particulate organic matter synthesized in the trophogenic zone is decomposed in the water column before reaching the sediments. Rates of algal decomposition are in the range of 0.5% (winter) to ca. 5% per day (summer).
      ii. In shallower and more productive lakes, the amount of organic matter sedimenting out of the trophogenic zone increases, the residence time in the water column decreases (shorter settling distance), and the close proximity of cold, anaerobic hypolimnetic conditions results in large amounts of partially decomposed organic seston reaching the sediments.
      iii. A portion of the organic matter of surficial sediments is resuspended into the water column during periods of water circulation and undergoes further degradation in the pelagic water.
8. Measurements of productivity of natural bacterioplankton communities over an annual period are relatively few.
   a. Annual production of bacterioplankton is usually less than the estimated annual production of phytoplankton and is greatly subsidized by the organic carbon of allochthonous and littoral sources.
   b. Bacterioplankton productivity as a percentage of all decomposition in lakes increases in deeper lakes that have a longer residence time for degradation of sedimenting particulate organic matter, relatively small allochthonous loading of dissolved organic matter, and large pelagic areas in comparison to those of the littoral.
9. Chemosynthetic bacterial metabolism, in which CO_2 is assimilated in the presence of utilizable dissolved organic substrates, is significant only in strata of steep redox gradients, such as between an anaerobic hypolimnion and the aerobic trophogenic zone. Bacterial chemosynthesis is usually insignificant in aerobic waters and is normally very low in oligotrophic and moderately productive lakes.

18

LAND–WATER INTERFACES: LARGER PLANTS

The land–water interface zone is difficult to define precisely. Physically, the boundaries of the intersection of land and water where the soils are water-saturated or water-influenced are highly variable over time as water levels fluctuate. The positions of the land–water boundaries vary in a fractal dimension depending on the scale of examination. At a large scale, for example from an aerial photograph, the boundary may seem quite distinct. However, as we have seen from the discussions of morphometry, slope gradients are often very small along the perimeters of lake and river valleys. Hence, small fluctuations in water levels with changes in precipitation can result in appreciable lateral movements of the interface boundaries. Among natural lakes, water level fluctuations are less than in manipulated water levels of reservoirs, but water levels in natural lakes can easily fluctuate annually in the range of a meter or more.

The gentle, small slopes along most land–water margins are conducive to sedimentation, and organic matter accumulations are common. Hydraulic conductivities of water movements through sediments of wetland and littoral areas are large. During active growth of floating and emergent plants in littoral areas, losses of water to the air by evapotranspiration contribute to a marked imbibition of water from upgradient soils. Thus, the land–water boundary is functionally and often physically diffuse; its position is constantly changing.

A very large percentage of the nonmarine biosphere is flat with very low relief profiles. Runoff from upland areas enters recipient river channels and lakes by gravitational flow. Lateral flooding of littoral wetland and shoreline areas are common as inputs from precipitation exceed temporarily the export of water downstream. Such inundated areas are often, particularly in the tropics, much larger, by many times, than the area of the lakes and rivers per se. Inundated and saturated soils almost immediately become anoxic and change to reducing conditions. Under reducing conditions, nutrient fluxes, rates of decomposition, and plant metabolism and community interactions are all altered markedly. These reducing conditions are chemically hostile to most higher plants. Those plants that have adapted to the reducing conditions have great competitive advantage because of the high availability of essential resources (nutrients, water) and reduced competition from less adapted biota that cannot survive under these conditions. As a result, the productivity of wetland and littoral plant communities includes the highest rates of organic matter synthesis in the biosphere (Fig. 8-1). A primary characteristic of wetlands and littoral areas of the land–water interface regions is that the water table is near, at, or above the hydrosoil (sediment) surface. These hydrological conditions are primary drivers of the coupled interactions of nutrient fluxes in the sediments and all aspects of plant physiology, growth, and productivity.

The littoral region is an interface zone between the land of the drainage basin and the open water of lakes and some streams. The size of the littoral zone in relation to the size of the pelagic region varies greatly among lakes and depends on the geomorphology of the basin and the rates of sedimentation that have occurred since the inception of the lake. Most lakes of the world are relatively small in area and shallow (Fig. 3-2). In such lakes, the littoral flora contributes significantly to the productivity and may dominate and regulate the metabolism of the entire lake ecosystem.

The general zonation of the littoral region has been introduced earlier (Chap. 8; Fig. 8-2). The physiological and ecological adaptations of freshwater aquatic angiosperms influence their distribution and often permit very high productivity. These plants also provide excellent habitats for photosynthetic and heterotrophic microflora, as well as many zooplankton (Chap. 19) and larger invertebrates (Chap. 22). In addition, the wetland and littoral flora synthesize large quantities of organic matter, most of which falls to and accumulates on and in the sediments of the littoral areas before degradation occurs (Chap. 23).

I. AQUATIC MACROPHYTES OF THE LITTORAL ZONE AND WETLANDS

A. Habitat Classification of Aquatic Macrophytes

In any inquiry into the botanical aspects of lakes, separate from the phytoplankton, one is confronted with an array of rather arbitrary definitions of the sessile flora of aquatic systems (Arber, 1920; Gessner, 1955, 1959; Sculthorpe, 1967; cf. Hutchinson, 1975). Many definitions are based on reproductive characteristics in relation to the aquatic stages in the life history of groups of angiosperms. From a strictly botanical viewpoint, such definitions are quite satisfactory. From an ecological standpoint, however, such species-orientated categorizations are unrealistic and ignore major system interrelationships. Alternatively, words such as *hydrophytes* are ambiguous; even though the term *hydrophyte* generally refers to vascular aquatic plants, it can include any form of aquatic plant. The term *aquatic macrophyte*, as it is commonly used, including in this work, refers to the macroscopic forms of aquatic vegetation and encompasses macroalgae (e.g., the alga *Cladophora*, the stoneworts such as *Chara*), the few species of mosses and ferns adapted to the aquatic habitat, as well as true angiosperms. Division on the basis of size is admittedly also arbitrary but, as discussed further on, when combined with the definition of the attached microflora, it permits a meaningful separation of primary producers in the littoral zone.

Numerous lines of evidence indicate that aquatic angiosperms originated on the land. Adaptation and specialization to the aquatic habitat have been achieved by only a few angiosperms ($<1\%$) and pteridophytes ($<2\%$). Consequently, the richness of plant species in aquatic habitats is relatively low compared with that of most terrestrial communities. The highly variable, changing habitat, particularly among floodplain inhabitants, results in low speciation and demands plastic adaptability (Henderson *et al.*, 1998). Many aquatic angiosperms possess relics of their terrestrial heritage, such as a cuticle, stomata, and a lignified xylem tracheal structure. Most are rooted, but a few species float freely in the water. Aquatic macrophytes have evolved from many diverse groups and often demonstrate extreme plasticity in structure and morphology in relation to changing environmental conditions. These factors, in combination with the very heterogeneous conditions of their littoral habitat, make difficult the precise ecological classification of this group into growth forms.

Numerous classification systems have been proposed and used. The primary groups of aquatic angiosperms, rooted and nonrooted, have been subdivided according to types of foliage and inflorescence and whether these organs are emergent, floating on the water surface, or submersed (Arber, 1920). Differences in the extent of emergence or submergence and the manner of attachment or rooting to the substratum have led to the creation of numerous complex subclassifications and a corresponding terminology (cf. Hejný, 1960; Luther, 1949; den Hartog and Segal, 1964). Although useful for certain phytosociological analyses, this terminology is cumbersome, and lines of demarcation are rarely distinct.

The following simple classification of aquatic macrophytes, after Arber (1920) and Sculthorpe (1967), is based on attachment but has also proven useful in morphological, physiological, and ecological studies.

1. Aquatic Macrophytes Attached to the Substratum

1. *Emergent Macrophytes.* These plants occur on water-saturated or submersed soils, from the point at which the water table is about 0.5 m below the soil surface to where the sediment is covered with approximately 1.5 m of water; they are primarily rhizomatous or cormous perennials (e.g., *Glyceria*, *Eleocharis*, *Phragmites*, *Scirpus*, *Typha*, and *Zizania*). In heterophyllous[1] species, submersed and/or floating leaves precede mature aerial leaves; many species may exist as (usually sterile) submersed forms; all produce aerial reproductive organs.

2. *Floating-Leaved Macrophytes.* These aquatic plants are primarily angiosperms that occur attached to submersed sediments at water depths from about 0.5 to 3 m. In heterophyllous species, submersed leaves precede or accompany the floating leaves; reproductive organs are floating or aerial; floating leaves are on long, flexible petioles (e.g., the water lilies *Nuphar* and *Nymphaea*) or on short petioles from long ascending stems (e.g., *Brasenia* and *Potamogeton natans*).

[1] Heterophylly refers to vegetative polymorphism where the same plant can develop morphologically very different leaves or other plant parts when submersed or aerial (see pp. 532ff).

3. *Submersed Macrophytes.* These plants comprise a few pteridophytes (e.g., the quillwort *Isoetes*), numerous mosses and charophytes (stonewort algae *Chara*, *Nitella*), and many angiosperms. They occur at all depths within the photic zone, but vascular angiosperms occur only to about 10 m (1 atm hydrostatic pressure). Leaf morphology is highly variable, from finely divided to broad; reproductive organs are aerial, floating, or submersed.

2. Freely Floating Macrophytes

An extremely diverse group, freely floating macrophytes are typically not rooted to the substratum, but live unattached within or upon the water. They are diverse in form and habit, ranging from large plants with rosettes of aerial and/or floating leaves and well-developed submersed roots (e.g., *Eichhornia*, *Trapa*, and *Hydrocharis*), to minute surface-floating or submersed plants with few or no roots (e.g., Lemnaceae, *Azolla*, and *Salvinia*). Reproductive organs are floating or aerial (e.g., aquatic *Utricularia*) but rarely submersed (e.g., *Ceratophyllum*).

Because of the major impact of the littoral vegetation on the metabolism of most lake and river ecosystems, a brief résumé of the main aspects of their morphological and physiological adaptations is required in order to appreciate their role in the productivity of fresh waters. Excellent reference works include Arber (1920), Gessner (1955, 1959), Sculthorpe (1967), and Hutchinson (1975).

II. AQUATIC PLANT CHARACTERISTICS

A. Emergent Flora

Aerial stems and leaves of emergent macrophytes possess many similarities in both morphology and physiology to related terrestrial plants. The emergent monocotyledons, such as *Phragmites* and the cattail *Typha*, produce erect, approximately linear leaves from an extensive anchoring system of rhizomes and roots. Epidermal cells are elongated parallel to the long axis of the leaf, which allows flexibility for bending. The cell walls are heavily thickened with cellulose, which provides the necessary rigidity. The mesophyll is generally undifferentiated and contains large gas spaces (lacunae) traversed at intervals by diaphragms that are porous to gases but not to water. These lacunae are separated from each other by thin walls of parenchyma cells. The anatomy of vascular bundles is similar to that of typical terrestrial plants: The xylem consists of scattered tracheids and parenchyma cells; the phloem, which consists of sieve tubes, companion, and parenchyma cells, is ensheathed with supporting sclerenchyma fibers. Emergent dicots produce erect, leafy stems, which show great anatomical differentiation. The mesophyll tissue of leaves is divided into typically dicotyledonous upper palisade and lower spongy layers.

The root and rhizome systems of these plants exist in permanently anaerobic sediments and must obtain oxygen from the aerial organs for sustained development. Similarly, the young foliage under water must be capable of respiring anaerobically for a brief period until the aerial habitat is reached, since the oxygen content of the water is extremely low in comparison to that of the air. Once the foliage has emerged into the aerial habitat, the intercellular gas channels and lacunae increase in size, thus facilitating gaseous exchange between the rooting tissues and the atmosphere.

TABLE 18-1 Vein Length Densities (per Unit Leaf Area, cm cm^{-2}) in Different Plant Types[a]

Aquatic plants		*Terrestrial plants*	
Floating leaves		*Rosa canina*	108
Nelumbo		*Acer campestris*	102
Aerial leaf		*Fraxinus excelsior*	88
Edge	104	*Quercus cerris*	86
Middle	91	*Salix rubrum*	52
Floating leaf			
Edge	68		
Middle	71		
Nymphaea mexicana	44		
Salvinia auriculata	27		
Submersed leaves			
Potamogeton praelongus	14		
Nuphar luteum	7		

[a] After Gessner (1959).

B. Floating-Leaved Macrophytes

Macrophytes that are attached to the substratum and possess leaves that float on the water surface are nearly all angiosperms, most conspicuously represented by the ubiquitous water lilies. The surface of the water is a habitat subjected to severe mechanical stresses from wind and water movements. Adaptations to these stresses by floating-leaved macrophytes include the tendency towards peltate leaves that are strong, leathery, and circular in shape with an entire margin. The leaves usually have hydrophobic surfaces and long, pliable petioles. This similar suite of adaptations is exhibited by floating-leaved macrophytes from many taxonomically unrelated groups. In spite of these adaptations, severe winds and water movements restrict these macrophytes to relatively sheltered habitats in which there is little water movement.

Floating leaves exhibit well-developed dorsoventral organization, in which the mesophyll usually is differentiated into an upper photosynthetic palisade tissue and an extensive lacunate tissue. Localized masses of spongy tissue aid buoyancy and, in combination with vascular tissues, offer resistance to tearing. The venous network of the leaves of aquatic macrophytes is much less extensive than among terrestrial plants and is least developed among submersed angiosperms (Table 18-1).

Positioning of leaf surfaces parallel to the water surface creates a vigorous competition for space to expose maximum leaf area to incident light. Leaf growth some distance from the root or rhizome system is accommodated by long, very pliable petioles. Between 10 cm and 4 m water depth, a complete proportionality is found between water depth and the length of the leaf petioles of water lilies. Petioles are about 20 cm longer than the water depth, which permits leaves to remain on the surface among undulating waves and small changes in water level. Stems of flowers, separate from the leaves in many species, grow to the surface in the same way. If gaseous exchange to the leaf surface is restricted experimentally, elongation of the petioles continues, and the leaves become aerial instead of floating. Neither oxygen nor carbon dioxide availability is involved in the control mechanisms of petiole elongation. Rather, petiole or stem growth is arrested at the surface when the leaves begin to lose ethylene to the atmosphere, a product that triggers production of the growth hormone gibberellic acid (see p. 532). Under crowded conditions, elongation frequently continues until leaves extend somewhat above the water surface, which reduces leaf overlap and increases light acquisition.

The presence and abundance of stomata vary among different species of floating-leaved angiosperms. The most common pattern is a restriction of stomata to the upper-leaf epidermis. A few stomata are found on the undersides of leaves; these are considered relict features and have no known function.

The epidermis of the leaf underside of many floating-leaved and submersed angiosperms contains groups of many smaller cells nested around a pore about 0.05 μm in diameter; an extremely thin cutin lamella lies over the pore. These "organs," termed *hydropoten*, assume a three-celled lenticular form in some water lilies or shieldlike hairs in certain submersed angiosperms. The hydropoten function in ion absorption in both submersed and floating leaves in contact with water by an active mechanism analogous to that of root absorption (Lüttge, 1964). Absorbed ions are then translocated to veins of the mesophyll.

C. Submersed Macrophytes

The submersed macrophytes are a heterogeneous group of plants that include (a) filamentous algae (e.g., *Cladophora*) that under certain conditions develop into profuse mats and may become loosely attached to the substrata of the littoral zone, (b) certain macroalgae (e.g., the calcareous stoneworts, *Charales*) that can dominate littoral macrophyte communities in hardwater lakes, (c) mosses that are occasionally the major macroflora of soft-water lakes and streams, (d) a few totally submersed nonvascular plants (e.g., the lycopsid quillwort *Isoetes*) that are most abundant in clear, softwater lakes, and (e) vascular submersed macrophytes of slightly <20 diverse families, mostly monocotyledons.

Among the vascular submersed macrophytes, numerous morphological and physiological modifications are found that allows existence in a totally aqueous environment. Stems, petioles, and leaves usually contain little lignin, even in vascular tissues. Sclerenchyma and collenchyma are usually absent. No secondary growth occurs, and no cambium can be recognized. Conditions of reduced illumination under the water are reflected in numerous characteristics: an extremely thin cuticle, leaves only a few cell layers in thickness, and an increase in the number of chloroplasts in epidermal tissue. Leaves tend to be much more divided and reticulated than those of terrestrial or other aquatic plants. The vascular system is greatly reduced, and all major conducting vessels have been reduced or lost from the stems. As a rule, conducting bundles have coalesced to axial vascular bundles in both mono- and dicotyledonous submersed species. Phloem and xylem are quite functional but difficult to differentiate, since phloem and woody parenchyma are nearly absent.

Leaves of submersed macrophytes occur in three main types: entire, fenestrated, and dissected. The entire leaf form is the most common throughout all groups and habitats. Entire leaves are often elongated to ribbonlike and filiform (threadlike) morphology in which length greatly exceeds width, even among more lanceolate-type leaves. Elongated, pliable leaves resist tearing in moving water, maximally utilize the reduced available light, increase the ratio of surface area to volume, and therefore presumably increase the efficiency of gaseous exchange and nutrient absorption. Fenestrated leaves are rare among submergents and occur only among a few tropical monocots. The adaptive advantages of such perforated, lacelike foliage are unclear. Dissected leaves are common among submersed dicotyledons. The most common form is extreme dissection with segments in whorls radiating from the petiole. Both the fenestrated and dissected leaf forms greatly increase the surface-area-to-volume ratios.

D. Freely Floating Macrophytes

The freely floating macrophytes, which occur submersed or on the surface, also exhibit great diversity in morphology and habitat. Several of these plants, notably *Lemna*, *Pistia*, *Salvinia*, *Eichhornia*, and *Trapa*, develop so profusely in some waterways and lakes that they inhibit the commercial use of these systems (Hillman, 1961; Moore, 1969; Mitchell and Thomas, 1972). The very high production and prolific development of these surface-inhabiting plants can result in excessive loading of organic matter and nearly total attenuation of light below the surface. As a result, periods of severe reduction or depletion of dissolved oxygen can result with large losses of invertebrates and fish.

The most elaborate of the free-floating life forms consists of a rosette of aerial and surface-floating leaves, a greatly condensed stem, and pendulous submersed roots. Most rosette species are perennials and are free-floating throughout growth, except for initial stages of seedling development. Among some freely floating groups, there is a strong tendency for reduction from the rosette habit. In the duckweeds (Lemnaceae), extreme reduction is seen in the trend toward elimination of roots and lack of separation between stem and leaf. Similarly, in the submersed free-floating "carnivorous" *Utricularia*, stems and leaves are not differentiated, and vegetative organs are greatly modified for entrapment of microfauna, especially rotifers. Some species of *Lemna* and *Utricularia* are, like the ubiquitous *Ceratophyllum*, submersed during vegetative growth and rootless. Freely floating macrophytes generally are restricted to sheltered habitats and slow-flowing rivers. Their nutrient absorption is completely from the water; most of these macrophytes are found in waters rich in dissolved nutrients.

Most floating plants possess little lignified tissue. The rigidity and buoyancy of the leaves are maintained by the turgor of living cells and extensively developed lacunate mesophyll tissue (often >70% gas by volume). Vascular tissues of the leaves are very poorly differentiated; in most, the protoxylem is represented by a lacuna. Vegetative propagation by the production of lateral stolons that develop into new rosettes is a common mode of rapid proliferation in this group. All freely floating rosette plants form well-developed adventitious roots, lateral roots, and epidermal hairs. The root system of the water hyacinth (*Eichhornia*), for example, represents 20–50% of the plant biomass.

E. Heterophylly: Vegetative Polymorphism

Although vegetative polymorphism or *heterophylly* is not unique to submersed macrophytes, extremes in foliar plasticity are particularly conspicuous among aquatic macrophytes that normally grow in shallow, submersed habitats that undergo fluctuations in water level. Such species can readily produce floating or aerial foliage. A marked polymorphism of leaves may often be found on the same stem or petiole. There is a tendency to shift from finely divided submersed leaves in the submersed habitat to more coalesced entire leaves when leaves of the same plant are surface-floating or aerial. Leaf form and anatomy can vary widely with age, water depth, current velocities, nutrient supplies, light intensity, day length, temperature, and other factors.

Heterophylly is particularly conspicuous among such common genera as *Ranunculus*, *Callitriche*, *Hippuris*, and *Potamogeton* and creates chaos when the taxonomy is based on leaf morphology. For example, leaf morphogenesis of the amphibious buttercup *Ranunculus* responds conspicuously to submergence and water temperature (Fig. 18-1). The lobate tripartite aerial leaf of the semiaquatic phase is approximated in the submersed phase at high temperatures. Both lobe numbers and blade lengths increase conspicuously with growth at decreasing temperatures. Analogous morphological differentiation has been shown to be directly related to carbon dioxide concentrations in at least three taxonomically distinct amphibious species (Bristow, 1969). Under conditions of high aqueous CO_2 concentrations, the plants developed dissected submersed leaves, whereas when the CO_2 content was reduced, lobate aerial foliage developed rapidly. These plants were unable to utilize bicarbonate as a source of inorganic carbon for photosynthesis. Among water lilies, continued elongation of petiole length occurs under conditions of low CO_2 of the water until the leaves reach the relatively high levels of atmospheric CO_2 (Gessner, 1959).

Transitions between submersed and aerial-type leaves of *Hippuris* are reversible by altering the ratio of red (660 nm) and far red (730 nm) light, which implies direct phytochromal control (Bodkin *et al.*, 1980; cf. Stross, 1981). A low ratio of red to far red light, such as occurs in very shallow water (far red light is selectively attenuated with increasing depth), induces aerial leaf formation on submersed shoots.

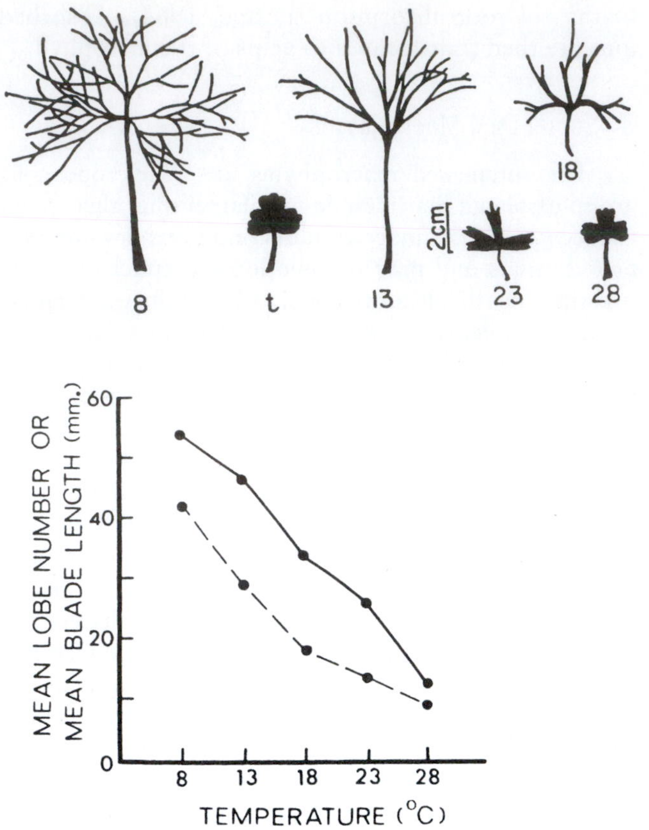

FIGURE 18-1 *Upper*: Silhouettes of leaves of a single clone of *Ranunculus flabellaris* grown in the semiaquatic phase (*t*) and submersed at temperatures of 8, 13, 18, 23, and 28°C. *Lower*: Mean blade length (———) and mean lobe number (- - -) in relation to temperature. (Modified from Johnson, M. P.: Temperature dependent leaf morphogenesis in *Ranunctilus flabellaris*. *Nature (London)* *214*:1354–1355, 1967.)

Ethylene is a plant hormone influencing many plant processes such as growth promotion, development of aerenchyma, abscission, and senescence. Ethylene biosynthesis involves multistep enzyme pathways in which methionine is converted to ethylene (e.g., Fluhr and Mattoo, 1996). When leaves of several aquatic angiosperms are submersed, ethylene concentrations reach high levels in the internal intercellular lacunae (Musgrave *et al.*, 1972; Musgrave and Walters, 1973; Cohen and Kende, 1986). These high levels of ethylene enhance the sensitivity of the tissue to the hormone gibberellic acid and consequently promote petiole elongation. When the tissues reach the water surface, gaseous contact is made between the intercellular air and the atmosphere, and the accumulated ethylene rapidly dissipates into the atmosphere. Gibberellic acid activity and cell elongation then decline. The plant growth regulator abscisic acid also induces the formation of floating leaves in totally submersed germinating tubers of *Potamogeton nodosus*

(Anderson, 1978) and of *Ranunculus flabellaris* (Young and Horton, 1985).

Numerous other factors influence leaf form among aquatic macrophytes (cf. Sculthorpe, 1967; Gaudet, 1968; Hutchinson, 1975). Generalizations are difficult to make, however, because among some plants polymorphism occurs irrespective of external factors; in such cases, polymorphism is distinguishable at the earliest primordial stages and is apparently determined either genetically or by the nutritional status of the shoot meristem.

F. Reproduction, Dispersal, and Dormancy

Many alterations in reproductive strategies have accompanied the invasion of the aquatic environment by higher plants of several unrelated terrestrial plant lineages. Most aquatic plants have simply modified the reproductive features of their terrestrial ancestors (Sculthorpe, 1967; Hutchinson, 1975; Grace, 1993; Philbrick and Les, 1996). Aquatic plants have adopted vegetative and clonal reproduction as a major mechanism for population growth and dispersal. Sexual reproduction and genetic recombination have been retained, although they are commonly subordinate to vegetative expansion and dispersal.

1. *Asexual Clonal Reproduction.*

Asexual reproduction includes both vegetative reproduction and seed production without fertilization (agamospermy). The latter process is very poorly understood in aquatic plants (Les, 1988). Population expansion of all three major types of aquatic plants—emergent, floating-leaved, and submersed—is largely by vegetative, clonal growth by both (a) the rapid development of aboveground tissues, particularly with apical meristematic growth, followed by fragmentation, and (b) expansion horizontally by creeping stems (e.g., rhizomes), modified shoot bases (corms and small bulbs), and root suckers and tubers. Clonal reproduction performs several functions (Grace, 1993): (a) numerical increase, (b) dispersal, (c) acquisition of resources, (d) storage, (e) protection, and (f) anchorage.

Another important vegetative propagule in aquatic plants is the *turion*. These morphological variations are found among temperate aquatic macrophytes that overwinter by the formation of *winter buds (turions)*. In the fall, many common macrophytes (e.g., *Utricularia*, *Myriophyllum*, and *Potamogeton*) form masses of dormant vegetative buds of aborted leaves with very short internodes in the axils of lower leaves (Weber and Noodén, 1974, 1976a,b; Aiken and Walz, 1979; Winston and Gorham, 1979; Sastroutomo, 1980a,b, 1981). These turions separate from the mother plant and sink or float some distance away and serve as a means of vegetative propagation (Weber, 1972). Many submersed plants persist for years by this means, without a sexual cycle. Turion formation is absent among the same species in the tropics but can be easily induced when plants near maturity are exposed to low water temperatures (<10–15°C) and short daily photoperiods (e.g., Van *et al.*, 1978; Spencer *et al.*, 1994; Netherland, 1997) or long daily photoperiods (e.g., Chambers *et al.*, 1985). Other winter buds of less modified leaves break off of the mother plant and these fragments disperse and subsequently can colonize in a new habitat.

Vegetative propagules are dispersed by many vectors. Mechanisms include both abiotic vectors, such as with water movements or strong winds for surface-floating macrophytes, and biota, such as amphibians, reptiles, birds, and mammals, particularly humans. Humans have dispersed many aquatic plants either intentionally (rice, water cress) or unintentionally (water hyacinth, *Myriophyllum, Hydrilla*). The latter three species have expanded so vigorously in some waters that their excessive development has altered the hydrology and chemistry to such an extent that the waters are unsuitable for many human uses.

2. *Sexual Reproduction.*

Sexual reproduction tends to be reduced in aquatic plants simply because of the efficacy of clonal and other vegetative reproduction. Some of the most effective competitors and most productive of aquatic plants, such as the cattail (*Typha*) and rushes (*Juncus*), combine very intensive clonal continuous growth (see later discussion) with moderate-to-intensive production of seeds for longer-distance dispersal. For example, the emergent rush *Juncus effusus* can devote as much as <1–7% of its annual production to seed production (several million very small seeds m^{-2}) even though germination and seedling mortality in the immediate vicinity of the mother plants is nearly 100% from autotoxicity from organic compounds released and from shading (Ervin and Wetzel, 2000).

Where flowering and sexual reproduction is retained in submersed aquatic plants, aerial flowering dominates on stems or flower stalks that project across the water surface. Among dioecious species in which only one sex occurs in an area, vegetative means are the only means of immediate reproduction (Haynes, 1988). Monoecious species or species with perfect flowers often produce viable seeds but they function largely in dispersal rather than as a means of increasing the immediate population size. Pollination occurs by insect or wind vectors. Water pollination (hydrophily), in which waterborne pollen is released and captured on the surface of water or under water, is rare and is found in <5% of aquatic species (Cox, 1993). Self-pollination is known to occur among aquatic species, as occurs in many terrestrial plants, but can be interpreted as adaptive where ecological conditions deter cross-pollination. Even though sexual reproduction is reduced or sometimes absent for long periods among submersed aquatic plants, genetic homogeneity from vegetative reproduction can be ameliorated by immigration of genetically distinct vegetative propagules from other habitats (Les, 1991).

G. Transpiration and Evapotranspiratory Water Losses

Evaporative losses from lakes and wetlands are greatly modified by the transpiration from emergent and floating-leaved aquatic plants. Rates of transpiration and evaporative losses to the atmosphere vary with an array of physical (e.g., wind velocity, humidity, and temperature) and metabolic parameters and structural characteristics of different plant species (Brezny *et al.*, 1973; Bernatowicz *et al.*, 1976; Boyd, 1987; Jones, 1992). For example, in a temperate-lake littoral, a distinct seasonality of evapotranspiration is common, and greatly accelerated rates of water loss from the lake habitat commonly occur in the summer period (cf. Table 4-3). In a humid tropical environment, floating plants and floating-leaved but rooted plants enhanced water losses from surface waters, often by a factor of 2 (Table 18-2), particularly during wet seasons (Rao, 1988).

The epidermal cuticle of emergent and floating leaves is generally well developed. Cuticular transpira-

TABLE 18-2 Representative Rates of Evapotranspiration (E_t) by Aquatic Plants and Comparison to Rates of Evaporation from Open Lake Water (E_o)[a]

Species	*mm d*$^{-1}$	E_t/E_o	*Reference*
Emergent			
Typha domingensis	2.7–4.7	1.3	Glenn *et al.* (1995); Abtew (1996)
Typha latifolia	4–12	1.41–1.84	Snyder and Boyd (1987)
	4.8	3.7–12.5	Price (1994)
Carex lurida	4.0–6.3	1.33	Boyd (1987)
Panicum regidulum	5.5–7.5	1.58	Boyd (1987)
Rice (*Oryza sativa*)	6–13		Humphreys *et al.* (1994)
Myriophyllum aquaticum	0.2–1.0		Sytsma and Anderson (1993)
Juncus effusus	3.8–8.0	1.52	Boyd (1987)
Justicia americana	2.2–6.4	1.17	Boyd (1987)
Alternanthera philoxeroides	4.0–6.3	1.26	Boyd (1987)
Willow carr (*Salix* spp.), Czech Republic	2.3–3.7		Pribán and Ondok (1986)
Sedge-grass marsh (*Carex, Calamagrostis, Glyceria*), Czech Republic	2.2–4.5		Pribán and Ondok (1985, 1986)
Lakeshore marsh (*Sagittaria, Pontederia, Panicum, Hibiscus* dominating), FL	0.5–1.0	0.35–1.2	Dolan *et al.* (1984)
Carex-dominated marsh, Ontario subarctic	2.6–3.1	0.74–1.02	Lafleur (1990)
Floodplain forest, FL	5.57		Brown (1981)
Reed (*Phragmites*) swamp, Czech Republic	1.4–6.9	1.03	Šmíd (1975)
Arctic wetland, Canada	4.5 (2.2–7.3)		Roulet and Woo (1988)
Quaking Fen, Netherlands			Koerselman and Beltman (1988)
Typha latifolia	0.9–4.7	1.87	
Carex diandra	1.1–3.9	1.68	
Carex acutiformis/Sphagnum	1.0–3.7	1.65	
Floating-leaved rooted			
Nymphaea lotus	2.5–6.0	0.82–1.35	Rao (1988)
Floating, not rooted			
Eichhornia crassipes	3.8–10.5	1.30–1.96	Rao (1988)
	6–11	1.45–2.02	Snyder and Boyd (1987)
		2.67	Lallana *et al.* (1987)
Salvinia molesta	2.1–6.8	0.96–1.39	Rao (1988)
Pistia stratiotes	19.9	1.07	Brezny *et al.* (1973)
Azolla caroliniana	7.1	0.95	Lallana *et al.* (1987)
Lemna spp.		1.03	Brown (1981)

[a] Modifed from Wetzel (1999a).

TABLE 18-3 Comparison of the Ratio of Maximal to Minimal Cuticular Transpiration among Aquatic Macrophytes and Terrestrial Flora[a]

Aquatic flora		*Terrestrial flora*	
Potamogeton natans	1.4	*Quercus ilex*	4.6
Eichhornia crassipes	1.5	*Syringa vulgaris*	4.9
Alisma plantago	1.7	*Viola tricolor*	7.6
Nymphaea marliacea	3.2	*Laurus nobilis*	8.4

[a] Modified from Gessner (1959).

tion is much lower in these plants than in land plants (Table 18-3). Cuticular transpiration is generally greater from floating-leaved macrophytes than from aerial foliage of emergent macrophytes. As a result, water losses by transpiration to the atmosphere by emergent macrophytes occurs largely via the stomata.

Rates of transpiration by emergent macrophytes are extremely high and result in an efflux of water vapor from the leaves that is much greater than evaporation from an equivalent area of water (Table 8-2). Evaporation from the water surface is lower when the water surface is densely covered by macrophytes, such as water lilies or duckweeds (*Lemna*). These plants restrict the contact of moving air with the water surface. However, this effect is counterbalanced by plant transpiration rates that increase markedly when the vegetation is exposed to moderate wind velocities (e.g., van der Weert and Kamerling, 1974; Benton *et al.*, 1978). Evapotranspiration rates of aquatic plants generally increase with increasing wind velocity up to modest speeds (ca. <2 m s^{-1}) and with decreasing relative humidity (Gessner, 1959; Rao, 1988). Evapotranspiration rates are generally positively correlated with temperatures and solar irradiance on a seasonal basis and furthermore increase with increasing rates of photosynthesis up to a maximum rate that differs among species of aquatic plants. Transpiration is usually lower in the basal leaves than in apical foliage and results in a general tendency toward xerophytic properties with increasing age of the leaves. Stomatal conductance and evapotranspiration rates decline as photosynthetic efficiencies are reduced during midday periods of intense irradiance and maximum diurnal temperatures (e.g., Bernatowicz *et al.*, 1976; Królikowska, 1978; Giurgevich and Dunn, 1982; Jones and Muthuri, 1984; Mann and Wetzel, 1999). These midday depressions in evapotranspiration suppress overall water losses to a greater extent than is counterbalanced by increased rates of evaporation from open water surfaces with the higher temperatures and lower relative humidity values that occur at midday.

Stomata of aquatic macrophytes generally do not close to nearly the extent observed in terrestrial flora during the light period, and higher transpiration rates occur in aquatic macrophytes than are found among land plants. In both aquatic plants and terrestrial flora, however, stomata prevent desiccation when water availability declines. Many aquatic plants with emergent and floating leaves have developed means of increasing surface area and evapotranspiration. Such structures include papillae, either single-celled or multicellular extensions of the epidermis, or perforations into the leaf blade, termed *stomatodes*. Many aquatic angiosperms, both emergents and submersed plants, exude water (guttation) from enlarged ends of veins (hydathodes) around the margins of their leaves.

Among aquatic macrophytes, population density is important in the regulation of transpiration rates. Evapotranspirational water losses by emergent macrophytes in the littoral zone of lakes and wetlands can be so effective that the plants can significantly reduce the water levels of the surrounding terrestrial area and can result in diminished growth of nearby terrestrial plants. During the macrophyte growing season, evapotranspiration by plants can cause a cone of depression such that water will seep to the cone for replacement from the lake or stream.

It is common among emergent macrophytes of large river systems, the borders of reservoirs, and bank riverine wetlands to periodically become inundated with rising water levels. In large river systems, such as the Amazon, herbaceous vegetation can be inundated for several months (Junk and Piedade, 1997). Increases in water level, to the point where large portions of or the entire emergent plant are submerged, result in a varied number of morphological and physiological changes. Many species can survive submergence for long periods of time, but then growth is greatly reduced because of conditions of diminished light and oxygen. Leaves are generally reduced in size and tend to elongate and lose rigidity. Mesophyll tissue often is reduced, and there is a corresponding increase in spongy tissue and intercellular air spaces. The xylem and lignified portions of the phloem fibers also generally decrease, concomitant with an increase in stem diameter and lacunal air spaces.

III. METABOLISM BY AQUATIC MACROPHYTES

A. Aeration, Gas Fluxes, and Metabolism within Higher Aquatic Plants

Clearly, the rooting tissues of aquatic plants nearly always exist in anaerobic organic sediments that contain high concentrations of relatively toxic organic solutes (e.g., volatile fatty acids) and gaseous substances (e.g., hydrogen sulfide) of fermentation. Movements of gases within and through the plant parts are essential to avoid toxicity effects and to maintain active photosynthetic and respiratory metabolism. The movements of oxygen, carbon dioxide, and other gases within and through the interconnecting aerenchymatous (lacunal) and nonaerenchymatous intercellular cortical gas spaces of shoots and roots of aquatic plants have been the subjects of intensive study in the past decade. Under the intensely and more severely reducing conditions of organic sediments, the gas spaces of the aerenchymatous tissues increase in size and can be induced to increase in size upon exposure to increasing concentrations of fermentative endproducts such as volatile fatty acids, alcohols, and H_2S (e.g., Penhale and Wetzel, 1983; Armstrong *et al.*, 1991a). Littoral and wetland plants that are tolerant of continuously saturated sediments or periodic flooding of the soils adjacent to lakes and rivers combine a range of both metabolic and morphoanatomical adaptations to enable them to tolerate seasonally flooded sediments. For example, the roots of some flood-tolerant species, when flooded seasonally, stop producing high amounts of ethanol and lactate and divert part of the pyruvate produced in respiration to nontoxic malate (Joly, 1994). The amount of pyruvate converted to malate is proportional to the amount of oxygen diffusing from the aerial parts to the roots.

Several mechanisms of gas fluxes are operationally specific to emergent and floating-leaved macrophytes and others are specific to submersed macrophytes (Blom *et al.*, 1990; Armstrong *et al.*, 1991b, 1992, 1996; Grosse *et al.*, 1996):

1. Mechanisms among Plants with Aerial Leaves

1. ***Diffusion-Dependent Throughflow Convections.*** A pressurization differential results from a Knudsen-diffusion of gases from the environment into the emergent shoot and root system. Gases enter newly emergent leaves, such as of floating-leaved water lilies, through minute pores that separate the aerenchyma of the leaves from the ambient atmosphere. A pressurized ventilation is created by the Knudsen-diffusion by both the physical effects of hygrometric diffusion (humidity-induced) and thermal transpiration (thermal osmosis) (Grosse, 1996). In mature leaves, these pores are larger and gas flows through them freely. Air moves into young, influx leaves along humidity and temperature gradients, causing pressurization in the aerenchyma, flows through the continuous intercellular space system, and exits from the mature, efflux leaves to the atmosphere (Fig. 18-2). In some cases, some of the gas exits through rhizomes to shoots of nearby plants.

2. ***Nonthroughflow Convections.*** A longitudinal gradient of gas velocity occurs through the plant in response to changes in concentrations of gases resulting from the reciprocal exchange of metabolic uptake or release of gases, especially carbon dioxide and oxygen. An excellent example is seen in rice, where bulk flows of gases occur into deepwater rice plants (Raskin and Knede, 1983, 1985; Stünzi and Kende, 1989). Gas flows in response to reduced pressure (suction) set up within the plant by the solubilization of respiratory CO_2 into surrounding water and gas films adhering to the leaves (Fig. 18-2). Oxygen diffusing from the atmosphere to submersed parts of the plant is not replaced by a reciprocating diffusion of CO_2 from below, and a sucking of air into the plant occurs alongside the normal diffusive intake of oxygen. The uptake of air into submerged organs of the rice plant is driven by the difference in the solubility of O_2 and CO_2 in water. The conduction of gases through the internal gas spaces of the leaf is negligible compared to the conduction of gases through the external air layers.

3. ***Venturi-induced Throughflow Convections.*** Venturi-induced pressure differentials are created when winds blow over the long (to >2 m) and persistent, dead, often broken, culms of emergent plants. Air is drawn out of these old shoots, which remain connected to the gas-space system of the underground rhizome network, by the Venturi effect and is replaced by an inflow form the atmosphere into other culms broken much closer to the sediment or water surface.

Internal pressurization and convective throughflow of gases are common attributes of wetland plants with cylindrical culms or linear leaves and among floating-leaved rooted macrophytes (e.g., Brix *et al.*, 1992; Armstrong *et al.*, 1996). Convective flowthrough mechanisms are much more efficient than nonthroughflow convection from the atmosphere. A distinct diurnal fluctuation in pressurization and convective flow is commonly observed, where rates are highest in the afternoon and lowest at night. The diurnal responses are thermo-osmotic pressurization differences regulated by changes in solar heating of tissues and alterations in humidity (cf. Dacey, 1981; Brix, 1988; Armstrong and Armstrong, 1990; Brix *et al.*, 1992).

The rate at which oxygen is transported through the lacunal system is influenced by the rate of net photosynthetic oxygen production in the foliage, which undergoes diurnal fluctuation, by the respiratory demands for oxygen of the root system, and by the resistance to diffusion afforded by diaphragms of the lacunae. Experimental analyses have demonstrated that roots and rhizomes can tolerate appreciable periods—as long as

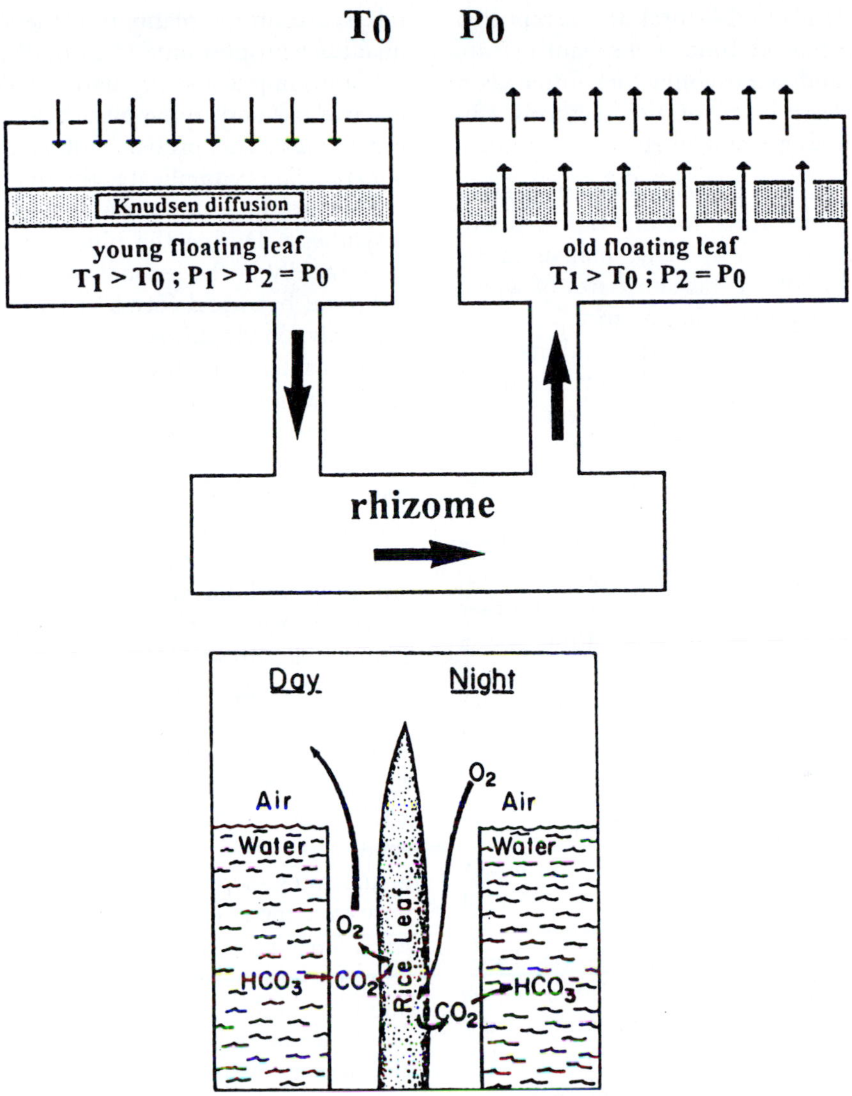

FIGURE 18-2 *Upper*: Convective gas throughflow in floating-leaved aquatic macrophytes resulting form thermal transpiration (thermal osmosis) and humidity-induced diffusion, where P = pressure in the young influx leaf (1), an old efflux leaf (2), and atmosphere (0), and T = temperature inside the leave and ambient air (0). (After Grosse, 1996.) *Lower*: Movement of gases along the exterior of leaves of deepwater rice. In darkness, respiration replaces relatively water-insoluble O_2 with CO_2. Rapid solubilization of CO_2 in water causes a decrease in the pressure inside the air layers and leads to movement of air from the atmosphere into the air layers. In light, photosynthetic CO_2 is absorbed from the water through the large liquid–gas interface provided by the air layers. Photosynthetic O_2 is expelled through the air layers into the abovewater atmosphere. (After Raskin and Kende, 1983.)

a month—of very low oxygen supply or anaerobiosis without any apparent harmful effects.

2. Methane Diffusion, Transport, and Evasion to the Atmosphere

The importance of methane (CH_4) release from inland aquatic plants of wetlands and littoral areas of lakes and rivers has been a subject of intense interest as a potential source of CH_4 additions to atmospheric gases that enhance retention of heat from solar irradiance ("greenhouse effect"). As the transport processes of CH_4 from bacterial production in anaerobic sediments through emergent and rooted floating-leaved aquatic plants were evaluated, a prodigious literature has appeared on this subject, particularly in relation to deepwater rice cultivation and the evasion of large amounts of CH_4 to the atmosphere by this aquatic plant. Among studies of CH_4 releases from aquatic plants of natural wetlands and littoral areas, reported evasion rates and purported control mechanisms are enormously variable and often conflicting, often as a result of the noncomparability of the methods employed. A few generalities are emerging:

1. Methane production is a microbiological process and results from anaerobic carbon mineralization in sediments. As discussed later (Chap. 21), methanogenic bacteria utilize a number of organic substrates of which acetate and hydrogen are of major importance.

2. Within the sediments, CH_4 diffuses along concentration gradients and can be oxidized by microbiological processes of methane-oxidizing bacteria (Chap. 21) under microaerobic conditions. The rate of methane oxidation increases proportionally to the rate of methane production.

3. Alternatively, in the root–rhizome rhizosphere of aquatic plants, some CH_4 diffuses into the intercellular gas spaces and can be transported along with other gases through the ventilation system to the atmosphere. This evasion to the atmosphere can be extensive in both floating-leaved rooted plants, such as the water lilies (e.g., Dacey and Klug, 1979), and emergent macrophytes (e.g., Segers, 1998).

4. Rates of potential methane production have been determined in a large number of studies in various natural wetlands and rice paddies. Rates are highly variable with environmental conditions within a range of 10^{-2}–10^{-3} μmol m^{-3} s^{-1} (reviews of Sebacher *et al.*, 1985; Schütz, *et al.*, 1989; Whiting and Chanton, 1993; Wang *et al.*, 1996; Hosono and Nouchi, 1997; Segers, 1998). Some of the variation is real and is associated with many differences in environmental factors regulating CH_4 production (e.g., temperature, pH, gas pressures, substrate availability, and nutrient availability, competition with alternate electron acceptors), microbial oxidation, and diffusion into the plants and their gaseous transport mechanisms. Changes in water level and inundation can rapidly alter conditions of anoxia and redox. Some of the variation is also attributable to large differences in the experimental methods used for assaying evasion rates and their control mechanisms. In general, however, CH_4 evasion to the atmosphere is positively correlated with net production of wetland vegetation.

5. On a diurnal basis, the highest emission rates of CH_4 are found in daytime, usually soon after sunrise, among many, but not all, emergent macrophytes examined (e.g., Miura *et al.*, 1992; Whiting and Chanton, 1996; van der Nat, *et al.*, 1998; many others).

3. Gaseous Transport in Submersed Macrophytes

It has been assumed, incorrectly, that as photosynthesis proceeds in submersed macrophytes, oxygen is released from the plants into the surrounding water in quantities proportional to rates of photosynthesis. Such a relationship is approximately accurate for micro- and macroalgae and apparently also for the nonvascular bryophytes. The presence of large lacunae in higher aquatic plants complicates the relationship, and oxygen released to the water variably lags behind rates of photosynthesis. During the photoperiod, oxygen accumulates rapidly in the lacunal intercellular atmosphere on a diurnal basis and forms an internal positive pressure and only slowly diffuses out into the surrounding water (Hartman and Brown, 1966; Sorrell, 1991). This pressurization assists with the internal mass flow of gases among plant parts. Consequently, diffusion rates of oxygen into the water are not directly correlated with the intensity of photosynthesis, and a large portion of the oxygen in the intercellular lacunae is utilized for respiration during both the photoperiod and in darkness, without affecting the oxygen concentrations of the surrounding water medium (e.g., Sorrell and Dromgoole, 1989). Although the release of bubbles, largely oxygen, from submersed plants can be significant during periods of intensive photosynthesis in warm water, this release should not be used as a measure of photosynthesis, as has been done (e.g., Odum, 1957, among others) because it can represent only a small proportion of the oxygen actually produced and actual rates of photosynthesis. At high rates of water flow over the plants, exchange of oxygen with the water is fairly rapid (e.g., Moeslund *et al.*, 1981) but still represents only a small portion of that produced by photosynthesis.

Most of the oxygen is retained within the plant and used for respiration, particularly by the rooting tissues, especially at low light intensities. Appreciable amounts of oxygen are released from roots of submersed aquatic macrophytes into the surrounding rhizosphere (e.g, Sand-Jensen *et al.*, 1982; Zhang *et al.*, 1998). Although much of the oxygen is utilized by bacteria, some reacts with ferrous iron to form a highly insoluble iron oxide that precipitates on the roots as an iron plaque. In addition to direct chemical reactions to form $Fe(OH)_3$, Fe-oxidizing bacteria are common in the rhizosphere of aquatic plant roots (Emerson *et al.*, 1999). Iron reduction then occurs in the anaerobic sediments immediately beyond the oxidized rhizosphere mediated largely by iron-reducing bacteria (cf. Chaps. 14 and 21). Such iron reduction can consume much of the organic carbon substrates by dissimilatory reduction of Fe^{+3} compounds.

The magnitude of the intercellular lacunal system is highly variable among species but is almost universally extensive and constitutes a major portion (often >70%) of the total plant volume. The fragility of the lacunal tissue and the possibility of its becoming filled with water if the plant were damaged at any point is avoided by a number of types of lateral plates and watertight diaphragms, permeable to gases, that interrupt the lacunae at intervals (Arber, 1920; Sculthorpe, 1967; Sorrell and Dromgoole, 1996). Although experimental evidence is meager, it is known that much of the oxygen produced during photosynthesis in submersed angiosperms and retained in the lacunal system diffuses from the leaves through the petioles and stems to underground root and rhizome systems, where respiratory demands are high. Some oxygen is released from the roots into the immediate (usually <0.5 mm) rhizosphere. Radial oxygen loss to the rhizosphere varies by over an order of magnitude (0.01–0.2 μg cm^{-2} root min^{-1}) among aquatic plant species.

The gases within the intercellular lacunae of submersed angiosperms diffuse along gas–partial-pressure gradients. In some emergent and floating-leaved plants (e.g., the yellow water lily *Nuphar*), however, evidence indicates that the internal gas spaces function as a pressurized flowthrough system (Dacey, 1981). Ambient air enters the youngest emergent leaves against a small gas-pressure gradient as a result of physical processes driven by solar-mediated gradients in temperature (thermal transpiration) and water vapor (hygrometric pressure) between the atmosphere and the lacunae. The lacunal gas spaces are continuous through young emergent leaves, petioles, rhizomes, and petioles of the older emergent leaves. The older leaves vent the elevated pressure generated by the younger leaves. The resulting flowthrough ventilation system accelerates both the rate of oxygen supply from the atmosphere to the root tissue and the rate of CO_2 and methane (Dacey and Klug, 1979) transport from the roots to the atmosphere.

The movement of oxygen (from the atmosphere in emergent and floating-leaved plants or from sites of oxygen production in submersed plants) to roots is essential to prevent the accumulation of toxic end products either from metabolism of the plant (e.g., ethanol produced during glycolysis) or from those that diffuse into the plant from the anaerobic rhizosphere (e.g., hydrogen sulfide). Of the two types of aerenchyma, schizogenous aerenchyma forms constitutively in some aquatic plants by specific patterns of cell separation and cell expansions to create intercellular spaces that fill with gases from surrounding cells (Raven, 1996). In contrast, lysigenous aerenchyma arises from the spatially selective death of grown cells. Any process that restricts root aeration stimulates production and accumulation of the plant hormone ethylene. The additional ethylene accelerates aerenchyma formation by a series of sensing and transduction processes that initiate cell death and degeneration and create the intercellular gas lacunae (Jackson and Armstrong, 1999).

Wetland and submersed plants often increase the volume of the lacunal gas system when sediments become more reducing (e.g., Katayama, 1961; Armstrong, 1978; Penhale and Wetzel, 1982; Jackson and Armstrong, 1999). In addition, a number of aquatic plants have developed metabolic adaptations to tolerate periods of anaerobic conditions: Diverse nontoxic end products (malate and shikimic acid) can be produced during glycolysis (Crawford, 1978; Penhale and Wetzel, 1982; Sale and Wetzel, 1982; McKee *et al.*, 1989), or ethanol can be released to the environment (Bertani *et al.*, 1980).

Bacterial end products of fermentation (H_2S and volatile fatty acids) in anoxic, reducing sediments are toxic to many plants (e.g., Koch and Mendelssohn, 1989). For example, H_2S suppresses alcohol dehydrogenase, an enzyme that catalyzes alcoholic fermentation and suppresses nutrient uptake. Some oxygen diffuses from the roots and forms an oxidized microzone in the surrounding sediments (reviewed by Armstrong, 1978; Sorrell and Dromgoole, 1989; Jespersen *et al.*, 1998). This oxidized microzone suppresses methanogenesis and reduces the toxicity of fermentation products but simultaneously reduces the availability of certain nutrients, such as iron and manganese, by the formation of oxidized precipitates in the microzone on the surfaces of the roots (Tessenow and Baynes, 1978; Crowder and Macfie, 1986; Caffrey and Kemp, 1991). Iron "plaques" are very commonly heavily deposited on the surfaces of roots, and such deposition is distinctly seasonal, reaching maximum deposition in midsummer. Certain sulfur bacteria (*Beggiatoa*) of the plant rhizosphere can oxidize hydrogen sulfide to sulfur granules and consequently lower H_2S concentrations adjacent to root tissues (Joshi and Hollis, 1977).

Photosynthetic carbon fixation of aquatic macrophytes is offset by growth, losses of CO_2 from mitochondrial respiration and photorespiration, and secretion of soluble organic compounds. Photosynthetic efficiency is influenced directly by the rates of these dynamic, constantly changing processes. In calculations of primary productivity, the respiration rate has usually been assumed to be the same in light as in the dark. Physiological evidence, however, indicates that such an assumption is erroneous. Mitochondrial (dark) respiration may be inhibited in the light in some plants, perhaps by suppression of glycolysis (Jackson and Volk, 1970). Efficient refixation of respired CO_2 in the light restricts the loss of respiratory CO_2 from submersed aquatic plants (Søndergaard, 1979; Søndergaard and Wetzel, 1980; Wetzel *et al.*, 1984a; Laing and Browse, 1985; Boston and Adams, 1986).

Photorespiration, well-known in terrestrial plants (Goldsworthy, 1970; Jackson and Volk, 1970; Hatch *et al.*, 1971), enhances the loss of photosynthetically fixed carbon as CO_2 and reduces photosynthetic efficiency of aquatic macrophytes. In this process, CO_2 is generated in the light from glycolic acid, a direct product of C_3–Calvin-cycle photosynthesis. The rate of this reaction is influenced by, and is proportional to, oxygen concentration, light intensity, and temperature. Glycolate metabolism is also enhanced when low CO_2 limits photosynthesis. The rate at which photorespired CO_2 is lost from the plant depends on the efficiency of CO_2 refixation in the intercellular spaces of the plant.

Terrestrial plants in which all cells photosynthesize by the C_3–Calvin cycle can lose up to 50% of fixed carbon immediately by photorespiration, depending on environmental conditions. In contrast, little or no CO_2 is lost in the light from plants in which photosynthesis proceeds through the C_4–β-carboxylation pathway in mesophyll cells. These plants efficiently refix CO_2 both from dark respiration and from photorespiration in the C_3 bundle sheath cells.

C_3 and C_4 plants can be distinguished by the following combination of characteristics, any one of which is often reasonable evidence for making the distinction (Black, 1971): (a) C_4 plants have highly developed bundle sheath cells in leaf cross sections, with unusually high concentrations of organelles and starch accumulation; C_3 leaves lack this differentiation. (b) Photosynthesis is difficult to light-saturate in C_4 plants, while in C_3 plants, saturation illuminance is in the range of 200–900 μmol quanta m^{-2} s^{-1}. (c) Photosynthetic temperature optima are 30–40°C in C_4 plants and 10–25°C in C_3 plants. (d) Photosynthetic CO_2 compensation points are low for C_4 plants (0–10 ppm CO_2) and high for C_3 plants (30–70 ppm CO_2). (e) Response of apparent photosynthesis to O_2 concentration is not detectable in C_4 plants and shows increasing inhibition >1% O_2 in C_3 plants. Response of photorespiration (CO_2 release in light) to O_2 concentration is not detectable in C_4 plants, but shows increasing enhancement with increasing O_2 in C_3 plants. (f) Glycolate synthesis and glycolate oxidase activity are low in C_4 plants as compared to C_3 plants.

Emergent and floating plants are partially exposed to environmental conditions similar to those of terrestrial plants. Possessing largely C_3 photosynthetic metabolism, photorespiration can be extensive (Longstreth, 1989). High CO_2 concentrations are common in the intercellular spaces of many of these emergent species, enhanced by high concentrations moving from roots and rhizomes to leaves through internal gas channels and lacunae. This CO_2 enrichment, in addition to common, relatively vertical leaf orientation with a high leaf-area index (leaf area per unit surface area) and relatively unlimited water supply, results in high productivity.

The C_4 photosynthetic system is likely of adaptive value in many emergent hydrophytes, particularly in regions of high temperature and high light intensity or in situations in which the salt content of the environment adversely affects internal CO_2 and water balance (Hatch *et al.*, 1971). In the latter context, *Spartina*, the dominant plant of the salt marshes of many marine coastal regions, and papyrus (*Cyperus papyrus*) of the tropics are C_4 plants (Black, 1971; Jones and Milburn, 1978). There is some evidence that *Typha latifolia* is a C_3 plant with relatively low rates of photorespiration (McNaughton, 1966a, 1969; McNaughton and Fullem, 1969); limiting mechanisms are not known in this case.

Submersed macrophytes are exposed to lower concentrations of ambient oxygen, lower levels of light, and lower summer temperatures than are terrestrial and emergent aquatic plants. Furthermore, submersed and emergent aquatic plants are rarely ever subjected to the water stress to which terrestrial C_4 plants of arid environments have become adapted. The much slower diffusion of CO_2 in water than in air and the presence of massive internal gas lacunae can retard loss of CO_2 from submersed angiosperms and facilitate refixation of respired CO_2 (Carr, 1969; Hough and Wetzel, 1972; Hough, 1974; Søndergaard, 1979; Søndergaard and Wetzel, 1980; Wetzel *et al.*, 1984a). In a submersed species of *Scirpus*, for example, 30–40% of the carbon dioxide for photosynthesis was CO_2 released from photorespiration and refixed.

In softwater lakes, concentrations of total inorganic carbon in the water are very low; certain softwater submersed angiosperms and ferns (e.g., *Juncus bulbosus, Littorella, Lobelia, and Isoetes*) utilize CO_2 of the sediment–interstitial water to supplement CO_2 assimilated from the water (Wium-Andersen, 1971; Sondergaard and Sand-Jensen, 1979a; Wetzel *et al.*, 1984b; Boston *et al.*, 1989). Carbon entry per unit root surface area can be several times more rapid than entry per unit shoot surface area at equal external CO_2 concentrations (Boston *et al.*, 1987). Sediment-derived CO_2 can make up 65–95% of the total carbon assimilated in these plants from very soft waters. Uptake of CO_2 by root tissue for photosynthetic fixation by submersed plants of hardwater lakes, in which concentrations of inorganic carbon were high, could not be demonstrated (Beer and Wetzel, 1981; Loczy *et al.*, 1983; Steinberg and Melzer, 1983).

The photosynthetic CO_2 compensation points of submersed angiosperms are variable from moderate to high in comparisons to terrestrial plants (Brown *et al.*, 1974; Helder *et al.*, 1974; Van *et al.*, 1976; Lloyd *et al.*, 1977; Hough and Wetzel, 1978; Bowes *et al.*, 1979; Maberly, 1985; Owttrim and Colman, 1989). In addition, photorespiration is enhanced under conditions of increasing concentrations of dissolved oxygen and with increasing temperatures within the normal limnological range. Nearly all submersed angiosperms possess the C_3–Calvin-type of photosynthetic pathway (Hough and Wetzel, 1977; Winter, 1978; Browse *et al.*, 1979; Valanne *et al.*, 1982; critically discussed in Beer and Wetzel, 1980, 1982; Spencer and Bowes, 1990).

Reduced CO_2 compensation points can result from the presence of oxygen-insensitive phosphoenolpyruvate carboxylase (PEPcase), characteristic of C_4-type metabolism, in the cytoplasm of certain submersed plants. Oxygen-sensitive ribulosebisphosphate carboxylase-oxygenase (RuBPcase) is found in the chloroplasts (Bowes, 1987). The presence of appreciable amounts of PEPcase permits the fixation of CO_2 from both photorespiration and mitochondrial respiration, thereby reducing losses from the cells. The PEPcase activity increases among certain submersed plants with low CO_2 compensation points and results in elevated CO_2 concentrations around RuBPcase in the chloroplasts, which increases the CO_2/O_2 ratio and thus reduces photorespiration. The enzymes are separated intracellularly, even though submersed angiosperms do not exhibit anatomical differentiation,[2] as is common in C_4 plants (Hough and Wetzel, 1977).

In contrast, in the submersed macrophyte *Hydrilla*, a highly competitive, aggressive plant in the subtropics and tropics, no bundle sheath cells are evident. The PEPcase is located in the cytosol where bicarbonate is fixed, and four-carbon acids occur that are transported to the chloroplast in the same cell and decarboxylated to CO_2 and pyruvate (Bowes and Salvucci, 1984; Spencer *et al.*, 1996). This gas-exchange pattern, analogous to that of terrestrial C_4 plants, yields low rates of photorespiration, low CO_2 compensation points, and low O_2 sensitivity of photosynthesis. This plasticity of C_4-like photosynthetic metabolism in *Hydrilla* allows the plant to optimize carbon gain as the inorganic carbon availability in productive littoral areas changes rapidly over a daily period (Spencer *et al.*, 1994).

B. Water Movements, Boundary Layers, and Diffusion along Submersed Surfaces

When an aquatic plant attached to a substratum is exposed to the hydrodynamic forces of moving water flow, the net force exerted on the plant thallus is a vector resultant of two orthogonally opposed forces: (a) the lift force, which operates perpendicular to the direction of flow, and (b) two horizontal force components, the pressure force and the acceleration force (Nicklas, 1994). Most species of submersed plants cope with these forces by bending easily, thereby reducing drag. The flexure permits plants to absorb large amounts of energy without breaking. Submersed and emergent aquatic plants are both structurally flexible but employ different strategies. In submersed plants, structural simplicity and elasticity maximize flexibility, whereas in some emergent plants a twisted cantilevered beam structure allows severe bending without buckling (Rowlatt and Morshead, 1992; Biehle *et al.*, 1998).

Despite these adaptive mechanisms to the pressures of movement of a dense medium such as water against plant tissues, most aquatic plants cannot survive in habitats that are exposed to severe wind-generated waves. Most freely floating macrophytes are absent in such areas, and submersed species do not grow well, if at all, in shallow waters. Submersed plants are usually found at depths of greater than a meter or more, where the erosive actions on the sediments and stress on the plant morphology are ameliorated (cf. Chap. 7). Experimentally increased exposure to wave action commonly leads to reduced biomass production, tiller formation, and growth among even well-adapted emergent macrophytes (e.g., Coops *et al.*, 1994). Marked fluctuations in water level, as is common in reservoirs, can induce increases in wave disruption of the macrophytes directly as well as sediment accumulations and result in decreases in both production and biodiversity of littoral plants and associated biota (e.g., Memezes *et al.*, 1993).

Low-velocity currents impinging on leaf surfaces increase photosynthetic rates among submersed angiosperms by reducing the stagnant boundary layer and increasing diffusion rates (e.g., Barth, 1957; Westlake, 1967). Increasing flow rates to velocities above ca. 15 mm s^{-1}, however, can decrease photosynthetic rates precipitously (e.g., Madsen and Sondergaard, 1983). Although flows commonly affect river macrophytes, among most lake, pond, and many river habitats flows within submersed macrophyte beds are greatly attenuated within a few centimeters of the outer plant-bed boundaries (Losee and Wetzel, 1993; Sand-Jensen and Mebus, 1996; Stevens and Hurd, 1997). Laminar-flow boundary layer thickness at the surfaces of submersed leaves are very large (several mm to cm), often greater than macrophyte leaf thicknesses themselves. The presence of epiphytes and attached debris alters the flows and diffusion rates because of roughness and turbulence at the microlevel (cf. Riber and Wetzel, 1987;

[2] Kranz anatomy, where CO_2 is fixed in mesophyll cells with few or no chloroplasts with PEPcase and transferred in organic acids to chloroplasts located in bundle sheath cells surrounding vascular tissue.

Koch, 1994). The boundary layer is furthermore highly dynamic at times, such as when a passing wave temporarily increases water movement and decreases boundary layer thickness. Clearly, however, in lakes and many stream conditions, the acquisition of gases and dissolved substances from the water depends on diffusion, which is generally very slow.

C. Carbon Assimilation among Aquatic Macrophytes

Atmospheric carbon dioxide is clearly a significant source of inorganic carbon among emergent (Brix, 1990; Singer *et al.*, 1994) and floating macrophytes. Carbon dioxide in the internal spaces of the aerenchyma, however, is often highly concentrated despite convectional flowthrough mechanisms, discussed earlier, for the internal gases, particularly at night (e.g., Constable *et al.*, 1992; Li and Jones, 1995). For example, in the emergent papyrus wetland species with C_4 photosynthesis, it was estimated that about half of the CO_2 of the intercellular spaces emanates from respiration and is rapidly recycled by photosynthesis.

A few species of freely floating angiosperms, such as some duckweeds (*Lemna*), utilize both atmospheric and aqueous carbon sources (Wohler, 1966; Wetzel and Manny, 1972; Ultsch and Anthony, 1973; Filbin and Hough, 1985). Some data on sources of CO_2 fixation by floating-leaved but rooted macrophytes, such as the water lilies, indicate that atmospheric CO_2 comprised only about half of the total carbon fixed (Hough and Filbin, 1990). The remainder was obtained from the water. Heterophyllous aquatic macrophytes are similarly adapted to use both atmospheric CO_2 by the aerial foliage and aqueous CO_2–HCO_3 by the submersed foliage. Clearly, however, the utilization of atmospheric CO_2 is the predominant *external* source of inorganic carbon uptake under most circumstances among amphibious and emergent species (cf. discussion of Maberly and Spence, 1989; Sand-Jensen *et al.*, 1992; Constable and Longstreth, 1994; Frost-Christensen and Sand-Jensen, 1995).

The availability of free carbon dioxide is always low in water because rates of diffusion in water are 4 orders of magnitude slower than in air (cf. Chap. 11) and CO_2 is not very soluble in water. As a result, CO_2 from aqueous solution can be readily reduced or depleted in microzones near leaves or cells during intensive photosynthetic assimilation. Morphological adaptations that increase absorption among submersed leaves, stems, and some petioles include (a) thin tissues (1–3 cell layers), (b) reduced cuticle development, (c) extreme reduction or elimination of mesophyll, and (d) dense concentration of chloroplasts in epidermal cells. These adaptations all increase utilization and exchange of gases dissolved in water. Massive intercellular gas spaces in leaves, stems, and petioles facilitate rapid internal diffusion among plant parts, including the rooting tissues.

In contrast, concentrations of bicarbonate of natural hard waters can be several orders of magnitude greater than those of CO_2. Because of the slow rates of diffusion and relatively long diffusion pathways from the medium through the unstirred layer to sites of biochemical assimilation, the ability to assimilate bicarbonate ions can be regarded as an adaptive advantage to those plants that possess this capability. The transport of bicarbonate and the removal of the CO_2 from the HCO_3^- within cells, however, has an energetic cost (Raven and Lucas, 1985; Eighmy *et al.*, 1991).

The question of limitation of photosynthesis by the low solubility and slow diffusion of CO_2 in water has received much attention, especially in relation to the alternative utilization of the relatively abundant bicarbonate ions. In large submersed angiosperms and algae, the diffusion path through the cells of the thallus or leaf is long (up to 50 μm). In addition, an effective boundary layer gradient zone of at least 10 μm to several millimeters of unstirred water occurs at the cellular surfaces (Losee and Wetzel, 1993). Moreover, diffusion rates through the larger cells are much slower than among microalgae (Steemann Nielsen, 1947; Raven, 1970; Smith and Walker, 1980). Ions and gas molecules move only by diffusion through the boundary layer. Water movements over submersed surfaces reduce the boundary layer thickness but do not eliminate it, regardless of current velocities.

In algae, especially larger macroalgae, the ability to utilize bicarbonate ions in photosynthesis is widespread but not universal. The extensive work on this subject among mosses and submersed angiosperms by Steemann Nielsen (1944, 1947), Ruttner (1947, 1948, 1960), and others (e.g., Bain and Proctor, 1980) has shown that the freshwater red alga *Batrachospermum* and all assayed genera of freshwater submersed mosses utilize only free CO_2 and cannot assimilate bicarbonate. These latter plants are restricted almost universally to soft waters of relatively low pH or to streams in which CO_2 concentrations are relatively high.

Different strategies have evolved in submersed macrophytes that can be viewed as both physiological and exploitative mechanisms to ameliorate the constraints of limited carbon availability (cf. reviews of Maberly and Spence, 1983; Madsen and Sand-Jensen, 1991):

a. Many species utilize primarily or only CO_2 and necessarily are restricted to habitats with multiple sources of CO_2 (e.g., Wetzel, 1969; Brown *et al.*, 1974; Van *et al.*, 1976; Moeller, 1978; Winter, 1978; Kadono, 1980).

b. Utilization of HCO_3^- under natural conditions has been reported for a number of freshwater angiosperms (e.g., Raven, 1970; Helder and Zanstra, 1977; Lucas *et al.*, 1978; Browse *et al.*, 1979; Prins, *et al.*, 1980; Beer and Wetzel, 1981; Lucas, 1983).

No other area of aquatic plant physiology has received greater study than has the assimilation of inorganic carbon as CO_2 and bicarbonate ions by submersed plants. The ability of submersed macrophytes and algae to assimilate bicarbonate ions as a carbon source supplementary to carbon dioxide has been demonstrated in a number of investigations. Space does not permit detailed discussion of the long history of remarkable experimental studies as well as conflicting results of some of these studies (cf. Raven, 1970, 1984; Helder and Zanstra, 1977; Lucas *et al.*, 1978; Browse *et al.*, 1979; Helder *et al.*, 1980; Prins *et al.*, 1980; Lucas, 1983; Wetzel and Grace, 1983; Lucas and Berry, 1985; Bowes, 1987; Prins and Elzenga, 1989; Madsen and Sand-Jensen, 1991). Variations in physiological mechanisms of bicarbonate uptake are largely though not completely understood. Three general mechanisms exist:

1. A polar differential, established by spatial separation of HCO_3^- influx from resulting OH^- efflux, common in submersed angiosperms and Characeae that assimilate bicarbonate.
   a. Carbon dioxide assimilation occurs in all cell layers.
   b. Bicarbonate ions are utilized only by cells of the lower epidermis.
   c. Carbon dioxide (both as free CO_2 and as CO_2 derived from the dehydration of bicarbonate) is fixed, and hydroxyl ions equivalent to the amount of carbon dioxide derived from bicarbonate pass out through the adaxial (upper) leaf surface.
   d. Cations, primarily calcium, equivalent to the quantity of bicarbonate taken in through the abaxial (lower) leaf surface, are transported from the abaxial to the adaxial leaf surface, thereby achieving stoichiometry and charge balance in all compartments.
   e. The passage of calcium and hydroxyl ions to the adaxial leaf surface results in a high pH on this surface and usually results in precipitation of $CaCO_3$ on that surface. The carbonate deposits encrusting the upper surfaces of the submersed leaves of macrophytes in calcareous hard waters often exceed the weight of the plant material (Wetzel, 1960). Among emergent and floating macrophytes that utilize atmospheric carbon dioxide, encrustation on submersed plant portions was found to be highly variable and proportional to the extent of development of epiphytic algae. Deposits on submersed species were much heavier and were correlated directly with morphological variations in surface area and the ability to utilize bicarbonate ions.
2. HCO_3^-–H^+ cotransport with subsequent removal of CO_2, particularly among the Characeae macroalgae.
3. Extracellular conversion of HCO_3^- to CO_2 by the release of carbonic anhydrase.

Some plants, such as certain species of *Ranunculus*, combine conversion of bicarbonate to CO_2 with carbonic anhydrase as well as with a photodependent increase in plasma membrane H^+ release (Rascio *et al.*, 1999).

D. CAM Metabolism

Crassulacean acid metabolism (CAM) is a CO_2-concentrating mechanism in which carbon is incorporated into malate (a characteristic of C_4 plants) and stored temporarily as a stable product rather than as an intermediary carbon source. Because of the restricted availability of inorganic carbon in softwater environments, some submersed angiosperms accumulate malate in the light, which may function as an ion to balance excess cation uptake (Browse *et al.*, 1980). However, malate may also accumulate in the dark for subsequent decarboxylation and refixation of CO_2 in the light, complementary to fixation of exogenous carbon (Beer and Wetzel, 1981; Keeley, 1981, 1982, 1998; Madsen, 1987).

Aquatic CAM plants inhabit sites where photosynthesis is potentially limited by carbon. CAM submersed plants have been found in a number of common genera: *Isoëtes* (Lycophyta), *Sagittaria, Vallisneria, Crassula*, and *Littorella*. Some of these plants occupy fertile shallow temporary ponds that experience extreme diurnal fluctuations in carbon availability. Elevated nighttime CO_2 concentrations are assimilated and fixed into malate for use in the subsequent day period. Some of the CAM submersed plants live in oligotrophic softwater lakes were carbon limitation prevails. Where CAM is a dominant mechanism for acquisition of CO_2, as in some species of *Isoëtes* and *Littorella*, CAM can contribute to over half of the annual gain for plant photosynthesis (e.g., Boston and Adams, 1986).

E. Daily Rates of Photosynthesis among Submersed Macrophytes

Rates of photosynthetic carbon fixation in submersed macrophytes are generally correlated with the intensity of solar radiation on a daily basis. However, a midday and afternoon depression of carbon fixation is often observed, so those highest rates of photosynthesis are skewed toward morning (Wetzel, 1965b; Sculthorpe, 1967; Goulder, 1970; Hough, 1974, 1979). *In situ* experiments with the submergent *Najas* showed that the afternoon decrease in net photosynthesis was correlated with a proportional increase in photorespiration. The results of numerous studies suggest that photorespiration may increase through the day with increasing light intensity, with increasing oxygen tension of photosynthetic origin, increasing temperature, and possibly decreasing CO_2 availability (Fig. 18-3). Conditions conducive to accelerated photorespiration can develop in the littoral zone, particularly among dense macrophytic populations during calm weather with little turbulence. An increase in photorespiration would reduce net photosynthesis. Simultaneously, however, the concentrations of available inorganic carbon decrease, and certainly this reduced availability could contribute equally to the observed afternoon reduction in net photosynthesis (e.g., Jones *et al.*, 1996).

With cessation of photosynthesis during darkness, internal oxygen content decreases rapidly and carbon dioxide increases; then respiration is limited by the rate at which oxygen diffuses from the water to the cells. Dark respiration rates increase with rising temperatures and are generally, but not in all cases, enhanced by high concentrations of oxygen in the surrounding water, especially if the water is agitated slightly by low-velocity currents (Pannier, 1957, 1958; Owens and Maris, 1964; McIntire, 1966; Westlake, 1967; McDonnell, 1971; Hough and Wetzel, 1972; Stanley, 1972; Prins and Wolff, 1974). During the daytime, the effect of oxygen on photorespiration and dark respiration would further increase the release of CO_2.

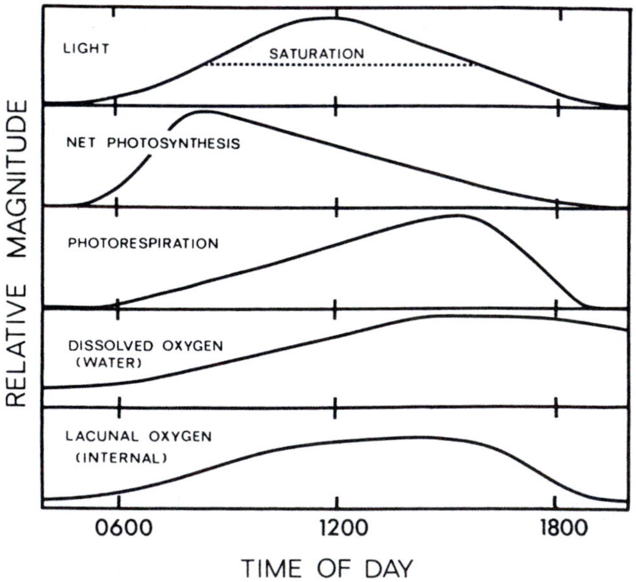

FIGURE 18-3 Commonly observed relative diurnal variations in light intensity, dissolved oxygen concentrations of the water and the internal lacunal spaces of submersed aquatic plants, net photosynthesis of submersed macrophyte (and some phytoplankton) populations, and proposed concurrent variations in photorespiration. (Modified from Hough, 1974.)

Physiological evidence indicates that most freshwater submersed angiosperms possess C_3–Calvin-type photosynthetic metabolism and high rates of photorespiration, which collectively reduce rates of net photosynthesis. Efficient internal recycling of respired CO_2, utilization of bicarbonate as well as dissolved CO_2, and augmentation of exogenous inorganic CO_2 by decarboxylation of malate formed during darkness are mechanisms by which certain submersed species adapt to the generally restricted availability of CO_2. There is no question that the evolution of an extensive internal lacunal gas system in submersed angiosperms constitutes an array of interrelated morphological adaptations related to (a) enhanced efficiency of carbon fixation, (b) flexibility to withstand water movements, and (c) positive buoyancy and positioning of leaves toward greater available light.

Little is known of the seasonal variations in respiration and photorespiration among submersed macrophytes. Photorespiration and dark respiration were found to be 10-fold greater during the fall than during summer as the annual submergent *Najas* was entering senescence following seed set in late summer (Hough, 1974). Photorespiration in the submersed perennial *Scirpus subterminalis* was relatively low during summer and increased under the ice in late winter.

Many submersed aquatic angiosperms of the temperate zone perennate beneath ice in lakes and ponds. Although some species lose their leaves and are dormant throughout the winter, many species retain leaf tissue and photosynthetic capacity beneath ice (Boylen and Sheldon, 1976; Best, 1986; Best and Visser, 1987). Submersed species that remain physiologically active beneath ice experience attenuated irradiance and cold nonfreezing temperatures. Analyses of submersed *Ceratophyllum demersum* found optimum temperatures for net photosynthesis of winter and summer plants were 5 and 30°C, respectively (Spencer and Wetzel, 1993). Dark respiration of winter plants was over 300% greater than that of summer plants. Reduced Rubisco carbon-fixing enzyme activity and increased dark respiration interacted to reduce net photosynthesis and increase CO_2 and light compensation points of winter plants.

F. Secretion of Dissolved Organic Matter

The loss of recently synthesized organic carbon and nitrogen as secreted dissolved organic compounds has been demonstrated for a few submersed and freely floating angiosperms (Wetzel, 1969; Hough and Wetzel, 1972, 1975; Wetzel and Manny, 1972; Hough, 1979; Søndergaard, 1981, 1990; Nalewajko and Godmaire, 1993). The organic compounds released are largely of low molecular weight (ca. 200 D). Well known among the planktonic algae (see Chap. 15), this loss of organic carbon during active photosynthesis in submersed macrophytes can represent a significant reduction in photosynthetic efficiency. Secretion rates of dissolved organic carbon vary greatly (0.05 to over 100% of carbon photosynthetically fixed) under an array of experimental conditions of varying light and ionic composition of the medium. Most values, however, including those from *in situ* analyses, were between 1 to 10% of photosynthetically fixed carbon, increasing somewhat during the course of the daylight period, but the maximum secretion percentage of total carbon fixed occurred at the lowest rates of photosynthesis.

The secretion discussed here refers to that released extracellularly during active photosynthesis, rather than soluble organic compounds released during senescence and early stages of decomposition. The secretion may represent an incomplete adaptation of submersed macrophytes to a totally aqueous medium. Functionally, however, the secretion of labile organic compounds by submersed macrophytes enhances the development of a highly productive epiphytic community of microflora (Wetzel and Allen, 1970; Wetzel, 1993a). Nutrient interactions, both inorganic and organic, take place between the submersed macrophyte and epiphytic algal and bacterial populations (cf. Allen, 1971; Allanson, 1973; Wetzel, 1996; Chap. 19). These interactions are lacking or greatly reduced in the diluted planktonic regime. This apparent small decrease in photosynthetic efficiency by the loss of dissolved organic carbon could be viewed as a symbiotic interaction between the macrophyte and its epiphytic bacteria and algae.

The release of organic acids by roots for the solubilization and mobilization by complexation of nutrients from soils is widely known among terrestrial plants (e.g., Uren and Reisenauer, 1988). A similar mechanism has been demonstrated in rice (Saleque and Kirk, 1995) and is likely among many aquatic plants under natural conditions. Interstitial dissolved organic carbon was much higher in sediments with growing emergent macrophytes than in unvegetated sediments, and bacterial productivity was shown to be appreciably greater utilizing dissolved organic matter from the vegetated sediments (Mann and Wetzel, 1999b). Analogous enhanced bacterial communities have been found in the rhizosphere of marine angiosperms (Donnelly and Herbert, 1999).

G. Absorption of Ions and Nutrition in Aquatic Plants

The functional significance of the rooting systems in aquatic macrophytes, primarily angiosperms, has been a long-standing question in limnology (reviewed in Gessner, 1959; Wetzel, 1964; Sculthorpe, 1967; Bristow, 1975; Hutchinson, 1975; Agami and Waisel, 1986). Two views emerge, each with supporting evidence: (a) that the rooting systems function in the absorption of nutrients from the substratum, and (b) that the roots function merely as organs of attachment, particularly among submersed plants. An overwhelming accumulation of evidence now demonstrates that rooting tissues are essential for primary acquisition of nutrients for aquatic plants, including those that are submersed.

The roots of most aquatic plants, including certain submersed plants that have reduced rooting tissues, possess root hairs (Shannon, 1953) or lateral root projections (Cumbus and Robinson, 1977) that increase the absorptive capacities of nutrients from the interstitial water of sediments. In addition, a large number of emergent, floating-leaved and submersed aquatic macrophytes possess vesicular–arbuscular mycorrhizae (e.g., Søndergaard and Laegaard, 1977; Tanner and Clayton, 1985; Nakatsubo *et al.*, 1994; Stenlund and Charvat, 1994; Wetzel and van der Valk, 1996). Despite the anoxia of sediments, the release of oxygen from the intercellular spaces of the plant roots into the immediate microrhizosphere is presumably sufficient to allow reasonable infection and support of the mycorrhizae. Increases in symbiotic mycorrhizal infections are common under conditions of reduced nutrient availability or other stress factors (increased salinity and reduced pH). These root extensions likely also function in the enhanced release of extracellular enzymes to increase the availability of growth factors.

1. Rooted Emergent and Floating Aquatic Plants

Among emergent and floating-leaved angiosperms with active transpiration-mediated root-pressure systems, nutrient absorption and translocation from the roots to the foliage are clearly operational. Most, but not all, emergent plants are rooted in the sediments. Where emergent plants occur in large floating mats, nutrients are essentially all absorbed from the water through the rooting tissues.

Among emergent aquatic macrophytes, phosphorus requirements are met by uptake of inorganic and organic phosphorus compounds from the sediments (e.g., Islam *et al.*, 1979; Atwell *et al.*, 1980). Detailed studies of the growth and reproduction of the cattail (*Typha latifolia*) in marshes of differing successional maturity demonstrated that phosphorus limitations affect not only overall growth but also the allocation of photosynthate to sexual and vegetative reproduction, early spring mortality, and competitive success (Grace and Wetzel, 1981a). Elevated phosphorus concentrations of the hydrosoils is a primary factor influencing rapid cattail (*Typha* spp.) growth and invasive competition with native emergent macrophyte species in the Everglades of Florida (Newman *et al.*, 1998). In another example, about half of the ammonium assimilated by the roots of rice is rapidly translocated to the shoots (Wang *et al.*, 1993). Fertilization of sediments with nitrogen and phosphorus can result in major increases in the aboveground production (e.g., Granéli, 1985; Bornkamm and Raghi-Atri, 1986). For example, enhanced nitrogen as ammonia and phosphate of sediment interstitial water resulted in marked increases in net annual aboveground production of *Sparganium* (+57%) and *Typha* (+19%) (Neely and Davis, 1985).

Although aquatic plants must possess a well-developed aeration system of rooting tissues that are confluent with aerial leaves, anaerobic sediments of nearly all aquatic ecosystems are highly reducing and consequently contain relatively large quantities of nutrients in available form. Concentrations in sediments are usually several orders of magnitude greater than those in the overlying water. Clearly, the sediment and associated organic detritus (e.g., Morris and Lajtha, 1986) are the primary sources of nutrients for aquatic plants.

As occurs among terrestrial plants, much (often >90%) of the nutrients used, recycled during growth, and stored in aboveground tissues of aquatic plants is translocated back to and stored within rooting tissues below ground (e.g., Granéli *et al.*, 1992; DeLucia and Schlesinger, 1995). The release of ions during senescence and decay may be into the water or to the sediment (cf. Denny, 1987), but in both cases the vigorously growing attached microbial community sequesters most of the nutrients being released (Godshalk and Wetzel, 1978; cf. Chap. 19). Most of the aquatic plant detritus accumulates on the sediments and becomes incorporated into the sediments. As a consequence, the sediments effectively become a nutrient sink.

2. Floating Aquatic Plants

Floating vegetation occurs in many forms. Floating swamps of tropical regions and large river deltas, such as the delta of the Danube, can be very large mats of floating root and rhizome mats often many hectares in size. The mats or rafts of emergent vegetation may be attached to, and grow out from, beds of rooted vegetation along the shores of lakes and rivers or be freely floating (Denny, 1985, 1987). Emergent plants, such as *Cyperus papyrus, Phragmites,* and *Typha*, often occur with floating plants, such as the water hyacinth *Eichhornia.* The floating mats of dense aerial biomass overlie a lower heterotrophic zone of rhizomes and partially decomposed organic matter from the prolific photosynthetic productivity of the aquatic plants. The water under these mats is commonly shallow and often anoxic extending down to the sediments. Large amounts of nutrients are stored in the plant tissues and detrital organic matter, and a tight internal cycling occurs from the living foliage and the supporting organic matter and water beneath it (e.g., Gaudet, 1982).

Most of the nutrients in a bed of nonrooted, freely floating higher aquatic plants are acquired by roots from the water below and around the bed. Pendulant roots assimilate ions directly from the water. Translocation is mediated by active root-pressure kinetics via intense evapotranspiration. Macroalgae, liverworts, mosses, ferns, and other rootless macrophytes that float on the surface or in the water or are variously loosely attached to the substratum presumably obtain nutrients by foliar absorption rather than by rhizoidal structures. An exception is the rhizoid-bearing macroalga *Chara*, which absorbs phosphorus equally well in all parts; phosphorus absorbed by the rhizoids is translocated to other parts of the plant (Littlefield and Forsberg, 1965). In a floating-leaved water lily (*Nuphar luteum*), phosphorus absorption rates differed with the absorbing tissue (roots > submersed leaves > floating leaves) (Twilley *et al.*, 1977). About four times more phosphorus moved from the roots to leaves than from the leaves to the roots during summer.

Special mention should be made of the heterosporous aquatic fern *Azolla*, whose sporophytes float freely on the water surface in shallow quiet waters, wetlands, and particularly in rice fields. This small (1–3 cm in diameter), prolific fern contains throughout its life cycle a symbiont, a heterocyst-forming, nitrogen-fixing cyanobacterium *Anabaena azollae* (Lumpkin and Plucknett, 1980; Ito and Watanabe, 1983; Peters and Meeks, 1989; Wagner, 1997). The nitrogen-fixing symbiont can provide the total nitrogen requirements for the fern and continues to fix nitrogen regardless of assimilation of ammonium or nitrate by the fern from the water. The *Azolla–Anabaena* symbioses are used as a major nitrogen fertilizer in lowland rice cultivation in Asia and other regions. It has been estimated that as much as a quarter of the total human nitrogen consumption is obtained from the *Azolla–Anabaena* source assimilated by rice.

3. Submersed Macrophytes

In rooted submersed plants, a positive absorptive capacity is present and guttation always occurs in small quantities, increasing from root tip to the basal portions of the plant and increasing from the apical parts to the base (reviewed by Stocking, 1956). Water exits via hydathodes on the upper portions of many submergents, and nutrients can enter apically through the numerous porous hydropoten. The extensive development of roots and root hairs among aquatic plants (>95% of 200 species of 105 genera and 54 families) emphasizes the dependency of these plants on roots for absorption of solutes as well as for anchorage (Shannon, 1953).

Submersed angiosperms have a reduced but well-structured vascular tissue containing xylem, phloem, and an endodermis with casparian strips important in the regulated flow of ions and water (Pedersen, 1993; Pedersen and Sand-Jensen, 1993). Acropetal water transport occurs from roots to leaves at low rates, but at rates markedly faster than is possible by passive diffusion. Transport depends upon light and presumably on photosynthetic products and energy. This transport clearly functions as a translocation system for nutrients taken up from the nutrient-rich sediments.

The absorption of phosphorus by roots of submersed plants both from the water and from the sediments is well known (cf. Chap. 13, and Gessner, 1959; Schwoerbel and Tillmanns, 1964a; Bristow, 1975; Denny, 1980; Vermaak *et al.*, 1982; Chambers *et al.*, 1989; Gunnison and Barko, 1989). Among most submersed macrophytes, phosphate absorption rates by foliage are proportional to and dependent upon the concentrations in the water. Very large quantities of several mg liter^{-1} are rapidly assimilated in excess of requirements until concentrations in the water are reduced to about 10 μg liter^{-1}. In most of these studies, however, discrimination between uptake and retention of phosphate by the attached periphyton is not differentiated from that taken into the macrophyte tissue itself. Where experiments are designed to specifically examine these pathways (e.g., Moeller *et al.*, 1988), clearly most of the phosphorus acquisition by the macrophyte was from roots, and uptake from the water was largely into epiphytic microorganisms and predominantly independently cycled in the epiphytic periphyton (cf. Chap. 19). Therefore, it is not clear in many of the older studies whether the nutrients from the water are being incorporated by the submersed plants or largely by the attached periphyton. The attached microorganisms are likely most important, particularly under conditions of low nutrient concentrations in the water.

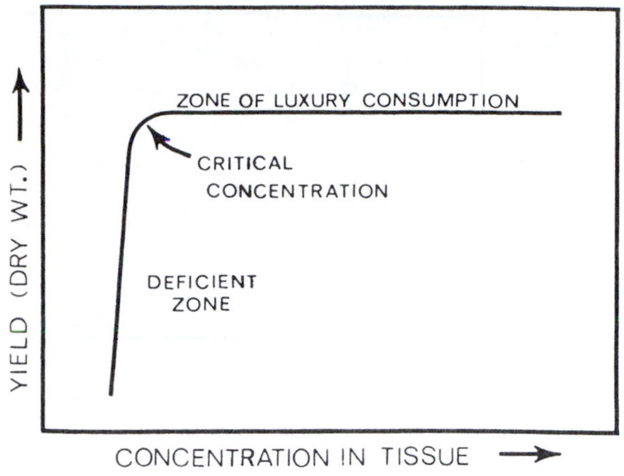

FIGURE 18-4 General relationship between aquatic plant biomass and the concentration of an essential element in plant tissue. (Modified from Gerloff, 1969.)

Luxury consumption of nutrients under high concentrations in the water, known especially for phosphorus but also for nitrogen, have led to the application of elemental analysis of tissues for evaluating nutrient supplies for algae and aquatic angiosperms, a technique long used in agriculture (Gerloff, 1969; Fitzgerald, 1972; Schmitt and Adams, 1981). Tissue analysis is based on the assumption that the concentration of an element in an organism varies over a wide range in response to the concentration in the environment and that over part of this range the yield (biomass) of the organisms is related to the tissue content of the element. Tissue analysis depends on establishing, for each species, the critical concentration for each element of interest, that concentration which is just adequate for maximum growth (Fig. 18-4). Below this point in the "deficient zone," plant yield is dependent on the supply of the element, but the elemental concentration in the organism changes little. Above this point, in the "zone of luxury consumption," the content of the element of the organism increases but plant growth does not. The method assumes that below a critical concentration, plant growth is being limited by the supply of that element. Such an assumption is true in a general way and is readily demonstrable in crop plants; some data additionally suggest a similar relationship among algae and aquatic plants grown hydroponically and axenically in completely aqueous media (e.g., Colman *et al.*, 1987). The correlation between nutrient concentration of the water and tissue content among submersed plants rooted in sediments, however, is less clear. When a macronutrient is distinctly limiting to growth of submersed macrophytes, however, the differences are conspicuously evident in tissue concentrations. For example, growth and biomass of submersed macrophytes

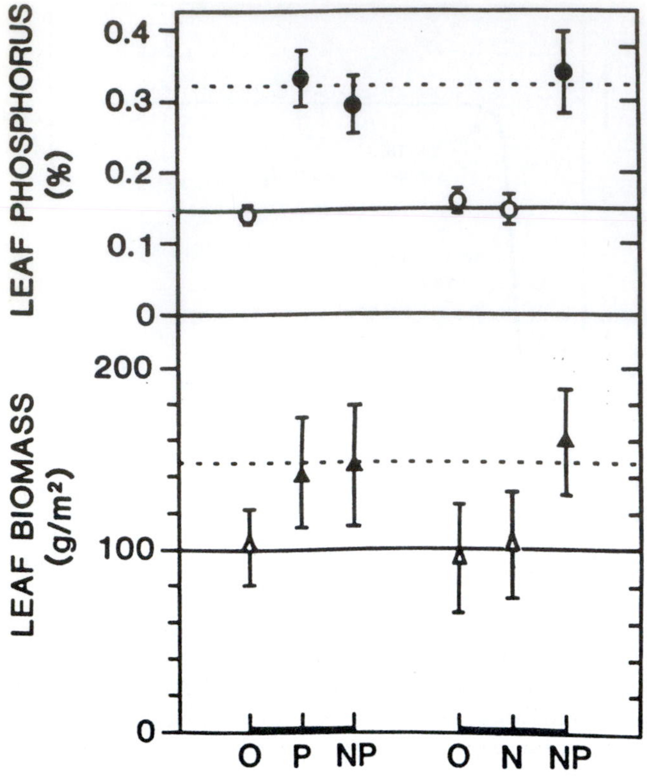

FIGURE 18-5 Response of the dominant submersed macrophyte *Scirpus subterminalis* to 1 year of inorganic phosphorus and nitrogen enrichments, separately and in combination, to sediments of hardwater Lawrence Lake, Michigan. *Upper*: P concentration in leaf tissue (% carbonate- and ash-free dry wt). *Lower*: Leaf biomass (g ash-free dry wt m^{-2}). Means with 95% confidence intervals are compared for three sites with a parallel P-alone enrichment and for two sites with N-alone enrichment (0 = control, NP = mixed fertilizer, N = +nitrogen, P = +phosphorus). (From data of Moeller *et al.*, 1998.)

increased significantly with enrichments of phosphate released experimentally in littoral sediments, but not with inorganic nitrogen, of an oligotrophic hardwater lake of southwestern Michigan (Fig. 18-5). The nutrient supply to the tissues, however, was from the sediment interstitial waters and not the overlying water. Root growth is directly enhanced in submersed *Myriophyllum* when nutrient concentrations of nitrogen and phosphorus in the water surrounding the foliage is reduced (Mantai and Newton, 1982).

Growth of the prolific nonrooted submersed angiosperm *Ceratophyllum demersum* is often limited by the ambient phosphorus concentration of lake water during summer. Winter-collected plants under the ice had 54 and 35% more phosphorus and nitrogen, respectively, than summer-collected plants (Spencer and Wetzel, 1993). Physiologically active perennation beneath ice enabled *C. demersum* to accumulate phosphorus during the winter when it was most abundant. Partial uncoupling of phosphorus acquisition from utilization may reduce phosphorus limitation upon growth during the summer when phosphorus concentration of the water is seasonally the lowest.

Experimental analyses have demonstrated that most rooted submersed angiosperms obtain most of their phosphorus from the interstitial water of the sediments (Bristow and Whitcombe, 1971; Schults and Malueg, 1971; DeMarte and Hartman, 1974; Bole and Allan, 1978; Welsh and Denny, 1979; Barko and Smart, 1980; Carignan and Kalff, 1980; Gabrielson *et al.*, 1984; Moeller *et al.*, 1988; Chambers *et al.*, 1989). Although some evidence is contradictory (e.g., Seadler and Alldridge, 1977; Swanepoel and Vermaak, 1977), most studies show an active absorption by the roots and translocation to the leaves. When concentrations of phosphate and sulfate of the water are low, absorption of these nutrients by the leaves of *Elodea* is limited by the rate of diffusion through the concentration gradient of the microlayer surrounding the leaves (Jeschke and Simonis, 1965). Absorption of phosphate is limited by the process of active uptake at intermediate exogenous concentrations and by the rate of diffusive influx into the leaves at high external concentrations (Grunsfelder and Simonis, 1973; Grunsfelder, 1974). Much of the phosphorus uptake from the water by the macrophyte–epiphyte community is utilized and recycled by the epiphytic algae (e.g., Howard-Williams, 1981; Pelton *et al.*, 1998). Detailed studies of the transport of phosphorus through the epiphytic algae and bacteria (Moeller *et al.*, 1988) have demonstrated that most of the phosphorus is actively recycled within the periphyton and that most of the phosphorus utilized by the submersed macrophytes is absorbed by the roots and translocated to the foliage. These studies indicate that many earlier studies that attributed high phosphorus uptake by the macrophytes from the water are in part simply a reflection of the inability to separate the uptake by epiphytic microbiota from the true assimilation from the water by the submersed angiosperm. High phosphorus concentrations in water frequently encourage the prolific development of epiphytic algae, to the detriment of submersed macrophytes by light attenuation (see Chap. 19).

Rates of nitrate assimilation by the foliage of several submersed macrophytes are considerably less than are rates of ammonia assimilation, especially at high pH values (Schwoerbel and Tillmanns, 1964b,c, 1977; Toetz, 1973b, 1974; Cole and Toetz, 1975; Nichols and Keeney, 1976; Holst and Yopp, 1979; Best, 1980; Madsen and Baattrup-Pedersen, 1995). High nitrogen concentrations are required by many submersed plants to maintain efficient carbon assimilation and carbon-concentrating mechanisms. Ammonium ions are readily absorbed by roots and translocated to apical tissues.

Furthermore, nitrogen from N_2-fixing bacteria of the sediments and plant rhizosphere can serve as a major nitrogen source for aquatic plants (e.g., Bristow, 1974a; Tjepkema and Evans, 1976; Purchase, 1977; Kana and Tjepkema, 1978; Ogan, 1979; Watanabe *et al.*, 1979; Blotnick *et al.*, 1980). Methanotrophic bacteria of anoxic sediments oxidize ammonium ions in the rhizosphere of aquatic plants and the nitrate is utilized by the macrophytes (Bodelier and Frenzel, 1999).

Among shallow lakes, phosphorus tends to be in abundance and is largely actively recycled to the sediments from aquatic plants and attached microflora. Under conditions of high production of organic matter and abundant anaerobic sediments and detritus, however, denitrification can result in major losses of nitrogen and contribute to nitrogen limitation (e.g., Lijklema, 1994). Such low levels of nitrogen can be correlated with lower productivity of submersed and emergent aquatic plants and with greater biodiversity than occurs at higher concentrations of nutrients (e.g., Gough and Grace, 1998; Moss, 1998).

Considerable information exists on the inorganic chemical composition of aquatic plants (cf. review of Hutchinson, 1975; Cowgill, 1989). There are large variations in chemical composition, even among closely related species and from site to site in the same species. Metals and alkalis are absorbed from the water by leaves and the interstitial water of sediment by roots (e.g., DeMarte and Hartman, 1974; Marquenie-van der Werff and Ernst, 1979), and many elements are concentrated in the macrophytes in great excess of metabolic requirements. As a result, submersed and emergent plants are often suggested to be scavengers of contaminants in surface waters (e.g., Wolverton and McDonald, 1978, 1979)

In summary, ion absorption in submersed macrophytes occurs both from the water by foliage and from the sediments by root and rhizoid systems. Under most circumstances, even in nutrient-rich waters, roots are the dominant sites of nutrient uptake and assimilation for aquatic plants. Translocation occurs in both directions, with the dominant acropetal translocation from the roots to the leaves. Because roots serve as a primary site of nutrient absorption from the sediments, the value of tissue analyses of element concentrations in aquatic macrophytes is suspect as an index of the fertility of the lake or river water. Nutrient content of the water can be quite unrelated to plant growth of those species having ready access to the abundant nutrient supply in the sediments. The method is, however, applicable to analyses of nutrient availability from the sediments or rootless submersed macrophytes. Also, as was indicated earlier (Chap. 13), the rooted plants can function as a "pump" of nutrients from the sediments; some of those nutrients can then be lost to the water during both active growth and decomposition. Most of the nutrients released from senescing tissues, however, is sequestered by attached microbial communities, intensively retained, and recycled (cf. Chap. 19).

IV. RATES OF PHOTOSYNTHESIS AND DEPTH DISTRIBUTION OF MACROPHYTES

A. Light and Temperature

Emergent macrophytes are among the most productive natural plant communities in the biosphere. The high rates of emergent macrophyte production have been attributed, among many factors, to a constant water supply and high nutrient concentrations of sediment interstitial waters. The leaf and growth morphology and minimal development of canopy structure of emergent macrophytes minimize shading and maximizes inception of light. In general, emergent and floating-leaved macrophytes exhibit sun plant characteristics and are capable of utilizing full sun with minimal photoinhibition at high irradiance levels.

Although macrophyte productivity varies with seasonal changes in temperature and irradiance within natural environments, the efficiency of photosynthetic carbon fixation and respiration/photorespiration vary markedly with light and temperature. For example, the overall photosynthetic efficiency of an emergent rush, *Juncus*, determined over a wide range of *in situ* irradiance and temperature levels, was 1.43 ± 0.03% (± SE, $n = 3974$) (Mann and Wetzel, 1999). Average photosynthetic efficiency was highest (4.5%) at the lowest irradiance and decreased with increasing irradiance levels (Fig. 18-6). Light saturation occurred at relatively low light levels (ca. 750 μmol quanta m^{-2} s^{-1}) and efficiency of light utilization decreased to about 0.5% at full sunlight without any apparent photoinhibition. The optimal temperature range for maximum average net photosynthesis was between 35 to 40°C for this species, but the photosynthetic efficiency decreased markedly at the higher temperatures (Fig. 18-6) as rates of respiration and photorespiration increased. Similar temperature optima have been found with other emergent and freely floating plant species (e.g., Wedge and Burris, 1982; Sale and Orr, 1987). Distinct photoinhibition by normal midsummer insolation has been found, however, among littoral bryophytes, such as *Sphagnum* mosses (Murray *et al.*, 1993). Transpirational cooling of the leaves and vertical orientation can reduce the temperature of the leaves several degrees below ambient temperature and improve efficiencies of photosynthesis (e.g., Pearcy *et al.*, 1974).

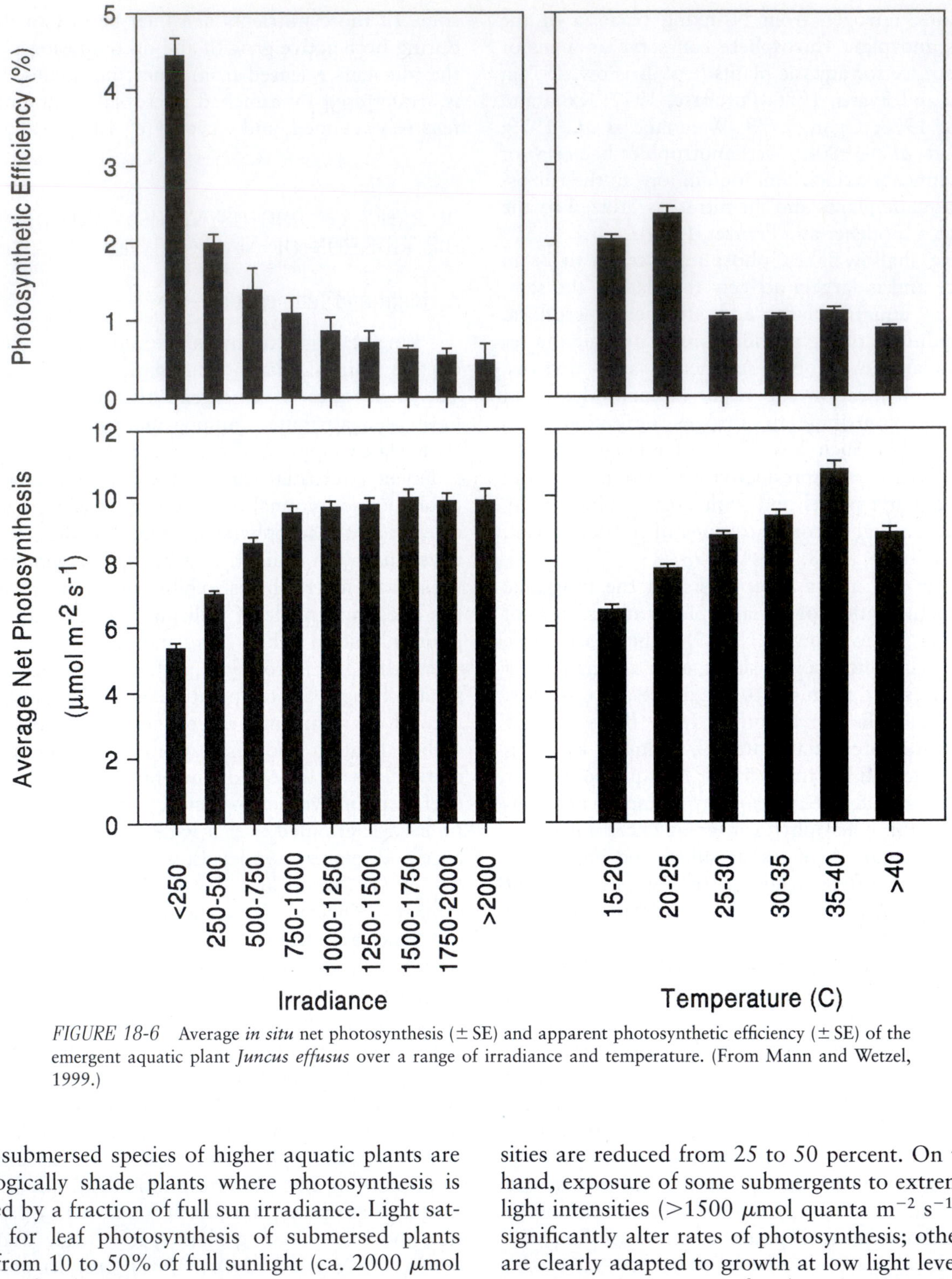

FIGURE 18-6 Average *in situ* net photosynthesis (± SE) and apparent photosynthetic efficiency (± SE) of the emergent aquatic plant *Juncus effusus* over a range of irradiance and temperature. (From Mann and Wetzel, 1999.)

All submersed species of higher aquatic plants are physiologically shade plants where photosynthesis is saturated by a fraction of full sun irradiance. Light saturation for leaf photosynthesis of submersed plants ranges from 10 to 50% of full sunlight (ca. 2000 μmol quanta m^{-2} s^{-1}) (Spencer and Bowes, 1990). The inhibitory effects of high light intensities in the surface waters, commonly observed among planktonic algae, are not generally observed among submersed macrophytes. Results from the few studies on *in situ* rates of photosynthesis in relation to light indicate a tendency toward increased photosynthesis at 1–2 m below the surface on bright, clear days, where surface light intensities are reduced from 25 to 50 percent. On the other hand, exposure of some submergents to extremely high light intensities (>1500 μmol quanta m^{-2} s^{-1}) did not significantly alter rates of photosynthesis; other species are clearly adapted to growth at low light levels as low as 20 μmol quanta m^{-2} s^{-1} or less (Gessner, 1955; Bowes *et al.*, 1977; Barko *et al.*, 1982; Wetzel *et al.*, 1984; Sand-Jensen and Madsen, 1991; Blanch *et al.*, 1998).

Many submersed macrophytes exhibit distinct morphological variations in relation to light intensity. Shade-adapted leaves are finely divided, whereas on the same plant growing at higher light intensities (shallower depths) leaves can be larger and much more lobate. Photosynthetic rates can be significantly higher near the surface in cases where overlying foliage of dense populations shades the lower leaves. Shade-adapted plants have also been shown to be sensitive to ultraviolet light, which is rapidly removed below the first meter of water, whereas photosynthetic rates of plants adapted to high light intensities were unaffected by ultraviolet light that is found in surface layers of water.

Gessner (1955) attempted to clarify the many physiological reactions of submersed macrophytes to light by grouping the plants into adaptation types either genetically fixed or phenotypically plastic. (a) A strictly shade-adapted form, such as the fenestrated tropical *Aponogeton*, is physiologically active only at low light intensities. (b) Numerous submergents are strictly light-adapted and require high light intensities for optimal metabolism. (c) Some submersed macrophytes exhibit a broad adaptation to light intensities, including some (d) that adapt to light or shade conditions depending on the area of development, (e) others that are generally tolerant of high light but develop best in weak light, and (f) still others that are shade-adapted but photosynthesize optimally at intermediate littoral light intensities. While such a scheme does not explain the physiological and morphological bases underlying such adaptations, it does demonstrate the extreme plasticity and adaptability of submersed macrophytes to the highly variable underwater light conditions.

Some insight into the mechanisms of photosynthetic adaptation of submersed macrophytes has been gained by the investigations of Spence and Chrystal (1970a) on two species of *Potamogeton*. The rates of net oxygen production at various irradiances, dark respiratory oxygen uptake, chlorophyll content, specific leaf area (cm^2 leaf area per mg leaf dry weight), and leaf thickness were measured for a shallowwater, "sun-adapted" species (*P. polygonifolius*) and a deepwater, "shade-adapted" species (*P. obtusifolius*). From estimates of relative rates of photosynthesis and respiration per leaf area and per pigment content of the species, and sun and shade leaves of each species, it was concluded that (a) surface leaf area increases and respiration and leaf thickness decrease with depth and with light reduction, and (b) higher net photosynthetic capacity per unit area of shade-adapted leaves and shade species of *Potamogeton* at low irradiances (approximately 1% of summer daylight) is achieved by lowered respiration per area. The latter may result from a reduction in leaf weight per unit area. Potential complications introduced by photorespiration, discussed earlier, were not considered in these studies.

Other studies on the photosynthetic capacities and compensation points of submersed plants growing at low light intensities indicate marked variability. Compensation points commonly occur at 1–3% of full sunlight (Wilkinson, 1963; Spence, 1976, 1982; Bowes *et al.*, 1977; Agami *et al.*, 1980; Moeller, 1980). However, considerable metabolic adaptation is found. For example, in an oligotrophic Danish lake, *Littorella* occurred between 0 and 2 m, and *Isoëtes* grew at between 2- and 4.5-m depths (Sand-Jensen, 1978). Light-saturated photosynthesis of *Littorella* was higher than that of *Isoëtes*, and the difference increased with rising temperatures within the range of 5–20°C. At low irradiance, photosynthetic rates of *Isoëtes* were similar to those of *Littorella*, despite greater chlorophyll content per unit biomass in *Isoëtes*. Respiration rates were lower in *Isoëtes* than in *Littorella*.

Two other common competing submersed angiosperms, *Myriophyllum spicatum* and *Vallisneria americana*, exhibit different growth forms; most of the biomass of *Myriophyllum* is near the water surface, whereas most of the leaf biomass of *Vallisneria* is near the sediment (Titus and Adams, 1979). *Vallisneria* was shade-adapted, and photosynthetic fixation rates were much more efficient than those of *Myriophyllum* at low light intensities. Optimum temperatures for photosynthesis (33°C) were nearly identical for the two species, but at 10°C, the photosynthetic rates of *Myriophyllum* were nearly twice those of *Vallisneria*. Similar light and temperature interactions have been found among several other submersed angiosperms (Barko and Smart, 1981).

The heat capacity of water buffers the water temperatures to which submersed plants are exposed, usually within a range of ca. 2–40°C. Active growth can occur at low temperatures near zero, including under ice cover, although growth is slow and near maintenance levels (e.g., Hough and Wetzel, 1975; Boylen and Sheldon, 1976; Spencer and Wetzel, 1993). Net photosynthetic rates of submersed plants under ice are 50% greater than those of summer-grown plants measured at similar low temperatures. For example, optimum temperatures for net photosynthesis of winter-adapted plants of the submersed non-rooted *Ceratophyllum* that remained physiologically active under ice were 5°C and for summer-adapted plants 30°C. Dark respiration rates of winter plants were many times greater (to >300%) than those of summer plants, which contributed to reduced net photosynthesis.

Temperature optima among summer-adapted submersed macrophytes are in the range of 25–37°C (Spencer and Bowes, 1990). Photosynthesis-to-respiration ratios decline with increasing temperatures, particularly above 20°C, although significant species variations occur (Barko and Smart, 1981). Some prolific, invasive, and highly competitive species, such as *Hydrilla*, have a greater capacity for growth at higher temperatures than most other submersed species.

The metalimnion of stratified lakes varies in depth and thickness but commonly in small lakes begins at a depth of 3–4 m and extends to 6–8 m. Water temperatures at the depths of the lower metalimnion and upper hypolimnion are commonly in the range of 5–12°C throughout the year in the temperate zone (e.g., Dale, 1986). As a result, if light adequate to support net photosynthesis of submersed macrophytes reaches these depths, as it does in many lakes of low-to-modest productivity, the cold temperatures suppress maximum rates of photosynthesis and limit the maximum depth distribution of aquatic macrophytes.

Any additional factors that influence attenuation of light under water can influence the growth, survival, and distribution of submersed vegetation. Floating macrophytes can form dense surface canopies that severely shade underlying submersed vegetation (e.g., Janes *et al.*, 1996). Growth of epiphytic microorganisms on the surfaces of submersed macrophytes can severely shade and even kill the macrophytes (e.g., Sand-Jensen, 1990), a subject discussed at length in the ensuing chapter (Chap. 19).

B. Hydrostatic Pressure

Normal growth can be abruptly inhibited when certain submersed angiosperms are exposed to moderate increases in hydrostatic pressure (Fig. 18-7). Leaves become shorter, stems are much thinner and possess greatly reduced lacunal intercellular spaces, and growth of adventitious roots is inhibited (Ferling, 1957; Gessner, 1961; Hutchinson, 1975). Increased pressure of as little as 0.5 atm above air pressure at the surface of the lake, equivalent to about 5-m depth, also induces increased growth of internodes, similar to the effects of low light intensities. Increased pressure also can inhibit flower formation; most submersed angiosperms produce aerial flowers that are wind or insect pollinated.

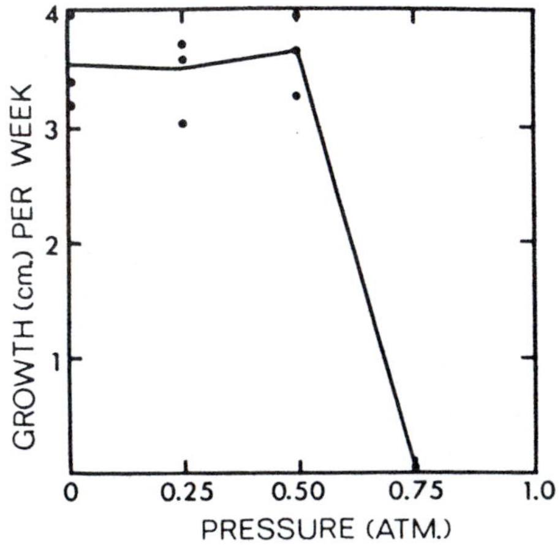

FIGURE 18-7 Average growth (cm) per week of *Hippurus* in relation to hydrostatic pressure. (Modified from Gessner, 1955.)

Growth inhibition is not usually proportional to pressure, and many species exhibit an irreversible "all-or-none" effect at an increase of about 1 atm of hydrostatic pressure above surface atmospheric pressure (approximately 10-m depth equivalency). Little can be said about the physiological effects of increased pressure. One effect is the inhibition of movement of intercellular gases to the root system. Short-term effects include increased secretion of soluble organic matter by excessive hydrostatic pressure. Such increased cellular permeability may lead to losses of nutrients and hormonal growth substances. Golubić (1963) demonstrated in a very clear, transparent lake that a hydrostatic pressure of 0.8 atm limited photosynthesis of submersed angiosperms. The macroalgae *Chara* and *Nitella*, lacking intercellular gas systems, are unaffected by pressure differences.

Growth inhibition by hydrostatic pressure is apparently not universal among all submersed angiosperms. For example, marine angiosperms grow commonly at depths of several atmospheres of hydrostatic pressure in clear waters, and rates of photosynthesis are unaffected in these plants by pressure changes (Payne, Beer, Wetzel, and Iverson, unpublished). Pressure effects (to 2.3 atm, corresponding to a depth of 13.3 m) on growth of the freshwater angiosperm *Hippuris* could be overcome by warm temperatures (>15°C) and high light intensities (>100 μmol quanta m^{-2} s^{-1}; Bodkin *et al.*, 1980). The physiological ramifications of various pressures remain obscure, but it is likely that hydrostatic pressure interacts, if at all, with other parameters, especially light, to restrict the distribution of freshwater angiosperms to depths of less than 10 m.[3] Shoots of *Myriophyllum spicatum* and other species are not affected by increased hydrostatic pressure (Dale, 1981, 1984; Payne 1982). Payne (1982) demonstrated that a lacunar arch system in stems of *Myriophyllum* expands the lacunae against external pressure by increasing cellular turgor pressure.

[3] *Potamogeton strictus* has been reported to a depth of 11 m in Lake Titicaca, Peru–Bolivia (Hutchinson, 1975). However, at an altitude of 3815 m, pressure at 11 m would be equivalent to a depth of 7.5 m in a lake at sea level. *Elodea canadensis* has been reported at a depth of 12 m in Lake George, New York (elevation approximately 100 m) (Sheldon and Boylen, 1977), although the range of this species in this lake is predominantly between 1 and 7 m.

Older stems and roots were weaker and were limited by low hydrostatic pressures.

C. Senescence

Most aquatic plant species grow more or less continuously, with losses of members of the population or parts of the same plant through senescence and death. A majority of aquatic plants grow perennially, with a constant turnover of population members that are senescing as new cohorts are emerging and growing. Some submersed plants grow actively at the apical tip and lower leaves senesce and slough off of the stem or petiole.

Although cell death is programmed genetically, the control agents and mechanisms for suppression homologs among aquatic plants are essentially unknown. Clearly, restricted light availability is involved in senescence induction along spatial dimensions in lakes (e.g., Rorslett, 1985) but the physiological mechanisms are unclear. During senescence, chlorophyll loss accelerates along with proteolysis and lipid oxidation, and carboxylation enzyme activities decrease (Jana and Choudhuri, 1982; Philosoph-Hadas *et al.*, 1994). Production of ethylene increases during senescence, which can enhance certain senescence-associated processes. Cell membrane integrity collapses during senescence with major losses of soluble cellular constituents. This released dissolved organic matter, including important organic nitrogen and phosphorus compounds, can be readily assimilated by attached microbial communities (Chap. 19).

V. PRIMARY PRODUCTIVITY OF MACROPHYTES

A. Changes in Biomass and Cohort Structure and Their Dynamics

Macrophyte productivity is most commonly evaluated by measuring changes in biomass (Westlake, 1965b). The initial biomass of seeds of annual macrophytes is negligible. The biomass of macrophytes typically increases in a sigmoid fashion during the growing season (Fig. 18-8). In this idealized example, gross productivity reaches a plateau and later declines in older tissues; net productivity decreases and becomes negative because respiration continues to increase. Maximum biomass, the maximum cumulative net production, is reached when the current daily net productivity becomes zero. Subsequently, both the biomass and cumulative net production decrease, and terminal net production is equal to the quantity of material that has not been respired by the time the whole plant is dead. Although numerous variations are found among natural populations, this general pattern of growth and metabolism prevails among *annual* (growth beginning from germinating seeds) macrophyte species in temperate and subtropical communities.

If the initial biomass (e.g., seeds of an annual plant) is negligible, or if the plant dies before the seasonal maximum is reached, and losses other than respiration are negligible, as is commonly the case, the seasonal maximum biomass is equal to the maximum annual net

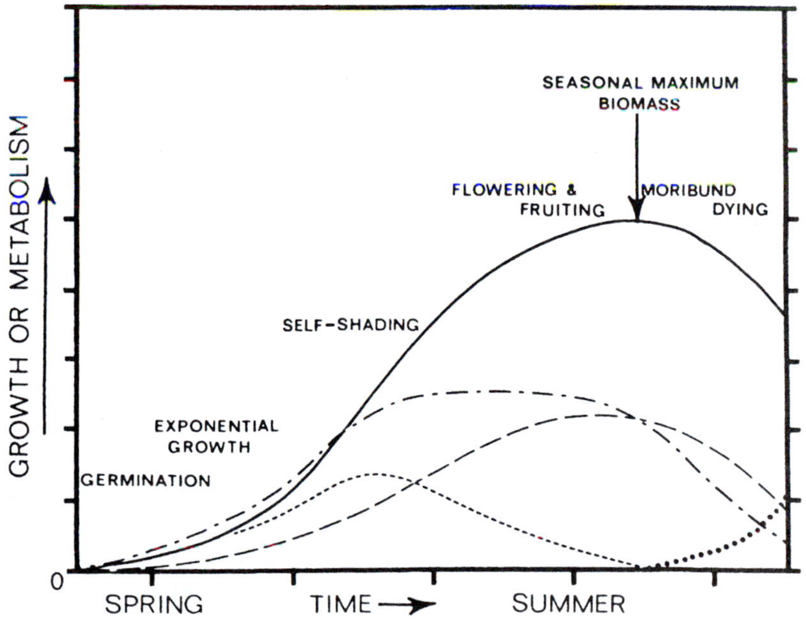

FIGURE 18-8 Generalized growth and metabolic patterns for a typical annual aquatic macrophyte. —— = total biomass;- – – – – – = current gross productivity; - - - - - = current net productivity; ——— = current respiration rate; .... = death losses. (After Westlake, 1965b.)

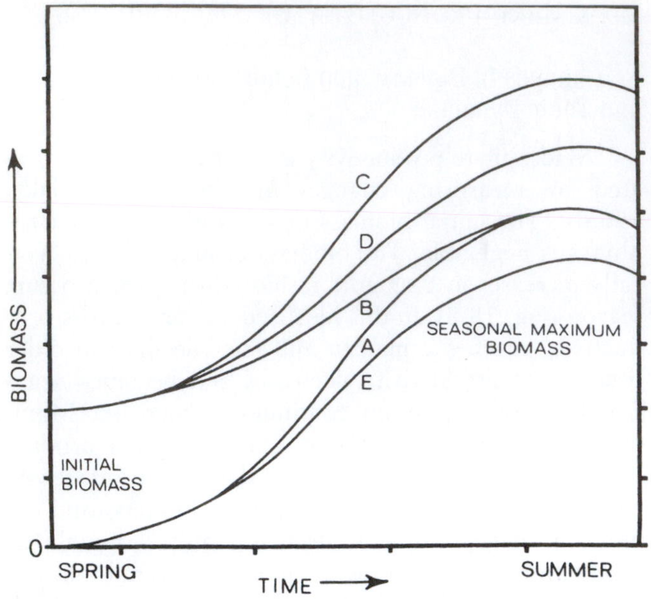

FIGURE 18-9 Types of growth curves among aquatic macrophytes. (*a*) True annual or a plant with manifest annual regrowth; (*b*) plant with obscured annual regrowth; (*c*) plant with spring biomass persisting until seasonal maximum; (*d*) plant with only part of spring biomass persisting until seasonal maximum; (*e*) plant with annual regrowth and losses from current year's biomass before seasonal maximum. (After Westlake, 1965b.)

production (Fig. 18-9, curves A and B) (Westlake, 1965b). This annual regrowth is characteristic of many plants that die down to a small quantity of perennating organs, but the extent of regrowth becomes less obvious when much of the plant survives until the next spring growing season. If the initial biomass persists without appreciable losses until after the seasonal maximum, the annual net production may be obtained as the difference between the final and initial biomass values for curve C (Fig. 18-9).

Most wetland and aquatic macrophytes are herbaceous perennials with massive carryover of perennating rooting tissues from one cohort of emergent foliage to the next. Most commonly, a variable proportion of the initial biomass is lost with senescence and decay (curve D). The maximum seasonal biomass in the latter case, as seen, for example, in populations of the macroalga *Chara*, is commonly only a few and highly variable percent (10–80%) of the annual net production. When losses of the current season's production are appreciable, the estimation of production from changes in biomass is complicated (Fig. 18-9, curve E), and can only be evaluated by detailed analyses of changes in the population demographics and biomass. Plant biomass remains more or less constant in extreme cases, such as in subtemperate and tropical communities, despite continuous growth. Very large amounts (50–90%) of the annual productivity would be missed by analyzing only maximum biomass at any given time.

Cohorts of a species population are constantly growing, senescing, and decaying back and being replaced simultaneously by new cohorts. In populations with a high plant mortality (e.g., in some reedswamp-emergent macrophytes or among populations with a high turnover of leaves) (Westlake, 1966; Bernard and Solsky, 1977; Dickerman and Wetzel, 1985; Dickerman *et al.*, 1986; Tsuchiya *et al.*, 1993; Wetzel and Howe, 1999), much of the production does not survive to be measured as terminal maximum biomass of the population. It is extraordinarily difficult to measure the productivity of these constantly emerging cohorts of new populations and to evaluate the losses occurring with senescing cohorts of shoots and roots in order to evaluate the composite population productivity. For example, continuous changes in the population dynamics, above- and belowground biomass, growth rates, and production of the emergent rush *Juncus effusus* were evaluated over an annual period in a subtemperate riparian littoral area (Wetzel and Howe, 1999). Shoots emerged continuously at all times of the year, and the number of ramets (shoots) increased four times from summer to winter (12,000 to >30,000 shoots m^{-2}) (Fig. 18-10). Individual culms (leaves) in November–December, however, were generally smaller (41–52%) than those in summer. Maximum growth of individual new culms ranged from a maximum of 6.4 mg day^{-1} in June (40–50 days to maximum biomass) to 1.1 mg day^{-1} in November (70–110 days to maximum biomass in winter). Continuous recruitment and gradual culm senescence resulted in numerous multiple cohorts on an annual basis. Annual root production was about 42% of shoot production. Average annual production estimates of combined above- and belowground components of this ubiquitous emergent species were nearly 10-g ash-free dry mass m^{-2} per year (Table 18-4). The extremely high production rates of this species result from the continuous growth, while simultaneously shoot densities changed seasonally as the inverse of biomass. These processes maximized photosynthetic carbon fixation under the mild climatic conditions of the winter and spring seasons. This dynamic growth strategy is likely common among herbaceous perennial emergent aquatic plants in mild climate regions.

These studies indicate the major importance of evaluating what percentage of total production is allocated to belowground tissues. For example, in the common sedge *Carex rostrata*, 88% of the total annual production was allocated to belowground tissues (Saarinen, 1996). Among total plant production allocated to rhizomes, coarse roots, and fine roots, 79% was directed to fine-root production (cf. Table 18-8).

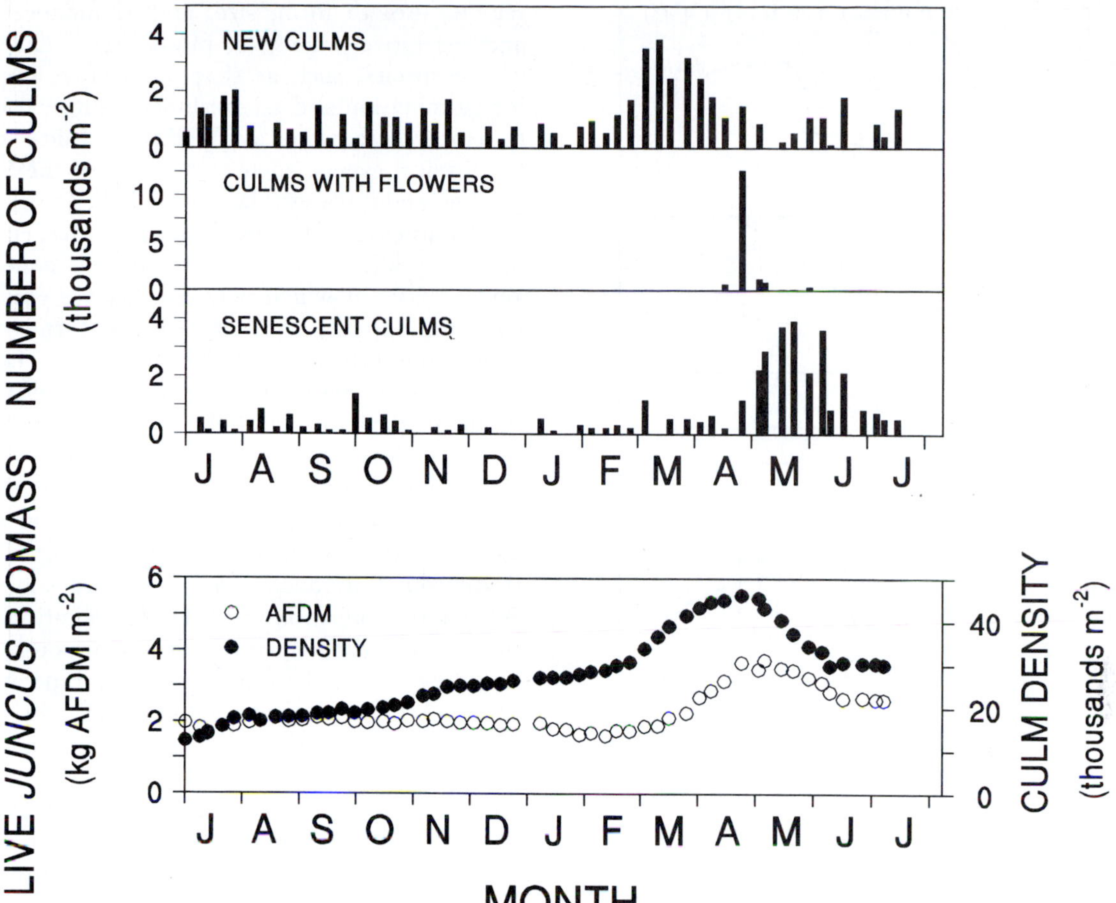

FIGURE 18-10 Example of number of new photosynthetically active culms, culms with flowers, and senescent culms of *Juncus effusus* formed each week per square meter over an annual period (*upper*) and total culm density (thousands m^{-2}) and biomass of photosynthetically active plants (kg AFDM m^{-2}) in Talladega Wetland Ecosystem, Alabama (*lower*). (Extracted from Wetzel and Howe, 1999.)

In more temperate regions with shorter growing seasons, the same multiple cohort strategy occurs among these plants but with reduced cohort turnover numbers. In a cattail (*Typha latifolia*) population in southern Michigan, 19% of the shoots had completely senesced prior to maximum aboveground biomass, and 95% of these "prematurely" senesced shoots had entirely disappeared from the population before the maximum biomass occurred (Fig. 18-11) (Dickerman and Wetzel, 1985). Other significant biomass losses resulted from high leaf turnover. On the average, 50 percent of the total number of leaves produced by a cattail shoot

TABLE 18-4 Production of the Emergent Macrophyte *Juncus effusus* among Different Plots of the Talladega Wetland Ecosystem, Alabama, and for the Entire Wetland Where Colonized by *Juncus* (64.8% Cover)[a]

	Annual production within plots				*Annual production within wetland*			
	(g AFDM m^{-2} per yr)			*(g C m^{-2} per yr)*	*(g AFDM m^{-2} per yr)*			*(g C m^{-2} per yr)*
	Aboveground	*Belowground*	*Total*	*Total*	*Aboveground*	*Belowground*	*Total*	*Total*
Plot 1	6052	2542	8594	3996	3922	1647	5569	2590
Plot 2	10,299	4325	14,624	6800	6674	2803	9476	4406
Plot 4	4422	1857	6279	2920	2865	1203	4069	1892
Total mean	6924	2908	9832	4572	4487	1884	6371	2963

[a] From Wetzel and Howe (1999).

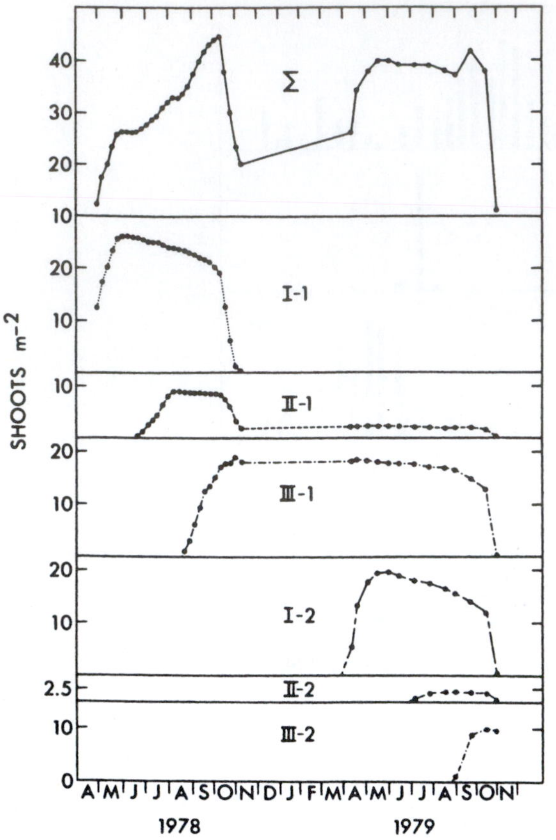

FIGURE 18-11 Shoot density of the composite population (Σ) and individual cohorts (roman numerals) during the first (1) and second (2) years of analysis of the common cattail (*Typha latifolia*), Lawrence Lake, Michigan. (After Dickerman and Wetzel, 1985.)

had undergone senescence and abscission by the time maximum aboveground biomass had been attained. As a result, the composite population experiences continual senescence after the first cohort reaches maturity and starts to decline. Further, early shoot mortality can also account for a significant loss of population biomass.

Emergence of cohorts throughout the growing season (Fig. 18-11) frequently more than offsets shoot mortality; a constant or increasing shoot density may help exclude potential competitor species. Density decreases with increasing accumulations of litter, increases when the area is flooded by rising water levels and increased inorganic nutrients, and decreases with grazing of buds and moderate increases in salinity (Rudescu *et al.*, 1965; Björk, 1967; Haslam, 1971a; Rodewald-Rudescu, 1974). High light, warmer temperatures, and higher water levels promote growth in dense, monospecific stands. Among mixed stands of emergents, the final balance of dominance is determined by a combination of (a) physical factors and inherent properties of the plants, such as soil and water régime, growth form, size, and abundance of sexual and vegetative organs, (b) physiological characteristics of the species, such as shade tolerance, requirements for germination and establishment, survival in relation to temperature, water availability, and litter cover, and (c) physical events, such as opening of the litter mats, grazing, and diseases (Haslam, 1971b, 1973).

Floating-leaved rooted macrophytes, particularly the water lilies, also exhibit numerous cohorts on an annual basis in which new cohorts are emerging and senescing rapidly, with an average turnover rate of aboveground foliage of 4–8 times a year (Table 18-5). Leaf longevity averages about 30 days and increases to about 50 days during cooler seasons. Senescence and sinking of leaves are essentially continuous throughout the growing season and, for example, averaged an annual loss rate of 5.2 leaves m^{-2} day^{-1} in a wetland pond at a latitude of 33°N (Carter *et al.*, 2000). True aboveground production rates require summation of the cohort production rates. That integration yields annual values nearly an order of magnitude higher than would be obtained from aerial instantaneous biomass estimates during the growing season (Table 18-5).

Similar situations have been found among most submersed macrophytes that have been examined in detail (Borutskii, 1950; Rich *et al.*, 1971; Best and Dassen, 1987). In the nonrooted submersed *Ceratophyllum demersum*, for example, production was over three times higher than that estimated from the difference between seasonal maximum and minimum biomass. Demographic methods developed for analyses of fish benthic invertebrate populations subject to high mortality have been applied successfully to dense macrophyte populations, in which, for example among *Glyceria* and *Typha* populations, net production was some 3–4 times greater than seasonal maximum biomass (Mathews and Westlake, 1969; Wetzel and Pickard, 1996; cf. Dickerman *et al.*, 1986).

B. Disease, Herbivory, and Other Mortality of Aquatic Plants

Pathogenic decimation of entire macrophyte populations is rare; natural diseases usually only cause gaps in existing populations, which result in a more heterogeneous distribution of individuals (see, e.g., Klötzli, 1971; cf. Zettler and Freeman, 1972; Freeman, 1977). Losses of plant biomass from damage from water movement is generally not severe in lakes and ponds, but can be a major mortality loss factor in streams and rivers. The few macrophytes growing in running waters are well adapted to moderate to even torrential river flows (cf. Gessner, 1955; Haslam, 1978). Many macrophyte fragments and seeds can be moved downstream

TABLE 18-5 Aboveground Production and Turnover Rates of Rooted Floating-Leaved Macrophytes from Various Sources

Floating-leaved macrophyte	*Location*	*Habitat*	*Biomass*[a]	*Annual production*[a]	*Turnover rate* (yr^{-1})	*Leaf longevity (days)*	*Reference*
Nymphaea tetragona	Japan	Pond	133–173.2 g DM m^{-2} (leaf and petiole)	515–559 g DM m^{-2}	6.7	31	Kunii and Aramaki (1992)
Nymphaea alba	Netherlands	Pond	137–154 g AFDM m^{-2}				Van Der Velde and Peelen-Bexkens (1983)
Nymphaea candida			59				
Nuphar lutea			88–182 (leaf)				
Nymphaea alba	Netherlands	Lake	139 g AFDM m^{-2}	319–348 g AFDM m^{-2}	3.6–3.8	46–49	Kok *et al.* (1990)
Nuphar lutea			50–162 (leaf)	108 (min)–447 (max)	3.8–4.4	38–50	
Nelumbo lutea	Illinois	River	337 g AFDM m^{-2} (leaf and petiole)				Grubaugh *et al.* (1986)
Nymphoides peltata	Netherlands	River	26.35 g AFDM m^{-2} (leaf)	480 g AFDM m^{-2} 560 g DM m^{-2}	7.3–8.6	23	Brock *et al.* (1983a)
	Japan	Lake	47.6 g DM m^{-2}	182 g DM m^{-2}	3.4	28	Tsuchiya *et al.* (1990)
Nymphaea odorata	Alabama	Wetland	20.6 g AFDM m^{-2}	251 g AFDM m^{-2}	6.6	32.6	Carter *et al.* (2000)
		Pond	23.7 g DM m^{-2} (leaf)	285.7 g DM m^{-2}			
Brasenia schreberi	Japan	Pond	52.5–80.5 g DM m^{-2} (leaf)	155–244 g DM m^{-2}	6.0–6.3	25–28	Kunii (1999)

[a] AFDM = ash-free dry mass; DM = dry mass.

during normal and spate river flows (Sand-Jensen *et al.*, 1999). The drifting shoots in particular can become caught on projections and serve as a means of recruitment and expansion of new colonies. Marginal bank and floodplain vegetation, however, can be vulnerable to strong spates and flood conditions, and large quantities of particulate organic matter is often moved downstream and heterogeneously distributed within the river channels as well as laterally on the flood plain (e.g., Furch and Junk, 1997). Most of the subsequent decomposition of the plant organic matter is microbial, particularly after the flood waters recede and conditions shift back to semiterrestrial phases.

Herbivory of aquatic plants is very poorly studied. Losses from grazing before the seasonal maximum biomass are generally considered small among temperate annual macrophytes (Westlake, 1965b; Sculthorpe, 1967). Although many animals, especially immature insects, eat live aquatic macrophytes, few cause extensive damage (see especially the extensive review of Gaevskaya, 1966). Although the percentage loss by animal ingestion or destructive maceration is set in the range of 25% for the *aboveground instantaneous biomass* of freshwater vascular macrophytes, and somewhat higher among nonvascular macrophytes (Lodge *et al.*, 1998), the percentage of loss of *total above- and belowground production* of aquatic plants by animal herbivory is nearly always unknown and certainly is lower.

Nearly all studies examine the losses to aboveground biomass, often during episodic events (e.g., intensive feeding during bird migrations), without evaluations of the recovery of subsequent cohort production of the plants or the impacts of herbivory on an annual basis. In the few cases where this evaluation is done correctly, herbivory of the total annual net production of the aquatic plants is usually low (<5%; e.g., Carter *et al.*, 2000). As a result, many of the quantitative values in the literature are certainly overestimates of true loss of organic matter by herbivory. Nonetheless, these data place in perspective some of the relative losses than can occur by animal herbivory.

Earlier assumptions that herbivory on aquatic plants was unimportant evolved because many plant tissues are difficult to ingest and of low nutritional quality; such assumptions are not valid (cf. reviews of Lodge 1991; Newman, 1991; Lodge *et al.*, 1998). Herbivore consumption of aboveground freshwater macrophyte biomass is similar to that among terrestrial vascular plants. About 25% of aboveground tissue of freshwater macrophytes is lost on the average to all animal herbivory, and this would be reduced by at least half or more of that value when one considers that aboveground production is about half of the total production. Moreover, total aquatic plant production is often made up of multiple cohorts, some of which may not be preyed upon as severely at certain times of the year. Certainly, consumption by herbivory of aboveground biomass of freshwater vascular plants is less than that of the attached microalgae of periphyton and of macroalgae growing on submersed portions. A distinction is needs to be made consumption of organic matter production and secondary effects that partial consumption of tissues can have on the plant biology. For example, damage to internal vascular tissues may affect plant metabolism severely even though the consumption of plant tissue is low.

Herbivory of aboveground vascular plant tissues by aquatic insects and snails is low in comparison to potential losses to crayfish crustaceans (e.g., Bolser and Hay, 1998; Bolser *et al.*, 1998), herbivorous fish specialists, mammals, and aquatic birds under certain conditions. Snails and many immature aquatic insects feed largely on detritus and periphyton rather than the macrophyte tissue (e.g., Kolodziejczyk, 1984). Mammals, particularly the muskrat (*Ondatra*) and nutria (*Myocastor*) commonly impact aboveground macrophyte biomass at levels intermediate between insects and the other vertebrates, birds and specialized fish. Herbivory of the leaf surface area of submersed plants ranged from 0% to nearly 100% in a variety of different studies (Lodge, 1991) but was usually at the low portion of that range and somewhat higher in streams (ca. 5%) than in lakes (ca. 2%) for a species of *Potamogeton* (Jacobsen and Sand-Jensen, 1995). Great seasonal and interannual variability in herbivory on submersed species was found (Jacobsen and Sand-Jensen, 1994). In vast areas of wetlands and littoral areas, however, reductions of total biomass by herbivory of these animals is much lower, particularly of the most productive emergent macrophytes.

Many classes of plant compounds are known to function as plant deterrents for herbivory (Otto, 1983; Ostrofsky and Zettler, 1986; Suren and Lake, 1989; Lodge, 1991; Newman, 1991; Bolser *et al.*, 1998). Direct chemical interactions are poorly known among aquatic plants. For example, when the high nitrogen-containing leaves of watercress (*Nasturtium officinale*) are damaged by feeding by caddisfly larvae, snails, or amphipods, myrosinase-mediated hydrolysis of phenylethyl glucosinolate forms a toxic end product, phenylethyl isothiocyanate, that inhibits further herbivory (Newman *et al.*, 1992, 1996). Components of the glucosinolate–myrosinase defensive system have been shown to be induced by the ubiquitous defensive signaling molecule jasmonic acid (Farmer *et al.*, 1998). Herbivores then shift to low nitrogen-containing senescent leaf tissues that also contain much less of the toxic compound. The freshwater orchid *Habenaria* synthesizes a defensive compound, habenariol, that is a significant deterrent against herbivory by the crayfish *Procambarus* (Wilson *et al.*, 1999). Similar mechanisms to suppress herbivory are likely common among the higher aquatic plants (Bolser *et al.*, 1998; Amudhan *et al.*, 1999). The metabolic costs of such chemical antiherbivore defenses versus potential gains in reproductive fitness are very poorly understood among freshwater aquatic plants.

Fish, birds, and a few mammals can, under exceptional circumstances, cause extensive damage or consume significant portions of the annual production of macrophytes. Some herbivorous fish, such as the grass carp (*Ctenopharyngodon idella*), can decimate submersed and floating macrophytes (e.g., Sutton, 1977; van der Zweerde, 1990). These fish have been used as a management tool to control aquatic macrophytes in semitropical and tropical fresh waters. Under natural conditions, values of grazing loss, however, usually range from about 0.5–8% of the total. Under most circumstances, most of macrophyte production is consumed by bacteria and fungi when the plants undergo senescence (see Chap. 21). Moreover, it must be recalled from an ecosystem perspective that even if the organic matter of the macrophytes is ingested, assimilation efficiencies are modest, usually much <50% in the animal digestive system, and much of the plant material is returned to the aquatic ecosystem as detrital particulate and dissolved organic matter. Most of the detrital particulate organic matter and all of the dissolved organic matter is largely metabolized subsequently by microbiota.

C. Belowground Biomass and Growth

The roots or rhizoids of submersed freshwater macrophytes generally constitute a small but significant proportion of the total plant biomass (Table 18-6). Among floating-leaved and emergent macrophytes, however, the extensive system of roots, rootstocks, and rhizomes constitutes a major, often dominating, portion of the total biomass. Not only are the respiratory demands on organic carbon and energy of the rooting tissues very high but these demands change under different growth and environmental conditions. For example, rhizome respiration rates of the reed *Phragmites* were found to be inversely related to rhizome age but directly related to temperature change seasonally (Cízková and Bauer, 1998). In another example, much greater amounts of the total aquatic plant productivity are often devoted to root growth under conditions of nutrient stress or limitation (e.g., Grace and Wetzel, 1981a). In general, allocation to root and rhizome production is lower under more reducing conditions (Day and Megonigal, 1993). Rates of decomposition of rooting tissues are similarly markedly reduced under more reducing conditions of saturated sediments.

Clearly, accurate analyses of productivity must include an evaluation of the dynamics of growth of the rooting system. Measurements of only the aboveground foliage ("standing crop," see definitions in Chap. 8) are inadequate and can be quite misleading

TABLE 18-6 Percentage of Total Biomass Found in Underground Tissues (Roots and Rhizomes) of Mature Aquatic Macrophytes[a]

Type and species	*% of total biomass*
Submersed	
Chara	<10
Ceratophyllum demersum[b]	<5
Elodea canadensis	2–6
Myriophyllum sp.	11
Potamogeton richardsonii	12
P. perfoliatus	39
P. lucens	49
P. pectinatus	17.5
Heteranthera dubia	19
Vallisneria americana	48
Isoetes locustris	25–40
Lobelia dortmanna	21–26
Littorella uniflora	46–55
Thalassia testudinum (marine)	75–85
Floating	
Eichhornia crassipes (submersed roots)	10–48
Floating-leaved	
Nuphar lutea and *N. pumilum*	50–80
N. variegatum	68
Nymphaea candida	48–80
Emergent	
Zizania aquatica	7–8
Cyperus fuscus	7–8
Alisma plantago-aquatica	40
Equisetum fluviatile	40–83
Pontederia cordata	67
Glyceria maxima	>30–67
Phragmites communis	>36–96
Schoenoplectus lacustris	>46–90
Sparganium fallax	40
S. mucronatus	>75
Sparganium spp.	>25–66
Typha angustifolia	>32–67
Typha hybrid	64
Typha latifolia	43–67
Eleocharis rostellata	46.5
Carex rostrata	20–60

[a] After Westlake (1965b, 1968); Nicholson and Best (1974); Kansanen and Niemi (1974); Ozimek *et al.* (1976); Fiala (1976, 1978); Szczepańska (1976); Grace and Wetzel (1981a); Han (1985); Seischab *et al.* (1985); Ozimek *et al.* (1986); Hwang *et al.* (1996); Saarinen (1998).
[b] Usually not rooted in sediments.

TABLE 18-7 Average and Range of Chemical Composition and Caloric Values of the Three Primary Types of Aquatic Plants[a]

			% of organic matter				
Ecological type	*Water (%)*	*Ash (% of dry weight)*	*Protein*	*Lipids (ether extract)*	*Crude fiber*	*Carbohydrate*	*Caloric value (cal g⁻¹)*
Emergent	79	12	13	2.1	32	50	4480
	(70–85)	(5–25)	(3.5–26.7)	(0.4–7.6)	(17.6–45.4)	(23–69)	(4207–4987)
Plants with	82	16	26.5	4.0	27	42	4770
floating leaves	(80–85)	(10–25)	(18-44)	(2.8–5.7)	(14–38)	(31–64)	(4560–5140)
Submersed	88	21	22	2.2	27	51	4580
	(85–92)	(9–25)	(7.5–34.7)	(4.4–5.1)	(14.0–61.6)	(20–70)	(4165–5201)
Average all groups	83	18	19	2.4	29	50	4570

[a] From Straškraba (1968), after numerous sources, and Esteves (1979b).

when evaluating true ecosystem constructs. The biomass of underground organs is difficult to measusre but can be estimated by randomized coring and quadrat excavation (Fiala *et al.*, 1968; Westlake, 1968; Fiala, 1971, 1973, 1978). A continuing problem is differentiating between living and moribund rooting tissues. Growth patterns and discrimination of annual growth among perennial root systems can be determined using a combination of tagging and growth experiments in which biomass increments of different age classes are analyzed (e.g., Ervin and Wetzel, 1998). The ratio of biomass of underground parts to above-sediment parts is relatively constant among certain species at maturity in a particular habitat (e.g., Szczepański, 1969) but usually increases markedly as the growing season progresses (e.g., Fiala, 1976, 1978; Szczepański, 1976; Granéli *et al.*, 1992; Ervin and Wetzel, 1998). Photosynthetic allocation to above- and belowground tissues can also be evaluated by *in situ* pulse labeling of photosynthetic tissue with radioactive CO_2 and determination of the allocation to above- and belowground plant parts over time (Grace and Wetzel, 1981a; cf. also Tietema, 1980).

D. Organic Chemical Composition of Aquatic Macrophytes

An extensive collation of existing data on the general organic chemical composition of aquatic macrophytes (Straškraba, 1968; Hutchinson, 1975; Muztar *et al.*, 1978a,b) provides useful information on the relative potential contribution of these plants to production and decomposition processes in aquatic ecosystems. Morphological and physiological differences among the different plant types are reflected in a higher content of water, ash, and protein and a lower content of fiber in submersed as compared to emergent macrophytes (Table 18-7). Carbohydrate and fat content is similar within the two groups. Floating-leaved macrophytes exhibit approximately intermediate composition values but contain significantly more lipids, which is reflected in somewhat higher average caloric values. The caloric values of aquatic macrophytes (Table 18-7) are approximately 20% lower than the mean for phytoplankton (approximately 6000 cal g^{-1} organic matter).

E. Comparative Macrophyte Productivity among Habitats

A comparison of productivity of any biotic group among freshwater ecosystems is difficult because of large variations in analytical and sampling techniques employed and frequency of sampling. Variability is especially great for evaluations of macrophyte productivity. Sampling of total biomass, including the rooting systems, has been done seriously only in recent years. Species distribution and production depend on an array of physical and chemical characteristics of both the water and the sediment. The result is extreme heterogeneity in distribution and productivity, both spatially and temporally. Further, results are commonly biased toward the growing season, which is often much shorter than a year. The annual time period is the only meaningful interval for quantitative comparisons of material and energy fluxes at population, community, and ecosystem levels (cf. Wetzel, 1995). Despite these difficulties, approximate comparisons of macrophyte productivity are possible (see especially Westlake, 1963, 1982; Sculthorpe, 1967; Květ, 1971). The few exemplary figures given in Table 18-8 indicate the general range of aboveground maximum annual biomass and productivity among macrophyte communities. Submersed macrophytes generally exhibit conspicuously lower productivity than plants of floating or emergent

TABLE 18-8 Representative Estimations of Annual Maximum Biomass and Productivity of Aquatic Macrophytes

Type and lake	*Seasonal maximum biomass or aboveground biomass (g dry m^{-2})*	*Productivity*[a] *(g m^{-2} yr^{-1})*	*Source*
Submergents dominating			
Trout L., WI (softwater)	0.07	—	Wilson (1941)
Sweeney L., WI (softwater)	1.73	—	Wilson (1937)
Weber L., WI (softwater)	16.8	—	Potzger and Engel (1942)
Mirror L., NH (*Lobelia dortmanna*)	80	55	Moeller (1978a)
L. Opinicon, Ontario	496	—	Crowder *et al.* (1977)
West Blue L., Manitoba	—	6.4	Love and Robinson, (1977)
Lake, northern WI (Two small isoetid species)	5–22	12–65	Boston and Adams (1987)
L. Baciver, Spain (*Isoëtes lacustris*)	25–60	75–226	Gacia and Bellasteros (1994)
Lowes L., Scotland (dystrophic)	32	—	Spence *et al.* (1971)
Spiggie L., Scotland (dystrophic)	100	—	
L. Mendota, WI (hardwater)	202	—	Rickett (1921)
Lawrence L., MI (hardwater)			
Submersed *Scirpus subterminalis*	338	565	Rich, Wetzel, and Thuy (1971)
Chara	110	155	
Annuals	130	199	
Croispol, Scotland	400	—	Spence *et al.* (1971)
Borax L., CA (saline lake, *Ruppia*)	60	64	Wetzel (1964)
Danish lowland stream (*Sparganium emersum*)	234	536	Nielsen *et al.* (1985)
Czech fish pond (*Elodea canadensis*)	450	500	Pokorný *et al.* (1984)
Lake Kariba, Zimbabwe (*Lagarosiphon ilicifolius*)	100	599	Machena *et al.* (1990)
Lake Marion, SC (*Egeria canadensis*)	77–374	—	Getsinger and Dillon (1984)
Floating marsh, Louisiana			
(*Panicum* dominated)	1160	1960	Sasser and Gosselink (1984)
(*Phragmites australis*)	—	2300	
River Ivel, England (*Berula*; very fertile)	500	—	Edwards and Owens (1960)
River Test, England (*Ranunculus*)	100–400	—	Owens and Edwards (1961)
River Yare, England (*Potamogeton*)	380	—	Owens and Edwards (1962)
New Zealand (fertile, *Nasturtium*)	1000	2200	Howard-Williams *et al.* (1982)
Saline channels PR (*Thalassia*)	700–7300	—	Burkholder *et al.* (1959)
Floating (see also Table 18-5)			
New Orleans, LA (*Eichhornia*)	630–1472	1500–4400	Penfound and Earle (1948)
Emergents dominating			
Ladoga L., Russia	0.4–10.7	2.4	Raspopov (1971)
Onega L., Russia	0.1–33	0.6	Raspopov *et al.* (1977)
Kubenskoje L., Russia	—	29	
Voze L., Russia	—	70	
Laca L., Russia	—	120	
Hymenjaure, Sweden (68°N)			
Equisetum fluviatile	10–21	16.8	Solander (1983)
Carex rostrata	12–25	14.0	
Czechoslovakian ponds			
Phragmites australis	2980	—	Dykyjová (1971)
Typha angustifolia	4040	—	
Scirpus lacustris	3000	—	
Southern Swedish lake, *Phragmites australis*	2380	—	Björk (1967)
L. Arreso, Denmark (eutrophic)	2900–9900	—	Andersen (1976)
Temperate wetlands, N. America *Zizania aquatica*	—	630–1450	Whigham and Simpson (1977)
Alberta oxbow lakes	465	—	Van der Valk and Bliss (1971)

(*continues*)

TABLE 18-8 (Continued)

Type and lake	*Seasonal maximum biomass or aboveground biomass (g dry m^{-2})*	*Productivity*[a] *(g m^{-2} yr^{-1})*	*Source*
Michigan Hollow, NY *Carex lacustris*			
Aboveground	1145	965	Bernard and Solsky
Belowground	575	208	(1977)
Blanket bog, England (*Sphagnum*)			
On hummocks	—	180	Clymo and Reddaway
In pools	—	290	(1971, 1974)
On "lawns"	—	340	
Polish lakes (emergent species)	440–830	—	Szczepańska (1973)
Minnesota wetlands (*Carex*)	850	738	Bernard (1973)
Surlingham Broad, England (*Glyceria, Typha*, and *Phragmites*)	800–1100	—	Buttery and Lambert (1965)
Opatoviky Pond, Czechoslovakia (*Phragmites*)			
Aerial	1100–2200	—	Dykyjová and Hradecká
Belowground	6000–8560	—	(1973)
Cedar Creek, MN (*Typha*)	4640	2500	Bray *et al.* (1959)
Lawrence L., MI (*Typha latifolia*, aboveground)	770	6000	Dickerman and Wetzel (1982a,b)
Swamp, Western Australia *Phragmites australis*			
Aboveground	9890	12898	Hocking (1989)
Belowground	16767	6338	
Fen, Finland *Carex rostrata*			
Shoots	185	176	Saarinen (1996)
Rhizomes	185	113	
Coarse roots	135	82	
Fine roots	1785	1053	
L. Naivasha, Kenya *Cyperus papyrus*			
Aboveground	1484	4180	Jones and Muturi (1997)
Belowground	2126	3160	
Mirrool Creek, Australia			
Typha orientalis, aboveground	2600	4379	Robert and Ganf (1986)
Carex wetlands	650–1950	1700	Bernard *et al.* (1988)
Tidal marsh, Quebec (*Scirpus americanus*; above- and belowground)	150–1500	38–216	Giroux and Bédard (1988)
Riverine wetland, AL (*Juncus effusus*)			Wetzel and Howe (1999)
Aboveground	1206	6924	
Belowground	505	2908	
Amazon flood plain (C_4 grass, *Echinocholoa polystachya*)	8000	9900	Piedade *et al.* (1991)
Texas pond (*Typha angustifolia*)			
Aerial	2727	2727	Hill (1987)
Belowground	708	2410	
Tropical swamp, Kenya			
Cyperus papyrus (C_4)	2050	—	Jones (1988)
Cyperus latifolius (C_4)	2170	—	
Typha domingensis (C_3)	1350	—	
Wetland, RI (*Juncus militaris*)	371	436	Hoagland and Killingbeck (1985)

[a] Where maximum seasonal biomass values are higher than productivity estimates, the biomass values include a carryover of biomass from the previous season.

TABLE 18-9 Generalized Summary Ranges of Most Frequently Encountered Production Rates of Both Above- and Belowground Components of Aquatic Plants in Aquatic Habitats[a]

	Seasonal maximum biomass (g dry m⁻²)	*Production rate (g m⁻² yr⁻¹)*
By plant type:		
Emergent	3980	1000–10,000
Floating-leaved rooted	850–1750	100–560
Freely floating	150 (Lemnaceae)	300
	1275 (*Eichhornia*)	4000–5000
Submersed		
Nutrient-poor	5–140	5–385
Enriched	65–700	200–1500
Bryophytes	15–200	40–400
By habitats:		
Marsh and swamp communities		
Aboveground		1500–7000
Belowground		150–3000
Total		1650–10,000
Peatbog and wet tundra communities		
Aboveground		300–1000
Belowground		70–1400
Total		370–2400
River communities		
Temperate rivers		
Emergent		320–3700
Submersed		8–400
Submersed, polluted		1,160
Tropical river		
Amazon "floating meadow"		2430–4050

[a] Integrated from Junk (1970), Esteves (1979a), Hejný *et al.* (1981), Westlake (1982), Bradbury and Grace (1983), Rodgers *et al.* (1983), Grigal (1985), Robert and Ganf (1986), Madsen and Adams (1988), Wallén *et al.* (1988), Boston *et al.* (1989), Thormann and Bayley (1997), and Wetzel and Howe (1999).

communities. Table 18-9 attempts to summarize many data that include both the above- and belowground production rates. Clearly, production is much higher, nearly double among the emergent plants, than is indicated by the aboveground production rates alone.

Annual net primary productivity (usually only aboveground) of submersed macrophytes ranges from 5 to 10 metric tons (mt) per hectare (500–1000 g m^{-2} yr^{-1}), the lower values being more common in oligotrophic and moderately fertile waters and the higher values in fertile lakes (Table 18-10). Greater macrophyte productivity occurs among submersed communities in subtropical regions, from 10 to as high as 60 mt ha^{-1} yr^{-1} in the marine *Thalassia* community of Puerto Rico. Few accurate data are available for productivities of floating or floating-leaved macrophytes, although it is apparent that this community reaches levels of 11–33 mt ha^{-1} yr^{-1} in warmer climates and can easily achieve greater productivity under tropical conditions (Westlake, 1963).

Numerous investigations of the productivity of emergent macrophytes of fertile reedswamps and marshes indicate very high values, even though the range is quite large. Temperate-zone emergents can attain productivity levels of 20–45 mt ha^{-1} yr^{-1} and, in subtropical and tropical regions, can reach 40 to at least 90 mt ha^{-1} yr^{-1}. Inspection of the annual net productivity of the plant communities (Tables 18-9 and 18-10) shows that the emergent reed communities are as productive as terrestrial communities. Under fertile conditions with a continuous growing season, emergent aquatic macrophyte communities can be more productive than terrestrial communities. In the temperate zones, the organic productivity of emergent communities is comparable to that of perennial agricultural crops and coniferous forests and appreciably greater than that of deciduous forests. In tropical zones, emergent communities are among the most productive of all plant communities.

By comparison, the submersed macrophytes, with the possible exception of tropical marine angiosperms, are relatively unproductive and are only slightly above the composite areal productivity of the least productive component, the phytoplankton. In spite of the number

TABLE 18-10 Annual Net Primary Productivity of Aquatic Communities on Fertile Sites in Comparison to That of Other Communities[a]

Type of ecosystem	*Approximate organic (dry) production (mt ha^{-1} yr^{-1})*[b]	*Range (mt ha^{-1} yr^{-1})*
Marine phytoplankton	2	1–4.5
Lake phytoplankton[a]	2	1–9
Freshwater submersed macrophytes		
Temperate	6	5–10
Tropical	17	12–20
Marine submersed macrophytes		
Temperate	29	25–35
Tropical	35	30–60
Marine emergent macrophytes (salt marsh)	30	25–85
Freshwater emergent macrophytes		
Temperate (largely C_3 plants)	38	30–70
Tropical (esp. C_4 plants)	75	60–90
Desert, arid	1	0–2
Temperate forest		
Deciduous	12	9–15
Coniferous	28	21–35
Temperate herbs	20	15–25
Temperate annuals	22	19–25
Tropical annuals	30	24–36
Rain forest	50	40–60

[a] After Westlake (1963, 1965b, 1982). More recent values given for phytoplankton are likely too high by nearly an order of magnitude and are not cited here.
[b] mt = metric tons. Values × 100 = g m^{-2} yr^{-1} (approximately) and × 50 = g C m^{-2} yr^{-1} (approximately).

of adaptations to aqueous conditions found among submersed macrophytes, the severe reductions in rates of diffusion and spectrally selective attenuation of light by the water preclude intensive productivity. In certain respects, the submersed habitat is less severe than the terrestrial (e.g., in terms of water availability and thermal stability). These ameliorated parameters permit some growth of perennial populations of submersed macrophytes to continue all year (Rich *et al.*, 1971; Hough and Wetzel, 1975; Boylen and Sheldon, 1976; Stickey *et al.*, 1978). However, these advantages are not sufficient to offset the strictures of being totally submersed (Table 18-11).

It should be emphasized in comparative analyses of productivity of macrophyte communities, as done briefly here (Table 18-11) and by others (e.g., Keefe, 1972), that geographically diverse ecosystems are not fully distinguished. The importance of geographic and genetic differences within a species may be reflected in their productivity and other growth characteristics, which vary in a given environment from population to population as well as among populations from one environment to another. These multidimensional variations among ecotypes of the same species are well characterized in detailed studies of *Typha* populations in which phenotypic adaptations to numerous environmental parameters, especially temperature, photoperiod, light availability, and nutrients, are seen reflected in marked differences in productivity (McNaughton, 1966b, 1970; Grace and Wetzel, 1981a–c, 1982a,b).

F. Productivity and Competition among Aquatic Macrophytes within Habitats

Competition with plant communities of a particular habitat can occur by indirect, exploitative competition for resources, such as light or nutrients, or direct interference among competing individuals (Tilman, 1988). The latter interference usually occurs from allelochemical interactions and suppression. Rarely have both exploitative and interference competition been studied simultaneously among aquatic plants. The literature on the qualitative phytosociological composition of aquatic macrophyte communities is prodigious (e.g., Hejný, 1960; Hutchinson, 1975). Although these analyses of changes in dominance yield much information on

TABLE 18-11 Comparative Characteristics of Larger Aquatic Plant Communities among River, Reservoir, and Natural Lake Ecosystems[a]

Characteristic	*Rivers*	*Reservoirs*	*Natural lakes*
Littoral zone/wetland	Major land–water floodplain interface of rivers, particularly lowland rivers of low gradient	Irregular and limited by severe water level fluctuations	Major land–water interface component of most lakes, particularly the dominant small and shallow lakes
Macrophyte community structure	Dominated by wetland macrophytes of flood plains and temporary islands of braided river channels and floating and emergent plants of large rivers; submersed macrophytes generally limited to specialized forms and to backwater areas	Wetland macrophytes often limited to riverine portions of reservoir; submersed macrophytes often limited to shallow embayments; massive development of floating macrophytes common under eutrophic conditions	A three-community gradient from emergent to floating-leaved but rooted to submersed macrophytes is common in littoral areas; macro- and microalgae often occur in the littoriprofundal zone below the lower submersed plant zone
Nutrient acquisition	Largely by rooting tissues in wetland and littoral areas; some nutrient acquisition from water by in-stream macrophytes	Mostly from sediments by rooting tissues	Nearly all from sediment if rooted; in competition with epiphytic periphyton from the water
Light acquisition	Low and restrictive in low-order streams in forested regions and by turbidity in large-order rivers	Commonly restricted to shallow areas by high and irregular turbidity	Not restricted in wetland emergent and floating-leaved macrophytes; along depth gradient among submersed macrophytes; restrictive by epiphytic periphyton and by dense phytoplankton in eutrophic lakes
Macrophyte biomass and production	High in floodplain areas; low in low-order canopied streams, increasing in mid-order (3–5) streams, and low to very low within large rivers	Low to moderate, largely in riverine wetland areas; floating-leaved macrophytes and floating islands can be very productive	High to extremely high, particularly by emergent macrophytes and some floating-leaved communities; submersed plant productivity reduced by limitations of light and gaseous exchange rates

[a] Adopted from the syntheses on rivers by Wetzel and Ward (1992) and on reservoirs and lakes by Wetzel (1990a,b).

the floristic structure of the communities, the criteria used (e.g., percentage areal cover) do not yield much insight into the causal mechanisms underlying competitive interactions.

Clonal growth is common among higher aquatic plants, and such vegetative growth is relied upon heavily for maintaining large populations. Clonal growth and other types of vegetative dispersion and expansion allow rapid, relatively complete expansion into favorable habitats. Seasonal changes in species dominance (e.g., percentage areal cover) are commonly observed among submersed, floating-leaved, and emergent communities, but a clear tendency to form extensive monospecific populations is also commonly observed.

An example of competitive interactions has been observed between two emergent reeds, *Glyceria* and *Phragmites* (Buttery *et al.*, 1965a,b). *Glyceria* grows rapidly in the spring, and and under optimal growing conditions it can completely suppress *Phragmites* by producing dense foliage that effectively shades its slower-growing competitor. When sediments are water-saturated and anaerobic, however, *Glyceria* growth is reduced and *Phragmites*, with taller and more erect foliage that absorbs more incident light, can increase effectively and eventually exclude *Glyceria*. Experimental manipulation of nutrient supply had no effect on this competitive interaction.

Two species of cattail, *Typha latifolia* and *T. angustifolia*, commonly segregate along a gradient of increasing water depth, with *T. latifolia* restricted to depths of <80 cm and *T. angustifolia* to depths >15 cm (Grace and Wetzel, 1981b,c). In these two species, habitat partitioning resulted from differences in morphology; *T. latifolia* was prevented from growing in deep water because of the higher metabolic cost of producing broader leaves but was able to outcompete *T. angustifolia* for light in shallow water owing to its greater leaf surface area. This competitive separation can persist for long periods of time (decades) in the same habitat (Grace and Wetzel, 1998).

Among submersed macrophytes of calcareous hardwater lakes, the ratio of organic matter to calcium carbonate of the sediments has been implicated in the competition between the macroalga *Chara* and rooted angiosperms (Wohlschlag, 1950). Higher organic matter content of manipulated lake plots favored the production of rooted angiosperms. Other experimental and correlational *in situ* studies indicate that, in general, growth and productivity of both submersed and emergent macrophytes increase on organic-rich sediments (Szczepańska and Szczepański, 1976; Barko and Smart, 1979; Sand-Jensen and Søndergaard, 1979b; Aiken and Picard, 1980; Grace and Wetzel, 1981a, 1982a).

The seasonal growth patterns of aquatic macrophytes can be exemplified by the common emergents *Phragmites* and *Typha*, a submersed annual *Myriophyllum*, and submersed perennial macrophytes in a hardwater lake. Many other examples exist and variations from these typical temperate zone patterns can be found. Growth and production analyses of emergent macrophyte communities are commonly based on measurements of total plant biomass, biomass of component species (or plant parts), or leaf area index. The leaf area index (LAI) is simply the ratio of leaf area per unit ground area (Květ, 1971). The reedswamp plants exemplified in Figure 18-12 demonstrate the marked differences in growth rates found among perennial emergent plant species during a summer growth season. In *Phragmites*, maximum biomass (*W*) occurred in midsummer, whereas the maximum biomass of *Typha*

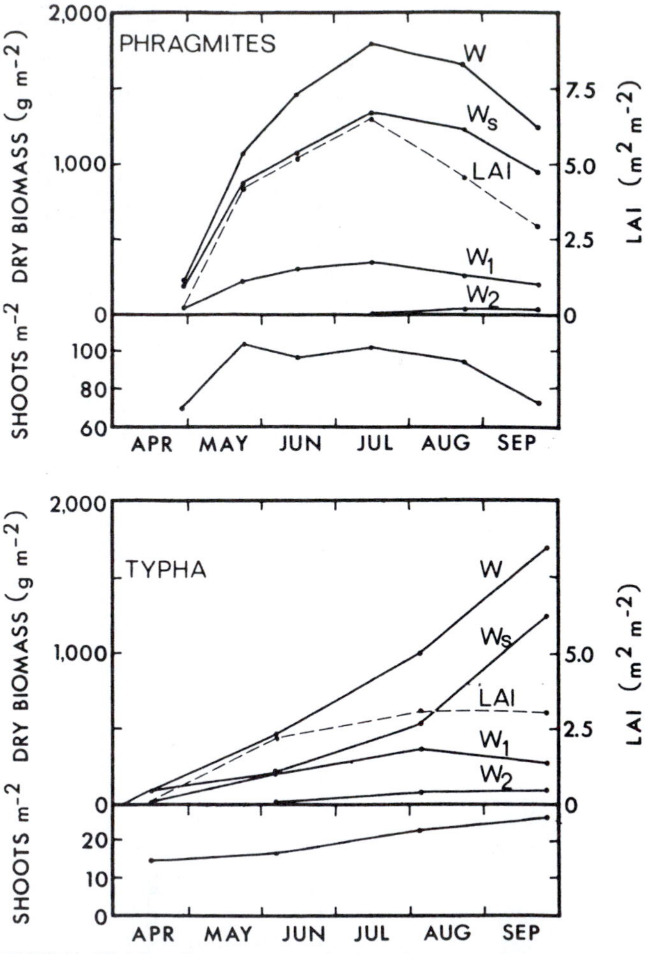

FIGURE 18-12 Changes in total plant biomass (dry, *W*); stem, sheaths, and stubble (W_s); leaf laminae (W_1); inflorescence (W_2); leaf area index (*LAI*); and the stand densities (shoot m^{-2}) of *Phragmites communis* and *Typha latifolia*, southern Czechoslovakia. (Modified from Květ, Svoboda, and Fiala, 1969.)

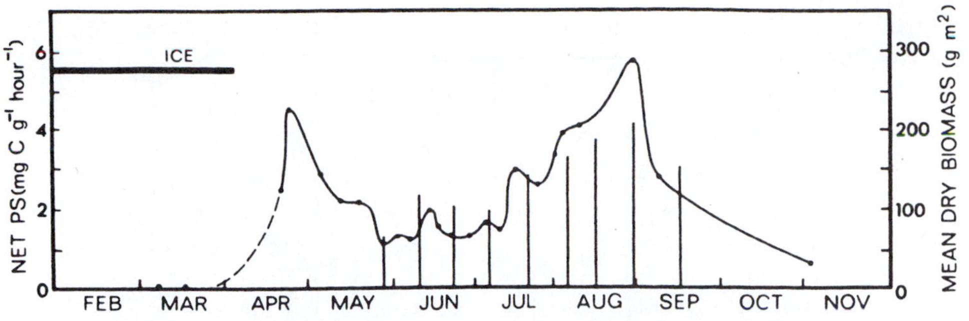

FIGURE 18-13 Rates of net photosynthesis (mg C gram dry weight^{-1} h^{-1}) of 15-cm tip sections of *Myriophyllum spicatum* (curved line) and mean ash-free organic weight (g m^{-2}) (vertical bars) in eutrophic Lake Wingra, Wisconsin, 1971. (From data of Adams and McCracken, 1974.)

increased over the entire growing season with little difference between the maximum values (1810 and 1620 g m^{-2}, respectively) for these two species. The average rate of increase in shoot biomass was 9.4 g m^{-2} day^{-1} in the cattail population and nearly 20 g m^{-2} day^{-1} in the *Phragmites* population. In both species, stem biomass (W_s) formed an increasingly greater proportion of the total (W), with a concomitant decrease in the proportion of leaf biomass (W_l). In *Phragmites*, however, both stem and leaf biomass decreased in the latter half of the growing season, whereas they continued to increase in *Typha.* It must be recalled, however, that these plant species experience constant turnover of cohorts of new shoots as well as senescence of portions of the earlier cohorts. As a result, the observed biomass maxima seasonally are generally marked underestimates of plant productivity. The leaf area index followed the changes in leaf biomass, and, coupled with changes in shoot density, it is clear that shoots of *Typha* that emerged later in the growing season had a better chance to survive in sparse stands than in the dense *Phragmites* stand. In the dense *Phragmites* stand, shorter shoots of *Typha* were handicapped by competition for light. Monodominant stands of emergent macrophytes (e.g., *Phragmites* and *Typha*) are common; invasion by other species can be resisted for long periods of time.

The seasonal succession of macrophyte communities on the flood plains of very large tropical river ecosystems is strongly coupled to the extent (water depth) and duration of seasonal flooding. A pronounced seasonal succession occurs with an alternation between aquatic and terrestrial conditions over large wetland areas of the flood plains (e.g., Prado *et al.*, 1994). On sites temporarily inundated for a period of several months, a succession occurred from purely terrestrial species to rapidly growing, short-lived annual aquatic species and perennial species that are able to tolerate periods of dryness by dormancy.

The littoral community of eutrophic Lake Wingra, Wisconsin, is dominated by the submersed annual macrophyte *Myriophyllum* (Adams and McCracken, 1974; Adams and Prentki, 1982). *In situ* measurements of photosynthetic rates demonstrated a rapidly formed spring maximum, followed by a marked summer reduction and subsequent maximum in late summer (Fig. 18-13). The spring maximum of photosynthetic rates did not coincide with the rates of biomass accumulation, a phenomenon similarly observed for the single submergent (*Ruppia maritima)* of a saline lake (Wetzel, 1964, and other studies). The biomass maxima occurred at the time of flowering in late summer (see review of *Myriophyllum* by Grace and Wetzel, 1978). A cumulative amount of organic matter approximately equal to 1.6 times the maximum biomass of *Myriophyllum* was lost by nighttime respiration and sloughing of leaves. Within the *Myriophyllum* community, most of the total biomass, leaf surface area, and photosynthetic activity occurred in the upper portion of the water column. Light reduction is definitely a major factor in the marked decrease in photosynthetic rates with depth observed by many investigators. Surface photoinhibition, so frequently found in phytoplankton photosynthesis, is seen only rarely among submersed macrophytes. Other factors, such as temperature and light reduction, greater age of tissue, and encrustations of epiphytic algae and carbonates on older leaf surfaces, contribute to lower rates of photosynthesis in deeper strata.

The importance of light to submersed macrophytes is further indicated by the depth distribution of biomass of the dominating perennial macrophyte *Scirpus subterminalis* (76% of total biomass) in oligotrophic Lawrence Lake, Michigan (Fig. 18-14). Two annual maxima are evident from two peaks at different depths. The larger fall maximum consisted of a major peak at 2 m in September, and an additional peak occurred at 4 m in midsummer when insolation was greatest. A fall

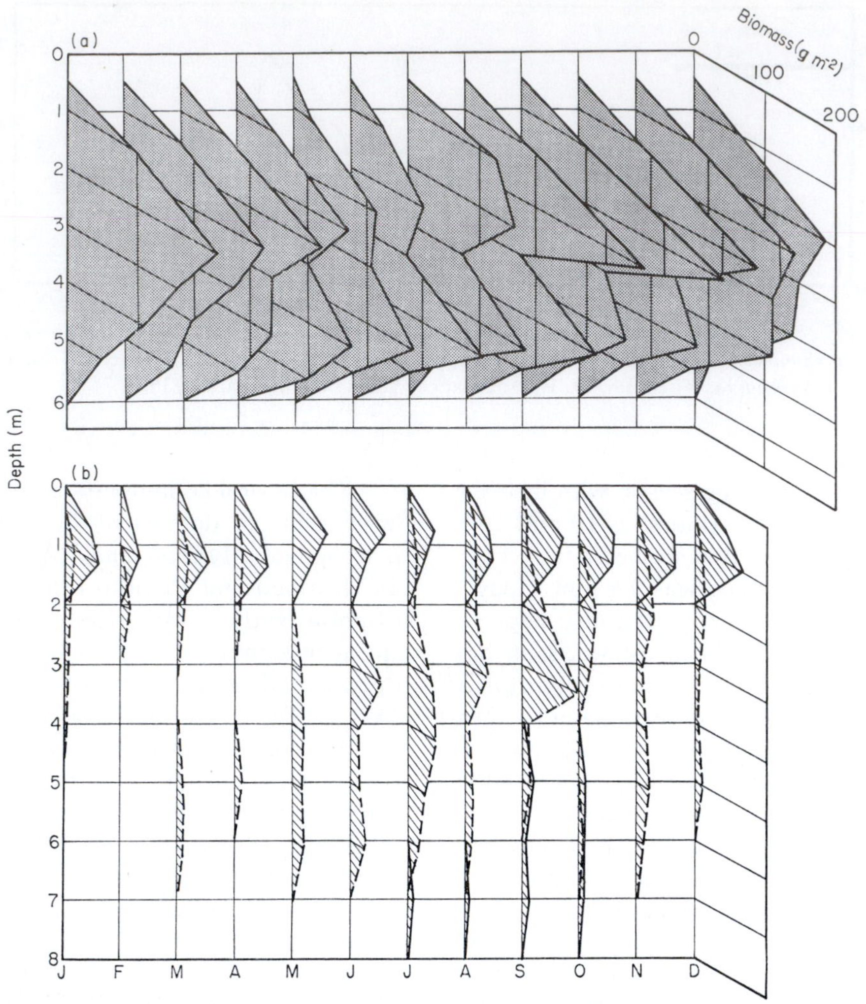

FIGURE 18-14 Annual biomass (g m^{-2} ash-free dry organic weight) depth distribution of (a) *Scirpus subterminalis*, and (b) Characeae (——) and annual macrophytes (- - - - -), Lawrence Lake, Michigan. (After Rich, P. H., Wetzel, R. G., and Thuy, N. V.: Distribution, production and role of aquatic macrophytes in a southern Michigan marl lake. *Freshwater Biol.* *1*:3–21, 1971.)

maximum of the macroalgae of the Characeae (16% of total biomass) was evident even though its seasonal biomass was relatively constant; the Characeae of this lake is similarly best described as a perennial population (Figs. 18-14 and 18-15). The submersed annual macrophytes (8% of total macrophytic biomass) of Lawrence Lake were dominated by two species of *Potamogeton*. These macrophytes exhibited two maxima and died back to perenniating rooting tissues in the autumn (Figs. 18-14 and 18-15).

The relatively constant biomass of perennial macrophytes throughout the year presented in this example is probably a much more common characteristic of lakes than is generally recognized. Turnover rates (ratio of production to biomass) of emergent and submergent *annual* plants are relatively low (1.02–2 times maximum biomass on an annual basis). Among perennial populations, turnover rates are much higher (range, 1.5–8 times the maximum biomass). Even among oligotrophic lakes, the productivity of perennial populations of submersed macrophytes can approach or exceed that of the commonly more productive annual macrophytes (Table 18-12). Comparisons of the relative contribution of the macrophytes to the primary productivity of attached algae and phytoplankton of lakes will be deferred until the conclusion of the discussion of littoral microalgae (Chap. 19).

The fate of organic matter produced by aquatic macrophytes as they undergo senescence, death, and decomposition is treated separately with a discussion of the metabolic processes in sediments (cf. Chaps. 21 and 23). From the perspective of the ecosystem, the

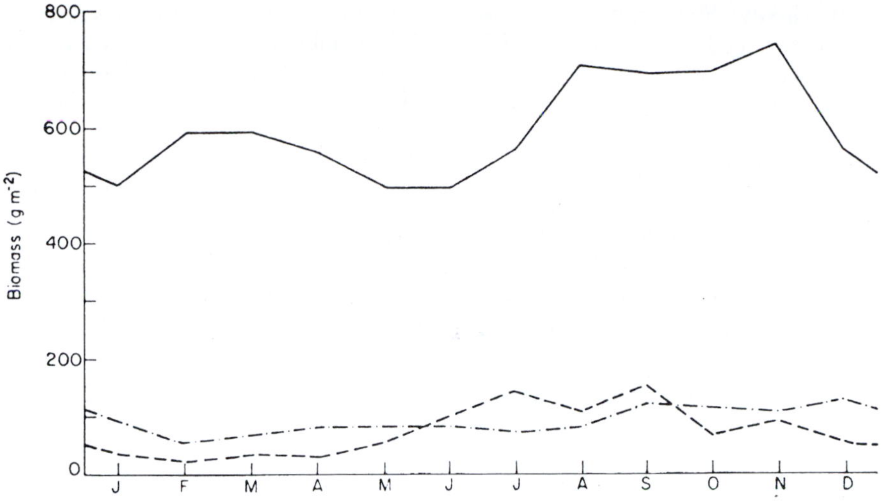

FIGURE 18-15 Annual biomass (g m^{-2} ash-free dry organic weight) of the important submersed macrophytic groups summed with respect to depth, Lawrence Lake, Michigan. —— = *Scirpus subterminalis* - · - · - · = Characeae, - - - - = annuals. (After Rich, P. H., Wetzel, R. G., and Thuy, N.V.: Distribution, production and role of aquatic macrophytes in a southern Michigan marl lake. *Freshwater Biol. 1*:3–21, 1971.)

productivity of the wetland and littoral interface regions between the land and the water usually exceeds the losses by decomposition. The factors that impede the decomposition of the accumulated organic matter, such as anoxia and labile organic nutrient availability (e.g., Wetzel, 1979; Ostendorp, 1992; Xiong and Nilsson, 1997), are simultaneously those that enhance the growth of macrophytes, particularly emergent macrophytes. Particulate organic matter tends to remain in and near sites of production in the land–water interface zones. What moves and is transported is colloidal and dissolved organic matter, from which more labile, readily decomposable organic compounds are degraded by attached microbial communities on the living and detrital particulate matter (Wetzel, 1992). The more recalcitrant dissolved organic compounds pass downgradient to the open waters of the stream or lake (cf. Chap. 23).

G. Interference Competition among Aquatic Plants

Allelochemical interactions and interspecific competition occur when an organism releases a chemical compound into the environment and this compound impacts, either positively or more commonly negatively, another organism (Rice, 1987; Einhellig, 1995). Extracts from live or dead tissues of a large number of aquatic plants have been found to inhibit growth or affect in some way the development of other species of aquatic macrophytes (Schröder, 1987; Gopal and Goel, 1993; Elakovich and Wooten, 1995; Inderjit, 1996). Of the many functional groups of allelochemicals, alkaloids, fatty acids, flavonoids, phenolic compounds, quinones, and terpenoids are most common. Many of the allelochemical compounds found in nature are secondary metabolites that are released during or after decomposition of structural plant tissues. Responses in

TABLE 18-12 Comparison of Annual Primary Productivity of Submersed Macrophytes in Several Lake Systems

Lake	*g C m^{-2} (of lake) yr^{-1}*	*Remarks*
Wingra, WI	117	*Myriophyllum*; eutrophic lake (Adams and McCracken, 1974)
Lawrence, MI	87.9	Perennials; oligotrophic lake (Rich, Wetzel, and Thuy, 1971)
Narock, Russia	59.4	Largely perennial charophytes; mesotrophic lake (Winberg *et al.*, 1972)
Marion, British Columbia	50.8	Annuals; mesotrophic lake (Davies, 1970)
Kalgaard, Denmark	21.0	Perennial isoetids; oligotrophic, softwater lake (Søndergaard and Sand-Jensen, 1978)
Mirror, NH	1.4	Perennials; oligotrophic, softwater lake (Moeller, 1975)
Borax, CA	1.2	Single annual plant *Ruppia*; saline lake (Wetzel, 1964)

affected species vary markedly. For example, certain phenolic compounds inhibit seed germination, while others complex proteins and enzymes and alter nutrient uptake rates (e.g., Blum, 1993; Wetzel, 1993; Lemke *et al.*, 1995; Inderjit, 1996). Allelochemicals can impact several processes of target organisms: cell division and elongation, cell membrane permeability, respiratory metabolism, organic synthesis, and complexation with hormones or enzymes. In some cases, allelochemicals released from aquatic plant tissues exhibit autotoxicity on the same plant at certain stages of its life history. For example, organic compounds released from the emergent rush *Juncus effusus* did not affect growth of several sympatric aquatic plant species or its own seed germination (Ervin and Wetzel, 1999). *Juncus* seedling development, however, was strongly suppressed by exudates in the nearby vicinity (<0.5 m) of the clonally reproducing mother plants.

H. Inhibition of Phytoplankton, Attached Algae, and Zooplankton by Submersed Macrophytes

A number of investigations have demonstrated a distinct inhibition of phytoplankton in the littoral zones containing heavy stands of emergent and submersed macrophytes (e.g., Schreiter, 1928; Hasler and Jones, 1949; Hogetsu *et al.*, 1960; Stangenberg, 1968; Goulder, 1969; Dokulil, 1973). Circumstantially, these investigations implied that organic compounds excreted by the macrophytes inhibited the growth of phytoplankton. However, competitive reduction of nutrients, light, and other substances by the macrophytes as factors inhibiting phytoplanktonic growth was not satisfactorily ruled out in these studies.

A detailed study of the influence of four species of submersed macrophytes on phytoplankton within the stands showed that decreases in rates of phytoplankton photosynthesis were caused by shading (Brandl *et al.*, 1970). Average reductions over three years in the rates of phytoplankton production among the plant stands as compared to rates in the open water were *Potamogeton pectinatus*, 60%; *P. lucens*, 46%; *Batrachium aquatile*, 40%; and *Elodea canadensis*, 12%. Phytoplankton production was markedly reduced when the phytoplankton were taken from the pelagic zone and incubated among the macrophytes and increased greatly when littoral phytoplanktonic algae were incubated at the same depth in the open water (Table 18-13).

The presence of membrane-filterable (1.2-μm pore size) organic substances capable of inhibiting the photosynthesis of phytoplankton could not be confirmed. However, increasing the pH of the water, as occurs with a reduction in available free CO_2 in dense stands of actively photosynthesizing submersed macrophytes (Fig. 18-16), reduced the rates of phytoplankton productivity. This decrease could be reversed completely by small additions of CO_2. Brammer (1979) similarly found that the submersed macrophyte *Stratiotes* inhibited the growth of contiguous populations of phytoplankton. Competition for nutrients and reductions in the concentrations of ions in the littoral water by the macrophytes were likely responsible for the decline of the phytoplankton.

Dense stands of macrophytes can markedly alter other immediate environmental conditions in comparison to those of the open water. Under floating macrophytes (e.g., duckweed and hyacinths), water temperature increases, pH values decline with increased CO_2 levels, and oxygen concentrations decrease (Ultsch, 1973; Dale and Gillespie, 1976; Rai and Munshi, 1979; Schreiner, 1980). In reed *(Phragmites)* stands, conductance and calcium and bicarbonate concentra-

TABLE 18-13 Changes in Rates of Primary Production of Phytoplankton from the Pelagic Zone Incubated among Submerged Macrophyte Stands of the Littoral and Littoral Phytoplankton Incubated in the Pelagic Zone, Czechoslovakian Shallow Lakes[a]

Submersed macrophyte species dominating littoral area	*Percentage reduction in littoral zone*	*Percentage increase in pelagic zone*
Potamogeton pectinatus	56	250
Potamogeton lucens	46	270
Batrachium aquatile	39	230
Elodea canadensis	27	500

[a] From data of Brandl, Brandlová, and Poštolková (1970)

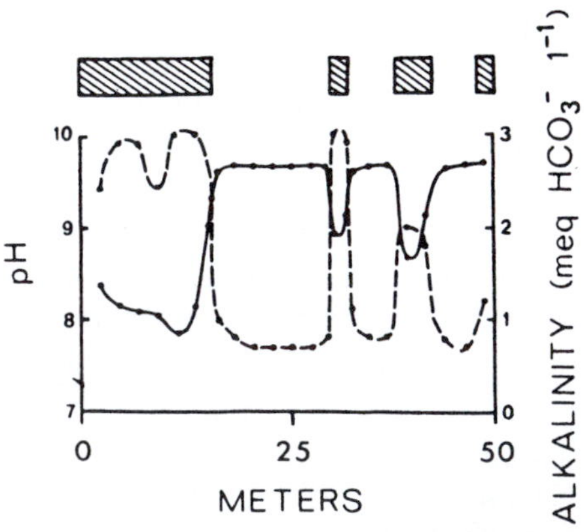

FIGURE 18-16 Changes in the pH (- - -) and alkalinity (——) in the surface water along a transect across Sangwin Pond, in the afternoon (June) through intermittent areas of dense cover of the submersed macrophyte *Ceratophyllum demersum* (indicated by hatching). (Modified from Goulder, 1969.)

tions can be higher than in the open water (Planter, 1970). In all cases, light is severely and selectively attenuated by the macrovegetation (Gessner, 1955; Dykyjová and Květ, 1978) and can result in the nearly complete removal of light and suppression of attached algal photosynthesis at the base of dense emergent macrophyte stands (Grimshaw *et al.*, 1997).

Extracts from several species of the submersed plant *Myriophyllum* were found to strongly inhibit cyanobacteria and to a lesser extent algae. Detailed chemical analyses indicated that the substances released were a number of phenolic compounds, including a potent ester, galloyl glycoside (Saito *et al.*, 1989; Gross and Sütfeld, 1994; Gross *et al.*, 1996). This polyphenolic compound, found to occur in concentrations as high as 1.5% dry weight of *Myriophyllum*, complexes with and inactivates extracellular enzymes. A large number of humic acids and other humic substances released from decaying higher aquatic plants were demonstrated similarly to noncompetitively inhibit extracellular and membrane-bound enzymes (Wetzel, 1991, 1993; Kim and Wetzel, 1993; Boavida and Wetzel, 1998). Other phenolic compounds with antialgal activity are also released from the freely floating water lettuce *Pistia* and seedlings of the water lily *Nuphar lutea* (Sütfeld, 1993; Dellagreca *et al.*, 1998). Several sulfur-releasing and related compounds (methylthio dithiolane and trithiane) occur in and are likely released by several species of the macroalgae *Chara* and the submersed angiosperm *Ceratophyllum* (Wium-Andersen, 1987). These compounds distinctly suppress microbial and insect activities.

Although much of the apparent inhibition of phytoplankton growth among dense stands of macrophytes is likely related to competition for light and nutrients, the production of biotically inhibitory or stimulatory organic substances cannot be totally excluded. For example, the presence of macrophytes favors the dominance of some cyanobacteria (Guseva and Goncharova, 1965) but is inhibitory to algal groups (Kogan and Chinnova, 1972). Macrophytes may influence the competition of cyanobacteria with algal groups by cyanobacterial heterotrophic utilization of certain excreted organic compounds. Zooplankton of many groups are inhibited strongly or repelled by substances secreted into the water by submersed angiosperms and macroalgae (much of the scattered literature is reviewed by Pennak, 1973, and Dorgelo and Koning, 1980). For example, *Daphnia* is strongly repelled by water containing *Elodea*, *Myriophyllum*, or *Nitella*. The magnitude of the negative responses varied with the plant species and was more pronounced at higher temperatures. Sexually mature *Daphnia* responded more rapidly than immature instars. The inhibitory or insecticidal effects of organic compounds secreted by macrophytes, especially the macroalgae Characeae, on mosquito larvae have been reported widely (and critically reviewed by Hutchinson, 1975). The active compounds are probably of the general allyl sulfide group, analogous to those derived enzymatically from garlic oil.

Despite the abundant evidence for intensive competition for light and nutrients among the macrophytes, attached microflora, and phytoplankton and the potential for release of organic suppressants from the larger aquatic plants, phytoplankton production can still be appreciable in nutrient-rich littoral and wetland areas (e.g., Robarts *et al.*, 1995). Clearly, phytoplankton production of shallow water bodies can constitute a significant part of the collective primary production.

I. Retention and Release of Nutrients

The surface area of wetland and littoral plants and associated detrital particulate organic matter increases greatly in the transition from emergent to submersed macrophytes. As a result, colonization of surfaces by epiphytic algae and bacteria increases along the littoral gradient from emergent to submersed macrophytes (Chap. 19). Both the macrophytes and epiphytic microflora utilize and release inorganic nutrients and dissolved organic compounds (Wetzel and Allen, 1970; Allanson, 1973; Mickle and Wetzel, 1978a,b, 1979; Wetzel, 1990, 1996). The photosynthetic productivity of epiphytic algae can be very high, often exceeding that of the submersed macrophytes (Chap. 19).

As water of the drainage basin containing inorganic nutrients and organic compounds passes through the complex vegetation and their epiphytes, marked alterations can occur in the chemical composition of the influent water. Submersed vegetation–epiphytic complexes can be extremely effective in removing such inorganic nutrients as Ca, K, combined N, and P of inflowing water, either by direct assimilation or via adsorption onto inorganic precipitates (such as $CaCO_3$) induced by photosynthetic activity (Mickle and Wetzel, 1978a). Labile organic compounds of low molecular weight (<1000 D) can be effectively removed by the epiphytic microflora. The fate of high-molecular-weight compounds (>1000–30,000 D), which dominate from terrestrial and wetland sources, is variable; they may be selectively removed or, alternately, pass through relatively unaffected, and their concentration may increase owing to release from the macrophytes themselves (Mickle and Wetzel, 1978b, 1979; Murray and Hodson, 1986; Wetzel *et al.*, 1995; Wetzel, 2000d).

During senescence of macrophytes, particularly emergent macrophytes, marked translocation of nutrients occurs from the dying tissues. Nutrient enrichment of plants tends to accelerate translocation to rooting tissues and retention within the plants (e.g., Davis, 1991). Accompanying senescence and after the death of the macrophytes, large releases of inorganic nutrients and dissolved organic matter can occur (Chap. 21). The release rates are governed by the population growth dynamics of the macrophytes, their mortality during and following the growth period, and the conditions prevailing at the time of death and decomposition. Release rates are also influenced by the type of vegetation; decomposition and release of nutrients and organic compounds are much faster from submersed and floating-leaved macrophytes than from emergent macrophytes (Chap. 21). Some of the release is from immediate autolysis of cells during senescence in addition to simple physical leaching. For example, dissolved organic matter releases are high during early stages of decomposition, and the loss rates of elements are commonly K > Na > P > Mg > C > N > Ca > Fe among submersed and floating-leaved macrophytes (Brock, 1984; Brock *et al.*, 1985; Ochiai and Nakajima, 1994). Depending upon when these nutrients and organic compounds enter the influent water, and how their concentrations are selectively modified and reduced by the metabolic sieve of the submersed macrophyte–epiphyte complex, the inputs can have marked influence on phytoplanktonic growth in the recipient stream or lake (Chap. 15) as well as the bacterioplankton (e.g., Murray and Hodson, 1986). For example, large, decaying stands of the submersed macrophyte *Myriophyllum* in a reservoir in Indiana contributed about 2% of the annual nitrogen and perhaps 18% of the potential phosphorus loading in comparison to allochthonous loadings (Landers, 1982). It should be recalled from earlier discussions that phytoplanktonic responses may be stimulated or inhibited by the inorganic–organic loadings, since the stimulatory effects of some inorganic nutrients can be counteracted by the inhibitory effects of certain organic compounds of macrophyte origin.

The release of dissolved organic matter and nutrients tends to be greatest per unit of macrophyte biomass immediately at senescence and during the early stages (initial few hours) of decomposition (e.g., Helbing *et al.*, 1986; Howard-Williams *et al.*, 1988; Moran and Hodson, 1989; Menéndez and Forés, 1998). The efficiency of bacterial utilization of these organic and inorganic compounds by bacterioplankton is high (30 to >50%) (e.g., Findlay *et al.*, 1986b; Kairesalo *et al.*, 1992; Mann and Wetzel, 1996), but under natural conditions most of such release is utilized directly by the epiphytic microbial community attached to the surfaces of the plants and its detritus (cf. Chap. 19). The dissolved organic matter released from living macrophyte leaves was used much more efficiently than that released from senescent leaves (Mann and Wetzel, 1996), but the humic substances are nonetheless utilized with moderate efficiency (Kairesalo *et al.*, 1992; Bicudo *et al.*, 1998).

VI. SUMMARY

1. The size of the littoral zone of lakes varies greatly in relation to the size of the open-water pelagic region. Similarly, the floodplain–littoral regions of streams and rivers are very dynamic and constantly changing in relation to hydrograph variations. Since most lakes of the world are small in area and shallow, and the floodplain-saturated bank littoral areas of streams are very large in relation to water of the channels, the wetland and littoral flora and its epiphytes are commonly a dominant source of synthesized organic matter to recipient waters.
2. Four groups of aquatic macrophytes can be distinguished on the basis of morphology and physiology; each of the groups dominates a major region of the littoral zone (Fig. 8-2):
   a. *Emergent* macrophytes grow on water-saturated or submersed soils from where the water table is about 0.5 m below the soil surface (supralittoral) to where the sediment is covered with approximately 1.5 m of water (upper littoral).
   b. *Floating-leaved* macrophytes are rooted in submersed sediments in the middle littoral zone (water depths of approximately 0.5–3 m) and possess either floating or slightly aerial leaves.
   c. *Submersed* macrophytes occur at all depths within the photic zone. Vascular angiosperms occur only to about 10 m (1 atm hydrostatic pressure) within the lower littoral (infralittoral), and nonvascular macrophytes (e.g., macroalgae) occur to the lower limit of the photic zone (deepest boundary of the littoriprofundal).
   d. *Freely floating* macrophytes are not rooted to the substratum; they float freely on or in the water and are usually restricted to nonturbulent, protected areas.
3. Emergent macrophytes produce erect, linear leaves from an extensive rhizome/root base and are physiologically similar to terrestrial plants.
   a. Rooting tissues grow in a water-saturated, anaerobic substratum and must obtain oxygen for respiration from aerial organs. The plants develop extensive intercellular gas lacunae that permit internal gas movement and exchange.

b. Rates of transpiration by emergent and floating macrophytes are often extremely high and result in water losses to the atmosphere that are greater than evaporation from an equivalent area of water. Vegetative evapotranspiratory losses of water can be sufficiently large to appreciably reduce water levels of the fresh waters and surrounding terrestrial areas.

4. Submersed angiosperms possess numerous morphological and physiological adaptations to their aqueous environments.
   a. Leaves are only a few cells in thickness, and photosynthetic pigments are concentrated in epidermal tissue. Leaves tend to be much more divided, with greater surface-to-volume ratios, than are leaves of other aquatic or terrestrial plants. Leaf morphology maximizes exposure of cellular surfaces to reduced light, gas, and nutrient availability under water.
   b. Many submersed angiosperms exhibit extreme vegetative polymorphism *(heterophylly)* in leaf structure, often on the same stem or petiole. Morphology commonly shifts from finely divided submersed leaves to more coalesced, entire shapes in surface-floating or aerial leaves. Heterophyllous growth is regulated by interactions of temperature, CO_2, and hormonal action mediated by ethylene concentrations and light.
5. Vegetative and clonal reproduction is a major mechanism for population growth and dispersal. Although sexual reproduction and genetic recombination has been retained, it is commonly subordinate to vegetative reproduction.
   a. Vegetative, clonal growth occurs by (i) fragmentation and dispersal of aboveground tissues, (ii) winter buds (turions), and (iii) horizontal expansion by rhizomes, tubers, corms, and bulbs.
   b. Many of the most effectively competitive aquatic plants combine continuous clonal growth with production of seeds for long-distance dispersal.
   c. Aerial flowering dominates among even submersed aquatic plants; pollination occurs by insect or wind vectors.
6. Movement of gases within and through plant parts are essential to alleviate toxicity effects of fermentation products (e.g., volatile fatty acids and hydrogen sulfide) in anaerobic sediments and to maintain active respiratory and photosynthetic metabolism.
   a. Oxygen, carbon dioxide, and other gases move within and through interconnecting aerenchymatous and intercellular cortical gas spaces of shoots and roots.
   b. Gas flux mechanisms among aquatic plants with aerial leaves include:
      i. Diffusion-dependent throughflow convections based on pressurization differentials
      ii. Nonthroughflow convections based on differential gas concentrations from metabolic uptake or release of gases
      iii. Venturi-induced throughflow convections from wind pressure differentials
      iv. Very large amounts of carbon dioxide and methane pass from sediments through rooting tissues into the intercellular gas system and to the atmosphere
   c. Within intercellular lacunae of submersed plants, oxygen from photosynthesis diffuses to sites of respiration, particularly to rooting tissues. Some oxygen diffuses into the microrhizosphere and reacts chemically and is utilized by microaerophyllic bacteria. Carbon dioxide from respiration diffuses to photosynthetic tissues and can augment CO_2 and bicarbonate sources from the water for photosynthesis.
7. Carbon assimilation is more difficult in submersed macrophytes than in emergent and floating-leaved plants that have access to atmospheric CO_2, primarily because rates of diffusion of gases are several orders of magnitude slower in water than in air. Nearly all submersed species possess C_3-type photosynthetic metabolism with relatively high rates of photorespiration. Because of chronic potential for inorganic carbon limitations to photosynthesis and growth, numerous mechanisms have evolved for conservation and recycling of CO_2:
   a. The cuticle of submersed angiosperms is reduced or absent, photosynthetic tissue is thin (1–3 cells thick), and chloroplasts are densely distributed in outer epidermal cells—all enhance exchange of gases and nutrients that are dissolved in the water.
   b. Extensive intercellular gas lacunae (often >70% of plant volume) in leaves, stems, petioles, and roots facilitate rapid diffusion of internal gases. Microzones at the cellular surface of leaves rapidly develop in quiescent littoral waters; these microzones set up gradients of reduced gas availability.
   c. Certain species of submersed macrophytes assimilate and subsequently decarboxylate bicarbonate ions. This assimilation represents a physiological adaptation to limited availability of free CO_2.
   d. Most submersed angiosperms possess C_3–Calvin-type photosynthetic metabolism with high rates of photorespiration. CO_2 from

mitochondrial respiration of foliage and rooting tissue, photorespiration, and, in a few softwater plants, CO_2 from sediments, augment CO_2 and/or HCO_3^- assimilated from the water. Some CO_2 from respiration is efficiently recycled in the internal gas lacunae of leaves.

e. Certain submersed macrophytes assimilate CO_2 in periods of darkness and relative abundance, store it as malate, and utilize the carbon in the subsequent daylight period of photosynthesis.

f. Many submersed macrophytes have low CO_2 compensation points and relatively high photosynthetic efficiency.

8. Nutrients are assimilated from the sediments by emergent and rooted floating-leaved macrophytes and translocated to their shoots and from the water in freely floating macrophytes.
   a. Most submersed rooted macrophytes obtain most of their nutrients from the interstitial water of sediments, where concentrations are much greater than in the overlying water. When nutrient concentrations of the water are high, a greater proportion is assimilated from the water.
   b. Some nutrients are released to the water by submersed macrophytes during active growth; large amounts can be released upon senescence and death, although most are immediately sequestered by attached microbial communities.
   c. Translocation of nutrients occurs in both directions with dominant acropetal translocation from the roots to the leaves.
9. Light availability is a major factor regulating the growth and competitive interactions of aquatic macrophytes.
   a. Light saturation occurs at relatively modest irradiances (ca. 750 μmol quanta m^{-2} s^{-1}) with efficiencies of light utilization at about 0.5% at full sunlight without apparent photoinhibition among emergent macrophytes. Photosynthetic efficiencies decrease markedly at higher temperatures as rates of respiration and photorespiration increase.
   b. Submersed macrophytes are distinctly shade-adapted with high concentrations of pigments and low-light compensation points of photosynthesis, commonly at 1–3% of full sunlight. Considerable photosynthetic and respiratory adaptation to low light and differing temperatures is found among submersed species.
   c. Among submersed macrophytes, light attenuation rapidly limits the vertical distribution of plant growth. Light availability and interspecific shading can be major factors influencing the competitive success of one species over another.
   d. Hydrostatic pressure likely interacts with low light intensities to limit the distribution of freshwater angiosperms to depths of $<$10 m.
10. The productivity of aquatic macrophytes has been evaluated from changes in biomass over time and, in a few cases, by *in situ* determinations of rates of photosynthesis.
    a. Most aquatic macrophytes are perennial, with varying degrees of biomass turnover during growth. As a result of continuous partial mortality and populations with high turnover of leaves, much of the annual production does not survive to be measured as seasonal maximum biomass. Multiple cohorts or continuous replacement growth and senescence prevail among most emergent, floating-leaved, and submersed macrophytes. In many cases, cohort longevity is in the range of 30–40 days.
    b. Roots and rhizomes of submersed macrophytes make up a lesser portion (1–40%) of total plant biomass than is the case for floating-leaved (30–70%) or emergent (30–95%) macrophytes.
    c. Emergent macrophytes contain a much higher content of structural tissue than do submersed or floating-leaved macrophytes. This structural tissue is of a chemical composition that is relatively recalcitrant to rapid microbial decomposition.
    d. Growth of aquatic macrophytes is generally higher on organic-rich sediments than it is on sandy sediments.
    e. On a unit-area basis, the net primary productivity of aquatic macrophytes is among the highest of any community in the biosphere (Table 18-10). Emergent macrophyte productivity is the highest (1000–10,000 g AFDW m^{-2} yr^{-1}); freely floating macrophytes is intermediate (300–5000 g m^{-2} yr^{-1}), floating-leaved rooted macrophyte productivity is modest (100–500 g m^{-2} yr^{-1}), and submersed macrophytic productivity is considerably less (5–1500 g m^{-2} yr^{-1}). Nearly all of the macrophyte productivities exceed those of phytoplankton (25–200 g AFDW or 50–450 g C m^{-2} yr^{-1}).
    f. Seasonal variations in aquatic macrophyte productivity are great. In the temperate zone, productivity of aquatic macrophytes commonly follows increasing insolation and temperature; some submergents persist under ice cover, although net productivity is low or zero. In the tropics, growth can be nearly continuous.
11. In addition to natural, genetically programmed senescence and death, cued by environmental

parameters, mortality can occur by disease, herbivory, and other factors.

a. Pathogenic decimation of entire macrophyte populations is rare; disease usually increases gaps and community heterogeneity.
b. Herbivory of aboveground vascular plant tissues by aquatic insects and snails is low in comparison to potential losses to crayfish crustaceans, herbivorous fish specialists, mammals, and aquatic birds under certain conditions. The percentage loss by animal ingestion or maceration is in the range of 25% for the *aboveground instantaneous biomass* of freshwater vascular macrophytes and somewhat higher among nonvascular macrophytes. However, the percentage loss of *total* above- and belowground *production* of aquatic plants by animal herbivory is usually unknown and certainly is lower (5–10%).
c. Many classes of plant compounds function as plant deterrents for herbivory and are known to be operational in many aquatic plants.
d. Losses of plant biomass by damage from water movement are often severe, particularly periodically in rivers and streams.

12. Interspecific competition among macrophyte species is poorly understood. Compounds released from living or dead tissues of many aquatic plants can inhibit growth and development of other aquatic macrophytes or attached microbiota.
13. Phytoplankton productivity is generally lower in littoral zones containing stands of aquatic macrophytes. Experimental evidence indicates that this phenomenon results largely from a competition for nutrients (including carbon) by submersed macrophytes and by a reduction of light by macrophyte foliage.
14. Wetland and littoral regions of freshwater ecosystems are commonly intensely metabolically active owing to the presence of aquatic macrophytes.
    a. The macrophytes are frequently the primary source of organic matter to both lakes and rivers.
    b. The synthesis of organic matter by emergent vegetation and its decomposition in and on wetland sediments result in high loading of dissolved organic matter and inorganic nutrients to recipient lakes. The timing of loading is strongly influenced by the stages of wetland plant growth, senescence, and decomposition.
    c. As water laden with nutrients and dissolved organic compounds passes to the lake, the chemical loading is greatly modified by the metabolic activity of littoral submersed macrophytes and their epiphytic microbiota. The littoral macrophyte–epiphyte complex can function as a selective metabolic sieve.

19

LAND–WATER INTERFACES: ATTACHED MICROORGANISMS, LITTORAL ALGAE, AND ZOOPLANKTON

Microbiota grow upon and attach to any surface immersed or growing in water. The succession of microorganisms upon a surface is complex and involves adsorption of organic substances to the surfaces and attachment of microorganisms in sequence. Usually bacteria attach first followed by algae, cyanobacteria, and protists. Extracellular mucilaginous materials, also referred to as exopolysaccharides or exopolymer secretions (EPS), often occur as coatings around individual microbial cells or projections from cells. Ultimately this material forms a matrix inhabited by a variety of microorganisms, particularly bacteria, algae, protists, and fungi (e.g., Fletcher and Marshall, 1982).

Excretion of exopolymer fibrils by bacteria is an important initial phase of attachment of microbes to surfaces (Fletcher and Floodgate, 1973; van Loosdrecht *et al.,* 1990; Brading *et al.,* 1995). A substantial portion of any attached aquatic microbial community will be composed of nonliving, mucilaginous materials. Most of this material is believed to be mucopolysaccharide, although the exact composition and texture of it varies with environmental conditions as well as the nature and condition of the organisms that secrete it (e.g., Sutherland, 1985; Hoagland *et al.,* 1993).

Much excellent study has been directed to biofilms attached to surfaces of technical devices (e.g., pipes,

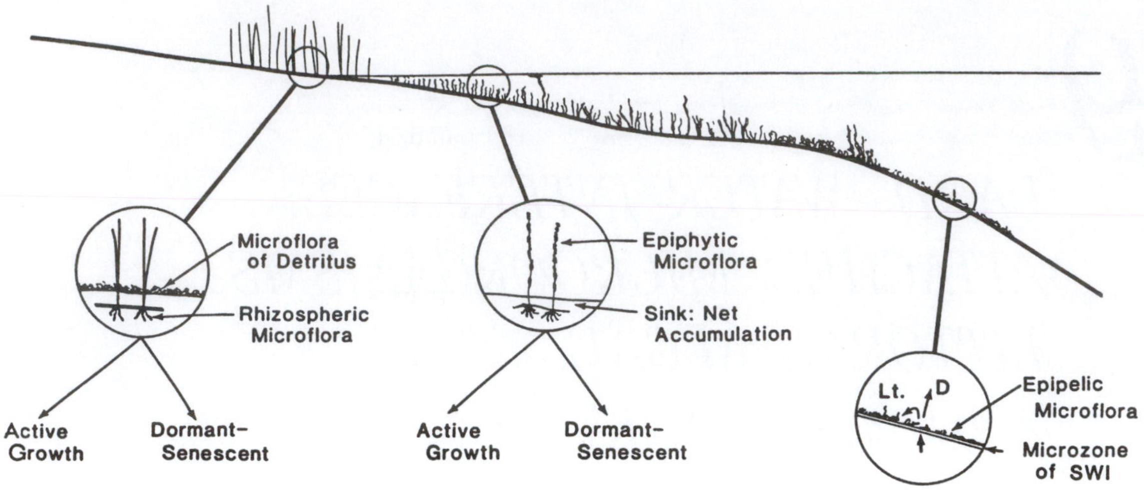

FIGURE 19-1 Primary habitat and attached algal community components along a wetland–littoral gradient of standing waters and backwater areas of river ecosystems. *SWI* = sediment–water interface; *D* = dark; *Lt* = light. (Modified slightly from Wetzel, 1990a.)

ship hulls, and teeth) that convey or are immersed in water (e.g., Characklis and Wilderer, 1989; Characklis and Marshall, 1990). These studies correctly focus largely on heterotrophic attached microorganisms, especially bacteria, because growth occurs on most of these anthropogenic surfaces in darkness. In natural ecosystems, however, algal photosynthesis is irrevocably coupled to heterotrophic attached microbial metabolism. As a result, mixed autotrophic–heterotrophic assemblages form much more complex and metabolically dynamic communities than those that are only heterotrophic.

I. ATTACHED MICROBES AND ALGAE OF LITTORAL REGIONS

A. General Habitat Characteristics

Benthic algae and other autotrophic microbes, such as the cyanobacteria, are generally associated with or attached to substrata. These substrata are living or dead, organic or inorganic, and in various combinations. Because photosynthesis is the only significant process for synthesis of organic matter among these organisms, light is mandatory within both the macrohabitats and the microhabitats of the communities in sufficient quantities to allow net synthesis of ATP. The land–water interface zone is a gradient that extends from detrital masses in hydrosoils and pools of wetlands through littoral zone regions of emergent, floating-leaved, and submersed macrophyte communities to sediments extending into deeper waters well beyond the vascular macrophytes (littoriprofundal). Available light shifts from abundant in wetland and shallow nontree-canopied littoral areas to low and highly variable seasonally in deeper areas as the densities of seston change in the overlying waters. Within this zone of adequate light, substrata diversity is highly variable but can be organized into general zones in relation to potential substratum nutrient and other environmental characteristics (Fig. 19-1).

1. Emergent Plant Zones

These zones consist of saturated hydrosoils and pools of upland wetlands and flood plains among emergent macrophytes that are made up of largely of standing and collapsed dead macrophytes and dense aggregations of particulate detritus among the understory. Although many critical plant nutrients are retained within living plant tissues and translocated to rooting tissues at the end of growth periods of individual plant cohorts, soluble organic matter and organic nutrients, particularly organic phosphorus, are released from the highly productive emergent macrophytes during and following active growth, senescence, and decomposition (Chaps. 18 and 21). Many of these herbaceous perennials have more or less continuous leaf or culm turnover of many cohorts per year in both northern and southern regions. As a consequence, relatively constant partial senescence of leaves and decomposition occur among macrophytes with release of cellular organic matter and nutrients.

The high production of the emergent macrophytes in excess of decomposition results in high detrital organic matter accumulation at the sediment–water interface. This loading, which contains relatively high proportions of moderately recalcitrant structural tissues from emergent plants, shrubs, and trees contributes to high substrata surface area for microbial colonization in relatively shallow water of high light and temperatures. The high loading of organic matter results in predominantly anaerobic conditions with increased solubility of critical nutrients, particularly P and N, in the interstitial waters near and at the sediment–water interface (cf. Chaps. 12 and 13). Much of the combined nitrogen as nitrate is commonly denitrified by facultative anaerobic bacteria in such detrital–sediment masses and released to the atmosphere as N_2 (Triska and Oremland, 1981; Bowden, 1987).

2. Submersed Plant Zones

As we have summarized in the Chap. 18, submersed macrophytes of lakes, ponds, rivers, and floodplain pools generally possess morphology of thin, finely divided, and reticulated leaves. The resulting increased surface area of leaves markedly enhances interceptions of light and the exchange of gases with those of the water. This greatly increased ratio of surface area to volume results in enormously increased substrata for colonization by epiphytic algae, cyanobacteria, and other microbes and protists. For example, the leaf surface area available for colonization by epiphytic algae on the submersed linear-leaved macrophyte *Scirpus subterminalis* averaged 24 m^2 of leaf area projecting upward three-dimensionally above every square meter of bottom in the moderately developed littoral zone of a lake in southwestern Michigan (Burkholder and Wetzel, 1989). The extensive area of these macrophytic surfaces projects myriad diverse microhabitats with attendant attached algal communities upward into littoral environments. In this spatial habitat in the water column, relative abundance occurs of light, dissolved gases from photosynthesis (O_2) and decomposition and respiration (CO_2), and nutrients diffusing from the high decomposition in interstitial waters of detritus and sediments and imported from upland allochthonous sources. Related to the very large diversity of microhabitats in this dynamic and highly productive region, most (>80%) of freshwater species of algae and cyanobacteria are attached, sessile forms (Round, 1981; Wetzel, 1999b).

3. Sediment Substrata

The benthic habitat in both standing and running fresh waters can be a physically hostile yet nutritionally advantageous environment. Epilithic and epipsammic microbes, attached to rocks and sand grains, respectively, are common in streams and wave-turbulent areas of lakes. Active water movements offer advantages for nutrient and gas exchanges but also expose the communities to molar action by particles such as sand and potential burial by sediments with reduction or removal of light (Meadows and Anderson, 1966; Moss and Round, 1967). Most lakes and ponds are small with modest fetch and wind-induced water movements. Water turbulence also decreases rapidly with increasing depth (cf. Chap. 7; also Imberger and Patterson, 1990), and submersed macrophyte communities greatly attenuate flows in both lakes and streams (cf. Chap. 18). As a result, even loosely aggregated organic-rich sediments can serve as substrata for algae and cyanobacteria attached to the surface of these sediments (epipelic) when they receive adequate light. The unstable nature of such sediments and frequently high rates of deposition from settling seston increase the risk of epipelic algal burial and light reduction. The high nutrient availability from interstitial waters of sediments, however, is a distinct advantage. Many epipelic algal species migrate, often in rhythmic fashion, to compensate for shifting light attenuation by sediment (cf. Round, 1981).

Algae growing adnate to unconsolidated, shifting sediments are constantly in danger of being removed from light, which is attenuated to zero in a few millimeters.[1] Whether epipelic algae supplement photosynthetic growth by heterotrophic utilization of organic substrates is unclear. Circumstantial evidence of the high concentrations of organic substrates in the interstitial waters of organic-rich sediments and many epipelic algae growing under very low light conditions is pervasive that facultative heterotrophy may occur among some benthic algae (e.g., Gaines and Elbrächter, 1987; Tuchman, 1996). In spite of this physiological potential, an important adaptation to the rigors of this habitat is an ability to move vertically within the sediments in response to light availability.

Persistent diurnal vertical migration rhythms have been shown in epipelic diatoms, flagellates, and cyanobacteria from flowing waters and shallow lakes (Round and Happey, 1965; Round and Eaton, 1966). Cell numbers on the sediment surface start to increase before dawn and reach a maximum about midmorning. Thereafter, surface cell numbers decrease and reach a minimum before the onset of darkness. This rhythm persists for a limited time in continuous darkness or continuous light, although the synchrony is lost more rapidly under continuous light conditions.

[1] The attenuation of light by sediments is highly dependent upon the type of sediment. Highly organic sediments attenuate light much more rapidly than does sand. For example, Palmer and Round (1965) found that radiation was reduced to about 1% of the incident value under about 3 mm of organic-rich sediment. Similar results were found by Wasmund and Kowalczewski (1982).

Studies of the diurnal rhythms of freshwater epipelic algae have shown that the maxima of cell emergence in the surface sediment layers and photosynthetic capacity coincide, although the increase in photosynthesis preceded cell emergence (Brown *et al.*, 1972). The photosynthetic rhythm persisted at light intensities well below saturation; minimum photosynthetic values were found in midafternoon, after which the rate increased again, a rise that preceded the reemergence of cells. The maximum photosynthetic rate occurred when maximum cell numbers were present on the sediment surface.

4. Metaphyton

Loosely aggregated algae and cyanobacteria, the metaphyton, in inundated water of wetlands and littoral areas of many lakes, ponds, and floodplain areas of rivers are neither strictly attached to substrata nor truly suspended. Metaphyton communities originate from fragmentation of dense epiphytic communities that aggregate and become clumped and loosely attached in littoral areas by wind-induced water movements and then can form dense microbial accumulations with intense internal nutrient recycling. The productivity and collective metabolism of metaphyton can be very high and radically alter the nutrient cycling of littoral areas.

In contrast to the immense amount of study on the systematics, physiology, and ecology of the phytoplanktonic algae of aquatic systems, there is a dearth of information on the algae attached to substrata or loosely aggregated in the littoral regions of lakes or shallow zones of streams and rivers. Although the taxa of these largely sessile algae are somewhat better understood, information is sparse on their geographical distribution, seasonal population dynamics, utilization of microhabitats, responses to parameters of water movement or water and substratum chemistry, or on their interactions with other organisms.

The extreme heterogeneity in distribution of the algae across an exceptionally variegated spectrum of microhabitats subjects attached organisms to much more variable environmental physicochemical and biotic parameters than usually occur in the open water. Spatial and temporal heterogeneity of phytoplankton in pelagic habitats is minor in comparison to that of the attached algae. Nonetheless, despite the numerous problems associated with obtaining quantitative information, indications are that the algal populations specialized to these habitats sustain high rates of productivity. Our poor understanding of the complex interactions between the sessile flora and their substrata, and the contributions of the attached microorganisms to the total system productivity, represents a major void in contemporary limnology that warrants intensified study.

B. Terminology and Zonation

The limnological terminology applied to attached algae is complex. These terms and the general zonation of littoral algae have been introduced earlier (Chap. 8; Fig 8-3, pp. 131–133).

C. General Distribution of Littoral Algae

Floristic surveys of the distribution of algae among subcommunities of the littoral zone result in long lists of species. Observed different associations reflect acute heterogeneity in parameters of light, temperature, nutrient availability, water movement, substratum, grazing pressure, and other factors. Detailed floristic analyses indicate specific characteristics of dominant taxa and designate the range of probable interactions among species, thereby providing a foundation upon which experimental approaches can be designed (Round, 1964a; Biggs, 1996; Stevenson, 1996a). The initial evaluations often lead to recognition of algal associations in various habitats, thereby enabling quantitative sampling and the separation of dominant and casual species. These evaluations are difficult to perform, especially among algae of the sediments, but have been done in a number detailed analyses of littoral communities.

The attached algae often dominate algal biomass in small streams and shallow lakes. Probably >90% of all algal species grow attached to a substratum; these species include practically all of the pennate diatoms and a majority of the Conjugales, Cyanophyta, Euglenophyta, Xanthophyceae, and Chrysophaceae (Round, 1964a; Biggs, 1996; Lowe, 1996).

Epilithic and epiphytic communities have many algal species in common, but there is relatively little interchange of species between those of the mud-living epipelic habitat and the other substrata types. Epipelic algae are largely motile; their motility is essential to enable them to move to the surface after any disturbance of the sediments. Algal species adnate to sediments of streams and rivers with moderate to fast flow are almost exclusively motile.

Epipsammic algae, consisting largely of small diatoms and cyanobacteria attached more or less firmly to crevices in the surface of sand grains and rock surfaces, exhibit an extremely variegated, heterogeneous distribution (Fig. 19-2). These algae tend to be less motile than epipelic algae, which usually grow on finer, organic sediments (Round, 1965a; Meadows and Anderson, 1966, 1968; Miller *et al.*, 1987). In the

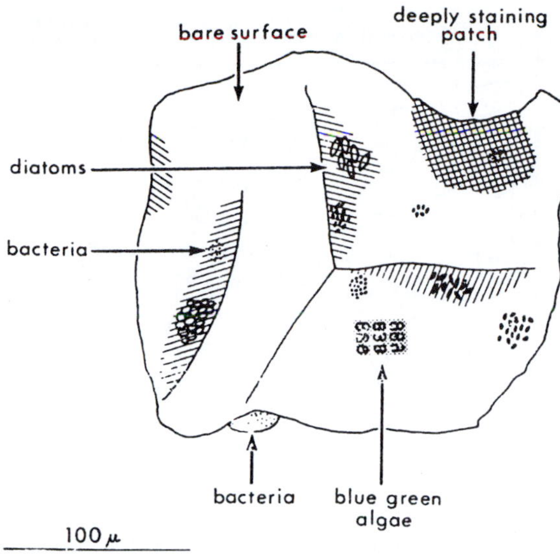

FIGURE 19-2 Diagrammatic representation of typical localized distribution of microorganisms and staining (carbol fuchsin) patches of organic materials on the surface of a sand grain. (From Meadows, P. S., and Anderson, J. G.: Micro-organisms attached to marine and freshwater sand grains. *Nature (London) 212*:1059–1060, 1966.)

shallows of a temperate English lake, Moss and Round (1967) found that both epipelic and epipsammic algae showed marked population maxima during the spring. The biomass of the epipsammic algae of the surface 5 cm of sediments was consistently greater than that of the epipelic algae. Algal populations in interstitial water of a sandy beach in a southern Michigan lake were also large and exhibited a spring maximum (Davies, 1971). Percentage water content in sand sediments was an important determinant of algal composition between the saturated zone of the water's edge and higher sand beaches subject to periods of desiccation. Within the sand, light is attenuated rapidly and is essentially extinguished within half a centimeter. Yet viable epipsammic algae have been found attached to sand grains as deep as 20 cm in sandy beach zones exposed to heavy wave action. It is believed that wave action is adequate to mix the sand grains and return attached algae to a lighted zone sufficiently often to permit active, although low, photosynthesis (Steele and Baird, 1968).

1. Attached Lichens of the Eulittoral

Lakes and streams with rocky shores and moderately stable water levels may have an abundant and diverse eulittoral crustose lichen flora. Largely studied among waters of northern Europe, a characteristic stratified zonation is commonly observed (Santesson, 1939; Hutchinson, 1975). Just above the lowest summer water level, grey–black *Verrucaria* spp. occur epilithically, being submersed some 80–90% of the year. Above this zone, some 20–30 cm above the lowest water level, several lichens, dominated by blackish-brown *Staurothele,* occur and are submersed well over half of the year, mostly in winter. A higher association, covered by water for 25–50% of the year, is dominated by light-brown or grey *Lecanora* spp. Above this zone at about the highest water level, the lichens shift to foliose species and associations. This distinct community demarcation, called the *lichen line* (*Flechten linie*; Naumann, 1931), represents the most probable highest water level in the immediate historic past.

Distribution of species within streams can be delimited into four distinct zones in the British Isles (James *et al.,* 1977; Gilbert, 1996): (1) The fluvial submersed zone involves about 15, mainly pyrenocarpous, species; (2) the fluvial mesic zone contains the richest aquatic lichen flora, with about 20 species and approximately 60% cover; (3) the fluvial xeric zone is lichen-depauperate, being scoured by spates about once a month; and (4) the lichen flora of the fluvial terrestrial zone is influenced only by nutrient loading from the stream water through the substrata. Along a stream gradient, the lichen flora varies from species-rich headwater flow with a cover of 80% of exposed rock to more erosive areas subjected to scour and substrata instability where species diversity was low and cover of bedrock was <1%.

Very little is known about the physiological characteristics and constraints of these semiaquatic lichens, although some of the ecophysiological factors that account for the delimited distribution of them have been investigated (Ried, 1960a,b). Practically no species tolerates continuous submersion. A notable exception is *Hydrothyria venosa,* a submersed species of the Collemataceae characterized by hyphae that ramify among the gelatinous matrix of single cells of the phycobiont *Nostoc* (Russell, 1856). This rheobiont lichen is rather commonly distributed on stones in relatively quiescent mountain streams with its thallus often in dense tufts to 10 cm or more in height. As in terrestrial environments, semiaquatic and submersed lichens generally occupy harsh, austere habitats that are physically stable and relatively free of nutrient enrichments and siltation (Gilbert, 1996). Stable springheads of small streams are common habitats in which interspecific competition of lichens is low. It is suspected that aquatic lichens usually possess slow growth rates and contribute relatively small percentages to overall material and energy fluxes of freshwater ecosystems.

The mineral content and pH of the water influence the lichen flora as instanced by impoverishment of communities on siliceous rocks on lakesides where inflow streams are basic in composition. Lake and stream lichens are also vulnerable to contamination by inorganic fertilizers in runoff and seepage. Hypereutrophication encourages cyanobacteria and green algae that can outcompete lichen communities. Long-term studies of aquatic lichen communities are of value in monitoring lowering or rising water levels that result from human-induced alterations and diversions of water (Seaward, 1996). River channel capacity can also be determined by lichen zonation (Gregory, 1976).

D. General Structure of Periphyton Communities

Problems of exchange of nutrients, gases, and metabolic products from within the periphyton communities and the overlying water involve problems of diffusion across the periphyton community boundaries and the external medium. Slow diffusion into attached microbial communities is particularly acute in relatively thick communities, where high productivity is maintained by intensive internal recycling of nutrients, including carbon, among mutualistic microbiota (Wetzel, 1992, 1993a). Abiotic adsorption of particulate and dissolved organic compounds, as well as incorporation of inorganic nutrients and metals into the mucilaginous matrix, can provide a concentration mechanism (Lock, 1981, 1982; Lock *et al.,* 1984; Roemer *et al.,* 1984; Beveridge and Graham, 1991; Freeman and Lock, 1995). Such concentration could enhance diffusion within the community matrix to the organisms. This mechanism likely not only stimulates the metabolism of the community but in the case of certain cations, such as calcium, magnesium, iron, and manganese, affect the physical properties of the mucilage matrix, such as hydrophobicity (e.g., Freeman *et al.,* 1995; Lemke *et al.,* 1995) and possibly texture. The extracellular matrix can also provide sites for attachment for extracellular enzymes, such as phosphatases and proteases, that are critical in rendering nutrients and carbon available to microorganisms (Wetzel, 1990a, 1991a).

Developing periphyton is three-dimensionally enmeshed with hydrated glycocalyx and other mucopolysaccharide materials, secreted by bacteria and algae, that both sequester ions and isolate the microorganisms from the water column (Wetzel and Allen, 1970; Allen, 1971; cf. review of Burkholder, 1996). Accumulated detrital particles and precipitated calcium carbonate crystals, as well as the glycocalyx materials, enhance nutrient enrichment by adsorbing phosphorus, ammonium, and various organic substances (Fig. 19-3). Nutrients may be assimilated, sometimes with the aid of excreted enzymes such as phosphatases (*), utilized, and released via leaching, excretions, secretions, or cell lysis by the benthic algae (A) (Fig. 19-3). Epiphytic microfauna and macrofauna can release phosphatases and nutrients during feeding

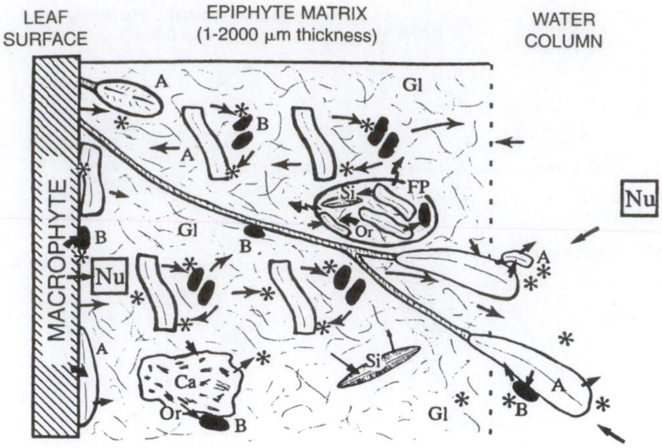

FIGURE 19-3 Diagram of the supply of nutrients (*Nu*, such as organic and inorganic forms of carbon, phosphorus, and nitrogen) to benthic algae and other microbes within a periphyton matrix on a substratum, in this case a submersed macrophyte leaf. Arrows indicate routes of nutrient cycling form the substratum and the water column, which represent the two major external sources of nutrient supply to the periphytic matrix. (See text for explanation of relationships and symbols.) (From Burkholder, 1996.)

and excretion (e.g., fecal pellets (FP) with viable algae and bacteria that have become nutrient-enriched during passage through the animal gut, nutrient waste products, and dead microflora and other organic detrital particles).

Other nutrient-sequestering or -releasing materials include inorganic calcium carbonate (Ca) that is precipitated by photosynthetic processes and siliceous frustules of dead diatoms (Si), organic debris (e.g., recently dead algae and animals, Or), and microbially derived hydrated glycocalyx/mucopolysaccharides (Gl) (Fig. 19-3). The microflora and fauna can also be colonized and attacked by fungi, especially when moribund, and this process would result in nutrient remineralization or release. Some algae and bacteria are directly adnate to and in contact with the leaf or other substratum surface. Other microflora are loosely attached in the overstory matrix, and some have stalks or other attachment structures to the substratum (Fig. 19-4).

Insight into the community structure of epiphytic microflora of submersed macrophytes can be gained from electron photomicrographs[2] of a typical association (Fig. 19-4). The surface structure of the epiphytic community (Fig. 19-4A) indicates the manner in which the loosely woven diatomaceous component is held in place by a stranded matrix formed largely from gelatinous stalks of mucoid substances (Fig. 19-4B), which

[2] The scanning electron photomicrogrraphs were kindly provided by Dr. B. R. Allanson University, South Africa; some are from his investigations reported in Allanson (1973).

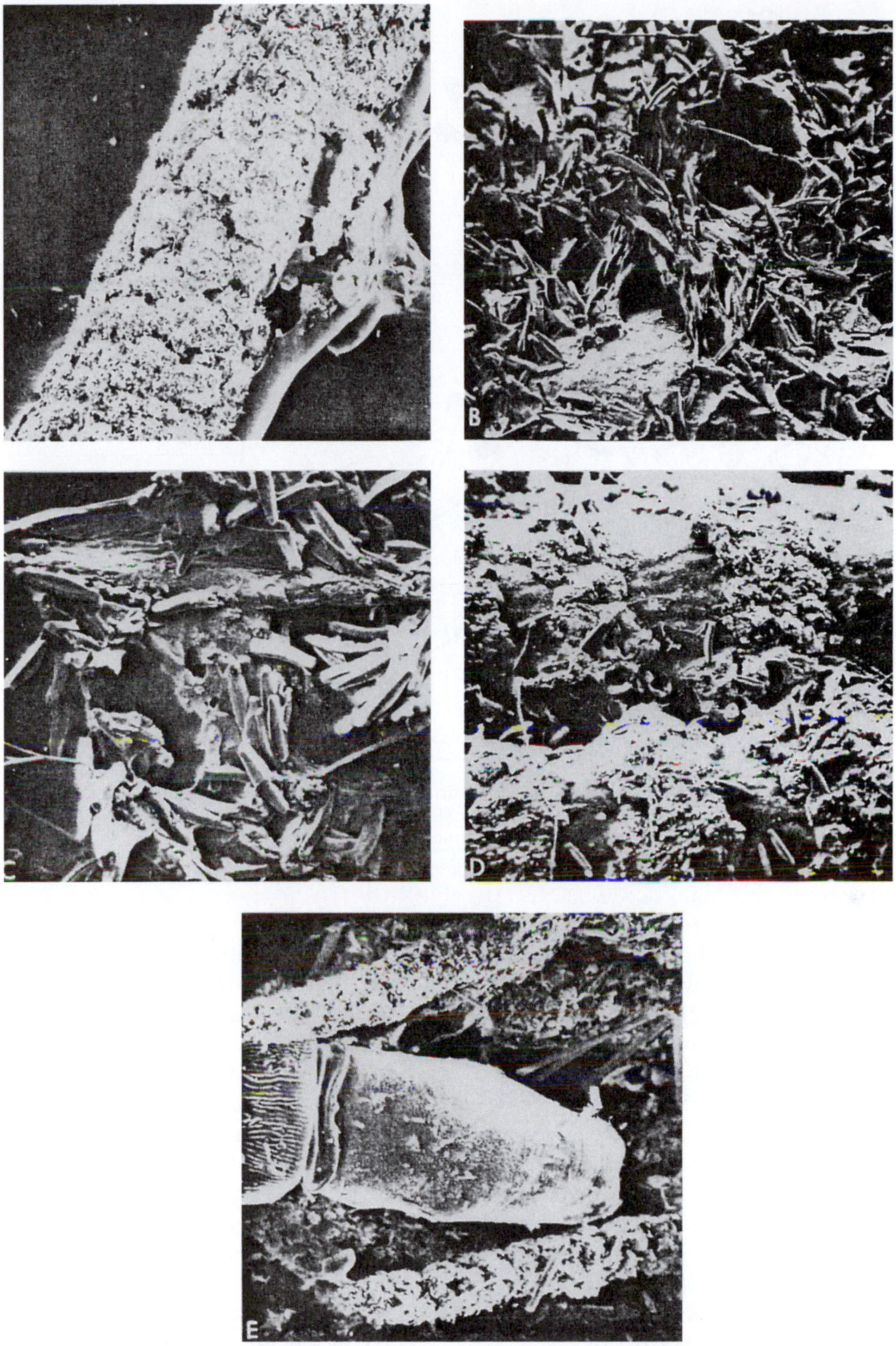

FIGURE 19-4 The structure of a microfloral community upon the macroalga *Chara* from Wytham Pond, Oxford, England. (*A*), Surface view of microflora on a stem (×75); (*B*), The epiphytic community after exposure to pH 4.5 to remove calcite deposits, showing the dominant diatom *Achnanthes minutissima* and mucoid membranes attached to the wall of the host plant (×500); (*C*), Intimate association between the diatom component and supporting mucoid matrix, calcite deposits removed (×1000); (*D*), transverse feeding tracks left by nymphs of the baetid mayfly *Cloeon dipterum* (×500); (*E*), a chironomid larva, found enclosed in fronds of *Chara*, which may have been feeding in this position (×90). (Photographs courtesy of B. R. Allanson, 1973. From Allanson, B. R.: Fine structure of the periphyton of *Chara* sp. and *Potamogeton natans* from Wytham Pond, Oxford, and its significance to the macrophyte periphyton metabolic model of R. G. Wetzel and H. L. Allen. *Freshwater Biol.* *3*:535–541, 1973.)

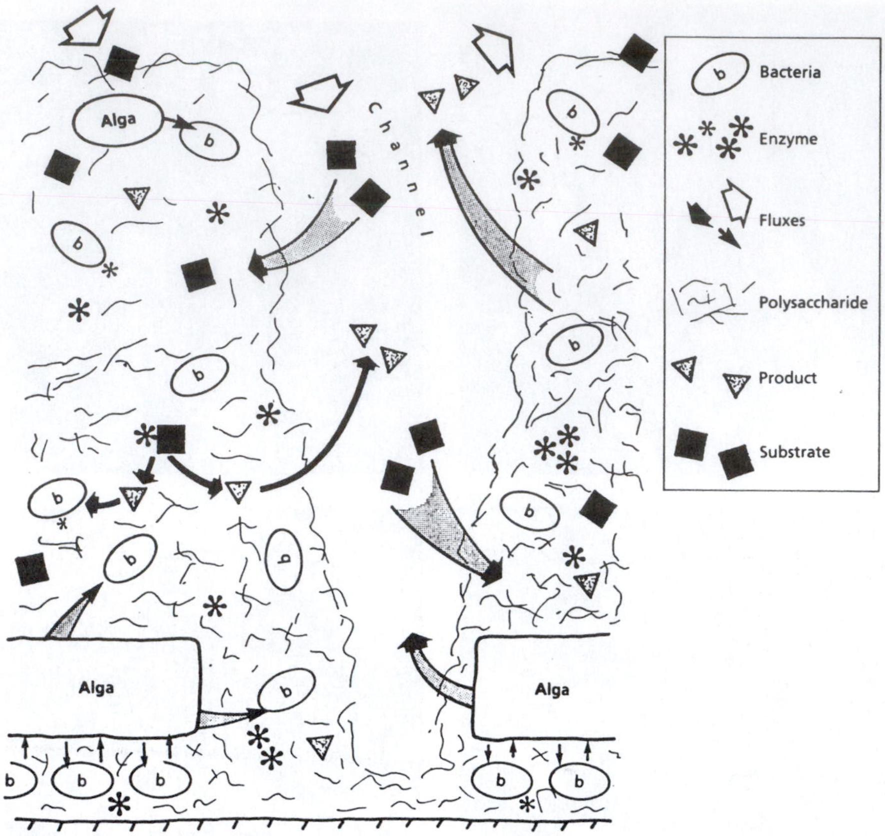

FIGURE 19-5 Conceptual model of the structure and metabolic fluxes within periphyton communities and from or into microchannels of the polysaccharide matrix and the overlying water. (See text; modified slightly from Lock, 1993.)

in places extends as a fragmented membrane above the deposits of calcium carbonate covering the cell wall (Allanson, 1973; Roos *et al.*, 1981). While some algae (diatoms and cyanobacteria) and heterotrophic bacteria are loosely associated within and on the calcite and mucoid complex, others penetrate through the complex to the macrophyte itself and attach themselves directly along its entire surface or by means of simple or branched mucilagenous stalks. Much of the matrix is associated with the mucopeptide cell walls of the epiphytic bacteria, algae, and cyanobacteria, an association that is lost in the techniques used to prepare these scanning electron photographs; this matrix can be seen, however, in transmission electron microscopy.

Most of this relatively rich community of microflora epiphytic upon macrophytes is destined to become detritus. Massive accumulation rates on the supporting plants *Chara* and *Potamogeton* (Allanson, 1973) and differences seen in the fine structure of the epiphytic-carbonate-mucoid complex between parts of the plant as it grows and provides new surfaces for colonization indicate that little of the community is utilized by grazing animals. In places, however, evidence for grazing occurs (Fig. 19-4 D,E).

Periphyton structure is envisaged to consist of microbes embedded within the polysaccharide matrix made from material secreted by bacteria and algae (Fig. 19-5). The polysaccharide matrix is potentially permeated by microchannels that may act as conduits in which diffusion of organic and inorganic nutrients is faster than through the polysaccharide matrix (Lock, 1993; Costerton *et al.*, 1994; DeBeer *et al.*, 1994; Wimpenny and Colasanti, 1997). The size of these microchannels varies greatly but is commonly in the range of 5–50 μm, and they permeate the microbiota and matrix in a dendritic, likely fractal pattern (Fig. 19-6). The amount of the polysaccharide matrix and likely the size of these microchannels is influenced by external factors. For example, microchannel size is increased in the presence of high dissolved humic substances and ultraviolet irradiance (Wetzel *et al.*, 1997, unpublished).

Fluxes of nutrients and dissolved gases from and to the overlying water and among microbial components within the periphyton would be faster in the liquid of

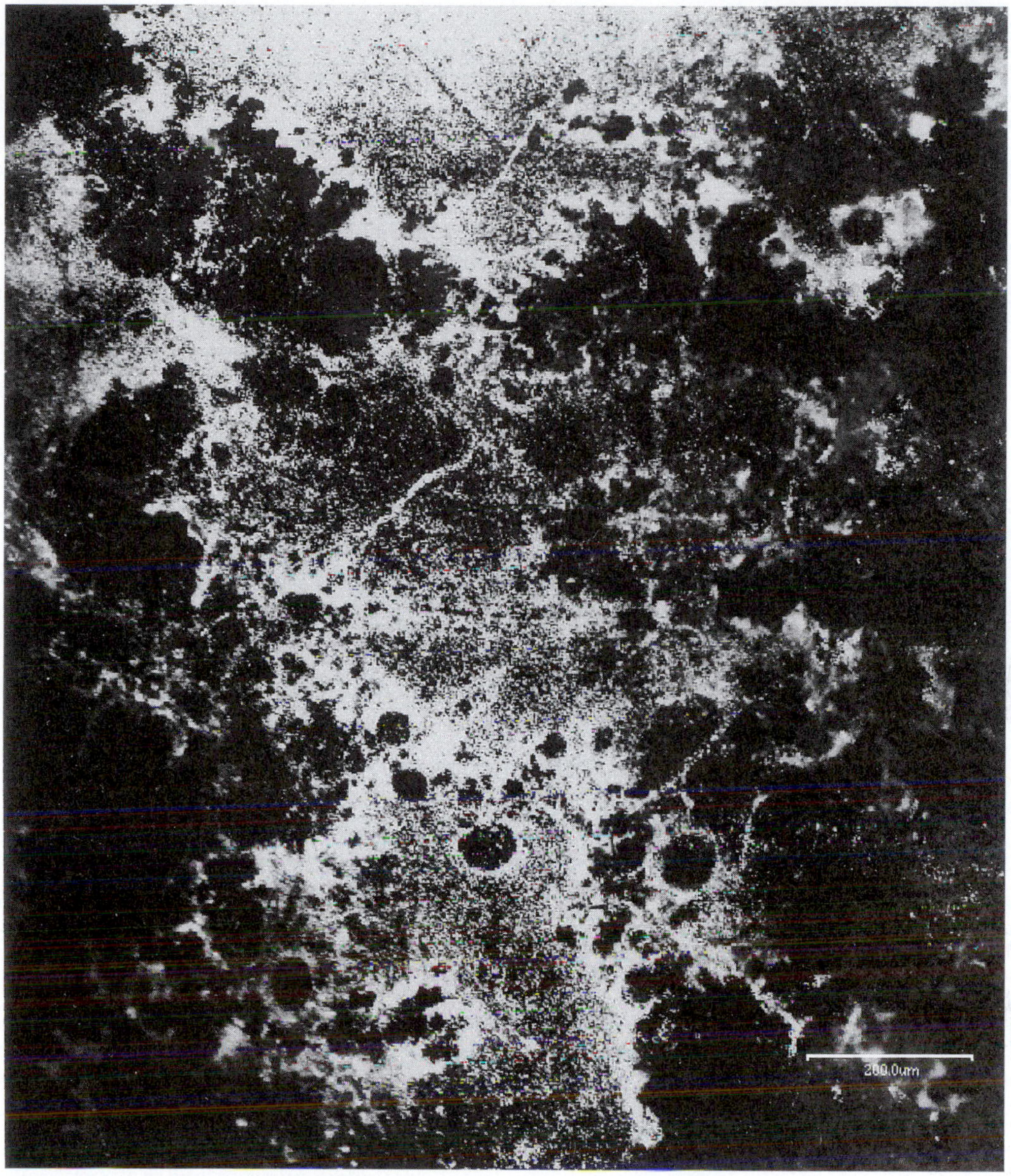

FIGURE 19-6 Fluorescent microspheres dispersed within microchannels of a lentic periphyton community at one focal plane about 75 μm below the surface–overlying water boundary. Black areas are bacterial and algal cell clusters within extracellular polysaccharide matrix. (E. Espeland and R. G. Wetzel, unpublished.)

the microchannels than through the matrix itself. Photosynthetic products of algae released intercellularly can be used by bacteria or may diffuse into the channels and to the overlying water. Reciprocally, bacteria secrete organic micronutrients (e.g., vitamins), CO_2, and enzymes intercellularly that may persist in the matrix for some time before being assimilated by algae or other microbes or diffusing from the matrix. Membrane-bound or -released extracellular enzymes may hydrolyze dissolved organic matter (DOM) released from other microbes, especially algae, or from DOM or particulate organic matter (POM) adsorbed from the overlying water. These hydrolyzed products can then be assimilated by microbes, adsorbed by the polysaccharide matrix, or be diffused into the microchannels or overlying water.

The polysaccharide matrix permeated with microchannels forms a physical medium in which diffusion is much slower than in the overlying water. A conceptual model, depicted in Figure 19-7, indicates the periphyton matrix attached to and overlying the substratum (Riber and Wetzel, 1987). In the interstitial water of the matrix, the concentration of a nutrient (C_m) will be the net effect of numerous factors,

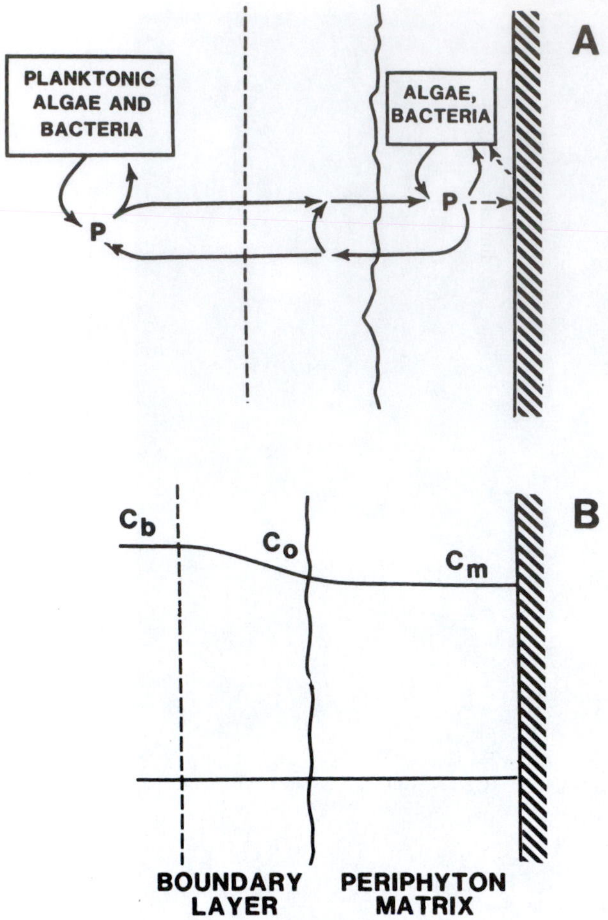

FIGURE 19-7 A conceptual model of nutrient (*P*) exchange between the organism in a periphyton matrix, the boundary layer, and the bulk water in which the nutrient availability is affected by exchange with planktonic organisms (*a, upper*). Nutrient concentration gradient (——) between the bulk-water concentration (C_b) and the periphyton matrix concentration (C_m). The concentration at the matrix surface is C_o (*b, lower*). (Modified from Riber and Wetzel, 1987.)

including uptake and release by algae and bacteria and adsorption by particles. C_m will vary somewhat as a function of periphyton depth due to biological stratification and exchange processes. The concentration at the matrix surface is denoted C_o. In the bulk water, the nutrient or gas concentration (C_b) is correspondingly determined by the activity of the plankton. The exchange of nutrient or gas between the two systems will be through the diffusive boundary layer. The mass flux through the boundary layer, *N*, is given by the general equation

$$N = h(C_b - C_o)$$

where *h* is the boundary-layer mass transfer coefficient, which depends on hydrodynamic conditions, including currents and turbulence in the bulk water (Chap. 7), and the geometry of the substratum and periphyton community structure. In some cases, *h* can be estimated by hydrodynamic calculations (Skelland, 1974). Since the net flux depends on the difference between bulk water and matrix concentrations, flux may go in either direction and can change rapidly with differences in metabolic conditions. When both systems are limited by nutrients and the concentrations are low, the net flux will be small or even zero.

The structure of the periphyton community can also influence transfer flux rates. Because of the uneven development and distribution of periphytic microbiota, some groups of organisms project upward and contain valleys in between. This increased roughness can theoretically increase turbulence as water flow velocities increase over the communities (e.g., Nikora *et al.,* 1997). However, at low flow rates, much of the water mass passes over the community with little turbulence between the high and low points of the community (Riber and Wetzel, 1987).

E. Growth-Regulating Resources

A number of consumable resource requirements influence the metabolism, growth, production, and exploitative competition among periphytic communities. Important distinctions are needed among (a) external resources in the overlying and surrounding medium, such as light and nutrients, (b) internal resources that are obtained from abiotic or biotic sources within the substrata on which the periphyton communities are attached, and (c) internal recycling within periphyton among algae and associated microbiota, as summarized later (pp. 593–594). All of these parameters are summarized in a comparative manner in Table 19-1.

1. Habitat Space

The availability of space for colonization is essential and markedly affects the composite productivity of the periphytic communities. As will be emphasized in the discussion on microspatial community structure and on comparative productivity later, the periphytic communities attached to surfaces (e.g., rock, and sediments) that are relatively fixed in area available for colonization and development can rapidly be limited by space when other growing conditions are good. Although these communities can project somewhat above the surface, and certain stalked and filamentous algal forms do so (e.g., Hoagland *et al.,* 1982), the communities readily become compacted and stratified with steep gradients of light attenuation, redox, and accumulated products of fermentation (Wetzel, 1993a). In contrast, periphyton growing epiphytically on submersed macrophytes have manifold increased surface

TABLE 19-1 Summary of Common Responses of Epilithic, Epipelic, and Epiphytic Autotrophic Microbiota to Dominant Physical, Chemical, and Biotic Resource Parameters[a]

Environmental parameter	*Epilithic*	*Epipelic*	*Epiphytic*
Physical			
Water movement	High in rivers and upper littoral areas; molar abrasion high	High → low in depth gradient; decreasing with depth	High exchange in water column; low exchange within dense submersed macrophtye stands
Substrata	Stable at low-to-moderate water velocities	Unstable; subject to frequent disturbance	Stable in position, but changing with growth and senescence; often increasing seasonally
Temperature	Often, marked seasonal and diurnal fluctuations	Decreasing with increasing depth; higher and variable in epilimnion, low and more stable in metalimnion	Commonly high in shallow areas and changing widely seasonally
Light			
a. Quantity	Steep seasonal changes, especially low in canopied streams; steep exponential declines with depth	High → low in depth gradient; rapid decrease within sediments	Moderate to high; decreasing with depth but often increasing seasonally as submersed macrophytes develop upward into water column
b. Quality	Selective attenuation by water column	Steep selective attenuation of red and blue spectra within periphyton and within sediments	Selective attenuation in water column and by pigments within periphyton communities
Habitat available	Moderate; sorting by wave action and flowing water; sets particle size and surface area	Low; essentially two-dimensional	High, particularly among submersed macrophytes; projection into water column, three-dimensional
Chemical-biotic			
General conditions	Usually oxic; low nutrients	Reducing but diurnal oxic; high nutrients	Oxic, with diurnal variations
Inorganic nutrients			
a. Phosphorus	Low, often restrictive	Very high from interstitial waters	Low, variable; some from macrophyte
b. Nitrogen	Low, often restrictive	Very high from interstitial waters	Low, variable; some from macrophyte
c. CO_2 (HCO_3^-)	Usually adequate to high	Very high from sediment metabolism	Low to high; much internal recycling
d. Others	Usually adequate	High	Low; likely much internal recycling
Organic substrates/nutrients			
a. Organic micronutrients	Low, but variable with season in lakes and floodplain inundation in rivers	Very high from sediment metabolism	Low, high microbial competition, much internal recycling
b. Heterotrophy	Algal heterotrophy low or absent; bacterial heterotrophy low to moderate	Algal heterotrophy potentially moderate, likely low; bacterial heterotrophy very high	Algal heterotrophy low; bacterial heterotrophy high
Primary productivity	Low to moderate	Low	High
Symbiotic interactions among microbes and substrata	Low	Moderate to high	High to very high

[a] Greatly modified from Wetzel (1983b).

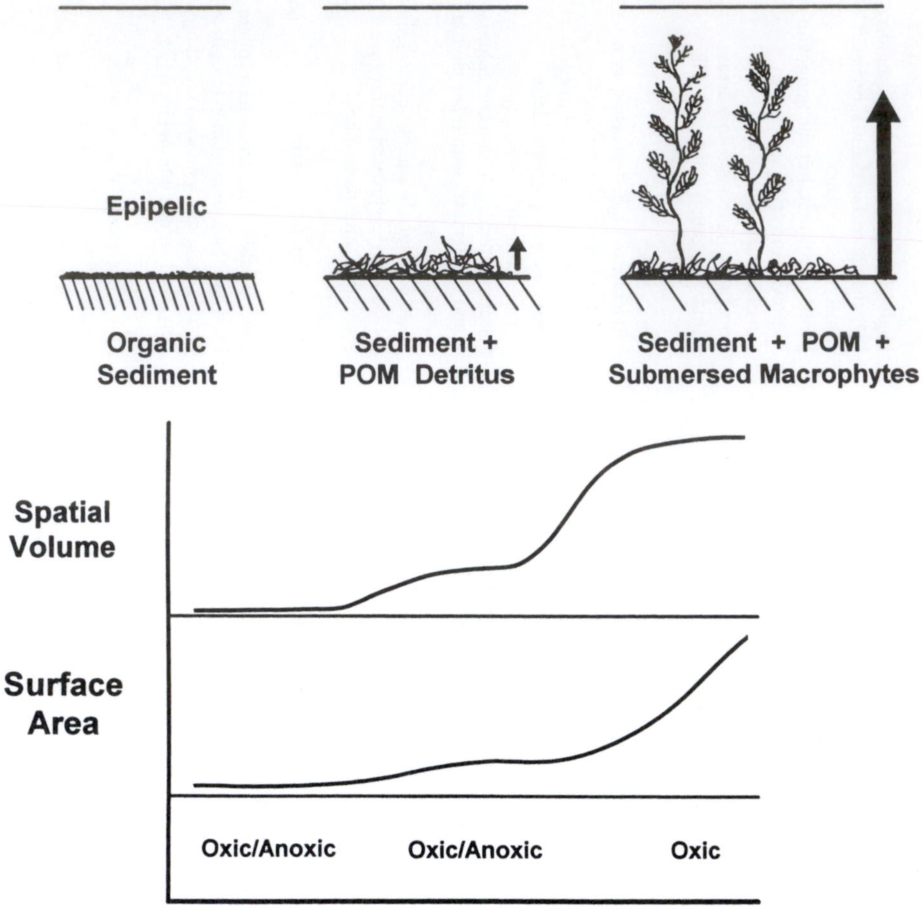

FIGURE 19-8 Transition of substrata surface area for attachment of periphyton from a predominantly two-dimensional spatial habitat through increasing surface area of particulate organic matter (POM) detritus to a fully three-dimensional spatial habitat of surface area of submersed aquatic plants.

area as the substrata becomes three-dimensional and projects upward into the water column and enhanced light field (Fig. 19-8). Not only does the surface area available for colonization increase exponentially but the spatial volume increases greatly with much greater exposure of the communities to largely oxic water column conditions (Fig. 19-8).

Analogous conditions occur in running waters. Although submersed macrophytes are much less well developed in streams and rivers (cf. Chap. 18), large woody debris often constitutes significant particulate organic debris (cf. Chap. 21) and can provide appreciable surface area for colonization by periphyton. Large woody debris tends to be moved downstream during flooding events considerably less than does fine particulate matter. In addition, stones and other particles tend to organize in clusters during high water velocity events and flooding in streams where a large anchoring stone (an obstacle clast) and smaller stones lying against the upstream face of the clast (the stoss) provide a hydraulic shelter for accumulation of smaller particles behind in the wake tail (Fig. 19-9). The surface area of the smaller rocks is increased considerably in comparison to large rock surfaces and shifts upward in a modest three-dimensional manner. Although the abrasive and export actions of flooding events often greatly reduce the abundance of attached algae, these microform bed clusters of hydraulic refuges can provide increased substrata area and support appreciably greater algal periphyton (e.g., Biggs *et al.*, 1997; Francoeur *et al.*, 1998). Among epipelic algal communities on organic sediments, however, a very small size of sediment grains, particularly below 125 μm, reduces the sites for microalgal community development (Cahoon *et al.*, 1999). With the shift toward a two-dimensional configuration, algal biomass is reduced and declines with the increasing percentage of very fine particles.

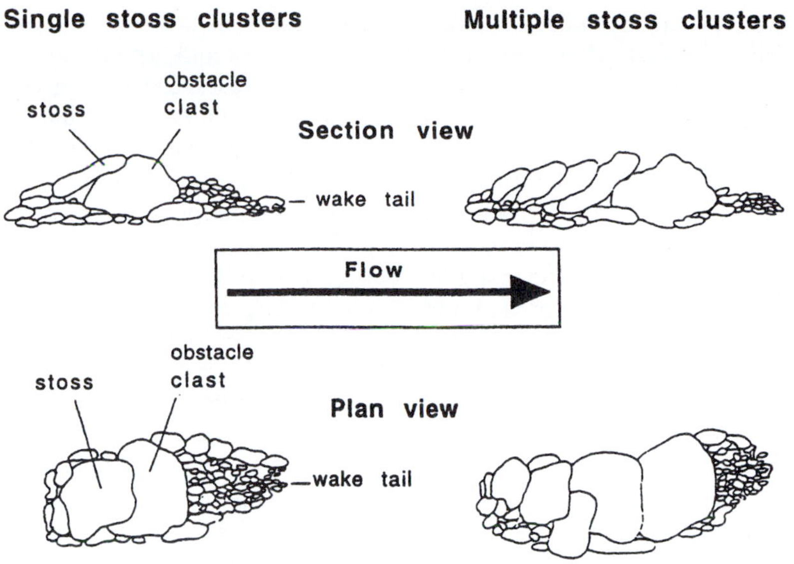

FIGURE 19-9 Microform bed clusters of sediment particles downstream of a large anchoring obstacle clast and smaller stones (the stoss) lying against the upstream face of the clast. (From Francoeur *et al.*, 1998.)

2. Nutrient Availability and Utilization from the Overlying Water

The physiological requirements and growth responses of attached algae are similar to those of phytoplankton, already discussed at length (cf. Chap. 15). Nutrients can be obtained by diffusion into the communities from the overlying water or from below from the substratum, which may be living or senescing (e.g., macrophytes), dead (organic detritus), or relatively inert, such as rock. Nutrient availability is determined by hydraulic conditions at the boundary layer that influence the thickness of the diffusive layer, the concentrations of compounds dissolved in the bulk environment of the surrounding water, and the metabolic activity of the attached organisms for utilization or production of the compounds.

At the macrogradient level of influence of nutrient concentrations of the overlying water on the development and production of periphyton communities, numerous studies have demonstrated that enrichments of nutrients, particularly P and N, in the overlying water commonly result in enhanced growth of attached algae on many different substrata. Examples are many. Growth responses of natural epiphytic communities of submersed macrophytes of an oligotrophic lake of Michigan were compared to that of another area of the littoral in which nutrients were administered to simulate release from macrophyte tissue (Moeller *et al.*, 1998). Epiphytic growth increased immediately where phosphate was released above the sediment, and after 10 weeks a 40-fold increase of epiphytic algal biomass occurred at the center of the plot as compared to biomass immediately outside of the m^2-plots. Whole-lake fertilization experiments with P and N in a shallow, subarctic Swedish lake resulted in massive increases in phytoplankton (Björk-Ramberg and Ånell, 1985). The resulting competition for light resulted in precipitous declines to less than half of the periphyton primary production to the total lake photosynthetic production. In unfertilized lakes, high periphyton productivity was maintained and contributed 70–83% of the total lake primary production. In habitats with relatively abundant nutrient concentrations, dense phytoplankton can shade periphyton development severely as the attached communities shift from nutrient to light limitations (e.g., Hansson, 1992).

Nutrient enrichment of the overlying water can also lead to marked changes in species composition of periphyton communities. For example, nitrogen enrichment led not only to an increase in epilithic algal abundance but to a shift from a diatom/N_2-fixing cyanobacteria-dominated community to one dominated by diatoms (Hawes and Smith, 1992). Where combined nitrogen insufficiency is maintained, N_2 fixation by attached cyanobacteria appears to be a successful strategy for survival in N-deficient environments on both a sustained and on a seasonal basis (Reuter *et al.*, 1983, 1986). However, species composition of algal periphyton is often relatively unresponsive to small changes in the nutrient concentrations and dynamics in the overlying water (e.g., Cattaneo, 1987; Pringle, 1987; Riber and Wetzel, 1987; Fairchild and Sherman, 1993; Neiderhauser and Schanz, 1993). Such recycling allows a short-term (days to weeks) functional separation of the metabolism within the periphyton from the nutrient concentrations in the overlying water, which indicates the importance of intensive recycling of nutrients within the microbial community. Although that separation cannot persist indefinitely, it represents a marked adaptive advantage in which inputs from the overlying water can be directed largely to new net growth and production to augment the recycling occurring within the community between algae and bacteria (Wetzel, 1993a; see later discussion). If enhanced nutrient concentrations are sustained over long periods of time, distinct species change in the attached algal communities. For example, along a long gradient of nutrient enrichment as a point source to a wetland of the Everglades, periphytic algal growth decreased with greater distances from the point source of nutrients and eutrophic diatom species were replaced by oligotrophic diatom species as the total phosphorus concentrations decreased below ca. 15 μg liter^{-1} (McCormick *et al.*, 1996, 1998). Because of light limitations by the dense macrophytes under the eutrophic conditions of the gradient, periphyton production was greater in the oligotrophic sites despite the reduced phosphorus availability.

Rates of diffusion from the overlying water to the periphyton communities are influenced by the thickness of the boundary layer as well as solute or gas concentration gradients. Because the thickness of the boundary layer, within which fluxes occur by diffusion, decreases with increasing current flow across the surface, the boundary layer is less in rivers than in lakes. Regardless of flows, nonturbulent boundary layer thicknesses cannot be reduced below ca. 10 μm (Raven, 1970; Smith and Walker, 1980; Koch, 1990). In protected areas, such as in backwater areas, refuges, or within aquatic plant beds of both standing and flowing waters, very much lower water current velocities and much greater boundary layer thicknesses (10^2–10^5 μm) occur (e.g., Losee and Wetzel, 1993; Sand-Jensen and Mebus, 1996; Stevens and Hurd, 1997). In addition, the thicknesses of the microbial communities of standing waters are very large, often several millimeters or even centimeters, than in streams. Water movements reduce the thicknesses appreciably by the sloughing off of cells and export downstream (cf. Stevenson, 1996b).

Within limits, increasing current enhances nutrient uptake rates and as a result often also photosynthesis, respiration, cell division, biomass, and productivity (cf. many references cited in Stevenson, 1996b). Obviously, water movements across periphyton interfaces would replenish nutrient concentrations in microzones within the boundary layers where they have been reduced by metabolic uptake and chemical reactions. The advantage accrued by increasing water velocities is highly variable, but the saturating current velocity for algal assemblages in many flowing habitats is in the range of 4 to perhaps 20 cm s^{-1}. As periphyton densities and community thicknesses increase, collective metabolism per unit area and growth rates decline, and nutrient demand increases. Depending on the architecture of the attached communities, reliance on the nutrients from the overlying water stabilizes and there is generally a shift to greater internal recycling of nutrients and gases within the periphyton community (Fig. 19-10) (Wetzel, 1993a). As communities develop and increase in thickness, instability can occur and cells and portions of the community may slough off. Alternatively, very dense biomass development of filamentous algae, such as *Cladophora*, requires water currents, either unidirectional in streams or multidirectional as in wave-turbulent areas of lakes, presumably because of the inadequacy of internal recycling within the community.

With increasing current velocities, shear stress resulting in drag on periphyton causes losses and export of cells downstream (cf. Stevenson, 1996b). In addition, at high current velocities and small boundary layers, the metabolism of attached algae can be reduced for reasons not well understood. Regenerated nutrients and extracellular enzymes may be lost across the interface to the moving water before they can be effectively utilized.

Attached benthic algal and microbial communities themselves, as a whole, are not porous, but rather are relatively impervious to throughflow. In contrast, many of the substrata upon which the algae grow, such as sand and organic particles, are very porous to intrusions from surface flows and from groundwater sources. The scale of this movement, however, is much larger than that of the metabolic environment of the individual cells of the attached microbial communities. An important distinction is that although the substrata may be porous, the periphyton communities are not; within the periphyton, fluxes of nutrients are by diffusion along concentration-mediated gradients. Those gradients are highly dynamic as mediated by the photosynthetic and respiratory metabolism of the microbiota and, in the case of macrophytes, the living substrata upon which the attached community may be growing.

3. Nutrient Interactions of Attached Algae with Substrata

Attached microbial communities are well positioned to utilize nutrients and organic compounds released from the substratum upon which they are growing as well as from the water column. Marked differences in the species and quantities of epiphyte communities have been observed on natural as opposed to artificial plants in oligotrophic and mesotrophic habitats (Cattaneo, 1978; Cattaneo and Kalff, 1978; Eminson and Moss, 1980; Gough and Gough, 1981; Burkholder and Wetzel, 1989a,). Artificial nutrient

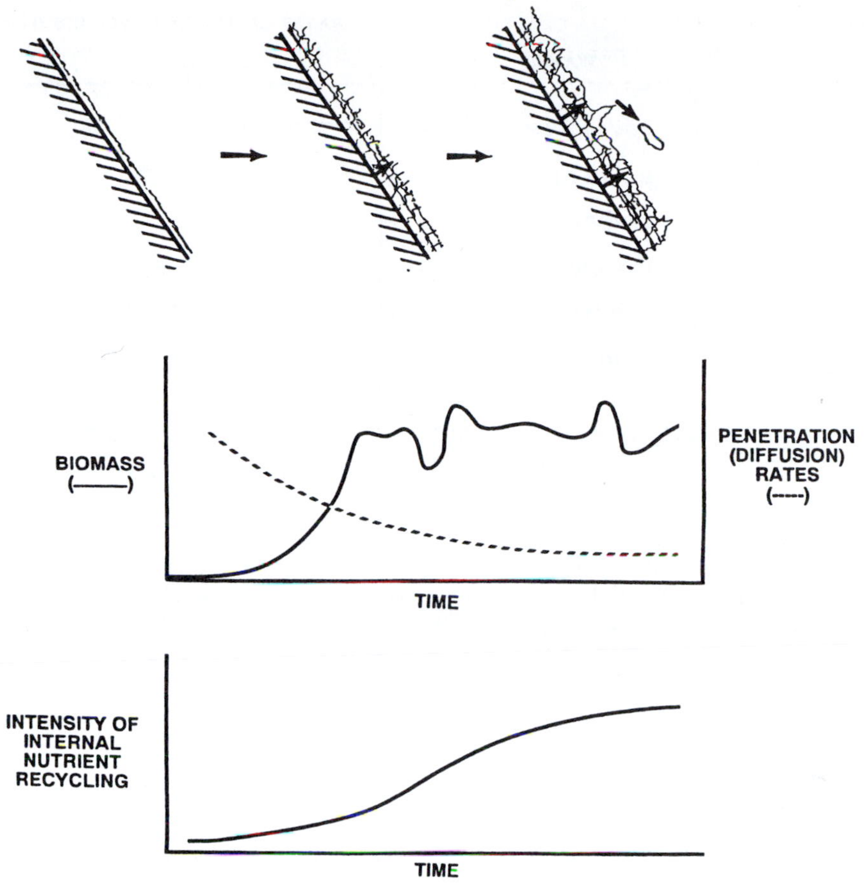

FIGURE 19-10 Successional development of, and increased nutrient recycling within, microbial communities on substrata (see text). (From Wetzel, 1993a.)

enrichment by experimentally following biomass differences on nutrient-diffusing artificial substrata in comparison to nutrient enrichments of the overlying water column have demonstrated that nutrients from the substratum are effective in increasing growth as well as altering community species composition (Fairchild *et al.,* 1985; Carrick and Lowe, 1988; Pringle, 1990). Although the capability for nutrient utilization is obvious from these studies, the question still remains about the utilization rates and significance of nutrient acquisition from the substrata under natural conditions.

Despite the early negativism that periphytic communities were obtaining essentially nothing nutritionally from the substrata, modern experimental evidences indicated quite the contrary—many periphytic communities are highly dependent upon the substrata for mineral and organic nutrients, particularly in oligotrophic and mesotrophic environments where the greatest periphyton productivity occurs. The influence of the supporting macrophytes in determining the epiphytic algal community is greatest in infertile waters. As the surrounding water becomes more fertile, external environmental factors influence epiphytic growth more than do inputs from the host macrophyte.

These relationships are well demonstrated among epiphytic algae and bacteria. In addition to the very large surface areas for colonization and retention of algae and other microbes, submersed macrophytes also provide significant sources of nutrients from both living as well as senescing tissues (Harlin, 1973; Moeller *et al.,* 1988; Burkholder and Wetzel, 1990; Burkholder *et al.,* 1990; Burkholder, 1996; Wetzel, 1996). The nutrients from the substrata internally supplement those diffusing into the attached communities from the water within and passing through the littoral zone. Complex nutrient interactions exist among algal species. For example, certain algal species growing adnate to the macrophyte can obtain >60% of their phosphorus from the macrophytes (Fig. 19-11). Algae near the outer portion of the community obtained less phosphorus from the macrophyte, but nonetheless significant quantities emanated from the plant. Even when phosphorus concentrations are very high in the water, some phosphorus, and presumably other nutrients, is

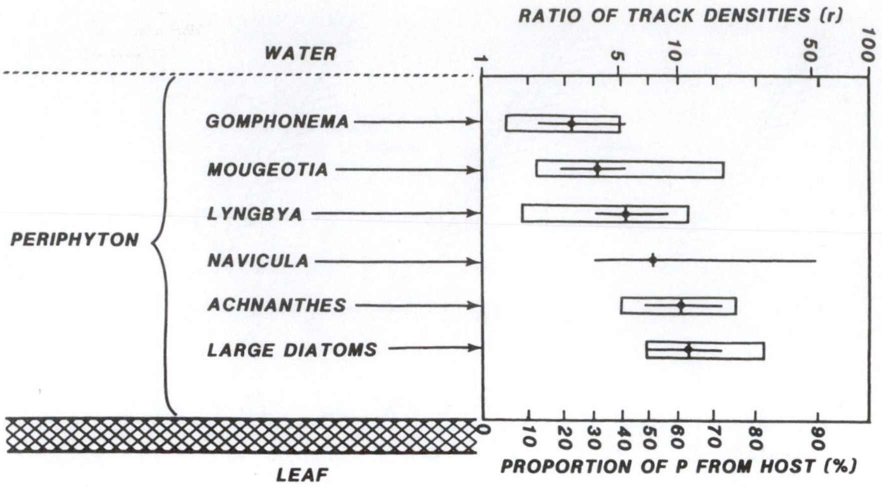

FIGURE 19-11 Sources of phosphorus within an epiphytic community on the submersed macrophyte *Najas flexilis*. Epiphytic algal taxa are arranged according to increasing proportion of phosphorus received from the host plant. The relation between *r* (*upper*) and the proportion (*lower*) is implicit in the dual scaling. A 90% confidence interval was calculated from raw data (open box) and corrected data (bar). (From Wetzel, 1990a, based on data from Moeller *et al.*, 1988.)

obtained from the macrophyte, simply because diffusion rates from the water into and within the complex epiphytic community are too slow to meet high metabolic demands. Little phosphorus incorporated into the periphyton passed to the macrophyte (Moeller *et al.*, 1988).

An additional example is the previously discussed (see Chap. 14) results of Jørgenson (1957), who found that diatoms epiphytic on the reed *Phragmites* competed with plankton for silica of the water early in the major growing season. Later in summer and fall, apparent utilization by the diatoms of silica from the *Phragmites* stems gave them a competitive advantage over planktonic forms (see Fig. 14-23). Certain other littoral algae also have high requirements for silicon (e.g., Moore and Traquair, 1976; Hooper-Reid and Robinson, 1978). Spring and autumnal population maxima are common in epiphytic diatom communities on submersed macrophytes and have been correlated with nutrient availability (e.g., Siver, 1978). Benthic microalgae are also important regulators of silica fluxes; epipelic diatom utilization can reduce fluxes to the overlying water and potentially influence phytoplankton growth and species composition (Sigmon and Cahoon, 1997).

As macrophytic tissue senesces, much of the phosphorus is translocated to rooting tissues. However, an inward swelling and disorganization of the epidermal cells of the macrorphyte occur (Rogers and Breen, 1981). Bacterial degradation penetrates and slowly degrades epidermal cell walls. This loss of cellular integrity of the leaves results in leaching from the leaves. Nutrients and dissolved organic compounds are readily utilized by periphytic microflora, which tend to develop profusely during autumn and winter periods in temperate regions. For example, more labile amino acids were readily assimilated and retained within periphyton communities to a greater extent (ca. 43%) than were more recalcitrant dissolved organic substrates leached from decaying emergent macrophytes (3–4%) (Bicudo *et al.*, 1998).

Microdistribution of species and groups of epilithic algae is correlated directly with differences in microdistributions of rock facies and differences in solubility of specific elements from the rock (Smith *et al.*, 1992). Mineral nutrient (Fe, Si, and trace elements) can leach from the rock substratum and be utilized by the attached microflora with alterations in microdistributions of algal species. The distribution of epipsammic algae is influenced and their productivity often increased by nutrient content and microscale differences in diffusion and microflows among interstitial waters through sandy sediments, as well as sand grain morphology and topography (Krejci and Lowe, 1986, 1987). In both streams and lakes, hyporheic and groundwater flows can enhance nutrient availability and productivity of attached algae, often in very patchy areas of the benthic materials where local variations of inflows enter the stream or lake (e.g., Hagerthey and Kerfoot, 1998).

Epipelic algae growing upon organic sediments often develop into dense communities, both loosely attached and in dense, matlike aggregations. Often these communities are several millimeters in thickness. Even in relatively sparsely developed communities on sediments, photosynthesis of the epipelic algae and community respiration can markedly affect nutrient fluxes from the sediments to the overlying water (reviewed by Burkholder, 1996; Wetzel, 1996). Fluxes are affected by metabolic alterations of redox and coupled chemical mobilities and by direct assimilation and utilization. The importance of epipelic photosynthesis, oxygen, and pH in a microzone at the sediment–water interface and bacteria to the mobility of phosphorus from the sediments and to nitrogen fluxes from the sediments was discussed in detail earlier (Fig. 13-8; pp. 215, 253, and 602). The production of oxygen and penetration of oxygen into the sediments can result in supersaturated oxygen conditions of interstitial water and rapid shifts, in a matter of minutes, in redox potentials between highly reducing to highly oxidized conditions. Fluxes of phosphorus from the sediments to the overlying water are nearly totally suppressed under these oxidizing conditions, as well as from microbial utilization (Carlton and Wetzel, 1988). Such marked diurnal fluctuations in oxygen and other chemical microgradients have been shown in various epipelic loosely attached algae and mat-type algal communities, (e.g., Jørgensen *et al.,* 1979, 1983; Hansson, 1989). Attached benthic algae also scavenge nitrogen as NH_4—N from interstitial waters of sediments (Jansson, 1980).

It should be recalled that the exchange of nutrients and metabolic products is not all from the macrophyte or detritus to the attached periphyton community. Much discussion (summarized later) has evolved on the negative effects of dense periphyton communities on the physiology of the supporting macrophytes, such as light reduction and especially inhibition by competition for nutrients and gases from the surrounding water. As we have seen in the discussion of the macrophytes (Chap. 18), the rooted macrophytes rely heavily on nutrient acquisition from the sediments and on internal recycling of oxygen and CO_2 within the intercellular gas channels. Nonetheless, it is likely that gases, particularly CO_2, produced within the periphyton communities diffuse to the macrophyte for assimilation. Similarly, it is probable that certain nutrients recycling among periphyton constituents also diffuse along concentration gradients to the macrophytic cells at the interface with the periphyton. Other than the studies of Moeller *et al.* (1988), there are few experimental demonstrations of such fluxes, which is a fertile area of investigation.

4. Nutrient Recycling within the Periphyton Community

The relatively significant hydraulic isolation by diffusive boundary layers of periphytic communities from the overlying water, as well as the close physical proximity of their cells, indicates a tight physiological coupling between attached algae and bacteria. Much of the evidence for this interdependent metabolic relationship is circumstantial. Dissolved organic matter released from stream benthic algae during photosynthesis was taken up by periphytic bacteria, and attached bacterial productivity was positively correlated with the release of dissolved organic matter from benthic algae (Haack and McFeters, 1982; Kaplan and Bott, 1989). The productivity of bacterial periphyton may depend to a greater extent upon external dissolved organic substrates in early stages of colonization and then shift to a greater reliance on organic substrates from algal production as the communities thicken and become more complex in structure (cf. Sobczak, 1996). More direct experiments in which photosynthesis of attached algae was inhibited under controlled conditions demonstrated that periphytic bacterial productivity depended on algal photosynthesis (Neely and Wetzel, 1995). Either algal exudates of organic matter stimulated bacterial production or photosynthetic oxygen increased aerobic bacterial activity, but in either or both cases a direct metabolic coupling was indicated.

Attached bacterial metabolism may also enhance algal metabolism and growth in periphyton, as has been shown often in mixed cultures, for a number of potential reasons (e.g., Lange, 1971; Haines and Guillard, 1974; Mouget *et al.,* 1995). Enhancement of algal growth can result from bacterial reduction of oxygen tension that could reduce photorespiration or from increased production of CO_2 or growth factors such as vitamins that the algae cannot produce. Phosphorus uptake kinetics of periphyton, for example, were found to be acutely limited by boundary-layer mass transfer and a power function of flow velocity (Riber and Wetzel, 1987). Kinetic calculations based on turnover measurements indicated that internal recycling of phosphorus and recycling from the boundary layer, rather than external uptake, accounted for most phosphate turnover within intact periphyton. Under optimal conditions, it was estimated that the turnover time of phosphorus was so rapid that phosphorus was recycled between algae and bacteria every 15 s. Similarly, denitrification rates by denitrifying bacteria within epiphytic periphyton are high in the dark and on senescent macrophyte tissues but inhibited by oxygen produced by the periphytic algae or the living macrophyte tissues (Eriksson and Weisner, 1996). Analogous results were found for the effects of light in epipelic periphyton, where photosynthetic oxygen of algal photosynthesis reduced the rate of denitrification within the microbial community precipitously (Risgaard-Petersen *et al.,* 1994).

Recently, efforts have been made to evaluate enzyme activities produced by periphyton communities as an estimate of potential capacities to hydrolyze organic substrates for specific nutrients, such as phosphorus or nitrogen, and energy. Many assays estimate enzyme activity by enzymatic degradation of a nonfluorescent substrate into a highly fluorescent product in the water surrounding the periphyton (e.g., Chappell and Goulder, 1992, 1994; Jones and Lock, 1993; Scholz and Boon, 1994). Because enzymes can occur and function within and throughout the attached microbial community, their synthesis and use can be only partially related to enzyme concentrations in the overlying water. Moreover, the enzymes may be hydrolyzing organic substrates of the substratum to which they are attached. The lack of explicit evaluation of sources and of where the enzymes are being produced and utilized by the use by the attached microbial communities confounds how they are functioning. Recent substrate assays allow evaluation of individual insoluble fluorescent products at the cellular levels for specific enzyme activities such as phosphatase and protease (Huang *et al.*, 1998; Francoeur and Wetzel, 2000).

Negative interactions can also occur. For example, competition for the same nutrients, well known among the plankton, likely operated among attached microbiota as well. Algae and bacteria may produce substances that are inhibitory to the growth of the competitors. Because of the large number of species present in periphyton communities, it is probable that both positive and negative responses are occurring simultaneously in the same community.

Aquatic fungi, primarily hyphomycetes, are abundant constituents of periphyton on and within submersed leaves, wood, and other organic detritus (e.g., Bärlocher, 1992; Shearer, 1992). Fungal growth and productivity usually far exceed that of attached bacteria on these substrata (cf. Chap. 21). However, hyphomycetes are rare on nonorganic substrata such as rocks. Even though fungi utilize simple, soluble organic molecules, they likely compete ineffectively with bacteria that have high affinities for these substrates (Suberkropp and Klug, 1976). However, there is some evidence for mutualistic fungal–bacterial interactions even at the species-specific level (Bengtsson, 1992).

5. Light Availability and Utilization

Light attenuation in relation to periphyton photosynthesis must be examined at two spatial scales, namely, with depth in the water column before reaching the periphyton and at the microscale within periphyton communities. The selectively spectral diminution of light with depth has been discussed at length in Chapter 5 (pp. 56–60), where it was shown that dissolved organic matter selectively removes large portions of the ultraviolet and blue portions of the spectrum and the water molecules per se selectively attenuate the infrared and red portions of the spectrum. Particulate matter such as phytoplankton is relatively unselective but can attenuate total photosynthetically active radiation appreciably when in high densities.

In very clear lakes, light can penetrate well below the common lower limit (ca. 10 m) of growth of vascular aquatic plants and support appreciable periphyton communities on rocks, sediments, and particulate detritus. For example, in Lake Tahoe of California—Nevada light adequate to support photosynthesis of epilithic periphyton can penetrate to 60 m (Loeb *et al.*, 1983). Although littoral attached communities may require modest light (ca. 200 μmol m^{-2} s^{-1}) for photosynthetic light saturation (e.g., Turner *et al.*, 1983; Hill, 1996), net photosynthesis occurs at much lower intensities, such as in the range of 5–25 μmol m^{-2} s^{-1} (Wetzel *et al.*, 1984; Lorenz *et al.*, 1991). As discussed in the descriptions of community and productivity dynamics later, depth and turbidity-associated alteration of light is clearly a factor regulating development of periphyton. Experimental analyses of light interactions under natural conditions are few but corroboratory (e.g., Hudon and Bourget, 1983; Marks and Low, 1993).

Light obviously affects rates of photosynthesis and thus influences primarily growth. Often, estimates of algal periphyton are made by evaluating biomass, commonly as chlorophyll *a* per unit area. Instantaneous biomass does not necessarily estimate growth rates over time, however, where mortality from disturbance, grazing, pathogens, and other factors is significant. In headwater streams where streambank vegetation is abundant and shades the channel, light reduction can markedly suppress periphyton photosynthesis. Algal biomass is often 5-fold times higher in unshaded sites in comparison to those with full tree canopy, particularly when grazing pressure is low or moderate (many references cited in Hill, 1996). In both lakes and streams, light- or nutrient-enhanced primary production that is not expressed in increased periphyton biomass when grazing is heavy is instead reflected in increased grazer growth or densities (Hill *et al.*, 1995; Moeller *et al.*, 1998).

Light is attenuated very rapidly within periphyton communities, but the community structure is heterogeneous and spatially variable on substrata. As a result, light gaps occur, and variable light attenuation occurs among cell aggregations, calcium carbonate crystals, and weakly pigmented algal cells and through dead diatom frustules that act as light conduits (Losee and Wetzel, 1983). Different species dominating the epiphytic community can also influence the nature of cellular structure and relative penetration of light into the communities as a whole (Dodds, 1992).

In addition to reductions in quantity, epiphytic microbiota can also modify severely light quantity and quality reaching the supporting macrophyte. It was demonstrated experimentally that it is primarily the photosynthetic pigments of algae and cyanobacteria that attenuate selectively light in the red (ca. 675 nm) and particularly the blue (ca. 430 nm) portions of the spectrum (Losee and Wetzel, 1983; Jørgensen and Des Marais, 1988; Plough *et al.,* 1993). Heterotrophic bacteria, dead diatom frustules, and carbonate crystals were nonselective. The recent development of fiber-optic microprobes (100 to <50 μm in size) now allow detailed examination of light quantity and quality within periphytic communities with minimal disturbance (e.g., Jørgensen and Des Marais, 1988; Dodds, 1989; Plough *et al.,* 1993; Kuhl and Jørgensen, 1994). Backscattering of light within these communities increases the total scalar irradiance to which embedded algal cells are exposed.

Algae of periphyton communities are relatively well protected by carotenoid pigments against high light intensities, and little photoinhibition has been demonstrated (cf. review of Hill, 1996). Algal productivity and photosynthetic oxygen production of periphyton were significantly reduced by small enhancements of ultraviolet irradiance, which in turn reduced the bacterial productivity coupled to rates of algal photosynthesis (Kahn and Wetzel, 1999). Microscale reductions in water level of only a few millimeters, as is common in shallow littoral and wetland areas on a daily basis, can result in marked increases in UV irradiance intensities and alterations in periphytic metabolism. There is some evidence (Bothwell *et al.,* 1993; Francoeur and Lowe, 1998) for long-term adaptation by periphytic diatoms in streams, however, where initially inhibitory effects of UV on growth were reversed after several weeks of species succession.

F. Growth-Regulating Mortality and Losses

In addition to growth regulation and exploitation competition for resources, a number of processes contribute to mortality, losses, and interference competition: predatory mortality, largely by grazing invertebrates and fishes; natural physiological senescence, death, and population turnover; diseases and viral mortality; and physical disturbances from water movements and substrata instabilities. Each is briefly summarized here and in Table 19-1.

1. Grazing Mortality of Periphyton

Ingestion of microbial communities attached to surfaces has many ramifications for nutrient cycling, productivity, and energy flows in inland waters. Several aspects need to be considered, namely: (1) direct ingestion and mortality of periphyton; (2) release of incompletely metabolized dissolved and particulate organic detritus by the grazers; (3) effects of grazing mortality on the relative productivity of the periphyton communities and the substrata (e.g., living macrophytes) supporting the communities; and (4) nutrition and energy derived by the grazing animals (see Chap. 22).

The growth and productivity of the sessile attached microbiota, particularly epiphytes on submersed portions of macrophytes, can be affected by grazing activities of animals (e.g., insect larvae, snails, crayfish, tadpoles, and certain phytophagous fishes) (Flint and Goldman, 1975; Mason and Bryant, 1975; Bowen, 1978; Eichenberger and Schlatter, 1978; Kitchell *et al.,* 1978; Hunter, 1980; Hagashi *et al.,* 1981; Power *et al.,* 1985; Lamberti *et al.,* 1987, 1992; Brönmark, 1989; Feminella *et al.,* 1989; Feminella and Resh, 1991; Steinman *et al.,* 1995; Steinman, 1996). In certain cases, when the epiphytic algal biomass is low, growth and productivity per unit area of the attached algal and bacterial community can be stimulated (turnover rates increased) by moderate levels of grazing. Moderate grazing could improve access of algae to nutrient and light resources from the overlying water and enhance algal growth as a result. For example, in grazing of stream periphyton communities by snails, the physical disruption and release of nutrients by ingestion and partial digestion can accelerate nutrient regeneration to the overlying water that is normally retained and recycled within the periphyton community. Similar results were found among grazing oligochaetes and snails on epiphytic periphyton of a lake (Kairesalo and Koskimies, 1987). From a quarter to over half of the phosphorus ingested as food was released from the animals through defecation and excretion.

Heavy grazing pressures often result in reductions in overall periphytic biomass or productivity under the commonly used experimental conditions of high grazer density without natural predation pressures upon the herbivores. However, many factors influence the attached microbial–grazer interactions (cf. Wetzel, 1983b; Steinman, 1996), including (a) herbivore type; (b) periphyton density and stage of development; (c) nutrient availability (concentrations, gradients, and water movements at the interface); (d) light quantity and quality; and (e) past disturbance frequency. Clearly, increased nutrient availability and light commonly have positive effects and grazing has negative effects on biomass and productivity of periphyton (e.g., Hart and Robinson, 1990; Rosemond, 1993; Rosemond *et al.,* 1993; Moeller *et al.,* 1998). Compensatory responses of both periphyton and grazing communities are marked and rapid, and as a result of these variable impacts, patterns resulting from the effects of herbivory are difficult to delineate. Extreme care is needed in extrapolation of grazing studies under intensively controlled laboratory conditions to field situations (cf., e.g., Fuller *et al.,* 1998). Although the instantaneous biomass of periphyton may be reduced by grazing, often the productivity of the periphyton community is enhanced by the constant cropping of portions of the attached communities. Little if any effects of macroinvertebrate grazing could be detected on the detachment and redistribution of attached bacteria (Leff *et al.,* 1994).

Very little evidence exists for true species-specific selectivity of attached microflora by grazers. Some experimental studies suggest that algivorous protozoan herbivores and oligochaete worms select prey diatoms on the basis of size and certain species (Bowker *et al.,* 1985; McCormick, 1991). Grazing macroinvertebrates are generally omnivorous and randomly ingest algae, bacteria, particulate detritus, and inorganic deposits (e.g., $CaCO_3$). The structure of benthic microbial communities, although highly variable, can influence grazing efficacy. For example, prostrate species are attached adnate to the substratum, upright or stalked species form a middle level, and filamentous species often emerge from the community to an upper level (Hoagland *et al.,* 1982). Mayfly larvae with gathering collector-feeding structures tend to feed along the outer layers among the loosely attached portions of the periphyton. Snails and caddisfly larvae use rasping and scraping mouth parts and are adapted to feed among the low-profile, adnate, attached periphyton. As a result, larger diatoms (>20 μm) may be more effectively removed by these herbivores than small green algae and cyanobacteria (Cuker, 1983). Herbivore abundance may not be in synchrony with seasonal variations in periphyton production, which may result in apparent minimal grazing impacts. Low-to-moderate grazing usually results in no appreciable changes in attached algal diversity, but intense grazing pressure can reduce diversity (e.g., Lowe and Hunter, 1988; Mulholland *et al.,* 1991). Certain fatty acid components of attached algae are distinctly inhibitory to grazing invertebrates (e.g., LaLonde *et al.,* 1979), and fatty acid composition of the periphyton can be altered by grazing (Steinman *et al.,* 1987). Whether this response is an allelochemical induction, as occurs among the phytoplankton (cf. Chaps. 15 and 16), is unclear.

There is some evidence that grazing mortality of periphyton is a greater percentage of total death and turnover of attached communities in stream ecosystems than in standing waters. Most of the microflora attached to macrophytes and other substrata is not consumed by animals, but instead enters detrital decomposition pathways within the ecosystem (cf. Chap. 23).

2. Disturbance

Disturbance to the habitat and general milieu of periphyton communities is relative in relation to physical disruptions, such as by water movements, and external impacts such as chemical alteration (e.g., acidification) or pollution. The latter topic is a subject of its own and is extensively treated in pollution and water quality literature. For example, much of the saprobic system of evaluating water quality is based on indicator species of attached algae and other microbiota under different conditions of pollution (e.g., Sládeček, 1973).

In lakes with relatively stable water levels, periphyton is well adapted to living in shallow areas of turbulence from wave actions. When periphyton that developed in relatively quiescent, nonturbulent habitats are subjected to unusual water turbulence, such as that associated with an episodic storm event, loosely attached microbiota can be dispersed. Stalked diatoms and similar microbes attach tightly to substrata but are sufficiently flexible to tolerate appreciable water movement (Cattaneo, 1990; Hoagland and Peterson, 1990).

In reservoirs, and to a certain extent also in wetlands, water levels can fluctuate widely. Obviously, the lowering of the water level and desiccation of periphyton communities suppresses growth and increases mortality. Although many diatoms and cyanobacteria are capable of surviving long periods of desiccation, others such as desmids cannot (cf. review of Goldsborough and Robinson, 1996). Fluctuating water levels in reservoirs often expose unconsolidated sediments to wave turbulence and result in resuspension and high turbidity. Similarly in shallow ponds and wetlands, high wind activities can increase turbidity and reduce light penetration. Suppression of photosynthesis and growth of periphyton accompanies such light reductions. If turbidity is short-lived, however, nutrient availability can be increased from the sediment disturbance and result in a positive effect on growth. Conversely, epidemic developments of the exotic zebra mussel (*Dreissena polymorpha*) in lakes can reduce planktonic turbidity by feeding activities and result in improved underwater light distribution (Nalepa and Schloesser, 1993). Increased periphyton growth can occur both in response to enhanced light but also because of the greatly increased surface area for colonization by the profuse development of mollusk shells.

In river ecosystems, the physical effects of turbulence and drag on cells of periphyton can be severe in removal of loosely attached cells and exporting them downstream and in disturbing substrata (Peterson *et al.,* 1994; cf. review of Stevenson, 1996b). Although attached microbial communities are well adapted to water movements, particularly up to about 15 cm s^{-1}, at sudden spates of greater velocities, the export of cells increases precipitously. Conversely, immigration of cells from upstream and colonization on substrata decrease as current velocities increase.

Complex indirect effects on periphyton communities of fluctuations in current velocities in rivers occur (Stevenson, 1996b). The size of sediment particles moved is proportional to current velocities (cf. pp. 97ff), and increasing currents can move substratum and attached microbes with devastating effects from abrasion and redistribution to a less suitable habitat downstream (buried, deeper or shaded areas, et cetera). The time between spate events of sufficient magnitude to move substrata is important. If sufficiently long to allow reasonable development of the microbial communities prior to the next major disturbance, overall community productivity can be quite high. Frequent or unusually high hydrographic events can markedly suppress periphyton productivity.

Changes in current velocities, particularly strong episodic events, can also modify the distribution and movement of herbivorous benthic invertebrates and fish, particularly in central areas of stream channels of fast currents (Poff and Ward, 1992; Hart and Finelli, 1999). Any disturbance of the sediments, such as from nesting of insects or fish, similarly affects the development of periphyton and alters the spatial distribution of attached microbiota in streams.

3. Other Losses

Losses of natural communities of attached microbiota by natural senescence and death, by parasitism particularly by viruses, and by disease are poorly known. As among the phytoplankton and bacterioplankton, nonpredatory losses are certainly significant and likely constitute over half of the natural mortality. This area is a fertile field for study.

G. Community Analyses and Seasonal Dynamics

The development of attached microbial communities on a new surface passes through several stages. Firstly, accrual occurs with colonization of the substratum and subsequent growth in approximately an exponential sigmoid curve (Fig. 19-10). The duration of the accrual phase toward maximum biomass varies with conditions, of course, but is commonly in the range of one to 2 weeks under natural conditions. A shift then occurs to a predominance of internal nutrient recycling and a greater balance between growth and loss processes by death from parasitism, disease, and grazing as well as sloughing and export of cells from the community (Fig. 19-10). Growth is approximately balanced where high productivity is maintained photosynthetically by internal recycling of nutrients and gases and utilization of external resources from the overlying water and counterbalanced by losses and disturbances.

Quantitative evaluations of attached benthic algal populations are much more difficult than qualitative floristic analyses. Recognition of the various habitats and their associations is a problem common to both analyses which is accentuated by ineffective sampling techniques that are difficult at best. It is difficult to separate epipelic algae from sediments, but separation is necessary for accurate enumeration. Phototactic responses of the algae can assist in this separation; alternatively, one can use other biomass methods such as pigment content (correcting for pigment degradation products common to sediments). Quantitative sampling of epilithic and epiphytic algal/cyanobacterial populations is even more difficult, but a number of devices and procedures have been devised (see the reviews of Sládečková, 1962; Lowe and Laliberte, 1996; Stevenson, 1996; Wetzel and Likens, 2000) that permit some degree of quantification. The problems of quantitative sampling of attached algae are further compounded by the settling of casual species originating from the plankton and metaphyton into the benthic populations, especially during winter. Only detailed sampling of the different communities combined with metabolic assays of growth can resolve this problem.

1. Use of Artificial Substrata

Because of extreme habitat and community heterogeneity, artificial substrata of uniform composition and colonizable area are commonly used to estimate colonization and growth characteristics of attached algae. In addition to examination of microbiota on natural substrata, artificial substrata that have been used include glass slides, glass fiber filters, concrete blocks, rocks and rock plates, sand, various plastic and metal plates, and wood.

There are several restrictions inherent in the use of artificial substrata. Marked differences in colonization rates and biomass accrual have been found in relation to the position of the supported substrata and their location within the aquatic system. For example, vertically positioned plates on which sedimenting organisms from the plankton or metaphyton could not accumulate contained less biomass but more realistically simulated population characteristics found on natural substrata than did those positioned horizontally (e.g., Castenholz, 1961; Pieczyńska and Spodniewska, 1963). The influence of water movement or simply the process of retrieving the slides following incubation can lead to significant losses of attached microflora. Some organisms will not adhere to artificial substrata such as glass until early successional forms have colonized the substratum, which may or may not occur within the period of incubation (e.g., Blinn *et al.*, 1980). Apparently glass is not seriously selective for the attachment of diatoms (Patrick *et al.*, 1954), although diatoms can utilize some of the silica from the glass. Grazing can lead to alterations of attached communities that may influence the results, depending on whether the questions being addressed concern specific attached flora or the entire attached community.

Some investigators have positioned artificial substrata in the pelagic zone of lakes or main flow of streams quite removed from sites of natural substrata. Although these analyses may have some value in a relative sense, such as in evaluating responses to inorganic, organic, or acidic toxicants in different water types or venues (Genter, 1996; Hoaglund *et al.*, 1996; Planas, 1996), they are of relative or little value in evaluating natural attached communities.

The most serious criticism that can be directed against studies using artificial substrata centers on the implicit assumption that any metabolite, inorganic or organic, of the living or nonliving substrata has no appreciable effects on the metabolism and growth of the attached community. Similarly, it is assumed that the metabolism of the attached microfloral community has no reciprocal effects upon the substratum. Recent evidence for such synergistic effects is overwhelming that much more than a passive relationship exists among the attached microbiota (see discussion on pp. 590–594).

With these thoughts in mind, the difficulties of quantitatively evaluating natural population growth on natural substrata become apparent. Even though the mutual metabolic relationships are real, especially among epiphytic algae and the supporting macrophytes, measuring the effects of the heterogeneous attached populations on the highly variable substrata often exceeds the capacities of existing methods. The problem is a technical one that must be approached without the use of artificial substrata. Alternatively, when used critically, artificial substrata can be a meaningful tool for the approximate estimation of biomass accrual of many attached microorganisms. Much of our existing information on these communities is based on analyses using artificial substrata. Only a few investigations have addressed the *in situ* rates of algal productivity by means of metabolic techniques.

2. Examples of Population Fluctuations

The detailed investigations of the population dynamics of epipelic algae of several lakes of the English Lake District (Round, 1957a–c, 1960, 1961a,b) serve to illustrate several characteristics of these populations, as well as the potential spatial, temporal, chemical, and biotic factors affecting growth. Diatoms completely dominated the epipelic community, and patterns of distribution could be discerned on the basis of general chemical features of the sediments and overlying water. Other algal groups were less well represented. The Volvocales were very sparse, and the Chlorococcales were represented only by the genera *Pediastrum* and *Scenedesmus*. Although not abundant, the desmid flora was richer in species than other groups. Euglenoids were much more common on more organic-rich sediments. Little correlation was found between the growth cycles of the epipelic and planktonic cyanobacteria, and in some lakes, a conspicuous development occurred on the sediments but not in the plankton. Epipelic cyanobacteria developed equally well in alkaline, neutral, and slightly acidic waters.

In general, a larger diatom biomass corresponded to a higher organic-matter content of the sediments, and growth was better on sediments with more leachable silica. Moderate organic-matter content of the sediments favored cyanobacterial populations, while low populations were associated with sediments of very high and very low organic-matter content. Populations of the poorly represented Chlorophyceae and the flagellated forms were distinctly greater on organic-rich sediments. Although little correlation was found between population numbers and sediment content of sodium, potassium, iron, or manganese, the diatom, cyanobacterial, and flagellated algal populations were directly correlated with calcium content. Algal growth was also greater on sediments with high phosphate content.

Although population numbers of epipelic algae exhibited large variations from lake to lake, and between sites within lakes, and with depth, certain consistent patterns emerged. The main growth period of epipelic diatoms and cyanobacteria of lakes of this moderate temperate region[3] was in the spring, commencing about February (Fig. 19-12). Except for the wave-active shallow zone (0–1-m depth), population numbers were relatively constant on sediments between 1- to 6-m depths; below this depth, numbers decreased precipitously to practically nil at depths greater than 8 m. The growth of the dominant benthic diatom, *Tabellaria flocculosa,* was shown experimentally to be largely related to light intensity, which decreased rapidly with increasing depth (Cannon *et al.*, 1961; Hillbricht-Ilkowska *et al.*, 1972). Strictly phototrophic diatoms are more sensitive to light than cyanobacteria, whose

[3] Ice cover is rare and usually short-lived, typically only a few days in duration.

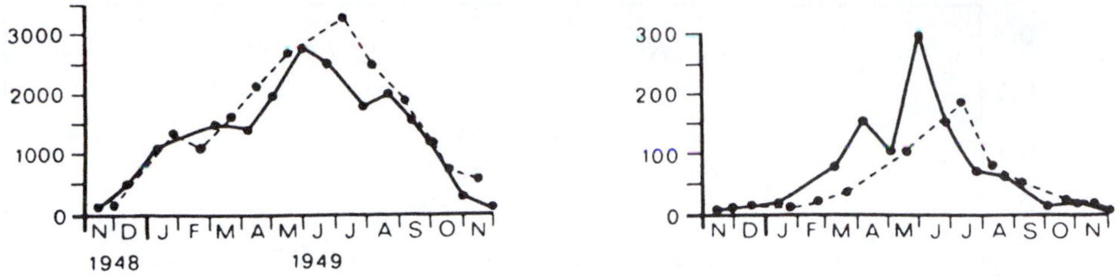

FIGURE 19-12 Seasonal cell counts of epipelic diatoms (*left*) and cyanobacteria (*right*), in which all the depth stations are added together for each sampling date in the photosynthetic zone (1–6 m) of Lake Windermere (———) and Blelham Tarn (- - - -), England. (Modified from Round, 1961b.)

depth distribution usually extends considerably deeper than that of the diatoms. The general seasonal biomass of epipelic algal populations in the photosynthetically illuminated zone generally followed the curves of incident light and water temperature.

In streams, light availability frequently limits benthic algal growth (e.g., Whitton, 1975; Wetzel, 1975b; Marker, 1976; Aizaki, 1978; Biggs, 1996; Hill, 1996). Light availability can be altered markedly by seasonal changes in the leaf canopy of deciduous trees adjacent to the stream. The succession of species within the overall annual cycle of epipelic algae is a much more elusive and complex matter. Among planktonic populations of algae, discussed in some detail in Chapter 15, the changes in physical and chemical parameters, particularly those associated with periods of circulation in stratified lakes, are very large and can influence dominance and competitive abilities of the species. The sudden collapse of populations at certain times of the year is related, in part, to these "shock" events of disturbance.[4] Epipelic populations tend to occur largely above the metalimnion, so the effects of disruption of stratification may be less on these populations than among the plankton. The seasonal succession of attached algal species is, however, just as pronounced as succession among the planktonic species (Round, 1972).

No temperate epipelic algal species is known to persist as a numerically dominant species over an entire year. Specific epipelic algal populations tend to increase quite rapidly and then decline equally rapidly. Often, several successive species associations tend to dominate during an annual cycle, each phasing in to the next from live cells that have remained in the habitat for some time after cell division has ceased. The substrata provide innumerable microhabitats, certainly many more than exist under planktonic conditions of the pelagic zone. Therefore, one would expect a large number of coexisting species in the attached algal community at almost any time of the year. Sampling techniques that tend to aggregate algae from many different microhabitats can obscure this distribution.

Annual and seasonal cycles of epipelic algae are marked and complex (e.g., Round, 1964a; Ilmavirta, 1977; Sheath and Hellebust, 1978; Romagoux, 1979). Diatom growth in lakes often commences in early spring and reaches maximum cell numbers in April to May in the temperate zone, after which a decline occurs in midsummer prior to a smaller autumnal peak that is over by November. Although this pattern is similar to that of planktonic species, the epipelic algal pattern occurs more slowly than that found among planktonic algae. On shallow sediments, the epipelic algal populations tend to exhibit early winter and spring maxima, followed by a later midsummer maximum.

Spring maxima are common among attached algal populations of streams, although these populations are subject to irregular decimation and disturbance by moving substrata during spates (see, e.g., Douglas, 1958; Sand-Jensen *et al.*, 1988). In woodland streams, the seasonally changing light passing through the terrestrial canopy can shift the productivity pattern from one that is autotrophic in the spring to one of essentially heterotrophic dominance later in the season (Kobayasi, 1961; Biggs, 1996; and also discussion in Chap. 22). That seasonal pattern may also reflect a seasonality in herbivory by aquatic invertebrates, which tends in the temperate zone to be greater in the autumn and early winter. Where disturbance is common, particularly from spates and flood events, seasonal patterns with a spring maximum are frequently interrupted. Immediately after a flood event, periphyton may recover rapidly and achieve maximum biomass during a window of time when invertebrate grazer densities are low because invertebrate reproduction and recolonization is much slower (Grimm and Fisher, 1989; Power, 1992; Peterson, 1996).

The seasonal population dynamics of epiphytic algae growing on submersed portions of macrophytes are

[4] A concept discussed by Round (1971).

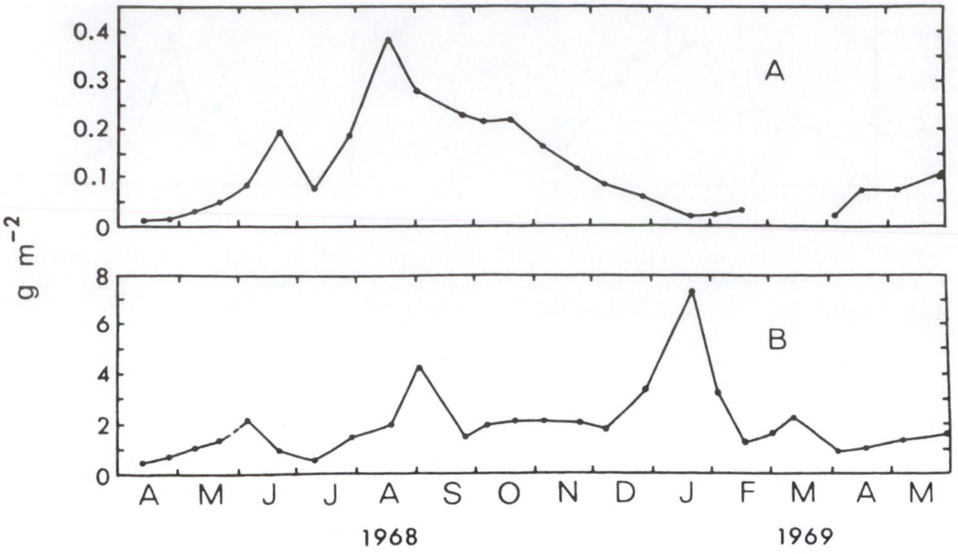

FIGURE 19-13 Chlorophyll *a* concentrations, extrapolated upward to g m^{-2} of the littoral zone on the basis of estimated surface areas of the macrophytes, of algae attached to artificial substrates in dense stands of the littoral zone dominated by (*A*), the emergent bulrush *Scirpus acutus*, and (*B*), the submergents *Najas flexilis* and *Chara*, Lawrence Lake, Michigan. Note the marked differences in the scales of the ordinate in g m^{-2}. (After Allen, 1971.)

much more involved because the surface area available for colonization and growth is continually changing as the macrophytes grow, senesce, and enter detrital phases. Computations of the population changes of the total epiphytic algal community are rare, but changes obviously are quite different among different supporting macrophyte communities (Kowalczewski, 1975a,b; Cattaneo and Kalff, 1980; Müller, 1996). For example, changes in the biomass of attached algae associated with a zone of the emergent bulrush *Scirpus acutus* were more than an order of magnitude lower and out of phase with those of algae attached to substrates in a shallow zone dominated by submersed plants (*Najas flexilis* and *Chara*) (Fig. 19-13). Detailed analyses were made of the population and community dynamics of epiphytic algae and cyanobacteria on leaves of the submersed sedge *Scirpus subterminalis* in a hardwater temperate lake (Burkholder and Wetzel, 1989a). Epiphytic diatoms comprised a mean of 90% of the total epiphytic algal biovolume, whereas cyanobacteria contributed a mean of 88% of the total algal–cyanobacterial cells. Epiphytic cyanobacteria were a negligible portion of the algal biomass except during autumn, when they contributed 40% of the total. However, cyanobacteria were estimated to produce >50% of the total algal carbon (Fig. 19-14).

Part of the reason that the productivity of epiphytes is so high is their persistence within the water column throughout much of the year (Fig. 19-15). Most submersed macrophytes are perennial plants; many species grow or persist in a dormant or "evergreen" condition for much of the year (cf. Chap. 18). The above-sediment biomass of submersed macrophytes commonly reaches a maximum in the spring–early summer period and gradually declines as resources are diverted to rooting tissues. An appreciable macrophyte biomass and surface area can persist all year in subtemperate regions and even under ice in temperate regions.

Epiphytic biomass and productivity tend to be relatively constant throughout the year (Fig. 19-15). Competition among epiphytic algae and phytoplankton for nutrients can reduce epiphytic development in the midyear period. This late spring maximum decline potentially results from increasing nutrient limitations, temperature, and grazing pressure. Epiphytic growth, particularly by cyanobacteria, often increases markedly in late summer and autumn. At this time, lake mixing occurs during autumn, nutrient supplies increase in the water, temperatures decline, macrophyte senescence accelerates, and epiphytic diatom/cyanobacteria populations increase (Fig. 19-15). Diatom productivity is particularly high in winter and spring periods.

Most of the nutrients of actively growing macrophytes are obtained from the sediments. As was demonstrated, most (95–99%) of phosphorus, for example, is obtained from sediment interstitial water and is retained within the plant and recycled repeatedly (cf. Chap. 13, Fig. 13-9). Nutrients from littoral water are very actively assimilated by the loosely attached epiphytic periphyton, incorporated into the periphyton,

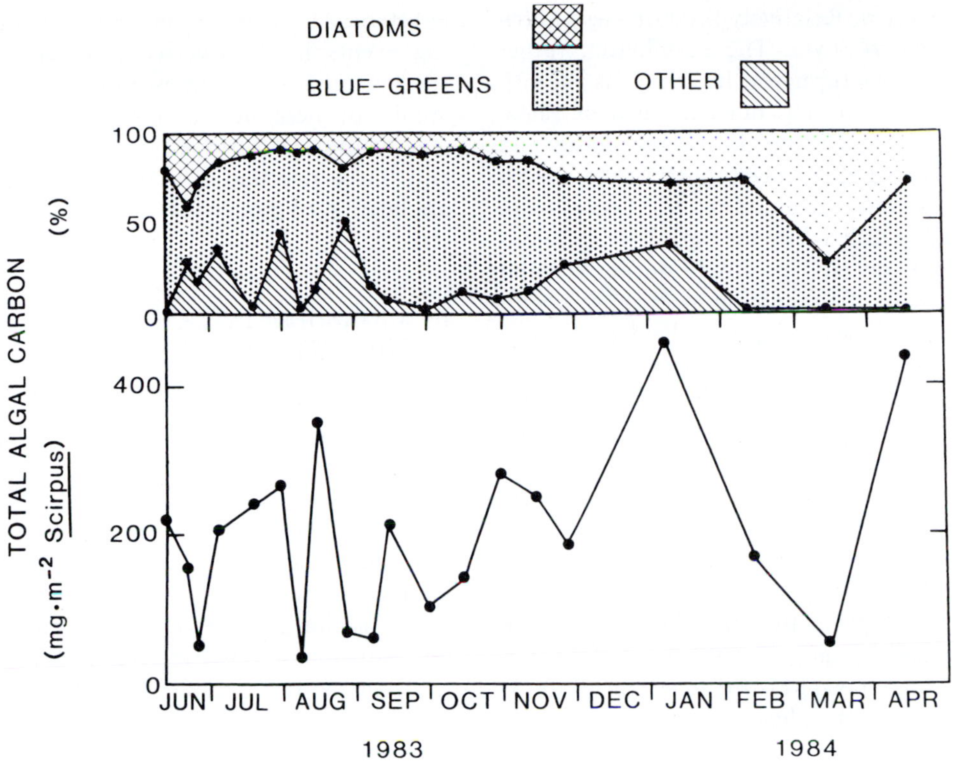

FIGURE 19-14 Organic carbon content of epiphytic algae and cyanobacteria (blue–greens) on apical portions of the submersed sedge *Scirpus subterminalis* in Lawrence Lake, southwestern Michigan. (From Burkholder and Wetzel, 1989a.)

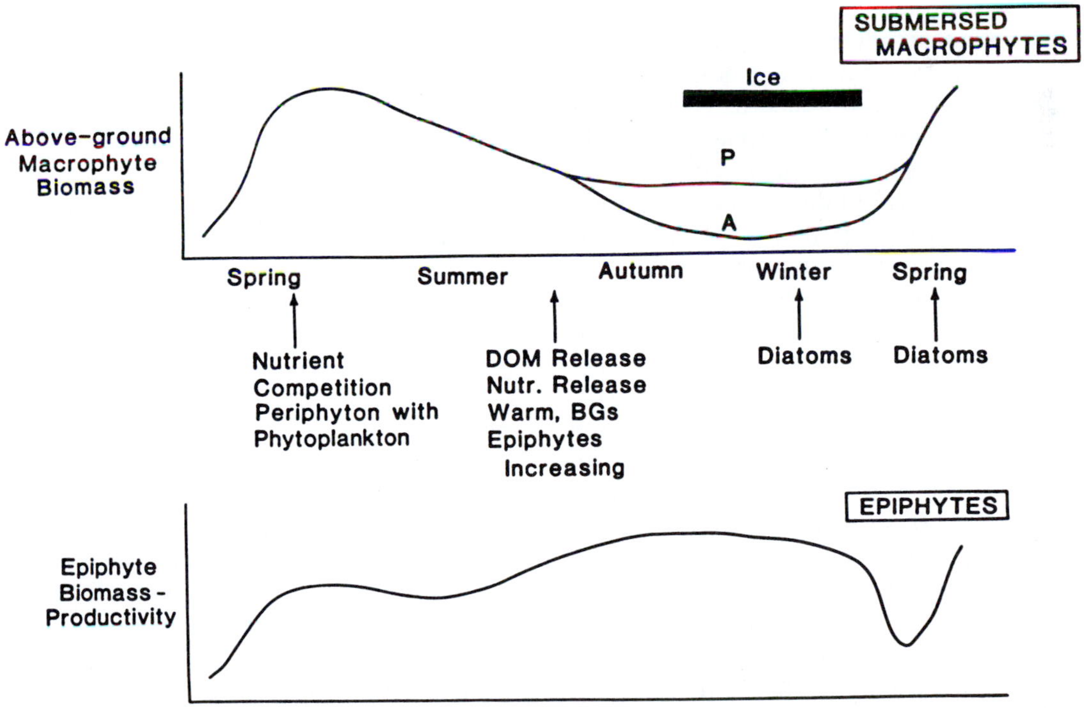

FIGURE 19-15 Generalized relationship of epiphytic biomass and productivity in relation to variations in aboveground biomass of submersed macrophytes (see text for discussion). *P* = perennial and *A* = annual plants; *BG* = cyanobacteria. (From Wetzel, 1990a.)

and intensively recycled. Relatively little of these nutrients pass to the macrophyte. The periphyton, rather than the submersed macrophytes, functions as the primary scavenger for a limiting nutrient such as phosphorus from the water. During dormancy phases and senescence of aboveground foliage, much of the critical nutrients is translocated to rooting tissues. Losses from the leaves are readily utilized by the periphytic community, which tends to develop profusely during autumn and winter periods. As the senescing macrophytes with their epiphyte communities collapse to the detrital mass near the sediments, much of the nutrients is incorporated into the sediments and is actively retained and recycled by the sediment epipelic microflora (Fig. 13-9). The fluxes from the sediments to overlying water are actively influenced by photosynthesis of epipelic algae within the microzone at the sediment–water interface and by the bacterial metabolism below the interface (see Chaps. 12 and 13).

The general assumption that the chemistry of the overlying water has a direct controlling effect on epipelic algae has not been verified experimentally. If buried in the sediments, the algae may also be exposed not only to lower intensities of light but also to anaerobic conditions within the sediments. Many benthic algae can survive anaerobiosis for several days and may retain considerable photosynthetic potential in the absence of oxygen (Moss, 1977). For those species adapted to epipelic existence, the proximity to the totally different chemical milieu of the interstitial water within and immediately adjacent to the sediment–water interface can give them a distinct advantage that is quite unrelated to chemical events occurring in the overlying water. For example, the biomass of epipelic algae commonly increases as the organic carbon, nitrogen, and phosphorus content of the sediments increases (Skorik *et al.*, 1972). A large portion of the ammonia nitrogen of interstitial water of sediments was found to be assimilated by benthic algae (Jansson, 1980). Dissolved organic nitrogen compounds released to the overlying water originated largely from the algae. Nitrogen fixation by epipelic, as well as epiphytic, cyanobacteria and heterotrophic bacteria is common (Finke and Seeley, 1978; Moeller and Roskowski, 1978) and, as indicated earlier (Chap. 12), can constitute a significant input of combined nitrogen to some lakes. Yet the epipelic forms are not completely exempt from the characteristics and events of the overlying water. For example, a high content of dissolved organic matter or a seasonally variable development of planktonic algae or of macrophytic vegetation populations would markedly affect the quality and quantity of light reaching the sediments, the thermal régime of the sediments, water movements near and along the sediments, et cetera. Although it is intuitively apparent that changing events in the overlying water can influence the species succession and productivity of epipelic algae, studies of these interactions are essentially nonexistent. The whole area of investigation is uncharted and difficult but is one that is approachable by the systematic application of modern research techniques.

II. METABOLIC INTERACTIONS IN THE LITTORAL REGIONS

A. Relationships between Algae and Bacteria of Periphyton

The macrophytes release large quantities of both inorganic compounds (e.g., oxygen, carbon dioxide, phosphates, silica, et cetera) and organic compounds secreted during active photosynthesis and released by autolysis during senescence into the immediate proximity of adnate epiphytic microflora and littoral waters (e.g., Howard-Williams *et al.*, 1978; Siver, 1980; cf. Chap. 18). As we have seen, this proximity alone is conducive to transfer and exchange of nutrients (PO_4 and CO_2) between the epiphyte and the supporting plant. In hard waters, precipitation of calcium carbonate is induced by photosynthetic activity (Chap. 11) to form a matrix intermixed with the microflora and mucoid substances (predominantly polysaccharides and peptides). The epiphytic bacteria actively utilize, in part, organic substrates released by the macrophytes and algae. Dissolved organic compounds not utilized by this association and not adsorbed within or to surfaces of precipitating calcium carbonate, enter the pool of littoral dissolved organic matter for further degradation by bacteria (Fig. 19-16). In turn, epiphytic bacteria produce CO_2 and certain organic micronutrients (e.g., vitamin B_{12}) (cf. Wetzel, 1969, 1972), that can serve as growth factors for the macrophytes or algae, or both. Photoassimilative active uptake of labile organic substrates has been demonstrated for a submersed angiosperm in axenic culture and for a species of the macroalga *Nitella* (Smith, 1967; Wetzel, unpublished).

There is evidence to support the conclusion that most phytoplanktonic algae cannot actively augment photosynthesis by heterotrophic utilization of organic substrates when in competition with bacteria that actively assimilate the same substrate (e.g., Mayfield and Inniss, 1978; cf. Chaps. 15 and 17). The immediate juxtaposition of attached algae to the metabolically active area of a living plant and bacterial concentrations can give them a distinct competitive advantage over their planktonic counterparts living in a habitat where

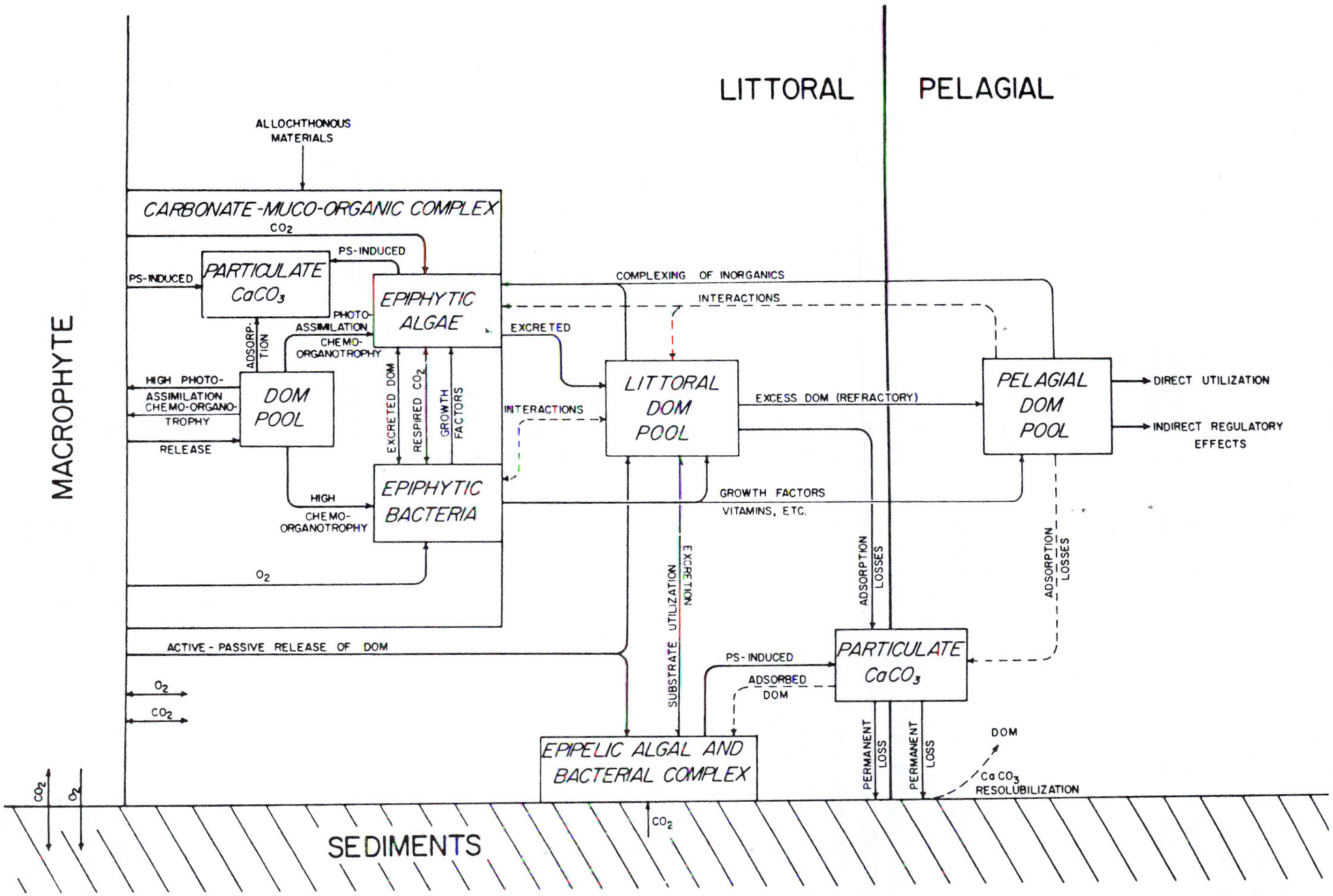

FIGURE 19-16 Diagrammatic representation of major metabolic pathways and interactions in the littoral zone of a typical lake; see text for discussion. *DOM* = dissolved organic matter; *PS* = photosynthesis. (Modified from Wetzel and Allen, 1970.)

nutrients are much more dispersed (see, e.g., the studies of Allen, 1971).

Analogous interactions certainly exist within the epipelic associations of algae and bacteria. Unfortunately, even less is known about the qualitative and quantitative metabolic interrelationships of these attached forms and their substrate than is known about the epiphytic microflora. Potential sources of inorganic nutrients and labile dissolved organic substrates are likely to be greater under reducing conditions of the sediment–water interface than at the surfaces of macrophytes.

Dissolved organic substrates emanating from littoral sources or from the drainage basin are degraded, flocculated, and adsorbed as they travel to the pelagic zone (cf. Chaps. 18, 20, and 23). The rates and kinds of "processing" that occur en route are highly variable and depend upon the dynamics of seasonal and individual differences in the physical and chemical parameters of lakes. The littoral region can be viewed as functioning as an effective but selective metabolic sieve or sink for certain dissolved organic compounds and inorganic nutrients from littoral and external sources toward the pelagic zone (e.g., Howard-Williams and Lenton, 1975; Klopatek, 1975; Lee *et al.,* 1975; Gaudet, 1977, 1978, 1979; Howard-Williams and Howard-Williams, 1978; Mickle and Wetzel, 1978a, b, 1979; Prentki *et al.,* 1978; Stewart and Wetzel, 1981a; Wetzel, 1990a, 1992). During this transport, the more easily degradable organic substrates tend to be decomposed in the littoral zone and compounds of greater recalcitrance (i.e., slower turnover rates) tend to move toward the open water. Therefore, even though the amounts, rates of utilization, and fate of inorganic and organic compounds released from the littoral zone vary seasonally and differ from lake to lake, their direct and indirect effects on the metabolism of the pelagic microflora can be great in many lakes.

B. Effects of Light Availability

Within dense stands of emergent macrophytes, the rapid attenuation of light plays a major role in the reduction of photosynthesis of littoral algae. For example, in reed stands of *Phragmites,* Straškraba and Pieczyńska (1970) found a rectilinear relationship between the percentage transmission of light through the emergent parts of the reed and shoot density. As much as 96% of the incoming radiation was removed above the water, whereas the density of the stand had little further influence on the light availability under water. The rather low light intensities occurring in reed stands, especially those of an average or high density, substantially lowered photosynthesis and biomass of phytoplankton entering the stands from the pelagic region. Photosynthetic activity of attached algae is similarly reduced in these dense stands. Cutting experiments showed a large increase in photosynthetic activity of each algal group, and the increase was directly proportional to the percentage improvement of light conditions (Table 19-2). In less dense stands of emergent plants, as, for example, in low-growing sedges, light reduction is not as serious, but a strong limitation of photosynthesis of phytoplankton still has been observed (Straškraba, 1963). Some decompositional organic compounds released from the plants were also believed to contribute to this reduction.

TABLE 19-2 Effect of Increased Light Availability on Photosynthetic Activity of Phytoplankton and Attached Algae in a Dense Stand of *Phragmites communis* before and after Removal of Plant Shoots, Smyslov Pond, Czechoslovakia, in September[a]

	With Phragmites		*Phragmites cut*	
	0 cm	*30 cm*	*0 cm*	*30 cm*
Phytoplankton (mg O_2 liter^{-1} day^{-1})				
Gross production	0.14	0.06	7.9	3.0
Respiration	0.17	0.78	2.6	1.5
Attached algae				
Gross production (mg O_2 dm^{-2} day^{-1})	3.1		8.7	

[a] From data of Straškraba and Pieczyńska (1970).

Another striking example of the importance of the shading of light from periphyton by emergent macrophytes was determined in the Everglade wetlands of Florida (Grimshaw *et al.,* 1997). Photosynthetically active radiation reaching the periphyton communities was reduced by only 35% in the more open sloughs of the emergent sawgrass (*Cladium mariscus*) but by 85% or more in dense stands of cattail (*Typha domingensis*). Photosynthetic rates of periphytic algae and cyanobacteria in sawgrass habitats were reduced by about 30% from those in openwater slough habitats; the net production rates of periphyton communities in the cattail communities were reduced by 80 ± 8% of rates in the open communities.

Light reduction by a seasonal development of an emergent macrophyte canopy can cause a seasonal reduction in the epiphytic periphyton as well. For example, the productivity of epiphytic algae on *Equisetum* was very high in early summer but declined during the summer as shading by the macrophytes increased (Kairesalo, 1984). Herbivory by snails also increased as the summer season progressed.

Light availability can also be reduced by abiotic factors at intermittent intervals and have marked effects on periphyton productivity. For example, algal biomass epiphytic on floating meadows of aquatic plants in an Amazon floodplain lake was severely reduced by inorganic turbidity imported by river water (Engle and Melack, 1993). Nonetheless, high turbidity impacted composite phytoplankton growth more severely, and as a result epiphytic algal productivity per unit area was much higher than that of the phytoplankton.

C. Attached Algae as a Source of Phytoplankton

The concept that littoral benthic algae can give rise to plankton algae emerged many years ago. It was assumed that plankton algae formed resting spores that overwinter in or upon the sediments and that under conditions favorable to germination and population development, algae on the sediments served as the inoculum for the phytoplankton. In moderate-size to large lakes, there is no evidence for any conspicuous buildup of algae on the sediments before a plankton bloom, but rather only afterward when the dead plankton settle out onto the sediment (Round, 1964).

An alternative concept is that planktonic populations are always present in the water, although in extremely low concentrations. These few remnants of past populations can then serve as the inoculum for a rapid population increase under favorable growing conditions. This situation seems to be the case for most species.[5] In one of the better studies of algae, resting spores of the diatom *Asterionella formosa* were not observed, and no evidence exists that appreciable numbers originate from shallow areas (Lund, 1949). Massive population developments more commonly originate from exponential growth of live cells present in the pelagic waters.

Some phytoplankton do settle out during periods of reduced turbulence and remain in a resting state on the sediments until water movements are again sufficient to reintroduce the resting cells back into the plankton.[6] *Aulacoseira*[7] *italica* is a large diatom that clearly does enter a resting stage on the sediments, particularly under ice cover and during summer stratification (Lund, 1954, 1955). There is no evidence, however, for active growth on the sediments, which are often below the photic zone. Resuspension of filaments during autumnal and spring circulation serves as a large inoculum for the rapidly developing pelagic population. Similar results have been found for cyanobacteria (Preston *et al.*, 1980; Fallon and Brock, 1981; cf. Chap. 15).

In shallow lakes, ponds, and backwater areas of streams and wetland, the separation between epipelic or epiphytic algae and littoral metaphyton is less distinct. It is common to find strictly benthic algae or predominantly planktonic forms intermixed, especially following irregular storm turbulence (see, e.g., Lehn, 1968; Moss, 1969a; Moss and Abdel Karim, 1969; Brown and Austin, 1973a). In shallow lakes in which thermal stratification is irregular and intermittent, shifts in the percentage composition of largely planktonic or of epipelic algae can be rapid. For example, in a pond in southern England, nonmotile species dominated planktonic populations only when the water column was relatively turbulent. Motile species persisted in the water column, and the largest biomass of any species occurred when the water column was stratified. In the shallow Obersee of Lake Constance, southern Germany, littoral diatoms and algae of the metaphyton are transported to the open water, especially during spring storms (Müller, 1967).

III. PRODUCTIVITY OF LITTORAL ALGAE

A. Quantitative Evaluations

Quantitative evaluation of attached algae is basic to analyses of population changes and productivity but is confounded seriously by the simultaneous collection of much debris and dead cells in various stages of decomposition. Identification, enumeration, and estimation of biovolume, with all proper precautions and statistical evaluations, have a number of advantages for detailed analyses (Lund and Talling, 1957; Lund *et al.*, 1958; Vollenweider, 1969b; Wetzel and Likens, 1991, 2000; American Public Health Association, 1998). Although developed for plankton, these methods of analysis usually are equally applicable to studies of littoral algae. Species enumeration can be used to estimate productivity when coupled to meaningful biomass parameters such as cell volumes or carbon at the species level (cf. Chap. 8). Cellular constituents such as pigment concentrations can be used effectively as an estimate of attached algal biomass (e.g., Szczepański and Szczepańska, 1966). It is particularly imperative, however, to correct for pigment degradation products (Wetzel and Westlake, 1969; Wetzel and Likens, 2000) and not to overextend these estimates, which vary

[5] Termed *holoplankton* in older literature, in contrast to meroplankton that have resting stages in sediments and are not observed in the open water for long periods of time (i.e., they are planktonic only at certain times in their life history).

[6] See page 359.

[7] Formerly Melosira.

significantly in response to an array of physiological conditions.

B. Biomass Measurements

To a limited extent, a temporal series of biomass measurements permit estimates of rates of primary production. These estimates generally underestimate actual productivity because of losses via excretion of organic compounds, respiration, mortality, decomposition, grazing, et cetera. The observed rates also are influenced by the rate of colonization of new propagula, which is generally faster when conditions for growth are better. Hence, the observed accumulation of biomass is a composite of colonization and production. The climax accumulations of biomass on substrata yield little insight into the production dynamics that have occurred before the cumulative maximum in balance between production and losses.

The rate of community turnover is an average of several variable species rates and is difficult to determine from biomass analyses. Subjective errors also occur in determinations of attached algae of lakes and streams; important factors include the frequency of sampling and estimation of the degree of colonization that represents a climax stage of the community (Sládeček and Sládečková, 1964; American Public Health Association, 1998). Apparently higher turnover rates result as the sampling interval is decreased, especially when the intervals are shorter than a month; sampling frequency must be compatible with reproductive rates of dominant algal populations.

Productivity values of littoral algae are difficult to compare from one lake to another, or even seasonally within a lake, because of the diversity of techniques employed. Very few *in situ* metabolic measurements have been made on these communities. As mentioned earlier, many analyses have employed artificial substrata to reduce the natural heterogeneity of the algae. The resemblance of benthic algae on artificial substrates to those on adjacent natural substances is highly variable and deviates significantly, both qualitatively and quantitatively, with seemingly minor differences in substrata, environmental differences in microhabitats, and particularly in rates and duration of colonization. The discrepancies found between populations on natural and artificial substrata are sufficient to necessitate a critical evaluation of each study in which artificial substrata is used (see, e.g., Tippett, 1970; Warren and Davis, 1971). The extent of compromise must be determined in the application of any ecological method, and this problem becomes especially acute among littoral communities.

The admixture of estimates of productivity values in the ensuing discussion should be viewed as representing the range encountered under natural conditions. Although no attempt has been made to summarize all of the literature, which is quite diverse and of variable precision, the examples are presented to point out primary characteristics and emphasize the major contributions that the attached flora can make to freshwater ecosystems.

C. Spatial and Temporal Variations

The rates of primary production of attached algae are obviously dependent upon the substrate area available for colonization within the zone of adequate light. In some lakes, the littoral substratum may be relatively uniform for much of the perimeter, but in most cases, the variation is great. Moreover, living macrophytic substrata are constantly changing, especially in the more seasonal temperate regions. An example of this spatial heterogeneity in productivity can be seen in a 2-yr analysis of littoral components—macrophytes, littoral phytoplankton, and attached, largely epiphytic, algae—in Mikolajskie Lake, northern Poland (Pieczyńska, 1965, 1968; Kowalczewski, 1965; Pieczyńska and Szczepańska, 1966). Simultaneous measurements in midsummer along numerous points of the littoral zone showed much variation, which was accentuated during sporadic mass appearances of sessile cyanobacterium *Gloeotrichia* and the filamentous green alga *Cladophora* (Fig. 19-17). Similar spatial variations in numbers and species assemblages have been demonstrated in the littoral of a eutrophic lake in British Columbia (Brown and Austin, 1973b).

Seasonal fluctuations in primary productivity of attached algae are also quite variable, and are analogous to changes in biomass discussed earlier. In general, estimates of biomass accrual, such as chlorophyll, and

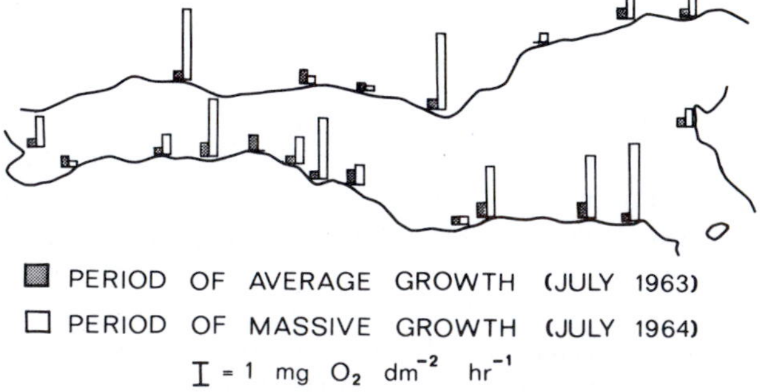

FIGURE 19-17 Spatial variations in the simultaneously measured rates of primary production of attached algae in different areas of the littoral zone of Mikolajskie Lake, central Poland. (After data of Pieczyńska and Szczepańska, 1966.)

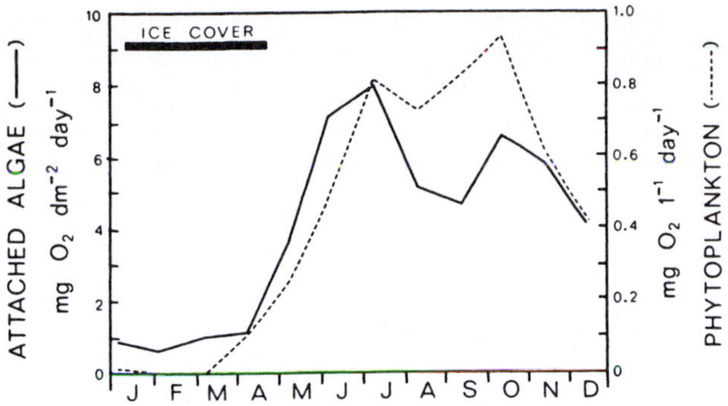

FIGURE 19-18 Rates of primary production of attached algae and phytoplankton in the littoral zone dominated by the emergent macrophyte *Phragmites*. (Modified from Pieczyńska and Szczepańska, 1966.)

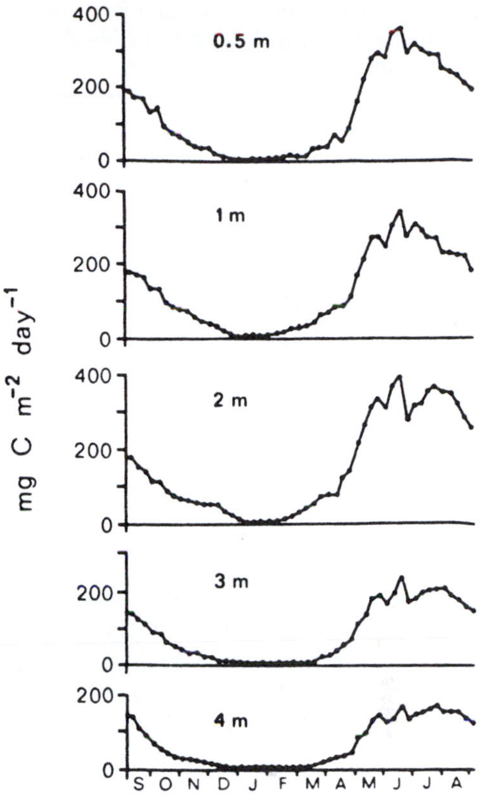

FIGURE 19-19 Weekly mean values of gross primary production rates of epipelic algae in shallow Marion Lake, a composite from five littoral stations, southern British Columbia. (Redrawn from Gruendling, 1971.)

in situ measurements of production rates by carbon uptake or oxygen production are fairly similar, particularly during periods of active growth (see Pieczyńska and Szczepańska, 1966; Allen, 1971; Hickman, 1971a; Hunding and Hargrave, 1973; Goldsborough and Brown, 1991; Robinson *et al.*, 1997). Reductions in temperature and light under winter conditions clearly are the dominant causal mechanisms limiting photosynthesis both seasonally and vertically with depth (see Figs. 19-18 and 19-19). The productivity of attached algae collected under ice and exposed to increased light was found to increase greatly and approached values obtained under summer conditions. In the example from Mikolajskie Lake, productivity in the spring increased rapidly in both the attached algae and littoral phytoplankton, decreased in midsummer during maximum development of the littoral macrophytes, and then increased in the fall with a bloom of diatoms. Where submersed macrophyte biomass persists in 'evergreen' form under ice, and that ice is not covered by dense light-reducing snow for long periods, attached algal productivity, particularly by diatoms, can be very high (Fig. 19-15).

Photosynthetic enhancement of epipelic algae was shown experimentally when sediment samples were moved to shallower depths with improved light conditions (Table 19-3). Hunding (1971) has shown that benthic diatoms exhibited light-saturated photosynthesis over a wide range of light intensities. During summer, light-saturated photosynthesis even at very high irradiances showed no photoinhibition up to 38 klux (= 742 μmol quanta m^{-2} s^{-1}). The light-saturated

TABLE 19-3 Enhancement of Photosynthesis of the Epipelic Algal Community of Marion Lake, British Columbia, in June by Moving Intact Sediments from Different Depths to 0.5 m[a]

Sample depth (m)	*Solar radiation (g cal cm*$^{-2}$ *h*$^{-1}$*)*	*Temperature* (°C)	*Incubated* in situ *at depth (ml* O_2 *m*$^{-2}$ *h*$^{-1}$*)*	*Incubated at 0.5 m depth (ml* O_2 *m*$^{-2}$ *h*$^{-1}$*)*
0.5	32.6	23.0	19.9	19.9
1.0	23.3	22.0	22.7	26.4
2.0	14.4	17.0	19.4	28.8
3.0	10.6	14.0	14.5	19.4
4.0	5.5	13.0	2.4	5.7

[a] From data of Gruendling (1971).

TABLE 19-4 Annual Mean Biomass and Rates of Primary Production of Epipelic Algae of Abbot's Pond, Southern England[a]

Year	*Littoral depth category (m)*	*Mean biomass (mg chlorophyll* a m^{-2})	*Mean rate of primary production* ($mg\ C\ m^{-2}\ h^{-1}$)
1966	0–1.0	11.7	5.50
	1.0–2.5	7.3	2.89
	2.5–4.0	4.0	0.56
1967	0–1.0	9.09	4.13
	1.0–2.5	1.38	0.75
	2.5–4.0	1.47	0.17
1968	0–1.0	13.6	6.23
	1.0–2.5	2.14	0.70
	2.5–4.0	1.47	0.23

[a] From data of Moss (1969b) and Hickman (1971a).

photosynthetic value decreased in autumn, indicating that maximum photosynthetic rates are probably maintained under reduced prevailing light conditions at that time of year. Similar depth relationships can be seen in comparisons of pigment biomass and primary productivity of epipelic algae in a small English lake (Table 19-4) and in a clear, deep alpine lake (Capblancq, 1973).

Within sediments, the rates of carbon fixation and the chlorophyll concentrations of epipelic algae are attenuated very rapidly with depths of a few millimeters (Table 19-5), whereas significant rates and pigment concentrations of epipsammic algae penetrate somewhat deeper. As discussed earlier, the light-mediated metabolism of these algae is probably intermittent and occurs predominantly when wave action circulates sand grains up near to the sediment surface. Biomass

TABLE 19-5 Relative Rates of Carbon Uptake by Epipelic and Epipsammic Algae at Different Depths within Sediment, Expressed as a Percentage of the Maximum Value, Shear Water, England[a]

	% of maximum values			
	Epipelic algae	*Epipsammic algae*		
Depth (cm)	*17 Mar. 68*	*23 Nov. 67*	*17 Mar. 68*	*9 Sept. 68*
0–1.0	100	100	100	100
1.0–2.0	11.6	57.4	42.1	6.6
2.0–3.0	0	33.4	5.7	0.2
3.0–4.0	0	30.5	3.1	1.2
4.0–5.0	0	14.3	2.5	0.3

[a] All samples incubated at the same light conditions. From data of Hickman and Round (1970).

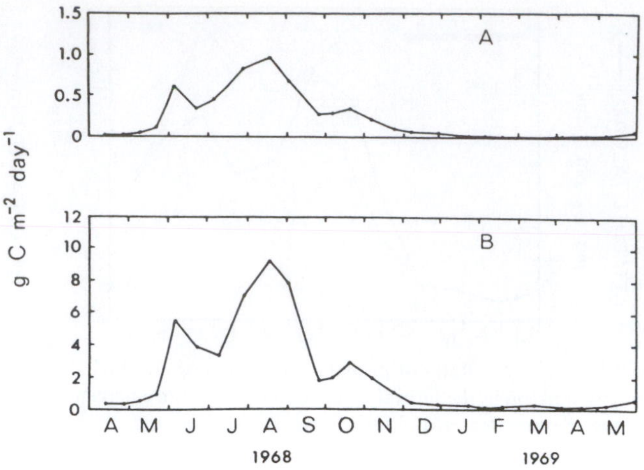

FIGURE 19-20 *In situ* primary production of attached algae in g C m^{-2} of the littoral zone day^{-1} among (*A*) emergent *Scirpus acutus*, and (*B*) submersed macrophytes *Najas flexilis* and *Chara*, Lawrence Lake, Michigan. Note marked differences in ordinate scales. (Modified from Allen, 1971.)

and rates of production of epipsammic algae were consistently greater throughout the year than those of epipelic algae in the same lake system.

Seasonal cycles of primary productivity of epiphytic algae differ considerably in relation to the type of supporting macrophyte. In the Lawrence Lake example shown in Figure 19-20, the annual maxima occurred in midsummer when macrophyte biomass, and hence substrata for colonization and growth, was greatest. At this time conservative estimates exceeded 8 g C m^{-2} littoral zone day^{-1} of carbon fixed by epiphytic algae in this oligotrophic lake. Even though the rates of production per substratum area were similar among the emergent bulrush *Scirpus acutus* site and that of the much more dissected submersed macrophytes *Najas* and *Chara*, the much greater area available for colonization in the submersed macrophyte site increased the rates of production by epiphytic algae in the littoral zone area dominated by the two more dissected macrophytes. In both cases, annual maxima were found in August near the sediments at the basal portions of the macrophytes. Although the winter rates of carbon fixation by attached algae among the *Najas–Chara* were 100–200 times greater than those among the *Scirpus*, the values per substrate area among the more open *Scirpus* littoral were consistently much higher, ranging from double to an order of magnitude higher per area of plant. Causes for this difference are unknown, but the more open *Scirpus* bed had greater light and water movement than occurred in the dense *Chara* beds. Furthermore, as discussed in Chapter 18, the macroalga *Chara* is known to release sulfur-containing organic compounds that are antimicrobial.

TABLE 19-6 Rate of Production of Attached Algae and Estimated Turnover Rates of the Algal Populations of Several Exemplary Aquatic Systems

Lake	*Type of algae*	*Average biomass* ($g\ dry\ m^{-2}$)	*Average annual estimated net productivity* ($mg\ dry\ m^{-2}\ day^{-1}$)	*Estimated turnover time* (*days*)	*Source*
Sodon Lake, MI	Glass slides, horizontal, open water	1.23	37.5	32.8	Newcombe (1950)
Falls Lake, WA	Glass slides, horizontal, at sediment interface (0.4 m)	3.24	148	21.8	Castenholz (1960)
Alkali Lake, WA	Glass slides, horizontal, at sediment interface (0.4 m)	2.94	131	22.4	Castenholz (1960)
Sedlice Reservoir, Czechoslovakia	Glass slides, vertical, open water, several depths	12.7	213	59.6	Sládeček and Sládečková (1963)
Lake Glubokoye, Russia	Epiphytic on *Equisetum*, biomass, and O_2 production	18.7	128.8	21.7	Assman (1953)
Silver Springs, FL	Epiphytic on *Sagittaria*, biomass, and O_2 production	177	12,300	14.4	Odum (1957)
Laboratory analyses, diatoms of Lake Balaton, Hungary	Epilithic diatoms, biomass, and O_2 production	—	—	Approx. 1 (optimal conditions)	Felföldy (1961)
Jeziorak and Tynwald, northern Poland	Glass slides; in reed zones	—	—	4.5–16	Bohr and Luscinska (1975)
Raspberry Pond, Manitoba	Epiphytic on rods in dystrophic pond; ^{14}C methods	—	1700	—	Goldsborough and Brown (1991)
Crescent Pond, Manitoba	Epiphytic algae on macrophytes; ^{14}C methods	40	147	—	Hooper and Robinson (1976, 1978)
Suomunjärvi, Finland	Epipelic algae of an oligotrophic lake; ^{14}C methods	—	1.1	—	Sarsa (1979)
Pääjärvi, Finland	Epipelic algae, ^{14}C methods	—	60	—	Kairesalo (1976, 1977)
	Epipelic algae, ^{14}C methods	—	6	—	
Changing, Antarctica	Epipelic algae on rocky debris; O_2 methods	—	18	—	Priddle (1980)
Sombre, Antarctica	Epipelic algae on rocky debris; O_2 methods	—	49	—	Priddle (1980)
Tundra, Ponds, Alaska	Epipelic algae of shallow ponds; ^{14}C methods	—	22–55	—	Stanley (1976)
Tamagawa River, Japan	Glass slides				Aizaki (1979)
	Bacteria	—	—	0.13–0.42	
	Algae	—	—	0.5–1.2	

In an oligotrophic lake in New York, photosynthetic rates of epiphytic algae were approximately 5% of the photosynthetic rates of the supporting macrophytes (Sheldon and Boylen, 1975). Epiphytic algal densities increased with the seasonal growth patterns of the macrophytes; maximal algal colonization of leaves of the plants, however, was similar per unit time. Epiphytic algal productivity from older leaves, lower on the macrophyte, was as much as 10-fold greater than that of epiphytes from younger top leaves. Similar results were found in the more detailed studies of Burkholder and Wetzel (1989a).

Rates of epiphytic bacterial utilization of simple organic substrates (glucose and acetate) followed first-order active transport kinetics (Allen, 1971). These rates were nearly an order of magnitude greater per unit area in the submersed macrophyte littoral region than among the emergent plants. This difference is potentially related to the greater quantities of organic substrates released by the submersed plants than by emergent macrophytes.

A comparison of production rates of attached algae among lakes and climatic regimes is difficult because of the variety of methods that have been used. However, there is a clear tendency for the rates of primary productivity of the attached algae to increase with decreasing latitude and longer growing season (Tables 19-6 and 19-7). The turnover times of the attached algal populations also tend, not unexpectedly, to decrease under the better growing conditions. The highest production values recorded are from subtropical springs in which constant temperatures, high flow, and adequate nutrient supplies combine to enhance high rates of production. A similar relationship is seen among streams (cf. Wetzel, 1975b; Aizaki, 1978; Wetzel and Ward, 1992), although available data are few and very difficult to compare in a satisfactory manner.

A comparison of the rates of in-stream primary productivity of attached algae among intensively studied sites from many geographical and geological regions of North America showed rates to be similar (Table 19-8). Concentrations of chlorophyll *a* of benthic microbiota spanned an order of magnitude (10–100 mg m^{-2}) with little geographical differentiation within a seasonal range of ca. 1–300 mg m^{-2}. Highest amounts were found in open plains or desert regions of high productivity.

Gross primary productivity of within-stream benthic algae (Table 19-9), as well as community respiration (e.g., consumption of dissolved oxygen in 24 h), commonly increases with downstream direction (e.g.,

TABLE 19-7 Comparisons of Net Rates of Production of Attached Algae, Expressed as Organic Matter Accrual over a 10-Day Period, of Several Different Fresh Waters[a]

Lake	*Mean net production rate (mg org. mat. dm^{-2} day^{-1})*	*Depth of maximal net production accrual*	*Remarks*
Sodon Lake, MI	0.5	0.2 m	Horizontal glass slides, 3 summer months; Newcombe (1950)
Falls Lake, WA	1.28	0.4 m	Horizontal glass slides, 22 months; Castenholz (1960)
Lenore Lake, WA	1.00	0.4 m	17 months
Alkali Lake, WA	1.76	0.4 m	12 months
Soap Lake, WA	1.67	0.4 m	17 months
Walnut Lake, MI	1.69	0.6 m	Horizontal glass slides, 2 summer months; Newcombe (1950)
Sedlice Reservoir, Czechoslovakia	2.13	3.0 m	Vertical slides, 10 months; Sládeček and Sládečková (1964)
Lake Tiberias, Israel	2.20	1.2 m	Vertical glass slides, 14 months; Dor (1970)
Shallow ponds, central MI	3.63	—	Glass slides, 2 summer months; Knight *et al.* (1962)
Borax Lake, CA	14.63	0.2 m	^{14}C methods, 12 months, *in situ* on epilithic algae; Wetzel (1964)
Red Cedar River, central MI	21.2	—	Plexiglas plates, summer months; King and Ball (1966)
Tiberias Hot Springs, Israel	73.0	0.1 m	Horizontal glass slides, 2 months; Dor (1970)
Silver Springs, FL	96.7	—	O_2 methods, epiphytic on *Sagittaria*, 12 months; Odum (1957)

[a] Techniques employed vary widely and only an approximate comparison is possible.

TABLE 19-8 Mean Annual Net Primary Productivity of In-Stream Community Producers of North American Streams

	Mean annual net community primary productivity (mg carbon m^{-2} day^{-1})
Eastern deciduous forest stream, coastal climate, Pennsylvania[a]	−27.0–246.8 ($\bar{x}_4$ 127.4)
Mesic hardwood forest stream, continental climate, Michigan[a]	−55.9–486.4 ($\bar{x}_4$ 207.3)
Cool, arid climate stream, much precipitation as snowfall, coniferous vegetation on north-facing slopes, sagebrush on south-facing slopes, Idaho[a]	48.2–524.7 ($\bar{x}_7$ 415.2)
Northern cool-desert stream open, no canopy, Idaho[b]	
Autochthonous	
Macrophytes	47.2–147.6 ($\bar{x}_3$ 93.4)
Periphyton	465.6–1748.0 ($\bar{x}_3$ 1004)
Allochthonous particulate organic matter	0.7–16.9 ($\bar{x}_3$ 7.0)
Coniferous forest stream, coastal climate, Oregon[a]	−21.0–93.8 ($\bar{x}_4$ 45.1)

[a] Extracted from data of Bott (1983).
[b] Calculated from Minshall (1978) assuming 1 kcal = 4.6 g organic matter and 46.5% carbon in organic matter of aquatic plants (Westlake, 1965).

TABLE 19-9 Examples of Rates of In-Stream Primary Production of Benthic Algae

Site	*Mean primary productivity (mg carbon m^{-2} day^{-1})*	*Technique*	*Source*
Direct estimates:			
Morgans Creek, KY	0.7–4	Biomass change	Minshall (1967)
Walker Branch, TN	8–11	Biomass change	Elwood and Nelson (1972)
Glade Branch, VA	10	^{14}C	Hornick *et al.* (1981)
Piney Branch, VA	11	^{14}C	Hornick *et al.* (1981)
Guys Run, VA	18	^{14}C	Hornick *et al.* (1981)
Berry Creek, OR	99	Oxygen exchange in recirculating chamber	Reese (1967)
Yellow Creek, AL	49.3	^{14}C	Lay and Ward (1987)
Little Schultz Creek, AL	91.2	^{14}C	Lay and Ward (1987)
Amazon River, Brazil	465–1040	^{14}C	Putz (1997)
Whole community estimates:			
New Hope Creek, NC[a]	330–1430	Diurnal oxygen curve	Hall (1972)
Riera Major, Spain	137	Diurnal oxygen, metabolism chamber	Guasch and Sabater (1998)
La Solana, Spain	430	Diurnal oxygen, metabolism chamber	Guasch and Sabater (1998)
Ober Water, England	103	Biomass change and estimated generation times	Shamsudin and Sleigh (1994)
River Itchen, England	822	Biomass change and estimated generation times	Shamsudin and Sleigh (1994)
Deep Creek, ID	1180	Oxygen exchange, upstream/downstream	Minshall (1978)
Catahoula Creek, MS	1850	Oxygen exchange, upstream/downstream	de la Cruz and Post (1977)
White Clay Creek, PA	2220	O_2/CO_2 exchange	Bott *et al.* (1978)

[a] Enriched by fertilizer and sewage.

Bott *et al.*, 1985). Photosynthetic efficiencies of benthic algae are generally below 4%, with averages near 1%, and decline downstream with increasing light availability (Fig. 15-20).

D. Production Rates of Attached Algae versus Phytoplankton and Macrophytes

Few detailed studies exist in which *in situ* measurements of productivity of attached algae, phytoplankton, and macrophytes have been made simultaneously (Table 19-10). However, these studies are important in that they demonstrate the magnitude of effect that littoral productivity can have on lake systems. The range is great. In the large, shallow Borax Lake in northern California, epilithic algae are a major source of organic matter. Because the littoral area is small, when productivity per unit area is expanded for the whole lake surface, the attached algal contribution is reduced but still constitutes nearly one-half of total primary productivity. In Marion Lake, the primarily epipelic algae dominate the lake system. In Lake Wingra, Wisconsin, the submersed macrophytes are major components of the total primary productivity. Littoral attached algae, largely epiphytic, of Lawrence Lake constitute nearly three-fourths of the primary lake productivity, in spite of the limited littoral area of this small, moderately deep lake (Table 19-10). In contrast, the littoral contribution to the total productivity of the two Ontario lakes constitutes only a few percent because the morphometry and geology of these lakes do not lend themselves to extensive littoral development.

A few other, less complete studies that do not permit full comparisons should be mentioned to relate the relative rates of photosynthesis of attached algae to those of phytoplankton. Slow-growing submerged mosses, almost exclusively *Marsupella aquatica*, cover some 40% of the bottom from between 2 to 35 m of clear, low mountain Lake Latnajaure in Swedish Lapland (latitude 68°). This large, deep ($\bar{z}$ = 16.5 m, z_m = 43.5 m) lake is covered with over a meter of ice for about 10 months of the year and has a mean annual water temperature of 2°C. During the period of investigation in August and September, the average primary productivity for the whole lake was between 3.5 and 4.0 kg C per day (Bodin and Nauwerck, 1969). The average turnover time of the biomass of the perennial moss was about 30 y, between extremes of about 15 y at a depth of 5 m and 125 y at depths greater than 30 m. Of the total primary productivity of the lake, the phytoplankton were responsible for 60%, the moss for 20%, the epipelic diatoms for 15%, and the epiphytic algae on the moss for 5%. Similar results have been found in Char Lake in the Canadian Arctic at latitude 75° N (Kalff and Wetzel, unpublished).

Evaluation of the productivity of phytoplankton and attached algae of a large, deep alpine lake of the central Pyrenees Range also indicated the major contribution of the macroalga *Nitella* and of epilithic algae to the total primary productivity (Capblancq, 1973). The attached algae of the littoral zone (0–6 m) and *Nitella* (6–19 m) formed a biomass some 140 times greater than the mean biomass of the phytoplankton during summer, and *Nitella* constituted about 80% of the total benthic algae. Based on *in situ* productivity measurements, the contribution of benthic algae to the total primary productivity of this lake was estimated to be 30%.

The work of Straškraba (1963) on two large, shallow fishponds of southern Czechoslovakia deserves particular mention. Based on extensive annual cycles of the nitrogen content and biomass dynamics of all plant and animal components of these shallow lakes, it was determined that about 73% of the primary productivity resulted from largely emergent macrophytes, 20% from attached, mainly epiphytic algae, and 7% from phytoplankton. It was shown that <10% of the primary productivity was utilized by higher trophic levels.

Studies evaluating the relative photosynthetic rates of algae of the littoral in comparison to the rates of phytoplankton are few. But the suggestion that the most productive site is epiphytic on macrophytes repeatedly appears, as shown, for example, Table 19-11 for a shallow English lake, Lawrence Lake, discussed earlier (Table 19-10), and in the littoral of Mikolajskie Lake (Table 19-12). Among the most detailed and instructive studies of the relative contributions of the attached algae of different substrata is from long-term analyses of periphyton of the extensive wetland on the shores of Lake Manitoba (Robinson *et al.*, 1997). Epiphytic and particularly metaphytic algae totally dominated the littoral algal productivity (Table 19-13), and production rates declined markedly as mean water levels were raised by half a meter.

These results support earlier discussion on the importance of factors of light, which is often coupled to water levels in shallow areas, wave action–sediment movement, and inorganic–organic nutrient exchanges between macrophytes and attached algae. It is apparent that growth conditions are often better in the littoral attached habitat than in the planktonic habitat. The significance of littoral productivity to a lake system depends to a great extent on the physical conditions of the lake morphometry, light availability, and characteristics of the substrata available for algal attachment and growth.

TABLE 19-10 Examples of Annual Net Productivity of Phytoplankton, Littoral Algae, and Macrophytes of Several lakes in Which Productivity Estimates of Attached Algae Were Made on Natural Substrata

Lake	*Area (ha)*	*Mean depth (m)*	*Annual mean* ($mg\ C\ m^{-2}\ day^{-1}$)	*Annual mean* ($kg\ C\ lake^{-1}\ day^{-1}$)	$kg\ C\ ha^{-1}$ *of lake surface* yr^{-1}	(%)	*Remarks*
Borax, CA	39.8	<0.5					Saline lake; benthic algae, primarily epilithic, some epiphytic and metaphyton; single macrophyte species *Ruppia maritima*; ^{14}C methods for all components (Wetzel, 1964)
Phytoplankton			249.3	101.0	926	(56.8)	
Littoral algae			731.5	75.5	692	(42.5)	
Macrophytes			76.5	1.36	12	(0.7)	
					1630		
Marion, British Columbia	13.3	2.2					Softwater, oligotrophic lake; benthic algae, primarily epipelic; O_2 techniques, from which net production was estimated (Efford, 1967; Hargrave, 1969; Gruendling, 1971)
Phytoplankton			21.9	0.29	8	(1.6)	
Littoral algae			109.6	11.3	310	(62.2)	
Macrophytes			49.3	6.5	180	(36.1)	
					498		
Lake 239, Ontario	56.1	10.5					Softwater, oligotrophic lake; probably underestimates since winter production is not included; benthic algae, primarily epilithic, macrophytes probably insignificant; CO_2 utilization methods (Schindler *et al.*, 1973)
Phytoplankton					823	(99.0)	
Littoral algae					8.1	(1.0)	
Macrophytes					N.D.		
					ca. 831		
Lake 240, Ontario	44.1	6.1					(Same as for Lake 239)
Phytoplankton					501	(98.2)	
Littoral algae					9.0	(1.8)	
Macrophytes					N.D.		
					ca. 510		
Lawrence, MI	5.0	5.9					Hardwater, oligotrophic marl lake; benthic algae are primarily epiphytic on sparse submersed macrophytes; ^{14}C methods (Wetzel *et al.*, 1972, and refinements of analyses of epiphytic algae in Burkholder and Wetzel, 1983b)
Phytoplankton			118.9	5.90	430.8	(13.2)	
Littoral algae (<1 m)			2001	15.10	1102.4	(33.7)	
Littoral algae (1–5 m)			500	16.35	1193.6	(36.6)	
Macrophytes			240.8	7.40	540.2	(16.5)	
				44.75	3267.0	100.0	

(*continues*)

TABLE 19-10 (*Continued*)

Lake	Area (*ha*)	Mean depth (*m*)	Annual mean (*mg C* m^{-2} day^{-1})	Annual mean (*kg C* $lake^{-1}$ day^{-1})	kg C ha^{-1} of lake surface yr^{-1}		Remarks
Wingra, WI	139.6	ca. 2				(%)	Large, shallow, hardwater eutrophic lake; large littoral zone with dominant submersed macrophyte *Myriophyllum* and metaphytic mats of macroalga *Oedogonium*; ^{14}C methods for all components; mostly only summer values (McCracken *et al.*, 1974; Adams and McCracken, 1974; J. F. Koonce, personal communication)
Phytoplankton			1200	1675	4380	(78.6)	
Metaphyton	(Summer, 1971)		3.0	4.2	11.	(0.4)	
(*Oedogonium*)	(Summer, 1972)		5.5	7.6	19.9		
Macrophytes			320.5	447	1170	(21.0)	
					5581		
Neusiedlersee, Austria	30,000	1.3					Large, shallow, 66% in *Phragmites* stands; various methods, mostly annual values (Khondker and Dokulil, 1988)
Phytoplankton (open lake)			235				
Phytoplankton (littoral)			20				
Epiphytic algae			11.6				
Epipelic algae			50				
Emergent *Phragmites*			2740				
Submersed macrophytes			254				
Algal, Antarctica	—	<0.5					Shallow oligotrophic lake; dense epipelic algal mats; ^{14}C techniques (Goldman *et al.*, 1972)
Epipelic algae			468				
Phytoplankton			8.2				
Kalgaard, Denmark	10.5	4.7					Small oligotrophic, softwater lake; ^{14}C techniques for phytoplankton and epiphytes; biomass of macrophytes (Søndergaard and Sand-Jensen, 1978)
Phytoplankton					241	(52.3)	
Littoral algae					5	(1.1)	
Macrophytes					215	(46.6)	
Watts, Antarctica	38	ca. 5					Shallow, oligotrophic, continental lake; ^{14}C techniques (Heath, 1988)
Epipelic algae			15.1				
Phytoplankton			27.7				
Laca, Russia	16,600	—					Large, shallow, eutrophic lake; macrophytes colonizing 48% of lake area; biomass methods (Raspopov, 1979)
Phytoplankton					996	(36.9)	
Littoral algae					725	(26.9)	
Macrophytes					979	(36.3)	
Eagle, CA	12,150	7.0					Large, subalpine, hardwater, eutrophic lake; oxygen and biomass techniques (Huntsinger and Maslin, 1976)
Phytoplankton			356	50,860	1168	(85.8)	
Epilithic algae			274	3800	95	(7.0)	
Epiphytic algae			1153	1919	47	(3.5)	
Macrophytes (emergent)			1249	2690	51	(3.8)	

TABLE 19-11 Mean Rates of Primary Production of Phytoplankton in Relation to Those of Epipelic Algae and Algae Epiphytic on the Horsetail (*Equisetum fluviatile* L.)

Lake	*Phytoplankton* (*mg C m*$^{-2}$ *h*$^{-1}$)	*Epipelic algae* (*mg C m*$^{-2}$ *h*$^{-1}$)	*Epiphytic algae* (*mg C m*$^{-2}$ *of substratum h*$^{-1}$)
Eutrophic, shallow; Priddy Pool, England[a]	1.55	1.71	63.9
Oligotrophic, deep, with limited littoral area; Lake Pääjärvi, Finland[b]			
0–1 m	5.35	3.68	3.6
2–4 m	3.85	1.68	—

[a] Mean annual values; based on data of Hickman (1971b).
[b] Mean summer values; Kairesalo (1980b).

E. Changing Littoral Productivity and Eutrophication

As lakes receive increasingly large loads of nutrients per unit volume per time, there is a strong tendency for phytoplankton to increase to the maximum capacity within existing limitations of temperature and available light (cf. Chap. 15). However, it is imperative that eutrophication of aquatic systems is not viewed in the restricted sense of phytoplanktonic productivity. Within obvious geomorphological restrictions on wetland and littoral development at the land–water interface regions, the common situation is for littoral productivity to play a major role in early and in final stages of increasing fertility of the lake system as a whole. Exceptions certainly exist, but these conditions of major littoral production are clearly widespread because a large percentage of lakes of the world are small in area and shallow.

A common natural or human-induced disturbance to lake and reservoir ecosystems involves progressive increases in loading of nutrients. These loadings, in excess of losses to sites of temporary or permanent inactivation such as sediments, result in enhanced nutrient availability for phytoplankton and other autotrophs with increased rates of growth and productivity. If nutrient loadings increase to an oligotrophic lake, increased productivity is rapid. Similarly, if the disturbance is brief (i.e., the duration of increased nutrient loading is relatively short), nutrient cycling is rapid, the ecosystem will recover rapidly, and productivity will be reduced proportionally to the load reduction.

As nutrient loading increases, particularly among shallow lakes that predominate globally (Wetzel, 1990), a marked shift in the productivity occurs from the pelagic to attached surfaces associated with living aquatic plants and the particulate detritus of senescing macrophyte biomass (Fig. 19-21). Under these conditions, the primary productivity and biomass of the lake ecosystem increase greatly. Habitat diversity among the massively dissected surfaces of submersed aquatic plants and particulate detritus of the sediments increase exponentially. As nutrient loading increases further, phytoplanktonic productivity per unit volume increases, but self-shading by the algae restricts the depth of the trophogenic zone. Within a given latitude and climate, maximum growth of phytoplankton in fertile waters is determined by light reduction induced by self-shading; growth can be increased beyond these limits only by increasing turbulence and light availability to levels greater than those normally found under natural conditions (Wetzel, 1966b). Phytoplanktonic productivity per unit area declines precipitously under these eutrophic conditions and is usually accompanied by a marked decrease in planktonic biodiversity as well.

This light limitation, usually associated with intense phytoplanktonic and epiphytic algal productivity,

TABLE 19-12 Percentage Contribution of Various Producers to the Annual Net Primary Productivity per m^2 in the Littoral Zone, Mikolajskie Lake, Poland[a]

Zone/component	*Percentage contribution*
Eulittoral	
Macrophytes	28
Planktonic	10
Metaphyton	21
Attached algae	41
Littoral overgrown with emergent vegetation	
Macrophytes	57.2
Planktonic	19.6
Metaphyton	0.1
Attached algae	23.1

[a] After Pieczyńska (1970).

TABLE 19-13 Comparison of Primary Productivity of Attached and Planktonic Algae of a Wetland of Manitoba under Different Water-Level Regimes[a]

		Mean daily productivity (mg C m^{-2} wetland area)			
	Mean algal biomass (mg chl. a m^{-2} wetland area)	*Low water level*	*Medium water level*	*High water level*	*Percentage contribution (mean all water levels)*
Phytoplankton	7	169	209	225	7.3
Epipelic	4	32	47	15	1.0
Epiphytic	67	1003	813	353	22.2
Metaphyton	530	2863	2372	1338	69.5
Total m^{-2}		4067	3441	1931	100.0

[a] Ice-free period (5–6 months); mean of five years. Low water = 7–28-cm depth; medium water = 37–58-cm depth; high water = 67–88-cm depth. Extracted from data in Robinson *et al.* (1997).

thus extends to submersed aquatic plants. Light attenuation is critical. Dense phytoplankton populations alone are sufficient to attenuate light to a point where it is inadequate to support growth of submersed macrophytes (e.g., Mulligan *et al.*, 1976; Jupp and Spence, 1977). Reductions in phytoplankton densities and improved light conditions (e.g., by reductions in nutrient or turbidity loading or by selective fish predation on zooplankton, which in turn alters phytoplankton composition; Chaps. 15 and 16; Leah *et al.*, 1980) can result in recolonization by submersed macrophytes. In addition, increased nutrient loading can increase epiphytic growth on macrophytes to the point of severely reducing the amount of light reaching the supporting

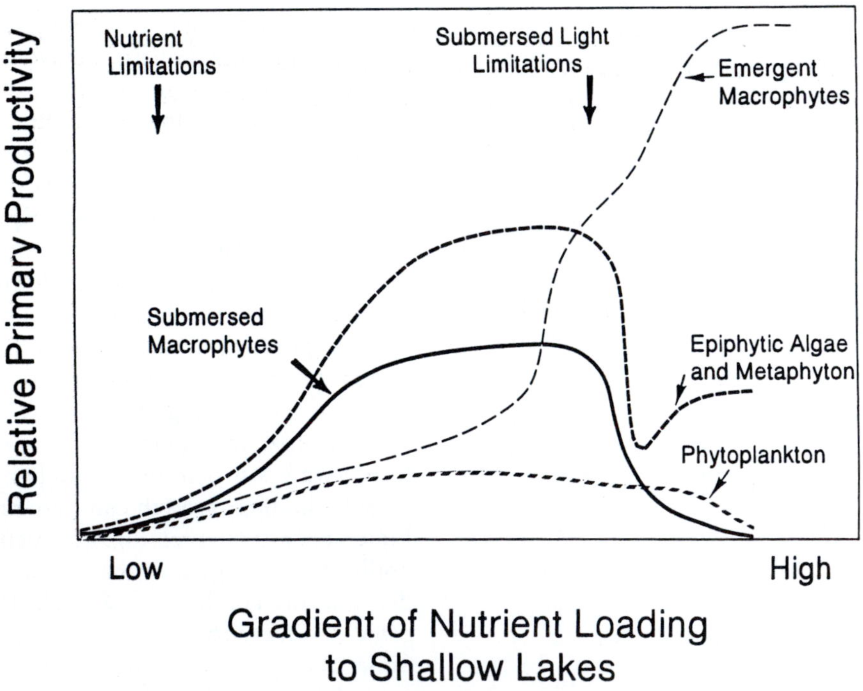

FIGURE 19-21 Relative changes in primary productivity of phytoplankton, macrophytes, and attached microflora along a gradient of nutrient loading to a spectrum of lake ecosystems. (From Wetzel, 1999a, expanded and modified from Wetzel and Hough, 1973.)

macrophytes (e.g., Sand-Jensen, 1977; Phillips *et al.*, 1978; Tsuchiya and Iwaka, 1979; Sand-Jensen and Søndergaard, 1981).

Submersed light limitation thus can reduce or eliminate submersed macrophyte growth and much of the attendant surfaces for microbial growth. The elimination of attached algal photosynthesis by attached microbiota decouples the metabolic mutualism to the attached heterotrophs and decreases collective productivity markedly (Wetzel, 1993a; Neely and Wetzel, 1995). The losses of attached microbial communities and their coupled metabolism cause a massive reduction in the capacities of the land–water interface regions of freshwater ecosystems to retain loaded nutrients and dissolved organic compounds.

Continual high nutrient loading and hypereutrophic phytoplanktonic conditions generate large areas of anaerobic reducing conditions and lower rates of decomposition of organic matter. Production exceeding decomposition leads to rapid sedimentation and the generation of increased shallow habitat conducive to colonization by emergent macrophytes (Fig. 19-21). The high productivity of emergent macrophytes increases the proportion of lignocellulose supporting tissues that are relatively recalcitrant to rapid decomposition, particularly under reducing conditions (Wetzel, 1979).

As the emergent vegetation assumes greater dominance in a lake ecosystem, and eventually covers a majority of the lake basin (see Chap. 24), an exceedingly productive combination of littoral macrophytes and attendant microflora develops. Attached, largely epiphytic algae, eulittoral algae, and metaphyton develop in association with the emergent flora. Natural changes in this general sequence are usually slow and may extend over centuries and millennia, depending on the basin morphometry. The process can be accelerated greatly by increased nutrient loading, either artificially (e.g., Hasler, 1947; Smith, 1969) or more gradually (e.g., Mattern, 1970). The collective result is markedly increased total ecosystem productivity, largely because of the high emergent macrophyte production.

The ratio of the interface zones to the pelagic permits one to estimate the importance of the littoral in relation to the pelagic zone. Most of the lakes of the world are small, and their morphometry is such that the ratio of the pelagic zone (P) to the colonizable littoral zone (L) is small (Fig. 19-22). On a global basis, the littoral zone is clearly dominant over the pelagic (low P/L ratios) among most standing-water and river ecosystems. Addition of the wetland components to the littoral zone results in low ratios ($P/W + L$) for nearly all standing-water ecosystems (Fig. 19-22). In these cases, there is little question that production of macrophytic vegetation, and importantly, its attendant microfloral community, has major impacts on the lake ecosystem.

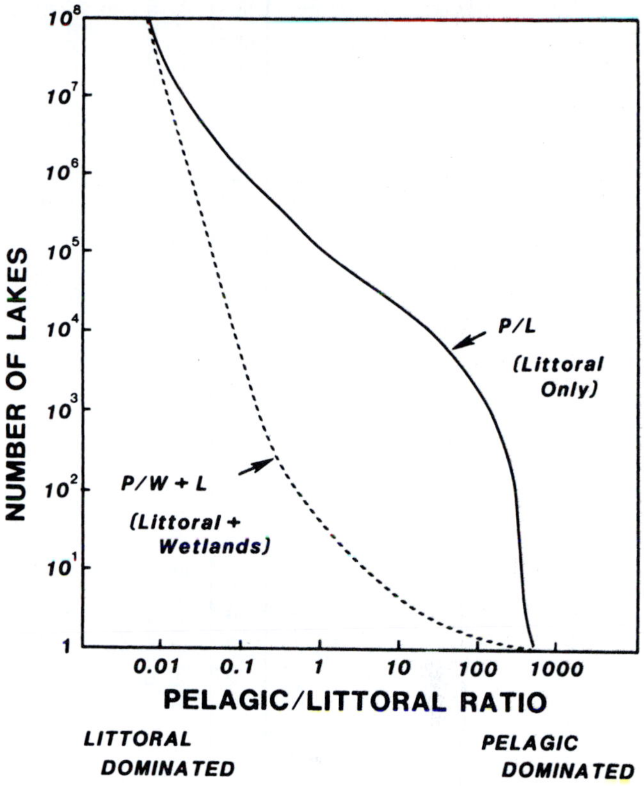

FIGURE 19-22 Number of lakes of the world in relation to the ratios of pelagic to littoral and wetland regions. (From Wetzel, 1990a.)

The distinction between what constitutes allochthonous production (e.g., eulittoral, terrestrial, et cetera) and what is part of true autochthonous lake production is an extremely artificial one and has, indeed, led to an artificial treatment of mechanisms contributing to and influencing ecosystem metabolism. All components of the drainage basin are influential in regulating lake metabolism. Integrated data on the functional impact of these components on the entire system are essentially nonexistent. The concept of a lake as a microcosm must finally be laid to rest.

IV. PERIPHYTON AMONG AQUATIC ECOSYSTEMS

Comparison of the periphyton communities and their productivity among river, reservoir, lake, and wetland/littoral ecosystems is difficult because of the high

TABLE 19-14 Comparative Characteristics among Periphyton Communities of River, Reservoir, and Natural Lake Ecosystems and Wetland Interface Regions

Characteristic	*Rivers*	*Reservoirs*	*Natural lakes*	*Wetland/littoral areas*
Periphytic algal diversity	Low in-channel, moderate in floodplain wetlands and backwater areas	Low, often restricted by severe water-level fluctuations and turbidity; macrophyte habitat often very low	Low in epilithic, moderate in epipelic, and very high in epiphytic communities	Very high among great habitat diversity of living and dead vegetation and particulate detritus
Periphytic algal biomass and production	Low in-channel, especially in low-order, canopied streams; increasing among mid-order (3–5) streams; low in large rivers; moderate to high in floodplain wetlands and backwaters	Low, often restricted by limited substrata in photic zones and low submersed macrophytes	High on two-dimensional surfaces (sediments, rocks) to very high on three-dimensional substrata (submersed plants)	Very high on abundant livingand detrital surfaces among shallow waters
Herbivory pressures on periphyton	Moderate to high seasonally in middle-order streams	Low, restricted availability	Low to moderate, distinctly seasonal; largely a production-detrital system dominated by microbial heterotrophy	Low, largely a production-detrital system dominated by microbial heterotrophy
Attached/littoral "zooplankton"	Low, restricted to floodplain and backwater areas; dominated by *r*-selected species that are adapted to frequent disturbances	Very low, restricted stable habitat and macrophytic refuges	High, large-sized among macrophyte habitats offering predation refuges and abundant food	Nearly all sessile, attached forms; predation relatively low in comparison to more open lake littoral regions

variability among habitats and their dynamic nature. Attached algal diversity is relatively low within the high-velocity portions of river channels but is considerably greater among the greater, somewhat more stable floodplain and backwater areas (Table 19-14). The high and variable turbidity and water-level fluctuations in many reservoirs preclude the development of moderately stable and diverse substrate habitat. In contrast, habitat diversity and attached algal diversity is extremely high among the littoral areas, particularly among submersed macrophytes and in wetland areas of abundant particulate detritus.

Attached algal and cyanobacterial productivity and biomass development is moderately high in mid-order (3–5) streams and in backwater areas but low in light-restricted, canopied low-order streams and in more turbid large rivers (Table 19-14). The productivity of phytoplankton, periphyton, and macrophytes becomes much more complicated in the highly dynamic large rivers of the tropics, such as the Amazon and Paraná/Paraguary river ecosystems. During the seasonal flood pulse, large fluctuations occur in periphyton development and productivity in relation to currents within backwater pools and ponds, loss of vegetation and associated substrata, and turbidity imported with the river water (Nieff, 1990; Junk, 1997). Limited habitat, substrata, and light often preclude extensive attached algal developments in reservoir ecosystems. Productivity and high biomass turnover occur in littoral and wetland areas, particularly among the high surface areas in well-lighted regions of submersed macrophytes and particulate detrital masses.

Herbivory by immature aquatic insects, other invertebrates, and certain fish can impose significant mortality on periphyton communities of streams at certain times of the year. In standing waters, production by periphyton is often greater, and herbivory is appreciably less than in many streams. In many productive littoral and wetland areas, the periphyton community is largely a production–detrital system dominated by microbial utilization of the attached autotrophic production.

V. LITTORAL ZOOPLANKTON COMMUNITIES

Littoral zooplankton communities are made up of a diverse assemblage of protozoans and other protists, rotifers, and microcrustaceans. Many of these animals are sessile upon the sediments or macrophytes and are not truly planktonic, or are only intermittently planktonic. The number of species within littoral macrophyte zones is generally larger among all groups of organisms than in the open-water zone of lakes, ponds, or rivers (e.g., Pennak, 1966).

Littoral zooplankton have been studied much less intensively than have pelagic forms. Many of the life history, reproductive, and population characteristics of pelagic zooplankton, discussed at length in Chapter 16, are equally applicable to littoral-inhabiting zooplankton. A few conspicuous characteristics of the littoral fauna are noteworthy.

Microcrustacean communities have been found to fall into three general groups in the littoral zone: (a) highly "plant-associated" and infrequently found away from aquatic macrophytes, (b) "free-swimming" among the large plants, and (c) sessile and living mainly in littoral sediments (e.g., Pennak, 1966; Whiteside *et al.*, 1978, Fairchild, 1981). Truly planktonic species are common in areas within littoral zones devoid of vegetation but are rare within areas densely colonized by macrophytes. Among the large plants, free zooplankton are predominantly plant browsers or sediment-inhabiting species that are temporarily swimming or open-water species that enter vegetated areas and develop under conditions of reduced predation pressures from fish.

Plant-associated microcrustacean taxa are often closely associated with macrophyte species or morphologically similar plant types (Quade, 1969, 1971) and are distributed according to available surface area (Fairchild, 1981). Numbers of Cladocera feeding on detrital periphyton were significantly correlated with epiphytic diatom density, whereas filter-feeding Cladocera attached to the plants were not. Similar animal–plant associations have been found among littoral rotifers (Edmondson, 1944, 1945; Wallace, 1978). Few benthic or epiphytic microcrustacea enter the water column, even at night (Paterson, 1993).

Among the common littoral microcrustaceans are the chydorids; most chydorid species increase in numbers during spring and autumn and decrease during midsummer (Goulden, 1971; Keen, 1973; Whiteside, 1974, 1988; Vuille, 1991). In a detailed study of the population dynamics of four species of a hardwater lake in southern Michigan (Keen, 1973), for example, *Chydorus sphaericus* reached its maximum population density in the spring, declined to a low level in the summer, and then increased to a smaller fall peak preceding a winter plateau (Fig. 19-23). It is common for *Chydorus* to become abundant in the open water in summer, coincident with the minimal littoral populations. As indicated in the upper portion of the figure, production of males and ephippial females was very small in this littorally perennial species. In contrast, *Graptoleberis*, *Acroperus*, and *Camptocercus* are aestival and after a winter absence appeared in spring, from ephippial eggs, and attained maximal densities in late summer and autumn (Fig. 19-23). The latter maxima

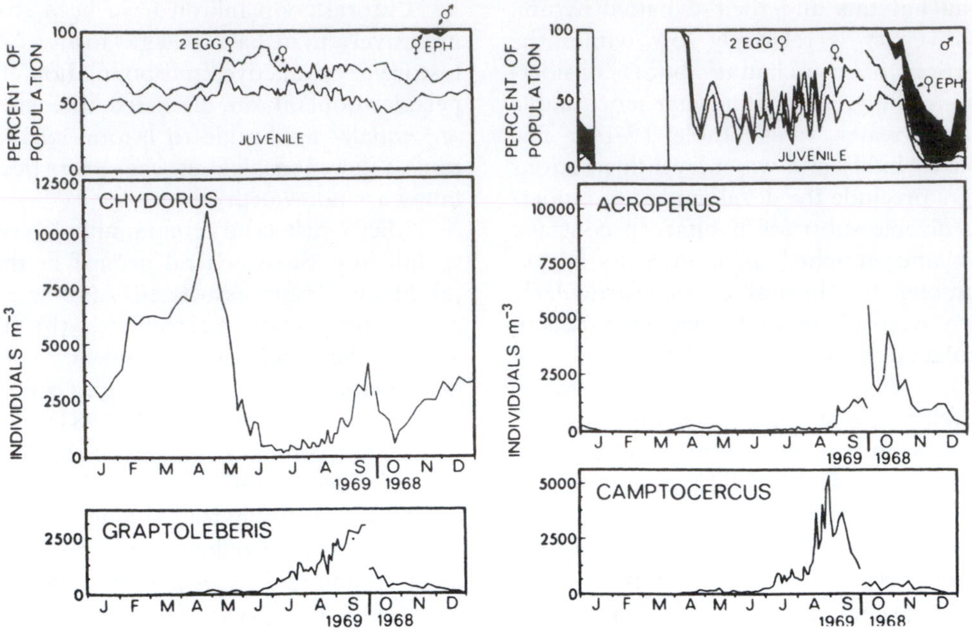

FIGURE 19-23 Annual population densities of the four major species of littoral chydorid cladocerans among submersed beds of *Scirpus subterminalis,* west littoral, Lawrence Lake, Michigan. Upper panels show percentage of the population of *Chydorus* and *Acroperus* seasonally in juveniles, females, females with eggs, males, and ephippial females. (Redrawn from Keen, 1973.)

terminated with the massive development of ephippial females and males prior to winter. Natural death and emigration were not clear but assumed to be negligible in this and several similar studies. High summer mortality was associated with active predation by small fishes (<4 cm in length) and nymphs of dragonflies (Odonata). In another analysis (Goulden, 1971), the midsummer population minima of chydorids were largely caused by predaceous tanypodine midge larvae. The midges were rare or absent in spring and autumn when the chydorids were most abundant.

In contrast, temporal and spatial correlations of distribution patterns of chironomid larvae of littoral sediments and chydorid cladocerans in the sandy littoral areas of a mesotrophic lake were very weak (van de Bund *et al.,* 1995). Alterations in species abundances need not always be associated with predation, however. For example, positively phototactic swarming of the cladoceran *Bosmina* and movement of large portions of the populations from littoral areas to the pelagic zone have been documented (Kairesalo and Penttilä, 1990).

Although the general relationships between temperature and egg and other developmental characteristics of littoral and benthic cladocerans and copepods are similar to those discussed earlier for planktonic zooplankton (Chap. 16), several differences have been found (e.g., Bottrell, 1975a,b; Sarvala, 1979). For example, egg development time is increased markedly with decreasing temperature, but among many species of cladocerans and copepods, eggs of sessile species take longer to develop than those of planktonic species, regardless of temperature. Instar duration and frequency of molting also tend to be greater among littoral species than among planktonic species. Diurnal migration has also been observed among zooplankton living among macrophytes. During the day, many populations, particularly copepods, aggregate among macrophytes near the sediments (Szlauer, 1963; Kairesalo, 1980a). Portions of the populations may then migrate to surface water strata during darkness.

Shallow lakes are often not sufficiently deep to stratify thermally and are therefore unable to provide a hypolimnetic refuge for zooplankton. In such waters, distribution of zooplankton is highly variable between littoral areas colonized by submersed aquatic plants and the open pelagic region. As indicated earlier, in littoral areas with well-developed aquatic macrophytes, the cladocerans *Alona, Chydorus, Diaphanosoma, Bosmina, Ceriodaphnia* and cyclopoid copepods are usually more abundant per area of habitat, often by a factor of 10, than in the pelagic areas (Vuille, 1991; Paterson, 1993; Lauridsen and Buenk, 1996; Lauridsen *et al.,* 1996). Some cladocerans, such as *Daphnia magna*, exhibited large diurnal changes, with highest populations during the daytime among the submersed macrophytes and movements into littoral boundary areas at night. Some species of cladocerans avoid submersed macrophytes, particularly during active photosynthesis periods, for unknown reasons (e.g., Dorgelo and Heykoop, 1985).

Many of the observed differences among the abundance of rotifers, cladocerans, and copepods within plant beds and the open water likely reflect variations in fish predation pressure (cf. further discussion in Chap. 20). The success of the habitat as a refuge for zooplankton clearly is related to plant density. Additionally, the size of zooplankton among submersed littoral vegetation tends to be larger than the size of zooplankton of the pelagic zone, indicative of selective predation upon larger forms. *Daphnia* species are commonly reduced to very low population levels in open water but can persist amongst extensive macrophyte stands after its elimination from open-water habitat. *Daphnia* is commonly succeeded by other cladocerans, all of which maintain a high grazing pressure on phytoplankton in the littoral areas (e.g., Stansfield *et al.,* 1997).

Although grazing pressures from small (0+-age fry) zooplanktivorous fish are high, plant refuges are significant. Based on correlational data from many lakes, it is clear that predatory control of zooplankton by planktivorous fish is reduced markedly in lakes rich in submersed macrophytes, where at least ca. 15–20% of the water volume is occupied by submersed vegetation (Jeppesen *et al.,* 1997; Persson and Crowder, 1998). Similar results can be demonstrated experimentally in controlled enclosure experiments in shallow lakes (e.g., Jeppesen *et al.,* 1998). The large plants serve as a refuge for pelagic cladocerans and encourage predatory fish at the expense of prey fish that tend to avoid dense littoral areas. Small planktivorous fish use macrophytes as a defense mechanism against predators and can compete effectively with predatory fish, particularly in the juvenile stages.

VI. SUMMARY

1. Benthic algae and cyanobacteria are generally associated with or attached to substrata in land–water interface habitats. These substrata are living or dead, organic or inorganic, and in various combinations within the plants, detrital masses, and sediments within emergent and submersed plant zones and with sediments below macrophyte colonization. Extremely diverse microhabitats occur in littoral areas among substrates of sand, rock, organic sediments, and macrophytes.
2. Limnological terminology applied to attached microbiota refers to generic types of substrata to which groups of organisms are associated (cf. Fig. 8-3, p. 131ff).
3. Attached algae of littoral communities often form the dominant algal biomass in streams and shallow lakes. Over 80% of all algal species likely grow attached in submersed habitats.
   a. Diatoms dominate on all substrata among moderately productive aquatic ecosystems.
   b. Sessile cyanobacteria and flagellates increase in abundance in organically enriched lakes and certain regions of streams.
4. Developing periphyton are three-dimensionally enmeshed with hydrated mucopolysaccharide materials, secreted by bacteria and algae, that both sequester ions and isolate the microorganisms from the water column (Fig. 19-3). Accumulated detrital particles and precipitated carbonates adsorb nutrient ions and organic compounds. Some microbiota are loosely attached and project within and above the general community aggregations. The polysaccharide matrix can contain microchannels that likely function as conduits that allow accelerated diffusion within the communities.
5. Resources regulating growth include both external resources in the overlying medium, such as light and nutrients in the surrounding water, and internal resources, such as nutrients that are obtained from the substrata to which the periphyton communities are attached.
   a. Availability of habitat space markedly affects the composite productivity of the periphytic communities. Some habitats, such as many sediments, are approximately two-dimensional and provide much less surface area and have much lower periphyton productivity than a three-dimensional habitat, such as submersed macrophytes. The epiphytic communities have manifold greater productivity as the spatial volume of the habitat projects upward into the water column and enhanced light field (see Fig. 19-8).
   b. Enrichments of nutrients, particularly phosphorus and nitrogen, in the overlying water commonly result in shifts in species composition and increases in growth and biomass of attached algae on substrata. Within a specific range of current velocities, the boundary layer thickness between the periphyton and the overlying water is reduced with increasing current velocities and results in enhanced nutrient exchange with the overlying water.
   c. As communities develop and increase in thickness, instability can occur and cells and portions of the community may slough off.

6. Many periphytic communities are dependent upon the substrata for mineral and organic nutrients, particularly from submersed macrophytes and in oligotrophic and mesotrophic environments where the greatest periphyton productivity occurs. As macrophytes senesce and organic matter decomposes, a large portion of the nutrients and organic substrates released are sequestered and intensively recycled among the periphytic microbiota. This retention is particularly effective in epiphytic and epipelic periphyton.
7. A tight, mutualistic, metabolic coupling exists between attached algae and bacteria. For example, bacterial production is increased by algal exudates of organic substrates and oxygen. Enhanced algal growth can occur from increased availability of CO_2 and growth factors such as vitamins and reduced oxygen tension. Nutrients are certainly recycled and exchanged rapidly among the periphyton microbiota.
8. Light availability influences periphyton production both spatially along depth gradients and seasonally. Even though benthic algae can adapt with increased pigment concentrations to very low light intensities, light is often restrictive not only with increasing depth but seasonally, particularly among low-order woodland streams that are heavily canopied with terrestrial vegetation. Light is attenuated selectively within periphyton as well, and excessive developments of periphyton can reduce light reaching sufficiently to kill the supporting macrophytes and hence habitat space for colonization.
9. Mortality losses among periphyton communities are poorly known despite much study.
   a. Heavy grazing of periphyton by animals (insect larvae, crayfish, and certain fishes) can result in reductions of periphytic biomass and productivity. Although most periphyton are not grazed by animals, attached microbial–grazer interactions and mortality vary with herbivore type, nutrient and light availability for compensatory growth, and disturbance frequency. There is some evidence that grazing mortality of periphyton is greater in stream ecosystems than in standing waters.
   b. Disturbance to the habitat of periphyton communities can be severe from physical disruptions of substrata, such as during river flooding, and from shifts in water level, particularly among shallow habitats. Algae living on loose sediments that are subject to disturbance and displacement by water movements and animals often exhibit an ability to migrate vertically in response to changes in light. This vertical migration is phased with diurnal rhythms in cell division and photosynthetic capacity.
   c. Losses of natural periphyton communities by natural senescence and death, parasitism particularly by viruses, and disease are poorly known, but it is apparent that most of the periphyton production enters the pool of detrital organic matter in both lake and river ecosystems.
10. In temperate regions, the seasonal community biomass of attached algae commonly follows seasonal changes in incident light and temperature.
    a. Epipelic diatoms frequently exhibit population maxima in the spring and in the autumn.
    b. Seasonal population dynamics of epiphytic algae are more variable and change as the surface area available for colonization of the supporting macrophytes changes seasonally.
11. Within the streams and rivers and the littoral zone of standing waters, photosynthesis of attached algae is influenced to a great extent by light availability, although compensatory mechanisms exist that permit appreciable active growth at low levels of irradiance.
    a. The biomass of epipelic algae commonly increases with increasing nutrient content of the sediments.
    b. Epipelic algae can utilize significant quantities of phosphorus and combined nitrogen in the interstitial water of sediments and alter, by photosynthetically changing sediment redox, the flux of nutrients from the sediments to overlying water in the littoral zone.
12. Where comparisons of growth among algal populations associated with different substrates are possible, they indicate that photosynthesis is commonly higher for algae attached to sand grains than it is for those associated with organic sediments. Algae epiphytic on macrophytes are often much more productive than algae associated with sediments.
13. Productivity of littoral algae is not much greater on a cellular basis than that of phytoplankton, and turnover rates tend to be somewhat slower.
    a. The large surface area available for colonization, however, particularly on submersed macrophytes, can result in very high contributions by attached littoral algae to the total primary productivity of many freshwater ecosystems.
    b. Spatial heterogeneity in rates of production by attached algae is usually very large because of the great variability in littoral habitats within freshwater ecosystems.

c. Community rates of primary production of attached algae commonly follow the annual insolation and temperature curves, with maxima occurring in midsummer.

d. Most lakes of the world are shallow and possess large littoral zones. The productivity of the littoral algae and macrophytes tends to increase in relation to that of the phytoplankton as lakes become enriched. In lakes with a modest development of submersed macrophytes, epiphytic productivity can be very high and dominate total in-lake photosynthetic productivity. Under eutrophic conditions, excessive growth of phytoplankton and epiphytic algae can reduce light sufficiently to cause the decline of submersed macrophytes and attached algae; a consequent overall reduction in total productivity within the lake will then result.

14. Many of the protozoan, other protistan, rotifer, and microcrustacean zooplankton of the littoral zone are sessile on the sediments or macrophytes and are not truly (or are only intermittently) planktonic.

a. Plant-associated zooplankton are commonly sessile on specific macrophyte types. Population densities of animals feeding on epiphytic microflora and detritus are positively correlated with epiphytic algal density.

b. Many of the life history, reproductive, and population characteristics of littoral-inhabiting zooplankton are similar to those of their pelagic counterparts.

15. In shallow lakes with a well-developed submersed aquatic plant littoral area, the littoral habitat can function as a refuge for zooplankton from fish and invertebrate predation that prevails in the open water.

a. Population numbers of common littoral microcrustaceans increase during spring and summer and decrease during summer. Midsummer population minima often result from intense predation by insect larvae and small fishes.

b. Predatory control of zooplankton by planktivorous fish is reduced markedly where minimally 15–20% of the water volume is occupied by submersed vegetation.

c. Zooplankton grazing can impact phytoplankton within the littoral and contribute to reduced phytoplankton and improved water transparency.

20

SHALLOW LAKES AND PONDS

Study of lakes has concentrated almost totally on moderately large lakes in regions of recent geological glacial activity where such waters abound. Although the volume of water in large lakes is appreciable as a percentage of the total fresh water, they represent a very small percentage of the total number of lakes (Wetzel, 1990a). Much less spectacular but important are the millions of shallow lakes, impoundments, and ponds of only a few meters of depth (cf. Fig. 3-2). Many human activities are associated with and dependent upon shallow waters and wetlands. In addition, maximum biodiversity of freshwater ecosystems occurs where wetland and littoral habitat heterogeneity interfaces with pelagic regions (Wetzel, 1999b). In many shallow lakes and ponds, the littoral structure and its productivity completely dominates the ecosystem.

I. ORIGINS AND DISTRIBUTION

Shallow lakes and ponds occur in abundance in lowland areas of very gentle relief. As discussed in Chapter 3, these shallow basins can originate from processes that have caused modest depressions in the landscape. Certainly some processes were major geological disturbances such as glacial movements. Other depressions were formed in relatively flat regions from altered river courses or wind deflation. Hundreds of thousands of small, often temporary, ponds and lakes occur in the flood plains of major river ecosystems. These floodplain impoundments are especially common adjacent to arctic and tropical rivers where large areas of the land are flooded seasonally for several months. Humans have created millions of shallow lakes and ponds, either fortuitously or intentionally, as they modify the landscape by construction and agriculture. The shallow waters may be used for agricultural, industrial, water storage, or aesthetic purposes.

II. CHARACTERISTICS

Such lowland areas tend to accumulate terrestrial organic matter and nutrients, and appreciable amounts of these substances can be transported to shallow lakes. Being of relatively small volume, the loading of nutrients per unit volume can be high. Wetland and littoral macrovegetation that have much higher rates of organic production than do phytoplankton per unit area often dominate production of organic matter within shallow lakes. Hence, the topography of the landscape of shallow lakes and the resulting morphometry of the lake basins tend to be highly irregular and heterogeneous. This heterogeneity leads to large variability in the distribution and productivity of higher vegetation and of the abundant microbiota associated with the surfaces of these larger plants and their particulate detritus. As a result, the growth and interactive metabolic properties of shallow lakes tend to be most variable, from which generalizations are more difficult than is the case with large lakes.

The importance of loading of nutrients to phytoplankton productivity of lakes, particularly phosphorus and nitrogen, has been discussed in detail (cf. Chaps. 12 and 13). Thermal stratification in these larger lakes is important for losses of phosphorus to the sediments. It was also emphasized that these relationships of nutrient losses and retention in the sediments deteriorate when applied to many shallow lakes. Shallow lakes and ponds tend to have a low percentage, if any, of total water volume in a thermally distinct hypolimnion. In the temperate region, thermal stratification for any appreciable length of time (months) is poor if basin depths are <5–7 m. As a result, water commonly circulates for long periods or continuously in shallow lakes.

Not only are the loadings of nutrients proportionally higher in shallow lakes than in deep ones but the losses of nutrients to potential depositories, such as the sediments or outflow, are lower and the rates of nutrient recycling faster in shallow lakes. As a result, factors other than nutrients, such as light availability, frequently regulate both photosynthetic productivity and the growth of organisms that depend on that productivity. Nutrient supply does not always influence phytoplankton productivity in shallow lakes in the manner discussed for large lakes because of the complex competitive interactions of macrophytes, attached microbiota, and phytoplankton.

III. INVASION AND GROWTH OF MACROPHYTES

Earlier discussion of the competition between submersed macrophytes, epiphytic periphyton, and phytoplankton was focused on large lakes with a large depth gradient and thermal stratification. In shallow lakes, the littoral can and often does extend over the entire lake or pond basin. A "***shallow lake***" or "***pond***" is usually defined as a permanent standing body of water that is sufficiently shallow to allow light penetration to the bottom sediments adequate to potentially support photosynthesis of higher aquatic plants over the entire basin. Turbidity from abiotic or biotic sources may prevent light from reaching the sediments, but the lakes or ponds are sufficiently shallow for this potential condition to occur. Commonly, macrophytes colonize only a portion of the ponds and lakes because of the frequent periodicity of high turbidity.

The type of macrophyte colonization is also highly variable. The common depth distribution and zonation of macrophytes in large lakes, as discussed earlier (Chap. 18) can apply similarly in shallow lakes but usually on a greatly contracted depth gradient, simply because light is often rapidly attenuated with depth in shallow lakes. Emergent macrophytes are largely limited to depths <1 m and rooted floating-leaved macrophytes to depths of <2–3 m. Submersed macrophytes can form a dominant component of shallow lakes and ponds and markedly influence the biological community structure of these lakes. Much of the following discussion is directed at recent intensive studies that have examined the shifts between dominance by submersed macrophytes to dominance by phytoplankton and what effects these shifts have on overall ecosystem productivity.

IV. SHIFTS BETWEEN MACROPHYTE AND PHYTOPLANKTON DOMINANCE

Nutrient loadings of phosphorus and nitrogen are still important, particularly at the lower concentrations, in shallow lakes. The resulting dominance of photosynthetic communities utilizing these nutrients, however, is coupled to the timing of competition between the macrophyte versus algal communities. Plant-dominated and phytoplankton algal-dominated alternative equilibria can exist over a wide range of nutrient concentrations (Scheffer *et al.*, 1993; Scheffer, 1998). Submersed macrophyte dominance is the prevalent condition, as in large lakes, when nutrient concentrations of the water are low (e.g., 25 μg total P liter^{-1}) with a high N:P ratio (>>10:1). In hardwater, calcareous shallow lakes, charophyte macroalgae are common (e.g., Wetzel, 1973; Blindow, 1992) under oligotrophic conditions. Isoetid submersed plants (e.g., *Isöetes, Littorella*, and *Lobelia*) are common in more acidic oligotrophic waters. These submersed plants utilize nutrients from the sediments, are small and growing close to the sediments, and are vulnerable to shading. Shading by phytoplankton communities, however, is generally modest at these nutrient levels.

As phosphorus and nitrogen levels increase, however, this plant community is replaced with larger submersed plants, such as *Myriophyllum* and a number of *Potamogeton* species, as well as increasing densities of water lilies (Nymphaeaceae). The larger submersed plants are much more productive and still derive most of their nutrients from the sediments (e.g., Barko and Smart, 1980; Denny, 1980; Moeller *et al.*, 1988). Although total phosphorus levels may increase further from external loadings, nitrogen is commonly limiting to both submersed macrophytes and phytoplankton because of heavy biological utilization and from losses by denitrification. Further increased loading of phosphorus and nitrogen can often result in a further shift in the submersed plant community to tall species (e.g., *Myriophyllum spicatum*, certain *Potamogeton* spp., *Ceratophyllum*) as well as water lilies. Although the phytoplankton communities expand and proliferate under conditions of increasing nutrients, they can coexist with certain aggressive, eutrophic submersed macrophytes that grow rapidly and concentrate photosynthetic tissues near the surface.

A. Macrophyte Dominance

Often, however, this submersed macrophyte–phytoplankton "equilibrium" over a wide range of total phosphorus concentrations (<10 to >1000 μg liter^{-1}) rapidly shifts to an alternative condition of dominance by phytoplankton with elimination of submersed macrophytes (Balls *et al.*, 1989; Jeppesen *et al.*, 1991). This shift can occur regardless of the water concentrations of total phosphorus, and obviously other factors than nutrient changes can cause this shift from the submersed plant-dominated state to one of phytoplankton dominance. In shallow lakes and ponds, submersed macrophyte dominance can be maintained by several mechanisms:

1. Geomorphic conditions, such as surrounding hills or trees, that minimize wave action and currents that would resuspend sediments and increase nonbiotic turbidity.

2. Sequestering of phosphorus and nitrogen by submersed macrophytes in excess of nutritional requirements, thereby reducing availability to phytoplankton and periphyton (e.g., Moeller *et al.*, 1988; Stephen *et al.*, 1998). Except for very oligotrophic environments and very hard waters, phosphorus is relatively adequate for submersed macrophytes, particularly from the sediments. Within submersed plant beds, anoxic areas occur near the sediments, which leads to appreciable bacterial denitrification and losses of nitrogen to the atmosphere. Because of the marked uptake of combined nitrogen by submersed macrophytes and especially epiphytic periphyton communities, nitrogen availability can become restricted to littoral phytoplankton (e.g., Fitzgerald, 1969; Howard-Williams, 1981; Van Donk *et al.*, 1993).

3. Submersed macrophytes that provide physical refuges for large zooplankton, such as *Daphnia* and other crustaceans that effectively graze on and reduce phytoplankton, from predation by fish (e.g., Timms and Moss, 1984; Schriver *et al.*, 1995). Small fish commonly inhabit littoral areas as a refuge from predation by piscivorous fish (e.g., Gerking, 1962; Venugopal and Winfield, 1993). Dense submersed vegetation, however, and reduction of light and oxygen in the lower portion of plant beds (cf. Fig. 9-8) can restrict small fish to water at the edges and above the plant beds. These refuges for zooplankton that feed upon and at times control phytoplankton densities are important to the regulation of littoral phytoplankton communities and their biomass (Irvine *et al.*, 1990; Stephan *et al.*, 1998). Clearly, some of the zooplankton migrate upward and out of the submersed plant protective habitat for feeding at night when the visually mediated predation pressure from fish is low. Lateral horizontal migration diurnally by littoral zooplankton (Davies, 1985) is less clear.

4. The large surface area of submersed macrophytes that provides habitat for prolific development of epiphytic periphyton. That development increases with increased nutrient loading and can suppress light availability to and growth of the macrophytes (e.g., Sand-Jensen and Sondergaard, 1981; Losee and Wetzel, 1983). Insect larvae, such as chironomids, and snails can be effective in grazing periphyton and reducing shading to the supporting plants. Again, habitat refuges for the macroinvertebrates from fish predation are abundant within the macrophytes (e.g., Brönmark, 1989; Moeller *et al.*, 1998).

5. Release of organic compounds that suppress the growth of phytoplankton and periphyton (e.g., Wuim-Andersen *et al.*, 1982; Gross and Sütfeld, 1994). This mechanism is constantly cited in the literature but is very poorly studied and is largely suppositional. When these compounds are released, their efficacy can be modified by interaction with natural organic compounds and bacterial metabolism. Suppression of phytoplankton growth by these compounds of macrophytes has not been demonstrated (e.g., Forsberg *et al.*, 1990).

6. The intense metabolism of the submersed macrophytes and epiphytic microbiota, as well as the large deposits of decomposing organic detritus, that frequently causes marked diurnal fluctuations of dissolved oxygen, pH, and other parameters in the littoral waters. Some of these conditions can suppress intrusions of fish and predation on the grazing zooplankton and macroinvertebrates.

B. Phytoplankton Dominance

In shallow lakes and ponds, the shift to phytoplankton-dominated conditions where submersed macrophytes are suppressed or eliminated can result from several mechanisms:

1. The development of prolific phytoplankton communities that compete with submersed plants mainly for light but also nutrients such as inorganic carbon. This condition is particularly effective for the

development of small algal and cyanobacterial species with a high capacity for light absorption and low light compensation points. With the reduction of grazing pressures by large zooplankton, discussed further on, the development of phytoplankton early in the season in nutrient-rich shallow waters of temperate regions can attenuate light markedly and suppress macrophyte development and growth from overwintering propagules or seeds.

2. If the small lakes and ponds are exposed to wind action, the common mixing of the water with resuspension of surficial sediments in the open habitat. As a result, abiogenic and biogenic turbidity increases with a marked reduction of light penetration. The loose, flocculent sediment is also a poor substratum for macrophyte colonization.

3. The absence of submersed macrophytes, which greatly reduces refuge habitat for large zooplankton that are effective in phytoplankton grazing. If a large proportion of small zooplanktivorous fish dominates the fish communities because of a small population of piscivorous fish, the grazing pressure on phytoplankton by large cladoceran and copepod zooplankton is reduced. As a result, the grazing mortality of phytoplankton is reduced. Many piscivorous fish are poorly adapted to lakes without macrophyte habitats and plant-associated food refuges for small fish and macroinvertebrates. Consumption of the larger fish, including cannibalism on their own species, skews the fish to smaller, effective zooplanktivorous fish. Gradually, piscivorous fish reproduction and recruitment can decline.

4. Once established and dominating under conditions of high nutrient loads, dense phytoplankton communities that tend to persist in spite of changing nutrient availability. As we have seen in the discussion of changes in phytoplankton populations seasonally in large lakes (Chap. 15), similar changes occur in shallow lakes and ponds seasonally. At very high phosphorus and combined nitrogen concentrations in the water, green algae can effectively compete with nitrogen-fixing cyanobacteria (e.g., Jeppesen *et al.*, 1994), a well-known and effectively demonstrated sequence in large lakes (cf. Chap. 15). Once the cyanobacterial populations are established, most species are largely inedible and poorly grazed (Gliwicz, 1990; Lampert, 1994).

C. Other Control Mechanisms

It should be noted that a number of other factors could induce the rapid transition from submersed-macrophyte domination to domination by phytoplankton. The introduction of certain large fish and other animals can cause the shift in conditions. For example, the sucking detrital feeding activities of the common carp (*Cyprinus carpio*, Cyprinidae) and the benthivorous bream (*Abramis brama*) remove large portions of sediments as they search for food. The sediment disturbance can markedly increase turbidity in shallow lakes and littoral areas in proportion to the biomass of benthivorous fish (Fletcher *et al.*, 1985; Breukelaar *et al.*, 1994). The large grass carp (*Ctemopharyngodon idella*, Cyprinidae), if sufficiently abundant, can consume submersed vegetation voraciously and completely denude shallow waters of these plants (Chilton and Muoneke, 1992) and often, though not always, result in large increases in phytoplankton biomass (Gasaway and Drda, 1978; Terrell, 1982; Richard *et al.*, 1984).

A number of other vertebrates can impact aquatic vegetation. Herbivorous waterfowl, such as the large Canada goose (*Branta canadensis*) and the mammals nutria (*Myocaster coypus*) and muskrat (*Ondotra zibethica*) are voracious herbivores that consume and disturb large quantities of macrovegetation (e.g, Manny *et al.*, 1994; Taylor *et al.*, 1997). Humans influence submersed vegetation either directly by application of herbicides purposefully for control or accidentally in runoff, by mechanical cutting, or from damage from boat motors or indirectly by excessive fishing removal of piscivorous fish. All of these factors, as well as others, such as selective winterkill of piscivorous fish, may impact the submersed vegetation or phytoplankton abundance sufficiently to induce a shift to dominance by phytoplankton.

V. TEMPORARY PONDS, POOLS, AND STREAMS

Small depressions in impermeable or semipermeable terrain can be filled with water during certain times of the year. Such temporary waters may be astatic seasonally or possess dry periods for longer intervals (Decksbach, 1929). The duration of flooding varies greatly and results in many variations in temporary or semipermanent shallow waters that last only a few weeks or, alternatively, pools that are dry for only short periods over an annual period.

In temperate regions, in spring melting snow and rain fill depressions of these temporary ***vernal pools,*** and these pools remain dry for the remainder of the year (cf. Wiggins *et al.*, 1980; Williams, 1987). Water levels subsequently decline, and the pools are generally without water for many months. The duration of the wet and dry phases varies widely, and the pools can often be inundated again in the autumn (termed *autumnal pools*) and remain inundated throughout the autumn and through the subsequent spring period. Other pools are inundated for a period of one year and may enter a drought period for months or years before being reflooded.

The marginal vegetation growing adjacent to, or vegetation including trees growing within, temporary pools of course influences the loading of organic matter to the pool ecosystem. Often, the predominant organic loading occurs as dissolved organic matter leached from the surrounding vegetation and soils. Particulate organic matter is frequently also predominantly allochthonous via leaf fall from trees and wetland vegetation. When exposed to appreciable light, autochthonous supplements are common from opportunist macrophytes and attached and planktonic algae/cyanobacteria. As was emphasized in the detailed treatment of the ecology and metabolism of littoral and floodplain environments (Chaps. 18 and 19) and the sediments (Chap. 21), the plankton component contributes only a small portion of the overall pool metabolism (cf., e.g., Cole and Fisher, 1978; Hobbie, 1980). Benthic microbial metabolism totally dominates in these ecosystems.

Temporary pools are essentially closed waters with influents largely from precipitation and runoff. Water losses by evaporation result in the concentration of imported salinity. In arid regions, concentrations of salts can reach extraordinary levels prior to drying and further constrain organisms to a few tolerant species (cf. Chap. 10; W. D. Williams, 1996). In addition, the annual disturbance imposed by the drying period allows more accelerated aerobic degradation and oxidation of organic matter and release of nutrients than would occur in many permanently inundated ponds. As a result, upon reflooding, nutrient availability is high with increased initial productivity.

Organisms living in such irregular, transient habitats must have many structural, behavioral, and physiological adaptations for surviving or avoiding drought. Organisms adapted to temporary pools are few but are a consistent and predictable assemblage. Only a few species of each major freshwater group have adapted to these habitats. Different animal communities can be distinguished on the basis of survival during drought periods and seasonal patterns of recruitment (Fernando and Galbraith, 1973; Wiggins *et al.*, 1980; D. D.Williams, 1987; Neckles *et al.*, 1990; Jeffries, 1994):

a. Animals that are permanent residents capable only of passive dispersal that aestivate and overwinter in the dry pool basin, either as stages resistant to desiccation [certain Turbellaria (cysts and egg cocoons), Bryozoa (statoblasts), Anostraca, Cladocera (ephippia), Copepoda (eggs), and Ostracoda] or while protected, largely as resistant eggs, in the sediments (Oligochaeta, Hirudinoidea, Decopoda, and molluscs).

b. Animals capable of some dispersal (certain Ephemeroptera, Coleoptera, Trichoptera, and Diptera) that oviposit on water in spring and then aestivate and overwinter in the dry basin in various stages of the life cycle.

c. Animals (certain Odonata, Trichoptera, and Diptera) that oviposit as summer recruits in the sediments of the pool basin after surface water has disappeared. Eggs or larvae then overwinter until the next inundation period.

d. Animals (certain Ephemeroptera, Odonata, Hemiptera, Coleoptera, Diptera, and amphibians) that have well-developed powers of dispersal that leave the disappearing pool and pass the dry phases in permanent waters. Some of these animals subsequently return to oviposit in the temporary pool in the following spring.

A requisite for all organisms, and particularly for larger animals, in temporary pools is a rapid rate of development during the wet phase. Development and locating a mate must be completed in a few weeks, often accelerated by rapid changes in physical (e.g., increasing temperatures and ultraviolet intensities), chemical (increasing salinity and increasing dissolved organic matter), and biological (increasing susceptibility to predation) conditions in the receding pools. Additionally, species of temporary pools have a marked seasonality in their life cycles that is coupled to the probability of inundation of temporary pools. Artificial elimination of the flooding–drying periodicity can alter or eliminate needed cues for oviposition, embryonic development, and hatching among dominant taxa.

Temporary or intermittent streams differ markedly from temporary pools in both physical and biological terms. Many organisms, such as benthic invertebrates, are capable of movement and many clearly seek refuge in the interstitial waters of the hyporheic zone as drying of the stream occurs progressively to the sediments and below the surface of the sediments (Williams and Hynes, 1974, 1977). Because of common groundwater persistence in the hyporheic zone, organisms inhabiting these regions are not well adapted to the drought and desiccation of temporary pools.

It may be noted that the water column and part of the sediments freeze solid during winter in certain shallow ponds (e.g., Hobbie, 1980). These ponds contain water at all times of the year and thus would be considered permanent water bodies. Because biological activity essentially ceases during the winter period, these waters have been termed ***aestival ponds*** (Welch, 1952). Organisms of aestival ponds are adapted to tolerate interruption of development by freezing until subsequent thawing periods. The physiological rigors of freezing without desiccation are less severe than being exposed to a drying temporary pool where both freezing and desiccation can occur (Daborn, 1974).

VI. SUMMARY

1. Millions of shallow lakes and ponds occur in lowland areas of gentle relief and in floodplain regions adjacent to river ecosystems. These shallow waters are rarely thermally stratified and if so, only briefly; water circulation is frequent and often continuous.
2. Loadings of nutrients are proportionally higher, losses are lower, and rates of nutrient recycling are faster in shallow lakes than in deep ones.
3. Complex competitive interactions of macrophytes, attached microbiota, and phytoplankton occur in shallow lakes and ponds.
   a. Light penetration to much of the sediment surface can support invasion and growth of macrophytes over much of the basin.
   b. Submersed macrophyte dominance prevails when nutrient concentrations of the water are low with a high N:P ratio ($\gg$10:1). As nitrogen and phosphorus concentrations increase, transitions occur in submersed aquatic plant communities.
   c. Submersed macrophyte dominance can be maintained by mechanisms (e.g., abiogenic turbidity, sequestering of nutrients by macrophytes, and large zooplankton herbivory) that suppress the development of profuse phytoplankton communities.
   d. A shift to phytoplankton-dominated conditions, where submersed macrophytes are suppressed or eliminated, can result from prolific development of algae or cyanobacteria adapted to competition for light as nutrient levels increase. The phytoplankton increase turbidity, reduce light for submersed macrophytes, and reduce the grazing pressure, because of food particle size, of large zooplankton on phytoplankton.
4. Temporary ponds and pools are shallow waters that are flooded periodically, largely from precipitation and immediate terrestrial runoff, and dry for periods of the year.
   a. Sources of organic matter are largely allochthonous, and benthic microbial metabolism totally dominates in these ecosystems.
   b. Losses of water are largely by evaporation and result in marked, rapid changes in ionic composition, nutrient availability, and desiccation.
   c. The few organisms adapted to living in these irregular, transient habitats have many structural, behavioral, and physiological adaptations for surviving (e.g., aestivation) or avoiding (e.g., dispersal) drought and desiccation. Very rapid rates of development during the wet phase are essential.

21 SEDIMENTS AND MICROFLORA

Many characteristics of sediments of freshwater ecosystems have been analyzed. Thus, sediments of the littoral zone of lakes and of running waters are sorted along a gradient by particle size as a result of water movements and differences in water velocity. Particle size of both inorganic and organic sediments is of major importance to the distribution and growth of many benthic invertebrates (cf. Cummins, 1962) and microbial metabolism. Because of the overriding importance of microbial metabolism in mineralization of organic matter, in biogeochemical cycling of nutrients and in the amount of matter deposited without appreciable further degradation in permanent sediments, the organic composition of sediments in lake systems has received much study.

I. GENERAL COMPOSITION OF SEDIMENTS

Sediments consist of three primary components: (a) organic matter in various stages of decomposition; (b) particulate mineral matter, including clays, carbonates, and nonclay silicates; and (c) an inorganic component of biogenic origin (e.g., diatom frustules and certain forms of calcium carbonate) (cf. Jones and Bowser, 1978; Kelts and Hsu, 1978). In many respects, the organic sediments of lakes are similar to the uppermost A_0 horizon in terrestrial soils (Hansen, 1959a). Heterogeneous humus in particulate form can be divided into two main types, acid humus and neutral humus (sapropel[1]). From a colloid-chemistry viewpoint, acid humus is largely an unsaturated sol whose particles have negatively charged surfaces, while neutral humus is largely a gel in which anions are adsorbed to the surfaces of the humus particles. Acid humus prevails in peat bog systems. Decomposing peat plant materials, such as the moss *Sphagnum*, form unsaturated humus colloids (dopplerite). Aggregations of black, gelatinous dopplerite are very soluble in water unless

[1] True sapropel is bluish-black, contains much hydrogen sulfide (H_2S) and methane (CH_4), and is deposited under anaerobic-reducing conditions.

dried; upon drying, they become insoluble. The nitrogen content of this humic material is very low (≈0–2%).

A. Dy and Gyttja

Two words of Scandinavian origin are widely used in limnology to describe the general character of organic sediments. The terms *dy* (pronounced "de", nasally) and *gyttja* (pronounced "yit-ja") were introduced in the middle of the nineteenth century by von Post (Hansen, 1959a,b). Gyttja is a coprogenous sediment containing the remains of all particulate organic matter, inorganic precipitations, and minerogenic matter. In a fresh state, gyttja is very soft and hydrous, with a dark greenish-gray to black color; it is never brown. In a dry state, some gyttjas are hard and black, while others are more friable and lighter in color, depending upon the main constituent. The organic carbon content of gyttja is <50%. Dy[2] is a gyttja mixed with unsaturated humus colloids. Fresh dy is soft, hydrous, and brown in color. In a dry condition, dy is very hard and dark brown. The organic carbon content of dy and peat is >50%.

Numerous comparative studies of the organic components in sediments have demonstrated that the protein and nitrogen content of acidic humus is low. The C:N ratio of *Sphagnum* peat is about 35; that of pure dopplerite varies from 46 to 52. Hansen concluded that if the C:N is <10, the humus is neutral humus and the sediment is gyttja. If the C:N >10, the gyttja is mixed with acid humus and the sediment is a dy.

The sources of particulate organic matter and their respective rates of decomposition during sedimentation in the pelagic zone were discussed elsewhere (Chaps. 17 and 23). In large oligotrophic as well as large eutrophic lakes in which phytoplanktonic productivity dominates, humic inputs to the sediments will be relatively low. Gyttja sediments would be anticipated in these lakes, and this is generally borne out by empirical observations (cf., e.g., Hansen, 1961). Dy sediments are found in some small lakes dominated by littoral productivity, especially by acid-producing *Sphagnum* mosses, or allochthonous inputs of humic organic matter of terrestrial and wetland plant origins. All sediments contain some humic matter, and there is a broad transition between gyttja and dy.

Comparisons of the changes in chemical composition of pelagic seston while it is sedimenting in the water column were discussed in Chapter 17. Diagenesis of this organic matter is faster within the water column than once it is buried in the sediments. Within the sediments, decomposition rates of organic matter follow the general sequence: carbohydrates-amino acids-amino sugars > humic compounds > lipids (e.g., Kayama *et al.*, 1973; Kemp and Johnston, 1979).

Examination of lipids within sediments showed that bound lipids are relatively stable in comparison to free lipids (Cranwell, 1981; Robinson *et al.*, 1987; Ogura, 1990). Short-chain and unsaturated[3] compounds are readily and preferentially utilized by sediment bacteria. The total lipid content of the sediments is related more to chemical composition than to nutrient availability; total lipid content tends to increase where the biodegradability of the total organic compounds of the sediments decrease.

B. Humic Compounds Originating from Wetland and Littoral Flora

Products formed by microbial degradation of lignin can polymerize as humic compounds (Flaig, 1964; Larson and Hufnal, 1980; Wang *et al.*, 1980). As a result, it would be expected that the wetland and littoral macrophytes, and particularly the highly productive and more highly lignified emergent flora, would be a greater source of humic compounds than submersed hydrophytes or planktonic microflora. Analyses of the successive degradation of macroflora cast some light on the decomposition products and the rates of their formation.

The carbon and nitrogen contents of fractions of (1) fresh, living plant material of *Stratiotes aloides* were compared with those of (2) semidecomposed, dead plant material that was morphologically distinct but undergoing intermediate phases of decomposition, (3) young, thin sapropel formed by partial mineralization, and (4) old, thick, stratified sapropel resulting from nearly complete mineralization and humification (Úlehlová, 1970, 1971, 1976). Results (Table 21-1) showed that the carbon and nitrogen content decreased markedly with progressive decomposition to sapropels, while little change occurred in the C:N ratio. Free humic substances exhibited a loss of carbon in the sapropel phases of decomposition; this loss occurred primarily in the fulvic acid fraction. In the chemically bound humic fractions, carbon content increased with progressive degradation; again, humic acid content increased at the expense of the fulvic acid component. Nitrogen content increased in the semidecomposed stage, and the quantity of old sapropel decreased in the bound humic fractions. Most of the nitrogen was found in the fulvic acid fractions of the humic substances. These results suggest that free humic

[2]Also termed *tyrfopel* in older literature, meaning a *fine-grained peat*.

[3]Alkyl groups containing carbon–carbon double bonds.

TABLE 21-1 Carbon, Nitrogen, and C:N Ratios among Fractions of Decomposing *Stratiotes* Organic Matter in Lake Venematen, Netherlands (See Text)[a]

Carbon nitrogen, and C:N ratio	*Total organic matter*	*Humic organic matter*	*Residual organic matter*	*Free humic substances*	*Free humic acids*	*Free fulvic acids*	*Sorbed humic acids*	*Bound humic substances*	*Bound humic acids*	*Bound fulvic acids*
Carbon										
Fresh *Stratiotes*	57.0	40.4	16.7	20.6	6.5	14.1	8.9	10.8	4.3	6.5
Semidecomposed plant material	60.6	38.6	22.0	20.6	10.2	10.4	3.1	14.9	9.3	5.6
Young, thin sapropel	34.0	34.1	—	14.4	10.7	3.8	3.7	16.0	16.0	0.00
Old, thick sapropel	34.6	34.6	—	14.7	10.3	4.4	3.9	16.0	15.2	0.7
Nitrogen										
Fresh *Stratiotes*	2.6	2.3	0.2	1.1	0.1	0.9	0.8	0.5	0.3	0.2
Semidecomposed	3.0	1.4	1.6	0.8	0.2	0.6	0.2	0.4	0.3	0.1
Thin sapropel	1.3	0.6	0.7	0.2	0.1	0.06	0.3	0.2	0.2	0.00
Thick sapropel	1.7	0.6	1.1	0.1	0.1	0.04	0.2	0.2	0.2	0.03
C:N										
Fresh *Stratiotes*	22.3	17.5	69.4	19.7	46.7	15.5	11.7	21.7	14.3	32.7
Semidecomposed	20.5	28.0	14.0	27.2	56.4	18.0	17.2	33.9	29.0	46.9
Thin sapropel	25.5	55.8	—	72.3	76.1	63.2	13.9	106.6	106.6	—
Thick sapropel	20.5	58.6	—	105.1	103.2	109.8	16.1	76.1	84.5	24.3

[a] After data of Úlehlová (1971).

substances present in the early stages of decomposition were microbiologically transformed to chemically bound humic substances in later stages of decomposition. As the nitrogen of humic compounds is selectively removed, C:N ratios increase markedly. Slower rates of degradation occur when nitrogen levels decline sufficiently, which then permit some net accumulation of organic matter in the sediments.

II. RESUSPENSION AND REDEPOSITION OF SEDIMENTS

A conspicuous feature of the annual cycle of sedimentation rates of seston among dimictic lakes is a marked increase during periods of spring and autumnal circulation (Fig. 21-1). Much of this greater rate of settling particulate matter results from resuspension of flocculent surface sediments by currents moving across the sediments. The extent of disturbance and sediment resuspension into the water column, transport to different locations, and redeposition is of course variable among lakes of differing morphology, exposure to wind, and other features.

Sediments are resuspended when the bottom shear exceeds a critical shear stress for the sediment materials. Critical shear stress is a function of properties of the sediments, such as water content and grain size, and must be sufficient to disrupt the cohesion of the particles and the forces of gravity. Resuspension occurs during (a) continuous mixing throughout the water column to the bottom sediments, (b) intermittent complete mixing, particularly in monomictic and dimictic lakes with full circulation following periods of stratification, (c) intermittent complete mixing of the epilimnion that intersects shallow sediments (many lakes have over half of their volume in the epilimnion, the base of which is in contact with sediments), and (d) peripheral wave intersection with sediments in littoral areas where the water is shallower than half the wave length (see Chap. 7) (Hilton *et al.*, 1986; Evans, 1994; Bloesch, 1995).

Resuspension occurs episodically in both shallow regions and in deepwater sediments primarily as a function of depth and the distribution of shear stresses as regulated by a combination of wind speed, wave height coupled with effective lake fetch, duration of the disturbance, and lake basin morphometry. In the zone of erosion, material of a size available for resuspension is usually quite small and is moved to the water column over relatively short time periods (Godshalk and Wetzel, 1984; Evans, 1994). The areal extent of resuspension increases with wind speed and is greatly increased during times of complete overturn and circulation of the water column. Deepwater sediments of many lakes are disturbed to depths of several millimeters to several centimeters usually one or more times a year (e.g., Davis, 1968, 1973; Wetzel *et al.*, 1972; Pennington, 1974; White and Wetzel, 1975). In calcareous

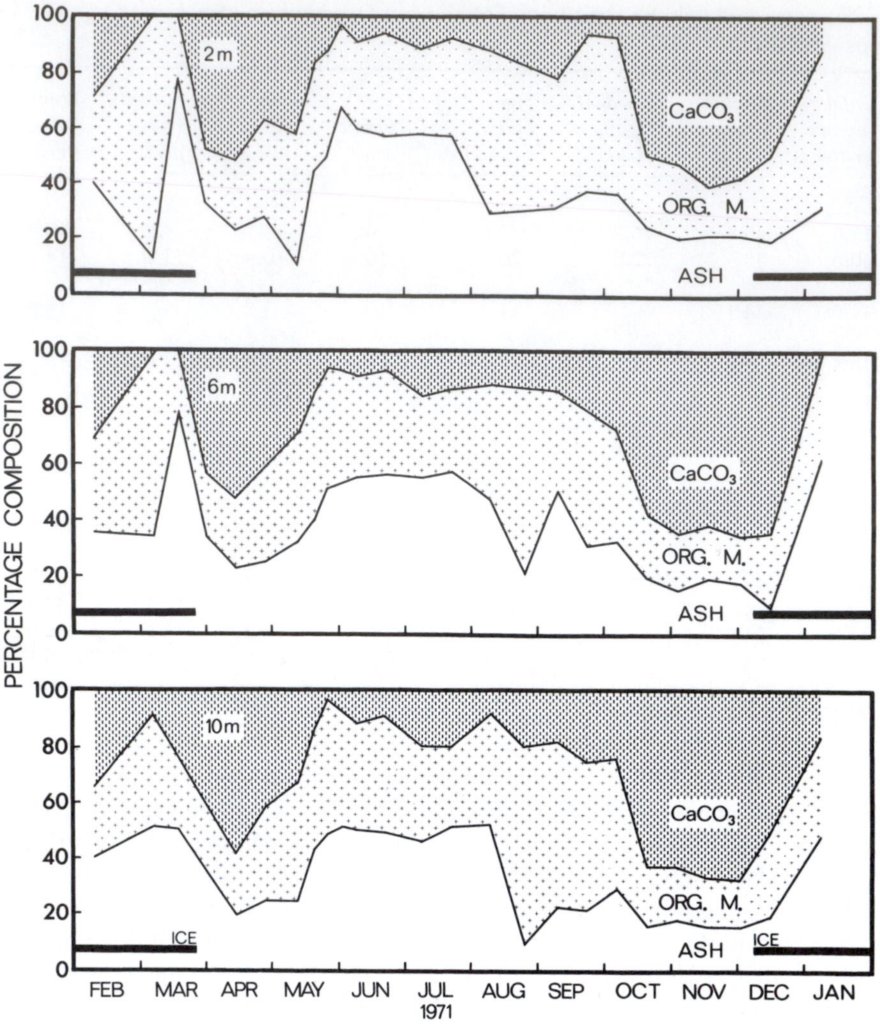

FIGURE 21-1 Seasonal changes in the percentage composition of calcium carbonate, particulate organic matter, and ash sedimenting to 2-, 6-, and 10-m depths. Lawrence Lake, Michigan, 1971. (From White and Wetzel, unpublished.)

lakes, as exemplified in Figure 21-1, resuspended calcium carbonate forms a major constituent of the seston during periods of circulation. This type of maximum resuspension during and at the end of water column circulation is common (e.g., Cobelas *et al.*, 1993), and the increased turbidity has been referred to in local vernacular as the "lake is cleaning itself." This fine-grained particulate matter is swept into the water column where it is subject to additional degradation by bacteria under aerobic conditions both during circulation as well as when it resettles to the bottom.

Differential horizontal sedimentation of particulate matter has been demonstrated in studies of the movement, deposition, and redeposition of pollen grains of different sizes and densities (Davis and Brubaker, 1973). Larger particles with rapid sinking rates are deposited fairly evenly onto the sediments throughout the lake basin. Smaller-sized particles with slower sinking rates in water are kept in suspension in the turbulent waters of the epilimnion, are carried across the lake in wind-mediated water currents, and are deposited preferentially onto littoral sediments. Later, particularly during fall circulation, these fine particles can be resuspended in the littoral areas, mixed in the lake water, and redeposited over the entire basin. Larger particles of organic matter or seston are not extensively resuspended and transported from the littoral to the center of the lake, although they are frequently shifted about in the littoral regions (Wetzel *et al.*, 1972; Godshalk and Wetzel, 1984; Hamilton and Mitchell, 1996). Aquatic plants, particularly emergent aquatic plants, disperse wave energy and ameliorate erosive effects and sediment resuspension (e.g., Moeller and Wetzel, 1988; James and Barko, 1994). The result is a size-selective

TABLE 21-2 Rates of Resuspension of Sediments of Lakes Based on Annual Measurements

Lake	Resuspension flux (g m^{-2} day^{-1})	Proportion of resuspended sediments of total settling particulate matter (%)	Source
Blue Chalk, Ontario, Canada	0.85	90	Dillon *et al.* 1990
Eau Galle, Wisconsin, USA	2.5	30–50	James and Barko (1993)
Hallwil, Switzerland	0–1.6	15–21	Bloesch and Uehlinger (1986)
Lough Neagh, Northern Ireland	21	90	Flower (1991)
Päijänne, Finland	0.6–1.4	51–79	Kansanen *et al.* (1991)
Wingra, Wisconsin, USA	11	82	Gasith (1975, 1976)
Ontario, USA–Canada	0.5	10–20	Rosa *et al.* (1983)
Erken, Sweden	1–47	54–80	Weyhenmeyer *et al.* (1995)

sorting and transport of the finest-sized particles from the littoral, which sediment into the quiescent waters below the metalimnion (cf. also Sebestyén, 1949). This resuspension and redistribution of organic matter has important consequences for the rates of decomposition of particulate organic matter deposited in littoral areas from either littoral production or allochthonous matter brought to the lake basin (cf. Chaps. 18 and 23). Resuspension of organic as well as inorganic particles can appreciably reduce light availability for photosynthesis and production of organic matter. For example, in a shallow lake, attenuation of light by resuspended particles was found to be directly proportional to wind velocity to the third power (Hellström, 1991) and the amount of organic matter in the resuspended particles (Blom *et al.*, 1994). In such lakes, total phosphorus content and other nutrients of the overlying water increased markedly during resuspension of the sediments, and phosphorus inputs to the water can be totally dominated by resuspension events (Kristensen *et al.*, 1992; Newman and Reddy, 1992; Reddy *et al.*, 1996).

Clearly, resuspension is greater in shallow lakes, even those that stratify thermally, than in deeper lakes. Although the data available on resuspended sediments as a proportion of the total settling particulate matter are few and variable because of differing techniques used for estimates, the conclusion is similar (Evans, 1994; Kozerski, 1994; Bloesch, 1995). The net sedimentation is small relative to the combined (gross) sedimentation of net sedimentation plus resuspension (Table 21-2). The percentage of suspended solids in the water column at any time originating from resuspension may exceed concentrations of new material by 10–20 times.

As the water levels in lake or reservoir basins change appreciably, the erosive energy of wave action is moved to different sediment strata, sometimes below the growth of littoral communities. Wind-mediated water movements can then cause appreciable resuspension of sediments and increases in nutrient loading to the water (e.g., Gibson and Guillot, 1997; Vernieu, 1997). Sediment resuspension is not limited to surface water movements. A zone of high shear and intense mixing can occur where internal seiches intersect the sediments and internal waves on the seiches break against the sediments and may be circulated within the metalimnion or hypolimnion (Gloor *et al.*, 1994; Pierson and Weyhenmeyer, 1994; Chikita *et al.*, 1995). Concentrations of resuspended particles, largely organic in nature, adjacent to the sediments have been observed to be several times greater than in the water directly above the benthic boundary layer.

III. MICROFLORA OF SEDIMENTS AND RATES OF DECOMPOSITION

A conspicuous feature of microbial populations in lakes is the great increase in numbers in the transition from the overlying water to the diffuse, uncompacted zone of the surficial sediments (Fig. 21-2). Bacteria increase about 3–5 orders of magnitude from the water to the surface sediments and decrease rapidly with increasing depth in the sediments. Expressed in numbers per gram dry weight of sediment, bacterial populations in the surficial sediments can reach 6–7 orders of magnitude greater than in an equivalent weight of overlying water (Table 21-3). Saprophytic bacteria decrease much more rapidly than do total bacteria with increasing depth below the sediment–water interface, which suggests a reduction of readily assimilable organic substrates below the interface.

Bacteria within the surface sediments are unevenly distributed over the lake basin as well (Henrici and McCoy, 1938; Steinberg, 1980a). When the littoral zone is covered with a well-developed macrophyte

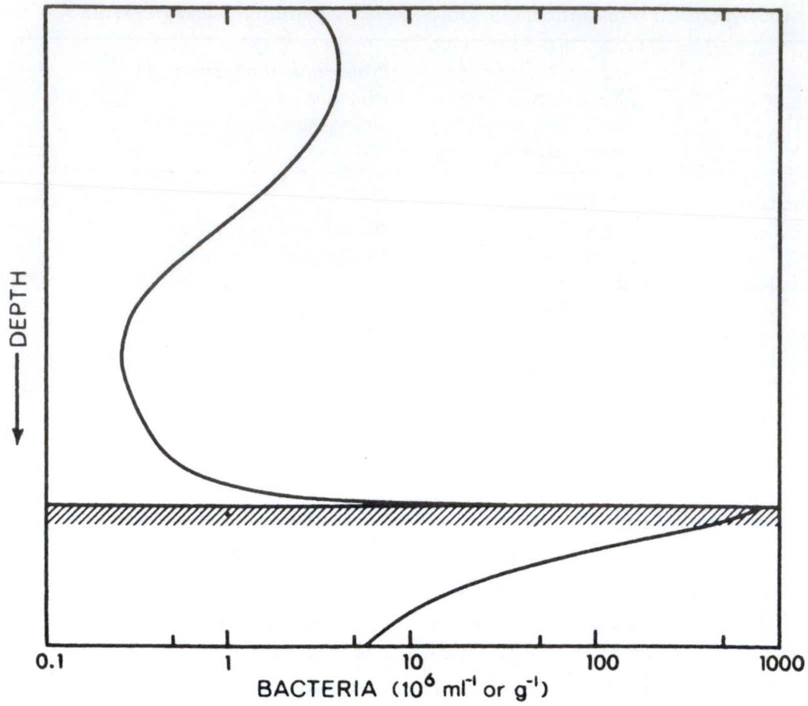

FIGURE 21-2 Generalized relative vertical distribution of bacterial numbers in water and sediments of a moderately productive lake. Depth scale in the water relative in meters and in the sediments in centimeters.

community, bacterial numbers of littoral sediments are much higher by several orders of magnitude than those found in profundal sediments of deepwater areas. Numbers of bacteria in sandy, organic-poor, wave-swept littoral sediments are lower than those in profundal sediments (e.g., Jones, 1980). In eutrophic Blelham Tarn, for example, aerobic respiration and nitrate reduction in the sediments of the littoral were greater than those of the profundal zone, whereas methanogenesis and sulfate reduction were greater in the profundal than in sediments of the littoral zone (Jones and Simon, 1981).

A more detailed analysis of bacterial concentrations in sediments within the macrovegetated littoral zone as compared with those of sediments free of vegetation showed similar relationships (Table 21-4), but with marked shifts seasonally in relation to growth of the macrovegetation. Nitrogen-fixing bacteria *Azotobacter* and the obligate anaerobe *Clostridium* were more abundant in sediments with macrovegetation than in sediments without vegetation or in the overlying water. Similar relationships were found among populations of hydrocarbon- and hydrogen-oxidizing bacteria, cellulose-decomposing bacteria and actino-

TABLE 21-3 Bacterial Numbers (Direct Enumeration) in the Surficial Layer of Sediments of Several Lakes of the Moscow Region, Russia[a]

Lake	Ash content of sediments	Bacterial numbers (10^6 g^{-1}) Fresh sediment	Dry sediment	Organic substances of bacteria in % of organic matter of sediments
Beloe (Kosino)	54.1	2326	54,219	7.8
B. Medvezh'e	58.3	1905	38,253	6.0
Chernoe (Kosino)	49.3	1285	35,109	4.5
M. Medvezh'e	47.3	1624	32,032	4.0
Sviatoe (Kosino)	18.6	922	29,790	2.3
Krugloe	79.2	1110	7991	2.5
Gab (Karelian region)	81.9	1883	15,680	5.8

[a] Modified from Kuznetsov (1970) after Khartulari.

TABLE 21-4 Comparison of Microbial Populations (Thousands per Gram of Sediment) of the Sediments among Emergent Bulrush Vegetation (A) and of Open Water Free of Bulrushes (B) in Lake Mekhteb, Russia[a]

	Total saprophytes		*Actinomycetes*		*Fungi*		*Clostridium pasteurianum*		*Hydrogen oxidizing*		*Methane forming*		*Sulfate reducing*	
Period	A	B	A	B	A	B	A	B	A	B	A	B	A	B
Dec–Jan: no bulrushes	590	790	0	80	0	30	10	10	1	10	0	1	0.1	2
Feb: young bulrushes	1800	2700	200	0	400	400	1	10	1	10	40	40	8	1
Mar: young bulrushes	4400	4800	400	300	0	0	1	10	10	10	5	240	40	10
Apr: developing stands	5300	3100	800	0	0	300	100	100	10	1	1500	140	50	1300
May:	1800	1800	1000	100	0	300	1000	10	1000	10	2000	15,000	100	100
June: mature bulrushes	7900	4800	2900	300	200	0	10	10	1000	10	2000	15,000	400	1200
July:	60,200	3800	0	0	100	200	100	10	10	100	1200	2000	240,000	5000
Aug: declining bulrushes	16,400	5400	1500	0	1000	0	100	100	100	10	1800	3000	1400	1500
Sep:	3800	2700	300	0	2800	0	10	10	10	100	170	13	1	1
Nov:	230	80	0	0	0	9	1	1	10	1	70	10	0	0

[a] From data of Aliverdieva-Gamidova (1969).

mycetes, and fungi. Large numbers of methane-forming and sulfate-reducing bacteria developed under anaerobic conditions in sediments that contain large quantities of decomposition products of higher aquatic plants.

Bacterial numbers and metabolic activities of the deepwater profundal sediments show a weak correspondence with the productivity of the lakes (Table 21-5), as one would anticipate the loading of organic matter and organic nutrients bound within the seston to be approximately proportional to lake productivity. Bacterial numbers in the open water tend to undergo rapid fluctuations and because of nutrient limitations and mortality from grazing and parasitism do not correlate well with lake productivity (cf. Chap. 17; Jones, 1979; Jones *et al.*, 1979). In extremely oligotrophic lakes, total numbers of bacteria of the sediments can be less than the cumulative total of the entire overlying water, which indicates that in these lakes most of the readily decomposable organic matter has been mineralized before the particulate residues reach the sediments (cf. Chap. 17; Bengtsson *et al.*, 1977; Mothes, 1981). Hence, strong positive correlations between bacterial abundance and organic matter in pelagic sediments would not be anticipated (cf. Bird and Duarte, 1989), but certainly increases in littoral areas.

A. Bacterial Activity of Sediments

The relationship of increased bacterial populations and metabolic activity to greater organic matter in sediments may appear obvious, but *in situ* measurements are very few. A significant correlation ($r = 0.93$; $p = 0.01$)

TABLE 21-5 Comparison of Relative Bacterial Numbers (Plate Culture Techniques) of the Profundal Sediments among Several Wisconsin Lakes of Differing Productivity[a]

Lake	*Type*	*Type of sediment*	*Average bacteria per cm³ of surface sediments*	*Total bacteria per cm³ of first 18 cm of sediment*
Crystal	Oligotrophic	Gyttja	2160	38,880
Weber	Oligotrophic	Gyttja	2350	42,300
Big Muskellunge	Mesotrophic	Gyttja	10,930	196,740
Trout	Oligotrophic	Gyttja	29,790	536,220
Little John	Mesotrophic	Gyttja	39,050	642,900
Mary	Dystrophic	Dy	39,450	710,100
Helmet	Dystrophic	Dy	120,300	2,165,400
Alexander	Eutrophic	Marl gyttja	144,240	2,599,320
Mendota	Eutrophic	Marl gyttja	609,300	>10,000,000

[a] After data of Henrici and McCoy (1938). The results must be treated in relative terms only because the methods employed likely estimate <5% of bacteria present.

between dehydrogenase activity and organic content of surface sediments has been found in a eutrophic reservoir (Lenhard *et al.*, 1962). These correlations, however, would not be expected to hold in organic-rich sediments in which other conditions, such as acidity as occurs in bogs, depress microbial activity or in sediments where much of the organic matter is chemically recalcitrant.

Data on oxygen consumption by the benthic community of bacteria, algae, and micro- and macrofauna offer some insight into sediment respiration rates, provided that oxygen utilization by these community components can be appropriately partitioned and corrected for abiotic oxygen uptake (Bowman and Delfino, 1980). Under aerobic conditions of shallow ($\bar{z}$ = 2.4 m) Marion Lake, Canada, respiratory oxygen consumption by sediment-dwelling bacteria over an annual period was found to be related primarily to temperature (Fig. 21-3; also Granéli, 1978). The proportion of community respiration resulting from bacterial respiration varied with season (Fig. 21-3) and was lowest during maximal community respiration during the summer period of higher temperatures. This percentage certainly increases under other conditions among lakes; in deep-water sediments, for example, nearly all metabolism is by heterotrophic microbes.

Microbes associated with particulate organic detritus consumed up to 3 orders of magnitude more oxygen per unit dry weight than did those of sand (Hargrave, 1972). Uptake rates were inversely related to particle size and directly related to particle organic content of carbon and nitrogen.

In shallow lakes, water turbulence can frequently disturb and mix sediments to appreciable depths (e.g., to 10 cm; Viner, 1975). After brief reoxygenation by turbulence or the respiratory or feeding activities of benthic animals, community respiration and oxygen consumption rates of highly organic sediments quickly (hours to a few days) return to values similar to those of undisturbed sediments of similar organic content and composition (Hargrave, 1975; Viner, 1975).

Assays of the heterotrophic activity of benthic microorganisms of shallow water sediments by the uptake of labeled glucose, acetate, and glycine demonstrated that the greatest activity occurred in the summer months when the water temperatures exceeded 10°C (K. J. Hall *et al.*, 1972; Toerien and Cavari, 1982). Availability of substrates and slow diffusion rates exerted some control over the uptake and utilization of dissolved organic compounds of the interstitial water. The fraction of the substrate respired as CO_2 varied with the amended compound and was highest (average 63%) with glycine in comparison to glucose (22%) and acetate (13%). Most of the heterotrophic activity

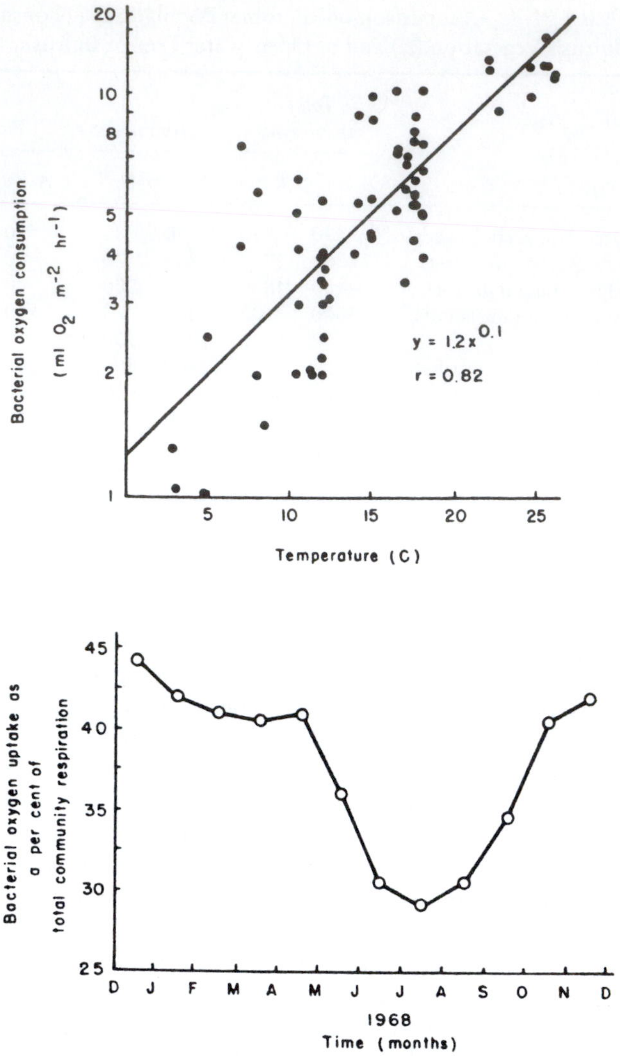

FIGURE 21-3 *Upper*: Relationship of temperature and sediment bacterial respiration; *lower*: bacterial respiration as a proportion of total community oxygen consumption in shallow-water Marion Lake sediments during 1968. (After Hargrave, B. T.: Epibenthic algae production and community respiration in the sediments of Marion Lake. *J. Fish. Res. Bd. Canada* 26:2003–2026, 1969.)

occurred in the upper sediment layers, the sites of highest bacterial numbers in the sediments.

Glucose uptake rates by bacteria of surficial sediments of a small, eutrophic alpine lake were greater in littoral areas than in profundal sediments (Steinberg, 1978b). Similar results were found in a hypereutrophic lake in southern Michigan (King and Klug, 1982b). The turnover times (T_t) of glucose were much more rapid in the anaerobic sediments (0.3–4 h) than in the overlying water column (20–40 h). About 40% of the methane carbon generated in the sediments emanated from the added glucose. Turnover rates were faster in the littoral sediments than in the profundal sediments.

A number of studies have indicated enhanced growth of bacteria in sediments at temperatures greater than usually occur in profundal sediments of temperate lakes (e.g., Inniss and Mayfield, 1978a,b, 1979; Tison and Pope, 1980; Tison *et al.*, 1980). Appreciable adaptation to cold temperatures apparently occurs by selection of bacterial species and by a physiological tolerance of temperature changes by individual species (e.g., Boylen and Brock, 1973; Leduc and Ferroni, 1979).

IV. ANAEROBIC DECOMPOSITION IN SEDIMENTS

Oxygen serves as the universal hydrogen acceptor for biochemical reactions of microbes under aerobic conditions. Under anaerobic conditions, however, the relationships are far more complicated, for various other substances and intermediate metabolic organic compounds become electron (hydrogen) acceptors. Often, the same compound can serve as an electron acceptor or donor, depending on the environmental conditions.

As was described earlier for the water column, when oxygen is utilized more rapidly than it is generated by photosynthetic processes or intrusion from other habitats, anaerobic microbial processes utilize a variety of inorganic and organic compounds as alternative electron acceptors. Specific anaerobic processes occur largely in a fixed sequence reduction of nitrate, manganese, iron, sulfate, and carbon dioxide utilizing organic substrates in methanogenesis (Table 21-6). The occurrence of specific anaerobic processes is dependent on the presence of the appropriate electron acceptor as well as competitive electron acceptors (e.g., review of Laanbroek, 1990). During the succession of anaerobic oxidation processes, the redox potential of the sediments decreases as a result of the reduced products formed (Table 21-6).

Each redox couple has a characteristic redox potential at pH 7. The potentially available maximum amount of energy generated in a redox reaction, assuming H_2 as the electron donor, decreases in the sequence from oxygen to carbon dioxide (Table 21-6). Generally, physiological adaptations have evolved to optimize the energy yield from the total oxidation of organic matter.

Some electron-acceptor compounds, such as nitrate and ferric oxide, can be reduced by the same bacteria. These microbes, however, have a preference to utilize nitrate in the reduction of organic compounds and are approximately mutually exclusive to utilize nitrate until the supply is exhausted before utilizing iron oxides. The reductions of ferric oxide, sulfate, and carbon dioxide are also mutually exclusive but vary because the reduction is performed by physiologically different types of bacteria with different affinities for electron donors such as acetate and hydrogen (Lovley and Klug, 1983b; Newell, 1984; Ward and Winfrey, 1985; Lovley, 1987; Laanbroek, 1990). Ferric oxide reduction occurs preferentially to the reduction of sulfate and carbon dioxide owing to a stronger affinity of ferric oxide–reducing bacteria for these electron donors than the other types of anaerobic bacteria. Similarly, sulfate-reducing bacteria are better scavengers for acetate and hydrogen than are the methanogenic bacteria because methanogens have the lowest affinities for these fermentation products.

The dynamics of nitrogen metabolism and exchanges in sediments is complicated because of numerous pools for nitrogen and the many simultaneously regulating roles of microorganisms on the nitrogen cycle. If light reaches the sediments, the processes and fluxes are further complicated by nitrogen assimilation and/or fixation, storage, and release as organic nitrogen. Because many sediments receive sufficient light to support an intercoupled benthic microalgae and

TABLE 21-6 Processes of Oxidants and Reductants and Approximate Redox Potentials (E_0) Resulting from Reduced Products Formed from Bacteria in Sediments during a Succession of Anaerobic Oxidation Processes[a]

Process	*Substrates*	*Products*	$\Delta G_0'$ (*kcal mole*$^{-1}$)	E_0 (*mV*)
Disappearance of oxygen	$O_2 + H_2$	$2H_2O$	−56.6	+330
Disappearance of nitrate	$2NO_3^- + 5H_2 + 2H^+$	$N_2 + 6H_2O$	−53.3	+220
	$NO_3^- + 4H_2 + 2H^+$	$NH_4^+ + 3H_2O$	−35.8	
Appearance of manganous ions	$MnO_2 + H_2 + 2H^+$	$Mn^{2+} + 2H_2O$	−37.5	+200
Appearance of ferrous ions	$2Fe(OH)_3 + H_2 + 4H^+$	$2Fe^{2+} + 6H_2O$	−10.5	+120
Disappearance of sulfate	$SO_4^{2-} + 4H_2$	$S^{2-} + 4H_2O$	−9.1	−150
Appearance of methane	$CO_2 + 4H_2$	$CH_4 + 2H_2O$	−7.9	−250

[a] Extracted from review of Laanbroek (1990) after many sources.

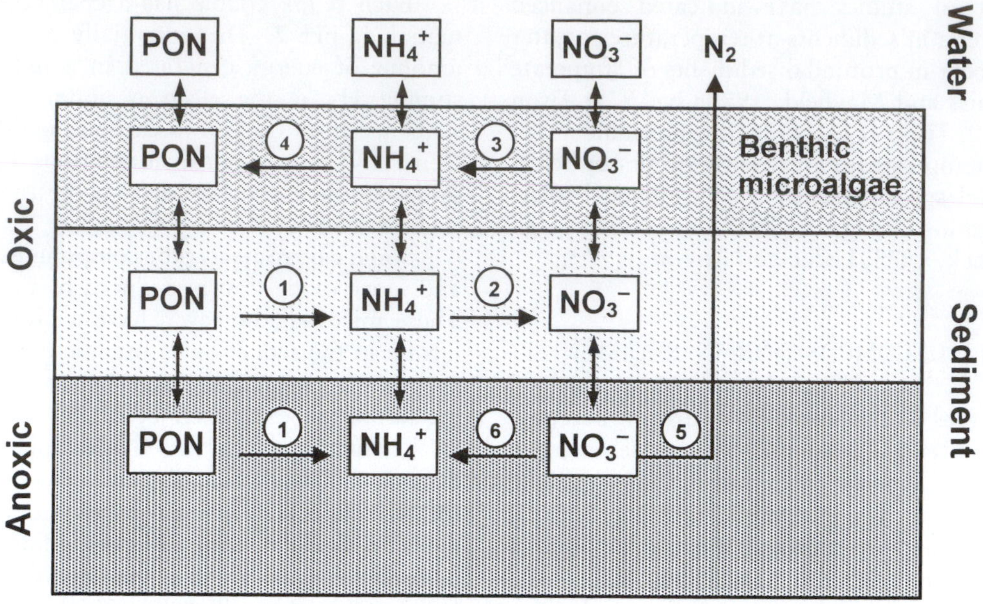

FIGURE 21-4 Vertical profile through sediment colonized by benthic microalgae, separated into an oxic layer and an anoxic layer (although when dark, the oxic layer usually rapidly becomes anoxic until the next period of light). PON = particulate organic nitrogen; 1 = mineralization; 2 = nitrification; 3 = assimilatory nitrate reduction; 4 = assimilation; 5 = denitrification; 6 = dissimlative nitrate reduction to ammonium (DNRA). (Modified from Rysgaard *et al.*, 1993.)

bacterial community on the surface of the sediments (cf. Chap. 19), it is included in the general model presented in Figure 21-4. Nitrogen transformations include assimilation, mineralization, nitrification, and denitrification in sediments (e.g., Rysgaard *et al.*, 1993). Particulate organic nitrogen is mineralized to NH_4^+ in both the oxic and anoxic strata in the sediments. Ammonium can then be oxidized to NO_3^- by nitrifying bacteria in the oxic layer of the sediment, assimilated by benthic algae and other microbes, or diffuse out of the sediment. The NO_3^- of nitrification can under anaerobic conditions be reduced to N_2 by denitrifying bacteria and lost from the environment, or it can be reduced to NH_4^+ by fermentative bacteria or diffuse out of the water. Nitrate and NH_4^+ in the overlying water can also diffuse into the sediment and be cycled as discussed.

As organic matter concentrates in the sediments, decomposition processes release ammonium (NH_4^+) that can be adsorbed to sediment particles (ca. 90% can be adsorbed; cf. Verdouw *et al.*, 1985), assimilated and metabolized directly by biota in the sediments, or nitrified bacterially to nitrate. In addition, NO_3^- can diffuse from the overlying water into the sediments. Denitrification in aquatic sediments, as in anoxic water strata overlying sediments, is a major potential loss of nitrogen from aquatic ecosystems, as it converts nitrate to N_2 that can evade to the overlying water and the atmosphere. A large percentage of the nitrate produced in the sediments is denitrified, and in many freshwater sediments, most (75–>95%) of the organic nitrogen mineralized in the sediments is lost through nitrification and denitrification (Seitzinger, 1990). Some of the highest rates of denitrification have been found in eutrophic river sediments. When benthic algae and cyanobacteria are present and active, much of the NO_3^- and NH_4^+ is assimilated by these microphytes, even for many hours (25–60 h) after darkness (Rysgaard *et al.*, 1993). Under steady-state conditions in the dark, however, as would be the case in profundal sediments, nearly all (≈100%) of the total NO_3^- uptake was by bacterial denitrification.

A. Methanogenesis

Degradation of large quantities of organic matter occurs under anaerobic conditions through methanogenesis in the sediments of lakes, ponds, wetlands, and streams, as well as in waste-treatment facilities (oxidation ponds, anaerobic lagoons, and septic tanks). In the degradation process, organic matter is converted quantitatively to methane and carbon dioxide in two stages (Fig. 21-5). In the first stage, a heterogeneous group of facultative and obligate anaerobic bacteria, termed *acid formers*, converts proteins, carbohydrates, and fats primarily into fatty acids by hydrolysis and fermentation (Deyl, 1961; McCarty, 1964; Zeikus, 1977; Nedwell,

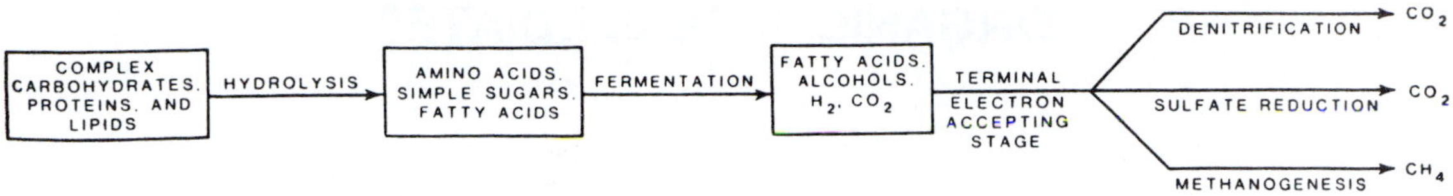

FIGURE 21-5 Stages of anaerobic metabolism of complex organic compounds. (Modified from Wolfe, 1971; Molongoski, 1978; Zehnder, 1978.)

1984). The methane-producing bacteria then utilize the organic acids, converting them to carbon dioxide and methane. Certain alcohols from carbohydrate fermentation can also be converted to methane and carbon dioxide by methane-producing bacteria.

Utilization of organic compounds in suspension or solution by acid-forming bacteria in the first stage results only in the synthesis of bacterial cells and the generation of end-product organic compounds, such as organic acids. Removal of oxidizable organic compounds occurs in the second stage and is directly proportional to the quantity of methane produced. During most anaerobic metabolism, the formation of hydrogen occurs (see later discussion); alternatively, the reduction of inorganic electron acceptors, such as nitrate, nitrite, oxidized iron, and sulfate takes place (described in detail in Chaps. 12 and 14).

The first stage of acid formation results from hydrolysis and fermentation (McCarty, 1964). Proteins are first enzymatically hydrolyzed to polypeptides and then to simple amino acids. Complex carbohydrates such as starch and cellulose are hydrolyzed to simple sugars, while fats and oils are hydrolyzed to glycerol and fatty acids. The amino acids, simple sugars, and glycerol formed by hydrolysis are soluble and are fermented by acid-forming bacteria. In the absence of oxygen, one portion of the organic molecule is oxidized, while another portion of the same molecule (or sometimes another compound) is reduced. These fermentative energy-yielding oxidation–reduction reactions produce reduced (saturated) fatty-acid compounds and oxidized carbon as CO_2. Ammonia is produced as an end product during amino-acid fermentation.

The methane-producing bacteria are strictly anaerobic and consist of four major genera: rod-shaped nonsporulating *Methanobacterium*, rod-shaped sporulating *Methanobacillus*, and the spherical *Methanococcus* and *Methanosarcina*. Species are differentiated on the basis of the substrates they are capable of using.

Methane is formed by two major processes. In the first, carbon dioxide serves as a hydrogen acceptor and is reduced to methane by enzymatic addition of hydrogen derived from the organic acids:

$$CO_2 + 8H \longrightarrow CH_4 + 2H_2O$$

In the second process, acetate, which is a major intermediate produced by the fermentation of complex organic compounds, is converted to CO_2 and methane. In this reaction, the carbon of the methane originates from the methyl carbon of acetate:

$$^*CH_3COOH \longrightarrow {}^*CH_4 + CO_2$$

Methane produced during the fermentation of mixed organic compounds originates largely (approximately 70%) via the intermediate, acetate. Much of the remainder of the methane results from reduction of CO_2. Propionic acid, an intermediate compound produced in significant quantities from protein, methanol, methylamines, carbohydrate, glycerol, and other compounds, also can serve as a substrate for methane production (e.g., Naguib, 1982; Lovley and Klug, 1983a). Longer-chain acids are of less importance. Acetate and hydrogen are, however, the major methane precursors.

The concentrations of acetate, H_2, and CO_2, the major *in situ* substrates for methanogenesis, are variable and dependent upon the inputs and rates of fermentation of settling organic matter. Acetate is the preferred substrate for methanogenesis at low partial pressures of hydrogen; as the partial pressure of hydrogen increases, bicarbonate is the preferred substrate (e.g., Winfrey *et al.*, 1977; Strayer and Tiedje, 1978a; Jones *et al.*, 1982; Lovley and Klug, 1982; Lovley *et al.*, 1982; Nedwell, 1984; Rothfuss and Conrad, 1993; Conrad, 1999). Hydrogen should theoretically account for about one-third of total methanogenesis when carbohydrates or similar forms of organic matter are degraded (Fig. 21-6). Many hypolimnetic temperatures are always cold ($<12°C$), which slows methanogenesis for thermodynamic reasons. Increases of temperature, however, can shift the degradation pathway and increase the role of H_2 as an intermediate substance (Schulz and Conrad, 1996). In methanogenic environments, H_2 is rapidly cycled, and its concentration is the result of simultaneous production by fermentation and syntrophic bacteria and consumption by methanogenic bacteria.

Nitrate, nitrite, oxidized iron, and sulfate in sediment in interstitial water inhibit methanogenesis either directly by metabolic inhibition of methanogens or

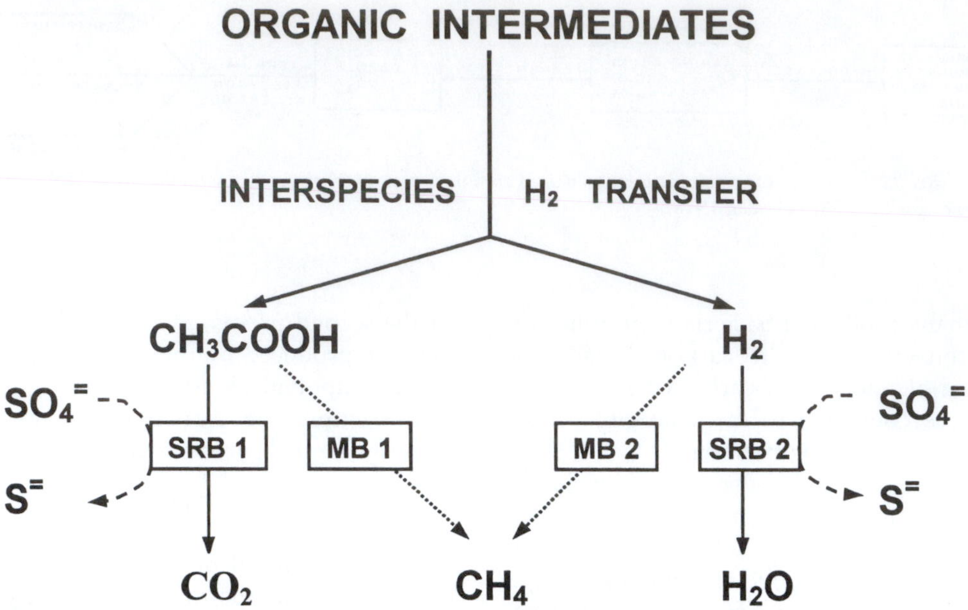

FIGURE 21-6 Competition for H_2 and acetate between methanogenic bacteria (MB) and sulfate-reducing bacteria (SRB). In the presence of high sulfate concentrations, MB are outcompeted for resource by SRB (*broken lines*). MB-1 = Acetate-utilizing MB; MB-2 = H_2 + CO_2-utilizing MB; SRB-1 = acetate-utilizing SRB; SRB-2 = H_2-utilizing SRB. (Modified from Nedwell, 1984.)

indirectly by channeling electron flow from methane carbon precursors to these alternate electron acceptors (Fig. 21-6, Table 21-6) (cf., Chap. 12 and 14, and, e.g., Chen *et al.*, 1972; MacGregor and Keeney, 1973; Cappenberg, 1974, 1975; Winfrey and Zeikus, 1977, 1979; Zeikus, 1977; Zehnder, 1978; Ward and Winfrey, 1985; Roden and Wetzel, 1996). The degree of inhibition depends upon the relative concentrations of alternate electron acceptors. Evidence suggests that certain sulfate-reducing bacteria can function commensally with methanogens: Methanogens ferment acetate released by the sulfate-reducing bacteria (Cappenberg, 1975). When sulfate is not limiting, the lower half-saturation constant of sulfate-reducing bacteria enables them to inhibit methane production by lowering the hydrogen partial pressure below levels that methanogenic bacteria can effectively utilize (Lovley *et al.*, 1982; Robinson and Tiedje, 1984; Ward and Winfrey, 1985; Conrad *et al.*, 1987). However, methanogenic bacteria can coexist with sulfate-reducing bacteria in the presence of sulfate (Fig. 21-6); the outcome of competition at any time is a function of the rate of hydrogen production, the relative population sizes, and sulfate availability. In low-sulfate environments methanogens are more competitive, whereas in high-sulfate environments, sulfate reducers are more competitive, in part because sulfate reducers have a greater affinity for acetate and H_2 (Conrad, 1999). In methanogenic environments, H_2 is rapidly turned over, and its concentration is the result of simultaneous production by fermenting bacteria and consumption by the methanogenic bacteria. The steady-state concentration observed in most methanogenic environments is close to the thermodynamic equilibrium of H_2-dependent methanogenesis to a boundary of ΔG_0 of ca. -96 kcal mol^{-1}.

Rates of sulfate reduction in pelagic sediments generally increase with greater productivity of phytoplankton in lakes (Table 21-7). The intensive sulfate-reduction processes, however, remove sulfate supplies as rapidly as they penetrate sediments from the overlying water, and exceed availability, particularly during stratification where sulfate is eliminated completely from the hypolimnion (Chap. 14). Methanogenesis then dominates in sediments receiving high amounts of organic matter.

B. Methane Oxidation

Methane can be oxidized abiologically by oxidation by hydroxyl radical and by aerobic bacterial oxidation. On a global basis, bacterial oxidation likely exceeds abiological conversions and substantially reduces methane emissions to the atmosphere (King and Blackburn, 1996). Data from many diverse aquatic ecosystems indicate that as much as 90% of the methane produced in sediments is oxidized as it passes into oxic zones within the sediments or water.

Although the production of methane in sediments can be very intense and reach as much as 85% of the

TABLE 21-7 Rates of Sulfate Reduction in Pelagic Sediments of Lakes of Differing Productivity

Lake	*Trophic state*	*Sulfate reduction mmol m^{-2} day^{-1}*	*Source*
Stechlin, Germany	Oligotrophic	500–1500	Babenzien *et al.* (1991)
Fuchskuhle, Germany	Acidotrophic	20–2000	
Washington, Washington, USA	Mesotrophic	117	Kuivila *et al.* (1989)
Constance, Germany	Eutrophic	1000–3000	Bak (1988)
Dagow, Germany	Eutrophic	1500–25,000	Babenzien *et al.* (1991)
Vechten, Netherlands	Eutrophic	3600	Hordijk *et al.* (1985)
Wintergreen, Michigan, USA	Hypereutrophic	7500	King and Klug (1982a)

total gas volume formed in the deposits, little of the methane escapes to the atmosphere in low to moderately productive lakes, owing to the presence and activity of methane-oxidizing bacteria in the sediments and overlying water (ZoBell, 1964). Methane oxidation rates are much higher per unit volume in sediments or surface flocs than in the water column (Lidstrom and Somers, 1984; Kuivila *et al.*, 1988; Frenzel *et al.*, 1990; reviewed in King and Blackburn, 1996). In profundal lake sediments, 50–>90% of methane flux is consumed by methane oxidation. Variables include the extent and rates of advection and diffusion of CH_4 and O_2, both of which vary markedly with seasonal stratification, lake circulation, and sediment resuspension. Consumption of oxygen diffusing into sediments usually exceeds inputs within the first 2 mm, so that much of the methane oxidation is maximized in this area where the two gradients meet. Under some circumstances, it is possible that ammonia-oxidizing bacteria can also oxidize CH_4, although it is unclear that this metabolism supports growth (cf. King and Blackburn, 1996). Very high concentrations of ammonia (>5–20 mM) can inhibit methane oxidation in the surface layers of sediments (Bosse *et al.*, 1993).

Methane oxidation is much more variable and ranges from <10–>90% in shallow water and wetland sediments. As discussed earlier, photosynthesis by benthic algae growing on the sediments in shallow water can shift oxygen concentrations within the surficial sediments (1–4 mm) from anaerobic conditions during darkness to supersaturated concentrations during the daylight (cf. Fig. 13-8 and pp. 253). During the transition from light to dark periods, methane flux increases accompanied by a decrease in methane oxidation (King, 1990a). The processes reverse under lighted conditions.

Emergent and floating-leaved but rooted aquatic plants are major conduits for methane exchange with the atmosphere (e.g., Dacey and Klug, 1979; Sebacher *et al.*, 1985; Schütz *et al.*, 1989; King, 1990b; Whiting and Chanton, 1992). Clearly, methane diffuses into the roots of aquatic plants—appreciable amounts (up to 95%) of the CH_4 is oxidized there or in the oxidized microrhizosphere around the roots (e.g., Schütz *et al.*, 1989; King *et al.*, 1990; Bosse and Frenzel, 1997; Calhoun and King, 1997), but collectively large amounts can be transported in the internal ventilation gaseous system of the plants and expelled to the atmosphere. Results generally indicate that >90% of the methane evasion to the atmosphere from wetlands and littoral areas is released from the plants rather than from direct ebullition. That gas flux to the atmosphere is coupled, in part, to rates of metabolism of the plants, increases with increasing emergent plant productivity, and commonly decreases as stomatal conductance decreases (Csermák *et al.*, 1992; Morrissey *et al.*, 1993; Whiting and Chanton, 1993). This release appears to be greater during vigorous plant growth than later during leaf senescence (Stanley *et al.*, 1998). Herbivory on the aquatic plant leaves had little effect on methane release. Although methane evasion to the atmosphere from the pelagic region of lakes is relatively low, these rates increase in shallow lakes and in the littoral areas (Table 21-8). Methane emissions from natural and cultivated (particularly rice) wetlands are appreciable (ca. 170–200 Tg yr^{-1}) on a global basis (Aselmann and Crutzen, 1989; Bouwman, 1991; Bartlet and Harriss, 1993; Sass and Fisher, 1996).

Methane concentrations in rivers can also be significant, and in large rivers the waters are nearly always supersaturated with methane (10–500 nM). Much of the methane evades directly to the atmosphere, with only about 25% being utilized by methane oxidizers (Lilley *et al.*, 1996). Undisturbed forested streams and rivers can contain high natural levels of CH_4 not attributable to pollution (de Angelis and Lilley, 1987; Jones *et al.*, 1995; Jones and Mulholland, 1998). Production can occur in the stream sediments as well as be imported by lateral diffusion and runoff from saturated forest and fertilized agricultural soils. Methane production is commonly greatest in bank sediments that are relatively isolated hydrologically and

TABLE 21-8 Examples of Methane Evasion to the Atmosphere from Openwater Areas and from Macrophyte-Covered Littoral and Wetland Areas[a]

Region/habitat	*Mean pelagic* mmol $CH_4\ m^{-2}\ day^{-1}$	*Mean littoral* mmol $CH_4\ m^{-2}\ day^{-1}$	*Source*
Arctic and boreal			
Small lakes	1.3	—	Whalen and Reeburgh (1990)
Average Arctic wetlands (>50°N)	—	7.8	Crill (1996)
Average Boreal wetlands (45–60°N)		6–21	
Temperate			
Florida lakes and wetlands	2.1	0.5–7	Barber *et al.* (1988)
Freshwater lakes, CA	10.1	—	Miller and Oremland (1988)
Saline lakes, CA	0.3	—	
Small lakes, CO	1.6	—	Smith and Lewis (1992)
Red Rock Lake, CO	4	12.6	
Duck Lake, MI	—	10	Dacey and Klug (1979)
Lake Kasumigaura, Japan	1.9	—	Nakamura *et al.* (1999)
Priest Pot, England			
Diffusion	0.4	—	Casper *et al.* (1999)
Ebullition	12.0	—	
Three rivers, OR	0.7	—	De Angelis and Lilley (1987)
Constructed wetlands, Netherlands	—	13–17	van der Nat and Middelburg (1998)
North temperate peatlands	—	0.2–41.5	Harriss *et al.* (1985)
German wetlands	—	0–10	Pfeiffer (1994)
Phragmites wetland, Nebraska	—	22.2	Kim *et al.* (1998)
Australian floodplain wetlands	—	<0.2–66	Boon and Sorrell (1995)
Mississippi River deltaic plain wetlands	—	9–57	Alford *et al.* (1997)
Average temperate wetlands (20–45°N)	—	5.4–6.2	Crill (1996)
Tropical			
Amazon, low water	4.6	12.6	Smith & Lewis (1992)
Amazon, high water	1.7	14.4	Bartlett *et al.* (1990)
Orinoco River, Venezuela			
High water	1.3	1.1	Smith and Lewis (1992)
Falling water	3.1	12.2	
Amazon floodplain lake, Brazil	0.25–4	0.6–12.4	Wassman and Thein (1996)
Average tropical wetlands (~3°S)	—	10–13.5	Crill (1996)

[a] Many additional values of methane evasion exist in a large literature on this subject (e.g., reviews of Shotyk (1989), Chanton and Dacey (1991), Harriss *et al.* (1991), Bartlett and Harriss (1993), Bubier *et al.* (1993), Sass and Fisher (1996), Javitt *et al.* (1997), and Segers (1998).

lowest in hyporheic and parafluvial sediments that are interactive with the surface stream water. In large rivers, increased areas of anoxia in sediments and reduced hydrologic exchange with surface water can enhance concentrations of methane.

Methane oxidizers are widely distributed in most natural waters (cf. Higgins *et al.*, 1981; Saralov *et al.*, 1984). Among the best-known of many species of methane-oxidizing bacteria is *Methanomonas methanica*, which, in addition to methane, can oxidize numerous other hydrocarbon compounds. All members of this group appear to be strict aerobes, although they tolerate microaerobic conditions. The overall empirical equation for methane oxidation is

$$5CH_4 + 8O_2 \longrightarrow 2(CH_2O) + 3CO_2 + 8H_2O$$

with a free-energy yield of − 195 kcal per mol.

The distribution of methane-oxidizing bacteria in dimictic lakes indicates that these forms are present throughout the year in significant numbers. The methane oxidizers increase during summer stratification, the highest numbers ($>5 \times 10^2$ ml^{-1}) occurring in metalimnetic and hypolimnetic waters with reduced oxygen concentrations and maximum concentrations of methane (Cappenberg, 1972). During summer stratification, the highest rates of methane oxidation have been consistently found in the lower metalimnion, where oxygen concentrations are reduced to approximately 1 mg $liter^{-1}$ or less (Rudd *et al.*, 1974, 1976; Rudd and Hamilton, 1975; Jannasch, 1975; Panganiban *et al.*, 1979). During the summer, when combined inorganic nitrogen sources in the epilimnion were low in the eutrophic lakes studied, the methane-oxidizing bacteria fixed molecular nitrogen. Methane oxidation in strata above the metalimnion declined by as much as 40%, because N_2 fixation can be inhibited by the higher concentrations of oxygen occurring there. High rates of methane oxidation, amounting to as much as 95% of the annual quantity of methane oxidized (Rudd and Hamilton, 1975; Heyer and Suckow, 1985), have been found in spring and fall periods of lake circulation. During the periods of lake turnover, combined inorganic nitrogen is available and the dependency of methane oxidation on N_2 fixation found in summer months is alleviated. In shallow but eutrophic lakes that do not stratify and are oxic continuously, methane oxidation is still the dominant methane sink. For example, in a large, eutrophic lake in Japan, methane oxidation consumed an annual average of 74% of dissolved methane in the water column (Utsumi *et al.*, 1998). Under ice cover, high rates of methane oxidation can occur throughout the water column of eutrophic and protected dystrophic lakes, possibly lowering oxygen concentrations or even causing total anoxia (cf. Chap. 9).

Oxidation of methane to carbon dioxide in the absence of oxygen has been observed but only at the sediment surface (Panganiban *et al.*, 1979). The methane-oxidizing organisms and the metabolic pathways involved are unclear but are likely associated with the activity of sulfate-reducing bacteria (cf. Reeburgh and Heggie, 1977; Rudd and Taylor, 1980).

C. Methane Cycling in Lakes

Several recent investigations have attempted to quantify methane cycling in eutrophic lakes in terms of overall carbon cycling and to evaluate its effects on the entire lake ecosystem (reviewed in Rudd and Taylor, 1980). For example, the annual methane cycle was evaluated in Lake 227, a lake in western Ontario rendered eutrophic by experimental fertilization (Rudd and Hamilton, 1978). Lake 227 is small (5 ha.), moderately deep (z_m = 10 m), and during the investigation circulated only for about six weeks in the fall before ice formation. During summer stratification, most methane produced in hypolimnetic sediments accumulated in the hypolimnion. Vertical diffusion of methane from the hypolimnion to the metalimnion was very slow and amounted to 11% of the methane produced in the hypolimnion. Nearly all of the methane that diffused into the metalimnion was converted to bacterial-cell carbon and CO_2 by methane-oxidizing bacteria. Small amounts of methane observed in the epilimnion during summer emanated largely from methanogenesis in epilimnetic sediments. Epilimnetic methane slowly evaded to the atmosphere, since little was oxidized because of low combined inorganic nitrogen concentrations and the inhibitory effects of high oxygen on N_2-fixation by methane-oxidizing bacteria.

Oxidation rates of methane in Lake 227 increased abruptly during fall circulation as hypolimnetic water rich in accumulated methane and inorganic nitrogen was mixed and oxygenated. About 40% of the methane that was transported into the upper strata escaped to the atmosphere. During fall circulation, however, the rates of methanogenesis in the surficial sediments were markedly reduced as the availability of alternative, more efficient electron acceptors, such as sulfate, became available in the sediments for reduction (e.g., Phelps and Zeikus, 1985). Under winter ice cover, methane production in hypolimnetic sediments was similar to that during the summer. Only about 10% of the methane was consumed by methane-oxidizing bacteria at the aerobic–anaerobic interface; the remainder accumulated and extended into the summer hypolimnion, since this lake did not circulate completely in the spring following ice loss.

In Lake 227, the annual methane production, oxidation, and evasion were compared to the annual phytoplankton primary production and total carbon inputs from runoff and the atmosphere. This comparison showed that the amounts of CO_2 and bacterial-cell material produced by methane oxidizers were insignificant in terms of carbon flow in comparison to primary production (Table 21-9). Methane cycling was much more significant, however, when compared with the total carbon input into the lake. During 1974, the amount of carbon regenerated as methane was equivalent to about 55 percent of the total carbon input into the lake (Table 21-9). Methane-oxidizing bacteria recycled about two-thirds of the methane carbon. Similar values

TABLE 21-9 Annual Rates of Methane Production, Methane Oxidation, Phytoplanktonic Primary Production, and Total Carbon Inputs from the Atmosphere and Runoff in Lake 227, Ontario[a]

	g C m^{-2} of lake surface[b]
CH_4 production	18
Total CH_4 oxidation	12
CH_4-oxidizing bacterial production	6
CO_2 production by CH_4-oxidizing bacteria	6
Phytoplanktonic primary production	138
Total carbon inputs	33

[a] After Rudd and Hamilton (1978).

[b] Primary production and total carbon inputs occurred mostly during the summer. The methane production and oxidation year included the succeeding winter since regeneration of methane carbon fixed during the previous summer would be occurring. The large difference between the amount of carbon entering the lake and the amount fixed by phytoplankton presumably resulted from rapid aerobic recycling of carbon within the epilimnion during the summer months.

TABLE 21-10 Relative Amounts of Particulate Organic Carbon Converted to Methane Carbon in Profundal Sediments and Losses of Methane Carbon by Evasion to the Atmosphere in Several Eutrophic Lakes

Lake	*Period*	*% of input C converted to CH_4*[a]	*% of input C lost from lake as CH_4–C*
Lake 227, Ontario (Rudd and Hamilton, 1978)	Annual	55	<10
Wintergreen Lake, MI (Strayer and Tiedje, 1978; Molongoski and Klug, 1980a)	Summer, 1976	34	14
	Summer, 1977	44	20
Third Sister Lake, MI (Robertson, 1979)	Summer	36	6
Frains Lake, MI (Robertson, 1979)	Summer	59	28
Lake Mendota, WI (Fallon *et al.*, 1980)	Summer	54	ca. 5
Lake Washington, WA (Kuivila *et al.*, 1988)	Feb., Oct.	20	ca. 2

[a] Values of particulate loading are based on phytoplanktonic sestonic sedimentation only; inputs of organic carbon from the wetlands and littoral productivity, which are extensive in many of these lakes and could be transported along the sediments to hypolimnetic areas and missed in pelagic sedimentation traps, were not evaluated. Therefore, the methanogenetic values could be overestimates.

have been found in a number of other eutrophic and hypereutrophic lakes during summer periods of thermal stratification (Table 21-10). A general direct relationship has been found between the rates of particulate organic carbon sedimentation and rates of methane release from *pelagic* sediments in eutrophic lakes (Robertson, 1979; Casper, 1992). In oligotrophic lakes where phytoplankton sedimentation to pelagic sediments is low, however, as well as in eutrophic lakes with well-developed littoral areas, methane production rates of the sediments in the littoral zone were found to be much higher than those of the profundal zone (Casper, 1996; Rolletschek, 1997).

The terminal processes involved in the decomposition of organic carbon compounds and electron flow in sediments can be divided into two groups (Fig. 21-7; Table 21-6): (a) those processes requiring electron acceptors (e.g., O_2, NO_3^-, Fe^{+3}, and SO_4^-,) generated or regenerated externally (i.e., from aerobic overlying water) to the zone of anaerobic metabolism in the sediments,

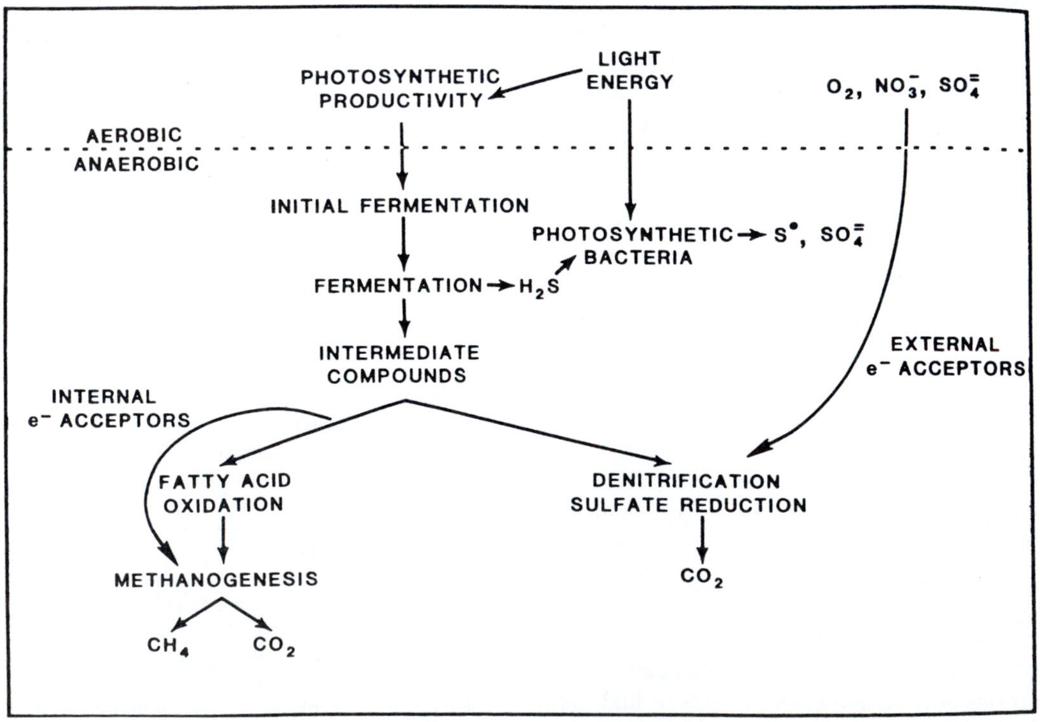

FIGURE 21-7 Relationship of main terminal degradation processes to the available electron acceptors in lake sediments. (Modified from Klug *et al.*, 1982.)

and (b) those processes utilizing electron acceptors (e.g., HCO_3^-), generated or regenerated internally within the zone of anaerobic metabolism (Klug *et al.*, 1982). Denitrification and sulfate reduction are examples of microbial processes dependent upon external electron acceptors. Fermentation and methanogenesis, in contrast, depend upon internally generated electron acceptors.

D. Hydrogen Metabolism

Molecular hydrogen is formed in the sediments, under anaerobic conditions, by fermentative degradation of carbohydrates, cellulose, and hemicelluloses to fatty acids, especially acetic acid. Hydrogen and CO_2 are generated by *Clostridium* from simple carbohydrates and from cellulose by *Achromobacter* and *Bacillus*. Formate serves as the carbon source for hydrogen formation in some species of *Bacterium*.

In sediments, only microorganisms catalyze the oxidation of H_2 coupled to the reduction of nitrate, Mn(IV), Fe(III), sulfate, or carbon dioxide. H_2 concentrations are dependent on the physiological characteristics of the microorganisms consuming the H_2 and utilizing H_2 to maintain lower concentrations in proportion to the availability of electron acceptors (Lovley and Goodwin, 1988). As discussed earlier (p. 639), each terminal electron-accepting reaction has a unique range of H_2 concentrations associated with it, greatest with methanogenesis and least with nitrate reduction. Because hydrogen is oxidized by several pathways, little reaches the surface and escapes to the atmosphere as bubbles. Certain strictly anaerobic sulfate-reducing bacteria of the genus *Desulfovibrio* oxidize hydrogen while utilizing CO_2 to synthesize at least part of their cell substance (ZoBell, 1973):

$$H_2SO_4 + 4H_2 \longrightarrow H_2S + 4H_2O$$
$$(\Delta G_0' = -73.1 \text{ kcal mol}^{-1})$$

Such sulfate reducers were found to fix an average of 0.24 g CO_2–C per gram of H_2 oxidized, and nearly all of the CO_2–C was accounted for as bacteria biomass. Oxygen for the oxidation is derived from the sulfate. Certain species of all of the major genera of the methane-forming bacteria can oxidize molecular hydrogen.

E. Carbon Monoxide Metabolism

Carbon monoxide results from certain microbial fermentations, but even under exceptionally favorable conditions, it generally does not exceed 3% of the gases formed by bacterial fermentation of organic wastes (ZoBell, 1964, 1973). Carbon monoxide can also be generated in surface waters by ultraviolet photolysis of humic substances (e.g., review of Moran and Zepp, 1997). The chemical reactivity of CO and the relative ease with which various bacteria reduce it to methane or oxidize it to CO_2 probably account for its absence or very low concentrations in most environments.

Aerobic CO-oxidizing bacteria are distributed widely, especially in organic-rich sediments. Species of *Carboxydomonas*, *Hydrogenomonas*, *Bacillus*, and certain methane-oxidizing bacteria oxidize CO (Hubley *et al.*, 1974):

$$CO + \tfrac{1}{2}O_2 \longrightarrow CO_2 \qquad (\Delta G_0' = -66 \text{ kcal mol}^{-1})$$

Additionally, several aerobic bacteria in mixed populations quantitatively convert carbon monoxide into methane in the presence of hydrogen:

$$CO + 3H_2 \longrightarrow CH_4 + H_2O$$
$$(\Delta G_0' = -46 \text{ kcal mol}^{-1})$$

In the absence of H_2, some anaerobic bacteria (*Methanosarcina*) produce CO_2 and methane:

$$4CO + 2H_2O \longrightarrow 3CO_2 + CH_4$$

F. Effects of Hydrostatic Pressure on Gaseous Metabolism in Sediments

Analyses of the distribution of the gases contained in lake sediments and water show that the total quantity of nitrogen, methane, and hydrogen in lake sediments increases with water depth (Table 21-11; cf. also

TABLE 21-11 Average Content of Gases in Sediments of Several Japanese Lakes in Relation to Gas and Hydrostatic Pressures[a]

Lake	Depth (m)	*ml gas liter^{-1} of interstitial water of sediments*					Average total gas pressure (atm)	Hydrostatic pressure the of bottom (atm)
		CO_2	O_2	N_2	CH_4	H_2		
Nakatsuna-ko	12	234	0	12	72	7	2.68	2.08
Kizaki-ko	29	137	0	17	110	17	4.13	3.71
Aoki-ko	56	89	0	18	161	24	5.14	6.3

[a] After Koyama (1964).

Fig. 12-1). The total pressure of the gases was usually slightly higher than the corresponding hydrostatic pressure (Koyama, 1964).

Experimental investigation of the metabolism of lake sediments under normal pressure demonstrated the following sequence (Koyama, 1955): (a) CO_2 production began immediately at high rates; (b) as the sediments became anaerobic, hydrogen was evolved, followed by increasing rates of methane production. The primary reaction of methane formation was the reduction of CO_2 by hydrogen and hydrogen-donor compounds. Hydrostatic pressures that would be encountered in the sediments of most lakes do not greatly alter bacterial production of gases. Methane production decreased slightly as pressures were increased from 0 to 50 atmospheres (atm) and then increased up to 300 atm, after which gradual decreases in methane production were observed.

G. Production of Gases in Sediments, Ebullition, and Relationships to Lake Productivity

Lakes and rivers are collectors of organic carbon. Total organic carbon within these ecosystems results from autochthonous production augmented by imported allochthonous organic carbon from wetland and terrestrial sources. All of the organic carbon is not mineralized because much is deposited within low redox conditions within sediments. As will be emphasized in the concluding chapters, most of the intense microbial respiration and fermentation is associated with the sediments. As a result, large quantities of CO_2 and CH_4 evade from the sediments. Carbon dioxide is only moderately soluble in water (cf. Chap. 11), and much more is produced by decomposition than can be refixed photosynthetically within the limitations of nutrient and light availability and herbivory. Consequently, large quantities of CO_2 evade to the atmosphere (Table 21-12). Marked seasonality occurs in the evasion of CO_2 to the atmosphere, with the greatest amounts following ice cover during lake circulation and the lowest during periods of highest photosynthesis (e.g., Otsuki and Wetzel, 1974; Anderson *et al.*, 1999). But for most lakes and rivers, a large net evasion of CO_2 to the atmosphere prevails.

Concentrations of fermentation gases within the water column of most productive lakes are very low in

TABLE 21-12 Rates of CO_2 Release from Various Inland Water Habitats

Habitat	*Conditions*	*Temperature range (°C)*	*g CO_2–C m^{-2} day^{-1}*		*Source*
			Average	*Range*	
Lakes					
Voeux, Quebec	Summer	—	0.16	—	Duchemin *et al.* (1999)
Two reservoirs, Quebec	Summer	—	2.5	—	
25 lakes of Arctic AK, USA	Annual	2–15	0.92	−0.24–2.63	Kling *et al.* (1992)
4 rivers of Arctic AK, USA	Annual	2–15	0.31	—	
Two seepage oligotrophic lakes,WI, USA	Annual	2–20	0.004	−0.24–0.72	Riera *et al.* (1999)
Two seepage bog lakes,WI, USA	Annual	2–20	0.51	0.3–2.9	
Williams Lake, MN, USA	Ice-loss day, April	4	2.6	—	Anderson *et al.* (1999)
	April	5	0.3	—	
Littoral/wetland					
Forested wetland, Denmark	Vegetation + litter	0.7–14	0.45	0.11–1.26	Paludan and Blicher-Mathiesen (1996)
Fen, Finland	Vegetation only	0–18	0.52	0.43–10.8	Martikainen *et al.* (1995)
Bog, Finland	Vegetation only	0–19	0.46	0.74–7.7	
Bog, MD, USA	Vegetation + litter	—	0.78	0.03–3.0	Yavitt (1994)
Fen, Finland	Vegetation only	5–20	0.79	0.18–10.8	Silvola *et al.* (1996)
Marsh, WV, USA	Vegetation + litter	—	1.13	0.03–4.1	Yavitt (1994)
Swamp, WV, USA	Vegetation + litter	—	1.76	0.00–6.2	
Forested swamp, AL, USA	Sediment + litter	7–20	1.24	0.12–2.4	Roden and Wetzel (1996)
Forested swamp, AL, USA	Standing dead litter	3–36	2.33	1.37–3.35	Kuehn and Suberkropp (1998b)
Bog, WV, USA	Vegetation + litter	−4–22	1.92	0.05–7.2	Yavitt *et al.* (1993)
Fen, Arctic Ontario, Canada	Vegetation + litter	18–23	5.67	1.73–10.4	Whiting (1994)
Bog, Arctic Ontario, Canada	Vegetation + litter	11–33	9.25	0.84–22.5	
Bog, MN, USA	Vegetation only	7–20	9.80	2.70–16.7	Kim and Verma (1992)

the surface layers and markedly increase in the hypolimnion and near the sediments. If production rates and pressure conditions are such (cf. p. 152ff.) as to permit bubble formation, the gases can rise rapidly to overlying strata. Often these reactive gases are redissolved and metabolized by microflora en route; much of the methane ebullition is redissolved and utilized by methane-oxidizing bacteria.

Bubble release from the water surface occurs in shallow aquatic situations, such as marshes and littoral areas, where intense anaerobic fermentation in the surficial sediments occurs (e.g., Koyama, 1963; King and Wiebe, 1978; Barber and Ensign, 1979; Svensson and Rosswall, 1984). Although methane emissions to the atmosphere from wetlands is largely (ca. 95% or more) through the internal ventilation system of emergent aquatic plants, methane is still released in significant quantities by ebullition from shallow eutrophic waters, particularly from nonvegetated sediments. Not unexpectedly, rates of evasion of methane to the atmosphere are roughly an order of magnitude greater in water above the sediments in shallow littoral areas than from the water column of pelagic areas (Table 21-8). Despite high release rates of CO_2 simultaneously from shallow waters and wetland, methane evasion could account for 50% or more of the carbon flux to the atmosphere. Measurements given in Table 21-8 are approximate, and data exhibit large variability. Some of that variability results from spatial heterogeneity in sediments, vegetation, temperature, and hydrology (e.g., Bartlett *et al.*, 1989), and that from day to day may result from natural variations in ebullition rates, such as marked increases from lakes associated with decreases in air pressure (Mattson and Likens, 1990) or water depth and hydrostatic pressure overlying sediments (Chanton *et al.*, 1989). Also, in polymictic, shallow, eutrophic lakes, ebullition of methane to the atmosphere is highly irregular in relation to mixing events and disturbance of sediments (Takita and Sakamoto, 1993). Ebullition in these waters increases with increasing temperatures in the sediments and overlying water during the active growing season.

As is the case with CO_2, methane also can accumulate greatly under ice (e.g., Rossolimo, 1935; Smith and Lewis, 1992) and evade to the atmosphere following ice loss and lake circulation at much greater rates than during other seasons. Rates of evasion from aquatic plants also vary markedly diurnally and seasonally with changes in physiology during the life cycles of the plants (e.g., Kim *et al.*, 1998).

As lakes become very eutrophic, the anaerobic fermentation of organic matter in the sediments can reach levels sufficient to result in the steady ebullition of gases. For example, Strayer and Tiedje (1978) found that an average of 21 mmole m^{-2} day^{-1} of methane left the anaerobic sediments by ebullition in hypereutrophic Wintergreen Lake, Michigan, during the period of late May through August. Ohle (1958) demonstrated by both chemical analyses and sonar echo sounding that, under stratified conditions in very productive lakes, ebullition is sufficient to induce "methane convection" currents by which sediment particles and dissolved substances of the hypolimnion can be transported to the trophogenic zone.

The return of nutrients from the sediment zone to the zone of primary production represents an ***internal fertilization*** of the system. The ebullition of gases from the sediments is closely related to lake productivity (Ohle, 1978). When the oxygen content of the deeper water is exhausted, ebullition of fermentation gases increases. These gases further reduce the oxygen content of water strata nearer the surface through bacterial oxidation and cyclically generate conditions for further intensification of gas generation and ebullition. The result can be observed in a jumplike increase in eutrophication *(rasante Seen-Eutrophierung)* (Fig. 21-8). The accelerated biogenic response intensifies production and inputs of organic matter to the sedimentary fermentation system and enhances rates of generation of gases.

Ohle (1958) further proposed a symbiotic relationship among the bacterial populations that utilize and generate fermentation gases. Sulfate-reducing bacteria of the sediments can utilize methane as an excellent

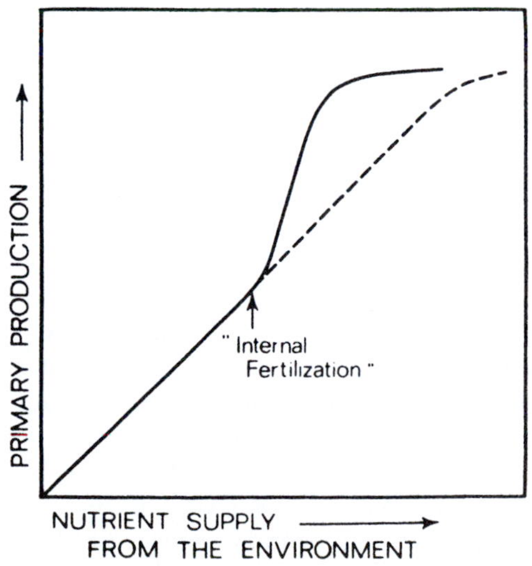

FIGURE 21-8 Accelerated primary production of lakes resulting from "internal fertilization," in part enhanced by ebullition of gases from sediments, during the progression of increasing nutrient inputs from the surrounding environment. (After Ohle, 1958.)

TABLE 21-13 Selected Estimates of Rates of Bacterial Production within Sediments of Pelagic and Running Water Habitats and within Littoral and Wetland Habitats

Habitat	*Measured sediment depth (cm)*	*Period of measurements*	*Bacterial production of 0–1 cm layer (mg C cm^{-2} day^{-1})*	*Source*
Eutrophic ponds	0–1	Summer	250–500	Moriarty (1986)
Saline pond	0–1	Summer	1200	Kemp (1988)
Ogeechee River, GA, USA	0–1	Jan.–Dec.	0.24–15	Findlay *et al.* (1986a)
Black Creek, GA, USA	0–1	Jan.–Dec.	0.2–26	
Black Cr.–Ogeechee R. backwaters	0–1	Jan.–Dec.	3.1–530	
Ogeechee River, GA, USA	0–20	Jan.–Dec.	49(3–110)	Meyer (1988)
Southern Ontario streams, Canada	0–2	May–Aug.	215–1700	Hudson *et al.* (1992)
Hudson River Estuary, NY, USA	0–1	June–Nov.	2–230	Austin and Findlay (1989)
Breitenbach, Germany	0–7	Jan.–Dec.	63(27–96)	Marxsen (1999)
14 lakes, Québec, Canada	0–2	Summer, 1 day	0.001–0.018	Gasol *et al.* (1993)
Aquatic plant detritus of littoral sediments	—	Summer	0.0001–0.002	Moran and Hodson (1989a)
Hyporheic sediments, Flathead River, MT, USA	—	Jan.–Dec.	0.01–0.166	Ellis *et al.* (1998)
Hyporheic sediments, Maple River, MI, USA	10 and 50	Jan.–Dec.	0.17–760	Hendricks (1996)
Epilithic bacteria, 5 Ontario lakes, Canada	<1	Summer	0.003–0.122	Hudson *et al.* (1990)
Hyporheic sediments, 4 Ontario streams, Canada	0–2	Summer	0.04–0.34	Hudson *et al.* (1992)

energy and carbon source, but they are strict anaerobes. Their generation of hydrogen sulfide, even though inhibitory to many bacteria, serves as an efficient means to maintain anoxia against intrusions of aerated water.

H. Rates of Bacterial Production and Mortality

Representative estimates of rates of bacterial production on and within the surficial sediments of lakes and streams (Table 21-13) demonstrate the extremely high production rates. In comparison to those rates of production of bacterioplankton within the integrated water column above the sediments, rates of bacterial production in sediments are several orders of magnitude greater. As will be emphasized in the concluding integrating chapter on carbon fluxes and budgets (cf. Chap. 23), decomposition in the sediments is where most degradation occurs in both lakes and streams. Decomposition in the sediments, largely by bacteria, results in a major source of oxidized carbon and contributes to the massive net evasion of CO_2 to the atmosphere from lake and river ecosystems. Lakes and rivers are basically sedimentary repositories for reduced organic carbon; metabolism within these ecosystems results in most being returned to the atmosphere as CO_2.

The highest bacterial carbon production is always found in the upper layers of natural sediments where concentrations of particulate organic matter are also usually highest in the vertical profile of sediments (e.g., Marxsen, 1996). The distribution of bacterial production over depth within the sediments is generally positively correlated with particulate organic matter content. In many studies of bacterial production, only the upper centimeter is evaluated. This stratum often contains most of the bacterial production, but in loose, particularly sandy, porous sediments as in many streams, a significant portion of total sediment bacterial production can extend lower. For example, in a small first-order stream (Marxsen, 1999), bacterial production was significant to much greater depths, even though those of the surficial centimeter usually dominated.

Bacterial production constitutes a large fraction of total secondary production in most aquatic sediments and benthic heterotrophy (cf. review of Kemp, 1990). Feeding by benthic macrofauna is largely restricted to surface sediment, and the macrofauna in most sediments consume directly only a small fraction (<10%) of bacterial production. Estimates of the fraction of bacterial production consumed by meiofauna are more variable but is also largely restricted to surface sediment and considered relatively small. An exception may be among the nematodes, many of which are largely bacterivorous and commonly occur in concentrations of ca. 10^6 m^{-2}. Microflagellates are abundant, particularly in organic-rich sediments, and are more likely to be important bacterivores. Ciliate bacterivory may be more important in sandy, aerobic sediments, but ciliates are unlikely major consumers of bacteria in organic-rich sediments. The overall trophic efficiency of transfers of organic carbon from bacteria to macrofauna via meiofaunal and protistan links is quite low (Kemp, 1990). Less than 20% of this rich bacterial production would be expected to enter macrofaunal food webs. Most of the bacterial organic carbon is respired and mineralized to CO_2 within the microbial food web.

I. Biotic Alterations of Sediment Redox

Although the direct consumption of bacteria by benthic deposit-feeding macroinvertebrates is a quite small portion of the whole, their feeding activities can significantly alter the gradients within the sediments by physical disturbance (Krantzberg, 1985; Koyama, 1993; Graf and Rosenberg, 1997). As a result, bacterial production can be increased within sediments and result in enhanced generation of respiratory gases. For example, feeding activities of amphipods, oligochaetes, and immature chironomid dipterans at high densities can reduce the steepness of redox gradients of sediments, increase bacterial production, and accelerate oxygen consumption from the overlying water in surficial sediments by factors in the range of 1.4–>4 times (Granéli, 1979; Andersen and Jensen, 1991; van de Bund *et al.*, 1994; Svensson and Leonardson, 1996).

Earlier discussion (Chap. 18) emphasized the major intercellular ventilation system of rooted aquatic plants by which gases are transported and exchanged to meet the respiratory demands of rooting tissues living in anoxic sediments. Oxygen is released from the roots and rhizomes of emergent, floating-leaved, and submersed aquatic plants (e.g., Carpenter *et al.*, 1988; Chen and Barko, 1988; Jaynes and Carpenter, 1988; Wetzel, 1990c; Boon and Sorrell, 1991; Armstrong *et al.*, 1992; Flessa, 1994; Christensen *et al.*, 1997). Appreciable quantities of oxygen are released into the rhizosphere and this oxidation can alter markedly the redox conditions in the immediate (usually <0.5 mm) microrhizosphere.

Fluxes of gases through the plants are mediated by both physical processes, some coupled with photosynthesis. Emergent and floating-leaved aquatic plants in contact with the air have mechanisms (Chap. 18) that allow moderate exchange of internal gases with the atmosphere during periods of darkness. In contrast, the release of oxygen into the rhizosphere from rooting tissues of submersed higher plants is light-dependent, and the release of oxygen from roots is appreciably reduced during darkness. The fluctuating redox conditions in sediments permeated with roots of higher aquatic plants leads to steep gradients within the sediments of increased redox, reduced nutrient availability, and increased organic substrates released by the plants.

Earlier discussion (Chap. 19) also evaluated the importance of attached algae on sediments in relation to their ability to influence nutrient fluxes into and from sediments by direct utilization as well as shifts in redox by the production of dissolved oxygen in excess of that consumed by heterotrophic biota. Photosynthesis of the epipelic algae and cyanobacteria can rapidly shift redox conditions at the sediment–water interface from anoxia to supersaturated concentrations that penetrate several millimeters into the sediment (e.g., Carlton and Wetzel, 1988; Wiltshire, 1993; Berninger and Huettel, 1997). Slow horizontal water movements over the epipelic microbial community at the sediment–water interface enhances the diffusion of photosynthetically produced oxygen out of the sediments.

V. LITTORAL DECOMPOSITION AND MICROBIAL METABOLISM

Evaluations of the horizontal distribution patterns of bacterioplankton in lakes demonstrate that bacterial densities are much greater in the littoral zone than in the open water (e.g., Mühlhauser, 1990). Although comprehensive studies of these spatial distributions are few, examples from both small and large lakes are consistent. The quantity of total bacterioplankton in the littoral zone of Lake Balaton, Hungary, determined by direct enumeration, was found to be 2–3 times greater than that of the open water (Oláh, 1969a,b). Saprophytic bacteria[4] of the littoral zone were 40–120 times more abundant than in the open water. Observed differences were greatest during the spring period of active growth of the emergent macrophyte reed zone (*Scirpus* and *Phragmites*) and in the autumn as portions of the littoral foliage senesced. High numbers of bacteria from the littoral zone extended out into the open water for considerable distances (1500 m) in this large lake during the spring, but they were restricted largely to the macrophyte belt during the summer.

Similarly, the bacterioplankton activity of large, shallow Lake Mekhtev, Russia, was strongly influenced by the aquatic macrophytes and associated sessile

[4] See footnote, p. 493.

TABLE 21-14 Distribution and Activity of Bacterioplankton in Water among Emergent Bulrush Vegetation and in Open Water without Macrovegetation in Lake Mekhteb, Dagestan, Russia[a]

		Water among macrophytes			*Open water*		
Period	*Water temperature (°C)*	*Total bacteria* ($\times 10^6$ ml^{-1})	*Generation time (hr)*	*Total saprophytes* ($\times 10^3$ ml^{-1})	*Total bacteria* ($\times 10^6$ ml^{-1})	*Generation time (hr)*	*Total saprophytes* ($\times 10^3$ ml^{-1})
Dec.–Jan.: no bulrushes	0–2	0.55	19–17	5–12	0.55	19–27	5–12
Feb.–Mar.: young bulrushes	3–7	1.4	11–16	28–39	1.0	17	22–103
Apr.–May: developing stands	17–23	1.3–1.8	10–13	11–31	1.4	10–12	14–21
June: mature bulrushes	26	3.3	5.2	40	3.0	7.1	31
July	28	5.3	4.4	232	2.5	6.5	28
Aug.: declining bulrushes	21	4.3	7.6	178	3.1	6.3	81
Oct.	11	1.5	11	21	1.1	23	17
Nov.	3	0.85	21	0.2	0.27	18	0.4

[a] After data of Aliverdieva-Gamidova (1969).

algal flora (Aliverdieva-Gamidova, 1969). Although temperature was strongly correlated directly to bacterial growth on a seasonal basis, and maximal populations and shortest generation times occurred in midsummer (Table 21-14), bacterioplankton were clearly more abundant and on the average grew faster in water among macrophyte stands than in open water lacking littoral flora. Although these studies indicate that large amounts of organic substrates are being released by the macrophytes and their epiphytic microflora during growth and senescence, measurements of actual transfer rates were not available in this study. As we have seen in the detailed discussions of Chapter 19, there is a strong tendency for conservation and retention of released organic substrates and nutrients by the epiphytic microflora on living, senescing, and detrital particulate matter of the macrophytes. This capacity for microbial retention and recycling can be overwhelmed with massive production of macrophytes under hypereutrophic conditions.

When the macrophyte-derived dissolved organic matter (DOM) passes through attached microflora, some of the DOM remains in dissolved and colloidal form and is transported within and out of the littoral areas. Some of the DOM is mineralized by littoral bacterioplankton and some aggregates into large colloidal masses and particles. Experiments with DOM from different macrophytes showed some 23% of the dissolved organic carbon remained in solution, 58% was mineralized by bacteria, and 19% formed aggregates (Alber and Valiela, 1994). The aggregates had low C:N ratios (4–12, depending on the source species) and contained large densities of active bacteria. These values of mineralization are similar to results of other studies (Findlay *et al.*, 1986b; Mann and Wetzel, 1996; Waichman, 1996; Bicudo *et al.*, 1998).

A. Losses of Macrophytic Particulate Organic Matter and Benthic Sediments

Relatively little, usually <5%, of the very high total (above- and belowground) annual organic productivity of wetland and littoral macrophytes is consumed by grazing animals. Most of the high net primary production senesces undisturbed and enters the detrital pool as dissolved and particulate organic matter (Polunin, 1984). Early stages of decomposition are dominated by leaching of soluble organic matter from the senescing tissues. Some of this degradation and leaching can occur while the senescing tissues are standing dead among living emergent macrophytes or after the tissues have fallen to the water. Earlier discussion indicated that the microbial utilization and degradation of dissolved organic matter released extracellularly during active plant growth and by autolysis was rapid and relatively complete (cf. Table 17-5).

In contrast, the degradation of the particulate organic matter of the littoral zone is much slower. Once in the water, microbial colonization, mostly bacteria, is rapid in a succession of forms. These bacteria initially utilize primarily released soluble organic matter (Mann and Wetzel, 1996), and the release of organic substrates influences bacterial growth and activity on leaf litter to a greater extent than does the process of fragmentation of the litter and associated increases in surface area available for bacterial colonization (Gunnarsson *et al.*, 1988). Studies are few and generally employ

TABLE 21-15 Percentage Loss of Dry Organic Matter of Aquatic Macrophytes over Long Periods of Time

	Loss of dry weight								
Plant type	*7–14 days*	*4 weeks*	*6 weeks*	*9 weeks*	*12 weeks*	*15–16 weeks*	*24 weeks*	*43 weeks*	*Source*
Emergent									
Juncus effusus	5	—	—	—	30	60	65	—	Boyd (1971)
Typha latifolia	5	—	—	—	25	30	45	—	Boyd (1970)
Spartina alterniflora	2	—	—	—	10	30	50	85	Burkholder and Bornside (1957)
Carex gracilis	—	42	42	55	—	—	—	—	Koreliakova (1959)
Sparganium simplex (overwintered)	22	—	48	—	58	100	—	—	Koreliakova (1958)
Salix leaves	17–56	—	—	—	—	—	—	—	Pieczyńska (1972)
Phragmites communis									
Overwintered plants	40	—	42	—	50	53	—	—	Koreliakova (1958)
Fresh plants	—	43	48	63	—	—	—	—	Koreliakova (1959)
Scirpus acutus									
Spring–summer	10–12	23	26	—	35	—	38	—	Godshalk (1977)
Fall–winter	5–10	15	18	—	—	18	—	—	
Echinochloa polystachya	20–60	40–75	50–80	60–85	—	65–95	—	—	Pompêo and Henary (1998)
Floating-leaved									
Lemnaceae	20	—	—	—	90	—	—	—	Laube and Wohler (1973)
Brasenia schreiberi									
Leaves in surface water	(2)	—	—	—	45	—	50	95	Kormondy (1968)
Leaves at sediment	20	—	—	—	20	—	45	90	
Nymphaea odorata									
Leaves in surface water	(15)	—	—	—	45	—	70	98	
Leaves at sediment	(20)	—	—	—	58	—	75	95	
Polygonum amphibium									
Overwintered plants	25	—	35	—	35	37	—	—	Koreliakova (1958, 1959)
Fresh plants	—	42	46	55	—	—	—	—	
Nuphar variegatum									
Spring–summer	23–53	70	80	—	88	—	90	—	Godshalk (1977)
Fall–winter	10–25	30	38	—	47	53	—	—	
Salvinia auriculata	32–34	—	—	45	—	47	50	90	Howard-Williams and Junk (1976)
Eichhornia crassipes	47–53	—	—	63	—	80	80	90	
	10–20	35	40	50	60	75	—	—	Gaur *et al.* (1992)
Submersed									
Elodea nutallii	5–30	10–80	15–85	15–90	25–95	30–100	—	—	Ochiai and Nakajima (1994)
Potamogeton lucens	6–92	—	—	—	—	—	—	—	Pieczyńska (1972), cf. also Bastardo (1979)
P. perfoliatus	6–95	—	—	—	—	—	—	—	
Scirpus subterminalis									
Spring–summer	4–10	33	55	—	62	—	70	—	Godshalk (1977)
Fall–winter	5–10	—	15	—	—	18	21	—	
Najas flexilis									
Spring–summer	15–33	55	85	—	95	—	99	—	
Fall–winter	10–15	20	25	—	—	25	—	—	
Myriophyllum heterophyllum									
Spring–summer	28–40	55	75	—	88	—	95	—	
Fall–winter	12–20	30	35	—	—	38	—	—	
Paspalum repens	57	—	—	—	—	80	90	95	Howard-Williams and Junk (1976)
Leersia hexandra	47	—	—	—	—	62	72	95	

approximate techniques (e.g., following the rates of weight loss over periods of time while being incubated in litter bags under experimental or *in situ* situations) (cf. Boulton and Boon, 1991, for discussion of methodological problems).

Emergent macrophytic vegetation generally exhibits very slow rates of degradation in comparison to those of floating-leaved and submersed aquatic plants (Table 21-15). Part of this greater resistance to decomposition is related to the greater quantity of lignified tissue found in emergent flora (Sculthorpe, 1967; cf. Chap. 18). With adaptation to a largely or totally submersed existence, a conspicuous reduction in lignification of supportive and conducting tissues is found in floating and submersed vegetation. Lignin, consisting of aromatic rings with side chains and -OH and -OCH_3 groups, exhibits a high degree of chemical stability. Catabolism of lignin by oxidative depolymerization is dominated by fungi, especially the Hymenomycetales (Basidiomycetes), which possess the inductive exoenzymes for this cleavage (Trojanowski, 1969). Bacteria are also active in the decomposition of lignocellulose macromolecules, often in concert with fungi, as side chains or monomers are detached from the lignin by fungi (e.g., Hackett *et al.*, 1977, Healy *et al.*, 1980; Benner *et al.*, 1984) or ultraviolet photolysis of partially degraded products. Large amounts of paracoumaric esters and lignin derivatives are found to persist during decomposition of emergent macrophytes under different environmental conditions (Boon *et al.*, 1982) and can further retard decay of emergent plants in the littoral zone of lakes. Under conditions of aerobic decomposition, appreciable amounts (to 30%) of the nitrogen can occur in nonprotein nitrogen compounds (Odum *et al.*, 1979). Much of this nitrogen is apparently chitinaceous and of fungal origin; these compounds are also relatively recalcitrant to rapid decay.

As the macrophytes are partially decomposed, appreciable quantities of dissolved organic compounds of relatively high molecular weight (1000–50000 D) are released. Bacterial degradation of these lignocellulose-derived dissolved organic compounds can be appreciable. For example, bacterial mineralization of a macrophyte DOC-pool was initially rapid (12% per day) but slowed to ca. 30% after 30 days (Moran and Hodson, 1989). Some of these macrophyte organic products are resistant to decay, but it can be greatly accelerated by exposure to natural UV light (Wetzel *et al.*, 1995, Wetzel, 2000c).

Planktonic fungi found in the open water likely originate mainly from wetland and littoral areas (Willoughby, 1965; Novotny and Tews, 1975), although some (e.g., many chytrid species) are highly specialized parasites of planktonic algae (cf. Chap. 15). Fungal propagules are transported in the water directly or in association with inflowing particulate plant debris. Little is known about their metabolic activities once the material becomes sestonic in the pelagic zone.

The sources of accumulated organic matter in the littoral zone are variable among lakes in relation to basin morphometry and production by the littoral algal and macrophyte components. In most situations, a majority of the particulate organic matter originates from the macrophytes (approximately 90%), most conspicuously from the emergent forms. Lesser quantities are derived from attached algae, loosely attached littoral algae, and windrowed planktonic algae. Rates of decomposition of the littoral particulate organic matter are quite variable in relation to conditions of accumulation (Table 21-16). In the example from Mikolajskie Lake, decomposition rates were lowest among plant material in the emergent part of the eulittoral zone on

TABLE 21-16 Decomposition Rates of Macrophyte Particulate Organic Matter in Various Habitats of the Littoral Zone of Mikolajskie Lake, Poland, 1967[a]

Plant material	*Period*	*Emergent, 1 m from shore line*	*Reed heaps on the shore line*	*Small pools on shore partly isolated from lake*	*In water, 0.5 m depth 2 m from shore line*
Emergent					
Phragmites communis leaves	Mid-June	4	25	43	21
	Late July	4	34	57	14
	Early Sept.	6	27	38	24
Salix leaves	Mid-June	5	35	37	18
	Late July	7	28	56	25
	Early Sept.	4	30	39	17
Submersed					
Potamogeton lucens	Mid-June	3	35	89	—
	Late July	5	41	92	5
	Early Sept.	6	—	77	4
Potamogeton perfoliatus	Mid-June	1	38	83	7
	Late July	6	40	95	6
	Early Sept.	7	50	70	8

[a] After Pieczynska (1970). Percentage losses of dry weight after 10 days of *in situ* exposure.

the surface of the sediments. Senescence and death of submersed species in the submersed littoral site was slow and delayed decomposition; when dead and decaying submersed plants occurred in pools and among reed heaps, decomposition was very rapid. Filamentous algae and accumulations of planktonic cyanobacteria in the littoral decayed much more rapidly (>95% in 3–10 days) under comparable conditions (Pieczyńska, 1970).

B. Degradation Patterns of Littoral Flora

Lignified tissue of emergent hydrophytes is degraded slowly by a succession of fungal flora. For example, it has been observed that the mycoflora of the cattail *Typha latifolia* passes through early stages that are similar to those of terrestrial plants, with primary colonization by leaf-surface fungi, followed by a secondary phase in which many species of *Leptosphaeria* dominate moribund leaves (Pugh and Mulder, 1971). The succession is associated with aging of plant material and substrate changes. Final stages of decomposition are accompanied by a dominance of fungi predaceous on nematodes. Similar successional patterns have been found on analogous substrates (Sparrow, 1968; Mason, 1976). Fungi generally very poorly colonize senescent or dead roots and rhizomes.

Lignin is not biodegraded anaerobically to any appreciable extent (cf. excellent review of Kirk and Farrell, 1987). As a result, much of the decomposition, fragmentation, photoalteration, and other degradation processes of the lignin components of the large lignocellulose mass of higher plants occurs under aerobic or intermittently aerobic conditions. Once permanently interred in anaerobic sediments, lignin is extremely recalcitrant to biomineralization.

Leaves of the emergent rush *Juncus effusus*, a plant common to many littoral areas and wetlands of lakes and streams, senesce from the leaf tip to the base at an exponential rate (90–225 days), the rate of which is greater with increasing temperatures seasonally (Kuehn and Suberkropp, 1998a). The leaves of *Juncus* remain standing while dead, a feature common among emergent macrophytes. Even though fungal biomass constitutes 3–8% of total detrital mass, decomposition is slow ($k = 0.40$ yr^{-1}), and the senescent leaves lose about half of their biomass in two years. Availability of water was a major factor affecting rates of fungal respiration in standing dead plant litter of this emergent macrophyte (Kuehn and Suberkropp, 1998b). Rates of CO_2 evolution from the plant litter increased precipitously in the evening with increasing relative humidity (>90%) and plant water potentials (>−1.0 Mpa). Fungal respiratory rates were manifoldly higher during night and early morning hours than during the daytime on clear days.

The movement of fine organic detritus particles of decomposing macrophytes from the littoral zone to the open water can occur by water movements and currents returning from the littoral to the pelagic zones, both along the surface and near the sediments (cf. Schroder, 1973, 1975; Gaudet, 1976; Kistritz, 1978; Rho and Gunner, 1978). In addition to these fine particles, nutrients, particularly phosphorus, are transported from the littoral flora to the open water. The maximum phosphorus inputs from the littoral occur in the summer period when the phosphate concentrations in the open lake water are minimal. The prevailing situation, however, is for the littoral community of macrophytes and microbial communities attached to living and detrital particulate matter to conserve and retain nutrients, especially soluble components released from the living and senescing macrophyte organic matter (cf. Wetzel, 1990a, and summary in Chaps. 18 and 19).

Very fine-sized (<100 μm) **particles of the emergent reed *Phragmites*** were studied as they underwent decomposition in lake water (Oláh, 1972). The successional patterns of bacterial populations in the water containing leached dissolved organic matter from the particles and on the particles correlate with progressive utilization of organic substances (Fig. 21-9). Within a few hours, large rods dominated the liquid phase, presumably utilizing the leached dissolved organic substrates of the detritus. A large population of small cocci

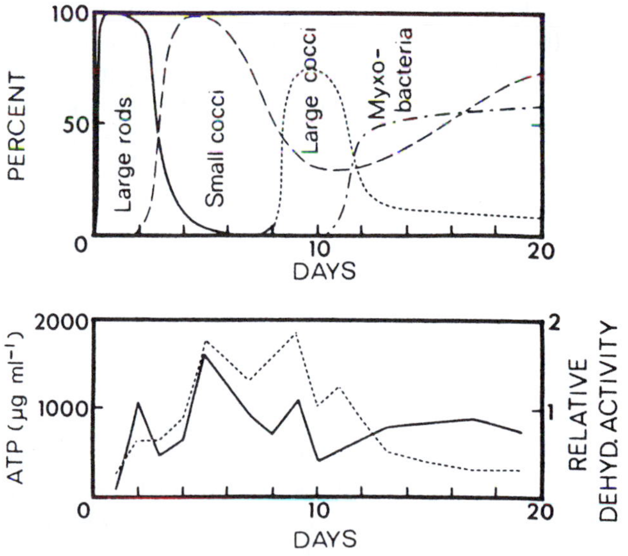

FIGURE 21-9 *Upper*: Bacterial populations; *lower*: activity associated with decomposition of fine particles of *Phragmites* detritus. —— = ATP biomass carbon; - - - - = relative dehydrogenase activity. (Drawn from data of Oláh, 1972.)

succeeded these bacteria; a few days later, the cocci were displaced by large cocci. Both of the coccal types were associated with the liquid phase and populated the surfaces of the small *Phragmites* fragments. Myxobacteria, well-known decomposers of living and dead particulate organic matter (Ruschke, 1968), colonized both the large cocci and the residual, stabilized detrital particles. Maximal bacterial biomass, measured as ATP content, coincided with the maximal development of the bacterial population succession and corresponded to the time of the mass development of small cocci (Fig. 21-9). Bacterial metabolism, determined by dehydrogenase activity, behaved similarly and decreased after about 10 days. The decrease was related to a stabilizing condition of decreasing quantities of easily assimilable organic substances in the residual particulate detritus. Analogous sequences of microbial succession were found during the decomposition of emergent reeds (*Typha*, *Phragmites*, and *Sparganium*), floating-leaved plants, and submersed macroflora (*Ceratophyllum, Myriophyllum, Najas,* and *Potamogeton*) (Gorbunov, 1953; Krasheninnikova, 1958; Casper *et al.*, 1988). Degradation rates were greater for the submersed flora under similar conditions than for emergent flora.

Intensive investigation of the decomposition of the particulate organic matter from several species of emergent, floating-leaved, and submersed macrophytes and of dissolved organic matter leached from them was undertaken under different conditions of temperature and oxygen availability (Godshalk and Wetzel, 1978a–c). Resistance of particulate tissue to microbial decomposition was strongly influenced by the chemical composition of the plant species. The floating-leaved species decomposed faster than the submersed plants, which decomposed more rapidly than the emergent species. Decay rates were related to initial nitrogen and fiber contents, with high-nitrogen, low-fiber plants decomposing most rapidly. Unlike the leached dissolved organic matter of the plants (whose rate of decomposition was found to be mostly influenced by oxygen and nutrient availability), primarily temperature, tissue nitrogen, and fiber content regulated the rate of conversion of particulate organic matter to carbon dioxide and/or dissolved organic compounds.

The relative rates of decomposition, k, of aquatic macrophytes have been variously defined mathematically (see Godshalk and Wetzel, 1978b; Carpenter, 1980; Belova, 1992) and are controlled by factors of the plant itself and the environment within which it is decaying (Godshalk and Wetzel, 1978c):

$$k \propto \frac{T \times O \times N}{R \times S}$$

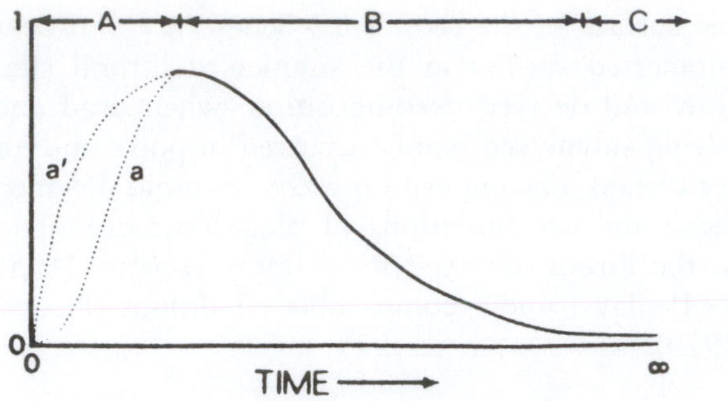

FIGURE 21-10 Generalized sequence of decay rates of dissolved and particulate organic matter with three phases of decay (see text). The logistic S-curve (line a) is found among plants beginning to decompose under natural conditions where decay is initially slow and is followed by increasing cellular senescence. Line a′ is found in laboratory studies using predried plant material at the start of experimental assays (cf. Rogers and Breen, 1982). (After Godshalk and Wetzel, 1978c.)

where T = temperature, O = dissolved oxygen, N = mineral nutrients necessary for microbial proliferation (e.g., Waite and Kurucz, 1977; Almazan and Boyd, 1978; Andersen, 1978; Carpenter and Adams, 1979), R = initial tissue recalcitrance, and S = particle size (i.e., particle volume/surface area). The relative rate of decomposition, k, can be summarized graphically (Fig. 21-10) versus undefined units of time; in this case, k is defined as the amount of detrital dissolved and particulate organic carbon metabolized (e.g., POM → DOM, POM → microbial cellular material, DOM → CO_2, et cetera) per unit time. Although decomposition is a continuous process, the *rate* of decomposition varies with time, and can conceptually be viewed as three distinct phases.

Phase A (Fig. 21-10) is a period of increasing biomass loss from leaching and/or autolytic production of dissolved organic matter (DOM). Much of this DOM is rapidly (hours) metabolized, and the quantity and quality of DOM that persists are influenced by temperature and oxygen availability. Following the maximum rate of weight loss, decay rates decrease to varying degrees during the next phase (B, Fig. 21-10) of decomposition, which can last from several days to many months. Microbial activity and nitrogen content of the particulate detritus increase in the initial phases; factors controlling decay rates have their greatest influence in this phase. The relative resistance of the remaining particulate organic matter continually increases as the more labile compounds are metabolized. In phase C, the rate of decomposition of the residual resistant particulate matter approaches an asymptotic limit of zero and is limited mostly by the high recalcitrance of the particulate detritus. Some of this matter can enter the anaerobic sediments and become buried; if so, decay rates become so slow that, for all practical purposes, decomposition ceases.

If particulate detritus from aquatic plants is colonized by epiphytic algae as well as bacteria and fungi on the sediments, there is a potential for a synergistic coupling between the photosynthesis and metabolic products of the algae, enhancing the metabolism of the heterotrophic microbiota that could influence and increase degradation of the macrophyte tissue. For example, the decomposition of cattail (*Typha latifolia*) detritus was accelerated moderately (5–10%) by the positive interactions between epiphytic algae and heterotrophic decomposers (Neely, 1994; Neely and Wetzel, 1997).

VI. DEGRADATION OF PARTICULATE ORGANIC MATTER IN SEDIMENTS OF RUNNING WATERS

An enormous literature exists on the inputs of particulate organic matter into stream and river ecosystems and the microbial degradation and utilization of this organic matter once it is in the sediments. Disproportionately intensive study and evaluation have resulted because of (a) the major loading of particulate organic matter seasonally in many low-order streams from leaf abscission of canopy vegetation in temperate regions, (b) the major importance of the macroinvertebrates in the food-web structure and energetics between the microbes and higher trophic levels, and (c) the many highly specialized feeding strategies that have evolved among immature and adult macroinvertebrates that spend much of their life histories in these habitats. Because terrestrial vegetation along river courses shades many low-order streams and results in low rates of photosynthesis within these streams (cf. Chaps. 15, 18, and 19), heterotrophic metabolism is highly dependent on dissolved and particulate organic matter that falls into the water channels or is imported with ground water and surface water. Similarly, large rivers are dominated by decomposition of imported organic matter because photosynthetic production is often limited by reduced light availability as a result of high turbidity.

Many river ecosystems of low elevational areas flood with predictable periodicity, as is common in many tropical and subtropical areas. Flood plains are periodically inundated by the lateral overflow of rivers or shallow riverine lakes. In many backwater areas, wetlands are particularly effective in sediment accretion, where prolific macrophyte growth both collimates and disperses flow as well as contributes directly through high productivity of organic matter (e.g., Rybak, 1980; Phillips, 1989; James and Barko, 1990; Feijoó *et al.*, 1994; Callaway *et al.*, 1996). Depending on the severity of flooding events, the wetlands typically serve as both short-term sediment sinks and long-term sediment storage sites.

These flood plains can be classified according to amplitude, frequency, predictability, and the source of flooding (Junk *et al.*, 1989; Junk, 1997a,b). For example, flood plains of large rivers have a predictable monomodal flood pulse of large amplitude and long duration (months) (e.g., the Amazon and Orinoco rivers of South America). Flood Plains of small creeks and streams are characterized by unpredictable polymodal flood pulses of highly variable amplitudes and duration. During these flooding events, large amounts of dissolved and particulate organic matter of various sizes are imported to the river channel sediments.

Detrital particulate organic matter associated with sediments in streams and rivers is commonly separated into size classes: (a) coarse particulate organic matter (CPOM, >1 mm), which consists of large woody debris, leaves, needles, and foliage of terrestrial and aquatic vegetation, and animal parts, and (b) fine particulate organic matter (FPOM, range, 0.5 μm–1 mm), which consists of fragmentation pieces of CPOM, imported terrestrial and wetland organic matter, and aggregated dissolved organic matter. Dissolved and colloidal organic matter is of a size <0.5 μm.

Much study has addressed the rates of decomposition of CPOM detritus of stream sediments mostly derived from terrestrial leaves that fall into small forested streams in the autumn of temperate regions. The rate of loss of mass of CPOM in streams follows approximately a simple exponential model (Petersen and Cummins, 1974; Webster and Benfield, 1986):

$$W_t = W_0 e^{-kt}$$

where

W_t = ash-free dry mass at time t in days
W_0 = initial ash-free dry mass
k = rate coefficient of breakdown loss rate of leaf mass vs. time, day^{-1}
t = elapsed time (days).

This model has been expanded to include a second independent variable, degree days or accumulated temperature, which improved data analyses slightly and is useful for examining differences influenced by temperature among different sites and seasons (Hanson *et al.*, 1984; McArthur *et al.*, 1988). The improvement by this modification over the simple negative exponential model is very small, however (Campbell *et al.*, 1994).

Allochthonous CPOM is often the largest source of particulate organic matter inputs to many streams. For example, in several Appalachian Mountain streams, $>86\%$ of the particulate organic matter loading was as CPOM (Wallace *et al.*, 1995). Movement of CPOM is often restricted by debris dams, where the CPOM is caught and accumulates against relatively immovable objects of the substratum or projections from the banks (e.g., Webster *et al.*, 1994). When discharge is at low base levels, CPOM settles relatively randomly onto the surface of the sediment (Jones and Smock, 1991; Shaddon *et al.*, 1992). During high discharge, however, backwater areas and debris dams can become primary retainers of CPOM. Because of the reduction of retention devices in higher-order streams, less CPOM is retained on the sediments and a greater percentage becomes suspended (Wallace *et al.*, 1982).

Because water is moving past the detrital CPOM on the sediments, fragments of FPOM derived from partial fungal degradation and the feeding activities of detritivorous invertebrates can be exported downstream from the CPOM. In many streams and rivers, particularly in low elevational gradient waters, most export of particulate organic matter is as FPOM. For example in several Appalachian Mountain streams, >96% of the particulate organic matter was exported as FPOM (Wallace *et al.*, 1995). Hence, the observed loss rates in streams are a combination of decomposition (respiratory CO_2 and CH_4 to the atmosphere), fragmentation, and export losses of dissolved (DOM) and fine particulate (FPOM) organic matter.

Very little is known about the fungal and bacterial decomposition of FPOM. In small headwater Appalachian streams, bacterial productivity was greatest per ash-free dry mass in the 106–250 μm fraction of FPOM, but differences were small (Peters *et al.*, 1987, 1989). The proportion of more recalcitrant organic compounds (lignin and cellulose) increased per unit organic mass as the size of the FPOM particles decreased in the range of 500–10 μm. Although productivity was positively correlated with temperature, no changes in bacterial productivity associated with FPOM were induced by amendments of nitrogen or phosphorus in these river ecosystems.

Rates of decomposition of CPOM in streams vary greatly with the chemical composition of the plant material, environmental variables, particularly temperature and nutrients, and the feeding activities of benthic invertebrates. After the initial leaching of large quantities of dissolved organic matter, the rates of decomposition of leaves from nonwoody plants are much faster (mean half-life of ca. 65 days; range, 15–200 days; breakdown rates, 0.3–4% day^{-1}) than rates of leaves of woody plants (mean half-life of ca. 200 days; range, 70–450 days; breakdown rates 0.2–1% day^{-1}) (Petersen and Cummins, 1974; Webster and Benfield, 1986). Nutrient availability is critical for microbial degradation of leaf materials. For example, decomposition rates and fungal productivity of leaves in moderately productive streams were doubled by enrichments of dissolved nitrogen and phosphorus (Meyer and Johnson, 1983; Suberkropp, 1995; Suberkropp and Chauvet, 1995).

As particulate organic matter falls into streams or is imported with runoff water from the flood plains, it is transported downstream for varying distances before being lodged in debris dams or deposited on or in sediments. Leaching of DOM occurs rapidly (e.g., Table 17-9) and within the first few days most loss, commonly 25% of initial ash-free dry mass of small particles, of immediate, nonbiologically mediated leaching occurs. Of the DOM leached from these materials, the more labile components are very rapidly metabolized by attached bacteria and bacterioplankton, whereas the more recalcitrant DOM compounds are metabolized more slowly. For example, carbon and nitrogen labile compounds of leachate from hickory and maple leaves in a small hardwater stream exhibited a half-life decomposition rate of 2 days, whereas bacterial decomposition of the more recalcitrant polyphenolic fraction had a half-life of 80 days (Wetzel and Manny, 1972b).

Leaf litter, fungal biomass, and fungal production per area of sediments is highly seasonal in woodland streams. For example, in a small woodland stream in Tennessee, leaf litter ranged from < 5 g AFDM m^{-2} in summer to 249 g m^{-2} in November after primary leaf fall (Suberkropp, 1997). Fungal growth rates averaged 2.6% day^{-1} over the year. Annual production of leaf-decaying fungi was 34 g m^{-2} (ca. 17 g C m^{-2}) within a daily production rate range of 0.006 (summer) to 0.49 g m^{-2} day^{-1} (November maximum), with an annual production-to-biomass ratio (*P*/*B*) of 8.2.

If the POM remains senescent as leaves on trees before abscission or is standing dead aerially as is common among emergent aquatic plants or was deposited on terrestrial soils before being transported to the river, the organic matter would be exposed to long periods of leaching by precipitation. Losses of labile dissolved organic matter by leaching of aerial senescent tissue can be large (e.g., Mann and Wetzel, 1996). Once in the stream or river, colonization of the POM by microbes proceeds rapidly by bacteria growing on the surfaces and by fungal hyphae growing on the surface and internally. Fungal growth has been observed to be at least 10 times greater than that of bacteria on the surfaces of newly introduced leaves (Bengtsson, 1992). Rates of decay of the POM vary greatly, in part related to the nutrient, especially nitrogen, content of the tissues (e.g., Kaushik and Hynes, 1971) and of the water. Decay rates are lower among POM sources of higher lignin content and polyphenolic compounds, some of which can competitively inhibit enzyme activities. Nitrogen content of decaying POM commonly increases as a percentage of ash-free dry mass, in part because nitrogen compounds bound to structural tissue compounds and contained in humic compounds decomposing at slower rates than other constituents and in part because of microbial biomass increases (Suberkropp *et al.*, 1976). Cellulose, hemicellulose, and lipids are mineralized at rates about the same as POM mass, whereas rates of lignin decomposition were slower and increased as a percentage of the residual organic mass. Under well-oxygenated conditions, most of the decomposition of POM in streams is by fungi, primarily hyphomycetes, during the early stages of decomposition over several months (Kaushik and Hynes, 1971; Bärlocher, 1982; Suberkropp, 1992). Bacterial degradation of these recalcitrant plant tissues is relatively minor during this period but increases and usually dominates in the later stages of the POM decomposition processes.

The heavy colonization of detrital POM by fungi and bacteria results in alterations in the texture and physical integrity of the POM tissues, as well as increases in the nutritional quality of the collective detrital mass and attached microbiota. Fungal metabolism renders leaf organic matter more palatable and nutritious for invertebrates by catalysis of leaf substrates to digestible subunits and by providing increased microbial biomass (Bärlocher, 1985). Some 5–10% of the total leaf litter mass can be of fungal hyphae biomass (Suberkropp, 1997). Detritivorous invertebrates feed actively on this detrital POM, and in the process, fragment the CPOM into smaller pieces. The maceration process reduces particle sizes and increases surface areas for microbial enzymatic activity and simultaneously nutrient-enriched feces. Both processes increase the rates of microbial decomposition of the POM. If macroinvertebrate feeding on decomposing POM is restricted by exclosures, such as in litter bags, or removed from the environment by selective poisons, rates of decomposition of the POM decrease significantly (Table 21-17). Decomposition of POM is clearly accelerated (5–25%) by the presence of macroinvertebrate detritivores (Petersen and Cummins, 1974; Polunin, 1984; Mulholland *et al.*, 1985b). Even though the efficiency of degradation within the invertebrate gut is low (usually much $< 50\%$), this direct heterotrophic utilization is a significant part of the whole. The maceration and fragmentation of the POM indirectly accelerates microbial rates of degradation, generation of FPOM, and leaching of DOM from the residual POM.

Decomposition of POM can be markedly suppressed if the conditions of decomposition shift to anoxia (Pattee and Chergui, 1994). Some bacterial degradation of lignified wood occurs under anaerobic conditions, but rates are very slow (Holt and Jones, 1983). As a result, the rates of decomposition slow greatly in backwater areas of streams and can lead to accelerated sedimentation in these areas.

If one examines the inputs and transformations of allochthonous organic matter, very few studies have dealt with the complete dissolved and particulate organic loadings and exports. In forested headwater streams of low stream order, leaves of terrestrial plants can constitute a major (e.g., ca. 70%, Iversen *et al.*, 1982) input of allochthonous *particulate* organic matter. The proportion of inputs into the stream sediments of coarse particulate organic matter (CPOM) of the total tends to increase as the size (width) of the streams decrease. The duration of retention of the CPOM in the stream sediments depends on the number of snags, the sinuosity of the river channels, and particularly the changes in the water discharge. Similarly, the retention capacity of fine POM in the sediments varies with the physical composition of sediment, and retention increases with roughness and substratum complexity (Webster *et al.*, 1987). FPOM is frequently deposited, resuspended, transported, and redeposited in a series of advection–dispersion movements governed by discharge and leading to constant resuspension and further export (Cushing *et al.*, 1993). The export downstream of FPOM increases with increasing water discharge and invertebrate feeding activities (Wallace *et al.*, 1991). In

TABLE 21-17 Decomposition Rates (Half-life, Days) of Dead Terrestrial Leaves in a Small Woodland Stream in Which Invertebrate Detritivores Had Been Removed[a]

Leaf species	*Reference stream (days)*	*Treated stream (days)*	*Increase in decay rate by invertebrates (%)*
Dogwood (*Cornus florida*)	41.0	65.4	160
Red maple (*Acer rubrum*)	50.2	135.9	271
White oak (*Quercus alba*)	64.2	173.3	270
Rhododendron (*Rhododendron maxima*)	128.4	577.6	450

[a] After data of Wallace *et al.* (1982). The stream section treated with insecticide removed about 90% of the macroinvertebrates.

larger streams and rivers, most of the allochthonous organic matter inputs to the river sediments are as dissolved organic matter retained by flocculation and bacterial uptake and fine particulate organic matter. Aquatic plant and terrestrial vegetation inundated by flooding can form a major source of organic matter in tropical river ecosystems (e.g., Walker, 1992). Outputs are largely as evasion of carbon gases (mainly CO_2, CH_4) by biological heterotrophic metabolism and photolysis and downstream export of dissolved organic matter and fine particulate organic matter.

Woody debris can constitute a major source of particulate organic matter in sediments of small streams and riverine wetlands. In rivers of very low elevational gradient, large woody debris can constitute the most stable portion of the river sediments and, in contrast to finer fragments, is seldom moved downstream except during uncommon floods (e.g., Wallace and Benke, 1984). Ranging in size from branches to large tree trunks, the rates of decomposition are generally very slow. Obviously, the wood tissue contains a high content of cellulose and lignin, is poorly penetrable to water, nutrients, and microbes and invertebrates, and has only a small surface area exposed to the enzymatic activities of microbes (Harmon *et al.*, 1986; Ward *et al.*, 1994). Nonetheless, wood serves as a persistent substratum that is actively degraded by many types of fungi in addition to aquatic hyphomycetes (Shearer, 1992). The low rate of decomposition is related to low surface-to-volume ratio and high density, which limits oxygen diffusion and microbial penetration into water-saturated substrates.

Depending upon the type of wood, small pieces <1 cm in diameter could require a decade for degradation if not fragmented by molar action of turbulence, ca. 50 years for branches 5–10 cm in diameter, and 100–300 years for larger tree trunks (Anderson *et al.*, 1978; Maser and Sedell, 1994). Examination of the rates of benthic respiration by consumption of dissolved oxygen per unit area per time by different types of CPOM revealed that microbial respiration associated with woody sticks nearly always made up <10% of the total (e.g., Tank *et al.*, 1993). Most of the benthic respiration was associated with smaller detritus, particularly from decaying leaves. Frequently, woody debris is partially buried in sediment by river turbulence during spates. Channel width and sinuosity are dominant factors that control the importation, storage, and effects on hydrology within the stream of woody debris (Nakamura and Swanson, 1994). The amount of coarse woody debris and the number of pool-forming pieces of wood are relatively high in wide, sinuous reaches of the river, particularly in association with a braided channel pattern. In wetlands, as areas are inundated (e.g., by beaver activity), trees are killed and fall into the sediments. Under anoxic conditions in the sediments, rates of decomposition are slowed even further to rates that are ecologically negligible.

VII. DEGRADATION OF DISSOLVED ORGANIC MATTER IN SEDIMENTS OF RUNNING WATERS

In addition to water flows in the open stream channel, water also flows through sediment interstices. In this subsurface region of sediments in streams and rivers, water exchanges with the overlying water of the channel. Flow also moves laterally into bank and floodplain sediments and then can reenter the flowing water of the stream channel at some point further downstream. This hyporheic zone is a transitional zone between surface water in the stream and ground water in the deeper zone in alluvium that holds ground water; the hyporheic zone connects these surficial sediments with water held and moving in surrounding floodplain sediments.

Bacterial mineralization processes, both aerobic and anaerobic, frequently occur as a result of intrusion of water both from the overlying channel water and with the lateral surficial and hyporheic water contained within the flood plains (cf. Chap. 18). The division between the hyporheic zone and the ground water is the interface region where <10% of the subsurface water originated from the channel water (Triska *et al.*, 1989). Because light penetrates poorly if at all into the hyporheic zone and is inadequate to support appreciable benthic photosynthesis, any biotic activity within the hyporheic zone must be supported by imported organic matter in the form of fine particulate or, predominantly, as dissolved organic matter.

If the sediments of a stream system contain abundant hydrous aluminum and iron metal oxides, an appreciable amount (ca. 40%) of dissolved organic carbon (DOC) can be removed from solution by sorption onto the oxides as they precipitate (McKnight *et al.*, 1992). DOC with greater contents of aromatic moieties, carboxylic acid groups, and amino acid residues were sorbed preferentially over DOC with fulvic and hydrophylic acid components.

From the perspective of bacterial activities within the sediments of streams, the sediments fall into two primary groups: (a) depositional in debris dam and backwater areas of very slow flows, as was discussed briefly earlier, and (b) porous sediments within which channel water can enter and leave by advection and connect with underlying ground water, if present. Although highly variable, with specific substrata and hydrologic conditions, advective flows commonly carry water into surficial sediments during which some chemical and mineralization can result in zones of microbial enrichment and increased bacterial productivity (review of Hendricks, 1993, 1996; cf. Chaps. 12 and 13). Nutrients and dissolved organic matter of interstitial water are enriched in comparison to those of surface water, often from water passing from lateral floodplain soils into the hyporheic zone (e.g., Valett *et al.*, 1990; Wainright *et al.*, 1992). Dissolved organic carbon is enriched in the upwelling zones of the sediments, emanating in part from enhanced anaerobic processes of denitrification, nitrate reduction, ammonification, sulfate reduction, and methanogenesis. Aerobic processes of respiration nitrification, sulfide oxidation, and methane oxidation are much more prevalent in the downwelling zone within the hyporheic sediments. Total heterotrophic bacterial productivity is appreciably greater in downwelling zones than in upwelling areas of the hyporheic sediments.

Rates of nitrification were found to be an order of magnitude greater in downwelling regions of the hyporheic sediments of streams in comparison to those in upwelling regions (Jones *et al.*, 1995a). When the sediments were disturbed, as following spates and flood-scouring events, nitrification rates were reduced appreciably and gradually increased as the microbial communities reestablished in subsequent succession. Because oxygen penetration can extend throughout surficial and deeper sediments of the hyporheic zone of the main channels of streams, denitrification rates are relatively low in these areas (Duff and Triska, 1990). Denitrification within sediments increased with increasing distance from the stream channel into hydrosoils of wetland and vegetated bank areas.

Population turnover times are rapid among natural communities of bacteria attached to sediments of streams. For example, in four small, temperate streams, estimated bacterial generations ranged from 67 to 375 per year (Bott and Kaplan, 1985). As appears to be true among the bacterioplankton, many of the attached bacteria of sediments are dead or dormant. In these four small streams mentioned, <25% of the bacteria were actively respiring.

Natural variations in loadings of dissolved organic matter to streams are large in response to precipitation events, snow melt runoff, deciduous leaf fall, and other seasonal changes in the flood plain and streams. Sediment microflora retain an ability to readily metabolize different sources of DOM that reach the communities at irregular intervals (Kaplan and Bott, 1985). In order to utilize naturally occurring pulses of carbon and energy, sediment bacteria must be enzymatically prepared and induced when the pulses occur. Within limits, bacterial metabolism is then roughly proportional to the availability of organic substrates.

Dissolved organic matter concentrations tend to be appreciably lower in ground waters and the hyporheic zone of sediments as the distance from a stream or river channel per se increases. In addition, the bioavailability of the organic carbon tends to become more recalcitrant to utilization by bacteria the greater the distance within the hyporheic zone from the main river channel (Ellis *et al.*, 1998). Bacterial production tends to be low in the lateral hyporheic zone and alluvial aquifers and decrease further with increasing distance from the river channel. Closer to the stream channel or within the stream sediments, however, bacterial utilization of DOC from terrestrial and floodplain wetland sources is much more rapid, even though rates differ from these with DOC from various higher plant species and sources (e.g., McArthur and Marzolf, 1986; Meyer *et al.*, 1987; Benner *et al.*, 1995; Bano *et al.*, 1997). Much of the DOC occurs as humic substances (60–80%), but even these relatively recalcitrant compounds are metabolized at rates of ca. 0.5% per day, with bacterial growth efficiencies of ca. 20–30% or more. Higher bacterial growth efficiencies often coincide with high water periods, which suggests that substrates imported from floodplain regions have increased bioavailability. In contrast, simple carbohydrates and amino acid compounds are rapidly immobilized by attached bacterial assimilation and a portion is respired (Fiebig, 1992; Marxsen and Fiebig, 1993). For example, in sandy sediments of a first-order stream, 99% of dissolved amino acids in ground water were immobilized, with some 14–25% respired.

As in lake sediments, by far the greatest bacterial production and metabolism occur in the sediments in comparison to that of the bacterioplankton in the overlying water. For example, bacterial respiration of the sediments of a low-gradient blackwater river was estimated to be about 97% of the total and at least 80% of that occurred in the sediments of the main channel rather than in backwater areas (Edwards *et al.*, 1990). Similarly, in a first-order blackwater stream, 70% of the respiration of the stream occurred in the hyporheic zone, and <1% occurred in the water column (Fuss and Smock, 1996). Many studies of benthic respiration rates, usually based on rates of oxygen consumption per unit area, have examined variations under different sediment conditions and spatial and temporal variability. Most of the findings would be anticipated from what is known from more detailed studies in lakes. For example, respiration per unit volume of hyporheic sediment was inversely related to the diameter of the sediment particles, which indicates that respiration is affected by substratum surface area and area available for microbial colonization (Jones, 1995). Respiration per unit of surface area on sediments, however, was positively correlated with sediment particle diameter, which indicated greater metabolic activity of microbes on larger particles and that conditions conducive to bacterial metabolism can be reduced among very fine particle-sized sediments.

The ratios between aerobic and anaerobic decomposition in the sediments of running waters vary greatly. Clearly, most stream and river ecosystems are predominantly heterotrophic and metabolize largely allochthonous dissolved and particulate organic matter. Bacterial metabolism within the sediments is influenced by such a large array of rapidly changing physical, chemical, and biotic factors that changes in metabolism also can shift rapidly both spatially and temporally within a river as the hydrology, loading of organic matter, and other parameters change. Rapid changes in nutrient and dissolved organic matter content within the sediments occur, often associated with changes in particle size. Large-scale interannual changes in discharge flow act as a general set-point mechanism that defines the prevalent particle-size composition and chemistry of the river sediments (Chambers *et al.*, 1992; Wondzell and Swanson, 1996). Flow changes of short duration have a less predictable effect on the chemistry of the sediments. Even though the overall pattern of subsurface flow changes little over the course of a year, the relative flux of advected channel water and ground water change among seasons and during storms.

Some generality about sediment metabolism in rivers is commonly observed, however. First, the consistent release of much greater amounts of CO_2 from benthic respiration than the amount of oxygen consumed in benthic respiration indicates, as in lake sediments, a predominance of anaerobic metabolism utilizing alternative electron acceptors (Dahm *et al.*, 1991). The ratio of aerobic to anaerobic metabolism in the sediments tends to increase with increasing size and stream order (e.g., Starzecka and Bednarz, 1994). Even if the metabolism of the surface microbial community is highly productive photosynthetically and appears as an autotrophic system, when the community metabolism of the hyporheic sediments and their large heterotrophic metabolism of allochthonous organic matter are included, as they must be in any three-dimensional evaluation of an ecosystem (Grimm and Fisher, 1984; Dahm *et al.*, 1987), the stream ecosystems are predominantly heterotrophic. Second, most of the benthic metabolism is microbial. For example, ca. 96% of the benthic community respiration of hyporheic sediments of a third-order stream was associated with organisms of a size $<100\ \mu m$ (Pusch and Schwoerbel, 1994).

VIII. SUMMARY

1. Sediments of inland waters are sites of major biological activity.
   a. Lake sediments are the major sites of microbial degradation of detrital organic matter and biogeochemical recycling of nutrients.
   b. Sediments of running waters and shallow areas of lakes are size-sorted in relation to velocity of water movements. Inorganic and organic particle size markedly influences the distribution and growth of many benthic invertebrates.
2. Sediments are composed of organic matter in various stages of decomposition, particulate mineral matter, and an inorganic component of biogenic origin (e.g., diatom frustules, photosynthetically induced precipitation of $CaCO_3$).
3. Particulate organic matter formed in the pelagic zone is decomposed more readily and completely during sedimentation than is the more recalcitrant structural tissue of littoral flora. As senescing littoral plants are decomposed and fragmented by successive communities of fungi and bacteria, redistribution often occurs. The partially degraded material is displaced widely among both littoral and pelagic sediments.
4. Resuspension of surficial sediments (0.5 to several cm) occurs episodically in both shallow and deepwater sediments, most frequently in lakes during complete water circulation (overturn) after periods of stratification. Resuspended particulate matter constitutes a major, often dominant, portion of the total sedimenting materials that reach the bottom on an annual basis.
5. Bacterial populations and metabolic activity in surficial sediments are several orders of magnitude larger than those in the overlying water column.
   a. Bacterial populations and metabolic activity decrease precipitously from the sediment–water interface to lower depths within the sediments.
   b. Bacterial numbers and metabolic activities of deepwater profundal sediments are often positively correlated with lake productivity and the quantity of particulate organic matter reaching the sediments. Bacterial activities are usually much greater in sediments of the littoral areas than in profundal sediments.
   c. Respiratory rates of sediment-dwelling bacteria are commonly inversely related to the surface area of particles (particle size) and directly related to organic nitrogen content.
6. Most sediments contain significant amounts of organic matter, and oxygen intrusions are limited. Bacterial metabolism rapidly produces anoxic,

reducing conditions. Under anaerobic conditions, rates of mineralization are slower than under aerobic conditions.

a. Under anoxic conditions, various substances and intermediate metabolic compounds function as electron acceptors and are reduced to yield energy instead of molecular oxygen. Reduced products formed by bacteria in sediments occur along a succession of anaerobic oxidation processes of reduction of nitrate, manganese and iron oxides, and sulfate (Table 21-6). Often, the same compound can serve as a hydrogen acceptor or donor, depending on the environmental conditions.
b. Large amounts of organic matter are anaerobically degraded by methane fermentation. Organic matter is converted to methane and CO_2 in two stages (Fig. 21-5).
c. Facultative and obligate anaerobic bacteria ('acid formers') convert proteins, carbohydrates, and fats primarily to fatty acids (especially acetate) by hydrolysis and fermentation.
d. Obligately anaerobic methanogenic bacteria then convert the organic acids to CH_4 and CO_2; alternatively, they can reduce CO_2 to CH_4 by enzymatic addition of hydrogen derived from the organic acids.

7. Methane generation can be very intense in anaerobic sediments. Methane diffuses into anaerobic overlying hypolimnetic waters of productive lakes. At the aerobic–anaerobic water interface (sediment–water interface or in the metalimnion overlying an anoxic hypolimnion), much of the methane is converted to CO_2 by aerobic methane-oxidizing bacteria.
   a. During summer stratification, the highest rates of methane oxidation often occur in the lower metalimnion, where oxygen concentrations are reduced (<1 mg liter^{-1}). For the lake as a whole, the total quantity of methane oxidized is often low during summer, however, because the bacteria are limited by the available combined inorganic nitrogen, and because oxygen present in the epilimnion is inhibitory to N_2 fixation. Nonetheless, little dissolved methane is found in the epilimnion, and relatively small amounts evade to the atmosphere at this time.
   b. High rates of methane oxidation (amounting to as much as 95% of the annual oxidation) occur in periods of lake circulation, especially during fall turnover. Considerable amounts of methane can diffuse and escape to the atmosphere during these periods.
   c. Under ice cover, high rates of methane oxidation can occur throughout the water column of eutrophic lakes. Methane oxidation under ice cover can contribute to reduced levels of dissolved oxygen or even total lake anoxia.
   d. Appreciable methane passes through emergent and floating-leaved aquatic plants and evades to the atmosphere. Methane emissions from natural and cultivated wetlands are appreciable on a global basis (ca. 20%).
   e. The amounts of CO_2 and bacterial-cell material production are usually insignificant in comparison to primary production. In relation to the total carbon inputs from the atmosphere and runoff, however, methane cycling can be significant in eutrophic and hypereutrophic lakes as well as in streams and rivers. An amount of organic carbon regenerated as methane equivalent to as much as 60% of the total carbon inputs has been found in several very productive lakes; methane-oxidizing bacteria can recycle about two-thirds of this carbon.
8. As lakes become very productive, anaerobic fermentation processes can evolve large quantities of gases, mainly CO_2, CH_4, and H_2. The quantities produced can exceed microbial oxidizing capacities and hydrostatic pressure constraints, which allows the formation of gas bubbles that rise to the lake surface and escape to the atmosphere.
   a. In relatively shallow hypereutrophic lakes, a steady ebullition of gases can result.
   b. Rising gas bubbles can generate small currents that are effective in redistributing hypolimnetic nutrients and degradation products to overlying water strata. These ebullition currents can enhance nutrient availability and primary productivity in the epilimnion and thus represent a positively reinforcing, internal fertilization process.
9. Most of the very high bacterial production of sediments is mineralized and respired as CO_2 and CH_4 within the microbial food web and evades to the atmosphere. Less than 20% of this rich bacterial production would be expected to reach higher trophic levels in either lake or river ecosystems.
10. The terrestrial vegetation adjacent to lakes and rivers, wetland vegetation, and higher aquatic plants of littoral and floodplain zones constitute a large source of organic matter.
    a. Senescent plant tissues decay during initial stages largely by fungal degradation.
    b. Within water, an appreciable portion (25% or more) of the tissue is leached as dissolved organic matter, some of which aggregates back to

particulate organic matter physically or by sediment microbial assimilation.

c. Particulate organic matter forms a major organic loading to sediments. Decomposition rates are initially rapid but decline with time as more labile organic constituents are decomposed. Rates vary greatly with differences in chemical composition of the particulate organic matter, particle size, nutrient availability, and temperature, among other factors.

d. In woodland streams, leaf fall from terrestrial vegetation can constitute a major seasonal input of particulate and dissolved organic matter. In larger rivers, external dissolved and particulate organic matter is derived largely from the flood plains and ground waters from adjacent terrestrial sources.

e. Loss rates in streams result from a combination of decomposition with evasion of respiratory CO_2 and CH_4 to the atmosphere, fragmentation of particulate organic matter by turbulence and macroinvertebrates, and export losses of dissolved and fine particulate organic matter. Fragmentation of coarse particulate organic matter by feeding activities of immature macroinvertebrates can increase markedly rates of microbial degradation.

11. In sediments of both lakes and rivers, much greater amounts of CO_2 are released from benthic respiration than the amount of oxygen consumed, which indicates a prevalence of anaerobic metabolism utilizing alternative electron acceptors. Most of the benthic metabolism is microbial, associated with organisms of a size $<100\ \mu m$.

22

BENTHIC ANIMALS AND FISH COMMUNITIES

I. BENTHIC ANIMAL COMMUNITIES

The distribution, abundance, and productivity of benthic organisms are determined by several ecological processes: (a) the historical events that have allowed or prevented a species from reaching a habitat, (b) the physiological limitations of the species at all stages of the life cycle, (c) the availability of energy resources, and (d) the ability of the species to tolerate competition, predation, and parasitism (Reynoldson, 1983; Hutchinson, 1996). We should examine briefly characteristics of the diversity of benthic invertebrates that inhabit inland waters.

Analyses of the complex interrelationships among the benthic animals of fresh waters have focused to a large extent on descriptions of species and their distributions within lakes and streams in relation to environmental variables. Although such analyses are essential to initial evaluations of the communities, physiologically oriented experimental analyses of regulating environmental parameters have not been utilized among benthic communities to nearly the extent that they have been in studies of planktonic communities. The population, productivity, and trophic interrelationships of the benthic fauna are poorly understood in lakes; they are somewhat better known in running waters (e.g., Hynes, 1970).

The distribution of the diverse fauna within lakes and streams is extremely heterogeneous, which is in part a product of variable requirements for feeding, growth, and reproduction. These requirements are influenced strongly by changes in the substratum and overlying water on a seasonal basis (e.g., changes in oxygen content and in the inputs of living and dead organic matter for food). The benthic organisms either possess adaptive mechanisms to cope with these changes, enter relatively dormant stages until more physiologically amenable conditions return, move, or die. The adaptive capabilities of the benthic animals to the dynamics of environmental parameters are basic to their distribution, growth and productivity, and reproductive potential.

Several major problems must be overcome in order to analyze benthic animal communities effectively. First is the difficulty of obtaining quantitative samples. Substrata heterogeneity leads to a patchy, nonrandom distribution that requires extensive replicated sampling according to procedures that depend on the organisms and bottom substrata involved (cf. detailed discussions of sampling methods in Edmondson and Winberg, 1971; Holme and McIntyre, 1971; Brinkhurst, 1974; Elliott, 1977; Benke, 1984; Downing, 1984; Peckarsky, 1984a; Wetzel and Likens, 2000). Organisms must be separated from samples obtained with benthic grabs, dredges, or cores, which retrieve organisms together with the substrate in which they live. The taxonomy of many animal groups is confusing to the nonspecialist; some groups are still very incompletely described. Emigration and immigration of members of the populations of certain groups, especially among the insects, necessitates more elaborate sampling methods. In spite of these problems, careful and detailed analyses of some populations provide insight into the controlling environmental and biotic interactions within benthic animal communities.

Benthic animals are extremely diverse and are represented by nearly all phyla from protozoans through large macroinvertebrates and vertebrates. This fact, together with their heterogeneous characteristics of habitats, feeding, growth, reproduction, mortality, and behavior, makes it exceedingly difficult to treat these animals in an integrated, functional manner. In the brief space available, we can only point out some of the major groups, giving examples of their population and production dynamics in relation to freshwater ecosystems. Important characteristics of the sediments have been discussed elsewhere in relation to oxygen (Chap. 9), redox (Chap. 14), microbial metabolism of organic matter (Chap. 21), and organic composition of different sediments (Chaps. 21 and 23).

A. Protozoa and Other Protists

Perhaps the least understood groups of benthic animals that occur in massive numbers on and in surficial sediments are the protozoa and related protists and the ostracod crustaceans. General characteristics of the protozoans and related protista and particularly their many modes of nutrition were discussed earlier among the planktonic populations (cf. Chap. 16, pp. 396–412; cf. also the excellent summary of Taylor and Sanders, 1991).

In spite of the abundant information that we have on protozoan morphology, physiology, genetics, and behavior, very little is known concerning their population dynamics and contributions to productivity in the sediments. Temperature, availability of dissolved oxygen, and characteristics of the substratum markedly influence the distribution and productivity of protozoa. Most protozoa are attached to substrata, and they are particularly abundant in habitats of active oxidative decomposition. Ciliate protozoans are particularly important in the metabolism of dissolved and particulate organic matter of sewage-treatment facilities and organically polluted streams (Cairns, 1974), and their community composition has been used extensively as part of a system for classifying the water quality of aquatic habitats (Sládeček, 1973; Foissner, 1988).

The diversity of species of protozoa, their possession of wide ranges of tolerance to environmental extremes and varied feeding capabilities (including algae, bacteria, particulate detritus, and other protozoa and protists), and their large population densities on aerobic, organic-rich sediments, all point to a significant metabolic role in freshwater ecosystems. Because of their low collective community biomass, it generally has been assumed that their contributions are small. However, their short generation times and rapid turnover under the optimal trophic conditions of the sediments indicate that the protozoa contribute appreciably to the degradation processes in lakes, just as they do in the pelagic and littoral zones.

Flagellates clearly dominate in many benthic communities, and have been found to range from 100 to 180,000 cells cm^{-3} of sediment, whereas ciliate numbers ranged from 26 to 11,000 cells ml^{-1} (Gasol, 1993), although much higher estimates were found in sediments of small temperate streams (Bott and Kaplan, 1989). Most of the protozoans occur in the upper 1 cm of sediment. The number of species within sediments of a lake can be very large (50–150; e.g., Webb, 1961; Laminger, 1973; Finlay *et al.*, 1979; Finlay, 1990). Although coexistence in the same habitat is common, microspatial differences in distribution exist, and species separation by differences in food utilization, reproduction, and timing of population development within community succession on substrata are also common (Goulder, 1971, 1974a,b; Taylor and Berger, 1980; Jax, 1997).

Many factors affect the distribution and growth of free-living protozoa (cf. reviews of Noland and

Gojdics, 1967; Laybourn-Parry, 1984, 1992; Taylor and Sanders, 1991), but dissolved oxygen is of paramount importance. The protists are sufficiently small that cellular exchange by diffusion is adequate without the necessity of organelles to increase surface area. Few species can tolerate the anaerobic conditions of the sediments for any appreciable period. Microaerophilic protists, however, can tolerate very low oxygen tensions, and these habitats can provide a refuge from competition and predation by larger animals. Protozoa are distributed in environments of widely ranging temperatures (0–50°C), but with a Q_{10} of about 2.0, rates of growth are most efficient below 25°C.

The distribution and abundance of chlorophyll-bearing forms, often microaerophilic ciliates with endosymbiotic green algae, are regulated to a large extent in ways analogous to the regulation of algae, such as by inorganic and organic nutrient availability and by light. Among heterotrophic protozoa, the distribution and quality of dissolved and particulate organic matter for food are important. Although the nutritional requirements of only a few species are known, there is a tendency for species living in unpolluted waters to feed on microalgae, while those of more polluted waters ingest predominantly bacteria. The significance of vitamins and other growth substances to protozoan population dynamics is just beginning to be appreciated.

Most species of protozoans are intolerant of low oxygen concentrations and reducing conditions in sediments, but some species tolerate reducing conditions much better than others (e.g., Goulder, 1971, 1980a; Laminger, 1973; Finlay, 1980). As a result, protozoan species are often segregated vertically in lake sediments along a depth gradient. These depth distributions may change seasonally, owing to changes in oxygen concentrations of overlying water. Many species migrate to sediments in shallower areas as anoxia develops in the hypolimnion, or the populations gradually assume a different depth distribution as reproductive rates change. Some species are negatively phototactic and avoid high light intensities within epilimnetic sediments.

Many of the benthic protistan populations exhibit pronounced summer maxima (Finlay, 1980, 1990; Bark, 1981; Rogerson, 1981). These changes have been positively correlated with summer increases in temperature, day length, and abundance of organic matter and microflora serving as food in the surficial sediments. Other studies found maxima in winter with decreases in nanoflagellates in the summer and autumn (Starink *et al.*, 1996). The importance of predation upon the benthic flagellates in summer by other meiobenthic invertebrates is unclear. As oxygen is depleted in the hypolimnion, many populations migrate upward to metalimnetic and epilimnetic regions; this migration also contributes to seasonal increases in the protozoan populations in these areas.

Although competition among protistan species has been demonstrated in cultures, most of the evidence for competitive species interactions is indirect under natural conditions (e.g., Cairns and Yongue, 1977). Both heterotrophic flagellates and benthic ciliates are estimated to consume only a small fraction (<2–5%) of the daily bacterial production in sediments (Kemp, 1988; Alongi, 1991; Starink *et al.*, 1994) and consume many other food materials, such as microalgae, particulate organic detritus, and other protists. Benthic bacterivory was in the range of 10–70 bacteria protist^{-1} h^{-1}, and somewhat higher for ciliates. Similarly, herbivory of benthic algivorous ciliates on attached diatoms was estimated to be <5% (Balczon and Pratt, 1996).

Predation upon protists is common by other protozoan species and by micrometazoa, particularly rotifers and some cladocerans and copepods. For example, experimental introduction of micrometazoa reduced the abundance of larger protists, mostly ciliates and certain algae (McCormick and Cairns, 1991). Benthic heterotrophic flagellate abundance increased under these conditions, in part related to declines in ciliates that prey and compete with the smaller flagellates. As was discussed among the planktonic protists (Chap. 16), changes in body size, feeding behaviors, and growth rates also occur among benthic protists, although much less is known about these interactions.

B. Porifera—Sponges

Few sponges occur in fresh waters; most species are marine. Although some species endemic to Lake Baikal grow to magnificent lobed structures of about half a meter in size, most are small, inconspicuous, and morphologically variable. The flagellated chambers create weak currents with water that enters through many pores (ostia), passes through the internal canals and chambers, where bacteria and particulate detrital material are trapped for intracellular digestion, and then collects and leaves through larger channels and openings (oscula) (reviewed in Frost, 1980, 1987, 1991). Filtration of water for feeding can be high (ca. 6 ml h^{-1} per mg dry mass of tissue). Zoochlorellae, ingested algae that persist inter- and intracellularly in the tissue, can grow and metabolize appreciably within certain sponges (Gilbert and Allen, 1973). The photosynthate produced can contribute significantly to the energy budget of the host, in the range of 2–22% of the animal respiration (Frost and Williamson, 1980; Sand-Jensen and Pedersen, 1994; Frost *et al.*, 1997). Rarely do sponges occur at depths in excess of the

Secchi disc transparency (Hutchinson, 1993), a depth equivalent of ca. 10–15% of surface illuminance. Although such symbiotic relationships between the organisms are common, the zoochlorellae are not essential to the survival of the sponge, at least not for *Spongilla* and a number of other genera.

Sponge growth occurs by proliferation of the tissue over the substratum with the addition of new chambers. Growth rates and energy demands generally increase progressively seasonally and reach a maximum in midsummer (e.g., Melão and Rocha, 1998). In mild climates, sponge populations can persist for many years without appreciable change (e.g., Pronzato and Manconi, 1995) until a major disturbance occurs, such a flooding. Under unfavorable conditions, such as in autumn in the temperate zone, tissue is reduced by partial or total deterioration. Highly resistant resting structures termed *gemmules*, usually less than a millimeter in diameter, are formed in many species (Gilbert, 1975; Gilbert and Simpson, 1976a). Some species, particularly those in sublittoral environments of muted seasonality, do not form gemmules; instead, sexual reproduction continues into the autumn (Gilbert and Allen, 1973; Simpson and Gilbert, 1974; Gilbert and Hadzisce, 1975, 1977). Major predation upon sponges by other animals is rare. Spicules and likely organic compounds can function as deterrents to predators. A number of animals, particularly certain immature insects, live within sponges, perhaps to find a refuge from predation. The relationships have been shown to be obligatory in only a few cases.

Gemmules contain undifferentiated mesenchymal cells and have an outer layer of compacted siliceous spicules within a tough sclerotized membrane. They can withstand freezing and desiccation. After a diapause of one to several months at colder temperatures, germination of the gemmules is usually associated with warmer temperatures >15°C in the early spring (Harrison, 1974; Gilbert *et al.*, 1975). Motile larvae move to appropriate substrata, attach, and develop. Syngamic sexual reproduction also occurs in warmer periods of the year. Sponges are usually dioecious, and the sex ratio is about 1:1 in natural populations. A common freshwater sponge, *Spongilla lacustris,* exhibits a type of alternate hermaphroditism, in which a sponge may be exclusively male or female during the period of sexual reproduction 1 year and the opposite sex the next year (Gilbert and Simpson, 1976b). This form of sexuality may facilitate larval production and thus dispersal following colonization of a new habitat.

Sponges are among the few animals requiring large amounts of silica for spicule development. Spicules occur in many types and species-specific forms and are bound together by collagens to form an inorganic skeleton. Sponges are restricted largely to waters of moderate silica content (>0.5 mg liter^{-1}) (Jewell, 1935). Species distribution is also directly correlated with the calcium content of the water but declines in very hard waters (Jewell, 1939; Strekal and McDiffett, 1974; Hutchinson, 1993); although causal mechanisms for this relationship remain obscure, they are probably related to food abundance rather than to a direct demand for calcium.

Freshwater sponges, which usually occur only in relatively clear, unproductive waters, rarely become a major component of benthic communities. Their significance in benthic productivity is minor in most situation, but has been found to constitute a significant portion of the total benthic productivity in certain conditions [e.g., 40% in the River Thames (Mann *et al.*, 1970) and in a *Sphagnum* bog pond (Frost, 1978)].

C. Cnideria (Coelenterata)

The hydroids and "jellyfish" of the class Hydrozoa belong to a predominately marine group that is poorly represented in fresh waters. The common *Hydra* occurs only as the tentacled polyp stage, while the rare *Craspedacusta* or freshwater jellyfish undergoes metagenesis with alternation between medusoid and polyp stages that is common to the marine Hydrozoa. During the life cycle of *Craspeducusta*, both medusa and polyp states feed on a variety of zooplankton, but because of low densities, they likely do not influence zooplankton densities (DeVries, 1992).

Epidermal cells contain nematocysts or stinging cells that function in capture of prey; food particles are then moved to the mouth orifice by the tentacles. The hydroids are sessile and are almost exclusively carnivorous on zooplankters. One species of green hydra (*Chlorohydra*) contains unicellular green algae in a symbiotic relationship, and the algae live intracellularly in gastrodermal cells. About 10% of the carbon photosynthetically fixed by the algae is released and assimilated by the *Hydra* (Muscatine and Lenhoff, 1963). This photosynthate is clearly of nutritional significance to the *Hydra* (Pardy and White, 1977; Phipps and Pardy, 1982).

Development in quiet waters is rapid in spring, and greatest densities usually are found in early summer and late summer (Miller, 1936; Cuker and Mozley, 1981). Growth rates decrease markedly when populations are crowded (Thorp and Barthalmus, 1975). Hydroid densities usually do not become great, however, and their contribution to total benthic productivity is usually, although with poor foundation, considered negligible. Under some conditions, populations of *Hydra* can become very large (to 30,000 m^{-2} among submersed macrophytes) and significantly reduce population sizes of small zooplankton by predation (Cuker and Mozley, 1981) and larval fish by both ingestion or mortality after being stung by nematocysts (Elliott *et al.*, 1997).

D. Turbellarian Flatworms

The dominant free-living members of the Platyhelminthes are the turbellarian flatworms. These organisms are acoelomate with only a single opening to the digestive tract (they lack an anus). Most members of the group, the tapeworms (Cestoidea) and the flukes (Trematoda), are entirely parasitic. Many free-living flatworms are small (<1 mm), but the triclads usually exceed 10 mm in length. Flatworms generally require a firm substratum and possess abundant cilia that assist them in movement over the substrata of shallow lakes and streams. When the animal encounters small invertebrates or detrital organic matter, the ventral pharynx is extruded through the mouth to incorporate the material (Young, 1973). Some of the rhabdocoels contain symbiotic zoochlorellae in their parenchyma and cells of the gastrodermis.

Reproduction is either asexual, by budding or fission (Pattee and Persat, 1978), or sexual. In the latter case, sex organs usually develop over winter and egg capsules are released in the late winter or spring in permanent bodies of water. Nearly all flatworms are hermaphroditic, with both sex organs occurring in the same individual. The eggs are encapsulated or sometimes enclosed in a stalked shell. Winter eggs are heavily encapsulated and are able to withstand low temperatures, but do not survive well to desiccation. Development is strongly temperature dependent in both the egg and immature stages; no larval stages are found in most species (e.g., Pattee, 1975; Folsom and Clifford, 1978). A few species are univoltine, but most species are multivoltine, with the number of generations correlated with the availability of habitat and food (cf. especially Harrmann, 1985).

Most flatworms, especially the triclads, are negatively phototactic and occur in shaded areas among debris, rocky substrata, and macrophytes. Among four species of lake-dwelling triclads studied extensively by Reynoldson (1966), all occupied the same general substratum in the shallow littoral zone and fed on the prey organisms there. Feeding mechanisms of all the species were similar, and a wide overlap existed in the type and size of prey eaten by young and adults of each species. The species populations exhibited restricted fluctuations in numbers and lived under conditions of food shortage for long periods of the year. Flatworms are able to avoid starvation because they can resorb and then regenerate tissues by reductions in size and metabolic rate (Calow, 1977) and because of low predation pressure by other organisms.

The distribution and abundance of individual triclad species are determined primarily by interspecific competition for food (Reynoldson and Bellamy, 1971, 1975; Reynoldson and Sefton, 1976; Reynoldson and Piearce, 1979; Adams, 1980; Reynoldson *et al.*, 1981; Young, 1981; Reynoldson, 1983; Armitage and Young, 1990a). All studies of lake populations indicate that they are food-limited and that severe intraspecific competition occurs. Despite the great overlap of food prey species, each triclad has been found to feed on a particular kind of prey to a greater extent than others. Thus, the competition depends largely on the pattern of distribution and abundance of important types of prey, such as gastropod mollusks, chironomids, stonefly and mayfly nymphs, amphipods, and oligochaetes, each of which represents the dominant prey of particular triclad species.

The coexistence of several species of triclads as well as leeches in lacustrine habitats is common even though both triclads and leeches are food limited and there is much overlap in their diets (e.g., Seaby *et al.*, 1996). Partitioning of food resources apparently occurs because the leeches are more efficient in capturing prey food items than are triclads. In contrast, the triclads can exploit damaged prey, including incapacitated, live, or recently dead animals that are leaking body fluids, better than do leeches. In contrast, the microturbellarians feed on bacteria, attached algae, protozoa, rotifers, and very small cladocerans, copepods, and oligochaetes (cf. Heitkamp, 1982). Predator–prey relations also exist among carnivorous microturbellarian species. Stream triclads reduce potential competition by spatial separation along the stream based on different responses to water temperature and stream gradient. Some evidence exists for interspecific competition for food among stream populations as well (Reynoldson, 1983).

Some species, particularly of the tropical planktonic turbellarian *Mesostoma*, can reach densities of 1000 m^{-3} and are occasionally significant (to 4 individuals day^{-1}) predators of cladoceran zooplankton, especially *Daphnia* and *Ceriodaphnia* (Rocha *et al.*, 1990). These flatworms produce and release a toxin that functions as a neurotoxin when cladoceran zooplankton encounter mucus released by the flatworm (Dumont and Carels, 1987). Capture and ingestion of the prey is assisted by this mechanism.

Although a number of invertebrates and vertebrates may consume triclads, free-living flatworms constitute a minor component of the diet of their predators (Davies and Reynoldson, 1971; Reynoldson, 1983).

Triclads are preyed upon only by a few predators such as dragonfly nymphs and newts. The chemical protection afforded triclads by rhabdites is apparently effective. The incidence of sporozoan, ciliate, nematode, and trematode parasitism is high, but mortality effects appear to be low (Armitage and Young, 1990b).

The abundance of these triclad species has been directly correlated with the general productivity of the waters and amount of food available in the littoral zone, which in turn has been directly related to concentrations of calcium and total dissolved matter (Reynoldson, 1966, 1981). As a result, unproductive lakes receiving low nutrient loading support smaller populations and fewer species, owing to competition for food, than lakes of higher productivity, in which both the diversity and abundance of food species increase. As food supply is reduced, short-term reproductive effort generally increases (Callow and Woollhead, 1977).

Similar relationships have been found among stream-dwelling flatworms (Chandler, 1966). Shallow streams exhibit wide fluctuations in temperature, which influence triclad feeding activity, reproductive capacity, and acclimation capacity. The distribution of individual species is widely separated as a result. Additionally, triclad abundance is inversely related to velocity of water flow. The shifting substrates of rapidly flowing streams and wave-swept littoral zones create an unfavorable habitat for flatworms and many additional benthic animals. Although adults may be able to tolerate the mechanical abrasion associated with shifting substrata, eggs or immature stages often cannot.

Turbellarian abundance in lakes is variable within a range of ca. several hundred to 80,000 m^{-2} (Heitkamp, 1982). For example, in mesotrophic Mirror Lake, New Hampshire, mean turbellarian densities were 27,000 m^{-2}, and decreased from 40,000 m^{-2} in shallow littoral areas to nearly zero in the profundal zone (Strayer, 1985). Although the turbellarians constitute a significant portion of the total meiobenthos productivity, their significance to the entire energy budget of lakes is less clear. Densities in streams are generally lower, with an upper end of the range <5000 m^{-2} where predation, particularly of triclads by a few species of stonefly nymphs and caddisfly larvae, is more effective than in lakes.

E. Gastrotricha

The gastrotrichs are a very small group of diminutive (50–800 μm), pseudocoelomate animals of or closely related to the nematodes of the Aschelminthes. Gastrotrichs occur in high densities in both aerobic and anaerobic sediments and in the periphyton of littoral areas of lakes and streams and in the hyporheic sediments of running waters. The gastrotrichs commonly attach the posterior end of their spindle-shaped body with adhesive tubes to a substratum. Rows of ventral cilia allow the animals to move by gliding.

Gastrotrichs feed on bacteria, algae, protozoans, and small detrital particles. The digestive system is a simple, relatively undifferentiated tubular gut with an anterior mouth and posterior dorsal anus. Little is understood of predation upon gastrotrichs by known predators of amoebae, cnidarians, and midge larvae.

Upon hatching from eggs, young gastrotrichs contain developing parthenogenetic eggs that can develop rapidly (e.g., 2 days), are quickly laid, and hatch (e.g., 1 day) under favorable conditions (Strayer and Hummon, 1991). This parthenogenesis allows rapid population development under favorable environmental conditions, with growth rates of 0.1–0.6 per day. Some of these gastrotrichs age longer and develop into hermaphrodites, and meiotic sexual eggs are produced with genetic recombination. A second type of parthenogenetic egg, a resting egg, is thick-shelled, very resistant to freezing and desiccation, and likely aids in dispersal.

Among the few population analyses, seasonal trends in gastrotrich abundance show species differences among the same sediment habitats. Generally, abundances are greatest in the spring for several species (range 10–50 cm^{-2}) on organic-rich lake sediments; other species reach greater population densities in the winter months (Kisielewska, 1982; Strayer and Hummon, 1991). Gastrotrich densities can be extremely high, within a reported range of 10,000–2,600,000 m^{-2}, and are generally more abundant in the littoral zone (Nesteruk, 1993, 1996). Densities within the sediments are concentrated within the uppermost 2 cm, although they have been found at much deeper depths, 10–15 cm, and among many sediment grain sizes, particularly more coarse particles (>0.5 mm) (Strayer, 1985; Nesteruk, 1991; Fregni *et al.*, 1998). Strayer estimated the secondary production of gastrotrichs in Mirror Lake, New Hampshire, at ca. 100 mg DM m^{-2} yr^{-1}, which was <1% of the total production of the zoobenthic community in this lake.

F. Nematoda—Roundworms

The Nematoda or roundworms, a class of the pseudocoelomate phylum Aschelminthes, are a significant component of the benthic fauna. Although most of the many nematode species are parasitic, the many free-living species are distributed widely in all types of freshwater habitats. The taxonomy of this group is difficult, incomplete, and has contributed to the relative paucity of understanding of the ecology and productivity of nematodes.

The external proteinaceous cuticle of nematodes is noncellular and layered and is molted four times during their life (Malakhov, 1994). This ridged cuticle is interfaced with longitudinal muscle cells that permit active, snakelike movements along and through sediments. Nearly all free-living freshwater nematodes are <1 cm in length.

Feeding habits are very diverse among the nematodes. Some members are strictly detritivores, feeding solely on dead plant or animal particulate matter, or both. Some species are bacterivorous and are prolific on surfaces among organically polluted saprobic conditions (Schiemer, 1983). Others are herbivorous and have specialized mouth parts for chewing living or dead plant material or for piercing and sucking cytoplasm from plants (Prejs, 1977a). Mouth parts are more specialized among carnivorous nematodes, enabling them to seize, rasp, and macerate small prey animals such as other nematodes, protozoans, gastrotrichs, tardigrades, oligochaetes, and small immature insects.

Reproduction is variable. Some species are parthenogenetic or hermaphroditic. Most species, however, are dioecious and syngamic, with separate females and males (Malakhov, 1994). Fertilized eggs, usually laid outside of the body, develop rapidly (1–10 days) and hatch into young, which are nearly fully developed except for the reproductive organs. Reproduction and egg laying by adults are more or less continuous. Little is known of the fecundity of free-living nematodes under natural conditions, but it is apparent that much of their productivity is diverted into reproduction in a normal life cycle. The reproductive phase is extended under habitat conditions of high and predictable food supply (Schiemer, 1983). Several hundred eggs with a biomass several times greater than that of the producing female are apparently common. The duration of the life cycle is also variable, from about 2–40 days. Most species, however, require 20–30 days for development and reproduction.

Nematode populations of the sediments of a typical dimictic, relatively oligotrophic mountainous lake were concentrated at depths of 3–4 cm in the sediments (Bretschko, 1973). Few penetrated below 6 cm into the sediments. Similarly, Traunspurger and Drews (1996) found most (89%) nematodes in the first 5 cm of sediment, although considerable spatial separation was observed among different species vertically within the sediments. Although the four molting instars could not be separated, analyses of the seasonal population dynamics of the 10 species demonstrated three generations per year. The first hatched in September–October and matured in November–December just after ice cover formed. The second generation was the most productive and reached maximum abundance and biomass in January (Table 22-1). The maximum relative abundance of adults of this generation occurred in April. Offspring of the winter generation formed the third generation, with a maximum in biomass and abundance in May (about 50% of that peak in January). As the ice cover was lost in May and the lake circulated, the entire community of nematode populations collapsed for unknown reasons. Over a 2-year period, the average population density was 235,000 m^{-2} under ice cover but only 60,000 m^{-2} during the ice-free period.

The greatest densities of nematodes are found in oligotrophic lakes (Prejs, 1977b,c). Nematode densities decrease in profundal sediments of eutrophic and dystrophic lakes; presumably low oxygen concentrations are a primary limitation in these lakes. Some species tolerate anoxia, at least for short periods, of the hypolimnion in mesotrophic lakes (e.g., Strayer, 1985).

Most studies have examined community dynamics of nematodes in profundal sediments, where densities can range between 30,000 and 750,000 m^{-2} (Traunspurger, 1996). In the littoral zone and in wetlands, however, nematodes can reach very high densities among the attached algae of dense emergent and submersed macrophytes. In a detailed study of the seasonal dynamics of nematode populations in the littoral zones of several lakes, Pieczyńska (1959, 1964) found that maxima occurred in May and June, when the populations numbered over one million individuals per square meter of substrate surface, a level brought about by mass reproduction of the dominant species. The same species reproduced throughout the year, although not as intensely. During the other seasons of the year, the nematodes occurred in lower, more or less constant densities. Variations in the development of the spring maximum from year-to-year were correlated with warming temperatures of the water and the duration of ice cover in the littoral. Species of *Prochromadorella* and *Punctodora* dominated in littoral substrata, especially among shoots of higher aquatic plants, to which they attached readily by the secretion of an adhesive substance from their caudal glands and by their ability to become entangled in colonies of littoral algae. Littoral ice cover was found to be especially destructive to the substrata and nematode populations. Wave action, in contrast, caused little mortality and was found to facilitate dispersal and subsequent rapid colonization of newly growing algal and macrophytic surfaces. The densities of nematodes were correlated directly with the development and density of attached algae.

Nematode species distribution and biomass production were strongly correlated with the type of substratum (Table 22-1). Over 80% of production occurred in the deeper zone of deep, fine sediments where gravel was scarce and boulders were absent within an oligotrophic mountainous lake. Within this zone, 90%

TABLE 22-1 Production of Nematodes Seasonally and along Depth Gradients in Two Lakes of Austria[a]

Substratum zone and its area	*Seasonal period (kg DW per area)* Oct.–Jan.	Feb.–May	June–Sep.	Oct.–Dec.	*Percentage of total production*
Vorderer Finstertalersee, high mountain lake, oligotrophic					
Littoral-slope zone, fine silt; no macrophytes 58,868 m^2	1.24	0.16	0.34	1.74	13
Current-steep slope, reduced silt covering gravel and boulders; 47,213 m^2	0.24	0.26	0.32	0.82	6
Flat-depth zone, below 20 m, deep fine mud, gravel rare, boulders absent; 51,520 m^2	8.64	1.56	0.44	10.64	81
Entire lake, 157,600 m^2	10.16	1.96	1.08	13.20	100

	Annual production kg ha^{-1}	kg total $area^{-1}$	
Piburger See, softwater, alpine lake mesotrophic			
Upper border (6–9 m) of profundal zone; 10,752 m^2	0.120	0.130	3.2
Stratum (9–13 m); 16,721 m^2	2.320	3.875	96.3
Lower border (13–13.5 m); 1938 m^2	0.103	0.020	0.5
Total Area (6–13.5 m); 29,411 m^2	2.543	4.025	100

[a] Summarized from data of Bretschko (1973, 1984) and Pehofer (1989). Dry weight estimated from fresh-weight values given by a factor of 20%; totals do not sum parts exactly because of rounding. The Finstertalersee has subsequently been modified extensively by damming of the valley and inundation for a reservoir.

of the biomass production was by one species of *Tobrilus*. The reasons for the observed high productivity in early winter are unclear but presumably were related to good food conditions that followed the ice-free period of higher primary productivity and to reduced predation by fishes and possibly by certain predacious cyclopoid copepods. Similar distributions and production of nematodes in distinct sediment habitats were found in a more productive lake (Table 22-1). Annual production-to-biomass ratios of three dominant nematode species, constituting most of the nematode production, were in the range of 4–7.5 (Table 22-2).

TABLE 22-2 Total Annual Production by Three Dominant Species of Profundal Nematodes and Their Production (*P*)-to-Biomass (*B*) Ratios[a]

Nematode species	*Production* ($mg\ m^{-2}\ yr^{-1}$)	*Annual P/B ratio*	*Source*
High mountain lake, oligotrophic			
Monhystera stagnalis (semivoltine)			Bretschko (1984)
1968/1969	31.2	ca. 4	
1969/1970	11.7		
Ethmolaimus pratensis (semivoltine)	83.3	5.5	
Mesotrophic, softwater alpine lake			Pehofer (1989)
Tobrilis gracilis	758	4.6	
Monhystera stagnalis	205–304	7.1–7.4	
Ethmolaimus pratensis	110	6.5	
Total mean (6–13.5 m sediments)	700	—	

[a] Extracted from data of Bretschko (1984) and Pehofer (1989).

Clearly, the small size of the nematodes belies their high production rates—in terms of densities per unit area and production, the nematodes are a major component of the benthic invertebrates.

G. Nematomorph—Horsehair Worms

Horsehair worms (phylum Nematomorpha; Gordian worms) are common in many fresh waters. Horsehair worms, evolutionarily distinct from other animals, are free-living as adults but as larvae are obligate parasites of insects and other invertebrates (Poinar, 1991). Larvae readily parasitize immature caddisflies, dragonflies, crustaceans, and leeches, although host specificity is not well documented. Often, sterility or death of the host is induced, and heavy parasitization may affect the population dynamics of the host species.

Adult free-living horsehair worms are common among littoral vegetation of shallow lakes, streams, and wetlands and can attain 1 m in length but are usually <3 mm in diameter throughout their body length. These worms are dioecious, and males are smaller than females. Sexual reproduction occurs during the warmer periods of the year. Eggs are laid in prodigious numbers (millions) in long gelatinous strings among vegetation and particulate detritus. Incubation is temperature-dependent and varies between 15 to 80 days. After hatching, larvae encyst on substrata and may be subsequently ingested by host animals. Internal metamorphosis occurs over several weeks, and juvenile worms emerge from hosts to mature. Emerging horsehair worms require water for maturation.

Very little is known about the feeding habits, population dynamics, and general ecology of these worms. Although the free-living nematomorphs may have relatively little impact on material and energy fluxes within most habitats, their commonly destructive parasitism in larval and juvenile stages could alter the population dynamics of host animals.

H. Bryozoans

The colonial bryozoan members of the primarily marine Ectoprocta are rarely of quantitative importance in freshwater ecosystems. These sessile forms, however, occasionally form massive colonies that can become conspicuous members of shallow eutrophic lakes and open areas of swamps for brief periods. The microscopic zooids possess ciliated tentacular crowns that project into the water and create water currents. Particles as large as rotifers, very small microcrustaceans, and small algae (<20 μm) are directed into the lophophore and mouth (Kamiński, 1984). Many zooids aggregate in colonial masses as a single structure. Some bryozoans contain many algae throughout the matrix that are attached to surfaces enclosed by the colonies (Joo *et al.*, 1992). Cyanobacteria dominated (90%) on surfaces enclosed by the colonies but were minor components of the attached algae on surfaces not colonized by bryozoans.

Although hermaphroditic sexual reproduction occurs in summer, bryozoan zooids are capable of asexual reproduction by budding new individuals from the midventral wall of the mother zooid. Budding permits rapid proliferation on the outer edge of the colonies, which occasionally form amorphous ball-like masses one-quarter of a meter in diameter. These colonies are common in small farm ponds in water less than a meter in depth, and their growth is influenced by fish predation (Dendy, 1963). When protected from fish predation, colonies were branched and more productive; when exposed to fish predation, unbranched, cropped colonies persisted. A large number of invertebrate animals, from protozoans to large insect larvae, particularly chironomids, live within the colonies of bryozoans. Predators of the bryozoans consume these coinhabitants as well as the bryozoans.

Under unfavorable environmental conditions, many bryozoans encapsulate dormant buds, termed *statoblasts*. Statoblasts have a thin outer chitinous shell that encloses yolk material and germinal tissue and can remain dormant for years under desiccation and freezing conditions. Statoblast viability, however, has been observed to be low (<10%) (Brown, 1933; Smyth and Reynolds, 1995). Statoblasts, particularly floating types, assist in dispersal; their production is approximately proportional to the size of the colony (Karlson, 1992).

Extensive studies of 12 species of bryozoans in 122 lakes, streams, and shallow waters of Michigan showed correlations of species distribution with several general limnological parameters such as pH (7–9), current, and substratum type (Bushnell, 1966; cf. the detailed review of Bushnell, 1974; Ricciardi and Reiswig, 1994). Some bryozoan species were associated with particular macrophyte substrata, and others were distributed more ubiquitously. Live colonies were found under ice at water temperatures <2°C. Growth was often erratic, but luxuriant colonies of *Plumatella* were found to grow rapidly, doubling in 3.4–4.9 days in summer and in 4.0–7.4 days in the spring. Growth ceased when temperatures decreased <9°C. Most mature colonies broke up and died after a period of growth, for reasons not completely understood. Predation on actively growing colonies was often very high, especially by caddisfly larvae and snails. *Plumatella* longevity ranged from 4 to 53 days; the death rate of individuals of the colony (the polypides) was highest

(80%) between 21 and 36 days after statoblast germination, and few lived longer than six weeks.

I. Aquatic Worms—Oligochaetes and Leeches

Two major groups of the annelids, or segmented worms, are represented in fresh waters. The first of these, the Oligochaeta or aquatic "earthworms," often form a major component of the benthic fauna, particularly of lakes. The other group, Hirudinea or leeches, is a diverse class of annelids of much biological interest. Although the significance of leeches as a component of total benthic animal productivity is unclear at this time, their predatory habits can materially influence the population dynamics of other benthic organisms, especially oligochaetes. By far the most comprehensive study of the aquatic oligochaetes is the work of Brinkhurst and Jamieson (1971). While largely a global taxonomic work, much of the presently rather limited knowledge of oligochaete anatomy, embryology, distribution, and ecology is reviewed in this book (cf. Learner *et al.*, 1978; Brinkhurst and Cook, 1980). A third minor group, the Branchiobdellida, consists of leech-like ectosymbionts living only on freshwater crustaceans, primarily on astacid crayfish (Brinkhurst and Gelder, 1991). This latter group is of much evolutionary interest but is a minor component of freshwater ecosystems and will not be discussed further here.

1. Oligochaete Worms

Oligochaetes are typically segmented, bilaterally symmetrical, hermaphroditic annelids with an anterior ventral mouth and a posterior anus. Size ranges from <1 mm to about 40 cm (maximum >200 cm), but most freshwater forms are <5 cm in length. Each segment, except for the first and a few terminal segments, bears four bundles of setae. Swimming occurs by rapid rhythmic movements in helical body wave cycles (to 12 s^{-1}) that pass from the anterior to posterier ends of the worm, thus propelling the worm forward (Drewes and Fourtner, 1993). Sexually mature Oligochaeta possess an anteriorly located thickened region in the body wall that is glandular and produces a cocoon at oviposition. Sexual reproduction predominates in many species under poor environmental conditions (Loden, 1981). Asexual reproduction is more common during favorable environmental conditions (e.g., McElhone, 1978).

Much of the information on oligochaete populations focuses on geographical distribution, habitat selection, and effects of organic pollution. Several species usually are found in each habitat, and these species differ little between streams and lakes of the same region. The number of species is often greater in larger lakes, perhaps because of the greater number of different microhabitats. Some species are restricted to relatively oligotrophic waters, while others are distributed widely in lakes of greatly differing productivity, from oligotrophic to extremely eutrophic (see, e.g., Milbrink, 1973a,b, 1978; Lang and Lang-Dobler, 1980; Särkkä and Aho, 1980). Obviously, a number of interacting parameters influence species distribution within the physiological tolerances of individual species. Some of these factors are discussed here.

The distribution of oligochaetes in sediments at different depths within a lake also indicates that a large number of species coexist in the same substrata. Individual species, however, do reach maximal abundance at different depths (Fig. 22-1). Although differences in the composition of the sediments were not analyzed critically in this study of a eutrophic lake, a number of studies suggest that the particulate composition and organic content of the sediments are correlated with the distribution and abundance of oligochaetes (Timm, 1962a,b). In a large, oligotrophic Italian lake, the number of oligochaetes was generally less in sediments of relatively large particle size (0.11–0.12 mm) than in more organic-rich sediments of finer particle size (0.07–0.08 mm) (Della Croce, 1955). A similar relationship has been found in relation to the total organic content of the sediments (Brinkhurst, 1967; Wachs, 1967). Experimental studies on the ubiquitously distributed *Tubifex tubifex* verified these general relationships. Nutrition and the availability of food are primary factors influencing distribution and abundance of oligochaetes.

Oligochaetes ingest surficial sediments containing organic matter of autochthonous and allochthonous origins colonized with bacteria and other microorganisms. Some species actively graze attached microorganisms growing epiphytically on macrophytic vegetation. Other oligochaetes actively ingest epipelic algae (e.g., Moore, 1978, 1981). Detailed analyses of the gut contents and feeding habits of oligochaetes living in the same habitat indicate that resource partitioning occurs by differences in worm morphology and by selective feeding on food items, particularly algae (Streit, 1977; McElhone, 1979). In studies of sediment feeding by this group (reviewed in Brinkhurst and Jamieson, 1971), it has been found that the organic carbon, caloric, and total nitrogen contents of the ingested sediments were only reduced slightly. However, critical experiments with mixed species cultures failed to demonstrate any reduction in the percentage of organic matter, nitrogen, or caloric content in feces as opposed to that in sediment provided as food (Brinkhurst *et al.*, 1972; Brinkhurst and Austin, 1979). These and other experimental studies on the utilization of bacterial species in sediments indicated that one species of

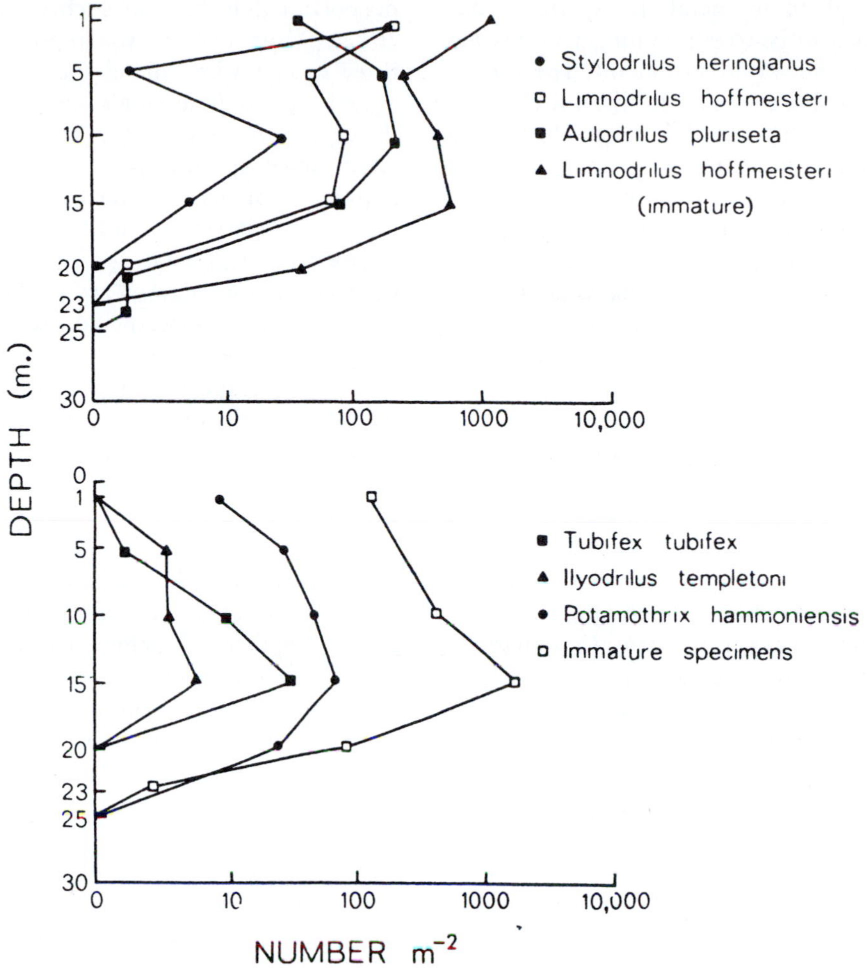

FIGURE 22-1 Distribution of various species of oligochaetes in relation to depth in hypereutrophic Rostherne Mere, southern England. The lake was devoid of oxygen below a depth of 25 m. (Drawn from data of Brinkhurst and Walsh, 1967.)

oligochaete may be able to selectively utilize different microbial components of the sediment (Brinkhurst and Chua, 1969; Wavre and Brinkhurst, 1971; Harper *et al.*, 1981).

As lakes and streams become organically polluted, it is common to find an abundance of tubificid oligochaetes (Brinkhurst and Cook, 1974; Milbrink, 1980). Associated with such organic enrichment is the acute reduction or elimination of dissolved oxygen. Acute reduction of oxygen is lethal to a majority of benthic animals. The number of species of tubificid worms also decreases with higher productivity and increased organic pollution. However, if some oxygen is available periodically and if toxic products of anaerobic metabolism do not accumulate, the combined conditions of a rich food supply and freedom from competing benthic animals permit rapid growth of oligochaetes adapted to these conditions.

Many oligochaetes, especially the tubificids, burrow headfirst into the sediments but leave their caudal ends, which contain most of their gills or respiratory appendages, projecting and undulating in the water above the sediment–water interface. The caudal respiratory movements increase in frequency and vigor as the oxygen content of the water decreases. In general, tubificids are able to adjust their respiration to decreasing oxygen concentration down to a critical level in the range of 10–15% saturation, values below which hemoglobin can no longer be loaded with oxygen. Below this critical level, feeding and defecation are then greatly reduced or cease (cf. Brandt, 1978). Many tubificids can tolerate anaerobic conditions for at least a month or longer but survive only if they are exposed intermittently to some oxygen, since they cannot respire anaerobically for any appreciable length of time. Water movement is apparently an important factor in removal of toxic metabolic wastes, since anaerobic conditions can be tolerated longer in moving situations than when the water is stagnant. The species of oligochaetes, both those inhabiting littoral areas as well as profundal sediments, exhibit relatively specific distributions in waters of differing productivity, organic loading, and tolerances to low dissolved oxygen. Therefore, the distribution of oligochaetes along gradients of oligotrophy to eutrophy and waters of high organic pollution (e.g., receiving pulp mill wastes) have been used as indicators of trophy and eutrophication. Such indicator organisms may be of some value at the extreme ends of the trophy spectrum but must be used with caution and in conjunction with other parameters because the distribution ranges of the species are quite large (cf. Särkkä, 1987).

Reproduction in oligochaetes varies seasonally, and even when more or less continuous, some periodicity in breeding intensity is common. Oligochaetes lack discrete age classes and production is often approximately continuous. The limited data from natural populations suggest that maturation requires as long as a year in some species and up to 2–4 years in others. Variation is great, however, within the same species, and even from site to site in the same body of water. Breeding intensifies in late winter and spring as temperatures increase above about 10°C. Most adult oligochaetes die after sexual reproduction. Selective predation by other animals as a factor in species coexistence has been studied very little in natural populations. However, it is clear that selective breeding and metabolic adaptations by oligochaetes permit cohabitation. For example, when three species were combined, the population respiration of mixed cultures decreased about 35%, while increasing the population density of a single species proportionately had no such effect (Brinkhurst *et al.*, 1972). More energy was used for growth when the three species were mixed than when they were maintained separately in single-species culture, and the rate of respiration decreased in the mixed populations. Isolated single species spent more time burrowing and searching for their preferred food, which is feces of other oligochaete species, than they did in feeding (Chua and Brinkhurst, 1973; Brinkhurst, 1974). Such interspecific interactions are probably causal factors in the frequently observed clumped distribution of several species of oligochaetes.

In the deep profundal sediments of the Danish eutrophic Lake Esrom, some tubificid *Potamothrix hammoniensis* can first reproduce at 3 years of age, but most do not until they are 4 years old (Jónasson and Thorhauge, 1972). The lifespan of this species can reach 5 or more years. Mature *Potamothrix* in this lake laid eggs in May–July, and this process was related to moderate increases in sediment temperatures. From 2 to 13 embryos hatched from each cocoon about 2 months later, in August. Growth was most rapid in the autumn and spring, and no weight increase occurred during the summer period of stratification. Numbers of this species varied between 10,000 and 25,000 m^{-2} over a 6-yr period. In spite of these large numbers, biomass varied between 0.75 and 1.80 g dry weight m^{-2} but usually was >1.0 g m^{-2}. These values are similar to those observed in other lakes (Table 22-3).

Similar or higher densities ($>30{,}000$ m^{-2}) of aquatic oligochaetes were found in the littoral zone of mesotrophic Lake Esrom in Denmark (Lindegaard *et al.*, 1990, 1995). Despite the relatively high annual net production of the two dominant species (*Stylodrilus heringianus*: 3.1 g AFDM m^{-2} and $P/\bar{B}$ of 3.2; and

TABLE 22-3 Average Population Densities and Biomass of 29 Species of Oligochaetes of Chud-Pskov Lake, Estonia[a]

	Average population density (no. m^{-2})	*Percent of total density of benthic animals in zone*	*Biomass[b] (g m^{-2})*	*Percent of total biomass of benthic animals in zone*
Total lake	370–1400	—	—	—
June	859	47	1.307	23
Fine-grained silt of profundal	287	58	0.433	44
Silt with fibrous particulate detritus	1452	62	2.776	16
Sandy with little silt; no vegetation	1015	22	1.409	36
Upper littoral	1185	20	0.133	6

[a] From data of Timms (1962a).
[b] Assumed to be wet weight; methods not stated (see Jónasson and Thorhauge (1972) for discussion of errors involved with preserved versus fresh wet weights).

Marionina southerni: 1.0 g AFDM m^{-2} and $P/\overline{B}$ of 2.5) and moderate net production efficiencies of ca. 43 and 26%, respectively, they constituted only a small portion (<1%) of the estimated mean zoobenthic production in the littoral zone from 0- to 2-m depth in Lake Esrom. Mortality in the Lake Esrom populations was clearly related to natural death following reproduction. However, predation by the dipteran larvae of *Chironomus anthracinus*, studied in great detail in this lake, also caused significant reduction of the oligochaete populations. The reproduction of the oligochaete populations was depressed severely every second year as a result of the 2-yr life cycle of the *Chironomus* larvae (see also Loden, 1974).

Few other estimates of production rates of oligochaete populations exist (Brinkhurst, 1980). Detailed evaluations of the egg development times, mortality, cohort structure, and population dynamics of two dominant species of a reservoir in northern Italy demonstrated marked differences between the seasonal population characteristics of these oligochaetes (Bonomi, 1979; Adreani *et al.*, 1981). *Tubifex* had higher fecundity and shorter generation times and was nearly twice as productive as *Limnodrilus* (Table 22-4). However, the annual production-to-biomass ratios (turnover times) of the two species were very similar.

Oligochaete species frequently segregate vertically within the sediments. Naidid oligochaetes are concentrated at the sediment–water interface, rarely >2–4 cm below the surface (Milbrink, 1973c). Tubificid oligochaetes are most dense between 2 to 4 cm of sediment depth and occasionally penetrate as deep as 15 cm. The movements of benthic organisms within the sediments are important because their activity can disrupt temporarily the oxidized microzone of the sediment–water interface and thereby alter rates of chemical exchange between the sediments and overlying water. Experiments with tubificids have shown that their activity can alter the stratigraphy of surface sediments (Davis, 1974; McCall and Fisher, 1980), which is of interest in micropaleontological work (Chap. 24). As was discussed in Chapter 21, it appears that the intensive microbial activity at the sediment–water interface is sufficient to reestablish redox conditions almost immediately after periodic disturbances. However, if oligochaete populations such as *Tubifex* are present in high densities (e.g., 50,000 m^{-2}) in sediments, water exchanges can double rates of oxygen consumption, increase denitrification of water-phase nitrate by three times, and greatly increase ammonium ion efflux from the sediments (Pelegri and Blackburn, 1995).

2. Hirudinea—Leeches

Excellent general summaries on the structure, physiology, and ecology of leeches are given by Mann (1962), Sawyer (1986), and Davies (1991). Many leeches are ectoparasites that intermittently consume blood and body fluids of vertebrates greatly in excess of their own body weight. Leeches then enter a period of fasting, during which the consumed food is utilized over a period that can be as long as 200 days, with progressive loss of body weight. Other leeches are predatory on invertebrates such as oligochaetes, amphipods, snails, and chironomid larvae and consume their prey entirely (Elliott, 1973a; Green, 1974; Davies and Everett, 1975; Bradley and Reynolds, 1987; Proctor and Young, 1987; Toman and Dall, 1997).

Most predaceous leeches have an annual or semiannual life cycle, breed once, and then die. Reproduction is usually initiated in the spring or early summer and apparently is governed by temperature, the density of the populations, and age. Many leeches breed at 1 year of age and die after breeding, although some pass through two or more generations in 1 year (Tillman and Barnes, 1973; Davies and Reynoldson, 1975; Davies, 1978; Malecha, 1984). Other species live 2 years, overwinter and breed in the second, and die soon thereafter (Elliott, 1973b). High postreproductive mortality is associated with both significant metabolic

TABLE 22-4 Average Biomass, Annual Production, and Annual Production/Biomass Ratios of Oligochaete Populations in Pietra del Pertusillo Reservoir, Italy[a]

		Immature (to 1 mm)	*Young*	*Mature adults*	*Egg-bearing adults*	*Total*
Production (P) (g m^{-2} yr^{-1})	*Tubifex*	18.89	21.31	6.46	2.26	48.92
	Limnodrilus	4.47	14.24	10.31	−1.50	27.53
Mean biomass ($\overline{B}$) (g m^{-2})	*Tubifex*	0.69	3.59	1.13	3.91	9.32
	Limnodrilus	1.07	1.78	0.87	2.44	6.15
Annual P/B	*Tubifex*	27.4	5.9	5.7	0.6	5.25
	Limnodrilus	4.2	8.0	11.8	—	4.48

[a] After Adreani *et al.* (1981). Biomass estimates based on wet (formalin) weights.

TABLE 22-5 Life History and Mortality of Two Common Freshwater Leeches in England[a]

	Glossiphonia complanata	*Erpobdella octoculata*
Breeding period	Mar.–May	June–Aug.
Average number of young per parent	26	23.5
Mortality in first 6–9 months (%)	97	91
Proportion of year–group breeding at 1 year old (%)	70 (or less)	87
Mortality from 6 to 18 months (%)	33	49
Proportion of year-group breeding at 2 years old (%)	100	100
Mortality from 18 to 30 months (%)	84–88	69
Proportion of total population surviving to 3 years (%)	5–6	4

[a] After Mann (1962), from several sources.

costs incurred during gametogenesis and the high energetic costs of reproductive output (Davies and Dratnal, 1996). All leeches are hermaphroditic and commonly exhibit protandry, but gametogenesis is not always by the same individual (cosexual). After fertilization, eggs are deposited into a cocoon that is usually subsequently laid on or attached to substrata.

Predation on flaccid cocoons by invertebrates, particularly beetles, and fish can be appreciable but usually is not considered as regulatory to population dynamics of leeches as is mortality of the young (Young, 1988). In the first 3 months after hatching there is heavy mortality of the young (approximately 95% see Table 22-5) from predators (carnivorous invertebrates, particularly odonates and beetles, amphibia, many fish, and birds). Thereafter mortality lessens but is still severe (e.g., Spelling and Young, 1987; Young, 1987). Swimming activity of leeches in search of food is much greater during, but not restricted to, darkness, presumably to minimize access by visual predators (Davies *et al.*, 1996; Angstadt and Moore, 1997) or to follow the diurnal vertical migration of pelagic amphipod prey (Blinn and Davies, 1990). When population densities become high, population size can be regulated by high mortality of the eggs, which are consumed by adult leeches (Elliott, 1973a). There is also considerable evidence for common food limitation that can result in a high mortality (to 98%) of recruited young (Martin *et al.*, 1994; Young *et al.*, 1995). The young are unable to locate damaged food in a spatially heterogeneous environment because of a poorly developed chemosensory system. If low food conditions persist for long periods of time, energy partitioned to growth and reproduction declines as leeches allocate more energy to lipid storage, which increases survival (Smith and Davies, 1995).

Parasitism of leeches and mortality associated with parasites is poorly understood. Prevalence levels of microsporidian and trematode metacercariae parasites are low (<10%) (Spelling and Young, 1986a,b). High levels of trematode infestation reduced egg production slightly (<10%) and were not considered to regulate lacustrine leech populations significantly.

Among blood-consuming leeches, the time required for individuals to attain maturity is dependent upon the number and frequency of blood meals that are consumed (Davies and Wilkialis, 1980; Wilkialis and Davies, 1980). A seasonal migration has been found in some species, in which populations move into shallow water areas, where encounters with vertebrates would be more probable, during the ice-free period and then back into deepwater zones during the period of ice cover. Leeches are able to tolerate considerable periods of anoxia but the duration is strongly influenced by temperature (>50 days at 5°C, <10 days at 20°C) (Davies, 1991). Leeches move to shallower regions as oxygen is depleted in the hypolimnion during summer stratification (e.g., Peterson, 1983).

The abundance of leeches is highly variable among different habitats of lakes and streams. A general direct correlation exists between leech abundance and lake productivity. This relationship probably is associated with the increasing diversity of substrate types among the macrophytes and sediments, with correspondingly greater amounts of invertebrate food sources for the predacious leeches and birds and other vertebrates for the blood-consuming leeches (Sander and Wilkialis, 1972). Growth and production rates are quite variable in the littoral areas, however, because of the heterogeneity of habitats and associated prey items. In general, however, leech productivity is highest in the upper littoral zone reaches. For example, in the most detailed study of the growth and production rates of 10 leech species of Lake Esrom, estimates of mean annual production among five littoral locations varied between 288 to 1721 mg AFDW m^{-2} yr^{-1} (Dall, 1987). Both

TABLE 22-6 Mean Annual Abundance, Net Production (*NP*), and *P/B* Ratios of Leeches in the Littoral of Lake Esrom, Denmark[a]

	Mean density (no. m^{-2})		*Mean annual NP (mg AFDW m^{-2} yr^{-1})*		*P/B ratios*	
Species	*0.3–0.5 m*	*2 m*	*0.3–0.5 m*	*2 m*	*1979–1980*	*1975–1976*
Erpobdella octoculata	175.8	79.7	464	175	2.5–4.1	5.0
E. testacea	61.2	1.0	250	5	1.7–2.3	2.4
Helobdella stagnalis	113.3	61.5	105	42	2.6–3.2	2.4
Glossiphonia complanata	38.8	15.3	80	40	2.1–2.7	2.0
G. concolor	47.7	1.0	100	5	1.4–2.2	3.0
G. heteroclita	35.3	0.4	11	8	2.6–3.2	3.1
Total	472.1	158.9	1010	275		

[a] From data of Dall (1987).

abundance and production rates of individual species of leeches were manifold higher in the shallow areas than in midlittoral reaches (Table 22-6).

Estimates of the production rates of leeches of running waters are rare. The figures cited in Table 22-7 are for moderately productive stream waters and emphasize some of the population characteristics. Production of the *Erpobdella* leech species varied considerably between years and year-classes. Production was always greatest in the first year of the life cycle, even though the mean biomass was very similar in both years of the life cycle. The rate of production was most rapid during March to July, after which production declined sharply. Feeding rates are usually maximal in the spring and summer and decline sharply in the autumn.

Overwintering leeches of the species *Helobdella stagnalis* were found to reproduce in May in a shallow, eutrophic reservoir in South Wales (Learner and Potter, 1974; cf. Davies and Reynoldson, 1975). These postreproductive adults then died. The spring generation grew rapidly, and about one-third of this generation reproduced in July. The nonbreeding portion of the spring generation overwintered with the summer generation; both reproduced at the same rate in the following spring. The mean biomass was 96 mg DW m^{-2} in 1970 and 393 mg DW m^{-2} in 1971. Production was

TABLE 22-7 Production, Mean Biomass, and the Ratio of Production:Biomass (*P/B*) for Populations of the Leeches in Running Waters

Population	*Mean production (g wet weight m^{-2} yr^{-1})*	*Mean biomass (g wet weight m^{-2})*	*P/B ratio*
Erpobdella octoculata[a]			
1964 year-class			
Second year	8.25	10.35	0.80
1965 year-class			
First year	23.42	10.02	2.34
Second year	8.52	11.99	0.71
Both years	31.94	11.01	2.90
1966 year-class			
First year	12.34	4.61	2.68
Second year	4.60	5.29	0.87
Both years	16.94	4.95	3.42
1967 year-class			
First year	6.54	2.47	2.65
Annual estimates			
1966	33.46	20.96	1.60
1967	19.77	14.36	1.38
1968	11.63	7.82	1.49
Helobdella stagnalis[b]	7.56	1.43	5.29

[a] After data of Elliott (1973b) from an English stream.
[b] After data of Murphy and Learner (1982) from an organically polluted river in southern Wales.

estimated to be 310 mg DW m^{-2} yr^{-1} in 1970 and 1190 mg DW m^{-2} yr^{-1} in 1971, with production-to-biomass ratios of 3.2 and 3.0 in the respective years, similar to those found elsewhere (Table 22-6).

J. Tardigrades—Water Bears

The Tardigrada, or water bears, are a group of small (100–500 μm) arthropod-like organisms with a bilaterally symmetrical body and four pairs of legs that usually terminate in claws. Although most tardigrades inhabit wet terrestrial habitats, a number of species occur in fresh waters. The origins and systematics of the Tardigrada are unclear because they share traits of deuterostomes, pseudocoelomates, and coelomate protostomes, but they likely are intermediary forms along the evolutionary progression from the annelids to the arthropods (Kinchin, 1994). The tardigrades have been separated into an individual phylum with detailed systematic analyses (Ramazzotti, 1962, 1972; Ramazzotti and Maucci, 1983).

Some tardigrades are bisexual and fertilization usually occurs within the female (reviewed by Nelson, 1991; Kinchin, 1994). Many tardigrades are hermaphroditic, with male and female gamete production by the same individual. Parthenogenesis is very common among tardigrades. Highly ornamented eggs gestate on substrata or are held within cast exuvia of the adults; development times vary widely between 5 and 40 days. Periodic molting, which requires 5–10 days, continues throughout the life of tardigrades. The pharyngeal apparatus is expulsed periodically, during which time feeding ceases, and the buccal apparatus is regenerated. The claws are also reformed during each molt.

Very little is known about the population dynamics and productivity of freshwater tardigrades. Some direct correlations have been observed between tardigrade population fluctuations and attached bacteria and algae, and inversely with prey nematodes (Kinchin, 1994). No detailed evaluations of the comparative productivity of tardigrades exist for freshwater ecosystems. Although it is improbable that these interesting animals constitute a major benthic component of ponds, lakes, and rivers in comparison to other benthic organisms, conclusions are unwise on the basis of existing information.

K. Water Mites

Water mites are a collective group of animals emanating from five unrelated groups; the Hydrachnida form the largest and most diverse group of freshwater Acari (Gledhill, 1985; Smith and Cook, 1991). The group is extremely diverse, with >5000 species, and is often found in abundances of >5000 m^{-2} and consisting of >50 species in both lake littoral areas and stream riffles and in interstitial waters of the hyporheic sediments (to 50 cm of sediments).

Their complex development begins with hexapod larvae that emerge from eggs and become ectoparasites on insects in or close to the water, from which they extract fluids while being passively transported and dispersed. Upon release from the host, the mites commonly enter a resting stage during which development occurs to form an octopod, actively swimming deutonymph stage. Although predaceously feeding and growing at this stage, the mites are not sexually mature. The mites then enter a second resting stage from which, after further development and metamorphosis, they emerge as active, mature adults. After mating and fertilization, eggs are laid in masses in a gelatinous matrix attached to substrata.

Analyses of dominant water mite populations within a complex community of many species in a eutrophic lake in the Netherlands indicated relatively constant populations densities of up to 1000 m^{-2} throughout the year, with the highest densities in shallow littoral areas (Davids *et al.*, 1994). Nymphs were abundant from spring through late autumn and declined precipitously in the winter. The growth rates of nymphs and adult mites increased with rising temperatures, and the duration of resting stages decreased with increasing temperatures (Butler and Burns, 1995).

The parasitic larvae and predaceous deutonymphs and adults of water mites have direct effects on the size and structure of insect populations in certain habitats. Larval water mites often parasitize 20–50% of natural populations of aquatic insects (Smith, 1988). This parasitism impairs the vitality, growth, mobility, and fecundity of host insects. Deutonymphs and adults of free-living species of mites are voracious predators of eggs of insects and fish, larvae of many diptera and other insects, ostracod, cladoceran, and copepod crustaceans (Böttger, 1970; Gledhill, 1985). Little is known of predation upon mites. Although many species are found in fish, others are specifically rejected by fish as distasteful. Practically no studies treat comprehensively the production of water mites.

L. Ostracods

The ostracods are small, bivalved crustaceans usually <1 mm in size that are widespread in nearly every aquatic habitat. Their size, distribution in the surface sediments with few planktonic species and difficult taxonomy have all contributed to a very poor understanding of their ecology, population dynamics, and productivity in fresh waters. Their valves, which superficially resemble clam shells, are held apart when undisturbed. Cladoceran-like appendages protrude between the open valves and are used for locomotion. Most ostracods move about on the sediments by beating movements of the antennae and the caudal ramus. Some species (e.g., *Cypridopsis*) are associated with periphyton on macrophytes and particulate detritus and feed voraciously on attached microflora (Mallwitz, 1984; Roca and Danielopol, 1991).

Ostracods are omnivorous and, similar to many cladoceran microcrustacea, they feed on bacteria, algae, detritus, and other microorganisms by means of filtration. Grazing rates increase in a linear function with rising food concentrations and then decrease precipitously at high food concentrations (Grant *et al.*, 1983). Very little is known of the effects of ostracods on the turnover and metabolism of benthic microflora. The large population numbers of ostracods, commonly in excess of 100,000 m^{-2}, suggest that their role in the metabolism of surficial sediments could be considerable.

Ostracod reproduction is usually parthenogenetic for much of their life cycle. In some species, males have not been found; in others, sexual reproduction is common. The time required for egg development is variable, from days to months, and is strongly temperature dependent. The larva hatches as a nauplius with a reduced number of appendages, and then undergoes a series of growth and molting stages; usually eight molts are needed to reach the ninth, adult stage, during which morphology becomes more complex and appendages develop.

Little is known of the population dynamics of ostracod populations, although many exhibit distinct seasonal periodicity. Some species exhibit a single generation per year; others two or three per year (e.g., Mallwitz, 1984). The detailed study of the reproductive potential and life history of a species of *Darwinula* in a deep, dimictic, temperate lake illustrates some of the characteristics of this group (McGregor, 1969). *D. stevensoni* was abundant throughout the littoral sediments of this mesotrophic lake and reached a maximum density at a depth of 6 m throughout the year. Most individuals occurred between depths of 3 and 6 m, and densities decreased markedly between 6 and 9 m. Only a few individuals were found below a depth of 9 m, and none <12 m. More than 95% of the adults and juveniles present in the sediments occurred in the upper 5 cm.

The reproductive period of this *Darwinula* species in this Michigan lake began in May and was effectively completed by October (Fig. 22-2). The number of young per individual increased from a maximum of 3 in May to 15 in August. Although the reproductive potential of a given individual was correlated strongly with temperature and varied somewhat with depth, most adults produced only one brood per year of about 11 young per individual at depths of 3–6 m (13 at 9 m).

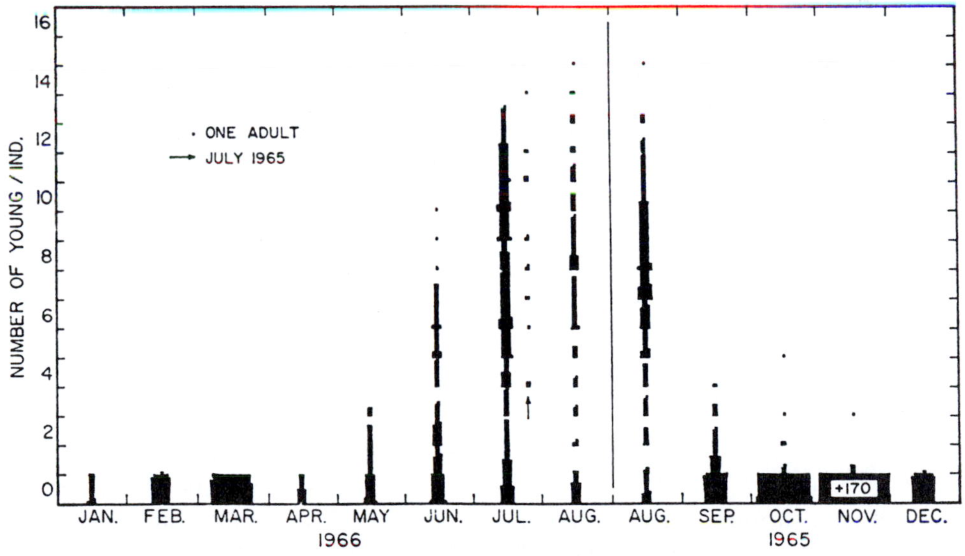

FIGURE 22-2 Annual reproductive cycle and reproductive potential of the benthic ostracod *Darwinula stevensoni* in Gull Lake, southwestern Michigan. (From McGregor, D. L.: The reproductive potential, life history and parasitism of the freshwater ostracod *Darwinula stevensoni* (Brady and Robertson). *In* Neale, J. W. (ed.): *The Taxonomy, Morphology and Ecology of Recent Ostracoda*, Edinburgh, Oliver & Boyd, 1969.)

Eggs of *Darwinula* are released into and develop within the carapace of the parent (McGregor, 1969). Juveniles produced by first and second reproductive season adults are released during the summer and early autumn and overwinter as juveniles. These young mature and reproduce the following summer. Surviving members of this age class overwinter as adults during the second winter and reproduce again the following summer. Most adults of the second reproductive year die following the release of young and are replaced by overwintering juveniles entering their first reproductive period. Thus, a nearly complete turnover of the adult portion of the population occurs each year. Similar population dynamics have been found with other lacustrine ostracod species (e.g., Strayer, 1988a).

Although much of the *Darwinula* population was parasitized by ectoparasitic rotifers, this did not appear to affect the reproductive potential. The population consequences of predation by other benthic invertebrates and bottom-feeding fish are unclear in this and other species. Ostracods alter their feeding and locomotion behaviors in response to chemical compounds released from fish predators and thereby reduce predation efficacy (Uiblein *et al.*, 1996).

Evaluations of densities of ostracod populations are few and range from an average of 57 m^{-2} in oligotrophic Great Slave Lake in the Canadian subarctic (Tressler, 1957) to about 10,000 m^{-2} of the *Darwinula* in mesotrophic Gull Lake. Much higher population densities of >50,000 m^{-2} are common in more productive lakes (e.g., Mallwitz, 1984), but densities decrease, and species diversity is reduced, in hypereutrophic lakes (e.g., Delorme, 1978). Given the common univoltine life cycle, low turnover rates, and long periods of nongrowing adulthood, productivity is likely moderate to low in comparison to other benthic invertebrates. Much greater study is needed.

M. Malacostracean Crustacea

Of the many species of the malacostracean crustaceans, four groups have distinct freshwater representatives of considerable interest and importance. All malacostraceans have a definite and fixed number of body segments.

1. Mysids–Opossum Shrimps

The mysids, or opossum shrimps, are morphologically similar to crayfish; they can attain a maximum length of about 3 cm. Their appendages, however, are elongated, contain abundant setae, and generally are greatly modified for active swimming. Their mode of feeding is by filtration of small zooplankton, phytoplankton, and particulate detritus via setose appendages, particularly two pairs of maxillipeds.

Aggregations of food particles are carried anteriorly to the mouth by water currents set up by movements of the appendages. Gut analyses of *Mysis relicta* showed that at night adults were voracious predators, feeding on *Daphnia*, other cladocerans, and the rotifer *Kellicottia* during their vertical migration (Lasenby and Langford, 1973). *Mysis* had a purported assimilation efficiency of about 85% when feeding on *Daphnia*, but this efficiency dropped to zero when the animal fed on debris from mosses growing in the lake. Several studies have shown that mysids are size-selective in their predation of zooplankton (Bowers and Grossnickle, 1978; Cooper and Goldman, 1980; Langeland, 1981; Murtaugh, 1981a,b). Large mysids prefer large-sized prey (*Daphnia* and *Epischura*) consistently, while small mysids select the smallest zooplankton available. The bivoltine life history of mysids in many lakes results in seasonal variations in population size structure (Murtaugh, 1983). These size changes influence the amount and quality of predation inflicted upon its zooplankton prey. Mysids will consume algae and detritus but are preferentially carnivores.

Freshwater mysids consume prey over a broader size range than most other invertebrate predators. When mysids are introduced into lake ecosystems as a food source for fish, however, size-selective predation pressure can lead to marked alterations in the zooplankton community (e.g., Richards *et al.*, 1975; Langeland, 1981; Cooper and Goldman, 1982; Nero and Sprules, 1986; Northcote, 1991). Selective predation on larger microcrustaceans by mysids can lead to the enhanced development of smaller species because of relaxed competition for food resources. Marked reductions or elimination of some large zooplankton species can thus result, contrary to intentions, in reductions of abundance and growth of some fishes.

The freshwater mysids are distinctly stenothermal, cold-water forms and in stratified lakes are restricted to hypolimnetic strata of <15°C. Reproduction occurs only during the colder periods of autumn, winter, and early spring. There are as many as 40 eggs per clutch, and the eggs are kept within the female brood pouch (marsupium). After hatching, juveniles are retained within the female for considerable periods of time (up to 3 months); hence, the name *opossum shrimps*.

A comparison of the life history, growth, and respiration of *Mysis relicta* in an arctic lake (Char Lake, Northwest Territories, Canada) and a temperate lake (Stony Lake) of southern Ontario is most instructive (Lasenby and Langford, 1972). In the arctic lake, *Mysis* takes 2 years to reach maturity, while those of the temperate lake mature in 1 year. *Mysis* from both lakes exhibited a trimodal distribution of size classes. In the arctic lake, the population consisted of 8- to-9-month-old immature mysids, 20-month-old adult males and females, and 32-month-old females. In contrast, in the temperate lake, the cohorts consisted of 2-month-old immatures, 1-year-old females, and a few 2-year-old females. Respiration measurements showed no differences between *Mysis* from the two lakes over their environmental temperature range. Over a 2-yr-period from hatching to maturity, the energy used by the arctic population in growth and reproduction was approximately 68% of that used by the temperate mysids in growth and reproduction over a 1-yr period from hatching to maturity (Table 22-8). Of much interest is the finding that the amount of energy (200 cal) used by each mysid for growth and respiration in order to become a reproducing adult was about the same for the arctic population in 2 years as for the temperate population in 1 year. Similarly, size–frequency distributions of *Mysis relicta* of the Superior, Huron, and Ontario basins of the Great Lakes showed differences in life cycles in the three lakes (Carpenter *et al.*, 1974). In the colder Lake Superior, the generation time appeared to be 2 years, while in lakes Huron and Ontario the generations matured in 18 months. When lake productivity is relatively high, the life cycle of mysids takes from 1 to 2 years, but when productivity is low, it may last up to 4 years (Morgan, 1981). Similar variation has been found in Scandinavian lakes (e.g., Kjellberg *et al.*, 1991).

TABLE 22–8 Energy Budgets in Calories per Individual Female *Mysis relicta* from Two Canadian Lakes, Assuming an Assimilation Efficiency of 85%[a]

Energy budget	Arctic Char Lake	Temperate Stony Lake
Ingestion[b]		
First year	62	245
Second year	180	510
Total	242	755
Growth		
First year	10	39
Second year	18	39
Total	28	78
Reproduction		
First year	0	14
Second year	8	14
Total	8	28
Respiration		
First year	43	156
Second year	127	376
Total	170	532
Egestion		
First year	9	36
Second year	27	81
Total	36	117

[a] Based on data of Lasenby and Langford (1972).
[b] Based on ingestion = growth + reproduction + respiration + egestion, when $0.85I = G + R$. An assimilation efficiency of 85% is likely an overestimate under natural conditions.

Lacking gills, mysids respire by exchange through a thin carapace. The high respiratory demands for oxygen and the requirement for cold temperatures restrict their distribution to oligotrophic lakes in which the oxygen content of the hypolimnion is not reduced appreciably.

Members of the two major genera, *Mysis* and *Neomysis,* all exhibit distinct, rapid diurnal migrations. During the day, the mysids are on the sediments or in the strata immediately overlying the sediments (Beeton, 1960; Herman, 1963). *Mysis relicta* ascends when surface light intensities decrease to about 3–0.2 μmol quanta m^{-2} s^{-1} (160–10 lux) and migrates into or below the metalimnion, in which temperatures are well below 15°C. This species rarely migrates into warmer surface waters. Descent occurred in populations of the Great Lakes when surface light intensities were increasing from 0.0002 to 0.002 μmol quanta m^{-2} s^{-1} (0.01–0.1 lux). Rates of the upward and downward migrations were similar at 0.5–0.8 m per minute over distances of at least 75 m and represent the fastest rates known among small invertebrates (see Table 16-16). *Mysis relicta* is very well adapted to the photic environment of these relatively clear, oligotrophic lakes in which light in the range of 490–540 nm penetrates the deepest. The peak photosensitivity of *Mysis* is to light at 515 nm, with a threshold response of at least 1.8×10^{-6} μmol quanta m^{-2} s^{-1} (0.00009 lux) (Beeton, 1959). This low irradiance is well within the light intensities at 100 m in lakes of low extinction coefficients even late in the daylight period. Moonlight can inhibit ascent of *Mysis* at night (Bowers and Grossnickle, 1978). Langsby and Langford (1972) indicate that the respiratory expenditure of energy for the daily migration is very small, about 3% of the total over a lifespan of 2 years.

Despite the avoidance of lighted zones, predation on mysids by planktivorous fish can be very high (e.g., Chicbu and Sibley, 1998). The mysid population structure can be appreciably decreased during abundant year classes of these fish species.

2. Isopods—Sowbugs

The isopods or sowbugs are largely marine or terrestrial; occasionally isopods become a significant part of the benthic community of lakes and streams. These small organisms (<2 cm) are flattened dorsoventrally and are well adapted to living on substrata exposed to water movements, as in streams or the littoral of lakes. Their seven pairs of walking legs are well developed. The first is modified for grasping particles. Isopods are omnivorous feeders on both plant and animal matter (Marcus *et al.*, 1978; Willoughby and Marcus, 1979), and for one species (*Lirceus*) both fungi and especially bacteria were determined to be variable and usually minor sources of organic carbon (Findlay *et al.*, 1984, 1986). Isopods are generally not active predators on other benthic fauna.

Reproduction is more common during periods of warmer temperatures but otherwise is similar to that of the mysids. The number of eggs per female is fairly high (to several hundred), and both eggs and young are retained in the brood pouch for about a month. Relatively little is known about their life cycle and population dynamics; generation time is about 8–12 months but quite variable. In an English lake, gravid females of *Asellus* were present in late winter–early spring and again during late summer. These females gave rise to spring and autumn cohorts (Adcock, 1979). Large *Asellus* of the spring cohort reproduced in the autumn, and the remainder in the following spring. The bimodal population growth characteristics showed that growth was faster, and productivity greater, in the overwintering spring cohort than was the case among the smaller individuals of the autumn cohort. Annual production (P) of the population was 3.01 g m^{-2} yr^{-1} from a mean biomass ($\bar{B}$) of 0.76 g m^{-2}. The $P/\bar{B}$ ratio of 4.0 was double that found for the same species in colder Swedish lakes (Andersson, 1969). Reynoldson (1961) has shown a strong direct correlation between the distribution of *Asellus* and the concentrations of calcium and total dissolved matter. This isopod was usually absent in soft, unproductive waters (<5 mg Ca liter^{-1}; <70 mg liter^{-1} of total dissolved matter). Isopods are common among the benthic invertebrates of saline inland waters and can tolerate high salinity (see, e.g., Eriksen, 1968; Ellis and Williams, 1970).

3. Decapods—Crayfish and Shrimps

The decapod Crustacea, of which the lobster is a familiar example, are largely marine. The few freshwater crayfish and shrimps are characterized by their approximately cylindrical body, heavily sclerotized translucent shell, and laterally compressed rostrum. Their 19 pairs of appendages include well-developed antennae and five pairs of large walking legs, the first three of which are clawed and the first greatly enlarged with a strong pincer claw used for crushing food. While the few shrimps are more or less continuous swimmers, the crayfish move slowly over the sediments and can move backward quickly by flexing the abdomen. Crayfish are omnivorous but herbivorous on algae and larger aquatic plants; occasionally, they are scavengers (Momot, 1967a, 1995; Jones and Momot, 1981). Some river species (*Orconectes*) acquire about two-thirds of their growth and production from allochthonous carbon sources and about a third from benthic invertebrates (Whitledge and Rabeni, 1997). Generally, however, crayfish can ingest large amounts of herbaceous and detrital materials while searching for and ingesting animal protein, but growth rates are best on a mixture of animal protein with other diets. Crayfish can become major competitors with fish for benthic invertebrate food sources. As sources of animal protein are exhausted, crayfish become facultative herbivores. Crayfish shift rapidly between herbivore/carnivore to scavenger/detritivore in response to food availability, a strategy that allows crayfish to maintain high production despite fluctuations in food resources.

Recently hatched juveniles of crayfish exploit lake littoral and stream riffle areas for food and shelter (Gowing and Momot, 1979; Momot 1984, 1995). Rapid growth of young of the year juveniles compensates somewhat for the high rates of predation, largely from fish. Commonly, larger juveniles migrate to deeper water as they develop and return to littoral and riffle areas as adults. As adults, much of the energy is diverted from growth to reproduction among several cohorts per year.

The lifespan of crayfish is generally fixed, although rapid growth and shorter lifespan occur among crayfish in lower latitudes (Momot, 1984). Longevity varies up to 16 years among the astacid crayfish, and commonly sexual maturity occurs during the third year and many become part of the breeding population during the fourth year. Crayfish occurring at higher latitudes and colder environments usually live longer and mature later (4–16 years). Development is more rapid among the cambarid crayfish and breeding can occur in the first year but mostly occurs in the second year (cf. Momot, 1984; Hobbs, 1991, for general details of life histories). Most species expend about 3–10 times more of ingested energy for maintenance than is expended for growth. Crayfish tend to have a high juvenile mortality with reduced adult mortality, and size-specific predation by fish on crayfish species of different sizes can lead to the eventual elimination of one of the species (DiDonato and Lodge, 1993).

Because crayfish are of some minor economic importance, considerable examination of disease and parasitic mortality has occurred (cf. Lodge and Hill, 1994; Momot, 1997). Although many bacterial and fungal disease organisms attack crayfish, many are nonlethal. Some fungal endemic and epidemic outbreaks have

TABLE 22-9 Successive Estimates of the Annual Biomass (in kg) of the Crayfish Populations of West Lost Lake, Michigan[a]

Age group	Sex	Summer 1962	Total ♂♂ and ♀♀	Spring 1963	Total ♂♂ and ♀♀	Summer 1963	Total ♂♂ and ♀♀
0	♂♂	8.7	—	6.2	—	4.2	—
0	♀♀	9.7	18.4	6.2	12.4	5.2	9.4
I (2nd year)	♂♂	36.2	—	23.7	—	41.9	—
I	♀♀	24.5	60.7	18.9	42.6	27.8	69.7
II (3rd year)	♂♂	49.7	—	25.9	—	39.0	—
II	♀♀	8.0	57.7	16.5	42.4	10.0	49.0
III (4th year)	♂♂	9.8	—	25.7	—	7.7	—
III	♀♀	3.7	13.5	1.6	27.3	1.8	9.5
Total, kg		150.3		124.7		137.6	
kg ha^{-1}		100.3		88.1		91.8	

[a] Data of Momot (1967a) with corrections for computational errors that appeared in the original publication (Momot, personal communication). West Lost Lake has an area of 1.5 ha and a maximum depth of 13.4 m (mean depth, 8.5 m).

eliminated species populations with nearly total mortality. Under natural conditions, however, evidence suggests that predation mortality, largely by fish, of the young stages is most severe.

The population dynamics and productivity of a common crayfish of temperate lakes have been studied in some detail (Momot, 1967a) and are exemplary for this group. In the crayfish population of a hardwater lake, the age structure exhibited a year-class fluctuation, and males had a higher growth rate than females. After the second year, mortality rates for females were greater than for males. Both sexes matured after the midsummer molting in the second year, then mated until early autumn. Eggs were laid the following spring. The average number of eggs carried by females in the spring, a measure of the reproductive capacity, was 58% of the eggs carried in the preceding fall. Within the maximum lifespan of 3 years of this species, *Orconectes virilis*, 2-year-old females produced most (92.5%) of the eggs. Young remained in shallow water after hatching in spring or early summer, but adult females molted and then migrated to deeper waters (3–7 m), where most remained all summer. Adult males also migrated, somewhat later than females, to deeper water. Population size appeared to be regulated by natural mortality at molting and cannibalism of both males and females on each other at high population densities. Predation by fish was not an important population control mechanism in this and other lakes (Momot, 1967b; Momot and Gowing, 1977a; Gowing and Momot, 1979) but can be particularly important in stream environments, as in pools, where refuge is difficult (e.g., Englund, 1999). A number of pollutants of lakes and streams, particularly those that reduce oxygen content or alter the substratum, cause changes in the population distribution or seasonal migrations (Hobbs and Hall, 1974). The successive year-class estimates of the crayfish biomass of this hardwater lake are given in Table 22-9. Seasonal changes in biomass were small because of compensatory changes among age groups. The estimated net production of the crayfish population was 12.7 g m^{-2} between the summer of 1962 and that of 1963.[1] These values are about intermediate within the range of biomass values found among crayfish populations studied in other waters (cf. Table 22-10).

In many lakes, crayfish can dominate the annual production of the benthic animal biomass, at times reaching 100–140 g m^{-2} and densities of 15 m^{-2} (Momot *et al.*, 1978). The rate of population growth, reproductive capacity, age at maturity, and lifespan are primarily density-dependent. Populations often regulate their development by control of brood stock size. Highest production rates are achieved by species with low mortality rates that coincide with the period of greatest growth (early age/size groups) (Momot, 1984). The annual production of crayfish populations ranges from one to several hundred g m^{-2} yr^{-1} in natural environments, but despite high biomass, low growth rates result in modest to low production (Table 22-10). In high-energy, high-nutrient environments or under artificial conditions where predation is minimized, much higher values can be obtained among species with short life cycles, high fecundity, and fast growth. Turnover

[1] A corrected value; see footnote to Table 22-9.

TABLE 22-10 Comparison of the Annual Production (P), Mean Biomass ($P/\bar{B}$) and Production/Biomass Ratios for Crayfish and Shrimp Populations[a]

Habitat/species	*Production* ($g\ m^{-2}\ yr^{-1}$)	*Mean biomass* ($g\ m^{-2}$)	$P/\bar{B}$ *ratio*
Crayfish			
Ontario (small lakes) *Orconectes virilis*	2.3–4.9	2.3–3.5	1.2–1.9
Michigan (small lakes) *Orconectes virilis*	1.7–14.2	3.2–14.3	0.5–1.5
Michigan (river) *Orconectes propinquus*	41.5	44.0	0.9
Oregon (river) *Pacifastacus leniusculus*	13.7	14.3	0.9
Lithuania (lake) *Astacus astacus*	4.2	6.0	0.7
England (stream) *Austropotamobius pallipes*	17.7	31.9–49.5	0.3–0.4
North Carolina (stream) *Cambarus bartonii*	0.96	1.67	0.6
Australia (ponds) *Cherax destructor*	67	32	2.1
Australia (ponds) *Cherax tenuimanus*	190–610	71–366	2.7–5.2
Portugal (river) *Procambarus clarkii*	27.2	5.4	5.0
California (river) *Pacifastacus leniusculus*	44.4	62.2	0.7
Shrimp			
Lake Sibaya, South Africa *Caridina nilotica*	24.4	2.69	9

[a] Modified from Momot (1984), after numerous sources, Hart (1981), Huryn and Wallace (1987), and Anastácio and Marques (1995).

rates, estimated from the ratios of production to mean biomass, are mostly around 1, but vary from 0.3 to 7 (Table 22-10).

The feeding and predatory activities of crayfish can have complex, multitrophic-level effects on food webs of streams and lake littoral areas where dense populations develop. For example, dense crayfish can significantly reduce other benthic invertebrates that graze on periphyton, and the reduction of grazing pressure can result in marked increases in periphyton primary production (Charlebois and Lamberti, 1996). In other cases, large attached algae, such as *Cladophora*, can be fed upon directly by the crayfish, which results in altered habitat for benthic invertebrates (Creed, 1994). The grazing activities of crayfish on submersed macrophytes and seedling stages of emergent and floating-leaved macrophytes, either directly or by damaging the plants during predation on macrophyte-associated invertebrate prey, can suppress macrophyte abundance appreciably (Lodge and Lorman, 1987; Nyström and Strand, 1996; Nyström *et al.*, 1996). At very high densities of crayfish, some species of macrophytes can be eliminated. There is some circumstantial evidence for macrophyte species selectivity by crayfish.

4. Amphipoda—Scuds

The amphipods or scuds, also chiefly marine, are represented in fresh waters by a few important species. With the exception of *Monoporeia*,[2] discussed further on, the amphipods live on the sediments. Most species are small (5–20 mm), with a laterally compressed, many-segmented body. At the base of their seven pairs of thoracic legs many have gills that, like the lateral gills on some species, are exposed to currents of water

[2] Previously *Pontoporeia affinis* (see Bousfield, 1989). Also *Hyalella montezuma*, endemic to a spring in Arizona, is a multivoltine pelagic amphipod that migrates similarly to *Monopeoreia* but has modified swimming appendages that allow it to filter-feed pelagically (Blinn and Johnson, 1982; Dehdashii and Blinn, 1991). This species of *Hyalella* has higher production rates than most other amphipods and $P/\bar{B}$ ratios between 3.7 and 5.5.

created by the beating of pleopod appendages on the abdomen. Amphipods have high oxygen requirements and are usually restricted to waters of high dissolved oxygen concentrations (Franke, 1977). Amphipods are generally omnivorous substrate feeders that consume bacteria, algae, fungi, and animal and plant remains (Minckley and Cole, 1963; Marzolf, 1965a; Hargrave, 1970a; Bärlocher and Kendrick, 1975; Kostalos and Seymour, 1976; Moore, 1977). Several amphipods possess digestive enzymes (cellulases, glucanase, and chitinase) that are capable of degrading plant and fungal cell walls (Chamier and Willoughby, 1986). Some amphipods, particularly *Gammarus* spp., are likely much more predacious on living animals and cannibalistic than previously believed (cf. MacNeil *et al.*, 1997).

In a detailed study of the population dynamics of the amphipod *Hyalella azteca* in a eutrophic lake in southeastern Michigan, Cooper (1965) was able to estimate production rates and biomass turnover of the natural populations (Table 22-11). The amphipod populations showed positive intrinsic rates of increase at temperatures above 10°C, with maxima between 20 to 25°C. Reproduction resumed again when water temperatures reached 20°C in the spring. Estimates of mortality rates based on size and age structures of the population indicated that a size-specific mortality factor was operating on the large, reproductively mature amphipods during the summer months. It was established that the highly size-specific feeding behavior of the yellow perch accounted for most of the mortality in the adult amphipod size class. A similar situation was found in the benthic invertebrate populations of shallow experimental ponds (Hall *et al.*, 1970), in which the fish (*Lepomis*) preyed upon larger particles, selectively removing the larger adult stages of *Hyalella* and the terminal aquatic stages of many benthic insects. There is some evidence that heavy predation pressure by fish can lead to genetic selection for smaller-sized adult amphipods (Strong, 1972; Wellborn, 1994). *Gammarus* species are also dominant macroinvertebrate prey of many fish both as a seasonal food source and as a year-round staple, but fish also feed selectively on larger stages and sizes (MacNeil *et al.*, 1999). Cannibalism is common within *Gammarus* spp., and intraguild predation by closely related but competing amphipod species and other invertebrates (e.g., flatworms, stoneflies, and crayfish) is also common. The impact of nonpiscean predators, particularly macroinvertebrates, on *Gammarus* may frequently be stronger than that of fish, amphibians, and water birds.

In spite of the relatively low biomass levels of the *Hyalella* populations, rapid growth rates and short generation times (33 days to maturation at 25°C; 98 days at 15°C) of the animals result in a mean production rate of about 0.13 kg ha^{-1} day^{-1} (0.013 g m^{-2} day^{-1}), which is significant in comparison to other components of the benthic fauna. The observed turnover rate of about 3% day^{-1} is close to the maximum of 4% day^{-1}, estimated from experimental studies, that the amphipod population could maintain over the course of the summer at temperatures above 20°C with any degree of population stability. Similar production rates were found for amphipods of temperate streams (Welton, 1979; Waters and Hokenstrom, 1980). For example, productivity of *Gammarus pseudolimnaeus* in a river in Ontario was 2.94 g DM m^{-2} yr^{-1}, and with a relatively modest mean biomass (0.63 g DM m^{-2}) exhibited a moderately high turnover rate ($P/\bar{B}$ of 4.65) (Marchant and Hynes, 1981).

Extensive studies of *Hyalella azteca* in shallow (z_m = 5 m), riverine Marion Lake, British Columbia, demonstrated that the growth, density, and body size of this deposit feeder depended on the quantity of epipelic algae and sediment microflora (Hargrave, 1970a–c). The highest concentrations of sedimentary chlorophyll and microflora, as well as the lowest

TABLE 22-11 Estimated Mean Production Rates and Numerical Turnover of the Amphipod *Hyalella azteca* Population from Four Stations of Sugarloaf Lake, Southeastern Michigan, May to October[a]

Age group	*Number of animals* m^{-2}	*Mean dry weight* (μg)	*Numerical yield* m^{-2} day^{-1}	*Mean population density* (mg m^{-2})	*Percent numerical turnover* day^{-1}
11–13 antennal segments	1259.4	15.3	39.8	19.3	3.2
14–16 antennal segments	1383.0	35.6	24.8	49.2	1.8
New adults	2044.6	133.3	71.8	272.5	3.5
Old adults	183.8	200.0	4.8	36.8	2.6
Total	4870.8	—	141.2	—	2.9

[a] Compiled from data of Cooper (1965).

concentrations of nondigestible lignin-like organic material, occurred in the upper 2 cm of sediment, which was also the limit of the vertical distribution of the *Hyalella* in the sediments. Increased growth of *Hyalella* during June was independent of temperature in Marion Lake but was correlated closely with increased rates of epipelic primary productivity. Egg production began in May as growth rates increased, and the maximum density of *Hyalella* was reached in August.

Bacteria and algae, except for cyanobacteria, were assimilated with an efficiency of about 50%; cellulose and lignin-like compounds were not assimilated at all by *Hyalella* (Hargrave, 1970a). In contrast, the amphipod *Gammarus* could digest and metabolize about 30–40% of the cellulose ingested (Chamier, 1991). Overall, sediment and associated microflora were assimilated with a 6–15% efficiency. In the estimated energy budget for adult *Hyalella* at 15°C in Marion Lake, 49% of calories assimilated were respired, 36% were lost as soluble excretory products, and 15% were accumulated as growth, egg production, and molts (Table 22-12). The fecal pellets produced were colonized rapidly by heterotrophic microorganisms, and the dissolved organic compounds excreted by *Hyalella* significantly increased the rate of recolonization of the fecal material (Hargrave, 1970c). Epipelic algal production was stimulated somewhat by the feeding activities of the amphipod at natural densities, even though <10% of the daily microfloral production was required to supply the energy necessary for observed rates of amphipod growth, respiration, and egg production. The mechanism of stimulation of algal productivity on the sediments is unclear. It may emanate directly from inorganic and organic nutrients from the amphipod excretions, indirectly from increased metabolism of the heterotrophic microorganisms of the sediments, or both. Alternatively, the mechanical disruption of the oxidized microzone of the sediment–water interface may be instrumental in encouraging release of certain reduced, and more soluble, nutrients into the proximity of the epipelic algae. Although these compounds would not migrate far into overlying aerobic water before being oxidized and their solubility reduced, the algae directly at the interface could benefit from such disturbance.

TABLE 22-12 The Estimated Energy Budget of Adult Amphipod *Hyalella azteca* of 700 μg Dry Weight at 15°C Feeding on the Surface Sediment of Marion Lake, British Columbia[a]

Energy	*Calories amphipod*$^{-1}$ *hr*$^{-1}$	*% of total*
Ingestion	0.0525	50.7
Production (growth, molts, and egg production)	0.0012	1.2
Respiration	0.0039	3.8
Egestion	0.0430	41.5
Excretion (soluble organic matter)	0.0029	2.8

[a] From data of Hargrave (1971). The mean caloric value of adult *Hyalella* was 3850 cal g^{-1} ash-free dry weight; therefore, a 700-μg amphipod contains 2.7 calories.

Fungi colonizing autumn-shed tree leaves are the preferred food of many woodland stream-dwelling amphipods (Bärlocher and Kendrick, 1975; Kostalos and Seymour, 1976; Sutcliffe *et al.*, 1981). Less than 20% of the ingested leaf material was assimilated, whereas 75% or more of the fungi of decomposing leaves were assimilated in some cases. Assimilation efficiencies vary greatly with the amounts of microbiota attached to and within detrital leaf materials from different plant species. As a result, growth rates of the amphipods feeding on detritus from leaves, such as oaks that are much more resistant to decay, were slower than those feeding on microflora and detritus of more labile leaves, such as from elm or aquatic plants.

Monoporeia affinis is exceptional among the amphipods in that it migrates extensively into the pelagic zone at night. The behavior and habitat requirements of *Monoporeia* are similar to those of *Mysis*, discussed earlier. *Monoporeia* is restricted to cold, relatively oligotrophic waters and migrates only into the upper hypolimnion and metalimnion, in which temperatures are usually <15°C (Marzolf, 1965b; Wells, 1968). Only a portion of the predominantly benthic population migrates into the water. During daytime, *Monoporeia* is largely on and in the sediments or in the immediately overlying water.

Monoporeia is a burrowing amphipod and a deposit feeder on detritus and algae of surficial sediments (Johnson, 1988). This amphipod is the predominant macrobenthic invertebrate of the upper Laurentian Great Lakes. In Lake Michigan, *Monoporeia* showed no significant correlation with depth, particle size, or organic matter content of the sediments (Marzolf, 1965a). Distribution of *Monoporeia* was, however, positively correlated with the number of bacteria in the sediments, which was correlated directly with larger brood sizes, smaller sizes at maturity and hence more rapid growth rates, and higher production rates (Siegfried, 1985). In experiments with different-sized sediment particles, *Monoporeia* selected sediments with a particle size <0.05 mm. Such sediments contained more bacteria and organic matter than coarser sediments. Extensive recent experiments, however, indicate that juvenile *Monoporeia* acquire only small portions

(<10%) of their carbon and energy requirements from bacteria, and much of the energy is derived from lipids of algal origins that reach the sediments (Goedkoop and Johnson, 1994). Productivity can be significant. For example, *Monoporeia hoyi* of Lake Huron averaged 1.15 g DM m^{-2} yr^{-1}, and with a mean biomass of 0.82 g DM m^{-2} had a modest $P/\overline{B}$ ratio of 1.4 (Johnson, 1988).

Because bacteria and detrital algae of the sediments are the predominant food source of *Monoporeia*, and there is no evidence for planktonic feeding, the adaptive significance of the nocturnal vertical migration of a small portion of the population is unclear. Predation can be high at night, such as by the predatory leech *Erpobdella*, which migrates nocturnally in synchrony with *Monoporeia* and feeds vigorously on the amphipod (Blinn and Davies, 1990). Marzolf (1965b) suggests that the migrations have adaptive value in maintaining genetic continuity across otherwise isolated benthic populations. Migration encourages dispersal and, possibly, copulation with individuals from populations from other areas within the community.

N. Molluscs

The freshwater Mollusca are separable into two distinct groups, the univalve snails (Gastropoda) and the bivalve clams and mussels (Bivalvia or Pelecypoda). Molluscs are soft-bodied and unsegmented with a body organized into a muscular foot, a head region, a visceral mass, and a fleshy mantle that secretes a shell of proteinaceous and crystalline calcium carbonate materials.

1. Gastropoda—Snails

Among the gastropod snails, the univalve shells generally are spirally coiled, while those of the few freshwater limpets are conically shaped. The gastropod mollusc head is distinct, with a pair of contractile tentacles and a ventral mouth with a chitinous, sclerotized jaw and a chitinous internal radula containing numerous transverse teeth. The radula is extended from the mouth and moves back and forth rapidly, scraping and macerating food particles. Respiration in the snails occurs by gills in many aquatic forms (Prosobranchia) and by pulmonary cavities or "lungs" in the pulmonate snails. Some of the most common pulmonate snails, however, are able to stay submersed for a very long time or for their entire life cycle without filling their pulmonary cavities with air. Cutaneous respiration through the body membranes is common to all freshwater snails (Ghiretti, 1966). Even though the lung is retained in pulmonate freshwater snails, it is of secondary function and is used only when the oxygen concentration drops to low values. Then, the snails come to the surface to breathe air while moving or floating. Other pulmonates have water-filled lungs through which gaseous exchange occurs, much the same as it does with a gill. Locomotion in snails is by muscular movements of the ventral surface of the body, the "foot."

The gastropods are separated into two distinct groups (Brown, 1991; Hutchinson, 1993). The subclass Prosobranchia has a gill and a flexible or calcareous operculum that is pulled into the shell aperture to protect the snail. The Pulmonata have a modified portion of the mantle cavity as a lung and lack an operculum. The prosobranch snails are usually dioecious, but there are many variations and a few species are hermaphroditic. The pulmonates are all monoecious. Nearly all gastropods live in standing waters; only one genus (*Acochlidium*) inhabits running water.

Gastropods feed by rasping movements of the radula against a substratum; retractors pull food particles back into the mouth and grind them against the roof of the mouth. Most snails rely on algae, detrital particles, and bacteria of the periphyton on submersed substrata. Claims of food selectivity by snails are frequently made and, although most are poorly substantiated by essential rigorous evaluations of chemical cues and receptors, likely occurs among microhabitats containing different dominant algal sources (cf. Kolodziejczyk, 1984; Lodge, 1985, 1986; Brendelberger, 1997a). Direct effects of grazing on living macrophytes are probably of minor importance, although a few snails consume submersed macrophytes in sufficient quantities to alter growth of the higher plants (Sheldon, 1987; Brönmark, 1989). However, snails can have a significant indirect effect on macrophytes by reducing the detrimental effects of epiphytic microbiota that shade and compete for nutrients from the water. Gastropods possess cellulases that assist in degradation of algae and detrital particles from plant sources (Brendelberger, 1997b). Some prosobranch gastropods can trap food particles from their respiratory current in mucus, which is then ingested.

The efficiency of assimilation (expressed as the percentage of ingested organic carbon that is assimilated) of the stream snail *Potamopyrgus* varied from 3.7 to 9.0%, depending on the type of sediment used as food (Heywood and Edwards, 1962). The average was 4% for adult snails, in which about 7 μg of organic carbon was assimilated per snail per day. Diatoms, a primary food of this snail, are crushed between sand grains, which are ingested with the food, and macerated by the strongly cuticularized stomach walls (Frenzel, 1979). In contrast, the tropical snail *Pila*, feeding on the macrophyte *Ceratophyllum*, had much higher assimilation rates, possibly >50% (Vivekanandan *et al.*, 1974).[3] Utilization efficiency of the plant food decreased as the supply of food increased. Assimilation efficiencies of diatoms, green algae, and cyanobacteria ingested by several gastropods varied between 30 and 73% and increased slightly with increased temperatures (Streit, 1975a,b; Brendelberger, 1997). Ingested cyanobacteria (*Anabaena*) were not assimilated by *Ancylus*; similarly, the green alga *Chlorella* was not assimilated by the tropical snail *Bulinus* (van Aardt and Wolmarans, 1981).

Pulmonate snails are moderately tolerant of desiccation and low levels of dissolved oxygen through the exchange of oxygen across membranes and anaerobic metabolism (McMahon, 1983). Because of the requirements for appreciable calcium for shell generation, most molluscs are found in waters with a dissolved calcium concentration >5 mg liter^{-1}. Above this concentration, little relationship exists between the chemical parameters and gastropod diversity, and habitat substratum and food availability free from major or frequent disturbance becomes important (Lodge *et al.*, 1987). Interspecific competition in natural communities of freshwater snails has not been demonstrated well and remains unclear; food and niche overlap is relatively narrow in highly heterogeneous and productive habitats (Brown, 1982, 1991; Strong *et al.*, 1984).

Predation on snails can markedly affect snail abundance. Certain fish adapted to crush snails, especially certain sunfish centrarchids, and crayfish are particularly effective in consuming snails that are not able to find cryptic refuge among aquatic vegetation (e.g., Vermeij and Covich, 1978; Brown and DeVries, 1985). Variations in thickness of snail shells, either within a species or between different species, can selectively alter predation vulnerability, as the energetic costs to predators increase as shell thickness increases (Stein *et al.*, 1984). Snails can crawl above the water line for several hours to avoid predation, particularly from crayfish (e.g., Alexander and Covich, 1991) or to deposit eggs on moist soil less vulnerable to invertebrate predation. Very little is known of the effects of parasites on snail dynamics and productivity. Clearly, parasites alter snail growth, and infections can often accelerate resource allocation to reproduction (Brown, 1991).

Many examples of population dynamics of snails exist in the literature, in part because of the relative ease of sampling and following changes in population structures over time. There is unquestionable correlative and experimental evidence to demonstrate that increased food resources commonly lead to enhanced growth, fecundity, and productivity and reduced generation times (e.g., Bruky, 1971; Brown, 1985; Moeller *et al.*, 1998). There is also evidence that restricted food availability can be commonly more important to limitation of somatic and reproductive growth than is predation (Osenberg, 1989; Jokela, 1996). A few examples illustrate the general sequences in the population development and differences in spatial and temporal separation.

The life cycle of freshwater snails in temperate regions, particularly the smaller species, tends to be annual (Harman, 1974). One reproductive period may occur in the spring or fall, or two or more reproductive periods may occur throughout the summer, during which the original cohort is replaced or supplemented. Some species overwinter as juveniles or adults and reach maturity the following spring or summer. Larger species tend to have life cycles that extend over 2, 3, or even 4 years. A few examples of the growth and population characteristics of some freshwater snails and clams illustrate these variations.

In the common pond snail *Physa gyrina* of the midtemperate region, copulation occurred in April (DeWitt, 1954). Egg masses adhere to substrata, and the number of eggs laid, about 100, varies with the size of the snail. After about a week of embryological development at 20–30°C, the young snails hatch and grow very rapidly, attaining most of their adult size of approximately 12 mm in about 8 weeks. Beyond this period, growth is very slow for the remainder of the 12- to 13-month lifespan. Growth in *Physa* is not determinate, but continues throughout the life of the individual.

Analyses of the population-size classes of the pulmonate snail *Lymnaea* that lives along the shore line in littoral regions of fresh waters showed a similar pattern, in which growth of overwintering individuals was linear but slow and then very rapid in the spring and early summer (McCraw, 1961, 1970). Little growth occurred before mid-April or after early July, when most of the production is shunted into reproductive development. Small individuals overwinter, and almost all of the spring and summer population is derived from snails ovipositing in later summer or autumn of the previous year. Although changes in population densities of *Lymnaea stagnalis* did not affect rates of food consumption, egg production under experimental conditions was significantly lower at high population densities (Mooij-Vogelaar *et al.*, 1973). Eisenberg (1966) found in both natural and experimental populations of *Lymnaea elodes* that both the quantity and quality of food limited the population densities. Adult fecundity

[3] The methods employed in measurements of changes in plant matter are questionable, and the reported values of loss most likely are overestimates.

TABLE 22–13 Life History Characteristics of Bivalve Mollusca[a]

Trait	*Unionacea*	*Sphaeriidae*	*Corbicula fluminae*	*Dreissena polymorpha*
Lifespan	<6–>100 yr	<1–5 yr	1–5 yr	4–7 yr
Age at maturity	6–12 yr	>0.2–<1.0 yr	0.25–0.75 yr	1–2 yr
Reproduction mode	Gonochoristic (two sexes)	Hermaphorditic	Hermaphroditic	Gonochoristic
Fecundity (young/average adult/breeding season)	200,000–17 × 10⁶	1–135	35,000	30,000–40,000
Growth rate	Rapid prior to maturity	Slow relative to other bivalves	Rapid throughout life	Rapid throughout life
Turnover time	1800–2850 days	Commonly <80 days	73–91 days	50–870 days

[a] Extracted from McMahon (1991).

increased some 25-fold and numbers of young increased 4- to 9-fold when abundant food of high quality was available. Mortality among natural populations of young snails, caused especially by dipteran larvae, was very high (93–98%), but food limitations on fecundity were considered to be the primary mechanism affecting the final population structure.

Comparative life cycle patterns of 13 species of snails were evaluated in the littoral community of a small, shallow eutrophic lake in Connecticut (Jokinen, 1985). The timing and duration of oviposition and juvenile recruitment varied considerably (Fig. 22-3). Four species had a well-defined annual life cycle, three species had almost continual recruitment from spring through autumn, and six species had two or three well-defined reproductive periods per year. Closely related species tended to differ the most in their reproductive periods or duration of oviposition. The division of the reproductive season (April through November) in this lake likely reduced interspecific competition and allowed snails to avail themselves of differences in seasonal variations in food resources.

2. Bivalvia—Clams and Mussels

The body of clams or mussels is enclosed in two symmetrical, opposing shell valves and lacks a head, tentacles, eyes, jaws, and a radula. The body is enclosed by membranous tissue, the mantle, which secretes the shell valves. At the posterior end of the clam, the mantle has two openings, the siphons. The lower ciliated siphon draws water into the body cavity. Thus, flow aerates the gills and carries in food particles. Food particles are removed from the water by filtration through the gills and cilia. Water then exits from the upper siphon. The muscular foot can be extended from the valves in front of the clam, implanted in the sediment, and then contracted to draw the animal forward. The food of the clams consists primarily of particulate detritus, microzooplankters, and phytoplankton (see reviews of Fuller, 1974; McMahon, 1991). Clams utilize living phytoplankton; living and active diatoms apparently pass through the alimentary canal unaffected. Some bivalve species supplement filter feeding by consuming organic detritus and associated microbiota of sediments by pedal feeding. Burrowing activities disturb particles that are then directed by cilia toward the mantle cavity, where they are drawn in. As much as half or more of the total organic carbon assimilated may arise from pedal deposit feeding mechanisms (Burky, 1983).

Bivalve molluscs tend to be long-lived and expend appreciable energy in high reproductive rates (Table 22-13). Nonpredatory mortality is known to occur but is poorly understood. Predatory losses can be very high by some large invertebrates (e.g., crayfish) and many vertebrates, particularly by shore birds and ducks, by fish, especially molluscivorous fish such as the carp (*Cyprinus carpio*), channel catfish (*Ictalurus punctatus*), and drum (*Aplodinotus grunniens*), and many semiaquatic mammals (Fuller, 1974). Two exotic species (Asian clam *Corbicula fluminae* and the zebra mussel *Dreissena polymorpha*) introduced recently into North America have caused massive destruction of habitat, reduced biodiversity of other molluscs, and appreciable economic damage. As a result, there has been a spate of recent bivalve literature on the ecology and control of these two species (cf. separate reviews, e.g., McMahon, 1991; Nalepa and Schloesser, 1993).

Filtration rates generally decrease with increasing particle concentration and decreasing temperatures. For example, the filtration rates and ingestion rates of a filter-feeding mussel (*Dreissena*) remained constant when the mussel was fed diatoms (*Nitzschia*) until high food concentrations were reached, above which rates declined (Walz, 1978a). The growth rate increased with rising food concentrations until a plateau was reached at ca. 2 mg C liter^{-1} and achieved a maximum assimilation efficiency of 40% (Walz, 1978b). In both experimental and field studies, high concentrations of inorganic suspended sediment, particularly above 1 mg liter^{-1},

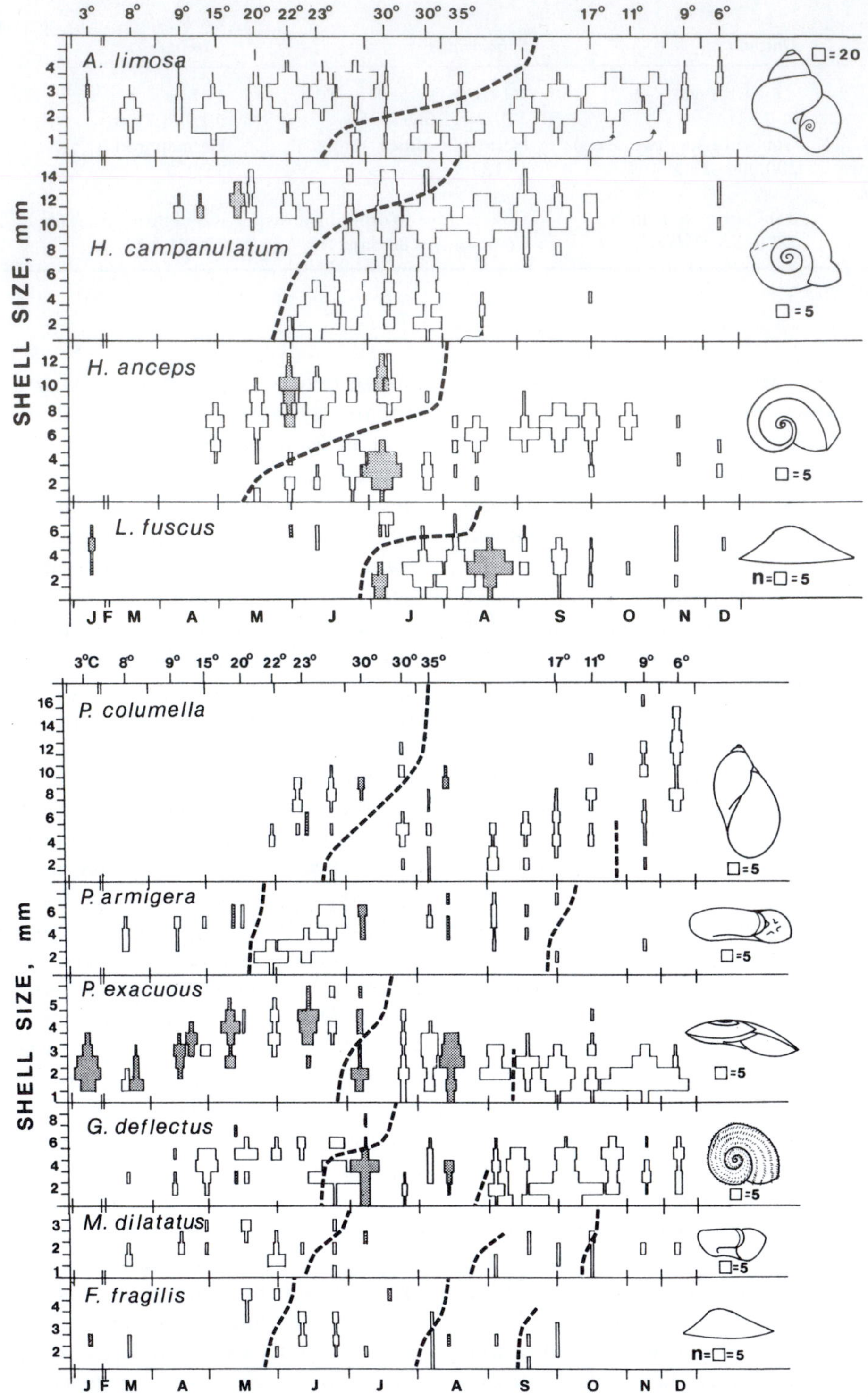

FIGURE 22-3 Life history patterns of snail populations with annual (*upper*) and with more than one (*lower*) generation produced per year in Roseland Lake, Connecticut, 1979–1980. Dashed lines separate generations. Shaded data bars from 1980 were added only when necessary to supplement 1979 data. (From Jokinen, 1985.)

severely reduce filtration rates, ingestion rates, and assimilation efficiencies (e.g., Madon *et al.*, 1998).

Bivalve molluscs can achieve high densities. For example, unionaceans of 10–30 individuals m^{-2} are not unusual. With filtering rates of ca. 300 ml g DM^{-1} h^{-1}, filtering rates in the range of 15–120 liters m^{-2} d^{-1} can remove appreciable amounts of seston and phytoplankton. The lower limit for filtration is about 0.5–1 μm, and as a result many bacteria are not removed and can be stimulated by nutrient cycling from the mortality of phytoplankton (Cotner *et al.*, 1995). Although these filtering rates are prodigious, normal turnover rates and replacement rates of phytoplankton can easily compensate for the losses from bivalve filtering. For example, seston was reduced by less than half of 1% of the total removed by planktivorous feeders in a eutrophic lake as a result of filtration by dense unionacean populations (Kasprzak, 1986). Seasonal energy budgets indicate that much >90% of energy consumption is expended in metabolic costs (Stoeckmann and Garton, 1997). In Lake Erie, with densities between 10,000 and 50,000 mussels m^{-2}, the zebra mussels potentially ingest between 10 and 50% of the summer primary production of phytoplankton.

When bivalve populations reach extreme densities, commonly exceeding 50,000 per m^{-2}, however, as in the case of *Corbicula* and *Dreissena*, many impacts on the river or lake ecosystem occur. For example, densities of unionid clams can decline markedly (90–100%) as a result of competition for food as well as by fouling as the mussels overgrow the larger clams (Ricciardi *et al.*, 1996; Strayer and Smith, 1996; Pace, 1998). Major reductions of phytoplankton densities and chlorophyll concentrations as well as microzooplankton (ciliates, rotifers, and copepod nauplii) can result. Some of the seston reductions from mussel filtration have been remarkable. For example, with the invasion of *Dreissena* into nutrient-enriched river and lake ecosystems, instantaneous phytoplankton biomass decreased by 85% or more in only a few years (e.g., Reeders *et al.*, 1989; Nicholls and Hopkins, 1993; Fahnenstiel *et al.*, 1995; Effler *et al.*, 1996; Caraco *et al.*, 1997). The increased transparency and transmittance of light can shift photosynthetic productivity from phytoplankton dominance in eutrophic environments to benthic algal proliferation (e.g., Lowe and Pillsbury, 1995). Not only is the nutrient turnover and accumulation in the sediment communities greater under such high mussel densities but the enhanced food availability at the sediments can increase other benthic macroinvertebrates as well, in spite of greater competition for habitat space (Stewart and Haynes, 1994). As the organic matter is removed and concentrated into pseudofeces that are directed toward the sediments, large quantities of organic nutrients are removed as well. Marked reductions of biogenic turbidity can result, which in turn reduce the rates of seston decomposition in the water column and rates of consumption of dissolved oxygen. No significant effects on the nutrients of the overlying water column are apparent from the intense benthic feeding and nutrient sequestering.

Freshwater bivalve molluscs are nearly all ovoviviparous and brood embryos through early development stages in the gill. Fully formed juveniles are released from the exhalant siphon in sphaeriids. The sperm of male clams is shed into the water and drawn into the female mantle cavity by the normal water current of the siphon system. Fertilized ova are incubated in portions of the gills, but mortality is very high from bacteria and protozoans. An important exception is the zebra mussel (*Dreissena polymorpha*), which releases both sperm and eggs externally, followed by external fertilization and development of a free-swimming veliger larval stage. The primary stimulus for reproduction appears to be the temperature above a critical level (McMahon, 1991).

In mussels of the Unionacea, the mature larvae hatch after periods of a few months to a year to temporarily free-living forms, called ***glochidia***, which have two chitinous valves bearing ventral hooks in some species. Glochidia development in temperate lakes occurred during the summer and was completed by autumn (Jokela *et al.*, 1991). Glochidia are stored in the gill blades until spring and are then discharged in large numbers, fall to the sediments, and die within a few days if they do not come into contact with a suitable fish host. If they come into contact with a fish, the glochidia attach to surface membranes or to the gills; some species of clams are specific to a particular species of fish. During this parasitic stage, which lasts from about 10–30 days, internal development occurs, although there is little change in size while the glochidia are encysted on the host fish. Juveniles then break encystment and fall to the sediments. Development ensues if substrata and food conditions are within tolerable limits. Large clams live 15 years or longer, while the smaller "fingernail clams" (Sphaeriidae) have a longevity of about a year or slightly longer. Mortality is exceedingly high in the egg, larval–glochidial, and juvenile stages. The adult populations and their distribution can be modified greatly by fish predation, certain birds, and a few mammals such as the muskrat (cf. Fuller, 1974).

Many freshwater bivalve molluscs can tolerate hypoxia or even anaerobic conditions, such as occurs in the hypolimnia of eutrophic lakes. Anaerobic molluscan metabolic pathways degrade glycogen and amino acids, particularly aspartate, to alanine and succinate, and then to volatile fatty acids (de Zwann, 1983). Some 5–6 moles of ATP are produced per mole of glucose, which is considerably higher than glycolysis at 2 moles ATP per mole of glucose catabolized. Acidic anaerobic end products accumulate under these conditions, however, and these are neutralized by erosion of the calcareous shells. Many species of both sphaerid and unionacean molluscs can tolerate long periods (months or even years) of desiccation in air as water levels change.

Study of the distribution of five species of sphaeriid clams of the common genus *Pisidium* showed that four coexisted among the more heterogeneous substrata of Lunzer Untersee, Austria (Hadl, 1972). The profundal sediments were dominated by a single species. The littoral species were exposed to large seasonal fluctuations in temperature and were gravid in spring and released young in the summer, during which rapid growth occurred. The new generation overwintered to reproduce the following spring. In contrast the deepwater species, which were exposed to relatively constant temperatures throughout the year, exhibited no definite reproductive cycle. Embryos were found throughout the year in mature females. The larger clams (e.g., the *Anodonta*), migrate seasonally on the sediments to different water depths (Burla, 1971). Species of this genus were found to move up to the upper littoral in spring and early summer, where they remained until late in the year. In autumn and early winter, most returned to deeper water. These movements were not correlated with temperature but rather with photoperiod, which may be related more to food availability. Although growth is reduced and development times lengthened at the colder temperatures of deeper waters (e.g., Hanson *et al.*, 1988a; Holopainen and Jónasson, 1989), these effects could be ameliorated by migration seasonally to areas of warmer water and greater food abundance.

3. Mollusc Production Rates

Production rates and turnover values of freshwater molluscs have been studied frequently. Despite low instantaneous biomass, the rapid growth and short life cycles of pulmonate snails yield relatively high production rates and short turnover times (Table 22-14). In contrast, long life cycles result in low production rates in spite of large biomass. Production-to-biomass ratios tend to be low (0.5–2) in molluscs but under exceptionally productive conditions can exceed 5 (Table 22-14). For example, detailed population and growth analyses of the snail *Lymnaea catascopium* in a lake in Québec showed considerable variability in the littoral areas (Pinel-Alloul, 1978). The mean biomass of this

TABLE 22–14 Examples of Production, Mean Biomass, and Ratios of Production:Biomass (*P/B*) for Populations of Gastropod and Bivalve Molluscs

Species	*Habitat*	*Mean productivity* (*g AFDM* m^{-2} yr^{-1})	*Mean biomass* (*g AFDM* m^{-2})	*P/B ratio*	*Source*
Gastropoda					
Elimia cahawbensis	Alabama stream	0.5	2	0.25	Richardson *et al.* (1988)
E. clara		1.5	5	0.3	
Lymnaea catascopium	Lake St. Louis, Québec	6.6–17.8	3.4–7.4	1.9–2.4	Pinel-Alloul (1978)
Viviparus subpurpureus	Louisiana bayous	20	10	2.0	Richardson and Brown (1989)
Campeloma decisum		40	20	2.0	
Pulmonate snails[a]	Many	4.36 ± 1.8	2.05 ± 1.0	2.1	Russell-Hunter and Buckley (1983)
Prosobranch snails[a]	Many	2.72 ± 0.4	9.71 ± 3.8	0.3	
Bivalvia					
Elliptio complanata	Mirror Lake, New Hampshire	0.06	0.52	0.12	Strayer *et al.* (1981)
Anodonta piscinalis	Lake Mikolajskie, Poland	2.2	6.6	0.35	Lewandowski and Stańczykowska (1975)
A. grandis	Narrow Lake, Alberta	1.2	4.5	0.27	Hanson *et al.* (1988b)
	Lake 377, Ontario	0.06	0.36	0.17	Huebner *et al.* (1990)
Dreissena polymorpha	Lake Constance, Germany	10.7	1.6	6.7	Walz (1978c)
Pisidium variabile	Ponds, southwest Ohio	1.8	0.54	3.3	Way (1988)
P. compressum		4.2	1.09	3.8	
Corbicula fluminea	Ogeechee River, Georgia				
	River sections	1.2–1.7	2.1–3.4	0.5–0.6	Stites *et al.* (1995)
	Population centers	9.5–22.9	10.3–12.9	0.9–1.8	

[a] Averages ± SD of 4–10 species and using 47.8% C content of AFDM (Benke *et al.*, 1999).

species varied between 3.4 and 7.4 g m^{-2}; production rates ranged between 6.6 and 17.8 g m^{-2} yr^{-1}. Turnover ratios ($P/\overline{B}$), however, were relatively constant at approximately 2 among the different populations and were found to be similar to those of other snail populations (e.g., Browne, 1978; Holopainen, 1979; Frenzel, 1980). Of two annual cohorts of a snail (*Potamopyrgus*) in a southcentral European lake, mean annual biomass was 8.8 g fresh weight m^{-2} (ca. 1.5 g AFDM m^{-2}), which resulted in production to mean biomass ratios of 2.5 and 3.7 for the first and second cohorts, respectively.

In a detailed study of the benthic fauna of a eutrophic, lowland lake in southern Norway, Ökland (1964) studied the mollusc fauna in particular (13 species of gastropod snails and 1 dominant bivalve clam, *Anodonta piscinalis*). The average quantitative data of Table 22-15 are included to point out a typical distribution with depth of the sediments. The greatest quantity of gastropods occurred in the zone of submersed plants at a depth of 1.5 m. The bivalve *Anodonta* was restricted to the littoral, and all mollusks were intolerant of the reduced oxygen concentrations of the deep strata during both summer and winter periods of stratification. The *Anodonta* constituted a trivial proportion of the numerical population density of the total benthic fauna (Table 22-15). Because of their large size and heavy shell, however, their mass can completely bias the quantitative biomass comparisons. Such data, while instructive in a comparative way, are meaningless for productivity estimates unless the rates of turnover of each of the components constituting the categories are known. The much smaller and more numerous nonmolluscan animals have a small instantaneous biomass but in general a higher turnover rate (shorter turnover time) (Table 22-14). The gastropod snails reached maximum numbers and biomass in the fall (October). The sphaeriid clams attained maximum biomass per area in early summer (June) prior to maximum densities in later summer (mid-August).

II. AQUATIC INSECTS

Insects are very abundant and diverse; most are terrestrial. Of those that are aquatic, nearly all are from fresh water. Some orders are entirely aquatic; others inhabit fresh waters only during certain life stages (cf. Hutchinson, 1981, 1993). Most insects are benthic, living on or burrowing into sediments or on macrophytic vegetation and plant detritus. The characteristics of the insects are well known, and no attempt will be made here to summarize the group differences except to point out salient features of the life cycles that are important to feeding and reproduction as related to benthic productivity and distribution.

The chitinous exoskeleton of arthropods necessitates molting for continued growth. The true bugs (Hemiptera), dragonflies (Odonata), stoneflies (Plecoptera), and mayflies (Ephemeroptera) are orders of winged insects that undergo gradual metamorphosis. In these orders, the young are referred to as *nymphs*; their wings develop as external pads, and the organisms increase in size with each molt. The other orders, including the flies (Diptera), caddisflies (Trichoptera), the alderflies and dobsonflies (Megaloptera), beetles (Coleoptera), a few species of moths (Lepidoptera), and spongeflies (Neuroptera) undergo

TABLE 22-15 Average Numerical Density and Wet Weight Biomass of Mollusks of Eutrophic Lake Borrevann, Norway, in Relation to Other Benthic Macroinvertebrates at Different Depths[a]

	Anodonta piscinalis		*Other mollusks*		*Other benthic fauna*	
Zone and depth	*Ind.* m^{-2} *(%)*	*g* m^{-2} *(%)*	*Ind.* m^{-2} *(%)*	*g* m^{-2} *(%)*	*Ind.* m^{-2} *(%)*	*g* m^{-2} *(%)*
Littoral						
0.2 m	— (—)	— (—)	121 (5)	3.3 (18)	2422 (95)	14.6 (82)
1.5 m	11 (—)	186 (80)	2041 (57)	28.7 (12)	2705 (43)	17.0 (7)
2 m	20 (—)	361 (94)	1182 (26)	6.6 (2)	3398 (74)	14.6 (4)
3 m	31 (2)	1185 (99)	93 (5)	0.3 (—)	1648 (93)	6.1 (1)
Sublittoral						
5 m	24 (2)	500 (99)	12 (1)	0.2 (—)	1242 (97)	5.4 (1)
6 m	11 (1)	73 (96)	5 (1)	— (—)	785 (98)	2.8 (4)
7 m	— (—)	— (—)	— (—)	— (—)	810 (100)	3.3 (100)
Profundal						
10 m	— (—)	— (—)	4 (—)	— (—)	1143 (100)	7.0 (100)
15 m	— (—)	— (—)	— (—)	— (—)	1598 (100)	5.9 (100)

[a] Wet-weight biomass can be compared in relative distribution only. Extracted from data of Ökland (1964).

complete metamorphosis. In this development, the wing pads develop internally in early *larval* instars and then evert to the outside in the preadult instar *pupal* stage. In nearly all of the important aquatic insects, only the immature stages live in the water; the adults and, in some groups, the pupae are terrestrial (cf. Hutchinson, 1981, 1993). Only some of the beetles and hemipterans have adapted to the point where both the adults and immature stages can live in the water.

A. Life History Characteristics

A brief summary of the major life history characteristics follows. In generalizing interrelationships among so diverse a group as the insects, less attention is given to taxonomic characteristics and more attention is given to functional relations of feeding, reproduction, and productivity within aquatic systems.

1. Dragonflies and Damselflies

After fertilization, the eggs of dragonflies and damselflies (Odonata) are deposited into the water, onto substrata in or near the water, or into submersed parts of macrovegetation. Egg development, as in most invertebrates, is temperature dependent; nymphs hatch without diapause after about 2–5 weeks of egg development. Growth in the nymphal stages is quite variable, especially in relation to temperature and food supply. Within a range of about 6 weeks to 3 years, 9 to about 16 instar moltings occur. The mature nymphs often leave the water on some emergent substratum as aerial adults. Of the three primary patterns of life history, spring emergence is common, and univoltine and hemivoltine species usually emerge in summer. The odonate nymphs are almost entirely littoral in habitat, living among macrovegetation and littoral sediments and burrowing into surficial sediments. The nymphs have fairly high respiratory demands and oxygen requirements.

2. Mayflies

The mayflies (Ephemeroptera) are almost totally aquatic (Edmunds *et al.*, 1976; Hutchinson, 1993). The adult longevity is very brief (3–4 days), during which no feeding occurs. After mating in flight, fertilized eggs are laid in the water or on submersed objects; the time to hatching varies from a few days to many weeks. Parthenogenesis is widespread, as well as egg diapause. Growth is relatively rapid, from <1 mm in length at hatching to about 2 cm in length in the nymphal stages of many species. From 9 to over 50 instars of molting occur over an average life cycle of about one year. A few species live as nymphs for 2 years or longer.

All mayflies are restricted to waters of relatively high oxygen content but are widely distributed even in waters with moderate organic loading (Roback, 1974). Respiratory exchange occurs through the cuticle in early instars, but later instars possess respiratory gills and lamellae that induce respiratory currents. Some species can regulate the respiratory movements of gills in response to changing oxygen concentrations (Eriksen, 1963a). Mayfly nymphs are characteristic of shallow streams and littoral areas of lakes and are distributed widely. However, many species are restricted to specific substrata of macrophytes, sediments of waveswept or moving stream areas, or sediments of specific-sized particles (Eriksen, 1963b).

3. Stoneflies

The stoneflies (Plecoptera) are terrestrial as adults, but in the nymphal stages they are strictly aquatic, and most are restricted to flowing waters of relatively high oxygen concentrations. Fertile eggs, laid over or in the water, require 2–3 weeks for hatching in many species and several months among some larger forms. The nymphal instars, from 12 to over 33 moltings, occur in 1–3 years. Nymphs tend to be predominately herbivores and detritivores, and a few become carnivorous in later instars.

4. Hemiptera

The Hemiptera, or true bugs, are essentially terrestrial; a few are semi-aquatic and very few species have adapted to submersed conditions. Most hemipterans overwinter as adults in moist sediments or vegetation (cf. the detailed synthesis of Hutchinson, 1993). Eggs are laid on semi-aquatic substrata or in aquatic macrophytes and develop rapidly in 1–4 weeks. Nymphs also develop rapidly in 1–2 months, commonly with five instar stages, and generally have a 1-yr life cycle. Few hemipterans are truly aquatic, with cutaneous respiration; most are dependent on atmospheric oxygen for respiration. They inhabit substrata either near or above the water, possess adaptations for obtaining air while under water, or carry entrapped pockets of air with them (under hemelytra of the adult or on hydrofuge surfaces of the thorax) on underwater excursions. Several large families of these hemipterans are adapted to stride over the water surface without disrupting the surface tension (see p. 13).

5. Diptera

The most important group of the aquatic insects with complete morphogenesis includes the flies, midges, and mosquitoes (Diptera). The dipterans are commonly dominant components of benthic invertebrate communities in many standing and running waters, and the chironomid larvae are ubiquitous, as will be seen in some of the examples that follow. Much variability is found in the morphology, reproductive biology, and respiration of dipteran larvae. Adults are essentially never aquatic, but most of their lives are spent as immature forms in fresh waters. The larval stage, with about three or four growth molts, extends from several weeks to at least 2 years in different species, and many overwinter in the larval stage. Most species have one generation per year, some have two per year, and a few of the species studied have a 2-yr life cycle (to 7 years in arctic ponds; Butler, 1982). Most larvae respire cutaneously or by means of "blood gills," which also function in ionic regulation. Fewer larvae rely on air and possess various adaptive structures to obtain air above the water or from the internal lacunae of aquatic angiosperms. Some chironomid larvae possess a type of hemoglobin in their blood that functions efficiently at low oxygen concentrations.

6. Caddisflies and Moths

The caddisflies or Trichoptera generally have a 1-yr cycle (Wiggins, 1977). Adults emerge in the warmer periods of the year, often from overlapping cohorts, from May to October. Eggs are dropped or placed on vegetation or laid under water on submersed substrata and develop in about 1–3 weeks. Many caddisfly larvae build beautifully intricate cases from substrate particles of sand, small stones, leaf fragments, and the like and are highly specific to types of substratum (cf. Cummins, 1964; Cummins and Lauff, 1969; Mackay and Wiggins, 1979; Wallace and Merritt, 1980). Some construct a net that traps microorganisms and detrital particles in flowing water. After 5–7 larval instars, pupation occurs under water within a cocoon. The pupal stage generally lasts only a few weeks, after which the pupa leaves the cocoon, moves to an aerial substratum, and emerges as an adult.

Few species of the moths (Lepidoptera) have aquatic larval stages; most aquatic moth species belong to the family *Pyralididae*. Many characteristics of the life history of the "aquatic caterpillars" are similar to those of the closely related caddisflies.

7. Spongeflies, Alderflies, and Dobsonflies

The spongeflies (Neuroptera) and alderflies and dobsonflies (Megaloptera) are closely related and often are grouped together under the Neuroptera. Both groups are primarily terrestrial for all of their life cycle. Of the few freshwater species, however, some megalopteran larvae are so large that they may represent a significant portion of the biomass of some benthic communities. Eggs generally are laid on aerial substrates overhanging the water. After a rapid development period (1–2 weeks), the larvae drop into the water and feed actively. The aquatic neuropterans (Sisyridae) are restricted to and parasitic on freshwater sponges and can have multiple generations annually. The megalopterans have numerous molting instars and most of their life cycle of 1–3 years is spent in the larval stage. The pupal stage occurs in soil out of water and lasts up to one month. The megalopteran larvae are vicious predators on other insects, and some can reach a size of several centimeters.

8. Aquatic Beetles

Such great diversity exists among the aquatic beetles (Coleoptera) that even broad generalization is difficult. Generally, the beetles adapted for larval and adult existence in water occur among the more phylogenetically primitive coleopteran groups. The life cycles of many (Gyrinidae, Haliplidae, Dytiscidae, and Hydrophilidae) are annual, with 3–8 larval instars. Eggs laid on or in macrophytes or sediments hatch in 1–3 weeks. Larval development is variable (1–8 months), and pupation takes place on some nearby terrestrial or aerial substratum. Overwintering generally occurs in the adult stage. Nearly all adult beetles are dependent on atmospheric oxygen and carry an air supply with them on the ventral side of their bodies or are variously adapted to obtain air from the surface or from aquatic angiosperms. Some larvae have tracheal gills to obtain oxygen directly from the water. Although beetles are generally omnivorous, their feeding habits often change markedly, with some exceptions, between the larval stages (herbivorous) to adults (omnivorous, largely detritus, insect eggs, algae).

B. Environmental Adaptations

Upon their invasion of inland waters, aquatic insects encountered the advantages of smaller fluctuations in temperature on a daily and an annual basis but also the disadvantages of decreased oxygen, greater stresses of osmoregulation, and restrictions to movements by the much more dense water medium. Aquatic insects adapted to life under water by essentially continuing to breathe air and developing special means for continuing respiration under water under greatly reduced oxygen availability (Eriksen *et al.*, 1992). Some insects make frequent excursions to the surface to acquire air directly or carry air stores under water in cavities (plastron) or among pubescent regions. Some immature insects extend tubular structures to the surface to acquire air. A few immature insects can penetrate into the aerenchymatous tissues of higher aquatic plants and remove oxygen.

Adaptations also include cutaneous exchange of dissolved oxygen and carbon dioxide by increased cuticular permeability, but this change requires osmoregulatory mechanisms to retain internal salt concentrations in a hypertonic hemolymph against the surrounding dilute water of low ionic strength (Ward, 1992). Ions are commonly reabsorbed in the hindgut before urine and feces are eliminated. In saline lake environments, the opposite problem of the need to conserve water arises. Aquatic insects of saline waters commonly osmoregulate by drinking saline water and excreting rectal fluids that are very hypertonic to the hemolymph (Stobbart and Shaw, 1974).

A number of immature stages of aquatic insects, especially among the larvae, develop tracheal gills. Gill surface area is large and increases the gaseous exchanges between hemolymph and the surrounding water. As dissolved oxygen concentrations of the water are reduced, body or appendage movements increase to move water or move within water in order to reduce concentration gradients and enhance exchange.

Some *Chironomus* larvae excrete lactic acid produced in anaerobic metabolism and avoid oxygen debt as a result. At very low concentrations of dissolved oxygen (e.g., 0.5 mg O_2 liter^{-1}), for example, the total energy production of *C. anthracinus* was only 20% of the rate at saturation and more than one-third was used in anaerobic degradation of glycogen (Hamberger *et al.*, 1994). This rate corresponded to a daily loss of ca. 5% of the body reserves. Long-term oxygen deficiency, however, induces a further suppression of energy metabolism. This process can permit these larvae to remain in oxygen-deficient water for several months. Thus, diurnal and especially seasonal changes in lactate and glycogen content of *Chaoborus* are seen in response to migrations diurnally from and into oxygen-deficient waters and as hypolimnetic waters become oxygen-reduced or anoxic seasonally (Franke, 1987a). Hemoglobin does occur in a few chironomids and hemipterans but is used more in buoyancy control than as an oxygen transport pigment during hypoxia.

Clearly, environmental temperature is critical to the ecology and evolution of aquatic insects as in all other organisms and has received exceptionally thorough study (cf. reviews of Ward and Stanford, 1982; Sweeney, 1984; Ward, 1992). Seasonal and diurnal temperature fluctuations that characterize most natural standing and running waters have been summarized in detail (cf. Chap. 6). The seasonal rate of temperature change in streams is usually greater than in lakes, although the annual temperature range is usually less than in lakes. Diurnal changes in temperature in streams are often greater than in the pelagic portions of lakes, although littoral thermal excursions on a diurnal basis can be broad. Thermal stratification is a dominant feature of most lakes of any appreciable depth (e.g., >5–7 m) but is essentially absent in natural streams and rivers.

Among aquatic habitats of temperate and high latitudes, the period of time above the developmental threshold of the organisms under evaluation is the *growing season*. Within this growing season, there may be a higher temperature threshold, the maturation threshold, which must be exceeded for a specified time period to complete larval development. The absolute values and duration of the minimum temperatures to which aquatic insects are exposed are important, but aquatic insects respond in development and maturation to the summation of thermal units ("degree days," see Baskerville and Emin, 1969). Entire orders and other major taxa of aquatic insects evolved in cool habitats, and members of these groups now inhabiting warmer waters are likely later derivatives of cool-adapted ancestral lines.

Nearly all facets of the life history and the distribution of aquatic insects are influenced by temperature. Temperature is a dominant factor influencing the distribution, diversity, and abundance patterns over elevational gradients and downstream along watercourses. Egg development time is inversely proportional (hyperbolic) to temperature, although hatching success is often optimal at temperatures lower than those temperatures that yield most rapid egg development. Fecundity in aquatic insects is directly proportional to adult female size. Diapause dormancy is common as a response, particularly among stream insects, to avoid warm periods, quite in opposition to what is commonly the case among terrestrial insects.

Development time among aquatic insects is variable in response to temperature. Some insects have a distinct temperature threshold for growth and for maturation, whereas others respond to thermal summation. Many species of aquatic insects of both lakes and streams grow and remain active at temperature near 0°C. Fluctuating temperatures, such as is commonly experienced in small streams and littoral areas, often stimulate growth rates appreciably over those rates found under constant temperature conditions. The large thermal differences observed among habitats at different latitudes or altitudes are not reflected as much in differences in growth rates but rather in differences in ***voltinism,*** the number of generations per year. The number of generations generally increases with higher thermal summation on an annual basis. The number of generations per year among aquatic insects is commonly greater in tropical waters of higher temperatures than in temperate regions (Talling and Lemoalle, 1998). Optimal thermal régimes likely exist for individual species, and deviations into warmer (southern in Northern Hemisphere) or cooler (northern) waters adversely affect fitness by decreasing body size and fecundity (Vannote and Sweeney, 1980).

The substrata of streams and the sediments of lake ecosystems have been discussed at some length in other chapters. A large diversity exists of substrata types, composition (inorganic to organic, ranging from gyttja to large woody tree debris), and especially particle sizes, from silt to boulders. Certainly this range of substrata types results in large differences in stability, particularly in relation to water movements, food organisms associated with the substrata, habitat refuges from predation, and many other factors. However, the velocity and characteristics of water movement in the littoral regions of lakes and in streams are dominant regulatory mechanisms of the development, feeding, growth, and reproductive capacities of aquatic insects.

As we have seen earlier (Chaps. 7 and 18), water movements and currents are steeply attenuated within aquatic vegetation of littoral areas of lakes and within beds of aquatic plants of rivers. However, where current flows are appreciable, there is a tendency for the morphology of aquatic insects associated with substrata to be dorsoventrally flattened, with a smooth dorsum. Although this morphology and behavior may appear to be an adaptation to reduce drag within the boundary layer, a convex surface would generate lift (Statzner and Holm, 1989; Vogel, 1994). Nevertheless, the habitat is clearly in the lower boundary layer of decreased drag. Clinging adaptations (suckers, hooks, spines, glands with adhesives, or "silk") allow exploitation of this habitat. In general, however, insects have developed behavioral patterns to avoid current.

C. Trophic Mechanisms and Food Types

The great diversity of food ingested by the aquatic insects and their various feeding mechanisms can be organized within a functional framework combining feeding mechanisms with the particle size of the dominant food utilized (Table 22-16). Distinctions can be made between herbivory, defined as the ingestion of

TABLE 22-16 A General Categorization of Trophic Mechanisms and Food Types of Aquatic Insects[a]

General category based on feeding mechanism	*General particle size range of food (μm)*	*Subdivision based on feeding mechanisms*	*Subdivision based on dominant food*	*Aquatic insect taxa containing predominant examples*
Shredders	$>10^3$	Chewers and miners of live macrophytes	Herbivores: living vascular plant tissue	Trichoptera (Phryganeidae, Leptoceridae) Lepidoptera Coleoptera (Chrysomelidae) Diptera (Chironomidae, Ephydridae)
		Chewers and miners of coarse particulate organic matter	Detritivores (large particle detritivores): decomposing vascular plant tissue	Plecoptera (Filipalpia) Trichoptera (Limnephilidae, Lepidostomatidae) Diptera (Tipulidae, Chironomidae)
		Gougers of wood	Wood	Trichoptera Coleoptera Diptera
Collectors	$<10^3$	Filter or suspension feeders[b]	Herbivore-detritivores: living algal cells, decomposing fine particulate organic matter	Ephemeroptera (Siphlonuridae) Trichoptera (Philopotamidae, Psychomyiidae, Hydropsychidae, Brachycentridae) Lepidoptera Diptera (Simuliidae, Chironomidae, Culicidae)
		Sediment or deposit (surface) feeders	Detritivores (fine particle detritivores): decomposing fine particulate organic matter	Ephemeroptera (Caenidae, Ephemeridae, Leptophlebiidae, Baetidae, Ephmerellidae, Heptageniidae) Trichoptera Lepidoptera Hemiptera (Gerridae) Coleoptera (Hydrophilidae) Diptera (Chironomidae, Ceratopogonidae)

(continues)

TABLE 22-16 (*Continued*)

General category based on feeding mechanism	*General particle size range of food (μm)*	*Subdivision based on feeding mechanisms*	*Subdivision based on dominant food*	*Aquatic insect taxa containing predominant examples*
Scrapers	$<10^3$	Mineral scrapers	Herbivores: algae and associated attached microflora to living and nonliving substrates	Ephemeroptera (Heptageniidae, Baetidae, Ephemerellidae) Trichoptera (Glossosomatidae, Helicopsychidae, Molannidae, Odontoceridae, Goreridae) Lepidoptera Coleoptera (Elmidae, Psephenidae) Diptera (Chironomidae, Tabanidae)
		Organic scrapers	Herbivores: algae and associated attached microflora	Ephemeroptera (Caenidae, Leptophlebiidae, Heptageniidae, Baetidae) Hemiptera (Corixidae) Trichoptera (Leptoceridae) Diptera (Chironomidae)
Predators	$>10^3$	Swallowers	Carnivores: ingest whole animals (or parts)	Odonata Plecoptera (Setipalpia) Megaloptera Trichoptera (Rhyacophilidae, Polycentropidae, Hydropsychidae) Coleoptera (Dytiscidae, Gyrinnidae) Diptera (Chironomidae)
		Piercers	Carnivores: cell and tissue fluids	Hemiptera (Belastomatidae, Nepidae, Notonectidae, Naucoridae) Diptera (Rhagionidae)
Macrophyte piercers	$>10^2$–$>10^3$	Herbivore piercers	Living vascular hydrophyte cell and tissue fluids	Neuroptera Trichoptera

[a] Nonparisitic nutrition of immature and adult stages of insects that occur in the water. Modified from Cummins (1973) and Cummins and Merritt (1996).
[b] See also Wallace and Merritt (1980).

living vascular plant tissue or algae; carnivory, the ingestion of living animal tissue; and detritivory, the intake of nonliving particulate organic matter and the microorganisms associated with it (Cummins, 1973; Hutchinson, 1993). The sources of this organic matter originate from (a) sedimenting phytoplankton, (b) benthic aquatic macrophytes, (c) attached microbial communities, and (d) terrestrial arboreal and wetland vegetation and animal matter, such as ants and other insects, that falls, is blown, or is washed into the water. The organic matter in any of these categories can be ingested either in particulate form (by swallowing, biting, or chewing) or in dissolved form (by piercing or sucking). Although the association of aquatic insects with a particular substratum usually is related directly to feeding on that substratum or on the associated microflora, this is not always the case. For example, the insect may use aquatic angiosperms as a source of air or as a site for reproduction or protection from predation. However, the relationship between insects and certain substrates (e.g., angiosperms) is often highly specialized and even species-specific (cf. the compilations of McGaha, 1952, and especially of Gaevskaia, 1966). Other insect groups, especially large ones such as the dipteran Chironomidae, show great diversity in feeding mechanisms and substrata ingested (Table 22-16). Most aquatic insects tend to be nonselective in their food habits, but clearly the abundance, biomass, and diversity of invertebrates is greatest among submersed macrophyte habitats (e.g., Dvořák, 1987).

Chemical communication is likely common among freshwater benthic animals (Dodson *et al.*, 1994). Prey species use chemical signals to modify their morphological development, life history strategies, feeding, and predator avoidance behavior. Such chemical signals can be used in dark or turbid environments, likely at the species level of distinction, by animals that do not have image-forming eyes.

Classification of the functional feeding groups among invertebrates, as has been done in Table 22-16 for aquatic insects, describes the morphological and behavioral capacities of these organisms to consume available food resources (see particularly the reviews of Cummins, 1973; Anderson *et al.*, 1978; Anderson and Sedell, 1979; Wallace and Merritt, 1980; Hutchinson, 1993; Wallace and Webster, 1996). Facultative-feeding invertebrates ingest a wider array of substrates (have a wider niche breadth) and inhabit a greater diversity of stream and lake habitats than more specialized feeders (Cummins and Klug, 1979).

The ingestion of food bears little relation to rates of assimilation. In streams and many shallow lakes and reservoirs, much of the organic matter consumed by benthic invertebrates and other feeders is in the form of dead particulate organic matter from higher plants. This particulate detritus contains a large amount of lignins and celluloses from plant structural tissues in various stages of slow decomposition. The rate of decomposition and abundance of associated decomposing microflora are related to the chemical composition and restricted accessibility to microbial enzymes and the general insolubility of the structural organic compounds.

Assimilation rates may not differ with variations in the caloric content of food material. For example, fortuitous ingestion of inorganic sediment by fine-particle detritivores obviously has low nutritive value, but coatings of organic compounds and attached microflora may be of considerable value. Alternatively, small inorganic particles may aid in digestion by means of mechanical disruption of ingested food cells within the gut. Little is understood about the digestive capabilities or efficiencies of food utilization by aquatic insects (see reviews of Cummins, 1973; Anderson *et al.*, 1978; Cummins and Klug, 1979; Benke and Wallace, 1980; Lamberti and Moore, 1984). Assimilation efficiencies by benthic consumers of plant and detrital particles are nearly always <50%, and most are within the range of 5–30%. Assimilation efficiencies are much lower (ca. 10%) for fine detritus and vascular plant detritus but increases to ca. 30% for algae. Large quantities of viable algae and other microorganisms can be found in gut contents and feces, which suggests incomplete digestion.

Dead plant material entering aquatic ecosystems is rapidly colonized by microbes, and degradation is dominated particularly by aquatic hyphomycetes and other microorganisms (Chaps. 18 and 21). Microbial catalysis of indigestible plant substances into digestible subunits by microbial enzymes and the addition of relatively easily digestible microbial biomass are clearly an important coupling between these recalcitrant organic materials and biotic utilization (Bärlocher, 1985). Different species of invertebrates vary in their abilities to utilize the microbial production and their biomass.

Much of the material ingested by litter and deposit-feeding invertebrates probably is indigestible and must be degraded by the enzymatic activity of symbiotic microflora and fauna of the gut. A number of carbohydrases have been found in the digestive tracts of several species of Trichoptera, the carnivorous amphipod *Gammarus*, and two species of sediment-feeding dipteran *Chironomus* (Bjarnov, 1972). All species degraded mono- and disaccharides, and species differences were found only in polysaccharide degradation. Cellulose was found to be degraded only by the amphipod, and chitinase activity was found only in carnivorous species and in one detritivorous caddisfly. Most aquatic insects have little enzymatic activity toward cellulose, however. Those species possessing digestive cellulase can live in a wide range of habitats because they can utilize a greater array of detritus and algal types than can organisms lacking this physiological adaptation. A number of dipteran and trichopteran larvae have high proteolytic activity for efficient digestion of bacteria and metazoans (e.g., Martin *et al.*, 1981a,b). Despite consumption of detritus laden with microbes, microbial biomass is usually <10% of the total detrital mass (see Table 21-3). In spite of relatively high assimilation efficiencies of microbes (e.g., Martin and Kukor, 1984), acquired organic carbon and nitrogen from these sources are often inadequate (usually <25%) to support the growth and reproduction of the insects. Appreciable sources must come from the particulate organic matter itself (e.g., 73–89% attributable to leaf organic matter in the growth of the dipteran larvae *Tipula*; Lawson *et al.*, 1984).

Macroinvertebrates utilize various feeding modes to compensate for the differing dietary sufficiency of ingested substrates (cf. reviews of Anderson and Cummins, 1979; Anderson and Sedell, 1979; Cummins and Klug, 1979). Shredders and some collectors feed preferentially on particulate organic detritus colonized by microorganisms and utilize the associated microorganisms and partially hydrolyzed substrates. Collectors, scrapers, and facultative shredders may increase consumption of low-quality food (e.g., resistant detrital plant parts) to compensate for the lower nutritional benefit. In general, gut retention time increases inversely with feeding rate. Feeding and ingestion rates are commonly slower with longer gut retention times as the quality of food declines.

A number of immature freshwater insects that consume relatively resistant plant materials and particulate organic detritus may harbor microbial symbionts in their alimentary tracts (Cummins and Klug, 1979). These animals are analogous to ruminant mammals, which ingest food of high C:N ratios (>17) and absorb through their gut wall materials with lower C:N ratios (<17) produced by intestinal microbiota that degrade polymers and supply essential amino acids.

Aquatic insect predators feed heavily upon other benthic prey. Most studies indicate that apparent prey selectivity is likely more a function of prey vulnerability to capture than of selection by the predator (Peckarsky, 1984b). Assimilation efficiencies of ingested animal prey and bacteria by aquatic insects are generally higher than those for plant and detrital materials, but data are few (Martin and Kukor, 1984). Aquatic insects have evolved a variety of defensive mechanisms and escape tactics by which predation is avoided. Primary defenses, such as morphological and chemical defenses, operate regardless of the presence of a predator (Edmunds, 1974). Primary defenses among aquatic insects include

1. Living in habitat refuges that decrease the chances of the prey encountering predators. Such refuges may be physical (crevices or holes) or temporal (e.g., limiting feeding activity and movements from physical refuges to periods of darkness).
2. Crypsis is any adaptation that causes an animal to be less conspicuous (e.g., visually, or chemically) to the predators. Morphological and behavioral modifications, color similarity to the habitat, and similar methods reduce susceptibility to predation.
3. Some prey contain chemical compounds that are apparently distasteful to predators and induce avoidance.

Secondary defense involves active predator avoidance behavior after the prey has detected the presence of the predator or the initiation of an attack by the predator. Prey can withdraw and escape by fleeing movements upon encountering a predator. Dispersal-related downstream drifting movements of aquatic insects, discussed later, is most active at night when visual predation by fish is reduced. Feigning death or attack postures can deter predators.

Despite the general opportunistic feeding behavior and lack of selectivity of food items, many benthic animals do not eat the same food throughout their development. Size is important at both the upper end of a range where whole particles cannot be ingested ("gape size limitation") and at the lower end of a range where the size of food particle ingested is too small to compensate for the energy expended in acquisition. For example, predatory benthic invertebrates often begin as herbivores or detritivores but shift to carnivores as they mature, and diet can shift with season, locality, and sex (Malmqvist *et al.*, 1991; Giller and Sangpradub, 1993).

Most particulate organic matter that enters lakes and streams from allochthonous and autochthonous sources is not eaten by animals and is either eventually degraded to CO_2 by microflora attached to the particles or stored for various periods of time in the sediments. The feeding and shredding activities of benthic invertebrates, however, can accelerate reduction of particle size and increase surface areas exposed to the microbiota and thereby stimulate microbial degradation rates.

The biomass of aquatic insects is relatively constant when they are supplied with consistent and abundant food supplies of similar caloric and protein content throughout the year (Cummins, 1973; Anderson and Cummins, 1979). The rate of biomass turnover in insects is controlled primarily by temperature and is mediated mainly by the positive relations among temperature, feeding rate, and respiration. The ratios of feeding and respiration to growth are relatively constant. In general, streams exhibit less seasonal fluctuation in temperature than do the littoral regions of lakes. Much of the feeding and growth of aquatic insects living in streams occurs during the fall and winter, even in the temperate zones. Assimilation efficiencies seem to be fairly constant over the broad range of foods consumed by insects. There is a suggestion (Welch, 1968), based on limited and variable data, that this efficiency[4] is somewhat higher in carnivores than in herbivores and detritivores. However, the net or tissue growth efficiencies of carnivores[5] tend to be lower than those of herbivores and detritivores.

III. LITTORAL AND PROFUNDAL BENTHIC COMMUNITIES OF LAKES

Some of the important relationships among the benthic macroinvertebrates can be seen by a comparison of the benthic fauna of the littoral with that of the

[4] Assimilation efficiency is defined here as assimilation/ingestion, or growth plus respiration/ingestion (see Chap. 8).
[5] Defined as growth/assimilation.

profundal. As lakes become more productive and the hypolimnetic water strata undergo periods of oxygen reduction and increases in the metabolic products of microbial decomposition, the number of animals adapted to these conditions decreases precipitously. Those adapted to these conditions are exposed to relatively homogeneous conditions of temperature and substratum throughout the year. Additionally, both competitive pressures for available resources and predation are reduced considerably. Therefore, a commonly observed community structure consists of a rich fauna with high oxygen demands in the littoral zone above the metalimnion. Substratum heterogeneity is much greater in the littoral, and species diversity and competitive interactions are more complex. By contrast, the profundal zone is more homogeneous and becomes more so as lakes become more productive and species diversity decreases correspondingly (cf. Jónasson, 1969, 1978, 1984).

The production of phytoplankton and macrophytes is a major determinant of the subsequent conditions of food supply, oxygen, ionic composition, pH, and numerous other factors that delimit the range and competitive abilities of benthic fauna (cf. general reviews of Macan, 1961; Brinkhurst, 1974; Jónasson, 1978; Ward, 1992; Hutchinson, 1993). The effects are seen not only in the qualitative distribution but also in the quantitative aspects. Two maxima of abundance are typically observed with depth, one in the littoral zone and another in the profundal. Examples of this distribution are very numerous. The data presented in Table 22-17 show these two maxima both in numbers and biomass, as well as the reduction in species diversity in the progression from the littoral to the profundal zones. The same type of distribution is illustrated by the example from Lake Washington (Fig. 22-4). If lakes become extremely enriched, to the point that the population densities of the phytoplankton and epiphytes become so great that they shade out the submersed macrovegetation, then the habitat diversity of the littoral decreases (cf. Fig. 26-1). Correspondingly, the diversity and quantity of littoral macroinvertebrates decrease, and often a maximum in animal biomass can be observed only in the lower profundal zone. With further increases in eutrophication and lengthening of the period of hypolimnetic oxygen reduction and associated chemical changes, the rates of respiratory activity of the adapted benthic animals are reduced. Rates of growth and survival also decline, and some insect larvae increase their life cycles from 1 to 2 years. As hypolimnetic strata of hypereutrophic waters undergo extreme eutrophicational or pollutional loading of organic matter, essentially all of the aquatic insects may be eliminated. Practically the only group of benthic fauna adapted to conditions of extremely high organic loading is the oligochaete annelids. Variations in the vertical distribution of biomass of benthic invertebrates are discussed in some detail by Deevey (1941), Brinkhurst (1974), and Jónasson and Lindegaard (1979).

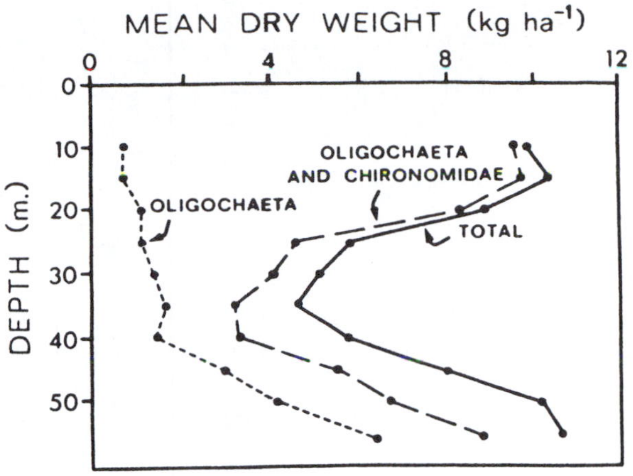

FIGURE 22-4 Vertical distribution of the mean biomass of oligochaetes, of oligochaetes and chironomids, and of the total benthic macroinvertebrates of Lake Washington, Washington. (Redrawn after Thut, 1969.)

The changes in species composition, especially among the dipteran larvae, and the quantity of benthic fauna in the profundal zone of lakes as they become more eutrophic have been subjects of intensive study since shortly after the turn of the century. Prompted by the perceptive work of Thienemann and other German workers, especially Lenz and Lundbeck, an array of typological schemes of lake classifications was developed on the basis of indicator species of dominant benthic fauna. Particular emphasis was given to the distribution of species within the ubiquitous dipteran family of midges (Chironomidae) in relation to the oxygen content of the hypolimnion. Much of the early work on the classification of lakes in relation to the Chironomidae is reviewed in the book by Thienemann (1954; see also the critical discussion by Deevey, 1941; Brinkhurst, 1974). Although these analyses did much to stimulate study of the benthic fauna, such classification systems are of limited value since they do not consider fully the complex interactions of morphometry, the chemical differences among sediments and water, the biotic interactions such as predation, and an array of other factors.

As lakes become more eutrophic, shifts occur in percentage composition of the two dominant groups of benthic animals in the profundal zone of lakes, the Chironomidae and the oligochaetes. The few examples given in Table 22-18, which could be expanded greatly, simply show the general reduction in the number of

TABLE 22-17 Average Numerical Density (no. m^{-2}) and Dry Weight Biomass (g m^{-2})[a] of the Principal Groups of Benthic Invertebrates at Different Depths of Eutrophic Lake Borrevann, Southern Norway[b]

	Littoral								*Sublittoral*						*Profundal*			
	0.2 m		*1.5 m*		*2 m*		*3 m*		*5 m*		*6 m*		*7 m*		*10 m*		*15 m*	
Benthic invertebrates	*no. m^{-2}*	*g m^{-2}*	*no. m^{-2}*	*g m^{-2}*	*no. m^{-2}*	*g m^{-2}*	*no. m^{-2}*	*g m^{-2}*	*no. m^{-2}*	*g m^{-2}*	*no. m^{-2}*	*g m^{-2}*	*no. m^{-2}*	*g m^{-2}*	*no. m^{-2}*	*g m^{-2}*	*no. m^{-2}*	*g m^{-2}*
Oligochaeta	114	0.12	1370	1.78	1652	1.68	897	0.90	252	0.32	280	0.28	245	0.16	378	0.28	406	0.14
Hirudina	126	0.58	163	0.42	368	0.44	63	0.10	2	0.004	—	—	—	—	—	—	—	—
Ephemeroptera	1416	1.0	150	0.22	18	0.008	9	0.002	—	—	—	—	—	—	—	—	—	—
Trichoptera	391	1.04	203	0.40	112	0.012	193	0.06	80	0.04	10	0.004	—	—	—	—	—	—
Chaoborus	—	—	—	—	2	0.002	3	0.002	—	—	5	0.004	60	0.02	172	0.1	958	0.6
Chironomidae	279	0.10	643	0.24	1138	0.50	463	0.14	900	0.70	490	0.28	505	0.50	575	1.0	299	0.44
Gastropoda	105	0.64	1964	5.58	902	1.04	60	0.04	6	0.04	—	—	—	—	2	—	—	—
Other Groups	112	0.10	253	0.50	388	0.46	53	0.04	14	0.02	5	0.004	—	—	20	0.02	5	0.002
Total	2543	3.58	4746	9.14	4580	1.28	1741	1.28	1254	1.12	790	0.57	810	0.68	1147	1.4	1668	1.18

[a] Original values, given in wet weight, were converted to an estimated dry masses by ×0.2 (cf. Winberg, 1971).
[b] From data of Ökland (1964); bivalve mollusk *Anodonta* excluded.

TABLE 22-18 Comparison of the Relative Composition of the Dominant Benthic Macroinvertebrates of Several Lakes of Differing Productivity Based on Other Criteria[a]

	Percentages				
Lake	*Chironomidae*	*Oligochaeta*	*Sphaeridae*	*Others*	*Source*
Oligotrophic					
Convict, CA	65.3	30.8	0.4	3.5	Reimers *et al.* (1955)
Bright Dot, CA	77.5	3.1	19.1	0.3	
Dorothy, CA	69.5	23.3	3.5	3.7	
Constance, CA	56.9	20.5	20.5	2.1	
Cultus, British Columbia	65.0	24.0	—	—	Ricker (1952)
Lake Ontario	1.8	6.4	3.4	88.4[b]	Johnson and Brinkhurst (1971)
Lake Erie (1929–1930)	10	1	2	87[c]	Wright (1955)
Eutrophic					
Washington, WA	43	51	3	3	Thut (1969)
Lake Erie (1958)	27	60	5	8	Beeton (1961)
Glenora Bay, Lake Ontario	42.3	29.4	6.2	22.1	Johnson and Brinkhurst (1971)

[a] Data are only approximately comparable because of different methods employed.
[b] Mostly amphipods.
[c] Mostly the mayfly *Hexagenia*.

chironomids and other benthic animals and a concomitant increase in oligochaete worms among more productive lakes. The transition can occur during natural development of a particular lake and can be accelerated greatly by the eutrophicational activities of humans, as was shown conspicuously in the case of Lake Erie (Table 22-18) in a period of <30 years (see the general review of this subject by Wiederholm, 1980).

A. Benthic Fauna of Lake Esrom

Certainly one of the foremost analyses of the benthic fauna can be found in the work of Jónasson (1972, 1978, 1984, 1990; Dall *et al.*, 1984). Environmental factors, especially the phytoplanktonic productivity of the moderately large (17.3 km^2) and deep (22 m), eutrophic Lake Esrom, a dimictic lake in Denmark, were studied in relation to the food, growth, life cycles, and population dynamics of three profundal detritivores and two carnivores. Because of the completeness of these studies, it is instructive to discuss them in some detail.

The dipteran detritivore *Chironomus anthracinus* was found to feed at the sediment surface. Its growth was limited to two very short periods, one in spring during the phytoplanktonic maximum when the hypolimnion was oxygen-rich, and the other after autumnal circulation when oxygen was available again but food production was declining. Growth continued during winter as long as the lake was ice free but essentially ceased under ice cover. During the summer, growth stopped when the oxygen concentrations of the hypolimnion were reduced slightly below 1 mg O_2 liter^{-1}. Biomass, protein content, and fat content of larvae of this species were positively correlated seasonally with the influx of oxygen and phytoplanktonic detritus to the hypolimnion.

Continuous quantitative data extending over two decades showed that deep (20-m) populations of *C. anthracinus* exhibited a 2-yr life cycle. The 2-yr cycle is maintained by alternate years of high initial population densities in which a large percentage of the population does not emerge as adults in the first year. These larvae prevent a new generation from becoming established because their feeding inflicts severe mortality on new eggs laid on the sediments. As a result, at any given time the population derives from a single generation of eggs. In alternate years, when the density of larvae declines below 2000 m^{-2} after the first year's emergence of adults, a new generation is able to establish itself because now the larvae are not able to feed over the whole of the sediment surface and most eggs survive. During these years of recruitment of a new generation, the larval population is a mixture of two generations (Fig. 22-5).

The larvae of this species in the metalimnetic sediments (11 and 14 m) grew longer because of greater oxygen concentrations, and each individual attained greater biomass. There was more exposure to fish predation in the shallower metalimnetic sediments, however, and population numbers were lower than in the profundal zone. Those at the 11-m depth had a 1-yr life cycle; the transitional zone between 1- and 2-yr life cycles was between 14- and 17-m sediments. In sediments at the latter depth, recruitment of larvae every year was not successful (Fig. 22-5). At 14 m, the

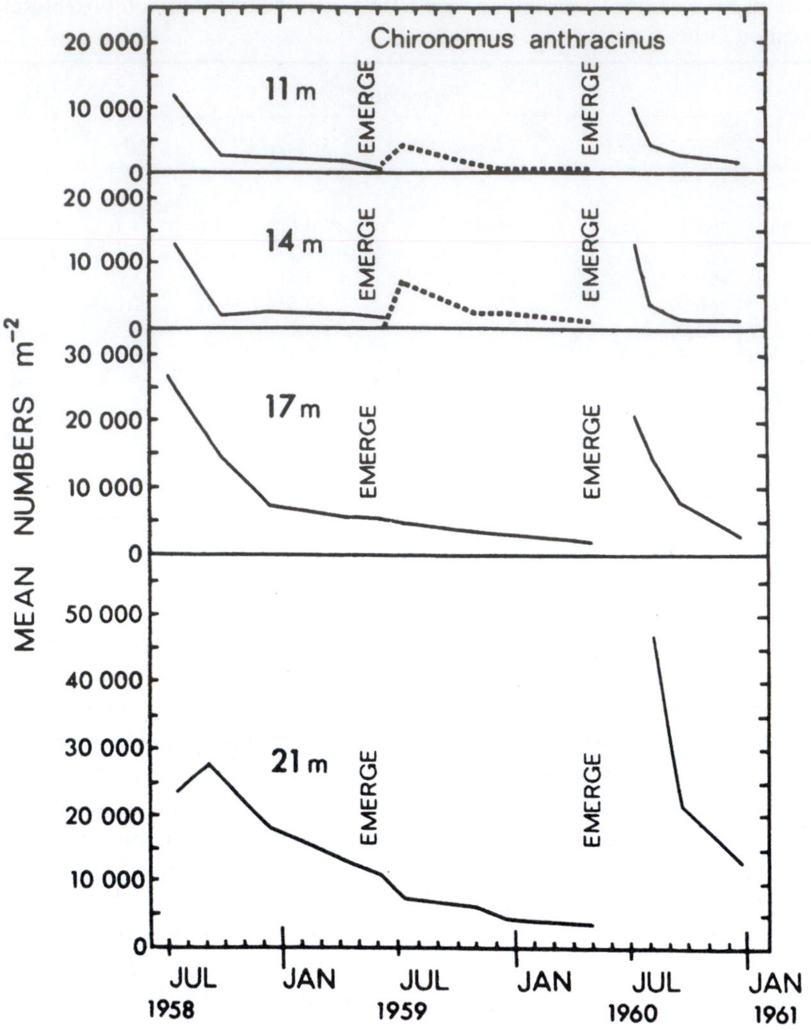

FIGURE 22-5 Seasonal fluctuations in populations of the midge larvae *Chironomus anthracinus* in sediments of different depths of Lake Esrom, Denmark. 1-yr life cycles were found at 11- and 14-m, 2-yr cycles at 17- and 21-m depths. (Modified from Jónasson, 1972.)

larvae succeeded in having a yearly recruitment in some years. Population size increased with depth, probably as a result of reduced fish predation. Larvae of the carnivorous phantom midge *Chaoborus flavicans* and of the midge *Procladius pectinatus* had similar growth patterns and consistently exhibited a 1-yr life cycle. The detritivorous tubificid oligochaete *Ilyodrilus hammoniensis* lived at least 5 years and did not begin to reproduce until after 3 years; the detritivorous small clam *Pisidium* also exhibited a long life cycle.

Detailed analyses of the population dynamics, survivorship, emergence, and respiration permitted an evaluation of the net production of the profundal macroinvertebrates of eutrophic Lake Esrom (Table 22-19). A large portion (89%) of the production was by the three benthic detritivores, mostly by the *Chironomus* species. Jónasson (1984, 1990) was able to extend these data, making a number of reasonable assumptions, to an energy flow budget. In this moderately eutrophic lake, an appreciable portion (perhaps 6%) of the phytoplanktonic primary production reaches the profundal sediments and, while undergoing decomposition, is utilized by profundal invertebrate fauna along with the microflora (Table 22-16).

Despite the relatively high mean annual densities of zoobenthos of Lake Esrom (65,000–200,000 individuals m^{-2}), about half of the animal production arises from the zooplankton and half from the zoobenthos (Table 22-20). In contrast, in the shallow, benthos-rich Lake Mývatn, Iceland, benthos accounts for the primary animal energy flow (ca. 21 g C m^{-2} yr^{-1}), which is some five times greater than that of the zooplankton (ca. 4.3 g C m^{-2} yr^{-1}) (Jónasson, 1979; Lindegaard, 1994). In further contrast, the zooplankton production (ca. 7.5 g C m^{-2} yr^{-1}) of deep, subarctic, oligotrophic Lake Thingvallavatn, Iceland, is four times greater than

TABLE 22-19 Estimates (g C m^{-2} yr^{-1}) of the Average Annual Production, Mortality, Respiration and Emergence of the Dominant Profundal Benthic Fauna of Lake Esrom, Denmark[a]

Species	*Respiration (minimal)*	*Mortality*	*Emergence*	*Net production*	$P/\bar{B}$
Dipterans					
Chironomus anthracinus	36.18	3.59	3.94	7.53	2.1
Chaoborus flavicans	5.12	0.26	1.16	1.44	1.7
Procladius pectinatus	0.73	0.13	0.17	0.29	1.9
Oligochaetes					
Potamothrix hammoniensis	0.73	—	—	0.60	0.7
Mollusks					
Pisidium casertanum/spp.	0.48	—	—	0.21	1.0
Total	43.24	3.98	5.27	10.07	

[a] Composite from data of Jónasson (1972) and Jónasson *et al.* (1990). Original data expressed in kcal m^{-2} yr^{-1} was converted by 1 g C ≈ 10 kcal.

that of the zoobenthos (ca. 1.9 g C m^{-2} yr^{-1}) (Jónasson and Lindegaard, 1988; Jónasson, 1992).

Studies of Lake Esrom for nearly a century have demonstrated the negative effects of the gradual eutrophication process on the benthic invertebrates (Jónasson, 1984; Lindegaard *et al.*, 1993). For many decades until the 1980s, Lake Esrom received sewage from nearby communities. Increased algal productivity and increased duration of hypolimnetic oxygen depletion caused marked declines in benthic invertebrates.

TABLE 22-20 Estimates of Biomass and Energy Flows among Components of Eutrophic Lake Esrom, Denmark, with Emphasis on the Benthic Invertebrates[a]

Component	*Mean biomass*[b]		*Production*		
	g *DM* m^{-2}	g *C* m^{-2}	(g *C* m^{-2} yr^{-1})	*P/B ratio*	*Percentage*
Inputs:					
Allochthonous organic matter[c]	9.0	4.5	—	—	—
Phytoplankton	5.8	2.9	170	58.6	65.6
Emergent macrophytes (0–3 m)	?	?	5	—	1.9
Submersed macrophytes[d] (2–6 m)	?	?	7	—	2.7
Epiphytic algae (2–6 m)	6.0	3.0	7	14.0	2.7
Benthic algae (0–3 m)	29.0	19.5	35	13.1	13.5
Zooplankton	4.0	2.0	18.6	9.3	7.2
Zoobenthos					
Shallow littoral (0–3 m)	1.4	0.7	2.6	3.64	1.0
Lower littoral (3–6 m)	12.4	6.2	4.3	0.71	1.7
Sublittoral (6–12 m)	9.8	4.9	3.2	0.65	1.2
Profundal (12–22 m)	10.6	5.3	6.4	1.21	2.5
					100.0
Outputs:					
Respiration by producers			72.2		24.4
Bacterial respiration in water			62.0		20.9
Secondary respiration			90.9		30.7
Microbial respiration in sediment[d]			13.4		4.5
Chemical oxidation in sediment			10.2		3.4
Secondary production			35.1		11.9
Emergence			0.4		0.14
Accumulation in sediment			8.4		2.8
Outflow (POM and DOM)[d]			2.4		0.8
					99.5

[a] From data of Jónasson (1984) and Jónasson *et al.* (1990).
[b] Mean biomass per average lake m^2, expanded over the entire lake area.
[c] Estimated from river inputs only.
[d] Likely underestimates.

The declines of profundal fauna lowered net zoobenthos production to about 30% of the levels in the 1950s. With improved sewage treatment and reduced nutrient loading to Lake Esrom, partial recovery has occurred, with reduced algal production and increased profundal production back to about 80% of the levels in the 1950s.

Food supply is of major importance in the regulation of the population dynamics of detritivore species (cf. also Iyengar *et al.*, 1963), especially in the profundal zone, in which predation shifts from other animals to cannibalism of eggs. Among carnivorous chironomid larvae, the young (first instar) are largely detritivores, but beyond the second instar, they shift to feeding on small chironomid larvae, small crustacea, and some plant food. Although *Procladius* are carnivorous, their diet is highly variable and opportunistic (Kajak and Dusoge, 1970). Mature larvae consume about 10% of their body weight per day and have a relatively high degree of utilization of animal food. Abundant food availability increases rates of growth and development greatly.

The Chironomidae form the dominant macroinvertebrates of all arctic lakes studied (Hobbie, 1973; Butler, 1982). It is noteworthy that those few studied have 2- to 7-yr cycles and are detritivores. The carnivorous *Procladius* is also usually present.

B. Phantom Midges

In addition to the chironomid larvae, oligochaetes, and the small clam *Pisidium*, another major component of the profundal zone of lakes is the phantom midge *Chaoborus*. Extensive studies of *C. flavicans* demonstrate some of the major characteristics of this group (Parma, 1971). This midge dominates the benthic fauna of the profundal zone of Lake Vechten, a small (5 ha), moderately deep (12 m) lake in the Netherlands. Lake Vechten is thermally stratified from April to November and mixes all winter (very brief period of ice cover). The lake is moderately eutrophic, and the hypolimnion becomes anaerobic during stratification.

The lifespan of an adult phantom midge is very brief (<6 days). Eggs (about 500 in number) are laid on the water in rafts and most (97%) hatch in 2–4 days. The larvae can develop to the fourth instar in 6 to 8 weeks. The first and second instars are always limnetic and positively phototactic, and they develop rapidly in a few weeks. The third instar, mostly limnetic but also occurring in the sediments, is of much longer duration; the larvae can overwinter in this stage (Stahl, 1966; Roth, 1968). After a variable period of up to several months, ecdysis to the fourth instar occurs; this instar is limnetic much of the time.

The fourth instar of many species of *Chaoborus* undergoes strong diurnal vertical migrations (see, e.g., Northcote, 1964; Malueg and Hasler, 1966; LaRow, 1968; Goldspink and Scott, 1971). A typical rhythm is movement from the sediments or from their daytime depth to the surface strata at sunset; they stay there until about sunrise, then descend. The initiating cue is light intensity and the orienting cue is gravity (Swift and Forward, 1988). Light intensities are variable at the initiation of and during upward migration, but the rapid movement is correlated with the timing of a relative light change (ca. 1.7×10^{-3} s^{-1}) (Haney *et al.*, 1990). Rates of ascent and descent are about the same at 4–6 m h^{-1}. Migration does not occur as actively at temperatures <5°C. When the larvae are burrowed into the sediments, they push up to the surface of the mud as if to sample the light conditions. If light levels are sufficiently low, they emerge from the sediments and begin the migration (LaRow, 1969). Experiments showed that the migration was influenced in part by the oxygen tension of the water near the sediments (LaRow, 1970). When the oxygen concentrations were high, most of the populations stayed in the sediments; when there was <1 mg O_2 $liter^{-1}$, about one-third of the population migrated into the water above the sediments and about one-third migrated to the surface during the night. Migration varies with season, being highest in midsummer and lowest in winter. There is evidence that *Chaoborus* also migrates horizontally at rapid rates (ca. 20 m h^{-1}) and can move into littoral areas at night (Franke, 1983, 1987b; Voss and Mumm, 1999).

Since the early instars are small and very transparent, predation pressure probably is low. With increasing size, predation pressure in the open water becomes greater and results in a more benthic habit. Oxygen stresses in the hypolimnion impose a demand for oxygen that can be alleviated by moving to the upper strata, but this activity is best accomplished at night when predation pressure from fish is reduced. Fish predation pressure is greater upon the larger third and fourth instar larvae of *Chaoborus* than on the smaller instars. Smaller species, or species in which the larval eye size and visibility to fish is reduced, have less pronounced vertical migration than larger species. Some highly visible species are eliminated in lakes by fish predation and do not occur when fish are present or introduced (Macan, 1977; Northcote *et al.*, 1978; Stenson, 1978a,b; von Ende, 1979; von Ende and Dempsey, 1981).

Pupation takes from 1 to 2 weeks and occurs from May to October in *C. flavicans*. The pupae migrate daily. Pupation and emergence as adults result in a reduction in the total benthic population during summer. Emergence has been correlated with lunar periodicity, beginning shortly after full moon and ending at the following new moon (e.g., Hare and Carter, 1986; Akeret, 1993). The highest densities of larvae, 1400–1800 m^{-2} in Lake Vechten, occurred in November. Mortality in the winter was high (65%) due to reductions in food supply, and a total mortality of about 97% occurred before the fourth larval stage was reached. The annual net production of the benthic part of the population was estimated to be 0.8–1.25 g m^{-2} yr^{-1} (8–12.5 kg DM ha^{-1}) for *C. flavicans* in this lake.

Production by *Chaoborus* larvae can be very high. The respiration rates of the larvae, when corrected for body size and temperature, were very low by comparison with the rates for most aquatic insects (Cressa and Lewis, 1986). As a result, growth efficiencies are high (40–75%). For example, in the eutrophic Socuy Reservoir, northwestern Venezuela, the largest portion of biomass and production resulted from the older instars (Table 22-21). Instar IV represented some 82% of the biomass and 61% of production. The marked increases in the *P*/*B* values in the older instars resulted in high average renewal rates (Table 22-21), but these values are similar among many other populations of *Chaoborus* in tropical eutrophic lakes (López and Cressa, 1996).

Chaoborus larvae are omnivorous and shift predominantly to predation on pelagic zooplankton and certain small benthic animals in later instars. All instars ingest large, flagellated phytoplankton, such as dinoflagellates and rotifers, but only instars III and IV fed on crustaceans, such as *Daphnia, Bosmina*, and *Diaphanosoma* (Hare and Carter, 1987; Moore, 1988). In the larger instars, the daily ration of food, amounting to about one crustacean zooplankter per hour, is variable but can amount to approximately 10% of their body weight (Kajak and Ranke-Rybicka, 1970). It was estimated that *Chaoborus* could consume about 12% of the macrozooplanktonic biomass on the average and a greater percentage at certain times of the year, such as late in the period of summer stratification. Predation efficiency on rotifers increased in successive larger instars (I–III) of *Chaoborus* but was lower in comparison with other invertebrate predators (Moore and Gilbert, 1987; Swift, 1992). Ingestion rates of rotifers by older instars of *Chaoborus* (III and IV) were among the highest reported for invertebrate predators on rotifers. Predation upon rotifers, however, was found to be inadequate to significantly affect natural rotifer community dynamics (Havens, 1990). Rotifer reproductive capacity during midsummer in eutrophic lakes was capable of exceeding predatory losses.

TABLE 22-21 Production, Biomass, and $P/\overline{B}$ of *Chaoborus* Larvae in Socuy Reservoir, Venezuela[a]

Instar	*Production* (*mg* m^{-2})	*Biomass* (*mg* m^{-2} day^{-1})	$P/\overline{B}$
I	0.80	2.13	2.7
II	5.16	16.07	3.1
III	8.02	85.11	10.6
IV	21.92	468.07	21.4
Total/mean	35.90	571.38	15.9

[a] From data of López and Cressa (1996).

Predatory larvae of *Chaoborus* release a chemical compound (kairomone) into the water that induces antipredatory morphological changes in daphnid zooplankton (e.g., Hebert and Grewe, 1985; Tollrain and von Elert, 1994). The compounds, characterized as low-molecular-weight hydroxy-carboxylic acids, induce adults to develop a large helmet and pronounced spines along the carapace margin. Conversely, *Chaoborus* movements and activity increase markedly in the presence of chemical compounds released by prey (Berendonk and O'Brien, 1996). Chemical compounds released from planktivorous fish increase sensitivity to low light intensities and accelerated vertical migration rates to refuges in the sediments (Dawidowicz *et al.*, 1990; Tjossem, 1990).

It has been demonstrated further that *C. flavicans* is conspicuously selective in the food it consumes (Swuste *et al.*, 1973). Among zooplankton, copepods were actively selected over the cladoceran *Daphnia*; the pelagic ostracod *Cypria* was attacked but then immediately rejected. When given the choice between benthic oligochaetes and zooplankton, the oligochaetes were taken and predation on *Daphnia* ceased. Similar results have been found in a number of other studies (Swift, 1976; Lewis, 1978b; Smyly, 1980; Vinyard and Menger, 1980; Winner and Greber, 1980; Chimney *et al.*, 1981; Pastorok, 1981; Christoffersen, 1990; Hanazato, 1990; Lüning-Krizan, 1997). Differences in prey size and shape, as well as in swimming, escape, and vertical migration behavior, influence the selection of prey by different instars of *Chaoborus*.

The coexistence of 2 or more species of *Chaoborus* without any apparent great differences in their characteristics or requirements that would reduce competition has been noted often (Stahl, 1966a,b). Previously, it was assumed that the abundance of food precluded serious competition. Roth (1968), however, observed severe competitive interactions among three species of *Chaoborus* in the same lake. Slight differences in morphology of the larvae and their distribution, especially during nocturnal migration, were adequate to permit the two quantitatively minor species to coexist with the dominant species. Similarly, two coexisting species of *Chaoborus*, the dominant profundal fauna of Lake George in tropical Africa and in other lakes, were found to have interspecific differences in the mean number of eggs per adult female, the nature of egg batches, the morphology of mouth parts, as well as differences in the vertical distribution of the larvae and seasonal occurrence of the adults (McGowan, 1974; Sardella and Carter, 1983).

IV. STREAM BENTHIC COMMUNITIES

The macroinvertebrates (>0.5 mm), particularly the aquatic insects, have received intensive study in running water ecosystems. There is reasonable evidence that many orders and other major taxa of aquatic insects evolved in cool running waters prior to expanding to warmer running waters and lake environments (cf. Ward, 1992). The cool flowing waters, where dissolved oxygen concentrations are high and flow minimizes distances for transport to cell membranes, are logical environments to develop respiratory mechanisms and adaptations for extracting dissolved oxygen from the water. These invertebrates feed largely on algae and other attached microbiota and serve as primary food resources for fish and other vertebrates. Invertebrate turnover rates are intermediate between the microbes and the larger predators (Cummins, 1992). Because of their importance for fish populations and as indicators of water quality (e.g., Holsenhoff, 1987; Waters, 1988; Johnson *et al.*, 1993; Kerans and Karr, 1994), macroinvertebrates have been used widely in stream assessments.

More recent analyses of stream benthic communities have shifted toward analyses of changes in population and community dynamics and production in response to environmental variables. As among all aquatic organisms, environmental parameters influencing growth and reproduction include (a) physiological constraints of parameters such as temperature, dissolved oxygen, and osmoregulation, (b) constraints of food acquisition and food quality, (c) biological interactions of predation and intra- and interspecific competition, and (d) physical constraints of changing habitats within stream ecosystems. This same set of factors, as well as mobility (drift, swimming, crawling, and flight), affects the rates of colonization of new or rewatered channels, rivers recovering from pollution, and unstable rivers with fluctuating discharge (Mackay, 1992). Insects are secondarily adapted to life in water and are dependent on the terrestrial environment for part of their life cycle (e.g., Wallace and Anderson, 1996). As a result, aquatic insects are prevalent in shallow standing waters and streams, decrease in large, deep lakes, and are nearly absent in the open ocean.

Habitats are generally separable into ***riffle*** areas of fast-flowing water characterized by sediment erosion and into *pools* of depositional habitats (pools, backwaters, and side channels). Flow of water is a dominant parameter in running water ecosystems that has led to major morphological and behavioral adaptations of invertebrates in relation to movements, attachment, concealment, feeding, and reproduction (cf. Cummins, 1992; Vogel, 1994). Flow and its velocity fields affect interactively the stream habitat, dispersal of organisms, their acquisition of essential resources, intra- and interspecific competition, and predation efficacy. If benthic invertebrates live on the surface of the substrata, the organisms usually reside within the roughness layer, where the unique arrangement of sediment particles produces strongly sheared and highly three-dimensional flow patterns (Nowell and Jumars, 1984; Bouckaert and Davis, 1998; Hart and Finelli, 1999). Flow velocities within the roughness layer are unpredictable based solely on knowledge of the overlying flow within the logarithmic boundary flow layer. Shear stress is highest in the vertical plane in the wake of boulders, and as a result, turbulence intensity and kinetic energy are greatest in the wake region. Benthic macroinvertebrate fauna is significantly richer and more abundant in the wakes than in the front of boulders. Current velocities decline markedly within sediments, and often benthic invertebrates penetrate well within the hyporheic interstitial spaces among sand and gravel substrata.

It is common to find the maximum density of macroinvertebrates, particularly oligochaetes, chironomids, amphipods, and microcrustaceans (especially cyclopoid copepods and ostracods), at a depth of between 5 and 20 cm within porous sediments (D. D. Williams, 1984; Brunke and Gonser, 1999; Malard *et al.*, 1999). The depth to which hyporheic invertebrates penetrate sediments is likely restricted by the availability of food resources, which generally decrease as the ratio of particulate organic carbon to total fine particles declines. Although bacterial biomass can be appreciable in areas of upwelling in the hyporheic zone (cf. Chap. 21), food resources are generally low within the hyporheic zone, and some of the invertebrates near the surface migrate diurnally, usually at night, to the surface areas for feeding. Most invertebrate movements within the hyporheic zone are over longer periods (days or seasonally). The hyporheic zone can function as a refuge from predation during normal flow conditions, although during severe spates the surficial sediment substrata can be eroded severely with a very high loss (50–90%) of invertebrates (e.g., Palmer *et al.*, 1992). Some benthic microinvertebrates inhabit the hyporheic zone many meters from the stream channel and likely feed on microbes that are utilizing dissolved organic substrates that are being transported from upland regions through the hyporheic zone to the stream channel.

A generalized model of the shifts in the relative abundances of invertebrate functional groups along a river tributary ecosystem from head waters to mouth was set forth in the brief introduction to the River Continuum Concept in Chapter 8 (pp. 140–142) (cf. Cummins, 1992). The concept was originally developed among small streams of woodland areas where the influence of the terrestrial vegetation was high. Headwater streams of low (1–3) order, supplied largely from groundwater sources, were described as heavily canopied with trees and light limited (Fig. 8-8). Leaf litter from these trees formed a major input of particulate and dissolved organic matter. In middle reaches of streams where tributaries increase stream order (3–5), the water is commonly still clear and shallow, but now the tree canopy does not cover the streams. Consequently, macrophytes and attached algae reach maximum development and enhance food resources for benthic invertebrates (cf. Fig. 15-20 and 15-21). Fluctuations in discharge and temperature are generally larger as the influence of tributary inflows becomes more significant. In the lower reaches of larger rivers with increasing stream order (>6), turbidity increases and water transparency decreases (Fig. 15-20). Primary productivity within the river decreases with increasing reliance on allochthonous sources of organic matter (Figs. 8-9 and 15-21). The River Continuum Concept was later broadened to accommodate regional differences in hydrology, climate, tributaries, location-specific geology and lithology, vegetation, and human-induced factors to encompass broader spatial and temporal scales (Minshall *et al.*, 1983, 1985; Cummins, 1992).

Flooding events are distinct disturbances in small rivers. In large floodplain rivers, however, flooding is less of a disturbance to invertebrates because it provides a periodic habitat expansion in two dimensions (Benke, 1993b). The great majority of invertebrate production in medium-sized to large rivers (>6th order) is by collector organisms (gathering collectors in sand and mud and filtering collectors on snags) that primarily consume amorphous detrital particles and microbes that are flushed from floodplain detritus and soils. Flooding increases snag habitat in the main channel in the vertical direction and inundates large floodplain areas in the horizontal dimension, often 10–100 times greater than the area of the main channel. The flood plain provides invertebrate habitat in benthic zones of heavy organic deposition and on the surfaces of wood, both living and dead, and other wetland vegetation. Therefore, unlike the River Continuum Concept for streams, high invertebrate production and drift levels are dependent on an abundance of snag habitats, the occurrence of natural flooding regimes, and mobilization of organic matter among these habitats (Table 22-22).

It should be noted in the application of the River Continuum Concept (e.g., Figs. 8-8 and 8-9) that the ratio of gross photosynthesis to total ecosystem

TABLE 22-22 Selected Ecosystem and Zoobenthos Community Characteristics for Small-to-Medium-Sized Streams and Floodplain Rivers (e.g., >6th Order)[a]

	Small-to-medium-sized streams	*Floodplain rivers*
Riparian zone	Narrow	Broad floodplain swamps
Channel habitats	Rocky (gravel to bedrock)	Sand, snags
Slackwater habitats	Pools	Backwaters (sloughs)
In-channel system metabolism	Heterotrophic to autotrophic	Heterotrophic
Effects of floods	Catastrophic, brief	Beneficial, prolonged
Floodplain fauna (aquatic)	None on surface	Extensive, year-round
Functional groups (aquatic)		
Channel	Shredders, scrapers	Gathering and filtering collectors
Flood plain		Shredders; gathering collectors
Invertebrate production	Low to moderate (4–25 g DM m^{-2} yr^{-1})	Moderate to high (>25 g DM m^{-2} yr^{-1})
Invertebrate drift	Short distances and times; origin from benthic habitats	Long distances and times; origin from snags
Invertebrate food	Microbially colonized riparian litterfall, periphyton	Wetland and swamp-derived DOM, FPOM, microbes

[a] Modified from Benke (1993b).

TABLE 22-23 Comparison of Invertebrate Feeding Group Ratios and Stream Ecosystem Parameters for the Kalamazoo River, Southeastern Michigan[a]

	Stream order			
Parameter	*1*	*2*	*3*	*5*
Stream width (m)	1	5	10	45
Trophic status	Heterotrophic	Autotrophic	Heterotrophic	Autotrophic
P/R ratio	0.47	1.13	0.90	1.23
Transport, CPOM / FPOM	0.022	0.016	0.019	0.022
Storage, CPOM / FPOM	0.36	0.11	0.15	0.10
POM, storage/transport	0.10	0.16	0.23	0.16
Mean annual				
Invertebrates $m^{-2} \times 10^3$	19.6	15.0	63.6	41.7
Shredders/total collectors	0.22	0.003	0.002	0.001
Filtering collectors/gathering collectors	0.67	0.42	0.45	1.50
Scrapers/shredders	0.18	12.23	3.99	16.91
Scrapers/(shredders + total collectors)	0.08	0.24	0.11	0.05

[a] All invertebrate data used were means of fall–winter and spring–summer densities m^{-2} of individuals >0.5 mm (extracted from Cummins *et al.*, 1981). P = gross annual primary production; R = annual community respiration, POM = particulate organic matter (CPOM = coarse POM and FPOM = fine POM).

respiration (P/R) indicates whether a stream ecosystem is a net producer or consumer of organic matter but does not indicate the extent to which stream consumers are supported by autochthonous or allochthonous organic matter (Rosenfeld and Mackay, 1987). The transition between dominance by autotrophy and heterotrophy in streams is not properly expressed by $P/R = 1$. Separation of respiration of autochthonous and allochthonous organic matter cannot be measured readily, but estimates of the fraction of net primary production respired in a particular reach of the river improves the assessment of relative P/R of the organic sources of the entire heterotrophic community of the ecosystem (Meyer, 1989).

Shifts also occur in the relative abundance of feeding benthic invertebrates as food resources change over the longitudinal profile along the River Continuum Concept. In Figures 8-8 and 8-9 (pp. 140–141), the patterns reflect (a) the importance of particulate organic matter (leaf litter) inputs, commonly maximized in the head waters, which influences the relative density of shredder insects; (b) increases of scraper insect types where light and nutrients favor increased production of attached algae and cyanobacteria, normally in wide, shallow midreaches of rivers; (c) an intensified linkage between the abundance of collector feeding types and fine particulate organic matter, either in the head waters related to litter degradation or in the lower reaches as a result of importation from upstream tributaries and floodplain scouring, and (d) the fairly constant relative abundance (ca. 10%) of predators in all reaches of the river ecosystem.

An example of the shifts in feeding strategies and types by benthic invertebrates, dominated by aquatic insects, is well illustrated in the comparison of invertebrate feeding group ratios along the Kalamazoo River basin in southwestern Michigan (Table 22-23). The ratio of gross primary production (P) to community respiration (R) can be used as an index (P/R) of the relative amounts of autotrophic and heterotrophic metabolism in the water of the stream/river channel. Particulate organic matter (POM) shifts from a dominance of coarse (>1 mm) to fine (<1 mm) size fractions in storage, and organic matter is slowly decomposed in the progression downstream by the actions of heterotrophic microbiota and feeding activities of benthic invertebrates (Table 22-23). In addition, inputs of coarse POM decrease with increasing river size as the relative width of the riparian zone is reduced in proportion to river width and volume. The ratio of shredding to collector feeding types of invertebrates is highest in the headwater regions of high coarse POM, and decrease as the POM shifts to fine size categories (Table 22-23 and Fig. 8-9). Disturbance of the drainage basin, however, such as by clear-cutting of forest or alteration of riparian vegetation, results in less production of fine POM and less retention of organic matter by streams (Dudgeon, 1988; Golladay *et al.*, 1989). These declines in fine POM inputs and in retention by higher velocity, more erosive water flows can result in appreciable reductions in invertebrate production. In contrast, when a river is impounded ("regulated") in a reservoir, the marked reductions in flow and alterations of substrata result in quantitative reductions in total macroinvertebrate community abundance and biomass (e.g., Ward and Voelz, 1988; Gumiero and Salmoiraghi, 1998). Many zoobenthic species are unable to adjust to sublethal modifications of environmental conditions, such as substrata, oxygenation, et cetera, induced by stream regulation.

A. Movement and Drift of Lotic Benthic Invertebrates

Many invertebrates move about localized substrata habitats on a diurnal basis, often hiding in refuges areas during daylight periods and emerging to upper surfaces at night for increased feeding activities with reduced predatory pressures (e.g., Glozier and Culp, 1989). Immature insects also move large distances (several to tens of meters) both upstream and downstream over periods of days to weeks. Dispersal between streams and rivers can also occur in the adult stages by aerial movements.

The drag forces of flowing water, particularly during periods of high flow, can dislodge invertebrates from the substrata with which they are associated and transport them downstream. ***Drift*** is the downstream transport by water currents of organisms normally living in or on substrata. As a result the benthic invertebrates are constantly being redistributed. The extent of this drift dispersal leading to population expansion through movement downstream and colonization of new habitats varies greatly among species. Some species rarely drift at all, whereas others have a marked propensity to drift. For example, some 18% of the invertebrate taxa of the Rhine River of France constituted 98% of the total drifting organisms (Cellot, 1996). Scraper feeding types of insects tend to cling and rarely drift, whereas many gathering-collector insects are swimmers and exhibit directed drift behavior (Wilzbach *et al.*, 1988).

Thus, several types of drift occur. Constant low densities of ***continuous drift*** occur by simple accidental displacement of organisms from the substrata (Waters, 1972; Brittain and Eikeland, 1988). ***Catastrophic drift*** can result in high-density drift movements as a result of major physical (e.g., high flow) or chemical (e.g., pollution) disturbances.

A third drift category is *behavioral drift* that results from distinct diurnal patterns of behavioral activity by which immature insects move at specific times to avoid predators and other stresses. In both temperate and tropical regions, a general pattern of maximum drift occurs during darkness, often with a maximum just after sunset, followed by low drift during daylight in a distinct circadian rhythm (Fig. 22-6, *upper*). Dominant drifting insects include particularly many of the Ephemeroptera, simulid Diptera, and many Trichoptera and Plecoptera. Because of differences in activity associated with life histories, maximum drift occurs in the summer, with reduced drift in spring and autumn periods, and least in winter (Fig. 22-6, *middle*). Drift is least among the most immature instars, and commonly nocturnal drift activity increases among the larger instars (Fig. 22-6, *lower*). Certainly many factors contribute to the diurnal periodicity of drift activity (e.g., light, including moonlight, current velocity, substratum type, turbidity, pollution, food availability, and predation intensity). Clearly, however, the behavioral drift periodicity is not the sole result of passive release of benthos from the substrata and movement by currents downstream. Waters (1972) and Allan (1978, 1995) present compelling evidence that increased insect activity for foraging and increased drifting at night are adaptive responses to reduced risk of visual predation by fish. Invertebrate predation, such as by large stonefly larvae, on drifting invertebrates, however, can be very high during darkness. Increased drift can function as a predatory avoidance mechanism. When predatory fish and large stonefly larvae are absent in streams of certain areas, this behavioral drift periodicity is weakened or absent (e.g., Malmqvist and Sjöström, 1987; Malmqvist, 1988; Lancaster, 1990; Flecker, 1992).

In the main channel of large rivers (>6th order), drift densities of invertebrates are extremely high and are dominated by snag-dwelling aquatic insects and microcrustaceans from swamps and wetlands (Benke, 1993b). Drift distances, drift times, and the percentage of organisms drifting are substantially greater than is the case for invertebrate populations in smaller streams (Table 22-22).

Continual depauperation of benthic invertebrates by accidental or behavioral drift is apparently compensated for, in part, by recolonization of upstream populations by adults migrating in an upstream direction and oviposition in headwater regions. Some upstream movements occur within the streams as well (e.g., Söderström, 1987). Although production in excess of carrying capacity of certain regions of streams would not require compensatory upstream recolonization (Waters, 1961), drift has been demonstrated to be density-dependent. Infrequent dispersal by randomly flying adults coupled with density dependence does allow population persistence (Anholt, 1995). The marked diurnal periodicity of insect drift is certainly coupled to increased susceptibility of dislodgment by current when actively feeding on upper surfaces of substrata during darkness periods of reduced risk of predation.

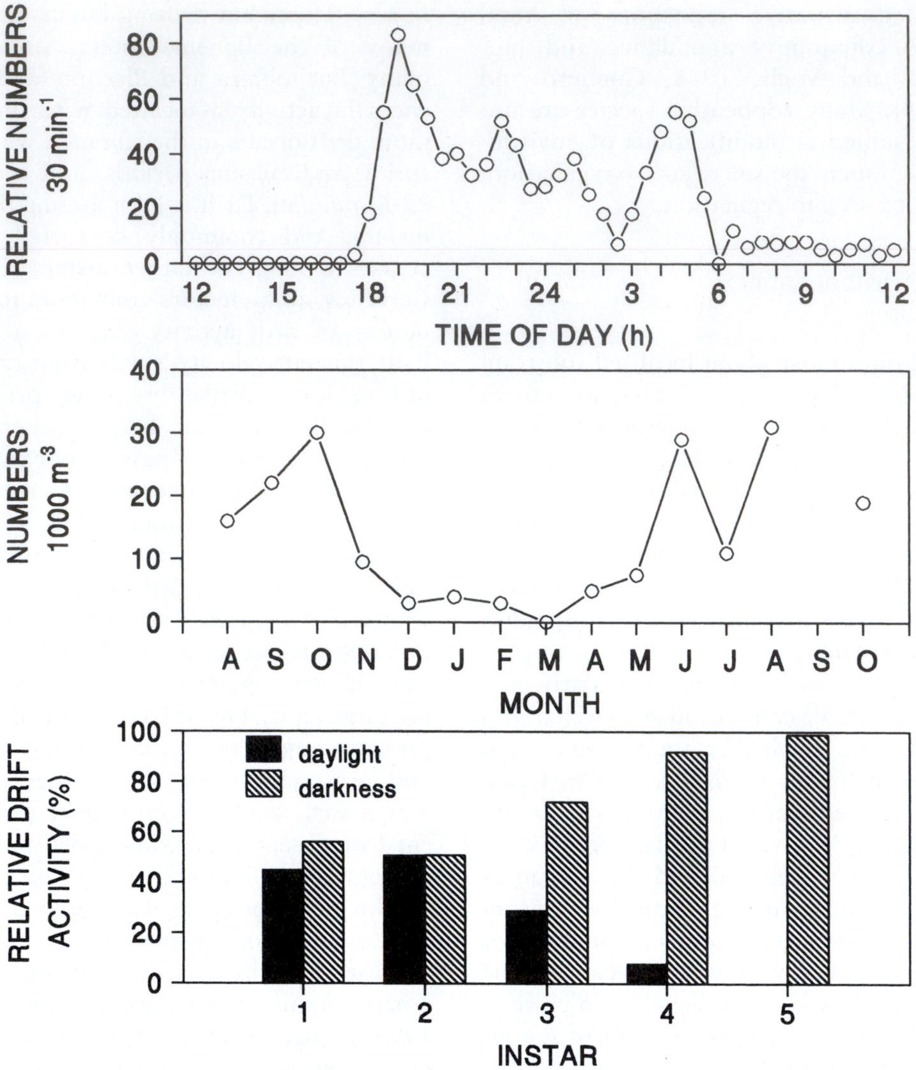

FIGURE 22-6 Drift patterns among certain stream invertebrates. *Upper:* Diurnal variations in comparative 30-min drift catches of the mayfly *Baetis rhodani* in a small stream in July. Ambient light was below 3 μmol quanta m^{-2} s^{-1} at 17.30 h and above this value at 06.00 h. (Drawn from data of Elliott, 1969.) *Middle panel*: Seasonal variations in drift intensity, with maxima in the summer and a minimum in winter, and (*lower panel*) changes in drift intensity with size, where larger instars became increasingly nocturnal in their drift activity, of the larvae of the caddisfly *Rhyacophila nubila* in a small stream. (Drawn from data of Fjellheim, 1980.)

V. RESERVOIR BENTHIC COMMUNITIES

Because soils flooded during the formation of reservoirs are relatively rich in nutrients, phytoplanktonic productivity is high. Although nutrients tend to decline during several years following flooding, usually nutrient loading from the drainage basin is adequate to maintain moderate-to-high primary production. Early high productivity of benthic invertebrates following inundation of the land is often associated with utilization of inundated herbaceous and woody terrestrial vegetation (e.g., Aggus, 1971). When trees and other vegetation are removed before inundation, benthic invertebrate abundance and production can be appreciably reduced (e.g., Applegate and Mullan, 1967).

Fluctuating water levels can alter severely the habitat and survival of benthic invertebrates of reservoirs. Macrovegetation, particularly submersed vegetation, does not survive well under marked fluctuations of water level. As a result, major habitat for benthic invertebrates is almost completely eliminated from many reservoirs. Changes in water level can also expose sediments to freezing conditions, and sandy sediments are more susceptible to freezing in winter than are organic-rich and stony detrital sediments (Palomärki and Koskenniemi, 1993). Such freezing of the sediments can eliminate over three-quarters of the invertebrate fauna from the exposed sediments.

In contrast, inundation of lowland riverine environments, as is common by the construction of low broad dams by the beaver (*Castor canadensis*) can greatly enhance the development of benthic invertebrates (e.g., Naiman *et al.*, 1984). Sediment accumulation increases markedly from the riverine erosive habitat to greatly enlarged (10- to 20-fold) depositional, organic-rich sediment that is readily colonized by highly productive aquatic macrophytes and periphyton. Increased nutrient loads are leached from the inundated surrounding land, in a manner similar to the well-known increased nutrient loading and productivity that follows periodic drawdown and desiccation of sediments of reservoirs (e.g., Neese, 1946; Baxter, 1977; Duthie and Ostrofsky, 1982). Essentially, the streams are shifted to highly productive wetland ecosystems, in which the productivity of benthic invertebrates increases markedly.

Increased productivity and organic loading to reservoirs can induce hypolimnetic anoxia and related toxic products of fermentative metabolism. The disappearance of many benthic invertebrate taxa and reductions in biomass and production of remaining invertebrates are found commonly in eutrophic reservoirs (Siegfried, 1984; Popp and Hoagland, 1995).

The regulation and suppression of flow in the stream or river below the dam can severely alter the microhabitats utilized by invertebrates downstream from the reservoir. Certain taxa, such as simulid dipterans, can be eliminated by reduced and altered flows, and in general the biomass of invertebrates declines under regulated flows (Gumiero and Salmoiraghi, 1998).

VI. PRODUCTION OF INVERTEBRATE BENTHIC FAUNA

Secondary production of invertebrate benthic fauna can be defined as the living organic matter or heterotrophic biomass produced by the animal populations during (through) an interval of time (Benke, 1984, 1993a). Therefore, secondary production is the flow rate of biomass produced, regardless of its fate (e.g., respiration, loss to predators, or emergence), and its units are biomass or energy per unit area per unit time (cf. Table 8-1, p. 144). As will be emphasized in Chapter 23, organic matter that enters an ecosystem and does not leave it physically (e.g., outflow or emergence of insects) must (a) be destroyed by respiration to CO_2 or CH_4; (b) be stored essentially permanently, such as in the sediments, or (c) be destroyed by physical processes such as photolysis of both particulate and dissolved organic matter with conversion to CO_2. Of course, because organic carbon is inefficiently recycled by consumers, the summed production of consumers exceeds the organic inputs to the ecosystem without violation of the thermodynamic principles (Strayer, 1988b). The secondary production of eukaryotic organisms, specifically the benthic invertebrates in this case, constitutes a small but important portion of the total heterotrophic production.

Measures of aquatic animal production are largely population-oriented, in part because of the methodology used for analyses of changes in complex life cycles. Determination of invertebrate and fish production at the population level usually measures growth and population survivorship of individual species. Single community-level measures for total animal production do not exist, although attempts have been made in that direction (cf. Hynes and Coleman, 1968; Benke, 1993a). Because production is a flow (flux) and biomass is a storage, the ratio of production to biomass ($P/\bar{B}$) is a rate in units of inverse time (Table 8-1). In invertebrates and fish, the time over which a cohort grows and completes development depends on the species and its environmental conditions. This time is quite variable (weeks or years), and cohort production must be known for accurate estimates of production. As discussed earlier among the zooplankton, the relationship of production to mean biomass provides important insights into the production process of different species, feeding types, and modes of reproduction. The extreme heterogeneity in distribution and the complex population dynamics of the benthic invertebrate fauna of lakes and streams have impeded detailed analyses of their productivity. Moreover, mortality from natural causes and by predation is complex and related to changes in the types and intensity of predation, since the macroinvertebrates change their size and distribution during their life cycles. Several aspects of these predation-related changes have already been discussed, especially among the oligochaetes, chironomids, and the phantom midge *Chaoborus*.

When a well-defined cohort structure exists in the population structure, production can be calculated between sampling intervals on the basis of instantaneous growth or size increment summation methods (cf. Benke, 1984, 1993a; Wetzel and Likens, 2000). When cohorts cannot be recognized, corrections are needed for development times that are greater than or less than a year based on changes in size distributions. Many of the existing quantitative data on the benthic fauna are expressed on the basis of biomass per unit area. A comparative set of examples of the average biomass among

TABLE 22-24 Comparisons of the Biomass of Benthic Macroinvertebrates in Lakes[a]

Lake	*Wet weight* (*kg ha*$^{-1}$)	*Dry weight* (*kg ha*$^{-1}$)	*Notes and source*
Great Slave, N.W.T.	20.0	4.0	Rawson (1955)
Athabaska, Alberta-Saskatchewan	(32.8)	6.6	
Minnewanka, Alberta	(36.0)	7.2	
Simcoe, Ontario	(99.2)	19.8	
Waskesiu, Saskatchewan	(198.0)	39.6	
Wyland, IN			
Mean 1955	—	8.05	Gerking (1962)
Mean 1956	—	8.60	
July	56.0	11.2	
August	31.6	6.3	
75 Finnish lakes	23.6	(4.7)	Deevey (1941) and Hayes (1957) after numerous sources
5 Swedish lakes	31.1	(6.2)	
3 New Brunswick lakes	25.4	(5.1)	
10 Lakes of Russia	41.3	(8.3)	
38 Lakes of USA, mostly of Connecticut and New York	87.2 (10–348)	(17.4) (2–70)	
13 Lakes of northern Canada	88.9	(17.8)	
43 European alpine lakes	76.1	(15.2)	
64 Lakes of northern Germany	115.0	(23)	(includes some mollusks)
19 Eutrophic Polish lakes	28 (2–56)	5.6 (0.4–11.2)	Pieczyńska *et al.* (1963)
15 Pond-type Polish lakes	101 (20–370)	20.2 (4–74)	
Dystrophic Store Gribsø, Denmark	94	(18.8)	Berg and Peterson (1956)
Weber, WI			
1940	553	(110.6)	Juday (1942)
1941	147	(29.4)	
Nebish, WI			
1940	122	(24.4)	
1941	590	(118)	
Shallow fish-culture ponds, Czechoslovakia	222–272	(44.4–54.4)	Lellák (1961)
Parvin Reservoir, CO	582	116.4	Buscemi (1961)
Green, WI			
0–1 m	36.6	7.3	Juday (1924)
0–10 m	82.2	16.4	
10–20 m	166.0	33.2	
20–40 m	171.1	34.2	
40–66 m	149.6	29.9	
Mendota, WI			
0–7 m	433.9	86.8	Juday (1942)
>20 m	696.8	139.4	
Average for lake	414	(82.8)	Juday (1921)
Memphremagog, Quebec–Vermont			Dermott *et al.* (1977)
North basin (mesotrophic)	—	21.9	
South basin (eutrophic)	—	43.2	

[a] Values are only approximately comparable because of the differences in methods and frequency and duration of sampling; all exclude mollusks unless stated to the contrary. Values in parentheses were estimated using the relationship: dry weight equals 20% of wet weight. Although Thut (1969) recommended 12.5%, Winberg *et al.* (1971), Gerking (1962), and others found about 20% (range, 15–25%) more realistic.

many lakes is set out in Table 22-24; the number of examples of this nature could be expanded many times over. Such data, however, are of quite limited value owing to great differences in sampling times and in sampling frequency and duration. As has been pointed out several times in the foregoing discussion, the summer period of low population densities and high predation pressure is often the poorest time for analyses. Many studies are conducted only during the ice-free period. In some streams and especially in sediments from deeper parts of lakes, the maximum period of growth often occurs during the winter months in the temperate zone. More importantly, in most of these studies data are insufficient on size–frequency distribution and other population characteristics to permit an evaluation of turnover rate and production rates. Animals with small biomass but high turnover rates (shorter generation times) can obviously be more productive and contribute as much or more to the system than can animals with high biomass but low turnover rates.

Thus, average data such as are summarized in Table 22-24 are not production estimates and must be viewed with caution. One not-too-surprising trend is observed, namely, that in general, the biomass of benthic fauna increases as the productivity of the ecosystem increases. The biomass of benthic animals of oligotrophic lakes in northern latitudes and alpine regions is quantitatively less than that of lowland lakes in temperate and tropical regions. Within some rather obvious limits, as food supply from autochthonous productivity and the diversity of littoral habitats increases, the productivity of benthic animals increases with greater eutrophication. If the organic loading becomes so great that the conditions of hypolimnion are intolerable to even the most adapted fauna, as occurs in extremely hypereutrophic lakes, productivity of the animals decreases. Generalizations based only on biomass data weaken rapidly beyond this level of discussion.

Changes in the total amount of biomass of benthic fauna with enrichment have been demonstrated conspicuously in the Lake Esrom example discussed earlier and many times in fertilization experiments. For example, high phosphate and nitrogen fertilization of fish culture ponds resulted in a 42% greater yield of benthic invertebrates, and 3.3 times greater yield of zooplankton, than yields of unfertilized ponds (Ball, 1949). In a similar study on ponds in Michigan, fertilization increased biomass of benthic invertebrates by about 75% and increased food organisms utilized by the common bluegill sunfish (*Lepomis*) by about 70%. (Patriarche and Ball, 1949). However, in a more detailed analysis of fertilization relationships in these shallow waters, Hall *et al.* (1970) demonstrated the complex interactions between the shifting development of macrophytic vegetation and increasing densities of phytoplankton, and the effects of these substrata on species composition of benthic animals and on their predation by fish. Although increases in productivity of benthic invertebrates were found in more nutrient-rich waters, in large part the production was compensated for by shifts in species composition owing to changing substrata and increases in mortality as a result of greater fish production and predation.

In the foregoing discussion, examples of estimates of the production rates of benthic macrofauna have been given for individual groups or for specific regions of streams or lakes, such as the profundal zone. Most benthic secondary production studies deal with only one to a few species simultaneously. Studies of the rates of production of the total invertebrate benthic fauna of whole-lake systems, for which sufficient detail is available to calculate turnover rates over the entire year, are much less common. The data in Table 22-25 for lakes and reservoirs and Table 22-26 for streams and rivers can be treated only as estimates, but nonetheless show the general relationships.

1. In plots of the frequency of occurrence of the annual production values among different benthic invertebrate groups (annelids, mollusks, crustaceans, and insects) in many lakes and rivers, the most common production values were between 1 and 10 g DM m^{-2} yr^{-1} (10 and 100 kg DM ha^{-1} yr^{-1}) for all groups (Waters, 1977; Benke, 1984, 1993a). Annual production of a benthic community in the 50 g DM m^{-2} yr^{-1} range is high in both lentic and lotic ecosystems.
   a. Production in running waters is usually at least two times greater than in standing waters. Highest production values were found among the filter feeders that receive food particles largely delivered by current (Fig. 22-7). Lower production values were found among gathering collectors and scrapers.
   b. No obvious relationship is apparent between discharge and between temperature and production, although total production of primary benthic consumers (that consume flow-transported particles) increases directly with discharge in the range of 0.1–2 m^3 s^{-1} (Benke, 1993a).
   c. Benthic secondary production of pollution-tolerant species is usually larger, commonly doubled, in organically enriched environments of both lakes and streams.
   d. Benthic production is generally higher in unshaded than in shaded reaches of streams

TABLE 22-25 Estimates of Annual Production (Dry Weight) of Total Benthic Invertebrates in Lakes and Reservoirs[a]

Benthic organisms	*Production* ($g\ m^{-2}\ yr^{-1}$)	*Lake*	*Source*
All benthic animals	2.95	Bodensee, Germany	Streit and Schroder (1978)
	8.4	Wyland Lake, IN	Gerking (1962)
	15.2	Marion Lake, British Columbia	Hall and Hyatt (1974)
	28.3	Mikolajskie Lake, Poland	Kajak and Rybak (1966)
	36.7	Taltowisko Lake, Poland	
	50.7	Bay of Quinte, Lake Ontario	Johnson and Brinkhurst (1971)
Herbivores-detritivores	0.37	Lake Krivoe, Russia	Alimov *et al.* (1970)
Carnivores	0.08		
Herbivores-detritivores	0.90	Lake Krugloe, Russia	
Carnivores	0.14		
Herbivores	4.10	Volchja Reservoir, Russia	Pidgaiko *et al.* (1970)
Predators	0.61		
Herbivores-detritivores	5.71	Lake Manitoba, Manitoba	Tudorancea *et al.* (1979)
Carnivores	1.01		
Herbivores-detritivores	21.44	River Thames, England	Mann *et al.* (1970)
Predators	3.48		
Nonpredators	1.81	Average of three lakes, Russia	Winberg *et al.* (1970)
Predators	0.56		
Nonpredators	0.47	Lake Zelenetzkoye, Russia (subarctic)	Winberg *et al.* (1973)
Predators	0.02		
Nonpredators	34.1	Kiev Reservoir, Russia	Winberg *et al.* (1970)
Nonpredators	20.4	Five Polish lakes	Kajak *et al.* (1970)
Chironomids ($\bar{x}$ for 4 yr, annual)			
Tokunagayusurika akamusi	3.7	Lake Kasumigaura, Japan	Iwakuma (1987)
Chironomus plumosus	3.2		
Clinotanypus sugiyamai	0.052		
Procladius culiciformis	0.006		

[a] Based on the approximate conversion coefficients for invertebrates of Waters (1977): 1 g dry weight = 5 g wet weight = 5 kcal = 0.9 g ash-free dry weight = 0.5 g carbon.

(Hopkins, 1976), related in part to higher autochthonous primary production. Annual benthic macroinvertebrate production commonly increases with stream size, with highest values in 6th- and 7th-order rivers (Grubaugh *et al.*, 1997).

2. The productivity of predatory benthic fauna was consistently found to be much lower than that of the nonpredatory benthic fauna (e.g., Fig. 22-7). For example, productivity estimates of the nonpredatory benthic fauna were found to be 5–35 times lower than those of nonpredatory zooplankton (Winberg, 1970).
3. In general, the productivity of the nonpredatory benthic fauna accounts for <10% of the phytoplanktonic primary productivity and decreases appreciably as the productivity of macrophytes increases and the loading of allochthonous organic matter to the lake increases in proportion to phytoplanktonic productivity.
4. In shallow lakes with relatively low mean depth (2–7 m) or where littoral zone development is extensive, zoobenthos production is commonly higher by 2–5 times than zooplankton production (Strayer and Likens, 1986; Jónasson, 1992; James *et al.*, 1998). In larger, deep lakes zooplankton production commonly equals or exceeds that of the zoobenthos, perhaps related to food limitations in the profundal zone (see Table 22-27).

Annual turnover rates for benthic invertebrates, calculated as the ratio of annual production to mean biomass, vary considerably (see reviews of Waters, 1977; Benke, 1993a). Most values, however, follow the general rule of larger $P/\bar{B}$ ratios with increasing number of generations per year (Table 22-28). Among benthos of lakes, cohort $P/\bar{B}$ ratios show considerable constancy between 3 and 5, irrespective of the length of cohort life, similar to the annual $P/\bar{B}$ ratios for univoltine species. The relationship of the annual $P/\bar{B}$ ratio to voltinism is directly attributable to water temperature among benthic invertebrates, since water temperature is clearly a dominant factor, in addition to food

TABLE 22-26 Total Invertebrate Production (or a High Fraction of Production) of Selected Streams and Rivers[a]

Water/location	*Production* ($g\ m^{-2}\ yr^{-1}$)	*P/B*	*Discharge* ($m^3\ s^{-1}$)	*Temperature* (°C)	*Source*
C III, NC	13.99	5.0	0.001	12.9	Lugthart and Wallace (1992)
Rold Kilde, Denmark	7.953	2.7	0.003	7.4	Iverson (1988)
Bear Brook, NH	4.100	2.7	0.020	7.0	Fisher and Likens (1973)
Snively, WA	14.143	19.4	0.040	10.8	Gaines *et al.* (1992)
Sycamore, AZ	120.90	117.4	0.050	17.0	Jackson and Fisher (1986)
Hinau, New Zealand					
Shaded	8.40	7.1	0.1	11.0	Hopkins (1976)
Open canopy	32.53	7.2	0.1	11.0	
Organic enrichment	73.69	7.2	0.1	11.0	
Estaragne, France	6.236	4.3	0.1	1.6	Lavendier and Décamps (1984)
North Branch, MN	22.516	5.0	0.2	9.5	Krueger and Waters (1983)
Caribou, MN	5.551	4.2	0.3	8.5	
Douglas, WA	23.185	13.2	0.6	13.0	Gaines *et al.* (1992)
Cedar, SC	3.054	8.8	1.2	17.6	Smock *et al.* (1985)
Ilm, Germany	43.170	5.1	5.0	9.0	Flössner (1982)
Oconee, GA	41.597	11.1	8.1	16.6	Nelson and Scott (1962)
Thames, England	27.780	0.9	37.9	11.2	Berrie (1972)
Satilla, GA					
Sand	21.025	163.4	66.0	19.0	Benke *et al.* (1984)
Mud	17.934	29.1	66.0	19.0	
On snags/wood	64.826	15.6	66.0	19.0	
St. Lawrence, USA–Canada	14.889	1.8	6739.0	—	Edwards *et al.* (1989)

[a] Selected from the larger analysis of Benke (1993).

TABLE 22-27 Relative Production of Zooplankton and Zoobenthos in Lakes, Expressed as a Percentage of Approximate Net Organic Carbon Inputs to the Lakes

Lake	*Net organic C input*[a] ($g\ C\ m^{-2}\ yr^{-1}$)	*Production (% of net organic C inputs)*		*Reference*
		Zooplankton	*Zoobenthos*	
Findley, WA, USA	12	4	6	Wissmar and Wetzel (1978)
Tundra Pond, AK, USA	26[b]	0.8	7	Hobbie (1980)
Mirror, NH, USA	49	5	12	Strayer and Likens (1986)
Pääjärvi, Finland	60	12	3	Sarvala *et al.* (1981)
Rybinsk Reservoir, Russia	93[c]	2[c]	0.3[c]	Sorokin (1979)
Marion, British Columbia, Canada	110	0.9	3	Wissmar and Wetzel (1978)
Red, Russia	140[c]	7[c]	1.4[c]	Andronikova *et al.* (1972)
Esrom, Denmark	160[b]	7	6	Jónasson (1972)
Naroch, Russia	160[b,c]	5[c]	0.8[c]	Winberg *et al.* (1972)
Dalnee, Russia	260[c]	22[c]	1[c]	Sorokin (1979)
Mikolajskie, Poland	260	21	2	Kajak (1978)
Kiev Reservoir, Russia	280[c]	7	6	Gak *et al.* (1972)
Mývatn, Iceland	330[b]	1.3	6	Jónasson (1981)
Wingra, WI, USA	610	4	0.4	Wissmar and Wetzel (1978)

[a] Primary production + allochthonous inputs—losses by outflow. From some sources, kcal $m^{-2}\ yr^{-1}$ were converted to g C $m^{-2}\ yr^{-1}$ by dividing by 10.
[b] Not including allochthonous inputs.
[c] Vegetative season only (May–Oct.).

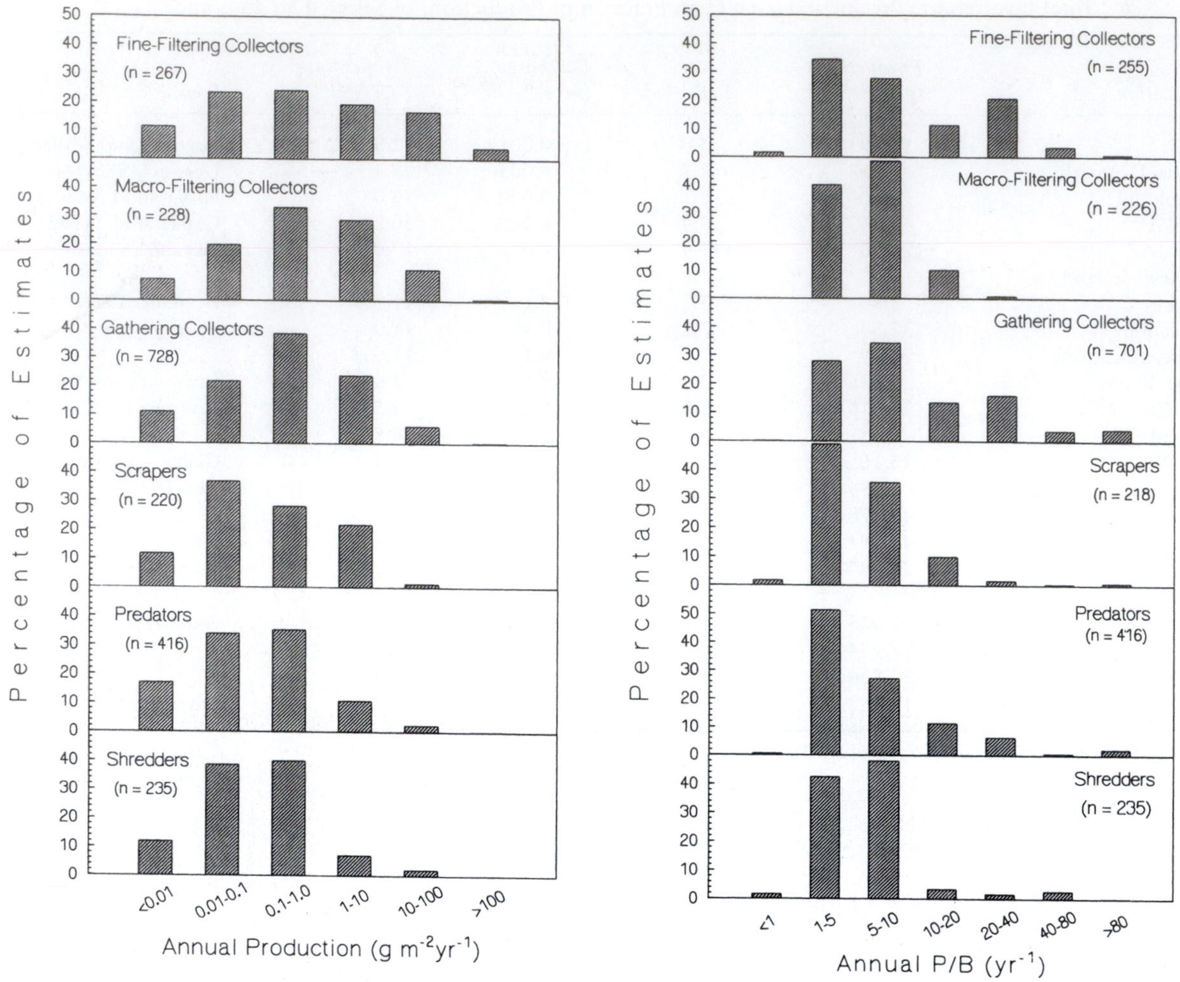

FIGURE 22-7 Distribution of production estimates (*left*) and annual $P/\overline{B}$ ratios (*right*) for various functional feeding groups of stream invertebrates. (After Benke, 1993a.)

resource availability of adequate quality, affecting the number of generations per year. $P/\overline{B}$ ratios have been empirically approximated by $T^2/10$, where T = mean temperature (°C) at the bottom of lakes (Johnson and Brinkhurst, 1971).

TABLE 22-28 Ranges and Average Production/Biomass Ratios of Benthic Invertebrates[a]

Generations of life cycle	*Average P/B ratio*	*P/B ratio range*
Multivoltine (2–3 generations per year)	8.3	3–16
Bivoltine	5.4	2–9
Univoltine	4.5	<1–9
Hemivoltine (2–3-yr lifespan)	2.2	<1–4
Long-lived mollusks	1.7	<1–5

[a] After discussion of Waters (1977).

Most studies of zoobenthos production report only that attributable to macroinvertebrates (>0.5 mm). Where the meiofauna (all metazoans <0.5 mm) are also analyzed, about half of the zoobenthic assimilation of carbon results from the meiofauna (Strayer and Likens, 1986). As a result, most of the cited values for zoobenthos production are conservative underestimates.

Among stream and river benthos, annual $P/\overline{B}$ values for the fine-filtering and gathering collectors is broad (Fig. 22-7) and is likely related to size variation among this group (e.g., large bivalves versus small dipterans) (Benke, 1993a). Macrofiltering collectors are capable of high production but have high biomass and modest $P/\overline{B}$ ratios. Scrapers are of relatively large size and have modest $P/\overline{B}$ ratios. Over 75% of stream predators have $P/\overline{B} < 10$, with relatively long development times and large body size. The lowest $P/\overline{B}$ ratios are found among the shredders, with $P/\overline{B}$ ratios <10 among 90% of the estimates.

VII. FISH PREDATION AND OTHER MORTALITY OF BENTHIC FAUNA

The subject of fish biology and the fish fauna of lakes and streams is an enormous field to which much attention has been devoted. The diversity in the morphology, physiology, behavior, reproductive capacity, growth, and feeding of freshwater fishes is a field of study that has been treated in detail in many recent works. This separation of the treatment of fish biology, ecology, and population dynamics with limnology is perhaps justifiable from the standpoint of giving most appropriate attention to these organisms of major economic importance and because the subject matter is so extensive. The schism is unfortunate in many other ways, however, particularly because fish behavior and feeding can alter the trophic structure and productivity of lakes and rivers appreciably.

As already discussed in the coupling of fish with zooplankton, fish are an integral component of freshwater ecosystems. Although the amount of energy and carbon flux to the fish trophic level is exceedingly small in most aquatic systems, fish are an operational component within the ecosystem. The impact of fish on the operation of the ecosystem in terms of carbon flux and nutrient regeneration at times can be quite significant, as has been discussed repeatedly in nearly every foregoing chapter. For example, the shift of fish species feeding on larger-sized food organisms to planktivorous species can have marked effects on zooplanktonic composition and productivity. This change can in turn influence the species composition of phytoplankton and facets of productivity at the primary level.

Another conspicuous example occurs during eutrophication of lake ecosystems, in which fish species change dramatically from salmonid and coregonid species of quite stringent low thermal and high oxygen requirements to warmwater species that are increasingly tolerant of eutrophic conditions. The warmwater fish are restricted to the epilimnion during summer stratification. Certain fish, such as the carp, have omnivorous feeding habits and can be very effective in modifying the littoral substrata to the point where many submersed macrophytes are eliminated (cf. Chaps. 18 and 20). Such feeding activities often can so disturb the sediments that turbidity is increased and transparency and phytoplanktonic productivity decline.

In general, evidence suggests that the proportion of zoobenthic production utilized by invertebrate predators is likely at least as great, if not greater, than that of vertebrate predators in many streams and lakes (Strayer and Likens, 1986; Wallace and Webster, 1996). For example, in Mirror Lake, New Hampshire, predation by invertebrates accounted for 80% of the fate of zoobenthic production, versus 15% by predation by fish and the remainder as insect emergence. In Lake Pääjärvi, Finland, some 50% of zoobenthic production was removed by invertebrate predation, 40% by fish predation, and the remainder in insect emergence (Sarvala *et al.*, 1981).

Nonetheless, a large number of studies have indicated the major effects of fish predation on the population dynamics and productivity of benthic macroinvertebrates, as has been discussed already in this and previous chapters. Many other examples exist (e.g., Post and Cucin, 1984). In Third Sister Lake, Michigan, the littoral benthic fauna increased to a maximum concentration in early winter (Ball and Hayne, 1952). The invertebrate populations declined steadily after the formation of ice cover until spring loss of ice, when an abrupt reduction occurred as adult insects emerged. During summer stratification, the populations and their biomass progressively increased from this annual spring low to the autumn and winter maxima. Experimental removal of all of the fish from the lake not only resulted in a large increase in the biomass and numbers of benthic fauna but made the low populations characteristic of early summer much less pronounced than under conditions of normal fish predation. Experimental manipulation of fish populations in half-hectare ponds showed that the presence of fish predation resulted in a very significant increase in the rate of production of benthic invertebrate fauna and a reduction in their biomass (Hayne and Ball, 1956). In the absence of fish, the apparent production rate of benthic fauna decreased and stabilized at a higher level of benthic faunal biomass relative to ponds in which fish predation occurred.

Although there are great differences in the qualitative composition of invertebrate fauna and seasonal succession of fauna among littoral stands of emergent vegetation, little difference is found between the quantity of macrofauna and the production of higher emergent vegetation in productive lakes or ponds (e.g., Dvořák, 1970). The presence or absence of submersed vegetation, however, results in large differences. These plants can in turn be influenced markedly by fish and waterfowl. A number of studies have experimentally excluded the feeding and disturbance activities of fish and waterfowl from the littoral vegetation (see, e.g., Korínková, 1967; Kajak, 1970). The biomass anc production rates of benthic macrofauna were severel reduced (usually by >50%) when the populations wer exposed to fish (carp) and birds (especially ducks). Th reductions resulted from increased predation pressu and from losses of submersed vegetation (the latt provided both refugia and food). Although mortality benthic invertebrates can be very high from fish prec tion (at least half), availability of adequate food sup

TABLE 22-29 Estimated Rates of Production of Profundal Macroinvertebrates of Lake Beloie, Russia, over an Annual Period, 1935–1936, and the Fate of the Fauna within the Lake[a]

	kg (dry weight) lake^{-1}	*Percentage*
Net production	4010	100
Emerging insects	256	6
Predation mortality (mainly by fish)	542	14
Natural mortality and decomposition	2213	55
Surviving population at end of year	999	25

[a] Generated from data of Borutskii (1939).

is usually the dominant factor controlling the population dynamics and productivity of benthic invertebrates (see Lellák, 1965; Hall *et al.*, 1970).

Studies of the profundal macroinvertebrate fauna–for example, in the early detailed analyses by Borutskii (1939)—showed that predation mortality, chiefly by fish, was relatively low (Table 22-29). Only a small percentage of the profundal macroinvertebrates lived to the adult stage. His studies indicated that the turnover rate was about two times the mean biomass, which is in general agreement with other studies (production rates equal about two to five times the mean annual biomass). There is evidence that the turnover values are considerably higher in the littoral zone (4–8 times the mean annual biomass) (see, e.g., Anderson and Hooper, 1956).

In a detailed study by Gerking (1962), the production rates of zooplankton and macroinvertebrate fauna were determined both directly and by measuring utilization by the dominant predator, the bluegill sunfish (*Lepomis macrochirus*). Population dynamics and mortality analyses showed that the dense fish population grew slowly in this eutrophic Indiana lake because individuals became overcrowded and undernourished. The diet of the fish was varied, but the dominant prey were midges (45%) and *Daphnia* (26%) among all size classes of fish (Table 22-30). Each year the diet shifted from a high proportion of *Daphnia* in July to a high proportion of midge larvae in August as the cladoceran populations decreased drastically during the summer. The estimated fraction of midge production that was consumed by fishes during the summer was 0.50 (Table 22-30), which indicates a high cropping efficiency of the benthic fauna, much higher than "losses" through emergence as adults.

It is often assumed that density-dependent competition for food resources becomes inconsequential when predation suppresses consumers (e.g., Hairston *et al.*, 1960; Carpenter and Kitchell, 1993; many others). Clearly, however, resource limitations and predation operate simultaneously on populations, and metabolic limitation is essential for thermodynamic and ecosystem stability (e.g., Wetzel, 1995; Osenberg and Mittelbach, 1996). In many situations, small insectivorous fish can influence benthic insect populations markedly at certain times of the year, but careful analyses indicate that seasonal limitations of detritus and associated microbes can influence insect growth and production more than predation (e.g., Batzer, 1998). Complex interactions of life history sequences, food availability and omnivory, food quality changes, prey refugia, and a number of other factors in addition to predation all influence growth and production of benthic invertebrates.

The production rate of the bluegill fish was estimated by multiplying the instantaneous rate of growth by the average summer biomass and adding the recruitment to this value (Table 22-30). The minimum summer production of benthic fauna was computed as the amount necessary to replace the losses by predation. Midges and other dipteran larvae accounted for more than one-half of the production of food that was derived from the bottom fauna. Therefore, it was estimated that minimum food production was 4 to 5 times the production of bluegills during the summer, and probably near 10 times bluegill production on an annual basis. As is common among fish of the temperate zone, both reproduction and growth of the bluegill sunfish are limited to the spring, summer, and early autumn. Spawning occurs for about a month in the spring, depending on the temperature. Growth occurs for about 5 summer months, during which time nearly all of the annual production occurs, about 625 kg ha^{-1} in this example (Table 22-30). Growth and feeding decreased abruptly in the autumn and remained low for the remainder of the year. In contrast, the production of benthic fauna continued for much of the year and increased greatly in the fall, under reduced predation pressure, until about the time ice cover formed. On an annual basis, the benthic faunal production is much larger (approximately 5–10 times) than that utilized by fish.

A similar situation in found in running water ecosystems. In studies of a stream in New Zealand, Allen (1951) found turnover rates by $P/\overline{B}$ ratios of benthic invertebrates to be much higher that reported previously.[6] Allen estimated that fish required many times more benthic prey biomass than was available at any instant in time, a discrepancy later termed the

[6] Subsequently corrected by Gerking (1962) and reduced by a factor of 2–3, but the values still demonstrated the large discrepancy.

TABLE 22-30 Relationships between the Bluegill Sunfish (*Lepomis macrochirus*) and Its Food Supply of Benthic Fauna and Zooplankton in Wyland Lake, Indiana[a]

Bluegill sunfish	*Weight (kg ha⁻¹)*		*Food supply*	*Weight (kg ha⁻¹)*	
	Protein	*Live weight*		*Protein*	*Live weight*
Biomass			July biomass		
June	17.0	98	Benthic fauna	6.1	56
Average	12.2	72	Zooplankton	9.2	191
			Total	15.3	247
Percent diet constituents			August biomass		
Benthic fauna	—	74	Benthic fauna	3.4	33
Zooplankton	—	26	Zooplankton	1.1	23
Midges	—	45	Total	4.5	56
Monthly summer food turnover			July-August decline		
Benthic fauna	6.2	64	Benthic fauna	2.7	24
Zooplankton	2.2	23	Zooplankton	8.1	168
Total	8.4	87	Total	10.8	192
Annual production			Minimum summer		
Growth	6.5	37	production rate		
Recruitment	9.1	54	Benthic fauna	17.8	196
Total	15.6	91	Midges	9.2	71

	Production turnover ratio	
	Protein	*Live weight*
Monthly summer food turnover compared with		
July biomass		
Benthic fauna	0.99	1.14
Zooplankton	0.24	0.12
July–August decline		
Benthic fauna	2.30	2.67
Zooplankton	0.27	0.14
Summer efficiencies based on total production		
Production biomass/intake	0.15	0.09
Production fish/production food	0.23	0.19
Intake by fish/production food		
Benthic animals	0.48	0.47
Midges	0.50	0.55

[a] Modified from Gerking (1962). Mahon (1976) reevaluated the fish production data of Gerking's study and found that production values of the younger bluegills were underestimated (corrected to approximately 625 kg ha^{-1} yr^{-1} total annual production). Most of these younger-aged fish would prey more heavily on zooplankton than upon benthic organisms.

Allen paradox (Hynes, 1970). Much of the discrepancy is related to the multiple generations per year of the prey species, where high prey turnover sometimes required by predators is easily obtainable, with an annual P/B of 30–40, by Chironomidae and other fast-growing invertebrates in warm environments (Benke, 1979, 1993; Waters, 1979). Such high annual $P/\bar{B}$ values are commonly found in both streams and small lakes among the benthos.

When discussing production of invertebrates within the stream and river channels, it is important to note that appreciable numbers and biomass of terrestrial arthropods enter streams, particularly during the summer (Mason and MacDonald, 1982; Cloe and Garman, 1996). Terrestrial arthropods represent an important energetic subsidy to stream fish during the period of low aquatic macroinvertebrate availability. Values of terrestrial arthropod inputs between 50 and 200 mg DM m^{-2} day^{-1} during late spring and summer periods are common. High mortality, particularly in the early stages of aquatic insects, is often attributed to predation, starvation, or mechanical (accidental) death (e.g., from molar abrasion of sediments in flowing waters). It is apparent that fish and other invertebrates can consume a major portion of benthic prey production in habitats, such as productive streams, where refuges are limited. However, there are many other sources of mortality of benthic invertebrates. Significant mortality occurs even when all of these causal mechanisms are eliminated. Protozoan, fungal, and anaerobic bacterial pathogens are common among invertebrates and likely result in a constant attrition of the diverse benthic communities (e.g., Cummins and Wilzbach, 1988).

Although much of the food demand by fish is apparently derived from surficial macroinvertebrates, which had high (6.9) $P/\overline{B}$ ratios, other prey sources with lower $P/\overline{B}$ ratios (hyporheic and terrestrial macroinvertebrates) were also important (Huryn, 1996). Where predation by fish is reduced or absent, invertebrate production is controlled to a greater extent by resource availability, and $P/\overline{B}$ ratios are lower (Huryn, 1998).

VIII. FISH PRODUCTION RATES

Although an enormous literature exists on the yield of fish per given area of freshwater habitat, relatively few analyses exist of their growth and mortality over annual periods. Consequently, only a few reasonable estimates of production rates can be made. Growth is highly variable among different species of fish. In general, the instantaneous growth is correlated inversely with fish size. At greater sizes, most of the food energy is utilized for maintenance; in young fish, a larger proportion of the energy intake is diverted into growth. Hence, the ratio of production to biomass usually decreases with age and greater size. Reproductive effort by sexually mature members of a fish population can represent a high proportion of the annual fish production.

Summary estimates of net production rates of various fish groups set forth in Table 22-31 demonstrate the general ranges encountered. Production rates are considerably higher in tropical waters than they are in temperate fresh waters, where growth is restricted to about half of the year. In fertile standing waters, the production rates of herbivorous fish in the tropics can reach several hundred g m^{-2} yr^{-1} (Chapman, 1967, 1978). In standing waters of temperate regions, in which a single species often predominates, the range of rates is from 1 to about 20 g m^{-2} yr^{-1}. Production rates in streams are usually higher than in standing waters (Table 22-31) and average about 50 g m^{-2} yr^{-1} in temperate areas.

Annual $P/\overline{B}$ ratios for fishes are generally lower than is the case among most invertebrates as a result of their longer lifespans (several years). Relatively high $P/\overline{B}$ ratios are found in the sculpins, for example, which have shorter lifespans (Table 22-31), because the younger, faster-growing stages contribute disproportionately to the population productivity. Annual $P/\overline{B}$ ratios of a population in an expanding or colonizing stage, such as in a new reservoir or in a stream recovering from flood damage, will be high, because growth is high relative to mortality (Waters, 1977). A population that is overcrowded and stunted, as is commonly the case with sunfish populations in temperate lakes, have lower annual $P/\overline{B}$ ratios, slower growth, and longer lifespans. A very approximate estimate of annual production can be obtained if the annual mean biomass of zooplankton is multiplied by 15–20, of benthic animals by 6–8, and of fish by 0.5–1.0 (1.2 for stream salmonids).

TABLE 22-31 Estimates of Rates of Production of Fish from Various Fresh Waters[a]

Fish group/communities	*Annual production ($kg\ ha^{-1}\ yr^{-1}$)*[b]	*$P/\overline{B}$ ratios*
Salmonidae (trout, salmon)		
From standing waters	0.21–66	0.62–2.0
From streams	11–300	1.0–5.0
Cottidae (sculpins)	8.0–431	1.2–5.5
Percidae (perch family)	0.91–52	0.35–2.4
Esocidae (pike family)	0.75–14.2	0.32–0.7
Cyprinidae (minnow and carp family)	0.1–915	0.18–1.94
Others	40–625	
Total fish fauna, multispecies		
Temperate zone	90–1980	
Tropical zone	1306–3468	
Planktivores (Russian lakes)	9–24	0.7–0.8
Benthivores (Russian lakes)	13–60	0.4–0.5
Piscivores (Russian lakes)	9–14	0.3–0.4

[a] After data from numerous sources, particularly Chapman (1967, 1978), Winberg *et al.* (1970), Waters (1977), and Morgan *et al.* (1980). Values are expressed as wet weight; approximate conversion values for fishes: 1 g wet weight = 0.2 g dry weight, or approximately 1 kcal.
[b] $kg\ ha^{-1}\ yr^{-1} = 0.1\ g\ m^{-2}\ yr^{-1}$.

IX. INVERTEBRATES AND FISH AMONG AQUATIC ECOSYSTEMS

The diversity of invertebrates associated with benthic substrata and their productivity is enormous among river, reservoir, lake, and wetland/littoral ecosystems. Because of the unusually extensive study and understanding of benthic invertebrates, particularly insects, of streams, some comparative distinctions are possible between small-to-medium-sized streams (1st–5th order) and larger floodplain rivers (6th or greater order). An overarching emergent property is

TABLE 22-32 Comparative Characteristics of Zoobenthos and Fish Communities among River, Reservoir, and Natural Lake Ecosystems[a]

	Rivers			
	Small-to-medium-sized streams	*Floodplain rivers*	*Reservoirs*	*Natural lakes*
Benthic invertebrate habitat	Narrow riparian zone; rocky (gravel to bedrock) and flooding brief but catastrophic	Broad floodplain swamps; sand, snags, and backwater sloughs; flooding prolonged and beneficial	Large, astatic littoral sediments, with large, irregular water-level fluctuations; sediment loading high and gradation high	Shoreline development low, water levels stable; sediment loading low; gradation slow
Zoobenthos community structure and feeding	Diverse with shredders and scrapers dominant in channel; absent on surface of flood plains. Food often dominated by microbially colonized litterfall and periphyton	Very diverse, with gathering and filtering collectors in channel and shredders and gathering collectors in flood plains. Food dominated by wetland and floodplain DOM, FPOM, and microbes	Low diversity with minimal and irregular littoral zone; largely gatherers	Moderate in profundal to very high diversity in littoral and marginal wetlands; decreasing diversity in profundal of eutrophic lakes; diverse feeding especially in littoral areas
Zoobenthos production	Low to moderate (4–25 g DM m^{-2} yr^{-1})	Moderate to high (>25 g DM m^{-2} yr^{-1})	Low generally and dominance of rapid growth (r) selection; high production common during initial flooding of terrestrial vegetation	Moderate in mesotrophic lakes, decreasing with major reductions as hypolimnetic oxygen depletion occurs; increasing markedly with greater littoral and wetland development
Fish communities	Warm and cold water species; numerous adaptations to turbulent environment	Largely warm water species; numerous adaptations to a turbid environment	Predominantly warm water species composition; differences often related to initial stocking; spawning success variable and low with low water levels; egg mortality increases with siltation; larval success reduced with less refugia	Warm and cold water species composition; spawning success good; egg mortality lower, larval success good
Fish production	Moderate to high, average ca. 50 g DM m^{-2} yr^{-1}	Moderate to high	Initially (5–20 years) high, then decreasing	Productivity moderate

[a] Assembled from the syntheses of Ryder and Psendorfer (1980), Wetzel (1990b), and many sources cited in those papers and the present chapter.

clearly that zoobenthos community structure is more diverse, more robust in responsiveness to disturbances (resilient), and more productive when the river or lake ecosystem includes an extensive littoral and associated wetland region (Table 22-32). That generalization can apply as well to fish communities.

The summary synthesis of the benthic invertebrates in this chapter indicates the remarkable variation in adaptive characteristics that emerge among the many organisms utilizing these habitats on surfaces (inert, organic, living, dead, decomposing detritus, et cetera). Strategies to acquire as much of the available organic resources as possible for growth and reproduction are played against minimizing expenditures of energy used in competitive interactions with other species, all done within environmental constraints on the individual physiological capacities of each species.

X. SUMMARY

1. The distribution, abundance, and productivity of benthic organisms are determined by several ecological processes: (a) the historical events that have

allowed or prevented a species from reaching a habitat; (b) the physiological limitations of the species at all stages of the life cycle; (c) the availability of energy resources; and (d) the ability of the species to tolerate competition, predation, parasitism, and natural mortality.

2. The benthic animals of inland waters are extremely diverse. Representatives of nearly every animal phylum occur in or are associated with the sediments of lakes and streams. This diversity and extreme heterogeneity in distribution, modes of feeding, reproduction, and morphological and behavioral characteristics make generalizations difficult for the entire benthic animal community. Nonetheless, complex general patterns of coexistence, interrelationships, and community productivity emerge. Invertebrate and vertebrate (primarily fish) predation upon benthic animals is an important regulator of spatial and temporal population structure and dynamics.
3. Most freshwater protozoa and other protists are attached to benthic substrata. Few protozoans tolerate low dissolved-oxygen concentrations; most inhabit surficial sediments and migrate to shallower water when dissolved oxygen of deeper strata declines in stratified productive lakes.
   a. Protistan species are segregated by differences in food utilization, reproduction, and timing of population development. Heterotrophic flagellates dominate in many benthic communities and can exceed 2×10^9 cells m^{-2}.
   b. Many protozoan and other protistan populations exhibit summer maxima; their population densities are positively correlated with summer increases in temperature and with the abundance of microbial food items of the surficial sediments. Consumption of total available benthic microbial production is low (usually <5%), however.
   c. Little is known of natural protozoan productivity in fresh waters. Although instantaneous protistan community biomass is low, rates of turnover (number of generations) are high. Benthic protistan productivity, analogously to protozooplanktonic communities, is likely much higher than is generally recognized and often dominates benthic productivity.
4. Freshwater sponges are rarely abundant in freshwater ecosystems.
   a. Sponges require large amounts of silica for the development of spicules that are bound by collagens to form an inorganic skeleton, which can function as a deterrent to predators.
   b. The contribution of sponges to total benthic productivity is usually minor.
5. The hydroid coelenterates are not common in fresh waters and rarely is their productivity a significant portion of the whole.
   a. Development is rapid in spring, and population maxima occur in early and late summer.
   b. Occasionally, dense populations of *Hydra* can impart significant predation pressures upon populations of small zooplankton and mortality to larval fish.
6. The turbellarian flatworms are the only important free-living members of the Platyhelminthes in fresh waters. Most are restricted to quiescent shallow areas of lakes and low-velocity streams where water movement is reduced.
   a. Flatworms prey upon other small invertebrates, and species segregate by the dominant type of prey consumed and by differences in reproductive capacities. Distribution and abundance are determined primarily by interspecific competition for food.
   b. Abundances in excess of 50,000 m^{-2} are common and are greater in littoral and wetland areas than among profundal sediments.
   c. Flatworm productivity is correlated directly with the general fertility and productivity of fresh waters.
7. Free-living nematodes, or roundworms, are widely distributed in fresh waters and can constitute a significant component of the benthic fauna. Nematode feeding habits are diverse: some species are strict herbivores, others are strict carnivores on other small animals, and still others are detritivores on dead particulate organic matter. A large proportion of production is expended in reproduction, inasmuch as mortality is high in the egg and juvenile stages.
   a. In the temperate zone, population dynamics of many nematode species exhibit three generations per year, with maximum production during winter and spring periods with reduced fish predation and increased food abundance.
   b. Highest densities (to >750,000 m^{-2}) are among deep, fine, organic-rich sediments. The greatest productivity of nematodes is commonly found among littoral substrata of productive lakes, in wetlands, and in floodplain areas of streams.
8. Horsehair worms (Nematomorpha) are free living as adults and parasitic on invertebrates as larvae. Little is known about the feeding habits, population dynamics, or productivity of these modest (to 1 m in length) animals, although they likely have

little impact on the material and energy fluxes within most habitats.

9. Colonial bryozoans are rarely quantitatively important in inland waters but occasionally form massive colonies in shallow waters and wetlands for brief periods.
   a. A number of invertebrates live commensally within bryozoan colonies.
   b. Bryozoans and these coinhabitants are heavily preyed upon by fish.
10. Two major groups of aquatic annelids, or segmented worms, form significant components of the benthic fauna.
   a. Oligochaete worms are diverse and occur in a spectrum of fresh waters, from unproductive to extremely eutrophic lakes and rivers.
      i. A large number of species of oligochaetes coexist in sediments of lakes, but species abundance patterns differ at different water depths, since the particle size and organic content of the sediment change with depth. Resource partitioning occurs by differences in worm morphology and selective feeding on microbiota.
      ii. As lakes become organically polluted and dissolved oxygen concentrations become reduced or are eliminated, an abundance of tubificid oligochaetes is commonly found concomitant with a precipitous reduction and exclusion of most other benthic animals. As long as some oxygen is periodically available, and toxic products of anaerobic sedimentary metabolism do not accumulate, the rich food supply and freedom from competing benthic animals and predators permit rapid growth.
      iii. Oligochaetes lack discrete age classes and production is approximately continuous. Maturation requires 1–4 years, with much variation in relation to optimality of growth conditions.
      iv. Oligochaete densities can be very large (>50,000 per m^2). Productivity can vary greatly from year to year because of changes in mortality associated with population dynamics of major long-lived predators (e.g., chironomid midge larvae).
   b. Leeches are primarily ectoparasites that intermittently consume vertebrate blood. A few species are predators on other invertebrates and consume their prey entirely.
      i. Leech abundance is highly variable but generally increases in more productive fresh waters.
      ii. Mortality is extremely high (ca. 95%), particularly among eggs and the young, from predation and food availability.
      iii. Production rates of leeches are greatest in the first year of the typical 2-yr life cycle. Rates of growth and production are most rapid during spring and early summer and are manifoldly higher in shallow areas.
11. The tardigrades, or water bears, are small, arthropod-like organisms of the littoral areas of lakes and streams. Tardigrades are of much evolutionary interest but little is known of their population dynamics and comparative productivity.
12. Water mites (Hydrachnida) are extremely speciose and are often found in abundance (>5000 m^{-2}) in littoral areas of lakes and stream backwater areas and the hyporheic zone of stream sediments.
   a. Larval water mites often parasitize up to 50% of aquatic insects, and adults are voracious predators of eggs, larvae, and smaller invertebrates.
   b. Little is known of the productivity of water mites.
13. The bivalved microcrustacean ostracods are widespread in fresh waters. Because of their small size (<1 mm), occurrence in surficial sediments, and difficult taxonomy, little is known of ostracod ecology, population dynamics, or productivity,
   a. Ostracods occur in the surficial sediments (0–5 cm), where they feed by filtration on bacteria, algae, detritus, and other microorganisms.
   b. Reproduction is parthenogenetic for much of the ostracod life cycle. Egg development is variable and is strongly temperature-dependent. One to three generations occur per year.
   c. Little is known about the effects of invertebrate or fish predation on ostracod population dynamics.
   d. Ostracod densities increase in more productive lakes (to >50,000 m^{-2}). Little is known of their productivity rates or of their roles in benthic metabolism; both are probably more significant than is presently realized.
14. Representatives of four groups of malocostracean crustaceans can form major components of the benthic fauna of some inland waters.
   a. The mysids, or opossum shrimps, feed by filtration on small zooplankton, phytoplankton, and particulate detritus.
      i. All mysids exhibit rapid diurnal migrations from the sediments into the metalimnion or lower hypolimnion (waters of <15°C) at night.

ii. Mysids are size-selective predators; their introduction into a lake can lead to marked alterations in the composition and productivity of the zooplankton community.
iii. When lake productivity is high, the mysid life cycle is 1–2 years in duration; when productivity or temperatures are low, mysids may require up to 4 years to complete their life cycle.

b. The isopods, or sowbugs, occasionally become a significant part of the benthic animal community of lakes and streams, commonly in turbulent littoral and stream habitats.
i. Isopod populations commonly have two distinct cohorts (spring and autumn).
ii. Productivity is greatest in the spring from the overwintering cohort.

c. The decapod crustacean, or freshwater crayfish and shrimps, is primarily herbivorous on algae and larger aquatic plants.
i. Crayfish ingest large amounts of herbaceous and detrital materials while searching for and ingesting animal protein.
ii. Longevity varies up to 16 years, although sexual maturity often occurs between the first and fourth years. Mortality from predation by fish is low where refuges are abundant but can be high in stream environments.
iii. Despite the long life cycle of crayfish (to 3–4 years), populations often reach high densities (e.g., 15 m^{-2}). Annual production is moderate, however, because of low growth rates.

d. The amphipods, or scuds, are generally omnivorous substrate feeders on bacteria, algae, fungi, and particulate organic detritus; they are restricted to well-oxygenated waters.
i. Size-selective fish predation commonly severely reduces the larger-sized adult amphipods.
ii. The amphipod *Monoporeia* burrows into the sediments and is a surficial deposit feeder during the daylight period. At night, *Monoporeia* migrates into the upper hypolimnion and metalimnion (<15°C), similarly to the mysids. In contrast to the mysids, however *Monoporeia* does not feed on plankton. The adaptive value of its nocturnal migration is unclear.

15. The freshwater mollusks consist of univalve snails (Gastropoda) and bivalve clams and mussels (Pelecypoda).
a. The snails are grazers by rasping movements on attached microorganisms and detrital particles; efficiency of assimilation is highly variable, but mostly in the range of 5–50%. Clams and mussels are filter feeders on particulate detritus and microzooplankton of or near the sediments.
b. Mortality is very high among young snails and bivalve molluscs, particularly by insect, crayfish, and fish predation, but food limitations on fecundity are often a primary mechanism affecting growth and population structure.
c. In spite of the large biomass of mollusks, the life cycle is long (1–4 years in snails; 1–15 years in clams) and productivity is moderate to relatively low. Bivalve molluscs expend much energy in respiration and high reproductive rates. Egg and juvenile mortality is commonly very high in both groups.
d. When bivalve populations reach extreme densities (>50,000 m^{-2}), such as in the case of *Dreissena* and *Corbicula* when introduced exotically into new habitats, filtering activities can ingest between 10 and 50% of summer phytoplanktonic production, creating marked reductions in turbidity and increased water transparency, as well as indirect effects of food competition among other organisms.

16. Aquatic insects are extremely diverse. Some orders of insects are entirely aquatic; others inhabit fresh waters only during certain life stages.
a. Most insects are benthic, living on or burrowing into sediments or on macrophytic vegetation and plant detritus.
b. In nearly all aquatic insects only the immature stages, particularly the nymph and larval stages, live in the water; in nearly all groups the adults, and in some groups the pupae, are terrestrial.
c. Aquatic insects adapt to low dissolved oxygen contentin water by acquiring air from the surface or taking air below water, increasing cuticular permeability to gases, developing tracheal gills, or, in some chironomid dipterans, developing the capacity for anaerobic metabolism.
d. Temperature is a dominant factor influencing the distribution, diversity, and abundance of aquatic insects along elevational gradients and along river courses. The number of generations (voltinism) generally increases with high annual thermal summation.
e. The velocity and characteristics of water movements in the littoral regions of lakes and in streams are dominant regulatory mechanisms of the development, feeding, growth, and reproductive capacities of aquatic insects.
f. Feeding mechanisms and types of food ingested by insects are extremely varied.

i. Organic matter is ingested either in particulate form (by swallowing, biting, or chewing) or in dissolved form (by piercing or sucking) (see Table 22-16).
ii. The association of aquatic insects with particular substrata is usually directly related to feeding on that substratum or on its associated attached microflora.
iii. Most aquatic insects tend to be nonselective in their food habits; a few species feed specifically on a given species of food substrate. Facultative-feeding invertebrates ingest a wider array of food substrates and inhabit a greater diversity of stream and lake habitats than do more specialized feeders.
iv. Assimilation efficiencies by benthic consumers of plant and detrital particles are <50% and most are within 5–30%; efficiencies of ingested animal prey and bacteria are generally higher than of plant and detrital materials. Microbial biomass is usually <10% of total detrital mass.
v. Some immature detritivorous insects harbor microbial symbionts in their alimentary tracts that degrade relatively resistant plant detritus. Species possessing these symbionts or digestive cellulase can utilize a wider range of habitats because they can use a greater array of detritus and algal types than can organisms that lack these adaptations.

g. The biomass of aquatic insects is relatively constant if food supplies of similar nutritional content are available. Insect biomass turnover is controlled primarily by water temperature and is governed by the positive relation between temperature, food availability, feeding rate, and respiration. Abundant food availability increases rates of growth and development. The growth efficiency of carnivores tends to be less than that of herbivores or detritivores.

17. The benthic community structure in lakes usually consists of a rich fauna with high oxygen demands in the littoral zone above the metalimnion. Heterogeneity of substrata is great in the littoral; benthic animal species diversity is greater in the littoral than in the more homogeneous profundal zone. As lakes become more productive, the number of benthic animals adapted to hypolimnetic conditions of reduced oxygen and increased decompositional endproducts declines.
    a. Two maxima in abundance and biomass of benthic animals are often observed—one in the littoral zone, the other in the lower profundal zone.
    b. As lakes become more fertile, submersed macrovegetation can be eliminated as a result of light attenuation. Maximum abundance and biomass of benthic animals may then shift to the profundal zone.
    c. With further eutrophication and intensive organic matter decomposition in the profundal zone, much of the benthic fauna of the profundal zone can be eliminated.
    d. As lakes become more tutrophic, a shift occurs in the percentage composition of two dominant benthic groups, with a decrease in the dipteran chironomid larvae and an increase in the more tolerant oligochaete worms (e.g., tubificids).
    e. The dipteran phantom midge *Chaoborus* is another major component of the profundal benthic fauna of lakes. *Chaoborus* larvae migrate into the open water at night and prey heavily on zooplankton. Feeding on limnetic zooplankton by *Chaoborus* is highly selective.

18. Benthic invertebrates of streams feed largely on algae and other attached microbiota and serve as primary food resources for fish and other vertebrates.
    a. Environmental parameters influencing growth and reproduction include:
        i. Physiological constraints of temperature, dissolved oxygen, and salinity-osmoregulation,
        ii. Constraints of food acquisition and food quality,
        iii. Intra- and interspecific competition and predation, and
        iv. Changing habitats in the substrata and associated food by physical variations in hydrology.
    b. The depth to which invertebrates penetrate sediments is influenced by the availability of attached food resources and decreases as particulate organic carbon among the particles declines.
    c. The relative abundances of invertebrate functional groups shift along a river tributary ecosystem from head waters to the mouth as the food resources change over the length of the river (see Figs. 8-8 and 8-9). Particulate organic matter and shredder insects are commonly maximized in head waters; scraper insect types that utilize increased production of attached algae and cyanobacteria are dominant in the shallow midreaches of rivers.
    d. Benthic organisms can be transported downstream in drift by water currents. Some species rarely drift, whereas others have a marked

propensity. Behavioral drift results from distinct diurnal patterns where maximum drift occurs during darkness when predation is reduced.

19. Composite densities of benthic invertebrates are often lowest during summer, especially among insect-dominated communities, both in streams and in the profundal zone of lakes. In general, biomass and productivity of benthic fauna increase as the overall fertility and productivity of lakes and streams increase.
    a. Mortality of benthic invertebrates from fish predation commonly increases with rising ecosystem productivity, in part offsetting enhanced benthic productivity.
    b. The most common production values among different benthic invertebrate communities (annelids, mollusks, crustaceans, and insects) are between 1 and 10 g DM m^{-2} yr^{-1} (10 and 100 kg DM ha^{-1} yr^{-1}). The annual production of a benthic community in the range of 50 g DM m^{-2} yr^{-1} is high in both lake and stream ecosystems.
    c. Production in running waters is usually at least two times greater than in standing waters.
    d. The productivity of nonpredatory benthic animals (herbivores and detritivores) is at least 5–10 times greater than that of predaceous (carnivorous) benthic animals.
    e. In shallow lakes with relatively low mean depths (2–7 m), zoobenthos production is commonly higher by 2–5 times than is zooplankton production. In larger, deep lakes, zooplankton production commonly equals or exceeds that of the zoobenthos.
    f. Most of the values given for zoobenthos production are conservative underestimates because about half of the total zoobenthos production occurs from the meiofauna (all metazoans <0.5 mm), which are rarely quantified.
    g. Ratios of annual production (P) to mean biomass ($\overline{B}$) for benthic invertebrates vary considerably, but in general $P/\overline{B}$ ratios increase with increasing number of generations per year (voltinism) (see Table 22-28). The relation of this annual $P/\overline{B}$ ratio to voltinism is, in part, related to water temperature, since temperature affects the number of generations in many of the dominant benthic organisms.
    h. Mortality of zoobenthic production by invertebrate predators is usually as great, if not greater, than that of vertebrate predators in many lakes and streams.

20. Although growth among different fish species is highly variable, growth rate and size are generally inversely correlated. In young fish, more energy is diverted into growth, and in larger fish, more food energy is utilized for maintenance.
    a. High production-to-biomass ratios of certain groups of benthic invertebrates are common in both streams and small lakes. The high prey turnover allows relatively high predator growth.
    b. Production rates of fish are considerably higher in tropical fresh waters than in temperate waters, where growth is restricted to about half of the year. In temperate regions, production rates are usually higher in streams than they are in standing waters.
    c. Annual production-to-mean-biomass ratios are higher among fish with shorter life cycles than they are among fish species that have slower growth and longer lifespans. Similarly, in cases where young stages contribute disproportionately to population productivity, annual $P/\overline{B}$ ratios are greater. In general, annual $P/\overline{B}$ ratios are highest among planktivorous fishes, lower for benthic feeders, and least for carnivores.

23

DETRITUS: ORGANIC CARBON CYCLING AND ECOSYSTEM METABOLISM

I. OVERVIEW OF ORGANIC TRANSFERS AND USES

The trophic structure of aquatic ecosystems has been dominated by evaluations of the rates of energy fixation by primary producers of pelagic communities and of transfer efficiencies of this algal–cyanobacterial energy by feeding to higher trophic levels. Original promulgations of energy flow in communities emphasized restrictions to trophic complexity by limitations of energy transfers among trophic levels (Lindeman, 1942; Hutchinson, 1959; Wetzel, 1995). These relationships of pelagic trophic structure and energy fluxes, however, consider solely particulate organic carbon—the ingestion and utilization of particulate organic matter (POM).

Flows of energy within the trophic structures specifically addressed predation (ingestion of particles of organic matter). As we have discussed in detail in Chapters 16 and 22, decades of ensuing studies of feeding relationships have greatly increased our understanding. In particular, studies addressed variations in size of ingested POM, morphological aspects of ingestion (e.g., filtration and gape), and avoidance of ingestion (e.g., transparency/visibility, interference by cellular or body projections). Many additional excellent studies addressed behavioral capabilities of organisms for movements within the pelagic zone in relation to refuges or escape from predators, nutritional differences in particulate food, and other factors. Despite recognition that many organisms ingest variable amounts of particulate detritus (dead organic matter) and that particulate detritus usually dominates over living POM of the plankton (e.g., Saunders, 1972), quantitative measures of consumption, assimilation, and residual egestion of detrital POM and its associated microbes remain practically unknown.

From the inception of evolving ecological constructs, early trophic dynamics, particularly among lakes, emphasized integration and interdependence among biotic components internally, but these dynamics were entirely predation based (see Chap. 16). As the metabolism of community components was analyzed with increasing accuracy, however, flux pathways and rates of transfer of organic carbon demonstrated a number of complexities and inconsistencies that could not be explained within the conventional food-web paradigms. Early carbon budgets demonstrated the importance of allochthonous dissolved organic carbon (DOC) from terrestrial, wetland, and littoral production. For example, the total annual budget of carbon fluxes of several lakes demonstrated that most of the DOC was relatively recalcitrant DOC derived from structural tissues of terrestrial and higher aquatic plants, even though higher aquatic plants colonized only small portions of the benthic area (discussed in greater detail later). Importantly, most of the heterotrophic respiration of organic matter occurred in the sediments, with a massive net evasion of CO_2 to the atmosphere.

Predation by ingestion of living particulate organic matter is not the prevailing cause of mortality of organisms. In addition, and also of great ecosystem significance, assimilation efficiencies of ingested food are modest at best under natural conditions (usually $\ll$50%), and much of the ingested organic matter is released or egested as both soluble and particulate detrital organic matter. Furthermore, commonly >90% of the total organic matter produced within the ecosystems or imported to the ecosystems are metabolized but are never consumed by particulate-ingesting metazoans. Recognition of this reality has been agonizingly slow, or, alternatively, it is acknowledged but completely ignored as the predation emphasis continues. Effective management of aquatic ecosystems is difficult and certainly imprecise if most of the metabolism, energetic fluxes, and control mechanisms of that microbial metabolism are poorly understood and separated from higher trophic levels.

The evaluation of organic carbon budgets of rivers, particularly among low- and medium-order streams, demonstrated that running waters are heterotrophic and allochthonously mediated ecosystems. These evaluations have assisted in erosion of the autochthonous dogma of lakes. Nonetheless, despite organic carbon budgets that indicate up to 99% of organic matter fluxes within aquatic ecosystems are detrital-based, the predation-based paradigms continue to prevail as the primary constructs of ecosystem operations, whereas in reality they are the minority. Essentially, all inland water ecosystems are microbially based heterotrophic ecosystems in which heterotrophic utilization of organic matter within lakes and streams exceeds—usually greatly exceeds—autochthonous autotrophic production.

A general consensus has emerged concerning the functions of dissolved (DOM) and particulate (POM) organic matter in aquatic ecosystems (Wetzel, 2000a). Two important changes in thought have arisen. Sealing a lake or river off from its drainage basin would drastically alter metabolism and biogeochemical cycling within it. Although lakes and rivers are frequently referred to as ecosystems, important influences of the drainage basin, particularly imported detrital organic matter, intimately link lakes and rivers with their surrounding terrestrial and wetland environments. A lake and associated inlet rivers themselves are only individual components of a larger landscape unit, the lake ecosystem, which must include the entire drainage basin of the lake per se. Clearly, inland water systems are being treated as integrated ecosystems with the increasing and full appreciation that the metabolism of lakes and rivers is dependent to a major extent on metabolism in the adjacent drainage basin.

Second, commonality of function is found amidst the plethora of individual habitat and biotic diversity (Wetzel, 1995, 2000a,b). This important merger is especially evident in the recognition that processes are functionally similar in freshwater and marine ecosystems. Differences between these ecosystems are apparent because of differences in the sources of organic matter, which affect the rates of utilization, not the process of utilization. The rates of utilization differ because of the dominance of more recalcitrant organic substrates in inland water ecosystems, as discussed here.

II. DEAD ORGANIC MATTER: THE CENTRAL ROLE OF DETRITUS

A. Carbon Content and General C:N Ratios

Nearly all of the organic carbon of natural waters consists of dissolved organic carbon (DOC) and dead particulate organic carbon (POC). DOC has been separated from POC by many techniques (e.g., sedimentation, centrifugation, and filtration). However, DOC is defined arbitrarily in most studies by the practical necessity of fractionation of POC from DOC by filtration at the 0.5-μm size level; hence, DOC concentrations frequently include a significant colloidal fraction in addition to truly dissolved organic carbon (e.g., Lock *et al.*, 1977; Gustafsson and Gschwend, 1997). Living POC of the biota constitutes a very small fraction of the total POC. The metabolism of the biota, however, creates a series of reversible fluxes between the dissolved and particulate phases of detrital carbon.

Detrital DOC and POC have long been known to exceed by many times the amount of organic carbon present as living material in the form of bacteria, plankton, flora, and fauna (Birge and Juday, 1926, 1934; Saunders, 1969; Wetzel *et al.*, 1972). An example from a lake in southeastern Michigan demonstrates the general relationships found among many lakes and rivers (see Fig. 17–9). The amount of carbon in higher organisms is even smaller than the smallest amount indicated in Figure 17–9, and is trivial ($\ll$1%) in comparison to the whole amount of carbon in the lake. In some lakes that receive large amounts of allochthonous dissolved organic matter, such as large, deep Loch Ness of Scotland with soft water of low alkalinity, total carbon is dominated by DOC and constitutes some 70% of the total (Jones *et al.*, 1997). These relationships are instantaneous measures of chemical mass, however, and, as indicated earlier, bear little relationship to the metabolic fluxes of carbon.

In spite of the relatively crude techniques for separation of particulate from dissolved organic matter and the approximate chemical analyses available to earlier workers on the subject, the massive data of Birge and Juday (1926, 1934) on a large number of Wisconsin lakes provide an initial orientation to general relationships. Seston consisting of living and mostly dead particulate organic matter was separated by centrifugation from colloidal organic matter and true DOC. The organic matter was analyzed for proteinaceous nitrogen and other nitrogen-containing compounds, lipids present in ether extracts, and the residual carbohydrate. The range of total organic carbon content of natural open waters is generally in the range of 1–30 mg C liter^{-1}. Higher values usually are encountered in very productive habitats, such as shallow waters of wetlands, and under organically polluted conditions. The average values from over 500 Wisconsin lakes were: dissolved organic matter 15.2 mg liter^{-1}, and particulate organic matter 1.4 mg liter^{-1}, a ratio of 11:1 (cf. Meybeck, 1982; Wetzel, 1984; world DOC average of rivers is 5.8 mg C liter^{-1} and of all surface waters 6.5 mg C liter^{-1}). Variations among different habitats, however, are very large (Table 23-1). The average composition of the dissolved organic matter (detailed in Table 12-8 of the earlier discussion on dissolved organic nitrogen) was crude protein 15.6%, lipid material 0.7%, and carbohydrate 83.7%.

Stokes's law states that spherical particles of a given density sediment through water at a rate directly proportional to the square of their radii. Both POC and DOC move with water, but nondecomposed POC ultimately will be deposited at the bottom of static water. If associated with inorganic particulate matter, DOC will also sediment, for example, when it is adsorbed to clay, organic, or $CaCO_3$ particles. These factors result in a displacement of some terrestrial production to the littoral zone of lakes with surface runoff and ground water. Littoral and pelagic production can also be largely displaced as detritus to the benthic sediments of lakes.

TABLE 23-1 Median Organic Carbon Content *of the Water* of Natural Aquatic Ecosystems[a]

Habitat	*Total organic carbon (mg liter^{-1})*	*Dissolved organic carbon (mg liter^{-1})*	*Particulate organic carbon (POC)*	*DOC:POC ratio*
Open ocean	0.5	0.45	0.05	9:1
Ground water	0.7	0.65	0.05	13:1
Precipitation	1.1	1.0	0.1	10:1
Oligotrophic lakes	2.2	2.0	0.2	10:1
Rivers	7.0	5.0	2.0	3:1
Eutrophic lakes	12.0	10.3	1.7	6:1
Wetlands–marshes	17.0	15.3	1.7	9:1
Bog water	33.0	30.3	2.7	11:1
World average, surface waters	—	6.5	0.6	10:1

[a] Extracted from Thurman (1985), Wetzel (1984), Hessen and Tranvik (1998), and Keskitalo and Eloranta (1999).

In terms of energetics of the ecosystem, the medium of exchange is detritus; functionally, POC and DOC are equivalent. These two detrital fractions are merely arbitrary divisions within a smooth continuum of large particles to small molecules (Rich and Wetzel, 1978). Although the specific nature of the detritus modifies the way in which it is utilized and its effects upon the environment, detritus inevitably carries energy from its point of origin to its place of transformation.

B. Composition of Organic Matter

Before discussing our limited knowledge of the utilization and degradation of detrital DOM and POM in river and lake ecosystems, a brief consideration of the heterogeneous composition of such materials must be presented. Detailed early reviews, including Vallentyne (1957) on the distribution and composition of organic matter in lakes, Breger *et al.* (1963) on organic geochemistry of all major components, and Konanova (1966) on soil organic matter, are especially recommended. Numerous compendia have appeared on organic matter in natural waters (e.g., the lengthy summations of pelagic and littoral organic compounds of Datsko, 1959; Duursma, 1961; Hood, 1970; Khailov, 1971; and particularly Thurman, 1985. Freshwater symposia include, for example, Maistrenko, 1965; Shtegman, 1969; Trifonova *et al.*, 1969; and Faust and Hunter, 1971. Humic substances in water are reviewed comprehensively by Swain, 1963; Haworth, 1971; Schnitzer and Khan, 1972; Gjessing, 1976; Stevenson, 1982; Aiken *et al.*, 1985; Hayes *et al.*, 1989; Suffett and MacCarthy, 1989; Purdue and Gjessing, 1990; Drozd *et al.*, 1997; Keskitalo and Eloranta, 1999; and especially Hessen and Tranvik, 1998.

The organic matter of soils and waters can be viewed as a mixture of plant, microbial, and animal products in various stages of decomposition; it consists of compounds synthesized biologically and chemically from degradation products and of microorganisms and their remains in various stages of decomposition. This complex system may be simplified by separation into two categories: nonhumic and humic substances (Aiken *et al.*, 1985; Thurman, 1985).

a. ***Nonhumic substances*** are a class of compounds that includes carbohydrates, proteins, peptides, amino acids, fats, waxes, resins, pigments, and other low-molecular-weight organic substances (see discussion in Chap. 17). These substances generally are labile, that is, relatively easily utilized and degraded by hydrolytic enzymes produced by microorganisms, and often exhibit rapid flux rates. Because of relatively rapid rates of utilization and turnover, the instantaneous concentrations of nonhumic substances in water are usually low.

b. ***Humic substances*** form most (70–80%) of the organic matter of soils and waters. Humic substances are naturally occurring, biogenic heterogeneous organic substances that are generally dark-colored (yellow to black), recalcitrant to biological degradation, and high in molecular weight (from hundreds to many thousands of daltons). Humic substances are formed largely as a result of microbial activity on plant material, but further polymerization can occur abiotically (cf. Larson and Hufnal, 1980; Haslam, 1998). The resulting compounds are relatively resistant to further microbial degradation and tend to have low turnover rates in aquatic systems.

1. Humic Substances

Humic substances are traditionally separated into three categories, each of which is defined in terms of acid-base–solubility characteristics of the materials when extracted into an alkaline solution: (1) ***Humic acids*** precipitate in water upon acidification (pH 2), but are soluble at greater pH; (2) ***fulvic acids*** are soluble at any pH value; and (3) ***humin*** is insoluble in water and in dilute solutions at any pH value. These three fractions have certain structural similarities but differ from one another in molecular weight and functional group content. Fulvic acids are of lower molecular weight than humic acid or humin fractions and have a higher proportion of oxygen-containing functional groups. This fractionation scheme is arbitrary; the individual fractions are all heterogeneous mixtures of various substances (sugars, carbohydrates, phenols, proteins, et cetera), whose specific compositions vary with the source material and its state of degradation.

High-molecular-weight humic materials exhibit a colloidal structure that is important in the physical behavior of humic solutions. The colloidal particles (approximately 0.02 μm in size) exhibit Brownian movement and can be fractionated by ultrafiltration, dialysis, gel permeation chromatography, ultracentrifugation, or electrophoresis. The molecular weight of humic substances influences their proton and metal binding characteristics, partitioning behavior, and adsorption properties (Gustafsson and Gschwend, 1997; Cabaniss *et al.*, 2000). Average molecular-weight (MW) distributions of aquatic fulvic acids vary from 2.7 to 3.7 kD and include low-MW fractions of 1110 D and a high-MW fraction of ca. 4890 D. Less than 10% of dissolved humic substances were found to be >5 kD. The low-MW fractions are more hydrophilic, exhibit faster diffusion and increased mobility, and are commonly more bioavailable. In contrast, the high-MW components have increased uptake of hydrophobic organic compounds, greater aromatic components, decreased mobility and diffusion, and are usually appreciably less susceptible to biological utilization. Molecular-size fractionation helps little in evaluation of bioavailability of different compounds, particularly above the size of effective cellular membrane permeability of ca. 500 D to transport of organic molecules. Bonding structure, particularly of the carboxyl group (C = O) and of aromatic components, and availability of sites for enzyme hydrolysis are of greater importance.

A basic characteristic of the humic colloids, as well as of truly dissolved humic and fulvic acid fractions, is their association with organic and inorganic materials via adsorption or peptization. The importance of the association of humic materials with Fe, P, Ca, and other ions has already been indicated in previous chapters. The pH and ionic strength of the solution governs both the extent of condensation and colloidal size, as well as the availability of adsorption sites.

Colloidal humic materials provide a very large surface area that is suitable for the adsorption of both inorganic and organic materials (e.g., Poirrier *et al.*, 1972; Stevenson, 1972; Seitz, 1982). Consequently, humic materials can fundamentally alter the availability of required or toxic metals and organic substances to aquatic biota. The binding of herbicides or pesticides (e.g., DDT) to humic materials, for example, can significantly change transport, fate, and bioaccumulation characteristics of these toxicants. Humic constituents can also link normally thermodynamically discrete redox couples by functioning as broad redox potential "flavoproteins" (Zimmerman, 1981). This feature can facilitate the oxidation or reduction of a number of biologically important substances (e.g., Miles and Brezonik, 1981; Francko and Heath, 1982; Gustafsson and Gschwend, 1997).

Humic material of the series humic acid–hymatomelanic acid (an alcohol-soluble component of humic acid)–fulvic acid–humin consists of polymeric micelles whose basic structure is aromatic rings that are variously bridged into condensed form by —O—, —NH—, —CH_2—, or —S— linkages (Swain, 1963; Schnitzer and Khan, 1972). Attached hydroxy and carboxyl groups provide acidity, hydrophilic properties, base-exchange capacity, and tanninlike character (e.g. they react with proteins). Fulvic acid is the least polymerized of the fractions, while humin is the most condensed humic component. Fulvic acid is highly oxidized, stable, and water soluble (Schnitzer, 1971). About 60% of the weight of the fulvic acid fraction is composed of functional groups such as carboxyls, hydroxyls, and carbonyls attached to a predominantly aromatic nucleus ring structure. Fulvic acid is a naturally occurring metal complexing agent that can bring di- and trivalent metal ions into stable solution from practically insoluble hydroxides and oxides. Carboxylic acid groups are of major importance in natural organic matter because they dominate the acidity and contribute to the solubility of the compounds.

Recently, structural information of humic substances has been increasingly examined with ^{13}C nuclear magnetic resonance (NMR) spectroscopy, which yields structural information of mixtures of organic compounds. Major bands in the spectra indicate functional groups (Table 23-2) and allow examination of variations in relative intensities of different groups in relation to differences in sources and biogeochemical transformations. For example, one could examine increases in aromatic carbon content (110–160 ppm) as dissolved organic matter from plant sources moves downgradient through a stream and wetland to a lake.

Analyses of chemical constituents of humic and fulvic acids can be further examined by pyrolytic thermal degradation of bonds and analysis of the fragmented chemical products by gas chromatography and mass spectrometry. These properties and proportions of various functional constituents, such as carboxyl, phenol, and methoxyl components, allows reconstruction of potential source materials. That is, the chemical composition of the fragments allows reconstruction of

TABLE 23-2 Spectral Regions for Different Functional Group Carbon Atoms in ^{13}C-NMR Spectra of Plant and Microorganism Tissue and Natural Organic Substances[a]

Designation	*Spectral regions (ppm)*	*Functional group carbons*
AL-I	0–62	Aliphatic carbons
AL-II	62–90	Alcoholic carbons in carbohydrates and other aliphatic alcohols and ethers
AL-III	90–110	Anomeric carbons in carbohydrates and some aromatic and olefinic carbons
AR	110–160	Aromatic and olefinic carbons
C-I	160–190	Carboxylate and amide carbons
C-II	190–230	Carbonyl carbons in aldehydes and ketones

[a] After Wershaw (1992). Chemical shifts are given in parts per million from tetramethylsilane (TMS) resonance at 0 ppm.

source materials and interpretation of potential sources of the organic compounds.

Humic substances contain a variable amount of nitrogen (1–6%), which is largely hydrolyzable to amino compounds. Most of the combined amino acid nitrogen occurs in amino acid–peptide linkages. The remainder is in the form of amino sugars, ammonia, and amine linkages.

2. Formation of Humic Compounds

Cellulose and lignin in plant structural materials account for the largest portion of organic material and humic compound formation. In spite of the diversity of vegetation types, a general similarity of the constituent functional groups is found during decomposition of detrital organic materials (McKnight and Aiken, 1998). The formation of humic substances occurs during the degradation of plant material; both microbiological and abiotic processes contribute to its production. Carbohydrates serve as the main microbial source of energy and carbon in the intercellular synthesis of protein and hemicellulose.

Lignin contains abundant aromatic rings and no nitrogen. Degradation of lignin in plant structural material is carried out by a limited number of microorganisms, primarily fungi under aerobic conditions. Polyphenolic lignin of higher plant tissues is modified during this degradation, and humiclike substances of high molecular weight result (Zeikus, 1981; McKnight and Aiken, 1998). By means of extracellular enzymes and abiotic processes, decomposition products are further partially degraded and oxidized, which results in soluble macromolecules enriched with carboxylic and other oxidized functional groups. Further degradation of these compounds results in the generation of humic and fulvic acids and various products of decomposition (e.g., fats, amino compounds, CO_2, CO, H_2, CH_4, N_2, NH_3, and H_2S). Additional polymerization of the decomposition products can occur on the surfaces of clay particles (Wang *et al.*, 1978; Larson and Hufnal, 1980; Steinberg and Münster, 1985). Autolysis of the microorganisms themselves represents a major source of humiclike substances.

It is clear that the fungi play a major role in the degradation of plant material and the synthesis of humic substances. Fungi, as well as cellulolytic actinomycetes and myxobacteria, enzymatically oxidize lignin-derived polyphenols, which condense with proteinaceous degradation products to form humic acids. Humic substances of terrestrial plant and soil systems form largely from cellulose and lignin. Mainly fungi degrade celluloses, hemicelluloses, and lignin of needles and other products of conifers. This coniferous litter is easily leached, especially of organic acids (Nykvist, 1963). Litter of deciduous vegetation, however, is decomposed more readily by fungi and bacteria to lignoprotein complexes. When allochthonous inputs of particulate organic matter from higher plants are large, as in the case of many river systems and many lakes with extensive littoral vegetation, fungi clearly dominate the degradation of lignified and cellulolytic tissue for subsequent bacterial decomposition. In the open water of lakes, in which autotrophic production by algae results in plant constituents composed mainly of hemicelluloses, proteins, and carbohydrates, cellulolytic activity is reduced. As would be anticipated, fungal populations are well developed in sediment accumulations, in littoral regions of higher plant densities, and in streams receiving large amounts of terrestrial plant debris. Sparrow (1968) and Suberkropp (1992) provide excellent reviews of the distribution and general ecology of the freshwater fungi.

The phenols, quinones, phenolic carboxylic acids, and related compounds, common constituents of humic substances, can be inhibitory to bacterial fermentation and to fungi. These inhibitory compounds, and associated acidity and resultant low redox potentials, frequently lead to accumulations of humic substances undergoing very slow rates of degradation. The antiseptic qualities of humic substances from plant remains result in accumulations in topographical depressions—for example, in the peat of bogs and lake sediments—over long periods of time without significant degradation. The superb preservation of human bodies placed into peat bogs in Europe over 2000 years ago (Glob, 1969) attests to the effectiveness of these combined properties and conditions.

C. Flocculation of Dissolved Organic Matter—Shift to Particles

Dissolved organic compounds, particularly in the colloidal size range of ca. 1 nm to 1 μm, can aggregate by a number of different processes. Flocculation, a generic term covering all aggregation processes, refers to the formation of a particle larger than 1 μm on which gravity dominates over colloidal interactions (Gregory, 1989; Buffle and Leppard, 1995). Hydrophilic colloids are thermodynamically stable and can be induced to aggregate and precipitate only by changing solvency conditions, such as by temperature changes. Hydrophobic colloids consist of substances of low solubility that exist in a finely divided state in water and remain kinetically stable by interparticle repulsion. This repulsion is usually electrical and most aqueous colloids are charged.

Flocculation occurs when particles collide and can adhere. An attractive energy from van der Waals forces is balanced by repulsion from negative electrical charges borne by most natural colloids. In many cases, hydrophobic particles will have a hydrophilic surface in the sense that it can be wetted by water. Hydrogen-bonded water molecules, for instance, hydrate insoluble oxides. Increased inorganic ionic salinity can give a lower colloidal stability for hydrophobic charge-stabilized colloids by reduction of both the effective surface potential and the extent of the diffuse layer surrounding the compounds. Aluminum and iron salts, commonly used in water treatment facilities, remove dissolved organic matter, especially high-molecular-weight humic substances, in which a colloidal metal-humate complex flocculates to a precipitate. Multivalent cations (Fe^{3+}) are more effective in flocculating ability than mono- and divalent cations (Na^{+} and Ca^{2+}) (Liu and Shei, 1991). An increase in temperature will also increase coagulation. The smallest colloids (<100 nm) coagulate into aggregates in the size range of 0.1–10 μm. Particles larger than 10 μm rapidly settle by sedimentation.

Many different types of aggregations are found, varying with source materials and conditions of flocculation. Colloidal constituents include biological debris, particularly from algae and cyanobacteria (Strycek *et al.*, 1992), iron oxyhydroxides, calcium carbonates, clays, amorphous silica, polysaccharides and other fibrillar material, gel-like organic material, and soil-derived fulvic and humic compounds (Buffle and Leppard, 1995). The proportion of organic matter increases with decreasing sizes of the colloidal material. Loose aggregates (flocs) can be formed by binding of several colloids and bridge together with other living or dead cells or particles. The structure, size, density, and formation rates of these flocs depend on the macromolecules, their relative length compared to colloid size, and relative concentrations of each.

A number of physical processes increase colloidal and particle collisions and probability of adhesion (particle stickiness) (O'Melia and Tiller, 1993; Kepkay, 1994): (a) Brownian motion, (b) shear from laminar or turbulent flows, (c) differential settling of large particles that collide with smaller particles, (d) diffusive capture of smaller particles within the boundary layer of larger particles, (e) aggregation of colloids (surface coagulation) at the gas–water interface by rising bubbles, (f) scavenging of small colloids by filtration as water passes through rapidly sinking porous macroaggregates ("lake snow"), and (g) bacterial motility in which swimming bacteria (to 30 $\mu m\ s^{-1}$; Lauffenberger, 1983) collide with colloids by interception rather than by diffusion. Aggregation into macroparticles can then occur with the formation of organic precipitates (lake snow).

The amorphous particles of organic coagulated precipitates contain appreciably less recalcitrant organic matter than do morphologically distinguishable particles (range of 10–100 μm) that are clearly plant fragments (Bowen, 1979, 1984). The colloidal organic matter, some of which is certainly in microaggregated particles, is clearly susceptible to appreciable degradation by bacteria (see Middelboe and Søndergaard, 1995). The amorphous particles are also more digestible when treated with simple digestive enzymes than are morphous particles and support better growth of detritivous animals. The role of these amorphous aggregations in the energetics of higher trophic levels remains unclear but is likely considerably underestimated. In some cases, higher animals, such as black fly larvae, appear to be able to induce aggregation of amorphous particulate organic material by passing water containing DOM in sufficient quantities to support growth (Ciborowski *et al.*, 1997).

III. ALLOCHTHONOUS ORGANIC MATTER

The sources and composition of organic matter are diverse and poorly understood. Production of dissolved and particulate organic carbon is a result of autotrophic synthesis and partial heterotrophic metabolism. Instantaneous measurements of the chemical mass of DOC and POC, however, are highly biased toward recalcitrant[1] compounds, that is, compounds that are relatively stable chemically, of low solubility, and slowly degraded microbially. These recalcitrant components of detrital organic carbon persist in inland waters for longer periods of time than do more labile organic compounds. The readily utilizable labile components cycle rapidly at low equilibrium concentrations but can represent major carbon pathways and energy fluxes. Moreover, in streams and most lakes, much of the detrital organic carbon, mostly dissolved, originates from terrestrial, wetland, and littoral-zone plant sources.

Most lakes are small, with a high proportion of their surface area being littoral zone. The flood plain and streamway adjacent to river channels also constitute large areas of littoral and wetland areas of high productivity of aquatic plants and terrestrial vegetation that are tolerant of wet soils. Allochthonous and littoral sources of dissolved and particulate detrital carbon form major inputs to both the lacustrine and river ecosystems, and these sources of organic compounds

[1] The word *recalcitrance* connotes contumacious reduction in capabilities for microbial degradation, usually related to the chemical structure of the organic matter (Alexander, 1965b) and is used in preference to *refractory*. Although interpretations vary, the term *refractory* often implies chemical resistance and nondegradability, which is not true in these natural ecosystems.

and energy markedly influence metabolism of organisms in the open-water in-lake, and in-stream communities. In streams, dissolved organic carbon is decomposed by both the benthic microflora and the planktonic bacteria (Cummins *et al.*, 1972; Wetzel and Manny, 1972b; Dahm, 1981). In the relatively static waters of lakes, gravity acts as an important selective agent by which sedimentation displaces a major portion of carbon and its metabolism to the sediments (Vallentyne, 1962; Wetzel *et al.*, 1972). To adequately assess the importance of the complex carbon cycle that is central to both the structure and function of lakes and streams, one must know of the productivity of all components of the ecosystem and have an understanding of the origins and metabolism of detrital organic carbon from all sources.

The importance of these and related studies to the present discussion is their relationship to the amount of allochthonous organic carbon entering most lakes from terrestrial, wetland, and littoral areas and to the qualitative changes in composition of this organic matter en route, usually in streams and ground water and then often littoral zones, to the pelagic zone. For example, humic and fulvic acids constituted over 60% of the riverine dissolved organic carbon (DOC) in the Amazon River, where fulvic-to-humic-acid ratios in the mainstem river averaged 4.7 (Ertel *et al.*, 1986). All of the dissolved humic and fulvic acids had clearly recognizable lignin components at levels (8 and 3% of the carbon, respectively) suggesting a predominantly allochthonous source from higher plant tissues. Fulvic acid substances had consistently lower lignin levels, lower lignin phenol methoxylation, higher acid:aldehyde ratios, and higher C:N ratios than coexisting humic acids—all indicative of greater aerobic degradation rates of the fulvic acid fractions. At least 40% of the total fluxes of carbon were associated with dissolved humic substances in the Amazon River. This situation is likely common among most stream and river ecosystems.

As the total organic matter of lake water increases, the percentage in the dissolved fraction increases disproportionately to that of the particulate fraction. The nitrogen content of the organic matter decreases progressively with increasing DOC concentration and similarly declines as the proportion of allochthonous to autochthonous organic matter increases. As a result, the organic C:N ratios increase—a reflection of an increased input of organic compounds low in nitrogen as well as decreased decomposition rates of the more recalcitrant organic compounds.

Dissolved organic matter of allochthonous origin contains a very low percentage of nitrogen (C:N about 50:1), while that produced autochthonously within the lake by algae and some macrophytes has a much higher initial nitrogen content (C:N about 12:1). Much of the DOC originating allochthonously from terrestrial and marsh plants is totally dominated by fulvic and humic acids that originate from the structural tissues of higher plants. Numerous studies have shown a nearly linear relationship between color units (Chap. 5) and dissolved organic matter or carbon. That color is associated with the chromophoric molecules of largely humic and fulvic acids of the colloidal or high-molecular-weight fractions of truly dissolved organic carbon (cf. Black and Christman, 1963; Stewart and Wetzel, 1981a).

Chromophoric dissolved organic matter (CDOM) is an optical definition for that portion of DOM that is able to absorb solar radiation. The molecular components, ***chromophores***, absorb UV radiation, and conjugation of chromophores results in absorption of solar radiation >290 nm. The concentration of CDOM, as indicated by the color of water, often correlates with the concentration of DOC (Juday and Birge, 1933; Birge and Juday, 1934), which indicates that CDOM is a major constituent of total DOC. Humic substances constitute a major portion of CDOM.

As CDOM absorbs UV and visible radiation, the molecules are electronically excited, which may lead to photochemical reactions that can modify the bonding structure of the macromolecules. Some of the energy is released as heat and some as emission of photons in fluorescence. Quantum yields for fluorescence are ca. 1% and have been found to vary little (<2.5-fold) in natural waters (Green and Blough, 1994). In some indirect photochemical reactions, the primary absorbing CDOM transfers energy to another molecule, such as oxygen, before returning to its ground state. The CDOM-initiated photochemical reactions in surface waters can mineralize organic compounds with the generation of CO_2, CO, NH_4, and PO_4. Photochemical reactions can covert DOC to dissolved inorganic carbon rapidly (Miller and Zepp, 1995) with simultaneous loss in absorption of UV and visible light by CDOM and in the quantum yields for photochemical reactions (Lean, 1998; Andrews *et al.*, 2000). In clear lakes with low DOC concentrations, DOC is not as strong a predictor of the attenuation coefficients for UV radiation. The photochemical reactions can also modify DOM by partial photodegradation with a reduction of molecular weight and an increase in bioavailability of the organic compounds.

The colloidal fraction of measured "dissolved" organic matter was often considered small (Krough and Lange, 1932), but because many humic substances can be large macromolecules and occur in aggregations, the colloids can constitute a significant fraction in waters

TABLE 23-3 Annual Average Particulate, Colloidal, and Dissolved Organic Fractions of Lake Furesø, Denmark[a]

Organic fraction[b]	*Total dry weight*[c] *(mg liter^{-1})*	*Organic weight*[c] *(mg liter^{-1})*	*Crude protein (%)*	*Lipid material (%)*	*Carbohydrate (%)*
Particulate	2.1	1.56	53.2	9.9	36.9
Colloidal	3.6	0.67	41.5	11.7	46.8
Dissolved	Approx. 70	8.8	37.5	—	62.5

[a] After Krogh and Lange (1932).
[b] Colloidal and dissolved fractions separated by ultrafiltration.
[c] Based on a limited number of largely surface and near-sediment samples.

with moderate-to-high concentrations of dissolved organic matter (see p. 734). Nitrogen content decreases and carbon content increases in the progression from particulate to colloidal to truly dissolved organic matter fractions (Table 23-3). The colloidal fraction is high in lakes rich in DOC, such as bog waters, and in hard waters in which organic compounds are adsorbed onto carbonate particles (cf. Ohle, 1934b; White, 1974; White and Wetzel, 1975).

Both allochthonous and autochthonous sources of detrital particulate organic matter (POM) and dissolved organic matter (DOM), therefore, constitute variable inputs to aquatic ecosystems. As is emphasized repeatedly in this synthesis, POM tends to remain near sites of production. Lateral transport of POM is small, relative to DOM transport, in terrestrial soils, and in land–water interface zones such as flood plains, wetlands, and littoral areas. As we have seen earlier (Chap. 21), even in streams, large POM is often retained (snags and debris dams) and transport is mainly via DOM and fine POM. Transport of DOM is the primary movement of allochthonous organic carbon and energy to all aquatic ecosystems (Wetzel, 1993, 1995).

Authochthonous sources within the lake or river supplement the allochthonous inputs and include

1. Littoral photosynthetic sources of POM and of DOM by active secretion, decomposition, and lysis of the macrophytes and attached algae and cyanobacteria
2. Primary producers of the open-water zone, primarily the algal and cyanobacterial phytoplankton. Under some circumstances, sulfur photosynthetic and chemosynthetic bacteria are also significant sources of organic carbon.

Rapid transformations between POC and DOC by heterotrophic microflora progressively degrade organic matter to CO_2 and heat. The amount of organic carbon utilized and transformed by animals is a quantitatively small portion (<10%) of that of the whole ecosystem. Most of the heterotrophic metabolism, which is almost entirely microbial, occurs both in the open water and in the sediments. Because so much detrital organic matter from pelagic and littoral sources is displaced to the sediments, much heterotrophic decomposition occurs in benthic sediments. The benthic region is the dominant site of organic carbon transformation to CO_2 in most lake ecosystems, especially as depth and volume decrease, in most reservoir ecosystems, and in all river ecosystems.

A. Allochthonous Organic Matter from Terrestrial Sources and within Streams

Terrestrial plants form much of the allochthonous organic matter of aquatic ecosystems. The organic carbon of residues of plant structural tissues is transformed variously by microbial utilization and degradation both at the sites of growth and while the organic matter is being transported by runoff ground water and vadose-surface water. Most (>95%) of the organic matter produced in the terrestrial portions of the drainage basin remains at the sites of production and is largely decomposed (Fisher and Likens, 1977; Dosskey and Bertsch, 1994). Little is stored permanently. Export of nondecomposed organic matter is largely as dissolved organic compounds in runoff and ground water. There is large variability in dissolved organic carbon (DOC) concentrations in terrestrial soils along this flow path (Cronan and Aiken, 1985; McDowell and Likens, 1988; Mulholland *et al.*, 1990; Nelson *et al.*, 1993; Hope *et al.*, 1994; Webster *et al.*, 1995; Kalbitz *et al.*, 2000; Magill and Aber, 2000).

When anthropogenic inputs of organic matter are low to rivers, concentrations of inorganic nitrogen are usually low and dissolved organic nitrogen (DON) is the primary form of nitrogen (Hedin *et al.*, 1995; Stepanauskas *et al.*, 1999, 2000). Much of that DON (to 70%) is bioavailable. Spring flooding from snowmelt is frequently a major hydrological event, and in some boreal rivers >50% of the annual water discharge may occur in a few weeks. During spring floods,

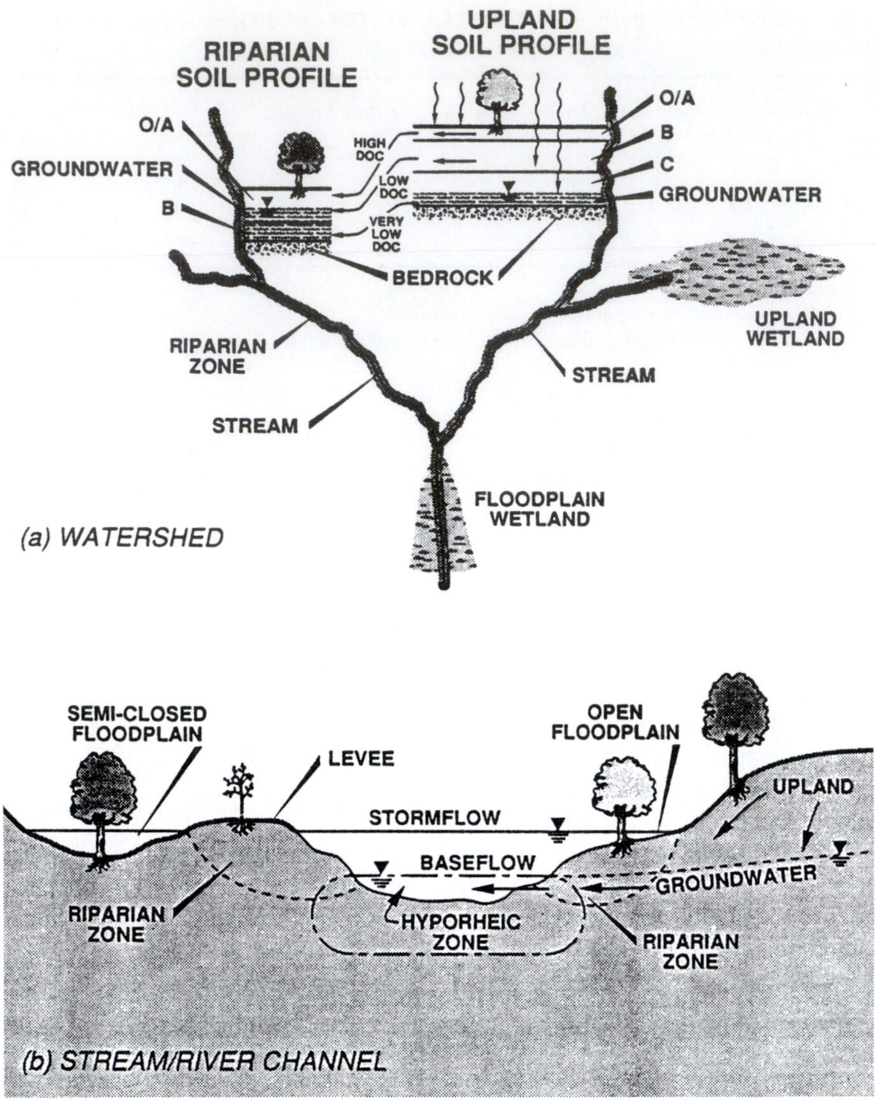

FIGURE 23-1 Potential dissolved organic carbon sources and hydrologic linkages with the stream/river at (a) the drainage basin (watershed) and (b) floodplain/channel spatial scales. (Extracted from Mulholland *et al.*, 1990.)

DON fluxes may exceed the baseflow fluxes by several orders of magnitude. During the flooding period, DON bioavailability increases markedly and has been correlated directly with increased concentrations of amino acids and related compounds.

Highest DOC concentrations occur in surface soil horizons (O and A) as a result of inputs from throughfall of the vegetation canopy and decomposition of surface POM (Fig. 23-1a), and tend to be much greater in forest soils among coniferous vegetation than among hardwood trees. Many humic substances of plant origins form microaggregates with clay-sized particles, which tend to reduce alteration by microbes (Clapp and Hayes, 1999). Removal of terrestrial vegetation, such as by clear cutting of forests of a drainage basin or removal of riparian vegetation for agriculture, reduces the inputs of organic matter to the soils and reduces the release of DOC in runoff to streams (Meyer and Tate, 1983; Delong and Brusven, 1994; Guyot and Wasson, 1994).

Seasonally, the concentration of DOC in soil leachates is inversely correlated with temperature. Additionally, warm temperate and tropical communities are more productive than those of temperate regions, but oxidation of organic matter of forest origin is rapid and reduces DOC concentrations in leachates. Hence, the mean DOC concentrations of streams among the tiaga is much greater (ca. 10 mg liter^{-1}) than from other upland habits (see Table 23-1), exceeded only by the DOC of swamps and wetlands (Thurman, 1985).

Concentrations of DOC decline steeply in lower soil horizons (B and C), primarily because of adsorption and coprecipitation in mineral soils with iron and aluminum sesquioxides (McDowell and Wood, 1984; David and Vance, 1991; Kaiser and Zech, 1998). Rates of microbial respiration in the upper soil horizons, especially B, indicate constant renewal of adsorption sites occurs as adsorbed DOC is utilized. Where soils, such as tropical podsols, do not have an iron-rich clay horizon, large quantities of DOC, particularly hydrophobic DOC, can be released. It is also probable that organic acids of root exudates from terrestrial vegetation can alter stereochemical hydrophobic arrangements within micellelike conformations of humic substances (Piccolo *et al.*, 1996; Nardi *et al.*, 2000). Organic compounds adsorbed to the soil can then be mobilized and released.

As DOC moves laterally within soils in gravity flows along decreasing elevation levels and reaches the floodplain and riparian soils, further changes occur (Fig. 23-1b). The DOC moving through soil particles is subject to continual microbial degradation and becomes progressively more recalcitrant. For example, over half of the DOC leached from soils was found to be readily biodegradable by soil microbes (Baker *et al.*, 2000). Indeed, the concentrations of DOC are reduced appreciably by microbial degradation. Despite this utilization in soil and groundwater flows, concentrations of dissolved organic carbon and nitrogen are often significantly higher in ground water than in surface water (Ford and Naiman, 1989).

During baseflow conditions of low-flow inflows, microbial activity of flood plain and subsurface ground water can be limited by DOC availability. DOC concentrations of subsurface flows tend to be positively correlated with discharge within the river channels and decrease with greater distances from the head of entry places, such as the head of gravel bars (Vervier and Naiman, 1992). DOC from these soil and groundwater sources is augmented by formation of DOC in small streams by both abiotic processes, such as leaching from particulate materials, and biotic, soluble processes and leaches rapidly (<24 h) from detrital particulate terrestrial vegetation.

During low-flow periods, DOC concentrations in the soils increase. Subsequent rapid declines in DOC concentrations are roughly proportional to increased water percolation through the soils such as is released from rapid snowmelt and from precipitation events (see p. 747) (Lewis and Saunders, 1989; Ciaio and McDiffett, 1990; Clair *et al.*, 1994; Hornberger *et al.*, 1994; Jones, et al., 1996). Maximum DOC concentrations appear in receiving streams and rivers soon thereafter and rapidly decline as the rivers return to base flows. As the area of flood plains increases in lowland river ecosystems, stream DOC concentrations often increase during high flows as DOC-rich water originating in wetlands is transported to the channel (e.g., Dalva and Moore, 1991).

The DOM input from terrestrial organic matter to streams results from direct leaching from living vegetation or from soluble compounds carried in runoff from dead plant material in various stages of decomposition (Fig. 23-2). POM, again mostly of plant origin, can fall directly into stream water from overhanging tree canopies, be transported by runoff water, particularly from flood plains, or be windblown into the stream. Foliage from trees and ground vegetation can provide very significant inputs of organic matter to streams, both as POM and as leached DOM from the dead POM. The variability in the ratios of DOC:POC in streams and rivers is very large (range between 0.09 to 70; Moeller *et al.*, 1979).

Total transport of organic carbon in the world's rivers from land masses to the ocean in both dissolved and particulate forms is ca. $0.37-0.41 \times 10^{15}$ g C yr^{-1} (Schlesinger and Melack, 1981; Meybeck, 1993c). Although this organic carbon transport is a small flux in the global carbon cycle, fluxes from land can be significant, though highly variable. Transport losses range from <1 g C m^{-2} yr^{-1} in grassland ecosystems to 10 g C m^{-2} yr^{-1} from some cultivated forests. Wetlands tend to release appreciably greater amounts of organic carbon (to ca. 20 g C m^{-2} yr^{-1}) (Table 23-4).

1. Transformations

The large (coarse) as well as fine particulate matter of terrestrial vegetation entering streams is leached of a significant portion of its organic content as dissolved compounds (Fig. 23-2). The amount and degradability varies with the plant species. For example, slower uptake by stream bacteria of DOC in an old-growth forest could be a result of a more limited supply of labile DOC as well as a greater concentration of inhibitory compounds such as polyphenolic and terpene compounds in the stream water (Dahm, 1984). Much of this DOM leachate, however, is metabolized very rapidly. As a result, bacterial populations in the water, attached on sediments, and in the hyporheic zone within the sediments increase markedly in response to the DOM loading. For example, in large experimental streams, leaf leachate was demonstrated to have a bacteriologically labile dissolved organic carbon (DOC) fraction that was decomposed rapidly ($T_{1/2} = 2$ days) and a recalcitrant DOC fraction ($T_{1/2} = 80$ days) (Wetzel and Manny, 1972b). Most of the recalcitrant dissolved organic nitrogen compounds persisted in a relatively unmodified state for at least 24 days. As the DOC mixtures from natural sources are degraded,

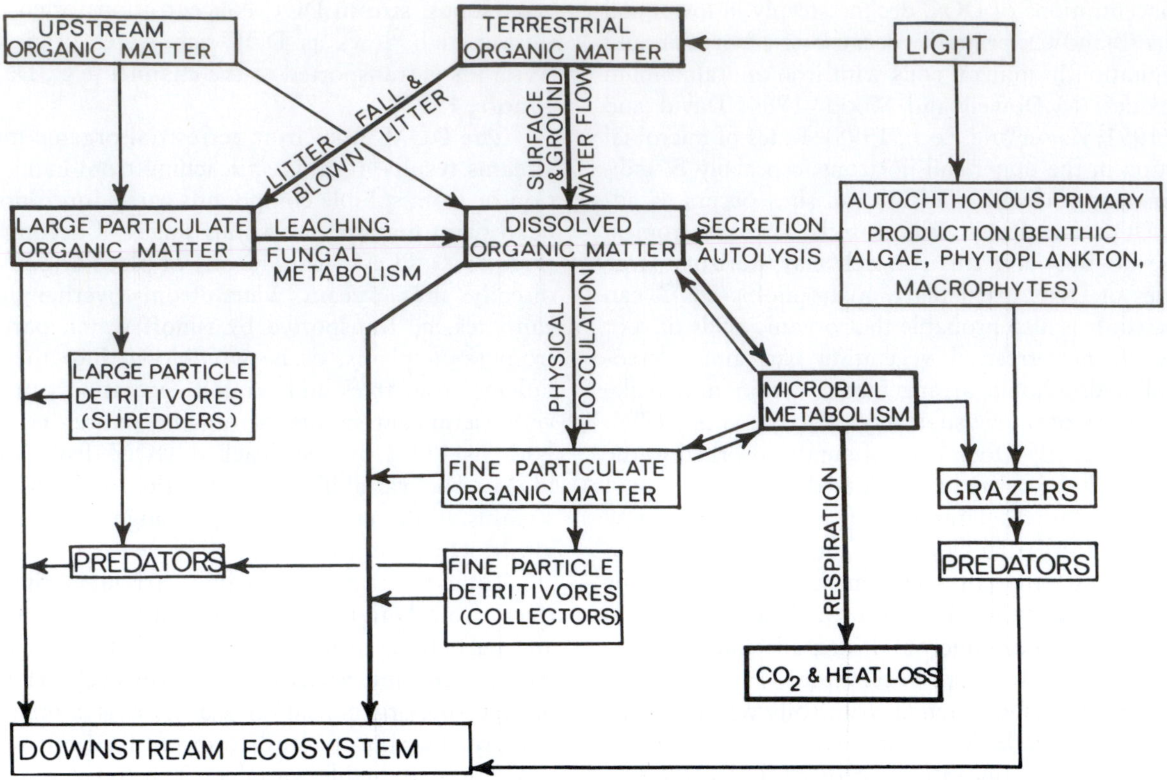

FIGURE 23-2 Simplified compartment model of the structure of an idealized stream ecosystem. Heavier lines indicate dominant transport and metabolic pathways of organic matter. (Composite of modified figures after Fisher and Likens, 1973; Cummins *et al.*, 1973.)

TABLE 23-4 Annual Riverine DOC Flux from Different Soils and Vegetation Habitat[a]

Biome	*Soil C:N*	*Observed DOC flux* ($g\ C\ m^{-2}\ yr^{-1}$)[b]
Cool grasslands	13.5	0.386
Tropical savannah	13.6	1.090
Taiga	13.8	0.700
Siberian steppe	14.7	1.290
Warm deciduous forests	15.3	1.410
Warm mixed woodlands	16.7	1.714
Cool deciduous forests	17.1	1.927
Warm conifer forests	21.0	3.684
Cool conifer forests	21.0	4.226
Northern mixed forests	23.2	5.260
Heath/moorlands	24.6	5.650
Tropical forests	25.0	6.336
Boreal/peat mix	25.7	6.349
Peatlands	30.1	8.567
Swamp forests	32.4	9.913

[a] From data reviewed by Aitkenhead and McDowell (2000).
[b] $\times 10 = kg\ C\ ha^{-1}\ yr^{-1}$.

largely by benthic microbes, aliphatic compounds such as organic acids are being released as degradation products along with CO_2 and CH_4 (Schindler and Krabbenhoft, 1998). These gaseous products either oxidize or evade to the atmosphere.

A quantity of the dissolved leachate precipitates to particulate form (Lush and Hynes, 1973); the rate of precipitation and the size of the resulting particles depend upon leaf species and water chemistry. As will be discussed later, a portion of the DOC can be photolysed upon exposure to sunlight to simple organic substrates readily utilizable by bacteria or to CO_2.

Aggregations of large POM, such as leaf packs trapped among stream sediments, rocks, and large woody debris undergo colonization by fungi in complex successional patterns within the detrital microhabitats (cf. detailed discussion in Chap. 21). As the resistant plant material is degraded, solubilized products of decomposition are utilized by bacteria living in highly stratified populations in the steep redox gradients of the compacted plant material (Fig. 23-2). The detrital material and its associated microflora serve as a major nutritive source for numerous aquatic invertebrates, especially the immature insect fauna (Kaushik and Hynes,

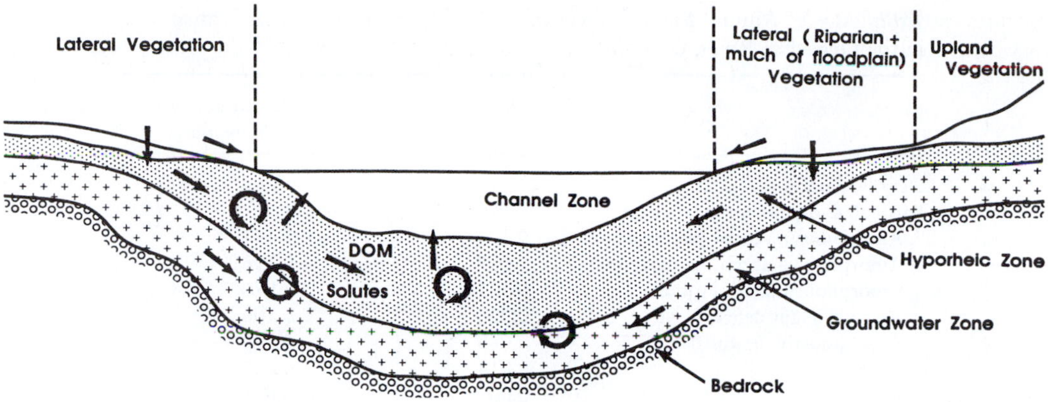

FIGURE 23-3 Conceptualization of lateral and vertical boundaries of running water ecosystems. The stream ecosystem boundary is defined as the hyporheic/groundwater interface and thereby includes a substantial volume beneath and lateral to the main channel. Vegetation rooted in the hyporheic zone is therefore part of the stream ecosystem primary production. Arrows indicate flow pathways of dissolved organic matter and inorganic solutes derived from plants and plant detritus within the stream ecosystem and flood plain. (From Wetzel and Ward, 1992; extensively modified from Triska *et al.*, 1989.)

1971; Cummins, 1973; Cummins *et al.*, 1973; Iversen, 1973; Anderson and Sedell, 1979) (see Chap. 22). There is no question that the shredding, collecting, and grazing activities of aquatic insects play a role in quickening the reduction of the size of POM and increasing the surface area for subsequent microbial degradation. It is also clear that a major portion of the animal nutrition is obtained from the microflora attached to the resistant detrital POM. Only a few animals are adapted with symbiotic gut microflora to utilize the particulate plant material directly (see Chap. 22).

Most of the lateral streamside and floodplain vegetation, however, will fall to the ground or hydrosoils beside the main channel and be incorporated into subsurface soil regions, including lateral hyporheic zones. Both organic matter stored and/or produced within the main channel, as well as that which feeds into the hyporheic zone from lateral regions, can fuel microbially active subsurface regions, where both aerobic and anaerobic processes further transform the detrital organic material (Fig. 23-3). Some of the dissolved organic and solute material eventually reemerges in the main channel, where it can serve as nutrient sources to within-channel microbial communities (e.g., Crocker and Meyer, 1987; Dahm *et al.*, 1987; Triska *et al.*, 1989; Coleman and Dahm, 1990; DeAngelis *et al.*, 1990; Leff and Meyer, 1991; Kaplan and Newbold, 1993).

There are few detailed analyses of the metabolism of organic carbon in streams. However, there is sufficient evidence to indicate that

1. Allochthonous inputs of terrestrial organic matter, in the form of detrital DOM and POM, commonly form the dominant source of material and energy for stream and river ecosystems (Tables 23-5 and 23-6). Much of that DOC is released from soils into ground water and from anaerobic processes in adjoining wetlands (e.g., Dahm *et al.*, 1987; Eckhardt and Moore, 1990; Dillon and Molot, 1997b).

2. From 10 to 25 times more organic matter occurs as DOM than occurs in particulate form on an annual basis (e.g., Table 23-5).

3. Groundwater inputs of DOM are the major source to upland streams, and most of the water reaching the streams passes through soil strata and is exposed to microbial degradation (e.g., Wallis *et al.*, 1981; Rutherford and Hynes, 1987).

4. The rates of decomposition of DOM are rapid (days) in comparison to those for much of the POM (leaves in weeks and woody material in years).

5. Bacteria rapidly metabolize the labile components of DOC and DON, but more recalcitrant components are decomposed at slower rates and are exported downstream (Wetzel and Manny, 1972; McDowell and Fisher, 1976; Leff and Meyer, 1991; Perdue, 1998; Findlay and Sinsabaugh, 1999; Stepanauskas *et al.*, 1999, 2000). Proteinaceous moieties in DOM are preferentially degraded. DOM with higher percentages of aromatic carbon and carboxyl (COOH) content are less bioavailable.

6. Despite variations related to the source materials of the DOC, bacterial utilization and growth efficiencies can be high (Kaplan and Bott, 1983; Findlay *et al.*, 1986; Edwards and Meyer, 1987).

7. Much of the bacterial respiration of DOM occurs in the hyporheic zone of the sediments or along the surface of the sediments. Community respiration of

TABLE 23-5 Annual Mean Concentrations of Organic Matter in Transport in the Ogeechee River, Georgia[a]

	Mean (mg AFDW liter^{-1})		*Percentage of total organic matter*	
Dissolved organic matter		25.4		96.32
Particulate organic matter				
Amorphous material (bacteria)	0.301		1.141	
Amorphous material (protozoans)	0.039		0.148	
Amorphous material (other)	0.521		1.976	
Vascular plant detritus	0.028		0.106	
Algae (mostly diatoms)	0.060		0.228	
Fungi	0.014		0.053	
Animals	0.002		0.008	
Total particulate organic matter		0.97		3.68
Total organic matter in transport		26.37		100.00

[a] Modified from Benke and Meyer (1988). AFDW = ash-free dry weight.

hyporheic sediments can be very high with as much as 60% degradation (McDowell, 1985; Hedin, 1990; Fiebig and Lock, 1991; Findlay and Sobczak, 1996; Pusch, 1996).

8. Up to half of the DOC of stream water can be metabolized by microbes attached to sediment particles from interstitial water flowing through the hyporheic zones of porous sediments of streams (e.g., Crocker and Meyer, 1987; Findlay *et al.*, 1993).

9. Rates of physical and microbial degradation of DOM, particularly humic substances, can be greatly accelerated by exposure to sunlight (see later discussion). In the absence of light, abiotic removal of DOC is slight (e.g., Kuserk *et al.*, 1984).

10. Large POM is decomposed slowly and has a longer retention time within a particular reach of the stream (Chap. 21).

Total organic carbon turnover length is the average downstream distance traveled by a carbon atom in a fixed or reduced (organic) form, calculated as the ratio of the downstream transport of organic carbon (per unit stream width) to benthic respiration (per unit area). Therefore, turnover length measures the rate at which organic material is lost from streams relative to the rate at which it is used (Webster *et al.*, 1995). Turnover length increases downstream and therefore with increasing stream order and average discharge. The bioavailability of riverine dissolved organic matter (DOM) appears to be greater under low discharge conditions and decreases with distance downstream. The chemical composition of the DOM tends to become more recalcitrant with greater time in the river (Leff and Meyer, 1991; Sun *et al.*, 1997). Downstream decreases in bioavailability were attributable primarily to selective degradation of aliphatic carbon in the riverine DOM.

Estimations of the rates and importance of autochthonous primary production in streams by the attached benthic algae, lotic phytoplankton, and larger aquatic plants are very difficult (reviewed by Wetzel, 1975b; Minshall, 1978; Bott, 1982; Wetzel and Ward, 1992) (cf. Chaps. 15, 18, and 19). Overwhelming evidence indicates that terrestrial photosynthesis and importation of this organic matter to stream ecosystems are the primary carbon and energy sources of these systems; that is, that streams are largely heterotrophic. Such is indeed the case in heavily canopied woodland and forested streams in which autochthonous primary production is very low or negligible.

In noncanopied streams and in rivers as they increase in size and decrease in velocity of flow, the significance of primary productivity of lotic phytoplanktonic and attached algae and macrophytes increases (Fig. 23-4). Animal consumers within the floodplain habitats of large rivers depend heavily on algal organic matter from both phytoplankton and periphyton for

TABLE 23-6 Estimates of Inputs of Particulate Organic Matter to a 135-km Reach of the New River (North Carolina, Virginia, West Virginia)[a]

Source	*Input (mT AFDW yr^{-1})*	*Percentage of total input*
Allochthonous		
Upstream and tributary	5893	53.8
Within study area	64	0.5
Autochthonous		
Periphyton	3570	32.6
Aquatic macrophytes	1435	13.1
Total particulate organic matter input	10962	

[a] From Hill and Webster (1983). AFDW = ash-free dry weight.

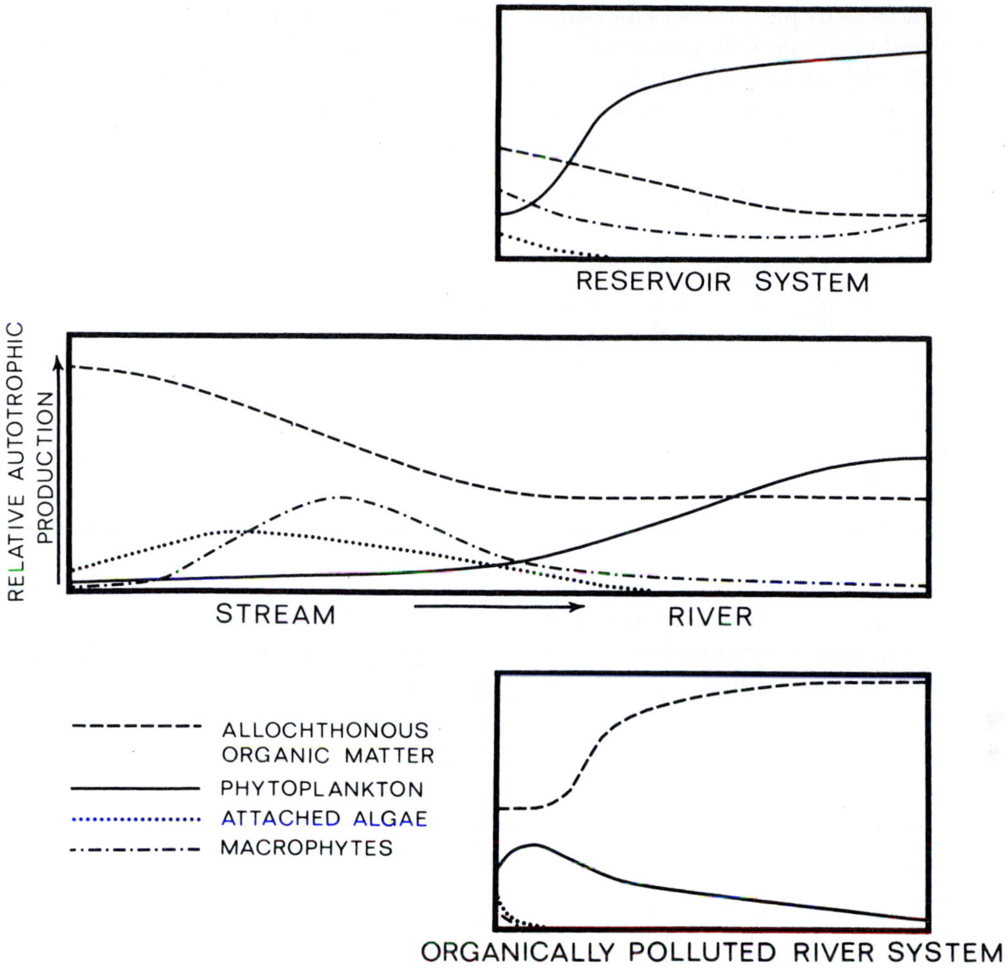

FIGURE 23-4 Generalized scheme of the relative contributions of allochthonous organic matter and autotrophic production by attached algae, phytoplankton, and aquatic macrophytes in the transition of a stream to a river system, if it is impounded where velocity of flow is reduced or practically eliminated or if it is organically polluted. (From Wetzel, R. G.: Primary production. *In* Whitton, B. (ed.): *River Ecology*. Oxford, Blackwell Sci. Publs., 1975.)

growth (Hamilton and Lewis, 1992; Hamilton *et al.*, 1992). However, the heterotrophic metabolism of these river ecosystems is totally dominated by microbial metabolism. Running waters can clearly vary in their proportion of heterotrophic and autotrophic metabolism. The relative significance of these types of metabolism is dynamic and varies considerably at local levels within a river system and seasonally with shifts in many physicochemical parameters and shifts in loading with allochthonous (natural or artificially by pollution) organic matter. On an annual basis, however, all streams and rivers are net heterotrophic ecosystems (Wetzel, 1995, 2000a).

In many rivers, particularly those of low elevational gradients, flooding events are important for redistribution of both particulate and dissolved organic matter. Inundation of the flood plains of rivers can increase the loading of organic matter from the decaying vegetation of the flood plain. Flooding behavior is very variable in relation to climate, precipitation periodicity and intensity, and extent of water-level regulation by reservoir damming. Examination of the extent and duration of floodplain inundation and flooding periodicity of an unregulated sixth-order river in the southeastern United States over a six-decade period indicated that >50% of the flood plain was inundated 15% of the time (54 day yr^{-1}). During 50% inundation, system width exceeded channel width by 19 times (Benke *et al.*, 2000). Over 50 to 100% of the flood plain was inundated for several months during the winter–spring period of high precipitation. Floods of 50% inundation of the flood plain had a duration of at least 30 days.

The flood pulse for this forested floodplain river was less predictable, and floods tended to be of shorter duration than those of large tropical rivers. Flooding in this river was correlated primarily with seasonal differences in temperature and plant evapotranspiration. Thus, increases in temperature with decreases in precipitation will lead to large decreases in export of dissolved organic substances (Clair and Ehrman, 1996). Evapotranspiration is linked to precipitation in determining the river discharge because even large increases in precipitation can lead to decreasing discharge when accompanied by higher temperatures. In contrast, in tropical rivers, where annual patterns of flooding are driven by large seasonal variations in precipitation, flood plains can be inundated for four or more months (Junk, 1997a,b; Lewis *et al.*, 2000).

The primary productivity within reservoirs is extremely variable because of individual characteristics of morphometry, seasonal changes in water-retention times, and human manipulations of water level and retention. In general, nutrient loading and trophic state are higher at the river end of reservoirs than at the deeper, dammed end. Although phytoplanktonic productivity per unit volume may be higher at the river end than at the dammed end, increased inorganic turbidity and light reduction at the river end can cause a reduction of the depth of the trophogenic zone and a decrease in total productivity (e.g., Fig. 23-4). The primary productivity per unit area in some reservoirs can be approximately the same over the length of the reservoir. In these cases, the higher trophic state and volumetric productivity at the inflow end of the reservoirs shift to lower epilimnetic trophic states, lower volumetric productivity, but increased depth of the trophogenic zone at the dammed end.

2. General Metabolism of Organic Matter in Streams

Therefore, in summary, much of the dissolved organic matter in streams originates from lignin and cellulose and related structural precursor compounds of higher plants. These substances are abundantly produced in the true lake and river ecosystems—that is, the ecosystem includes the drainage basin and organic matter produced photosynthetically within it. The productivity of terrestrial vegetation and aquatic plants associated with the land–water interface region is manifoldly (usually several orders of magnitude) greater than that of algae. Organic substances from higher plant tissues are abundant, chemically complex, and relatively recalcitrant to rapid biological degradation (Thurman, 1985; Haslam, 1998). During oxidative and anaerobic degradation, these compounds are modified by microbial activities in detrital masses, including standing dead tissues that can remain in an oxidative aerial environment for months or years. Much of the dissolved organic compounds released from partial decomposition of the plant tissues and from associated microbial degradation products are leached and partially degraded en route toward recipient lakes and streams. That microbial modification continues during partial decomposition in terrestrial soils and hydrosoils of wetlands. Once within land–water interface regions, water containing dissolved organic compounds moves, often diffusely, through dense aggregations of living emergent and submersed aquatic plants and massive amounts of particulate, largely plant derived, detritus (extensively reviewed in Wetzel, 1990a). The enormous surface areas of these habitats support large aggregations of rapidly growing microbial communities. During transport through these microbial metabolic sieves associated with wetland and littoral areas, appreciable further selective degradation of more labile dissolved organic constituents occurs before final movement into the receiving lake body or river channel per se.

Although difficult to generalize, a common pattern is that DOM entering rivers is quite different among different sources. For example, in the Amazon ecosystem, heavy storm precipitation events transport relatively labile DOM, whereas DOM from wetlands and ground water tends to be hydrophobic and recalcitrant from selective biological utilization of the more labile components (McClain and Richey, 1996). DOM from litterfall leachates tends to be relatively labile.

As will be emphasized further on, the trophic dynamic structure of aquatic ecosystems depends operationally on a dynamic detrital structure (Wetzel *et al.*, 1972; Wetzel and Rich, 1973; Rich and Wetzel, 1978; Wetzel, 1995). From the standpoint of carbon fluxes, most energy and organic carbon of the systems is dead, of autotrophic origin, and undergoing microbial degradation, the rates of which are variable and serially decrease with increasing recalcitrance of the organic matter. In lakes, reservoirs, and rivers, detrital organic matter is the main supportive metabolic base of carbon and energy. Since most lakes of the world are small to very small, much of the autochthonous production of detrital matter originates from benthic littoral and wetland vegetation, which augments the loadings of dissolved organic matter from terrestrial origins. Similarly, detrital heterotrophic metabolism dominates in streams, where usually the major sources are allochthonously derived terrestrial plant material. Common to both lake and stream ecosystems is the dominance of detrital metabolism, which gives the ecosystems a fundamental metabolic stability. The trophic structure above the producer–decomposer level, with all of its complexities of population fluctuations, metabolism, and behavior, has a relatively small impact on the total carbon flux of the system. The detrital system provides stability to streams; the slower, relatively consistent degradation of dissolved and particulate detritus by microorganisms underlies the more sporadic autochthonous metabolism that responds rapidly to, and depends to a greater extent on, environmental fluctuations. Autochthonous primary production is often small and variable, but in combination with allochthonous, autotrophically produced detritus, drives the lotic system. The functional operation of lentic and lotic systems converges at this point of similarity in detrital metabolism.

B. Allochthonous Organic Matter Received by Lakes via Streams and Rivers

1. Dissolved Organic Matter

The DOM of surface runoff is composed of relatively recalcitrant organic compounds resistant to rapid microbial degradation. The amounts of DOC and POC reaching a lake and the chemical composition of these organic compounds change seasonally with the volume of flow in relation to duration of retention in the stream, the growth and decay cycles of the terrestrial and wetland vegetation through which runoff flows, and other factors, especially climatic variations.

An example is seen from quantitative analyses of dynamics of influxes of allochthonous detrital soluble and particulate organic carbon to a small temperate lake over an annual period in relation to their fate within and losses from the lake (Wetzel and Otsuki, 1974; Wetzel, 1989). Detrital organic carbon influxes were determined in water from a primary inlet stream, both before and after its flow had traversed a wetland adjacent to the lake, and from a second inlet stream, both at its headwaters and after the stream water passed through the wetland. Similarly, measurements were made of organic content of ground water where it entered the lake and at the outlet. Concentrations of organic carbon in inflows and outflows were converted to values of total carbon loading and outflow using a detailed annual water budget (Fig. 23-5). The DOC and the dissolved organic nitrogen (DON) increased significantly (×2) during the active growing season as the inlet water of the streams passed through the

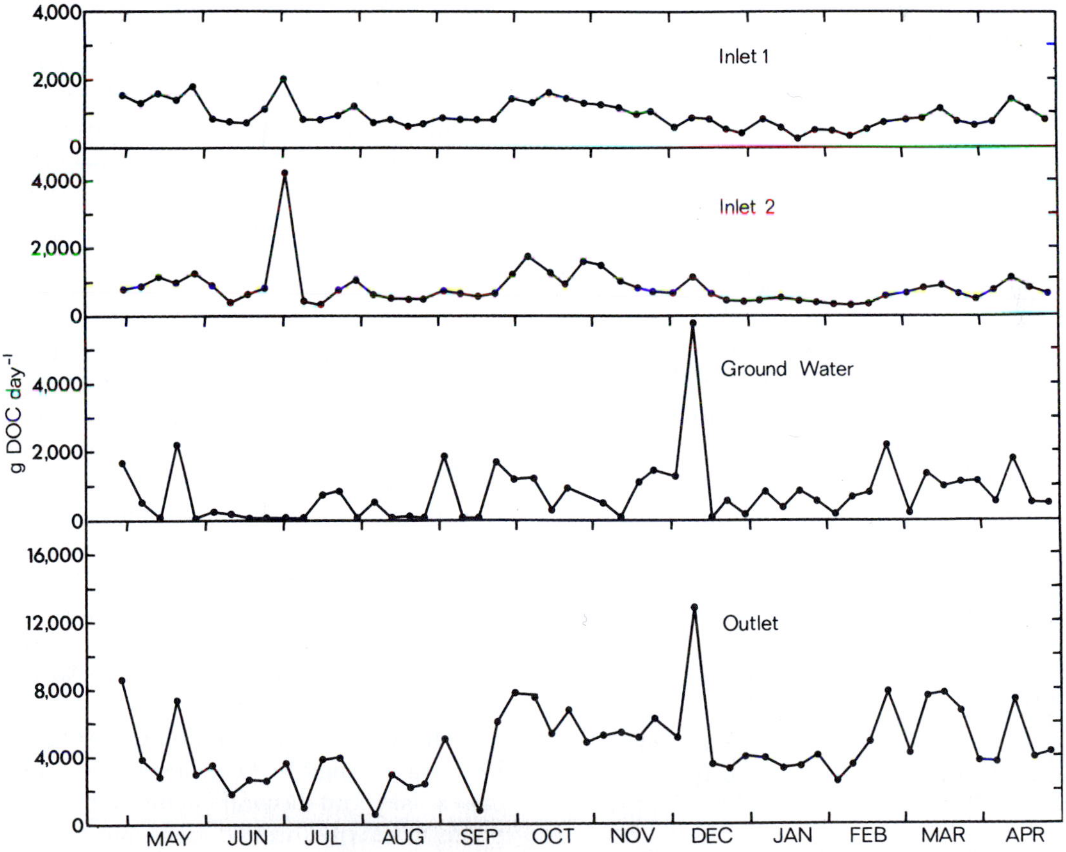

FIGURE 23-5 Dissolved organic carbon entering Lawrence Lake, Michigan, via ground water and the two stream inlets and leaving via the lake outlet. (From Wetzel, R. G., and Otsuki, A.: Allochthonous organic carbon of a marl lake. *Arch. Hydrobiol.* 73:31–56, 1974.)

marshes prior to entering the lake (Wetzel, 1989). The dissolved organic C:N ratios (DOC-C:DON-N) varied between 15 and 150, with an annual average of ca. 30 over the year of study. In addition, the DOC:DON ratios decreased during the active growing season (April–October) as one progressed from the upper wetlands toward the wetlands of the littoral of the lake (Wetzel, 1989). Similar results were found in larger streams as they passed through extensive marsh systems (Manny and Wetzel, 1973; Grieve, 1994).

The DOC:DON ratios decreased during the active growing season of the wetland from values twofold higher during the colder and winter periods of the year (Wetzel, 1989). Clearly, the wetland plants and attendant microflora of the sediments and detrital particulate organic matter served as a distinct source for dissolved organic nitrogen during much of the year but particularly during the active growing season of the wetland ecosystem. Much of the DON-N was associated with the humic compounds. For the smaller amounts of more labile DON-N associated with proteinaceous and amino compounds, degradation rates were rapid. For example, mineralization of protein-N by the wetland sediment microflora had turnover rates of 13–69 h between May and October (Cunningham and Wetzel, 1989). Utilization of amino-N was 9–57 times more rapid than that of protein-N, with turnover times of 1.6–3.7 h in the wetland sediments over the growing season.

The lowest inputs of DOC occurred during the summer, when there was little precipitation, temperatures and bacterial metabolism were higher, and vegetation was growing rapidly (Wetzel, 1989). During the autumn and early winter, DOC increased markedly. The DOC content of the ground water was very low, but because of the large volume of this inflow, it constituted one-third of the annual influx of allochthonous DOC (Table 23-7). A net loss of 15 g DOC m^{-2} $year^{-1}$ from the outflow of this lake was observed, which represented a major pathway of organic carbon removal from the lake. Spectrophotometric analyses (absorption of ultraviolet light and fluorescence) of water flowing into and out of the lake indicated that a high percentage of the DOC was composed of humic compounds and yellow organic acids.

TABLE 23-7 Annual Dissolved Organic Carbon Entering Allochthonously and Leaving Lawrence Lake, 1971–1972[a]

DOC	*g C m^{-2} yr^{-1}*
Influxes of DOC	
Inlet 1	7.00
Inlet 2	7.94
Ground water and seepage	6.01
Total	20.95
Outflow DOC	35.82

[a] From Wetzel, R. G., and Otsuki, A.: Allochthonous organic carbon of a marl lake. *Arch. Hydrobiol.* 73:31–56, 1974.

Analogous situations exist in large rivers, particularly those with extensive flood plains. For example, in Australian floodplain rivers, in-stream primary productivity is trivial in comparison to that of riparian floodplain sources of organic matter (Robertson *et al.*, 1999). Riverine forests have primary production of ca. 600 g C m^{-2} yr^{-1}, whereas those of floodplain aquatic macrophytes and periphyton exceed 2500 g C m^{-2} yr^{-1} and 620 g C m^{-2} yr^{-1}, respectively. Large pools of particulate organic carbon as detritus exist as litter (>500 g C m^{-2}) and coarse woody debris (ca. 6000 g C m^{-2}). Bacterial decomposition of particulate and dissolved organic matter of the sediments is very high, with respiration and methanogenesis generating and releasing ca. 1 g C m^{-2} day^{-1}. Very large quantities of DOC are released plant from litter and exported to the river channels. Similar production, utilization, and transport of organic matter is found in other large rivers (e.g., Hedges *et al.*, 1994; Lewis *et al.*, 2000). For example, in the Amazon River, highly degraded leaf material is solubilized during degradation and then partitioned between soil minerals and water during transport to the river, which results in suspended fine particulate organic materials of soil detritus origin that are nitrogen-enriched coexisting with dissolved organic substances that are nitrogen-poor.

2. Particulate Organic Matter

Concentrations of POC in the influxes, in water within the lake, and in the outflow water were consistently about 10% that of the DOC. Inputs of POC were minimal during the summer months, when marsh vegetation was growing rapidly, but increased markedly during periods of high autumnal rainfall and spring runoff (Fig. 23-6). Groundwater POC was very low, but on an annual basis constituted over one-fourth of the allochthonous POC input (Table 23-8). Allochthonous POC was small in relation to the quantities of littoral and pelagic organic carbon synthesized within the lake and represented <5% of the gross inputs of POC to the lake.

The quantity of windblown POM to Lawrence Lake was negligible. Leaf litterfall and material blown into a lake can become significant in small lakes or ponds in heavily forested areas, but generally this input is small (Table 23-9) in comparison to other sources of organic loading from external sources. Domestic and industrial POM pollution of lakes by organic wastes is

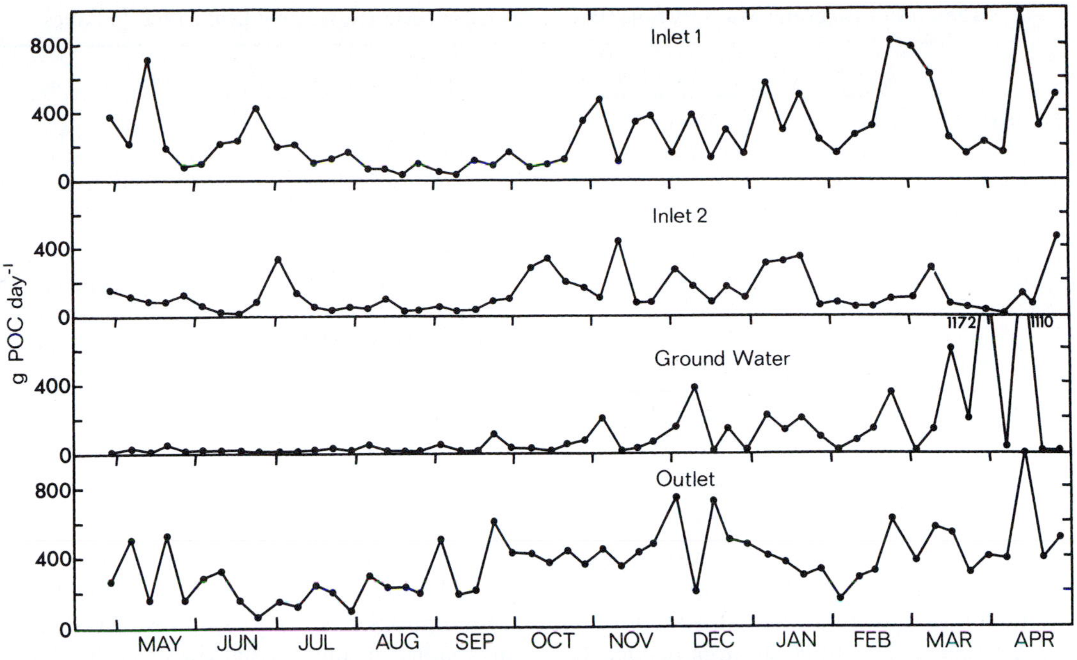

FIGURE 23-6 Particulate organic carbon of allochthonous inputs entering and leaving via the lake outflow, Lawrence Lake, Michigan. Note that the POC values are an order of magnitude lower than those of DOC in Figure 23-5. (From Wetzel, R. G., and Otsuki, A.: Allochthonous organic carbon of a marl lake. *Arch Hydrobiol.* 73:31–56, 1974.)

an exceedingly individualistic parameter that lends itself poorly to generalization. Pollution of this nature is normally transported to lakes and reservoirs via stream and river inflows rather than via groundwater sources. The types of organic pollution, partial degradation en route, and effects on stream properties are treated excellently by Hynes (1960, 1970). As pointed out by Pieczyńska (1972; cf. also Rai and Hill, 1981), the degradation of organic matter by microflora in lakes that receive a high proportion of allochthonous organic matter is generally more variable and requires a greater diversity of metabolic pathways than microbial degradation in lakes that receive POM predominantly from autotrophic sources. It should be stressed, however, that many lakes derive much of their organic carbon and many nutrients from terrestrial and wetland sources of the surrounding drainage basin. The magnitude of allochthonous inputs is difficult to measure and, more importantly, the metabolic role of allochthonous materials within the lake is difficult to separate from that of autochthonous inputs. Despite much recent progress in understanding, this area is in great need of further investigation.

TABLE 23-8 Annual Particulate Organic Carbon Entering Allochthonously and Leaving Lawrence Lake, Michigan, 1971–1972[a]

POC	$g\ C\ m^{-2}\ yr^{-1}$
Influxes of POC	
Inlet 1	1.99
Inlet 2	0.962
Ground water and seepage	1.15
Windblown POC	0.01
Total	4.10
Outflow POC	2.75

[a] From Wetzel, R. G., and Otsuki, A.: Allochthonous organic carbon of a marl lake. *Arch. Hydrobiol.* 73:31–56 1974.

Leaf fall into streams, particularly among forest-canopied small streams (first to fourth order), is, of course, a major input of organic matter, as has been discussed repeatedly earlier (cf. Chaps. 8, 18, 21, and 22). Mean DOC concentrations in water of small streams is directly related to the mass of leaf litter (Mulholland, 1997; Meyer *et al.*, 1998). In-stream generation of DOC from leaf litter trapped and stored among stream sediments contributes about 20% of the daily DOC exported from many forested headwater streams. In the temperate zone, this source of DOC is very seasonal—greatest during autumn and winter and least during spring and summer. The remainder of DOC emanates largely from the floodplain vegetation and its degradation and from terrestrial sources.

TABLE 23-9 Estimates of Airborne Transport of Particulate Organic Matter, Largely Leaf Litter, to Lakes[a]

Location	*Lake area (ha)*	*% as leaves*	*Total airborne litter to lake*	*Source*
Mikolajskie, Poland	469	100	8 kg leaves ha^{-1} yr^{-1}	Szczepański (1965)
Wingra, WI	130	>80	23 kg litter ha^{-1} $autumn^{-1}$	Gasith and Hasler (1976)
Glenwild, NJ	44	66	176 kg litter ha^{-1} $autumn^{-1}$	M. Sebetich (pers. comm).
Mirror, NH	15	62	113 kg litter ha^{-1} yr^{-1}	Gosz *et al.* (1972); Jordan and Likens (1975)
Frongoch, Wales	7.2	95	14 kg litter ha^{-1} yr^{-1}	Hanlon (1981)
Lawrence, MI	4.9	<1	<2 kg litter ha^{-1} yr^{-1} of 89 kg POM[a] ha^{-1} yr^{-1} of allochthonous POM entering lake	Wetzel and Otsuki (1974)

[a] POM = particulate organic matter (>0.5 μm).

The implications of the preceding discussion are that a large portion of the particulate and dissolved detrital organic matter that enters lakes from allochthonous sources has undergone microbial stripping of more labile compounds. The bulk of organic detrital carbon is in the form of humic and other substances that are relatively resistant to further microbial degradation. It is then necessary to superimpose upon this variable allochthonous input of detrital organic matter the organic matter that is produced in the body of water itself by phytoplankton of the open water and that produced in the littoral zone.

3. Total Energy Transfer Rates

In previous discussions, we have systematically scrutinized the autochthonous synthesis of organic matter in both the open water (Chap. 15) and the littoral regions (Chaps. 18 and 19). The utilization of portions of this organic matter by planktonic and benthic animals was considered in Chapters 16 and 22. The complexities of microbial degradation of the dissolved and particulate organic matter in the open water and the sediments were treated in Chapters 17 and 21, respectively. These analyses provide the parts that can be fitted into an integrated whole that quantitatively combines the pathways of flow among producer, consumer, and decomposer components.

Historically, integration of the ecosystem components has been attempted in many different ways. Early attempts coupled the plant and animal components qualitatively as complex food webs, in which the feeding or trophic relationships among organisms were interconnected. Although these analyses provide descriptors of the biotic components, the food-production and food-consumption processes are very dynamic and constantly change with differing environmental conditions and during the life histories of the organisms. So many organisms are involved that it is exceedingly difficult to quantify, in the words of Hutchinson, the roles of even the major actors of the ecological play.

Alternatively, the energy content of organisms and communities has been examined. Present technology is incapable of measuring directly the rates of energy transfer among individual organisms and communities.[2] If the energy content of all organisms and their energy expenditures for all behavioral, maintenance, and reproductive costs were known, however, it would be possible to estimate the energy transfer among biotic components. The dearth of bioenergetic knowledge for many of the organisms occurring in fresh waters means that only very approximate estimates are possible at this time.

Quantitative carbon fluxes among communities of freshwater ecosystems have been used with some success. The underlying premise of this approach is that the synthesis of organic matter begins with the fixation (primarily photosynthetic) of CO_2. Much of this organic carbon is variously cycled within the ecosystem before being respired (primarily via decomposition) and returned as CO_2. Despite the limitations of this approach, discussed later, much functional information on the operation of freshwater ecosystems has emerged from quantitative analyses of rates of carbon cycling and what parameters regulate these rates.

We initiated discussion throughout previous chapters with the quantitative aspects of productivity and the dispersion of gross productivity in metabolic respiration, growth, and reproduction to obtain estimates of net productivity. These values were expressed in terms

[2] In the future, it should be possible theoretically to quantify directly energy fluxes from a source of molecular mass in one organism to other organisms that are physiologically similar. In addition, it should be possible to evaluate energy dissipation routes and transfer rates into reproduction, growth, and losses by direct measures of molecular energy fluxes.

of carbon wherever possible for comparative purposes under differing physical, chemical, and biotic (e.g., predation) environmental conditions. The biotic components were coupled qualitatively earlier, as, for example, in Figure 17–1. It is now instructive to examine the mass and flux of organic carbon among the components in as quantitative a manner as possible in representative lakes, reservoirs, and rivers. None of these ecosystem analyses is complete, but the analyses do provide insight into the relative importance of the different communities and processes. Additionally, they emphasize areas in need of intensive further investigation in order to quantitatively evaluate control mechanisms of those energy fluxes.

IV. DISTRIBUTION OF ORGANIC CARBON

A. Dissolved of Organic Carbon (DOC)

A conspicuous feature of the depth–time distribution of total dissolved organic carbon in a lake (e.g., Fig. 23-7) is the absence of strong vertical stratification or seasonal fluctuations, despite the parallel thermal and density stratification and seasonal pulses of metabolic activity. It is apparent that the DOC pool consists primarily of carbon compounds relatively recalcitrant to rapid bacterial decomposition. Inputs of these dissolved organic substrates are approximately equal to their slow rates of microbial degradation. Highest concentrations occur during summer stratification in the epilimnion and consistently fluctuate to a greater extent in the upper strata, especially near the surface, than do concentrations in the hypolimnion. It is evident from earlier discussion (see Chaps. 15, 18, and 19) that extracellular release of DOC by living and senescing phytoplanktonic algae and littoral algae and macrophytes is associated with a portion of the higher epilimnetic concentrations of DOC. Decomposition of largely labile secreted organic compounds often is very rapid (e.g., <48 h), and their dynamics would not be delineated by the sampling frequency (weekly) employed for this generalized picture of the DOC pool. In addition, an appreciable input of allochthonous DOC commonly enters the upper strata during the period of summer stratification. In addition to the data presented here, the relative constancy of the DOC pool spatially and temporally was also found in many other waters (e.g., Fukushima *et al.*, 1996).

The constancy of the DOC pool is reflected further in annual average values per square meter of water column and for the whole lake calculated as the depthwise integration of the DOC concentrations and volume of water of each meter of water stratum (Table 23-10). Much of the seasonal change is observed in the surface waters, subject to greatest insolation (Hessen *et al.*, 1997). DOC concentrations at lower depths are relatively constant. Horizontal variations of DOC within this lake were very small (<10% deviation from the mean of the central depression). In some lakes, however, and especially in reservoirs, point–source inputs of DOC, such as river influxes high in DOC, would cause greater horizontal variations within the basin and irregular temporal fluctuations.

Therefore, the major sources to the DOC pool are (1) photosynthetic inputs of the littoral and pelagic flora

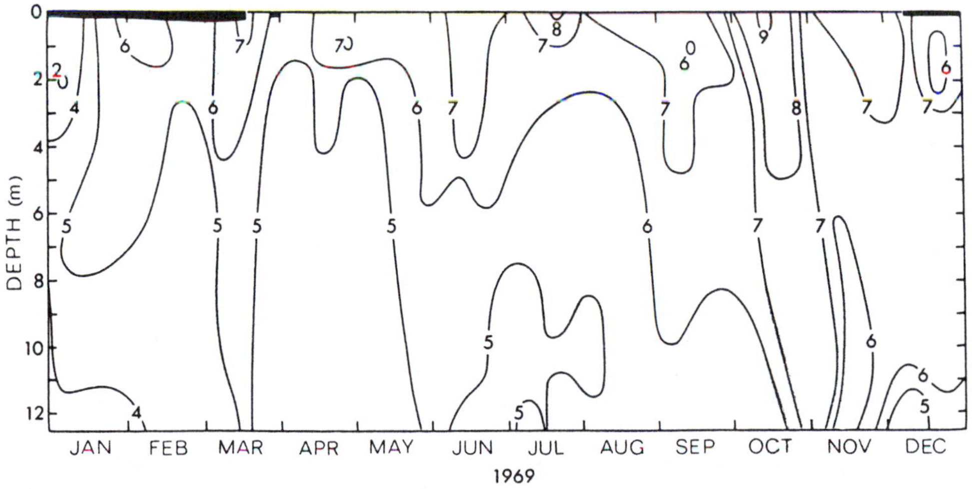

FIGURE 23-7 Depth–time diagram of isopleths of dissolved organic carbon (mg C liter^{-1}), Lawrence Lake, Michigan, 1969. (From Wetzel, R. G., Rich, P. H., Miller, M. C., and Allen, H. L.: Metabolism of dissolved and particulate detrital carbon in a temperate hard-water lake. *Mem. Ist. Ital. Idrobiol.* *29*(Suppl.):185–243, 1972.)

TABLE 23-10 Dissolved Organic Carbon of Lawrence Lake, Michigan, 1968–1979[a]

Years	*Mean g C* m^{-2}	*Mean kg C* $lake^{-1}$
1968	30.93	1535
1969	34.73	1724
1970	27.80	1380
1971	26.71	1326
1972	20.19	1002
1973	25.15	1248
1974	23.25	1153
1975	25.87	1284
1976	25.47	1264
1977	21.54	1069
1978	22.12	1098
1979	28.70	1424
Average, 12 yr	26.04	1292

[a] Data of Wetzel (unpublished).

added to the pool through secretions and lysis of cellular contents; (2) allochthonous DOC, composed largely of humic substances originating from terrestrial and wetland/littoral higher plant tissues that are recalcitrant to rapid bacterial degradation; and (3) bacterial degradation and chemosynthesis of organic matter with subsequent release of DOC. Very much smaller amounts, usually quantitatively negligible but qualitatively of potential importance, include (4) DOC from excretions of secondary producers, zooplankton and higher animals, that have ingested living or detrital organic matter.

Phytoplanktonic productivity and allochthonous sources from the drainage basin are the primary sources of the DOC pool of oligotrophic waters. In moderately large or very large bodies of water, phytoplanktonic photosynthesis can constitute a major contributor to the DOC pool. However, most lakes are small, and the ratio of littoral to pelagic photosynthetic productivity greatly increases as the mean depth of the basin decreases.

With increasing fertility, the relative contributions to system productivity by phytoplankton, submerged and emergent littoral/wetland macrophytes, and eulittoral algae are altered (Chaps. 18 and 19; see Fig. 19-21). The shift in dominance with respect to the productivity of these major groups results in differential contributions to the DOC pool (Fig. 23-8). If the drainage basin is not manipulated extensively by human activities, allochthonous inputs of DOC to a lake are defined by characteristics of its drainage basin and are relatively constant on a long-term annual basis; their relative contribution to the ecosystem consequently decreases as autochthonous sources increase

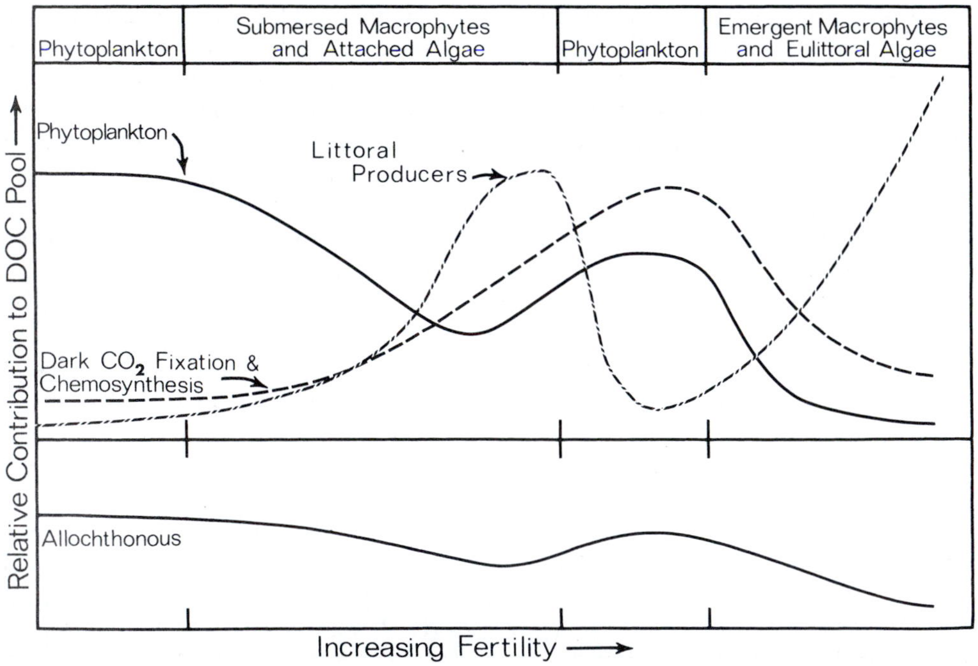

FIGURE 23-8 Generalized relative contributions of phytoplankton, littoral producers, dark and chemosynthetic CO_2 fixation, and allochthonous sources of dissolved organic carbon (DOC) to lakes of increasing fertility in the progression of dominating pelagic and littoral flora. Upper panel of generalized changes in dominating productivity by autotrophic flora refers to those changes detailed in Figure 19-21. In the final stages, highly productive emergent plants subsume all other inputs of organic matter (cf. Chap. 25).

(Fig. 23-8). A partial dissolved organic carbon budget of acidified Lake Gårdsjön, Sweden, demonstrated that over half of the DOC entered the lake from precipitation, surface runoff, and seepage (Broberg and Persson, 1984). Nearly half of the dissolved organic carbon was respired, photooxidized, or precipitated and sedimented. High-molecular-weight DOC compounds tend to prevail in the open water, with long chemical turnover times (Cole *et al.*, 1984). In reservoir systems, allochthonous inputs of DOC can exceed by several times the amounts of POC and DOC produced autochthonously (e.g., Romanenko, 1967).

Temporal variations in DOC among streams are large and often irregular. Clearly, most of the DOC in rivers is usually dominated by recalcitrant organic compounds of terrestrial and floodplain origins (Meyer, 1986; Moore, 1987; Burney, 1990; Grieve, 1990). DOC commonly reaches maximum values during low-flow summer and autumn periods, likely the result of concentration effects during summer periods of low precipitation and high evapotranspiration, as well as increased rates of decomposition with higher temperatures. In larger rivers, concentrations of DOC and total export of DOC is coupled to high-flow periods when flood plains are inundated, often in the winter. In very large rivers, primary production by phytoplankton and periphyton may contribute appreciably to the DOC during low-flow periods. Production is rapidly reduced as a result of turbidity and light attenuation during high-flow periods. In tropical rivers, DOC concentrations are much more variable owing to reduced seasonality of temperature, variable precipitation and flooding periods and the duration of flooding of the flood plain. Although the initial flush of DOC from the land or wetlands can be large following snowmelt or a large precipitation event, the sustained DOC concentrations in the runoff is often rapidly reduced by the dilution with large amounts of water.

Human disturbances can markedly alter releases of DOC to streams. Nonseasonal inputs of domestic and industrial organic wastes can radically alter discharge-corrected loads of DOC. Following clear-cutting of forest within the drainage basins of small streams, DOC concentrations of streams were reduced with little downstream accumulation (Kaplan *et al.*, 1980; Meyer and Tate, 1983; Grieve, 1990). Streams within undisturbed drainage basins had much higher levels of DOC with summer maxima and concentrations increasing downstream. In steep-drainage river basins, seasonal differences in DOC levels were small among streams draining forested and deforested sections of the drainage basin.

Therefore, in summary, a large portion, usually >90%, of the organic matter imported from allochthonous and littoral/wetland sources to these aquatic ecosystems is predominantly in dissolved or colloidal form. Although a portion of the dissolved organic compounds may aggregate and shift to a particulate and hence gravitoidal form that may sediment out of the water, most of the imported dissolved organic matter is dispersed within the water and is moved about with the hydrodynamic movements within the water body. That dispersion of DOM is important owing to its retention in zones of utilization or modification by physical processes (e.g., photolysis) within the aquatic ecosystems, which may not be the case with particulate organic matter subject to gravitoidal sedimentation. Great seasonal and daily variations can occur in those hydrodynamics, particularly in streams and rivers. In lakes, daily and seasonal variations are generally appreciably slower and smaller than in rivers. Retention times, which are important for effective utilization by enzymatic hydrolysis, tend to be longer in lakes than in streams.

Furthermore, massive microbial degradation of DOM occurs within lakes and streams despite the general recalcitrance of the organic matter. Many examples were detailed earlier (Chaps. 17 and 21) and are being further verified experimentally (e.g., Wehr *et al.*, 1999). Because of limited accessibility of large portions of these predominantly humic and fulvic acid molecules to enzymatic hydrolysis, degradation rates are slow. Thus, many of these commonly acidic macromolecules have long turnover times and relatively long residence times once within the lakes or streams. Dissolved macromolecules are often of considerable age (years) but are mixed with variable and rapidly changing inputs of younger humic and nonhumic substances of more recent synthesis. Recent studies are indicating that humic substances, particularly relatively recalcitrant fulvic acids, are generated by algae and contribute to the multitude of diverse compounds that constitute the collective dissolved organic matter (McKnight *et al.*, 1991, 1994). This source is potentially important in littoral areas, where attached and sessile algal productivity is often several orders of magnitude greater than that of phytoplankton (e.g., Wetzel, 1990a, 1996). Because of the recalcitrance of these dominating dissolved organic compounds, the dissolved organic matter can reside *within* lakes and rivers for long periods of time (months or years).

Strong budgetary evidence in early lake studies indicated both that, despite continual loading, the dissolved organic matter did not accumulate and losses by aggregation and sedimentation were small. However, large quantities of CO_2 evade to the atmosphere, even among oligotrophic lakes where rates of production and heterotrophic microbial respiration are low (e.g., Otsuki and Wetzel, 1974; Kling *et al.*, 1992;

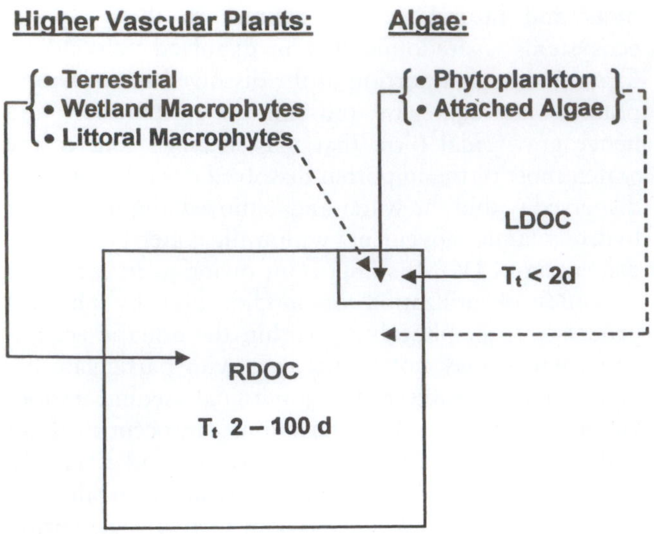

FIGURE 23-9 Common sources of organic carbon and rates of mineralization of labile (LDOC) and recalcitrant (RDOC) dissolved organic carbon in lakes and streams. (After Wetzel, 2000a.)

recently verified globally, cf. Cole *et al.*, 1994; Dillon and Molot, 1997a). Even though at certain times of the year, CO_2 (levels may indicate a small invasion) during periods of high primary productivity, total CO_2 evasion to the atmosphere is always in excess, usually greatly in excess, of autochthonous photosynthetic organic carbon production by phytoplankton on an annual basis.

For decades in aquatic ecology, the apparent chemical recalcitrance of humic substances that dominated the instantaneous bulk DOC of lakes and streams led to the belief that these compounds were poorly used by microbiota. Recent evidence clearly shows that bacterial production depends to a great extent upon humic materials from allochthonous sources and is stimulated by high-flow events that deliver humic organic influxes into the lakes and streams (e.g., Geller, 1985; Meyer *et al.*, 1987; Gremm and Kaplan, 1988; Kairesalo *et al.*, 1992; Straškrabová *et al.*, 1993; Volk *et al.*, 1997; Tranvik, 1998; Bergström and Jansson, 2000; cf. Chap. 17). Utilization of the imported DOC is enhanced by the longer retention times and exposure to higher temperatures and to light. To be sure, loss rates are slow, but consistently in the range of 0.5–2% per day under many different environmental conditions (Fig. 23-9). Often, rates of degradation of total dissolved organic matter are faster than 2% per day.

B. Particulate Organic Carbon (POC)

Examination of the general spatial and temporal distribution of pelagic POC of lakes of varying degrees of productivity permits a few generalizations. In oligotrophic to moderately productive lakes, the observed depth–time distribution of pelagic POC follows the productivity and biomass distribution of phytoplankton rather closely during stratified periods. When Lawrence Lake, for example, is stratified, the POC maxima lags behind highs in productivity from several days to about two weeks, especially when sedimentation of plankton is slowed in thermal density gradients of the metalimnion (Fig. 23-10). During the period of ice cover, one can frequently observe reduced primary production, lower inputs of allochthonous POC, and

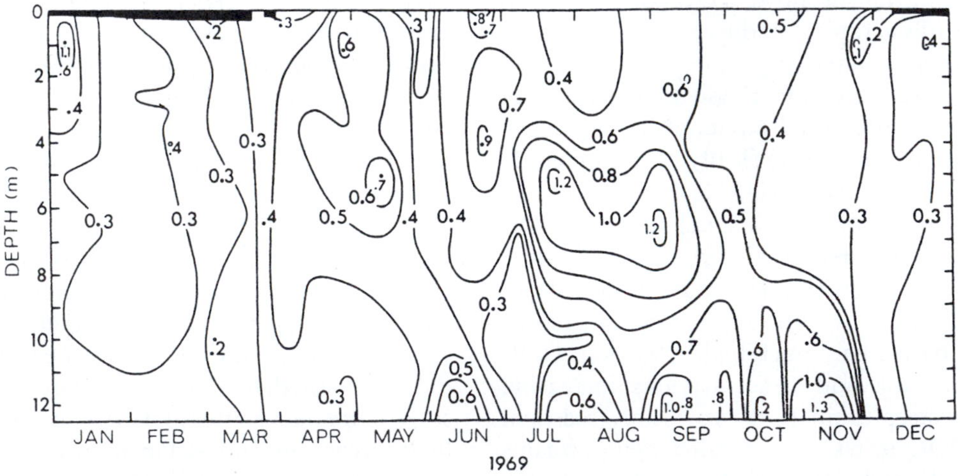

FIGURE 23-10 Depth–time distribution of pelagic particulate organic carbon (mg C liter^{-1}), Lawrence Lake, Michigan, 1969. (From Wetzel, R. G., Rich, P. H., Miller, M. C., and Allen, H. L.: Metabolism of dissolved and particulate detrital carbon in a temperate hard-water lake. *Mem. Ist. Ital. Idrobiol.* *29*(Suppl.):185–243, 1972.)

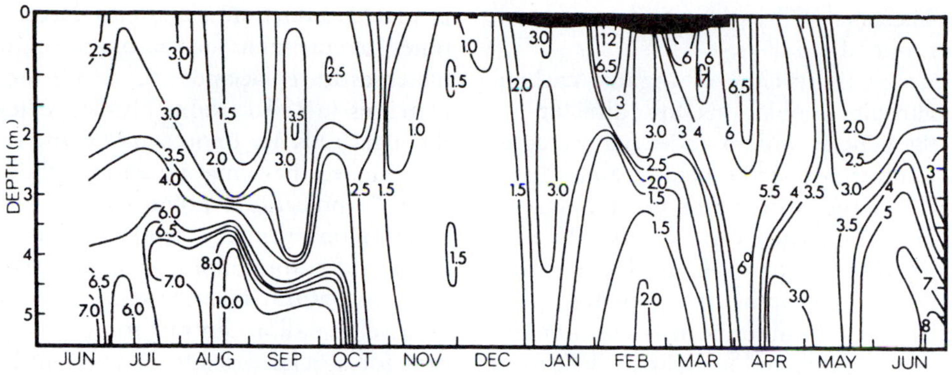

FIGURE 23-11 Depth–time distribution of pelagic particulate organic carbon (mg C liter^{-1}) of hypereutrophic Wintergreen Lake, Michigan, 1971–1972. (From Wetzel *et al.*, unpublished.)

a gradual reduction in concentrations of POC. An increase in POC generally is observed during periods of circulation when surficial sediments are disturbed and resuspended into the water column. In less productive lakes, hypolimnetic POC generally does not increase greatly unless the lower hypolimnion becomes anaerobic—for example, in the latter portions of summer stratification when specialized bacterial populations or algae (e.g., euglenophytes) may develop in profusion. In hypereutrophic lakes that receive large inputs of planktonic and littoral POC, the hypolimnia are rapidly rendered anoxic (as exemplified by Wintergreen Lake, Fig. 23-11) and bacterial productivity contributes to marked increases in POC.

The amounts of pelagic POC of a lake are relatively constant from year to year (Table 23-11), as long as the lake system is not disturbed by external influences such as enrichment from human activities. The ratio of DOC to POC is rather constant at about 10:1 in most unproductive to moderately productive lakes but is lower (ca. 5:1 to 3:1) and more variable in streams (e.g., Moeller *et al.*, 1979; cf. Table 23-1). Deviations from this 10:1 ratio with depth and season are small in less productive waters (cf., e.g., Figs. 23-7 and 23-10). As lakes become more eutrophic, the DOC:POC ratio fluctuates greatly with season and depth. In the Wintergreen Lake example (Fig. 23-11), the annual average is about 5:1; but during periods of intensive algal (epilimnetic) and bacterial (metalimnetic and hypolimnetic) growth, the ratio decreases to 1:1 or less and then increases to about 10:1 during the fall period of circulation (see also Weinmann, 1970; Pendl and Stewart, 1986). In riverine lakes and in reservoirs with high allochthonous loading of POC, DOC:POC ratios fluctuate and have lower than the global average values (ca. 4:l; Rai and Hill, 1981).

TABLE 23-11 Particulate Organic Carbon of the Pelagic Zone of Lawrence Lake, Michigan, 1968–1981[a]

Years	*Mean g C^{-2}*	*Mean kg C lake^{-1}*
1968	2.03	101.0
1969	2.65	131.8
1970	2.18	108.4
1971	2.10	104.2
1972	1.73	86.0
1973	1.91	94.6
1974	2.06	102.2
1975	2.14	106.0
1976	2.25	111.8
1977	2.07	102.7
1978	2.72	135.2
1979	2.38	118.2
1980	2.42	120.1
1981	1.84	91.5
Average, 14 yr	2.18	108.1

[a] Data of Wetzel (unpublished).

In a productive lake in Michigan, Saunders (1972) found that the particulate organic detritus constituted from 1.3 to 16.9 times the phytoplanktonic biomass and made up >50% of the seston. The remainder of the seston was dominated by inorganic matter, such as particulate $CaCO_3$ and silica.

Estimates of the replacement of algal cell carbon in the pelagic POC have been made from measurements of the biomass of algalcell carbon and the rates of net primary production (Miller, 1972). Compensating for respiratory loss of carbon, the daily net accumulation of POC can be estimated from net primary production. The total suspended epilimnetic POC of Lawrence Lake had an average replacement time (turnover) of 40.7 days (range, 8.1–544 days). In the POC pool, a mean of 83 mg C m^{-2} (range, 30–241 mg C m^{-2}) was algal cell carbon that was replaced by primary productivity in 1.1 days (range, 0.30–2.55 days) during the ice-free seasons. The algal cell carbon had an annual mean replacement time of 3.6 days.

Generalizations on the cycling of organic carbon among phytoplanktonic populations are difficult to make because of the paucity of data under natural conditions. POC concentrations of the pelagic zones usually are considerably larger (double or greater) in eutrophic lakes than in infertile waters (Fig. 23-12). Differences in concentrations of DOC are less marked in the transition from oligotrophic to highly eutrophic waters. Algal cell carbon commonly increases nonlinearly with increasing fertility, and a slight tendency towards algal cells of greater size is found in eutrophic lakes. Replacement times of algal cell carbon by net primary production are usually larger (slower turnover times) in eutrophic than in oligotrophic waters, but the increases in replacement times are not proportional to increases in cell carbon. Hence, in less fertile waters, the algal cells are photosynthesizing more per unit cell carbon; i.e., they have a greater carbon flux per cell.

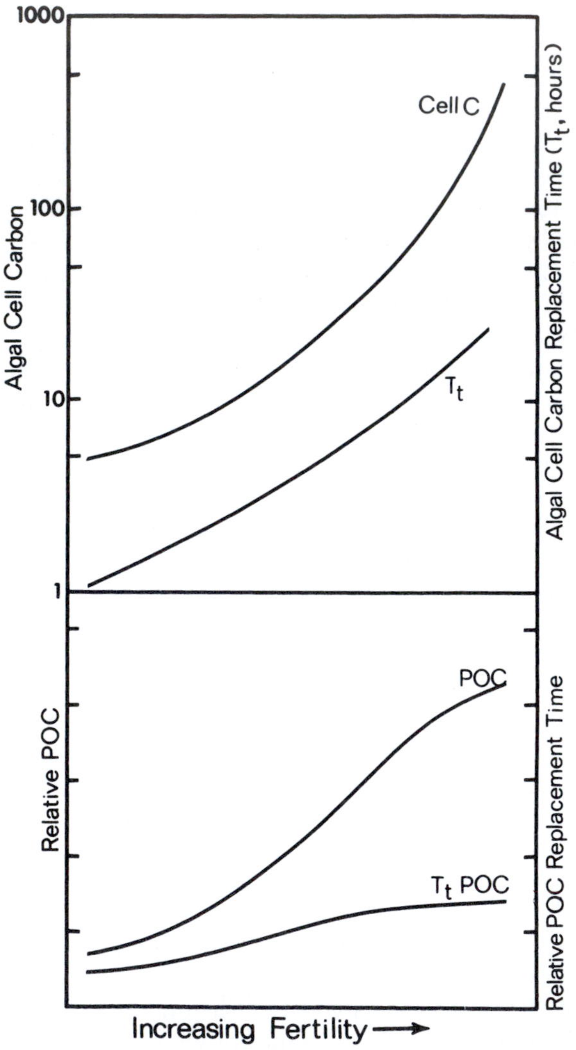

FIGURE 23-12 Generalized relationship of pelagic algal cell carbon and particulate organic carbon (POC) and their relative replacement times (T_t) in lakes of increasing fertility. (From Wetzel, R. G., and Rich, P. H.: Carbon in freshwater systems. *In* Woodwell, G. M., and Pecan, E. V. (eds.): *Carbon and the Biosphere*. Springfield, Va., National Information Service, 1973.)

All indications point to the importance of autochthonous primary production by the phytoplanktonic and littoral flora as major contributors to the POC of natural lake systems. Allochthonous POC is relatively small in contrast to major inputs of DOC, except for special cases (e.g., in small lakes or ponds in heavily forested areas or reservoirs that are small in volume in relation to inputs and flowthrough). The sources of POC shift in their relative contributions to the total POC pool as the aquatic ecosystems frequently progress through stages of increasing fertility when exposed to enhanced nutrient loading or as seen among a series of lakes at different stages of development (Fig. 23-13). Generalizations are again problematic because of the high degree of lake individuality. The importance of the littoral components to the production of POC in a given series of lakes increases greatly and changes markedly in the transition from nutrient-limited conditions of oligotrophic lakes to light limitations imposed by biogenic turbidity, to dominance by emergent macrophytes and associated littoral microflora.

C. Chemosynthesis and Nutrient Regeneration

In many cases, only low intensities of light reach the sediments within the pelagic zone; this fact, combined with the continual sedimentation of predominantly dead POC, ensures that benthic metabolism is primarily heterotrophic and detrital (Rich and Wetzel, 1978). The low rate of diffusion of oxygen into saturated sediments and its rapid utilization there mean that much metabolism is anaerobic, even when hypolimnetic oxygen depletion does not occur. Hutchinson (1941) termed this metabolism "pelometabolism" and describes its importance relative to "hydrometabolism" that occurs in the free water of the lake. Carbon dioxide produced by anaerobic detrital pelometabolism represents the escape of the oxidized product of an oxidation–reduction reaction (Fig. 23-14). The continued production of CO_2 during anaerobiosis indicates that the respiratory quotient (CO_2 release/O_2 uptake) of benthic metabolism is >1 and that alternate electron acceptors are being reduced instead of molecular oxygen. Thus, alternate electron acceptors are receiving energy originally captured during photosynthesis and transferred to carbon as water is oxidized to oxygen and CO_2 is reduced to carbohydrate (cf. Chap. 21).

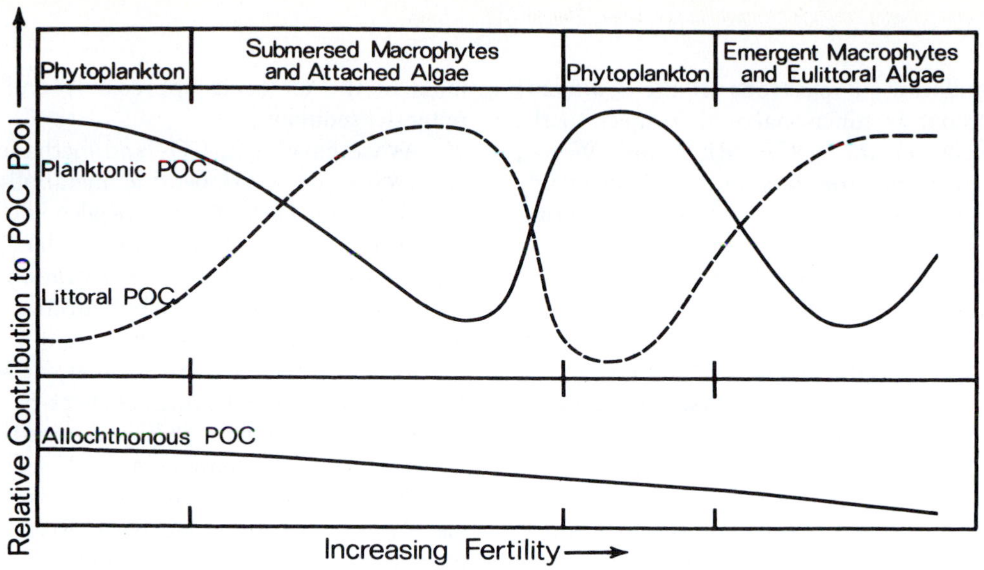

FIGURE 23-13 Generalized relative contributions of planktonic, littoral, and allochthonous sources of particulate organic carbon (POC) to lakes of increasing fertility as the dominance of pelagic and littoral primary productivity shifts (as detailed in Fig. 19-21). (From Wetzel, R. G., and Rich, P. H.: Carbon in freshwater systems. *In* Woodwell, G. M., and Pecan, E. V. (eds.): *Carbon and the Biosphere.* Springfield, Va., National Information Service, 1973.)

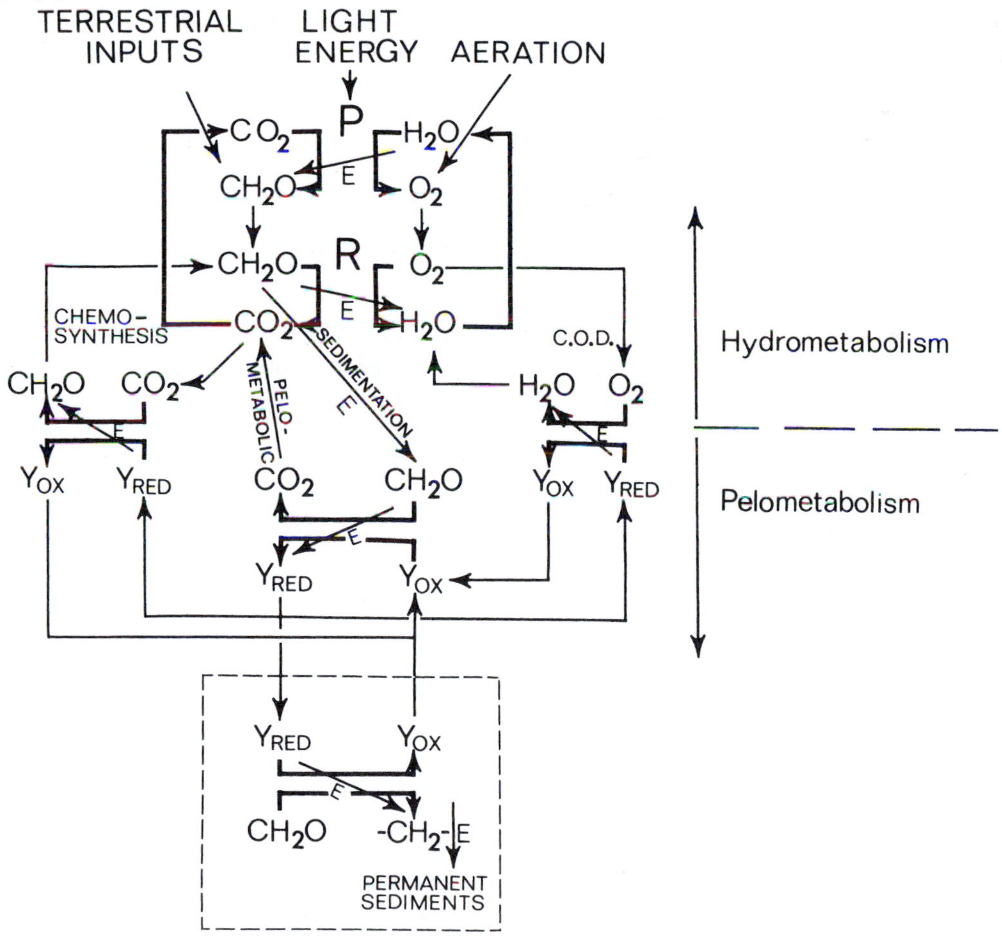

FIGURE 23-14 Oxidation–reduction reactions associated with benthic detrital electron flux from sediments. *P* = photosynthesis; *R* = respiration; *E* = electrons; red = reduced; ox = oxidized; C.O.D. = chemical oxygen demand. (Modified from Rich and Wetzel, 1978.)

This accumulation of electrons in the sediments must be interpreted as functional with respect to the ecosystem (Wetzel *et al.*, 1972; Rich and Wetzel, 1978). The ***detrital electron flux*** into and ultimately out of the sediments originates from the same photosynthetic energy source as does a predator's food and is a much larger proportion of the total source. Presumably, evolution, or at least integration, of the ecosystem has incorporated this energy flow into its overall function to the same extent as the more recognized flow through predator–prey systems. Reduced ions, radicals, and molecules are generally more soluble than their oxidized form; as a result, the reducing conditions favor an increase in the net rate of diffusion out of the sediments. Upon entry of these products into the oxidized layers of the lake, further oxidation may proceed either chemically or biologically. If the original detrital reduction involved a nutrient (sulfate, nitrate, or ferric phosphate) or a fermentation product, the detrital energy has effected both nutrient regeneration and translocation. If the subsequent "hydrometabolic" oxidation is biological (chemosynthesis), the detrital energy has been reintroduced into the biota. The energy thus transferred must be considered heterotrophic rather than synthetic in that it is originally derived from primary production (cf. Sorokin, 1964, 1965).

In order to further understand the consequences of the benthic detrital electron flux, some basic considerations of the possible alternatives of benthic metabolism are worth evaluating. Material enters the sediments by sedimentation because it is particulate—that is, it is relatively insoluble. If it remains insoluble it will enter the permanent sediments of the lake and decrease the volume of the lake—that is, reduce the lifespan of the lake. In order to escape the sediments, it must become more soluble, gaseous, or both.

Presumably, cellulose and similar materials constitute much of the organic material that enters the sediments. The intermediary metabolic products of such compounds are increasingly soluble and culminate in carbon dioxide that is both gaseous and highly soluble. Very reduced carbon compounds—for example, methane—are also gaseous and volatile, soluble, or both. The loss of dissolved organic carbon from the lake sediments is not well understood, but organic acids, acetic acid in particular, are significant products (Chap. 21). A carbohydrate (e.g., cellulose) requires one molecule of oxygen for each carbon atom in order to be oxidized to CO_2 and water. The removal of a molecule of oxygen will release two carbon atoms when the product is methane. However, oxygen represents the heaviest atom in cellulose, and the simple removal of oxygen via the production of CO_2 reduces the mass of the molecule appreciably and results in less potential sediment.

As we have seen, CO_2 and methane are the major gases escaping from benthic metabolism. Highly reduced carbon compounds are also well-known products of long-term sedimentation—that is, coal and petroleum—and increasing concentrations of carbon, in excess of that found in carbohydrates, are measurable in lake sediments over much shorter time periods (Shiegl, 1972). Given this format, at least two additional ecosystem-level consequences of the benthic detrital electron flux may be hypothesized.

The redox potential in lake sediments brought about by anaerobic metabolism represents a measure of electron activity that has been dissociated from biochemical (enzymatic) "constraints" or reaction specificity (Rich and Wetzel, 1978). Consequently, these electrons may diffuse out of the sediments in association with a number of inorganic and organic compounds, as has already been described. The exhaustion of a particular class of electron acceptors causes the benthic redox potential to become more negative relative to molecular oxygen; this continues until new acceptors precipitate back into the sediments following hydrometabolic oxidation as highly insoluble compounds (e.g., FeS or CuS). On the other hand, the carbon and oxygen in the form of CO_2 are the relatively massive products of the oxidized component of the anaerobic oxidation–reduction reactions. Thus, the net diffusion of reduced compounds out of the sediments is controlled by redox potential, and the net diffusion controls the export of mass—that is, carbon and oxygen—from the lake sediments. In this manner, the lifespan of the lake is affected directly by reduction of mass through losses of large quantities of gases of carbon and oxygen. The rate of the detrital electron flux is fundamental to eutrophication rates of lake systems.

The importance of detrital electron flux in benthic metabolism is reflected further by the changes in ratios of CO_2 evolution to consumption of molecular oxygen in hypolimnia over an annual period. The oxygen uptake by benthic sediments in lakes generally is considered to exceed the concomitant evolution of CO_2. Oxidations of proteins and fats result in respiratory quotients ($RQ = CO_2$ released/O_2 consumed) of less than unity, as has been demonstrated often by caloric combustions of benthic organisms. An RQ value of 0.85, generally accepted as an average value (Ohle, 1952; Hutchinson, 1957), has been used in estimates of lake productivity by hypolimnetic CO_2 accumulation (Chap. 11). Anaerobic bacterial fermentative metabolism produces excess CO_2, a positive CO_2 anomaly, and volatile organic compounds, such as methane, that diffuse out of the sediments.

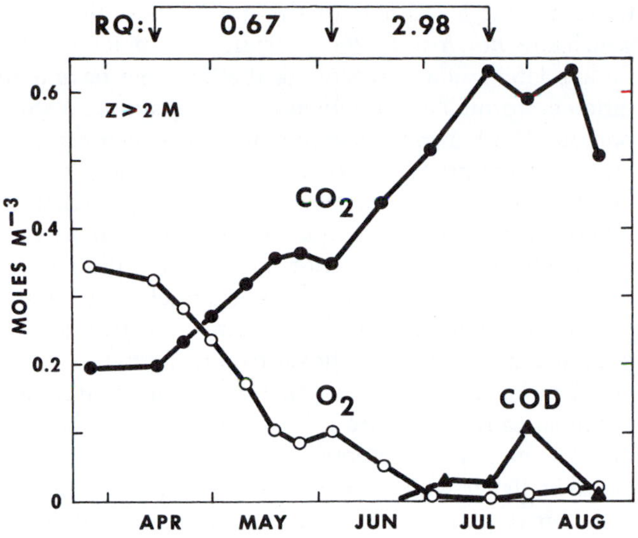

FIGURE 23-15 Changes in the hypolimnetic concentrations of CO_2, O_2, and chemical oxygen demand (COD) and the benthic respiratory quotients, Dunham Pond, Connecticut. (From data of Rich, 1975.)

Lake sediments have an "oxygen debt" that is indicated by low and negative redox potentials and nonbiological chemical uptake of molecular oxygen by reduced substrates that were formed under anaerobic conditions (Rich, 1975). This suggests that microbial metabolism involving glycolysis and reactions in the tricarboxylic acid cycle are continuing, while oxidative reactions using molecular oxygen as the terminal electron acceptor (i.e., the respiratory cytochrome system) are impeded by limited O_2 diffusion (Wetzel *et al.*, 1972). Viewing the *RQ* concept at the benthic community level, which would include the anaerobic bacteria that utilize electron acceptors other than molecular O_2, the benthic respiratory quotients would be greater than those of organisms undergoing aerobic metabolism. Additionally, the *in situ RQ* values would be expected to vary seasonally with water circulation periods at overturn and changes in redox gradients in response to varying rates of oxygen diffusion. Such was the case in Dunham Pond, Connecticut, in which Rich (1975) found during stratification that respiratory quotients of the hypolimnion varied inversely with the availability of oxygen (Fig. 23-15) from <1 after spring circulation and oxygen renewal. The respiratory quotients increased, however, to nearly 3 under anoxic conditions during the latter portion of summer stratification. Similar results have been found in a number of other lakes (Rich and Devol, 1978; Rich, 1980; Mattson and Likens, 1993). Under anaerobic conditions, high respiratory quotients are the result of the oxidation of organic carbon to CO_2 during the reduction of alternate electron acceptors, which then appear as oxidizable substrates. Reductions in sediment mass also result from this process, because soluble and gaseous products are formed that increase diffusion of reduced matter out of the sediments. Further, the removal of oxygen (reduction) from permanently sedimented carbon compounds represents the removal of more than half their mass.

To conclude, the alternate role of detrital electron flux suggests that existing lakes can be viewed as temporarily sedimenting energy rather than mass, which permits an extended existence. Simultaneously, benthic detrital electron flux tends to close the carbon and particularly the oxygen cycles of lakes.

V. DETRITUS: ORGANIC MATTER AS A COMPONENT OF THE ECOSYSTEM

Studies that address detrital origins and its metabolism directly, including the nonplanktonic and terrestrial components of the lacustrine ecosystem, are few. The heterogeneity and diversity of lacustrine detritus are reflected in the compound nature of most lake ecosystems. The role of allochthonous inputs in the metabolism and trophy of streams has been emphasized by numerous investigations (Ross, 1963; Hynes, 1963; Darnell, 1964; Cummins, 1973). Similarly, pelagic metabolism can be strongly influenced by edaphic factors and terrestrial metabolism in small lakes with high drainage-area-to-volume ratios. The importance of very high wetland and littoral productivity and extensive loading of organic matter and nutrients from the drainage basin disputes Forbes's (1887) statement that a lake. . . "is an islet of older, lower life in the midst of the higher, more recent life of the surrounding region." Further, it is simply untrue, as Shelford (1918) proposed, that "one could probably remove all the larger plants from a lake and substitute glass structure of the same form and surface texture without greatly affecting the immediate food relations." Most lakes are small with high shoreline-to-surface-area ratios, and pelagic metabolism is modified by littoral metabolic activities and inputs. Moreover, sedimentation of organic matter from both littoral and pelagic sources in the relatively static waters of lakes results in a displacement of much of the lake's metabolism to the sediments. Prerequisite to a study of aquatic ecosystem structure is a complete representation of the productivity inputs of all components. The metabolism of detrital organic matter results in a complex carbon cycle that dominates both the structure and function of lake and river ecosystems.

A. Definitions of Detritus and its Functions

Organic detritus or "biodetritus" was described by Odum and de la Cruz (1963) as dead particulate organic matter inhabited by decomposer microorganisms. The existence and importance of detritus in many habitats have since become the subject of a large and widespread literature. However, the position of detritus in relation to the trophic dynamic concept of Lindeman (Chap. 8) was not clarified until recently. Balogh (1958) demonstrated that detritus, as egested material, can constitute an important fraction of total metabolism in several terrestrial soil and litter communities. On the other hand, Odum (1962, 1963) and others have emphasized that detritus originating as ungrazed primary production supports a "detritus food chain" that is essentially parallel to the conventional "grazer food chain" at succeeding trophic levels.

The importance of the divergence between the concepts of the grazer food chain and the detritus food chain has been emphasized and clarified by Wetzel and co-workers (1972; Rich and Wetzel, 1978; Rich, 1984). Part of the difficulty stems from the concept of ecological efficiencies as it grew out of the original statement of the trophic dynamic model by Lindeman (1942; cf. Kozlovsky, 1968). Lindeman's trophic dynamic concept indeed offers explanations for Eltonian pyramids of numbers and biomass. The postulated trophic levels estimated predation and concluded that energy transmitted by predation influences the amount of biomass at each trophic level. In this way, productivity included biomass and turnover dynamics by respiration, predation, and nonpredatory losses. Lindeman estimated predation or trophic transfers at every trophic level simply by biomass plus respiration at each trophic level. Thus, his corrected productivity was the sum of biomass, respiration, and inferred predation and decomposition. Decomposition was egestion (defecation) by a predator or ingested but unassimilated prey by simple difference (decomposition of trophic level n material was egestion by trophic level $n + 1$). Other nonpredatory losses, such as prey not killed or prey killed or that died but were not eaten, were ignored as trivial, which they are not.

Efficiency is a ratio of product over reactant, or, in the case of an ecological efficiency, the productivity of a predator over predation upon its prey. Lindeman defined trophic efficiency as the ratio between assimilation by one trophic level and the assimilation of the preceding trophic level (i.e., Λ_n/Λ_{n-1}). This formulation recognizes neither the existence of material egested or otherwise lost by trophic level n nor postassimilatory, nonpredatory losses at Λ_{n-1}, which may be lost to those trophic levels and to the grazer food chain but which *are not lost to the ecosystem as a whole*. Although later studies on ecological efficiencies have used different formulae (e.g., Ingestion$_n$/Ingestion$_{n-1}$) (Slobodkin, 1960, 1962), discrimination between egestion and other nonpredatory losses from respiration is commonly not done (Burns, 1989; Hairston and Hairston, 1993). Therefore, most empirically derived aquatic ecological efficiencies that do not specify respiration are in reality agricultural or grazer food chain efficiencies and not applicable to complete ecosystems in which detrital organic matter is the significant trophic pathway.

A more realistic system operation is integrated diagrammatically in Figure 23-16. Here autochthonous organic matter represents all carbon fixed by autotrophs living within the lake or river ecosystem, minus their respiratory losses. The actual immediate form of organic matter productivity may be living cellular material (primary POM), which may become detrital secondarily, or dead particulate or dissolved organic carbon, or both.

A three-phase system of organic carbon is operational: dead organic matter, both particulate and dissolved (Fig. 23-16, *left*); living particulate organic material (*right*), which has the potential of entering either the POC or DOC pool upon death and whose metabolism may either create or destroy all three phases; and CO_2. Nonliving equilibrium forces are also present (e.g., photooxidation and hydrolysis of organic compounds). As is discussed later, these processes potentially represent turnover rates that can equal biological metabolism and certainly can enhance rates of microbial utilization. Therefore, metabolism represents the organizing force in the structure of the organic system, and it is metabolism that determines to a large extent the kinds and magnitudes of transformations that occur between the oxidized and reduced forms of carbon. The top of the diagram represents the zone of reduced carbon inputs, the middle portion the zone of oxidation to CO_2, and the lower portion represents losses of organic matter from boundaries of the system. The two parallel dynamic pools of organic carbon (dead, *left*; living, *right*) originate photosynthetically and are depleted by oxidation to CO_2 and exports. Although bacteria are shown in association with detritus, they function as biota. Odum's detritus food chain is really a link between the two pools in the direction of the biota. From the standpoint of the ecosystem, however, microbial and photochemical utilization of organic matter is a major utilization of organic matter that totally dwarfs utilization by higher organisms. Moreover, nonpredatory losses are in reality "nonpredatory productivity"—that is, productivity that is available to organisms other than predators, namely bacteria.

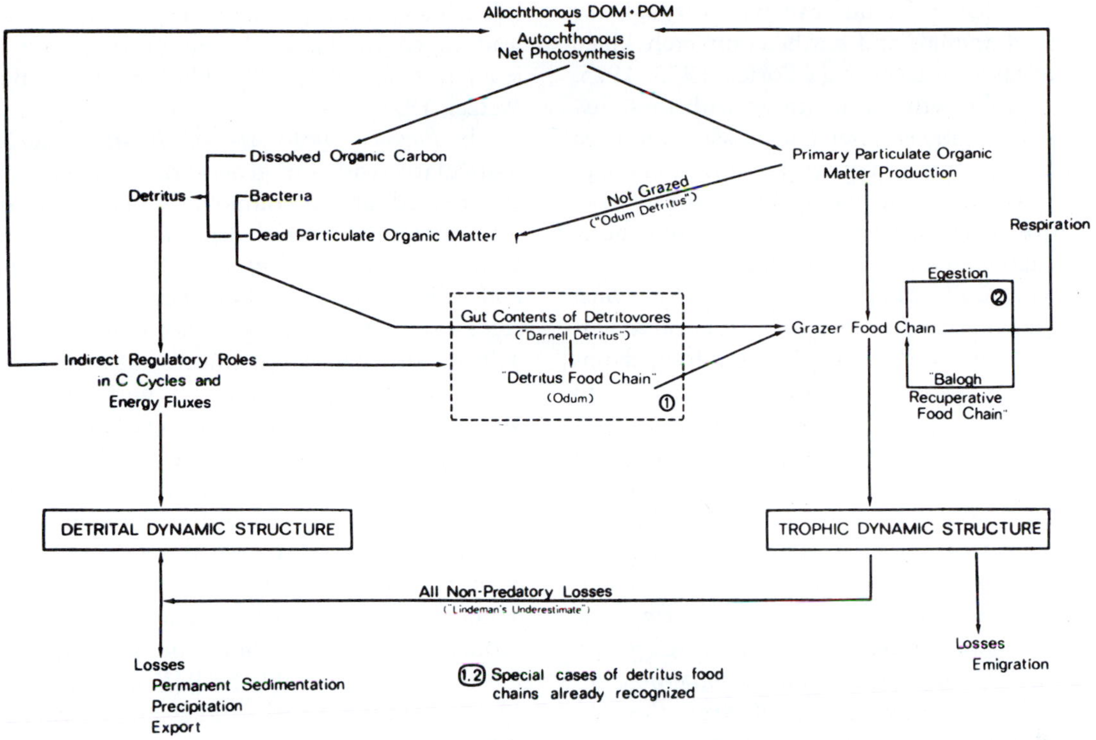

FIGURE 23-16 Generalized integration of the trophic and detrital dynamics of aquatic ecosystems. (From Wetzel *et al.*, 1972.)

A major point to be emphasized is that frequently >90% of the organic matter entering the system from photosynthetic sources is not utilized in the grazer and detrital food chains. Among studies of the trophic dynamic food cycle relationships, lack of appropriate recognition of nonpredatory losses from autotrophs ("Odum's detritus") and nonpredatory losses from heterotrophs ("Lindeman's underestimate"; see Fig. 23-16) represents a major void in contemporary system analyses.[3] As noted earlier (Table 16-12), Richman (1958) and Slobodkin (1962) demonstrated that a large amount of ingestion is not assimilated and that egestion can be 80% or more of what is ingested, instead of the 10–20% estimated by Lindeman. Ingestion and the percentage of egestion increase markedly with increasing food availability.

Detritus and associated terms have been redefined (Wetzel *et al.*, 1972; Rich and Wetzel, 1978) in an attempt to make current terminology consistent with the ecosystem concept and to avoid perpetuating the oversight of nonpredatory pathways. The new definitions do not contradict the old, but are defined more broadly so as to avoid some of the ambiguity inherent in the concepts of the ecosystem and trophic-level efficiencies espoused by earlier investigators and perpetuated by many recent ones.

***Detritus** consists of organic carbon lost by nonpredatory means from any trophic level (includes egestion, excretion, secretion, and so forth) or inputs from sources external to the ecosystem that enter and cycle in the system (allochthonous organic carbon).* This definition removes the highly arbitrary "particulate" restriction from existing definitions of detritus. In terms of the ecosystem, there is no energetic difference between DOC lost from a phytoplankter or feces or other exudates lost from an animal. Rates of utilization may differ, but functionally the organic carbon and energy are the same. Detritus is all dead organic carbon, distinguishable from living organic and inorganic carbon.

Second, the bacterial component is not combined with detritus ("biodetritus" = POM plus bacteria of Odum and de la Cruz, 1963) because this interferes with the general applicability of the term to situations in which detritus is not simply ingested. Such cases include the use of detrital energy in the regeneration of nutrients such as CO_2, N, and P; algal heterotrophy; losses by adsorption; flocculation; precipitation with $CaCO_3$; chelation of elements, and so forth. In some

[3] Notable exceptions are the theoretical analyses and models of Patten (1985), who has shown the important thermodynamic relationships of detritus and how the ecosystems must operate.

cases, plant and algal materials can pass through the digestive tracts of animals and not be completely killed before being released in feces (e.g., Porter, 1973, 1975; Velimirov, 1991). Growth can continue with these living cells or plant fragments after gut passage and even be enhanced by exposure to concentrated nutrient supplies en route. This material would be excluded from the definition of detritus. Detritus is dead, and therefore living egested material is not detritus.

The ***detritus food chain*** *is any route by which chemical energy contained within detrital organic carbon becomes available to the biota.* Detrital food chains must include the cycling of detrital organic carbon, both dissolved and particulate, to the biota by direct heterotrophy of DOC, chemoorganotrophy, or absorption and ingestion. The definition of "detritus food chain" emphasizes the actual trophic linkage between the nonliving detritus and living organisms and recognizes the metabolic activities of bacteria attached to detrital substrates as a trophic transfer. *The special case of a detrital food chain in which detrital energy is subsequently transferred by noncarbon substrates in an anaerobic environment has been termed* ***detrital electron flux***. As discussed in the previous section, such energy may reenter the biota (chemosynthesis) or mediate chemical or physical phenomena, or both, such as increasing the availability of inorganic nutrients in an anaerobic hypolimnion or in the sediments. The term *flux* specifies the flow of electrons, not carbon, to alternate electron acceptors in the absence of molecular oxygen.

Detritus, as a component of the environment*, can also affect facets of the chemical and physical milieu without clear, defined energetic transformations of the detritus itself.* In this situation, reference is made to effects of detritus that do not involve oxidation–reduction reactions. Examples include several indirect effects by which detrital organic carbon can influence and regulate the total energy and carbon flux of an aquatic system. Adsorption onto the surfaces of and coprecipitation of dissolved organic compounds with inorganic particulate matter (clays, $CaCO_3$, and others), and complexing of inorganic nutrients by dissolved organic substances (e.g., humic compounds) are examples of this interaction.

Physical processes, such as partial or complete photochemical modification of organic macromolecules, can result in major alterations in biological availability of portions of complex, heterogeneous dissolved organic compounds. These processes can affect the decomposition, utilization, and effects of dissolved organic matter in many ways. Examples include:

a. *Alterations of enzymatic accessibility by the macromolecules.* In particular, cross-linking of polypeptide chains with polyphenolic humic substances can lead to enzymatic inhibition or reduction of activity (e.g., Wetzel, 1991, 1993; Haslam, 1998; Boavida and Wetzel, 1999).

b. *Partial photolysis of humic macromolecules*, particularly with the generation of volatile fatty acids and related simple compounds that serve as excellent substrates for bacterial degradation (e.g., Stewart and Wetzel, 1981; Wetzel *et al.*, 1995; Moran and Covert, 2001; discussed in detail in Chap. 17). It is important to recognize that of the total photolytic irradiance, about half of the partial photolysis of organic substrates results from UV-B (285–300 nm) and UV-A (320–400 nm) irradiance. Transmittance and photolytic activity from UV-B and UV-A is restricted largely to the surface waters. In contrast, photosynthetically active radiation (PAR, 400–720 nm), although much weaker energetically than UV, penetrates into water to much greater depths. Although photolysis of organic compounds is appreciably less than that induced by UV at surface waters, the photolytic generation of simple substrates is appreciable by PAR as well as by UV (Fig. 23-17). Results of such studies are indicating that over half of labile organic substrates are generated by PAR irradiance of dissolved humic substances.

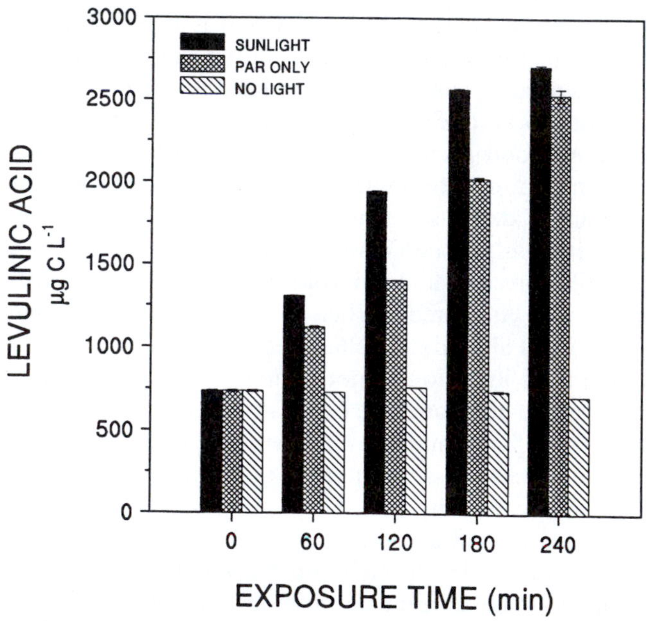

FIGURE 23-17 Net generation of levulinic acid from the partial photolysis of sterile whole leachate (0.2-μm pore-size filtrate) of the emergent macrophyte *Juncus effusus* (10 mg DOC liter^{-1}) after four weeks of microbial decomposition at 20°C in the dark. Exposed to full natural sunlight of UV-B, UV-A, and photosynthetically active radiation (PAR) (13.05 mol m^{-2} over the 4-h period), to PAR only, and incubated simultaneously in the dark. Error bars = SD, n = 3. (Modified from Wetzel, 2000d.)

If the photodegradation continues to the point that most of the chromophoric humic compounds are photolysed and utilized biologically or converted to CO_2, the residual nonchromophoric DOC appears to be fairly resistant to appreciable biological degradation. For example, in very shallow wetland pools and saline lakes, chromophoric DOC can be reduced appreciably (Tranvik and Kokalj, 1998; Arts *et al.*, 2000). In many shallow saline lakes, the nonchromophoric DOC concentrations can be very high (to 30 mg C liter^{-1} or more). The specific nature of the recalcitrance of nonchromophoric DOC to bacterial degradation is unclear. Exposure of dissolved humic substances to strong UV-C (254 nm) reduced subsequent growth of bacteria (Lund and Hongve, 1994). These authors suggested that radicals induced by radiation mediated the negative growth effects. However, most of the identified nonabsorbing photodegradation products are good substrates for bacteria (e.g., Wetzel *et al.*, 1995).

c. *Partial photodegradation of dissolved organic nitrogen and phosphorus compounds* to release inorganic nutrient compounds such as nitrate, ammonia, and phosphate (e.g., Manny *et al.*, 1971; review of Moran and Zepp, 1997).

d. *Complete photolysis of humic substances to CO and CO_2*, with some dissociation to dissolved inorganic bicarbonate as well as evasion of some CO_2 to the atmosphere. Photochemical oxidation by sunlight of natural dissolved organic carbon compounds to both carbon monoxide (CO) and dissolved inorganic carbon (DIC) has been known for some time (e.g., Miller and Zepp, 1995). Depending on the dissociation conditions, some excess CO_2 will evade to the atmosphere. Early studies on the photolytic degradation of dissolved organic matter suggested that the dominant component of solar irradiance was UV-B and UV-A and that PAR above 400 nm was of little consequence. Many of these studies, however, were not performed under sterile conditions, and as a result findings were confounded by nearly instantaneous microbial utilization of the organic compounds generated with rapid degradation and generation of CO_2. Moreover, many of the DOM sources of these studies had been exposed to natural sunlight for long (e.g., weeks) and noncomparable periods of light. Contemporary research is indicating that although UV-B and UV-A are significant and can contribute to more than half of photochemical mineralization, photosynthetically active radiation (PAR, 400–720 nm) is also a major photolytic agent (Vähätalo *et al.*, 2000; Wetzel, 2000c,d). For example, from nearly 200 separate photolytic experiments on DOM from different waters and plant sources under different conditions of decomposition, the UV-B portion of the spectrum was always most effective in complete photodegradation to CO_2, but UV-A was also highly effective, with small differences from the photolytic capacities of UV-B (Fig. 23-18). PAR is also highly effective in photolytic degradation of DOM to CO_2, and about one-quarter (23.4%) of the collective photolysis can be attributed to the largely blue portion of the PAR spectrum (400–700 nm) and ca. 68% to UV-A (320–400 nm) (Fig. 23-19).

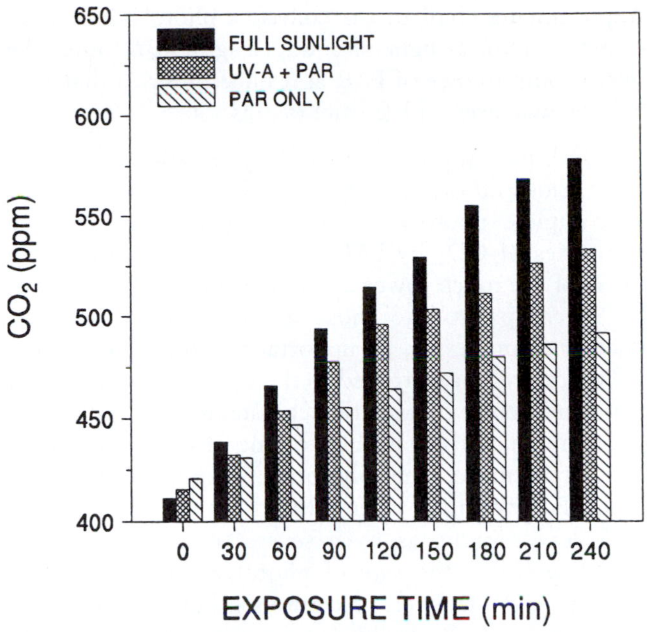

FIGURE 23-18 Photolytic degradation of sterile dissolved organic matter (leachate from *Juncus effusus*, 0.2-μm pore-size filtrate) to CO_2 under replicated, aseptic conditions exposed to full sunlight (UV-B, UV-A, and PAR) of 15.53 mol m^{-2} over the 4-h period, to UV-A + PAR, and PAR only. (From Wetzel, 2000a.)

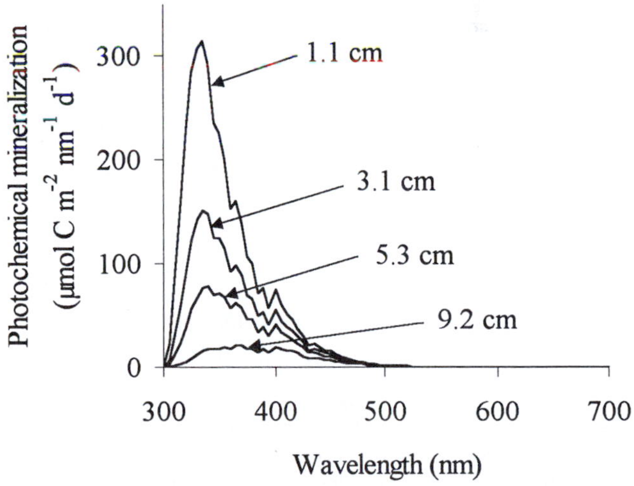

FIGURE 23-19 Photochemical mineralization of chromophoric dissolved organic matter at different depths in humic lake Valkea-Kotinen, Finland, 26 July 1994, indicating the increasing proportion of mineralization with greater depth resulting from irradiance in the visible range. (Modified from Vähätalo *et al.*, 2000.)

Approximately half of the collective photolysis was induced by natural light of a wavelength >360 nm. The relative importance of PAR as a mineralizer to that proportion increases with greater depths.

Both partial photolysis to the generation of volatile fatty acids and other simple organic substrates, as well as complete photolysis with the generation of large quantities of CO_2 by PAR, are important findings because of the much lower extinction rates of PAR in water in comparison to those of ultraviolet irradiance. Photolytic processes, so important to nutrient cycling, are therefore not restricted to the uppermost strata of a few centimeters of aquatic ecosystems, but rather can affect much of the variable volume of the photic zone.

Furthermore, the photolytic responses are very rapid. For example, when natural sunlight was attenuated very rapidly, as by a severe thunderstorm (Fig. 23-20, *middle*), the rate of photolytic degradation of DOC to CO_2 declined precipitously, but the photolytic capacities of PAR declined more rapidly than did the effects of UV-B (Fig. 23-20, *lower*). The precise chemical degradation processes involved in photolysis of dissolved macromolecules, particularly the helical humic substances, represent a major void in our understanding. Interdisciplinary collaboration among chemists and biologists is essential to progress effectively in our understanding of biological implications of organic molecular structure.

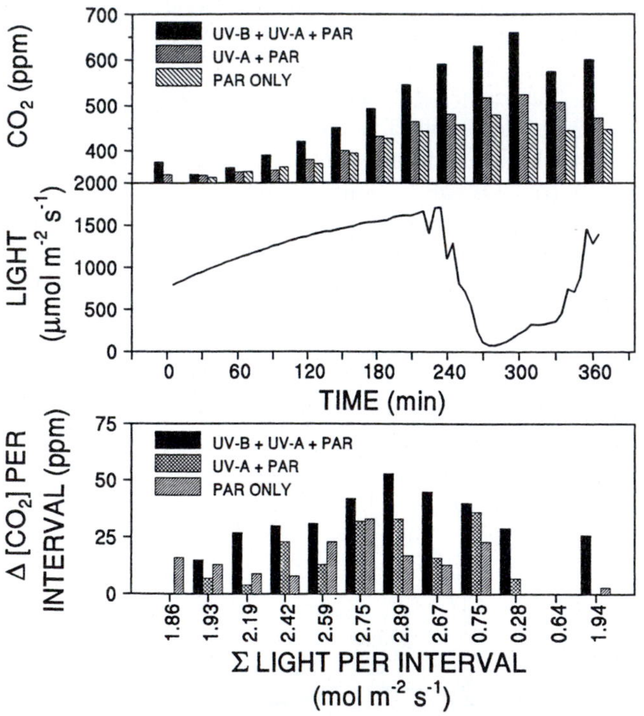

FIGURE 23-20 Photolytic degradation of sterile dissolved organic matter (aquatic plant leachate, *Juncus effusus*, 0.2-μm pore-size filtrate after 20 weeks of microbial decomposition) to CO_2 under replicated, aseptic conditions exposed to full sunlight (UV-B + UV-A + PAR), UV-A + PAR only, and PAR only (*upper*). A heavy rainstorm occurred during the incubations, which reduced light severely for an hour (*middle*). The net change in CO_2 production per amount of light received per interval under these conditions (*lower*). (From Wetzel, 2000a.)

Less direct but important biogeochemical interactions of dissolved organic matter in aquatic systems are also important but poorly studied at the ecosystem level. Natural dissolved organic substances, particularly humic compounds, in aquatic ecosystems can interact with other important metabolic components. For example, dissolved organic compounds can:

a. *Interact with inorganic compounds,* particularly in complexation reactions such as chelation, as discussed earlier (Chap. 14; see also Perdue, 1998). Depending on the concentration ratios of the complexing DOM to inorganic elements, the mode of organic complexation, biological availability, and, in some cases, elemental toxicity can be increased or decreased.

b. *Interact with other organic compounds*, such as peptidization, and alter biological susceptibility to enzymatic hydrolysis. Bonding of proteins, glycoproteins, and carbohydrates to polyphenolic humic macromolecules is a common process and results in macromolecules of considerable size (1500–90,000 D) (Münster, 1985). Membrane properties, such as lipid hydrophobicity, can be altered by such interactions with humic substances and in turn affect enzyme hydrolysis rates and nutrient transport mechanisms (e.g., Lemke *et al.*, 1995, 1998). In a most interesting interaction, humic substances can complex with proteins, particularly enzymes both freely soluble and membrane-bound, with noncompetitive inhibition (Wetzel, 1992, 1993; Münster and De Haan, 1998). Enzymes can be stored for long periods (days or weeks) in this complexed, inactive state, be redistributed in the ecosystem with water parcel movements, and be reactivated by partial photolytic cleavage by ultraviolet irradiance (Fig. 23-21) (Wetzel, 1991, 1995, 2000c; Boavida and Wetzel, 1999).

In addition, dissolved humic substances can sorb hydrophobic organic chemicals and thereby reduce their bioconcentration and toxicity (Kukkonen, 1999; Steinberg *et al.*, 2000). The binding capacity is linearly related to the hydrophobicity of the contaminant. The detoxifying capacity of humic substances is directly correlated with the quotient between aromatic and aliphatic carbons in the humic substances.

c. *Alter chemical properties such as redox and pH*; for example, exposure of natural dissolved organic matter to UV of sunlight can result in a photochemical

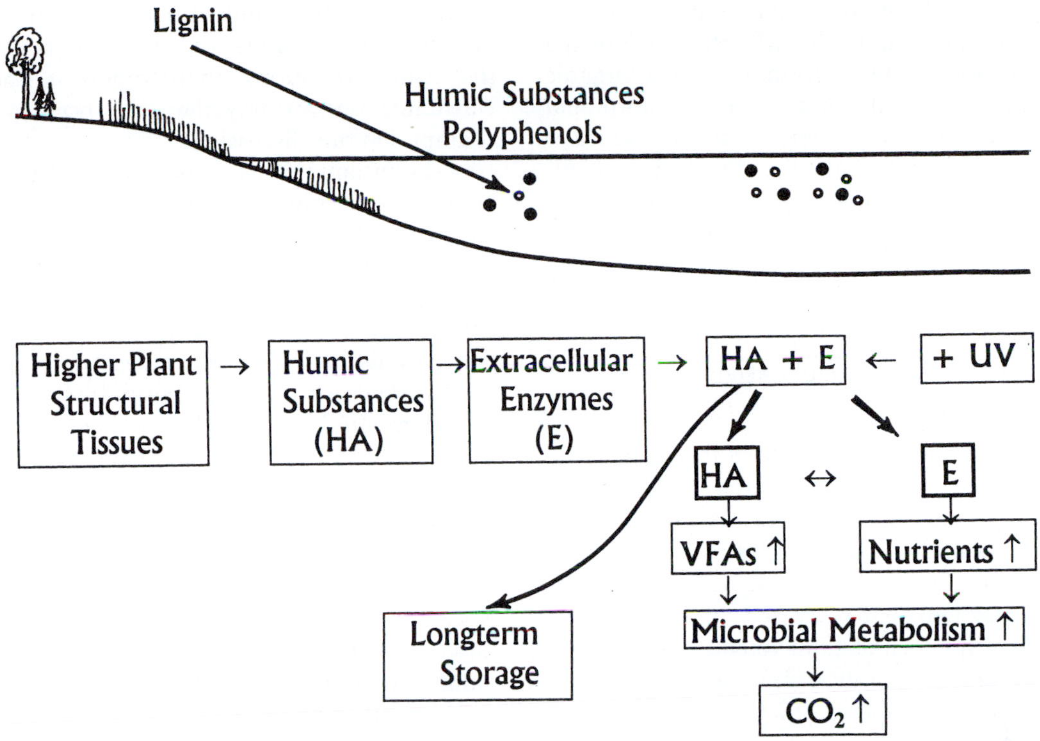

FIGURE 23-21 Potential interactive pathways and processes of humic substances emanating from decomposition products of higher plant tissues with extracellular and surface-bound enzymes and photolytic reactions, particularly with UV irradiance. Humic acid–enzyme complexes can be stable for long periods (weeks or months) and subsequently reactivated upon exposure to weak UV light. Further photolysis can cleave simple compounds from the macromolecules for subsequent utilization. (From Wetzel, 2000a.)

formation of reduced oxygen species, particularly hydrogen peroxide (H_2O_2) (Cooper *et al.*, 1988; Scully *et al.*, 1994). H_2O_2, with a half-life of several hours in natural waters, can radically alter redox cycling of metals (Moffet and Zika, 1987).

Humic substances may buffer against acidification but may also add natural acidity to surface waters when in very high concentrations in waters of low ionic strength. Organic carboxylic and phenolic acids provide a pH buffer when the pH is between 4 and 5 and the carbonate buffering capacity is absent (Kullberg *et al.*, 1993; Lydersen, 1998). Organic acids can modify the acidity of surface waters and depress water from 0.5 to 2.5 pH units when the acid-neutralizing capacity is in the range of 0–50 μeq liter^{-1}. Thus, a predominance of humic acids can result in an organic acidity that can influence, and at times exceed, inorganically derived acidity form natural or anthropogenic sources (cf. Chap. 11). However, usually organic anions represent <10% of the total anions and contribute little to acidity except in waters of extremely low ionic strength (Henriksen *et al.*, 1988; Urban *et al.*, 1988; Driscoll *et al.*, 1989; David *et al.*, 1992). Strong acids, particularly H_2SO_4, can exceed the buffering capacities of organic acids and shift surface water acidification by organic acids to total dominance by strong acids. High production of organic acids in peat wetlands and similar environments are often neutralized within the fens and bogs.

d. *Microbially reduced humic substances can*, upon entering less-reduced zones of sediments, *serve as electron donors* for the microbial reduction of several environmentally significant electron acceptors (Lovley *et al.*, 1999). Once microbially reduced, humic substances can transfer electrons to various Fe(III) or Mn(IV) oxide forms abiotically and recycle the humic compounds to the oxidized form, which can then accept more electrons from the humic compound-reducing microorganisms.

e. *Altered physical properties* such as selective modifications of light penetration. The well-known selective attenuation of light by chromophoric dissolved organic matter (cf. Chap. 5) can further modify biogeochemical cycling in numerous ways. Increasing light extinction can result in increased temperatures of surface strata with steep thermal discontinuities in the metalimnion and strong isolation of the colder hypolimnetic waters. The mixed layer depth decreases, which results in reduced total heat content of the lake or reservoir (Hocking and Straškraba, 1999). Modifications of the spectral penetration of light from dissolved organic matter can alter rates of photosynthesis, hormonal activities, and migratory distribution and reproductive behaviors (see Chaps. 15 and 16). Absorption of ultraviolet irradiance by humic substances can protect organisms from genetic damage as well as modify macromolecules and enhance bioavailability of organic substrates.

B. Detrital Dynamic Structure of Carbon in Lakes and Reservoirs

There have been very few attempts to examine the dynamics of detrital organic carbon on a functional basis. Such studies require measurements of all nonpredatory losses of organic carbon from any trophic level within the ecosystem as well as inputs from sources external to the ecosystem that enter and cycle in the system. Obviously, a number of simultaneously interacting components are operational. It is clear, nonetheless, that most of the organic metabolism of lake systems involves the detrital dynamic structure and that an understanding of this functional system is a prerequisite for meaningful evaluation of control mechanisms of the whole ecosystem. Stated another way, we must understand carbon flux rates among all components at the same time that we seek an understanding of parameters or human disturbances important in regulating metabolism.

Quantitative analyses of the major organic carbon transformations in lake systems have shown the following general characteristics: (1) The central pool of organic carbon is in the dissolved form; (2) Three major sources of particulate organic carbon occur: allochthonous, and two distinct major zones of autochthonous carbon flux, the littoral and the pelagic of lakes or the channel water of rivers; (3) Allochthonous inputs from the drainage basin and exports from the lake or river occur largely as dissolved organic carbon and represent a major flow of carbon through the metabolism of the system; and (4) Detrital metabolism occurs principally in the benthic region, where a majority of POC is decomposed in many lakes and rivers, and the pelagic area during sedimentation in lakes. Secondary metabolism thus is displaced from sites of production and input, by sedimentation in the case of POC and by aggregation and coprecipitation with inorganic matter in the case of DOC.

Examples in support of these generalizations can be obtained from analyses of an arctic tundra pond and of the pelagic zone of a eutrophic temperate lake for a single day, of a large reservoir over a 30-day period of the summer, and of a softwater and a hardwater temperate lake over an annual period. Because of the complexity of these studies and the insights they offer into the central questions on metabolism, these analyses warrant closer scrutiny than has been given to other examples in this discussion.

The instantaneous carbon mass of the major components and transfer rates of carbon between them were evaluated for a typical midsummer day in a small arctic tundra pond by Hobbie *et al.* (1972, as modified and corrected in Hobbie 1980). These small ponds cover the northern tundra by the thousands, lying in depressions between low ridges formed above ice wedges; beneath them lies permafrost, which prevents any outflow. Most of the ponds are very shallow (<40 cm) and are frozen solid for nine months of the year. The tundra vegetation surrounding the ponds is dominated by two sedges (*Carex aquatilis* and *Eriophorum angustifolium*) and a grass (*Dupontia fischeri*). The dominant herbivore of the area is the brown lemming, which clips this vegetation close to the ground. Litter from these clippings and lemming feces form a significant source of particulate and dissolved matter reaching the ponds.

The DOC pool represented most of the total carbon present in the aqueous portion of the system because the inorganic carbon content was quite low (Fig. 23-22). Although the primary productivity on an annual basis was very low, 50% of the total was contributed by the macrophytes, 47% by the benthic algae, and 3% by the phytoplankton. The most active transfer path was the movement of detrital organic carbon from the aquatic plants to the sediments, where decomposition resulted in major respiratory loss as CO_2. The flux rates were much smaller in the water than in the sediments, and the transfer of organic carbon between POC and the zooplanktonic crustacea was small but significant. The photooxidation of dissolved organic compounds to CO_2, although small, is significant in this shallow habitat, as has been indicated for certain streams and lakes (cf. Gjessing, 1970; Gjessing and Gjerdahl, 1970; Stewart and Wetzel, 1981a; Wetzel, 2000a).

The organic carbon cycle of a shallow (1.5 m) tundra lake of the Northwest Territories of Canada was analyzed for five weeks of the open-water period by examination of CO_2 fluxes through benthic respiration and anaerobic decomposition, photosynthesis of benthic and phytoplankton communities, and gas exchange at the air–water interface (Ramlal *et al.*, 1994). The annual carbon budget estimate for the lake indicated that 50% of the carbon was produced by the benthic community, 20% by phytoplankton, and 30% by allochthonous organic matter. Benthic respiration dominated heterotrophic utilization of the organic matter.

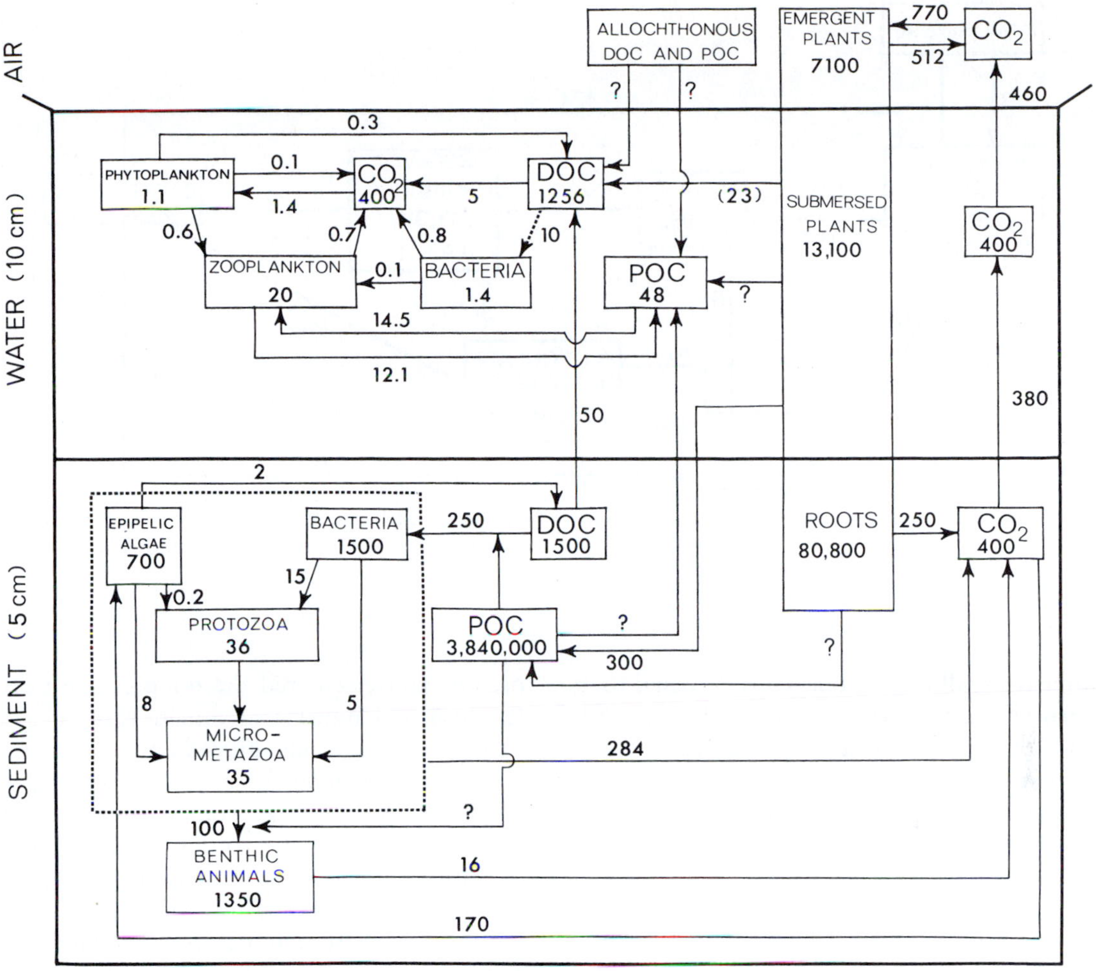

PERMAFROST

FIGURE 23-22 Carbon flow among components of a tundra pond ecosystem representative of approximately average midsummer conditions at coastal Barrow, Alaska (lat. 71°N). Data from Pond B on 12 July 1971 are used in the water portion of the diagram and those for Pond J on the same date for the sediment portion. Numbers in boxes refer to organic carbon mass (mg C m^{-2}), while arrows indicate the transfer rates (mg C m^{-2} day^{-1}). DOC = dissolved organic carbon; POC = particulate organic carbon. (Modified from Hobbie *et al.*, 1972, as corrected by Hobbie, 1980.)

The carbon flux in the pelagic zone at midsummer in the surface water of eutrophic Frains Lake, southeastern Michigan (Saunders, 1972a) demonstrates an analogous relationship (Fig. 23-23). Within a 24-h period, fluctuations in DOC ranged from two to six times more than DOC released extracellularly by the phytoplanktonic algae. Particulate organic detritus, ordinarily five to ten times greater than that of the phytoplankton, constituted a much larger potential source of DOC than living phytoplankton and was dispersed both in the photic and aphotic zones. Similarly, detrital particulate carbon was, at this time and depth, much more important energetically than phytoplankton as a food source to the crustacean zooplankton. Clearly, the organic detrital matter is quantitatively a major constituent of this system, in terms of both mass and metabolism. Its importance increases markedly from the photic to the tropholytic zone.

An approximate organic carbon budget for a shallow hypereutrophic lake exhibited very high productivity of phytoplankton, submersed macrophytes (predominantly *Myriophyllum spicatum*), and epiphytic algae (Table 23-12). Benthic respiration dominated in this shallow lake, in which the organic sediments were resuspended constantly during the ice-free period.

Another example of the relationships among pools of organic matter, living components, and fluxes among components emerges from investigations of the pelagic zone of Dalnee Lake, eastern Russia (Sorokin and Paveljeva, 1972). This lake receives a major influx of

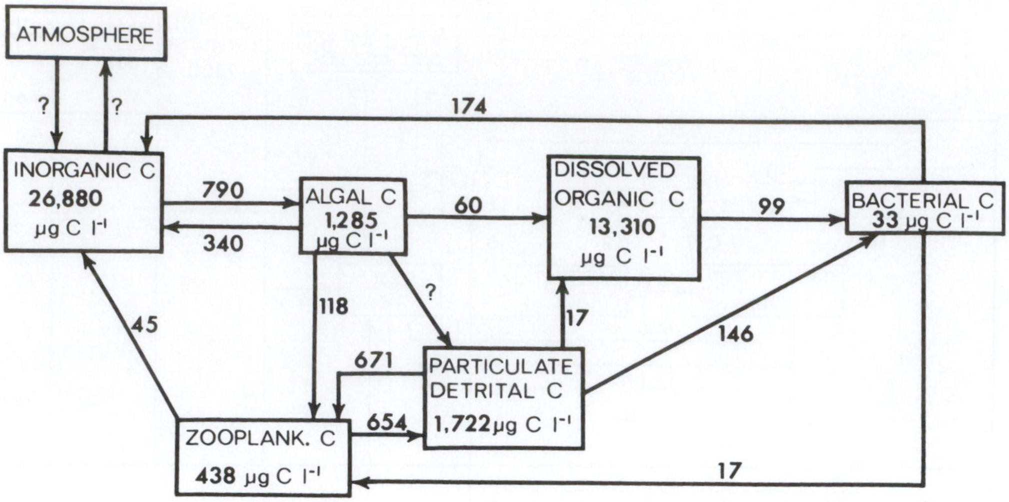

FIGURE 23-23 Pelagic carbon flow at 1-m depth in Frains Lake, Michigan, on 22 July 1968. Numbers on arrows refer to transfer rates in μg C liter^{-1} 24 h^{-1}; numbers in blocks refer to carbon concentration in μg C liter^{-1} at sunrise. (Modified from Saunders, 1972a.)

organic matter from allochthonous sources, equal to at least one-third of the total organic input (Fig. 23-24). During the month of July, the average flux rates of organic matter, expressed as calories m^{-2} per 30 days, indicated that the planktonic bacterial populations were consumed mainly by protozoa. Most of the protozoan organic matter entered the nonpredatory organic matter pool and underwent mineralization. Zooplankton and higher organisms utilized only a very small portion of the total organic base of detritus and microflora. About one-third of the zooplanktonic utilization of the microflora was by the rotifer *Asplanchna*, which was predatory on protozoa at this time of year. It was estimated that well over 70% of the organic matter entering the system from autochthonous photosynthetic and allochthonous sources became part of the nonpredatory organic matter pool and was subsequently decomposed or lost from the system by sedimentation and outflow without entering the animal components of the system.

TABLE 23-12 Annual Organic Carbon Budget for Lake Wingra, Wisconsin[a]

Components	*g C m^{-2} yr^{-1}*	*% of subtotal*
Inputs		
Autochthonous		
Phytoplankton	430	62.2
Macrophytes	120	17.4
Attached algae	61.9	9.0
Allocthonous		
DOC in runoff (urban)	18	2.6
DOC in ground water	3.1	0.4
DOC in precipitation	2.8	0.4
POC in runoff	51	7.4
Airborne litter	1.4	0.2
Dustfall	2.8	0.4
	691	100.0
Outputs		
Respiration	500	82.2
Methane loss to atmosphere	6.1	1.0
Permanent sedimentation	71	11.7
Outflow		
DOC	21	3.5
POC	8.2	1.3
Ground water	1.4	0.2
Insect emergence	0.5	0.1
	608.2	100.0

[a] Extracted from estimates of Adams and Prentki (1982).

A detailed evaluation of the carbon budgets of a whole-lake system at meter depth intervals over several years was undertaken on a hardwater lake in southwestern Michigan (Wetzel *et al.*, 1972). In this system, organic carbon inputs and utilization were evaluated over an annual period for allochthonous inputs, exports, and dynamics in the littoral, pelagic, and benthic zones (Fig. 23-25). Even though the littoral flora of this lake is not well developed in comparison to most lakes of similar morphometry, the single largest flow of organic carbon originated as primary particulate production of the littoral zone (Table 23-13). This material was transported to the deeper sediments, where it was supplemented by additional particulate organic carbon sedimenting from the pelagic zone (Table 23-14). Allochthonous POC input to the sediments was a fraction of the POC sedimenting in the pelagic zone and is included there. Benthic microbial metabolism of the sediments released a major amount of the sedimented

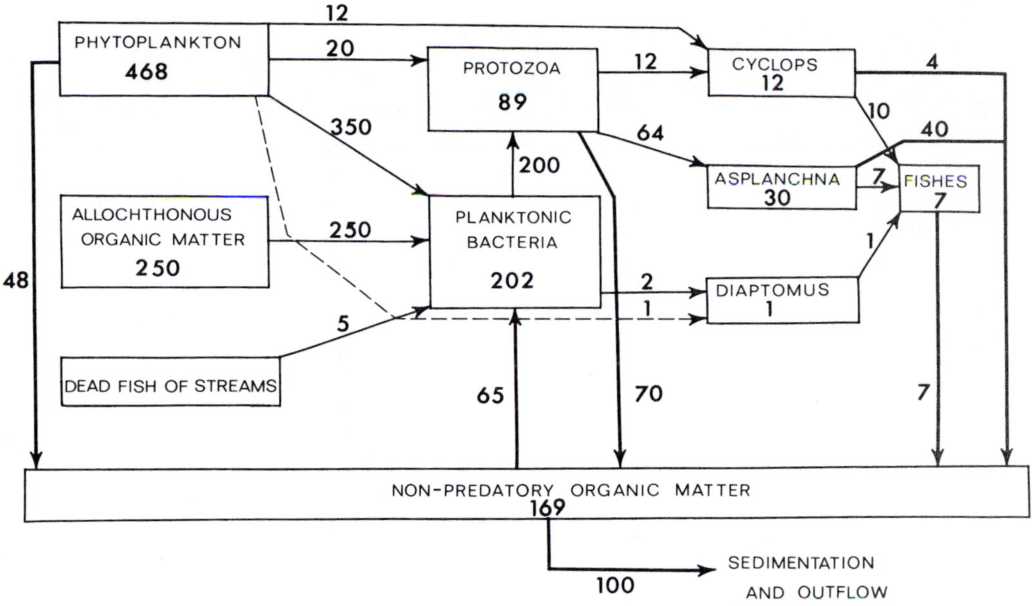

FIGURE 23-24 Generalized relationships of the average energy flow in the pelagic zone of Dalnee Lake, Kamchatka, Russia, during the month of July. Data expressed in calories m^{-2} per 30 days. (From data of Sorokin and Paveljeva, 1972.)

organic carbon to the dissolved CO_2—HCO_3^-—CO_3^- pool in respiration (Table 23-15). CO_2 released by microbial respiration was the primary flow of carbon to the photic zone, and much of it evaded to the atmosphere. A much smaller amount of benthic organic carbon was directly utilized by the pelagic community, and approximately an equivalent amount was lost by permanent sedimentation. Utilization of benthic detrital carbon (resuspension of Fig. 23-25) was almost equivalent to sedimentation losses from the pelagic zone and represented an important contribution to pelagic productivity of material originating from the littoral zone.

TABLE 23-13 Littoral Carbon Budget, Lawrence Lake, Michigan[a]

Carbon	*g C m^{-2} yr^{-1}*
Inputs	
Macrophytes	87.9
Epiphytic algae	37.9
Epipelic algae	2.0
Heterotrophy	2.8
Resuspension	0.0
Total	130.6
Outputs	
Secretion	5.5
Gross productivity	125.1
Benthic respiration	117.5
Net productivity	+7.6

[a] From Wetzel *et al.* (1972). Estimates are based on a lakewide areal basis rather than on an arbitrarily defined littoral zone. Values for epiphytic algal productivity are larger based on a subsequent detailed analysis (Burkholder and Wetzel, 1989a), but the conservative value here is used for illustration of the points discussed.

An integration of these budgets of inputs and outputs permits an evaluation of the annual carbon fluxes in a whole-lake system in which the allochthonous, littoral, benthic, and pelagic organic carbon flows are considered (Table 23-16). The whole-lake budget is within 5% of being balanced, which is very good considering the magnitude of variances involved in the analyses under *in situ* conditions. Analyses of zooplanktonic grazing by serially increasing animal concentrations to 200-fold *in situ* (Wetzel *et al.*, 1972) and *in situ* measurements of grazing of phytoplankton (Haney, Wetzel, and Manny, unpublished; Crumpton and Wetzel, 1982) yielded maximal values of 25% (nighttime) and <5% (daytime) of the epilimnetic volume grazed per day during the period of maximum development of zooplanktonic populations. At other times of the year, the values were lower and therefore were considered negligible in this lake. Without question, zooplanktonic grazing can cause marked alterations in algal species composition and succession within the phytoplanktonic community at certain times of the year (see critical analyses of Lawrence Lake algal populations in Crumpton and Wetzel, 1982; Taylor

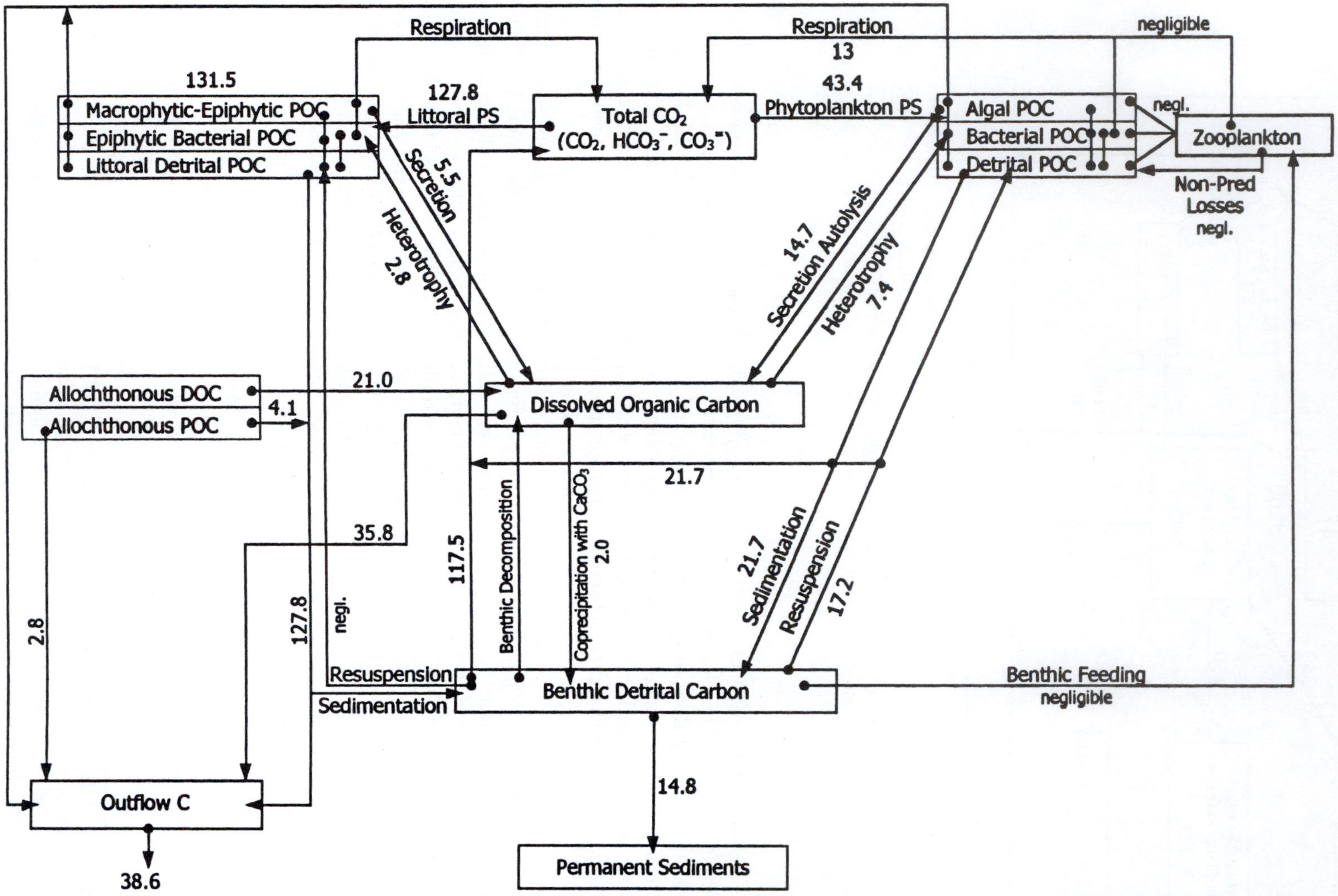

FIGURE 23-25 Detrital structure and flux of organic carbon (g C m^{-2} y^{-1}) of Lawrence Lake, southwestern Michigan. DOC = dissolved organic carbon; POC = particulate organic carbon; PS = photosynthesis. (After Wetzel *et al.*, 1972.)

TABLE 23-14 Pelagic Organic Carbon Budget, Lawrence Lake, Michigan[a]

Pelagic dissolved organic carbon	*g C m^{-2} yr^{-1}*	*Pelagic particulate organic carbon*	*g C m^{-2} yr^{-1}*
Inputs		Inputs	
Algal secretion } Algal autolysis }	14.7	Phytoplankton	43.4
		Resuspension	17.2
Littoral secretion:		Allochthonous	4.1
Macrophytes	3.5	Bacteria:	
Epiphytic algae	1.9	Chemosynthesis	7.1
Epipelic algae	0.1	Heterotrophy of DOC	7.4
Allochthonous DOC	21.0	Total	79.2
Total	41.2	Losses	
Losses		Outflow POC	2.8
Outflow	35.8	Sedimentation	21.7
Coprecipitation with $CaCO_3$	2.0	Total	24.5
Total	37.8	Gross production	+54.7
Gross production	+3.4	Respiration	
Respiration		Algae	13.0
Bacteria (estimated as 50% of DOC production)	−20.6	Sedimenting POC	21.7
		Total	−34.7
Net production	−17.2	Net production	+20.0

Total pelagic net production (DOC + POC) = +2.8 g C m^{-2} yr^{-1}

[a] From Wetzel, R. G., *et al.*: Metabolism of dissolved and particulate detrital carbon in a temperate hardwater lake. *Mem. Ist. Ital. Idrobiol. 29*(Suppl.):185–243, 1972.

and Wetzel, 1988). These analyses, as well as others (e.g., Jewson *et al.*, 1981), however, demonstrated the importance of algal losses by sedimentation out of the photic zone; on an annual basis, sedimentation losses were much greater than losses from grazing. Sedimenting algae, as well as detritus egested by zooplankton, enter the detrital portion of the carbon cycle.

Analyses of the dynamics of the annual carbon fluxes in Lawrence Lake, of which only summary values are given in Tables 23-13 through 23-16, demonstrate the major inputs from the littoral flora and the dominance of respiration in the sediments, where this

TABLE 23-15 Annual Benthic Particulate Carbon Budget of Lawrence Lake, Michigan[a]

Particulate carbon	*g C m^{-2} yr^{-1}*
Inputs	
Submersed macrophytes	+87.9
Epiphytic algae	+37.9
Epipelic algae	+2.0
Windblown POC	+0.0
Precipitation of DOC with $CaCO_3$	+2.0
Sedimentation	+21.7
Outputs	
Benthic respiration	−117.5
Permanent sedimentation	−14.8
Balance (utilization)	+19.2

[a] After Wetzel *et al.* (1972).

TABLE 23-16 Total Annual Budget of Carbon Fluxes for Lawrence Lake, Michigan[a]

Components	*g C m^{-2} yr^{-1}*	*Percentage*
Inputs		
Autochthonous		
Phytoplankton	43.4	19.1
Submersed macrophytes	87.9	38.8
Epiphytic algae	37.9	16.8
Epipelic algae	2.0	0.9
Algal secretion and autolysis	14.7	6.5
Littoral plant secretion	5.5	2.5
Heterotrophy	2.8	1.2
Dark CO_2 fixation	7.1	3.1
Allochthonous		
Stream and groundwater DOC	21.0	9.3
Stream POC	4.1	1.8
Shoreline litter	0.01	0.0
	226.4	100.0
Outputs		
Respiration		
Benthic respiration	117.5	54.6
Bacterial respiration of DOC	20.6	9.6
Bacterial respiration of POC	8.6	4.0
Algal respiration	13.0	6.1
Permanent sedimentation	14.8	6.9
Coprecipitation of DOC with $CaCO_3$	2.0	1.0
Outflow		
Dissolved	35.8	16.5
Particulate	2.8	1.3
	215.1	100.0

[a] Derived from data of Wetzel *et al.* (1972); Otsuki and Wetzel (1974); Burkholder and Wetzel (1989).

TABLE 23-17 Annual Organic Carbon Fluxes in Mirror Lake, New Hampshire[a]

Components	*g C m^{-2} yr^{-1}*	*Percentage of subtotal (%)*
Inputs		
Autochthonous		
Phytoplankton	56.5	69.5
Epilithic algae	2.2	2.7
Epipelic algae	0.6	0.7
Epiphytic algae	0.06	0.07
Macrophytes	2.5	3.1
Dark CO_2 fixation	2.1	2.6
Allochthonous		
With precipitation	1.4	1.7
Shoreline litter	4.3	5.3
Stream DOC	10.5	12.9
Stream POC	1.15	1.4
	81.31	100.0
Outputs		
Respiration		
Phytoplankton	19.1	23.5
Zooplankton	12.0	14.8
Macrophytes	1.0	1.2
Attached algae	1.16	1.4
Benthic invertebrates	2.8	3.4
Fish	0.2	0.2
Sediment bacteria	17.3	21.3
Planktonic bacteria	4.9	6.0
Permanent sedimentation	10.7[b]	13.2
Outflow		
Dissolved	10.87	13.4
Particulate	0.78	1.0
Insect emergence	0.5	0.6
	81.31	100.0

[a] From unpublished data of Jordan, M., Likens, G. E., and Petersen, B. (1982); Likens (1985).
[b] Estimated by difference.

material is decomposed. When macrophytes are minor contributors to the total autochthonous organic inputs, as is the case in Mirror Lake, New Hampshire (Table 23-17), the respiratory metabolism of the sediments is still a major site of decomposition. In Mirror Lake (Table 23-17), phytoplankton dominate the inputs of organic carbon; the output percentages of Mirror Lake are nearly identical to those of Lawrence Lake (Table 23-16), in which the littoral inputs dominate. The importance of dissolved organic detritus in the inputs from allochthonous sources and in the outflow is in the same relationship to the whole in both lakes. The decomposition of photosynthetically fixed carbon by microflora of the littoral and the sediments, which probably dominates in most lakes of the world since a vast majority of lakes are small to very small, relegates the role of animals as "decomposers" to a lesser category, and the burden of heterotrophic metabolism is shifted to the microflora. In large lakes in which the inputs from littoral sources are proportionately low or in hypereutrophic lakes in which phytoplanktonic densities reduce light to the point of excluding much of the littoral flora, animals may assume a somewhat greater importance in the degradation of POC. The bulk of organic carbon is dissolved, however, and decomposition of that organic matter is almost completely microbial and by physical photolysis.

C. Detrital Dynamic Structure in River Ecosystems

The structure of biomass or carbon fluxes in flowing waters is much less well documented. From a detailed budget of the average daily biomass flow in a temperate woodland stream, it was estimated that 25–30% of the particulate biomass input flowed through and was decomposed by the benthic animals, assuming an average assimilation rate of 40% for nonpredators (Cummins, 1972). Allochthonous and autochthonous inputs of dissolved organic matter, which constitute at least 5 times and probably 10 times that of the POC, were not considered in this budget.

The dissolved organic matter, as in lakes, drives the metabolism of streams and rivers. It is common for >90% of the annual organic carbon movement in streams and rivers to be in the dissolved form (e.g., Table 23-18). Experimental analyses in artificial streams have demonstrated that a large portion (>50%) of this DOM can be rapidly degraded (24–48 h) by planktonic and attached bacteria under optimal conditions. Under natural conditions, some of the DOC being transported is photolysed by natural sunlight either completely to CO_2 or partially with the generation of substrates available for bacterial degrada-

TABLE 23-18 Annual Mean Concentrations of Organic Matter in Transport in the Fifth-to Sixth-Order Ogeechee River, Georgia[a]

	Mean (mg AFDW $liter^{-1}$)		*Total organic matter (%)*	
Dissolved organic matter		25.4		96.3
Particulate organic matter				
Amorphous material (bacteria)	0.30		1.14	
Amorphous material (protozoans)	0.04		0.15	
Amorphous material (other)	0.52		1.98	
Vascular plant detritus	0.03		0.11	
Algae (mostly diatoms)	0.06		0.23	
Fungi	0.014		0.05	
Animals	0.002		0.01	
Total particulate organic matter		0.97		3.7
Total organic matter in transport		26.37		100.0

[a] From data of Benke and Meyer (1988).

TABLE 23-19 Annual Organic Carbon Budget Estimates in a 500-m Segment of the Kogesawa River, Uratakao, Japan[a]

	Organic matter *kg C* yr^{-1}	*Percent* (%)
Inputs		
Primary production	100	5.1
Litterfall into stream	90	4.6
Lateral movement to stream	38	1.9
Fine particulate organic matter	480	24.5
Dissolved organic matter	1170	59.8
Groundwater inflow, DOC	80	4.1
Subtotals	1958	100.0
Outputs:		
Community respiration	77	4.1
Fine particulate organic matter	620	32.7
Dissolved organic matter	1200	63.2
Subtotal	1897	100.0

[a] Extracted from data of Yasuda *et al.* (1989).

tion. The amount of DOC that is converted to POM via flocculation, especially in hardwater streams as discussed earlier, or that enters bacterial biomass that is then partially degraded by animals is unclear. This area is of great interest and is in need of further research. Less than 1% of detrital organic inputs was estimated to be metabolized by macroinvertebrate fauna of a mountain stream in New England (Fisher and Likens, 1973). The role of animals in comparison to microflora in decomposition of POM varies widely, and may be even more variable in streams than in lakes, but is certainly small.

These relationships are exemplified in an annual organic carbon budget for a section of a small river in Japan (Table 23-19). Dissolved organic matter and fine particulate organic matter made up >88% of the inputs of organic carbon. Total community respiration constituted for <5% utilization of the organic matter loading. The remainder was either stored briefly or largely exported as fine POC and mostly as DOC.

It is of the utmost importance to emphasize that these examples of the functional structure of lake and stream ecosystems are only preludes to the important underlying questions of what parameters control the observed structural integrity of the system. Analyses of structure and dynamics of ecosystems are a prerequisite to analyses of factors regulating the observed and changing structure. But such analyses must not become an end in themselves. Sufficient underlying similarities between the structure and dynamics of ecosystems probably exist that will permit generalizations to emerge with less rigorous analyses. As a result, more effort could be devoted to experimental evaluation of control mechanisms. It is critical that the regulatory mechanisms of natural ecosystems be understood before human perturbations of the systems can be effectively evaluated and minimized. Meaningful applied research and wise management of aquatic resources cannot be undertaken effectively without a basic understanding of functional control mechanisms.

D. Death and Organic Cycling

Therefore, in summary, the observed heterotrophic biotic productivity of most lakes and rivers, as estimated in the detailed organic carbon budgets just discussed, could not possibly be supported by the autochthonous organic carbon generated by pelagic primary productivity within these waters. The aquatic heterotrophic productivity must be supplemented by external allochthonous organic matter imported from terrestrial and land–water interface regions. From the standpoint of the lake or a river per se, bounded by the shore line, the lake or river is a net heterotrophic ecosystem that decomposes much more organic matter than is produced within those boundaries.

Furthermore, the annual time period is the only interval of relevance in comparative analyses of productivity and utilization of organic matter (Wetzel, 1995). A significant portion of the total autochthonous annual photosynthetic production, even in the pelagic regions, occurs during cold, low-light periods when consumption by zooplankton and higher animals is reduced or virtually absent. Although that noningested production may be of little immediate consequence to the higher animal heterotrophs, it is of major importance and usually dominates energy flows within the pelagic region. Furthermore, nonpredatory heterotrophic metabolism completely dominates energy flows of the whole-lake or river ecosystem. Large amounts of dissolved and noningested particulate organic matter enter the detrital pool for the dominant microbial heterotrophy.

Utilization of organic matter by metazoan predation, however, represents only a small portion of the total organic carbon pool of the ecosystem and a very small source of DOM. Material and energy fluxes of egested particulate and soluble organic matter within the pelagic zone and the rest of aquatic ecosystems by nonmetazoan heterotrophs are largely ignored or discussed in terms of conversion back to particulate organic matter of sufficient size for ingestion by predation of metazoans (Wetzel, 1995). The common thesis that living net production in organisms, particularly algae and microbes as well as metazoans, that die largely by ingestive predation is not only unsubstantiated but also unreasonable. Predation by animals functions in the facilitation of detritus turnover, expediting the transfer rates of a small portion of organic matter by microbes in the conversion to CO_2.

Many if not most organisms—particularly bacteria, fungi, algae, as well as higher organisms—simply mature physiologically, senesce, and die, and then enter the combined particulate and dissolved organic detrital pool for utilization by microbial heterotrophs. Biochemical death of many cells and organisms is genetically programmed and fixed within relatively narrow physiological sequences. Turnover of dissolved organic matter occurs many times, with large conversions of portions of the organic matter to CO_2 heterotrophically, before predation, with much slower generation times, can respond to changes in particulate organic matter.

Senescence and mortality of bacteria, algae, and other microbes from viral infections is large (40 to >50%). Clearly, viral mortality leads to major evasions of dissolved organic matter as cellular integrity is compromised (e.g., Suttle, 1994; Fuhrman, 2000; Wommack and Colwell, 2000). Average virus-to-bacteria ratios of 50:1 have been observed with a high direct correlation between abundance of viruses and bacteria (Thingstad, 2000; Wetzel, 2000b). Most estimates indicate viral phages lyse about one-fourth of the bacterial production per hour (Heidal and Bratbak, 1991). Phage-induced mortality of bacterioplankton in Lake Konstanz, Germany, for example, was estimated to vary between 1 and 24% of total mortality over an annual period (Simon *et al.*, 1998). Other viral mortality estimates of bacterioplankton and phytoplankton are in a similar or higher range (11–66%) for lakes, coastal waters, and rivers (e.g., Suttle, 1994; Fuhrman and Noble, 1995; Mathias *et al.*, 1995). One can view the viral mortality as an important component of the microbial loop that accelerates the dissipation of organic carbon to CO_2.

Loss of cellular integrity upon senescence or viral parasitism results in a rapid large loss of soluble cellular organic matter, often exceeding 40% of the total organic carbon within the first 24 h (e.g., Krause, 1962; Otsuki and Wetzel, 1974b). DOM lost at this stage often consists of simple, nonstructural organic compounds of relatively high energy and availability to microbes. Similarly, ingestion of bacteria and pico- and ultraplankton by protists, which purportedly although erroneously was proposed to "short circuit" metabolism in the "microbial loop", is a major diversion of organic carbon away from animal higher trophic levels. A much more realistic food-web ecosystem structure properly places appropriate metabolic emphasis on the microbial communities from viruses through protists (Fig. 23-26). Clearly, the metabolism and energetic fluxes associated with dead detrital organic matter as DOM dominates in all these ecosystems. Among some communities, such as higher aquatic plants and their attached periphytic microbial

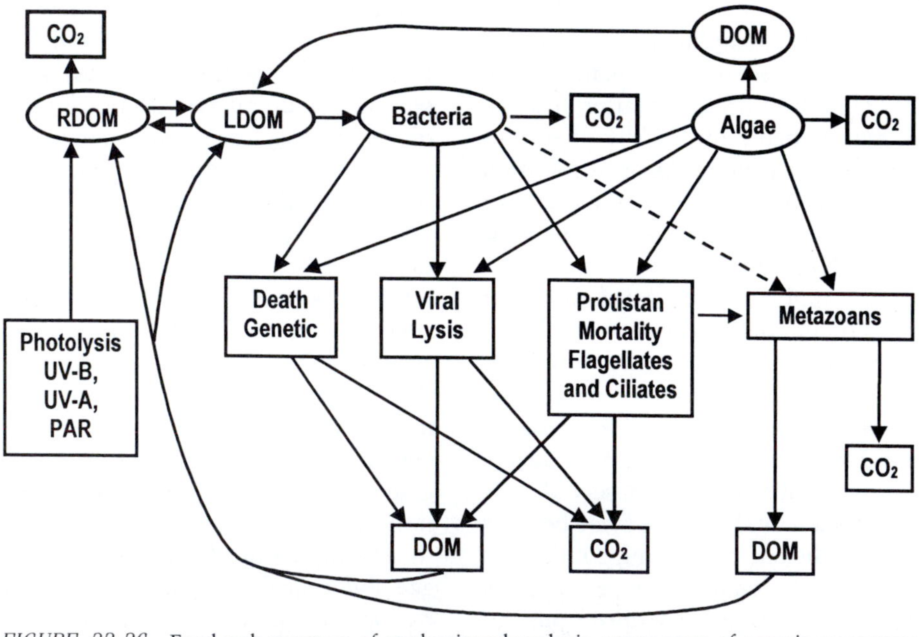

FIGURE 23-26 Food-web structure of predominantly pelagic components of aquatic ecosystems that reflects the importance of the microbiota, the microbial loop, and physical processes by which most of the organic matter inputs are metabolized heterotrophically or degraded *by physical processes*. DOM = dissolved organic matter, with labile (LDOM) and recalcitrant (RDOM) components.

TABLE 23-20 Net Ecosystem Production of Organic Carbon, Lawrence Lake, Michigan[a]

	Dissolved organic carbon	Particulate organic carbon	Total ($g\ C\ m^{-2}\ yr^{-1}$)
Inputs			
Inlet 1	7.0	2.0	9.0
Inlet 2	7.9	1.0	8.9
Ground water	6.0	1.2	7.2
Total	20.9	4.2	25.1
Outputs			
Outflow	35.8	2.8	38.6
Sedimentation	—	14.8	14.8
Total	35.8	17.6	53.4
Net ecosystem production (NEP): $+28.3\ g\ C\ m^{-2}\ yr^{-1}$			

[a] After Wetzel *et al.* (1972).

complex, most (>90%) of the total organic matter production (foliage, sloughed tissues, excreted DOM, rooting tissues, and others) never enters a metazoan digestive tract, analogous to the situation in terrestrial ecosystems (Wetzel, 1990a, 1995). Functionally, organic carbon cycling in terrestrial and inland aquatic ecosystems is similar.

VI. NET ECOSYSTEM PRODUCTION

The dissolved organic carbon pool of lakes receives inputs from both pelagic and littoral photosynthesis. In the Lawrence Lake example, an equivalent amount was received allochthonously from the drainage basin. The major loss of dissolved organic carbon through outflow represented most of the net ecosystem production of Lawrence Lake relative to the surrounding environment (Table 23-20). This estimate of *net ecosystem production* yields essentially a measure of export from the system, for example, by outflow, sedimentation, and emigration.

Woodwell and Whittaker (1968) originally defined net ecosystem production (NEP) as

$$\text{NEP} = GP - [RS_{(A)} + RS_{(H)}]$$

when GP = gross production
$RS_{(A)}$ = respiration of autotrophs
$RS_{(H)}$ = respiration of heterotrophs

or simply NEP = (gross production)–(respiration of the ecosystem). The essential feature of NEP is that it accounts for importation and losses via respiration. The original expression was modified by Woodwell *et al.* (1972) to account for the import and export of organic material:

$$NEP = (GP + NP_{\text{in}}) - [RS_{(A)} + RS_{(H)} + NP_{\text{out}}]$$

when NP_{in} = net production imported from another ecosystem
NP_{out} = net production exported

Both formulations interpret *NEP* as the positive or negative increment of organic matter, either as living biota or dead organic storage, after total respiration within the ecosystem. The latter definition of *NEP*, which specifically subtracts organic exports, limits application of the term to only that material which remains inside ecosystem boundaries.

VII. BIOTIC STABILITY AND SUCCESSION OF PRODUCTIVITY

The mechanisms and couplings of detrital metabolism, as elaborated in the three-phase system (allochthonous, littoral, and pelagic) and exemplified in the dynamics of whole-lake ecosystems, as exemplified in Lawrence and Mirror lakes, provide a fundamental stability to the entire biotic dynamics of lakes and streams. Living components of the system generally undergo rapid oscillations of productivity in an opportunistic series of competitive responses to changes in availability of constraining nutritional and physical factors governing their growth. However, the overall metabolism of both lakes and streams is dependent to a major extent on the detritus components of dead DOC and POC that form a primary source of utilizable energy.

Much of the present emphasis on DOM of aquatic ecosystems correctly addresses how this abundant reservoir of recalcitrant dissolved organic carbon can be utilized in microbial metabolism and respiration and can influence microbial nutrient regeneration. The commonly observed incomplete photolysis of DOC is critical to accelerated utilization of these macromolecules

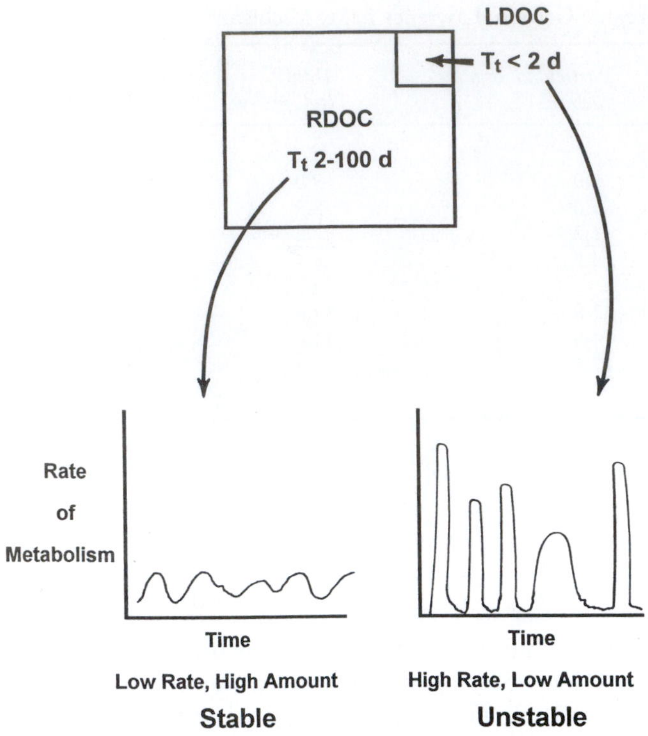

FIGURE 23-27 Relative rates of metabolism of the large, dominating pool of recalcitrant dissolved organic carbon (RDOC) vs. the small instantaneous pool of labile dissolved organic carbon compounds that are largely under the supply control of highly ephemeral algal communities. (From Wetzel, 2000a.)

but is clearly not mandatory. Portions of the complex DOM pools, including fractions of humic and fulvic acid compounds, are degraded, but total degradation rates are clearly slow. Such chemical organic recalcitrance of DOM is instrumental in providing a thermodynamic stability to metabolism within lake, reservoir, wetland–littoral land–water interface, and river ecosystems (Wetzel, 1983, 1984, 1992, 1995, 2000a). The chemical recalcitrance is truly a "brake" on ecosystem metabolism, and that brake is critical for the maintenance of stability.

Most of the detrital organic pool, both in particulate and dissolved phases, of inland aquatic ecosystems consists of residual organic compounds of plant structural tissues. The more labile organic constituents of complex dissolved and particulate organic matter are commonly hydrolysed and metabolized more rapidly than more recalcitrant organic compounds that are less accessible enzymatically (Fig. 23-27). The result is a general increase in concentration of the more recalcitrant compounds, commonly exceeding 80% of the total, with slower rates of metabolism and turnover. These recalcitrant compounds, however, are metabolized at rates slowed and regulated in large part by their molecular complexity and bonding structure.[4]

In every detailed *annual* organic carbon budget of lake and river ecosystems, the heterotrophic metabolism of the ecosystem cannot be supported by organic matter generated by phytoplankton. At least severalfold support of the total metabolism is from organic subsidies from the land–water interface communities and allochthonous production. From the standpoint of metabolic stability, it is particularly important that most of the organic carbon is dissolved and relatively recalcitrant, which ameliorates the violent oscillations so characteristic of the pelagic particulate components of the ecosystem. The relatively slow utilization rate of this detrital reservoir gives stability to aquatic systems, tiding the system over during periods of low detrital carbon inputs and recharging the reserve during excessive inputs.

In addition, much of the particulate organic matter formed in the dominating land–water interface regions of lake and river ecosystems is displaced to reducing, anoxic environments of the littoral and profundal sediments. The dissolved organic carbon, largely of higher plant origins, provides the stability and is the currency for the quantitatively more important detrital pathways in aquatic ecosystems (Fig. 23-28). The same underpinnings of that stability prevail in terrestrial ecosystems and likely much of the marine ecosystem. The trophic dynamic structure is almost totally dependent energetically upon the detrital dynamic structure. The manner in which it is manifested differs in fine detail among systems, but functionally, it must be the same in all ecosystems.

Detritus includes nonliving particulate, colloidal, and dissolved organic matter, and metabolically, size only affects rates of hydrolytic attack (Wetzel, 1995). Inland aquatic ecosystems collect organic matter, particularly in dissolved forms, from terrestrial, wetland, and littoral sources in quantities that supplement if not exceed those produced autochthonously. Rates of utilization of that organic matter are slowed by a combination of chemical recalcitrance as well as displacement to anoxic environments. As a result, inland aquatic ecosystems are heterotrophic and functionally, detrital bowls, not algal bowls.

A. Human Influences

In long-term evolutionary scales, humans now have the abilities to intervene rapidly in this interdependent relationship and alter the stability of the rates

[4] On the basis of disequilibrium thermodynamics, the chemical recalcitrance functions to regulate and balance dissipation of energy (entropy) at a high level, which is needed to maintain stability.

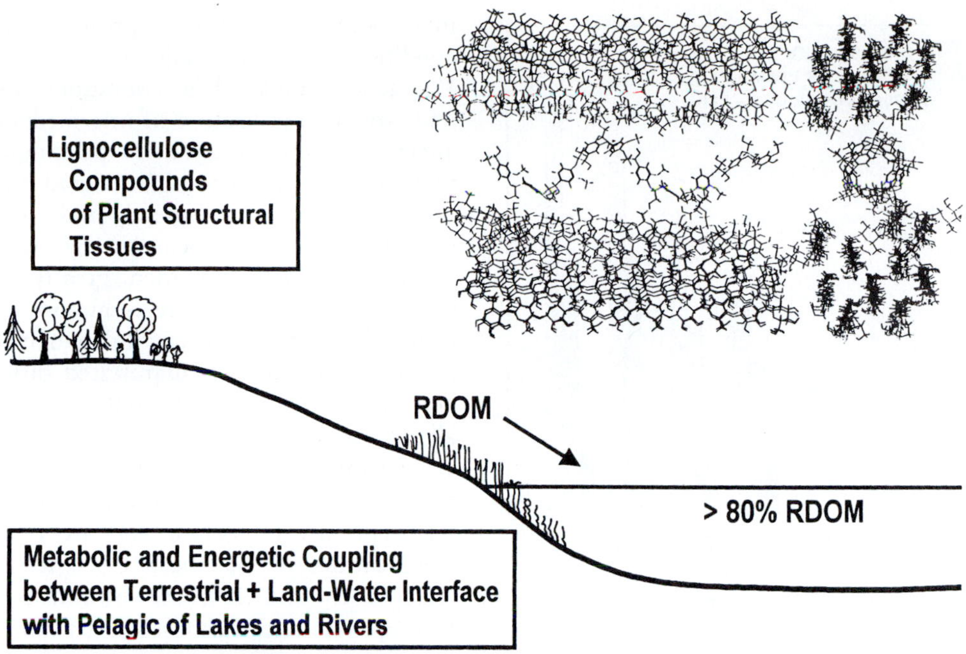

FIGURE 23-28 Lignocellulose compounds of higher plant structural tissues from terrestrial and land–water interface regions form a major source of recalcitrant dissolved organic matter (RDOM) within aquatic ecosystems and a major metabolic coupling between the drainage basin and aquatic ecosystems. (From Wetzel, 2000a.)

of metabolism of organic matter. For example, reduction of ozone in the stratosphere and associated increased UV irradiance could lead to accelerated photolytic degradation of macromolecules of DOM by both abiotic[5] and biotic pathways to CO_2. The photolytic enhancement of substrates for bacterial metabolism by UV photolysis of recalcitrant humic substances can result in accelerated rates of biogeochemical cycling of nutrients and stimulated productivity of the ecosystems. In addition to decreasing the metabolic stability of the lakes and streams, the enhanced microbial respiration will certainly lead to enhanced generation of CO_2 and evasion to the atmosphere.

Moreover, as the concentrations of CO_2 in the atmosphere increase, largely from anthropogenic combustion of fossil fuels, plant growth commonly accelerates. For example, CO_2 concentrations will double to ca. 720 ppm yet in the present century. Experimental doubling of CO_2 leads to increased rates of photosynthesis and growth by about 20–35% above rates at ambient CO_2 concentrations. The increased growth also commonly leads to nitrogen limitations. As a result, C:N ratios increase and commonly structural tissues, particularly lignin content, increase. As these plants grown in enriched CO_2 decay, DOM leached into the surface waters contain larger amounts of dissolved humic substances. Photolysis of this DOM from those plants grown under enriched atmospheric CO_2 resulted in a conspicuous increase in CO_2 release among both terrestrial plants (Fig. 23-29) and emergent aquatic plants (Fig. 23-30). Clearly, photodegradation of dissolved organic substrates is a major process that can alter rates of biogeochemical cycling in aquatic ecosystems.

The stability is afforded in part by the recalcitrant chemical structure of the organic substrates and in part by the displacement of much of the organic matter to anoxic environments. When oxygen is absent or its diffusion rate into a habitat is insufficient to fulfill the requirements of aerobic microflora, the rate of carbon mineralization is slower and the number of microbial cells formed per unit of substrate is less than when oxygen is abundant (Alexander, 1971; J. B. Hall, 1971). Complex and often simple organic compounds, such as fatty acids and amines, can accumulate when the quantity of readily fermentable carbon is large, as in the sediments. The low redox conditions lead to the generation of anaerobic fermentative products, including gases (CO_2, CH_4, H_2, and H_2S), that simultaneously inhibit

[5]Although the contribution of UV-B to photochemical mineralization of DOC is small in comparison to UV-A and blue portions of PAR (see Fig. 23-19), UV-B has high impact for photochemical damage to DNA. Therefore, the implications of increased UV-B are likely higher for availability and metabolism of organic substrates and for reproduction of biological organisms than for abiotic photochemical reactions that convert DOC to CO_2.

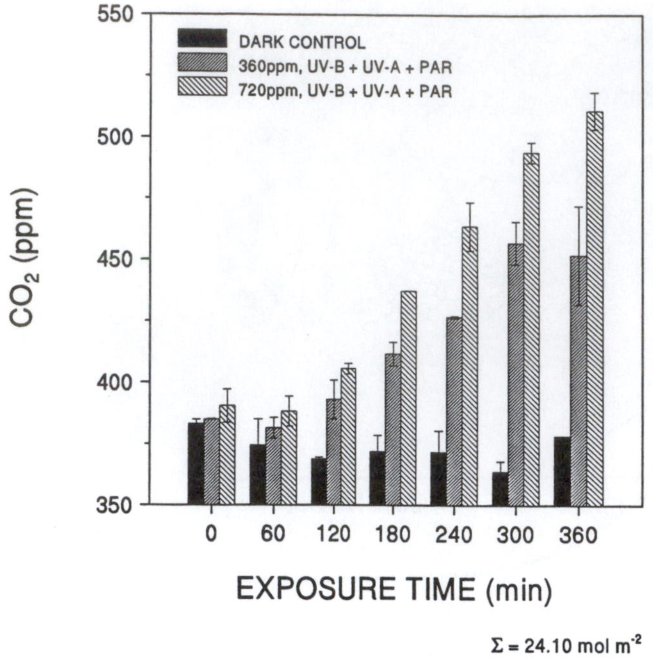

FIGURE 23-29 Release of CO_2 above saturation by complete photolytic degradation of sterile leachate from leaves of the riverine popular tree (*Populus tremuloides*), grown under ambient atmospheric CO_2 and doubled CO_2 concentrations, after one week of microbial decomposition. (From Wetzel, 2000a.)

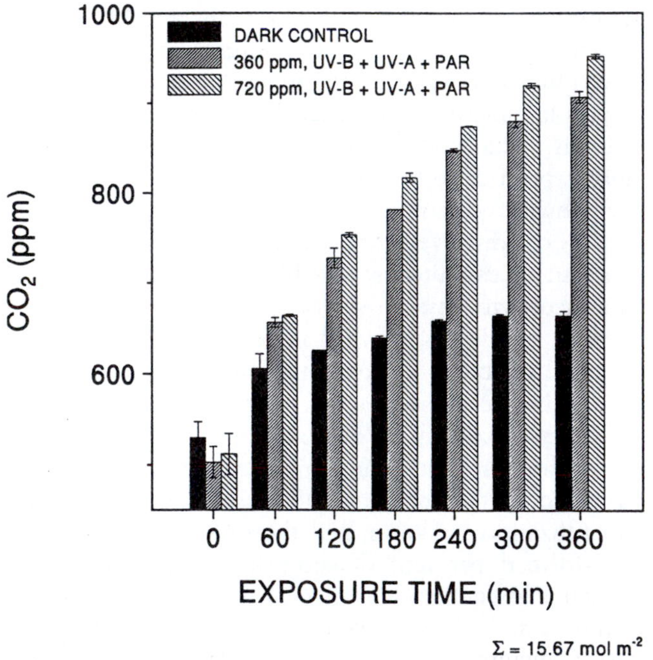

FIGURE 23-30 Release of CO_2 above saturation by complete photolytic degradation of sterile leachate from leaves of the cattail (*Typha latifolia*), grown under ambient atmospheric CO_2 and doubled CO_2 concentrations, after one week of microbial decomposition. (From Wetzel, 2000a.)

many forms of microbial growth and under aerobic conditions serve as a major sink for molecular oxygen. Degradation under these conditions is slowed and limited largely to utilization of intermediate compounds of anaerobic microflora. When oxic conditions are in close juxtaposition to anaerobic degradation, intermediate products are more completely and efficiently degraded. Inhibitory conditions caused by low E_h and toxic gaseous products are then greatly diminished. The trophic dynamic structure is almost totally dependent energetically upon the detrital dynamic structure. The manner in which it is manifested differs in fine detail among systems, but functionally, it must be the same in all ecosystems.

Within the operation of the detrital dynamic structure, the sources, quantities, and composition of detrital inputs are important and are related to lake eutrophication. As has been discussed in detail earlier (Chaps. 18 and 19), the littoral plant complex of macrophytes and their associated microflora dominate autochthonous productivity of many lake and most river ecosystems. The contributions by phytoplankton, submersed and emergent macrophytes, and attached and eulittoral algae change with increasing fertility. Dominance by the sessile flora is usually the case, except for certain extremely oligotrophic lakes, lakes of very large area and volume, or hypereutrophic lakes, where biogenic turbidity severely attenuates submersed light conditions (see Fig. 19–21).

The detrital dynamic structure functionally controls metabolism in a majority of lakes, in part because nutrient regeneration is controlled by detrital dynamics. Wetland and littoral plants are implicated as a major metabolic force driving detrital dynamics in most lakes, where they are dominant sources of allochthonous and autochthonous organic matter. Moreover, the detrital inputs of particulate and dissolved organic substrates vary with the flora adapted to individual lake conditions. For example, in hardwater marl lakes, DOC interacts with inorganic components such as $CaCO_3$, which effectively suppresses subsequent pelo- and hydrometabolism. These suppressing mechanisms can be loaded to saturation, resulting in rapid transition to highly eutrophic conditions, geologically speaking. In other cases, the succession of certain macrophytes can lead to a community dominated by *Sphagnum* and other mosses that function as effective ion exchangers with simultaneous release of refractory organic compounds (cf. Chap. 25), creating conditions in which detrital metabolism is slowed and permitting accumulation of organic matter in particulate form. Detrital metabolism and the rates and cycling of dissolved organic matter underlie the entire metabolic and trophic structures of all ecosystems.

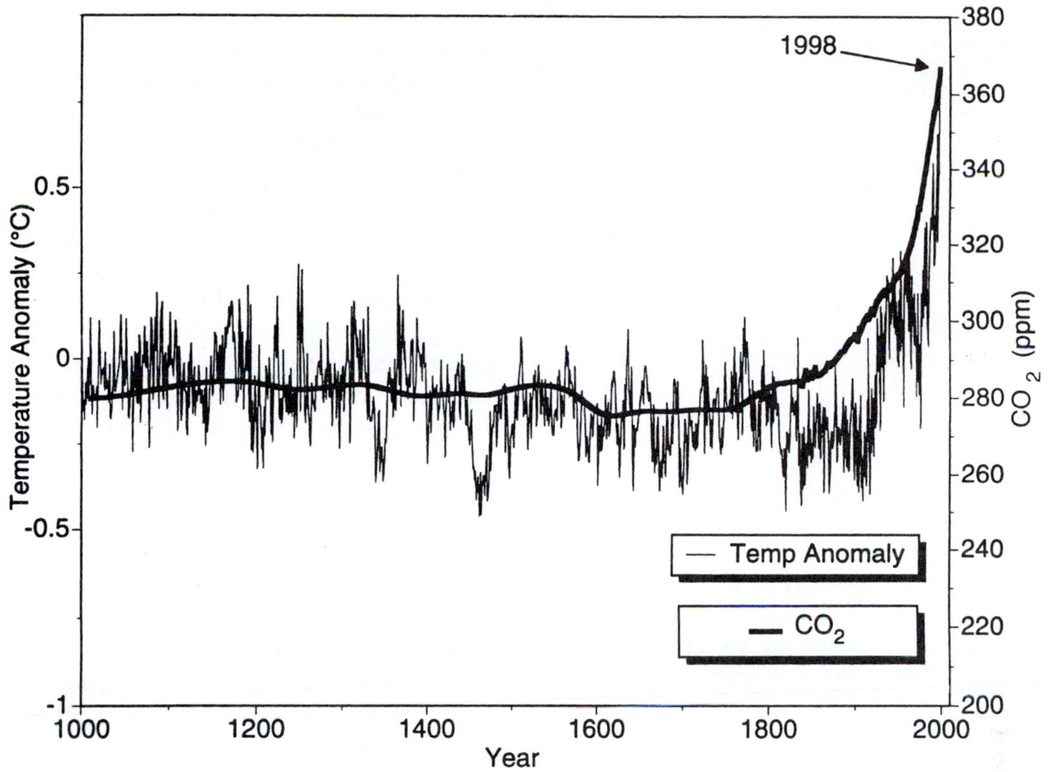

FIGURE 23-31 Long-term determinations of CO_2 concentrations of the atmosphere, directly for the past half century and from paleorecords earlier and average paleo-temperature estimates of deviations from the average over the past 1000 years. (Courtesy of Dr. James A. Teeri, adapted from data of M. E. Mann, 1999. Geophysical Research Letters, 26, 759–762.)

VIII. SYNERGIES AMONG DISSOLVED ORGANIC MATTER, SUNLIGHT, CLIMATIC WARMING, ENHANCED ATMOSPHERIC CO_2, AND ACIDIFICATION

Reduction of ozone in the stratosphere, in part associated with releases of chlorofluorohydrocarbon compounds used in refrigeration and propellant devices, has resulted in large increases in the amounts of UV-B and, to a lesser extent, UV-A reaching the surface of the Earth (Kerr and McElroy, 1993; Madronich, 1994). The increases vary with atmospheric conditions but are in the range of 1% per decade near the equator and >13% per decade at latitudes greater than >45°.

Atmospheric concentrations of CO_2 have progressively increased markedly in the past century, largely as a result of human combustion of fossil fuels (Fig. 23-31) and are anticipated to double to approximately 720 ppm in the present century. The increases in CO_2 concentrations can stimulate plant productivity, as discussed earlier, but additionally result in a "greenhouse effect," in which back radiation of heat from the Earth is reduced. As a result, a gradual warming of the average temperatures occurs that is directly correlated with rising CO_2 concentrations (Fig. 23-31).

The primary thesis of this chapter is that dissolved organic matter is a primary component of aquatic ecosystems that regulates biogeochemical cycling and provides a fundamental thermodynamic stability to lakes and streams. Alterations of the atmospheric and climatic environment by human activities can impact the biogeochemistry of organic matter cycling and flux rates in aquatic ecosystems. A number of processes are already evident:

1. *Reduction in DOC Concentrations.* A general trend has been observed among long-term data sets that DOC concentrations are declining gradually (Schindler *et al.*, 1991; Schindler and Curtis, 1997). Particularly in oligotrophic lakes where DOC concentrations are often low, UV light penetrates to depths of several meters and can negatively influence organisms by genetic damage, diverting production to increased synthesis of protective pigments, and other ways. In lakes of alpine regions or in high latitudes where higher plant source materials are few, allochthonous loading

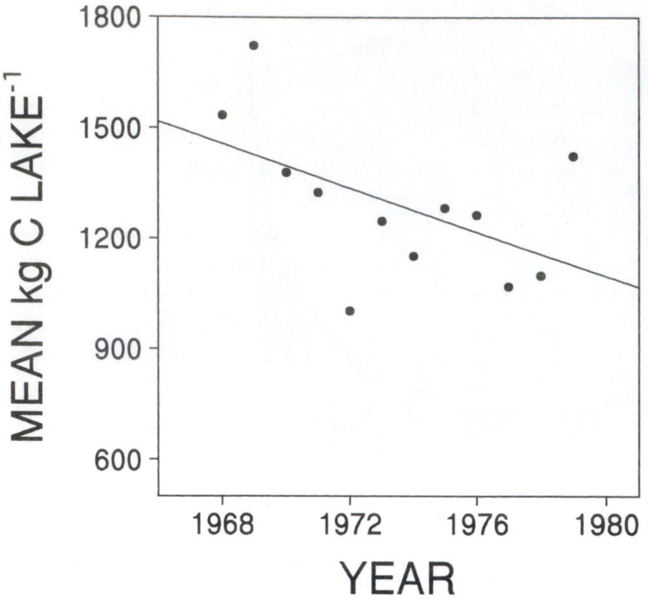

FIGURE 23-32 Decline ($r^2 = 0.28$) of mean annual dissolved organic carbon in Lawrence Lake, southwestern Michigan, determined by analyses at each meter depth at weekly or biweekly intervals over the 12-yr period and weighted for volume in each meter stratum of water. (Data of Wetzel, unpublished.)

of chromophoric organic matter is low. Organisms in such lakes can be exposed to high intensities of ultraviolet irradiance (McKnight *et al.*, 1997). Even in lakes with higher concentrations of DOC, the long-term trends are often toward slowly decreasing concentrations of DOC (Fig. 23-32).

2. *Climatic Warming and Droughts.* There is little question that both temperature and carbon dioxide concentrations of the atmosphere are increasing (Fig. 23-31). Rising temperature has also influenced precipitation patterns and particularly has led to large regions in which rainfall and snow accumulations have been reduced. Droughts are a cumulative result of numerous meteorological factors affecting precipitation, evapotranspiration, and losses. Droughts usually do not become severe until after long periods of deficient rainfall and unrestrained water use.

DOC in some lakes has declined appreciably over the last quarter century coincident with substantial warming (Fig. 23-32 and in the Ontario lakes, Schindler *et al.*, 1992). Reduced precipitation and increased evapotranspiration in the drainage basin result in reduced stream flows and lower DOC loading to the stream and lake. Transparency of lake water to UV photolysis increases under these conditions. Similar reductions in DOC in streams have been observed (e.g., Dillon and Molot, 1997). The decrease in annual DOC yields of streams occurs in spite of higher concentrations in storm flows following periods of prolonged drought (Schindler *et al.*, 1992; Hinton *et al.*, 1997).

3. *Effects of Acidification.* As poorly buffered softwater lakes are acidified (cf. Chap. 11), the concentrations of DOC decline and transparency of the water increases (Driscoll and van Dreason, 1993; Dillon and Molot, 1997; Gjessing *et al.*, 1998). The fate of the DOC upon acidification is unclear. Some DOC may be mineralized and evade to the atmosphere, and some may flocculate and sediment. Under drought conditions, reoxidation of previously deposited reduced sulfur in wetland and littoral sediments can yield strong acid pulses during intermittent rainfall events following drought periods (Bayley *et al.*, 1992).

4. *Increased Rates of Bacterial Metabolism and Nutrient Recycling.* As discussed earlier in this chapter, the increased penetration of UV-B, UV-A, and PAR irradiance can result in the acceleration of partial photolysis of organic macromolecules. Simple organic substrates generated by this process accelerate bacterial metabolism, simultaneously generating CO_2 and enhancing nutrient regeneration. The decrease in the metabolic buffering capacity from DOC results in reduced stability of the ecosystem.

5. *Increased Rates of CO_2 Evasion to the Atmosphere.* The photolytic mineralization of organic substances to CO_2 directly by UV-B, UV-A, and PAR irradiance increases as the amounts of DOC decline. With the decline of chromophoric DOC, penetration of light increases in a positive feedback mechanism.

IX. SUMMARY

1. Nearly all of the organic carbon of natural waters consists of dissolved organic carbon (DOC) and dead particulate organic carbon (POC). The ratio of DOC to POC is commonly between 6:1 and 10:1, both in lacustrine and running water ecosystems (cf. Table 23-1).
2. Living POC of the biota constitutes a very small portion of the total POC; most POC and all DOC is dead detrital organic matter. The metabolism of the biota, however, is important in mediating reversible fluxes of carbon between the dissolved and particulate phases of organic carbon.
3. Organic matter consists almost totally of a mixture of plant, microbial, and animal products in various stages of decomposition.
   a. *Nonhumic substances* are low-molecular-weight organic substances, such as carbohydrates, proteins, amino acids, fats, waxes, resins, pigments, et cetera. These substances are relatively labile, are easily degraded by microorganisms (hours),

and exhibit rapid flux rates. Instantaneous concentrations are relatively low.

b. *Humic substances* are naturally occurring heterogeneous organic substances recalcitrant to rapid biological utilization (days to weeks) and are high in molecular weight (most 700–5000 D). Humic substances are derived largely by microbial degradation from cellulose and lignin of plant structural materials and constitute 70–80% of the organic matter of water and soils.

c. Fungi and certain specialized bacteria degrade plant materials and form humic substances.

d. Dissolved organic carbon moves with the water and will sediment if it adsorbs to inorganic or organic particulate matter or if it undergoes polymerization to particulate form. Only a small percentage of DOC aggregates, flocculates, and shifts to colloidal and particulate states; these organic particles are of relatively high nutritive value.

4. The distribution of DOC within lakes and streams is relatively constant, with changes in season and in depth.

5. Particulate organic matter tends to remain near sites of production. In the soils and hydrosoils, POC undergoes degradation, with the release of DOC of partial degradation by microbes.

6. DOC of allochthonous origin (terrestrial and wetland marsh sources) constitutes the primary source of external organic carbon loading to fresh waters. This DOC is extensively modified by microbial metabolism in soils, streams, wetlands, and littoral areas as it flows toward recipient lakes.

   a. Dissolved organic matter (DOM) that enters lakes via surface runoff or groundwater seepage contains a low percentage of nitrogen (organic C:N approximately 50:1). The high C:N ratio results from a large proportion of structural compounds from terrestrial and higher aquatic plants and from the selective decomposition of more labile compounds by microflora as the DOM is transported by water flow.

   b. DOM derived from algal and cyanobacterial production within water bodies has a higher initial nitrogen content, with a C:N of about 12:1.

   c. The molecules of many humic compounds of DOM absorb solar radiation >290 nm. Photochemical reactions from absorption of irradiance by these chromophoric compounds (CDOM) can modify the compounds to render them more bioavailable to bacteria and, in some cases, degrade them entirely to CO_2.

7. There are three major sources of particulate organic carbon (POC):

   a. Allochthonous sources of POC, which consist primarily of plant materials. This POC is transported by water drainage and by air movements to recipient streams or lakes. Rates of decomposition of POC in recipient waters vary greatly with the organic composition of the material but are usually much slower (e.g., leaves in weeks; woody material in years) than those of DOC (days to weeks).

   b. Most of the POC of streams and rivers is produced in the soils of the drainage basin and in the floodplain areas adjacent to the channels. POC inputs can be seasonal, particularly among heavily forest-canopied streams (cf. Chap. 21).

   c. Much of the POC in lakes originates from autochthonous production in (i) the pelagic zone and (ii) the littoral zone. Because most lakes are small and shallow, much of the POC originates in littoral areas from aquatic macrophytes and associated attached algae. In reservoirs, the proportion of autochthonous POC production to allochthonous POC loading is more equal, although it can be quite variable (particularly seasonally).

8. The spatial and temporal distribution of pelagic particulate organic carbon is commonly closely correlated with the productivity and biomass distribution of phytoplankton and bacteria during periods of thermal stratification. The inputs from littoral productivity to the POC pool of lakes increase in relative importance in the transition (e.g., in a series of lakes) from the nutrient-limited conditions of oligotrophic lakes to more fertile conditions of eutrophic lakes (see Fig. 23-13).

9. Much functional insight into the operation of freshwater ecosystems has been obtained from quantitative analyses of the dynamics of carbon cycling among the biotic communities and the inorganic and organic carbon reservoirs. From several examples of organic carbon budgets discussed in detail, those general properties discussed earlier were verified. In addition:

   a. Detrital particulate organic carbon from allochthonous, littoral, and pelagic components is often metabolized under anoxic conditions largely in the sediments. Detrital dissolved organic matter is dominated by recalcitrant organic substances, largely of allochthonous and littoral sources, that chemically suppress rapid degradation. These degradability characteristics provide a fundamental thermodynamic stability to ecosystems. Many components of the biotic (living), trophic dynamic structure of freshwater ecosystems generally undergo rapid oscillations in relation to an array of dynamic factors governing growth and reproduction, while detrital metabolism is slower and more evenly sustained by a much larger organic reserve.

   b. Increases in autochthonous organic carbon inputs in relation to allochthonous inputs result in progressively greater decomposition in the sediments than occurs in the water column.
   c. The trophic dynamic structure of freshwater ecosystems is almost totally dependent for its energy upon the detrital dynamic structure. In many lakes and reservoirs, the littoral and allochthonous components are the dominant sources of synthesized organic carbon; these two components also dominate inputs of detrital organic carbon and are implicated as the major driving force of the ecosystems.

10. The major sites of detrital metabolism are the benthic region, where most particulate organic carbon is decomposed in many lakes, and the pelagic region largely of dissolved organic carbon. Considerable secondary metabolism is thus displaced from sites of production to sites of decomposition by sedimentation in the case of POC and by aggregation and coprecipitation with particulate matter in the case of DOC.
11. Detrital organic matter functions in several pivotal roles:
    a. Although the specific nature of detritus molecules modifies their behavior and effects upon the environment, detritus inevitably carries most of the energy of the ecosystem from its points of origin to its places of transformation.
    b. Essentially, all lakes and streams are heterotrophic net CO_2 producers with massive evasion of CO_2 to the atmosphere in excess of autochthonous production.
    c. Most benthic metabolism is heterotrophic, microbial, and anaerobic, even when hypolimnetic oxygen depletion does not occur. The continued production of CO_2 by anaerobic benthic metabolism, and release of this oxidized product (synthesized using alternate electron acceptors) to the overlying water, results in benthic and hypolimnetic respiratory quotients (CO_2 released/O_2 uptake) >1.
    d. The reduction of alternate electron acceptors (nitrate, iron and manganese oxides, sulfate, or various organic substrates) instead of molecular oxygen results in an accumulation of electrons in the sediments and the receipt of energy originally transferred during photosynthesis (where CO_2 was reduced to carbohydrate and water was oxidized to oxygen). This flow, termed *detrital electron flux*, into and ultimately out of the sediments (i) derives its energy from photosynthesis, just as a predatory animal does from its food, and (ii) is a much larger proportion of the total energy inputs and transfer processes because most organic matter becomes detrital in the ecosystem without ever passing through an animal. Thus, the ecosystems incorporate detrital energy flow into their overall functioning to an extent that is similar to, and usually much greater than, that occurring through the well-recognized energy flow of predator–prey processes.
    e. From a functional standpoint of ecosystem operation, *detritus* consists of all organic carbon (DOC and POC) lost by nonpredatory means from any trophic level (by egestion, excretion, secretion, autolysis, et cetera.) or inputs from sources external to the ecosystem that enter and cycle within the system (allochthonous organic carbon). The *detritus food chain* is any pathway by which the chemical energy contained within detrital organic carbon compounds becomes available to the biota.
    f. Recent experiments demonstrate microbial degradation of DOC manifoldly greater than could be derived from autochthonous primary production and that utilization of this recalcitrant allochthonous and littoral DOC is mediated by partial photolysis by UV-B, UV-A, and photosynthetically active (PAR) radiation.
    g. Relatively small portions of the allochthonous and autochthonous organic matter reach higher trophic levels. Degradation of allochthonous and autochthonous organic matter is largely by physical photolytic and by microbial processes.
    h. A functional commonality exists in the way organic matter is utilized; differences emerge in sources of organic matter, which affect the rates of utilization, not the process of utilization. Freshwater ecosystems are heterotrophic and are functionally detrital bowls, not algal bowls.
12. Natural dissolved organic substances, particularly humic substances, can interact with other important metabolic components.
    a. These compounds can complex, often by chelation, with inorganic elements and alter biological availability.

b. These compounds can complex with other organic compounds, such as enzymes, and reduce their availability and utilization by biota.
c. Humic substances can buffer against acidification but also may add acidity when in high concentrations to surface waters of very low ionic strength.
d. Microbially reduced humic substances can serve as electron donors and a redox shunt between oxidized and reduced compounds under anaerobic conditions.
e. Dissolved organic substances can modify light penetration severely and thereby affect photoreception, organism behavior, and thermal cycles and coupled stratification.

24 PAST PRODUCTIVITY: PALEOLIMNOLOGY

In the past, alterations of aquatic ecosystems have been brought about by changes in climatic conditions; more recently, changes have been greatly accelerated by human activities. These changes have affected conditions of the drainage basins, water budgets, nutrient loading and budgets, and the resultant productivity of fresh waters.

Lakes function as natural traps for sediment. Once materials entering or produced in the lake settle to the lake bottom, energy levels and hydrodynamics are usually insufficient to transport materials out of the basin. Records of climatic and human-induced changes in the environment have been left in the sediments of lakes and certain reservoirs. These records represent relatively static derivatives of the dynamic ecosystems that are constantly changing over short-term (e.g., seasonal) and long-term periods. Paleolimnology focuses on the interpretation of sedimentary sequences and on diagenetic processes that can alter that record. The ultimate goal is to gain insights into past conditions, productivity, and changes in regulatory parameters that have caused aquatic ecosystems to enter a different stage of productivity. In addition, insights can be gained toward a prediction of the trajectory that an ecosystem may take in future time (Deevey, 1984).

A lake is at the receiving, downgradient end of a drainage basin. Lake sediment is derived from both external and internal sources. Allochthonous materials are transported by water movements from its surrounding drainage basin or enter from the air as particulate precipitates (e.g., leaf fall or dust). Autochthonous materials settle from materials generated within the lake (dead organic matter, chemical precipitates such as carbonates, and siliceous diatom frustules). Organic accumulation within the sediments of a lake results from a positive balance between production and loading to the sediments and diagenesis within the sediments. It is generally assumed that the rate of change of organic deposition provides a record of past productivity levels. Such assumptions, however, are based on the premise that the rates of diagenesis have been constant or can be estimated and that most of the organic matter has been produced autochthonously within the lake basin.

Most of the bulk of lake sediments is mineral detritus released and imported from the drainage basin. This proportion increases with decreasing size and volume of lakes and is very high in reservoirs, where river influents are a much larger proportion of the total reservoir volume than is the case in most natural lakes.

Biota and biotic structure within the lake are derived biogeochemically largely from these materials. The sediments provide a partial record of both the biota within the lake and chemical and biotic materials imported from the drainage basin. Some changes occurring in the drainage basin but recorded by mineral, chemical, or biotic (e.g., pollen) importations to the sediments can be very useful (Binford *et al.,* 1983). For example, climatic changes, recorded directly by changing mineralogy or indirectly by changing terrestrial plant pollen assemblages found in the lake sediments, are important in the interpretation of environmental conditions to which lake biota were exposed, parts of which may also be preserved in the sediment.

I. STRATIGRAPHY AND GEOCHEMISTRY

A. Sedimentary Record

The fossil record of sediments includes both biochemical substances produced by organisms, or resulting from their degradation, and morphological remnants of specific organisms. The subject is large and the literature is voluminous. Particularly instructive reviews on paleolimnology, from which much of the following résumé was drawn, include Vallentyne, 1960; Frey, 1964, 1969a,b, 1974, 1988; Swain, 1965; Juse, 1966; Korde, 1966; Binford *et al.,* 1983; Smol, 1990, 1992; and Anderson, 1993.

Sediments that accumulate in lake basins originate from numerous source materials. The composition is influenced to a great extent by the geomorphology of the lake basin and the drainage basin. Although many lakes were formed between the intervals of cyclic glaciation and deglaciation, most lakes of earlier glacial periods have been obliterated by succeeding glaciations, by filling with sedimentary materials or by draining. It is only because of the very recent retreat of the last major glaciation phase that so many lakes exist at the present time. With the notable exceptions of some very ancient lakes, particularly in rift areas in Africa and Asia (Livingstone, 1975), most lakes are very young ($<$20,000 years).

The primary sedimented materials are controlled by regional geology and climate and are modified by biological processes of both the drainage basin of the lake and the lake itself. Water movements such as wave action and currents sort particulate matter according to its size and density and the energy available for displacement. Movement of particles to the central depression of a lake basin that leads to sediment focusing (cf. Chap. 21, pp. 633–635) in the lake results from three major processes (Hilton, 1985): (a) peripheral wave action that generates turbulence and resuspends sediments in littoral areas that are subsequently redeposited in deeper areas; (b) intermittent, random redistribution of sediments of the entire basin when the lake is not stratified, particularly during autumnal overturn; and (c) movement of sedimentary material parallel to a sloping bed (sliding) or movement initiated by a rotational failure of the sediment (slumping), resulting in material being transported by flows from shallow regions to deeper waters. Little sediment will accumulate on slopes $>$14%, and $<$4% slope has little effect (Håkanson, 1981). The morphometry of the lake basin dictates the dominant process of sediment redistribution.

The result is a general gradient from coarser particles near shore regions to finer particles in sediments under deeper waters. This gradient can be disrupted by various mechanisms that often are related to the morphology of the lake basin. For example, deposits of sediments in littoral regions or river deltas can accumulate until they become gravitationally unstable and slump in landslide fashion or move as turbidity flows to deeper portions of the basin where they come to overlie finer, often younger, sediments.

B. Dating of Sediments

Several methods have been used to determine the age of sedimented materials in lake deposits. In postglacial lakes, the most important method has been to evaluate the age of organic material by measuring its content of radioactive carbon (^{14}C). Radiocarbon is formed in the upper atmosphere by the reaction of nitrogen (^{14}N) with neutrons (n): $^{14}N + n \rightarrow {}^{14}C + {}^{1}H$. The production of ^{14}C has been nearly constant over time but has varied because of changes in the supply of neutrons produced by cosmic radiation. Neutron quantity is thus dependent upon the intensity of cosmic radiation. The cosmic-ray flux in the upper atmosphere is influenced by (a) the intensity of the Earth's magnetic field and (b) short-term changes in solar wind magnetic properties (Ralph and Michael, 1974; Stuiver and Quay, 1979). When the Earth's geomagnetic field is strong, fewer cosmic rays reach the upper atmosphere; when it is weak, the inverse occurs. These geomagnetic changes cause about an 8% change in atmospheric ^{14}C over a period of approximately 6000 years. The short-term solar magnetic variations of ^{14}C production occur in cycles of a few hundred years or less and superimpose small deviations upon the long-term geomagnetic trends. Radioactive carbon decays back to ^{14}N with the emission of a β^- particle and has a half-life of 5568 years.

The ^{14}C of the atmosphere is a very small part of Earth's CO_2 reservoir (approximately 1 in 10^{12} parts).

For purposes of ^{14}C dating, it is assumed that ^{14}C produced in the atmosphere is in rapid equilibrium with these CO_2 reservoirs and that the balance among the carbon reservoirs has been constant over time. During photosynthesis, $^{14}CO_2$ is photosynthetically incorporated into organic matter in proportion to its availability in the atmosphere. When the organisms die and are interred in sediments, the ^{14}C contained in the organic matter continues to decay back to ^{14}N with the emission of a β^- particle. The half-life of ^{14}C disintegration is constant, and when the residual specific activity of ^{14}C in the carbon of old organic matter is accurately radioassayed, it provides an estimate of the age of the organic matter. The decay rate of ^{14}C should permit age determinations back to a limit of about 75,000 years. Technical detection limitations of assaying for negative beta radiation of ^{14}C above background radiation usually limit ^{14}C-age determinations to 40,000 years, although techniques have improved (e.g., Grootes, 1978; Stuiver, 1978b).

Much effort has been devoted to evaluating the accuracy of ^{14}C dating of organic deposits, including potential errors caused by variations of ^{14}C production within the atmosphere and ^{14}C incorporation into plant organic matter (e.g., reviews of Ralph and Michael, 1974; Krishnaswami and Lal, 1978; Stuiver, 1978a; Stuiver and Quay, 1979). Recent changes in past atmospheric ^{14}C levels can be determined by measuring the present-day ^{14}C activity of wood laid down in trees (e.g., Douglas fir, Sequoia redwoods, and bristlecone pine) and accurately dated by counting tree rings. Calibrations of ^{14}C dating measurements of older samples have been made by careful comparisons with other dating techniques ($^{230}Th/^{234}U$ ratios, thermoluminescence, magnetic dating). In hardwater lakes containing appreciable quantities of geologically old, ^{14}C-deficient carbonates, inorganic carbon fixed in photosynthesis is not in equilibrium with atmospheric ^{14}C activity, but includes carbon dissolved from carbonates of the drainage basin. This old ^{14}C-deficient carbon dilutes contemporary carbon originating from atmospheric sources. Organic matter from sediments of calcareous lakes thus gives dates that are spuriously old. Corrections are only approximate, and often the radiocarbon-dating values from these lakes must be compared with independent chronological measures (MacDonald *et al.*, 1991).

The dating of very recent sediments can be problematic with ^{14}C methodology. More recent dates of sediment deposits have been obtained by analyses of lead-210 (^{210}Pb). Radium-226 in soils decays to radon-222, which escapes to the atmosphere, where it decays to ^{210}Pb. This ^{210}Pb enters a lake via precipitation and is eventually incorporated into the sediments; it is termed "unsupported" ^{210}Pb, since it was produced from radium located outside the sediments. Dating by ^{210}Pb relies on the difference in the total ^{210}Pb in the sediment and the "supported" ^{210}Pb resulting from the presence of ^{226}Ra and produced within the sediments. The difference is the concentration of "unsupported" ^{210}Pb resulting from atmospheric deposition. "Unsupported" ^{210}Pb will decrease with depth in the sediments because of radioactive decay, provided the input of ^{210}Pb to a lake is constant, its residence time is constant, and there is no significant migration within the sediment (Robbins and Edgington, 1975; Pennington *et al.*, 1976; Oldfield and Appleby, 1984; El-Daoushy, 1988; Turner and Delorme, 1996). The age of the sediment at any particular depth can be calculated from its activity of unsupported ^{210}Pb relative to that at the surface, since the radioactive half-life (22.26 years) is constant. Because of the short half-life of ^{210}Pb, this dating technique is limited to about 150 years. Caution is needed in interpretations because of differential rates of accumulation and of remobilization of Pb in the early diagenesis of sediments, particularly in relation to differences in iron and manganese cycling within the sediments (Appleby and Oldfield, 1978; H. E. Evans, 1984; Oldfield and Appleby, 1984; Benoit and Hemond, 1988; Binford and Brenner, 1988; R. D. Evans, 1991; Wan, 1993).

The distribution in sediments of cesium-137, which has been present in the atmosphere since 1954 because of atomic bomb testing, has been a useful method of dating sediments deposited within the last 50 years. More recently (1986), the nuclear accident at Chernobyl created a new maximum in ^{137}Cs that has been used as a marker for short-term accretion rate and transfer coefficient estimates in sediments (Mitchell *et al.*, 1983; Appleby *et al.*, 1993; Callaway *et al.*, 1996). This method assumes that ^{137}Cs fallout with rain became attached to particles and that these particles were quickly (<1 year) transported from the drainage basin to lake deposits (Ritchie *et al.*, 1973). ^{137}Cs falling directly upon the lake surface is adsorbed onto suspended particulate matter and sedimented. There is evidence (Davis *et al.*, 1984) that downward molecular diffusion and adsorption of ^{137}Cs and biological recycling can greatly confound the application of this method for dating recent sediments.

In both the ^{210}Pb and ^{137}Cs dating techniques for recently deposited sediments, it is assumed that disturbance of the sediment stratigraphy is small. In some cases, disturbance by water movements and biological redistribution of sediments by benthic macroinvertebrates (cf. Chap. 21) can be sufficient to obscure dating chronology (e.g., Robbins *et al.*, 1977; McCall and Fischer, 1980). In other cases, careful interpretation of recent dating analyses, combined with other paleolimnological data, provides much insight into reconstruction of recent lake events (e.g., Pennington *et al.*, 1976; von Damm *et al.*, 1979).

After sediments are deposited in a lake, properties of magnetic intensity and geomagnetic declination (direction) are retained in the sediments as ***remnant magnetism***. These magnetic characteristics reflect past, relatively short (approximately 1000 years) variations in the direction of the Earth's magnetic field. Much of the magnetic remanence is carried by hematite (Fe_2O_3) (Mackereth, 1971, Creer *et al.*, 1972). Initial studies carried out in Lake Windermere, England, showed that the direction of the horizontal magnetization oscillates about a mean direction with an amplitude of approximately $\pm 20°$ and a frequency of 2700 years. Remnant magnetism was similarly demonstrated in sediments of a number of other lakes, although with somewhat different periods of oscillation (e.g., Yaskawa *et al.*, 1973; Tolonen *et al.*, 1975; Creer *et al.*, 1976; Oldfield *et al.*, 1978a,b; Thompson *et al.*, 1980). This method assists in rapidly dating sediments by nondestructive methods once the detailed chronology has been established within a given lake or continental region (Creer, 1982).

A further magnetic property of sediments, their ***magnetic susceptibility***, is useful in evaluating probable sources of sedimented materials in lakes (Thompson *et al.*, 1975; Oldfield *et al.*, 1978, 1983; Thompson, 1978; Appleby *et al.*, 1985). Magnetic susceptibility is simply the magnetizability of sediment (i.e., its degree of attraction to a magnet). The most magnetically susceptible sediments contain the highest amount of allochthonous inorganic material washed into the lake from the drainage basin. From the paleomagnetic secular variations the chronology of sedimentation rates can be coupled to deforestation, afforestation, and agricultural changes in the drainage basin. Magnetic susceptibility also assists in evaluating spatial differences in stratigraphy within a lake basin. These differences result from variations in the deposition in different places, including the slumping of deposits from shallower areas over deeper deposits.

Many recent lake sediment profiles contain atmospherically derived fly ash and various particles from industrial processes. These particles include a magnetic fraction that can be characterized and approximately quantified by magnetic remanences preserved in the sediments (Oldfield and Richardson, 1990). Sediment profiles from lakes in industrially developed regions show widespread increases, steeply so in recent decades, in magnetite and hematite deposition beginning from the mid-nineteenth century onwards.

Airborne ***soot spherules*** are small (ca. 30 μm) carbonaceous particles generated during coal and oil combustion and released into the atmosphere. Soot particles are preserved in lake sediments, and the content in the sedimentary record is proportional to the history and intensity of coal and oil combustion (Renberg and Wik, 1984). Soot particle deposition increased in the middle of the nineteenth century and has risen progressively since, with some optimistic suggestions for reduced loading coincident with emission controls in recent times.

Some lake deposits exhibit distinct ***laminations*** in which temporal differences in the composition and quantities of suspended matter laid down create fine alternations of light- and dark-colored layers. When a pair of laminae (a dark and a light) represent a time period of one year, the lamination is called a ***varve*** (DeGeer, 1912; Nipkow, 1920; Sturm, 1979; Renberg and Segerström, 1981). Varves are generally restricted to sediments of lakes that have not been physically disturbed during deposition. In such lakes, varves provide a means of dating sediments at annual intervals and can be used to interpret former conditions of climate, mineralogy of the drainage basin, water level, the succession of certain organisms, and trophic conditions as evaluated by other sediment parameters and from biotic remains (e.g., Merkt, 1971; Renberg, 1978; Satake and Saijo, 1978; Sturm and Matter, 1978; Renberg and Segerström, 1981; Edmondson, 1991b).

Whether the deposition of sediments occurs in distinct laminations or varves depends upon the timing of lake stratification in relation to the timing of inputs of suspended matter from sources to the basin (from river runoff, deposition from the atmosphere, autochthonous production of organic matter or inorganic precipitates (e.g., $CaCO_3$), resuspension of deposited matter, or importation of suspended matter from ground water) (Sturm, 1979). Often, inputs are not sufficiently coordinated with quiescent stratified conditions; as a result, deposition is chaotic and distinct annual laminations are not formed. Bimodal sedimentary varves occur if the seasonal (discontinuous) influx of suspended matter coincides with thermal stratification of the water column in annually stratified lakes and no appreciable bioturbation of the sediments erases the pattern subsequently.

In glacial environments, relatively coarse sediments are brought to recipient lakes with the inflow of meltwater in the spring and summer and are deposited during the period of thermal stratification. Finer sediments are deposited over these during the winter when the lake is ice covered. Similar bimodal deposition is often found in temperate lakes, particularly in meromictic lakes in which the sediments are not disturbed by water circulation (e.g., Tippett, 1964; Brunskill, 1969; Digerfeldt *et al.*, 1975; Ludlam, 1976; Simola, 1977). In these cases, records of seasonal differences can be found in the deposition of diatoms, tree pollen, clay, iron oxides, and/or calcium carbonate.

C. Inorganic Chemistry

External source materials in dissolved and particulate forms that leave the drainage basin and enter the recipient lake basin are influenced markedly by the terrestrial vegetative cover. Numerous examples were discussed quantitatively in the preceding chapters and need not be reiterated here. The long-term effects on the productivity of lakes are influenced by climatic variations as well as by human-induced changes in the vegetation and weathering processes (e.g., Mackereth, 1965, 1966; Bradbury *et al.,* 1975; McColl and Grigal, 1975; Dean and Gorham, 1976; Likens *et al.,* 1977; Engstrom and Wright, 1984; Likens and Bormann, 1995). An example of the effect of climate on weathering was demonstrated in some endorheic lakes of East Africa (Hecky and Kilham, 1972). Lakes in humid areas that had high weathering rates of the drainage basin received greater inputs of silica and were largely bicarbonate dominated. In more arid regions, sodium chloride originating from precipitation dominated the inorganic composition, and inputs of silica were low. The chemistry and climate changes of these endorheic lakes were reflected in the sedimentary record as well as in the extant diatom populations.

Chemical constituents of the sediments have been used variously to interpret rates of both limnological activities of the lake as well as changes in climate and alterations of the drainage basin (Engstrom and Wright, 1984). For example, the relative intensity of soil erosion is generally reflected in the concentrations of elements primarily associated with clastic minerals (Na, K, and Mg). Estimates of soil weathering based on the level of these elements in the mineral matter (inorganic ash) may be confounded by variations in sedimentation of biogenic silica and autochthonous biochemically precipitated oxides. Phosphorus occurs in lake sediments as sorbed components of amorphous iron oxides, in discrete mineral phases, and as organically bound phosphorus (Chap. 13). The inorganic phosphorus–iron complex is largely responsible for the exchange of dissolved phosphorus in many sediments. If iron content and other sedimentological conditions remain constant, a record of changing phosphorus levels in the water may be preserved in the sediments. As discussed earlier (Chap. 13), the retention of phosphorus in lake sediments is affected by redox conditions, sediment mixing, and pH. Thus, because of the changes in retention capacities of the sediments for phosphorus, only long-term major changes or abrupt pollutional loading alterations are conspicuous.

The analyses assume that the chemical chronology represents, at least on a long-term basis, the composite changes in inputs to the lake and metabolic transformations that have occurred within the lake. For example, comparison of the chemical composition of numerous elements, mostly biologically nonessential, in lake sediments as well as in rock and soils of the drainage basins permits an effective evaluation of the chronology of erosion and rates of leaching (Cowgill, Hutchinson, and collaborators, 1963, 1966a,b, 1970, 1973). Inputs of extremely inert, conservative elements, such as titanium, permit another evaluation of erosion rates of the lake basins. Depending on the geology of the area, other elements may indicate disturbances of drainage basins. In Laguna de Petenxil, a small lake in Guatemala, sediment concentrations of calcium, strontium, potassium, and, to some extent, sodium increased during periods of active Mayan agricultural activity in the drainage basin. Manganese, iron, and phosphorus followed the fluctuations of agriculturally determined alkalies and alkaline earths in an analogous way (cf. Deevey *et al.,* 1979).

Detailed investigations of the paleolimnology of a small, closed lake of volcanic origin in Italy, Lago di Monterosi, demonstrated the marked effects of alterations to the drainage basin area by the construction of a road alongside the lake in early historical times (Hutchinson *et al.,* 1970). The lake was quite productive shortly after its formation for a period of about 2000 years, probably related to initial easy leaching of materials rich in potassium and of fairly high phosphorus content. By 20,000 years BP[1], sedimentation rates were very low. Low rates of deposition of inorganic constituents and organic matter, indicative of very low productivity, continued until at least about 5000 years BP and the Roman period. This interval was marked by a period of cold, dry conditions following the last glacial retreat from the area. A slight increase in productivity was discernible during the latter portion of this interval and a more recent period, most likely related to slight climatic amelioration and some human activity. A very large increase in nutrients and productivity occurred slightly over 2000 years BP (Fig. 24-1). The changes in the sedimentary records coincided rather precisely with the major disturbance of the drainage basin by the construction of a Roman road, the Via Cassia, in about 171 BC. A great increase in inputs of phosphorus and alkaline earths, particularly calcium, resulted, and the lake rapidly became eutrophic, causing the deposition of much organic matter. The productivity of the lake since the period of major disturbance has declined as land uses in the area have changed.

[1] BP = before present.

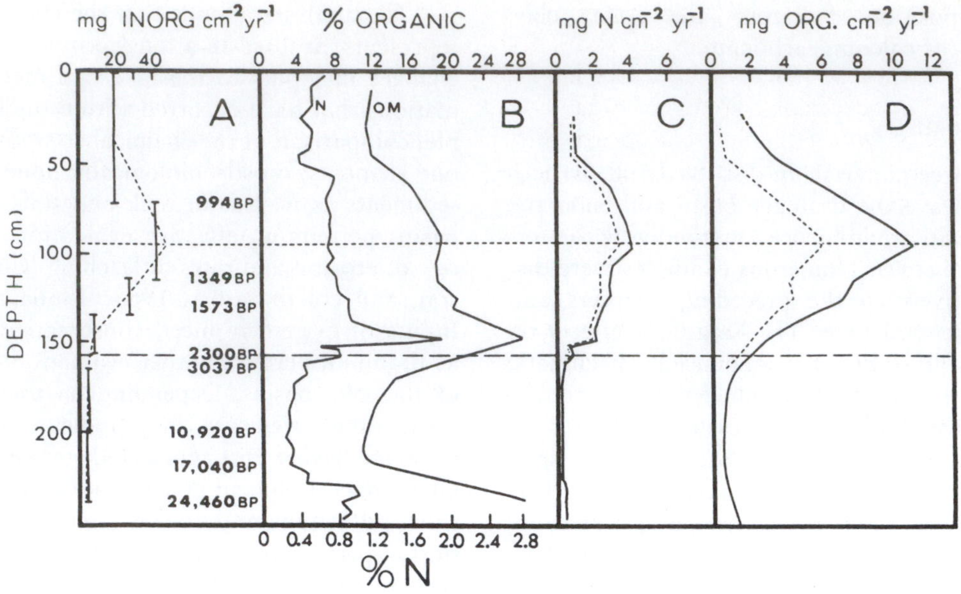

FIGURE 24-1 Stratigraphy of organic matter and nitrogen in the sediments of Lago di Monterosi, Italy. (a) Estimation of sedimentation rates as a function of sediment depth and radiocarbon dates; (b) ignitable organic matter and nitrogen as percentages of dry sediment; (c) estimated rate of deposition of nitrogen, the broken line being corrected for organic nitrogen brought in by erosion; and (d) estimated rate of deposition of organic matter, the broken line being corrected for organic matter brought in by erosion. (Redrawn from Hutchinson *et al.*, 1970.)

Sedimentary records of such disturbances in the drainage basin are common. For example, the productivity of Grosser Plöner See in northern Germany was increased markedly in the early thirteenth century by the construction of a mill dam that raised the water depth a few meters and flooded portions of the surrounding catchment basin (Ohle, 1972, 1979; Averdieck, 1978). In more recent times in the New World, the total clearance of forests over large portions of North America has led to marked changes, usually increases, in the productivity of lakes, which recorded in the chemical stratigraphy of sediments deposited during the last two centuries. Often, the increased leaching and nutrient loading leads to a marked acceleration of productivity. In contrast, when this increased leaching occurs in calcareous regions, extreme loading of calcium and carbonates can result in a negative effect on productivity owing to carbonate interactions and sedimentation of essential inorganic and organic nutrients (Wetzel, 1970; Manny *et al.*, 1978; Wetzel and Manny, 1978; cf. Livingstone, 1957; Livingstone and Boykin, 1962).

The paleolimnological conditions and productivity of a eutrophic hardwater lake in southern Michigan were reconstructed by combined analyses of the sedimentation rates of organic carbon, nutrients, fossil pigments, pollen, and diatoms (Manny *et al.*, 1978). During the late glacial period (14,075 to 10,200 years BP), algal productivity was low under boreal conditions of low temperatures and competition for nutrients, particularly phosphorus, which was coprecipitated with $CaCO_3$. A long period of relatively constant but slowly increasing productivity followed from about 10,200 to 5900 years BP. During this period, the surrounding vegetation shifted from pine to oak dominance as the climate ameliorated. From about 5900 to 140 years BP, several lines of evidence indicated a gradual increase in areas suitable for the growth of submersed littoral macrophytes, until littoral productivity likely exceeded that of the phytoplankton. Alternating shifts between littoral algal and macrophytic dominance and then to phytoplanktonic dominance occurred repeatedly, which suggested fluctuations in water levels in response to alternating moist and dry periods. During this period, evidence also indicated that the hypolimnion became anoxic each year. As the land surrounding the lake was deforested by humans who settled into the region approximately 140 years BP, sedimentation rates increased sevenfold. The increase in sedimentation rates occurred concomitant with markedly increased sediment concentrations of carbonates and phosphorus and decreased organic matter, nitrogen, and plant pigments. Littoral diatoms increased over planktonic diatoms during this period. In 1926 (some 75 years BP), a bird sanctuary with extensive waterfowl populations was established on the lake in conjunction with a reforestation program designed to stabilize eroding soils around the lake. These actions rapidly resulted in reduced carbonate loading and markedly increased concentrations of organic nitrogen, pigments, organic matter, and particularly phosphorus. During this period, diatoms of the sediments were dominated by eutrophic, planktonic species. The hypereutrophic conditions continue to the present time, with exceedingly high sedimentation rates (approximately 6 mm yr^{-1}).

The old concept that lakes progress from states of oligotrophy to eutrophy is not necessarily universal (Deevey, 1955; Hutchinson, 1973). As has been demonstrated repeatedly from chemical analyses of sediment stratigraphy, coupled with other indices of past productivity, lake productivity responds to changing nutrient incomes, climate, and morphometry. Accumulation rates of nitrogen, phosphorus, and organic matter in sediments are commonly high in early postglacial times, a period of high availability of nutrients. Accumulation rates typically then decline as leaching gradually impoverishes the drainage basin. In recent times, increases or decreases in lake productivity are usually directly related to disturbances caused by human activities.

D. Organic Constituents

Examination of the organic compounds in sediments must address the problem of origin: Did the organic compound originate within or outside of the lake basin? Furthermore, many organic compounds undergo degradation both during and after sedimentation. The extent of preservation is not always uniform with geologically changing conditions at and within the sediments. However, certain organic constituents of sediments, especially pigment degradation products and certain lipids, are relatively stable and very helpful in the interpretation of past episodes of productivity, particularly when coupled with other indices.

Total amino acid, carbohydrate, and chlorophyllous and flavinoid pigment residues in lake and bog sediments typically increase toward the surface (Swain, 1965). Carotenoid pigments also commonly increase upward in more recent lake sediments but decrease markedly in the developmental transition from a lake to a bog (see discussions further on). Maxima of many organic compounds occur just beneath the surface of the sediment, probably associated both with increased microbial metabolism at the interface and with redox-mediated adsorption phenomena.

Much information is available on organic compounds and their distribution in sediments (Vallentyne, 1957b, 1960, 1969; Cranwell, 1976; Philp *et al.*, 1976). With improved analytical technology in the last decade, a shift has occurred from simple measurements, such as the amount and distribution of total organic matter, carbon, or nitrogen, to molecular characterization of individual compounds. Many organic compounds are not very stable and tend to be hydrolyzed enzymatically and chemically; they may then be metabolized microbially. It is often very difficult to distinguish the time, origin, and mechanism of synthesis of many organic compounds. Significant progress has been made in the association of specific compounds with specific organisms, particularly with certain pigment products (see the following discussion) and with lipid derivatives (e.g., Mermoud *et al.*, 1981, 1982; Cranwell, 1990; Buchholz *et al.*, 1993).

The distribution patterns of fatty acids, hydroxyacids, and total neutral lipid constituents of sediments allow identification of biological sources of a significant percentage (ca. 50%) of the lipids (Cranwell, 1973, 1990; Robinson *et al.*, 1987). For example, analyses of carbon-chain lengths of *n*-alkanes derived from plant lipids showed that sediment could be characterized by the type of humic material from which it was derived. In higher plants, *n*-alkanes consist of largely odd-numbered carbon chain lengths of C_{23} to C_{35}, whereas lower plants contain *n*-alkanes of both odd- and even-numbered carbon chains. Humic derivatives in sediments from acidic peat areas around a lake contained predominantly C_{31} *n*-alkanes, whereas those from deciduous forested areas had mostly C_{27} and C_{29} *n*-alkanes. Investigation of monocarboxylic unsaturated fatty acid components in lake sediments, soil, and peat indicated that autochthonous material was the main source of C_{12}–C_{18} *n*-alkanoic acids, while terrestrial sources provided the C_{22}–C_{28} *n*-alkanoic acids (Cranwell, 1974, 1979, 1990). The relative contribution of these two sources could be related to the trophic conditions (greater C_{12}–C_{18} acid content in the sediment of productive lakes) and to the rate of erosion of the drainage basin. High abundance of branched/cyclic alkanoic acids, typically occurring in lipopolysaccharides of gram-negative bacteria, was correlated with a large autochthonous contribution from algae and bacteria, whereas a low abundance was correlated with a high terrestrial contribution to the *n*-alkanoic acids. High amounts of alkanoic acids were also found in productive, eutrophic lake sediments that were strongly reducing.

These and similar organic chemical approaches hold much potential and deserve intensified investigation. Problems associated with diagenetic changes in organic compounds after deposition, about which little is known, are serious because these transformations can be confused with those resulting from succession or other changes in the organisms that synthesize the compounds. Controlled experimental analyses, however, can provide insight into the mechanisms and rates of diagenesis involved.

II. BIOLOGICAL INDICATORS

A. Morphological Remains

Paleolimnology has a long history that began with analyses of morphological remains of aquatic and terrestrial organisms. Preservation of dead organisms in the sediments is typically incomplete, and the extent of preservation depends both upon the type of organism and the environmental conditions that prevail at the time of sedimentation. Pollen and spores of terrestrial plants are most abundant, as are frustules of diatoms and chrysomonad cysts (Frey, 1974, 1988). Remnants of some other groups of algae can occur, but many are not represented. The Cladocera and chironomid midges have the most abundant and diversified of animal remains, although all groups of animals are represented to some extent. Many remains can be matched to species, particularly among the diatoms, desmids, Cladocera, ostracods, and beetles. When only a stage in the life cycle is represented (e.g., cysts of chrysomonads or resting eggs of rotifers), features allowing specific identification are lacking and differentiation often is possible only to the point of the genus or family level.

Of the dominant organisms found in sediments, each group consists of two separate communities, one in the warm littoral zone and the other pelagic. As we have seen with the focusing effects of sediment particles and organic matter, remains of inshore organisms are moved offshore by wind-generated water movements and currents. The amount of such movement and displacement varies widely with different drainage basin and lake basin characteristics.

B. Pollen and Spores

Plant pollen and spores are produced in great abundance, but only a few of those produced fulfill their reproductive function. As pollen types are dispersed, they are well mixed by atmospheric turbulence in proportion to the quantity and composition of the parent vegetation in and surrounding lakes. Most pollen grains have a heavy protective layer (exine) and are resistant to decay if deposited in nonoxidizing sites, such as occur in the sediments of most standing waters. The taxonomy of pollen, based largely on the exine structure and sculpture, is relatively well known for vegetation of many parts of the world. Since pollen is relatively abundant in sediments, small aliquots of sediment are sufficient for analysis. As a result, analyses can be performed at very close intervals within the stratigraphy of sedimentary deposits. The pollen assemblage in sediments of known age, which can be determined by various dating techniques, provides an index of vegetation in and surrounding the lake in the past and also reflects changes that have occurred in the vegetation and drainage basin through time.

A number of detailed works are available on the principles of pollen analysis, including pollen identification and interpretation of pollen data. A good summary of these analyses is given in Birks and Birks (1980). Although many earlier analyses depicted the stratigraphy of pollen as relative percentages of the different pollen types, much greater interpretative power is gained from determinations of the absolute pollen frequency (number of pollen grains cm^{-3}). Then, with determinations of the sediment accumulation rates obtained from dating measurements, the absolute pollen accumulation rate (grains cm^{-3} yr^{-1}) can be determined for the different types or species.

Absolute pollen accumulation rates (pollen influx), unlike percentages, provide independent rates for each species in proportion to its densities and its past population dynamics. Careful interpretation is required, however, because pollen influx into lakes and their sediments can vary with factors not associated with the population and community dynamics of the surrounding vegetation. For example, some lakes are more efficient than others in collecting pollen because of differences in their morphology, their position within the landscape, and the quantities of pollen influx received from inflowing streams versus that received from the atmosphere (Davis *et al.,* 1973, 1984; Pennington, 1973, 1979; Bonny, 1976).

Within a lake, sediments and pollen can be resuspended and moved from shallow to deep parts of the basin (Lundqvist, 1927; Davis, 1968, 1973). This process of sediment focusing often can result in a greater net accumulation of sediment in the deeper parts of the lake basin (e.g., Lehman, 1975; Likens and Davis, 1975; Davis and Ford, 1982; Davis *et al.,* 1984). Additionally, the rate of sediment focusing can differ in different areas of a lake basin, and analyses of several sediment cores are required to resolve spatial and temporal patterns of deposition. The intensity of sediment focusing can change over time. For example, focusing can decrease as the basin fills. This reduction may be erroneously interpreted as a decrease in apparent rates of accumulation per unit area (when measured at a single point), because more recent deposits are being spread more thinly over a greater area than was the case previously.

The stratigraphy of pollen and spores provides a chronology of the vegetation from which presumptive evidence for climatic changes can be obtained. These vegetational changes of the landscape and catchment area provide insight into temperature and rainfall changes, soil development, and changes in the drainage basin caused by human activities. Rapid changes in vegetational succession, such as caused by catastrophic forest pathogen outbreaks in the past or recently by human introduction of exotic pathogens, is readily seen in the pollen record in laminated lake sediments (e.g., Allison *et al.,* 1986). Knowledge of the detailed vegetational history of the drainage basin is indispensable for the interpretation of changes in the developmental stratigraphy of the lake through its sediments.

Nearly all detailed paleolimnological analyses augment their interpretative power by means of pollen diagrams. Hence, the number of studies on pollen stratigraphy is very large. A few examples will demonstrate their usefulness.

In a classical study of the now meromictic Längsee, Austria, Frey (1955) demonstrated that permanent meromixis was well established and intensified at least 2000 years ago when humans moved into the area and began clearing forest vegetation from the drainage basin for agriculture. Pollen stratigraphy showed a marked decrease in the dominant forest species and a sharp increase in pollen of smaller species, two cultivated plants, and a number of agricultural weeds. Evidence also was found for markedly increased erosion of fine clay particles that probably were sufficient to increase further the density of lower water strata and subsequently maintain meromixis biogenically (cf. biogenic meromixis, Chap. 6). Subsequent detailed analyses of the complete chemical and fossil stratigraphy, with the loss of most benthic species, indicated that meromictic conditions began about 15,000 BP with a combination of postglacial climate changes, poor inflow system, and high carbonate content (Harmsworth, 1984; Löffler, 1997).

Numerous other examples exist in which the pollen stratigraphy reflects how sensitively lake metabolism responds to changes in the drainage basin. The response of the Italian Lago di Monterosi to construction of a Roman road, discussed earlier, also can be seen in the abrupt changes in pollen deposition from surrounding terrestrial vegetation. More recent changes in lake drainage basins resulting from agricultural clearing of land are seen in many parts of North America. As the land was denuded of trees, erosion and leaching of ions was increased markedly, usually resulting in accelerated nutrient loading and productivity. This sedimentary horizon is demonstrated by many chemical and biotic changes, the most conspicuous of which is the sudden rise in pollen of the agricultural weed *Ambrosia,* or common ragweed. The importance of the erosion rate in the drainage basin also was shown to be of major importance in the stratigraphy of lake sediments in the English Lake District (Tutin, 1969; Pennington, 1981). Pollen spectra throughout late and postglacial times correspond closely with the chemical history of the drainage basin. During the late glacial and early postglacial period, the erosion rates were controlled by low temperatures. In the midpostglacial period of deciduous forest, erosion was at a minimum. Within the last 5000 years, variation in or destruction of the deciduous forests surrounding many lakes in the English Lake District is evident from pollen stratigraphy and increased rates of erosional deposition.

C. Algal Remains

Several types of algae are well preserved in lake sediments. By far the most common algal microfossils are the siliceous frustules of diatoms (Round, 1964b; Juse, 1966; Korde, 1966; Bradbury, 1975). Diatoms are useful paleoecological indicators because their remains commonly occur in abundance, are often well preserved (see following discussion), and most can be identified to species from frustule morphology. Additionally, the physiological and ecological characteristics of many diatoms are distinct and well known, which permits reconstruction of probable past lake conditions from past diatom communities.

Diatom profiles in sediments are complicated by differential rates of dissolution of the frustules during and after sedimentation, which raises questions about how rigorously the remains found in sediments reflect the quantitative composition of producing populations. The few studies directed toward this question are conflicting, but it is apparent that dissolution rates vary among individual lake systems in relation to their chemical and morphometrically related history (cf. Chap. 14). For example, it has been found that frustules sedimented in the littoral and shallower profundal zones dissolved at higher rates than those sedimenting in the deep profundal zone (Tessenow, 1966). Under mildly acidic conditions, as in the hypolimnia of some meromictic lakes, dissolution of diatom frustules can be higher than in less acidic sediments (e.g., Meriläinen, 1969, 1971). Diffusive dissolved silica fluxes from sediments to the overlying water were proportional to the concentrations of biogenic silica in the sediments and increased with biogenic concentrations to ca. 100 mg g^{-1} of sediments, above which fluxes were limited by diffusive transfer rates (Conley and Schelske, 1989).

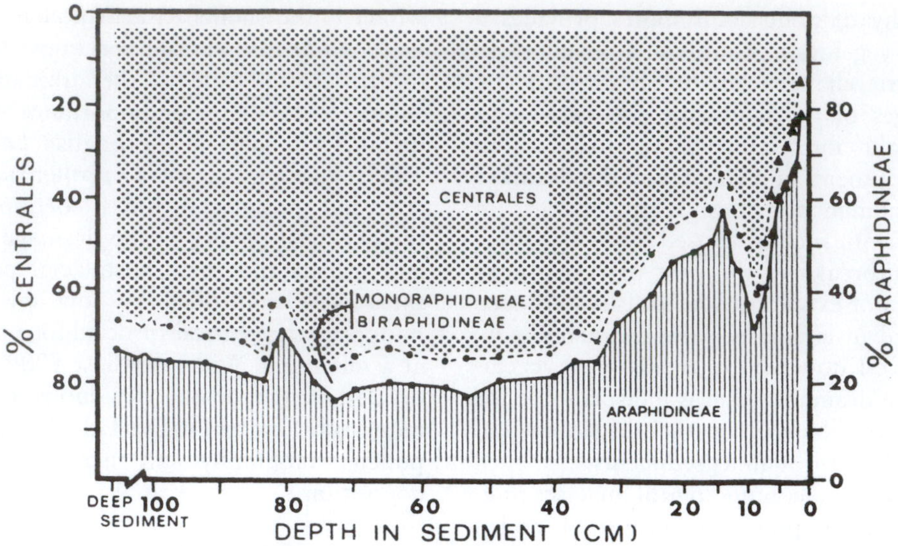

FIGURE 24-2 Changes in the percentage composition of the centric and araphid pennate diatoms in the surficial sediments of Lake Washington. The changes around 80 cm are associated with a clay turbidite layer peculiar to this core but not generally present throughout the lake. (Redrawn and modified from Stockner and Benson, 1967.)

Diatoms have been used extensively to assess environmental changes. Because much is known about how many diatom species respond quickly to changes in habitat characteristics, particularly their chemical environment, the diverse diatom communities contain much ecological information (e.g., Dixit *et al.*, 1992; Smol *et al.*, 1998; Stoermer and Smol, 1999). The identification of changes in species composition of diatom communities preserved in sediments can be used to infer long-trend ecological changes resulting from lake eutrophication, lake acidification, and climatic change.

In very general terms, centric diatoms are associated with more oligotrophic waters, while planktonic pennate diatoms tend to be more characteristic of eutrophic waters. The ratio of these two indicator groups of diatoms in the stratigraphic record can offer some help in interpreting the chronology of positive or negative changes in productivity. Around Lake Washington in Seattle, early development of the human population extended over the period from 1850 to the early 1900s (Edmondson, 1991a). From about 1910 to the mid-1920s, raw sewage entered the lake in increasing amounts. Between 1926 and 1941, much of the primary effluent was diverted away from the lake. In 1941, with progressively increasing regional human population expansion, large amounts of secondarily treated sewage entered the lake. These effluents of treated sewage continued to enter the lake until the mid-1960s, when all sewage was diverted to Puget Sound. Examination of the diatoms preserved in the recent sediments indicated detectable responses to the known sequence of enrichment, reduction, and renewed enrichment (Stockner and Benson, 1967). Major changes in the diatom community occurred (Figs. 24-2 and 24-3), related largely to reciprocal fluctuations in the abundance of the centric *Aulacoseira* (formerly *Melosira*) *italica* and the pennate *Fragilaria crotonensis*. The relative composition of the diatoms deposited, as well as species diversity redundancy values, was constant for the period prior to enrichment. Algal dom-

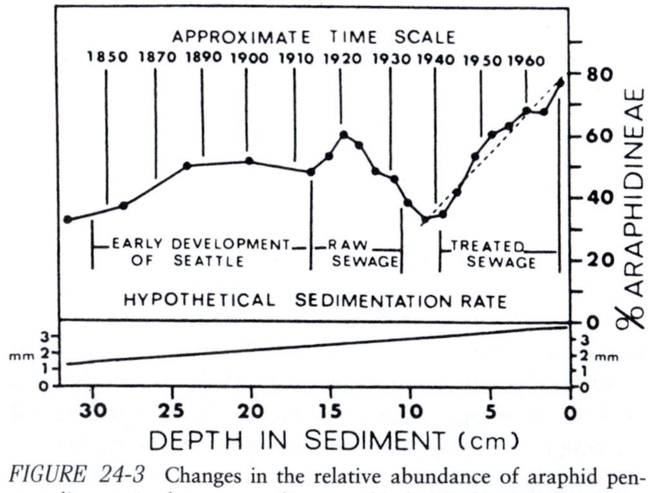

FIGURE 24-3 Changes in the relative abundance of araphid pennate diatoms in the recent sediments of Lake Washington in relation to the developmental history of the region surrounding the lake. The time scale is derived from an estimated average sedimentation rate of 2.5 mm per year over the upper 30 cm. (Redrawn and modified from Stockner and Benson, 1967.)

inance could be correlated with the pattern of sewage discharge into the lake. Many of the species changed in proportion in accordance with their generally known nutritional behavior as indicator species. It was concluded that the use of indicator groups (in this case, the ratio of centric to araphid pennate diatoms) was more reliable than the use of single indicator species.

The increased ratio of Araphidineae (e.g., *Fragilaria, Asterionella, and Synedra*) to centric diatoms (A/C) has been successfully applied to a number of other recent sediments as an indicator of eutrophication of lakes (e.g., Stockner, 1971, 1972, 1975; Bailey and Davis, 1978; Brugam, 1978; Culver *et al.,* 1981). It is clear, however, that high A/C indices of eutrophic conditions are not universal and are characteristic only in certain lakes of moderate total phosphorus concentrations and low alkalinities (Hall and Smol, 1999). High ratios of Araphidineae to centric diatoms are associated with low-alkalinity eutrophic lakes (Brugam and Patterson, 1983). In other lakes, even though araphidinate diatoms such as *Fragilaria* commonly increase in response to human settlement and increased nutrient loading, certain centric diatoms [particularly *Stephanodiscus* and *Aulacoseira* (*Melosira*)] increased even more vigorously (Bradbury and Megard, 1972; Haworth, 1972; Bradbury, 1975; Bradbury and Winter, 1976; Battarbee, 1978; Koivo and Ritchie, 1978; Brugam, 1979). Eutrophic lakes dominated by the centric diatom *Stephanodiscus* are nearly always of high alkalinity (>1.5 meq liter^{-1}). Although the changes in species composition of diatom communities can provide some insights into changing trophic conditions, correlations of diatom abundances with phosphorus concentrations are poor, and as a result past diatom-inferred total phosphorus predictions are poor (e.g., Fritz *et al.,* 1993; Hall *et al.,* 1997). Caution is clearly warranted in any use of simple A/C ratios or of indicatory species as an index of trophic conditions.

Experimentally determined tolerances of diatoms to ranges of alkalinity and other chemical parameters have permitted reconstructions of past chemical conditions in lakes based on past diatom communities. For example, in certain shallow African lakes, wide fluctuations in total ion content, lake morphometry, and paleoclimate have been inferred from stratigraphic changes in diatom communities (Richardson *et al.,* 1978). A significant linear relationship was found between the biogenic silica concentration and diatom biovolume abundance in sediment cores, from which estimates of past changes in diatom abundance could be inferred (Conley, 1988). In several of the Laurentian Great Lakes, sedimentary stratigraphy of diatoms, silica, and phosphorus concentrations have provided evidence for water column reduction in silica as a response of increased diatom production to phosphorus enrichment and decreased diatom production resulting from silica reduction (Schelske *et al.,* 1986; Schelske, 1988, 1999). Rate constants of pelagic cycling of phosphorus are higher than those for silica recycling. As a consequence, relatively small levels of phosphorus enrichment can increase diatom production and sedimentation that eventually causes silica reduction and Si-limited diatom production in the water mass.

The predominant environmental habitats of many diatoms are well understood. Therefore, stratigraphic separation of diatom communities into littoral (epiphytic or benthic) versus planktonic-dwelling species can yield further insight into past conditions of lake morphometry, water-level fluctuations, and shifts between littoral versus planktonic dominance of primary productivity (e.g., Round, 1961c; Evans, 1970; Manny *et al.,* 1978; Wolin and Duthie, 1999).

Sedimentary diatoms have been used to evaluate rates of lake acidification, associated changes in water chemistry, and recovery. To reconstruct past estimates of lakewater pH, present-day pH optima of surface-sediment diatom communities are applied to down-core fossil diatom assemblages preserved in dated lake sediments (Battarbee *et al.,* 1999). These separations are based on known correlations of diatom species with different conditions of pH, acid-neutralizing capacities, monomeric aluminum concentrations, and dissolved organic carbon concentrations, in addition to the inferences about diatom assemblages in terms of nutrient (P and N) concentrations discussed earlier. For example, in analyses of 37 Adirondack lakes of New York, diatom stratigraphies indicated that many lakes have acidified considerably, some by over two pH units, since the 1850s (Kingston and Birks, 1990; Smol *et al.,* 1998). Many other examples exist of the reconstruction of historical lake pH from alkaliphilous to acidophilous sedimentary diatom remains (e.g., Del Prete and Schofield, 1981; Davis and Anderson, 1985; Birks *et al.,* 1990). Highly acidified lakes often have chronically toxic concentrations of solubilized monomeric aluminum. Changes in dissolved organic carbon from acidification were small and indicated no definite pattern. Acidification had occurred primarily because of atmospheric deposition of mineral acid pollution in liquid (rain or snow) and particulate (dust) precipitation (Battarbee and Charles, 1994). Weighted-averaging models have also been developed to infer lakewater total phosphorus in the low range (<25 μg liter^{-1}) from diatom assemblages preserved in sediments (e.g., Hall and Smol, 1992). Using a number of assumptions, estimates of the onset, rate, and magnitude of lake eutrophication can be inferred in response to natural processes and human disturbances.

Other algae or parts of algae are preserved less completely than diatoms. Certain green algae (e.g., *Pediastrum*), scales and cysts of chrysophytes, cysts of dinoflagellates, and heterocysts of cyanobacteria are preserved in lake sediments (e.g., Leventhal, 1970; Brugam, 1978; Crisman, 1978a; Elner and Happey-Wood, 1978; Munch, 1980; Smol, 1980). The stratigraphy of these algae, and of diatoms as well, can be useful in interpreting past conditions when coupled with changes in other parameters recorded in the sedimentary record. However, less is known of the physiology and ecology of these other algae than is known of many diatoms, and as a result interpretation is often less rigorous than it is with changes in diatom stratigraphy.

The chrysophycean algae produce resting stages in the form of endogenously formed silicious cysts (statospores). The cysts are spherical or oval and often the walls are ornamented, which, along with other features, allows fairly specific taxonomic identification (Cronberg, 1986; Smol, 1988). The cells of chrysophytes of the family Mallomonadaceae are covered with loosely attached silica scales, some of which are also of taxonomic value. Upon death, scales and cysts are often deposited on sediments. Chrysophytes are primarily planktonic and have been used as indicators of water quality. Some six different assemblages of chrysophyte species, based largely on cysts, can be differentiated on the basis of water quality characteristics of pH, conductivity, and chemical constituents (Rybak, 1986; Siver, 1993; Duff *et al.*, 1997).

Mallomonadaceae species, identified from scales, have been used widely as indicators from which changes in lake development could be interpreted (e.g., Battarbee *et al.*, 1980; Munch, 1980; Smol, 1980). Changes in species composition of *Synura* and *Mallomonas* were seen in response to alterations of water quality, particularly changes in the drainage basin (e.g., road construction) and eutrophication. Differences in chrysophycean species were also shown in response to onset of meromixis (Smol *et al.*, 1983) and acidification (Smol *et al.*, 1984; Marchetto and Lami, 1994; Duff *et al.*, 1997) and recovery from acidification (Dixit *et al.*, 1989). Because chrysophytes tend to be associated with relatively oligotrophic lakes, declines are commonly associated with eutrophic phases of the development of a lake. As a result, a ratio of fossil eutrophic diatom frustules to chrysophycean cysts has been suggested as a potentially useful index of trophic status in temperate lakes (Smol, 1985).

D. Plant Macrofossils

Remnants of large plants are frequently preserved in lake sediments. Plant macrofossils usually consist of fruits, seeds, megaspores, and fragments of leaves, rhizomes, flowers, and woody tissue (see reviews of Dilcher, 1974; Birks, 1980; Birks and Birks, 1980; Collinson, 1988; Hannon and Gaillard, 1997). Seeds and other reproductive parts are often well preserved and identifiable to species. Such macrofossils supplement the pollen record. Many pollen grains cannot be identified to the species level, and some pollen types are poorly preserved. In addition, some common plants, particularly larger aquatic plants, produce so little pollen that they are not routinely represented in pollen counts. Because of their larger size, plant macrofossils are usually deposited close to their sites of origin; in contrast, pollen is easily and widely dispersed. The limited transport of plant macrofossils can result in heterogeneous distributions within sediments in different areas of a lake. Moreover, because the quantity of plant macrofossils per volume of sediment is usually very low, one must examine large quantities of sediment in order to acquire significant fossils.

Plant macrofossils preserve well in anaerobic sediments, although little is known of long-term diagenesis of vascular plant organic matter in lake sediments. Analyses of biogeochemical differences in contemporary tissues of white spruce (*Picea glauca*) with those preserved 10,000 years in lake sediments showed small changes (Meyers *et al.*, 1995). Hydrocarbon and total alcohol concentrations and distributions showed little diagenetic change. Cellulose components decreased relative to lignin, but C:N ratios were not significantly different.

Quantitative analyses of the stratigraphy of plant macrofossils assist in the reconstruction of past lake development and changes in primary productivity (see the Wintergreen Lake example discussed earlier). Changes in lake levels, water chemistry, and regional climate have been evaluated, in part, from plant remains and inferences from known physiological and ecological characteristics and tolerance ranges of the plants (e.g., Watts and Winter, 1966; Birks, 1980; Dearing, 1997; Hannon and Gaillard, 1997). The strongest lines of sedimentary evidence for lake-level change are the advance and retreat of littoral vegetation deduced from analyses of macrophyte fossils and coarse organic matter. Macrofossils can also provide insight into the rates and types of plant succession that have occurred in and surrounding lakes.

E. Animal Remains

Nearly all groups of animals leave at least some identifiable morphological remains in lake sediments. At the microscopic level, the most abundant animal remains in sediments are those of ostracods, Cladocera, and midges, and under favorable circumstances, shells or cases of testaceous rhizopods, spicules of sponges, egg cocoons of neorhabdocoele Turbellaria, resting eggs of rotifers, bryozoan statoblasts, copepod spermatophores, and oribatid mites can also be found (Frey, 1964, 1969, 1976, 1988; Crisman, 1978b; Walker, 1993). At the macroscopic level, mollusk and beetle remains dominate. The remains of larger organisms (e.g., fish scales and bones) are generally less abundant and require larger volumes of sediments for the recovery of significant numbers. Therefore, their stratigraphy must be based on coarser intervals than that of microscopic fossils.

Numbers of species or groups are often expressed in relative percentages of the total assemblage or, preferably, are related to the weight of organic matter or some other volumetric mass parameter of sediment. Varying rates of sedimentation can be determined from radiocarbon dating and certain other stratigraphic criteria that permit expression as an absolute rate of accumulation, which can result in quite different interpretations than those based on percentage composition (cf. Davis and Deevey, 1964).

1. Midges

The nonbiting midges, or Chironomidae, are of particular interest in microfossil analyses of lake sediments because they provide insights into the ecological conditions of earlier periods of lake development (Walker, 1987; Hofmann, 1988). Some species of chironomids (*Chironomus*) are commonly distributed in the profundal zone of eutrophic lakes, which often experience low dissolved-oxygen concentrations at the end of the stratified period. Others (e.g., *Tanytarsus lugens*) are restricted to oligotrophic lakes, which commonly have high dissolved oxygen in the hypolimnion at the end of the stratified period. Therefore, if a lake becomes more eutrophic with decreasing oxygen content of the hypolimnion, one might expect, for example, a succession of profundal midge populations from those that require high oxygen to those that tolerate decreasing oxygen concentrations. It is important to note, however, that the profundal midges respond to the concentrations of dissolved oxygen, which may result from a combination of decreased volume of the hypolimnion through accumulation of sediments as well as increases in nutrient-controlled productivity.

In detailed studies of the stratigraphy of chironomid and *Chaoborus* midge faunas of several lakes in northern Indiana, Stahl (1959) demonstrated such a succession in midge populations and offered strong evidence for the transition from aerobic to anoxic hypolimnetic conditions among the more productive lakes. Additional evidence has been found for historical changes in eutrophication of lakes from marked changes in benthic invertebrate remains in sediments (e.g., Stahl, 1969; Carter, 1977; Hofman, 1978b; Wiederholm and Eriksson, 1979).

Changes in chironomid community stratigraphy have been used as an additional indicator of the effects of acidification of softwater lakes. For example, marked species alterations occurred as an oligotrophic lake shifted to a *Sphagnum*-dominated acidic, ombrotrophic lake (Walker and Paterson, 1983). Similar marked changes in profundal chironomid communities occur in lakes that have been acidified from acidic precipitation (Henrikson *et al.*, 1982; Uutala, 1990).

2. Cladocerans

Exoskeletons of cladoceran zooplankton preserve reasonably well in lake sediments. This group has been studied in the fossil record in great detail. Analyses of past populations also have been made in which information theory at the community level has been applied, thereby refining interpretations of responses to changing ecological conditions.

Species of chydorid Cladocera are largely littoral and occur on a variety of substrata. Considerable evidence exists, however, that the littoral faunal remains are redistributed by currents within lake basins, where they become integrated with remains of planktonic Cladocera (Mueller, 1964; DeCosta, 1968; Goulden, 1969a; Frey, 1988). The populations recorded in profundal sediments, therefore, offer a reasonable integration of community dynamics over habitats and seasons. Changes with time or among lakes can then reflect changes in the communities in response to such varying biotic conditions as shifts in food quantity and quality, competition, or predators (Frey, 1976, 1988; Hofmann, 1978a,b, 1986; Crisman and Whitehead, 1978).

There are many examples of the usefulness of cladoceran remains in interpreting past conditions (Frey, 1969, 1974; Birks and Birks, 1980; Hofmann, 1986). Changes in species diversity and equitability of chydorid communities, as pioneered by Goulden (1969b), have been used to assess the stability and responses of the communities to disturbances. Equilibria of the population associations have been shown to be disrupted by the establishment of competing species or by agricultural disturbances, climatic alterations, and shifts in the ratio of lake productivity contributed by littoral and phytoplanktonic sources (Whiteside, 1969). The ratio between planktonic and littoral cladoceran fossil remains also is useful in interpreting past changes in water level and the extent of littoral development. For example, in many glacial lakes of Europe and

North America, *Bosmina* is often the most abundant zooplankton group. *B. longispina* is often found early in the lake development and is a species unable to survive at high levels of phytoplankton productivity and conditions associated with that productivity. That *Bosmina* species is often replaced later in the development of the lake by *B. longirostris,* a species common to more eutrophic conditions (Boucherle and Züllig, 1983; Hofmann, 1986).

Changes in morphology of fossil cladocerans have been associated with known induction of such changes by invertebrate predators (Brugam and Speziale, 1983). Changes in piscivore fish that affect planktivore predatory efficiency on large cladocerans, such as *Daphnia,* appear in the fossil records with the increased abundance of *Bosmina* and decreased *Daphnia.* If fish community structure is altered by natural processes (e.g., winterkill) or pollution (e.g., lake acidification), the events can be reconstructed from the cladoceran fossil records (Nilssen and Sandøy, 1990). Fossil zooplankton are likely more sensitive indicators of predation than of lake trophic status.

3. Ostracods

Paleolimnological interpretation of fossil ostracod communities is poor because of limited modern ecological understanding at the species level. Although correlations have been made among ostracod distributions and limnological parameters (Delorme, 1971; Binford, 1982; Carbonel *et al.,* 1988), these data allow only limited estimates of past chemical conditions based on fossil ostracod stratigraphy. Ostracods are good indicators of differences in salinity and certain alkaline earth elements.

4. Molluscs

Fossil molluscs have been studied extensively and allow selective interpretations of past changes in water level, salinity, and trophic conditions (Frey, 1964). However, freshwater molluscs are relatively poor indicators of past paleoclimatic fluctuations and lake conditions because of their generally broad environmental tolerances (Ouellet, 1975). In some cases, however, temporal changes in the molluscan fauna provide useful interpretations of postglacial conditions of lakewater chemistry and water-level fluctuations (e.g., Covich, 1976; Ohlhorst *et al.,* 1977).

F. Pigments

Fossil pigments of biota have been investigated extensively as organic constituents of sedimentary stratigraphy (cf. Vallentyne, 1957b, 1960; Gorham, 1960; Brown, 1969; Sanger, 1988). Upon plant senescence and death, the photosynthetic pigments undergo molecular transformations in which ions (such as the coordinately bound magnesium atom of chlorophylls) and side groups (long-chain terpene alcohol, phytol) are lost progressively during physical or biological degradation. There is a tendency for the degradation products to increase in stability—that is, decrease in solubility and increase in relative resistance to further microbial and physical decomposition. Chlorophyll *a,* for example, degrades to pheophytin *a* with the loss of magnesium; pheophytin *a* then degrades to pheophorbide *a* with the loss of the phytyl group. The alternative sequence in which the phytyl group is lost first leads to the formation of the intermediate product, chlorophyllide *a.*

The number of pigments found in sediments is large. Nondegraded chlorophylls are rare, but pheophytins, chlorophyllides, pheophorbides, and bacteriochlorophyll degradation products have been identified and quantified. Additionally, a number of carotenoids have been found, some of which are highly specific to groups of organisms such as the cyanobacteria or certain families of organisms. Carotenoids commonly present in sediments are separated into (a) epiphasic (hydrocarbon carotenes), soluble in nonaqueous organic solvents (e.g., β-carotene); and (b) hypophasic, many oxygenated xanthophylls in hydroxy, keto, epoxy, or methoxy groups (e.g., lutein, violaxanthin, and zeaxanthin). As discussed earlier (Chap. 15), there is a general positive correlation between long-term growth and biomass of photosynthetic microbes and their community concentrations of pigments. Total carotenoid concentrations were found to be the most strongly correlated to other indices of lake trophic status (Sanger, 1988; Hilton *et al.,* 1991). If pigments are reasonably preserved by the time they are interred in sediments, the stratigraphic distribution of pigments would offer the possibility of identification and estimation of past plant populations as well as of environmental conditions of the water and sediment at the time of their deposition.

In the 1950s, Vallentyne and colleagues pioneered studies of fossil pigments in sediments of many different lake ecosystems. These data, when correlated with the stratigraphy of microfossils of organisms, allow interpretation of the ecological records of the various lakes. The potential information that pigments can provide, with critical interpretation, is exemplified in analyses of Bethany Bog, Connecticut (Vallentyne, 1956). Myxoxanthin, a carotenoid specific to the cyanobacteria, was present only in the sediments of the eutrophic phase of the lake and was not found in the more recent sediments of the lake as it entered its bog phase.

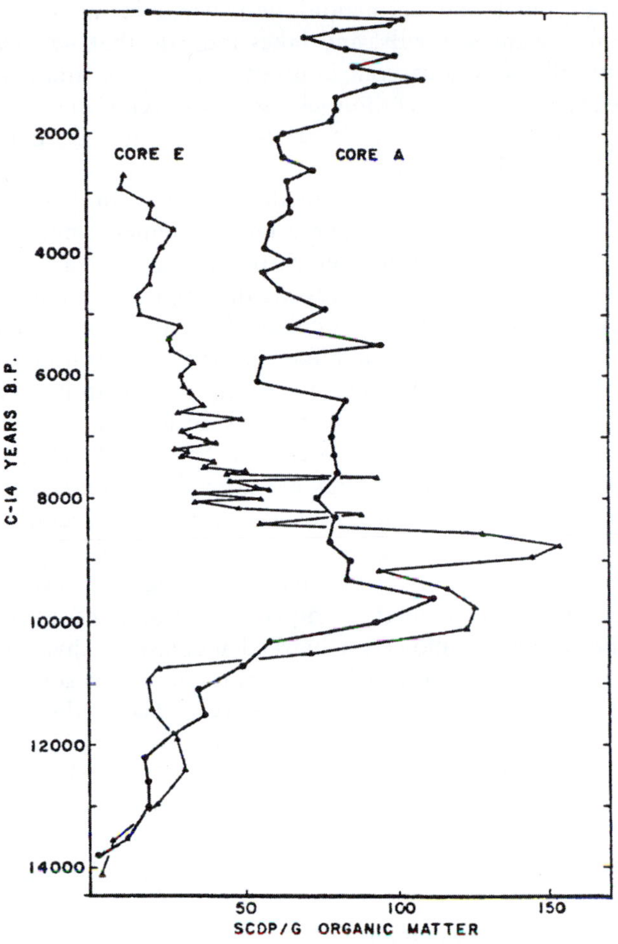

FIGURE 24-4 Sedimentary chlorophyll degradation products (SCDP) per gram organic matter in two cores from Pretty Lake, northeastern Indiana. Core A was from the deepwater central depression of the lake at a water depth of 25 m; Core E was from a shallow marl lakemount near shore that ceased accretion of sediment 2740 years BP because of water movements at a water depth of about 1 m below the surface. (From Wetzel, 1970, as modified by Frey, D. G.: Paleolimnology. *Mitteilungen Internat. Verein. Limnol.* *20*:95–123, 1974.)

A number of analyses on several different types of lakes have indicated general correlations between sedimentary pigments and productivity. Most studies show maximum concentrations of pigments in early postglacial times after a period of tundra conditions associated with retreating glaciation (Fig. 24-4). The maximum concentrations are generally attributed to phytoplankton productivity stimulated by early leaching of nutrients from the drainage basin during the ameliorated thermal period following deglaciation, which is characterized by transitions from tundra to coniferous and then deciduous vegetation. This initial surge in pigment degradation products usually is followed by an intermediate period of relative stability or slowly declining rates of pigment deposition, with minor fluctuations associated with variations in climate, rainfall, and nutrient conditions. Subsequent pigment concentrations, assumed by inference to emanate from phytoplanktonic productivity, are variable but commonly have been lower until very recent times. This reduction has been associated with alterations within the drainage basin that increase the ratio of allochthonously derived organic matter (including pigment degradation products) to autochthonous organic matter sedimenting from production within the lake (Fogg and Belcher, 1961).

The rapid eutrophication of many lakes in recent times, often within the last century, generally is reflected in large increases in the pigment concentrations of recent sediments. This recent increase often coincides with renewed leaching of nutrients from the drainage basin as the land was deforested for agriculture. In highly calcareous drainage basins, deforestation can accelerate inputs of calcium and bicarbonate, which interact with numerous inorganic and organic factors to suppress productivity. The inverse relationship between $CaCO_3$ and pigment stratigraphy has been demonstrated in several hardwater lakes (Wetzel, 1970; Manny *et al.*, 1978; Wetzel and Manny, 1978). The abrupt decline in pigment concentrations and other measures of algal abundance and production within 200 years BP (Fig. 24-4, upper portion of Core A) coincided precisely with deforestation, massive increases in $CaCO_3$, and the initiation of *Ambrosia* (ragweed) pollen associated with agriculture.

As discussed above, morphologically recognizable parts of a number of algae (diatoms, certain desmids and green algae, and cysts of some dinoflagellates and chrysophytes) are well preserved in freshwater sediments. Parts of other algae are preserved less well or not at all. Therefore, if pigments that are preserved are restricted to specific taxonomic groups that are characteristic of relatively distinct ecological conditions, their stratigraphy can be most helpful in interpretation of past conditions in the ontogeny of a lake.

Chlorophyllous pigments are cosmopolitan in distribution among plants. Certain algal and bacterial carotenoids and bacteriochlorophyllous degradation products of purple photosynthetic and green sulfur bacteria have been found in sediment stratigraphy and associated with events that led to eutrophication and meromixis (e.g., Brown, 1968; Czeczuga and Czerpak, 1968; Züllig, 1982, 1989; Guilizzoni *et al.*, 1986; Sabater and Haworth, 1995; Hodgson *et al.*, 1998). Myxoxanthin, myxoxanthophyll, oscillaxanthin, and other carotenoids have been used in this manner to infer the time of invasion and development of algae and cyanobacteria (Züllig, 1961, 1981, 1982, 1989; Brown and Coleman, 1963; Griffiths *et al.*, 1969; Griffiths, 1978). For example, quantitative measurements of myxoxanthophyll in sediment cores of many Swiss lakes provided a clear record of the first appearance of cyanobacteria, in the absence of morphological fossils, during the marked recent eutrophicational changes in some of these lakes (Züllig, 1961, 1982). The appearance of oscillaxanthin, which is specific to cyanobacteria of the family Oscillatoriaceae, in recent sediments of Lake Washington in Seattle was related directly to the development of dense populations of this microbial group as the lake became excessively eutrophic owing to enrichment with sewage (Griffiths *et al.*, 1969; Griffiths and Edmondson, 1975).

Interpretation of fossil pigments as a measure of qualitative and quantitative changes of former microbial populations must be done critically. Several assumptions are often made that are never completely valid and can vary with specific conditions of individual lake systems (cf. reviews of Moss, 1968; Brown, 1969; Wetzel, 1970; Swain, 1985; Hendry *et al.*, 1987; Sanger, 1988). A number of complex abiotic and biotic interactions alter the diagenesis of pigments before and after arriving to the sediments.

1. *Importation of Allochthonous and Littoral Pigments to Pelagic Sediments.* The extent to which pigment products of terrestrial and wetland origin dilute those formed within the lake is not always clear. Chlorophyllous pigments decompose readily to pheopigments in soils and in senescent leaves (Hoyt, 1966). The relative allochthonous input of inorganic and organic matter is largely a function of morphometry of the lake basin in relation to its drainage basin and the rate of erosion (cf. Mackereth, 1966). Although allochthonous and littoral inputs of *particulate* organic matter can be quite significant in small, shallow lakes, the distribution and decomposition rates of chlorophyll and carotenoid derivatives in woodland soils, swamps, ponds, and lakes indicate that *particulate* allochthonous contributions usually are small in relation to autochthonous sources (Gorham and Sanger, 1964, 1972, 1975, 1976; Sanger and Gorham, 1970, 1973).

Sanger and Gorham introduced a pigment diversity index, based on the enumeration of pigment spots on two-dimensional thin-layer chromatograms of acetone extracts, as a means of assessing the origins of sedimentary organic matter. Particulate detritus of terrestrial origin contains three dominant pigments—pheophytin *a*, β-carotene, and lutein—even though quantities of these pigments in leaves of various deciduous trees are very low at the time of abscission (Sanger, 1971). The carotenoids in aerobic soils are degraded more rapidly than pheophytins. Therefore, the terrestrial vegetation has a low pigment diversity, which decreases further on degradation in soils. Examinations of pigment diversities of upland vegetation, aquatic macrophytes, algae, and sediments showed a marked increase in diversity with the transition to lakes and autochthonous sources (Table 24-1). The pigment diversity in lake sediments was greater than that of other materials as a result of decomposition of the phytoplankton. Thus, the pigment diversity index can be viewed as primarily a measure of pigment diagenesis. The initially high pigment diversity of algae is further increased during decomposition (cf. Daley, 1973).

The ranges commonly observed were 7–8 for upland vegetation, 12–15 for aquatic macrophytes, 10–21 for algae, and 24–47 for lake sediments. The highest diversity was found under ideal conditions for preservation of pigments in sediments of a meromictic lake (Table 24-1). The results indicate that during oligotrophic conditions, much of the organic matter preserved in sediments is allochthonous in origin, while under eutrophic conditions pigments of autochthonous

TABLE 24-1 Pigment Maxima and Diversity Indices in Woodland Soil Detritus, Swamp Peats, and Sediments of a Series of Dimictic Lakes and a Meromictic Lake[a]

	Relative units per gram organic matter			
System	*Chlorophyll derivatives*	*Epiphasic (xanthophyll) carotenoids*	*Hypophasic (carotenes) carotenoids*	*Pigment diversity*
Woodland maximum	1.3	0.2	0.4	10
Swamp maximum	12.6	2.4	5.0	23
Dimictic lakes, minimum	1.1	0.4	0.8	24
Dimictic lakes, maximum	16.3	26.6	31.4	47
Meromictic lake, maximum	59.2	40.5	64.7	50

[a] Modified from Sanger and Gorham (1973) and Gorham and Sanger (1972).

origins can exceed those sedimented from external sources. Nonplanktonic inputs of pigments found in sediment traps, determined by the difference between the total trapped particulate pigments and that estimated in the phytoplankton, were found to be more than half of the total in midsummer in three northern Wisconsin lakes (Carpenter *et al.,* 1986). Sources of nonplanktonic inputs were unknown but believed to be largely from littoral periphyton and the resuspension and lateral transport of littoral sediments, where pigment concentrations can be very high (cf. Chap. 19 and Cyr, 1998).

2. *Proportionality between Planktonic Pigment Concentrations and Productivity.* Another assumption generally made is that the pigment concentrations of living algal populations represent a reasonable relative estimate of algal productivity, so that the sedimentary record reflects past levels of algal productivity. Although chlorophylls have been used widely as a measure of contemporary algal productivity (Chaps. 15 and 19), the pigment content of living populations is influenced by numerous environmental factors and can vary greatly under different conditions of growth. Over time, the succession of species and changes in nutrient status may result in changes in the ratio of pigment to organic matter in sedimenting materials. In a general and relative way, however, pigment concentrations do reflect contemporary microbial photosynthetic productivity. If nonplanktonic inputs of particulate pigments originate from littoral production, then that is part of the total lake productivity. However, it is probable that littoral inputs are highly variable under different environmental conditions affecting lateral displacement to the pelagic zone.

3. *Differential Degradation of Pigments.* Differential degradation of pigments during and after sedimentation is a more serious problem that is difficult to resolve. If pigment diagenesis is quite variable, it is difficult to distinguish between high productivity and rapid rates of degradation or low productivity and slow rates of diagenesis. Detailed experimental investigations of the destruction of chlorophyll and the formation of derivatives from lacustrine pigments demonstrated that several factors are involved (Daley, 1973; Daley and Brown, 1973). Photooxidative destruction of chlorophyll occurred in senescent phytoplanktonic cells and at an accelerated rate in cells lysed artificially, by bacteria or by a virus. Chlorophyll *a* was degraded faster than chlorophyll *b* (cf. Brown *et al.,* 1977). Enzyme-mediated chlorophyll degradation could not be detected. In prolonged darkness, reduced destruction was observed, but aphotic pigment degradation in the hypolimnion was still significant (Leavitt and Carpenter, 1990a). Decay of carotenoids ranged from 0 to an observed maximum of 8.7% day^{-1} and chlorophylls from 0 to 12% day^{-1}. Preservation varies greatly among different carotenoids, and preservation is relatively high for secondary carotenoids (e.g., β-carotene and diatoxanthin) (Hurley and Armstrong, 1991). From a comparison of available information, it was concluded that photooxidation is a principal cause of chlorophyll destruction *in situ*.

4. *Conditions for Preservation: Oxidation, Acidity, and Temperature.* The processes of destruction of pigments and derivative formation are unrelated. Pheophytins and pheophorbides accumulated only in lysed cells, and concentrations of these two derivatives increased in the presence of dilute acids and oxygen. Derivative formation is strictly chemical and governed by oxygen concentration and pH. The rate of derivative formation decreases precipitously at pH values above neutrality. Although most inland waters have a pH above neutrality, pH tends to decrease in the hypolimnion of stratified waters near the sediments where these materials are interred. As would be anticipated, preservation is increased and optimized in sediments of meromictic lakes that are anoxic, dark, cold, and not resuspended into the water column during mixing events.

The low temperatures found in hypolimnetic waters favor preservation of pigments. Thus one would anticipate that the best conditions for preservation occur in the anoxic, cold, and relatively dark monimolimnia of meromictic lakes (cf. Table 24-1). In more typical lakes, however, the effects of circulation should be kept in mind. Pigments deposited in organic matter of surficial sediments may be resuspended during circulation of the water column and thereby reexposed to oxygenated, lighted conditions. Temperatures are relatively low in hypolimnia during circulation periods in the temperate regions and during winter. Hence, the conditions for chemically mediated formation of pigment derivatives may be fairly similar over much of the history of the lake development. As the basin fills in and becomes too shallow to permit stratification or becomes meromictic, however, conditions for preservation and diagenesis can change markedly (Daley *et al.,* 1977).

5. *Ingestion by Zooplankton and Microherbivores.* Ingestion of algae by herbivores, such as *Daphnia* and the chrysophycean flagellate *Ochromonas,* resulted in the destruction of chlorophyll and the formation of pheophytins and pheophorbides (Daley, 1973; Daley and Brown, 1973; Carpenter *et al.,* 1986; Leavitt and Carpenter, 1990b). Pigment sedimentation increased significantly with the mean size of cladocerans and omnivorous copepods, presumably related to concentration of algal detrital particles in fecal pellets that sediment relatively rapidly into aphotic, cooler,

and potentially anoxic strata of the hypolimnion. Fecal pellets that are disrupted during sedimentation to a very small size (<30 μm) or released by microzooplankton and small (<20–35 μm) are likely to be degraded significantly by photooxidation (Soohoo and Keifer, 1982a,b).

Therefore, it must be recognized that use of concentrations of different pigment constituents as paleolimnological indices requires critical evaluation. The methods have merit in interpreting past lacustrine events but should complement morphological and other reconstructions (e.g., Manny *et al.*, 1978; Bradbury *et al.*, 1981; Schmidt *et al.*, 1998) rather than serve as a sole paleolimnological approach.

III. SEDIMENTARY RECORD AND LAKE ONTOGENY

As one examines the historical development of lake ecosystems, the chemical and biotic records within the sediments provide historical records of change. The cumulative organic matter retained in the sediments follows a general sigmoid increase, somewhat analogous to primary ecological succession on land (Fig. 24-5). The initial gradual eutrophication implies intensified recycling of nutrients as increasingly anoxic hypolimnetic conditions occur and are associated with increasing accumulation of organic matter. The following pseudoequilibrium in the trophic state is approximately set and maintained by the intensity of nutrient loading and internal recycling. Major disturbances, such as human-induced eutrophication, accelerate nutrient loading by disturbance of the drainage basins. Associated with that disturbance is a marked increase (1–5 orders of magnitude) in erosion and deposition of inorganic matter (Binford *et al.*, 1983). Accelerated nutrient and nonnutrient inputs into lakes tend to have opposing effects on the pelagic productivity. In the next Chapter 25, we will see how sedimentation and paludification, terminal organic production processes that overwhelm pelagic production, can result from a shift to predominant accumulation of organic sediments associated with higher plants that are more resistant to rapid decomposition. Assuming that diagenesis within anoxic organic sediments is relatively constant over millennial time periods, the rates of pelagic organic accumulation respond to disturbances. The ecosystem exhibits appreciable resilience to minor disturbances that are relatively unsustained, from which recovery can be relatively complete (Fig. 24-5). Major disturbances, however, such as those associated with many cases of human eutrophication, result in major accelerations in organic accumulations and reductions in pelagic biotic diversity.

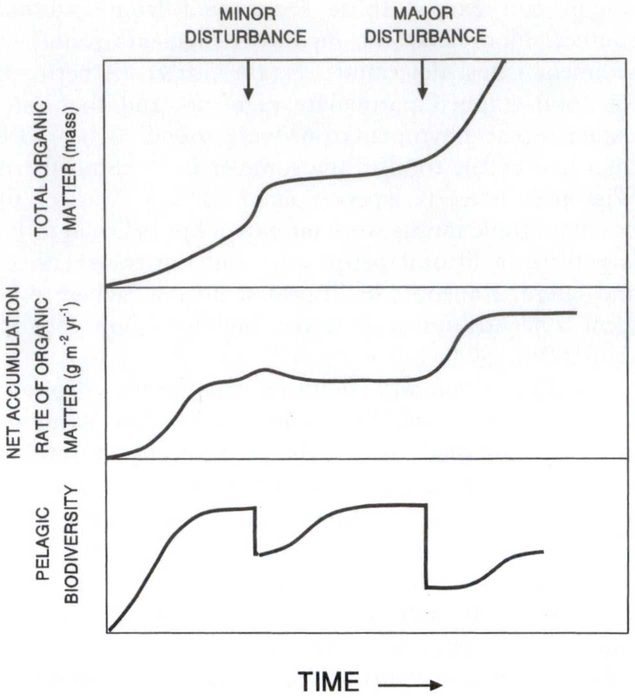

FIGURE 24-5 Trajectories inferred from sedimentary profiles for the cumulative total organic matter in lake sediments, net rate of accumulation of organic sediment, and biotic diversity within the pelagic zone. A major disturbance, as indicated by cultural eutrophication, that can result from major alteration of the drainage basin with accelerated inputs of nutrients is depicted as a continuation of earlier Holocene eutrophication. (Modified extensively from Binford *et al.*, 1983.)

IV. SUMMARY

1. Paleolimnology allows reconstruction of past lake communities, lake biogeochemistry, and lake development (ontogeny).
   a. Disturbance to a lake ecosystem often is derived from changes in the drainage basin. Materials within sediments include both remains of production from the drainage basin (erosional inputs, pollen, and other organic matter from vegetation), the atmosphere (metals, nutrients, organic compounds, and particles), and the biota within the lake.
   b. Changes in preserved lake biota (algae, zooplankton, certain benthic invertebrates, and macrophytes) can then be correlated with changes in external inputs and geochemical remnants.

c. Changes recorded in lake sediments prior to human interference and alterations to the drainage basins and compared to contemporary records allow interpretation of the responses of the lake communities to natural and human-induced environmental changes (e.g., eutrophication, acidification, and climate change).

2. As in all paleontological records, gaps occur and the record is incomplete. Variable inputs from the drainage basin, redistribution of sediments within the lake basin, and differential preservation of interred remains demand critical interpretation of the record. Nonetheless, the amount of information in the sediments is large. Accurate interpretation of the historical records depends, however, upon a good understanding of the biology and regulatory physicochemical processes of contemporary ecosystems.

3. Evaluations of past events in the historical development of a lake require accurate determinations of the rates of sedimentation over time. Several methods have been used to determine the age of sedimented materials.
   a. Radioactive carbon-14 is being continuously generated from atmospheric nitrogen by cosmic radiation. This ^{14}C is also in equilibrium with CO_2 in the water. During photosynthesis, a small amount of the radioactive ^{14}C is incorporated proportionally into new organic matter. When organic matter is interred in the sediments, the ^{14}C decays back to ^{14}N at a constant, known rate. Accurate radioassay of the amount of ^{14}C remaining in old organic matter thereby provides an estimate of its age. Numerous variables in the synthesis of ^{14}C and its incorporation into organic matter require careful time corrections in order to provide accurate dates. The ^{14}C dating technique permits age estimates over the past 40,000 years or somewhat longer.
   b. The age of recent sediments (to 150 years) can be estimated by the content of more short-lived isotopes (e.g., lead-210).
   c. As sediments are deposited in a lake, they retain remnant properties of magnetic declination and intensity in minerals such as hematite (F_2O_3). The remnant magnetism is caused by Earth's magnetic field, which oscillates about a mean direction with a constant frequency at a given location (approximately 2700 years).
   d. Some lake deposits are laid down in distinct laminae. When annual laminations are distinguishable, they are called *varves*. Under ideal depositional circumstances, as is found in certain meromictic lakes and in lakes where annually pulsed loading (e.g., glacially fed streams) occurs during thermal stratification, varves permit direct age determinations.

4. Paleolimnological analyses of many chemical and biological remains in sediments indicate that the old concept that most temperate lakes progress from oligotrophy to eutrophy is not universal, or is applicable only to very recent times.
   a. In glaciated regions, nutrient loading and productivity were commonly very low for long periods following the inception of lakes under the cold boreal conditions after glaciation. But the transition period from tundra to coniferous to deciduous forested conditions, with its ameliorated temperatures, is characterized by high lake productivity. Productivity often declined and stabilized as land surrounding lakes became impoverished of nutrients by leaching of the soils and as nutrients were retained by forested vegetation surrounding lakes.
   b. As the vegetation of the drainage basin was disturbed by human activities for agriculture and other purposes, nutrient loading commonly increased, resulting in enhanced productivity.
   c. In calcareous regions, drainage basin disturbances increased cation and carbonate loading to recipient lakes, which counteracted the effects of simultaneous loading of important nutrients (e.g., phosphorus) and thereby reduced productivity.

5. Sediments of lakes contain numerous organic compounds of biological origins produced either autochthonously or within the drainage basin and transported to the lake.
   a. Some compounds (e.g., lipid derivatives) are specifically characteristic not only of terrestrial plants but also of algae and bacteria. Analyses of the proportions of these compounds in sediments provide insight into the differences in the proportion of allochthonous versus autochthonous sources of the organic matter found in sediments.
   b. Degradation products of plant pigments are relatively stable after the death of the organisms that synthesized these compounds. Certain pigment degradation products preserve well in sediments, and their composition and remnant concentrations in sediments have been used to interpret past changes in algal and bacterial communities and productivity.
      i. Certain carotenoids are synthesized only by specific algal or bacterial groups. The stratigraphy and quantities of these compounds have permitted inference about the time of

invasion and development of certain algae, cyanobacteria, and other photosynthetic bacteria that do not leave morphologically recognizable fossils in the sediments.

ii. In conjunction with other chemical and biological characteristics, the quantitative stratigraphy of chlorophyllous and carotenoid degradation products in sediments has been used in a number of paleolimnological studies to corroborate past general productivity. Critical chromatographic discrimination of the many pigment degradation products is necessary for the accurate evaluation of those compounds that originated from autochthonous primary productivity. In many large lakes, evidence indicates that much of the remnant pigment products originated from autochthonous primary producers (pelagic and littoral).

6. Morphological remains of many plants and animals are fossilized in lake sediments. Because most lakes are young (<20,000 years since inception), if the contemporary physiology and ecology of the organisms are known, inferences can be made about past ecological conditions.
   a. Plant pollen and spores are commonly deposited into lakes in proportion to changes in the composition and quantity of parent plants in or surrounding lakes. Most pollen grains preserve well once they have settled into anoxic reducing sediments, and they can be identified, sometimes to species. Absolute pollen accumulation rates provide independent measures of each species in proportion to its density and past population dynamics.
      i. The stratigraphy of pollen and spores provides presumptive evidence of the chronology for vegetational changes in the drainage basin related to changes in temperature, rainfall, soil development, and disturbances caused by the activities of humans.
      ii. Pollen stratigraphy is used in all comprehensive studies to augment other paleolimnological analyses. Changes in the plant communities of the drainage basin markedly influence the nutrient loading characteristics and are reflected in subsequent chemical and biological indicators of lake productivity.
   b. Although the siliceous frustules of diatoms are the most common algal microfossils preserved in lake sediments, some other algae or algal parts (e.g., cysts or scales of chrysophytes or dinoflagellates) are also found. Because diatoms generally preserve well, can be identified to species from the frustules, and much is known of their physiological ecology, their stratigraphy has been studied extensively in conjunction with other paleolimnological parameters.
      i. Changes in diatom communities through the history of lake development have been related particularly to probable past changes in nutrient chemistry and salinity, and by inference to alterations in lake level and climate.
      ii. Changes in ratios of littoral-dwelling to planktonic diatoms have also been used to evaluate past changes in water levels and littoral development.
   c. Plant macrofossils (fruits, seeds, megaspores, and tissue fragments) are frequently found in sediments but are less abundant than are microfossils. Plant macrofossils assist in reconstruction of past changes in plant succession, water levels, water chemistry, and regional climate from known physiological and ecological characteristics of the plants.
   d. Nearly all animals of inland waters leave some identifiable remains fossilized in lake sediments. The most widely studied and abundant fossils are remains of Cladocera, ostracods, and midges, many of which can be identified to species from exoskeleton and mouth-part remnants.
      i. Study of past midge community structure has proven particularly useful in the reconstruction of the historical transitions of increased productivity and resulting depletion of oxygen from the hypolimnia of lakes in more productive stages.
      ii. Based upon known population dynamics and factors controlling community composition and species diversity, past cladoceran communities provide presumptive evidence for past environmental conditions, including food quantity and quality, competition, and selective predation. Ratios of past littoral versus planktonic cladocera can be useful in interpreting previous changes in lakewater levels and littoral development.

25

THE ONTOGENY OF INLAND AQUATIC ECOSYSTEMS

The successional development or ontogeny of inland aquatic ecosystems is governed by many interacting causal mechanisms that affect the rates of autotrophic productivity and utilization of that organic matter. Nearly all of the preceding discussion of this book has been directed to these regulating mechanisms and their integration. As we have seen, a number of complex factors influence productivity at all levels within inland waters. Those mechanisms control from the base by limitations of resources needed for production and from predation that limits growth and reproductive capacities.

Among the multitude of dynamic factors that regulate the productivity of individual ecosystems at any given time, a fundamental property prevails: The ecosystems are constantly changing, succeeding, and evolving, largely as a result of sedimentation. For example, lakes are formed, usually by catastrophic geological events, and proceed through a series of stages to eventual filling. Often lakes are obliterated as lakes per se and are incorporated into the terrestrial landscape. As we have seen, particularly in Chapter 24, the rate at which lakes fill varies greatly. Filling by inorganic materials is governed to a great extent by the basin morphometry, characteristics of the drainage basin, and climatic factors. Organic loading is influenced to a greater extent by factors regulating autotrophic productivity and the decomposition rates of the organic matter produced. Rivers also are influenced by sedimentation and are constantly maturing as the channels are eroded and aggraded, with appreciable migration and filling of the channels within flood plains.

Although the general ontogeny of lakes is from low to higher productivity, we have seen many cases (Chap. 24) in which past productivity was higher in the earlier developmental history of lakes than in subsequent periods. In very recent times, however, many lakes are experiencing accelerated productivity or *eutrophication*, often as a direct result of increased loading of nutrients (see particularly Chap. 13). The enhanced productivity is desired and is encouraged by certain human cultures to increase, despite very low conversion rates of energy, production of protein for consumption. In Western cultures, eutrophication is held less desirable, at least at the present time, because lakes are used primarily for water supply and recreational purposes, and these uses are more compatible with oligotrophic or mesotrophic characteristics.

Throughout the previous discussions, I have attempted to integrate the drainage basin, river metabolism, and littoral and wetland productivity into the overall functions of aquatic ecosystem functions. In the late stages of lake ecosystem ontogeny, autochthonous productivity shifts from phytoplanktonic prevalence to total dominance by littoral components. In shallow lake basins, which predominate the Earth, and in lake systems in which sedimentation has reduced water depth sufficiently for littoral dominance, rates of lake ontogeny accelerate greatly. The ensuing brief discussion attempts to summarize and integrate the earlier treatments.

I. SUCCESSIONAL DEVELOPMENT OF AQUATIC ECOSYSTEMS

Succession is a concept developed over a century ago largely from phytosociology and plant ecology, including some plant communities of ponds. However, lakes have not been merged well into conceptual successional theory but have been organized under the concept of eutrophication (Rodhe, 1969; Likens, 1970).

A. River Ecosystems

Successional theory has been applied only moderately well to stream ecosystems (Fisher, 1990). ***Succession*** can be defined simply as changes that occur on a site or in a community after a disturbance. Usually the disturbance induces habitat change or causes substantial mortality. Primary succession occurs after a disturbance that is so intense that no trace of the previous community remains. Secondary succession from disturbances that are not as severe as to remove the entire habitat and organisms is more common.

Successional theory is difficult to apply in spatially heterogeneous, hierarchically organized ecosystems like streams. Although succession in streams has been studied at many scales from microdistributions to drainage basin levels, much attention has been directed to longitudinal patterns of functional succession under the river continuum concept discussed earlier (see Chaps. 8 and 22), which is likely a product of disturbance. Succession occurs almost totally as secondary succession in many streams. Streams are spatially heterogeneous. After a disturbance, rates of succession are modified by the extent of spatial heterogeneity, and as a result successional theory is difficult to apply. Succession is really a part of ecosystem stability (see Chap. 23, pp. 775–778), and there is not great support for its separate application to streams.

B. Lake Ecosystems

The biogeochemical context, with its quantitative flux approach, provides insight into the mechanisms controlling contemporary productivity. Aquatic ecosystems usually are not in steady state, except in a restricted, short-term sense. The differences among aquatic ecosystems, as is evident from comparative studies, are obviously related to an array of dynamic factors that change over time. Short-term changes, keyed to diurnal and seasonal fluctuations, are reasonably repetitious and can be accommodated within a longer-term steady state. The preceding summary of paleolimnology (Chap. 24), however, should make it clear that the long-term progression cannot be adequately viewed as a steady-state process.

Aquatic ecosystems change. Erosion is a dominant geomorphological feature in running water systems. Lotic metabolism changes with progressive physical and chemical changes that occur in the headwater regions. Geomorphology also is critical to the ontogeny of lake ecosystems. Geochemical inputs to lake basins, coupled with morphometric characteristics of the lake basin depression in the landscape and changes in both the morphology and drainage patterns, influence lake productivity.

A primary characteristic of this ontogeny is the process of sedimentation, which gradually alters river valleys and fills lake basins. Inorganic inputs to the sediments vary greatly with the parent materials of the surrounding catchment area. Sedimentation is generally quite slow, and most organic inputs are mineralized prior to and soon after initial sedimentation. The ratio of organic to inorganic deposition can shift markedly in response to changes in the nutrient incomes, drainage basin characteristics, and basin morphometry. A time is reached at which the rate of organic deposition exceeds the capacity for its decomposition. Sedimentation rates can increase very rapidly once such a combination of characteristics is reached and can result in an accelerated conversion of the lake basin toward a terrestrial or bog landscape. Although the precise mechanisms involved in this ontogeny are highly individualistic, certain similarities among lakes emerge.

Inputs of organic matter are crucial in determining ecosystem metabolism and ontogeny. The organic productivity of rivers and lakes is fundamentally heterotrophic and is based on utilization of both autotrophic productivity and imported organic matter. Both production and utilization of the organic matter are controlled to a great extent by regulating factors of inorganic and organic biogeochemical cycling.

The boundary of the lake per se can usually be delineated relatively easily by the shore line and the supralittoral area, but in many cases, it is much more diffuse. From a functional standpoint, both in terms of the pelagic region and of the entire lake, saturated wetland and littoral areas that surround many lakes constitute a major source of inorganic and organic inputs that usually radically alter metabolic processes within the lake or stream and its ontogeny. Separation of these massive sources of organic matter from consideration of the metabolic processes that occur in the open-water "lake" is functionally incorrect. Similarly, one cannot ignore the sensitivity of the lake per se to differences in the types, quantities, and timing of allochthonous inputs of nutrients and organic materials.

Rivers and especially lakes cannot be treated as separate microcosms. While the importance of allochthonous inputs has long been known, especially from British work, there remains an erroneous, persistent tendency to treat the pelagic community as an isolated, nonintegrated component of the lake ecosystem. Dissolved organic matter is important in changing the trophic states of lakes by accelerating the cyclic regeneration of nutrients through bacterial metabolism, which both accelerates the decomposition of organic matter and increases the recycling and availability of inorganic nutrients required for photosynthesis.

Major causal pathways governing the eutrophicational ontogeny of lakes are based on interacting mechanisms, mainly nutrients and their availability, that regulate autotrophic metabolism. In many cases, the following discussion is applicable to certain lakes since their inception. In other cases, the pathways described are applicable only to recent times, such as since the recent disturbances by humans or those changes that could be anticipated with changes induced by human activities that affect the loading rates of nutrients and organic matter.

C. Increasing Eutrophication

Low rates of productivity of oligotrophic lakes are frequently determined by low inputs of inorganic nutrients from external sources. Morphometric characteristics of relatively large size and depth that yield high ratios of hypolimnion to epilimnion volume are commonly characteristic of oligotrophic lakes and affect nutrient cycling. The low production of organic matter, resultant low rates of decomposition, and oxidizing hypolimnetic conditions result in relatively low rates of nutrient release from the sediments in a cyclical causal system (Fig. 25-1). The importance of phosphorus and nitrogen limitations to maintenance of low productivity in a large percentage of lakes of the world was discussed in detail earlier (see particularly Chap. 13, pp. 279–286).

Oligotrophic conditions can be maintained by a combination of properties, including:

1. Low external loading of nutrients, particularly phosphorus and nitrogen
2. Geomorphic conditions of many oligotrophic lakes that often preclude the extensive

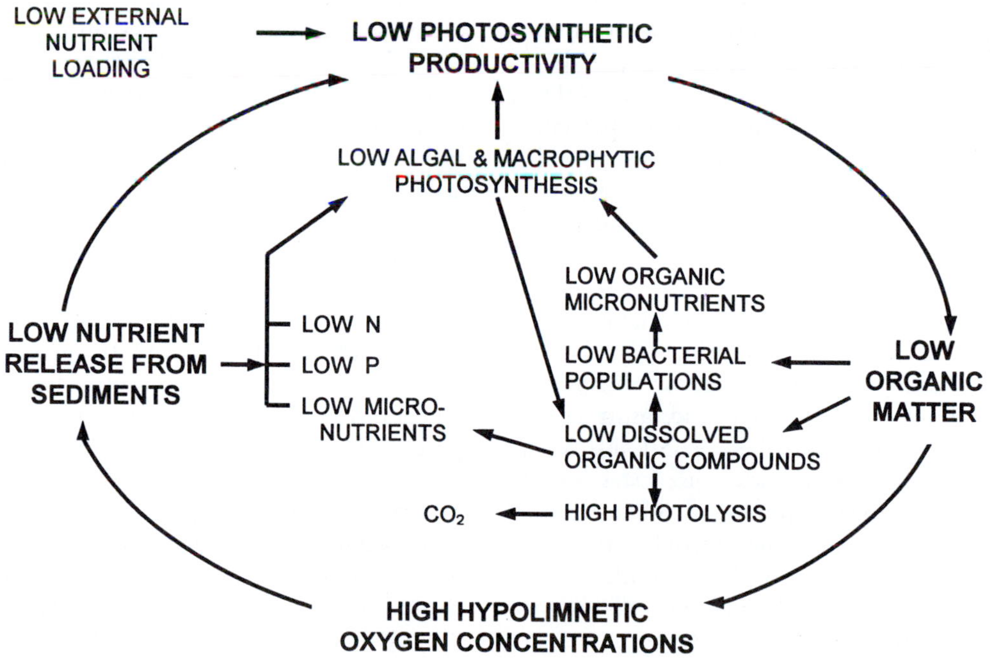

FIGURE 25-1 Major inorganic and organic interactions influencing the metabolism of phytoplankton of oligotrophic lakes. (Extensively modified from Wetzel and Allen, 1970.)

development of the highly productive wetland and littoral communities of higher plants and attached microflora.

3. The low production of phytoplankton, resulting from many interacting factors:
   a. Low loading of nutrients from external sources
   b. The larger size and depth of oligotrophic lakes and their deep mixed layers
   c. Carbon to phosphorus (C:P) ratios that tend to rise when more light is available, which encourages development of more efficient phytoplankton species; in turn, higher C:P ratios increase zooplankton grazing efficiencies and growth rates (Sterner *et al.*, 1998)
4. The decomposition of much of the organic matter production of the lake under oxic conditions and long sedimentation pathways; and the low organic loading to the sediments in both the littoral and pelagic regions
5. Low dissolved organic matter because of low autotrophic production, and the high transparency of the water, which increases the rates and depths of photolytic degradation of humic substances imported from the drainage basin
6. Low bacterial and protistan degradation of algae and other particulate organic matter because of limitations of nutrients and dissolved organic matter
7. Correspondingly limited synthesis of organic micronutrients, which are essential to most planktonic algae. Complexing of essential inorganic micronutrients, particularly iron, by dissolved organic compounds, by which solubility and physiological availability could be partially maintained under oxidizing conditions, would be less effective under oligotrophic conditions of low organic matter and high rates of degradation of available substrates.

The circumstances of the oligotrophic lakes of the Precambrian Shield region of Ontario, Canada, are ideal for demonstrating the importance of phosphorus limitation and rapid recovery from eutrophic conditions once the inputs are controlled. The Lake Washington and Swiss lake examples, discussed earlier (Chap. 24), are analogous examples of the success in the abatement of phytoplanktonic eutrophication that is possible once control by phosphorus-loading measures are instituted. In spite of the obvious caution that must be exercised in making generalizations about such complex and varied systems as lakes, it is important to emphasize once more that phosphorus abatement will not return all lakes to preenrichment conditions. As discussed in Chapter 13, recycling of phosphorus from internal stores in the sediments can sustain high rates of production for many years after influent loading has been reduced. The importance of phosphorus demand and supply for plant growth is such that its reduction in inputs to fresh waters is the first place to begin and likely to succeed in most cases.

Further, if phosphorus inputs to surface water are to be reduced most effectively, point sources should be eliminated as rapidly as possible. For example, the millions of tons of phosphate still being introduced into surface waters by synthetic detergents can be eliminated relatively easily by means of phosphate-free substitutes of equal cleaning efficacy. In addition, technologically available methods exist for nearly complete phosphate removal during wastewater treatment. The scientific basis for the importance of phosphate and nitrogen in eutrophication is so overwhelming that an international resolution was ratified at the 19th International Congress of Theoretical and Applied Limnology in 1974 (Wetzel, 1975a). This resolution emphasized the critical role of phosphorus in the rapid eutrophication of inland waters and the need to control the addition of this element to any inland water by any means available. In addition to full secondary, and in many cases tertiary, treatment of sewage, methods of control include (a) restrictions on the use of cleaning products that contain phosphates or other ecologically harmful substances, (b) removal of phosphate at sewage treatment facilities discharging effluents into such water, and (c) control of drainage from feedlots, agricultural areas, septic tanks, and other diffuse sources of nutrients. Control measures for nitrogen should also be considered in basins in which there is evidence that such controls are appropriate. Implementation of such control measures is socially complex, politically controversial, but technologically attainable and economically prudent.

Under eutrophic conditions, the loading rates of phosphorus and nitrogen, as well as of other nutrients of less acute demand, are relatively high (Fig. 25-2). As the rates of photosynthetic productivity increase during eutrophication, the cyclic interactions involving regeneration of inorganic nutrients and organic compounds increase in intensity. Planktonic productivity increases markedly per unit volume of water and results in increasing light-limited compression of the trophogenic zone. Reduction in depth of the trophogenic zone continues with intensification of eutrophication, largely as a result of biogenic turbidity of the phytoplankton. Eventually, planktonic populations impose self-shading to such an extent that further increases are not possible under natural conditions of

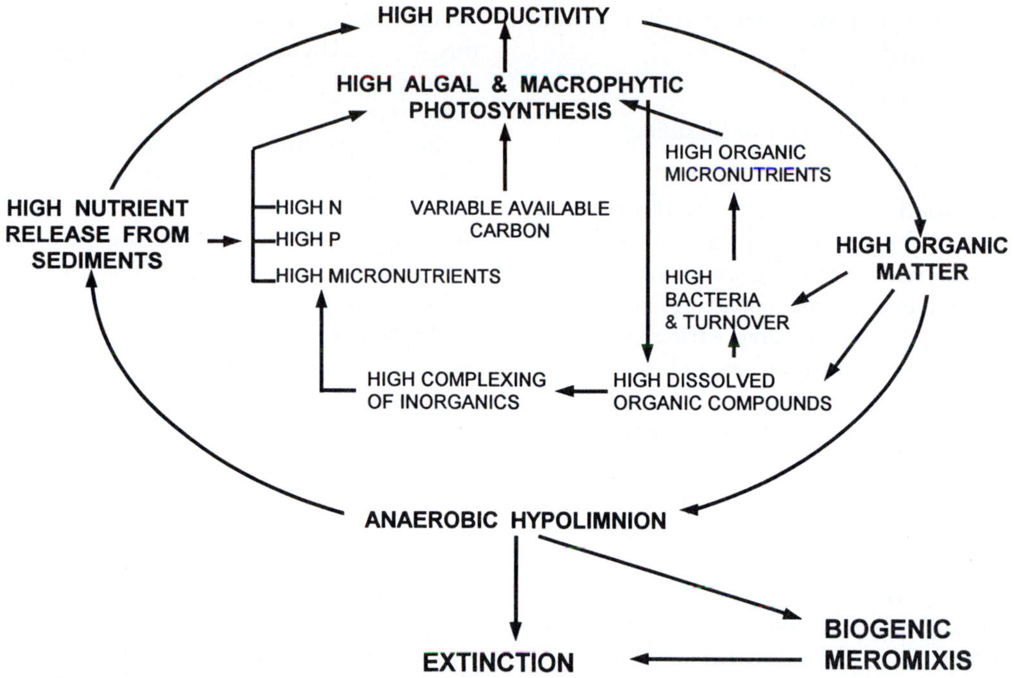

FIGURE 25-2 Major inorganic and organic interactions influencing the metabolism of phytoplankton of eutrophic lakes. (Extensively modified from Wetzel and Allen, 1970.)

incident solar irradiance (Fig. 25-3). The rate of planktonic production then reaches a plateau, with a gradual progression toward eventual extinction through sedimentation of organic matter that accrues faster than decomposition can remove it. Under certain conditions of lake basin morphometry, meteorology, and productivity, it is not rare for mesotrophic to eutrophic lakes of small surface area and moderate depth to undergo biogenically induced meromixis (Chap. 6). A reduction, usually temporary, in productivity can

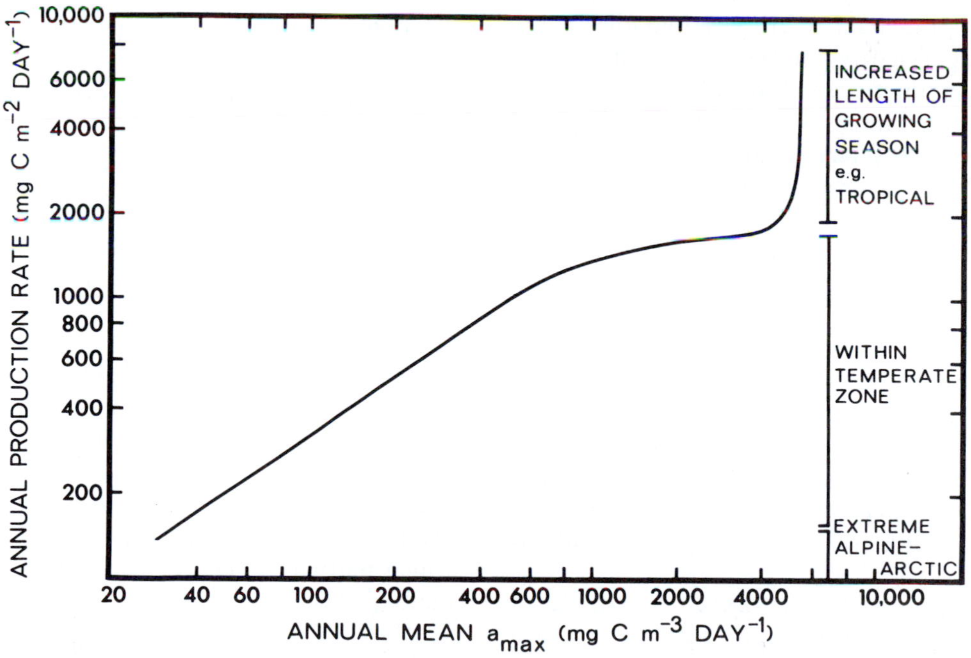

FIGURE 25-3 General relationship between the annual mean volumetric production rate at the depth of optimal growth and the annual mean areal production rates of phytoplankton. (After Wetzel, 1966a, and others.)

occur when redistribution of nutrient-rich hypolimnetic waters is prevented.

D. Food-Chain Length in Relation to Changing Ecosystem Productivity

A community food web or chain includes the feeding relations among the organisms found in a defined habitat. Organisms with similar groups of prey and similar predators have been combined into aggregated trophic species or levels. As indicated earlier (Chap. 8), the "top" species is a predator that has no higher predator, intermediate trophic species are those organisms that are both a predator and a prey, and a basal trophic species is a prey that has no prey (Cohen and Briand, 1984; Cohen and Luczak, 1992). A trophic link is any feeding relation between two trophic species in a community web. The mean number of trophic links between any two categories of trophic species (basal, intermediate, and top) is proportional to the geometric mean number of species in the categories joined.

The food chain is a sequence of trophic links in which energy starts at a species that eats no other species in the web and ends at a species that is eaten by no other species in the web (Briand and Cohen, 1987). The length of a food chain is the number of trophic links it comprises. The mean chain length of a food web is the arithmetic average of the lengths of all chains in the web. Many studies have correlated food-chain length of pelagic organisms to various environmental parameters. For example, based on energetic relationships, the food-chain length was hypothesized to be limited by inefficiencies of energy transmission by predation in relation to the minimal energy requirements of predators (Hutchinson, 1959). This hypothesis predicted that food-chain length should be longer in ecosystems with higher primary productivity of phytoplankton. Some support for this direct relationship has emerged (cf. Pimm, 1982; Jenkins *et al.*, 1992; Persson *et al.*, 1992; Vander Zanden *et al.*, 1999). However, examination of many community food webs of lakes indicated that the average and maximal lengths of food chains of higher organisms (phytoplankton to fish) are independent of primary productivity (Briand and Cohen, 1987; Hairston and Hairston, 1993; Post *et al.*, 2000). Food-chain length was found to be directly correlated with "ecosystem size" (lake volume) among many lakes, but not with resource (phytoplankton) availability.

As Hutchinson (1959) so effectively noted, ecosystem productivity certainly constrains food-chain length. It must be recalled, however, that higher trophic levels consume heterotrophically a very small percentage of the productivity, particularly in more productive pelagic waters (Wetzel, 1995; Chaps. 16 and 23 of this work). The microbial communities metabolize heterotrophically most of the organic matter in lake ecosystems. Hence, it is not surprising to find in the microbial loop with protists that resource availability and basal productivity was directly correlated with population-level responses and food-chain length (Kaunzinger and Morin, 1998). Except under conditions of very low productivity or in small, shallow lakes where the habitats are much more heterogeneous and varied than in the pelagic of large lakes, one would anticipate that productivity is quite adequate to support higher trophic levels and that other factors dominate in regulation of food-chain length. Greater volume among natural lakes, particularly in regions analyzed by Post *et al.* (2000), is often inversely correlated with primary productivity but can obviously be limited by many other factors quite unrelated to basin morphometry (cf. Chaps. 3 and 15).

The common separation and emphasis of the higher trophic of the pelagic of lakes from the rest of the lake ecosystem provides for interesting theoretical analyses. However, these studies are frequently naive because of their decoupling and false assumptions of relative independence from the energetic fluxes of the entire lake as an integrated ecosystem. It also should be recalled once again that many of these studies are based on feeding relationships during summer periods of maximum growth and productivity of higher trophic levels. These trophic relationships are totally different and many feeding couplings are completely absent during large portions of the annual cycle.

E. Differences among Hardwater Calcareous Lakes

The ontogeny of an oligotrophic lake toward eutrophic conditions can be altered markedly by natural inputs of carbonates and associated cations from surrounding sedimentary bedrock and glacial till. Calcareous hardwater lakes are common to large regions of the world, and in many regions of the temperate zone a quarter of the lakes contain very hard water with large amounts of precipitated $CaCO_3$. In such lakes, oligotrophic states are often maintained by high calcareous inputs sustained over long periods of time. In these waters, reduced productivity is maintained by decreased nutrient availability rather than the type of oligotrophy persistent in softwater lakes of low productivity, in which nutrient inputs to the systems are deficient and limiting at times.

Among both calcareous and noncalcareous lakes, early postglacial eutrophication generally proceeded rapidly as the regional climate ameliorated. Initial high rates of eutrophication were favored by inputs of

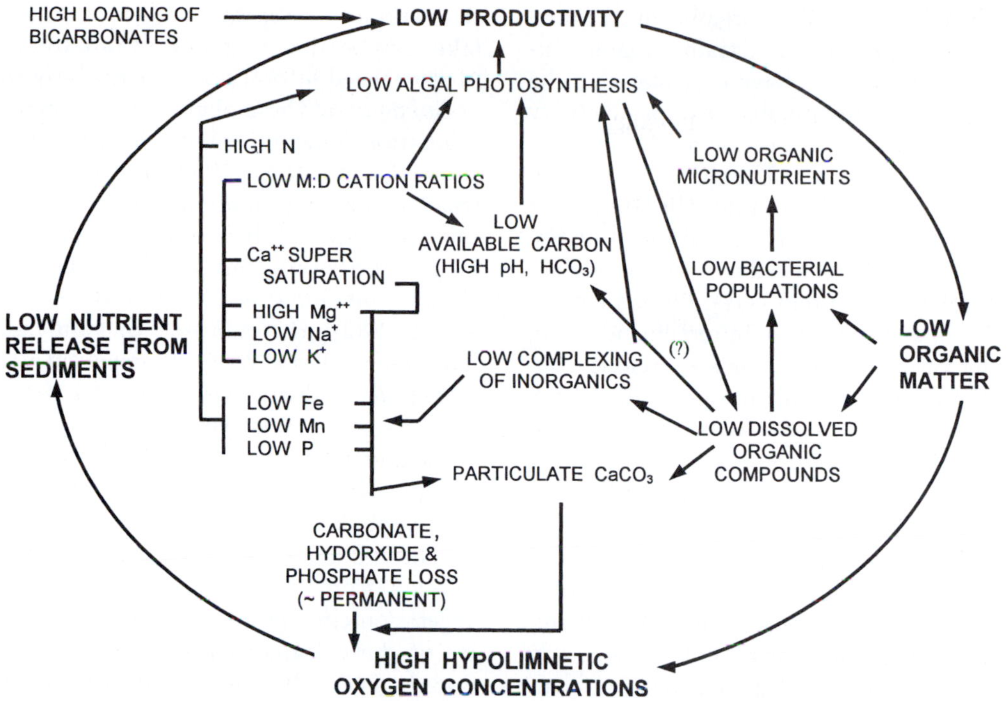

FIGURE 25-4 Major inorganic and organic interactions influencing the metabolism of producers in hardwater calcareous lakes. (Extensively modified from Wetzel, 1972.)

readily leached nutrients that entered lakes postglacially from adjacent terrestrial areas. In calcareous regions, however, high inputs of carbonates can be maintained over long periods of time (millennia). Under excessively buffered bicarbonate conditions, productivity is suppressed by an array of inorganic-organic interactions affecting the metabolism of the micro- and macroflora (Fig. 25-4). Major interactions have been discussed in earlier chapters and need be mentioned only briefly here for continuity.

Phosphate and essential inorganic micronutrients, particularly iron and manganese, form highly insoluble compounds and precipitate from the trophogenic zone of hardwater lakes (Wetzel, 1972). Combined nitrogen compounds tend to occur in high concentrations as a result of the relatively high inputs common in calcareous regions and the low rates of biotic utilization within the lake. In extremely calcareous lakes, it is not rare to find inorganic nitrogen concentrations that approach levels considered toxic ($>$10 mg NO_3 liter^{-1}) for human consumption. Calcium and magnesium concentrations can be exceedingly high; calcium is often in apparent supersaturation and approaches or exceeds 100 mg Ca^{++} liter^{-1}. Intensive biogenically induced decalcification is characteristic of these hard waters, during which massive amounts of particulate carbonate precipitate from the trophogenic zone. The sedimenting particulate and colloidal $CaCO_3$ adsorbs inorganic nutrients and certain labile and recalcitrant dissolved organic materials and effectively removes them from the trophogenic zone. The major monovalent cations, sodium and potassium, commonly occur at low concentrations. Low levels of sodium, an essential element of certain cyanobacteria, have been implicated as a contributory factor in reducing the growth of heterocystous cyanobacteria in calcareous lakes (Ward and Wetzel, 1975).

Although inorganic carbon concentrations occur at very high levels, the low availability of free CO_2 can reduce photosynthesis of certain plants under stagnated conditions when the pH is excessively high (see Chap. 11). Certain species of natural phytoplanktonic algae of calcareous lakes are capable of significantly ($<$10%) augmenting photosynthesis by photoheterotrophy of simple organic substrates and compounds produced extracellularly by aquatic macrophytes (Chap. 15).

The low photosynthetic productivity of calcareous lakes contributes to reduced inputs of dissolved organic compounds, such as lytic and extracellular losses from phytoplankton and macrovegetation. This situation is compounded by effective adsorptive losses of dissolved organic matter to monocarbonates as they precipitate. Bacterial rates of metabolism are low because of the low concentrations and turnover of easily decomposable organic compounds, which can lead further to concomitant reduced synthesis of organic micronutrients. Reduced concentrations of certain organic micronutrients such as vitamin B_{12} have been shown to be contributory to reduced phytoplanktonic productivity in calcareous lakes; a portion can be metabolically inactivated by adsorption to $CaCO_3$. The more reactive organic substrates are either utilized rapidly by bacteria or are inactivated by adsorption to sedimenting $CaCO_3$. The loss of organic compounds reduces the complexation capacity of the water, so that certain inorganic nutrients cannot be maintained in solution. In addition, the selective removal of the more labile organic constituents leads to a relative accumulation of more recalcitrant dissolved organic matter. It is common to find very hard lakes with a stained coloration (brown) and high concentrations of dissolved humic materials; these compounds degrade slowly and may inhibit additional epilimnetic decalcification by interfering with the formation of $CaCO_3$ particles, thereby sustaining an oligotrophic condition. A portion of the dissolved organic matter is degraded to CO_2 by solar photolysis.

Factors that reduce the buffering capacity and carbonate reservoirs of calcareous lakes promote eutrophication in these ecosystems. Two obvious means of altering the major controlling aspects of calcareous inputs are reductions in or depletion of bicarbonate and cation sources in the drainage basin or increases in the loading of dissolved organic matter. At some point in this transition, the inhibitory effects of the calcareous state on nutrient availability are reduced and the stimulatory effects (both direct and indirect) of dissolved organic matter increase. At that point, eutrophication can proceed relatively rapidly. Morphometric changes that accompany eutrophication-induced increases in sedimentation rates and infilling also have significant effects on stratification patterns, hypolimnetic capacity of maintaining oxidizing conditions, and nutrient recycling. These changes accelerate as productivity increases and operate to further enhance eutrophication.

Where large areas of sedimentary rock have been glaciated, it is common to find former very calcareous lakes underlying bog lakes[1] or bogs (discussed in some detail further on). It is apparent that numerous marl lakes have shifted relatively rapidly, in the range of a millennium, from highly calcareous, alkaline conditions to a very acidic, organic-rich state, markedly depleted in divalent cations and bicarbonate. In partially closed basins of moderate to small size that have shifed to dominance of water inputs from precipitation rather than surface flows, the ontogeny of initially calcareous lakes can be altered at a rapid rate by the development of specialized littoral flora, particularly the mosses such as *Sphagnum*. These plants and associated encroaching vegetation function as particularly effective cationic sieves (Clymo, 1963, 1964). It is not uncommon to find small pioneering hummocks of *Sphagnum* growing in the wetland and back-littoral reaches of calcareous lakes directly on a thin organic deposit overlying sediments containing >50% of $CaCO_3$ (see, e.g., Glime *et al.*, 1982). The eulittoral of highly stained alkaline hardwater lakes is often interspersed with stretches of *Sphagnum* in dense mats. These alkaline bog lakes represent a transitional stage between certain calcareous lakes and true acidic *Sphagnum* bogs. The ionic exchange mechanisms of these mosses, accompanied by the simultaneous release of organic acids (e.g., polyuronic acids), are effective in reducing the cation influxes from surface sources. As the littoral sieving vegetation circumscribes the basin, the buffering capacity of the calcareous lake system is progressively reduced. The subsequent vegetative development then can proceed relatively rapidly, in a classical pattern of bog succession (discussed further next) via accelerated accumulation of organic matter under acidic, reducing conditions that does not favor rapid decomposition.

[1] Bog lakes have been variously described; they commonly refer to strongly basic alkaline lakes surrounded by an extensive acid-forming bog mat (see Welch, 1936).

II. DYSTROPHY AND BOG ECOSYSTEMS

Early workers formulating relationships in studies of comparative regional limnology were acutely aware of the important differences among lakes in regard to the proportion of allochthonous and autochthonous organic matter.[2] ***Trophy*** of a lake refers to the rate at which organic matter is supplied by or to the lake per unit time.[3] Trophy, then, is an expression of the combined effects of organic matter supplied to the lake.

As has already been indicated (Chap. 23), under natural conditions, the relatively resistant humic substances of largely terrestrial plant origin represent the most common component of allochthonous organic

[2] Birge and Juday (1927), for example, in their extensive studies on the origin and supplies of dissolved organic matter, differentiated between autotrophic lakes dominated by autotrophic inputs and allotrophic lakes that received a majority of their soluble organic matter from the drainage basin.

[3] A concise review of the development of the trophic concept is given by Rodhe (1969). A number of other terms have been used to indicate the direction of the state of trophy, for example, *eutrophication* in the broad sense of an increase in productivity (also *auxotrophication*; Thunmark, 1948; Lillieroth, 1950), which refers to an increase in trophic standard. The contradictory term *oligotrophication* (also *meiotrophication*; Quennerstedt, 1955) refers to a decrease in trophic standard.

matter. Lakes that receive large amounts of their organic matter supply from allochthonous sources commonly are heavily stained and have been referred to as "brown-water lakes." These lakes were termed ***dystrophic***, in reference to their high content of humic organic matter.

The productivity of most dystrophic lakes classically has been described as low. However, more detailed examinations indicate the contrary. Phytoplankton biomass (chlorophyll concentrations) is significantly higher in colored ("brown-water") lakes than in clear lakes. Phytoplanktonic primary productivity was higher in dystrophic lakes when expressed per volume of epilimnion (Jasser, 1997; Nürnberg and Shaw, 1998). Annual integral primary productivity expressed on an areal basis was somewhat smaller in colored lakes, probably as a result of shallower phototrophic depths in these lakes, than in clear lakes.[4] It must be emphasized further, however, that this moderate productivity criterion refers to the planktonic productivity and, as it was developed, ignored the littoral plant components of the lake system. As we shall see, in dystrophic lakes that develop bog flora, the littoral plants completely dominate the metabolism of these lake ecosystems as sources of dissolved and particulate organic matter and energy.

Annual integral productivity of the epilimnetic bacterioplankton was also found to be much higher (four times) in dystrophic lakes than in clear lakes (Nürnberg and Shaw, 1998; see also Satoh and Abe, 1987). Also in the hypolimnia of organically stained dystrophic lakes, hypolimnetic bacterial production is commonly higher than epilimnetic production, particularly in anoxic hypolimnia that are frequent in dystrophic lakes. Even though these DOC compounds are relatively recalcitrant, photolytic degradation certainly assists in enhanced availability for bacteria (see Chap. 23).

As a consequence of uncritical use of the terms associated with dystrophy, bogs, bog lakes, and dystrophic lakes have been loosely, and incorrectly, equated. Much of the unfortunate confusion emanates from inappropriate generalizations about ontogeny of lake systems from specific lake investigations. For example, the famous successional scheme put forth by Lindeman for Cedar Bog Lake, Minnesota, has been more than once uncritically proposed in general ecology texts as the universal situation. In this way, the erroneous concept that all lakes become bogs and then land may become widely accepted. Although some

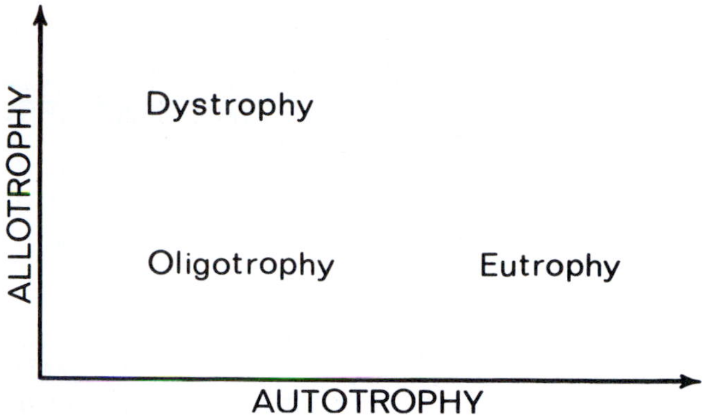

FIGURE 25-5 Classical trophic types of lakes based on the rate of supply of organic matter from autotrophic and allochthonous sources. In the original development of these concepts, only phytoplanktonic productivity was considered. (Modified from Rodhe, 1969.)

lakes do progress through this sequence, it is far from the rule. A large, heavily stained Finnish lake on a primary bedrock basin with no characteristic bog vegetation can be correctly termed dystrophic just as effectively as the *open pelagic water* of a quaking *Sphagnum* bog. *Trophy* refers to the rate of supply of organic matter, and *dystrophy* denotes a high loading of allochthonous organic matter (Fig. 24-5). Dystrophy is a subset of trophy (oligotrophy to eutrophy), rather than a parallel concept.

These examples emphasize an important point. As developed originally and as largely used today, the trophic concept refers to the pelagic-zone-planktonic portion of the lake ecosystem. The differentiation was between the rate of organic matter input to the system from autotrophic phytoplanktonic sources and from allochthonous sources of the catchment basin (allotrophy) (Fig. 25-5). The littoral flora and its often dominating supply of autochthonous organic matter to the system, was, and usually still is, ignored. Although conceptually it is easy to place the wetland and littoral productivity within such trophic schemes, there are few quantitative evaluations of its contribution. Sufficient information is available, however, to indicate that it cannot be ignored in most cases.

The photosynthetic rates of phytoplankton are moderate to high on a volumetric basis in highly stained open basins that receive copious quantities of dissolved organic matter from allochthonous, wetland, and littoral vegetation sources (Fig. 25-6). The abundance of dissolved humic materials in these systems is conducive to high bacterial metabolism, in part certainly stimulated by solar photolysis of the complex humic macromolecules (cf. Chap. 23). Nutrient regeneration rates are generally high under such conditions. Additionally, monovalent-to-divalent cation ratios tend

[4] High productivity of phytoplankton in heavily stained dystrophic lakes is known, especially in lowland areas of Scandinavia. Järnefelt (1925) called these lakes *mixotrophic*, but introduction of another term in this conceptual discussion is unnecessary and confusing.

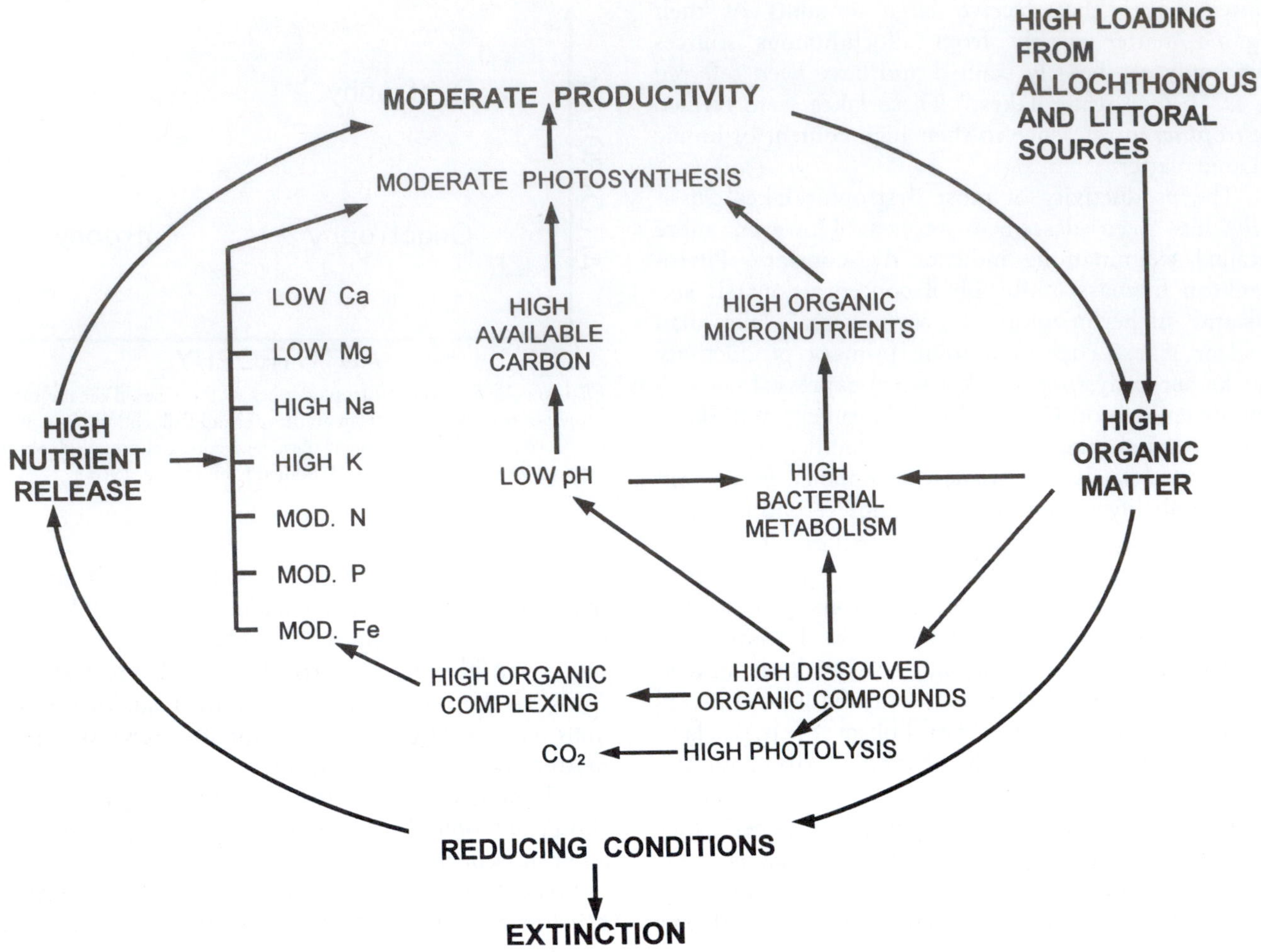

FIGURE 25-6 Dominant inorganic and organic interactions influencing the metabolism of phytoplankton in the pelagic waters of dystrophic and bog lakes. (Extensively modified from Wetzel and Allen, 1970.)

to be high because of low inputs from the drainage basin or because of effective cation exchange mechanisms of the littoral flora, particularly if dominated by bryophytes.

The relationships among the major causal mechanisms regulating the phytoplanktonic trophic states in the four main types of lakes may be further connected in an overall sequence of development (Fig. 25-7). This scheme of lake ontogeny emphasizes the importance of changes in functional relationships and environmental parameters that regulate algal, and to some extent, macrophytic productivity. The designated pathways of ontogeny (Fig. 25-7) can obviously vary greatly, depending upon the geomorphology of the specific ecosystem and the macroclimate of the area. The rate at which a lake proceeds to extinction via sedimentation is determined both by interactions that occur within the pelagic zone and by certain aspects of geomorphology of the drainage basin. Finally, as has been stressed repeatedly throughout this text, one must have an appreciation for the importance of littoral macrovegetation in the culmination of this sequence.

A. Littoral Development

Shallow water bodies, correctly termed *lakes* but often referred to in older literature as *ponds*, are usually characterized by an abundance of aquatic macrovegetation and associated microflora attached to all surfaces (see Chap. 20). It is not unusual for the vegetation to extend over the entire basin, provided the depth does not exceed plant tolerances, primarily for adequate light (cf. Chaps. 18 and 20). Shallow lakes include basins that were never preceded by a larger, deeper lake, as well as those basins that represent a terminal stage in the extinction of deeper lakes. From a hydrological viewpoint, the shallow water bodies can be separated into those that are permanent, containing

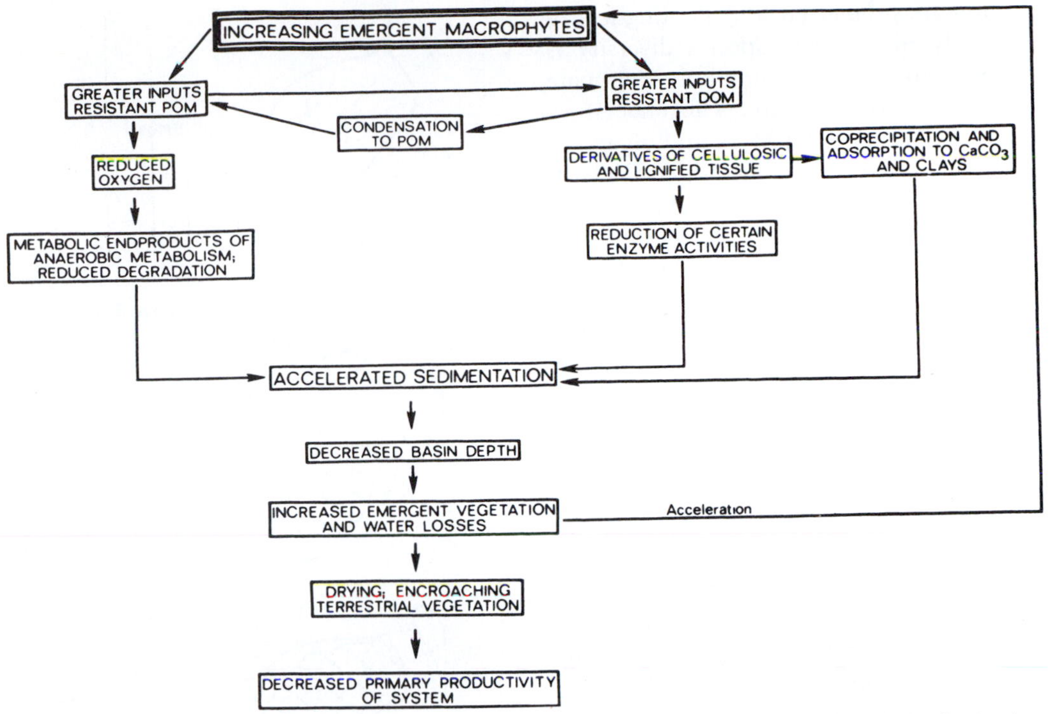

FIGURE 25-10 Relationships of increased loading of relatively resistant organic matter from dominating emergent littoral and wetland flora to controls of decomposition rates, acceleration of emergent macrophyte development, and the ontogeny of lake systems. POM and DOM = particulate and dissolved organic matter, respectively. (From Wetzel, 1979.)

ground water. The ionic composition of the water of these "rheophilous mires"[5] is dominated by calcium and bicarbonate. Allochthonous organic matter inputs are high and result in accrual of peat within the central portion of the basin (Stage 2). In some cases, this peat is sufficiently buoyant to float as a mat. The main flow of water tends to be channelized to peripheral areas. In the third transitional stage of mire development,[6] continued accrual of peat deposits diverts inflow from the basin. Water supply is derived largely from the immediate catchment of the mire and from direct precipitation on the surface. Ionic composition usually is dominated by calcium and sulphate. Further accrual of peat results in large areas of the mire surface that are not affected directly by moving water (Stage 4), but are inundated periodically when the water level rises during heavy rainfall.

Ombrogenous or ombrophilous (ombro = rain) mires[7] (Stage 5) are no longer subject to the influence of flowing ground water. Nearly all of the water, which is dominated by sulphate and hydrogen ions, is received from direct rainfall. The surface of the mire system is too high to be materially influenced by fluctuations in the level of ground water. The change in ionic composition of water from dominance by calcium and bicarbonate to that of sulphate and hydrogen ions is a conspicuous feature characterizing these mire systems (Fig. 25-11). The loss of ions from surface water supplies results in very low salinities and greatly reduced buffering capacity. Under these conditions, relatively small additions of acids can result in a considerable reduction in pH.

The origin of acidity and control of cation exchanges of mire systems result from a combination of factors. The increase in acidity and reduction of cations is particularly marked in communities dominated by the mosses of the genus *Sphagnum*. Water surrounding *Sphagnum*, a dominant genus of mosses over immense areas of the temperate, boreal, and subarctic regions of the world (see e.g., Sjörs, 1961), has a pH usually below 4.5 and sometimes below 3.0. Much of the acidity of the aquatic environment surrounding *Sphagnum* can be attributed to cation exchange (Anschütz and Gessner, 1954; Clymo, 1963, 1967; Brehm, 1970; Glime *et al.*, 1982; and others). *Sphagnum* and other mosses behave as cation exchangers even when dead because of their high concentrations (up to 30% of the plant's dry weight) of polymerized unesterified

[5] Synonymous with terms "low moor" and *Niedermoore*.
[6] *Ubergangsmoore* of older German literature.
[7] Termed "high moors" or *Hochmoore* in early works.

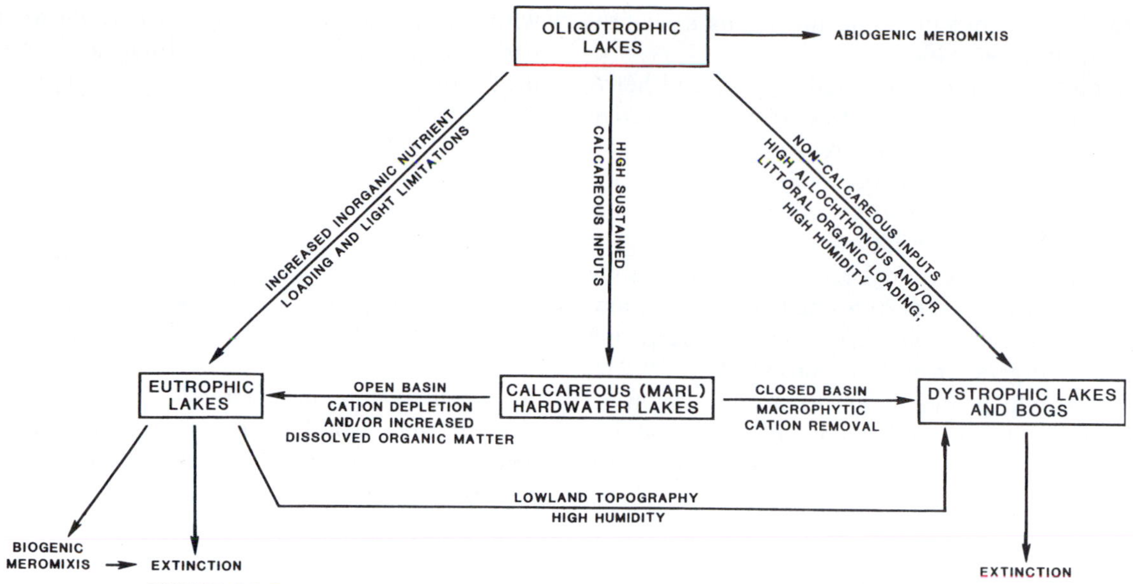

FIGURE 25-7 Potential ontogeny of the four main types of lakes. See text for discussion.

some water at all times of the year, and those that are temporary, in which the basin periodically becomes dry (Chap. 20). It is apparent that a nearly infinite variety of conditions lead to the development and persistence of small, shallow lakes in relation to seasonal and long-term climatic changes in the water budget. Outflow (surface and subsurface), evaporation, and retention, including water retained by large accumulations of organic matter of plant origin, counterbalance inflow and precipitation.

Excellent reviews and interpretation of the widely dispersed literature on peat-producing ecosystems or ***mires*** are given by Gorham (1957) and in the books by Kulczynski (1949), Moore and Bellamy (1974), Kivinen *et al.* (1979), Crum (1988), Gottlich (1990), and Feehan and O'Donovan (1996). The terminology and synonymy of this area of study are complex; only a brief résumé of major features of the systems is given here. It should be noted that *mire* is a collective term which includes both ***bog*** ("moss" in British literature) and ***fen***, differentiated according to water and nutrient sources, as well as by rather subtle floristic variations. "Mire" is equivalent to the Swedish ***myr*** and the German ***Moor***. All mires are in areas where the groundwater table is permanently at or near the surface of the ground. Mires are commonly separated into ***minerotrophic mires*** (fens) in which water and nutrients are supplied from ground water, surface sources, and atmospheric precipitation. The water of fens tends to be base-rich (pH > 5.5) and commonly supports herbaceous vegetation. In ***ombrotrophic mires*** (bogs), water-table recharge and nutrient inputs originate mainly from atmospheric precipitation. Water is of very low salinity and tends to be acidic (pH < 5.5). A ***carr*** refers to a mire or portion of a mire that is dominated by woody shrubs or trees among the herbaceous vegetation on a relatively stable peat mass (e.g., a fen carr could be dominated by trees, such as species of *Alnus*, *Fraxinus*, and *Betula*, with a dense understory of fen herbaceous plants). The terminology and characteristics of these gradients in peatlands and associated waters are discussed more extensively in Damman (1986), Bridgham *et al.* (1996), and particularly Wheeler and Proctor (2000).

The terminal stages of the transition from lake to terrestrial ecosystems are characterized by an accumulation of organic matter in excess of degradation. Partially decayed organic matter, mainly of plant origin and termed ***peat***, accumulates in aquatic ecosystems under a wide variety of conditions in all except the driest macroclimatic regions of the world. Mire ecosystems form (1) in basins or depressions (primary mire systems); (2) beyond the physical confines of the basin or depression (secondary), the peat itself acting as a reservoir and increasing the surface retention of water; and (3) above the physical limits of the ground water (tertiary), the peat functioning as a reservoir holding a volume of water by capillarity above the level of the main groundwater mass of the region (Moore and Bellamy, 1974). In addition, large quantities of trapped gases, particularly methane, occur in the deeper deposits of peat (Brown *et al.*, 1989). Entrapped gases could reduce the hydraulic conductivity in the lower layers of the peat by blocking the soil pore spaces, preventing fluid movement, and generating an elevated water table. This last mire system obtains much of its

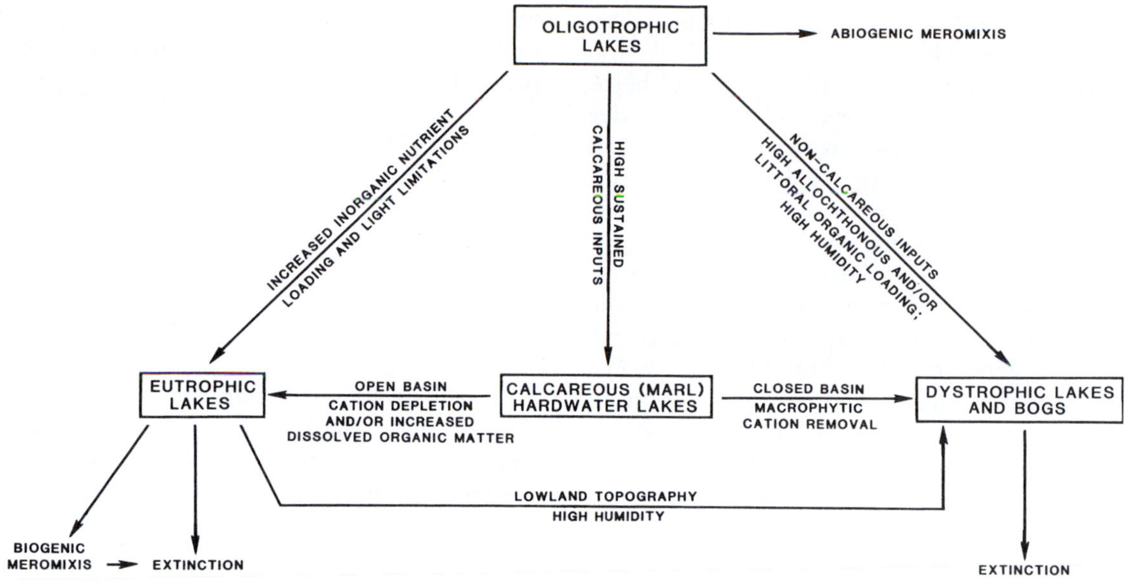

FIGURE 25-7 Potential ontogeny of the four main types of lakes. See text for discussion.

some water at all times of the year, and those that are temporary, in which the basin periodically becomes dry (Chap. 20). It is apparent that a nearly infinite variety of conditions lead to the development and persistence of small, shallow lakes in relation to seasonal and long-term climatic changes in the water budget. Outflow (surface and subsurface), evaporation, and retention, including water retained by large accumulations of organic matter of plant origin, counterbalance inflow and precipitation.

Excellent reviews and interpretation of the widely dispersed literature on peat-producing ecosystems or ***mires*** are given by Gorham (1957) and in the books by Kulczynski (1949), Moore and Bellamy (1974), Kivinen *et al.* (1979), Crum (1988), Gottlich (1990), and Feehan and O'Donovan (1996). The terminology and synonymy of this area of study are complex; only a brief résumé of major features of the systems is given here. It should be noted that *mire* is a collective term which includes both ***bog*** ("moss" in British literature) and ***fen***, differentiated according to water and nutrient sources, as well as by rather subtle floristic variations. "Mire" is equivalent to the Swedish ***myr*** and the German ***Moor***. All mires are in areas where the groundwater table is permanently at or near the surface of the ground. Mires are commonly separated into ***minerotrophic mires*** (fens) in which water and nutrients are supplied from ground water, surface sources, and atmospheric precipitation. The water of fens tends to be base-rich (pH >5.5) and commonly supports herbaceous vegetation. In ***ombrotrophic mires*** (bogs), water-table recharge and nutrient inputs originate mainly from atmospheric precipitation. Water is of very low salinity and tends to be acidic (pH <5.5). A ***carr*** refers to a mire or portion of a mire that is dominated by woody shrubs or trees among the herbaceous vegetation on a relatively stable peat mass (e.g., a fen carr could be dominated by trees, such as species of *Alnus, Fraxinus*, and *Betula*, with a dense understory of fen herbaceous plants). The terminology and characteristics of these gradients in peatlands and associated waters are discussed more extensively in Damman (1986), Bridgham *et al.* (1996), and particularly Wheeler and Proctor (2000).

The terminal stages of the transition from lake to terrestrial ecosystems are characterized by an accumulation of organic matter in excess of degradation. Partially decayed organic matter, mainly of plant origin and termed ***peat***, accumulates in aquatic ecosystems under a wide variety of conditions in all except the driest macroclimatic regions of the world. Mire ecosystems form (1) in basins or depressions (primary mire systems); (2) beyond the physical confines of the basin or depression (secondary), the peat itself acting as a reservoir and increasing the surface retention of water; and (3) above the physical limits of the ground water (tertiary), the peat functioning as a reservoir holding a volume of water by capillarity above the level of the main groundwater mass of the region (Moore and Bellamy, 1974). In addition, large quantities of trapped gases, particularly methane, occur in the deeper deposits of peat (Brown *et al.*, 1989). Entrapped gases could reduce the hydraulic conductivity in the lower layers of the peat by blocking the soil pore spaces, preventing fluid movement, and generating an elevated water table. This last mire system obtains much of its water directly from precipitation and forms a slightly raised (perched) water table.

Although dense stands of small mosses and herbs are the dominant components of contemporary active peat formations over immense areas of subarctic and temperate regions, shallow depressions of lake systems commonly gain excessive organic matter deposits through submersed, floating-leaved, and emergent macrophytes. The successional sequence may or may not terminate in moss and associated bog vegetation. Many highly fertile eutrophic lakes develop increasing amounts of emergent littoral vegetation (Fig. 19-21), to the point at which organic matter accumulation increases the level of flora and sediments above the water table, which allows terrestrial vegetation to encroach. A common successional pattern associated with the ontogeny of shallow basins follows along the pathway of excessive macrophytic and sessile algal production:

Macrovegetation and associated microflora ⟶ Reed swamps ↓ Marshes and fens ↗ Moss vegetation ⟶ Bog systems
Marshes and fens ↘ Terrestrial vegetation

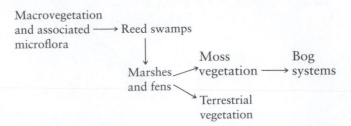

Reed swamps are characterized by closing vegetation of the littoral, often polycorm, tall graminoid emergent plants—for example, *Phragmites, Cyperus papyrus, Scirpus*, and tall *Carex* species—with only

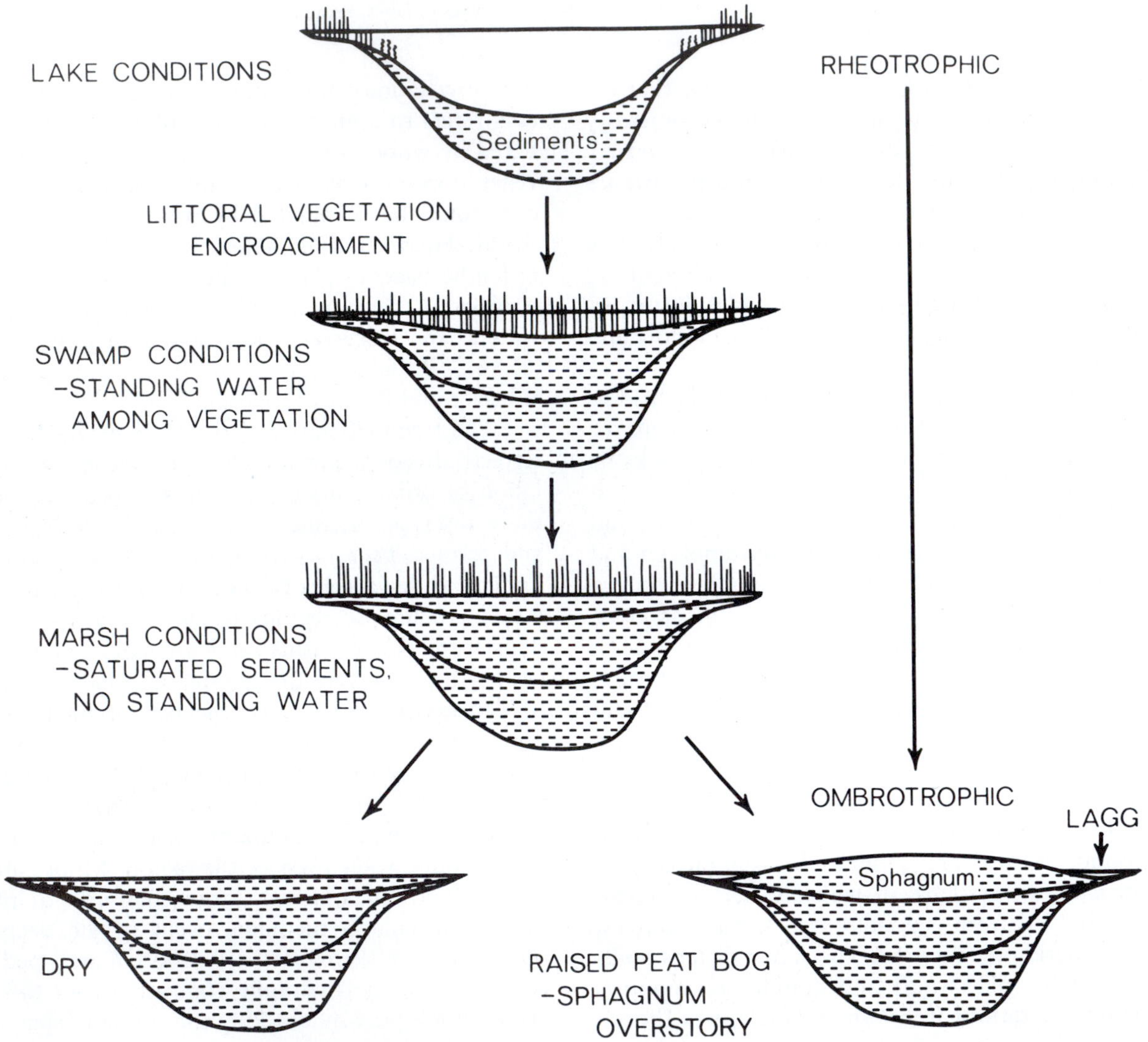

FIGURE 25-8 Frequently observed ontogeny of shallow lake systems through swamp and marsh stages to terrestrial conditions or to raised peat bogs. See text for discussion.

occasional mixtures of submersed and floating-leaved macrophytes. Usually macrophyte species diversity is low, and two herbaceous layers commonly dominate within the main vegetative structure. The understory can contain main species of bryophytes and algae. As organic accumulations increase with gradual displacement of the standing water, the water-logged peat habitat often is characterized by moderate-sized graminoid vegetation and small herbs in two or three layers. This vegetation is referred to as *marsh* or *fen* in areas of slightly flowing surficial water, or *bogs* in areas of subsurface seepage drainage. The distinction between the terms *marsh* and *fen* is largely made on the basis of phytosociological differences in floristic associations. As marshes and fens "dry" and depositions of organic matter exceed the mean water table, very tall terrestrial dicotyledonous vegetation increasingly colonizes and develops over the basin and displaces the emergent aquatic macrophytes. In nutrient-poor areas that are humid and receive abundant rainfall, marshes and fens may possess well-developed bryophyte floras. In many instances, mosses (especially species of *Sphagnum*) dominate the system and can lead to the formation of bogs (e.g., Glime *et al.*, 1982).

Early distinctions of various bog systems and types were made mostly on the basis of the primary, secondary, and tertiary stages of mire development just discussed in relation to water sources and floristic differences. Classical hydrosere succession hypothesized that infilling of a shallow lake by organic sediments produces a sequence of vegetation communities beginning with a marsh community of aquatic plants such as sedges, often followed by a fen or bog community composed mainly of *Sphagnum* mosses, sedges, and ericaceous shrubs, and culminating in a mature upland or mesic "climax" forest (Fig. 25-8) (reviewed in Klinger, 1996). Succession to fens and terrestrialization can proceed as set forth in this model. The ontogeny of mire systems is generally interpreted with respect to a combination of hydrological, phytosociological, and chemical processes.

A sequence under ideal geomorphological and hydrological conditions illustrating common stages in the succession of mire systems in shallow surface depressions is portrayed in Figure 25-9 (Moore and Bellamy, 1974, after Kulczynski). Water supply to the model lake system of Figure 25-9 consists of direct rainfall, runoff, and seepage from the immediate catchment of the lake basin, and a continuous flow of ground water that enters by an inflow stream of sufficient volume to affect the whole lake. The depth of the open water decreases continually because of the sedimentation of organic matter. As the basin progressively fills, the littoral vegetation simultaneously advances toward the center of the

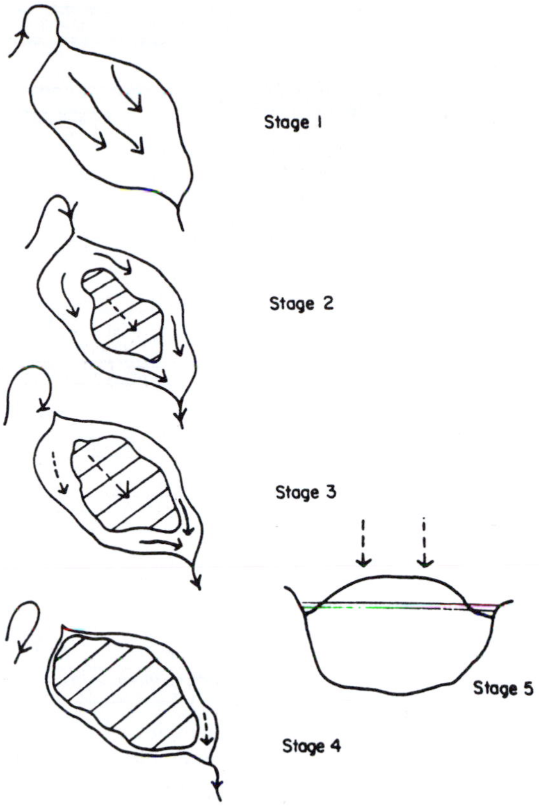

FIGURE 25-9 Idealized successional stages in mire systems from a small, shallow drainage lake (1), through a seepage system (2, 3), to a raised bog deriving most of its water supply from precipitation directly on its surface (4, 5). Hatched area = peat accumulation. (From Moore, P. D., and Bellamy, D. J.: *Peatlands*. London, Paul Elek Scientific Books Ltd., 1974.)

lake basin. Eventually, the entire area is covered with what is loosely referred to as ***swamp*** conditions, with standing water occurring among the vegetation. Production of littoral vegetation increases because of the high productivity of emergent macrophytes, and sediments eventually reach the surface of the original lake. This stage can be referred to as a ***marsh***, for the sediments are continually saturated but there is little, if any, standing water among the vegetation. Under these conditions, decomposition proceeds rapidly. However, for a number of reasons (discussed in detail in Chap. 21 and summarized in Fig. 25-10), inputs of relatively recalcitrant organic matter from the emergent macrophytes occur even more rapidly. The mean pH decreases substantially, and concentrations of dissolved humic materials in the water increase. Changes in the dominant vegetation also occur, and acidophilic flora (especially sedges and grasses) becomes more abundant.

When these mire systems initially develop in a shallow drainage lake (Fig. 25-9), the system is under the influence of continuously or intermittently flowing

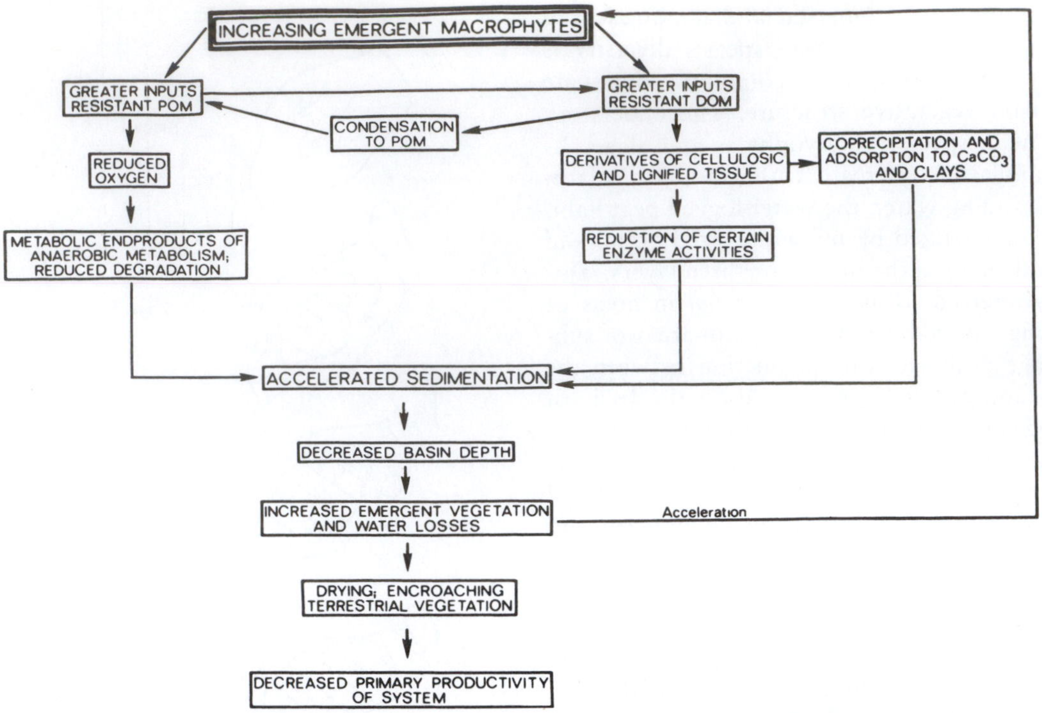

FIGURE 25-10 Relationships of increased loading of relatively resistant organic matter from dominating emergent littoral and wetland flora to controls of decomposition rates, acceleration of emergent macrophyte development, and the ontogeny of lake systems. POM and DOM = particulate and dissolved organic matter, respectively. (From Wetzel, 1979.)

ground water. The ionic composition of the water of these "rheophilous mires"[5] is dominated by calcium and bicarbonate. Allochthonous organic matter inputs are high and result in accrual of peat within the central portion of the basin (Stage 2). In some cases, this peat is sufficiently buoyant to float as a mat. The main flow of water tends to be channelized to peripheral areas. In the third transitional stage of mire development,[6] continued accrual of peat deposits diverts inflow from the basin. Water supply is derived largely from the immediate catchment of the mire and from direct precipitation on the surface. Ionic composition usually is dominated by calcium and sulphate. Further accrual of peat results in large areas of the mire surface that are not affected directly by moving water (Stage 4), but are inundated periodically when the water level rises during heavy rainfall.

Ombrogenous or ombrophilous (ombro = rain) mires[7] (Stage 5) are no longer subject to the influence of flowing ground water. Nearly all of the water, which is dominated by sulphate and hydrogen ions, is received from direct rainfall. The surface of the mire system is too high to be materially influenced by fluctuations in the level of ground water. The change in ionic composition of water from dominance by calcium and bicarbonate to that of sulphate and hydrogen ions is a conspicuous feature characterizing these mire systems (Fig. 25-11). The loss of ions from surface water supplies results in very low salinities and greatly reduced buffering capacity. Under these conditions, relatively small additions of acids can result in a considerable reduction in pH.

The origin of acidity and control of cation exchanges of mire systems result from a combination of factors. The increase in acidity and reduction of cations is particularly marked in communities dominated by the mosses of the genus *Sphagnum*. Water surrounding *Sphagnum*, a dominant genus of mosses over immense areas of the temperate, boreal, and subarctic regions of the world (see e.g., Sjörs, 1961), has a pH usually below 4.5 and sometimes below 3.0. Much of the acidity of the aquatic environment surrounding *Sphagnum* can be attributed to cation exchange (Anschütz and Gessner, 1954; Clymo, 1963, 1967; Brehm, 1970; Glime *et al.*, 1982; and others). *Sphagnum* and other mosses behave as cation exchangers even when dead because of their high concentrations (up to 30% of the plant's dry weight) of polymerized unesterified

[5] Synonymous with terms "low moor" and *Niedermoore*.
[6] *Ubergangsmoore* of older German literature.
[7] Termed "high moors" or *Hochmoore* in early works.

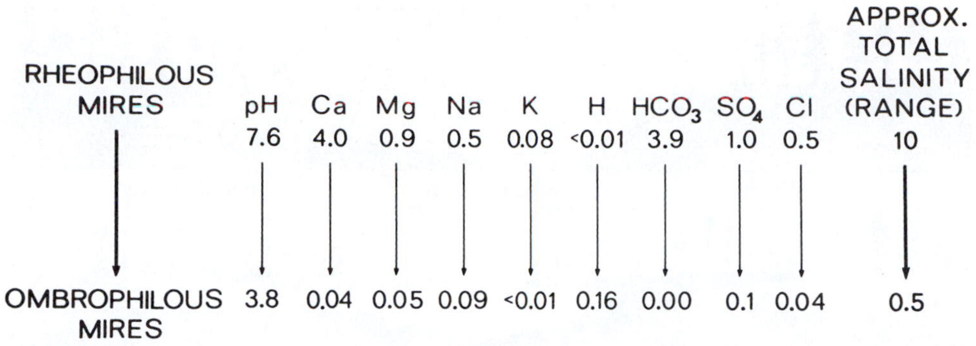

RHEOPHILOUS MIRES
pH Ca Mg Na K H HCO3 SO4 Cl
7.6 4.0 0.9 0.5 0.08 <0.01 3.9 1.0 0.5
APPROX. TOTAL SALINITY (RANGE)
10
OMBROPHILOUS MIRES
3.8 0.04 0.05 0.09 <0.01 0.16 0.00 0.1 0.04
0.5

FIGURE 25-11 General shifts in ionic composition (milliequivalents liter^{-1}) of the water in the transition from rheophilous to ombrophilous mire systems of Western European mires. (From average values of Bellamy and of Sjörs given in Moore and Bellamy, 1974.)

uronic acids.[8] Although small amounts of free organic acids are released from *Sphagnum*, evidence indicates that most of the acidity of *Sphagnum*-dominated mire systems originates from the release of carboxyl-associated H^+ ions by freshly produced plant growth. The hydrogen ions are exchanged for nutritionally important cations that occur in rain or ground water. Some of the acidity of bog pools may be derived from H^+ ions released during the metabolic activities of anaerobic sulfur-metabolizing bacteria (Clymo, 1965, Gorham, 1966), but this source of acidity is clearly minor.

B. Bogs and Quaking Bogs

Littoral vegetation as the dominant source of organic matter produced in excess of decomposition is clearly critical in the senescence of a lake. Bogs are acidic, dystrophic aquatic ecosystems that possess an exceptionally characteristic littoral flora. Bogs do not represent an ontogenetic stage through which all lakes must pass in their route to extinction. The development of a bog system requires climatic conditions of abundant precipitation and relatively high humidity over much of the annual cycle. The littoral vegetation of bogs completely dominates key characteristics of the system: acidic pH, heavy staining resulting from the presence of large amounts of dissolved and colloidal humic materials, and low concentrations of nutritionally important cations.

Commonly, the transition of a small lake to a bog and a terrestrial ecosystem occurs sequentially, as depicted in Figure 25-12. A *Sphagnum* bog climax model of hydrosere succession proceeds from two directions from the margin of a lake or a pond: (a) inward toward the lake or pond center following the classical hydrosere succession (i.e., terrestrialization), and (b) in the other direction outward from the margin into the surrounding forests involving the successional conversion of upland forest to bog (i.e., paludification—swamping or inundation that results in saturated hydrosoils, which retards the decomposition of organic matter) (Godwin, 1956; Klinger, 1996). Water levels of the original water body increase in height because of impeded drainage from peat accumulation (Heinselman, 1963) and the capillarity of the moss vegetation. In advanced stages of succession, the ombrotrophic bogs develop perched water tables that are decoupled from the groundwater table.

The formation of a bog depends greatly on the prevailing climate. In moderately dry regions, the transition is from marsh conditions to terrestrial vegetation, without the characteristic development of mosses. Over much of the temperate zone, however, conditions are sufficiently moist to encourage the colonization and eventual dominance of *Sphagnum* and other mosses among the often predominantly sedge flora. Through a succession of mosses, especially *Sphagnum* species, cations decrease and acidity increases, which effectively decreases rates of decomposition (Chap. 21) and accelerates net accumulation of organic matter. The accumulation of particulate organic matter in the understory, continued growth of plants at the surface, and excellent capillarity of the dead and living mass of mosses raise the water level within the mat above the original level of the lake. In this way, the vegetative mat can develop slightly (usually <50 cm) above the water table; hence, the term *raised bog*.[9]

[8] Similar to sugars, but the sixth carbon is part of a carboxyl group. These long-chain molecules may be mixed polymers containing both sugars and uronic acids.

[9] Raised bogs occur in depressions, often in former lake basins, where losses of water from runoff are not as great as in more upland areas. In areas of high rainfall (>130 cm yr^{-1}) and low evaporation, bog vegetation can extend onto low-lying landscapes that have slopes of >15°, forming ***blanket bogs***. Blanket bogs cover large regions, notably in northern Great Britain, western Norway, and boreal continental areas in general (Sjörs, 1961).

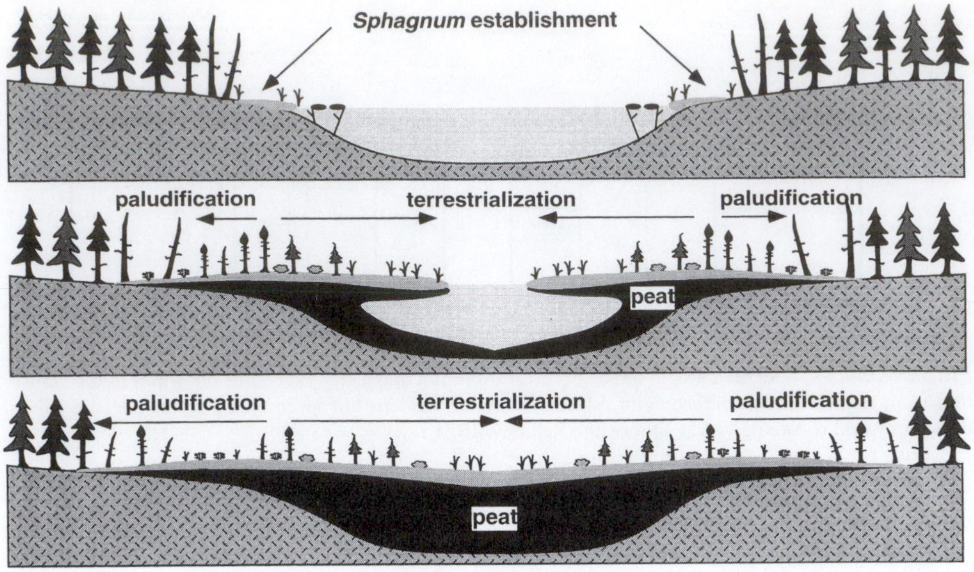

FIGURE 25-12 Bog climax model of succession whereby development of climax bog communities occurs via both terrestrialization and paludification beginning at the margins of a water body. Key stages in the succession involve (*upper*) establishment of peat-forming (e.g., *Sphagnum*) mosses along shore lines; (*middle*) peat accumulation via both terrestrialization and paludification, resulting in slight impoundment and a rise in water levels; and (*lower*) continued terrestrialization and paludification, resulting in the formation of an ombrotrophic bog. (From Klinger, L. F. 1996, reproduced by permission of the Regents of the University of Colorado. The myth of the classic hydosere model of bog succession. Arctic Alpine Res, 28:1–9.)

Moatlike areas of shallow water characteristically surround the central mat of peat; these areas, which are called *lagg zones*, are the remnants of flowing ground water whose path has been diverted around the central peat mat (cf. Fig. 25-13). Organic matter accumulates more slowly in the lagg zones than it does in the raised portions of the bog because of greater movements of water and better oxidizing conditions. Vegetation of the lagg zone is commonly dominated by small graminoid species, bryophytes, and foliaceous liverworts.

A dominant feature during the succession of lakes to bogs is a shift in the origin of nutrients. Initially, most of the nutrients are derived from soil in the drainage basin; in later successional stages, most of the nutrients enter the bog ecosystem via atmospheric precipitation and particulate fallout. In mire systems in general, the terms ***rheotrophic*** and ***ombrotrophic*** have been introduced[10] to designate systems whose nutrients are derived predominantly from surface water/ground water or from atmospheric sources, respectively. These two terms emphasize the interplay of geomorphological, chemical, climatic, and biotic factors that are involved in the succession of wetlands and of bogs. In regions where the ionic composition of rainfall is strongly influenced by oceanic contributions (cf. Chap. 10), such as in western Britain and much of Ireland, ombrotrophic nutrition is only slightly less effective than weak rheotrophic nutrition, and little difference in vegetation results.

Ombrotrophic bogs occur in cool temperate regions (45–65° lat.) with annual precipitation exceeding evaporation and a summer precipitation deficit <100–150 mm (Proctor, 1995). Raised bogs occupy flat sites (filled lake basins, flood plains, and coastal flats), whereas ombrotrophic bogs growing on sloped land (*blanket bogs*) occur only where the summer precipitation deficit is ca. <25 mm. Strong-acid anions in equivalents are less than the total cations, and this "anion deficit" is made up by organic anions of the dissolved organic compounds. Bog plants typically grow slowly and are adapted to low nutrient supply.

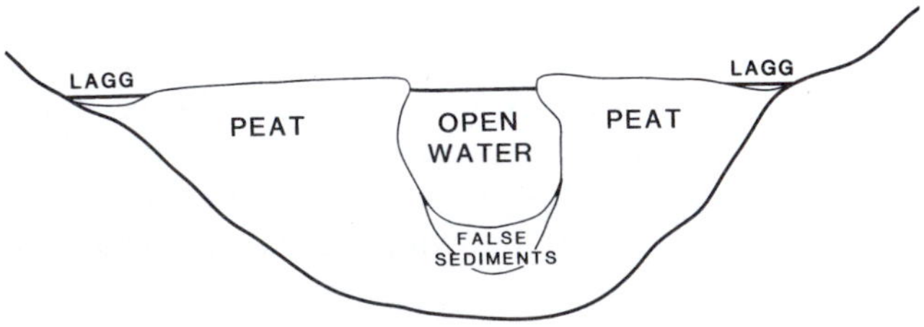

FIGURE 25-13 An idealized quaking *Sphagnum* bog in which the mat encroachment toward the center of the basin overlies littoral peat accumulations and much of the open water.

[10] Originally proposed as "minerotrophic" and "ombrotrophic" by DuRietz (Sjörs, 1961). A detailed discussion of the interrelationships and terminology is given by Moore and Bellamy (1974).

Nitrogen is commonly limiting in unpolluted areas, and phosphorus is limiting in regions of high atmospheric nitrogen deposition.

Quaking bogs represent a particular development of bogs within a relatively deep lake basin of small surface area. The development of *Sphagnum* mosses occurs early in the succession of littoral flora and in geological context leads to a very rapid accumulation of organic matter in the littoral areas (Figs. 25-12 and 25-13). In more advanced conditions, the littoral vegetation encroaches more rapidly on the open water than deposition of peat deposits occurs beneath it. This growth results in a thick mat (several meters) that floats above the open water and sediments consisting of loosely aggregated, flocculant organic matter. Growth proceeds all along the periphery toward the lake center, and eventually covers the entire surface.

A marked feature of quaking bogs is the concentric zonation of plant communities. *Sphagnum* is most often one of several pioneering mosses and dominates the floating mat formation closest to the center of the lake. Within the open water of the bog, certain submersed angiosperms (e.g., *Potamogeton* and *Utricularia* spp.) and floating-leaved macrophytes (e.g., *Nuphar* and *Nymphaea*) may be abundant. Lakeward portions of the mat may contain sedges (especially *Carex* spp.) and low shrubs such as leather leaf (*Chamaedaphne*) and blueberry (*Vaccinium*). Farther landward, tall shrubs and, finally, bog trees, such as black spruce and tamarack (*Larex*), are common (Gates, 1942).

The older portions of the mat are grounded firmly in the basin by underlying deposits of peat. The floating mat, growing considerably above the level of water-saturated flocculent organic matter, often is attached only by the vegetative mass and can be displaced by the added weight of a person and "quake." The sensation is that of walking on an immense floating, saturated sponge. The sedimentation of much particulate, littorally derived organic matter in the open water areas, results in loosely aggregated, flocculent sediment that is not compacted sufficiently to support small weights (e.g., an anchor), often for many meters; hence, the term *false bottom* or *false sediments* (Fig. 25-13).

Quantitative information on the productivity and population dynamics of biota of the pelagic and profundal regions of bog lakes and quaking bogs is extremely sparse. Much more knowledge exists on qualitative aspects simply because of the extremely diverse and interesting biota that are adapted to the extreme conditions of bogs. The microflora of the plankton and of small pools among hummocks of *Sphagnum* and larger plants are characterized by a great diversity of algal species, although few are ubiquitous. Most abundant are desmids, often represented by several hundred species. Species of cyanobacteria, chrysomonads, dinophyceans, and diatoms are adapted to bog conditions but seldom are found in abundance. The microfauna is dominated by testaceous rhizopod protozoans and other protists and rotifers; very few cladoceran and copepod zooplankters are tolerant of the bog milieu.

Larger animals have not adapted well to the extreme acidity and low salinity of bog waters. Species diversity of larger animals is low and entire groups are lacking or poorly represented. For example, in acidic, quaking *Sphagnum* bogs, sponges, coelenterates, ostracods, hydracarnid mites, oligochaetes, Ephemeroptera, Malacostraca, mollusks, nematodes, flatworms, and fish are completely absent or very poorly represented.

The genesis and ontogeny of bog systems can be viewed as a transitional succession of lake basins of greatly differing geomorphology to wetlands in which organic matter of macrophytic vegetation accumulates in excess of complete degradation. The causal processes regulating productivity can be effectively viewed from a short-term biogeochemical standpoint as operating in a reasonably steady-state system. However, the very nature of excessive accumulation of organic matter, the successional pathways of which are profoundly influenced by climatic and hydrological factors, indicates that the systems are not balanced even over very brief geological time periods.

III. SUMMARY

1. The interacting mechanisms regulating autotrophic productivity follow several pathways of development (ontogeny) as lake basins undergo succession and are ultimately obliterated as aquatic ecosystems and transformed into terrestrial ecosystems. Many rate-regulating mechanisms are influenced by the general transition from a dominance of allochthonous loading of nutrients and organic matter to accelerated loading of organic matter from littoral and wetland productivity.
   a. In the early stages of lake ontogeny, particularly among lakes formed by glaciation, autochthonous and allochthonous productivity is severely limited by climatic conditions, especially cold temperatures.
   b. In the intermediate stages of ontogeny, autochthonous autotrophy by phytoplanktonic production is supplemented with some contributions by littoral production.
   c. In the later and terminal stages, however, a rapidly accelerating shift occurs to a total dominance of productivity by littoral and wetland components.

d. The planktonic-dominated stages of lake development are relatively lengthy; during these stages, organic sedimentation and deposition are nearly balanced by decomposition.
e. In shallow basins and in lakes in which gradual sedimentation has reduced the depth sufficiently, massive production by littoral plants and wetland *emergent* macrophytes dominates inputs of photosynthetically accrued organic matter.
   i. The increasing contribution of derivatives of lignified and cellulosic tissues results in reduced rates of, and less complete, decomposition, and
   ii. Leads to accelerated rates of sedimentation and favors conditions conducive to enhanced emergent macrophyte development in a cyclically reinforcing process (cf. Fig. 25-10).

2. Increased inorganic nutrient loading, particularly of phosphorus and combined nitrogen, is fundamental to initial eutrophication and to maintenance of high sustained productivity by the phytoplankton community.
   a. Low rates of productivity in oligotrophic lakes are maintained to a large extent by low inputs of inorganic nutrients from external sources. Low production of organic matter, concomitant low rates of decomposition, and oxidizing hypolimnetic conditions result in low rates of nutrient release from the sediments in a cyclical causal system (Fig. 25-1). Dissolved organic compounds from both internal and external sources are usually low, in part as a result of solar photolysis, which results in limited availability of organic micronutrients and reduced complexing capability for essential inorganic micronutrients.
   b. Under eutrophic conditions, the loading rates of inorganic nutrients, especially of phosphorus and combined nitrogen, are relatively high (Fig. 25-2). As rates of photosynthetic productivity and organic loading to lower strata increase, nutrients are released from sediments into anoxic hypolimnia, increasing recycling rates.
   c. Phytoplanktonic productivity of eutrophic lakes increases markedly. In increasingly eutrophic lakes, dense algal communities reduce light penetration and thereby compress the depth of the trophogenic zone. Light limitations caused by self-shading set an upper boundary to the phytoplanktonic photosynthetic productivity, beyond which further increases are not possible, regardless of increased nutrient availability (Fig. 25-3). Further increases in photosynthetic productivity are possible only by extending the length of the growing season (e.g., in the tropics) or by increasing turbulence and frequency of exposure to available light (as in artificially mixed sewage lagoons).
   d. In calcareous hardwater lakes, the availability of certain organic micronutrients (such as vitamins) and inorganic nutrients (particularly phosphorus and inorganic micronutrients) can be suppressed by inactivation, that is, chemical competition or sedimentation with inorganic particulate materials (e.g., coprecipitation with $CaCO_3$) (Fig. 25-4). Maintenance of reduced productivity in calcareous lakes by lowered nutrient availability depends upon high carbonate loading from the drainage basin. Calcareous loading may be reduced through long-term leaching or overcome by increased loading of dissolved organic matter. The result is a relatively rapid transition to eutrophic conditions. Under certain conditions, high cationic (e.g., Ca or Mg) loading also can be counterbalanced and reduced by cationic exchange mechanisms of littoral-wetland plants (e.g., bryophytes), which may result in the transition of calcareous lakes to acidic bogs.

3. Many lakes receive large amounts of dissolved and particulate organic matter from allochthonous and surrounding wetland-littoral sources. These lakes are commonly heavily stained as a result of the abundance of dissolved humic compounds of plant origin; such lakes are termed *dystrophic*.
   a. Phytoplanktonic productivity is usually moderately high in dystrophic lakes (Fig. 25-6), and productivity of the surrounding littoral vegetation is moderate to very high.
   b. The surrounding macrovegetation can function as effective nutrient scavengers and frequently reduces the nutrient loading that reaches the open water. In some cases, particularly in bogs, the vegetation can effectively shift the lake per se from an open to a closed basin, so that nutrient inputs to the lake are mainly from atmospheric precipitation.

4. Terminal stages in the successional sequence of lake ontogeny are dependent upon the types of macrovegetation in the littoral zone and wetland areas surrounding the basin. The development of the vegetation is, in turn, strongly influenced by prevailing climatic conditions (rainfall and humidity) and the geomorphology of the region.
   a. Many fertile eutrophic lakes develop ever-increasing amounts of emergent littoral vegetation until littoral vegetation encroaches over the entire lake basin (Fig. 25-8). The resultant ***swamp*** conditions, with standing water among the emergent vegetation, gradually succeed to ***marsh*** conditions, with little or no standing

water over the water-saturated sediments. Evapotranspiration from the vegetation can gradually exceed water income, permitting invasion of a terrestrial flora. Ecosystem productivity decreases in the transition to terrestrial conditions.

b. Under climatic conditions of high rainfall and humidity, various types of mire ecosystems develop; in these systems, partially decayed plant organic matter (peat) accumulates in abundance.
   i. *Rheophilous mires* can develop from shallow drainage lake basins in which large amounts of organic matter accumulate in the depression.
   ii. The development of climax bog communities occurs via both terrestrialization and by paludification (inundation), beginning at the margins of a water body. Key stages of the succession involve establishment of peat-forming mosses, peat accumulation, and progressive terrestrialization and paludification (Fig. 25-12).
   iii. Eventually, the mire is no longer under the influence of flowing ground water. In these *ombrophilous mires*, water and nutrients are received primarily from rainfall. The salinity, buffering capacity, and pH of the water are low.
   iv. The acidity of the water often increases during mire development. Mosses, particularly *Sphagnum*, often develop in profusion in *bog* mires. Mosses have effective cation-exchange mechanisms, in which divalent cations are retained with a commensurate release of H^+ ions and organic acids. *Sphagnum* and other acidophilic plants can develop above the mean water table (raised bog) yet still acquire sufficient water by capillarity.
   v. Bog mats and underlying peat accumulations can gradually encroach upon, and eventually eliminate, the open water of the lake remnant within the bog.
   vi. Individual bog succession proceeds at rates and along developmental pathways that are controlled by a complex interplay of geomorphological, chemical, climatic, and biotic factors.

26

INLAND WATERS: UNDERSTANDING IS ESSENTIAL FOR THE FUTURE

I. WATER QUALITY IS ESSENTIAL; WATER QUALITY IS BIOLOGICAL

Water is an essential physiological requirement of humans for survival and for provision of food and basic living needs. Even with the catastrophic dangers that are upon human society because of excessive population growth, availability of fresh water could be reasonably satisfactory for modest standards of living if essential characteristics of water quality, allocation, and prudent use were instituted and practiced.

Water problems are not only of hydrological availability. We have great quantities of water in the oceans and it is close to human concentrations, as about 50% of humans are aggregated in coastal regions. It requires large amounts of energy to desalinate and clean marine water to a quality acceptable for human use. That energy requirement is too great at the present time simply because water is used and misused on such a massive scale without regard for its true economic value to societies. Moreover, much water is needed inland for agriculture and industry, and distribution costs are simply too great at present rates of water use and waste.

A false idea has emerged in modern society that water is a basic human right. Humans do not have a right to water but rather have a responsibility for wise and optimal use of available water. Responsibility is an obligation for human survival and realization of a reasonable standard of living. If that responsibility is given to society and its leaders, then it is essential that humans understand the fundamentals of water science for its responsible use and the effective management of water resources for both hydrological availability and acceptable water quality.

Contemporary management of freshwater resources is often performed in erratic and irrational ways. The rationale behind many practices is commonly purely hydrological, without the slightest interest in or appreciation of the biotic interactions of the water bodies themselves or the long-term related effects on the climate of large regional areas and, in the end, on the biosphere. Water quality is biologically mediated. Aquatic ecosystems have changed slowly over time. Adjustments have occurred internally as organisms adapted to these changes. The result is a complex integration of the biological components within environmental variability. Certainly one can poison a water body to such an extreme extent that biotic metabolism is eliminated, but that is not usually the situation. Inland aquatic ecosystems are biological systems. Understanding biogeochemical cycling and metabolic regulation of productivity is essential for effective management and use of fresh waters. The biological metabolism of aquatic ecosystems *does* matter and must be considered and understood in all the manipulative measures taken in management.

A. "End of Science" Nonsense and Effects on the Discipline

A strange theme that irrupts periodically and is especially prevalent at the present time is that "all major functions of the natural world are known" and that we "know enough science to understand and manage our world." That is to say that we understand the mechanisms controlling the chemical and biological processes of freshwater ecosystems so that we can manage, use, and restore damaged ecosystems effectively. Such statements are irrevocably and absolutely nonsense. Limnology has just recently emerged from descriptive stages and is just beginning to evaluate controlling mechanisms that regulate and produce the observed biological and chemical results. Statements to the effect that we know enough to manage lakes and rivers are tantamount to saying that we know enough at present to control and manage human cancer and enough to restore health—again a nonsense statement. Excising an organ or irradiating cells to delay the spread of cancer are not solutions to cancer. Causality must be delineated and understood.

Only feeble insights exist regarding control mechanisms in lake, reservoir, and river ecosystems. Fortunately, improved or completely changed insights of previous conceptions of how these ecosystems work are being gained. For example, less than a decade ago we were aware of the abundance of humic substances dissolved in lake and river waters, but prevailing dogma assumed that most dissolved organic matter was recalcitrant to enzymatic hydrolysis and relatively unavailable for use by bacteria and other organisms. It is now known from extensive recent experimental work that both natural ultraviolet light as well as visible light induce major photolytic changes in the complex organic molecules and generate large quantities of readily utilizable substrates for microbial metabolism (cf. Chap. 23). This important process makes available a huge reservoir of organic carbon and energy that was largely ignored previously. Moreover, much of that source of organic matter and energy is derived from structural tissues of higher plants of external and littoral sources that were previously treated as unimportant. These data also explain in part why good, quantitative budgets of organic carbon fluxes would not balance without including this massive external subsidy of energy. The entire picture of how lakes and rivers operate is undergoing changes, and there will be many analogous important discoveries in the future. These changes in understanding must modify how we manage lakes and rivers.

II. BIODIVERSITY OF INLAND WATERS

The biodiversity of most microbial, plant, and animal groups of stream, lake, and wetland ecosystems is very poorly known. Moreover, it is likely erroneous to believe that the biodiversity of many taxa in temperate ecosystems is appreciably less than in tropical inland aquatic ecosystems. For example, the rivers and streams of Alabama, a "hot spot" within the United States, contain 43% of all gill-breathing snails, 52% of freshwater aquatic or semiaquatic turtles, 60% of mussels, and 38% of freshwater fishes of North America (Lydeard and Mayden, 1995). Over half of the species in some of the groups mentioned are either threatened or endangered under the U.S. Endangered Species Act of 1973.

Over a decade has passed since the perceptive insights and concerns of E. O. Wilson (1985, 1988, 1992) brought the problems of diminishing global biodiversity into overdue prominence. Rates of loss of biodiversity vary both because of differences in intensities of disturbances to the habitats as well as the number of species inhabiting particular areas. Because of the extremely high biodiversity in many tropical forested areas that are very vulnerable to catastrophic alterations by human activities, great attention has been given to these critical ecosystems or "hot spots" (Myers 1988; Wilson, 1992). Essentially, all of the emphasized biodiversity crises have focused on terrestrial ecosystems, particularly in tropical rainforests, which are under intensive siege.

A. Water Quality Alterations as Disturbances to Biodiversity

The fresh waters of the world are collectively experiencing markedly accelerating rates of degradation.

These major sources of disturbance impact the biodiversity of fresh waters in many ways. Direct chemical toxicants released into surface and ground waters are common. Many forms of heavy metals, inorganic reducing agents, and organic compounds enter the environment and eventually fresh waters. Although some are inactivated, such as by chemical precipitation, or oxidized, many have long residue resident times (see, e.g., Francko and Wetzel, 1983). Despite dispersion and dilution in the aquatic environment, bioconcentration of both metals and organic compounds is common, by which toxicity can be increased exponentially. The biological effects of many of these compounds are unknown. Elemental and compound toxicity is increased markedly when the pollutant substances are radioactive or acidic. Increased acidity in poorly buffered fresh waters can increase the bioavailability and potential toxicity of metals, such as aluminum, that are not normally abundant in soluble, reactive states (e.g., Hall *et al.,* 1985; Havas and Likens, 1985; Cronan and Schofield, 1990).

Plant nutrients, particularly phosphorus and nitrogen, can lead to well-understood enrichments of plant and other organic productivity in fresh waters (cf. Chaps. 13, 15, and 25). This eutrophication process often results in enhanced rates of decomposition and in chemical conditions that greatly reduce or eliminate suitable habitat for many species of plants and animals. Similar excessive loading of organic matter and enhanced rates of degradation result from organic sewage from human populations, industry, and agriculture. A further common pollutant that reduces habitat availability markedly is the suspension of finely divided organic and inert inorganic matter. Erosion and transport of such suspensions are increasing as large areas of forest and interface buffer zones of former wetlands between land and water are eliminated, primarily for agricultural expansions. As a result, flow patterns of surface waters are often altered and benthic habitats of surface waters can be obliterated by sedimentation.

Changes in biodiversity in freshwater ecosystems can arise from many disturbances. The introduction of certain competitively superior species can result in marked losses of biodiversity. The infamous example of the introduction of the Nile perch into Lake Victoria of eastern Africa resulted in the extinction of over 200 species of its cichlid fish taxa in two decades (Kaufman, 1992; Lowe-McConnell, 1994, 1996; Goldschmidt, 1996). Many other examples exist of introductions of exotic plant and animal species that resulted either in direct destruction of prey species or indirect alteration of habitats required by many species. Dense floating macrophyte communities and other eutrophication-associated excessive plant productivity often result in deoxygenation and reduction in habitat and elimination of many plant and animal species.

Technological and social responses to degradation of water, and the rates at which changes can be implemented, vary under different societal structures (Francko and Wetzel, 1983; Wetzel, 1992a). Effective utilization of extant freshwater resources is complicated by distributions of humans and their exploitations in regions low in water availability. Accelerating anthropogenically induced changes in hydrological patterns associated with flood controls and water supply reservoirs and with ongoing and impending climatic changes further obfuscate efficient utilization practices. Certain societies can cope with pollution and water availability constraints and even reduce freshwater degradation with habitat restoration. In most of the world, however, human population growth continues without any significant reduction of rates. Until human growth is stabilized, either by intelligence or catastrophes, further losses and degradation of fresh waters can be controlled only partially on a global basis (Wetzel, 1992a). It is sadly remarkable that organisms as intelligent as humans can be so stupid in unrestrained population growth and unsustainably exploiting and destroying the environment essential for survival. Other animals of far less advanced intelligence have evolved the recognition that it is essential not to foul their living quarters. Control and reversal of degradation requires a proper economic valuation of fresh waters for efficient, conserved utilization of water supplies for agricultural, industrial, and residential purposes.

B. Shifting Resilience, Stability, and Biodiversity of Freshwater Ecosystems

Many definitions of stability exist. A central feature of stability is the ability of a property (nutrient concentration, population, and community) of an ecosystem to return toward a steady-state equilibrium following a disturbance. Resilience, or relative stability, is a measure of the rate at which the property or system approaches steady state following the disturbance. The return time (T_R) to equilibrium can be formalized mathematically by integration of the asymptotic curve of return to equilibrium (e.g., DeAngelis, 1980, 1992). Resilience is approximately the inverse of the return time ($1/T_R$). As a rate of recovery, resilience is a function of the turnover time of limiting resources, such as a limiting nutrient, and thus approximately the inverse of turnover time ($1/T_{res}$) (DeAngelis, 1992). A faster rate of nutrient input per unit biomass, for example, will generally decrease the nutrient turnover time and thus increase the rate at which a system can recover from a disturbance. Therefore, any mechanism by which the rates of nutrient turnover and recycling are increased should generally result in a reduction in resilience to disturbances.

Many theoretical ecologists have evaluated resilience and its couplings to the structure of ecological food webs. Contemporary views of ecosystem stability focus on the rates of recovery (i.e., resiliency) after perturbations (Pimm, 1984; DeAngelis, 1992) or the strengths of trophic interactions in community food webs (de Ruiter *et al.,* 1995). Largely based on terrestrial studies, the Eltonian diversity–stability hypothesis suggested that, because of the many different traits of multiple species, more diverse ecosystems will likely have species that will survive and expand during and following an environmental disturbance and compensate for those species that are reduced by the perturbation (supporting evidence reviewed by Tilman, 1996). Thus, a more species-diverse ecosystem should be able to be more resilient to disturbances than a less biodiverse system.

Because physiological differences and tolerances among species can be small, the individual interactive strengths among some species in an ecosystem can become saturating at high biodiversity. As a result, a point is reached where increasing species may be functionally redundant and have reduced individual impact on ecosystem processes (e.g., Vitousek and Hooper, 1993). On the basis of both theoretical and experimental grounds, only a small fraction of species manipulations have strong influences on food-web structure (de Ruiter *et al.,* 1995). This species-redundancy hypothesis implies an appreciable functional resiliency in which the ecosystem can compensate in collective metabolism and biogeochemical cycling when disturbed. Thus, although population dynamics become progressively less stable as the biodiversity and number of competing species increases (May, 1973; Tilman, 1996), biodiversity can enhance the resiliency of many community and ecosystem processes in terms of the rate at which the system metabolism returns to equilibrium states following a perturbation.

Dead organic matter or detritus clearly functions in the regulation of nutrient dynamics of ecosystems and, as a result, has manifold implications for ecosystem stability (Wetzel, 1983, 1984, 1995; DeAngelis 1992). Nine essential points impact biodiversity and ecosystem stability. Some of these points are well supported by empirical data; some are hypothetical but probable in most freshwater aquatic ecosystems.

1. Most organic matter of aquatic ecosystems is dead (detritus) and most is in dissolved or in colloidal form. Dissolved organic matter is energetically and functionally detritus (Chap. 23). This source of organic matter is physically unavailable to the metazoan food web and can enter the metazoan food web primarily only through bacteria.
2. Most of the instantaneous mass of dissolved organic matter (soluble detritus) present in lake, pond, and river ecosystems originates from the structural tissues of higher plants. These higher plants occur in the littoral, adjacent wetland and floodplain regions, and upland terrestrial areas. Upon partial decomposition of these plants, the relatively recalcitrant macromolecules from structural tissues are imported in soluble or colloidal form to the aquatic systems.
3. Once within the aquatic ecosystem, the recalcitrant DOC is metabolized, although slowly (ca. 0.5–2% per day) as mediated by a combination of hydrolytic enzymatic and physical (e.g., photolysis, Wetzel *et al.,* 1995) processes.
4. Bacterial, viral, and much of protistan heterotrophic metabolism (respiration) represents a major output of carbon, largely as CO_2, from the aquatic ecosystem. Although this output is a loss of carbon from the ecosystem, it is not an energetic loss from the ecosystem. Nonpredatory metabolism and death (biochemical senescence, or lysis) of prokaryotic and protistan heterotrophs dominate. Although the organic carbon packaged in these microbial heterotrophs is largely not available or little used for ingestion by higher trophic levels, this nonmetazoan pathway is a dominant energetic and carbon flow within the entire ecosystem (pelagic, benthic, and littoral) and dominates nutrient regeneration from organic carbon substrates. Even ignoring the benthic portions of lakes where most of organic carbon metabolism occurs, and considering only the pelagic metabolism of lakes, for example, most of the organic carbon entering the system does not reach higher trophic levels (Wetzel *et al.,* 1972; Wetzel, 1983a, 1984, 1995; Gaedke and Straile, 1994; Gaedke *et al.,* 1996).
5. There is very little storage capacity for organic carbon within the higher trophic levels, and low system residence times results in rapid cycling of carbon and nutrients within the higher food-web components. Such rapid cycling and recycling results in a reduction in the resilience of the higher trophic levels of the ecosystem.
6. Most of the storage of organic carbon occurs in the dissolved organic carbon (DOC) compartment in the open water and in the particulate organic carbon (POC) and DOC compartments of the sediments. In both of these compart-

ments, the soluble organic carbon of the pelagic of lakes or running water of streams and the DOC and POC of the sediments, the cycling of carbon is slowed. That rate of cycling is slowed in the pelagic by the recalcitrant chemical composition of the dissolved organic carbon emanating largely from higher plants and in the sediments additionally by the anoxic conditions that prevail almost universally among aquatic sediments. The reduced rates of cycling and recycling result in an inherent increase in the resilience stability of the ecosystem.

7. Any process that changes the sources and hence chemical composition of soluble organic matter and their rates of loading to or within aquatic ecosystems will affect the rates of nutrient recycling. Reduced percentage composition of recalcitrant dissolved organic matter will increase the rates of nutrient recycling and decrease the resilience of the system to return to an equilibrium state following a disturbance.
8. Pivotal to these arguments is the coupling of the chemical type of organic matter loaded to the aquatic ecosystems. Shifts from organic matter inputs dominated by in-lake (autochthonous) algal production to organic matter loadings dominated by littoral/wetland and external (allochthonous) organic matter, largely in the form of imported dissolved organic macromolecules, could result in decreased rates of degradation of total organic matter. As a result, the shift would be toward decreased rates of nutrient and carbon recycling and increased resilience to disturbance.
9. These arguments address total energetic fluxes of the aquatic ecosystem, of which only a small portion consists of the metazoan higher food web. Much, and often most, of the energetic and material (carbon and nutrients) fluxes never enter the metazoan food web, as was long ago demonstrated quantitatively (Chap. 23) and theoretically (Patten, 1985, 1995). These arguments, however, do not imply in any way that the microbial metabolism pathways are not coupled with higher freshwater food webs:
   a. Rates of nutrient cycling and recycling *in the entire aquatic ecosystem* are governed largely by bacterial, fungal, and protistan metabolism either (i) directly by hydrolytic degradation of *dissolved* organic substrates or (ii) indirectly by modifying the retention (e.g., adsorption to particles) and movements of nutrients as a consequence of the release of metabolic products into, and alteration of the redox of, the environment. Higher trophic levels, in particular herbivorous zooplankton, can have a significant effect on nutrient recycling by partial heterotrophic decomposition of particulate organic matter, which can decrease nutrient turnover time in the food web by creating short circuits in nutrients directly back to the nutrient pool without delays of storage in particulate detritus (DeAngelis, 1992). An important distinction made in the present text (Wetzel, 1984, 1995) is that most of the organic detritus is in dissolved form and includes major loading sources from allochthonous and littoral sources. This point differs from that of others, such as DeAngelis's (1992) excellent synthesis, which largely restricts the detrital compartment in aquatic ecosystems to that particulate detritus emanating from the metazoan food web.
   b. Any factor that influences the rates of nutrient and carbon cycling will influence the resilience of the ecosystem to disturbances. Changing sources of organic matter, as discussed later, and hence bacterial metabolism and nutrient cycling thus change resilience and biotic stability.
   c. A wealth of limnological data from a spectrum of hundreds of lake ecosystems suggests that with a shift in nutrient loadings, concomitant shifts in the development of photosynthetic producers and loadings of organic matter occur. During the common sequential development of lake ecosystems over time (reviewed in Chaps. 23 and 25), shifts in the ratios of higher vegetation versus algal dominance can occur. Increased relative organic loadings from higher vegetation result in proportionally greater loadings of recalcitrant DOC, which can suppress nutrient cycling and increase resilience of the ecosystem.
   d. In addition, the development of higher vegetation in littoral and wetland combinations increases enormously the habitat heterogeneity, commonly by a factor of 10, 100, or more. Species diversity nearly always increases under these circumstances by at least an order of magnitude among nearly all major groups of organisms, particularly among the lower phyla.

Resilience or relative stability in an aquatic ecosystem is governed by *process rates,* that is, the rates at which energy and materials are transferred among biotic components (Wetzel, 1999b). Because so much of the organic matter of aquatic ecosystems is detrital, most of the assimilation and decomposition metabolism is microbial (bacterial, fungal, or protistan). Each of the species constituting the biotic groups has a physiological range for growth, competition, and reproduction within the environmental constraints. Some of these ranges differ with developmental or temporal changes in life histories of the different species, which can affect species responses to disturbances. Resilience occurs at the community level, however, from an ecosystem perspective within the medley of combined species process-response capabilities.

Physical and chemical environmental conditions, mediated by a host of biogeochemical and climatic/meteorological factors, regulate the bounds of these material and energy flux rates. The average range of these environmental conditions is reasonably finite for different ecosystems and habitats within the ecosystems. Major deviations of natural ranges of these environmental conditions from the average can arise from human or abnormal natural disturbances. The collective, integrated ecosystem responds to major disturbances by shifting dominant metabolism from one species to alternative species as the physiological tolerances of certain species are exceeded and they succumb. In ecosystems with greater species diversity, physiological and behavioral differences among many species may be small. Within limits, productivity and energy flux rates may increase with mild disturbances that remove certain species, because greater species biodiversity can result in increased interspecific competition and cause a reduction of community productivity and biomass. Given sufficient information, one can model the differences in physiological ranges and abilities of the individual species to tolerate change and still maintain stable productivity and reproductive capacity.

Greater biodiversity may have a greater collective effect by improving the ability for ecosystems to recover from large disturbances. Reduced biodiversity increases vulnerability by reducing the total collective physiological tolerances of the community to large habitat changes/disturbances. Recovery after a major or catastrophic disturbance would be slower, with a reduced aggregation of residual physiological ranges within the remaining species. Recovery then must depend to a greater extent on slower fortuitous methods, such as importation of species rather than generation from residual surviving species or slow recolonization processes such as from remnants in resting stages or seed banks. In some cases, such as in many ponds, streams, and reservoirs in clay-rich regions where high turbidity often occurs with successive rain events, total submersed photosynthetic productivity is intermittently but repeatedly suppressed; biodiversity is likely also suppressed and restricted to *r*-strategist species with high reproductive potential to utilize periods between turbidity disturbance events.

C. Extent and Timing of Disturbances and Shifting Stability

As implied in the earlier introductory comments, disturbance to freshwater ecosystems can occur in many forms and to different extents. Certain perturbations can be catastrophic, such as overwhelming a lake or stream with an organic or inorganic poison in which most of the biota are eliminated (e.g., Hanazato, 1994; or copper poisoning of an entire lake, Corbella *et al.*, 1958; Bonacina *et al.*, 1973). However, many disturbances are more gradual over long periods of time (months or years), such as nutrient enrichment, or irregularly episodic and often of short duration (days or weeks), such as severe flooding and scouring of a section of a river. Several examples are worthy of brief discussion to point out couplings of biodiversity to ecosystem stability (resilience) and the type and extent of disturbances.

The ensuing hypothetical model of the responses of the photosynthetic productivity of lake ecosystems to changes in nutrient loadings from the drainage basins (Fig. 26-1, *upper panel*), originally set forth by Wetzel and Hough (1973) and variously improved (Wetzel, 1983a; Carpenter *et al.*, 1998), has been abundantly verified by nutrient and comparative primary productivity data of phytoplankton, attached algae, and macrophytes from dozens of lakes in different stages of successional development. The differences in plant productivity result in very different types of chemical composition of organic matter loaded to lakes, especially that associated with the structural tissues of higher plants. Because of the large amounts and relative chemical recalcitrance of dissolved and particulate organic detrital sources from higher plants, heterotrophic utilization of this pool is slowed (Chap. 23). The resulting large but slow metabolism of organic detritus provides an inherent ecosystem stability that energetically buffers the reserves from rapid exploitation and nutrient recycling. Alterations to the quality and quantities of dissolved organic matter entering the aquatic ecosystem from the drainage basins will influence this inherent chemically mediated metabolic stability. In this context, forested watersheds, undisturbed riparian flood plains, and wetland/littoral zones are important to the metabolic stability within the lake and stream ecosystems. Alterations to these organic loading sources will be translated into altered stability within the lakes and streams. Stated another way, the organic carbon cou-

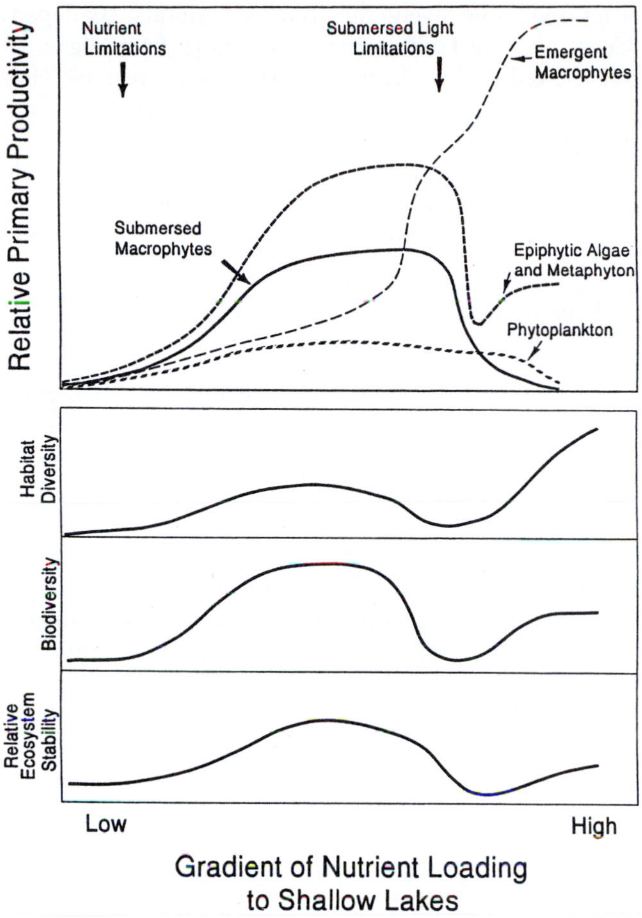

FIGURE 26-1 Relative changes in primary productivity of phytoplankton, macrophytes, and attached microflora and habitat diversity, species diversity, and ecosystem stability along a gradient of nutrient loading to lake ecosystems (see text for discussion). (From Wetzel, 1999b.)

plings between the drainage basin and land–water interface zones are metabolically as important to lake and stream energy and carbon flux stability as are the traditionally studied nutrient (N, P, Si) loading relationships to food-web productivity. Indeed, the predominating recalcitrant chemical quality of the organic reserves retards the rates of utilization and related nutrient recycling within the lakes and streams. As the sources of the organic matter change, the chemical recalcitrance to degradation can change. Alterations of *rates* of utilization, nutrient recycling, and energetic resilience can ensue as a result.

These quantitative differences in photosynthetic organic matter sources in relation to nutrient loadings from the drainage basin can be integrated with hypothetical differences in biodiversity. These relationships can be coupled to lakes of different developmental stages among a series of, for example, a hundred lakes along the trophic gradient.

A common disturbance to lake and reservoir ecosystems involves progressive increases in loading of nutrients. These loadings, in excess of losses to sites of temporary or permanent inactivation, such as to the sediments, result in enhanced nutrient availability for phytoplankton and other autotrophs and increased rates of growth and productivity. If nutrient loadings increase to an oligotrophic lake, increased productivity is rapid. Similarly, if the disturbance is brief (i.e., the duration of increased nutrient loading is relatively short), nutrient cycling is rapid, the ecosystem will recover rapidly, and productivity would be reduced proportionally to the load reduction. The return time of such an oligotrophic ecosystem is high, but the resiliency is low. Species diversity of the plankton tends to be high in oligotrophic lakes, but because of limited habitats, particularly associated with littoral areas and surfaces, composite biodiversity of the lake ecosystem would be low (Fig. 26-1).

As nutrient loading increases, particularly among shallow lakes that predominate globally (Wetzel, 1990a), a marked shift in productivity occurs from the pelagic to the attached surfaces associated with living aquatic plants and the particulate detritus of senescing macrophyte biomass (Fig. 26-1). Under these conditions, primary productivity and biomass increase greatly. Habitat diversity among the massively dissected surfaces of submersed aquatic plants increases exponentially, and biodiversity among attached biota and associated metaphyton would increase by at least an order of magnitude. It is likely that among all autotrophic and heterotrophic microbial groups, as well as most of the smaller metazoans, >90% of the species are sessile in association with surfaces.

As nutrient loading increases further, phytoplanktonic productivity per unit volume increases, but self-shading by the algae restricts the depth of the photogenic zone. Phytoplanktonic productivity per unit area declines precipitously under these eutrophic conditions and is usually accompanied by a marked decrease in planktonic biodiversity as well (Fig. 26-1). Submersed light limitations also reduce or eliminate submersed macrophyte growth and attendant surfaces for microbial growth. The elimination of photosynthesis of attached microbiota decouples the metabolic mutualism to the attached heterotrophs and decreases collective productivity markedly (Wetzel, 1993a; Neely and Wetzel, 1995). Biodiversity of the attached biota would decline precipitously. The losses of attached microbial communities and their coupled metabolism cause a massive reduction in the capacities of the lake ecosystems to retain loaded nutrients and dissolved organic compounds. Because of the great accumulation of nutrients in the sediments from a largely planktonic eutrophic system, the time for recovery from a reduction of the loading (reduced disturbance) would be long and not proportional to nutrient load reductions. Under these conditions, after the loading perturbations end, the internal nutrient loading and recycling could be greater than external nutrient loading. With increased nutrient recycling, resiliency (relative stability) would be reduced.

Continual high nutrient loadings and hypereutrophic phytoplanktonic conditions generate large areas of anaerobic reducing conditions and lower rates of decomposition of organic matter. Production exceeding decomposition leads to rapid sedimentation and generation of increased shallow habitat conducive to colonization by emergent macrophytes. The high productivity of emergent macrophytes increases the proportion of lignocellulose supporting tissues that are relatively recalcitrant to rapid decomposition, particularly under reducing conditions. The collective result is markedly increased productivity and habitat diversity (Wetzel, 1979, 1990a). However, although habitat diversity is high among very shallow wetland-dominated waters, environmental fluctuations are also more extreme than in continually submersed habitats. Quite different community structures occur and biodiversity is likely tempered as a result, though comparative data to support this statement are few.

D. Biodiversity and Nutrient Recycling (Turnover Rates)

Fragments of data from many descriptive and experimental studies suggest that *greater biodiversity results in a greater efficiency of utilization of nutrients.* Stated quite simply, greater biodiversity likely results in a greater physiological diversity to cope with natural vagaries in environmental parameters. Competition for resources—nutrients in this specific case—is intensified. The efficiency of nutrient retention in an ecosystem would be maximized under conditions of greatest *microbial* community diversity. That microbial diversity (bacterial, algae, fungi, and protists) is maximized in the attached communities where habitat diversity and microenvironmental differences are the greatest. The physical constraints of boundary layers and mucopolysaccharide matrices in which the microbiota live mandate ionic and gaseous movements predominately by diffusion to organisms living within the attached communities (Riber and Wetzel, 1987; Losee and Wetzel, 1993). Nutrients, once acquired, are intensively recycled among the attached biota and conserved. Resources from external sources can then be utilized largely for new growth and reproduction (Wetzel, 1993a, 1996). Nutrient retention is very high within the microcommunities and, collectively, the attached habitats. As dependence on more labile organic substrates from algae increases, reduced resilience to disturbances results. That reduction in resilience accelerates when external nutrient loadings to the ecosystems push the productivity to hypereutrophic phytoplanktonic densities that shade out submersed macrophytes and, by elimination of substrata, the attached microflora.

E. Predation-Induced Shifts in Energy Fluxes and Biodiversity

As we have seen (Chaps. 16 and 20), in certain moderately productive lakes, piscivore consumption of planktivorous fishes can lead to sporadically enhanced development of cladoceran zooplankton. The high grazing rates of the cladocerans can result in selective reduction of larger algae for brief-to-moderate intervals of time. The effects of these changes among the larger forms are, however, poorly translated to smaller microbiota, and only minor changes are seen at the microalgal and bacterial levels (e.g., Pace, 1993). For example, removal of portions of the larger algae can decrease competition for nutrient and light resources. Often, the smaller forms with shorter generation times increase in productivity. Nutrient recycling likely also increases, particularly as decomposition of these algae and metabolically coupled bacteria is accelerated by protistian microconsumers (heteroflagellates, ciliates, and related organisms) and tightly retained among the microbiota. Little is known concerning changes in biodiversity under these conditions, but it appears that biodiversity among the microbiota would increase and likely more than compensate for the losses of larger forms by the crustacean grazing. Once again, a shift in energy fluxes occurs, here from larger planktonic forms to smaller forms with much higher turnover rates and increased rates of resource turnover. Compensatory mechanisms likely also appear in composite biodiversity. Resiliency within the pelagic food web likely declines as a result.

III. RIVER REGULATION

A. Riparian-Littoral-River Interface

Flooding adds dissolved and particulate organic matter and mineral nutrients to both the aquatic and terrestrial portions of the river ecosystems. The river ecosystem includes both the channel and the flood plain, both of which are dependent on erosion and sedimentation associated with lateral migration. Inundation results in silts and nutrients that replenish the floodplain pools and flooding of terrestrial mineral and organic matter releases nutrients to the water. Primary production increases markedly, usually far in excess of losses by decomposition. After the flood maximum is reached, decomposition processes usually exceed production.

Many of the organisms, particularly plants, are highly adapted to such inundation gradients and flooding disturbances with rapid growth and reproduction (*r*-strategy traits). Some species, such as the red gum and cypress trees, are totally dependent upon fluvial flooding. Many animals (invertebrates, fish, birds, and mammals) are adapted to the flood cycle and the high plant and microbial productivity associated with it.

Migrations of the river channel occur across the flood plain as alluvial rivers erode and deposit sediment along and above their banks. During more active flows, meandering can lead to rapid avulsions of parts of the flood plain. Spatial and temporal variations in lateral migrations vary greatly (0–800 m yr^{-1}, Hooke, 1980). Periodic disturbances by riverine erosion and deposition are essential to the successional processes in floodplain forests and their associated high habitat and biological diversity (Johnson, 1992; Marston *et al.*, 1995). In addition, the timing and duration of flooding as an agent of erosion and deposition are important in the regulation of riparian plant communities, particularly the forests (Scott *et al.*, 1997; Auble and Scott, 1998; Johnson, 1998). The natural periodic movement of sediment and redeposition within the flood plain are essential to the biota at all levels—primary production, secondary (animal) production, and critical decomposition cycles for nutrient regeneration. The floodplain habitats must constantly be reset by natural disturbances mediated by variations in river discharge.

Fluvial processes create a pattern of physical features that balance successional processes of sedimentation and terrestrialization with scour and channel migration. Periodic rejuvenation of the flood plain by erosion and deposition is essential to sustain the health of the ecosystem (Bayley, 1995). Floodplain patches or habitats are defined both by type (e.g., sand bars, levees, cutoff channels, and backswamps) as well as by age, frequency, and duration of connectivity by flooding and channel migration. Different habitat types and their changing arrangements relate to the channel pattern (meandering, braided, or anastomosed) and mobility over time scales of many years. The diversity of vegetation development of flood plains is directly related to the rejuvenation and succession generated by channel erosion (Petts, 1996).

A complex of flowing channels, backwater areas, lakes, springs, wetlands, and terrestrial patches characterizes floodplain rivers. These diverse patch and habitat areas exhibit a high degree of connectivity, particularly during flooding conditions. Primary process variables, such as diurnal and annual temperature ranges, channel stability, and flood régime and predictability, tend to increase downstream among large rivers (Fig. 26-2). Coarse particulate matter (CPOM) and the importance of the floodplain biotic communities also increase in large floodplain rivers. The interactions of these factors contribute to the high biodiversity found among large floodplain rivers (Fig. 26-2).

Most river flood plains function as storage reservoirs for flows at or below the 40-yr return-period events (Bhowmik and Demissee, 1982). At higher flows, the storage advantage is reduced, as the river and flood plain function more as a conveyance channel. Flooding disturbances, however, function positively because floods can "reset" floodplain patches, destroy floodplain forest, flush pools, and rejuvenate successions.

In large floodplain rivers, the ecological integrity of the river is dependent upon the predictable connection to a diverse range of floodplain habitats. Most (to 90%) of animal biomass in large rivers derives directly or indirectly from production within the flood plain and not from downstream transport of organic matter (Junk *et al.*, 1989; Junk, 1997a,b). Most of that production of organic matter and energy is derived from plant and microbial production within the flood plain, including herbaceous vegetation that shows high productivity during the dry season and abundant aquatic plant growth during the wet season. The high range of habitats created and rejuvenated by channel instability and flooding in a highly predictable manner results in an extremely high productivity and biodiversity of the flood plain. Any reduction in the floodplain habitats and inundation periodicities results in a precipitous decrease in the biodiversity of these ecosystems.

B. Dam Impoundments and Channelization

About two-thirds of the fresh water flowing to the oceans is obstructed by approximately 40,000 large dams (>15 m in height) and more than 800,000 smaller dams (McCully, 1996). In addition, many more rivers are constrained by levees or dikes. Such hydrological alterations are undertaken to hold water for agricultural, industrial, and domestic purposes, for hydroelectricity generation, and for flood protection (Nilsson and Berggren, 2000). Regulation of river flows by dams and the controlled release of water downstream and by channelization reduce or eliminate the linkage between the river and its floodplain margins. The fluvial processes of flooding, erosion, and deposition determine land areas within these lateral habitats adjacent to the river channel. These abiotic variables tend to be relatively regular and predictable on an annual basis. Water quality is governed by

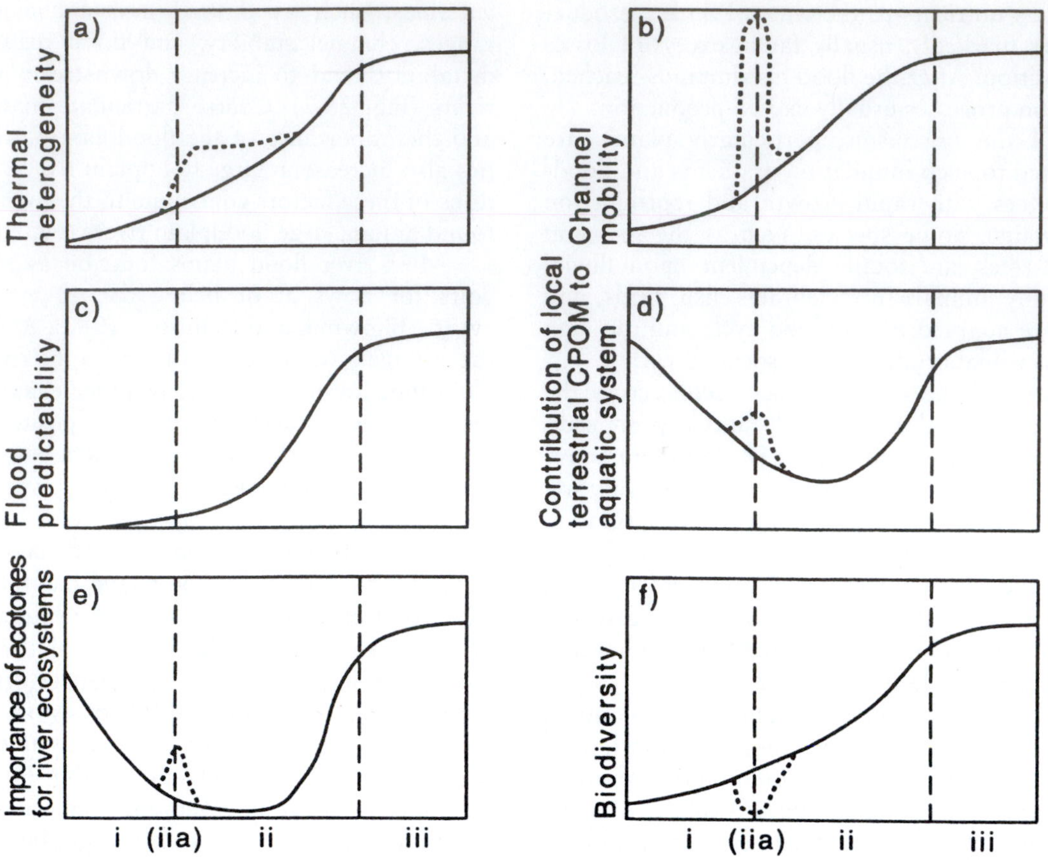

FIGURE 26-2 Downstream patterns (low to high) along an idealized river contrasting the large floodplain river (iii) with the middle river (ii) including braided sections (iia) and headwater streams (i). (From Petts, G. E. 1996, Sustaining the ecological integrity of large floodplain rivers. In M. G. Anderson, D. E. Walling, and P. D. Bates, eds. Floodplain Processes. John Wiley & Sons, Limited, Chichester. Pp. 535–551. Reproduced with permission.)

biologically mediated processes of uptake, retention, transformation, and regulation of physical retentive processes on particles. The biotic communities are governed by dynamic interactions of fluvial and terrestrial processes (current, nutrient loading, physiological adaptations to flooding and desiccation, et cetera).

In regulated rivers, floods are controlled and channel morphology is stabilized (Shields *et al.,* 2000). Reservoirs that reduce the frequency and duration of high flows typically reduce later migration rates by factors of 3–6. Following flow regulation of the Missouri River downstream from Fort Peck Dam in Montana for flood control and hydropower production, for example, lateral migration of the river channel was severely attenuated. The mean rate of channel centerline migration declined from 6.6 to 1.8 m yr^{-1} after impoundment. The primary effects of the isolation of natural flood plains and channelization effectively remove the resetting function. Secondary succession then ensues, with a cascade of responses as terrestrialization progresses closer to the river per se. Production in the controlled flood plain and particularly the entire river ecosystem declines precipitously (Table 26-1). The biodiversity of the river ecosystem also declines markedly, often in the range of <10% of the preregulation conditions. The great biodiversity of floodplain river ecosystems is a direct result of the high spatial and temporal heterogeneity (Ward *et al.,* 1999). That heterogeneity is a direct result of the patterns of disturbance created by flooding and channel migration.

Without question, large dams and river diversions destroy aquatic habitat, contribute substantially to the destruction of fisheries, the extinction of species, and overall loss of ecosystem services on which the human economy depends (McCully, 1996; Postel, 1998; Pringle *et al.,* 2000; Rosenberg *et al.,* 2000). Hydrological alterations induce habitat fragmentation, deterioration of irrigated terrestrial environments and associated surface waters, and dewatering of the rivers from diversions. Hydrological modifications have resulted in regional bi-

TABLE 26-1 Contrasts between Natural and Regulated (Canalized and Embanked) Large Floodplain Rivers Having Late Spring/Early Summer Flood in Temperate Regions[a]

	Natural	*Regulated*
Productive processes	High P/R ratio due to high *in situ* production on the inundated flood plain High CPOM/FPOM ratio—floodplain contributes wood, bark, leaves, fruit, flowers, and terrestrial arthropods. High productivity of herbaceous vegetation during dry phase and hydrophytes during wet phase	Low P/R ratio due to processing of material from upstream and low *in situ* production Low CPOM/FPOM ratio—dependent upon downstream flux of organic matter from head waters
Dominant role of floods	Resource: life cycles of many aquatic and terrestrial organisms are adapted to the flood régime	Disturbance: *frequent* flushing "resets" physical and biological environments and limits successions
Dominant role of main channel	Routeway for gaining access to adult feeding areas, nurseries, spawning grounds, and refuges	Supports whole life cycle
Bulk of riverine animal biomass derives from:	Production within flood plain	Downstream transport of organic matter from head waters
Biodiversity	Very high	Limited by homogeneous substratum (bed and banks), turbidity, and oxygen deficits
Sensitivity of biota to headwater catchment "events" (major floods or pollution incidents)	Low	High

[a] After Petts (1996). P = production; R = respiration; CPOM = coarse particulate organic matter; FPOM = fine particulate organic matter.

otic impoverishment, ranging from reduced population abundance and biodiversity to range fragmentation and increases in exotic and lake-adapted taxa.

C. Climate Changes: Effects on Rivers

Global warming is clearly occurring and is directly coincident with increases in the CO_2 concentration in the atmosphere. CO_2 concentrations are clearly increasing (Fig. 23-31), largely from combustion of fossil fuels, and will more than double in the present century. Global warming is also clearly occurring in part related to the "greenhouse effect" of reduced radiative thermal loss of heat. The combined effects, based on several climatic models (Cushing, 1997; Lettenmaier *et al.*, 1999; Magnuson *et al.*, 2000), are predicted to result in an increase in temperature between 1–3°C in the present century, with perhaps a greater warming in the central parts of continents. Summer precipitation is anticipated to decrease appreciably, with marked increases in evapotranspiration, in some continental regions and increase in others. Downstream flows are anticipated to fluctuate to a much greater extent relative to base flow on an annual basis than was the case previously.

Snowmelt is a dominant climate-controlled factor regulating the hydrology of many large river ecosystems. Increasing temperatures and early seasonal melting result in rapid alterations of the rates of snowmelt, discharge, and seasonality of streamflow hydrographs (Murdoch *et al.*, 2000). Spring snowmelt maxima are reduced and average winter flows increase. Of greater importance than the seasonal shape of the hydrographs, however, are changes in precipitation that are directly coupled to annual total runoff volumes. In general, climatic change models predict that the increased warming will result in decreased precipitation and additionally a decrease in frequency but with more intense precipitation events.

Climate warming effects that generate reduced river flows may be favorable for urban and agricultural flood control but would be detrimental for all other ecosystem use objectives (water supply, navigation, hydropower, recreation, and most importantly biodiversity). In particular, major declines in habitat quality of semipermanent wetlands of the flood plains will occur as the differences between flooding and drying events increase (Poiani and Johnson, 1991; Grimm, 1993; Poiani *et al.*, 1996). Nearly balanced cover:water ratios will shift to closed basins with no open-water areas, with elimination of many fish and wildfowl populations. Clearly, human-induced changes in discharge patterns will be exacerbated by additional progressive climatic changes.

IV. RESTORATION OF AQUATIC ECOSYSTEMS

Aquatic ecosystems are damaged in a number of ways. Alterations of water quality by pollution can occur in many different forms; all decrease how effectively the aquatic ecosystems can be used for human activities. The most common type of degradation is contamination of the water and biota by inorganic and organic pollutants. In the case of toxic substances, remedial measures are difficult or impractical once the substances are dispersed (Wetzel, 1992a). Indeed, dispersion and dilution are unfortunately relied upon as a common corrective, remedial measure. Once the loading of the pollutant to the aquatic ecosystem has ceased or been reduced appreciably, natural water renewal rates are relied upon to dilute the contaminants to acceptable levels. Lakes are more susceptible to function as sinks for pollutants than are rivers, which means that pollutants tend to accumulate and potentially increase in toxicity with time.

The need for management of lakes and rivers reflects an inability of the ecosystems to operate in self-sustaining ways because of interference or damage that exceed the capacities of the ecosystem for self-repair (Moss, 1999). Most demands for management are a result of disturbances by human activities. Impediments to successful management include (a) the basic complexity and stochasticity of natural ecosystems, which may change or be altered faster than they can be understood and managed; (b) an ecosystem scale required in evaluation of appropriate methods for effective management, a scale that does not lend itself readily to rigorous scientific evaluation methods and experimentation of potential effects; and (c) political impediments, often large, such that if a portion of a society should choose to exploit a resource in an unsustainable way, scientific understanding is of little assistance.

Four major types of degradation of surface waters have occurred in recent times. Nutrient loadings, particularly of phosphorus and nitrogen, have resulted in nutrient enrichments and excessive growth of algae and macrophytic plants. This accelerated ***eutrophication*** process has led to a concert of remedial techniques to mitigate the problems and return the lake or reservoir to some state of improved, lower productivity. A second major degradation involves soil erosion, primarily from agriculture, leading to ***siltation,*** loss of volume in the aquatic habitat, and pesticide contamination. A third major type of degradation results from excessive loading of hydrogen ions associated largely with strong acids. These acids result from gases (e.g., SO_2, and NO_x) from fossil fuel combustion products that are dissolved in precipitation or adsorbed to particles that are deposited from the atmosphere onto surface waters or the land (see Chaps. 12 and 14; also Schindler, 1988b). The resulting ***acidification*** of poorly buffered waters not only increases acidity and alters osmoregulatory capacities of organisms but alters the solubility and bioavailability of many metals and other ions of variable toxicity. A fourth type of degradation includes the introduction of ***toxic materials*** such as heavy metals, chlorinated hydrocarbons, and radioactive materials. The sources of all of these pollutants are often diffuse (nonpoint), which makes their control very difficult.

When one attempts to restore a disturbed aquatic ecosystem to some former state or condition, the question arises as to what are the objectives of such recovery. The objective may be the complex task of restoring specific habitats for important native species or, more commonly, to restore community functions such as trophic structure (Palmer *et al.,* 1997). On the basis of much study of community development in streams and rivers, there is reasonable probability that natural succession processes can be exploited to accelerate restoration of disturbed ecosystems.

A. Lake Management and Restoration

Lake management and restoration have focused upon problems particularly associated with excessive nutrient loading and poor land management. The subject has been treated in detail in several recent monographs (Ryding and Rast, 1989; Klapper, 1991; Cairns *et al.,* 1992; Welch, 1992; Cooke *et al.,* 1993; Eiseltová, 1994; Kortmann and Rich, 1994; Geller *et al.,* 1998). Restoration of acidified surface waters involves complex socioeconomic issues to reduce atmospheric pollution. Temporary mitigation of surface water acidity can be accomplished by chemical treatment (Olem, 1991). Only a brief summary of management and restoration is presented here.

1. Eutrophication

Eutrophication is a process leading to increased biological productivity and decreased basin volume from the excessive addition of dissolved and particulate inorganic and organic materials to lakes and reservoirs (Cooke *et al.,* 1993). Although much of the earlier emphasis on eutrophication was on the basis of increased growth of phytoplanktonic algae (e.g., Likens, 1972), a broader context in which rapid alterations of the basin morphometry occurs by siltation is more realistic. The increased littoral areas lead to rapid increases in the development of aquatic macrophytes of high productivity and accelerated internal loadings of organic matter (Wetzel, 1979). In addition, this increased proximity of the sediments with the productive trophogenic zone as lakes become shallower can result in enhanced exchange of nutrients that are remineralized in the sediments.

The cause of the enhanced productivity in lakes and reservoirs is clearly most frequently enhanced phosphorus availability (cf. Chap.13). Although other elements, particularly nitrogen (e.g., Suttle and Harrison, 1988; Dodds *et al.,* 1989) can become the dominant limiting nutrient for phytoplanktonic growth in certain alpine waters when phosphorus supplies are sufficient. On a global basis, phosphorus is the foremost nutrient limitation to autochthonous photosynthesis on a sustained long-term basis. Lake restoration efforts have, as a result, been directed largely toward reducing the loading of phosphorus, and to a lesser extent nitrogen, to the surface waters by advanced wastewater treatment, diversion, land management, or reducing the phosphorus load in wastewater by restricting the phosphorus content of detergents. If releases of sediment phosphorus stores become significant, in part stimulated by low redox potentials that have resulted from organic matter loading, production, and deposition (cf. Chap. 13), attempts are often made to control the availability or recycling of nutrients by physical or chemical methods within the lake or reservoir.

Prevention of nutrient pollution and eutrophication is clearly the most prudent long-term solution. Often, reductions in nutrient loadings require evaluation of diffuse and point sources within the drainage basin and a systematic multifaceted program of reduction and control (Cooke *et al.,* 1993). Once the loadings of nutrients and water are known, a nutrient budget can be used to evaluate the dynamics of annual inputs and outputs of substances. The budget permits the estimation of changes in algal biomass that could result from increases or decreases of nutrient loadings, water residence time, or increased mean depth as may occur from sediment removal. Various phosphorus and water input–output models have been developed that rather accurately predict what loading rates are and what effects reductions would have on phytoplankton biomass (see Chap. 13 and Dillon and Rigler, 1974; Vollenweider, 1975, 1976; Uttormark and Hutchins, 1980; Reckhow and Chapra, 1983; Cooke *et al.,* 1993). Phosphorus release from sediments can be very high in lakes with anoxic hypolimnia, which can delay the predicted reduction in phytoplankton biomass following reductions in external nutrient loading (Marsden, 1989). Internal loading can be estimated and incorporated into nutrient budget data and models (e.g., Nurnberg, 1984; Chapra and Canale, 1991; Rossi and Premazzi, 1991). Restoration of lakes from excessive productivity from eutrophication is accomplished by (a) nutrient loading reduction and (b) control or removal of plant biomass.

2. Nutrient Control

Eight common methods are used, singly or in combination, to reduce nutrient availability or suitable habitat for photosynthesis (Klapper, 1991; Cooke *et al.,* 1993).

1. *Nutrient Removal by Advanced Treatment and Land Management.* Regulation and reduction in external loadings of nutrients to eutrophic lakes is the best of long-term corrective measures, without which treatments within the lakes will have minimal lasting results. Advanced wastewater treatment is only practical where used water is collected in a wastewater system where phosphorus of the water can then be removed to quantities that will not appreciably alter lake productivity when the treated water is returned to the stream, reservoir, or lake. Lake quality, as measured by biomass of phytoplankton or transparency, is often a function of the absolute equilibrium concentration of lake phosphorus, not the proportional change in concentration. Agricultural land-management practices, including manure storage and changes in tillage, can also reduce nutrient loading.

2. *Nutrient Diversion.* Occasionally diversion of major external nutrient loadings has been adequate to restore a eutrophic lake—Lake Washington in Seattle is a well-documented example (Edmondson, 1991a). In situations where the primary source or sources of nutrient loadings are defined, the waste water or storm water may be diverted without appreciable alteration of lake or reservoir hydrology. Diversion is most likely to be successful in rapidly flushed lakes (low retention time) and where internal nutrient loading from sediments is small.

3. *Hypolimnetic Withdrawal* Nutrient-enriched hypolimnetic water is removed by large-scale siphoning, pumping, or deepwater discharge at the dam. Nutrient-rich water, rather than nutrient-poor epilimnetic water, is discharged. Residence time of hypolimnetic water is reduced and the oxygen content of water overlying sediments may be increased with a concomitant decrease in internal loading of phosphorus from the sediments. If designed and regulated carefully, it is possible to have hypolimnetic removal without thermal destratification.

4. *Dilution and Flushing.* If a major source of nutrient-poor water is available, addition of such water to an enriched lake or reservoir may be adequate to dilute nutrient levels adequately to suppress algal productivity. If sufficiently large quantities can be added, the water alone can be adequate to flush out algal cells and maintain lower productivity by dilution. Lower algal biomass can be maintained even with nutrient-rich water if flow is adequate to wash out the algae faster than

algal growth rates. Both methods require the availability of a suitable water supply, particularly during seasons of high productivity. These methods are sometimes used in coordinated management with induced lake circulation, chemical precipitation, and biological manipulations.

5. *Phosphorus Precipitation and Inactivation.* Recovery from eutrophication (oligotrophication), in which essential nutrients are reduced to growth-limiting concentrations, can be a slow process if lake water renewal times are long and if nutrient releases from sediments and entrainments of nutrient-rich hypolimnetic water into the epilimnion are large. Phosphorus is adsorbed to and/or coprecipitated with clay and carbonate particles (cf. Chap. 13). A similar process can be induced by adding aluminum compounds to the water and the sediments. Aluminum sulfate (alum) or sodium aluminate or both form a precipitate of aluminum phosphate or a colloid of aluminum hydroxide, depending upon the alkalinity and pH conditions. In both cases, phosphorus is bound and scavenged to the sediments without aluminum toxicity if circumneutral pH is maintained. Further additions of these aluminum compounds to the sediments inactivate and retard phosphorus migration into the overlying water.

6. *Sediment Oxidation.* High bacterial respiration of organic matter in sediments results in anaerobic conditions in nearly all sediments. These reducing conditions lead to reduced iron and the release of associated phosphate into interstitial sediments and overlying waters (see Chap. 13). Additions of nitrate as an alternative electron acceptor for oxygen delays the reduction of iron and release of phosphate. Artificial addition of nitrate to stimulate denitrification and oxidation of organic matter, ferric chloride to remove hydrogen sulfide and precipitate phosphorus, and lime to raise the somewhat pH of the hypolimnetic water and/or of sediments can result in inactivation of phosphorus release from sediments.

7. *Aeration.* Two methods have been employed to aerate water of stratified lakes and reservoirs—artificial circulation (destratification) and hypolimnetic aeration. Oxygenation can induce precipitation of phosphorus, iron, and manganese and lower ammonium content and increase pH. Artificial water circulation by pumps, jets, and bubbled air or other gases destratifies the water strata, making the lake isothermal and increasing water temperatures, particularly of the former hypolimnion of smaller volume. This destratification reduces the competitive advantages of cyanobacteria. In deep lakes, circulation may light-limit the phytoplankton. Hypolimnetic aeration with contained aeration and flow devices increases hypolimnetic oxygen content without destratifying the water column or warming the hypolimnion. Internal loading of phosphorus often declines from renewed aerobic conditions at the sediment–water interface if adequate iron is present. Otherwise, iron may be added to the hypolimnetic aeration devices.

8. *Sediment Removal.* Shallow eutrophic lakes can often be restored by removal of sediments. This technique simultaneously removes large quantities of nutrients stored in the sediments, increases the mean depth of the basin, and can remove sedimented toxic substances. Often, the habitat for nuisance development of macrophytes is reduced by the deepening of the basin, which can result in light limitations.

9. *Food-Web Manipulations.* The management of food-web structure by manipulation of animal herbivory, predation, and nutrient recycling by animals can influence the quantitative and qualitative composition of algae on a seasonal basis (see critical review in Chaps. 16 and 20). Nutrient concentrations and loading rates set general limits on the types of phytoplankton and their productivities. Alterations of zooplanktonic herbivore types and densities can alter algal biomass and species composition to levels below those anticipated by the nutrient loadings to the lake ecosystem. The phytoplankton productivity is usually rapidly replaced by smaller algal species that may be less obtrusive for human uses of the water bodies.

Browsing of littoral sediments of benthivorous fishes and their digestive activities can add to internal loading of nutrients in shallow lakes and reservoirs (LaMarra, 1975; Brabrand *et al.*, 1990). This enhancement can be suppressed if intensive fish bioturbation activities resuspend fine particled sediments sufficiently to maintain turbid conditions of the water (Hosper and Jagtman, 1990). Fish removal, coupled with reduced nutrient loading, may "switch" the lake to a clear state dominated by submersed macrophytes.

Natural factors or manipulative procedures that significantly increase and maintain strong populations of piscivorous fish should reduce populations of planktivorous fish. This reduction should lead to enhanced survival of large-bodied zooplankton species and to a reduction in phytoplankton biomass and improved water transparency. It should be emphasized that these animal community-induced changes in algal mass do not really alter nutrient content *within the ecosystem*, but rather displace nutrient availability temporarily to areas (tropholytic zone) where it is less immediately available photosynthetically. Downward-cascading (Paine, 1980) piscivore influences to lower trophic levels have less effect on algal biomass in productive lakes where nutrient regulation of phytoplanktonic biomass is large relative to influences of the feeding interactions of piscivore through planktivore to zooplankton communities (McQueen *et al.*, 1989).

Food-web management to improve resulting algal community structure and biomass levels remains experimental (Crowder *et al.*, 1988). Many confounding interactions, such as variations in mixing events, climate, zooplanktonic refuges, and long-term fish reproductive successes, limit applications of fish manipulations as predictive management tools. Food-web manipulation, however, is not a control method but offers some potential as a supplementary procedure to manipulate the types of phytoplankton at certain times of the year in low to moderately productive lakes (see pp. 462–466).

3. Control of Macrophyte Biomass

Excessive growth of macrophytes, particularly of certain exotic nuisance species such as the water hyacinth (*Eichhornia crassipes*), hydrilla (*Hydrilla verticillata*), and the Eurasian watermilfoil (*Myriophyllum spicatum*), can curtail or eliminate the use of lakes, reservoirs, and river ecosystems by humans. Entire disciplines have emerged that are directed to the control and management of aquatic macrophytes (e.g., Gopal, 1987; Gangstad, 1988; Pieterse and Murphy, 1990). The ensuing few paragraphs summarize common approaches to the control and management of excessive development of aquatic macrophytes (see Cooke *et al.*, 1993).

1. *Drawdown of Water Level.* The drawdown of water level is a multipurpose restoration and management technique for ponds and reservoirs to control certain aquatic plants and to modify habitats for the management of fish populations. Exposure of aquatic macrophytes to dry or freezing conditions for adequate (i.e., species-specific and variable) periods of time can kill the plants. Negative effects also occur (e.g., losses of or damage to important adjacent wetlands and losses of important benthic communities).

2. *Mechanical Removal.* Aquatic macrophyte communities that have developed to nuisance densities have been controlled by numerous manual and mechanical methods. Mechanized devices effectively drag, dredge, and till the sediments by mowing and tillage machinery or by suction and diver-operated dredging equipment. Harvested plants are preferably removed from the water, thereby simultaneously removing appreciable amounts of nutrients.

3. *Plant and Sediment Shading.* Sediments and submersed macrophytes can be covered with opaque sheeting or dense screen materials that are impenetrable by plants growing from root systems. These methods are limited to small areas of intensive uses where macrophytes are not desired.

4. *Biological Controls.* Introduction of phytophagous insects and fish or plant pathogens such as fungi and viruses have been used as species-specific control agents to reduce the success and biomass of targeted macrophytes. The objectives are to establish an equilibrium between the control organism and its target plant at an acceptable level of plant biomass (reviewed in Cooke *et al.*, 1993).

5. *Chemical Controls.* A number of general and moderately selective herbicides have been effectively applied to the control of aquatic macrophytes (Gangstad, 1988a,b; Westerdahl and Getsinger, 1988). Such compounds are often used in concert with other control methods.

B. River Management and Restoration

A fundamental aspect of both river management and restoration is to protect or restore the riparian floodplain areas of streams and rivers. In many cases, streams and rivers have been channelized by straightening and deepening the channel, particularly in lowland agricultural areas to increase drainage. Accompanying the loss of stream length is a loss of riffles and pools, a loss of riparian flood plains and wetlands and a loss of riparian vegetation. As a result, hydraulic head is increased when energy dissipation is lost. Increased flows reduce habitat diversity, particularly in the sediments and adjacent wetland, with a catastrophic reduction in habitat diversity and biodiversity (e.g., Petersen *et al.*, 1992). Loss of riparian flood plains and wetlands decreases water tables, increases rates of water runoff, enhances nutrient losses from adjacent land, and increases scouring and sediment yield to the channel. Water retention capacities are greatly reduced, and as a result, during high precipitation events, flooding is very common and much more severe than would be the case if portions of the riparian and floodplain environments were retained. It is estimated that drainage basins comprised of 5–10% wetlands are capable of providing a 50% reduction in peak flood period compared to those drainage basins that have none.

The relative stability of flowing water ecosystems and communities living within them is indicated by their resistance to disturbance as well as their rate of recovery from disturbance (Webster *et al.*, 1983; Yount and Niemi, 1990). After a disturbance, such as a flooding event, a poisoning phenomenon, or alteration of channel stability, lotic communities and ecosystems recovered relatively rapidly because (a) life histories of communities are commonly adapted for rapid recolonization and reestablishment of the disturbed areas, (b) unaffected internal refugia as well as unaffected upstream and downstream populations can serve as seed inocula for reestablishment, (c) flushing characteristics assist in dilution and replacement of polluted regions, and (d) streams and rivers have frequent natural disturbances and are adapted to human-induced perturbations. Major alterations, such as channelization, that severely alter sediment and flow characteristics will result in much longer recovery times.

Several management and restoration measures for streams and rivers are commonly used (see Gore, 1985; Peterson *et al.*, 1990, 1992; Osborne and Kovacic, 1993; Large and Petts, 1994; Eiseltová and Biggs, 1995; Muhar *et al.*, 1995; Boon *et al.*, 2000; Petts *et al.*, 2000). A means of river restoration is to allow the river channel to migrate laterally with periodic inundation of the flood plain above river bankfull.

1. Buffer Strips

Plant and microbial communities on a buffer strip of land, ca. 10- to 30-m wide, along each side of the river channel function to stabilize banks and as a metabolic filter for nutrients and other substances potentially released from land. These narrow strips of vegetated riparian land can retain and reduce nutrient loadings to the channel by 65–100%. Hence, buffer zones and riparian strips function as barriers to eutrophication of streams, aid in channel stability, and have great ecological influence by provision of habitat patch variability. Variations in natural revegetation of the buffer strips are advantageous to enhance biodiversity and habitat differences. Restoration methods are also used to encourage development of wetlands at juncture points of small tributaries or drainage ditches to the main channel.

2. Reduction of Channel Slide–Slope

Reduction of the slope gradient at the edge of the channel, particularly when dredging is periodically required, will encourage channel meandering and floodplain revegetation.

3. Channel Migration and Floodplain Development

Reversion of the channel to natural migrations and meandering morphology increases channel length, alters path morphology, and greatly increases habitat diversity. The erosional and depositional process (see pp. 97–102) will seek a meander frequency of about five to seven channel widths, dissipates flow energy, and allows for greatest habitat variability. Because of the greater time that water spends in contact with sediment surfaces, nutrient retention and spiraling properties are optimized.

C. Acidification

Mitigation of the acidification of lakes and rivers requires long-term reductions in atmospheric pollution and involves complex socioeconomic issues. Reduction at the sources of acidic precipitation is clearly the best alternative strategy for restoration of acidified lakes and streams (Schindler *et al.*, 1991; Schindler, 1997). Acidified lakes and streams begin to recover rather quickly once acid inputs are reduced, assisted by alkalinity producing anaerobic bacterial processes in the systems (see earlier discussion, p. 197). Depletion of base cations in soils impairs the ability of catchment soils to neutralize acid (Likens *et al.*, 1996). Short-term restoration of acidified surface waters has been accomplished by the addition of alkaline materials to the affected lake, stream, or drainage basin (Olem, 1991). Although several substances have been used, additions of limestone ($CaCO_3$) are most common and generally increase nutrient cycling, decomposition, and productivity of phytoplankton and other biota. These restoration procedures are never ideal in their effectiveness for recovery, often of very short duration before repetition of treatment is needed, and are extraordinarily expensive.

V. EPILOGUE

In the comparative approach taken in the foregoing pages, a concerted attempt was made to integrate the massive number of dynamic variables that govern and effect the resultant biotic productivity and conditions. Dominant factors regulating the differing productivity of the ecosystems of lakes, reservoirs, and rivers were contrasted. The underlying rationale was to emphasize the basic control mechanisms that operate on the progressive changes that lead to enrichment and increases in productivity per given quantity of water and why the resulting medley of organisms function together. Only by understanding these regulatory mechanisms can we understand what is likely to happen as aquatic ecosystems are variously disturbed by human activities.

Our understanding of the operation of freshwater ecosystems has come a great distance in the last century. However, if we really scrutinize the existing information in an unbiased way, the only possible conclusion is that we remain quite ignorant of this beautifully complex integration in aquatic ecosystems. The accrual of needed information must be greatly accelerated in order to understand how these systems function before they are irreversibly altered by human activities. As has been demonstrated so well in the human medical profession, there is no alternative to devoting a greater percentage of our intellectual and financial resources to understanding biotic regulation, in this case of freshwater quality. Everything allocated to this critical resource, the foremost resource of our survival, thus far has been utter tokenism. We have passed the point at which we can continue to take unabatedly without casting back some comprehensive understanding and wise use in return. Nature is remarkably resilient to human insults. Yet, humans must learn what are nature's dynamic capacities, because excessive violation without harmony will only unleash her intolerable vengeance. The very survival of humankind depends on our understanding and wise use of our finite freshwater resources.

VI. SUMMARY

1. Water quality is biologically mediated and is the result of a complex integration of biological components within environmental variability. Understanding that mediation is essential to effective water management.
2. Species diversity of lakes, streams, and wetlands is poorly understood but is certainly equal to or greater than that in terrestrial environments. Biodiversity is highest in the land–water interface regions where habitat diversity is remarkably large.
3. Biodiversity decreases precipitously as salinity increases. Saline lakes are species-poor, although reduced competition often allows exploitation of resources with reduced competition and extensive development and high productivity of a few species.
4. Biodiversity within freshwater ecosystems is examined in relation to a gradient of loadings of nutrients and of the chemical composition of organic matter to these ecosystems (Fig. 26-1). These changes are also examined in relation to hypothesized shifts in ecosystem resilience or relative stability—that is, the rate at which a property or the ecosystem returns to a quasi-steady state following a perturbation. Increased rates of nutrient recycling tend to decrease resilience.
5. Regulation of biotic regeneration and recycling of nutrients, and thus resilience to disturbances, is markedly influenced by the sources of organic substrates.
   a. From an entire ecosystem perspective, much of the organic carbon entering aquatic ecosystems is respired by microbial heterotrophs without entering the metazoan food web.
   b. Dissolved organic carbon emanating from decomposing higher plant tissues buffers the rates of nutrient recycling and increases resilience of the ecosystem. Shifts in the ratios of higher vegetation vs. algal dominance in the development of aquatic ecosystems therefore alter relative stability.
6. Maximum metabolic stability of aquatic ecosystems occurs where dissolved organic matter loadings derived from higher plant tissues from land–water interface and terrestrial sources are high. As a result, a concordance of biodiversity and relative stability occur.
7. Major disruptions of river ecosystems occur by regulation of flows, particularly by impoundment in reservoirs and channelization.
   a. Periodic disturbance by flooding and inundation of the flood plain is critical to functions of nutrient retention and cycling, maintenance of habitat diversity and productivity, and high biodiversity.
   b. Nearly all major rivers and many tributaries are impounded with dams, and the channels are often constrained by levees or dikes to retain water for agricultural, industrial, and domestic purposes, hydroelectricity, and for flood protection.
   c. Regulation of river flows by dams and controlled release of water and channelization reduce or eliminate the essential linkages between the river and its floodplain margins.
   d. Removal of the flooding and lateral channel migration effectively eliminates the resetting functions and degrades the rivers and eliminates most of their biological functions.
8. Five major types of degradation of surface waters have occurred in recent times. Various remediation methods are summarized.
   e. Eutrophication from nutrient enrichments, which causes excessive growth of algae and macrophytic plants. Thirteen common methods for remediation to reduce nutrient availability or habitat for photosynthesis of phytoplankton and macrophytes within the contaminated waters are summarized.
   f. Siltation as a result of soil erosion, particularly from agriculture.
   g. Acidification of waters from excessive loading of hydrogen ions from strong acids from ions in acidic atmospheric precipitation derived from human combustion products of petroleum. Remediation of acidified waters is difficult and expensive.
   h. Diffuse introduction of toxic materials (e.g., heavy metals, chlorinated hydrocarbons, and radioisotopes). Remediation, if possible at all once contaminants are dispersed, is often ineffective and manifoldly (at least 10 times) more expensive than removing the contamination at the sources where they are concentrated.

References Cited

van Aardt, W. J. and C. T. Wolmarans. 1981. Evidence for non-assimilation of Chlorella by the African freshwater snail *Bulinus* (*Physopsis*) *globosus*. S. African J. Sci. 77:319–320.

Aaronson, W., S. W. Dhawale, N. J. Patni, B. DeAngelis, O. Frank, and H. Baker. 1977. The cell content and secretion of water-soluble vitamins by several freshwater algae. Arch. Microbiol. *112*:57–59.

Abella, C., E. Montesinos, and R. Guerrero. 1980. Field studies on the competition between purple and green sulfur bacteria for available light (Lake Siso, Spain). Developments Hydrobiol. *3*:173–181.

Åberg, B. and W. Rodhe. 1942. über die Milieufaktoren in einigen südschwedischen Seen. Symbol. Bot. Upsalien. *5*(*3*), 256 pp.

Abrahams, M. and M. Kattenfeld. 1997. The role of turbidity as a constraint on predator-prey interactions in aquatic environments. Behav. Ecol. Sociobiol. *40*:169–174.

Abrew, W. 1996. Evapotranspiration measurements and modeling for three wetland systems in south Florida. Water Resour. Bull. *32*:465–473.

Adam, N. K. 1937. A rapid method for determining the lowering of tension of exposed water surfaces, with some observations on

Note: A few discrepancies occur in the transliteration of Slavic languages. When the work of an author is translated into another language and I have cited only one of his or her works, that citation is recorded as printed even if the transliteration is incorrect. When more than one work of the same author is cited but some are from the mother language, the name is cited consistently as it should be transliterated. An exception is Vinberg in the Russian, because of his European ancestry, the author used Winberg in several important writings in the English. Conforming to the transliteration in this case could lead to confusion if the references were sought.

Further discrepancies appear as abbreviations for certain journal names have evolved over time. The original definitions and abbreviations of journals have been retained, but are later cited to what they have been changed (e.g., Amer. is now Am., Mich. is now MI, Water is now Wat., etc.).

Placing some names in alphabetical order can be confusing. The surname (family) name should of course have priority, and usually does (e.g., de Angelis is under A). Where the foreign article de or preposition van, among others, is part of the name, it is organized under the first letter (e.g., DeAngelis is under D).

the surface tension of the sea and of inland waters. Proc. Roy. Soc. London (Ser. B) *122*:134–139.

Adams, D. G. and N. G. Carr. 1981. The developmental biology of heterocyst and akinete formation in cyanobacteria. Crit. Rev. Microbiol. *9*:45–100.

Adams, J. 1980. The role of competition in the population dynamics of a freshwater flatworm, *Bdellocephala punctata* (Turbellaria, Tricladida). J. Anim. Ecol. *49*:565–579.

Adams, M. S. and M. D. McCracken. 1974. Seasonal production of the *Myriophyllum* component of the littoral of Lake Wingra, Wisconsin. J. Ecol. *62*:457–465.

Adams, M. S. and R. T. Prenkti. 1982. Biology, metabolism and function of littoral submersed weedbeds of Lake Wingra, Wisconsin, USA: A summary and review. Arch. Hydrobiol./Suppl. *62*:333–409.

Adams, W. A. 1978. Effects of ice cover on the solar radiation regime in Canadian lakes. Verh. Internat. Verein. Limnol. *20*:141–149.

Adams, W. A., W. P. Adams, P. A. Flavelle, and N. T. Roulet. 1984. Variability of light beneath a modified portion of the snow and ice cover of a lake. Verh. Internat. Verein. Limnol. *22*:65–71.

Adams, W. P. 1981. Snow and ice on lakes. *In* D. M. Gray and D. H. Male, eds. Handbook of Snow: Principles, Processes, Management, Use. Pergamon Press, New York. pp. 437–474.

Adams, W. P. and D. C. Lasenby. 1978. The role of ice and snow in lake heat budgets. Limnol. Oceanogr. *23*:1025–1028.

Adams, W. P. and N. T. Roulet. 1982. Areal differentiation of land and lake snowcover in a small sub-arctic drainage basin. Nordic Hydrol. *13*:139–156.

Adams, W. P. and T. D. Prowse. 1978. Observations on special characteristics of lake snowcover. Proc. Eastern Snow Conf. *1978*:117–128.

Adams, W. P., M. C. English, and D. C. Lasenby. 1979. Snow and ice in the phosphorus budget of a lake in south central Ontario. Water Res. *13*:213–215.

Adcock, J. A. 1979. Energetics of a population of the isopod *Asellus aquaticus*: Life history and production. Freshwat. Biol. *9*:343–355.

Adeniji, H. A. 1978. Diurnal vertical distribution of zooplankton during stratification in Kainji Lake, Nigeria. Verh. Internat. Verein. Limnol. *20*:1677–1683.

Adreani, L., C. Bonacina, and G. Bonomi. 1981. Production and population dynamics in profundal lacustrine Oligochaeta. Verh. Internat. Verein. Limnol. *21*:967–974.

Adrian, R. 1987. Viability of phytoplankton in fecal pellets of two cyclopoid copepods. Arch. Hydrobiol. *110*:321–330.

Agami, M. and Y. Waisel. 1986. The ecophysiology of roots of submersed vascular plants. Physiol. Veg. *24*:607–624.

Agami, M., S. Beer, and Y. Waisel. 1980. Growth and photosynthesis of *Najas marina* L. as affected by light intensity. Aquat. Bot. *9*:285–289.

Aggus, L. R. 1971. Summer benthos in newly flooded areas of Beaver Reservoir during the second and third years of filling, 1965–1966. *In* G. E. Hall, ed. Reservoir Fisheries and Limnology. Spec. Publ. Am. Fish. Soc. *8*:139–152.

Ahl, T. 1975. Effects of man-induced and natural loadings of phosphorus and nitrogen on the large Swedish lakes. Verh. Internat. Verein. Limnol. *19*:1125–1132.

Ahl, T. 1980. Variability in ionic composition of Swedish lakes and rivers. Arch. Hydrobiol. *89*:5–16.

Ahlgren, G. 1993. Seasonal variation of fatty acid content in natural phytoplankton in two eutrophic lakes. A factor controlling zooplankton species? Verh. Internat. Verein. Limnol. *25*:144–149.

Ahlgren, G., L. Lundstedt, M. Brett, and C. Forsberg. 1990. Lipid composition and food quality of some freshwater phytoplankton for cladoceran zooplankters. J. Plankton Res. *12*:809–818.

Ahlgren, G., I.-B. Gustafsson, and M. Boberg. 1992. Fatty acid content and chemical composition of freshwater microalgae. J. Phycol. *28*:37–50.

Ahlgren, G., L. Sonesten, M. Boberg, and I.-B. Gustafsson. 1996. Fatty acid content of some freshwater fish in lakes of different trophic levels—a bottom-up effect? Ecol. Freshwat. Fish *5*:15–27.

Ahlgren, G., W. Goedkoop, H. Markensten, L. Sonesten, and M. Boberg. 1997. Seasonal variations in food quality for pelagic and benthic invertebrates in Lake Erken—the role of fatty acids. Freshwat. Biol. *38*:555–570.

Ahlgren, G., K. Zeipel, and I.-B. Gustafsson. 1998. Phosphorus limitation effects on the fatty acid content and nutritional quality of a green alga and a diatom. Verh. Internat. Verein. Limnol. *26*:1659–1664.

Aiken, G. R., D. M. McKnight, R. L. Wershaw, and P. MacCarthy, eds. 1985. Humic Substances in Soil, Sediment, and Water—Geochemistry, Isolation, and Characterization. John Wiley & Sons, New York. 692 pp.

Aiken, S. G. and K. F. Walz. 1979. Turions of *Myriophyllum exalbescens*. Aquat. Bot. *6*:357–363.

Aiken, S. G. and R. R. Picard. 1980. The influence of substrate on the growth and morphology of *Myriophyllum exalbescens* and *Myriophyllum spicatum*. Can. J. Bot. *58*:1111–1118.

Aitkenhead, J. A. and W. H. McDowell. 2000. Soil C:N ratio as a predictor of annual riverine DOC flux at local and global scales. Global Biogeochem. Cycles *14*:127–138.

Aizaki, M. 1978. Seasonal changes in standing crop and production of periphyton in the Tamagawa River. Jap. J. Ecol. *28*:123–134.

Aizaki, M. 1979. Growth rates of microorganisms in a periphyton community. Jap. J. Limnol. *40*:1–10.

Akeret, B. 1993. Lunar periodicity of the emergence of *Chaoborus flavicans*. Verh. Internat. Verein. Limnol. *25*:625–626.

Alber, M. and I. Valiela. 1994. Production of microbial organic aggregates from macrophyte-derived dissolved organic material. Limnol. Oceanogr. *39*:37–50.

Albers, J. 1963. Interaction of Color. Yale Univ. Press, New Haven. 75 pp.

Albrecht, M.-L. 1964. Die Lichtdurchlässigkeit von Eis und Schnee und ihre Bedeutung für die Sauerstoffproduktion im Wasser. Deutsche Fischerei-Zeitung *11*:371–376.

Alexander, J. E., Jr. and A P. Covich. 1991. Predation risk and avoidance behavior in two freshwater snails. Biol. Bull. *180*:387–393.

Alexander, M. 1961. Introduction to Soil Microbiology. John Wiley & Sons, New York. 472 pp.

Alexander, M. 1965a. Nitrification. Agronomy *10*:307–343.

Alexander, M. 1965b. Biodegradation: Problems of molecular recalcitrance and microbial fallibility. Adv. Appl. Microbiol. *7*:35–80.

Alexander, M. 1971. Microbial Ecology. John Wiley & Sons, New York. 511 pp.

Alexander, M. 1975. Environmental and microbiological problems arising from recalcitrant molecules. Microb. Ecol. *2*:17–27.

Alexander, V., D. W. Stanley, R. J. Daley, and C. P. McRoy. 1980. Primary producers. *In* J. E. Hobbie, ed. Limnology of Tundra Ponds. Dowden, Hutchinson & Ross, Inc., Stroudsburg. pp. 179–250.

Alford, D. P., R. D. Delaune, and C. W. Lindau. 1997. Methane flux from Mississippi River deltaic plain wetlands. Biogeochemistry *37*:227–236.

Ali, W., C. R. O'Melia, and J. K. Edzwald. 1985. Colloidal stability of particles in lakes: Measurement and significance. Wat. Sci. Tech. *17*:701–712.

Alimov, A. F., *et al.* 1970. Biological productivity of lakes Krivoe and Krugloe. *In* Z. Kajak and A. Hillbricht-Ilkowska, eds. Productivity Problems of Freshwaters. PWN Polish Scientific Publishers, Warsaw. pp. 39–56.

Aliverdieva-Gamidova, L. A. 1969. Microbiological processes in Lake Mekhteb. Mikrobiologiya *38*:1096–1100. (Translated into English, Consultants Bureau, New York, 1970.)

Allan, J. D. 1978. Trout predation and the size composition of stream drift. Limnol. Oceanogr. *23*:1231–1237.

Allan, J. D. 1995. Stream Ecology: Structure and Function of Running Waters. Chapman & Hall, London. 388 pp.

Allanson, B. R. 1973. The fine structure of the periphyton of *Chara sp.* and *Potamogeton natans* from Wytham Pond, Oxford, and its significance to the macrophyte-periphyton metabolic model of R. G. Wetzel and H. L. Allen. Freshwat. Biol. *3*:535–541.

Allen, E. D. and D. H. N. Spence. 1981. The differential ability of aquatic plants to utilize the inorganic carbon supply in fresh waters. New Phytol. *87*:269–283.

Allen, H. L. 1969. Chemo-organotrophic utilization of dissolved organic compounds by planktic algae and bacteria in a pond. Int. Rev. ges. Hydrobiol. *54*:1–33.

Allen, H. L. 1971. Primary productivity, chemo-organotrophy, and nutritional interactions of epiphytic algae and bacteria on macrophytes in the littoral of a lake. Ecol. Monogr. *41*:97–127.

Allen, H. L. 1972. Phytoplankton photosynthesis, micronutrient interactions, and inorganic carbon availability in a soft-water Vermont lake. *In* G. E. Likens, ed. Nutrients and Eutrophication: The Limiting-Nutrient Controversy. Special Symposium. Amer. Soc. Limnol. Oceanogr. *1*:63–83.

Allen, H. L. 1976. Dissolved organic matter in lakewater: Characteristics of molecular weight size fractions and ecological implications. Oikos *27*:64–70.

Allen, K. R. 1951. The Horokiwi Stream. A study of a trout population. Fish. Bull. N. Z. Mar. Dept. *10*:1–238.

Allen, M. B. 1952. The cultivation of Myxophyceae. Arch. Mikrobiol. *17*:34–53.

Allen, M. B. and D. I. Arnon. 1955. Studies on nitrogen-fixing blue-green algae. II. The sodium requirement of *Anabaena cylindrica*. Physiol. Plant *8*:653–660.

Aller, R. C. 1982. The effects of macrobenthos on chemical properties of marine sediment and overlying water. *In* P. L. McCall and M. J. S. Tevesz, eds. Animal-sediment relations: The biogenic alteration of sediments. Plenum Press, New York. pp. 53–102.

Allgeier, R. J., B. C. Hafford, and C. Juday. 1941. Oxidation-reduction potentials and pH of lake waters and of lake sediments. Trans. Wis. Acad. Sci. Arts Lett. *33*:115–133.

Allison, T. D., R. E. Moeller, and M. B. Davis. 1986. Pollen in laminated sediments provides evidence for a mid-Holocene forest pathogen outbreak. Ecology *67*:1101–1105.

Allott, N. A. 1986. Temperature, oxygen and heat-budgets of six small western Irish lakes. Freshwat. Biol. *16*:145–154.

Almazan, G. and C. E. Boyd. 1978. Effects of nitrogen levels on rates of oxygen consumption during decay of aquatic plants. Aquat. Bot. *5*:119–126.

Almer, B., W. Dickson, C. Ekströum, and E. Hörnström. 1978. Sulfur pollution and the aquaticst ecosystem. *In* J. O. Nriagu, ed. Sulfur in the Environment. II. Ecological Impacts. John Wiley & Sons, New York. pp. 271–311.

Alongi, D. M. 1991. Flagellates of benthic communities: Characteristics and methods of study. *In* D. J. Patterson and J. Larsen, eds. The Biology of Free-living Heterotrophic Flagellates. Clarendon Press, Oxford. pp. 57–75.

Alsterberg, G. 1927. Die Sauerstoffschichtung der Seen. Bot. Not. *1927*:255–274.

Amblard, C. 1988. Seasonal succession and strategies of phytoplankton development in two lakes of different trophic states. J. Plankton Res. *10*:1189–1208.

Amblard, C., S. Rachiq, and G. Bourdier. 1992. Photolithotrophy, photoheterotrophy and chemoheterotrophy during spring phytoplankton development (Lake Pavin). Microb. Ecol. *24*:109–123.

Ambrosetti, W. and L. Barbanti. 1993. Heat content and vertical mixing in Lake Orta. Mem. Ist. Ital. Idrobiol. *51*:1–10.

American Public Health Association. 1998. Periphyton. *In* Standard Methods for the Examination of Water and Wastewater. 20th Edition. American Public Health Association, Washington, DC. 1514 pp.

Ammerman, J. W. and F. Azam. 1985. Bacterial 5′-nucleotidase in aquatic ecosystems: A novel mechanism of phosphorus regeneration. Science *227*:1338–1340.

Ammerman, J. W. and F. Azam. 1991a. Bacterial 5′-nucleotidase activity in estuarine and coastal marine waters: Characterization of enzyme activity. Limnol. Oceanogr. *36*:1427–1436.

Ammerman, J. W. and F. Azam. 1991b. Bacterial 5′-nucleotidase activity in estuarine and coastal marine waters: Role in phosphorus regeneration. Limnol. Oceanogr. *36*:1437–1447.

Amon, R. M. W. and R. Benner. 1996a. Bacterial utilization of different size classes of dissolved organic matter. Limnol. Oceanogr. *41*:41–51.

Amon, R. M. W. and R. Benner. 1996b. Photochemical and microbial consumption of dissolved organic carbon and dissolved oxygen in the Amazon River system. Geochim. Cosmochim. Acta *60*:1783–1792.

Amudhan, S., U. P. Rao, and J. S. Bentur. 1999. Total phenol profile in some rice varieties in relation to infestation by Asian rice gall midge *Orseolia oryzae* (Wood-Mason). Curr. Sci. *76*:1577–1580.

Anastácio, P. M. and J. C. Marques. 1995. Population biology and production of the red swamp crayfish *Procambarus clarkii* (Girard) in the Lower Mondego River Valley, Portugal. J. Crustacean Biol. *15*:156–168.

Andersen, F. Ø. 1976. Primary production in a shallow water lake with special reference to a reed swamp. Oikos *27*:243–250.

Andersen, F. Ø. 1978. Effects of nutrient level on the decomposition of *Phragmites communis* Trin. Arch. Hydrobiol. *84*:42–54.

Andersen, F. Ø. and H. S. Jensen. 1991. The influence of chironomids on decomposition of organic matter and nutrient exchange in a lake sediment. Verh. Internat. Verein. Limnol. *24*:3051–3055.

Andersen, F. Ø. and J. S. Jensen. 1992. Regeneration of inorganic phosphorus and nitrogen from decomposition of seston in a freshwater sediment. Hydrobiologia *228*:71–81.

Andersen, J. M. 1975. Influence of pH on release of phosphorus from lake sediments. Arch. Hydrobiol. *76*:411–419.

Andersen, J. M. 1977. Rates of denitrification of undisturbed sediment from six lakes as a function of nitrate concentration, oxygen and temperature. Arch. Hydrobiol. *80*:147–159.

Andersen, J. M. 1982. Effect of nitrate concentrations in lake water on phosphate release from the sediment. Water Res. *16*:1119–1126.

Andersen, O. K., J. C. Goldman, D. A. Caron, and M. R. Dennett. 1986. Nutrient cycling in a microflagellate food chain. III. Phosphorus dynamics. Mar. Ecol. Progr. Ser. *31*:47–55.

Anderson, D. E., R. G. Striegl, D. I. Stannard, C. M. Michmerhuizen, T. A. McConnaughey, and J. W. LaBaugh. 1999. Estimating lake-atmosphere CO_2 exchange. Limnol. Oceanogr. *44*:988–1001.

Anderson, D. V. 1961. A note on the morphology of the basins of the Great Lakes. J. Fish. Res. Board Can. *18*:273–277.

Anderson, G. and R. J. Hance. 1963. Investigation of an organic phosphorus component of fulvic acid. Plant Soil *19*:296–303.

Anderson, G. C. 1958. Some limnological features of a shallow saline meromictic lake. Limnol. Oceanogr. *3*:259–270.

Anderson, L. W. J. 1978. Abscisic acid induces formation of floating leaves in the heterophyllous aquatic angiosperm *Potamogeton nodusus*. Science *210*:1135–1138.

Anderson, L. W. J. and B. M. Sweeney. 1977. Diel changes in sedimentation characteristics of *Ditylum brightwelli*: Changes in cellular lipid and effects of respiratory inhibitors and ion-transport modifiers. Limnol. Oceanogr. *22*:539–552.

Anderson, M. P. and J. A. Munter. 1981. Seasonal reversals of groundwater flow around lakes and the relevance to stagnation points and lake budgets. Water Resour. Res. *17*:1139–1150.

Anderson, N. H. and J. R. Sedell. 1979. Detritus processing by macroinvertebrates in stream ecosystems. Ann. Rev. Entomol. *24*:351–377.

Anderson, N. H. and K. W. Cummins. 1979. Influences of diet on the life histories of aquatic insects. J. Fish. Res. Board Can. *36*:335–342.

Anderson, N. H., J. R. Sedell, L. M. Roberts, and F. J. Triska. 1978. The role of aquatic invertebrates in processing of wood debris in coniferous forest streams. Amer. Midland Nat. *100*:64–82.

Anderson, N. J. 1993. Natural versus anthropogenic change in lakes: The role of the sediment record. Trends Ecol. Evol. *8*:356–361.

Anderson, R. A. 1998. What to do with protists? Aust. Syst. Bot. *11*:185–201.

Anderson, R. F. and S. L. Schiff. 1987. Alkalinity generation and the fate of sulfur in lake sediments. Can. J. Fish. Aquat. Sci. *44*(Suppl. 1):188–193.

Anderson, R. O. and F. F. Hooper. 1956. Seasonal abundance and production of littoral bottom fauna in a southern Michigan lake. Trans. Amer. Microsc. Soc. *75*:259–270.

Anderson, R. S. 1979. Predator-prey relationships and predation rates for crustacean zooplankters from some lakes in western Canada. Can. J. Zool. *48*:1229–1240.

Andersson, A., U. Larsson, and Å. Hagström. 1986. Size-selective grazing by a microflagellate on pelagic bacteria. Mar. Ecol. Progr. Ser. *33*:51–57.

Andersson, E. 1969. Life cycle and growth of *Asellus aquaticus* (L.). Rept. Inst. Freshwat. Res. Drottningholm *49*:5–26.

Andersson, G., W. Granéli, and J. Stenson. 1988. The influence of animals on phosphorus cycling in lake ecosystems. Hydrobiologia *170*:267–284.

Andrew, T. E. and A. G. Fitzsimons. 1992. Seasonality, population dynamics and production of planktonic rotifers in Lough Neagh, Northern Ireland. Hydrobiologia *246*:147–164.

Andrews, S. S., S. Caron, and O. C. Zafiriou. 2000. Photochemical oxygen consumption in marine waters: A major sink for colored dissolved organic matter? Limnol. Oceanogr. *45*:267–277.

Andronikova, I. N., V. G. Drabkova, K. N. Kuzmenko, N. F. Michailova, and E. A. Stravinskaya. 1970. Biological productivity of the main communities of the Red Lake. *In* Z. Kajak and A. Hillbright-Ilkowska, eds. Productivity Problems of Freshwaters. PWN Polish Scientific Publishers, Warsaw. pp. 57–71.

de Angelis, M. A. and M. D. Lilley. 1987. Methane in surface waters of Oregon estuaries and rivers. Limnol. Oceanogr. *32*:716–722.

Angstadt, J. D. and W. H. Moore. 1997. A circadian rhythm of swimming behavior in a predatory leech of the family Erpobdellidae. Am. Midland Nat. *137*:165–172.

Anholt, B. R. 1995. Density dependence resolves the stream drift paradox. Ecology *76*:2235–2239.

Anschütz, I. and F. Gessner. 1954. Der Ioneanaustausch bei Torfmoosen (*Sphagnum*). Flora *141*:178–236.

Anthony, E. H. and F. R. Hayes. 1964. Lake water and sediment. VII. Chemical and optical properties of water in relation to the bacterial counts in the sediments of twenty-five North American lakes. Limnol. Oceanogr. *9*:35–41.

Appleby, P. G. and F. Oldfield. 1978. The calculation of lead–210 dates assuming a constant rate of supply of unsupported ^{210}Pb to the sediment. Catena *5*:1–8.

Appleby, P. G., J. A. Dearing, and F. Oldfield. 1985. Magnetic studies of erosion in a Scottish lake catchment. 1. Core chronology and correlation. Limnol. Oceanogr. *30*:1144–1153.

Appleby, P. G., N. Richardson, and J. T. Smith. 1993. The use of radionuclide records from Chernobyl and weapons test fallout for assessing the reliability of Pb^{210} in dating very recent sediments. Verh. Internat. Verein. Limnol. *25*:266–269.

Applegate, R. L. and J. W. Mullan. 1967. Food of young largemouth bass (*Micropterus salmoides*), in a new and an old reservoir. Trans. Am. Fish. Soc. *96*:74–77.

Arber, A. 1920. Water Plants. A Study of Aquatic Angiosperms. Cambridge University Press, Cambridge. 436 pp.

Armitage, M. J. and J. O. Young. 1990a. The realized food niches of three species of stream-dwelling triclads (Turbellaria). Freshwat. Biol. *24*:93–100.

Armitage, M. J. and J. O. Young. 1990b. A field and laboratory study of the parasites of the triclad *Phagocata vitta* (Dugàs). Freshwat. Biol. *24*:101–107.

Armstrong, J. and W. Armstrong. 1990. Light-enhanced convective throughflow increases oxygenation in rhizomes and rhizosphere of *Phragmites australis* (Cav.) Trin. ex Steud. New Phytol. *114*:121–128.

Armstrong, J., W. Armstrong, and P. M. Beckett. 1992. *Phragmites australis*: Venturi- and humidity-induced pressure flows enhance rhizome aeration and rhizosphere oxidation. New Phytol. *120*:197–207.

Armstrong, W. 1978. Root aeration in the wetland condition. *In* D. D. Hook and R. M. M. Crawford, eds. Plant Life in Anaerobic Environments. Ann Arbor Science Publishers, Inc., Ann Arbor. pp. 269–297.

Armstrong, W., S. H. F. W. Justin, P. M. Beckett, and S. Lythe. 1991a. Root adaptation to soil waterlogging. Aquat. Bot. *39*:57–73.

Armstrong, W., J. Armstrong, P. M. Beckett, and S. H. F. W. Justin. 1991b. Convective gas-flows in wetland plant aeration. *In* M. B. Jackson, D. D. Davies, and H. Lambers, eds. Plant Life under Oxygen Deprivation. SPB Academic Publ., B.V., The Hague. pp. 283–302.

Armstrong, W., J. Armstrong, and P. M. Beckett. 1996. Pressurised ventilation in emergent macrophytes: The mechanism and mathematical modelling of humidity-induced convection. Aquat. Bot. *54*:121–135.

Arndt, H. 1993. Rotifers as predators on components of the microbial web (bacteria, heterotrophic flagellates, ciliates)—a review. Hydrobiologia *255/256*:231–246.

Arndt, H., D. Dietrich, B. Auer, E.-J. Cleven, T. Gräfenhan, M. Weitere, and A. P. Mylnikov. 2000. Functional diversity of heterotrophic flagellates in aquatic ecosystems. *In* B. S. C. Leadbeater and J. C. Green, eds. The Flagellates. Taylor and Francis, London (In press.)

Arnold, D. E. 1971. Ingestion, assimilation, survival and reproduction by *Daphnia pulex* fed seven species of blue-green algae. Limnol. Oceanogr. *16*:906–920.

Arruda, J. A., G. R. Marzolf, and R. T. Faulk. 1983. The role of suspended sediments in the nutrition of zooplankton in turbid reservoirs. Ecology *64*:1225–1235.

Arts, M. T. 1999. Lipids in freshwater zooplankton: Selected ecological and physiological aspects. *In* M. T. Arts and B. C. Wainman, eds. Lipids in Freshwater Ecosystems. Springer-Verlag, New York. pp. 71–90.

Arts, M. T. and H. Rai. 1997. Effects of enhanced ultraviolet-B radiation on the production of lipid, polysaccharide and protein in three freshwater algal species. Freshwat. Biol. *38*:597–610.

Arts, M. T., H. Rai, and V. P. Tumber. 2000. Effects of artificial UV-A and UV-B radiation on carbon allocation in *Synechococcus elongatus* (cyanobacterium) and *Nitzschia palea* (diatom). Verh. Internat. Verein. Limnol. *27* (In press).

Arts, M. T., M. S. Evans, and R. D. Robarts. 1992. Seasonal patterns of total and energy reserve lipids of dominant zooplanktonic crustaceans from a hyper-eutrophic lake. Oecologia *90*:560–571.

Arts, M. T., R. D. Robarts, F. Kasai, M. J. Waiser, V. P. Tumber, A. J. Plante, H. Rai, and H. J. de Lange. 2000. The attenuation of ultraviolet radiation in high dissolved organic carbon waters of wetlands and lakes on the northern Great Plains. Limnol. Oceanogr. *45*:292–299.

Arts, M. T., R. D. Robarts, and M. S. Evans. 1993. Energy reserve lipids of zooplanktonic crustaceans from an oligotrophic saline lake in relation to food resources and temperature. Can. J. Fish. Aquat. Sci. *50*:2404–2420.

Arts, M. T., R. D. Robarts, and M. S. Evans. 1997. Seasonal changes in particulate and dissolved lipids in a eutrophic prairie lake. Freshwat. Biol. *38*:525–537.

Aruga, Y. 1965. Ecological studies of photosynthesis and matter production of phytoplankton. II. Photosynthesis of algae in relation to light intensity and temperature. Bot. Mag. Tokyo *78*:360–365.

Arvola, L. 1984. Diel variation in primary production and the vertical distribution of phytoplankton in a polyhumic lake. Arch. Hydrobiol. *101*:503–519.

Arvola, L. and P. Kankaala. 1989. Winter and spring variability in phyto- and bacterioplankton in lakes with different water colour. Aqua Fennica *19*:29–39.

Aselmann, I. and P. J. Crutzen. 1989. Global distribution of natural freshwater wetlands and rice paddies, their net primary productivity, seasonality and possible methane emissions. J. Atm. Chem. *8*:307–358.

Assman, A. V. 1953. Rol vodoroslevykh obrastanii v obrazovanii organicheskogo veshchestva v Glubokom Ozere. Trudy Vsesoiuznogo Gidrobiol. Obshchestva *5*:138–157.

Atkinson, K. M., J. M. Elliott, D. G. George, J. G. Jones, E. Y. Haworth, S. I. Heaney, C. A. Mills, C. S. Reynolds, and J. F. Talling. 1986. A general assessment of environmental and biological features of Windermere and their susceptibility to change. Rept. Freshwater Biological Association, Ambleside, UK. (Cited in Reynolds, 1997.)

Attridge, E. M. and P. Rowell. 1997. Growth, heterocyst differentiation and nitrogenase activity in the cyanobacteria *Anabaena variabilis* and *Anabaena cylindrica* in response to molybdenum and vanadium. New Phytol. *135*:518–526.

Atwell, B. J., M. T. Veerkamp, B. Stuiver, and P. J. C. Kuiper. 1980. The uptake of phosphate by *Carex* species from oligotrophic to eutrophic swamp habitats. Physiol. Plant. *49*:487–494.

Auble, G. T. and M. L. Scott. 1998. Fluvial disturbance patches and cottonwood recruitment along the upper Missouri River, Montana. Wetlands *18*:546–556.

Aumen, N. G., P. J. Bottomley, and S. V. Gregory. 1985. Nitrogen dynamics in stream wood samples incubated with [^{14}C]lignocellulose and potassium [^{15}N]nitrate. Appl. Environ. Microbiol. *49*:1119–1123.

Austin, H. K. and S. E. G. Findlay. 1989. Benthic bacterial biomass and production in the Hudson River Estuary. Microb. Ecol. *18*:105–116.

Averdieck, F.-R. 1978. Palynologischer Beitrag zur Entwicklungsgeschichte des Grossen Plöner Sees und der Vegetation seiner Umgebung. Arch. Hydrobiol. *83*:1–46.

Axler, R. P., R. M. Gersberg, and C. R. Goldman. 1980. Stimulation of nitrate uptake and photosynthesis by molybdenum in Castle Lake, California. Can. J. Fish. Aquat. Sci. *37*:707–712.

Ayers, J. C., D. C. Chandler, G. H. Lauff, C. F. Powers, and E. B. Henson. 1958. Currents and Water Masses of Lake Michigan. Publ. Great Lakes Res. Div., Univ. Mich., 3, 169 pp.

Azam, F., T. Fenchel, J. G. Field, J. S. Gray, L.-A. Meyer-Reil, and F. Thingstad. 1983. The ecological role of water-column microbes in the sea. Mar. Ecol. Prog. Ser. *10*:257–263.

Van Baalen, C. 1968. The effects of ultraviolet irradiation on a coccoid blue-green alga: Survival, photosynthesis, and photoreactivation. Plant Physiol. *43*:1689–1695.

Baas Becking, L. G. M., I. R. Kaplan, and D. Moore. 1960. Limits of the natural environment in terms of pH and oxidation-reduction potentials. J. Geol. *68*:243–284.

Babenzien, H.-D., P. Casper, and C. Babenzien. 1991. Bacterial processes in sediments: Comparative studies in the Lake Stechlin area. Verh. Internat. Verein. Limnol. *24*:2613–2617.

Baccini, P. 1976. Untersuchungen über den Schwermetallhaushalt in Seen. Schweiz. Z. Hydrol. *38*:121–158.

Baccini, P. and T. Joller. 1981. Transport processes of copper and zinc in a highly eutrophic and meromictic lake. Schweiz. Z. Hydrol. *43*:176–199.

Bachmann, R. W. 1963. Zinc-65 in studies of the freshwater zinc cycle. *In* V. Schultz and A. W. Klement, Jr., eds. Radioecology. First National Symposium on Radioecology. Reinhold Publishing Corp., New York. pp. 485–496.

Bachmann, R. W. 1984. Calculation of phosphorus and nitrogen loadings to natural and artificial lakes. Verh. Internat. Verein. Limnol. *22*:239–243.

Bachmann, R. W. and C. R. Goldman. 1965. Hypolimnetic heating in Castle Lake, California. Limnol. Oceanogr. *10*:233–239.

Bailey, J. H. and R. B. Davis. 1978. Quantitative comparisons of lake water quality and surface sediment diatom assemblages in Maine, U.S.A. lakes. Verh. Internat. Verein. Limnol. *20*:531.

Bailey-Watts, A. E., A. Kirika, L. May, and D. H. Jones. 1990. Changes in phytoplankton over various time scales in a shallow, eutrophic: The Loch Leven experience with special reference to the influence of flushing rate. Freshwat. Biol. *23*:85–111.

Bain, J. T. and M. C. F. Proctor. 1980. The requirement of aquatic bryophytes for free CO_2 as an inorganic carbon source: Some experimental evidence. New Phytol. *86*:393–400.

Baines, S. B. and M. L. Pace. 1991. The production of dissolved organic matter by phytoplankton and its importance to bacteria: Patterns across marine and freshwater systems. Limnol. Oceanogr. *36*:1078–1090.

Bak, F. 1988. Sulfatreduzierende Bakterien und ihre Aktivität im Litoralsediment der Unteren Güll (Überlinger See). Thesis, Univ. Konstanz, Germany. 123 pp. (Not seen in original; cited in Babenzien *et al.*, 1991.)

Bak, F. and N. Pfennig. 1991. Microbial sulfate reduction in littoral sediment of Lake Constance. FEMS Microbiol. Ecol. *85*:31–42.

Baker, A. L., K. K. Baker, and P. A. Tyler. 1985. Fine-layer depth relationships of lakewater chemistry, planktonic algae and photosynthetic bacteria in meromictic Lake Fidler, Tasmania. Freshwat. Biol. *15*:735–747.

Baker, L. A., P. L. Brezonik, and C. D. Pollman. 1986. Model of internal alkalinity generation: Sulfate retention component. Water Air Soil Pollut. *31*:89–94.

Baker, L. A., A. T. Herlihy, P. R. Kaufmann, and J. M. Eilers. 1991. Acidic lakes and streams in the United States: The role of acidic deposition. Science *252*:1151–1154.

Baker, M. A., H. M. Valett, and C. N. Dahm. 2000. Organic carbon retention and metabolism in a shallow groundwater ecosystem. Ecology (In press).

Baker, R. L. 1979. Birth rate of planktonic rotifers in relation to food concentration in a shallow, eutrophic lake in western Canada. Can. J. Zool. *57*:1206–1214.

Balczon, J. M. and J. R. Pratt. 1996. The functional responses of two benthic algivorous ciliated protozoa with differing feeding strategies. Microb. Ecol. *31*:209–224.

Baldi, F. 1997. Microbial transformation of mercury species and their importance in the biogeochemical cycle of mercury. *In* A. Sigel and H. Sigel, eds. Mercury and Its Effects on Environment and Biology. M. Dekker, Inc., New York. pp. 213–257.

Balikungeri, A., D. Robin, and W. Haerdi. 1985. Manganese in natural waters. II. Evidence for microbiological oxidation of Mn(II). Toxicol. Environ. Chem. *9*:309–325.

Balistrieri, L. S., J. W. Murray, and B. Paul. 1992. The cycling of iron and manganese in the water column of Lake Sammamish, Washington. Limnol. Oceanogr. *37*:510–528.

Ball, R. C. 1949. Experimental use of fertilizer in the production of fish-food organisms and fish. Tech. Bull. Mich. State Univ. Agricult. Exp. Stat., Sec. Zool. *210*, 28 pp.

Ball, R. C. and D. W. Hayne. 1952. Effects of the removal of the fish population on the fish-food organisms of a lake. Ecology *33*:41–48.

Ball, R. C. and F. F. Hooper. 1963. Translocation of phosphorus in a trout stream ecosystem. *In* V. Schultz and A. W. Klement, Jr., eds. Radioecology, First National Symposium on Radioecology. Reinhold Publishing Corp., New York. pp. 217–228.

Ball, R. C., T. A. Wojtalik, and F. F. Hooper. 1963a. Upstream dispersion of radiophosphorus in a Michigan trout stream. Pap. Mich. Acad. Sci. Arts Lett. *48*:57–64.

Balls, H., B. Moss, and K. Irvine. 1980. The loss of submerged plants with eutrophication. I. Experimental design, water chemistry, aquatic plant and phytoplakton biomass in experiments carried out in ponds in the Norfolk Broadland. Freshwat. Biol. *22*:71–87.

Balogh, J. 1958. On some problems of production biology. Acta Zool. Acad. Sci. Hungaricae *4*:89–114.

Bamforth, S. S. 1958. Ecological studies on the planktonic protozoa of a small artificial pond. Limnol. Oceanogr. *3*:398–412.

Bandurski, R. S. 1965. Biological reduction of sulfate and nitrate. *In* J. Bonner and J. E. Varner, eds. Plant Chemistry. Academic Press, New York. pp. 467–490.

Bannerman, R. T., D. E. Armstrong, G. C. Holdren, and R. F. Harris. 1974. Phosphorus mobility in Lake Ontario sediments. Proc. Conf. Great Lakes Res. *17*:158–178.

Bano, N., M. A. Moran, and R. E. Hodson. 1997. Bacterial utilization of dissolved humic substances from a freshwater swamp. Aquat. Microb. Ecol. *12*:233–238.

Barber, L. E. and J. C. Ensign. 1979. Methane formation and release in a small Wisconsin lake. Geomicrobiol. J. *1*:341–353.

Barber, T. R., R. A. Burke, Jr., and W. M. Sackett. 1988. Diffusive flux of methane from warm wetlands. Global Biogeochem. Cycles *2*:411–425.

Barbiero, R. P. and E. B. Welch. 1992. Contribution of benthic blue-green algal recruitment to lake populations and phosphorus translocation. Freshwat. Biol. *27*:249–260.

Barcina, I., B. Ayo, A. Muela, L. Egea, and J. Iriberri. 1991. Predation rates of flagellate and ciliated protozoa on bacterioplankton in a river. FEMS Microbiol. Ecol. *85*:141–150.

Barica, J. 1970. Untersuchungen über den Stickstoff-Kreislauf des Titisees und seiner Quellen. Arch. Hydrobiol./Suppl. *38*:212–235.

Barica, J. 1984. Empirical models for prediction of algal blooms and collapses, winter oxygen depletion and a freeze-out effect in lakes: Summary and verification. Verh. Internat. Verein. Limnol. *22*:309–319.

Barica, J. and F. A. J. Armstrong. 1971. Contribution by snow to the nutrient budget of some small northwest Ontario lakes. Limnol. Oceanogr. *16*:891–899.

Barica, J. and J. A. Mathias. 1979. Oxygen depletion and winterkill risk is small prairie lakes under extended ice cover. J. Fish. Res. Board Can. *36*:980–986.

Bark, A. W. 1981. The temporal and spatial distribution of planktonic and benthic protozoan communities in a small productive lake. Hydrobiologia *85*:239–255.

Bark, A. W. 1985. Studies on ciliated protozoa in eutrophic lakes. I. Seasonal distribution in relation to thermal stratification and hypolimnetic anoxia. Hydrobiologia *124*:167–176.

Bark, A. W. and J. M. Watts. 1984. A comparison of the growth characteristics and spatial distribution of hypolimnetic ciliates in a small lake and an artificial lake ecosystem. J. Gen. Microbiol. *130*:3113–3122.

Barko, J. W. 1982. Influence of potassium source (sediment vs. open water) and sediment composition on the growth and nutrition of a submersed freshwater macrophyte (*Hydrilla verticillata* (L.f.) Royle). Aquat. Bot. *12*:157–172.

Barko, J. W. and R. M. Smart. 1979. The nutritional ecology *Cyperus esculentus*, an emergent aquatic plant, grown on different sediments. Aquat. Bot. *6*:13–28.

Barko, J. W. and R. M. Smart. 1980. Mobilization of sediment phosphorus by submersed freshwater macrophytes. Freshw. Biol. *10*:229–238.

Barko, J. W. and R. M. Smart. 1981. Comparative influences of light and temperature on the growth and metabolism of selected submersed freshwater macrophytes. Ecol. Monogr. *51*:219–235.

Barko, J. W. and R. M. Smart. 1981. Sediment-based nutrient of submersed macrophytes. Aquat. Bot. *10*:339–352.

Barko, J. W., D. G. Hardin, and M. S. Matthews. 1982. Growth and morphology of submersed freshwater macrophytes in relation to light and temperature. Can. J. Bot. *60*:877–887.

Barko, J. W., P. G. Murphy, and R. G. Wetzel. 1977. An investigation of primary production and ecosystem metabolism in a Lake Michigan dune pond. Arch. Hydrobiol. *81*:155–187.

Bärlocher, F. 1982. On the ecology of Ingoldian fungi. BioScience *32*:581–586.

Bärlocher, F. 1985. The role of fungi in the nutrition of stream invertebrates. Bot. J. Linn. Soc. *91*:83–94.

Bärlocher, F. 1992. Community organization. *In* F. Bärlocher, ed. The Ecology of Aquatic Hyphomycetes. Springer-Verlag, New York. pp. 38–76.

Bärlocher, F. 1995. The role of fungi in the nutrition of stream invertebrates. Bot. J. Linn. Soc. *91*:83–94.

Bärlocher, F. and B. Kendrick. 1975. Assimilation efficiency of *Gammarus pseudolimnaeus* (Amphipoda) feeding on fungal mycelium or autumn-shed leaves. Oikos *26*:55–59.

Barlow, J. P. and J. W. Bishop. 1965. Phosphate regeneration by zooplankton in Cayuga Lake. Limnol. Oceanogr. *10*(Suppl.): R15-R25.

Barnhisel, D. R. 1991. The caudal appendage of the cladoceran *Bythotrephes cederstroemi* as defense against young fish. J. Plankton Res. *13*:529–537.

Barrett, P. H. 1957. Potassium concentrations in fertilized trout lakes. Limnol. Oceanogr. *2*:287–294.

Bartell, S. M. and J. F. Kitchell. 1978. Seasonal impact of planktivory on phosphorus release by Lake Wingra zooplankton. Verh. Internat. Verein. Limnol. *20*:466–474.

Barth, H. 1957. Aufnahme und Abgabe von CO_2 and O_2 bei submersen Wasserpflanzen. Gewässer Abwässer *4*(17/18): 18–81.

Bartlett, D. S., K. B. Bartlett, J. M. Hartman, R. C. Harriss, D. I. Sebacher, R. Pelletier-Travis, D. D. Dow, and D. P. Brannon. 1989. Methane emissions from the Florida Everglades: Patterns of variability in a regional wetland ecosystem. Global Biogeochem. Cycles *3*:363–374.

Bartlett, K. B. and R. C. Harriss. 1993. Review and assessment of methane emissions from wetlands. Chemosphere *26*:261–320.

Baskerville, G. L. and P. Emin. 1969. Rapid estimation of heat accumulation from maximum and minimum temperatures. Ecology *50*:514–517.

Bastardo, H. 1979. Laboratory studies on decomposition of littoral plants. Polskie Arch. Hydrobiol. *26*:267–299.

Basu, B. K. and F. R. Pick. 1996. Factors regulating phytoplankton and zooplankton biomass in temperate rivers. Limnol. Oceanogr. *41*:1572–1577.

Basu, B. K. and F. R. Pick. 1997. Phytoplankton and zooplankton development in a lowland, temperate river. J. Plankton Res. *19*:237–253.

Battarbee, R. W. 1978. Relative composition, concentration, and calculated influx of diatoms from a sediment core from Lough Erne, Northern Ireland. Polskie Arch. Hydrobiol. *25*:9–16.

Battarbee, R. W. and D. F. Charles. 1994. Lake acidification and the role of paleolimnology. *In* C. E. W. Steinberg and R. F. Wright, eds. Acidification of Freshwater Ecosystems: Implications for the Future. John Wiley & Sons, Chichester. pp. 51–65.

Battarbee, R. W., D. F. Charles, S. S. Dixit, and I. Renberg. 1999. Diatoms as indicators of surface water acidity. *In* E. F. Stoermer and J. P. Smol, eds. The Diatoms: Applications for the Environmental and Earth Sciences. Cambridge Univ. Press, Cambridge. pp. 85–127.

Batzer, D. P. 1998. Trophic interactions among detritus, benthic midges, and predatory fish in a freshwater marsh. Ecology *79*:1688–1698.

Baudo, R., R. de Bernardi, E. Soldavini, B. Locht, and H. Muntau. 1981. Spatial and temporal variations of metal concentrations in plankton of Lago Maggiore and Lago Mergozzo. Mem. Ist. Ital. Idrobiol. *38*:79–100.

Baumeister, W. 1943. Der Einfluss des Bors auf die Photosynthese und Atmung submerser Pflanzen. Jb. wiss. Bot. *91*:242–277.

Baxter, R. M. 1977. Environmental effects of dams and impoundments. Annu. Rev. Ecol. Syst. *8*:255–283.

Baxter, R. M. 1985. Environmental effects of reservoirs. *In* D. Gunnison, ed. Microbial processes in reservoirs. Developments in Hydrobiology 27. Dr. W. Junk Publ., Dordrecht. pp. 1–26.

Baxter, R. M., R. B. Wood, and M. V. Prosser. 1973. The probable occurrence of hydroxylamine in the water of an Ethiopian lake. Limnol. Oceanogr. *18*:470–472.

Bayley, P. B. 1995. Understanding large river-floodplain ecosystems. BioScience *45*:153–158.

Bayley, S. E., D. W. Schindler, B. R. Parker, M. P. Stainton, and K. G. Beaty. 1992. Effect of forest fire and drought on acidity of a base-poor boreal forest stream: Similarities between climatic warming and acidic precipitation. Biogeochemistry *17*:191–204.

Bayliss, P., J. C. Ellis-Evans, and J. Laybourn-Parry. 1997. Temporal patterns of primary production in a large ultra-oligotrophic Antarctic freshwater lake. Polar Biol. *18*:363–370.

Bayly, I. A. E. and W. D. Williams. 1973. Inland Waters and Their Ecology. Longman Australia Pty. Ltd., Victoria. 316 pp.

Bazin, M. and G. W. Saunders. 1971. The hypolimnetic oxygen deficit as an index of eutrophication in Douglas Lake, Michigan. Mich. Academician *3*(Pt. 4):91–106.

Beadle, L. C. 1943. Osmotic regulation and the faunas of inland waters. Biol. Rev. *18*:172–183.

Beadle, L. C. 1957. Comparative physiology: osmotic and ionic regulation in aquatic animals. Ann. Rev. Physiol. *19*:329–358.

Beadle, L. C. 1959. Osmotic and ionic regulation in relation to the classification of brackish and inland saline waters. Arch. Oceanogr. Limnol. *11*(Suppl.):143–151.

Beadle, L. C. 1969. Osmotic regulation and the adaptation of freshwater animals to inland saline waters. Verh. Internat. Verein. Limnol. *17*:421–429.

Beadle, L. C. 1974. The Inland Waters of Tropical Africa: An Introduction to Tropical Limnology. Longman Group Ltd., London. 365 pp.

Beaulac, M. N. and K. H. Reckhow. 1982. An examination of land use-nutrient export relationships. Water Resour. Bull. *18*:1013–1024.

Beaumont, P. 1975. Hydrology. *In* B. A. Whitton, ed. River Ecology. Blackwell Sci. Publ., Oxford. pp. 1–38.

Beaver, J. R. and T. L. Crisman. 1982. The trophic response of ciliated protozoans in freshwater lakes. Limnol. Oceanogr. *27*:246–253.

Beaver, J. R. and T. L. Crisman. 1989. The role of ciliated protozoans in pelagic freshwater ecosystems. Microb. Ecol. *17*:111–136.

Becher, G., G. Oestvold, P. Paus, and H. M. Seip. 1983. Complexation of copper by aquatic humic matter studied by reversed-phase liquid chromatography and atomic absorption spectroscopy. Chemosphere *12*:1209–1215.

Beer, S. and R. G. Wetzel. 1981. Photosynthetic carbon metabolism in the submerged aquatic angiosperm *Scirpus subterminalis*. Plant Sci. Lett. *21*:199–207.

Beer, S. and R. G. Wetzel. 1982. Photosynthesis in submersed macrophytes of a temperate lake. Plant Physiol. *70*:488–492.

Beeton, A. M. 1958. Relationship between Secchi disc readings and light penetration in Lake Huron. Trans. Amer. Fish. Soc. *87*(1957):73–79.

Beeton, A. M. 1959. Photoreception in the opossum shrimp, *Mysis relicta* Lovén. Biol. Bull. *116*:204–216.

Beeton, A. M. 1960. The vertical migration of Mysis relicta in lakes Huron and Michigan. J. Fish. Res. Board Can. *17*:517–539.

Beeton, A. M. 1961. Environmental changes in Lake Erie. Trans. Amer. Fish. Soc. *90*:153–159.

Behre, K. 1956. Die Algenbesiedlung einiger Seen um Bremen und Bremerhaven. Ver. Inst. Meeresforsch. Bremerhaven *4*:221–283.

Bell, E. M. and J. Laybourn-Perry. 1999. Annual plankton dynamics in an Antarctic saline lake. Freshwat. Biol. *41*:507–519.

Bell, J. G. and J. R. Dick. 1993. The appearance of rods in the eyes of herring and increased didocosahexaenoyl molecular species of phospholipids. J. Mar. Biol. Assoc. U.K. *73*:679–688.

Bell, J. G., D. R. Tocher, F. M. Macdonald, and J. R. Sargent. 1994. Effects of diets rich in linoleic (18:1n–6) and -linolenic (18:3n–3) acids on the growth, lipid class and fatty acid compositions and eicosanoid production in juvenile turbot (*Scopthalmus maximus* L.). Fish Physiol. Biochem. *13*:105–118.

Bell, J. G., R. S. Batty, J. R. Dick, K. Fretwell, J. C. Navarro, and J. R. Sargent. 1995. Dietary deficiency of docosahexaenoic acid impairs vision at low light intensities in juvenile herring (*Clupea harengus* L.). Lipids *26*:565–573.

Bell, R. T., G. M. Ahlgren, and I. Ahlgren. 1983. Estimating bacterioplankton production by measuring [^{3}H]thymidine incorporation in a eutrophic Swedish lake. Appl. Environ. Microbiol. *45*:1709–1721.

Bell, R. T. and L. Tranvik. 1993. Impact of acidification and liming on the microbial ecology of lakes. Ambio *22*:325–330.

Bell, W. H. 1980. Bacterial utilization of algal extracellular products. 1. The kinetic approach. Limnol. Oceanogr. *25*:1007–1020.

Bell, W. H. and E. Sakshaug. 1980. Bacterial utilization of algal extracellular products. 2. A kinetic study of natural populations. Limnol. Oceanogr. *25*:1021–1033.

Belova, M. A. 1992. Bacterial destruction of organic substance of higher aquatic plants in lakes. Mikrobiologii *61*:261–268. (In Russian).

Bencala, K. E. and D. M. McKnight. 1987. Identifying in-stream variability: Sampling iron in an acidic stream. *In* R. C. Averett and D. M. McKnight, eds. Chemical quality of water and the hydrologic cycle. Lewis Publishing, Chelsea, MI. pp. 255–269.

Benderliev, K. M. and N. I. Ivanova. 1994. High-affinity siderophore-mediated iron-transport system in the green alga *Scenedesmus incrassatulus*. Planta *193*:163–166.

Bengtsson, G. 1992. Interactions between fungi, bacteria and beech leaves in a stream microcosm. Oecologia *89*:542–549.

Bengtsson, L. 1986. Dispersion in ice-covered lakes. Nord. Hydrol. *17*:151–170.

Bengtsson, L. 1996. Mixing in ice-covered lakes. Hydrobiologia *322*:91–97.

Bengtsson, L. and T. Svensson. 1996. Thermal regime of ice covered Swedish lakes. Nord. Hydrol. *27*:39–56.

Bengtsson, L., T. Brorson, S. Fleischer, and C. Aström. 1977. Beschaffenheitsänderungen des Sestons in Seen nach dem Sedimentieren. Acta Hydrochim. Hydrobiol. *5*:153–165.

Benke, A. C. 1979. A modification of the Hynes method for estimating secondary production with particular significance for multivoltine populations. Limnol. Oceanogr. *24*:168–171.

Benke, A. C. 1984. Secondary production in aquatic insects. *In* V. H. Resh and D. M. Rosenberg, eds. The Ecology of Aquatic Insects. Praeger, New York. pp. 289–322.

Benke, A. C. 1993a. Concepts and patterns of invertebrate production in running waters. Verh. Internat. Verein. Limnol. *25*:15–38.

Benke, A. C. 1993b. Invertebrate production dynamics of large rivers—deviations from the stream paradigm. *In* S. W. Hamilton, E. W. Chester, and A. F. Scott, eds. Proceedings 5th Annual Symposium on the Natural History of Lower Tennessee and Cumberland River Valleys. Austin Peay State Univ., Clarksville, TN. pp. 1–15.

Benke, A. C. and J. B. Wallace. 1980. Trophic basis of production among net-spinning caddisflies in a southern Appalachian stream. Ecology *61*:108–118.

Benke, A. C., T. C. Van Arsdall, D. M. Gillespie, and F. K. Parrish. 1984. Invertebrate productivity in a subtropical blackwater river: The importance of habitat and life history. Ecol. Monogr. *54*:25–63.

Benke, A. C., A. D. Huryn, L. A. Smock, and J. B. Wallace. 1999. Length-mass relationships for freshwater macroinvertebrates in North America with particular reference to the southeastern United States. J. N. Am. Benthol. Soc. *18*:308–343.

Benke, A. C., I. Chaubey, G. M. Ward, and E. L. Dunn. 2000. Flood pulse dynamics of an unregulated river-floodplain ecosystem in the southeastern USA coastal plain. Ecology (In press).

Benner, R., A. E. Maccubbin, and R. E. Hodson. 1984. Anaerobic biodegradation of the lignin and polysaccharide components of lignocellulose and synthetic lignin by sediment microflora. Appl. Environ. Microbiol. *47*:998–1004.

Benner, R., S. Opsahl, G. Chin-Leo, J. E. Richey, and B. R. Forsberg. 1995. Bacterial carbon metabolism in the Amazon River system. Limnol. Oceanogr. *40*:1262–1270.

Bennett, S. J., R. W. Sanders, and K. G. Porter. 1990. Heterotrophic, autotrophic, and mixotrophic nanoflagellates: Seasonal abundances and bacterivory in a eutrophic lake. Limnol. Oceanogr. *35*:1821–1832.

Benoit, G. and H. F. Hemond. 1988. Comment on "Dilution of ^{210}Pb by organic sedimentation in lakes of different trophic states, and application to studies of sediment-water interaction" (Binford and Brenner). Limnol. Oceanogr. *33*:299–304.

Benoit, R. J. 1957. Preliminary observations on cobalt and vitamin B_{12} in fresh water. Limnol. Oceanogr. *2*:233–240.

Benoit, R. J. 1969. Geochemistry of eutrophication. *In* Eutrophication: Causes, Consequences, Correctives. National Academy of Sciences, Washington, DC. pp. 614–630.

Benson, B. B. and D. Krause, Jr. 1980. The concentration and isotopic fractionation of gases dissolved in freshwater in equilibrium with the atmosphere. 1. Oxygen. Limnol. Oceanogr. *25*:662–671.

Bentley, J. A. 1958a. Role of plant hormones in algal metabolism and ecology. Nature *181*:1499–1502.

Bentley, J. A. 1958b. The naturally-occurring auxins and inhibitors. Ann. Rev. Plant Physiol. *9*:47–80.

Benton, A. R., Jr., W. P. James, and J. W. Rouse, Jr. 1978. Evapotranspiration from water hyacinth (*Eichhornia crassipes* (Mart). Solms) in Texas reservoirs. Water Resour. Bull. *14*:919–930.

Bentzen, E. and W. D. Taylor. 1991. Estimating organic P utilization by freshwater plankton using [^{32}P]ATP. J. Plankton Res. *13*:1223–1238.

Berberovic, R., K. Bikar, and W. Geller. 1990. Seasonal variability of the embryonic development time of three planktonic crustaceans: Dependence on temperature, adult size, and egg weight. Hydrobiologia *203*:127–136.

Berendonk, T. U. and W. J. O'Brien. 1996. Movement response of *Chaoborus* to chemicals from a predator and prey. Limnol. Oceanogr. *41*:1829–1832.

Berg, K. 1938. Studies on the bottom animals of Esrom Lake. K. Danske Vidensk. Selsk. Skr. Nat. Mat. Afd. 9, 8, 255 pp.

Berg, K. and I. C. Petersen. 1956. Studies on the humic, acid Lake Gribsß. Folia Limnol. Scandinavica 8, 273 pp.

Berger, F. 1955. Die Dichte natürlicher Wässer und die Konzentrations-Stabilität in Seen. Arch. Hydrobiol./Suppl. *22*:286–294.

Bergh, Ø., K. Y. Børsheim, G. Bratbak, and M. Heldal. 1989. High abundance of viruses found in aquatic environments. Nature *340*:467–468.

Bergmann, M. A. and H. E. Welch. 1985. Spring meltwater mixing in small Arctic lakes. Can. J. Fish. Aquat. Sci. *42*:1789–1798.

Bergstein, T., Y. Henis, and B. Z. Cavari. 1981. Nitrogen fixation by the photosynthetic sulfur bacterium *Chlorobium phaeobacteroides* from Lake Kinneret. Appl. Environ. Microbiol. *41*:542–544.

Bergstrom, A.-K. and M. Jansson. 2000. Bacterioplankton production in humic lake Örträsket in relation to input of bacterial cells and input of allochthonous organic carbon. Microb. Ecol. *39*:101–115.

Berman, M. S. and S. Richman. 1974. The feeding behavior of *Daphnia pulex* from Lake Winnebago, Wisconsin. Limnol. Oceanogr. *19*:105–109.

Berman, T. 1970. Alkaline phosphatases and phosphorus availability in Lake Kinneret. Limnol. Oceanogr. *15*:663–674.

Berman, T. 1976. Release of dissolved organic matter by photosynthesizing algae in Lake Kinneret, Israel. Freshwat. Biol. *6*:13–18.

Berman, T., B. F. Sheer, E. Sherr, D. Wynne, and J. J. McCarthy. 1984. The characteristics of ammonium and nitrate uptake by phytoplankton in Lake Kinneret. Limnol. Oceanogr. *29*:287–297.

Berman, T., B. Kaplan, S. Chava, R. Parparova, and A. Nishri. 1993.

Effects of iron and chelation on Lake Kinneret bacteria. Microb. Ecol. *26*:1–8.

Bern, L. 1987. Zooplankton grazing on [methyl-^{3}H]thymidine-labelled natural particle assemblages: Determination of filtering rates and food selectivity. Freshwat. Biol. *17*:151–159.

Bernard, J. M. 1973. Production ecology of wetland sedges: The genus *Carex*. Pol. Arch. Hydrobiol. *20*:207–214.

Bernard, J. M. and B. A. Solsky. 1977. Nutrient cycling in a *Carex lacustris* wetland. Can. J. Bot. *55*:630–638.

Bernard, J. M., D. Solander, and J. Květ. 1988. Production and nutrient dynamics in *Carex* wetlands. Aquat. Bot. *30*:125–147.

Bernatowicz, S., S. Leszczynski, and S. Tyczynska. 1976. The influence of transpiration by emergent plants on the water balance in lakes. Aquat. Bot. *2*:275–288.

Berner, E. K. and R. A. Berner. 1987. The Global Water Cycle: Geochemistry and Environment. Prentice-Hall, Englewood Cliffs, NJ. 397 pp.

Berninger, U.-G. and M. Huettel. 1997. Impact of flow on oxygen dynamics in photosynthetically active sediments. Aquat. Microb. Ecol. *12*:291–302.

Berninger, U.-G., B. J. Finlay, and P. Kuuppo-Leinikki. 1991. Protozoan control of bacterial abundances in freshwater (*sic*). Limnol. Oceanogr. *36*:139–147.

Berrie, A. D. 1972. Productivity of the River Thames at Reading. Symp. Zool. Soc. Lond. *29*:69–86.

Bertani, A., I. Brambilla, and F. Menegus. 1980. Effect of anaerobiosis on rice seedlings: Growth, metabolic rate, and fate of fermentation products. J. Exp. Bot. *31*:325–331.

Bertoni, R. and C. Callieri. 1989. Organic matter and decomposers in Lago Maggiore: A pluriannual study. Mem. Ist. Ital. Idrobiol. *46*:145–172.

Berzinš, B. 1958. Ein planktologisches Querprofil. Rept. Inst. Freshwat. Res. Drottningholm *39*:5–22.

Berzinš, B. and B. Pejler. 1989. Rotifer occurrence in relation to temperature. Hydrobiologia *175*:223–231.

Best, E. P. H. 1980. Effect of nitrogen on the growth and nitrogenous compounds of *Ceratophyllum demersum*. Aquat. Bot. *8*:197–206.

Best, E. P. H. 1986. Photosynthetic characteristics of the submerged macrophyte *Ceratophyllum demersum*. Plant Physiol. *84*:502–510.

Best, J. L. and A. G. Roy. 1991. Mixing-layer distortion at the confluence of channels of different depth. Nature *350*:411–413.

Best, E. P. H. and J. H. A. Dassen. 1987. Biomass, stand area, primary production characteristics and oxygen regime of the *Ceratophyllum demersum* L. population in Lake Vechten, The Netherlands. Arch. Hydrobiol./Suppl. *76*:347–367.

Best, E. P. H. and H. W. C. Visser. 1987. Seasonal growth of the submerged macrophyte *Ceratophyllum demersum* L. in mesotrphic Lake Vechten in relation to insolation, temperature and reserve carbohydrates. Hydrobiologia *148*:231–243.

Bettoli, P. W., M. F. Cichra, and W. J. Clark. 1985. Phytoplankton community structure and dynamics in Lake Conroe, Texas. Texas J. Sci. *36*:221–233.

Beveridge, T. J. and L. L. Graham. 1991. Surface layers of bacteria. Microbiol. Rev. *55*:684–705.

Bhowmik, N. G. and M. Demissee. 1982. Carrying capacity of floodplains. Proc. Am. Soc. Chem. Eng., J. Hydraulics Div. *108*(HY3):443–452.

Bickford, E. D. and S. Dunn. 1972. Lighting for Plant Growth. Kent State University Press, Kent, OH. 221 pp.

Bicudo, D. C., A. K. Ward, and R. G. Wetzel. 1998. Fluxes of dissolved organic carbon within attached aquatic microbiota. Verh. Internat. Verein. Limnol. *26*:1608–1613.

Biehle, G., T. Speck, and H.-C. Spatz. 1998. Hydrodynamics and biomechanics of the submerged water moss *Fontinalis antipyretica*—a comparison of specimens from habitats with different flow velocities. Bot. Acta *111*:42–50.

Bienfang, P. K. 1979. A new phytoplankton sinking rate method suitable for field use. Deep-Sea Res. *26*:719–729.

Biggar, J. W. and R. B. Corey. 1969. Agricultural drainage and eutrophication. *In* Eutrophication: Causes, Consequences, Correctives. National Academy of Sciences, Washington, DC. pp. 404–445.

Biggs. B. J. F. 1996. Patterns in benthic algae of streams. *In* R. J. Stevenson, M. L. Bothwell, and R. L. Lowe, eds. Algal Ecology: Freshwater Benthic Ecosystems. Academic Press, San Diego. pp. 31–56.

Biggs, B. J. F., M. J. Duncan, S. N. Francoeur, and W. D. Meyer. 1997. Physical characterisation of microform bed cluster refugia in 12 headwater streams, New Zealand. N. Z. J. Mar. Freshwat. Res. *31*:413–422.

Bilby, R. E. and G. E. Likens. 1980. Importance of organic debris dams in the structure and function of stream ecosystems. Ecology *61*:1107–1113.

Billaud, V. A. 1968. Nitrogen fixation and the utilization of other inorganic nitrogen sources in a subarctic lake. J. Fish. Res. Board Can. *25*:2101–2110.

Bilsky, L. J., ed. 1980. Historical Ecology. Essays on Environment and Social Change. Kennikat Press, Port Washington, NY. 195 pp.

Binford, M. W. 1982. Ecological history of Lake Valencia, Venezuela: Interpretation of animal microfossils and some chemical, physical, and geological features. Ecol. Monogr. *52*:307–333.

Binford, M. W. and M. Brenner. 1988. Reply to comment by Benoit and Hemond. Limnol. Oceanogr. *33*:304–310.

Binford, M. W., E. S. Deevey, and T. L. Crisman. 1983. Paleolimnology: An historical perspective on lacustrine ecosystem. Annu. Rev. Ecol. Syst. *14*:255–286.

Bird, D. F. and C. M. Duarte. 1989. Bacteria-organic matter relationship in sediments: A case of spurious correlation. Can. J. Fish. Aquat. Sci. *46*:904–908.

Bird, D. F. and J. Kalff. 1986. Bacterial grazing by planktonic lake algae. Science *231*:493–495.

Birge, E. A. 1915. The heat budgets of American and European lakes. Trans. Wis. Acad. Sci. Arts Lett. *18*(Pt. 1):166–213.

Birge, E. A. 1916. The work of the wind in warming a lake. Trans. Wis. Acad. Sci. Arts Lett. *18*(Pt. 2):341–391.

Birge, E. A. and C. Juday. 1911. The inland lakes of Wisconsin. The dissolved gases of the water and their biological significance. Bull. Wis. Geol. Nat. Hist. Survey 22, Sci. Ser. 7, 259 pp.

Birge, E. A. and C. Juday. 1914. A limnological study of the Finger Lakes of New York. Bull. U.S. Bur. Fish. *32*:525–609.

Birge, E. A. and C. Juday. 1926. Organic content of lake water. Bull. U.S. Bur. Fish *42*:185–205.

Birge, E. A. and C. Juday. 1927. The organic content of the water of small lakes. Proc. Amer. Phil. Soc. *66*:357–372.

Birge, E. A. and C. Juday. 1934. Particulate and dissolved organic matter in inland lakes. Ecol. Monogr. *4*:440–474.

Birge, E. A., C. Juday, and H. W. March. 1927. The temperature of the bottom deposits of Lake Mendota. A chapter in the heat exchanges in the lake. Trans. Wis. Acad. Sci. Arts Lett. *23*:187–231.

Birks, H. H. 1980. Plant macrofossils in quaternary lake sediments. Arch. Hydrobiol. Ergeb. Limnol. *15*, 60 pp.

Birks, H. J. B. and H. H. Birks. 1980. Quaternary Palaeoecology. Edward Arnold Publ., London. 289 pp.

Birks, H. J. B., J. M. Line, S. Juggins, A. C. Stevenson, and C. J. F. ter

Braak. 1990. Diatoms and pH reconstruction. Phil. Trans. R. Soc. Lond. B *327*:263–278.

Birmingham, M. W., R. W. Bachmann, and W. G. Crumpton. 1994. Nitrate uptake by stream sediments: The influence of sediment character. Verh. Internat. Verein. Limnol. *25*:1467–1470.

Bjarnov, N. 1972. Carbohydrases in *Chironomus*, *Gammarus* and some Trichoptera larvae. Oikos *23*:261–263.

Bjerke, G., A. H. Erlandsen, and K. Vennerød. 1979. The temperature of maximum density. Temperature profiles and circulation in deep, temperate lakes. Arch. Hydrobiol. *86*:87–94.

Björk, S. 1967. Ecologic investigations of *Phragmites communis*. Studies in theoretic and applied limnology. Folia Limnol. Scand. *14*, 248 pp.

Björk-Ramberg, S. 1985. Uptake of phosphate and inorganic nitrogen by a sediment-algal system in a subarctic lake. Freshwat. Biol. *15*:175–183.

Björk-Ramberg, S. and C. Ånell. 1985. Production and chlorophyll concentration of epipelic and epilithic algae in fertilized and nonfertilized subarctic lakes. Hydrobiologia *126*:213–219.

Black, A. P. and R. F. Christman. 1968. Chemical characteristics of fulvic acids. J. Amer. Water Works Assoc. *55*:897–912.

Black, A. R. and S. I. Dodson. 1990. Demographic costs of *Chaoborus*-induced phenotypic plasticity in *Daphnia pulex*. Oecologia *83*:117–122.

Black, C. C. 1971. Ecological implications of dividing plants into groups with distinct photosynthetic production capacities. Adv. Ecol. Res. *7*:87–114.

Black, R. W. 1980a. The nature and causes of cyclomorphosis in a species of the *Bosmina longirostris* complex. Ecology *61*:1122–1132.

Black, R. W. 1980b. The genetic component of cyclomorphosis in Bosmina. *In* W. C. Kerfoot, ed. Evolution and Ecology Zooplankton Communities. Univ. Press New England, Hanover, NH. pp. 456–469.

Blanch, S. J., G. G. Ganf, and K. F. Walker. 1998. Growth and recruitment in *Vallisneria americana* as related to average irradiance in the water column. Aquat. Bot. *61*:181–205.

Bland, R. D. and A. J. Brook. 1974. The spatial distribution of desmids in lakes in northern Minnesota, U.S.A. Freshwat. Biol. *4*:543–556.

Blanton, J. O. 1973. Vertical entrainment into the epilimnia of stratified lakes. Limnol. Oceanogr. *18*:697–704.

Blazka, P., Z. Brandl, and L. Procházková. 1982. Oxygen consumption and ammonia and phosphate excretion in pond zooplankton. Limnol. Oceanogr. *27*:294–303.

Blevins, D. G. and K. M. Lukaszewski. 1998. Boron in plant structure and function. Annu. Rev. Plant Physiol. Mol. Biol. *49*:481–500.

Blindow, I. 1992. Decline of charophytes during eutrophication: Comparison with angiosperms. Freshwat. Biol. *28*:15–27.

Blinn, D. W. 1993. Diatom community structure along physicochemical gradients in saline lakes. Ecology *74*:1246–1263.

Blinn, D. W. and K. S. Button. 1973. The effect of temperature on parasitism of *Pandorina sp.* by *Dangeardia mammillata* B. Schröder in an Arizona mountain lake. J. Phycol. *9*:323–326.

Blinn, D. W. and D. B. Johnson. 1982. Filter-feeding of *Hyalella montezuma*, an unusual behavior for a freshwater amphipod. Freshwat. Invertebr. Biol. *5*:48–52.

Blinn, D. W. and R. W. Davies. 1990. Concomitant diel vertical migration of a predatory leech and its amphipod prey. Freshwat. Biol. *24*:401–407.

Blinn, D. W., A. Fredericksen, and V. Korte. 1980. Colonization rates and community structure of diatoms on three different rock substrata in a lotic system. Br. Phycol. J. *15*:303–310.

Bloem, J. and M.-J. B. Bär-Gilissen. 1989. Bacterial activity and protozoan grazing potential in a stratified lake. Limnol. Oceanogr. *34*:297–309.

Bloem, J., F. M. Ellenbroek, M.-J. B. Bär-Gilissen, and T. E. Cappenberg. 1989. Protozoan grazing and bacterial production in stratified lake Vechten estimated with fluorescently labeled bacteria and by thymidine incorporation. Appl. Environ. Microbiol. *55*:1787–1795.

Bloesch, J. 1995. Mechanisms, measurement and importance of sediment resuspension in lakes. Mar. Freshwat. Res. *46*:295–304.

Bloesch, J. and N. M. Burns. 1980. A critical review of sedimentation trap technique. Schweiz. Z. Hydrol. *42*:15–55.

Bloesch, J. and U. Uehlinger. 1986. Horizontal sedimentation differences in a eutrophic Swiss lake. Limnol. Oceanogr. *31*:1094–1109.

Blom, C. W. P. M., G. M. Bögemann, P. Laan, A. J. M. van der Sman, H. M. van de Steeg, and L. A. C. J. Voesenek. 1990. Adaptations to flooding in plants from river areas. Aquat. Bot. *38*:29–47.

Blom, G., E. H. S. van Duin, and L. Lijklema. 1994. Sediment resuspension and light conditions in some shallow Dutch lakes. Wat. Sci. Tech. *30*:243–252.

Blomqvist, P., A. Pettersson, and P. Hyenstrand. 1994. Ammonium-nitrogen: A key regulatory factor causing dominance of non-nitrogen-fixing cyanobacteria in aquatic systems. Arch. Hydrobiol. *132*:141–164.

Bloomfield, J. A., ed. 1978. Lakes of New York State. Vol. I. Ecology of the Finger Lakes. Academic Press, New York. 499 pp.

Blotnick, J. R., J. Rho, and H. B. Gunner. 1980. Ecological characteristics of the rhizosphere microflora of *Myriophyllum heterophyllum*. J. Environ. Qual. *9*:207–210.

Blough, N. V., O. C. Zafiriou, and J. Bonilla. 1993. Optical absorption spectra of waters from the Orinoco River outflow: Terrestrial input of colored organic matter to the Caribbean. J. Geophys. Res. *98*:2271–2278.

Blum, U. 1996. Allelopathic interactions involving phenolic acids. J. Nematol. *28*:129–132.

Boavida, M. J. and R. T. Heath. 1984. Are the phosphatases released by *Daphnia magna* components of its food? Limnol. Oceanogr. *29*:641–645.

Boavida, M. J. and R. T. Marques. 1995. Low activity of alkaline phosphatase in two eutrophic reservoirs. Hydrobiologia *297*:11–16.

Boavida, M. J. and R. G. Wetzel. 1998. Inhibition of phosphatase activity by dissolved humic substances and hydrolytic reactivation by natural UV. Freshwat. Biol. *40*:285–293.

Bodelier, P. L. E. and P. Frenzel. 1999. Contribution of methanotrophic and nitrifying bacteria to CH_4 and NH_4 oxidation in the rhizosphere of rice plants as determined by new methods of discrimination. Appl. Environ. Microbiol. *65*:1826–1833.

Bodin, K. and A. Nauwerck. 1969. Produktionsbiologische Studien über die Moosvegetation eines klaren Gebirgssees. Schweiz. Z. Hydrol. *30*:318–352.

Bodkin, P. C., D. H. N. Spence, and D. C. Weeks. 1980. Photoreversible control of heterophylly in *Hippuris vulgaris* L. New Phytol. *84*:533–542.

Bodkin, P. C., U. Posluszny, and H. M. Dale. 1980. Light and pressure in two freshwater lakes and their influence on the growth, morphology and depth limits of *Hippuris vulgaris*. Freshwat. Biol. *10*:545–552.

Boers, J. J. and J. C. H. Carter. 1978. The life history of *Cyclops scutifer* Sars (Copepoda: Cyclopoida) in a small lake of the Matamek River System, Quebec. Can. J. Zool. *56*:2603–2607.

Boers, P. C. M. 1991. The influence of pH on phosphate release from lake sediments. Water Res. *25*:309–311.

Boers, P. C. M. and O. van Hese. 1988. Phosphorus release from the peaty sediments of the Loosdrecht lakes (The Netherlands). Water Res. *22*:355–363.

Boersma, M. and J. Vijverberg. 1996. Food effects on life history traits and seasonal dynamics of *Ceriodaphnia pulchella*. Freshwat. Biol. *35*:25–34.

Boersma, M., O. F. R. van Tongeren, and W. M. Mooij. 1996. Seasonal patterns in the mortality of *Daphnia* species in a shallow lake. Can. J. Fish. Aquat. Sci. *53*:18–28.

Bogdan, K. G. and J. J. Gilbert. 1984. Body size and food size in freshwater zooplankton. Proc. Natl. Acad. Sci. USA *81*:6427–6431.

Bogdan, K. G., J. J. Gilbert, and P. L. Starkweather. 1980. In situ clearance rates of planktonic rotifers. Hydrobiologia *73*:73–77.

Bohl, E. 1980. Diel pattern of pelagic distribution and feeding in planktivorous fish. Oecologia *44*:368–375.

Bohle-Carbonell, M. 1986. Currents in Lake Geneva. Limnol. Oceanogr. *31*:1255–1266.

Bohr, R. and M. Luscinska. 1975. The primary production of the periphyton association *Oedogonio-Epithemietum litoralae*. Verh. Internat. Verein. Limnol. *19*:1309–1312.

Bole, J. B. and J. R. Allan. 1978. Uptake of phosphorus from sediment by aquatic plants, *Myriophyllum spicatum* and *Hydrilla verticillata*. Water Res. *12*:353–358.

Bolsenga, S. J., C. E. Herdendorf, and D. C. Norton. 1991. Spectral transmittance of lake ice from 400–850 nm. Hydrobiologia *218*:15–25.

Bolsenga, S. J., M. Evans, H. A. Vanderploeg, and D. G. Norton. 1996. PAR transmittance through thick, clear freshwater ice. Hydrobiologia *330*:227–230.

Bolser, R. C. and M. E. Hay. 1998. A field test of inducible resistance to specialist and generalist herbivores using the water lily *Nuphar luteum*. Oecologia *116*:143–153.

Bolser, R. C., M. E. Hay, N. Lindquist, W. Fenical, and D. Wilson. 1998. Chemical defenses of freshwater macrophytes against crayfish herbivory. J. Chem. Ecol. *24*:1639–1658.

Bonacina, C., G. Bonomi, and D. Ruggiu. 1973. Reduction of the industrial pollution of Lake Orta (N. Italy): An attempt to evaluate its consequences. Mem. Ist. Ital. Idrobiol. *30*:149–168.

Bond, C. E. 1996. Biology of Fishes. 2nd ed. Saunders College Publ., Philadelphia.

Boney, A. D. 1981. Mucilage: The ubiquitous algal attribute. Br. Phycol. J. *16*:115–132.

Bonny, A. P. 1976. Recruitment of pollen to the seston and sediment of some Lake District lakes. J. Ecol. *64*:859–887.

Bonomi, G. 1979. Ponderal production of *Tubifex tubifex* Müller and *Limnodrilus hoffmeisteri* Claparède (Oligochaeta, Tubificidae), benthic cohabitants of an artificial lake. Boll. Zool. *46*:153–161.

Booker, M. J. and A. E. Walsby. 1979. The relative form resistance of straight and helical blue-green algal filaments. Br. Phycol. J. *14*:141–150.

Booker, M. J. and A. E. Walsby. 1981. Bloom formation and stratification by a planktonic blue-green alga in an experimental water column. Br. Phycol. J. *16*:411–421.

Boon, J. J., R. G. Wetzel, and G. L. Godshalk. 1982. Pyrolysis mass spectrometry of some *Scirpus* species and their decomposition products. Limnol. Oceanogr. *27*:839–848.

Boon, P. I. 1991. Bacterial assemblages in rivers and billabongs of southeastern Australia. Microb. Ecol. *22*:27–52.

Boon, P. I. 1994. Interactions between suspended solids and phosphatase activity in turbid rivers of south-eastern Australia. Verh. Internat. Verein. Limnol. *25*:1827–1833.

Boon, P. I. and B. K. Sorrell. 1991. Biogeochemistry of billabong sediments. I. The effect of macrophytes. Freshwat. Biol. *26*:209–226.

Boon, P. I. and B. K. Sorrell. 1995. Methane fluxes from an Australian floodplain wetland: The importance of emergent macrophytes. J. N. Am. Benthol. Soc. *14*:582–598.

Boon, P. I., P. Virtue, and P. D. Nichols. 1996. Microbial consortia in wetland sediments: A biomarker analysis of effects of hydrological regime, vegetation and season on benthic microbes. Mar. Freshwat. Res. *47*:27–41.

Boon, P. J., B. R. Davies, and G. E. Petts, eds. 2000. Global Perspectives on River Conservation: Science, Policy and Practice. J. Wiley & Sons, Chichester. 548 pp.

Boon, P. J., P. Calow, and G. E. Petts, eds. 1992. River Conservation and Management. J. Wiley & Sons, Chichester. 470 pp.

Borass, M. E., K. W. Estep, P. W. Johnson, and J. McN. Sieburth. 1988. Phagotrophic phototrophs: The ecological significance of mixotrophy. J. Protozool. *35*:249–252.

Borchardt, M. A. 1996. Nutrients. *In* R. J. Stevenson, M. L. Bothwell, and R. L. Lowe, eds. Algal Ecology: Freshwater Benthic Ecosystems. Academic Press, San Diego. pp. 183–227.

Børcheim, K. Y. and Y. Olsen. 1984. Grazing activities by *Daphnia pulex* on natural populations of bacteria and algae. Verh. Internat. Verein. Limnol. *22*:644–648.

Boring, L. R., W. T. Swank, J. B. Waide, and G. S. Henderson. 1988. Sources, fates, and impacts of nitrogen inputs to terrestrial ecosystems: Review and synthesis. Biogeochemistry *6*:119–159.

Bornkamm, R. and F. Raghi-Atri. 1986. Über die Wirkung unterschiedlicher Gaben von Stickstoff und Phosphor auf die Entwicklung von *Phragmites australis* (Cav.) Trin. ex Steudel. Arch. Hydrobiol. *105*:423–441.

Borutskii, E. V. 1939. Dynamics of the total benthic biomass in the profundal of Lake Beloie. Proc. Kossino Limnol. Stat. Hydrometeorol. Serv. USSR *22*:196–218. (Translated by M. Ovchynnyk, Michigan State University.)

Borutskii, E. V. 1950. Dinamika organicheskogo veshchestva v vodoeme. Trudy Vses. Gidrobiol. Obshchestva 2:43–68.

Bosch, F. van den and B. Santer. 1993. Cannibalism in *Cyclops abyssorum*. Oikos *67*:19–28.

Bosse, U. and P. Frenzel. 1997. Activity and distribution of methane-oxidizing bacteria in flooded rice soil microcosms and in rice plants (*Oryza sativa*). Appl. Environ. Microbiol. *63*:1199–1207.

Bosse, U., P. Frenzel, and R. Conrad. 1993. Inhibition of methane oxidation by ammonium in the surface layer of a littoral sediment. FEMS Microbiol. Ecol. *13*:123–134.

Bosselmann, S. 1975. Production of *Eudiaptomus graciloides* in Lake Esrom, 1970. Arch. Hydrobiol. *76*:43–64.

Bosselmann, S. 1979a. Population dynamics of *Keratella cochlearis* in Lake Esrom. Arch. Hydrobiol. *87*:152–165.

Bosselmann, S. 1979b. Production of *Keratella cochlearis* in Lake Esrom. Arch. Hydrobiol. *87*:304–313.

Boston, H. L. and M. S. Adams. 1986. The contribution of crassulacean acid metabolism to the annual productivity of two aquatic vascular plants. Oecologia *68*:615–622.

Boston, H. L. and M. S. Adams. 1987. Productivity, growth and photosynthesis of two small 'isoetid' plants, *Littorella uniflora* and *Isoetes macrospora*. J. Ecol. *75*:333–350.

Boston, H. L., M. S. Adams, and T. P. Pienkowski. 1987. Utilization of sediment CO_2 by selected North American isoetids. Ann. Bot. *60*:485–494.

Boston, H. L., M. S. Adams, and J. D. Madsen. 1989. Photosynthetic strategies and productivity in aquatic systems. Aquat. Bot. *34*:27–57.

Boström, B. 1981. Factors controlling the seasonal variation of nitrate in Lake Erken. Int. Revue ges. Hydrobiol. *66*:821–835.

Boström, B., J. M. Andersen, S. Fleischer, and J. Jansson. 1988. Exchange of phosphorus across the sediment-water interface. Hydrobiologia *170*:229–244.

Boström, B., M. Jansson, and C. Forsberg. 1982. Phosphorus release from lake sediments. Arch. Hydrobiol. Beih. Ergebn. Limnol. *18*:5–59.

Botan, E. A., J. J. Miller, and H. Kleerekoper. 1960. A study of the microbiological decomposition of nitrogenous matter in fresh water. Arch. Hydrobiol. *56*:334–354.

Botelho, C. M. S., R. A. R. Boaventura, M. L. S. S. Goncalves, and L. Sigg. 1994. Interactions of lead(II) with natural river water. Part II. Particulate matter. Sci. Total Environ. *151*:101–112.

Bothwell, M. L., D. Sherbot, A. C. Roberge, and R. J. Daley. 1993. Influence of natural ultraviolet radiation on lotic periphytic diatom community growth, biomass accrual, and species composition: Short-term versus long-term effects. J. Phycol. *29*:24–35.

Bott, T. L. 1983. Primary productivity in steams. *In* J. R. Barnes and G. W. Minshall, eds. Stream Ecology Application and Testing of General Ecological Theory. Plenum Press, New York. pp. 29–53.

Bott, T. L. and L. A. Kaplan. 1985. Bacterial biomass, metabolic state, and activity in stream sediments: Relation to environmental variables and multiple assay comparisons. Appl. Environ. Microbiol. *50*:508–522.

Bott, T. L. and L. A. Kaplan. 1989. Densities of benthic protozoa and nematodes in a Piedmont stream. J. N. Am. Benthol. Soc. *8*:187–196.

Bott, T. L., J. T. Brock, C. E. Cushing, S. V. Gregory, D. King, and R. C. Petersen. 1978. A comparison of methods for measuring primary productivity and community respiration in streams. Hydrobiologia *60*:3–12.

Bott, T. L., J. T. Brock, C. S. Dunn, R. J. Naiman, R. W. Ovink, and R. C. Petersen. 1985. Benthic community metabolism in four temperate stream systems: An inter-biome comparison and evaluation of the river continuum concept. Hydrobiologia *123*:3–45.

Böttger, K. 1970. Die Ernährungsweise der Wassermilben (Hydrachnellae, Acari). Int. Revue ges. Hydrobiol. *55*:895–912.

Bottrell, H. H. 1975a. The relationship between temperature and duration of egg development in some epiphytic Cladocera and Copepoda from the River Thames, Reading, with a discussion of temperature functions. Oecologia *18*:63–84.

Bottrell, H. H. 1975b. Generation time, length of life, instar duration and frequency of moulting, and their relationship to temperature in eight species of Cladocera from the River Thames, Reading. Oecologia *19*:129–140.

Bottrell, H. H., A. Duncan, Z. M. Gliwicz, E. Grygierek, A. Herzig, A. Hillbricht-Ilkowska, H. Kurasawa, P. Larsson, and T. Weglenska. 1976. A review of some problems in zooplankton production studies. Norw. J. Zool. *24*:419–456.

Boucherie, M. M. and H. Züllig. 1983. Cladoceran remains as evidence of change in trophic state in three Swiss lakes. Hydrobiologia *103*:141–146.

Bouckaert, F. W. and J. Davis. 1998. Microflow regimes and the distribution of macroinvertebrates around stream boulders. Freshwat. Biol. *40*:77–86.

Boulton, A. J. and P. I. Boon. 1991. A review of methodology used to measure leaf litter decomposition in lotic environments: Time to turnover an old leaf? Aust. J. Mar. Freshwat. Res. *42*:1–43.

Bousfield, E. L. 1989. Revised morphological relationships within the amphipod genera *Pontoporeia* and *Gammaracanthus* and the 'glacial relict' significance of their postglacial distributions. Can. J. Fish. Aquat. Sci. *46*:1714–1725.

Bouwman, A. F. 1991. Agronomic aspects of wetland rice cultivation and associated methane emissions. Biogeochemistry *15*:65–88.

Bouwmeester, R. J. B. and P. L. G. Vlek. 1981. Rate control of ammonia volatilization from rice paddies. Atmospheric Environ. *15*:131–140.

Bowden, W. B. 1987. The biogeochemistry of nitrogen in freshwater wetlands. Biogeochemistry *4*:313–348.

Bowen, S. H. 1978. Benthic diatom distribution and grazing by *Sarotherodon mossambicus* in Lake Sibaya, south Africa. Freshwat. Biol. *8*:449–453.

Bowen, S. H. 1979. Determinants of the chemical composition of periphytic detrital aggregate in a tropical lake (Lake Valencia, Venezuela). Arch. Hydrobiol. *87*:166–177.

Bowen, S. H. 1984. Evidence of a detritus food chain based on consumption of organic precipitates. Bull. Mar. Sci. *35*:440–448.

Bowers, J. A. and N. E. Grossnickle. 1978. The herbivorous habits of *Mysis relicta* in Lake Michigan. Limnol. Oceanogr. *23*:767–776.

Bowes, G. 1987. Aquatic plant photosynthesis: Strategies that enhance carbon gain. *In* R. M. M. Crawford, ed. Plant Life in Aquatic and Amphibious Habitats. Blackwell Scientific Publs., Oxford. pp. 77–96.

Bowes, G., A. S. Holaday, and W. T. Haller. 1979. Seasonal variation in the biomass, tuber density, and photosynthetic metabolism of *Hydrilla* in three Florida lakes. J. Aquat. Plant Manage. *17*:61–65.

Bowes, G., T. K. Van, L. A. Garrard, and W. T. Haller. 1977. Adaptation to low light levels by *Hydrilla*. J. Aquat. Plant Manage. *15*:32–35.

Bowie, I. S. and P. A. Gillespie. 1976. Microbial parameters and trophic status of ten New Zealand lakes. N. Z. J. Mar. Freshw. Res. *10*:343–354.

Bowker, D. W., M. T. Wareham, and M. A. Learner. 1985. The ingestion and assimilation of algae by *Nais elinguis* (Oligochaeta: Naididae). Oecologia *67*:282–285.

Bowling, L. C. 1990. Heat contents, thermal stabilities and Birgean wind work in dystrophic Tasmanian lakes and reservoirs. Aust. J. Mar. Freshwat. Res. *41*:429–441.

Bowman, G. T. and J. J. Delfino. 1980. Sediment oxygen demand techniques: A review and comparison of laboratory and in situ systems. Water Res. *14*:491–499.

Boyd, C. E. 1970. Losses of mineral nutrients during decomposition of *Typha latifolia*. Arch. Hydrobiol. *66*:511–517.

Boyd, C. E. 1971. The dynamics of dry matter and chemical substances in a *Juncus effusus* population. Amer. Midland Nat. *86*:28–45.

Boyd, C. E. 1987. Evapotranspiration/evaporation (E/E_o) ratios for aquatic plants. J. Aquat. Plant Manage. *25*:1–3.

Boyle, E. A. 1979. Copper in natural waters. *In* J. O. Nriagu, ed. Copper in the Environment. I. Ecological Cycling. John Wiley & Sons, New York. pp. 77–88.

Boylen, C. W. and R. B. Sheldon. 1976. Submergent macrophytes: Growth under winter ice cover. Science *194*:841–842.

Boylen, C. W. and T. D. Brock. 1973. Bacterial decomposition processes in Lake Wingra sediments during winter. Limnol. Oceanogr. *18*:628–634.

Brabrand, Å., B. A. Faafeng, and J. P. M. Nilssen. 1990. Relative importance of phosphorus supply to phytoplankton production: Fish excretion versus external loading. Can. J. Fish. Aquat. Sci. *47*:364–372.

Brabrand, Å., B. A. Faafeng, T. Källqvist, and J. P. Nilssen. 1983. Biological control of undesirable cyanobacteria in culturally eutrophic lakes. Oecologia *60*:1–5.

Bradbury, I. K. and J. Grace. 1983. Primary production in wetlands. *In* A. J. P. Gore, ed. Mires: Swamp, Bog, Fen and Moor. A. General Studies. Elsevier Scientific Publ., Amsterdam. pp. 285–310.

Bradbury, J. P. 1975. Diatom stratigraphy and human settlement in Minnesota. Spec. Pap. Geol. Soc. Amer. *171*. 74 pp.

Bradbury, J. P. and R. O. Megard. 1972. Stratigraphic record of pollution in Shagawa Lake, northeastern Minnesota. Bull. Geol. Soc. Amer. *85*:2639–2648.

Bradbury, J. P. and T. C. Winter. 1976. Areal distribution and stratigraphy of diatoms in the sediments of Lake Sallie, Minnesota. Ecology *57*:1005–1014.

Bradbury, J. P., B. Leyden, M. Salgado-Labouriau, W. M. Lewis, Jr., C. Schubert, M. W. Binford, D. G. Frey, D. R. Whitehead, and F. H. Weibezahn. 1981. Late Quaternary environmental history of Lake Valencia, Venezuela. Science *214*:1299–1305.

Bradbury, J. P., S. J. Tarapchak, J. C. B. Waddington, and R. F. Wright. 1975. The impact of a forest fire on a wilderness lake in northeastern Minnesota. Verh. Internat. Verein. Limnol. *19*:875–883.

Bradford, G. R., F. I. Bair, and V. Hunsaker. 1968. Trace and major element content of 170 high Sierra lakes in California. Limnol. Oceanogr. *13*:526–530.

Brading, M. G., J. Jass, and H. M. Lappin-Scott. 1995. Dynamics of bacterial biofilm formation. *In* H. M. Lappin-Scott and J. W. Costerton, eds. Microbial Biofilms. Cambridge University Press, Cambridge. pp. 46–63.

Bradley, M. D. K. and J. D. Reynolds. 1987. Diet of the leeches *Erpobdella octoculata* (L) and *Helobdella stagnalis* (L) in a lotic habitat subject to organic pollution. Freshwat. Biol. *18*:267–275.

Bragg, A. N. 1960. An ecological study of the protozoa of Crystal Lake, Norman, Oklahoma. Wasmann J. Biol. *18*:37–85.

Brakke, D. F., A. Henriksen, and S. A. Norton. 1987. The relative importance of acidity sources for humic lakes in Norway. Nature *329*:432–434.

Brammer, E. S. 1979. Exclusion of phytoplankton in the proximity of dominant water-soldier (*Stratiotes aloides*). Freshwat. Biol. *9*:233–249.

Brammer, E. S. and R. G. Wetzel. 1984. Uptake and release of K^+, Na^+ and Ca^{2+} by the water soldier, *Stratiotes aloides* L. Aquat. Bot. *19*:119–130.

Brandl, Z., J. Brandlová, and M. Poàtolková. 1970. The influence of submerged vegetation on the photosynthesis of phytoplankton in ponds. Rozpravy Ceskosl. Akad. Ved, Rada Matem. Prír. Ved. *80*(*6*):33–62.

Brandt, E. 1978. Anpassungen von *Tubifex tubifex* Müller (Annelida, Oligochaeta) an die Temperatur, den Sauerstoffgehalt und den Ernährungszustand. Arch. Hydrobiol. *84*:302–338.

Branstrator, D. K. and J. T. Lehman. 1991. Invertebrate predation in Lake Michigan: Regulation of *Bosmina longirostris* by *Leptodora kindtii*. Limnol. Oceanogr. *36*:483–495.

Bratbak, G. 1985. Bacterial biovolume and biomass estimations. Appl. Environ. Microbiol. *49*:1488–1493.

Bratbak, G., F. Thingstad, and M. Heldal. 1994. Viruses and the microbial loop. Microb. Ecol. *28*:209–221.

Bray, J. R., D. B. Lawrence, and L. C. Pearson. 1959. Primary production in some Minnesota terrestrial communities for 1957. Oikos *10*:38–49.

Breger, I. A., ed. 1963. Organic Geochemistry. Macmillan Company, New York. 658 pp.

Brehm, J. 1967. Untersuchungen über den Aminosäure-Haushalt holsteinischer Gewässer, insbesondere des Pluss-Sees. Arch. Hydrobiol./Suppl. *32*:313–435.

Brehm, K. 1979. Kationenaustausch bei Hochmoorsphagnen: Die Wirkung von an den Austauscher gebundenen Kationen in Kulturversuchen. Beitr. Biol. Pflanzen *47*:91–116.

Brendelberger, H. 1997a. Contrasting feeding strategies of two freshwater gastropods, *Radix peregra* (Lymnaeidae) and *Bithynia tentaculata* (Bithyniidae). Arch. Hydrobiol. *140*:1–21.

Brendelberger, H. 1997b. Determination of digestive enzyme kinetics: A new method to define trophic niches in freshwater snails. Oecologia *109*:34–40.

Brendelberger, H. 1991. Filter mesh size of cladocerans predicts retention efficiency for bacteria. Limnol. Oceanogr. *36*:884–894.

Brendelberger, H. and W. Geller. 1985. Variability of filter structures in eight *Daphnia* species: Mesh sizes and filtering areas. J. Plankton Res. *7*:473–486.

Brendelberger, H., M. Herbeck, H. Lang, and W. Lampert. 1986. *Daphnia*'s filters are not solid walls. Arch. Hydrobiol. *107*:197–202.

Bretschko, G. 1973. Benthos production of a high- mountain lake: Nematoda. Verh. Internat. Verein. Limnol. *18*:1421–1428.

Bretschko, G. 1984. Free-living nematodes of a high mountain lake (Vorderer Finstertaler See, Tyrol, Austria, 2237 m asl). I. *Monhystera* cf.*stagnalis* and *Ethmolaimus pratensis*. Arch. Hydrobiol. *101*:39–72.

Brett, M. T. and D. C. Müller-Navarra. 1997. The role of highly unsaturated fatty acids in aquatic foodweb processes. Freshwat. Biol. *38*:483–499.

Brett, M. T., K. Wiackowski, F. S. Lubnow, A. Mueller-Solger, J. J. Elser, and C. R. Goldman. 1994. Species-dependent effects of zooplankton on planktonic ecosystem processes in Castle Lake, California. Ecology *75*:2243–2254.

Breukelaar, A. W., E. H. R. R. Lammens, J. G. P. K Breteler, and I Tátrai. 1994. Effects of benthivorous bream (*Abramis brama*) and carp (*Cyprinus carpio*) on sediment resuspension and concentrations of nutrients and chlorophyll *a*. Freshwat. Biol. *32*:113–121.

Brezny, O., I. Mehta, and R. K. Sharma. 1973. Studies on evapotranspiration of some aquatic weeds. Weed Sci. *21*:197–204.

Brezonik, P. L. 1994. Chemical kinetics and process dynamics in aquatic systems. Lewis Publishers, Boca Raton. 754 pp.

Brezonik, P. L. and C. L. Harper. 1969. Nitrogen fixation in some aquatic lacustrine environments. Science *164*:1277–1279.

Brezonik, P. L. and G. F. Lee. 1968. Denitrification as a nitrogen sink in Lake Mendota, Wisconsin. Environ. Sci. Technol. *2*:120–125.

Brezonik, P. L., J. J. Delfino, and G. F. Lee. 1969. Chemistry of N and Mn in Cox Hollow Lake, Wisc., following destratification. J. Sanit. Eng. Div. Proc. Amer. Soc. Civil Eng. *SA–5*:929–940.

Briand, F. and J. E. Cohen. 1987. Environmental correlates of food chain length. Science *238*:956–960.

Bridgham, S. D., J. Pastor, J. A. Janssens, C. Chapin, and T. J. Malterer. 1996. Multiple limiting gradients in peatlands: A call for a new paradigm. Wetlands *16*:45–65.

Bright, T., F. Ferrari, D. Martin, and G. A. Franceschini. 1972. Effects of a total solar eclipse on the vertical distribution of certain oceanic zooplankters. Limnol. Oceanogr. *17*:296–301.

Brinkhurst, R. O. 1967. The distribution of aquatic oligochaetes in Saginaw Bay, Lake Huron. Limnol. Oceanogr. *12*:137–143.

Brinkhurst, R. O. 1974a. Factors mediating interspecific aggregation of tubificid oligochaetes. J. Fish. Res. Board Can. *31*:460–462.

Brinkhurst, R. O. 1974b. The Benthos of Lakes. St. Martin's Press, New York. 190 pp.

Brinkhurst, R. O. 1980. The production biology of the Tubificidae (Oligochaeta). *In* R. O. Brinkhurst and D. G. Cook, eds. Aquatic Oligochaete Biology. Plenum Press, New York. pp. 205–209.

Brinkhurst, R. O. and B. G. M. Jamieson. 1971. Aquatic Oligochaeta of the World. University of Toronto Press, Toronto. 860 pp.

Brinkhurst, R. O. and B. Walsh. 1967. Rostherne Mere, England, a further instance of guanotrophy. J. Fish. Res. Board Can. *24*:1299–1309.

Brinkhurst, R. O. and D. G. Cook. 1974. Aquatic earthworms (Annelida: Oligochaeta). *In* C. W. Hart, Jr. and S. L. H. Fuller, eds. Pollution Ecology of Freshwater Invertebrates. Academic Press, New York. pp. 143–156.

Brinkhurst, R. O. and D. G. Cook, eds. 1980. Aquatic Oligochaete Biology, Plenum Press, New York.

Brinkhurst, R. O. and K. E. Chua. 1969. Preliminary investigation of the exploitation of some potential nutritional resources by three sympatric tubificid oligochaetes. J. Fish. Res. Board Can. *26*:2659–2668.

Brinkhurst, R. O. and M. J. Austin. 1979. Assimilation by aquatic Oligochaeta. Int. Rev. ges. Hydrobiol. *64*:245–250.

Brinkhurst, R. O. and S. R. Gelder. 1991. Annelica: Oligochaeta and Branchiobdellida. *In* J. H. Thorp and A. P. Covich, eds. Ecology and Classification of North American Freshwater Invertebrates. Academic Press, San Diego. pp. 401–435.

Brinkhurst, R. O., K. E. Chua, and N. K. Kaushik. 1972. Interspecific interactions and selective feeding by tubificid oligochaetes. Limnol. Oceanogr. *17*:122–133.

Bristow, J. M. 1969. The effects of carbon dioxide on the growth and development of amphibious plants. Can. J. Bot. *47*:1803–1807.

Bristow, J. M. 1974. Nitrogen fixation in the rhizosphere of freshwater angiosperms. Can. J. Bot. *52*:217–221.

Bristow, J. M. 1975. The structure and function of roots in aquatic vascular plants. *In* J. G. Torrey and D. Clarkson, eds. The Development and Function of Roots. Academic Press, London. pp. 221–236.

Bristow, J. M. and M. Whitcombe. 1971. The role of roots in the nutrition of aquatic vascular plants. Am. J. Bot. *58*:8–13.

Brittain, J. E. and T. J. Eikeland. 1988. Invertebrate drift—a review. Hydrobiologia *166*:77–93.

Brix, H. 1988. Light-dependent variations in the composition of the internal atmosphere of *Phragmites australis* (Cav.) Trin. ex Steudel. Aquat. Bot. *30*:319–329.

Brix, H. 1990. Uptake and photosynthetic utilization of sediment-derived carbon by *Phragmites australis* (Cav.) Trin. ex Steudel. Aquat. Bot. *38*:377–389.

Brix, H. and J. E. Lyngby. 1985. Uptake and translocation of phosphorus in eelgrass (*Zostera marina*). Mar. Biol. *90*:111–116.

Brix, H., B. K. Sorrell, and P. T. Orr. 1992. Internal pressurization and convective gas flow in some emergent freshwater macrophytes. Limnol. Oceanogr. *37*:1420–1433.

Broberg, O. and G. Persson. 1984. External budgets for phosphorus, nitrogen and dissolved organic carbon for the acidified Lake Gårdsjön. Arch. Hydrobiol. *99*:160–175.

Broch, E. S. and W. Yake. 1969. A modification of Maucha's ionic diagram to include ionic concentrations. Limnol. Oceanogr. *14*:933–935.

Brock, T. C. M. 1984. Aspects of the decomposition of *Nymphoides peltata* (Gmel.) O. Kuntze (Menyanthaceae). Aquat. Bot. *19*:131–156.

Brock, T. C. M., G. H. P. Arts, L. L. M. Goossen, and A. H. M. Rutenfrans. 1983a. Structure and annual biomass production of *Nymphoides peltata* (Gmel.) O. Kuntze (Menyanthaceae). Aquat. Bot. *17*:167–188.

Brock, T. C. M., M. C. M. Bongaerts, G. J. M. A. Heijnen, and J. H. F. G. Heijthuijsen. 1983b. Nitrogen and phosphorus accumulation and cycling by *Nymphoides peltata* (Gmel.) O. Kuntze (Menyanthaceae). Aquat. Bot. *17*:189–214.

Brock, T. C. M., M. J. H. de Lyon, E. M. J. M. van Laar, and E. M. M. van Loon. 1985. Field studies on the breakdown of *Nuphar lutea* (L.) Sm. (Nymphaeaceae), and a comparison of three mathematical models for organic weight loss. Aquat. Bot. *21*:1–22.

Brock, T. D. 1966. Principles of Microbial Ecology. Prentice-Hall, Inc., Englewood Cliffs, NJ. 306 pp.

Brock, T. D. and J. Clyne. 1984. Significance of algal excretory products for growth in epilimnetic bacteria. Appl. Environ. Microbiol. *47*:731–734.

Broecker, W. S. 1965. An application of natural radon to problems in ocean circulation. *In* Diffusion in Oceans and Freshwaters. Symposium Proceedings, New York, Lamont Geological Observatory, Columbia University. pp. 116–145.

Broecker, W. S. 1973. Factors controlling CO_2 content in the oceans and atmosphere. *In* G. M. Woodwell and E. V. Pecan, eds. Carbon and the Biosphere. Brookhaven, N.Y., Proc. Brookhaven Symp. in Biol. 24. Tech. Information Center, U.S. Atomic Energy Commission CONF–720510. pp. 32–50.

Broecker, W. S. and T.-H. Peng. 1971. The vertical distribution of radon in the Bonex area. Earth Planet Sci. Lett. *11*:99–108.

Broecker, W. S., J. Cromwell, and Y. H. Li. 1968. Rates of vertical eddy diffusion near the ocean floor based on measurement of the distribution of excess ^{222}Rn. Earth Planet. Sci. Lett. *5*:101–105.

Broecker, W. S., T. Takahashi, H. J. Simpson, and T.-H. Peng. 1979. Fate of fossil fuel carbon dioxide and the global carbon budget. Science *206*:409–418.

Brönmark, C. 1989. Interactions between epiphytes, macrophytes and freshwater snails: A review. J. Molluscan Stud. *55*:299–311.

Bronowski, J. 1956. Science and human values. Harper & Brothers Publs., New York. 94 pp.

Brook, A. J. 1981. The Biology of Desmids. Univ. California Press, Berkeley. 276 pp.

Brooks, A. S. and B. G. Torke. 1977. Vertical and seasonal distribution of chlorophyll *a* in Lake Michigan. J. Fish. Res. Board Can. *34*:2280–2287.

Brooks, J. L. 1947. Turbulence as an environmental determinant of relative growth in *Daphnia*. Proc. Nat. Acad. Sci. USA *33*:141–148.

Brooks, J. L. 1964. The relationship between the vertical distribution and seasonal variation of limnetic species of *Daphnia*. Verh. Internat. Verein. Limnol. *15*:684–690.

Brooks, J. L. 1965. Predation and relative helmet size in cyclomorphic *Daphnia*. Proc. Nat. Acad. Sci. USA *53*:119–126.

Brooks, J. L. 1966. Cyclomorphosis, turbulence and overwintering in Daphnia. Verh. Internat. Verein. Limnol. *16*:1653–1659.

Brooks, J. L. 1968. The effects of prey size selection by lake planktivores. Syst. Zool. *17*:273–291.

Brooks, J. L. and G. E. Hutchinson. 1950. On the rate of passive sinking in *Daphnia*. Proc. Nat. Acad. Sci. USA *36*:272–277.

Brooks, J. L. and S. I. Dodson. 1965. Predation, body size, and composition of plankton. Science *150*:28–35.

Browder, J. A., P. J. Gleason, and D. R. Swift. 1994. Periphyton in the Everglades: Spatial variation, environmental correlates, and ecological implications. *In* S. M. Davis and J. C. Ogden, eds. Everglades: The ecosystem and its restoration. St. Lucie Press, Delray Beach, FL. pp. 379–418.

Brown, A., S. P. Mathur, and D. J. Kushner. 1989. An ombrotrophic bog as a methane reservoir. Global Biogeochem. Cycles *3*:205–213.

Brown, B. E., J. L. Fassbender, and R. Winkler. 1992. Carbonate production and sediment transport in a marl lake of southeastern Wisconsin. Limnol. Oceanogr. *37*:184–191.

Brown, C. J. D. 1933. A limnological study of certain fresh-water Polyzoa with special reference to their statoblasts. Trans. Am. Microscop. Soc. *52*:271–316.

Brown, D. H., C. E. Gibby, and M. Hickman. 1972. Photosynthetic rhythms in epipelic algal populations. Brit. Phycol. Bull. 7:37–44.

Brown, G. W. 1969. Predicting temperatures of small streams. Water Resour. Res. *5*:68–75.

Brown, J. M. A., F. I. Dromgoole, M. W. Towsey, and J. Browse. 1974. Photosynthesis and photorespiration in aquatic macrophytes. *In* R. L. Bieleski, A. R. Ferguson, and M. M. Cresswell, eds. Mechanisms of Regulation of Plant Growth. Bull. 12, Royal Soc. New Zealand, Wellington. pp. 243–249.

Brown, K. M. 1982. Resource overlap and competition in pond snails: An experimental analysis. Ecology *63*:412–422.

Brown, K. M. 1985. Interspecific life history variation in a pond snail: The roles of population divergence and phenotypic plasticity. Evolution *39*:387–395.

Brown, K. M. 1991. Mollusca: Gastropoda. *In* J. H. Thorp and A. P. Covich, eds. Ecology and Classification of North American Freshwater Invertebrates. Academic Press, San Diego. pp. 285–314.

Brown, K. M. and D. R. DeVries. 1985. Predation and the distribution and abundance of a pulmonate pond snail. Oecologia *66*:93–99.

Brown, S. 1981. A comparison of the structure, primary productivity, and transpiration of cypress ecosystems in Florida. Ecol. Monogr. *51*:403–427.

Brown, S. and B. Coleman. 1963. Oscillaxanthin in lake sediments. Limnol. Oceanogr. *8*:352–353.

Brown, S. R. 1968. Bacterial carotenoids from freshwater sediments. Limnol. Oceanogr. *13*:233–241.

Brown, S. R. 1969. Paleolimnological evidence from fossil pigments. Mitt. Internat. Verein. Limnol. *17*:95–103.

Brown, S. R., R. J. Daley, and R. N. McNeely. 1977. Composition and stratigraphy of the fossil phorbin derivatives of Little Round Lake, Ontario. Limnol. Oceanogr. *22*:336–348.

Brown, S.-D. and A. P. Austin. 1973a. Diatom succession and interaction in littoral periphyton and plankton. Hydrobiologia *43*:333–356.

Brown, S.-D. and A. P. Austin. 1973b. Spatial and temporal variation in periphyton and physicochemical conditions in the littoral of a lake. Arch. Hydrobiol. *71*:183–232.

Browne, R. A. 1978. Growth, mortality, fecundity, biomass and productivity of four lake populations of the prosobranch snail, *Viviparus georgianus*. Ecology *59*:742–750.

Brownlee, B. G. and T. P. Murphy. 1983. Nitrogen fixation and phosphorus turnover in a hypertrophic prairie lake. Can. J. Fish. Aquat. Sci. *40*:1853–1860.

Browse, J. A., F. I. Dromgoole, and J. M. A. Brown. 1979. Photosynthesis in the aquatic macrophyte *Egeria densa*. III. Gas exchange studies. Aust. J. Plant Physiol. *6*:499–512.

Browse, J. A., J. M. A. Brown, and F. I. Dromgoole. 1979. Photosynthesis in the aquatic macrophyte *Egeria densa*. II. Effects of inorganic carbon conditions on ^{14}C fixation. Aust. J. Plant Physiol. *6*:1–9.

Browse, J. A., J. M. A. Brown, and F. I. Dromgoole. 1980. Malate synthesis and metabolism during photosynthesis in *Egeria densa* Planch. Aquat. Bot. *8*:295–305.

Brugam, R. B. 1978. Human disturbance and the historical development of Linsley Pond. Ecology *59*:19–36.

Brugam, R. B. and B. J. Speziale. 1983. Human disturbance and the paleolimnological record of change in the zooplankton community of Lake Harriet, Minnesota. Ecology *64*:578–591.

Brugam, R. B. and C. Patterson. 1983. The A/C (Araphidineae/Centrales) ratio in high and low alkalinity lakes in eastern Minnesota. Freshwat. Biol. *13*:47–55.

Brugam, R. G. 1979. A re-evaluation of the Araphidineae/Centrales index as an indicator of lake trophic status. Freshwat. Biol. *9*:451–460.

Van Bruggen, J. J. A., C. K. Stumm, and G. D. Vogels. 1983. Symbiosis of methanogenic bacteria and sapropelic protozoa. Arch. Mikrobiol. *136*:89–96.

Bruning, K. 1991a. Infection of the diatom *Asterionella* by a chytrid. I. Effects of light on reproduction and infectivity of the parasite. J. Plankton Res. *13*:103–117.

Bruning, K. 1991b. Infection of the diatom *Asterionella* by a chytrid. II. Effects of light on survival and epidemic development of the parasite. J. Plankton Res. *13*:119–129.

Bruning, K. and J. Ringelberg. 1987. The influence of phosphorus limitation of the diatom *Asterionella formosa* on the zoospore production of its fungal parasite *Rhizophydium planktonicum*. Hydrobiol. Bull. *21*:49–54.

Brunke, M. and T. Gonser. 1999. Hyporheic invertebrates—the clinal nature of interstitial communities structured by hydrological exchange and environmental gradients. J. N. Am. Benthol. Soc. *18*:344–362.

Brunskill, G. J. 1969. Fayetteville Green Lake, New York. II. Precipitation and sedimentation of calcite in a meromictic lake with laminated sediments. Limnol. Oceanogr. *14*:830–847.

Brunskill, G. J. and S. D. Ludlam. 1969. Fayetteville Green Lake, New York. I. Physical and chemical limnology. Limnol. Oceanogr. *14*:817–829.

Brylinsky, M. 1980. Estimating the productivity of lakes and reservoirs. *In* E. D. Le Cren and R. H. Lowe-McConnell, eds. The Functioning of Freshwater Ecosystems. Cambridge Univ. Press, Cambridge. pp. 411–453.

Bryson, R. A. and R. A. Ragotzkie. 1960. On internal waves in lakes. Limnol. Oceanogr. *5*:397–408.

Bryson, R. A. and T. J. Murray. 1977. Climates of hunger. Mankind and the world's changing weather. Univ. Wisconsin Press, Madison. 171 pp.

Bubier, J. L., T. R. Moore, and N. T. Roulet. 1993. Methane emissions from wetlands in the midboreal region of northern Ontario, Canada. Ecology *74*:2240–2254.

Buchholz, B., E. Laczko, N. Pfennig, M. Rohmer, and S. Neunlist. 1993. Hopanoids of a recent sediment from Lake Constance as eutrophication markers. FEMS Microbiol. Ecol. *102*:217–223.

Budd, K. and G. W. Kerson. 1987. Uptake of phosphate by two cyanophytes: Cation effects and energetics. Can. J. Bot. *65*:1901–1907.

Buffle, J. and G. G. Leppard. 1995. Characterization of aquatic colloids and macromolecules. 1. Structure and behavior of colloidal material. Environ. Sci. Technol. *29*:2169–2175.

Bühlmann, B., P. Bossard, and U. Uehlinger. 1987. The influence of longwave ultraviolet radiation (u.v.-A) on the photosynthetic activity (^{14}C-assimilation) of phytoplankton. J. Plankton Res. *9*:935–943.

van de Bund, W., W. Goedkoop, and R. K. Johnson. 1994. Effects of deposit-feeder activity on bacterial production and abundance in profundal lake sediment. J. N. Am. Benthol. Soc. *13*:532–539.

van de Bund, W. J., C. Davids, and S. J. H. Spaas. 1995. Seasonal dynamics and spatial distribution of chydorid cladocerans in relation to chironomid larvae in the sandy littoral zone of an oligo-mesotrophic lake. Hydrobiologia *299*:125–138.

Bunner, H. C. and K. Halcrow. 1977. Experimental induction of the production of ephippia by *Daphnia magna* Straus (Cladocera). Crustaceana *32*:77–86.

Buresh, R. J., M. E. Casselman, and W. H. Patrick, Jr. 1980. Nitrogen fixation in flooded soil systems: A review. Adv. Agronomy *33*:149–192.

Burgis, M. J. 1970. The effect of temperature on the development time of eggs of *Thermocyclops* sp., a tropical cyclopoid copepod from Lake George, Uganda. Limnol. Oceanogr. *15*:742–747.

Burkholder, J. M. 1996. Interactions of benthic algae with their substrata. *In* R. J. Stevenson, M. L. Bothwell, and R. L. Lowe, eds. Algal Ecology: Freshwater Benthic Ecosystems. Academic Press, San Diego. pp. 253–297.

Burkholder, J. M. and R. G. Wetzel. 1989a. Epiphytic microalgae on natural substrata in a hardwater lake: Seasonal dynamics of community structure, biomass and ATP content. Arch. Hydrobiol./Suppl. *83*:1–56.

Burkholder, J. M. and R. G. Wetzel. 1989b. Microbial colonization on natural and artificial macrophytes in a phosphorus-limited, hardwater lake. J. Phycol. *25*:55–65.

Burkholder, J. M. and R. G. Wetzel. 1990. Epiphytic alkaline phosphatase on natural and artificial plants in an oligotrophic lake: Re-evaluation of the role of macrophytes as a phosphorus source for epiphytes. Limnol. Oceanogr. *35*:736–747.

Burkholder, J. M., R. G. Wetzel, and K. L. Klomparens. 1990. Direct comparison of phosphate uptake by adnate and loosely attached microalgae within an intact biofilm matrix. Appl. Environ. Microbiol. *56*:2882–2890.

Burkholder, P. R. and G. H. Bornside. 1957. Decomposition of marsh grass by aerobic marine bacteria. Bull. Torr. Bot. Club *84*:366–383.

Burkholder, P. R., L. M. Burkholder, and J. A. Rivero. 1959. Some chemical constituents of turtlegrass, *Thalassia testudinum*. Bull. Torr. Bot. Club *86*:88–93.

Burky, A. J. 1971. Biomass turnover, respiration and interpopulation variation in the stream limpet *Ferrissia rivularis*. Ecol. Monogr. *41*:235–251.

Burky, A. J. 1983. Physiological ecology of freshwater bivalves. *In* W. D. Russell-Hunter, ed. The Mollusca. Vol. 6. Ecology. Academic Press, New York. pp. 281–327.

Burla, H. 1971. Gerichtete Ortsveränderung bei Muscheln der Gattung Anodonta im Zürichsee. Vierteljahrsschr. Naturf. Gesellschaft Zürich *116*:181–194.

Burney, C. M. 1990. Seasonal and diel changes in dissolved and particulate organic material. *In* R. S. Wotton, ed. The Biology of Particles in Aquatic Systems. CRC Press, Boca Raton. pp. 83–116.

Burnison, B. K. and R. Y. Morita. 1973. Competitive inhibition for amino acid uptake by the indigenous microflora of Upper Klamath Lake. Appl. Microbiol. *25*:103–106.

Burns, C. W. 1968a. The relationship between body size of filter-feeding Cladocera and the maximum size of particle ingested. Limnol. Oceanogr. *13*:675–678.

Burns, C. W. 1968b. Direct observations of mechanisms regulating feeding behavior of *Daphnia* in lakewater. Int. Rev. ges. Hydrobiol. *53*:83–100.

Burns, C. W. 1969a. Particle size and sedimentation in the feeding behavior of two species of *Daphnia*. Limnol. Oceanogr. *14*:392–402.

Burns, C. W. 1969b. Relation between filtering rate, temperature, and body size in four species of *Daphnia*. Limnol. Oceanogr. *14*:696–700.

Burns, C. W. 1979. Population dynamics and production of *Boeckella dilatata* (Copepoda: Calanoida) in Lake Hayes, New Zealand. Arch. Hydrobiol./Suppl. *54*:409–465.

Burns, C. W. 1985. Fungal parasitism in a copepod population: The effects of *Aphanomyces* on the population dynamics of *Boeckella dilatata* Sars. J. Plankton Res. *7*:201–205.

Burns, C. W. and F. H. Rigler. 1967. Comparison of filtering rates of *Daphnia rosea* in lake water and in suspensions of yeast. Limnol. Oceanogr. *12*:492–502.

Burns, C. W. and J. J. Gilbert. 1993. Predation on ciliates by freshwater calanoid copepods: Rates of predation and relative vulnerabilities of prey. Freshwat. Biol. *30*:377–393.

Burns, C. W. and Z. Xu. 1990. Utilization of colonial cyanobacteria and algae by freshwater calanoid copepods: Survivorship and reproduction of adult *Boeckella* spp. Arch. Hydrobiol. *117*:257–270.

Burns, D. A. 1998. Retention of NO_3^- in an upland stream environment: A mass balance approach. Biogeochemistry *40*:73–96.

Burns, N. M. 1995. Using hypolimnetic dissolved oxygen depletion rates for monitoring lakes. N. A. J. Mar. Freshwat. Res. *29*:1–11.

Burns, N. M. and C. Ross. 1971. Nutrient relationships in a stratified eutrophic lake. Proc. Conf. Great Lakes Res., Int. Assoc. Great Lakes Res. *14*:749–760.

Burns, N. M. and J. O. Nriagu. 1976. Forms of iron and manganese in Lake Erie waters. J. Fish. Res. Board Can. *33*:463–470.

Burns, T. P. 1989. Lindeman's contradiction and the trophic structure of ecosystems. Ecology *70*:1355–1362.

Burris, R. H., F. J. Eppling, H. B. Wahlin, and P. W. Wilson. 1943. Detection of nitrogen-fixation with isotopic nitrogen. J. Biol. Chem. *148*:349–357.

Burton, J. D., T. M. Leatherland, and P. S. Liss. 1970. The reactivity of dissolved silicon in some natural waters. Limnol. Oceanogr. *15*:473–476.

Burton, T. M., M. P. Oemke, and J. M. Molloy. 1991. Contrasting effects of nitrogen and phosphorus additions on epilithic algae in a hardwater and a softwater stream in northern Michigan. Verh. Internat. Verein. Limnol. *24*:1644–1653.

Buscemi, P. A. 1958. Littoral oxygen depletion produced by a cover of *Elodea canadensis*. Oikos *9*:239–245.

Buscemi, P. A. 1961. Ecology of the bottom fauna of Parvin Lake, Colorado. Trans. Amer. Microsc. Soc. *80*:266–307.

Bushnell, J. H. 1974. Bryozoans (Ectoprocta). *In* C. W. Hart, Jr. and S. L. H. Fuller, eds. Pollution Ecology of Freshwater Invertebrates. Academic Press, New York. pp. 157–194.

Bushnell, J. H., Jr. 1966. Environmental relations of Michigan Ectoprocta, and dynamics of natural populations of *Plumatella repens*. Ecol. Monogr. *36*:95–123.

Butler, M. G. 1982. A 7-year life cycle for two *Chironomus* species in arctic Alaskan tundra ponds (Diptera: Chironomidae). Can. J. Zool. *60*:58–70.

Butler, M. I. and C. W. Burns. 1995. Effects of temperature and food level on growth and development of a planktonic water mite. Hydrobiologia *308*:153–165.

Butlin, K. R. 1953. The bacterial sulphur cycle. Research *6*:184–191.

Buttery, B. R., W. T. William, and J. M. Lambert. 1965. Competition between *Glyceria maxima* and *Phragmites communis* in the region of Surlingham Broad. II. The fen gradient. J. Ecol. *53*:183–195.

Buttery, B. R. and J. M. Lambert. 1965. Competition between *Glyceria maxima* and *Phragmites communis* in the region of Surlingham Broad. I. The competition mechanism. J. Ecol. *53*:163–181.

Button, D. K. 1978. On the theory of control of microbial growth kinetics by limiting nutrient concentrations. Deep-Sea Res. *25*:1163–1177.

Byron, E. R. 1982. The adaptive significance of calanoid copepod pigmentation: A comparative and experimental analysis. Ecology *63*:1871–1886.

Cabaniss, S. E., Q. Zhou, P. A. Maurice, Y.-P. Chin, and G. R. Aiken. 2000. A log-normal distribution model for the molecular weight of aquatic fulvic acids. Environ. Sci. Technol. *34*:1103–1109.

Caffrey, J. M. and W. M. Kemp. 1991. Seasonal and spatial patterns of oxygen production, respiration and root-rhizome release in *Potamogeton perfoliatus* L. and *Zostera marina* L. Aquat. Bot. *40*:109–128.

Cahoon, L. B., J. E. Nearhoff, and C. L. Tilton. 1999. Sediment grain size effect on benthic microalgal biomass in shallow aquatic ecosystems. Estuaries *22*:735–741.

Cai, Y., R. Jaffé, and R. Jones. 1997. Ethylmercury in the soils and sediments of the Florida everglades. Environ. Sci. Technol. *31*:302–305.

Cairns, J., Jr. 1974. Protozoans (Protozoa). *In* C. W. Hart, Jr. and S. L. H. Fuller, eds. Pollution Ecology of Freshwater Invertebrates. Academic Press, New York. pp. 1–28.

Cairns, J., Jr. and W. H. Yongue. 1977. Factors affecting the number of species in freshwater protozoan communities. *In* J. Cairns, Jr., ed. Aquatic Microbial Communities. Garland, New York. pp. 257–303.

Cairns, J., Jr. and 14 other authors. 1992. Restoration of Aquatic Ecosystems: Science, Technology and Public Policy. National Academy Press, Washington, DC. 552 pp.

Caldwell, D. E. and J. M. Tiedje. 1975. The structure of anaerobic bacterial communities in the hypolimnia of several Michigan lakes. Can. J. Microbiol. *21*:377–385.

Caldwell, D. R., J. M. Brubaker, and V. T. Neal. 1978. Thermal microstructure on a lake slope. Limnol. Oceanogr. *23*:372–374.

Calhoun, A. and G. M. King. 1997. Regulation of root-associated methanotrophy by oxygen availability in the rhizosphere of two aquatic macrophytes. Appl. Environ. Microbiol. *63*: 3051–3058.

Callaway, J. C., J. A. Nyman, and R. D. DeLaune. 1996. Sediment accretion in coastal wetlands: A review and a simulation model of processes. Curr. Topics Wetland Biogeochem. 2:2–23.

Callaway, J. C., R. D. DeLaune, and W. H. Patrick, Jr. 1996. Chernobyl ^{137}Cs used to determine sediment accretion rates at selected northern European coastal wetlands. Limnol. Oceanogr. *41*:444–450.

Calow, P. 1977. The joint effect of temperature and starvation on the metabolism of triclads. Oikos *29*:87–92.

Calow, P. and A. S. Woollhead. 1977. The relationship between ration, reproductive effort and age-specific mortality in the evolution of life-history strategies—some observations on freshwater triclads. J. Anim. Ecol. *46*:765–781.

Calow, P. and G. E. Petts, eds. 1992. The Rivers Handbook. Vol. 1. Hydrological and Ecological Principles. Blackwell Scientific Publications, Oxford. 526 pp.

Camacho, A. and E. Vicente. 1998. Carbon photoassimilation by sharply stratified phototrophic communities at the chemocline of Lake Arcas (Spain). FEMS Microbiol. Ecol. *25*:11–22.

Camarero, L. and J. Catalan. 1991. Horizontal heterogeneity of phytoplankton in a small high mountain lake. Verh. Internat. Verein. Limnol. *24*:1005–1010.

Campbell, I. C., G. M. Enierga, L. Fuchshauber, and K. R. James. 1994. An evaluation of the two variable model for stream litter processing using data from southeastern Australia: How important is temperature? Verh. Internat. Verein. Limnol. *25*:1837–1840.

Campbell, P. and T. Torgersen. 1980. Maintenance of iron meromixis by iron redeposition in a rapidly flushed monimolimnion. Can. J. Fish. Aquat. Sci. *37*:1303–1313.

Canfield, D. E., Jr. and R. W. Bachmann. 1981. Prediction of total phosphorus concentrations, chlorophyll a, and Secchi depths in natural and artificial lakes. Can. J. Fish. Aquat. Sci. *38*:414–423.

Canfield, D. E., Jr., R. W. Bachmann, and M. V. Hoyer. 1983. Freeze-out of salts in hard-water lakes. Limnol. Oceanogr. *28*:970–977.

Canfield, T. J. and J. R. Jones. 1996. Zooplankton abundance, biomass, and size-distribution in selected midwestern waterbodies and relation with trophic state. J. Freshwat. Ecol. *11*:171–181.

Cannon, D., J. W. G. Lund, and J. Sieminska. 1961. The growth of *Tabellaria flocculosa* (Roth) Kütz. var. *flocculosa* (Roth) Knuds. under natural conditions of light and temperature. J. Ecol. *49*:277–287.

Canter, H. M. 1973. A new primitive protozoan devouring centric diatoms in the plankton. Zool. J. Linn. Soc. *42*:63–83.

Canter, H. M. 1979. Fungal and protozoan parasites and their importance in the ecology of phytoplankton. Freshwat. Biol. Assoc. Ann. Rep. *47*:43–50.

Canter, H. M. and G. H. M. Jaworski. 1979. The occurrence of a hypersensitive reaction in the planktonic diatom *Asterionella formosa* Hassall parasitized by the chytrid *Rhizophydium planktonicum* Canter Emend., in culture. New Phytol. *82*:187–206.

Canter, H. M. and J. W. G. Lund. 1948. Studies on plankton parasites. I. Fluctuations in the numbers of *Asterionella formosa* Hass. in relation to fungal epidemics. New Phytol. *47*:238–261.

Canter, H. M. and J. W. G. Lund. 1968. The importance of Protozoa in controlling the abundance of planktonic algae in lakes. Proc. Linn. Soc. Lond. *179*:203–219.

Canter, H. M. and J. W. G. Lund. 1969. The parasitism of planktonic desmids by fungi. Österr. Bot. Z. *116*:351–377.

Canter, H. M. and S. I. Heaney. 1984. Observations on zoosporic fungi of *Ceratium* spp. in lakes of the English Lake District; importance for phytoplankton population dynamics. New Phytol. *97*:601–612.

Capblancq, J. 1973. Phytobenthos et productivité primaire d'un lac de haute montagne dans les Pyrénées centrales. Ann. Limnol. *9*:193–230.

Cappenberg, T. E. 1972. Ecological observations on heterotrophic, methane oxidizing and sulfate reducing bacteria in a pond. Hydrobiologia *40*:471–485.

Cappenberg, T. E. 1974. Interrelations between sulfate-reducing and methane-producing bacteira in bottom deposits of a fresh-water lake. II. Inhibition experiments. Antonie van Leeuwenhoek *40*:297–306.

Cappenberg, T. E. 1975. A study of mixed continuous cultures of sulfate-reducing and methane-producing bacteria. Microb. Ecol. *2*:60–72.

Capriulo, G. M., E. B. Sherr, and B. F. Sherr. 1991. Trophic behaviour and related community feeding activities of heterotrophic marine protists. *In* P. C. Reid, C. M. Turley, and Ph. H. Burkill, eds. Protozoa and Their Role in Marine Processes. Springer-Verlag, Berlin. pp. 219–265.

Caraco, N., J. Cole, and G. E. Likens. 1990. A comparison of phosphorus immobilization in sediments of freshwater and coastal marine systems. Biogeochemistry 9:277–290.

Caraco, N., J. J. Cole, and G. E. Likens. 1989. Evidence for sulphate-controlled phosphorous release from sediments of aquatic systems. Nature *341*:316–318.

Caraco, N. F., J. J. Cole, P. A. Raymond, D. L. Strayer, M. L. Pace, S. E. G. Findlay, and D. T. Fischer. 1997. Zebra mussel invasion in a large, turbid river: Phytoplankton response to increased grazing. Ecology. *78*:588–602.

Carbonel, P., J.-P. Colin, D. L. Danielopol, H. Löffler, and I. Neustrueva. 1988. Paleoecology of limnic ostracodes: A review of some major topics. Palaeogeogr. Palaeoclimatol. Palaeoecol. *62*:413–461.

Carignan, R. 1985. Quantitative importance of alkalinity flux from the sediments of acid lakes. Nature *317*:158–160.

Carignan, R. and J. Kalff. 1980. Phosphorus sources for aquatic weeds: Water or sediments? Science *207*:987–989.

Carignan, R. and J. Kalff. 1982. Phosphorus release by submerged macrophytes: Significance to epiphyton and phytoplankton. Limnol. Oceanogr. *27*:419–427.

Carignan, R. and R. J. Flett. 1981. Postdepositional mobility of phosphorus in lake sediments. Limnol. Oceanogr. *26*:361–366.

Carling, P. A. 1992. In-stream hydraulics and sediment transport. *In* P. Calow and G. E. Petts, eds. The Rivers Handbook. I. Hydrological and Ecological Principles. Blackwell Sci. Publs., Oxford. pp. 101–125.

Carlough, L. A. and J. L. Meyer. 1989. Protozoans in two southeastern blackwater rivers and their importance to trophic transfer. Limnol. Oceanogr. *34*:163–177.

Carlough, L. A. and J. L. Meyer. 1990. Rates of protozoan bacterivory in three habitats of a southeastern blackwater river. J. N. Am. Benthol. Soc. *9*:45–53.

Carlough, L. A. and J. L. Meyer. 1991. Bacterivory by sestonic protists in a southeastern blackwater river. Limnol. Oceanogr. *36*:873–883.

Carlson, R. E. 1977. A trophic state index for lakes. Limnol. Oceanogr. *22*:361–369.

Carlson, R. E. 1980. More complications in the chlorophyll-Secchi disk relationship. Limnol. Oceanogr. *25*:378–382.

Carlson, R. E. 1992. Expanding the trophic state concept to identify non-nutrient limited lakes and reservoirs. *In* Proceedings of a National Conference on Enhancing the States' Lake Management Programs. Monitoring and Lake Impact Assessment. Chicago. pp. 59–71.

Carlson, R. E. and J. Simpson. 1996. Trophic state. *In* A Coordinator's Guide to Volunteer Lake Monitoring Methods. North American Lake Management Society. pp. 7-1–7-20.

Carlton, R. G. and R. G. Wetzel. 1987. Distributions and fates of oxygen in periphyton communities. Can. J. Bot. *65*:1031–1037.

Carlton, R. G. and R. G. Wetzel. 1988. Phosphorus flux from lake sediments: Effects of epipelic algal oxygen production. Limnol. Oceanogr. *33*:562–570.

Carlucci, A. F. and P. M. Bowes. 1970. Production of vitamin B_{12}, thiamine, and biotin by phytoplankton. J. Phycol. *6*:351–357.

Carmack, E. C., R. C. Wiegand, R. J. Daley, C. B. J. Gray, S. Jasper, and C. H. Pharo. 1986. Mechanisms influencing the circulation and distribution of water mass in a medium residence-time lake. Limnol. Oceanogr. *31*:249–265.

Carmichael, W. W. 1986. Algal toxins. Adv. Bot. Res. *12*:47–101.

Carmichael, W. W. 1997. The cyanotoxins. Adv. Bot. Res. *27*:211–256.

Caron, D. A. 1991. Evolving role of protozoa in aquatic nutrient cycles. *In* Reid, P. C., C. M. Turley, and P. H. Burkill, eds. Protozoa and Their Role in Marine Processes. Springer-Verlag, Berlin. pp. 387–415.

Caron, D. A., J. C. Goldman, O. K. Andersen, and M. R. Dennett. 1985. Nutrient cycling in a microflagellate food chain. II. Population dynamics and carbon cycling. Mar. Ecol. Prog. Ser. *24*:243–254.

Caron, D. A., J. C. Goldman, and M. R. Dennett. 1988. Experimental demonstration of the roles of bacteria and bacterivorous protozoa in plankton nutrient cycles. Hydrobiologia *159*:27–40.

Caron, D. A., K. G. Porter, and R. W. Sanders. 1990. Carbon, nitrogen, and phosphorus budgets for the mixotrophic phytoflagellate *Poterioochromonas malhamensis* (Chrysophyceae) during bacterial ingestion. Limnol. Oceanogr. *35*:433–443.

Caron, F. and J. R. Kramer. 1994. Formation of volatile sulfides in freshwater environments. Sci. Total Environ. *153*:177–194.

Carpenter, G. F., E. L. Mansey, and N. H. F. Watson. 1974. Abundance and life history of *Mysis relicta* in the St. Lawrence Great Lakes. J. Fish. Res. Board Can. *31*:319–325.

Carpenter, S. R. 1980. Enrichment of Lake Wingra, Wisconsin, by submersed macrophyte decay. Ecology *61*:1145–1155.

Carpenter, S. R. and J. F. Kitchell, eds. 1989. The Trophic Cascade in Lakes. Cambridge Univ. Press, Cambridge. 385 pp.

Carpenter, S. R. and M. S. Adams. 1979. Effects of nutrients and temperature on decomposition of *Myriophyllum spicatum* L. in a hard-water eutrophic lake. Limnol. Oceanogr. *24*:520–528.

Carpenter, S. R., D. L. Christensen, J. J. Cole, K. L. Cottingham, X. He, J. R. Hodgson, J. F. Kitchell, S. E. Knight, M. L. Pace, D. M. Post, D. E.Schindler, and N. Voichick. 1995. Biological control of eutrophication in lakes. Environ. Sci. Technol. *29*:784–786.

Carpenter, S. R., E. van Donk, and R. G. Wetzel. 1998. Nutrient loading gradient in shallow lakes. *In* E. Jeppesen, Ma. Søndergaard, Mo. Søndergaard, and K. Christoffersen, eds. The Structuring Role of Submerged Macrophytes in Lakes. Springer-Verlag, New York. pp. 393–396.

Carpenter, S. R., J. F. Kitchell, and J. R. Hodgson. 1985. Cascading trophic interactions and lake productivity. BioScience *35*:634–639.

Carpenter, S. R., J. J. Elser, and K. M. Olson. 1983. Effects of roots of *Myriophyllum verticillatum* L. on sediment redox conditions. Aquat. Bot. *17*:243–249.

Carpenter, S. R., M. M. Elser, and J. J. Elser. 1986. Chlorophyll production, degradation, and sedimentation: Implications for paleolimnology. Limnol. Oceanogr. *31*:112–124.

Carr, J. F. 1962. Dissolved oxygen in Lake Erie, past and present. Publ. Great Lakes Res. Div., Univ. Mich. *9*:1–14.

Carr, J. L. 1969. The primary productivity and physiology of *Ceratophyllum demersum*. II. Micro primary productivity, pH, and the P/R ratio. Austral. J. Mar. Freshwat. Res. *20*:127–142.

Carrias, J.-F., C. Amblard, and G. Bourdier. 1994. Vertical and temporal heterogeneity of planktonic ciliated protozoa in a humic lake. J. Plankton Res. *16*:471–485.

Carrias, J.-F., C. Amblard, and G. Bourdier. 1996. Protistan bacterivory in an oligomesotrophic lake: Importance of attached ciliates and flagellates. Microb. Ecol. *31*:249–268.

Carrick, H. J. and R. L. Lowe. 1988. Response of Lake Michigan benthic algae to an *in situ* enrichment with Si, N, and P. Can. J. Fish. Aquat. Sci. *45*:271–279.

Carrick, H. J., G. L. Fahenstiel, E. F. Stoermer, and R. G. Wetzel. 1991. The importance of zooplankton-protozoan trophic couplings in Lake Michigan. Limnol. Oceanogr. *36*:1335–1345.

Carrick, H. J., G. L. Fahnenstiel, and W. D. Taylor. 1992. Growth and production of planktonic protozoa in Lake Michigan: In situ versus in vitro comparisons and importance to food web dynamics. Limnol. Oceanogr. *37*:1221–1235.

Carrillo, P., I. Reche, and L. Cruz-Pizarro. 1996. Intraspecific stoichiometric variability and the ratio of nitrogen to phosphorus resupplied by zooplankton. Freshwat. Biol. *36*:363–374.

Carroll, D. 1962. Rainwater as a chemical agent of geologic processes—a review. U.S. Geol. Surv. Water-Supply Pap. 1535-G, 18 pp.

Carter, C. E. 1977. The recent history of the chironomid fauna of Lough Neagh, from the analysis of remains in sediment cores. Freshwat. Biol. *7*:415–423.

Carter, J. C. H. 1974. Life cycles of three limnetic copepods in a beaver pond. J. Fish. Res. Board Can. *31*:421–434.

Carter, S., G. M. Ward, R. G. Wetzel, and A. C. Benke. 2000. Growth, production, and senescence of *Nymphaea odorata* Aiton in a southeastern (U.S.A.) wetland. Aquat. Bot. (Submitted).

Carvalho, G. R. and R. N. Hughes. 1983. The effect of food availability, female culture-density and photoperiod on ephippia production in *Daphnia magna* Straus (Crustacea: Cladocera). Freshwat. Biol. *13*:37–46.

Casper, P. 1992. Methane production in lakes of different trophic state. Arch. Hydrobiol. Beih. Ergebn. Limnol. *37*:149–154.

Casper, P. 1996. Methane production in littoral and profundal sediments of an oligotrophic and a eutrophic lake. Arch. Hydrobiol. Spec. Issues Advanc. Limnol. *48*:253–259.

Casper, P., C. Babenzien, and G. Proft. 1988. Untersuchungen zum mikrobiellen Abbau von pflanzlichem Detritus in Seen. Limnologica *19*:147–159.

Casper, P., S. C. Maberly, G. H. Hall, and B. J. Finlay. 2000. Fluxes of methane and carbon dioxide from a small productive lake to the atmosphere. Biogeochemistry *49*:1–19.

Casselman, J. M. and H. H. Harvey. 1975. Selective fish mortality resulting fromm low winter oxygen. Verh. Internat. Verein. Limnol. *19*:2418–2429.

Castenholz, R. W. 1960. Seasonal changes in the attached algae of freshwater and saline lakes in the Lower Grand Coulee, Washington. Limnol. Oceanogr. *5*:1–28.

Castenholz, R. W. 1961. An evaluation of a submerged glass method of estimating production of attached algae. Verh. Internat. Verein. Limnol. *14*:155–159.

Castenholz, R. W. 1969. Thermophilic blue-green algae and the thermal environment. Bacteriol. Rev. *33*:476–504.

Caswell, H. 1972. On instantaneous and finite birth rates. Limnol. Oceanogr. *17*:787–791.

Catalan, J. and L. Camarero. 1993. Seasonal changes in alkalinity and pH in two Pyrenean lakes of very different water residence time. Verh. Internat. Verein. Limnol. *25*:749–753.

Cattaneo, A. 1978. The microdistribution of eiphytes on the leaves of natural and artificial macrophytes. Br. Phycol. J. *13*:183–188.

Cattaneo, A. 1987. Periphyton in lakes of different trophy. Can. J. Fish. Aquat. Sci. *44*:296–303.

Cattaneo, A. 1990. The effect of fetch on periphyton spatial variation. Hydrobiologia *206*:1–10.

Cattaneo, A. and J. Kalff. 1978. Seasonal changes in the epiphyte community of natural and artifical macrophytes in Lake Memphremagog (Que. & Vt.). Hydrobiologia *60*:135–144.

Cattaneo, A. and J. Kalff. 1979. Primary production of algae growing on natural and artificial aquatic plants: A study of interactions between epiphytes and their substrate. Limnol. Oceanogr. *24*:1031–1037.

Cattaneo, A. and J. Kalff. 1980. The relative contribution of aquatic macrophytes and their epiphytes to the production of macrophyte beds. Limnol. Oceanogr. *25*:280–289.

Cavari, B. and N. Grossowicz. 1977. Seasonal distribution of vitamin B_{12} in Lake Kinneret. Appl. Environ. Microbiol. *34*:120–124.

Cavari, B. Z. 1977. Nitrification potential and factors governing the rate of nitrification in Lake Kinneret. Oikos *28*:285–290.

Cavari, B. Z., G. Phelps, and O. Hadas. 1978. Glucose concentrations and heterotrophic activity in Lake Kinneret. Verh. Internat. Verein. Limnol. *20*:2249–2254.

Cellot, B. 1996. Influence of side-arms on aquatic macroinvertebrate drift in the main channel of a large river. Freshwat. Biol. *35*:149–164.

Cembella, A. D., N. J. Antia, and P. J. Harrison. 1983. The utilization of inorganic and organic phosphorous compounds as nutrients by eukaryotic microalgae: A multidisciplinary perspective. Part 1. CRC Crit. Rev. Microbiol. *10*:317–391.

Cembella, A. D., N. J. Antia, and P. J. Harrison. 1984. The utilization of inorganic and organic phosphorous compounds as nutrients by eukaryotic microalgae: A multidisciplinary perspective. Part 2. CRC Crit. Rev. Microbiol. *11*:13–81.

Chamberlain, W. M. 1968. A preliminary investigation of the nature and importance of soluble organic phosphorus in the phosphorus cycle of lakes. Ph.D. Diss., University of Toronto, Ontario. 232 pp.

Chambers, P. A., D. H. N. Spence, and D. C. Weeks. 1985. Photocontrol of turion formation by *Potamogeton crispus* L. in the laboratory and natural water. New Phytol. *99*:183–194.

Chambers, P. A., E. E. Prepas, and K. Gibson. 1992. Temporal and spatial dynamics in riverbed chemistry: The influence of flow and sediment composition. Can. J. Fish. Aquat. Sci. *49*:2128–2140.

Chambers, P. A., E. E. Prepas, M. L. Bothwell, and H. R. Hamilton. 1989. Roots versus shoots in nutrient uptake by aquatic macrophytes in flowing waters. Can. J. Fish. Aquat. Sci. *46*:435–439.

Chamier, A.-C. 1991. Cellulose digestion and metabolism in the freshwater amphipod *Gammarus pseudolimnaeus* Bousfield. Freshwat. Biol. *25*:33–40.

Chamier, A.-C. and L. G. Willoughby. 1986. The role of fungi in the diet of the amphipod *Gammarus pulex* (L.): An enzymatic study. Freshwat. Biol. *16*:197–208.

Chan, Y. K. and N. E. R. Campbell. 1978. Phytoplankton uptake and excretion of assimilated nitrate in a small Canadian Shield lake. Appl. Environ. Microbiol. *35*:1052–1060.

Chan, Y. K. and N. E. R. Campbell. 1980. Denitrification in Lake 227 during summer stratification. Can. J. Fish. Aquat. Sci. *37*:506–512.

Chan, Y. K. and R. Knowles. 1979. Measurement of denitrification in two freshwater sediments by an in situ acetylene inhibition method. Appl. Environ. Microbiol. *37*:1067–1072.

Chandler, C. M. 1966. Environmental factors affecting the local distribution and abundance of four species of stream-dwelling triclads. Invest. Indiana Lakes Streams *7*:1–56.

Chandler, D. C. 1937. Fate of typical lake plankton in streams. Ecol. Monogr. *7*:445–479.

Chang, C. C. Y., J. S. Kuwabara, and S. P. Pasilis. 1992. Phosphate and iron limitation of phytoplankton biomass in Lake Tahoe. Can. J. Fish. Aquat. Sci. *49*:1206–1215.

Chang, T.-P. 1980. Mucilage sheath as a barrier to carbon uptake in a cyanophyte, *Oscillatoria rubescens* D. C. Arch. Hydrobiol. *88*:128–133.

Chang, T.-P. 1981. Excretion and DOC utilization by *Oscillatoria rubescens* D. C. and its accompanying micro-organisms. Arch. Hydrobiol. *91*:509–520.

Chanton, J. P. and J. W. H. Dacey. 1991. Effects of vegetation on methane flux, reservoirs, and carbon isotopic composition. *In* Tracer Gas Emissions by Plants. Academic Press, San Diego. pp. 65–92.

Chanton, J. P., C. S. Martens, and C. A. Kelley. 1989. Gas transport from methane-saturated, tidal freshwater and wetland sediments. Limnol. Oceanogr. *34*:807–819.

Chapin, J. D. and P. D. Uttormark. 1973. Atmospheric Contributions of Nitrogen and Phosphorus. Tech. Rep. Wat. Resources Ctr. Univ. Wis. 73–2, 35 pp.

Chapelle, F. H., S. K. Haack, P. Adriaens, M. A. Henry, and P. M. Bradley. 1996. Comparison of E_h and H_2 measurements for delineating redox processes in a contaminated aquifer. Environ. Sci. Technol. *30*:3565–3569.

Chapman, D. W. 1967. Production in fish populations. *In* S. D. Gerking, ed. The Biological Basis of Freshwater Fish Production. John Wiley & Sons, New York. pp. 3–29.

Chapman, D. W. 1978. Production in fish populations. *In* S. D. Gerking, ed. Ecology of Freshwater Fish Production. John Wiley & Sons, New York. pp. 5–25.

Chappell, K. R. and R. Goulder. 1992. Epilithic extracellular enzyme activity in acid and calcareous headstreams. Arch. Hydrobiol. *125*:129–148.

Chappell, K. R. and R. Goulder. 1994. Seasonal variation of epiphytic extracellular enzyme activity on two freshwater plants, *Phragmites australis* and *Elodea candensis*. Arch. Hydrobiol. *132*:237–253.

Chapra, S. C. and K. H. Reckhow. 1979. Expressing the phosphorus loading concept in probabilistic terms. J. Fish Res. Board Can. *36*:225–229.

Chapra, S. C. and K. H. Reckhow. 1983. Engineering approaches for lake management. II. Mechanistic modelling. Ann Arbor Science-Butterworth Publishers, Boston.

Chapra, S. C and R. P. Canale. 1991. Long-term phenomenological model of phosphorus and oxygen for stratified lakes. Water Res. *25*:707–715.

Characklis, W. G. and P. A. Wilderer, eds. 1989. Structure and Function of Biofilms. Johnn Wiley & Sons, Chichester.

Characklis, W. G. and K. C. Marshall, eds. 1990. Biofilms. John Wiley & Sons, New York.

Charlebois, P. M. and G. A. Lamberti. 1996. Invading crayfish in a Michigan stream: Direct and indirect effects on periphyton and macroinvertebrates. J. N. Am. Benthol. Soc. *15*:551–563.

Charlton, M. N. 1980. Hypolimnion oxygen consumption in lakes: Discussion of productivity and morphometry effects. Can. J. Fish. Aquat. Sci. *37*:1531–1539.

Chatarpaul, L., J. B. Robinson, and N. K. Kaushik. 1980. Effects of tubificid worms on denitrification and nitrification in stream sediment. Can. J. Fish. Aquat. Sci. *37*:656–663.

Chatwin, P. D. and C. M. Allen. 1985. Mathematical models of dispersion in rivers and estuaries. Ann. Rev. Fluid Mechan. *17*:119–149.

Chave, K. E. 1965. Carbonates: Association with organic matter in surface seawater. Science *148*:1723–1724.

Chave, K. E. and E. Suess. 1970. Calcium carbonate saturation in seawater: Effects of dissolved organic matter. Limnol. Oceanogr. *15*:633–637.

Chawla, V. K. and Y. K. Chau. 1969. Trace elements in Lake Erie. Proc. Conf. Great Lakes Res., Int. Assoc. Great Lakes Res. *12*:760–765.

Chen, K. Y. and J. C. Morris. 1972. Kinetics of oxidation of aqueous sulfide by O_2. Environ. Sci. Technol. *6*:529–537.

Chen, R. L. and J. W. Barko. 1988. Effects of freshwater macrophytes on sediment chemistry. J. Freshwat. Ecol. *4*:279–289.

Chen, R. L., D. R. Keeney, D. A. Graetz, and A. J. Holding. 1972. Denitrification and nitrate reduction in lake sediments. J. Environ. Quality *1*:158–162.

Chen, R. L., D. R. Keeney, and J. G. Konrad. 1972. Nitrification in lake sediments. J. Environ. Quality *1*:151–154.

Chen, R. L., D. R. Keeney, J. G. Konrad, A. J. Holding, and D. A. Graetz. 1972. Gas production in sediments of Lake Mendota, Wisconsin. J. Environ. Quality *1*:155–158.

Chigbu, P. and T. H. Sibley. 1998. Predation by longfin smelt (*Spirinchus thaleichthys*) on the mysid *Neomysis mercedis* in Lake Washington. Freshwat. Biol. *40*:295–304.

Chikita, K., K. Sakata, and S. Hino. 1995. Transportation of suspended sediment slowly settling in a caldera lake. Jpn. J. Limnol. *56*:245–257.

Chikita, K., M. Hattori, and E. Hagiwara. 1985. A field study on river-induced currents—An intermountain lake, Lake Okotanpe, Hokkaido. Jpn. J. Limnol. *46*:256–267.

Chikita, K., Y. Hosogaya, and S. Natsume. 1993. The characteristics of internal waves in a caldera lake introduced from field measurements: Lake Kuttara, Hokkaido. Jpn. J. Limnol. *54*:213–224.

Chilton, E. W. and M. I. Muoneke. Biology and management of grass carp (*Ctenopharyngodon idella*, Cyprinidae) for vegetation control: A North American perspective. Rev. Fish Biol. Fisheries 2:283–320.

Chimney, M. J., R. W. Winner, and S. K. Seilkop. 1981. Prey utilization by *Chaoborus punctipennis* Say in a small, eutrophic reservoir. Hydrobiologia *85*:193–199.

Cho, J.-C., M. W. Kim, D.-H. Lee, and S.-J. Kim. 1997. Response of bacterial communities to changes in composition of extracellular organic carbon from phytoplankton in Daechung Reservoir (Korea). Arch. Hydrobiol. *138*:559–576.

Chorley, R. J. 1978. The hillslope hydrological cycle. *In* M. J. Kirkby, ed. Hillslope hydrology. John Wiley & Sons, Chichester. pp. 1–42.

Chow, V. T., ed. 1964. Handbook of Applied Hydrology. McGraw-Hill Book Co., New York.

Chow-Fraser, P. and H. C. Duthie. 1983. An interpretation of phosphorus loadings in dystrophic lakes. Arch. Hydrobiol. 97:109–121.

Chow-Fraser, P. and W. G. Sprules. 1986. Inhibitory effect of *Anabaena* sp. on in situ filtering rate of *Daphnia*. Can. J. Zool. *64*:1831–1834.

Christensen, K. K., F. Ø. Andersen, and H. S. Jensen. 1997. Comparison of iron, manganese, and phosphorus retention in freshwater littoral sediment with growth of *Littorella uniflora* and benthic microalgae. Biogeochemistry *38*:149–171.

Christensen, P. B., L. P. Nielsen, J. Sørensen, and N. P. Revsbech. 1990. Denitrification in nitrate-rich streams: Diurnal and seasonal variation related to benthic oxygen metabolism. Limnol. Oceanogr. *35*:640–651.

Christoffersen, K. 1990. Evaluation of *Chaoborus* predation on natural populations of herbivorous zooplankton in a eutrophic lake. Hydrobiologia *200/201*:459–466.

Christoffersen, K., B. Riemann, A. Klysner, and M. Søndergaard. 1993. Potential role of fish predation and natural populations of zooplankton in structuring a plankton community in eutrophic lake water. Limnol. Oceanogr. *38*:561–573.

Chróst, R. J. 1991. Environmental control of the synthesis and activity of aquatic microbial ectoenzymes. *In* R. J. Chróst, ed. Microbial Enzymes in Aquatic Environments. Springer-Verlag, New York. pp. 29–59.

Chróst, R. J. and A. J. Gajewski. 1995. Microbial utilization of lipids in lake water. FEMS Microbiol. Ecol. *18*:45–50.

Chróst, R. J. and H. Rai. 1994. Bacterial secondary production. *In* J. Overbeck and R. J. Chróst, eds. Microbial Ecology of Lake Plußsee. Springer-Verlag, New York. pp. 92–117.

Chróst, R. J. and J. Overbeck. 1987. Kinetics of alkaline phosphatase activity and phosphorus availability for phytoplankton and bacterioplankton in Lake Plußsee (North German eutrophic lake). Microb. Ecol. *13*:229–248.

Chrzanowski, T. H. 1985. Seasonality, abundance, and biomass of bacteria in a southwestern reservoir. Hydrobiologia *127*: 117–123.

Chrzanowski, T. H. and J. G. Hubbard. 1988. Primary and bacterial secondary production in a southwestern reservoir. Appl. Environ. Microbiol. *54*:661–669.

Chrzanowski, T. H. and J. G. Hubbard. 1989. Bacterial utilization of algal extracellular products in a southwestern reservoir. Hydrobiologia *179*:61–71.

Chrzanowski, T. H. and K. Šimek. 1990. Prey-size selection by freshwater flagellated protozoa. Limnol. Oceanogr. *35*:1429–1436.

Chrzanowski, T. H. and K. Šimek. 1993. Bacterial growth and losses due to bacterivory in a mesotrophic lake. J. Plankton Res. *15*:771–785.

Chrzanowski, T. H. and M. Kyle. 1996. Ratios of carbon, nitrogen and phosphorus in *Pseudomonas fluorescens* as a model for bacterial element ratios and nutrient regeneration. Aquat. Microb. Ecol. *10*:115–122.

Chrzanowski, T. H., M. Kyle, J. J. Elser, and R. W. Sterner. 1996. Element ratios and growth dynamics of bacteria in an oligotrophic Canadian shield lake. Aquat. Microb. Ecol. *11*:119–125.

Chrzanowski, T. H., R. D. Crotty, and J. G. Hubbard. 1988. Seasonal variation in cell volume of epilimnetic bacteria. Microb. Ecol. *16*:155–163.

Chu, S. P. 1942. The influence of the mineral composition of the medium on the growth of planktonic algae. I. Methods and culture media. J. Ecol. *30*:284–325.

Chu, S. P. 1943. The influence of the mineral composition of the medium on the growth of planktonic algae. II. The influence of the concentration of inorganic nitrogen and phosphate phosphorus. J. Ecol. *31*:109–148.

Chua, K. E. and R. O. Brinkhurst. 1973. Evidence of interspecific interactions in the respiration of tubificid oligochaetes. J. Fish. Res. Board Can. *30*:617–622.

Ciaio, C.-J. and W. F. McDiffett. 1990. Dissolved organic carbon dynamics in a small stream. J. Freshwat. Ecol. *5*:383–390.

Ciborowski, J. J. H., D. A. Craig, and K. M. Fry. 1997. Dissolved organic matter as a food for black fly larvae (Diptera: Simuliidae). J. N. Am. Benthol. Soc. *16*:771–780.

Cízková, H. and V. Bauer. 1998. Rhizome respiration of *Phragmites australis*: Effect of rhizome age, temperature, and nutrient status of the habitat. Aquat. Bot. *61*:239–253.

Clair, T. A. and J. M. Ehrman. 1996. Variations in discharge and dissolved organic carbon and nitrogen export from terrestrial basins with changes in climate: A neural network approach. Limnol. Oceanogr. *41*:921–927.

Clair, T. A., T. L. Pollock, and J. M. Ehrman. 1994. Exports of carbon and nitrogen from river basins in Canada's Atlantic Provinces. Global Biogeochem. Cycles *8*:441–450.

Clapp, C. E. and M. H. B. Hayes. 1999. Characterization of humic substances isolated from clay- and silt-sized fractions of a corn residue-amended agricultural soil. Soil Sci. *164*:899–913.

Clark, C. 1967. Population Growth and Land Use. Macmillan Press Ltd., London. 406 pp.

Clarke, F. W. 1924. The Data of Geochemistry. 5th ed. Bull. U.S. Geol. Surv. *770*, 841 pp.

Clarke, G. L. 1939. The utilization of solar energy by aquatic organisms. *In* Problems in lake biology. Publ. Amer. Assoc. Adv. Sci. *10*:27–38.

van Cleemput, O. 1994. Biogeochemistry of nitrous oxide in wetlands. Curr. Topics Wetland Biogeochem. *1*:3–14.

Clémencon, H. 1963. Verbreitung einiger B-Vitamine in Kleingewässern aus der Umgebung von Bern und Untersuchungen über die Abgabe von B-Vitamine durch Algen. Schweiz. Z. Hydrol. *25*:157–165.

Clesceri, N. L., S. J. Curran, and R. I. Sedlak. 1986. Nutrient loads to Wisconsin lakes. Part 1. Nitrogen and phosphorus export coefficients. Water Resour. Bull. *22*:983–990.

Cloe, W. W. and G. C. Garman. 1996. The energetic importance of terrestial arthropod inputs to three warm-water streams. Freshwat. Biol. *36*:105–114.

Clymo, R. S. 1963. Ion exchange in *Sphagnum* and its relation to bog ecology. Ann. Bot., N.S. 27:309–324.

Clymo, R. S. 1964. The origin of acidity in *Sphagnum* bogs. Bryologist *67*:427–431.

Clymo, R. S. 1965. Experiments on breakdown of *Sphagnum* in two bogs. J. Ecol. *53*:747–758.

Clymo, R. S. 1967. Control of cation concentrations, and in particular of pH, in *Sphagnum* dominated communities. *In* H. L. Golterman and R. S. Clymo, eds. Chemical Environment in the Aquatic Habitat. N. V. Noord-Hollandsche Uitgevers Maatschappij, Amsterdam. pp. 273–284.

Clymo, R. S. and E. J. F. Reddaway. 1971. Productivity of *Sphagnum* (bog moss) and peat accumulation. Hidrobiologia (Romania) *12*:181–192.

Clymo, R. S. and E. J. F. Reddaway. 1974. Growth rate of *Sphagnum rubellum* Wils, on pennine blanket bog. J. Ecol. *62*:181–192.

Coats, R. N., R. L. Leonard, and C. R. Goldman. 1976. Nitrogen uptake and release in a forested watershed, Lake Tahoe basin, California. Ecology *57*:995–1004.

Cobelas, M. A. 1992. Temperature and heat in a hypertrophic, gravel-pit lake. Arch. Hydrobiol. *125*:279–294.

Cobelas, M. A., J. L. V. Díaz, and A. R. Olmo. 1993. Settling seston in a hypertrophic lake. Arch. Hydrobiol. *127*:327–343.

Coesel, P. F. M. 1983. The significance of desmids as indicators of the trophic status of freshwaters (*sic*). Schweiz. Z. Hydrol. *45*:388–393.

Coffin, C. C., F. R. Hayes, L. H. Jodrey, and S. G. Whiteway. 1949. Exchange of materials in a lake as studied by the addition of radioactive phosphorus. Can. J. Res. (Ser. D) *27*:207–222.

Cohen, E. and H. Kende. 1986. The effect of submergence, ethylene and gibberellin on polyamines and their biosynthetic enzymes in deepwater-rice internodes. Planta *169*:498–504.

Cohen, J. E. and F. Briand. 1984. Trophic links of community food webs. Proc. Natl. Acad. Sci. USA *81*:4105–4109.

Cohen, J. E. and T. Luczak. 1992. Trophic levels in community food webs. Evol. Ecol. *6*:73–89.

Cole, B. S. and D. W. Toetz. 1975. Utilization of sedimentary ammonia by *Potamogeton nodosus* and *Scirpus*. Verh. Internat. Verein. Limnol. *19*:2765–2772.

Cole, J. and S. G. Fisher. 1978. Annual metabolism of a temporary pond ecosystem. Am. Midland Nat. *100*:15–22.

Cole, J. J. and M. L. Pace. 1995. Bacterial secondary production in oxic and anoxic freshwaters (*sic*). Limnol. Oceanogr. *40*:1019–1027.

Cole, J. J., G. E. Likens, and J. E. Hobbie. 1984. Decomposition of planktonic algae in an oligotrophic lake. Oikos *42*:257–266.

Cole, J. J., M. L. Pace, N. F. Caraco, and G. S. Steinhart. 1993. Bacterial biomass and cell size distributions in lakes: More and larger cells in anoxic waters. Limnol. Oceanogr. *38*:1627–1632.

Cole, J. J., N. F. Caraco, and G. E. Likens. 1990. Short-range atmospheric transport: A significant source of phosphorus to an oligotrophic lake. Limnol. Oceanogr. *35*:1230–1237.

Cole, J. J., N. F. Caraco, G. W. Kling, and T. K. Kratz. 1994. Carbon dioxide supersaturation in the surface waters of lakes. Science *265*:1568–1570.

Cole, J. J., R. W. Howarth, S. S. Nolan, and R. Marino. 1986. Sulfate inhibition of molybdate assimilation by planktonic algae and bacteria: Some implications for the aquatic nitrogen cycle. Biogeochemistry 2:179–196.

Cole, J. J., S. Findlay, and M. L. Pace. 1988. Bacterial production in fresh and saltwater ecosystems: A cross-system overview. Mar. Ecol. Prog. Ser. *43*:1–10.

Cole, J. J., W. H. McDowell, and G. E. Likens. 1984. Sources and molecular weight of "dissolved" organic carbon in an oligotrophic lake. Oikos *42*:1–9.

Cole, T. M. and H. H. Hannan. 1990. Dissolved oxygen dynamics. *In* K. W. Thornton, B. L. Kimmel, and F. E. Payne, eds. Reservoir Limnology: Ecological Perspectives. J. Wiley & Sons, New York. pp. 71–107.

Coleman, R. L. and C. N. Dahm. 1990. Stream geomorphology: Effects on periphyton standing crop and primary production. J. N. Am. Benthol. Soc. *9*:293–302.

Collienne, R. H. 1983. Photoreduction of iron in the epilimnion of acidic lakes. Limnol. Oceanogr. *28*:83–100.

Collins, M. L. P. and C. C. Remsen. 1991. The purple phototrophic bacteria. *In* J. F. Stolz, ed. Structure of Phototrophic Prokaryotes. CRC Press, Boca Raton. pp. 49–77.

Collinson, M. E. 1988. Freshwater macrophytes in palaeolimnology. Palaeogeogr. Palaeoclimat. Palaeoecol. *62*:317–342.

Colman, J. A. and D. E. Armstrong. 1983. Horizontal diffusivity in a small, ice-covered lake. Limnol. Oceanogr. *28*:1020–1026.

Colman, J. A. and D. E. Armstrong. 1987. Vertical eddy diffusivity determined with ^{222}Rn in the benthic boundary layer of ice-covered lakes. Limnol. Oceanogr. *32*:577–590.

Colman, J. A., K. Sorsa, J. P. Hoffmann, C. S. Smith, and J. H. Andrews. 1987. Yield- and photosynthesis-derived critical concentrations of tissue phosphorus and their significance for growth of Eurasian water milfoil, *Myriophyllum spicatum* L. Aquat. Bot. *29*:111–122.

Comita, G. W. 1964. The energy budget of *Diaptomus siciloides*, Lilljeborg. Verh. Internat. Verein. Limnol. *15*:646–653.

Comita, G. W. 1972. The seasonal zooplankton cycles, production and transformations of energy in Severson Lake, Minnesota. Arch. Hydrobiol. *70*:14–66.

Comita, G. W. and G. C. Anderson. 1959. The seasonal development of a population of *Diaptomus ashlandi* Marsh, and related phytoplankton cycles in Lake Washington. Limnol. Oceanogr. *4*:37–52.

Confer, J. L. 1971. Intrazooplankton predation by *Mesocyclops edax* at natural prey densities. Limnol. Oceanogr. *16*:663–666.

Confer, J. L. 1972. Interrelations among plankton, attached algae, and the phosphorus cycle in artificial open systems. Ecol. Monogr. *42*:1–23.

Confer, J. L. and G. Applegate. 1979. Size-selective predation by zooplankton. Amer. Midland Nat. *102*:378–383.

Confer, J. L. and J. M. Cooley. 1977. Copepod instar survival and predation by zooplankton. J. Fish. Res. Board Can. *34*:703–706.

Conley, D. J. 1988. Biogenic silica as an estimate of siliceous microfossil abundance in Great Lakes sediments. Biogeochemistry *6*:161–179.

Conley, D. J. and C. L. Schelske. 1989. Processes controlling the benthic regeneration and sedimentary accumulation of biogenic silica in Lake Michigan. Arch. Hydrobiol. *116*:23–43.

Conley, D. J. and C. L. Schelske. 1993. Potential role of sponge spicules in influencing the silicon biogeochemistry of Florida lakes. Can. J. Fish. Aquat. Sci. *50*:296–302.

Conley, D. J., C. L. Schelske, and E. F. Stoermer. 1993. Modification of the biogeochemical cycle of silica with eutrophication. Mar. Ecol. Prog. Ser. *101*:179–192.

Conrad, H. M. and P. Saltman. 1962. Growth substances. *In* R. A. Lewin, ed. Physiology and Biochemistry of Algae. Academic Press, New York. pp. 663–671.

Conrad, R. 1999. Contribution of hydrogen to methane production and control of hydrogen concentrations in methanogenic soils and sediments. FEMS Microbiol. Ecol. *28*:193–202.

Conrad, R., F. S. Lupton, and J. G. Zeikus. 1987. Hydrogen metabolism and sulfate-dependent inhibition of methanogenesis in a eutrophic lake sediment (Lake Mendota). FEMS Microbiol. Ecol. *45*:107–115.

Constable, J. V. H. and D. J. Longstreth. 1994. Aerenchyma carbon dioxide can be assimilated in *Typha latifolia* L. leaves. Plant Physiol. *106*:1065–1072.

Constable, J. V. H., J. B. Grace, and D. J. Longstreth. 1992. High carbon dioxide concentrations in aerenchyma of *Typha latifolia*. Am. J. Bot. *79*:415–418.

Conway, E. J. 1942. Mean geochemical data in relation to oceanic evolution. Proc. Royal Irish Acad. (Ser. B) *48*:119–159.

Conway, H. L, J. I. Parker, E. M. Yaguchi, and D. L. Mellinger. 1977. biological utilization and regeneration of silicon in Lake Michigan. J. Fish. Res. Board Can. *34*:537–544.

Cook, R. B. 1984. Distributions of ferrous iron and sulfide in an anoxic hypolimnion. Can. J. Fish. Aquat. Sci. *41*:286–293.

Cook, R. B., C. A. Kelly, D. W. Schindler, and M. A. Turner. 1986. Mechanisms of hydrogen ion neutralization in an experimentally acidified lake. Limnol. Oceanogr. *31*:134–148.

Cooke, G. D., E. B. Welch, S. A. Peterson, and P. R. Newroth. 1993. Restoration and Management of Lakes and Reservoirs. 2nd ed. Lewis Publishers, Chelsea, MI. 548 pp.

Cooke, G. D., M. R. McComas, D. W. Waller, and R. H. Kennedy. 1977. The occurrence of internal phosphorus loading in two small, eutrophic glacial lakes in northeastern Ohio. Hydrobiologia *56*:129–135.

Cooke, G. W. and R. J. B. Williams. 1973. Significance of man-made sources of phosphorus: Fertilizers and farming. Water Res. *7*:19–33.

Cooke, J. G. and R. E. White. 1987a. The effect of nitrate in stream water on the relationship between denitrification and nitrification in a stream-sediment microcosm. Freshwat. Biol. *18*:213–226.

Cooke, J. G. and R. E. White. 1987b. Spatial deistribution of denitrifying activity in a stream draining an agricultural catchment. Freshwat. Biol. *18*:509–519.

Cooke, W. B. 1956. Colonization of artificial bare areas by microorganisms. Bot. Rev. *22*:613–638.

Cooney, J. D. and C. W. Gehrs. 1980. The relationship between egg size and naupliar size in the calanoid copepod *Diaptomus clavipes* Schacht. Limnol. Oceanogr. *25*:549–552.

Cooper, C. F. 1969. Nutrient output from managed forests. *In* Eutrophication: Causes, Consequences, Correctives. National Academy of Sciences, Washington, DC. pp. 446–463.

Cooper, S. D. and C. R. Goldman. 1980. Opossum shrimp (*Mysis relicta*) predation on zooplankton. Can. J. Fish. Aquat. Sci. *37*:909–919.

Cooper, S. D. and C. R. Goldman. 1982. Environmental factors affecting predation rates of *Mysis relicta*. Can. J. Fish. Aquat. Sci. *39*:203–208.

Cooper, K. L., K. D. Hyatt, and D. P. Rankin. 1992. Life history and production of *Neomysis mercedis* in two British Columbia coastal lakes. Hydrobiologia *230*:9–30.

Cooper, W. E. 1965. Dynamics and production of a natural population of a fresh-water amphipod, *Hyalella azteca*. Ecol. Monogr. *35*:377–394.

Cooper, W. J., R. G. Zika, R. G. Petasne, and J. M. C. Plane. 1988. Photochemical formation of H_2O_2 in natural waters exposed to sunlight. Environ. Sci. Technol. *22*:1156–1160.

Coops, H., N. Geilen, and G. van der Velde. 1994. Distribution and growth of the helophyte species *Phragmites australis* and *Scirpus lacustris* in water depth gradients in relation to wave exposure. Aquat. Bot. *48*:273–284.

Corbella, C., V. Tonolli, and L. Tonolli. 1958. I sedimenti del Lago d'Orta, testimoni di una disastrosa polluzione cupro-ammonicale. Mem. Ist. Ital. Idrobiol. *10*:9–50.

Cornett, R. J. and F. H. Rigler. 1979. Hypolimnetic oxygen deficits: Their prediction and interpretation. Science *205*:580–581

Cornett, R. J. and F. H. Rigler. 1980. The areal hypolimnetic oxygen deficit: An empirical test of the model. Limnol. Oceanogr. *25*:672–679.

Correll, D. L. 1998. The role of phosphorus in the eutrophication of receiving waters: A review. J. Environ. Qual. *27*:261–266.

Costa, R. R. and K. W. Cummins. 1972. The contribution of *Leptodora* and other zooplankton to the diet of various fish. Amer. Midland Nat. *87*:559–564.

Cotner, J. B. and R. G. Wetzel. 1991a. Bacterial phosphatase from different habitats in a small, hardwater lake. *In* R. J. Chróst, ed. Microbial Enzymes in Aquatic Environments. Springer-Verlag, New York. pp. 187–205.

Cotner, J. B. and R. G. Wetzel. 1991b. 5′-nucleotidase activity in a eutrophic lake and an oligotrophic lake. Appl. Environ. Microbiol. *57*:1306–1312.

Cotner, J. B. and R. G. Wetzel. 1992. Uptake of dissolved inorganic and organic phosphorus compounds by phytoplankton and bacterioplankton. Limnol. Oceanogr. *37*:232–243.

Cotner, J. B., Jr. and R. T. Heath. 1988. Potential phosphate release from phosphomonoesters by acid phosphatase in a bog lake. Arch. Hydrobiol. *111*:329–338.

Cotner, J. B. and R. T. Heath. 1990. Iron redox effects on photosensitive phosphorus release from dissolved humic materials. Limnol. Oceanogr. *35*:1175–1181.

Cotner, J. B., W. S. Gardner, J. R. Johnson, R. H. Sada, J. F. Cavaletto, and R. T. Heath. 1995. Effects of zebra mussels (*Dreissena polymorpha*) on bacterioplankton: Evidence for both size-selective consumption and growth stimulation. J. Great Lakes Res. *21*:517–528.

Coughlan, M. P. 1971. The role of iron in microbial metabolism. Sci. Progress, Oxford *59*:1–23.

Coulombe, A. M. and G. G. C. Robinson. 1981. Collapsing *Aphanizomenon flos-aquae* blooms: Possible contributions of photooxidation, O_2 toxicity, and cyanophages. Can. J. Bot. *59*:1277–1284.

Coulson, K. L. 1975. Solar and terrestrial radiation. Methods and measurements. Academic Press, New York. 322 pp.

Coulter, G. W., ed. l991. Lake Tanganyika and its Life. Oxford Univ. Press, Oxford. 354 pp.

Coveney, M. F. 1982. Bacterial uptake of photosynthetic carbon from freshwater phytoplankton. Oikos *38*:8–20.

Coveney, M. F. and R. G. Wetzel. 1989. Bacterial metabolism of algal extracellular carbon. Hydrobiologia *173*:141–149.

Coveney, M. F. and R. G. Wetzel. 1992. Effects of nutrients on specific growth rate of bacterioplankton in oligotrophic lake water cultures. Appl. Environ. Microbiol. *58*:150–156.

Coveney, M. F. and R. G. Wetzel. 1995. Biomass, production, and specific growth rate of bacterioplankton and coupling to phytoplankton in an oligotrophic lake. Limnol. Oceanogr. *40*:1187–1200.

Coveney, M. F., G. Cronberg, M. Enell, K. Larsson, and L. Olofsson. 1977. Phytoplankton, zooplankton and bacteria—standing crop and production relationships in a eutrophic lake. Oikos *29*:5–21.

Covich, A. 1976. Recent changes in molluscan species diversity of a large tropical lake (Lago de Peten, Guatemala). Limnol. Oceanogr. *21*:51–59.

Cowen, W. F. and G. F. Lee. 1973. Leaves as source of phosphorus. Environ. Sci. Technol. *7*:853–854.

Cowgill, U. M. 1976. The hydrogeochemistry of Linsley Pond, North Branford, Connecticut. IV. Minor constituents by optical emission spectroscopy. Arch. Hydrobiol. *78*:279–309.

Cowgill, U. M. 1977a. The hydrogeochemistry of Linsley Pond, North Branford, Connecticut. VI. Summary, Part 1: Geochemical relationships within zones of the lake. Arch. Hydrobiol./Suppl. *50*:401–438.

Cowgill, U. M. 1977b. The hydrogeochemistry of Linsley Pond, North Branford, Connecticut. VI. Summary, Part 2: Geochemical relationships between zones of the lake. Arch. Hydrobiol./Suppl. *53*:1–47.

Cowgill, U. M. 1977c. The molybdenum cycle in Linsley Pond, North Branford, Connecticut. *In* W. R. Chappell and K. K. Petersen, eds. Molybdenum in the Environment. II. M. Dekker, New York. pp. 705–723.

Cowgill, U. M. 1989. The chemical and minerological content of the plants of the Lake Huleh Preserve, Israel. Phil. Trans. R. Soc. Lond. *B326*:59–118.

Cowgill, U. M. and G. E. Hutchinson. 1963. The history of a pond in Guatemala. Arch. Hydrobiol. *62*:355–372.

Cowgill, U. M. and G. E. Hutchinson. 1966. El Bajo de Santa Fé. Trans. Amer. Phil. Soc., N.S. *53*(7), 51 pp.

Cowgill, U. M., C. E. Goulden, G. E. Hutchinson, R. Patrick, A. A. Racek, and M. Tsukada. 1966. The History of Laguna de Petenxil. A Small Lake in Northern Guatemala. Mem. Conn. Acad. Arts Sci. *17*, 126 pp.

Cox, P. A. 1993. Water-pollination in plants. Sci. Amer. *269*:68–74.

Craik, A. D. D. 1977. The generation of Langmuir circulations by an instability mechanism. J. Fluid Mech. *81*:209–223.

Craik, A. D. D. and S. Leibovich. 1976. A rational model for Langmuir circulations. J. Fluid Mech. *73*:410–426.

Cranwell, P. A. 1973. Chain-length distribution of *n*-alkanes from lake sediments in relation to post-glacial environmental change. Freshwat. Biol. *3*:259–265.

Cranwell, P. A. 1974. Monocarboxylic acids in lake sediments: Indicators, derived from terrestrial and aquatic biota, of paleoenvironmental trophic levels. Chem. Geol. *14*:1–14.

Cranwell, P. A. 1976. Organic geochemistry of lake sediments. *In* J. O. Nriagu, ed. Environmental Biogeochemistry. I. Carbon, Nitrogen, Phosphorus, Sulfur and Selenium Cycles. Ann Arbor Sci. Publs., Inc., Ann Arbor, MI. pp. 75–88.

Cranwell, P. A. 1979. Decomposition of aquatic biota and sediment formation: Bound lipids in algal detritus and lake sediments. Freshwat. Biol. *9*:305–313.

Cranwell, P. A. 1981. Geochemistry of lipids in lacustrine sediments. Rep. Freshwat. Biol. Assoc. U.K. *49*:45–56.

Cranwell, P. A. 1990. Paleolimnological studies using sequential lipid extraction from recent lacustrine sediment: Recognition of source organisms from biomarkers. Hydrobiologia *214*:293–303.

Crawford, R. L., L. Johnson, and M. Martinson. 1982. Bacterial enrichments in surface films of freshwater lakes. J. Great Lakes Res. *8*:323–325.

Crawford, R. M. M. 1978. Metabolic adaptations to anoxia. *In* D. D. Hook and R. M. M. Crawford, eds. Plant Life in Anaerobic Environments. Ann Arbor Science Publ. Inc., Ann Arbor, MI. pp. 119–136.

Creed, R. P., Jr. 1994. Direct and indirect effects of crayfish grazing in a stream community. Ecology *75*:2091–2103.

Creer, K. M. 1982. Lake sediments as recorders of geomagnetic field variations—applications to dating post-Glacial sediments. Hydrobiologia *92*:587–596.

Creer, K. M., D. L. Gross, and J. A. Lineback. 1976. Origin of regional geomagnetic variations recorded by Wisconsinan and Holocene sediments from Lake Michigan, U.S.A., and Lake Windermere, England. Geol. Soc. Amer. Bull. *87*:531–540.

Creer, K. M., R. Thompson, and L. Molyneux. 1972. Geomagnetic secular variation recorded in the stable magnetic remanence of recent sediments. Earth Planetary Sci. Lett. *14*:115–127.

Cressa, C. and W. M. Lewis, Jr. 1986. Ecological energetics of *Chaoborus* in a tropical lake. Oecologia *70*:326–331.

Crill, P. M. 1996. Latitudinal differences in methane fluxes from natural wetlands. Mitt. Internat. Verein. Limnol. *25*:163–171.

Crisman, T. L. 1978a. Algal remains in Minnesota lake types: A comparison of modern and late-glacial distributions. Verh. Internat. Verein. Limnol. *20*:445–451.

Crisman, T. L. 1978b. Reconstruction of past lacustrine environments based on the remains of aquatic invertebrates. *In* D. Walker and J. C. Guppy, eds. Biology and Quaternary Environments. Australian Academy of Sciences, Canberra. pp. 69–101.

Crisman, T. L. 1980. Algal control through trophic-level interactions: A subtropical perspective. U.S.E.P.A. Workshop on Algal Management and Control. 20 pp.

Crisman, T. L. and D. R. Whitehead. 1978. Paleolimnological studies on small New England (U.S.A.) ponds. Part II. Cladoceran community responses to trophic oscillations. Polskie Arch. Hydrobiol. *25*:75–86.

Crisman, T. L., J. R. Beaver, and J. S. Bays. 1981. Examination of the relative impact of microzooplankton and macrozooplankton on bacteria in Florida lakes. Verh. Internat. Verein. Limnol. *21*:359–362.

Crisman, T. L., P. Scheuerman, R. W. Bienert, J. R. Beaver, and J. S. Bays. 1984. A preliminary characterization of bacterioplankton seasonality in subtropical Florida lakes. Verh. Internat. Verein. Limnol. *22*:620–626.

Crisp, D. T. 1990. Water temperature in a stream gravel bed and implications for salmonid incubation. Freshwat. Biol. *23*:601–612.

Crisp, D. T. and G. Howson. 1982. Effect of air temperature upon mean water temperature in streams in the north Pennines and English Lake District. Freshwat. Biol. *12*:359–367.

Crisp, D. T., A. M. Matthews, and D. F. Westlake. 1982. The temperatures of nine flowing waters in southern England. Hydrobiologia *89*:193–204.

Crittenden, R. N. 1981. Morphological characteristics and dimensions of the filter structures from three species of *Daphnia* (Cladocera). Crustaceana *41*:233–248.

Crocker, M. T. and J. L. Meyer. 1987. Interstitial dissolved organic carbon in sediments of a southern Appalachian headwater stream. J. N. Am. Benthol. Soc. *6*:159–167.

Cronan, C. S. and C. L. Schofield. 1979. Aluminum leaching response to acid precipitation: Effects on high-elevation watersheds in the Northeast. Science *204*:304–306.

Cronan, C. S. and C. L. Schofield. 1990. Relationships between aqueous aluminum and acidic deposition in forested watersheds of North America and Northern Europe. Environ. Sci. Technol. *24*:1100–1105.

Cronan, C. S. and G. R. Aiken. 1985. Chemistry and transport of soluble humic substances in forested watersheds of the Adirondack Park, New York. Geochem. Cosmochem. Acta *49*:1697–1705.

Cronan, C. S., W. J. Walker, and P. R. Bloom. 1986. Predicting aqueous aluminum concentrations in natural waters. Nature *324*: 140–143.

Cronberg, G. 1986. Chrysophycean cysts and scales in lake sediments: A review. *In* J. Kristiansen and R. A. Andersen, eds. Chrysophytes: Aspects and Problems. Cambridge Univ. Press, Cambridge. pp. 281–315.

Crowder, A. A. and S. M. Macfie. 1986. Seasonal deposition of ferric hydroxide plaque on roots of wetland plants. Can. J. Bot. *64*:2120–2124.

Crowder, A. A., J. M. Bristow, M. R. King, and S. Vanderkloet. 1977. Distribution, seasonality, and biomass of aquatic macrophytes in Lake Opinicon (Eastern Ontario). Naturaliste Can. *104*:441–456.

Crowder, C. B., R. W. Drenner, W. C. Kerfoot, D. J. McQueen, E. L. Mills, U. Sommer, C. N. Spencer, and J. J. Vanni. 1988. Food web interactions in lakes. *In* S. R. Carpenter, ed. Complex Interactions in Lake Communities. Springer-Verlag, New York. pp. 141–160.

Crowder, L. B. and W. E. Cooper. 1982. Habitat structural complexity and the interaction between bluegills and their prey. Ecology *63*:1802–1813.

Crum, H. 1988. A Focus on Peatlands and Peat Mosses. Univ. Michigan Press, Ann Arbor. 306 pp.

Crumpton, W. G. and R. G. Wetzel. 1982. Effects of differential growth and mortality in the seasonal succession of phytoplankton populations in Lawrence Lake, Michigan. Ecology *63*:1729–1739.

de la Cruz, A. and A. H. Post. 1977. Production and transport of organic matter in a woodland stream. Arch. Hydrobiol. *80*: 227–238.

Cruz-Pizarro, L. 1978. Comparative vertical zonation and diurnal migration among Crustacea and Rotifera in the small mountain lake LaCaldera (Granada, Spain). Verh. Internat. Verein. Limnol. *20*:1026–1032.

Csermák, K., A. Csermák, and F. Máté. 1992. Methanbildung im Sediment des Balaton (Plattensee, Ungarn). Limnologica *22*:277–282.

Cuffney, T. F., J. B. Wallace, and G. J. Lugthart. 1990. Experimental evidence quantifying the role of benthic invertebrates in organic matter dynamics of headwater streams. Freshwat. Biol. *23*:281–299.

Cuker, B. E. 1983. Grazing and nutrient interactions in controlling the activity and composition of the epilithic algal community of an arctic lake. Limnol. Oceanogr. *28*:133–141.

Cuker, B. E. and L. Hudson, Jr. 1992. Type of suspended clay influences zooplankton response to phosphorus loading. Limnol. Oceanogr. *37*:566–576.

Cuker, B. E. and S. C. Mozley. 1981. Summer population fluctuations, feeding, and growth of *Hydra* in an arctic lake. Limnol. Oceanogr. *26*:697–708.

Cuker, B. E., P. T. Gama, and J. M. Burkholder. 1990. Type of suspended clay influences lake productivity and phytoplankton community in response to phosphorus loading. Limnol. Oceanogr. *35*:830–839.

Cullen, J. J. 1982. The deep chlorophyll maximum: Comparing vertical profiles of chlorophyll *a*. Can. J. Fish. Aquat. Sci. *39*: 791–803.

Cullen, J. J. and P. J. Neale. 1994. Ultraviolet radiation, ozone depletion, and marine photosynthesis. Photosyn. Res. *39*:303–320.

Culver, D. A., R. M. Vaga, C. S. Munch, and S. M. Harris. 1981. Paleoecology of Hall Lake, Washington: A history of meromixis and disturbance. Ecology *62*:848–863.

Cumbus, I. P. and L. W. Robinson. 1977. The function of root systems in mineral nutrition of watercress (*Rorippa nasturtium-aquaticum* (L.) Hayek). Plant Soil *47*:395–406.

Cummins, K. W. 1962. An evaluation of some techniques for the collection and analysis of benthic samples with special emphasis on lotic waters. Amer. Midland Nat. *76*:477–504.

Cummins, K. W. 1964. Factors limiting the microdistribution of larvae of the caddisflies *Pycnopsyche lepida* (Hagen) and *Pycnopsyche guttifer* (Walker) in a Michigan stream (Trichoptera: Limnephilidae). Ecol. Monogr. *34*:271–295.

Cummins, K. W. 1972. Predicting variations in energy flow through a semi-controlled lotic ecosystem. Tech. Rept. Inst. Water Res. Mich. State Univ. *19*, 21 pp.

Cummins, K. W. 1973. Trophic relations of aquatic insects. Ann. Rev. Entomol. *18*:183–206.

Cummins, K. W. 1992. Invertebrates. *In* P. Calow and G. E. Petts., eds. The Rivers Handbook. I. Hydrological and Ecological Principles. Blackwell Scientific Publs., Oxford. pp. 234–250.

Cummins, K. W. and G. H. Lauff. 1969. The influence of substrate particle size on the microdistribution of stream macrobenthos. Hydrobiologia *34*:145–181.

Cummins, K. W. and J. C. Wuycheck. 1971. Caloric equivalents for investigations in ecological energetics. Mitt. Internat. Verein. Limnol. *18*, 158 pp.

Cummins, K. W. and M. A. Wilzbach. 1988. Do pathogens regulate stream invertebrate populations? Verh. Internat. Verein. Limnol. *23*:1232–1243.

Cummins, K. W. and M. J. Klug. 1979. Feeding ecology of stream invertebrates. Ann. Rev. Ecol. Syst. *10*:147–172.

Cummins, K. W. and R. W. Merritt. 1996. Ecology and distribution of aquatic insects. *In* R. W. Merritt and K. W. Cummins, eds. An Introduction to the Aquatic Insects of North America. 3rd ed. Kendall/Hunt Publ. Co., Dubuque, IA. pp. 74–86.

Cummins, K. W., C. E. Cushing, and G. W. Minshall. 1995. Introduction: An overview of stream ecosystems. *In* C. E. Cushing, K. W. Cummins, and G. W. Minshall, eds. River and Stream Ecosystems. Elsevier, Amsterdam. pp. 1–8.

Cummins, K. W., M. J. Klug, G. M. Ward, G. L. Spengler, R. W. Speaker, R. W. Ovink, D. C. Mahan, and R. C. Petersen. 1981. Trends in particulate organic matter fluxes, community processes and macroinvertebrate functional groups along a Great Lakes Drainage Basin river continuum. Verh. Internat. Verein. Limnol. *21*:841–849.

Cummins, K. W., M. J. Klug, R. G. Wetzel, R. C. Petersen, K. F. Suberkropp, B. A. Manny, J. C. Wuycheck, and F. O. Howard. 1972. Organic enrichment with leaf leachate in experimental lotic ecosystems. BioScience *22*:719–722.

Cummins, K. W., R. C. Petersen, F. O. Howard, J. C. Wuycheck, and V. I. Holt. 1973. The utilization of leaf litter by stream detritivores. Ecology *54*:336–345.

Cummins, K. W., R. R. Costa, R. E. Rowe, G. A. Moshiri, R. M. Scanlon, and R. K. Zajdel. 1969. Ecological energetics of a natural population of the predaceous zooplankter *Leptodora kindtii* Focke (Cladocera). Oikos *20*:189–223.

Cunningham, H. W. and R. G. Wetzel. 1989. Kinetic analysis of protein degradation by a freshwater wetland sediment community. Appl. Environ. Microbiol. *55*:1963–1967.

Cunningham, L. 1972. Vertical migrations of *Daphnia* and copepods under the ice. Limnol. Oceanogr. *17*:301–303.

Curds, C. R. 1982. The ecology and role of protozoa in aerobic sewage treatment processes. Ann. Rev. Microbiol. *36*:27–46.

Currie, D. J. 1990. Large-scale variability and interactions among phytoplankton, bacterioplankton, and phosphorus. Limnol. Oceanogr. *35*:1437–1455.

Currie, D. J. and J. Kalff. 1984a. The relative importance of bacterioplankton and phytoplankton in phosphorus uptake in freshwater (*sic*). Limnol. Oceanogr. *29*:311–324.

Currie, D. J. and J. Kalff. 1984b. Can bacteria outcompete phhytoplankton for phosphorus: A chemostat test. Microb. Ecol. *10*:205–216.

Currie, D. J. and J. Kalff. 1984c. A comparison of the abilities of freshwater algae and bacteria to acquire and retain phosphorus. Limnol. Oceanogr. *29*:298–310.

Cushing, C. E., ed. 1997. Freshwater Ecosystems and Climate Change in North America: A Regional Assessment. John Wiley & Sons, Chichester. 262 pp.

Cushing, C. E., C. D. McIntire, K. W. Cummins, G. W. Minshall, R. C. Petersen, J. R. Sedell, and R. L. Vannote. 1983. Relationships among chemical, physical, and biological indices along river continua based on multivariate analyses. Arch. Hydrobiol. *98*:317–326.

Cushing, C. E., G. W. Minshall, and J. D. Newbold. 1993. Transport dynamics of fine particulate organic matter in two Idaho streams. Limnol. Oceanogr. *38*:1101–1115.

Cyr, H. 1998. How does the vertical distribution of chlorophyll vary in littoral sediments of small lakes? Freshwat. Biol. *40*:25–40.

Czeczuga, B. 1959. On oxygen minimum and maximum in the metalimnion of Rajgród lakes. Acta Hydrobiol. *1*:109–122.

Czeczuga, B. 1968a. An attempt to determine the primary production of the green sulphur bacteria, *Chlorobium limicola* Nads, (Chlorobacteriaceae). Hydrobiologia *31*:317–333.

Czeczuga, B. 1968b. Primary production of the green hydrosulphuric bacteria, *Chlorobium limicola* Nads. (Chlorobacteriaceae). Photosynthetica *2*:11–15.

Czeczuga, B. 1968c. Primary production of the purple sulphuric bacteria, *Thiopedia rosea* Winogr. (Thiorhodaceae). Photosynthetica *2*:161–166.

Czeczuga, B. and E. Bobiatynska-ksok. 1970. The extent of consumption of the energy contained in the food suspension by *Ceriodaphnia reticulata* (Jurine). *In* Z. Kajak and A. Hillbricht-Ilkowska, eds. Productivity Problems of Freshwaters. PWN Polish Scientific Publishers, Warsaw. pp. 739–748.

Czeczuga, B. and R. Czerpak. 1967. Obserwacje nad bakterioplanktonem jezior leginskich. Zesz. Nauk WSR Olszt. *23*:35–44.

Czeczuga, B. and R. Czerpak. 1968. Investigations on vegetable pigments in post-glacial bed sediments of lakes. Schweiz. A. Hydrol. *30*:217–231.

Daan, N. and J. Ringelberg. 1969. Further studies on the positive and negative phototactic reaction of *Daphnia magna* Straus. Netherlands. J. Zool. *19*:525–540.

Daborn, G. R. 1974. Biological features of an aestival pond in western Canada. Hydrobiologia *44*:287–299.

Dacey, J. W. H. 1981. Pressurized ventilation in the yellow waterlily. Ecology *62*:1137–1147.

Dacey, J. W. H. and M. J. Klug. 1979. Methane efflux from lake sediments through water lilies. Science *203*:1253–1254.

Daft, N. J., J. Begg, and W. D. P. Stewart. 1970. A virus of blue-green algae from freshwater habitats in Scotland. New Phytol. *69*:1029–1038.

Dahm, C. N. 1981. Pathways and mechanisms for removal of dissolved organic carbon from leaf leachate in streams. Can. J. Fish. Aquat. Sci. *38*:68–76.

Dahm, C. N. 1984. Uptake of dissolved organic carbon in mountain streams. Verh. Internat. Verein. Limnol. *22*:1842–1846.

Dahm, C. N., E. H. Trotter, and J. R. Sedell. 1987. Role of anaerobic zones and processes in stream ecosystem productivity. *In* R. C. Averett and D. M. McKnight, eds. Chemical Quality of Water and the Hydrologic Cycle. Lewis Publishers, Chelsea, MI. pp. 157–178.

Dahm, C. N., D. L. Carr, and R. L. Coleman. 1991. Anaerobic carbon cycling in stream ecosystems. Verh. Internat. Verein. Limnol. *24*:1600–1604.

Dahms, H.-U. 1995. Dormancy in the Copepoda—an overview. Hydrobiologia *306*:199–211.

Daisley, K. W. 1969. Monthly survey of vitamin B_{12} concentrations in some waters of the English Lake District. Limnol. Oceanogr. *14*:224–228.

Dale, H. M. 1981. Hydrostatic pressure as the controlling factor in the depth distribution of Eurasian watermilfoil, *Myriophyllum spicatum* L. Hydrobiologia *79*:239–241.

Dale, H. M. 1984. Hydrostatic pressure and aquatic plant growth: A laboratory study. Hydrobiologia *111*:193–200.

Dale, H. M. 1986. Temperature and light: The determining factors in maximum depth distribution of aquatic macrophytes in Ontario, Canada. Hydrobiologica *133*:73–77.

Dale, H. M. and T. Gillespie. 1977. Diurnal fluctuations of temperature near the bottom of shallow water bodies as affected by solar radiation, bottom color and water circulation. Hydrobiologia *55*:87–92.

Dale, H. M. and T. Gillespie. 1976. The influence of floating vascular plants on the diurnal fluctuations of temperature near the water surface in early spring. Hydrobiologia *49*:245–256.

Daley, R. J. 1973. Experimental characterization of lacustrine chlorophyll diagenesis. II. Bacterial, viral and herbivore grazing effects. Arch. Hydrobiol. *72*:409–439.

Daley, R. J. and S. R. Brown. 1973. Experimental characterization of lacustrine chlorophyll diagenesis. I. Physiological and environmental effects. Arch. Hydrobiol. *72*:277–304.

Daley, R. J., S. R. Brown, and R. N. McNeely. 1977. Chromatographic and SCDP measurements of fossil phorbins and the postglacial history of Little Round Lake, Ontario. Limnol. Oceanogr. *22*:349–360.

Dall, P. C. 1987. The ecology of the littoral leech fauna (Hirudinea) in Lake Esrom, Denmark. Arch. Hydrobiol./Suppl. *76*:256–313.

Dall, P. C., C. Lindegaard, E. Jónsson, G. Jónsson, and P. M. Jónasson. 1984. Invertebrate communities and their environment in the exposed littoral zone of Lake Esrom, Denmark. Arch. Hydrobiol./Suppl. *69*:477–524.

Dalva, M. and T. R. Moore. 1991. Sources and sinks of dissolved organic carbon in a forested swamp catchment. Biogeochemistry *15*:1–19.

Von Damm, K. L., L. K. Benninger, and K. K. Turekian. 1979. The ^{210}Pb chronology of a core from Mirror Lake, New Hampshire. Limnol. Oceanogr. *24*:434–439.

Damman, A. W. H. 1986. Hydrology, development, and biogeochemistry of ombrogenous peat bogs with special reference to nutrient relocation in a western Newfoundland bog. Can. J. Bot. *64*:384–353.

Danell, K. and K. Sjöberg. 1979. Decomposition of *Carex* and *Equisetum* in a northern Swedish lake: Dry weight loss and colonization by macro-invertebrates. J. Ecol. *67*:191–200.

Danen-Louwerse, H., L. Lijklema, and M. Coenraats. 1993. Iron content of sediment and phosphate adsorption properties. Hydrobiologia *253*:311–317.

Danforth, W. F. 1962. Substrate assimilation and heterotrophy. *In* R. A. Lewin, ed. Physiology and Biochemistry of Algae. Academic Press, New York. pp. 99–123.

Darnell, R. M. 1964. Organic detritus in relation to secondary production in aquatic communities. Verh. Internat. Verein. Limnol. *15*:462–470.

Datsko, V. G. 1959. Organischeskoe Veshchestvo v Vodakh Iuzhnykh Morei SSSR. Izdatel'stvo Akademii Nauk SSSR, Moscow. 271 pp.

David, M. B. and G. F. Vance. 1991. Chemical character and origin of organic acids in streams and seepage lakes of central Maine. Biogeochemistry *12*:17–41.

David, M. B. and M. J. Mitchell. 1985. Sulfur constituents and cycling in waters, seston, and sediments of an oligotrophic lake. Limnol. Oceanogr. *30*:1196–1207.

David, M. B., G. F. Vance, and J. S. Kahl. 1992. Chemistry of dissolved organic carbon and organic acids in two streams draining forested watersheds. Water Resour. Res. *28*:389–396.

Davids, C., E. H. T. Winkel, and C. J. de Groot. 1994. Temporal and spatial patterns of water mites in Lake Maarsseveen I. Netherlands J. Aquat. Ecol. *28*:11–17.

Davies, G. S. 1970. Productivity of macrophytes in Marion Lake, British Columbia. J. Fish. Res. Board Can. *27*:71–81.

Davies, J. 1985. Evidence for a diurnal horizontal migration in *Daphnia hyalina*. Hydrobiologia *120*:103–106.

Davies, R. W. 1978. Reproductive strategies shown by freshwater Hirudinoidea. Verh. Internat. Verein. Limnol. *20*:2378–2381.

Davies, R. W. 1991. Annelida: Leeches, polychaetes, and acanthobdellids. *In* J. H. Thorp and A. P. Covich, eds. Ecology and Classification of North American Freshwater Invertebrates. Academic Press, San Diego. pp. 437–479.

Davies, R. W. and E. Dratnal. 1996. Differences in energy allocation during growth and reproduction by semelparous and iteroparous *Nephelopsis obscura* (Erpobdellidae). Arch. Hydrobiol. *138*:45–55.

Davies, R. W. and J. Wilkialis. 1980. The population ecology of the leech (Hirudinoidea: Glossiphoniidae) *Theromyzon rude*. Can. J. Zool. *58*:913–916.

Davies, R. W. and R. P. Everett. 1975. The feeding of four species of freshwater Hirudinoidea in southern Alberta. Verh. Internat. Verein. Limnol. *19*:2816–2827.

Davies, R. W. and T. B. Reynoldson. 1971. The incidence and intensity of predation on lake-dwelling triclads in the field. J. Anim. Ecol. *40*:191–214.

Davies, R. W. and T. B. Reynoldson. 1975. Life history of *Helobdella stagnalis* (L.) in Alberta. Verh. Internat. Verein. Limnol. *19*:2828–2839.

Davies, R. W., E. Dratnal, and L. R. Linton. 1996. Activity and foraging behaviour in the predatory freshwater leech *Nephelopsis obscura* (Erpobdellidae). Funct. Ecol. *10*:51–54.

Davies, W. 1971. The phytopsammon of a sandy beach transect. Amer. Midland Nat. *86*:292–308.

Davis, C. C. 1963. On questions of production and productivity in ecology. Arch. Hydrobiol. *59*:145–161.

Davis, J. A. and L. A. Barmuta. 1989. An ecologically useful classification of mean and near-bed flows in streams and rivers. Freshwat. Biol. *21*:271–282.

Davis, J. A. and R. Gloor. 1981. Adsorption of dissolved organics in lake water by aluminum oxide. Effect of molecular weight. Environ. Sci. Technol. *15*:1223–1229.

Davis, M. B. 1968. Pollen grains in lake sediments: Redeposition caused by seasonal water circulation. Science *162*:796–799.

Davis, M. B. 1973. Redeposition of pollen grains in lake sediment. Limnol. Oceanogr. *18*:44–52.

Davis, M. B. and E. S. Deevey, Jr. 1964. Pollen accumulation rates: Estimates from late-glacial sediment of Rogers Lake. Science *145*:1293–1295.

Davis, M. B. and L. B. Brubaker. 1973. Differential sedimentation of pollen grains in lakes. Limnol. Oceanogr. *18*:635–646.

Davis, M. B. and M. S. J. Ford. 1982. Sediment focusing in Mirror Lake, New Hampshire. Limnol. Oceanogr. *27*:137–150.

Davis, M. B., L. B. Brubaker, and T. Webb. 1973. Calibration of absolute pollen influx. *In* H. J. B. Birks and R. G. West, eds. Quaternary Plant Ecology. Blackwell Sci. Publs., Oxford. pp. 9–25.

Davis, M. B., R. E. Moeller, and J. Ford. 1984. Sediment focusing and pollen influx. *In* E. Y. Haworth and J. W. G. Lund, eds. Lake Sediments and Environmental History. Univ. Minnesota Press, Minneapolis. pp. 261–293.

Davis, R. B. 1974. Stratigraphic effects of tubificids in profundal lake sediments. Limnol. Oceanogr. *19*:466–488.

Davis, R. B. and D. S. Anderson. 1985. Methods of pH calibration of sedimentary diatom remains for reconstructing history of pH in lakes. Hydrobiologia *120*:69–87.

Davis, R. B., C. T. Hess, S. A. Norton, D. W. Hanson, K. D. Hoagland, and D. S. Anderson. 1983. ^{137}Cs and ^{210}Pb dating of sediments from soft-water lakes in New England (U.S.A.) and Scandinavia, a failure of ^{137}Cs dating. Chem. Geol. *44*:151–185.

Davis, R. B., D. L. Thurlow, and F. E. Brewster. 1975. Effects of burrowing tubificid worms on the exchange of phosphorus between lake sediment and overlying water. Verh. Internat. Verein. Limnol. *19*:382–394.

Davis, S. M. 1991. Growth, decomposition, and nutrient retention of *Cladium jamaicense* Crantz and *Typha domingensis* Pers. in the Florida Everglades. Aquat. Bot. *40*:203–224.

Davis, S. N. 1964. Silica in streams and ground water. Am. J. Sci. *262*:870–891.

Davis, S. N. and R. J. M. DeWiest. 1966. Hydrogeology. John Wiley & Sons, New York. 463 pp.

Davis, W. M. 1883. On the classification of lake basins. Proc. Boston Soc. Nat. Hist. *21*(1880–1882):315–381.

Davis, W. M. 1933. The lakes of California. Calif. J. Mines Geology *29*:175–236.

Davison, I. R. 1991. Environmental effects on algal photosynthesis: Temperature. J. Phycol. *27*:2–8.

Davison, W. 1987. Internal elemental cycles affecting the long-term alkalinity status of lakes: Implications for lake restoration. Schweiz. Z. Hydrol. *49*:186–201.

Davison, W. 1991. The solubility of iron sulphides in synthetic and natural waters at ambient temperature. Aquat. Sci. *53*:309–329.

Davison, W. and B. J. Finlay. 1990. Ferrous iron and phototrophy as alternative sinks for sulphide in the anoxic hypolimnia of two adjacent lakes. J. Ecol. *74*:663–673.

Davison, W. and R. De Vitre. 1992. Iron particles in freshwater. *In* J. Buffle and H. P. van Leeuwen, eds. Environmental Particles. Vol. 1. Lewis Publishers, Boca Raton. pp. 315–355.

Davison, W. and S. I. Heaney. 1978. Ferrous iron-sulfide interactions in anoxic hypolimnetic waters. Limnol. Oceanogr. *23*:1194–1200.

Davison, W., S. I. Heaney, J. F. Talling, and E. Rigg. 1980. Seasonal transformations and movements of iron in a productive English lake with deep-water anoxia. Schweiz. Z. Hydrol. *42*:196–224.

Dawidowicz, P. and C. J. Loose. 1992a. Cost of swimming by *Daphnia* during diel vertical migration. Limnol. Oceanogr. *37*:665–669.

Dawidowicz, P. and C. J. Loose. 1992b. Metabolic costs during predator-induced diel vertical migration of *Daphnia*. Limnol. Oceanogr. *37*:1589–1595.

Dawidowicz, P., J. Pijanowska, and K. Ciechomski. 1990. Vertical migration of *Chaoborus* larvae is induced by the presence of fish. Limnol. Oceanogr. *35*:1631–1637.

Dawson, F. H. 1978. The seasonal effects of aquatic plant growth on the flow of water in a stream. Proc. EWRS Symp. Aquat. Weeds *5*:1–6.

Day, F. P., Jr. and J. P. Megonigal. 1993. The relationship between variable hydroperiod, production allocation, and belowground organic turnover in forested wetlands. Wetlands *13*:115–121.

Dean, J. V. and D. D. Biesboer. 1985. Loss and uptake of ^{15}N-ammonium in submerged soils of a cattail marsh. Am. J. Bot. *72*:1197–1203.

Dean, W. E. and E. Gorham. 1976. Major chemical and mineral components of profundal surface sediments in Minnesota lakes. Limnol. Oceanogr. *21*:259–284.

DeAngelis, D. L. 1980. Energy flow, nutrient cycling, and ecosystem resilience. Ecology *51*:764–771.

DeAngelis, D. L. 1992. Dynamics of nutrient cycling and food webs. Chapman & Hall, London. 270 pp.

DeAngelis, D. L., P. J. Mulholland, J. W. Elwood, A. V. Palumbo, and A. D. Steinman. 1990. Biogeochemical cycling constraints on stream ecosystem recovery. Environ. Manage. *14*:685–697.

Dearing, J. A. 1997. Sedimentary indicators of lake-level changes in the humic temperate zone: A. critical review. J. Paleolimnol. *18*:1–14.

DeBiase, A. E., R. W. Sanders, and K. G. Porter. 1990. Relative nutritional value of ciliate protozoa and algae as food for *Daphnia*. Microb. Ecol. *19*:199–210.

Debroas, D. 1998. Decomposition of protein compounds in an eutrophic lake: Spatial and temporal distribution of exopeptidase and endopeptidase activities in various size fractions. Hydrobiologia *382*:161–173.

Decksbach, N. K. 1929. Zur Klassification der Gewässer vom astatischen Typus. Arch. Hydrobiol. *20*:399–406.

DeCosta, J. 1968. Species diversity of chydorid fossil communities in the Mississippi Valley. Hydrobiologia *32*:497–512.

Deevey, E. S., Jr. 1984. Stress, strain, and stability of lacustrine ecosystems. *In* E. Y. Haworth and J. W. G. Lund, eds. Lake Sediments and Environmental History. Univ. Minnesota Press, Minneapolis. pp. 203–229.

Deevey, E. S., D. S. Rice, P. M. Rice, H. H. Vaughan, M. Brenner, and M. S. Flannery. 1979. Mayan urbanism: Impact on a tropical karst environment. Science *206*:298–306.

Deevey, E. S., Jr. 1941. Limnological studies in Connecticut. VI. The quantity and composition of the bottom fauna of thirty-six Connecticut and New York lakes. Ecol. Monogr. *11*:414–455.

Deevey, E. S., Jr. 1955. The obliteration of the hypolimnion. Mem. Ist. Ital. Idrobiol. *8*(Suppl.):9–38.

DeGeer, G. 1912. A geochronology of the last 12,000 years. Proc. 11th Int. Geol. Congr., Stockholm. Part 1, pp. 241–253.

Dehdashti, B. and D. W. Blinn. 1991. Population dynamics and production of the pelagic amphipod *Hyalella montezuma* in a thermally constant system. Freshwat. Biol. *25*:131–141.

Delfino, J. J. and F. G. Lee. 1971. Variation of manganese, dissolved oxygen and related chemical parameters in the bottom waters of Lake Mendota, Wisconsin. Water Res. *5*:1207–1217.

Delfino, J. J. and G. F. Lee. 1968. Chemistry of manganese in Lake Mendota, Wisconsin. Environ. Sci. Technol. *2*:1094–1100.

Della Croce, N. 1955. The conditions of sedimentation and their relations with Oligochaeta populations of Lake Maggiore. Mem. Ist. Ital. Idrobiol. *8*(Suppl.):39–62.

Dellagreca, M., A. Fiorentino, P. Monaco, G. Pinto, A. Pollio, L. Previtera, and A. Zarrelli. 1998. Structural characterization and antialgal activity of compounds from *Pistia stratiotes* exudates. Allelopathy J. *5*:53–66.

Delong, M. D. and M. A. Brusven. 1994. Allochthonous input of organic matter from different riparian habitats of an agriculturally impacted stream. Environ. Manage. *18*:59–71.

Delorme, L. D. 1978. Distribution of freshwater ostracodes in Lake Erie. J. Great Lakes Res. *4*:216–220.

Del Prete, A. and C. Schofield. 1981. The utility of diatom analyses of lake sediments for evaluating acid precipitation effects on dilute lakes. Arch. Hydrobiol. *91*:332–340.

DeLucia, E. H. and W. H. Schlesinger. 1995. Photosynthetic rates and nutrient-use efficiency among evergreen and deciduous shrubs in Okefenokee Swamp. Int. J. Plant Sci. *156*:19–28.

DeMarte, J. A. and R. T. Hartman. 1974. Studies on absorption of ^{32}P, ^{59}Fe, and ^{45}Ca by water-milfoil (*Myriophyllum exalbescens* Fernald). Ecology *55*:188–194.

DeMelo, R., R. France, and D. J. McQueen. 1992. Biomanipulation: Hit or myth? Limnol. Oceanogr. *37*:192–207.

Demers, C. L. and R. W. Sage, Jr. 1990. Effects of road deicing salt on chloride levels in four Adirondak streams. Water Air Soil Poll. *49*:369–373.

DeMontigny, C. and Y. T. Prairie. 1993. The relative importance of biological and chemical processes in the release of phosphorous from a highly organic sediment. Hydrobiologia *253*:141–150.

DeMott, W. R. 1986. The role of taste in food selection by freshwater zooplankton. Oecologia *69*:334–340.

DeMott, W. R. and D. C. Müller-Navarra. 1997. The importance of highly unsaturated fatty acids in zooplankton nutrition: Evidence from experiments with *Daphnia*, a cyanobacterium and lipid emulsions. Freshwat. Biol. *38*:649–664.

Demuth, J., H. Neve, and K.-P. Witzel. 1993. Morphological diversity of bacteriophage populations in lake Plußsee, studied by direct electron microscopy. Appl. Environ. Microbiol. *59*:3378–3384.

den Hartog, C. and S. Segal. 1964. A new classification of the water-plant communities. Acta Bot. Nerl. *13*:367–393.

Dendy, J. S. 1963. Observations on bryozoan ecology in farm ponds. Limnol. Oceanogr. *8*:478–482.

Deng, Y. and W. Stumm. 1993. Kinetics of redox cycling of iron coupled with fulvic acid. Aquat. Sci. *55*:103–111.

Denny, M. W. 1988. Biology and the mechanics of the wave-swept environment. Princeton Univ. Press, Princeton, NJ. 329 pp.

Denny, P. 1972. Sites of nutrient absorption in aquatic macrophytes. J. Ecol. *60*:819–829.

Denny, P. 1980. Solute movement in submerged angiosperms. Biol. Rev. *50*:65–92.

Denny, P., ed. 1985. The Ecology and Management of African Wetland Vegetation. Dr. W. Junk, Publ., The Hague. 344 pp.

Denny, P. 1987. Mineral cycling by wetland plants—a review. Arch. Hydrobiol Beih. Ergebn. Limnol. *27*:1–25.

DeNoyelles, F., Jr. and W. J. O'Brien. 1978. Phytoplankton succession in nutrient enriched experimental ponds as related to changing carbon, nitrogen and phosphorus conditions. Arch. Hydrobiol. *84*:137–165.

Dermott, R. M., J. Kalff, W. C. Leggett, and J. Spence. 1977. Production of *Chironomus*, *Procladius*, and *Chaoborus* at different levels of phytoplankton biomass in Make Memphremagog, Quebec-Vermont. J. Fish. Res. Board Can. *34*:2001–2007.

Desortová, B. 1981. Relationship between chlorophyll-*a* concentration and phytoplankton biomass in several reservoirs in Czechoslovakia. Int. Rev. ges. Hydrobiol. *66*:153–169.

De Souza-Lima, Y. and J.-C. Romano. 1983. Ecological aspects of the surface microlayer. 1. ATP, ADP, AMP contents, and energy charge ratios of microplanktonic communities. J. Exp. Mar. Biol. Ecol. *70*:107–122.

Desvilettes, C., G. Bourdier, C. Amblard, and B. Barth. 1997. Use of fatty acids for the assessment of zooplankton grazing on bacteria, protozoans and microalgae. Freshwat. Biol. *38*: 629–637.

Detenbeck, N. E. and P. L. Brezonik. 1991. Phosphorus sorption by sediments from a soft-water seepage lake. 2. Effects of pH and sediment composition. Environ. Sci. Technol. *25*:403–409.

Dévai, I., L. Felföldy, I. Wittner, and S. Plósz. 1988. Detection of phosphine: New aspects of the phosphorus cycle in the hydrosphere. Nature *333*:343–345.

DeVries, D. R. 1992. Freshwater jellyfish *Craspedacusta sowerbyi*: A summary of its life history, ecology, and distribution. J. Freshwat. Ecol. *7*:7–16.

DeWitt, R. M. 1954. Reproduction, embryonic development, and growth in the pond snail, *Physa gyrina* Say. Trans. Amer. Micros. Soc. *73*:124–137.

Deyl, Z. 1961. Anaerobic fermentations. Sci. Pap. Inst. Chem. Technol. Praque, Technol. Water *5*(2):131–234.

Diana, J. S. 1995. Biology and Ecology of Fishes. Cooper Publishing Group LLC, Carmel, IN. 441 pp.

Dias, F. and C. Kharif. 1999. Nonlinear gravity and capillary-gravity waves. Annu. Rev. Fluid Mech. *31*:301–346.

Diaz, M. M. and F. L. Pedrozo. 1993. Seasonal succession of phytoplankton in a small Andean patagonian lake (Rep. Argentina) and some considerations about the PEG Model. Arch. Hydrobiol. *127*:167–184.

Diaz, O. A., K. R. Reddy, and P. A. Moore, Jr. 1994. Solubility of inorganic phosphorus in stream water as influenced by pH and calcium concentration. Wat. Res. *28*:1755–1763.

Dice, L. R. 1914. The factors determining the vertical movements of *Daphnia*. J. Anim. Behavior *4*:229–265.

Dickerman, J. A. and R. G. Wetzel. 1985. Clonal growth in *Typha latifolia*: Population dynamics and demography of the ramets. J. Ecol. *73*:535–552.

Dickerman, J. A., A. J. Stewart, and R. G. Wetzel. 1986. Estimates of net annual aboveground production: Sensitivity of sampling frequency. Ecology *67*:650–659.

Dickerman, M. D. and J. S. Hartman. 1979. A rationale for the subclassification of biogenic meromictic lakes. Int. Revue ges. Hydrobiol. *64*:189–192.

DiCola, G., M. Dilingenti, and I. Guerri. 1977. Identification of the thermal eddy diffusivity in a stratified lake. Mem. Ist. Ital. Idrobiol. *34*:175–185.

DiDonato, G. T. and D. M. Lodge. 1993. Species replacements among *Orconectes* crayfishes in Wisconsin lakes: The role of predation by fish. Can. J. Fish. Aquat. Sci. *50*:1484–1488.

Dietz, A. S., L. J. Albright, and T. Tuominen. 1976. Heterotrophic activities of bacterioneuston and bacterioplankton. Can. J. Microbiol. *22*:1699–1709.

Digerfeldt, G., R. W. Battarbee, and L. Bengtsson. 1975. Report on annually laminated sediment in Lake Järlasjön. Geol. Fören. Förhandl. Stockholm 97:29–40.

Dilcher, D. L. 1974. Approaches to the identification of angiosperm leaf remains. Bot. Rev. *40*:1–157.

Dillon, P. J. 1974. The prediction of phosphorus and chlorophyll concentrations in lakes. Ph.D. Dissertation, Univ. Toronto. 330 pp.

Dillon, P. J. and F. H. Rigler. 1974. A test of a simple nutrient budget model predicting the phosphorus concentration in lake water. J. Fish. Res. Board Can. *31*:1771–1778.

Dillon, P. J. and F. H. Rigler. 1974. The phosphorus-chlorophyll relationship in lakes. Limnol. Oceanogr. *19*:767–773.

Dillon, P. J. and L. A. Molot. 1997a. Dissolved organic and inorganic carbon mass balances in central Ontario lakes. Biogeochemistry *36*:29–42.

Dillon, P. J. and L. A. Molot. 1997b. Effect of landscape form on export of dissolved organic carbon, iron, and phosphorus from forested stream catchments. Water Resour. Res. *33*:2591–2600.

Dillon, P. J. and W. B. Kirchner. 1975. The effects of geology and land use on the export of phosphorus from watersheds. Water Res. *9*:135–148.

Dillon, P. J., H. E. Evans, and R. Girard. 1997. Hypolimnetic alkalinity generation in two dilute, oligotrophic lakes in Ontario, Canada. Water Air Soil Poll. *99*:373–380.

Dillon, P. J., N. D. Yan, and H. H. Harvey. 1983. Acidic deposition: Effects on aquatic ecosystems. CRC Crit. Rev. Environ. Control *13*:167–194.

Dillon, P. J., R. D. Evans, and L. A. Molot. 1990. Retention and resuspension of phosphorus, nitrogen, and iron in a central Ontario lake. Can. J. Fish. Aquat. Sci. *47*:1269–1274.

Dinsdale, M. T. and A. E. Walsby. 1972. The interrelations of cell turgor pressure, gas-vacuolation, and buoyancy in a blue-green alga. J. Exp. Bot. *23*:561–570.

Dixit, S. S., A. S. Dixit, and J. P. Smol. 1989. Lake acidification recovery can be monitored using chrysophycean microfossils. Can. J. Fish. Aquat. Sci. *46*:1309–1312.

Dixit, S. S., J. P. Smol., J. C. Kingston, and D. F. Charles. 1992. Diatoms: Powerful indicators of environmental changes. Environ. Sci. Technol. *26*:23–33.

Dodds, W. K. 1989. Microscale vertical profiles of N_2 fixation, photosynthesis, O_2, chlorophyll *a* and light in a cyanobacterial assemblage. Appl. Environ. Microbiol. *55*:882–886.

Dodds, W. K. 1992. A modified fiber-optic microprobe to measure spherically integrated photosynthetic photon flux density: Characterization of periphyton photosynthesis-irradiance patterns. Limnol. Oceanogr. *37*:871–878.

Dodds, W. K. and R. D. Jones. 1987. Potential rates of nitrification and denitrification in an oligotrophic freshwater sediment system. Microb. Ecol. *14*:91–100.

Dodds, W. K. and R. W. Castenholz. 1988. The nitrogen budget of an oligotrophic cold water pond. Arch. Hydrobiol./Suppl. *79*:343–362.

Dodds, W. K., B. K. Ellis, and J. C. Priscu. 1991. Zooplankton induced decrease in inorganic phosphorus uptake by plankton in an oligotrophic lake. Hydrobiologia *211*:253–259.

Dodds, W. K., J. R. Jones, and E. B. Welch. 1998. Suggested classification of stream trophic state: Distributions of temperate stream types by chlorophyll, total nitrogen, and phosphorus. Wat. Res. *32*:1455–1462.

Dodds, W. K., K. R. Johnson, and J. C. Priscu. 1989. Simultaneous nitrogen and phosphorus deficiency in natural phytoplankton assemblages: Theory, empirical evidence, and implications for lake management. Lake Reservoir Manage. *5*:21–26.

Dodson, S. I. 1970. Complementary feeding niches sustained by size-selective predation. Limnol. Oceanogr. *15*:131–137.

Dodson, S. I. 1972. Mortality in a population of *Daphnia rosea*. Ecology *53*:1011–1023.

Dodson, S. I. 1974a. Zooplankton competition and predation: An experimental test of the size-efficiency hypothesis. Ecology *55*:605–613.

Dodson, S. I. 1974b. Adaptive change in plankton morphology in response to size-selective predation: A new hypothesis of cyclomorphosis. Limnol. Oceanogr. *19*:721–729.

Dodson, S. I. 1989. Cyclomorphosis in *Daphnia galeata mendotae* Birge and *D. retrocurva* Forbes as a predator-induced response. Freshwat. Biol. *19*:109–114.

Dodson, S. 1990. Predicting diel vertical migration of zooplankton. Limnol. Oceanogr. *35*:1195–1200.

Dodson, S. I., T. A. Crowl, B. L. Peckarsky, L. B. Kats, A. P. Covich, and J. M. Culp. 1994. Non-visual communication in freshwater benthos: An overview. J. N. Am. Benthol. Soc. *13*:268–282.

Döhler, G. and K.-R. Przybylla. 1973. Einfluss der Temperatur auf die Lichtatmung der Blaualge *Anacystis nidulans*. Planta *110*:153–158.

Dokulil, M. 1973. Planktonic primary production within the Phragmites community of Lake Neusiedlersee (Austria). Pol. Arch. Hydrobiol. *20*:175–180.

Dokulil, M. 1988. Seasonal and spatial distribution of cryptophycean species in the deep, stratifying, alpine lake Mondsee and their role in the food web. Hydrobiologia *161*:185–201.

Dokulil, M. and C. Skolaut. 1986. Succession of phytoplankton in a deep stratifying lake: Mondsee, Austria. Hydrobiologia *138*:9–24.

Dolan, T. J., A. J. Hermann, S. E. Bayley, and J. Zoltek, Jr. 1984. Evapotranspiration of a Florida, U.S.A., freshwater wetland. J. Hydrol. *74*:355–371.

Domogalla, B. P. and E. B. Fred. 1926. Ammonia and nitrate studies of lakes near Madison, Wisconsin. J. Amer. Soc. Agronomy *18*:897–911.

Domogalla, B. P., E. B. Fred, and W. H. Peterson. 1926. Seasonal variations in the ammonia and nitrate content of lake waters. J. Amer. Water Works Assoc. *15*:369–385.

Van Donk, E. 1989. The role of fungal parasites in phytoplankton succession. *In* U. Sommer, ed. Plankton Ecology: Succession in Plankton Communities. Springer-Verlag, Berlin. pp. 171–194.

Van Donk, E. and A. Otte. 1996. Effects of grazing by fish and waterfowl on the biomass and species composition of submerged macrophytes. Hydrobiologia *340*:285–290.

Van Donk, E. and D. O. Hessen. 1993. Grazing resistance in nutrient-stressed phytoplankton. Oecologia *93*:508–511.

Van Donk, E. and K. Bruning. 1995. Effects of fungal parasites on planktonic algae and the role of environmental factors in the fungus-alga relationship. *In* W. Wiessner, E. Schnepf, and R. C. Starr, eds. Algae, Environment and Human Affairs. Biopress Ltd., Bristol. pp. 223–234.

Van Donk, E., B. A. Faafeng, D. O. Hessen, and T. Källqvist. 1993. Use of immobilized algae for estimating bioavailable phosphorus released by zooplankton. J. Plankton Res. *15*:761–769.

Van Donk, E., R. D. Gulati, A. Iedema, and J. T. Meulemans. 1993. Macrophyte-related shifts in the nitrogen and phosphorus contents of the different trophic levels in a biomanipulated shallow lake. Hydrobiologia *251*:19–26.

Donnelly, A. P. and R. A. Herbert. 1999. Bacterial interactions in the rhizosphere of seagrass communities in shallow coastal lagoons. J. App. Microbiol. *85(*Suppl.):151S–160S.

Dor, I. 1970. Production rate of the periphyton in Lake Tiberias as measured by the glass-slide method. Israel J. Bot. *19*:1–15.

Dorgelo, J. and K. Koning. 1980. Avoidance of macrophytes and additional notes on avoidance of the shore by '*Acanthodiaptomus denticornis*' (Wierzejski, 1887) from Lake Pavin (Auvergne, France). Hydrobiol. Bull. *14*:196–208.

Dorgelo, J. and M. Heykoop. 1985. Avoidance of macrophytes by *Daphnia longispina*. Verh. Internat. Verein. Limnol. *22*:3369–3372.

Dosskey, M. G. and P. M. Bertsch. 1994. Forest sources and pathways of organic matter transport to a blackwater stream: A hydrological approach. Biogeochemistry *24*:1–19.

Doucette, G. J., N. M. Price, and P. J. Harrison. 1987. Effects of selenium deficiency on the morphology and ultrastructure of the coastal marine diatom *Thalassiosira pseudonana* (Bacillariophyceae). J. Phycol. *23*:9–17.

Douglas, B. 1958. The ecology of the attached diatoms and other algae in a small stony stream. J. Ecol. *46*:295–322.

Downes, M. T. and H. W. Paerl. 1978. Separation of two dissolved reactive phosphorus fractions in lakewater. J. Fish. Res. Board Can. *35*:1636–1639.

Downing, J. A. 1984. Sampling the benthos of standing waters. *In* J. A. Downing and F. H. Rigler, eds. A Manual on Methods for the Assessment of Secondary Productivity in Fresh Waters. 2nd ed. Blackwell Scientific Publs., Oxford. pp. 87–130.

Downing, J. A. and F. H. Rigler, eds. 1984. A Manual on Methods for the Assessment of Secondary Productivity in Fresh Waters. 2nd ed. Blackwell Scientific Publications, Oxford. 501 pp.

Doyle, R. D. and T. R. Fisher. 1994. Nitrogen fixation by periphyton and plankton on the Amazon floodplain at Lake Calado. Biogeochemistry *26*:41–66.

Doyle, R. W. 1968a. The origin of the ferrous ion-ferric oxide Nernst potential in environments containing dissolved ferrous iron. Am. J. Sci. *265*:840–859.

Doyle, R. W. 1968b. Identification and solubility of iron sulfide in anaerobic lake sediment. Am. J. Sci. *266*:980–994.

Drago, E. C. 1989. Thermal summer characteristics of lakes and ponds on Deception Island, Antarctica. Hydrobiologia *184*:51–60.

Drake, J. C. and S. I. Heaney. 1987. Occurrence of phosphorus and its potential remobilization in the littoral sediments of a productive English lake. Freshwat. Biol. *17*:513–523.

Drewes, C. D. and C. R. Fourtner. 1993. Helical swimming in a freshwater oligochaete. Biol. Bull. *185*:1–9.

Driscoll, C. T. and R. Van Dreason. 1993. Seasonal and long-term temporal patterns in the chemistry of Adirondack lakes. Water Air Soil Poll. *67*:319–344.

Driscoll, C. T., C. Yan, C. L. Schofield, R. Munson, and J. Holsapple. 1994. The mercury cycle and fish in the Adirondack lakes. Environ. Sci. Technol. *28*:136A–143A.

Driscoll, C. T., J. Holsapple, C. L. Schofield, and R. Munson. 1998. The chemistry and transport of mercury in a small wetland in the Adirondack region of New York, USA. Biogeochemistry *40*:137–146.

Driscoll, C. T., R. D. Fuller, and W. D. Schecher. 1989. The role of organic acids in the acidification of surface waters in the Eastern U.S. Water Air Soil Poll. *43*:21–40.

Driscoll, C. T., V. Blette, C. Yan, C. L. Schofield, R. Munson, and J. Holsapple. 1995. The role of dissolved organic carbon in the chemistry and bioavailability of mercury in remote Adirondack lakes. Water Air Soil Poll. *80*:499–508.

Droop, M. R. 1973. Some thoughts on nutrient limitation in algae. J. Phycol. *9*:264–272.

Droop, M. R. 1974a. Heterotrophy of carbon. *In* W. D. P. Stewart, ed. Algal Physiology and Biochemistry. Univ. of California Press, Berkeley. pp. 530–559.

Droop, M. R. 1974b. The nutrient status of algal cells in continuous culture. J. Mar. Biol. Assoc. U. K. *54*:825–855.

Droop, M. R. 1977. An approach to quantitative nutrition of phytoplankton. J. Protozool. *24*:528–532.

Drozd, J., S. S. Gonet, S. Senesi, and J. Weber, eds. 1997. The Role of Humic Substances in the Ecosystems and in Environmental Protection. Polish Soc. Humic Substances, Wroclaw, Poland. 1002 pp.

Drummond, A. J. 1971. Recent measurements of the solar radiation incident on the atmosphere. *In* Space Research XI. Akademie-Verlag, Berlin. pp. 681–693.

Dubay, C. I. and G. M. Simmons, Jr. 1979. The contribution of macrophytes to the metalimnetic oxygen maximum in a montane, oligotrophic lake. Amer. Midland Nat. *101*:108–117.

Duchemin, E., M. Lucotte, and R. Canuel. 1999. Comparison of static chamber and thin boundary layer equation methods for measuring greenhouse gas emissions from large water bodies. Environ. Sci. Technol. *33*:350–357.

Dubinina, G. A., V. M. Gorlenko, and J. I. Suleimanov. 1973. A study of microorganisms involved in the circulation of manganese, iron, and sulfur in meromictic Lake Gek-Gel'. Mikrobiologiya *42*:918–924.

Dubinsky, Z. and T. Berman 1981b. Photosynthetic efficiencies in aquatic ecosystems. Verh. Internat. Verein. Limnol. *21*:237–243.

Dubinsky, Z. and T. Berman. 1976. Light utilization efficiencies of phytoplankton in Lake Kinneret (Sea of Galilee). Limnol. Oceanogr. *21*:226–230.

Dubinsky, Z. and T. Berman. 1981a. Light utilization by phytoplankton in Lake Kinneret (Israel). Limnol. Oceanogr. *26*:660–670.

Dudel, G. and J.-G. Kohl. 1991. Contribution of dinitrogen fixation and denitrification to the N-budget of a shallow lake. Verh. Internat. Verein. Limnol. *24*:884–888.

Dudgeon, D. 1988. The influence of riparian vegetation on macroinvertebrate community structure in four Hong Kong streams. J. Zool. Lond. *216*:609–627.

Duff, J. H. and F. J. Triska. 1990. Denitrification in sediments from the hyporheic zone adjacent to a small forested stream. Can. J. Fish. Aquat. Sci. *47*:1140–1147.

Duff, K. E., B. A. Zeeb, and J. P. Smol. 1997. Chrysophyte cyst biogeographical and ecological distributions: A synthesis. J. Biogeogr. *24*:791–812.

Dugdale, R. C. and V. A. Dugdale. 1961. Sources of phosphorus and nitrogen for lakes on Afognak Island. Limnol. *6*:13–23.

Dugdale, V. A. and R. C. Dugdale. 1962. Nitrogen metabolism in lakes. II. Role of nitrogen fixation in Sanctuary Lake, Pennsylvania. Limnol. Oceanogr. *7*:170–177.

Dumont, H. J. 1972. The biological cycle of molybdenum in relation to primary production and waterbloom formation in a eutrophic pond. Verh. Internat. Verein. Limnol. *18*:84–92.

Dumont, H. J. and I. Carels. 1987. Flatworm predator (*Mesostoma* cf. *lingua*) releases a toxin to catch planktonic prey (*Daphnia magna*). Limnol. Oceanogr. *32*:699–702.

Dumont, H. J. and J. Green, eds. 1980. Rotatoria. Hydrobiologia *73*:1–262. (Also as Developments in Hydrobiology, Vol. 1.)

Duncan, A. 1989. Food limitation and body size in the life cycles of planktonic rotifers and cladocerans. Hydrobiologia *186/187*: 11–28.

Dunne, T. 1978. Field studies of hillslope flow processes. *In* M. J. Kirkby, ed. Hillslope Hydrology. John Wiley & Sons, New York. pp. 227–293.

Dunne, T. and R. D. Black. 1970a. An experimental investigation of runoff production in permeable soils. Water Resour. Res. *6*:478–490.

Dunne, T. and R. D. Black. 1970b. Partial area contributions to storm runoff in a small New England watershed. Water Resour. Res. *6*:1296–1311.

Dunne, T. and R. D. Black. 1971. Runoff processes during snowmelt. Water Resour. Res. *7*:1160–1172.

Duong, T. P. 1972. Nitrogen fixation and productivity in a eutrophic hard-water lakes: *In situ* and laboratory studies. Ph.D. Disser., Michigan State University. 241 pp.

Durum, W. H. and J. Haffty. 1961. Occurrence of minor elements in water. Circ. U.S. Geol. Surv. *445*, 11 pp.

Dussart, B. 1966. Limnologie. L'Etude des Eaux Continentales. Gauthier-Villars, Paris. 677 pp.

Dussart, B. H., K. F. Lagler, P. A. Larkin, T. Scudder, K. Szesztay, and G. F. White. 1972. Man-made lakes as modified ecosystems. SCOPE Rept. 2. Int. Council Sci. Unions, Paris. 76 pp.

Dustin, N. M., B. H. Wilkinson, and R. M. Owen. 1986. Littlefield Lake, Michigan: Carbonate budget of Holocene sedimentation in a temperate-region lacustrine system. Limnol. Oceanogr. *31*:1301–1311.

Duston, N. M., R. M. Owen, and B. H. Wilkinson. 1986. Water chemistry and sedimentological observations in Littlefield Lake, Michigan: Implications for lacustrine marl deposition. Environ. Geol. Water Sci. *8*:229–236.

Duthie, H. C. and M. L. Ostrofsky. 1982. Use of phosphorus budget models in reservoir management. Can. Wat. Resour. J. *7*: 337–347.

Duthie, H. C. and V. M. Stout. 1986. Phytoplankton periodicity of the Waitaki lakes, New Zealand. Hydrobiologia *138*:221–236.

Dutton, J. A. and R. A. Bryson. 1962. Heat flux in Lake Mendota. Limnol. Oceanogr. *7*:80–97.

Duursma, E. K. 1961. Dissolved organic carbon, nitrogen and phosphorus in the sea. Netherlands J. Mar. Res. *1*:1–148.

Duursma, E. K. 1967. The mobility of compounds in sediments in relation to exchange between bottom and supernatant water. *In* H. L. Golterman and R. S. Clymo, eds. Chemical Environment in the Aquatic Habitat. N.V. Noord-Hollandsche Uitgevers Maatsahappij, Amsterdam. pp. 288–296.

Dvorák, J. 1970. A quantitative study on the macrofauna of stands of emergent vegetation in a carp pond of south-west Bohemia. Rozpravy Ceskosl. Akad. Ved, Rada Matem. Prírod. Ved *80*(6):63–110.

Dvorák, J. 1987. Production-ecological relationships between aquatic vascular plants and invertebrates in shallow waters and wetlands—a review. Arch. Hydrobiol. Beih. Ergebn. Limnol. *27*:181–184.

Dykyjová, D. 1971. Production, vertical structure and light profiles in littoral stands of reed-bed species. Hidrobiologia (Romania) *12*:361–376.

Dykyjová, D. and D. Hradecká. 1973. Productivity of reed-bed stands in relation to the ecotype, microclimate and trophic conditions of the habitat. Pol. Arch. Hydrobiol. *20*:111–119.

Dykyjová, D. and J. Květ, eds. 1978. Pond Littoral Ecosystems. Springer-Verlag, New York. 464 pp.

Eaton, K. A. 1983. The life history and production of *Chaoborus punctipennis* (Diptera: Chaoboridae) in Lake Norman, North Carolina, U.S.A. Hydrobiologia *106*:247–252.

Eberly, W. R. 1959. The metalimnetic oxygen maximum in Myers Lakes. Invest. Indiana Lakes Streams *5*:1–46.

Eberly, W. R. 1963. Oxygen production in some northern Indiana lakes. Proc. Indust. Wastes Conf., Purdue Univ. *17*:733–747.

Eberly, W. R. 1964. Further studies on the metalimnetic oxygen maximum, with special reference to its occurrence throughout the world. Invest. Indiana Lakes Streams 6:103–139.

Eccleston-Parry, J. D. and B. S. C. Leadbeater. 1995. Regeneration of phosphorus and nitrogen by four species of heterotrophic nanoflagellates feeding on three nutritional states of a single bacterial strain. Appl. Environ. Microbiol. *61*:1033–1038.

Eckerrot, Å. and K. Pettersson. 1993. Pore water phosphorus and iron concentrations in a shallow, eutrophic lake—indications of bacterial regulation. Hydrobiologia *253*:165–177.

Eckert, W. and H. G. Trüper. 1993. Microbial-related redox changes in a subtropical lake. I. *In situ* monitoring of the annual redox cycle. Biogeochemistry *21*:1–19.

Eckhardt, B. W. and T. R. Moore. Controls on dissolved organic carbon concentrations in streams, southern Québec. Can. J. Fish. Aquat. Sci. *47*:1537–1544.

Edlund, M. B. and E. F. Stoermer. 1997. Ecological, evolutionary, and systematic significance of diatom life histories. J. Phycol. *33*:897–918.

Edmondson, W. T. 1944. Ecological studies of sessile Rotatoria. Part I. Factors affecting distribution. Ecol. Monogr. *14*:31–66.

Edmondson, W. T. 1945. Ecological studies of sessile Rotatoria. Part II. Dynamics of populations and social structures. Ecol. Monogr. *15*:141–172.

Edmondson, W. T. 1946. Factors in the dynamics of rotifer populations. Ecol. Monogr. *16*:357–372.

Edmondson, W. T. 1948. Ecological applications of Lansing's physiological work on longevity in Rotatoria. Science *108*:123–126.

Edmondson, W. T. 1960. Reproductive rates of rotifers in natural populations. Mem. Ist. Ital. Idrobiol. *12*:21–77.

Edmondson, W. T. 1961. Changes in Lake Washington following an increase in the nutrient income. Verh. Internat. Verein. Limnol. *14*:167–175.

Edmondson, W. T. 1965. Reproductive rate of planktonic rotifers as related to food and temperature in nature. Ecol. Monogr. *35*:61–111.

Edmondson, W. T. 1966. Changes in the oxygen deficit of Lake Washington. Verh. Internat. Verein. Limnol. *16*:153–158.

Edmondson, W. T. 1968. A graphical model for evaluating the use of egg ratio for measuring birth and death rates. Oecologia *1*:1–37.

Edmondson, W. T. 1969. Eutrophication in North America. *In* Eutrophication: Causes, Consequences, Correctives. National Academy of Sciences, Washington, DC. pp. 124–149.

Edmondson, W. T. 1972. Nutrients and phytoplankton in Lake Washington. *In* G. E. Likens, ed. Nutrients and Eutrophication: The Limiting-Nutrient Controversy. Special Symposium, Amer. Soc. Limnol. Oceanogr. *1*:172–193.

Edmondson, W. T. 1974. Secondary productivity. Mitt. Internat. Verein. Limnol. *20*:229–272.

Edmondson, W. T. 1991a. The Uses of Ecology: Lake Washington and Beyond. Univ. Washington Press, Seattle. 329 pp.

Edmondson, W. T. 1991b. Sedimentary record of changes in the condition of Lake Washington. Limnol. Oceanogr. *36*:1031–1044.

Edmondson, W. T. and G. G. Winberg, eds. 1971. A Manual on Methods for the Assessment of Secondary Productivity in Fresh Waters. Int. Biol. Program Handbook, 17. Blackwell Scientific Publications, Oxford. 358 pp.

Edmondson, W. T. and J. T. Lehman. 1981. The effect of changes in the nutrient income on the condition of Lake Washington. Limnol. Oceanogr. *26*:1–29.

Edmunds, G. F., Jr., S. L. Jensen, and L. Berner. 1976. The mayflies of North and Central America. Univ. Minnesota Press, Minneapolis. 330 pp.

Edmunds, M. 1974. Defense in Animals. A Survey of Anti-predator Defenses. Longman Group Ltd., New York. 357 pp.

Edwards, A. M. C. and P. S. Liss. 1973. Evidence for buffering of dissolved silicon in fresh waters. Nature *243*:341–342.

Edwards, C. J., P. L. Hudson, W. G. Duffy, S. J. Nepszy, C. D. McNabb, R. C. Haas, C. R. Liston, B. Manny, and W.-D. N. Busch. 1989. Hydrological, morphometrical, and biological characteristics of the connecting rivers of the International Great Lakes: A review. *In* D. P. Dodge, ed. Proceedings of the International Large River Symposium. Can. Spec. Publ. Fish. Aquat. Sci. *106*:240–264.

Edwards, R. T. 1987. Sestonic bacteria as a food source for filtering invertebrates in two southeastern blackwater rivers. Limnol. Oceanogr. *32*:221–234.

Edwards, R. T. and J. L. Meyer. 1987. Metabolism of a sub-tropical low gradient blackwater river. Freshwat. Biol. *17*:251–264.

Edwards, R. T., J. L. Meyer, and S. E. G. Findlay. 1990. The relative contribution of benthic and suspended bacteria to system biomass, production, and metabolism in a low-gradient blackwater river. J. N. Am. Benthol. Soc. *9*:216–228.

Edwards, R. W. and M. Owens. 1960. The effects of plants on river conditions. I. Summer crops and estimates of net productivity of macrophytes in a chalk stream. J. Ecol. 48:151–160.

Effler, S. W. and D. L. Johnson. 1987. Calcium carbonate precipitation and turbidity measurements in Otisco Lake, New York. Water Resour. Bull. *23*:73–79.

Effler, S. W., C. M. Brooks, K. Whitehead, B. Wagner, S. M. Doerr, M. Perkins, C. A. Siegfried, L. Walrath, and R. P. Canale. 1996. Impact of zebra mussel invasion on river water quality. Water Environ. Res. *68*:205–214.

Effler, S. W., M. G. Perkins, H. Greer, and D. L. Johnson. 1987. Effect of "whiting" on optical properties and turbidity in Owasco Lake, New York. Water Resour. Bull. *23*:189–196.

Effler, S. W., M. T. Auer, N. Johnson, M. Penn, and H. C. Rowell. 1996. Sediments. *In* S. W. Effler, ed. Limnological and Engineering Analysis of a Polluted Urban Lake. Springer-Verlag, New York. pp. 600–666.

Efford, I. E. 1967. Temporal and spatial difference in phytoplankton productivity in Marion Lake, British Columbia. J. Fish. Res. Board Can. *24*:2283–2307.

Eggleton, F. E. 1956. Limnology of a meromictic, interglacial, plunge-basin lake. Trans. Amer. Microsc. Soc. *75*:334–378.

Egli, T. 1995. The ecological and physiological significance of the growth of heterotrophic microorganisms with mixtures of substrates. Adv. Microb. Ecol. *14*:305–386.

Egloff, D. A. and D. S. Palmer. 1971. Size relations of the filtering area of two *Daphnia* species. Limnol. Oceanogr. *16*:900–905.

Ehrenreich, A. and F. Widdel. 1994. Anaerobic oxidation of ferrous iron by purple bacteria, a new type of phototrophic metabolism. Appl. Environ. Microbiol. *60*:4517–4526.

Eichenberger, E. and A. Schlatter. 1978. Effect of herbivorous insects on the production of benthic algal vegetation in outdoor channels. Verh. Internat. Verein. Limnol. *20*:1806–1810.

Eighmy, T. T., L. S. Jahnke, and W. R. Fagerberg. 1991. Studies of *Elodea nuttallii* grown under photorespiratory conditons. II.

Evidence for bicarbonate active transport. Plant Cell Environ. *14*:157–165.

Eilers, J. M., D. F. Brakke, and A. Henriksen. 1992. The inapplicability of the Gibbs model of world water chemistry for dilute lakes. Limnol. Oceanogr. *37*:1335–1337.

Einhellig, F. A. 1995. Allelopathy: Current status and future goals. *In* Inderjit, K. M. M. Dakshini, and F. A. Einhellig, eds. Allelopathy: Organisms, Processes, and Applications. American Chemical Society, Washington. pp. 1–23.

Einsele, W. 1936. Über die Beziehungen des Eisenkreislaufs zum Phosphatkreislauf im eutrophen See. Arch. Hydrobiol. *29*:664–686.

Einsele, W. 1941. Die Umsetzung von zugeführtem, anorganischen Phosphat im eutrophen See und ihre Rüchwirkungen auf seinen Gesamthaushalt. Zeitsch. f. Fischerei *39*:407–488.

Einsele, W. 1960. Die Strömungsgeschwindigkeit als beherrschender Faktor bei der limnologischen Gestaltung der Gewasser. Osterreichs Fischerei 2:1–40.

Eiseltová, M., ed. 1994. Restoration of Lake Ecosystems: A Holistic Approach. Publ. Int. Waterflow Wetlands Res. Bureau, Gloucester, UK 32. 182 pp.

Eiseltová, M. and J. Biggs, eds. 1995. Restoration of Stream Ecosystems: An Integrated Catchment Approach. Publ. Int. Waterflowl Wetlands Res. Bureau, Gloucester, UK 37. 170 pp.

Eisenberg, D. and W. Kauzmann. 1969. The Structure and Properties of Water. Oxford University Press, New York. 296 pp.

Eisenberg, R. M. 1966. The regulation of density in a natural population of the pond snail, *Lymnaea elodes*. Ecology *47*:889–906.

Eklund, H. 1963. Fresh water: Temperature of maximum density calculated from compressibility. Science *142*:1457–1458.

Eklund, H. 1965. Stability of lakes near the temperature maximum density. Science 149:632–633.

Elakovich, S. D. and J. W. Wooten. 1995. Allelopathic, herbaceous, vascular hydrophytes. *In* Inderjit, K. M. M. Dakshini, and F. A. Einhellig, eds. Allelopathy: Organisms, Processes, and Applications. American Chemical Society, Washington. pp. 58–73.

El-Daoushy, F. 1988. A summary on the lead–210 cycle in nature and related applications in Scandinavia. Environ. Internat. *14*:305–319.

Elert, E. V. and C. J. Loose. 1996. Predator-induced diel vertical migration in *Daphnia*: Enrichment and preliminary chemical characterization of a kairomone exuded by fish. J. Chem. Ecol. *22*:885–895.

Elgmork, K. 1959. Seasonal occurrence of *Cyclops strenuus strenuus* in relation to environment in small water bodies in southern Norway. Folia Limnol. Scandinavica *11*, 196 pp.

Elgmork, K. 1980. Evolutionary aspects of diapause in freshwater copepods. *In* W. C. Kerfoot, ed. Evolution and Ecology of Zooplankton Communities. Univ. Press of New England, Hanover. pp. 411–417.

Elgmork, K. 1985. Prolonged life cycles in the planktonic copepod *Cyclops scutifer* Sars. Verh. Internat. Verein. Limnol. 22:3154–3158.

Elgmork, K. 1990. Coexistence with similar life cycles in two species of freshwater copepods (Crustacea). Hydrobiologia *208*: 187–199.

Elgmork, K. and A. Langeland. 1980. *Cyclops scutifer* Sars: One and two-year life cycles with diapause in the meromictic Lake Blankvatn. Arch. Hydrobiol. *88*:178–201.

Elgmork, K. and J. P. Nilssen. 1978. Equivalence of copepod and insect diapause. Verh. Internat. Verein. Limnol. *20*:2511–2517.

Elliott, J. I. 1977. Seasonal changes in the abundance and distribution of planktonic rotifers in Grasmere (English Lake District). Freshwat. Biol. *7*:147–166.

Elliott, J. K., J. M. Elliott, and W. C. Leggett. 1997. Predation by *Hydra* on larval fish: Field and laboratory experiments with bluegill (*Lepomis macrochirus*). Limnol. Oceanogr. *42*: 1416–1423.

Elliott, J. M. 1969. Diel periodicity in invertebrate drift and the effect of different sampling periods. Oikos *20*:524–528.

Elliott, J. M. 1973a. The diel activity pattern, drifting and food of the leech *Erpobdella octoculata* (L.) (Hirudinea: Erpobdellidae) in a Lake District stream. J. Anim. Ecol. *42*:449–459.

Elliott, J. M. 1973b. The life cycle and production of the leech *Erpobdella octoculata* (L.) (Hirudinea: Erpobdellidae) in a Lake District stream. J. Anim. Ecol. *42*:435–448.

Elliott, J. M. 1977. Some methods for the statistical analysis of samples of benthic invertebrates. 2nd ed. Sci. Publ. Freshwat. Biol. Assoc. *25*, 156 pp.

Ellis, B. K. and J. A. Stanford. 1982. Comparative photoheterotrophy, chemoheterotrophy, and photolithotrophy in a eutrophic reservoir and an oligotrophic lake. Limnol. Oceanogr. *27*:440–454.

Ellis, B. K., J. A. Stanford, and J. V. Ward. 1998. Microbial assemblages and production in alluvial aquifers of the Flathead River, Montana, USA. J. N. Am. Benthol. Soc. *17*:382–402.

Ellis, C. R., H. G. Stefan, and R. Gu. 1991. Water temperature dynamics and heat transfer beneath the ice cover of a lake. Limnol. Oceanogr. *36*:324–335.

Ellis, J. and S. Kanamori. 1973. An evaluation of the Miller method for dissolved oxygen analysis. Limnol. Oceanogr. *18*: 1002–1005.

Ellis, P. and W. D. Williams. 1970. The biology of *Haloniscus searlei* Chilton, an oniscoid isopod living in Australian salt lakes. Austral. J. Mar. Freshw. Res. *21*:51–69.

Elmore, J. L. 1983. The influence of temperature on egg development times of three species of *Diaptomus* from subtropical Florida. Am. Midland Nat. *109*:300–308.

Elner, J. K. and C. M. Happey-Wood. 1978. Diatom and chrysophycean cyst profiles in sediment cores from two linked but contrasting Welch lakes. Brit. Phycol. J. *13*:341–360.

Elser, J. J. and N. B. George. 1993. The stoichiometry of N and P in the pelagic zone of Castle Lake, California. J. Plankton Res. *15*:977–992.

Elser, J. J. and S. R. Carpenter. 1988. Predation-driven dynamics of zooplankton and phytoplankton communities in a whole-lake experiment. Oecologia *76*:148–154.

Elser, J. J., E. R. Marzolf, and C. R. Goldman. 1990. Phosphorus and nitrogen limitation of phytoplankton growth in the freshwaters (*sic*) of North America: A review and critique of experimental enrichments. Can. J. Fish. Aquat. Sci. *47*:1468–1477.

Elser, J. J., L. B. Stabler, and R. P. Hassett. 1995a. Nutrient limitation of bacterial growth and rates of bacterivory in lakes and oceans: A comparative study. Aquat. Microb. Ecol. *9*:105–110.

Elser, J. J., N. C. Goff, N. A. MacKay, A. L. St. Amand, M. M. Elser, and S. R. Carpenter. 1987. Species-specific algal responses to zooplankton: Experimental and field observations in three nutrient-limited lakes. J. Plankton Res. *9*:699–717.

Elser, J. J., T. H. Chrzanowski, R. W. Sterner, J. H. Schampel, and D. K. Foster. 1995b. Elemental ratios and the uptake and release of nutrients by phytoplankton and bacteria in three lakes of the Canadian Shield. Microb. Ecol. *29*:145–162.

Elster, H.-J. 1954a. Einige Gedanken zur Systematik, Terminologie und Zielsetzung der dynamischen Limnologie. Arch. Hydrobiol./Suppl. *20*:487–523.

Elster, H.-J. 1954b. über die Populationsdynamik von *Eudiaptomus gracilis* Sars und *Heterocope borealis* Fischer in Bodensee-Obersee. Arch. Hydrobiol./Suppl. *20*:546–614.

Elster, H.-J. and M. Stepánek. 1967. Eine neue Modifikation der Secchischeibe. Arch. Hydrobiol./Suppl. *33*:101–106.

Elwood, J. W. and D. Nelson. 1972. Periphyton production and grazing rates in a stream measured with a P^{32} material balance method. Oikos *23*:295–303.

Emerson, D. and N. P. Revsbech. 1994. Investigation of an iron-oxidizing microbial mat community located near Aarhus, Denmark: Field studies. Appl. Environ. Microbiol. *60*: 4022–4031.

Emerson, D., J. V. Weiss, and J. P. Megonigal. 1999. Iron-oxidizing bacteria are associated with ferric hydroxide precipitates (Fe-plaque) of wetland plants. Appl. Environ. Microbiol. *65*:2758–2761.

Emerson, K. R. C. Russo, R. E. Lund, and R. V. Thurston. 1975. Aqueous ammonia equilibrium calculations: Effect of pH and temperature. J. Fish. Res. Board Can. *32*:2379–2383.

Emerson, S., W. Broecker, and D. W. Schindler. 1973. Gas-exchange rates in a small lake as determined by the radon method. J. Fish. Res. Board Can. *30*:1475–1484.

Emerson, S. 1975. Chemically enhanced CO_2 gas exchange in a eutrophic lake: A general model. Limnol. Oceanogr. *20*:743–753.

Eminson, D. and B. Moss. 1980. The composition and ecology of periphyton communities in freshwaters. I. The influence of host type and external environment on community composition. Br. Phycol. J. *15*:429–446.

von Ende, C. N. 1979. Fish predation, interspecific predation, and the distribution of two *Chaoborus* species. Ecology *60*:119–128.

von Ende, C. N. and D. O. Dempsey. 1981. Apparent exclusion of the cladoceran *Bosmina longirostris* by the invertebrate predator *Chaoborus americanus*. Am. Midland Nat. *105*:240–248.

von Engeln, C. D. 1961. The Finger Lakes Region: Its Origin and Nature. Cornell University Press, Ithaca, NY. 156 pp.

Engle, D. L. and J. M. Melack. 1993. Consequences of riverine flooding for seston and the periphyton of floating meadows in an Amazon floodplain lake. Limnol. Oceanogr. *38*:1500–1520.

Englert, J. P. and K. M. Stewart. 1983. Natural short-circuiting of inflow to outflow through Silver Lake, New York. Water Resour. Res. *19*:529–537.

English, M. C. 1978. The magnitude and significance of the terrestrial snowpack and white ice contribution to the phosphorus budget of a lake in the Canadian Shield Region of the Kawartha lakes. Proc. Eastern Snow Conf. *35*:173–189.

Englund, G. 1999. Effects of fish on the local abundance of crayfish in stream pools. Oikos *87*:48–56.

Engstrom, D. R. and H. E. Wright, Jr. 1984. Chemical stratigraphy of lake sediments as a record of environmental change. *In* E. Y. Haworth and J. W. G. Lund, eds. Lake Sediments and Environmental History. Univ. Minnesota Press, Minneapolis. pp. 11–67.

Enright, J. T. 1977a. Problems in estimating copepod velocity. Limnol. Oceanogr. *22*:160–162.

Enright, J. T. 1977b. Diurnal vertical migration: Adaptive significance and timing. Part 1. Selective advantage: A metabolic model. Limnol. Oceanogr. *22*:856–872.

Epp, R. W. and W. M. Lewis, Jr. 1984. Cost and speed of locomotion for rotifers. Oecologia *61*:289–292.

Eppley, R. W. 1962. Major cations. *In* R. A. Lewin, ed. Physiology and Biochemistry of Algae. Academic Press, New York. pp. 255–266.

Eppley, R. W. and F. M. Macias. 1963. Role of the alga *Chlamydomonas mundana* in anaerobic waste stabilization lagoons. Limnol. Oceanogr. *8*:411–416.

Eppley, R. W. and W. H. Thomas. 1969. Comparison of half-saturation constants for growth and nitrate uptake of marine phytoplankton. J. Phycol. *5*:375–379.

Epstein, E. 1965. Mineral metabolism. *In* J. Bonner and J. E. Varner, eds. Plant Biochemistry. Academic Press, New York. pp. 438–466.

Eriksen, C. H. 1963a. Respiratory regulation in *Ephemera simulans* Walker and *Hexagenia limbata* (Serville) (Ephemeroptera). J. Exp. Biol. *40*:455–467.

Eriksen, C. H. 1963b. The relation of oxygen consumption to substrate particle size in two burrowing mayflies. J. Exp. Biol. *40*:447–453.

Eriksen, C. H. 1968. Aspects of the limno-ecology of *Corophium spinicorne* Stimpson (Amphipoda) and *Gnorimosphaeroma oregonensis* (Dana) (Isopoda). Crustaceana *14*:1–12.

Eriksen, C. H., G. A. Lamberti, and V. H. Resh. 1992. Aquatic insect respiration. *In* R. W. Merritt and K. W. Cummins, eds. An Introduction to the Aquatic Insects of North America. Kendall/Hunt, Dubuque, IA. pp. 29–39.

Erikson, R., K. Vammen, A. Zelaya, and R. T. Bell. 1998. Distribution and dynamics of bacterioplankton production in a polymictic tropical lake (Lago Xolotlán, Nicaragua). Hydrobiologia *382*:27–39.

Eriksson, C. and C. Pedrós-Alió. 1990. Selenium as a nutrient for freshwater bacterioplankton and its interactions with phosphorus. Can. J. Microbiol. *36*:475–483.

Eriksson, P. G. and S. E. B. Weisner. 1996. Functional differences in epiphytic microbial communities in nutrient-rich freshwater ecosystems: An assay of denitrifying capacity. Freshwat. Biol. *36*:555–562.

Ertel, J. R., J. I. Hedges, A. H. Devol, J. R. Richey, and M. N. G. Ribeiro. 1986. Dissolved humic substances of the Amazon River system. Limnol. Oceanogr. *31*:739–754.

Ervin, G. N. and R. G. Wetzel. 1997. Shoot:root dynamics during growth stages of the aquatic rush *Juncus effusus* L. Aquat. Bot. *59*:63–73.

Ervin, G. N. and R. G. Wetzel. 2000. Allelochemical autotoxicity in the emergent wetland macrophyte *Juncus effusus* (Juncaceae). Am. J. Bot. *87*:853–860.

Espie, G. S., A. G. Miller, R. A. Kandasamy, and D. T. Canvin. 1991. Active HCO_3^- transport in cyanobacteria. Can. J. Bot. *69*:936–944.

Esteves, F. A. 1979a. Die Bedeutung der aquatischen Makrophyten für den Stoffhaushalt des Schöhsees. I. Die Produktion an Biomasse. Arch. Hydrobiol./Suppl. *57*:117–143.

Esteves, F. A. 1979b. Die Bedeutung der aquatischen Makrophyten für den Stoffhaushalt des Schöhsees. II. Die organischen Hauptbestandteile und der Energiegehalt der aquatischen Makrophyten. Arch. Hydrobiol./Supp. *57*:144–187.

Eugster, H. P. and L. A. Hardie. 1978. Saline lakes. *In* A. Lerman, ed. Lakes: Chemistry, Geology, Physics. Springer-Verlag, New York. pp. 237–293.

Evans, G. H. 1970. Pollen and diatom analyses of late-Quaternary deposits in the Blelham Basin, North Lancashire. New Phytol. *69*:821–874.

Evans, H. E. 1984. A test of the ^{210}Pb dating method: A comparison of the CRS and CIC models. Verh. Internat. Verein. Limnol. *22*:338–344.

Evans, J. H. 1961. Growth of Lake Victoria phytoplankton in enriched cultures. Nature *189*:417.

Evans, J. H. 1962. The distribution of phytoplankton in some central East African waters. Hydrobiologia *19*:299–315.

Evans, M. S., M. T. Arts, and R. D. Robarts. 1996. Algal productivity, algal biomass, and zooplankton biomass in a phosphorus-rich, saline lake: Deviations from regression model predictions. Can. J. Fish. Aquat. Sci. *53*:1048–1060.

Evans, R. D. 1991. The impact of sediment focusing on total residual ^{210}Pb: Implications for choice of a dating model. Verh. Internat. Verein. Limnol. *24*:2335–2339.

Evans, R. D. 1994. Empirical evidence of the importance of sediment resuspension in lakes. Hydrobiologia *284*:5–12.

Fahnenstiel, G. L., L. Sicko-Goad, D. Scavia, and E. F. Stoermer. 1986. Importance of picoplankton in Lake Superior. Can. J. Fish. Aquat. Sci. *43*:235–240.

Fahnenstiel, G. L., T. B. Bridgeman, G. A. Lang, M. J. McCormick, and T. F. Nalepa. 1995. Phytoplankton productivity in Saginaw Bay, Lake Huron: Effects of zebra mussel (*Dreissena polymorpha*) colonization. J. Great Lakes Res. *21*:465–475.

Fairchild, G. W. 1981. Movement and microdistribution of *Sida crystallina* and other littoral microcrustacea. Ecology *62*:1341–1352.

Fairchild, G. W. and J. W. Sherman. 1993. Algal periphyton response to acidity and nutrients in softwater lakes: Lake comparison vs. nutrient enrichment approaches. J. North Am. Benthol. Soc. *12*:157–167.

Fairchild, W., R. L. Lowe, and W. B. Richardson. 1985. Algal periphyton growth on nutrient-diffusing substrates: An *in situ* bioassay. Ecology *66*:465–472.

Falkner, G., F. Horner, and W. Simonis. 1980. The regulation of the energy-dependent phosphate uptake by blue-green alga *Anacystis nidulans*. Planta *149*:138–143.

Falkowski, P. G. and J. A. Raven. 1997. Aquatic Photosynthesis. Blackwell Sciences, Oxford. 375 pp.

Faller, A. J. 1969. The generation of Langmuir circulations by the eddy pressure of surface waves. Limnol. Oceanogr. *14*:504–513.

Faller, A. J. and E. A. Caponi. 1978. Laboratory studies of wind-driven Langmuir circulations. J. Geophys. Res. *83*:3617–3633.

Fallon, R. D. and T. D. Brock. 1981. Overwintering of *Microcystis* in Lake Mendota. Freshwat. Biol. *11*:217–226.

Fallon, R. D., S. Harrits, R. S. Hanson, and T. D. Brock. 1980. The role of methane in internal carbon cycling in Lake Mendota during summer stratification. Limnol. Oceanogr. *25*:357–360.

Fang, X. and H. G. Stefan. 1996. Dynamics of heat exchange between sediment and water in a lake. Wat. Resour. Res. *32*:1719–1727.

Farmer, E. E., H. Weber, and S. Vollenweider. 1998. Fatty acid signaling in *Arabidopsis*. Planta *206*:167–174.

Faust, B. C. 1994. A review of the photochemical redox reactions of iron(III) species in atmospheric, oceanic, and surface waters: Influences on geochemical cycles and oxidant formation. *In* G. R. Helz, R. G. Zepp, and D. G. Crosby, eds. Aquatic and Surface Photochemistry. Lewis Publishers, Boca Raton. pp. 3–37.

Faust, S. J. and J. V. Hunter, eds. 1971. Organic Compounds in Aquatic Environments. Marcel Dekker, Inc., New York. 638 pp.

Fay, P. 1992. Oxygen relations of nitrogen fixation in cyanobacteria. Microbiol. Rev. *56*:340–373.

Feehan, J. and G. O'Donovan. 1996.The Bogs of Ireland: An Introduction to the Natural, Cultural and Industrial Heritage of Irish Peatlands. University College, Dublin. 518 pp.

Feijoó, C. S., G. A. Ferreyra, N. M. Tur, and F. R. Momo. 1994. Influence of a macrophyte bed on sediment deposition in a small plain stream. Verh. Internat. Verein. Limnol. *25*:1888–1892.

Felföldy, L. J. M. 1960. Experiments on the carbonate assimilation of some unicellular algae by Ruttner's conductiometric method. Act. Biol. Hung. *11*:67–75.

Felföldy, L. J. M. 1961. On the chlorophyll content and biological productivity of periphytic diatom communities on the stony shores of Lake Balaton. Ann. Biol. Tihany *28*:99–104.

Felip, M., M. L. Pace, and J. J. Cole. 1996. Regulation of planktonic bacterial growth rates: The effects of temperature and resources. Microb. Ecol. *31*:15–28.

Feminella, J. W. and V. H. Resh. 1991. Herbivorous caddisflies, macroalgae, and epilithic microalgae: Dynamic interactions in a stream grazing system. Oecologia *87*:247–256.

Feminella, J. W., M. E. Power, and V. H. Resh. 1989. Periphyton responses to invertebrate grazing and riparian canopy in three northern California coastal streams. Freshwat. Biol. *22*: 445–457.

Fenchel, T. 1986. The ecology of heterotrophic microflagellates. Adv. Microb. Ecol. *9*:57–97.

Fenchel, T. 1987. Ecology of Protozoa: The Biology of Free-living Phagotrophic Protists. Springer-Verlag, Berlin. 197 pp.

Fenchel, T. 1991. Flagellate design and function. *In* D. J. Patterson and J. Larsen, eds. The Biology of Free-living Heterotrophic Flagellates. Claredon Press, Oxford. pp. 7–19.

Fenchel, T. and T. H. Blackburn. 1979. Bacteria and mineral cycling. Academic Press, New York. 225 pp.

Fennessy, M. S. and J. K. Cronk. 1997. The effectiveness and restortion potential of riparian ecotones for the management of nonpoint pollution, particularly nitrate. Crit. Rev. Environ. Sci. Technol. *27*:285–317.

Ferling, E. 1957. Die Wirkungen des erhöhten hydrostatischen Druckes auf Wachstum und Differenzierung submerser Blütenpflanzen. Planta *49*:235–270.

Fernando, C. H. 1994. Zooplankton, fish and fisheries in tropical freshwaters (*sic*). Hydrobiologia *272*:105–123.

Fernando, C. H. and D. Galbraith. 1973. Seasonality and dynamics of aquatic insects colonizing small habitats. Verh. Internat. Verein. Limnol. *18*:1564–1575.

Ferrante, J. G. 1976. The role of zooplankton in the intrabiocoenotic phosphorus cycle and factors affecting phosphorus excretion in a lake. Hydrobiologia *49*:203–214

Ferrante, J. G. and J. I. Parker. 1978. The influence of planktonic and benthic crustaceans on silicon cycling in Lake Michigan, U.S.A. Verh. Internat. Verein. Limnol. *20*:324–328.

Ferreira, F. and N. A. Straus. 1994. Iron deprivation in cyanobacteria. J. Appl. Phycol. *6*:199–210.

Ferrier-Pagès, C. and F. Rassoulzadegan. 1994. N remineralization in planktonic protozoa. Limnol. Oceanogr. *39*:411–419.

Ferrier-Pagès, C., M. Karner, and F. Rassoulzadegan. 1998. Release of dissolved amino acids by flagellates and ciliates grazing on bacteria. Oceanol. Acta *21*:485–494.

Feth, J. H. 1971. Mechanisms controlling world water chemistry: Evaporation-crystallization process. Science *172*:870.

Feuillade, M., P. Defour, and J. Feuillade. 1988. Organic carbon release by phytoplankton and bacterial assimilation. Schweiz. Z. Hydrol. *50*:115–135.

Fiala, K. 1971. Seasonal changes in the growth of clones of *Typha latifolia* L. in natural conditions. Folia Geobot. Phytotax. Praha *6*:255–270.

Fiala, K. 1973. Growth and production of underground organs of *Typha angustifolia* L., *Typha latifolia* L. and *Phragmites communis* Trin. Pol. Arch. Hydrobiol. *20*:59–66.

Fiala, K. 1976. Underground organs of *Phragmites communis*, their growth, biomass and net production. Folia Geobot. Phytotas. Praha *11*:225–259.

Fiala, K. 1978. Underground organs of *Typha angustifolia* and *Typha latifolia,* their growth, propagation and production. Acta Sci. Nat. Brno *12*(8):1–43.

Fiala, K., D. Dykyjová, J. Květ, and J. Svoboda. 1968. Methods of assessing rhizome and root production in reed-bed stands. *In* Methods of Productivity Studies in Root Systems and Rhizosphere Organisms. Publishing House Nauka, Leningrad. pp. 36–47.

Ficke, E. R. and J. F. Ficke. 1977. Ice on Rivers and Lakes: A Bibliographic Essay. U.S. Geol. Surv. Water-Resources Invest. Report *75–95*. 173 pp.

Fiebig, D. M. 1992. Fates of dissolved free amino acids in groundwater discharged through stream bed sediments. Hydrobiologia *235/236*:311–319.

Fiebig, D. M. and M. A. Lock. 1991. Immobilization of dissolved organic matter from groundwater discharging through the stream bed. Freshwat. Biol. *26*:45–55.

Filbin, G. J. and R. A. Hough. 1985. Photosynthesis, photorespiration and productivity in *Lemna minor* L. Limnol. Oceanogr. *30*:322–334.

Findenegg, I. 1935. Limnologische Untersuchungen im Kärntner Seengebiete. Ein Beitrag zur Kenntnis des Stoffhaushaltes in Alpenseen. Int. Rev. ges. Hydrobiol. *32*:369–423.

Findenegg, I. 1937. Holomiktische und meromiktische Seen. Int. Rev. ges. Hydrobiol. *35*:586–610.

Findenegg, I. 1943. Untersuchungen über die Ökologie und die Produktionsverhältnisse des Planktons im Kärntner Seengebiete. Int. Rev. ges. Hydrobiol. *43*:368–429.

Findenegg, I. 1965. Relationship between standing crop and primary productivity. Mem. Ist. Ital. Idrobiol. *18*(Suppl.):271–289.

Findlay, S. and R. L. Sinsabaugh. 1999. Unravelling the sources and bioavailability of dissolved organic matter in lotic aquatic ecosystems. Mar. Freshwat. Res. *50*:781–790.

Findlay, S. and W. V. Sobczak. 1996. Variability in removal of dissolved organic carbon in hyporheic sediments. J. N. Am. Benthol. Soc. *15*:35–41.

Findlay, S., D. Strayer, C. Goumbala, and K. Gould. 1993. Metabolism of streamwater dissolved organic carbon in the shallow hyporheic zone. Limnol. Oceanogr. *38*:1493–1499.

Findlay, S., J. L. Meyer, and P. J. Smith. 1984. Significance of bacterial biomass in the nutrition of a freshwater isopod (*Lirceus* sp.). Oecologia *63*:38–42.

Findlay, S., J. L. Meyer, and P. J. Smith. 1986a. Contribution of fungal biomass to the diet of a freshwater isopod (*Lirceus* sp.). Freshwat. Biol. *16*:377–385.

Findlay, S., J. L. Meyer, and R. Risley. 1986b. Benthic bacterial biomass and production in two blackwater rivers. Can. J. Fish. Aquat. Sci. *43*:1271–1276.

Findlay, S., L. Carlough, M. T. Crocker, H. K. Gill, J. L. Meyer, and P. J. Smith. 1986c. Bacterial growth on macrophyte leachate and fate of bacterial production. Limnol. Oceanogr. *31*:1335–1341.

Finke, L. R. and H. W. Seeley, Jr. 1978. Nitrogen fixation (acetylene reduction) by epiphytes of freshwater macrophytes. Appl. Environ. Microbiol. *36*:129–138.

Finlay, B. J. 1980. Temporal and vertical distribution of ciliophoran communities in the benthos of a small eutrophic lock with particular reference to the redox profile. Freshwat. Biol. *10*:15–34.

Finlay, B. J. 1981. Oxygen availability and seasonal migrations of ciliated protozoa in a freshwater lake. J. Gen. Microbiol. *123*:173–178.

Finlay, B. J. 1990. Physiological ecology of free-living protozoa. Adv. Microb. Ecol. *11*:1–35.

Finlay, B., P. Bannister, and J. Stewart. 1979. Temporal variation in benthic ciliates and the application of association analysis. Freshwat. Biol. *9*:45–53.

Finlay, B. J., T. Fenchel, and S. Gardener. 1986. Oxygen perception and O_2 toxicity in the freshwater ciliated protozoan *Loxodes*. J. Protozool. *33*:157–165.

Firth, P. and S. G. Fisher, eds. 1992. Global Climate Change and Freshwater Ecosystems. Springer-Verlag, New York. 321 pp.

Fischer, Z. 1970. Elements of energy balance in grass carp *Ctenopharyngodon idella* Val. Pol. Arch. Hydrobiol. *17*: 421–434.

Fish, G. R. 1956. Chemical factors limiting growth of phytoplankton in Lake Victoria. East African Agricult. J. *21*:152–158.

Fisher, D. W., A. W. Gambell, G. E. Likens, and F. H. Bormann. 1968. Atmospheric contributions to water quality of streams in the Hubbard Brook Experimental Forest, New Hampshire. Water Resour. Res. *4*:1115–1126.

Fisher, S. G. 1977. Organic matter processing by a stream-segment ecosystem: Fort River, Massachusetts, U.S.A. Int. Rev. ges. Hydrobiol. *62*:701–727.

Fisher, S. G. 1990. Recovery processes in lotic ecosystems: Limits of successional theory. Environ. Manage. *14*:725–736.

Fisher, S. G. and G. E. Likens. 1973. Energy flow in Bear Brook, New Hampshire: An integrative approach to stream ecosystem metabolism. Ecol. Monogr. *43*:421–439.

Fitzgerald, G. P. 1969. Some factors in the competition or antagonism among bacteria, algae and aquatic weeds. J. Phycol. *5*:351–359.

Fitzgerald, G. P. 1972. Bioassay analysis of nutrient availability. *In* H. E. Allen, and J. R. Kramer, eds. Nutrients in Natural Waters. John Wiley & Sons, New York. pp. 147–169.

Fjellheim, A. 1980. Differences in drifting of larval stages of *Ryacophila nubila* (Trichoptera). Holarctic Ecol. *3*:99–103.

Fjerdingstad, E. 1979. Sulfur Bacteria. Tech. Publ. 650, Am. Soc. Testing Materials, Philadelphia. 121 pp.

Flaig, W. 1964. Effects of micro-organisms in the transformation of lignin to humic substances. Geochim. Cosmochim. Acta *28*:1523–1535.

Flecker, A. S. 1992. Fish predation and the evolution of invertebrate drift periodicity: Evidence from neotropical streams. Ecology *73*:438–448.

Fleischer, S. 1978. Evidence for the anaerobic release of phosphorus from lake sediments as a biological process. Naturwissenschaften *65*:109.

Fleischer, S. 1986. Aerobic uptake of Fe(III)-precipitated phosphorus by microorganisms. Arch. Hydrobiol. *107*:269–277.

Flessa, H. 1994. Plant-induced changes in the redox potential of the rhizospheres of the submerged vascular macrophytes *Myriophyllum verticillatum* L. and *Ranunculus circinatus* L. Aquat. Bot. *47*:119–129.

Fletcher, A. R., A. K. Morison, and D. J. Hume. 1985. Effects of carp, *Cyprinus carpio* L., on communities of aquatic vegetation and turbidity of waterbodies in the Lower Goulburn River Basin. Aust. J. Mar. Freshwat. Res. *36*:311–327.

Fletcher, M. and G. D. Floodgate. 1973. An electron-microscopic demonstration of an acidic polysaccharide involved in the adhesion of a marine bacterium to solid surfaces. J. Gen. Microbiol. *74*:325–334.

Fletcher, M. and K. C. Marshall. 1982. Are solid surfaces of ecological significance to aquatic bacteria? Adv. Microb. Ecol. *6*:199–236.

Flett, R. J., D. W. Schindler, R. D. Hamilton, and N. E. R. Campbell. 1980. Nitrogen fixation in Canadian Precambrian Shield lakes. Can. J. Fish. Aquat. Sci. *37*:494–505.

Flint, R. W. and C. R. Goldman. 1975. The effects of a benthic grazer on the primary productivity of the littoral zone of Lake Tahoe. Limnol. Oceanogr. *20*:935–944.

Flint, R. W. and C. R. Goldman. 1977. Crayfish growth in Lake Tahoe: Effects of habitat variation. J. Fish. Res. Board Can. *34*:155–159.

Flohn, H. 1973. Der Wasserhaushalt der Erde Schwankungen und Eingriffe. Naturwissenschaften *60*:340–348.

Florin, M.-B. and H. E. Wright, Jr. 1969. Diatom evidence for the persistence of stagnant glacial ice in Minnesota. Bull. Geol. Soc. Amer. *80*:695–704.

Flössner, D. 1982. Untersuchungen zur Biomasse und Produktion des Makrozoobenthos der Ilm und mittleren Saale. Limnologica *14*:297–327.

Flower, R. J. 1991. Field calibration and performance of sediment traps in a eutrophic holomictic lake. J. Paleolimnol. *5*:175–188.

Fluhr, R. and A. K. Mattoo. 1996. Ethylene – biosynthesis and perception. Crit. Rev. Plant Sci. *15*:479–523.

Fogg, G. E. 1963. The role of algae in organic production in aquatic environments. Brit. Phycol. Bull. *2*:195–205.

Fogg, G. E. 1965. Algal Cultures and Phytoplankton Ecology. University of Wisconsin Press, Madison. 126 pp.

Fogg, G. E. 1971a. Nitrogen fixation in lakes. Plant Soil (Spec. Vol.) *1971*:393–401.

Fogg, G. E. 1971b. Extracellular products of algae in freshwater. Arch. Hydrobiol. Beih. Ergebn. Limnol. *5*, 25 pp.

Fogg, G. E. 1974. Nitrogen fixation. *In* W. D. P. Stewart, ed. Algal Physiology and Biochemistry. Univ. California Press, Berkeley. pp. 560–582.

Fogg, G. E. 1983. The ecological significance of extracellular products of phytoplankton photosynthesis. Bot. Mar. *26*:3–14.

Fogg, G. E. 1991. The phytoplanktonic ways of life. New Phytol. *118*:191–232.

Fogg, G. E. 1995. Some comments on picoplankton and its importance in the pelagic ecosystem. Aquat. Microb. Ecol. *9*:33–39.

Fogg, G. E. and A. E. Walsby. 1971. Buoyancy regulation and the growth of planktonic blue-green algae. Mitt. Internat. Verein. Limnol. *19*:182–188.

Fogg, G. E. and D. F. Westlake. 1955. The importance of extracellular products of algae in freshwater. Verh. Internat. Verein. Limnol. *12*:219–232.

Fogg, G. E. and J. H. Belcher. 1961. Pigments from the bottom deposits of an English lake. New Phytol. *60*:129–142.

Fogg, G. E., W. D. P. Stewart, P. Fay, and A. E. Walsby. 1973. The Blue-Green Algae. Academic Press, New York. 459 pp.

Foissner, W. 1988. Taxonomic and nomenclatural revision of Sládeček's list of ciliates (Protozoa: Ciliophora) as indicators of water quality. Hydrobiologia *166*:1–64.

Foissner, W., H. Berger, and F. Kohmann. 1992. Taxonomische und ökologische Revision der Ciliaten des Saprobiensystems. II. Peritrichia, Heterotrichida, Odontostomatida. Bayerisches Landesamt Wasserwirtschaft, München. 502 pp.

Foissner, W., H. Berger, and F. Kohmann. 1994. Taxonomische und ökologisches Revision der Ciliaten des Saprobiensystems. III. Hymenostomata, Prostomatida, Nassulida. Bayerisches Landesamt Wasserwirtschaft, München. 548 pp.

Foissner, W., H. Berger, H. Blatterer, and F. Kohmann. 1995. Taxonomische und ökologische Revision der Ciliaten des Saprobiensystems. IV. Gymnostomatea, *Loxodes*, Suctoria. Bayerisches Landesamt Wasserwirtschaft, München. 540 pp.

Foissner, W., H. Blatterer, H. Berger, and F. Kohmann. 1991. Taxonomische und ökologische Revision der Ciliaten des Saprobiensystems. I. Cyrtophorida, Oligotrichida, Hypotrichia, Colpodea. Bayerisches Landesamt Wasserwirtschaft, München. 478 pp.

Folsom, T. C. and H. F. Clifford. 1978. The population biology of *Dugesia tigrina* (Platyhelminthes: Turbellaria) in a thermally enriched Alberta, Canada lake. Ecology *59*:966–975.

Folt, C. and C. R. Goldman. 1981. Allelopathy between zooplankton: A mechanism for interference competition. Science *213*:1133–1135.

Forbes, S. A. 1887. The lake as a microcosm. Bull. Peoria (Ill.) Sci. Assoc. 1887. Reprinted in Bull. Ill. Nat. Hist. Surv. *15*:537–550 (1925).

Ford, T. E. and R. J. Naiman. 1989. Groundwater-surface water relationships in boreal forest watersheds: Dissolved organic carbon and inorganic nutrient dynamics. Can. J. Fish. Aquat. Sci. *46*:41–49.

Forel, F.-A. 1892. Le Léman: Monographie Limnologique. Tome I. Géographie, Hydrographie, Géologie, Climatologie, Hydrologie, Lausanne, F. Rouge, 543 pp. (Reprinted Genève, Slatkine Reprints, 1969.)

Forel, F.-A. 1895. Le Léman: Monographie Limnologique. Tome II. Mécanique, Hydraulique, Thermique, Optique, Acoustique, Chemie, Lausanne, F. Rouge, 651 pp. (Reprinted Genève, Slatkine Reprints, 1969.)

Forel, F.-A. 1904. Le Léman: Monographie Limnologique. Tome III. Biologie, Histoire, Navigation, Peche. Lausanne, F. Rouge, 715 pp. (Reprinted Genève, Slatkine Reprints, 1969.)

Forsberg, C. and S. O. Ryding. 1980. Eutrophication parameters and trophic state indices in 30 Swedish waste-receiving lakes. Arch. Hydrobiol. *89*:189–207.

Forsberg, C., S. Kleiven, and T. Willén. 1990. Absence of allelopathic effects of *Chara* on phytoplankton in situ. Aquat. Bot. *38*:289–294.

Forsyth, D. J. and M. R. James. 1984. Zooplankton grazing on lake bacterioplankton and phytoplankton. J. Plankton Res. *6*:803–810.

Forsyth, D. J. and M. R. James. 1991. Population dynamics and production of zooplankton in eutrophic Lake Okaro, North Island, New Zealand. Arch. Hydrobiol. *120*:287–314.

Fott, B. 1954. Ein interessanter Fall der Neustonbildung und deren Bedeutung für die Produktionsbiologie des Teiches. Preslia *26*:95–104.

Fox, L. E. 1993. The chemistry of aquatic phosphate: Inorganic processes in rivers. Hydrobiologia *253*:1–16.

Foy, R. H. 1986. Suppression of phosphorus release from lake sediments by the addition of nitrate. Water Res. *20*:1345–1351.

Francisco, D. E., R. A. Mah, and A. C. Rabin. 1973. Acridine orange-epifluorescence technique for counting bacteria in natural waters. Trans. Amer. Microsc. Soc. *92*:416–421.

Francko, D. A. 1986. Epilimnetic phosphorus cycling: Influence of humic materials and iron on coexisting major mechanisms. Can. J. Fish. Aquat. Sci. *43*:302–310.

Francko, D. A. and R. G. Wetzel. 1980. Cyclic adenosine-3′:5′-monophosphate: Production and extracellular release from green and blue-green algae. Physiol. Plant. *49*:65–67.

Francko, D. A. and R. G. Wetzel. 1981. Dynamics of cellular and extracellular cAMP in *Anabaena flos-aquae* (Cyanophyta): Intrinsic culture variability and correlation with metabolic variables. J. Phycol. *17*:129–134.

Francko, D. A. and R. G. Wetzel. 1982. The isolation of cyclic adenosine 3′:5′-monophosphate (cAMP) from lakes of differing trophic status: Correlation with planktonic metabolic variables. Limnol. Oceanogr. *27*:27–38.

Francko, D. A. and R. G. Wetzel. 1983. To Quench Our Thirst. The Present and Future Status of Freshwater Resources of the United States. University of Michigan Press, Ann Arbor. 153 pp.

Francko, D. A. and R. T. Heath. 1979. Functionally distinct classes of complex phosphorus compounds in lake water. Limnol. Oceanogr. *24*:463–473.

Francko, D. A. and R. T. Heath. 1982. UV-sensitive complex phosphorus: Association with dissolved humic material and iron in a bog lake. Limnol. Oceanogr. *27*:564–569.

Francoeur, S. N. and R. G. Wetzel. 2000. Detecting protease activity in biofilms. Appl. Environ. Microbiol. (Submitted.)

Francoeur, S. N. and R. L. Lowe. 1998. Effects of ambient ultraviolet radiation on littoral periphyton: Biomass accrual and taxon-specific responses. J. Freshwat. Ecol. *13*:29–37.

Francoeur, S. N., B. J. F. Biggs, and R. L. Lowe. 1998. Microform bed clusters as refugia for periphyton in a flood-prone headwater stream. N. Z. J. Mar. Freshwat. Res. *32*:363–374.

Frank, P. W. 1952. A laboratory study of intraspecific and interspecific competition in *Daphnia pulicaria* (Forbes) and *Simocephalus vetulus* O. F. Müller. Physiol. Zool. *25*:178–204.

Frank, P. W. 1957. Coactions in laboratory populations of two species of *Daphnia*. Ecology *38*:510–519.

Franke, C. 1983. Transversal migration of 4th instar larvae and pupae of *Chaoborus flavicans* (Meigen, 1818) (Diptera, Chaoboridae) in Lake Heiligensee (Berlin-West). Arch. Hydrobiol. *99*:93–105.

Franke, C. 1987a. Diurnal and seasonal change of lactate, glucose, and glycogen in larvae and pupae of *Chaoborus flavicans* (Diptera, Chaoboridae). Arch. Hydrobiol. *110*:565–577.

Franke, C. 1987b. Detection of transversal migration of larvae of *Chaoborus flavicans* (Diptera, Chaoboridae) by the use of a sonar system. Arch. Hydrobiol. *109*:355–366.

Franke, U. 1977. Experimentelle Untersuchungen zur Respiration von *Gammarus fossarum* Koch 1835 (Crustacea-Amphipoda) in Abhängigkeit von Temperatur, Sauerstoffkonzentration und Wasserbewegung. Arch. Hydrobiol./Suppl. *48*:369–411.

Freeman, C. and M. A. Lock. 1995. The biofilm polysaccharide matrix: A buffer against changing organic substrate supply? Limnol. Oceanogr. *40*:273–278.

Freeman, C., P. J. Chapman, K. Gilman, M. A. Lock, B. Reynolds, and H. S. Wheater. 1995. Ion exchange mechanisms and the entrapment of nutrients by river biofilms. Hydrobiologia *297*:61–65.

Freeman, T. E. 1977. Biological control of aquatic weeds with plant pathogens. Aquat. Bot. *3*:175–184.

Freeze, R. A. and J. A. Cherry. 1979. Groundwater. Prentice-Hall, Inc., Englewood Cliffs, NJ. 604 pp.

Fregni, E., M. Balsamo, and P. Tongiorgi. 1998. Interstitial gastrotrichs from lotic Italian fresh waters. Hydrobiologia *368*:175–187.

Frenzel, P. 1979. Untersuchungen zur Biologie und Populationsdynamik von *Potamopyrgus jenkinsi* (Smith) (Gastropoda: Prosobranchia) im Litoral des Bodensees. Arch. Hydrobiol. *85*:448–464.

Frenzel, P. 1980. Die Produktion von *Potamopyrgus jenkinsi* Smith (Gastropoda, Prosobranchia) im Bodensee. Hydrobiologia *74*:141–144.

Frenzel, P., B. Thebrath, and R. Conrad. 1990. Oxidation of methane in the oxic surface layer of a deep lake sediment (Lake Constance). FEMS Microbiol. Ecol. *73*:149–158.

Frevert, T. 1979a. Phosphorus and iron concentrations in the interstitial water and dry substance of sediments of Lake Constance (Obersee). I. General discussion. Arch. Hydrobiol./Suppl. *55*:298–323.

Frevert, T. 1979b. The pE redox concept in natural sediment-water systems; its role in controlling phosphorus release from lake sediments. Arch. Hydrobiol./Suppl. *55*:278–297.

Frevert, T. 1980. Dissolved oxygen dependent phosphorus release from profundal sediments of Lake Constance (Obersee). Hydrobiologia *74*:17–28.

Frey, D. G. 1955a. Längsee: A history of meromixis. Mem. Ist. Ital. Idrobiol. *8*(Suppl.):141–164.

Frey, D. G. 1955b. Distributional ecology of the cisco (*Coregonus artedii*) in Indiana. Invest. Indiana Lakes Streams *4*:177–228.

Frey, D. G. 1964. Remains of animals in Quaternary lake and bog sediments and their interpretation. Arch. Hydrobiol. Beih. Ergebn. Limnol. 2, 114 pp.

Frey, D. G. 1969a. The rationale of paleolimnology. Mitt. Internat. Verein. Limnol. *17*:7–18.

Frey, D. G. 1969b. Evidence for eutrophication from remains of organisms in sediments. *In* Eutrophication: Causes, Consequences, Correctives. National Academy of Sciences, Washington, DC. pp. 594–613.

Frey, D. G. 1974. Paleolimnology. Mitt. Internat. Verein. Limnol. *20*:95–123.

Frey, D. G. 1976. Interpretation of Quaternary paleoecology from Cladocera and midges, and prognosis regarding usability of other organisms. Can. J. Zool. 54:2208–2226.

Frey, D. G. 1988. Littoral and offshore communities of diatoms, cladocerans and dipterous larvae, and their interpretation in paleolimnology. J. Paleolimnol. *1*:179–191.

Friedrich, C. G. 1998. Physiology and genetics of sulfur-oxidizing bacteria. Adv. Microb. Physiol. *39*:235–289.

Friedrich, G. and M. Viehweg. 1984. Recent developments of the phytoplankton and its activity in the Lower Rhine. Verh. Internat. Verein. Limnol. *22*:2029–2035.

Frimmel, F. H. and R. F. Christman, eds. 1988. Humic substances and their role in the environment. John Wiley & Sons, Chichester. 340 pp.

Frink, C. R. 1967. Nutrient budget: Rational analysis of eutrophication in a Connecticut lake. Environ. Sci. Technol. *1*:425–428.

Fritz, S. C., J. C. Kingston, and D. R. Engstrom. 1993. Quantitative trophic reconstruction from sedimentary diatom assemblages: A cautionary tale. Freshwat. Biol. *30*:1–23.

Frost, T. M. 1978. Impact of the freshwater sponge *Spongilla lacustris* on a *Sphagnum* bog-pond. Verh. Internat. Verein. Limnol. *20*:2368–2371.

Frost, T. M. 1980. Selection in sponge feeding processes. *In* D. C. Smith and Y. Tiffon, eds. Nutrition in Lower Metazoa. Pergamon, Oxford. pp. 33–44.

Frost, T. M. 1987. Porifera. *In* T. J. Pandian and J. F. Vernberg, eds. Animal Energetics. Vol. 1. Academic Press, New York. pp. 27–53.

Frost, T. M. 1991. Porifera. *In* J. H. Thorp and A. P. Covich, eds. Ecology and Classification of North American Freshwater Invertebrates. Academic Press, San Diego. pp. 95–124.

Frost, T. M., L. E. Graham, J. E. Elisa, M. J. Haase, D. W. Kretchmer, and J. A. Kranzfelder. 1997. A yellow-green algal symbiont in the freshwater sponge, *Corvomeyenia everetti*: Convergent evolution of symbiotic associations. Freshwat. Biol. *38*: 395–399.

Frost-Christensen, H. and K. Sand-Jensen. 1995. Comparative kinetics of photosynthesis in floating and submerged *Potamogeton* leaves. Aquat. Bot. *51*:121–134.

Fryer, G. 1954. Contributions to our knowledge of the biology and systematics of the freshwater Copepoda. Schweiz. Z. Hydrol. *16*:64–77.

Fryer, G. 1957a. The food of some freshwater cyclopoid copepods and its ecological significance. J. Anim. Ecol. *26*:263–286.

Fryer, G. 1957b. The feeding mechanism of some freshwater cyclopoid copepods. Proc. Zool. Soc. London *129*:1–25.

Fryer, G. 1996. Diapause, a potent force in the evolution of freshwater crustaceans. Hydrobiologia *320*:1–14.

Fryer, G. and W. J. P. Smyly. 1954. Some remarks on the resting stages of some freshwater cyclopoid and harpacticoid copepods. Ann. Mag. Nat. Hist. Ser.*12*:65–72.

Fuhrman, J. A. 2000. Impact of viruses on bacterial processes. *In* D. L. Kirchman, ed. Microbial Ecology of the Oceans. Wiley-Liss, Inc., New York. pp. 327–350.

Fuhrman, J. A. and R. T. Noble. 1995. Viruses and protists cause similar bacterial mortality in coastal seawater. Limnol. Oceanogr. *40*:1236–1242.

Fuhs, G. W., S. D. Demmerle, E. Canelli, and M. Chen. 1972. Characterization of phosphorus-limited plankton algae (with reflections on the limiting-nutrient concept). *In* G. E. Likens, ed. Nutrients and Eutrophication: The Limiting-Nutrient Controversy. Special Symposium, Amer. Soc. Limnol. Oceanogr. *1*: 113–133.

Fukami, K., B. Meier, and J. Overbeck. 1991. Vertical and temporal changes in bacterial production and its consumption by heterotrophic nanoflagellates in a north German eutrophic lake. Arch. Hydrobiol. *122*:129–145.

Fukuhara, H. and K. Yasuda. 1985. Phosphorus excretion by some zoobenthos in a eutrophic freshwater lake and its temperature dependency. Jpn. J. Limnol. *46*:287–296.

Fukuhara, H. and K. Yasuda. 1989. Ammonium excretion by some freshwater zoobenthos from a eutrophic lake. Hydrobiologia *173*:1–8.

Fukuhara, H. and M. Sakamoto. 1987. Enhancement of inorganic nitrogen and phosphate release from lake sediment by tubificid worms and chironomid larvae. Oikos *48*:312–320.

Fukushima, T., J. Park, A. Imai, and K. Matsushige. 1996. Dissolved organic carbon in a eutrophic lake: Dynamics, biodegradability and origin. Aquat. Sci. *58*:139–157.

Fuller, R. L., C. Ribble, A. Kelley, and E. Gaenzie. 1998. Impact of stream grazers on periphyton communities: A laboratory and field manipulation. J. Freshwat. Ecol. *13*:105–114.

Fuller, S. L. H. 1974. Clams and mussels (Mollusca: Bivalvia). *In* C. W. Hart, Jr. and S. L. H. Fuller, eds. Pollution Ecology of Freshwater Invertebrates. Academic Press, New York. pp. 215–273.

Fulton, R. S. 1988. Grazing on filamentous algae by herbivorous zooplankton. Freshwat. Biol. *20*:263–271.

Furch, K. and W. J. Junk. 1997. The chemical composition, food value, and decomposition of herbaceous plants, leaves, and leaf litter of floodplain forests. *In* W. J. Junk, ed. The Central Amazon Floodplain: Ecology of a Pulsing System. Springer-Verlag, Berlin. pp. 187–205.

Fuss, C. L. and L. A. Smock. 1996. Spatial and temporal variation of microbial respiration rates in a blackwater stream. Freshwat. Biol. *36*:339–349.

Gabriel, W., B. E. Taylor, and S. Kirsch-Prokosch. 1987. Cladoceran birth and death rates estimates: Experimental comparisons of egg-ratio methods. Freshwat. Biol. *18*:361–372.

Gabriel, W. and B. Thomas. 1988a. Vertical migration of zooplankton as an evolutionarily stable strategy. Am. Nat. *132*:199–216.

Gabriel, W. and B. Thomas. 1988b. The influence of food availability, predation risk, and metabolic costs on the evolutionary stability of diel vertical migration in zooplankton. Verh. Internat. Verein. Limnol. *23*:807–811.

Gabrielson, J. O., M. A. Perkins, and E. B. Welch. 1984. The uptake, translocation and release of phosphorus by *Elodea densa*. Hydrobiologia *111*:43–48.

Gächter, R. and J. Bloesch. 1985. Seasonal and vertical variation in the C:P ratio of suspended and settling seston of lakes. Hydrobiologia *128*:193–200.

Gachter, R. and J. S. Meyer. 1993. The role of microorganisms in mobilization and fixation of phosphorous in sediments. Hydrobiologia *253*:103–121.

Gacia, E. and E. Ballesteros. 1994. Production of *Isoetes lacustris* in a Pyrenean lake: Seasonality and ecological factors involved in the growing period. Aquat. Bot. *48*:77–89.

Gaedeke, A. and U. Sommer. 1986. The influence of the frequency of periodic disturbances on the maintenance of phytoplankton diversity. Oecologia *71*:25–28.

Gaedke, U. and D. Straile. 1994. Seasonal changes in the quantitative importance of protozoans in a large lake—An ecosystem approach using mass-balanced carbon flow diagrams. Mar. Microb. Food Webs *8*:163–188.

Gaedke, U., D. Straile, and C. Pahl-Wostl. 1996. Trophic structure and carbon flow dynamics in the pelagic community of a large lake. *In* G. A. Polis and K. O. Winemiller, eds. Food Webs: Integration of Patterns and Dynamics. Chapman & Hall, New York. pp. 60–71.

Gaevskaia, N. S. 1966. Rol' vysshikh vodnykh rastenii v pitanii zhivotnykh presnykh vodoemov. Moskva, Izdatel'stvo Nauka, 327 pp. (Translated into English, Nat. Lending Libr. Sci. Technol., Yorkshire, England, 1969, as: The Role of Higher Aquatic Plants in the Nutrition of the Animals of Fresh-Water Basins.)

Gaines, G. and M. Elbrachter. 1987. Heterotrophic nutrition. *In* F. J. R. Taylor, eds. The Biology of Dinoflagellates. Blackwell Scientific Publications, Oxford, pp. 224–268.

Gaines, W. L., C. E. Cushing, and S. D. Smith. 1992. Secondary production estimates of benthic insects in three cold desert streams. Great Basin Nat. *52*:11–24.

Gajewski, A. J. and R. J. Chróst. 1995. Production and enzymatic decomposition of organic matter by microplankton in a eutrophic lake. J. Plankton Res. *17*:709–728.

Gak, D. Z. 1959. Fiziologicheskaia aktivnost' i sistematicheskoe polozhenie mobilizuiushchikh fosfor mikroorganizmov, vydelennykh iz vodoemov pribaltiki. Mikrobiologiya, *28*:551–556.

Gak, D. Z. 1963. Vertikal'noe raspredelenie mobilizuiushchikh fosfor bakterii v gruntakh Latviiskikh vodoemov. Mikrobiologiya *32*:838–842.

Gak, D. Z., V. V. Gurvich, I. L. Korelyakova, L. E. Kostikova, N. A. Kostantinova, G. A. Olivari, A. D. Priimachenko, Y. Y. Tseeb, K. S. Vladimirova, and L. N. Zimbalevskaya. 1970. Productivity of aquatic organism communities of different trophic levels in Kiev Reservoir. *In:* Z. Kajak and A. Hillbricht-Ilkowska, eds. Productivity Problems of Freshwaters. PWN Polish Scientific Publishers, Warszawa. pp. 447–455.

Galbraith, M. G., Jr. 1967. Size-selective predation on *Daphnia* by rainbow trout and yellow perch. Trans. Amer. Fish. Soc. *96*:1–10.

Gallepp, G. W. 1979. Chironomid influence on phosphorus release in sediment-water microcosms. Ecology *60*:547–556.

Gallepp, G. W., J. F. Kitchell, and S. M. Bartell. 1978. Phosphorus release from lake sediments as affected by chironomids. Verh. Internat. Verein. Limnol. *20*:458–465.

Gallon, J. R., T. A. LaRue, and W. G. W. Kurz. 1974. Photosynthesis and nitrogenase activity in the blue-green alga *Gloeocapsa*. Can. J. Microbiol. *20*:1633–1637.

Galloway, J. N. 1998. The global nitrogen cycle: Changes and consequences. Environ. Poll. *102*(S1):15–24.

Galloway, J. N. and D. M. Whelpdale. 1980. An atmospheric sulfur budget for eastern North America. Atmospheric Environ. *14*:409–417.

Galloway, J. N., G. E. Likens, and M. E. Hawley. 1984. Acid precipitation: Natural versus anthropogenic components. Science *226*:829–831.

Gálvez, J. A., F. X. Niell, and J. Lucena. 1989. Seston vertical flux model for a eutrophic reservoir. Arch. Hydrobiol. Beih. Ergebn. Limnol. *33*:9–18.

Gálvez, J. A., F. X. Niell, and J. Lucena. 1991. C:N:P ratio of settling seston in a eutrophic reservoir. Verh. Internat. Verein. Limnol. *24*:1390–1395.

Gálvez, J. A., F. X. Niell, and J. Lucena. 1993. Sinking velocities of principal phytoplankton species in a stratified reservoir: Ecological implications. Verh. Internat. Verein. Limnol. *25*: 1228–1231.

Gambrell, R. P. 1994. Trace and toxic metals in wetlands—A review. J. Environ. Qual. *23*:883–891.

Ganf, G. G. 1974a. Incident solar irradiance and underwater light penetration as factors controlling the chlorophyll *a* content of a shallow equatorial lake (Lake George, Uganda). J. Ecol. *62*:593–609.

Ganf, G. G. 1974b. Diurnal mixing and the vertical distribution of phytoplankton in a shallow equatorial lake (Lake George, Uganda). J. Ecol. *62*:611–629.

Ganf, G. G. 1974c. Phytoplankton biomass and distribution in a shallow eutrophic lake (Lake George, Uganda). Oecologia *16*:9–29.

Ganf, G. G. and R. J. Shiel. 1985. Particle capture by *Daphnia carinata*. Aust. J. Mar. Freshw. Res. *36*:371–381.

Ganf, G. G. and R. L. Oliver. 1982. Vertical separation of light and available nutrients as a factor causing replacement of green algae by blue-green algae in the plankton of a stratified lake. J. Ecol. *70*:829–844.

Gangstad, E. O. 1988a. Freshwater Vegetation Management. Thomas Publications, Fresno, CA. 380 pp.

Gangstad, E. O. 1988b. Resource Recreation Management. Thomson Publications, Fresno, CA. 296 pp.

Garcia-Gil, L. J., L. Sala-Genoher, J. V. Esteva, and C. A. Abella. 1990. Distribution of iron in Lake Banyoles in relation to the ecology of purple and green sulfur bacteria. Hydrobiologia *192*:259–270.

Gardner, W. S. and G. F. Lee. 1975. The role of amino acids in the nitrogen cycle of Lake Mendota. Limnol. Oceanogr. *20*:379–388.

Gardner, W. S., J. F. Chandler, and G. A. Laird. 1989. Organic nitrogen mineralization and substrate limitation of bacteria in Lake Michigan. Limnol. Oceanogr. *34*:478–485.

Gardner, W. S., J. F. Chandler, G. A. Laird, and H. J. Carrick. 1987. Sources and fate of dissolved free amino acids in epilimetic Lake Michigan water. Limnol. Oceanogr. *32*:1353–1362.

Garrels, R. M. 1965. Silica: Role in the buffering of natural waters. Science *148*:69.

Gasaway, R. D. and T. F. Drda. 1978. Effects of grass carp introduction on macrophytic vegetation and chlorophyll content of phytoplankton in four Florida lakes. Florida Sci. *41*:101–109.

Gasith, A. 1975. Tripton sedimentation in eutrophic lakes—simple correction for the resuspended matter. Verh. Internat. Verein. Limnol. *19*:116–122.

Gasith, A. 1976. Seston dynamics and tripton sedimentation in the pelagic zone of a shallow eutrophic lake. Hydrobiologia *51*: 225–231.

Gasith, A. and A. D. Hasler. 1976. Airborne litterfall as a source of organic matter in lakes. Limnol. Oceanogr. *21*:253–258.

Gasol, J. M. 1993. Benthic flagellates and ciliates in fine freshwater sediments: Calibration of a live counting procedure and estimation of their abundances. Microb. Ecol. *25*:247–262.

Gasol, J. M., A. M. Simons, and J. Kalff. 1995. Patterns in the top-down versus bottom-up regulation of heterotrophic nanoflagellates in temperate lakes. J. Plankton Res. *17*:1879–1903.

Gasol, J. M., B. Sander, and M. Schallenberg. 1993. Production of bacteria in freshwater sediments: Comparison of different cell-specific measurements to mineralization rates. Verh. Internat. Verein. Limnol. *25*:325–330.

Gasol, J. M., J. Garcia-Cantizano, R. Massana, F. Peters, R. Guerrero, and C. Pedrós-Alió. 1991. Diel changes in the microstratification of the metalimnetic community in Lake Cisó. Hydrobiologia *211*:227–240.

Gassmann, G. 1994. Phosphine in the fluvial and marine hydrosphere. Mar. Chem. *45*:197–205.

Gassmann, G. and F. Schorn. 1993. Phosphine from harbor surface sediments. Naturwissenschaften *80*:78–80.

Gat, J. R. and M. Shatkay. 1991. Gas exchange with saline waters. Limnol. Oceanogr. *36*:988–997.

Gates, D. M. 1962. Energy Exchange in the Biosphere. Harper & Row, New York. 151 pp.

Gates, D. M. 1980. Biophysical Ecology. Springer-Verlag, New York. 611 pp.

Gates, D. M. 1993. Climate change and its biological consequences. Sinauer Associates, Inc., Sunderland, MA. 280 pp.

Gates, F. C. 1942. The bogs of northern Lower Michigan. Ecol. Monogr. *12*:213–254.

Gates, M. A. 1984. Quantitative importance of ciliates in the planktonic biomass of lake ecosystems. Hydrobiologia *108*:233–238.

Gates, M. A. and U. T. Lewg. 1984. Contribution of ciliated protozoa to the planktonic biomass in a series of Ontario lakes: Quantitative estimates and dynamical relationships. J. Plankton Res. *6*:443–456.

Gaudet, J. J. 1968. The correlation of physiological differences and various leaf forms of an aquatic plant. Physiol. Plant *21*:594–601.

Gaudet, J. J. 1976. Nutrient relationships in the detritus of a tropical swamp. Arch. Hydrobiol. *78*:213–239.

Gaudet, J. J. 1977. Uptake, accumulation, and loss of nutrients by papyrus in tropical swamps. Ecology *58*:415–422.

Gaudet, J. J. 1978. Effect of a tropical swamp on water quality. Verh. Internat. Verein. Limnol. *20*:2202–2206.

Gaudet, J. J. 1979. Seasonal changes in nutrients in a tropical swamp: North Swamp, Lake Naivasha, Kenya. J. Ecol. *67*: 953–981.

Gaudet, J. J. 1982. Nutrient dynamics of papyrus swamps. *In* B. Gopal, R. E. Turner, R. G. Wetzel, and D. F. Whigham, eds. Wetlands: Ecology and Management. Internat. Sci. Publ., Jaipur, India. pp. 305–319.

Gaudet, J. M. and A. G. Roy. 1995. Effect of bed morphology on flow mixing length at river confluences. Nature *373*:138–139.

Gaur, S., P. K. Singhal, and S. K. Hasija. 1992. Relative contributions of bacteria and fungi to water hyacinth decomposition. Aquat. Bot. *43*:1–15.

Gebre-Mariam, Z. and W. D. Taylor. 1990. Heterotrophic bacterioplankton production and grazing mortality rates in an Ethiopian rift-valley lake (Awassa). Freshwat. Biol. *22*:369–381.

Geertz-Hansen, O., M. Olesen, P. K. Bjørnsen, J. B. Larsen, and B. Riemann. 1987. Zooplankton consumption of bacteria in a eutrophic lake and in experimental enclosures. Arch. Hydrobiol. *110*:553–563.

Geesey, G. G. and J. W. Costerton. 1979. Microbiology of a northern river: Bacterial distribution and relationship to suspended sediment and organic carbon. Can. J. Microbiol. *25*:1058–1062.

Geiling, W. T. and R. S. Campbell. 1972. The effect of temperature on the development rate of the major life stages of *Diaptomus pallidus* Herrick. Limnol. Oceanogr. *17*:304–307.

Geller, A. 1985. Degradation and formation of refractory DOM by bacteria during simultaneous growth on labile substrates and persistent lake water constituents. Schweiz. Z. Hydrol. *47*:27–44.

Geller, A. 1986. Comparison of mechanisms enhancing biodegradability of refractory lake water constituents. Limnol. Oceanogr. *31*:755–764.

Geller, W. 1975. Die Nahrungsaufnahme von *Daphnia pulex* in Abhängigkeit von der Futterkonzentration, der Temperatur, der Körpergrösse und dem Hungerzustand der Tiere. Arch. Hydrobiol./Suppl. *48*:47–107.

Geller, W. 1985. Production, food utilization and losses of two coexisting, ecologically different *Daphnia* species. Arch. Hydrobiol. Beih. Ergebn. Limnol. *21*:67–79.

Geller, W. 1986. Diurnal vertical migration of zooplankton in a temperate great lake (L. Constance): A starvation avoidance mechanism? Arch. Hydrobiol./Suppl. *74*:1–60.

Geller, W. 1987. On estimating the age and the development time of *Daphnia* as a function of body size and temperature. J. Plankton Res. *9*:1225–1230.

Geller, W. and H. Müller. 1981. The filtration apparatus of Cladocera: Filter mesh-sizes and their implications on food selectivity. Oecologia *49*:316–321.

Geller, W., H. Klapper, and W. Salomons, eds. 1998. Acidic Mining Lakes: Acid Mine Drainage, Limnology and Reclamation. Springer-Verlag, Berlin. 435 pp.

Gensemer, R. W. and R. C. Playle. 1999. The bioavailability and toxicity of aluminum in aquatic environments. Crit. Rev. Environ. Sci. Technol. *29*:315–450.

Genter, R. B. 1996. Ecotoxicology of inorganic chemical stress. *In* R. J. Stevenson, M. L. Bothwell, and R. L. Lowe, eds. Algal Ecology: Freshwater Benthic Ecosystems. Academic Press, San Diego. pp. 403–468.

George, D. G. 1976. Life cycle and production of *Cyclops vicinus* in a shallow eutrophic reservoir. Oikos *27*:101–110.

George, D. G. and R. W. Edwards. 1973. *Daphnia* distribution within Langmuir circulations. Limnol. Oceanogr. *18*:798–800.

George, D. G. and R. W. Edwards. 1974. Population dynamics and production of *Daphnia hyalina* in a eutrophic reservoir. Freshwat. Biol. *4*:445–465.

George, D. G. and S. I. Heaney. 1978. Factors influencing the spatial distribution of phytoplankton in a small productive lake. J. Ecol. *66*:133–155.

George, M. G. and C. H. Fernando. 1969. Seasonal distribution and vertical migration of planktonic rotifers in two lakes of eastern Canada. Verh. Internat. Verein. Limnol. *17*:817–829.

George, M. G. and C. H. Fernando. 1970. Diurnal migration in three species of rotifers in Sunfish Lake, Ontario. Limnol. Oceanogr. *15*:218–223.

Gerking, S. D. 1962. Production and food utilization in a population of bluegill sunfish. Ecol. Monogr. *32*:31–78.

Gerking, S. D. 1994. Feeding Ecology of Fish. Academic Press, San Diego. 416 pp.

Gerletti, M. 1968. Dark bottle measurements in primary productivity studies. Mem. Ist. Ital. Idrobiol. *23*:197–208.

Gerloff, G. C. 1963. Comparative mineral nutrition of plants. Ann. Rev. Plant Physiol. *14*:107–123.

Gerloff, G. C. 1969. Evaluating nutrient supplies for the growth of aquatic plants in natural waters. *In* Eutrophication: Causes, Consequences, Correctives. National Academy of Sciences, Washington, DC. pp. 537–555.

Gerloff, G. C. and F. Skoog. 1957. Availability of iron and manganese in southern Wisconsin lakes for the growth of *Microcystis aeruginosa*. Ecology *38*:551–556.

Gerloff, G. C., G. P. Fitzgerald, and F. Skoog. 1952. The mineral nutrition of *Microcystis aeruginosa*. Am. J. Bot. *39*:26–32.

Gerritsen, J. and K. G. Porter. 1982. The role of surface chemistry in filter feeding by zooplankton. Science *216*:1225–1227.

Gerritsen, J., K. G. Porter, and J. R. Strickler. 1988. Not by sieving alone: Observations of suspension feeding in *Daphnia*. Bull. Mar. Sci. *43*:366–376.

Gervais, F. 1997. Diel vertical migration of *Cryptomonas* and *Chromatium* in the deep chlorophyll maximum of a eutrophic lake. J. Plankton Res. *19*:533–550.

Gessner, F. 1955. Hydrobotanik. Die Physiologischen Grundlagen der Pflanzenverbreitung im Wasser. I. Energiehaushalt. VEB Deutscher Verlag der Wissenschaften, Berlin. 701 pp.

Gessner, F. 1961. Hydrostatischer Druck und Pflanzenwachstum. *In* W. Ruhland, ed. Handbuch der Pflanzenphysiologie. Vol. 16. Springer-Verlag, Heidelberg. pp. 668–690.

Gessner, F. and A. Diehl. 1951. Die Wirkung natürlicher Ultraviolettstrahlung auf die Chlorophyllzerstörung von Planktonalgen. Arch. Mikrobiol. *15*:439–453.

Gest, H., A. San Pietro, and L. P. Vernon, eds. 1963. Bacterial Photosynthesis. Antioch Press, Yellow Springs, OH. 523 pp.

Getsinger, K. D. and C. R. Dillon. 1984. Quiescence, growth and senescence of *Egeria densa* in Lake Marion. Aquat. Bot. *20*:329–338.

Ghiorse, W. C. 1984. Biology of iron- and manganese-depositing bacteria. Ann. Rev. Microbiol. *38*:515–550.

Ghiretti, F. 1966. Respiration. *In* Wilbur, K. M. and C. M. Yonge, eds. Physiology of Mollusca, Vol. 2. Academic Press, New York. pp. 175–208.

Gibbs, M. M. 1992. Influence of hypolimnetic stirring and underflow on the limnology of Lake Rotoiti, New Zealand. N. Z. J. Mar. Freshwat. Res. *26*:453–463.

Gibbs, R. J. 1970. Mechanisms controlling world water chemistry. Science *170*:1088–1090.

Gibbs, R. J. 1973. Mechanisms of trace metal transport in rivers. Science *180*:71–73.

Gibbs, R. J. 1992. A reply to the comment of Eilers et al. Limnol. Oceanogr. *37*:1338–1339.

Giblin, A. E., G. E. Likens, D. White, and R. W. Howarth. 1990. Sulfur storage and alkalinity generation in New England lake sediments. Limnol. Oceanogr. *35*:852–869.

Gibson, C. E. 1981. Silica budgets and the ecology of planktonic diatoms in a unstratified lake (Lough Neagh, N. Ireland). Int. Revue ges. Hydrobiol. *66*:641–664.

Gibson, C. E. 1984. Sinking rates of planktonic diatoms in an unstratified lake: A comparison of field and laboratory observations. Freshwat. Biol. *14*:631–638.

Gibson, C. E. and J. Guillot. 1997. Sedimentation in a large lake: The importance of fluctuations in water level. Freshwat. Biol. *37*:597–604.

Gibson, C. E., Y. Wu, and D. Pinkerton. 1995. Substance budgets of an upland catchment: The significance of atmospheric phosphorus inputs. Freshwat. Biol. *33*:385–392.

Gibson, C. E., Y. Wu, S. J. Smith, and S. A. Wolfe-Murphy. 1995. Synoptic limnology of a diverse geological region: Catchment and water chemistry. Hydrobiologia *306*:213–227.

Gilbert, J. J. 1967a. Control of sexuality in the rotifer *Asplanchna brightwelli* by dietary lipids of plant origin. Proc. Nat. Acad. Sci. USA *57*:1218–1225.

Gilbert, J. J. 1967b. *Asplanchna* and postero-lateral spine production in *Brachionus calyciflorus*. Arch. Hydrobiol. *64*:1–62.

Gilbert, J. J. 1968. Dietary control of sexuality in the rotifer *Asplanchna brightwelli* Gosse. Physiol. Zool. *41*:14–43.

Gilbert, J. J. 1972. α-tocopherol in males of the rotifer *Asplanchna sieboldi*: Its metabolism and its distribution in the testis and rudimentary gut. J. Exp. Zool. *181*:117–128.

Gilbert, J. J. 1973. Induction and ecological significance of gigantism in the rotifer *Asplanchna sieboldi*. Science *181*:63–66.

Gilbert, J. J. 1975a. Polymorphism and sexuality in the rotifer *Asplanchna*, with special reference to the effects of prey-type and clonal variation. Arch. Hydrobiol. *74*:442–483.

Gilbert, J. J. 1975b. Field experiments on gemmulation in the freshwater sponge Spongilla lacustris. Trans. Amer. Microsc. Soc. *94*:347–356.

Gilbert, J. J. 1976. Polymorphism in the rotifer *Asplanchna sieboldi*: Biomass, growth, and reproductive rate of the saccate and campanulate morphotypes. Ecology *57*:542–551.

Gilbert, J. J. 1977. A note on the effect of cold shock on mictic-female production in *Brachionus calyciflorus*. Arch. Hydrobiol. Beih. Ergebn. Limnol. *8*:158–160.

Gilbert, J. J. 1980a. Female polymorphism and sexual reproduction in the rotifer *Asplanchna*: Evolution of their relationship and control by dietary tocopherol. Am. Nat. *116*:409–431.

Gilbert, J. J. 1980b. Further observations on developmental polymorphism and its evolution in the rotifer *Brachionus calyciflorus*. Freshwat. Biol. *10*:281–294.

Gilbert, J. J. 1980c. Feeding in the rotifer *Asplanchna*: Behavior, cannibalism, selectivity, prey defenses, and impact on rotifer communities. *In* W. C. Kerfoot, ed. Evolution and Ecology of Zooplanktonic Communities. Univ. Press New England, Hanover, NH. pp. 158–172.

Gilbert, J. J. 1988. Suppression of rotifer populations by *Daphnia*: A review of the evidence, the mechanisms, and the effects on zooplankton community structure. Limnol. Oceanogr. *33*: 1286–1303.

Gilbert, J. J. 1993. Rotifera. *In* K. G. Adiyodi and R. G. Adiyodi, eds. Reproductive Biology of Invertebrates. 5. Sexual Differentiation and Behaviour. Oxford IBH Publ., New Delhi. pp. 115–136.

Gilbert, J. J. 1995a. Rotifera. *In* K. G. Adiyodi and K. G. Adiyodi, eds. Reproductive Biology of Invertebrates. 6A. Asexual Propagation and Reproductive Strategies. Oxford IBH Publ., New Delhi. pp. 231–263.

Gilbert, J. J. 1995b. Structure, development and induction of a new diapause stage in rotifers. Freshwat. Biol. *34*:263–270.

Gilbert, J. J. and C. E. Williamson. 1978. Predator-prey behavior and its effect on rotifer survival in associations of *Mesocyclops edax*, *Asplanchna girodi*, *Polyarthra vulgaris*, and *Keratella cochlearis*. Oecologia *37*:13–22.

Gilbert, J. J. and C. W. Birky, Jr. 1971. Sensitivity and specificity of the *Asplanchna* response to dietary α-tocopherol. J. Nutrition *101*:113–126.

Gilbert, J. J. and D. E. Schreiber. 1995. Induction of diapausing amictic eggs in *Synchaeta pectinata*. Hydrobiologia *313/314*: 345–350.

Gilbert, J. J. and G. A. Thompson, Jr. 1968. Alpha tocopherol control of sexuality and polymorphism in the rotifer *Asplanchna*. Science *159*:734–736.

Gilbert, J. J. and H. L. Allen. 1973. Studies on the physiology of the green freshwater sponge *Spongilla lacustris*: Primary productivity, organic matter, and chlorophyll content. Verh. Internat. Verein. Limnol. *18*:1413–1420.

Gilbert, J. J. and J. K. Waage. 1967. *Asplanchna*, *Asplanchna*-substance, and posterolateral spine length variation of the rotifer *Brachionus calyciflorus* in a natural environment. Ecology *48*:1027–1031.

Gilbert, J. J. and J. R. Litton, Jr. 1975. Dietary tocopherol and sexual reproduction in the rotifers *Brachionus calyciflorus* and *Asplanchna sieboldi*. J. Exp. Zool. *194*:485–494.

Gilbert, J. J. and P. L. Starkweather. 1977. Feeding in the rotifer *Brachionus calyciflorus*. I. Regulatory mechanisms. Oecologia *28*:125–131.

Gilbert, J. J. and P. L. Starkweather. 1978. Feeding in the rotifer *Brachionus calyciflorus*. III. Direct observations on the effects of food type, food density, change in food type, and starvation on the incidence of pseudotrochal screening. Verh. Internat. Verein. Limnol. *20*:2382–2388.

Gilbert, J. J. and R. S. Stemberger. 1984. *Asplanchna*-induced polymorphism in the rotifer *Keratella slacki*. Limnol. Oceanogr. *29*:1309–1316.

Gilbert, J. J. and R. S. Stemberger. 1985. Control of *Keratella* populations by interference competition from *Daphnia*. Limnol. Oceanogr. *30*:180–188.

Gilbert, J. J. and S. Hadèiàce. 1975. Sexual reproduction in the freshwater sponge *Ochridaspongia rotunda*. Verh. Internat. Verein. Limnol. *19*:2785–2792.

Gilbert, J. J. and S. Hadèiàce. 1977. Life cycle of the freshwater sponge *Ochridaspongia rotunda* Arndt. Arch. Hydrobiol. *79*:285–318.

Gilbert, J. J. and T. L. Simpson. 1976a. Gemmule polymorphism in the freshwater sponge *Spongilla lacustris*. Arch. Hydrobiol. *78*:268–277.

Gilbert, J. J. and T. L. Simpson. 1976b. Sex reversal in a freshwater sponge. J. Exp. Zool. *195*:145–151.

Gilbert, J. J., T. L. Simpson, and G. S. de Nagy. 1975. Field experiments on egg production in the fresh-water sponge *Spongilla lacustris*. Hydrobiologia *46*:17–27.

Gilbert, O. L. 1996. The lichen vegetation of chalk and limestone streams in Britain. Lichenologist *28*:145–159.

Giller, P. S. and B. Malmqvist. 1998. The Biology of Streams and Rivers. Oxford Univ. Press, New York. 296 pp.

Giller, P. S. and N. Sangpradub. 1993. Predatory foraging behaviour and activity patterns of larvae of two species of limnephilid cased caddis. Oikos *67*:351–357.

Gillespie, P. A. and M. J. Spencer. 1980. Seasonal variation of heterotrophic potential in Lake Rotoroa. N. Z. J. Mar. Freshw. Res. *14*:15–21.

Ginzburg, B., I. Chalifa, F. J. Gun, I. Dor, O. Hadas, and O. Lev. 1998. DMS formation by dimethylsulfoniopropionate route in freshwater (*sic*). Environ. Sci. Technol. *32*:2130–2136.

Giroux, J.-F. and J. Bédard. 1988. Above- and below-ground macrophyte production in *Scirpus* tidal marshes of the St. Lawrence estuary, Quebec. Can. J. Bot. *66*:955–962.

Giurgevich, J. R. and E. L. Dunn. 1982. Seasonal patterns of daily net photosynthesis, transpiration and net primary productivity of *Juncus roemerianus* and *Spartina alterniflora* in a Georgia salt marsh. Oecologia *52*:404–410.

Gjessing, E. T. 1964. Ferrous iron in water. Limnol. Oceanogr. *9*: 272–274.

Gjessing E. T. 1970. Reduction of aquatic humus in streams. Vatten *26*:14–23.

Gjessing, E. T. 1976. Physical and Chemical Characteristics of Aquatic Humus. Ann Arbor Science Publs., Ann Arbor, MI. 120 pp.

Gjessing, E. T. and T. Gjerdahl. 1970. Influence of ultra-violet radiation on aquatic humus. Vatten *26*:144–145.

Gjessing, E. T., G. Riise, and E. Lydersen. 1998. Acid rain and natural organic matter (NOM). Acta Hydrochim. Hydrobiol. *26*: 131–136.

Gladyshev, M. I. and K. G. Malyshevskiy. 1982. The concept of "neuston." Hydrobiol. J. *18*:7–10.

Gledhill, T. Water mites—predators and parasites. Annu. Rep. Freshwat. Biol. Assoc. U.K. *53*:45–59.

Glenn, E., T. L. Thompson, R. Frye, J. Riley, and D. Baumgartner. 1995. Effects of salinity on growth and evapotranspiration of *Typha domingensis* Pers. Aquat. Bot. *52*:75–91.

Glime, J. M, R. G. Wetzel, and B. J. Kennedy. 1982. The effects of bryophytes on succession from alkaline marsh to *Sphagnum* bog. Amer. Midland Nat. *108*:209–223.

Glindemann, D., U. Stottmeister, and A. Bergmann. 1996. Free phosphine from the anaerobic biosphere. Environ. Sci. Pollut. Res. *3*:17–19.

Gliwicz, Z. M. 1979. Metalimnetic gradients and trophic state of lake epilimnia. Mem. Ist. Ital. Idrobiol. *37*:121–143.

Gliwicz, Z. M. 1980. Filtering rates, food size selection, and feeding rates in cladocerans—another aspect of interspecific competition in filter-feeding zooplankton. *In* W. C. Kerfoot, ed. Evolution and Ecology of Zooplankton Communities. Univ. Press New England, Hanover, NH. pp. 282–291.

Gliwicz, Z. M. 1986a. Predation and the evolution of vertical migration in zooplankton. Nature *320*:746–748.

Gliwicz, Z. M. 1986b. A lunar cycle in zooplankton. Ecology *67*: 883–897.

Gliwicz, Z. M. 1986c. Suspended clay concentration controlled by filter-feeding zooplankton in a tropical reservoir. Nature *323*: 330–332.

Gliwicz, Z. M. 1990a. Food thresholds and body size in cladocerans. Nature *343*:638–640.

Gliwicz, Z. M. 1990b. Why do cladocerans fail to control algal blooms? Hydrobiologia *200/201*:83–97.

Gliwicz, Z. M. and A. Prejs. 1977. Can planktivorous fish keep in check planktonic crustacean populations? A test of size-efficiency hypothesis in typical Polish lakes. Ekol. Polska *25*:567–591.

Gliwicz, Z. M. and E. Siedlar. 1980. Food size limitation and algae interfering with food collection in *Daphnia*. Arch. Hydrobiol. *88*:155–177.

Gliwicz, Z. M. and W. Lampert. 1990. Food thresholds in *Daphnia* species in the absence and presence of blue-green filaments. Ecology *71*:691–702.

Gliwicz, Z. M., P. Dawidowicz, and J. Pijanowska. 1985. Structure of phytoplankton in rivers and in lakes of northeastern Poland. Ekol. Polska *33*:537–546.

Glob, P. V. 1969. The Bog People. Iron-Age Man Preserved. Cornell University Press, Ithaca, NY. 200 pp.

Gloor, M., A. Wüest, and M. Münnich. 1994. Benthic boundary mixing and resuspension induced by internal seiches. Hydrobiologia *284*:59–68.

Glooschenko, W. A., J. E. Moore, and R. A. Vollenweider. 1973. Chlorophyll *a* distribution in Lake Huron and its relationship to primary production. Proc. Conf. Grat Lakes Res., Int. Assoc. Great Lakes Res. *16*:40–49.

Glooschenko, W. A., J. E. Moore, M. Munawata, and R. A. Vollenweider. 1974a. Primary production in lakes Ontario and Erie: A comparative study. J. Fish. Res. Board Can. *31*:253–263.

Glooschenko, W. A., J. E. Moore, M. Munawata, and R. A. Vollenweider. 1974b. Spatial and temporal distribution of chlorophyll *a* and pheopigments in surface waters of Lake Erie. J. Fish. Res. Board Can. *31*:265–274.

Glozier, N. E. and J. M. Culp. 1989. Experimental investigations of diel vertical movements by lotic mayflies over substrate surfaces. Freshwat. Biol. *21*:253–260.

Gocke, K. 1970. Untersuchungen über Abgabe und Aufnahme von Aminosäuren und Polypeptiden durch Planktonorganismen. Arch. Hydrobiol. *67*:285–367.

Godbout, L. and H. B. N. Hynes. 1982. The three dimensional distribution of the fauna in a single riffle in a stream in Ontario. Hydrobiologia *97*:87–96.

Gode, P. and J. Overbeck. 1972. Untersuchungen zur heterotrophen Nitrifikation im See. Z. Allg. Mikrobiol. *12*:567–574.

Godlewska-Lipowa, W. A. 1975. Ecosystem of the Mikolajskie Lake. The role of heterotrophic bacteria in the pelagial. Pol. Arch. Hydrobiol. *22*:79–87.

Godlewska-Lipowa, W. A. 1976. Bacteria as indicators of the degree of eutrophication and degradation of lakes. Pol. Arch. Hydrobiol. *23*:341–356.

Godshalk, G. L. 1977. Decomposition of aquatic plants in lakes. Ph.D. Dissertation, Michigan State University. 139 pp. + 340 figs.

Godshalk, G. L. and R. G. Wetzel. 1978a. Decomposition of aquatic angiosperms. I. Dissolved components. Aquat. Bot. *5*:281–300.

Godshalk, G. L. and R. G. Wetzel. 1978b. Decomposition of aquatic angiosperms. II. Particulate components. Aquat. Bot. *5*: 301–327.

Godshalk, G. L. and R. G. Wetzel. 1978c. Decomposition of aquatic angiosperms. III. *Zostera marina* L. and a conceptual model of decomposition. Aquat. Bot. *5*:329–354.

Godshalk, G. L. and R. G. Wetzel. 1978d. Decomposition in the littoral zone of lakes. *In* R. E. Good, D. F. Whigham, and R. L. Simpson, eds. Freshwater Wetlands: Ecological Processes and Management Potential. Academic Press, New York. pp. 131–144.

Godshalk, G. L. and R. G. Wetzel. 1984. Accumulation of sediment organic matter in a hardwater lake with reference to lake ontogeny. Bull. Mar. Sci. *35*:576–586.

Godward, M. B. E. 1962. Invisible radiations. *In* R. A. Lewis, ed. Physiology and Biochemistry of Algae. Academic Press, New York. pp. 551–566.

Godwin, H. 1956. The History of the British Flora. Cambridge Univ. Press., London. 384 pp.

Goedkoop, W. and R. K. Johnson. 1994. Exploitation of sediment bacterial carbon by juveniles of the amphipod *Monoporeia affinis*. Freshwat. Biol. *32*:553–563.

Goering, J. J. and J. C. Neess. 1964. Nitrogen fixation in two Wisconsin lakes. Limnol. Oceanogr. *9*:530–539.

Goering, J. J. and V. A. Dugdale. 1966. Estimates of the rates of denitrification in a subarctic lake. Limnol. Oceanogr. *11*:113–117.

Goes, J. I., N. Handa, S. Taguchi, and T. Hama. 1994. Effects of UV-B radiation on the fatty acid composition of the marine phytoplankton *Tetraselmis* sp.: Relationship to cellular pigments. Mar. Ecol. Prog. Ser. *114*:259–274.

Golachowska, J. B. 1971. The pathways of phosphorus in lake water. Pol. Arch. Hydrobiol. *18*:325–245.

Goldacre, R. J. 1949. Surface films on natural bodies of water. J. Anim. Ecol. *18*:36–39.

Golden, S. S., M. Ishiura, C. H. Johnson, and T. Kondo. 1997. Cyanobacterial circadian rhythms. Annu. Rev. Plant Physiol. Plant Mol. Biol. *48*:327–354.

Goldman, C. R. 1960. Molybdenum as a factor limiting primary productivity in Castle Lake, California. Science *132*:1016–1017.

Goldman, C. R. 1960. Primary productivity and limiting factors in three lakes of the Alaska Peninsula. Ecol. Monogr. *30*:207–230.

Goldman, C. R. 1961. The contribution of alder trees (*Alnus tenuifolia*) to the primary productivity of Castle Lake, California. Ecology *42*:282–288.

Goldman, C. R. 1970. Antarctic freshwater ecosystems. *In* M. W. Holdgate, ed. Antarctic Ecology. Academic Press, New York. pp. 609–627.

Goldman, C. R. 1972. The role of minor nutrients in limiting the productivity of aquatic ecosystems. *In* G. E. Likens, ed. Nutrients and Eutrophication: The Limiting-Nutrient Controversy. Special Symposium, Amer. Soc. Limnol. Oceanogr. *1*:21–38.

Goldman, C. R. and R. G. Wetzel. 1963. A study of the primary productivity of Clear Lake, Lake County, California. Ecology *44*:283–294.

Goldman, C. R., D. T. Mason, and B. J. B. Wood. 1963. Light injury and inhibition in Antarctic freshwater phytoplankton. Limnol. Oceanogr. *8*:313–322.

Goldman, C. R., D. T. Mason, and B. J. B. Wood. 1972. Comparative study of the limnology of two small lakes on Ross Island, Antarctica. *In* G. A. Llano, ed. Antarctic Terrestrial Biology. American Geophysical Union, Washington. pp. 1–50.

Goldman, J. C. 1977. Steady state growth of phytoplankton in continuous culture: Comparison of internal and external nutrient equations. J. Phycol. *13*:251–258.

Goldman, J. C. and S. J. Graham. 1980. Inorganic carbon limitation and chemical composition of two freshwater green microalgae. Appl. Environ. Microbiol. *41*:60–70.

Goldman, J. C., D. B. Porcella, E. J. Middlebrooks, and D. F. Toerien. 1972. The effect of carbon on algal growth—its relationship to eutrophication. Water Res. *6*:637–679.

Goldman, J. C., J. J. McCarthy, and D. G. Peavey. 1979. Growth rate influence on the chemical composition of phytoplankton in oceanic waters. Nature *279*:210–214.

Goldsborough, L. G. and D. J. Brown. 1991. Periphyton production in a small, dystrophic pond on the Canadian Precambrian Shield. Verh. Internat. Verein Limnol. *24*:1497–1502.

Goldsborough, L. G. and G. G. C. Robinson. 1996. Pattern in wetlands. *In* R. J. Stevenson, M. L. Bothwell, and R. L. Lowe, eds.

Algal Ecology: Freshwater Benthic Ecosystems. Academic Press, San Diego. pp. 77–117.

Goldschmidt, T. 1996. Darwin's Dreampond: Drama in Lake Victoria. MIT Press, Cambridge, MA. 274 pp.

Goldspink, C. R. and D. B. C. Scott. 1971. Vertical migration of *Chaoborus flavicans* in a Scottish loch. Freshwat. Biol. *1*:411–421.

Goldsworthy, A. 1970. Photorespiration. Bot. Rev. *36*:321–340.

Golladay, S. W., J. R. Webster, and E. F. Benfield. 1989. Changes in stream benthic organic matter following watershed disturbance. Holarctic Ecol. *12*:96–105.

Golterman, H. L. 1960. Studies on the cycle of elements in fresh water. Acta Bot. Neerlandica *9*:1–58.

Golterman, H. L., ed. 1969. Methods for Chemical Analysis of Fresh Waters. Int. Biol. Program Handbook 8. Blackwell Scientific Publications, Oxford. 172 pp.

Golterman, H. L. 1971. The determination of mineralization losses in correlation with the estimation of net primary production with the oxygen method and chemical inhibitors. Freshwat. Biol. *1*:249–256.

Golterman, H. L. 1982. Loading concentration models for phosphate in shallow lakes. Hydrobiologia *91*:169–174.

Golterman, H. L. 1995. The role of the iron hydroxide-phosphate-sulphide system in the phosphate exchange between sediments and overlying water. Hydrobiologia *297*:43–54.

Golterman, H. L. and M. L. Meyer. 1985. The geochemistry of two hard water rivers, the Rhine and the Rhone: Part 1: Presentation and screening of data. Hydrobiologia *126*:3–10.

Golterman, H. L., C. C. Bakels, and J. Jakobs-Mögelin. 1969. Availability of mud phosphates for the growth of algae. Verh. Internat. Verein. Limnol. *17*:467–479.

Golubić, S. 1963. Hydrostatischer Druck, Licht und submerse Vegetation im Vrana-See. Int. Rev. ges. Hydrobiol. *48*:1–7.

Golubić, S. 1967. Algenvegetation der Felsen. Eine ökologische Algenstudie im dinarischen Karstgebiet. Die Binnengewässer *23*. 183 pp.

Golubić, S. 1969. Cyclic and noncyclic mechanisms in the formation of travertine. Verh. Internat. Verein. Limnol. *17*:956–961.

Golubić, S. 1973. The relationship between blue-green algae and carbonate deposits. *In* N. G. Carr and B. A. Whitton, eds. The Biology of Blue-Green Algae. University of California Press, Berkeley. pp. 434–472.

Goma, R. H., M. Aizaki, T. Fukushima, and A. Otsuki. 1996. Significance of zooplankton grazing activity as a sources of dissolved organic nitrogen, urea and dissolved free amino acids in a eutrophic shallow lake: Experiments using outdoor continuous flow pond systems. Jpn. J. Limnol. *57*:1–13.

González, J. M. 1999. Bacterivory rate estimates and fraction of active bacterivores in natural protist assemblages from aquatic systems. Appl. Environ. Microbiol. *65*:1463–1469.

González, J. M., E. B. Sherr, and B. F. Sherr. 1990. Size-selective grazing on bacteria by natural assemblages of estuarine flagellates and ciliates. Appl. Environ. Microbiol. *56*:583–589.

González, M. J. and T. M. Frost. 1992. Food limitation and seasonal population declines of rotifers. Oecologia *89*:560–566.

Goodwin, T. W. 1974. Carotenoids and biliproteins. *In* W. D. P. Stewart, ed. Algal Physiology and Biochemistry. University of California Press, Berkeley. pp. 176–205.

Gopal, B. 1987. Water Hyacinth. Elsevier, Amsterdam. 471 pp.

Gopal, B. and U. Goel. 1993. Competition and allelopathy in aquatic plant communities. Bot. Rev. *59*:155–210.

Gophen, M. and W. Geller. 1984. Filter mesh size and food particle uptake by *Daphnia*. Oecologia *64*:408–412.

Gophen, M, B. Z. Cavari, and T. Berman. 1974. Zooplankton feeding on differentially labelled algae and bacteria. Nature *247*:393–394.

Gorbunov, K. V. 1953. Raspad ostatkov vysshikh vodnykh rastenii i ego ekologicheskaia rol' v vodoemakh nizhnei zony Del 'ty Volgi. Trudy Vses. Gidrobiol. Obshschestva *5*:158–202.

Gordon, N. D., T. A. McMahon, and B. L. Finlayson. 1992. Stream Hydrology: An Introduction for Ecologists. John Wiley & Sons, Chichester. 526 pp.

Gore, J. A. 1985. Restoration of Rivers and Streams. Butterworths, Boston. 320 pp.

Gorham, E. 1955. On some factors affecting the chemical composition of Swedish fresh waters. Geochim. Cosmochim. Acta *7*:129–150.

Gorham, E. 1956. On the chemical composition of some waters from the Moor House Nature Reserve. J. Ecol. *44*:377–382.

Gorham, E. 1957. The development of peat lands. Quart. Rev. Biol. *32*:145–166.

Gorham, E. 1958. Observations on the formation and breakdown of the oxidized microzone at the mud surface in lakes. Limnol. Oceanogr. *3*:291–298.

Gorham, E. 1960. Chlorophyll derivatives in the surface muds from the English lakes. Limnol. Oceanogr. *5*:29–33.

Gorham, E. 1961. Factors influencing supply of major ions to inland waters, with special reference to the atmosphere. Bull. Geol. Soc. Amer. *72*:795–840.

Gorham, E. 1964. Morphometric control of annual heat budgets in temperate lakes. Limnol. Oceanogr. *9*:525–529.

Gorham, E. and F. M. Boyce. 1989. Influence of lake surface area and depth upon thermal stratification and the depth of the summer thermocline. J. Great Lakes Res. *15*:233–245.

Gorham, E. and J. E. Sanger. 1972. Fossil pigments in the surface sediments of a meromictic lake. Limnol. Oceanogr. *17*:618–622.

Gorham, E. and J. E. Sanger. 1975. Fossil pigments in Minnesota lake sediments and their bearing upon the balance between terrestrial and aquatic inputs to sedimentary organic matter. Verh. Internat. Verein. Limnol. *19*:2267–2273.

Gorham, E. and J. E. Sanger. 1976. Fossilized pigments as stratigraphic indicators of cultural eutrophication in Shagawa Lake, northeastern Minnesota. Geol. Soc. Amer. Bull. *87*: 1638–1642.

Gorham, E. and J. Sanger. 1964. Chlorophyll derivatives in woodland, swamp, and pond soils of Cedar Creek Natural History Area, Minnesota, U.S.A. *In* Recent Researches in the Fields of Hydrosphere, Atmosphere and Nuclear Geochemistry. Mauzen Co., Ltd., Tokyo. pp. 1–12.

Gorham, E., J. K. Underwood, F. B. Martin, and J. G. Ogden. 1986. Natural and anthropogenic causes of lake acidification in Nova Scotia. Nature *324*:451–453.

Gorham, E., W. E. Dean, and J. E. Sanger. 1983. The chemical composition of lakes in the north-central United States. Limnol. Oceanogr. *28*:287–301.

Gorlenko, V. M., M. B. Vainshtein, and E. N. Chebotarev. 1980. Bacteria of sulfur and iron cycles in the low-sulfate meromictic Lake Kuznechikha. Mikrobiologiya *49*:804–812.

Gosz, J. R., G. E. Likens, and F. H. Bormann. 1972. Nutrient content of litter fall on the Hubbard Brook Experimental Forest, New Hampshire. Ecology *53*:769–784.

Göttlich, K., ed. 1990. Moor- und Torfkunde. E. Schweizerbart'sche Verlagsbuchhandlung, Stuttgart. 529 pp.

Gough, L. and J. B. Grace. 1998. Herbivore effects on plant species density at varying productivity levels. Ecology *79*:1586–1594.

Gough, S. B. and L. P. Gough. 1981. Comment on "Primary production of algae growing on natural and artificial aquatic plants: A study of interactions between epiphytes and their substrate" (Cattaneo and Kalff). Limnol. Oceanogr. *26*:987–989.

Goulden, C. E. 1969a. Interpretative studies of cladoceran microfossils in lake sediments. Mitt. Internat. Verein. Limnol. *17*:43–55.

Goulden, C. E. 1969b. Temporal changes in diversity. *In* G. M. Woodwell and H. H. Smith, eds. Diversity and Stability in Ecological Systems. Brookhaven Symposia in Biology *22*:96–102.

Goulden, C. E. 1971. Environmental control of the abundance and distribution of the chydorid Cladocera. Limnol. Oceanogr. *16*:320–331.

Goulden, C. E. and A. R. Place. 1990. Fatty acid synthesis and accumulation rates in daphnids. J. Exp. Zool. *256*:168–178.

Goulden, C. E. and A. R. Place. 1993. Lipid accumulation and allocation in daphnid Cladocera. Bull. Mar. Sci. *53*:106–114.

Goulden, C. E. and L. Henry. 1985. Lipid energy reserves and their role in Cladocera. *In* D. G. Meyers and J. R. Strickler, eds. Trophic Interactions within Aquatic Ecosystems. Westview Press, Boulder, CO. pp. 167–185.

Goulden, C. E. and L. Hornig. 1980. Population oscillations and energy reserves in plankton Cladocera and their consequences to competition. Proc. Nat. Acad. Sci. USA *77*:1716–1720.

Goulden, C. E. and L. L. Henry. 1981. Lipid energy reserves and their role in Cladocera. Spec. Symposium, Amer. Assoc. Advancement Science. pp. 25.

Goulden, C. E., L. Hornig, and C. Wilson. 1978. Why do large zooplankton species dominate? Verh. Internat. Verein. Limnol. *20*:2457–2460.

Goulden, C. E., R. E. Moeller, J. N. McNair, and A. R. Place. 1999. Lipid dietary dependencies in zooplankton. *In* M. T. Arts and B. C. Wainman, eds. Lipids in Freshwater Ecosystems. Springer-Verlag, New York. pp. 91–108.

Goulder, R. 1969. Interactions between the rates of production of a freshwater macrophyte and phytoplankton in a pond. Oikos *20*:300–309.

Goulder, R. 1970. Day-time variations in the rates of production by two natural communities of submerged freshwater macrophytes. J. Ecol. *58*:521–528.

Goulder, R. 1972. Grazing by the ciliated protozoan *Loxodes magnus* on the alga *Scenedesmus* in a eutrophic pond. Oikos *23*:109–115.

Goulder, R. 1974a. The seasonal an spatial distribution of some benthic ciliated Protozoa in Esthwaite Water. Freshwat. Biol. *4*:127–147.

Goulder, R. 1974b. Relationships between natural populations of the ciliated protozoa *Loxodes magnus* Stokes and *L. striatus* Penard. J. Ecol. *43*:429–438.

Goulder, R. 1975. The effects of photosynthetically raised pH and light on some ciliated protozoa in a eutrophic pond. Freshwat. Biol. *5*:127–147.

Goulder, R. 1980a. The ecology of two species of primitive ciliated protozoa commonly found in standing fresh waters (*Loxodes magnus* Stokes and *L. striatus* Penard). Hydrobiologia *72*: 131–158.

Goulder, R. 1980b. Seasonal variation in heterotrophic activity and population density of planktonic bacteria in a clean river. J. Ecol. *68*:349–363.

Gouy, J.-L., P. Bergé, and L. Labroue. 1984. *Gallionella ferruginea*, facteur de dénitrification dans les eaux pauvres en matière organique. C. R. Acad. Sc. Paris *298*:153–156.

Govindjee and B. Z. Braun. 1974. Light absorption, emission and photosynthesis. *In* W. D. P. Stewart, ed. Algal Physiology and Biochemistry. University of California Press, Berkeley. pp. 346–390.

Gower, A. M. 1980. Ecological effects of changes in water quality. *In* Gower, A. M., ed. Water quality in catchment ecosystems. J. Wiley & Sons, Chichester. pp. 145–171.

Gowing, H. and W. T. Momot. 1979. Impact of brook trout (*Salvelinus fontinalis*) predation on the crayfish *Orconectes virilis* in three Michigan lakes. J. Fish. Res. Board Can. *36*:1191–1196.

Grace, J. B. 1993. The adaptive significance of clonal reproduction in angiosperms: An aquatic perspective. Aquat. Bot. *44*:159–180.

Grace, J. B. and R. G. Wetzel. 1978. The production biology of Eurasian waterfilfoil (*Myriophyllum spicatum* L.): A review. J. Aquat. Plant Manage. *16*:1–11.

Grace, J. B. and R. G. Wetzel. 1981a. Phenotypic and genotypic components of growth and reproduction in *Typha latifolia*: Experimental studies in marshes of differing successional maturity. Ecology *62*:789–801.

Grace, J. B. and R. G. Wetzel. 1981b. Effects of size and growth rate on vegetative reproduction in *Typha*. Oecologia *50*:158–161.

Grace, J. B. and R. G. Wetzel. 1981c. Habitat partitioning and competitive displacement in cattails (*Typha*): Experimental field studies. Am. Nat. *118*:463–474.

Grace, J. B. and R. G. Wetzel. 1982a. Niche differentiation between two rhizomatous plant species: *Typha latifolia* and *Typha angustifolia*. Can. J. Bot. *60*:46–57.

Grace, J. B. and R. G. Wetzel. 1982b. Variations in growth and reproduction within populations of two rhizomatous plant species: *Typha latifolia* and *Typha angustifolia*. Oecologia *53*:258–263.

Grace, J. B. and R. G. Wetzel. 1998. Long-term dynamics of *Typha* populations. Aquat. Bot. *61*:137–146.

Graf, G. and R. Rosenberg. 1997. Bioresuspension and biodeposition: A review. J. Mar. Syst. *11*:269–278.

Graham, W. F., S. R. Piotrowicz, and R. A. Duce. 1979. The sea as a source of atmospheric phosphorus. Mar. Chem. *7*:325–342.

Granberg, K. and H. Harjula. 1982. On the relation of chlorophyll a to phytoplankton biomass in some Finnish freshwater lakes. Arch. Hydrobiol. Beih. Ergebn. Limnol. *16*:63–75.

Granéli, W. 1978. Sediment oxygen uptake in south Swedish lakes. Oikos *30*:7–16.

Granéli, W. 1979. The influence of *Chironomus plumosus* larvae on the oxygen uptake of sediment. Arch. Hydrobiol. *87*:385–403.

Granéli, W. 1985. Biomass response after nutrient addition to natural stands of reed, *Phragmites australis*. Verh. Internat. Verein. Limnol. *22*:2956–2961.

Granéli, W. and D. Solander. 1988. Influence of aquatic macrophytes on phosphorus cycling in lakes. Hydrobiologia *170*:245–266.

Granéli, W., S. E. B. Weisner, and M. D. Sytsma. 1992. Rhizome dynamics and resource storage in *Phragmites australis*. Wetlands Ecol. Manage. *1*:239–247.

Grant, G. E. , F. J. Swanson, and M. G. Wolman. 1990. Pattern and origin of stepped-bed morphology in high-gradient streams, western Cascades, Oregon. Geol. Soc. Am. Bull. *102*:340–352.

Grant, I. F., E. A. Egan, and M. Alexander. 1983. Measurement of rates of grazing of the ostracod *Cyprinotus carolinensis* on blue-green algae. Hydrobiologia *106*:199–208.

Grant, J. W. G. and I. A. E. Bayly. 1981. Predator induction of crests in morphs of the *Daphnia carinata* King complex. Limnol. Oceanogr. *26*:201–208.

Green, E. J. and D. E. Carritt. 1967. New tables for oxygen saturation of seawater. J. Mar. Res. *25*:140–147.

Green, J. 1993. Zooplankton associations in East African lakes spanning a wide salinity range. Hydrobiologia *267*:249–256.

Green, J. 1994. The temperate-tropical gradient of planktonic Protozoa and Rotifera. Hydrobiologia *272*:13–26.

Green, J. and O. B. Lan. 1974. *Asplanchna* and the spines of *Brachionus calyciflorus* in two Javanese sewage ponds. Freshwat. Biol. *4*:223–226.

Green, J. D. 1976. Population dynamics and production of the calanoid copepod *Calamoecia lacasi* in a northern New Zealand lake. Arch. Hydrobiol./Suppl. *50*:313–400.

Green, R. F. 1980. A note on *K*-selection. Am. Nat. *116*:291–296.

Green, S. A. and N. V. Blough. 1994. Optical absorption and fluorescence properties of chromophoric dissolved organic matter in natural waters. Limnol. Oceanogr. *39*:1903–1916.

Greenbank, J. 1945. Limnological conditions in ice-covered lakes, especially as related to winterkill of fish. Ecol. Monogr. *15*: 343–392.

Greene, K. L. 1974. Experiments and observations on the feeding behavior of the freshwater leech *Erpobdella octoculata* (L.) (Hirudinea: Erpobdellidae). Arch. Hydrobiol. *74*:87–99.

Greeney, W. J., D. A. Bella, and H. C. Curl, Jr. 1973. A theoretical approach to interspecific competition in phytoplankton communities. Am. Nat. *107*:405–425.

Greenwood, P. H. 1974. The cichlid fishes of L. Victoria, East Africa: The biology and evolution of a species flock. Bull. Br. Mus. (Nat. Hist.), Zool., *6*(Suppl.):1–134.

Gregory, J. 1989. Fundamentals of flocculation. Crit. Rev. Environ. Control *19*:185–230.

Gregory, K. J. 1976. Lichens and the determination of river channel capacity. Earth Surface Processes *1*:273–285.

Gregory, K. J. and D. E. Walling. 1973. Drainage basin form and process. Edward Arnold, London.

Gregory, S. V. 1978. Phosphorus dynamics on organic and inorganic substrates in streams. Verh. Internat. Verein. Limnol. *20*:1340–1346.

Gremm, T. J. and L. A. Kaplan. 1998. Dissolved carbohydrate concentration, composition, and bioavailability to microbial heterotrophs in stream water. Acta Hydrochim. Hydrobiol. *26*: 167–171.

Grey, D. M., ed. 1970. Handbook on the Principles of Hydrology. National Research Council of Canada, Ottawa. 676 pp.

Grieco, E. and R. Desrochers. 1978. Production de vitamin B_{12} par une algue bleue. Can. J. Microbiol. *24*:1562–1566.

Grieve, I. C. 1990. Seasonal, hydrological, and land management factors controlling dissolved organic carbon concentrations in the Loch Fleet catchments, southwest Scotland. Hydrol. Processes *4*:231–239.

Grieve, I. C. 1994. Dissolved organic carbon dynamics in two streams draining forested catchments at Loch Ard, Scotland. Hydrol. Processes *8*:457–464.

Griffith, E. J., A. Beeton, J. M. Spencer, and D. T. Mitchell. 1973. Environmental Phosphorus Handbook. John Wiley & Sons, New York. 718 pp.

Griffiths, M. 1978. Specific blue-green algal carotenoids in sediments of Esthwaite Water. Limnol. Oceanogr. *23*:777–784.

Griffiths, M. and W. T. Edmondson. 1975. Burial of oscillaxanthin in the sediment of Lake Washington. Limnol. Oceanogr. *20*: 945–952.

Griffiths, M., P. S. Perrott, and W. T. Edmondson. 1969. Oscillaxanthin in the sediment of Lake Washington. Limnol. Oceanogr. *14*:317–326.

Grigal, D. F. 1985. *Sphagnum* production in forested bogs of northern Minnesota. Can. J. Bot. *63*:1204–1207.

Grim, J. 1952. Vermehrungsleistungen planktischer Algenpopulationen in Gleichgewichtsperioden. Arch. Hydrobiol./Suppl. *20*: 238–260.

Grimm, N. B. 1987. Nitrogen dynamics during succession in a desert stream. Ecology *68*:1157–1170.

Grimm, N. B. 1993. Implications of climate change for stream communities. *In* P. M. Kareiva, J. G. Kingsolver, and R. B. Huey, eds. Biotic Interactions and Global Change. Sinauer Associates Inc., Sunderland, MA. pp. 293–314.

Grimm, N. B. and S. G. Fisher. 1984. Exchange between interstitial and surface waters: Implications for stream metabolism and nutrient cycling. Hydrobiologia *111*:219–228.

Grimm, N. B. and S. G. Fisher. 1986. Nitrogen limitation in a Sonoran Desert stream. J. N. Am. Benthol. Soc. *5*:2–15.

Grimm, N. B. and S. G. Fisher. 1989. Stability of periphyton and macroinvertebrates to disturbance by flash floods in a desert stream. J. North Am. Benthol. Soc. *8*:292–307.

Grimm, N. B., H. M. Valett, E. H. Stanley, and S. G. Fisher. 1991. The contribution of the hyporheic zone to the stability of a desert stream. Verh. Internat. Verein. Limnol. *24*:1595–1599.

Grimshaw, H. J., R. G. Wetzel, M. Brandenburg, K. Segerblom, L. J. Wenkert, G. A. Marsh, W. Charnetzky, J. E. Haky, and C. Carraher. 1997. Shading of periphyton communities by wetland emergent macrophytes: Decoupling of algal photosynthesis from microbial nutrient retention. Arch. Hydrobiol. *139*:17–27.

Grobler, D. C. and E. Daves. 1981. Sediments as a source of phosphate: A study of 38 impoundments. Water S.A. *7*:54–60.

Grobler, D. C. and M. J. Silberbauer. 1985. The combined effect of geology, phosphate sources and runoff on phosphate export from drainage basins. Water Res. *19*:975–981.

Groffman, P. M. 1994. Denitrification in freshwater wetlands. Curr. Topics Wetland Biogeochem. *1*:15–35.

Grootes, P. M. 1978. Carbon–14 time scale extended: Comparison of chronologies. Science *200*:11–21.

Gross, E. M. and R. Sütfeld. 1994. Polyphenols with algicidal activity in the submersed macrophyte *Myriophyllum spicatum* L. Acta Horticulturae *381*:710–714.

Gross, E. M., C. P. Wolk, and F. Jüttner. 1991. Fischerellin, a new allelochemical from the freshwater cyanobacterium *Fischerella musicola*. J. Phycol. *27*:686–692.

Gross, E. M., H. Meyer, and G. Schilling. 1996. Release and ecological impact of algicidal hydrolysable polyphenols in *Myriophyllum spicatum*. Phytochemistry *41*:133–138.

Grosse, W. 1996. The mechanism of thermal transpiration (= thermal osmosis). Aquat. Bot. *54*:101–110.

Grosse, W., J. Armstrong, and W. Armstrong. 1996. A history of pressurised gas-flow studies in plants. Aquat. Bot. *54*:87–100.

Groth, P. 1971. Untersuchungen über einige Spurenelemente in Seen. Arch. Hydrobiol. *68*:305–375.

Grubaugh, J. W., J. B. Wallace, and E. S. Houston. 1997. Production of benthic macroinvertebrate communities along a southern Appalachian river continuum. Freshwat. Biol. *37*:581–596.

Grubaugh, J. W., R. V. Anderson, D. M. Day, K. S. Lubinski, and R. E. Sparks. 1986. Production and fate of organic material from *Sagittaria latifolia* and *Nelumbo lutea* on Pool 19, Mississippi River. J. Freshwat. Ecol. *3*:477–484.

Gruendling, G. K. 1971. Ecology of the epipelic algal communities in Marion Lake, British Columbia. J. Phycol. *7*:239–249.

Grünsfelder, M. and W. Simonis. 1973. Aktive und inaktive Phosphataufnahme in Blattzellen von *Elodea densa* bei hohen Phosphat-Aussenkonzentrationen. Planta *115*:173–186.

Guasch, H. and S. Sabater. 1998. Estimation of the annual primary production of stream epilithic biofilms based on photosynthesis-irradiance relations. Arch. Hydrobiol. *141*:469–481.

Güde, H. 1986. Loss processes influencing growth of planktonic bacterial populations in Lake Constance. J. Plankton Res. *8*:795–810.

Güde, H. 1989. The role of grazing on bacteria in plankton succession. *In* U. Sommer, ed. Plankton Ecology: Succession in Plankton Communities. Springer-Verlag, New York. pp. 337–364.

Güde, H. 1991. Participation of baterioplankton in epilimnetic phosphorus cycles of Lake Constance. Verh. Internat. Verein. Limnol. *24*:816–820.

Güde, H., B. Haibel, and H. Müller. 1985. Development of planktonic bacterial populations in a water column of Lake Constance (Bodensee-Öbersee). Arch. Hydrobiol. *105*:59–77.

Guerinot, M. L. 1994. Microbial iron transport. Annu. Rev. Microbiol. *48*:743–772.

Guerrero, R., E. Montesinos, C. Pedrós-Alió, I. Esteve, J. Mas, H. Gemerden, P. A. G. Hofman, and J. F. Bakker. 1985. Phototrophic sulfur bacteria in two Spanish lakes: Vertical distribution and limiting factors. Limnol. Oceanogr. *30*:919–931.

Guerrero, R., E. Montesinos, I. Esteve, and C. Abella. 1980. Physiological adaptation and growth of purple and green sulfur bacteria in a meromictic lake (Vila) as compared to a holomictic lake (Siso). Developments Hydrobiol. *3*:161–171.

Guhl, B. E., B. J. Finlay, and B. Schink. 1996. Comparison of ciliate communities in the anoxic hypolimnia of three lakes: General features and the influence of lake characteristics. J. Plankton Res. *18*:335–353.

Guilizzoni, P., A. Lami, D. Ruggiu, and G. Bonomi. 1986. Stratigraphy of specific algal and bacterial carotenoids in the sediments of Lake Varese (N. Italy). Hydrobiologia *143*:321–325.

Gulati, R. D. 1978. Vertical changes in the filtering, feeding and assimilation rates of dominant zooplankters in a stratified lake. Verh. Internat. Verein. Limnol. *20*:950–956.

Gulati, R. D., J. Ejsmont-Karabin, J. Rooth, and K. Siewertsen. 1989. A laboratory study of phosphorus and nitrogen excretion of *Euchlanis dilatata lucksiana*. Hydrobiologia *186/187*:347–354.

Gulati, R. D., K. Siewertsen, and L. Van Liere. 1991. Carbon and phosphorus relationships of zooplankton and its seston food in Loosdrecht lakes. Mem. Ist. Ital. Idrobiol. *48*:279–298.

Gumiero, B. and G. Salmoiraghi. 1998. Influence of an impoundment on benthic macroinvertebrate habitat utilization. Verh. Internat. Verein. Limnol. *26*:2063–2069.

Gunnarsson, T., P. Sundin, and A. Tunlid. 1988. Importance of leaf litter fragmentation for bacterial growth. Oikos *52*:303–308.

Gunnison, D. and J. W. Barko. 1989. The rhizosphere ecology of submersed macrophytes. Water Resour. Bull. *25*:193–201.

Gunnison, D. and M. Alexander. 1975. Resistance and susceptibility of algae to decomposition by natural microbial communities. Limnol. Oceanogr. *20*:64–70.

Gusev, M. V. and K. A. Nikitina. 1974. A study of the death of blue-green algae under conditions of darkness. (Trans. Consultants Bureau, Plenum Publ. Corp.) Mikrobiologiya *43*:333–337.

Guseva, K. A. and S. P. Goncharova. 1965. O vliianii vysshei vodnoi rastitel 'nosti na razvitie planktonnykh sinezelenykh vodoroslei. *In* Ekologiia i Fiziologiia Sinezelenykh Vodoroslei. Leningrad, pp. 230–234.

Gustafsson, Ö. and P. M. Gschwend. 1997. Aquatic colloids: Concepts, definitions, and current challenges. Limnol. Oceanogr. *42*:519–528.

Guyot, J. L. and J. G. Wasson. 1994. Regional pattern of riverine dissolved organic carbon in the Amazon drainage basin of Bolivia. Limnol. Oceanogr. *39*:452–458.

Haack, T. K. and G. A. McFeters. 1982. Nutritional relationships among microorganisms in an epilithic biofilm community. Microb. Ecol. *8*:115–126.

de Haan, H. 1972. Some structural and ecological studies on soluble humic compounds from Tjeukemeer. Verh. Internat. Verein. Limnol. *18*:685–695.

de Haan, H. 1974. Effect of a fulvic acid fraction on the growth of a *Pseudomonas* from Tjeukemeer (The Netherlands). Freshwat. Biol. *4*:301–309.

de Haan, H., and T. de Boer. 1986. Geochemical aspects of aqueous iron, phosphorus and dissolved organic carbon in the humic Lake Tjeukemeer, The Netherlands. Freshwat. Biol. *16*:661–672.

de Haan, H., J. Voerman, T. de Boer, J. R. Moed, J. Schrotenboer, and H. L. Hoogveld. 1990. Trace metal chemistry of a Dutch reservoir, the Tjeukemeer. Freshwat. Biol. *24*:391–400.

de Haan, H., M. J. W. Veldhuis, and J. R. Moed. 1985. Availability of dissolved iron from Tjeukemeer, The Netherlands, for iron-limited growing *Scenedesmus quadricauda*. Water Res. *19*:235–239.

de Haan, H., R. I. Jones, and K. Salonen. 1990. Abiotic transformations of iron and phosphate in humic lake water revealed by double-isotope labeling and gel filtration. Limnol. Oceanogr. *35*:491–497.

Hackett, W. F., W. J. Connor, T. K. Kirk, and J. G. Zeikus. 1977. Microbial decomposition of synthetic ^{14}C-labeled lignins in nature: Lignin biodegradation in a variety of natural materials. Appl. Environ. Microbiol. *33*:43–51.

Hadas, O. and T. Berman. 1998. Seasonal abundance and vertical distribution of Protozoa (flagellates, ciliates) and bacteria in Lake Kinneret, Israel. Aquat. Microb. Ecol. *14*:161–170.

Hadas, O., N. Malinsky-Rushansky, R. Pinkas, and T. E. Cappenberg. 1998. Grazing on autotrophic and heterotrophic picoplankton by ciliates isolated from Lake Kinneret, Israel. J. Plankton Res. *20*:1435–1448.

Hadl, G. 1972. Zur Ökologie und Biologie der Pisidien (Bivalvia: Sphaeriidae) im Lunzer Untersee. Sitzungsber. Österr. Akad. Wissensch. Math.-nat. Kl., Abt. I *180*:317–338.

Hagedorn, H. 1971. Experimentelle Untersuchungen über den Einfluss des Thiamins auf die natürliche Algenpopulation des Pelagials. Arch. Hydrobiol. *68*:382–399.

Hagerthey, S. E. and W. C. Kerfoot. 1998. Groundwater flow influences the biomass and nutrient ratios of epibenthic algae in a north temperate seepage lake. Limnol. Oceanogr. *43*:1227–1242.

Hagmann, L. and F. Jüttner. 1996. Fischerellin A, a novel photosystem-II-inhibiting allelochemical of the cyanobacterium *Fischerella musicola* with antifungal and herbicidal activity. Tetrahedron Lett. *37*:6539–6542.

Haines, D. A. and R. A. Bryson. 1961. An empirical study of wind factor in Lake Mendota. Limnol. Oceanogr. *6*:356–364.

Haines, K. G. and R. R. L. Guillard. 1974. Growth of vitamin B_{12}-requiring marine diatoms in mixed laboratory cultures with vitamin B_{12}-producing marine bacteria. J. Phycol. *10*:245–252.

Hairston, N. G., Jr. 1976. Photoprotection by carotenoid pigments in the copepod *Diaptomus nevadensis*. Proc. Natl. Acad. Sci. USA *73*:971–974.

Hairston, N. G., Jr. 1980. The vertical distribution of diaptomid copepods in relation to body pigmentation. *In* W. C. Kerfoot, ed. Evolution and Ecology of Zooplankton Communities. Univ. Press New England, Hanover, NH. pp. 98–110.

Hairston, N. G., Jr. 1981. The interaction of salinity, predators, light and copepod color. Hydrobiologia *81*:151–158.

Hairston, N. G., Jr. 1987. Diapause as a predator-avoidance adaptation. *In* W. C. Kerfoot and A. Sih, eds. Predation: Direct and Indirect Impacts on Aquatic Communities. Univ. Press New England, Hanover, NH. pp. 281–290.

Hairston, N. G., Jr. 1996. Zooplankton egg banks as biotic reservoirs in changing environments. Limnol. Oceanogr. *41*:1087–1092.

Hairston, N. G., Jr. and C. E. Cáceres. 1996. Distribution of crustacean diapause: Micro- and macroevolutionary pattern and process. Hydrobiologia *320*:27–44.

Hairston, N. G., Jr. and N. G. Hairston, Sr. 1993. Cause-effect relationships in energy flow, trophic structure, and interspecific interactions. Am. Nat. *142*:379–411.

Hairston, N. G., Jr. and R. A. van Brunt. 1994. Diapause dynamics of two diaptomid copepod species in a large lake. Hydrobiologia *292/293*:209–218.

Hairston, N. G., Jr. and W. R. Munns, Jr. 1984. The timing of copepod diapause as an evolutionarily stable strategy. Am. Nat. *123*:733–751.

Hairston, N. G., F. E. Smith, and L. B. Slobodkin. 1960. Community structure, population control, and competition. Am. Nat. *94*:421–425.

Hairston, N. G., Jr., R. A. van Brunt, C. M. Kearns, and D. R. Engstrom. 1995. Age and suvivorship of diapausing eggs in a sediment egg bank. Ecology *76*:1706–1711.

Håkanson, L. 1981. A Manual of Lake Morphometry. Springer-Verlag, New York. 78 pp.

Halbach, U. 1970. Einfluß der Temperatur auf die Populationsdynamik des planktischen Rädertieres *Brachionus calyciflorus* Pallas. Oecologia *4*:176–204.

Halbach, U. and G. Halbach-Keup. 1974. Quantitative Beziehungen zwischen Phytoplankton und der Populationsdynamik des Rotators *Brachionus calyciflorus* Pallas. Befunde aus Laboratoriumsexperimenten und Freilanduntersuchungen. Arch. Hydrobiol. *73*:273–309.

Halemejko, G. Z. and R. J. Chróst. 1986. Enzymatic hydrolysis of proteinaceous particulate and dissolved material in an eutrophic lake. Arch. Hydrobiol. *107*:1–21.

Hall, C. A. S. 1972. Migration and metabolism in a temperate stream ecosystem. Ecology *53*:585–604.

Hall, D. J. 1964. An experimental approach to the dynamics of a natural population of *Daphnia galeata mendotae*. Ecology *45*:94–112.

Hall, D. J. and T. J. Ehlinger. 1989. Perturbation, planktivory, and pelagic community structure: The consequence of winterkill in a small lake. Can. J. Fish. Aquat. Sci. *46*:2203–2209.

Hall, D. J., E. E. Werner, J. F. Gilliam, G. G. Mittelbach, D. Howard, C. G. Doner, J. A. Dickerman, and A. J. Stewart. 1979. Diel foraging behavior and prey selection in the golden shiner (*Notemigonus crysoleucas*). J. Fish. Res. Board Can. *36*:1029–1039.

Hall, D. J., S. T. Threlkeld, C. W. Burns, and P. H. Crowley. 1976. The size-efficiency hypothesis and the size structure of zooplankton communities. Ann. Rev. Ecol. Syst. 7:177–208.

Hall, D. J., W. E. Cooper, and E. E. Werner. 1970. An experimental approach to the production dynamics and structure of freshwater animal communities. Limnol. Oceanogr. *15*:839–928.

Hall, J. B. 1971. Evolution of the prokaryotes. J. Theor. Biol. *30*:429–454.

Hall, K. J. and K. D. Hyatt. 1974. Marion Lake (IBP)—from bacteria to fish. J. Fish. Res. Board Can. *31*:893–911.

Hall, K. J. and T. G. Northcote. 1990. Production and decomposition processes in a saline meromictic lake. Hydrobiologia *197*:115–128.

Hall, K. J., P. M. Kleiber, and I. Yesaki. 1972. Heterotrophic uptake of organic solutes by microorganisms in the sediment. Mem. Ist. Ital. Idrobiol. 29(Suppl.):441–471.

Hall, R. I. and J. P. Smol. 1992. A weighted-averaging regression and calibration model for inferring total phosphorus concentration from diatoms in British Columbia (Canada) lakes. Freshwat. Biol. *27*:417–434.

Hall, R. I., P. R. Leavitt, J. P. Smol, and N. Zirnhelt. 1997. Comparison of diatoms, fossil pigments and historical records as measures of lake eutrophication. Freshwat. Biol. *38*:401–417.

Hall, R. J., C. T. Driscoll, G. E. Likens, and J. M. Pratt. 1985. Physical, chemical, and biological consequences of episodic aluminum additions to a stream. Limnol. Oceanogr. *30*:212–220.

Hall, R. L. and J. P. Smol. 1999. Diatoms as indicators of lake eutrophication. *In* E. F. Stoermer and J. P. Smol, eds. The Diatoms: Applications for the Environmental and Earth Sciences. Cambridge Univ. Press, Cambridge. pp. 128–168.

Halldal, P. 1967. Ultraviolet action spectra in algology. A review. Photochem. Photobiol. *6*:445–460.

Hallegraeff, G. M. and J. Ringelberg. 1978. Characterization of species diversity of phytoplankton assemblages by dominance-diversity curves. Verh. Internat. Verein. Limnol. *20*:939–949.

Halmann, M. and A. Elgavish. 1975. The role of phosphate in eutrophication. Stimulation of plankton growth and residence times of inorganic phosphate in Lake Kinneret water. Verh. Internat. Verein. Limnol. *19*:1351–1356.

Hama, T. and N. Handa. 1980. Molecular weight distribution and characterization of dissolved organic matter from lake waters. Arch. Hydrobiol. *90*:106–120.

Hamblin, P. F. 1982. On the free surface oscillations of Lake Ontario. Limnol. Oceanogr. *27*:1039–1049.

Hambright, K. D. and R. O. Hall. 1992. Differential zooplankton feeding behaviors, selectivities, and community impacts of two planktivorous fishes. Environ. Biol. Fishes *35*:401–411.

Hamburger, K. 1986. Energy flow in the populations of *Eudiaptomus graciloides* and *Daphnia galeata* in Lake Esrom. Arch. Hydrobiol. *105*:517–530.

Hamburger, K., P. C. Dall, and C. Lindegaard. 1994. Energy metabolism of *Chironomus anthracinus* (Diptera: Chironomidae) from the profundal zone of Lake Esrom, Denmark, as a function of body size, temperature and oxygen concentration. Hydrobiologia *294*:43–50.

Hamilton, D. P. and S. F. Mitchell. 1996. An empirical model for sediment resuspension in shallow lakes. Hydrobiologia *317*:209–220.

Hamilton, E. I. 1994. The geobiochemistry of cobalt. Sci. Total Environ. *150*:7–39.

Hamilton, S. K. and W. M. Lewis, Jr. 1990. Basin morphology in relation to chemical and ecological characteristics of lakes on the Orinoco River floodplain, Venezuela. Arch. Hydrobiol. *119*:393–425.

Hamilton, S. K. and W. M. Lewis, Jr. 1992. Stable carbon and nitrogen isotopes in algae and detritus from the Orinoco River floodplain, Venezuela. Geochim. Cosmochim. Acta *56*: 4237–4246.

Hamilton, S. K., J. M. Melack, M. F. Goodchild, and W. M. Lewis, Jr. 1992. Estimation of the fractal dimension of terrain from lake size dimensions. *In* P.A. Carling and G. E. Petts, eds. Lowland Floodplain Rivers: Geomorphological Perspectives. John Wiley & Sons Ltd., London. pp. 145–163.

Hamilton, S. K., S. J. Sippel, and J. M. Melack. 1995. Oxygen depletion and carbon dioxide and methane production in waters of the Pantanal wetland of Brazil. Biogeochemistry *30*:115–141.

Hamilton, S. K., S. J. Sippel, D. F. Calheiros, and J. M. Melack. 1997. An anoxic event and other biogeochemical effects of the Pantanal wetland on the Paraguay River. Limnol. Oceanogr. *42*:257–272.

Hamilton, S. K., W. M. Lewis, Jr., and S. J. Sippel. 1992. Energy sources for aquatic animals in the Orinoco River floodplain: Evidence from stable isotopes. Oecologia *89*:324–330.

Hamilton, W. A. 1985. Sulphate-reducing bacteria and anaerobic corrosion. Ann. Rev. Microbiol. *39*:195–217.

Hamilton-Taylor, J. and E. B. Morris. 1985. The dynamics of iron and manganese in the surface sediments of a seasonally anoxic lake. Arch. Hydrobiol./Suppl. *72*:135–165.

Hammer, U. T. 1978. The saline lakes of Saskatchewan. III. Chemical characterization. Int. Revue ges. Hydrobiol. *63*:311–335.

Hammer, U. T. 1986. Saline Lake Ecosystems of the World. Junk Publishers, Dordrecht. 616 pp.

Hammer, U. T. 1993. Zooplankton distribution and abundance in saline lakes of Alberta and Saskatchewan, Canada. Int. J. Salt Lake Res. *2*:111–132.

Hammer, U. T. and W. W. Sawchyn. 1968. Seasonal succession and congeneric associations of *Diaptomus* spp. (Copepoda) in some Saskatchewan ponds. Limnol. Oceanogr. *13*:476–484.

Hammer, U. T., J. Shamess, and R. C. Haynes. 1983. The distribution and abundance of algae in saline lakes of Saskatchewan, Canada. Hydrobiologia *105*:1–26.

Han, J.-S. 1985. Net primary production in a marsh. Mich. Bot. *24*:55–62.

Hanazato, T. 1990a. A comparison between predation effects on zooplankton communities by *Neomysis* and *Chaoborus*. Hydrobiologia *198*:33–40.

Hanazato, T. 1990b. Induction of helmet development by a *Chaoborus* factor in *Daphnia ambigua* during juvenile stages. J. Plankton Res. *12*:1287–1294.

Hanazato, T. 1991. Induction of development of high helmets by a *Chaoborus*-released chemical in *Daphnia galeata*. Arch. Hydrobiol. *122*:167–175.

Hanazato, T. 1994. Stability and diversity of a zooplankton community in experimental ponds. *In* M. Yasuno and M. M. Watanabe, eds. Biodiversity: Its Complexity and Role. Global Environmental Forum, Tokyo. pp. 177–186.

Hanazato, T. 1995. Life history responses of two *Daphnia* species of different sizes against a fish kairomone. Jpn. J. Limnol. *56*:27–32.

Haney, J. F. 1971. An *in situ* method for the measurement of zooplankton grazing rates. Limnol. Oceanogr. *16*:970–977.

Haney, J. F. 1973. An *in situ* examination of the grazing activities of natural zooplankton communities. Arch. Hydrobiol. *72*:87–132.

Haney, J. F. and D. J. Hall. 1975. Diel vertical migration and filter-feeding activities of *Daphnia*. Arch. Hydrobiol. *75*:413–441.

Haney, J. F., A. Craggy, K. Kimball, and F. Weeks. 1990. Light control of evening vertical migrations by *Chaoborus punctipennis* larvae. Limnol. Oceanogr. *35*:1068–1078.

Haney, J. F., D. J. Forsyth, and M. R. James. 1994. Inhibition of zooplankton filtering rates by dissolved inhibitors produced by naturally occurring cyanobacteria. Arch. Hydrobiol. *132*:1–13.

Haney, J. F., J. J. Sasner, and M. Ikawa. 1995. Effects of products released by *Aphanizomenon flos-aquae* and purified saxitoxin on the movements of *Daphnia carinata* feeding appendages. Limnol. Oceanogr. *40*:263–272.

Hanisch, K., B. Schweitzer, and M. Simon. 1996. Use of dissolved carbohydrates by planktonic bacteria in a mesotrophic lake. Microb. Ecol. *31*:41–55.

Hanlon, R. D. G. 1981. Allochthonous plant litter as a source of organic material in an oligotrophic lake (Llyn Frongoch). Hydrobiologia *80*:257–261.

Hannon, G. E. and M.-J. Gaillard. 1997. The plant-macrofossil record of past lake-level changes. J. Paleolimnol. *18*:15–28.

Hansen, A-M. and E. Jeppesen. 1992. Life cycle of *Cyclops vicinus* in relation to food availability, predation, diapause and temperature. J. Plankton Res. *14*:591–605.

Hansen, K. 1959a. Sediments from Danish lakes. J. Sediment. Petrol. *29*:38–46.

Hansen, K. 1959b. The terms gyttja and dy. Hydrobiologia *13*:309–315.

Hansen, K. 1961. Lake types and lake sediments. Verh. Internat. Verein. Limnol. *14*:285–290.

Hansen, L., G. F. Krog, and M. Søndergaard. 1986. Decomposition of lake phytoplankton. 1. Dynamics of short-term decomposition. Oikos *46*:37–44.

Hanson, B. J., K. W. Cummins, J. R. Barnes, and M. W. Carter. 1984. Leaf litter processing in aquatic systems: A two variable model. Hydrobiologia *111*:21–29.

Hanson, J. M., W. C. Mackay, and E. E. Prepas. 1988a. The effects of water depth and density on the growth of a unionid clam. Freshwat. Biol. *19*:345–355.

Hanson, J. M., W. C. Mackay, and E. E. Prepas. 1988b. Population size, growth, and production of a unionid clam, *Anodonta grandis simpsoniana*, in a small, deep Boreal Forest lake in central Alberta. Can. J. Zool. *66*:247–253.

Hanson, M. A., M. G. Butler, J. L. Richardson, and J. L. Arndt. 1990. Indirect effects of fish predation on calcite supersaturation, precipitation and turbidity in a shallow prairie lake. Freshwat. Biol. *24*:547–556.

Hansson, L.-A. 1989. The influence of a periphytic biolayer on phosphorus exchange between substrate and water. Arch. Hydrobiol. *115*:21–26.

Hansson, L.-A. 1992. Factors regulating periphytic algal biomass. Limnol. Oceanogr. *33*:121–128.

Hansson, L.-A., E. Bergman, and G. Cronberg. 1998a. Size structure and succession in phytoplankton communities: The impact of interactions between herbivory and predation. Oikos *81*:337–345.

Hansson, L.-A., J. Annadotter, E. Bergman, S. F. Hamrin, E. Jeppesen, T. Kairesalo, E. Luokkanen, P.-Å. Nilsson, M. Søndergaard, and J. Strand. 1998b. Biomanipulation as an application of food-chain theory: Constraints, synthesis, and recommendations for temperate lakes. Ecosystems *1*:558–575.

Hanuàová, J. 1962. Ein Beitrag zum Studium des Schwefelkreislaufes während der Sommerstagnation in der Talsperre Sedlice. Sci. Pap. Inst. Chem. Technol. Prague Technol. Water *6*:177–191.

Happey-Wood, C. 1976. Vertical migration patterns in phytoplankton of mixed species composition. Br. Phycol. J. *11*:355–369.

Happey-Wood, C. M. 1991. Temporal and spatial patterns in the distribution and abundance of pico, nano and microphytoplankton in an upland lake. Freshwat. Biol. *26*:453–480.

Happey-Wood, C. M. and A. H. Lund. 1994. Production of new organic carbon and its distribution between autotrophic picoplankton, bacteria, extracellular organic carbon and phytoplankton in an upland lake. Freshwat. Biol. *31*:1–18.

Harbison, G. R. and V. L. McAlister. 1980. Fact and artifact in copepod feeding experiments. Limnol. Oceanogr. *25*:971–981.

Hardman, Y. 1941. The surface tension of Wisconsin lake waters. Trans. Wis. Acad. Sci. Arts Lett. *33*:395–404.

Hardman, Y. and A. T. Henrici. 1939. Studies of freshwater bacteria. V. The distribution of *Siderocapsa treubii* in some lakes and streams. J. Bact. *37*:97–104.

Hardy, J. T. 1973. Phytoneuston ecology of a temperate marine lagoon. Limnol. Oceanogr. *18*:525–533.

Hardy, R. W. F., R. C. Burns, and R. D. Holsten. 1973. Applications of the acetylene-ethylene assay for measurement of nitrogen fixation. Soil Biol. Biochem. *5*:47–81.

Hare, L. and J. C. H. Carter. 1986. The benthos of a natural West African lake, with emphasis on the diel migrations and lunar and seasonal periodicities of the *Chaoborus* populations (Diptera, Chaoboridae). Freshwat. Biol. *16*:759–780.

Hare, L. and J. C. H. Carter. 1987. Zooplankton populations and the diets of three *Chaoborus* species (Diptera, Chaoboridae) in a tropical lake. Freshwat. Biol. *17*:275–290.

Hargrave, B. T. 1969. Epibenthic algal production and community respiration in the sediments of Marion Lake. J. Fish. Res. Board Can. *26*:2003–2026.

Hargrave, B. T. 1970a. The utilization of benthic microflora by *Hyalella azteca* (Amphipoda). J. Anim. Ecol. *39*:427–437.

Hargrave, B. T. 1970b. Distribution, growth and seasonal abundance of *Hyalella azteca* (Amphipoda) in relation to sediment microflora. J. Fish. Res. Board Can. *27*:685–699.

Hargrave, B. T. 1971. An energy budget for a deposit-feeding amphipod. Limnol. Oceanogr. *16*:99–103.

Hargrave, B. T. 1972a. A comparison of sediment oxygen uptake, hypolimnetic oxygen deficit and primary production in Lake Esrom, Denmark. Verh. Internat. Verein. Limnol. *18*:134–139.

Hargrave, B. T. 1972b. Oxidation-reduction potentials, oxygen concentration and oxygen uptake of profundal sediments in a eutrophic lake. Oikos *23*:167–177.

Hargrave, B. T. 1972c. Aerobic decomposition of sediment and detritus as a function of particle surface area and organic content. Limnol. Oceanogr. *17*:583–596.

Hargrave, B. T. 1975. Stability in structure and function of the mud-water interface. Verh. Internat. Verein. Limnol. *19*:1073–1079.

Hargrave, B. T. and G. H. Geen. 1968. Phosphorus excretion by zooplankton. Limnol. Oceanogr. *13*:332–342.

Harlin, M. M. 1973. Transfer of products between epiphytic marine algae and host plants. J. Phycol. *9*:243–248.

Harman, W. N. 1974. Snails (Mollusca: Gastropoda). *In* C. W. Hart, Jr. and S. L. H. Fuller, eds. Pollution Ecology of Freshwater Invertebrates. Academic Press, New York. pp. 275–312.

Harmon, M. E., J. F. Franklin, F. J. Swanson, P. Sollins, S. V. Gregory, J. D. Lattin, N. H. Anderson, S. P. Cline, N. G. Aumen, J. R. Sedell, G. W. Lienkaemper, K. Cromack, Jr., and K. W. Cummins. 1986. Ecology of coarse woody debris in temperate ecosystems. Adv. Ecol. Res. *15*:133–302.

Harmsworth, R. V. 1984. Längsee: A geochemical history of meromixis. Hydrobiologia *108*:219–231.

Harper, D. M. 1992. Eutrophication of Freshwaters (*sic*): Principles, Problems and Restoration. Chapman and Hall, London.

Harper, R. M., J. C. Fry, and M. A. Learner. 1981. A bacteriological investigation to elucidate the feeding biology of *Nais variabilis* (Oligochaeta: Naididae). Freshwat. Biol. *11*:227–236.

Harris, G. P. 1986. Phytoplankton ecology: Structure, function and fluctuation. Chapman and Hall, Ltd., London. 384 pp.

Harris, G. P. 1999. Comparison of the biogeochemistry of lakes and estuaries: Ecosystem processes, functional groups, hysteresis effects and interactions between macro- and microbiology. Mar. Freshwat. Res. *50*:791–811.

Harris, G. P. and J. N. A. Lott. 1973. Observations of Langmuir circulations in Lake Ontario. Limnol. Oceanogr. *18*:584–589.

Harris, G. P., S. I. Heaney, and J. F. Talling. 1979. Physiological and environmental constraints in the ecology of the planktonic dinoflagellate *Ceratium hirundinella*. Freshwat. Biol. *9*:413–428.

Harrison, F. W. 1974. Sponges (Porifera: Spongillidae). *In* C. W. Hart, Jr. and S. L. H. Fuller, eds. Pollution Ecology of Freshwater Invertebrates. Academic Press, New York. pp. 29–66.

Harrison, M. J., R. T. Wright, and R. Y. Morita. 1971. Method for measuring mineralization in lake sediments. Appl. Microbiol. *21*:698–702.

Harriss, R., K. Bartlett, S. Frolking, and P. Crill. 1991. Methane emissions from northern high-latitude wetlands. *In* Biogeocehmistry of Global Change: Radiatively Active Trace Gases. Chapman & Hall, New York. pp. 449–486.

Harriss, R. C. 1967. Silica and chloride in interstitial waters of river and lake sediments. Limnol. Oceanogr. *12*:8–12.

Harriss, R. C., E. Gorham, D. I. Sebacher, K. B. Bartlett, and P. A. Flebbe. 1985. Methane flux from northern peatlands. Nature *315*:652–654.

Hart, B. T. 1982. Uptake of trace metals by sediments and suspended particulates: A review. Hydrobiologia *91*:299–313.

Hart, B. T. and S. H. R. Davies. 1981. Trace metal speciation in three Victorian lakes. Aust. J. Mar. Freshwat. Res. *32*:175–189.

Hart, B. T., P. Freeman, I. D. McKelvie, S. Pearse, and D. G. Ross. 1991. Phosphorus spiralling in Myrtle Creek, Victoria, Australia. Verh. Internat. Verein. Limnol. *24*:2065–2070.

Hart, D. D. and C. T. Robinson. 1990. Resource limitation in a stream community: Phosphorus enrichment effects on periphyton and grazers. Ecology *71*:1494–1502.

Hart, D. D. and C. M. Finelli. 1999. Physical-biological coupling in streams: The pervasive effects of flow on benthic organisms. Annu. Rev. Ecol. Syst. *30*:363–395.

Hart, R. C. 1977. Feeding rhythmicity in a migratory copepod (*Pseudodiaptomus hessei* (Mrázek)). Freshwat. Biol. *7*:1–8.

Hart, R. C. 1981. Population dynamics and production of the tropical freshwater shrimp *Caridina nilotica* (Decopoda: Atyidae) in the littoral of Lake Sibaya. Freshwat. Biol. *11*:531–547.

Hart, R. C. 1986. Aspects of the feeding ecology of turbid water zooplankton. In situ studies of community filtration rates in silt-laden Lake le Roux, Orange River, South Africa. J. Plankton Res. *8*:401–426.

Hart, R. C. 1987. Population dynamics and production of five crustacean zooplankters in a subtropical reservoir during years of contrasting turbidity. Freshwat. Biol. *18*:287–318.

Hart, R. C. 1990. Copepod post-embryonic durations: Pattern, conformity, and predictability. The realities of isochronal and equiproportional development, and trends in the copepodid-naupliar duration ratio. Hydrobiologia *206*:175–206.

Hart, R. C. and A. C. Jarvis. 1993. *In situ* determinations of bacterial selectivity and filtration rates by five cladoceran zooplankters in a hypertrophic subtropical reservoir. J. Plankton Res. *15*:295–315.

Hart, R. C. and B. R. Allanson. 1976. The distribution and diel vertical migration of *Pseudodiaptomus hessei* (Mrázek) (Calanoida: Copepoda) in a subtropical lake in southern Africa. Freshwat. Biol. *6*:183–198.

Harter, R. D. 1968. Adsorption of phosphorus by lake sediment. Proc. Soil Sci. Soc. Amer. *32*:514–518.

Hartikainen, H., M. Pitkänen, T. Kairesalo, and L. Tuominen. 1996. Co-occurrence and potential chemical competition of phosphorus and silicon in lake sediment. Water Res. *30*:2472–2478.

Hartman, R. T. and D. L. Brown. 1967. Changes in internal atmosphere of submersed vascular hydrophytes in relation to photosynthesis. Ecology *48*:252–258.

Hartmann, H. J. 1985. Feeding of *Daphnia pulicaria* and *Diaptomus ashlandi* on mixtures of unicellular and filamentous algae. Verh. Internat. Verein. Limnol. *22*:3178–3183.

Harvey, H. H. and J. F. Coombs. Physical and chemical limnology of the lakes of Manitoulin Island. J. Fish. Res. Board Can. *28*:1883–1897.

Haselkorn, R. and W. J. Buikema. 1992. Nitrogen fixation in cyanobacteria. *In* G. Stacey, R. H. Burris, and H. J. Evans, eds. Biological nitrogen fixation. Chapman & Hall, New York. pp. 166–190.

Haslam, E. 1998. Practical polyphenolics: From structure to molecular recognition and physiological action. Cambridge Univ. Press, Cambridge. 422 pp.

Haslam, S. M. 1971a. Community regulation in *Phragmites communis* Trin. I. Monodominant strands. J. Ecol. *59*:65–73.

Haslam, S. M. 1971b. Community regulation in *Phragmites communis* Trin. II. Mixed strands. J. Ecol. *59*:75–88.

Haslam, S. M. 1973. Some aspects of the life history and autecology of *Phragmites communis* Trin. A review. Pol. Arch. Hydrobiol. *20*:79–100.

Haslam, S. M. 1978. River Plants: The Macrophytic Vegetation of Watercourses. Cambridge Univ. Press, Cambridge. 396 pp.

Hasler, A. D. 1947. Eutrophication of lakes by domestic drainage. Ecology *28*:383–395.

Hasler, A. D. and E. Jones. 1949. Demonstration of the antagonistic action of large aquatic plants on algae and rotifers. Ecology *30*:359–364.

Hasler, A. D. and W. G. Einsele. 1948. Fertilization for increasing productivity of natural inland waters. Trans. N. Amer. Wildl. Conf. *13*:527–555.

Hatch, M. D., C. B. Osmond, and R. O. Slatyer, eds. 1971. Photosynthesis and Photorespiration. John Wiley & Sons, New York. 565 pp.

Haury, L. and D. Weihs. 1976. Energetically efficient swimming behavior of negatively buoyant zooplankton. Limnol. Oceanogr. *21*:797–803.

Hausinger, R. P. 1994. Nickel enzymes in microbes. Sci. Total Environ. *148*:157–166.

Havas, M. and G. E. Likens. 1985. Toxicity of aluminum and hydrogen ions to *Daphnia catawba*, *Holopedium gibberum*, *Chaoborus punctipennis*, and *Chironomus anthrocinus* from Mirror Lake, New Hampshire. Can. J. Zool. *63*:1114–1119.

Havel, J. E. and S. I. Dodson. 1984. *Chaoborus* predation on typical and spined morphs of *Daphnia pulex*: Behavioral observations. Limnol. Oceanogr. *29*:487–494.

Havel, J. E. and S. I. Dodson. 1985. Environmental cues for cyclomorphosis in *Daphnia retrocurva* Forbes. Freshwat. Biol. *15*:469–478.

Havens, K. E. 1990. *Chaoborus* predation and zooplankton community structure in a rotifer-dominated lake. Hydrobiologia *198*:215–226.

Havens, K. and J. DeCosta. 1985. An analysis of selective herbivory in an acid lake and its importance in controlling phytoplankton community structure. J. Plankton Res. 7:207–222.

Hawes, I. H. and R. Smith. 1992. Effect of localised nutrient enrichment on the shallow epilithic periphyton of oligotrophic Lake Taupo. N. Z. J. Mar. Freshwat. Res. *27*:365–372.

Haworth, E. Y. 1972. The recent diatom history of Loch Leven, Kinross. Freshwat. Biol. *2*:131–141.

Haworth, R. D. 1971. The chemical nature of humic acid. Soil Sci. *111*:71–79.

Hayes, F. R. 1955. The effect of bacteria on the exchange of radiophosphorus at the mud-water interface. Verh. Internat. Verein. Limnol. *12*:111–116.

Hayes, F. R. 1957. On the variation in bottom fauna and fish yield in relation to trophic level and lake dimensions. J. Fish. Res. Board Can. *14*:1–32.

Hayes, F. R. 1964. The mud-water interface. Oceanogr. Mar. Biol. Rev. *2*:121–145.

Hayes, F. R. and C. C. Coffin. 1951. Radioactive phosphorus and the exchange of lake nutrients. Endeavor *10*:78–81.

Hayes, F. R. and E. H. Anthony. 1959. Lake water and sediment. VI. The standing crop of bacteria in lake sediments and its place in the classification of lakes. Limnol. Oceanogr. *4*:299–315.

Hayes, F. R. and E. H. Anthony. 1964. Productive capacity of North American lakes as related to the quantity and the trophic level of fish, the lake dimensions, and the water chemistry. Trans. Amer. Fish. Soc. *93*:53–57.

Hayes, F. R. and J. E. Phillips. 1958. Lake water and sediment. IV. Radiophosphorus equilibrium with mud, plants, and bacteria under oxidized and reduced conditions. Limnol. Oceanogr. *3*:459–475.

Hayes, F. R. and M. A. MacAulay. 1959. Lake water and sediment. V. Oxygen consumed in water over sediment cores. Limnol. Oceanogr. *4*:291–298.

Hayes, F. R., B. L. Reid, and M. L. Cameron. 1958. Lake water and sediment. II. Oxidation-reduction relations at the mud-water interface. Limnol. Oceanogr. *3*:308–317.

Hayes, F. R., J. A. McCarter, M. L. Cameron, and D. A. Livingstone. 1952. On the kinetics of phosphorus exchange in lakes. J. Ecol. *40*:202–216.

Hayes, M. H. B., P. MacCarthy, R. L. Malcolm, and R. S. Swift, eds. 1989. Humic Substances II: In Search of Structure. John Wiley & Sons, Chichester. 764 pp.

Hayne, D. W. and R. C. Ball. 1956. Benthic productivity as influenced by fish predation. Limnol. Oceanogr. *1*:162–175.

Haynes, R. R. 1988. Reproductive biology of selected aquatic plants. Ann. Missouri Bot. Gard. *75*:805–810.

Hazelwood, D. H. 1966. Illumination and turbulence effects on relative growth in *Daphnia*. Limnol. Oceanogr. *11*:212–216.

Healey, F. P. and L. L. Hendzel. 1975. Effect of phosphorus deficiency on two algae growing in chemostats. J. Phycol. *11*:303–309.

Healey, F. P. and L. L. Hendzel. 1980. Physiological indicators of nutrient deficiency in lake phytoplankton. Can. J. Fish. Aquat. Sci. *37*:442–453.

Healy, J. B., Jr., L. Y. Young, and M. Reinhard. 1980. Methanogenic decomposition of ferulic acid, a model lignin derivative. Appl. Environ. Microbiol. *39*:436–444.

Heaney, S. I. 1976. Temporal and spatial distribution of the dinoflagellate *Ceratium hirundinella* O. F. Müller within a small productive lake. Freshwat. Biol. *6*:531–542.

Heaney, S. I. and J. F. Talling. 1980a. Dynamic aspects of dinoflagellate distribution patterns in a small productive lake. J. Ecol. *68*:75–94.

Heaney, S. I. and J. F. Talling. 1980b. *Ceratium hirundinella*—ecology of a complex, mobile, and successful plant. Ann. Rept. Freshwat. Biol. Assoc. U.K. *48*:26–40.

Heaney, S. I., D. V. Chapman, and H. R. Morison. 1983. The role of the cyst stage in the seasonal growth of the dinoflagellate *Ceratium hirundinella* within a small productive lake. Br. Phycol. J. *18*:47–59.

Heath, C. W. 1988. Annual primary productivity of an Antarctic continental lake: Phytoplankton and benthic algal mat production strategies. Hydrobiologia *165*:77–87.

Heath, R. T. and G. D. Cooke. 1975. The significance of alkaline phosphatase in a eutrophic lake. Verh. Internat. Verein. Limnol. *19*:959–965.

Hebert, P. D. N. 1978a. Cyclomorphosis in natural populations of *Daphnia cephalata* King. Freshwat. Biol. *8*:79–90.

Hebert, P. D. N. 1978b. The adaptive significance of cyclomorphosis in *Daphnia*: More possibilities. Freshwat. Biol. *8*:313–320.

Hebert, P. D. N. and C. J. Emery. 1990. The adaptive significance of cuticular pigmentation in *Daphnia*. Functional Ecol. *4*:703–710.

Hebert, P. D. N. and P. M. Grewe. 1985. *Chaoborus*-induced shifts in the morphology of *Daphnia ambigua*. Limnol. Oceanogr. *30*:1291–1297.

Hecky, R. E. and H. J. Kling. 1981. The phytoplankton and protozooplankton of the euphotic zone of Lake Tanganyika: Species composition, biomass, chlorophyll content, and spatio-temporal distribution. Limnol. Oceanogr. *26*:548–564.

Hecky, R. E. and P. Kilham. 1973. Diatoms in alkaline, saline lakes: Ecology and geochemical implications. Limnol. Oceanogr. *18*:53–71.

Hecky, R. E., H. J. Kling, and G. J. Brunskill. 1986. Seasonality of phytoplankton in relation to silicon cycling and interstitial water circulation in large, shallow lakes of central Canada. Hydrobiologia *138*:117–126.

Hecky, R. E., P. Campbell, and L. L. Hendzel. 1993. The stoichiometry of carbon, nitrogen, and phosphorus in particulate matter of lakes and oceans. Limnol. Oceanogr. *38*:709–724.

Hedges, J. I., G. L. Cowie, J. E. Richey, P. D. Quay, R. Benner, M. Strom, and B. R. Forsberg. 1994. Origins and processing of organic matter in the Amazon River as indicated by carbohydrates and amino acids. Limnol. Oceanogr. *39*:743–761.

Hedin, L. L. 1990. Factors controlling sediment community respiration in a woodland stream ecosystem. Oikos *57*:94–105.

Hedin, L. O., J. J. Armesto, and A. H. Johnson. 1995. Patterns of nutrient loss from unpolluted, old-growth temperate forests: Evaluation of biogeochemical theory. Ecology *76*:493–509.

Hegel, G. W. F. 1807. Phänomenologie des Geistes. System der Wissenschaft. J. A. Goebhardt, Bamberg u. Würzburg. p. 21.

Hegi, H.-R. 1976. Schwermetalle (Fe, Mn, Cd, Cr, Cu, Pb, Zn) im

Pelagial des Bodensees (Obersee und Untersee) und des Greifensees. Schweiz. Z. Hydrol. *38*:35–47.

Heidal, M. and G. Bratbak. 1991. Production and decay of viruses in aquatic environments. Mar. Ecol. Progr. Ser. *72*:205–212.

Heinselman, M. L. 1963. Forest sites, bog processes, and peatland types in the glacial Lake Agassiz region, Minnesota. Ecol. Monogr. *33*:327–374.

Heisey, D. and K. G. Porter. 1977. The effect of ambient oxygen concentration on filtering and respiration rates of *Daphnia galeata mendotae* and *Daphnia magna*. Limnol. Oceanogr. *22*:839–845.

Heitkamp, U. 1982. Untersuchungen zur Biologie, Ökologie und Systematik limnischer Turbellarien periodischer und perennierender Kleingewässer Sudniedersachsens. Arch. Hydrobiol./Suppl. *64*: 65–188.

Hejný, S. 1960. Ökologische Charakteristik der Wasser- und Sumpfpflanzen in den slowakischen Tiefebenen (Donau- und Theissegebiet). Verlag Slowakischen Akad. Wissenschaften, Bratislava. 487 pp.

Hejný, S., J. Květ, and D. Dykyjová. 1981. Survey of biomass and net production of higher plant communities in fishponds. Folia Geobot. Phytotax., Praha *16*:73–94.

Helbing, U. W., F. A. Esteves, M. M. Tilzer, and H. H. Stabel. 1986. Influência dos produtos de decomposição da macrófita aquática *Nymphoides indica* (L.) O. Kuntze, na composição química da água da Represa do Lobo (Broa)—São Paulo. Acta Limnol. Brasil. *1*:611–637.

Heldal, M. and G. Bratbak. 1991. Production and decay of viruses in aquatic environments. Mar. Ecol. Progr. Ser. *72*:205–212.

Helder, R. J. and P. E. Zanstra. 1977. Changes of the pH at the upper and lower surface of bicarbonate assimilating leaves of *Potamogeton lucens* L. Proc. Koninklijke Nederl. Akad. Wetenschappen, Ser. C, Biol. Med. Sci. *80*:421–436.

Helder, R. J., H. B. A. Prins, and J. Schuurmans. 1974. Photorespiration in leaves of *Valisneria spiralis*, Proc. Koninklijke Nederl. Akad. Wetenschappen, Ser. C, Biol. Med. Sci. 77:338–344.

Helder, R. J., J. Boerma, and P. E. Zanstra. 1980. Uptake of carbon dioxide and bicarbonate by leaves of *Potamogeton lucens* L. Proc. Koninklijke Nederl. Akad. Wetenschappen, Ser. C, Biol. Med. Sci. *83*:151–166.

Hellebust, J. A. 1974. Extracellular products. *In* W. D. P. Stewart, ed. Algal Physiology and Biochemistry. University of California Press, Berkeley. pp. 838–863.

Heller, M. D. 1977. The phased division of the freshwater dinoflagellate *Ceratium hirundinella* and its use as a method of assessing growth in natural populations. Freshwat. Biol. *7*:527–533.

Hellsten, M. E. and J. A. E. Stenson. 1995. Cyclomorphosis in a population of *Bosmina coregoni*. Hydrobiologia *312*:1–9.

Hellström, T. 1991. The effect of resuspension on algal production in a shallow lake. Hydrobiologia *213*:183–190.

Hellström, T. 1996. An empirical study of nitrogen dynamics in lakes. Wat. Environ. Res. *68*:55–65.

Helmer, E. H., N. R. Urban, and S. J. Eisenreich. 1990. Aluminum geochemistry in peatland waters. Biogeochemistry *9*:247–276.

Hem, J. D. 1960a. Restraints on dissolved ferrous iron imposed by bicarbonate, redox potential, and pH. *In* Chemistry of Iron in Natural Water. U.S. Geol. Surv. Water-Supply Pap. *1459-B*:33–55.

Hem, J. D. 1960b. Complexes of ferrous iron with tannic acid. *In* Chemistry of Iron in Natural Water. U.S. Geol. Surv. Water-Supply Pap. *1459-D*:75–94.

Hem, J. D. 1960c. Some chemical relationships among sulfur species and dissolved ferrous iron. *In* Chemistry of Iron in Natural Water. U.S. Geol. Surv. Water-Supply Pap. *1459-C*:57–73.

Hem, J. D. 1963. Chemical equilibria and rates of manganese oxidation. *In* Chemistry of Manganese in Natural Water. U.S. Geol. Surv. Water-Survey Pap. *1667-A*, 64 pp.

Hem, J. D. 1964. Deposition and solution of manganese oxides. *In* Chemistry of Manganese in Natural Water. U.S. Geol. Surv. Water-Survey Pap. *1667-B*, 42 pp.

Hem, J. D. and M. W. Skougstad. 1960. Coprecipitation effects in solutions containing ferrous, ferric, and cupric ions. *In* Chemistry of Iron in Natural Water. U.S. Geol. Surv. Water-Survey Pap. *1459-E*:95–110.

Hem, J. D., and W. H. Cropper. 1959. Survey of ferrous-ferric chemical equilibria and redox potentials. *In* Chemistry of Iron in Natural Water. U.S. Geol. Surv. Water-Survey Pap. *1459-A*:1–30.

Hemond, H. F. 1994. Role of organic acids in acidification of fresh waters. *In* C. E. W. Steinberg and R. F. Wright, eds. Acidification of Freshwater Ecosystems: Implications for the Future. John Wiley & Sons, Chichester. pp. 103–115.

Henderson, P. A., W. D. Hamilton, and W.G. R. Crampton. 1998. Evolution and diversity in Amazonian floodplain communities. *In* D. M. Newbery, H. H. T. Prins, and N. D. Brown, eds. Dynamics of Tropical Communities. Blackwell Science, Oxford. pp. 385–419.

Hendricks, S. P. 1993. Microbial ecology of the hyporheic zone: A perspective integrating hydrology and biology. J. N. Am. Benthol. Soc. *12*:70–78.

Hendricks, S. P. 1996. Bacterial biomass, activity, and production within the hyporheic zone of a north-temperate stream. Arch. Hydrobiol. *136*:467–487.

Hendricks, S. P. and D. S. White. 1991. Physicochemical patterns within a hyporheic zone of a northern Michigan river, with comments on surface water patterns. Can. J. Fish. Aquat. Sci. *48*:1645–1654.

Hendry, G. A. F., J. D. Houghton, and S. B. Brown. 1987. The degradation of chlorophyll—a biological enigma. New Phytol. *107*:255–302.

Henrici, A. T. and E. McCoy. 1938. The distribution of heterotrophic bacteria in the bottom deposits of some lakes. Trans. Wis. Acad. Sci. Arts Lett. *31*:323–361.

Henriksen, A., D. F. Brakke, and S. A. Norton. 1988. Total organic carbon concentrations in acidic lakes in southern Norway. Environ. Sci. Technol. *22*:1103–1005.

Henriksen, K., M. B. Rasmussen, and A. Jensen. 1983. Effect of bioturbation on microbial nitrogen transformations in the sediment and fluxes of ammonium and nitrate to the overlying water. Environ. Biogeochem. Ecol. Bull. (Stockholm) *35*:193–205.

Henrikson, L., J. B. Olofsson, and H. G. Oscarson. 1982. The impact of acidification on Chironomidae (Diptera) as indicated by subfossil stratification. Hydrobiologia *86*:223–229.

Henry, E. A., L. J. Dodge-Murphy, G. N. Bigham, S. M. Klein, and C. C. Gilmour. 1995. Total mercury and methylmercury mass balance in an alkaline, hypereutrophic urban lake (Onondaga Lake, NY). Water Air Soil Poll. *80*:509–518.

Henry, R., K. Hino, J. G. Gentil, and J. G. Tundisi. 1985. Primary production and effects of enrichment with nitrate and phosphate on phytoplankton in the Barra Bonita Reservoir (State of Sao Paulo, Brazil). Int. Revue ges. Hydrobiol. *70*:561–573.

Hepher, B. 1966. Some aspects of the phosphorus cycle in fish ponds. Verh. Internat. Verein. Limnol. *16*:1293–1297.

Herbst, V. and J. Overbeck. 1978. Metabolic coupling between the alga *Oscillatoria redekei* and accompanying bacteria. Naturwissenschaften *65*:598.

Herdendorf, C. E. 1984. Inventory of the morphometric and limnologic characteristics of the large lakes of the world. Tech. Bull. Ohio Sea Grant OHSU-TB–17. 78 PP.

Hering, J. G. and F. M. M. Morel. 1988. Humic acid complexation of calcium and copper. Environ. Sci. Technol. *22*:1234–1237.

Herlihy, A. T. and A. L. Mills. 1985. Sulfate reduction in freshwater sediments receiving acid mine drainage. Appl. Environ. Microbiol. *49*:179–186.

Herman, S. S. 1963. Vertical migration of the opossum shrimp, *Neomysis americana* Smith. Limnol. Oceanogr. *8*:228–238.

Hermansson, M. and B. Dahlbäck. Bacterial activity at the air/water interface. Microb. Ecol. *9*:317–328.

Herndl, G. J., A. Brugger, S. Hager, E. Kaiser, I. Obernosterer, B. Reitner, and D. Slezak. 1997. Role of ultraviolet-B radiation on bacterioplankton and the availability of dissolved organic matter. Plant Ecol. *128*:42–51.

Herndl, G. J., G. Müller-Niklas, and J. Frick. 1993. Major role of ultraviolet-B in controlling bacterioplankton growth in the surface layer of the ocean. Nature *361*:717–719.

Heron, J. 1961. The seasonal variation of phosphate, silicate, and nitrate in waters of the English Lake District. Limnol. Oceanogr. *6*:338–346.

Herrmann, J. 1985. Reproductive strategies in *Dendrocoelum lacteum* (Turbellaria)—comparisons between Swedish and British populations. Verh. Internat. Verein. Limnol. *22*:2974–2978.

Herzig, A. 1980. Ten years quantitative date on a population of *Rhinoglena fertöensis* (Brachionidae, Monogononta). Hydrobiologia *73*:161–167.

Herzig, A. 1985. Resting eggs—a significant stage in the life cycle of crustaceans *Leptodora kindti* and *Bythotrephes longimanus*. Verh. Internat. Verein. Limnol. *22*:3088–3098.

Hessen, D. O. 1985a. Filtering structures and particle size selection in coexisting Cladocera. Oecologia *66*:368–372.

Hessen, D. O. 1985b. Selective zooplankton predation by pre-adult roach (*Rutilus rutilus*): The size-selective hypothesis versus the visibility-selective hypothesis. Hydrobiologia *124*:73–79.

Hessen, D. O. 1985c. The relation between bacterial carbon and dissolved humic compounds in oligotrophic lakes. FEMS Microbiol. Ecol. *31*:215–223.

Hessen, D. O. 1993. DNA-damage and pigmentation in alpine and arctic zooplankton as bioindicators of UV-radiation. Verh. Internat. Verein. Limnol. *25*:482–486.

Hessen, D. O. 1994. *Daphnia* responses to UV-light. Arch. Hydrobiol. Beih. Ergebn. Limnol. *43*:185–195.

Hessen, D. O. and E. van Donk. 1993. Morphological changes in *Scenedesmus* induced by substances released from *Daphnia*. Arch. Hydrobiol. *127*:129–140.

Hessen, D. O. and L. J. Tranvik, eds. 1998. Aquatic Humic Substances: Ecology and Biogeochemistry. Springer-Verlag, Berlin. 346 pp.

Hessen, D. O. and T. Andersen. 1990. Bacteria as a source of phosphorus for zooplankton. Hydrobiologia *206*:217–223.

Hessen, D. O., B. J. Faafeng, and T. Andersen. 1995. Replacement of herbivore zooplankton species along gradients of ecosystem productivity and fish predation pressure. Can. J. Fish. Aquat. Sci. *52*:733–742.

Hessen, D. O., E. T. Gjessing, J. Knulst, and E. Fjeld. 1997. TOC fluctuations in a humic lake as related to catchment acidification, season and climate. Biogeochemistry *36*:139–151.

Hessen, D. O., H. J. de Lange, and E. van Donk. 1997. UV-induced changes in phytoplankton cells and its effects on grazers. Freshwat. Biol. *38*:513–524.

Hessen, D. O., T. Andersen, and A. Lyche. 1990. Carbon metabolism in a humic lake: Pool sizes and cycling through zooplankton. Limnol. Oceanogr. *35*:84–99.

Hesslein, R. and P. Quay. 1973. Vertical eddy diffusion studies in the thermocline of a small stratified lake. J. Fish. Res. Board Can. *30*:1491–1500.

Heyer, J. and R. Suckow. 1985. Ökologische Untersuchungen der Methanoxydation in einem sauren Moorsee. Limnologica *16*:247–266.

Heyman, U. and P. Blomqvist. 1984. Diurnal variations in phytoplankton cell numbers and primary productivity in Siggeforasjön. Arch. Hydrobiol. *100*:219–233.

Heywood, J. and R. W. Edwards. 1962. Some aspects of the ecology of *Potamopyrgus jenkinsi* Smith. J. Anim. Ecol. *31*:239–250.

Hickey, J. R. and others. 1980. Initial solar irradiance determinations from Nimbus 7 cavity radiometer measurements. Science *208*:281–283.

Hickman, M. 1971a. Standing crops and primary productivity of the epipelon of two small ponds in North Somerset, U.K. Oecologia *6*:238–253.

Hickman, M. 1971b. The standing crop and primary productivity of the epiphyton attached to *Equisetum fluviatile* L. in Priddy Pool, North Somerset. Brit. Phycol. J. *6*:51–59.

Hickman, M. and F. E. Round. 1970. Primary production and standing crops of epipsammic and epipelic algae. Brit. Phycol. J. *5*:247–255.

Hietz, P. 1992. Decomposition and nutrient dynamics of reed (*Phragmites australis* (Cav.) Trin. ex Steud.) litter in Lake Neusiedl, Austria. Aquat. Bot. *43*:211–230.

Higashi, M, T. Miura, K. Tanimizu, and Y. Iwasa. 1981. Effect of the feeding activity of snails on the biomass and productivity of an algal community attached to a reed stem. Verh. Internat. Verein. Limnol. *21*:590–595.

Higgins, I. J., D. J. Best, R. C. Hammond, and D. Scott. 1981. Methane-oxidizing microorganisms. Microbiol. Rev. *45*:556–590.

Hill, B. H. 1987. *Typha* productivity in a Texas Pond: Implications for energy and nutrient dynamics in freshwater wetlands. Aquat. Bot. *27*:385–394.

Hill, S. 1992. Physiology of nitrogen fixation in free-living heterotrophs. *In* G. Stacey, R. H. Burris, and H. J. Evans, eds. Biological nitrogen fixation. Chapman & Hall, New York. pp. 87–134.

Hill, W. 1996. Effects of light. *In* R. J. Stevenson, M. L. Bothwell, and R. L. Lowe, eds. Algal Ecology: Freshwater Benthic Ecosystems. Academic Press, New York. pp. 121–148.

Hill, W. R., M. G. Ryon, and E. M. Schilling. 1995. Light limitation in a stream ecosystem: Responses by primary producers and consumers. Ecology *76*:1297–1309.

Hillbricht-Ilkowska, A. 1983. Response of planktonic rotifers to the eutrophication process and to the autumnal shift of blooms in Lake Biwa, Japan. II. Changes in fecundity and turnover time of the dominant species. Jpn. J. Limnol. *44*:107–115.

Hillbricht-Ilkowska, A. and W. Lawacz. 1983. Biotic structure and processes in the lake system of R. Jorka watershed (Masurian lakeland, Poland). Ekol. Pol. *31*:539–585.

Hillbricht-Ilkowska, A., A. Kowalczewski, and I. Spodniewska. 1972. Field experiment on the factors controlling primary production of the lake plankton and periphyton. Ekol. Polska *20*:315–326.

Hillbricht-Ilkowska, A., I. Spodniewska, T. Weglenska, and A. Karabin. 1970. The seasonal variation of some ecological efficiencies and production rates in the plankton community of several Polish lakes of different trophy. *In* Z. Kajak and A. Hillbricht-Ilkowska, eds. Productivity Problems of Freshwaters. PWN Polish Scientific Publishers, Warsaw. pp. 111–127.

Hillebrand, H., C.-D. Dürselen, D. Kirschtel, U. Pollingher, and T. Zohary. 1999. Biovolume calculation for pelagic and benthic microalgae. J. Phycol. *35*:403–424.

Hillman, T. J. 1986. Billabongs. *In* P. de Deckker and W. D. Williams, eds. Limnology in Australia. W. Junk Publ., Dordrecht. pp. 457–470.

Hillman, W. S. 1961. The Lemnaceae, or duckweeds. Bot. Rev. *27*:221–287.

Hilsenhoff, W. L. 1987. An improved biotic index of organic stream pollution. Great Lakes Entomol. *20*:31–39.

Hilton, J. 1985. A conceptual framework for predicting the occurrence of sediment focusing and sediment redistribution in small lakes. Limnol. Oceanogr. *30*:1131–1143.

Hilton, J., J. P. Lishman, and P. V. Allen. 1986. The dominant processes of sediment distribution and focusing in a small, eutrophic, monomictic lake. Limnol. Oceanogr. *31*:125–133.

Hilton, J., J. P. Lishman, T. R. Carrick, and P. V. Allen. 1991. An assessment of the sources of error in estimations of bulk sedimentary pigment concentrations and its implications for trophic status assessment. Hydrobiologia *218*:247–254.

Hines, M. E. 1996. Emissions of sulfur gases from wetlands. Mitt. Internat. Verein. Limnol. *25*:153–161.

Hino, S. 1991. A large biomass and the survival of phytoplankton beneath the ice layer of Lake Akan. Jpn. J. Limnol. *52*:153–160.

Hinton, M. J., S. L. Schiff, and M. C. English. 1997. The significance of storms for the concentration and export of dissolved organic carbon from two Precambrian Shield catchments. Biogeochemistry *36*:67–88.

Hirata, T. and K. Muraoka. 1984. Internal seiche and waves in Lake Chuzenji (Japan). Verh. Internat. Verein. Limnol. *22*:91–96.

Hoagland, K. D. and C. G. Peterson. 1990. Effects of light and wave disturbance on vertical zonation of attached microalgae in a large reservoir. J. Phycol. *26*:450–457.

Hoagland, K. D., J. P. Carder, and R. L. Spawn. 1996. Effects of organic toxic substances. *In* R. J. Stevenson, M. L. Bothwell, and R. L. Lowe, eds. Algal Ecology: Freshwater Benthic Ecosystems. Academic Press, San Diego. pp. 469–496.

Hoagland, K. D., J. R. Rosowski, M. R. Gretz, and S. C. Roemer. 1993. Diatom extracellular polymeric substances: Function, fine structure, chemistry, and physiology. J. Phycol. *29*:537–566.

Hoagland, K. D., S. C. Roemer, and J. R. Rosowski. 1982. Colonization and community structure of two periphyton assemblages, with emphasis on the diatoms (Bacillariophyceae). Am. J. Bot. *69*:188–213.

Hobbie, J. E. 1961. Summer temperatures in Lake Schrader, Alaska. Limnol. Oceanogr. *6*:326–329.

Hobbie, J. E. 1964. Carbon-14 measurements of primary production in two arctic Alaskan lakes. Verh. Internat. Verein. Limnol. *15*:360–364.

Hobbie, J. E. 1967. Glucose and acetate in freshwater: Concentrations and turnover rates. *In* H. L. Golterman and R. S. Clymo, eds. Chemical Environment in the Aquatic Habitat. N. V. Noord-Hollandsche Uitgevers Mattschappij, Amsterdam. pp. 245–251.

Hobbie, J. E. 1971. Heterotrophic bacteria in aquatic ecosystems; Some results of studies with organic radioisotopes. *In* J. Cairns, Jr., ed. The Structure and Function of Fresh-Water Microbial Communities. Res. Div. Monogr. 3, Virginia Polytechnic Inst., pp. 181–194.

Hobbie, J. E. 1973. Arctic limnology: A review. *In* M. E. Britton, ed. Alaskan Arctic Tundra. Tech. Paper No. 25, Arctic Inst. of North America, pp. 127–168.

Hobbie, J. E., ed. 1980. Limnology of Tundra Ponds. Dowden, Hutchinson, & Ross, Inc., Stroudsburg, PA. 514 pp.

Hobbie, J. E. and C. C. Crawford. 1969. Bacterial uptake of organic substrate: New methods of study and application to eutrophication. Verh. Internat. Verein. Limnol. *17*:725–730.

Hobbie, J. E. and G. E. Likens. 1973. Output of phosphorus, dissolved organic carbon and fine particulate carbon from Hubbard Brook watersheds. Limnol. Oceanogr. *18*:734–742.

Hobbie, J. E. and R. T. Wright. 1965. Competition between planktonic bacteria and algae for organic solutes. Mem. Ist. Ital. Idrobiol. *18*(Suppl.):175–185.

Hobbie, J. E., C. C. Crawford, and K. L. Webb. 1968. Amino acid flux in an estuary. Science *159*:1463–1464.

Hobbie, J. E., M. Bahr, and P. A. Rublee. 1999. Controls on microbial food webs in oligotrophic arctic lakes. Arch. Hydrobiol. Spec. Issues Advanc. Limnol. *54*:61–76.

Hobbie, J. E., T. L. Corliss, and B. J. Peterson. 1983. Seasonal patterns of bacterial abundance in an arctic lake. Arctic Alpine Res. *15*:253–259.

Hobbie, J. E., R. J. Barsdate, V. Alexander, D. W. Stanley, C. P. McRoy, R. G. Stross, D. A. Bierle, R. D. Dillon, and M. C. Miller. 1972. Carbon Flux Through a Tundra Pond Ecosystem at Barrow, Alaska. U.S. Tundra Biome Report *72–1*, 28 pp.

Hobbs, H. H. 1991. Decapoda. *In* J. H. Thorp and A. P. Covich, eds. Ecology and Classification of North American Freshwater Invertebrates. Academic Press, San Diego. pp. 823–858.

Hobbs, H. H., Jr. and E. T. Hall, Jr. 1974. Crayfishes (Decapoda: Astacidae). *In* C. W. Hart, Jr. and S. L. H. Fuller, eds. Pollution Ecology of Freshwater Invertebrates. Academic Press, New York. pp. 195–214.

Hobbs, P. V. 1974. Ice physics. Oxford Univ. Press, London. 837 pp.

Hochmüller, K. and H. Simoneth. 1980a. Zur Situation bei der Angabe von Ergebnissen im Bereich der Wasserchemie und -technologie. Teil I. Das Val wurde nicht ersatzlos gestrichen. Z. Wasser Abwasser Forsch. *13*:153–158.

Hochmüller, K. and H. Simoneth. 1980b. Zur Situation bei der Angabe von Ergebnissen im Bereich der Wasserchemie und -technologie. Teil II. Gesetzliche Einheiten, Grössen, Terminologie. Z. Wasser Abwasser Forsch. *13*:227–232.

Hocking, G. C. and M. Straškraba. 1999. The effect of light extinction on thermal stratification in reservoirs and lakes. Internat. Rev. Hydrobiol. *84*:535–556.

Hocking, P. J. 1989. Seasonal dynamics of production, and nutrient accumulation and cycling by *Phragmites australis* (Cav.) Trin. ex Stuedel in a nutrient-enriched swamp in inland Australia. I. Whole plants. Aust. J. Mar. Freshwat. Res. *40*:421–444.

Hodgson, D. A., S. W. Wright, P. A. Tyler, and N. Davies. 1998. Analysis of fossil pigments from algae and bacteria in meromictic Lake Fidler, Tasmania, and its application to lake management. J. Paleolimnol. *19*:1–22.

Hodoki, Y. and Y. Watanabe. 1998, Attenuation of solar ultraviolet radiation in eutrophic freshwater lakes and ponds. Jpn. J. Limnol. *59*:27–37.

Hodson, P. V., U. Borgmann, and H. Shear. 1979. Toxicity of copper to aquatic biota. *In* J. O. Nriagu, ed. Copper in the Environment. II. Health Effects. John Wiley & Sons, New York. pp. 307–372.

van den Hoek, D. G. Mann, and H. M. Jahns. 1995. Algae: An Introduction to Phycology. Cambridge Univ. Press, Cambridge. 627 pp.

Hoenicke, R. 1984. The effects of a fungal infection of *Diaptomus novamexicanus* eggs on the zooplankton community structure of Castle Lake, California. Verh. Internat. Verein. Limnol. *22*:573–577.

Hoffman, F. A. and J. W. Bishop. 1994. Impacts of a phosphate detergent ban on concentrations of phosphorus in the James River, Virginia. Water Res. *28*:1239–1240.

Hoffman, R. W., C. R. Goldman, S. Paulson, and G. R. Winters. 1981. Aquatic impacts of deicing salts in the central Sierra Nevada Mountains, California. Water Resour. Bull. *17*:280–285.

Hofmann, W. 1978a. *Bosmina* (*Eubosmina*) populations of Grosser Plöner See and Schöhsee lakes during late-glacial and postglacial times. Pol. Arch. Hydrobiol. *25*:167–176.

Hofmann, W. 1978b. Analysis of animal microfossils from the Grosser Segeberger See (F.R.G.). Arch. Hydrobiol. *82*:316–346.

Hofmann, W. 1982. On the coexistence of two pelagic *Filinia* species (Rotatoria) in Lake Plußsee. I. Dynamics of abundance and dispersion. Arch. Hydrobiol. *95*:125–137.

Hofmann, W. 1986. Developmental history of the Großer Plöner See and the Schöhsee (north Germany): Cladoceran analysis, with special reference to eutrophication. Arch. Hydrobiol./Suppl.*74*: 259–287.

Hofmann, W. 1988. The significance of chironomid analysis (Insecta: Diptera) for paleolimnological research. Palaeogeogr. Palaeoclimat. Palaeoecol. 62:501–509.

Hogeland, A. M. and K. T. Killingbeck. 1985. Biomass, productivity and life history traits of *Juncus militaris* Bigel. in two Rhode Island (U.S.A.) freshwater wetlands. Aquat. Bot. *22*:335–346.

Hogetsu, K., Y. Okanishi, and H. Sugawara. 1960. Studies on the antagonistic relationship between phytoplankton and rooted aquatic plants. Jpn. J. Limnol. *21*:124–130.

Holden, A. V. 1961. The removal of dissolved phosphate from lake waters by bottom deposits. Verh. Internat. Verein. Limnol. *14*:247–251.

Holdren, G. C., Jr., D. E. Armstrong, and R. F. Harris. 1977. Interstitial inorganic phosphorus concentrations in lakes Mendota and Wingra. Water Res. *11*:1041–1047.

Höll, K. 1972. Water: Examination, Assessment, Conditioning, Chemistry, Bacteriology, Biology. Walder de Gruyter, Berlin. 389 pp.

Hollibaugh, J. T. 1979. Metabolic adaptation in natural bacterial populations supplemented with selected amino acids. Estuarine Coastal Mar. Sci. 9:215–230.

Holm, N. P. and D. E. Armstrong. 1981. Role of nutrient limitation and competition in controlling the populations of *Asterionella formosa* and *Microcystis aeruginosa* in semicontinuous culture. Limnol. Oceanogr. *26*:622–634.

Holm, N. P. and J. Shapiro. 1984. An examination of lipid reserves and the nutritional status of *Daphnia pulex* fed *Aphanizomenon flos-aquae*. Limnol. Oceanogr. *29*:1137–1140.

Holm, N. P., G. G. Ganf, and J. Shapiro. 1983. Feeding and assimilation rates of *Daphnia pulex* fed *Aphanizomenon flos-aquae*. Limnol. Oceanogr. *28*:677–687.

Holm-Hansen, O. 1968. Ecology, physiology and biochemistry of bluegreen algae. Ann. Rev. Microbiol. *22*:47–70.

Holme, N. A. and A. D. McIntyre, eds. 1971. Methods for the Study of Marine Benthos. Int. Biol. Program Handbook 16. Blackwell Scientific Publications, Oxford. 334 pp.

Holmes, R. M., S. G. Fisher, and N. B. Grimm. 1994. Parafluvial nitrogen dynamics in a desert stream ecosystem. J. N. Am. Benthol. Soc. *13*:468–478.

Holmgren, S. K. 1984. Experimental lake fertilization in the Kuokkel Area, northern Sweden. Phytoplankton biomass and algal composition in natural and fertilized subarctic lakes. Int. Revue ges. Hydrobiol. *69*:781–817.

Holopainen, I. J. 1979. Population dynamics and production of *Pisidium* species (Bivalvia, Sphaeriidae) in the oligotrophic and mesohumic Lake Pääjarvi, southern Finland. Arch. Hydrobiol./Suppl. *54*:466–508.

Holopainen, I. J. and P. M. Jónasson. 1989. Reproduction of *Pisidium* (Bivalvia, Sphaeriidae) at different depths in Lake Esrom, Denmark. Arch. Hydrobiol. *116*:85–95.

Holst, R. W. and J. H. Yopp. 1979. Comparative utilization of inorganic and organic compounds as sole nitrogen sources by the submergent duckweed, *Lemna trisulca* L. Biol. Plantarum (Praha) *21*:245–252.

Holt, D. M. and E. B. G. Jones. 1983. Bacterial degradation of lignified wood cell walls in anaerobic aquatic habitats. Appl. Environ. Microbiol. *46*:722–727.

Holzmann, R. 1993. Seasonal fluctuations in the diversity and compositional stability of phytoplankton communities in small lakes in upper Bavaria. Hydrobiologia *249*:101–109.

Hongve, D. 1980. Chemical stratification and stability of meromictic lakes in the upper Romerike district. Schweiz. Z. Hydrol. *42*:171–195.

Hongve, D. 1993. Total and reactive aluminum concentrations in non-turbid Norwegian surface waters. Verh. Internat. Verein. Limnol. *25*:133–136.

Hongve, D. and G. Åkesson. 1996. Spectrophotometric determination of water colour in Hazen Units. Wat. Res. *30*:2771–2775.

Hood, D. W., ed. 1970. Organic matter in natural waters. Occas. Publ. Inst. Mar. Sci. Univ. Alaska *1*, 625 pp.

Hooke, J. M. 1980. Magnitude and distribution of rates of river bank erosion. Earth Surface Processes *5*:143–157.

Hooper, F. F. and A. M. Elliott. 1953. Release of inorganic phosphorus from extracts of lake mud by protozoa. Trans. Amer. Microsc. Soc. *72*:276–281.

Hooper, N. M. and G. G. C. Robinson. 1976. Primary production of epiphytic algae in a marsh pond. Can. J. Bot. *54*:2810–2815.

Hooper, R. P. and C. A. Shoemaker. 1985. Aluminum mobilization in an acidic headwater stream: Temporal variation and mineral dissolution disequilibria. Science *229*:463–465.

Hooper-Reid, N. M. and G. G. C. Robinson. 1978. Seasonal dynamics of epiphytic algal growth in a marsh pond: Productivity, standing crop, and community composition. Can. J. Bot. *56*:2434–2440.

Hope, B. K. 1994. A global biogeochemical budget for vanadium. Sci. Total Environ. *141*:1–10.

Hope, D., M. F. Billett, and M. S. Cresser. 1994. A review of the export of carbon in river water: Fluxes and processes. Environ. Poll. *84*:301–324.

Hopkins, C. L. 1976. Estimate of biological production in some stream invertebrates. N. Z. J. Mar. Freshwat. Res. *10*:629–640.

Hoppe, H.-G. 1983. Significance of exoenzymatic activities in the ecology of brackish water: Measurements by means of methylumbelliferyl substrates. Mar. Ecol. Progr. Ser. *11*:299–308.

Hoppe, H.-G. 1991. Microbial extracellular enzyme activity: A new key parameter in aquatic ecology. *In* R. J. Chróst, ed. Microbial Enzymes in Aquatic Environments. Springer-Verlag, New York. pp. 60–83.

Hordijk, K. A., C. P. M. M. Haagenars, and T. E. Cappenberg. 1985. Kinetic studies of bacterial sulfate reduction in freshwater sediments by high-pressure liquid chromatography and microdistillation. Appl. Environ. Microbiol. *49*:434–440.

Horie, S. 1962. Morphometric features and the classification of all the lakes in Japan. Mem. College Sci. Univ. Kyoto (Ser. B) *29*:191–262.

Horn, W., C. H. Mortimer, and D. J. Schwab. 1986. Wind-induced internal seiches in Lake Zurich observed and modeled. Limnol. Oceanogr. *31*:1232–1254.

Hornberger, G. M., K. E. Bencala, and D. M. McKnight. 1994. Hydrological controls on dissolved organic carbon during snowmelt in the Snake River near Montezuma, Colorado. Biogeochemistry *25*:147–165.

Horne, A. J. 1979. Nitrogen fixation in Clear Lake, California. 4. Diel studies on *Aphanizomenon* and *Anabaena* blooms. Limnol. Oceanogr. *24*:329–341.

Horne, A. J. and C. R. Goldman. 1972. Nitrogen fixation in Clear Lake, California. I. Seasonal variation and the role of heterocysts. Limnol. Oceanogr. *17*:678–692.

Horne, A. J. and D. L. Galat. 1985. Nitrogen fixation in an

oligotrophic, saline desert lake: Pyramid Lake, Nevada. Limnol. Oceanogr. *30*:1229–1239.

Horne, A. J. and G. E. Fogg. 1970. Nitrogen fixation in some English lakes. Proc. Roy. Soc. London (Ser. B) *175*:351–366.

Horne, A. J. and J. W. Carmiggelt. 1975. Algal nitrogen fixation in Californian streams: Seasonal cycles. Freshwat. Biol. *5*:461–470.

Horne, A. J., J. C. Sandusky and C. J. W. Carmiggelt. 1979. Nitrogen fixation in Clear Lake, California. 3. Repetitive synoptic sampling of the spring *Aphanizomenon* blooms. Limnol. Oceanogr. *24*:316–328.

Horne, A. J., J. E. Dillard, D. K. Fujita, and C. R. Goldman. 1972. Nitrogen fixation in Clear Lake, California. II. Synoptic studies on the autumn *Anabaena* bloom. Limnol. Oceanogr. *17*:693–703.

Horne, R. A., ed. 1972. Water and Aqueous Solutions. Structure, Thermodynamics, and Transport Processes. Wiley-Interscience, New York. 837 pp.

Hornick, L. E., J. R. Webster, and E. F. Benfield. 1981. Periphyton production in an Appalachian mountain trout stream. Am. Midland Nat. *106*:22–36.

Horvath, R. S. 1972. Microbial co-metabolism and the degradation of organic compounds in nature. Bacteriol. Rev. *36*:146–155.

Hosono, T. and I. Nouchi. 1997. Effect of gas pressure in the root and stem base zone on methane transport through rice bodies. Plant Soil *195*:65–73.

Hosper, S. H. and E. Jagtman. 1990. Biomanipulation additional control to nutrient control for restoration of shallow lakes in The Netherlands. Hydrobiologia *200*:523–534.

Hough, J. L. 1958. Geology of the Great Lakes. University of Illinois Press, Urbana. 313 pp.

Hough, R. A. 1979. Photosynthesis, respiration, and organic carbon release in *Elodea canadensis* Michx. Aquatic Bot. 7:1–11.

Hough, R. A. 1974. Photorespiration and productivity in submersed aquatic vascular plants. Limnol. Oceanogr. *19*:912–927.

Hough, R. A. and G. J. Filbin. 1990. Photosynthesis, photorespiration, and organic carbon release in *Nymphaea tuberosa* Paine and *Nuphar variegatum* Engelm. J. Freshwat. Ecol. *5*:307–312.

Hough, R. A. and R. G. Wetzel. 1972. A ^{14}C-assay for photorespiration in aquatic plants. Plant Physiol. *49*:987–990.

Hough, R. A. and R. G. Wetzel. 1975. The release of dissolved organic carbon from submersed aquatic macrophytes: Diel, seasonal, and community relationships. Verh. Internat. Verein. Limnol. *19*:939–948.

Hough, R. A. and R. G. Wetzel. 1977. Photosynthetic pathways of some aquatic plants. Aquatic Bot. *3*:297–313.

Hough, R. A. and R. G. Wetzel. 1978. Photorespiration and CO_2 compensation point in *Najas flexilis*. Limnol. Oceanogr. *23*:719–724.

Houle, D. and R. Carignan. 1995. Role of SO_4 adsorption and desorption in the long-term S budget of a coniferous catchment on the Canadian Shield. Biogeochemistry *28*:161–182.

House, W. A. 1984. The kinetics of calcite precipitation and related processes. Ann. Rept. Freshwat. Biol. Assoc. *52*:75–90.

House, W. A. and L. Donaldson. 1986. Adsorption and coprecipitation of phosphate on calcite. J. Colloid Interface Sci. *112*: 309–324.

House, W. A., H. Casey, L. Donaldson, and S. Smith. 1986. Factors affecting the coprecipitation of inorganic phosphate with calcite in hardwaters. I. Laboratory studies. Water Res. *20*:917–922.

Howard, H. H. and S. W. Chisholm. 1975. Seasonal variation of manganese in a eutrophic lake. Amer. Midland Nat. *93*:188–197.

Howard-Williams, C. 1981. Studies on the ability of a *Potamogeton pectinatus* community to remove dissolved nitrogen and phosphorus compounds from lake water. J. Appl. Ecol. *18*:619–637.

Howard-Williams, C. and B. R. Allanson. 1981. Phosphorus cycling in a dense *Potamogeton pectinatus* L. bed. Oecologia *49*:56–66.

Howard-Williams, C. and B. R. Davies. 1978. The influence of periphyton on the surface structure of a *Potamogeton pectinatus* L. leaf (an hypothesis). Aquatic Bot. *5*:87–91.

Howard-Williams, C. and G. M. Lenton. 1975. The role of the littoral zone in the functioning of a shallow tropical lake ecosystem. Freshwat. Biol. *5*:445–459.

Howard-Williams, C. and W. Howard-Williams. 1978. Nutrient leaching from the swamp vegetation of Lake Chilwa, a shallow African lake. Aquatic Bot. *4*:257–267.

Howard-Williams, C. and W. J. Junk. 1976. The decomposition of aquatic macrophytes in the floating meadows of a central Amazonian Várzea Lake. Biogeographica 7:115–123.

Howard-Williams, C., J. Davies, and S. Pickmere 1982. The dynamics of growth, the effects of changing area and nitrate uptake by watercress *Nasturtium officinale* R. Br. in a New Zealand stream. J. Appl. Ecol. *19*:589–601.

Howard-Williams, C., S. Pickmere, and J. Davies. 1988. The effect of nutrients on aquatic plant decomposition rates. Verh. Internat. Verein. Limnol. *23*:1973–1978.

Howarth, R. W., R. Marino, and J. J. Cole. 1988b. Nitrogen fixation in freshwater, estuarine, and marine ecosystems. 2. Biogeochemical controls. Limnol. Oceanogr. *33*:688–701.

Howarth, R. W., R. Marino, J. Lane, and J. J. Cole. 1988a. Nitrogen fixation in freshwater, estuarine, and marine ecosystems. 1. Rates and importance. Limnol. Oceanogr. *33*:669–687.

Hoyer, M. V. and J. R. Jones. 1983. Factors affecting the relation between phosphorus and chlorophyll *a* in midwestern reservoirs. Can. J. Fish. Aquat. Sci. *40*:192–199.

Hoyt, P. B. 1966. Chlorophyll-type compounds in soil. II. Their decomposition. Plant Soil *25*:313–328.

Hrbáček, J. 1958. Typologie und Produktivität der teichartigen Gewässer. Verh. Internat. Verein. Limnol. *13*:394–399.

Hrbáček, J. 1959. Circulation of water as a main factor influencing the development of helmets in *Daphnia cucullata* Sars. Hydrobiologia *13*:170–185.

Hrbáček, J. 1962. Species composition and the amount of the zooplankton in relation to the fish stock. Rozpravy Ceskosl. Akad. Ved, Rada Matem. Prír. Ved *72*(10):1–114.

Hrbáček, J. and M. Dvoráková-Novotná. 1965. Plankton of four backwaters related to their size and fish stock. Rozpravy Ceskosl. Akad. Ved, Rada Matem. Prír. Ved *75*(13):1–65.

Hrbáček, J. and M. Straškraba. 1966. Horizontal and vertical distribution of temperature, oxygen, pH and water movements in Slapy Reservoir (1958–1960). Hydrobiol. Stud. *1*:7–40.

Hrbáček, J., M. Dvorákova, V. Korínek, and L. Procházkóva. 1961. Demonstration of the effect of the fish stock on the species composition of zooplankton and the intensity of metabolism of the whole plankton association. Verh. Internat. Verein. Limnol. *14*:192–195.

Huang, C.-T., K. D. Xu, G. A. McFeters, and P. S. Stewart. 1998. Spatial patterns of alkaline phosphatase expression within bacterial colonies and biofilms in response to phosphate starvation. Appl. Environ. Microbiol. *64*:1526–1531.

Hubley, J. H., J. R. Mitton, and J. F. Wilkinson. 1974. The oxidation of carbon monoxide by methane-oxidizing bacteria. Arch. Microbiol. *95*:365–368.

Hudec, P. P. and P. Sonnenfeld. 1974. Hot brines on Los Roques, Venezuela. Science *185*:440–442.

Hudon, C. and E. Bourget. 1983. The effect of light on the vertical structure of epibenthic algal communities. Bot. Mar. *26*:317–330.

Hudson, J. J., J. C. Roff, and B. K. Burnison. 1990. Measuring epilithic bacteria production in streams. Can. J. Fish. Aquat. Sci. *47*:1813–1820.

Hudson, J. J., J. C. Roff, and B. K. Burnison. 1992. Bacterial productivity in forested and open streams in southern Ontario. Can. J. Fish. Aquat. Sci. *49*:2412–2422.

Huebner, J. D., D. F. Malley, and K. Donkersloot. 1990. Population ecology of the freshwater mussel *Anodonta grandis grandis* in a Precambrian Shield lake. Can. J. Zool. *68*:1931–1941.

Hühnerfuss, H., W. Walter, and G. Kruspe. 1977. On the variability of surface tension with mean wind speed. J. Phys. Oceanogr. *7*:567–571.

Hulbert, M. H. and S. Krawiec. 1977. Cometabolism: A critique. J. Theor. Biol. *69*:287–291.

Humphreys, E., W. S. Meyer, S. A. Prathapar, and D. J. Smith. 1994. Estimation of evapotranspiration from rice in southern New South Wales: A review. Aust. J. Exp. Agricul. *34*:1069–1078.

Hunding, C. 1971. Production of benthic microalgae in the littoral zone of a eutrophic lake. Oikos 22:389–397.

Hunding, C. and B. T. Hargrave. 1973. A comparison of benthic microalgal production measured by C^{14} and oxygen methods. J. Fish. Res. Board Can. *30*:309–312.

Hunter, R. D. 1980. Effects of grazing on the quantity and quality of freshwater Aufwuchs. Hydrobiologia *69*:251–259.

Hunter, S. H. and L. Provasoli. 1964. Nutrition of algae. Ann. Rev. Plant Physiol. *15*:37–56.

Huntsinger, K. R. G. and P. E. Maslin. 1976. Contribution of phytoplankton, periphyton, and macrophytes to primary production in Eagle Lake, California. Calif. Fish. Game *62*:187–194.

Hurley, J. P. and D. E. Armstrong. 1991. Pigment preservation in lake sediments: A comparison of sedimentary environments in Trout Lake, Wisconsin. Can. J. Fish. Aquat. Sci. *48*:472–486.

Hurley, J. P., D. E. Armstrong, G. J. Kenoyer, and C. J. Bowser. 1985. Ground water as a silica source for diatom production in a precipitation-dominated lake. Science *227*:1576–1578.

Huryn, A. D. 1996. An appraisal of the Allen paradox in a New Zealand trout stream. Limnol. Oceanogr. *41*:243–252.

Huryn, A. D. 1998. Ecosystem-level evidence for top-down and bottom-up control of production in a grassland stream system. Oecologia *115*:173–183.

Huryn, A. D. and J. B. Wallace. 1987. Production and litter processing by crayfish in an Appalachian mountain stream. Freshwat. Biol. *18*:277–286.

Hutchinson, G. E. 1937. A contribution to the limnology of arid regions primarily founded on observations made in the Lahontan Basin. Trans. Conn. Acad. Arts Sci. *33*:47–132.

Hutchinson, G. E. 1938. Chemical stratification and lake morphology. Proc. Nat. Acad. Sci. USA *24*:63–69.

Hutchinson, G. E. 1941. Limnological studies in Connecticut. IV. The mechanisms of intermediary metabolism in stratified lakes. Ecol. Monogr. *11*:21–60.

Hutchinson, G. E. 1944a. Limnological studies in Connecticut: VII. A critical examination of the supposed relationship between phytoplankton periodicity and chemical changes in lake water. Ecology *25*:3–26.

Hutchinson, G. E. 1944b. Nitrogen in the biogeochemistry of the atmosphere. Am. Scientist *32*:178–195.

Hutchinson, G. E. 1950. The biogeochemistry of vertebrate excretion. Bull. Amer. Mus. Nat. Hist. *96*, 554 pp.

Hutchinson, G. E. 1957a. A Treatise on Limnology. I. Geography, Physics, and Chemistry. John Wiley & Sons, New York. 1015 pp.

Hutchinson, G. E. 1957b. Concluding remarks. Cold Spring Harbor Symp. Quant. Biol. *22*:415–427.

Hutchinson, G. E. 1959. Homage to Santa Rosalia, or why are there so many kinds of animals? Am. Nat. *93*:145–159.

Hutchinson, G. E. 1961. The paradox of the plankton. Am. Nat. *95*:137–146.

Hutchinson, G. E. 1964. The lacustrine microcosm reconsidered. Am. Sci. *52*:334–341.

Hutchinson, G. E. 1967. A Treatise on Limnology. II. Introduction to Lake Biology and the Limnoplankton. John Wiley & Sons, New York. 1115 pp.

Hutchinson, G. E. 1973. Eutrophication. The scientific background of a contemporary practical problem. Am. Sci. *61*:269–279.

Hutchinson, G. E. 1974. De rebus planktonicis. Limnol. Oceanogr. *19*:360–361.

Hutchinson, G. E. 1975. A Treatise on Limnology. III. Limnological botany. John Wiley & Sons, New York. 660 pp.

Hutchinson, G. E. 1978. An Introduction to Population Ecology. Yale Univ. Press, New Haven. 260 pp.

Hutchinson, G. E. 1981. Thoughts on aquatic insects. BioScience *31*:495–500.

Hutchinson, G. E. 1993. A Treatise on Limnology. Vol. IV. The Zoobenthos. John Wiley & Sons, New York. 944 pp.

Hutchinson, G. E. and H. Löffler. 1956. The thermal classification of lakes. Proc. Nat. Acad. Sci. USA *42*:84–86.

Hutchinson, G. E. and U. M. Cowgill. 1973. The waters of Merom: A study of Lake Heleh. III. The major chemical constituents of a 54 m core. Arch. Hydrobiol. *72*:145–185.

Hutchinson, G. E. and V. T. Bowen. 1947. A direct demonstration of the phosphorus cycle in a small lake. Proc. Nat. Acad. Sci. USA *33*:148–153.

Hutchinson, G. E. and V. T. Bowen. 1950. Limnological studies in Connecticut. IX. A quantitative radiochemical study of the phosphorus cycle in Linsley Pond. Ecology *31*:194–293.

Hutchinson, G. E., E. S. Deevey, Jr., and A. Wollack. 1939. The oxidation-reduction potentials of lake waters and their ecological significance. Proc. Nat. Acad. Sci. USA *25*:87–90.

Hutchinson, G. E., *et al.* 1970. Ianula: An account of the history and development of the Lago di Monterosi, Latium, Italy. Trans. Amer. Phil. Soc. N. S. *60*(4), 178 pp.

Hwang, S.-J. and R. T. Heath. 1997. Bacterial productivity and protistan bacterivory in coastal and offshore communities of Lake Erie. Can. J. Fish. Aquat. Sci. *54*:788–799.

Hwang, Y.-H., C.-W. Fan, and M-H. Yin. 1996. Primary production and chemical composition of emergent aquatic macrophytes, *Schoenoplectus mucronatus* ssp. *robustus* and *Sparganium fallax*, in Lake Yuan-Yang, Taiwan. Biol. Bull. Acad. Sin. *37*:265–273.

Hyman, L. H. 1951. The Invertebrates: Acanthocephala, Aschelminthes, and Entoprocta. The Pseudocoelomate Bilateria, Vol. III. McGraw-Hill Book Co., New York. 572 pp.

Hynes, H. B. N. 1960. The Biology of Polluted Waters. Liverpool University Press, Liverpool. 202 pp.

Hynes, H. B. N. 1963. Imported organic matter and secondary productivity in streams. Proc. 16th Int. Congr. Zool. *4*:324–329.

Hynes, H. B. N. 1970. The Ecology of Running Waters. University of Toronto Press, Toronto. 555 pp.

Hynes, H. B. N. and B. J. Greib. 1970. Movement of phosphate and other ions from and through lake muds. J. Fish. Res. Board Can. *27*:653–668.

Hynes, H. B. N. and M. J. Coleman. 1968. A simple method of assessing the annual production of stream benthos. Limnol. Oceanogr. *13*:569–573.

Ichimura, S., N. Takase, and H. Seki. 1981. Growth and filament length of *Oscillatoria mougeotii* under metalimnetic environment. Arch. Hydrobiol. *91*:276–286.

Idso, S. B. 1973. On the concept of lake stability. Limnol. Oceanogr. *18*:681–683.

Idso, S. B. and R. G. Gilbert. 1974. On the universality of the Poole and Atkins Secchi desk-light extinction equation. J. Appl. Ecol. *11*:399–401.

Illies, J. 1953. Die Besiedlung der Fulda (insbes. das Benthos der Salmonidenregion) nach dem jetzigen Stand der Untersuchung. Berl. Limnol. Flußstat. Freudenthal *5*:1–28.

Ilmavirta, V. 1975. Diel periodicity in the phytoplankton community of the oligotrophic lake Pääjärvi, southern Finland. II. Late summer phytoplankton biomass. Ann. Bot. Fennici *12*:37–44.

Ilmavirta, V. 1983. The role of flagellated phytoplankton in chains of small brown-water lakes in southern Finland. Ann. Bot. Fennici *20*:187–195.

Ilmavirta, V., R. I. Jones, and T. Kairesalo. 1977. The structure and photosynthetic activity of pelagial and littoral plankton communities in Lake Pääjärvi, southern Finland. Ann. Bot. Fennici *14*:7–16.

Imberger, J. 1985. The diurnal mixed layer. Limnol. Oceanogr. *30*:737–770.

Imberger, J. and J. C. Patterson. 1990. Physical limnology. Adv. Appl. Mechanics *27*:303–475.

Imberger, J. and P. F. Hamblin. 1982. Dynamics of lakes, reservoirs, and cooling ponds. Ann. Rev. Fluid Mech. *14*:153–187.

Imboden, D. M. and S. Emerson. 1978. Natural radon and phosphorus as limnologic tracers: Horizontal and vertical eddy diffusion in Greifensee. Limnol. Oceanogr. *23*:77–90.

Imboden, D. M., U. Lemmin, T. Joller, and M. Schurter. 1983. Mixing processes in lakes: Mechanisms and ecological relevance. Schweiz. Z. Hydrol. *45*:11–44.

Incoll, L. D., S. P. Long, and M. R. Ashmore. 1977. SI units in publications in plant science. Current Adv. Plant Sci. *9*:331–343.

Inderjit. 1996. Plant phenolics in allelopathy. Bot. Rev. *62*:186–202.

Infante, A. 1973. Untersuchungen über die Ausnutzbarkeit verschiedener Algen durch das Zooplankton. Arch. Hydrobiol./Suppl. *42*:340–405.

Infante, A. and A. H. Litt. 1985. Differences between two species of *Daphnia* in the use of 10 species of algae in Lake Washington. Limnol. Oceanogr. *30*:1053–1059.

Ingram, L. O., C. Van Baalen, and J. A. Calder. 1973b. Role of reduced exogenous organic compounds in the physiology of the blue-green bacteria (algae): Photoheterotrophic growth of an "autotrophic" blue-green bacterium. J. Bacteriol. *114*:701–705.

Ingram, L. O., J. A. Calder, C. Van Baalen, F. E. Plucker, and P. L. Parker. 1973a. Role of reduced exogenous organic compounds in the physiology of the blue-green bacteria (algae): Photoheterotrophic growth of a "heterotrophic" blue-green bacterium. J. Bacteriol. *114*:695–700.

Ingvorsen, K., J. G. Zeikus, and T. D. Brock. 1981. Dynamics of bacterial sulfate reduction in a eutrophic lake. Appl. Environ. Microbiol. *42*:1029–1036.

Inniss, W. E. and C. I. Mayfield. 1978a. Growth rates of psychrotrophic sediment microorganisms. Water Res. *12*:231–236.

Inniss, W. E. and C. I. Mayfield. 1978b. Psychrotrophic bacteria in sediments from the Great Lakes. Water Res. *12*:237–241.

Inniss, W. E. and C. I. Mayfield. 1979. Seasonal variation of psychrotrophic bacteria in sediment from Lake Ontario. Water Res. *13*:481–484.

International Association of Limnology. 1959. Symposium on the classification of brackish waters. Arch. Oceanogr. Limnol. *11*(Suppl.):1–248.

Iriberri, J., B. Avo, I. Artolozaga, I. Barcina, and L. Egea. 1994. Grazing on allochthonous vs. autochthonous bacteria in river water. Lett. Appl. Microbiol. *18*:12–14.

Iriberri, J., B. Ayo, E. Santamaria, I. Barcina, and L. Egea. 1995. Influence of bacterial density and water temperature on the grazing activity of two freshwater ciliates. Freshwat. Biol. *33*:223–231.

Irion, G. and U. Förstner. 1975. Chemismus und Mineralbestand amazonischer See-Tone. Naturwissenschaften *62*:179.

Irvine, K., H. Balls, and B. Moss. 1990. The entomostracan and rotifer communities associated with submerged plants in the Norfolk Broadland—effects of plant biomass and species composition. Int. Revue ges. Hydrobiol. *75*:121–141.

Isirimah, N. O., D. R. Keeney, and E. H. Dettmann. 1976. Nitrogen cycling in Lake Wingra. J. Environ. Qual. *5*:182–188.

Islam, A, R. Mandal, and K. T. Osman. 1979. Direct absorption of organic phosphate by rice and jute plants. Plant Soil *53*:49–54.

Ito, O. and I. Watanabe. 1983. The relationship between combined nitrogen uptakes and nitrogen fixation in *Azolla-Anabaena* symbiosis. New Phytol. *95*:647–654.

Iversen, H. W. 1952. Laboratory study of breakers. *In* Gravity Waves. Circ. U.S. Bur. Standards *521*:9–32.

Iversen, T. M. 1973. Decomposition of autumn-shed beech leaves in a springbrook and its significance for the fauna. Arch. Hydrobiol. *72*:305–312.

Iversen, T. M. 1988. Secondary production and trophic relationships in a spring invertebrate community. Limnol. Oceanogr. *33*:582–592.

Iversen, T. M., J. Thorup, and J. Skriver. 1982. Inputs and transformation of allochthonous particulate organic matter in a headwater stream. Holarct. Ecol. *5*:10–19.

Iwakuma, T. 1987. Density, biomass, and production of Chironomidae (Diptera) in Lake Kasumigaura during 1982–1986. Jpn. J. Limnol. *48*:S59-S75.

Iwakuma, T. and K. Shibata. 1989. Production ecology of phyto- and zooplankton in a eutrophic pond dominated by *Chaoborus flavicans* (Diptera: Chaoboridae). Ecol. Res. *4*:31–53.

Iyengar, V. K. S., D. M. Davies, and H. Kleerekoper. 1963. Some relationships between Chironomidae and their substrate in nine freshwater lakes of southern Ontario, Canada. Arch. Hydrobiol. *59*:289–310.

Jack, J. D. and J. J. Gilbert. 1993a. Susceptibilities of different-sized ciliates to direct suppression by small and large cladocerans. Freshwat. Biol. *29*:19–29.

Jack, J. D. and J. J. Gilbert. 1993b. The effect of suspended clay on ciliate population growth rates. Freshwat. Biol. *29*:385–394.

Jack, J. D. and J. J. Gilbert. 1994. Effects of *Daphnia* on microzooplankton communities. J. Plankton Res. *16*:1499–1512.

Jack, J. D. and J. J. Gilbert. 1997. Effects of metazoan predators on ciliates in freshwater plankton communities. J. Euk. Microbiol. *44*:194–199.

Jackson, D. and V. Sládeček. 1970. Algal viruses—eutrophication control potential Yale Sci. Mag. *44*:16–21.

Jackson, D. A., H. H. Harvey, and K. M. Somers. 1990. Ratios in aquatic sciences: Statistical shortcomings with mean depth and the morphoedaphic index. Can. J. Fish. Aquat. Sci. *47*:1788–1795.

Jackson, J. K. and S. G. Fisher. 1986. Secondary production, emergence, and export of aquatic insects of a Sonoran Desert stream. Ecology *67*:629–638.

Jackson, M. B. and W. Armstrong. 1999. Formation of aerenchyma and the processes of plant ventilation in relation to soil flooding and submergence. Plant Biol. *1*:274–287.

Jackson, T. A. and D. W. Schindler. 1975. The bio-geochemistry of phosphorus in an experimental lake environment: Evidence for the formation of humic-metal-phosphate complexes. Verh. Internat. Verein. Limnol. *19*:211–221.

Jackson, T. A. and R. E. Hecky. 1980. Depression of primary productivity by humic matter in lake and reservoir waters of the boreal forest zone. Can. J. Fish. Aquat. Sci. *37*:2300–2317.

Jackson, T. A., G. Kipphut, R. H. Hesslein, and D. W. Schindler. 1980. Experimental study of trace metal chemistry in soft-water lakes at different pH levels. Can. J. Fish. Aquat. Sci. *37*:387–402.

Jackson, W. A. and R. J. Volk. 1970. Photorespiration. Ann. Rev. Plant Physiol. *21*:385–432.

Jacobs, J. 1961. Cyclomorphosis in *Daphnia galeata mendotae*, a case of environmentally controlled allometry. Arch. Hydrobiol. *58*:7–71.

Jacobs, J. 1962. Light and turbulence as co-determinants of relative growth rates in cyclomorphic *Daphnia*. Int. Rev. ges. Hydrobiol. *47*:146–156.

Jacobs, J. 1966. Predation and rate of evolution in cyclomorphic *Daphnia*. Verh. Internat. Verein. Limnol. *16*:1645–1652.

Jacobs, J. 1967. Untersuchungen zur Funktion und Evolution der Zyklomorphose bei *Daphnia*, mit besonderer Berücksichtigung der Selektion durch Fische. Arch. Hydrobiol. *62*: 467–541.

Jacobs, J. 1980. Environmental control of cladoceran cyclomorphosis via target-specific growth factors in the animal. *In* W. C. Kerfoot, ed. Evolution and Ecology of Zooplankton Communities. Univ. Press New England, Hanover, NH. pp. 429–437.

Jacobs, J. 1987. Cyclomorphosis in *Daphnia*. Mem. Ist. Ital. Idrobiol. *45*:325–352.

Jacobsen, D. and K. Sand-Jensen. 1994. Invertebrate herbivory on the submerged macrophyte *Potamogeton perfoliatus* in a Danish stream. Freshwat. Biol. *31*:43–52.

Jacobsen, D. and K. Sand-Jensen. 1995. Variability in invertebrate herbivory on the submerged macrophyte *Potamogeton perfoliatus*. Freshwat. Biol. *34*:357–365.

Jacobsen, T. R. and G. W. Comita. 1976. Ammonia-nitrogen excretion in *Daphnia pulex*. Hydrobiologia *51*:195–200.

Jäger, P and J. Röhrs. 1990. Phosphorfällung über Calciumcarbonat im eutrophen Wallersee (Salzburger Alpenvorland, Österreich). Int. Revue ges. Hydrobiol. *75*:153–173.

James, H. R., with E. A. Birge. 1938. A laboratory study of the absorption of light by lake waters. Trans. Wis. Acad. Sci. Arts Lett. *31*:1–154.

James, M. R., C. W. Burns, and D. J. Forsyth. 1995. Pelagic ciliated protozoa in two monomictic, southern temperate lakes of contrasting trophic state: Seasonal distribution and abundance. J. Plankton Res. *17*:1479–1500.

James, M. R., M. Weatherhead, C. Stanger, and E. Graynoth. 1998. Macroinvertebrate distribution in the littoral zone of Lake Coleridge, South Island, New Zealand—effects of habitat stability, wind exposure, and macrophytes. N. Z. J. Mar. Freshwat. Res. *32*:287–305.

James, P. W., D. L. Hawksworth, and F. Rose. 1977. Lichen communities in the British Isles: A preliminary conspectus. *In* M. R. D. Seaward, ed. Lichen Ecology. Academic Press, London. pp. 295–413.

James, W. F. and J. W. Barko. 1990. Macrophyte influences on the zonation of sediment accretion and composition in a north-temperate reservoir. Arch. Hydrobiol. *120*:129–142.

James, W. F. and J. W. Barko. 1991. Littoral-pelagic phosphorus dynamics during nighttime convective circulation. Limnol. Oceanogr. *36*:949–960.

James, W. F. and J. W. Barko. 1993. Sediment resuspension, redeposition, and focusing in a small dimictic reservoir. Can. J. Fish. Aquat. Sci. *50*:1023–1028.

James, W. F. and J. W. Barko. 1994. Macrophyte influences on sediment resuspension and export in a shallow impoundment. Lake Reserv. Manage. *10*:95–102.

James, W. F., R. H. Kennedy, and R. F. Gaugush. 1990. Effects of large-scale metalimnetic migration events on phosphorus dynamics in a north-temperate reservoir. Can. J. Fish. Aquat. Sci. *47*:156–162.

Jana, S. and M. A. Choudhuri. 1982. Changes in the activities of ribulose 1,5-bisphosphate and phosphoenolpyruvate carboxylases in submersed aquatic angiosperms during aging. Plant Physiol. *70*:1125–1127.

Janes, R. A., J. W. Eaton, and K. Hardwick. 1996. The effects of floating mats of *Azolla filiculoides* Lam.and *Lemna minuta* Kunth on the growth of submerged macrophytes. Hydrobiologia *340*:23–26.

Jannasch, H. W. 1970. Threshold concentration of carbon sources limiting bacterial growth in seawater. *In* D. W. Hood, ed. Symposium on Organic matter in Natural Waters. Occas. Publ. Univ. Alaska Mar. Sci. *1*:321–330.

Jannasch, H. W. 1975. Methane oxidation in Lake Kivu (central Africa). Limnol. Oceanogr. *20*:860–864.

Jannasch, H. W. and G. E. Jones. 1959. Bacterial populations in sea water as determined by different methods of enumeration. Limnol. Oceanogr. *4*:128–139.

Janssen, J. 1980. Alewives (*Alosa pseudoharengus*) and ciscoes (*Coregonus artedii*) as selective and non-selective planktivores. *In* W. C. Kerfoot, ed. Evolution and Ecology of Zooplankton communities. Univ. Press New England, Hanover, NH. pp. 580–586.

Jansson, M. 1976. Phosphatases in lakewater: Characterization of enzymes from phytoplankton and zooplankton by gel filtration. Science *194*:320–321.

Jansson, M. 1980. Role of benthic algae in transport of nitrogen from sediment to lake water in a shallow clearwater lake. Arch. Hydrobiol. *89*:101–109.

Jansson, M. 1981. Induction of high phosphatase activity by aluminum in acid lakes. Arch. Hydrobiol. *93*:32–44.

Jansson, M. 1984. Experimental lake fertilization: Turnover of nitrogen and phosphorus in stratified and non-stratified subarctic lakes in northern Sweden. Verh. Internat. Verein. Limnol. *22*:708–711.

Jansson, M., G. Persson, and O. Broberg. 1986. Phosphorus in acidified lakes: The example of Lake Gårdsjön, Sweden. Hydrobiologia *139*:81–96.

Jansson, M., H. Olsson, and K. Pettersson. 1988. Phosphatases: Origin, characteristics and function in lakes. Hydrobiologia *170*:157–175.

Jansson, M., H. Olsson, and O. Broberg. 1981. Characterization of acid phosphatases in the acidified Lake Gärdsjön, Sweden. Arch. Hydrobiol. 92:377–395.

Jansson, M., P. Blomqvist, A. Jonsson, and A.-K. Bergström. 1996. Nutrient limitation of bacterioplankton, autotrophic and mixotrophic phytoplankton, and heterotrophic nanoflagellates in Lake Örträsket. Limnol. Oceanogr. *41*:1552–1559.

Janus, L. L. and R. A. Vollenweider. 1984. Phosphorus residence time in relation to trophic conditions in lakes. Verh. Internat. Verein. Limnol. 22:179–184.

Jaquet, J.-M., G. Nembrini, J. Garcia, and J.-P. Vernet. 1982. The manganese cycle in Lac Léman, Switzerland: The role of *Metallogenium*. Hydrobiologia *91*:323–340.

Järnefelt, H. 1925. Zur Limnologie einiger Gewässer Finlands. Ann. Soc. Zool.-Bot. Fennicae Vanamo 2:185–352.

Jarvis, N. L. 1967. Adsorption of surface-active material at the sea-air interface. Limnol. Oceanogr. *12*:213–221.

Jarvis, N. L., W. D. Garrett, M. A. Scheiman, and C. O. Timmons. 1967. Surface chemical characterization of surface-active material in seawater. Limnol. Oceanogr. *12*:88–96.

Jassby, A. and T. Powell. 1975. Vertical patterns of eddy diffusion during stratification in Castle Lake, California. Limnol. Oceanogr. *20*:530–543.

Jassby, A. D. 1975. The ecological significance of sinking to planktonic bacteria. Can. J. Microbiol. *21*:270–274.

Jassby, A. D. and C. R. Goldman. 1974. A quantitative measure of succession rate and its application to the phytoplankton of lakes. Am. Nat. *108*:688–693.

Jassby, A. D., J. E. Reuter, R. P. Axler, C. R. Goldman, and S. H. Hackley. 1994. Atmospheric deposition of nitrogen and phosphorus in the annual nutrient load of Lake Tahoe (California-Nevada). Wat. Resour. Res. *30*:2207–2216.

Jasser, I. 1997. The dynamics and importance of picoplankton in shallow, dystrophic lake in comparison with surface waters of two deep lakes with contrasting trophic status. Hydrobiologia *342/343*:87–93.

Jaworski, G. H. M., J. F. Talling, and S. I. Heaney. 1981. The influence of carbon dioxide-depletion on growth and sinking rate of two planktonic diatoms in culture. Br. Phycol. J. *16*:395–410.

Jax, K. 1997. On functional attributes of testate amoebae in the succession of freshwater Aufwuchs. Europ. J. Protistol. *33*:219–226.

Jaynes, M. L. and S. R. Carpenter. 1986. Effects of vascular and nonvascular macrophytes on sediment redox and solute dynamics. Ecology *67*:875–882.

Jeffries, M. 1994. Invertebrate communities and turnover in wetland ponds affected by drought. Freshwat. Biol. *32*:603–612.

Jenkins, B., R. L. Kitching, and S. L. Pimm. 1992. Productivity, disturbance and food web structure at a local spatial scale in experimental container habitats. Oikos *65*:249–255.

Jensen, J. P., E. Jeppesen, K. Olrik, and P. Kristensen. 1994. Impact of nutrients and physical factors on the shift from cyanobacterial to chlorophyte dominance in shallow Danish lakes. Can. J. Fish. Aquat. Sci. *51*:1692–1699.

Jensen, K. H., P. J. Jakobsen, and O. T. Kleiven. 1998. Fish kairomone regulation of internal swarm structure in *Daphnia pulex* (Cladocera: Crustacea). Hydrobiologia *368*:123–127.

Jensen, M. L. and N. Nakai. 1961. Sources and isotopic composition of atmospheric sulfur. Science *134*:2102–2104.

Jeppesen, E., J. P. Jensen, M. Søndergaard, T. Lauridsen, L. J. Pedersen, and L. Jensen. 1997. Top-down control in freshwater lakes: The role of nutrient state, submerged macrophytes and water depth. Hydrobiologia *342/343*:151–164.

Jeppesen, E., M. Erlandsen, and M. Søndergaard. 1997. Can simple empirical equations describe the seasonal dynamics of bacterioplankton in lakes: An eight-year study in shallow hypertrophic and biologically highly dynamic Lake Søbygård, Denmark. Microb. Ecol. *34*:11–26.

Jeppesen, E., M. Søndergaard, J. P. Jensen, M. Mortensen, and O. Sortkjær. 1996. Fish-induced changes in zooplankton grazing on phytoplankton and bacterioplankton: A long-term study in shallow hypertrophic Lake Søbygaard. J. Plankton Res. *18*:1605–1625.

Jeppesen, E., Ma. Søndergaard, Mo. Søndergaard, K. Christoffersen, J. Theil-Nielsen, K. Jürgens, S. Bosselmann, and L. Schlüter. 2000. Cascading trophic interactions in the littoral zone of a shallow lake. Freshwat. Biol. (In press).

Jeppesen, E., P. Kristensen, J. P. Jensen, M. Søndergaard, E. Mortensen, and T. Lauridsen. 1991. Recovery resilience following a reduction in external phosphorus loading of shallow, eutrophic Danish lakes: Duration, regulating factors and methods for overcoming resilience. Mem. Ist. Ital. Idrobiol. *48*:127–148.

Jeppesen, E., T. L. Lauridsen, T. Kairesalo, and M. Perrow. 1998. Impact of submerged macrophytes on fish-zooplankton interactions in lakes. *In* E. Jeppesen, Ma. Søndergaard, Mo. Søndergaard, and K. Christoffersen, eds. The Structuring Role of Submersed Macrophytes in Lakes. Springer-Verlag, Berlin. pp. 91–114.

Jeschke, W. D. and W. Simonis. 1965. Über die Aufnahme von Phosphat- und Sulfationen durch Blätter von *Elodea densa* und ihre Beeinflussung durch Licht, Temperatur und Aussenkonzentration. Planta *67*:6–32.

Jespersen, D. N., B. K. Sorrell, and H. Brix. 1998. Growth and root oxygen release by *Typha latifolia* and its effects on sediment methanogenesis. Aquat. Bot. *61*:165–180.

Jewell, M. E. 1935. An ecological study of the freshwater sponges of northern Wisconsin. Ecol. Monogr. *5*:461–504.

Jewell, M. E. 1939. An ecological study of the freshwater sponges of Wisconsin. II. The influence of calcium. Ecology *20*:11–28.

Jewell, W. J. and P. L. McCarty. 1971. Aerobic decomposition of algae. Environ. Sci. Technol. *5*:1023–1031.

Jewson, D. H. 1984. Comparison of scalar and cosine instruments for measuring photosynthetically available radiation in L. Neagh, N. Ireland. Verh. Internat. Verein. Limnol. *22*:77–81.

Jewson, D. H., B. H. Rippey, and W. K. Gilmore. 1981. Loss rates from sedimentation, parasitism, and grazing during the growth, nutrient limitation, and dormancy of a diatom crop. Limnol. Oceanogr. *26*:1045–1056.

Jobling, M. 1994. Fish Bioenergetics. Chapman & Hall, London.

Johannes, R. E. 1964a. Uptake and release of phosphorus by a benthic marine amphipod. Limnol. Oceanogr. *9*:235–242.

Johannes, R. E. 1964b. Uptake and release of dissolved organic phosphorus by representatives of a coastal marine ecosystem. Limnol. Oceanogr. *9*:224–234.

Johannes, R. E. 1964c. Phosphorus excretion and body size in marine animals: Microzooplankton and nutrient regeneration. Science *146*:923–924.

Johannsson, O. E. 1995. Response of *Mysis relicta* population dynamics and productivity to spatial and seasonal gradients in Lake Ontario. Can. J. Fish. Aquat. Sci. *52*:1509–1522.

Johansson, J.-Å. 1983. Seasonal development of bacterioplankton in two forest lakes in central Sweden. Hydrobiologia *101*:71–83.

Johnson, C. E., T. G. Siccama, C. T. Driscoll, G. E. Likens, and R. E. Moeller. 1995. Changes in lead biogeochemistry in response to decreasing atmospheric inputs. Ecol. Appl. *5*:813–822.

Johnson, L. 1964. Temperature regime of deep lakes. Science *144*:1336–1337.

Johnson, L. 1966. Temperature of maximum density of fresh water and its effect on circulation in Great Bear Lake. J. Fish. Res. Board Can. *23*:963–973.

Johnson, M. G. 1988. Production by the amphipod *Pontoporeia hoyi* in South Bay, Lake Huron. Can. J. Fish. Aquat. Sci. *45*:617–624.

Johnson, M. G. and R. O. Brinkhurst. 1971. Production of benthic macroinvertebrates of Bay of Quinte and Lake Ontario. J. Fish. Res. Board Can. *28*:1699–1714.

Johnson, M. P. 1967. Temperature dependent leaf morphogenesis in *Ranunculus flabellaris*. Nature *214*:1354–1355.

Johnson, N. M. and D. H. Merritt. 1979. Convective and advective circulation of Lake Powell, Utah-Arizona during 1972–1975. Water Resour. Res. *15*:873–884.

Johnson, N. M., J. S. Eaton, and J. E. Richey. 1978. Analysis of five North American lake ecosystems. II. Thermal energy and mechanical stability. Verh. Internat. Verein. Limnol. *20*:562–567.

Johnson, N. M., R. C. Reynolds, and G. E. Likens. 1972. Atmospheric sulfur: Its effect on the chemical weathering of New England. Science *177*:514–516.

Johnson, R. K. 1987. The life history, production and food habits of *Pontoporeia affinis* Lindström (Crustacea: Amphipoda) in mesotrophic Lake Erken. Hydrobiologia *144*:277–283.

Johnson, R. K., T. Wiederholm, and D. M. Rosenberg. 1993. Freshwater biomonitoring using individual organisms, populations, and species assemblages of benthic macroinvertebrates. *In* D. M. Rosenberg and V. H. Resh, eds. Freshwater Biomonitoring and Benthic Macroinvertebrates. Chapman & Hall, New York. pp. 40–158.

Johnson, W. C. 1992. Dams and riparian forests: Case study from the upper Missouri River. Rivers *3*:229–242.

Johnson, W. C. 1998. Adjustment of riparian vegetation to river regulation in the Great Plains, USA. Wetlands *18*:608–618.

Johnston, W. R., F. Ittihadieh, R. M. Daum, and A. F. Pillsbury. 1965. Nitrogen and phosphorus in tile drainage effluent. Proc. Soil Sci. Soc. Amer. *29*:287–289.

Johnstone, I. M., D. Harding, and R. D. S. Robertson. 1988. A geothermal density current. Verh. Internat. Verein. Limnol. *23*:96–102.

Jokela, J. 1996. Within-season reproductive and somatic energy allocation in a freshwater clam, *Anodonta piscinalis*. Oecologia *105*:167–174.

Jokela, J., E. T. Valtonen, and M. Lappalainen. 1991. Development of glochidia of *Anodonta piscinalis* and their infection of fish in a small lake in northern Finland. Arch. Hydrobiol. *120*:345–355.

Jokinen, E. H. 1985. Comparative life history patterns within a littoral zone snail community. Verh. Internat. Verein. Limnol. *22*:3292–3299.

Joly, C. A. 1994. Flooding tolerance: A reinterpretation of Crawford's metabolic theory. Proc. Roy. Soc. Edinburgh *102B*:343–354.

Jónasson, P. M. 1969. Bottom fauna and eutrophication. *In* Eutrophication: Causes, Consequences, Correctives. National Academy of Sciences, Washington, DC. pp. 274–305.

Jónasson, P. M. 1972. Ecology and production of the profundal benthos in relation to phytoplankton in Lake Esrom. Oikos *14*(Suppl.), 148 pp.

Jónasson, P. M. 1978. Zoobenthos of lakes. Verh. Internat. Verein. Limnol. *20*:13–37.

Jónasson, P. M. 1979. The Lake Mývatn ecosystem, Iceland. Oikos *32*:289–305.

Jónasson, P. M. 1981. Energy flow in a subarctic, eutrophic lake. Verh. Internat. Verein. Limnol. *21*:389–393.

Jónasson, P. M. 1984a. The ecosystem of eutrophic Lake Esrom. *In* F. B.Taub, ed. Lakes and Reservoirs. Ecosystems of the World, Vol. 23. Elsevier Science Publishers, Amsterdam. pp. 177–204.

Jónasson, P. M. 1984b. Decline of zoobenthos through five decades of eutrophication in Lake Esrom. Verh. Internat. Verein. Limnol. *22*:800–804.

Jónasson, P. M. 1992. The ecosystem of Thingvallavatn: A synthesis. Oikos *64*:405–434.

Jónasson, P. M. and C. Lindegaard. 1979. Zoobenthos and its contribution to the metabolism of shallow lakes. Arch. Hydrobiol. Beih. Ergebn. Limnol. *13*:162–180.

Jónasson, P. M. and C. Lindegaard. 1988. Ecosystem studies of North Atlantic Ridge lakes. Verh. Internat. Verein. Limnol. *23*:394–402.

Jónasson, P. M. and F. Thorhauge. 1972. Life cycle of *Potamothrix hammoniensis* (Tubifidcidae) in the profundal of a eutrophic lake. Oikos *23*:151–158.

Jónasson, P. M. and H. Mathiesen. 1959. Measurements of primary production in two Danish eutrophic lakes. Esrom Sø and Furesø. Oikos *10*:137–167.

Jónasson, P. M., C. Lindegaard, and K. Hamburger. 1990. Energy budget of Lake Esrom, Denmark. Verh. Internat. Verein. Limnol. *24*:632–640.

Jones, A. K. and R. C. Cannon. 1986. The release of micro-algal photosynthate and associated bacterial uptake and heterotrophic growth. Br. Phycol. J. *21*:341–358.

Jones, B. F. and C. J. Bowser. 1978. The mineralogy and related chemistry of lake sediments. *In* A. Lerman, ed. Lakes: Chemistry, Geology, Physics. Springer-Verlag, New York. pp. 179–235.

Jones, F. E. 1992. Evaporation of Water. With Emphasis on Applications and Measurements. Lewis Publishers, Chelsea, MI. 188 pp.

Jones, J. B., Jr. 1995. Factors controlling hyporheic respiration in a desert stream. Freshwat. Biol. *34*:91–99.

Jones, J. B., Jr. and L. A. Smock. 1991. Transport and retention of particulate organic matter in two low-gradient headwater streams. J. N. Am. Benthol. Soc. *10*:115–126.

Jones, J. B., Jr. and P. J. Mulholland. 1998. Methane input and evasion in a hardwood forest stream: Effects of subsurface flow from shallow and deep pathways. Limnol. Oceanogr. *43*:1243–1250.

Jones, J. B., Jr., S. G. Fisher, and N. B. Grimm. 1995a. Nitrification in the hyporheic zone of a desert stream ecosystem. J. N. Am. Benthol. Soc. *14*:249–258.

Jones, J. B., Jr., R. M. Holmes, S. G. Fisher, N. B. Grimm, and D. M. Greene. 1995b. Methanogenesis in Arizona, USA dryland stream. Biogeochemistry *31*:155–173.

Jones, J. G. 1972. Studies on freshwater bacteria: Association with algae and alkaline phosphatase activity. J. Ecol. *60*:59–75.

Jones, J. G. 1976. The microbiology and decomposition of seston in open water and experimental enclosures in a productive lake. J. Ecol. *64*:241–278.

Jones, J. G. 1977. The effect of environmental factors on estimated viable and total populations of planktonic bacteria in lakes and experimental enclosures. Freshwat. Biol. *7*:67–91.

Jones, J. G. 1978. The distribution of some freshwater planktonic bacteria in two stratified eutrophic lakes. Freshwat. Biol. *8*:127–140.

Jones, J. G. 1979a. Microbial nitrate reduction in freshwater sediments. J. Gen. Microbiol. *115*:27–35.

Jones, J. G. 1980. Some differences in the microbiology of profundal and littoral lake sediments. J. Gen. Microbiol. *117*:285–292.

Jones, J. G. 1986. Iron transformations by freshwater bacteria. Adv. Microb. Ecol. *9*:149–185.

Jones, J. G. and B. M. Simon. 1980. Variability in microbiological data from a stratified eutrophic lake. J. Appl. Bacteriol. *49*:127–135.

Jones, J. G. and B. M. Simon. 1981. Differences in microbial decomposition processes in profundal and littoral lake sediments, with particular reference to the nitrogen cycle. J. Gen. Microbiol. *123*:297–312.

Jones, J. G., B. M. Simon, and R. W. Horsley. 1982. Microbiological sources of ammonia in freshwater lake sediments. J. Gen. Microbiol. *128*:2823–2831.

Jones, J. G., B. M. Simon, and S. Gardener. 1982. Factors affecting methanogenesis and associated anaerobic processes in the sediments of a stratified eutrophic lake. J. Gen. Microbiol. *128*:1–11.

Jones, J. G., B. M. Simon, and J. V. Roscoe. 1982. Microbiological sources of sulphide in freshwater lake sediments. J. Gen. Microbiol. *128*:2833–2839.

Jones, J. G., M. J. L. G. Orlandi, and B. M. Simon. 1979. A microbiological study of sediments from the Cumbrian lakes. J. Gen. Microbiol. *115*:37–48.

Jones, J. I., K. Hardwick, and J. W. Eaton. 1996. Diurnal carbon restrictions on the photosynthesis of dense stands of *Elodea nuttallii* (Planch.) St. John. Hydrobiologia *340*:11–16.

Jones, J. R. 1979b. Microbial activity in lake sediments with particular reference to electrode potential gradients. J. Gen. Microbiol. *115*:19–26.

Jones, J. R. and R. W. Bachmann. 1976. Prediction of phosphorus and chlorophyll levels in lakes. J. Water Poll. Control Fed. *48*:2176–2182.

Jones, J. R., M. M. Smart, and J. N. Sebaugh. 1984. Factors related to algal biomass in Missouri Ozark streams. Verh. Internat. Verein. Limnol. *22*:1867–1875.

Jones, M. B. 1988. Photosynthetic responses of C_3 and C_4 wetland species in a tropical swamp. J. Ecol. *76*:253–262.

Jones, M. B. and F. M. Muthuri. 1984. The diurnal course of plant water potential, stomatal conductance and transpiration in a papyrus (*Cyperus papyrus* L.) canopy. Oecologia *63*:252–255.

Jones, M. B. and F. M. Muthuri. 1997. Standing biomass and carbon distribution in a papyrus (*Cyperus papyrus* L.) swamp on Lake Naivasha, Kenya. J. Tropical Ecol. *13*:347–356.

Jones, M. R. and T. R. Milburn. 1978. Photosynthesis in papyrus (*Cyperus papyrus* L.). Photosynthetica *12*:197–199.

Jones, P. D. and W. T. Momot. 1981. Crayfish productivity, allochthony, and basin morphometry. Can. J. Fish. Aquat. Sci. *38*:175–183.

Jones, R. I. 1990. Phosphorus transformations in the epilimnion of humic lakes: Biological uptake of phosphate. Freshwat. Biol. *23*:323–337.

Jones, R. I. 1991. Advantages of diurnal vertical migrations to phytoplankton in sharply stratified, humic forest lakes. Arch. Hydrobiol. *120*:257–266.

Jones, R. I. and K. Salonen. 1985. The importance of bacterial utilization of released phytoplankton photosynthate in two humic forest lakes in southern Finland. Holarctic Ecol. *8*:133–140.

Jones, R. I. and V. Ilmavirta. 1978. A diurnal study of the phytoplankton in the eutrophic Lake Lovojärvi, southern Finland. Arch. Hydrobiol. *83*:494–514.

Jones, R. I., A. S. Fulcher, J. K. U. Jayakody, J. Laybourn-Parry, A. J. Shine, M. C. Walton, and J. M. Young. 1995. The horizontal distribution of plankton in a deep, oligotrophic lake—Loch Ness, Scotland. Freshwat. Biol. *33*:161–170.

Jones, R. I., J. Laybourn-Parry, M. C. Walton, and J. M. Young. 1997. The forms and distribution of carbon in a deep, oligotrophic lake (Loch Ness, Scotland). Verh. Internat. Verein. Limnol. *26*:330–334.

Jones, R. I., K. Salonen, and H. de Haan. 1988. Phosphorus transformations in the epilimnion of humic lakes: Abiotic interactions between dissolved humic materials and phosphate. Freshwat. Biol. *19*:357–369.

Jones, R. I., P. J. Shaw, and H. de Haan. 1993. Effects of dissolved humic substances on the speciation of iron and phosphate at different pH and ionic strength. Environ. Sci. Technol. *27*:1052–1059.

Jones, S. E. and M. A. Lock. 1993. Seasonal determinations of extracellular hydrolytic activities in heterotrophic and mixed heterotrophic/autotrophic biofilms from two contrasting rivers. Hydrobiologia *257*:1–16.

Jones, S. W. and R. Goulder. 1973. Swimming speeds of some ciliated Protozoa from a eutrophic pond. Naturalist (Hull, England) (No. 924): 33–35.

Jonsson, A. and M. Jansson. 1997. Sedimentation and mineralisation of organic carbon, nitrogen and phosphorus in a large humic lake, northern Sweden. Arch. Hydrobiol. *141*:45–65.

Joo, G.-J., A. K. Ward, and G. M. Ward. 1992. Ecology of *Pectinatella magnifica* (Bryozoa) in an Alabama oxbow lake: Colony growth and association with algae. J. N. Am. Benthol. Soc. *11*:324–333.

Jordan, M. and G. E. Likens. 1975. An organic carbon budget for an oligotrophic lake in New Hampshire, U.S.A. Verh. Internat. Verein. Limnol. *19*:994–1003.

Jordan, M. J. and G. E. Likens. 1980. Measurement of planktonic bacterial production in an oligotrophic lake. Limnol. Oceanogr. *25*:719–732.

Jordan, T. E. and D. E. Weller. 1996. Human contributions to terrestrial nitrogen flux. BioScience *46*:655–664.

Jørgensen, B. B. 1983. The microbial sulfur cycle. *In* W. E. Krumbein, ed. Microbial Geochemistry. Blackwell Sci. Publs., Oxford. pp. 91–124.

Jørgensen, B. B. 1990. The sulfur cycle of freshwater sediments: Role of thiosulfate. Limnol. Oceanogr. *35*:1329–1342.

Jørgensen, B. B. and D. J. Des Marais. 1988. Optical properties of benthic photosynthetic communities: Fiber-optic studies of cyanobacterial mats. Limnol. Oceanogr. *33*:99–113.

Jørgensen, B. B., N. P. Revsbech, T. H. Blackburn, and Y. Cohen. 1979. Diurnal cycle of oxygen and sulfide microgradients and microbial photosynthesis in a cyanobacterial mat sediment. Appl. Environ. Microbiol. *38*:46–58.

Jørgensen, B. B., N. P. Revsbech, and Y. Cohen. 1983. Photosynthesis and structure of benthic microbial mats: Microelectrode and SEM studies of four cyanobacterial communities. Limnol. Oceanogr. *28*:1075–1093.

Jørgensen, E. G. 1957. Diatom periodicity and silicon assimilation. Dansk Bot. Arkiv. *18*(1), 54 pp.

Jørgensen, E. G. 1964. Adaptation to different light intensities in the diatom *Cyclotella meneghiniana* Kütz. Physiol. Plant. *17*:136–145.

Jørgensen, E. G. 1968. The adaptation of plankton algae. II. Aspects of the temperature adaptation of *Skeletonema costatum*. Physiol. Plant. *21*:423–427.

Jørgensen, E. G. 1969. The adaptation of plankton algae. IV. Light adaptation in different algal species. Physiol. Plant. *22*:1307–1315.

Joseph, I. N. and C. Bravo. 1990. Solubilización de los silicatos por microorganismos. Ciencias Biológicas (Cuba) *23*:1–5.

Joshi, M. M. and J. P. Hollis. 1977. Interaction of *Beggiatoa* and rice plant: Detoxification of hydrogen sulfide in the rice rhizosphere. Science *195*:179–180.

Juday, C. 1921. Quantitative studies of the bottom fauna in the deeper waters of Lake Mendota. Trans. Wis. Acad. Sci. Arts Lett. *20*:461–493.

Juday, C. 1924. The productivity of Green Lake, Wisconsin. Verh. Internat. Verein. Limnol. *2*:357–360.

Juday, C. 1940. The annual energy budget of an inland lake. Ecology *21*:438–450.

Juday, C. 1942. The summer standing crop of plants and animals in four Wisconsin lakes. Trans. Wis. Acad. Sci. Arts Lett. *34*:103–135.

Juday, C. and E. A. Birge. 1931. A second report on the phosphorus content of Wisconsin lake waters. Trans. Wis. Acad. Sci. Arts Lett. *26*:353–382.

Juday, C. and E. A. Birge. 1932. Dissolved oxygen and oxygen consumed in the lake waters of northeastern Wisconsin. Trans. Wis. Acad. Sci. Arts Lett. *27*:415–486.

Juday, C. and E. A. Birge. 1933. The transparency, the color and the specific conductance of the lake waters of northeastern Wisconsin. Trans. Wis. Acad. Sci. Arts Lett. *28*:205–259.

Juday, C., E. A. Birge, and V. W. Meloche. 1938. Mineral content of the lake waters of northeastern Wisconsin. Trans. Wis. Acad. Sci. Arts Lett. *31*:223–276.

Juday, C., E. A. Birge, G. I. Kemmerer, and R. J. Robinson. 1927. Phosphorus content of lake waters in northeastern Wisconsin. Trans. Wis. Acad. Sci. Arts Lett. *23*:233–248.

Judd, J. H. 1970. Lake stratification caused by runoff from street deicing. Water Res. *4*:521–532.

Jun, S.-H. and J. O. Bae. 1998. Distribution of dissolved copper species in Nakdong River water. Jpn. J. Limnol. *59*:457–464. (In Japanese.)

Jungmann, D., M. Henning, and F. Jüttner. 1991. Are the same compounds in *Microcystis* responsible for toxicity to *Daphnia* and inhibition of its filtering rate? Int. Rev. ges. Hydrobiol. *76*:47–56.

Junk, W. 1970. Investigations on the ecology and production biology of the 'floating meadows' (Paspalo-Echinochloctum) on the middle Amazon. Part 1. The floating vegetation and its ecology. Amazoniana 2:449–495.

Junk, W. J., ed. 1997a. The Central Amazon Floodplain: Ecology of a Pulsing System. Springer-Verlag, New York. 525 pp.

Junk, W. J. 1997b. General aspects of floodplain ecology with special reference to Amazonian floodplains. *In* W. J. Junk, ed. The Central Amazon Floodplain: Ecology of a Pulsing System. Springer-Verlag, Berlin. pp. 3–20.

Junk, W. J. and B. A. Robertson. 1997. Aquatic invertebrates. *In* W. Junk, ed. The Central Amazon Floodplain Ecology of a Pulsing System. Springer-Verlag, Berlin. pp. 279–298.

Junk, W. J. and M. T. F. Piedade. 1997. Plant life in the floodplain with special reference to herbaceous plants. *In* W. J. Junk, ed. The Central Amazon Floodplain: Ecology of a Pulsing System. Springer-Verlag, Berlin. pp. 147–185.

Junk, W. J., P. B. Bayley, and R. E. Sparks. 1989. The flood pulse concept in river-floodplain systems. *In* D. P. Dodge, ed. Proceedings of the International Large River Symposium. Can. Spec. Publ. Fish. Aquat. Sci. *106*:110–127.

Jupp, B. P. and D. H. N. Spence. 1977. Limitations on macrophytes in a eutrophic lake, Loch Leven. I. Effects of phytoplankton. J. Ecol. *65*:175–186.

Jürgens, K. 1994. Impact of *Daphnia* on planktonic microbial food webs—a review. Mar. Microb. Food Webs *8*:295–324.

Jürgens, K. and H. Güde. 1990. Incorporation and release of phosphorus by planktonic bacteria and phagotrophic flagellates. Mar. Ecol. Prog. Ser. *59*:271–284.

Jürgens, K. and H. Güde. 1994. The potential importance of grazing-resistant bacteria in planktonic systems. Mar. Ecol. Prog. Ser. *112*:169–188.

Jürgens, K. and G. Stolpe. 1995. Seasonal dynamics of crustacean zooplankton, heterotrophic nanoflagellates and bacteria in a shallow, eutrophic lake. Freshwat. Biol. *33*:27–38.

Jürgens, K., H. Arndt, and H. Zimmermann. 1997. Impact of metazoan and protozoan grazers on bacterial biomass distribution in microcosm experiments. Aquat. Microb. Ecol. *12*:131–138.

Jürgens, K., H. Arndt, and K. O. Rothhaupt. 1994. Zooplankton-mediated changes of bacterial community structure. Microb. Ecol. *27*:27–42.

Jürgens, K., J. M. Gasol, R. Massana, and C. Pedrós-Alió. 1994. Control of heterotrophic bacteria and protozoans by *Daphnia pulex* in the epilimnion of Lake Cisó. Arch. Hydrobiol. *131*:55–78.

Jürgens, K., S. A. Wickham, K. O. Rothhaupt, and B. Santer. 1996. Feeding rates of macro- and microzooplankton on heterotrophic nanoflagellates. Limnol. Oceanogr. *41*:1833–1839.

Juse, A. 1966. Diatomeen in Seesedimenten. Arch. Hydrobiol. Beih. Ergebn. Limnol. *4*, 32 pp.

Jüttner, F. 1981. Biologically active compounds released during algal blooms. Verh. Internat. Verein. Limnol. *21*:227–230.

Jüttner, F. 1995. Physiology and biochemistry of odorous compounds from freshwater cyanobacteria and algae. Wat. Sci. Technol. *31*:69–78.

Jüttner, F. and H. Faul. 1984. Organic activators and inhibitors of algal growth in water of a eutrophic shallow lake. Arch. Hydrobiol. *102*:21–30.

Jüttner, F. and R. Friz. 1974. Excretion products of *Ochromonas* with special reference to pyrrolidone carboxylic acid. Arch. Microbiol. *96*:223–232.

Kadlec, R. H. and R. L. Knight. 1996. Treatment Wetlands. CRC Press, Inc., Boca Raton, FL. 893 pp.

Kadono, Y. 1980. Photosynthetic carbon sources in some *Potamogeton* species. Bot. Mag. Tokyo *93*:185–193.

Kahn, W. E. and R. G. Wetzel. 1999. Effects of microscale water level fluctuations and altered ultraviolet radiation on periphytic microbiota. Microb. Ecol. *38*:253–263.

Kairesalo, T. 1976. Measurements of production of epilithiphyton and littoral plankton in Lake Pääjärvi, southern Finland. Ann. Bot. Fennici *13*:114–118.

Kairesalo, T. 1977. On the production ecology of epipelic algae and littoral plankton communities in Lake Pääjärvi, southern Finland. Ann. Bot. Fennici *14*:82–88.

Kairesalo, T. 1980a. Diurnal fluctuations within a littoral plankton community in oligotrophic Lake Pääjärvi, southern Finland. Freshwat. Biol. *10*:533–537.

Kairesalo, T. 1980b. Comparison of in situ photosynthetic activity of epiphytic, epipelic and planktonic algal communities in an oligotrophic lake, southern Finland. J. Phycol. *16*:57–62.

Kairesalo, T. 1984. The seasonal succession of epiphytic communities within an *Equisetum fluviatile* L. stand in Lake Pääjärvi, southern Finland. Int. Rev. ges. Hydrobiol. *69*:475–505.

Kairesalo, T. and I. Koskimies. 1987. Grazing by oligochaetes and snails on epiphytes. Freshwat. Biol. *17*:317–324.

Kairesalo, T. and S. Penttilä. 1990. Effect of light and water flow on the spatial distribution of littoral *Bosmina longispina* Leydig (Cladocera). Verh. Internat. Verein. Limnol. *24*:682–687.

Kairesalo, T. and T. Matilainen. 1988. The importance of low flow rates to the phosphorus flux between littoral and pelagial zones. Verh. Internat. Verein. Limnol. *23*:2210–2215.

Kairesalo, T. and T. Seppälä. 1987. Phosphorus flux through a littoral ecosystem: The importance of cladoceran zooplankton and young fish. Int. Rev. ges. Hydrobiol. *72*:385–403.

Kairesalo, T., A. Lehtovaara, and P. Saukkonen. 1992. Littoral-pelagial interchange and the decomposition of dissolved organic matter in a polyhumic lake. Hydrobiologia *229*: 199–224.

Kairesalo, T., S. Laine, T. Malinen, M. Suoraniemi, and J. Keto. 1999. Life of Lake Vesijärvi. City of Lahti, Finland. 92 pp.

Kaiser, K. and W. Zech. 1998. Rates of dissolved organic matter release and sorption in forest soils. Soil Sci. *163*:714–725.

Kajak, Z. 1970a. Some remarks on the necessities and prospects of the studies on biological production of freshwater ecosystems. Pol. Arch. Hydrobiol. *17*:43–54.

Kajak, Z. 1970b. Analysis of the influence of fish on benthos by the method of enclosures. *In* Z. Kajak and A. Hillbricht-Ilkowska, eds. Productivity Problems of Freshwaters. PWN Polish Scientific Publishers, Warsaw. pp. 781–793.

Kajak, Z. 1978. The characteristics of a temperate eutrophic, dimictic lake (Lake Mikolajskie, northern Poland). Int. Rev. ges. Hydrobiol. *63*:451–480.

Kajak, Z. and B. Ranke-Rybicka. 1970. Feeding and production efficiency of *Chaoborus flavicans* Meigen (Diptera, Culicidae) larvae in eutrophic and dystrophic lake. Pol. Arch. Hydrobiol. *17*:225–232.

Kajak, Z. and J. I. Rybak. 1966. Production and some trophic dependences in benthos against primary production and zooplankton production of several Masurian lakes. Verh. Internat. Verein. Limnol. *16*:441–451.

Kajak, Z. and K. Dusoge. 1970. Production efficiency of *Procladius choreus* MG (Chironomidae, Diptera) and its dependence on the trophic conditions. Pol. Arch. Hydrobiol. *17*:217–224.

Kajak, Z., A. Hillbricht-Ilkowska, and E. Pieczyńska. 1970. The production processes in several Polish lakes. *In* Z. Kajak and A. Hillbricht-Ilkowska, eds. Productivity Problems of Freshwaters. PWN Polish Scientific Publishers, Warsaw. pp. 129–147.

Kajosaari, E. 1966. Estimation of the detention period of a lake. Verh. Internat. Verein. Limnol. *16*:139–143.

Kalbitz, K., S. Solinger, J.-H. Park, B. Michalzik, and E. Matzner. 2000. Controls on the dynamics of dissolved organic matter in soils: A review. Soil Sci. *165*:277–304.

Kalff, J. and H. E. Welch. 1974. Phytoplankton production in Char Lake, a natural polar lake, and in Meretta Lake, a polluted polar lake, Cornwallis Island, Northwest Territories. J. Fish. Res. Board Can. *31*:621–636.

Kalinin, G. P. and V. D. Bykov. 1969. The world's water resources, present and future. Impact of Science on Technology *19*:135–150.

Kalk, M., A. J. McLachlan, and C. Howard-Williams. 1979. Lake Chilwa: Studies of change in a tropical ecosystem. Junk BV Publishers, The Hague. 462 pp.

Kaminski, M. 1984. Food composition of three bryozoan species (Bryozoa, Phylactolaemata) in a mesotrophic lake. Pol. Arch. Hydrobiol. *31*:45–53.

Kamiyama, K., S. Okuda, and A. Kawai. 1977. Studies on the release of ammonium nitrogen from the bottom sediments in freshwater regions. II. Ammonium nitrogen in dissolved and absorbed form in the sediments. Jpn. J. Limnol. *38*:100–106.

Kamp-Nielsen, L., H. Mejer, and S. E. Jørgensen. 1982. Modelling the influence of bioturbation on the vertical distribution of sedimentary phosphorus in L. Esrom. Hydrobiologia *91*:197–206.

Kana, T. M. and J. D. Tjepkema. 1978. Nitrogen fixation associated with *Scripus atrovirens* and other nonnodulated plants in Massachusetts. Can. J. Bot. *56*:2636–2640.

Kankaala, P. 1988. The relative importance of algae and bacteria as food for *Daphnia longispina* (Cladocera) in a polyhumic lake. Freshwat. Biol. *19*:285–296.

Kansanen, A. and R. Niemi. 1974. On the production ecology of isoetids, especially *Isoëtes lacustris* and *Lobelia dortmanna*, in Lake Pääjärvi, southern Finland. Ann. Bot. Fennici *11*:178–187.

Kansanen, P. H., T. Jaakkola, S. Kulmala, and R. Suutarinen. 1991. Sedimentation and distribution of gamma-emitting radionuclides in bottom sediments of southern Lake Päijänne, Finland, after the Chernobyl accident. Hydrobiologia *222*:121–140.

Kaplan, L A. and J. D. Newbold. 1993. Biogeochemistry of dissolved organic carbon entering streams. *In* T. E. Ford, ed. Aquatic Microbiology: An Ecologial Approach. Blackwell Sci. Publ., Oxford. pp. 139–165.

Kaplan, L. A. and T. L. Bott. 1983. Microbial heterotrophic utilization of dissolved organic matter in a piedmont stream. Freshwat. Biol. *13*:363–377.

Kaplan, L. A. and T. L. Bott. 1985. Acclimation of stream-bed heterotrophic microflora: Metabolic responses to dissolved organic matter. Freshwat. Biol. *15*:479–492.

Kaplan, L. A. and T. L. Bott. 1989. Diel fluctuations in bacterial activity on streambed substrata during vernal algal blooms: Effects of temperature, water chemistry, and habitat. Limnol. Oceanogr. *34*:718–733.

Kaplan, W. A. and S. C. Wofsy. 1985. The biogeochemistry of nitrous oxide: A review. Adv. Aquat. Microbiol. *3*:181–206.

Kapustina, L. L. 1996. Bacterioplankton response to eutrophication in Lake Ladoga. Hydrobiologia *322*:17–22.

Karcher, F. H. 1939. Untersuchungen über den Stickstoff-haushalt in ostpreussischen Waldseen. Arch. Hydrobiol. *35*:177–266.

Karentz, D., M. L. Bothwell, R. B. Coffin, A. Hanson, G. J. Herndl, S. S. Kilham, M. P. Lesser, M. Lindell, R. E. Moeller, D. P. Morris, P. J. Neale, R. W. Sanders, C. S. Weiler, and R. G. Wetzel. 1994. Impact of UV-B radiation on pelagic freshwater ecosystems: Report of working group on bacteria and phytoplankton. Arch. Hydrobiol. Beih. Ergebn. Limnol. *43*:31–69.

Karl, D. M. 1993. Microbial RNA and DNA synthesis derived from the assimilation of [2,^{3}H]-adenine. *In* P. F. Kemp, B. F. Sherr, E. B. Sherr, and J. J. Cole, eds. Handbook of Methods in Aquatic Microbial Ecology. Lewis Publishers, Boca Raton. pp. 471–481.

Karlson, R. H. 1992. Divergent dispersal strategies in the freshwater bryozoan *Plumatella repens*: Ramet size effects on statoblast numbers. Oecologia *89*:407–411.

Karp-Boss, L., E. Boss, and P. A. Jumars. 1996. Nutrient fluxes to planktonic osmotrophs in the presence of fluid motion. Oceanogr. Mar. Biol. Annu. Rev. *34*:71–107.

Kashiwada, K., A. Kanazawa, and S. Tachibanazono. 1963. Studies on organic compounds in natural water. II. On the seasonal variations in the content of nicotinic acid, pantothenic acid, biotin, folic acid and vitamin B_{12} in the water of the Lake Kasumigaura. (In Japanese.) Mem. Fac. Fish., Kagoshima Univ. *12*:153–157.

Kasprzak, K. 1986. Role of the Unionidae and Sphaeriidae (Mollusca, Bivalvia) in the eutrophic Lake Zbechy and its outflow. Int. Rev. ges. Hydrobiol. *71*:315–334.

Kasprzak, P., V. Vyhnálek, and M. Straškraba. 1986. Feeding and food selection in *Daphnia pulicaria* (Crustacea: Cladocera). Limnologica *17*:309–323.

Kasuga, S. and A. Otsuki. 1984. Phosphorus release by stirring up sediments and mysids feeding activities. Res. Rep. Natl. Inst. Environ. Stud., Jpn. *51*:141–155.

Katayama, T. 1961. Studies on the intercellular spaces in rice. Crop Sci. Soc. Japan Proc. *29*:229–233.

Kato, K. and M. Sakamoto. 1979. Vertical distribution of carbohydrate utilizing bacteria in Lake Kizaki. Jpn. J. Limnol. *40*:211–214.

Kato, K. and M. Sakamoto. 1981. Vertical distribution of free-living and attached heterotrophic bacteria in Lake Kizaki. Jpn. J. Limnol. *42*:154–159.

Kato, K. and M. Sakamoto. 1983. The function of the free-living bacterial fraction in the organic matter metabolism of a mesotrophic lake. Arch. Hydrobiol. *97*:289–302.

Kaufman, L. 1992. Catastrophic change in species-rich freshwater ecosystems: The lessons of Lake Victoria. BioScience *42*:846–858.

Kaunzinger, C. M. K. and P. J. Morin. 1998. Productivity controls food-chain properties in microbial communities. Nature *395*:495–497.

Kaushik, N. K. and H. B. N. Hynes. 1971. The fate of the dead leaves that fall into streams. Arch. Hydrobiol. *68*:465–515.

Keating, K. I. 1977. Allelopathic influence on blue-green bloom sequence in a eutrophic lake. Science *196*:885–887.

Keating, K. I. and B. C. Dagbusan. 1984. Effect of selenium deficiency on cuticle integrity in the Cladocera (Crustacea). Proc. Natl. Acad. Sci. USA *81*:3433–3437.

Keefe, C. W. 1972. Marsh production: A summary of the literature. Contr. Mar. Sci. Univ. Texas *16*:163–181.

Keeley, J. E. 1981. *Isoetes howellii*: A submerged aquatic CAM plant? Am. J. Bot. *68*:420–424.

Keeley, J. E. 1982. Distribution of diurnal acid metabolism in the genus *Isoetes*. Am. J. Bot. *69*:254–257.

Keeley, J. E. 1998. CAM photosynthesis in submerged aquatic plants. Bot. Rev. *64*:121–175.

Keen, R. 1973. A probabilistic approach to the dynamics of natural populations of the Chydoridae (Cladocera, Crustacea). Ecology *54*:524–534.

Keen, R. and R. Nassar. 1981. Confidence intervals for birth and death rates estimated with the egg-ratio technique for natural populations of zooplankton. Limnol. Oceanogr. *26*:131–142.

Keen, W. H. and J. Gagliardi. 1981. Effect of brown bullheads on release of phosphorus in sediment and water systems. Prog. Fish-Cult. *43*:183–185.

Keeney, D. R. 1972. The fate of nitrogen in aquatic ecosystems. Literature Rev., *3*, Water Resources Center, University of Wisconsin, 59 pp.

Keeney, D. R. 1973. The nitrogen cycle in sediment-water systems. J. Environ. Quality *2*:15–29.

Keeney, D. R., J. G. Konrad, and G. Chesters. 1970. Nitrogen distribution in some Wisconsin lake sediments. J. Water Poll. Control Fed. *42*:411–417.

Keeney, D. R., R. L. Chen, and D. A. Graetz. 1971. Importance of denitrification and nitrate reduction in sediments to the nitrogen budgets of lakes. Nature *233*:66–67.

Kelderman, P., H. J. Lindeboom, and J. Klein. 1988. Light dependent sediment-water exchange of dissolved reactive phosphorus and silicon in a producing microflora mat. Hydrobiologia *159*:137–147.

Kellerhals, R. and M. Church. 1989. The morphology of large rivers: Characterization and management. *In* D. P. Dodge, ed. Proceedings of the International Large River Symposium. Can. Spec. Publ. Fish. Aquatic Sci. *106*:31–48.

Kellerhals, R., M. Church, and D. I. Bray. 1976. Classification and analysis of river processes. J. Amer. Soc. Civil Eng. Hydraulics Div. *102*:813–829.

Kellogg, W. W., R. D. Cadle, E. R. Allen, A. L. Lazrus, and E. A. Martel. 1972. The sulfur cycle. Science *175*:587–596.

Kelly, C. A., J. W. M. Rudd, R. H. Hesslein, D. W. Schindler, P. J. Dillon, C. T. Driscoll, S. A. Gherini, and R. E. Hecky. 1987. Prediction of biological acid neutralization in acid-sensitive lakes. Biogeochemistry *3*:129–140.

Kelly, D. P. 1990. Physiology and biochemistry of unicellular sulfur bacteria. *In* H. G. Schlegel and B. Bowien, eds. Autotrophic Bacteria. Springer-Verlag, Berlin. pp. 193–217.

Kelso, B. H. L., R. V. Smith, R. J. Laughlin, and S. D. Lennox. 1997. Dissimilatory nitrate reduction in anaerobic sediments leading to river nitrite accumulation. Appl. Environ. Microbiol. *63*:4679–4685.

Kelso, B. H. L., R. V. Smith, and R. J. Laughlin. 1999. Effects of carbon substrates on nitrite accumulation in freshwater sediments. Appl. Environ. Microbiol. *65*:61–66.

Kelts, K. and K. J. Hsü. 1978. Freshwater carbonate sedimentation. *In* A. Lerman, ed. Lakes: Chemistry, Geology, Physics. Springer-Verlag, New York, pp. 295–323.

Kemp, A. L. W. and L. M. Johnston. 1979. Diagenesis of organic matter in the sediments of lakes Ontario, Erie, and Huron. J. Great Lakes Res. *5*:1–10.

Kemp, P. F. 1988. Bacterivory by benthic ciliates: Significance as a carbon source and impact on sediment bacteria. Mar. Ecol. Progr. Ser. *49*:163–169.

Kemp, P. F. 1990. The fate of benthic bacterial production. Rev. Aquat. Sci. *2*:109–124.

Kemp, P. F., B. F. Sherr, E. B. Sherr, and J. J. Cole, eds. 1993. Handbook of Methods in Aquatic Microbial Ecology. Lewis Publishers, Boca Raton. 777 pp.

Kenney, B. C. 1991. Under-ice circulation and the residence time of a shallow bay. Can. J. Fish. Aquat. Sci. *48*:152–162.

Kepkay, P. E. 1994. Particle aggregation and the biological reactivity of colloids. Mar. Ecol. Prog. Ser. *109*:293–304.

Kerans, B. L. and J. R. Karr. 1994. A benthic index of biotic integrity (B-IBI) for rivers of the Tennessee Valley. Ecol. Appl. *4*:768–785.

Kerfoot, W. C. 1978. Combat between predatory copepods and their prey: *Cyclops*, *Epischura*, and *Bosmina*. Limnol. Oceanogr. *23*:1089–1102.

Kerfoot, W. C. 1980b. Perspectives on cyclomorphosis: Separation of phenotypes and genotypes. *In* W. C. Kerfoot, ed. Evolution and Ecology of Zooplankton Communities. Univ. Press New England, Hanover, NH. pp. 470–496.

Kerfoot, W. C. and R. A. Pastorok. 1978. Survival versus competition: Evolutionary compromises and diversity in the zooplankton. Verh. Internat. Verein. Limnol. *20*:362–374.

Kerfoot, W. C., ed. 1980a. Evolution and Ecology of Zooplankton Communities. Univ. Press New England, Hanover, NH. 793 pp.

Kerekes, J., S. Beauchamp, R. Tordon, C. Tremblay, and T. Pollock. 1986. Organic versus anthropogenic acidity in tributaries of the Kejimkujik watersheds in western Nova Scotia. Water Air Soil Poll. *31*:165–173.

Kern, D. M. 1960. The hydration of carbon dioxide. J. Chem. Education *37*:14–23.

Kerr, J. B. and C. T. McElroy. 1993. Evidence for large upward trends of ultraviolet-B radiation linked to ozone depletion. Science *262*:1032–1034.

Kerr, P. C., D. L. Brockway, D. F. Paris, and J. T. Barnett, Jr. 1972. The interrelation of carbon and phosphorus in regulating heterotrophic and autotrophic populations in an aquatic ecosystem, Shriner's Pond. *In* G. E. Likens, ed. Nutrients and Eutrophication: The Limiting-Nutrient Controversy. Special Symposium, Amer. Soc. Limnol. Oceanogr. *1*:41–62.

Kerry, A., D. E. Laudenbach, and C. G. Trick. 1988. Influence of iron limitation and nitrogen source on growth and siderophore production by cyanobacteria. J. Phycol. *24*:566–571.

Kersting, K. and W. Holterman. 1973. The feeding behaviour of *Daphnia magna*, studies with the Coulter Counter. Verh. Internat. Verein. Limnol. *18*:1434–1440.

Keskitalo, J. and P. Eloranta, eds. 1999. Limnology of Humic Waters. Backhuys Publishers, Leiden. 284 pp.

van Kessel, J. F. 1978. The relation between redox potential and denitrification in a water-sediment system. Water Res. *12*:285–290.

Keup, L. E. 1968. Phosphorus in flowing waters. Water Res. 2: 373–386.

Khailov, K. M. 1971. Ekologicheskii metabolizm v more. Izdatel 'stvo Naukova Dumka, Kiev. 252 pp.

Khondker, M. and M. Dokulil. 1988. Seasonality, biomass and primary productivity of epipelic algae in a shallow lake (Neusiedlersee, Austria). Acta Hydrochim. Hydrobiol. *16*: 499–515.

Kibby, H. V. 1971. Energetics and population dynamics of *Diaptomus gracilis*. Ecol. Monogr. *41*:311–327.

Kibby, H. V. and F. H. Rigler. 1973. Filtering rates of *Limnocalanus*. Verh. Internat. Verein. Limnol. *18*:1457–1461.

Kiene, R. P. 1996. Microbial cycling of organosulfur gases in marine and freshwater environments. Mitt. Internat. Verein. Limnol. *25*:137–151.

Kiene, R. P. and M. E. Hines. 1995. Microbial formation of dimethyl sulfide in anoxic *Sphagnum* peat. Appl. Environ. Microbiol. *61*:2720–2726.

Kilham, P. 1971. A hypothesis concerning silica and the freshwater planktonic diatoms. Limnol. Oceanogr. *16*:10–18.

Kilham, P. 1975. Mechanisms controlling world water chemistry based on data for African lakes and rivers. Proc. Int. Symp. Geochemistry of Natural Waters. Burlington, Ontario. 5 pp.

Kilham, P. 1981. Pelagic bacteria: Extreme abundances in African saline lakes. Naturwissenschaften *67*:380–381.

Kilham, P. 1982. The effect of hippopotamuses on potassium and phosphate ion concentrations in an African lake. Amer. Midland Nat. *108*:202–205.

Kilham, P. 1982. Acid precipitation: Its role in the alkalization of a lake in Michigan. Limnol. Oceanogr. *27*:856–867.

Kilham, P. 1984. Sulfate in African inland waters: Sulfate to chloride ratios. Verh. Internat. Verein. Limnol. *22*:296–302.

Kilham, P. 1990. Mechanisms controlling the chemical composition of lakes and rivers: Data from Africa. Limnol. Oceanogr. *35*:80–83.

Kilham, P. and D. Tilman. 1979. The importance of resource competition and nutrient gradients for phytoplankton ecology. Arch. Hydrobiol. Beih. Ergebn. Limnol. *13*:100–119.

Kilham, P. and D. Titman. 1976. Some biological effects of atmospheric inputs to lakes: Nutrient ratios and competitive interactions between phytoplankton. J. Great Lakes Res. *2*(Suppl.):187–191.

Kilham, P. and R. E. Hecky. 1988. Comparative ecology of marine and freshwater phytoplankton. Limnol. Oceanogr. *33*:776–795.

Kilham, P. and S. S. Kilham. 1980. The evolutionary ecology of phytoplankton. *In* I. Morris, ed. The Physiological Ecology of Phytoplankton. Univ. California Press, Berkeley. pp. 571–597.

Kilham, S. S. 1975. Kinetics of silicon-limited growth in the freshwater diatom *Asterionella formosa*. J. Phycol. *11*:396–399.

Kilham, S. S. 1978. Nutrient kinetics of freshwater planktonic algae using batch and semicontinuous methods. Mitt. Internat. Verein. Limnol. *21*:147–157.

Kilham, S. S. 1986. Dynamics of Lake Michigan natural phytoplankton communities in continuous cultures along a Si:P loading gradient. Can. J. Fish. Aquat. Sci. *43*:351–360.

Kilham, S. S. 1990. Relationship of phytoplankton and nutrients to stoichiometric measures. *In* M. M. Tilzer and C. Serruya, eds. Large lakes: Ecological structure and function. Springer-Verlag, New York. pp. 403–414.

Kilham, S. S. and P. Kilham. 1978. Natural community bioassays: Predictions of results based on nutrient physiology and competition. Verh. Internat. Verein. Limnol. *20*:68–74.

Kilham, S. S. and P. Kilham. 1990. Tropical limnology: Do African lakes violate the "first law" of limnology? Verh. Internat. Verein. Limnol. *24*:68–72.

Kilham, S. S., D. A. Kreeger, C. E. Goulden, and S. G. Lynn. 1997. Effects of nutrient limitation on biochemical constituents of *Ankistrodesmus falcatus*. Freshwat. Biol. *38*:591–596.

Kim, B. and R. G. Wetzel. 1993. The effect of dissolved humic substances on the alkaline phosphatase and growth of microalgae. Verh. Internat. Verein. Limnol. *25*:129–132.

Kim, D.-S. and Y. Watanabe. 1993. The effect of long wave ultraviolet radiation (UV-A) on the photosynthetic activity of natural population of freshwater phytoplankton. Ecol. Res. *8*:225–234.

Kim, D.-S. and Y. Watanabe. 1994. Inhibition of growth and photosynthesis of freshwater phytoplankton by ultraviolet A (UVA) radiation and subsequent recovery from stress. J. Plankton Res. *16*:1645–1654.

Kim, J. and S. B. Verma. 1992. Soil surface CO_2 flux in a Minnesota peatland. Biogeochemistry *18*:37–51.

Kim, J., S. B. Verma, and D. P. Billesbach. 1998. Seasonal variation in methane emission from a temperate *Phragmites*-dominated marsh: Effect of growth stage and plant-mediated transport. Global Change Biol. *5*:433–440.

Kimball, K. D. 1973. Seasonal fluctuations of ionic copper in Knights Pond, Massachusetts. Limnol. Oceanogr. *18*:169–172.

Kimmel, B. L., O. T. Lind, and L. J. Paulson. 1990. Reservoir primary production. *In* K. W. Thornton, B. L. Kimmel, and F. E. Payne, eds. Reservoir Limnology: Ecological Perspectives. John Wiley & Sons, New York. pp. 133–193.

Kinchin, I. M. 1994. The biology of tardigrades. Portland Press Ltd., London. 186 pp.

King, C. E. 1967. Food, age, and the dynamics of a laboratory population of rotifers. Ecology *48*:111–128.

King, C. H., R. W. Sanders, E. B. Shotts, and K. G. Porter. 1991. Differential survival of bacteria ingested by zooplankton from a stratified eutrophic lake. Limnol. Oceanogr. *36*:829–845.

King, C. R. 1979. Secondary productivity of the North Pine Dam. Tech. Pap. Australian Water Resour. Council No. 39. 75 pp.

King, C. R. and J. G. Greenwood. 1992a. The productivity and carbon budget of a natural population of *Daphnia lumholtzi* Sars. Hydrobiologia *231*:197–207.

King, C. R. and J. G. Greenwood. 1992b. The seasonal population changes and carbon budget of the calanoid copepod *Boeckella minuta* Sars in a newly formed sub-tropical reservoir. J. Plankton Res. *14*:329–342.

King, D. L. and R. C. Ball. 1966. A qualitative and quantitative measure of *Aufwuchs* production. Trans. Amer. Microsc. Soc. *85*:232–240.

King, G. M. 1990a. Regulation by light of methane emission from a Danish wetland. Nature *345*:513–515.

King, G. M. 1990b. Dynamics and controls of methane oxidation in a Danish wetland sediment. FEMS Microbiol. Ecol. *74*:309–324.

King, G. M. and M. J. Klug. 1980. Sulfhydrolase activity in sediments of Wintergreen Lake, Kalamazoo County, Michigan. Appl. Environ. Microbiol. *39*:950–956.

King, G. M. and M. J. Klug. 1982a. Comparative aspects of sulfur mineralization in sediments of a eutrophic lake basin. Appl. Environ. Microbiol. *43*:1406–1412.

King, G. M. and M. H. Klug. 1982b. Glucose metabolism in sediments of a eutrophic lake: Tracer analysis of uptake and product formation. Appl. Environ. Microbiol. *44*:1308–1317.

King, G. M. and T. H. Blackburn. 1996. Controls of methane oxidation in sediments. Mitt. Internat. Verein. Limnol. *25*:25–38.

King, G. M. and W. J. Wiebe. 1978. Methane release from soils of a Georgia salt marsh. Geochim. Cosmochim. Acta *42*:343–348.

King, G. M., P. Roslev, and H. Skovgaard. 1990. Distribution and rate of methane oxidation in sediments of the Florida Everglades. Appl. Environ. Microbiol. *56*:2902–2911.

Kingsland, S. 1982. The refractory model: The logistic curve and the history of population ecology. Quart. Rev. Biol. *57*:29–52.

Kingston, J. C. and H. J. B. Birks. 1990. Dissolved organic carbon reconstructions from diatom assemblages in PIRLA project lakes, North America. Phil. Trans. R. Soc. Lond. B *327*:279–288.

Kinsman, R., B. W. Ibelings, and A. E. Walsby. 1991. Gas vesicle collapse by turgor pressure and its role in buoyancy regulation by *Anabaena flos-aquae*. J. Gen. Microbiol. *137*:1171–1178.

Kirby, M. J., ed. 1978. Hillslope Hydrology. John Wiley & Sons, New York. 389 pp.

Kirchman, D. 1983. The production of bacteria attached to particles suspended in a freshwater pond. Limnol. Oceanogr. *28*:858–872.

Kirchner, W. B. and P. J. Dillon. 1975. An empirical method of estimating the retention of phosphorus in lakes. Water Resour. Res. *11*:182–183.

Kirk, G. J. D. and J. B. Bajita. 1995. Root-induced iron oxidation, pH changes and zinc solubilization in the rhizosphere of lowland rice. New Phytol. *131*:129–137.

Kirk, J. T. O. 1983. Light and photosynthesis in aquatic ecosystems. Cambridge Univ. Press, Cambridge, England. 401 pp.

Kirk, J. T. O. 1985. Effects of suspensoids (turbidity) on penetration of solar radiation in aquatic ecosystems. Hydrobiologia *125*: 195–208.

Kirk, J. T. O. 1994. Optics of UV-B radiation in natural waters. Arch. Hydrobiol. Beih. Ergebn. Limnol. *43*:1–16.

Kirk, K. L. 1991. Inorganic particles alter competition in grazing plankton: The role of selective feeding. Ecology *72*:915–923.

Kirk, K. L. 1992. Effects of suspended clay on *Daphnia* body growth and fitness. Freshwat. Biol. *28*:103–109.

Kirk, K. L. 1997. Life-history responses to variable environments: Starvation and reproduction in planktonic rotifers. Ecology *78*:434–441.

Kirk, K. L. and J. J. Gilbert. 1990. Suspended clay and the population dynamics of planktonic rotifers and cladocerans. Ecology *71*:1741–1755.

Kirk, T. K. and R. L. Farrell, 1987. Enzymatic "combustion": The microbial degradation of lignin. Ann. Rev. Microbiol. *41*: 465–505.

Kirkland, D. W., J. P. Bradbury, and W. E. Dean. 1983. The heliothermic lake—a direct method of collecting and storing solar energy. Arch. Hydrobiol./Suppl. *65*:1–60.

Kisielewska, G. 1982. Gastrotricha of two complexes of peat hags near Siedlce. Fragmenta Faunistica *27*:39–57.

Kisker, C., H. Schindelin, and D. C. Rees. 1997. Molybdenum-cofactor-containing enzymes: Structure and mechanism. Annu. Rev. Biochem. *66*:233–267.

Kistritz, R. U. 1978. Recycling of nutrients in an enclosed aquatic community of decomposing macrophytes (*Myriophyllum spicatum*). Oikos *30*:561–569.

Kitchell, J. F., R. A. Stein, and B. Kneèevic. 1978. Utilization of filamentous algae by fishes in Skadar Lake, Yugoslavia. Verh. Internat. Verein. Limnol. *20*:2159–2165.

Kivinen, E., L. Heikurainen, and Pakarin, eds. 1979. Classification of Peat and Peatlands. International Peat Society, Helsinki. 367 pp.

Kjeldsen, K. 1994. The relationship between phosphorus and peak biomass of benthic algae in small lowland streams. Verh. Internat. Verein. Limnol. *25*:1530–1533.

Kjellberg, G., D. O. Hessen, and J. P. Nilssen. 1991. Life history, growth and production of *Mysis relicta* in the large fiord-type Lake Mjøsa, Norway. Freshwat. Biol. *26*:165–173.

Kjensmo, J. 1962. Some extreme features of the iron metabolism in lakes. Schweiz. Z. Hydrol. *24*:244–252.

Kjensmo, J. 1967. The development and some main features of "iron-meromictic" soft water lakes. Arch. Hydrobiol./Suppl. *32*:137–312.

Kjensmo, J. 1968. Iron as the primary factor rendering lakes meromictic, and related problems. Mitt. Internat. Verein. Limnol. *14*:83–93.

Kjensmo, J. 1970. The redox potentials in small oligo and meromictic lakes. Nordic Hydrol. *1*:56–65.

Klapper, H. 1991. Control of Eutrophication in Inland Waters. Ellis Horwood, New York. 337 pp.

Klein, R. M. 1978. Plants and near-ultraviolet radiation. Bot. Rev. *44*:1–127.

Kleiner, J. 1988. Coprecipitation of phosphate with calcite in lake water: A laboratory experiment modelling phosphorus removal with calcite in Lake Constance. Water Res. *22*:1259–1265.

Kleiner, J. 1990. Calcite precipitation—regulating mechanisms in hardwater lakes. Verh. Internat. Verein. Limnol. *24*:136–139.

Kleiven, O. T., P. Larsson, and A. Hobæk. 1996. Direct distributional response in *Daphnia pulex* to a predator kairomone. J. Plankton Res. *18*:1341–1348.

Klekowski, R. Z. 1970. Bioenergetic budgets and their application for estimation of production efficiency. Pol. Arch. Hydrobiol. *17*:55–80.

Klekowski, R. Z. and E. A. Shushkina. 1966. Ernährung, Atmung, Wachstum und Energie-Umformung in *Macrocyclops albidus* (Jurine). Verh. Internat. Verein. Limnol. *16*:399–418.

Klekowski, R. Z., E. Fischer, Z. Fischer, M. B. Ivanova, T. Prus, E. A. Shushkina, T. Stachurska, Z. Stepien, and H. Zyromska-Rudzka. 1970. Energy budgets and energy transformation efficiencies of several animal species of different feeding types. *In* Z. Kajak and A. Hillbricht-Ilkowska, eds. Productivity Problems of Freshwaters. PWN Polish Scientific Publishers, Warsaw. pp. 749–763.

Klemer, A. R. 1978. Nitrogen limitation of growth and gas vacuolation in *Oscillatoria rubescens*. Verh. Internat. Verein. Limnol. *20*:2293–2297.

Klemer, A. R. 1986. Nutrient-induced migrations of blue-green algae (cyanobacteria). *In* M. A. Rankin, ed. Migration: Mechanisms and Adaptive Significance. Contribut. Mar. Sci. Suppl. *27*:153–165.

Klemer, A. R., J. Feuillade, and M. Feuillade. 1982. Cyanobacterial blooms: Carbon and nitrogen limitation have opposite effects on the buoyancy of *Oscillatoria*. Science *215*:1629–1631.

Kling, G. W., G. W. Kipphut, and M. C. Miller. 1992. The flux of CO_2 and CH_4 from lakes and rivers in arctic Alaska. Hydrobiologia *240*:23–36.

Kling, G. W., M. A. Clark, H. R. Compton, J. D. Devine, W. C. Evans, A. M. Humphrey, E. J. Koenigsberg, J. P. Lockwood, M. L. Tuttle, and G. N. Wagner. 1987. The 1986 Lake Nyos gas disaster in Cameroon, West Africa. Science *236*:169–175.

Kling, G. W., M. L. Tuttle, and W. C. Evans. 1989. The evolution of thermal structure and water chemistry in Lake Nyos. J. Volcan. Geotherm. Res. *39*:151–165.

Kling, G. W., W. C. Evans, and M. L. Tuttle. 1991a. A comparative view of lakes Nyos and Monoun, Cameroon, West Africa. Verh. Internat. Verein. Limnol. *24*:1102–1105.

Kling, G. W., A. E. Giblin, B. Fry, and B. J. Peterson. 1991b. The role of seasonal turnover in lake alkalinity dynamics. Limnol. Oceanogr. *36*:106–122.

Klingensmith, K. M. and V. Alexander. 1983. Sediment nitrification, denitrification, and nitrous oxide production in a deep arctic lake. Appl. Environ. Microbiol. *46*:1084–1092.

Klinger, L. F. 1996. The myth of the classic hydrosere model of bog succession. Arctic Alpine Res. *28*:1–9.

Klopatek, J. M. 1975. The role of emergent macrophytes in mineral cycling in a freshwater marsh. *In* F. G. Howell, J. B. Gentry, and M. H. Smith, eds. Mineral Cycling in Southeastern Ecosystems. U.S. Energy Res. Development Admin., Washington, DC. pp. 367–393.

Klotz, R. L. 1985. Factors controlling phosphorus limitation in stream sediments. Limnol. Oceanogr. *30*:543–553.

Klötzli, F. 1971. Biogenous influence on aquatic macrophytes, especially *Phragmites communis*. Hidrobiologia (Romania) *12*:107–111.

Klug, M. J., G. M. King, R. L. Smith, D. R. Lovley, and J. W. H. Dacey. 1982. Comparative aspects of anaerobic carbon and electron flow in freshwater sediments. (Manuscript.)

Knight, A., R. C. Ball, and F. F. Hooper. 1962. Some estimates of primary production rates in Michigan ponds. Pap. Mich. Acad. Sci. Arts Lett. *47*:219–233.

Knight, D. W. and K. Shiono. 1996. River channel and floodplain hydraulics. *In* M. G. Anderson, D. E. Walling, and P. D. Bates, eds. Floodplain Processes. John Wiley & Sons, Chichester. pp. 139–181.

Knoechel, R. and J. Kalff. 1975. Algal sedimentation: The cause of a diatom-blue-green succession. Verh. Internat. Verein. Limnol. *19*:745–754.

Knoechel, R. and L. B. Holtby. 1986. Cladoceran filtering rate:body length relationships for bacterial and large algal particles. Limnol. Oceanogr. *31*:195–200.

Kobayasi, H. 1961. Productivity in sessile algal community of Japanese mountain river. Bot. Mag. Tokyo *74*:331–341.

Kobayashi, T., R. J. Shiel, and P. Gibbs. 1998a. Size structure of river zooplankton: Seasonal variation, overall pattern and functional aspect. Mar. Freshwat. Res. *49*:547–552.

Kobayashi, T., R. J. Shiel, P. Gibbs, and P. I. Dixon. 1998b. Freshwater zooplankton in the Hawkesbury-Nepean River: Comparison of community structure with other rivers. Hydrobiologia *377*:133–145.

Koch, E. W. 1994. Hydrodynamics, diffusion-boundary layers and photosynthesis of the seagrasses *Thalassia testudium* and *Cymodocea nodosa*. Mar. Biol. *118*:767–776.

Koch, M. S. and I. A. Mendelssohn. 1989. Sulphide as a soil phytotoxin: Differential responses in two marsh species. J. Ecol. *77*:565–578.

Kocsis, O., B. Mathis, M. Gloor, M. Schurter, and A. Wüest. 1998. Enhanced mixing in narrows: A case study at the Mainau sill (Lake Constance). Aquat. Sci. *60*:236–252.

Kodomari, S. 1984. Studies on the internal wave in small lakes. II. Effect of the lake basin shape on the internal wave. Jpn. J. Limnol. *45*:269–278.

Koehl, M. A. R. and J. R. Strickler. 1981. Copepod feeding currents: Food capture at low Reynolds number. Limnol. Oceanogr. *26*:1062–1073.

Koenings, J. P. 1976. In situ experiments on the dissolved and colloidal state of iron in an acid bog lake. Limnol. Oceanogr. *21*:674–683.

Koenings, J. P. and F. F. Hooper. 1976. The influence of colloidal organic matter on iron and iron-phosphorus cycling in an acid bog lake. Limnol. Oceanogr. *21*:684–696.

Koerselman, W. and B. Beltman. 1988. Evapotranspiration from fens in relation to Penman's potential free water evaporation (E_o) and pan evaporation. Aquat. Bot. *31*:307–320.

Kogan, Sh. I. and G. A. Chinnova. 1972. Relations between *Ceratophyllum demersum* (L.) and some blue-green algae. Hydrobiol. J. (USSR; Translation Ser.) *8*:14–25.

Kohlenbrander, G. J. 1972. The eutrophication of surface water by agriculture and the urban population. Stickstoff *13*:56–67.

Köhler, J. 1994. Dynamics of phytoplankton in the lowland River Spree (Germany). Verh. Internat. Verein. Limnol. *25*:1590–1594.

Koivo, L. K. and J. C. Ritchie. 1978. Modern diatom assemblages from lake sediments in the boreal-arctic transition region near the Mackenzie Delta, N.W.T., Canada. Can. J. Bot. *56*:1010–1020.

Kok, C. J., G. van der Welde, and K. M. Landsberger. 1990. Production, nutrient dynamics and initial decomposition of floating leaves of *Nymphaea alba* L. and *Nuphar lutea* (L.) Sm. (Nymphaeaceae) in alkaline and acid waters. Biogeochemistry *11*:235–250.

Kolodziejczyk, A. 1984. Occurrence of gastropoda in the lake littoral and their role in the production and transformation of detritus. II. Ecological activity of snails. Ekol. Polska *32*:469–492.

Konda, T. 1984. Seasonal variations in four bacterial size fractions from a hypertrophic pond in Tokyo, Japan. Int. Rev. ges. Hydrobiol. *69*:843–858.

Konda, T., S. Takii, M. Fukui, Y. Kusuoka, G. I. Matsumoto, and T. Torii. 1994. Vertical distribution of bacterial population in Lake Fryxell, an Antarctic lake. Jpn. J. Limnol. *55*:185–192.

Kondrat 'eva, E. N. 1965. Photosynthetic Bacteria. Moscow, Izdatel 'stvo Akademii Nauk SSSR 1963. (Translated into English, Israel Program for Scientific Translations, Jerusalem.) 243 pp.

Kononova, M. M. 1966. Soil Organic Matter. Its Nature, Its Role in Soil Formation and in Soil Fertility. 2nd ed. Pergamon Press, Oxford. 544 pp.

Konopka, A. 1981. Influence of temperature, oxygen, and pH on a metalimnetic population of *Oscillatoria rubescens*. Appl. Environ. Microbiol. *42*:102–108.

Konopka, A. 1982. Buoyancy regulation and vertical migration by *Oscillatoria rubescens* in Crooked Lake, Indiana. Br. Phycol. J. *17*:427–442.

Konrad, J. G., D. R. Keeney, G. Chesters, and K.-L. Chen. 1970. Nitrogen and carbon distribution in sediment cores of selected Wisconsin lakes. J. Water Poll. Control Fed. *42*:2094–2101.

Korde, N. W. 1966. Algenreste in Seesedimenten. Zur Entwicklungsgeschichte der Seen und umliegenden Landschaften. Arch. Hydrobiol. Beih. Ergebn. Limnol. *3*, 38 pp.

Koreliakova, I. L. 1958. Nekotorye nabluideniia nad raspodom perezimovavshei pribrezhno-vodnoi rastitel'nosti Rybinskogo Vodokhranilishcha. Bull. Inst. Biol. Vodokhranilishch *1*:22–25.

Koreliakova, I. L. 1959. O raslade skoshennoi pribrezhno-vodnoi rastitel'nosti. Bull. Inst. Biol. Vodochranilishch *3*:13–16.

Korínková, J. 1967. Relations between predation pressure of carp, submerged plant development and littoral bottom-fauna of Pond Smyslov. Rozpravy Ceskosl. Akad. Ved, Rada Matem. Prírod. Ved 77(11):35–62.

Kormondy, E. J. 1968. Weight loss of cellulose and aquatic macrophytes in a Carolina bay. Limnol. Oceanogr. *13*:522–526.

Kortmann, R. W. 1980. Benthic and atmospheric contributions to the nutrient budgets of a soft-water lake. Limnol. Oceanogr. *25*:229–239.

Kortmann, R. W. and P. H. Rich. 1994. Lake ecosystem energetics: The missing management link. Lake Reserv. Manage. *8*:77–97.

Koschel, R. 1990. Pelagic calcite precipitation and trophic state of hardwater lakes. Arch. Hydrobiol. Beih. Ergebn. Limnol. *33*:713–722.

Koschel, R. 1997. Structure and function of pelagic calcite precipitation in lake ecosystems. Verh. Internat. Verein. Limnol. *26*:343–349.

Koschel, R., J. Benndorf, G. Proft, and F. Recknagel. 1983. Calcite precipitation as a natural control mechanism of eutrophication. Arch. Hydrobiol. *98*:380–408.

Koshinsky, G. D. 1970. The morphometry of shield lakes in Saskatchewan. Limnol. Oceanogr. *15*:695–701.

Kostalos, M. and R. L. Seymour. 1976. Role of microbial enriched detritus in the nutrition of *Gammarus minus* (Amphipoda). Oikos *27*:512–516.

Kowalczewski, A. 1965. Changes in periphyton biomass of Mikolajskie Lake. Bull. Acad. Polon. Sci. (Cl. II) *13*:395–398.

Kowalczewski, A. 1975a. Algal primary production in the zone of submerged vegetation of a eutrophic lake. Verh. Internat. Verein. Limnol. *19*:1305–1308.

Kowalczewski, A. 1975b. Periphyton primary production in the zone of submerged vegetation of Mikolajskie Lake. Ekol. Polska *23*:509–543.

Kowalczewski, A. and J. I. Rybak. 1981. Atmospheric fallout as a source of phosphorus for Lake Warniak. Ekol. Polska *29*:63–71.

Koyama, T. 1955. Gaseous metabolism in lake muds and paddy soils. J. Earth Sci. Nagoya Univ. *3*:65–76.

Koyama, T. 1963. Gaseous metabolism in lake sediments and paddy soils and the production of atmospheric methane and hydrogen. J. Geophys. Res. *68*:3971–3973.

Koyama, T. 1964. Gaseous metabolism in lake sediments and paddy soils. *In* U. Colombo and G. D. Hobson, eds. Advances in Organic Geochemistry. Macmillan Co., New York. pp. 363–375.

Koyama, T. 1993. Zoobenthos effects on the gaseous metabolism in lake sediments. Verh. Internat. Verein. Limnol. *25*:827–831.

Koyama, T., M. Nakaido, T. Tomino, and H. Hayakawa. 1973. Decomposition of organic matter in lake sediments. *In* E. Ingerson, ed. Proceedings of Symposium on Hydrogeochemistry and Biogeochemistry. Clarke Company, Washington, DC. pp. 512–535.

Kozerski, H.-P. 1994. Possibilities and limitations of sediment traps to measure sedimentation and resuspension. Hydrobiologia *284*:93–100.

Kozhov, M. 1963. Lake Baikal and Its Life. W. Junk Publishers, The Hague. 344 pp.

Kozhova, O. M. and L. R. Izmest'eva, eds. 1998. Lake Baikal. Evolution and Biodiversity. Backhuys Publishers, Leiden. 447 pp.

Kozlovsky, D. G. 1968. A critical evaluation of the trophic level concept. I. Ecological efficiencies. Ecology *49*:48–60.

Kózminski, Z. and J. Wisniewski. 1935. Über die Vorfrühlingthermik der Wigry-Seen. Arch. Hydrobiol. *28*:198–235.

Krabbenhoft, D. P., M. P. Anderson, and C. J. Bowser. 1990. Estimating groundwater exchange with lakes. 2. Calibration of a three-dimensional, solute transport model to a stable isotope plume. Water Resour. Res. *26*:2455–2462.

Krambeck, C. 1978. Changes in planktonic microbial populations—an analysis by scanning electron microscopy. Verh. Internat. Verein. Limnol. *20*:2255–2259.

Krambeck, H.-J. 1978. A numerical-topographical model of Lake Grosser Plöner See and its application to the calculation of seiches. Arch. Hydrobiol. *97*:262–273.

Kramer, J. R. 1978. Acid precipitation. *In* J. O. Nriagu, ed. Sulfur in the Environment. I. The Atmospheric Cycle. John Wiley & Sons, New York. pp. 325–370.

Krantzberg, G. 1985. The influence of bioturbation on physical, chemical and biological parameters in aquatic environments: A review. Environ. Poll. (Ser. A) *39*:99–122.

Krasheninnikova, S. A. 1958. Mikrobiologicheskie protsessy raslada vodnoi rastitel'nosti v litorali Rybinskogo Vodokhanilishcha. Bull. Inst. Biol. Vodokhranilishch 2:3–6.

Kratz, T. K., R. B. Cook, C. J. Bowser, and P. L. Brezonik. 1987. Winter and spring pH depressions in northern Wisconsin lakes caused by increases in pCO_2. Can. J. Fish. Aquat. Sci. *44*:1082–1088.

Kratz, W. A. and J. Myers. 1955. Nutrition and growth of several blue-green algae. Am. J. Bot. *42*:282–287.

Krause, H. R. 1961. Einige Bemerkungen über den postmortalen Abbau von Süsswasser-Zooplankton unter Laboratoriums- und Freilandbedingungen. Arch. Hydrobiol. *57*:539–543.

Krause, H. R. 1962. Investigation of the Decomposition of Organic Matter in Natural Waters. FAO Fish. Biol. Report *34*(FB/R34), 19 pp.

Krause, H. R. 1964. Zur Chemie und Biochemie der Zersetzung von Süsswasserorganismen, unter besonderer Berücksichtigung des Abbaues der organischen Phosphorkomponenten. Verh. Internat. Verein. Limnol. *15*:549–561.

Krause, H. R., L. Möchel, and M. Stegmann. 1961. Organische Säuren als gelöste Intermediärprodukte des postmortalen Abbaues von Süsswasser-Zooplankton. Naturwissenschaften *48*:434–435.

Krauskopf, K. B. 1956. Dissolution and precipitation of silica at low temperatures. Geochim. Cosmochim. Acta *10*:1–26.

Krauss, R. W. 1958. Physiology of the fresh-water algae. Ann. Rev. Plant Physiol. *9*:207–244.

Krauss, R. W. 1962. Inhibitors. *In* R. A. Lewin, ed. Physiology and Biochemistry of Algae. Academic Press, New York. pp. 673–685.

Kreeger, D. A., C. E. Goulden, S. S. Kilham, S. G. Lynn, S. Datta, and S. J. Interlandi. 1997. Seasonal changes in the biochemistry of lake seston. Freshwat. Biol. *38*:535–554.

Krejci, M. E. and R. L. Lowe. 1986. Importance of sand grain minerology and topography in determining micro-spatial distribution of epipsammic diatoms. J. N. Am. Benthol. Soc. *5*:211–220.

Krejci, M. E. and R. L. Lowe. 1987. Spatial and temporal variation of epipsammic diatoms in a spring-fed brook. J. Phycol. *23*:585–590.

Kretsinger, R. H. 1979. The informational role of calcium in cytosol. *In* P. Greengard and G. A. Robison, eds. Advances in Cyclic Nucleotide Research. Vol. 11. Raven Press, New York. pp. 2–26.

Kring, R. L. and W. J. O'Brien. 1976. Effect of varying oxygen concentrations on the filtering rate of *Daphnia pulex*. Ecology *57*:808–814.

Krishnaswami, S. and D. Lal. 1978. Radionuclide limnochronology. *In* A. Lerman, ed. Lakes: Chemistry, Geology, Physics. Springer Verlag, New York. pp. 153–177.

Kriss, A. E. and R. Tomson. 1973. Origin of the warm water near the bottom of Lake Vanda in the Antarctic (25.5–27°): Microbiological data. Mikrobiologiya *42*:942–943.

Kristensen, P., M. Søndergaard, and E. Jeppesen. 1992. Resuspension in a shallow eutrophic lake. Hydrobiologia *228*:101–109.

Krogh, A. 1939. Osmotic Regulation in Aquatic Animals. Cambridge University Press, Cambridge. 242 pp.

Krogh, A. and E. Lange. 1932. Quantitative Untersuchungen über Plankton, Kolloide und gelöste organische und anorganische Substanzen in dem Füresee. Int. Rev. ges. Hydrobiol. *26*:20–53.

Królikowska, J. 1978. The transpiration of helophytes. Ekol. Polska *26*:193–212.

Krueger, C. C. and T. F. Waters. 1983. Annual production of macroinvertebrates in three streams of different water quality. Ecology *64*:840–850.

Küchler-Krischun, J. and J. Kleiner. 1990. Heterogeneously nucleated calcite precipitation in Lake Constance. A short time resolution study. Aquatic Sci. *52*:176–197.

Kudoh, S. and M. Takahashi. 1990. Fungal control of population changes of the planktonic diatom *Asterionella formosa* in a shallow eutrophic lake. J. Phycol. *26*:239–244.

Kudoh, S. and M. Takahashi. 1992. An experimental test of host population size control by fungal parasitism in the planktonic diatom *Asterionella formosa* using mesocosms in a natural lake. Arch. Hydrobiol. *124*:293–307.

Kuehn, K. A. and K. Suberkropp. 1998a. Decomposition of standing litter of the freshwater emergent macrophyte *Juncus effusus*. Freshwat. Biol. *40*:717–727.

Kuehn, K. A. and K. Suberkropp. 1998b. Diel fluctuations in rates of CO_2 evolution from standing dead leaf litter of the emergent macrophyte *Juncus effusus*. Aquat. Microb. Ecol. *14*:171–182.

Kuenen, J. G., L. A. Robertson, and H. V. Gemerden. 1985. Microbial interactions among aerobic and anaerobic sulfur-oxidizing bacteria. Adv. Microb. Ecol. *8*:1–58.

Kuenzler, E. J. 1970. Dissolved organic phosphorus excretion by marine phytoplankton. J. Phycol. *6*:7–13.

Kuhl, A. 1962. Inorganic phosphorus uptake and metabolism. *In* R. A. Lewin, ed. Physiology and Biochemistry of Algae. Academic Press, New York. pp. 211–229.

Kuhl, A. 1974. Phosphorus. *In* W. D. P. Stewart, ed. Algal Physiology and Biochemistry. Univ. California Press, Berkeley. pp. 636–654.

Kuhl, M. and B. B. Jørgensen. 1994. The light field of microbenthic communities: Radiance distribution and microscale optics of sandy coastal sediments. Limnol. Oceanogr. *39*:1368–1398.

Kuivila, K. M., J. W. Murray, A. H. Devol, M. E. Lidstrom, and C. E. Reimers. 1988. Methane cycling in the sediments of Lake Washington. Limnol. Oceanogr. *33*:571–581.

organic carbon along the Ogeechee River. Limnol. Oceanogr. 36:315–323.

Leff, L. G., J. V. McArthur, J. L. Meyer, and L. J. Shimkets. 1994. Effect of macroinvertebrates on detachment of bacteria from biofilms in stream microcosms. J. N. Am. Benthol. Soc. 13:74–79.

Lehman, J. T. 1975. Reconstructing the rate of accumulation of lake sediment: The effect of sediment focusing. Quat. Res. 5:541–550.

Lehman, J. T. 1976. Ecological and nutritional studies on *Dinobryon* Ehrenb.: Seasonal periodicity and the phosphate toxicity problem. Limnol. Oceanogr. 21:646–658.

Lehman, J. T. 1977. On calculating drag characteristics for decelerating zooplankton. Limnol. Oceanogr. 22:170–172.

Lehman, J. T. 1979. Physical and chemical factors affecting the seasonal abundance of *Asterionella formosa* Hass. in a small temperate lake. Arch. Hydrobiol. 87:274–303.

Lehman, J. T. 1980a. Release and cycling of nutrients between planktonic algae and herbivores. Limnol. Oceanogr. 25:620–632.

Lehman, J. T. 1980b. Nutrient recycling as an interface between algae and grazers in freshwater communities. *In* W. C. Kerfoot, ed. Evolution and Ecology of Zooplankton Communities. Univ. Press New England, Hanover, NH. pp. 251–263.

Lehman, J. T. 1986. The goal of understanding in limnology. Limnol. Oceanogr. 31:1160–1166.

Lehman, J. T. 1988. Ecological principles affecting community structure and secondary production by zooplankton in marine and freshwater environments. Limnol. Oceanogr. 33:931–945.

Lehman, J. T. 1991. Causes and consequences of cladoceran dynamics in Lake Michigan: Implications of species invasion by *Bythotrephes*. J. Great Lakes Res. 17:437–445.

Lehman, J. T. and C. D. Sandgren. 1978. Documenting a seasonal change from phosphorus to nitrogen limitation in a small temperate lake, and its impact on the population dynamics of *Asterionella*. Verh. Internat. Verein. Limnol. 20:375–380.

Lehman, J. T. and D. K. Branstrator. 1994. Nutrient dynamics and turnover rates of phosphate and sulfate in Lake Victoria, East Africa. Limnol. Oceanogr. 39:227–233.

Lehman, J. T. and D. Scavia. 1982. Microscale patchiness of nutrients in plankton communities. Science 216:729–730

Lehman, J. T. and T. Naumoski. 1985. Content and turnover rates of phosphorus in *Daphnia pulex*: Effect of food quality. Hydrobiologia 128:119–125.

Lehmusluoto, P., *et al.* 1999. Limnology in Indonesia: From the Legacy of the Past to the Prospects for the Future. *In* R. G. Wetzel and B. Gopal, eds. Limnology of Developing Countries. Vol. 2. Societas Internationalis Limnologiae. pp. 119–234.

Lehn, H. 1965. Zur Durchsichtigkeitsmessung im Bodensee. Schrift. Ver. Geschichte Bodensees Umgebung 83:32–44.

Lehn, H. 1968. Litorale Aufwuchsalgen im Pelagial des Bodensees. Beitr. Naturk. Forsch. Südw.-Dlt. 27:97–100.

Lei, C.-H. and K. B. Armitage. 1980. Ecological energetics of a *Daphnia ambigua* population. Hydrobiologia 70:133–143.

Leibovich, S. 1977. Convective instability of stably stratified water in the ocean. J. Fluid Mech. 82:561–581.

Leifer, A. 1988. The kinetics of environmental aquatic photochemistry. Amer. Chem. Soc., York, PA. 304 pp.

Leighly, J. 1942. Effects of the Great Lakes on the annual march of air temperature in their vicinity. Pap. Mich. Acad. Sci. Arts Lett. 21:377–414.

Lelieveld, J., G.-J. Roelofs, L. Ganzeveld, J. Feichter, and H. Rodhe. 1997. Terrestrial sources and distribution of atmospheric sulphur. Phil. Trans. R. Soc. Lond. B 352:149–158.

Lellák, J. 1961. Zur Benthosproduktion und ihrer Dynamik in drei böhmischen Teichen. Verh. Internat. Verein. Limnol. 14:213–219.

Lellák, J. 1965. The food supply as a factor regulating the population dynamics of bottom animals. Mitt. Internat. Verein. Limnol. 13:128–138.

Lemke, M. J., P. F. Churchill, and R. G. Wetzel. 1995. Effect of substrate and cell surface hydrophobicity on phosphate utilization in bacteria. Appl. Environ. Microbiol. 61:913–919.

Lemke, M. J., P. F. Churchill, and R. G. Wetzel. 1998. Humic acid interaction with extracellular layers of wetland bacteria. Verh. Internat. Verein. Limnol. 26:1621–1624.

Lemly, A. D. and J. F. Dimmick. 1982. Phytoplankton communities in the littoral zone of lakes: Observations on structure and dynamics in oligotrophic and eutrophic systems. Oecologia 54:359–369.

Lemmin, U. J. 1989. Dynamics of horizontal turbulent mixing in a nearshore zone of Lake Geneva. Limnol. Oceanogr. 34:420–434.

Lenhard, G., W. R. Ross, and A. du Plooy. 1962. A study of methods for the classification of bottom deposits of natural waters. Hydrobiologia 20:223–240.

Lenz, P. H., J. M. Melack, B. Robertson, and E. A. Hardy. 1986. Ammonium and phosphate regeneration by the zooplankton of an Amazon floodplain lake. Freshwat. Biol. 16:821–830.

Leopold, L. B., M. G. Wolman, and J. P. Miller. 1964. Fluvial Processes in Geomorphology. W. H. Freeman and Co., San Francisco. 522 pp.

Leppard, G. G., J. Buffle, R. R. de Vitre, and D. Perret. 1988. The ultrastructure and physical characteristics of a distinctive colloidal iron particulate isolated from a small eutrophic lake. Arch. Hydrobiol. 113:405–424.

Lepš, J., M. Straškraba, B. Desortova, and L. Prochazková. 1990. Annual cycles of plankton species composition and physical chemical conditions in Slapy Reservoir detected by multivariate statistics. Arch. Hydrobiol. Beih. Ergebn. Limnol. 33:933–945.

Lerman, A. and M. Stiller. 1969. Vertical eddy diffusion in Lake Tiberias. Verh. Internat. Verein. Limnol. 17:323–333.

Les, D. H. 1988. Breeding systems, population structure, and evolution in hydrophilous angiosperms. Ann. Missouri Bot. Gard. 75:819–835.

Les, D. H. 1991. Genetic diversity in the monoecious hydrophile *Ceratophyllum* (Ceratophyllaceae). Am. J. Bot. 78:1070–1082.

Lesack, L. F. W. and J. M. Melack. 1991. The deposition, composition, and potential sources of major ionic solutes in rain of the central Amazon Basin. Water Resour. Res. 27:2953–2977.

Letey, J., J. W. Blair, D. Devitt, L. J. Lund, and P. Nash. 1977. Nitrate-nitrogen in effluent from agricultural tile drains in California. Hilgardia 45:289–319.

Lettenmaier, D. P., A. W. Wood, R. N. Palmer, E. F. Wood, and E. Z. Stakhiv. 1999. Water resources implications of global warming: A U.S. regional perspective. Climatic Change 43:537–579.

Leventhal, E. A. 1970. The Chrysomonadina. *In* G. E. Hutchinson, ed. Ianula: An Account of the History and Development of the Lago di Monterosi, Latium, Italy. Trans. Amer. Philos. Soc. 60(Pt. 4):123–142.

Levine, S. 1975. Orthophosphate concentration and flux within the epilimnia of two Canadian Shield lakes. Verh. Internat. Verein. Limnol. 19:624–629.

Levine, S. N. and D. W. Schindler. 1992. Modification of the N:P ratio in lakes by in situ processes. Limnol. Oceanogr. 37:917–935.

Levine, S. N. and W. M. Lewis, Jr. 1984. Diel variation of nitrogen fixation in Lake Valencia, Venezuela. Limnol. Oceanogr. 29:887–893.

Levine, S. N. and W. M. Lewis, Jr. 1985. The horizontal heterogeneity of nitrogen fixation in Lake Valencia, Venezuela. Limnol. Oceanogr. 30:1240–1245.

Lewandowski, K. and A. Stanczykowska. 1975. The occurrence and

Kuivila, K. M., J. W. Murray, A. H. Devol, and P. C. Novelli. 1989. Methane production, sulfate reduction and competition for substrates in the sediments of Lake Washington. Geochim. Cosmochim. Acta 53:409–416.

Kukkonen, J. V. K. 1999. Toxicity and bioavailability of contaminants. *In* J. Keskitalo and P. Eloranta, eds. Limnology of Humic Waters. Backhuys Publishers, Leiden. pp. 117–129.

Kulczynski, S. 1949. Peat Bogs of Polesie. Mémoires de l'Académie Polonaise des Sciences et des Lettres, Cl. Sci. Math. Natur., Ser. B, 15. 355 pp. + 46 pl.

Kullberg, A., K. H. Bishop, A. Hargeby, M. Jansson, and R. C. Petersen, Jr. 1993. The ecological significance of dissolved organic carbon in acidified waters. Ambio 22:331–337.

Kümmel, R. 1981. Zur Phosphateliminierung durch Fällung mit Calciumionen. Acta Hydrochim. Hydrobiol. 9:585–588.

Kunii, H. 1999. Annual and seasonal variations in net production, biomass and life span of floating leaves in *Brasenia schreberi* J. F. Gmel. Jpn. J. Limnol. 60:281–289.

Kunii, H. and M. Aramaki. 1992. Annual net production and life span of floating leaves in *Nymphaea tetragona* Georgi: A comparison with other floating-leaved macrophytes. Hydrobiologia 242:185–193.

Kurata, A., C. Saraceni, and H. Kadota. 1979a. The status of B group vitamins in macrophyte and pelagic zones of Lake Biwa. Mem. Ist. Ital. Idrobiol. 37:63–85.

Kurata, A., C. Saraceni, and H. Kadota. 1979b. Diurnal changes of concentration in water of B group vitamins in macrophyte and pelagic zones of Lake Biwa. Mem. Ist. Ital. Idrobiol. 37:87–103.

Kurata, A., C. Saraceni, D. Ruggiu, M. Nakanishi, U. Melchiorri-Santolini, and H. Kadota. 1976. Relationship between B group vitamins and primary production and phytoplankton population in Lake Mergozzo (Northern Italy). Mem. Ist. Ital. Idrobiol. 33:257–284.

Kurmayer, R. and F. Jüttner. 1999. Strategies for the co-existence of zooplankton with the toxic cyanobacterium *Planktothrix rubescens* in Lake Zürich. J. Plankton Res. 21:659–683.

Kuserk, F. T., L. A. Kaplan, and T. L. Bott. 1984. In situ measures of dissolved organic carbon flux in a rural stream. Can. J. Fish. Aquat. Sci. 41:964–973.

Kussatz, C., A. Gnauck, W. Jorga, H.-G. Mayer, L. Schürmann, and G. Weise. 1984. Untersuchungen zur Phosphataufnahme durch Unterwasserpflanzen. Acta Hydrochim. Hydrobiol. 12:659–677.

Kuuppo-Leinikki, P. and H. Kuosa. 1990. Estimation of flagellate grazing on bacteria by size fractionation in the northern Baltic Sea. Arch. Hydrobiol. Beih. Ergebn. Limnol. 34:283–290.

Kuznetsov, S. I. 1935. Microbiological researches in the study of the oxygenous regimen of lakes. Verh. Internat. Verein. Limnol. 7:562–582.

Kuznetsov, S. I. 1959. Die Rolle der Mikroorganismen im Stoffkreislauf der Seen. Berlin, VEB Deutsch. Verlag Wissenschaften, 301 pp.

Kuznetsov, S. I. 1964. Biogeochemistry of sulphur. *In* Lo zolfo in agricoltura. Simposio Int. Agrochimica (Palermo, Italy) 5:312–330.

Kuznetsov, S. I. 1968. Recent studies on the role of microorganisms in the cycling of substances in lakes. Limnol. Oceanogr. 13:211–224.

Kuznetsov, S. I. 1970. Mikroflora ozer i ee geokhimicheskaya deyatel'nost'. (Microflora of Lakes and Their Geochemical Activities.) (In Russian.) Izdatel'stvo Nauka, Leningrad. 440 pp.

Kuznetsov, S. I. and E. M. Khartulari. 1941. Mikrobiologicheskaia kharakteristika protsessov anaerobnogo raspada organicheskogo veshchestva ila Belogo Ozera v Kosine. (Microbiological characteristics of the process of anaerobic decomposition of organic substances of sediments of Beloye Lake in Kosine.) Mikrobiologiya 10:834–849.

Kuznetsov, S. I. and G. S. Karzinkin. 1931. Direct method for the quantitative study of bacteria in water and some considerations on causes which produce a zone of oxygen-minimum in Lake Glubokoje. Zbl. Bakt., Ser. II 83:169–174.

Kuznetsov, S. I. and V. I. Romanenko. 1963. Mikrobiologicheskoe izuchenie viutrennikh yodoemov. Laboratornoe rukobodstvo. Izdatel'stvo Akademii Nauk SSSR, Moscow. 129 pp.

Kuznetsov, S. I., G. A. Dubinina, and N. A. Lapteva. 1979. Biology of oligotrophic bacteria. Ann. Rev. Microbiol. 33:377–387.

Květ, J. 1971. Growth analysis approach to the production ecology of reedswamp plant communities. Hidrobiologia (Romania) 12:15–40.

Květ, J., J. Svoboda, and K. Fiala. 1969. Canopy development in stands of *Typha latifolia* L. and *Phragmites communis* Trin. in South Moravia. Hidrobiologia (Romania) 10:63–75.

Laanbroek, H. J. 1990. Bacteria cycling of minerals that affect plant growth in waterlogged soils: A review. Aquat. Bot. 38:109–125.

Laane, R. W. P. M., W. W. C. Gieskes, G. W. Kraay, and A. Eversdijk. 1985. Oxygen consumption from natural waters by photo-oxidizing processes. Netherlands J. Sea Res. 19:125–128.

LaBaugh, J. W., D. O. Rosenberry, and T. C. Winter. 1995. Groundwater contribution to the water and chemical budgets of Williams Lake, Minnesota, 1980–1991. Can. J. Fish. Aquat. Sci. 52:754–767.

Lafleur, P. M. 1990. Evapotranspiration from sedge-dominated wetland surfaces. Aquat. Bot. 37:341–353.

Lagler, K. F., J. E. Bardach, and R. R. Miller. 1962. Ichthyology. John Wiley & Sons, New York. 545 pp.

Laing, W. A. and J. Browse. 1985. A dynamic model for photosynthesis by an aquatic plant, *Egeria densa*. Plant Cell Environ. 8:639–649.

Lair, N. 1990. Effects of invertebrate predation on the seasonal succession of a zooplankton community: A two year study in Lake Aydat, France. Hydrobiologia 198:1–12.

Lal, R. 1998. Soil erosion impact on agronomic productivity and environment quality. Crit. Rev. Pl. Sci. 17:319–464.

Lallana, V. H., R. A. Sabattini, and M. C. Lallana. 1987. Evapotranspiration from *Eichhornia crassipes, Pistia stratiotes, Salvinia herzogii* and *Azolla caroliniana* during summer in Argentina. J. Aquat. Pl. Manage. 25:48–50.

LaLonde, R. T., C. D. Morris, C. F. Wong, L. C. Gardner, D. J. Eckert, D. R. King, and R. H. Zimmerman. 1979. Response of *Aedes triseriatus* larvae to fatty acids of Cladophora. J. Chem. Ecol. 5:371–381.

Lam, C. W. Y., W. F. Vincent, and W. B. Silvester. 1979. Nitrogenase activity and estimates of nitrogen fixation by freshwater benthic blue-green algae. N. Z. J. Mar. Freshwat. Res. 13:187–192.

LaMarra, V. J., Jr. 1975. Digestive activities of carp as a major contributor to the nutrient loading of lakes. Verh. Internat. Verein. Limnol. 19:2461–2468.

Lamberti, G. A. and J. W. Moore. 1984. Aquatic insects as primary consumers. *In* V. H. Resh and D. M. Rosenberg, eds. The Ecology of Aquatic Insects. Praeger, New York. pp. 164–195.

Lamberti, G. A., L. R. Ashkenas, S. V. Gregory, and A. D. Steinman. 1987. Effects of three herbivores on periphyton communities in laboratory streams. J. N. Am. Benthol. Soc. 6:92–104.

Lamberti, G. A., S. V. Gregory, C. P. Hawkins, R. C. Wildman, L. R. Ashkenas, and D. M. DeNicola. 1992. Plant-herbivore interactions in streams near Mount St. Helens. Freshwat. Biol. 27:237–247.

Laminger, H. 1973. Untersuchungen über Abundanz und Biomasse der sedimentbewohnenden Testaceen (Protozoa, Rhizopoda) in einem Hochgebirgssee (Vorderer Finstertaler See, Kühtai, Tirol). Int. Rev. ges. Hydrobiol. 58:543–568.

Kuivila, K. M., J. W. Murray, A. H. Devol, and P. C. Novelli. 1989. Methane production, sulfate reduction and competition for substrates in the sediments of Lake Washington. Geochim. Cosmochim. Acta *53*:409–416.

Kukkonen, J. V. K. 1999. Toxicity and bioavailability of contaminants. *In* J. Keskitalo and P. Eloranta, eds. Limnology of Humic Waters. Backhuys Publishers, Leiden. pp. 117–129.

Kulczynski, S. 1949. Peat Bogs of Polesie. Mémoires de l'Académie Polonaise des Sciences et des Lettres, Cl. Sci. Math. Natur., Ser. B, 15. 355 pp. + 46 pl.

Kullberg, A., K. H. Bishop, A. Hargeby, M. Jansson, and R. C. Petersen, Jr. 1993. The ecological significance of dissolved organic carbon in acidified waters. Ambio 22:331–337.

Kümmel, R. 1981. Zur Phosphateliminierung durch Fällung mit Calciumionen. Acta Hydrochim. Hydrobiol. *9*:585–588.

Kunii, H. 1999. Annual and seasonal variations in net production, biomass and life span of floating leaves in *Brasenia schreberi* J. F. Gmel. Jpn. J. Limnol. *60*:281–289.

Kunii, H. and M. Aramaki. 1992. Annual net production and life span of floating leaves in *Nymphaea tetragona* Georgi: A comparison with other floating-leaved macrophytes. Hydrobiologia *242*:185–193.

Kurata, A., C. Saraceni, and H. Kadota. 1979a. The status of B group vitamins in macrophyte and pelagic zones of Lake Biwa. Mem. Ist. Ital. Idrobiol. *37*:63–85.

Kurata, A., C. Saraceni, and H. Kadota. 1979b. Diurnal changes of concentration in water of B group vitamins in macrophyte and pelagic zones of Lake Biwa. Mem. Ist. Ital. Idrobiol. *37*:87–103.

Kurata, A., C. Saraceni, D. Ruggiu, M. Nakanishi, U. Melchiorri-Santolini, and H. Kadota. 1976. Relationship between B group vitamins and primary production and phytoplankton population in Lake Mergozzo (Northern Italy). Mem. Ist. Ital. Idrobiol. *33*:257–284.

Kurmayer, R. and F. Jüttner. 1999. Strategies for the co-existence of zooplankton with the toxic cyanobacterium *Planktothrix rubescens* in Lake Zürich. J. Plankton Res. *21*:659–683.

Kuserk, F. T., L. A. Kaplan, and T. L. Bott. 1984. In situ measures of dissolved organic carbon flux in a rural stream. Can. J. Fish. Aquat. Sci. *41*:964–973.

Kussatz, C., A. Gnauck, W. Jorga, H.-G. Mayer, L. Schürmann, and G. Weise. 1984. Untersuchungen zur Phosphataufnahme durch Unterwasserpflanzen. Acta Hydrochim. Hydrobiol. *12*:659–677.

Kuuppo-Leinikki, P. and H. Kuosa. 1990. Estimation of flagellate grazing on bacteria by size fractionation in the northern Baltic Sea. Arch. Hydrobiol. Beih. Ergebn. Limnol. *34*:283–290.

Kuznetsov, S. I. 1935. Microbiological researches in the study of the oxygenous regimen of lakes. Verh. Internat. Verein. Limnol. *7*:562–582.

Kuznetsov, S. I. 1959. Die Rolle der Mikroorganismen im Stoffkreislauf der Seen. Berlin, VEB Deutsch. Verlag Wissenschaften, 301 pp.

Kuznetsov, S. I. 1964. Biogeochemistry of sulphur. *In* Lo zolfo in agricoltura. Simposio Int. Agrochimica (Palermo, Italy) *5*:312–330.

Kuznetsov, S. I. 1968. Recent studies on the role of microorganisms in the cycling of substances in lakes. Limnol. Oceanogr. *13*:211–224.

Kuznetsov, S. I. 1970. Mikroflora ozer i ee geokhimicheskaya deyatel'nost'. (Microflora of Lakes and Their Geochemical Activities.) (In Russian.) Izdatel'stvo Nauka, Leningrad. 440 pp.

Kuznetsov, S. I. and E. M. Khartulari. 1941. Mikrobiologicheskaia kharakteristika protsessov anaerobnogo raspada organicheskogo veshchestva ila Belogo Ozera v Kosine. (Microbiological characteristics of the process of anaerobic decomposition of organic substances of sediments of Beloye Lake in Kosine.) Mikrobiologiya *10*:834–849.

Kuznetsov, S. I. and G. S. Karzinkin. 1931. Direct method for the quantitative study of bacteria in water and some considerations on causes which produce a zone of oxygen-minimum in Lake Glubokoje. Zbl. Bakt., Ser. II *83*:169–174.

Kuznetsov, S. I. and V. I. Romanenko. 1963. Mikrobiologicheskoe izuchenie viutrennikh yodoemov. Laboratornoe rukobodstvo. Izdatel'stvo Akademii Nauk SSSR, Moscow. 129 pp.

Kuznetsov, S. I., G. A. Dubinina, and N. A. Lapteva. 1979. Biology of oligotrophic bacteria. Ann. Rev. Microbiol. *33*:377–387.

Květ, J. 1971. Growth analysis approach to the production ecology of reedswamp plant communities. Hidrobiologia (Romania) *12*:15–40.

Květ, J., J. Svoboda, and K. Fiala. 1969. Canopy development in stands of *Typha latifolia* L. and *Phragmites communis* Trin. in South Moravia. Hidrobiologia (Romania) *10*:63–75.

Laanbroek, H. J. 1990. Bacteria cycling of minerals that affect plant growth in waterlogged soils: A review. Aquat. Bot. *38*:109–125.

Laane, R. W. P. M., W. W. C. Gieskes, G. W. Kraay, and A. Eversdijk. 1985. Oxygen consumption from natural waters by photo-oxidizing processes. Netherlands J. Sea Res. *19*:125–128.

LaBaugh, J. W., D. O. Rosenberry, and T. C. Winter. 1995. Groundwater contribution to the water and chemical budgets of Williams Lake, Minnesota, 1980–1991. Can. J. Fish. Aquat. Sci. *52*:754–767.

Lafleur, P. M. 1990. Evapotranspiration from sedge-dominated wetland surfaces. Aquat. Bot. *37*:341–353.

Lagler, K. F., J. E. Bardach, and R. R. Miller. 1962. Ichthyology. John Wiley & Sons, New York. 545 pp.

Laing, W. A. and J. Browse. 1985. A dynamic model for photosynthesis by an aquatic plant, *Egeria densa*. Plant Cell Environ. *8*:639–649.

Lair, N. 1990. Effects of invertebrate predation on the seasonal succession of a zooplankton community: A two year study in Lake Aydat, France. Hydrobiologia *198*:1–12.

Lal, R. 1998. Soil erosion impact on agronomic productivity and environment quality. Crit. Rev. Pl. Sci. *17*:319–464.

Lallana, V. H., R. A. Sabattini, and M. C. Lallana. 1987. Evapotranspiration from *Eichhornia crassipes, Pistia stratiotes, Salvinia herzogii* and *Azolla caroliniana* during summer in Argentina. J. Aquat. Pl. Manage. 25:48–50.

LaLonde, R. T., C. D. Morris, C. F. Wong, L. C. Gardner, D. J. Eckert, D. R. King, and R. H. Zimmerman. 1979. Response of *Aedes triseriatus* larvae to fatty acids of Cladophora. J. Chem. Ecol. *5*:371–381.

Lam, C. W. Y., W. F. Vincent, and W. B. Silvester. 1979. Nitrogenase activity and estimates of nitrogen fixation by freshwater benthic blue-green algae. N. Z. J. Mar. Freshwat. Res. *13*:187–192.

LaMarra, V. J., Jr. 1975. Digestive activities of carp as a major contributor to the nutrient loading of lakes. Verh. Internat. Verein. Limnol. *19*:2461–2468.

Lamberti, G. A. and J. W. Moore. 1984. Aquatic insects as primary consumers. *In* V. H. Resh and D. M. Rosenberg, eds. The Ecology of Aquatic Insects. Praeger, New York. pp. 164–195.

Lamberti, G. A., L. R. Ashkenas, S. V. Gregory, and A. D. Steinman. 1987. Effects of three herbivores on periphyton communities in laboratory streams. J. N. Am. Benthol. Soc. *6*:92–104.

Lamberti, G. A., S. V. Gregory, C. P. Hawkins, R. C. Wildman, L. R. Ashkenas, and D. M. DeNicola. 1992. Plant-herbivore interactions in streams near Mount St. Helens. Freshwat. Biol. *27*:237–247.

Laminger, H. 1973. Untersuchungen über Abundanz und Biomasse der sedimentbewohnenden Testaceen (Protozoa, Rhizopoda) in einem Hochgebirgssee (Vorderer Finstertaler See, Kühtai, Tirol). Int. Rev. ges. Hydrobiol. *58*:543–568.

Lampert, W. 1977a. Studies on the carbon balance of *Daphnia pulex* de Geer as related to environmental conditions. II. The dependence of carbon assimilation on animal size, temperature, food concentration and diet species. Arch. Hydrobiol./Suppl. *48*:310–335.

Lampert, W. 1977b. Studies on the carbon balance of *Daphnia pulex* de Geer as related to environmental conditions. III. Production and production efficiency. Arch. Hydrobiol./Suppl. *48*:336–360.

Lampert, W. 1978a. Climatic conditions and planktonic interactions as factors controlling the regular succession of spring algal bloom and extremely clear water in Lake Constance. Verh. Internat. Verein. Limnol. *20*:969–974.

Lampert, W. 1978b. Release of dissolved organic carbon by grazing zooplankton. Limnol. Oceanogr. *23*:831–834.

Lampert, W. 1981. Inhibiting and toxic effects of blue-green algae on *Daphnia*. Int. Rev. ges. Hydrobiol. *66*:285–298.

Lampert, W. 1982. Further studies on the inhibitory effect of the toxic blue-green *Microcystis aeruginosa* on the filtering rate of zooplankton. Arch. Hydrobiol. *95*:207–220.

Lampert, W. 1986. Response of the respiratory rate of *Daphnia magna* to changing food conditions. Oecologia *70*:495–501.

Lampert, W. 1989. The adaptive significance of diel vertical migration of zooplankton. Functional Ecol. *3*:21–27.

Lampert, W. 1993. Ultimate causes of diel vertical migration of zooplankton: New evidence for the predator-avoidance hypothesis. Arch. Hydrobiol. Beih. Ergebn. Limnol. *39*:79–88.

Lampert, W. and H. Brendelberger. 1996. Strategies of phenotypic low-food adaptation in *Daphnia*: Filter screens, mesh sizes, and appendage beat rates. Limnol. Oceanogr. *41*:216–223.

Lampert, W. and H. G. Wolf. 1986. Cyclomorphosis in *Daphnia cucullata*: Morphometric and population genetic analyses. J. Plankton Res. *8*:289–303.

Lampert, W. and R. Bohrer. 1984. Effect of food availability on the respiratory quotient of *Daphnia magna*. Comp. Biochem. Physiol. *78A*:221–223.

Lancaster, J. 1990. Predation and drift of lotic macroinvertebrates during colonization. Oecologia *85*:48–56.

Lancaster, J. and A. G. Hildrew. 1993. Characterizing in-stream flow refugia. Can. J. Fish. Aquat. Sci. *50*:1663–1675.

Lancaster, J. and A. L. Robertson. 1995. Microcrustacean prey and macroinvertebrate predators in a stream food web. Freshwat. Biol. *34*:123–134.

Landers, D. H. 1982. Effects of naturally senescing aquatic macrophytes on nutrient chemistry and chlorophyll *a* of surrounding waters. Limnol. Oceanogr. *27*:428–439.

Landner, L. and T. Larsson. 1973. Indications of disturbances in the nitrification process in a heavily nitrogen-polluted water body. Ambio 2:154–157.

Landrum, P. F. and S. W. Fisher. 1999. Influence of lipids on the bioaccumulation and trophic transfer of organic contaminants in aquatic systems. *In* M. T. Arts and B. C. Wainman, eds. Lipids in Freshwater Ecosystems. Springer-Verlag, New York. pp. 203–234.

Lane, L. S. 1977. Microbial community fluctuations in a meromictic, Antarctic lake. Hydrobiologia *55*:187–190.

Lang, C. and B. Lang-Dobler. 1980. Structure of tubificid and lumbriculid worm communities, and three indices of trophy based upon these communities, as descriptors of eutrophication level of Lake Geneva (Switzerland). *In* R. O. Brinkhurst and D. G. Cook, eds. Aquatic Oligochaete Biology. Plenum Press, New York. pp. 457–470.

Lange, W. 1971. Enhancement of algal growth in Cyanophyta-bacteria systems by carbonaceous compounds. Can. J. Microbiol. *17*:303–314.

Langeland, A. 1981. Decreased zooplankton density in two Norwegian lakes caused by predation of recently introduced *Mysis relicta*. Verh. Internat. Verein. Limnol. *21*:926–937.

Langeland, A. 1988. Decreased zooplankton density in a mountain lake resulting from predation by recently introduced *Mysis relicta*. Verh. Internat. Verein. Limnol. *23*:419–429.

Langford, R. R. and E. G. Jermolajev. 1966. Direct effect of wind on plankton distribution. Verh. Internat. Verein. Limnol. *16*:188–193.

Langmuir, I. 1938. Surface motion of water induced by wind. Science *87*:119–123.

Lansactôha, F. A., A. F. Lima, S. M. Thomaz, and M. C. Roberto. 1993. Zooplâncton de uma planície de inundação do Rio Paraná. II. Variação sazonal e influência dos níveis fluviométricos sobre a comunidade. Acta Limnol. Brasiliensia *4*:42–55.

Large, A. R. G. and G. E. Petts. 1994. Rehabilitation of river margins. *In* P. Calow and G. E. Petts, eds. The Rivers Handbook. 2. Hydrological and Ecological Principles. Blackwell Scientific Publs., Oxford. pp. 401–418.

LaRow, E. J. 1968. A persistent diurnal rhythm in *Chaoborus* larvae. I. The nature of the rhythmicity. Limnol. Oceanogr. *13*:250–256.

LaRow, E. J. 1969. A persistent diurnal rhythm in *Chaoborus* larvae. II. Ecological significance. Limnol. Oceanogr. *14*:213–218.

LaRow, E. J. 1970. The effect of oxygen tension on the vertical migration of *Chaoborus* larvae. Limnol. Oceanogr. *15*:357–362.

LaRow, E. J. 1975. Secondary productivity of *Leptodora kindtii* in Lake George, N. Y. Am. Midland Nat. *94*:120–126.

Larsen, D. P. and H. T. Mercier. 1976. Phosphorus retention capacity of lakes. J. Fish. Res. Board Can. *33*:1742–1750.

Larsen, D. P., K. W. Malueg, D. W. Schults, and R. M. Brice. 1975. Response of eutrophic Shagawa Lake, Minnesota, U.S.A., to point-source phosphorus reduction. Verh. Internat. Verein. Limnol. *19*:884–892.

Larson, G. L. 1989. Geographical distribution, morphology and water quality of caldera lakes: A review. Hydrobiologia *171*:23–32.

Larson, R. A. and J. M. Hufnal, Jr. 1980. Oxidative polymerization of dissolved phenols by soluble and insoluble inorganic species. Limnol. Oceanogr. *25*:505–512.

Larsson, U. and A. Hagström. 1979. Phytoplankton exudate release as an energy source for the growth of pelagic bacteria. Mar. Biol. *52*:199–206.

Lasenby, D. C. 1975. Development of oxygen deficits in 14 southern Ontario lakes. Limnol. Oceanogr. *20*:993–999.

Lasenby, D. C. and R. R. Langford. 1972. Growth, life history, and respiration of *Mysis relicta* in an arctic and temperate lake. J. Fish. Res. Board. Can. *29*:1701–1708.

Lasenby, D. C. and R. R. Langford. 1973. Feeding and assimilation of *Mysis relicta*. Limnol. Oceanogr. *18*:280–285.

Lastein, E. 1976. Recent sedimentation and resuspension of organic matter in eutrophic Lake Esrom, Denmark. Oikos *27*:44–49.

Latimer, J. R. 1972. Radiation measurement. Tech. Manual Ser. Int. Field Year Great Lakes 2, 53 pp.

Laube, H. R. and J. R. Wohler. 1973. Studies on the decomposition of a duckweed (Lemnaceae) community. Bull. Torr. Bot. Club. *100*:238–240.

Läuchli, A. 1993. Selenium in plants: Uptake, functions, and environmental toxicity. Bot. Acta *106*:455–468.

Lauffenberger, D. A. 1983. Effects of cell motility properties on cell populations in ecosystems. *In* H. W. Blanch, E. T. Papoutsakis, and G. Stephanopoulos, eds. Foundations of Biochemical Engineering: Kinetics and Thermodynamics in Biological Systems. American Chemical Society, Washington, DC. pp. 266–292.

Laurent, M. and J. Badia. 1973 Étude comparative du cycle biologique de l'azote dans deux étangs. Ann. Hydrobiol. *4*:77–102.

Lauridsen, T. L. and I. Buenk. 1996. Diel changes in the horizontal distribution of zooplankton in the littoral zone of two shallow eutrophic lakes. Arch. Hydrobiol. *137*:161–176.

Lauwers, A. M. and W. Heinen. 1974. Biol-degradation and utilization of silica and quartz. Arch. Microbiol. *95*:67–78.

Lavandier, P. 1990. Dynamics of bacterioplankton in a mesotrophic French reservoir (Pareloup). Hydrobiologica *207*:79–86.

Lavandier, P. and H. Décamps. 1984. Estaragne. *In* B. A. Whitton, ed. Ecology of European Rivers. Blackwell Scientific Publ., Oxford. pp. 237–264.

Lawacz, W. 1969. The characteristics of sinking materials and the formation of bottom deposits in a eutrophic lake. Mitt. Internat. Verein. Limnol. *17*:319–331.

Lawrence, G. B., R. D. Fuller, and C. T. Driscoll. 1986. Spatial relationships of aluminum chemistry in the streams of the Hubbard Brook Experimental Forest, New Hampshire. Biogeochemistry *2*:115–135.

Lawson, D. L., M. J. Klug, and R. W. Merritt. 1984. The influence of physical, chemical and microbiological characteristics of decomposing leaves on the growth of the detritivore *Tipula abdominalis* (Diptera: Tipulidae). Can. J. Zool. *62*:2339–2343.

Laxen, D. P. H. 1984. Cadmium in freshwaters (*sic*): Concentrations and chemistry. Freshwat. Biol. *14*:587–595.

Laxhuber, R. 1987. Abundance and distribution of pelagic rotifers in a cold, deep oligotrophic alpine lake (Königssee). Hydrobiologia *147*:189–196.

Lay, J. A. and A. K. Ward. 1987. Algal community dynamics in two streams associated with different geological regions in the southeastern United States. Arch. Hydrobiol. *108*:305–324.

Laybourn-Parry, J. 1984. A Functional Biology of Free-Living Protozoa. Univ. California Press, Berkeley.

Laybourn-Parry, J. 1992. Protozoan Plankton Ecology. Chapman & Hall, London. 231 pp.

Laybourn-Parry, J. 1994. Seasonal successions of protozooplankton in freshwater ecosystems of different latitudes. Mar. Microbial Food Webs *8*:145–162.

Laybourn-Parry, J., A. Rogerson, and D. W. Crawford. 1992. Temporal patterns of protozooplankton abundance in the Clyde and Loch Striven. Estuar. Coastal Shelf Sci. *35*:533-543.

Laybourn-Parry, J., B. Baldock, and J. C. Kingsmill-Robinson. 1980. Respiration studies on two small freshwater amoebae. Microb. Ecol. *6*:209–216.

Laybourn-Parry, J., J. Olver, A. Rogerson, and P. L. Duverge. 1990a. The temporal pattern of protozooplankton abundance in a eutrophic temperate lake. Hydrobiologia *203*:99–110.

Laybourn-Parry, J., J. Olver, and S. Rees. 1990b. The hypolimnetic protozoan plankton of a eutrophic lake. Hydrobiologia *203*:111–119.

Laybourn-Parry, J., M. Walton, J. Young, R. I. Jones, and A. Shine. 1994. Protozooplankton and bacterioplankton in a large oligotrophic lake—Loch Ness, Scotland. J. Plankton Res. *16*:1655–1670.

Laybourn-Parry, J., P. Bayliss, and J. C. Ellis-Evans. 1995. The dynamics of heterotrophic nanoflagellates and bacterioplankton in a large ultra-oligotrophic Antarctic lake. J. Plankton Res. *17*:1835–1850.

Leach, J. H. 1975. Seston composition in the Point Pelee Area of Lake Erie. (Unpublished manuscript.)

Leah, R. T., B. Moss, and D. E. Forrest. 1980. The role of predation in causing major changes in the limnology of a hyper-eutrophic lake. Int. Rev. ges. Hydrobiol. *65*:223–247.

Lean, D. 1998. Attenuation of solar radiation in humic waters. *In* D. O. Hessen and L. J. Tranvik, eds. Aquatic Humic Substances: Ecology and Biogeochemistry. Springer-Verlag, Berlin. pp. 109–124.

Lean, D. R. S. 1973a. Phosphorus dynamics in lake water. Science *179*:678–680.

Lean, D. R. S. 1973b. Movements of phosphorus between its biologically important forms in lake water. J. Fish. Res. Board Can. *30*:1525–1536.

Lean, D. R. S. and C. Nalewajko. 1976. Phosphate exchange and organic phosphorus excretion by freshwater algae. J. Fish. Res. Board Can. *33*:1312–1323.

Lean, D. R. S. and F. H. Rigler. 1974. A test of the hypothesis that abiotic phosphate complexing influences phosphorus kinetics in epilimnetic lake water. Limnol. Oceanogr. *19*:784–788.

Lean, D. R. S., C. F.-H. Liao, T. P. Murphy, and D. S. Painter. 1978. The importance of nitrogen fixation in lakes. *In* Environmental Role of Nitrogen-Fixing Blue-Green Algae and Asymbiotic Bacteria. Ecol. Bull. (Stockholm) *26*:41–51.

Lean, D. R. S., T. P. Murphy, and F. R. Pick. 1982. Photosynthetic response of lake plankton to combined nitrogen enrichment. J. Phycol. *18*:509–521.

Learner, M. A. and D. W. B. Potter. 1974. Life-history and production of the leech *Helobdella stagnalis* (L.) (Hirudinea) in a shallow eutrophic reservoir in South Wales. J. Anim. Ecol. *43*:199–208.

Learner, M. A., G. Lochhead, and B. D. Hughes. 1978. A review of the biology of British Naididae (Oligochaeta) with emphasis on the lotic environment. Freshwat. Biol. *8*:357–375.

Leavitt, P. R. and S. R. Carpenter. 1990a. Aphotic pigment degradation in the hypolimnion: Implications for sedimentation studies and paleolimnology. Limnol. Oceangr. *35*:520–534.

Leavitt, P. R. and S. R. Carpenter. 1990b. Regulation of pigment sedimentation by photo-oxidation and herbivore grazing. Can. J. Fish. Aquat. Sci. *47*:1166–1176.

Lebaron, P., N. Parthuisot, and P. Catala. 1998. Comparison of blue nucleic acid dyes for flow cytometric enumeration of bacteria in aquatic systems. Appl. Environ. Microbiol. *64*:1725–1730.

Leckie, J. O. and J. A. Davis, III. 1979. Aqueous environmental chemistry of copper. *In* J. O. Nriagu, ed. Copper in the Environment. I. Ecological Cycling. John Wiley & Sons, New York. pp. 89–121.

LeCren, E. D. 1958. Observations on the growth of perch (*Perca fluviatalis* L.) over twenty-two years with special reference to the effects of temperature and changes in population density. J. Anim. Ecol. *27*:287–334.

Leduc, L. G. and G. D. Ferroni. 1979. Quantitative ecology of psychrophilic bacteria in an aquatic environment and characterization of heterotrophic bacteria from permanently cold sediments. Can. J. Microbiol. *25*:1433–1442.

Lee, G. F., E. Bentley, and R. Amundson. 1975. Effects of marshes on water quality. *In* A. D. Hasler, ed. Coupling of Land and Water Systems. Springer-Verlag, New York. pp. 105–127.

Lee, J. G., S. B. Roberts, and F. M. M. Morel. 1995. Cadmium: A nutrient for the marine diatom *Thalassiosira weissflogii*. Limnol. Oceanogr. *40*:1056–1063.

Lee, K. 1983. Vanadium in the aquatic ecosystem. *In* J. O. Nriagu, ed. Aquatic Toxicology. John Wiley & Sons, New York. pp. 155–187.

Lee, R. E. and P. Kugrens. 1992. Relationship between the flagellates and the ciliates. Microbiol. Rev. *56*:529–542.

Lefèvre, M. 1964. Extracellular products of algae. *In* D. F. Jackson, ed. Algae and Man. Plenum Press, New York. pp. 337–367.

Leff, L. G. and J. L. Meyer. 1991. Biological availability of dissolved organic carbon along the Ogeechee River. Limnol. Oceanogr. *36*:315–323.

Leff, L. G., J. V. McArthur, J. L. Meyer, and L. J. Shimkets. 1994. Effect of macroinvertebrates on detachment of bacteria from biofilms in stream microcosms. J. N. Am. Benthol. Soc. *13*:74–79.

Lehman, J. T. 1975. Reconstructing the rate of accumulation of lake sediment: The effect of sediment focusing. Quat. Res. *5*:541–550.

Lehman, J. T. 1976. Ecological and nutritional studies on *Dinobryon* Ehrenb.: Seasonal periodicity and the phosphate toxicity problem. Limnol. Oceanogr. *21*:646–658.

Lehman, J. T. 1977. On calculating drag characteristics for decelerating zooplankton. Limnol. Oceanogr. 22:170–172.

Lehman, J. T. 1979. Physical and chemical factors affecting the seasonal abundance of *Asterionella formosa* Hass. in a small temperate lake. Arch. Hydrobiol. *87*:274–303.

Lehman, J. T. 1980a. Release and cycling of nutrients between planktonic algae and herbivores. Limnol. Oceanogr. *25*:620–632.

Lehman, J. T. 1980b. Nutrient recycling as an interface between algae and grazers in freshwater communities. *In* W. C. Kerfoot, ed. Evolution and Ecology of Zooplankton Communities. Univ. Press New England, Hanover, NH. pp. 251–263.

Lehman, J. T. 1986. The goal of understanding in limnology. Limnol. Oceanogr. *31*:1160–1166.

Lehman, J. T. 1988. Ecological principles affecting community structure and secondary production by zooplankton in marine and freshwater environments. Limnol. Oceanogr. *33*:931–945.

Lehman, J. T. 1991. Causes and consequences of cladoceran dynamics in Lake Michigan: Implications of species invasion by *Bythotrephes*. J. Great Lakes Res. *17*:437–445.

Lehman, J. T. and C. D. Sandgren. 1978. Documenting a seasonal change from phosphorus to nitrogen limitation in a small temperate lake, and its impact on the population dynamics of *Asterionella*. Verh. Internat. Verein. Limnol. *20*:375–380.

Lehman, J. T. and D. K. Branstrator. 1994. Nutrient dynamics and turnover rates of phosphate and sulfate in Lake Victoria, East Africa. Limnol. Oceanogr. *39*:227–233.

Lehman, J. T. and D. Scavia. 1982. Microscale patchiness of nutrients in plankton communities. Science *216*:729–730

Lehman, J. T. and T. Naumoski. 1985. Content and turnover rates of phosphorus in *Daphnia pulex*: Effect of food quality. Hydrobiologia *128*:119–125.

Lehmusluoto, P., *et al.* 1999. Limnology in Indonesia: From the Legacy of the Past to the Prospects for the Future. *In* R. G. Wetzel and B. Gopal, eds. Limnology of Developing Countries. Vol. 2. Societas Internationalis Limnologiae. pp. 119–234.

Lehn, H. 1965. Zur Durchsichtigkeitsmessung im Bodensee. Schrift. Ver. Geschichte Bodensees Umgebung *83*:32–44.

Lehn, H. 1968. Litorale Aufwuchsalgen im Pelagial des Bodensee. Beitr. Naturk. Forsch. Südw.-Dlt. *27*:97–100.

Lei, C.-H. and K. B. Armitage. 1980. Ecological energetics of a *Daphnia ambigua* population. Hydrobiologia *70*:133–143.

Leibovich, S. 1977. Convective instability of stably stratified water in the ocean. J. Fluid Mech. *82*:561–581.

Leifer, A. 1988. The kinetics of environmental aquatic photochemistry. Amer. Chem. Soc., York, PA. 304 pp.

Leighly, J. 1942. Effects of the Great Lakes on the annual march of air temperature in their vicinity. Pap. Mich. Acad. Sci. Arts Lett. *21*:377–414.

Lelieveld, J., G.-J. Roelofs, L. Ganzeveld, J. Feichter, and H. Rodhe. 1997. Terrestrial sources and distribution of atmospheric sulphur. Phil. Trans. R. Soc. Lond. B *352*:149–158.

Lellák, J. 1961. Zur Benthosproduktion und ihrer Dynamik in drei böhmischen Teichen. Verh. Internat. Verein. Limnol. *14*:213–219.

Lellák, J. 1965. The food supply as a factor regulating the population dynamics of bottom animals. Mitt. Internat. Verein. Limnol. *13*:128–138.

Lemke, M. J., P. F. Churchill, and R. G. Wetzel. 1995. Effect of substrate and cell surface hydrophobicity on phosphate utilization in bacteria. Appl. Environ. Microbiol. *61*:913–919.

Lemke, M. J., P. F. Churchill, and R. G. Wetzel. 1998. Humic acid interaction with extracellular layers of wetland bacteria. Verh. Internat. Verein. Limnol. *26*:1621–1624.

Lemly, A. D. and J. F. Dimmick. 1982. Phytoplankton communities in the littoral zone of lakes: Observations on structure and dynamics in oligotrophic and eutrophic systems. Oecologia *54*:359–369.

Lemmin, U. J. 1989. Dynamics of horizontal turbulent mixing in a nearshore zone of Lake Geneva. Limnol. Oceanogr. *34*:420–434.

Lenhard, G., W. R. Ross, and A. du Plooy. 1962. A study of methods for the classification of bottom deposits of natural waters. Hydrobiologia *20*:223–240.

Lenz, P. H., J. M. Melack, B. Robertson, and E. A. Hardy. 1986. Ammonium and phosphate regeneration by the zooplankton of an Amazon floodplain lake. Freshwat. Biol. *16*:821–830.

Leopold, L. B., M. G. Wolman, and J. P. Miller. 1964. Fluvial Processes in Geomorphology. W. H. Freeman and Co., San Francisco. 522 pp.

Leppard, G. G., J. Buffle, R. R. de Vitre, and D. Perret. 1988. The ultrastructure and physical characteristics of a distinctive colloidal iron particulate isolated from a small eutrophic lake. Arch. Hydrobiol. *113*:405–424.

Lepš, J., M. Straškraba, B. Desortova, and L. Prochazková. 1990. Annual cycles of plankton species composition and physical chemical conditions in Slapy Reservoir detected by multivariate statistics. Arch. Hydrobiol. Beih. Ergebn. Limnol. *33*:933–945.

Lerman, A. and M. Stiller. 1969. Vertical eddy diffusion in Lake Tiberias. Verh. Internat. Verein. Limnol. *17*:323–333.

Les, D. H. 1988. Breeding systems, population structure, and evolution in hydrophilous angiosperms. Ann. Missouri Bot. Gard. *75*:819–835.

Les, D. H. 1991. Genetic diversity in the monoecious hydrophile *Ceratophyllum* (Ceratophyllaceae). Am. J. Bot. *78*:1070–1082.

Lesack, L. F. W. and J. M. Melack. 1991. The deposition, composition, and potential sources of major ionic solutes in rain of the central Amazon Basin. Water Resour. Res. *27*:2953–2977.

Letey, J., J. W. Blair, D. Devitt, L. J. Lund, and P. Nash. 1977. Nitrate-nitrogen in effluent from agricultural tile drains in California. Hilgardia *45*:289–319.

Lettenmaier, D. P., A. W. Wood, R. N. Palmer, E. F. Wood, and E. Z. Stakhiv. 1999. Water resources implications of global warming: A. U.S. regional perspective. Climatic Change *43*:537–579.

Leventhal, E. A. 1970. The Chrysomonadina. *In* G. E. Hutchinson, ed. Ianula: An Account of the History and Development of the Lago di Monterosi, Latium, Italy. Trans. Amer. Philos. Soc. *60*(Pt. 4):123–142.

Levine, S. 1975. Orthophosphate concentration and flux within the epilimnia of two Canadian Shield lakes. Verh. Internat. Verein. Limnol. *19*:624–629.

Levine, S. N. and D. W. Schindler. 1992. Modification of the N:P ratio in lakes by in situ processes. Limnol. Oceanogr. *37*:917–935.

Levine, S. N. and W. M. Lewis, Jr. 1984. Diel variation of nitrogen fixation in Lake Valencia, Venezuela. Limnol. Oceanogr. *29*:887–893.

Levine, S. N. and W. M. Lewis, Jr. 1985. The horizontal heterogeneity of nitrogen fixation in Lake Valencia, Venezuela. Limnol. Oceanogr. *30*:1240–1245.

Lewandowski, K. and A. Stanczykowska. 1975. The occurrence and role of bivalves of the family Unionidae in Mikolajskie Lake. Ekol. Pol. *23*:317–334.

Lewin, J. C. 1962. Silicification. *In* R. A. Lewin, ed. Physiology and Biochemistry of Algae. Academic Press, New York. pp. 445–455.

Lewis, W. M., Jr. 1973. The thermal regime of Lake Lanao (Philippines) and its theoretical implications for tropical lakes. Limnol. Oceanogr. *18*:200–217.

Lewis, W. M., Jr. 1974. Primary production in the plankton community of a tropical lake. Ecol. Monogr. *44*:377–409.

Lewis, W. M., Jr. 1975. A theoretical comparison of the attenuation of light energy and quanta in waters of divergent optical properties. Arch. Hydrobiol. *75*:285–296.

Lewis, W. M., Jr. 1977a. Ecological significance of the shapes of abundance-frequency distributions for coexisting phytoplankton species. Ecology *58*:850–859.

Lewis, W. M., Jr. 1977b. Feeding selectivity of a tropical *Chaoborus* population. Freshwat. Biol. 7:311–325.

Lewis, W. M., Jr. 1978a. A compositional, phytogeographical, and elementary community structural analysis of the phytoplankton in a tropical lake. J. Ecol. *66*:213–226.

Lewis, W. M., Jr. 1978b. Dynamics and succession of the phytoplankton in a tropical lake: Lake Lanao, Philippines. J. Ecol. *66*:849–880.

Lewis, W. M., Jr. 1979. Zooplankton Community Analysis: Studies on a Tropical System. Springer-Verlag, New York. 163 pp.

Lewis, W. M., Jr. 1983a. A revised classification of lakes based on mixing. Can. J. Fish. Aquat. Sci. *40*:1779–1787.

Lewis, W. M., Jr. 1983b. Temperature, heat, and mixing in Lake Valencia, Venezuela. Limnol. Oceanogr. *28*:273–286.

Lewis, W. M., Jr. 1985. Protozoan abundance in the plankton of two tropical lakes. Arch. Hydrobiol. *104*:337–343.

Lewis, W. M., Jr. 1986. Nitrogen and phosphorus runoff losses from a nutrient-poor tropical moist forest. Ecology *67*:1275–1282.

Lewis, W. M., Jr. 1987. Tropical limnology. Annu. Rev. Ecol. Syst. *18*:159–184.

Lewis, W. M., Jr. 1996. Tropical lakes: How latitude makes a difference. *In* F. Schiemer and K. T. Boland, eds. Perspectives in Tropical Limnology. SPB Academic Publishing BV, Amsterdam. pp. 43–64.

Lewis, W. M., Jr. and J. F. Saunders. 1989. Concentration and transport of dissolved and suspended substances in the Orinoco River. Biogeochemistry *7*:203–240.

Lewis, W. M., Jr., and S. N. Levine. 1984. The light response of nitrogen fixation in Lake Valencia, Venezuela. Limnol. Oceanogr. *29*:894–900.

Lewis, W. M., Jr., M. C. Grant, and S. K. Hamilton. 1985. Evidence that filterable phosphorus is a significant atmospheric link in the phosphorus cycle. Oikos *45*:428–432.

Lewis, W. M., Jr., S. K. Hamilton, M. A. Lasi, M. Rodríguez, and J. F. Saunders. 2000. Ecological determinism on the Orinoco floodplain. BioScience *50*:681–692.

Lewis, W. M., Jr., T. Frost, and D. Morris. 1986. Studies of planktonic bacteria in Lake Valencia, Venezuela. Arch. Hydrobiol. *106*:289–305.

Li, M. and M. B. Jones. 1995. CO_2 and O_2 transport in the aerenchyma of *Cyperus papyrus* L. Aquat. Bot. *52*:93–106.

Li, W. C., D. E. Armstrong, J. D. H. Williams, R. F. Harris, and J. K. Syers. 1972. Rate and extent of inorganic phosphate exchange in lake sediments. Proc. Soil Sci. Soc. Amer. *36*:279–285.

Liao, C. F.-H. and D. R. S. Lean. 1978. Nitrogen transformations within the trophogenic zone of lakes. J. Fish. Res. Board Can. *35*:1102–1108.

Lidstrom, M. E. and L. Somers. 1984. Seasonal study of methane oxidation in Lake Washington. Appl. Environ. Microbiol. *47*:1255–1260.

Lighthill, J. 1978. Waves in fluids. Cambridge University Press, Cambridge.

Lijklema, L. 1994. Nutrient dynamics in shallow lakes: Effects of changes in loading and role of sediment-water interactions. Hydrobiologia *275/276*:335–348.

Likens, G. E. 1970. Eutrophication and aquatic ecosystems. *In* G. E. Likens, ed. Nutrients and Eutrophication. Amer. Soc. Limnol. Oceanogr. Spec. Symp. 1. pp. 3–13.

Likens, G. E., ed. 1972. Nutrient and Eutrophication: The Limiting-Nutrient Controversy. Special Publ. Amer. Soc. Limnol. Oceanogr. *1*. 328 pp.

Likens, G. E. 1975. Primary production of inland aquatic ecosystems. *In* H. Lieth and R. W. Whittaker, eds. The Primary Productivity of the Biosphere. Springer-Verlag, New York. pp. 185–202.

Likens, G. E., 1984. Beyond the shore line: A watershed-ecosystem approach. Verh. Internat. Verein. Limnol. *22*:1–22.

Likens, G. E., ed. 1985. An ecosystem approach to aquatic ecology: Mirror Lake and its environment. Springer-Verlag, New York. 516 pp.

Likens, G. E. and A. D. Hasler. 1960. Movement of radiosodium in a chemically stratified lake. Science *131*:1676–1677.

Likens, G. E. and A. D. Hasler. 1962. Movements of radiosodium (Na^{24}) within an ice-covered lake. Limnol. Oceanogr. *7*:48–56.

Likens, G. E. and F. H. Bormann. 1972. Nutrient cycling in ecosystems. *In* J. A. Weins, ed. Ecosystem Structure and Function. Oregon State University Press, Corvallis. pp. 25–67.

Likens, G. E. and F. H. Bormann. 1995. Biogeochemistry of a forested ecosystem. 2nd Edition. Springer-Verlag, New York. 159 pp.

Likens, G. E. and M. B. Davis. 1975. Post-glacial history of Mirror Lake and its watershed in New Hampshire, U.S.A.: An initial report. Verh. Internat. Verein. Limnol. *19*:982–993.

Likens, G. E. and N. M. Johnson. 1969. Measurement and analysis of the annual heat budget for the sediments in two Wisconsin lakes. Limnol. Oceanogr. *14*:115–135.

Likens, G. E. and P. L. Johnson. 1966. A chemically stratified lake in Alaska. Science *153*:875–877.

Likens, G. E. and R. A. Ragotzkie. 1965. Vertical water motions in a small ice-covered lake. J. Geophys. Res. *70*:2333–2344.

Likens, G. E. and R. A. Ragotzkie. 1966. Rotary circulation of water in an ice-covered lake. Verh. Internat. Verein. Limnol. *16*:126–133.

Likens, G. E., C. T. Driscoll, D. C. Buso, T. G. Siccama, C. E. Johnson, G. M. Lovett, D. F. Ryan, T. Fahey, and W. A. Reiners. 1994. The biogeochemistry of potassium at Hubbard Brook. Biogeochemistry *25*:61–125.

Likens, G. E., C. T. Driscoll, and D. C. Buso. 1996. Long-term effects of acid rain: Response and recovery of a forest ecosystem. Science *272*:244–246.

Likens, G. E., F. H. Bormann, and N. M. Johnson. 1972. Acid rain. Environment *14*:33–40.

Likens, G. E., F. H. Bormann, N. M. Johnson, and R. S. Pierce. 1967. The calcium, magnesium, potassium and sodium budgets for a small forested ecosystem. Ecology *48*:772–785.

Likens, G. E., F. H. Bormann, N. M. Johnson, D. W. Fisher, and R. S. Pierce. 1970. Effects of forest cutting and herbicide treatment on nutrient budgets in the Hubbard Brook watershed-ecosystem. Ecol. Monogr. *40*:23–47.

Likens, G. E., F. H. Bormann, R. S. Pierce, J. S. Eaton, and N. M. Johnson. 1977. Biogeochemistry of a forested ecosystem. Springer-Verlag, New York. 146 pp.

Likens, G. E., R. F. Wright, J. N. Galloway, and T. J. Butler. 1979. Acid rain. Sci. Amer. *251*(4):43–51.

Likens, G. E., J. S. Eaton, N. M. Johnson, and R. S. Pierce. 1985. Flux and balance of water and chemicals. *In* G. E. Likens, ed. An Ecosystem Approach to Aquatic Ecology. Mirror Lake and Its Environment. Springer-Verlag, New York. pp. 135–155.

Lilley, M. D., M. A. de Angelis, and E. J. Olson. 1996. Methane concentrations and estimated fluxes from Pacific Northwest rivers. Mitt. Internat. Verein. Limnol. *25*:187–196.

Lillieroth, S. 1950. Über Folgen kulturbedingter Wasserstandsenkungen für Makrophyten- und Planktongemeinschaften in seichten Seen des südschwedischen Oligotrophiegebietes. Acta Limnol. 3. 188 pp.

Lin, C.-J. and S. O. Pehkonen. 1999. The chemistry of atmospheric mercury: A review. Atmosph. Environ. *33*:2067–2079.

Lind, O. T. 1978. Interdepression differences in the hypolimnetic areal relative oxygen deficits of Douglas Lake, Michigan. Verh. Internat. Verein. Limnol. *20*:2689–2696.

Lind, O. T. 1986. The effect of non-algal turbidity on the relationship of Secchi depth to chlorophyll *a*. Hydrobiologia *140*:27–35.

Lind, O. T. 1987. Spatial and temporal variation in hypolimnetic oxygen deficits of a multidepression lake. Limnol. Oceanogr. *32*:740–744.

Lind, O. T. and L. Dávalos-Lind. 1991. Association of turbidity and organic carbon with bacterial abundance and cell size in a large, turbid, tropical lake. Limnol. Oceanogr. *36*:1200–1208.

Lindegaard, C. 1994. The role of zoobenthos in energy flow in two shallow lakes. Hydrobiologia *275/276*:313–322.

Lindegaard, C. and P. M. Jónasson. 1979. Abundance, population dynamics and production of zoobenthos in Lake Myvatn, Iceland. Oikos *32*:202–227.

Lindegaard, C., K. Hamburger, and P. C. Dall. 1990. Population dynamics and energy budget of *Stylodrilus heringianus* Clap. (Lumbriculidae, Oligochaeta) in the shallow littoral of Lake Esrom, Denmark. Verh. Internat. Verein. Limnol. *24*:626–631.

Lindegaard, C., K. Hamburger, and P. C. Dall. 1994. Population dynamics and energy budget of *Marionina southerni* (Cernosvitov) (Enchytraeidae, Oligochaeta) in the shallow littoral of Lake Esrom, Denmark. Hydrobiologia *278*:291–301.

Lindegaard, C., P. C. Dall, and S. B. Hansen. 1993. Natural and imposed variability in the profundal fauna of Lake Esrom, Denmark. Verh. Internat. Verein. Limnol. *25*:576–581.

Lindell, M. J., W. Granéli, and L. J. Tranvik. 1995. Enhanced bacterial growth in response to photochemical transformation of dissolved organic matter. Limnol. Oceanogr. *40*:195–199.

Lindell, M. J., W. Granéli, and L. J. Tranvik. 1996. Effects of sunlight on bacterial growth in lakes of different humic content. Aquat. Microb. Ecol. *11*:135–141.

Lindeman, R. L. 1942. The trophic-dynamic aspect of ecology. Ecology *23*:399–418.

Lindholm, T. 1985. *Mesodinium rubrum*—A unique photosynthetic ciliate. Adv. Aquat. Microbiol. *3*:1–48.

Lindström, K. 1980. *Peridinium cinctum* bioassays of Se in Lake Erken. Arch. Hydrobiol. *89*:110–117.

Lindström, K. 1983. Selenium as a growth factor for plankton algae in laboratory experiments and in some Swedish lakes. Hydrobiologia *101*:35–48.

Lindström, K. 1985. Selenium requirement of the dinoflagellate *Peridinopsis Borgei* (Lemm). Int. Rev. ges. Hydrobiol. *70*:77–85.

Lindström, K. and W. Rodhe. 1978. Selenium as a micronutrient for the dinoflagellate *Peridinium cinctum* fa. westii. Mitt. Internat. Verein. Limnol. *21*:168–173.

Lingeman, R, B. J. G. Flik, and J. Ringelberg. 1975. Stability of the oxygen stratification in a eutrophic lake. Verh. Internat. Verein. Limnol. *19*:1193–1201.

Lipschultz, F., S. C. Wofsy, and L. E. Fox. 1986. Nitrogen metabolism of the eutrophic Delaware River ecosystem. Limnol. Oceanogr. *31*:701–716.

Littlefield, L. and C. Forsberg. 1965. Absorption and translocation of phosphorus–32 by *Chara globularis* Thuill. Physiol. Plant. *18*:291–296.

Liu, X, and P. Shei. 1991. Reduction of dissolved organic carbon in Donghu Lake through flocculation and bacterial utilization. Ann. Rept. Freshwat. Ecol. Biotechnol. China *1991*:67–73.

Livingston, D., A. Pentecost, and B. A. Whitton. 1984. Diel variations in nitrogen and carbon dioxide fixation by the blue-green alga *Rivularia* in an upland stream. Phycologia *23*:125–133.

Livingstone, D. A. 1954. On the orientation of lake basins. Am. J. Sci. *252*:547–554.

Livingstone, D. A. 1957. On the sigmoid growth phase in the history of Linsley Pond. Am. J. Sci. *255*:364–373.

Livingstone, D. A. 1963a. Chemical composition of rivers and lakes. Chap. G. Data of Geochemistry. 6th Edition. Prof. Pap. U.S. Geol. Surv. *440-G*, 64 pp.

Livingstone, D. A. 1963b. Alaska, Yukon, Northwest Territories, and Greenland. *In* D. G. Frey, ed. Limnology in North America. University of Wisconsin Press, Madison. pp. 559–574.

Livingstone, D. A. 1975. Late Quaternary climatic change in Africa. Ann. Rev. Ecol. Syst. *6*:249–280.

Livingstone, D. A. and J. C. Boykin. 1962. Vertical distribution of phosphorus in Linsley Pond mud. Limnol. Oceanogr. *7*:57–62.

Livingstone, D. A., K. Bryan, Jr., and R. G. Leahy. 1958. Effects of an arctic environment on the origin and development of freshwater lakes. Limnol. Oceanogr. *3*:192–214.

Livingstone, D. and G. H. M. Jaworski. 1980. The viability of akinetes of blue-green algae recovered from the sediments of Rostherne Mere. Br. Phycol. J. *15*:357–364.

Livingstone, D. M. 1991. The diel oxygen cycle in three subalpine Swiss streams. Arch. Hydrobiol. *120*:457–479.

Lloyd, N. D. H., D. T. Canvin, and J. M. Bristow. 1977. Photosynthesis and photorespiration in submerged aquatic vascular plants. Can. J. Bot. *55*:3001–3005.

Lock, M. A. 1981. River epilithon—a light and organic energy transducer. *In* M. A. Lock and D. D. Williams, eds. Perspectives in Running Water Ecology. Plenum Press, New York. pp. 3–40.

Lock, M. A. 1990. The dynamics of dissolved and particulate organic material over the substratum of water bodies. *In* R. S. Wotton, ed. The Biology of Particles in Aquatic systems. CRC Press, Boca Raton. pp. 117–144.

Lock, M. A. 1993. Attached microbial communities in rivers. *In* T. E. Ford, ed. Aquatic Microbiology. Blackwell Scientific, Oxford. pp. 113–138.

Lock, M. A., P. M. Wallis, and H.B. N. Hynes. 1977. Colloidal organic carbon in running waters. Oikos *29*:1–4.

Lock, M. A., R. R. Wallace, J. W. Costerton, R. M. Ventullo, and S. E. Charlton. 1984. River epilithon: Toward a structural-functional model. Oikos *42*:10–22.

Loczy, S., R. Carignan, and D. Planas. 1983. The role of roots in carbon uptake by the submersed macrophytes *Myriophyllum spicatum, Vallisneria americana,* and *Heteranthera dubia.* Hydrobiologia *98*:3–7.

Loden, M. S. 1974. Predation by chironomid (Diptera) larvae on oligochaetes. Limnol. Oceanogr. *19*:156–159.

Loden, M. S. 1981. Reproductive ecology of Naididae (Oligochaeta). Hydrobiologia *83*:115–123.

Lodge, D. M. 1985. Macrophyte-gastropod associations: Observations and experiments on macrophyte choice by gastropods. Freshwat. Biol. *15*:695–708.

Lodge, D. M. 1986. Selective grazing on periphyton: A determinant of freshwater gastropod microdistributions. Freshwat. Biol. *16*:831–841.

Lodge, D. M. 1991. Herbivory on freshwater macrophytes. Aquat. Bot. *41*:195–224.

Lodge, D. M. and A. M. Hill. 1994. Factors governing species composition, population size, and productivity of cool-water crayfishes. Nordic J. Freshwat. Res. *69*:111–136.

Lodge, D. M. and J. G. Lorman. 1987. Reductions in submersed macrophyte biomass and species richness by the crayfish *Orconectes rusticus*. Can. J. Fish. Aquat. Sci. *44*:591–597.

Lodge, D. M., G. Cronin, E. van Donk, and A. J. Froelich. 1998. Impact of herbivory on plant standing crop: Comparisons among biomes, between vascular and nonvascular plants, and among freshwater herbivore taxa. *In* E. Jeppesen, Ma. Søndergaard, Mo. Søndergaard, and K. Christoffersen, eds. The Structuring Role of Submerged Macrophytes in Lakes. Springer-Verlag, New York. pp. 149–174.

Lodge, D. M., K. M. Brown, S. P. Klosiewski, R. A. Stein, A. P. Covich, B. K. Leathers, and C. Brönmark. 1987. Distribution of freshwater snails: Spatial scale and the relative importance of physicochemical and biotic factors. Am. Malacol. Bull. *5*:73–84.

Loeb, S. L., J. E. Reuter, and C. R. Goldman. 1983. Littoral zone production of oligotrophic lakes. The contributions of phytoplankton and periphyton. *In* R. G. Wetzel, ed. Periphyton of Freshwater Ecosystems. W. Junk Publ., The Hague. pp. 161–167.

Loeblich, A. R. and L. A. Loeblich. 1984. Dinoflagellate cysts. *In* D. L. Spector, ed. Dinoflagellates. Academic Press, Orlando. pp. 443–480.

Loehr, R. C., C. S. Martin, and W. Rast, eds. 1980. Phosphorus Management Strategies for Lakes. Ann Arbor Science Publ., Inc., Ann Arbor. 490 pp.

Löffler, H. 1997. Längsee: A history of meromixis; 40 years laters: Homage to Dr. D. G. Frey. Verh. Internat. Verein. Limnol. *26*:829–832.

Logan, T. J. 1982. Mechanisms for release of sediment-bound phosphate to water and the effects of agricultural land management on fluvial transport of particulate and dissolved phosphate. Hydrobiologia *92*:519–530.

Lohuis, D., V. W. Meloche, and C. Juday. 1938. Sodium and potassium content of Wisconsin lake waters and their residues. Trans. Wis. Acad. Sci. Arts Lett. *31*:285–304.

Long, E. T. and G. D. Cooke. 1978. Phosphorus variability in three streams during storm events: Chemical analysis vs. algal assay. Mitt. Internat. Verein. Limnol *21*:441–452.

Long, S. P., S. Humphries, and P. G. Falkowski. 1994. Photoinhibition of photosynthesis in nature. Annu. Rev. Plant Physiol. Plant Mol. Biol. *45*:633–662.

Longstreth, D. J. 1989. Photosynthesis and photorespiration in freshwater emergent and floating plants. Aquat. Bot. *34*:287–299.

van Loosdrecht, M. C. M., J. Lyklema, W. Norder, and A. J. B. Zehnder. 1990. Influence of interfaces on microbial activity. Microbial. Rev. *54*:75–87.

Loose, C. J. 1993. Lack of endogenous rhythmicity in *Daphnia* diel vertical migration. Limnol. Oceanogr. *38*:1837–1841.

Loose, C. J. and P. Dawidowicz. 1994. Trade-offs in diel vertical migration by zooplankton: The costs of predator avoidance. Ecology *75*:2255–2263.

López, C. and C. Cressa. 1996. Ecological studies on a *Chaoborus* larvae population in a tropical reservoir (Socuy Reservoir, Venezuela). Arch. Hydrobiol. *136*:421–431.

Loranger, T. J. and D. F. Brakke. 1988. Birgean heat budgets and rates of heat uptake in two monomictic lakes. Hydrobiologia *160*:123–127.

Lorch, D. W. 1978. Desmids and heavy metals. II. Manganese: Uptake and influence on growth and morphogenesis of selected species. Arch. Hydrobiol. *84*:166–179.

Lorenz, R. C., M. E. Monaco, and C. E. Herdendorf. 1991. Minimum light requirements for substrate colonization by *Cladophora glomerata*. J. Great Lakes Res. *17*:536–542.

Lorieri, D. and H. Elsenbeer. 1997. Aluminum, iron and manganese in near-surface waters of a tropical rainforest ecosystem. Sci. Total Environ. *205*:13–23.

Losee, R. F. and R. G. Wetzel. 1983. Selective light attenuation by the periphyton complex. *In* R. G. Wetzel, ed. Periphyton of Freshwater Ecosystems. Developments in Hydrobiology 17. W. Junk Publishers, The Hague. pp. 89–96.

Losee, R. F. and R. G. Wetzel. 1993. Littoral flow rates within and around submersed macrophyte communities. Freshwat. Biol. *29*:7–17.

Loucks, O. L. and W. E. Odum. 1978. Analysis of five North American lake ecosystems. I. A strategy for comparison. Verh. Internat. Verein. Limnol. *20*:556–561.

Love, R. J. R. and G. G. C. Robinson. 1977. The primary productivity of submerged macrophytes in West Blue Lake, Manitoba. Can. J. Bot. *55*:118–127.

Lovley, D. R. 1987. Organic matter mineralization with the reduction of ferric iron: A review. Geomicrobiol. J. *5*:375–399.

Lovley, D. R. 1991. Dissimilatory Fe(III) and Mn(IV) reduction. Microbiol. Rev. *55*:259–287.

Lovley, D. R. 1995. Microbial reduction of iron, manganese, and other metals. Adv. Agronomy *54*:175–231.

Lovley, D. R. and M. J. Klug. 1982. Intermediary metabolism of organic matter in the sediments of a eutrophic lake. Appl. Environ. Microbiol. *43*:552–560.

Lovley, D. R. and M. J. Klug. 1983a. Methanogenesis from methanol and methylamines and acetogenesis from hydrogen and carbon dioxide in the sediments of a eutrophic lake. Appl. Environ. Microbiol. *45*:1310–1315.

Lovley, D. R. and M. J. Klug. 1983b. Sulfate reducers can outcompete methanogens at freshwater sulfate concentrations. Appl. Environ. Microbiol. *45*:187–192.

Lovley, D. R. and S. Goodwin. 1988. Hydrogen concentrations as an indicator of the predominant terminal electron-accepting reactions in aquatic sediments. Geochim. Cosmochim. Acta *52*:2993–3003.

Lovley, D. R., D. F. Dwyer, and M. J. Klug. 1982. Kinetics analysis of competition between sulfate and reducers and methanogens for hydrogen in sediments. Appl. Environ. Microbiol. *43*:1373–1379.

Lovley, D. R., J. D. Coates, E. L. Blunt-Harris, E. J. P. Phillips, and J. C. Woodward. 1996. Humic substances as electron acceptors for microbial respiration. Nature *382*:445–448.

Lovley, D. R., J. L. Fraga, E. L. Blunt-Harris, L. A. Hayes, E. J. P. Phillips, and J. D. Coates. 1998. Humic substances as a mediator for microbially catalyzed metal reduction. Acta Hydrochim. Hydrobiol. *26*:152–156.

Lovley, D. R., J. L. Fraga, J. D. Coates, and E. L. Blunt-Harris. 1999. Humics as an electron donor for anaerobic respiration. Environ. Microbiol. *1*:89–98.

Lowe, R. L. 1996. Periphyton patterns in lakes. *In* R. J. Stevenson, M. L. Bothwell, and R. L. Lowe, eds. Algal Ecology: Freshwater Benthic Ecosystems. Academic Press, San Diego. pp. 57–76.

Lowe, R. L. and G. D. Laliberte. 1996. Benthic stream algae: Distribution and structure. *In* F. R. Hauer and G. A. Lamberti, eds. Methods in Stream Ecology. Academic Press, San Diego. pp. 269–293.

Lowe, R. L. and R. D. Hunter. 1988. Effects of grazing by *Physa integra* on periphyton community structure. J. N. Am. Benthol. Soc. *7*:29–36.

Lowe, R. L. and R. W. Pillsbury. 1995. Shifts in benthic algal community structure and function following the appearance of zebra mussels (*Dreissena polymorpha*) in Saginaw Bay, Lake Huron. J. Great Lakes Res. *21*:558–566.

Lowe-McConnell, R. H. 1975. Fish Communities in Tropical Fresh Waters, Their Distribution, Ecology and Evolution. Longman, London. 375 pp.

Lowe-McConnell, R. H. 1987. Ecological Studies in Tropical Fish Communities. Cambridge Univ. Press, Cambridge. 382 pp.

Lowe-McConnell, R. 1996. Fish communities in the African Great Lakes. Environ. Biol. Fishes *45*:219–235.

Lowe-McConnell, R., F. C. Roest, G. Ntakimazi, and L. Risch. 1994. The African great lakes. *In* Biological Diversity in African Fresh- and Brackish Water Fishes. Ann. Mus. r. Afr. Centr. Zool. *275*:87–94.

Lucas, A. 1996. Bioenergetics of Aquatic Animals. Taylor & Francis, London. 169 pp.

Lucas, W. J. 1983. Photosynthetic assimilation of exogenous HCO_3^- by aquatic plants. Annu. Rev. Plant Physiol. *34*:71–104.

Lucas, W. J. and J. A. Berry, eds. 1985. Inorganic carbon uptake by aquatic photosynthetic organisms. Amer. Soc. Plant Physiologists, Rockville, MD. 494 pp.

Lucas, W. J., M. T. Tyree, and A. Petrov. 1978. Characterization of photosynthetic ^{14}carbon assimilation by *Potamogeton lucens* L. J. Exp. Bot. *29*:1409–1421.

Ludlam, S. D. 1976. Laminated sediments in holomictic Berkshire lakes. Limnol. Oceanogr. *21*:743–746.

Luecke, C. and W. J. O'Brien. 1981. Phototoxicity and fish predation: Selective factors in color morphs in *Heterocope*. Limnol. Oceanogr. *26*:454–460.

Luecke, C. and W. J. O'Brien. 1983. Photoprotective pigments in a pond morph of *Daphnia middendorffiana*. Arctic *36*:365–368.

Lueschow, L. A., J. M. Helm, D. R. Winter, and G. W. Karl. 1970. Trophic nature of selected Wisconsin lakes. Trans. Wis. Acad. Sci. Arts Lett. *58*:237–264.

Lugthart, G. J. and J. B. Wallace. 1992. Effects of disturbance on benthic functional structure and production in mountain streams. J. N. Am. Benthol. Soc. *11*:138–164.

Lumpkin, T. A. and D. L. Plucknett. 1980. *Azolla*: Botany, physiology, and use as a green manure. Econ. Bot. *34*:111–153.

Lund, J. W. G. 1949. Studies on *Asterionella*. I. The origin and nature of the cells producing seasonal maxima. J. Ecol. *37*:389–419.

Lund, J. W. G. 1950. Studies on *Asterionella formosa* Hass. II. Nutrient depletion and the spring maximum (Parts I and II). J. Ecol. *38*:1–35.

Lund, J. W. G. 1954. The seasonal cycle of the plankton diatom, *Melosira italica* (Ehr.). Kütz. subsp. *subarctica* O. Müll. J. Ecol. *42*:151–179.

Lund, J. W. G. 1955. Further observations on the seasonal cycle of *Melosira italica* (Ehr.). Kütz. subsp. *subarctica* O. Müll. J. Ecol. *43*:90–102.

Lund, J. W. G. 1959. Buoyancy in relation to the ecology of the freshwater phytoplankton. Brit. Phycol. Bull. *1*:1–17.

Lund, J. W. G. 1964. Primary production and periodicity of phytoplankton. Verh. Internat. Verein. Limnol. *15*:37–56.

Lund, J. W. G. 1965. The ecology of the freshwater phytoplankton. Biol. Rev. *40*:231–293.

Lund, J. W. G. and J. F. Talling. 1957. Botanical limnological methods with special reference to the algae. Bot. Rev. *23*:489–583.

Lund, J. W. G., C. Kipling, and E. D. LeCren. 1958. The inverted microscope method of estimating algal numbers and the statistical basis of estimations by counting. Hydrobiologia *11*:143–170.

Lund, J. W. G., F. J. H. Mackereth, and C. H. Mortimer. 1963. Changes in depth and time of certain chemical and physical conditions and of the standing crop of *Asterionella formosa* Hass. in the North Basin of Windermere in 1947. Phil. Trans. Roy. Soc. London (Ser. B) *246*:255–290.

Lund, V. and D. Hongve. 1994. Ultraviolet irradiated water containing humic substances inhibits bacterial metabolism. Wat. Res. *28*:1111–1116.

Lundgren, A. 1978. Nitrogen fixation induced by phosphorus fertilization of a subarctic lake. *In* Environmental Role of Nitrogen-Fixing Blue-Green Algae and Asynbiotic Bacteria. Ecol. Bull. (Stockholm) *26*:52–59.

Lundqvist, G. 1927. Bodenablagerungen und Entwicklungstypen der Seen. Die Binnengewässer *2*:124 pp.

Lüning-Krizan, J. 1997. Selective feeding of third- and fourth-instar larvae of *Chaoborus flavicans* in the field. Arch. Hydrobiol. *140*:347–365.

Lunte, C. C. and C. Luecke. 1990. Trophic interactions of *Leptodora* in Lake Mendota. Limnol. Oceanogr. *35*:1091–1100.

Lürling, M. and E. van Donk. 1996. Zooplankton-induced unicell-colony transformation in *Scenedesmus acutus* and its effect on growth of herbivore *Daphnia*. Oecologia *108*:432–437.

Lürling, M. and E. van Donk. 1997. Morphological changes in *Scenedesmus* induced by infochemicals released in situ from zooplankton grazers. Limnol. Oceanogr. *42*:783–788.

Lush, D. L. and H. B. N. Hynes. 1973. The formation of particles in freshwater leachates of dead leaves. Limnol. Oceanogr. *18*:968–977.

Lüttge, U. 1964. Mikroautoradiographische Untersuchungen über die Funktion der Hydropoten von *Nymphaea*. Protoplasma *59*:157–162.

Lvovitch, M. I. 1973. The global water balance. Trans. Amer. Geophys. Union *54*:28–42.

Lyche, A., T. Andersen, K. Christoffersen, D. O. Hessen, P. H. B. Hansen, and A. Klysner. 1996. Mesocosm tracer studies. 1. Zooplankton as sources and sinks in the pelagic phosphorus cycle of a mesotrophic lake. Limnol. Oceanogr. *41*:460–474.

Lyche, A., T. Andersen, K. Christoffersen, D. O. Hessen, P. H. B. Hansen, and A. Klysner. 1996. Mesocosm tracer studies. 2. The fate of primary production and the role of consumers in the pelagic carbon cycle of a mesotrophic lake. Limnol. Oceanogr. *41*:475–487.

Lydeard, C. and R. L. Mayden. 1995. A diverse and endangered aquatic ecosystem of the southeast United States. Conservat. Biol. *9*:800–805.

Lydersen, E. 1998. Humus and acidification. *In* D. O. Hessen and L. J. Tranvik, eds. Aquatic Humic Substances: Ecology and Biogeochemistry. Springer-Verlag, Berlin. pp. 63–92.

Lynch, M. 1979. Predation, competition, and zooplankton community structure: An experimental study. Limnol. Oceanogr. *24*:253–272.

Lynch, M. 1982. How well does the Edmondson-Paloheimo model approximate instantaneous birth rates? Ecology *63*:12–18.

Maberly, S. C. 1985. Photosynthesis by *Fontinalis antipyretica*. I. Interaction between photon irradiance, concentration of carbon dioxide and temperature. New Phytol. *100*:127–140.

Maberly, S. C. and D. H. N. Spence. 1983. Photosynthetic inorganic carbon use by freshwater plants. J. Ecol. *71*:705–724.

Maberly, S. C. and D. H. N. Spence. 1989. Photosynthesis and photorespiration in freshwater organisms: Amphibious plants. Aquat. Bot. *34*:267–286.

Macan, T. T. 1961. Factors that limit the range of freshwater animals. Biol. Rev. *36*:151–198.

Macan, T. T. 1970. Biological Studies of the English Lakes. American Elsevier Publishing Co., Inc., New York. 260 pp.

Macan, T. T. 1977. The influence of predation on the composition of freshwater animal communities. Biol. Rev. *52*:45–70.

MacArthur, J. W. and W. H. T. Baillie. 1929. Metabolic activity and duration of life. I. Influence of temperature on longevity in *Daphnia magna*. J. Exp. Zool. *53*:221–242.

MacArthur, R. H. and E. O. Wilson. 1967. The Theory of Island Biogeography. Princeton Univ. Press, Princeton, NJ. 203 pp.

MacDonald, G. M., R. P. Beukens, and W. E. Kieser. 1991. Radiocarbon dating of limnic sediments: A comparative analysis and discussion. Ecology *73*:1150–1155.

Macek, M., K. Simek, J. Pernthaler, V. Vyhnálek, and R. Psenner. 1996. Growth rates of dominant planktonic ciliates in two freshwater bodies of different trophic degree. J. Plankton Res. *18*:463–481.

MacFayden, A. 1948. The meaning of productivity in biological systems. J. Anim. Ecol. *17*:75–80.

MacFayden, A. 1950. Biologische Produktivität. Arch. Hydrobiol. *43*:166–170.

MacGregor, A. N. and D. R. Keeney. 1973a. Denitrification in lake sediments. Environ. Lett. *5*:175–181.

MacGregor, A. N. and D. R. Keeney. 1973b. Methane formation by lake sediments during in vitro incubation. Water Resour. Bull. *9*:1153–1158.

Machena, C., N. Kautsky, and G. Lindmark. 1990. Growth and production of *Lagarosiphon ilicifolius* in Lake Kariba—a man-made tropical lake. Aquat. Bot. *37*:1–15.

Machácek, J. 1995. Inducibility of life history changes by fish kairomone in various developmental stages of *Daphnia*. J. Plankton Res. *17*:1513–1520.

Machata-Wenninger, C. and G. A. Janauer. 1991. The measurement of current velocities in macrophyte beds. Aquat. Bot. *39*: 221–230.

Machta, L. 1973. Prediction of CO_2 in the atmosphere. *In* G. M. Woodwell and E. V. Pecan, eds. Carbon and the Biosphere. Brookhaven, N Y, Proc. Brookhaven Symp. in Biol. 24. Tech. Information Center, U.S. Atomic Energy Commission CONF–720510, pp. 21–31.

Maciolek, J. A. 1954. Artificial fertilization of lakes and ponds. A review of the literature. USFWS, Spec. Sci. Rep. Fish. *113*, 41 pp.

MacIsaac, H. J. and J. J. Gilbert. 1991. Discrimination between exploitative and interference competition between cladocera and *Keratella cochlearis*. Ecology *72*:924–937.

Mackay, R. J. 1992. Colonization by lotic macroinvertebrates: A review of processes and patterns. Can. J. Fish. Aquat. Sci. *49*:617–628.

Mackay, R. J. and G. B. Wiggins. 1979. Ecological diversity in Trichoptera. Ann. Rev. Entomol. *24*:185–208.

Mackenzie, F. T. and R. M. Garrels. 1965. Silicates: Reactivity with sea water. Science *150*:57–58.

Mackenzie, F. T., R. M. Garrels, O. P. Bricker, and F. Bickley. 1967. Silica in sea water: Control by silica minerals. Science *155*:1404–1405.

Mackereth, F. J. H. 1965. Chemical investigation of lake sediments and their interpretation. Proc. Roy. Soc. (Ser. B) *161*:295–309.

Mackereth, F. J. H. 1953. Phosphorus utilization by *Asterionella formosa* Hass. J. Exp. Bot. *4*:296–313.

Mackereth, F. J. H. 1966. Some chemical observations on postglacial lake sediments. Phil. Trans. Roy. Soc. London (Ser. B) *250*:165–213.

Mackereth, F. J. H. 1971. On the variation in direction of the horizontal component of remanent magnetisation in lake sediments. Earth Planetary Sci. Lett. *12*:332–338.

MacNeil, C., J. T. A. Dick, and R. W. Elwood. 1997. The trophic ecology of freshwater *Gammarus* spp. (Crustacea: Amphipoda): Problems and perspectives concerning the functional feeding group concept. Biol. Rev.*72*:349–364.

MacNeil, C., J. T. A. Dick, and R. W. Elwood. 1999. The dynamics of predation on *Gammarus* spp. (Crustacea: Amphipoda). Biol. Rev. *74*:375–395.

Macpherson, L. B., N. R. Sinclair, and F. R. Hayes. 1958. Lake water and sediment. III. The effect of pH on the partition of inorganic phosphate between water and oxidized mud or its ash. Limnol. Oceanogr. *3*:318–326.

Madeira, P. T., A. S. Brooks, and D. B. Seale. 1982. Excretion of total phosphorus, dissolved reactive phosphorus, ammonia, and urea by Lake Michigan *Mysis relicta*. Hydrobiologia *93*:145–154.

Madigan, M. T. 1988. Microbiology, physiology, and ecology of phototrophic bacteria. *In* A. J. B. Zehnder, ed. Biology of Anaerobic Microorganisms. John Wiley & Sons, New York. pp. 39–111.

Madon, S. P., D. W. Schneider, J. A. Stoeckel, and R. E. Sparks. 1998. Effects of inorganic sediment and food concentrations on energetic processes of the zebra mussel, *Dreissena polymorpha*: Implications for growth in turbid rivers. Can. J. Fish. Aquat. Sci. *55*:401–413.

Madronich, S. 1994. Increases in biologically damaging UV-B radiation due to stratospheric ozone reductions: A brief review. Arch. Hydrobiol. Beih. Ergebn. Limnol. *43*:17–30.

Madsen, J. D. and M. S. Adams. 1988. The seasonal biomass and productivity of the submerged macrophytes in a polluted Wisconsin stream. Freshwat. Biol. *20*:41–50.

Madsen, T. V. 1987. Interactions between internal and external CO_2 pools in the photosynthesis of the aquatic CAM plants *Littorella uniflora* (L.) Aschers and *Isoetes lacustris* L. New Phytol. *106*:35–50.

Madsen, T. V. and A. Baattrup-Pedersen. 1995. Regulation of growth and photosynthetic performance in *Elodea canadensis* in response to inorganic nitrogen. Funct. Ecol. *9*:239–247.

Madsen, T. V. and E. Warncke. 1983. Velocities of currents around and within submerged aquatic vegetation. Arch. Hydrobiol. *97*:389–394.

Madsen, T. V. and K. Sand-Jensen. 1991. Photosynthetic carbon assimilation in aquatic macrophytes. Aquat. Bot. *41*:5–40.

Madsen, T. V. and M. Søndergaard. 1983. The effects of current velocity on the photosynthesis of *Callitriche stagnalis* Scop. Aquat. Bot. *15*:187–193.

Maeda, O. and H. Tomioka. 1977. Vertical distributions of particulate protein and nucleic acids in lakes in central Japan in late summer. Jpn. J. Limnol. *38*:109–155. (In Japanese.)

Maeda, O. and S. Ichimura. 1973. On the high density of a phytoplankton population found in a lake under ice. Int. Rev. ges. Hydrobiol. *58*:673–685.

Maeso, E. S., F. F. Pinas, M. G. Gonzalez, and E. F. Valiente. 1987. Sodium requirement for photosynthesis and its relationship with dinitrogen fixation and the external CO_2 concentration in cyanobacteria. Plant Physiol. *85*:585–587.

Magill, A. H. and J. D. Aber. 2000. Dissolved organic carbon and nitrogen relationships in forest litter as affected by nitrogen deposition. Soil Biol. Biochem. *32*:603–613.

Magnien, R. E. and J. J. Gilbert. 1983. Diel cycles of reproduction and vertical migration in the rotifer *Keratella crassa* and their influence on the estimation of population dynamics. Limnol. Oceanogr. *28*:957–969.

Magnuson, J. J. and 13 others. 2000. Historical trends in lake and river ice cover in the Northern Hemisphere. Science *289*:1743–1746.

Mague, T. H., E. Friberg, D. H. Hughes, and I. Morris. 1980. Extracellular release of carbon by marine phytoplankton: A physiological approach. Limnol. Oceanogr. *25*:262–279.

Magyar, B., H. C. Moor, and L. Sigg. 1993. Vertical distribution and transport of molybdenum in a lake with a seasonally anoxic hypolimnion. Limnol. Oceanogr. *38*:521–531.

Mahon, R. 1976. A second look at bluegill production in Wyland Lake, Indiana. Environ. Biol. Fish. *1*:85–86.

Maier, G. 1989a. The effect of temperature on the development times of eggs, naupliar and copepodite stages of five species of cyclopoid copepods. Hydrobiologia *184*:79–88.

Maier, G. 1989b. Variable life cycles in the freshwater copepod *Cyclops vicinus* (Uljanin 1875): Support for the predator avoidance hypothesis? Arch. Hydrobiol. *115*:203–219.

Maier, G. 1990a. Reproduction and life history of a central European, planktonic *Diacyclops bicuspidatus* population. Arch. Hydobiol. *117*:485–495.

Maier, G. 1990b. The life history of two copepods with special reference to *Eudiaptomus vulgaris* Schmeil, 1898. Crustaceana *59*:204–212.

Maier, G. 1990c. Coexistence of the predatory cyclopoids *Acanthocyclops robustus* (Sars) and *Mesocyclops leuckarti* (Claus) in a small eutrophic lake. Hydrobiologia *198*:185–203.

Maier, G. 1990d. Spatial distribution of resting stages, rate of emergence from diapause and times to adulthood and to the appearance of the first clutch in 3 species of cyclopoid copepods. Hydrobiologia *206*:11–18.

Maier, G. 1994. Patterns of life history among cyclopoid copepods of central Europe. Freshwat. Biol. *31*:77–86.

Maier, G. 1996. Copepod communities in lakes of varying trophic degree. Arch. Hydrobiol. *136*:455–465.

Maiss, M., J. Ilmberger, and K. O. Münnich. 1994. Vertical mixing in Überlingersee (Lake Constance) traced by SF_6 and heat. Aquat. Sci. *56*:329–347.

Maistrenko, Iu. G. 1965. Organicheskoe Veshchestvo Vody i Donnykh Otlozhenii Rek i Vodoemov Ukrainy (Basseiny Dnepra i Dunaia). Inst. Gidrobiologii, Kiev. 239 pp.

Makarewicz, J. C. and G. E. Likens. 1975. Niche analysis of a zooplankton community. Science *190*:1000–1003.

Makarewicz, J. C. and G. E. Likens. 1979. Structure and function of the zooplankton community of Mirror Lake, New Hampshire. Ecol. Monogr. *49*:109–127.

Maki, A. W., D. B. Porcella, and R. H. Wendt. 1984. The impact of detergent phosphorus bans on receiving water quality. Water Res. *18*:893–903.

Malakhov, V. V. 1994. Nematodes: Structure, Development, Classification, and Phylogeny. Smithsonian Inst. Press., Washington, DC. 286 pp.

Malard, F., J. V. Ward, and C. T. Robinson. 2000. An expanded perspective of the hyporheic zone. Verh. Internat. Verein. Limnol. *27* (In press.)

Malcolm, R. L. 1990. The uniqueness of humic substances in each of soil, stream and marine environments. Anal. Chem. Acta *232*:19–30.

Malcolm, R. L. and W. H. Durum. 1976. Organic carbon and nitrogen concentration and annual organic carbon load for six selected rivers of the U.S.A. U.S. Geol. Surv. Water-Supp. Pap. 1817-F. 21 pp.

Malecha, J. 1984. Cycle biologique de l'hirudinée rhynchobdelle *Piscicola geometra* L. Hydrobiologia *118*:237–243.

Mallwitz, J. 1984. Untersuchungen zur Ökologie litoraler Ostracoden im Schmal- und Lüttauersee (Schleswig-Holstein, BRD). Arch. Hydrobiol. *100*:311–339.

Malm, J. 1995. Spring circulation associated with the thermal bar in large temperate lakes. Nordic Hydrol. *26*:331–358.

Malm, J. 1998. Bottom buoyancy layer in an ice-covered lake. Water Resour. Res. *34*:2981–2993.

Malm, J. 1999. Some properties of currents and mixing in a shallow ice-covered lake. Water Resour. Res. *35*:221–232.

Malm, J., L. Bengtsson, A. Terzhevic, P. Boyarinov, A. Glinsky, N. Palshin, and M. Petrov. 1998. Field study on currents in a shallow, ice-covered lake. Limnol. Oceanogr. *43*:1669–1679.

Malmqvist, B. 1988. Downstream drift in Madeiran levadas: Tests of hypotheses relating to the influence of predators in the drift of insects. Aquat. Insects *10*:141–152.

Malmqvist, B. and P. Sjöström. 1987. Stream drift as a consequence of disturbance by invertebrate predators. Field and laboratory experiments. Oecologia *74*:396–403.

Malmqvist, B., P. Sjöström, and K. Frick. 1991. The diet of two species of *Isoperla* (Plecoptera: Perlodidae) in relation to season, site and sympatry. Hydrobiologia *213*:191–203.

Malone, B. J. and D. J. McQueen. 1983. Horizontal patchiness in zooplankton populations in two Ontario kettle lakes. Hydrobiologia 99:101–124.

Malovitskaia, L. M. and Ju. Sorokin. 1961. Eksperimental'noe issledovanie pitaniia *Diaptomus* (Crustacea, Copepoda) s pomoshch'iu C^{14}. Trudy Inst. Biol. Vodokhranilishch *4*:262–272.

Maltchik, L., S. Molla, C. Casado, and C. Montes. 1994. Measurement of nutrient spiralling in a Mediterranean stream: Comparison of two extreme hydrological periods. Arch. Hydrobiol. *130*:215–227.

Malueg, K. W. and A. D. Hasler. 1966. Echo sounder studies on diel vertical movements of *Chaoborus* larvae in Wisconsin (U.S.A.) lakes. Verh. Internat. Verein. Limnol. *16*:1697–1708.

Malueg, K. W., D. P. Larsen, D. W. Schults, and H. T. Mercier. 1975. A six year water, phosphorus, and nitrogen budget for Shagawa Lake, Minnesota. J. Environ. Qual. *4*:236–242.

Maly, E. J. and M. P. Maly. 1974. Dietary differences between two co-occurring calanoid copepod species. Oecologia *17*:325–333.

Mancuso, C. A., P. D. Franzmann, H. R. Burton, and P. D. Nichols. 1990. Microbial community structure and biomass estimate of a methanogenic Antarctic lake ecosystem as determined by phospholipid analyses. Microb. Ecol. *19*:73–95.

Mann, C. J. and R. G. Wetzel. 1995. Dissolved organic carbon and its utilization in a riverine wetland ecosystem. Biogeochemistry *31*:91–120.

Mann, C. J. and R. G. Wetzel. 1996. Loading and bacterial utilization of dissolved organic carbon from emergent macrophytes. Aquat. Bot. *53*:61–72.

Mann, C. J. and R. G.Wetzel. 1999a. Photosynthesis and stomatal conductance of *Juncus effusus* in a temperate wetland ecosystem. Aquat. Bot. *63*:127–144.

Mann, C. J. and R. G. Wetzel. 1999b. Effects of the emergent macrophyte *Juncus effusus* L. on the chemical composition of interstitial water and bacterial productivity. Biogeochemistry *48*:307–322.

Mann, K. H. 1962. Leeches (Hirudinea). Their Structure, Physiology, Ecology and Embryology. Pergamon Press, Oxford. 201 pp.

Manny, B. A. 1972a. Seasonal changes in dissolved organic nitrogen in six Michigan lakes. Verh. Internat. Verein. Limnol. *18*:147–156.

Manny, B. A 1972b. Seasonal changes in organic nitrogen content of net- and nannophytoplankton in two hardwater lakes. Arch. Hydrobiol. *71*:103–123.

Manny, B. A. and R. G. Wetzel. 1973. Diurnal changes in dissolved organic and inorganic carbon and nitrogen in a hardwater stream. Freshwat. Biol. *3*:31–43.

Manny, B. A. and R. G. Wetzel. 1982. Allochthonous dissolved organic and inorganic nitrogen budget of a marl lake. (Unpublished manuscript.)

Manny, B. A., M. C. Miller, and R. G. Wetzel. 1971. Ultraviolet combustion of dissolved organic nitrogen compounds in lake waters. Limnol. Oceanogr. *16*:71–85.

Manny, B. A., R. G. Wetzel, and R. E. Bailey. 1978. Paleolimnological sedimentation of organic carbon, nitrogen, phosphorus, fossil pigments, pollen, and diatoms in a hypereutrophic, hardwater lake: A case history of eutrophication. Pol. Arch. Hydrobiol. *25*:243–267.

Manny, B. A., R. G. Wetzel, and W. C. Johnson. 1975. Annual contribution of carbon, nitrogen, and phosphorus to a hard-water lake by migrant Canada geese. Verh. Internat. Verein. Limnol. *19*:949–951.

Manny, B. A., W. C. Johnson, and R. G. Wetzel. 1994. Nutrient additions by waterfowl to lakes and reservoirs: Predicting their effects on productivity and water quality. Hydrobiologia *279/280*:121–132.

Mantai, K. E. and M. E. Newton. 1982. Root growth in *Myriophyllum*: A specific plant response to nutrient availability? Aquat. Bot. *13*:45–55.

Marchant, R. and H. B. N. Hynes. 1981. The distribution and production of *Gammarus pseudolumnaeus* (Crustacea: Amphipoda) along a reach of the Credit River, Ontario. Freshwat. Biol. *11*:169–182.

Marchetto, A. and A. Lami. 1994. Reconstruction of pH by chrysophycean scales in some lakes of the southern Alps. Hydrobiologia *274*:83–90.

Marcus, J. H., D. W. Sutcliffe, and L. G. Willoughby. 1978. Feeding and growth of *Asellus aquaticus* (Isopoda) on food items from the littoral of Windermere, including green leaves of *Elodea canadensis*. Freshwat. Biol. *8*:505–519.

Marcus, N. H. and J. Schmidt-Gengenbach. 1986. Recruitment of individuals into the plankton: The importance of bioturbation. Limnol. Oceanogr. *31*:206–210.

Margalef, R. 1955. Temperature and morphology in freshwater organisms. Verh. Internat. Verein. Limnol. *12*:507–514.

Margalef, R. 1969. Size of centric diatoms as an ecological indicator. Mitt. Internat. Verein. Limnol. *17*:202–210.

Margalef, R. 1983. Ecologia. Ediciones Omega, Barcelona. 1010 pp.

Marino, R., R. W. Howarth, J. Shamess, and E. Prepas. 1990. Molybdenum and sulfate as controls on the abundance of nitrogen-fixing cyanobacteria in saline lakes in Alberta. Limnol. Oceanogr. *35*:245–259.

Marion, L., P. Clergeau, L. Brient, and G. Bertru. 1994. The importance of avian-contributed nitrogen (N) and phosphorus (P) to Lake Grand-Lieu, France. Hydrobiologia *279/280*:133–147.

Marker, A. F. H. 1976. The benthic algae of some streams in southern England. II. The primary production of the epilithon in a small chalkstream. J. Ecol. *64*:359–373.

Marks, J. C. and R. L. Lowe. 1993. Interactive effects of nutrient availability and light levels on the periphyton composition of a large oligotrophic lake. Can. J. Fish. Aquat. Sci. *50*:1270–1278.

Marnette, E. C. L., C. A. Hordijk, N. van Breemen, and T. E. Cappenberg. 1992. Sulfate reduction and S-oxidation in a moorland pool sediment. Biogeochemistry *17*:123–143.

Marquenie-van der Werff, M. and W. H. O. Ernst. 1979. Kinetics of copper and zinc uptake by leaves and roots of aquatic plant, *Elodea muttallii*. Z. Pflanzenphysiol. *92*:1–10.

Marsden, M. W. 1989. Lake restoration by reducing external phosphorus loading: The influence of sediment phosphorus release. Freshwat. Biol. *21*:139–162.

Marshall, E. J. P. and D. F. Westlake. 1990. Water velocities around water plants in chalk streams. Folia Geobot. Phytotax. Praha *25*:279–289.

Marston, R. A., J. Girel, G. Pautou, H. Piegay, J. P. Bravard, and C. Arneson. 1995. Channel metamorphosis, floodplain disturbance, and vegetation development: Ain River, France. Geomorphology *13*:121–131.

Martin, A. J., R. M. H. Seaby, and J. O. Young. 1994. Food limitation in lake-dwelling leeches: Field experiments. J. Anim. Ecol. *63*:93–100.

Martin, J. H., G. A. Knauer, and A. R. Flegal. 1980a. Cadmium in natural waters. *In* J. O. Nriagu, ed. Cadmium in the Environment. Part I. Ecological Cycling. John Wiley & Sons, New York. pp. 141–145.

Martin, J. H., G. A. Knauer, and A. R. Flegal. 1980b. Distribution of zinc in natural waters. *In* J. O. Nriagu, ed. Zinc in the Environment. I. Ecological Cycling. John Wiley & Sons, New York. pp. 193–197.

Martin, M. M. and J. J. Kukor. 1984. Role of mycophagy and bacteriophagy in invertebrate nutrition. *In* M. J. Klug and C. A. Reddy, eds. Current Perspectives in Microbial Ecology. Amer. Soc. Microbiology, Washington, DC. pp. 257–263.

Martin, M. M., J. S. Martin, J. J. Kukor, and R. W. Merritt. 1981a. The digestive enzymes of detritus-feeding stonefly nymphs (Plecoptera: Pteronarcyidae). Can. J. Zool. *59*:1947–1951.

Martin, M. M., J. J. Kukor, J. S. Martin, D. L. Lawson, and R. W. Merritt. 1981b. Digestive enzymes of larvae of three species of caddisflies (Trichoptera). Insect Biochem. *11*:501–505.

Martin, S. 1981. Frazil ice in rivers and oceans. Ann. Rev. Fluid Mech. *13*:379–397.

Marx, J. L. 1980. Calmodulin: A protein for all seasons. Science *208*:274–276.

Marxsen, J. 1980. Untersuchungen zur Ökologie der Bakterien in der fließenden Welle von Bächen. III. Aufnahme gelöster organischer Substanzen. Arch. Hydrobiol./Suppl. *58*:207–272.

Marxsen, J. 1988. Evaluation of the importance of bacteria in the carbon flow of a small open grassland stream, the Breitenbach. Arch. Hydrobiol. *111*:339–350.

Marxsen, J. 1996. Measurement of bacterial production in streambed sediments via leucine incorporation. FEMS Microbiol. Ecol. *21*:313–325.

Marxsen, J. 1999. Importance of bacterial production in the carbon flow of an upland stream, the Breitenbach. Arch. Hydrobiol. Spec. Issues Advanc. Limnol, *54*:135–145.

Marxsen, J. and D. M. Fiebig. 1993. Use of perfused cores for evaluating extracellular enzyme activity in stream-bed sediments. FEMS Microbiol. Ecol. *13*:1–12.

Marzolf, G. R. 1965a. Substrate relations of the burrowing amphipod *Pontoporeia affinis* in Lake Michigan. Ecology *46*: 579–592.

Marzolf, G. R. 1965b. Vertical migration of *Pontoporeia affinis* (Amphipoda) in Lake Michigan. Publ. Great Lakes Res. Div., Univ. Mich. *13*:133–140.

Marzolf, G. R. 1990. Reservoirs as environments for zooplankton. *In* K. W. Thornton, B. L. Kimmel, and F. E. Payne, eds. Reservoir Limnology: Ecological Perspectives. Wiley-Interscience, New York. pp. 195–208.

Maser, C. and J. R. Sedell. 1994. From the forest to the sea: The ecology of wood in streams, rivers, estuaries, and oceans. St. Lucie Press, Delray Beach, FL. 200 pp.

Mason, C. F. 1976. Relative importance of fungi and bacteria in the decomposition of *Phragmites* leaves. Hydrobiologia *51*:65–69.

Mason, C. F. and M. M. Abdul-Hussein. 1991. Population dynamics and production of *Daphnia hyalina* and *Bosmina longirostris* in a shallow, eutrophic reservoir. Freshwat. Biol. *25*:243–260.

Mason, C. F. and R. J. Bryant. 1975. Periphyton production and grazing by chironomids in Alderfen Broad, Norfolk. Freshwat. Biol. *5*:271–277.

Mason, C. F. and S. M. MacDonald. 1982. The input of terrestrial invertebrates from tree canopies to a stream. Freshwat. Biol. *12*:305–311.

Mason, M. A. 1952. Some observations of breaking waves. *In* Gravity Waves. Circ. U.S. Bur. Standards *521*:215–220.

Mason, R. P., F. M. M. Morel, and H. F. Hemond. 1995. The role of microorganisms in elemental mercury formation in natural waters. Water Air Soil Poll. *80*:775–787.

Mather, M. E., M. J. Vanni, T. E. Wissing, S. A. Davis, and M. H. Schaus. 1995. Regeneration of nitrogen and phosphorus by bluegill and gizzard shad: Effect of feeding history. Can. J. Fish. Aquat. Sci. *52*:2327–2338.

Mathes, J. and H. Arndt. 1995. Annual cycle of protozooplankton (ciliates, flagellates and sarcodines) in relation to phyto- and metazooplankton in Lake Neumühler See (Mecklenburg, Germany). Arch. Hydrobiol. *134*:337–358.

Mathews, C. P. and D. F. Westlake. 1969. Estimation of production by populations of higher plants subject to high mortality. Oikos *20*:156–160.

Mathias, C. B., A. K. T. Kirschner, and B. Velimirov. 1995. Seasonal variations of virus abundance and viral control of the bacterial production in a backwater system of the Danube River. Appl. Environ. Microbiol. *61*:3734–3740.

Matsuyama, M. 1973. Organic substances in sediment and settling matter during spring in meromictic Lake Suigetsu. J. Oceanogr. Soc. Japan *29*:53–60.

Mattern, H. 1970. Beobachtungen über die Algenflora im Uferbereich des Bodensees (Überlinger See und Gnadensee). Arch. Hydrobiol./Suppl. *37*:1–163.

Matthews, P. C. and S. I. Heaney. 1987. Solar heating and its influence on mixing in ice-covered lakes. Freshwat. Biol. *18*:135–149.

Matthews, W. J. 1998. Patterns in Freshwater Fish Ecology. International Thomson Publ., New York. 756 pp.

Mattson, M. D. and G. E. Likens. 1990. Air pressure and methane fluxes. Nature *347*:718–719.

Mattson, M. D. and G. E. Likens. 1993. Redox reactions of organic matter decomposition in a soft water lake. Biogeochemistry *19*:149–172.

Matveev, V. P. 1964. O vertikal'nom raspredelenii temperatury v donnykh otlozheniyakh Ozer Dolgogo (Pitkayarvi) i Volochaevskogo (Vuotyarvi). (On the vertical distribution of temperature in the bottom deposits of Lake Dolgom (Pitkayarvi) and Volochaevskom (Vuotyarvi). *In* Ozera Karel'skogo Ieresheika. Izdatel'stvo Nauka, Moscow. pp. 45–50.

Maucha, R. 1932. Hydrochemische Methoden in der Limnologie. Die Binnengewässer *12*, 173 pp.

May, L. 1980. On the ecology of *Notholca squamula* Müller in Loc Leven, Kinross, Scotland. Hydrobiologia *73*:177–180.

May, R. M. 1973. Stability and Complexity in Model Ecosystems. Princeton University Press, Princeton, NJ.

Mayfield, C. I. and W. E. Inniss. 1978. Interactions between freshwater bacteria and *Ankistrodesmus braunii* in batch and continuous culture. Microb. Ecol. *4*:331–344.

Mazumder, A., W. D. Taylor, D. J. McQueen, and D. R. S. Lean. 1989. Effects of fertilization and planktivorous fish on epilimnetic phosphorus and phosphorus sedimentation in large enclosures. Can. J. Fish. Aquat. Sci. *46*:1735–1742.

Mazumder, A., D. J. McQueen, W. D. Taylor, D. R. S. Lean, and M. D. Dickman. 1990a. Micro- and mesozooplankton grazing on natural pico- and nanoplankton in contrasting plankton communities produced by planktivore manipulation and fertilization. Arch. Hydrobiol. *118*:257–282.

Mazumder, A., W. D. Taylor, D. J. McQueen, D. R. S. Lean, and N. R. Lafontaine. 1990b. A comparison of lakes and lake enclosures with contrasting abundances of planktivorous fish. J. Plankton Res. *12*:109–124.

Mazumder, A., W. D. Taylor, D. J. McQueen, and D. R. S. Lean. 1990c. Effects of fish and plankton on lake temperature and mixing depth. Science *247*:312–315.

Mazumder, A., W. D. Taylor, D. R. S. Lean, and D. J. McQueen. 1992. Partitioning and fluxes of phosphorus: Mechanisms regulating the size-distribution and biomass of plankton. Arch. Hydrobiol. Beih. Ergebn. Limnol. *35*:121–143.

McArthur, J. V. and G. R. Marzolf. 1986. Interactions of the bacterial assemblages of a prairie stream with dissolved organic carbon from riparian vegetation. Hydrobiologia *134*:193–199.

McArthur, J. V., J. R. Barnes, B. J. Hansen, and L. G. Leff. 1988. Seasonal dynamics of leaf litter breakdown in a Utah alpine stream. J. N. Am. Benthol. Soc. *7*:44–50.

McCall, P. L. and J. B. Fisher. 1980. Effects of tubificid oligochaetes on physical and chemical properties of Lake Erie sediments. *In* R. O. Brinkhurst and D. G. Cook, eds. Aquatic Oligochaete Biology. Plenum Press, New York. pp. 253–317.

McCarter, J. A., F. R. Hayes, L. H. Jodrey, and M. L. Cameron. 1952. Movements of materials in the hypolimnion of a lake as studied by the addition of radioactive phosphorus. Can. J. Zool. *30*:128–133.

McCarty, P. L. 1964. The methane fermentation. *In* H. Heukelekian and N. C. Dondero, eds. Principles and Applications in Aquatic Microbiology. John Wiley & Sons, New York. pp. 314–343.

McClain, M. E. and J. E. Richey. 1996. Regional-scale linkages of terrestrial and lotic ecosystems in the Amazon basin: A conceptual model for organic matter. Arch. Hydrobiol./Suppl. *113*:111–125.

McClain, M. E., J. E. Richey, and T. P. Pimentel. 1994. Groundwater nitrogen dynamics at the terrestrial-lotic interface of a small catchment in the central Amazon basin. Biogeochemistry *27*:113–127.

McColl, J. G. and D. F. Grigal. 1975. Forest fires: Effects on phosphorus movement to lakes. Science *188*:1109–1111.

McConnaughey, T. A. 1994. Calcification, photosynthesis, and global carbon cycles. Bull. Inst. Oceanogr. Monaco *13*:137–161.

McCormick, P. V. 1991. Lotic protistan herbivore selectivity and its potential impact on benthic algal assemblages. J. N. Am. Benthol. Soc. *10*:238–250.

McCormick, P. V., and J. Cairns, Jr. 1991. Effects of micrometazoa on the protistan assemblage of a littoral food web. Freshwat. Biol. *26*:111–119.

McCormick, P. V., P. S. Rawlik, K. Lurding, E. P. Smith, and F. J. Sklar. 1996. Periphyton-water quality relationships along a nutrient gradient in the northern Florida Everglades. J. N. Am. Benthol. Soc. *15*:433–449.

McCormick, P. V., R. B. E. Shuford, J. G. Backus, and W. C. Kennedy. 1998. Spatial and seasonal patterns of periphyton biomass and productivity in the northern Everglades, Florida, U.S.A. Hydrobiologia *362*:185–208.

McCracken, M. D., T. D. Gustafson, and M. S. Adams. 1974. Productivity of *Oedogonium* in Lake Wingra, Wisconsin. Amer. Midland Nat. *92*:247–254.

McCraw, B. M. 1961. Life history and growth of the snail *Lymnaea humilis* Say. Trans. Amer. Microsc. Soc. *80*:16–27.

McCraw, B. M. 1970. Aspects of the growth of the snail *Lymnaea palustris* (Müller). Malacologia *10*:399–413.

McCully, P. 1996. Silenced Rivers. The Ecology and Politics of Large Dams. Zed Books, London. 350 pp.

McDonnell, A. J. 1971. Variations in oxygen consumption by aquatic macrophytes in a changing environment. Proc. Conf. Great Lakes Res., Int. Asso. Great Lakes Res. *14*:52–58.

McDowell, W. H. 1985. Kinetics and mechanisms of dissolved organic carbon retention in a headwater stream. Biogeochemistry *1*:329–352.

McDowell, W. H. and G. E. Likens. 1988. Origin, composition, and flux of dissolved organic carbon in the Hubbard Brook Valley. Ecol. Monogr. *58*:177–195.

McDowell, W. H. and S. G. Fisher. 1976. Autumnal processing of dissolved organic matter in a small woodland stream ecosystem. Ecology *57*:561–569.

McDowell, W. H. and T. Wood. 1984. Podzolization: Soil processes control dissolved organic carbon concentrations in stream water. Soil Sci. *137*:23–32.

McElhone, M. J. 1978. A population study of littoral dwelling Naididae (Oligochaeta) in an shallow mesotrophic lake in North Wales. J. Anim. Ecol. *47*:615–626.

McElhone, M. J. 1979. A comparison of the gut contents of two coexisting lake-dwelling Naididae (Oligochaeta), *Nais pseudobtusa* and *Chaetogaster diastrophus*. Freshwat. Biol. *9*:199–204.

McGaha, V. J. 1952. The limnological relations of insects to certain aquatic flowering plants. Trans. Am. Microsc. Soc. *71*:355–381.

McGowan, L. M. 1974. Ecological studies on *Chaoborus* (Diptera, Chaoboridae) in Lake George, Uganda. Freshwat. Biol. *4*:483–505.

McGregor, D. L. 1969. The reproductive potential, life history and parasitism of the freshwater ostracod *Darwinula stevensoni* (Brady and Robertson). *In* J. W. Neale, ed. The Taxonomy, Morphology and Ecology of Recent Ostracoda. Oliver & Boyd, Edinburgh. pp. 194–221.

McIntire, C. D. 1966. Some factors affecting respiration of periphyton communities in lotic environments. Ecology *47*:918–930.

McIntosh, R. P. 1980. The background and some current problems of theoretical ecology. Synthese *43*:195–255.

McKay, C. P., G. D. Clow, R. A. Wharton, and S. W. Squyres. 1985. Thickness of ice on perennially frozen lakes. Nature *313*:561–562.

McKeague, J. A. and M. G. Cline. 1963a. Silica in soil solutions. I. The form and concentration of dissolved silica in aqueous extracts of some soils. Can. J. Soil Sci. *43*:70–82.

McKeague, J. A. and M. G. Cline. 1963b. Silica in soil solutions. II. the adsorption of monosilicic acid by soil and by other substances. Can. J. Soil Sci. *43*:83–96.

McKee, K. L., I. A. Mendelssohn, and D. M. Burdick. 1989. Effect of long-term flooding on root metabolic response in five freshwater marsh plant species. Can. J. Bot. *67*:3446–3452.

McKinley, K. R. 1977. Light-mediated uptake of ^{3}H-glucose in a small hard-water lake. Ecology *58*:1356–1365.

McKnight, D. 1981. Chemical and biological processes controlling the response of a freshwater ecosystem to copper stress: A field study of the $CuSO_4$ treatment of Mill Pond Reservoir, Burlington, Massachusetts. Limnol. Oceanogr. *26*:518–531.

McKnight, D. M. 1979. Release of weak and strong copper-complexing agents by algae. Limnol. Oceanogr. *24*:823–837.

McKnight, D. M. and G. R. Aiken. 1998. Sources and age of aquatic humus. *In* D. O. Hessen and L. J. Tranvik, eds. Aquatic Humic Substances: Ecology and Biogeochemistry. Springer-Verlag, New York. pp. 9–39.

McKnight, D. M., B. A. Kimball, and K. E. Bencala. 1988. Iron photoreduction and oxidation in an acidic mountain stream. Science *240*:637–640.

McKnight, D. M., E. D. Andrews, R. L. Smith, and R. Dufford. 1994. Aquatic fulvic acids in algal-rich Antarctic ponds. Limnol. Oceanogr. *39*:1972–1979.

McKnight, D. M., G. R. Aiken, and R. L. Smith. 1991. Aquatic fulvic acids in microbially based ecosystems: Results from two Antarctic desert lakes. Limnol. Oceanogr. *36*:998–1006.

McKnight, D. M., K. E. Bencala, G. W. Zellweger, G. R. Aiken, G. L. Feder, and K. A. Thorn. 1992. Sorption of dissolved organic carbon by hydrous aluminum and iron oxides occurring at the confluence of Deer Creek with the Snake River, Summit County, Colorado. Environ. Sci. Technol. *26*:1388–1396.

McKnight, D. M., R. Harnish, R. L. Wershaw, J. S. Baron, and S. Schiff. 1997. Chemical characteristics of particulate, colloidal, and dissolved organic material in Loch Vale watershed, Rocky Mountain National Park. Biogeochemistry *36*:99–124.

McKnight, D. M., R. L. Smith, R. A. Harnish, C. L. Miller, and K. E. Bencala. 1993. Seasonal relationships between planktonic microorganisms and dissolved organic material in an alpine stream. Biogeochemistry *21*:39–59.

McLaren, I. A. 1963. Effects of temperature on growth of zooplankton, and the adaptive value of vertical migration. J. Fish. Res. Board Can. *20*:685–727.

McLaren, I. A. 1974. Demographic strategy of vertical migration by a marine copepod. Am. Nat. *108*:91–102.

McLaren, I. A. and C. J. Corkett. 1984. Singular, mass-specific *P/B* ratios cannot be used to estimate copepod production. Can. J. Fish. Aquat. Sci. *41*:828–830.

McMahon, J. W. 1965. Some physical factors influencing the feeding behavior of *Daphnia magna* Straus. Can. J. Zool. *43*:603–612.

McMahon, J. W. 1969. The annual and diurnal variation in the vertical distribution of acid-soluble ferrous and total iron in a small dimictic lake. Limnol. Oceanogr. *14*:357–367.

McMahon, J. W. and F. H. Rigler. 1963. Mechanisms regulating the feeding rate of *Daphnia magna* Straus. Can. J. Zool. *14*:321–332.

McMahon, J. W. and F. H. Rigler. 1965. Feeding rate of *Daphnia magna* Straus in different foods labeled with radioactive phosphorus. Limnol. Oceanogr. *10*:105–113.

McMahon, R. F. 1983. Physiological ecology of freshwater pulmonates. *In* W. D. Russell-Hunter, ed. The Mollusca. Vol. 6. Ecology. Academic Press, Orlando. pp. 359–430.

McMahon, R. F. 1991. Mollusca: Bivalvia. *In* J. H. Thorp and A. P. Covich, eds. Ecology and Classification of North American Freshwater Invertebrates. Academic Press, San Diego. pp. 315–399.

McNaught, D. C. 1966. Depth control by planktonic cladocerans in Lake Michigan. Publ. Great Lakes Res. Div., Univ. Mich. *15*:98–108.

McNaught, D. C. 1978. spatial heterogeneity and niche differentiation in zooplankton of Lake Huron. Verh. Internat. Verein. Limnol. *20*:341–346.

McNaught, D. C. and A. D. Hasler. 1961. Surface schooling and feeding behavior in the white bass, *Roccus chrysops* (Rafinesque), in Lake Mendota. Limnol. Oceanogr. *6*:53–60.

McNaught, D. C. and A. D. Hasler. 1964. Rate of movement of populations of *Daphnia* in relation to changes in light intensity. J. Fish. Res. Board Can. *21*:291–318.

McNaught, D. C. and A. D. Hasler. 1966. Photoenvironments of planktonic Crustacea in Lake Michigan. Verh. Internat. Verein. Limnol. *16*:194–203.

McNaught, D. C., D. Griesmer, and M. Kennedy. 1980. Resource characteristics modifying selective grazing by copepods. *In* W. C. Kerfoot, ed. Evolution and Ecology of Zooplankton Communities. Univ. Press New England, Hanover, NH. pp 292–298.

McNaughton S. J. 1966b. Ecotype function in the *Typha* community-type. Ecol. Monogr. *36*:297–325.

McNaughton, S. J. 1966a. Light stimulated oxygen uptake and glycolic acid oxidase in *Typha latifolia* L. leaf discs. Science *211*:1197–1198.

McNaughton, S. J. 1969. Genetic and environmental control of glycolic acid oxidase activity in ecotypic populations of *Typha latifolia*. Am. J. Bot. *56*:37–41.

McNaughton, S. J. 1970. Fitness sets for *Typha*. Am. Nat. *104*:337–341.

McNaughton, S. J. and L. W. Fullem. 1969. Photosynthesis and photorespiration in *Typha latifolia*. Plant Physiol. *45*:703–707.

McQueen, D. J. 1969. Reduction of zooplankton standing stocks by predaceous *Cyclops bicuspidatus thomasi* in Marion Lake, British Columbia. J. Fish. Res. Board Can. *26*:1605–1618.

McQueen, D. J. 1970. Grazing rates and food selection in *Diaptomus oregonensis* (Copepoda) from Marion Lake, British Columbia. J. Fish. Res. Board Can. *27*:13–20.

McQueen, D. J. 1990. Manipulating lake community structure: Where do we go from here? Freshwat. Biol. *23*:613–620.

McQueen, D. J., J. R. Post, and E. L. Mills. 1986. Trophic relationships in freshwater pelagic ecosystems. Can. J. Fish. Aquat. Sci. *43*:1571–1581.

McQueen, D. J., M. R. S. Johannes, J. R. Post, T. J. Stewart, and D. R. S. Lean. 1989. Bottom-up and top-down impacts on freshwater pelagic community structure. Ecol. Monogr. *59*:289–309.

McQueen, D. J., R. France, and C. Kraft. 1992. Confounded impacts of planktivorous fish on freshwater biomanipulations. Arch. Hydrobiol. *125*:1–24.

McRae, G. and C. J. Edwards. 1994. Thermal characteristics of Wisconsin headwater streams occupied by beaver: Implications for brook trout habitat. Trans. Amer. Fish. Soc. *123*:641–656.

McRoy, C. P., R. J. Barsdate, and M. Nebert. 1972. Phosphorus cycling in an eelgrass (*Zostera marina* L.) ecosystem. Limnol. Oceanogr. *17*:58–67.

Meadows, P. S. and J. G. Anderson. 1966. Microorganisms attached to marine and freshwater sand grains. Nature *212*:1059–1060.

Meadows, P. S. and J. G. Anderson. 1968. Microorganisms attached to marine sand grains. J. Mar. Biol. Assoc. U.K. *48*:161–175.

Medlin, L. K., D. M. Williams, and P. A. Sims. 1993. The evolution of the diatoms (Bacillariophyta). I. Origin of the group and assessment of the monophyly of its major divisions. Eur. J. Phycol. *28*:261–275.

Medlin, L. K., W. H. C. F. Kooistra, D. Potter, G. W. Saunders, and R. A. Andersen. 1997. Phylogenetic relationships of the 'golden algae' (haptophytes, heterokont chromophytes) and their plastids. Pl. Syst. Evol. *11*(Suppl.):187–219.

Meeks, J. C. 1974. Chlorophylls. *In* W. D. P. Stewart, ed. Algal Physiology and Biochemistry. University of California Press, Berkeley. pp. 161–175.

de Meester, L. and H. de Jager. 1993. Hatching of *Daphnia* sexual eggs. I. Intraspecific differences in the hatching responses of *D. magna* eggs. Freshwat. Biol. *30*:219–226.

Meffert, M.-E. and H. Zimmermann-Telschow. 1979. Net release of nitrogenous compounds by axenic and bacteria-containing cultures of *Oscillatoria raedekei* (Cyanophyta). Arch. Hydrobiol. *87*:125–138.

Meffert, M.-E. and J. Overbeck. 1979. Regulation of bacterial growth by algal release products. Arch. Hydrobiol. *87*:118–121.

Megard, R. O. 1972. Phytoplankton, photosynthesis, and phosphorus in Lake Minnetonka, Minnesota. Limnol. Oceanogr. *17*:68–87.

Meinzer, O. E., ed. 1942. Hydrology. McGraw-Hill Book Co., New York. 712 pp.

Meisch, H.-U., H. Benzschawel and H.-J. Bielig. 1977. The role of vanadium in green plants. II. Vanadium in green algae—two sites of action. Arch. Microbiol. *114*:67–70.

Melack, J. M. 1979. Temporal variability of phytoplankton in tropical lakes. Oecologia *44*:1–7.

Melack, J. M. and T. R. Fisher. 1983. Diel oxygen variations and their ecological implications in Amazon floodplain lakes. Arch. Hydrobiol. *98*:422–442.

Melão, M. G. G. and O. Rocha. 1998. Growth rates and energy budget of *Metania spinata* (Carter 1881) (Porifera, Metaniidae) in Lagoa Dourada, Brazil. Verh. Internat. Verein. Limnol. *26*:2098–2102.

Menéndez, M. and E. Forès. 1998. Early decomposition of *Ruppia cirrhosa* (Petagna) Grande and *Potamogeton pectinatus* L. leaves. Oecologia Aquat. *11*:73–86.

Menezes, C. F. S., F. A. Esteves, and A. M. Anesio. 1993. Influência da variação artificial do nível d'água da Represa do Lobo (SP) sobre a biomassa e produtividade de *Nymphoides indica* (L.) O. Kuntze e *Pontederia cordata* L. Acta Limnol. Brasiliensia *6*:163–172.

Mengis, M., R. Gächter, and B. Wehrli. 1996. Nitrous oxide emissions to the atmosphere from an artificially oxygenated lake. Limnol. Oceanogr. *41*:548–553.

Mengis, M., R. Gächter, and B. Wehrli. 1997. Sources and sinks of nitrous oxide (N_2O) in deep lakes. Biogeochemistry *38*:281–301.

Meon, B. and F. Jüttner. 1999. Concentrations and dynamics of free mono- and oligosaccharides in a shallow eutrophic lake measured by thermospray mass spectrometry. Aquat. Microb. Ecol. *16*:281–293.

Meriläinen, J. 1969. Distribution of diatom frustules in recent sediments of some meromictic lakes. Mitt. Internat. Verein. Limnol. *17*:186–192.

Meriläinen, J. 1971. The recent sedimentation of diatom frustules in four meromictic lakes. Ann. Bot. Fenn. *8*:160–176.

Merkt, J. 1971. Zuverlässige Auszählungen von Jahresschichten in Seesedimenten mit Hilfe von Gross-Cünnschliffen. Arch. Hydrobiol. *69*:145–154.

Mermoud, F., F. O. Gülaçar, S. Siles, B. Chassaing, and A. Buchs. 1982. 4-Methylsterols in recent lacustrine sediments: Terrestrial, planktonic or some other origin? Chemosphere *11*:557–567.

Mermoud, F., O. Clerc, F. O. Gülacar, and A. Buchs. 1981. Analyse des acides gras et des stérols dans le plancton du Lac Léman. Arch. Sc. Genève *34*:367–382.

Merrell, J. R. and D. K. Stoecker. 1998. Differential grazing on protozoan microplankton by developmental stages of the calanoid copepod *Eurytemora affinis* Poppe. J. Plankton Res. *20*:289–304.

Messina, D. S. and A. L. Baker. 1982. Interspecific growth regulation in species succession through vitamin B_{12} competitive inhibition. J. Plankton Res. *4*:41–46.

Meybeck, M. 1981. Pathways of major elements from land to ocean through rivers. *In* River Inputs to Ocean Systems. UNEP/Unesco Report. pp. 18–30.

Meybeck, M. 1982. Carbon, nitrogen, and phosphorus transport by world rivers. Am. J. Sci. *282*:401–450.

Meybeck, M. 1993a. Riverine transport of atmospheric carbon: Sources, global typology and budget. Water Air Soil Poll. *70*:443–463.

Meybeck, M. 1993b. Natural sources of C, N, P and S. *In* R. Wollast, F. T. Mackenzie, and L. Chou, eds. Interactions of C, N, P and S biogeochemical cycles and global change, Springer-Verlag, Berlin. pp. 163–193.

Meybeck, M. and R. Helmer. 1989. The quality of rivers: From pristine stage to global pollution. Palaeogergr. Palaeoclimatol. Palaeoecol. *75*:283–309.

Meyer, J. L. 1979. The role of sediments and bryophytes in phosphorus dynamics in a headwater stream ecosystem. Limnol. Oceanogr. *24*:365–375.

Meyer, J. L. 1980. Dynamics of phosphorus and organic matter during leaf decomposition in a forest stream. Oikos *34*:44–53.

Meyer, J. L. 1986. Dissolved organic carbon dynamics in two subtropical blackwater rivers. Arch. Hydrobiol. *108*:119–134.

Meyer, J. L. 1988. Benthic bacterial biomass and production in a black-water river. Verh. Internat. Verein. Limnol. *23*:1832–1838.

Meyer, J. L. 1989. Can P/R ratio be used to assess the food base of stream ecosystems? A comment on Rosenfeld and Mackay (1987). Oikos *54*:119–121.

Meyer, J. L. and C. Johnson. 1983. The influence of elevated nitrate concentration on rate of leaf decomposition in a stream. Freshwat. Biol. *13*:177–183.

Meyer, J. L. and G. E. Likens. 1979. Transport and transformation of phosphorus in a forest stream ecosystem. Ecology *60*:1255–1269.

Meyer, J. L. and C. M. Tate. 1983. The effects of watershed disturbance on dissolved organic carbon dynamics of a stream. Ecology *64*:33–44.

Meyer, J. L., J. B. Wallace, and S. L. Eggert. 1998. Leaf litter as a source of dissolved organic carbon in streams. Ecosystems *1*:240–249.

Meyer, J. L., R. T. Edwards, and R. Risley. 1987. Bacterial growth on dissolved organic carbon from a blackwater river. Microb. Ecol. *13*:13–29.

Meyers, P. A. and J. G. Quinn. 1971. Interaction between fatty acids and calcite in seawater. Limnol. Oceanogr. *16*:922–997.

Meyers, P. A., M. J. Leenheer, B. J. Eadie, and S. J. Maule. 1984. Organic geochemistry of suspended and settling particulate matter in Lake Michigan. Geochim. Cosmochim. Acta *48*:443–452.

Meyers, P. A., M. J. Leenheer, and R. A. Bourbonniere. 1995. Diagenesis of vascular plant organic matter components during burial in lake sediments. Aquat. Geochem. *1*:35–52.

Mickle, A. M. and R. G. Wetzel. 1978a. Effectiveness of submersed angiosperm-epiphyte complexes on exchange of nutrients and organic carbon in littoral systems. I. Inorganic nutrients. Aquat. Bot. *4*:303–316.

Mickle, A. M. and R. G. Wetzel 1978b. Effectiveness of submersed angiosperm-epiphyte complexes on exchange of nutrients and organic carbon in littoral systems. II. Dissolved organic carbon. Aquat. Bot. *4*:317–329.

Mickle, A. M. and R. G. Wetzel. 1979. Effectiveness of submersed angiosperm-epiphyte complexes on exchange of nutrients and organic carbon in littoral systems. III. Refractory organic carbon. Aquat. Bot. *6*:339–355.

Middelboe, M. and M. Søndergaard. 1995. Concentration and bacterial utilization of sub-micron particles and dissolved organic carbon in lakes and a coastal area. Arch. Hydrobiol. *133*:129–147.

Milbrink, G. 1973a. Communities of Oligochaeta as indicators of the water quality in Lake Hjälmaren. Zoon *1*:77–88.

Milbrink, G. 1973b. On the use of indicator communities of Tubificidae and some Lumbriculidae in the assessment of water pollution in Swedish lakes. Zoon *1*:125–139.

Milbrink, G. 1973c. On the vertical distribution of oligochaetes in lake sediments. Rep. Inst. Freshw. Res. Drottningholm *53*:34–50.

Milbrink, G. 1978. Indicator communities of oligochaetes in Scandinavian lakes. Verh. Internat. Verein. Limnol. *20*:2406–2411.

Milbrink, G. 1980. Oligochaete communities in pollution biology: The European situation with special reference to Scandinavia. *In* R. O. Brinkhurst and D. G. Cook, eds. Aquatic Oligochaete Biology. Plenum Press, New York. pp. 433–455.

Miles, C. J. and P. L. Brezonik. 1981. Oxygen consumption in humic-colored waters by a photochemical ferrous-ferric catalytic cycle. Environ. Sci. Technol. *15*:1089–1095.

Miller, A. G., G. S. Espie, and D. T. Canvin. 1990. Physiological aspects of CO_2 and HCO_3^- transport by cyanobacteria: A review. Can. J. Bot. *68*:1291–1302.

Miller, A. G., G. S. Espie, and D. T. Canvin. 1991. Active CO_2 transport in cyanobacteria. Can. J. Bot. *69*:925–935.

Miller, A. R., R. L. Lowe, and J. T. Rotenberry. 1987. Succession of diatom communities on sand grains. J. Ecol. *75*:693–709.

Miller, D. E. 1936. A limnological study of *Pelmatohydra* with special reference to their quantitative seasonal distribution. Trans. Am. Microsc. Soc. *55*:123–193.

Miller, J. C. 1987. Evidence for the use of non-detrital dissolved organic matter by microheterotrophs on plant detritus in a woodland stream. Freshwat. Biol. *18*:483–494.

Miller, J. R. and J. B. Ritter. 1996. An examination of the Rosgen classification of natural rivers. Catena *27*:295–299.

Miller, L. G. and R. S. Oremland. 1988. Methane efflux from the pelagic regions of four lakes. Global Biogeochem. Cycles *2*:269–277.

Miller, M. C. 1972. The carbon cycle in the epilimnion of two Michigan lakes. Ph.D. Diss., Michigan State University, 214 pp.

Miller, R. W. 1991. Molybdenum nitrogenase. *In* M. J. Dilworth and A. R. Glenn, eds. Biology and biochemistry of nitrogen fixation. Elsevier, Amsterdam. pp. 9–36.

Miller, W. L. and R. G. Zepp. 1995. Photochemical production of dissolved inorganic carbon from terrestrial organic matter: Significance to the oceanic carbon cycle. Geophys. Res. Lett. *22*:417–420.

Millero, F. J., W. Yao, and J. Aicher. 1995. The speciation of Fe(II) and Fe(III) in natural waters. Mar. Chem. *50*:21–39.

Mills, A. L. and M. Alexander. 1974. Microbial decomposition of species of freshwater planktonic algae. J. Environ. Quality *3*:423–428.

Mills, E. L. and J. L. Forney. 1983. Impact on *Daphnia pulex* of predation by young yellow perch in Oneida Lake, New York. Trans. Am. Fish. Soc. *112*:154–161.

Mills, E. L. and R. T. Oglesby. 1971. Five trace elements and vitamin B_{12} in Cayuga Lake, New York. Proc. Conf. Great Lakes Res., Int. Assoc. Great Lakes Res. *14*:256–267.

Minckley, W. L. 1963. The ecology of a spring stream Doe Run, Meade County, Kentucky. Wildl. Monogr. 11. 124 pp.

Minckley, W. L. and G. A. Cole. 1963. Ecological and morphological studies on gammarid amphipods (*Gammarus* spp.) in spring-fed streams of northern Kentucky. Occas. Pap. Adams Center Ecol. Stud. *10*, 35 pp.

Minder, L. 1922. Über biogene Entkalkung im Zürichsee. Verh. Internat. Verein. Limnol. *1*:20–32.

Minder, L. 1923. Studien über den Sauerstoffgehalt des Zürichsees. Arch. Hydrobiol./Suppl. *3*:197–155.

Minshall, G. W. 1967. Role of allochthonous detritus in the trophic structure of a woodland springbrook community. Ecology *48*:139–149.

Minshall, G. W. 1978. Autotrophy in stream ecosystems. BioScience *28*:767–771.

Minshall, G. W. 1988. Stream ecosystem theory: A global perspective. J. N. Am. Benthol. Soc. *7*:263–288.

Minshall, G. W., R. C. Petersen, K. W. Cummins, T. L. Bott, J. R. Sedell, C. E. Cushing, and R. L. Vannote. 1983. Interbiome comparison of stream ecosystem dynamics. Ecol. Monogr. *53*:1–25.

Minshall, G. W., K. W. Cummins, R. C. Petersen, C. E. Cushing, D. A. Bruns, J. R. Sedell, and R. L. Vannote. 1985. Developments in stream ecosystem theory. Can. J. Fish. Aquat. Sci. *42*:1045–1055.

Miskimmin, B. M., J. W. M. Rudd, and C. A. Kelly. 1992. Influence of dissolved organic carbon, pH, and microbial respiration rates on mercury methylation and demethylation in lake water. Can. J. Fish. Aquat. Sci. *49*:17–22.

Mitamura, O. and Y. Saijo. 1980. Urea supply from decomposition and excretion of zooplankton. J. Oceanogr. Soc. Jpn. *36*:121–125.

Mitamura, O. and Y. Saijo. 1986. Urea metabolism and its significance in the nitrogen cycle in the euphotic layer of Lake Biwa. I. In situ measurement of nitrogen assimilation and urea decomposition. Arch. Hydrobiol. *107*:23–51.

Mitchell, B. D. 1978. Cyclomorphosis in *Daphnia carinata* King (Crustacea: Cladocera) from two adjacent sewage lagoons in South Australia. Aust. J. Mar. Freshwat. Res. *29*:565–576.

Mitchell, D. S. and P. A. Thomas. 1972. Ecology of water weeds in the neotropics: An ecological survey of the aquatic weeds *Eichhornia crassipedes* and *Salvinia* species, and their natural enemies in the neotropics. Tech. Pap. in Hydrology, UNESCO 12, 50 pp.

Mitchell, J. K., S. Mostaghimi, D. S. Freeny, and J. R. McHenry. 1983. Sediment deposition estimation from cesium–137 measurements. Wat. Resour. Bull. *19*:549–555.

Mitchell, M. J., D. H. Landers, and D. F. Brodowski. 1981. Sulfur constituents of sediments and their relationship to lake acidification. Water Air Soil Poll. *16*:351–359.

Mitchell, M. J., M. B. David, and A. J. Uutala. 1985. Sulfur distribution in lake sediment profiles as an index of historical depositional patterns. Hydrobiologia *121*:121–127.

Mitchell, M. J., D. H. Landers, D. F. Brodowski, G. B. Lawrence, and M. B. David. 1984. Organic and inorganic sulfur constituents of the sediments in three New York lakes: Effect of site, sediment depth and season. Water Air Soil Poll. *21*:231–245.

Mitchell, S. F. and C. W. Burns. 1979. Oxygen consumption in the epilimnia and hypolimnia of two eutrophic, warm-monomictic lakes. N. Z. J. Mar. Freshw. Res. *13*:427–441.

Mittelbach, G. G. 1981. Foraging efficiency and body size: A study of optimal diet and habitat use by bluegills. Ecology *62*:1370–1386.

Miura, Y., A. Watanabe, M. Kimura, and S. Kuwatsuka. 1992. Methane emission from paddy field. 2. Main route of methane transfer through rice plant, and temperature and light effects on diurnal variation of methane emission. Environ. Sci. *5*:187–193.

Miyajima, T. 1992a. Biological manganese oxidation in a lake. I. Occurrence and distribution of *Metallogenium* sp. and its kinetic properties. Arch. Hydrobiol. *124*:317–335.

Miyajima, T. 1992b. Recycling of nitrogen and phosphorus from the particulate organic matter associated with the proliferation of bacteria and microflagelates. Jpn. J. Limnol. *53*:133–138.

Moaledj, K. and J. Overbeck. 1980. Studies on uptake kinetics of oligocarbophilic bacteria. Arch. Hydrobiol. *89*:303–312.

Mobley, C. D. 1994. Light and Water: Radiative Transfer in Natural Waters. Academic Press, San Diego. 583 pp.

Moeller, J. R., G. W. Minshall, K. W. Cummins, R. C. Petersen, C. E. Cushing, J. R. Sedell, R. A. Larson, and R. L. Vannote. 1979. Transport of dissolved organic carbon in streams of differing physiographic characteristics. Org. Geochem. *1*:139–150.

Moeller, R. E. 1975. Hydrophyte biomass and community structure in a small, oligotrophic New Hampshire lake. Verh. Internat. Verein. Limnol. *19*:1005–1012.

Moeller, R. E. 1978a. Seasonal changes in biomass, tissue chemistry, and net production of the evergreen hydrophyte, *Lobelia dortmanna*. Can. J. Bot. *56*:1425–1433.

Moeller, R. E. 1978b. Carbon uptake by the submerged hydrophyte *Utricularia purpurea*. Aquat. Bot. *5*:209–216.

Moeller, R. E. 1980. The temperature-determined growing season of a submerged hydrophyte: Tissue chemistry and biomass turnover of *Utricularia purpurea*. Freshwat. Biol. *10*:391–400.

Moeller, R. E. 1994. Contribution of ultraviolet radiation (UV-A, UV-B) to photoinhibition of epilimnetic phytoplankton in lakes of differing UV transparency. Arch. Hydrobiol. Beih. Ergebn. Limnol. *43*:157–170.

Moeller, R. E. and J. P. Roskoski. 1978. Nitrogen-fixation in the littoral benthos of an oligotrophic lake. Hydrobiologia *60*:13–16.

Moeller, R. E. and R. G.Wetzel. 1988. Littoral vs. profundal components of sediment accumulation: Contrasting roles as phosphorus sinks. Verh. Internat. Verein. Limnol. *23*:386–393.

Moeller, R. G., J. M. Burkholder, and R. G. Wetzel. 1988. Significance of sedimentary phosphorus to a submersed freshwater macrophyte (*Najas flexilis*) and its algal epiphytes. Aquat. Bot. *32*:261–281.

Moeller, R. E., R. G. Wetzel, and C. W. Osenberg. 1998. Concordance of phosphorus limitation in lakes: Bacterioplankton, phytoplankton, epiphyte-snail consumers, and rooted macrophytes. *In* E. Jeppesen, Ma. Søndergaard, Mo. Søndergaard, and K. Christoffersen, eds. The Structuring Role of Submerged Macrophytes in Lakes. Springer-Verlag, New York. pp. 318–325.

Moeslund, B., M. G. Kelly, and N. Thyssen. 1981. Storage of carbon and transport of oxygen in river macrophytes: Mass-balance, and the measurement of primary productivity in rivers. Arch. Hydrobiol. *93*:45–51.

Moffet, J. W. and R. G. Zika. 1987. Reaction kinetics of hydrogen peroxide with copper and iron in seawater. Environ. Sci. Technol. *21*:804–810.

Moll, R. and M. Brahce. 1986. Seasonal and spatial distribution of bacteria, chlorophyll and nutrients in nearshore Lake Michigan. J. Great Lakes Res. *12*:52–62.

Molongoski, J. J. 1978. Sedimentation and anaerobic metabolism of particulate organic matter in the sediments of a hypereutrophic lake. Ph.D. Dissertation, Michigan State University. 143 pp.

Molongoski, J. J. and M. J. Klug. 1980a. Anaerobic metabolism of particulate organic matter in the sediments of a hypereutrophic lake. Freshwat. Biol. *10*:507–518.

Molongoski, J. J. and M. J. Klug. 1980b. Quantification and characterization of sedimenting particulate organic matter in a shallow hypereutrophic lake. Freshwat. Biol. *10*:497–506.

Molot, L. A. and P. J. Dillon. 1993. Nitrogen mass balances and denitrification rates in central Ontario lakes. Biogeochemistry *20*:195–212.

Molot, L. A., P. J. Dillon, and B. D. LaZerte. 1989. Factors affecting alkalinity concentrations of streamwater during snowmelt in central Ontario. Can. J. Fish. Aquat. Sci. *46*:1658–1666.

Momot, W. T. 1967a. Population dynamics and productivity of the crayfish., *Orconectes virilis*, in a marl lake. Amer. Midland Nat. *78*:55–81.

Momot, W. T. 1967b. Effects of brook trout predation on a crayfish population. Trans. Amer. Fish. Soc. *96*:202–209.

Momot, W. T. 1978. Annual production and production/biomass ratios of the crayfish, *Orconectes virilis*, in two northern Ontario lakes. Trans. Amer. Fish. Soc. *107*:776–784.

Momot, W. T. 1984. Crayfish production: A reflection of community energetics. J. Crustacean Biol. *4*:35–54.

Momot, W. T. 1995. Redefining the role of crayfish in aquatic ecosystems. Rev. Fisheries Sci. *3*:33–63.

Momot, W. T., ed. 1997. Freshwater Crayfish 11. Int. Assoc. Astacology, Louisiana State Univ., Baton Rouge. 674 pp.

Momot, W. T. and H. Gowing. 1977a. Response of the crayfish *Orconectes virilis* to exploitation. J. Fish. Res. Board Can. *34*:1212–1219.

Momot, W. T. and H. Gowing. 1977b. Production and population dynamics of the crayfish *Orconectes virilis* in three Michigan lakes. J. Fish. Res. Board Can. *34*:2041–2055.

Momot, W. T., H. Gowing, and P. D. Jones. 1978. The dynamics of crayfish and their role in ecosystems. Am. Midland Nat. *99*:10–35.

Monahan, E. C. 1969. Fresh water whitecaps. J. Atm. Sci. *26*:1026–1029.

Monakov, A. B. and Ju. I. Sorokin. 1961. Kolichestvenn'ie dann'ie o pitanii Dafnii. Trudy Inst. Biol. Vodokhranilishch *4*:251–261.

Monson, B. A. and P. L. Brezonik. 1998. Seasonal patterns of mercury species in water and plankton from softwater lakes in northeastern Minnesota. Biogeochemistry *40*:147–162.

Mooij-Vogelaar, J. W., J. C. Jager, and W. J. van der Steen. 1973. Effects of density levels, and changes in density levels on reproduction, feeding and growth in the pond snail *Lymnaea stagnalis* (L.). Proc. Nederl. Akad. Wetensc. Amsterdam (Ser. C) *76*:245–256.

Moore, A. W. 1969. Azolla: Biology and agronomic significance. Bot. Rev. *35*:17–34.

Moore, J. W. 1977. Importance of algae in the diet of subarctic populations of *Gammarus lacustris* and *Pontoporeia affnis*. Can. J. Zool. *55*:637–641.

Moore, J. W. 1978. Importance of algae in the diet of the oligochaetes *Lumbriculus variegatus* (Müller) and *Rhyacodrilus sodalis* (Eisen). Oecologia *35*:357–363.

Moore, J. W. 1981. Inter-species variability in the consumption of algae by oligochaetes. Hydrobiologia *83*:241–244.

Moore, L. F. and J. A. Traquair. 1976. Silicon, a required nutrient for *Cladophora glomerata* (L) Kütz. (Chlorophyta). Planta *128*:179–182.

Moore, M. V. 1988. Differential use of food resources by the instars of *Chaoborus punctipennis*. Freshwat. Biol. *19*:249–268.

Moore, M. V. and J. J. Gilbert. 1987. Age-specific *Chaoborus* predation on rotifer prey. Freshwat. Biol. *17*:233–236.

Moore, M. V., N. D. Yan, and T. Pawson. 1994. Omnivory of the larval phantom midge (*Chaoborus* spp.) and its potential significance for freshwater planktonic food webs. Can. J. Zool. *72*:2055–2065.

Moore, P. D. and D. J. Bellamy. 1974. Peatlands. Elek Science, London. 221 pp.

Moore, T. R. 1987. Patterns of dissolved organic matter in sub-Arctic peatlands. Earth Surf. Processes Landforms *12*:387–397.

Moran, M. A. and J. S. Covert. 2001. Photochemically-mediated linkages between dissolved organic matter and bacterioplankton. *In* S. Findlay and R. Sinsabaugh, eds. Integrating Approaches to Microbial-Dissolved Organic Matter Trophic Linkages. Academic Press, San Diego. (In press.)

Moran, M. A. and R. E. Hodson. 1989a. Bacterial secondary production on vascular plant detritus: Relationships to detritus composition and degradation rate. Appl. Environ. Microbiol. *55*:2178–2189.

Moran, M. A. and R. E. Hodson. 1989b. Formation and bacterial utilization of dissolved organic carbon derived from detrital lignocellulose. Limnol. Oceanogr. *34*:1034–1047.

Moran, M. A. and R. G. Zepp. 1997. Role of photoreactions in the formation of biologically labile compounds from dissolved organic matter. Limnol. Oceanogr. *42*:1307–1316.

Morel, A. 1978. Available, usable, and stored radiant energy in relation to marine photosynthesis. Deep-Sea Res. *25*:673–688.

Morel, A. and R. C. Smith. 1974. Relation between total quanta and total energy for aquatic photosynthesis. Limnol. Oceanogr. *19*:591–600.

Morel, F. M. M. and J. G. Hering. 1993. Principles and Applications of Aquatic Chemistry. John Wiley & Sons, New York. 374 pp.

Morgan, A. E. 1971. Dams and other disasters. Porter Sargent Publ., Boston. 422 pp.

Morgan, M. D. 1981. Abundance, life history, and growth of introduced populations of the opossum shrimp (*Mysis relicta*) in subalpine California lakes. Can. J. Fish. Aquat. Sci. *38*:989–993.

Morgan, M. D. 1991. Sources of stream acidity in the New Jersey Pinelands. Verh. Internat. Verein. Limnol. *24*:1707–1710.

Morgan, N. C., T. Backiel, G. Bretschko, A. Duncan, A. Hillbricht-Ilkowska, Z. Kajak, J. F. Kitchell, P. Larsson, C. Lévêque, A. Nauwerck, F. Schiemer, and J. E. Thorpe. 1980. Secondary production. *In* E. D. LeCren and R. H. Lowe-McConnell, eds. The Functioning of Freshwater Ecosystems. Cambridge Univ. Press, Cambridge. pp. 247–340.

Moriarty, D. J. W. 1986. Measurement of microbial growth rates in aquatic systems using rates of nucleic acid synthesis. Adv. Microb. Ecol. *9*:245–292.

Moriarty, D. J. W. and R. T. Bell. 1993. Bacterial growth and starvation in aquatic environments. *In* S. Kjelleberg, ed. Starvation in Bacteria. Plenum Press, New York. pp. 25–53.

Morikawa, M, Y. Fukuo, and F. Hirao. 1959. Limnological researches in Lake Biwa near the mouth of the River Ado. I. Density distribution of the lake water off the mouth of the river. (In Japanese.) Jpn. J. Limnol. *20*:10–20.

Morisawa, M. 1968. Streams: Their dynamics and morphology. McGraw-Hill Book Co., New York. 175 pp.

Morita, R. Y. 1982. Starvation survival of heterotrophs. Adv. Microb. Ecol. *6*:171–198.

Morowitz, H. J. 1980. The dimensionality of niche space. J. Theor. Biol. *86*:259–263.

Morris, D. P. and W. M. Lewis, Jr. 1992. Nutrient limitation of bacterioplankton growth in Lake Dillon, Colorado. Limnol. Oceanogr. *37*:1179–1192.

Morris, D. P., H. Zagarese, C. E. Williamson, E. G. Balseiro, B. R. Hargreaves, B. Modenutti, R. Moeller, and C. Queimalinos. 1995. The attenuation of solar UV radiation in lakes and the role of dissolved organic carbon. Limnol. Oceanogr. *40*:1381–1391.

Morris, I. 1967. An Introduction to the Algae. Hutchinson University Library, London. 189 pp.

Morris, J. C. and W. Stumm. 1967. Redox equilibria and measurements of potentials in the aquatic environment. Advances in Chemistry Series *67*:270–285.

Morris, J. T. 1991. Effects of nitrogen loading on wetland ecosystems with particular reference to atmospheric deposition. Annu. Rev. Ecol. Syst. *22*:257–279.

Morris, J. T. and K. Lajtha. 1986. Decomposition and nutrient dynamics of litter from four species of freshwater emergent macrophytes. Hydrobiologia *131*:215–223.

Morrissey, L. A., D. B. Zobel, and G. P. Livingston. 1993. Significance of stomatal control on methane release from *Carex*-dominated wetlands. Chemosphere *26*:339–355.

Mortimer, C. H. 1941. The exchange of dissolved substances between mud and water in lakes (Parts I and II). J. Ecol. *29*:280–329.

Mortimer, C. H. 1942. The exchange of dissolved substances between mud and water in lakes (Parts III, IV, summary, and references). J. Ecol. 30:147–201.

Mortimer, C. H. 1951. The use of models in the study of water movement in stratified lakes. Verh. Internat. Verein. Limnol. *11*:254–260.

Mortimer, C. H. 1952. Water movements in lakes during summer stratification; evidence from the distribution of temperature in Windermere. Proc. Roy. Soc. London (Ser. B) *236*:355–404.

Mortimer, C. H. 1953. The resonant response of stratified lakes to wind. Schweiz. Z. Hydrol. *15*:94–151.

Mortimer, C. H. 1954. Models of the flow-pattern in lakes. Weather *9*:177–184.

Mortimer, C. H. 1955. Some effects of the earth's rotation on water movements in stratified lakes. Verh. Internat. Verein. Limnol. *12*:66–77.

Mortimer, C. H. 1961. Motion in thermoclines. Verh. Internat. Verein. Limnol. *14*:79–83.

Mortimer, C. H. 1963. Frontiers in physical limnology with particular reference to long waves in rotating basins. Publ. Great Lakes Res. Div., Univ. Mich. *10*:9–42.

Mortimer, C. H. 1965. Spectra of long surface waves and tides in Lake Michigan and at Green Bay, Wisconsin. Publ. Great Lakes Res. Div., Univ. Mich. *13*:304–325.

Mortimer, C. H. 1971. Chemical exchanges between sediments and water in the Great Lakes—speculations on probable regulatory mechanisms. Limnol. Oceanogr. *16*:387–404.

Mortimer, C. H. 1971. Large-Scale Oscillatory Motions and Seasonal Temperature Changes in Lake Michigan and Lake Ontario. Spec. Rept. No. 12, Center for Great Lakes Studies, University of Wisconsin-Milwaukee. Part I, Text, 111 pp. and Part II, Illustrations, 106 pp.

Mortimer, C. H. 1974. Lake hydrodynamics. Mitt. Internat. Verein. Limnol. *20*:124–197.

Mortimer, C. H. 1975. Substantive corrections to SIL Communications (IVL Mitteilungen numbers 6 and 20.) Verh. Internat. Verein. Limnol. *19*:60–72.

Mortimer, C. H. 1981. The oxygen content of air-saturated fresh waters over ranges of temperature and atmospheric pressure of limnological interest. Mitt. Internat. Verein. Limnol. *22*: 23 pp.

Mortimer, C. H. and C. F. Hickling. 1954. Fertilizers in fishponds. Fish. Publ. U.K. Colonial Office, London *5*, 155 pp.

Mortimer, C. H. and F. J. H. Mackereth. 1958. Convection and its consequences in ice-covered lakes. Verh. Internat. Verein. Limnol. *13*:923–932.

Mortimer, C. H., D. C. McNaught, and K. M. Stewart. 1968. Short internal waves near their high-frequency limit in central Lake Michigan. Proc. Conf. Great Lakes Res., Int. Assoc. Great Lakes Res. *11*:454–469.

Morton, S. D. and G. F. Lee. 1968. Calcium carbonate equilibria in lakes. J. Chem. Educ. *45*:511–513.

Mortonson, J. A. and A. S. Brooks. 1980. Occurrence of a deep nitrite maximum in Lake Michigan. Can. J. Fish. Aquat. Sci. *37*:1025–1027.

Moshiri, G. A., K. W. Cummins, and R. R. Costa. 1969. Respiratory energy expenditure by the predaceous zooplankter *Leptodora kindtii* (Focke) (Crustacea: Cladocera). Limnol. Oceanogr. *14*:475–484.

Moskalenko, B. K. and K. K. Votinsev. 1970. Biological productivity and balance of organic substance and energy in Lake Baikal. *In* Z. Kajak and A. Hillbricht-Ilkowska, eds. Productivity Problems of Freshwaters. PWN Polish Scientific Publishers, Warsaw. pp. 207–226.

Moss, B. 1968. Studies on the degradation of chlorophyll *a* and carotenoids in freshwaters. New Phytol. *67*:49–59.

Moss, B. 1969a. Vertical heterogeneity in the water column of Abbot's Pond. II. The influence of physical and chemical conditions on the spatial and temporal distribution of the phytoplankton and of a community of epipelic algae. J. Ecol. *57*:397–414.

Moss, B. 1969b. Algae of two Somersetshire pools: Standing crops of phytoplankton and epipelic algae as measured by cell numbers and chlorophyll *a*. J. Phycol. *5*:158–168.

Moss, B. 1972a. Studies on Gull Lake, Michigan. I. Seasonal and depth distribution of phytoplankton. Freshwat. Biol. *2*:289–307.

Moss, B. 1972b. Studies on Gull Lake, Michigan. II. Eutrophication—evidence and prognosis. Freshwat. Biol. *2*:309–320.

Moss, B. 1972c. The influence of environmental factors on the distribution of freshwater algae: An experimental study. I. Introduction and the influence of calcium concentration. J. Ecol. *60*:917–932.

Moss, B. 1973a. The influence of environmental factors on the distribution of freshwater algae: An experimental study. II. The role of pH and the carbon dioxide-bicarbonate system. J. Ecol. *61*:157–177.

Moss, B. 1973b. The influence of environmental factors on the distribution of freshwater algae: An experimental study. III. Effects of temperature, vitamin requirements and inorganic nitrogen compounds on growth. J. Ecol. *61*:179–192.

Moss, B. 1973c. The influence of environmental factors on the distribution of freshwater algae: An experimental study. IV. Growth of test species in natural lake waters, and conclusion. J. Ecol. *61*:193–211.

Moss, B. 1973d. Diversity in fresh-water phytoplankton. Am. Midland Nat. *90*:341–355.

Moss, B. 1977. Adaptations of epipelic and epipsammic freshwater algae. Oecologia *28*:103–108.

Moss, B. 1998. The E numbers of eutrophication—errors, ecosystem effects, economics, eventualities, and environment and education. Wat. Sci. Tech. *37*:75–84.

Moss, B. 1999. Ecological challenges for lake management. Hydrobiologia *395/396*:3–11.

Moss, B. and A. G. Abdel Karim. 1969. Phytoplankton associations in two pools and their relationships with associated benthic flora. Hydrobiologia *33*:587–600.

Moss, B. and F. E. Round. 1967. Observations on standing crops of epipelic and epipsammic algal communities in Shear Water, Wilts. Brit. Phycol. Bull. *3*:241–248.

Moss, B. and H. Balls. 1989. Phytoplankton distribution in a floodplain lake and river system. II. Seasonal changes in the phytoplankton communities and their control by hydrology and nutrient availability. J. Plankton Res. *11*:839–867.

Moss, B. and J. Moss. 1969. Aspects of the limnology of an endorheic African lake (L. Chilwa, Malawi). Ecology *50*:109–118.

Moss, B., R. G. Wetzel, and G. H. Lauff. 1980. Studies on Gull Lake, Michigan. III. Annual productivity and phytoplankton changes between 1979 and 1974. Freshwat. Biol. *10*:113–121.

Mothes, G. 1981. Sedimentation und Stoffbilanzen in Seen des Stechlinseegebiets. Limnologica *13*:147–194.

Mouget, J.-L., A. Dakhama, M. C. Lavoie, and J. de la Noue. 1995. Algal growth enhancement by bacteria: Is consumption of oxygen involved? FEMS Microbiol. Ecol. *18*:35–44.

Mourelatos, S., C. Rougier, and R. Pourriot. 1989. Diel patterns of zooplankton grazing in a shallow lake. J. Plankton Res. *11*:1021–1035.

Moyle, P. M. and J. J. Cech, Jr. 1988. Fishes—An Introduction to Ichthyology. 2nd editor. Prentice-Hall, Englewood Cliffs, NJ.

Moyo, S. M. 1991. Cyanobacterial nitrogen fixation in Lake Kariba, Zimbabwe. Verh. Internat. Verein. Limnol. *24*:1123–1127.

Mueller, W. P. 1964. The distribution of cladoceran remains in surficial sediments from three northern Indiana lakes. Invest. Indiana Lakes Streams 6:1–63.

Muhar, S., S. Schmutz, and M. Jungwirth. 1995. River restoration concepts—goals and perspectives. Hydrobiologia *303*:183–194.

Mühlhauser, H. A. 1990. Organic matter mineralization in the littoral and open lake sediments of the shallow Neusiedlersee, Österreich. Acta Hydrochim. Hydrobiol. *18*:421–431.

Mulholland, P. J. 1981. Organic carbon flow in a swamp-stream ecosystem. Ecol. Monogr. *51*:307–322.

Mulholland, P. J. 1992. Regulation of nutrient concentrations in a temperate forest stream: Roles of upland, riparian, and instream processes. Limnol. Oceanogr. *37*:1512–1526.

Mulholland, P. J. 1997. Dissolved organic matter concentrations and flux in streams. J. N. Am. Benthol. Soc. *16*:131–140.

Mulholland, P. J., A. D. Steinman, and J. W. Elwood. 1990. Measurement of phosphorus uptake length in streams: Comparison of radiotracer and stable PO_4 releases. Can. J. Fish. Aquat. Sci. *47*:2351–2357.

Mulholland, P. J., A. D. Steinman, A. V. Palumbo, J. W. Elwood, and D. B. Kirschtel. 1991. Role of nutrient cycling and herbivory in regulating periphyton communities in laboratory streams. Ecology *72*:966–982.

Mulholland, P. J., C. N. Dahm, M. B. David, D. M. Di Toro, T. R. Fisher, H. F. Hemond, I. Kögel-Knabner, M. H. Meybeck, J. L. Meyer, and J. R. Sedell. 1990. What are the temporal and spatial variations of organic acids at the ecosystem level. *In* E. M. Perdue and E. T. Gjessing, eds. Organic Acids in Aquatic Ecosystems. John Wiley & Sons, Chichester. pp. 315–329.

Mulholland, P. J., J. D. Newbold, J. W. Elwood, and C. L. Hom. 1983. The effect of grazing intensity on phosphorus spiralling in autotrophic streams. Oecologia *58*:358–366.

Mulholland, P. J., J. D. Newbold, J. W. Elwood, and L. A. Ferren. 1985a. Phosphorus spiralling in a woodland stream: Seasonal variations. Ecology *66*:1012–1023.

Mulholland, P. J., J. W. Elwood, J. D. Newbold, and L. A. Ferren. 1985b. Effect of a leaf-shredding invertebrate on organic matter dynamics and phosphorus spiralling in heterotrophic laboratory streams. Oecologia *66*:199–206.

Mulholland, P. J., J. W. Elwood, J. D. Newbold, J. R. Webster, L. A. Ferren, and R. E. Perkins. 1984. Phosphorus uptake by decomposing leaf detritus: Effect of microbial biomass and activity. Verh. Internat. Verein. Limnol. *22*:1899–1905.

Mulholland, P. J., R. A. Minear, and J. W. Elwood. 1988. Production of soluble, high molecular weight phosphorus and its subsequent uptake by stream detritus. Verh. Internat. Verein. Limnol. *23*:1190–1197.

Müller, H. 1967. Eine neue qualitative Bestandsaufnahme des Phytoplanktons des Bodensee-Obersees mit besonderer Berücksichtigung der tychoplanktischen Diatomeen. Arch. Hydrobiol./Suppl. *33*:206–236.

Müller, H. 1989. The relative importance of different ciliate taxa in the pelagic food web of Lake Constance. Microb. Ecol. *18*:261–273.

Müller, H. and W. Geller. 1993. Maximum growth rates of aquatic ciliated protozoa: The dependence on body size and temperature reconsidered. Arch. Hydrobiol. *126*:315–327.

Müller, H., A. Schöne, R. M. Pinto-Coelho, A. Schweizer, and T. Weisse. 1991. Seasonal succession of ciliates in Lake Constance. Microb. Ecol. *21*:119–138.

Müller, J. and G. Weise. 1987. Oxygen budget of a river rich in submerged macrophytes (River Zschopau in the south of the GDR). Int. Rev. ges. Hydrobiol. *72*:653–667.

Müller, K. 1951. Fisch und Fischregionen der Fulda. Berl. Limnol. Flußstat. Freudenthal *2*:18–23.

Müller, U. 1996. Production rates of epiphytic algae in a eutrophic lake. Hydrobiologia *330*:37–45.

Müller-Haeckel, A. 1965. Tagesperiodik des Siliziumgehaltes in einem Fliessegewässer. Oikos *16*:232–233.

Müller-Navarra, D. 1995. Evidence that a highly unsaturated fatty acid limits *Daphnia* growth in nature. Arch. Hydrobiol. *132*:297–307.

Mulligan, H. F., A. Baranowski, and R. Johnson. 1976. Nitrogen and phosphorus fertilization of aquatic vascular plants and algae in replicated ponds. I. Initial response to fertilization. Hydrobiologia *48*:109–116.

Mullin, M. M., P. R. Sloan, and R. W. Eppley. 1966. Relationship between carbon content, cell volume, and area in phytoplankton. Limnol. Oceanogr. *11*:307–311.

Munch, C. S. 1980. Fossil diatoms and scales of Chrysophyceae in the recent history of Hall lake, Washington. Freshwat. Biol. *10*:61–66.

Munn, N. and E. Prepas. 1986. Seasonal dynamics of phosphorus partitioning and export in two streams in Alberta, Canada. Can. J. Fish. Aquat. Sci. *43*:2464–2471.

Münster, U. 1984. Distribution, dynamic and structure of free dissolved carbohydrates in the Plußsee, a north German eutrophic lake. Verh. Internat. Verein. Limnol. *22*:929–935.

Münster, U. 1985. Investigations about structure, distribution and dynamics of different organic substrates in the DOM of Lake Plußsee. Arch. Hydrobiol./Suppl. *70*:429–480.

Münster, U. 1991. Extracellular enzyme activity in eutrophic and polyhumic lakes. *In* R. J. Chróst, ed. Microbial Enzymes in Aquatic Environments. Springer-Verlag, New York. pp. 96–122.

Münster, U. 1993. Concentrations and fluxes of organic carbon substrates in the aquatic environment. Antonie van Leeuwenhoek *63*:243–274.

Münster, U. 1994. Studies on phosphatase activities in humic lakes. Environ. Int. *20*:49–59.

Münster, U. and H. De Haan. 1998. The role of microbial extracellular enzymes in the transformation of dissolved organic matter in humic waters. *In* D. O. Hessen and L. J. Tranvik, eds. Aquatic Humic Substances: Ecology and Biogeochemistry. Springer-Verlag, Berlin. pp. 199–257.

Münster, U. and R. J. Chróst. 1990. Origin, composition and microbial utilization of dissolved organic matter. *In* J. Overbeck and R. J. Chróst, eds. Aquatic Microbial Ecology: Biochemical and Molecular Approaches. Springer-Verlag, New York. pp. 8–46.

Münster, U., E. Heikkinen, M. Likolammi, M. Järvinen, K. Salonen, and H. De Haan. 1999. Utilisation of polymeric and monomeric aromatic and amino acid carbon in a humic boreal forest lake. Arch. Hydrobiol. Spec. Issues Advanc. Limnol. *54*:105–134.

Murakami, T., C. Isaji, N. Kuroda, K. Yoshida, and H. Haga. 1992. Potamoplanktonic diatoms in the Nagara River: Flora, population dynamics and influences on water quality. Jpn. J. Limnol. *53*:1–12.

Murdoch, P. S., J. S. Baron, and T. L. Miller. 2000. Potential effects of climate change on surface-water quality in North America. J. Amer. Water Resour. Assoc. *36*:347–366.

Murphy, D. H. and B. H. Wilkinson. 1980. Carbonate deposition and facies distribution in a central Michigan marl lake. Sedimentology *27*:123–135.

Murphy, P. M. and M. A. Learner. 1982. The life history and production of the leech *Helobdella stagnalis* (Hirudinea: Glossiphonidae) in the River Ely, South Wales. Freshwat. Biol. *12*:321–329.

Murphy, T. P., D. R. S. Lean, and C. Nalewajko. 1976. Blue-green algae: Their excretion of iron-selective chelators enables them to dominate other algae. Science *192*:900–902.

Murphy, T. P., K. J. Hall, and I. Yesaki. 1983. Coprecipitation of phosphate with calcite in a naturally eutrophic lake. Limnol. Oceanogr. *28*:58–69.

Murray, K. J., J. D. Tenhunen, and R. S. Nowak. 1993. Photoinhibition as a control on photosynthesis and production of *Sphagnum* mosses. Oecologia *96*:200–207.

Murray, R. E. and R. E. Hodson. 1985. Annual cycle of bacterial secondary production in five aquatic habitats of the Okefenokee swamp ecosystem. Appl. Environ. Microbiol. *49*:650–655.

Murray, R. E. and R. E. Hodson. 1986. Influence of macrophyte decomposition on growth rate and community structure of

Okefenokee Swamp bacterioplankton. Appl. Environ. Microbiol. *51*:293–301.

Murtaugh, P. A. 1981a. Selective predation by *Neomysis mercedis* in Lake Washington. Limnol. Oceanogr. *26*:445–453.

Murtaugh, P. A. 1981b. Size-selective predation on *Daphnia* by *Neomysis mercedis*. Ecology *62*:894–900.

Murtaugh, P. A. 1981c. Inferring properties of mysid predation from injuries to *Daphnia*. Limnol. Oceanogr. *26*:811–821.

Murtaugh, P. A. 1983. Mysid life history and seasonal variation in predation pressure on zooplankton. Can. J. Fish. Aquat. Sci. *40*:1968–1974.

Muscatine, L. and H. M. Lenhoff. 1963. Symbiosis: On the role of algae symbiotic with hydra. Science *142*:956–958.

Musgrave, A. and J. Walters. 1973. Ethylene-stimulated growth and auxin transport in *Ranunculus sceleratus* petioles. New Phytol. *72*:783–789.

Musgrave, A., M. B. Jackson, and E. Ling. 1972. Callitriche stem elongation is controlled by ethylene and gibberellin. Nature *238*:93–96.

Muztar, A. J., S. J. Slinger, and J. H. Burton. 1978a. Chemical composition of aquatic macrophytes. I. Investigation of organic constituents and nutritional potential. Can. J. Plant Sci. *58*:829–841.

Muztar, A. J., S. J. Slinger, and J. H. Burton. 1978b. The chemical composition of aquatic macrophytes. II. Amino acid composition of the protein and non-protein fractions. Can. J. Plant Sci. *58*:843–849.

Myer, G. E. 1969. A field study of Langmuir circulations. Proc. Conf. Great Lakes Res., Int. Assoc. Great Lakes Res. *12*:652–663.

Myers, N. 1988. Threatened biotas: "Hot spots" in tropical forests. Environmentalist *8*:187–208.

Nagasawa, M. 1969. On the dichotomous microstratification of pH in a lake. Jpn. J. Limnol. *20*:75–79.

Nagata, T. 1984a. Bacterioplankton in Lake Biwa: Annual fluctuations of bacterial numbers and their possible relationship with environmental variables. Jpn. J. Limnol. *45*:126–133.

Nagata, T. 1984b. Production rate of planktonic bacteria in the north basin of Lake Biwa, Japan. Appl. Environ. Microbiol. *53*:2872–2882.

Nagata, T. 1988. The microflagellate-picoplankton food linkage in the water column of Lake Biwa. Limnol. Oceanogr. *33*:504–517.

Nagata, T. and D. L. Kirchman. 1997. Roles of submicron particles and colloids in microbial food webs and biogeochemical cycles within marine environments. Adv. Microb. Ecol. *15*:81–103.

Nägeli, A. and F. Schanz. 1990. Planktoneustonic algae in the surface films of Lake Zürich: Occurrence and dependence on phytoplanktonic succession. Aquat. Sci. *52*:269–286.

Nägeli, A. and F. Schanz. 1991. The influence of extracellular algal products on the surface tension of water. Int. Rev. ges. Hydrobiol. *76*:89–103.

Naguib, M. 1982. Methanogenese im Sediment der Binnengewässer. 1. Methanol als dominanter Methan-"Precursor" im sediment eines eutrophen Sees. Arch. Hydrobiol. *95*:317–329.

Naiman, R. J. 1983. The annual pattern and spatial distribution of aquatic oxygen metabolism in boreal forest watersheds. Ecol. Monogr. *53*:73–94.

Naiman, R. J. and J. M. Melillo. 1984. Nitrogen budget of a subarctic stream altered by beaver (*Castor canadensis*). Oecologia (Berlin) *62*:150–155.

Naiman, R. J., D. M. McDowell, and B. S. Farr. 1984. The influence of beaver (*Castor canadensis*) on the production dynamics of aquatic insects. Verh. Internat. Verein. Limnol. *22*:1801–1810.

Nakamura, F. and F. J. Swanson. 1994. Distribution of coarse woody debris in a mountain stream, western Cascade Range, Oregon. Can. J. For. Res. *24*:2395–2403.

Nakamura, T., Y. Nojin, M. Utsumi, T. Nozawa, and A. Otsuki. 1999. Methane emission to the atmosphere and cycling in a shallow eutrophic lake. Arch. Hydrobiol. *144*:383–407.

Nakano, S. 1994a. Rates and ratios of nitrogen and phosphorus released by a bacterivorous flagellate. Jpn. J. Limnol. *55*:115–123.

Nakano, S. 1994b. Estimation of phosphorus release rate by bacterivorous flagellates in Lake Biwa. Jpn. J. Limnol. *55*:201–211.

Nakano, S. 1994c. Carbon:nitrogen:phosphorus ratios and nutrient regeneration of a heterotrophic flagellate fed on bacteria with different elemental ratios. Arch. Hydrobiol. *129*:257–271.

Nakatsubo, T., M. Katiyu, N. Nakagoshi, and T. Horikoshi. 1994. Distribution of vesicular-arbuscular mycorrhizae in plants growing in a river floodplain. Bull. Jpn. Soc. Microb. Ecol. *9*:109–117.

Nakazawa, S. 1973. Artificial induction of lake balls. Naturwissenschaften *60*:481.

Nalepa, T. F. and D. W. Schloesser. 1993. Zebra Mussels: Biology, Impacts and Control. Lewsih Publishers, Boca Raton.

Nalepa, T. F., W. S. Gardner, and J. M. Malczyk. 1983. Phosphorus release by three kinds of benthic invertebrates: Effect of substrate and water medium. Can. J. Fish. Aquat. Sci. *40*:810–813.

Nalewajko, C. 1977. Extracellular release in freshwater algae and bacteria: Extracellular products of algae as a source of carbon for heterotrophs. *In* J. Cairns, Jr., ed. Aquatic Microbial Communities. Garland Publ., Inc., New York. pp. 589–624.

Nalewajko, C. and H. Godmaire. 1993. Extracellular products of *Myriophyllum spicatum* L. as a function of growth phase and diel cycle. Arch. Hydrobiol. *127*:345–356.

Nalewajko, C, K. Lee, and P. Fay. 1980. Significance of algal extracellular products to bacteria in lakes and in cultures. Microb. Ecol. *6*:199–207.

Nalewajko, C. and D. R. S. Lean. 1972. Growth and excretion in planktonic algae and bacteria. J. Phycol. *8*:361–366.

Nalewajko, C., K. Lee, and T. R. Jack. 1995. Effects of vanadium on freshwater phytoplankton photosynthesis. Water Air Soil Poll. *81*:93–105.

Nanazato, T. and M. Yasuno. 1985. Population dynamics and production of cladoceran zooplankton in the highly eutrophic Lake Kasumigaura. Hydrobiologia *124*:13–22.

Nancollas, G. H. and M. M. Reddy. 1971. The crystallization of calcium carbonate. II. Calcite growth mechanism. J. Colloid Interface Sci. *37*:824–830.

Napolitano, G. E. 1999. Fatty acids as trophic and chemical markers in freshwater ecosystems. *In* M. T. Arts and B. C. Wainman, eds. Lipids in Freshwater Ecosystems. Springer-Verlag, New York. pp. 21–44.

Napolitano, G. E. and D. S. Cicerone. 1999. Lipids in water-surface microlayers and foams. *In* M. T. Arts and B. C. Wainman, eds. Lipids in Freshwater Ecosystems. Springer-Verlag, New York. pp. 235–262.

Nardi, S., G. Concheri, D. Pizzeghello, A. Sturaro, R. Rella, and G. Parvoli. 2000. Soil organic matter mobilization by root exudates. Chemosphere *41*:653–658.

van der Nat, F.-J. W. A. and J. J. Middelburg. 1998. Effects of two common macrophytes on methane dynamics in freshwater sediments. Biogeochemistry *43*:79–104.

van der Nat, F.-J., J. J. Middelburg, D. van Meteren, and A. Wielemakers. 1998. Diel methane emission patterns from *Scirpus lacustris* and *Phragmites australis*. Biogeochemistry *41*:1–22.

National Academy of Sciences. 1969. Eutrophication: Causes, Consequences, Correctives. National Academy of Sciences, Washington, DC. 661 pp.

National Academy of Sciences. 1996. Freshwater Ecosystems: Revitalizing Educational Programs in Limnology. National Academy Press, Washington, DC. 364 pp.

Naumann, E. 1917. Beiträge zur Kenntnis des Teichnannoplanktons. II. Über das Neuston des Süßwassers. Biologisches Zentralblatt *37*:98–106.

Naumann, E. 1919. Några synpunkter angående limnoplanktons ökologi med särskild hänsyn till fytoplankton. Svensk Bot. Tidskr. *13*:129–163. (English translat. by the Freshwater Biological Association, No. 49.)

Naumann, E. 1929. The scope and chief problems of regional limnology. Int. Rev. ges. Hydrobiol. *22*:423–444.

Naumann, E. 1931. Limnologische Terminologie. Handbuch der biologischen Arbeitsmethoden, Abt. IX, Teil 8. Urban & Schwarzenberg, Berlin. 776 pp.

Naumann, E. 1932. Grundzüge der regionalen Limnologie. Die Binnengewässer *11*, 176 pp.

Nauwerck, A. 1959. Zur Bestimmung der Filterierrate limnischer Planktontiere. Arch. Hydrobiol. Suppl. *25*:83–101.

Nauwerck, A. 1963. Die Beziehungen zwischen Zooplankton und Phytoplankton im See Erken. Symbol. Bot. Upsalien. *17*(5), 163 pp.

Neal, J. T., ed. 1975. Playas and dried lakes. Occurrence and development. Dowden, Hutchinson & Ross, Stroudsburg, PA. 411 pp.

Neale, P. J., S. I. Heaney, and G. H. M. Jaworski. 1991a. Responses to high irradiance contribute to the decline of the spring diatom maximum. Limnol. Oceanogr. *36*:761–768.

Neale, P. J., J. F. Talling, S. I. Heaney, C. S. Reynolds, and J. W. G. Lund. 1991b. Long time series from the English Lake district: Irradiance-dependent phytoplankton dynamics during the spring maximum. Limnol. Oceanogr. *36*:751–760.

Nealson, K. H. and D. Saffarini. 1994. Iron and manganese in anaerobic respiration: Environmental significance, physiology, and regulation. Annu. Rev. Microbiol. *48*:311–343.

Nebaeus, M. 1984. Algal water-blooms under ice-cover. Verh. Internat. Verein. Limnol. *22*:719–724.

Neckles, H. A., H. R. Murkin, and J. A. Cooper. 1990. Influences of seasonal flooding on macroinvertebrate abundance in wetland habitats. Freshwat. Biol. *23*:311–322.

Nedoma, J., J. Vrba, J. Hejzlar, K. Šimek, and V. Straškrabová. 1994. *N*-acetylglucosamine dynamics in freshwater environments: Concentration of amino sugars, extracellular enzyme activities, and microbial uptake. Limnol. Oceanogr. *39*:1088–1100.

Nedwell, D. B. 1984. The input and mineralization of organic carbon in anaerobic aquatic sediments. Adv. Microb. Ecol. *7*:93–131.

Neely, R. K. 1994. Evidence for positive interactions between epiphytic algae and heterotrophic decomposers during the decomposition of *Typha latifolia*. Arch. Hydrobiol. *129*:443–457.

Neely, R. K. and C. B. Davis. 1985. Nitrogen and phosphorus fertilization of *Sparganium eurycarpum* Engelm. and *Typha glauca* Godr. stands. I. Emergent plant production. Aquat. Bot. *22*:347–361.

Neely, R. K. and R. G. Wetzel. 1995. Simultaneous use of ^{14}C and ^{3}H to determine autotrophic production and bacterial protein production in periphyton. Microb. Ecol. *30*:227–237.

Neely, R. K. and R. G. Wetzel. 1997. Autumnal production by bacteria and autotrophs attached to *Typha latifolia* L. detritus. J. Freshwat. Ecol. *12*:253–267.

Neess, J. C. 1946. Developmental status of pond fertilization in Central Europe. Trans. Am. Fish. Soc. *76*:335–358.

Neiderhauser, P. and F. Schanz. 1993. Effects of nutrient (N, P, C) enrichment upon the littoral diatom community of an oligotrophic high-mountain lake. Hydrobiologia *269/270*:453–462.

Neiff, J. J. 1990. Aspects of primary productivity in the lower Parana and Paraguay riverine system. Acta Limnol. Brasil. *3*:77–113.

Neill, W. E. 1975. Experimental studies of microcrustacean competition, community composition and efficiency of resource utilization. Ecology *56*:809–826.

Neill, W. E. 1984. Regulation of rotifer densities by crustacean zooplankton in an oligotrophic montane lake in British Columbia. Oecologia *61*:175–181.

Neill, W. E. 1990. Induced vertical migration in copepods as a defence against invertebrate predation. Nature *345*:524–526.

Neilson, A. H. and R. A. Lewin. 1974. The uptake and utilization of organic carbon by algae: An essay in comparative biochemistry. Phycologia *13*:227–264.

Nekrasova, V. K., L. M. Gerasimenko, and A. K. Ramanova. 1984. Precipitation of calcium carbonate in the presence of cyanobacteria. Mikrobiologiya *53*:833–836.

Nelson, D. J. and D. C. Scott. 1962. Role of detritus in the productivity of a rock-outcrop community in a Piedmont stream. Limnol. Oceanogr. *7*:396–413.

Nelson, D. R. 1991. Tardigrada. *In* J. H. Thorp and A. P. Covich, eds. Ecology and Classification of North American Freshwater Invertebrates. Academic Press, San Diego. pp. 501–521.

Nelson, P. N., J. A. Baldock, and J. M. Oades. 1993. Concentration and composition of dissolved organic carbon in streams in relation to catchment soil properties. Biogeochemistry *19*:27–50.

Nero, R. W. and W. G. Sprules. 1986. Zooplankton species abundance and biomass in relation to occurrence of *Mysis relicta* (Malacostraca: Mysidacea). Can. J. Fish. Aquat. Sci. *43*:420–434.

Nesteruk, T. 1991. Vertical distribution of Gastrotricha in organic bottom sediment of inland water bodies. Acta Hydrobiol. *33*:253–264.

Nesteruk, T. 1993. A comparison of values of freshwater Gastrotricha densities determined by various methods. Acta Hydrobiol. *35*:321–328.

Nesteruk, T. 1996. Density and biomass of Gastrotricha in sediments of different types of standing waters. Hydrobiologia *324*:205–208.

Netherland, M. D. 1997. Turion ecology of *Hydrilla*. J. Aquat. Plant Manage. *35*:1–10.

Neumann, J. 1959. Maximum depth and average depth of lakes. J. Fish. Res. Board Can. *16*:923–927.

Newbold, J. D. 1992. Cycles and spirals of nutrients. *In* P. Calow, and G. E. Petts, eds. The rivers handbook. I. Hydrological and ecological principles. Blackwell Sci. Publs., Oxford. pp. 379–408.

Newbold, J. D., J. W. Elwood, R. V. O'Neill, and W. VanWinkle. 1981. Measuring nutrient spiralling in streams. Can. J. Fish. Aquat. Sci. *38*:860–863.

Newbold, J. D., J. W. Elwood, R. V. O'Neill, and A. L. Sheldon. 1983a. Phosporus dynamics in a woodland stream ecosystem: A study of nutrient spiralling. Ecology *64*:1249–1265.

Newbold, J. D., J. W. Elwood, M. S. Schulze, R. W. Stark, and J. C. Barmeier. 1983b. Continuous ammonium enrichment of a woodland stream: Uptake kinetics, leaf decomposition, and nitrification. Freshwat. Biol. *13*:193–204.

Newcombe, C. L. 1950. A quantitative study of attachment materials in Sodon Lake, Michigan. Ecology *31*:204–215.

Newman, E. I. 1995. Phosphorus inputs to terrestrial ecosystems. J. Ecol. *83*:713–726.

Newman, R. M. 1991. Herbivory and detritivory on freshwater macrophytes by invertebrates. J. N. Am. Benthol. Soc. *10*:89–114.

Newman, R. M., W. C. Kerfoot, and Z. Hanscom. 1986. Watercress allelochemical defends high-nitrogen foliage against consumption: Effects on freshwater invertebrate herbivores. Ecology *77*:2312–2323.

Newman, R. M., Z. Hanscom, and W. C. Kerfoot. 1992. The watercress glucosinolate-myrosinase system: A feeding deterrent to caddisflies, snails and amphipods. Oecologia *92*:1–7.

Newman, S. and K. R. Reddy. 1992. Sediment resuspension effects on alkaline phosphatase activity. Hydrobiologia *245*:75–86.

Newman, S., F. J. Aldridge, E. J. Phlips, and K. R. Reddy. 1994. Assessment of phosphorus availability for natural phytoplankton populations from a hypereutrophic lake. Arch. Hydrobiol. *130*:409–427.

Newman, S., J. Schuette, J. B. Grace, K. Ratchey, T. Fontaine, K. R. Reddy, and M. Pietrucha. 1998. Factors influencing cattail abundance in the northern Everglades. Aquat. Bot. *60*:265–280.

Nicholls, K. H. and G. J. Hopkins. 1993. Recent changes in Lake Erie (North Shore) phytoplankton: Cumulative impacts of phosphorus loading reductions and the zebra mussel introduction. J. Great Lakes Res. *19*:637–647.

Nichols, D. S. and D. R. Keeney. 1973. Nitrogen and phosphorus release from decaying water milfoil. Hydrobiologia *42*:509–525.

Nichols, D. S. and D. R. Keeney. 1976. Nitrogen nutrition of *Myriophyllum spicatum*: Uptake and translocation of ^{15}N by shoots and roots. Freshwat. Biol. *6*:145–154.

Nicholson, S. A. and D. G. Best. 1974. Root:shoot and leaf area relationships of macrophyte communities in Chautauqua Lake, New York. Bull. Torrey Bot. Club *101*:96–100.

Nielsen, L. W., K. Nielsen, and K. Sand-Jensen. 1985. High rates of production and mortality of submerged *Sparganium emersum* Rehman during its short growth season in a eutrophic Danish stream. Aquat. Bot. *22*:325–334.

Nielsen, L. P., P. B. Christensen, N. P. Revsbech, and J. Sorensen. 1990. Denitrification and photosynthesis in stream sediment studied with microsensor and whole-core techniques. Limnol. Oceanogr. *35*:1135–1144.

Nieuwenhuyse, E. E. V. and J. R. Jones. 1996. Phosphorus-chlorophyll relationship in temperate streams and its variation with stream catchment area. Can. J. Fish. Aquat. Sci. *53*:99–105.

Niewolak, S. 1970. Seasonal changes of nitrogen-fixing and nitrifying and denitrifying bacteria in the bottom deposits of Ilawa lakes. Pol. Arch. Hydrobiol. *17*:89–103.

Niewolak, S. 1972. Fixation of atmospheric nitrogen by *Azotobacter* sp., and other heterotrophic oligonitrophilous bacteria in the Ilawa lakes. Acta Hydrobiol. *14*:287–305.

Niewolak, S. 1974. Distribution of microorganisms in the waters of the Kortowskie Lake. Pol. Arch. Hydrobiol. *21*:315–333.

Niklas, C. J. 1994. Aquatic plants. *In* K. J. Niklas, ed. Plant Allometry: The Scaling of Form and Process. Univ. Chicago Press, Chicago. pp. 60–122.

Nikolsky, G. V. 1963. The Ecology of Fishes. Academic Press, London.

Nikora, V. I., D. G. Goring, and B. J. F. Biggs. 1997. On stream periphyton-turbulence interactions. N. Z. J. Mar. Freshwat. Res. *31*:435–448.

Nilssen, J. P. 1978. On the evolution of life histories of limnetic cyclopoid copepods. Mem. Ist. Ital. Idrobiol. *36*:193–214.

Nilssen, J. P. and K. Elgmork. 1977. *Cyclops abyssorum*: Life cycle dynamics and habitat selection. Mem. Ist. Ital. Idrobiol. *34*:197–238.

Nilssen, J. P. and S. Sandøy. 1990. Recent lake acidification and cladoceran dynamics: Surface sediment and core analyses from lakes in Norway, Scotland and Sweden. Phil. Trans. R. Soc. Lond. B *327*:299–309.

Nilsson, C. and K. Berggren. 2000. Alterations of riparian ecosystems caused by river regulation. BioScience *50*:783–792.

Nilsson, N.-A. and B. Pejler. 1973. On the relation between fish fauna and zooplankton composition in North Swedish lakes. Rep. Inst. Freshw. Res. Drottningholm *53*:51–77.

Nipkow, F. 1920. Vorläufige Mitteilungen über Untersuchungen des Schlammabsatzes im Zürichsee. Z. Hydrol. *1*:1–27.

Nisbet, B. 1984. Nutrition and Feeding Strategies in Protozoa. Croom Helm, London.

Nishijima, T., R. Shiozaki, and Y. Hata. 1979. Production of vitamin B_{12}, thiamine, and biotin by freshwater phytoplankton. Bull. Japan. Soc. Sci. Fish. *45*:199–204.

Nishijima, T. and Y. Hata. 1977. Distribution of thiamine, biotin, and vitamin B_{12} in Lake Kojima. I. Distribution in the lake water. Bull. Japan. Soc. Sci. Fish. *43*:1403–1410.

Nishijima, T. and Y. Hata. 1978. Distribution of thiamine, biotin, and vitamin B_{12} in Lake Kojima. II. Distribution in the bottom sediments. Bull. Japan. Soc. Sci. Fish. *44*:815–818.

Nishimura, M., S. Nakaya, and K. Tanaka. 1973. Boron in the atmosphere and precipitation: Is the sea the source of atmospheric boron? *In* Proc. Symposium on Hydrogeochemistry and Biogeochemistry. I. Hydrogeochemistry. Clarke Company, Washington, DC. pp. 547–557.

Nix, J. 1981. Contribution of hypolimnetic water on metalimnetic dissolved oxygen minima in a reservoir. Wat. Resour. Res. *17*:329–332.

Noble, R. T. and J. A. Fuhrman. 1998. Use of SYBR Green I for rapid epifluorescence counts of marine viruses and bacteria. Aquat. Microb. Ecol. *14*:113–118.

Noble, V. E. 1961. Measurement of horizontal diffusion in the Great Lakes. Publ. Great Lakes Res. Div., Univ. Mich. *7*:85–95.

Noble, V. E. 1967. Evidences of geostrophically defined circulation in Lake Michigan. Proc. Conf. Great Lakes Res., Int. Assoc. Great Lakes Res. *10*:289–298.

Noland, L. E. and M. Gojdics. 1967. Ecology of Free-Living Protozoa. *In* T. Chen, ed. Research in Protozoology. Vol. 2. Pergamon Press, Oxford. pp. 215–266.

Nolen, S. L., J. Wilhm, and G. Howick. 1985. Factors influencing inorganic turbidity in a great plains reservoir. Hydrobiologia *123*:109–117.

Nordlie, F. G. 1972. Thermal stratification and annual heat budget of a Florida sinkhole lake. Hydrobiologia *40*:183–200.

Norkrans, B. 1980. Surface microlayers in aquatic environments. Adv. Microb. Ecol. *4*:51–85.

Northcote, T. G. 1964. Use of a high-frequency echo sounder to record distribution and migration of *Chaoborus* larvae. Limnol. Oceanogr. *9*:87–91.

Northcote, T. G. 1991. Success, problems, and control of introduced mysid populations in lakes and reservoirs. Amer. Fish. Soc. Symposium *9*:5–16.

Northcote, T. G., C. J. Walters, and J. M. B. Hume. 1978. Initial impacts of experimental fish introductions on the macrozooplankton of small oligotrophic lakes. Verh. Internat. Verein. Limnol. *20*:2003–2012.

Novac, J. T. and D. E. Brune. 1985. Inorganic carbon limited growth kinetics of some freshwater algae. Water Res. *19*:215–225.

Novotná, M. and V. Korínek. 1966. Effect of the fishstock on the quantity and species composition of the plankton of two backwaters. Hydrobiol. Stud. *1*:297–322.

Novotny, I. H. and L. L. Tews. 1975. Lentic moulds of southern Lake Winnebago. Trans. Br. Mycol. Soc. *65*:433–441.

Nowell, A. R. M. and P. A. Jumars. 1984. Flow environments of aquatic benthos. Annu. Rev. Ecol. Syst. *15*:303–328.

Nriagu, J. O. 1968. Sulfur metabolism and sedimentary environment: Lake Mendota, Wisconsin. Limnol. Oceanogr. *13*:430–439.

Nriagu, J. O. 1978. Dissolved silica in pore waters of lakes Ontario, Erie, and Superior sediments. Limnol. Oceanogr. *23*:53–67.

Nriagu, J. O. 1979. Copper in the atmosphere and precipitation. *In* J. O. Nriagu, ed. Copper in the Environment. I. Ecological Cycling. John Wiley & Sons, New York. pp. 43–75.

Nriagu, J. O. 1980. Cadmium in the atmosphere and in precipitation. *In* J. O. Nriagu, ed. Cadmium in the Environment. Part I. Ecological Cycling. John Wiley & Sons, New York. pp. 71–114.

Nriagu, J. O. 1996. A history of global metal pollution. Science *272*:223–224.

Nriagu, J. O. and C. I. Davidson. 1980. Zinc in the atmosphere. *In* J. O. Nriagu, ed. Zinc in the Environment. I. Ecology Cycling. John Wiley & Sons, New York. pp. 113–159.

Nriagu, J. O. and J. D. Hem. 1978. Chemistry of pollutant sulfur in natural waters. *In* J. O. Nriagu, ed. Sulfur in the Environment. II. Ecological Impacts. John Wiley & Sons, New York. pp. 211–270.

Nriagu, J. O., G. Lawson, H. K. T. Wong, and J. M. Azcue. 1993. A protocol for minimizing contamination in the analysis of trace metals in Great Lakes waters. J. Great Lakes Res. *19*: 175–182.

Nriagu, J. O., G. Lawson, H. K. T. Wong, and V. Cheam. 1996. Dissolved trace metals in lakes Superior, Erie, and Ontario. Environ. Sci. Technol. *30*:178–187.

Nürnberg, G. K. 1984. The prediction of internal phosphorus load in lakes with anoxic hypolimnia. Limnol. Oceanogr. *29*:111–124.

Nürnberg, G. K. 1985. Availability of phosphorus upwelling from iron-rich anoxic hypolimnia. Arch. Hydrobiol. *104*:459–476.

Nürnberg, G. K. and M. Shaw. 1998. Productivity of clear and humic lakes: Nutrients, phytoplankton, bacteria. Hydrobiologia *382*:97–112.

Nürnberg, G. K. and P. J. Dillon. 1993. Iron budgets in temperate lakes. Can. J. Fish. Aquat. Sci. *50*:1728–1737.

Nygaard, G. 1938. Hydrobiologische Studien über dänische Teiche und Seen. 1. Teil: Chemisch-Physikalische Untersuchungen und Plankton-wägungen. Arch. Hydrobiol. *32*:523–692.

Nygaard, G. 1949. Hydrobiological studies on some Danish ponds and lakes. Part II. The quotient hypothesis and some new or little known phytoplankton organisms. Kongel. Danske Vidensk. Selskab Biol. Skrift. *7*(1), 293 pp.

Nygaard, G. 1955. On the productivity of five Danish waters. Verh. Internat. Verein. Limnol. *12*:123–133.

Nykvist, N. 1963. Leaching and decomposition of water-soluble organic substances from different types of leaf and needle litter. Stud. Forest. Suecica *3*, 31 pp.

Nyström, P. and J. A. Strand. 1996. Grazing by a native and an exotic crayfish on aquatic macrophytes. Freshwat. Biol. *36*:673–682.

Nyström, P., C. Brönmark, and W. Granéli. 1996. Patterns in benthic food webs: A role for omnivorous crayfish? Freshwat. Biol. *36*:631–646.

Oborn, E. T. 1960a. Iron content of selected water and land plants. *In* Chemistry of Iron in Natural Water. U.S. Geol. Surv. Water-Supply Pap. *1459-G*:191–211.

Oborn, E. T. 1960b. A survey of pertinent biochemical literature. *In* Chemistry of Iron in Natural Water. U.S. Geol. Surv. Water-Supply Pap. *1459-F*:111–190.

Oborn, E. T. 1964. Intercellular and extracellular concentration of manganese and other elements by aquatic organisms. *In* Chemistry of Manganese in Natural Water. U.S. Geol. Surv. Water-Supply Pap. *1667-C*, 18 pp.

Oborn, E. T. and J. D. Hem. 1962. Some effects of the larger types of aquatic vegetation on iron content of water. *In* Chemistry of Iron in Natural Water. U.S. Geol. Surv. Water-Supply Pap. *1459-I*:237–268.

O'Brien, W. J. 1987. Planktivory by freshwater fish: Thrust and parry in the pelagia. *In* W. C. Kerfoot and A. Sih, eds. Prediation: Direct and Indirect Impacts on Aquatic Communities. Univ. New England Press, Hanover, NH. pp. 3–16.

O'Brien, W. J. and F. deNoyelles. 1974. Filtering rate of *Ceriodaphnia reticulata* in pond waters of varying phytoplankton concentrations. Am. Midland Nat. *91*:509–512.

O'Brien, W. J., N. A. Slade, and G. L. Vinyard. 1976. Apparent size as the determinant of prey selection by bluegill sunfish (*Lepomis macrochirus*). Ecology *57*:1304–1310.

Ochiai, M. and T. Nakajima. 1994. Decomposition of aquatic plant *Elodea nutallii* from Lake Biwa. Verh. Internat. Verein. Limnol. *25*:2276–2278.

Ochs, C. A., J. J. Cole, and G. E. Likens. 1995. Population dynamics of bacterioplankton in an oligotrophic lake. J. Plankton Res. *17*:365–391.

Odum, E. P. 1962. Relationships between structure and function in the ecosystem. Jpn. J. Ecol. *12*:108–118.

Odum, E. P. 1963. Primary and secondary energy flow in relation to ecosystem structure. Proc. 16th Int. Congr. Zool. *4*:336–338.

Odum, E. P. 1971. Fundamentals of Ecology, 3rd ed. W.B. Saunders Co., Philadelphia. 574 pp.

Odum, E. P. and A. A. de la Cruz. 1963. Detritus as a major component of ecosystems. Bull. Amer. Inst. Biol. Sci. *13*:39–40.

Odum, H. T. 1957. Trophic structure and productivity of Silver Springs, Florida. Ecol. Monogr. *27*:55–112.

Odum, W. E., P. W. Kirk, and J. C. Zieman. 1979. Non-protein nitrogen compounds associated with particles of vascular plant detritus. Oikos *32*:363–367.

Ogan, M. T. 1979. Potential for nitrogen fixation in the rhizosphere and habitat of natural stands of the wild rice *Zizania aquatica*. Can. J. Bot. *57*:1285–1291.

Ogawa, R. E. and J. F. Carr. 1969. The influence of nitrogen on heterocyst production in blue-green algae. Limnol. Oceanogr. *14*:342–351.

Oglesby, R. T. and W. R. Schaffner. 1978. Phosphorus loadings to lakes and some of their responses. Part 2. Regression models of summer phytoplankton standing crops, winter total P, and transparency of New York lakes with phosphorus loadings. Limnol. Oceanogr. *23*:135–145.

Oglesby, R. T., C. A. Carlson, and J. A. McCann, eds. 1972. River Ecology and Man. Academic Press, New York. 465 pp.

Ogura, K. 1990. Lipid compounds in lake sediments. Verh. Internat. Verein. Limnol. *24*:274–278.

Ohle, W. 1934a. Chemische und physikalische Untersuchungen norddeutscher Seen. Arch. Hydrobiol. *26*:386–464 and 584–658.

Ohle, W. 1934b. Über organische Stoffe in Binnenseen. Verh. Internat. Verein. Limnol. *6*:249–262.

Ohle, W. 1938. Zur Vervolkommnung der hydrochemischen Analyse. III. Die Phosphorbestimmung. Angew. Chem. *51*:906–911.

Ohle, W. 1952. Die hypolimnische Kohlendioxyd-Akkumulation als produktionsbiologischer Indikator. Arch. Hydrobiol. *46*:153–285.

Ohle, W. 1953. Die chemisch und elektrochemische Bestimmung des molekular gelösten Sauerstoffs der Binnengewässer. Mitt. Internat. Verein. Limnol. *3*:44 pp.

Ohle, W. 1954. Sulfat als "Katalysator" des limnischen Stoffkreislaufes. Vom Wasser *21*:13–32.

Ohle, W. 1955a. Ionenaustausch der Gewässersedimente. Mem. Ist. Ital. Idrobiol. *8*(Suppl.):221–245.

Ohle, W. 1955b. Beiträge zur Produktionsbiologie der Gewässer. Arch. Hydrobiol./Suppl. *22*:456–479.

Ohle, W. 1956. Bioactivity, production, and energy utilization of lakes. Limnol. Oceanogr. *1*:139–149.

Ohle, W. 1958a. Die Stoffwechseldynamik der Seen in Abhängigkeit von der Gasausscheidung ihres Schlammes. Vom Wasser *25*:127–149.

Ohle, W. 1958b. Typologische Kennzeichnung der Gewässer auf Grund ihrer Bioaktivität. Verh. Internat. Verein. Limnol. *13*:196–211.

Ohle, W. 1962. Der Stoffhaushalt der Seen als Grundlage einer allgemeinen Stoffwechseldynamik der Gewässer. Kieler Meeresforschungen *18*:107–120.

Ohle, W. 1964. Kolloidkomplexe als Kationen- und Anionenaustauscher in Binnengewässern. Vom Wasser *30*(1963):50–64.

Ohle, W. 1965. Nährstoffanreicherung der Gewässer durch Düngemittel und Meliorationen. Münchner Beiträge *12*:54–83.

Ohle, W. 1972. Die Sedimente des Grossen Plöner Sees als Dokumente der Zivilisation. Jahrb. Heimatkunde Plön *2*:7–27.

Ohle, W. 1978. Ebullition of gases from sediment, conditions, and relationship to primary production of lakes. Verh. Internat. Verein. Limnol. *20*:957–962.

Ohle, W. 1979. Ontogeny of the lake Grosser Plöner See. *In* S. Horie, ed. Paleolimnology of Lake Biwa and the Japanese Pleistocene *7*:3–33.

Ohlhorst, S., G. E. Hutchinson, and J. G. J. Kuiper. 1977. The waters of Meron: A study of Lake Huleh. V. Temporal changes in the molluscan fauna. Arch. Hydrobiol. *80*:1–19.

Ohmori, M. and A. Hattori. 1972. Effect of nitrate on nitrogen-fixation by the blue-green alga *Anabaena cylindrica*. Plant Cell Physiol. *13*:589–599.

Ohwada, K. and N. Taga. 1972. Vitamin B_{12}, thiamine, and biotin in Lake Sagami. Limnol. Oceanogr. *17*:315–320.

Ohwada, K., M. Otsuhata, and N. Taga. 1972. Seasonal cycles of vitamin B_{12}, thiamine and biotin in the surface water of Lake Tsukui. Bull. Japan. Soc. Sci. Fish. *38*:817–823.

Ojala, A., P. Kankaala, T. Kairesalo, and K. Salonen. 1995. Growth of *Daphnia longispina* L. in a polyhumic lake under various availabilities of algal, bacterial and detrital food. Hydrobiologia *315*:119–134.

Ökland, J. 1964. The eutrophic Lake Borrevann (Norway)—an ecological study on shore and bottom fauna with special reference to gastropods, including a hydrographic survey. Folia Limnol. Scandinavica *13*, 337 pp.

Oláh, J. 1969a. The quantity, vertical and horizontal distribution of the total bacterioplankton of Lake Balaton in 1966/167. Annal. Biol. Tihany *36*:185–195.

Oláh, J. 1969b. A quantitative study of the saprophytic and total bacterioplankton in the open water and the littoral zone of Lake Balaton in 1968. Annal. Biol. Tihany *36*:197–212.

Oláh, J. 1972. Leaching, colonization and stabilization during detritus formation. Mem. Ist. Ital. Idrobiol. *29*(Suppl.):105–127.

Oldfield, F. and N. Richardson. 1990. Lake sediment magnetism and atmospheric deposition. Phil. Trans. R. Soc. Lond. B *327*:325–330.

Oldfield, F. and P. G. Appleby. 1984. Empirical testing of ^{210}Pb-dating models for lake sediments. *In* E. Y. Haworth and J. W. G. Lund, eds. Lake Sediments and Environmental History. Univ. Minnesota Press, Minneapolis. pp. 93–124.

Oldfield, F., C. Barnosky, E. B. Leopold, and J. P. Smith. 1983. Mineral magnetic studies of lake sediments: A brief review. Hydrobiologia *103*:37–44.

Oldfield, F., R. Thompson, and K. E. Barber. 1978b. Changing atmospheric fallout of magnetic particles recorded in recent ombrotrophic peat sections. Science *199*:679–680.

Oldfield, F., J. A. Dearing, R. Thompson, and S. Garret-Jones. 1978a. Some magnetic properties of lake sediments and their possible links with erosion rates. Polskie Arch. Hydrobiol. *25*:321–331.

Olem, H. 1991. Liming Acidic Surface Waters. Lewis Publishers, Chelsea, MI. 331 pp.

Oleson, D. J. and J. C. Makarewicz. 1990. Effect of sodium and nitrate on growth of *Anabaena flos-aquae* (Cyanophyta). J. Phycol. *26*:593–595.

Olila, O. G. and K. R. Reddy. 1993. Phosphorus sorption characteristics of sediments in shallow eutrophic lakes of Florida. Arch. Hydrobiol. *129*:45–65.

Oliveira, L. and N. J. Antia. 1986a. Nickel ion requirements for autotrophic growth of several marine microalgae with urea serving as nitrogen source. Can. J. Fish. Aquat. Sci. *43*:2427–2433.

Oliveira, L. and N. J. Antia. 1986b. Some observations on the urea-degrading enzyme of the diatom *Cyclotella cryptica* and the role of nickel in its production. J. Plankton Res. *8*:235–242.

Olsen, S. 1958a. Phosphate adsorption and isotopic exchange in lake muds. Experiments with P32; Preliminary report. Verh. Internat. Verein. Limnol. *13*:915–922.

Olsen, S. 1958b. Fosfatbalancen mellem bund og vand i Furesø. Forsøg med radioaktivt fosfor. Folia Limnol. Scandinavica *10*:39–96.

Olsen, S. 1964. Phosphate equilibrium between reduced sediments and water. Laboratory experiments with radioactive phosphorus. Verh. Internat. Verein. Limnol. *15*:333–341.

Olsen, Y. 1999. Lipids and essential fatty acids in aquatic food webs: What can freshwater ecologists learn from mariculture? *In* M. T. Arts and B. C. Wainman, eds. Lipids in Freshwater Ecosystems. Springer-Verlag, New York. pp. 161–202.

Olsen, Y. and K. Østgaard. 1985. Estimating release rates of phosphorus from zooplankton: Model and experimental verification. Limnol. Oceanogr. *30*:844–852.

Olsen, Y., A. Jensen, H. Reinertsen, K. Y. Børsheim, M. Heldal, and A. Langeland. 1986a. Dependence of the rate of release of phosphorus by zooplankton on the P:C ratio in the food supply, as calculated by a recycling model. Limnol. Oceanogr. *31*:34–44.

Olsen, Y., K. M. Varum, and A. Jensen. 1986b. Some characteristics of the carbon compounds released by *Daphnia*. J. Plankton Res. *8*:505–517.

Olson, F. C. W. 1960. A system of morphometry. Int. Hydrogr. Rev. *37*:147–155.

Olson, F. C. W. and T. Ichiye. 1959. Horizontal diffusion. Science *130*:1255.

Olsson, H. and M. Jansson. 1984. Stability of dissolved ^{32}P-labelled phosphorus compounds in lake water and algal cultures—resistance to enzymatic treatment and algal uptake. Verh. Internat. Verein. Limnol. *22*:200–204.

O'Melia, C. R. 1998. Coagulation and sedimentation in lakes, reservoirs and water treatment plants. Wat. Sci. Tech. *37*:129–135.

O'Melia, C. R. and C. L. Tiller. 1993. Physicochemical aggregation and deposition in aquatic environments. *In* J. Buffle and H. P. van Leeuwen, eds. Environmental Particles. Vol. 2. Lewis Publishers, Ann Arbor. pp. 353–385.

Ondok, J. P. 1973a. Photosynthetically active radiation in a stand of *Phragmites communis* Trin. I. Distribution of irradiance and foliage structure. Photosynthetica 7:8–11.

Ondok, J. P. 1973b. Photosynthetically active radiation in a stand of *Phragmites communis* Trin. III. Distribution of irradiance on sunlit foliage area. Photosynthetica 7:311–319.

Ondok, J. P. 1978. Radiation climate in fishpond littoral plant communities. *In* D. Dykyjová and J. Květ, eds. Pond Littoral Ecosystems. Springer-Verlag, Berlin. pp. 113–125.

O'Neill, J. G. and J. F. Wilkinson. 1977. Oxidation of ammonia by methane-oxidizing bacteria and the effects of ammonia on methane oxidation. J. Gen. Microbiol. *100*:407–412.

Ooms-Wilms, A. L., G. Postema, and R. D. Gulati. 1995. Evaluation of bacterivory of Rotifera based on measurements of in situ ingestion of fluorescent particles, including some comparisons with Cladocera. J. Plankton Res. *17*:1057–1077.

Orcutt, J. D., Jr. and K. G. Porter. 1983. Diel vertical migration by zooplankton: Constant and fluctuating temperature effects on life history parameters of *Daphnia*. Limnol. Oceanogr. *28*:720–730.

Orcutt, J. D., Jr. and M. L. Pace. 1984. Seasonal dynamics of rotifer and crustacean zooplankton populations in a eutrophic, monomictic lake with a note on rotifer sampling techniques. Hydrobiologia *119*:73–80.

Orlov, D. S., I. A. Pivovarova, and N. I. Gorbunov. 1973. Interaction of humic substances with minerals and the nature of their bond—a review. Agrokhimiya *1973*(9):140–153. (Translat. in Soviet Soil Science *5*:568–581.)

Ortiz, J. L. and R. P. Martinez. 1984. Applicability of the OECD eutrophication models to Spanish reservoirs. Verh. Internat. Verein. Limnol. *22*:1521–1535.

Osborne, L. L. and D. A. Kovacic. 1993. Riparian vegetated buffer strips in water-quality restoration and stream management. Freshwat. Biol. *29*:243–258.

Osenberg, C. W. 1989. Resource limitation, competition and the influence of life history in a freshwater snail community. Oecologia *79*:512–519.

Osenberg, C. W. and G. G. Mittelbach. 1996. The relative importance of resource limitation and predatory limitation in food chains. *In* G. A. Polis and K. O. Winemuller, eds. Food Webs: Integration of Patterns and Dynamics. Chapman and Hall, New York. pp. 134–148.

Ostendorp, W. 1992. Sedimente und Sedimentbildung in Seeuferröhrichten des Bodensee-Untersees. Limnologia *22*:16–33.

Ostendorp, W. and T. Frevert. 1979. Untersuchungen zur Manganfreisetzung und zum Mangangehalt der Sedimentoberschicht im Bodensee. Arch. Hydrobiol./Suppl. *55*:255–277.

Ostrofsky, M. L. and E. R. Zettler. 1986. Chemical defences in aquatic plants. J. Ecol. *74*:279–287.

Oswald, G. K. A. and G. de Q. Robin. 1973. Lakes beneath the Antarctic ice sheet. Nature *245*:251–254.

Otsuki, A. and T. Hanya. 1968. On the production of dissolved nitrogen-rich organic matter. Limnol. Oceanogr. *13*:183–185.

Otsuki, A. and R. G. Wetzel. 1972. Coprecipitation of phosphate with carbonates in a marl lake. Limnol. Oceanogr. *17*:763–767.

Otsuki, A. and R. G. Wetzel. 1973. Interaction of yellow organic acids with calcium carbonate in freshwater. Limnol. Oceanogr. *18*:490–493.

Otsuki, A. and R. G. Wetzel. 1974a. Calcium and total alkalinity budgets and calcium carbonate precipitation of a small hardwater lake. Arch. Hydrobiol. *73*:14–30.

Otsuki, A. and R. G. Wetzel. 1974b. Release of dissolved organic matter by autolysis of a submersed macrophyte, *Scirpus subterminalis*. Limnol. Oceanogr. *19*:842–845.

Otsuki, A. and T. Hanya. 1972a. Production of dissolved organic matter from dead green algal cells. I. Aerobic microbial decomposition. Limnol. Oceanogr. *17*:248–257.

Otsuki, A. and T. Hanya. 1972b. Production of dissolved organic matter from dead green algal cells. II. Anaerobic microbial decomposition. Limnol. Oceanogr. *17*:258–264.

Ottesen Hansen, N.-E. 1978a. Effects of boundary layers on mixing in small lakes. *In* W. H. Graf and C. H. Mortimer, eds. Hydrodynamics of Lakes. Elsevier Sci. Publ. Co., Amsterdam. pp. 341–356.

Ottesen Hansen, N.-E. 1978b. Mixing processes in lakes. Nordic Hydrol. *9*:57–74.

Otto, C. 1983. Adaptations to benthic freshwater herbivory. *In* R. G. Wetzel, ed. Periphyton of Freshwater Ecosystems. W. Junk Publ., The Hague. pp. 199–205.

Otto, G. and J. Benndorf. 1971. Über den Einfluss des physiologischen Zustandes sedimentierender Phytoplankter auf die Abbauvorgänge während der Sedimentation. Limnologica *8*:365–370.

Oude, P. J. den and R. D. Gulati. 1988. Phosphorus and nitrogen excretion rates of zooplankton from the eutrophic Loosdrecht lakes, with notes on other P sources for phytoplankton requirements. Hydrobiologia *169*:379–390.

Ouellet, M. H. 1975. Paleoclimatological implications of a Late-Quaternary molluscan fauna from Atkins Lake, Ontario. Verh. Internat. Verein. Limnol. *19*:2251–2258.

Overbeck, H.-J. 1968. Bakterien im Gewässer-Ein Beispiel für die Gegenwärtig Entwicklung der Limnologie. Mitt. Max-Planck-Ges. *3*:165–182.

Overbeck, J. 1962. Untersuchungen zum Phosphathaushalt von Grünalgen. III. Das Verhalten der Zellfraktionen von *Scenedesmus quadricauda* (Turp.) Bréb. im Tagescyclus unter verschiedenen Belichtungsbedingungen und bei verschidenen Phosphatverbindungen. Arch. Mikrobiol. *41*:11–26.

Overbeck, J. 1963. Untersuchungen zum Phosphathaushalt von Grünalgen. VI. Ein Beitrag zum Polyphosphatstoffwechsel des Phytoplanktons. Ber. Deut. Bot. Gesellsch. *76*:276–286.

Overbeck, J. 1965. Primärproduktion und Gewässerbakterien. Naturwissenschaften *51*:145.

Overbeck, J. 1968. Prinzipielles zum Vorkommen der Bakterien im See. Mitt. Internat. Verein. Limnol. *14*:134–144.

Overbeck, J. 1975. Distribution pattern of uptake kinetic responses in a stratified eutrophic lake. Verh. Internat. Verein. Limnol. *19*:2600–2615.

Overbeck, J. 1994. Heterotrophic potential of bacteria. *In* J. Overbeck and R. J. Chróst, eds. Microbial Ecology of Lake Plußsee. Springer-Verlag, New York. pp. 192–200.

Overbeck, J. and H.-D. Babenzien. 1964. Bakterien und Phytoplankton eines Kleingewässers im Jahreszyklus. Z. Allg. Mikrobiol. *4*:59–76.

Overbeck, J. and D. Tóth. 1978. Einfluss des Phosphatgehalts auf die Glucoseaufnahme bei Bakterien. Arch. Hydrobiol. *82*:114–122.

Overmann, J., J. T. Beatty, K. J. Hall, N. Pfennig, and T. G. Northcote. 1991. Characterization of a dense, purple sulfur bacterial layer in a meromictic salt lake. Limnol. Oceanogr. *36*:846–859.

Owens, M. and P. J. Maris. 1964. Some factors affecting the respiration of some aquatic plants. Hydrobiologia *23*:533–543.

Owens, M. and R. W. Edwards. 1961. The effects of plants on river conditions. II. Further crop studies and estimates of net productivity of macrophytes in a chalk stream. J. Ecol. *49*:119–126.

Owens, M. and R. W. Edwards. 1962. The effects of plants on river conditions. III. Crop studies and estimates of net productivity of macrophytes in four streams in southern England. J. Ecol. *50*:157–162.

Ownbey, C. R. and D. A. Kee. 1967. Chlorides in Lake Erie. Proc. Conf. Great Lakes Res., Int. Assoc. Great Lakes Res. *10*:382–389.

Owttrim, G. W. and B. Colman. 1989. Measurement of the photorespiratory activity of the submerged aquatic plant *Myriophyllum spicatum* L. Plant Cell Environ. *12*:805–811.

Ozimek, T., A. Prejs, and K. Prejs. 1976. Biomass and distribution of underground parts of *Potamogeton perfoliatus* L. and *P. lucens* L. in Mikolajskie Lake, Poland. Aquat. Bot. *2*:309–316.

Ozimek, T., K. Prejs, and A. Prejs. 1986. Biomass and growth rate of *Potamogeton pectinatus* L. in lakes of different trophic state. Ekologia Polska *34*:125–131.

Pace, M. L. 1982. Planktonic ciliates: Their distribution, abundance and relationship to microbial resources in a monomictic lake. Can. J. Fish. Aquat. Sci. *39*:1106–1116.

Pace, M. L. 1988. Bacterial mortality and the fate of bacterial production. Hydrobiologia *159*:41–49.

Pace, M. L. 1993. Heterotrophic microbial processes. *In* S. R.Carpenter and J. F. Kitchell, eds. The Trophic Cascade in Lakes. Cambridge University Press, New York. pp. 252–277.

Pace, M. L. and D. Vaqué. 1994. The importance of *Daphnia* in determining mortality rates of protozoans and rotifers in lakes. Limnol. Oceanogr. *39*:985–996.

Pace, M. L. and J. D. Orcutt, Jr. 1981. The relative importance of protozoans, rotifers, and crustaceans in a freshwater zooplankton community. Limnol. Oceanogr. *26*:822–830.

Pace, M. L. and J. J. Cole. 1996. Regulation of bacteria by resources and predation tested in whole-lake experiments. Limnol. Oceanogr. *41*:1448–1460.

Pace, M. L., G. B. McManus, and S. E. G. Findlay. 1990. Planktonic community structure determines the fate of bacterial production in a temperate lake. Limnol. Oceanogr. *35*:795–808.

Pace, M. L., S. E. G. Findlay, and D. Fischer. 1998. Effects of an invasive bivalve on the zooplankton community of the Hudson River. Freshwat. Biol. *39*:103–116.

Pace, M. L., S. E. G. Findlay, and D. Lints. 1992. Zooplankton in advective environments: The Hudson River community and a comparative analysis. Can. J. Fish. Aquat. Sci. *49*:1060–1069.

Paerl, H. W. 1975. Microbial attachment to particles in marine and freshwater populations. Microbial Ecol. *2*:73–83.

Paerl, H. W. 1978. Role of heterotrophic bacteria in promoting N_2 fixation by *Anabaena* in aquatic habitats. Microbial Ecol. *4*:215–231.

Paerl, H. W. 1979. Optimization of carbon dioxide and nitrogen fixation by the blue-green alga *Anabaena* in freshwater blooms. Oecologia *38*:275–290.

Paerl, H. W. 1980. Ecological rationale for H_2 metabolism during aquatic blooms of the cyanobacterium *Anabaena*. Oecologia *47*:43–45.

Paerl, H. W. 1985. Microzone formation: Its role in the enhancement of aquatic N_2 fixation. Limnol. Oceanogr. *30*:1246–1252.

Paerl, H. W. and D. R. S. Lean. 1976. Visual observations of phosphorus movement between algae, bacteria, and abiotic particles in lake water. J. Fish. Res. Board Can. *33*:2805–2813.

Paerl, H. W. and J. L. Pinckney. 1996. A mini-review of microbial consortia: Their roles in aquatic production and biogeochemical cycling. Microb. Ecol. *31*:225–247.

Paerl, H. W. and L. E. Prufert. 1987. Oxygen-poor microzones as potential sites of microbial N_2 fixation in nitrogen-depleted aerobic marine waters. Appl. Environ. Microbiol. *53*:1078–1087.

Paerl, H. W. and M. T. Downes. 1978. Biological availability of low versus high molecular weight reactive phosphorus. J. Fish. Res. Board Can. *35*:1639–1643.

Paerl, H. W. and P. E. Kellar. 1979. Nitrogen-fixing *Anabaena*: Physiological adaptations instrumental in maintaining surface blooms. Science *204*:620–622.

Paine, R. T. 1980. Food webs: Linkage, interaction strength and community infrastructure. J. Anim. Ecol. *49*:667–685.

Pallesen, L., P. M. Berthouex, and K. Booman. 1985. Environmental intervention analysis: Wisconsin's ban on phosphate detergents. Water Res. *19*:353–362.

Palmer, F. E., R. D. Methot, Jr., and J. T. Staley. 1976. Patchiness in the distribution of planktonic heterotrophic bacteria in lakes. Appl. Environ. Microbiol. *31*:1003–1005.

Palmer, J. D. and F. E. Round. 1965. Persistent, vertical-migration rhythms in benthic microflora. I. The effect of light and temperature on the rhythmic behaviour of *Euglena obtusa*. J. Mar. Biol. Assoc. U.K. *45*:567–582.

Palmer, M. A., A. E. Bely, and K. E. Berg. 1992. Response of invertebrates to lotic disturbance: A test of the hyporheic refuge hypothesis. Oecologia *89*:182–194.

Palmer, M. A., R. F. Ambrose, and N. L. Poff. 1997. Ecological theory and community restoration ecology. Restoration Ecol. *5*:291–300.

Palmer, R. W. and J. H. O'Keeffe. Temperature characteristics of an impounded river. Arch. Hydrobiol. *116*:471–485.

Paloheimo, J. E. 1974. Calculation of instantaneous birth rate. Limnol. Oceanogr. *19*:692–694.

Palomäki, R. and E. Koskenniemi. 1993. Effects of bottom freezing on macrozoobenthos in regulated Lake Pyhäjärvi. Arch. Hydrobiol. *128*:73–90.

Paludan, C. and G. Blicher-Mathiesen. 1996. Losses of inorganic carbon and nitrous oxide from a temperate freshwater wetland in relation to nitrate loading. Biogeochemistry *35*:305–326.

Panganiban, A. T., Jr., T. E. Patt, W. Hart, and R. S. Hanson. 1979. Oxidation of methane in the absence of oxygen in lake water samples. Appl. Environ. Microbiol. *37*:303–309.

Pannier, F. 1957. El consumo de oxigeno de plantas acuaticas en relacion a distintas concentraciones de oxigeno. Parte I. Acta Cient. Venez. *8*:148–161.

Pannier, F. 1958. El consumo de oxigeno de plantas acuaticas en relacion a distintas concentraciones de oxigeno. Parte II. Acta Cient. Venez. *9*:2–13.

Pardy, R. L. and B. N. White. 1977. Metabolic relationships between green *Hydra* and its symbiotic algae. Biol. Bull. *153*:228–236.

Parejko, K. and S. Dodson. 1990. Progress towards characterization of a predator/prey kairomone: *Daphnia pulex* and *Chaoborus americanus*. Hydrobiologia *198*:51–59.

Park, J.-C., M. Aizaki, T. Fukushima, and A. Otsuki. 1997. Production of labile and refractory dissolved organic carbon by zooplankton excretion: An experimental study using large outdoor continuous flow-through ponds. Can. J. Fish. Aquat. Sci. *54*:434–443.

Parker, B. C. and M. A. Wachtel. 1971. Seasonal distribution of cobalamins, biotin and niacin in rainwater. *In* J. Cairns, Jr., ed. The structure and function of freshwater microbial communities. Res. Div. Monogr., Virginia Polytech. Inst. *3*:195–207.

Parker, M. 1977. Vitamin B_{12} in Lake Washington, USA: Concentration and rate of uptake. Limnol. Oceanogr. *22*:527–538.

Parker, M. and A. D. Hasler. 1969. Studies on the distribution of cobalt in lakes. Limnol. Oceanogr. *14*:229–241.

Parker, R. A. and D. H. Hazelwood. 1962. Some possible effects of trace elements on fresh-water microcrustacean populations. Limnol. Oceanogr. 7:344–347.

Parkin, T. B. and T. D. Brock. 1990. Photosynthetic bacterial production in lakes: The effects of light intensity. Limnol. Oceanogr. *25*:711–718.

Parma, S. 1971. *Chaoborus flavicans* (Meigan) (Diptera, Chaoboridae): An autecological study. Ph.D. Diss., University of Groningen. Rotterdam, Bronder-Offset n.v., 128 pp.

Parrish, C. C. 1999. Determination of total lipid, lipid classes, and fatty acids in aquatic samples. *In* M. T. Arts and B. C. Wainman, eds. Lipids in Freshwater Ecosystems. Springer-Verlag, New York. pp. 4–20.

Parrish, C. C. and P. J. Wangersky. 1987. Particulate and dissolved lipid classes in cultures of *Phaeodactylum tricornutum* grown in cage culture turbidostats with a range of nitrogen supply rates. Mar. Ecol. Prog. Ser. *35*:119–128.

Parsons, T. R. and J. D. H. Strickland. 1962. On the production of particulate organic carbon by heterotrophic processes in sea water. Deep-Sea Res. *8*:211–222.

Pasciak, W. J. and J. Gavis. 1974. Transport limitation of nutrient uptake in phytoplankton. Limnol. Oceanogr. *19*:881–888.

Pastorok, R. A. 1981. Prey vulnerability and size selection by *Chaoborus* larvae. Ecology *62*:1311–1324.

Patalas, K. 1961. Wind- und morphologiebedingte Wasserbewegungstypen als bestimmender Faktor für die Intensität des Stoffkreislaufes in nordpolnischen Seen. Verh. Internat. Verein. Limnol. *14*:59–64.

Paterson, M. 1993. The distribution of microcrustacea in the littoral zone of a freshwater lake. Hydrobiologia *263*:173–183.

Patriarche, M. H. and R. C. Ball. 1949. An analysis of the bottom fauna production in fertilized and unfertilized ponds and its utilization by young-of-the-year fish. Tech. Bull. Mich. State Univ. Agricult. Exp. Stat., Sec. Zool. *207*, 35 pp.

Patrick, R. 1978. Effects of trace metals in the aquatic ecosystem. Amer. Sci. *66*:185–191.

Patrick, R. 1988. Importance of diversity in the functioning and structure of riverine communities. Limnol. Oceanogr. *33*:1304–1307.

Patrick, R., B. Crum, and J. Coles. 1969. Temperature and manganese as determining factors in the presence of diatom or blue-green algal floras in streams. Proc. Nat. Acad. Sci. USA *64*:472–478.

Patrick, R., M. H. Hohn, and J. H. Wallace. 1954. A new method for determining the pattern of the diatom flora. Notulae Naturae Acad. Nat. Sci. Philadelphia, *259*, 12 pp.

Pattee, E. 1975. Température stable et température fluctuante. I. Etude comparative de leurs effets sur le développement de certaines Planaires. Verh. Internat. Verein. Limnol. *19*:2795–2802.

Pattee, E. and H. Chergui. 1994. On the incomplete breakdown of submerged leaves. Verh. Internat. Verein. Limnol. *25*:1545–1548.

Pattee, E. and H. Persat. 1978. Contribution of asexual reproduction to the distribution of triclad flatworms. Verh. Internat. Verein. Limnol. *29*:2372–2377.

Patten, B. C. 1985. Energy cycling in the ecosystem. Ecol. Model. *28*:1–71.

Patten, B. C. 1995. Network integration of ecological extremal principles: Exergy, emergy, power, ascendency, and indirect effects. Ecol. Model. *79*:79–84.

Patterson, D. J. 1992. Free-living Freshwater Protozoa: A Color Guide. CRC Press Inc., Boca Raton. 223 pp.

Patterson, D. J. and J. Larsen, eds. 1991. The Biology of Free-living Heterotrophic Flagellates. Oxford University Press, Oxford. 502 pp.

Paul, B. J. and H. C. Duthie. 1989. Nutrient cycling in the epilithon of running waters. Can. J. Bot. *67*:2302–2309.

Pauli, H.-R. 1991. Estimates of rotifer productivity in Lake Constance: A comparison of methods. Verh. Internat. Verein. Limnol. *24*:850–853.

Payne, F. C. 1982. Influence of hydrostatic pressure on gas balance and lacunar structure in *Myriophyllum spicatum* L. Ph.D. Dissertation, Michigan State Univ. 109 pp.

Payne, W. J. and W. J. Wiebe. 1978. Growth yield and efficiency in chemosynthetic microorganisms. Annu. Rev. Microbiol. *32*:155–183.

Pearcy, R. W., J. A. Berry, and B. Bartholomew. 1974. Field photosynthetic performance and leaf temperatures of *Phragmites communis* under summer conditions in Death Valley, California. Photosynthetica *8*:104–108.

Pearsall, W. H. 1922. A suggestion as to factors influencing the distribution of free-floating vegetation. J. Ecol. *9*:241–253.

Pearsall, W. H. 1932. Phytoplankton in the English lakes. II. The composition of the phytoplankton in relation to dissolved substances. J. Ecol. *20*:241–262.

Pechan'-Finenko, G. A. 1971. Effectiveness of the assimilation of food by plankton crustaceans. (Translat. Consultants Bureau.) Ekologia *2*:64–72.

Pechlaner, R. 1970. The phytoplankton spring outburst and its conditions in Lake Erken (Sweden). Limnol. Oceanogr. *15*:113–130.

Pechlaner, R. 1971. Factors that control the production rate and biomass of phytoplankton in high-mountain lakes. Mitt. Internat. Verein. Limnol. *19*:125–145.

Pechlaner, R., G. Bretschko, P. Gollmann, H. Pfeifer, M. Tilzer, and H. P. Weissenbach. 1970. The production processes in two high-mountain lakes (Vorderer and Hinterer Finstertaler See, Kühtai, Austria). *In* Z. Kajak and A. Hillbricht-Ilkowska, eds. Productivity Problems of Freshwaters. PWN Polish Scientific Publishers, Warsaw. pp. 239–269.

Peck, H. D. 1993. Bioenergetic strategies of the sulfate-reducing bacteria. *In* J. M. Odom and R. Singleton, eds. The Sulfate-Reducing Bacteria: Contemporary Perspectives. Springer-Verlag, New York. pp. 41–76.

Peckarsky, B. L. 1984a. Sampling of stream benthos. *In* J. A. Downing and F. H. Rigler, eds. A Manual on Methods for the Assessment of Secondary Productivity in Fresh Waters. 2nd Edition. Blackwell Scientific Publs., Oxford. pp. 131–160.

Peckarsky, B. L. 1984b. Predator-prey interactions among aquatic insects. *In* V. H. Resh and D. M. Rosenberg, eds. The Ecology of Aquatic Insects. Praeger, New York. pp. 196–254.

Pedersen, O. 1993. Long-distance water transport in aquatic plants. Plant Physiol. *103*:1369–1375.

Pedersen, O. and K. Sand-Jensen. 1993. Water transport in submerged macrophytes. Aquat. Bot. *44*:385–406.

Pedrós-Alió, C. and T. D. Brock. 1982. Assessing biomass and production of bacteria in eutrophic Lake Mendota, Wisconsin. Appl. Environ. Microbiol. *44*:203–218.

Pedrós-Alió, C. and T. D. Brock. 1983. The impact of zooplankton feeding on the epilimnetic bacteria of a eutrophic lake. Freshwat. Biol. *13*:227–239.

Pedrós-Alió, C., R. Massana, M. Latasa, J. García-Cantizano, and J. M. Gasol. 1995. Predation by ciliates on a metalimnetic *Cryptomonas* population: Feeding rates, impact and effects of vertical migration. J. Plankton Res. *17*:2131–2154.

Pehofer, H. E. 1989. Spatial distribution of the nematode fauna and production of three nematodes (*Tobrilus gracilis, Monhystera stagnalis, Ethmolaimus pratensis*) in the profundal of Piberger See (Austria, 913 m a.s.l.). Int. Rev. ges. Hydrobiol. *74*:135–168.

Pelegri, S. P. and T. H. Blackburn. 1995. Effects of *Tubifex tubifex* (Oligochaeta: Tubificidae) on N-mineralization in freshwater sediments, measured with ^{15}N isotopes. Aquat. Microb. Ecol. *9*:289–294.

Pelton, D. K., S. N. Levine, and M. Braner. 1998. Measurements of phosphorus uptake by macrophytes and epiphytes from the LaPlatte River (VT) using ^{32}P in stream microcosms. Freshwat. Biol. *39*:285–299.

Pendl, M. P. and K. M. Stewart. 1986. Variations in carbon fractions within a dimictic and a meromictic basin of the Junius Ponds, New York. Freshwat. Biol. *16*:539–555.

Penfound, W. T. and T. T. Earle. 1948. The biology of the water hyacinth. Ecol. Monogr. *18*:447–472.

Penhale, P. A. and R. G. Wetzel. 1983. Structural and functional adaptations of eelgrass (*Zostera marina* L.) to the anaerobic sediment environment. Can. J. Bot. *61*:1421–1428.

Pennak, R. W. 1940. Ecology of the microscopic Metazoa inhabiting the sandy beaches of some Wisconsin lakes. Ecol. Monogr. *10*:537–615.

Pennak, R. W. 1944. Diurnal movements of zooplankton organisms in some Colorado mountain lakes. Ecology *25*:387–403.

Pennak, R. W. 1966. Structure of zooplankton populations in the littoral macrophyte zone of some Colorado lakes. Trans. Amer. Microsc. Soc. *85*:329–349.

Pennak, R. W. 1973. Some evidence for aquatic macrophytes as repellents for a limnetic species of *Daphnia*. Int. Rev. ges. Hydrobiol. *58*:569–576.

Pennak, R. W. 1978. Freshwater Invertebrates of the United States, 2nd Edition. Wiley-Interscience, New York. 803 pp.

Pennington, W. 1973. Absolute pollen frequencies in the sediments of lakes of different morphometry. *In* H. J. B. Birks and R. G. West, eds. Quaternary Plant Ecology. Blackwell Sci. Publs., Oxford. pp. 79–104.

Pennington, W. 1974. Seston and sediment formation in five Lake District lakes. J. Ecol. *62*:215–251.

Pennington, W. 1979. The origin of pollen in lake sediments: An enclosed lake compared with one receiving inflow streams. New Phytol. *83*:189–213.

Pennington, W. 1981. Records of a lake's life in time: The sediments. Hydrobiologia *79*:197–219.

Pennington, W., R. S. Cambray, J. D. Eakins, and D. D. Harkness. 1976. Radionuclide dating of the recent sediments of Blelham Tarn. Freshwat. Biol. 6:317–331.

Pentecost, A. 1984. Observations on a bloom of the neuston alga, *Nautococcus pyriformis*, from southern England with an explanation of the floatation mechanism. Br. Phycol. J. *19*:227–232.

Penttinen, S., A. Kostamo, and J. V. K. Kukkonen. 1998. Combined effects of dissolved organic material and water hardness on toxicity of cadmium to *Daphnia magna*. Environ. Toxicol. Chem. *17*:2498–2503.

Perdue, E. M. 1998. Chemical compositon, structure, and metal binding properties. *In* D. O. Hessen and L. J. Tranvik, eds. Aquatic Humic Substances: Ecology and Biogeochemistry. Springer-Verlag, Berlin. pp. 41–61.

Perdue, E. M. and E. T. Gjessing, eds. 1990. Organic acids in aquatic ecosystems. John Wiley & Sons, New York. 345 pp.

Pérez-Fuentetaja, A., D. J. McQueen, and C. W. Ramcharan. 1996. Predator-induced bottom-up effects in oligotrophic systems. Hydrobiologia *317*:163–176.

Perfil'ev, B. V. and D. R. Gabe. 1969. Capillary Methods of Investigating Micro-Organisms. (Translat. of 1961 edition by J. M. Shewan.) University of Toronto Press, Toronto. 627 pp.

Perry, S. A., W. B. Perry, and G. M. Simmons, Jr. 1990. Bacterioplankton and phytoplankton populations in a rapidly-flushed eutrophic reservoir. Int. Rev. ges. Hydrobiol. *75*:27–44.

Persson, G. 1985. Community grazing and the regulation of in situ clearance and feeding rates of planktonic crustaceans in lakes in the Kuokkel area, northern Sweden. Arch. Hydrobiol./Suppl. *70*:197–238.

Persson, L. 1999. Trophic cascades: Abiding heterogeneity and the trophic level concept at the end of the road. Oikos *85*:385–397.

Persson, L. and L. B. Crowder. 1998. Fish-habitat interactions mediated via ontogenetic niche shifts. *In* E. Jeppesen, Ma. Søndergaard, Mo. Søndergaard, and K. Christoffersen, eds. The Structuring Role of Submerged Macrophytes in Lakes. Springer-Verlag, New York. pp. 3–23.

Persson, L, S. Diehl, L. Johansson, G. Anderson, and S. F. Hamrin. 1992. Trophic interactions in temperate lake ecosystems: A test of food chain theory. Am. Nat. *140*:59–84.

Peters, F. 1994. Prediction of planktonic protistan grazing rates. Limnol. Oceanogr. *39*:195–206.

Peters, G. A. and J. C. Meeks. 1989. The *Azolla-Anabaena* symbiosis: Basic biology. Ann. Rev. Plant Physiol. Plant Mol. Biol. *40*:193–210.

Peters, G. T., J. R. Webster, and E. F. Benfield. 1987. Microbial activity associated with seston in headwater streams: Effects of nitrogen, phosphorus and temperature. Freshwat. Biol. *18*:405–413.

Peters, G. T., E. F. Benfield, and J. R. Webster. 1989. Chemical composition and microbial activity of seston in a southern Appalachian headwater stream. J. N. Am. Benthol. Soc. *8*:74–84.

Peters, R. and D. Lean. 1973. The characterization of soluble phosphorus released by limnetic zooplankton. Limnol. Oceanogr. *18*:270–279.

Peters, R. H. 1975. Orthophosphate turnover in central European lakes. Mem. Ist. Ital. Idrobiol. *32*:297–311.

Peters, R. H. 1979. Concentrations and kinetics of phosphorus fractions along the trophic gradient of Lake Memphremagog. J. Fish. Res. Board Can. *36*:970–979.

Peters, R. H. and J. A. Downing. 1984. Empirical analysis of zooplankton filtering and feeding rates. Limnol. Oceanogr. *29*:763–784.

Peters, R. H. and S. MacIntyre. 1976. Orthophosphate turnover in East African lakes. Oecologia *25*:313–319.

Petersen, R. C. and K. W. Cummins. 1974. Leaf processing in a woodland stream. Freshwat. Biol. *4*:343–368.

Petersen, R. C., L. B.-M. Petersen, and J. Lacoursiére. 1992. A building-block model for stream restoration. *In* P. J. Boon, P. Calow, and G. E. Petts, eds. River Conservation and Management. John Wiley and Sons, Chichester. pp. 293–309.

Petersen, R. C., L. B.-M. Petersen, and J. O. Lacoursiére. 1990. Restoration of lowland streams: The building block. Vatten *46*:244–249.

Peterson, B. J. 1980. Aquatic primary productivity and the $^{14}C-CO_2$ method: A history of the productivity problem. Ann. Rev. Ecol. Syst. *11*:359–385.

Peterson, B. J. and 10 others. 1985. Transformation of a tundra river from heterotrophy to autotrophy by addition of phosphorus. Science *229*:1383–1386.

Peterson, C. G. 1996. Response of benthic algal communities to natural physical disturbance. *In* R. J. Stevenson, M. L. Bothwell, and R. L. Lowe, eds. Algal Ecology: Freshwater Benthic Ecosystems. Academic Press, San Diego. pp. 375–402.

Peterson, C. G., A. C. Weibel, N. B. Grimm, and S. G. Fisher. 1994. Mechanisms of benthic algal recovery following spates: Comparison of simulated and natural events. Oecologia *98*:280–290.

Peterson, D. L. 1983. Life cycle and reproduction of *Nephelopsis obscura* Verrill (Hirudinea: Erpobdellidae) in permanent ponds of northwestern Minnesota. Freshwat. Invertebr. Biol. *2*:165–172.

Peterson, F. 1983. Population dynamics and production of *Daphnia galeata* (Crustacea, Cladocera) in Lake Esrom. Holarct. Ecol. *6*:285–294.

Peterson, H. G., F. P. Healey, and R. Wagemann. 1984. Metal toxicity to algae: A highly pH dependent phenomenon. Can. J. Fish. Aquat. Sci. *41*:974–979.

Peterson, W. H., E. B. Fred, and B. P. Domogalla. 1925. The occurrence of amino acids and other organic nitrogen compounds in lake water. J. Biol. Chem. *23*:287–295.

Pettersson, K. 1980. Alkaline phosphatase activity and algal surplus phosphorus as phosphorus-deficiency indicators in Lake Erken. Arch. Hydrobiol. *89*:54–87.

Petts, G. E. 1996. Sustaining the ecological integrity of large floodplain rivers. *In* M. G. Anderson, D. E. Walling, and P. D. Bates, eds. Floodplain Processes. John Wiley & Sons, Chichester. pp. 535–551.

Petts, G. E., R. Sparks, and I. Campbell. 2000. River restoration in developed economies. *In* P. J. Boon, B. R. Davies, and G. E.

Petts, eds. Global perspectives on River Conservation: Science, Policy and Practices. John Wiley & Sons, Chichester. pp. 493–508.

Pfeiffer, E.-M. 1994. Methane fluxes in natural wetlands (marsh and moor) in northern Germany. Curr. Topics Wetland Biogeochem. *1*:36–47.

Pfennig, N. 1967. Photosynthetic bacteria. Ann. Rev. Microbiol. *21*:285–324.

Phelps, T. J. and J. G. Zeikus. 1985. Effect of fall turnover on terminal carbon metabolism in Lake Mendota sediments. Appl. Environ. Microbiol. *50*:1285–1291.

Philbrick, C. T. and D. H. Les. 1996. Evolution of aquatic angiosperm reproductive systems. BioScience *46*:813–826.

Phillips, G. L., D. Eminson, and B. Moss. 1978. A mechanism to account for macrophyte decline in progressively eutrophicated freshwaters. Aquat. Bot. *4*:103–126.

Phillips, J. D. 1989. Fluvial sediment storage in wetlands. Water Resour. Bull. *25*:867–873.

Philosoph-Hadas, S., S. Meir, and N. Aharoni. 1994. Role of ethylene in senescence of watercress leaves. Physiol. Plant. *90*:553–559.

Philp, R. P., J. R. Maxwell, and G. Eglinton. 1976. Environmental organic geochemistry of aquatic sediments. Sci. Prog. Oxford *63*:521–545.

Phipps, D. W., Jr. and R. L. Pardy. 1982. Host enhancement of symbiont photosynthesis in the hydra-algae symbiosis. Biol. Bull. *162*:83–94.

Pia, J. 1933. Kohlensäure und Kalk. Die Binnengewässer *13*, 183 pp.

Piccolo, A., S. Nardi, and G. Concheri. 1996. Macromolecular changes of humic substances induced by interaction with organic acids. Eur. J. Soil Sci. *47*:319–328.

Pidgaiko, M. L., B. G. Grin, L. A. Kititsina, L. G. Lenchina, M. F. Polivannaya, O. A. Sergeeva, and T. A. Vinogradskaya. 1970. Biological productivity of Kurakhov's Power Station cooling reservoir. *In* Z. Kajak and A. Hillbricht-Ilkowska, eds. Productivity Problems of Freshwaters. PWN Polish Sci. Publs., Warsaw. pp. 477–491.

Pieczyńska, E. 1959. Character of the occurrence of free-living Nematoda in various types of periphyton in Lake Tajty. Ekol. Polska (Ser. A) 7:317–337. (Polish; English summary.)

Pieczyńska, E. 1964. Investigations on colonization of new substrates by nematodes (Nematoda) and some other periphyton organisms. Ekol. Polska (Ser. A) *12*:185–234.

Pieczyńska, E. 1965. Variations in the primary production of plankton and periphyton in the littoral zone of lakes. Bull. Acad. Polon. Sci. Cl. II, *13*:219–225.

Pieczyńska, E. 1968. Dependence of the primary production of periphyton upon the substrate area suitable for colonization. Bull. Acad. Polon. Sci. Cl. II, *16*:165–169.

Pieczyńska, E. 1970. Production and decomposition in the eulittoral zone of lakes. *In* Z. Kajak and A. Hillbricht-Ilkowska, eds. Productivity Problems of Freshwaters. PWN Polish Scientific Publishers, Warsaw. pp. 271–285.

Pieczyńska, E. 1972a. Ecology of the eulittoral zone of lakes. Ekol. Polska *20*:637–732.

Pieczyńska, E. 1972b. Rola materii allochtonicznej w jeziorach. Wiadomosci Ekologiczne *18*:131–140.

Pieczyńska, E. and I. Spodniewska. 1963. Occurrence and colonization of periphyton organisms in accordance with the type of substrate. Ekol. Polska (Ser. A) *11*:533–545.

Pieczyńska, E. and W. Szczepańska. 1966. Primary production in the littoral of several Masurian lakes. Verh. Internat. Verein. Limnol. *16*:372–379.

Pieczyńska, E., E. Pieczyński, T. Prus, and K. Tarwid. 1963. The biomass of the bottom fauna of 42 lakes in the Wegorzewo District. Ekol. Polska (Ser. A) *11*:495–502.

Piedade, M. T. F., W. J. Junk, and S. P. Long. 1991. The productivity of the C_4 grass *Echinochloa polystachya* on the Amazon floodplain. Ecology *72*:1456–1463.

Pielou, E. C. 1977. Mathematical Ecology. John Wiley & Sons, New York. 385 pp.

Pielou, E. C. 1981. The usefulness of ecological models: A stocktaking. Quart. Rev. Biol. *56*:17–31.

Pielou, E. C. 1998. Fresh Water. Univ. Chicago Press, Chicago. 275 pp.

Pierson, D. C. and G. A. Weyhenmeyer. 1994. High resolution measurements of sediment resuspension above an accumulation bottom in a stratified lake. Hydrobiologia *284*:43–57.

Pieterse, A. H. and K. J. Murphy, eds. 1990. Aquatic Weeds: The Ecology and Management of Nuisance Aquatic Vegetation. Oxford Univ. Press, Oxford. 593 pp.

Pijanowska, J. and G. Stolpe. 1996. Summer diapause in *Daphnia* as a reaction to the presence of fish. J. Plankton Res. *18*:1407–1412.

Pimentel, G. C. and A. L. McClellan. 1960. The Hydrogen Bond. W. H. Freeman and Co., San Francisco. 475 pp.

Pimm, S. L. 1984. The complexity and stability of ecosystems. Nature *307*:321–326.

Pinay, G., H. Décamps, E. Chauvet, and E. Fustec. 1990. Functions of ecotones in fluvial systems. *In* R. J. Naiman and H. Décamps, eds. The Ecology and Management of Aquatic-Terrestrial Ecotones. Parthenon Publishing Group, Carnforth, UK. pp. 141–169.

Pinel-Alloul, B. 1978. Ecologie des populations de *Lymnaea catascopium catascopium* (Mollusques, Gastéropodes, Pulmonées) du lac St-Louis, près de Montréal, Québec. Verh. Internat. Verein. Limnol. *20*:2412–2426.

Pinel-Alloul, B., E. Magnin, and G. Codin-Blumer. 1982. Effets de la mise en eau du réservoir Desaulniers (Territoire de la Baie de James) sur le zooplancton d'une rivière et d'une tourbière reticulée. Hydrobiologia *86*:271–296.

Pithart, D. 1997. Diurnal vertical migration study during a winter bloom of Cryptophyceae in a floodplain pool. Int. Rev. ges. Hydrobiol. *82*:33–46.

Pivovarov, A. A. 1973. Thermal Conditions in Freezing Lakes and Rivers. Israel Program for Scientific Translations of 1972 Russian edition. John Wiley & Sons, New York. 136 pp.

Planas, D. 1996. Acidification effects. *In* R. J. Stevenson, M. L. Bothwell, and R. L. Lowe, eds. Algal Ecology: Freshwater Benthic Ecosystems. Academic Press, San Diego. pp. 497–530.

Planas, D. and R. E. Hecky. 1984. Comparison of phosphorus turnover times in northern Manitoba reservoirs with lakes of the Experimental Lakes Area. Can. J. Fish. Aquat. Sci. *41*:605–612.

Planter, M. 1970. Physico-chemical properties of the water of reedbelts in Mikolajskie, Taltowisko, and Sniardwy lakes. Pol. Arch. Hydrobiol. *17*:337–356.

Ploug, H., C. Lassen, and B. B. Jørgensen. 1993. Action spectra of microalgal photosynthesis and depth distribution of spectral scalar irradiance in a coastal marine sediment of Limfjorden, Denmark. FEMS Miocrobiol. Ecol. *102*:261–270.

Poff, N. L. and J. V. Ward. 1992. Heterogeneous currents and algal resources mediate *in situ* foraging activity of a mobile stream grazer. Oikos *65*:465–478.

Poiani, K. A. and W. C. Johnson. 1991. Global warming and prairie wetlands. BioScience *41*:611–618.

Poiani, K. A., W. C. Johnson, G. A. Swanson, and T. C. Winter. 1996. Climate change and northern prairie wetlands: Simulations of long-term dynamics. Limnol. Oceanogr.*41*:871–881.

Poindexter, J. S. 1981. Oligotrophy: Fast and famine existence. Adv. Microb. Ecol. *5*:63–89.

Poirrier, M. A., B. R. Bordelon, and J. L. Laseter. 1972. Adsorption and concentration of dissolved carbon-14 DDT by coloring colloids in surface waters. Environ. Sci. Technol. *6*:1033–1035.

Pokorný, J., J. Květ, J. P. Ondok, Z. Toul, and I. Ostrý. 1984. Production-ecological analysis of a plant community dominated by *Elodea canadensis* Michx. Aquat. Bot. *19*:363–392.

Pokorný, J., L. Hammer, and J. P. Ondok. 1987. Oxygen budget in the reed belt and open water of a shallow lake. Arch. Hydrobiol. Beih. Ergebn. Limnol. *27*:185–201.

Polishchuk, L. V. and A. M. Ghilarov. 1981. Comparison of two approaches used to calculate zooplankton mortality. Limnol. Oceanogr. *26*:1162–1168.

Pollingher, U., B. Kaplan, and T. Berman. 1995. The impact of iron and chelators on Lake Kinneret phytoplankton. J. Plankton Res. *17*:1977–1992.

Pollingher, U., H. R. Bürgi, and H. Ambühl. 1993. The cysts of *Ceratium hirundinella*: Their dynamics and role within a eutrophic lake (Lake Sempach, Switzerland). Aquatic Sci. *5*:10–18.

Poltz, J. 1972. Untersuchungen über das Vorkommen und den Abbau von Fetten und Fettsäuren in See. Arch. Hydrobiol./Suppl. *40*:315–399.

Polunin, N. V. C. 1984. The decomposition of emergent macrophytes in fresh water. Adv. Ecol. Res. *14*:115–166.

Pomeroy, L. R. 1974. The ocean's food web: A changing paradigm. BioScience *24*:499–504.

Pomeroy, L. R. and W. J. Wiebe. 1988. Energetics of microbial food webs. Hydrobiologia *159*:7–18.

Pomeroy, L. R., E. E. Smith, and C. M. Grant. 1965. The exchange of phosphate between estuarine water and sediments. Limnol. Oceanogr. *10*:167–172.

Pompêo, M. L. M. and R. Henry. 1998. Decomposition of macrophyte *Echinochloa polystachya* (H.B.K.) Hitchcock, in a Brazilian reservoir (Paranapanema River mouth). Verh. Internat. Verein. Limnol. *26*:1871–1875.

Poole, H. H. and W. R. G. Atkins. 1929 Photo-electric measurements of submarine illumination throughout the year. J. Mar. Biol. Assoc. U.K. *16*:297–324.

Popp, A. and K. D. Hoagland. 1995. Changes in benthic community composition in response to reservoir aging. Hydrobiologia *306*:159–171.

Porter, K. G. 1973. Selective grazing and differential digestion of algae by zooplankton. Nature *244*:179–180.

Porter, K. G. 1975. Viable gut passage of gelatinous green algae ingested by *Daphnia*. Verh. Internat. Verein. Limnol. *19*: 2840–2850.

Porter, K. G. 1976. Enhancement of algal growth and productivity by grazing zooplankton. Science *192*:1332–1334.

Porter, K. G. 1988. Phagotrophic phytoflagellates in microbial food webs. Hydrobiologia *159*:89–97.

Porter, K. G. and J. D. Orcutt, Jr. 1980. Nutritional adequacy, manageability, and toxicity as factors that determine the food quality of green and blue-green algae for *Daphnia*. *In* W. C. Kerfoot, ed. Evolution and Ecology of Zooplankton Communities. Univ. Press New England, Hanover, NH. pp. 268–281.

Porter, K. G. and Y. S. Feig. 1980. The use of DAPI for identifying and counting aquatic microflora. Limnol. Oceanogr. 25:943–948.

Porter, K. G., M. L. Pace, and J. F. Battey. 1979. Ciliate protozoans as links in freshwater planktonic food chains. Nature *277*:563–565.

Portnoy, J. W. and M. A. Soukup. 1990. Gull contributions of phosphorus and nitrogen to a Cape Cod kettle pond. Hydrobiologia *202*:61–69.

Posch, T. and H. Arndt. 1996. Uptake of sub-micrometre- and micrometre-sized detrital particles by bacterivorous and omnivorous ciliates. Aquat. Microb. Ecol. *10*:45–53.

Post, D. M., M. L. Pace, and N. G. Hairston, Jr. 2000. Ecosystem size determines food-chain length in lakes. Nature *405*:1047–1049.

Post, D. M., S. R. Carpenter, D. L. Christensen, K. L. Cottingham, J. F. Kitchell, D. E. Schindler, and J. R. Hodgson. 1997. Seasonal effects of variable recruitment of a dominant piscivore on pelagic food web structure. Limnol. Oceanogr. *42*:722–729.

Post, J. R. and D. Cucin. 1984. Changes in the benthic community of a small precambrian lake following the introduction of yellow perch (*Perca flavescens*). Can. J. Fish. Aquat. Sci. *41*:1496–1501.

Postel, S. L. 1998. Water for food production: Will there be enough in 2025? BioScience *48*:629–637.

Postgate, J. R. 1976. Death in macrobes and microbes. *In* T. R. G. Gray and J. R. Postgate, eds. The Survival of Vegetative Microbes. Cambridge Univ. Press, Cambridge. pp. 1–18.

Postolková, M. 1967. Comparison of the zooplankton amount and primary production of the fenced and unfenced littoral regions of Smyslov Pond. Rozpravy Ceskosl. Akad. Ved, Rada Matem. Prír. Ved 77(11):63–79.

Potts, W. T. W. and G. Parry. 1964. Osmotic and Ionic Regulation in Animals. Pergamon Press, Oxford. 423 pp.

Potzger, J. E. and W. A. van Engel. 1942. Study of the rooted aquatic vegetation of Weber Lake, Vilas County, Wisconsin. Trans. Wis. Acad. Sci. Arts Lett. *34*:149–166.

Poulícková, A. 1993. Ecological study of seasonal maxima of centric diatoms. Algological Stud. *68*:85–106.

Pourriot, R. 1965. Recherches sur l'écologie des rotifères. Vie et Milieu (suppl.) *21*, 224 pp.

Pourriot, R. 1974. Relations prédateur-proie chez les rotifères: influence du prédateur (*Asplanchna brightwelli*) sur la morphologie de la proie (*Brachionus bidentata*). Ann. Hydrobiol. *5*:43–55.

Pourriot, R. and P. Clément. 1975. Influence de la durée de l'éclairement quotidien sur le taux de femelles mictiques chez *Notommata copeus* Ehr. (Rotifère). Oecologia *22*:67–77.

Pourriot, R. and T. W. Snell. 1983. Resting eggs in rotifers. Hydrobiologia *104*:213–224.

Pourriot, R., C. Rougier, and A. Miquelis. 1997. Origin and development of river zooplankton: Example of the Marne. Hydrobiologia *345*:143–148.

Povoledo, D. 1961. Ulteriori studi sulle sostanze organiche disciolte nell'acqua del Lago Maggiore: Frazionamento e separazione delle proteine dai peptidi e dagli aminoacidi liberi. Mem. Ist. Ital. Idrobiol. *13*:203–222.

Powell, T. and A. Jassby. 1974. The estimation of vertical eddy diffusivities below the thermocline in lakes. Water Resour. Res. *10*:191–198.

Powell, T., M. H. Kirkish, P. J. Neale, and P. J. Richardson. 1984. The diurnal cycle of stratification in Lake Titicaca: Eddy diffusion. Verh. Internat. Verein. Limnol. *22*:1237–1243.

Power, M. E. 1992. Hydrological and trophic controls of seasonal algal blooms in northern California rivers. Arch. Hydrobiol. *125*:375–410.

Power, M. E., W. J. Matthews, and A. J. Stewart. 1985. Grazing minnows, piscivorous bass, and stream algae: Dynamics of a strong interaction. Ecology *66*:1448–1456.

Prado, A. L., C. W. Heckman, and F. R. Martins. 1994. The seasonal succession of biotic communities in wetlands of the tropical wet-and-dry climatic zone. II. The aquatic macrophyte vegetation in the Pantanal of Mato Grosso, Brazil. Int. Revue ges. Hydrobiol. *79*:569–589.

Preissler, K. 1977. Do rotifers show "avoidance of the shore"? Oecologia *27*:253–260.

Prejs, K. 1977a. The nematodes of the root region of aquatic macrophytes, with special consideration of nematode groupings penetrating the tissues of roots and rhizomes. Ekol. Polska *25*:5–20.

Prejs, K. 1977b. The species diversity, numbers and biomass of benthic nematodes in central part of lakes with different trophy. Ekol. Polska *25*:31–44.

Prejs, K. 1977c. The littoral and profundal benthic nematodes of lakes with different trophy. Ekol. Polska *25*:21–30.

Prentki, R. T., T. D. Gustafson, and M. S. Adams. 1978. Nutrient movements in lakeshore marshes. *In* R. E. Good, D. F. Whigham, and R. L. Simpson, eds. Freshwater Wetlands: Ecological Processes and Management Potential. Academic Press, New York. pp. 169–194.

Prepas, E. E. 1983. Orthophosphate turnover time in shallow productive lakes. Can. J. Fish. Aquat. Sci. *40*:1412–1418.

Prepas, E. E. and F. H. Rigler. 1982. Improvements in quantifying the phosphorus concentration in lake water. Can. J. Fish. Aquat. Sci. *39*:822–829.

Preston, T., W. D. P. Stewart, and C. S. Reynolds. 1980. Bloom-forming cyanobacterium *Microcystis aeruginosa* overwinters on sediment surface. Nature *288*:365–367.

Prézelin, B. B. and B. M. Sweeney. 1978. Photoadaptation of photosynthesis in *Gonyaulax polyedra*. Mar. Biol. *48*:27–35.

Pribán, K. and J. P. Ondok. 1985. Heat balance components and evapotranspiration from a sedge-grass marsh. Fol. Geobot. Phytotax. Praha *20*:41–56.

Pribán, K. and J. P. Ondok. 1986. Evapotranspiration of a willow carr in summer. Aquat. Bot. *25*:203–216.

Price, N. M. and F. M. M. Morel. 1990. Cadmium and cobalt substitution for zinc in a marine diatom. Nature *344*:658–660.

Price, N. M. and F. M. M. Morel. 1991. Colimitation of phytoplankton growth by nickel and nitrogen. Limnol. Oceanogr. *36*:1071–1077.

Priddle, J. 1980. The production ecology of benthic plants in some Antarctic lakes. I. In situ production studies. J. Ecol. *68*:141–153.

Priddle, J. and R. B. Heywood. 1980. Evolution of Antarctic lake ecosystems. Biol. J. Linn. Soc. *14*:51–66.

Priest, F. G. 1984. Extracellular Enzymes. Van Nostrand/Reinhold, Wokingham. 79 pp.

Pringle, C. M. 1987. Effects of water and substratum nutrient supplies on lotic periphyton growth: An integrated bioassay. Can. J. Fish. Aquat. Sci. *44*:619–629.

Pringle, C. M. 1990. Nutrient spatial heterogeneity: Effects on community structure, physiognomy, and diversity of stream algae. Ecology *71*:905–920.

Pringle, C. M., M. C. Freeman, and B. J. Freeman. 2000. Regional effects of hydrologic alterations on riverine macrobiota in the New World: Tropical-temperate comparisons. BioScience *50*:807–823.

Prins, H. B. A. and R. W. Wolff. 1974. Photorespiration in leaves of *Vallisneria spiralis*. The effect of oxygen on the carbon dioxide compensation point. Proc. Akad. van Wetensc. Amsterdam (Ser. C) *77*:239–245.

Prins, H. B. A. and J. T. M. Elzenga. 1989. Bicarbonate utilization: Function and metabolism. Aquat. Bot. *34*:59–83.

Prins, H. B. A., J. F. H. Snel, R. J. Helder, and P. E. Zanstra. 1980. Photosynthetic HCO_3^- utilization and OH^- excretion in aquatic angiosperms. Light-induced pH changes at the leaf surface. Plant Physiol. *66*:818–822.

Priscu, J. C. 1983. Suspensoid characteristics in subalpine Castle Lake, California. II. Optical properties. Arch. Hydrobiol. *97*:425–433.

Priscu, J. C. 1984. A comparison of nitrogen and carbon metabolism in the shallow and deep-water phytoplankton populations of a subalpine lake: Response to photosynthetic photon flux density. J. Plankton Res. *6*:733–749.

Priscu, J. C. 1997. The biogeochemistry of nitrous oxide in permanently ice-covered lakes of the McMurdo Dry Valleys, Antarctica. Global Change Biol. *3*:301–315.

Priscu, J. C. and C. R. Goldman. 1983. Seasonal dynamics of the deep-chlorophyll maximum in Castle Lake, California. Can. J. Fish. Aquat. Sci. *40*:208–214.

Priscu, J. C. and C. R. Goldman. 1984. The effect of temperature on photosynthetic and respiratory electron transport system activity in the shallow and deep-living phytoplankton of a subalpine lake. Freshwat. Biol. *14*:143–155.

Priscu, J. C., W. F. Vincent, and C. Howard-Williams. 1989. Inorganic nitrogen uptake and regeneration in perennially ice-covered lakes Fryxell and Vanda, Antarctica. J. Plankton Res. *11*:335–351.

Proctor, M. C. F. 1995. The ombrogenous bog environment. *In* B. D. Wheeler, S. C. Shaw, W. J. Fojt, and R. A. Robertson, eds. Restoration of Temperate Wetlands. John Wiley & Sons, Chichester. pp. 287–303.

Proctor, R. M. and J. O. Young. 1987. The life history, diet and migration of a lake-dwelling population of the leech *Alboglossiphonia heteroclita* (L.). Hydrobiologia *150*:133–139.

Proctor, V. W. 1957. Studies of algal antibiosis using *Haematococcus* and *Chlamydomonas*. Limnol. Oceanogr. *2*:125–139.

Pronzato, R. and R. Manconi. 1995. Long-term dynamics of a freshwater sponge population. Freshwat. Biol. *33*:485–495.

Provasoli, L. 1958. Nutrition and ecology of protozoa and algae. Ann. Rev. Microbiol. *12*:279–308.

Provasoli, L. 1963. Organic regulation of phytoplankton fertility. *In* M. N. Hill, ed. The Sea. Vol. 2. Interscience, New York. pp. 165–219.

Provasoli, L., J. J. A. McLaughlin, and I. J. Pintner. 1954. Relative and limiting concentrations of major mineral constituents for the growth of algal flagellates. Trans. N.Y. Acad. Sci. (Ser. 2) *16*:412–417.

Prowse, T. D. 1994. Environmental significance of ice to streamflow in cold regions. Freshwat. Biol. *32*:241–259.

Psenner, R. 1984. The proportion of empneuston and total atmospheric inputs of carbon, nitrogen and phosphorus in the nutrient budget of a small mesotrophic lake (Piburger See, Austria). Int. Rev. ges. Hydrobiol. *69*:23–39.

Pugh, G. J. F. and J. L. Mulder. 1971. Mycoflora associated with *Typha latifolia*. Trans. Br. Mycol. Soc. *57*:273–282.

Purchase, B. S. 1977. Nitrogen fixation associated with *Eichhornia crassipes*. Plant Soil *46*:283–286.

Pusch, M. 1996. The metabolism of organic matter in the hyporheic zone of a mountain stream, and its spatial distribution. Hydrobiologia *323*:107–118.

Pusch, M. and J. Schwoerbel. 1994. Community respiration in hyporheic sediments of a mountain stream (Steina, Black Forest). Arch. Hydrobiol. *130*:35–52.

Putnam, H. D. and T. A. Olson. 1961. Studies on the Productivity and Plankton of Lake Superior. Rep. School Public Health, University of Minnesota, 24 pp.

Putz, R. 1997. Periphyton communities in Amazonian black- and whitewater habitats: Community structure, biomass and productivity. Aquat. Sci. *59*:74–93

Pyke, G. H., H. R. Pulliam, and E. L. Charnov. 1977. Optimal foraging: A selective review of theory and tests. Quart. Rev. Biol. *52*:137–154.

Qian, J., H. B. Xue, L. Sigg, and A. Albrecht. 1998. Complexation of cobalt by natural ligands in freshwater. Environ. Sci. Technol. *32*:2043–2050.

Quade, H. W. 1969. Cladoceran faunas associated with aquatic macrophytes in some lakes in northwestern Minnesota. Ecology *50*:170–179.

Quade, H. W. 1971. Niche specificity of littoral Cladocera: Habitat. Trans. Amer. Microsc. Soc. *90*:104–105.

Quennerstedt, N. 1955. Diatoméerna i Långans Sjövetetation. Acta Phytogeographica Suecica *36*. 208 pp.

Quigley, M. A. and J. A. Robbins. 1984. Silica regeneration processes in nearshore southern Lake Michigan. J. Great Lakes Res. *10*:383–392.

Quirós, R. and S. Cuch. 1990. The fisheries and limnology of the lower Plata Basin. *In* D. P. Dodge, ed. International Large River Symposium. Can. Spec. Publ. Fish. Aquat. Sci. *106*:429–443.

Radek, R. and K. Hausmann. 1994. Endocytosis, digestion, and defecation in flagellates. Acta Protozool. *33*:127–147.

Ragotzkie, R. A. 1978. Heat budgets of lake. *In* A. Lerman, ed. Lakes: Chemistry, Geology, Physics. Springer-Verlag, New York. pp. 1–19.

Rahm, L. 1985. The thermally forced circulation in a small, ice-covered lake. Limnol. Oceanogr. *30*:1122–1128.

Rai, D. N. and J. D Munshi. 1979. The influence of thick floating vegetation (water hyacinth: *Eichhornia crassipes*) on the physico-chemical environment of a freshwater wetland. Hydrobiologia *62*:65–69.

Rai, H. 1978. Utilizacào de glicose por bactérias heterotróficas no ecossistema lacustre de Amazonia Central. Acta Amazonica *8*:225–232.

Rai, H. 1979. Microbiology of central Amazon lakes. Amazoniana *6*:583–599.

Rai, H. and G. Hill. 1981. Bacterial biodynamics in Lago Tupé, a central Amazonian black water "Ria Lake." Arch. Hydrobiol./ Suppl. *58*:420–468.

Rai, H. and T. R. Jacobsen. 1993. Dissolved alkaline phosphatase activity (APA) and the contribution of APA by size fractionated plankton in Lake Schöhsee. Verh. Internat. Verein. Limnol. *25*:164–169.

Rai, H., M. T. Arts, B. C. Wainman, N. Dockal, and H. J. Krambeck. 1997. Lipid production in natural phytoplankton communities in a small freshwater Baltic lake, Lake Schöhsee, Germany. Freshwat. Biol. *38*:581–590.

Raidt, H. and R. Koschel. 1988. Morphology of calcite crystals in hardwater lakes. Limnologica (Berlin) *19*:3–12.

Raidt, H. and R. Koschel. 1993. Variable morphology of calcite crystals in hardwater lakes. Limnologica *23*:85–89.

Raikow, D. F., S. A. Grubbs, and K. W. Cummins. 1995. Debris dam dynamics and coarse particulate organic matter retention in an Appalachian mountain stream. J. N. Am. Benthol. Soc. 14:535–546.

Rainwater, F. H. and L. L. Thatcher. 1960. Methods for collection and analysis of water samples. U.S. Geol. Surv. Water-Supply Pap. *1454*, 301 pp.

Ralph, E. K. and H. N. Michael. 1974. Twenty-five years of radiocarbon dating. Am. Sci. *62*:553–560.

Ramaiah, N. 1995. Summer abundance and activities of bacteria in the freshwater lakes of Schirmacher Oasis, Antarctica. Polar Biol. *15*:547–553.

Ramazzotti, G. 1962. Il Phylum Tardigrada. Mem. Ist. Ital. Idrobiol. *16*:1–595.

Ramazzotti, G. 1972. Il Phylum Tardigrada (Seconda edizione aggiornata). Mem. Ist. Ital. Idrobiol. *28*:1–732.

Ramazzotti, G. and W. Maucci. 1983. Il Phylum Tardigrada (III edizione reveduta e aggiornata). Mem. Ist. Ital. Idrobiol. *41*:1–1012.

Ramberg, L. 1987. Phytoplankton succession in the Sanyati basin, Lake Kariba. Hydrobiologia *153*:193–202.

Ramcharan, C. W., D. J. McQueen, E. Demers, S. A. Popiel, A. M. Rocchi, N. D. Yan, A. H. Wong, and K. D. Hughes. 1995. A comparative approach to determining the role of fish predation in structuring limnetic ecosystems. Arch. Hydrobiol. *133*:389–416.

Ramcharan, C. W., R. L. France, and D. J. McQueen. 1996. Multiple effects of planktivorous fish on algae through a pelagic trophic cascade. Can. J. Fish. Aquat. Sci. *53*:2819–2828.

Ramlal, P. S., R. H. Hesslein, R. E.Hecky, E. J. Fee, J. W. M. Rudd, and S. J. Guildford. 1994. The organic carbon budget of a shallow Arctic tundra lake on the Tuktoyaktuk Peninsula, N.W.T., Canada. Biogeochemistry *24*:145–172.

Ramsey, W. L. 1962a. Bubble growth from dissolved oxygen near the sea surface. Limnol. Oceanogr. *7*:1–7.

Ramsey, W. L. 1962b. Dissolved oxygen in shallow nearshore water and its relation to possible bubble formation. Limnol. Oceanogr. *7*:453–461.

Rao, A. S. 1988. Evapotranspiration rates of *Eichhornia crassipes* (Mart.) Solms, *Salvinia molesta* D. S. Mitchell and *Nymphaea lotus* (L.) Willd. Linn. In a humid tropical climate. Aquat. Bot. *30*:215–222.

Rao, I. M. 1997. The role of phosphorus in photosynthesis. *In* M. Pessarakli, ed. Handbook of Photosynthesis. M. Dekker, Inc., New York. pp. 173–194.

Rao, S. S. and B. K. Burnison. 1976. Bacterial distributions in Lake Erie (1967, 1970). J. Fish. Res. Board Can. *33*:574–580.

Rascio, N., F. Cuccato, F. Dalla Vecchia, N. LaRocca, and W. Larcher. 1999. Structural and functional features of the leaves of *Ranunculus trichophyllus* Chaix., a freshwater submerged macrophyte. Plant Cell Environ. *22*:205–212.

Raskin, I. and H. Kende. 1983. How does deep water rice solve its aeration problem. Plant Physiol. *72*:447–454.

Raskin, I. and H. Kende. 1985. Mechanism of aeration in rice. Science *228*:327–329.

Raspopov, I. M. 1971. Litoralvegatation der Onega-und Ladogaseen. Hidrobiologia (Romania) *12*:241–247.

Raspopov, I. M. 1979. Vegetation der grossen seichten Seen im Nordwesten der UdSSR und ihre Produktion. Arch. Hydrobiol. *86*:242–253.

Raspopov, I. M., V. A. Ekzercev, and I. L. Koreljakova. 1977. Production by freshwater vascular plant (macrophyte) communities of lakes and reservoirs in the European part of the U.S.S.R. Folia Geobot. Phytotax. Praha *12*:113–120.

Raspor, B. 1980. Distribution and speciation of cadmium in natural waters. *In* J. O. Nriagu, ed. Cadmium in the Environment. Part I. Ecological Cycling. John Wiley & Sons, New York. pp. 147–236.

Rast, W. and G. F. Lee. 1978. Summary analysis of the North American (U.S. portion) OECD Eutrophication Project: Nutrient loading-lake response relationships and trophic state indices. U.S. Environm. Protection Agency Rept. EPA-600/3–78–008, 455 pp.

Rast, W., R. A. Jones, and G. F. Lee. 1983. Predictive capability of U.S. OECD phosphorus loading-eutrophication response models. J. Water Poll. Contr. Fed. *55*:990–1003.

Rasumov, A. S. 1962. Mikrobial'nyi plankton vody. Trudy Vsesoyusnogo Gidrobiol. Obshch. *12*:60–190.

Raven, J. A. 1970. Exogenous inorganic carbon sources in plant photosynthesis. Biol. Rev. *45*:167–221.

Raven, J. A. 1984. Energetics and Transport in Aquatic Plants. A. R. Liss, Inc., New York. 587 pp.

Raven, J. A. 1995. Photosynthetic and non-photosynthetic roles of carbonic anhydrase in algae and cyanobacteria. Phycologia *34*:93–101.

Raven, J. A. 1996. Into the voids: The distribution, function, development and maintenance of gas spaces in plants. Ann. Bot. *78*:137–142.

Raven, J. A. and K. Richardson. 1984. Dinophyte flagella: A cost-benefit analysis. New Phytol. *98*:259–276.

Raven, J. A. and W. J. Lucas. 1985. Energy costs of carbon acquisition. *In* W. J. Lucas, and J. A. Berry, eds. Inorganic Carbon Uptake by Aquatic Photosynthetic Organisms. Amer. Soc. Plant Physiol., Beltsville, MD. pp. 305–324.

Ravera, O. and M. C. Gatti. 1993. Release of carbon, nitrogen and hydrogen from dead zooplankton. Verh. Internat. Verein. Limnol. *25*:766–769.

Rawson, D. S. 1939. Some physical and chemical factors in the metabolism of lakes. *In* Problems of Lake Biology. Publ. Amer. Assoc. Adv. Sci. *10*:9–26.

Rawson, D. S. 1952. Mean depth and the fish production of large lakes. Ecology *33*:515–521.

Rawson, D. S. 1955. Morphometry as a dominant factor in the productivity of large lakes. Verh. Internat. Verein. Limnol. *12*:164–175.

Rawson, D. S. 1956. Algal indicators of trophic lake types. Limnol. Oceanogr. *1*:18–25.

Rawson, D. S. and J. E. Moore. 1944. The saline lakes of Saskatchewan. Can. J. Res. (Sec. D), *22*:141–201.

Rebsdorf, A., N. Thyssen, and M. Erlandsen. 1991. Regional and temporal variation in pH, alkalinity and carbon dioxide in Danish streams, related to soil type and land use. Freshwat. Biol. *25*:419–435.

Reckhow, K. H. 1979. Empirical lake models for phosphorus: Development, applications, limitations and uncertainty. *In* D. Scavia and A. Robertson, eds. Perspectives on Lake Ecosystem Modeling. Ann Arbor Sci. Publs., Ann Arbor, MI. pp. 193–221.

Reckhow, K. H. and S. C. Chapra. 1983. Engineering Approaches for Lake Management. I. Data Analysis and Empirical Modeling. Butterworths, Boston. 340 pp.

Reddy, K. R. and W. H. Patrick. 1984. Nitrogen transformations and loss in flooded soils and sediments. CRC Crit. Rev. Environ. Control *13*:273–309.

Reddy, K. R., M. M. Fisher, and D. Ivanoff. 1996. Resuspension and diffusive flux of nitrogen and phsophorus in a hypereutrophic lake. J. Environ. Qual. *25*:363–371.

Redfield, G. W. and C. R. Goldman. 1980. Diel vertical migration by males, females, copepodids and nauplii in a limnetic population of *Diaptomus* (Copepoda). Hydrobiologia *74*:241–248.

Redfield, G. W. and W. F. Vincent. 1979. Stages of infection and ecological effects of a fungal epidemic on the eggs of a limnetic copepod. Freshwat. Biol. *9*:503–510.

Reeburgh, W. A. and D. T. Heggie. 1977. Microbial methane consumption reactions and their effect on methane distributions in freshwater and marine environments. Limnol. Oceanogr. *22*:1–9.

Reede, T. 1995. Life history shifts in response to different levels of fish kairomones in *Daphnia*. J. Plankton Res. *17*:1661–1667.

Reeders, H. H., A. B. D. Vaate, and F. J. Slim. 1989. The filtration rate of *Dreissena polymorpha* (Bivalvia) in three Dutch lakes with reference to biological water quality management. Freshwat. Biol. *22*:133–141.

Rees, T. A. V. 1984. Sodium dependent photosynthetic oxygen evolution in a marine diatom. J. Exp. Bot. *35*:332–337.

Reese, W. H. 1967. Physiological ecology and structure of benthic communities in a woodland stream. Ph.D. Dissertation. Oregon State University, Corvallis.

Reichardt, W., J. Overbeck, and L. Steubing. 1967. Free dissolved enzymes in lake waters. Nature *216*:1345–1347.

Reif, C. B. 1969. Temperature profiles and heat flow in sediments of Nuangola. Proc. Pennsylvania Acad. Sci. *43*:98–100.

Reif, C. B. and D. W. Tappa. 1966. Selective predation: Smelt and cladocerans in Harveys Lake. Limnol. Oceanogr. *11*:437–438.

Reif, C. B., B. B. Smith, and A. Case. 1983. The desmids and physical characteristics of 100 lakes in northeastern Pennsylvania. J. Freshwat. Ecol. *2*:25–36.

Reimers, N., J. A. Maciolek, and E. P. Pister. 1955. Limnological study of the lakes in Convict Creek Basin, Mono County, California. Fish. Bull. U.S. Fish. Wildl. Serv. *56*:437–503.

Remane, A. and C. Schlieper. 1971. Biology of brackish water. 2nd Edition. Die Binnengewässer *25*, 372 pp.

Rempel, R. S. and P. J. Colby. 1991. A statistically valid model of the morphoedaphic index. Can. J. Fish. Aquat. Sci. *48*:1937–1943.

Renberg, I. 1978. Palaeolimnology and varve counts of the annually laminated sediment of Lake Rudetjärn, Northern Sweden. Early Norrland *11*:63–92.

Renberg, I. and M. Wik. 1984. Dating recent lake sediments by soot particle counting. Verh. Internat. Verein. Limnol. *22*:712–718.

Renberg, I. and U. Segerström. 1981. Applications of varved lake sediments in palaeoenvironmental studies. Wahlenbergia *7*:125–133.

Renberg, I., M. W. Persson, and O. Emteryd. 1994. Pre-industrial atmospheric lead contamination detected in Swedish lake sediments. Nature *368*:323–326.

Repka, S., M. van der Vlies, and J. Vijverberg. 1999. Food quality of detritus derived from the filamentous cyanobacterium *Oscillatoria limnetica* for *Daphnia galeata*. J. Plankton Res. (In press).

Reuter, J. E., S. L. Loeb, and C. R. Goldman. 1983. Nitrogen fixation in periphyton of oligotrophic Lake Tahoe. *In* R. G. Wetzel, ed. Periphyton of Freshwater Ecosystems. W. Junk Publishers, The Hague. pp. 101–109.

Reuter, J. E., S. L. Loeb, and C. R. Goldman. 1986. Inorganic nitrogen uptake by epilithic periphyton in a N-deficient lake. Limnol. Oceanogr. *31*:149–160.

Revsbech, N. P. and B. B. Jørgensen. 1986. Microelectrodes: Their use in microbial ecology. Microb. Ecol. *9*:293–352.

Revsbech, N. P., B. B. Jørgensen, T. H. Blackburn, and Y. Cohen. 1983. Microelectrode studies of the photosynthesis and O_2, H_2S, and pH profiles of a microbial mat. Limnol. Oceanogr. *28*:1062–1074.

Reynolds, C. S. 1976a. Succession and vertical distribution of phytoplankton in response to thermal stratification in a lowland mere, with special reference to nutrient availability. J. Ecol. *64*:529–551.

Reynolds, C. S. 1976b. Sinking movements of phytoplankton indicated by a simple trapping method. I. A. *Fragilaria* population. Br. Phycol. J. *11*:279–291.

Reynolds, C. S. 1978. Stratification in natural populations of bloom-forming blue-green algae. Verh. Internat. Verein. Limnol. *20*:2285–2292.

Reynolds, C. S. 1979. The limnology of the eutrophic meres of the Shropshire-Cheshire Plain: A review. Field Stud. *5*:93–173.

Reynolds, C. S. 1984. Phytoplankton periodicity: The interactions of form, function and environmental variability. Freshwat. Biol. *14*:111–142.

Reynolds, C. S. 1988. Potamoplankton: Paradigms, paradoxes and prognoses. *In* F. E. Round, ed. Algae and the Aquatic Environment. Biopress Ltd., Bristol. pp. 285–311.

Reynolds, C. S. 1990. Temporal scales of variability in pelagic environments and the response of phytoplankton. Freshwat. Biol. *23*:25–53.

Reynolds, C. S. 1993. Scales of disturbance and their role in plankton ecology. Hydrobiologia *249*:157–171.

Reynolds, C. S. 1994. The long, the short and the stalled: On the attributes of phytoplankton selected by physical mixing in lakes and rivers. Hydrobiologia *289*:9–21.

Reynolds, C. S. 1997. Vegetation Processes in the Pelagic: A Model for Ecosystem Theory. Ecology Institute, Oldendorf/Luhe, Germany. 371 pp.

Reynolds, C. S. and A. E. Walsby. 1975. Waterblooms. Biol. Rev. *50*:437–481.

Reynolds, C. S. and J.-P. Descy. 1996. The production, biomass and structure of phytoplankton in large rivers. Arch. Hydrobiol. Suppl. *113*:161–187.

Reynolds, C. S., J.-P. Descy, and J. Padisák. 1994. Are phytoplankton dynamics in rivers so different from those in shallow lakes? Hydrobiologia *289*:1–7.

Reynolds, G. L. and J. Hamilton-Taylor. 1992. The role of planktonic algae in the cycling of Zn and Cu in a productive soft-water lake. Limnol. Oceanogr. *37*:1759–1769.

Reynoldson, T. B. 1961. Observations on the occurrence of *Asellus* (Isopoda, Crustacea) in some lakes of northern Britain. Verh. Internat. Verein. Limnol. *14*:988–994.

Reynoldson, T. B. 1966. The distribution and abundance of lake-dwelling triclads—towards a hypothesis. Adv. Ecol. Res. *3*:1–71.

Reynoldson, T. B. 1981. The ecology of the Turbellaria with special reference to the freshwater triclads. Hydrobiologia *84*:87–90.

Reynoldson, T. B. 1983. The population biology of Turbellaria with special reference to the freshwater triclads of the British Isles. Adv. Ecol. Res. *13*:235–326.

Reynoldson, T. B. and A. D. Sefton. 1976. The food of *Planaria torva* (Müller) (Turbellaria-Tricladida), a laboratory and field study. Freshwat. Biol. 6:383–393.

Reynoldson, T. B. and B. Piearce. 1979. Feeding on gastropods by lake-dwelling *Polycelis* in the absence and presence of *Dugesia polychroa* (Turbellaria, Tricladida). Freshwat. Biol. *9*:357–367.

Reynoldson, T. B. and L. S. Bellamy. 1971. Intraspecific competition in lake-dwelling triclads. A laboratory study. Oikos *22*:315–328.

Reynoldson, T. B. and P. Bellamy. 1975. Triclads (Turbellaria: Tricladida) as predators of lake-dwelling stonefly and mayfly nymphs. Freshwat. Biol. *5*:305–312.

Reynoldson, T. B., F. J. Gilliam, and R. M. Jaques. 1981. Competitive exclusion and co-existence in natural populations of *Polycelis nigra* and *P. tenuis* (Tricladida, Turbellaria). Arch. Hydrobiol. 92:71–113.

Rhee, G.-Y. 1972. Competition between an alga and an aquatic bacterium for phosphate. Limnol. Oceanogr. *17*:505–514.

Rhee, G.-Y. 1973. A continuous culture study of phosphate uptake, growth rate and polyphosphate in *Scenedesmus sp.* J. Phycol. *9*:495–506.

Rhee, G.-Y. 1974. Phosphate uptake under nitrate limitation by *Scenedesmus sp.* and its ecological implications. J. Phycol. *10*:470–475.

Rhee, G.-Y. 1978. Effects of N:P atomic ratios and nitrate limitation on algal growth, cell composition, and nitrate uptake. Limnol. Oceanogr. *23*:10–25.

Rhee, G.-Y. and I. J. Gotham. 1980. Optimum N:P ratios and coexistence of planktonic algae. J. Phycol. *16*:486–489.

Rhee, G.-Y. and I. J. Gotham. 1981a. The effect of environmental factors on phytoplankton growth: Temperature and interactions of temperature with nutrient limitation. Limnol. Oceanogr. *26*:635–648.

Rhee, G.-Y. and I. J. Gotham. 1981b. The effect of environmental factors on phytoplankton growth: Light and the interactions of light with nitrate limitation. Limnol. Oceanogr. *26*:649–659.

Rho, J. and H. B. Gunner. 1978. Microfloral response to aquatic weed decomposition. Water Res. *12*:165–170.

Riber, H. H. and R. G. Wetzel. 1987. Boundary-layer and internal diffusion effects on phosphorus fluxes in lake periphyton. Limnol. Oceanogr. *32*:1181–1194.

Ricciardi, A. and H. M. Reiswig. 1994. Taxonomy, distribution, and ecology of the freshwater bryozoans (Ectoprocta) of eastern Canada. Can. J. Zool. *72*:339–359.

Ricciardi, A., F. G. Whoriskey, and J. B. Rasmussen. 1996. Impact of the *Dreissena* invasion on native unionid bivalves in the upper St. Lawrence River. Can. J. Fish. Aquat. Sci. *53*:1434–1444.

Rice, E. L. 1987. Allelopathy: An overview. *In* G. R. Walker, ed. Allelochemicals: Role in Agriculture and Forestry. American Chemical Society, Washington. pp. 8–20.

Rice, E. L. and S. K. Pancholy. 1972. Inhibition of nitrification by climax ecosystems. Am. J. Bot. *59*:1033–1040.

Rice, E. L. and S. K. Pancholy. 1973. Inhibition of nitrification by climax ecosystems. II. Additional evidence and possible role of tannins. Am. J. Bot. *60*:691–702.

Rich, P. H. 1975. Benthic metabolism of a soft-water lake. Verh. Internat. Verein. Limnol. *19*:1023–1028.

Rich, P. H. 1980. Hypolimnetic metabolism in three Cape Cod lakes. Am. Midland Nat. 1 *104*:102–109.

Rich, P. H. 1984. Trophic vs. detrital energetics: Is detritus productive? Bull. Mar. Sci. *35*:312–317.

Rich, P. H. and A. H. Devol. 1978. Analysis of five North American lake ecosystems. VII. Sediment processing. Verh. Internat. Verein. Limnol. *20*:598–604.

Rich, P. H. and R. G. Wetzel. 1969. A simple, sensitive underwater photometer. Limnol. Oceanogr. *14*:611–613.

Rich, P. H. and R. G. Wetzel. 1978. Detritus in lake ecosystems. Amer. Nat. *112*:57–71.

Rich, P. H. and T. E. Murray. 1990. De-icing salts in an urban drainage basin. Verh. Internat. Verein. Limnol. *24*:162–165.

Rich, P. H., R. G. Wetzel, and N. V. Thuy. 1971. Distribution, production and role of aquatic macrophytes in a southern Michigan marl lake. Freshwat. Biol. *1*:3–21.

Richard, D. I., J. W. Small, Jr., and J. A. Osborne. 1984. Phytoplankton responses to reduction and elimination of submerged vegetation by herbicides and grass carp in four Florida lakes. Aquat. Bot. *20*:307–319.

Richards, K. S. 1982. Rivers: Form and process in alluvial channels. Methuen, London.

Richards, S. R., C. A. Kelly, and J. W. M. Rudd. 1991. Organic volatile sulfur in lakes of the Canadian Shield and its loss to the atmosphere. Limnol. Oceanogr. *36*:468–482.

Richards, S. R., J. W. M. Rudd, and C. A. Kelly. 1994. Organic volatile sulfur in lakes ranging in sulfate and dissolved salt concentrations over five orders of magnitude. Limnol. Oceanogr. *39*:562–572.

Richardson, J. L., T. J. Harvey, and S. A. Holdship. 1978. Diatom in the history of shallow East African lakes. Pol. Arch. Hydrobiol. *25*:341–353.

Richardson, L. F. 1925. Turbulence and vertical temperature difference near trees. Phil. Mag. *49*:81–90.

Richardson, T. D. and K. M. Brown. 1989. Secondary production of two subtropical snails (Prosobranchia: Viviparidae). J. N. Am. Benthol. Soc. *8*:229–236.

Richardson, T. D., J. F. Scheiring, and K. M. Brown. 1988. Secondary production of two lotic snails (Pleuroceridae: *Elimia*). J. N. Am. Benthol. Soc. 7:234–245.

Richardson, W. B. 1991. Seasonal dynamics, benthic habitat use, and drift of zooplankton in a small stream in southern Oklahoma, U.S.A. Can. J. Zool. *69*:748–756.

Richardson, W. B. 1992. Microcrustacea in flowing water: Experimental analysis of washout times and a field test. Freshwat. Biol. *28*:217–230.

Richerson, P. J., M. Lopez, and T. Coon. 1978. The deep chlorophyll maximum layer of Lake Tahoe. Verh. Internat. Verein. Limnol. *20*:426–433.

Richerson, P., R. Armstrong, and C. R. Goldman. 1970. Contemporaneous disequilibrium, a new hypothesis to explain the "Paradox of the Plankton." Proc. Nat. Acad. Sci. USA *67*: 1710–1714.

Richey, J. S., W. H. McDowell, and G. E. Likens. 1985. Nitrogen transformations in a small mountain stream. Hydrobiologia *124*:129–139.

Richman, S. 1958. The transformation of energy by *Daphnia pulex*. Ecol. Monogr. *28*:273–291.

Richman, S. 1966. The effect of phytoplankton concentration on the feeding rate of *Diaptomus oregonensis*. Verh. Internat. Verein. Limnol. *16*:392–398.

Richman, S., M. D. Bailiff, L. J. Mackey, and D. W. Bolgrien. 1984. Zooplankton standing stock, species composition and size distribution along a trophic gradient in Green Bay, Lake Michigan. Verh. Internat. Verein. Limnol. *22*:475–487.

Ricker, W. E. 1934. A critical discussion of various measures of oxygen saturation in lakes. Ecology *15*:348–363.

Ricker, W. E. 1937. Physical and chemical characteristics of Cultus Lake, British Columbia. J. Biol. Board Can. *3*(4):363–402.

Ricker, W. E. 1952. The benthos of Cultus Lake. J. Fish. Res. Board Can. *9*:204–212.

Rickett, H. W. 1921. A quantitative study of the larger aquatic plants of Lake Mendota, Wisconsin. Trans. Wis. Acad. Sci. Arts Lett. *20*:501–527.

Ricklefs, R. E. 1979. Ecology, 2nd Edition. Chiron Press, Inc., New York. 966 pp.

Riddolls, A. 1985. Aspects of nitrogen fixation in Lough Neagh. I. Acetylene reduction and the frequency of *Aphanizomenon flos-aquae* heterocysts. Freshwat. Biol. *15*:289–297.

Ried, A. 1960a. Stoffwechsel und Verbreitungsgrenzen von Flechten. I. Flechtenzonierung an Bachufern und ihre Beziehungen zur jarhlichen überflutungsdauer und zum Mikroklima. Flora, Jena *148*:612–638.

Ried, A. 1960b. Stoffwechsel und Verbreitungsgrenzen von Flechten. II. Wasser- und Assimilationshaushalt, Entquellungs- und Submersion-resistenz von Krustenflechten. Flora, Jena *149*: 345–385.

Riemann, B. 1978. Differentiation between heterotrophic and photosynthetic plankton by size fractionation, glucose uptake, ATP and chlorophyll content. Oikos *31*:358–367.

Riemann, B. 1983. Biomass and production of phyto- and bacterioplankton in eutrophic Lake Tystrup, Denmark. Freshwat. Biol. *13*:389–398.

Riemann, B. 1985. Potential importance of fish predation and zooplankton grazing on natural populations of freshwater bacteria. Appl. Environ. Microbiol. *50*:187–193.

Riemann, B., N. O. G. Jørgensen, W. Lampert, and J. A. Fuhrman. 1986. Zooplankton induced changes in dissolved free amino acids and in production rates of freshwater bacteria. Microb. Ecol. *12*:247–258.

Riera, J. L., J. E. Schindler, and T. K. Kratz. 1999. Seasonal dynamics of carbon dioxide and methane in two clear-water lakes and two bog lakes in northern Wisconsin, U.S.A. Can. J. Fish. Aquat. Sci. *56*:265–274.

Riessen, H. P. 1984. The other side of cyclomorphosis: Why *Daphnia* lose their helmets. Limnol. Oceanogr. *29*:1123–1127.

Rigby, C. H., S. R. Craig, and K. Budd. 1980. Phosphate uptake by *Synechococcus leopoliensis* (Cyanophyceae): Enhancement by calcium ions. J. Phycol. *16*:389–393.

Riggs, L. Q. and J. J. Gilbert. 1972. The labile period for a-tocopherol-induced mictic female and body wall outgrowth responses in embryos of the rotifer *Asplanchna sieboldi*. Int. Rev. ges. Hydrobiol. *57*:675–683.

Rigler, F. H. 1956. A tracer study of the phosphorus cycle in lake water. Ecology *37*:550–562.

Rigler, F. H. 1961. The uptake and release of inorganic phosphorus by *Daphnia magna* Straus. Limnol. Oceanogr. *6*:165–174.

Rigler, F. H. 1964a. The phosphorus fractions and the turnover time of inorganic phosphorus in different types of lakes. Limnol. Oceanogr. *9*:511–518

Rigler, F. H. 1964b. The contribution of zooplankton to the turnover of phosphorus in the epilimnion of lakes. Can. Fish Culturist *32*:3–9.

Rigler, F. H. 1971. Feeding rates: Zooplankton. *In* W. T. Edmondson and G. G. Winberg, eds. A Manual on Methods for the Assessment of Secondary Productivity in Fresh Waters. Int. Biol. Program Handbook 17. Blackwell Scientific Publs., Oxford. pp. 228–255.

Rigler, F. H. 1973. A dynamic view of the phosphorus cycle in lakes. *In* E. J. Griffith, A. Beeton, J. M. Spencer, and D. T. Mitchell, eds. Environmental Phosphorus Handbook. John Wiley & Sons, New York. pp. 539–572.

Rigler, F. H. and J. A. Downing. 1984. The calculation of secondary productivity. *In* J. A. Downing and F. H. Rigler, eds. A Manual on Methods for the Assessment of Secondary Productivity in Fresh Waters. 2nd Edition. Blackwell Scientific Publs., Oxford. pp. 19–58.

Rigler, F. H., M. E. MacCallum, and J. C. Roff. 1974. Production of zóoplankton in Char Lake. J. Fish. Res. Board Can. *31*:637–646.

Rijkeboer, M., F. de Bles, and H. J. Gons. 1991. Role of sestonic detritus as a P-buffer. Mem. Ist. Ital. Idrobiol. *48*:251–260.

van Rijn, J. and M. Shilo. 1983. Buoyancy regulation in a natural population of *Oscillatoria* spp. in fishponds. Limnol. Oceanogr. *28*:1034–1037.

Riley, G. A. 1939. Limnological studies in Connecticut. Part I. General limnological survey. Part II. The copper cycle. Ecol. Monogr. *9*:66–94.

Ringelberg, J. 1964. The positively phototactic reaction of *Daphnia magna* Straus: A contribution to the understanding of diurnal vertical migration. Netherlands. J. Sea Res. *2*:319–406.

Ringelberg, J. 1980. Aspects of red pigmentation in zooplankton, especially copepods. *In* W. C. Kerfoot, ed. Evolution and Ecology of Zooplankton Communities. Univ. Press New England, Hanover, NH. pp. 91–97.

Ringelberg, J. 1981. On the variation in carotenoid content of copepods. Limnol. Oceanogr. *26*:995–997.

Ringelberg, J. 1987. Light induced behaviour in *Daphnia*. Mem. Ist. Ital. Idrobiol. *45*:285–323.

Ringelberg, J. 1991a. Enhancement of the phototactic reaction in *Daphnia hyalina* by a chemical mediated by juvenile perch (*Perca fluviatilis*). J. Plankton Res. *13*:17–25.

Ringelberg, J. 1991b. A mechanism of predator-mediated induction of diel vertical migration in *Daphnia hyalina*. J. Plankton Res. *13*:83–89.

Ringelberg, J. and E. van Gool. 1995. Migrating *Daphnia* have a memory for fish kairomones. Mar. Fres. Behav. Physiol. *26*:249–257.

Ringelberg, J. and E. van Gool. 1998. Do bacteria, not fish, produce "fish kairomone"? J. Plankton Res. *20*:1847–1852.

Ringelberg, J., B. J. G. Flik, D. Lindenaar, and K. Royackers. 1991. Diel vertical migration of *Eudiaptomus gracilis* during a short summer period. Hydrobiol. Bull. *25*:77–84.

Rippey, B. 1983. A laboratory study of the silicon release process from a lake sediment (Lough Neagh, Northern Ireland). Arch. Hydrobiol. *96*:417–433.

Risgaard-Petersen, N., S. Rysgaard, L. P. Nielsen, and N. P. Revsbech. 1994. Diurnal variation of denitrification and nitrification in sediments colonized by benthic microphytes. Limnol. Oceanogr. *39*:573–579.

Ritchie, J. C., J. R. McHenry, and A. C. Gill. 1973. Dating recent reservoir sediments. Limnol. Oceanogr. *18*:254–263.

Rivkin, R. B. and E. Swift. 1980. Characterization of alkaline phosphatase and organic phosphorus utilization in the oceanic dinoflagellate *Pyrocystis noctiluca*. Mar. Biol. *61*:1–8.

Roback, S. S. 1974. Insects (Arthropoda: Insecta). *In* C. W. Hart, Jr. and S. L. H. Fuller, eds. Pollution Ecology of Freshwater Invertebrates. Academic Press, New York. pp. 313–376.

Robarts, R. D. 1979. Heterotrophic utilization of acetate and glucose in Swartvlei, South Africa. J. Limnol. Soc. South Africa *5*:84–88.

Robarts, R. D. and R. J. Wicks. 1990. Heterotrophic bacterial production and its dependence on autotrophic production in a hypertrophic African reservoir. Can. J. Fish. Aquat. Sci. *47*:1027–1037.

Robarts, R. D., D. B. Donald, and M. T. Arts. 1995. Phytoplankton primary production of three temporary northern prairie wetlands. Can. J. Fish. Aquat. Sci. *52*:897–902.

Robarts, R. D., M. T. Arts, M. S. Evans, and M. J. Waiser. 1994. The coupling of heterotrophic bacterial and phytoplankton production in a hypertrophic, shallow prairie lake. Can. J. Fish. Aquat. Sci. *51*:2219–2226.

Robbins, J. A. and E. Callender. 1975. Diagenesis of manganeses in Lake Michigan sediments. Am. J. Sci. 275:512–533.

Robbins, J. A. and D. N. Edgington. 1975. Determination of recent sedimentation rates in Lake Michigan using Pb–210 and Cs–137. Geochim. Cosmochim. Acta *39*:285–304.

Robbins, J. A., E. Landstrom, and M. Wahlgren. 1972. Tributary inputs of soluble trace metals to Lake Michigan. Proc. Conf. Great Lakes. Res., Int. Assoc. Great Lakes Res. *15*:270–290.

Robbins, J. A., J. R. Krezoski, and S. C. Mozley. 1977. Radioactivity in sediments of the Great Lakes: Post-depositional redistribution by deposit-feeding organisms. Earth Planetary Sci. Lett. *36*:325–333.

Robinson, G. G. C., S. E. Gurney, and L. G. Goldsborough. 1997. The primary productivity of benthic and planktonic algae in a prairie wetland under controlled water-level regimes. Wetlands *17*:182–194.

Robinson, J. A. and J. M. Tiedje. 1984. Competition between sulfate-reducing and methanogenic bacteria for H_2 under resting and growing conditions. Arch. Microbiol. *137*:26–32.

Robinson, N., P. A. Cranwell, and G. Eglinton. 1987. Sources of the lipids in the bottom sediments of an English oligo-mesotrophic lake. Freshwat. Biol. *17*:15–33.

Roberts, G. P. and P. W. Ludden. 1992. Nitrogen fixation by photosynthetic bacteria. *In* G. Stacey, R. H. Burris, and H. J. Evans, eds. Biological nitrogen fixation. Chapman & Hall, New York. pp. 135–165.

Roberts, J. and G. G. Ganf. 1986. Annual production of *Typha orientalis* Presl. in inland Australia. Aust. J. Mar. Freshw. Res. *37*:659–668.

Robertson, A. I., S. E. Bunn, P. I. Boon, and K. F. Walker. 1999. Sources, sinks and transformations of organic carbon in Australian floodplain rivers. Mar. Freshwat. Res. *50*:813–829.

Robertson, A. L. 1995. Secondary production of a community of benthic Chydoridae (Cladocera: Crustacea) in a large river, UK. Arch. Hydrobiol. *134*:425–440.

Robertson, A. L., J. Lancaster, and A. G. Hildrew. 1995. Stream hydraulics and the distribution of microcrustacea: A role for refugia? Freshwat. Biol. *33*:469–484.

Robertson, C. K. 1979. Quantitative comparison of the significance of methane in the carbon cycles of two small lakes. Arch. Hydrobiol. Beih. Ergebn. Limnol. *12*:123–135.

Robertson, J. D. 1941. The function and metabolism of calcium in the Invertebrata. Biol. Rev. *16*:106–133.

Robertson, R. N. 1960. Ion transport and respiration. Biol. Rev. *35*:231–264.

Roca, J. R. and D. L. Danielopol. 1991. Exploration of interstitial habitats by the phytophilous ostracod *Cypridopsis vidua* (O.F. Müller): Experimental evidence. Annls. Limnol. *27*:243–252.

Rocha, O. and T. Matsumura-Tundisi. 1984. Biomass and production of *Argyrodiaptomus furcatus*, a tropical calanoid copepod in Broa Reservoir, southern Brazil. Hydrobiologia *113*:307–311.

Rocha, O., T. Matsumura-Tundisi, J. G. Tundisi, and C. P. Fonseca. 1990. Predation on and by pelagic Turbellaria in some lakes in Brazil. Hydrobiologia *198*:91–101.

Rodda, J. C., R. A. Downing, and F. M. Law. 1976. Systematic Hydrology. Newnes-Butterworths, London. 399 pp.

Roden, E. E. and J. W. Edmonds. 1997. Phosphate mobilization in anaerobic sediments: Microbial Fe(III) oxide reduction versus iron-sulfide formation. Arch. Hydrobiol. *139*:347–378.

Roden, E. E. and R. G. Wetzel. 1996. Organic carbon oxidation and suppression of methane production by microbial Fe(III) oxide reduction in vegetated and unvegetated freshwater wetland sediments. Limnol. Oceanogr. *41*:1733–1748.

Roden, E. E. and R. G. Wetzel. 2000. Humic substances promote sediment Fe(III) oxide reduction. Appl. Environ. Microbiol. (In press).

Rodewald-Rudescu, L. 1974. Das Schilfrohr *Phragmites communis* Trinius. Die Binnengewässer *27*, 302 pp.

Rodgers, G. K. 1966. The thermal bar in Lake Ontario, spring 1965 and winter 1965–66. Publ. Great Lakes Res. Div., Univ. Mich. *15*:369–374.

Rodgers, J. H., Jr., M. E. McKivitt, D. O. Hammerlund, and K. L. Dickson. 1983. Primary production and decomposition of submergent and emergent aquatic plants of two Appalachian rivers. *In* T. D. Fontaine and S. M. Bartell, eds. Dynamics of Lotic Ecosystems. Ann Arbor Science, Ann Arbor, MI. pp. 283–301.

Rodhe, W. 1948. Environmental requirements of freshwater plankton algae. Experimental studies in the ecology of phytoplankton. Symbol. Bot. Upsalien. *10*(1), 149 pp.

Rodhe, W. 1949. The ionic composition of lake waters. Verh. Internat. Verein. Limnol. *10*:377–386.

Rodhe, W. 1951. Minor constituents in lake waters. Verh. Internat. Verein. Limnol. *11*:317–323.

Rodhe, W. 1955. Can plankton production proceed during winter darkness in subarctic lakes? Verh. Internat. Verein. Limnol. *12*:117–122.

Rodhe, W. 1958. Primärproduktion und Seetypen. Verh. Internat. Verein. Limnol. *13*:121–141.

Rodhe, W. 1969. Crystallization of eutrophication concepts in Northern Europe. *In* Eutrophication: Causes, Consequences, Correctives. Nat. Acad. Sciences, Washington, DC. pp. 50–64.

Rodhe, W. 1974. Plankton, planktic, planktonic. Limnol. Oceanogr. *19*:360.

Rodhe, W. 1975. The SIL founders and our fundament. Verh. Internat. Verein. Limnol. *19*:16–25.

Rodhe, W., J. E. Hobbie, and R. T. Wright. 1966. Phototrophy and heterotrophy in high mountain lakes. Verh. Internat. Verein. Limnol. *16*:302–313.

Rodhe, W., R. A. Vollenweider, and A. Nauwerck. 1958. The primary production and standing crop of phytoplankton. *In* A. A. Buzzati-Traverso, ed. Perspectives in Marine Biology. University of California Press, Berkeley. pp. 299–322.

Rodina, A. G. 1965. Metody Vodnoi Mikrobiologii. Izdatel'stvo Nauka, Moscow. 363 pp.

Rodina, A. G. 1972. Methods in Aquatic Microbiology. Translat. and rev. R. R. Colwell and M. S. Zambruski. University Park Press, Baltimore. 461 pp.

Rodusky, A. J. and K. E. Havens. 1996. The potential effects of a small *Chaoborus* species (*C. punctipennis*) on the zooplankton of a small eutrophic lake. Arch. Hydrobiol. *138*:11–31.

Roemer, S. C. and K. D. Hoagland. 1979. Seasonal attenuation of quantum irradiance (400–700 nm) in three Nebraska reservoirs. Hydrobiologia *63*:81–92.

Roemer, S. C., K. D. Hoagland, and J. R. Rosowski. 1984. Development of a freshwater periphyton community as influenced by diatom mucilages. Can. J. Bot. *62*:1799–1813.

Roff, J. C., J. T. Turner, M. K. Webber, and R. R. Hopcroft. 1995. Bacterivory by tropical copepod nauplii: Extent and possible significance. Aquat. Microb. Ecol. *9*:165–175.

Rogers, K. H. and C. M. Breen. 1981. Effects of epiphyton on *Potamogeton crispus* L. leaves. Microb. Ecol. *7*:351–363.

Rogers, K. H. and C. M. Breen. 1982. Decomposition of *Potamogeton crispus* L.: The effects of drying on the pattern of mass and nutrient loss. Aquat. Bot. *12*:1–12.

Rogerson, A. 1981. The ecological energetics of *Amoeba proteus* (Protozoa). Hydrobiologia *85*:117–128.

Rohlich, G. A. 1969. Engineering aspects of nutrient removal. *In* Eutrophication: Causes, Consequences, Correctives. Nat. Acad. Sciences, Washington, DC. pp. 371–382.

Roijackers, R. M. M. 1986. Development and succession of scale-bearing Chrysophyceae in two shallow freshwater bodies near Nijmegen, The Netherlands. *In* J. Kristiansen and R. A. Andersen, eds. Chrysophytes: Aspects and Problems. Cambridge Univ. Press, Cambridge. pp. 241–258.

Rolletschek, H. 1997. Temporal and spatial variations in methane cycling in Lake Müggelsee. Arch. Hydrobiol. *140*:195–206.

Romagoux, J.-C. 1979. Caracteristiques du microphytobenthos d'un lac volcanique meromitique (Lac Pavin, France). I. Biomasse chlorophyllienne et déterminisme du cycle annuel. Int. Rev. ges. Hydrobiol. *64*:303–343.

Romanenko, V. I. 1966. Microbiological processes in the formation and breakdown of organic matter in the Rybinsk Reservoir. Trudy Inst. Biol. Vnutrennikh Vod *13*:137–158. (Translated into English, Israel Program for Scientific Translations, Jerusalem, 1969.)

Romanenko, V. I. 1967. Sootnoshenie mezhdu fotosintezom fitoplanktona i destruktsiei organicheskogo veshchestva v vodokhranilishchakh. Trudy Inst. Biol. Vnutrennikh Vod *15*:61–74.

Romanenko, V. I. and E. G. Dobrynin. 1973. Potreblenie kisloroda, temnovaia assimiliatsiia CO_2 i intensivnost fotosinteza v natural'nykh i profil'trovannykh probakh vody. Mikrobiologiya *42*:573–575.

Romanenko, V. I. and S. I. Kuznetsov. 1974. Ekologiia Mikroorganizmov Presnykh Vodoemov. Laboratornoe Rukovodstvo. Izdatel'stvo Nauka, Leningrad. 194 pp.

Roos, P. J., A. F. Post, and J. M. Revier. 1981. Dynamics and architecture of reed periphyton. Verh. Internat. Verein. Limnol. *21*:948–953.

Rørslett, B. 1985. Death of submerged macrophytes—actual field observations and some implications. Aquat. Bot. *22*:7–19.

Rosa, F., J. G. Nriagu, and H. K. Wong. 1983. Particulate flux at the bottom of Lake Ontario. Chemosphere *12*:1345–1354.

Rosemond, A. D. 1993. Interactions among irradiance, nutrients, and herbivores constrain a stream algal community. Oecologia *94*:585–594.

Rosemond, A. D., P. J. Mulholland, and J. W. Elwood. 1993. Top-down and bottom-up control of stream periphyton: Effects of nutrients and herbivores. Ecology *74*:1264–1280.

Rosenberg, D. M., P. McCully, and C. M. Pringle. 2000. Global-scale environmental effects of hydrological alterations: Introduction. BioScience *50*:746–751.

Rosenberry, D. O., P. A. Bukaveckas, D. C. Buso, G. E. Likens, A. M. Shapiro, and T. C. Winter. 1999. Movement of road salt to a small New Hampshire lake. Water Air Soil Poll. *109*:179–206.

Rosenfeld, J. S. and R. J. Mackay. 1987. Assessing the food base of stream ecosystems: Alternatives to the P/R ratio. Oikos *50*:141–147.

Rosenstock, B. and M. Simon. 1993. Use of dissolved and combined and free amino acids by planktonic bacteria in Lake Constance. Limnol. Oceanogr. *38*:1521–1531.

Rosgen, D. L. 1994. A classification of natural rivers. Catena *22*:169–199.

Rosgen, D. L. 1996. Applied River Morphology. Wildlife Hydrology Books, Pagosa Springs, CO. 365 pp.

Ross, H. H. 1963. Stream communities and terrestrial biomes. Arch. Hydrobiol. *59*:235–242.

Rossi, G. and G. Premazzi. 1991. Delay in lake recovery caused by internal loading. Water Res. *25*:567–575.

Rossolimo, L. 1935. Die Boden-Gasausscheidung und das Sauerstoffregime der Seen. Verh. Internat. Verein. Limnol. *7*: 539–561.

Rosswall, T., ed. 1973. Modern Methods in the Study of Microbial Ecology. Bull. Ecological Research Committee, Swedish Natural Science Research Council *17*, 511 pp.

Roth, J. C. 1968. Benthic and limnetic distribution of three *Chaoborus* species in a southern Michigan lake (Diptera, Chaoboridae). Limnol. Oceanogr. *13*:242–249.

Roth, J. R., J. G. Lawrence, and T. A. Bobik. 1996. Cobalamin (coenzyme B_{12}): Synthesis and biological significance. Annu. Rev. Microbiol. *50*:137–181.

Rothfuss, F. and R. Conrad. 1993. Thermodynamics of methanogenic intermediary metabolism in littoral sediment of Lake Constance. FEMS Microbiol. Ecol. *12*:265–276.

Rothhaupt, K. O. 1990. Population growth rates of two closely related rotifer species: Effects of food quantity, particle size, and nutritional quality. Freshwat. Biol. *23*:561–570.

Rothhaupt, K. O. 1992. Stimulation of phosphorus-limited phytoplankton by bacterivorous flagellates in laboratory experiments. Limnol. Oceanogr. *37*:750–759.

Roulet, N. T. and M. Woo. 1988. Wetland and lake evaporation in the low Arctic. Arctic Alpine Res. *18*:195–200.

Roulet, N. T. and W. P. Adams. 1986. Spectral distribution of light under a subarctic winter lake cover. Hydrobiologia *134*: 89–95.

Round F. E. 1960. Studies on bottom-living algae in some lakes of the English Lake District. Part IV. The seasonal cycles of the Bacillariophyceae. J. Ecol. *48*:529–547.

Round F. E. 1961a. Studies on the bottom-living algae in some lakes of the English Lake District. Part V. The seasonal cycles of the Cyanophyceae. J. Ecol. *49*:31–38.

Round F. E. 1961b. Studies on bottom-living algae in some lakes of the English Lake District. Part VI. The effect of depth on the epipelic algal community. J. Ecol. *49*:245–254.

Round F. E. 1961c. The diatoms of a core from Esthwaite Water. New Phytol. *60*:43–59.

Round F. E. 1964a. The ecology of benthic algae. *In* D. F. Jackson, ed. Algae and Man. Plenum Press, New York. pp. 138–184.

Round F. E. 1964b. The diatom sequence in lake deposits: Some problems of interpretation. Verh. Internat. Verein. Limnol. *15*:1012–1020.

Round F. E. 1965. The epipsammon; a relatively unknown freshwater algal association. Brit. Phycol. Bull. *2*:456–462.

Round F. E. 1971. The growth and succession of algal populations in freshwaters. Mitt. Internat. Verein. Limnol. *19*:70–99.

Round F. E. 1972. Patterns of seasonal succession of freshwater epipelic algae. Brit. Phycol. J. *7*:213–220.

Round, F. E. 1957a. Studies on bottom-living algae in some lakes of the English Lake District. Part I. Some chemical features of the sediments related to algal productivities. J. Ecol. *45*:133–148.

Round, F. E. 1957b. Studies on bottom-living algae in some of the lakes of the English Lake District. Part II. The distribution of Bacillariophyceae on the sediments. J. Ecol. *45*:343–360.

Round, F. E. 1957c. Studies on bottom-living algae in some lakes of the English Lake District. Part III. The distribution on the sediments of algal groups other than the Bacillariophyceae. J. Ecol. *45*:649–664.

Round, F. E. 1981. The Ecology of Algae. Cambridge University Press, Cambridge.

Round, F. E. and C. M. Happey. 1965. Persistent, vertical-migration rhythms in benthic microflora. IV. A diurnal rhythm of the epipelic diatom association in non-tidal flowing waters. Brit. Phycol. Bull. *2*:463–471.

Round F. E. and J. W. Eaton. 1966. Persistent, vertical-migration rhythms in benthic microflora. III. The rhythm of epipelic algae in a freshwater pond. J. Ecol. *54*:609–615.

Rowell, P., W. James, W. L. Smith, L. L. Handley, and C. M. Scrimgeour. 1998. ^{15}N discrimination in molybdenum- and vanadium-grown N_2-fixing *Anabaena variabilis* and *Azotobacter vinelandii*. Soil Biol. Biochem *30*:2177–2180.

Rowlatt, U. and H. Morshead. 1992. Architecture of the leaf of the greater reed mace, *Typha latifolia* L. Bot. J. Linn. Soc. *110*:161–170.

Rublee, P. A. 1992. Community structure and bottom-up regulation of heterotrophic microplankton in arctic LTER lakes. Hydrobiologia *240*:133–141.

Ruck, P. 1965. The components of the visual system of a dragonfly. J. Gen. Physiol. *49*:289–307.

Rudd, J. W. M. and C. D. Taylor. 1980. Methane cycling in aquatic environments. Adv. Aquat. Microbiol. *2*:77–150.

Rudd, J. W. M. and R. D. Hamilton. 1975. Factors controlling rates of methane oxidation and the distribution of the methane oxidizers in a small stratified lake. Arch. Hydrobiol. *75*:522–538.

Rudd, J. W. M. and R. D. Hamilton. 1978. Methane cycling in a eutrophic shield lake and its effects on whole lake metabolism. Limnol. Oceanogr. *23*:337–348.

Rudd, J. W. M., A. Furutani, R. J. Flett, and R. D. Hamilton. 1976. Factors controlling methane oxidation in Shield lakes: The role of nitrogen fixation and oxygen. Limnol. Oceanogr. *21*:357–364.

Rudd, J. W. M., C. A. Kelly, and A. Furutani. 1986. The role of sulfate reduction in long term accumulation of organic and inorganic sulfur in lake sediments. Limnol. Oceanogr. *31*:1281–1291.

Rudd, J. W. M., C. A. Kelly, V. St. Louis, R. H. Hesslein, A. Furutani, and M. H. Holoka. 1986. Microbial consumption of nitric and sulfuric acids in acidified north temperate lakes. Limnol. Oceanogr. *31*:1267–1280.

Rudd, J. W. M., R. D. Hamilton, and N. E. R. Campbell. 1974. Measurement of microbial oxidation of methane in lake water. Limnol. Oceanogr. *19*:519–524.

Rudescu, L., C. Niculescu, and I. P. Chivu. 1965. Monografia Stufului din Delta Dunarii. Bucharest, Editura Acad. Rep. Soc. Romania, 542 pp.

Rudstam, L. G., R. C. Lathrop, and S. R. Carpenter. 1993. The rise and fall of a dominant planktivore: Direct and indirect effects on zooplankton. Ecology *74*:303–319.

Rueter, J. G. 1988. Iron stimulation of photosynthesis and nitrogen fixation in *Anabaena* 7120 and *Trichodesmium* (Cyanophyceae). J. Phycol. *24*:249–254.

Rueter, J. G. and D. R. Ades. 1987. The role of iron nutrition in photosynthesis and nitrogen assimilation in *Scenedesmus quadricauda* (Chlorophyceae). J. Phycol. *23*:452–457.

de Ruiter, P. C., A.-M. Neutel, and J. C. Moore. 1995. Energetics, patterns of interaction strengths, and stability in real ecosystems. Science *269*:1257–1260.

Ruschke, R. 1968. Die Bedeutung von Wassermyxobakterien für den Abbau organischen Materials. Mitt. Internat. Verein. Limnol. *14*:164–167.

Rusness, D. and R. H. Burris. 1970. Acetylene reduction (nitrogen fixation) in Wisconsin lakes. Limnol. Oceanogr. *15*:808–813.

Russell, J. L. 1856. *Hydrothyria venosa*: A new genus and species of the Collemaceae. Proc. Essex Inst. *1*:188–191.

Russell-Hunter, W. D. and D. E. Buckley. 1983. Actuarial bioenergetics of nonmarine molluscan productivity. *In* W. D. Russell-Hunter, ed. The Mollusca. Vol. 6. Ecology. Academic Press, Orlando. pp. 463–503.

Rutherford, J. C. 1994. River mixing. John Wiley & Sons, Chichester. 347 pp.

Rutherford, J. E. and H. B. N. Hynes. 1987. Dissolved organic carbon in streams and groundwater. Hydrobiologia *154*:33–48.

Ruttner, F. 1931. Hydrographische und hydrochemische Beobachtungen auf Java, Sumatra und Bali. Arch. Hydrobiol./Suppl. *8*:197–454.

Ruttner, F. 1933. Über metalimnische Sauerstoffminimum. Die Naturwissenschaften *1933*(21–23):401–404.

Ruttner, F. 1947. Zur Frage der Karbonatassimilation der Wasserpflanzen. I. Die Beiden Haupttypen der Kohlenstoffaufnahme. Öst. Bot. Z. *94*:265–294.

Ruttner, F. 1948. Zur Frage der Karbonatassimilation der Wasserpflanzen. II. Das Verhalten von *Elodea canadensis* und *Fontinalis antipyretica* in Lösungen von Natrium- bzw. Kaliumbikarbonat. Öst. Bot. Z. *95*:208–238.

Ruttner, F. 1960. über die Kohlenstoffaufnahme bei Algen aus der Rhodophyceen-Gattung *Batrachospermum*. Schweiz. Z. Hydrol. *22*:280–291.

Ruttner, F. 1963. Fundamentals of Limnology. (Translat. D. G. Frey and F. E. J. Fry.) University of Toronto Press, Toronto. 295 pp.

Ruttner-Kolisko, A. 1964. Über die labile Periode im Fortpflanzungszyklus der Rädertiere. Int. Rev. ges. Hydrobiol. *49*:473–482.

Ruttner-Kolisko, A. 1972. Rotatoria. *In* Das Zooplankton der Binnengewässer. 1. Teil. Die Binnengewässer *26*(Pt. 1):99–234.

Ruttner-Kolisko, A. 1975. The vertical distribution of planktonic rotifers in a small alpine lake with a sharp oxygen depletion (Lunzer Obersee). Verh. Internat. Verein. Limnol. *19*:1286–1294.

Ruttner-Kolisko, A. 1980. The abundance and distribution of *Filinia terminalis* in various types of lakes as related to temperature, oxygen, and food. Hydrobiologia *73*:169–175.

Rybak, J. I. 1980. The structure of littoral bottom deposits in several Masurian lakes. Bull. Acad. Polon. Sci., Cl. II, *28*:389–394.

Rybak, M. 1986. The chrysophycean paleocyst flora on the bottom sediments of Kortowskie Lake (Poland) and its ecological significance. Hydrobiologia *140*:67–84.

Ryder, R. A. 1965. A method for estimating the potential fish production of north-temperate lakes. Trans. Amer. Fish. Soc. *94*:214–218.

Ryder, R. A. 1982. The morphoedaphic index—use, abuse, and fundamental concepts. Trans. Amer. Fish. Soc. *111*:154–164.

Ryder, R. A. and J. Pesendorfer. 1989. Large rivers are more than flowing lakes: A comparative review. *In* D. P. Dodge, ed. Proceedings of the International Large River Symposium. Can. Spec. Publ. Fish. Aquat. Sci. *106*:65–85.

Ryding, S.-O. and W. Rast, eds. 1989. The Control of Eutrophication of Lakes and Reservoirs. Parthenon Publishing Group Ltd., Carnforth. 314 pp.

Ryhänen, R. 1968. Die Bedeutung der Humussubstanzen im Stoffhaushalt der Gewässer Finnlands. Mitt. Internat. Verein. Limnol. *14*:168–178.

Rysgaard, S., N. Risgaard-Petersen, L. P. Nielsen, and N. P. Revsbech. 1993. Nitrification and denitrification in lake and estuarine sediments measured by the ^{15}N dilution technique and isotope pairing. Appl. Environ. Microbiol. *59*:2093–2098.

Saarinen, T. 1996. Biomass and production of two vascular plants in a boreal mesotrophic fen. Can. J. Bot. *74*:934–938.

Saarinen, T. 1998. Internal C:N balance and biomass partitioning of *Carex rostrata* grown at three levels of nitrogen supply. Can. J. Bot. *76*:762–768.

Sabater, S. and E. Y. Haworth. 1995. An assessment of recent trophic changes in Windermere South Basin (England) based on diatom remains and fossil pigments. J. Paleolimnol. *14*:151–163.

Safferman, R. S. and M.-E. Morris. 1963. Algal virus: Isolation. Science *140*:679–680.

Sager, P. E. and A. D. Hasler. 1969. Species diversity in lacustrine phytoplankton. I. The components of the index of diversity from Shannon's formula. Amer. Nat. *102*:51–59.

Saito, K., M. Matumoto, T. Sekine, I. Murakoshi, N. Morisaki, and S. Iwasaki. 1989. Inhibitory substances from *Myriophyllum brasiliense* on growth of blue-green algae. J. Nat. Prod. *52*:1221–1226.

Sakamoto, M. 1966. Primary production of phytoplankton community in some Japanese lakes and its dependence on lake depth. Arch. Hydrobiol. *62*:1–28.

Sako, Y., Y. Ishida, H. Kadota, and Y. Hata. 1985. Excystment in the freshwater dinoflagellate *Peridinium cunningtonii*. Bull. Jpn. Soc. Sci. Fish. *51*:267–272.

Salas, H. J. and P. Martino. 1991. A simplified phosphorus trophic state model for warm-water tropical lakes. Water Res. *25*:341–350.

Sale, P. J. M. and P. T. Orr. 1987. Growth responses of *Typha orientalis* Presl to controlled temperatures and photoperiods. Aquat. Bot. *29*:227–243.

Sale, P. J. M. and R. G. Wetzel. 1983. Growth and metabolism of *Typha* species in relation to cutting treatments. Aquat. Bot. *15*:321–334.

Saleque, M. A. and G. J. D. Kirk. 1995. Root-induced solubilization of phosphate in the rhizosphere of lowland rice. New Phytol. *129*:325–336.

Salonen, K., R. I. Jones, and L. Arvola. 1984. Hypolimnetic phosphorus retrieval by diel vertical migrations of lake phytoplankton. Freshwat. Biol. *14*:431–438.

Salvucci, M. E. and G. Bowes. 1983. Two photosynthetic mechanisms mediating the low photorespiratory state in submersed aquatic angiosperms. Plant Physiol. *73*:488–496.

Sanders, R. W. 1991a. Trophic strategies among heterotrophic flagellates. *In* D. J. Patterson and J. Larsen, eds. The Biology of Free-living Heterotrophic Flagellates. Clarendon Press, Oxford. pp. 21–38.

Sanders, R. W. 1991b. Mixotrophic protists in marine and freshwater ecosystems. J. Protozool. *38*:76–81.

Sanders, R. W. and K. G. Porter. 1988. Phagotrophic phytoflagellates. Adv. Microb. Ecol. *10*:167–192.

Sanders, R. W., K. G. Porter, S. J. Bennett, and A. E. DeBiase. 1989. Seasonal patterns of bacterivory by flagellates, ciliates, rotifers, and cladocerans in a freshwater planktonic community. Limnol. Oceanogr. *34*:673–687.

Sanders, R. W., K. G. Porter, and D. A. Caron. 1990. Relationship between phototrophy and phagotrophy in the mixotrophic chrysophyte *Poterioochromonas malhamensis*. Microb. Ecol. *19*:97–109.

Sanders, R. W., C. E. Williamson, P. L. Stutzman, R. E. Moeller, C. E.Goulden, and R. Aoki-Goldsmith. 1996. Reproductive success of "herbivorous" zooplankton fed algal and nonalgal food resources. Limnol. Oceanogr. *41*:1295–1305.

Sandgren, C. D. 1986. Effects of environmental temperature on the vegetative growth and sexual life history of *Dinobryon cylindricum* Imhof. *In* J. Kristiansen and R. A. Andersen, eds. Chrysophytes: Aspects and Problems. Cambridge Univ. Press, Cambridge. pp. 207–225.

Sand-Jensen, K. 1977. Effect of epiphytes on eelgrass photosynthesis. Aquat. Bot. *3*:55–63.

Sand-Jensen, K. 1990. Epiphyte shading: Its role in resulting depth distribution of submerged aquatic macrophytes. Folia Geobot. Phytotax., Praha *25*:315–320.

Sand-Jensen, K. 1978. Metabolic adaptation and vertical zonation of *Littorella uniflora* (L.) Aschers. and *Isoetes lacustris* L. Aquat. Bot. *4*:1–10.

Sand-Jensen, K. and J. R. Mebus. Fine-scale patterns of water velocity within macrophyte patches in streams. Oikos *76*:169–180.

Sand-Jensen, K. and M. F. Pedersen. 1994. Photosynthesis by symbiotic algae in the freshwater sponge, *Spongilla lacustris*. Limnol. Oceanogr. *39*:551–561.

Sand-Jensen, K. and M. Søndergaard. 1979. Distribution and quantitative development of aquatic macrophytes in relation to sediment characteristics in oligotrophic Lake Kalgaard, Denmark. Freshwat. Biol. *9*:1–11.

Sand-Jensen, K. and M. Søndergaard. 1981. Phytoplankton and epiphyte development and their shading effect on submerged macrophytes in lakes of different nutrient status. Int. Rev. ges. Hydrobiol. *66*:529–552.

Sand-Jensen, K. and T. V. Madsen. 1991. Minimum light requirements of submerged freshwater macrophytes in laboratory growth experiments. J. Ecol. *79*:749–764.

Sand-Jensen, K., C. Prahl, and H. Stokholm. 1982. Oxygen release from roots of submerged aquatic macrophytes. Oikos *38*:349–354.

Sand-Jensen, K., E. Jeppesen, K. Nielsen, L. van der Bijl, L. Hjermind, L. W. Nielsen, and T. M. Iversen. 1989. Growth of macrophytes and ecosystem consequences in a lowland Danish stream. Freshwat. Biol. *22*:15–32.

Sand-Jensen, K., J. Møller, and B. H. Olesen. 1988. Biomass regulation of microbenthic algae in Danish lowland streams. Oikos *53*:332–340.

Sand-Jensen, K., K. Andersen, and T. Andersen. 1999. Dynamic properties of recruitment, expansion and mortality of macrophyte patches in streams. Internat. Rev. Hydrobiol. *84*:497–508.

Sand-Jensen, K., M. F. Pedersen, and S. L. Nielsen. 1992. Photosynthetic use of inorganic carbon among primary and secondary water plants in streams. Freshwat. Biol. *27*:283–293.

Sandercock, G. A. 1967. A study of selected mechanisms for the coexistence of *Diaptomus* spp. in Clarke Lake, Ontario. Limnol. Oceanogr. *12*:97–112.

Sandner, H. and J. Wilkialis. 1972. Leech communities (Hirudinea) in the Mazurian and Bialystok regions and the Pomeranian Lake District. Ekol. Polska *20*:345–365.

Sanger, J. E. 1971a. Identification and quantitative measurement of plant pigments in soil humus layers. Ecology *52*:959–963.

Sanger, J. E. 1971b. Quantitative investigations of leaf pigments from their inception in buds through autumn coloration to decomposition in falling leaves. Ecology *52*:1075–1089.

Sanger, J. E. 1988. Fossil pigments in paleoecology and paleolimnology. Palaeogeogr. Palaeoclimat. Palaeoecol. *62*:343–359.

Sanger, J. E. and E. Gorham. 1970. The diversity of pigments in lake sediments and its ecological significance. Limnol. Oceanogr. *15*:59–69.

Sanger, J. E. and E. Gorham. 1973. A comparison of the abundance and diversity of fossil pigments in wetland peats and woodland humus layers. Ecology *54*:605–611.

Santer, B. and W. Lampert. 1995. Summer diapause in cyclopoid copepods: Adaptive response to a food bottleneck? J. Anim. Ecol. *64*:600–613.

Santesson, R. 1939. über die Zonationsverhältnisse der lakustrinen Flechten einiger Seen im Anebodagebiet. Medd. från Lund Univ. Limnol. Inst. *1*:70 pp.

Saralov, A. I., I. N. Krylova, E. E. Saralova, and S. I. Kuznetsov. 1984. Distribution and species composition of methane-oxidizing bacteria in lake waters. Mikrobiologiya *53*:837–842.

Sardella, L. C. and J. C. H. Carter. 1983. Factors contributing to coexistence of *Chaoborus flavicans* and *C. punctipennis* (Diptera, Chaoboridae) in a small meromictic lake. Hydrobiologia *107*:155–164.

Särkkä, J. 1987. The occurrence of oligochaetes in lake chains receiving pulp mill waste and their relation to eutrophication on the trophic scale. Hydrobiologia *155*:259–266.

Särkkä, J. and J. Aho. 1980. Distribution of aquatic Oligochaeta in the Finnish Lake District. Freshwat. Biol. *10*:197–206.

Sarvala, J. 1979. Effect of temperature on the duration of egg, nauplius and copepodite development of some freshwater benthic Copepoda. Freshwat. Biol. *9*:515–534.

Sarvala, J., T. Kairesalo, I. Koskimies, A. Lehtovaara, J. Ruuhijärvi, and I. Vähä-Piikkiö. 1982. Carbon, phosporus and nitrogen budgets of the littoral *Equisetum* belt in an oligotrophic lake. Hydrobiologia *86*:41–53.

Sarvala, J., V. Ilmavirta, L. Paasivirta, and K. Salonen. 1981. The ecosystem of the oligotrophic Lake Pääjärvi. 3. Secondary production and an ecological energy budget of the lake. Verh. Internat. Verein. Limnol. *21*:454–459.

Sass, R. L. and F. M. Fisher, Jr. 1996. Methane from irrigated rice cultivation. Curr. Topics Wetland Biogeochem. *2*:24–39.

Sasser, C. E. and J. E. Gosselink. 1984. Vegetation and primary production in a floating freshwater marsh in Louisiana. Aquat. Bot. *20*:245–255.

Sastroutomo, S. S. 1980a. Environmental control of turion formation in curly pondweed (*Potamogeton crispus*). Physiol. Plant. *49*:261–264.

Sastroutomo, S. S. 1980. Dormancy and germination in axillary turions of *Hydrilla verticillata*. Bot. Mag. Tokyo *93*:265–273.

Sastroutomo, S. S. 1981. Turion formation, dormancy and germination of curly pondweed, *Potamogeton crispus* L. Aquat. Bot. *10*:161–173.

Satake, K. and Y. Saijo. 1978. Mechanism of lamination in bottom sediment of the strongly acid Lake Katanuma. Arch. Hydrobiol. *83*:429–442.

Satoh, Y. and H. Abe. 1987. Dissolved organic matter in colored water from mountain bog pools in Japan. II. Biological decomposability. Arch. Hydrobiol. *111*:25–35.

Sauberer, F. 1939. Beiträge zur Kenntnis des Lichtklimas einiger Alpenseen. Int. Rev. ges. Hydrobiol. *39*:20–55.

Sauberer, F. 1950. Die spektrale Strahlungsdurchlässigkeit des Eises. Wetter u. Leben *2*:193–197.

Sauberer, F. 1962. Empfehlungen für die Durchführung von Strahlungsmessungen an und in Gewässern. Mitt. Internat. Verein. Limnol. *11*, 77 pp.

Saunders, J. F. and W. M. Lewis, Jr. 1988. Dynamics and control mechanisms in a tropical zooplankton community (Lake Valencia, Venezuela). Ecol. Monogr. *58*:337–353.

Saunders, G. W. 1957. Interrelations of dissolved organic matter and phytoplankton. Bot. Rev. *23*:389–410.

Saunders, G. W. 1963. The biological characteristics of fresh water. Publ. Great Lakes Res. Div., Univ. Mich. *10*:245–257.

Saunders, G. W. 1969. Some aspects of feeding in zooplankton. *In* Eutrophication: Causes, Consequences, Correctives. Nat. Acad. Sciences, Washington, DC. pp. 556–573.

Saunders, G. W. 1971. Carbon flow in the aquatic system. *In* J. Cairns, Jr., ed. The Structure and Function of Fresh-Water Microbial Communities. Res. Div. Monogr. 3, Virginia Polytechnic Inst. pp. 31–45.

Saunders, G. W. 1972a. The transformation of artificial detritus in lake water. Mem. Ist. Ital. Idrobiol. *29*(Suppl.):261–288.

Saunders, G. W. 1972b. Summary of the general conclusions of the symposium. Mem. Ist. Ital. Idrobiol. *29*(Suppl.):533–540.

Saunders, G. W. 1972c. Potential heterotrophy in a natural population of *Oscillatoria agardhii* var. *isothrix* Skuja. Limnol. Oceanogr. *17*:704–711.

Saunders, G. W. and T. A. Storch. 1971. Couples oscillatory control mechanism in a planktonic system. Nature (New Biol.) *230*:58–60.

Saunders, G. W., K. W. Cummins, D. Z. Gak, E. Pieczyńska, V. Straškrabová, and R. G. Wetzel. 1980. Organic matter and decomposers. *In* E. D. Le Cren and R. H. Lowe-McConnell, eds. The Functioning of Freshwater Ecosystems. Cambridge Univ. Press, Cambridge. pp. 341–392.

Savostin, P. 1972. Microbial transformation of silicates. Z. Pflanzenernährung Bodenkunde *132*:37–45.

Sawyer, C. N. 1947. Fertilization of lakes by agricultural and urban drainage. J. New England Water Works Assoc. *61*:109–127.

Sawyer, R. T. 1986. Leech Biology and Behaviour. Clarendon Press, Oxford. 1065 pp.

Scavia, D. and G. A. Laird. 1987. Bacterioplankton in Lake Michigan: Dynamics, controls, and significance to carbon flux. Limnol. Oceanogr. *32*:1017–1033.

Schafran, G. C. and C. T. Driscoll. 1987. Spatial and temporal variations in aluminum chemistry of a dilute, acidic lake. Biogeochemistry *3*:105–119.

Schaller, T., H. C. Moor, and B. Wehrli. 1997. Reconstructing the iron cycle from the horizontal distribution of metals in the sediment of Baldeggersee. Aquat. Sci. *59*:326–344.

Schaus, M. H., M. J. Vanni, T. E. Wissing, M. T. Bremigan, J. E. Garvey, and R. A. Stein. 1997. Nitrogen and phosphorus excretion by detritivorous gizzard shad in a reservoir ecosystem. Limnol. Oceanogr. *42*:1386–1397.

Scheffer, M. 1998. Ecology of Shallow Lakes, Chapman and Hall, London.

Scheffer, M., S. H. Hosper, M. L. Meijer, B. Moss, and E. Jeppesen. 1993. Alternative equilibria in shallow lakes. Trends Ecol. Evol. *8*:275–279.

Schelske, C. L. 1962. Iron, organic matter, and other factors limiting primary productivity in a marl lake. Science *136*:45–46.

Schelske, C. L. 1985. Biogeochemical silica mass balances in Lake Michigan and Lake Superior. Biogeochemistry *1*:197–218.

Schelske, C. L. 1988. Historical trends in Lake Michigan silica concentrations. Int. Rev. ges. Hydrobiol. *73*:559–591.

Schelske, C. L. 1999. Diatoms as mediators of biogeochemical silica depletion in the Laurentian Great Lakes. *In* E. F. Stoermer and J. P. Smol, eds. The Diatoms: Applications for the Environmental and Earth Sciences. Cambridge Univ. Press, Cambridge. pp. 73–84.

Schelske, C. L. and E. F. Stoermer. 1971. Eutrophication, silica depletion, and predicted changes in algal quality in Lake Michigan. Science *173*:423–424.

Schelske, C. L. and E. F. Stoermer. 1972. Phosphorus, silica, and eutrophication of Lake Michigan. *In* G. E. Likens, ed. Nutrients and Eutrophication. Special Symposium, Amer. Soc. Limnol. Oceanogr. *1*:157–171.

Schelske, C. L., D. J. Conley, E. F. Stoermer, T. L. Newberry, and C. D. Campbell. 1986. Biogenic silica and phosphorus accumulation in sediments as indices of eutrophication in the Laurentian Great Lakes. Hydrobiologia *143*:79–86.

Schelske, C. L., F. F. Hooper, and E. J. Haertl. 1962. Responses of a marl lake to chelated iron and fertilizer. Ecology, *43*:646–653.

Schelske, C. L., H. Züllig, and M. Boucherle. 1987. Limnological investigation of biogenic silica sedimentation and silica biogeochemistry in Lake St. Moritz and Lake Zürich. Schweiz. Z. Hydrol. *49*:42–50.

Schelske, C. L., L. E. Feldt, M. A. Santiago, and E. F. Stoermer. 1972. Nutrient enrichment and its effects on phytoplankton production and species composition in Lake Superior. Proc. Conf. Great Lakes Res., Int. Assoc. Great Lakes Res. *15*:149–165.

Schenk, R. U. 1991. Alternative road deicer. *In* D. L. Wise, Y. A. Levendis, and M. Metghalchi, eds. Calcium magnesium acetate: An emerging bulk chemical for environmental applications. Elsevier, Amsterdam. pp. 37–48.

Schiegl, W. E. 1972. Deuterium content of peat as a paleoclimatic recorder. Science *175*:512–513.

Schiemer, F. 1983. Comparative aspects of food dependence and energetics of freeliving nematodes. Oikos *41*:32–42.

Schiff, S. L. and R. F. Anderson. 1986. Alkalinity production in epilimnetic sediments: Acidic and non-acidic lakes. Water Air Soil Pollut. *31*:941–948.

Schindler, D. E., J. F. Kitchell, X. He, S. R. Carpenter, J. R. Hodgson, and K. L. Cottingham. 1993. Food web structure and phosphorus cycling in lakes. Trans. Amer. Fish. Soc. *122*:756–772.

Schindler, D. W. 1968. Feeding, assimilation and respiration rates of *Daphnia magna* under various environmental conditions and their relation to production estimates. J. Anim. Ecol. *37*:369–385.

Schindler, D. W. 1970. Production of phytoplankton and zooplankton in Canadian Shield lakes. *In* Z. Kajak and A. Hillbricht-Ilkowska, eds. Productivity Problems of Freshwaters. PWN Polish Scientific Publishers, Warsaw. pp. 311–331.

Schindler, D. W. 1974. Eutrophication and recovery in experimental lakes: Implications for lake management. Science *184*:897–899.

Schindler, D. W. 1978. Predictive eutrophication models. Limnol. Oceanogr. *23*:1080–1081.

Schindler, D. W. 1986. The significance of in-lake production of alkalinity. Water Air Soil Pollut. *30*:931–944.

Schindler, D. W. 1988a. Confusion over the origin of alkalinity in lakes. Limnol. Oceanogr. *33*:1637–1640.

Schindler, D. W. 1988b. Effects of acid rain on freshwater ecosystems. Science *239*:149–157.

Schindler, D. W. 1997. Liming to restore acidified lakes and streams: A typical approach to restoring damaged ecosystems? Restoration Ecol. *5*:1–6.

Schindler, D. W. and P. J. Curtis. 1997. The role of DOC in protecting freshwaters subjected to climatic warming and acidification from UV exposure. Biogeochemistry *36*:1–8.

Schindler, D. W., E. J. Fee, and T. Ruszczynski. 1978. Phosphorus input and its consequences for phytoplankton standing crop and production in the Experimental Lakes Area and in similar lakes. J. Fish. Res. Board Can. *35*:190–196.

Schindler, D. W., G. J. Brunskill, S. Emerson, W. S. Broecker, and T.-H. Peng. 1972. Atmospheric carbon dioxide: Its role in maintaining phytoplankton standing crops. Science *177*:1192–1194.

Schindler, D. W., M. A. Turner, and R. H. Hesslein. 1985. Acidification and alkalization of lakes by experimental addition of nitrogen compounds. Biogeochemistry *1*:117–133.

Schindler, D. W., M. A. Turner, M. P. Stainton, and G. A. Linsey. 1986. Natural sources of acid neutralizing capacity in low alkalinity lakes of the Precambrian Shield. Science *232*:844–847.

Schindler, D. W., R. W. Newberry, K. G. Beaty, and P. Campbell. 1976. Natural water and chemical budgets for a small Precambrian lake basin in central Canada. J. Fish. Res. Board Can. *33*:2526–2543.

Schindler, D. W., S. E. Bayley, P. J. Curtis, B. R. Parker, M. P. Stainton, and C. A. Kelly. 1992. Natural and man-caused factors affecting the abundance and cycling of dissolved organic substances in Precambrian Shield lakes. Hydrobiologia *229*:1–21.

Schindler, D. W., T. W. Frost, K. H. Mills, P. S. S. Chang, I. J. Davies, D. L. Findlay, D. F. Malley, J. A. Shearer, M. A. Turner, P. J. Garrison, C. J. Watras, K. Webster, J. M. Gunn, P. L. Brezonik, and W. A. Swenson. 1991. Comparisons between experimentally- and atmospherically-acidified lakes during stress and recovery. Proc. Royal Soc. Edinburgh *97B*:193–226.

Schindler, D. W., V. E. Frost, and R. V. Schmidt. 1973. Production of epilithiphyton in two lakes of the Experimental Lakes Area, northwestern Ontario. J. Fish. Res. Board Can. *30*:1511–1524.

Schindler, J. E. and D. P. Krabbenhoft. 1998. The hyporheic zone as a source of dissolved organic carbon and carbon gases to a temperate forested stream. Biogeochemistry *43*:157–174.

Schlesinger, W. H. 1997. Biogeochemistry: An Analysis of Global Change. 2nd Edition. Academic Press, San Diego. 588 pp.

Schlesinger, W. H. and A. E. Hartley. 1992. A global budget for atmospheric NH_3. Biogeochemistry *15*:191–211.

Schlesinger, W. H. and J. M. Melack. 1981. Transport of organic carbon in the world's rivers. Tellus *33*:172–187.

Schlott-Idl, K. 1984a. Die räumlich und zeitliche Verteilung der pelagischem Ciliaten im Lunzer Untersee 1981/82. Arch. Hydrobiol. *101*:279–287.

Schlott-Idl, K. 1984b. Qualitative und quantitative Untersuchungen der pelagischen Ciliaten des Piburger Sees (Tirol, Österreich). Limnologica *15*:43–54.

Schmidt, R., S. Wunsam, U. Brosch, J. Fott, A. Lami, H. Löffler, A. Marchetto, H. W. Müller, M. Prazáková, and B. Schwaighofer. 1998. Late and post-glacial history of meromictic Längsee (Austria), in respect to climate change and anthropogenic impact. Aquat. Sci. *60*:56–88.

Schmidt, W. 1915. Über den Energie-gehalt der Seen. Mit Beispielen vom Lunzer Untersee nach Messungen mit einen enfachen Temperaturlot. Int. Rev. Hydrobiol. Suppl., *6*. (Not seen in original.)

Schmidt, W. 1928. Über Temperatur und Stabilitätsverhältnisse von Seen. Geographiska Annaler *10*:145–177.

Schmitt, M. R. and M. S. Adams. 1981. Dependence on rates of apparent photosynthesis on tissue phosphorus concentrations in *Myriophyllum spicatum* L. Aquat. Bot. *11*:379–387.

Schneller, M. V. 1955. Oxygen depletion in Salt Creek, Indiana. Invest. Indiana Lakes Streams *4*:163–175.

Schnitzer, M. 1971. Metal-organic matter interactions in soils and waters. *In* S. J. Faust and J. V. Hunter, eds. Organic Compounds in Aquatic Environments. Marcel Dekker, Inc., New York. pp. 297–315.

Schnitzer, M. and K. Ghosh. 1982. Characteristics of water-soluble fulvic acid-copper and fulvic acid-iron complexes. Soil Sci. *134*:354–363.

Schnitzer, M. and S. U. Khan. 1972. Humic Substances in the Environment. Marcel Dekker, Inc., New York. 327 pp.

Schoener, T. W. 1993. On the relative importance of direct versus indirect effects in ecological communities. *In* H. Kawanabe, J. E. Cohen, and K. Iwasaka, eds. Mutualism and Community Organization—Behavioural, Theoretical and Food-web Approaches. Oxford Univ. Press, Oxford. pp. 364–411.

Scholz, O. and P. I. Boon. 1993. Biofilm development and extracellular enzyme activities on wood in billabongs of south-eastern Australia. Freshwat. Biol. *30*:359–368.

Schönborn, W. 1962. Über Planktismus und Zyklomorphose bei *Difflugia limnetica* (Levander) Penard. Limnologica *1*:21–34.

Schreiner, S. P. 1980. Effect of water hyacinth on the physicochemistry of a south Georgia pond. J. Aquat. Plant Manage. *18*:9–12.

Schreiner, S. P. 1984. Particulate matter distribution and heating patterns in a small pond. Verh. Internat. Verein. Limnol. *22*:119–124.

Schreiter, T. 1928. Untersuchungen über den Einfluss einer Helodeawucherung auf das Netzplankton des Hirschberger Grossteiches in Böhmen in den Jahren 1921 bis 1925 incl. Sborník Vyzk. ìst. Zemed. RCS *61*: 98 pp.

Schriver, P., Bøgestrand, E. Jeppesen, and M. Søndergaard. 1995. Impact of submerged macrophytes on the interactions between fish, zooplankton and phytoplankton: Large scale enclosure experiments in a shallow lake. Freshwat. Biol. *33*:255–270.

Schröder, R. 1973. Die Freisetzung von Pflanzennährstoffen im Schilfgebiet und ihr Transport in das Freiwasser am Beispiel des Bodensee-Untersees. Arch. Hydrobiol. *71*:145–158.

Schröder, R. 1975. Release of plant nutrients from reed borders and their transport into the open waters of the Bodensee-Untersee. Symp. Biol. Hungarica *15*:21–27.

Schröder, R. 1987. Das Schilfsterben am Bodensee-Untersee Beobachtungen, Untersuchungen und Gegenmaßnahmen. Arch. Hydrobiol./Suppl. *76*:53–99.

Schroeder, L. A. 1981. Consumer growth efficiencies: Their limits and relationships to ecological energetics. J. Theor. Biol. *93*:805–828.

Schuldt, J. A. and A. E. Hershey. 1995. Effect of salmon carcass decomposition on Lake Superior tributary streams. J. N. Am. Benthol. Soc. *14*:259–268.

Schults, D. W. and K. W. Malueg. 1971. Uptake of radiophosphorus by rooted aquatic plants. Proc. Third Nat. Symp. Radioecology. pp. 417–424.

Schultz, D. M. and J. G. Quinn. 1973. Fatty acid composition of organic detritus from *Spartina alterniflora*. Estuarine and Coastal Mar. Sci. *1*:177–190.

Schulz, S. and R. Conrad. 1996. Influence of temperature on pathways to methane production in the permanently cold profundal sediment of Lake Constance. FEMS Microbiol. Ecol. *20*:1–14.

Schuman, G. E., R. E. Burwell, R. F. Priest, and R. C. Spomer. 1973. Nitrogen losses in surface runoff from agricultural watersheds on Missouri Valley loess. J. Environ. Qual. *2*:299–302.

Schütz, H., W. Seiler, and R. Conrad. 1989. Processes involved in formation and emission of methane in rice paddies. Biogeochemistry *7*:33–53.

Schuurkes, J. A. A. R. and C. J. Kok. 1988. *In vitro* studies on sulphate reduction and acidification in sediments of shallow soft water lakes. Freshwat. Biol. *19*:417–426.

Schuurkes, J. A. A. R., A. J. Kempers, and C. J. Kok. 1988. Aspects of biochemical sulphur conversions in sediments of a shallow soft water lake. J. Freshwat. Ecol. *4*:369–381.

Schwartz, S. S. 1984. Life history strategies in *Daphnia*: A review and predictions. Oikos *42*:114–122.

Schwartz, S. S. and G. N. Cameron. 1993. How do parasites cost their host? Preliminary answers from trematodes and *Daphnia obtusa*. Limnol. Oceanogr. *38*:602–612.

Schweizer, A. 1997. From littoral to pelagial: Comparing the distribution of phytoplankton and ciliated protozoa along a transect. J. Plankton Res. *19*:829–848.

Schwoerbel, J. and G. C. Tillmanns. 1964a. Konzentrationsabhängige Aufnahme von wasserlöslichem PO_4-P bei submersen Wasserpflanzen. Naturwissenschaften *51*:319–320.

Schwoerbel, J. and G. C. Tillmanns. 1964b. Untersuchungen über die Stoffwechseldynamik in Fliessgewässern. I. Die Rolle höherer Wasserpflanzen: *Callitriche hamulata* Kütz. Arch. Hydrobiol./Suppl. *28*:245–258.

Schwoerbel, J. and G. C. Tillmanns. 1964c. Untersuchungen über die Stoffwechseldynamik in Fliessgewässern. II. Experimentelle Untersuchungen über die Ammoniumaufnahme und pH-Änderung im Wasser durch *Callitriche hamulata* Kütz. und *Fontinalis antipyretica* L. Arch. Hydrobiol./Suppl. *28*: 259–267.

Schwoerbel, J. and G. C. Tillmanns. 1972. Ammonium-Adaptation bei submersen Phanerogamen in situ. Arch. Hydrobiol./Suppl. *42*:139–141.

Schwoerbel, J. and G. C. Tillmanns. 1977. Nitrataufnahme aus dem Wasser und Nitratreduktase-Aktivität bei *Fontinalis antipyretica* L. im Hell-Dunkel-Wechsel. Arch. Hydrobiol./Suppl. *48*: 412–423.

Scott, D. T., D. M. McKnight, E. L. Blunt-Harris, S. E. Kolesar, and D. R. Lovley. 1998. Quinone moieties act as electron acceptors in the reduction of humic substances by humics-reducing microorganisms. Environ. Sci. Technol. *32*:2984–2989.

Scott, J. T., G. E. Myer, R. Stewart, and E. G. Walther. 1969. On the mechanism of Langmuir circulations and their role in epilimnion mixing. Limnol. Oceanogr. *14*:493–503.

Scott, M. L., G. T. Auble, and J. M. Friedman. 1997. Flood dependency of cottonwood establishment along the Missouri River, Montana, USA. Ecological Applications *7*:677–690.

Scott, W. 1924. The diurnal oxygen pulse in Eagle (Winona) Lake. Proc. Indiana Acad. Sci. *33*(1923):311–314.

Scrimgeour, G. J., T. D. Prowse, J. M. Culp, and P. A. Chambers. 1994. Ecological effects of river ice break-up: A review and perspective. Freshwat. Biol. *32*:261–275.

Scully, N. M. and D. R. S. Lean. 1994. The attenuation of ultraviolet radiation in temperate lakes. Arch. Hydrobiol. Beih. Ergebn. Limnol. *43*:135–144.

Scully, N. M., D. R. S. Lean, D. J. McQueen, and W. J. Cooper. 1995. Photochemical formation of hydrogen peroxide in lakes: Effects of dissolved organic carbon and ultraviolet radiation. Can. J. Fish. Aquat. Sci. *52*:2675–2681.

Sculthorpe, C. D. 1967. The Biology of Aquatic Vascular Plants. St. Martin's Press, New York. 610 pp.

Seaby, R. M. H., A. J. Martin, and J. O. Young. 1996. Food partitioning by lake-dwelling triclads and glossiphoniid leeches: Field and laboratory experiments. Oecologia *106*:544–550.

Seadler, A. W. and N. A. Alldridge. 1977. The translocation of radioactive phosphorus by the aquatic vascular plant *Najas minor*. Ohio J. Sci. *77*:76–80.

Seaward, M. R. D. 1996. Lichens and the environment. *In* B. C. Sutton, ed. A Century of Mycology. Cambridge Univ. Press, Cambridge. pp. 293–320.

Sebacher, D. L., R. C. Harriss, and K. B. Bartlett. 1985. Methane emissions to the atmosphere through aquatic plants. J. Environ. Qual. *14*:40–46.

Sebestyén, O. 1949. Studies of detritus drifts in Lake Balaton. Verh. Internat. Verein. Limnol. *10*:414–419.

Sedell, J. R., G. H. Reeves, F. R. Hauer, J. A. Stanford, and C. P. Haukins. 1990. Role of refugia in recovery from disturbances: Modern fragmented and disconnected river systems. Environ. Management *14*:711–724.

Sederholm, H., A. Mauranen, and L. Montonen. 1973. Some observations on the microbial degradation of humous substances in water. Verh. Internat. Verein. Limnol. *18*:1301–1305.

Segers, R. 1998. Methane production and methane consumption: A review of processes underlying wetland methane fluxes. Biogeochemistry *41*:23–51.

Seghers, B. H. 1975. Role of gill rakers in size-selective predation by lake whitefish, *Coregonus clupeaformis* (Mitchell). Verh. Internat. Verein. Limnol. *19*:2401–2405.

Seitz, A. 1980. The coexistence of three species of *Daphnia* in the Klostersee. I. Field studies on the dynamics of reproduction. Oecologia *45*:117–130.

Seitz, W. R. 1981. Fluorescence methods for studying speciation of pollutants in water. Trends Anal. Chem. *1*:79–83.

Seitzinger, S. P. 1988. Denitrification in freshwater and coastal marine ecosystems: Ecological and geochemical significance. Limnol. Oceanogr. *33*:704–724.

Seitzinger, S. P. 1990. Denitrification in aquatic sediments. *In* N. P. Revsbech and J. Sørensen, eds. Denitrification in Soil and Sediment. Plenum Press, New York. pp. 301–322.

Selby, M. J. 1985a. Earth's changing surface: An introduction to geomorphology. Oxford Univ. Press, Oxford. 607 pp.

Selby, M. J. 1985b. River valleys. *In* Earth's Changing Surface: An Introduction to Geomorphology. Clarendon Press, Oxford. pp. 260–302.

Sell, D. W. 1982. Size-frequency estimates of secondary production by *Mysis relicta* in lakes Michigan and Huron. Hydrobiologia *93*:69–78.

Sen, Z. 1995. Applied hydrogeology for scientists and engineers. Lewis Publishers, Boca Raton. 444 pp.

Senft, W. H. 1978. Dependence of light-saturated rates of algal photosynthesis on intracellular concentrations of phosphorus. Limnol. Oceanogr. *23*:709–718.

Serruya, C., U. Pollingher, and M. Gophen. 1975. N and P distribution in Lake Kinneret (Israel) with emphasis on dissolved organic nitrogen. Oikos *26*:1–8.

Servais, P., G. Billen, and J. V. Rego. 1985. Rate of bacterial mortality in aquatic environments. Appl. Environ. Microbiol. *49*:1448–1454.

Seto, M., S. Nishida, and M. Yamamoto. 1982. Dissolved organic carbon as a controlling factor in oxygen consumption in natural and man-made waters. Jpn. J. Limnol. *43*:96–101.

Shaffer, P. W. and M. R. Church. 1989. Terrestrial and in-lake contributions to alkalinity budgets of drainage lakes: An assessment of regional differences. Can. J. Fish. Aquat. Sci. *46*:509–515.

Shamsudin, L. and M. A. Sleigh. 1994. Seasonal changes in composition and biomass of epilithic algal floras of a chalk stream and a soft water stream with estimates of production. Hydrobiologia *273*:131–146.

Shan, R. K. 1974. Reproduction in laboratory stocks of *Pleuroxus* (Chydoridae, Cladocera) under the influence of photoperiod and light intensity. Int. Rev. ges. Hydrobiol. *59*:643–666.

Shannon, C. E. and W. Weaver. 1949. The Mathematical Theory of Communication. Univ. Illinois Press, Urbana. 117 pp.

Shannon, E. L. 1953. The production of root hairs by aquatic plants. Am. Midland Nat. *50*:474–479.

Shapiro, J. 1957. Chemical and biological studies on the yellow organic acids of lake water. Limnol. Oceanogr. *2*:161–179.

Shapiro, J. 1960. The cause of a metalimnetic minimum of dissolved oxygen. Limnol. Oceanogr. *5*:216–227.

Shapiro, J. 1964. Effect of yellow organic acids on iron and other metals in water. J. Amer. Water Wks. Assoc. *56*:1062–1082.

Shapiro, J. 1966. The relation of humic color to iron in natural waters. Verh. Internat. Verein. Limnol. *16*:477–484.

Shapiro, J. 1969. Iron in natural waters—its characteristics and biological availability as determined with the ferrigram. Verh. Internat. Verein. Limnol. *17*:456–466.

Shapiro, J. and G. E. Glass. 1975. Synergistic effects of phosphate and manganese on growth of Lake Superior algae. Verh. Internat. Verein. Limnol. *19*:395–404.

Shapiro, J., V. Lamarra, and M. Lynch. 1975. Biomanipulation: An ecosystem approach to lake restoration. *In* P. L. Brezonik and J. L. Fox, eds. Water Quality Management through Biological Control. Univ. Florida, Gainesville. pp. 85–96.

Sharma, P. C. and M. C. Pant. 1984. An energy budget for *Simocephalus vetulus* (O. F. Muller) (Crustacea: Cladocera). Hydrobiologia *111*:37–42.

Shaw, P. J. 1994. The effect of pH, dissolved humic substances, and ionic composition on the transfer of iron and phosphate to particulate size fractions in epilimnetic lake water. Limnol. Oceanogr. *39*:1734–1743.

Shcherbakov, A. D. 1969. Quantity and biomass of protozoa in the plankton of a eutrophic lake. Hydrobiol. J. *5*:9–15.

Shear, H. and A. E.Walsby. 1975. An investigation into the possible light-shielding role of gas vacuoles in a planktonic blue-green alga. Br. Phycol. J. *10*:241–251.

Shearer, C. A. 1992. The role of woody debris. *In* F. Bärlocher, ed. The Ecology of Aquatic Hyphomycetes. Springer-Verlag, Berlin. pp. 77–98.

Sheath, R. G. 1986. Seasonality of phytoplankton in northern tundra ponds. Hydrobiologia *138*:75–83.

Sheath, R. G. and J. A. Hellebust. 1978. Comparison of algae in the euplankton, tychoplankton, and periphyton of a tundra pond. Can. J. Bot. *56*:1472–1483.

Sheen, J. 1990. Metabolic repression of transcription in higher plants. Plant Cell *2*:1027–1038.

Sheldon, R. B. and C. W. Boylen. 1975. Factors affecting the contribution by epiphytic algae to the primary productivity of an oligotrophic freshwater lake. Appl. Microbiol. *30*:657–667.

Sheldon, R. B. and C. W. Boylen. 1977. Maximum depth inhabited by aquatic vascular plants. Am. Midland Nat. *97*:248–254.

Sheldon, S. P. 1987. The effects of herbivorous snails on submerged macrophyte communities in Minnesota lakes. Ecology *68*:1920–1931.

Shelford, V. E. 1918. Conditions of existence. *In* H. B. Ward and G. C. Whipple, eds. Freshwater Biology. John Wiley & Sons, New York. pp. 21–60.

Sherr, E. B. and B. Sherr. 1988. Role of microbes in pelagic food webs: A revised concept. Limnol. Oceanogr. *33*:1225–1227.

Sherr, E. B. and B. F. Sherr. 1994. Bacterivory and herbivory: Key roles in phagotrophic protists in pelagic food webs. Microb. Ecol. *28*:223–235.

Sherwood, J. E., F. Stagnitti, M. J. Kokkinn, and W. D. Williams. 1991. Dissolved oxygen concentrations in hypersaline waters. Limnol. Oceanogr. *36*:235–250.

Sherwood, J. E., F. Stagnitti, M. J. Kokkinn, and W. D. Williams. 1992. A standard table for predicting equilibrium dissolved oxygen concentrations in salt lakes dominated by sodium chloride. Int. J. Salt Lake Res. *1*:1–6.

Shields, F. D., Jr., A. Simon, and L. J. Steffen. 2000. Reservoir effects on downstream river channel migration. Environ. Conserv. *27*:54–66.

Shiklomanov, I. A. 1990. Global water resources. Nature & Resour. *26*:34–43.

Shilo, M. 1970. Lysis of blue-green algae by myxobacter. J. Bacteriol. *104*:453–461.

Shilo, M. 1971. Biological agents which cause lysis of blue-green algae. Mitt. Internat. Verein. Limnol. *19*:206–213.

Shimaraev, M. N., N. G. Granin, and A. A. Zhdanov. 1993. Deep ventilation of Lake Baikal waters due to spring thermal bars. Limnol. Oceanogr. *38*:1068–1072.

Shoesmith, E. A. and A. J. Brook. 1983. Monovalent-divalent cation ratios and the occurrence of photoplankton, with special reference to the desmids. Freshwat. Biol. *13*:151–155.

Shotyk, W. 1989. An overview of the geochemistry of methane dynamics in mires. Internat. Peat J. *3*:25–44.

Shtegman, B. K., ed. 1966. Produtsirovanie i krugovorot organicheskogo veshchestva vo vnutrennikh vodoemakh. (Production and circulation of organic matter in inland waters.) (Translated into English, Israel Program for Scientific Translations, Jerusalem, 1969.) Trudy Inst. Biol. Vnutrennikh Vod *13*, 287 pp.

Shukla, S. S., J. K. Syers, J. D. H. Williams, D. E. Armstrong, and R. F. Harris. 1971. Sorption of inorganic phosphate by lake sediments. Proc. Soil Sci. Soc. Amer. *35*:244–249.

Shulman, M. D. and R. A. Bryson. 1961. The vertical variation of wind-driven currents in Lake Mendota. Limnol. Oceanogr. *6*:347–355.

Shumm, S. A. 1977. The Fluvial System. John Wiley & Sons, New York.

Shuter, B. J. and K. K. Ing. 1997. Factors affecting the production of zooplankton in lakes. Can. J. Fish. Aquat. Sci. *54*:359–377.

Sicko-Goad, L. and E. F. Stoermer. 1984. The need for uniform terminology concerning phytoplankton cell size fractions and examples of picoplankton from the Laurentian Great Lakes. J. Great Lakes Res. *10*:90–93.

Siderius, M., A. Musgrave, H. van den Ende, H. Koerten, P. Cambier, and P. van der Meer. 1996. *Chlamydomonas eugametos* (Chlorophyta) stores phosphate in polyphosphate bodies together with calcium. J. Phycol. *32*:402–409.

Siebeck, O. 1960. Untersuchungen über die Vertikalwanderung planktischer Crustaceen unter Berücksichtigung der Strahlungsverhältnisse. Int. Rev. ges. Hydrobiol. *45*:381–454.

Siebeck, O. 1968. "Uferflucht" und optische Orientierung pelagischer Crustaceen. Arch. Hydrobiol./Suppl. *35*:1–118.

Siebeck, O. and J. Ringelberg. 1969. Spatial orientation of planktonic crustaceans. I. The swimming behaviour in a horizontal plane. 2. The swimming behaviour in a vertical plane. Verh. Internat. Verein. Limnol. *17*:831–847.

Siebeck, O., T. L. Vail, C. E. Williamson, R. Vetter, D. Hessen, H. Zagarese, E. Little, E. Balseiro, B. Modenutti, J. Seva, and A. Shumate. 1994. Impact of UV-B radiation on zooplankton and fish in pelagic freshwater ecosystems. Arch. Hydrobiol. Beih. Ergebn. Limnol. *43*:101–114.

Sieburth, J. M., V. Smetacek, and J. Lenz. 1978. Pelagic ecosystem structure: Heterotrophic compartments of the plankton and their relationship to plankton size fractions. Limnol. Oceanogr. *23*:1256–1263.

Siegfried, C. A. 1984. The benthos of a eutrophic mountain reservoir: Influence of reservoir level on community composition, abundance, and production. Calif. Fish Game *70*:39–52.

Siegfried, C. A. 1985. Life history, population dynamics and production of *Pontoporeia hoyi* (Crustacea, Amphipoda) in relation to the trophic gradient of Lake George, New York. Hydrobiologia *122*:175–180.

Siegfried, C. A. 1991. The pelagic rotifer community of an acidic clearwater lake in the Adirondack mountains of New York State. Arch. Hydrobiol. *122*:441–462.

Siegfried, C. A. and M. E. Kopache. 1984. Zooplankton dynamics in a high mountain reservoir of southern California. Calif. Fish Game *70*:18–38.

Sieschab, F. K., J. M. Bernard, and K. Fiala. 1985. Above- and belowground standing crop partitioning of biomass by *Eleocharis rostellata* Torr. in the Byron-Bergen Swamp, Genesee County, New York. Am. Midland Nat. *114*:70–76.

Sigg, L., A. Kuhn, H. Xue, E. Kiefer, and D. Kistler. 1995. Cycles of trace elements (copper and zinc) in a eutrophic lake: Role of speciation and sedimentation. *In* C. P. Huang, C. R. O'Melia, and J. J. Morgan, eds. Aquatic Chemistry: Interfacial and Interspecies Processes. Amer. Chem. Soc., Washington, DC. pp. 177–194.

Sigmon, D. E. and L. B. Cahoon. 1997. Comparative effects of benthic microalgae and phytoplankton on dissolved silica fluxes. Aquat. Microb. Ecol. *13*:275–284.

Silver, C. S. and R. S. DeFries. 1990. One Earth, One Future: Our Changing Global Environment. Nat. Acad. Press, Washington, DC. 196 pp.

Silvola, J., J. Alm, U. Ahlholm, U. Nykänen, and P. J. Martikainen. 1996. CO_2 fluxes from peat in boreal mires under varying temperature and moisture conditions. J. Ecol. *84*:219–228.

Šimek, K. and T. H. Chrzanowski. 1992. Direct and indirect evidence of size-selective grazing on pelagic bacteria by freshwater nanoflagellates. Appl. Environ. Microbiol. *58*:3715–3720.

Šimek, K., J. Armengol, M. Comerma, J.-C. Garcia, T. H. Chrzanowski, M. Macek, J. Nedoma, and V. Straškrabová. 2000. Characteristics of protistan control of bacterial production in three reservoirs of different trophy. Int. Revue ges. Hydrobiol. (In press).

Šimek, K., J. Bobková, M. Macek, J. Nedoma, and R. Psenner. 1995. Ciliate grazing on picoplankton in a eutrophic reservoir during the summer phytoplankton maximum: A study at the species and community level. Limnol. Oceanogr. *40*:1077–1090.

Simó, R., R. de Wit, J. Villanueva, and J. Grimalt. 1993. Metabolism of volatile sulfur compounds in Lake Cisó. Verh. Internat. Verein. Limnol. *25*:743–746.

Simola, H. 1977. Diatom succession in the formation of annually laminated sediments in Lovojärvi, a small eutrophicated lake. Ann. Bot. Fennica *14*:143–148.

Simon, M. 1987. Biomass and production of small and large free-living and attached bacteria in Lake Constance. Limnol. Oceanogr. *32*:591–607.

Simon, M. 1988. Growth characteristics of small and large free-living and attached bacteria in Lake Constance. Microb. Ecol. *15*:151–163.

Simon, M. 1994. Diel variability of bacterioplankton biomass production and cell multiplication in Lake Constance. Arch. Hydrobiol. *130*:283–302.

Simon, M. 1998. Bacterioplankton dynamics in a large mesotrophic lake. II. Concentrations and turnover of dissolved amino acids. Arch. Hydrobiol. *144*:1–23.

Simon, M. and C. Wünsch. 1998. Temperature control of bacterioplankton growth in a temperate large lake. Aquat. Microb. Ecol. *16*:119–130.

Simon, M. and M. M. Tilzer. 1982. Bacterial decay of the autumnal phytoplankton in Lake Constance (Bodensee). Schweiz. Z. Hydrol. *44*:263–275.

Simon, M., C. Bunte, M. Schulz, M. Weiss, and C. Wünsch. 1998a. Bacterioplankton dynamics in Lake Constance (Bodensee): Substrate utilization, growth control, and long-term trends. Arch. Hydrobiol. Spec. Issues Advanc. Limnol. *53*:195–221.

Simon, M., M. M. Tilzer, and H. Müller. 1998b. Bacterioplankton dynamics in a large mesotrophic lake. I. Abundance, production and growth control. Arch. Hydrobiol. *143*:385–407.

Simpson, J. H. and J. D. Woods. 1970. Temperature microstructure in a freshwater thermocline. Nature *226*:832–834.

Simpson, T. L. and J. J. Gilbert. 1974. Gemmulation, gemmule hatching, and sexual reproduction in fresh-water sponges. II. Life cycle events in young, larva-produced sponges of *Spongilla lacustris* and an unidentified species. Trans. Amer. Microsc. Soc. *93*:39–45.

Singer, A., A. Eshel, M. Agami, and S. Beer. 1994. The contribution of aerenchymal CO_2 to the photosynthesis of emergent and submerged culms of *Scirpus lacustris* and *Cyperus papyrus*. Aquat. Bot. *49*:107–116.

Singleton, R. 1993. The sulfate-reducing bacteria: An overview. *In* J. M. Odom and R. Singleton, eds. The Sulfate-Reducing Bacteria: Contemporary Perspectives. Springer-Verlag, New York. pp. 1–20.

Siver, P. A. 1978. Development of diatom communities on *Potamogeton robbinsii* Oakes. Rhodora *80*:417–430.

Siver, P. A. 1980. Microattachment patterns of diatoms on leaves of *Potamogeton robbinsii* Oakes. Trans. Amer. Microsc. Soc. *99*:217–220.

Siver, P. A. 1993. Inferring the specific conductivity of lake water with scaled chrysophytes. Limnol. Oceanogr. *38*:1480–1492.

Siver, P. A. and J. S. Chock. 1986. Phytoplankton dynamics in a chrysophycean lake. *In* J. Kristiansen and R. A. Andersen, eds. Chrysophytes: Aspects and Problems. Cambridge Univ. Press, Cambridge. pp. 165–183.

Sjörs, H. 1961. Surface patterns in Boreal peatland. Endeavour *20*:217–224.

Skaggs, R. W., M. A. Brevé, and J. W. Gilliam. 1994. Hydrologic and water quality impacts of agricultural drainage. Crit. Rev. Environ. Sci. Technol. *24*:1–32.

Skelland, A. H. P. 1974. Diffusional mass transfer. Wiley-Interscience, New York.

Skogstad, A., L. Granskog, and D. Klaveness. 1987. Growth of freshwater ciliates offered planktonic algae as food. J. Plankton Res. *9*:503–512.

Skorik, L. V., K. S. Valdimirova, G. A. Yenaki, and I. K. Palamarchuk. 1972. Organic matter in Kiev Reservoir soils and its role in the development of benthic algae. Hydrobiol. J. *8*:63–66.

Slack, K. V. 1964. Effect of tree leaves on water quality in the Cacapon River, West Virginia. Prof. Pap. U.S. Geol. Surv. *475-D*:181–185.

Sládeček, V. 1973. System of water quality from the biological point of view. Arch. Hydrobiol. Ergebn. Limnol. *7*:1–218.

Sládeček, V. and A. Sládečková. 1963. Limnological study of the Reservoir Sedlice near Zeliv. XXIII. Periphyton production. Sbornik Vysoké úkoly Chem.-Technol. Praze, Technol. Vody *7*:77–133.

Sládeček, V. and A. Sládeček. 1964. Determination of the periphyton production by means of the glass slide method. Hydrobiologia *23*:125–158.

Sládeček, V. and A. Sládeček. 1998. Revision of polysaprobic indicators. Verh. Internat. Verein. Limnol. *26*:1277–1280.

Sládečeková, A. 1960. Limnological study of the Reservoir Sedlice near Zeliv. XI. Periphyton stratification during the first yearlong period (June 1957–July 1958). Sci. Pap. Inst. Chem. Technol. Prague, Faculty of Technol. Fuel Water *4*:143–261.

Sládečeková, A. 1962. Limnological investigation methods for the periphyton ("Aufwuchs") community. Bot. Rev. *28*:286–350.

Sleigh, M. 1989. Protozoa and Other Protists. E. Arnold/Hodder & Stoughton, London. 342 pp.

Sleigh, M. A., B. M. Baldock, and J. H. Baker. 1992. Protozoan communities in chalk streams. Hydrobiologia *248*:53–64.

Slobodkin, L. B. 1954. Population dynamics in *Daphnia obtusa* Kurz. Ecol. Monogr. *24*:69–88.

Slobodkin, L. B. 1960. Ecological energy relationships at the population level. Am. Nat. *94*:213–236.

Slobodkin, L. B. 1962. Energy in animal ecology. Adv. Ecol. Res. *1*:69–101.

Slusarczyk, M. 1995. Predator-induced diapause in *Daphnia*. Ecology 76:1008–1013.

Smayda, T. J. and B. J. Boleyn. 1965. Experimental observations on the flotation of marine diatoms. I. *Thalassiosira* cf. *nana*, *Thalassiosira rotula* and *Nitzschia seriata*. Limnol. Oceanogr. *10*:499–509.

Smayda, T. J. and B. J. Boleyn. 1966a. Experimental observations on the flotation of marine diatoms. II. *Skeletonema costatum* and *Rhizosolenia setigera*. Limnol. Oceanogr. *11*:18–34.

Smayda, T. J. and B. J. Boleyn. 1966b. Experimental observations on the flotation of marine diatoms. III. *Bacteriastrum hyalinum* and *Chaetoceros lauderi*. Limnol. Oceanogr. *11*:35–43.

Smetacek, V. S. 1985. Role of sinking in diatom life-history cycles: Ecological, evolutionary and geological significance. Mar. Biol. *84*:239–251.

Šmid, P. 1975. Evaporation from a reedswamp. J. Ecol. *66*:299–309.

Smith, B. P. 1988. Host-parasite interaction and impact of larval water mites on insects. Annu. Rev. Entomol. *33*:487–507.

Smith, C. S. and M. S. Adams. 1986. Phosphorus transfer from sediments by *Myriophyllum spicatum*. Limnol. Oceanogr. *31*:1312–1321.

Smith, D. E. C. and R. W. Davies. 1995. Effects of feeding frequencies on energy partitioning and the life history of the leech *Nephelopsis obscura*. J. N. Am. Benthol. Soc. *14*:563–576.

Smith, F. A. 1967. Rates of photosynthesis in Characean cells. I. Photosynthetic $^{14}CO_2$ fixation by *Nitella translucens*. J. Exp. Bot. *18*:509–517.

Smith, F. A. and N. A. Walker. 1980. Photosynthesis by aquatic plants: Effects of unstirred layers in relation to the assimilation of CO_2 and HCO_3^- and to carbon isotopic discrimination. New Phytol. *86*:245–259.

Smith, I. M. and D. R. Cook. 1991. Water mites. *In* J. H. Thorp and A. P. Covich, eds. Ecology and Classification of North American Freshwater Invertebrates. Academic Press, San Diego. pp. 523–592.

Smith, I. R. 1975. Turbulence in lakes and rivers. Sci. Publ. Freshwater Biol. Assoc. U.K. 29, 79 pp.

Smith, I. R. and I. J. Sinclair. 1972. Deep water waves in lakes. Freshwat. Biol. *2*:387–399.

Smith, J., A. K. Ward, and M. S. Stock. 1992. Quantitative estimation of epilithic algal patchiness caused by microtopographical irregularities of different rock types. Bull. North Am. Benthol. Soc. 9:148.

Smith, K. 1981. The prediction of river water temperatures. Hydrol. Sci. Bull. *26*:19–32.

Smith, L. K. and W. M. Lewis, Jr. 1992. Seasonality of methane emissions from five lakes and associated wetlands of the Colorado Rockies. Global Biogeochem. Cycles *6*:323–338.

Smith, M. W. 1969. Changes in environment and biota of a natural lake after fertilization. J. Fish. Res. Board Can. *26*:3101–3132.

Smith, R. C. and W. H. Wilson, Jr. 1972. Photon scalar irradiance. Appl. Optics *11*:934–938.

Smith, R. C., J. E. Tyler, and C. R. Goldman. 1973. Optical properties and color of Lake Tahoe and Crater Lake. Limnol. Oceanogr. *18*:189–199.

Smith, R. J. 1995. Calcium and bacteria. Adv. Microbial Physiol. *37*:83–133.

Smith, R. L. and M. J. Klug. 1981. Electron donors utilized by sulfate-reducing bacteria in eutrophic lake sediments. Appl. Environ. Microbiol. *42*:116–121.

Smith, R. L. and M. J. Klug. 1981. Reduction of sulfur compounds in sediments of a eutrophic lake basin. Appl. Environ. Microbiol. *41*:1230–1237.

Smith, V. H. 1983. Low nitrogen to phosphorus ratios favor dominance by blue-green algae in lake phytoplankton. Science *221*:669–671.

Smith, V. H. and J. Shapiro. 1981. Chlorophyll-phosphorus relations in individual lakes. Their importance to lake restoration strategies. Environ. Sci. Technol. *15*:444–451.

Smock, L. A., E. Gilinsky, and D. L. Stoneburner. 1985. Macroinvertebrate production in a southeastern United States stream. Ecology *66*:1491–1503.

Smol, J. P. 1980. Fossil synuracean (Chrysophyceae) scales in lake sediments: A new group of paleoindicators. Can. J. Bot. *48*:458–465.

Smol, J. P. 1985. The ratio of diatom frustules to chrysophycean statospores: A useful paleolimnological index. Hydrobiologia *123*:199–208.

Smol, J. P. 1988. Chrysophycean microfossils in paleolimnological studies. Palaeogeogr. Palaeoclimat. Palaeoecol. *62*:287–297.

Smol, J. P. 1990. Paleolimnology: Recent advances and future challenges. Mem. Ist. Ital. Idrobiol *47*:253–276.

Smol, J. P. 1992. Paleolimnology: An important tool for effective ecosystem management. J. Aquat. Ecosystem Health *1*:49–58.

Smol, J. P., B. F. Cumming, A. S. Dixit, and S. S. Dixit. 1998. Tracking recovery patterns in acidified lakes: A paleolimnological perspective. Restoration Ecol. *6*:318–326.

Smol, J. P., D. F. Charles, and D. R. Whitehead. 1984. Mallomonadacean (Chrysophycean) assemblages and their relationships with limnological characteristics in 38 Adirondack (New York) lakes. Can. J. Bot. *62*:911–923.

Smol, J. P., S. R. Brown, and R. N. McNeely. 1983. Cultural disturbances and trophic history of a small meromictic lake from central Canada. Hydrobiologia *103*:125–130.

Smyly, W. J. P. 1973. Bionomics of *Cyclops strenuus abyssorum* Sars (Copepoda: Cyclopoida). Oecologia *11*:163–186.

Smyly, W. J. P. 1980. Food and feeding of aquatic larvae of the midge *Chaoborus flavicans* (Meigen) (Diptera: Chaoboridae) in the laboratory. Hydrobiologia *70*:179–188.

Smyth, T. and J. D. Reynolds. 1995. Survival ability of statoblasts of freshwater Bryozoa found in Renvyle Lough, County Galway. Proc. Royal Irish Acad. *95B*:65–68.

Snaddon, C. D., B. A. Stewart, and B. R. Davies. 1992. The effect of discharge on leaf retention in two headwater streams. Arch. Hydrobiol. *125*:109–120.

Snell, T. W. 1980. Blue-green algae and selection in rotifer populations. Oecologia *46*:343–346.

Snowden, R. E. D. and B. D. Wheller. 1995. Chemical changes in selected wetland plant species with increasing Fe supply, with specific reference to root precipitates and Fe tolerance. New Phytol. *131*:503–520.

Snyder, R. L. and C. E. Boyd. 1987. Evapotranspiration by *Eichhorina crassipes* (Mart.) Solms and *Typha latifolia* L. Aquat. Bot. *27*:217–227.

Søballe, D. M. and B. L. Kimmel. 1987. A large-scale comparison of factors influencing phytoplankton abundance in rivers, lakes, and impoundments. Ecology *68*:1943–1954.

Sobczak, W. H. 1996. Epilithic bacterial responses to variations in algal biomass and labile dissolved organic carbon during biofilm colonization. J. N. Am. Benthol. Soc. *15*:143–154.

Soeder, C. J. 1965. Some aspects of phytoplankton growth and activity. Mem. Ist. Ital. Idrobiol. *18*(Suppl.):47–59.

Söderström, O. 1987. Upstream movements of invertebrates in running waters—a review. Arch. Hydrobiol. *111*:197–208.

Soeder, C. J. and G. Engelmann. 1984. Nickel requirement in *Chlorella emersonii*. Arch. Microbiol. *137*:85–87.

Sokolov, A. A. and T. G. Chapman, eds. 1974. Methods for Water Balance Computations. Unesco Press, Paris. 127 pp.

Sokolova, G. A. 1961. Rol' zhalezobakterii v dinamike zheleza v Glubokom Ozere. Trudy Vses. Gidrobiol. Obshch. *11*:5–11.

Solander, D. 1983. Biomass and shoot production of *Carex rostrata* and *Equisetum fluviatile* in unfertilized and fertilized subarctic lakes. Aquat. Bot. *15*:349–366.

Solski, A. 1962. Mineralizacja roslin wodnych. I. Uwalnianie fosforu i potasu przez wymywanie. Pol. Arch. Hydrobiol. *10*:167–196.

Sommaruga, R. and R. Psenner. 1995. Trophic interactions within the microbial food web in Piburger See (Austria). Arch. Hydrobiol. *132*:257–278.

Sommaruga, R. and R. Psenner. 1997. Ultraviolet radiation in a high mountain lake of the Austrian Alps: Air and underwater measurements. Photochem. Photobiol. *65*:957–963.

Sommaruga, R., I. Obernosterer, G. J. Herndl, and R. Psenner. 1997. Inhibitory effect of solar radiation on thymidine and leucine incorporation by freshwater and marine bacterioplankton. Appl. Environ. Microbiol. *63*:4178–4184.

Sommer, U. 1981. The role of *r*- and *K*-selection in the succession of phytoplankton in Lake Constance. Acta Oecologica Oecol. Gener. *2*:327–342.

Sommer, U. 1982. Vertical niche separation between two closely related planktonic flagellate species (*Rhodomonas lens* and *Rhodomonas minuta* v. *nannoplanctica*). J. Plankton Res. *4*:137–142.

Sommer, U. 1984a. Population dynamics of three planktonic diatoms in Lake Constance. Holarctic Ecol. *7*:257–261.

Sommer, U. 1984b. Sedimentation of principal phytoplankton species in Lake Constance. J. Plankton Res. *6*:1–14.

Sommer, U. 1985. Seasonal succession of phytoplankton in Lake Constance. BioScience *35*:351–357.

Sommer, U. 1986. Differential migration of Cryptophyceae in Lake Constance. *In* M. A. Rankin, ed. Migration: Mechanisms and Adaptive Significance. Contribut. Mar. Sci. *27*(Suppl.):166–175.

Sommer, U. 1988. Some size relationships in phytoflagellate motility. Hydrobiologia *161*:125–131.

Sommer, U. 1990. The role of competition for resources in phytoplankton succession. *In* U. Sommer, ed. Plankton ecology: Succession in plankton communities. Springer-Verlag, New York. pp. 57–106.

Sommer, U. 1991. A comparison of the Droop and the Monod models of nutrient limited growth applied to natural populations of phytoplankton. Functional Ecol. *5*:535–544.

Sommer, U. and H.-H. Stabel. 1983. Silicon consumption and population density changes of dominant planktonic diatoms in Lake Constance. J. Ecol. *71*:119–130.

Sommer, U., Z. M. Gliwicz, W. Lampert, and A. Duncan. 1986. The PEG-model of seasonal succession of planktonic events in fresh waters. Arch. Hydrobiol. *106*:455–471.

Sommers, L. E., R. F. Harris, J. D. H. Williams, D. E. Armstrong, and J. K. Syers. 1970. Determination of total organic phosphorus in lake sediments. Limnol. Oceanogr. *15*:301–304.

Søndergaard, M. 1979. Light and dark respiration and the effect of the lacunal system on refixation of CO_2 in submerged aquatic plants. Aquat. Bot. *6*:269–283.

Søndergaard, M. 1981. Kinetics of extracellular release of ^{14}C-labelled organic carbon by submerged macrophytes. Oikos *36*:331–347.

Søndergaard, M. 1990. The effect of environmental variables on release of extracellular organic carbon by freshwater macrophytes. Folia Geobot. Phytotax. Praha *25*:321–332.

Søndergaard, M. 1991. Phototrophic picoplankton in temperate lakes: Seasonal abundance and importance along a trophic gradient. Int. Rev. ges. Hydrobiol. *76*:505–522.

Søndergaard, M. 1993. Organic carbon pools in two Danish lakes: Flow of carbon to bacterioplankton. Verh. Internat. Verein. Limnol. *25*:593–598.

Søndergaard, M. and H.-H. Schierup. 1982. Release of extracellular organic carbon during a diatom bloom in Lake Mossø: Molecular weight fractionation. Freshwat. Biol. *12*:313–320.

Søndergaard, M. and K. Sand-Jensen. 1978. Total autotrophic production in oligotrophic Lake Kalgaard, Denmark. Verh. Internat. Verein. Limnol. *20*:667–673.

Søndergaard, M. and K. Sand-Jensen. 1979. Carbon uptake by leaves and roots of *Littorella uniflora* (L.) Aschers. Aquat. Bot. *6*:1–12.

Søndergaard, M. and R. G. Wetzel. 1980. Photorespiration and internal recycling of CO_2 in the submersed angiosperm *Scirpus subterminalis*. Can. J. Bot. *58*:591–598.

Søndergaard, M. and S. Laegaad. 1977. Vesicular-arbuscular mycorrhiza in some aquatic vascular plants. Nature *268*:232–233.

Soohoo, J. B. and D. A. Keifer. 1982a. Vertical distribution of phaeopigments. I. A simple grazing and photooxidative scheme for small particles. Deep Sea Res. *29*:1539–1552.

Soohoo, J. B. and D. A. Keifer. 1982b. Vertical distribution of phaeopigments. II. Rates of production and kinetics of photooxidation. Deep Sea Res. *29*:1553–1563.

Sorhannus, U., F. Gasse, R. Perasso, and A. B. Tourancheau. 1995. A preliminary phylogeny of diatoms based on 28S ribosomal RNA sequence data. Phycologia *34*:65–73.

Sorokin, Ju. I. 1964a. On the primary production and bacterial activities in the Black Sea. J. Cons. Int. Explor. Mer. *29*:41–60.

Sorokin, Ju. I. 1964b. On the trophic role of chemosynthesis in water bodies. Int. Rev. ges. Hydrobiol. *49*:307–324.

Sorokin, Ju. I. 1965. On the trophic role of chemosynthesis and bacterial biosynthesis in water bodies. Mem. Ist. Ital. Idrobiol. *18*(Suppl.):187–205.

Sorokin, Ju. I. 1966. Vzaimosviaz mikrobiologicheskikh protessov krugovorota sepy i ugleroda v meromikticheskom Ozere Belovod. *In* Plankton i Bentos. Trudy Inst. Biol. Vnutrennikh Vod *12*:332–355.

Sorokin, Ju. I. 1968. The use of ^{14}C in the study of nutrition of aquatic animals. Mitt. Internat. Verein. Limnol. *16*, 41 pp.

Sorokin, Ju. I. 1970. Interrelations between sulphur and carbon turnover in meromictic lakes. Arch. Hydrobiol. *66*:391–446.

Sorokin, Ju. I. and E. B. Paveljeva. 1972. On the quantitative characteristics of the pelagic ecosystem of Dalnee Lake (Kamchatka). Hydrobiologia *40*:519–552.

Sorokin, Y. 1979. On methodology of lake ecosystem studies. Arch. Hydrobiol. Beih. Ergebn. Limnol. *13*:225–233.

Sorokin, Y. I. and H. Kadota, eds. 1972. Techniques for the Assessment of Microbial Production and Decomposition in Fresh Waters. Int. Biol. Program Handbook 23. Blackwell Scientific Publs., Oxford. 112 pp.

Sorrell, B. K. 1991. Transient pressure gradients in the lacunar system of the submerged macrophyte *Egeria densa* Planch. Aquat. Bot. *39*:99–108.

Sorrell, B. K. and F. I. Dromgoole. 1989. Oxygen diffusion and dark respiration in aquatic macrophytes. Plant Cell Environ. *12*:293–299.

Sorrell, B. K. and F. I. Dromgoole. 1987. Oxygen transport in the submerged freshwater macrophyte *Egeria densa* Planch. I. Oxygen production, storage and release. Aquat. Bot. *28*:63–80.

Sorrell, B. K. and F. I. Dromgoole. 1996. Mechanical properties of the lacunar gas in *Egeria densa* Planch. shoots. Aquat. Bot. *53*:47–60.

Sorsa, K. 1979. Primary production of epipelic algae in Lake Suomunjärvi, Finnish North Karelia. Ann. Bot. Fennici *16*:351–366.

Spain, J. D., G. M. Wernert, and D. W. Hubbard. 1976. The structure of the spring thermal bar in Lake Superior, II. J. Great Lakes Res. *2*:296–306.

Spalding, R. F. and M. E. Exner. 1993. Occurrence of nitrate in groundwater—a review. J. Environ. Qual. *22*:392–402.

Sparrow, F. K., Jr. 1968. Ecology of freshwater fungi. *In* G. C. Ainsworth and A. S. Sussman, eds. The Fungi. Vol. III. The Fungal Population. Academic Press, New York. pp. 41–93.

Spelling, S. M. and J. O. Young. 1986. The population dynamics of metacercariae of *Apatemon gracilis* (Trematoda: Digenea) in three species of lake-dwelling leeches. Parasitology *93*:517–530.

Spelling, S. M. and J. O. Young. 1987a. Predation on lake-dwelling leeches (Annelida: Hirudinea): An evaluation by field experiment. J. Anim. Ecol. *56*:131–146.

Spelling, S. M. and J. O. Young. 1987b. A field study of the parasite *Nosema herpobdellae* Conet (Microspora) in *Erpobdella octoculata* (L.) (Hirudinea). Hydrobiologia *156*:91–93.

Spence, D. H. N. 1976. Light and plant response in fresh water. *In* G. C. Evans, R. Bainbridge, and O. Rackham, eds. Light as an Ecological Factor: II. Blackwell Scientific Publs., Oxford. pp. 93–133.

Spence, D. H. N. 1982. The zonation of plants in freshwater lakes. Adv. Ecol. Res. *12*:37–125.

Spence, D. H. N. and J. Chrystal. 1970a. Photosynthesis and zonation of freshwater macrophytes. I. Depth distribution and shade tolerance. New Phytol. *69*:205–215.

Spence, D. H. N. and J. Chrystal. 1970b. Photosynthesis and zonation of freshwater macrophytes. II. Adaptability of species of deep and shallow water. New Phytol. *69*:217–227

Spence, D. H. N., R. M. Campbell, and J. Chrystal. 1971. Productivity of submerged freshwater macrophytes. Hidrobiologia (Romania) *12*:169–176.

Spencer, C. N. and F. R. Hauer. 1991. Phosphorus and nitrogen dynamics in streams during a wildfire. J. N. Amer. Benthol. Soc. *10*:24–30.

Spencer, D. F., L. W. J. Anderson, and G. G. Ksander. 1994. Field and greenhouse investigations on winter bud production by *Potamogeton gramineus* L. Aquat. Bot. *48*:285–295.

Spencer, M. J. 1978. Microbial activity and biomass relationships in 26 oligotrophic to mesotrophic lakes in South Island, New Zealand. Ver. Internat. Verein. Limnol. *20*:1175–1181.

Spencer, W. and G. Bowes. 1990. Ecophysiology of the world's most troublesome aquatic weeds. *In* A. H. Pieterse and K. J. Murphy, eds. Aquatic Weeds: The Ecology and Management of Nuisance Aquatic Vegetation. Oxford Univ. Press, Oxford. pp. 39–73.

Spencer, W. E. and R. G. Wetzel. 1993. Acclimation of photosynthesis and dark respiration of a submersed angiosperm beneath ice in a temperate lake. Plant Physiol. *101*:985–991.

Spencer, W. E., J. Teeri, and R. G. Wetzel. 1994. Acclimation of photosynthetic phenotype to environmental heterogeneity. Ecology *75*:301–314.

Spencer, W. E., R. G. Wetzel, and J. Terri. 1996. Photosynthetic phenotype plasticity and the role of phosphoenolpyruvate carboxylase in *Hydrilla verticillata*. Plant Sci. *118*:1–9.

Sprules, W. G. 1972. Effects of size-selective predation and food competition on high altitude zooplankton communities. Ecology *53*:375–386.

Stabel, H.-H. 1986. Calcite precipitation in Lake Constance: Chemical equilibrium, sedimentation, and nucleation by algae. Limnol. Oceanogr. *31*:1081–1093.

Stabel, H.-H. 1989. Coupling of strontium and calcium cycles in Lake Constance. Hydrobiologia *176/177*:323–329.

Stabel, H.-H. and M. Geiger. 1985. Phosphorus adsorption to riverine suspended matter: Implications for the P-budget of Lake Constance. Water Res. *19*:1347–1352.

Stabel, H.-H., K. Moaledj, and J. Overbeck. 1979. On the degradation of dissolved organic molecules from Plusssee by oligocarbophilic bacteria. Arch. Hydrobiol. Beih. Ergebn. Limnol. *12*:95–104.

Stabell, T. 1996. Ciliate bacterivory in epilimnetic waters. Aquat. Microb. Ecol. *10*:265–272.

Stadelmann, P. 1971. Stickstoffkreislauf und Primärproduktion im mesotrophen Vierwaldstättersee (Horwer Bucht) und im eutrophen Rotsee, mit besonderer Berücksichtigung des Nitrats als limitierenden Faktors. Schweiz. Z. Hydrol. *33*:1–65.

Stahl, J. B. 1959. The developmental history of the chironomid and *Chaoborus* faunas of Myers Lake. Invest. Indiana Lakes Streams *5*:47–102.

Stahl, J. B. 1966a. The ecology of *Chaoborus* in Myers Lakes, Indiana. Limnol. Oceanogr. *11*:177–183.

Stahl, J. B. 1966b. Coexistence in Chaoborus and its ecological significance. Invest. Indiana Lakes Streams *7*:99–113.

Stahl, J. B. 1969. The uses of chironomids and other midges in interpreting lake histories. Mitt. Internat. Verein. Limnol. *17*:111–125.

Stallard, R. F. 1980. Major element geochemistry of the Amazon River system. Ph.D. Dissertation, Massachusetts Institute of Technology.

Stallard, R. F. and J. M. Edmond. 1981. Geochemistry of the Amazon. I. Precipitation chemistry and the marine contribution to the dissolved load at the time of peak discharge. J. Geophys. Res. *86*:9844–9858.

Stanford, J. A. and J. V. Ward. 1988. The hyporheic habitat of river ecosystems. Nature *335*:64–66.

Stangenberg, M. 1968. Bacteriostatic effects of some algae- and *Lemna minor* extracts. Hydrobiologia *32*:88–96.

Stangenberg-Oporowska, K. 1967. Sodium contents in carp-pond waters in Poland. Pol. Arch. Hydrobiol. *14*:11–17.

Stanier, R. Y. 1973. Autotrophy and heterotrophy in unicellular blue-green algae. *In* N. G. Carr and B. A. Whitton, eds. The Biology of Blue-Green Algae. University of California Press, Berkeley. pp. 501–518.

Stanier, R. Y. and G. Cohen-Bazire. 1977. Phototrophic prokaryotes: The cyanobacteria. Ann. Rev. Microbiol. *31*:225–274.

Stanley, D. W. 1976. Productivity of epipelic algae in tundra ponds and a lake near Barrow, Alaska. Ecology *57*:1015–1024.

Stanley, E. H., E. C. Barrett, and A. K. Ward. 1998. Methane efflux through *Nymphaea*: Potential effects of leaf damage and leaf age. Verh. Internat. Verein. Limnol. *26*:1882–1885.

Stanley, R. A. 1972. Photosynthesis in Eurasian watermilfoil (*Myriophyllum spicatum* L.) Plant Physiol. *50*:149–151.

Stansfield, J. H., M. R. Perrow, L. D. Tench, A. J. D. Jowitt, and A. A. L. Taylor. 1997. Submerged macrophytes as refuges for grazing Cladocera against fish predation: Observations on seasonal changes in relation to macrophyte cover and predation pressure. Hydrobiologia *342/343*:229–240.

Starink, M., I. N. Krylova, M.-J. Bär-Gilissen, R. P. M. Bak, and T. E. Cappenberg. 1994. Rates of benthic protozoan grazing on free and attached sediment bacteria measured with fluorescently stained sediment. Appl. Environ. Microbiol. *60*:2259–2264.

Starink, M., M.-J. Bär-Gilissen, R. P. M. Bak, and T. E. Cappenberg. 1996. Seasonal and spatial variations in heterotrophic nanoflagellate and bacteria abundances in sediments of a freshwater littoral zone. Limnol. Oceanogr. *41*:234–242.

Starkweather, P. L. 1975. Diel patterns of grazing in *Daphnia pulex* Leydig. Verh. Internat. Verein. Limnol. *19*:2851–2857.

Starkweather, P. L. 1980a. Behavioral determinants of diet quantity and diet quality in *Brachionus calyciflorus*. *In* W. C. Kerfoot, ed. Evolution and Ecology of Zooplankton Communities. Univ. Press New England, Hanover, NH. pp. 151–157.

Starkweather, P. L. 1980b. Aspects of the feeding behavior and trophic ecology of suspension-feeding rotifers. Hydrobiologia *73*:63–72.

Starkweather, P. L. 1990. Zooplankton community structure of high elevation lakes: Biogeographic and predator-prey interactions. Verh. Internat. Verein. Limnol. *24*:513–517.

Starkweather, P. L. and J. J. Gilbert. 1977. Feeding in the rotifer *Brachionus calyciflorus*. II. Effect of food density on feeding rates using *Euglena gracilis* and *Rhodotorula glutinis*. Oecologia 28:133–139.

Starkweather, P. L. and J. J. Gilbert. 1978a. Feeding in the rotifer *Brachionus calyciflorus*. IV. Selective feeding on tracer particles as a factor in trophic ecology and in situ technique. Verh. Internat. Verein. Limnol. *20*:2389–2394.

Starkweather, P. L. and J. J. Gilbert. 1978b. Radiotracer determination of feeding in *Brachionus calyciflorus*: The importance of gut passage times. Arch. Hydrobiol. Beih. Ergebn. Limnol. *8*:261–263.

Starkweather, P. L. and K. G. Bogdan. 1980. Detrital feeding in natural zooplankton communities: Discrimination between live and dead algal foods. Hydrobiologia *73*:83–85.

Starzecka, A. and T. Bednarz. 1994. Decomposition of organic matter in bottom sediments of a stream and the Dobczyce dam-reservoir in the area of the Wolnica Creek (southern Poland). Arch. Hydrobiol. *129*:327–337.

de Stasio, B. T., Jr. 1989. The seed bank of a freshwater crustacean: Copepodology for the plant ecologist. Ecology *70*:1377–1389.

Statzner, B. and B. Higler. 1986. Stream hydraulics as a major determinant of benthic invertebrate zonation patterns. Freshwat. Biol. *16*:127–139.

Statzner, B. and T. F. Holm. 1989. Morphological adaptation of shape to flow: Microcurrents around lotic macroinvertebrates with known Reynolds numbers at quasi-natural flow conditions. Oecologia *78*:145–157.

Stauffer, R. E. 1985. Relationships between phosphorus loading and trophic state in calcareous lakes of southeast Wisconsin. Limnol. Oceanogr. *30*:123–145.

Stauffer, R. E. l986. Cycling of manganese and iron in Lake Mendota, Wisconsin. Environ. Sci. Technol. *20*:449–457.

Stauffer, R. E. 1987. Effects of oxygen transport on the areal hypolimnetic oxygen deficit. Wat. Resour. Res. *23*:1887–1892.

Stauffer, R. E. and D. E. Armstrong. 1986. Cycling of iron, manganese, silica, phosphorus, calcium, and potassium in two stratified basin of Shagawa Lake, Minnesota. Geochim. Cosmochim. Acta *50*:215–229.

Steele, D. W. and J. M. Buttle. 1994. Sulphate dynamics in a northern wetland catchment during snowmelt. Biogeochemistry *27*:187–211.

Steele, J. H. and I. E. Baird. 1968. Production ecology of a sandy beach. Limnol. Oceanogr. *13*:14–25.

Steemann Nielsen, E. 1944. Dependence of freshwater plants on quantity of carbon dioxide and hydrogen ion concentration. Dansk Bot. Ark. *11*:1–25.

Steemann Nielsen, E. 1947. Photosynthesis of aquatic plants with special reference to the carbon sources. Dansk Bot. Ark. *12*:5–71.

Steemann Nielsen, E. 1962a. Inactivation of the photochemical mechanism in photosynthesis as a means to protect the cells against too high light intensities. Physiol. Plant *15*:161–171.

Steemann Nielsen, E. 1962b. On the maximum quantity of plankton chlorophyll per surface unit of a lake or the sea. Int. Rev. ges. Hydrobiol. *47*:333–338.

Steemann Nielsen, E. and E. G. Jørgensen. 1962. The physiological background for using chlorophyll measurements in hydrobiology and a theory explaining daily variations in chlorophyll concentration. Arch. Hydrobiol. *58*:349–357.

Steemann Nielsen, E. and E. G. Jørgensen. 1968a. The adaptation of plankton algae. I. General part. Physiol. Plant *21*:401–413.

Steemann Nielsen, E. and E. G. Jørgensen. 1968b. The adaptation of plankton algae. III. With special consideration of the importance in nature. Physiol. Plant. *21*:647–654.

Steemann Nielsen, E. and M. Willemoës. 1971. How to measure the illumination rate when investigating the rate of photosynthesis of unicellular algae under various light conditions. Int. Rev. ges. Hydrobiol. *56*:541–556.

Stein, J. R., ed. 1973. Handbook of Phycological Methods. Culture Methods and Growth Measurements. Cambridge Univ. Press, Cambridge. 448 pp.

Stein, R. A., C. G. Goodman, and E. A. Marschall. 1984. Using time and energetic measures of cost in estimating prey value for fish predators. Ecology *65*:702–715.

Steinberg, C. 1978a. Bakterien und ihre Aktivität in der Oberfläche von Profundalsedimenten des Walchensees (Oberbayern). Arch. Hydrobiol. *84*:29–41.

Steinberg, C. 1978b. Bacteria, bacterial activities, and uptake kinetics in surficial sediments of a small eutrophicated mountain lake. Verh. Internat. Verein. Limnol. *20*:2260–2263.

Steinberg, C. 1980. Species of dissolved metals derived from oligotrophic hard water. Water Res. *14*:1239–1250.

Steinberg, C. and A. Melzer. 1983. Aufnahme, Transport und Abgabe von Kohlenstoff durch submerse Makrophyten von Fliesswasserstandorten. Schweiz. Z. Hydrol. *45*:333–344.

Steinberg, C. and U. Münster. 1985. Geochemistry and ecological role of humic substances in lakewater. *In* G. R. Aiken, D. M. McKnight, R. L. Wershaw, and P. MacCarthy, eds. Humic Substances in Soil, Sediment, and Water: Geochemistry, Isolation, and Characterization. John Wiley & Sons, New York. pp. 105–145.

Steinberg, C. E. W., M. Haitzer, R. Brüggemann, I. V. Perminova, N. Y. Yashchenko, and V. S. Petrosyan. 2000. Toward a quantitative structure activity relationship (QSAR) of dissolved humic substances as detoxifying agents in freshwaters. Internat. Rev. Hydrobiol. *85*:253–266.

Steinman, A. D. 1996. Effects of grazers on freshwater benthic algae. *In* R. J. Stevenson, M. L. Bothwell, and R. L. Lowe, eds. Algal Ecology: Freshwater Benthic Ecosystems. Academic Press, San Diego. pp. 341–373.

Steinman, A. D. and G. A. Lamberti. 1996. Biomass and pigments of benthic algae. *In* F. R. Hauer and G. A. Lamberti, eds. Methods in Stream Ecology. Academic Press, San Diego. pp. 295–313.

Steinman, A. D., C. D. McIntire, and R. R. Lowry. 1987. Effect of herbivore type and density on chemical composition of algal assemblages in laboratory streams. J. N. Am. Benthol. Soc. *6*:189–197.

Steinman, A. D., P. J. Mulholland, and J. J. Beauchamp. 1995. Effects of biomass, light, and grazing on phosphorus cycling in stream periphyton communities. J. N. Am. Benthol. Soc. *14*:371–381.

Stemberger, R. S. and J. J. Gilbert. 1985. Body size, food concentration, and population growth in planktonic rotifers. Ecology *66*:1151–1159.

Stemberger, R. S. and J. J. Gilbert. 1987. Rotifer threshold food concentrations and the size-efficiency hypothesis. Ecology *68*:181–187.

Stemberger, R. S. and M. S. Evans. 1984. Rotifer seasonal succession and copepod predation in Lake Michigan. J. Great Lakes Res. *10*:417–428.

Stemler, A. and Govindjee. 1973. Bicarbonate ion as a critical factor in photosynthetic oxygen evolution. Plant Physiol. *51*:119–123.

Stenlund, D. L. and I. D. Charvat. 1994. Vesicular arbuscular mycorrhizae in floating wetland mat communities dominated by *Typha*. Mycorrhiza *4*:131–137.

Stenson, J. A. E. 1978a. Differential predation by fish on two species of *Chaoborus* (Diptera, Chaoboridae). Oikos *31*:98–101.

Stenson, J. A. E. 1978b. Relations between vertebrate and invertebrate zooplankton predators in some arctic lakes. Astarte *11*:21–26.

Stepanauskas, R., H. Laudon, and N. O. G. Jørgensen. 2000. High DON bioavailability in boreal streams during a spring flood. Limnol. Oceanogr. *45*:1298–1307.

Stepanauskas, R., L. J. Tranvik, and L. Leonardson. 1999. Bioavailability of wetland-derived DON to freshwater and marine bacterioplankton. Limnol. Oceanogr. *44*:1477–1485.

Štepánek, M. 1959. Limnological study of the reservoir Sedlice near Zeliv. IX. Transmission and transparency of water. Sci. Pap. Inst. Chem. Technol., Prague, Fac. Technol. Fuel Water *3*(Pt. 2):363–430.

Stephen, D., B. Moss, and G. Phillips. 1998. The relative importance of top-down and bottom-up control of phytoplankton in a shallow macrophyte-dominated lake. Freshwat. Biol. *39*:699–713.

Stephens, D. B. 1996. Vadose Zone Hydrology. Lewis Publishers, Boca Raton, FL.

Sterner, R. W. 1989. The role of grazers in phytoplankton succession. *In* U. Sommer, ed. Plankton Ecology: Succession in Plankton Communities. Springer-Verlag, Berlin. pp. 107–170.

Sterner, R. W. and D. O. Hessen. 1994. Algal nutrient limitation and the nutrition of aquatic herbivores. Annu. Rev. Ecol. Syst. *25*:1–29.

Sterner, R. W. and J. L. Robinson. 1994. Thresholds for growth in *Daphnia magna* with high and low phosphorus diets. Limnol. Oceanogr. *39*:1228–1232.

Sterner, R. W., J. Clasen, W. Lampert, and T. Weisse. 1998. Carbon:phosphorus stoichiometry and food chain production. Ecol. Lett. *1*:146–150.

Sterner, R. W., J. J. Elser, and D. O. Hessen. 1992. Stoichiometric relationships among producers, consumers and nutrient cycling in pelagic ecosystems. Biogeochemistry *17*:49–67.

Sterner, R. W., J. J. Elser, E. J. Fee, S. J. Guildford, and T. H. Chrzanowski. 1997. The light:nutrient ratio in lakes: The balance of energy and materials affects ecosystem structure and process. Am. Nat. *150*:663–684.

Sterner, R. W., T. H. Chrzanowski, J. J. Elser, and N. B. George. 1995. Sources of nitrogen and phosphorus supporting the growth of bacterio- and phytoplankton in an oligotrophic Canadian shield lake. Limnol. Oceanogr. *40*:242–249.

Sterzynski, W. 1979. Fecundity and body size of planktic rotifers in 30 Polish lakes of various trophic state. Ekol. Polska *27*:307–321.

Stevens, C. L. and C. L. Hurd. 1997. Boundary-layers around bladed aquatic macrophytes. Hydrobiologia *346*:119–128.

Stevens, C. L. and G. A. Lawrence. 1997. Estimation of wind-forced internal seiche amplitudes in lakes and reservoirs, with data from British Columbia, Canada. Aquat. Sci. *59*:115–134.

Stevens, R. J. and M. P. Parr. 1977. The significance of alkaline phosphatase activity in Lough Neagh. Freshwat. Biol. *7*:351–355.

Stevenson, F. J. 1972. Role and function of humus in soil with emphasis on adsorption of herbicides and chelation of micronutrients. BioScience *22*:643–650.

Stevenson, F. J. 1982. Humus Chemistry: Genesis, Composition, Reactions. John Wiley & Sons, New York. 443 pp.

Stevenson, L. H. 1978. A case for bacterial dormancy in aquatic systems. Microb. Ecol. *4*:127–133.

Stevenson, R. J. 1996a. An introduction to algal ecology in freshwater benthic habitats. *In* R. J. Stevenson, M. L. Bothwell, and R. L. Lowe, eds. Algal Ecology: Freshwater Benthic Ecosystems. Academic Press, San Diego. pp. 3–30.

Stevenson, R. J. 1996b. The stimulation and drag of current. *In* R. J. Stevenson, M. L. Bothwell, and R. L. Lowe, eds. Algal Ecology: Freshwater Benthic Ecosystems. Academic Press, San Diego. pp. 321–340.

Stewart, A. J. and R. G. Wetzel. 1981. Dissolved humic materials: Photodegradation, sediment effects, and reactivity with phosphate and calcium carbonate precipitation. Arch. Hydrobiol. *92*:265–286.

Stewart, A. J. and R. G. Wetzel. 1982a. Phytoplankton contribution to alkaline phosphatase activity. Arch. Hydrobiol. *93*:265–271.

Stewart, A. J. and R. G. Wetzel. 1982b. Influence of dissolved humic materials on carbon assimilation and alkaline phosphatase activity in natural algal-bacterial assemblages. Freshwat. Biol. *12*:369–380.

Stewart, A. J. and R. G. Wetzel. 1986. Cryptophytes and other microflagellates as couplers in planktonic community dynamics. Arch. Hydrobiol. *106*:1–19.

Stewart, K. M. 1973a. Detailed time variations in mean temperature and heat content of some Madison lakes. Limnol. Oceanogr. *18*:218–226.

Stewart, K. M. 1973b. Winter conditions in Lake Erie with reference to ice and thermal structure and comparison to lakes Winnebago (Wisconsin) and Mille lacs (Minnesota). Proc. Conf. Great Lakes Res. *16*:845–857.

Stewart, K. M. 1976. Oxygen deficits, clarity, and eutrophication in some Madison lakes. Int. Rev. ges. Hydrobiol. *61*:563–579.

Stewart, K. M. 1988. Tracing inflows in a physical model of Lake Constance. J. Great Lakes Res. *14*:466–478.

Stewart, K. M. 1993. Waves in some Madison lakes. Verh. Internat. Verein. Limnol. *25*:56–64.

Stewart, K. M. and B. E. Brockett. 1984. Transmission of light through ice and snow of Adirondack lakes, New York. Verh. Internat. Verein. Limnol. *22*:72–76.

Stewart, K. M. and E. Hollan. 1975. Meromixis in Ulmener Maar (Germany). Verh. Internat. Verein. Limnol. *19*:1211–1219.

Stewart, K. M. and E. Hollan. 1984. Physical model study of Lake Constance. Schweiz. Z. Hydrol. *46*:5–40.

Stewart, K. M. and R. K. Haugen. 1990. Influence of lake morphometry on ice dates. Verh. Internat. Verein. Limnol. *24*:122–127.

Stewart, R. and R. K. Schmitt. 1968. Wave interaction and Langmuir circulations. Proc. Conf. Great Lakes Res., Int. Assoc. Great Lakes Res. *11*:496–499.

Stewart, T. W. and J. M. Haynes. 1994. Benthic macroinvertebrate communities of southwestern Lake Ontario following invasion of *Dreissena*. J. Great Lakes Res. *20*:479–493.

Stewart, W. D. P. 1968. Nitrogen input into aquatic ecosystems. *In* D. F. Jackson, ed. Algae, Man, and the Environment. Syracuse University Press, Syracuse, NY. pp. 53–72.

Stewart, W. D. P. 1969. Biological and ecological aspects of nitrogen fixation by free-living miocroorganisms. Proc. Roy. Soc. London (Ser. B) *172*:367–388.

Stewart, W. D. P. 1973. Nitrogen fixation. *In* N. G. Carr and B. A. Whitton, eds. The Biology of the Blue-Green Algae. University of California Press, Berkeley. pp. 260–278.

Stibe, L. and S. Fleischer. 1991. Agricultural production methods—impact on drainage water nitrogen. Verh. Internat. Verein. Limnol. *24*:1749–1752.

Stich, H.-B. and W. Lampert. 1984. Growth and reproduction of migrating and non-migrating *Daphnia* species under simulated food and temperature conditions of diurnal vertical migration. Oecologia *61*:192–196.

Stigebrandt, A. 1978. Dynamics of an ice-covered lake with throughflow. Nord. Hydrol. *9*:219–244.

Stites, D. L., A. C. Benke, and D. M. Gillespie. 1995. Population dynamics, growth, and production of the Asiatic clam, *Corbicula fluminea*, in a blackwater river. Can. J. Fish. Aquat. Sci. *52*:425–437.

Stobbart, R. H. and J. Shaw. 1974. Salt and water balance; excretion. *In* M. Rockstein, ed. The Physiology of Insecta. V. 2nd Edition. Academic Press, New York. pp. 361–446.

Stöber, W. 1967. Formation of silicic acid in aqueous suspensions of different silica modifications. *In* Equilibrium concepts in natural water systems. Adv. Chem. Ser. *67*:161–182.

Stocking, C. R. 1956. Vascular conduction in submerged plants. *In* W. Ruhland, ed. Handbuch der Pflanzenphysiologie. Band 3. Pflanze und Wasser. Springer-Verlag, Berlin. pp. 587–595.

Stockner, J. G. 1971. Preliminary characterization of lakes in the Experimental Lakes Area, northwestern Ontario using diatom occurrence in sediments. J. Fish. Res. Board Can. *28*: 265–275.

Stockner, J. G. 1972. Paleolimnology as a means of assessing eutrophication. Verh. Internat. Verein. Limnol. *18*:1018–1030.

Stockner, J. G. 1975. Phytoplankton heterogeneity and paleolimnology of Babine Lake, British Columbia, Canada. Verh. Internat. Verein. Limnol. *19*:2236–2250.

Stockner, J. G. 1991. Autotrophic picoplankton in freshwater ecosystems: The view from the summit. Int. Rev. ges. Hydrobiol. *76*:483–492.

Stockner, J. G. and N. J. Antia. 1986. Algal picoplankton from marine and freshwater ecosystems: A multidisciplinary perspective. Can. J. Fish. Aquat. Sci. *43*:2472–2503.

Stockner, J. G. and W. W. Benson. 1967. The succession of diatom assemblages in the recent sediments of Lake Washington. Limnol. Oceanogr. *12*:513–532.

Stoecker, D. K. 1984. Particle production by planktonic ciliates. Limnol. Oceanogr. *29*:930–940.

Stoeckmann, A. M. and D. W. Garton. 1997. A seasonal energy budget for zebra mussels (*Dreissena polymorpha*) in western Lake Erie. Can. J. Fish. Aquat. Sci. *54*:2743–2751.

Stoermer, E. F. and J. P. Smol. 1999. The Diatoms: Applications for the Environmental and Earth Sciences. Cambridge Univ. Press, Cambridge. 469 pp.

Stoermer, E. F., B. G. Ladewski, and C. L. Schelske. 1978. Population responses of Lake Michigan phytoplankton to nitrogen and phosphorus enrichment. Hydrobiologia *57*:249–265.

Stolz, J. F. 1991. The ecology of phototrophic bacteria. *In* J. F. Stolz, ed. Structure of Phototrophic Prokaryotes. CRC Press, Boca Raton. pp. 105–123.

Stommel, H. 1949. Horizontal diffusion due to oceanic turbulence. J. Mar. Res. *8*:199–225.

Stout, G. E., ed. 1967. Isotope Techniques in the Hydrologic Cycle. Geophys. Monogr. Ser. 11. American Geophysical Union, Washington, DC. 199 pp.

Strahler, A. N. 1964. Quantitative geomorphology of drainage basins and channel networks. *In* V. T. Chow, ed. Handbook of applied hydrology. McGraw-Hill, New York. pp. 4–39–4–76.

Straile, D. 1998. Biomass allocation and carbon flow in the pelagic food web of Lake Constance. Arch. Hydrobiol. Spec. Issues Advanc. Limnol. *53*:545–563.

Strain, H. 1951. The pigments of algae. *In* G. M. Smith, ed. Manual of Phycology. Chronica Botanica Co., Waltham, MA. pp. 243–262.

Straškraba, M. 1963. Share of the littoral region in the productivity of two fishponds in southern Bohemia. Rozpravy Ceskosl. Akad. Ved, Rada Matem. Prír. Ved, *73*(13), 64 pp.

Straškraba, M. 1965. The effect of fish on the number of invertebrates in ponds and streams. Mitt. Internat. Verein. Limnol. *13*:106–127.

Straškraba, M. 1967. Quantitative study on the littoral zooplankton of the Poltruba Backwater with an attempt to disclose the effect on fish. Rozpravy Ceskosl. Akad. Ved Rada Matem. Prír. Ved, *77*(11):7–34.

Straškraba, M. 1968. Der Anteil der höheren Pflanzen an der Produktion der stehenden Gewässer. Mitt. Internat. Verein. Limnol. *14*:212–230.

Straškraba, M. 1996. Lake and reservoir management. Verh. Internat. Verein. Limnol. *26*:193–209.

Straškraba, M. and E. Pieczyńska. 1970. Field experiments on shading effect by emergents on littoral phytoplankton and periphyton production. Rozpravy Ceskosl. Akad. Ved Rada Matem. Prír. Ved *80*(6):7–32.

Straškraba, M., J. G. Tundisi, and A. Duncan. 1993. State-of-the-art of reservoir limnology and water quality management. *In* M. Straškraba, J. G. Tundisi, and A. Duncan, eds. Comparative reservoir limnology and water quality management. Kluwer Acad. Publ., Dordrecht. pp. 213–288.

Straškraba, M., J. Hrbáček, and P. Javornicky. 1973. Effect of an upstream reservoir on the stratification conditions in Slapy Reservoir. Hydrobiol. Studies *2*:7–82.

Straškraba, V. and J. Komárková. 1979. Seasonal changes of bacterioplankton in a reservoir related to algae. I. Numbers and biomass. Int. Rev. ges. Hydrobiol. *64*:285–302.

Straškrabová, V., J. Komárková, and V. Vyhnálek. 1993. Degradation of organic substances in reservoirs. Wat. Sci. Technol. *28*:95–104.

Strayer, D. 1985. The benthic micrometazoans of Mirror Lake, New Hampshire. Arch. Hydrobiol./Suppl.*72*:287–426.

Strayer, D. 1988a. Life history of a lacustrine ostracod. Hydrobiologia *160*:189–191.

Strayer, D. 1988b. On the limits to secondary production. Limnol. Oceanogr. *33*:1217–1220.

Strayer, D. and G. E. Likens. 1986. An energy budget for the zoobenthos of Mirror Lake, New Hampshire. Ecology *67*:303–313.

Strayer, D. L. and L. C. Smith. 1996. Relationships between zebra mussels (*Dreissena polymorpha*) and unionid clams during the early stages of the zebra mussel invasion of the Hudson River. Freshwat. Biol. *36*:771–779.

Strayer, D. L. and W. D. Hummon. 1991. Gastrotricha. *In* J. H. Thorp and A. P. Covich, eds. Ecology and Classification of North American Freshwater Invertebrates. Academic Press, San Diego. pp. 173–185.

Strayer, D. L., J. J. Cole, G. E. Likens, and D. C. Busco. 1981. Biomass and annual production of the freshwater mussel *Elliptio complanata* in an oligotrophic softwater lake. Freshwat. Biol. *11*:435–440.

Strayer, R. F. and J. M. Tiedje. 1978a. Kinetic parameters of the conversion of methane precursors to methane in a hypereutrophic lake sediment. Appl. Environ. Microbiol. *36*:330–340.

Strayer, R. F. and J. M. Tiedje. 1978b. In situ methane production in a small, hypereutrophic, hard-water lake: Loss of methane from sediments by vertical diffusion and ebullition. Limnol. Oceanogr. *23*:1201–1206.

Stream Solute Workshop. 1990. Concepts and methods for assessing solute dynamics in stream ecosystems. J. N. Am. Benthol. Soc. 9:95–119.

Streit, B. 1975a. Experimentelle Untersuchungen zum Stoffhaushalt von *Ancylus fluviatilis* (Gastropoda—Basommatophora). 1. Ingestion, Assimilation, Wachstum und Eiablage. Arch. Hydrobiol./Suppl. *47*:458–514.

Streit, B. 1975b. Experimentelle Untersuchungen zum Stoffhaushalt von *Ancylus fluviatilis* (Gastropoda—Basommatophora). 2. Untersuchungen über Einbau und Umsatz des Kohlenstoffs. Arch. Hydrobiol./Suppl. *48*:1–46.

Streit, B. 1977. Morphometric relationships and feeding habits of two species of *Chaetogaster*, *Ch. limnaei* and *Ch. diastrophus* (Oligochaeta). Arch. Hydrobiol./Suppl. *48*:424–437.

Streit, B. and P. Schröder. 1978. Dominierende Benthosinvertebraten in der Geröllbrandungszone des Bodensees: Phänologie, Nahrungsökologie und Biomasse. Arch. Hydrobiol./Suppl. *55*:211–234.

Strekal, T. A. and W. F. McDiffett. 1974. Factors affecting germination, growth, and distribution of the freshwater sponge, *Spongilla fragilis* Leidy (Porifera). Biol. Bull. *146*:267–278.

Strickland, J. D. H. 1958. Solar radiation penetrating the ocean. A review of requirements, data and methods of measurement, with particular reference to photosynthetic productivity. J. Fish. Res. Board Can. *15*:453–493.

Strickland, J. D. H. 1960. Measuring the production of marine phytoplankton. Bull. Fish. Res. Board Can. *122*, 172 pp.

Strickland, J. D. H. and T. R. Parsons. 1972. A practical handbook of seawater analysis. 2nd Edition. Bull. Fish. Res. Board Can. *167*, 310 pp.

Strickler, J. R. 1975. Swimming of planktonic *Cyclops* species (Copepoda, Crustacea): Pattern, movements and their control. *In* T.Y.-T. Wu, C. J. Brokaw, and C. Brennen, eds. Swimming and Flying in Nature, vol. 2. Plenum Press, New York. pp. 599–613.

Strickler, J. R. 1977. Observation of swimming performances of planktonic copepods. Limnol. Oceanogr. *22*:165–170.

Strickler, J. R. and S. Twombly. 1975. Reynolds number, diapause, and predatory copepods. Verh. Internat. Verein. Limnol. *19*:2943–2950.

Strøm, K. M. 1933. Nutrition of algae. Experiments upon: The feasibility of the Schreiber Method in fresh waters; the relative importance of iron and manganese in the nutritive medium; the nutritive substance given off by lake bottom muds. Arch. Hydrobiol. *25*:38–47.

Strøm, K. M. 1945. The temperature of maximum density in fresh waters. An attempt to determine its lowering with increased pressure from observations in deep lakes. Geofysiske Publikasjoner *16*(8), 14 pp.

Strøm, K. M. 1947. Correlation between C^{18} and pH in lakes. Nature *159*:782–783.

Strøm, K. M. 1955. Waters and sediments in the deep of lakes. Mem. Ist. Ital. Idrobiol. *8*(Suppl.):345–356.

Strome, D. J. and M. C. Miller. 1978. Photolytic changes in dissolved humic substances. Verh. Internat. Verein. Limnol. *20*:1248–1254.

Strong, D. R. 1992. Are trophic cascades all wet? Differentiation and donor-control in speciose ecosystems. Ecology *73*:747–754.

Strong, D. R., Jr. 1972. Life history variation among populations of an amphipod (*Hyalella azteca*). Ecology *53*:1103–1111.

Strong, D. R., Jr., D. Simberloff, L. G. Abele, and A. B. Thistle. 1984. Ecological Communities: Conceptual Issues and the Evidence. Princeton Univ. Press, Princeton, NJ.

Stross, R. G. 1966. Light and temperature requirements for diapause development and release in *Daphnia*. Ecology *47*:368–374.

Stross, R. G. 1981. Photomorphogenesis in underwater environments. *In* H. Smith, ed. Plants and the Daylight Spectrum. Academic Press, London. pp. 377–397.

Stross, R. G. and J. C. Hill. 1965. Diapause induction in *Daphnia* requires two stimuli. Science *150*:1462–1464.

Strum, M. 1979. Origin and composition of clastic varves. *In* C.

Schlüchter, ed. Moraines and Varves: Origin, Genesis, Classification. A. A. Balkema, Rotterdam. pp. 281–285.

Strum, M. and A. Matter. 1978. Turbidites and varves in Lake Brienz (Switzerland): Deposition of clastic detritus by density currents. *In* A. Matter and M. E. Tucker, eds. Modern and Ancient Lake Sediments. Spec. Publs. Int. Assoc. Sedimentol. 2:147–168.

Strycek, T., J. Acreman, A. Kerry, G. G. Leppard, M. V. Nermut, and D. J. Kushner. 1992. Extracellular fibril production by freshwater algae and cyanobacteria. Microb. Ecol. *23*:53–74.

Stuckey, R. L., J. R. Wehrmeister, and R. J. Bartolotta. 1978. Submersed aquatic vascular plants in ice-covered ponds of central Ohio. Rhodora *80*:575–580.

Stuiver, M. 1967. The sulfur cycle in lake waters during thermal stratification. Geochim. Cosmochim. Acta *31*:2151–2167.

Stuiver, M. 1968. Oxygen-18 content of atmospheric precipitation during last 11,000 years in the Great Lakes region. Science *162*:994–997.

Stuiver, M. 1978a. Radiocarbon timescale tested against magnetic and other dating methods. Nature *273*:271–274.

Stuiver, M. 1978b. Carbon-14 dating: A comparison of beta and ion counting. Science *202*:881–883.

Stuiver, M. and P. D. Quay. 1980. Changes in atmospheric carbon-14 attributed to a variable sun. Science *207*:11–19.

Stull, E. A., E. deAmezaga, and C. R. Goldman. 1973. The contribution of individual species of algae to primary productivity of Castle Lake, California. Verh. Internat. Verein. Limnol. *18*:1776–1783.

Stumm, W. 1966. Redox potential as an environmental parameter; conceptual significance and operational limitation. 3rd Int. Conf. on Water Poll. Res., Water Poll. Control Federation. Sec. 1, Paper 13, 16 pp.

Stumm, W. and F. G. Lee. 1960. The chemistry of aqueous iron. Schweiz. Z. Hydrol. *22*:295–319.

Stumm, W. and J. J. Morgan. 1981. Aquatic Chemistry. An Introduction Emphasizing Chemical Equilibria in Natural Waters. 2nd Edition. J. Wiley & Sons, New York. 780 pp.

Stumm, W. and J. J. Morgan. 1995. Aquatic chemistry: Chemical equilibria and rates in natural waters. 3rd Edition. J. Wiley & Sons, New York. 1040 pp.

Stumm, W. and J. O. Leckie. 1971. Phosphate exchange with sediments: Its role in the productivity of surface waters. Proc. Water Poll. Res. Conf. III, Art. 16, 16 pp.

Stünzi, J. T. and H. Kende. 1989. Gas composition in the internal air spaces of deepwater rice in relation to growth induced by submergence. Plant Cell Physiol. *30*:49–56.

Suberkropp, K. 1992. Aquatic hyphomycete communities. *In* G. C. Carroll and D. T. Wicklow, eds. The Fungal Community: Its Organization and Role in the Ecosystem. 2nd Edition. Marcel. Dekker, Inc., New York. pp. 729–747.

Suberkropp, K. 1995. The influence of nutrients on fungal growth, productivity, and sporulation during leaf breakdown in streams. Can. J. Bot. *73*(Suppl. 1):S1361-S1369.

Suberkropp, K. 1997. Annual production of leaf-decaying fungi in a woodland stream. Freshwat. Biol. *38*:169–178.

Suberkropp, K. and E. Chauvet. 1995. Regulation of leaf breakdown by fungi in streams: Influences of water chemistry. Ecology *76*:1433–1445.

Suberkropp, K. and M. J. Klug. 1976. Fungi and bacteria associated with leaves during processing in a woodland stream. Ecology *57*:707–719.

Suberkropp, K. and M. J. Klug. 1980. The maceration of deciduous leaf litter by aquatic hyphomycetes. Can. J. Bot. *58*:1025–1031.

Suberkropp, K., G. L. Godshalk, and M. J. Klug. 1976. Changes in the chemical composition of leaves during processing in a woodland stream. Ecology *57*:720–727.

Suberkropp, K., M. J. Klug, and K. W. Cummins. 1975. Community processing of leaf litter in woodland streams. Verh. Internat. Verein. Limnol. *19*:1653–1658.

Suess, E. 1968. Calcium carbonate interaction with organic compounds. Ph.D. Diss., Lehigh University, Bethlehem, PA.

Suess, E. 1970. Interaction of organic compounds with calcium carbonate. I. Association phenomena and geochemical implications. Geochim. Cosmochim. Acta 34:157–168.

Suffet, I. H. and P. MacCarthy, eds. 1989. Aquatic Humic Substances: Influence on Fate and Treatment of Pollutants. American Chemical Soc., Washington, DC. 864 pp.

Sugawa, A. 1987. On eddy diffusivity in the hypolimnion in Lake Kanna. Jpn. J. Limnol. *48*:9–17.

Sugawara, K. 1961. Na, Cl and Na/Cl in inland waters. Jpn. J. Limnol. *22*:49–65.

Sugawara, K., T. Koyama, and E. Kamata. 1957. Recovery of precipitated phosphate from lake muds related to sulfate reduction. Chem. Inst. Fac. Sci. Nagoya Univ. *5*:60–67.

Sugihara, G. 1980. Minimal community structure: An explanation of species abundance patterns. Am. Nat. *116*:770–787.

Sulzberger, B., H. Laubscher, and G. Karametaxas. 1994. Photoredox reactions at the surface of iron(III) (hydr)oxides. *In* G. R. Helz, R. G. Zepp, and D. G. Crosby, eds. Aquatic and Surface Photochemistry. Lewis Publishers, Boca Raton. pp. 53–73.

Sun, L., E. M. Perdue, J. L. Meyer, and J. Weis. 1997. Use of elemental composition to predict bioavailability of dissolved organic matter in a Georgia river. Limnol. Oceanogr. *42*:714–721.

Sundbom, M. and T. Vrede. 1997. Effects of fatty acid and phosphorus content of food on the growth, survival and reproduction of *Daphnia*. Freshwat. Biol. *38*:665–674.

Sundh, I. 1992. Biochemical composition of dissolved organic carbon derived from phytoplankton and used by heterotrophic bacteria. Appl. Environ. Microbiol. *58*:2938–2947.

Sundh, I. and R. T. Bell. 1992. Extracellular dissolved organic carbon released from phytoplankton as a source of carbon for heterotrophic bacteria in lakes of different humic content. Hydrobiologia *229*:93–106.

Suren, A. M. and P. S. Lake. 1989. Edibility of fresh and decomposing macrophytes to three species of freshwater invertebrate herbivores. Hydrobiologia *178*:165–178.

Sutcliffe, D. W. and T. R. Carrick. 1983a. Chemical composition of water-bodies in the English Lake District: Relationships between chloride and other major ions related to solid geology, and a tentative budget for Windermere. Freshwat. Biol. *13*:323–352.

Sutcliffe, D. W. and T. R. Carrick. 1983b. Relationships between chloride and major cations in precipitation and streamwaters in the Windermere catchment (English Lake District). Freshwat. Biol. *13*:415–441.

Sutcliffe, D. W., T. R. Carrick, and L. G. Willoughby. 1981. Effects of diet, body size, age and temperature on growth rates in the amphipod *Gammarus pulex*. Freshwat. Biol. 11:183–214.

Sütfeld, R. 1993. Exudation of UV-light absorbing natural products by seedlings of *Nuphar lutea*. Chemoecology *4*:108–114.

Sutherland, I. W. 1985. Biosynthesis and composition of gram-negative bacterial extracellular and wall polysaccharides. Ann. Rev. Microbiol. *39*:243–270.

Suttle, C. A. 1994. The significance of viruses to mortality in aquatic microbial communities. Microb. Ecol. *28*:237–243.

Suttle, C. A. and P. J. Harrison. 1988. Ammonium and phosphate uptakes rates, N:P supply ratios, and evidence for N and P limitation in some oligotrophic lakes. Limnol. Oceanogr. *33*:186–202.

Suttle, C. A., A. M. Chan, and M. T. Cottrell. 1990. Infection of phytoplankton by viruses and reduction of primary productivity. Nature *347*:467–469.

Suttle, C. A. and F. Chen. 1992. Mechanisms and rates of decay of marine viruses in seawater. Appl. Environ. Microbiol. *58*:3721–3729.

Sutton, D. L. 1977. Grass carp (*Ctenopharyngodon idella* Val.) in North America. Aquat. Bot. *3*:157–164.

Suzuki, M. T., E. B. Sherr, and B. F. Sherr. 1996. Estimation of ammonium regeneration efficiencies associated with bacterivory in pelagic food webs via a ^{15}N tracer method. J. Plankton Res. *18*:411–428.

Suzuki, Y., Y. Sugimura, and Y. Miyake. 1981. Selenium content and its chemical form in river waters of Japan. Jpn. J. Limnol. *42*:89–91.

Svensson, B. H. and T. Rosswall. 1984. In situ methane production from acid peat in plant communities with different moisture regimes in a subarctic mire. Oikos *43*:341–350.

Svensson, J. M. 1997. Influence of *Chironomus plumosus* larvae on ammonium flux and denitrification (measured by the acetylene blockage- and the isotope pairing-technique) in eutrophic lake sediment. Hydrobiologia *346*:157–168.

Svensson, J. M. and L. Leonardson. 1996. Effects of bioturbation by tube-dwelling chironomid larvae on oxygen uptake and denitrification in eutrophic lake sediments. Freshwat. Biol. *35*:289–300.

Swain, E. B. 1985. Measurement and interpretation of sedimentary pigments. Freshwat. Biol. *15*:53–75.

Swain, F. M. 1963. Geochemistry of humus. *In* I. A. Breger, ed. Organic Geochemistry. Macmillan Company, New York. pp. 87–147.

Swain, F. M. 1965. Geochemistry of some Quaternary lake sediments of North America. *In* H. E. Wright, Jr. and D. G. Frey, eds. The Quaternary of the United States. Princeton University Press, Princeton, NJ. pp. 765–781.

Swanepoel, J. H. and J. F. Vermaak. 1977. Preliminary results on the uptake and release of ^{32}P by *Potamogeton pectinatus*. J. Limnol. Soc. South Africa *3*:63–65.

Sweeney, B. W. 1984. Factors influencing life-history patterns of aquatic insects. *In* V. H. Resh and D. M. Rosenberg, eds. The Ecology of Aquatic Insects. Praeger Publishers, New York. pp. 56–100.

Swift, M. C. 1976. Energetics of vertical migration in *Chaoborus trivittatus* larvae. Ecology *57*:900–914.

Swift, M. C. 1992. Prey capture by the four larval instars of *Chaoborus crystallinus*. Limnol. Oceanogr. *37*:14–24.

Swift, M. C. and R. B. Forward, Jr. 1988. Absolute light intensity vs. rate of relative change in light intensity: The role of light in the vertical migration of *Chaoborus punctipennis* larvae. Bull. Mar. Sci. *43*:604–619.

Swift, M. C. and U. T. Hammer. 1979. Zooplankton population dynamics and *Diaptomus* production in Waldsea Lake, a saline meromictic lake in Saskatchewan. J. Fish. Res. Board Can. *36*:1431–1438.

Swüste, H. F. J., R. Cremer, and S. Parma. 1973. Selective predation by larvae of *Chaoborus flavicans* (Diptera, Chaoboridae). Verh. Internat. Verein. Limnol. *18*:1559–1563.

Symons, F. 1972. On the changes in the structure of two algal populations: Species diversity and stability. Hydrobiologia *40*:499–502.

Symposium on the Classification of Brackish Waters. 1959. Venice, 1958. Societas Internationalis Limnologiae. Arch. Oceanogr. Limnol. *11*(Suppl.), 248 pp.

Sytsma, M. D. and L. W. J. Anderson. 1993. Transpiration by an emergent macrophyte: Source of water and implications for nutrient supply. Hydrobiologia *271*:97–108.

Szczepańska, W. 1973. Production of helophytes in different types of lakes. Pol. Arch. Hydrobiol. *20*:51–57.

Szczepańska, W. 1976. Development of the underground parts of *Phragmites communis* Trin. and *Typha latifolia* L. Pol. Arch. Hydrobiol. *23*:227–232.

Szczepańska, W. and A. Szczepański. 1976. Growth of *Phragmites communis* Trin., *Typha latifolia* L., and *Typha angustifolia* L. in relation to the fertility of soils. Pol. Arch. Hydrobiol. *23*:233–248.

Szczepański, A. 1965. Deciduous leaves as a source of organic matter in lakes. Bull. Acad. Polon. Sci., Cl. II, *13*:215–217.

Szczepański, A. 1968. Scattering of light and visibility in water of different types of lakes. Pol. Arch. Hydrobiol. *15*:51–77.

Szczepański, A. 1969. Biomass of underground parts of the reed *Phragmites communis* Trin. Bull. Acad. Polon. Sci., Cl. II, *17*:245–246.

Szczepański, A. and W. Szczepańska. 1966. Primary production and its dependence on the quantity of periphyton. Bull. Acad. Polon. Sci., Cl. II, *14*:45–50.

Szestay, K. 1982. River basin development and water management. Water Qual. Bull. *7*:155–162.

Szilágyi, M. 1973. The redox properties and the determination of the normal potential of the peatwater system. Soil Sci. *115*:434–437.

Szlauer, L. 1963. Diurnal migrations of minute invertebrates inhabiting the zone of submerged hydrophytes in a lake. Schweiz. Z. Hydrol. *25*:56–64.

Szumiec, M. 1961. Eingangsmessungen der Strahlungsintensität in Teichen. Acta Hydrobiol. *3*:133–142.

Takahashi, M. and S. Ichimura. 1968. Vertical distribution and organic matter production of photosynthetic sulfur bacteria in Japanese lakes. Limnol. Oceanogr. *13*:644–655.

Takahashi, M. and Y. Saijo. 1981. Nitrogen metabolism in Lake Kizaki, Japan. II. Distribution and decomposition of organic nitrogen. Arch. Hydrobiol. *92*:359–376.

Takamura, N. and M. Yasuno. 1983. Food selection of the ciliated protozoa, *Condylostoma vorticella* (Ehrenberg) in Lake Kasumigaura. Jpn. J. Limnol. *44*:184–189.

Takamura, N. and M. Yasuno. 1984. Diurnal changes in the vertical distribution of phytoplankton in hypertrophic Lake Kasumigaura, Japan. Hydrobiologia *112*:53–60.

Takita, M. and M. Sakamoto. 1993. Methane flux in a shallow eutrophic lake. Verh. Internat. Verein. Limnol. *25*:822–826.

Tal, E. 1962. Preliminary study of fluctuations of vitamin B_{12} content in fish ponds in the Beith-Shean Valley. Bamidgeh (Israel) *14*:19–26.

Talling, J. F. 1958. The longitudinal succession of the water characteristics in the White Nile. Hydrobiologia *11*:73–89.

Talling, J. F. 1960. Self-shading effects in natural populations of a planktonic diatom. Wetter u. Leben *12*:235–242.

Talling, J. F. 1965. The photosynthetic activity of phytoplankton in East African lakes. Int. Rev. ges. Hydrobiol. *50*:1–32.

Talling, J. F. 1969. The incidence of vertical mixing, and some biological and chemical consequences, in tropical African lakes. Verh. Internat. Verein. Limnol. *17*:998–1012.

Talling, J. F. 1971. The underwater light climate as a controlling factor in the production ecology of freshwater phytoplankton. Mitt. Internat. Verein. Limnol. *19*:214–243.

Talling, J. F. 1973. The application of some electrochemical methods to the measurement of photosynthesis and respiration in fresh waters. Freshwat. Biol. *3*:335–362.

Talling, J. F. 1976. The depletion of carbon dioxide from lake water by phytoplankton. J. Ecol. *64*:79–121.

Talling, J. F. 1979. Factor interactions and implications for the prediction of lake metabolism. Arch. Hydrobiol. Beih. Ergebn. Limnol. *13*:96–109.

Talling, J. F. 1986. The seasonality of phytoplankton in African lakes. Hydrobiologia *138*:139–160.

Talling, J. F. and I. B. Talling. 1965. The chemical composition of African lake waters. Int. Rev. ges. Hydrobiol. *50*:421–463.

Talling, J. F., R. B. Wood, M. V. Prosser, and R. M. Baxter. 1973. The upper limit of photosynthetic productivity by phytoplankton: Evidence from Ethiopian soda lakes. Freshwat. Biol. *3*:53–76.

Tam, A. C. and C. K. N. Patel. 1979. Optical absorptions of light and heavy water by laser optoacoustic spectroscopy. Appl. Optics *18*:3348–3358.

Tamminen, T. 1989. Dissolved organic phosphorus regeneration by bacterioplankton: 5'-nucleotidase activity and subsequent phosphate uptake in a mesocosm enrichment experiment. Mar. Ecol. Prog. Ser. *58*:89–100.

Tan, T. L. 1973. Physiologie der Nitratreduktion bei *Pseudomonas aeruginosa*. Z. Allg. Mikrobiol. *13*:83–94

Tan, T. L. and J. Overbeck. 1973. Ökologische Untersuchungen über nitratreduzierende Bakterien im Wasser des Pluss-sees (Schleswig-Holstein). Z. Allg. Mikrobiol. *13*:71–82.

Tanaka, N., Y. Ueda, M. Onizawa, and H. Kadota. 1977. Bacterial populations in water masses of different organic matter concentrations in Lake Biwa. Jpn. J. Limnol. *38*:41–47.

Tanner, C. C. and J. S. Clayton. 1985. Effects of vesicular-arbuscular mycorrhizas on growth and nutrition of a submerged aquatic plant. Aquat. Bot. *22*:377–386.

Tank, J. L., J. R. Webster, and E. F. Benfield. 1993. Microbial respiration on decaying leaves and sticks in a southern Appalachian stream. J. N. Am. Benthol. Soc. *12*:394–405.

Tappa, D. W. 1965. The dynamics of the association of six limnetic species of *Daphnia* in Aziscoos Lake, Maine. Ecol. Monogr. *35*:395–423.

Tarapchak, S. J. and L. R. Herche. 1985. Perspectives in epilimnetic phosphorus cycling. *In* Z. Dubinsky and Y. Steinberger, eds. Environmental Quality and Ecosystem Stability. Vol. 3A/B. Bar-Ilan Univ. Press, Ramat-Gan, Israel. pp. 245–255.

Tarapchak, S. J. and L. R. Herche. 1986. Phosphate uptake by microorganisms in lake water: Deviations from simple Michaelis-Menten kinetics. Can. J. Fish. Aquat. Sci. *43*:319–328.

Tarapchak, S. J., S. M. Bigelow, and C. Rubitschun. 1982. Overestimation of orthophosphorus concentrations in surface waters of southern Lake Michigan: Effects of acid and ammonium molybdate. Can. J. Fish. Aquat. Sci. *39*:296–304.

Tate, C. M. 1990. Patterns and controls of nitrogen in tallgrass prairie streams. Ecology *71*:2007–2018.

Taub, F. B. and A. M. Dollar. 1968. The nutritional inadequacy of *Chlorella* and *Chlamydomonas* as food for *Daphnia pulex*. Limnol. Oceanogr. *13*:607–617.

Taylor, B. E. and M. Slatkin. 1981. Estimating birth and death rates of zooplankton. Limnol. Oceanogr. *26*:143–158.

Taylor, B. E., G. A. Wyngaard, and D. L. Mahoney. 1990. Hatching of *Diaptomus stagnalis* eggs from a temporary pond after a prolonged dry period. Arch. Hydrobiol. *117*:271–278.

Taylor, F. J. R. and U. Pollingher. 1987. Ecology of dinoflagellates. *In* F. J. R. Taylor, ed. The Biology of Flagellates. Blackwell Scientific Publs., Oxford. pp. 398–529.

Taylor, H. H. and J. M. Anstiss. 1999. Copper and haemocyanin dynamics in aquatic invertebrates. Mar. Freshwat. Res. *50*:907–931.

Taylor, K. L., J. B. Grace, and B. D. Marx. 1997. The effects of herbivory on neighbor interactions along a coastal marsh gradient. Am. J. Bot. *84*:709–715.

Taylor, M. and G. W. Stone. 1996. Beach-ridges: A review. J. Coastal Res. 12:612–621.

Taylor, W. D. 1984. Phosporus flux through epilimnetic zooplankton from Lake Ontario: Relationship with body size and significance to phytoplankton. Can. J. Fish. Aquat. Sci. *41*:1702–1712.

Taylor, W. D. and D. R. S. Lean. 1981. Radiotracer experiments on phosphorus uptake and release by limnetic microzooplankton. Can. J. Fish. Aquat. Sci. *38*:1316–1321.

Taylor, W. D. and D. R. S. Lean. 1991. Phosphorus pool sizes and fluxes in the epilimnion of a mesotrophic lake. Can. J. Fish. Aquat. Sci. *48*:1293–1301.

Taylor, W. D. and J. Berger. 1980. Microspatial heterogeneity in the distribution of ciliates in a small pond. Microb. Ecol. *6*:27–34.

Taylor, W. D. and R. G. Wetzel. 1984. Population dynamics of *Rhodomonas minuta* v. *nannoplanctica* Skuja (Cryptophyceae) in a hardwater lake. Verh. Internat. Verein. Limnol. *22*:536–541.

Taylor, W. D. and R. G. Wetzel. 1988. Phytoplankton community dynamics in Lawrence Lake of southwestern Michigan. Arch. Hydrobiol./Suppl. *81*:491–532.

Taylor, W. D. and R. W. Sanders. 1991. Protozoa. *In* J. H. Thorp and A. P. Covich, eds. Ecology and Classification of North American Freshwater Invertebrates. Academic Press, San Diego. pp. 37–93.

Taylor, W. D., J. W. Barko, and W. F. James. 1988. Contrasting diel patterns of vertical migration in the dinoflagellate *Ceratium hirundinella* in relation to phosphorus supply in a north temperate reservoir. Can. J. Fish. Aquat. Sci. *45*:1093–1098.

Taylor, W. E. and O. E. Johannsson. 1991. A comparison of estimates of productivity and consumption by zooplankton for planktonic ciliates in Lake Ontario. J. Plankton Res. *13*:363–372.

Teal, J. M. 1957. Community metabolism in a temperate cold spring. Ecol. Monogr. *27*:283–302.

Telesh, I. V. 1993. The effect of fish on planktonic rotifers. Hydrobiologia *255/256*:289–296.

Terai, H. 1987. Studies on denitrification in the water column of Lake Kizaki and Lake Fukami-Ike. Jpn. J. Limnol. *48*:257–264.

Terai, H., M. Yoh, and Y. Saijo. 1987. Denitrifying activity and population growth of denitrifying bacteria in Lake Fukami-Ike. Jpn. J. Limnol. *48*:211–218.

Terrell, T. T. 1982. Responses of plankton communities to the introduction of grass carp into some Georgia ponds. J. Freshwat. Ecol. *1*:395–406.

Tessenow, U. 1966. Untersuchungen über den Kieselsäurehaushalt der Binnengewässer. Arch. Hydrobiol./Suppl. *32*:1–136.

Tessenow, U. and Y. Baynes. 1978. Redoxchemische Einflüsse von *Isoëtes lacustris* L. im Litoralsediment des Feldsees (Hochschwarzwald). Arch. Hydrobiol. *82*:20–48.

Tessier, A. J. 1986. Comparative population regulation in two planktonic Cladocera (*Holopedium gibberum* and *Daphnia catawba*). Ecology *67*:285–302.

Tessier, A. J. and C. E. Goulden. 1982. Estimating food limitation in cladoceran populations. Limnol. Oceanogr. *27*:707–717.

Tessier, A. J., L. L. Henry, C. E. Goulden, and M. W. Durand. 1983. Starvation in *Daphnia*: Energy reserves and reproductive allocation. Limnol. Oceanogr. *28*:667–676.

Thibodeaux, L. J. and J. D. Boyle. 1987. Bedform-generated convective transport in bottom sediment. Nature *325*:341–343.

Thienemann, A. 1921. Seetypen. Naturwissenschaften *18*:1–3.

Thienemann, A. 1925. Die Binnengewässer Mitteleuropas. Eine limnologische Einführung. Die Binnengewässer, *1*, 255 pp.

Thienemann, A. 1926. Der Nahrungskreislauf im Wasser. Verh. Deutsch. Zool. Gesell. *31*:29–79.

Thienemann, A. 1927. Der Bau des Seebeckens in seiner Bedeutung für den Ablauf des Lebens im See. Verh. Zool.-Bot. Ges. Wien *77*:87–91.

Thienemann, A. 1928. Der Sauerstoff im eutrophen und oligotrophen See. Ein Beitrag zur Seetypenlehre. Die Binnengewässer *4*, 175 pp.

Thienemann, A. 1931. Der Productionsbegriff in der Biologie. Arch. Hydrobiol. *22*:616–622.

Thienemann, A. 1954. *Chironomus*. Leben, Verbreitung und wirtschaftliche Bedeutung der Chironomiden. Die Binnengewässer *20*, 834 pp.

Thingstad, T. F. 2000. Elements of a theory for the mechanisms controlling abundance, diversity, and biogeochemical role of lytic bacterial viruses in aquatic systems. Limnol. Oceanogr. *45*:1320–1328.

Thomas, E. A. 1951. Sturmeinfluss auf das Tiefenwasser des Zürichsees im Winter. Schweiz. Z. Hydrol. *13*:5–23.

Thomas, E. A. 1960. Sauerstoffminima und Stoffkreislaüfe im ufernahen Oberflächenwasser des Zürichsees (Cladophora- und Phragmites-Gürtel). Monatsbull. Schweiz. Ver. Gas- Wasserfachmännern *1960*(6):1–8.

Thomas, R. H. and A. J. Walsby. 1985. Buoyancy regulation in a strain of *Microcystis*. J. Gen. Microbiol. *131*:799–809.

Thomas, W. H., B. C. Cho, and F. Azam. 1991. Phytoplankton and bacterial production and biomass in subalpine Eastern Brook Lake, Sierra Nevada, California. I. Seasonal interrelationships between the two biotic groups. Arctic Alpine Res. *23*:287–295.

Thompson, A. 1986. Secondary flows and the pool-riffle unit: A case study of the processes of meander development. Earth Surf. Processes Landforms *11*:631–641.

Thompson, R. 1978. European paleomagnetic secular variation 13,000-B.P. Pol. Arch. Hydrobiol. *25*:413–418.

Thompson, R., J. Bloemendal, J. A. Dearing, F. Oldfield, T. A. Rummery, J. C. Stober, and G. M. Turner. 1980. Environmental applications of magnetic measurements. Science *207*:481–486.

Thompson, R., R. W. Battarbee, P. E. O'Sullivan, and F. Oldfield. 1975. Magnetic susceptibility of lake sediments. Limnol. Oceanogr. *20*:687–698.

Thormann, M. N. and S. E. Bayley. 1997. Aboveground net primary production along a bog-fen-marsh gradient in southern boreal Alberta, Canada. Ecoscience *4*:374–384.

Thornton, K. W. 1990. Perspectives on reservoir limnology. *In* K. W. Thornton, B. L. Kimmel, and F. E. Payne, eds. Reservoir Limnology: Ecological Perspectives. John Wiley & Sons, New York. pp. 1–13.

Thornton, K. W., B. L. Kimmel, and F. E. Payne, eds. 1991. Reservoir limnology: Ecological perspectives. John Wiley & Sons, New York. 246 pp.

Thorp, J. H. and G. T. Barthalmus. 1975. Effects of crowding on growth rate and symbiosis in green hydra. Ecology *56*:206–212.

Threlkeld, S. T. 1976. Starvation and the size structure of zooplankton communities. Freshwat. Biol. *6*:489–496.

Threlkeld, S. T. 1979a. The midsummer dynamics of two *Daphnia* species in Wintergreen Lake, Michigan. Ecology *60*:165–179.

Threlkeld, S. T. 1979b. Estimating cladoceran birth rates: The importance of egg mortality and the egg age distribution. Limnol. Oceanogr. *24*:601–612.

Threlkeld, S. T. 1985. Egg degeneration and mortality in cladoceran populations. Verh. Internat. Verein. Limnol. *22*:3083–3087.

Threlkeld, S. T. 1986. Life table responses and population dynamics of four cladoceran zooplankton during a reservoir flood. J. Plankton Res. *8*:639–647.

Threlkeld, S. T. 1987. *Daphnia* life history strategies and resource allocation patterns. Mem. Ist. Ital. Idrobiol. *45*:353–366.

Threlkeld, S. T. 1988. Planktivory and planktivore biomass effects on zooplankton, phytoplankton, and the trophic cascade. Limnol. Oceanogr. *33*:1362–1375.

Thunmark, S. 1945. Zur Soziologie des Süsswasserplanktons. Eine methodologisch-ökologische Studie. Folia Limnol. Scandin. *3*, 66 pp.

Thunmark, S. 1948. Sjöar och myrar i Lenhovda socken. Lenhovda. En värendssocken berättar. Moheda. 48 pp. (Unable to obtain original; discussed in Lillieroth, 1950.)

Thurman, E. M. 1985. Organic Geochemistry of Natural Waters. Martinus Nijhoff/W. Junk Publishers, Dordrecht. 497 pp.

Thut, R. N. 1969. A study of the profundal bottom fauna of Lake Washington. Ecol. Monogr. *39*:79–100.

Tibbles, B. J. 1997. Effects of temperature on the relative incorporation of leucine and thymidine by bacterioplankton and bacterial isolates. Aquat. Microb. Ecol. *11*:239–250.

Tice, R. H. 1968. Magnitude and frequency of floods in the United States. Part 1-B. North Atlantic Slope Basins, New York to York River. U.S. Geol. Surv. Water-Supply Pap. *1672*. 585 pp.

Tietema, T. 1980. Ecophysiology of the sand sedge *Carex arenaria* L. II. The distribution of ^{14}C assimilates. Acta Bot. Neerl. *29*:165–178.

Tillman, D. L. and J. R. Barnes. 1973. The reproductive biology of the leech *Helobdella stagnalis* (L.) in Utah Lake, Utah. Freshwat. Biol. *3*:137–145.

Tilman, D. 1977. Resource competition between phytoplanktonic algae: An experimental and theoretical approach. Ecology *58*:338–348.

Tilman, D. 1978. The role of nutrient competition in a predictive theory of phytoplankton population dynamics. Mitt. Internat. Verein. Limnol. *21*:585–592.

Tilman, D. 1980. Resources: A graphical-mechanistic approach to competition and predation. Amer. Nat. *116*:362–393.

Tilman, D. 1982. Resource competition and community structure. Princeton Univ. Press, Princeton, NJ. 296 pp.

Tilman, D. 1988. Plant Strategies and the Dynamics and Structure of Plant Communities. Princeton Univ. Press, Princeton, NJ. 360 pp.

Tilman, D. 1996. Biodiversity: Population versus ecosystem stability. Ecology *77*:350–363.

Tilman, D. and S. S. Kilham. 1976. Phosphate and silicate growth and uptake kinetics of the diatoms *Asterionella formosa* and *Cyclotella meneghiniana* in batch and semicontinuous culture. J. Phycol. *12*:375–383.

Tilman, D., R. Kiesling, R. Sterner, S. S. Kilham, and F. A. Johnson. 1986. Green, bluegreen and diatom algae: Taxonomic differences in competitive ability for phosphorus, silicon and nitrogen. Arch. Hydrobiol. *106*:473–485.

Tilman, D., S. S. Kilham, and P. Kilham. 1982. Phytoplankton community ecology: The role of limiting nutrients. Ann. Rev. Ecol. Syst. *13*:349–372.

Tilzer, M. 1972. Dynamik und Produktivität von Phytoplankton und pelagischen Bakterien in einem Hochgebirgssee (Vorderer Finstertaler See, Österreich). Arch. Hydrobiol./Suppl. *40*:201–273.

Tilzer, M. 1973. Diurnal periodicity in the phytoplankton assemblage of a high mountain lake. Limnol. Oceanogr. *18*:15–30.

Tilzer, M. M. 1984. Estimation of phytoplankton loss rates from daily photosynthetic rates and observed biomass changes in Lake Constance. J. Plankton Res. *6*:309–324.

Tilzer, M. M., C. R. Goldman, and E. de Amezaga. 1975. The efficiency of photosynthetic light energy utilization by lake phytoplankton. Verh. Internat. Verein. Limnol. *19*:800–807.

Tilzer, M. M., H. W. Paerl, and C. R. Goldman. 1977. Sustained viability of aphotic phytoplankton in Lake Tahoe (California–Nevada). Limnol. Oceanogr. *22*:84–91.

Timm, T. 1962a. Maloschetnikovie chervi Chudsko Pskovskogo Ozera. Gidrobiol. Issledovaniia *3*:106–108.

Timm, T. 1962b. O rasprostranenii maloshchetnikovikh chervei (Oligochaeta) v ozerakh Estonii. Gidrobiol. Issledovaniia *3*:162–168.

Timms, B. V. 1986. The coastal dune lakes of eastern Australia. *In* P. de Deckker, and W. D. Williams, eds. Limnology in Australia. W. Junk Publ., Dordrecht. pp. 421–432.

Timms, B. V. 1992. Lake geomorphology. Gleneagles Publishing, Glen Osmond, Australia. 180 pp.

Timms, R. M. and B. Moss. 1984. Prevention of growth of potentially dense phytoplankton populations by zooplankton grazing, in the presence of zooplanktivorous fish, in a shallow wetland ecosystem. Limnol. Oceanogr. *29*:472–486.

Timperley, M. H., R. J. Vigor-Brown, M. Kawashima, and M. Ishigami. 1985. Organic nitrogen compounds in atmospheric precipitation: Their chemistry and availability to phytoplankton. Can. J. Fish. Aquat. Sci. *42*:1171–1177.

Tippett, R. 1964. An investigation into the nature of the layering of deep-water sediments in two eastern Ontario lakes. Can. J. Bot. *42*:1693–1709.

Tippett, R. 1970. Artificial surfaces as a method of studying populations of benthic micro-algae in fresh water. Brit. Phycol. J. *5*:187–199.

Tipping, E. 1981. The adsorption of aquatic humic substances by iron oxides. Geochim. Cosmochim. Acta *45*:191–199.

Tipping, E. 1984. Temperature dependence of Mn(II)oxidation in lakewaters: A test of biological involvement. Geochim. Cosmochim. Acta *48*:1353–1356.

Tipping, E. and C. Woof. 1983. Elevated concentrations of humic substances in a seasonally anoxic hypolimnion: Evidence for co-accumulation with iron. Arch. Hydrobiol. *98*:137–145.

Tipping, E. and D. C. Higgins. 1982. The effect of adsorbed humic substances on the colloid stability of hematite particles. Colloids Surfaces *5*:85–92.

Tipping, E., C. Woof, and D. Cooke. 1981. Iron oxide from a seasonally anoxic lake. Geochim. Cosmochim. Acta *45*:1411–1419.

Tipping, E., C. Woof, and M. Ohnstad. 1982. Forms of iron in the oxygenated waters of Esthwaite Water, U.K. Hydrobiologia *92*:383–393.

Tipping, E., J. G. Jones, and C. Woof. 1985. Lacustrine manganese oxides: Mn oxidation states and relationships to "Mn-depositing bacteria," Arch. Hydrobiol. *105*:161–175.

Tirén, T. and K. Pettersson. 1985. The influence of nitrate on the phosphorus flux to and from oxygen depleted lake sediments. Hydrobiologia *120*:207–223.

Tirén, T., J. Thorin, and H. Nômmik. 1976. Denitrification measurements in lakes. Acta Agriculturae Scand. *26*:175–184.

Tirlapur, U., R. Scheuerlein, and D.-P. Häder. 1993. Motility and orientation of a dinoflagellate, *Gymnodinium*, impaired by solar and ultraviolet radiation. FEMS Microbiol. Ecol. *102*:167–174.

Tison, D. L. and D. H. Pope. 1980. Effect of temperature on mineralization by heterotrophic bacteria. Appl. Environ. Microbiol. *39*:584–587.

Tison, D. L., D. H. Pope, and C. W. Boylen. 1980. Influence of seasonal temperature on the temperature optima of bacteria in sediments of Lake George, New York. Appl. Environ. Microbiol. *39*:675–677.

Tison, D. L., F. E. Palmer, and J. T. Staley. 1977. Nitrogen fixation in lakes of the Lake Washington drainage basin. Water Res. *11*:843–847.

Titman, D. 1975. A fluorometric technique for measuring sinking rates of freshwater phytoplankton. Limnol. Oceanogr. *20*:869–875.

Titman, D. 1976. Ecological competition between algae: Experimental confirmation of resource-based theory. Science *192*:463–465.

Titman, D. and P. Kilham. 1976. Sinking in freshwater phytoplankton: Some ecological implications of cell nutrient status and physical mixing processes. Limnol. Oceanogr. *21*:409–417.

Titus, J. E. and M. S. Adams. 1979. Coexistence and the comparative light relations of the submersed macrophytes *Myriophyllum spicatum* L. and *Vallisneria americana* Michx. Oecologia *40*:273–286.

Tjepkema, J. D. and H. J. Evans. 1976. Nitrogen fixation associated with *Juncus balticus* and other plants of Oregon wetlands. Soil Biol. Biochem. *8*:505–509.

Tjossem, S. F. 1990. Effects of fish chemical cues on vertical migration behavior of *Chaoborus*. Limnol. Oceanogr. *35*:1456–1468.

Todd, D. K. 1980. Groundwater hydrology. 2nd Edition. John Wiley & Sons, New York. 535 pp.

Toerien, D. F. and B. Cavari. 1982. Effect of temperature on heterotrophic glucose uptake, mineralization, and turnover rates in lake sediments. Appl. Environ. Microbiol. *43*:1–5.

Toetz, D. and B. Cole. 1980. Ammonia mineralization and cycling in Shagawa Lake, Minnesota. Arch. Hydrobiol. *88*:9–23.

Toetz, D. W. 1973a. The limnology of nitrogen in an Oklahoma reservoir: Nitrogenase activity and related limnological factors. Am. Midland Nat. *89*:369–380.

Toetz, D. W. 1973b. The kinetics of NH_4 uptake by *Ceratophyllum*. Hydrobiologia *41*:275–290.

Toetz, D. W. 1974. Uptake and translocation of ammonia by freshwater hydrophytes. Ecology *55*:199–201.

Toetz, D., L. Varga, and B. Huss. 1977. Observations on uptake of nitrate and ammonia by reservoir phytoplankton. Arch. Hydrobiol. *79*:182–192.

Tollrian, R. 1990. Predator-induced formation in *Daphnia cucullata* (Sars). Arch. Hydrobiol. *119*:191–196.

Tollrian, R. 1994. Fish-kairomone induced morphological changes in *Daphnia lumholtzi* (Sars). Arch. Hydrobiol. *130*:69–75.

Tollrian, R. and E. von Elert. 1994. Enrichment and purification of *Chaoborus* kairomone from water: Further steps toward its chemical characterization. Limnol. Oceanogr. *39*:788–796.

Tolonen, K., A. Siiriänen, and R. Thompson. 1975. Prehistoric field erosion sediment in Lake Lojärvi, S. Finland and its palaeomagnetic dating. Ann. Bot. Fenn. *12*:161–164.

Toman, M. J. and P. C. Dall. 1997. The diet of *Erpobdella octoculata* (Hirudinea: Erpobdellidae) in two Danish lowland streams. Arch. Hydrobiol. *140*:549–563.

Toolan, T., J. D. Wehr, and S. Findlay. 1991. Inorganic phosphorus stimulation of bacterioplankton production in a meso-eutrophic lake. Appl. Environ. Microbiol. *57*:2074–2078.

Tranvik, L. and S. Kokalj. 1998. Decreased biodegradability of algal DOC due to interactive effects of UV radiation and humic matter. Aquat. Microb. Ecol. *14*:301–307.

Tranvik, L. J. 1988. Availability of dissolved organic carbon for planktonic bacteria in oligotrophic lakes of differing humic content. Microb. Ecol. *16*:311–322.

Tranvik, L. J. 1989. Bacterioplankton growth, grazing mortality and quantitative relationship to primary production in a humic and a clearwater lake. J. Plankton Res. *11*:985–1000.

Tranvik, L. J. 1990. Bacterioplankton growth on fractions of dissolved organic carbon of different molecular weights from humic and clear lakes. Appl. Environ. Microbiol. *56*:1672–1677.

Tranvik, L. J. 1998. Degradation of dissolved organic matter in humic waters by bacteria. *In* D. O. Hessen and L. J. Tranvik, eds. Aquatic Humic Substances: Ecology and Biogeochemistry. Springer-Verlag, Berlin. pp. 259–283.

Traunspurger, W. 1996. Distribution of benthic nematodes in the littoriprofundal and profundal of an oligotrophic lake (Königssee, National Park Berchtesgaden, FRG). Arch. Hydrobiol. *135*: 557–575.

Traunspurger, W. and C. Drews. 1996. Vertical distribution of benthic nematodes in an oligotrophic lake: Seasonality, species and age segregation. Hydrobiologia *331*:33–42.

Tressler, W. L. 1957. The Ostracoda of Great Slave Lake. J. Wash. Acad. Sci. *47*:415–423.

Trifonova, N. A., *et al.* 1969. Materialy k Soveshchaniiu po Prognozirovaniiu Soderzhaniia Biolennykh Elementov i Organicheskogo Veshchestva v Vodokhranilishchakh. Inst. Biol. Vnutrennikh Vod., Rybinsk. 175 pp.

Trimbee, A. M. and G. P. Harris. 1984. Phytoplankton population dynamics of a small reservoir: Use of sedimentation traps to quantify the loss of diatoms and recruitment of summer bloom-forming blue-green algae. J. Plankton Res. *6*:897–918.

Triska, F. J. and J. R. Sedell. 1976. Decomposition of four species of leaf litter in response to nitrate manipulation. Ecology *57*:783–792.

Triska, F. J. and R. S. Oremland. 1981. Denitrification assoicated with periphyton communities. Appl. Environ. Microbiol. *42*:745–748.

Triska, F. J., A. P. Jackman, J. H. Duff, and R. J. Avanzino. 1994. Ammonium sorption to channel and riparian sediments: A transient storage pool for dissolved inorganic nitrogen. Biogeochemistry *26*:67–83.

Triska, F. J., J. H. Duff, and R. J. Avanzino. 1993. Patterns of hydrological exchange and nutrient transformation in the hyporheic zone of a gravel-bottom stream: Examining terrestrial-aquatic linkages. Freshwat. Biol. *29*:259–274.

Triska, F. J., J. R. Sedell, and B. Buckley. 1975. The processing of conifer and hardwood leaves in two coniferous forest streams. II. Biochemical and nutrient changes. Verh. Internat. Verein. Limnol. *19*:1628–1639.

Triska, F. J., J. R. Sedell, K. Cromack, Jr., S. V. Gregory, and F. M. McCorison. 1984. Nitrogen budget for a small coniferous forest stream. Ecol. Monogr. *54*:119–140.

Triska, F. J., V. C. Kennedy, R. J. Avanzino, G. W. Zellweger, and K. E. Bencala. 1989. Retention and transport of nutrients in a third-order stream in northwestern California: Hyporheic processes. Ecology *70*:1893–1905.

Triska, F. J., V. C. Kennedy, R. J. Avanzino, G. W. Zellweger, and K. E. Bencala. 1990a. In situ retention-transport response to nitrate loading and storm discharge in a third-order stream. J. N. Am. Benthol. Soc. *9*:229–239.

Triska, F. J., J. H. Duff, and R. J. Avanzino. 1990b. Influence of exchange flow between the channel and hyporheic zone on nitrate production in a small mountain stream. Can. J. Fish. Aquat. Sci. *47*:2009–2111.

Trojanowski, J. 1969. Biological degradation of lignin. Int. Biodetn. Bull. *5*:119–124.

Truesdale, G. A., A. L. Downing, and G. F. Lowden. 1955. The solubility of oxygen in pure water and sea-water. J. Appl. Chem. *5*:53–62.

Trussell, R. P. 1972. The percent un-ionized ammonia in aqueous ammonia solutions at different pH levels and temperatures. J. Fish. Res. Board Can. *29*:1505–1507.

Tsuchiya, T. and H. Iwaki. 1979. Impact of nutrient enrichment in a waterchestnut ecosystem at Takahama-iri Bay of Lake Kasumigaura. II. Role of waterchestnut in primary productivity and nutrient uptake. Water Air Soil Poll. *12*:503–510.

Tsuchiya, T., S. Nohara, and T. Iwakuma. 1990. Net primary production of *Nymphoides peltata* (Gmel.) O. Kuntze growing on sandy sediment in Edosaki-iri Bay in Lake Kasumigaura, Japan. Jpn. J. Limnol. *51*:307–312.

Tsuchiya, T., A. Shinozuka, and I. Ikusima. 1993. Population dynamics, productivity and biomass allocation of *Zizania latifolia* in an aquatic-terrestrial ecotone. Ecol. Res. *8*:193–198.

Tuchman, N. C. 1996. The role of heterotrophy in algae. *In* R. J. Stevenson, M. L. Bothwell, and R. L. Lowe, eds. Algal Ecology: Freshwater Benthic Ecosystems. Academic Press, San Diego. pp. 299–319.

Tucker, A. 1957. The relation of phytoplankton periodicity to the nature of the physico-chemical environment with special reference to phosphorus. I. Morphometrical, physical and chemical conditions. II. Seasonal and vertical distribution of the phytoplankton in relation to the environment. Am. Midland Nat. *57*:300–370.

Tudorancea, C., R. H. Green, and J. Huebner. 1979. Structure, dynamics and production of the benthic fauna in Lake Manitoba. Hydrobiologia *64*:59–95.

Tuomi, P., T. Torsvik, M. Heldal, and G. Bratbak. 1997. Bacterial population dynamics in a meromictic lake. Appl. Environ. Microbiol. *63*:2181–2188.

Turner, A. M. and G. G. Mittelbach. 1990. Predator avoidance and community structure: Interactions among piscivores, planktivores, and plankton. Ecology *71*:2241–2254.

Turner, L. J. and L. D. Delorme. 1996. Assessment of ^{210}Pb data from Canadian lakes using the CIC and CRS models. Environ. Geol. *28*:78–87.

Turner, M. A., D. W. Schindler, and R. W. Graham. 1983. Photosynthesis-irradiance relationships of epilithic algae measured in the laboratory and *in situ*. *In* R. G. Wetzel, ed. Periphyton of Freshwater Ecosystems. W. Junk Publishers, The Hague. pp. 73–88.

Turner, R. R., E. A. Laws, and R. C. Harriss. 1983. Nutrient retention and transformation in relation to hydraulic flushing rate in a small impoundment. Freshwat. Biol. *13*:113–127.

Tuschall, J. R., Jr. and P. L. Brezonik. 1980. Characterization of organic nitrogen in natural waters: Its molecular size, protein content, and interactions with heavy metals. Limnol. Oceanogr. *25*:495–504.

Tutin, W. 1969. The usefulness of pollen analysis in interpretation of stratigraphic horizons, both Late-glacial and Post-glacial. Mitt. Internat. Verein. Limnol. *17*:154–164.

Twilley, R. R., M. M. Brinson, and G. J. Davis. 1977. Phosphorus absorption, translocation, and secretion in *Nuphar luteum*. Limnol. Oceanogr. *22*:1022–1032.

Twinch, A. J. and R. H. Peters. 1994. Phosphate exchange between littoral sediments and overlying water in an oligotrophic north-temperate lake. Can. J. Fish. Aquat. Sci. *41*:1609–1617.

Twiss, M. R. and P. G. C. Campbell. 1998. Trace metal cycling in the surface waters of Lake Erie: Linking ecological and geochemical fates. J. Great Lakes Res. *24*:791–807.

Twombly, S. and W. M. Lewis, Jr. 1989. Factors regulating cladoceran dynamics in a Venezuelan floodplain lake. J. Plankton Res. *11*:317–333.

Tyler, J. E. 1961a. Sun-altitude effect on the distribution of underwater light. Limnol. Oceanogr. *6*:24–25.

Tyler, J. E. 1961b. Scattering properties of distilled and natural waters. Limnol. Oceanogr. *6*:451–456.

Tyler, J. E. 1968. The Secchi disc. Limnol. Oceanogr. *13*:1–6.

Tyler, J. E. and R. C. Smith. 1970. Measurements of Spectral Irradiance Underwater. Gordan and Breach Science Publ., New York. 103 pp.

Uehlinger, U. 1991. Spatial and temporal variability of the periphyton biomass in a pre-alpine river (Necker, Switzerland). Arch. Hydrobiol. *123*:219–237.

Uehlinger, U. and J. Bloesch. 1987. Variation in the C:P ratio of suspended and settling seston and its significance for P uptake calculations. Freshwat. Biol. *17*:99–108.

Úesták, Z., J. Catsky, and P. G. Jarvis, eds. 1971. Plant Photosynthetic Production Manual of Methods. W. Junk N. V. Publishers, The Hague. 818 pp.

Uiblein, F., J. R. Roca, A. Baltanás, and D. L. Danielopol. 1996. Tradeoff between foraging and antipredator behaviour in a macrophyte dwelling ostracod. Arch. Hydrobiol. *137*:119–133.

Úlehlová, B. 1970. An ecological study of aquatic habitats in northwest Overijssel, The Netherlands. Acta Bot. Neerl. *19*:830–858.

Úlehlová, B. 1971. Decomposition and humification of plant material in the vegetation of *Stratiotes aloides* in NW Overijssel, Holland. Hidrobiologia (Romania) *12*:279–285.

Úlehlová, B. 1976. Microbial decomposers and decomposition processes in wetlands. Ceskoslovenská Akad. Ved Studie CSAV *17*. 112 pp.

Ultsch, G. R. 1973. The effects of water hyacinths (*Eichhornia crassipes*) on the microenvironment of aquatic communities. Arch. Hydrobiol. *72*:460–473.

Ultsch, G. R. and D. S. Anthony. 1973. The role of the aquatic exchange of carbon dioxide in the ecology of the water hyacinth (*Eichhornia crassipes*). Florida Sci. *36*:16–22

Urabe, J. 1993. N and P cycling coupled by grazers' activities: Food quality and nutrient release by zooplankton. Ecology *74*: 2337–2350.

Urabe, J. 1995. Direct and indirect effects of zooplankton on seston stoichiometry. Ecoscience *2*:286–296.

Urabe, J. and Y. Watanabe. 1991. Effect of food conditions on the bacterial feeding of *Daphnia galeata*. Hydrobiologia *225*: 121–128.

Urban, N. R. and P. L. Brezonik. 1993. Transformations of sulfur in sediment microcosms. Can. J. Fish. Aquat. Sci. *50*:1946–1960.

Urban, N. R., P. L. Brezonik, L. A. Baker, and L. A. Sherman. 1994. Sulfate reduction and diffusion in sediments of Little Rock Lake, Wisconsin. Limnol. Oceanogr. *39*:797–815.

Urban, N. R., S. E. Bayley, and S. J. Eisenreich. 1989. Export of dissolved organic carbon and acidity from peatlands. Wat. Resour. Res. *25*:1619–1628.

Uren, N. C. and H. M. Reisenauer. 1988. The role of root exudates in nutrient acquisition. Adv. Plant Nutr. *3*:79–114.

Utkilen, H. C., R. L. Oliver, and A. E. Walsby. 1985a. Buoyancy regulation in a red *Oscillatoria* unable to collapse gas vacuoles by turgor pressure. Arch. Hydrobiol. *102*:319–329.

Utkilen, H. C., O. M. Skulberg, and A. E. Walsby. 1985b. Buoyancy regulation and chromatic adaptation in planktonic *Oscillatoria* species: Alternative strategies for optimising light absorption in stratified lakes. Arch. Hydrobiol. *104*:407–417.

Utsumi, M., Y. Nojiri, T. Nakamura, T. Nozawa, A. Otsuki, and H. Seki. 1998. Oxidation of dissolved methane in a eutrophic, shallow lake: Lake Kasumigaura, Japan. Limnol. Oceanogr. *43*:471–480.

Uttormark, P. H. and M. L. Hutchins. 1980. Input/output models as decision aids for lake restoration. Water Res. Bull. *16*:494–500.

Uutala, A. J. 1990. *Chaoborus* (Diptera: Chaoboridae) mandibles—paleolimnological indicators of the historical status of fish populations in acid-sensitive lakes. J. Paleolimnol. *4*:139–151.

Vadstein, O., A. Jensen, Y. Olsen, and H. Reinertsen. 1988. Growth and phosphorus status of limnetic phytoplankton and bacteria. Limnol. Oceanogr. *33*:489–503.

Vadstein, O., B. O. Harkjerr, A. Jensen, Y. Olsen, and H. Reinertsen. 1989. Cycling of organic carbon in the photic zone of a eutrophic lake with special reference to the heterotrophic bacteria. Limnol. Oceanogr. *34*:840–855.

Vadstein, O., O. Brekke, T. Andersen, and Y. Olsen. 1995. Estimation of phosphorus release rates from natural zooplankton communities feeding on planktonic algae and bacteria. Limnol. Oceanogr. *40*:250–262.

Vadstein, O., Y. Olsen, H. Reinertsen, and A. Jensen. 1993. The role of planktonic bacteria in phosphorus cycling in lakes: Sink and link. Limnol. Oceanogr. *38*:1539–1544.

Vähätalo, A. V., M. Salkinoja-Salonen, P. Taalas, and K. Salonen. 2000. Spectrum of the quantum yield for photochemical mineralization of dissolved organic carbon in a humic lake. Limnol. Oceanogr. *45*:664–676.

Vaishampayan, A. 1983. Vanadium as a trace element in the blue-green alga, *Nostoc muscorum*: Influence on nitrogenase and nitrate reductase. New Phytol. *95*:55–60.

Valanne, N., E.-M. Aro, and E. Rintamäki. 1982. Leaf and chloroplast structure of two aquatic *Ranunculus* species. Aquat. Bot. *12*:13–22.

Valett, H. M., S. G. Fisher, and E. H. Stanley. 1990. Physical and chemical characteristics of the hyporheic zone of a Sonoran Desert stream. J. N. Am. Benthol. Soc. *9*:201–215.

Valett, H. M., S. G. Fisher, N. B. Grimm, and P. Camill. 1994. Vertical hydrologic exchange and ecological stability of a desert stream ecosystem. Ecology *75*:548–560.

Valiente, E. F. and M. C. Avendaño. 1993. Sodium-stimulation of phosphate uptake in the cyanobacterium *Anabaena* PCC 7119. Plant Cell Physiol. *34*:201–207.

van der Valk, A. G. and L. C. Bliss. 1971. Hydrarch succession and net primary production of Oxbow lakes in central Alberta. Can. J. Bot. *49*:1177–1199.

Vallentyne, J. R. 1956. Epiphasic carotenoids in post-glacial lake sediments. Limnol. Oceanogr. *1*:252–262.

Vallentyne, J. R. 1957a. Principles of modern limnology. Am. Sci. *45*:218–244.

Vallentyne, J. R. 1957b. The molecular nature of organic matter in lakes and oceans, with lesser reference to sewage and terrestrial soils. J. Fish. Res. Board Can. *14*:33–82.

Vallentyne, J. R. 1960. Fossil pigments. *In* M. B. Allen, ed. Comparative Biochemistry of Photoreactive Systems. Academic Press, New York. pp. 83–105.

Vallentyne, J. R. 1962. Solubility and the decomposition of organic matter in nature. Arch. Hydrobiol. *58*:423–434.

Vallentyne, J. R. 1967. A simplified model of a lake for instructional use. J. Fish. Res. Board Can. *24*:2473–2479.

Vallentyne, J. R. 1969. Sedimentary organic matter and paleolimnology. Mitt. Internat. Verein. Limnol. *17*:104–110.

Vallentyne, J. R. 1970. Phosphorus and the control of eutrophication. Can. Res. Development (May–June, 1970):36–43, 49.

Vallentyne, J. R. 1972. Freshwater supplies and pollution: Effects of the demophoric explosion on water and man. *In* N. Polunin, ed. The Environmental Future. Macmillan Press Ltd., London. pp. 181–211.

Vallentyne, J. R. 1974. The algal bowl—lakes and man. Misc. Special Publ. 22, Dept. of the Environment, Ottawa. 185 pp.

Vallentyne, J. R. 1988. First direction, then velocity. Ambio *17*:409.

Vandal, G. M., W. F. Fitzgerald, K. R. Rolfhus, and C. H. Lamborg. 1995. Modeling the elemental mercury cycle in Pallette Lake, Wisconsin, USA. Water Air Soil Poll. *80*:529–538.

Vanderploeg, H. A., W. S. Gardner, C. C. Parrish, J. R. Liebig, and J. F. Cavaletto. 1992. Lipids and the life-cycle strategy of a hypolimnetic copepod in Lake Michigan. Limnol. Oceanogr. *37*:413–424.

Van, T. K., W. T. Haller, and L. A. Garrard. 1978. The effect of daylength and temperature on *Hydrilla* growth and tuber production. J. Aquat. Plant Manage. *16*:57–59.

Vanni, M. J. 1987. Effects of food availability and fish predation on a zooplankton community. Ecol. Monogr. *57*:61–88.

Vanni, M. J. 1988. Freshwater zooplankton community structure: Introduction of large invertebrate predators and large herbivores to a small-species community. Can. J. Fish. Aquat. Sci. *45*:1758–1770.

Vanni, M. J. and D. L. Findlay. 1990. Trophic cascades and phytoplankton community structure. Ecology *71*:921–937.

Vanni, M. J., C. Luecke, J. F. Kitchell, and J. J. Magnuson. 1990. Effects of planktivorous fish mass mortality on the plankton community of Lake Mendota, Wisconsin: Implications for biomanipulation. Hydrobiologia *200/201*:329–336.

Vannote, R. L. and B. W. Sweeney. 1980. Geographic analysis of thermal equilibria: A conceptual model for evaluating the effect of natural and modified thermal regimes on aquatic insect communities. Am. Nat. *115*:667–695.

Vannote, R. L., G. W. Minshall, K. W. Cummins, J. R. Sedell, and C. E. Cushing. 1980. The river continuum concept. Can. J. Fish. Aquat. Sci. *37*:130–137.

Vareschi, E. 1987. Saline lake ecosystems. *In* Schulze, E. D. and H. Zwölfer, eds. Potentials and Limitations of Ecosystem Analysis. Springer-Verlag, Berlin. pp. 347–364.

Vareschi, E. and J. Jacobs. 1984. The ecology of Lake Nakuru (Kenya). V. Production and consumption of consumer organisms. Oecologia *61*:83–98.

Vargo, G. A. 1979. The contribution of ammonia excreted by zooplankton to phytoplankton production in Narragansett Bay. J. Plankton Res. *1*:75–84.

Vaux, W. G. 1962. Interchange of stream and intergravel water in a salmon spawning riffle. Spec. Sci. Publ. Fisheries, U.S. Fish Wildlife Serv. 405. 11 pp.

Velimirov, B. 1991. Detritus and the concept of non-predatory loss. Arch. Hydrobiol. *121*:1–20.

Venugopal, M. N. and I. J. Winfield. 1993. The distribution of juvenile fishes in a hypereutrophic pond: Can macrophytes potentially offer a refuge for zooplankton? J. Freshwat. Ecol. *8*:389–396.

Verber, J. L. 1964. Initial current studies in Lake Michigan. Limnol. Oceanogr. *9*:426–430.

Verber, J. L. 1966. Inertial currents in the Great Lakes. Publ. Great Lakes Res. Div., Univ. Mich. *15*:375–379.

Verdouw, H. and E. M. J. Dekkers. 1980. Iron and manganese in Lake Vechten (The Netherlands): Dynamics and role in the cycle of reducing power. Arch. Hydrobiol. *89*:509–532.

Verdouw, H., P. C. M. Boers, and E. M. J. Dekkers. 1985. The dynamics of ammonia in sediments and hypolimnion of Lake Vechten (The Netherlands). Arch. Hydrobiol. *105*:79–92.

Verduin, J. 1959. Photosynthesis by aquatic communities in northwestern Ohio. Ecology *40*:377–383.

Verduin, J. 1961. Separation rate and neighbor diffusivity. Science *134*:837–838.

Verhoff, F. H., D. A. Melfi, and S. M. Yaksich. 1982. An analysis of total phosphorus transport in river systems. Hydrobiologia *91*:241–252.

Verity, P. G. 1991. Feeding in planktonic protozoans: Evidence for non-random acquisition of prey. J. Protozool. *38*:69–76.

Vermaak, J. F., J. H. Swanepoel, and H. J. Schoonbee. 1982. The phosphorus cycle in Germiston Lake. III. Seasonal patterns in the absorption, translocation and release of phosphorus by *Potamogeton pectinatus* L. Water SA *8*:138–141.

Vermeij, G. J. and A. P. Covich. 1978. Coevolution of freshwater gastropods and their predators. Am. Nat. *112*:833–843.

Vernadskii, V. I. 1926. Biosfera. Leningrad. (Seen in reprinted version: Vernadskii, V. I. 1989. Biosfera i Noosfera. Akad. Nauk SSSR, Isdatel'stvo Nauka, Moskva. 261 pp.)

Vernieu, W. S. 1997. Effects of reservoir drawdown on resuspension of deltaic sediments in Lake Powell. J. Lake Reserv. Manage. *13*:67–78.

Vernon, L. P. 1964. Bacterial photosynthesis. Ann. Rev. Plant Physiol. *15*:73–100.

Verstreate, D. R., T. A. Storch, and V. L. Dunham. 1980. A comparison of the influence of iron on the growth and nitrate metabolism of *Anabaena* and *Scenedesmus*. Physiol. Plant. *50*:47–51.

Verta, M. 1984. The mercury cycle in lakes: Some new hypotheses. Aqua Fenn. *14*:215–221.

Vervier, P. and R. J. Naiman. 1992. Spatial and temporal fluctuations of dissolved organic carbon in subsurface flow of the Stillaguamish River (Washington, USA). Arch. Hydrobiol. *123*:401–412.

Verweij, W., H. de Haan, T. de Boer, and J. Voerman. 1989. Copper complexation in eutrophic and humic Lake Tjeukemeer, The Netherlands. Freshwat. Biol. *21*:427–436.

Vijverberg, J. 1977. Population structure, life histories and abundance of copepods in Tjeukemeer, The Netherlands. Freshwat. Biol. *7*:579–597.

Vinberg, G. G. 1934. K voprosu o metalimnianom minimume kisloroda. Trudy Limnol. Sta. Kosine *18*:137–142.

Vinberg, G. G. 1960. Pervichnaya Produktsiia Voedoemov. Izdatel'stvo Akademii Nauk, Minsk. 329 pp. (English translat. 1963. The Primary Production of Bodies of Water. U.S. Atomic Energy Comm., Div. Tech. Info. AEC-tr-5692, 601 pp.)

Vinberg, G. G. and V. P. Liakhnovich. 1965. Udobrenie prudov. (Fertilization of fish ponds.) Izdatel'stvo "Pishchevaia Promyshlennost," Moscow. 271 pp. (English translat. 1969. Fish. Res. Board Can. Translation Ser. No. 1339, 482 pp.)

Vinberg, G. G., *et al.* 1972. Biological productivity of different types of lakes. *In* Z. Kajak and A. Hillbricht-Ilkowska, eds. Productivity Problems of Freshwaters. PWN Polish Scientific Publishers, Warsaw. pp. 382–404.

Vincent, W. F. 1980. The physiological ecology of a *Scenedesmus* population in the hypolimnion of a hypertrophic pond. II. Heterotrophy. Br. Phycol. J. *15*:35–41.

Vincent, W. F. and C. R. Goldman. 1980. Evidence for algal heterotrophy in Lake Tahoe, California–Nevada. Limnol. Oceanogr. *25*:89–99.

Vincent, W. F. and S. Roy. 1993. Solar ultraviolet-B radiation and aquatic primary production: Damage, protection, and recovery. Environ. Rev. *1*:1–12.

Vincent, W. F., C. L. Vincent, M. T. Downes, and P. J. Richerson. 1985. Nitrate cycling in Lake Titicaca (Peru-Bolivia): The effects of high-altitude and tropicality. Freshwat. Biol. *15*:31–42.

Vincent, W. F., M. M. Gibbs, and R. H. Spigel. 1991. Eutrophication processes regulated by a plunging river inflow. Hydrobiologia *226*:51–63.

Viner, A. B. 1975. The sediments of Lake George (Uganda). I. Redox potentials, oxygen consumption and carbon dioxide output. Arch. Hydrobiol. *76*:181–197.

Viner, A. B. 1975. The supply of minerals to tropical rivers and lakes (Uganda). *In* A. D. Hasler, ed. Coupling of land and water systems. Springer-Verlag, New York. pp. 227–261.

Viner, A. B. 1985. Thermal stability and phytoplankton distribution. Hydrobiologia *125*:47–69.

Vinyard, G. L. and R. A. Menger. 1980. *Chaoborus americanus* predation on various zooplankters; functional response and behavioral observations. Oecologia *45*:90–93.

Visser, S. A. 1964a. Origin of nitrates in tropical rainwater. Nature *201*:35–36.

Visser, S. A. 1964b. Oxidation-reduction potentials and capillary activities of humic acids. Nature *204*:581.

Vitousek, P. M. and D. U. Hooper. 1993. Biological diversity and terrestrial ecosystem biogeochemistry. *In* E. D. Schulze and H. A. Mooney, eds. Biodiversity and Ecosystem Function. Springer-Verlag, Berlin. pp. 3–14.

Vitousek, P. M. and W. A. Reiners. 1975. Ecosystem succession and nutrient retention: A hypothesis. BioScience *25*:376–381.

Vivekanandan, E., M. A. Haniffa, T. J. Pandian, and R. Raghuraman. 1974. Studies on energy transformation in the freshwater snail *Pila globosa* 1. Influence of feeding rate. Freshwat. Biol. *4*:275–280.

Vlymen, W. J. 1970. Energy expenditure of swimming copepods. Limnol. Oceanogr. *15*:348–356.

Vogel, S. 1994. Life in moving fluids: The physical biology of flow. 2nd Edition. Princeton Univ. Press, Princeton, NJ. 467 pp.

Volk, C. J., C. B. Volk, and L. A. Kaplan. 1997. Chemical composition of biodegradable dissolved organic matter in streamwater. Limnol. Oceanogr. *42*:39–44.

Vollenweider, R. A. 1950. Ökologische Untersuchungen von planktischen Algen auf experimenteller Grundlage. Schweiz. Z. Hydrol. *12*:193–262.

Vollenweider, R. A. 1955. Ein Nomogramm zur Bestimmung des Transmissionskoeffizienten sowie einige Bemerkungen zur Methode seiner Berechnung in der Limnologie. Schweiz. Z. Hydrol. *17*:205–216.

Vollenweider, R. A. 1964. Über oligomiktische Verhältnisse des Lago Maggiore und einiger anderer insubrischer Seen. Mem. Ist. Ital. Idrobiol. *17*:191–206.

Vollenweider, R. A. 1965. Calculation models of photosynthesis-depth curves and some implications regarding day rate estimates in primary production measurements. Mem. Ist. Ital. Idrobiol. *18*(Suppl.):425–457.

Vollenweider, R. A. 1966. Advances in defining critical loading levels for phosphorus in lake eutrophication. Mem. Ist. Ital. Idrobiol. *33*:53–83.

Vollenweider, R. A. 1968. Scientific Fundamentals of the Eutrophication of Lakes and Flowing Waters, with Particular Reference to Nitrogen and Phosphorus as Factors in Eutrophication. Paris, Rep. Organisation for Economic Cooperation and Development. DAS/CSI/68.27, 192 pp.; Annex, 21 pp.; Bibliography, 61 pp.

Vollenweider, R. A. 1969a. Möglichkeiten und Grenzen elementarer Modelle der Stoffbilanz von Seen. Arch. Hydrobiol *66*:1–36.

Vollenweider, R. A., ed. 1969b. A Manual on Methods for Measuring Primary Production in Aquatic Environments. Int. Biol. Program Handbook 12. Blackwell Scientific Publications, Oxford. 213 pp.

Vollenweider, R. A. 1972. Input-Output Models. Mimeographed report. Can. Cent. Inland Waters, Burlington, Ontario. 40 pp.

Vollenweider, R. A. 1975. Input-output models, with special reference to the phosphorus loading concept in limnology. Schweiz. Z. Hydrol. *37*:53–84.

Vollenweider, R. A. 1976. Advances in defining critical loading levels for phosphorus in lake eutrophication. Mem. Ist. Ital. Idrobiol. *33*:53–83.

Vollenweider, R. A. 1979. Das Nährstoffbelastungskonzept als Grundlage für den externen Eingriff in den Eutrophierungsprozess stehender Gewässer und Talsperren. Z. Wasser-u. Abwasser-Forschung *12*:46–56.

Vollenweider, R. A. 1985. Elemental and biochemical composition of plankton biomass; some comments and explorations. Arch. Hydrobiol. *105*:11–29.

Vollenweider, R. A. 1990. Eutrophication: Conventional and non-conventional considerations and comments on selected topics. Mem. Ist. Ital. Idrobiol. *47*:77–134.

Vollenweider, R. A. and J. Kerekes. 1980. The loading concept as a basis for controlling eutrophication philosophy and preliminary results of the OECD Programme on eutrophication. Prog. Water Technol. *12*:5–18.

Vollenweider, R. A., M. Munawar, and P. Stadelmann. 1974. A comparative review of phytoplankton and primary production in the Laurentian Great Lakes. J. Fish. Res. Board Can. *31*:739–762.

Vollenweider, R. A., W. Rast, and J. Kerekes. 1980. The phosphorus loading concept and Great Lakes eutrophication. *In* R. C. Loehr, C. S. Martin, and W. Rast, eds. Phosphorus Management Strategies for Lakes. Ann Arbor Science Publs., Ann Arbor, MI. pp. 207–234.

Voss, S. and H. Mumm. 1999. Where to stay by night and day: Size-specific and seasonal differences in horizontal and vertical distribution of *Chaoborus flavicans* larvae. Freshwat. Biol. *42*:201–213.

Vrede, K. 1996. Regulation of bacterioplankton production and biomass in an oligotrophic clearwater lake—the importance of the phytoplankton community. J. Plankton Res. *18*:1009–1032.

Vuille, T. 1991. Abundance, standing crop and production of microcrustacean populations (Cladocera, Copepoda) in the littoral zone of Lake Biel, Switzerland. Arch. Hydrobiol. *123*:165–185.

Vuori, K.-M. 1995. Direct and indirect effects of iron on river ecosystems. Ann. Zool. Fennici *32*:317–329.

Vuorinen, I., M. Ketola, and M. Walls. 1989. Defensive spine formation in *Daphnia pulex* Leydig and induction by *Chaoborus crystallinus* De Geer. Limnol. Oceanogr. *34*:245–248.

Wachs, B. 1967. Die Oligochaeten-Fauna der Fliessgewässer unter besonderer Berücksichtigung der Beziehungen zwischen der Tubificiden-Besiedlung und dem Substrat. Arch. Hydrobiol. *63*:310–386.

Wagner, G. M. 1997. *Azolla*: A review of its biology and utilization. Bot. Rev. *63*:1–26.

Wagner-Döbler, I. 1990. Vertical migration of *Chaoborus flavicans* (Diptera, Chaoboridae): Control of onset of migration and migration velocity by environmental stimuli. Arch. Hydrobiol. *117*:279–307.

Waichman, A. V. 1996. Autotrophic carbon sources for heterotrophic bacterioplankton in a floodplain lake of central Amazon. Hydrobiologia *341*:27–36.

Wainman, B. C. and D. R. S. Lean. 1990. Seasonal trends in planktonic lipid content and lipid class. Verh. Internat. Verein. Limnol. *24*:416–419.

Wainman, B. C. and D. R. S. Lean. 1992. Carbon fixation into lipid in small freshwater lakes. Limnol. Oceanogr. *37*:956–965.

Wainman, B. C., D. J. McQueen, and D. R. S. Lean. 1993. Seasonal trends in zooplankton lipid concentration and class in freshwater lakes. J. Plankton Res. *15*:1319–1332.

Wainman, B. C., R. E. H. Smith, H. Rai, and J. A. Furgal. 1999. Irradiance and lipid production in natural algal populations. *In* M. T. Arts and B. C. Wainman, eds. Lipids in Freshwater Ecosystems. Springer-Verlag, New York. pp. 45–70.

Wainright, S. C., C. A. Couch, and J. L. Meyer. 1992. Fluxes of bacteria and organic matter into a blackwater river from river sediments and floodplain soils. Freshwat. Biol. *28*:37–48.

Waite, T. D. and C. Kurucz. 1977. The kinetics of bacterial degradation of the aquatic weed '*Hydrilla*' sp. Water Air Soil Poll. *7*:33–43.

Waite, T. D. and F. M. M. Morel. 1984. Photoreductive dissolution of colloidal iron oxides in natural waters. Environ. Sci. Technol. *18*:860–868.

Walker, I. 1992. The benthic litter habitat with its sediment load in the inundation forest of the central Amazonian blackwater river Tarumã Mirím. Amazoniana *12*:143–153.

Walker, I. R. 1987. Chironomidae (Diptera) in paleoecology. Quaternary Sci. Rev. *6*:29–40.

Walker, I. R. 1993. Paleolimnological biomonitoring using freshwater benthic macroinvertebrates. *In* D. M. Rosenberg and V. H. Resh, eds. Freshwater Biomonitoring and Benthic Macroinvertebrates. Chapman & Hall, New York. pp. 306–343.

Walker, I. R. and C. G. Paterson. 1983. Post-glacial chironomid succession in two small, humic lakes in the New Brunswick–Nova Scotia (Canada) border area. Freshwat. Invertebr. Biol. *2*:61–73.

Walker, K. F. and G. E. Likens. 1975. Meromixis and a reconsidered typology of lake circulation patterns. Verh. Internat. Verein. Limnol. *19*:442–458.

Walker, K. F., W. D. Williams, and U. T. Hammer. 1970. The Miller method for oxygen determination applied to saline waters. Limnol. Oceanogr. *15*:814–815.

Walker, T. A. 1980. A correction to the Poole and Atkins Secchi disc/light-attenuation formula. J. Mar. Biol. Ass. U.K. *60*:769–771.

Wallace, J. B. and A. C. Benke. 1984. Quantification of wood habitat in subtropical Coastal Plain streams. Can. J. Fish. Aquat. Sci. *41*:1643–1652.

Wallace, J. B. and J. R. Webster. 1996. The role of macroinvertebrates in stream ecosystem function. Annu. Rev. Entomol. *41*:115–139.

Wallace, J. B. and N. H. Anderson. 1996. Habitat, life history and behavioral adaptations of aquatic insects. *In* R. W. Merritt and K. W. Cummins, eds. An Introduction to the Aquatic Insects of North America. Kendall/Hunt, Dubuque, IA. pp. 41–73.

Wallace, J. B. and R. W. Merritt. 1980. Filter-feeding ecology of aquatic insects. Ann. Rev. Entomol. *25*:103–132.

Wallace, J. B., D. H. Ross, and J. L. Meyer. 1982. Seston and dissolved organic carbon dynamics in a southern Appalachian stream. Ecology *63*:824–838.

Wallace, J. B., J. R. Webster, and T. F. Cuffney. 1982. Stream detritus dynamics: Regulation by invertebrate consumers. Oecologia *53*:197–200.

Wallace, J. B., M. R. Whiles, S. Eggert, T. F. Cuffney, G. J. Lugthart, and K. Chung. 1995. Long-term dynamics of coarse particulate organic matter in three Appalachian Mountain streams. J. N. Am. Benthol. Soc.*14*:217–232.

Wallace, J. B., T. F. Cuffney, J. R. Webster, G. J. Lugthart, K. Chung, and B. S. Goldowitz. 1991. Export of fine organic particles from headwater streams: Effects of season, extreme discharges, and invertebrate manipulation. Limnol. Oceanogr. *36*:670–682.

Wallace, R. L. 1978. Substrate selection by larvae of the sessile rotifer *Ptygura beauchampi*. Ecology *59*:221–227.

Wallace, R. L. and T. W. Snell. 1991. Rotifera. *In* J. H. Thorp and A. P. Covich, eds. Ecology and Classification of North American Freshwater Invertebrates. Academic Press, San Diego. pp. 187–248.

Wallén, B., U. Falkengren-Grerup, and N. Malmer. 1988. Biomass, productivtiy and relative rate of photosynthesis of *Sphagnum* at different water levels on a south Swedish peat bog. Holarctic Ecol. *11*:70–76.

Walling, D. E. and B. W. Webb. 1986. Solutes in river systems. *In* S. T. Trudgill, ed. Solute Processes. John Wiley & Sons, Chichester. pp. 251–327.

Walling, D. E. and B. W. Webb. 1992. Water quality. I. Physical characteristics. *In* P. Calow and G. E. Petts, eds. The rivers Handbook. I. Hydrological and ecological principles. Blackwell Scientific Publs., Oxford. pp. 48–72.

Wallis, P. M., H. B. N. Hynes, and S. A. Telang. 1981. The importance of groundwater in the transportation of allochthonous dissolved organic matter to the streams draining a small mountain basin. Hydrobiologia *79*:77–90.

Walsby, A. E. 1972. Structure and function of gas vacuoles. Bacteriol. Rev. *36*:1–32.

Walsby, A. E. 1975. Gas vesicles. Ann. Rev. Plant Physiol. *26*:427–439.

Walsby, A. E. 1985. The permeability of heterocysts to the gases nitrogen and oxygen. Proc. R. Soc. London Ser. B *223*:177–196.

Walsby, A. E. 1994. Gas vesicles. Microbiol. Rev. *58*:94–144.

Walsby, A. E. and C. S. Reynolds. 1980. Sinking and floating. *In* I. Morris, ed. The Physiological Ecology of Phytoplankton. Blackwell Scientific Publ., Oxford. pp. 371–412.

Walsby, A. E. and M. J. Booker. 1980. Changes in buoyancy of a planktonic blue-green alga in response to light intensity. Br. Phycol. J. *15*:311–319.

Walsh, G. E. 1965a. Studies on dissolved carbohydrate in Cape Cod waters. I. General Survey. Limnol. Oceanogr. *10*:570–576.

Walsh, G. E. 1965b. Studies on dissolved carbohydrate in Cape Cod waters. II. Diurnal fluctuation in Oyster Pond. Limnol. Oceanogr. *10*:577–582.

Walsh, G. E. 1966. Studies on dissolved carbohydrate in Cape Cod waters. III. Seasonal variation in Oyster Pond and Wequaquet Lake, Massachusetts. Limnol. Oceanogr. *11*:249–256.

Walz, N. 1978a. The energy balance of the freshwater mussel *Dreissena polymorpha* Pallas in laboratory experiments and in Lake Constance. I. Pattern of activity, feeding and assimilation efficiency. Arch. Hydrobiol./Suppl. *55*:83–105.

Walz, N. 1978b. The energy balance of the freshwater mussel *Dreissena polymorpha* Pallas in laboratory experiments and in Lake Constance. III. Growth under standard conditions. Arch. Hydrobiol./Suppl. *55*:121–141.

Walz, N. 1978c. Die Produktion der *Dreissena*-Population und deren Bedeutung im Stoffkreislauf des Bodensees. Arch. Hydrobiol. *82*:482–499.

Walz, N. and F. Rothbucher. 1991. Effect of food concentration on body size, egg size, and population dynamics of *Brachionus angularis* (Rotatoria). Verh. Internat. Verein. Limnol. *24*:2750–2753.

Walz, N., H.-J. Elster, and M. Mezger. 1987. The development of the rotifer community structure in Lake Constance during its eutrophication. Arch. Hydrobiol./Suppl. *74*:452–487.

Wan, G. 1993. Remobilization of Pb during the early diagenesis in the lacustrine sediments. Verh. Internat. Verein. Limnol. *25*:242–248.

Wandschneider, K. 1979. Vertical distribution of phytoplankton during invetigations of a natural surface film. Mar. Biol. *52*:105–111.

Wang, K. S. and T. J. Chai. 1994. Reduction in omega–3 fatty acids by UV-B irradiation in microalgae. J. Appl. Phycol. *6*:415–421.

Wang, L. and J. C. Priscu. 1994. Influence of phytoplankton on the response of bacterioplankton growth to nutrient enrichment. Freshwat. Biol. *31*:183–190.

Wang, L., T. D. Miller, and J. C. Priscu. 1992. Bacterioplankton nutrient deficiency in a eutrophic lake. Arch. Hydrobiol. *125*:423–439.

Wang, M. Y., M. Y. Siddiqi, T. J. Ruth, and A. D. M. Glass. 1993. Ammonium uptake by rice roots. I. Fluxes and subcellular distribution of $^{13}NH_4^+$. Plant Physiol. *103*:1249–1258.

Wang, T. S. C., M.-M. Kao, and P. M. Huang. 1980. The effect of pH on the catalytic synthesis of humic substances by illite. Soil Sci. *129*:333–338.

Wang, T. S., S. W. Li, and Y. L. Ferng. 1978. Catalytic polymerization of phenolic compounds by clay minerals. Soil Sci. *126*:15–21.

Wang, Z. P., D. Zeng, and W. H. Patrick. 1996. Methane emissions from natural wetlands. Environ. Monit. Assess. *42*:143–161.

Wangersky, P. J. 1963. Manganese in ecology. *In* V. Schultz and A. W. Klement, Jr., eds. Radioecology. First National Symposium on Radioecology. Reinhold Publishing Corp., New York. pp. 499–508.

Wangersky, P. J. 1986. Biological control of trace metal residence time and speciation: A review and synthesis. Mar. Chem. *18*:269–297.

Ward, A. K. and R. G. Wetzel. 1975. Sodium: Some effects on blue-green algal growth. J. Phycol. *11*:357–363.

Ward, A. K. and R. G. Wetzel. 1980a. Interactions of light and nitrogen source among planktonic blue-green algae. Arch. Hydrobiol. *90*:1–25.

Ward, A. K. and R. G. Wetzel. 1980b. Photosynthetic responses of blue-green algal populations to variable light intensities. Arch. Hydrobiol. *90*:129–138.

Ward, D. M. and M. R. Winfrey. 1985. Interactions between methanogenic and sulfate-reducing bacteria in sediments. Adv. Aquat. Microbiol. *3*:141–179.

Ward, G. M., A. K. Ward, C. N. Dahm, and N. G. Aumen. 1994. Origin and formation of organic and inorganic particles in aquatic systems. *In* R. S. Wotton, ed. Particulate and Dissolved Material in Aquatic Systems. 2nd Edition. CRC Press, Boca Raton, FL. pp. 45–73.

Ward, J. V. 1985. Thermal characteristics of running waters. Hydrobiologia *125*:31–46.

Ward, J. V. 1992. Aquatic Insect Ecology. I. Biology and Habitat. John Wiley & Sons, New York. 438 pp.

Ward, J. V. and J. A. Stanford. 1982. Thermal responses in the evolutionary ecology of aquatic insects. Annu. Rev. Entomol. *27*:97–117.

Ward, J. V. and J. A. Stanford. 1995. The serial discontinuity concept: Extending the model to floodplain rivers. Regul. Rivers Res. Manage. *10*:159–168.

Ward, J. V. and N. J. Voelz. 1988. Downstream effects of a large, deep-release, high mountain reservoir on lotic zoobenthos. Verh. Internat. Verein. Limnol. *23*:1174–1178.

Ward, J. V., K. Tochner, and F. Schiemer. 1999. Biodiversity and floodplain river ecosystems: Ecotones and connectivity. Regul. Rivers Res. Manage. *15*:125–139.

Warren, C. E. and G. E. Davis. 1971. Laboratory stream research: Objectives, possibilities, and constraints. Ann. Rev. Ecol. Systematics *2*:111–114.

Wasmund, N. and A. Kowalczewski. 1982. Production and distribution of benthic microalgae in the littoral sediments of Mikolajskie Lake. Ekol. Polska *30*:287–301.

Wassmann, R. and U. G. Thein. 1996. Spatial and seasonal variation of methane emission from an Amazon floodplain lake. Mitt. Internat. Verein. Limnol. *25*:179–185.

Watanabe, I., W. L. Barraquio, M. R. de Guzman, and D. A. Cabrera. 1979. Nitrogen-fixing (acetylene reduction) activity and population of aerobic heterotrophic nitrogen-fixing bacteria associated with wetland rice. Appl. Environ. Microbiol. *37*:813–819.

Watanabe, Y. 1990. C:N:P ratios of size-fractionated seston and planktonic organisms in various trophic levels. Verh. Internat. Verein. Limnol. *24*:195–199.

Watanabe, Y. 1996. Limiting factors for bacterioplankton production in mesotrophic and hypereutrophic lakes: Estimation by [^{3}H]thymidine incorporation. Jpn. J. Limnol. *57*:107–117.

Waters, T. F. 1961. Standing crop and drift of stream bottom organisms. Ecology *42*:352–357.

Waters, T. F. 1972. The drift of stream insects. Annu. Rev. Entomol. *17*:253–272.

Waters, T. F. 1977. Secondary production in inland waters. Adv. Ecol. Res. *10*:91–164.

Waters, T. F. 1979. Influence of benthos life history upon the estimation of secondary production. J. Fish. Res. Board Can. *36*:1425–1430.

Waters, T. F. 1988. Fish production-benthos production relationships in trout streams. Polskie Arch. Hydrobiol. *35*:548–561.

Waters, T. F. and J. C. Hokenstrom. 1980. Annual production and drift of the stream amphipod *Gammarus pseudolimnaeus* in Valley Creek, Minnesota. Limnol. Oceanogr. *25*:700–710.

Watras, C. J. 1983. Reproductive cycles in diaptomid copepods: Effects of temperature, photocycle, and species on reproductive potential. Can. J. Fish. Aquat. Sci. *40*:1607–1613.

Watras, C. J., N. S. Bloom, S. A. Claas, K. A. Morrison, C. C. Gilmour, and S. R. Craig. 1995. Methylmercury production in the anoxic hypolimnion of a dimictic seepage lake. Water Air Soil Poll. *80*:735–745.

Watson, N. H. F. and B. N. Smallman. 1971. The role of photoperiod and temperature in the induction and termination of an arrested development in two species of freshwater cyclopid copepods. Can. J. Zool. *49*:855–862.

Watson, S., E. McCauley, and J. A. Downing. 1992. Sigmoid relationships between phosphorus, algal biomass, and algal community structure. Can. J. Fish. Aquat. Sci. *49*:2605–2610.

Watts, W. A. and T. C. Winter. 1966. Plant macrofossils from Kirchner Marsh, Minnesota—A paleoecology study. Geol. Soc. Amer. Bull. *77*:1339–1360.

Waughman, G. J. and D. J. Bellamy. 1980. Nitrogen fixation and the nitrogen balance in peatland ecosystems. Ecology *61*:1185–1198.

Wavre, M. and R. O. Brinkhurst. 1971. Interactions between some tubificid oligochaetes and bacteria found in the sediments of Toronto Harbour, Ontario. J. Fish. Res. Board Can. *28*:335–341.

Way, C. M. 1988. Seasonal allocation of energy to respiration, growth and reproduction in the freshwater clams, *Pisidium variabile* and *P. compressum* (Bivalvia: Pisidiidae). Freshwat. Biol. *19*:321–332.

Weast, R. C., ed. 1970. Handbook of Chemistry and Physics. 51st Edition. Chemical Rubber Co., Cleveland, OH. 2367 pp.

Weaver, C. I. and R. G. Wetzel. 1980. Carbonic anhydrase levels and internal lacunar CO_2 concentrations in aquatic macrophytes. Aquatic Bot. *8*:173–186.

Webb, B. W. and D. E. Walling. 1993. Temporal variability in the impact of river regulation on thermal regime and some biological implications. Freshwat. Biol. *29*:167–182.

Webb, M. G. 1961. The effects of thermal stratification on the distribution of benthic protozoa in a freshwater pond. J. Anim. Ecol. *30*:137–151.

Weber, C. A. 1907. Aufbau und Vegetation der Moore Norddeutschlands. Beibl. Bot. Jahrb. *90*:19–34.

Weber, J. A. 1972. The importance of turions in the propagation of *Myriophyllum exalbescens* (Haloragidaceae) in Douglas Lake, Michigan. Mich. Bot. *11*:115–121.

Weber, J. A. and L. D. Noodén. 1974. Turion formation and germination in *Myriophyllum verticillatum*: Phenology and its interpretation. Mich. Bot. *13*:151–158.

Weber, J. A. and L. D. Noodén. 1976a. Environmental and hormonal control of turion formation in *Myriophyllum verticillatum*. Plant Cell Physiol. *17*:721–731.

Weber, J. A. and L. D. Noodén. 1976b. Environmental and hormonal control of turion germination in *Myriophyllum verticillatum*. Am. J. Bot. *63*:936–944.

Webster, J. R. and B. C. Patten. 1979. Effects of watershed perturbation on stream potassium and calcium dynamics. Ecol. Monogr. *49*:51–72.

Webster, J. R. and E. F. Benfield. 1986. Vascular plant breakdown in freshwater ecosystems. Annu. Rev. Ecol. Syst. *17*:567–594.

Webster, J. R., A. P. Covich, J. L. Tank, and T. V. Crockett. 1994. Retention of coarse organic particles in streams in the southern Appalachian Mountains. J. N. Am. Benthol. Soc. *13*:140–150.

Webster, J. R., D. J. D'Angelo, and G. T. Peters. 1991. Nitrate and phosphate uptake in streams at Coweeta Hydrologic Laboratory. Verh. Internat. Verein. Limnol. *24*:1681–1686.

Webster, J. R., E. F. Benfield, S. W. Golladay, B. R. Hill, L. E. Hornick, R. F. Kazmierczak, Jr., and W. B. Perry. 1987. Experimental studies of physical factors affecting seston transport in streams. Limnol. Oceanogr. *32*:848–863.

Webster, J. R., J. B. Wallace, and E. F. Benfield. 1995. Organic processes in streams of the eastern United States. *In* C. E. Cushing, K. W. Cummins, and G. W. Minshall, eds. River and Stream Ecosystems. Elsevier, Amsterdam. pp. 117–187.

Webster, J. R., M. E. Gurtz, J. J. Hains, J. L. Meyer, W. T. Swank, J. B. Waide, and J. B. Wallace. 1983. Stability of stream ecosystems. *In* J. R. Barnes and G. W. Minshall, eds. Stream Ecology: Application and Testing of General Ecological Theory. Plenum Press, New York. pp. 355–395.

Wedge, R. M. and J. E. Burris. 1982. Effects of light and temperature on duckweed photosynthesis. Aquat. Bot. *13*:133–140.

Weers, P. M. M. and R. D. Gulati. 1997. Effect of the addition of polyunsaturated fatty acids to the diet on the growth and fecundity of *Daphnia galeata*. Freshwat. Biol. *38*:721–729.

Weert, R. van der, and G. E. Kamerling. 1974. Evapotranspiration of water hyacinth (*Eichhornia crassipes*). J. Hydrol. *22*:201–212.

Wehr, J. D. 1991. Nutrient and grazer-mediated effects on picoplankton and size structure in phytoplankton communities. Int. Rev. ges. Hydrobiol. *76*:643–656.

Wehr, J. D. and L. M. Brown. 1985. Selenium requirement of a bloom-forming planktonic alga from softwater and acidified lakes. Can. J. Fish. Aquat. Sci. *42*:1783–1788.

Wehr, J. D., J. Petersen, and S. Findlay. 1999. Influence of three contrasting detrital carbon sources on planktonic bacterial metabolism in a mesotrophic lake. Microb. Ecol. *37*:23–35.

Wehr, J. D., L. M. Brown, and K. O'Grady. 1987. Highly specialized nitrogen metabolism in a freshwater phytoplankter, *Chrysochromulina breviturrita*. Can. J. Fish. Aquat. Sci. *44*:736–742.

Weibel, S. R. 1969. Urban drainage as a factor in eutrophication. *In* Eutrophication: Causes, Consequences, Correctives. Nat. Acad. Sciences, Washington, DC. pp. 383–403.

Weidemann, A. D., T. T. Bannister, S. W. Effler, and D. L. Johnson. 1985. Particulate and optical properties during $CaCO_3$ precipitation in Otisco Lake. Limnol. Oceanogr. *30*:1078–1083.

Weiler, R. R. 1974. Exchange of carbon dioxide between the atmosphere and Lake Ontario. J. Fish. Res. Board Can. *31*:329–332.

Weinbauer, M. G., C. Beckmann, and M. G. Höfle. 1998. Utility of green fluorescent nucleic acid dyes and aluminum oxide membrane filters for rapid epifluorescence enumeration of soil and sediment bacteria. Appl. Environ. Microbiol. *64*:5000–5003.

Weinmann, G. 1970. Gelöste Kohlenhydrate und andere organische Stoffe in natürlichen Gewässern und in Kulturen von *Scenedesmus quadricauda*. Arch. Hydrobiol./Suppl. *37*:164–242.

Weiss, R. F. 1970. The solubility of nitrogen, oxygen and argon in water and seawater. Deep-Sea Res. *17*:721–735.

Weisse, T. 1990. Trophic interactions among heterotrophic microplankton, nanoplankton, and bacteria in Lake Constance. Hydrobiologia *191*:111–122.

Weisse, T. 1991. The annual cycle of heterotrophic freshwater nanoflagellates: Role of bottom-up versus top-down control. J. Plankton Res. *13*:167–185.

Weisse, T. 1997. Growth and production of heterotrophic nanoflagellates in a meso-eutrophic lake. J. Plankton Res. *19*:703–722.

Welch, E. B. 1992. Ecological Effects of Wastewater: Applied Limnology and Pollutant Effects. 2nd Edition. Chapman & Hall, London. 425 pp.

Welch, H. E. 1968. Relationships between assimilation efficiencies and growth efficiencies for aquatic consumers. Ecology *49*:755–759.

Welch, H. E. and M. A. Bergmann. 1985. Water circulation in small arctic lakes in winter. Can. J. Fish. Aquat. Sci. *42*:506–520.

Welch, P. S. 1936. Limnological investigation of a strongly basic bog lake surrounded by an extensive acid-forming bog mat. Pap. Mich. Acad. Sci. Arts Lett. *21*:727–751.

Welch, P. S. 1948. Limnological Methods. Blakiston Co., Philadelphia. 381 pp.

Welch, P. S. 1952. Limnology. 2nd Edition. McGraw-Hill Book Co., New York. 538 pp.

Welch, P. S. and F. E. Eggleton. 1932. Limnological investigations on northern Michigan lakes. II. A further study of depression individuality in Douglas Lake. Pap. Mich. Acad. Sci. Arts Lett. *15*(1931):491–508.

Welch, R. M. 1995. Micronutrient nutrition of plants. Crit. Rev. Plant Sci. *14*:49–82.

Van Der Welde, G. and P. M. M. Peelen-Bexkens. 1983. Production and biomass of floating leaves of three species of Nymphaeaceae in two Dutch waters. Proceedings of the International Symposium on Aquatic Macrophytes, Nijmegen. pp. 230–235.

Wellborn, G. A. 1994. Size-biased predation and prey life histories: A comparative study of freshwater amphipod populations. Ecology *75*:2104–2117.

Wells, L. 1960. Seasonal abundance and vertical movements of planktonic Crustacea in Lake Michigan. Fish. Bull. U.S. Fish. Wildl. Serv. *60*(172):343–369.

Wells, L. 1968. Daytime distribution of *Pontoporeia affinis* off bottom in Lake Michigan. Limnol. Oceanogr. *13*:703–705.

Welsh, R. P. H. and P. Denny. 1979. The translocation of ^{32}P in two submerged aquatic angiosperm species. New Phytol. *82*:645–656.

Welton, J. S. 1979. Life-history and production of the amphipod *Gammarus pulex* in a Dorset chalk stream. Freshwat. Biol. *9*:263–275.

Wentz, D. A. and G. F. Lee. 1969. Sedimentary phosphorus in lake cores—analytical procedure. Environ. Sci. Technol. *3*:750–754.

Werner, D., ed. 1977. The Biology of Diatoms. Univ. California Press, Berkeley. 498 pp.

Werner, E. E. 1974. The fish size, prey size, handling time relation in several sunfishes and some implications. J. Fish. Res. Board Can. *31*:1531–1536.

Werner, E. E. and D. J. Hall. 1974. Optimal foraging and the size selection of prey by the bluegill sunfish (*Lepomis macrochirus*). Ecology *55*:1042–1052.

Werner, E. E. and G. G. Mittelbach. 1981. Optimal foraging: Field tests of diet choice and habitat switching. Am. Zool. *21*:813–829.

Werner, E. E., G. G. Mittelbach, and D. J. Hall. 1981. The role of foraging profitability and experience in habitat use by the bluegill sunfish. Ecology *62*:116–125.

Wershaw, R. L. 1992. Membrane-micelle model for humus in soils and sediments and its relation to humification. U.S. Geol. Surv., Denver. Open-File Rep. *91–513*. 64 pp.

Westerdahl, H. E. and K. D. Getsinger. 1988. Aquatic Plant Identification and Herbicide Use Guide. Vols. 1 and 2. Technical Report A–88–9. U.S. Army Corps of Engineers, Vicksburg, MS.

Westlake, D. F. 1963. Comparisons of plant productivity. Biol. Rev. *38*:385–425.

Westlake, D. F. 1965a. Theoretical aspects of comparability of productivity data. Mem. Ist. Ital. Idrobiol. *18*(Suppl.):313–322.

Westlake, D. F. 1965b. Some basic data for investigations of the productivity of aquatic macrophytes. Mem. Ist. Ital. Idrobiol. *18*(Suppl.):229–248.

Westlake, D. F. 1965c. Some problems in the measurement of radiation under water: A review. Photochem. Photobiol. *4*:849–868.

Westlake, D. F. 1966. The biomass and productivity of *Glyceria maxima*. I. Seasonal changes in biomass. J. Ecol. *54*:745–753.

Westlake, D. F. 1967. Some effects of low-velocity currents on the metabolism of aquatic macrophytes. J. Exp. Bot. *18*:187–205.

Westlake, D. F. 1968. Methods used to determine the annual production of reedswamp plants with extensive rhizomes. *In* Methods of Productivity Studies in Root Systems and Rhizosphere Organisms. Publishing House Nauka, Leningrad. pp. 226–234.

Westlake, D. F. 1982. The primary productivity of water plants. *In* J. J. Symoens, S. S. Hooper, and P. Compère, eds. Studies On Aquatic Vascular Plants. Royal Botanical Society of Belgium, Brussels. pp. 165–180.

Westlake, D. F., *et al.* 1980. Primary production. *In* E. D. Le Cren and R. H. Lowe-McConnell, eds. The Functioning of Freshwater Ecosystems. Cambridge Univ. Press, Cambridge. pp. 141–246.

Wetzel, P. R. and A. G. van der Valk. 1996. Vesicular-arbuscular mycorrhizae in prairie pothole wetland vegetation in Iowa and North Dakota. Can. J. Bot. *74*:883–890.

Wetzel, R. G. 1960. Marl encrustation on hydrophytes in several Michigan lakes. Oikos *11*:223–236.

Wetzel, R. G. 1964. A comparative study of the primary productivity of higher aquatic plants, periphyton, and phytoplankton in a large, shallow lake. Int. Rev. ges. Hydrobiol. *49*:1–64.

Wetzel, R. G. 1965a. Nutritional aspects of algal productivity in marl lakes with particular reference to enrichment bioassays and their interpretation. Mem. Ist. Ital. Idrobiol. *18*(Suppl.): 137–157.

Wetzel, R. G. 1965b. Techniques and problems of primary productivity measurements in higher aquatic plants and periphyton. Mem. Ist. Ital. Idrobiol. *18*(Suppl.):249–267.

Wetzel, R. G. 1966a. Variations in productivity of Goose and hypereutrophic Sylvan lakes, Indiana. Invest. Indiana Lakes Streams 7:147–184.

Wetzel, R. G. 1966b. Productivity and nutrient relationships in marl lakes of northern Indiana. Verh. Internat. Verein. Limnol. *16*:321–332.

Wetzel, R. G. 1967. Dissolved organic compounds and their utilization in two marl lakes. *In* Problems of organic matter determination in freshwater. Hidrológiai Közlöny *47*:298–303.

Wetzel, R. G. 1968. Dissolved organic matter and phytoplanktonic productivity in marl lakes. Mitt. Internat. Verein. Limnol. *14*:261–270.

Wetzel, R. G. 1969. Factors influencing photosynthesis and excretion of dissolved organic matter by aquatic macrophytes in hard-water lakes. Verh. Internat. Verein. Limnol. *17*:72–85.

Wetzel, R. G. 1970. Recent and postglacial production rates of a marl lake. Limnol. Oceanogr. *15*:491–503.

Wetzel, R. G. 1972. The role of carbon in hard-water marl lakes. *In* G. E. Likens, ed. Nutrients and Eutrophication: The Limiting-Nutrient Controversy. Special Symposium, Amer. Soc. Limnol. Oceanogr. *1*:84–97.

Wetzel, R. G. 1973. Productivity investigations of interconnected lakes. I. The eight lakes of the Oliver and Walters chains, northeastern Indiana. Hydrobiol. Stud. *3*:91–143.

Wetzel, R. G. 1975a. General Secretary's Report—19th Congress of the Societas Internationalis Limnologiae. Verh. Internat. Verein. Limnol. *19*:3232–3292.

Wetzel, R. G. 1975b. Primary production. *In* B. A. Whitton, ed. River Ecology. Blackwell Scientific Publs., Oxford. pp. 230–247.

Wetzel, R. G. 1978. Foreword and introduction. *In* Good, R. E., D. F. Whigham, and R. L. Simpson, eds. Freshwater Wetlands: Ecological Processes and Management Potential. Academic Press, New York. pp. xiii-xvii.

Wetzel, R. G. 1979. The role of the littoral zone and detritus in lake metabolism. Arch. Hydrobiol. Beih. Ergebn. Limnol. *13*:145–161.

Wetzel, R. G. 1981. Longterm dissolved and particulate alkaline phosphatase activity in a hardwater lake in relation to lake stability and phosphorus enrichments. Verh. Internat. Verein. Limnol. *21*:337–349.

Wetzel, R. G. 1983a. Limnology. 2nd Edition. Saunders College Publishing, Philadelphia. 860 pp.

Wetzel, R. G. 1983b. Attached algal-substrata interactions: Fact or myth, and when and how? *In* R. G. Wetzel, ed. Periphyton in Freshwater Ecosystems. W. Junk Publishers, The Hague. pp. 207–215.

Wetzel, R. G., ed. 1983b. Periphyton of Aquatic Ecosystems. B. V. Junk Publishers, The Hague. 346 pp.

Wetzel, R. G. 1984. Detrital dissolved and particulate organic carbon functions in aquatic ecosystems. Bull. Mar. Sci. *35*:503–509.

Wetzel, R. G. 1989. Wetland and littoral interfaces of lakes: Productivity and nutrient regulation in the Lawrence Lake ecosystem. *In* R. R. Sharitz and J. W. Gibbons, eds. Freshwater Wetlands and Wildlife. U.S. Dept. Energy, Office Sci. Technical Information, Oak Ridge, TN. pp. 283–302.

Wetzel, R. G. 1990a. Land-water interfaces: Metabolic and limnological regulators. Verh. Internat. Verein. Limnol. *24*:6–24.

Wetzel, R. G. 1990b. Reservoir ecosystems: Conclusions and speculations. *In* K. W. Thornton, B. L. Kimmel, and F. E. Payne, eds. Reservoir Limnology: Ecological Perspectives. Wiley-Interscience, New York. pp. 227–238.

Wetzel, R. G. 1990c. Detritus, macrophytes and nutrient cycling in lakes. Mem. Ist. Ital. Idrobiol. *47*:233–249.

Wetzel, R. G. 1991. Extracellular enzymatic interactions in aquatic ecosystems: Storage, redistribution, and interspecific communication. *In* R. J. Chróst, ed. Microbial Enzymes in Aquatic Environments. Springer-Verlag, New York. pp. 6–28.

Wetzel, R. G. 1992a. Clean water: A fading resource. Hydrobiologia *243/244*:21–30.

Wetzel, R. G. 1992b. Gradient-dominated ecosystems: Sources and regulatory functions of dissolved organic matter in freshwater ecosystems. Hydrobiologia *229*:181–198.

Wetzel, R. G. 1993a. Microcommunities and microgradients: Linking nutrient regeneration, microbial mutualism, and high sustained aquatic primary production. Netherlands J. Aquat. Ecol. *27*:3–9.

Wetzel, R. G. 1993b. Humic compounds from wetlands: Complexation, inactivation, and reactivation of surface-bound and extracellular enzymes. Verh. Internat. Verein. Limnol. *25*:122–128.

Wetzel, R. G. 1995. Death, detritus, and energy flow in aquatic ecosystems. Freshwat. Biol. *33*:83–89.

Wetzel, R. G. 1996. Benthic algae and nutrient cycling in lentic freshwater ecosystems. *In* R. J. Stevenson, M. L. Bothwell, and R. L. Lowe, eds. Algal Ecology: Freshwater Benthic Ecosystems. Academic Press, San Diego. pp. 641–667.

Wetzel, R. G. 1999a. Plants and water in and adjacent to lakes. *In* A. J. Baird and R. L. Wilby, eds. Eco-hydrology: Plants and Water in Terrestrial and Aquatic Environments. Routledge, London. pp. 269–299.

Wetzel, R. G. 1999b. Biodiversity and shifting energetic stability within freshwater ecosystems. Arch. Hydrobiol. Spec. Issues Advanc. Limnol. *54*:19–32.

Wetzel, R. G. 1999c. Organic phosphorus mineralization in soils and sediments. *In* K. R. Reddy, G. A. O'Connor, and C. L. Schelske, eds. Phosphorus Biogeochemistry of Subtropical Ecosystems. CRC Press, Inc., Boca Raton, FL. pp. 225–245.

Wetzel, R. G. 2000a. Dissolved organic carbon: Detrital energetics, metabolic regulators, and drivers of ecosystem stability of aquatic ecosystems. *In* S. Findlay and R. Sinsabaugh, eds. Dissolved Organic Matter in Aquatic Ecosystems. Academic Press, San Diego. (In press.)

Wetzel, R. G. 2000b. Freshwater ecology: Changes, requirements, and future demands. Limnology *1*: 3–11.

Wetzel, R. G. 2000c. Natural photodegradation by UV-B of dissolved organic matter of different decomposing plant sources to readily degradable fatty acids. Verh. Internat. Verein. Limnol. 27 (In press.)

Wetzel, R. G. 2000d. Origins, fates, and ramifications of natural organic compounds of wetlands. *In* A. Clark, ed. Organic Compounds in Natural Ecosystems. (In press.)

Wetzel, R. G. and A. K. Ward. 1992. Primary production. *In* P. Calow and G. E. Petts, eds. Rivers Handbook. I. Hydrological and Ecological Principles. Blackwell Scientific Publs., Oxford. pp. 354–369.

Wetzel, R. G. and A. Otsuki. 1974. Allochthonous organic carbon of a marl lake. Arch. Hydrobiol. *73*:31–56.

Wetzel, R. G. and B. A. Manny. 1972a. Secretion of dissolved organic carbon and nitrogen by aquatic macrophytes. Verh. Internat. Verein. Limnol. *18*:162–170.

Wetzel, R. G. and B. A. Manny. 1972b. Decomposition of dissolved organic carbon and nitrogen compounds from leaves in an experimental hard-water stream. Limnol. Oceanogr. *17*:927–931.

Wetzel, R. G. and B. A. Manny. 1977. Seasonal changes in particulate and dissolved organic carbon and nitrogen in a hardwater stream. Arch. Hydrobiol. *80*:20–39.

Wetzel, R. G. and B. A. Manny. 1978. Postglacial rates of sedimentation, nutrient and fossil pigment deposition in a hardwater marl lake of Michigan. Pol. Arch. Hydrobiol. *25*:453–469.

Wetzel, R. G. and D. F. Westlake. 1969. Periphyton. *In* R. A. Vollenweider, ed. A Manual on Methods for Measuring Primary Production in Aquatic Environments. Int. Biol. Program Handbook 12. Blackwell Scientific Publications, Oxford. pp. 33–40.

Wetzel, R. G. and D. G. McGregor. 1968. Axenic culture and nutritional studies of aquatic macrophytes. Am. Midland Nat. *80*:52–64.

Wetzel, R. G. and D. Pickard. 1996. Application of secondary production methods to estimates of net aboveground primary production of emergent aquatic macrophytes. Aquat. Bot. *53*:109–120.

Wetzel, R. G. and G. E. Likens. 1991. Limnological Analyses. 2nd Edition. Springer-Verlag, New York. 391 pp.

Wetzel, R. G. and G. E. Likens. 2000. Limnological Analyses. 3rd Edition. Springer-Verlag, New York. 429 pp.

Wetzel, R. G. and H. L. Allen. 1970. Functions and interactions of dissolved organic matter and the littoral zone in lake metabolism and eutrophication. *In* Z. Kajak and A. Hillbricht-Ilkowska, eds. Productivity Problems of Freshwaters. PWN Polish Scientific Publishers, Warsaw. pp. 333–347.

Wetzel, R. G. and J. B. Grace. 1983. Atmospheric CO_2 enrichment effects on aquatic plants. *In* E. H. Lemon, ed. The Response of Plants to Rising Levels of Atmospheric Carbon Dioxide. Amer. Assoc. Advancement Sci., Washington, DC. pp. 223–280.

Wetzel, R. G. and M. J. Howe. 1999. High production in a herbaceous perennial plant achieved by continuous growth and synchronized population dynamics. Aquat. Bot. *64*:111–129.

Wetzel, R. G. and P. H. Rich. 1973. Carbon in freshwater systems. *In* G. M. Woodwell and E. V. Pecan, eds. Carbon and the Biosphere. Proc. Brookhaven Symp. in Biol. 24. Tech. Information Center, U.S. Atomic Energy Commission CONF–720510, Brookhaven, NY. pp. 241–263.

Wetzel, R. G. and R. A. Hough. 1973. Productivity and role of aquatic macrophytes in lakes: An assessment. Pol. Arch. Hydrobiol. *20*:9–19.

Wetzel, R. G., B. A. Manny, W. S. White, R. A. Hough, and K. R. McKinley. 1982. Wintergreen Lake: A study in hypereutrophy. (Unpublished manuscript.)

Wetzel, R. G., E. S. Brammer, and C. Forsberg. 1984. Photosynthesis of submersed macrophytes in acidified lakes. I. Carbon fluxes and recycling of CO_2 in *Juncus bulbosus* L. Aquat. Bot. *19*:329–342.

Wetzel, R. G., E. S. Brammer, K. Lindström, and C. Forsberg. 1985. Photosynthesis of submersed macrophytes in acidified lakes. II. Carbon limitations and utilization of benthic CO_2 sources. Aquatic Bot. *22*:107–120.

Wetzel, R. G., P. G. Hatcher, and T. S. Bianchi. 1995. Natural photolysis by ultraviolet irradiance of recalcitrant dissolved organic matter to simple substrates for rapid bacterial metabolism. Limnol. Oceanogr. *40*:1369–1380.

Wetzel, R. G., P. H. Rich, M. C. Miller, and H. L. Allen. 1972. Metabolism of dissolved and particulate detrital carbon in a temperate hard-water lake. Mem. Ist. Ital. Idrobiol. *29*(Suppl.): 185–243.

Weyhenmeyer, G. A., M. Meili, and D. C. Pierson. 1995. A simple method to quantify sources of settling particles in lakes: Resuspension versus new sedimentation of material from planktonic production. Mar. Freshwat. Res. *46*:223–231.

Whalen, S. C. and J. C. Cornwell. 1985. Nitrogen, phosphorus, and organic carbon cycling in an arctic lake. Can. J. Fish. Aquat. Sci. *42*:797–808.

Whalen, S. C. and W. S. Reeburgh. 1990. A methane flux transect along the trans-Alaska pipeline haul road. Tellus, Ser. B *42*:237–249.

Wharton, R. A., Jr., C. P. McKay, G. M. Simmons, Jr., and B. C. Parker. 1986. Oxygen budget of a perennially ice-covered Antarctic lake. Limnol. Oceanogr. *31*:437–443.

Wheeler, B. D. and M. C. F. Proctor. 2000. Ecological gradients, subdivisions and terminology of north-west European mires. J. Ecol. *88*:187–203.

Whigham, D. and R. Simpson. 1977. Growth, mortality, and biomass partitioning in freshwater tidal wetland populations of wild rice (*Zizania aquatica* var. *aquatica*). Bull. Torrey Bot. Club *104*:347–351.

Whipple, G. C. 1898. Classifications of lakes according to temperature. Am. Nat. *32*:25–33.

Whipple, G. C. 1927. The Microscopy of Drinking Water. 4th Edition. John Wiley & Sons, New York. 586 pp.

White, D. S. 1993. Perspectives on defining and delineating hyporheic zones. J. N. Am. Benthol. Soc. *12*:61–69.

White, D. S., C. H. Elzinga, and S. P. Hendricks. 1987. Temperature patterns within the hyporheic zone of a northern Michigan river. J. N. Am. Benthol. Soc. *6*:85–91.

White, E., G. Payne, S. Pickmere, and P. Woods. 1986. Nutrient demand and availability related to growth among natural assemblages of phytoplankton. N. Z. J. Mar. Freshwat. Res. *20*:199–208.

White, J. R. and C. T. Driscoll. 1987. Manganese cycling in an acidic Adirondack lake. Biogeochemistry *3*:87–103.

White, P. A., J. Kalff, J. B. Rasmussen, and J. M. Gasol. 1991. The effect of temperature and algal biomass on bacterial production and specific growth rate in freshwater and marine habitats. Microb. Ecol. *21*:99–118.

White, W. S. 1974. Role of calcium carbonate precipitation in lake metabolism. Ph.D. Diss., Michigan State University, East Lansing. 141 pp.

White, W. S. and R. G. Wetzel. 1975. Nitrogen, phosphorus, particulate and colloidal carbon content of sedimenting seston of a hard-water lake. Verh. Internat. Verein. Limnol. *19*:330–339.

White, W. S. and R. G. Wetzel. 1985. Association of vitamin B_{12} with calcium carbonate in hardwater lakes. Arch. Hydrobiol. *104*:305–309.

Whitehead, H. C. and J. H. Feth. 1961. Recent chemical analyses of waters from several closed-basin lakes and their tributaries in the western United States. Bull. Geol. Soc. Amer. *72*: 1421–1426.

Whiteside, M. 1988. 0+ fish as major factors affecting abundance patterns of littoral zooplankton. Verh. Internat. Verein. Limnol. *23*:1710–1714.

Whiteside, M. C. 1969. Chydorid (Cladocera) remains in surficial sediments of Danish lakes and their significance to paleolimnological interpretations. Mitt. Internat. Verein. Limnol. *17*:193–201.

Whiteside, M. C. 1974. Chydorid (Cladocera) ecology: Seasonal patterns and abundance of populations in Elk Lake, Minnesota. Ecology *55*:538–550.

Whiteside, M. C., J. B. Williams, and C. P. White. 1978. Seasonal bundance and pattern of chydorid Cladocera in mud and vegetative habitats. Ecology *59*:1177–1188.

Whiting, G. J. 1994. CO_2 exchange in the Hudson Bay lowlands: Community characteristics and multispectral reflectance properties. J. Geophys. Res. *99D*:1519–1528.

Whiting, G. J. and J. P. Chanton. 1992. Plant-dependent CH_4 emission in a subarctic Canadian fen. Global Biogeochem. Cycles *6*:225–231.

Whiting, G. J. and J. P. Chanton. 1993. Primary production control of methane emission from wetlands. Nature *364*:794–795.

Whiting, G. J. and J. P. Chanton. 1996. Control of the diurnal pattern of methane emission from emergent aquatic macrophytes by gas transport mechanisms. Aquat. Bot. *54*:237–253.

Whitledge, G. W. and C. F. Rabeni. 1997. Energy sources and ecological role of crayfishes in an Ozark stream: Insights from stable isotopes and gut analysis. Can. J. Fish. Aquat. Sci. *54*:2555–2563.

Whitney, L. V. 1937. Microstratification in the waters of inland lakes in summer. Science *85*:224–225.

Whitton, B. A. 1975. Algae. *In* B. A. Whitton, ed. River Ecology. Blackwell Scientific Publications, Oxford. pp. 81–105.

Whitton, B. A., ed. 1975. River Ecology. Univ. California Press, Berkeley. 725 pp.

Wickham, S. A. 1995a. *Cyclops* predation on ciliates: Species-specific differences and functional responses. J. Plankton Res. *17*:1633–1646.

Wickham, S. A. 1995b. Trophic relations between cyclopoid copepods and ciliated protists: Complex interactions link the microbial and classic food webs. Limnol. Oceanogr. *40*:1173–1181.

Wickham, S. A. and J. J. Gilbert. 1993. The comparative importance of competition and predation by *Daphnia* on ciliated protists. Arch. Hydrobiol. *126*:289–313.

Wiederholm, T. 1980. Use of benthos in lake monitoring. J. Water Poll. Control Fed. *52*:537–547.

Wiederholm, T. and L. Eriksson. 1979. Subfossil chironomids as evidence of eutrophication in Ekoln Bay, central Sweden. Hydrobiologia *62*:195–208.

Wierzbicka, M. 1966. Les résultats des recherches concernant l'état de repos (resting stage) des Cyclopoida. Verh. Internat. Verein. Limnol. *16*:592–599.

Wiessner, W. 1962. Inorganic micronutrients. *In* R. A. Lewin, ed. Physiology and Biochemistry of Algae. Academic Press, New York. pp. 267–286.

Wiggins, G. B. 1977. Larvae of the North American caddisfly genera. Univ. Toronto Press, Toronto.

Wiggins, G. B., R. J. Mackay, and I. M. Smith. 1980. Evolutionary and ecological strategies of animals in annual temporary pools. Arch. Hydrobiol./Suppl. *58*:97–206.

Wilhelm, C. 1990. The biochemistry and physiology of light-harvesting processes in chlorophyll *b*- and *c*-containing algae. Plant Physiol. Biochem. *28*:293–306.

Wilhelm, S. W. 1995. Ecology of iron-limited cyanobacteria: A review of physiological responses and implications for aquatic systems. Aquat. Microb. Ecol. *9*:295–303.

Wilkialis, J. and R. W. Davies. 1980. The population ecology of the leech (Hirudinoidea: Glossiphoniidae) *Theromyzon tessulatum*. Can. J. Zool. *58*:906–912.

Wilkins, A. S. 1972. Physiological factors in the regulation of alkaline phosphatase synthesis in *Escherichia coli*. J. Bacteriol. *110*:616–623.

Wilkinson, R. E. 1963. Effects of light intensity and temperature on the growth of waterstargrass, coontail, and duckweed. Weeds *11*:287–290.

Williams, D. D. 1984. The hyporheic zone as a habitat for aquatic insects and associated arthropods. *In* V. H. Resh and D. M. Rosenberg, eds. The Ecology of Aquatic Insects. Praeger Scientific, New York. pp. 430–455.

Williams, D. D. 1987. The Ecology of Temporary Waters. Croom Helm (Routledge, Chapman & Hall Ltd.), London. 205 pp.

Williams, D. D. and H. B. N. Hynes. 1974. The occurrence of benthos deep in the substratum of a stream. Freshwat. Biol. *4*:233–256.

Williams, D. D. and H. B. N. Hynes. 1977. The ecology of temporary streams. II. General remarks on temporary streams. Int. Rev. ges. Hydrobiol. *62*:53–61.

Williams, J. B. 1982. Temporal and spatial patterns of abundance of the Chydoridae (Cladocera) in Lake Itasca, Minnesota. Ecology *63*:345–353.

Williams, J. B. and J. E. Pinder. 1990. Ground water flow and runoff in a coastal plain stream. Water Resour. Bull. 26:343–352.

Williams, J. D. H. and T. Mayer. 1972. Effects of sediment diagenesis and regeneration of phosphorus with special reference to lakes Eire and Ontario. *In* H. E. Allen and J. R. Kramer, eds. Nutrients in Natural Waters. John Wiley & Sons, New York. pp. 281–315.

Williams, J. D. H., J. K. Syers, R. F. Harris, and D. E. Armstrong. 1970. Adsorption and desorption of inorganic phosphorus by lake sediments in a 0.1M NaCl system. Environ. Sci. Technol. *4*:517–519.

Williams, J. D. H., J. K. Syers, S. S. Shukla, R. F. Harris, and D. E. Armstrong. 1971a. Levels of inorganic and total phosphorus in lake sediments as related to other sediment parameters. Environ. Sci. Technol. *5*:1113–1120.

Williams, J. D. H., J. K. Syers, R. F. Harris, and D. E. Armstrong. 1971b. Fractionation of inorganic phosphate in calcareous lake sediments. Proc. Soil Sci. Soc. Amer. *35*:250–255.

Williams, J. D. H., J. K. Syers, D. E. Armstrong, and R. F. Harris. 1971c. Characterization of inorganic phosphate in noncalcareous lake sediments. Proc. Soil Sci. Soc. Amer. *35*:556–561.

Williams, W. D. 1986. Limnology, the study of inland waters: A comment on perceptions of studies of salt lakes, past and present. *In* P. de Deckker, and W. D. Williams, eds. Limnology in Australia. W. Junk Publ., Dordrecht. pp. 471–484.

Williams, W. D. 1988. Limnological imbalances: An antipodean viewpoint. Freshwat. Biol. *20*:407–420.

Williams, W. D. 1994. Definition and measurement of salinity in salt lakes. Int. J. Salt Lake Res. *3*:53–63.

Williams, W. D. 1996. The largest, highest and lowest lakes of the world: Saline lakes. Verh. Internat. Verein. Limnol. *26*:61–79.

Williams, W. D. and N. V. Aladin. 1991. The Aral Sea: Recent limnological changes and their conservation significance. Aquat. Conservation: Mar. Freshwat. Ecosystems *1*:3–23.

Williams, W. D. and H. F. Wan. 1972. Some distinctive features of Australian inland waters. Water Res. 6:829–836.

Williamson, C. E. 1980. The predatory behavior of *Mesocyclops edax*: Predator preferences, prey defenses, and starvation-induced changes. Limnol. Oceanogr. *25*:903–909.

Williamson, C. E. 1991. Copepoda. *In* J. H. Thorp and A. P. Covich, eds. Ecology and Classificaiton of North American Freshwater Invertebrates. Academic Press, San Diego. pp. 787–822.

Williamson, C. E., R. S. Stemberger, D. P. Morris, T. M. Frost, and S. G. Paulsen. 1996. Ultraviolet radiation in North American lakes: Attenuation estimates from DOC measurements and implications for plankton communities. Limnol. Oceanogr. *41*:1024–1034.

Willoughby, L. G. 1965. Some observations on the location of sites of fungal activity at Blelham Tarn. Hydrobiologia *25*:352–356.

Willoughby, L. G. 1974. Decomposition of litter in fresh water. *In* C. H. Dickinson and G. J. F. Pugh, eds. Biology of Plant Litter Decomposition. Vol. 2. Academic Press, New York. pp. 659–681.

Willoughby, L. G. and J. H. Marcus. 1979. Feeding and growth of the isopod *Asellus aquaticus* on actinomycetes, considered as model filamentous bacteria. Freshwat. Biol. *9*:441–449.

Willsky, G. R. 1990. Vanadium in the biosphere. *In* N. D. Chasteen, ed. Vanadium in Biological Systems: Physiology and Biochemistry. Kluwer Acad. Publ., Dordrecht. pp. 1–24.

Wilson, D. M., W. Fenical, M. Hay, N. Lindquist, and R. Bolser. 1999. Habenariol, a freshwater feeding deterrent from the aquatic orchid *Habenaria repens* (Orchidaceae). Phytochemistry *50*:1333–1336.

Wilson, E. O. 1985. The biological diversity crisis. BioScience *35*:700–706.

Wilson, E. O., ed. 1988. Biodiversity. Nat. Acad. Press, Washington, DC. 521 pp.

Wilson, E. O. 1992. The Diversity of Life. Harvard Univ. Press, Cambridge, MA. 424 pp.

Wilson, E. O. and W. H. Bossert. 1971. A primer of population biology. Sinauer Associates, Inc. Publs., Stamford, CT. 192 pp.

Wilson, L. R. 1937. A quantitative and ecological study of the larger aquatic plants of Sweeney Lake, Oneida County, Wisconsin. Bull. Torr. Bot. Club *64*:199–208.

Wilson, L. R. 1941. The larger aquatic vegetation of Trout Lake, Vilas County, Wisconsin. Trans. Wis. Acad. Sci. Arts Lett. *33*:135–146.

Wilson, M. S. and H. C. Yeatman. 1959. Free-living Copepoda. *In* W. T. Edmondson, ed. Fresh-Water Biology. 2nd Edition. John Wiley & Sons, Inc. New York. pp. 735–861.

Wiltshire, K. H. 1993. The effects of the photosynthetic production of microphytobenthos on the nutrient and oxygen status of sediment-water systems. Verh. Internat. Verein. Limnol. *25*:1141–1146.

Wilzbach, M. A., K. W. Cummins, and R. Knapp. 1988. Toward a functional classification of stream invertebrate drift. Verh. Internat. Verein. Limnol. 23:1244–1254.

Winberg, G. G. 1970. Some interim results of Soviet IBP investigations on lakes. *In* Z. Kajak and A. Hillbricht-Ilkowska, eds. Productivity Problems of Freshwaters. PWN Polish Scientific Publishers, Warsaw. pp. 363–381.

Winberg, G. G., ed. 1971. Methods for the Estimation of Production of Aquatic Animals. Academic Press, New York. 175 pp.

Winberg, G. G., V. A. Babitsky, S. I. Gavrilov, G. V. Bladky, I. S. Zakharenkov, R. Z. Kovalevskaya, T. M. Mikheeva, P. S. Nevyadomskaya, A. P. Ostapenya, P. G. Petrovich, J. S. Potaenko, and O. F. Yakushko. 1970. Biological productivity of different types of lakes. *In* Z. Kajak and A. Hillbricht-Ilkowska, eds. Productivity Problems of Freshwaters. PWN Polish Scientific Publishers, Warsaw. pp. 383–404.

Winberg, G. G., *et al.* 1971. Symbols, Units and Conversion Factors in Studies of Freshwater Productivity. Int. Biol. Program, Sec. PF-Productivity of Freshwaters, 23 pp.

Winberg, G. G., *et al.* 1973a. The progress and state of research on the metabolism, growth, nutrition, and production of freshwater invertebrate animals. Hydrobiol. J. *9*:77–84.

Winberg, G. G., *et al.* 1973b. Biological productivity of two subarctic lakes. Freshwat. Biol. *3*:177–197.

Winfrey, M. R. and J. G. Zeikus. 1977. Effect of sulfate on carbon and electron flow during microbial methanogenesis in freshwater sediments. Appl. Environ. Microbiol. *33*:275–281.

Winfrey, M. R. and J. G. Zeikus. 1979. Microbial methanogenesis and acetate metabolism in a meromictic lake. Appl. Environ. Microbiol. *37*:213–221.

Winfrey, M. R., D. R. Nelson, S. C. Klevickis, and J. G. Zeikus. 1977. Association of hydrogen metabolism with methanogenesis in Lake Mendota sediments. Appl. Environ. Microbiol. *33*:312–318.

Winner, R. W. and J. S. Greber. 1980. Prey selection by *Chaoborus punctipennis* under laboratory conditions. Hydrobiologia *68*:231–233.

Winston, R. D. and P. R. Gorham. 1979. Turions and dormancy states in *Utricularia vulgaris*. Can J. Bot. *57*:2740–2749.

Winter, K. 1978. Short-term fixation of ^{14}carbon by the submerged aquatic angiosperm *Potamogeton pectinatus*. J. Exp. Bot. *29*:1169–1172.

Winter, T. C. 1978. Ground-water component of lake water and nutrient budgets. Verh. Internat. Verein. Limnol. *20*:438–444.

Winter, T. C. 1981a. Effects of water-table configuration on seepage through lakebeds. Limnol. Oceanogr. *26*:925–934.

Winter, T. C. 1981b. Uncertainties in estimating the water balance of lakes. Water Resour. Bull. *17*:82–115.

Winter, T. C. 1985. Physiographic setting and geologic origin of Mirror Lake. *In* G. E. Likens, ed. An Ecosystem Approach to Aquatic Ecology: Mirror Lake and its Environment. Springer-Verlag, New York. pp. 40–53.

Winter, T. C. 1995. Hydrological processes and the water budget of lakes. *In* A. Lerman, D. Imboden, and J. Gat, eds. Physics and chemistry of lakes. Springer-Verlag, Berlin. pp. 37–62.

Winter, T. C., J. W. Harvey, O. L. Franke, and W. M. Alley. 1998. Ground Water and Surface Water: A Single Resource. U.S. Geol. Survey Circ. *1139*. 79 pp.

Wirick, C. D. 1989. Herbivores and the spatial distributions of the phytoplankton. II. Estimating grazing in planktonic environments. Int. Rev. ges. Hydrobiol. *74*:249–259.

Wiseman, S. W., G. H. M. Jaworski, and C. S. Reynolds. 1983. Variability in sinking rate of the freshwater diatom *Asterionella formosa* Hass.: The influence of the excess density of colonies. Br. Phycol. J. *18*:425–432.

Wisniewski, R. J. 1991. The role of benthic biota in the phosphorus flux through the sediment-water interface. Verh. Internat. Verein. Limnol. *24*:913–916.

Wissmar, R. C. and F. J. Swanson. 1990. Landscape disturbances and lotic ecotones. *In* R. J. Naiman and H. Décamps, eds. The Ecology and Management of Aquatic-Terrestrial Ecotones. Parthenon Publishing Group, Carnforth, UK. pp. 65–89.

Wissmar, R. C. and R. G. Wetzel. 1978. Analysis of five North American lake ecosystems. VI. Consumer community structure and production. Verh. Internat. Verein. Limnol. *20*:587–597.

Wissmar, R. C., M. D. Lilley, and M. deAngelis. 1987. Nitrous oxide release from aerobic riverine deposits. J. Freshwat. Ecol. *4*:209–218.

Witzel, K.-P., J. Demuth, and C. Schütt. 1994. Viruses. *In* J. Overbeck and R. J. Chróst, eds. Microbial Ecology of Lake Plußsee. Springer-Verlag, New York. pp. 270–286.

Wium-Andersen, S. 1987. Allelopathy among aquatic plants. Arch. Hydrobiol. Beih. Ergebn. Limnol. *27*:167–172.

Wium-Andersen, S., U. Anthoni, C. Christophersen, and G. Houen. 1982. Allelopathic effects on phytoplankton by substances isolated form aquatic macrophytes (Charales). Oikos *39*:187–190.

Wohler, J. R. 1966. Productivity of the duckweeds. M.Sc. Thesis, University of Pittsburgh. 69 pp.

Wohlschlag, D. E. 1950. Vegetation and invertebrate life in a marl lake. Invest. Indiana Lakes Streams *3*:321–372.

Wolfe, J. M. and E. L. Rice. 1979. Allelopathic interactions among algae. J. Chem. Ecol. *5*:533–542.

Wolfe, R. S. 1971. Microbial formation of methane. Adv. Microbiol. Physiol. *6*:107–145.

Wölfl, S. 1991. The pelagic copepod species in Lake Constance: Abundance, biomass, and secondary productivity. Verh. Internat. Verein. Limnol. *24*:854–857.

Wolin, J. A. and H. C. Duthie. 1999. Diatoms as indicators of water level in freshwater lakes. *In* E. F. Stoermer and J. P. Smol, eds. The Diatoms: Applications for the Environmental and Earth Sciences. Cambridge Univ. Pres, Cambridge. pp. 183–202.

Wolk, C. P. 1968. Movement of carbon from vegetative cells to heterocysts in *Anabaena cylindrica*. J. Bacteriol. *96*:2138–2143.

Wolk, C. P. 1973. Physiology and cytological chemistry of blue-green algae. Bacteriol. Rev. *37*:32–101.

Wolverton, B. C. and R. C. McDonald. 1978. Bioaccumulation and detection of trace levels of cadmium in aquatic systems by *Eichhornia crassipes*. Environ. Health Perspectives *27*:161–164.

Wolverton, B. C. and R. C. McDonald. 1979. Upgrading facultative wastewater lagoons with vascular aquatic plants. J. Water Poll. Control Fed. *51*:305–313.

Wommack, K. E. and R. R. Colwell. 2000. Virioplankton: Viruses in aquatic ecosystems. Microbiol. Molecul. Bio. Rev. *64*:69–114.

Wondzell, S. M. and F. J. Swanson. 1996. Seasonal and storm dynamics of the hyporheic zone of the 4th-order mountain stream. I. Hydrologic processes. J. N. Am. Benthol. Soc. *15*:3–19.

Wondzell, S. M. and F. J. Swanson. 1999. Floods, channel change, and the hyporheic zone. Water Resour. Res. *35*:555–567.

Wong, B. and F. J. Ward. 1972. Size selection of *Daphnia pulicaria* by yellow perch (*Perca flavescens*) fry in West Blue Lake, Manitoba. J. Fish. Res. Board Can. *29*:1761–1764.

Wong, P. T. S., C. I. Mayfield, and Y. K. Chau. 1980. Cadmium toxicity to phytoplankton and microorganisms. *In* J. O. Nriagu, ed. Cadmium in the Environment. Part I. Ecological Cycling. John Wiley & Sons, New York. pp. 571–585.

Wood, K. G. 1974. Carbon dixode diffusivity across the air-water interface. Arch. Hydrobiol. *73*:57–69.

Wood, K. G. 1977. Chemical enhancement of CO_2 flux across the air-water interface. Arch. Hydrobiol. *79*:103–110.

Woodwell, G. M. and R. H. Whittaker. 1968. Primary production in terrestrial ecosystems. Am. Zool. *8*:19–30.

Woodwell, G. M., P. H. Rich, and C. A. S. Hall. 1973. The carbon cycle of estuaries. *In* G. M. Woodwell and E. V. Pecan, eds. Carbon and the Biosphere. Proc. 24th Brookhaven Symposium in Biology. U.S. Atomic Energy Commission. Symp. Ser. CONF–720510, Brookhaven, NY. pp. 221–240.

Wootton, R. J. 1990. Ecology of Teleost Fishes. Chapman & Hall, London.

Worthington, E. B. 1931. Vertical movements of fresh-water macroplankton. Int. Rev. ges. Hydrobiol. *25*:394–436.

Wright, J. C. 1965. The population dynamics and production of *Daphnia* in Canyon Ferry Reservoir, Montana. Limnol. Oceanogr. *10*:583–590.

Wright, J., W. J. O'Brien, and G. L. Vinyard. 1980. Adaptive value of vertical migration: A simulation model argument for predation hypothesis. *In* W. C. Kerfoot, ed. Evolution and Ecology of Zooplankton Communities. Univ. Press New England, Hanover, NH. pp. 138–147.

Wright, J. F., A. C. Cameron, P. D. Hiley, and A. D. Berrie. 1982. Seasonal changes in biomass of macrophytes on shaded and unshaded sections of the River Lambourn, England. Freshwat. Biol. *12*:271–283.

Wright, R. T. 1964. Dynamics of a phytoplankton community in an ice-covered lake. Limnol. Oceanogr. *9*:163–178.

Wright, R. T. 1975. Studies on glycolic acid metabolism by freshwater bacteria. Limnol. Oceanogr. *20*:626–633.

Wright, R. T. 1978. Measurement and significance of specific activity in the heterotrophic bacteria of natural waters. Appl. Environ. Microbiol. *36*:297–305.

Wright, R. T. and J. E. Hobbie. 1966. Use of glucose and acetate by bacteria and algae in aquatic ecosystems. Ecology *47*:447–464.

Wright, S. 1955. Limnological Survey of Western Lake Erie. Spec. Sci. Rept. Fish. U.S. Fish Wildl. Serv., *139*, 341 pp.

Wu, Y. and C. E. Gibson. 1996. Mechanisms controlling the water chemistry of small lakes in Northern Ireland. Wat. Res. *30*:178–182.

Wuim-Andersen, S. 1971. Photosynthetic uptake of free CO_2 by the roots of *Lobelia dortmanna*. Physiol. Plant *25*:245–248.

Wunderlich, W. O. 1971. The dynamics of density-stratified reservoirs. *In* G. E. Hall, ed. Reservoir Fisheries and Limnology. Spec. Publ. 8, Amer. Fish. Soc., Washington, DC. pp. 219–231.

Wurtsbaugh, W. and H. Li. 1985. Diel migrations of a zooplanktivorous fish (*Menidia beryllina*) in relation to the distribution of its prey in a large eutrophic lake. Limnol. Oceanogr. *30*:565–576.

Wurtsbaugh, W. A. and A. J. Horne. 1983. Iron in eutrophic Clear Lake, California: Its importance for algal nitrogen fixation and growth. Can. J. Fish. Aquat. Sci. *40*:1419–1429.

Wurtsbaugh, W. A., W. F. Vincent, R. A. Tapia, C. L. Vincent, and P. J. Richerson. 1985. Nutrient limitation of algal growth and nitrogen fixation in a tropical alpine lake, Lake Titicaca (Peru/Bolivia). Freshwat. Biol. *15*:185–195.

Wylie, J. L. and D. J. Currie. 1991. The relative importance of bacteria and algae as food sources for crustacean zooplankton. Limnol. Oceanogr. *36*:708–728.

Wynne, D. and M. Gophen. 1981. Phosphatase activity in freshwater zooplankton. Oikos *37*:369–376.

Wynne, D. and G-Y. Rhee. 1988. Changes in alkaline phosphatase activity and phosphate uptake in P-limited phytoplankton, induced by light intensity and spectral quality. Hydrobiologia *160*:173–178.

Xiong, F., J. Komenda, J. Kopecky, and L. Nedbal. 1997. Strategies of ultraviolet-B protection in microscopic algae. Physiol. Plant. *100*:378–388.

Xiong, S. and C. Nilsson. 1997. Dynamics of leaf litter accumulation and its effects on riparian vegetation: A review. Bot. Rev. *63*:240–264.

Xue, H. B. and L. Sigg. 1993. Free cupric iron concentration and Cu(II) speciation in a eutrophic lake. Limnol. Oceanogr. *38*:1200–1213.

Xue, H.-B., R. Gächter, and L. Sigg. 1997. Comparison of Cu and Zn cycling in eutrophic lakes with oxic and anoxic hypolimnion. Aquat. Sci. *59*:176–189.

Yagi, A. 1988. Dissolved organic manganese in the anoxic hypolimnion of Lake Fukami-ike. Jpn. J. Limnol. *49*:149–156.

Yagi, A. 1990. Dissolved organic manganese in the interstitial water of Lake Fukami-ike. Jpn. J. Limnol. *51*:269–279.

Yagi, A. 1993. Manganese cycle in Lake Fukami-ike. Verh. Internat. Verein. Limnol. *25*:193–199.

Yaskawa, K., T. Nakajima, N. Kawai, M. Torii, N. Natsuhara, and S. Horie. 1973. Palaeomagnetism of a core from Lake Biwa (I). J. Geomag. Geoelectr. *25*:447–474.

Yasuda, T., H. Ichikawa, and N. Ogura. 1989. Organic carbon budget in a headwater stream at Uratakao. Jpn. J. Limnol. *50*:227–234.

Yavitt, J. B. 1994. Carbon dynamics in Appalachian peatlands of West Virginia and western Maryland. Water Air Soil Poll. 77:271–290.

Yavitt, J. B., C. J. Williams, and R. K. Wieder. 1997. Production of methane and carbon dioxide in peatland ecosystems across North America: Effects of temperature, aeration, and organic chemistry of peat. Geomicrobiol. J. *14*:299–314.

Yavitt, J. B., R. K. Wieder, and G. E. Lang. 1993. CO_2 and CH_4 dynamics of a *Sphagnum*-dominated peatland in West Virginia. Global Biogeochem. Cycles 7:259–274.

Yentsch, C. S. 1960. The influence of phytoplankton pigments on the color of sea water. Deep-Sea Res. 7:1–9.

Yin, X. and S. E. Nicholson. 1998. The water balance of Lake Victoria. Hydrol. Sci. J. *43*:789–811.

Yoh, M, H. Terai, and Y. Saijo. 1988. Nitrous oxide in freshwater lakes. Arch. Hydrobiol. *113*:273–294.

Yoh, M. 1992. Marked variation in lacustrine N_2O accumulation level and its mechanism. Jpn. J. Limnol. *53*:75–81.

Yoh, M., A. Yagi, and H. Terai. 1990. Significance of low-oxygen zone for nitrogen cycling in a freshwater lake: Production of N_2O by simultaneous denitrification and nitrification. Jpn. J. Limnol. *51*:163–171.

Yoshida, T., M. Aizaki, T. Asami, and N. Makishima. 1979. Biological nitrogen fixation and denitrification in Lake Kasumiga-ura. Jpn. J. Limnol. *40*:1–9. (In Japanese.)

Yoshimura, S. 1935. Relation between depth for maximum amount of excess oxygen during the summer stagnation period and the transparency of freshwater lakes of Japan. Proc. Imperial Acad. Japan *11*:356–358.

Yoshimura, S. 1936a. A contribution to the knowledge of deep water temperatures of Japanese lakes. I. Summer temperatures. Jpn. J. Astronomy Geophys. *13*:61–120.

Yoshimura, S. 1936b. Contributions to the knowledge of iron dissolved in the lake waters of Japan. Second report. Jpn. J. Geol. Geogr. *13*:39–56.

Young, J. O. 1973. The prey and predators of *Phaenocora typhlops* (Vejdovsky) (Turbellaria: Neorhabdocoela) living in a small pond. J. Anim. Ecol. *42*:637–643.

Young, J. O. 1981. A comparative study of the food niches of lake-dwelling triclads and leeches. Hydrobiologia *84*:91–102.

Young, J. O. 1987. Predation on leeches in a weedy pond. Freshwat. Biol. *17*:161–167.

Young, J. O. 1988. Intra- and interspecific predation on the cocoons of *Erpobdella octoculata* (L.) (Annelida: Hirudinea). Hydrobiologia *169*:85–89.

Young, J. O., R. M. H. Seaby, and A. J. Martin. 1995. Contrasting mortality in young freshwater leeches and triclads. Oecologia *101*:317–323.

Young, J. P. and R. F. Horton. 1985. Heterophylly in *Ranunculus flabellaris* Raf.: The effect of abscisic acid. Ann. Bot. *55*:899–902.

Yount, J. D. and G. J. Niemi. 1990. Recovery of lotic communities and ecosystems from disturbance—a narrative review of case studies. Environ. Manage. *14*:547–569.

Yurista, P. M. 1997. *Bythotrephes cederstroemi* diapausing egg distribution and abundance in Lake Michigan and the environmental cues for breaking diapause. J. Great Lakes Res. *23*:202–209.

Yusoff, F. M. and I. Patimah. 1994. A comparative study of phytoplankton populations in two Malaysian lakes. Mitt. Internat. Verein. Limnol. *24*:251–257.

Zafar, A. R. 1959. Taxonomy of lakes. Hydrobiologia *13*:287–299.

Zaffagnini, F. 1987. Reproduction in *Daphnia*. Mem. Ist. Ital. Idrobiol. *45*:245–284.

Zago, M. S. A. 1976. A preliminary investigation on the cyclomorphosis of *Daphnia gessneri* Herbst, 1967, in a Brazilian reservoir. Bolm. Zool. Univ. S. Paulo *1*:147–160.

Zaiss, U. 1984. Acetate, a key intermediate in the metabolism of anaerobic sediments containing sulfate. Arch. Hydrobiol. Beih. Ergebn. Limnol. *19*:215–221.

Zaiss, U. 1985. Physiologische und ökologische Untersuchungen zur Regulation der Phosphatspeicherung bei *Oscillatoria redekei*. II. Der Einfluss ökologischer Parameter auf die Regulation des Polyphosphatstoffwechsels. Arch. Hydrobiol./Suppl. *72*:166–219.

Zaitsev, Yu. P. 1970. Marine neustonology. Akad. Nauk Ukrainskoi SSR, Naukova Dumka, Kiev. 207 pp. (Translat. into English, Programs for Scientific Translation, Jerusalem, 1971.)

Vander Zanden, M. J., B. J. Shuter, N. Lester, and J. B. Rasmussen. 1999. Patterns of food chain length in lakes: A stable isotope study. Am. Nat. *154*:406–416.

Zaret, T. M. 1972a. Predator-prey interaction in a tropical lacustrine ecosystem. Ecology *53*:248–257.

Zaret, T. M. 1972b. Predators, invisible prey, and the nature of polymorphism in the Cladocera (Class Crustacea). Limnol. Oceanogr. *17*:171–184.

Zaret, T. M. 1975. Strategies for existence of zooplankton prey in homogeneous environments. Verh. Internat. Verein. Limnol. *19*:1484–1489.

Zaret, T. M. 1980. Predation and Freshwater Communities. Yale Univ. Press, New Haven. 1187 pp.

Zaret, T. M. and W. C. Kerfoot. 1975. Fish predation on *Bosmina longirostris*: Body-size selection versus visibility selection. Ecology *56*:232–237.

Zaret, T. M. and J. S. Suffern. 1976. Vertical migration in zooplankton as a predator avoidance mechanism. Limnol. Oceanogr. *21*:804–813.

Zehnder, A. J. B. 1978. Ecology of methane formation. *In* R. Mitchell, ed. Water Pollution Microbiology, Vol. 2. John Wiley & Sons, New York. pp. 349–376.

Zehr, J. P., S. G. Paulsen, R. P. Axler, and C. R. Goldman. 1988. Dynamics of dissolved organic nitrogen in subalpine Castle Lake, California. Hydrobiologia *157*:33–45.

Zeikus, J. G. 1977. Biology of methanogenic bacteria. Bacteriol. Rev. *41*:514–541.

Zeikus, J. G. 1981. Lignin metabolism and the carbon cycle. Polymer biosynthesis, biodegradation, and environmental recalcitrance. Adv. Microb. Ecol. *5*:211–243.

Zepp, R. G., G. L. Baughmann, and P. F. Schlotzhauer. 1981. Comparison of photochemical behavior of various humic substances in water. 1. Sunlight induced reactions of aquatic pollutants photosensitized by humic substances. Chemosphere *10*:109–117.

Zettler, F. W. and T. E. Freeman. 1972. Plant pathogens as biocontrols of aquatic weeds. Ann. Rev. Phytopath. *10*:455–470.

Zhang, X., F. Zhang, and D. Mao. 1998. Effect of iron plaque outside roots on nutrient uptake by rice (*Oryza sativa* L.). Zinc uptake by Fe-deficient rice. Plant Soil *202*:33–39.

Zhang, Y. 1996. Dynamics of CO_2-driven lake eruptions. Science *379*:57–59.

Zicker, E. L., K. C. Berger, and A. D. Hasler. 1956. Phosphorus release from bog lake muds. Limnol. Oceanogr. *1*:296–303.

Zimmerman, A. P. 1981. Electron intensity, the role of humic acids in extracellular electron transport and chemical determination of pE in natural waters. Hydrobiologia *78*:259–265.

Zimmermann, U., H. Müller, and T. Weisse. 1996. Seasonal and spatial variability of planktonic heliozoa in Lake Constance. Aquat. Microb. Ecol. *11*:21–29.

Zinder, S. H. and T. D. Brock. 1978. Microbial transformations of sulfur in the environment. *In* J. O. Nriagu, ed. Sulfur in the Environment. II. Ecological Impacts. John Wiley & Sons, New York. pp. 445–466.

Zmyslowska, I. and M. Sobierajska. 1977. Microbiological studies of the Kortowskie Lake. Pol. Arch. Hydrobiol. *24*:61–71.

ZoBell, C. E. 1964. Geochemical aspects of the microbial modification of carbon compounds. *In* U. Colombo and G. D. Hobson, eds. Advances in Organic Geochemistry. Macmillan Co., New York. pp. 339–356.

ZoBell, C. E. 1973. Microbial biogeochemistry of oxygen. *In* A. A. Imshenetskii, ed. Geokhimicheskaia Deiatel'nost' Mikroorganizmov v Vodoemakh i Mestorozhdeniiach Poleznykh Iskopaemykh. Tipografiia Izdatel'stva Sovetskoe Radio, Moscow. pp. 3–76.

Zohary, T. and R. D. Robarts. 1990. Hyperscums and the population dynamics of *Microcystis aeruginosa*. J. Plankton Res. *12*: 423–432.

Zohary, T., A. M. Pais-Madeira, R. Robarts, and K. D. Hambright. 1996. Interannual phytoplankton dynamics of a hypertrophic African lake. Arch. Hydrobiol. *136*:105–126.

Züllig, H. 1961. Die Bestimmung von Myxoxanthophyll in Bohrprofilen zum Nachweis vergangener Blaualgenentfaltungen. Verh. Internat. Verein. Limnol. *14*:263–270.

Züllig, H. 1981. On the use of carotenoid stratigraphy in lake sediments for detecting past developments of phytoplankton. Limnol. Oceanogr. *26*:970–976.

Züllig, H. 1982. Untersuchungen über die Stratigraphie von Carotinoiden im geschichteten Sediment von 10 Schweizer Seen zur Erkundung früherer Phytoplankton-Entfaltungen. Schweiz. Z. Hydrol. *44*:1–98.

Züllig, H. 1989. Role of carotenoids in lake sediments for reconstructing trophic history during the late Quaternary. J. Paleolimnol. 2:23–40.

Zumberge, J. H. 1952. The Lakes of Minnesota. Their Origin and Classification. University of Minnesota Press, Minneapolis. 99 pp.

van der Zweerde, W. 1990. Biological control of aquatic weeds by means of phytophagous fish. *In* A. H. Pieterse and K. J. Murphy, eds. Aquatic Weeds: The Ecology and Management of Nuisance Aquatic Vegetation. Oxford Univ. Press, Oxford. pp. 201–221.

Zygmuntowa, J. 1972. Occurrence of free amino acids in pond water. Acta Hydrobiol. *14*:317–325.

Appendix

Conversion Factors

Quantity	*Unit*	*Conversion factor*
Length	1 in.	= 25.4 mm
	1 ft	= 0.3048 m
	1 yd	= 0.9144 m
	1 fathom	= 1.8288 m
	1 chain	= 20.1168 m
	1 mile	= 1.60934 km
	1 International nautical mile	= 1.852 km
	1 U.K. nautical mile	= 1.85318 km
Area	1 in.^{2}	= 6.4516 cm^2
	1 ft^2	= 0.092903 m^2
	1 yd^2	= 0.836127 m^2
	1 acre	= 4046.86 m^2 = 0.404686 ha (hectare)
	1 sq. mile	= 2.58999 km^2 = 258.999 ha
Volume	1 U.K. minim	= 0.0591938 cm^3
	1 U.K. fluid drachm	= 3.55163 cm^3
	1 U.K. fluid ounce	= 28.4131 cm^3

(continued)

Quantity	Unit		Conversion factor
Volume (*continued*)	1 U.S. fluid ounce	=	29.5735 cm^3
	1 U.S. liquid pint	=	473.176 cm^3 = 0.4732 dm^3 (= liter)
	1 U.S. dry pint	=	550.610 cm^3 = 0.5506 dm^3
	1 Imperial pint	=	568.261 cm^3 = 0.5683 dm^3
	1 U.K. gallon	=	1.201 U.S. gallon
		=	4.54609 dm^3 (liter)
	1 U.S. gallon	=	0.833 U.K. gallon
		=	3.78541 dm^3 (liter)
	1 U.K. bu (bushel)	=	0.0363687 m^3 = 36.3687 dm^3
	1 U.S. bushel	=	0.0352391 m^3 = 35.2391 dm^3
	1 $in.^3$	=	16.3871 cm^3
	1 ft^3	=	0.0283168 m^3
	1 yd^3	=	0.764555 m^3
	1 board foot (timber)	=	0.00235974 m^3 = 2.35974 dm^3
	1 cord (timber)	=	3.62456 m^3
Moment of inertia	1 lb ft^2	=	0.0421401 kg m^2
	1 slug ft^2	=	1.35582 kg m^2
Mass	1 grain	=	0.0647989 g = 64.7989 mg
	1 dram (avoir.)	=	1.77185 g = 0.00177185 kg
	1 drachm (apoth.)	=	3.88793 g = 0.00388793 kg
	1 ounce (troy or apoth.)	=	31.1035 g = 0.0311035 kg
	1 oz (avoir.)	=	28.3495 g
	1 lb	=	0.45359237 kg
	1 slug	=	14.5939 kg
	1 sh cwt (U.S. hundredweight)	=	45.3592 kg
	1 cwt (U.K. hundredweight)	=	50.8023 kg
	1 U.K. ton	=	1016.05 kg
		=	1.01605 tonne
	1 short ton	=	2000 lb
		=	907.185 kg
		=	0.907 tonne
Mass per unit length	1 lb/yd	=	0.496055 kg/m
	1 U.K. ton/mile	=	0.631342 kg/m
	1 U.K. ton/1000 yd	=	1.11116 kg/m
	1 oz/in.	=	1.11612 kg/m = 11.1612 g/cm
	1 lb/ft	=	1.48816 kg/m
	1 lb/in.	=	17.8580 kg/m
Mass per unit area	1 lb/acre	=	0.112085 g/m^2 = 1.12085 × 10^{-4} kg/m^2
	1 U.K. cwt/acre	=	0.0125535 kg/m^2
	1 oz/yd^2	=	0.0339057 kg/m^2
	1 U.K. ton/acre	=	0.251071 kg/m^2
	1 oz/ft^2	=	0.305152 kg/m^2
	1 lb/ft^2	=	4.88243 kg/m^2
	1 $lb/in.^2$	=	703.070 kg/m^2
	1 U.K. $ton/mile^2$	=	0.392298 g/m^2 = 3.92298 × 10^{-4} kg/m^2
Density	1 lb/ft^3	=	16.0185 kg/m^3
	1 lb/U.K. gal	=	99.7763 kg/m^3 = 0.09978 kg/l
	1 lb/U.S. gal	=	119.826 kg/m^3 = 0.1198 kg/l
	1 $slug/ft^3$	=	515.379 kg/m^3
	1 ton/yd^3	=	1328.94 kg/m^3 = 1.32894 $tonne/m^3$
	1 $lb/in.^3$	=	27.6799 Mg/m^3 = 27.6799 g/cm^3
Specific volume	1 $in.^3/lb$	=	36.1273 cm^3/kg
	1 ft^3/lb	=	0.0624280 m^3/kg = 62.4280 dm3/kg
Velocity	1 in./min	=	0.42333 cm/s
	1 ft/min	=	0.00508 m/s = 0.3048 m/min

Quantity	*Unit*		*Conversion factor*
Velocity (*continued*)	1 ft/s	=	0.3048 m/s = 1.09728 km/h
	1 mile/h	=	1.60934 km/h = 0.44704 m/s
	1 U.K. knot	=	1.85318 km/h = 0.514773 m/s
	1 International knot	=	1.852 km/h = 0.514444 m/s
Acceleration	1 ft/s^2	=	0.3048 m/s^2
Mass flow rate	1 lb/h	=	0.125998 g/s = 1.25998 $\times$ 10^{-4} kg/s
	1 U.K. ton/h	=	0.282235 kg/s
Force or weight	1 dyne	=	10^{-5} N
	1 pdl (poundal)	=	0.138255 N
	1 ozf (ounce)	=	0.278014 N
	1 lbf	=	4.44822 N
	1 kgf	=	9.80665 N
	1 tonf	=	9.96402 kN
Force or weight per unit length	1 lbf/ft	=	14.5939 N/m
	1 lbf/in.	=	175.127 N/m = 0.175127 N/mm
	1 tonf/ft	=	32.6903 kN/m
Force (weight) per unit area or pressure or stress	1 pdl/ft^2	=	1.48816 N/m^2 or Pa (pascal)
	1 lbf/ft^2	=	47.8803 N/m^2
	1 mm Hg	=	133.322 N/m^2
	1 in. H_2O	=	249.089 N/m^2
	1 ft H_2O	=	2989.07 N/m^2 = 0.0298907 bar
	1 in. Hg	=	3386.39 N/m^2 = 0.0338639 bar
	1 lbf/in.^{2}	=	6.89476 kN/m^2 = 0.0689476 bar
	1 bar	=	10^5 N/m^2
	1 std. atmos.	=	101.325 kN/m^2 = 1.01325 bar
	1 tonf/ft^2	=	107.252 kN/m^2
	1 tonf/in.^{2}	=	15.4443 MN/m^2 = 1.54443 hectobar
Specific weight	1 lbf/ft^3	=	157.088 N/m^3
	1 lbf/U.K. gal	=	978.471 N/m^3
	1 tonf/yd^3	=	13.0324 kN/m^3
	1 lbf/in.^{3}	=	271.447 kN/m^3
Moment, torque, or couple	1 ozf in. (ounce-force inch)	=	0.00706155 N m
	1 pdl ft	=	0.0421401 N m
	1 lbf in.	=	0.112985 N m
	1 lbf ft	=	1.35582 N m
	1 tonf ft	=	3037.03 N m = 3.03703 kN m
Energy, heat, or work	1 erg	=	10^{-7} J
	1 hp h (horsepower hour)	=	2.68452 MJ
	1 thermie = 10^6 cal$_{IT}$	=	4.1855 MJ
	1 therm = 100000 Btu	=	105.506 MJ
	1 cal$_{IT}$	=	4.1868 J
	1 Btu	=	1.05506 kJ
	1 kWh	=	3.6 MJ
Power	1 hp = 550 ft lbf/s	=	0.745700 kW
	1 metric horsepower (ch, PS)	=	735.499 W
Specific heat capacity	1 Btu/lb °F 1 Chu/lb °C 1 cal/g °C	=	4.1868 kJ/kg K
Heat flow rate	1 Btu/h	=	0.293071 W
	1 kcal/h	=	1.163 W
	1 cal/s	=	4.1868 W

(continued)

Quantity	*Unit*		*Conversion factor*
Intensity of heat flow rate	1 Btu/ft^2 h	=	3.15459 W/m^2
Electric stress	1 kV/in.	=	0.0393701 kV/mm
Dynamic viscosity	1 lb/ft s	=	14.8816 poise = 1.48816 kg/m s
Kinematic viscosity	1 ft^2/s	=	929.03 stokes = 0.092903 m^2/s
Caloric value or specific enthalpy	1 Btu/ft^3	=	0.0372589 J/cm^3 = 37.2589 kJ/m^3
	1 Btu/lb	=	2.326 kJ/kg
	1 cal/g	=	4.1868 J/g
	1 $kcal/m^3$	=	4.1868 kJ/m^3
Specific entropy	1 Btu/lb °R	=	4.1868 kJ/kg K
Thermal conductivity	1 cal cm/cm^2 s °C	=	41.868 W/m K
	1 Btu ft/ft^2 h °F	=	1.73073 W/m K
Gas constant	1 ft lbf/lb °R	=	0.00538032 kJ/kg K
Plane angle	1 rad (radian)	=	57.2958°
	1 degree	=	0.0174533 rad = 1.1111 grade
	1 min	=	2.90888×10^{-4} rad = 0.0185 grade
	1 sec	=	4.84814×10^{-6} rad = 0.0003 grade
Radioactivity	1 Ci (curie)	=	37 GBq (becquerel)
	1 μCi	=	37 kBq

Interrelationships of Selected Units of Irradiance and Illuminance

	μmol quanta/m²-s	*J/m²-s*	*erg/cm²-s*	*gcal/cm²-min*	*lux[a] (illuminance unit)*
μmol quanta/m^2-s	1	1.20×10^{-1}	1.20×10^{2}	1.72×10^{-4}	~5.12×10^{1}
J/m^2-s (= W/m^2)	5.03	1	10^3	1.43×10^{-3}	~2.5×10^{2}
gcal/cm^2-min	5.83×10^{3}	6.98×10^{2}	6.98×10^{5}	1	~1.8×10^{5}
lux[a] or m-candle (= 1 lumen/m^2 or 0.0929 ft-candle)	~1.953×10^{-2}	~4.0×10^{-3}	~4.0	~5.7×10^{-6}	1

[a] Energy equivalents are given in terms of visible range in daylight (380–720 nm). 1 W/m^2 ≈ 4.6 μmol quanta/m^2-s; 1 lux × 0.01953 μmol quanta/m^2-s; 1 lux = 4.1×10^{-7} W/cm^2 = 6.0×10^{-6} gcal/cm^2-min; 1 lumen = 4.17×10^{-3} W (1 W = 240 lumens); 1 ft-candle = 4.6×10^{-6} W/cm^2 = 6.5×10^{-5} gcal/cm^2-min; 1 W/cm^2 = 2.4×10^{6} lux = 2.2×10^{5} ft-candles; and 1 gcal/cm^2-min = 1.6×10^{5} lux = 1.5×10^{4} ft-candles.

Index